The
Principles
of
Physiology

The Principles of Physiology

David Jensen, Ph.D.

Denver, Colorado

Illustrated by **Barbara Jensen**

APPLETON-CENTURY-CROFTS/New York
A Publishing Division of Prentice-Hall, Inc.

LIBRARY OF CONGRESS CATALOGING IN
PUBLICATION DATA

Jensen, David, 1926-
     The principles of physiology.

     Bibliography: p. 1257
     1. Physiology.   2. Physiology, Pathological.
I. Title.   [DNLM: 1.   Physiology.   QT4 J53p]
QP31.2.J46   599'.01   76-3683
ISBN 0-8385-7932-9

*Prentice-Hall International, Inc., London
Prentice-Hall of Australia, Pty. Ltd., Sydney
Prentice-Hall of India Private Limited, New Delhi
Prentice-Hall of Japan, Inc., Tokyo
Prentice-Hall of Southeast Asia (Pte.) Ltd., Singapore*

PRINTED IN THE UNITED STATES OF AMERICA

cover design: Judy Forster

Dedicated
To
PROFESSOR LESLIE L. BENNETT,
Mentor and Friend,
who introduced the author to the science of
Physiology some thirty years ago, an introduction
which has culminated in this work

Acknowledgments

The preparation of a book such as this would have been impossible without the splendid cooperation and unstinting support of many individuals. In particular, the author would like to express his deepest appreciation to the following persons whose valuable contributions have aided substantially and materially in the completion of this volume:

Prof. Richard A. Deitrich, Dr. Kenneth L. Goetz, Dr. Raymond L. Powis, Dr. Frederick A. Seydel, Prof. Harold Tarver, Miss Margaret A. Berry, and Mr. Michael Drebena all have read critically certain portions of the manuscript. Their many helpful and constructive suggestions have enhanced considerably the accuracy and clarity of the text.

Mrs. Joan W. Hamilton deserves special mention for her unselfish devotion to the preparation of the entire manuscript of this book for publication, as well as for her unfailing assistance in securing many of the references contained herein.

Mrs. Barbara Jensen has spent many hundreds of hours in drawing all of the figures which are published in this book; their accuracy and clarity speak clearly of her great skill and patience.

Mr. Floyd A. Holland has prepared the photographs from which the plates used in printing the illustrations were made. The quality of the final product is a tribute to his technical ability.

Mr. Phillip D. Jensen has devoted much time to the chore of proofreading this volume at various stages in its development. His valuable contribution is acknowledged with particular thanks.

Mr. Robert B. R. Gratiot, Mrs. Barbara Drebena, Miss Anita E. Jensen, and Miss Jean K. Lisle all have rendered a great service in helping to check the copy of this text as well as to assist in preparation of the index.

All too often a major contributor to the preparation of a book is somewhat neglected insofar as acknowledgment is concerned: the publisher. Therefore it is a distinct pleasure for the author to recognize the unfailing cooperation and invaluable material assistance which were provided by Appleton-Century-Crofts. Mr. David W. Stires, General Manager, as well as Mrs. Berta Steiner Rosenberg, Managing Editor, Mrs. Joan Caldwell, Marketing Manager, Miss Theresa Kornak, Assistant Editor, and Miss Laura Jane Bird, Art Director, are entitled to special and particular mention here.

Finally, this book could not have been written without the significant contributions made by innumerable investigators, authors, editors, and publishers in a wide variety of scientific disciplines. The bibliography of this textbook, as well as the legends to certain of the figures and tables, reflect with gratitude the profound debt of the author to these sources of information.

Preface

This textbook has been designed to present the basic concepts of physiology in a manner that will enable the medical, graduate, or other student to achieve a clear understanding of the subject as rapidly as possible. Mammalian, and in particular human, physiology has been stressed throughout wherever possible, although certain data obtained from lower organisms perforce have been included, since in many instances such animals provide "model systems" that are vital to the solution of many fundamental problems.

The material has been treated in depth, but without unnecessary minutae. The reader is referred to the works cited in the bibliography for such particulars as, for example, detailed experimental procedures.

Many examples of abnormal physiology are included in order to emphasize as well as to contrast the relationships between normal functions and clinically encountered derangements thereof.

The Principles of Physiology includes several unique features, which may be enumerated briefly:

As indicated in the Contents, the material has been divided into 16 chapters. The first of these consists exclusively of a review of the principal physical and chemical concepts that are of fundamental importance to an understanding of the material to follow. Also included in this chapter is a brief review of the pertinent features of organic chemistry.

In subsequent chapters, the material is presented on a scale of ascending structural as well as functional complexity. Prefatory to these discussions, relevant gross as well as microscopic anatomic features of the tissues and organs under consideration are reviewed briefly.

The final chapter consists of several essays in which a number of normal and pathophysiologic processes are discussed in some detail from the standpoint of the entire organism; ie, a holistic approach is followed.

All cross references within the text are given by citing specific page, figure, and/or table numbers. Thus no time will be lost by the reader who is seeking additional information related to the topic under discussion.

The illustrations are line drawings and conceptual diagrams which were designed specifically to complement the text. Where possible, many factual details have been summarized in tabular form for purposes of convenient reference and comparison.

The bibliography has been organized under two principal headings: *General References* includes works that contain useful discussions of topics covered in more than one chapter of this book, whereas *Specific References* indicates citations for particular topics that are discussed in individual chapters or which provide additional information on particular topics. No references to books or papers are made within the text proper. The interested reader can ascertain these readily by consulting the titles of the articles cited.

Every effort has been made to include pertinent current research data. A number of significant older works of major importance to the development of certain fields also are listed.

Many technical terms are defined within the text in relation to the topic under discussion. In this regard, words in everyday usage, but which have been adopted and endowed with special technical meanings by physiologists, also have been defined.

A comment on eponyms is in order. Usage of these denotations is as common in physiology as in any other branch of science. However, such names seldom give the uninitiated reader any indication of the anatomic fact, physiologic mechanism, or clinical situation under discussion. Consequently, eponyms have been given parenthetic treatment in this book immediately following the subject to which they refer. For example: The force of contraction of myocardial fibers is, within limits, directly proportional to the initial fiber length (Starling's law of the heart). It is suggested that the student take advantage of eponymic designations as mnemonic devices.

In the interests of precision and consistency, metric units have been used exclusively throughout. Conversion factors to other systems of measurement will be found in Chapter 1.

The science of physiology is a dynamic rather than a static discipline. Active research in many areas constantly is producing new information and data; older concepts and interpretations are being revised or modified continually. As a consequence, much controversy exists today among various schools of thought regarding the ultimate interpretation of many experimental findings. Obviously, it is impossible in this volume to present and discuss adequately the various aspects of such controversies, however interesting and important they may be. For these reasons, the material presented herein has been limited to those facts and concepts which appear to be reasonably well established and which are deemed most pertinent to a rational approach to the study and eventual practice of clinical medicine and other health related disciplines.

David Jensen
Denver, Colorado

Contents

The
Principles
of
Physiology

CHAPTER 1

Summary of Physical and Chemical Principles

All physiologic processes ultimately involve mechanisms that are based upon the physics and chemistry of biologic systems. Therefore, an understanding of such mechanisms in the human body depends in large measure upon a firm knowledge of the basic physical and chemical units, terms, and concepts that are commonly employed in physiology. For this reason, the present chapter is devoted exclusively to a review of this essential material.

At the outset, the reader should note clearly the distinction between *absolute units* and *derived* or *calculated units,* since many examples of both types of expression will be encountered throughout this text. Length and time, for example, are measured in absolute units; velocity, on the other hand, is a conceptual derivation involving both of these factors — ie, it is the distance (or length) traversed by a moving body per unit of time. Derived units enable one not only to obtain more information than is possible from empirical data expressed in absolute units, but also to calculate indirectly many important parameters of bodily functions which cannot be measured readily or directly (eg, cardiac output, p. 640).

STANDARD UNITS OF THE METRIC SYSTEM

Only metric units will be employed in this book; for conversion factors to other unit systems, see Table 1.1.

The *absolute* or *standard units* in the metric system are *length, mass,* and *time.* For this reason, the metric scale is sometimes referred to as the centimeter-gram-second (cgs) or meter-kilogram-second (mks) system. Both designations are used in physics as well as in biology; one should bear in mind, however, the different *magnitudes* involved — ie, centimeters and grams in the former, and meters and kilograms in the latter. Certain prefixes are applied in combination with these basic units

Table 1.1 CONVERSION FACTORS AMONG VARIOUS STANDARD UNITS[a]

A. Length

1 meter = 100 centimeters = 39.37 inches = 3.28 feet = 1.09 yards
1 centimeter = 10 millimeters = 0.39 inch
1 inch = 2.54 centimeters = 254 millimeters
1 kilometer = 1,000 meters = 0.62 mile (U.S., statute) = 3,280.80 feet
1 mile (U.S. statute) = 5,280 feet = 1.61 kilometers

B. Mass

1 kilogram = 1,000 grams = 2.20 pounds (avoirdupois) = 2.68 pounds (apothecary[b])
1 gram = 1,000 milligrams = 0.03 ounce (avoirdupois) = 0.03 ounce (apothecary)
1 pound (avoirdupois) = 16 ounces (avoirdupois) = 0.45 kilogram = 454.60 grams
1 pound (apothecary) = 12 ounces (apothecary) = 0.37 kilogram = 373.24 grams
1 ounce (avoirdupois) = 28.35 grams
1 ounce (apothecary) = 31.10 grams
1 grain = 65.80 milligrams

C. Volume

1 liter = 1,000 milliliters = 1.06 quart (U.S., fluid) = 0.88 quart (British or Imperial, fluid)
1 milliliter = 0.03 ounce (U.S., fluid) = 0.03 ounce (British, fluid) = 1.00 cubic centimeter
1 ounce (U.S., fluid) = 29.57 milliliters
1 ounce (British, fluid) = 28.41 milliliters
1 quart (U.S., fluid) = 0.25 gallon (U.S., fluid) = 0.95 liter = 946.36 milliliters
1 quart (British, fluid) = 0.25 gallon (British or Imperial, fluid) = 1.14 liters = 1,136.52 milliliters

[a] *All values in this table have been rounded off to two decimal places.*
[b] *Apothecary weight is the same as troy weight.*

in order to designate their multiples or fractions, as indicated in Table 1.2.

A number of conversion factors illustrating

Table 1.2 PREFIX NAMES OF MULTIPLES AND SUBMULTIPLES OF BASIC UNITS

PREFIX	ABBREVIATION OF PREFIX	FACTOR BY WHICH UNIT IS MULTIPLIED
tera-	T	10^{12}
giga-	G	10^{9}
mega-	M	10^{6}
kilo-	K	10^{3}
hecto-	h	10^{2}
deka-	da	10
deci-	d	10^{-1}
centi-	c	10^{-2}
milli-	m	10^{-3}
micro-	μ	10^{-6}
nano-	n	10^{-9}
pico-	p	10^{-12}
femto-	f	10^{-15}
atto-	a	10^{-18}

certain interrelationships among various units to be discussed below are summarized in Table 1.3.

Length

The unit of length is the standard meter (m). The centimeter (cm) is one-hundredth of the meter, and the millimeter (mm) equals 0.1 cm.

A unit of length commonly used in physiology and microscopic anatomy is the micron (μ) or micrometer (μm). The term "micron," denoted by the Greek letter mu, will be employed in this book. One micron equals 0.001 mm.

A number of multiples and submultiples of length are tabulated at the top of Figure 1.3.

Mass

The unit of mass (or, less precisely, weight) is the kilogram (kg). One gram (gm or gram mass) is one-thousandth of a kilogram, and one milligram (mg) is equal to 0.001 gm. (See also the section titled "Force.")

One gram mass is equal (approximately) to the weight of one cubic centimeter of pure water at its maximum density (about 4 C).

Time

The unit of time is the second (sec). One second is equal to 1/86,400 of a mean solar day.

Volume

The unit of volume is the liter (l). One milliliter (ml) is one-thousandth of a liter. For all practical purposes, one cubic centimeter (cc) is equivalent to one milliliter insofar as volume is concerned; however, the designation *ml* rather than *cc* is preferable and will be employed in this text.

Temperature

The unit of temperature is the degree Celsius or centigrade (C).

On the absolute, or Kelvin (K), temperature scale, 0 C equals 273 K. The magnitude of the units for temperature on the Celsius and Kelvin scales is identical; thus, 0 K, or absolute zero, equals − 273.2 C.

A comparison of the Fahrenheit and Celsius temperature scales, as well as the mathematical relationship between these two systems, is shown in Figure 1.1.

Table 1.3 CONVERSION FACTORS AMONG CERTAIN PHYSICAL UNITS

TO CONVERT	INTO	MULTIPLY BY
Atmospheres	Millimeters Hg	760
Atmospheres	Dynes/cm^2	1.01×10^6
Calories, gram (mean)	Joules/gm (mean)	4.186
Calories, kilogram (mean)	Joules (absolute)	4,186
Calories, kilogram (mean)	Calories, gram (mean)	0.001
Dynes/cm	Ergs/cm^2	1.0
Dynes/cm^2	Cm H_2O, 4 C	0.001
Dynes/cm^2	Gm/cm^2	0.001
Dynes/cm^2	Kg/m^2	0.01
Dynes/cm^2	Mm Hg, at 0 C	7.50×10^{-4}
Ergs	Kilogram−meters	1.02×10^{-8}
Ergs	Joules (absolute)	1×10^{-7}
Gravity, acceleration of	cm/sec^2	980.67 (or 981)
Joules	Ergs	1×10^7
Joules (absolute)	Kilogram−meters	0.102
Kilogram-meters	Joules (absolute)	9.81
Kilogram-meters/second	Watts (absolute)	9.81
Kilogram-meters	Ergs	9.81×10^7
Millimeters Hg, 0 C	Atmospheres	0.0013
Millimeters Hg, 0 C	pounds/ft^2	2.785
Millimeters Hg, 0 C	pounds/inch^2	0.0193

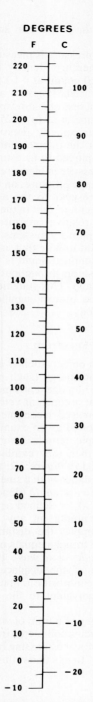

DEGREES

F   C

FIG. 1.1. Comparison of Fahrenheit and Celsius temperature scales. To convert degrees Fahrenheit (°F) into degrees Celsius (°C), subtract 32 and multiply by $\frac{5}{9}$. To convert °C into °F, multiply by $\frac{9}{5}$ and add 32.

Note that "temperature" is merely an arbitrary way of designating the relative heat or thermal energy content of a system measured on an equally arbitrary scale.

MISCELLANEOUS PHYSICAL UNITS AND CONCEPTS

Velocity

The unit for velocity is either the meter per second or the centimeter per second (m/sec or cm/sec). Both definitions of velocity are used in physiology, for various purposes, and it should be remembered that the magnitude of the former unit is a hundredfold greater than that of the latter.

Acceleration

The unit of acceleration is the centimeter per second per second (cm/sec²). The acceleration produced by the force of gravity is 980.67 cm/sec².

Force

The cgs unit of force is the dyne. One dyne is the force which, when acting on a *mass* of 1 gm, imparts to it an acceleration of 1 cm/sec². One gram *weight*, on the other hand, is the force with which the earth's gravitational field attracts 1 gm mass, and is approximately equal to 981 dynes. That is, the gram weight (or weight of a gram mass) is the cgs gravitational unit.

Work and Energy

The cgs unit for work is the erg. One erg is a force of 1 dyne acting through a distance of 1 cm; one joule is equal to 10⁷ ergs. Note that the joule is in mks units.

The most commonly used unit of work in physiology is the kilogram-meter (kg-m), which is defined as the amount of work expended in raising a mass of 1 kg to a height of 1 meter. This mks unit also may be defined as the work done by a force of 1 kg acting through a distance of 1 meter in the direction of the force.

Thus, work is measured by the product of a force acting upon a given mass and the distance through which the mass is moved. (Remember that weight, as defined above, actually is a force acting upon a specified mass.)

The term *energy* denotes the capacity for performing work. One may speak of mechanical, thermal (heat), electrical, or chemical energy. Generally, these fundamental types of energy can be interconverted readily. For example, in the muscles of the body, chemical energy is converted into a mechanical contraction with the liberation of heat as a by-product of the contractile process.

Energy also may be designated as *potential* or *kinetic*. In the former situation, the energy is indicated in a system at rest, whereas in the latter case the energy is being released. For example, a mass of 1 kg situated 1 meter above a tabletop has a potential energy of 1 kg-meter. If this mass is released suddenly, the potential energy is converted into kinetic energy as the body falls, and as the mass strikes the surface of the table, both the potential and the kinetic energy of the system becomes zero as equilibrium is reached. During the free fall of the body, 1 kg-meter of energy has been released.

Power

The cgs unit for power is the erg per second, whereas in mks units this becomes joules per second. One watt = 10^7 ergs/sec = 1 joule/sec; one kilowatt (kw) = 10^3 watts = 10^{10} ergs/sec or 10^3 joules/sec.

In other words, power is the time rate at which work is performed. If a quantity of work (W) is done in a time interval (t), the power (P), or rate of doing work, is given by the expression: $P = W/t$. The power is obtained in watts if W is expressed in joules (one joule = 10^7 ergs) and t is in seconds. Similarly, the power developed by an individual performing physical work can be expressed in terms of kilogram-meters per second.

Heat

The unit for heat is the calorie (cal), also called the gram-calorie or small calorie. One kilogram-calorie (Cal) equals 1,000 cal.

One gram calorie is defined as the quantity of heat energy necessary to raise the temperature of 1 gm of water 1 C, usually between the range from 14.5 to 15.5 C. Similarly, 1 Cal is the quantity of heat energy required to raise the temperature of 1 kg of water a like amount.

Although the joule generally is used in measuring mechanical work, note that calories may be converted readily into these units, as indicated in Table 1.3. In physiology, however, the kilogram-calorie or Cal is the unit of heat energy in common usage.

Density

The density of a substance is defined as the mass of that substance per unit volume (gm/ml). The term *specific gravity* refers to the ratio of the mass of a body to the mass of an equal volume of water at 4 C.

Pressure

The term *pressure* (P) signifies a force (F) applied to or distributed over a surface, and is measured as the force per unit area (A). Thus, $P = F/A$. The unit of pressure is the dyne per square centimeter (dyne/cm²).

Frequently, pressure is measured by the height of a column of mercury (density 13.6 gm/ml) or of water (density 1.0 gm/ml at 4 C) that is supported by the force exerted. Pressure also can be measured in atmospheres: one atmosphere (atm) = 760 mm Hg. See Figure 16.25 (p. 1225) for the relationship between atmospheric pressure and altitude, and Table 1.3 for the interrelationships among the various units employed to measure pressure.

Absolute pressure is measured with respect to zero pressure; gauge pressure, on the other hand, is measured with respect to the ambient pressure. This distinction is important in the determination of absolute versus relative pressures in the cardiovascular and respiratory systems.

It is important to note that pressure exerted at any point on a confined liquid is transmitted in an undiminished fashion throughout the mass of the liquid (Pascal's law). This statement of elementary physics is of the utmost significance to cardiovascular physiology.

The Frequency Spectrum

When a vibrating object (such as a tuning fork) or a source of radiant energy (such as the sun) emits waves through a homogeneous medium, the waves are propagated at a constant velocity (v), and this value can be expressed in centimeters or meters per second. Using sound as an example, if the source vibrates in a simple periodic fashion and the waves are transverse, then they exhibit the general appearance depicted in Figure 1.2; that is, the waves oscillate in accordance with a sine curve. As shown in this figure, the *wavelength* (denoted by the Greek symbol lambda, λ), is defined as the distance between identical points on any two consecutive waves. For example, the distance between two successive wave troughs equals one wavelength.

The *amplitude* of a wave is defined by the maximum value of the displacement (A) of the wave about a mean position, and the amplitude of the waves is equivalent to the amplitude of the emitting source.

The *frequency* of a train of waves is defined by the number of cycles completed in a unit of time (ie, by the number of waves passing by any given point per unit of time). In cgs units, this factor is expres-

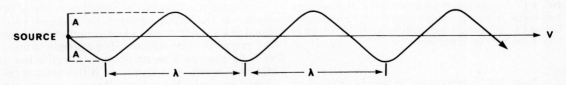

FIG. 1.2. Definition of wavelength (λ) as the distance between corresponding points on two consecutive waves emitted from a vibrating or radiant source. The amplitude (A) is the maximum displacement of the wave from a mean, and the velocity (V) of the wave is the distance that the wave front travels per unit time.

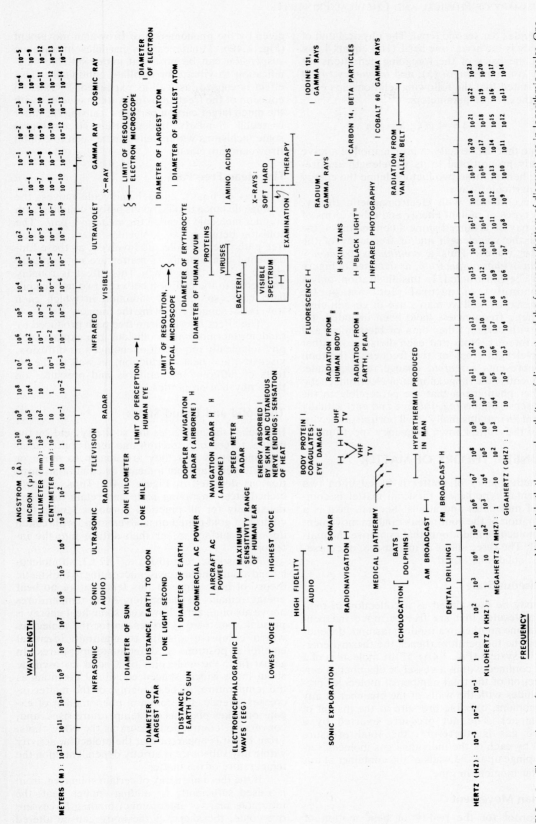

FIG. 1.3. The continuous frequency or energy spectrum. Note that both the wavelength (top of diagram) and the frequency (bottom of diagram) are in logarithmic scales. One millimicron (mμ) equals 10^{-3} μ.

Only certain arbitrarily selected, naturally occurring as well as artificially produced wave bands are shown, with emphasis on data of biomedical interest. Thus certain physical data pertaining to radio and other frequencies are included for comparative purposes. The practical rather than theoretical limits of resolution for optical and electron microscopes are indicated.

Note particularly that the visible and auditory spectra occupy an infinitely small fraction of the entire frequency spectrum.

sed as cycles per second (cps). The physical unit of frequency is the *hertz;* one hertz (Hz) equals 1 cps.

On the basis of the foregoing definitions for *velocity* (*v*), *wavelength* (λ), and *frequency* (*n*), it can be stated that the following relationships exist among these three parameters:

$$v = n\lambda \; ; n = v/\lambda \; ; \lambda = v/n$$

Note that the velocity of propagation of a wave is directly proportional to its wavelength and frequency; hence, these two factors define the energy of that particular wave.

In accordance with electromagnetic theory, various types of radiant energy and other forms of energy may be arranged along a continuum on the basis of the wavelength and/or frequency of the particular waves. Thus a *continuous frequency spectrum* or *energy spectrum,* as depicted in Figure 1.3, may be constructed. In this illustration, only a few examples of natural electromagnetic phenomena, as well as man's use of certain electromagnetic frequencies, have been included, together with certain other data of biomedical and general interest. Note that each division on either the wavelength (top) or the frequency (bottom) scale represents a *tenfold* change in magnitude. Consequently, the range of natural electromagnetic and other phenomena that are perceptible to the human sense organs (eg, the eye and ear) actually represent an infinitesimally small portion of the vast range of the entire known frequency spectrum.

THE KINETIC THEORY OF MATTER

The kinetic theory of matter is based upon two fundamental hypotheses: First, all matter is composed of molecules and atoms. Second, heat is a manifestation of the continuous random movement of the molecules and atoms that comprise the matter itself. The three states of matter are, of course, gaseous, solid, and liquid.

The Gaseous State

A gas may be considered as a collection of individual molecules that are free to move independently of one another in a random fashion, depending upon the temperature (hence the thermal energy) of the system (Fig. 1.4A). If the molecules of a gas are confined within a vessel or chamber, a certain fraction of the total number of molecules present collides with the walls of the chamber at any given moment, exerting pressure at the instant of their impact. The net pressure exerted by a confined gas is, therefore, the total pressure exerted by each of the individual gas molecules as they impinge upon the walls of the container at any particular moment in time.

Brownian Movement

Visual proof for the reality of heat motion of molecules, as postulated by the kinetic theory, is given by the phenomenon of Brownian movement (Fig. 1.4B). Minute carbon particles in aqueous suspension can be observed under suitable magnification to vibrate or oscillate continually. This effect is caused, as well as explained, by random collision of the freely moving water molecules with the much larger carbon particles, causing the latter to recoil. Similarly, under certain circumstances some inclusions within cells can be seen to exhibit Brownian movement.

The Mean Free Path

In a gas or liquid, as each molecule moves, the average distance that it travels before colliding with another molecule is called the *mean free path*. The collision between two molecules alters the mean free path of each molecule involved, as these fundamental constituents of matter are not perfectly elastic bodies (Fig. 1.4C). This effect is roughly analogous to the collision between two moving billiard balls, so that the direction in which each travels is changed following the impact.

The alterations in mean free path produced by random intermolecular collisions can explain such physiologically important phenomena as diffusion, viscosity, and heat conduction, as well as the effects of different temperatures and pressures on these physical properties.

The Solid and Liquid States

In a solid, the molecules are so firmly held by the mutual forces of attraction and repulsion of adjacent molecules that they are unable to move or oscillate very far from a mean, or equilibrial, position, as depicted in Figure 1.4D. Thus, a finely etched steel engraving plate can retain its shape indefinitely for all practical purposes, and a thin coating of gold plated on a silver spoon will remain upon the surface rather than diffuse into the interior.

At absolute zero (0 K, -273.2 C), all molecular and atomic motion ceases according to the kinetic theory of heat, since at this temperature no heat energy remains in the matter. At all temperatures above absolute zero, however, some heat energy is present; and there is no reason for the molecules which comprise a solid to maintain identical equilibrial positions at different temperatures. In actual fact, the molecules of a solid do oscillate about more widely spaced equilibrial positions as the temperature, or heat energy, of the matter increases. Hence, the common phenomenon of expansion takes place as the temperature rises, and, conversely, contraction occurs as the body cools. From this it is apparent that the molecular activity within any substance is strictly dependent upon the temperature of the matter.

If the thermal energy of certain solids (eg, iron) is raised sufficiently by adding enough heat, the intermolecular (or interatomic) binding forces are overcome, the shape of the body can be altered readily, and the substance is said to have become

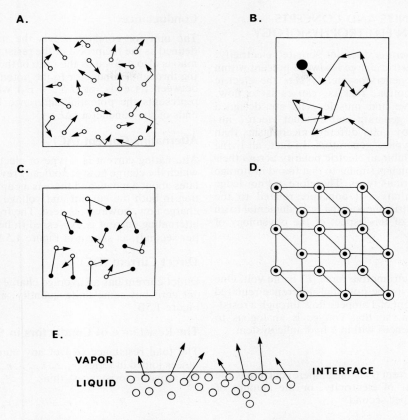

A.

B.

C.

D.

E.

VAPOR

LIQUID

INTERFACE

FIG. 1.4. Aspects of the kinetic theory of matter. **A.** In a gas, the individual atoms (or molecules) move in a random fashion and their kinetic energy is dependent upon the heat content of the system as measured by the temperature. The pressure exerted by a confined gas upon the walls of a container is proportional to the total number of random collisions between the gas particles and the container walls at any instant in time. **B.** Brownian movement provides visual proof of the random movement of the atoms (or molecules) that comprise a gas or liquid. Thus a minute carbon particle (dot) suspended in water is seen to oscillate or vibrate in a random fashion (pathway shown by arrows) when viewed under suitable magnification. This random motion is caused by elastic recoil from the unequal bombardment of the particle by water molecules. **C.** The mean free path of an atom (or molecule) of a gas or liquid is the average distance that each particle moves before it collides with another particle. As shown in the diagram, the elastic recoil between the particles causes them to change their direction of movement (arrows on black dots). Similarly, the particles bombarded (open circles) also change their direction of movement in response to the collisions (not shown). **D.** Schematic representation of the atoms (or molecules) in a solid. The individual particles (dots) are held in relatively fixed positions in the lattice by mutual forces of attraction and repulsion (depicted by lines) so that each particle can oscillate only to a very limited extent (circles) about a mean position. **E.** An atom (or molecule) can pass from the liquid to the vapor state (arrows) when the thermal energy of the system is elevated sufficiently.

plastic. Thus iron is more malleable, and hence is worked more readily, when red hot than when at ordinary temperatures. When still more heat is added to the plastic solid, the energy of the individual molecules finally becomes so great that their mutually attractive forces are completely overcome; the position of the molecules is constantly rearranged in a random manner, and the substance now has entered the liquid state.

When any liquid such as, for example, water or molten iron is heated sufficiently, a certain number of the molecules located at the interface between the liquid and the surrounding atmosphere become sufficiently agitated by thermal energy to pass from the liquid to the vapor (or gaseous) state (Fig. 1.4E). The quantity of heat necessary to produce

this effect is known as the *heat of vaporization* of the substance.

Similarly, under appropriate conditions, certain substances, such as ice or sulfur, are able to pass directly from the solid to the vapor state without passing through a liquid phase. This effect, known as *sublimation,* is the physical basis for the process whereby whole blood plasma is dried for future use without causing denaturation of the proteins. The plasma is frozen in a thin layer upon the inner surface of a suitable container, and, while frozen, it is subjected to a high vacuum. The water sublimes off, leaving the soluble plasma constituents behind in an unmodified state, to be reconstituted later by the addition of sterile distilled water.

PHYSICAL UNITS AND CONCEPTS EMPLOYED IN ELECTROPHYSIOLOGY

An electric generator does not "create" electricity. Rather, the electric charge is always present within the wire; the generator merely sets the electric charge into motion, that is, causes it to flow. Likewise, nerve and muscle cells are designed specifically to generate and conduct electric impulses, albeit by quite different mechanisms than are found in typical generators. In fact, all living cells at rest exhibit an electric polarity across their membranes which is similar to that found in storage batteries of various types. Therefore, a knowledge of the basic units and concepts related to the physics of electric phenomena is fundamental to an understanding of this aspect of the physiology of living systems.

Voltage

The unit of electromotive force (E) is the volt. One volt is defined as the potential difference required to make a current of 1 ampere flow through a resistance of 1 ohm, so that voltage is analogous to pressure differences within a hydraulic system.

Current

The unit for current (I) is the ampere; current is the rate of transfer of electricity, or the amount of charge moved per second.

Resistance

The unit for resistance (R) is the ohm. One ohm is that resistance through which a potential difference of 1 volt produces a current of 1 ampere.

Note that voltage, current, and resistance are wholly interdependent factors and that each of these variables is defined in terms of the other two.

The Relationship Among Current, Voltage, and Resistance (Ohm's Law)

Voltage (E), current (I), and resistance (R) are related mathematically in accordance with an expression known as Ohm's law:

$$E = IR; \text{ thus, } I = E/R, \text{ and } R = E/I$$

It is implicit in this expression that the voltage, or electromotive force E, represents the *potential difference* between both ends of the conductor ($E_1 - E_2$) or ΔE.

An easy way to remember Ohm's law is provided by the following diagram:

The unit covered by a finger is defined by the product or quotient of the remaining two units.

Conductance

The unit for conductance is the mho, which is defined as the reciprocal of the resistance, $1/R$. The mho is also defined as the ratio of the current flowing through a conductor to the potential difference between its ends, or ($I/E_1 - E_2$) where $E_1 - E_2$ represents the potential difference between both ends of the conductor (ΔE).

Alternating Current (AC)

Alternating current is a type of electric current in which the charge flow periodically reverses or oscillates about a mean, and there is a continuous variation in both the current and voltage. The average charge flow, however, is zero. The frequency of an alternating current is expressed in hertz (or cycles per second), as shown in Figure 1.5A.

Direct Current (DC)

Direct current has an average charge flow which is not zero but some finite quantity, as depicted in Figure 1.5B.

The Resistance of Conductors in Series

The total resistance (R_t) of any number of resistances joined in series ($r_1, r_2, r_3 \ldots r_n$) is the sum of the separate resistances, thus:

$$R_t = r_1 + r_2 + r_3 \ldots r_n$$

A schematic diagram of three resistances in series is given in Figure 1.6A.

The Resistance of Conductors in Parallel

The total resistance (R_t) of a number of conductors in parallel, whose individual resistances are $r_1, r_2, r_3 \ldots r_n$, is given by the expression:

$$1/R_t = 1/r_1 + 1/r_2 + 1/r_3 \ldots 1/r_n$$

For three terms, this expression becomes:

$$R_t = \frac{(r_1 r_2 r_3)}{(r_1 + r_2 + r_3)}$$

Three resistances in parallel are shown in Figure 1.6B.

Coulomb

The coulomb is the quantity unit of electricity (Q). Thus, one ampere can be defined as a current flow of 1 coulomb/sec.

The coulomb is defined as that quantity of electricity that will deposit a given amount of silver from a silver nitrate solution under specified conditions.

Farad

The farad is the practical unit of capacitance. The farad is equal to the capacitance (C) of a condensor

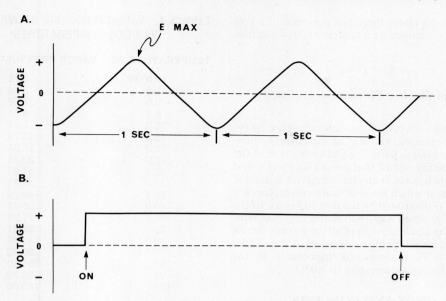

FIG. 1.5. **A.** Alternating current (AC) periodically reverses its charge flow about a mean (0); therefore, a continuous variation in current and voltage are present. If this reversal of charge occurs 120 times per second, then the frequency of the AC is 60 cycles per second (or hertz). (One Hertz = 1 Hz = 1 cps.) **B.** Direct current (DC) does not oscillate about a mean; rather, the average charge flow is some finite quantity, as depicted in the figure.

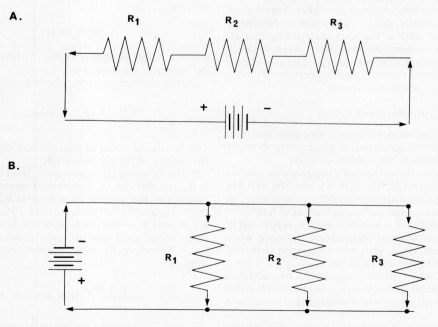

FIG. 1.6 **A.** Resistances (r_1, r_2, r_3) in series. The arrows indicate the direction of current flow to and from the current source (battery). The total resistance of the circuit is equal to the sum of the individual resistances ($R_T = r_1 + r_2 + r_3$). **B.** Resistances in parallel. The total resistance of this circuit (R_T) is equal to the sum of the reciprocals of the individual resistances, ie, $1/r_1 + 1/r_2 + 1/r_3$.

between whose plates there is a potential of 1 volt (V) when it is charged by 1 coulomb (Q) of electricity, thus:

$$C = Q/V$$

Behavior of Electric Currents in a Closed Circuit

In any closed network of conductors that are carrying electric currents, the sum of the currents flowing into any branch point is equal to the sum of the currents flowing out of that point (Kirchoff's first law). A branch point in an electric circuit is defined as any point at which three or more conductors are joined; this is illustrated by the dots in Figure 1.6B.

Stated another way, Kirchoff's first law indicates that the algebraic sum of all the electric forces flowing to a single point in a network is zero. This statement is of fundamental importance to the theory of electrocardiography (p. 603).

PHYSICAL UNITS AND CONCEPTS EMPLOYED IN RESPIRATORY PHYSIOLOGY

As noted earlier, a gas is a state of matter in which the molecules are unrestricted by cohesive or other binding forces. Therefore, gases have neither definite shape nor volume, and, unlike liquids, gases are readily compressible.

A vapor is the gaseous state of a substance which is usually a liquid or solid at room temperature. Thus, one speaks of water *vapor* and oxygen *gas*. The vapor pressure of a substance is that pressure exerted when a solid or liquid is in equilibrium with its own vapor, and it is a function of the type of substance as well as the temperature (Table 1.4). The pressure exerted by any gas or vapor is due to the kinetic energy of the molecules, and this in turn is directly proportional to the temperature; ie, it is a manifestation of thermal energy.

Standard Conditions for Gases

Since a gas has neither a definite shape nor volume, certain standard conditions are necessary in order to define precisely the volume of a gas.

Generally, the volume of a dry gas is calculated to a temperature of 0 C (273 K) and 760 mm Hg pressure (the atmospheric pressure at sea level, or 1 atm. These conditions are abbreviated STPD.

The volume of a gas may also be calculated at body temperature and the ambient pressure when the gas is saturated with water vapor at that temperature and pressure (BTPS).

Lastly, the volume of a gas may be corrected to the ambient temperature and pressure while saturated with water vapor under these conditions (ATPS).

These different methods for correcting a given volume of gas to a standard set of conditions are used for different purposes; the important fact is that, in order to define the volume of a gas accurately, those conditions must be specified clearly.

Table 1.4 VAPOR PRESSURE OF WATER AT VARIOUS TEMPERATURES[a]

TEMPERATURE, °C	VAPOR PRESSURE, mm Hg
−10.0	2.15
−5.0	3.16
0.0	4.58
5.0	6.54
10.0	9.21
15.0	12.79
20.0	17.54
25.0	23.76
30.0	31.82
35.0	42.18
37.5	48.36
40.0	55.32
45.0	71.88
50	92.51
60	149.38
70	233.70
80	355.10
90	525.76
100	760.00

[a]*These pressures are accurate for water in contact with its own vapor. Note that at its freezing point, water still exerts a significant, albeit slight, vapor pressure, and that at 100 C the vapor pressure equals atmospheric pressure at sea level. The latter conditions define the boiling point of pure water.*

The Equation of State of an Ideal Gas

The fundamental relationship among pressure (P), volume (V), and temperature (T) of an ideal gas is written:

$$PV = nRT,$$

where n is the number of moles of the gas and R is the universal gas constant.

It is obvious from the above expression that:

$$\frac{PV}{T} = nR = \text{a constant}$$

The numerical value of the constant depends upon the weight of the gas involved.

If n is set equal to one mole, then the constant is R, and this can be determined quantitatively by measuring the volume of a known weight of gas under standard conditions, ie, 273.2 K and 760 mm Hg pressure. Under such standard conditions, one mole of an ideal gas occupies a volume of 22.4 liters. Therefore, by substitution in the above equation:

$$R = \frac{P_0 V_0}{T_0} = \frac{1 \times 22.4}{273.2} \text{ liter atm per deg per mole}$$

$$= 0.082 \text{ liter} \times \text{atm} \times \text{deg}^{-1} \times \text{mole}^{-1}$$

Obviously, the numerical value for R depends upon the units employed for P, V, and T. If V is in milliliters, the value of R is 82.0. Using other units of pressure and volume, the following values are obtained:

$$R = 8.3 \times 10^7 \text{ ergs} \times \text{deg}^{-1} \times \text{mole}^{-1} \text{(cgs system)}$$
$$= 8.3 \text{ joules} \times \text{deg}^{-1} \times \text{mole}^{-1} \quad \text{(mks system)}$$
$$= 1.9 \text{ cal} \times \text{deg}^{-1} \times \text{mole}^{-1}$$

The units for R must be stated clearly in order to give any meaning to the gas constant.

Since $PV/T = $ a constant, it follows that the relationship between volume, pressure, and temperature also may be written:

$$\frac{P_1 V_1}{T_1} = \frac{P_2 V_2}{T_2}$$

This equation states that, at a constant temperature, the volume occupied by a fixed quantity of gas is inversely proportional to the pressure to which the gas is subjected (Boyle's law), and that at constant pressure, the volume occupied by a fixed mass of gas varies in direct proportion to the absolute temperature (Charles' law or Gay-Lussac's law).

This expression is reasonably accurate in calculations involving volumes of air, oxygen, or nitrogen at moderate pressures only. The equation of state of an ideal gas also is important in the calculation of osmotic pressure, since dilute osmotically active solutions behave in accordance with the gas laws (p. 25).

The Law of Partial Pressures

The total pressure exerted by a mixture of gases (P_t) is equal to the sum of the separate pressures $(P_1 + P_2 + P_3 \ldots P_n)$ which each gas would exert individually if it alone occupied the entire volume (V) (Dalton's law of gases). Thus:

$$P_t V = V (P_1 + P_2 + P_3 \ldots P_n)$$

The partial pressure (P_p) of any gas (G) in a mixture of several gases may be calculated readily from the total pressure of that mixture (P_t) and the percentage concentration $(\%G)$ of any particular gas in the mixture:

$$P_p = \frac{(P_t)(\%G)}{100}$$

In respiratory physiology, the vapor pressure of water within the lungs at body temperature (approximately 50 mm Hg) affects the actual respiratory gas mixture and is dealt with mathematically in accordance with Dalton's law.

The Solution of Gases in Liquids

The mass of a slightly soluble gas that dissolves in a given mass of liquid at a specific temperature is directly proportional to the partial pressure of that gas (Henry's law).

This law does not hold in two situations: first, if there is a chemical reaction between the gas phase and the liquid phase (or any substance dissolved in the liquid). For example, carbon dioxide in contact with a sodium hydroxide solution will be absorbed regardless of its partial pressure. Second, chemical combinations may be dependent upon the partial pressure, but *not* in direct proportion to this factor — for example, the combination of oxygen with hemoglobin.

Number of Molecules in a Gas or Other Substance

At the same pressure and temperature, equal volumes of different gases contain the same number of molecules (Avogadro's law). One mole (or grammolecular weight) of any compound contains the same number of molecules (Avogadro's number, N, equals 6.02×10^{23}).

A Note on Pressure and Tension

The words "pressure" and "tension" are used synonymously in many older texts of physiology with reference to the pressure or partial pressure of gases, especially when they are in solution. This terminology is ambiguous, since the term "tension" implies tautness, stretching, or strain, such as surface tension or muscle tension. The term "pressure," defined as a force per unit area, will be employed throughout this text, whether with reference to free gases or those dissolved in liquids.

Although a volume of gas dissolved in a liquid does not exert physical pressure in the same sense that it does when in the free state, it is treated mathematically as though it *were* in the free state under specific conditions, and thus was able to exert an independent pressure or partial pressure. It is also imperative to avoid confusing the *partial pressure* of a gas in the free state (or in solution) with the *concentration* (eg, milliliters of gas per 100 milliliters or volumes percent) of that gas in a mixture of gases or dissolved in solution.

MISCELLANEOUS CHEMICAL UNITS AND CONCEPTS

The Atom

An atom is defined as the smallest particle of any chemical element that can enter into a chemical combination (Appendix, Table 2, p. 1262). All chemical compounds are composed of atoms; the difference between various compounds is attributable to the nature, number, and arrangement of the constituent atoms.

The atom consists of a central *nucleus* and a number of negatively charged *electrons* that circle about the nucleus in orbits in a manner that is similar to the planets' circling about the sun. The important elemental constituents of the human body together with certain of their chemical properties are listed in Table 1.5.

Table 1.5 PRINCIPAL ELEMENTS ASSOCIATED WITH THE HUMAN BODY

ELEMENT	SYMBOL	ATOMIC NUMBER	ATOMIC WEIGHT[a]	ORBITAL ELECTRONS: SHELL 1 2 3 4 5
Hydrogen	H	1	1.00	1
Carbon[b]	C	6	12.01	2, 4
Nitrogen	N	7	14.01	2, 5
Oxygen	O	8	16.00	2, 6
Sodium	Na	11	23.00	2, 8, 1
Magnesium	Mg	12	24.31	2, 8, 2
Phosphorus	P	15	31.00	2, 8, 5
Sulfur	S	16	32.1	2, 8, 6
Chlorine	Cl	17	35.45	2, 8, 7
Potassium	K	19	39.10	2, 8, 8, 1
Calcium	Ca	20	40.10	2, 8, 8, 2
Manganese	Mn	25	54.93	2, 8, 13, 2
Iron	Fe	26	55.85	2, 8, 14, 2
Cobalt	Co	27	58.93	2, 8, 15, 2
Copper	Cu	29	63.54	2, 8, 18, 2
Zinc	Zn	30	65.37	2, 8, 18, 2
Iodine	I	53	126.90	2, 8, 18, 18, 7

[a] *Atomic weights have been rounded off to two decimal places. The variations in atomic weights are due to variations in the natural isotopic composition of the elements.*
[b] *The atomic weight of the most common isotope of carbon (12.0000) is the standard upon which all atomic weights are based.*

The nucleus contains the principal mass of the entire atom; hence, its mass is considered that of the atom itself. The nucleus consists of a number of positively charged particles known as *protons,* whose number always equals the *atomic number* (Z), as well as a number of particles having no charge, called *neutrons.* The number of neutrons present in any atom is equal to the difference between the atomic weight (A) and the atomic number (Z).

All molecular weights are measured in *daltons.* One dalton equals the mass of one hydrogen atom or 1.67×10^{-24} gm. The nucleus of ordinary hydrogen (H) consists of one proton and one neutron.

Atomic weights of the elements are based arbitrarily upon the atomic weight of the most common natural isotope of carbon, or 12.0000. According to this scale, the atomic weight of the commonest isotope of hydrogen is 1.0000, or one dalton = 1.67×10^{-24} gm.

The atomic nucleus carries an integral number of positive charges (ie, an integral number of protons), each having a charge of 1.6×10^{-19} coulomb; likewise, each orbital electron carries one equivalent negative charge of 1.6×10^{-19} coulomb. Since the number of orbital electrons is equal to the number of protons in the nucleus (ie, the number of electrons is equal to Z), the entire atom has a net charge of zero.

The electrons are arranged in successive shells about the nucleus, and each of these shells contains a particular number of electrons (Table 1.5; Appendix, Table 2, p. 1262). Therefore, the extranuclear structure of the atom is peculiar to each element, and this, in turn, confers upon each element its characteristic properties. The electrons that comprise the innermost shells are tightly bound to the nucleus, whereas those present in the outer shells are less tightly bound, so that the latter are responsible for the chemical properties of the element.

Isotopes are forms of a given element that have the same atomic number (Z) but different mass numbers. Isotopes of a particular element have the same number of protons in their nuclei, but differ as to the number of neutrons that are present in the nucleus. An isotope may be stable (ie, nonradioactive) or radioactive. In the latter instance, the isotope emits radiation (or energy), for example, in the form of electrons or gamma rays.

Valence

The valence of an atom is that property of an element that is measured by the ability of the atom to donate or accept electrons. The valence electrons of an atom are those that are gained, lost, or shared in chemical reactions, and these electrons are usually found in the outer shell of orbital electrons in the atom. For example, in the reaction, $NaCl \rightleftharpoons Na^+ + Cl^-$, when crystalline salt is dissolved in water, sodium loses one valence electron (e^-), becoming positively charged, whereas the chloride atom acquires this electron and becomes negatively charged. Thus both sodium and chloride have a valence of 1.

Equivalent Weight (or Combining Weight)

Atoms of elements that enter into chemical combination to form molecules or compounds always do so in quantities that bear a simple multiple relationship to one another, ie, in quantities that are proportional to their equivalent weights (Dalton's law).

The equivalent (or combining) weight of an element or a chemical radical is its atomic weight divided by its valence. Thus, the equivalent weight for oxygen is 16/2 or 8, and that for calcium is 40.1/2 or 20.05.

The Gram Atom

If the atomic weight of an element is expressed in grams, it is called the *gram-atomic weight* of that element. This is sometimes referred to as *gram atoms* of the element.

The *gram equivalent* is defined as the mass of a substance that displaces, or reacts with, 1.0 gm of hydrogen or 8.0 gm of oxygen.

The Mole and Molecular Weight

One mole is defined as the weight of a compound that is numerically equal to the molecular weight of that compound. A gram mole is the weight in grams equal to the molecular weight.

The molecular weight of a chemical compound is given by the sum of the atomic weights of all of the individual atoms in that compound; for example, the gram-molecular weight of common table salt (NaCl) is 23 + 35.5 = 55.5 daltons.

Concentration

The concentration of a substance is defined as the quantity of that substance (the solute) which is dissolved in a unit volume of solvent. The concentration of solutes commonly is expressed in three ways:

1. Mass Per Unit Volume: (a) Gram (or milligram) of solute per 100 ml solvent = a gram (or milligram) percent solution. (b) Gram (or milligram) of solute per 1,000 ml = a gram (or milligram) per liter solution. (c) If the solute is itself a liquid, the concentration may be expressed on a volume solute per volume of solvent. For example, 30 ml of absolute alcohol mixed with 70 ml of water yields a 30 percent solution of alcohol in water.

2. Molar and Molal Solutions: When the mass in grams of one mole (or one gram-molecular weight) of a compound is dissolved in 1 liter of solvent, then a one *molar* (1 M) solution results. A *molal* solution, on the other hand, contains one gram-molecular weight of solute dissolved in 1,000 gm of solvent. For all practical purposes, these two concentration units are identical.

3. Normal Solutions: A one normal (1 N) solution contains one gram-molecular weight of the solute divided by its hydrogen equivalent (or the valence of the substance) in 1 liter of solvent. That is, a 1 N solution contains one equivalent per liter, and equal volumes of two solutions with the same normality are said to be chemically equivalent to each other, ie, the elements in each solution combine with each other in equivalent or similar quantities.

In physiology, the concentrations of inorganic constituents in the body fluids usually are expressed in milliequivalents per liter (mEq/liter). The chemical equivalent of an element (ion) or compound in solution is obtained by dividing its concentration in grams per liter by the atomic (or molecular) weight and multiplying by the chemical valence. In order to obtain the result in milliequivalents per liter directly, instead of in equivalents per liter, milligrams rather than grams are substituted in the following expression:

$$mEq/liter = \frac{mg \text{ of substance/liter}}{atomic \text{ (or molecular) weight}} \times valence$$

For example, the blood calcium level normally is about 10 mg per 100 ml (or 100 mg/liter). The atomic weight of this element is 40.1, and the valence is 2 (Ca^{++}). Therefore, the calcium concentration is 100/40.1 × 2 = 4.99 mEq/liter, or 0.0049 Eq/liter.

Similarly, sodium is a monovalent cation (Na^+) that is present in blood serum at a concentration of approximately 325 mg/100 ml. Therefore, the concentration of sodium in the blood is 3,250/23 × 1 = 141 mEq/liter.

The concentrations of substances that are expressed in milliequivalents per liter may be converted readily into milligrams per 100 ml, or milligram percent, simply by rearranging the above expression:

$$mg/liter = \frac{mEq/liter \times atomic \text{ weight}}{valence}$$

The value obtained by use of this relationship is converted into milligram percent by dividing by 10.

Electrolytic Dissociation or Ionization (Arrhenius Theory)

If an acid, base, or salt is dissolved in water (or other solvent that permits dissociation), a portion of all of the solute is broken up into charged particles called *ions*. Some of these ions have a positive electric charge, and these migrate to the negative pole or cathode in an electric field; hence, they are termed *cations*. Concomitantly, an equivalent number of particles develop a negative charge, and thus migrate to the positive pole or anode in an electric field; hence, they are termed *anions*.

Cations are formed when an electrically neutral metallic element (atom) or molecule *loses* one or more negatively charged valence electrons; anions are formed when those electrons are *gained* by an element or chemical radical. (Compare oxidation and reduction below.)

Acids

An acid usually is considered to be a substance that dissociates upon solution in water to yield one or more hydrogen ions, H^+. However, in more general terms, an acid is defined as *any substance that can donate one or more protons,* also indicated by the symbol H^+ (Brønsted–Lowry concept).

The hydronium ion, H_3O^+, formed by the reaction of a proton with a water molecule, is the actual acid radical. Thus: $H^+ + H_2O \rightarrow H_3O^+$. However, by convention, the symbol H^+ is used to indicate an acid, although it is implicit in this notation that a hydronium ion is involved, because a proton cannot exist in the free state.

Bases

A base may be considered to be any substance that dissociates upon solution in water to yield one or more hydroxyl ions, OH^-. According to the Brønsted–Lowry concept however, a base is defined as *any substance that can accept one or more protons*. Hence, in the equation $H^+ + H_2O \rightarrow H_3O^+$, the water acts as a base in the formation of hydronium ion since it accepts a proton.

The Brønsted–Lowry concept of acids and bases is considerably broader than the older definitions for these substances, and thus is of considerable importance in biochemical reactions, as will be noted in subsequent discussions.

Salts

A normal salt is any compound that dissociates into ions (other than hydrogen or hydroxyl ions) upon solution in water. Salts are derived from acids by replacing all or part of the acid hydrogen by a metal or a radical that behaves as a metal.

Oxidation

The process of oxidation involves a *loss* of one or more electrons by an atom, ion, or compound so that the positive valence of the substance which lost the electrons is increased proportionally to the number of electrons involved.

In an electrolytic cell (battery), oxidation takes place at the anode, or positive terminal.

Reduction

The process of reduction involves a *gain* of one or more electrons by an atom, ion, or compound, so that the positive valence of the substance that gained the electrons is *reduced* in proportion to the number of electrons involved.

In an electrolytic cell, reduction takes place at the cathode, or negative terminal.

The processes of oxidation and reduction are wholly inseparable. Therefore, when one substance is oxidized, another is reduced simultaneously.

Oxidation–reduction reactions are of fundamental importance to energy transfer processes in living systems.

Catalysts and Enzymes

A catalyst is any substance whose presence accelerates the velocity of a chemical reaction. This effect usually is caused by the lowering of the free energy of activation required for that reaction to take place. Thus, a catalytic agent enables a reaction to proceed more rapidly and under milder conditions (eg, at a lower temperature and/or pressure) than would be possible in its absence. At the end of the reaction, the catalyst remains unchanged, either quantitatively or qualitatively.

Catalysts may be organic or inorganic substances. In plants and animals, *enzymes* are the principal organic catalysts. These substances consist of protein molecules of various degrees of complexity. Not only are enzymes synthesized by the living systems themselves, but they are also of fundamental importance in regulating the host of chemical reactions that constitute the properties of life itself within the same systems. This regulatory activity may range from highly specific (in which one enzyme regulates one reaction) to nonspecific (in which one enzyme regulates a number of reactions). Furthermore, the activity of particular enzymes may be manifest at either an intracellular or an extracellular level.

Certain important properties of these essential biocatalysts will be considered in more detail in Chapters 2 and 14 (pp. 44, 908).

Bonding Forces in Matter

There are three primary and two secondary physical forces which give cohesive or binding properties to matter, whether that matter be a solid, a liquid, or a gas. All of these forces are electric in nature, and are graded in strength from the strongest to the weakest as outlined below.

COVALENT BONDS. The covalent bond is formed when two electrons per valence bond are *shared* among the atoms that make up a compound. The formation of covalent compounds involves *no actual transfer* of electrons between the atoms, such as takes place during ionization or electrostatic bond formation.

At ordinary temperatures, covalent bonds are relatively strong; hence, the compounds formed by such bond forces are stable.

If the valence electrons are represented by dots, the diatomic molecules of the common elementary gases may be written as follows: oxygen, $:\!\ddot{O}\!:\ddot{O}\!:$; hydrogen, $H\!:\!H$; and nitrogen, $:\!N\!:\!N\!:$.

Similarly, water may be represented as $H\!:\!\ddot{O}\!:\!H$.

Covalent bonds also are denoted by one or more dashes, for example, C—C.

The covalent bond will be discussed further on page 26.

ELECTROSTATIC (ELECTROVALENT) BONDS. An electrostatic bond is formed when two or more atoms are held together by the actual *transfer of one or more valence electrons* from one of the atoms to the other. In the sodium chloride molecule, sodium yields an electron to the chlorine atom and the two ions are held rigidly together in the solid or crystalline state by an electrostatic attraction between the Na^+ and Cl^- by means of this electron in the crystal lattice.

Since NaCl is a highly polar (electrically charged) electrolyte — that is, the distribution of electrons is not uniform within the molecule — the

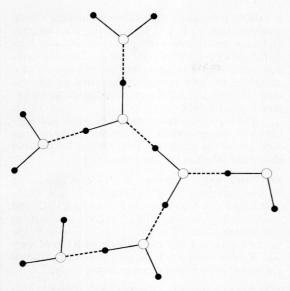

FIG. 1.7. Hydrogen bonds (dashed lines) are formed between individual water molecules by the attraction of the positively charged hydrogen atoms (dots) to the negatively charged, electron-dense oxygen atoms (open circles). See also Figure 1.9.

electrovalent bond can be dissociated readily in polar substances such as water. A nonelectrolyte such as glucose, on the other hand, is termed a nonpolar compound because it has a fairly uniform electric charge distribution about its atoms, and it does not ionize appreciably upon solution in water. Therefore, a glucose solution is a poor conductor of electricity, in contrast to a solution of a highly electrolytic compound such as salt.

HYDROGEN BONDS. The hydrogen bond is a relatively weak linkage that is formed whenever the hydrogen of one molecule is bound to a strong electronegative (electron-dense) atom of another element. Hydrogen can form only one covalent bond since it has only one orbital electron; however, as this atom is very small, its electrostatic field is very intense, and it will associate with such electron-dense atoms as oxygen and nitrogen. Thus, water molecules readily associate with each other by hydrogen bond formation, as illustrated by the dashed lines in Figure 1.7.

The hydrogen bond is weaker than either the covalent or the electrostatic bond.

HYDROPHOBIC INTERACTIONS (BONDS). In a hydrophobic interaction, the highly polar solvent (water) molecules repel the nonpolar (or uncharged) groups on other molecules, forcing them to orient themselves in juxtaposition to one another so that micellar structures result (Fig. 1.8).

Physiologically important examples of this type of bonding are found in the bimolecular lipid layers of the plasma membrane, as well as in fat-and-water emulsions.

VAN DER WAALS FORCES. The weak van der Waals forces are attractions that take place between electrically neutral atoms and molecules. These forces develop because of the electric polarization *induced* in each of the particles by the presence of other particles. Van der Waals forces are effective at relatively greater distances than either covalent or electrostatic bonds, and they are important to the biologic interactions of macromolecules such as proteins.

There are fundamental and special physical forces involved in the constitution and interactions of matter other than those discussed above. However, covalent, electrostatic, and hydrogen bonds are the principal forces involved in the constitution and functions of living systems.

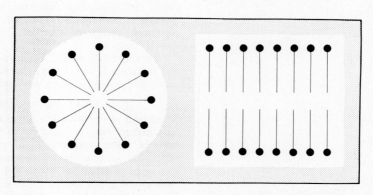

A.   **B.**

FIG. 1.8. Hydrophobic interactions result in the formation of micellar structures **(A)** or bimolecular layers **(B).** Such bonds are formed when charged water molecules (dots in shaded area) attract the polar groups of larger molecules (large dots) while simultaneously repelling the nonpolar groups of the same molecules (indicated by lines). The result is to force the molecules into a regular spherical (or micellar) orientation or into a bimolecular layer in which the free energy of the system is at a minimum.

PHYSICAL AND CHEMICAL PROPERTIES OF WATER

Although water is so commonplace in everyday life as to be taken almost for granted, a review of some of the elementary physical and chemical properties of this substance is necessary prior to the discussion of a number of processes which cannot take place except in aqueous systems and which are essential for the maintenance of life.

The water molecule (denoted by the symbols H_2O, HOH, or $H:\overset{..}{O}:H$) consists of two atoms of hydrogen, each bonded through a single valence electron to one atom of oxygen. Taking the oxygen atom as the apex, the two hydrogen atoms are bonded at an angle of about 104° with respect to one another, as shown in Figure 1.9. The distance between the oxygen and hydrogen nuclei is approximately 0.97 Å. Thus a single water molecule is an extremely small entity capable of rapid and continuous exchange among the various compartments within the body.

Physical Properties of Water

THE DIELECTRIC CONSTANT. Because of the unequal (asymmetric) distribution of electric charges within the water molecule (Fig. 1.9), water is said to possess a very strong permanent dipole moment; that is, it is a highly polar (or charged) molecule, and this property gives rise to a very high dielectric constant for pure water. The dielectric constant, in turn, is a measure of the capacity of a substance to store an electric potential (or electric charge) when the substance is subjected to an external electric field.

The high dielectric constant of water accounts for its enormous solvent and ionizing properties, as well as for its relatively poor ability to conduct electricity when in the pure state.

MELTING AND BOILING POINTS. Considering its low molecular weight (18.0), water is a stable compound that has relatively low melting (0 C) and relatively high boiling (100 C) points. The melting and boiling points of water, in fact, are used as the primary standards for comparison with the properties of other compounds, as they provide the physical basis for the Celsius temperature scale.

THERMAL (HEAT) CAPACITY. The term "thermal capacity" signifies the quantity of heat (in calories) required to raise the temperature of a given mass of water 1 C at 760 mm Hg barometric pressure. The heat capacity of air-free water is 1.0 cal/gm/°C. Thus, water serves as the primary standard for definition of the gram calorie in cgs units.

A biologically important consequence of the high thermal capacity of water is that it takes a relatively large quantity of heat to raise the temperature of a given mass of water appreciably. Conversely, cooling this same mass of water to its prior temperature requires the removal of a similar large amount of heat. Consequently, living organisms, including man, are protected to some extent from sudden and extreme fluctuations in environmental temperature.

On a vastly larger scale, the heat contained in the mass of water present in the world's oceans exerts an enormous influence upon the global climatic conditions. So much so, in fact, that one commonly refers to "tropical," "temperate," and "polar" regions.

In summary, water, because of its large thermal capacity, is an effective thermal buffer not only on an individual but on a worldwide scale.

THE LATENT HEAT OF VAPORIZATION. The latent heat of vaporization refers to the quantity of heat energy that is required to vaporize a given mass of water from the liquid to the vapor state *without a change in its temperature*. The unit employed to define the latent heat of vaporization is calories per gram (cal/gm). Thus, 1 gm of water at 100 C requires 0.54 Cal/gm (or 540 cal/gm) for conversion from the liquid to the vapor state, and a like amount of water at 40 C (about body temperature) requires 0.6 Cal/gm (600 cal/gm) to vaporize.

This property of water is of major significance in the maintenance of body temperature, since the evaporation of any quantity of water as sweat from

A.

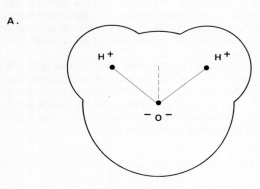

B.

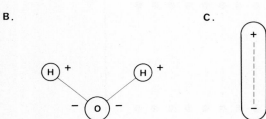

C.

FIG. 1.9. Diagrams indicating the strongly dipolar nature of the water molecule. **A.** The hydrogen atoms are shown bound to the oxygen atom at an angle of 104°. The nuclei of the hydrogen and oxygen atoms are separated by a distance of 0.965 Å (solid lines), and the large semicircles represent the approximate relative sizes of the atoms. The vertical dashed line represents the vector giving rise to the dipole moment. **B** and **C** illustrate two additional ways of portraying the water molecule.

the body surface is attended by a considerable heat loss (p. 1008).

DENSITY AND THE ANOMALOUS EXPANSION OF WATER. The density of water attains a maximum value of 1 gm/ml at about 4 C (actually, 3.98 C). Pure water at this temperature serves as the primary standard for density.

Normally, the density of matter is related inversely to the temperature (p. 4). That is, the density of most substances *decreases* as temperature *increases*. Hence, the expansion of water with a concomitant decrease in its density below 4 C is termed anomalous.

Since the density of pure (gas-free) water at 4 C is somewhat greater than at its freezing point (0 C), ice floats. Oceans and other large bodies of water generally do not freeze completely, a fact of major biologic significance when viewed on a worldwide scale.

WATER AS A PRIMARY STANDARD: SPECIFIC GRAVITY AND VISCOSITY. Water not only serves as the primary standard in the metric system for the definition of mass, volume, temperature, the calorie, and density, but it is also used as a reference standard in the determination of specific gravity and viscosity.

The term *specific gravity* is defined as the ratio of the mass of a body to the mass of an equal volume of water at 4 C (or other specified temperature).

The term *viscosity* is defined as the resistance to flow (or shear) when a force is applied to a liquid. The absolute unit for the measurement of viscosity is the poise, named after Poiseuille, who determined experimentally that the velocity of flow (V) of a liquid through a tube per unit of time (t) is directly proportional to the pressure differential between the two ends of the tube (p) and the fourth power of the radius of the tube (r^4), but inversely proportional to the length of the tube (l) and the viscosity (η). Mathematically, this relationship (Poiseuille's law) is expressed:

$$V/t = \frac{\pi p r^4}{8 l \eta}$$

The symbol π is a constant (3.1416), and when p is expressed in dynes per square centimeter, r and l in centimeters, and t in seconds, then the viscosity is expressed in dyne-seconds per square centimeter, or poises. The centipoise is 0.01 poise.

The viscosity of water at 20 C is 0.01002 poise, and this figure is used as an absolute value for the calibration instruments designed to measure viscosity (viscosimeters).

The Poiseuille equation may be simplified for practical use in understanding certain aspects of cardiovascular physiology, so that, by analogy to Ohm's law (p. 8), it becomes: $R = P/F$, where R is the resistance to flow, P is the pressure difference between both ends of a blood vessel (ΔP or $P_1 - P_2$), and F is the blood flow per unit of time. To further extend the analogy, this relationship can be remembered by the device:

It should be stressed that Poiseuille's law holds only for *streamline* and not for *turbulent* flow (p. 1008).

MISCELLANEOUS PHYSICAL PROPERTIES OF WATER.

Compressibility. Unlike gases, water is a fluid which is relatively incompressible; that is, its volume is not altered significantly upon the application of large external forces, and, as noted earlier, the pressure exerted at any point on a confined liquid is transmitted equally and undiminished in all directions throughout the liquid (Pascal's law). These facts are of significance in cardiovascular physiology.

The Coefficient of Thermal Expansion. Water possesses a low coefficient of thermal expansion. Hence, the volume of a particular mass of water does not increase appreciably with a moderate rise in temperature; thus living cells or tissues are not in danger of rupture when there are moderate temperature fluctuations.

Surface Tension. The property of surface tension exhibited by a liquid may be considered to be caused by an unequal attraction among the molecules lying at the surface and those lying deeper within the liquid (Fig. 1.10). Thus, at a water–air interface, the surface tension is manifest by physical properties resembling those of an elastic "skin." This is apparent when one observes soap bubbles or very small droplets of water or mercury; the bubbles or droplets assume the smallest possible surface area and thus become spherical — ie, the skin shrinks.

The surface tension of water is modified readily by solutes of various types, and this physical property assumes great importance in certain aspects of respiratory physiology.

Surface tension is measured in dynes per centimeter or ergs per square centimeter.

Chemical Properties of Water

Water is the most abundant constituent of all living systems, including man. Because of the unique physical and chemical properties of this liquid, it serves as a solvent for mineral ions and a multitude of other substances, as well as a dispersion medium for macromolecules and other colloidal material found in protoplasm.

Cellular water may be either free or bound. Free water amounts to about 95 percent of the total quantity present, and this fraction serves as the

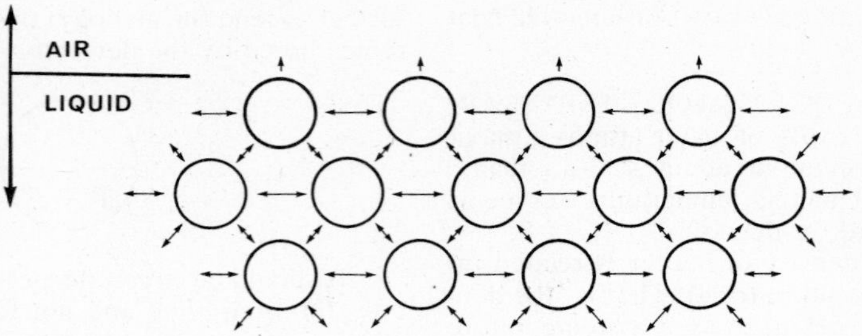

FIG. 1.10. Representation of the forces involved in surface tension at an air-water interface (horizontal line at left). Water molecules (open circles) at the surface of the liquid are bonded strongly to each other and to deeper-lying water molecules within the mass of the liquid (represented by light lines and arrows) so that the net force of attraction among the surface molecules is together and downward, rather than toward the gas phase (small vertical arrows above molecules at interface). This cohesive property of the molecules at the air-liquid interface results in the phenomenon of surface tension.

solvent and dispersion medium. The bound water (roughly 5 percent of the total) is held loosely within the structure of macromolecules, such as proteins and the mineral substance of bone (apatite), by hydrogen bonds and other electric forces. However, cellular water in both free and bound forms should not be considered to be sequestered in rigid compartments; rather, the water within each category is capable of rapid, dynamic exchange into the other.

There is also some evidence that water per se has a definite, although pliant, "structure" within the cell as compared with water found outside of living matter.

Water is indispensable for *all* metabolic activity in living organisms, and it may be formed chemically as the result of certain metabolic processes. Water may be considered to act as a solvent-catalyst in many biochemical processes, and it participates in many enzymatic reactions. This fact often is assumed tacitly rather than stated specifically.

Thus water, regardless of the level of organization with the organism at which it is considered, plays a dynamic and active role as a major participant in, as well as constituent of, living systems.

Following are some physicochemical properties of reactions involving aqueous systems.

THE COLLOIDAL STATE. A colloidal system is a heterogeneous mixture of at least two components. One component, known as the *dispersed phase,* is scattered (or suspended) among the molecules of the other component, which is called the *continuous phase,* or the *dispersion medium.* Sometimes colloids are referred to as a "fourth state of matter."

Colloidal dispersions are classified as such on the basis of particle size; the particle sizes which comprise the colloidal state range from about 1μ to 0.5μ. A colloid usually is stable in that the particulate matter (or dispersed phase) does not settle out of or separate from the continuous phase. The size of particles within a colloid is so large that diffusion of the dispersed phase proceeds very slowly; further-

more, the dispersed phase does not pass through an ultrafilter or dialyzing membrane. These properties of colloids have important physiologic consequences; eg, plasma proteins normally do not pass through the capillary membrane (an ultrafilter).

Two types of colloid are important physiologically: the *sol,* which is a dispersion of a solid in a liquid, and the *emulsion,* which is the dispersion of a liquid in a liquid. Water is the only dispersing medium in living systems.

Sols may be either hydrophilic (water attracting) or hydrophobic (water repelling), depending upon the electric charge on the particles as well as the dipolar nature of the water molecule. The plasma proteins are an excellent example of a hydrophilic colloid.

Emulsions are colloidal systems in which both the dispersed phase and the dispersion medium are liquids; the one in excess usually is considered as the dispersing medium. Oil in water and water in oil are two kinds of emulsion. When these substances are shaken together in various proportions, the emulsions so formed are not stable and eventually will separate into their individual components. An *emulsifying agent* must be added in order to stabilize the system. For example, the digestion and absorption of fat by the intestine is facilitated by emulsification: The dispersed fat droplets in the aqueous dispersing phase are stabilized by the presence of a small quantity of soap, which is formed by the alkaline digestive juice acting upon some of the fat that is present.

A colloidal dispersion of various materials, whether as a sol or as an emulsion, increases enormously the total surface area relative to the mass of the individual particles; this, in turn, greatly facilitates any chemical reactions that may be involved subsequently. Thus, emulsification of dietary fats exposes a relatively large surface area to enzymatic digestion.

ACIDS, BASES, AND NEUTRALIZATION. By definition, an acid is a substance that can supply a hydrogen ion (H_3O^+) or simply a proton (H^+), and a base is a substance that can accept a proton or yield a hydroxyl ion (OH⁻). An acid–base reaction, therefore, involves the transfer of a proton from an acid to a base according to the fundamental equation: acid $\rightleftharpoons$ base + proton. The strength of an acid is a quantitative measure of its tendency to lose protons, whereas the strength of a base is a quantitative measure of its tendency to accept protons (Brønsted–Lowry concept).

A *strong electrolyte* is an electrovalent compound which can dissociate completely into ions in aqueous solution. A *weak electrolyte*, on the other hand, does not dissociate completely under similar conditions. Thus a solution of hydrochloric acid in water is a strong acid as well as a strong electrolyte, but carbonic acid is a weak acid as well as a weak electrolyte.

It should be reiterated that the hydrogen ion is not the proton per se (H⁺). Actually, the hydrogen ion is the proton in combination with one or more molecules of water, or H_3O^+, and this hydronium radical is the acid. But as noted earlier, by convention, the hydronium ion (hydrogen ion) usually is denoted by the symbol H⁺.

A base is a compound containing a hydroxyl radical (−OH⁻) which dissociates in aqueous solution, giving the hydroxyl ion (OH⁻). Hence, the reaction between an acid (proton donor) and a base (proton acceptor) may be expressed by the equation:

$$H_3O^+ + OH^- \rightarrow 2H_2O$$

The ionization of water is indicated by the reverse of the above reaction:

$$2H_2O \rightarrow H_3O^+ + OH^-$$

This concept of an acid and a base applies *only* to aqueous systems, since *only* water can ionize exclusively into H⁺ and OH⁻ ions, albeit to a very limited extent. Hence water is a very weak acid, base, and electrolyte.

Neutralization is the reaction of an acid with a base to form water. For example, $HCl + NaOH \rightleftharpoons NaCl + H_2O \rightleftharpoons Na^+ + Cl^- + H_2O$.

Water can act *both* as an acid and a base according to the Brønsted–Lowry concept. Thus:

$$H_2O + H^+ \rightleftharpoons H_3O^+$$

$$\text{base} + \text{proton} \rightleftharpoons \text{acid}$$

The reversibility of these reactions is indicated by the double arrows.

Since a proton (H⁺) cannot exist in the free state in an aqueous solution, an acid can yield a proton only if a base is present to take it up. Therefore *two* acids and *two* bases must participate in every reaction involving a proton, according to the equation:

$$\text{acid}_1 + \text{base}_2 \rightleftharpoons \text{acid}_2 + \text{base}_1$$

HYDROLYSIS. As indicated in the previous section, water itself is very weakly ionized. According to the ionic theory of electrolytic dissociation, the properties and reactions of electrolytes in aqueous solution are the properties of the ions themselves, *except* when the solute ions react chemically with the ions of water. When the solute ions do react with the ions of water, the process is called *hydrolysis*.

Many biochemical reactions involving hydrolysis are important physiologically. For example, in carbohydrate, fat, and protein metabolism, the following hydrolyses take place:

a) Sucrose + $H_2O \rightarrow$ glucose + fructose

b) Fat + NaOH + $H_2O \rightarrow$ glycerol + Na salt (soap)

c) The hydrolysis of peptide linkages in a protein molecule involves one molecule of water for each peptide bond broken:

$$\underset{\underset{O}{\|}\,\underset{H}{|}}{R-C-N}-R^1 + H_2O \rightarrow \underset{\underset{O}{\|}}{R-C}-OH + R^1-NH_2$$

Inorganic hydrolytic reactions generally are reversible merely by changing the equilibrial conditions as discussed below. However, in biochemical systems, energy generally is required for the resynthesis of a compound following hydrolysis. For example, the re-formation of a peptide bond following its hydrolytic cleavage is not possible unless energy is supplied to the system.

THE LAW OF MASS ACTION OR THE LAW OF CONCENTRATION EFFECT (GULDBERG AND WAAGE). The velocity of a chemical reaction is defined in terms of the change in concentration of the reactant per unit of time, eg, Δ moles per liter per second. Three factors usually influence the rate of chemical and biochemical reactions: first, the concentration of the substances involved in the reaction (the reactants); second, the temperature at which the reaction takes place; and third, the presence of catalysts (eg, enzymes in biochemical systems).

The law of mass action deals with the first factor, namely, the concentration of the reactants. This law may be stated as follows: The rate of a simple chemical reaction at a constant temperature is directly proportional to the concentration of each reactant, raised to a power equal to the number of molecules of the reactants present in the equation for that particular reaction. The *order* of a reaction is defined by the number of concentration factors which enter into the equation for the rate.

The important fact to remember concerning the law of mass action is that the *velocity* of the reaction in either direction is directly proportional to the concentration of the substances participating in the reaction. Furthermore, at chemical equilibrium (ie, when the reaction is proceeding with equal velocity in *both* directions with no net change in the concentration of the reactants), an alteration in the concentration of any of the original reactants, or the products of the reaction, can shift once again the direction and quantitative balance of the reaction as well as its velocity.

This law of concentration effect has important and far-reaching consequences in many physiologic mechanisms. Hence, it has been stated in general terms prior to a consideration of ionic equilibria in general, and the concept of pH in particular.

IONIC EQUILIBRIA AND THE CONCEPT OF pH.   As defined above, a chemical equilibrium is that state of a chemical reaction in which the velocity of the reaction in both directions is equal. Any chemical system in equilibrium at a given temperature is characterized by an equilibrium constant (K).

Since electrolytes in aqueous solution do not follow the theoretical predictions of their behavior exactly (the so-called ideal situation), the term *activity* (a) is used to correct for the discrepancies between the predicted and the actual situation. Thus, the term activity represents the *effective* concentration of a substance as evaluated by its actual behavior in solution. As the concentration of the reactants increases, their activity decreases, indicating that fewer molecules are able to participate in the reaction. The activity factor is related to the molal concentration of the electrolyte (m) by a factor gamma (γ), which is called the *activity coefficient*; hence, $\gamma = a/m$. Thus, the activity coefficient is a number by which the molal concentration of the reactant must be multiplied in order to obtain the activity of that substance.

The Ionization Constant for a Weak Acid.   The ionization constant for a *weak* acid (K) may now be calculated, using acetic acid (HA) as an example. In aqueous solution, this compound dissociates according to the equation:

$$HA + H_2O \rightleftharpoons H_3O^+ + A^-$$

Note that dissociation of this acid is weak, as indicated by the small arrow to the right; ie, the equilibrium favors the undissociated state. Therefore,

$$K_{HA} = \frac{a\,H_3O^+ \times a\,A^-}{a\,H_2O \times a\,HA}$$

Since the activity of a solute in dilute aqueous solution in water is itself a constant for all practical purposes, and is equal to its concentration (1,000/18 = 55.5 molar), then the activity of the water may be replaced by another constant, denoted by k. Therefore,

$$K_{HA} = \frac{a\,H_3O^+ \times a\,A^-}{k_a HA}$$

And $K_{HA} \times k = K_a$, where K_a is the ionization constant for the weak acid, HA.

This equation may be written in a simplified form for weak acids in solutions that do not contain many other ions in terms of *concentration*, as indicated by the brackets in the following expression:

$$K_a = \frac{[H_3O^+][A^-]}{[HA]}$$

In this equation, the concentrations are expressed in molarity.

The Ionization Constant for a Weak Base.   Weak bases fall into two classes: first, bases that dissociate into a hydroxyl (OH^-) radical and a cation in aqueous solution; second, bases that ionize by reacting with water and thus acquire a proton, thereby producing OH^-.

For the first type of base, the general equilibrium equation is written:

$$BOH \rightleftharpoons B^+ + OH^-$$

Therefore,

$$K_b = \frac{a\,B^+ \times a\,OH^-}{a\,BOH}$$

In terms of concentration, this expression becomes:

$$K_b = \frac{[B^+][OH^-]}{[BOH]}$$

For the second type of base, which ionizes by reaction with water, ammonia (NH_3) may be used as the example:

$$NH_3 + H_2O \rightleftharpoons NH_4^+ + OH^-$$

The equilibrium constant (K_c) written in terms of concentration is:

$$K_c = \frac{[NH_4^+][OH^-]}{[NH_3][H_2O]}$$

As in the case of acetic acid, the concentration of water may be dealt with as a constant (k), so that:

$$K_c \times k = K_b$$

The Ionization Constant for Water.   The ion product for water (which ionizes *very* weakly in the pure state) is derived in a manner similar to that for acids and bases:

$$2H_2O \rightleftharpoons H_3O^+ + OH^-$$

So that:

$$K = \frac{a\,H_3O^+ \times a\,OH^-}{a^2 H_2O}$$

Since the actual quantity of water molecules that are ionized in any given mass of this liquid is so small, its activity may be considered to be constant (k). Therefore:

$$a\,H_3O^+ \times a\,OH^- = K \times k^2 = K_w$$

K_w is the ion product constant (or ion product) for water. In dilute solutions, the ion product may be expressed as concentration for water, as is done for acids and bases; thus:

$$K_w = [H_3O^+][OH^-]$$

This expression emphasizes one particularly important point: Although the *relative* concentrations for hydrogen and hydroxyl ions may be varied greatly, *their product always remains the same*.

The value of K_w has been determined by several methods, and at 25 C its value is 1.0×10^{-14}. The concentrations of hydrogen and hydroxyl ions are equal, since the ionization of one mole of water produces one mole of each ion. Therefore, at 25 C:

$$[H_3O^+] = [OH^-] = \frac{}{\sqrt{1 \times 10^{-14}}} = 1 \times 10^{-7} \text{mole/liter}$$

The fundamental significance of this relationship cannot be overemphasized, for it is the basis of the pH scale.

pH: The pH scale was first suggested by Sørensen as a convenient method for expressing the range of acidity (hydrogen ion concentration) and alkalinity encountered in living systems.

The pH of a solution is defined as the negative exponent of the hydrogen ion concentration (the *p* stands for "power" in a mathematical, not a physical sense). Thus, in aqueous solutions *only* the following relationship holds:

$$pH = -\log_{10} [H_3O^+] = \log_{10} [1/H_3O^+]$$

and

$$[H_3O^+] = 10^{-pH}$$

If the $[H_3O^+]$ is 1×10^{-5} mole/liter, the pH of the solution is $-\log_{10} 10^{-5}$ or 5. Conversely, if a solution has a pH of 9, then:

$$9 = -\log_{10} [H_3O^+] = \log_{10} [1/H_3O^+]$$

and

$$[H_3O^+] = 10^{-9}$$

The notation pOH to indicate alkalinity of a solution is rarely used; it is related to the hydroxyl ion concentration as follows:

$$pOH = -\log_{10} [OH^-] = \log_{10} [1/OH^-]$$

The relationship between pH and pOH may be shown simply as:

$$pH + pOH = 14$$

In a neutral solution at 25 C, pH = pOH = 7. If the pH is *less than* 7, the solution is acid; if the pH is *greater than* 7, it is basic (Fig. 1.11). The entire range of the pH scale is from 0 (strongly acid) to 14 (strongly alkaline).

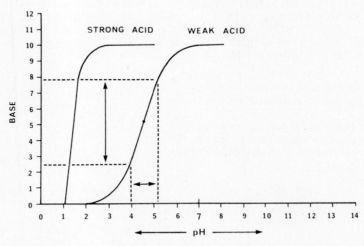

FIG. 1.11. Comparison of the titration curves of a strong and a weak acid against a strong base (ordinate in arbitrary units).

The strong acid has no reserve of protons (ie, is completely dissociated) so that the pH (abscissa) changes abruptly as the protons are used up suddenly during the titration.

The weak acid has a sigmoid titration curve. The pK is indicated by the dot on the curve at pH 4.5, and this pH represents the point at which one-half of the molecules are ionized, the other half un-ionized. The distance between the horizontal dashed lines (vertical arrow) indicates the region of maximum buffering capacity above and below the pK on the linear portion of the curve, and shows that the pH changes the least for the maximum amount of added titrant in this region (horizontal arrow).

Any aqueous solution is neutral at pH 7.0; values below 7.0 indicate acidity (arrow to left on abscissa) whereas values above 7.0 indicate alkalinity (arrow to right).

If a weak base is titrated against a strong acid, a curve similar to that shown in the figure would be obtained.

It is essential to realize that a change of *one* pH unit represents a *tenfold* change in the hydrogen ion concentration, because of the logarithmic relationship between the two scales.

The majority of physiologic processes take place in a pH range of about 6 to 8; therefore, such reactions occur in weakly acidic to weakly basic solutions.

THE HENDERSON–HASSELBALCH EQUATION AND BUFFERS. According to the Brønsted-Lowry concept, acids are proton donors, whereas bases are proton acceptors. Henceforth, this definition of an acid will be limited strictly to proton donors and acceptors found in living, hence aqueous, systems.

In the following discussion, the symbol HA will be used for the undissociated acid. It is important to reiterate that a *weak* acid is very *slightly* dissociated, so that relatively few protons are formed and most of the acid is present in solution in the *undissociated* form denoted by HA. Thus, in the reaction of HA with water:

$$HA + H_2O \rightleftharpoons H_3O^+ + A^-$$

The reaction is strongly to the *left,* as indicated by the large arrow. (It is important *not* to confuse the strength of an acid, as indicated by its degree of dissociation, with its concentration, usually expressed as moles per liter.)

According to the law of mass action (p. 19), one can calculate the velocities of the reactions in the equation above, both from left to right (v_1) and right to left (v_2), in the following way:

$$v_1 = k_1 [HA][H_2O] \quad \text{(left to right)}$$
$$v_2 = k_2 [H_3O^+][A^-] \quad \text{(right to left)}$$

In these equations, the velocity, v, is proportional to the rate constant, k, for each reaction multiplied by the molar concentrations of the reactants.

If a millimole of HA is added to a liter of water, an equilibrium is reached in which there is no net change in the concentration of reactants and $v_1 = v_2$.

Since
$$v_1 = k_1 [HA][H_2O]$$
$$v_2 = k_2 [H_3O^+][A^-]$$

and, at equilibrium,

$$v_1 = v_2$$

then, by substitution for v_1 and v_2, one obtains:

$$k_1 [HA][H_2O] = k_2 [H_3O^+][A^-]$$

Rearranging this equation yields the expression:

$$\frac{k_1}{k_2} = \frac{[H_3O^+][A^-]}{[H_2O][HA]}$$

The ratio k_1/k_2 can be expressed as a third constant, K:

$$K = \frac{k_1}{k_2}$$

K is called the *equilibrium constant* for the reaction.

By substitution of the factors for k_1 and k_2, we find that:

$$K = \frac{[H_3O^+][A^-]}{[HA][H_2O]}$$

For practical use in acid-base chemistry, this equation usually is written:

$$K' = \frac{[H^+][A^-]}{[HA]}$$

In this simplified expression, K' is the *apparent ionization* (or *dissociation*) *constant*. The symbol K' also includes the activity coefficient, and it removes the water term from the denominator on the right side of the equation. The shorthand symbol H^+ is substituted for H_3O^+ (the acid or hydronium ion, H_3O^+, is implicit in the symbol H^+ as used henceforth).

K' is not actually a constant but may be used as such when dealing with systems of *specifically defined* temperature and ionic strength, such as are encountered in the body fluids of living organisms.

K' is of immediate practical significance, as it indicates the relative strength of an acid. Thus K' for hydrochloric acid is very high, whereas K' for acetic acid is 10^{-5} moles/liter.

The Henderson–Hasselbalch equation is a very useful tool in the quantitative evaluation of buffers and buffer systems such as are found in the body. This equation is obtained by rearrangement of the expression

$$K' = \frac{[H^+][A^-]}{[HA]}$$

so that

$$[H^+] = K' \frac{[HA]}{[A^-]}$$

This equation now may be converted into the $\log_{10}$ form, so that it becomes:

$$\log_{10} [H^+] = \log_{10} K' + \log_{10} \frac{[HA]}{[A^-]}$$

Since pH is defined as $-\log_{10} [H^+]$, multiplying this expression by -1 yields:

$$-\log_{10} [H^+] = -\log_{10} K' + \log_{10} \frac{[A^-]}{[HA]}$$

Note that the right-hand term in this equation is reexpressed by *inverting* $-\log_{10} [HA^-]/[A^-]$, which equals $+\log_{10} [A^-]/[HA]$.

In a fashion analogous to the definition of pH, the term pK' is obtained, or p$K' = -\log K'$.

The Henderson-Hasselbalch equation is obtained by substituting pH and pK' in the expression given above:

$$pH = pK' + \log_{10}\frac{[A^-]}{[HA]}$$

This equation may also be written:

$$pH = pK' + \log_{10}\frac{[base]}{[acid]}$$

$$= pK' + \log_{10}\frac{[proton\ acceptor]}{[proton\ donor]}$$

The degree of ionization of any particular compound depends upon the pK' of that compound, as well as the pH of the solution in which the compound is dissolved. It is helpful to consider the dissociation constant of a substance as the pH at which there are equal proportions of ionized and nonionized molecules.

An aqueous buffer solution is defined simply as one that resists a change in pH (compared to water) when an acid or an alkali is added to it. The solutes present that are responsible for the resistance to alteration in pH are called the buffer system or simply, buffers. A buffer may be considered as a reservoir of proton donor (acid) or proton acceptor (base). Remember also that according to this definition of a base, hydroxyl (OH^-) radicals are not necessarily present, and the proton acceptor (base) also may be electrically neutral (eg, R—NH_2), or even positively charged (eg, $NH_3^+ + H_3O^+ \rightleftharpoons NH_4^+ + H_2O$).

Solutions of specific weak acids and bases — for example, H_2CO_3, H_3PO_4, and $NaHCO_3$, — when titrated against strong bases or acids, respectively, each have a definite pH range where their resistance to alteration of pH is at a maximum; that is, the change in pH is minimal for the addition of a maximum amount of titrant. This is called the *maximum buffering region* of the titration curve (Fig. 1.11).

The *maximum buffering capacity* of a buffer system is the *total quantity* (eg, in moles per liter) of acid (or base) that a given buffer system can neutralize without a significant change of pH. The pK', as defined above, is shown in Figure 1.11 as a dot on the titration curve for the weak acid and lies at pH 4.5.

The buffer systems in the human body are quite complex, and the regulation of blood pH at a value close to 7.4 is critical to the survival of the individual. Thorough comprehension of the concept of pH and the Henderson–Hasselbalch equation is fundamental to an understanding of the mechanisms that control the pH of the body fluids and their interrelationships.

THE IONIZATION OF AMPHOLYTES (AMPHOTERIC SUBSTANCES) AND THE ISOELECTRIC POINT. Ampholytes are substances capable of reacting chemically either as acids (proton donors) or as bases (proton acceptors). Consequently, these ions have very high dipole moments when doubly ionized; that is, at their *isoelectric point*. The isoelectric point also is defined as the pH at which an ampholyte does not migrate toward either the cathode or the anode in a DC electric field; the *net* electric charge on the molecules is zero.

There are many biologically important substances which are ampholytes, among them amino acids, peptides, and proteins.

The amino acid glycine may be used as an example for the reactions of an ampholyte:

(a) Glycine as a base, reacting with an acid:

$$\begin{array}{c} CH_2\!-\!NH_3^+ + H_3O^+ \rightleftharpoons CH_2\!-\!NH_3^+ + H_2O \\ | \qquad\qquad\qquad\qquad | \\ COO^- \qquad\qquad\qquad COOH \end{array}$$

(b) Glycine as an acid, reacting with a base:

$$\begin{array}{c} CH_2\!-\!NH_3^+ + OH^- \rightleftharpoons CH\!-\!NH_2 + H_2O \\ | \qquad\qquad\qquad\qquad | \\ COO^- \qquad\qquad\qquad COO^- \end{array}$$

The doubly ionized (charged) ampholytic form of glycine is shown in the equations above on the left. Every amphoteric substance exhibits this characteristic double dissociation at the pH corresponding to its isoelectric point.

DIFFUSION, OSMOSIS, AND OSMOTIC PRESSURE. Diffusion is the process whereby particles (such as ions and molecules) of liquids and gases mix as the result of their spontaneous movement. This movement is caused by the kinetic energy (thermal agitation) of the particles themselves.

Dissolved substances thus diffuse, or move *down*, their individual concentration gradients from regions of higher to regions of lower concentration (Fig. 1.12A).

The *rate* of diffusion of a solute depends upon its nature (eg, charge, particle size), the temperature, and the concentration of the solution, as well as the properties and nature of the solvent itself.

The term *osmosis* merely signifies the diffusion of *water* (the solvent per se) through a semipermeable membrane.

Osmotic pressure is a measure of the tendency of water to diffuse from a region of high concentration (of water, *not* solutes) to a region of lower concentration. Osmotic pressure is a real force and may be determined directly or, more usually, indirectly by measuring the pressure (eg, dynes per square centimeter) necessary to prevent the *osmot-*

ic flow (or diffusion) across a semipermeable membrane (Fig. 1.12B). A semipermeable membrane is an interface or boundary between two separate compartments which permits the free exchange of water molecules but *not* particles of the solute. The solute effectively reduces the *activity* (or concentration of free water), so that the compartment containing pure water has a higher *effective concentration* (activity) of water molecules than the compartment containing the solute. Water

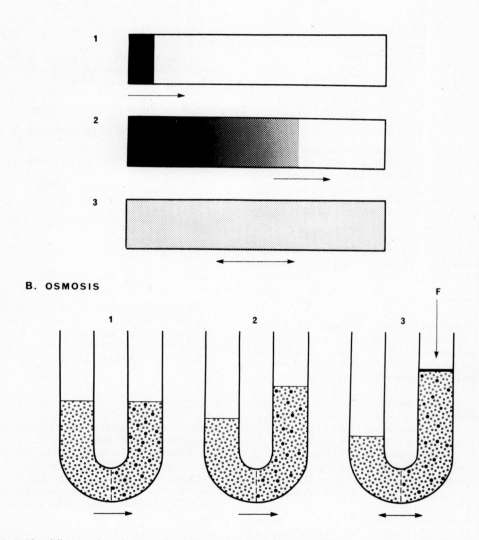

A. DIFFUSION

B. OSMOSIS

FIG. 1.12. A. The diffusion of a solute in a liquid down its concentration gradient is indicated by shading in the three boxes. In **A.1,** the solute has been added (dark bar). The solute molecules gradually pass down their concentration gradient with time, **A.2,** and eventually become distributed uniformly throughout the solvent at equilibrium, **A.3. B.** Osmosis is the diffusion of water molecules (open circles) down their own concentration gradient, ie, from the left- to the right-hand compartment, across a semipermeable membrane (vertical dashed line). The nonpermeable solute is indicated by solid dots in the right compartment. In **B.1,** pure water is in the left compartment whereas a solute is present in the right compartment, thus reducing the effective concentration of water in this part of the system, and thereby establishing a concentration gradient for the solvent between the two regions. Thus water molecules move from left to right across the semipermeable membrane, and the total volume in the right-hand compartment increases as shown in **B.2.** At osmotic equilibrium, **B.3,** equal numbers of water molecules pass in each direction across the membrane, and the activity of the water is equal in both compartments of the system.

In **B.3,** the force (*F,* vertical downward arrow) necessary to *prevent* the actual movement of water is a measure of the osmotic pressure.

diffuses down its own concentration gradient to dilute the solution in the other compartment; therefore, when physical equilibrium is reached, the number of water molecules passing through the membrane in each direction per unit time is equal. If the actual movement of the water across the membrane is prevented, as by a piston in a cylinder, then the force exerted by the piston at equilibrium is a measure of the osmotic pressure (Fig. 1.12B).

The osmotic pressure developed by a solution is directly related to, and dependent upon, the absolute temperature as well as the concentration of dissolved solute (ie, the total number of particles present, moles per liter) and *not* upon the size or chemical nature of the particles themselves. The solute may be an electrolyte (ie, ionized) or a nonelectrolyte. For example, equal volumes of equimolar concentrations of table salt (NaCl) and sucrose (cane sugar) have different osmotic properties because NaCl dissociates completely into Na^+ and Cl^- ions (two particles per molecule), whereas sucrose goes into solution without dissociation (one particle per molecule). Therefore, the saline solution has twice the osmotic activity of the sucrose solution, because the salt has effectively reduced the activity of the water twice as much as the sucrose; hence, the diffusion gradient between the water and NaCl solution compartments is twice that of sucrose.

Solutions that have the same osmotic properties are said to be osmotically equivalent to each other, or *osmolar*.

Colloids, such as plasma proteins, also exert a significant effect upon the osmotic pressure of a solution; consequently, they are very important physiologically. The concept of osmotic pressure tends to be confusing since one speaks of the osmotic pressure of a given solution in terms of the *solute* it contains, whereas, in actuality, one is dealing with the *activity* (or concentration) of *free water* in that solution relative to pure water itself. The term *osmotic activity* of a solution would be preferable, since this phrase implies specifically the concentration of free water per se that is available for diffusion in a given solution relative to pure water.

There are two important generalizations concerning the osmotic pressures of solutions of nonelectrolytes (van't Hoff).

First, the osmotic pressure (π) of a dilute solution of a nonelectrolyte is directly proportional to the concentration (C) and inversely proportional to the volume (V) of a solvent at a constant absolute temperature (T). Stated mathematically, this relationship becomes:

$$\pi = k_1C \text{ or } \pi V = k_1 \text{ (at constant temperature)}$$

This expression is analogous to Boyle's law for gases (p. 11). The symbol k_1 is a proportionality constant.

Second, the osmotic pressure of a dilute solution with a given concentration of solute varies in direct proportion to the absolute temperature, thus:

$$\pi = k_2T \text{ (at constant concentration or volume)}$$

This is similar to Charles' law for gases. Combining the two equations, we get:

$$\pi V = kT \text{ (when } V \text{ and } T \text{ both vary)}$$

This is the same form as that for the combined equation for the gas laws, $PV = RT$. Or, as expressed by van't Hoff, this relationship becomes $\pi V = RT$.

If the concentration of a solute is expressed in moles (m), then the osmotic pressure may be calculated from the equation:

$$\pi = mRT$$

In this expression, R is the universal gas constant.

THE GIBBS–DONNAN MEMBRANE EQUILIBRIUM. This principle of the Gibbs-Donnan membrane equilibrium forms an important part of certain physiologic mechanisms, for it defines the state of equilibrium which exists when solutions of two electrolytes with a common ion are separated by a membrane permeable to this common ion but *impermeable* to one of the other ions.

Suppose that a semipermeable membrane separates a solution of potassium chloride (KCl) and a solution of the potassium salt of a protein (P), the protein being the nonpermeable ion. If the protein solution is placed on the *inner* side of the membrane, and the potassium chloride on the *outer* side, diffusion of potassium and chloride ions takes place spontaneously, and this tends to restore a concentration equilibrium. That is, the K^+ and Cl^- diffuse freely to the inner side of the membrane until the ionic activity product of the potassium chloride inside (i) is the same as that on the outside (o). If the concentrations are made equal to the activities of the ions involved, then the following relationship holds at equilibrium:

$$[K^+]_i[Cl^-]_i = [K^+]_o[Cl^-]_o$$
$$\text{Inside} \qquad \text{Outside}$$

It is important to realize that at equilibrium the Gibbs–Donnan relationship defines *only* the unequal distribution of the *diffusible* ions across a semipermeable membrane. Significant electric and osmotic differences also develop between the two compartments as a result of this phenomenon; these effects will be discussed later.

REVIEW OF ORGANIC CHEMISTRY

In broadest terms, an organic compound may be defined as any substance that contains the element carbon (C), regardless of whether or not such a substance is a product of, or may be derived from, living matter. All other substances may be regarded as inorganic or mineral.

The organic compounds associated with protoplasm usually contain hydrogen and oxygen in addition to carbon. Other elemental constituents may be present as components of the molecular structure of specific compounds. For example, nitrogen (N), phosphorus (P), and sulfur (S) are of frequent occurrence (Table 1.5).

There are several compounds of great physiologic importance which contain carbon but which are considered to be *inorganic* substances because of their chemical and physical properties. Included in this category are carbon dioxide (CO_2), carbon monoxide (CO), the bicarbonate ion (HCO_3^-), and the carbonate ion ($CO_3^=$). The cyanide ion (CN^-) is also listed among the inorganic carbon-containing compounds.

The Physicochemical Properties of Carbon

Since the element carbon is of such enormous significance in the structural and functional constitution of all biologic systems, it is necessary to discuss certain of its basic properties.

THE COVALENT BOND AND THE TETRAVALENCE OF CARBON. Unlike atoms which dissociate in aqueous solution, producing charged particles (ions) by the actual loss or gain of electrons, the element carbon forms *nonionic covalent bonds by sharing electrons* (p. 14). The carbon atom has six electrons (the atomic number of carbon equals the nuclear charge, which is also six). Two of these electrons are unpaired and are present in separate orbitals surrounding the nucleus (Table 1.5). In accordance with the Pauli exclusion principle, only two electrons may occupy a given orbital, and these electrons must spin in opposite directions. This general and basic rule governs the configuration of electrons in atoms as well as in molecules. In the formation of a covalent bond between two atoms of carbon, each atom must become positioned so that its orbital overlaps that of the other carbon atom. Thus each orbital contains a single electron whose spin is opposite to that of the electron in the second orbital. The combination of the two orbitals causes formation of a *single* bond orbital which contains both of the electrons; each of these electrons has an opposite spin with respect to the other, in accordance with the Pauli exclusion principle. The greater the overlap between the orbitals of a given pair of carbon atoms, the greater the strength of the bond; in other words, the bond strength is directly proportional to the degree of orbital overlap.

Although the foregoing discussion might suggest that carbon forms only two bond orbitals with other atoms such as hydrogen, in actuality four bonds are produced, giving rise to, for example, methane (CH_4). Carbon has a valence of four and is *tetravalent*. The reasons for this are as follows: Since covalent bond formation results in the release or evolution of energy (ie, it is an *exergonic process*), the resulting new atomic configuration is far more stable than that present initially in the individual atoms. Since the process of bond formation is exergonic, therefore stabilizing, as many bonds as possible are formed in order that the most stable configuration of the atoms sharing the electrons be achieved. This process of multiple bond formation takes place regardless of the shape of the initial *atomic* orbitals in relation to the ultimate configuration of the same orbitals within the *molecule*. Thus carbon produces hybrid or mixed overlapping orbitals within organic molecules, such as methane. Consequently, a total of four of the six electrons present in the two outermost orbitals about the nucleus of the carbon atom always participate in the formation of four covalent bonds with similar or other atomic species.

In order to achieve a maximal overlap of the carbon atom with the orbital electrons of hydrogen in the formation of methane, the mixed (or hybrid) orbitals are oriented toward the corners of a regular tetrahedron, and thus the individual orbitals are separated as widely as possible, in accordance with the Pauli exclusion principle stated earlier (Fig. 1.13).

Several important general principles regarding the covalent bond derive from the facts stated above.

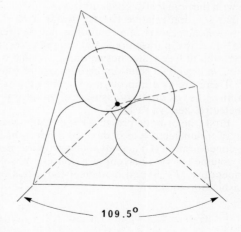

FIG. 1.13. Model of the methane molecule (CH_4). The carbon atom is indicated by a dot at the center of a regular tetrahedron, and the orbitals of the four hydrogen atoms are indicated by large circles.

First, because of the spatial arrangement of the hydrogen atoms in relation to the carbon atom in the methane molecule, characteristic *bond angles* are found among the atoms in this and all other organic compounds. In methane, the example cited, the angle between the adjacent C—H bonds is 109.5°, as shown in Figure 1.13. In contrast to the covalent bond, which exhibits marked *directional characteristics,* ionic (or electrovalent) bonds exhibit equal strength in all directions. That is, no spatial orientation is present.

Second, covalent bonds have characteristic *bond lengths.* And third, covalent bonds also have characteristic *bond energies.*

Therefore, covalent bonds within specific organic compounds are largely responsible for governing the molecular size and shape of those compounds as well as such physical properties as isomerism.

Several other atomic species of great biologic significance also possess the ability to form covalent bonds when present in their elemental state. These elements are hydrogen (H), oxygen (O), and nitrogen (N). By sharing electrons, as discussed above, these atoms exhibit covalences of 1, 2, and 3, respectively. It must be stressed that, in addition to covalent bond formation, H, O, and N are also capable of electron transfer; hence they can form electrovalent (ionic valence) bonds as well as covalent bonds.

CATENATION. Because of the repeated formation of stable covalent bonds, carbon atoms may become attached together in long chains or rings of various types and configurations. This process is known as *catenation.*

The tetravalent nature of carbon atoms and their ability to form stable covalent bonds with specific directional and dimensional characteristics, combined with the property of catenation, enables carbon to participate in the formation of a vast number of organic molecules. These molecules may exhibit highly specific physicochemical properties depending upon their individual three-dimensional characteristics. This basic fact is of the utmost importance to an understanding of many of the biochemical processes that take place in the human body.

The Schematic Representation of Valence States

The hydrogen molecule, H_2, represents the simplest example of a covalent molecule in which *one pair* of electrons is shared. The shared electrons may be depicted by dots, H:H (Lewis, 1916) in the electronic formula of this molecule, or by a dash, H—H (Kekulé, 1859).

Similarly, the chemical formula for methane

may be written as CH_4 and depicted as H:C̈:H or
$$\begin{array}{c} H \\ | \\ H\!-\!\ddot{C}\!-\!H \\ | \\ H \end{array}$$

H—C—H. Water, H_2O, becomes :Ö:H, and ammonia, NH_3, is expressed H:N̈:H.

A double bond between two atoms consists of *two pairs* of shared electrons and two orbitals for each atom. For example, the electronic formula of ethylene, C_2H_4, may be given as

$$\begin{array}{cc} H & H \\ \ddot{C}::\ddot{C} \\ H & H \end{array}\ ;$$

more commonly, this formula is written

$$\begin{array}{cc} H & H \\ \diagdown & \diagup \\ C\!=\!C \\ \diagup & \diagdown \\ H & H \end{array}$$

or $H_2C = CH_2$.

Occasionally, an atom may form covalent bonds which differ in number from the normal number, and the atom thereby completes its stable electronic configuration by *transferring electrons.* A biologically important example of this effect is in the formation of the ammonium ion, NH_4^+:

$$\left[\begin{array}{c} H \\ H\!:\!\ddot{N}\!:\!H \\ H \end{array}\right]^+$$

The nitrogen atom in this radical forms three covalent bonds and one electrovalent (ionic) bond with a total of four hydrogen atoms. In addition to nitrogen, oxygen (O) and sulfur (S) also form covalent bonds with carbon. Atoms such as these three are termed *heteroatoms,* since they possess one or more pairs of electrons which are not involved in covalent bonding and which also can participate in electrovalent (ionic) bonding as noted above.

Furthermore, since the unshared electrons have a negative field, organic compounds having heteroatoms (N, O, and S) act as bases, since they attract protons (Brønsted–Lowry concept). For example, ethylamine accepts a proton, thereby becoming a positively charged compound:

$$C_2H_5NH_2 + H^+ \rightleftharpoons C_2H_5NH_3^+$$

Many examples of this effect are encountered in biologic systems.

According to the generally accepted convention, a dash or a short line is used to represent every valence available to an element for chemical combinations:

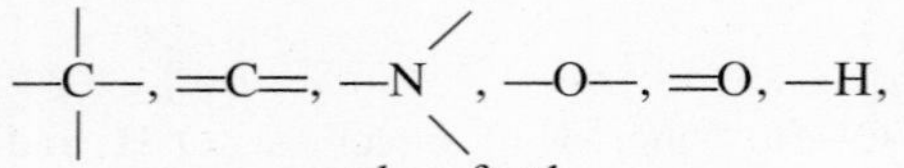

$$-\overset{|}{\underset{|}{C}}-, \quad =C=, \quad -N\overset{\diagup}{\diagdown}, \quad -O-, \quad =O, \quad -H,$$

and so forth.

Thus methane and carbon dioxide may be written

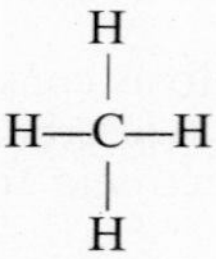

$$H-\overset{\overset{\textstyle H}{|}}{\underset{\underset{\textstyle H}{|}}{C}}-H \text{ and } O=C=O, \text{ respectively.}$$

In depicting more complex formulas such as are encountered in compounds of physiologic and biochemical interest, further abbreviation is accomplished by grouping certain elements together and by omitting certain of the Kekulé bonds. For example:

CH_4	$R-CH_3$	CH_3-CH_3
Methane	Methyl	Ethane

$R-CH_2-CH_3$	CO_2
Ethyl	Carbon Dioxide

The symbol R will be used henceforth to denote a *radical*, whose chemical nature will be specified for the particular organic compound under discussion. The Kekulé representation will be used for the structural formulas presented in this book, with due recognition of the fact that each dash signifies one pair of shared electrons.

Isomerism

As noted earlier, the tetravalent nature of carbon, combined with the ability of this element to form covalent bonds having specific directional and dimensional properties, gives rise to organic molecules which may exhibit a definite spatial, or three-dimensional, structure. In addition, the property of catenation contributes to the formation of such compounds. As a direct consequence of these facts, two compounds may have identical numbers of the same atoms of the same elements in their molecules (ie, their empirical or chemical formulas and their molecular weights are the same) yet differ markedly in the structural or spatial arrangement of their constituent atoms. This difference in structure may in turn give rise to one or more sharply differing chemical, physical, and biologic properties between the two molecules. This phenomenon is called *isomerism*. In other words, isomerism depends upon the fact that two compounds having an identical chemical *composition* exhibit a different spatial *configuration* of their individual atoms. The sequence in which the atoms are joined together in a molecule is termed the *constitution* of that molecule. Hence the terms ''composition,'' ''configuration,'' and ''constitution'' denote specific attributes of organic molecules and may not be used interchangeably.

Organic compounds exhibit two principal forms of isomerism: *structural isomerism* and *stereoisomerism*.

STRUCTURAL ISOMERISM. Structural isomerism, illustrated in Figure 1.14, results from a different configuration and constitution of the atoms of the individual isomeric forms of a compound, although their elemental composition is identical, as can be seen from the empirical (or chemical) formulas (depicted at the left of the figure).

When structurally isomeric compounds belong to different homologous series, as shown in Figure 1.14D, these substances are called *functional isomers*, since they possess different functional groups.

In all instances, structural isomers have different chemical and physical properties, including differences between their boiling and melting points, densities, rates of hydrolysis, and so forth. The physicochemical differences between the properties of given structural isomers may be slight, or they may be quite marked, such as is the case among functional isomers. Furthermore, profound differences in biologic activity may be observed between structural isomers, as, for example, between dimethyl ether and ethyl alcohol (ethanol).

STEREOISOMERISM. The term *stereoisomerism* denotes *spatial isomerism*. Stereoisomers of the same compound have the same elemental composition (chemical formula) and constitution of the atoms within their molecules, but differ markedly in the three-dimensional configuration of their atoms. Stereoisomers thus may exhibit one or more different chemical, physical, and/or biologic properties due to their configuration.

Two kinds of stereoisomerism are encountered: geometric *isomerism* and *optical isomerism*.

Geometric Isomerism. Geometric isomerism is found in molecules that have a double covalent bond. Since the double bond is far more rigid than a single bond, the atoms connected to the double bond cannot rotate freely, as is the case with a single bond connecting two carbon atoms in the skeleton of the molecule.

For example, the simple organic compound ethane can have two arrangements of its methyl (CH_3) groups because of rotation of one group with respect to the other about the single carbon-to-carbon bond, as depicted in Figure 1.15A. Thus, the ethane molecule can exhibit two different spatial conformations of its atoms.* In contrast to this arrangement, the presence of a double bond in the molecule, as, for example, in ethene (Fig. 1.15B), prevents rotation of the end groups unless the double bond is broken. Therefore, geometric

---

The term ''conformation'' signifies that different arrangements of the atoms within the molecule can take place without breaking any bonds. This term is applied principally to geometric isomers.

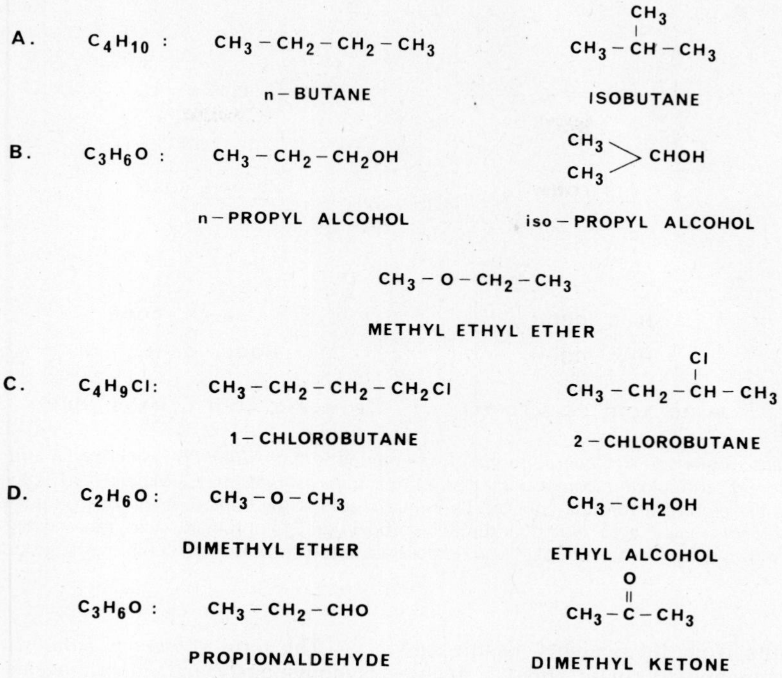

A. C_4H_{10} : $CH_3 - CH_2 - CH_2 - CH_3$

n-BUTANE

$$CH_3 - \overset{\overset{\displaystyle CH_3}{|}}{CH} - CH_3$$

ISOBUTANE

B. C_3H_6O : $CH_3 - CH_2 - CH_2OH$

n-PROPYL ALCOHOL

$$\overset{\displaystyle CH_3}{\underset{\displaystyle CH_3}{\Large >}} CHOH$$

iso-PROPYL ALCOHOL

$$CH_3 - O - CH_2 - CH_3$$

METHYL ETHYL ETHER

C. C_4H_9Cl: $CH_3 - CH_2 - CH_2 - CH_2Cl$

1-CHLOROBUTANE

$$CH_3 - CH_2 - \overset{\overset{\displaystyle Cl}{|}}{CH} - CH_3$$

2-CHLOROBUTANE

D. C_2H_6O : $CH_3 - O - CH_3$

DIMETHYL ETHER

$CH_3 - CH_2OH$

ETHYL ALCOHOL

C_3H_6O : $CH_3 - CH_2 - CHO$

PROPIONALDEHYDE

$$CH_3 - \overset{\overset{\displaystyle O}{\|}}{C} - CH_3$$

DIMETHYL KETONE

FIG. 1.14.   Some examples of structural isomerism. Empirical formulas are shown on the left in the figure.

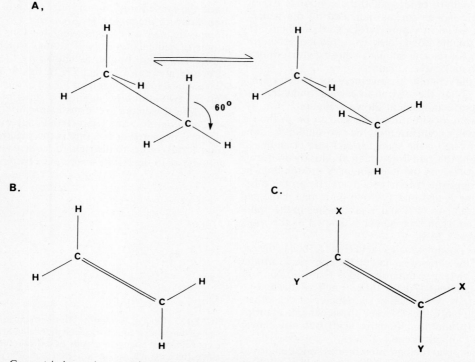

FIG. 1.15   Geometric isomerism. **A.** Ethane can exhibit two different spatial conformations because of free rotation of the methyl groups around the single covalent bond that joins them. **B.** Presence of the double bond in ethene prevents rotation of the end groups without rupturing this bond. **C.** General formula for compounds that exhibit geometric isomerism. Note that in addition to a double bond, different end groups (X and Y) must be present in the molecule if geometric isomers of the compounds are to be formed.

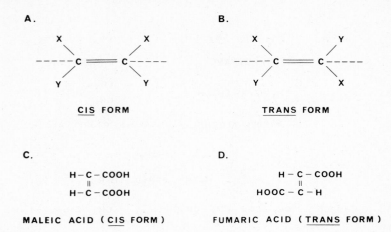

A.

CIS FORM

B.

TRANS FORM

C.

H – C – COOH
||
H – C – COOH

MALEIC ACID (CIS FORM)

D.

H – C – COOH
||
HOOC – C – H

FUMARIC ACID (TRANS FORM)

FIG. 1.16. *Cis-trans* isomerism. **A.** General formula for compounds that exhibit *cis* isomerism. X and Y indicate different end groups. **B.** General formula for compounds that exhibit *trans* isomerism. **C.** Maleic acid is the *cis* isomer of the compound having the empirical formula $C_4H_4O_4$. **D.** Fumaric acid is the *trans* form of the compound having the same chemical composition as maleic acid. Note that the two carboxyl groups of fumaric acid are arranged on opposite *(trans)* sides of the two carbons joined by the double bond, in contrast to the *cis* orientation of the same groups in maleic acid.

isomerism results from the inability of the end groups of a compound to rotate about a double bond without breaking the bond. In addition to this requirement, different chemical groups or atoms must be attached to each carbon of the double-bonded pair in order that the compounds exhibit geometric isomerism, as indicated by the general formula given in Figure 1.15C. Carbon rings containing double bonds also can produce geometric isomers, for reasons similar to those given for linear (or aliphatic) carbon chains.

If the spatial arrangement of the terminal groups about the carbon-to-carbon bond follows the general arrangement shown in Figure 1.16A, the configuration of the atoms is said to be in the *cis form.* In this configuration, the two similar X atoms or groups lie on the same (or *cis*) side of an imaginary plane (dashed line) which divides the molecule in half, whereas the two similar Y groups both lie on the opposite side of this plane. If, on the other hand, one X and one Y group lie in one plane and the second pair of X and Y groups lies in the opposite plane, then this isomer is known as the *trans form* (Fig. 1.16B).

For example, maleic acid (Fig. 1.16C) is the *cis* form of the compound having the empirical formula $C_4H_4O_4$, whereas fumaric acid is the *trans* form of this compound (Fig. 1.16D). Each of these two compounds is the geometric isomer of the other, and each has quite different physicochemical properties. In addition, fumaric acid, but not maleic acid, is biologically active.

Thus, two biologic membranes, one composed of *cis* and one of *trans* isomers, would exhibit totally different properties. In addition, a specific enzyme that acts upon one geometric isomer of a compound might be completely inactive insofar as the other isomer is concerned.

The term *geometric isomerism* is often used synonymously and interchangeably with the term *cis–trans isomerism.*

Optical Isomerism. The second major type of stereoisomerism, or optical isomerism, depends upon the presence of one or more asymmetric carbon atoms in the molecule. An *asymmetric carbon atom* is defined as a carbon which has four different atoms or radicals (groups of atoms) attached to it through covalent bonds, as shown by the general structure given in Figure 1.17. In this figure, note that each isomer is the mirror image of the other, so that the two structures cannot be superimposed.

Optical isomers are termed *enantiomorphs* (or *optical antipodes*) and characteristically exhibit the physical property of rotating plane polarized light (Fig. 1.18B) either to the right *(dextrorotatory compounds)* or to the left *(levorotatory compounds)*, as

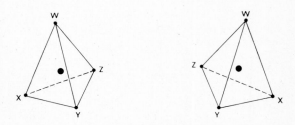

FIG. 1.17. Optical isomerism. The carbon atom (dot) in each of the general structures given in this figure is attached to four different chemical groups or radicals (W, X, Y, Z); hence, the carbon is said to be asymmetric. The presence of one (or more) asymmetric carbon atoms in a molecule is necessary for the formation of optical isomers (or enantiomorphs). Note that each structure in the figure is the mirror image of the other, and that the two compounds cannot be superimposed.

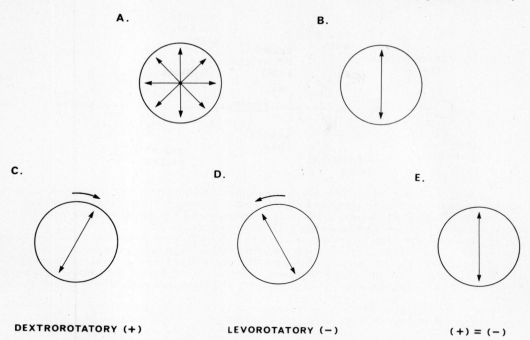

A. **B.** **C.** **D.** **E.**

DEXTROROTATORY (+) **LEVOROTATORY (−)** **(+) = (−)**

FIG. 1.18. In all of the examples depicted in this figure, the beam of light is oriented in a direction perpendicular to the surface of the page. **A.** The waves of ordinary light vibrate in all directions, as indicated by the arrows. **B.** Plane polarized light, in contrast to ordinary light, vibrates in only one plane as indicated by the single vertical arrow. **C.** Optical isomers that rotate plane polarized light to the right are termed dextrorotatory (+). **D.** Optical isomers that rotate plane polarized light to the left are termed levorotatory (−). **E.** A racemic mixture contains equal quantities of dextrorotatory (+) and levorotatory (−) isomers of the same optically active compound. Consequently, a racemic mixture produces no rotation of a beam of plane polarized light, as the rotation produced by each member of the isomeric pair is balanced exactly by the rotation in the opposite direction that is produced by the other member. For this reason, enantiomorphs sometimes are called optical antipodes. See also Figure 1.19.

depicted in Figures 1.18C and D, respectively. Optical isomers are mirror images of each other. Dextrorotatory compounds are indicated by a plus sign, whereas levorotatory compounds are indicated by a minus sign.

The actual degree of rotation of plane polarized light that is caused by an optically active compound, or enantiomorph, is determined under specific conditions by the use of an instrument called a polarimeter, which is shown in Figure 1.19 and explained in the legend to this figure.

Diastereoisomerism. As stated above, the phenomenon of enantiomorphism denotes a mirror-image relationship exhibited by the molecular structures of two stereoisomeric compounds, and the isomers of such a pair are optical antipodes. In contrast to this form of optical isomerism, the term *diastereoisomerism* sometimes is used to denote a type of optical isomerism between two compounds whose molecules contain more than one asymmetric carbon and do not exhibit a mirror-image relationship to each other. Thus, glucose and galactose (Fig. 1.20A) exhibit *diastereoisomerism,* in contrast to levolactic acid and dextrolactic acid (Fig. 1.20B) which are *enantiomorphs* (cf. epimers, below).

If a solution contains a mixture of dextrorotatory and levorotatory enantiomorphs of an optically active compound in equimolar amounts, then the optical rotation is zero, as the rotation of the polarized light caused by each of the compounds is opposite to and exactly balances that caused by the other (Fig. 1.18E). If the two constituents of such a *racemic* or DL *mixture,* as it is called, are separated physically (resolved) into each optically active constituent, then each isomer exhibits its characteristic ability to rotate polarized light either in a dextrorotatory or in a levorotatory direction. Compounds produced synthetically always are racemic mixtures, because the possibility for the formation of either the D or L isomer in a random fashion is identical under such conditions.

The spatial configurations of sugars, amino acids, and many other optically active substances of biologic importance may be related to the configuration of the dextro (D) and levo (L) forms of glyceraldehyde (Fig. 1.21), and these two isomers are used as references to which other compounds may be compared both structurally and optically. It so happens that L-glyceraldehyde is levorotatory, whereas D-glyceraldehyde is dextrorotatory; therefore, the geometric and optical

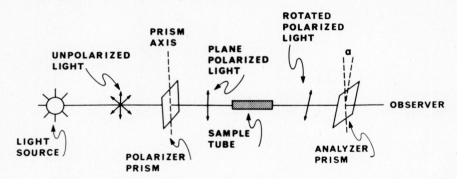

FIG. 1.19. Diagram of a polarimeter such as is used to measure quantitatively the optical activity of organic compounds. The polarized light beam is directed through a tube containing a known concentration of the compound under study which has been dissolved in an appropriate solvent, and the analyzer prism is rotated so that the angle (α) the compound rotates the light beam from the vertical is measured directly on a calibrated eyepice.

The specific rotation of the light beam that is produced by an optically active compound is determined from the following expression: $[\alpha]\lambda^t = \alpha/1\text{dm}$. In this equation, $[\alpha]\lambda^t$ is the specific rotation of a polarized light beam of wavelength λ at a known temperature, t. The factor α is the rotation of the light beam measured in degrees as it passes through a tube 1 dm in length that is filled with the optically active substance at a concentration of α gm/ml. By standardizing carefully the conditions under which the optical activity of various compounds is measured, comparable data for different substances can be obtained.

The rotation is termed dextrorotatory (+) if it is in a clockwise direction from the position of the observer, as shown in the figure above, and levorotatory (−) if in a counterclockwise direction.

properties of these two compounds coincide. However, the properties of geometric isomerism and optical isomerism are independent of each other in many compounds; hence, any given compound may be D (+) or (−) or L (+) or (−). The symbols D and L refer to the geometric isomerism of the compound compared to glyceraldehyde, whereas the plus and minus signs indicate the direction in which that compound rotates plane polarized light. For example, as shown in Figure 1.20B, D-lactic acid is

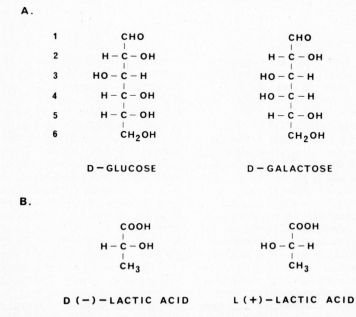

FIG. 1.20. **A.** Diastereoisomerism is the term sometimes used to distinguish between two compounds that contain more than one asymmetric carbon atom, and whose structures have no mirror-image relationship to one another. The numbering of the carbon atoms is indicated to the left of the glucose molecule, and carbon atoms 2, 3, 4, and 5 in this compound are asymmetric. Two sugars, such as glucose and galactose, which exhibit a difference in their configuration about only one carbon atom (C-4) are called epimers. **B.** D(−)-lactic acid and L(+)-lactic acid are examples of enantiomorphic compounds.

A. **B.**

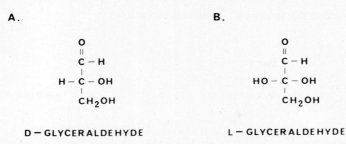

$$
\begin{array}{ccc}
& \text{O} & \\
& \| & \\
& \text{C} - \text{H} & \\
& | & \\
\text{H} - & \text{C} - \text{OH} & \\
& | & \\
& \text{CH}_2\text{OH} &
\end{array}
$$

D – GLYCERALDEHYDE

$$
\begin{array}{ccc}
& \text{O} & \\
& \| & \\
& \text{C} - \text{H} & \\
& | & \\
\text{HO} - & \text{C} - \text{OH} & \\
& | & \\
& \text{CH}_2\text{OH} &
\end{array}
$$

L – GLYCERALDEHYDE

FIG. 1.21. The configurations of sugars, amino acids, and other compounds are based upon the use of the D and L forms of glyceraldehyde (or glycerose) as reference standards.

levorotatory, whereas, L-lactic acid is dextrorotatory.

THE RULE OF *n.* The total number of possible stereoisomers that may be formed by any given compound is determined by the rule of *n:* If *n* represents the number of asymmetric carbon atoms in a molecule, then the number of isomers which may be formed by that compound is 2^n. For example, the glucose molecule ($C_6H_{12}O_6$) has four asymmetric carbon atoms, hence 2^4 or 16 stereoisomers of this compound are possible; eight of these compounds will be D-compounds and eight will be the mirror images of these, or L-compounds.

EPIMERS. If two *sugars* exhibit a structural difference from one another only in their configuration about a *single* carbon atom, then these sugars are termed *epimers* (Fig. 1.20A); D-glucose and D-galactose are examples of such an epimeric pair, which differ in their configuration about the number 4 carbon atom. The interconversion of epimers, or epimerization, within body tissues may be exemplified by the conversion of galactose to glucose within the liver; the specific enzyme which catalyzes this process is called an *epimerase.*

TAUTOMERISM. Occasionally, an organic compound may behave in chemical reactions or in living systems as though the atoms in its molecules were arranged in more than one configuration, and each configuration may be expressed by a different structural formula. This type of isomerism is known as *tautomerism* or *dynamic isomerism.* Each isomer is known as the tautomer of the other. Keto-enol isomerism is one example of tautomerism. For instance, as shown in Figure 1.22, enolpyruvic acid and ketopyruvic acid readily undergo spontaneous tautomerization.

The interconvertible keto and enol isomers may not necessarily be isolable, but chemically pyruvic acid behaves as though it underwent transformation to the enol configuration before undergoing further metabolic reactions.

Summary of Organic Nomenclature and Organic Reactions

ALIPHATIC COMPOUNDS. Hydrocarbons, as the name implies, are composed solely of carbon

and hydrogen, and these elements are joined together by covalent bonds in a wide variety of straight (normal, abbreviated *n-*) or branched (isomeric form, abbreviated *iso-*) chains. Organic substances having such configurations are known as *aliphatic compounds* (Fig. 1.23).

If only single covalent bonds are present in the skeleton of the molecule, the compound belongs to a family of organic substances known as the *alkanes* (also called the saturated hydrocabon series or paraffin hydrocarbons). Several examples of such compounds are depicted in Figure 1.23. The simplest compound in the aliphatic series is, of course, methane. A few common aliphatic compounds, together with the alkyl groups derived from them, are given in Table 1.6.

Aliphatic hydrocarbons that have double bonds in the molecule are unsaturated compounds or *alkenes;* several examples of such substances also are given in Figure 1.23.

Many derivatives of aliphatic compounds are of great physiologic and biochemical importance in the body. For example, the saturated and unsaturated fatty acids, as well as their metabolic degradation products such as β-hydroxybutyric acid, are compounds with a basic alkane or alkene structure.

Aliphatic compounds often are termed *acyclic compounds,* since they contain no closed rings of carbon atoms; this type of structure is in contrast to the various cyclic compounds discussed below.

CYCLIC COMPOUNDS. Organic substances in which the carbon atoms are joined together in a circular pattern are called *cyclic compounds* or *ring compounds;* such compounds must contain three or more carbon atoms. As depicted in Figure 1.24A

$$
\begin{array}{ccc}
\text{COOH} & & \text{COOH} \\
| & & | \\
\text{C} = \text{O} & \rightleftharpoons & \text{C} - \text{OH} \\
| & & \| \\
\text{CH}_3 & & \text{CH}_2
\end{array}
$$

KETOPYRUVIC ACID ENOLPYRUVIC ACID

FIG. 1.22. An example of tautomerism (or dynamic isomerism) is given by the spontaneous and reversible interconversion of ketopyruvic acid into enolpyruvic acid. See also Figure 1.25C.

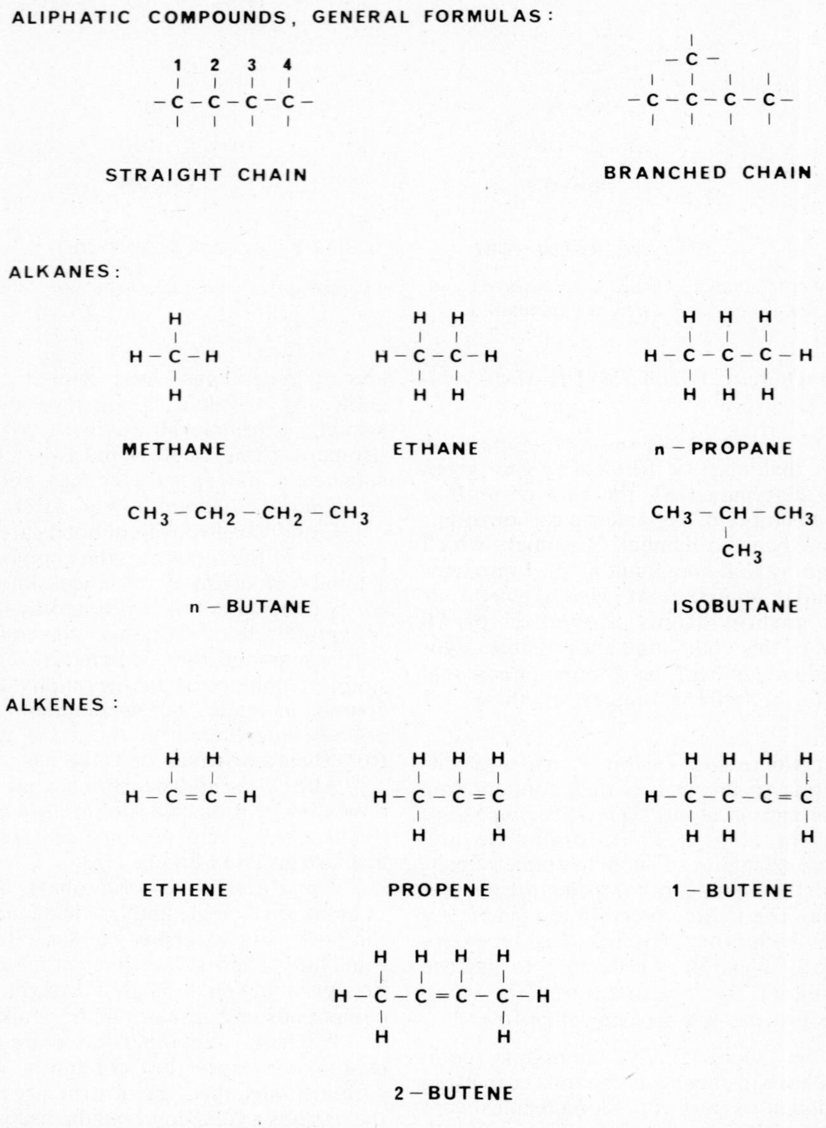

ALIPHATIC COMPOUNDS, GENERAL FORMULAS:

STRAIGHT CHAIN

BRANCHED CHAIN

ALKANES:

METHANE

ETHANE

n – PROPANE

$CH_3 - CH_2 - CH_2 - CH_3$

n – BUTANE

$CH_3 - CH - CH_3$
$| $
CH_3

ISOBUTANE

ALKENES:

ETHENE

PROPENE

1 – BUTENE

2 – BUTENE

FIG. 1.23. General formulas of aliphatic (straight chain) organic compounds as well as some specific examples of alkanes (saturated) and alkenes (unsaturated).

and B, ring compounds may be *aromatic* or *alicyclic*.

In aromatic compounds such as benzene, only carbon atoms are present in the ring; these differ from alicyclic compounds in being written as though alternating double and single bonds were present between the individual carbon atoms. Aromatic compounds are believed to exhibit the phenomenon of *resonance;* that is, two or more electronic structures may be written for the compound without changing the positions of the atoms in the molecule, as depicted in the hexagonal representations to the right in Figure 1.24A. In other words, the actual position of the shared electrons in the benzene molecule is a hybrid of the extremes

shown in the figure. The presence of hydrogen atoms is implicit in the abbreviated structural formulas.

Alicyclic compounds, like aromatic structures, have only carbon atoms present in their rings; but in contrast to aromatic compounds, *no* double bonds are present in the molecule. Therefore, the carbon atoms of alicyclic compounds are joined together by single covalent bonds, ie, by one pair of shared electrons. Alicyclic hydrocarbons are often called *cycloparaffins*. Several examples of this type of ring structure are given in Figure 1.24B.

In contrast to aromatic and alicyclic compounds, *heterocyclic rings* contain one or more atoms of another element in the ring in addition to

Table 1.6 COMMON ALKYL GROUPS

PARENT COMPOUND	NAME OF GROUP	STRUCTURAL FORMULA
Methane: CH_4	Methyl	CH_3
Ethane: CH_3—CH_3	Ethyl	CH_3—CH_2—
Propane: CH_3—CH_2—CH_3	*n*-Propyl	CH_3—CH_2—CH_2—
n-Butane: CH_3—CH_2—CH_2—CH_3	*n*-Butyl	CH_3—CH_2—CH_2—CH_2—

carbon (Fig. 1.25A). In compounds of biologic importance, these atoms are usually oxygen, nitrogen, or sulfur. Two specific groups of heterocyclic compounds having special biologic significance are derived from the parent compounds pyrimidine and purine (Fig. 1.25B), and these derived substances form essential constituents of the nucleic acids.

Pyrimidine derivatives exhibit keto-enol tautomerism. This phenomenon is illustrated in Figure 1.25C for the compound uracil.

A.

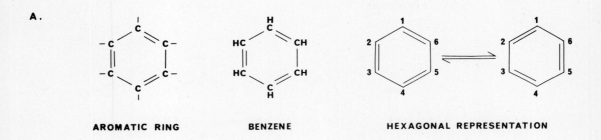

AROMATIC RING BENZENE HEXAGONAL REPRESENTATION

B.

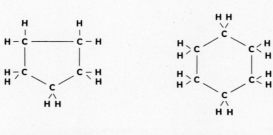

ALICYCLIC RINGS

CYCLOPENTANE CYCLOHEXANE

FIG. 1.24. Cyclic or ring compounds. **A.** Aromatic compounds, such as benzene, have only carbon atoms, three of which are joined by double bonds, present in the ring. Thus aromatic rings are unsaturated ring compounds. In the hexagonal representation to the right in the figure, note the numbering of the carbon atoms of the benzene molecule, as well as the spontaneous shift in the position of the double bonds, a phenomenon called resonance. **B.** Alicyclic compounds have only carbon atoms present in their ring structure, but in contrast to aromatic compounds, no double bonds are found in the ring; hence, alicyclic compounds are saturated. Compare with Figure 6.25, page 202.

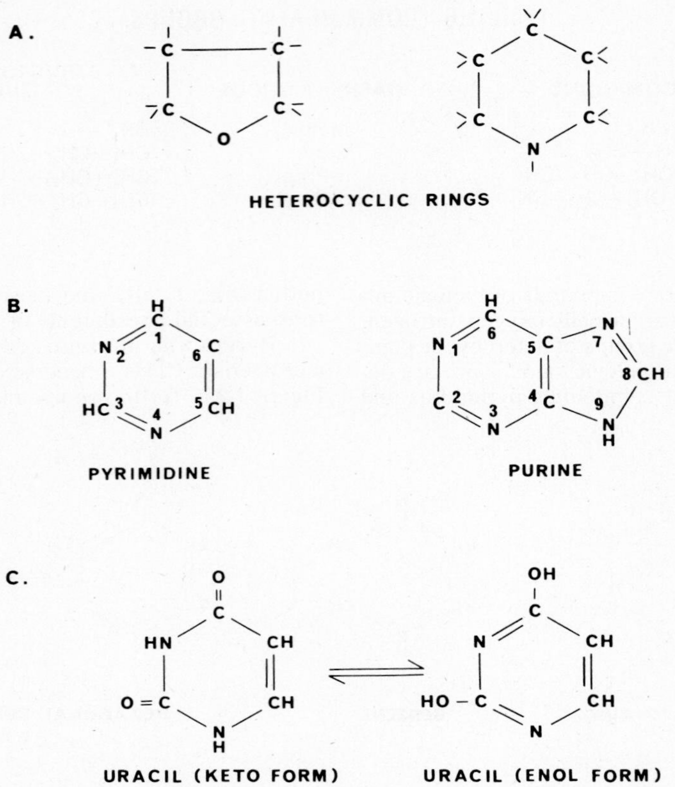

A.

HETEROCYCLIC RINGS

B.

PYRIMIDINE

PURINE

C.

URACIL (KETO FORM) ⟺ URACIL (ENOL FORM)

FIG. 1.25. General formulas of heterocyclic rings (**A**) as well as some specific examples of heterocyclic compounds (**B** and **C**). Note that heterocyclic rings contain an element other than carbon, and that double bonds may or may not be present.

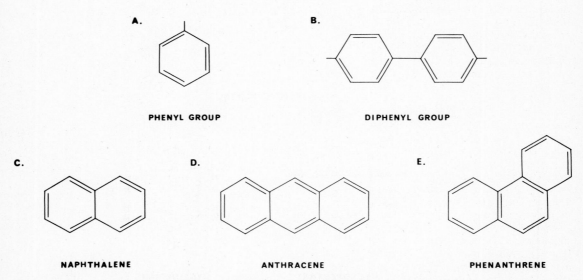

A.

PHENYL GROUP

B.

DIPHENYL GROUP

C.

NAPHTHALENE

D.

ANTHRACENE

E.

PHENANTHRENE

FIG. 1.26. Some aromatic radicals and compounds derived from the condensation of benzene molecules. **A.** The phenyl group, derived from benzene, is an important constituent of tyrosine and phenylalanine. **B.** The diphenyl group is an essential constituent of thyroxine and the catecholamines epinephrine and norepinephrine. **C.** Condensation of two benzene molecules yields naphthalene. **D.** Condensation of three benzene molecules in a linear position with respect to each other yields anthracene. **E.** Condensation of three benzene molecules in a nonlinear pattern yields phenanthrene. The phenanthrene nucleus provides the underlying structure of such physiologically important compounds as cholesterol, the steroid hormones of the adrenal cortex, and the male and female sex hormones.

A. BOAT **B. CHAIR**

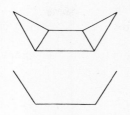 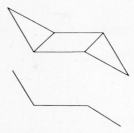

FIG. 1.27. Alicyclic compounds having five or more carbon atoms in their rings are aplanar. Consequently, the spatial arrangement of the atoms of the cyclohexane ring can assume one of two conformations, the so-called boat **(A)** or chair **(B)**. From a thermodynamic standpoint, the chair conformation is the most stable.

Examples of several complex aromatic rings formed by the condensation of benzene molecules are shown in Figure 1.26.

The *phenyl* group, derived from benzene (Fig. 1.26A), is an important constituent of several naturally occurring organic molecules. For example, two coupled phenyl groups (diphenyl, Fig. 1.26B) provide an essential feature of such physiologically active compounds as epinephrine, norepinephrine, and thyroxin, whereas a single phenyl group is a basic structural component of the amino acids phenylalanine and tyrosine.

Any alicyclic compound that contains five or more carbon atoms in its ring is *aplanar;* that is, the carbon atoms lie in different spatial planes. (This is in contrast to the carbon atoms of the benzene molecule, all of which lie in the same plane; hence, benzene is said to be a *planar* molecule.) Thus cyclohexane (Fig. 1.24B) can assume two spatial conformations: These are the so-called *boat* and *chair conformations* (Fig. 1.27A and B), and they represent more accurately the true three-dimensional shape of the cyclohexane molecule. The chair conformation of the cyclohexane molecule is more stable from a thermodynamic standpoint, and it is the form most commonly encountered in the biologically active groups of compounds known as *sterols.*

The complex alicyclic ring of perhydrocyclopentanophenanthrene (Fig. 1.28) is of importance here because this structure forms the parent, or *steroid,* nucleus of many physiologically active compounds. Note that the perhydrocyclopentanophenanthrene nucleus consists of three six-membered cyclohexane rings (A, B, and C in the figure) in the nonlinear or phenanthrene arrangement (Fig. 1.26E) attached to a five-membered terminal cyclopentane ring (Fig. 1.24B). The rings are fused as shown, and the 17 carbon atoms present in this structure are numbered as shown in the figure.

Many natural steroids have a side chain containing between 8 and 10 carbon atoms on position 17 as well as a functional alcoholic radical in the

molecule, and thus are *secondary alcohols,* called *sterols.*

A few examples of naturally occurring compounds with a perhydrocyclopentanophenanthrene nucleus are cholesterol, the adrenocortical hormones, the male and female sex hormones, and the cardiac glycosides. Sterols and steroids may be of either animal or plant origin.

Each of the four alicyclic rings present in the steroid nucleus is capable of assuming either the boat or the chair conformation (usually the latter). Consequently, the actual three-dimensional shape of individual bioactive steroids may be quite complex, and this physical property may assume major significance in determining the biologic activity of the particular molecule.

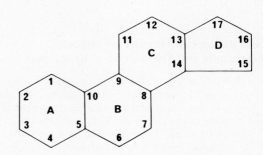

FIG. 1.28. The perhydrocyclopentanophenanthrene nucleus. The individual rings that comprise this chemical structure are denoted by letters (A, B, C, D), and the individual carbon atoms are indicated by numbers. This alicyclic ring can assume the boat and chair conformations indicated for cyclohexane in Figure 1.27; the latter is the form most commonly found in biologic material.

The perhydrocyclopentanophenanthrene ring is of great physiologic importance, as it is the parent structure of such compounds as cholesterol, the bile acids, and the adrenocortical steroids, as well as the male and female sex hormones. This ubiquitous structure, derived from phenanthrene (Fig. 1.26E), also provides the chemical foundation of a group of compounds found in plants known as cardiac glycosides, which are of considerable therapeutic significance in certain cardiac malfunctions.

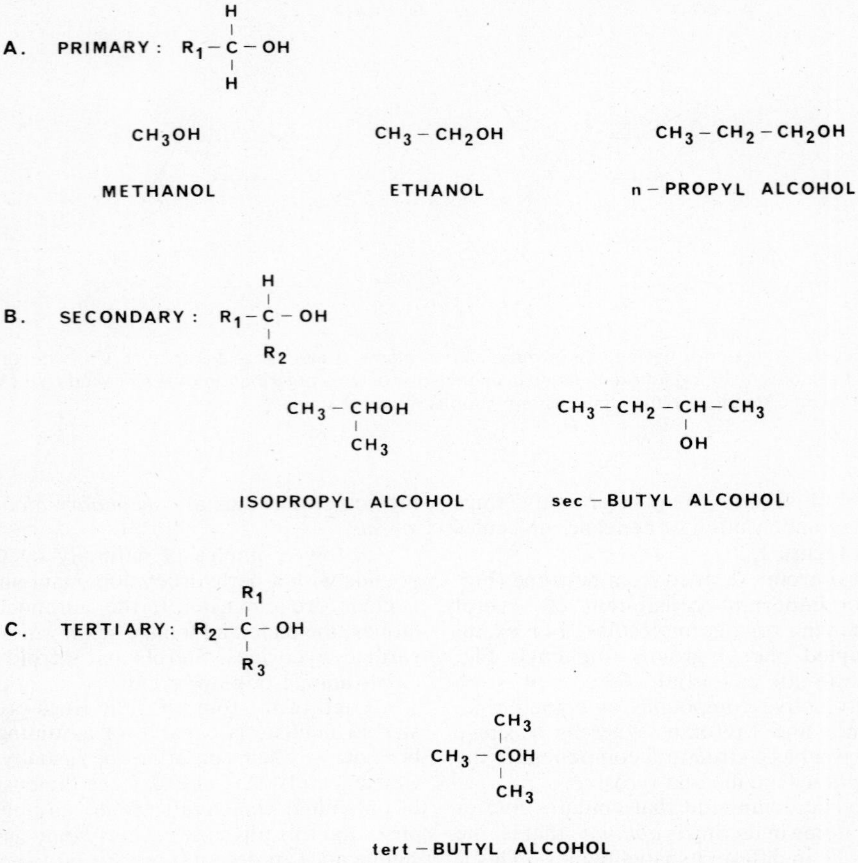

A.   PRIMARY: $R_1 - \overset{\overset{\displaystyle H}{|}}{\underset{\underset{\displaystyle H}{|}}{C}} - OH$

CH_3OH $CH_3 - CH_2OH$ $CH_3 - CH_2 - CH_2OH$

METHANOL   ETHANOL   n - PROPYL ALCOHOL

B.   SECONDARY: $R_1 - \overset{\overset{\displaystyle H}{|}}{\underset{\underset{\displaystyle R_2}{|}}{C}} - OH$

$CH_3 - \underset{\underset{\displaystyle CH_3}{|}}{CHOH}$ $CH_3 - CH_2 - \underset{\underset{\displaystyle OH}{|}}{CH} - CH_3$

ISOPROPYL ALCOHOL   sec - BUTYL ALCOHOL

C.   TERTIARY: $R_2 - \overset{\overset{\displaystyle R_1}{|}}{\underset{\underset{\displaystyle R_3}{|}}{C}} - OH$

$CH_3 - \overset{\overset{\displaystyle CH_3}{|}}{\underset{\underset{\displaystyle CH_3}{|}}{C}OH}$

tert - BUTYL ALCOHOL

FIG. 1.29. General formulas and examples of primary (**A**), secondary (**B**), and tertiary (**C**) alcohols.

Some Biologically Important Organic Compounds and Their General Reactions

ALCOHOLS AND PHENOLS. Organic compounds which contain the polar *hydroxyl group* (—OH)* as the functional (or reactive) portion of their molecules are called alcohols and phenols.

An *alcohol* may be considered as a derivative of an aliphatic hydrocarbon and has a nonpolar alkyl group in which one or more hydrogen atoms are replaced by the hydroxyl radical.

Phenols are compounds in which one or more hydrogen atoms attached to an aromatic ring *(aryl group)* are replaced by hydroxyl.

Customarily, alcohols are classified as *primary, secondary, or tertiary,* depending upon

The hydroxyl group (—OH) must not be confused with the electrovalent hydroxide ion (OH⁻), discussed on page 19. The hydroxyl group, as found in alcohols and phenols, is attached to the carbon atom by a covalent bond, and does not ionize when these compounds are in aqueous solution. Thus, aqueous solutions of alcohols and phenols do not exhibit alkaline properties.

whether the hydroxyl group is bound to a carbon atom attached to one, two, or three other carbon atoms, respectively. This classification of alcohols is illustrated by the general formulas together with appropriate examples of each type of alcohol in Figure 1.29.

Several phenols, or aromatic alcohols, are illustrated in Figure 1.30. In particular, catechol (1, 2-dihydroxybenzene) is of considerable importance in physiology, as this aromatic ring compound is an essential component of the catecholamine hormones epinephrine and norepinephrine.

Alcohols that have up to three carbon atoms are infinitely soluble in water; however, the water solubility of alcohols decreases in proportion to any further increase in the length of the alkyl chain. Stated another way, the solubility of an alcohol in water decreases as the nonpolar character of the molecule increases.

An alcohol may be *monohydric*, ie, have only one —OH group, or *polyhydric*, with more than one —OH group. The structural formulas of two common polyhydric alcohols, ethylene glycol and glycerol (or glycerin), are given in Figure 1.31. The

PHENOL CATECHOL RESORCINOL

HYDROXYQUINONE PYROGALLOL

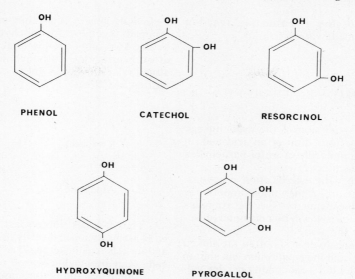

FIG. 1.30. Some examples of aromatic alcohols or phenols. In particular, catechol is physiologically important, as this compound is a basic structure in the constitution of the catecholamine hormones epinephrine and norepinephrine.

former compound is highly toxic in the body but frequently is used as an antifreeze liquid in the cooling systems of automobile engines; the latter is an essential chemical component of certain natural fatty substances, the glycerides.

Sugars are derivatives of polyhydric alcohols, and cyclic alcohols such as the sterols or inositol are examples of other naturally occurring alcohols.

Polyhydric alcohols are highly polar compounds, unlike their monohydric equivalents. Therefore, polyhydric alcohols are far more water soluble than monohydric alcohols with the same number of carbon atoms. Consequently, sugars are quite soluble in water.

Alcohols may undergo three general types of chemical reaction, each of which can take place in biochemical systems: *oxidation, esterification* and *hydrolysis,* and *thioether formation.*

Oxidation or Dehydrogenation. Primary alcohols may be oxidized (dehydrogenated) by suitable enzyme systems or by strong oxidizing agents to form *carboxylic acids* (p. 41), in accordance with the following general equation: $R{-}CH_2{-}OH + O_2 \rightarrow R{-}COOH + H_2O$.

The oxidation of a secondary alcohol, on the other hand yields a ketone:

$$\begin{matrix} R_1 \\ \diagdown \\ CH{-}OH \\ \diagup \\ R_2 \end{matrix} + O_2 \rightarrow \begin{matrix} R_1 \\ \diagdown \\ C{=}O \\ \diagup \\ R_2 \end{matrix} + H_2O$$

In contrast to primary and secondary alcohols, tertiary alcohols cannot be oxidized without rupture of a C—C bond. Consequently, tertiary alcohols do not undergo oxidation readily.

Esterification and Hydrolysis. When a primary, secondary, or tertiary alcohol reacts with either an organic or an inorganic acid, a molecule of water is split out, and the resulting compound is called an *ester*. The general reaction for esterification is:

$$R_1{-}\overset{\displaystyle O}{\overset{\|}{C}}{-}OH + HO{-}R_2 \rightarrow R_1{-}\overset{\displaystyle O}{\overset{\|}{C}}{-}O{-}R_2 + H_2O$$

Many lipids (fatty substances) contain carboxylic ester linkages. For example, simple lipids are esters of fatty acids with various alcohols, whereas *fats* are esters of fatty acids with the polyhydric alcohol glycerol.

As noted above, ester formation, or esterification, may also take place between inorganic acids and an alcohol. For example, certain esters formed between phosphoric acid (H_3PO_4) and various sugars are important in physiology and biochemistry.

$$\begin{matrix} CH_2-OH \\ | \\ CH_2-OH \end{matrix} \qquad\qquad \begin{matrix} CH_2-OH \\ | \\ CH_2-OH \\ | \\ CH_2-OH \end{matrix}$$

A. ETHYLENE GLYCOL B. GLYCEROL

FIG. 1.31. Two examples of polyhydric alcohols. **A.** Ethylene glycol commonly is used an antifreeze. **B.** Glycerol is an essential component of the group of dietary and body lipids known as neutral fats or glycerides.

The reverse of the esterification reaction shown above is termed *hydrolysis*.

$$R_1-\overset{\overset{\textstyle O}{\|}}{C}-O-R_2 + H_2O \rightarrow R_1-\overset{\overset{\textstyle O}{\|}}{C}-OH + HA-R_2$$

In this process, scission of the ester linkage takes place with the formation of an acid and an alcohol; note that a molecule of water participates in the hydrolytic reaction.

As is the case with esterification reactions, hydrolytic reactions are of the utmost importance in physiology and biochemistry; specific examples of both of these processes will be encountered frequently throughout this text.

Esterification and hydrolysis often are linked in biologic systems, and the equilibrium can shift in either direction under different conditions, depending upon the specific reactants and enzyme systems involved.

Ether and Thioether Formation. Additional Sulfur Analogs of Oxygen Compounds. Primary, secondary, and tertiary alcohols form *ethers* by replacement of the hydrogen atom of their —OH group with an alkyl group. The general formula for an ether may be written R_1-O-R_2. This ether linkage occurs rarely in living systems, however. Sulfur, on the other hand, belongs to the same group of elements as oxygen in the periodic table, and thus both elements form chemically similar compounds. Hence, *thioalcohols (mercaptans), thioesters,* and *thioethers* all are found in living systems. The general formulas of such compounds are:

$$R-CH_2-SH \qquad R_1-\overset{\overset{\textstyle O}{\|}}{C}-S-R_2$$

Thioalcohol (mercaptan)   Thioester

$$R_1-S-R_2$$

Thioether

The general chemical reactions of these thio- (or sulfur-containing) compounds in biologic systems are quite similar to those encountered for their oxygen analogs.

In addition to the three general groups of thio-compounds outlined above, sulfur forms *disulfide linkages* which are extremely important in protein structure but which have *no* oxygen counterpart. The general formula for the disulfide linkage is: $R_1-S-S-R_2$.

ALDEHYDES AND KETONES. The aldehydes and ketones both contain a strongly reducing carbon-oxygen bond known as a *carbonyl group*

(—C=O); hence, such compounds generally are grouped together. An aldehyde has one and a ketone two alkyl groups attached to the carbon atom bearing the carbonyl group. The general structural formulas of these compounds may be written as follows:

$$R-\overset{\overset{\textstyle H}{|}}{C}=O \qquad \begin{matrix} R_1 \\ \diagdown \\ C=O \\ \diagup \\ R_2 \end{matrix}$$

Aldehyde   Ketone

Thus ketones, in contrast to aldehydes, have *no* H atoms attached directly to the carbonyl group, and for this reason ketones are less reactive chemically than are aldehydes.

Aldehydes and ketones are of great biochemical significance; for example, sugars are not only polyhydric alcohols, but also are either aldehydes or ketones.

Some biochemically important reactions of aldehydes and ketones are the following.

Oxidation. Ketones do not oxidize readily, since they cannot lose oxygen without rupture of a C—C bond, a property similar to that exhibited by tertiary alcohols.

Aldehydes, in contrast to ketones, can undergo oxidation to the corresponding carboxylic acid, in accordance with the following general reaction:

$$R-\overset{\overset{\textstyle H}{|}}{C}=O + \tfrac{1}{2}O_2 \rightarrow R-COOH$$

Reduction. Reduction of an aldehyde produces the corresponding primary alcohol:

$$R-\overset{\overset{\textstyle H}{|}}{C}=O + 2H^+ \rightarrow R-CH_2OH$$

Reduction of a ketone yields the corresponding secondary alcohol:

$$\overset{\overset{\textstyle R_1}{|}}{\underset{\underset{\textstyle R_2}{|}}{C}}=O + 2H^+ \rightarrow R_1-\overset{\overset{\textstyle H}{|}}{\underset{\underset{\textstyle R_2}{|}}{C}}-OH$$

Hemiacetal and Acetal Formation. Aldehydes can react with either one or two of the hydroxyl groups of an alcohol, forming a hemiacetal or an acetal, respectively:

$$R_1\!\!-\!\overset{\overset{\displaystyle H}{|}}{C}\!\!=\!O + R_2OH \rightarrow R_1\!\!-\!\overset{\overset{\displaystyle H}{|}}{\underset{\underset{\displaystyle O\!\!-\!R_2}{|}}{C}}\!\!-\!OH$$

Aldehyde   Alcohol   Hemiacetal

$$R_1\!\!-\!\overset{\overset{\displaystyle H}{|}}{C}\!\!=\!O + 2R_2OH \rightarrow R_1\!\!-\!\overset{\overset{\displaystyle H}{|}}{\underset{\underset{\displaystyle O\!\!-\!R_2}{|}}{C}}\!\!-\!O\!\!-\!R_2 + H_2O$$

Aldehyde   Alcohol   Acetal

Note that the alcohol and carbonyl groups both may function as components of the same molecule. Thus, aldose (aldehyde) sugars are present in solution principally as *internal hemiacetals*.

Aldehydes also may react with thioalcohols (mercaptans) to form thiohemiacetals and thioacetals.

$$R_1\!\!-\!\overset{\overset{\displaystyle H}{|}}{C}\!\!=\!O + R_2\!\!-\!SH \rightarrow R_1\!\!-\!\overset{\overset{\displaystyle H}{|}}{\underset{\underset{\displaystyle S\!\!-\!R_2}{|}}{C}}\!\!-\!OH$$

Aldehyde   Thioalcohol   Thiohemiacetal

$$R_1\!\!-\!\overset{\overset{\displaystyle H}{|}}{C}\!\!=\!O + 2R_2SH \rightarrow R_1\!\!-\!\overset{\overset{\displaystyle H}{|}}{\underset{\underset{\displaystyle S\!\!-\!R_2}{|}}{C}}\!\!-\!S\!\!-\!R_2 + H_2O$$

Aldehyde   Thioalcohol   Thioacetal

In the enzymatic oxidation of aldehydes to acids, thiohemiacetals are present as enzyme-bound intermediates.

Aldol Condensation. Aldehydes and, to a more limited extent, ketones, under alkaline conditions, may undergo condensation between their α-carbon atoms and carbonyl groups. The compounds that can result from this condensation process are aldols, β-hydroxy aldehydes, or β-hydroxy ketones. For example, two molecules of acetaldehyde may condense to form one molecule of β-hydroxybutyraldehyde:

$$CH_3\!\!-\!\overset{\overset{\displaystyle H}{|}}{C}\!\!=\!O + CH_3\!\!-\!\overset{\overset{\displaystyle H}{|}}{C}\!\!=\!O + OH^- \rightarrow$$

$$CH_3\!\!-\!\overset{\overset{\displaystyle H}{|}}{\underset{\underset{\displaystyle OH}{|}}{C}}\!\!-\!CH_2\!\!-\!\overset{\overset{\displaystyle H}{|}}{C}\!\!=\!O$$

The β-hydroxy acids derived from the oxidation of aldehydes such as this are of great importance in the metabolism of fatty acids.

CARBOXYLIC ACIDS. Many organic compounds found in living systems contain a characteristic functional group of atoms known as *carboxyl group*. This name is derived from a contraction of the words *carbonyl* and *hydroxyl*, both of which groups form the carboxyl group.

The carboxyl group may be written —COOH
or $-\overset{\overset{\displaystyle O}{\|}}{C}\!\!-\!OH$.

Carboxylic acids typically are weak acids, and dissociate only partially in water, yielding a proton (H^+) and a carboxylate anion ($R\!\!-\!COO^-$). The negative charge of the anion is shared equally between the two oxygen atoms.

It is worth noting that the presence of the carbonyl group (—C=O) on the same carbon atom as the hydroxyl (—OH) group so modifies the reactivity of the hydroxyl group that dissociation of the hydrogen atom occurs with greater facility than is the case with an alcohol. Therefore, one major difference between carboxylic acids and alcohols is the much greater acidity of the former compared to the latter.

A number of physiologically important reactions of carboxylic acids include reduction, salt formation, amide formation, and acid anhydride formation, as well as ester and thioester formation.

Many biologically active organic acids contain two carboxyl groups; hence, they are known as *dicarboxylic acids*. Aspartic and glutamic acids, for example, are dicarboxylic compounds.

Reduction. The complete reduction of a carboxylic acid yields the corresponding primary alcohol in accordance with the following general reaction:

$$R\!\!-\!COOH + 4H^+ \rightarrow R\!\!-\!CH_2OH + H_2O$$

Salt Formation. Carboxylic acids react stoichiometrically, ie, equivalent for equivalent, with bases. The salts thus formed dissociate completely in aqueous solution.

Amide Formation. An amide is formed when a molecule of water is split off between a carboxylic acid and an amine. For example:

$$CH_3\!\!-\!\overset{\overset{\displaystyle O}{\|}}{C}\!\!-\!OH + NH_3 \rightarrow CH_3\!\!-\!\overset{\overset{\displaystyle O}{\|}}{C}\!\!-\!NH_2 + H_2O$$

Acetic acid   Ammonia   Acetamide

Peptides are amides of enormous biologic significance. The peptide linkage is formed between the amino (—NH₂) group of one amino acid and the carboxyl group (—COOH) of another, according to the following general reaction:

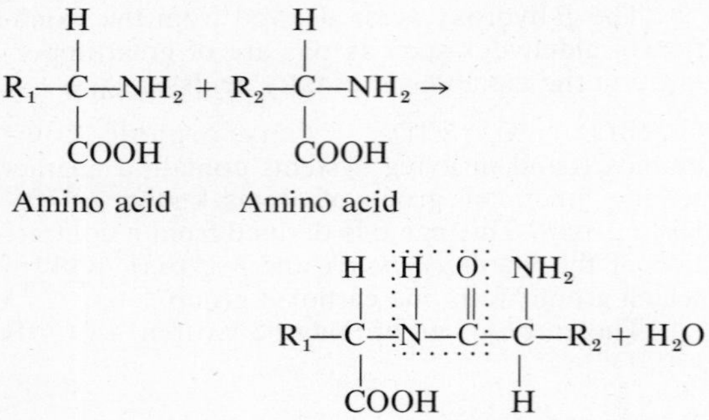

$$
\begin{array}{ccc}
\text{H} & & \text{H} \\
| & & | \\
R_1\text{—C—NH}_2 & + & R_2\text{—C—NH}_2 \rightarrow \\
| & & | \\
\text{COOH} & & \text{COOH} \\
\text{Amino acid} & & \text{Amino acid}
\end{array}
$$

$$
\begin{array}{c}
\quad\ \text{H}\ \ \text{H}\ \ \text{O}\ \ \text{NH}_2 \\
\quad\ |\quad |\quad ||\quad | \\
R_1\text{—C—N—C—C—}R_2 + \text{H}_2\text{O} \\
\quad\ |\qquad\quad | \\
\quad\ \text{COOH}\qquad \text{H}
\end{array}
$$

Dipeptide

The peptide linkage, shown in the box, provides the fundamental chemical union between amino acids in all peptides, polypeptides, and proteins found in living matter.

The relative strengths of carboxylic acids are expressed in terms of pKa values as discussed on page 23. Briefly, for the dissociation reaction

$$R\text{—COOH} \rightarrow R\text{—COO}^- + \text{H}^+$$

the dissociation constant (Ka) is defined by the expression:

$$Ka = \frac{[\text{RCOO}^-][\text{H}^+]}{[\text{RCOOH}]}$$

Acid Anhydride Formation. When the carboxyl groups of two carboxylic acid molecules react together, one molecule of water is split out, and an acid anhydride is formed:

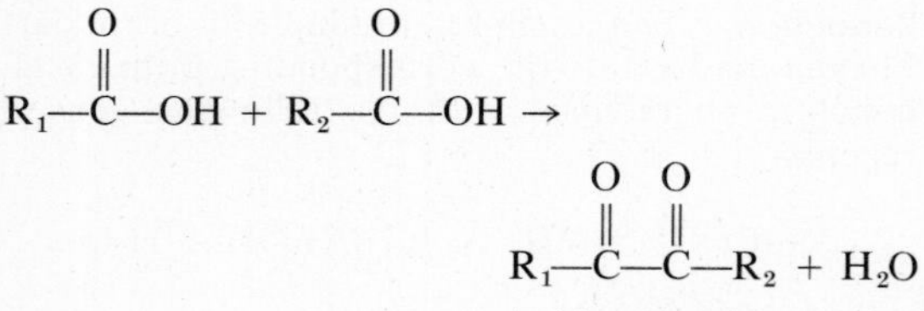

$$
\begin{array}{ccc}
\text{O} & & \text{O} \\
|| & & || \\
R_1\text{—C—OH} & + & R_2\text{—C—OH} \rightarrow
\end{array}
$$

$$
\begin{array}{c}
\text{O}\ \ \text{O} \\
||\ \ || \\
R_1\text{—C—C—}R_2 + \text{H}_2\text{O}
\end{array}
$$

Acid anhydride

If both molecules of acid involved in this reaction are the same, then a *symmetric anhydride* is formed, whereas if the acids are different (eg, R_1—COOH; R_2—COOH), a *mixed anhydride* is produced, as depicted in the above reaction.

In living systems, important anhydrides include those of phosphoric acid, such as is found in adenosine triphosphate.

Ester and Theioester Formation. See the discussion concerned with alcohols (p. 38).

AMINES. Derivatives of ammonia (NH_3) in which one, two, or three hydrogen atoms are replaced by alkyl or aryl groups are known as amines. If one hydrogen atom of the ammonia molecule is replaced by an alkyl or aryl radical, then the amine is classed as a *primary amine*. Similarly, if two hydrogen atoms are replaced, the compound is a *sec-*

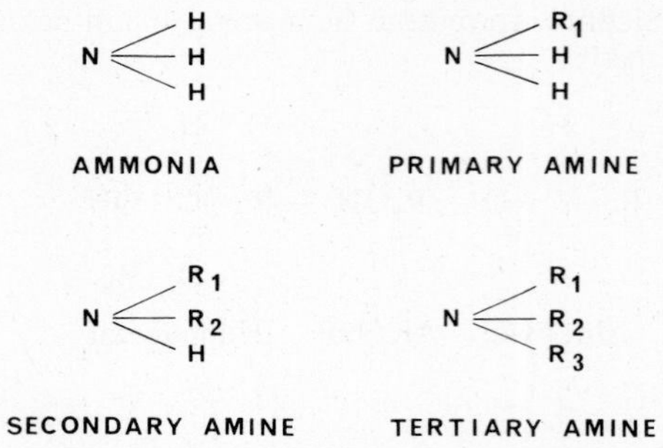

FIG. 1.32. The chemical relationship of primary, secondary, and tertiary amines to their parent compound, ammonia, are indicated. The symbols R_1, R_2, and R_3 denote substitution of one or more chemical groups or radicals for one or more hydrogen atoms.

ondary amine; and if all three of the hydrogen atoms are replaced by other groups, then the resulting compound is a *tertiary amine*. The general formulas of these three types of amines are shown in Figure 1.32.

In aqueous solution, ammonia is present as both charged and uncharged molecules:

$$\text{NH}_3 + \text{H}^+ \rightleftharpoons \text{NH}_4^+$$

Amines behave in an entirely similar manner:

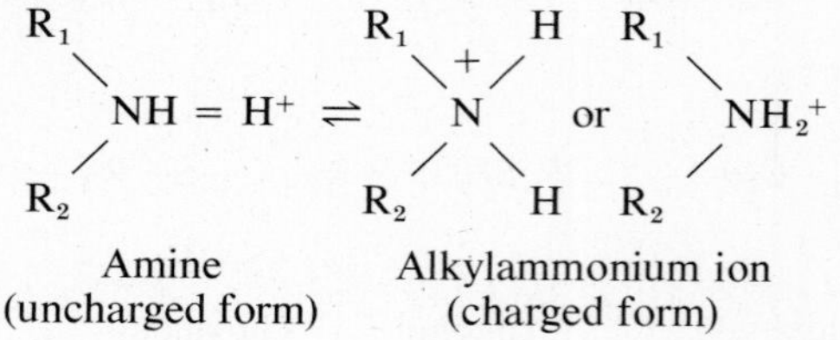

$$
\begin{array}{ccc}
R_1 & & R_1\ \ \text{H}\quad\ R_1 \\
\diagdown & & \diagdown\ +\ \diagup\quad\ \diagdown \\
\text{NH} = \text{H}^+ \rightleftharpoons & & \text{N}\quad \text{or} \quad \text{NH}_2^+ \\
\diagup & & \diagup\ \diagdown\quad\ \diagup \\
R_2 & & R_2\ \ \text{H}\quad\ R_2
\end{array}
$$

Amine   Alkylammonium ion
(uncharged form)   (charged form)

Amines are *proton acceptors* in the uncharged form; consequently, they act as bases under such conditions. When present in their charged form, however, amines can act as *proton donors,* and hence behave as acids. The relative strengths of different amines may be expressed by suitable pKa values, in a manner quite similar to that used for describing the relative strengths of carboxylic and other weak acids as discussed earlier (p. 23).

The pKa values for various aliphatic amines (Table 1.7) indicate that these compounds are weaker acids (or stronger bases!) than ammonia; but, most importantly, at pH 7.4 practically all of the aliphatic amines are in the undissociated or *charged form*. Since this is the pH associated with most body fluids, amines in the body fluids are associated with an anion such as chloride (Cl$^-$).

In contrast to aliphatic amines, the nitrogen atoms of aromatic amines such as the purines and pyrimidines are fairly strong acids. Consequently, at pH 7.4 most aromatic amines are present in the dissociated (or uncharged) form.

Table 1.7 ACID DISSOCIATION CONSTANTS OF CERTAIN AMINES

NAME	ACID FORM	pK_a
Ammonia	NH_4^+	9.26
Methylamine	$CH_3NH_3^+$	10.64
Dimethylamine	$(CH_3)_2NH_2^+$	10.72
Trimethylamine	$(CH_3)_3NH^+$	9.74

Note that the terms *dissociated* and *uncharged* are synonymous when used in connection with amines, in sharp contrast to the situation which obtains with other electrovalent compounds. This is shown by the relationship:

$$NH_3 + H^+ \rightleftharpoons NH_4^+$$

(Dissociated,   (Undissociated,
  uncharged   charged form)
    form)

Furthermore, note that ammonia (NH_3) can act as a *base* in accepting a proton to form an ammonium ion (NH_4^+), whereas, when the ammonium ion dissociates, it behaves as an *acid* by donating a proton.

Amide formation by amines was discussed earlier (p. 41).

Amines and their derivatives are extremely important in physiology and medicine. For example, such compounds participate in many biochemical reactions involving amino acids, lipids, and nucleic acids. In addition, many pharmacologically active compounds and drugs are amines.

CHAPTER 2

Introduction: Some General Concepts

The difference between a piece of stone and an atom is that an atom is highly organized, whereas the stone is not. The atom is a pattern, and the molecule is a pattern, and the crystal is a pattern; but the stone, although it is made up of these patterns, is just a mere confusion. It's only when life appears that you begin to get organization on a larger scale. Life takes the atoms and molecules and crystals; but, instead of making a mess of them like the stone, it combines them into new and more elaborate patterns of its own.

Aldous Huxley
Time Must Have a Stop

Physiology is that branch of the biologic sciences which deals with the functions that characterize living organisms, in contrast to the processes that occur in the inanimate world. Ultimately, physiology is a study of the controlled physicochemical mechanisms whereby living systems carry out the various processes that collectively are termed "life."

Philosophically, the science of physiology is not concerned particularly with *why* a given process takes place within a living system, but rather with *how* that process takes place. Therefore, questions appropriate to the province of physiology are: How does a nerve conduct impulses? How does a muscle perform mechanical work? How does the kidney produce urine? What are the mechanisms involved in the cardiorespiratory adaptations to exercise?

As stated in the quotation from Huxley which opens this chapter, the organization of atoms and molecules into elaborate structural patterns is a basic feature of all living matter; but this is only one aspect of the physiologic approach to the study of life processes. Not only is there a complex morphologic organization within a living organism that ranges from the gross down to the atomic level, but

there are also a number of critical functional attributes that serve to distinguish living from nonliving systems.

First, a living organism is discrete from its environment, and thus presents a physicochemical discontinuity from its surroundings which is maintained only by a continuous exchange or *turnover* of materials with that environment. In other words, a living organism is an unstable *open system* in a thermodynamic sense, and maintains its characteristic and unique structural and functional properties only by the continual expenditure of energy through a multitude of chemical reactions that are termed collectively *metabolism*. Thus, a living system is at all times in a state of *dynamic equilibrium,* in contrast to the static equilibrium exhibited by a crystal or a stone.

Second, it is implicit in the foregoing remarks that, in order for an organism to maintain its particular characteristics while acting as an open system, there must be an extremely sophisticated *control* exerted over the exchange of materials and energy, both within the living system and with the environment, so that the unique properties of the organism are maintained. In fact, a large part of the science of physiology is concerned with a study of the specific mechanisms that are responsible for regulating the constancy of material and energy exchange in order that the structural and functional properties of the organism be maintained at relatively stable levels.

Third, in higher organisms such as man, whose body contains many specialized tissues and organs that subserve particular functions, there must be effective mechanisms whereby the individual activities of the various structures are *coordinated* in an orderly fashion that is of benefit to the individual as a whole. Therefore, communications and bulk transport systems within the body are essential to effect these processes. Effective control of these internal processes is critical to survival.

Fourth, the organism must have means whereby it can react and adapt in an appropriate manner to the external environment itself.

Fifth, survival of the particular species of organism must be assured through suitable reproductive processes.

Certain aspects of the foregoing generalizations will be examined in greater detail in the sections to follow. However, it should be reiterated that a living system possesses an extremely high degree not only of structural but also of functional organization within its many components. Both of these characteristics are maintained only through a regulated exchange of materials and energy with the environment.

WATER: THE MATRIX OF LIFE

Without the presence of water, life as we know it would be impossible. The physicochemical properties of this deceptively simple and ubiquitous substance provide the milieu in which the myriad processes which are termed "life" take place. Therefore, water quite literally is the matrix in which the normal structural and functional attributes of life are manifest, and as such it is an indispensable component of protoplasmic systems. From a quantitative standpoint, water comprises the largest single constituent of such systems. Furthermore, water not only provides the matrix for the life processes but also participates actively in many of these processes.

MAN AS AN AQUATIC ANIMAL

Despite his terrestrial habitat, man is essentially an aquatic organism, a feature he shares with other land-dwelling animals. A moment's reflection will indicate the essential truth of this statement.

All living tissues in the human body that are exposed directly to the external environment must be bathed continuously in a film of aqueous fluids, however thin this film may be. The prime example of this fact is found in the alveoli of the lungs, where the actual exchange of atmospheric gases with those in the blood takes place. For respiratory exchange to occur, oxygen first must pass into solution within the fluid lining the alveolar sac, from whence it moves through the alveolar membrane to the blood within the pulmonary capillary. The maintenance of this miniscule volume of fluid is of critical importance to the survival and well-being of the individual.

Consider also the mucous membranes that are in direct contact with the external environment. These tissues also are bathed in various watery solutions. The lining of the nasal, respiratory, oral, gastrointestinal, and urogenital tracts may serve as examples. The corneas of the eyes and related ocular structures, as well as the bodily orifices, also are continually laved with fluids.

The remainder of the body surface that is exposed to the atmosphere is covered with the integument, a thin layer of keratinized and dead cells. Although the skin is far from a lifeless packaging material for the internal structures, at the environment–body interface, it does provide a relatively inert and impermeable barrier. In its deeper layers, however, the skin is a most complex and physiologically active structure. Here, too, an aqueous habitat is essential to the survival of its component cells.

Turning to the interior of the body, one can readily appreciate the fact that all cells, tissues, and organs must live in an aqueous medium whose complex and varied composition must be maintained within relatively narrow limits throughout the life of the individual.

These points have been emphasized because a large part of the science of human physiology deals explicitly with those mechanisms involved in the maintenance of both the quantity and the composition of the body fluids. Furthermore, these fluids subserve many vital bulk transport functions for various nutrients as well as waste products that are destined for elimination. The critical nature of the body fluids becomes even more evident in many pathologic situations encountered in clinical medicine.

Man, once again in common with other terrestrial organisms, has achieved independence from a strictly limited existence in an aquatic or marine habitat for a number of reasons — principally because a host of complex physiologic regulatory mechanisms have developed during the long and tortuous path of evolution that enable him to survive in, and adapt to, an essentially hostile environment in the atmosphere. One of the principal mechanisms, as just discussed, has been the development of a suitable "portable pond" within which his cells can function effectively. A second major development along this evolutionary path was the condition of homeothermy. This ability to regulate the body temperature independently from that of the external environment is found only in mammals and birds among the million-odd living species with which we share this planet, and the specific physicochemical properties of water itself play a fundamental role in this process.

THE BIOCHEMICAL AND MORPHOLOGIC ORGANIZATION OF LIVING SYSTEMS

In all living organisms, an ascending series of hierarchies of ever-increasing complexity is present within the biochemical and morphologic constitution of the living system (Fig. 2.1). As indicated by the arrows in this conceptual diagram, note that a potentiality for free exchange of substances among the various biochemical and morphologic compartments of the body is present, and that water is the fundamental transport medium involved in such exchange.

The important concepts to be obtained from Figure 2.1 may be summarized briefly as follows.

Atomic, Molecular, and Macromolecular Organization

At the atomic level are found the ionic forms of certain elements such as sodium, potassium, and chloride. Next are found relatively simple inorganic and organic molecules including water, amino acids, and simple sugars. These substances may be present as independent chemical entities or may be organized with varying degrees of intricacy into still larger and larger biochemical compounds with ever greater complexity, until the ultimate size is reached in such structures as bimolecular lipid layers and macromolecules like the proteins and nucleic acids. These macromolecules, in turn, may be polymerized so as to form structures that are

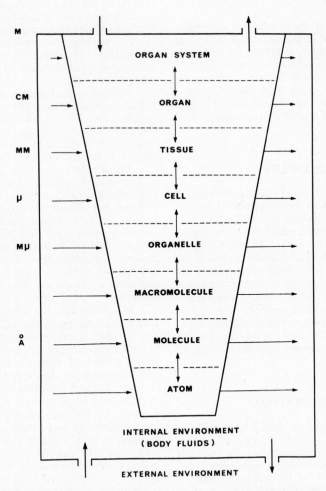

FIG. 2.1. Diagram showing the hierarchical levels of organization within a complex living system and their interrelationships with the external environment. Note that all levels of organization are connected with each other either directly or indirectly, actually or potentially (arrows). The inverted pyramid in the center of the figure denotes progressively increasing levels of physicochemical as well as morphologic complexity. The approximate magnitudes at each level are given by the scale on the left of the figure. The arrows at the top and bottom are indicative of the exchange of material, energy, and information between the body and the external environment.

visible under suitable magnification and which constitute the morphologic framework as well as certain organelles of the cell.

The Cell, Organelles, and the Cell Membrane

Within living systems, it is convenient as well as traditional to think of the *cell* as the basic structural and functional unit of living matter. A cell may be defined as a discrete mass of protoplasm (known as the *cytoplasm*) that is bounded by a *plasma (cell) membrane* and which can function as an independent unit. In the eukaryotic cells of higher organisms, a *nucleus* is the most obvious cytoplasmic organelle; but in addition to this structure, typical cells usually contain such organelles as mitochondria, ribosomes, lysosomes, Golgi apparatus, and inclusions of various types. These organelles are discussed in greater detail in Chapter 3 (p. 51).

The plasma membrane of cells in general has the important functional property of *selective permeability* to various ionic and molecular species. This membrane may be considered as a specialized boundary or interface between the cell interior and its immediate environment which has specific morphologic and physiologic properties (p. 71).

The cell membrane is of particular functional significance as it regulates the passage of materials into and out of the cell, whether this is accomplished by *passive transport* (p. 85) or by *active transport* (p. 92). One direct consequence of these transport mechanisms is that all living cells possess a measurable electric potential across their membranes at rest. By convention, the interior of the membrane is considered negative to the exterior under such conditions, and this typical polarity must be considered an integral part of the physical constitution of the living cell.

Tissue

Any aggregation of similar cells that performs special functions is called a *tissue*. Although normally a liquid, the blood may be considered as a complex fluid tissue.

Organ

The gross morphologic units of the body are called *organs*, and these structures may be composed of one or more kinds of tissue. An organ can subserve one or more basic physiologic roles in the body.

From an anatomic standpoint, an organ usually is considered to be a discrete mass of tissue (eg, the pancreas, liver, kidney), irrespective of its functional attributes.

Organ Systems

Groups of organs, which may or may not be related specifically on a purely anatomic basis but which are concerned with the performance of one or more general physiologic goals, constitute the top level in the morphologic hierarchy — the *organ system*.

The major organ systems of the body may be summarized on a functional basis: (1) locomotor; (2) nervous; (3) cardiovascular; (4) respiratory; (5) excretory; (6) digestive; (7) endocrine; (8) reproductive.

Magnitudes at the Various Levels of Organization

It is helpful to have a clear appreciation of the magnitudes involved at the various levels of biochemical and morphologic organization, as shown by the approximate dimensions on the left in Figure 2.1. The reader can relate these dimensions to any given physiologic process under discussion. For example, one is dealing in the size range of a few to several hundred angstrom units when one is considering events taking place at the atomic, molecular, or cell membrane level. Within the size ranges of individual cells and their components, the dimensions are in millimicrons or in microns. When one considers tissues, organs, and organ systems, however, the magnitudes are reckoned in terms of millimeters, centimeters, or even meters at the level of the entire body.

A similar broad range of magnitudes must also be considered when dealing with chemical quantities as well as the total masses and volumes of tissues and other substances involved in the bodily processes.

SOME ASPECTS OF BIOLOGIC CONTROL SYSTEMS

Since large areas of physiology are concerned intimately with the regulation of biologic processes at all levels of organization, it is imperative to consider several general concepts relating to the mechanisms involved in such control systems.

Every biologic system has inherent controls that either initiate or accelerate various processes under certain conditions, and that terminate or decelerate the same processes under other conditions. In other words, all biologic processes must be turned *on* or *off* under appropriate circumstances; otherwise, a chaotic situation would ensue that would be detrimental to survival of the cell or organism. These general statements are of fundamental significance to an understanding of bioregulatory processes.

A Note on Surfaces

The balanced exchange of materials and energy between an organism and its environment requires not only an exceedingly sophisticated control over the multitude of substances that enter and leave the body, but also a high degree of specialization of certain structures which are closely related to the exchange processes. Certain *surfaces* are of extreme importance in this regard, because these are the regions where the actual exchange takes place. Included in this category are the plasma membranes of cells as well as the epithelial linings of certain organs, including the lungs, gastrointestinal tract, kidneys, capillaries, and skin. All of these structures are semipermeable membranes that are specialized to allow certain substances to cross or be transported readily, whereas other materials are allowed such passage only with difficulty, or not at all.

The compartments separated by membraneous structures may be *internal* or *external*. For example, the capillary membrane separates the blood from the interstitial fluid, while the skin as well as the mucous membrane of the gastrointestinal tract separate the internal compartments of the body from the external environment.

Consequently, if a cell or an entire organism lacked a quantitative as well as qualitative control over the flow of materials across these various interfaces, then a physicochemical equilibrium would be reached shortly, and the organism would cease to be a living entity.

In the human body, the major exchanges of materials and energy take place principally in the lungs, gut, and kidneys. The skin, except for its vital role in thermal energy exchange, is a relatively minor region insofar as exchange of materials between the body and the environment is concerned.

The lungs, gut, and kidneys share two common features that render them able to perform their exchange functions effectively. First, these organs are arranged morphologically so that they have an enormous total surface area relative to their volume. Second, the surface membranes of these structures are in exceedingly close contact with their particular vascular beds, so that exchange of substances between the capillary blood and the other region served by these organs is accomplished with great facility.

At the molecular level in the hierarchy of organization, it is important to note that the process of catalysis by enzymes also takes place on surfaces. Thus, the proteins that constitute these organic molecules present specific surfaces upon which the actual catalytic processes take place.

Control of Biologic Processes at the Molecular Level

A living system, whether it be a single cell or an entire organism, must use energy continually; and that energy, derived from metabolic processes, is essential to maintain the morphologic and functional integrity of the system. Thus, the synthesis of compounds (anabolism) required for growth, repair, and maintenance of the bodily structures, as well as reproduction, requires energy expenditure; and this energy is derived from the closely controlled degradation (catabolism) of certain chemical compounds that contain high levels of potential chemical energy. In turn, the orderly, stepwise release of energy by sequences of specific reactions depends upon enzymatic catalysis, so that ultimately all biologic control rests upon events at the level of protein molecules themselves. How is this regulatory function accomplished?

Recent experimental work has established beyond reasonable doubt that protein molecules have the ability to change their shape under suitable external influences, and this shape change, in turn, provides the fundamental on-off control mechanism that underlies all biologic control systems. This relatively simple phenomenon also underlies the regulation of vast numbers of complex physiologic processes in the body.

Not only are enzymatic processes controlled by alterations in the conformation of flexible protein molecules by rotation about a single double bond within their structure (Fig. 1.15), but, for example, proteins in certain sensory receptors such as the eye, ear, taste buds, and olfactory tissues also are capable of deformation by appropriate stimuli, and thereby a characteristic response is elicited. Similarly, antibody proteins are able to undergo conformational changes in response to suitable agents, and thus the immune response is evoked.

Consequently, changes in the shape of a multitude of specific proteins induced by appropriate stimuli appear to be responsible for the ultimate control of all functional processes in living systems at the molecular level.

The Internal Environment and Homeostasis

Over a century ago, Claude Bernard advanced the concept that the blood mediated between the internal environment (*le milieu intérieur*) and the external environment (*le milieu extérieur*). According to this concept, the blood acted as the intermediate vehicle for the exchange of all nutrients and waste products between the cells of the body and the external environment. This fundamental concept still is valid and useful today.

About 50 years ago, Walter B. Cannon advanced the general concept of *homeostasis*. The term "homeostasis" may be defined as a tendency toward the maintenance of a relatively stable internal environment in the bodies of higher animals through a series of interacting physiologic and biochemical processes.

The science of physiology deals in large measure with the specific mechanisms which are responsible for maintaining the relative constancy of the internal milieu within limits that are compatible with life. Hence, these homeostatic mechanisms become operative whenever internal or external events produce a fluctuation in some vital parameter that exceeds certain critical limits.

Physiologic homeostasis enables systems of cells (organs) to maintain a constant structure and function. Homeostasis at this level of organization ultimately results from a dynamic equilibrium, or steady-state relationship, among the molecular constituents of cells. Consequently, most cellular components are being synthesized and broken down continually at rates that depend upon bodily requirements for the individual compounds. This unending synthesis and breakdown of the molecular components of living systems is called *turnover*. (Deoxyribonucleic acid is the exception to a general turnover of all molecular species in cells, as this substance is replicated only.)

HOMEOSTATIC REGULATION AT THE MOLECULAR LEVEL. At the molecular level, there are four general mechanisms whereby homeostasis is achieved:

a. Activity may be regulated through the *enzymes* themselves; ie, their particular activity toward specific substrates, their kinetic properties, and their tendency to be inhibited by the product(s) of their action.
b. Activity may be regulated as the consequence of the kinetic properties of *systems of enzymes.*
c. Activity may be regulated by *compartmentalization* of enzymes or enzyme systems within the cell or certain organelles therein.
d. Activity may be regulated by controlling the *rate at which the enzyme is synthesized,* as well as by the *structure of the enzyme* itself. These factors, in turn, are controlled by the genetic machinery of the cell.

Note that the first two of these regulatory mechanisms can be related directly to the conformational changes in the protein molecules that constitute the enzymes, as discussed in the previous section.

HOMEOSTATIC REGULATION AT THE LEVEL OF THE ENTIRE ORGANISM. Three major systems in the mammalian body underlie homeostatic mechanisms at the level of the body as a whole, and these systems coordinate the activities of many specialized tissues and organs as well as the composition of the internal environment itself:

a. The *nervous system* is responsible for the rapid transmission and integration of information among all regions of the body.
b. The *endocrine system,* composed of a number of discrete glands as well as certain specialized tissues, transmits information by means of chemical substances *(hormones).* In general, this chemical integration of various bodily functions is slower, albeit of longer duration, than that accomplished by the nervous system.
c. The *cardiovascular system* serves as the essential vehicle whereby the chemical messages that originate in the endocrine glands are conveyed in the blood to their appropriate sites of action *(target tissues).*

The actions of these three systems are not mutually exclusive; rather, they usually operate in a smoothly coordinated, integrated, and interdependent fashion to achieve a homeostatic balance among a considerable number of bodily functions above the molecular level of organization. Furthermore, the nervous, endocrine, and cardiovascular systems are of vital importance in the operation of the negative feedback mechanisms that underlie many homeostatic processes.

The Operation of Homeostatic Mechanisms: Feedback Control Systems

During recent years, it has become increasingly evident that many of the physiologic and biochemical regulatory mechanisms that achieve homeostasis in the human body are mediated through *negative feedback control systems* of various types. A simple negative feedback control system is shown in Figure 2.2, and its operation is explained in the legend to this figure.

This elementary scheme becomes more complex in the following situations:

1. The entire feedback system or any of its components may exhibit a nonlinear response.
2. A time lag or delay may exist in the response of any or all of the components involved in a feedback system, eg, in the control element (2) or feedback loop (4).
3. Several independent or interdependent factors may influence either the reference input (1) or the control element (2) as indicated by "disturbances" (vertical arrows at the top of the figure). The control element also may have to decide whether or not to respond to the input from the disturbances.
4. Several independent or interdependent feedback loops may impinge upon the controlled variable.

The negative feedback mechanisms of interest in physiology are activated in order to restore equilibrial conditions within definite limits (homeostasis) when a normal balance is disturbed. Once equilibrial conditions are restored, the stimulus that activated the feedback loop is removed, so that the system ceases to function until an appropriate stimulus initiates the feedback process once again. That is, negative feedback systems within the body normally are reversible and come into play only upon demand.

Feedback control systems are present at all levels of organization in living organisms, and these mechanisms, in turn, control a host of metabolic and other bodily functions. The endocrine control of blood sugar level, the regulation of arterial blood pressure, and the physiologic timing of ovulation in relation to the menstrual cycle are but three examples of the operation of such mechanisms in the human body.

In contrast to negative feedback systems, positive feedback mechanisms are typified by the following general properties: They are usually irreversible, progressive, self-perpetuating, and self-amplifying. Under physiologic conditions, positive feedback processes rarely are encountered in normal homeostasis and usually are one-way processes. Childbirth (parturition) is one example of a normal positive feedback process, and the act of

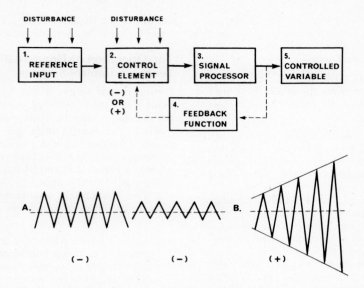

FIG. 2.2. Diagram of the essential elements of a simple feedback loop system such as is encountered in many physiologic processes. The reference input or set point (1) determines the level at which the control element (2) responds to alterations or fluctuations within the system as well as to external disturbances. The signal from the control element may or may not be amplified or otherwise treated by the signal processor (3). A portion of the output signal from the signal processor is fed back through a loop (4) into the control elements and may diminish in a controlled fashion (negative feedback, **A**), or augment in an uncontrolled fashion (positive feedback, **B**), the net or overall system response which is the controlled variable (5).

It is helpful to think of the negative feedback control of the air temperature within a refrigerator in regard to this concept. In the diagram at the left in **A,** the temperature fluctuates widely above and below the desired mean (dashed line) because the thermostat (which is both the reference input and the control element) is not sufficiently sensitive. The control element in this example actuates the refrigerator motor and compressor through a relay (signal processor), so that the air temperature is the controlled variable. Replacement of the thermostat with another, more sensitive regulator would result in a more precise control of the air temperature about the desired mean (right-hand diagram, **A**), and the overall system response now would be increased.

In **B,** positive feedback would result in an unrestricted variation in the overall system response (controlled variable) so that there would be a net increase in this factor with time (indicated by the diverging solid lines) and the system is uncontrolled. Thus the system response would fluctuate (oscillate) in an ever-widening pattern about the mean in an unregulated fashion. In the refrigerator, positive feedback would result in continuous operation of motor and compressor with a concomitant and uncontrolled drop in air temperature.

urination (micturition) is another. Usually, however, positive feedback mechanisms lead to severe dysfunction and even death, eg, terminal cardiac failure and irreversible shock.

Parenthetically, it should be noted that a large part of clinical medicine is devoted to restoring the normal homeostatic levels of various functions by ameliorating or correcting defects in particular negative feedback systems within the body, whether by drugs or other kinds of therapy.

CHAPTER 3

Morphologic and Physiologic Properties of Cells and Tissues

A *cell* is defined as the smallest discrete structural aggregate of living matter *(protoplasm)* which is capable of functioning as an independent unit under appropriate conditions, and which is bounded externally by a *semipermeable membrane*. In broadest terms, cells are classified as prokaryotic or eukaryotic.

From a structural viewpoint, *prokaryotic cells* are simple and quite small, and only a cell membrane is present. No membrane invests the nucleus of prokaryotic cells, and no membrane-limited organelles are present. Prokaryotes contain only one chromosome, and this structure consists merely of a single molecule of double-stranded deoxyribonucleic acid (DNA) which is coiled tightly in the *nuclear zone* of the cell. Examples of prokaryotic cells are eubacteria, blue-green algae, spirochetes, and rickettsiae. Presumably, prokaryotic cells were the first to develop in the course of biologic evolution.

The *eukaryotic cells,* found in all higher forms of life in both the plant and animal kingdoms, are considerably larger and far more complex than procaryotic cells. Eucaryotes contain a membrane-limited nucleus in addition to various membrane-limited organelles, such as the mitochondria, endoplasmic reticulum, lysosomes, and Golgi apparatus. The material within the eukaryote nucleus is divided into a few to many chromosomes, depending upon the species of organism, and these structures undergo mitosis during the process of cell division (p. 106). The diploid chromosome number $(2n)$ found in several plant and animal species is summarized in Table 3.1

Eukaryotic cells are found in protozoa, fungi, green plants, and most algae, as well as in all vertebrates, including man. Eukaryotic cells are believed to have developed more recently in the course of evolution than prokaryotic cells. Quite possibly, prokaryotic cells were the progenitors of eukaryotic cells.

Remarkably few basic cell types are found within the human body, considering the enormous structural and functional diversity that is encountered. Histologically, cells may be classified as muscular, nervous, epithelial, or supportive. Regardless of the extreme morphologic variation and specialization found within various cell populations (tissues), there are only these four distinct races of cells present in the human body.

The individual cells of the body are far too small to be seen and studied by the unaided eye, since their diameter is only a few microns at most. Consequently, in order to elucidate the structural and functional properties of cells, the use of various types of microscopy is required, together with the application of a variety of other highly sophisticated physical and chemical techniques.

The various hierarchies or organization found within a complex living system were discussed in Chapter 2 (p. 46) and depicted in Figure. 2.1. It is necessary to stress at this point that the laws or rules that apply to functions at one level of morphologic or biochemical organization do not necessarily apply at lower levels of organization. Furthermore, the entire cell or organism is considerably more than the sum of its constituent parts. For example, the physicochemical properties and biologic activity of a protein molecule differ considerably from the properties of the amino acids which comprise that protein, as well as the properties of the individual elements which comprise the amino acids themselves. Therefore, it is impossible to predict either the properties or behavior of large molecules from their composition alone. These general concepts also can be applied to the various structure constituents of cells and tissues.

Table 3.1 DIPLOID NUMBER OF CHROMOSOMES (2n) IN SOME COMMON PLANTS AND ANIMALS

PLANTS

Common Name	Scientific Name	Chromosomes
Cabbage	Brassica oleracea	18
Potato	Solanum tuberosum	48
Onion	Allium cepa	16
Tomato	Lycopersicum solanum	24
Tobacco	Nicotiana tabacum	48
Rice	Oryza sativa	24
Sugarcane	Saccharum officinarum	80
Cucumber	Cucumis sativus	14
Watermelon	Citrullus vulgaris	22
Oats	Avena sativa	42
Summer wheat	Triticum dicoccum	28
Bread wheat	Triticum vulgare	42
Coffee	Coffea arabica	44
Apple	Malus silvestris	34, 51
Cherry	Prunus cerasus	32
Plum	Prunus domestica	48
Eucalyptus	Eucalyptus spp.	22
Yellow pine	Pinus ponderosa	24
Black sorghum	Sorghum almum	40
Garden pea	Pisum sativum	14
Wood rush	Luzula purpurea	6

ANIMALS

Common Name	Scientific Name	Chromosomes
Spanish butterfly	Lysandra nivescens	380
Vinegar fly	Drosophila melanogaster	8
Housefly	Musca domestica	12
Mosquito	Culex pipiens	6
Honeybee	Apis mellifica	32, 16
Frog	Rana spp.	26
Toad	Bufo spp.	22
Chicken	Gallus domesticus	approximately 78
Turkey	Meleagris gallipavo	82
Duck	Anas platyrhyncha	80
Mouse	Mus musculus	40
Rabbit	Oryctolagus cuniculus	44
Albino rat	Rattus norvegicus	42
Common rat	Rattus rattus	42
Guinea pig	Cavia cobaya	64
Dog	Canis familiaris	78
Cat	Felis domestica	38
Horse	Equus caballus	66
Sheep	Ovis aries	54
Pig	Sus scrofa	40
Cow	Bos taurus	60
Rhesus monkey	Macaca mulata	42
Gorilla	Gorilla gorilla	48
Orangutan	Pongo pygmaeus	48
Chimpanzee	Pan troglodytes	48
Man	Homo sapiens	46

THE PHYSICOCHEMICAL COMPOSITION OF PROTOPLASM

Elemental Constituents

As summarized in Table 3.2, some 24 chemical elements have been shown to be normal con-stituents of protoplasm within animal cells. Of these two dozen elements, only four (carbon, hydrogen, oxygen, and nitrogen) constitute the principal mass of the four basic classes of organic compounds that make up the actual substance of living matter. These fundamental compounds are proteins, lipids, carbohydrates, and nucleic acids.

Since the major constituent of protoplasm is an

Table 3.2 THE CHEMICAL ELEMENTS ASSOCIATED WITH ANIMAL PROTOPLASM

ELEMENT	CHEMICAL SYMBOL	PERCENT ABUNDANCE IN HUMAN BODY
1. Carbon	C	9.5
2. Hydrogen	H	63.0
3. Oxygen	O	25.5
4. Nitrogen	N	1.4
5. Phosphorus	P	
6. Sulfur	S	
7. Sodium	Na	
8. Potassium	K	
9. Calcium	Ca	
10. Magnesium	Mg	
11. Chlorine	Cl	
12. Iron	Fe	
13. Iodine	I	
14. Copper	Cu	
15. Cobalt	Co	0.6 approximately
16. Manganese	Mn	
17. Zinc	Zn	
18. Fluorine	F	
19. Molybdenum	Mo	
20. Selenium	Se	
21. Chromium	Cr	
22. Tin	Sn	
23. Vanadium	V	
24. Silicon	Si	

aqueous matrix (between 70 and 85 percent of the total cell mass is water), it is hardly surprising that of the four elements listed above, hydrogen and oxygen predominate. About 89 percent of the total mass of atoms in the body belongs to these two atomic species.

Each of the remaining 20 elements listed in the table are found in the body in widely varying quantities, ranging from hundreds of grams or even kilograms (eg, sodium, calcium), to grams (eg, iron), milligrams (eg, iodine), or micrograms (eg, cobalt). Yet in toto, all of these 20 atomic species comprise only about 0.6 percent of the total mass of the human body.

Actually, all 92 naturally occurring elements may be detected in the protoplasm obtained from various organisms. The important question then becomes: Which elements are requisite to the normal functions of the organism, and which elements are present fortuitously? When a given element is found only in trace (microgram or picogram) quantities in the entire body, it is an exceedingly difficult technical problem to demonstrate conclusively any specific effects of that element on such processes as growth, development, reproduction, or other functions. For example, certain elements that are found in only minute total quantities in the whole body may perform definite and vital roles as constituents of, or cofactors for, the action of certain enzymes (eg, zinc, manganese). On the other hand, while the necessity for such trace elements as molybdenum, selenium, chromium, tin, and vanadium, is known, their specific functions have not yet been proved unequivocally in all higher species of animals.

It is sometimes useful to view the total quantity of an element or other substance in the body as representing a pool of that substance that is available for various metabolic processes. For example, the total quantity of sodium within the body is the pool for that element. The stability of the sodium pool, in turn, is dependent upon sodium loss through excretion compared to sodium intake in the diet over a given period of time. That is, the *turnover rate* of the sodium pool can be expressed as the fraction of the total quantity of sodium present that is excreted in the urine per unit of time. This type of data gives valuable information concerning the balance of certain bodily constituents in both health and disease. The general pool concept coupled to the measurement of the turnover rate is applicable with equal facility to organic compounds as well as to inorganic substances.

Gross analytic data of the sort presented in Table 3.2 are of strictly limited value, as they yield no specific information concerning the complex organization of the individual chemical elements into various organic compounds, as well as the structural and functional interrelationships among these compounds within living cells. Consequently, the material to be discussed in the remainder of this section will deal with the major ionic species of the body as well as with the assemblage of certain elements into those chemical compounds that are fundamental to the composition and functions of protoplasm.

Intracellular and Extracellular Electrolytes

A number of elements are present in solution as electrolytes, both within the aqueous protoplasmic matrix of the cell and in the extracellular fluids (Table 3.3). Note that the principal ions found in the body as a whole are Na^+, K^+, Cl^-, and HCO_3^-.

It is apparent from the analytic data summarized in Table 3.3 that the major intracellular ions are K^+, Mg^{++}, HCO_3^-, and protein$^-$, whereas the chief extracellular ions are Na^+, Cl^-, HCO_3^-, and protein$^-$. These fundamental differences between the composition of the intra- and extracellular fluids reflect the selective ability of the cell membrane to concentrate certain ions within the cell and to exclude others, thereby creating concentration differences between the cell interior and its immediate environment. This fact is of basic significance to many cellular functions. The mechanisms whereby these concentration differences are produced will be discussed later (p. 85).

The intra- and extracellular anions and cations collectively are essential to the normal functional integrity of the cell in a variety of ways, and several examples of this may be introduced here. First, certain ions are important in the regulation of *osmotic* pressure within the cell. Second, the *bioelectric properties* of all cells, in particular those of muscle and nerve (pp. 130, 162), are wholly dependent upon the participation of several ionic species. Third, the regulation of normal *acid-base balance* in cells (as well as within the body tissues as a whole) is carried out by various electrolytes. Fourth, certain cations function as essential components in a variety of *enzyme systems;* in fact, certain enzymes require specific metallic ions in trace quantities in order to function at all.

Certain of the elements summarized in Table 3.2 may be found only in specific cells and tissues. For example, iodine is concentrated only within the cells of the thyroid gland.

Consequently, electrolytes perform many specific and vital regulatory functions insofar as various physiologic or biochemical mechanisms within the cell are concerned. This statement also holds true for the body and its tissues as a whole, to the extent that a large part of the subject of physiology is concerned with a detailed consideration of the specific functions of electrolytes in the body under normal and abnormal conditions.

Molecules and Macromolecules: The Organic Constituents of Protoplasm

Carbon, hydrogen, oxygen, and nitrogen are the principal elements that comprise the vast array of organic molecules found in protoplasm. In addition to these fundamental atomic species, small quantities of other elements, such as phosphorus and sulfur, also are essential constituents of many organic compounds present in living systems.

Ultimately, the form and organization of the cell and its individual functional substructures (*organelles*) resides in several types of *macromolecule*. These cellular macromolecules, in turn, are built up from many small individual organic subunits (*monomers*) that are linked together (*polymerized*) by the repetitive formation of covalent bonds, so that the resulting compounds have an enormously high molecular weight. The principal classes of polymerized macromolecules found within the animal cell are *proteins, nucleic acids,* and *polysaccharides*. In animal tissues, however, polysaccharides are found chiefly as constituents of the extracellular matrix or as glycogen within the cell, rather than as major structural components of the cells theselves.

Although *lipids* do not in themselves form true macromolecules, they do form micellar structures (p. 15), and these compounds are of the utmost importance to the formation of specialized cellular and intracellular membranes. Consequently, lipids will be discussed in this section together with proteins, nucleic acids, and polysaccharides.

Many of the cellular structures rendered visible by light or electron microscopy reflect the presence of submicroscopic macromolecular components which are themselves too small to be seen individually; but when such submicroscopic mac-

Table 3.3 PRINCIPAL INTRACELLULAR AND EXTRACELLULAR ELECTROLYTES[a]

Intracellular Fluid (Muscle)

Cations	mEq/liter water
Na^+	10
K^+	148
Ca^{++}	2
Mg^{++}	40
Total	200

Anions	
HCO_3^-	8
HPO_4^{--}, PO_4^{---}, SO_4^{--}, Cl^-, organic acids$^-$	136
Protein$^-$	56
Total	200

Extracellular Fluid (Plasma)

Cations	
Na^+	142
K^+	5
Ca^{++}	5
Mg^{++}	3
Total	155

Anions	
Cl^-	103
HCO_3^-	27
HPO_4^{--}	2
SO_4^{--}	1
Organic acids$^-$	6
Protein$^-$	16
Total	155

[a]The values cited in this table are representative rather than absolute concentrations. The value for Ca^{++} in intracellular fluid is subject to error, owing to the difficulty of determining accurately the small quantity of this ion that is present.

romolecules are present in sufficiently large aggregates, they may be seen as typical morphologic features of the cell when suitably magnified. In addition, the technique of *x-ray diffraction* has proved to be a powerful tool for the study of macromolecules and their structural organization at levels far beyond the resolving capacity of the best available optical and electron microscopes. Therefore, our knowledge of the intimate structure of macromolecules and their actual or potential relationship to cellular structure is far greater than would be possible without the use of this technique in conjunction with light and electron microscopy.

PROTEINS AND PEPTIDES. Every protein is a high-molecular-weight polymer composed of a definite number of specific monomers, the *amino acids* (Table 14.10, p. 970), which are joined together in an exact sequence by *peptide linkages* (p. 42). Each type of protein molecule appears to be tailored to a precise biologic role, insofar as its molecular size, conformation, and composition are concerned. In all, about 20 individual amino acids are the fundamental subunits of natural proteins, although not necessarily all of these compounds are found within the structure of any particular protein molecule.

If a peptide linkage forms between only two amino acids, the resulting compound is known as a *dipeptide,* whereas, if three amino acids are involved, the compound is called a *tripeptide.* *Polypeptides,* on the other hand, contain a large number of amino acid residues but have a molecular weight considerably below that of small proteins.

All protein and peptide molecules are *amphoteric substances* (p. 23), because there is present a free acidic group at one end of the compound and a basic group at the other end. In addition, such molecules contain lateral radicals in their structure which can be either basic or acidic in character.

The free amino acids present within a cell may be derived from the degradation of cellular proteins or obtained by absorption from the extracellular fluid that surrounds the cell. Such free amino acids constitute the *amino acid pool* of the cell and are utilized for the synthesis of new proteins.

Since the distance between two adjacent peptide linkages is approximately 3.5 Å, a protein having a molecular weight of 30,000 daltons and consisting of 300 amino acid residues would be about 1,000 Å long, 10 Å wide, and 4.6 Å thick if fully extended. These facts emphasize the point that protein molecules can form extremely long chains (*fibrous proteins*). On the other hand, many protein molecules are folded into spherical or other conformations, so that the resulting molecule may be spherical in shape (*globular proteins*).

The peptide bond which links the individual amino acid residues together in the primary structure of a protein molecule is, of course, a covalent bond (p. 14). Similarly, covalent disulfide bonds can form between cystine residues (—S—S—).

Other (and weaker) bonding forces that are important in determining the three-dimensional structure of proteins are ionic (or electrostatic) bonds (p. 14), hydrogen bonds (p. 15), and van der Waal's forces (p. 15). Hence, the functions of specific proteins depend not only upon their particular amino acid composition, but also upon their conformation; and, as noted earlier, the conformation of many proteins is not rigid but rather is flexible, depending upon a number of factors (p. 48).

In general, proteins of various types constitute between 10 and 20 percent of the total cell mass, and a given cell may contain many hundreds of specific types of protein molecule when all of the enzymes present are enumerated.

A few examples of fibrous and globular proteins may be cited.

Fibrous Proteins. Collagen is a long-chain, fibrous protein that plays an essential role in the organization of tissues that have an extracellular framework, or *stroma.* Collagen is the most abundant protein in the body, and it is found primarily as insoluble extracellular fibers that account for a major proportion of the organic material in such structures as the tendons, skin, blood vessels, and bone.

Collagen is a unique protein molecule in a number of respects. First, collagen fibrils consist of three helically disposed chains of polypeptides (practically all other fibrous proteins have an even number of polypeptide chains). Second, the helical conformation of the polypeptides in collagen macromolecules is found in this protein alone (Fig. 3.1). Third, collagen has an especially high proline and glycine content. Fourth, the glycine residues are positioned at every third amino acid locus, a fact that accounts for a number of the peculiar molecular characteristics of collagen, including its special helical conformation.

Individual collagen molecules are capable of spontaneous assembly into macromolecular structures, or fibrils, which exhibit a characteristic transverse banding at intervals of 640 Å throughout their length. This fact is indicative of the repetitive pattern in the amino acid sequence in the individual collagen molecules, which in turn are linked together through aldehyde and other reactive groups.

Elastin is another fibrous protein which, as its name implies, possesses highly developed elastic properties. Collagen and elastin quite frequently are found in close association in the connective tissues of the body.

Certain fibrous proteins are formed in the body at sites remote from their ultimate locus of action and transported as precursors. For example, *fibrinogen* is the precursor of the structural protein of blood clots, *fibrin.* Fibrinogen is produced solely by the liver, from where it is transported in the bloodstream as a small, globular molecule. During clot formation, the enzyme thrombin catalyzes the hydrolysis of fibrinogen into fibrin monomers and small peptide fragments, which then polymerize

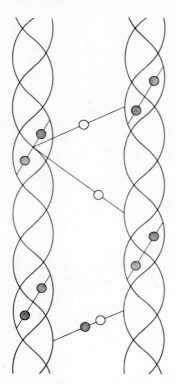

FIG. 3.1. The molecular organization of collagen. This fibrous protein consists of three helically wound polypeptide chains that are linked together by aldehyde (shaded circles) and other chemical groups (open circles). (After Tanzer. *Science* 180:561, 1973.)

into a network of fibrin macromolecules. Hence, the formation of a network of fibrous protein molecules from the globular fibrinogen underlies the blood-clotting mechanism.

Globular Proteins. Polypeptide chains that are folded into compact spheroid or globular shapes are known as globular proteins (Fig. 3.2A). Most of the globular proteins are water soluble and are able to diffuse readily; hence, such molecules generally have dynamic functions within living systems.

Approximately 1,000 individual enzymes are known currently, and the majority of these catalysts are globular proteins. Certain hormones, antibodies, hemoglobin, and serum albumin are further examples of globular proteins.

Some proteins are classified as having characteristics that lie between those of fibrous and globular proteins. Included in this category is myosin, an important contractile element of muscle cells.

Levels of Organization Within Protein Molecules. Just as different levels of organization are found within living systems, protein molecules also exhibit several intramolecular, structural hierarchies which are described by specific terms.

The term *primary structure* of a protein denotes the specific sequence of amino acid residues linked together by covalent (or chemical) bonds in a polypeptide chain, ie, this term refers to the "backbone" of the protein molecule.

The term *secondary structure* indicates either the extended or the helically coiled conformation of the polypeptide chains that constitute a protein molecule, especially as found in fibrous proteins.

The term *tertiary structure* signifies the folding of the polypeptide chains in order to form the definite compact structure that is found in globular proteins.

The terms "primary," "secondary," and "tertiary structure" refer only to a single polypeptide chain, whereas the term *quaternary structure* indicates the manner in which two or more polypeptide chains are arranged in space in an aggregated form.

In contrast to the primary structure of a polypeptide, which is determined exclusively by covalent bonds (either peptide linkages or disulfide bonds that link cystine residues), the secondary and tertiary structures are formed by weaker bonds, including electrostatic (ionic) forces, hydrogen bonds, and van der Waal's forces.

The general term *conformation* is applied to the combined secondary and tertiary structure of the polypeptide chain in proteins.

Proteins that have two or more polypeptide chains are known as *oligomeric proteins;* each chain is termed a *protomer.* For example, hemoglobin consists of four globular polypeptide chains that are fitted closely together and that form an extremely stable structure, although no covalent bonds are present among the protomers.

Generally, oligomeric proteins contain an even number of polypeptide chains. A notable exception is collagen.

Denaturation and Renaturation. The majority of protein molecules are capable of functioning biologically only within an extremely narrow range of temperature and pH. If the temperature and/or pH limits under which a given native protein can function are exceeded, a loss of the characteristic physicochemical and biologic properties of that protein can ensue. This change in properties is known as *denaturation.* A loss in solubility is the most obvious change that accompanies the denaturation of a globular protein. For example, heating an egg white produces a coagulated mass of insoluble denatured albumin. Similarly, heating an enzyme may cause it to lose its characteristic ability to catalyze a specific chemical reaction; hence, the biologic function of that enzyme is lost.

Denaturation does not rupture the covalent bonds in the primary structure of a protein molecule. Therefore, it appears that denaturation involves an unfolding of the molecule, which now assumes a random conformation, as shown in Figure 3.2B.

The process of denaturation has been demonstrated to be spontaneously reversible in many instances. Therefore, a denatured protein may return to its native folded state once again, provided

A.

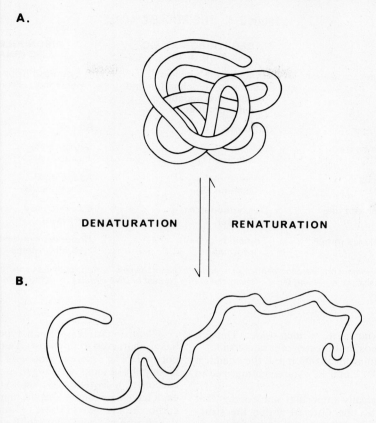

DENATURATION   RENATURATION

B.

FIG. 3.2. Structure of a native and a denatured globular protein. **A.** The polypeptide chains that compose the primary structure of a globular protein molecule in its native state are folded into a compact spheroid shape. **B.** Denaturation causes an unfolding of the polypeptide chains so that they assume a more random conformation; this change is accompanied by a loss of biologic activity. Given sufficient time, denaturation can be spontaneously reversed, as indicated by the arrows in the figure.

that the temperature and pH are compatible with the stability of the native conformation of the protein. This process is called *renaturation.* Because renaturation may take place at an extremely low rate, it was once thought that denaturation was an irreversible process.

Renaturation of a denatured enzyme can result in the return of the original catalytic (biologic) activity, after the native conformation of the protein is achieved by a refolding of the uncoiled molecule.

No input of chemical energy is necessary for renaturation to take place, ie, it is a spontaneous process. Since every protein molecule has a particular amino acid composition as well as a characteristic sequence in which those amino acids are arranged within the primary structure of the polypeptide, these properties underlie the conformation of that protein molecule in the native state. Consequently, it appears that renaturation is a process whereby the most stable biologic form of a protein molecule is achieved. In other words, this is the state in which the free energy of the molecule is at a minimum as represented by the particular conformation of the molecule.

NUCLEIC ACIDS. Nucleic acids are of the utmost biologic importance, for they underlie and regulate the synthesis of all the proteins found within living systems and are also responsible for the transmission of hereditary characteristics from generation to generation. Therefore, it is no exaggeration to say that the nucleic acids are the basic molecules of life itself, since all of the biochemical and physiologic characteristics of an organism, regardless of the level of organization, depend ultimately upon the specific structure and functions of these compounds.

Composition of Nucleic Acids. The cells of all living organisms contain two kinds of nucleic acid, *deoxyribonucleic acid* (DNA) and *ribonucleic acid* (RNA). A nucleic acid, in turn, consists of a high-molecular-weight polymer that is composed of a large number of monomers known as *nucleotides.* Each nucleotide is composed of a pentose sugar molecule (either deoxyribose or ribose), a molecule of a nitrogenous base (either a purine or a pyrimidine), and a phosphoric acid moiety. Within each nucleotide, the combination of a pentose sugar

Table 3.4 THE NUCLEIC ACIDS

	DEOXYRIBONUCLEIC ACID (DNA)	RIBONUCLEIC ACID (RNA)
Intracellular localization	Principally in nucleus; some in mitochondria and organelles capable of self-reproduction	In cytoplasm (ribosomes), nucleolus, chromosomes
Pentose sugar	Deoxyribose	Ribose
Pyrimidine bases	Cytosine (C) Thymine (T)	Cytosine (C) Uracil (U)
Purine bases	Adenine (A)[a] Guanine (G)	Adenine (A) Guanine (G)
Hydrolyzing enzyme	Deoxyribonuclease (DNase)	Ribonuclease (RNase)
Functional role in cell	Transmits genetic information	Underlies synthesis of all proteins

[a]*In DNA, the molar concentration of adenine is equal to that of thymine, and the guanine content is equal to that of cytosine, ie, A = T and G = C. Hence, the purine content of DNA equals that of the pyrimidines on a molar basis.*

with a base is known as a *nucleoside*. Thus, a nucleotide is a phosphoric ester of a nucleoside.

The intracellular localization and the chemical composition of DNA and RNA are summarized in Table 3.4.

A few biologically important nucleotides that are not specifically encountered within the structure of nucleic acids are summarized in Table 3.5.

Pentose Sugars. Each type of nucleic acid has a characteristic pentose. Deoxyribose is present in DNA, whereas ribose is present in RNA. Deoxyribose lacks an oxygen atom on the second carbon (2'). Structurally, deoxyribose and ribose each have a pentagon-shaped ring composed of five carbon atoms; two of these (3' and 5') are linked to the phosphoric acid and a third (1') is linked to the

Table 3.5 SOME BIOLOGICALLY IMPORTANT NUCLEOTIDES[a]

NAME	ABBREVIATION	ENZYMATIC AND OTHER FUNCTIONS
1. Adenosine triphosphate	ATP	Transphosphorylation reactions; principal compound involved in storage and utilization of chemical energy in the body
2. Guanosine triphosphate	GTP	Transphosphorylation reactions
3. Nicotinamide adenine dinucleotide (diphosphopyridine nucleotide)[b]	NAD (DPN)	Hydrogen carrier
4. Nicotinamide adenine dinucleotide phosphate (triphosphopyridine nucleotide)[b]	NADP (TPN)	Hydrogen carrier
5. Pyridoxal phosphate	—	Coenzyme for transaminases, amino acid decarboxylases, and other enzymes
6. Thiamine pyrophosphate	TPP	Oxidative decarboxylation, aldehyde carrier
7. Coenzyme A	CoA	Acetyl and other acyl group transfers; essential in fatty acid synthesis and oxidation
8. Flavin mononucleotide	FMN	Hydrogen carrier
9. Flavin adenine dinucleotide	FAD	Hydrogen carrier

[a]*The nucleotides listed in this table are not constituents of nucleic acids, but rather serve to illustrate the broad range of functions and compounds in which nucleotides are involved.*

[b]*The terms nicotinamide adenine dinucleotide (NAD) and nicotinamide adenine dinucleotide phosphate (NADP) are used preferentially to the terms diphosphopyridine nucleotide (DPN) and triphosphopyridine nucleotide (TPN), respectively.*

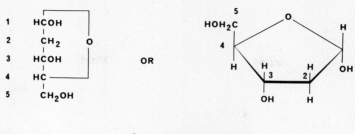

A. D – 2 – DEOXYRIBOSE

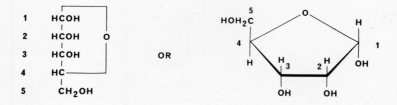

B. D – RIBOSE

FIG. 3.3. Two methods of depicting the structural formulas of deoxyribose **(A)** and ribose **(B)** illustrating the furanose ring structure of these compounds; ie, an oxygen bridge linking carbon atoms 1 and 4. The pentagonal representations shown on the right are a more accurate portrayal of these structures.

base. The prime sign (′) following the number of each carbon atom refers to the position in the sugar moiety to which either phosphoric acid or a base is linked in a nucleoside or its derivatives. (See Figure 3.7B.)

The structural formulas of deoxyribose and ribose are given in Figure 3.3.

Pyrimidine Bases. The principal pyrimidine bases found in nucleic acids are *cytosine* (C), *thymine* (T), and *uracil* (U). The structural formulas of these substances are shown in Figure 3.4. Cytosine is present in both DNA and RNA. Thymine is characteristically found in DNA, whereas uracil is characteristically present in RNA. Thus, DNA and RNA differ not only in the structure of their respective pentose sugars but also in their pyrimidine bases.

The biologic significance of pyrimidines is not limited solely to their role in the composition of nucleic acids. Several pyrimidine nucleotides are essential participants in carbohydrate and fat metabolism. For example, vitamin B_1 (thiamine, p. 878) is a pyrimidine derivative.

All of the pyrimidines exhibit keto-enol tautomerism (Fig. 1.22).

Purine Bases. The purine basis adenine (A) and guanine (G) are common to both DNA and RNA (Fig. 3.5).

In certain RNA molecules, especially *transfer RNA* (tRNA), a considerable proportion of the bases is methylated. That is, such RNA molecules contain methyladenine, methylguanine, or methylcytosine.

Uric acid and xanthine are two examples of additional purines that are biologically important.

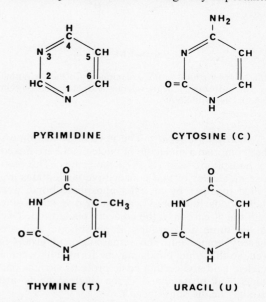

FIG. 3.4. Structural formulas of pyrimidine and of the three major pyrimidine bases found in nucleic acids. Cytosine occurs in both DNA and RNA. Thymine is characteristic of DNA, and uracil is characteristic of RNA.

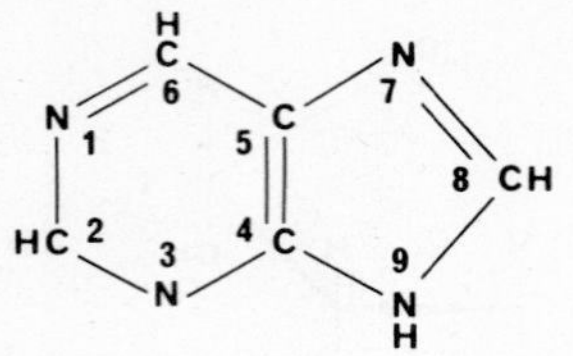

PURINE

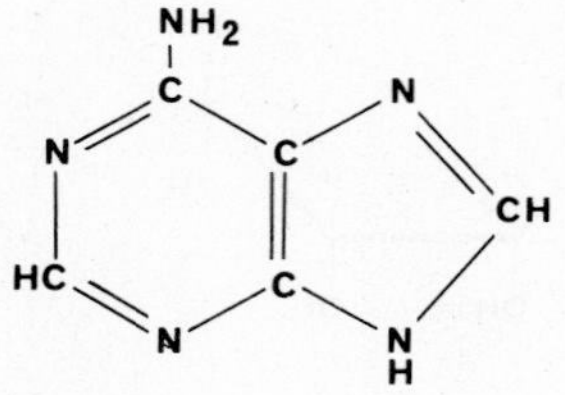

ADENINE (A)

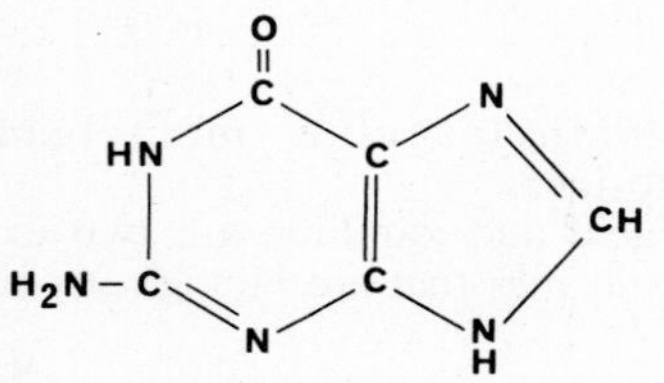

GUANINE (G)

FIG. 3.5.   Structural formulas of purine and of the two purine bases found in nucleic acids. Adenine and guanine are present in both DNA and RNA.

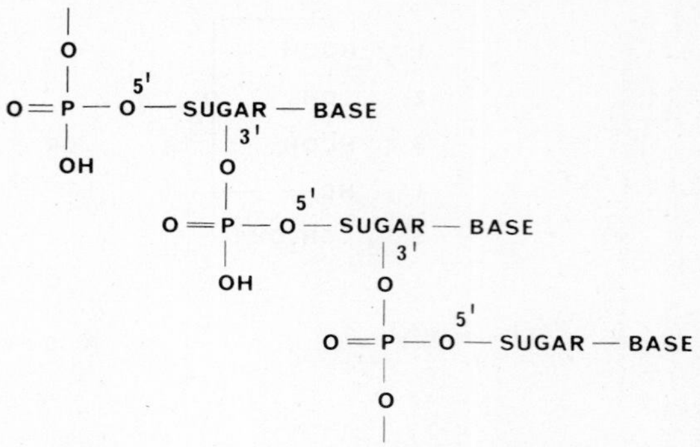

FIG. 3.6.   Chemical structure of a portion of a single polynucleotide strand, illustrating the participation of phosphoric acid as the phosphodiester bond that links the 5' carbon of one pentose sugar to the 3' carbon of the adjacent pentose sugar. See also Figure 3.7B.

The phosphodiester linkage found in polynucleotide strands chemically is analogous to the peptide linkage which links individual amino acids into polypeptide chains.

(After West, Todd, Mason and Van Bruggen. *Textbook of Biochemistry*, 4th ed. The Macmillan Co.)

Phosphoric Acid. The nucleotide monomers that comprise a molecule of nucleic acid are linked by phosphoric acid (H_3PO_4) molecules. The pentose molecules of two consecutive nucleotides are joined together by an ester-phosphate bond, and such bonds link the 3' carbon in one nucleoside with the 5' carbon in the adjacent nucleoside. Two of the three acid groups of phosphoric acid are thereby employed. The remaining acid group is free, so that ionic bonds may be formed between a nucleic acid and certain basic proteins such as histones and protamines. (Free nucleic acids are strongly acidic substances.)

The participation of phosphoric acid as the link between the successive pentose moieties of adjacent nucleotides in the structure of a nucleic acid is depicted in Figure 3.6.

The Structure of DNA. An understanding of the molecular structure of DNA is essential, as it explains the physicochemical and biologic properties of this substance — in particular, how DNA is duplicated (replicated) within the cell. On the basis of earlier work, in 1953 Watson and Crick proposed a model for the structure of DNA which has the following characteristics.

Each molecule of DNA is made up of two extremely long polydeoxyribonucleotide chains that are twisted about each other to form a double helix surrounding a central axis.

Each nucleoside within the two nucleotide chains is arranged so that it lies in a plane perpendicular to the polynucleotide chain, and the purine and pyrimidine bases within the nucleosides are planar (flat) molecules. The planes that contain the successive pairs of bases lie parallel to each other, and are 3.4 Å apart.

The two polynucleotide chains are linked together by hydrogen bonds between each pair of bases in the nucleosides, ie, the binding forces are interchain.

The two polynucleotide strands run in opposite directions; that is, they are antiparallel with respect to each other, and they are right-hand helices.

The two polynucleotide strands in the DNA molecule are *complementary,* and the pairing between the bases is highly specific. Consequently, because there is a fixed distance of 11 Å between the two sugar moieties on the opposite nucleotides, one purine base is able to pair only with one pyrimidine base, and adenine–thymine (A–T) and guanine–cytosine (G–C) pairs are the *only* possible combinations that can be formed.

The features of the Watson–Crick model for DNA summarized above are shown in Figure 3.7.

A.

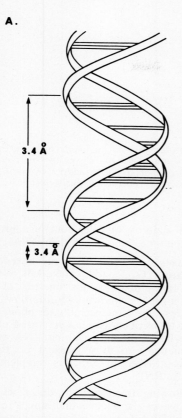

B.

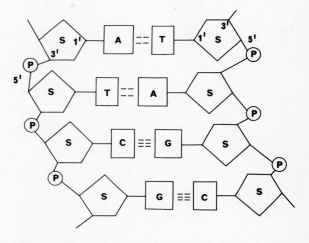

FIG. 3.7. The molecular pattern of DNA organization. **A.** The Watson–Crick model of DNA. The individual polydeoxyribonucleotide chains are twisted about each other to form a double helix around a central axis. The horizontal lines denote hydrogen bonds that link the base pairs of the individual deoxynucleotide strands.

FIG. 3.7. **B.** Small portion of a DNA molecule illustrating the 3′-5′ linkage between the pentose sugars (S) through a phosphoric acid molecule (P) as well as the hydrogen bonds (horizontal dashed lines) that connect the complementary bases of the two individual polydeoxyribonucleotide strands. Note that the base pair A–T has two such bonds, whereas the base pair C–G has three hydrogen bonds. (Data from De-Robertis, Nowinski, and Saez. *Cell Biology,* 5th ed, 1970. Saunders; White, Handler, and Smith. *Principles of Biochemistry,* 4th ed, 1968. The Blakiston Division, McGraw-Hill).

As indicated by the dashed lines in part B of this figure, two hydrogen bonds are formed between the A–T pair, whereas three hydrogen bonds are formed between the G–C pair; therefore A–C and G–T pairs *cannot* be formed.

Although the *pairs* of bases found at any particular locus on the polynucleotide chains must invariably be A–T or G–C, the sequence of the bases along the long axis of *one* polynucleotide chain can vary considerably. However, the bases on the opposite chain must be exactly complementary to those on the other chain, as indicated by this hypothetical example:

Chain 1: A C G T C A T G G
| | | | | | | | |
Chain 2: T G C A G T A C C

Long axis of molecule ⟷

During the process of DNA replication, the two polynucleotide chains dissociate by uncoiling from each other in a manner roughly analogous to opening a zipper, and each chain now serves as a template upon which two complementary DNA chains are produced (Fig. 3.8). In this manner, two DNA molecules are synthesized, each of which has an identical molecular composition to the other.

Variations in the sequence of the four bases along the DNA chain forms the basis for the transmission of all genetic information. DNA molecules are extremely long polymers with molecular weights ranging between 6×10^6 and 6×10^8 daltons, and many contain over 20,000 individual nucleotides. Although only four major nucleotides are present in the DNA chain, it may be readily appreciated that the possible number of combinations of these nucleotides along the DNA molecule is truly astronomic; and this fact underlies the enormous amount of coded information that is contained within a single DNA molecule.

Consequently DNA molecules are the basis upon which all genetic information is stored, expressed, and transmitted. In addition, the fact that nucleic acids can be altered structurally, thereby producing *mutations,* is of major phylogenetic importance and carries great clinical significance as far as hereditary diseases are concerned.

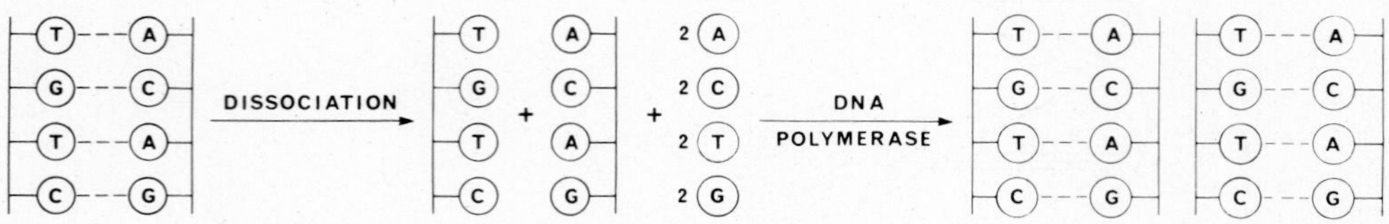

A.   **B.**   **C.**

FIG. 3.8. The replication of DNA. **A.** Segment of a DNA helix with the two deoxynucleotide strands connected by hydrogen bonds (single dashed lines) that link the base pairs. **B.** The double helix of the DNA has dissociated into two individual deoxypolyribonucleotide strands whose bases now are free to react with complementary purines and pyrimidines. **C.** Under the influence of the enzyme DNA polymerase, two new molecules of DNA are synthesized, each of which is identical to the parent molecule.

The Gene. A gene may be defined as an intracellular element that serves as a specific transmitter of a single hereditary characteristic. At the molecular level, a gene may be considered to be a portion of a DNA molecule which is arranged in a fixed position in a linear fashion along the DNA chain and which functions by controlling the synthesis of a specific polypeptide chain. That is, each gene controls the synthesis of a single protein, regardless of the functional role of that protein within the cell in particular or in the body in general. This statement explains the fundamental mechanism whereby the transmission and expression of genetic information takes place in all living systems.

Every cell contains a quantity of DNA that represents approximately 10^9 nucleotide pairs. One certain gene has been demonstrated to consist of some 74 nucleotide pairs; however, some genes are undoubtedly of much greater length.

RNA and Protein Synthesis. Three major types of ribonucleic acid are present in cells: messenger RNA (mRNA), ribosomal RNA (rRNA), and transfer RNA (tRNA). Each of these nucleic acids has a characteristic molecular weight and complement of bases.

These three kinds of RNA each are found in a number of molecular species. Thus, many hundreds of mRNA species are found; rRNA has 3 major species; and tRNA is present in about 60 species.

The majority of cells contain approximately five to ten times more RNA than DNA.

Messenger RNA. The only bases found in mRNA are A, G, C, and U (Table 3.4). Synthesis of mRNA takes place in the nucleus of the cell through a process known as *transcription*. During this process, the sequence of bases of *one* strand of DNA is transcribed into a *single* strand of mRNA, and the bases on the resulting RNA strand are complementary to those present on the parent DNA chain. Note that uracil is complementary to adenine in RNA.

Following transcription, the newly formed mRNA leaves the nucleus and moves to the ribosomes (p. 79), where the mRNA acts as a template for the specific and sequential organization of particular amino acids during protein synthesis. Groups of three nucleotides *(triplets),* known as *codons,* are located along the entire length of the mRNA molecule. The triplets stipulate the specific amino acid sequence in the polypeptide chain (Fig. 3.9). Thus, the structure of each of the thousands of individual enzymes and other proteins produced in living cells is encoded by a specific mRNA molecule or a portion of it, and the codons on the mRNA blueprint are complementary nucleotides to those present on the parent DNA molecule.

Transfer RNA. Compared to mRNA, tRNA molecules are relatively small. These nucleic acids

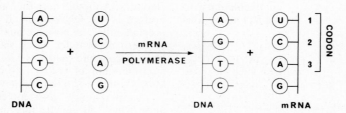

FIG. 3.9. The replication of mRNA. At the left of the figure, the double helix of the parent DNA molecule has uncoiled, thereby exposing a sequence of bases, A, G, T, C. These bases now are free to react with the complementary bases, U, C, A, and G. (Note that uracil replaces thymine in the synthesis of RNA). Under the influence of mRNA polymerase, a complementary sequence of bases is synthesized on the mRNA strand. The triplet of bases U–C–A forms a codon (indicated by the numbers) for the amino acid serine.

act as carriers or adaptors of particular amino acids during the process of protein synthesis on the ribosomes.

Twenty amino acids normally are found in proteins, and it appears that there is at least one specific tRNA present for each of these amino acids. Each tRNA adaptor molecule contains an *anticodon* — a nucleotide triplet that is *complementary* in sequence to the nucleotide present in the codon of the mRNA molecule which codes for the amino acid.

The first step of protein synthesis involves the activation of an amino acid to its aminoacyl form by a suitable enzyme. The amino acid then becomes attached to its particular tRNA molecule, forming an aminoacyl–tRNA complex.

Ribosomal RNA. About 65 percent of the total mass of the ribosomes is composed of rRNA; however, the specific role of this type of RNA in protein synthesis is unclear at present.

Summary of the Events During Protein Synthesis. Storage of genetic information and its replication within the cell are the responsibility of DNA molecules. RNA molecules, on the other hand, are critical to the actual synthesis of the specific polypeptide chains of particular proteins. The sequence of events underlying this process may be visualized as a flow of information.

$$\text{DNA} \xrightarrow{\text{Transcription}} \text{RNA} \xrightarrow{\text{Translation}} \text{Protein}$$

Transcription does not involve a change in the genetic code, whereas *translation* does involve such a change. Therefore, translation of the information carried by mRNA to the site of protein synthesis on the ribosome requires a "code book" so that each codon on the mRNA molecule can act as the template for a particular amino acid. The nucleotide sequences in mRNA that are responsible for coding specific amino acids are shown in Table 3.6. Note in this table that a sequence of three nucleotides are involved in coding for each amino acid, and that several different codons may be involved for a given amino acid, a phenomenon known as *degeneracy*.

Assembly of the polypeptide chain takes place as follows (Fig. 3.10): Particular tRNA molecules collide with their specific amino acids, thereby forming activated aminoacyl–tRNA molecules. These molecules, in turn, collide at random with the codons located on mRNA molecules that are attached to ribosomes. When the anticodon of an aminoacyl-tRNA molecule precisely matches the codon on the mRNA strand, the amino acid is attached to the growing peptide chain, the ribosome passes down one codon, and the process is repeated in a manner crudely analogous to the movement of a typewriter carriage as each key is struck for a different letter (arrow in Fig. 3.10D).

In this manner, the primary structure of the polypeptide chain is built up one amino acid at a time on the growing protein molecule.

Briefly, the sequence of bases on one strand of a double-stranded DNA molecule is copied in a complementary manner in the primary structure of an mRNA molecule. The mRNA molecule then provides a template upon which the primary structure of the growing protein molecule is constructed on the ribosome.

Table 3.6   KNOWN NUCLEOTIDE SEQUENCES IN mRNA CODONS

POSITION ONE	POSITION TWO				POSITION THREE
	U	C	A	G	
U	Phe	Ser	Tyr	Cys	U
	Phe	Ser	Tyr	Cys	C
	Leu	Ser	(CT)	(CT)	A
	Leu	Ser	(CT)	Trp	G
C	Leu	Pro	His	Arg	U
	Leu	Pro	His	Arg	C
	Leu	Pro	Gln	Arg	A
	Leu	Pro	Gln	Arg	G
A	Ile	Thr	Asn	Ser	U
	Ile	Thr	Asn	Ser	C
	Ile	Thr	Lys	Arg	A
Met (CI)	Thr	Lys	Arg	G
G	Val	Ala	Asp	Gly	U
	Val	Ala	Asp	Gly	C
	Val	Ala	Glu	Gly	A
Val (CI)	Ala	Glu	Gly	G

Note: U = uracil; C = cytosine; A = adenine; G = guanine. Position one indicates the initial nucleotide of the triplet bearing a 5'-OH or 5'-phosphate. The third nucleotide of each triplet has a 3'-phosphate that links to the next successive triplet in the mRNA molecule. The symbol (CI) indicates a chain-initiating codon for the amino (—NH$_2$) end of a peptide chain. The symbol (CT) indicates a chain-terminating codon for the carboxyl (—COOH) end of a peptide chain. The names of the individual amino acids are summarized in Table 14-10 (p. 970).

Mutations. DNA is replicated with an extremely high order of precision. This mechanism is not absolutely perfect, however, and errors or *mutations* can occur in this process as a consequence of various types of external influences. For example, mutations can be induced by high-energy radiation (x-rays or gamma rays) as well as by many chemicals (mustard gas, peroxides, epoxides, and nitrites, as well as some pyrimidines and purines).

If a mutation takes place at the level of the gene, it is called a *point mutation*. A point mutation can involve the addition, substitution, or deletion of a nucleotide. Hence such mutations may be likened to typographic errors in a printed message, and the resulting alteration in the mRNA codon produced by the altered DNA strand will, in turn, produce an altered polypeptide chain or protein molecule.

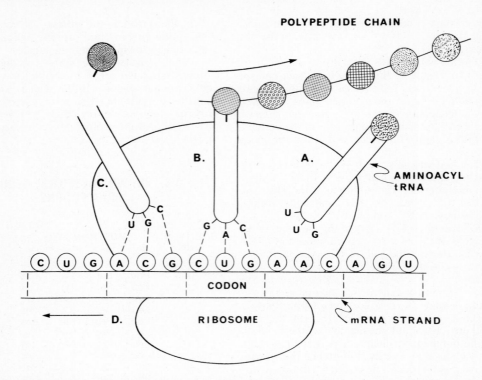

FIG. 3.10. The steps in protein synthesis. **A.** Specific tRNA molecules react with particular amino acids (shaded circles) and form activated aminoacyl tRNA molecules. **B.** The aminoacyl tRNA molecule bonds to the specific codon on an mRNA strand associated with a ribosome. When the anticodon of the aminoacyl tRNA matches the codon on the mRNA strand, the amino acid is attached to the elongating polypeptide chain (top of figure). **C.** The reactive site for amino acid binding on the tRNA molecule now is free to combine with its particular amino acid once again. **D.** The mRNA strand moves one codon to the left as indicated by the arrow, and the process is repeated with the next aminoacyl tRNA molecule.

In this manner an extremely long polypeptide chain can be synthesized one amino acid at a time with amazing rapidity. See also Figure 3.26.

Some point mutations are of minor significance as far as survival of the individual or the species is concerned; others, however, can produce serious defects. For example, the only difference between the hemoglobin in sickle cell anemia (p. 568) and normal hemoglobin lies in a single amino acid in the molecule; and this substitution, in turn, is the result of an error in a single codon in the gene responsible for the synthesis of hemoglobin.

CARBOHYDRATES. The carbohydrates are an important group of naturally occurring organic compounds that are distributed universally in both the animal and the plant kingdoms. The term "carbohydrate" indicates that these compounds are composed of carbon, oxygen, and hydrogen in the proportions $C_n(H_2O)_n$, where n equals three or more.

All organic nutrients utilized by living organisms are obtained ultimately from carbohydrates by means of photosynthesis, a process that takes place only in green plants. In addition, the chemical degradation of carbohydrates (catabolism) provides the major proportion of the energy required for the maintenance of life and for the performance of work by practically all living organisms. Carbohydrates also function as reservoirs of chemical energy and play structural roles as both intra- and extracellular components.

The carbohydrates that are of biologic significance are divided into three principal classes: *monosaccharides, oligosaccharides,* and *polysaccharides.* Generally, the monosaccharides have between four and eight carbon atoms, and only one ketone or aldehyde group is present in the molecule. Oligosaccharides are composed of only a few monosaccharide monomers that are linked together by glycosidic bonds. Carbohydrates that are characterized by a greater degree of polymerization than is found in oligosaccharides are termed polysaccharides. Extremely large polysaccharide molecules with molecular weights of several million daltons are often encountered.

Monosaccharides. The most common monosaccharide found in living material is the simple six-carbon sugar (or hexose) known as glucose (Fig. 3.11A, B, and C). This molecule contains four asymmetric carbon atoms, and thus there are $2^4 = 16$ optical isomers of this compound. Because there

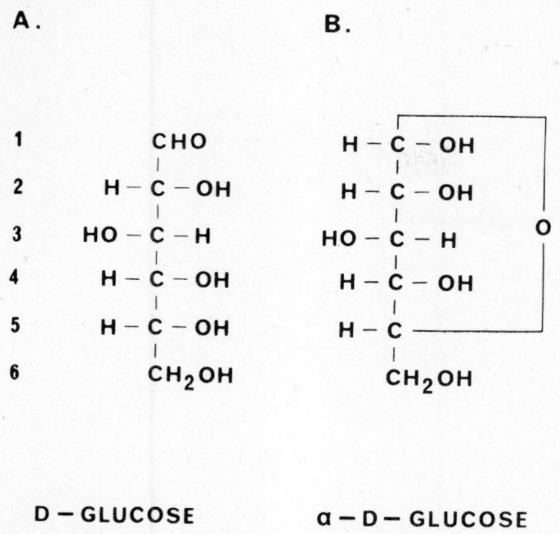

A. **B.**

1	CHO	H – C – OH
2	H – C – OH	H – C – OH
3	HO – C – H	HO – C – H
4	H – C – OH	H – C – OH
5	H – C – OH	H – C
6	CH₂OH	CH₂OH

D – GLUCOSE **α – D – GLUCOSE**

FIG. 3.11. A and B. Representations of the structural formula of glucose ($C_6H_{12}O_6$).

is an aldehyde group on C-1, glucose sometimes is referred to as an *aldose*.

Sugars that have a ketone group on C-2 are known as *ketoses*. The most abundant ketose sugar is fructose (Fig. 3.12).

Two additional examples of monosaccharides may be cited: These are the pentose (five-carbon) sugars deoxyribose and ribose, whose importance as constituents of the nucleic acids was discussed earlier (p. 59).

The ability to react to form glycosides is an extremely important property of the monosaccharides. This process involves the substitution of another compound or radical for the proton of the hydroxyl group on C-1 in the carbon chain of the

C. **D.**

HO – C – H	
H – C – OH	
HO – C – H	
H – C – OH	
H – C	
CH₂OH	

β – D – GLUCOSE **α – D – GLUCOSE**

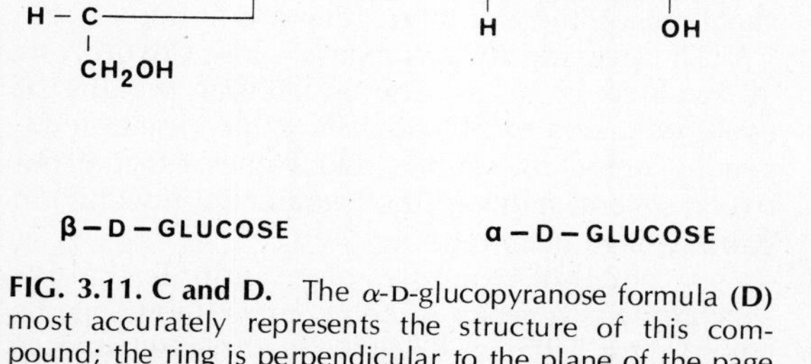

FIG. 3.11. C and D. The α-D-glucopyranose formula **(D)** most accurately represents the structure of this compound; the ring is perpendicular to the plane of the page as indicated by heavy shading on the bonds nearest to the reader (Haworth projection formula).

sugar. Thus, an acetal linkage joins the two compounds (R_1—O—R_2). Compounds that have one or more glycosidic linkages are of fundamental biochemical importance, as this is the chemical bond whereby the majority of oligosaccharides and all polysaccharides are formed. If a carbohydrate molecule is linked to a nonsugar moiety, the noncarbohydrate residue is called an *aglycone*.

In certain glycosides, the carbohydrate is linked to the aglycone through a nitrogen atom rather than through an oxygen atom (R_1—N—R_2); the resulting compounds are known as N-glycosides. This type of linkage is found in such compounds as DNA and RNA, as well as in adenosine triphosphate (ATP).

The glycosidic linkage is as fundamental in carbohydrate structure as is the peptide bond in proteins and the phosphodiester linkage in nucleic acids.

Glycosidic bonds commonly are found in the linkages between sugars and steroids. For example, the so-called cardiac glycosides contain steroids as the aglycone component of their molecules. Glycosides also are present in certain spices and as constituents of both plant and animal tissues.

Oligosaccharides. The oligosaccharides are subdivided into classes known as disaccharides, trisaccharides, tetrasaccharides, and so forth, depending upon whether the particular sugar is composed of two, three, four, or more carbohydrate monomers.

Oligosaccharides are sugars that are formed by the condensation of two or more monomers of monosaccharide through the loss of one or more water molecules. The empirical formula for a disaccharide is written $C_{12}H_{22}O_{11}$.

The most important oligosaccharides are sucrose (cane sugar), maltose, and lactose or milk sugar (Figs. 3.13A, B, and C).

Polysaccharides. From a quantitative standpoint, most of the carbohydrates found in natural sources are present in the form of polysaccharides. Functionally, such substances can be divided into two broad groups: Some polysaccharides (eg, cellulose) have primarily a structural role in the constitution of living systems; other polysaccharides serve principally as storage forms for nutrients (eg, starch and glycogen).

Polysaccharides are formed by the condensation of large numbers of monosaccharide monomers with a concomitant loss of water molecules. The empirical formula of these substances is $(C_6H_{10}O_5)_n$.

Hydrolysis of polysaccharides yields molecules of simple sugars which may be all of one type (in homopolysaccharides) or of two or more monomeric types (in heteropolysaccharides).

Cellulose. Cellulose is the most abundant structural polysaccharide in plants, and in fact the most abundant organic material found on the earth, since roughly 10^{11} tons of this material are synthe-

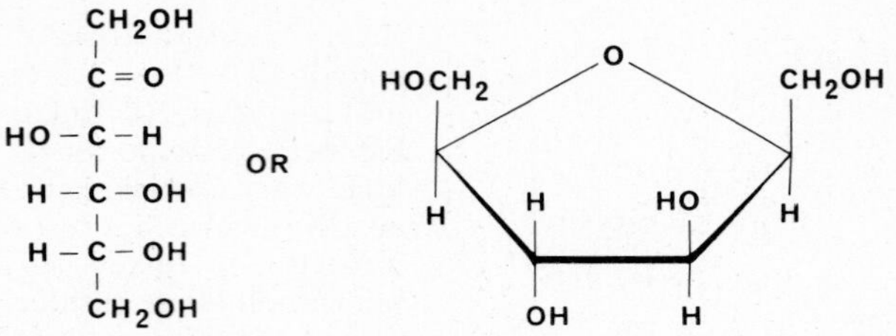

CH$_2$OH
|
C = O
|
HO – C – H
|
H – C – OH
|
H – C – OH
|
CH$_2$OH

OR

D – FRUCTOSE

FIG. 3.12. Two representations of the structural formula of the ketose sugar fructose.

sized each year. In plants, cellulose accounts for approximately 50 percent of the total carbon present and is the major structural component of such organisms.

Cellulose is composed of carbohydrate subunits known as cellobiose ($C_{12}H_{22}O_{11}$), which yield exclusively glucose monomers upon hydrolysis. Therefore, cellulose consists entirely of glucose molecules linked by β-(1→4)-glycosidic bonds.

Molecular weights of isolated cellulose preparations may vary from 50,000 to over 1,000,000 daltons. The techniques involved in isolation and purification, however, may cause some degradation of the native cellulose, so that the empirically determined weights may be far too low. In fact, cellulose is not present in the cell walls of plants as molecules per se, but rather as microfibrils that are several hundred angstroms in length. These microfibrils, in turn, are built up of many cellulose chains arranged in parallel arrays.

Starch. Starch is an example of a polysaccharide that serves as a nutritional reservoir for the plants in which it is formed. From a structural standpoint, starches are composed of long, unbranched chains of glucose monomers called *amyloses*, as well as branched chains called *amylopectins*. Amyloses are built up of repetitive α-(1→4)-glycosidic bonds between successive glucose molecules. The branch points in the amylopectin chains are formed by α-(1→6)-glycosidic linkages.

Starches exhibit considerable variation among their individual molecular weights; this may be partly a result of degradation during the isolation and purification processes. Molecular weights of several million daltons are encountered frequently for starch preparations.

Glycogen. Glycogen may be considered as the starch that is found in animal cells. This nutritional polysaccharide also is composed of long chains of glucose molecules that are highly branched and thus resemble amylopectin in structure.

Molecular weights of glycogen have been obtained that range from several hundred thousand to around a hundred million daltons.

Complex Polysaccharides. In addition to the homopolysaccharides composed exclusively of hexose monomers, discussed above, a number of complex or heteropolysaccharides are found in living systems. These substances play substantial roles in the molecular organization of cells and tissues, especially as intercellular substances. Such heteropolysaccharides may be found either in the free state or combined with various proteins.

The long molecular chains of the complex polysaccharides contain amino nitrogen (eg, glucosamine) that can additionally undergo acetylation (as in acetylglucosamine), or may have sulfuric or phosphoric acid substituents.

The major groups of complex polysaccharides found in the tissues of higher animals are acidic mucopolysaccharides, mucoproteins (mucoids), and glycoproteins.

Acidic Mucopolysaccharides. The acidic mucopolysaccharides contain sulfuric or other acids; consequently, they are strongly basophilic substances. Hyaluronic acid, for example, is composed of equimolar quantities of D-glucuronic acid and 2-acetamido-2-deoxy-D-glucose. Thus substance is distributed widely in living organisms, and is always present in the connective tissues of higher animals as well as in the synovial and vitreous fluids. Hyaluronic acid usually is found as a complex with protein.

Functionally, hyaluronic acid appears to bind water in the interstitial spaces and acts as a jellylike matrix which forms a supporting medium for the cells. Furthermore, this substance may provide the fluids present in the joints of the skeleton with shock-absorbing and lubricating properties.

The enzyme hyaluronidase causes hydrolysis of hyaluronic acid. The molecular weight of hyaluronic acid exhibits considerable variation depending upon its source, and values ranging between several hundred thousand and a few million daltons have been reported.

A second example of a complex polysaccharide is a group of compounds known as the chondroitin sulfates. Chemically these substances are identical with hyaluronic acid, except that D-galactosamine replaces D-glucosamine. Chondroitin sulfates are found in cartilage, skin, cornea, umbilical cord, and other tissues.

The molecular weight of the chondroitin sulfates is rather low for mucopolysaccharides in general, and ranges between 50,000 and 100,000 daltons.

Heparin is another example of a mucopolysaccharide. Its structure and biologic role in vivo have not yet been elaborated completely. It is clear, however, that this substance is a powerful anti-coagulant for the blood, and it appears to be synthesized and stored in the mast cells of mammalian tissues. Heparin occurs naturally in a firm complex with protein, and its molecular weight varies between 10,000 and 20,000 daltons, depending upon the preparation.

Mucoproteins and Glycoproteins. The mucoproteins and glycoproteins are complexes formed between acetylglucosamine or other carbohydrates and various proteins.

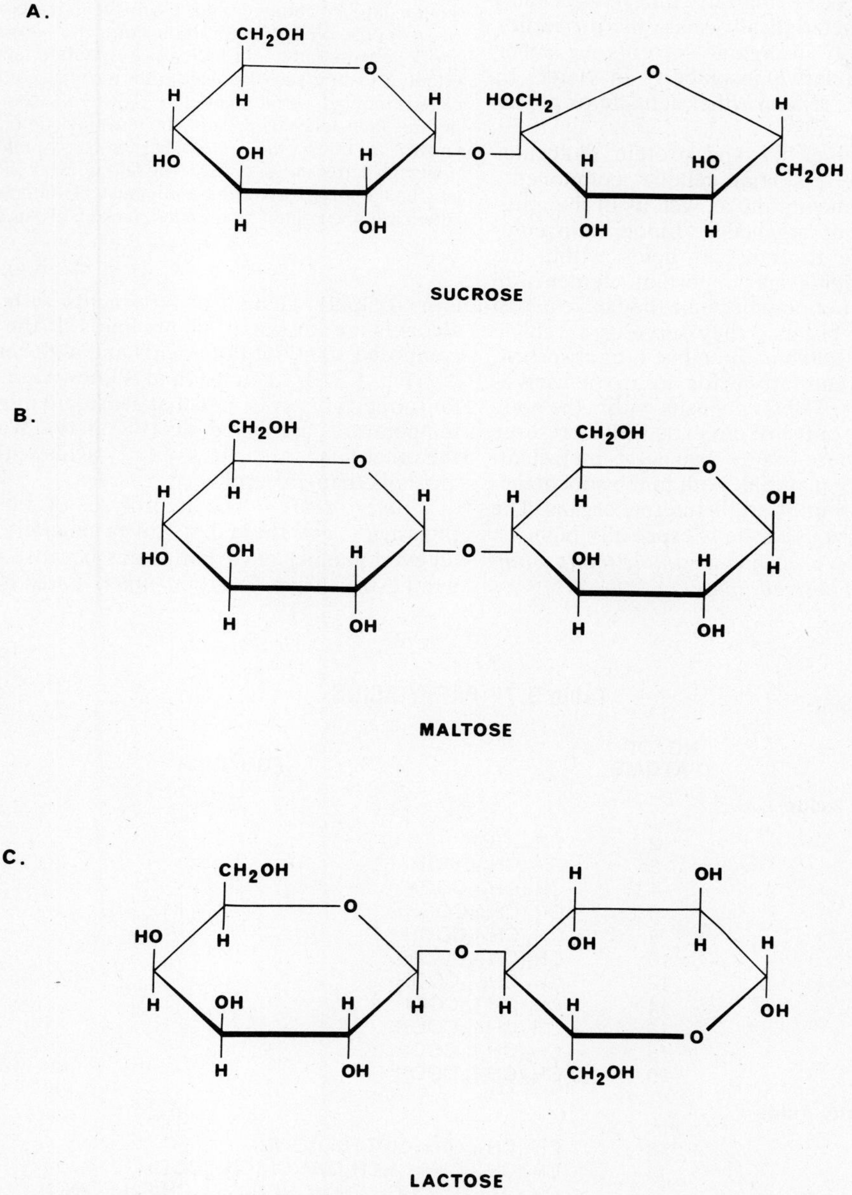

A.

SUCROSE

B.

MALTOSE

C.

LACTOSE

FIG. 3.13. Structural formulas of the three most common disaccharides. **A.** Sucrose (or cane sugar) consists of one molecule of glucose and one of fructose. **B.** Maltose consists of two molecules of glucose. **C.** Lactose (or milk sugar) consists of one molecule of glucose and one of galactose.

Mucoproteins are found, for example, in the saliva, the gastric mucosa, and ovomucoid. Important glycoproteins are ovalbumin and serum albumin.

In addition to the particular complex polysaccharides discussed above, such substances also are found in the materials responsible for specific blood groups, in the antigenic coatings of the encapsulated bacteria, and as components of bacterial cell walls.

LIPIDS. The lipids are a large and heterogeneous class of naturally occurring fatty and related compounds that characteristically share the properties of relative solubility in organic solvents (eg, ether or alcohol) and relative insolubility in water. In addition, lipids are related either actually or potentially to the fatty acids.

Combinations of fat and protein molecules (lipoproteins) are essential cellular components found in the cell membrane as well as in the mitochondria and other organelles. Lipoprotein complexes also serve to transport lipids within the blood. Dietary lipids are important elements in human nutrition, not only because they have a high energy value but because they serve as a vehicle whereby the fat-soluble vitamins are absorbed. Furthermore, certain fatty acids are in themselves essential nutrients. The fat deposits within the body are an important source of direct as well as reserve energy. Such adipose tissues also act as an insulating and cushioning material, both in the subcutaneous regions and around certain internal organs. The fat content of nervous tissue is especially high.

Lipids may be classified as *simple lipids, compound lipids,* and *derived lipids.*

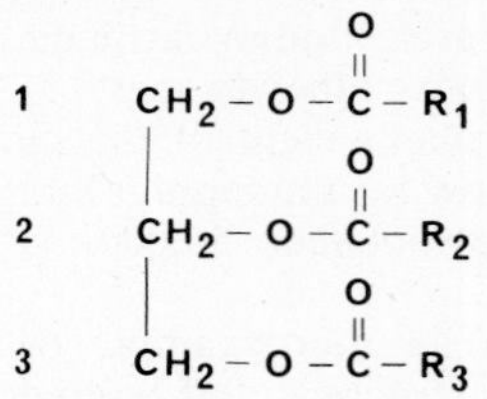

$$1 \quad CH_2-O-\overset{\overset{O}{\|}}{C}-R_1$$
$$2 \quad CH_2-O-\overset{\overset{O}{\|}}{C}-R_2$$
$$3 \quad CH_2-O-\overset{\overset{O}{\|}}{C}-R_3$$

TRIGLYCERIDE

FIG. 3.14. General structural formula of a triglyceride (or neutral fat). Positions 1 and 3 are the α carbon atoms, whereas position 2 is the β carbon atom. The symbols R_1, R_2, and R_3 denote fatty acid radicals. If only one fatty acid is esterified to the glycerol molecule, the resulting compound is called a monoglyceride (or monoacylglycerol), whereas if two or three fatty acids are present, the resulting compounds are known as diglycerides (diacylglycerols) or triglycerides (triacylglycerols), respectively. The fatty acids (R_1, R_2 and R_3) may be the same, or, more usually, may be different; in the latter case, the neutral fat is called a mixed glyceride.

Simple Lipids. Esters of fatty acids with various alcohols are known as simple lipids. If the ester is composed solely of fatty acids and a glycerol moiety (Fig. 1.31B), then the lipid is known as a neutral fat (or acylglycerol); fats that are liquid at ordinary temperatures are called oils (Fig. 3.14). Waxes, on the other hand, are esters of fatty acids with higher alcohols than glycerol.

Fatty Acids. The hydrolysis of fats yields fatty acids, and these derivatives generally contain an even number of carbon atoms because they are synthesized from 2-carbon units. Fatty acids are

Table 3.7 FATTY ACIDS

NAME	NO. OF C ATOMS	FORMULA
A. Saturated Fatty Acids		
Acetic	2	CH_3COOH
Propionic	3	$CH_3(CH_2)COOH$
Butyric (butanoic)	4	$CH_3(CH_2)_2COOH$
Caproic (hexanoic)	6	$CH_3(CH_2)_4COOH$
Caprylic (octanoic)	8	$CH_3(CH_2)_6COOH$
Decanoic (capric)	10	$CH_3(CH_2)_8COOH$
Lauric	12	$CH_3(CH_2)_{10}COOH$
Myristic	14	$CH_3(CH_2)_{12}COOH$
Palmitic[a]	16	$CH_3(CH_2)_{14}COOH$
Stearic[a]	18	$CH_3(CH_2)_{16}COOH$
Arachidic	20	$CH_3(CH_2)_{18}COOH$
B. Unsaturated Fatty Acids		
Oleic	18	$CH_3(CH_2)_7CH{=}CH(CH_2)_7COOH$
Linoleic[b]	18	$CH_3(CH_2)_4CH{=}CHCH_2CH{=}CH(CH_2)_7COOH$
Linolenic[b]	18	$CH_3CH_2CH{=}CHCH_2CH{=}CHCH_2CH{=}CH(CH_2)_7COOH$
Arachidonic[b]	20	$CH_3(CH_2)_4CH{=}CHCH_2CH{=}CHCH_2CH{=}CHCH_2CH{=}CH(CH_2)_3COOH$

[a]*Common in all animal and plant fats.*
[b]*Linoleic, linolenic, and arachidonic acids are the nutritionally essential fatty acids.*

straight-chain (ie, aliphatic) compounds. The chain may contain no double bonds, in which case it is said to be saturated, or it may contain one or more double bonds, in which case the compound is called an unsaturated fatty acid.

Saturated fatty acids are based upon the first member of the series, acetic acid (CH_3COOH), and have the general formula $C_nH_{2n+1}COOH$ (Table 3.7A.)

Monounsaturated fatty acids have the general formula $C_nH_{2n-1}COOH$. For example, oleic acid has one double bond between C-9 and C-10. Similarly, the general formula of polyunsaturated fatty acids with two double bonds is $C_nH_{2n-3}COOH$; for three double bonds, it is $C_nH_{2n-5}COOH$; and for four double bonds, $C_nH_{2n-7}COOH$.

Examples of the important unsaturated fatty acids are given in Table 3.7B.

The prostaglandins (p. 1173) are a group of substances that are of physiologic interest because of their biochemical and pharmacologic properties. These compounds are synthesized from arachidonic acid.

Alcohols. The alcohols commonly found in lipid molecules include glycerol and cholesterol, as well as higher alcohols.

If only one fatty acid is esterified to one OH radical of the glycerol moiety, the resulting fatty compound is known as a monoglyceride or monoacylglycerol. If two or three fatty acids are esterified to two or three hydroxyl radicals of glycerol, the resulting compounds are known as diglycerides (diacylglycerols) or triglycerides (triacylglycerols), respectively.

The neutral fats in the body generally contain a mixture of different fatty acids; hence, they are called mixed glycerides.

Compound Lipids. Esters of fatty acids that contain chemical groups or radicals in addition to fatty acids and an alcohol are known as compound lipids. Included in this category are phospholipids, cerebrosides (or glycolipids), and other compound lipids.

Phospholipids. The phospholipids contain a phosphoric acid residue as well as nitrogenous bases and other constituents, in addition to fatty acids and glycerol.

Several important types of phospholipids may be reviewed briefly.

The lecithins, also called phosphatidyl cholines, are composed of glycerol, fatty acids, phosphoric acid, and the nitrogenous base choline (Fig. 3.15A, B). Lecithins are distributed widely in the tissues of the body, where they have both structural and metabolic functions in the cells. For example, dipalmityl lecithin serves as a surface-active agent within the lungs, where it prevents collapse of the air sacs (alveoli) due to the effects of surface tension (p. 721).

The cephalins (phosphatidyl ethanolamines) are quite similar chemically to the lecithins, except

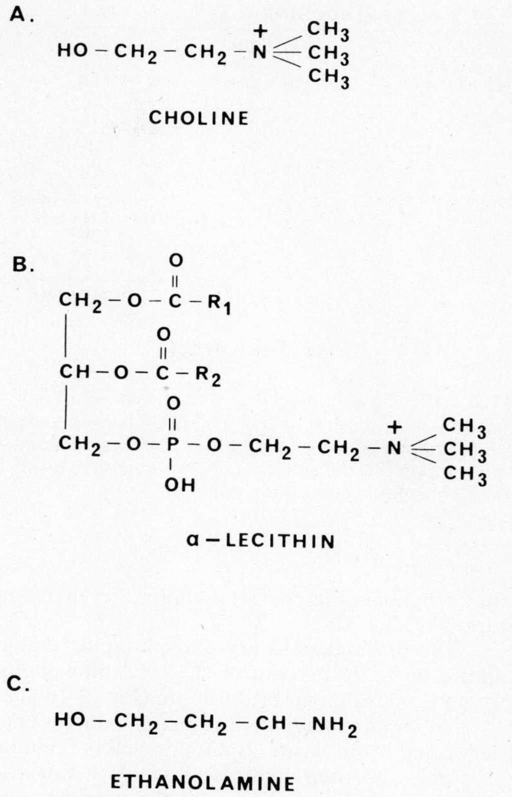

FIG. 3.15. **A.** The nitrogenous base choline is found characteristically in the group of phospholipids called lecithins. **B.** General formula of an α-lecithin. The symbols R_1 and R_2 indicate fatty acids esterified to the glycerol moiety. **C.** Ethanolamine instead of choline is the characteristic nitrogenous base found in the group of phospholipids known as cephalins; otherwise, their structure is quite similar to that of the lecithins **(B).**

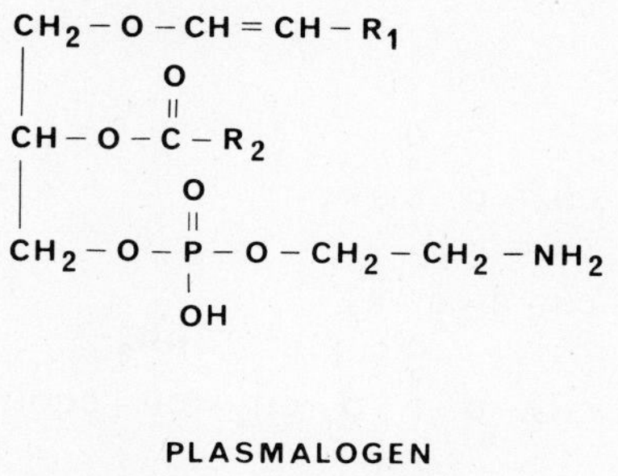

FIG. 3.16. Plasmalogens are important phospholipid constituents of brain and muscle. The symbols R_1 and R_2 denote fatty acids. Note that there is an ether linkage (—O—) on C-1 instead of an ester linkage. The aliphatic (or alkyl) chain of the fatty acid in this position typically is unsaturated rather than saturated, as indicated by the double bond. In this figure, the nitrogeneous constituent shown is ethanolamine; however, either choline or inositol may be substituted for the ethanolamine.

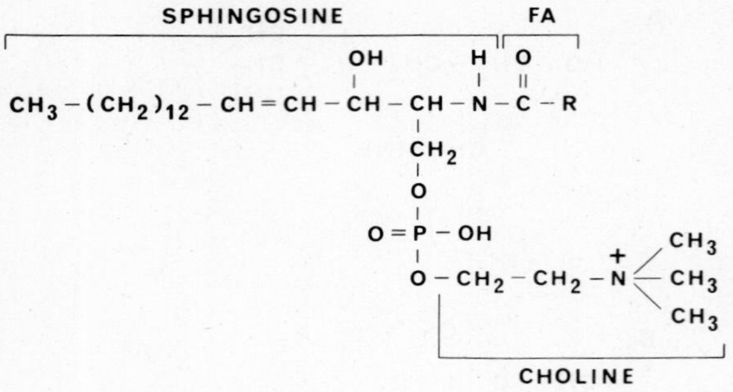

SPHINGOSINE   FA

$$CH_3 - (CH_2)_{12} - CH = CH - \underset{\underset{\displaystyle CH_2}{|}}{\underset{\displaystyle OH}{\overset{|}{CH}}} - \underset{\overset{|}{H}}{\overset{|}{CH}} - \underset{}{N} - \overset{\overset{\displaystyle O}{\|}}{C} - R$$

CH₂

O

O = P − OH

O − CH₂ − CH₂ − N⁺ ⟨ CH₃ CH₃ CH₃

CHOLINE

SPHINGOMYELIN

FIG. 3.17. Sphingomyelins are major phospholipid components of brain and neural tissue. Unlike other types of phospholipid, no glycerol is present. Rather, sphingomyelins are composed of the amino alcohol sphingosine, a fatty acid (FA), phosphoric acid, and choline.

that ethanolamine replaces choline in their structure (Fig. 3.15C).

The plasmalogens are phospholipids that constitute up to 10 percent of the total phospholipids found in muscle and brain tissue (Fig. 3.16).

The sphingomyelins are major constituents of nerve and brain tissue. Sphingomyelins contain no glycerol, and their hydrolysis yields a fatty acid, phosphoric acid, choline, and the complex amino alcohol sphingosine (Fig. 3.17).

Certain other phospholipids may contain an amino acid moiety rather than ethanolamine. For example, phosphatidyl serine is a cephalinlike phospholipid that contains the amino acid serine (Fig. 3.18).

Cerebrosides (Glycolipids). The cerebrosides contain galactose, phosphoric acid, and sphingosine as well as a high-molecular-weight fatty

acid; consequently, they may also be classified with the sphingomyelins. The cerebrosides commonly occur in many tissues in addition to the brain. In peripheral nerve fibers, these compounds are found in much higher concentration in medullated (myelinated) than in nonmedullated nerve fibers.

Derived Lipids. The compounds that may be obtained by hydrolysis from the various groups of lipids discussed above are referred to as derived lipids. Included in this category are saturated as well as unsaturated fatty acids, glycerol, steroids, alcohols, fatty aldehydes, ketone bodies, and sterols (Fig. 3.19).

Because they are uncharged molecules, the glycerides, cholesterol, and cholesterol esters are known as neutral lipids.

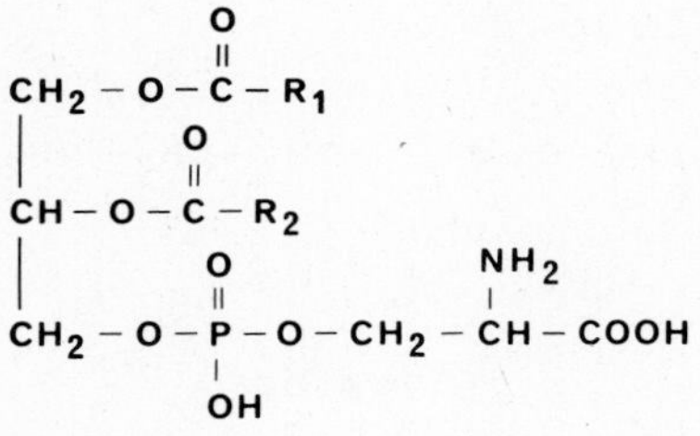

THE CELL

The cells which comprise the individual tissues of the human body exhibit a broad variety of morphologic, physiologic, and biochemical specializations; thus, there is no such thing as a "typical" cell. Nevertheless, most cells share a number of common features which have similar morphologic and functional properties regardless of the type of cell. Hence, it is possible to discuss in broad terms a number of the attributes that are shared by cells in general, and this approach will be followed in the following presentation.

From a physicochemical standpoint, protoplasm may be considered to be a dynamic multiphasic colloidal system. Therefore, the living matter that comprises a cell consists of numerous localized regions or structures (the dispersed phase) suspended in an aqueous medium (the continuous phase) that has some of the properties of a fibrous gel. In accordance with this concept, pro-

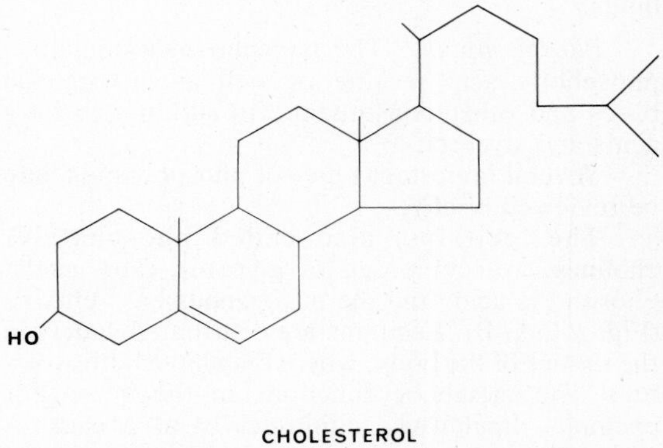

$$\underset{\displaystyle OH}{\underset{|}{\underset{\displaystyle CH_2 - O - P - O - CH_2 - \overset{\overset{\displaystyle NH_2}{|}}{CH} - COOH}{\overset{}{}}}}$$

CH₂ − O − C − R₁

CH − O − C − R₂

CH₂ − O − P − O − CH₂ − CH − COOH

3 − PHOSPHATIDYL SERINE

FIG. 3.18. Some phospholipids contain an amino acid as a nitrogenous group instead of choline or ethanolamine, for example, phosphatidylserine. The symbols R₁ and R₂ indicate fatty acids esterified to the glycerol moiety.

CHOLESTEROL

FIG. 3.19. Cholesterol is an example of a derived lipid. (Compare Figure 1.28.) Cholesterol is found in all cells as well as in the blood, but it is particularly abundant in neural tissue.

toplasm may be envisaged as a complex system rather than as a discrete entity.

As noted in the previous section, the principal macromolecules of the cell belong to three major classes — proteins, nucleic acids, and polysaccharides — and these compounds, together with highly organized laminar lipoprotein aggregates, form relatively large, specific, intracellular structures known as *organelles.* An organelle may be defined as a minute intracellular organ that has certain definite functions. These membrane-limited organelles tend to compartmentalize the functional activities of the cell to a certain extent, so that quite dissimilar and independent chemical processes can take place simultaneously and without interference from each other; thus the overall cellular machinery functions in a smoothly coordinated and highly regulated fashion with regard to both space and time.

The combined use of light and electron microscopy is indispensable to the resolution of many problems surrounding the morphologic characteristics of the cell and its components, but in general these visual techniques yield no clues as to the specific functional properties of the various subcellular particulates, structures, and organelles that may be seen. One technique that has proved to be invaluable in functional studies is the mechanical disruption of cells combined with subsequent fractionation of the subcellular components by means of differential ultracentrifugation. In this manner, relatively pure masses of individual organelles can be obtained in sufficient quantities to carry out a multitude of sophisticated biochemical and other types of study. Much of the information summarized in this section has been obtained by employing this technique in combination with light and electron microscopy.

In this presentation, the cell will be divided arbitrarily into two regions, the *cell membrane* and the *cytoplasm;* the latter will be considered together with its specific organelles and inclusions. The general structural aspects of the cell are shown in Figure 3.20.

The Cell Membrane

MORPHOLOGY.   All living cells are surrounded externally by an extremely thin limiting membrane known as the *plasmalemma, plasma membrane,* or *cell membrane.** This cell membrane generally is invisible under a light microscope but appears as a dense linear structure with a thickness ranging between 50 and 100 Å (average about 80 Å) when viewed in profile at moderate magnifications under an electron microscope. At much higher magnifications, the cell membrane in cross section is re-

solved into two dense layers, each about 25 Å thick, which are separated by a third lighter, intermediate layer approximately 30 Å in thickness. Although some asymmetry exists between the thickness of the outer (external) and inner (cytoplasmic) dense layers of the cell membrane, which probably reflects functional differences, all cell membranes have been shown to have the same fundamental *trilaminar structure,* which is often referred to as the *unit membrane* (Fig. 3.21).

There is no paucity of models that have been advanced to explain the physicochemical and biologic properties of the plasma membrane. In the classic view, set forth about 40 years ago by Davson and Danielli, the plasma membrane was envisaged as a bimolecular layer of lipids. The hydrophilic portions of the lipid molecules faced outward, as shown in Figure 3.21, whereas the hydrophobic tails on the molecules, the fatty acid chains, were apposed in the center of the membrane (p. 15).† In accordance with this model, layers of globular protein molecules were postulated to cover the lipids so that in effect the cell membrane was composed of a lipid bilayer sandwiched between two separate layers of protein. Insofar as its functions were concerned, the cell membrane was presumed at this time to be a relatively static morphologic feature of the cell that was responsible primarily for delimiting the cell and serving as a container for the cytoplasm and its components.

Although certain basic features of the Davson–Danielli model of the cell membrane are correct, a number of important details now are known that render this model untenable in detail. A current view of the functions and structures of the cell membrane may be summarized as follows.

(1) The cell membrane and the individual membranes that invest certain intracellular organelles have essentially the same structure, and both are composed of similar compounds, namely, proteins and lipids.

(2) As in the Davson–Danielli model, the structural framework or backbone of the membrane is a bilayer of amphipathic lipid molecules that are oriented as depicted in Figure 3.21. The cell membrane consists of phospholipids, glycolipids, and neutral lipids, particularly cholesterol. The membranes of the intracellular organelles consist principally of phospholipids.

Regardless of the type of membrane, the lipids and carbohydrates present are bound to protein molecules in the form of glyco- and/or lipoproteins.

There is no exchange of lipids across the bilayer, except for that which is catalyzed enzymatically. However, the lipid molecules *are* able to diffuse laterally to a limited extent within the membrane.

---

*The plasmalemma must not be confused with the *cell wall, a relatively gross structure composed of cellulose which invests plant cells.*

†*The term* amphipathic *is applied to substances which have strongly polar (hydrophilic) and nonpolar (hydrophobic) regions in the same molecule.*

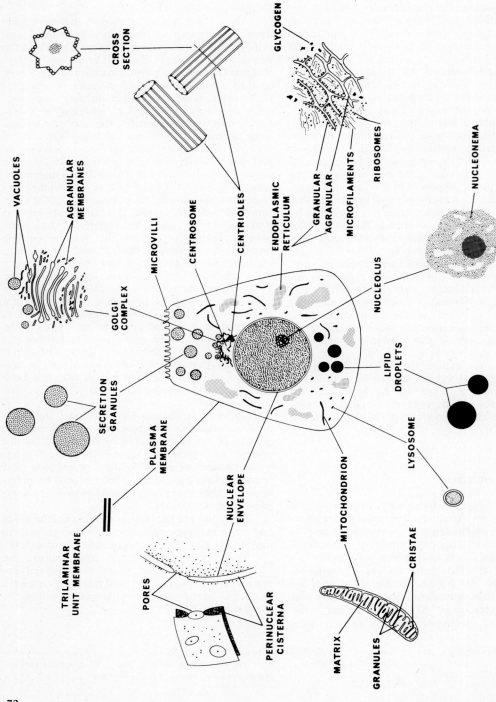

VACUOLES

AGRANULAR MEMBRANES

GOLGI COMPLEX

MICROVILLI

CENTROSOME

CENTRIOLES

CROSS SECTION

GLYCOGEN

ENDOPLASMIC RETICULUM

GRANULAR

AGRANULAR

MICROFILAMENTS

RIBOSOMES

NUCLEONEMA

NUCLEOLUS

SECRETION GRANULES

PLASMA MEMBRANE

LIPID DROPLETS

LYSOSOME

TRILAMINAR UNIT MEMBRANE

NUCLEAR ENVELOPE

PORES

PERINUCLEAR CISTERNA

MITOCHONDRION

MATRIX

GRANULES

CRISTAE

FIG. 3.20. A cell and its organelles and inclusions as observed by light microscopy (center of figure) and by electron microscopy (periphery of figure). (After Bloom and Fawcett. *A Textbook of Histology,* 9th ed., 1968, Saunders.)

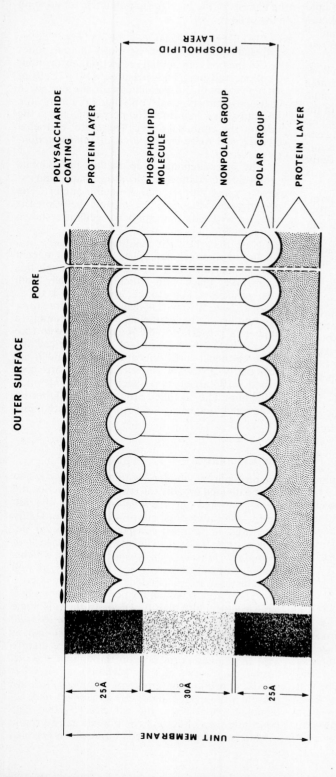

FIG. 3.21. The classic trilaminar unit membrane of the cell as revealed by electron microscopy. Compare Figure 3.22.

73

(3) The structural proteins of the membrane are a special class of heterogeneous globular molecules that are elaborated in the cytoplasm. The proteins are arranged in an amphipathic structure (Fig. 3.22). The molecules may be adsorbed upon the membrane surface, partially imbedded in the lipid matrix, or even extend through the entire thickness of the membrane, in which case carbohydrate is present upon the extracellular surface and the protein is asymmetric.

Thus, the membrane may be visualized as composed of a *mosaic* of different kinds of protein molecules that are imbedded in a partially fluid matrix of lipids. In other words, the structure of the membrane resembles that of a liquid crystal in certain respects, and the protein molecules are embedded either wholly or partially within a discontinuous fluid bilayer. Only a small proportion of the total lipid present reacts specifically with the proteins.

(4) Certain of the proteins can move into or out of the membrane to a limited extent, as well as laterally along the plane of the membrane, as indicated by the arrows in Figure 3.22. However, as is the case with the lipid molecules, there is no actual exchange of structural protein across the membrane.

(5) The proteins and glycoproteins have a number of particular roles in membranes. They can act as structural elements, function as enzymes for synthetic or degradative processes, operate as pumps for the transport of materials into and out of cells and their organelles, and behave as receptor sites for hormones, antigens, and drugs. In fact, it is the diversity of the proteins as well as their specific functions that gives each particular cell membrane its unique characteristics.

(6) In addition to its fundamental structure, the plasmalemma acts as though it contained a number of continuous *pores* scattered in a random fashion about its surface. These pores are approximately 7.5 Å in diameter, according to calculations based upon the relative permeability of the cell membrane to solutes of various molecular sizes (Fig. 3.21).

Based on the foregoing statements, it is clear that the membranes of the cell, both individually and collectively, are highly dynamic and functionally complex structures that perform a wide variety of physiologic operations which are critical to almost every phase of cellular activity.

GENERAL PHYSIOLOGIC PROPERTIES OF THE CELL MEMBRANE. It is essential to reiterate that most of the important physiologic processes that occur in living organisms take place on surfaces and at interfaces. The cell membrane as well as the membranes that invest certain intracellular organelles provide such surfaces. The cell membrane is the region where specific interactions occur between the cell interior and its immediate environment. The intracellular membranes serve to compartmentalize and thus localize ions, metabo-

lites, enzymes, and other protoplasmic constituents.

A number of significant membrane functions were presented in (5) above. Furthermore, the intramembraneous transport mechanisms underlie the genesis of the bioelectric potentials found in all living cells (p. 98).

SPECIALIZED JUNCTIONS BETWEEN CELLS. Epithelial cells have one or more diverse functions in the body, including protection, absorption, secretion, transport, and special sensation. Cells of this type maintain close contact with one another in sheets, so that materials entering or leaving the body must pass in a direction perpendicular to the plane of the epithelial sheet. Therefore, epithelial cells are said to be *morphologically and functionally polarized*.

The plasmalemma of epithelial cells exhibits certain characteristic specializations that permit extremely close lateral cell to cell contact. The specializations that may be present between the membranes of adjacent epithelial cells are shown in Figure 3.23.

Note that the tight junction obliterates completely the intercellular space between adjacent cells, so that diffusion of substances through this region takes place with difficulty, if at all. Such tight junctions sometimes are referred to as *nexuses*. Note also in this figure that the free *(apical)* border of the epithelial cell may contain a number of projections, the *microvilli*, that in toto vastly increase the total surface area of the cell and of the epithelial surface as a whole. Microvilli are particularly abundant on surfaces that are concerned with absorption, such as the intestinal epithelium. The so-called striated or brush border of the renal tubules also is composed of microvilli.

The Cytoplasm

The protoplasm contained within the plasmalemma may be divided broadly into two compartments: the *cytoplasm,* and the *nucleoplasm* within the nucleus. The latter organelle, together with its substructures, is the most conspicuous organized element within the cell. The cytoplasm consists of *ground substance* or *cytoplasmic matrix,* in which a number of more or less evident organelles, as well as inclusions, are suspended.

The cytoplasmic matrix per se lacks any obvious structure, although functionally there is some evidence that a phenomenon known as *metabolic channeling* is present in this part of the cell, and, as a consequence, many small molecules are distributed in a nonuniform fashion throughout the cell during the course of their metabolism. In other words, concentration gradients of metabolites are present within cells, and, since organelles may accumulate metabolites, the net effect is that high concentrations of certain compounds can be found in quite specific and localized intracellular regions.

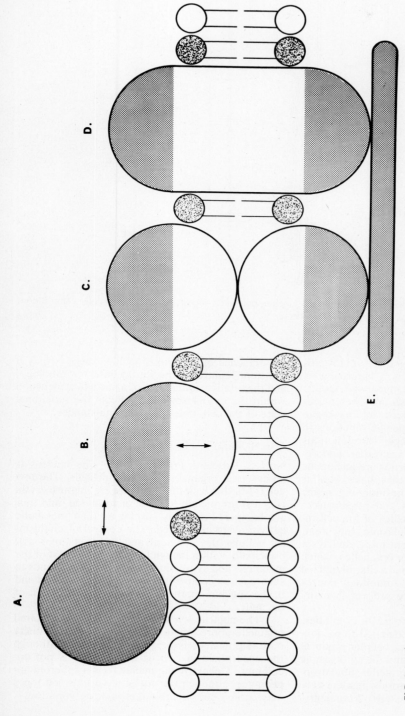

FIG. 3.22. The possible relationships of different protein molecules to the bimolecular lipid layer of the cell membrane. Shading indicates hydrophilic portions of molecules. **A.** The protein molecule adsorbed on the outer surface of the cell membrane can move back and forth to some extent (horizontal arrow). **B.** Amphipathic protein molecule with the hydrophilic region of the molecule on the outer surface, the hydrophobic region penetrating into the bimolecular lipid layer. The direction of movement of the molecule depending on the functional state of the cell is indicated by the vertical arrow. The immobilized polar groups of the lipid molecules immediately adjacent to the protein molecule are indicated by shading. **C.** Two amphipathic protein molecules with their hydrophobic portions apposed and their hydrophilic regions facing outward on opposite sides of the membrane. **D.** Single protein molecule with central hydrophilic portion and hydrophilic ends that passes entirely through membrane. **E.** Rodlike hydrophilic protein molecule linking other proteins on the inner surface of the membrane into aggregates so that the molecules potentially can move and function as units.

The specific types and orientations of various protein molecules lend functional and morphologic asymmetry to the membrane. The protein molecules are not static but rather mechanically and metabolically dynamic entities which confer the characteristic structural, enzymatic, transport, and receptor properties upon particular cell membranes. Compare Figure 3.25. (Data from Capaldi. *Sci Am* 230:26, 1974.)

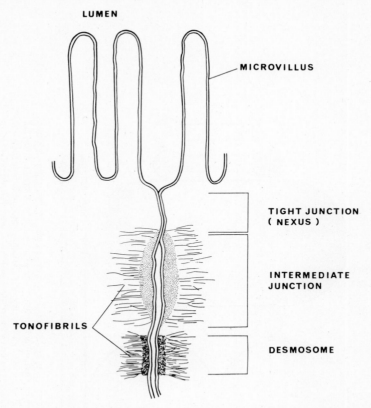

FIG. 3.23. The morphologic specializations present at the junction between two columnar epithelial cells. (After Bloom and Fawcett. *A Textbook of Histology,* 9th ed, 1968, Saunders.)

Thus the cytoplasmic matrix, together with its various organelles and inclusions, is a heterogenous mixture when viewed from a chemical standpoint.

As noted above, an *organelle* may be considered as any small, highly organized, internal organ of the cell, which carries out specific functions in cellular metabolism. In this category are functions such as digestion, respiration, excretion, and reproduction. Thus cellular organelles are analogous to entire anatomic systems that have similar physiologic roles at the level of the entire organism.

In addition to the nucleus and its substructures, the other principal organelles of the cell include the mitochondria, endoplasmic reticulum, ribosomes, Golgi apparatus, centrioles, and lysosomes. In addition to these major organelles, which are found in practically all animal cells, organized structures such as annulate lamellae, microfilaments, and microtubules are present in many cells.

Organelles are considered to be living entities, whereas cellular inclusions are deemed to be nonliving metabolites or accumulated cellular products that are found as aggregates within the cytoplasmic matrix. For example, glycogen granules, lipid droplets, secretory droplets, and various pigments are classified as intracellular inclusions. Frequently, the presence of inclusions such as these is only a transient feature of the intracellular morphology, and the relative abundance of a particular inclusion within the cytoplasm often reflects the functional state of the cell under given circumstances.

THE NUCLEUS AND NUCLEOLUS

Nuclear Structure and Functions. The nucleus is an essential organelle in nearly all cells. The few cell types that lack this structure (eg, mature erythrocytes) are unable to grow or to divide, and thus are hampered severely insofar as their metabolic activities are concerned. The control mechanisms for cellular development and function are found principally within the nucleus, whereas most of the metabolic and synthetic machinery that responds to such nuclear control is found in the cytoplasm and in the specific organelles present in this region of the cell.

The nucleus is a prominent, membrane-limited organelle approximately 5 μ in diameter. Usually only one nucleus is present in each cell, although the occurrence of two or more nuclei is not uncommon. Generally the nucleus has a spheroid appearance, although the nuclei of various cell types may exhibit considerable morphologic variation.

All nuclei contain DNA (p. 57), the genetic material of the cell, and, as noted earlier, this DNA specifically determines the morphologic and biochemical characteristics of every cell.

Nuclear DNA is not present in the free state; rather, it is found combined with histones and other structural proteins into a substance known as *chromatin*. Chromatin may be present in either of two interconvertible states depending upon its degree of dispersion or condensation. Dispersed chromatin stains weakly, if at all, whereas condensed chromatin stains deeply.* During cell division (mitosis, p. 109), the dispersed chromatin that was present in the nucleus during interphase undergoes condensation into a number of deep-staining, elongated, rodlike bodies known as *chromosomes* whose total number is a characteristic of every species of organism (Table 3.1).

The nucleus is clearly delineated from the ctyoplasm by a well-defined *nuclear membrane*. Electron microscopy has revealed that this structure consists of *two* parallel unit membranes which enclose a narrow space called the *perinuclear cisterna* (Fig. 3.20). At intervals over the entire surface of the nuclear membrane, these two unit membranes fuse; such regions are punctuated by small, circular nuclear pores, each of which is covered by a thin diaphragm (or *septum*) that is somewhat thinner than the usual unit membrane. The nuclear pores have an approximate diameter of 1,000 Å. Between 1,000 and 10,000 of these fenestrated openings are scattered over the entire surface of the nucleus.

The nuclear pores form channels whereby certain macromolecules that are synthesized within the nucleus but function in the cytoplasm pass from one compartment of the cell to the other. For example, ribonucleic acid (RNA), which is synthesized at intranuclear sites, enters the cytoplasm by means of these pores.

The external membrane of the nuclear envelope frequently is continuous with the canalicular system of the endoplasmic reticulum; hence, the perinuclear space is considered to be a cisterna. Consequently, the nuclear membrane itself sometimes is included as a part of this cytoplasmic system. Ribosomes often are found in close association with the outer or cytoplasmic surface of the nuclear membrane.

Nucleolar Structure and Function. The nucleolus is a discrete and prominent organelle that is located eccentrically within the nucleoplasm. Usually between one and four of these structures are found within the nucleus of a cell, and each nucleolus is composed of a dense branching and anastomosing strand, the *nucleolonema*. The nucleolus is not membrane limited (Fig. 3.20).

The nucleolus stains deeply with basic dyes as well as with a number of acidic dyes. The basophilic property is due to a high concentration

of RNA. Therefore, many investigators believe that the nucleolus is an important intermediate region through which these nucleic acids must pass after their synthesis in the nucleus before they migrate to their ultimate functional locus within the cytoplasm.

The nucleolus exhibits considerable variation in its size depending upon the cell type as well as the physiologic state of the cell. For example, the nucleolus is particularly large in cells that are engaged actively in protein synthesis.

THE MITOCHONDRIA

Structure. The mitochondria are minute, membrane-bounded cytoplasmic organelles that have an appearance ranging from small spheroid granules to ellipsoid rods, or even long filaments (Figs. 3.20, 3.24). In general, mitochondria are about 0.5 μ in diameter and between 2 to 4 μ in length. Within the living cell, they are moved above passively by the normal flow of the cytoplasm *(cytoplasmic streaming),* but mitochrondria also appear to be capable of independent active movement as well. The total number of mitochondria per cell averages between several hundred and a thousand.

Electron microscopy has revealed that mitochondria have a very complex and characteristic structure. An individual mitochondrion is encapsulated by two distinct unit membranes (Figs. 3.24, 3.25). The smooth outer limiting membrane is about 50 Å in thickness. The inner membrane has a similar thickness, but it forms narrow folds, the *cristae,* which project into the central region of the organelle. The inner membrane and the cristae contain morphologic subunits that can be visualized only at extremely high magnifications. These subunits undoubtedly represent highly organized arrays of specific enzymes. Between the individual cristae, the central cavity of the mitochondrion is occupied by a relatively dense granular mitochondrial matrix. Thus a mitochondrion is divided into an intercristal space in which the matrix is

---

The Feulgen reaction is a highly specific histochemical test for deoxyribonucleoprotein.

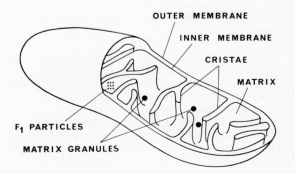

FIG. 3.24. The ultrastructure of a mitochondrion.

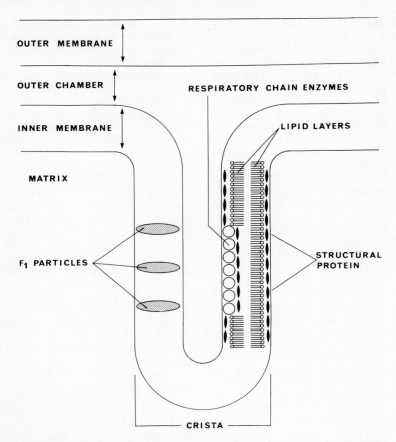

FIG. 3.25. Ultrastructure of a portion of a mitochondrion illustrating the location on the crista of the F_1 particles containing the phosphorylase enzymes that synthesize ATP (shown on left) as well as the location of the respiratory enzymes in relation to the protein and lipid components of the unit membrane (shown on right). (Data from DeRobertis, Nowinski, and Saez. *Cell Biology*, 5th ed, 1970, Saunders.)

present, as well as an outer chamber or membrane space which lies between the outer and inner membranes of the organelle.

The membrane space is quite narrow and appears empty in electron micrographs, whereas the intercristal space is occupied by a number of discrete and dense *matrix granules,* which have a diameter of around 400 Å. The presence within the matrix of deoxyribonucleoprotein filaments has also been demonstrated. Small ribonucleic acid granules also are found in this region of the mitochondria.

Functions.   The technique of fractional centrifugation mentioned earlier has permitted the isolation of mitochondria in sufficient quantities to allow characterization of many of the biochemical activities of these organelles.

The mitochondria function primarily as specific loci for the chemical reactions involved in intracellular energy production and energy transfer (energy *transduction*). Consequently, these organelles frequently are referred to as the "powerhouses of the cell."

The enzymes of the inner mitochondrial membrane are present in a highly organized, repetitive pattern of subunits that facilitate the sequential catalytic reactions carried out by these organelles (Fig. 3.25). Disruption of mitochondria that have been isolated by fractional centrifugation yields enzymes of two general classes: *membrane-bound enzymes* and *soluble enzymes.* The soluble enzymes represent those present in the matrix of the intact mitochrondrion. Generally speaking, the respiratory and phosphorylating enzymes are associated with the membranes, whereas most of the enzymes that function in the citric acid cycle, as well as those concerned with protein and lipid synthesis, are located in the matrix (Table 3.8). The respiratory enzymes of the citric acid cycle act in a specific sequence to oxidize the 2-carbon fragments generated by intermediary metabolism into carbon dioxide and water. During this oxidative process, electrons are transferred along the chain of respiratory enzymes (p. 931). Oxidative phosphorylation (p. 933) with a concomitant synthesis of the ubiquitous high-energy compound *adenosine triphosphate* (ATP, p. 104) from adenosine diphosphate (ADP) and inorganic phosphate is coupled closely to electron transport, and the phosphorylase enzymes that catalyze this process are found as F_1 subunits in the inner membrane (Fig. 3.25).

Table 3.8 ENZYMES OF MITOCHONDRIA

I. Membrane Bound

    A. Dehydrogenases for:
       1. Succinic acid
       2. Isocitric acid (?)
       3. Reduced adenine dinucleotide (NADH)
       4. β -hydroxybutyric acid
       5. α -glycerophosphate
       6. Choline
    B. Cytochromes
    C. Coenzyme Q
    D. Flavoproteins
    E. Enzymes of oxidative phosphorylation

II. Soluble

    A. Enzymes of the citric acid (Krebs) cycle, except certain dehydrogenases; eg, succinic acid (isocitric acid ?)
    B. Glutamic acid dehydrogenase
    C. Enzymes involved in protein synthesis
    D. Enzymes involved in lipid synthesis

The matrix granules are believed to be involved in regulating the ionic composition within the mitochondria. Divalent cations such as calcium and magnesium tend to accumulate inside mitochondria when they are present at sufficiently high concentrations in the external medium, and the matrix granules concomitantly become more electron dense and larger under such conditions. Presumably, the excess cations bind organically to the matrix granules, causing them to swell.

The mitochondrial DNA apparently constitutes an extranuclear genetic system within the cell which has limited functions. This DNA appears to be responsible for the protein synthesis necessary for the proliferation of mitochondria when it acts in conjunction with the mitochondrial RNA. However, it also appears that the nuclear and mitochondrial genetic systems are closely interrelated, so that both of these components are necessary to the formation of a full complement of proteins by the mitochondria. The DNA present in the mitochondria alone contains insufficient genetic information to enable all protein constituents to be synthesized by these organelles.

THE ENDOPLASMIC RETICULUM. The advent of electron microscopy has established clearly that the cytoplasm contains two distinct types of membrane-limited, interconnecting, fluid-filled canalicular systems: the granular or rough endoplasmic reticulum, and the agranular or smooth endoplasmic reticulum.

The Granular Endoplasmic Reticulum and the Ribosomes

Structure. When stained with basic dyes, the cytoplasm of many kinds of cell contains either diffuse or aggregated masses of strongly basophilic material, which was termed by light microscopists the chromidial substance or ergastoplasm. Later it was demonstrated that this basophilic cytoplasmic material was composed of ribonucleoproteins.

The application of electron microscopy to this cytologic problem revealed that most cells have a tridimensional, anastomosing network of channels, and this system was given the status of an intracellular organelle, the *endoplasmic reticulum*. Typically, this organelle consists of irregular branching tubules that frequently are continuous with flattened, *saccular* structures known as *cisternae* (Figs. 3.20, 3.22). Generally, the cisternae form lamellar systems of flat, parallel cavities, although these dilatations may occur singly. In addition to the tubular network and cisternae, certain isolated cytoplasmic vesicles also are considered to be part of the endoplasmic reticulum. There is a considerable morphologic variation in the extent to which the endoplasmic reticulum is developed among different cells, and even in the same cell type under various physiologic states.

The ergastoplasm of the classic microscopist now is known to correspond to the endoplasmic reticulum as revealed by electron microscopy. However, the basophilic reaction resides *not* in the reticulum of canaliculi per se, but rather in minute dense particles of ribonucleoprotein termed *ribosomes* (Fig. 3.26). The ribosomes are found in enormous numbers attached to the outer surface of the endoplasmic membrane, and they are quite uniform in size (120 Å to 150 Å in diameter), hence the designation *granular* or *rough endoplasmic reticulum*. Ribosomes also may occur as free structures that lie within the cytoplasmic matrix.

Regardless of whether they are free or attached to the limiting membrane of the endoplasmic reticulum, a number of ribosomes often are aggregated together in clusters known as *polysomes* or *polyribosomes*. Such aggregations of ribosomes may consist of between 3 and 30 individual particles held together by a slender filament approximately 12 Å in diameter. This filament is believed to be composed of a single macromolecular thread of mRNA (Figs. 3.10; 3.26B).

It can be shown by appropriate biochemical techniques that both the membrane-bound and the free ribosomes are composed of two dissimilar subunits held together by hydrogen bonds and magnesium ions (Fig. 3.26A). The nucleic acid within each subunit is associated with histones and other structural proteins. In the living cell, however, the two subunits are always combined into the single particulate mass known as the ribosome during protein synthesis, whereas they may dissociate at the end of this process.

A.

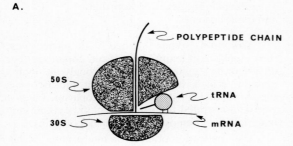

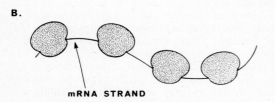

B.

FIG. 3.26. The ribosome. **A.** Ultrastructure of a single ribosome illustrating the two subunits classified according to their sedimentation properties in an ultracentrifuge (50 S and 30 S). The relationship among the mRNA strand, a tRNA molecule, and the polypeptide chain of a protein undergoing synthesis also are indicated schematically. Compare Figure 3.10. **B.** The polysome (or polyribosome) consisting of a number of individual ribosomes strung like beads along a single strand of mRNA.

Functions. The canalicular network of membrane-limited tubules and cisternae which constitute the granular endoplasmic reticulum provide intracellular channels whereby rapid diffusion of materials can take place within the cytoplasm. In addition, the limiting membrane of this organelle contains enzyme systems that play an important role in cellular metabolism.

The ribosomes associated with the endoplasmic reticulum as well as the free cytoplasmic ribosomes are the loci for *de novo* synthesis of protein within the cell (p. 63). The proteins that are elaborated by a cell for secretion are manufactured by the ribosomes attached to the endoplasmic reticulum. Following their synthesis, such proteins enter the lumen of the canaliculi as the first step in the secretory process (p. 1080; Fig. 15.27). The free cytoplasmic ribosomes and polysomes, on the other hand, are believed to elaborate the proteins required by the cell itself — that is, enzymes, structural proteins, and so forth.

Consequently, the ribosomes may be considered to be complex, highly ordered macromolecules that provide the machinery for the selection of particular molecules as well as their orderly assemblage into proteins during the translation phase of polypeptide synthesis.

The strand of messenger RNA that is found in association with the polysomes is synthesized within the nucleus, and this mRNA, in turn, transmits the detailed and encoded information to the ribosomes so that the specific amino acid sequence of particular proteins is established.

The Agranular Endoplasmic Reticulum

Structure. As implied by its name, the agranular or smooth endoplasmic reticulum lacks associated ribosomes (Fig. 3.20). In fact, the existence of such a branching and anastomosing tubular network within the cell was unsuspected by cytologists until it was revealed by electron microscopy. Hence, the smooth endoplasmic reticulum has no counterpart in conventional light microscopy.

The mere presence or absence of ribosomes is not the only feature that distinguishes the two forms of endoplasmic reticulum, even though both types are often continuous with each other. The agranular reticulum generally forms a tightly meshed three-dimensional network of tubules in which cisternae usually are absent.

Functions. In many cell types, either a granular or an agranular endoplasmic reticulum is predominant. For example, pancreatic acinar cells, whose secretion contains much protein, have an endoplasmic reticulum which is almost entirely the granular variety. In contrast, skeletal muscle cells contain principally agranular endoplasmic reticulum, which apparently is concerned with the release as well as the sequestration of calcium ions during the contractile process (Fig. 4.2, p. 117). Hepatic cells, on the other hand, contain both granular and agranular reticula in practically equal proportions.

Unlike the granular reticulum, which has been established unequivocally as a site of intracellular protein synthesis, no specific metabolic function has been ascribed as yet to the agranular reticulum. In several endocrine glands (eg, the adrenal glands, testes, and ovaries), the agranular reticulum apparently is involved in the biosynthesis of steroid hormones. In the liver, the agranular reticulum has been implicated as a site for lipid and cholesterol metabolism, as well as for reactions involving the hydroxylation of various compounds. The agranular reticulum probably serves also as a low-resistance intracellular transport system for various substances, a feature which it shares with the granular endoplasmic reticulum.

THE GOLGI APPARATUS

Structure. Usually invisible in the living cell, the *Golgi apparatus* or *complex* sometimes appears in ordinary histologic preparations as an unstained region in the vicinity of the nucleus, ie, as a negative image (Fig. 3.20, p. 72). When the tissue is impregnated with silver or osmium, however, the Golgi apparatus appears as a blackened juxtanuclear network.

Electron microscopy has elucidated the morphology of this organelle still further. At suitable magnifications, the Golgi apparatus appears in cross section as a concentric series of membrane-limited, curved, parallel cisternae or small sacs (saccules). The individual cisternae may exhibit dilatations at their ends, and numerous vesicles or vacuoles of various sizes are associated intimately with the convex outer surfaces of the laminae. The concave inner surface of the array of laminae often is associated with a number of larger vacuoles. Hence, the Golgi complex appears somewhat like a stack of dinner plates when viewed from the side.

Functions. The Golgi apparatus is believed to be the organelle within secretory cells that is responsible for concentrating and packaging the secretory product. Thus, the vacuoles mentioned above which are associated with the surface of the Golgi complex apparently represent the concentrated secretory product of the cell, and these vacuoles in turn become secretory granules or droplets (Fig. 3.20).

The Golgi apparatus *does not* have any significant role in the synthesis of protein-rich secretions; this function is performed by the granular endoplasmic reticulum. On the other hand, there is abundant evidence that the Golgi complex in certain cells is concerned with the synthesis of a type of secretory material that is rich in complex polysaccharides. In many nonsecretory cells, however, the function of the Golgi apparatus remains obscure.

THE LYSOSOMES

Structure. Lysosomes usually appear as dense, membrane-limited organelles that vary between 0.25 and 0.5 μ in diameter and exhibit considerable diversity in their morphology within a cell type as well as among different cell types. In electron micrographs, lysosomes appear as spheres or ovid bodies with a dense, homogeneous interior or as irregular structures that enclose material of varying electron densities. Lysosomes also may be present as globular organelles with a pale matrix. Some lysosomes may even exhibit a pattern of crystals oriented along the equator of the organelle and embedded in a matrix of lower electron density (Fig. 3.20).

Functions. Lysosomes contain a variety of potent hydrolytic enzymes that collectively are termed *acid hydrolases* and that are capable of degrading proteins, DNA, RNA, and certain carbohydrates. A number of the specific enzymes found within lysosomes are summarized in Figure 3.27 together with their substrates.

Although the functional role of the lysosomes in normal cellular activities is still under investigation, the concept has emerged that these organelles form a discontinuous intracellular digestive system. In a normal cell, the hydrolytic enzymes of the lysosomes are confined within a unit membrane so that autolysis of the cell due to their indiscriminate action is prevented. However, if the permeability of the lysosomal membrane is increased, or if the membrane is destroyed by the action of drugs or by other agents, then the enzymes are released and digestion or *lysis* of the cell can take place. Hence, the term "lysosome" is applied to these organelles.

Lysosomes are important to the digestion of worn out or damaged cells within the body, as well as to solubilization of their constituent molecules. The lysosomal enzymes are released when a cell dies.

Certain white blood cells, the granulocytes (p. 535), contain an abundance of lysosomes that are capable of digesting foreign particulate matter such as bacteria after they are ingested by the cell (see phagocytosis, p. 94). The bacteria are engulfed by the cell and sequestered within membrane-bounded phagocytic vacuoles *(phagosomes)*. Lysosomes then attach to, and subsequently fuse with, the membrane of the phagosome so that the hydrolytic enzymes are released; these kill and ultimately digest the bacteria contained within the phagocytic vacuole. Thus, lysosomes form an important cellular defense mechanism of the body against invasion by pathogenic microorganisms.

During the normal life of cells, certain organelles such as mitochondria may become damaged. The defective organelles fuse with lysosomes within the same cell, forming vacuoles known as *cytolysosomes* in which the aforementioned hydrolases are liberated and the damaged organelles disintegrated. These cytolysosomal vacuoles are termed *autophagic vacuoles* if morphologically recognizable fragments of the defective organelles are present. With advancing age, a pigment called *lipofuscin* accumulates intracellularly; this pigment presumably represents an accumulation of undigested remnants of incomplete lysosomal activity.

The examples cited above emphasize that substances undergoing digestion by the lysosomes may have either an exogenous source (ie, originate outside of the cell) or an endogenous source (ie, arise within the cell itself).

THE CENTROSOME AND CENTRIOLES

Structure. The term *centrosome* is used to describe a clear region of the cytoplasm in which a pair of short rods or granules known as *centrioles* are located (Fig. 3.20). Usually the centrosome is in close promixity to the nucleus and may occupy a slight indentation on the surface of this organelle. The centrosome frequently is partially surrounded by the Golgi complex, as shown in Figure 3.20. However, in certain cells, such as those of the intestinal epithelium, the centrioles are unrelated to a centrosome and appear within the apical cytoplasm just beneath the luminal surface.

Each centriole consists of a hollow cylinder 150 nm in diameter and between 300 and 500 nm in length; the cylinders are closed at one end but open at the other. Low-density cytoplasm occupies the central core of the centriole. When viewed in cross section, a centriole is circular in outline. The wall of the centriole consists of nine groups of longitudinally disposed, parallel subunits. Each of these

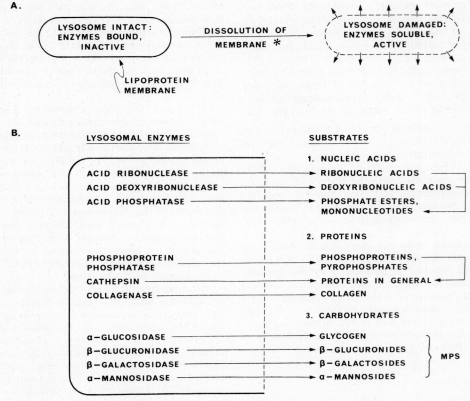

FIG. 3.27. The lysosome, an intracellular digestive organelle. **A.** In normal cells, the acid hydrolases present within the lysosome are sequestered within a membrane and hence are unable to function. Disruption of this membrane (*), as by freezing and thawing, mechanical processes (eg, ultrasound, grinding), enzymes, fat solvents, detergents, and the death of or damage to the cell per se, liberate the lysosomal enzymes so that they now are free to exert their hydrolytic effects upon other cellular organelles and inclusions. **B.** Listing of the lysosomal enzymes and the particular substrates they hydrolyze. (Data from DeRobertis, Nowinski, and Saez. *Cell Biology*, 5th ed, 1970, Saunders.)

subunits, in turn, consists of an array of three fused tubular elements (a *triplet*). The triplets are oriented as shown in Figure 3.20, and each one is connected to the adjacent triplet by a slender filament.

Centrioles generally occur in pairs known as *diplosomes*, and usually the long axes of each centriole of a diplosome are perpendicular to one another.

Both cilia (p. 152) and flagella (p. 153) originate from centrioles, and each of these motile cellular processes contains nine double fibrils (or tubules) which originate from the triplets within the cylindrical wall of the centriole. Each cilium or flagellum is anchored to the cell from which it originates by a structure known as a *basal body;* the basal body is identical morphologically to the centriole.

Centrioles are self-replicating organelles; however, they do not appear to undergo simple fission. After the centrioles reproduce early in cell division, the pairs migrate to opposite sides of the nucleus, where they function as the centers around which the formation of the microtubules of the mitotic spindle and asters takes place (Fig. 3.41).

Functions. The centrosome is the region of the cell that is concerned with certain processes involved in cell division. The centrioles are related to the movement of the chromosomes during mitosis, and they also appear to be regulatory centers for the rhythmic contractile activity of both cilia and flagella in nondividing cells that have such motile structures.

MICROBODIES. Certain spherical particulate microbodies that have been isolated from hepatic cells by fractional ultracentrifugation contain high concentrations of certain enzymes, including catalase, D-amino acid oxidase, and urate oxidase (uricase). Since these membrane-limited microbodies are able to catalyze the reduction of oxygen to hydrogen peroxide (H_2O_2), as well as the oxidation of hydrogen peroxide to water, these microbodies have been called *peroxisomes*. Presumably, the peroxisomes function in vivo in a manner similar to the lysosomes; however, their overall contribution to the economy of the cell is unclear at present.

The *synaptic vesicles* of nerve endings represent another highly specialized type of microbody which is of major significance in the junctional transmission of impulses between individual nerve cells.

MISCELLANEOUS CYTOPLASMIC ORGANELLES. In addition to the well-defined cytoplasmic organelles discussed up to this point, several other types of cytoplasmic organelle have been discerned in animal cells; these structures are annulate lamellae, microfilaments, and microtubules.

Annulate Lamellae.   These cytoplasmic organelles consist of cisternae arranged in closely aligned and highly organized parallel layers. The cisternae have small openings or fenestrations, each of which is covered by a thin membrane or septum quite similar to the pores in the nuclear membrane.

No specific function has been determined as yet for this organelle, which has been seen as a normal component in a large number of somatic cell types as well as in the germ cells of many vertebrate and invertebrate species.

Microfilaments.   The majority of cell types contain randomly occurring filaments between 30 and 60 Å in thickness and of indefinite length. These filaments may be organized into bundles that are resolved by light microscopy as *tonofibrils,* although the individual filaments that comprise these aggregations cannot be visualized except by the electron microscope.

Frequently, tonofibrils terminate in dense regions known as *desmosomes* that lie in close proximity to the cell membrane at specialized loci where adjacent cells are attached to each other (Fig. 3.23).

There is no evidence that the cytoplasmic filaments found in various cell types are identical insofar as their chemical composition and functions are concerned. The microfilaments of muscle cells are definitely contractile and are responsible, at least in part, for the ability of this type of cell to shorten markedly. On the other hand, the microfilaments found within the deep-lying epithelial cells of the skin may have a totally different physicochemical composition and thus may have merely a passive structural role in the organization of this tissue (see also p. 243).

Microtubules.   All nucleated cells contain microtubules, at least during certain phases of their existence. These organelles have a wall that is roughly 60 Å thick and an overall diameter of approximately 230 Å. The walls of microtubules are composed of high-molecular-weight polymers made up of subunits of a protein known as *tubulin.* Polymerization of the tubulin into elongated macromolecules takes place in the absence of calcium ions but requires the presence of magnesium ions as

well as an energy source such as adenosine triphosphate (ATP) or guanosine triphosphate (GTP).

Tubulin itself is composed of at least two different proteins, each having a molecular weight of about 55,000 daltons. Consequently, tubulin is a dimer with a molecular weight of around 110,000 daltons.

The microtubules of particular cell types, although morphologically and chemically similar in many respects, exhibit a wide diversity of functions. For example, microtubules are intimately involved in cell division (mitosis, p. 109), the transport of substances along the axons of nerves, the secretion of such hormones as thyroxine and insulin, and changes in the shape of cells. Microtubules also may participate as skeletal elements in maintaining the shape of certain cells, eg, the axons of nerve cells, and they provide the basic contractile elements in flagella and cilia (p. 152).

THE CYTOPLASMIC MATRIX AND PARTICULATE INCLUSIONS. The discussions of particular cytoplasmic organelles presented above have emphasized that the ultrastructure of protoplasm is not only highly complex but also highly ordered. In addition to the discrete cytoplasmic organelles, a considerable proportion of the total cell volume is occupied by a seemingly amorphous cytoplasmic fluid, or matrix. In this region, free enzymes, nutrients, metabolites, and various macromolecular products are found, either in solution or in suspension. Hence, it would be naive to assume that the aqueous cytoplasmic matrix is not as highly organized as are the visible organelles, merely because this level of organization is invisible with the techniques currently available for study.

The cytoplasmic matrix often contains a number of inclusions that are rendered visible to optical microscopy, generally after the application of special staining techniques. Such particulate inclusions are not considered to be part of the living cell mass, but at times they may contribute a significant fraction to the total volume of the cell. Included in this category are glycogen granules, lipid droplets, pigments of various types, secretion granules, and crystals.

Glycogen.   Animal cells are capable of storing glucose as a relatively compact, high-molecular-weight polymer known as *glycogen* or *animal starch* (p. 66). By the use of special staining techniques, glycogen may be seen within the cytoplasmic matrix either as diffuse particles or as coarse aggregates. The manner in which the tissue is preserved markedly affects the ultimate distribution pattern of intracellular glycogen.

From a physiologic standpoint, glycogen storage by cells in general, and hepatic cells in particular, is essential to the homeostatic regulation of the blood glucose level. Therefore, intracellular glycogen forms a highly labile reservoir of carbohydrate for the body as a whole.

Lipids. Frequently lipids are stored within cells as small, generally spherical, oil droplets. Oils are lipids that are fluid at body temperature because of a high degree of unsaturation in their molecules. Usually the oils present in cells are simple triglycerides (p. 69).

Generally, lipid droplets within the cytoplasm appear in routine histologic sections as clear, round areas, since the fatty material present within the cells has been extracted by the solvents (eg, alcohol) used for preparation of the specimen. Tissue sections that are prepared by freezing and subsequently stained with osmium clearly reveal the cytoplasmic lipid as black, spherical droplets of varying size (Fig. 3.20).

The intracellular lipids serve as a form of highly concentrated, stored energy. In particular, the cells of adipose tissue in a normal individual contain large quantities of lipids that can be mobilized to provide a supply of energy for the body tissues as a whole. In fact, the cytoplasm of such fat cells may be almost completely obscured by the presence of one or more large droplets of lipid, and, as a consequence, the nucleus is forced to occupy an eccentric position at one side of the cell. Lipids also serve as a source of short carbon chains that are employed by cells in general for the synthesis of their membranes and other structural elements.

Pigments

Melanocytes. Within the human body, cells known as *melanocytes* contain cytoplasmic granules of a dark brown or black pigment called *melanin*. The dense pigment granules found within such cells are termed *melanosomes*. Melanocytes are present in the basal layer of the epidermis (p. 996), in the pigmented retinal epithelium as well as in the iris of the eye (p. 468), and in certain brain cells (p. 287).

Lipofuscin. Lipofuscin is a tan or light brown pigment that forms prominent deposits within the cells of older persons. It is thought that lipofuscin represents an end product of the metabolic degradation of various substances by the lysosomes, and accumulates within the cells with increasing age.

Hemosiderin. Hemosiderin is a gold-brown, iron-containing pigment resulting from the degradation of hemoglobin, the respiratory pigment of the red blood cells (erythrocytes).

Mature red blood cells have a restricted lifespan in the circulation due partly to the fact that they lack a nucleus and partly to the mechanical damage they sustain in traversing the blood vessels. Therefore, ingestion and digestion of old and damaged erythrocytes are continuous processes within the body, and these processes take place in cells known as *phagocytes*. As a result, hemosiderin accumulates normally as large, irregular masses within the phogocytes themselves. Phagocytes are present in practically all tissues but are especially numerous in the spleen, liver, and bone marrow.

If a disease process is operative in which the rate of red cell destruction is abnormally high, as in acute malaria, then the phagocytes contain a significantly elevated quantity of hemosiderin, and this pigment can be detected readily by histochemical techniques specific for iron.

Electron micrographs of intracellular hemosiderin aggregates show granules approximately 90 Å in diameter; these granules represent particles of an iron-containing proteinaceous compound known as *ferritin*.

Secretion Granules and Crystals. The cytoplasm of active secretory cells contains an abundance of secretory granules or droplets of different sizes which can be visualized histologically by the use of appropriate techniques (Figs. 3.20, 3.33).

Microcrystalline deposits of various substances also may be present in the cytoplasm under normal as well as pathologic conditions. For example, crystalline granules of renin (p. 770) are normal cytologic components of the juxtaglomerular cells of the kidney (p. 770), whereas extensive deposits of glycogen are found in hepatic and splenic cells in glycogen storage disease (von Gierke's disease).

A crystal is defined as a body that is produced by the solidification of an element, compound, or isomorphous mixture which has a regularly repeating internal arrangement of its atoms, especially when natural external plane faces result from the rigid internal structure. The inorganic matrix of bone is the most extensive crystalline structure found in the body. A *liquid crystal,* on the other hand, behaves mechanically as though it were a viscous fluid, although the arrangement of its component atoms and molecules exhibits the characteristic orderly arrangement found in a solid crystal. In other words, mechanical deformation of a liquid crystal causes the regular atomic and molecular layers of which it is composed to slide past each other, unlike in a solid crystal, in which the components are locked together.

Many biologic structures normally exhibit the properties of liquid crystals. For example, the plasma membrane, the orderly lipid layers found in the myelin sheaths of nerves, and the rods and cones of the eye all have the characteristics of liquid crystals. There is also evidence that certain disease states may have abnormal liquid crystal formation as a component of their etiology. In sickle-cell anemia, during the process of sickling, the hemoglobin molecules form liquid crystals within the erythrocytes instead of remaining in solution as a random association of molecules (p. 568).

The study of liquid crystals in relation to the cell in particular and to biomedical problems in general is in its infancy. However, it is apparent that further research in this field could yield much information of fundamental significance to clinical medicine within the next decade.

GENERAL PHYSIOLOGIC PROPERTIES OF CELLS AND TISSUES

Throughout their life-span, all cells and tissues have in common a number of basic physiologic properties. These general properties and their underlying mechanisms are responsible for many of the specific functions of organs and organ systems within the body as a whole. Included in this category of general functions are the transport of materials, the genesis of bioelectric potentials, the storage, release, and utilization of energy, and cellular reproduction.

Transport of Materials Across the Cell Membrane

The transport of substances across the cell membrane is accomplished by two basic processes: passive transport and active transport. The principal difference between these two kinds of transport is implicit in the names that denote their mechanisms. Thus *passive transport* is merely the physical diffusion of a substance *down* its own concentration (or electrochemical) gradient (p. 23), whereas *active transport* requires the expenditure of energy in order to move a substance *against* its own concentration (or electrochemical) gradient. Stated another way, passive transport is the movement of a substance from a region with a high concentration of that substance to a region with a lower concentration of the same substance. Active transport is the reverse of this process; that is, energy must be supplied to the transport system in order to move the substance "uphill" against the "downhill" force of diffusion for that substance.

In addition to these two general mechanisms that govern the distribution of materials across the cell membrane, there are several other processes that are of importance to the transport of materials across both cell and tissue membranes. These are facilitated passive transport, the cellular processes of endocytosis and exocytosis, and the ultrafiltration of various substances across membrane sheets due to hydrostatic pressure differences.

PASSIVE TRANSPORT (DIFFUSION). The term *passive transport* is applied to the simple physical diffusion of substances in living systems. Because of the complex physicochemical makeup of cell membranes, the rate of simple diffusion across these membranes is modified significantly by several factors. Some of these factors depend upon the nature of the diffusing particles, while certain properties of the membranes themselves also contribute to alterations in the rate of diffusion. The major factors that contribute to changes in the diffusion rate are summarized in Table 3.9.

As discussed in Chapter 1 (p. 23), the diffusion of a substance in a nonliving system depends upon the random motion of the individual atoms and molecules within that system due to their kinetic energy. The kinetic energy, in turn, is directly proportional to the heat content of the system as

Table 3.9 FACTORS THAT MODIFY THE PASSIVE TRANSPORT OF SUBSTANCES ACROSS CELL MEMBRANES

A. Properties of the Diffusible Substance

1. Concentration gradients of the diffusible substance across the membrane
2. Permeability coefficient (or diffusion coefficient)[a]
3. Relative oil–water solubility (partition coefficient) of the diffusible substance
4. Effective diameter of the diffusible particles
5. Presence or absence of an electric charge on the diffusible particles

B. Properties of the Cell Membrane

1. Total surface area available for diffusion
2. Thickness and structure of the membrane[b]
3. Diameter of the water-filled membrane pores
4. Presence or absence of an electric charge on the membrane pores
5. Presence or absence of carrier molecules for particular diffusible substances

[a] *The rate factor in diffusion is taken into account as part of the permeability coefficient (P), and P has the dimensions of velocity, ie, centimeters per second.*

[b] *Note that the distance traveled by the solute particle is accounted for in the permeability coefficient. See text for further explanation.*

measured on the absolute temperature scale. These facts hold true whether the particles composing the system are in a gaseous or a liquid state. If such an artificial system is composed of an aqueous solution of a substance that is concentrated in one region and has a lower concentration in another region, then the particles of the substance will move (or *diffuse*) freely and spontaneously down its concentration gradient. This movement is due to the kinetic energy of the system resulting from the thermal agitation of the water and the solute molecules (Fig. 1.12A). At equilibrium, therefore, the solute particles will become uniformly distributed in a random fashion throughout the aqueous system, and the original concentration difference will have been abolished.

If an inert membrane which is freely permeable to the solute as well as to the solvent molecules is placed in the system so as to separate an aqueous compartment from a compartment that contains the concentrated solution, at equilibrium the outcome is still the same. That is, the concentration difference of solute across the membrane eventually disappears, and the solute molecules now are distributed at random throughout the system, regardless of the presence of the membrane.

In contrast to this artificial situation, the cell membrane of a living system imposes a semipermeable, or selective, barrier between the extracellular and the intracellular fluids so that the free diffusion of molecules and electrolytes into and out of the cell is considerably impeded. In fact, the cell membrane is relatively impermeable to all substances when compared with an aqueous layer of similar thickness. Consequently, the free diffusion of substances across the membrane of living cells is altered by membrane properties such as those listed in Table 3.9.

Viewed from a functional standpoint, the cell membrane may be considered as composed of two general regions through which substances may diffuse from the extracellular fluid in order to reach the cytoplasmic matrix or vice versa. By far the greatest proportion of the total area of the membrane consists of a bimolecular lipid leaflet; thus, the relative lipid-water solubility of a substance directly affects its diffusion rate through the membrane. The cell membrane also behaves as though it were traversed by a number of circular, continuous, water-filled pores whose diameter is estimated to range between 7 and 8 Å. As a consequence, the aqueous phase of the extracellular fluid is continuous in the regions of the pores with the aqueous phase of the cytoplasmic matrix. Thus, the cell membrane may be considered functionally as an extremely thin oil film which is pierced at intervals by minute water-filled openings, as depicted in Figure 3.21.

This simplistic concept of the cell membrane is useful in analyzing the general factors that govern the diffusion of substances into and out of cells.

The net movement of a substance through the cell membrane by diffusion is a spontaneous and passive event, and it should be stressed here that this process takes place in both directions simultaneously. The net passive transport, or *net flux,* of a substance in one direction or the other may be modified by, or dependent upon, other seemingly unrelated processes (eg, metabolism of the transported substance). The several factors governing diffusion across the cell membrane now may be considered individually.

The Concentration Gradient and the Permeability Coefficient. In physical terms, the major factors that determine the rate of diffusion are the concentration gradient of the diffusible substance across the membrane and the membrane thickness, as well as the total surface area that is available for diffusion. If a given cell has a constant surface area (A), then the rate of diffusion (Q) across the membrane is directly proportional to A as well as to the concentration difference of the diffusible substance $(C_1 - C_2)$ across the membrane, or $Q = PA(C_1 - C_2)$. In this expression, P signifies the *permeability* or *diffusion coefficient* for the diffusible substance and a particular membrane. Since the term P has the dimensions of velocity, or centimeters per second, it defines the transport rate of the solute through a membrane of given thickness when a unit concentration gradient of the solute is maintained across that membrane. The value of P, determined by calculation from empirical data using the equation presented above, is useful in expressing the relative permeability of a membrane to various materials. A few examples of P for common substances are given in Table 3.10. A high value for P indicates

Table 3.10 PERMEABILITY COEFFICIENTS FOR VARIOUS SUBSTANCES[a]

SUBSTANCE	PERMEABILITY COEFFICIENT (P)
Water	1×10^{-2}
Urea	1×10^{-4}
Cl^-	4×10^{-6}
K^+	2×10^{-6}
Na^+	2×10^{-8}
Protein (anion)	~ 0

[a]*A high permeability coefficient indicates that the substance is transported through the membrane with relative ease, whereas a low permeability coefficient indicates that the membrane is an effective barrier to transport of that substance. The values listed for water and urea were obtained on erythrocytes, whereas those for Cl^-, K^+, and Na^+ were obtained on skeletal muscle cells.*

that the substance is transported easily through the membrane (eg, water and urea), whereas a low value for P indicates that the membrane acts as an effective barrier to transport of that substance (eg, sodium). These values give a direct indication of the relative ease with which a particular substance diffuses through a given membrane. A high permeability coefficient indicates that a substance passes through the membrane with facility — as is true, for example, for water — whereas a low permeability coefficient, such as is found for sodium ions, indicates that the membrane is relatively impermeable to that substance.

The net *unidirectional transport* or *flux* for a given substance through the cell membrane at any moment in time reflects the difference between two oppositely directed diffusional fluxes. However, it is the net concentration difference between the individual concentrations of the substance on both sides of the membrane that is of major significance in determining the direction and magnitude of the transport by diffusion, as illustrated in Figure 3.28A, B, and C). Any factor or mechanism that enhances this concentration gradient directly increases the rate of diffusion, a fact of considerable importance in many physiologic processes.

In addition to the concentration gradient and total surface area available for diffusion, there are several additional factors that influence this process. These are the temperature, the thickness of the membrane (or length of the pathway) through which the diffusing substance must travel, and the *atomic* or *molecular weight* of the substance involved.

The rate of diffusion in an artificial physicochemical system is directly proportional to the absolute temperature, because the kinetic energy of the particles increases with their heat energy. Since the body temperature of mammals is closely regulated under normal circumstances, this factor is negligible insofar as its influence on the net diffusion rates of various substances in the human body is concerned.

The net rate of diffusion decreases as the distance through which a particle must travel in-

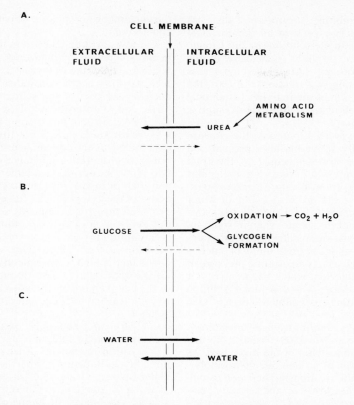

FIG. 3.28. Diffusion (or passive transport) of three different substances across the cell membrane. **A.** The net flux of urea, produced within the cell by amino acid metabolism, is toward the extracellular fluid owing to a steep concentration gradient toward the outside of the cell (heavy arrow to left). **B.** The net transport of glucose, in contrast to urea, is toward the intracellular fluid (heavy arrow to right). The continuous metabolic utilization of glucose as well as its intracellular conversion into glycogen maintains a high concentration gradient across the membrane for this compound. See also Figure 3.29. The light dashed arrows in both A and B indicate the minor, oppositely directed fluxes for urea and glucose, respectively. **C.** In contrast to both urea and glucose, the flux of water molecules is quite high and extremely rapid in both directions across the cell membrane (heavy arrows).

creases. In other words, the diffusion rate is inversely proportional to the distance the diffusible substance is transported. In the body, the distances through which substances move by simple diffusion alone are quite small (on the order of magnitude of angstroms or microns), and adequate concentration gradients are maintained by physical mixing and by metabolic reactions, so that where diffusion is the sole mechanism involved in the transport of a substance, the rate is quite adequate to meet the requirements of the cells and tissues. For example, the normal exchange of respiratory gases between the lungs and the blood takes place solely by passive transport.

Osmosis. The term *osmosis* merely signifies the passive transport (diffusion) of water per se. The diffusion of water is governed by the same physical processes that were described above for solutes; however, in comparison with other substances, water diffuses across the cell membrane at an extremely high rate. As shown in Table 3.10, the permeability coefficient for water through the erythrocyte membrane is 100 times faster than that for

urea, and about 1,000,000 times greater than that for sodium. Consequently, the water within an erythrocyte is exchanged completely with extracellular water within a second, although under normal circumstances the net volume of the cell remains constant. This constancy results from the operation of two precisely equal, but oppositely directed, fluxes of water across the membrane so that no net transport in either direction takes place (Fig. 3.28C). Even the slightest imbalance between these opposing fluxes, however, results in a net movement of water across the membrane, and the cell swells or shrinks, depending upon which flux is altered.

The net movement or diffusion of water caused by a concentration difference (or gradient) in the molecules of this liquid is called *osmosis*. The physicochemical basis for osmosis and the development of an osmotic pressure across a semipermeable membrane were discussed on page 23 and illustrated in Figure 1.12B. It should be reiterated here that the magnitude of the pressure difference developed by osmosis depends strictly upon the total number of nondiffusible solute particles on each side of the membrane, regardless of their nature, and this factor is expressed by the chemical activity of the solute in the solution (p. 20). Chemical activity, in turn, provides a measure of the potential osmotic pressure that will develop across a membrane separating two solutions. In turn, the total activity is the sum of the individual fractional contributions of each of the solutes that comprise each of the solutions.

The term *osmol* is used to denote the concentration of osmotically active particles present in a given solution. One osmol is equal to the total number of particles present in one gram-molecular weight of a solute per liter of water. Consequently, 180 grams of glucose dissolved in 1 liter of water exerts an osmotic pressure of 1 osmol, since this carbohydrate does not dissociate when in solution. If, however, the solute dissociates into, for example, two particles, then 1 gram mole of such a compound in solution exerts an osmotic pressure of 2 osmols. Sodium chloride, which has a molecular weight of 58.5, produces a two osmolar solution when 58.5 grams of this compound are dissolved in 1 liter of water.

Usually the *milliosmol* is used in physiology; one milliosmol is equal to 0.001 osmol. Normal extra- and intracellular fluids in the body have an osmotic concentration of around 300 milliosmols.

There is a definite relationship between osmolality, which is a *concentration,* and osmotic pressure, which is a *force.* At 38 C a solution with a concentration of one osmol will exert an osmotic pressure of 19,300 mm Hg. Consequently, one milliosmol is equivalent to 19.3 mm Hg at body temperature.

Any system, such as a cell, a mitochondrion, or other structure that is bounded by a semipermeable membrane can exhibit osmotic properties. Normally, in mammalian organisms the osmotic pressure of the blood and other extracellular fluids is regulated with extreme precision so that net exchanges of water between the intracellular and extracellular fluid compartments are minimal, despite enormous and rapid fluxes of water between these regions.

Oil-Water Solubility: The Partition Coefficient.   Another factor that is important to the diffusion of various substances across the cell membrane is their relative solubility in oil and water, as defined by the *partition coefficient.* The partition coefficient is obtained from the relative solubility of a substance in a two-phase system consisting of oil and water. At equilibrium, the partition coefficient, *B,* is equal to the relative concentration of the solute in each of the two phases:

$$B = \frac{C \text{ oil}}{C \text{ water}}$$

Hence, the partition coefficient defines the quantitative distribution of the substances between the oil and water phases when the system is at equilibrium.

A low partition coefficient indicates that a particular solute has a greater tendency to leave the oil phase and enter the water phase; and, conversely, a high partition coefficient indicates a greater tendency on the part of the solute to leave the water phase and to enter the oil phase.

Earlier in this discussion, the cell membrane was presented as a thin lipid film which is punctuated at infrequent intervals by water-filled pores. In order that a substance diffuse through this two-phase barrier, it must either enter the lipid film and then diffuse across this region to the aqueous phase of the cytoplasm, or traverse a pore directly, provided that it is in aqueous solution.

A few biologically important substances have a great affinity for the lipid layer, ie, they are said to be *hydrophobic,* since their oil–water partition coefficient is high. Consequently, these substances diffuse with relative ease through the lipid portion of the cell membrane. Included in this category are oxygen, carbon dioxide, and fatty acids, as well as alcohol.

Hydrophilic materials, on the other hand, are far less soluble in lipid than in water, and thus have low partition coefficients. Such materials encounter difficulty in diffusing across the lipid phase of the cell membrane; water itself is an excellent example of such a substance. Therefore, highly water-soluble substances, including electrolytes, tend to diffuse through the water-filled membrane pores or else are carried by means of an active transport mechanism (p. 92).

In general, the magnitude of the partition coefficient for a substance is directly proportional to the permeability coefficient for the same substance (Overton's rule). In other words, there is a direct relationship between lipid solubility and membrane permeability.

Diffusion Through the Membrane Pores: Particle Size and Hydrophilic Substances.   As noted above, the cell membrane behaves as though it contained continuous, water-filled pores, which have an approximate diameter of 7 to 8 Å, and which occupy a total area equal to about 0.0001 of the total surface area of the cell. These figures have been obtained from experiments performed chiefly on erythrocytes. Other cell types may possess membranes with quite different surface properties; for example, the chloride ion diffuses much more slowly through the neurilemma than through the erythrocyte membrane, as shown in Table 3.11. It is also important to realize that the permeability characteristics and pore sizes of a given type of cell membrane may be altered considerably under different functional states of the cell itself.

Despite the minute proportion of the total area of the cell membrane that is covered by aqueous pores, it is evident that small-diameter substances, including water, sodium, potassium, and chlorine can traverse these aqueous regions of the cell

Table 3.11 EFFECTIVE DIAMETER OF VARIOUS PARTICLES IN RELATION TO PORE DIAMETER[a]

SUBSTANCE	APPROXIMATE DIAMETER	RELATIVE DIFFUSION RATE[b]
Pore diameter	8.0 Å (average)	—
Water molecule	3.0	5×10^7
Urea molecule	3.6	4×10^7
Chloride ion, hydrated†	3.9	
Erythrocyte membrane		3.6×10^7
Neurilemma		2×10^2
Potassium ion, hydrated	4.0	1×10^2
Sodium ion, hydrated	5.1	1.0
Lactate ion	5.2	—
Glycerol molecule	6.2	—
Ribose molecule	7.4	—
Galactose molecule	8.4	—
Glucose molecule	8.6	—
Mannitol molecule	8.6	—
Sucrose	10.4	—
Lactose	10.8	—

[a]*Water molecules associated with ions form a hydration shell that surrounds each ion, and thus the effective or overall diameter of such particles includes not only the diameter of the ion per se, but also the thickness of the layer of water molecules. Thus, hydration renders the effective diameter of a sodium ion approximately 30 percent greater than that of a potassium ion, a factor that greatly retards diffusion of sodium through the membrane.*

[b]*These figures are representative and are not absolute values. For the most part, these values were obtained using the erythrocyte plasmalemma as the model system. Note the considerably lower diffusion rate of Cl⁻ through neurilemma compared with the membrane of red blood cells.*

membrane quite rapidly. This rapid diffusion of such substances through the cell membrane is related, to some extent, to the effective size of the individual particles in relation to the pore diameter, as shown in Table 3.11. For a series of related chemical elements or compounds there is, in general, an inverse relationship between the atomic or molecular weight and the permeability coefficient. Since the Brownian motion of a particle is related inversely to the particle size, then the rate of diffusion decreases as the particle size increases. Consequently, polar macromolecules such as proteins and nucleic acids have extremely small theoretically derived permeability constants, and mechanisms other than simple passive diffusion are operative within the cell in order to transport such substances across membrane barriers (p. 94).

Many substances of biologic importance, whether atoms or organic compounds, are ionized. Hence they possess an electric charge when dissolved in aqueous solution. In fact, water itself has an unequal distribution of electric charge on its molecule and therefore is a highly polar compound (p. 16). Consequently, an interaction between water molecules and other charged particles, eg, electrolytes, can take place. This interaction increases the *effective size* of the atom or molecule by the electric association of a layer of water

molecules with the particle known as the *hydration shell*. Polar compounds that dissociate in solution and thus readily attract water molecules are said to be hydrophilic, in contrast to nonpolar compounds, including lipids, which are hydrophobic. For example, the permeability coefficients for sodium and potassium ions are exceedingly small compared to those for water, since these highly polar ions are quite insoluble in the lipid phase of the cell membrane. Consequently, these ions are transported passively through the cell membrane at very low rates compared with urea or water.

Another factor that impedes the passive transport of electrolytes such as sodium and potassium through the aqueous pores is believed to reside within the molecular structure of the pores themselves. These pores are considered by some investigators to be lined by proteins whose terminal groups can dissociate and thereby produce a net positive charge within the pores. Such a positive electrostatic charge would, of course, tend to repel any cations entering the pore. Furthermore, the presence of adsorbed cations, such as calcium, on the lining of the pore would also contribute to this repellent effect, and this would tend to decrease still further the diffusion rate of cations across the membrane.

In marked contrast to cations, anions such as chloride diffuse far more readily across the cell membrane, since particles of this type are attracted rather than repelled by the fixed or adsorbed cationic charges that may be present within the pore.

The pores within certain membranes, such as those found on erythrocytes and nerve cells, are believed to lack significant numbers of fixed negative charges, whereas an abundance of fixed positive charges are present. Therefore, the diffusion rate of anions in general is greatly enhanced in such cells. In other words, the charge distribution within the pores tends to favor the passive transport of such ions as bicarbonate and chloride over that of sodium and potassium in the erythrocyte membrane. In the resting nerve cell, the passive transport of chloride ions is about twice that of potassium ions and roughly 150 times greater than that of sodium ions.

These general remarks indicate that a great deal of individual specificity exists with regard to the permeability of cell membranes of various types.

The overall permeability of the cell membrane and of the pores themselves may vary considerably in the same cell type under different circumstances. For example, an increased level of calcium ions in the extracellular fluid decreases the overall membrane permeability, whereas a reduced extracellular calcium ion level increases the permeability to a variety of substances. This property is of considerable importance to the maintenance of normal excitability of muscle and nerve cells. A second example of this effect is to be found in the action of antidiuretic hormone upon the pore size of the membranes that invest the epithelial cells within the renal tubules. Briefly, this hormone dilates or enlarges the membrane pores in these cells, so that water and other substances that have been filtered into the tubule can readily diffuse back into the bloodstream.

In conclusion, diffusion is a passive physical mechanism of critical importance to the rapid transport and distribution of many substances across cell membranes in the body. The magnitude of the concentration gradient of specific substances across the membrane is the principal factor that regulates the rate of this process. Another factor that can modify the diffusion rate to varying degrees is the permeability of the membrane to various specific solutes. This permeability is determined, in turn, by the relative oil–water solubility of the diffusing substance, the size of the membrane pores, and the net electric charge on both the pores and the diffusing particles. Finally, there is in general an inverse relationship between the ionic or molecular size and the ability of a substance to be transported passively across the cell membrane.

FACILITATED PASSIVE TRANSPORT. Many substances critical to normal cell functions, including sugars and amino acids, are capable of traversing the cell membrane from the extracellular fluid to the cytoplasm by simple diffusion, but only at very low rates, since the membrane is relatively impermeable to such substances. For example, glucose is quite insoluble in the lipid phase of the cell membrane, and its molecular size, as indicated in Table 3.11, is somewhat greater than the average pore diameter. Consequently, this important carbohydrate is transported into cells by a process known as *facilitated passive transport, facilitated diffusion*, or *carrier-mediated diffusion*. There is considerable experimental evidence that indicates that substances transported by this mechanism combine reversibly with specific carrier molecules that are confined to the cell membrane. The substance — carrier complex, which is relatively soluble in lipids, crosses the membrane proper by the process of simple diffusion, and the substance transported is then released from the carrier into the intracellular fluid on the cytoplasmic surface of the membrane, as shown in Figure 3.29. The net effect of this process is a marked acceleration in the rate of transport of the substance across the lipid phase of the cell membrane. The carrier molecule then is free to return to the external surface of the membrane, where the transport cycle is repeated. This process may be visualized as a shuttle system in which the carrier molecules diffuse back and forth across the membrane. The exact nature of the carrier molecules as well as their detailed modus operandi are still unknown.

The carrier substance for glucose is believed to be a protein molecule. Presumably, the attachment of glucose to this molecule lowers the thermal energy that is required for glucose to enter the lipid phase of the membrane from the aqueous extracellular phase. The concentration difference between the external and the internal surfaces of the membrane provides sufficient impetus for diffusion of the sugar-carrier complex across the membrane. Two important features of the facilitated passive transport mechanism should be emphasized. First, no energy is required to drive the system; and, second, the substance transported is diffused down its own concentration gradient, as is the situation with simple passive transport in the absence of a carrier. Nevertheless, facilitated transport by a car-

EXTRACELLULAR   CELL   INTRACELLULAR
FLUID   MEMBRANE   FLUID

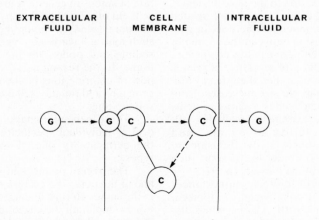

FIG. 3.29. The mechanism of facilitated passive transport. Since glucose (G) is transported into the cell by passive transport at very low rates (Fig. 3.28), the combination of glucose with a carrier molecule (C) within the cell membrane renders the glucose more soluble within this structure, and hence markedly increases the rate at which this sugar can diffuse into the cell. The reversible combination of a substance with a carrier followed by its subsequent diffusion across the membrane and release upon the opposite side is known as facilitated passive transport. Following release of the substance, the carrier molecule then is free to cross the membrane once again and there combine with another molecule of the substance transported. Thus a cyclic process ensues. It must be stressed that facilitated passive transport requires no expenditure of energy.

rier system differs in several important respects from simple passive diffusion.

Saturation.   In the absence of a carrier system, the net rate of transport of a substance by passive diffusion is directly proportional to the concentration difference, as shown by the examples given in Figure 3.28 A and B. Therefore, if the concentration difference doubles, then the rate of transport doubles.

In a carrier-mediated transport system, on the other hand, as the concentration of the substance is increased, an optimal level is reached at which the carrier molecules are *saturated,* as shown in Figure 3.30A. Increasing the concentration beyond this optimal level causes the transport rate to decline sharply or to reach a plateau, because all of the available carrier molecules are occupied by the substance. Therefore, a further increase in concentration of the substance transported will not increase the rate of transport significantly, and the carrier system now is operating at the maximum rate possible. Below this optimum concentration, of course, the rate of transport is proportional to

the concentration difference of the substance on both sides of the membrane, as is the case for passive diffusion.

Specificity.   One type of carrier molecule may transport only one kind of substance, whereas a second carrier molecule may transport only another kind of substance. In such a case, each carrier system exhibits *specificity* toward a given molecular type and will not transport other substances.

Specificity may be demonstrated by the addition of substance B to a transport system while substance A is being transported. If the addition of B does not alter the rate of carrier-mediated transport of A, then the two substances do not utilize the same carrier system, and the specificity of the substance-carrier is demonstrated for the materials involved.

The property of specificity in a carrier-mediated transport system is illustrated in Figure 3.30B. In actuality, the specificity is never absolute, and a given carrier may transport several chemically related substances but with a different

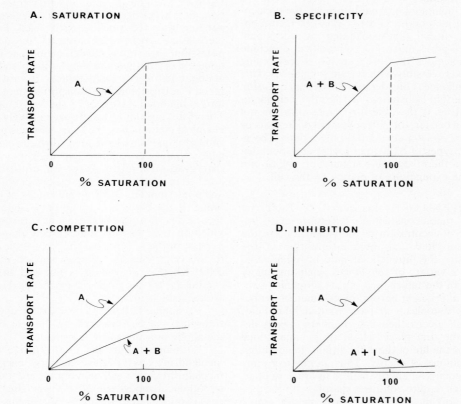

FIG. 3.30.   Characteristic properties of carrier-mediated transport systems. **A.** Saturation. The rate of transport of a substance A is directly proportional to the concentration difference of that substance across the membrane until some point is reached at which the carrier molecules become saturated, and at this point the transport rate declines sharply. **B.** Specificity. The presence of a substance B which does not utilize the same carrier molecules as substance A has no effect on the transport rate of A. **C.** Competition. The transport rate of substance A is decreased markedly by the presence of substance B which utilizes the same carrier molecule as substance A. **D.** Inhibition. If the carrier molecules for substance A are bound irreversibly to an inhibitory substance, I, then the transport rate of A becomes zero for all practical purposes.

efficiency for each one. For example, one carrier system may transport several sugars, whereas another will transport a number of amino acids.

Competition. If two substances, A and B, utilize the same carrier molecule for their transport through the cell membrane, then addition of B to the extracellular fluid will depress the net transport of A, as shown in Figure 3.30C. Since the total number of carrier molecules in the membrane is limited, B will occupy some of the carrier sites previously available to substance A. Thus, the two substances *compete* for the same loci on the carrier molecule. Likewise, the presence of A will reduce the transport of B, since competition is a mutual effect.

Inhibition. In inhibition, which actually is a special type of competition, an inhibitory substance, I, combines with the carrier molecule strongly and irreversibly. Consequently, the transport of substance A is inhibited completely, even though I is not transported at all but remains bound to the active sites on the carrier molecule (Fig. 3.30D).

In accordance with the facts presented above regarding facilitated passive transport, the concept of the cell membrane as a two-phase system consisting of an oil film perforated by occasional water-filled pores must now be modified to include carrier molecules that are capable of combining rapidly and reversibly with certain substances. This process enhances considerably the overall diffusion rate of certain substances across the cell membrane, but, as is the case for simple diffusion not involving a carrier, the net movement of substances takes place *down* a concentration gradient. Therefore, both types of diffusion process are spontaneous from a thermodynamic standpoint, and *do not* require the expenditure of energy by the cell.

ACTIVE TRANSPORT. Inspection of Table 3.3 will reveal that a considerable difference exists between the concentration of individual ions in the extracellular fluid and the concentration of the same ions in the intracellular fluid of living cells. Thus, for a variety of substances, including many not listed in the table, there exists a definite *electrochemical gradient* across the cell membrane between the cytoplasm and the extracellular fluid. Such uphill electrochemical gradients cannot develop in the first place, let alone be maintained throughout the life of the cell, without the continuous expenditure of energy. Thus, *active transport systems* require an energy source in order to operate, and this feature contrasts sharply with the passive transport mechanism which is merely simple physical diffusion of substances *down* their individual electrical and/or chemical gradients.

The quantity of energy required for the active transport of a substance is related directly to the *magnitude of the electrochemical gradient* that is involved. All living cells contain a variety of such active transport systems whereby particular substances are transported against their own electro-

chemical gradients. The energy required to transport, or *pump*, an ion against its own electrochemical gradient is derived principally from the hydrolysis of high-energy phosphate bonds in certain compounds, notably *adenosine triphosphate* (ATP, p. 105). In addition, certain active transport systems possibly may be driven by energy obtained either wholly or partially from direct oxidative processes involving the electron transport chain (p. 931).

Since active transport is a second type of carrier-mediated transport, the same criteria as define facilitated passive transport are involved. Namely, active transport systems exhibit the properties of *saturation, specificity, competition,* and *inhibition,* as discussed in the previous section and illustrated in Figure 3.30. In addition to these four general characteristics of carrier-mediated transport systems in general, an *energy source* also is a critical requisite of active transport mechanisms. Consequently, any factor that inhibits energy production and/or utilization within the cell also blocks active transport systems, since energy consumption and active transport are *coupled* inseparably together. Such blockage of an active transport system may be either reversible or irreversible, depending upon the blocking agent used. For example, the electrochemical gradients for sodium and potassium ions across the cell membrane are maintained in erythrocytes as long as the blood remains at about 38 C (body temperature). If, however, the blood is refrigerated, the transport system for these ions is inhibited (blocked) by the low temperature. Therefore, sodium now diffuses passively down its chemical gradient *into* the cells, whereas potassium similarly and simultaneously diffuses *out* of the cells into the extracellular fluid along its own chemical gradient. Ultimately, an equilibrium is reached in which the concentrations of both sodium and potassium are equal on both sides of the membrane. If the blood is warmed once again to 38 C, then active transport of these ions is resumed, provided that an energy source such as glucose is present in the plasma and that oxygen is available to the cells. Sodium is now pumped out of the cells, whereas potassium is pumped in, so that eventually the original electrochemical gradients are reestablished.

A similar but *irreversible* effect upon active transport can be observed by the use of certain metabolic poisons, such as cyanide.

Although the details of the active transport mechanism as well as the coupling of energy to this mechanism within cell membranes are unknown, a hypothetical model useful in visualizing this process is presented in Figure 3.31A. In this diagram, the carrier molecule is depicted as a rotating wheel, and the energy to drive this wheel (ATP) is indicated at the axle. Situated about the periphery of the rotating carrier molecule are specific chemical loci that form points of attachment (receptor sites) for the substance being transported. As the wheel turns, an enzyme causes the substance transported to be released from the carrier molecule at the

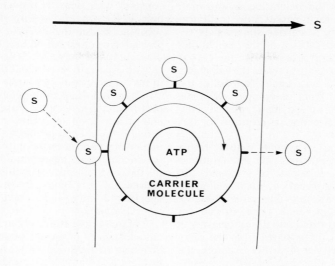

A.

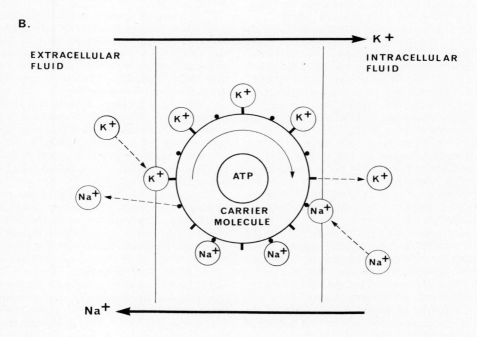

B.

FIG. 3.31. Active transport mechanisms. **A.** Hypothetical model of an active transport mechanism for a substance S. On the left of the figure, the substance is shown to combine with a receptor site upon a large carrier molecule (presumably a protein) located within the cell membrane (delimited by light vertical lines). The carrier molecule rotates in a clockwise direction, and the substance is released from the carrier molecule upon the opposite side of the membrane at the right. The receptor sites (black spokes) then are free to return to the other side of the membrane, and there become attached to another molecule of the substance transported. In sharp contrast to facilitated passive transport (Fig. 3.29), active transport mechanisms require the continual expenditure of energy, derived from the breakdown of ATP, in order to operate. Consequently, active transport systems can transport ions or molecules against concentration gradients; ie, from regions of low to regions of high concentrations. The net direction of flux of S is indicated by the heavy arrow directed toward the right at the top of the figure. **B.** Hypothetical model of the coupled sodium-potassium pump. The operation of this mechanism is identical to that described in A, except that the carrier molecule is envisaged as having two receptor sites, one for K^+ (spokes) and one for Na^+ (dots). Thus each time a potassium ion is transported against its concentration gradient to the intracellular fluid, a sodium ion is transported simultaneously against its own concentration gradient to the extracellular fluid. Thus K^+ and Na^+ are exchanged across the membrane on a one-for-one basis. The net fluxes of these ions across the cell membrane are indicated by the heavy arrows.

intracellular surface of the membrane, as shown in the figure, from where it passes into the intracellular fluid. The empty carrier site then is returned to the external surface of the membrane, where some of the energy used to turn the carrier molecule also is employed to activate the substance and/or the receptor site once again. Release of the transported substance from its complex with the carrier is believed to be an enzymatic process; consequently, no chemical energy is required for this particular step.

In certain instances, it appears that the transport of one substance across the membrane is *coupled tightly* to the transport of a second substance in the opposite direction. Such a coupled active transport mechanism is illustrated in Figure 3.31B. In this system, each time an ion (or molecule) of A is transported into the cytoplasm, an ion (or molecule) of B is extruded from the cytoplasm into the extracellular fluid. For example, the accumulation of each potassium ion within muscle and nerve cells at rest is accompanied by the simultaneous extrusion of a sodium ion (p. 131).

It should be reiterated that this model is purely hypothetical, and none of its components have been isolated or defined biochemically. Nevertheless, such a hypothetical presentation of the active transport mechanism is quite useful in interpreting a number of experimental facts relating to this process. Some investigators believe that the enzyme adenosine triphosphatase (ATPase) that is present in the membrane may act as the true carrier mechanism for the extrusion of sodium ions from the cell when it is coupled to its substrate, ATP. A similar mechanism that acts in the opposite direction also has been postulated to account for the active accumulation of potassium ions within the cell. Both of these mechanisms, however, are largely conjectural at present.

It might appear to the reader as though the continuous active transport of a substance across a cell membrane would give rise to a steadily increasing electrochemical gradient for that substance. Such an effect does not take place for two important reasons: First, as the concentration of the substance transported increases on one side of the membrane, the passive diffusion of that substance also increases, but in a direction *opposite* to that of active transport. Thus, a *steady state* is reached in which the active transport in one direction is balanced precisely by passive diffusion in the opposite direction across the membrane. Second, as the concentration of the transported substance increases on one side of the membrane, the probability that the carrier molecule will return to the opposite side of the membrane with fewer unoccupied receptor sites also increases. In other words, the equilibrium is shifted in a direction opposite to that in which active transport takes place and the carrier molecule tends to exhibit a certain degree of saturation.

Once an active transport system is functioning, the ratio of the extracellular concentration to the intracellular concentration of a substance may be constant with time. However, even small alterations in the *rate of pumping* of different individual substances can induce profound effects in the physiologic activity of different cells. For example, the activity of nerve (p. 164) and muscle (p. 131) cells is especially dependent upon such alterations in the transport rate of sodium and potassium ions.

It is important to remember that active transport mechanisms can be present in any membrane and at any level of organization within the *individual cellular compartments*. Thus, there is a particular mechanism in the mitochondrial membrane for concentrating magnesium ions. The membrane of the endoplasmic reticulum may also exhibit specialized properties; in general, however, the specialized intracellular membrane functions are *terra incognita*.

Many active transport mechanisms for different substances or groups of substances have been demonstrated. Amino acids, sugars, and ions such as Na^+, K^+, Ca^{++}, Mg^{++}, and I^- all may be transported actively.

All transport systems conceivably may have a similar fundamental mechanism that differs only in detail for specific substances. That is, the shape, size, or chemical nature of the receptor sites on the individual carrier molecules within the membrane may differ, although all of the carrier molecules themselves may be lipoproteins.

The active transport systems are given different names in various cell types. In the renal tubules, one speaks of active secretion and active reabsorption of various substances, whereas in nerve or muscle the active transport of sodium and potassium is carried out by the so-called *sodium–potassium pump* (p. 94).

In conclusion, the concept of the cell membrane as a functional system now can be modified still further. The membranes of the cell may be viewed as a *two-phase lipid–water system* in which the carrier molecules responsible for the *facilitated passive transport* of certain substances are found. In addition, the membrane structure contains highly organized *active transport carrier systems* (composed of lipoproteins?) together with certain *enzymes* and the *metabolic systems* essential for the provision of high-energy phosphate compounds necessary to drive such active transport mechanisms. Furthermore, certain unit membranes are postulated to contain *specific receptor sites* (enzymes or other proteins?) that are able to respond to specific chemical and other stimuli (see the next section).

The concept of the cell membrane as a complex structural mosaic was presented earlier in this chapter. This concept now may be amplified so that the cell membrane may be considered to be a dynamic functional mosaic as well.

ENDOCYTOSIS: PINOCYTOSIS AND PHAGOCYTOSIS. The cellular transport mechanisms discussed up to this point all share one feature in common: The substances transported are inherently permeable or are rendered relatively permeable, hence they are able to traverse the cellular barrier between the extra- and intracellular fluids by passive transport, facilitated passive transport, or active transport. In general, the atomic or molecular weights of such substances are relatively low. The permeability of the cell membrane to very large molecules or particles, on the other hand, is negligible. Yet it has been demonstrated repeatedly and unequivocally that native proteins and other macromolecules, such as DNA and viruses, are able to traverse various cellular membranes intact and without undergoing degradation into smaller components.

Endocytosis includes the processes of *pinocytosis* as well as *phagocytosis*. Pinocytosis involves the uptake by the cell of liquid-filled vacuoles (vesicles) in which large molecules may be in solution; hence, this process has been referred to as "drinking by cells." Phagocytosis is similar to pinocytosis, except that the material ingested by the cell is in particulate or solid form, suspended in the aqueous medium of the extracellular fluid. Therefore, phagocytosis can be designated by the phrase "eating by cells."

Endocytosis represents a specialized active transport mechanism; however, the carrier involved in this process is a portion of the cell membrane itself. Both pinocytosis and phagocytosis appear to have a similar underlying mechanism, as depicted in Figure 3.32. During the first step of pinocytosis (or phagocytosis), the material to be ingested is adsorbed upon the surface of the cell by electrostatic or other bonds, perhaps at specific receptor sites (Fig. 3.32A). Undoubtedly a chemical attraction between the substance being taken up by the cell and specific receptor molecules in the membrane plays a role in this process in some instances. Next, the cell membrane invaginates and a discrete membrane-limited vacuole forms in which the substance is carried into the cytoplasmic matrix (Fig. 3.32B, C, D). This *pinocytotic* (or *phagocytic*) *vacuole* shrinks in size and moves away from the cell membrane, which has resealed. In some unknown manner, the contents of the vacuole are liberated into the cytoplasm (Fig. 3.32F). Possibly the vacuolar membrane merely disintegrates, thereby liberating the transported materials, although this is conjectural.

Alternatively, as during the phagocytosis of cellular debris or invading microorganisms by the leukocytes and macrophages (p. 537), the phagocytic vacuole can fuse with a lysosome so that the vacuolar contents are digested intracellularly. This activity provides an important defense mechanism to the body against infectious and other types of microorganisms. Solid particulate matter (eg, carbon) that cannot be digested readily by the phagocytic cells can be stored for variable periods within the cytoplasm of the cell. When such inert particulate matter is liberated, as by the death of the phagocyte itself, then the particle is ingested once again by another phagocyte.

During both pinocytosis and phagocytosis, any extracellular fluid that is ingested together with the

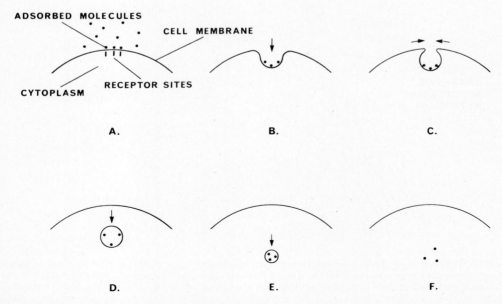

FIG. 3.32. The mechanism of endocytosis, whereby either particulate matter or large molecules (such as proteins) in solution are transported into the cell. Such transport of particulate matter is called phagocytosis, whereas the transport of large molecules is called pinocytosis. **A.** Large molecules or particulate matter attach to receptor sites on the cell membrane. **B** and **C.** The cell membrane invaginates. **D.** A membrane-limited vacuole containing the material undergoing transport is formed and then moves away from the cell surface and into the cytoplasm. **E.** The vacuole shrinks. **F.** The membrane presumably disintegrates, thereby liberating its contents into the cytoplasm.

substance or material transported appears to be taken up by the cell simply by chance rather than by necessity.

Endocytosis is a form of active transport that is less widespread than are the general permeability processes described earlier; the phenomena of pinocytosis and phagocytosis are exhibited by only a few cell types. Furthermore, in such cells endocytosis may take place only during limited periods of cellular life. Therefore, endocytosis should be considered as a specialized and auxiliary kind of active transport.

EXOCYTOSIS OR SECRETION. The term *secretion* denotes the active release of membrane-limited soluble products elaborated by the cell into the extracellular fluid. Therefore, secretion is *exocytosis*, in contrast with endocytosis, which takes place in the opposite direction.

The transport of secretory products to the exterior is shown in Figure 3.33, and the individual steps in this process are summarized in the footnote. The secretory vesicles (or secretion granules) found in the cytoplasm of active secretory cells represent a concentrated storage form of the products which are usually synthesized by the endoplasmic reticulum and which have been concentrated and packaged within a membrane by the Golgi apparatus. Upon receipt of an appropriate stimulus, the cell actively discharges these droplets into the extracellular fluid, presumably by a sort of reverse pinocytosis; that is, the membrane of the vesicle fuses with the plasmalemma, and the secretion is released.

The secreted products may contain protein, so that exocytosis provides a transport mechanism whereby these macromolecules are released in their biologically active (native) state from the cells that produce them for extracellular transport to their site of action, which may be quite remote from their intracellular source.

The process of secretion underlies a large number of vital physiologic processes, and many epithelial cells have specific secretory functions. For example, the exocrine tissue of the pancreas, the salivary galnds, and the gastrointestinal tract all secrete specialized products that are required for the normal digestion of foodstuffs. In addition, the transmission of impulses between contiguous nerve cells takes place by the extremely rapid secretion of chemical substances called *neurotransmitters* (p. 195).

TRANSPORT THROUGH CELLULAR LAYERS. In many tissues and organs of the body, bulk transport (or flow) of water and solutes on

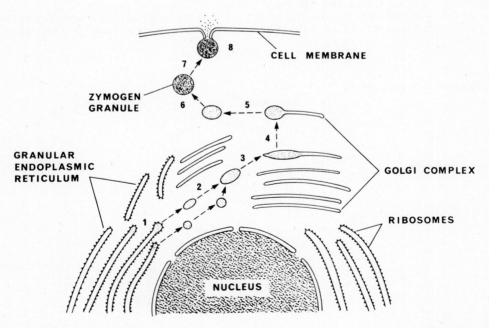

FIG. 3.33. Exocytosis, or secretion, provides an active transport mechanism for the release in an unchanged form of high-molecular-weight compounds elaborated by the cell. Protein-containing cellular products elaborated by the ribosomes of the granular endoplasmic reticulum enter the lumen of the tubules (shading, 1). Small membrane-limited vacuoles containing the secretion bud off from the edges of the cisternae of the reticular tubules (2), coalesce into larger droplets (3), and then fuse with the tubules of the Golgi complex (4). The secretory droplets or granules than bud off from the Golgi apparatus (5), and the vesicles pass toward the cell membrane as the secretion becomes more concentrated (6 and 7, darker shading). Lastly, the vacuolar membrane fuses with that of the cell and the contents of the vesicle are released outside of the cell (8). (Data from Bloom and Fawcett. *A Textbook of Histology,* 9th ed, 1968, Saunders; DeRobertis, Nowinski, and Saez. *Cell Biology,* 5th ed, 1970, Saunders.)

a large scale takes place across an entire layer or sheet of epithelial cells, rather than through only one cell membrane. Two general types of mechanism are involved in the bulk transport of water and other substances across epithelial membranes: *active transport* and *hydrostatic pressure differences* that are generated across the entire epithelial layer.

The operation of these transport mechanisms is essential to the normal functioning of such organs and structures as the kidneys, intestine, capillaries, exocrine glands, and choroid plexus of the brain.

Transport Across Epithelial Sheets. The mechanism whereby active transport of solutes takes place across an epithelial membrane, such as the brush border of the renal tubule, actually consists of three closely interrelated phases which involve both passive and active transport. This mechanism is portrayed in Figure 3.34. The three sequential steps are as follows:

(1) On the luminal (apical) surface of the epithelial sheet, the membranes of the individual cells are highly permeable to water and solutes *except* in those regions where the apices of the cells are in close apposition, the so-called tight junctions (Fig. 3.23). The cell membranes on the apical surface of the epithelium are folded into numerous projections known as the brush or striated border to the light microscopist. In fact, the brush border is composed of numerous microvilli which can be seen individually only with the electron microscope. This morphologic feature of certain epithelial cells is of considerable physiologic importance, as it vastly increases the total surface area of the membrane for transport of water and other substances. Since the apical membrane of the epithelial cells and the microvilli is quite permeable, water and other materials can diffuse freely and rapidly into the cells in this region.

(2) Once the solute has entered the cell, it is pumped into the intercellular space by an active transport mechanism. In particular, sodium ion is one of the most important solutes transported in this manner. As sodium is pumped out, chloride also moves passively into the intercellular space in order that electric neutrality be maintained.

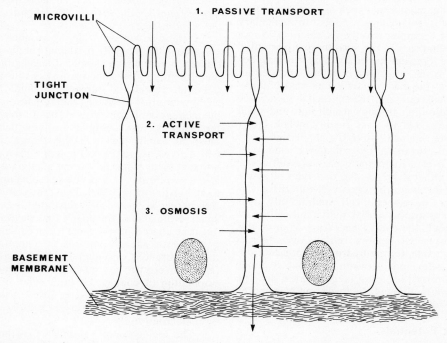

FIG. 3.34. Mechanism of transport of substances (eg, Na^+ and K^+) across a sheet of epithelial cells. The free apical surface of the cell membrane is freely permeable to water as well as to many solutes, except at the tight junctions between the cells (Fig. 3.23). The microvilli of the brush border also may contain hydrolytic enzymes that degrade certain compounds (eg, disaccharides into monosaccharides), thereby enhancing their transport greatly. This is particularly true of the microvilli of the intestinal epithelium, so that the absorption of nutrients from the intestinal lumen is enhanced significantly. Therefore ions and other substances can move into the cells in the apical region by means of passive transport (1). The microvilli in this region greatly increase the total surface area available for diffusion. The ions within the cytoplasm then are transported actively into the intercellular space beneath the tight junctions (2). This net transport increases the osmotic pressure of the intercellular space so that water now passes into the intercellular space by osmosis (3). The increased intercellular fluid volume elevates the intercellular hydrostatic pressure slightly above that in the basement membrane; therefore, the net flow of water and solutes is toward this region (vertical arrow, bottom of figure). (After Guyton. *Textbook of Medical Physiology,* 4th ed, 1971, Saunders.)

(3) As the net concentration of sodium and chloride ions within the extracellular fluid of the intercellular space increases, an osmotic gradient is established. Therefore, water diffuses passively out of the cells in order to maintain an osmotic equilibrium. The increased intercellular fluid volume is sufficient to cause a minute localized pressure differential that is adequate to cause a net movement of the fluid into the basement membrane from the intercellular region. From there, the excess fluid and solutes can move into other regions of the tissue, for example, the terminal blood and/or lymphatic capillaries.

When water moves between various compartments of the body under the influence of an osmotic pressure difference like that described above, movement of the solvent molecules themselves may cause some solute particles to move passively along with the water. This phenomenon is called *solvent drag.* The overall effects of this process in the body are negligible in most instances, however.

Hydrostatic Pressure Differences Across Epithelial Membranes: Ultrafiltration. In certain regions of the body, notably the capillaries, a large difference in hydrostatic pressure normally is present across the endothelial membrane, such as that comprising the walls of the capillary vessels. Intracapillary pressure, for example, may be as much as 25 mm Hg greater than that in the surrounding interstitial fluid, and the force developed by this pressure gradient causes a net transport of water and certain solutes across the membrane by a process of ultrafiltration. Consequently, the epithelial membrane acts as an ultrafilter through which ions and small molecules are able to pass readily, while larger molecules are retained within the capillaries.

The transport of various substances caused by the presence of a hydrostatic pressure difference across an epithelial membrane is a type of active transport. However, the energy required for this process is not developed locally and directly, as is the case for transport across individual cell membranes. Rather, the energy necessary to drive this process is derived from a remote and indirect source. For example, the hydrostatic pressure differential that develops between the capillary lumen and the interstitial fluid is generated by the rhythmic contractions of the heart and transmitted via the arterial system to the capillary bed.

Bioelectric Potentials

Due to the unique physicochemical properties of the plasmalemma, all living cells exhibit a difference in electric potential across this membrane, which is known as the *transmembrane resting potential,* the *membrane potential,* or simply the *rest potential.* * In turn, the rest potential of the cell

The term "rest" as used in this context merely signifies that the cell is not active insofar as the conduction of electric impulses is concerned, and does not imply that the cell is metabolically quiescent. Rather, a steady state is present with regard to its metabolic and other activities.

develops while it is in a steady state because of the specific differences in the composition of the extracellular and the intracellular fluids (Table 3.3). The concentration differences among the individual constituents of these compartments depends, in turn, upon two major factors: first, the relative rates of *passive diffusion* of ions across the cell membrane, which results directly from concentration differences of the ions on both sides of this barrier; and, second, the *active transport* of ions across the membrane. The net effect of these two factors is an unequal distribution or imbalance of electric charges on both sides of the membrane so that a rest potential develops, the outside of the membrane being positively charged with respect to the interior (Fig. 3.35A).

The fluid on each side of the membrane is an aqueous solution of electrolytes, which contains about 155 mEq per liter of cations and the same concentration of anions. However, an extremely small number of anions (negatively charged ions) are concentrated in excess upon the inner surface of the cell membrane, whereas an equal number of cations (positively charged ions) accumulate along the outer surface of this structure. The membrane potential thus develops as a direct consequence of this *unequal spatial distribution of charges across the membrane.*

It must be stressed that the number of cations and anions that actually participate in the genesis of the rest potential is a chemically negligible and minute fraction of the total quantity of ions present in the extracellular and intracellular fluids. Even the slight separation of such a relatively few ions across the membrane, however, is sufficient to give rise to the observed rest potentials of living cells. It also must be emphasized that the charge difference, which gives rise to the rest potential, is a phenomenon that takes place *only at the immediate surfaces of the membrane,* a distance spanning about 80 Å, as shown in Figure 3.35A. Thus, the actual distance the charges are physically separated is exceedingly small. Note also in this figure that the positive and negative charges responsible for producing the rest potential are aligned opposite to each other across the membrane. Consequently, there is a sharp difference in electric potential across this barrier. This orientation of charges on both sides of the plasmalemma is exactly the same process that occurs when an electric condenser (capacitor) is charged with electricity.

The dielectric or "insulator" within the cell membrane that is responsible for separating the charges is the bimolecular lipid layer. Since the ability of a condenser to store electric charges is inversely proportional to the distance between the plates, and since the cell membrane is only about 80 Å thick, it can be readily appreciated that the capacitance of the cell membrane is tremendous.

The resting transmembrane potentials of various cell types can be measured directly by inserting a suitable *microelectrode* into the cell. After proper amplification, the membrane potential is deter-

A.

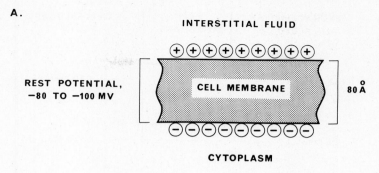

B.

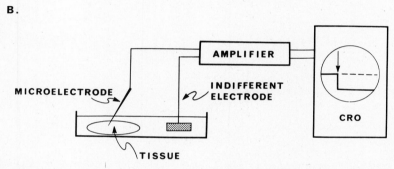

FIG. 3.35. **A.** The rest potential across the cell membrane (see text.) **B.** The technique used for measuring the rest potential of living cells. CRO = cathode-ray oscilloscope. When a cell is penetrated by the tip of the microelectrode (vertical arrow on CRO screen), the beam of the oscilloscope shows a sharp downward (negative) deflection from isopotential (dashed line) as indicated by the drop in the solid line on the CRO screen. The magnitude of the difference in millivolts between the two potentials is the rest potential. See also Figure 5.5, page 164.

mined with respect to the exterior of the cell by means of a galvanometer, such as a cathode-ray oscilloscope (Fig. 3.35B). Since the interior of the resting cell is negative with respect to the exterior, by convention such steady-state membrane potentials are written with a *minus sign* preceding them. The magnitude of actual rest potentials measured by such a technique varies considerably, and extreme values may range from -10 to -100 millivolts (mv) depending upon the cell under study. Usually, however, the rest potentials of mammalian cells are on the order of -80 to -95 mv.

The role of diffusion and of active transport in the development of the rest potential may now be assessed, together with other pertinent factors.

PASSIVE TRANSPORT: DIFFUSION POTENTIALS. Two conditions must be fulfilled in order that a potential difference develop across a membrane by passive diffusion. First, the membrane must be *semipermeable* so that ions with one charge can diffuse through the membrane with greater ease than ions with an opposite charge. Second, a *concentration (chemical) gradient* for the diffusible ions must exist across the membrane.

The operation of these factors is shown in Figure 3.36A. When a concentration difference exists across the membrane for a diffusible ion, the magnitude of the potential difference *at equilibrium* for that ion which is produced by diffusion alone is determined by the tendency of the ions to diffuse in one direction compared with their tendency to diffuse in the opposite direction.

The contribution of each ion to the overall membrane potential at equilibrium may now be calculated by using an expression known as the *Nernst equation:*

$$E_I = \frac{RT}{FZ_I} \ 1_n \ \frac{[I_{\text{out}}]}{[I_{\text{in}}]}$$

In this expression, E is the potential difference that develops across a membrane due to the presence of a diffusible ion, I; R is the universal gas constant; T is the absolute temperature; F is faradays (or number of coulombs per mol of charge); Z_I is the valence of the ion; and $[I_{\text{out}}]/[I_{\text{in}}]$ is the ratio of the concentrations of the ion on the outer and inner surfaces of the membrane, respectively.

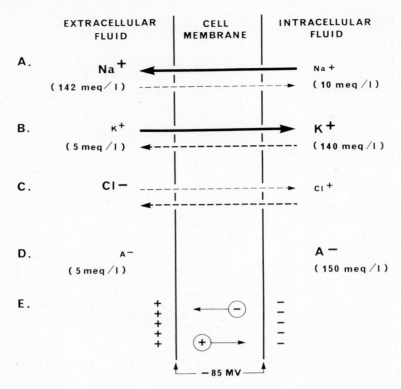

FIG. 3.36. Distribution and fluxes of ions across the steady-state cell membrane. **A.** The resting membrane has a slight but definite permeability to Na^+ such that this ion tends to diffuse passively down its electrochemical gradient into the cell (dashed arrow). However, sodium is pumped actively out of the cell against its electrochemical gradient, and thereby a steep concentration gradient for Na^+ is maintained across the membrane (heavy arrow) for this ion. **B.** The permeability of the resting membrane to K^+ is much higher than for Na^+; consequently, K^+ tends to diffuse outward down its concentration gradient and against its electric gradient (dashed arrow). K^+ is present at a high concentration within the cell because this ion is transported actively from the extracellular fluid into the cell against its chemical gradient, but down its electric gradient. **C.** The concentration of Cl^- is higher in the extracellular fluid than in the intracellular fluid. Furthermore, Cl^- has a relatively high permeability coefficient; therefore, this ion tends to diffuse down its chemical gradient into the cell (light dashed arrow). However, the intracellular fluid contains a high concentration of negatively charged, nondiffusible protein anions (A^-) as shown in **D.** Therefore, Cl^- diffuses passively out of the cell down this electric gradient (heavy dashed arrow). **E.** The net effect of the unequal distribution of the ions across the membrane as described above is that a slight excess of positive charges (K^+) is built up on the external surface of the membrane, whereas a slight excess of negative charges (A^-) is present upon the inner surface. Consequently, the resting membrane is electrically polarized because of this separation of charges, the outside being positive with respect to the interior. The tendency of this charge to become neutralized is indicated by the arrows within the membrane (E); but because of the extremely high dielectric constant of the resting membrane, as well as the electrochemical gradients for Na^+ and K^+ maintained by the activity of the sodium-potassium pump, such neutralization of the charge does not take place. See also Figure 6.37, page 217.

The Nernst equation may be simplified for practical use as follows:

$$E = 61 \log \frac{C_{in}}{C_{out}}$$

In this statement, E is the electromotive force or *equilibrium potential* in millivolts for a given ion, and C_{in} and C_{out} are the relative concentrations of the diffusible ion on the inner and outer surfaces of the membrane. The factor 61 represents the conversion of natural to base 10 logs, as well as the inclusion of the constants listed above in addition to a factor which renders this expression valid at mammalian body temperature (38 C). Therefore, if

the ratio for the concentration of an ion upon both sides of a membrane (C_{in}/C_{out}) is 10/1, then the diffusion potential at equilibrium for the ion as calculated by this expression is 61 mv. In this manner, the equilibrium potential for each ion that contributes to the rest potential can be calculated.

In a living cell, however, the membrane is permeable to several different ions, so that the potential in such a system now depends upon the *polarity* (or *valence state*) of each ion, in addition to its concentration difference across the membrane and its permeability.

Specifically, the major ions responsible for the development and maintenance of the rest potential in mammalian cells such as muscle or nerve are the cations Na$^+$ and K$^+$ as well as the Cl$^-$ anion. In addition, the intracellular compartment contains a quantity of nondiffusible protein anions, usually denoted A$^-$. The important properties of each of these ions in relation to the development of the rest potential now may be considered, together with the forces acting upon their movement.

As noted in Table 3.3, the chemical (concentration) gradient for sodium ions is *inward* across the cell membrane, and the electric gradient, as shown in Figure 3.36, is in the *same direction* for this ion. The membrane has a low but definite resting permeability to Na$^+$. Consequently, a sodium ion must be transported actively *out* of the cell, against its electrochemical gradient, as discussed below (Table 3.3; Fig. 3.36). This active transport is in addition to the passive diffusion forces acting upon the sodium ions.

In the case of potassium ions, the concentration gradient is *outward* but the electric gradient is *inward* (Table 3.3; Fig. 3.36). Since the net concentration of K$^+$ within the cell is much greater than it would be if only concentration and electric gradients were involved in its movement, than potassium ions must be accumulated within the cell by an active transport mechanism. Furthermore, the permeability coefficient for K$^+$ is considerably greater than that for Na$^+$, as shown in Table 3.10; hence, potassium can diffuse across the cell membrane far more readily than sodium, since the effective diameter of potassium ions is smaller than that of sodium ions.

The chloride ion, on the other hand, is more concentrated in the extracellular fluid and has a relatively high permeability coefficient (Table 3.10). Consequently, this anion tends to diffuse along its electrochemical gradient across the membrane *into* the intracellular fluid. The intracellular compartment, however, has a high concentration of negatively charged protein molecules which are nondiffusible; therefore, chloride ions are expelled passively from the cell down the electric gradient. An equilibrium ultimately develops for chloride ions, at which point the influx and efflux of this anion are equal, and the membrane potential at which this situation exists for Cl$^-$ is the equilibrium potential for this ion as defined by the Nernst equation.

The effects of active transport of sodium and potassium ions on the electrochemical gradients for these ions in relation to the membrane potential must now be considered.

ACTIVE TRANSPORT: THE SODIUM-POTASSIUM PUMP. A particular active transport system is present in all cells within the body which is responsible for the development and maintenance of the electrochemical gradients for sodium and potassium across the cell membrane (Table 3.3; Fig. 3.36). These electrochemical gradients, in turn, are an essential feature in the development of the membrane potential. This intramembranous pumping mechanism actively transports sodium *out* of the cell at rest, while simultaneously trans-porting potassium *into* the cell; consequently, the transport system involved is referred to as the *sodium-potassium pump* or, more generally, the *sodium pump*.

The sodium pump, in common with all other cellular active transport systems, requires a continual expenditure of metabolic energy in order to function, as discussed earlier (p. 92). As shown in Figure 3.31B, the transport of sodium out of the cell generally, but not invariably, is coupled to the transport of potassium into the cell. Therefore, extrusion of one sodium ion usually is accompanied by the inward transport of one potassium ion.

The molecular machinery for the sodium pump is located within the cell membrane and derives its energy from the hydrolysis of adenosine triphosphate (ATP). Therefore, the presence of molecular oxygen — and hence aerobic metabolism — is critical to the sustained activity of this mechanism.

The rate at which the sodium pump functions is proportional to the intracellular concentration of sodium ions. Thus, a *feedback mechanism* is present that governs the rate at which sodium is extruded from the cell, and this rate, in turn, is regulated by the intracellular sodium ion concentration itself.

Although the actual details of the mechanism whereby sodium and potassium ions are moved across the membrane are unknown at present, it is clear that a specific enzyme is involved in this process. This enzyme is activated by the presence of Na$^+$ and K$^+$, causing the hydrolysis of ATP to adenosine diphosphate (ADP), with a concomitant release of usable free energy (p. 104). The enzyme that appears to be responsible for sodium–potassium transport is called *sodium–potassium-activated adenosine triphosphatase* (or Na-K ATPase). This ATPase is a large lipoprotein that requires Mg^{++} for its action, and it has two specific loci or sites on its molecule, one with a high affinity for Na$^+$, the other for K$^+$.

In membranes derived from various cell types, the concentration of Na-K ATPase is directly proportional to the normal transport rate of Na$^+$ and K$^+$ by the membrane of the particular cell; hence, it is presumed that this enzyme has an intimate role in the transport mechanism.

The sodium pump not only plays a critical part in the development and maintenance of the electrochemical gradients of sodium and potassium that are required for normal bioelectric functions of the cell, but it is also critical to the normal homeostatic maintenance of *cell volume* and *pressure*. The efflux of Na$^+$ from the cell coupled to the influx of K$^+$ prevents the entrance of Na$^+$ and Cl$^-$ by diffusion into the cell down their individual chemical concentration gradients. In the absence of a sodium pump, therefore, these ions would move freely into the cell, and water also would enter by osmosis concomitantly with the ions, causing the cells to swell until their internal pressure balanced that of the interstitial fluid. Because the sodium pump functions actively to extrude sodium and to concentrate potassium, an osmotic equilibrium is reached, the membrane potential is maintained, and the intracellular chloride ion concentration remains low.

SUMMARY OF THE FACTORS RESPONSIBLE FOR PRODUCING THE REST POTENTIAL. As described earlier, the unequal net distribution of several ions across the cell membrane at rest determines both the *magnitude* and the *polarity* of the potential difference across the cell membrane at any given moment. The contribution of the diffusional and active transport mechanisms to this process may now be summarized and related to the individual ions which participate in this process (Fig. 3.37).

First, there is an *active transport mechanism* within the cell membrane that specifically and selectively transports sodium *out* of the cell, while simultaneously pumping potassium *into* the cell. Thus, in the resting cell, steep chemical and electric gradients for these ions are maintained across the membrane.

Second, the membrane *at rest* is far more permeable to the diffusion of potassium ions than to sodium ions. Potassium ions diffuse *out* of the cell with far greater facility than sodium ions diffuse *into* the cell.

Third, the interior of the cell contains a high concentration of nondiffusible anions. Of particular importance in this regard are *protein, organic phosphate,* and *organic sulfate* anions. These three factors are summarized in Figure 3.36, and each contributes to the membrane potential of a resting cell as follows.

As was stated on page 98, there is a very slight excess of positive charges upon the external surface of the membrane, whereas an excess of negative charges develops upon the inner surface of the membrane. This situation arises directly from the relative ease with which potassium ions diffuse across the resting membrane to the external surface, due to the steep electrochemical gradient of

this cation from the inside to the outside of the plasmalemma. Concomitantly, with the outward movement of potassium ions, a similar, very slight excess of negative charges builds up along the inner surface of the membrane, since the major intracellular anions are *nondiffusible* for all practical purposes. Thus a potential difference, the *transmembrane rest potential,* develops across the membrane. It must be reiterated that the total number of anions and cations responsible for this effect represents only a *minute fraction of the total quantity* of these substances that is present. The sodium influx that occurs simultaneously with development of the membrane potential due to an outward diffusion of potassium *does not* neutralize this potential, because the permeability of the membrane to sodium ions is quite small relative to the outward diffusion rate of potassium. Furthermore, steep electrochemical gradients for both Na^+ and K^+ are maintained continuously by the sodium pump.

Although chloride diffuses down its electrochemical gradient — that is, it tends to pass *into* the cell — any *net* movement of this anion is balanced by the membrane potential when the cell is in a steady state; in other words, the electric gradient alone is responsible for the distribution of Cl^- across the membrane of the resting cell.

The sodium-potassium pump functions solely to maintain the concentration gradients of sodium and potassium ions, because without these chemical gradients, there could be no membrane potential. The action of this pumping mechanism itself is electrically neutral, since for each sodium ion transported out of the cell, one potassium ion is transported in. If the pumping mechanism is inhibited, the ions gradually assume their equilibrium concentrations on each side of the membrane by passive diffusion and in accordance with their individual concentration gradients, so that at equilibrium any potential difference which remains across the membrane is governed strictly by *passive diffusion* of the ions.

At any point in time, the actual magnitude of the membrane potential is dependent upon the relative distribution of Na^+, K^+, and Cl^- across the membrane as well as upon the relative permeability of the membrane to these ions. The *Goldman constant-field equation* defines this magnitude quite accurately in living cells, which generally are not in a true steady-state equilibrium:

$$Em =$$

$$\frac{RT}{F} \ln \left(\frac{P_K^+ [K_o^+] + P_{Na}^+[Na_o^+] + P_{Cl}^-[Cl_i^-]}{P_K^+ [K_i^+] + P_{Na}^+[Na_i^+] + P_{Cl}^-[Cl_o^-]} \right)$$

In this expression, *Em* is the membrane potential in millivolts; *R* is the universal gas constant; *T* is the absolute temperature; *F* is faradays; and *P* refers to the individual permeabilities of the membrane to Na^+, K^+, and Cl^-. The brackets denote concentrations of the individual ions, while the letters *i* and *o*

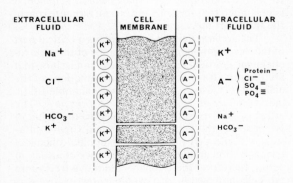

EXTRACELLULAR FLUID — CELL MEMBRANE — INTRACELLULAR FLUID

FIG. 3.37. Summary of the factors involved in producing the resting transmembrane potential. Note that the K^+ and A^- are aligned on opposite surfaces of the membrane only in the immediate vicinity of this structure.

signify the inner and outer compartments, respectively.

Like the Nernst equation, the Goldman field equation can be simplified by combining R, T, and F and by converting to base 10 logarithms, so that at 38 C it becomes:

$$Em =$$

$$61 \log \left(\frac{P_K^+ [K^+_o] + P_{Na}^+ [Na^+_o] + P_{Cl}^- [Cl^-_i]}{P_K^+ [K^+_i] + P_{Na}^+ [Na^+_i] + P_{Cl}^- [Cl^-_o]} \right)$$

It is especially important to note from this relationship that a *cation gradient from the inside to the outside of the cell causes the inside of the membrane to become electronegative with respect to the outside.* Since potassium ions exhibit such a gradient, the actual magnitude of the transmembrane potential depends primarily upon the concentration gradient of potassium, as expressed by the ratio K^+_o/K^+_i, as well as upon the relatively high permeability of the membrane to diffusion of this ion down its own chemical gradient. These facts also explain why a positively charged ion, potassium, generates a negative intracellular potential, a concept that causes considerable difficulty for many students.

Similarly, an anion gradient from the *outside* of the cell to the *inside* also produces electronegativity on the inside of the membrane. That is, in the case of an anion, the concentration gradient of a diffusible ion that produces electronegativity within the cell is directly *opposite* to that for a cation. And this is exactly the situation that exists for chloride ions.

In a resting cell, the membrane permeability to sodium ions is quite low compared with that for potassium and chloride ions. Therefore, Na^+ contributes very little to E_m, the net membrane potential. Changes in the external $[Na^+_o]$, or sodium ion concentration outside the cell, cause little alteration in the resting membrane potential, as would be predicted from the Goldman equation. However, a similar increase in the external $[K^+_o]$, or potassium ion concentration outside the cell, will *decrease* the membrane potential, whereas a reduction in the extracellular K^+ concentration will increase the magnitude of the rest potential.

In summary, the membrane potential of living cells results directly from an unequal distribution of cations and anions across the plasmalemma. There is a slight net excess of potassium ions immediately outside, and a similar excess of anions, principally chloride and protein, immediately inside the membrane barrier. In turn, this unequal spatial distribution of cations and anions, and hence of electric charges along the membrane surfaces, depends upon the relative permeability of the membrane to diffusion of the individual ions, as well as upon the concentration differences (or gradients) of those ions across the membrane surface. Finally, the development of these concentration differences for sodium and potassium is critically dependent upon a specific active transport mechanism which requires the continuous expenditure of metabolic energy.

The intracellular and extracellular fluids that are not *immediately adjacent* to the inner and outer surfaces of the membrane are electrically neutral, because of the mutual attraction of like numbers of positive and negative charges.

The foregoing discussion has introduced the genesis of the membrane rest potential in some detail, since a thorough understanding of this process is essential to a further consideration of the bioelectric properties of *muscle* (Chapter 4) and *nerve* (Chapter 5). Both of these tissues are irritable, or excitable; hence, each responds to appropriate chemical or physical stimuli by particular electric responses. In muscle, this response ultimately results in a physical shortening or *contraction* of the cells involved so that mechanical work is performed. In nerve, the response elicited upon such stimulation is the rapid conduction or transmission of an electric wave known as an *action potential* or *nerve impulse*. Consequently, the basic functions of both muscle and nerve cells are dependent upon the bioelectric potentials developed across their membranes during activity as well as at rest; and thus each of these cell types shares with the other certain common electric phenomena, such as the development and transmission of electric impulses in addition to the maintenance of steady-state resting potentials.

In conclusion, it should be reemphasized that up to this point only the bioelectric potential *developed and measured at rest across the membrane of a single cell* has been considered, and this relatively static voltage is a feature that appears to be present in all living cells. If, on the other hand, one records the voltage from the *surface* of a muscle or nerve cell, or a group of muscle or nerve cells during activity, then both the pattern and the magnitude of the electric events recorded is quite different. Thus, a clear distinction must be made henceforth between the *transmembrane potentials* of single cells and their relationship to the *surface potentials* which may be recorded from an active cell or, more generally, a tissue or an organ. Two examples of such surface recordings of electric events that are of major clinical importance are the *electroencephalograph* (p. 420) and the *electrocardiograph* (p. 600). The material presented in this section will serve as a foundation for the analysis of the bioelectric properties of muscle and nerve cells during states of activity.

General Features of Energy Storage, Release, and Utilization within the Cell

All manifestations of life require energy. The growth and differentiation of cells and tissues involve the synthesis of such essential compounds as nucleic acids and proteins, and in turn these synthetic reactions, among many others, require the use of energy. Maintenance of the body temperature in mammals depends upon controlled energy release.

The mechanical work performed by muscles, cilia, and ameboid cells necessitates energy expenditure. The production of electric impulses in nerve as well as brain, the performance of osmotic work, and the active transport of substances against electrochemical gradients also require the utilization of energy.

In living systems, the energy needed for all of the vital processes outlined above is derived from the oxidation of foods and is made available to the cells and tissues indirectly as *chemical energy* that is released by the degradation of certain energy-rich chemical compounds synthesized during oxidation of the nutrients. The specific chemical systems that provide energy for chemical, thermal, mechanical, electric, and osmotic work, in turn, are catalyzed and regulated by particular enzyme systems. Thus all aspects of the life processes are wholly dependent upon a continuous and closely regulated supply of energy in a form that the cells can utilize readily. As defined in Chapter 1 (p. 3), "energy" is a basic physical concept, and it is merely the capacity to perform work. Thus work and energy are equivalent in the physical sense and have the same dimensions.

The ultimate source of energy for all life processes is the radiation from the sun, and this radiant energy has its source in thermonuclear reactions. In this regard, all forms of energy can be interconverted from one form to another. Therefore, radiant energy is transformed into chemical energy by living systems (green plants), and the chemical energy thus stored can then be transduced or converted by various biologic mechanisms into mechanical, biosynthetic, and osmotic work.

The law of the conservation of energy states that all forms of energy may be converted from one into another, and that in the process energy is neither created nor destroyed. This law applies equally to living and to nonliving systems. It is important to realize, however, that during the conversion of chemical to other forms of energy within biologic systems, only a fraction of the total energy released during the process is available for the performance of any type of useful work. This fraction of the total energy that can be utilized is known as the *free energy* and is considered in greater detail in Chapter 14 (p. 919).

ENERGY STORAGE AND RELEASE. All cells derive their energy indirectly from the stepwise oxidation of one or more of the three major classes of nutrients — carbohydrates, fats, and proteins. The latter compounds enter the general metabolic pathways of the cell as amino acids and are oxidized together with carbohydrates and fats. Oxidation of these nutrients may take place in the presence of molecular oxygen, that is, under *aerobic* conditions, or in the absence of oxygen, in which case the process is *anaerobic* and is referred to as *glycolysis*. In either event, only a fraction of the total free energy released by oxidation of the nutrient is trapped chemically in a biologically usable form, and this energy is used to synthesize the so-called high-energy bonds of the mononucleotide ATP from ADP and inorganic phosphate. The chemical energy that is stored in the labile terminal phosphate group of ATP yields 7,000 cal (or 7 Cal) of utilizable free energy per mole during hydrolysis of this compound to ADP and inorganic phosphate (Fig. 3.38). Thus ATP is called a *high-energy phosphate compound,* although it is not unique in this respect, as discussed below (p. 106).

High-energy phosphate bonds always are denoted by the symbol ~P; however, this term actually means that the *difference in energy content* between the parent compound and the end products of its hydrolysis is relatively high, and *not* that the bond per se has a high energy content that is liberated when the compound is split.

ATP occupies an *intermediate position* in the thermodynamic scale of phosphate compounds, so that the ATP–ADP system serves as the key link between compounds that have a high phosphate group transfer energy potential and other compounds that have a lower phosphate group transfer energy potential. Therefore, ADP acts as the specific phosphate acceptor when phosphate groups are transferred to it from metabolic intermediates with a very high chemical potential energy. By the acceptance of the phosphate group, ADP is converted into ATP, and the synthesis of the terminal ~P bond of ATP results in the storage of 7 Cal of usable free energy that can be released immediately or stored until needed by the cell.*

Roughly 90 percent of the total ATP synthesis within the cell takes place in the mitochondria; the remainder is synthesized within the cytoplasmic matrix.

To summarize the foregoing remarks, the instantaneous hydrolysis of ATP into ADP yields considerably more utilizable free energy than is stored in the average covalent chemical bond. Therefore, ATP is the major source of usable chemical energy in living cells. The continuous resynthesis of ATP from ADP and inorganic phosphate in a controlled and cyclic fashion by means of the aerobic and anaerobic metabolic pathways assures the cell of an uninterrupted supply of this compound. ATP may be stored only in limited quantities within the cell, so that its continual resynthesis in adequate quantities is critical to survival.

ENERGY UTILIZATION. The precise mechanisms whereby the free energy derived from the chemical breakdown of ATP within the cell is linked or coupled to other intracellular mechanisms and processes are unknown. It is clear, however,

---

The figure 7 Cal/mole for the terminal phosphate bond has been determined under standard thermodynamic conditions in vitro. There is much evidence, however, that the intracellular release of free energy through hydrolysis of this bond in vivo can yield considerably more free energy, even up to 12 Cal/mole.

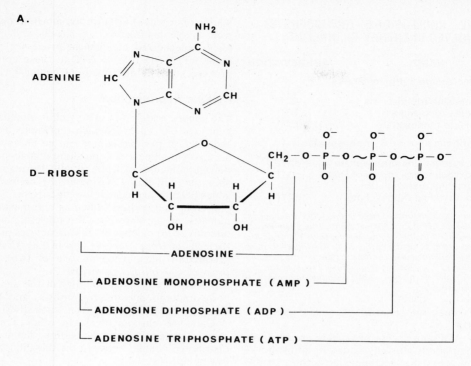

FIG. 3.38. A. Structural formula of the high-energy mononucleotide adenosine triphosphate (ATP). The two bonds which yield utilizable free energy during hydrolysis of this compound are denoted by the symbol ~. In living cells, very little free ATP is present; rather, this compound is found as a soluble and stable complex with divalent cations, particularly Mg^{++}. **B.** Reactions indicating the synthesis of ADP and AMP. P_i = inorganic phosphate. (After Lehninger. *Bioenergetics,* 1965, WA Benjamin Co.)

that ATP has three distinct and vital roles insofar as general cellular functions are concerned. These roles include the transport of various substances across the plasmalemma, participation in intracellular biosynthetic processes, and the performance of mechanical work by a number of specialized cells and tissues.

Membrane Transport. The role of ATP as the prime energy source in the active transport mechanisms for various electrolytes and other substances across the cell membrane was presented on page 92, and illustrated in Figures 3.31 through 3.34. In fact, the utilization of energy derived from the hydrolysis of ATP by the coupled sodium-potassium transport system is so great that a large proportion of the total energy expended in maintaining the basal metabolism of the entire body is employed for this one process alone (p. 989). Membrane transport is such an important function that active secretory cells may expend roughly one-third of their total energy output for this proc-

ess. Endocytosis and exocytosis also require ATP, but the actual participation of this compound in these processes is unclear.

Biosynthetic Reactions. ATP and ADP are found universally in all cells, and these compounds, together with adenosine monophosphate (AMP, Fig. 3.38), comprise the main pathway whereby phosphate group (hence energy) transfer takes place between the oxidative reactions that yield energy and the other processes in cells that require energy in order to take place at all. In addition to ATP, however, most cells also contain an abundance of other phosphate compounds that are quite similar to ATP in structure and functions but which act to *channel phosphate bond energy* into other biosynthetic pathways. The important high-energy phosphate compounds are summarized together with their abbreviations in Table 3.12. All of these phosphate compounds are nucleotides like ATP, and they all have approximately the same free energy of hydrolysis for their terminal phosphate

Table 3.12 NUCLEOSIDE 5'-TRIPHOSPHATES INVOLVED IN ENERGY CHANNELING[a]

NAME	ABBREVIATION
A. Ribonucleoside 5'-triphosphates	
1. Guanosine triphosphate	GTP
2. Uridine triphosphate	UTP
3. Cytidine triphosphate	CTP
B. Deoxyribonucleoside 5'-triphosphates	
1. Deoxyadenosine triphosphate	dATP
2. Deoxyguanosine triphosphate	dGTP
3. Thymidine triphosphate	TPP
4. Deoxycytidine triphosphate	dCTP

[a]*The high-energy compounds listed in this table are identical in structure with ATP (Fig. 3.38), except that the adenine ring of ATP is replaced by the purine guanine (Fig. 3.5) or by the pyrimidines uracil or cytosine (Fig. 3.4). The nomenclature and abbreviations for these compounds are similar to those employed for ATP. In the series of deoxyribonucleoside 5'-phosphates (part B above), 2-deoxyribose replaces the ribose moiety of ATP.*

groups as does the terminal phosphate group of ATP (7 Cal).

Only ADP can accept phosphate groups from other compounds during oxidative phosphorylation, a process that accompanies aerobic oxidation or glycolysis, to form ATP; however, the ATP–ADP system can, in turn, phosphorylate such compounds as guanosine diphosphate (GDP), uridine diphosphate (UDP), cytidine diphosphate (CDP), and so forth. A group of enzymes known as the *nucleoside diphosphokinases* catalyzes these reversible phosphorylation reactions, for example:

$$\text{ATP} + \text{GDP} \rightleftharpoons \text{ADP} + \text{GTP}$$
$$\text{ATP} + \text{UDP} \rightleftharpoons \text{ADP} + \text{UTP}$$
$$\text{ATP} + \text{CDP} \rightleftharpoons \text{ADP} + \text{CTP}$$

In turn, these phosphorylated nucleosides can channel the free energy contained in their terminal phosphate bonds into the biosynthetic pathways for a wide variety of fundamental biologic molecules. This channeling of energy is illustrated in Figure 3.39. It is clear from inspection of this figure, however, that ATP occupies the key position in all of these phosphate and energy transfer reactions, and that the other nucleotides involved are merely intermediates in the transfer of phosphate bond energy in the various biosynthetic reactions.

Cells in general may utilize up to 75 percent of the total quantity of ATP that they synthesize merely for the *de novo* synthesis of various compounds. Thus biosynthetic and transport mechanisms can account for all of the ATP utilization in many cells.

Mechanical Work. In addition to performing the work of biosynthesis and transport, a number of specialized cells also utilize the chemical energy derived from the hydrolysis of ATP to perform *mechanical work*. The contraction of skeletal, smooth, and cardiac muscle cells depends upon this source of energy. Furthermore, ameboid motion as well as the contractions of flagella and cilia are highly specialized mechanical processes that require ATP. The role of ATP in these contractile processes will be discussed in Chapter 4.

Cellular Reproduction

The genetic material of every cell is organized into genes (p. 62), each of which represents a specific sequence of pyrimidine and purine bases that are arranged in a linear fashion along double-stranded, helical molecules of DNA (p. 60). During cell division, the individual DNA molecules containing the genes become aggregated into microscopically visible, elongated bodies known as chromosomes. In humans, 46 chromosomes are present; this is the diploid (or $2n$) number of these structures present in the cells of *Homo sapiens* (Table 3.1). Cellular reproduction takes place by one of two different processes — mitosis or meiosis.

An overwhelming majority of the cells in the body undergo division by mitosis, and the two daughter cells that result from such division each contains 46 chromosomes. In other words, the genetic material contained within the nucleus is distributed equally between the daughter cells following mitosis.

Meiosis, in contrast, is a highly specialized process that involves *only* the germ cells, or *gametes,* of the body, the spermatozoa in the male and the ova in the female. If the gametes each had a diploid set of chromosomes, the resulting fertilized ovum *(zygote)* would have twice the normal diploid number of chromosomes. This outcome is avoided, however, because, during meiosis, each gamete undergoes *two* distinct nuclear divisions, the net result of which is the production of four daughter cells from every gamete, each of which contains a haploid (or n) set of chromosomes. For this reason, meiosis often is called *reduction division*. Essentially, therefore, meiosis reduces the chromosome number in half so that when a spermatozoon fertilizes an ovum, the resulting zygote contains a diploid number of chromosomes. In humans, the haploid number of chromosomes in the gametes that results from meiosis is, of course, 23.

The details of the process of meiosis are considered in context with the physiology of human reproduction (Chapter 15, p. 1130). The following discussion is devoted exclusively to cell division by the more general process of mitosis.

GENERAL ASPECTS OF MITOSIS. Since the body as a whole is a dynamic system, most of the cells that comprise its structure undergo the continual processes of growth and reproduction. Newly formed cells replace the older cells as they die and are degraded so that the total number of these morphologic elements present in the body remains about the same throughout the lifetime of a mature individual. Nerve cells are the most important exception to this generalization. Once the full complement of these elements is formed during

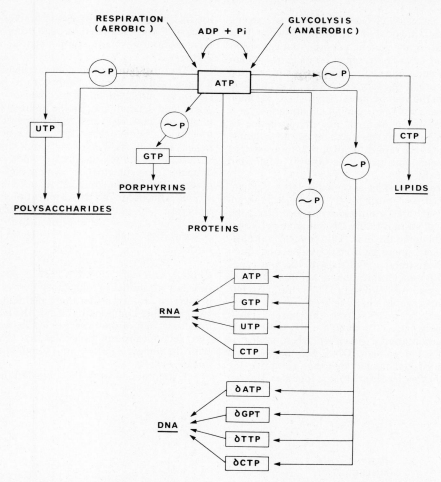

FIG. 3.39. The pathways whereby the phosphate bond energy of ATP is channeled into particular biosynthetic pathways (see text). (After Lehninger. *Bioenergetics,* 1965, WA Benjamin Co.)

early fetal life, nerve cells no longer are capable of reproduction, and those that die, regardless of the cause, are *not* replaced.

In tissue culture, the majority of mammalian cell types exhibit a life cycle that ranges between 10 and 30 hours under optimal conditions; that is, the period between mitoses (the *intermitotic interval,* or *interphase*) lasts for this length of time. Within the body, however, there appear to be inhibitory controls that regulate the rate of cell division. In general, therefore, cells in vivo have a much lower turnover rate than is found for similar cells that are cultured in vitro.

Individual cells within the body have considerably different turnover rates. For example, epithelial cells such as are found in the gut or germinal layers of the skin are replaced every few days. Mature erythrocytes survive for a few months within the circulation before they are degraded. Muscle or hepatic cells rarely undergo mitosis in normal individuals, so that these cell types have a turnover rate of several years. Nerve cells, on the

other hand, once formed, must survive for the entire lifetime of the individual.

The total number of cells in the human body has been estimated roughly at around 100 trillion (or 10^{14}).* It also has been estimated that the overall turnover rate of cells takes place at the astronomic rate of many billion per day throughout the lifetime of a normal individual. It is readily apparent from these rough approximations that the process of mitosis is an essential mechanism not only for the repair of injuries but also for the maintenance of a normal morphologic steady state within the tissues of the body as a whole.

If an opportunity is presented, the reader should not fail to observe for himself the process of mitosis under phase-contrast microscopy or as recorded by time-lapse cinephotomicrography. In no other way can a better appreciation of this dynamic cellular process be obtained.

*According to the American system, one trillion = 10^{12}, and one billion = 10^9.

DNA AND GENE REPLICATION. The net result of mitosis is an equal distribution between two daughter cells of the genetic material that is present in the parent cell. The replication of genes within the nucleus takes approximately one hour, and this process occurs at the submicroscopic level several hours before the visible onset of mitosis. Once gene replication has commenced, all of the genes in a dividing cell are replicated, not merely some of them. However, there may be a temporal sequence in which genes are replicated in various cells so that all of these elements are not necessarily duplicated simultaneously.

Gene replication is essentially DNA replication, as described on page 61. Briefly, this event takes place in three continuous stages. First, the two individual strands of the DNA double helix that constitute the gene unwind. This process can take place at an extremely high velocity. Second, each separate DNA strand combines with the four types of deoxyribose nucleotides that are *complementary* to those present on the parent DNA strand (p. 60). The specific gene is characterized by the linear sequence of the bases along the parent DNA strand, which acts as a template (p. 62). Third, bond formation in the daughter DNA strand is catalyzed by the enzyme DNA polymerase, and thus the individual nucleotides that comprise the daughter DNA molecule are linked together. Energy

for this biosynthetic process is derived ultimately from ATP and channeled through other appropriate high-energy phosphate compounds, as depicted in Figure 3.39.

The mechanism of gene replication in the somatic cells of mammals appears to be *semiconservative*, although other mechanisms have been postulated depending upon the mode of distribution of the parental DNA material among the daughter molecules. Semiconservative replication, as proposed originally by Watson and Crick, takes place by separation of the parent DNA strands so that each daughter molecule that is formed consists of one parental DNA strand attached to one daughter DNA strand. This mode of DNA replication is illustrated in Figure 3.40, and there is good experimental evidence to support the contention that semiconservative replication is operative in many types of cells.

CHROMOSOME REPLICATION. Although the detailed structure of chromosomes is unclear at present, it is known that these elements consist of two principal components. First, the genes, which are composed of double-stranded helices of DNA, and, second, protein molecules that are associated with the genes, resulting in a combination termed *nucleoprotein*.

Experimental work has revealed that all of the

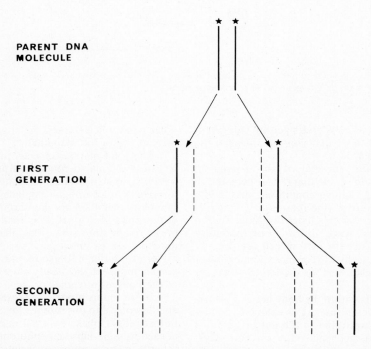

PARENT DNA
MOLECULE

FIRST
GENERATION

SECOND
GENERATION

FIG. 3.40.   Semiconservative replication of DNA in accordance with the Watson–Crick hypothesis. The parent polynucleotide chains of DNA (indicated by stars) separate, and each chain serves as a template for the synthesis of its complementary DNA strand (vertical dashed lines). Consequently, each daughter DNA molecule consists of one parent DNA strand together with one complementary DNA strand that has been synthesized *de novo*. Thus, as the process continues into further generations, the single DNA chains remain unchanged, as shown at the bottom of the figure.

DNA in a single chromosome is arranged in one enormously long double-helical molecule, and the individual genes within this macromolecular structure are attached end-to-end (ie, in a linear fashion) to each other. Although DNA macromolecules as found in the chromosomes are extremely long in relation to their diameter, it appears that the double helices of the individual DNA strands are compressed like springs and held in this folded position by linkages to adjacent protein molecules. These protein molecules, of course, have no role insofar as the genetic potency of the chromosomes is concerned, and they appear to function merely as physical adjuncts in the structure of the chromosomes.

Chromosomal replication follows directly as the natural consequence of replication of the individual DNA strands, as shown in Figure 3.41. The term *chromatid* is applied to each new daughter chromosome that is formed by replication of these nuclear elements.

MITOSIS. The microscopically visible physical process whereby a single cell divides into two daughter cells is called *mitosis*. After gene and chromosome replication have taken place at the submicroscopic level, as described above, mitosis usually occurs spontaneously within a few hours.

During interphase, the chromosomal material within the nucleus generally is in a highly dispersed state, and is known as *chromatin* because of its staining characteristics. However, there is much experimental evidence indicating that, even during interphase when the individual chromosomes cannot be recognized, the chromosomes, as well as the genes therein, retain their individual functional specificity and potency.

Following the typical mitotic division of a cell, there is an equal distribution of the nuclear material and of the protoplasm of the cell body between the two daughter cells. Nuclear division per se is termed *karyokinesis*, whereas cytoplasmic division is called *cytokinesis*.

In living cells, the process of mitosis is a smooth continuum. For the purpose of discussion, however, mitosis usually is divided arbitrarily into four major stages that occur in a sequential fashion. These stages are *prophase, metaphase, anaphase,* and *telophase* (Fig. 3.41). Many individual variations are evident in the details of mitosis as observed in the cells of various species; therefore, Figure 3.41 is merely illustrative of the general steps in this process.

Prophase. Observed in a living cell, the onset of *visible prophase* is heralded by the condensation of submicroscopic chromosomal material into minute granules of chromatin that are near the limit of resolution of the phase-contrast microscope. Subsequently these granules increase in size as well as number and become aggregated into larger elongated bodies that develop ultimately into pairs of chromosomes. The paired chromosomes are held together by a common structure, the *centromere* (Fig. 3.41B, C).

Meanwhile, the cell center divides, and, as prophase continues, the two cell centers move apart while the microtubules of the *mitotic spindle* develop between them, together with the *asters* which radiate outward from the cell centers.

During prophase, the nucleolus elongates and gradually becomes incorporated into one of the larger developing chromosomes (Fig. 3.41C).

The dissolution of the nuclear membrane signals the end of prophase.

Metaphase. During early metaphase *(prometaphase)*, the spindle enlarges and the centrioles appear clearly toward the opposite poles of the cell (Fig. 3.41D). The chromosomes now move into the equatorial plane of the spindle and become aligned at approximate right angles to the long axis of this organelle (Fig. 3.41E). Microtubules of the spindle apparatus become attached on opposite sides of each centromere.

In late metaphase, each chromosome becomes more clearly delineated into two daughter chromosomes, or *chromatids*.

Anaphase. During anaphase, which follows shortly after metaphase, the chromatids separate. The result is two sets of identical daughter chromosomes (Fig. 3.41F). The two individual sets of daughter chromosomes now move rapidly toward each pole of the spindle (Fig. 3.41G) and then gradually fuse into the two daughter nuclei.

During late anaphase, the spindle elongates and cytokinesis begins as the cell body constricts.

Telophase. As shown in Figure 3.41H, telophase is marked by the appearance of a nuclear membrane in each daughter cell. The fused chromatids disperse into chromatin, one or more nucleoli appear, and cytokinesis progresses as the daughter nuclei enlarge. The spindle fibers now disappear, forming the ephemeral *intermediate body of Flemming,* and the daughter cells once again enter an intermitotic state.

During telophase, cisternae of the granular endoplasmic reticulum accumulate around the daughter chromosomes, and the nuclear membranes of the daughter cells then are produced by these organelles.

Replication of the Centrioles and Other Organelles. The centrioles are replicated by simple fission; this process can take place early in prophase, or, as is shown in Figure 3.41, during a later stage of mitosis. The actual stage at which the formation of new diplosomes takes place depends upon the cell type involved.

During cytokinesis, organelles such as the mitochondria, ribosomes, plasma genes, and other organized particulates as well as the cytoplasm itself appear to be distributed in approximately equal quantities and in a random fashion between the daughter cells.

Forces Involved in Mitosis. There are two varieties of microtubule present in the mitotic spindle from a

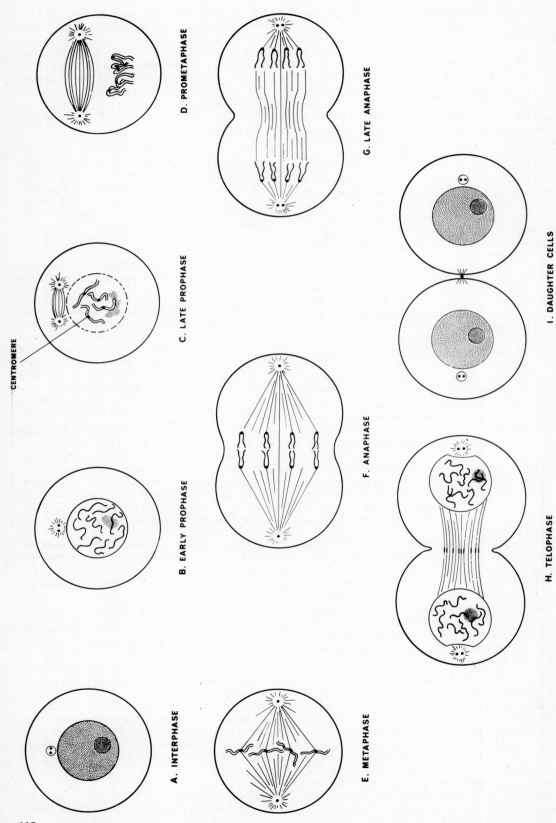

FIG. 3.41. Cell division or mitosis actually is a smoothly continuous process. For descriptive purposes, however, mitosis is divided arbitrarily into prophase, metaphase, anaphase, and telophase.

CENTROMERE

A. INTERPHASE

B. EARLY PROPHASE

C. LATE PROPHASE

D. PROMETAPHASE

E. METAPHASE

F. ANAPHASE

G. LATE ANAPHASE

H. TELOPHASE

I. DAUGHTER CELLS

morphologic standpoint: those that traverse the entire spindle from pole to pole, and those that are attached to the chromosomes. As described earlier (p. 83), the microtubules are composed of smaller protein subunits of tubulin, and the orderly, end-to-end polymerization of these subunits gives rise to the microtubules themselves.

Bascially, two hypothetical mechanisms have been proposed to account for centriolar and chromosomal migration during mitosis. According to one hypothesis, the orderly and progressive elongation of the microtubules in an outward direction from postulated organization centers pushes the centrioles and chromosomes apart and toward the opposite poles of the cell during mitosis. Calcium ions may have a regulatory role in this process, since a high concentration of this cation prevents the polymerization of tubulin monomers into microtubules. Alternatively, the polymerized tubulin molecules may contract actively through a kind of sliding filament mechanism similar to that involved in muscular contraction (p. 138). The energy source for this process could be ATP, although guanosine triphosphate (GTP) also has been implicated.

Perhaps both of these mechanisms are involved to some extent during various stages of mitosis; however, this is conjectural.

The forces involved in cytokinesis per se are completely unknown, although microfilaments apparently play a part in this process, as discussed on page 149.

CELLULAR GROWTH. The ultimate size reached by the daughter cells following mitosis depends almost exclusively on the quantity of nuclear DNA that is present. In turn, this DNA regulates the type of proteins that are synthesized by the cell. It is assumed here, of course, that the cell has an adequate supply (in a quantitative as well as a qualitative sense) of all of the substrates that are necessary for the myriad biosynthetic processes involved.

In the human body, the underlying mechanisms that actually control cellular reproduction and growth are unknown for the most part. Although most cells have a potential for continual growth and division as discussed earlier, the rates of these processes in vivo appear to be much suppressed in normal individuals. Presumably this retardant effect on unlimited growth in size *(hypertrophy)* as well as reproduction *(hyperplasia)* of the cells is caused by a number of inhibitory substances elaborated by the cells themselves. These chemical inhibitors may act in a kind of negative feedback mechanism at the cellular level in order to prevent the uncontrolled growth and multiplication of cells such as occurs in cancer. In fact, according to one hypothesis, it is precisely a lack of such negative feedback mechanisms that allow cancer cells to grow and reproduce in an unrestrained fashion. Normal cells also exhibit an unlimited growth and reproductive potential when they are cultured in

vitro under optimal conditions. Possible, this effect also is due to the removal of the appropriate inhibitory factors.

It is interesting to note that if a deficiency of certain cell types is created in vivo, then the remaining cells of that type will proliferate rapidly until a relatively normal complement of such elements is once again achieved. For example, surgical removal of most of the liver by performing a subtotal hepatectomy so stimulates the growth and reproduction of the remaining hepatic cells that an approximately normal quantity of liver tissue is restored within a few days. Once normal liver size has been attained, the rate of mitotic division by the cells returns to its low preoperative level.

CELLULAR DIFFERENTIATION. Differentiation is a specialized aspect of the growth and reproduction of cells whereby these elements assume particular morphologic and functional attributes during prenatal and postnatal life. The structural changes that occur in cells and tissues, as well as the regulation of these processes during growth and development, lie principally within the realm of embryology. One recent experiment is of considerable interest, however, because it has demonstrated certain general features regarding the role of the nucleus in the process of cell differentiation. This experiment was conducted as follows. The nucleus was removed from an epithelial cell of frog intestine, and this nucleus then was implanted into an enucleated frog ovum. Subsequent fertilization of ova containing an "intestinal nucleus" frequently resulted in the development of tadpoles, which subsequently underwent their usual metamorphosis into completely normal frogs.

This simple experiment demonstrates that, during the process of differentiation, there is apparently a selective repression of the different *operator genes* that normally regulate the function of the *structural genes*. In turn, the activity of the structural genes is manifest by the wide variety of morphologic and functional specializations that develop in cells and tissues as the individual matures. Furthermore, it also is evident from this experiment that even a fully mature and specialized cell nucleus, such as that found in an intestinal epithelial cell, still contains the entire complement of genetic patterns within its genes that are necessary to stimulate the growth, reproduction, and differentiation of a single cell into a complete and normal organism.

Although other biochemical substances, including the so-called *inducers,* are known to be involved in cell differentiation, the particular mechanisms responsible for the normal control of this process are poorly understood.

INTERCELLULAR TRANSMISSION OF INFORMATION

In the previous sections of this chapter, emphasis was placed upon the microscopic and submicroscopic components of cells, as well as on certain of the basic functional attributes of these fundamental units of living matter. However, it is necessary to emphasize at this juncture that isolated cells rarely are encountered in the human body; rather, cells that are designed to perform a common function generally are assembled into organized units or tissues which perform their functions in a smoothly coordinated and integrated manner. This association of many cells into larger groupings bound together into morphologic units by fibrous connective tissues, as well as by intercellular substances of various types, automatically increases the total mass of the tissues considerably. Therefore, effective means of *communication* among the individual cells within a tissue are critical to the survival of the organism as a whole. Furthermore, the distances between various tissues are so great that mechanisms to transmit information rapidly and effectively throughout the body are also essential.

The term *information* as used in this context signifies a *physiologic communication* among succeeding generations of a species as well as among the individual cells, tissues, organs, and organ systems of the body as a whole. Thus, the general mechanisms that underlie intercellular communication occur at all morphologic levels of organization and are of major significance to the functions of the body.

Fundamentally, there are three mechanisms whereby information is transmitted among and within complex living systems. These mechanisms are *genetic, nervous,* and *metabolic.*

Genetic information is transmitted from generation to generation of cells in higher organisms by means of sexual reproduction. Thus, when a male and female gamete fuse, their respective complements of DNA each have transmitted a like amount of genetic information to the succeeding generation. This genetic information, in turn, completely regulates the physicochemical attributes of the offspring. In addition, the occasional occurrence of mutations (p. 63) lends a certain flexibility to genetic communication between cells in the long-term sense, so that evolutionary processes, whether for better or worse insofar as the survival of the individual or of the species is concerned, can take place. Hence, the transmission of genetic information is a relatively pliant form of intercellular communication, since it may permit the gradual development of a population with greater survival potentialities in the face of marked environmental changes.

As might be anticipated from the foregoing remarks, intercellular transmission of information by the genetic route is a relatively slow process indeed. In marked contrast to genetic communication, the intercellular transmission of information by means of highly specialized *nerve cells (neurons)* is an extremely rapid process. Hence, the conduction and integration of signals that arise both within and without the body enable the individual organism and its component structures to function in a coordinated fashion.

The third mechanism for the intercellular transmission of information depends upon a number of specific chemical substances that are elaborated by certain cells and that are transported in the circulatory system to their target tissues or organs. These chemical messengers or *hormones* achieve their effects at a rate that is considerably slower than that of the nerves but infinitely greater than that of genetic communication between cells. The various hormones, in one way or another, have a profound regulatory effect upon *all* cellular metabolic processes. Therefore, hormones are indispensable chemical agents both in the orderly development of the individual from conception onwards and in many of the homeostatic mechanisms that are critical to survival.

CHAPTER 4

The Physiology of Contractile Cells and Tissues

The ability to move or to perform mechanical work is one of the most fundamental and obvious characteristics of living organisms. This physiologic property in turn depends upon the ability of specialized contractile cells and tissues to undergo a physical shortening and thickening along some linear axis. The contraction either may be spontaneous or may be in response to appropriate external stimuli; hence, contractile systems are said to be *excitable* or *irritable*. The ultimate physiologic goal of most contractile cells and tissues is the performance of mechanical work that is useful to the individual as a whole, either directly or indirectly.

In the body there are three major cell types that are highly specialized for contraction. These are *skeletal, cardiac,* and *smooth muscle* cells. The distinction between these contractile elements is based upon their morphologic and physiologic differences as well as upon their reactions to various pharmacologic agents. In the aggregate, the three types of muscle cell comprise about 50 percent of the total body mass in human beings.

The general property of contractility is by no means confined to muscle cells, although it is most evident in these elements. For example, during growth and differentiation of the embryo, many individual cells are capable of active, carefully regulated migration from their sites of origin to their permanent structural and functional loci within various tissues. Thus, such cells are said to be *motile,* at least during a portion of their life cycle, and this motility depends upon an inherent ability of the cells to contract actively. Since this type of cellular locomotion resembles that exhibited by ameba, it is called *ameboid motion.*

Even after embryonic cells have come to occupy a relatively sessile niche within the structure of an adult tissue, they still may undergo a type of limited contractile activity. Thus, changes in the shape of individual cells can influence profoundly the ultimate gross and microscopic structure of an organ.

In the adult body there are present a number of particular cells that still possess the latent capacity to wander as independent units among the tissues by ameboid motion, and usually such motility is eilicted in response to specific chemical stimuli.

A number of particular tissues contain specialized cells that have from a few to many actively contractile, threadlike organelles known as *cilia* (sing, *cilium*). These motile structures project from the surfaces of the individual cells and beat in a coordinated, rhythmic pattern. Since ciliated epithelial cells occupy fixed positions and are arranged in sheets, the net effect of ciliary motion is to move a layer of mucus slowly along the surface of the entire epithelial layer.

In contrast with ciliated epithelial cells, each mature male germ cell *(spermatozoon)* has but a single hairlike motile organelle called a *flagellum* (pl, flagella). Morphologically, flagella are similar to cilia; however, flagellar motion causes each spermatozoon to swim actively in a manner resembling the undulations of a tadpole.

It is obvious from this brief summary that the general property of contractility is found in a broad variety of cell types. Nevertheless, all contractile and motile cells, as well as cilia and flagella, appear to share one common feature: Contractile systems in general are able to convert or *transduce* chemical energy into mechanical energy. In other words, contraction leads to the performance of mechanical work, and contractile systems may be considered *chemomechanical transducers.*

In this chapter, emphasis is placed upon the morphologic, bioenergetic, and bioelectric processes that underlie the phenomenon of contractility, with particular reference to skeletal muscle. The physiologic mechanisms that initiate, regulate, and coordinate the activity of specific contractile tissues in vivo will be dealt with in subsequent chapters.

THE PHYSIOLOGY OF MUSCULAR CONTRACTION

Introductory Remarks

The spatial organization of muscle cells into larger groupings makes possible a vast number of intricate voluntary and involuntary movements which range from localized contractions within individual structures and organs to generalized activities on the part of the entire body.

Muscle cells or *fibers* can be divided into two broad classes, depending on their cytologic characteristics. These classes are *striated muscle* and *smooth muscle*. Striated muscle cells, in turn, may be subdivided into two additional categories: skeletal and cardiac muscle. Striated muscle fibers are characterized by the presence of transverse bands that repeat at regular intervals along the length of each fiber, whereas smooth muscle cells lack these transverse bands.

Muscle fibers are elongated structures that shorten and thicken during contraction in a linear direction along the axes of the fibers.

Skeletal muscle is innervated by cerebrospinal nerves so that contraction of this tissue in vivo is under *voluntary* control (p. 403). Cardiac muscle, which comprises the wall of the heart, has the property of inherent contractility (p. 143). However, the rhythmic contractions of the heart are governed by the autonomic nervous system, among other factors; therefore, fundamentally the contractions of cardiac muscle are *involuntary*.

The musculature of the viscera is composed principally of smooth muscle, and, like that of the heart, the activity of smooth muscle is both inherent and regulated by the autonomic nervous system (p. 270). Endocrine factors also are of major significance in controlling the activity of the smooth muscle in certain structures and organs.

The Structure and Ultrastructure of Skeletal Muscle

In few other tissues of the body is the structural organization of the cells so closely and obviously related to their function as in skeletal muscle (Fig. 4.1). This fact is clearly evident at all levels of organization, from the gross anatomic features down to the molecular morphology of the individual contractile elements. Consequently, a description of the structure of a muscle is also a description of the contractile machinery per se.

GENERAL PATTERNS OF ORGANIZATION. Skeletal muscle cells are multinucleate, elongated, cylindrical structures grouped closely together in parallel bundles or *fascicles* that are visible to the naked eye (Fig. 4.1A, B). The indi-

vidual fascicles are organized into a number of distinct patterns which may be discerned readily upon gross inspection. Thus, an anatomically distinct "muscle" is composed of a large number of fascicles bound together into a single anatomic and functional unit by connective tissue. Although this connective tissue forms a continuous framework or *stroma*, its various parts are designated arbitrarily by different terms (Fig. 4.1A). The connective tissue layer that invests a discrete muscle is called the *epimysium*. Each fascicle, in turn, is surrounded by a thin, collagenous wall or *septum* that is continuous with the epimysium, and collectively these septa are known as the *perimysium*. The individual muscle fibers are surrounded by an exceedingly delicate network of connective tissue called the *endomysium*.

The connective tissue stroma serves to bind the individual contractile elements of a muscle into an integrated, functional unit, although some freedom of movement between the individual fibers and fascicles is permitted.

Skeletal muscle cells are provided with rich capillary networks that arise from larger blood vessels entering the muscle in the septa. The capillaries are convoluted so that they can adapt their length to the functional state of the muscle. Thus, during contraction the capillaries become more tortuous, whereas they elongate as the muscle fibers lengthen during relaxation.

In addition to blood vessels, the septa also contain the larger branches of nerves which ramify and ultimately make anatomic and physiologic contact with the individual muscle fibers. For example, the terminal branches of motor nerves are connected to individual muscle fibers at highly specialized structures known as *motor end plates* or *neuromuscular junctions* (Fig. 4.1B; see also Fig. 6.29, p. 208).

In most muscles, the individual fibers are believed to be shorter than the length of the entire muscle. Therefore, one end of the fiber generally is joined to a connective tissue septum, whereas the other end is attached to a tendon at either the origin or the insertion of that particular muscle. Certain muscles (eg, the sartorius) do not taper at the ends, however, and the individual fibers appear to be as long as the entire muscle.

Individual muscle fibers exhibit considerable variation in diameter even in the same muscle, and may range in thickness from 10 to over 100 μ. The diameter of a fiber depends upon a number of factors. For example, the thickness of muscle fibers increases with age. In the adult, a further *compensatory hypertrophy* of individual fibers results from the performance of long-sustained, vigorous physical work. Conversely, a lack of exercise results in an *atrophy* of muscle fibers, so that their diameter shrinks.

CYTOLOGY. When a longitudinal section of a skeletal muscle is viewed under an optical microscope, each cell is seen to be enveloped in a delicate limiting membrane, known to the classic histologist as the *sarcolemma*. Actually, the sarcolemma is a compound structure, a fact that becomes evident when it is visualized under the electron microscope. Hence, the "sarcolemma" of the light microscopist actually consists of the cell membrane (plasmalemma), an external lipoprotein-polysaccharide coating, and a delicate reticulum of connective tissue fibers (the endomysium).

The plasmalemma is, of course, invisible under the best magnifications that can be achieved using a light microscope; however, the term "sarcolemma" currently is used to denote *only* the plasmalemma (or cell membrane) of muscle cells. Henceforth, the word "sarcolemma" as used in this text refers specifically to the cell membrane of muscle cells. The ultrastructure of the sarcolemma is quite similar to that of the unit membranes encountered in other types of cells (p. 71; Fig. 3.21, p. 73).

When seen in transverse section under the light microscope, the fibers of fixed skeletal muscle are arranged into irregular polygonal areas known as *Cohnheim's fields*. Actually, living muscle fibers are round in cross section, and the Cohnheim's fields represent fixation artifacts.

Skeletal muscle fibers are multinucleate, the individual nuclei being roughly elliptical structures whose long axes are oriented parallel to the fiber axis (Fig. 4.1C). The nuclei are situated immediately beneath the sarcolemma, and this feature distinguishes skeletal muscle fibers from cardiac and smooth muscle cells, whose nuclei are located centrally within each fiber.

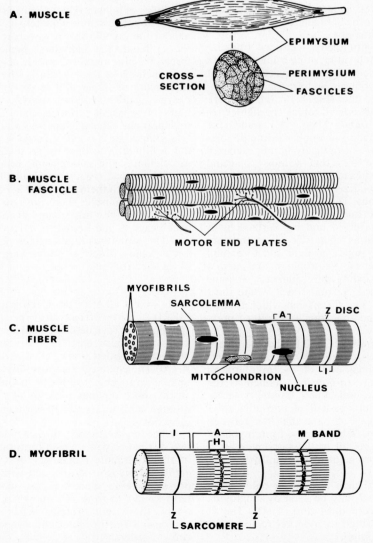

FIG. 4.1.   Gross and fine structure of skeletal muscle as revealed by optical microscopy.

The cytoplasm of muscle fibers *(sarcoplasm)* contains a Golgi complex that is located near one pole of each nucleus. The mitochondria *(sarcosomes)* also are located most abundantly near the nuclei and just beneath the sarcolemma, although these structures may be found deeper within the body of the cell. Glycogen and lipid droplets are present as inclusions within the sarcoplasmic matrix of each muscle cell. A pigment called myoglobin, which imparts a reddish color to muscle, is also found in the sarcoplasm. If the concentration of this pigment is low, the muscle is called *pale* or *phasic,* whereas, if the myoglobin concentration is high, the muscles are termed *red* or *tonic.*

The most obvious organelles found within the scanty cytoplasmic matrix of skeletal muscle fibers are the dense, tightly packed, parallel bundles of small fibers which bear regular cross-striations. These are the myofibrils (Fig. 4.1C, D). The myofibrils form the basis of the contractile machinery, and they are the smallest morphologic units that are visible under the light microscope, having a diameter of approximately 1 μ.

In stained, longitudinally sectioned preparations of skeletal muscle, alternating transverse dark and light bands are seen. These bands are apparent in the individual myofibrils and are also seen running across the entire fiber (Fig. 4.1C, D). The dark bands are called *A bands,* whereas the lighter regions are called *I bands.* The "A" stands for *anisotropic,* birefringent, or doubly refractile. The "I" denotes *isotropic* or optically homogeneous.

It is important to stress that the anisotropic and isotropic characteristics of striated muscle are seen only under plane polarized light and in *unstained* histologic preparations. The A or anisotropic bands appear as pale areas and the I or isotropic bands appear dark when viewed under polarized light. The anisotropic regions of muscle fibers thus are able to rotate plane polarized light and to exhibit birefringence, whereas the isotropic regions neither rotate the light nor exhibit birefringence. Staining the tissue renders the A bands dark, while the I bands remain light. The discussion which follows will consider only the appearance of *stained* preparations.

As shown in Figure 4.1C and D, the I bands are bisected by dark lines called the *Z bands* or *membranes.* The functional unit of the myofibril in striated muscle is the *sarcomere,* defined as the distance between two adjacent Z bands. A sarcomere thus consists of one-half of two adjacent I bands together with the A band that lies between them (Fig. 4.1D). The Z bands pass from myofibril to myofibril; therefore, all of the sarcomeres in a single muscle fiber are aligned structurally with respect to one another.

In the center of each A band, a pale region known as the *H band* may be seen by employing special histologic techniques (Fig. 4.1D). The H band, in turn, is bisected by a darker region in its center called the *M band.*

The transverse striations described above give skeletal and cardiac muscle fibers their characteristic striped appearance, and they also are indicative of a highly ordered structural pattern of the elements that are encountered at the submicroscopic level.

The myofibrils present in smooth muscle cells lack the typical cross-banding that is found in striated muscle. Consequently, smooth muscle sometimes is referred to as *plain muscle.*

The *sarcoplasmic reticulum* of skeletal muscle cells is smooth, and it consists of an elaborate but delicate network of membrane-limited tubules which surround the individual myofibrils (Fig. 4.2). This reticular system is similar to the smooth endoplasmic reticulum present in other types of cells; however, in muscle fibers the reticulum is structured in intimate relationship to the myofibrils, and a ribosomal component is absent.

Overlying the A bands, the reticular sarcotubules of human skeletal muscle fibers course in a direction that is approximately parallel to the long axis of the myofibrils. However, in the vicinity of the H band, the individual tubules anastomose freely, forming larger transverse channels. Near the A-I junctions of the myofibrils, the tubules join to form pairs of larger and more regular transverse channels called the *terminal cisternae.* Between the pairs of adjacent terminal cisternae lies a third membranous tubular element known as the *transverse tubule* or *T tubule.* As shown in Figure 4.2, the T tubule is formed by an invagination of the sarcolemma. The sarcoplasmic reticula of two adjacent cisternae together with the T tubule that is associated with them are called a *triad.* In mammalian skeletal muscle, there are two triads per sarcomere, and they are located at the junctions between each A and I band.

The structural relationships of the sarcoplasmic reticulum described above were clarified by use of the electron microscope.

THE MOLECULAR STRUCTURE OF THE MYOFIBRIL. The ultrastructure of individual myofibrils as revealed by electron microscopy indicates that these elements are composed of still smaller elongated units called *myofilaments* (Fig. 4.3A). The chemical composition, size, and regular spatial arrangement of these subunits are reflected in the more obvious patterns of cross-banding in striated muscle visualized by means of light microscopy.

The myofilaments, in turn, are composed of two distinct proteins which form the major portion of the contractile machinery of the muscle cell: *actin* and *myosin.* Associated with the actin, but present in lesser quantities, are two additional proteins known as *tropomyosin* and *troponin.* Collectively, these four proteins acting in concert are responsible for the contractile properties of muscle cells in general.

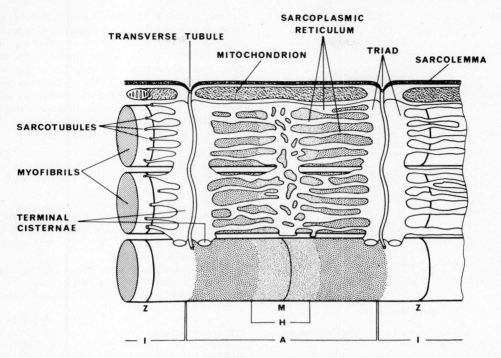

FIG. 4.2. The sarcoplasmic reticulum in relation to other physiologically important morphologic features of mammalian skeletal muscle.

As shown in Figure 4.3A, the sarcomere of a myofilament is composed of alternating thin and thick filaments which overlap to a certain extent. The Thin Filaments. The thin filaments of the myofibril consist principally of high-molecular-weight polymers composed of individual globular or G-actin molecules, and are called fibrillar or *F-actin filaments* (Fig. 4.3B). As shown in the figure, two individual F-actin filaments are twisted together into a helix. The thin filaments also con-

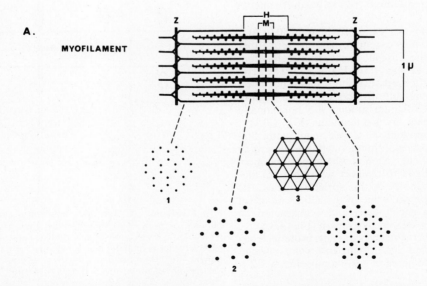

FIG. 4.3. Ultrastructure of skeletal muscle. **A.** Fine structure of a portion of a myofilament illustrating the relationship between the thick and thin filaments of one sarcomere in both longitudinal and cross section (1–4).

B.

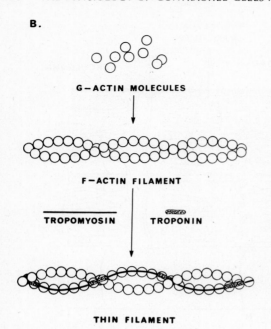

FIG. 4.3 (cont.)   **B.** Morphologic relationships among the proteins that comprise a thin filament.

tween the ends of the thin filaments delimits the H band. Consequently, in a contracted muscle, the H bands are either quite narrow or obliterated completely (Fig. 4.4B). In a stretched muscle, on the other hand, the H bands are quite broad (Fig. 4.4C).

The thin filaments exhibit a definite polarity that is conferred upon them by the polarity of their constituent G-actin molecules. This polarity is an essential feature of muscular contraction.

The Thick Filaments.   Each thick filament is composed of several hundred individual myosin molecules (Fig. 4.3A, C). These elements of the myofibril are about 160 Å in diameter, 1.5 μ in length, and lie parallel to each other about 450 Å apart. The myosin filaments themselves are somewhat thicker in the middle than at the ends, and they form the principal component of the A band. The length of the thick filaments determines the width of the A band itself.

The thick filaments are held in precise relationship to each other by cross-linkages that are aligned at the midpoint of the A band. These cross-linkages give rise to the M band (Fig. 4.3A-3).

As shown in Figure 4.3C, the individual myosin molecules that make up a thick filament consist of two portions. The longest part of the molecule is a thin, rodlike protein called *light*

tain the tropomyosin and troponin (Fig. 4.3B). Individual tropomyosin molecules are polymerized into long, exceedingly fine threads that follow along the surface of each F-actin strand; therefore, two such threads are present on each F-actin filament. Troponin molecules, on the other hand, have a roughly globular shape, and these proteins are found at intervals along each F-actin filament resting upon the individual tropomyosin filaments (Fig. 4.3B).

The thin filaments are approximately 50 to 70 Å in diameter, and they extend for about 1 μ in either direction from the Z band; hence, this region constitutes the I band (Fig. 4.3A; Fig. 4.4A). The actin filaments do not terminate at the A band, however, but extend for variable distances into this region, where they interdigitate among the hexagonally packed and thicker myosin filaments (Fig. 4.3A).

As they approach the Z line, each actin filament appears to be continuous with two diverging and thinner filaments, each of which runs obliquely through the Z band to connect with an actin filament on the other side of this structure. Thus, the actin filaments on each side of the Z line are offset with respect to each other, and a characteristic zigzag pattern is seen in this region (Fig. 4.3A).

The Z bands themselves are flat, proteinaceous structures that are 2.2 μ apart in *resting* muscle. Hence, the sarcomeres of resting muscle are 2.2 μ in length.

In a resting muscle, the actin filaments that penetrate an A band from opposite ends of the sarcomere do not meet, so that the distance be-

C.

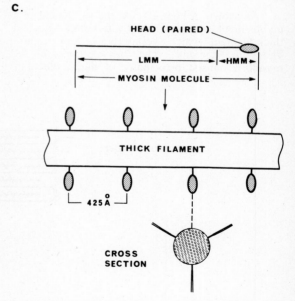

FIG. 4.3. (cont.)   **C.** The thick filament is composed of bundles of individual myosin molecules. Each myosin molecule, in turn, consists of a rodlike light meromyosin (LMM) component and a heavy meromyosin (HMM) component. The HMM has paired heads at its end which form the radially disposed cross-bridges, indicated by dots in part A of the figure. Adenosine triphosphatase activity is localized within the heads of the HMM molecules, which also form cross-linkages with the helical thin filaments during muscular contraction.

meromyosin (LMM); it is attached to a "head" that contains a pair of projections or cross-bridges and is composed of *heavy meromyosin molecules* (HMM), as shown in Figure 4.3C.

In the regions where the thick and thin filaments overlap at the ends of the A bands, the filaments are only about 100 to 200 Å apart. The cross-bridges extend radially in a spiral fashion from each myosin filament toward the actin filaments that surround it (Fig. 4.3A, C).

Heavy meromyosin exhibits *ATPase* activity, and thus can hydrolyze ATP, which is the immediate source of energy for muscular contraction. Furthermore, HMM also has the important physiologic ability to form transient cross-linkages with actin, and therein lies the basic mechanism of muscular contraction.

The Molecular Basis of Muscular Contraction. The overlapping and parallel arrays of alternating thick and thin protein filaments within every myofilament are so arranged that they are able to slide past each other without hindrance (Fig. 4.4). This sliding motion, in turn, is the fundamental mechanism that underlies all muscular contraction, and it will be discussed in greater detail on page 138 following a consideration of certain additional physiologic properties of contractile systems.

The Mechanical Properties of Skeletal Muscle

When a muscle shortens physically during the process of contraction, it applies tension at the points from which it originates and at which it is inserted on the bones. The bones, in turn, are so arranged that they form lever systems of various types that are actuated by particular muscles or groups of muscles. In this way, many intricate bodily positions and movements are possible.

Skeletal muscle can develop a maximal tension equal to between 3 and 4 kg per square centimeter of its cross-sectional area. Since the human quadriceps may have a cross-sectional area of approximately 100 cm², this muscle can exert a tension upon the patellar tendon of up to 400 kg. Furthermore, a muscle can shorten at rates up to ten times its own length per second. As a result of these two mechanical characteristics, humans can execute very powerful as well as rapid movements of many kinds.

The gross contractile properties of skeletal muscle are best analyzed in two stages: first, in terms of the mechanical properties of muscle in vitro; and, second, in terms of the events that take place in vivo and that result in smoothly coordinated muscular activity. At the outset, however, it is necessary to emphasize the difference between

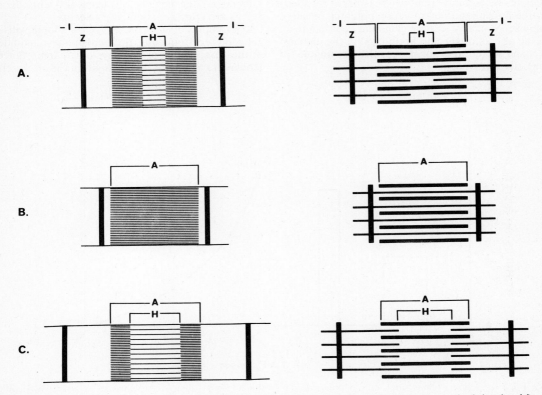

FIG. 4.4. The microscopic appearance of the cross-striations in one sarcomere of a skeletal muscle (left side of figure) together with the physical relationship between the thick and thin filaments (right side of figure). **A.** During relaxation of muscle. **B.** During contraction. **C.** During stretch. Note particularly how the appearance of the cross-striations is altered by these different physical states, depending upon the degree of interdigitation of the thick and thin filaments.

the mechanical and electric processes that take place in muscle. Normally, the excitation of a muscle that leads to its contraction is the result of electric events. That is, electric impulses *(action potentials)* are transmitted by motor nerves to motor end plates within the muscles (p. 208). The result of this excitation is a localized depolarization of the sarcolemma in the motor end plate region, which leads to the generation of a propagated action potential along each muscle fiber itself, and these electric events subsequently initiate the contractile process per se. Therefore, the electric and mechanical events that take place during muscular contraction operate in a sequential and interlocking fashion. The physiologic basis for the electric events is quite different from that for the contractile process, however, and each will be discussed independently.

CELLULAR EXCITABILITY AND THE LAW OF SPECIFIC ENERGIES. The protoplasm of all living cells is able to react to an appropriate environmental change by giving a suitable response. This fundamental characteristic of living systems that enables them to react to a particular environmental change or *stimulus* is called *excitability* or *irritability*. The property of excitability is most highly developed and obvious in muscular and neural tissues. Muscle responds to an adequate stimulus by a contraction, whereas nerve responds by the transmission of an action potential or nerve impulse. Similarly, certain glandular cells can react to the effects of specific hormones by the elaboration and secretion of substances that are unique for the cells involved.

Thus, it is evident that any given cell responds to a stimulus in a manner that is characteristic for that type of cell, regardless of the kind of stimulus that is applied to it. This generalization is sometimes called the *law of specific cellular energies.*

GENERAL PROPERTIES AND TYPES OF STIMULI. Every stimulus has two basic characteristics — *intensity* and *duration.* If the strength or intensity of a stimulus is too weak to evoke a reaction, then it is said to be *below threshold, subthreshold,* or *subliminal.* When the intensity of a particular stimulus is barely sufficient to evoke a response, then it is called *adequate, threshold, minimal,* or *liminal.* As the intensity of a stimulus is increased above threshold to suprathreshold levels, a point is reached beyond which the response of a tissue no longer is augmented; the response then is said to be *maximal.*

If a stimulus fails to act for a sufficient length of time on a cell or tissue, then obviously no response will be elicited, regardless of the intensity of the applied stimulus. Consequently, the intensity and duration of any stimulus are critical and interdependent factors, both of which are involved in the excitation of living systems.

Stimuli may be classified broadly into two categories, *physical* and *chemical.* Physical stimuli include *electric, mechanical, osmotic,* and *thermal* factors; a large number of chemical agents themselves are effective stimulatory agents.

Examples of all of these types of stimuli will be encountered frequently throughout the remainder of this text, and all are of physiologic importance. However, artificial electric stimulation is the

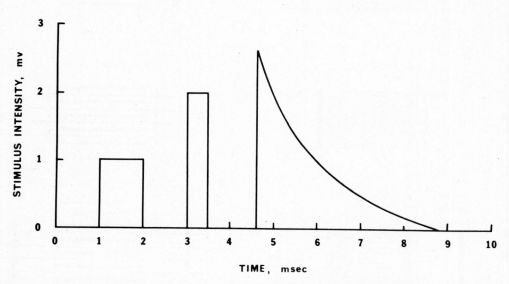

FIG. 4.5. The shapes of three single current pulses used for artificially stimulating excitable tissues. **A.** Square wave pulse of 1 mv intensity and 1 msec duration. **B.** Square wave pulse of 2 mv intensity and 0.5 msec duration. **C.** Pulse having an intensity of approximately 2.7 mv and duration of about 5 msec. Note the exponential decay.

Electronic stimulators used in physiologic research can generate an extremely broad variety of pulse shapes; the frequency, intensity, and duration of each pulse can be regulated with considerable precision.

method of choice for studying the physiologic properties of muscle and nerve, because the intensity, duration, shape of individual pulses, and rate of stimulation or *frequency* can be regulated closely, both individually and collectively (Fig. 4.5). Furthermore, electric stimulation produces the least damage to a tissue during extended periods of observation, and it approximates most closely the physiologic events in vivo.

On the basis of the foregoing general considerations, the properties of an adequate electric stimulus for skeletal muscle now may be defined more precisely. As depicted in Figure 4.6A, an electric potential having a magnitude of 8 v, acting for about 0.1 msec, is sufficient to excite the tissue; ie, this value is the *threshold* for stimulation. If, however, a much lower voltage — for example, 2 v — is permitted to act for a longer interval, about 0.7 msec, the current still reaches threshold, and the tissue once again responds by contraction (Fig. 4.6B). When the applied voltage is dropped still lower (to 1 v) and permitted to act for a still longer interval, the stimulus also is adequate to produce excitation (Fig. 4.6C). As shown in the figure, the term *rheobase* defines the minimum voltage acting for an extended period of time that is able to produce a response, whereas the term *chronaxie* is defined as a current that is twice the rheobase and acts for a shorter, finite interval.

Thus, by choosing different voltages as well as pulse durations, a strength-duration curve can be prepared for a muscle (or nerve) which relates these two factors insofar as effective stimulation of the tissue is concerned.

As mentioned earlier, regardless of the intensity of a stimulus, if it acts for too short a period, then no response is elicited. On the other hand, if the intensity is too low, then, regardless of the length of time that the stimulus acts, no response will ensue. These facts are evident from inspection of the curve given in Figure 4.6. The strength-duration curve thus defines the conditions under which a stimulus is effective on a given tissue under specified conditions.

MUSCULAR CONTRACTION IN VITRO

The All-or-None Law. Under given circumstances, a single muscle fiber that is activated by an adequate stimulus contracts maximally. This general statement is known as the *all-or-none law,* and it is implicit in this concept that a weak stimulus does not give rise to a weak contraction, nor does a strong stimulus evoke a stronger response. If a stimulus is able to excite a muscle cell at all, then during the resulting contraction the cell develops the maximum tension of which it is capable. This fact does not imply that a structurally discrete muscle, which is composed of many

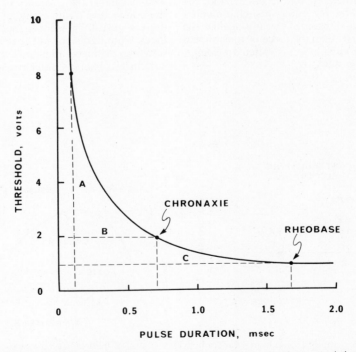

FIG. 4.6. Hypothetical strength-duration curve for the excitation of a tissue. A, The vertical dashed line indicates the minimal time required for a response when the stimulus intensity is at (or above) a certain value, eg, 8 v in the figure. B, The chronaxie is defined as the minimum time in which an electric current exactly twice the rheobase will excite the tissue. C, The rheobase is defined as the minimum voltage necessary to stimulate the tissue. The strength-duration curve for a given tissue under particular experimental conditions is obtained by varying stimulus intensity and pulse duration over a broad range of values for each of these variables.

thousands of individual fibers, is incapable of producing contractions of graded intensity. On the contrary, graded responses of this type are of the utmost physiologic importance, although each fiber within a given muscle always contracts in accordance with the all-or-none law.

The Single Muscle Twitch.   The basic physiology of muscular contraction is well exemplified by analysis of a simple muscle twitch. In the following discussion, it is assumed that both the stimulus and the response are maximal.

Isotonic and Isometric Contractions.   As depicted in Figure 4.7A, a muscle is arranged so that it can be stimulated electrically and the resulting contractions are recorded graphically on a moving strip of paper. The stimulus can be delivered either indirectly through the motor nerve or directly to the surface of the muscle itself.

As shown in Figure 4.7B, a single stimulus causes a single rapid contractile response which follows shortly after the development of an action potential. The time lag between the application of the stimulus and the mechanical response is called the *latent period*. The muscle contracts to a peak tension and then returns quickly to its former resting length, since obviously muscular contraction is a reversible process. This type of contraction is called *isotonic* because the tension on the muscle remains constant as its length changes and mechanical work is performed. If, on the other hand, the muscle is arranged for recording in such a manner that it is unable to shorten physically against the load imposed upon it, then the force or tension that develops following stimulation is called an *isometric contraction* (Fig. 4.8A, B).

There are several fundamental differences between isometric and isotonic contractions. First, when a muscle performs mechanical work during an isotonic contraction, the oxygen consumption — and hence energy utilization — is much greater than when it merely develops tension during an isometric contraction. This increased oxygen consumption during isotonic contraction is called the *Fenn effect*. Second, isometric contraction does not involve a physical shortening of the contractile elements; consequently, no sliding takes place between the thin and thick filaments, and therefore no external work is performed. Third, isotonic contraction involves the movement of a load. Such movement requires that the inertia of the object moved be overcome by an acceleration until a certain velocity is reached. The load also has a certain momentum after the contraction has ended. Consequently, an isotonic contraction generally persists for a longer period of time than does an isometric contraction.

Elastic Elements During Muscular Contraction.   During either an isotonic or an isometric contraction, a portion of the total work expended is utilized to deform or to stretch the noncontractile elements of the muscle. Thus, only part of the energy liberated during a contraction is employed to move a load or to develop tension in a muscle. The contractile elements also must act against the viscosity of the sarcoplasm and its components as well as against the elasticity of the sarcoplasm and the connective tissue elements, including the tendons. Figure 4.9 presents a model to illustrate these facts. Note that some of the resistances to contraction are arranged in series with the contractile structures, whereas other resistances are arranged

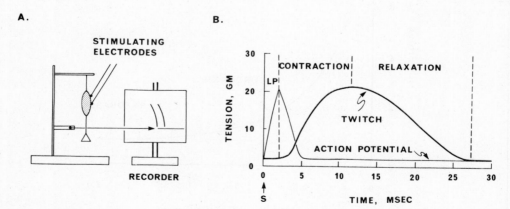

FIG. 4.7.   A. A simple technique used to make recordings (myograms) of the isotonic contractions of skeletal muscle in vitro. One end of the muscle is supported by an arm, and the other end is attached to a pivoted lever that writes upon a rotating drum. The muscle can be stretched prior to stimulation by suspending appropriate weights from the lever (small triangle in figure). Alternatively, muscular contractions can be recorded electronically by use of a strain gauge coupled to a suitable amplifier which drives the pen of an electronic chart recorder. **B.** Components of a single muscle twitch. S = stimulus. LP = latent period. The latent period is defined as the time between application of the stimulus and onset of the mechanical response. The action potential curve has been superimposed upon the twitch curve for comparative purposes.

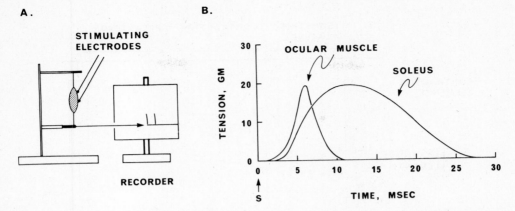

FIG. 4.8. **A.** Technique used for recording isometric contractions of skeletal muscle in vitro. The setup is quite similar to that described in the legend for Figure 4.7A, except that the muscle is restrained from shortening appreciably (black bar on lever). In practice, it is exceedingly difficult from a technical standpoint to secure a "pure" isometric contraction, since some mechanical shortening of the muscle, however slight, is inevitable. **B.** Comparison of the isometric contraction curves obtained from the application of one stimulus to two particular muscles. The marked temporal differences in their onset and duration are obvious. S = stimulus.

in parallel to the contractile elements. The mechanical events that take place during isotonic and isometric contractions in relation to the elastic elements of muscle can be diagrammed similarly (Figs. 4.10, 4.11). Note that during isometric contraction the elastic elements are stretched and tension is applied to the external load even though the muscle as a whole does not shorten physically.

The Active State and Heat Evolution During Muscular Contraction. It must be stressed that both the contraction and the relaxation of muscle are active processes, and that the latter is not merely a passive return to the resting state or length. The contractile elements are said to be in an *active state* as long as they are shortened, and they remain so until nearly the end of contraction (Fig. 4.12). Contraction and relaxation merely reflect an increase and decrease of the active state.

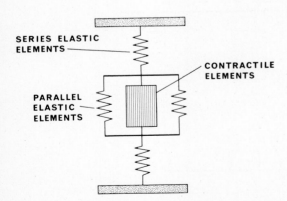

FIG. 4.9. The location of the noncontractile series and parallel elastic elements of skeletal muscle in relation to the contractile elements. (After Florey. *An Introduction to General and Comparative Physiology,* 1966, Saunders.)

The active state develops within a few milliseconds after application of an adequate stimulus, and it is accompanied by a sudden increase in heat production by the muscle. This *heat of activation* is independent of the load and it occurs under isometric as well as isotonic conditions. If the muscle is permitted to shorten (ie, an isotonic contraction ensues following stimulation), then the physical contraction is accompanied by a slowly rising additional increment of heat production called the *heat of contraction* or *heat of shortening.*

The heat of activation is developed concomitantly with a shortening of the contractile elements themselves, whereas the heat of contraction accompanies the performance of external work as the muscle shortens physically against an external load, after the series elastic elements of the muscle have been stretched.

The elevated heat production declines rapidly as the active state decreases. This heat is generated principally as a result of chemical events that accompany contraction; friction produces only an insignificant fraction of the total heat released during muscular contraction.

The heat of activation develops during the latent period and has its onset simultaneously with the shortening of the contractile elements. During a theoretically ideal isometric contraction, however, no mechanical work is performed other than that required initially to stretch the series elastic elements of the muscles. Therefore, practically all of the energy released during an isometric contraction is lost as heat energy. On the other hand, the total energy released during an isotonic contraction is the sum of the energy lost as heat plus the energy converted into mechanical work (p. 103).

Note in Figure 4.12 that the rate of development of maximal tension in a muscle during a twitch lags considerably behind the rate at which the ac-

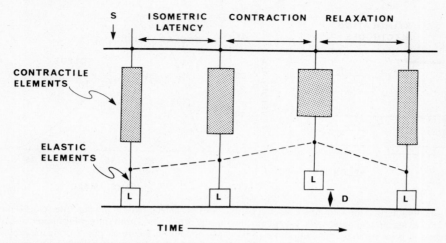

FIG. 4.10. Isotonic muscular contraction from a mechanical standpoint. S = stimulus; L = load; D = maximum distance the muscle has shortened at the peak of contraction (dashed line). During the period of isometric latency following application of the stimulus, the tension within the muscle is increasing as the contractile elements must deform the noncontractile (elastic) elements of the muscle (Fig. 4.9) before any external work can be performed upon the load. (After Florey. *An Introduction to General and Comparative Physiology,* 1966, Saunders.)

tive state develops. In this context, the term *tension* defines the force with which a muscle pulls upon a load.

Relationship Between Sarcomere Length and Force of Muscular Contraction; Free-Loading and After-Loading. As depicted in Figure 4.13, a definite relationship exists between the initial length of the individual sarcomeres in a muscle and the force which develops upon their contraction. Thus, within definite limits, the force of contraction is proportional to the initial fiber length, and maximal tension is developed when the individual sarcomeres are stretched to an optimal length between 2.0 and 2.2 μ (see also p. 1246). If the length of the sarcomere is less (or greater) than this optimal length, then the force of contraction is correspondingly diminished.

The general facts cited above reflect the events taking place at the molecular level. As indicated by the spatial relationship between the actin and myosin filaments shown in Figure 4.13, maximal tension is achieved in the sarcomere when the ends

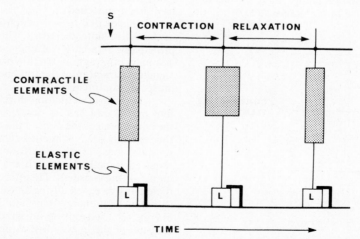

FIG. 4.11. Isometric muscular contraction from a mechanical standpoint. Symbols are the same as in Figure 4.10. The muscle is prevented from shortening physically by the clamp, but note particularly that tension is exerted upon the load nonetheless as the contractile elements stretch the elastic elements (center of figure). (After Florey. *An Introduction to General and Comparative Physiology,* 1966, Saunders.)

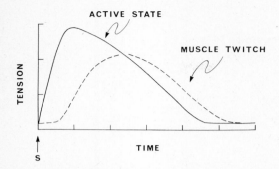

FIG. 4.12. The relationship between the active state of a skeletal muscle during contraction and the tension developed. Note that the active state persists until relaxation is nearly complete, as both contraction as well as relaxation are active processes.

of the actin filaments are closely apposed and neither widely separated nor overlapping each other.

A similar physiologic relationship is found among the contractile elements present in an entire muscle. However, because of the presence of a large amount of connective tissue, as well as the fact that all of the sarcomeres do not necessarily contract simultaneously, the length-tension curve can have a somewhat different form than that presented in Figure 4.13. Nonetheless, each skeletal muscle has an optimal stretched (ie, resting) length from which it can develop a maximum contractile tension (Fig. 4.14).

If a muscle is stretched passively (ie, develops a passive tension) by the application of an external load before it contracts, then it is said to be *free-loaded*. On the other hand, if a muscle is loaded with a weight in such a manner that it cannot lengthen physically (or develop a passive tension in response to the load) before it contracts, then it is said to be *after-loaded*. For example, if an arm is bent at right angles to the body and a 5-kg weight is placed in the palm of the hand, the biceps muscle is now free-loaded. If, however, the arm is supported by the surface of a table in a similar position so that the biceps is relaxed, any weight that is added to the palm of the hand will be supported by the table, and the biceps now is after-loaded. The weight is lifted vertically more easily from the stretched or free-loaded state of the muscle than from the after-loaded state, a fact that the reader can verify readily for himself. This simple demonstration will serve to illustrate the major point made in this section; namely, that muscle fibers contract most effectively when they are at an optimal length. The practical consequences of these facts are evident in everyday life.

Relationship Between Load and Velocity of Contraction. In general, the velocity of a contraction is inversely proportional to the load against which the muscle is performing work (Fig. 4.15). Thus, an unloaded muscle contracts quite rapidly, whereas the velocity decreases markedly as the load increases. Ultimately, a point is reached at which the load equals the maximal tension that the muscle can develop, the velocity of the contraction becomes zero, and the contraction now is isometric.

Summation of Contractions and Tetanus. The normal electric phenomena that precede contractile

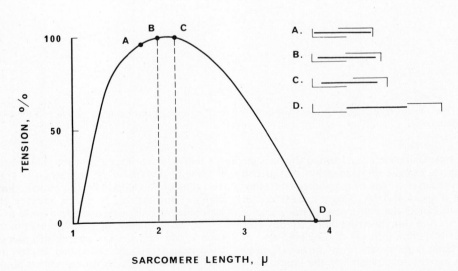

FIG. 4.13. The relationship between sarcomere length and tension developed during contraction of a skeletal muscle. Inset at right side of the figure indicates the sarcomere length as well as the degree of overlap of the thick and thin filaments for the similar points indicated upon the tension-sarcomere length curve (A, B, C, D). Note that peak tension develops when the sarcomere is stretched to between 2.0 and 2.2 μ (B, C), and that the tension declines sharply at either shorter or longer sarcomere lengths. (After Guyton. *Textbook of Medical Physiology*, 4th ed, 1971, Saunders.)

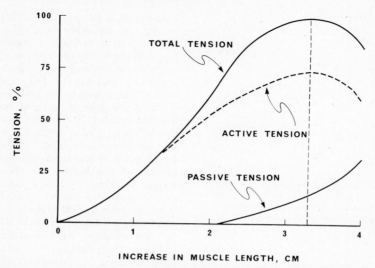

FIG. 4.14 The length-tension curve for an entire skeletal muscle has a somewhat different configuration than that for an individual sarcomere (Fig. 4.13). Nevertheless, each skeletal muscle has an optimal resting length from which it develops a maximal tension during contraction (dashed vertical line).

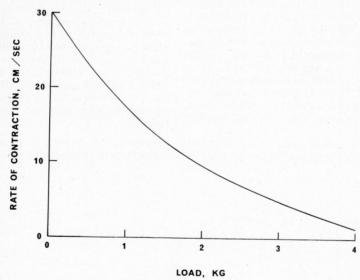

FIG. 4.15. The inverse relationship between velocity or rate of shortening of a skeletal muscle and the external load applied to the muscle.

activity in muscle are such that, during the passage of an action potential, depolarization of the cell momentarily causes a loss of irritability to further stimulation (Figs. 4.4, 4.20). This transient loss of irritability is called the *refractory period,* and it prevents a second stimulus from being effective for a very brief but finite interval of time (see also p. 133). The contractile mechanism, however, does not exhibit a refractory period. If the contraction that is initiated by the first stimulus is not complete before a second stimulus is applied after the refractory period has ended, then a second activation of the contractile machinery is initiated *before* relaxation from the first has taken place. Hence,

the individual contractions fuse. This phenomenon is called *summation of contractions* (Fig. 4.16). The tension developed by the muscle during summated contractions is much greater than that which develops during a single muscle twitch.

If the frequency of stimulation is increased to such a rate that the contractile mechanism is excited repeatedly, the individual contractions fuse into one continuous response called *tetanus* or a *tetanic contraction.** As shown in Figure 4.17,

---

*Physiologic tetanus must not be confused either with the disease of the same name caused by the bacterium Clostridium tetani *or with* tetany *(p. 1072).*

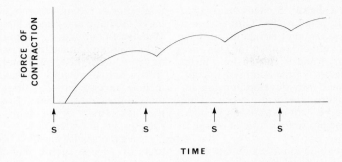

FIG. 4.16. Summation or fusion of individual contractions induced by repeated stimuli (S) applied to a skeletal muscle.

tetanus can be either complete or incomplete. When there are intervals during which an incomplete relaxation occurs between the individual contractions, then an *incomplete tetanus* is present. On the other hand, if no relaxation takes place between contractions due to a high rate of stimulation, a *complete tetanus* is present.

It can be appreciated readily from the discussion presented above that summation of the individual contractile responses of muscle is a smoothly graded effect that is dependent solely upon the rate at which the muscle is stimulated, and the response can vary from a single twitch to a complete tetanus.

The maximal tension produced during a complete tetanus is approximately four times greater than the same muscle can achieve during a single twitch, a fact that has profound physiologic significance (p. 128).

The frequency of stimulation at which summation of individual contractions takes place is governed strictly by the *twitch duration* of the particular muscle under study. For example, assuming that the twitch duration of a muscle is 10 msec, then stimulation at frequencies less than one every 10 msec (100/sec) will produce individual contractile responses punctuated by complete relaxation of the muscle, whereas frequencies higher than 100/sec will cause summation of the contractions.

If the tetanic stimulation of a muscle is continued for a sufficient length of time, *fatigue* sets in, and the muscle is no longer able to remain in a strongly contracted state (Fig. 4.17). Fatigue results from the inability of the metabolic processes and contractile machinery of the muscle to maintain a sustained work output indefinitely and at a high level. Consequently, if the blood supply to an actively contracting muscle in vivo is decreased, the resulting diminution in the supply of oxygen and other nutrients to the tissue, as well as the accumulation of waste metabolites, results in the rapid onset of fatigue. In vivo, however, additional factors are operative, so that fatigue of skeletal muscle is far more complex than these general remarks would indicate (p. 431).

In conclusion to this section, it is essential to stress that skeletal and cardiac muscles exhibit one basic physiologic difference insofar as their refractory periods are concerned. The refractory period of cardiac muscle is so prolonged that normally summation of contractions as well as tetanus *cannot* take place in this tissue — a fact of major physiologic importance (see also p. 143).

The Staircase Phenomenon or Treppe. When a number of maximal stimuli are delivered to an isolated skeletal muscle at a rate that is below the frequency required to tetanize, each succeeding

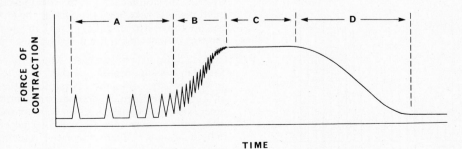

FIG. 4.17. The genesis of tetanus as the frequency of stimulation is increased. A, Single twitches that lead to summation as the rate of stimulation is increased progressively. B, Incomplete tetanus (or summation of individual contractions) is manifest by some relaxation between the individual twitches. C, Complete tetanus is evident when the individual twitches fuse smoothly into a single, sustained contractile response when the rate of stimulation becomes sufficiently high. D, If the stimulation is maintained at the same elevated frequency for a sufficiently long interval, the muscle no longer is able to develop peak tension and fatigue sets in, as indicated by the decline in contractile force with time.

contraction produces a slightly greater tension than the one preceding it (Fig. 4.18). Ultimately, a plateau in tension development is reached after a few contractions. This effect is called the *staircase phenomenon,* or *treppe,* and is caused by a little-understood "warming up" of the actomyosin complex so that its capacity to develop tension increases as the muscle performs work. This phenomenon also can be demonstrated in isolated cardiac muscle, but the physiologic role of treppe is obscure.

The staircase phenomenon as exhibited under particular circumstances by skeletal and cardiac muscle should not be confused with summation of contractions or tetanus.

MUSCULAR CONTRACTION IN VIVO. Simple muscle twitches rarely are encountered in the body, the notable exception being the muscular contraction that closes the eyelids during the act of blinking. Rather, the majority of bodily movements are a smooth combination of summated isotonic and isometric contractions of graded intensity. The general mechanical features of muscular activity in vivo, as well as the mechanisms which underlie this activity, will be considered in this section.

The Motor Unit. Practically every motor nerve or *motoneuron* that leaves the spinal cord branches near its terminations, and each branch of the single nerve fiber innervates a skeletal muscle fiber. All of the muscle fibers that are innervated by the branches of a single nerve fiber are called a *motor unit.* Thus, a motor unit may be considered as the functional unit of a muscle, even as the fascicle may be considered as the anatomic unit.

The number of muscle fibers which constitutes a motor unit can vary considerably. Generally speaking, the motor units of fast-acting muscles whose control is quite precise and whose movements are delicate include only a few fibers, whereas motor units that are responsible for more gross activities are comprised of a much greater number of fibers. For example, the ocular muscles

that control eye movements and the muscles that move the hand and fingers have approximately 5 to 10 muscle fibers per motor unit. Consequently, the ratio of nerve to muscle fibers is quite high in these tissues. In contrast, each motor unit in the large, slowly acting postural muscles of the back may include several hundred fibers, and the nerve-to-muscle-fiber ratio is much lower. On the average, the motor units of the body as a whole contain roughly 200 fibers each.

In general, the muscle fibers of adjacent motor units overlap in such a manner that the separate motor units function cooperatively rather than as independent entities (see also p. 396).

Isotonic and Isometric Contractions in Vivo. As remarked earlier, the muscular contractions encountered in the body generally are a graded combination of isotonic and isometric contractions. For example, in walking or running, the extensor muscles that keep the legs stiff and support the body undergo principally isometric contractions, whereas flexion of the legs is carried out mainly by isotonic contractions.

In this regard, it is of interest to note that specific muscles exhibit an isometric twitch duration in vitro that is consistent with their general functions in vivo.* For example, the ocular muscles have a twitch duration of about 10 msec, and the movement of these muscles is quite rapid in order to maintain fixation of the eyes on various stationary as well as moving objects. The gastrocnemius muscle has an intermediate twitch duration of about 30 msec; this muscle must contract with sufficient rapidity to enable the person to walk, run, or jump. The soleus, on the other hand, which functions principally in relatively slow reactions involving continued support of the body against the force of gravity, has a twitch duration of about 100 msec. A simultaneous combination of isotonic and isometric contractions thus enables an individual to perform an enormous variety of voluntary movements that are delicately graded, insofar as force and duration are concerned.

Since most voluntary activity in vivo is a combination of isotonic and isometric muscular contractions, it is necessary to examine the factors that are operative to produce graded mechanical responses of muscles in the individual as a whole.

Factors Underlying Graded Muscular Activity. Skeletal muscle, unlike cardiac muscle, is not spontaneously contractile. Therefore, the activity of skeletal muscle in the body depends strictly upon innervation by the nervous system, in particular, by motoneurons that arise from the spinal cord. The individual nerve fibers innervate specific motor units of particular muscles. Therefore, a graded

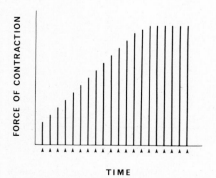

TIME

FIG. 4.18. The staircase phenomenon or treppe. Repeated maximal stimuli delivered to a muscle in vitro (arrowheads along abscissa) evoke a progressively greater contractile response until a plateau is reached.

Isometric contractions are recorded since they permit a more accurate comparison of the individual twitch durations, whereas the rate of isotonic contractions is dependent largely upon muscle shape, size, and load, among other factors.

response depends partially upon the total number of motor units that are activated in a specific muscle at any particular moment in time. Another factor that contributes to the graded response is the rate or frequency at which a motor nerve fires impulses. Lastly, the individual motor units are discharged asynchronously, an effect that produces a fusion of individual muscle fiber contractions so that the entire muscle contracts smoothly.

Recruitment of Motor Units. As shown in Figure 4.19, the force of contraction of a muscle is dependent upon the total number of individual motor units that are contracting simultaneously. The activation of progressively greater numbers of individual motor units in a muscle is called *recruitment, multiple motor unit summation,* or *spatial summation.*

In any muscle, the number of fibers present in each motor unit, as well as the sizes of the individual fibers, can vary considerably. Therefore, one motor unit not only can be larger but, as a consequence, can develop a considerably greater tension than other motor units in the same muscle. Furthermore, the smaller motor units are stimulated far more readily than the larger ones, because their nerve fibers have a greater inherent excitability. Thus, there is a gradation of contractile force during muscular contraction in the body because of the progressive recruitment of larger and larger motor units.

Temporal Summation. As the frequency of impulse discharge along a particular motoneuron increases progressively, the single contractions of the individual muscle fibers in the motor unit fuse or summate until finally the muscle is in a state of tetanic contraction (Fig. 4.16). This phenomenon is called *temporal summation* or *wave summation.*

Asynchronous Firing of Motor Units. Under physiologic conditions, spatial and temporal summation rarely occur independently of one another.

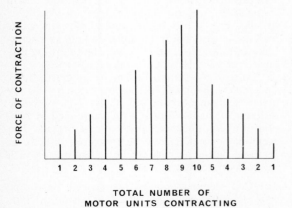

FIG. 4.19. The relationship between the force of contraction and the number of motor units that are active at any particular time. The increase in the force of contraction that develops as progressively more motor units are activated is called recruitment.

The spinal cord has inherent neural mechanisms that can increase the number of individual motor units firing at any time. In addition, these mechanisms can increase the frequency of impulse discharge to particular motor units. Therefore, the individual motor units fire *asynchronously.* That is, while some units are contracting, others are relaxing. This process repeats in such a way that weaker contractions of the entire muscle are smooth, rather than spasmodic.

Skeletal Muscle Tone. In the body, all of the skeletal muscles are normally and continually in a state of very slight partial contraction, even though they are at rest. This weak contraction is called *muscle tone.* Since nerve impulses are essential to the contraction of normal skeletal muscle, centers within the spinal cord are believed to discharge spontaneously the impulses responsible for producing this tone; but these tonic impulses are subject to modification by other nerve impulses that arise within the brain and within muscle spindles situated in the muscle itself. The muscle spindles are *sensory receptors* that are present in practically all skeletal muscles and that can detect with exquisite sensitivity the extent to which that particular muscle is contracted. The structure and function of the muscle spindles are discussed in detail in Chapter 6 (p. 181).

Denervation of Skeletal Muscle. Following section of a motor nerve, the skeletal muscle (or muscles) that it innervates becomes completely flaccid. Ultimately, the denervated muscle atrophies, but prior to this it becomes hyperexcitable and its sensitivity to acetylcholine increases (p. 226).

Shortly after denervation, the individual muscle fibers may undergo spontaneous abnormal contractions, a phenomenon known as *fibrillation.* If the motor nerve regenerates so that the denervated fibers are reinnervated, the fibrillations disappear.

Fibrillation takes place at the microscopic level; hence, it is not visible to the unaided eye. Fibrillation should not be confused with *fasciculation,* which is caused by the pathologic discharge of spinal motoneurons. Fasciculation is manifest by grossly visible, jerky contractions of bundles of muscle fibers which are seen as irregular ripplings of the skin above the affected muscle.

Physiologic Contracture. Extreme fatigue of a skeletal muscle can induce a prolonged shortening known as *physiologic contracture* (p. 230).

It should be mentioned that contracture and fibrillation are two instances in which muscle fibers contract *without* prior activation by nerve impulses.

Functional Types of Skeletal Muscles. The human body contains three distinct kinds of skeletal muscles: red, white, and intermediate. Each of these types of muscle has distinct functional and other characteristics.

Red Muscle. Red muscle fibers contain a high concentration of the pigment myoglobin, hence their common name. In addition, red muscles have less prominent cross-striations than the other types. Functionally, red muscles have a longer latent period than do white muscles, and they contract more slowly; they also are called *slow* or *tonic* muscles. The latter name is derived from the fact that red muscles are found at locations in the body where prolonged contractions are required in order for the muscles to be effective; for example, the muscles that maintain an erect posture are red muscles.

White Muscle. As implied by their name, white muscle fibers have a much lower myoglobin content than do red fibers. Functionally, white muscles have a short twitch duration. Consequently, the white fibers also are called *fast* or *phasic*. White muscles are specialized for the performance of delicate, precise movements that must be executed rapidly. For example, the extraocular muscles as well as certain of the muscles in the hand contain a large proportion of white fibers.

Intermediate Muscle. The intermediate fibers resemble the red fibers in that they contain a high myoglobin content and contract slowly; however, in other respects, they are similar to white fibers. Intermediate fibers are found in high concentration in the soleus muscle.

Every motor unit is composed of only one type of muscle fiber, as described above. Regardless of the mixture of fiber types in any particular muscle, each motoneuron innervates only one kind of muscle fiber.

Some of the physiologic and biochemical characteristics of particular muscles appear to be governed by the nerve itself. As noted earlier, denervation results in atrophy and a flaccid paralysis of skeletal muscle, and it can also result in fibrillation as well as hypersensitivity to acetylcholine. Thus, the motoneurons are said to perform a nutritional or *trophic* function with regard to skeletal muscle. In addition, however, the nerves themselves appear to determine, to a certain extent, the inherent character of a muscle. For example, if the nerves to a tonic and a phasic muscle are severed, crossed, and allowed to regenerate, then the tonic muscle acquires the physiologic characteristics of the phasic muscle and vice versa. It thus appears that the biochemical characteristics of muscles are regulated to some extent by their innervation. In this regard, there is abundant evidence that a number of chemical substances are transported physically along nerves and can enter muscle fibers directly from this source.

The Bioelectric Properties of Skeletal Muscle

Both muscle and nerve fibers are excitable tissues that are specialized for the rapid and repetitive generation of electric impulses, known as *action potentials*. In muscle fibers, the action potentials serve to initiate contractility, whereas in nerve fibers the action potentials underlie the rapid transmission of coded information throughout the body. Qualitatively, the bioelectric events in both muscle and nerve cells responsible for their property of irritability have a similar foundation in the ionic fluxes that take place across their respective membranes upon stimulation. Quantitatively, however, considerable differences exist between muscle and nerve in regard to the magnitude and temporal relationships of the bioelectric events following excitation.

This section will examine the ionic fluxes following stimulation that are responsible for producing an action potential in muscle and nerve; discussion of the transmission or *conduction* of these impulses will be deferred until Chapter 5. The electric events that are induced in skeletal muscle following its stimulation are identical, regardless of the site of application of the stimulus. That is, the stimulus may be applied directly to the entire muscle, to the muscle fiber, or, under physiologic conditions, through the motoneuron and its junction with the fiber.

GENERAL FEATURES OF THE ELECTRIC ACTIVITY IN MUSCLE. Skeletal muscle fibers have a resting membrane potential of about −90 mv. The magnitude of the action potential that develops following stimulation is roughly 120 mv; the duration of the action potential is about 2 to 4 msec; and the rate of propagation or transmission of the electric impulse in muscle is approximately 5 meters/sec. The absolute refractory period of muscle lasts for 1 to 3 msec, and the afterpotentials are relatively prolonged when compared to the other phases of the action potential.

Individual muscle fibers have slightly different thresholds of excitation. However, the electric properties of muscle fibers are so similar that no compound action potential, such as is found in nerve, is evident (p. 173).

THE REST POTENTIAL IN MUSCLE FIBERS. The physicochemical origin of the rest potential in living cells was presented in detail in Chapter 3 (p. 98), but it will be reviewed here since this material is essential to an understanding of excitability in both muscle and nerve; the following discussion is valid for both of these cell types.

At rest, the cell membrane of the muscle fiber develops a potential difference of about −90 mv across a distance of about 80 Å. This potential difference is caused as well as maintained by several physical, chemical, and biologic factors. The membrane potential can be measured directly at rest and during activity of the cell by means of a microelectrode (Fig. 3.35).

The Cell Membrane. The resting cell membrane itself is *semipermeable;* that is, it readily permits passage of certain ions with greater ease than it does others, while still other ions cannot pass this barrier at all. Because of this property, there is an unequal distribution of ions across this interface, in keeping with the Gibbs–Donnan effect.

Table 4.1[a] STEADY-STATE ION DISTRIBUTION AND EQUILIBRIUM POTENTIALS FOR MAMMALIAN SKELETAL MUSCLE FIBERS AND EXTRACELLULAR FLUID

	ION CONCENTRATIONS			
ION	**Extracellular Fluid mM/1**	**Intracellular Fluid mM/1**	$\dfrac{[I_o]}{[I_i]}$	**EQUILIBRIUM POTENTIAL[b] mv**
Cations				
Na^+	145	12	12.1	+66
K^+	4	155	1/39	−97
H^+	3.8×10^{-5}	13×10^{-5}	1/3.4	−32
pH	7.4	6.9	—	—
Miscellaneous	5	—	—	—
Anions				
Cl^-	120	4	30	−90
HCO_3^-	27	8	3.4	−32
A^-	—	155	—	—
Miscellaneous	7	—	—	—

[a]*Data from Ruch and Patton.* Physiology and Biophysics. *Saunders, 1965.*
[b]*The equilibrium potentials for each ion were calculated using the Nernst equation (p. 99). The net membrane potential is −90 mv.*

The Diffusion Potential. In a strictly physicochemical system that contains a nondiffusible anion (A^-) in one compartment, at equilibrium the product of the concentrations* of the diffusible ions on one side of the membrane is equal to the product of the concentrations of the diffusible ions on the other side of the membrane (Gibbs–Donnan effect, p. 25). For example, at equilibrium in such a nonliving system, $[K^+_i] [Cl^-_o] = [K^+_o] [Cl^-_i]$. In this expression, the symbol i stands for the inner compartment, whereas o stands for the outer compartment.

In a living system, on the other hand, the nondiffusible anions within the cell membrane are proteins and other large organic anions. Thus, under rest conditions (ie, at equilibrium), despite the fact that the *products* of the diffusible ions are equal on each side of the membrane, the total *sums* of the ionic activities on each side of the membrane are different. Therefore, $[K^+_i] + [Cl^-_i] + [A^-_i] \neq [K^+_o] + [Cl^-_o]$. Consequently, the total concentration of ions within the cell is somewhat greater than that on the outside. This concentration difference causes only a negligible difference in the osmotic pressure on each side of the membrane; but, more importantly, because of the *net* distribution of the total quantities of ions on each side of the membrane, at equilibrium there is an extremely slight excess of total *anions* within the cell. The basic law of electrochemical neutrality requires that the sum of the anions equal the sum of the cations on the *same side* of the membrane. This situation is attained at equilibrium; however, such a condition is not met *across* the membrane by the diffusible cations,

*Or, more strictly, the activities (p. 20).

since their products differ. Consequently, there will be an excess of cations outside (+) and of anions (−) inside of the cell. The net result is that a small electric potential difference develops across the membrane. This is called the *diffusion potential*.

In living cells the *net* contribution of the diffusion potential that develops from operation of the Gibbs–Donnan effect to the overall rest potential is quite small.

Membrane Permeability to Specific Ions. The steady-state cell membrane is impermeable to intracellular protein and organic ions for all practical purposes, so that these anions (A^-) are sequestered within the cell. This membrane also is slightly permeable to Na^+ and much more permeable to K^+ and Cl^-. In fact, the permeability of the resting cell membrane is about 100 times greater to K^+ than to Na^+.

As shown in Table 4.1, there are marked concentration differences across the membrane for particular ions; therefore, the forces that act upon the ions must be analyzed.

Physicochemical Gradients Across the Cell Membrane and the Origin of the Rest Potential. The data cited in Table 4.1 indicate that considerable differences in the concentrations of various ions are found on opposite sides of the membrane of living cells. In accordance with these data, and as indicated in the table, the equilibrium potential may be calculated and analyzed for each ion in a resting cell (p. 99). However, the cell membrane is *not* at equilibrium; furthermore, this structure contains a coupled sodium-potassium pump that actively extrudes Na^+ from the cell, while K^+ is accumulated simultaneously within the cell in a similar active fashion (p. 101). The marked concentration differences for these ions that develop across the membrane as a result of this metabolic pump, together with the individual diffusion characteristics of the same ions, are responsible for the genesis of the *rest potential* (p. 102).

Briefly, the high concentration of intracellular K^+ generated by the sodium-potassium pump and the relatively high permeability of the membrane to this ion result in an outward diffusion of some K^+, and a net accumulation of positive charges along the outer surface of the membrane takes place. Concomitantly, a slight excess of negative charges is built up along the inner surface of the membrane, and since this structure has an extremely high capacitance and impedance in the physical sense, the charges remain separate. The rest potential develops as a direct consequence of an electrochemical gradient for K^+ across the cell membrane; this gradient, as well as that for Na^+, is maintained solely by the operation of an active transport mechanism, the coupled sodium-potassium pump that is located within the cell membrane.

It must be reiterated that the total *quantities* of K^+ and A^- that are actually responsible for producing the measured potential difference across the cell membrane are extremely minute fractions of the total quantities of these ions that are present in solution within the intra- and extracellular fluids (p. 98). When a living cell is in a resting or *steady state*, there is no *net transport* (either a net influx or net efflux) of any ionic constituent across the membrane (p. 100). The membrane potential is a phenomenon that is associated strictly with the outer and inner surfaces of the membrane itself (p. 98).

During the steady state, ie, in a resting cell, the concentration or diffusion gradient for each ion is balanced exactly by an electric force. In addition to these purely physical forces, the active transport mechanism continually maintains the observed inward concentration gradient for sodium as well as the outward concentration gradient for potassium. Consequently, any factor that interferes with the usual operation of the sodium-potassium pump and/or increases the permeability of the steady-state membrane allows the ions to move readily down their respective chemical gradients. Such is the case when an adequate stimulus is applied to a muscle or nerve fiber and an all-or-none action potential develops as the consequence of the stimulus.

NORMAL FLUCTUATIONS IN THE REST PO-TENTIAL. The steady-state or resting cell membrane can exhibit minute fluctuations in its rest potential. However, these fluctuations are minimized as follows. If the rest potential is decreased by any stimulus, the electric gradient responsible for keeping potassium ions within the cell is decreased, so that more potassium diffuses out of the cell. As the potassium diffuses out, there is a concomitant net influx of chloride ions into the cell; therefore, the overall result of these two processes is an increase in the membrane potential to its pre-

vious resting level. Conversely, if the membrane potential increases (ie, the membrane *hyperpolarizes*), then potassium ions diffuse into the cell whereas chloride ions diffuse out.

These opposing processes take place in all living cells, and thus maintain the rest potential at a constant value within very narrow limits. In muscle and nerve cells, however, any stimulus that reduces the membrane potential to a sufficient extent triggers an action potential which is transmitted spontaneously along the membrane in an all-or-none fashion.

THE ACTION POTENTIAL. An effective stimulus causes a rapid and transient increase in sodium and potassium permeability of the membrane at the point of application of the stimulus. Thus the steady-state permeability of the membrane to these ions is disrupted momentarily as an action potential is generated, and the cell becomes *depolarized*.

As shown in Figure 4.20A, the action potential first is manifest by a relatively slow depolarization from the resting level. When the transmembrane voltage has decreased sufficiently, the rate of discharge of the fiber increases suddenly and markedly, indicated by the arrow in the figure. The point at which this change in the rate of discharge takes place is called the *firing level* or *threshold* for the particular fiber. Subsequently, the membrane potential falls to zero and then rapidly *overshoots* the zero potential level *(isopotential)*. As shown in Figure 4.20A, the total magnitude of the action potential is 120 mv, and the overshoot is 40 mv. During the overshoot, the polarity of the membrane reverses suddenly; that is, the outside of the membrane becomes negative with respect to the inside. After the peak of the action potential has been reached, the membrane *repolarizes,* quickly at first, then more gradually. Ultimately the steady-state resting potential develops once again across the membrane.

The initial steep rise and fall in potential are called the *spike.* The subsequent, more gradual fall in potential toward the resting level is labeled afterpotential 1 in Figure 4.20A. This is followed by a slower, even more gradual hyperpolarization to a negative voltage that actually is below the resting level (afterpotential 2).* The voltage then decreases slowly until the initial rest potential is reached once more, and the membrane now is able to respond to

---

The conventional nomenclature employed in describing the two afterpotentials that occur following the spike is somewhat confusing. The first afterpotential usually is called the negative afterpotential *or afterdepolarization, whereas the second afterpotential is called the* positive afterpotential, *or afterhyperpolarization, even though the latter actually is a more negative voltage than the former, as depicted in Figure 4.20A. For the sake of clarity, the terms afterpotential 1 and afterpotential 2 have been adopted for use in this text; but the reader must keep in mind that these denotations are synonymous with the* negative *and* positive afterpotentials, *respectively.*

A.

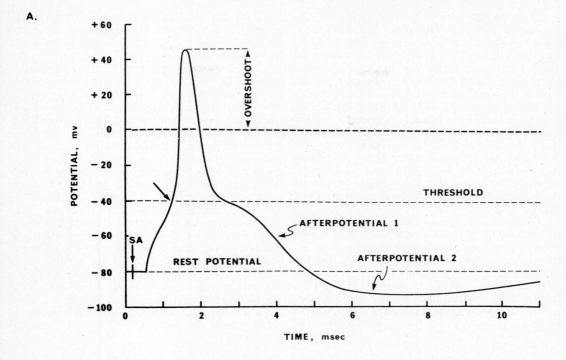

B.

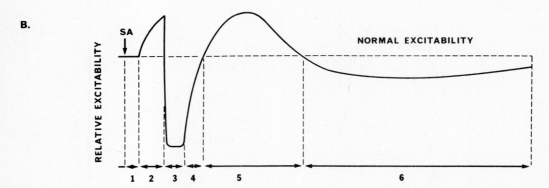

FIG. 4.20. **A.** Action potential of a single fiber recorded by a microelectrode. SA = stimulus artifact. Arrow at left indicates firing level of cell. **B.** The relative excitability of the fiber during the various phases of the action potential shown in **A.** 1. Normal excitability. 2. Increased excitability following application of the stimulus as the transmembrane potential approaches threshold to firing (−40 mv in A). 3. Absolute refractory period develops rapidly as cell fires. 4. Relative refractory period ensues as cell repolarizes. 5. Excitability rises above normal during first afterpotential. 6. Excitability falls below normal during the second afterpotential, then gradually returns to normal as the transmembrane potential returns to its steady state level.

another stimulus. This entire sequence of changes is called the *action potential.*

The electric events described above are observed by means of a microelectrode inserted within a single muscle fiber, as depicted in Figure 3.35B (p. 99), and recorded photographically using a camera attached to the screen of a cathode-ray oscilloscope.

For purposes of comparison, the relative excitability of the fiber during the course of an action potential is depicted in Figure 4.20B. The normal excitability of the cell is shown by the horizontal line, interval 1 in the figure. Note that during the initial depolarization toward the firing level or threshold following application of an adequate stimulus (Fig. 4.20A), the excitability of the cell increases simultaneously (interval 2). As the spike of the action potential develops, however, the excitability drops to a minimum value (interval 3). This period coincides with the *absolute refractory period* of the fiber, and during this interval a second applied stimulus, regardless of its intensity, will not excite the cell again. During interval 4, the cell repolarizes gradually so that a suprathreshold stimulus now may excite the fiber; consequently, this phase is called the *relative refractory period.* During the first afterpotential, the cell hypopolarizes slightly, and once again the excitability becomes greater than normal (interval 5). As the second afterpotential ensues, the excitability falls below the normal level for a second time; this phase coincides with a transient hyperpolarization of the cell beyond the resting level (interval 6). Gradually the transmembrane voltage — and thus the excitability of the cell — returns to its normal resting level.

If the initial stimulus is inadequate to excite the cell sufficiently to produce an all-or-none action potential (ie, it is subthreshold), there is some depolarization of the fiber toward threshold, but the transmembrane voltage rapidly returns to the resting level; that is, −80 mv, as shown in Figure 4.21. Consequently, the cell does not fire. On the other hand, if an electric current is passed into a fiber so that the transmembrane voltage is driven below the

level of the normal rest potential, then the cell membrane is said to be *hyperpolarized;* hence, the irritability is decreased and the membrane becomes more stable electrically (cf. p. 147).

As shown by the arrows in the upper part of Figure 4.22, the current flows from cathode to anode in the external circuit when an excitable tissue is stimulated electrically, and from anode to cathode within the tissue itself under the same conditions. As depicted by the curve and arrows in the lower part of the figure, the anodal current increases the transmembrane voltage (ie, causes hyperpolarization of the membrane) so that the irritability of the fiber is decreased in the vicinity of the anode. This phenomenon of decreased irritability under the anode is called *anelectrotonus.* Concomitantly, under the cathode, the rest potential is decreased toward threshold and the excitability of the fiber is increased in the phenomenon known as *catelectrotonus.* Therefore, by the application of anodal or cathodal currents to an excitable tissue, the rest potential can be changed artificially so as to increase (hyperpolarize) or decrease (hypopolarize) the electric stability of the membrane. In brief, anodal currents are *inward* and *hyperpolar-*

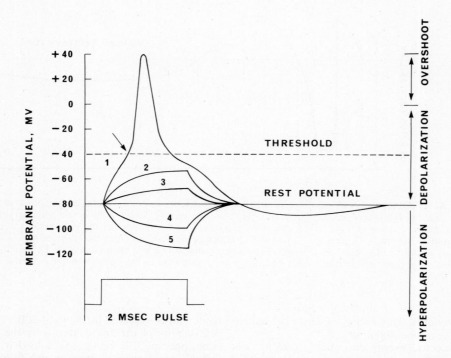

FIG. 4.21. Effect of depolarizing and hyperpolarizing currents upon the transmembrane potential of an excitable cell. In all instances the pulse duration is the same. 1. The intensity of the depolarizing current was sufficient to evoke an all-or-none action potential when the membrane potential reached threshold to firing (arrow). 2. and 3. The intensity of the depolarizing current was insufficient to drive the rest potential to the firing level (−40 mv). Therefore when the stimulus was turned off, the membrane spontaneously repolarized to its previous rest level (−80 mv). 4. and 5. Hyperpolarizing currents drive the membrane potential below its resting level and decrease still further the irritability of the fiber; ie, the membrane becomes more stable electrically. See also Figure 4.22.

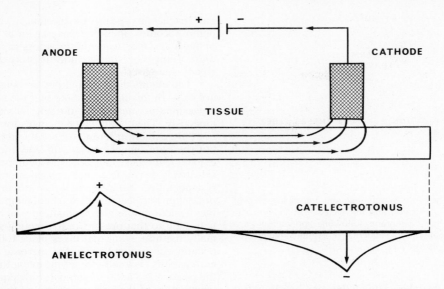

FIG. 4.22. Anelectrotonus and catelectrotonus. As shown in the upper part of this figure, anodal currents are inward and hyperpolarize the membrane. Cathodal currents, on the other hand, are outward and depolarize the membrane. As indicated in the lower part of this figure, anelectrotonus increases the transmembrane potential (upward sweep of curve and arrow), whereas catelectrotonus produces the opposite effect. The centers of the electrodes are indicated by plus and minus signs. Note also that the electrotonic potentials decrease exponentially with distance from the electrodes.

ize, whereas cathodal currents are *outward* and *hypopolarize.*

IONIC FLUXES DURING THE ACTION POTENTIAL. The ionic fluxes that underlie the action potential itself result directly from alterations in the permeability of the membrane of an excitable cell that are caused by an adequate stimulus. It will be recalled that the cell membrane at rest has a low permeability to sodium ions, whereas permeability to potassium ions is relatively high. Following the application of a threshold stimulus, this steady-state pattern changes dramatically, rapidly, and reversibly as follows.

A threshold stimulus causes a marked change in the permeability of the steady-state membrane so that sodium ions are now able to pass rapidly down their electrochemical gradient into the cell, thus neutralizing the resting transmembrane potential. In other words, the slight partial depolarization of the cell caused by a threshold stimulus becomes amplified spontaneously. In turn, the *sodium conductance,* which is a measure of the membrane permeability to sodium, is dependent upon the magnitude of the membrane potential. Consequently, at the normal rest potential, the sodium conductance is quite low. However, for unknown reasons, it increases markedly when the rest potential is decreased toward threshold. Hence, sodium ions enter the fiber at an increased rate, and, since they carry a positive charge across the membrane, the initial decrease in the rest potential caused by the stimulus is augmented. Concomitantly, sodium permeability increases still further as the transmembrane voltage falls, so that the process is self-amplifying (Fig. 4.23). The activity of the sodium pump apparently ceases during this process, even as the "channels" or "gates" through which the sodium can diffuse through the membrane into the cytoplasm open. Thus, the membrane depolarization itself reinforces the increased sodium conductance as excitation proceeds.

As is evident from inspection of Figure 4.20A, sodium ions do not pass into the cell merely until the transmembrane voltage reaches zero, or isopotential. Rather, there is an overshoot, so that the interior of the fiber becomes momentarily positive with respect to the exterior. This overshoot is caused by a permeability change of the membrane in such a way that a "negative resistance" is offered to the sodium ions throughout a definite range of their voltage/current relationship. Therefore, an excess of sodium ions is built up on the intracellular membrane surface, and the polarization of the membrane reverses in a rapidly transient fashion.

During the explosive, self-amplifying events described above, the membrane becomes nonexcitable, and this interval coincides with the absolute refractory period.

The voltage changes that occur during an action potential are reversible; that is, the membrane ultimately repolarizes to its resting value. The factors that are responsible for the regeneration of the steady-state membrane potential following depolarization are the conductances of potassium and chloride ions. The firing level or threshold of the membrane is reached following application of a stimulus when the rate of sodium influx is sufficient to balance the rate of potassium efflux as well as chloride influx. Thus, the membrane is put into an

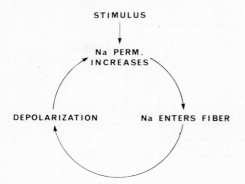

FIG. 4.23. The cyclic, progressive, self-amplifying effect of membrane deplarization and the increase in sodium permeability of the membrane following application of a threshold stimulus. This sequence of events sometimes is called the Hodgkin cycle.

electrically unstable state by the stimulus. If, at this point, the sodium current decreases slightly, the activity will decline and no all-or-none action potential will develop as the membrane repolarizes to its steady-state level. This is the cause of a *local* or *subthreshold response* (Fig. 4.21). On the other hand, if the stimulus is adequate and the threshold is exceeded, then the increase in the inward sodium current will cause the change in potential together with the rise in sodium conductance to drive the membrane potential rapidly toward the sodium equilibrium potential (Table 4.1). From a theoretical standpoint, the sodium influx could continue until the cytoplasmic surface of the membrane reached about +66 mv with respect to the exterior. This does not take place in excitable cells, however, because the increased sodium permeability is only a very brief and transient effect and also because there is a rapid increase in potassium conductance which begins to operate as the action poten-

tial reaches its peak. Furthermore, the sodium pump becomes operative once again at about this point, and collectively these factors rapidly accelerate return of the membrane potential to its steady-state level.

The events discussed above are shown in Figure 4.24. In this diagram, note that the sodium and potassium permeabilities do not increase simultaneously. The conductance changes for these ions not only exhibit different temporal characteristics, but they are out of phase with one another. It is evident from these facts that the membrane still has selectively permeable characteristics even when it is depolarized. Note also that the ratio of sodium to potassium permeability is reversed temporarily during the action potential, a situation which contrasts markedly to that present in the resting state. Normally the first event that takes place following application of a stimulus, ie, the increase in sodium permeability, is a self-amplifying event that leads to the peak of the action potential. Subsequently, the inactivation of the sodium pump and the increased potassium permeability are self-limiting events. The accelerated efflux of potassium ions down their chemical gradient produces a rapid decline of the potential within the cell to its steady-state level, and this, in turn, leads automatically to a restoration of the original ionic permeabilities. Therefore, the increase in potassium permeability during an action potential is a *negative feedback* process which turns itself off as it progresses. The cycle of activity is terminated rapidly and the normal excitability of the fiber is restored within a few milliseconds after application of a stimulus.

The membrane potential is not restored quickly, as described above, in all instances. In particular, the action potential duration of cardiac muscle fibers is quite prolonged, since the increase of potassium permeability shown in Figure 4.24 does not occur and sodium conductance is not turned off completely (Fig. 10.20, p. 596). There-

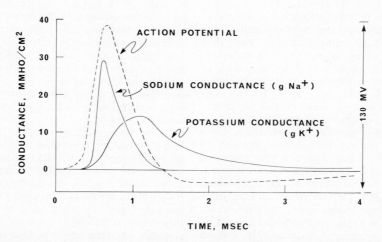

FIG. 4.24. The relationship between the action potential and the concomitant changes in sodium and potassium conductance.

fore, a long plateau appears on the cardiac fiber action potential; functionally, this permits the stimulus to act for a sufficient length of time to permit systolic contraction of all of the fibers within the heart.

ELECTROMYOGRAMS. Electromyography is the procedure whereby the activation of motor units in experimental animals and humans is studied. The electric activity is monitored by applying small metal electrodes to the skin over the muscle under study or by inserting needle electrodes directly into the muscle itself. The voltages produced by the muscle during its activity are amplified and then recorded by an ink stylus on a strip chart recorder or visualized on a cathode-ray oscilloscope tube and photographed. The records so obtained are called *electromyograms*.

It should be noted that the voltages obtained by electromyography are *surface potentials* rather than transmembrane potentials, as discussed on page 394.

Energy Metabolism and Muscular Contraction

The direct source of energy for muscular contraction is the high-energy compound ATP (p. 104). This compound is synthesized during the intermediate metabolism of carbohydrates and lipids, as discussed in detail in Chapter 14 (p. 907).

The overall metabolic reactions that produce ATP in muscle are summarized in Table 4.2. When each mole of ATP is hydrolyzed in vivo to ADP and phosphoric acid, it yields up to 12 Cal of utilizable free energy (reaction 1) that are used directly for muscular contraction and other processes within the fibers. Note also in this table that the aerobic oxidation of glucose, which is the principal normal source of energy for muscular contraction, yields considerably more energy in the form of ATP synthesized than does anaerobic glycolysis of an equivalent quantity of this carbohydrate.

Table 4.2 SOURCES OF ENERGY FOR MUSCULAR CONTRACTION

1. $ATP + H_2O \rightarrow ADP + H_3PO_4 + {\sim}12$ Cal

2. Glucose + 2ATP $\xrightarrow{\text{aerobic}}$ $6CO_2 + 6H_2O + 40ATP$

3. Glycogen + ATP $\xrightarrow{\text{aerobic}}$ $6CO_2 + 6H_2O + 40$ ATP

4. Glucose + 2ATP $\xrightarrow{\text{anaerobic}}$ 2 lactic acid + 4ATP

5. Glycogen + ATP $\xrightarrow{\text{anaerobic}}$ 2 lactic acid + 4ATP

6. Free fatty acids + ADP $\xrightarrow{\text{aerobic}}$ $CO_2 + H_2O + ATP$[a]

7. Phosphocreatine + ADP $\rightleftharpoons$ Creatine + ATP

[a]*The quantity of ATP synthesized during the oxidation of free fatty acids depends upon the length of the carbon chain of the particular acid undergoing oxidation.*

In muscle, there is a reservoir of another high-energy phosphate compound called *phosphocreatine,* which can yield its free energy during hydrolysis for the synthesis of ATP from ADP (reaction 7). At rest, some of the ATP that normally is present in muscle transfers its phosphate bond energy to creatine. Thus, a reserve supply of phosphocreatine is synthesized during rest so that during exercise the free energy liberated by the hydrolysis of this compound can be used to produce ATP from ADP and muscular contraction will be able to continue.

CARBOHYDRATE METABOLISM. The metabolic degradation (catabolism) of carbohydrate to carbon dioxide and water provides a major proportion of the total energy required by muscle for the synthesis of ATP and phosphocreatine. Therefore, a general summary of this process will be included here.

Glucose circulating in the blood enters body cells in general and muscle cells in particular, where it is degraded in a number of sequential chemical stages to pyruvic acid (Fig. 14.25, p. 938). Glycogen (p. 66), a polysaccharide that is especially abundant in muscle and hepatic cells, provides another important source of intracellular glucose.

In the presence of an adequate supply of molecular oxygen, the pyruvate formed by the catabolism of glucose enters the citric acid cycle (Fig. 14.29, p. 948), where it is metabolized to carbon dioxide and water. The enzymes of the respiratory chain are linked intimately and inseparably to this aerobic metabolic process (Fig. 14.20, p. 932). The net effect of the aerobic metabolism of glucose, therefore, is the gradual release of free energy, a large part of which is trapped in a useful form through the synthesis of high-energy phosphate bonds in ATP.

In contrast to the aerobic metabolism of glucose to carbon dioxide and water (Table 4.2, reactions 2 and 3), if the supply of molecular oxygen is insufficient, the pyruvic acid formed by the degradation of glucose does not enter the citric acid cycle. Rather, the pyruvate is reduced to lactic acid in a process known as *anaerobic glycolysis* (Table 4.2, reactions 4 and 5; Fig. 14.25, p. 938). Anaerobic glycolysis, like the aerobic oxidation of glucose, is associated with the net production of ATP. However, considerably fewer high-energy phosphate bonds are synthesized under anaerobic conditions than during the aerobic oxidation of glucose. Nonetheless, anaerobic glycolysis provides an important physiologic mechanism whereby muscular contraction in vivo can proceed for limited periods of time in the absence of oxygen (p. 1184).

FATTY ACID METABOLISM. Skeletal muscle also is able to extract free fatty acids (FFAs) from the bloodstream and oxidize them to carbon dioxide and water (p. 963). In fact, FFAs appear to be the major substrates for energy production in skeletal muscle during periods of rest as well as during recovery from vigorous muscular activity (Fig. 14.35, p. 964).

OXYGEN DEBT. The oxygen debt mechanism is considered in detail in Chapter 16 in relation to muscular exercise in vivo (p. 1184). Briefly, however, the salient features of this process may be reviewed as follows.

During the performance of muscular work, the blood vessels in the skeletal muscles dilate so that the blood flow to the contractile tissues increases up to a point. That is, within limits, oxygen consumption is proportional to the energy expended in the performance of mechanical work. Under such conditions, all of the energy requirements are met by the aerobic metabolic processes described above.

If, however, the level of muscular exertion becomes so great that the rate of energy utilization exceeds the rate of aerobic synthesis of ATP, then the phosphocreatine reservoir is used for resynthesis of ATP and ADP. As the available phosphocreatine supply is depleted, more phosphocreatine, hence ATP, is resynthesized through the anaerobic glycolysis of glucose to lactic acid. Eventually, the lactate accumulates at such levels in the exercising muscles that the pH declines and the enzymes responsible for the contractile processes are inhibited. However, for short periods of time, this anaerobic mechanism enables the individual to perform work at levels far beyond the capacity of the aerobic metabolic system to provide energy.

When the period of high-energy expenditure has ended, excess oxygen is consumed by the individual in order to oxidize the excess lactic acid as well as to replenish the ATP and phosphocreatine supplies. The quantity of excess oxygen that is required during the period of recovery from intense muscular exertion directly reflects the energy demands of the body above the capacity of the aerobic mechanisms to provide energy; this is called the *oxygen debt* that was incurred during the exercise.

HEAT PRODUCTION. From a thermodynamic standpoint, the total energy input to a muscle must equal its total energy output. The energy output is manifest as the mechanical work performed by the muscle, expressed as efficiency, as high-energy phosphate bonds synthesized, and as heat.

The net mechanical efficiency* developed by a skeletal muscle may range from 0 percent (during isometric contraction) up to about 50 percent (during isotonic contraction). Storage of chemical energy by the synthesis of high-energy bonds is a small fraction of the overall utilization of energy. Therefore, heat production represents a considerable proportion of the total energy output of a muscle. The heat evolved by a muscle under various conditions may be defined as follows.

Efficiency = mechanical work performed/total energy expenditure × 100.

Resting Heat. During periods of rest, the muscle continuously gives off heat as a consequence of exothermic basal metabolic processes; this is called the *resting heat* of the muscle.

Initial Heat. During contraction, an excess quantity of heat is given off above the resting heat. This extra energy release is called the *initial heat*. The initial heat is divided into the *activation heat* (or heat of activation) given off during contraction, and the *shortening heat* (or heat of shortening) liberated in quantitative proportion to the distance through which the muscle shortens. The shortening heat apparently is caused by the structural changes that accompany the physical contraction of muscle.

Recovery Heat. Following contraction of a muscle, heat production above the resting level may continue for up to 30 minutes; this is called the *recovery heat*. The recovery heat is yielded by exothermic metabolic reactions that return the muscle to its precontraction state. The recovery heat is about equal to the initial heat. In other words, the heat liberated during recovery is approximately the same as that released during contraction.

Relaxation Heat. Following an isotonic contraction, the restoration of a muscle to its resting length causes the evolution of excess heat above the recovery heat. This extra heat is called the *relaxation heat*. It is a reflection of the work that must be performed upon the muscle in order to return it to its precontraction length.

The Mechanism of Muscular Contraction and Excitation–Contraction Coupling

With the data currently available, it is possible to construct in some detail the events that take place at the molecular level during muscular contraction and to integrate these events with the process of excitation of the contractile machinery.

MUSCULAR CONTRACTION. The morphologic features essential to an understanding of muscular contraction may be reviewed briefly prior to a discussion of the contractile mechanism.

The major contractile elements of muscle are composed of two distinct kinds of proteins. Each of these proteins, in turn, is organized into elongated structures, the *thick filaments* and the *thin filaments* (Figs. 4.3, 4.25).

The thick filaments are composed chiefly of the protein *myosin,* and each of these filaments resembles a long rod with two globular heads at one end. The myosin molecules are so organized within each thick filament that the heads are oriented to face toward the Z bands of each sarcomere. The rodlike portions of the myosin molecules are arranged so that no projections are present in the center of each thick filament. In electron photomicrographs, the heads of the myosin molecules appear as crossbridges between the thick and thin filaments. These

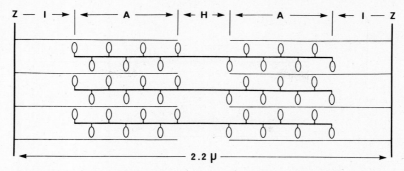

FIG. 4.25. The relationship between the thick and thin filaments of a resting sarcomere, which provide the molecular basis for muscular contraction. The heavy meromyosin heads which form cross bridges linking the myosin to the actin filaments during contraction are depicted as ellipses. The dimensions of the thick and thin filaments as well as the pertinent distances between these molecular components of the contractile mechanism are given in the text.

heads or cross-bridges are composed of heavy meromyosin, a protein subunit of the myosin molecule that has the ability to react with actin as well as to exert ATPase activity. Both of these properties are essential to muscular contraction. The rodlike portion of each myosin molecule is an important structural component. Each cigar-shaped thick filament is about 1.5 μ in length and 160 Å in diameter, and is made up of several hundred individual myosin molecules, from which the heads project at intervals in radially disposed groups of three (Fig. 4.25).

The thin filaments are composed of about 300 to 400 globular *actin* molecules polymerized into two long chains (Fig. 4.3B) roughly 1 μ in length and 50 to 70 Å in diameter. Each thin filament therefore consists of two single actin filaments that are twisted into a long helical structure like two intertwined strands of beads, each bead representing a single actin molecule. In living muscle, one end of each actin filament is attached to a Z membrane (Fig. 4.25).

The thin filaments also contain small quantities of two additional proteins, *troponin* and *tropomyosin,* whose functional roles in the contractile process will be discussed subsequently (p. 141). Quantitatively, however, of the three proteins, actin is present in the highest concentration in the thin filaments.

As is shown in Figures 4.3 and 4.25, the thick and thin filaments are interdigitated between one another in a parallel fashion along the fiber axis, so that in a resting myofilament the H band (Fig. 4.1) represents the smooth regions of the thick filaments. In Figure 4.25, it will be noted that the cross-bridges formed by the heads of the individual myosin molecules form the only link between the thin and thick filaments. Note also that the thick and thin filaments are separated where they overlap by a small distance (about 100 to 200 Å); therefore, these filaments are able to slide freely past one another. This sliding motion is the basis of the shortening that takes place during muscular contraction.

The key to the contractile mechanism of skeletal muscle is to be found in the alternating regular pattern of thick and thin filaments within each myofilament of a muscle fiber, and, in particular, the heavy meromyosin heads or cross-bridges on the myosin molecules that can form rapidly reversible cross-linkages between the individual myosin and actin molecules. Since the thick filaments are spaced at regular intervals between the thin filaments, each kind of filament can glide readily past the other, producing the mechanical shortening of contraction. The force that drives the sliding motion is derived from the cross-bridges on the thick filaments (Fig. 4.26). As shown in this figure, the cross-bridges are not static and rigid entities; rather, they are believed to be quite flexible and able to swivel or pivot freely at their point of attachment to the thick filament. Therefore, during contraction, the heads of the myosin molecules attach to the thin filament at a definite angle (Fig. 4.26B) and then pivot to a different angle, thereby drawing the thin filaments past the thick filaments (Fig. 4.26C).

As shown, the cross-bridges on opposite sides of the bare region of the thick filaments (ie, the H band) are polarized so that they pivot in opposite directions. Thus, the actin filaments are drawn together in a greater overlap than would be possible otherwise; in this way, the distance between the two Z bands decreases, and the sarcomere is shortened.

Each cross-bridge must act in a definite cyclic fashion in order to effect a meaningful shortening of the individual sarcomere. That is, the cross-bridge must attach at a particular angle to the thin filament, pivot through a specified distance, detach from the thin filament, and then reattach at a point farther along the actin filament. A single cycle such as that just described produces a movement of approximately 100 Å. If all of the cross-bridges in a muscle fiber completed only one cycle during a contraction, the muscle would be shortened about 1 percent of its resting length. Since muscles are able to shorten many times this minimal value

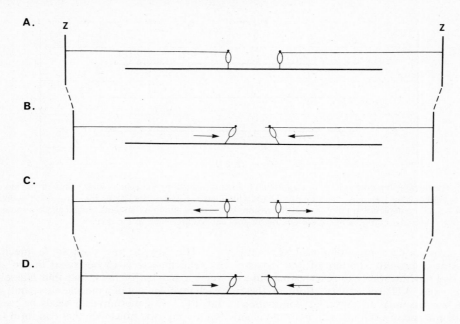

FIG. 4.26.   The molecular mechanism of muscular contraction. **A.** Heavy meromyosin heads projecting from the thick filaments attach to receptor sites on the thin filaments, as indicated by dots. **B.** The cross bridges thus formed on each side of the H band (Fig. 4.25) then swivel toward each other (arrows), thereby pulling the thin filaments past the thick filament. **C.** The cross bridges become detached from the actin filaments, swivel in the opposite direction (arrows), and then reattach to the next adjacent receptor sites on the actin filaments. **D.** The cross bridges swivel once again, and the entire process repeats over and over as the entire sarcomere shortens (dashed lines at both edges of figure).

(up to between 25 and 35 percent of their resting length), it is obvious that, during a maximal isotonic contraction, the cyclic formation of crossbridges must be repeated many times in rapid succession.

Figure 4.13 indicates that the maximum force of contraction is developed in a muscle when the resting sarcomere length is between 2.0 and 2.2 μ. During isometric or isotonic contraction, the force developed by the individual myofilaments themselves is proportional to the number of cross-linkages that form between the actin and myosin molecules. When a muscle is stretched so that the sarcomere length exceeds 2.2 μ,* the overlap between the thick and thin filaments is reduced, and the number of cross-linkages that can form is reduced to a similar extent. On the other hand, when the sarcomere length is less than its resting length, the actin filaments overlap to a greater extent at the outset of the contraction, and once again the total number of possible cross-linkages that can develop is reduced; hence, the contraction is less forceful.

The mechanism discussed above generally is called the *sliding filament* or *ratchet theory* of muscular contraction.

If a muscle is stretched to about three times its resting length (so that the length of the individual sarcomere exceeds about 6.6 μ), the muscle ruptures.

EXCITATION–CONTRACTION COUPLING. All of the myofibrils throughout a single muscle fiber are activated almost instantaneously following a localized depolarization of the sarcolemma at the motor end plate (or myoneural junction, p. 210) by an impulse from a motor nerve. The key morphologic unit involved in the rapid spread of excitation through a muscle fiber is the transverse tubule or T-tubule shown in Figure 4.2. As indicated in the footnote to this figure, the membrane of the T-tubule is continuous with the sarcolemma, a fact of considerable importance to the rapid spread of the electric excitation throughout the muscle fiber, and hence to excitation–contraction coupling.

It has long been known that calcium ions play an essential role in the contractile process. In muscle fibers at rest, most of the calcium present is sequestered within the membranes of the sarcoplasmic reticulum (Fig. 4.2). When an action potential passes down the sarcolemma, it also depolarizes the T-system, thereby causing a transient release of calcium ions from the adjacent cisternae and the sarcoplasmic reticulum proper. Presumably, this release is caused by altering the permeability of this membrane-limited organelle to calcium and/or by shutting off the active pump that transports calcium into the reticulum during rest. Hence, calcium ions both initiate and regulate the contractile process in vivo.

How are calcium ions responsible for accomplishing this regulatory process? The ultimate source of energy for muscular contraction is derived from the hydrolysis of ATP to ADP. The free energy liberated during this process causes the cross-bridges to attach to the actin filaments and then to pivot, thereby causing the thin filaments to slide past the thick filaments. The release of calcium ions from the terminal cisternae of the sarcoplasmic reticulum activates the ATPase present in the heavy meromyosin heads of the myosin molecules located on the thick filaments. Therefore, ATP is split to ADP with the concomitant release of free energy, which in turn powers attachment and swiveling movements of the cross-linkages.

Following contraction, the calcium is pumped rapidly out of the sarcoplasm by an active transport mechanism situated within the membranes of the sarcoplasmic reticulum. The calcium removal inhibits the further hydrolysis of ATP; this, in turn, stops the cyclic activity of the cross-bridges so that the muscle relaxes. In this regard, it should be mentioned that the process of relaxation is not entirely passive, and that there is a considerable body of experimental evidence which indicates that one or more "relaxing factors" are present within the sarcoplasmic reticulum. For example, in vitro, fragments of the sarcoplasmic reticulum contain vesicles that rapidly and reversibly bind calcium in the presence of ATP.

The release of calcium following muscular excitation as well as its subsequent removal from the sarcoplasm following contraction are extremely fast processes, each of which takes place in a few milliseconds. Thus, calcium ion concentration within the muscle fiber provides the key link between excitation and contraction. The next important consideration, however, is: How do calcium ions confer sensitivity upon the actin filaments so that myosin can form reversible cross-bridges with actin during contraction? The answer to this question lies in the proteins troponin and tropomyosin that are normal components of each actin filament.

Experimentally it has been demonstrated in vitro that, in the absence of troponin and tropomyosin, the contractile process is no longer regulated by calcium; therefore, ATP hydrolysis continues unabated until this compound is used up.

It is relevant at this point to examine further the steps that take place during the hydrolysis of ATP in muscle, and then to relate these steps to the roles of troponin and tropomyosin in producing sensitivity of the system to calcium itself. The chemical reactions that result in hydrolysis of ATP occur in four steps, and all of these take place on the heavy meromyosin heads of the myosin molecules extending from the thick filaments (Fig. 4.27).

First, an ATP molecule binds to a receptor site upon the surface of the heavy meromyosin. In normal muscle, this binding tendency is so great that

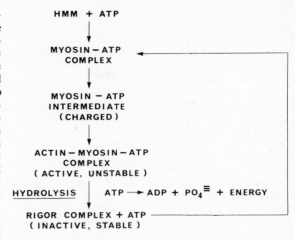

FIG. 4.27. Sequence of the biochemical events during muscular contraction. HMM = heavy meromyosin. (Data from Murray and Weber. *Sci Am* 230:58, 1974.)

practically every myosin head has an ATP molecule bound to it.

Second, the ATP–meromyosin complex is converted into an electrically charged intermediate form which has a great tendency to bind immediately to an actin molecule on the thin filament.

Third, the charged complex is degraded rapidly so that ATP is hydrolyzed and free energy is released as soon as the ATP–meromyosin complex becomes attached to an actin molecule. The charged complex, when in an isolated form — ie, without the presence of actin in vitro — is fairly stable. The free energy that is liberated during this step causes a pivoting of the cross-bridge to a new angle, and thus the thick filament is drawn along the thin filament and the muscle contracts.

Fourth, the cross-bridge is detached from the actin molecule, an event that can *only* take place *after* another ATP molecule has bound to the actin–myosin complex. The resulting complex of actin, myosin, and ATP now dissociates rapidly so that a free actin molecule and an uncharged myosin–ATP complex are formed once again and the cycle can repeat.

It is evident from the foregoing presentation that during the course of ATP hydrolysis, two types of actin–myosin complex are formed. One of these complexes has a high energy, the other a lower energy. The high-energy or *active complex* is produced when the intermediate, charged myosin–ATP complex combines with an actin molecule. This active complex is a very transient entity; within a few milliseconds, its component ATP undergoes hydrolysis to ADP and releases its energy, and the low-energy actin–myosin complex is formed. This low-energy actin–myosin complex remains stable until another molecule of ATP reacts with it to form a new myosin–ATP complex. The low-energy actin–myosin complex is so stable that unless another ATP molecule is immediately available for the cycle to continue, the actin remains permanently bound to the myosin, and these proteins will not dissociate but will remain joined together indefinitely. The extreme stability of the low-energy actin–myosin complex is responsible for the rigidity of the skeletal muscles in the state of *rigor mortis* that develops following death. As ATP gradually disappears from the muscles, progressively greater numbers of myosin molecules become irreversibly bound to actin molecules so that the thick and thin filaments become locked together until gross rigor of the muscle ensues.*

All of the processes described above are regulated by calcium ions. It has been demonstrated clearly that the calcium-sensitive step in the cycle outlined above, and shown in Figure 4.27, is the formation of a link between the charged form of myosin–ATP and the actin molecule (step two). Therefore, the formation of an active or charged myosin–actin–ATP complex is the stage that is controlled by calcium ions. Tropomyosin and troponin are essential for calcium to exert its regulatory function, and these proteins undoubtedly prevent the formation of the high-energy myosin–ATP–actin complex in the absence of calcium. This effect would inhibit contractile processes during periods of relaxation in a muscle. Furthermore, it is known that troponin actively binds calcium, and this, in turn, activates or "turns on" the thin filament so that an actin molecule can form a complex with myosin. However, low-energy-state actin–myosin complexes can be formed in the absence as well as in the presence of calcium; consequently, the formation of rigor complexes is independent of calcium ions. ATP itself also appears to exert a graded control on the effects of calcium ions during the contractile process, as well as upon the number of low-energy or rigor complexes that are present at any particular time. Thus, at high ATP concentrations, the regulation of muscular contractility by calcium is normal, whereas at lower ATP levels the loss of such control by calcium becomes progressively greater; ultimately, when ATP levels are very low and a large proportion of low-energy-state actin–myosin complexes are present, calcium has no effect on contractile activity. Since ATP acts together with myosin, and not upon tropomyosin or troponin directly, the effects of ATP upon the calcium regulation of contractile activity must be exerted indirectly. Experimental evidence indicates that the level of activity within the actin filaments is graded, and that the tropomyosin component serves to coordinate this activity within the actin chains of the thin filament. Troponin, on the other hand, apparently initiates the "information" carried throughout the actin filament through its sensitivity to calcium ions. Thus, each troponin molecule, together with its associated tropomyosin

molecule, serves to coordinate the activity of a segment of each thin filament consisting of seven actin molecules (Fig. 4.25).

Cardiac Muscle

The morphology and physiologic properties of cardiac muscle are dealt with fully in Chapter 10 (p. 577); however, a few outstanding features of this contractile tissue will be presented here in order to compare and contrast them with the similar properties of skeletal muscle.

STRUCTURE. The microscopically visible striations found in cardiac muscle fibers are similar to those encountered in skeletal muscle, and typical Z bands are present. The ultrastructural pattern of cardiac muscle likewise is similar to that of skeletal muscle, and the individual fibers branch and anastomose freely. But in contrast to skeletal muscle, whose individual fibers are present as a morphologic syncytium, cardiac muscle fibers each are surrounded completely by a cell membrane. At the points where the ends of the individual cardiac fibers are apposed, the two individual membranes fuse into a highly folded structure. These regions, known as *intercalated discs,* are always found at the Z bands, and they provide a strong connection among all of the individual fibers of the heart. Functionally, this structural pattern is important in transmitting the tension developed during contraction of individual fibers throughout the heart as a whole along the axes of the fibers.

The external cell membranes of single cardiac muscle fibers fuse together in the regions of the intercalated discs along the sides of the cells and thus provide bridges of low electric resistance for the rapid transmission of action potentials between fibers. It follows that cardiac muscle is a *functional syncytium,* although there are no structural or protoplasmic bridges among the cells.

The sarcoplasmic reticulum of cardiac muscle is not as highly developed as that found in skeletal muscle, and no transverse elements comparable to the terminal cisternae of the triads in skeletal muscle are present. The T-system is situated at the Z bands of the cardiac fibers, whereas, in mammalian skeletal muscle, these elements are located at the A-I junctions.

Cardiac muscle also contains an abundance of elongated mitochondria that are found in intimate contact with the myofibrils.

MECHANICAL PROPERTIES. Cardiac muscle begins to contract immediately following the onset of depolarization. The duration of the contractile response is about one and a half times longer than that of the action potential (Fig. 4.28). The response of each fiber is all-or-none.

The absolute refractory period lasts throughout contraction, or *systole,* and well into relaxation, or *diastole,* of the heart (Fig. 4.28). This important physiologic property of cardiac muscle prevents the development of tetanus from excessively high rates

---

*The low-energy-state actin–myosin complex also can be called the rigor complex in contrast to the high-energy or active state.

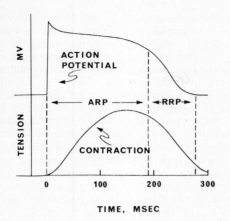

TIME, MSEC

FIG. 4.28. The temporal relationship between an action potential and the mechanical contraction of a cardiac muscle fiber. ARP = absolute refractory period; RRP = relative refractory period. Note the long plateau on the action potential curve which coincides with the depolarization of the fiber and gives rise to a similarly long ARP.

of stimulation, in contrast to skeletal muscle, in which tetanus is a normal component of its functional activity. In the heart, of course, such tetanic contraction would be fatal; therefore, the long absolute refractory period serves as a normal protective mechanism.

The proportional relationship between the initial fiber length and the tension developed during contraction of cardiac muscle fibers is like that seen in skeletal muscle (Fig. 4.13). Hence, there is a resting fiber length at which the tension developed by cardiac muscle fibers during contraction is maximal. In vivo, the initial fiber length prior to contraction is determined by the extent to which the heart is filled with blood during diastole, and the intraventricular pressure developed during systole is proportional to the total tension developed (see also p. 663). As diastolic filling increases, the force of contraction of the ventricles likewise is increased within definite limits (Starling's law of the heart, p. 660).

BIOLELECTRIC PROPERTIES. The rest potential of individual cardiac muscle fibers is around −80 mv. Stimulation causes the rapid development of a propagated action potential, and the initial depolarization is quite rapid, as in skeletal muscle and nerve (Fig. 4.28). Depolarization of a cardiac fiber thus takes about 2 msec.

Repolarization of the cardiac fiber proceeds quite slowly, however, and in mammalian hearts this phase may last for over 200 msec. Consequently, as noted above, the repolarization is not complete until the mechanical response is nearly over.

In cardiac muscle, there is an inverse relationship between the rate of contraction and the repolarization time. Functionally this is important because, when the heart rate accelerates in vivo,

the time required to recover full excitability decreases.

METABOLISM. Cardiac muscle differs quantitatively from skeletal muscle in several metabolic respects, although, in general, the biochemical pathways whereby energy is derived are the same for both types of tissue.

Under normal circumstances, cardiac muscle obtains its energy from aerobic rather than anaerobic metabolism. In fact, the heart cannot obtain enough energy from strictly anaerobic glycolysis to sustain its contractions for any length of time; ventricular failure ensues rapidly when the oxygen supply is severely curtailed.

Cardiac muscle that is contracting under basal conditions oxidizes roughly 35 percent carbohydrates, 5 percent ketones and amino acids, and 60 percent fatty acids. Thus, a very high percentage of the total energy utilized by the heart in a resting individual is derived from fat oxidation. The proportions of these nutrients employed by the heart in different physiologic and pathologic states can exhibit great variation. For example, in untreated diabetics the proportion of carbohydrate oxidized is reduced considerably, whereas that of fat is increased proportionately.

CONTRACTION AND EXCITATION–CONTRACTION COUPLING. Contraction in cardiac muscle is produced by a sliding filament mechanism identical to that found in skeletal muscle (p. 138).

The role of calcium in excitation–contraction coupling within cardiac muscle fibers also is similar to that already described in skeletal muscle (p. 140), except that, during relaxation or diastole, calcium ions are found primarily in the extracellular fluid. Therefore, in order to initiate the contractile process in cardiac muscle fibers, the calcium ions first must traverse the cell membrane from the extracellular to the intracellular fluid following electric excitation of the cell.

AUTOMATICITY. In sharp contrast to skeletal muscle, whose normal activity is strictly dependent upon impulses from motor nerves, cardiac muscle possesses the inherent property of *automaticity* or *autorhythmicity*. That is, all cardiac muscle fibers potentially or actually are able to contract spontaneously. This property of automaticity is most highly developed in certain specialized *pacemaker tissues* found within the heart, which regulate the rate of the entire organ (p. 584).

The fundamental property of autorhythmic cells which distinguishes them from all other types of excitable fiber is that they do not have a stable transmembrane resting potential. Therefore, immediately following the repolarization of an autorhythmic cell to a "resting" voltage, the membrane gradually leaks current so that the transmembrane voltage decreases spontaneously to the threshold value for the cell, at which point the cell fires an all-or-none action potential. The slow, spontaneous depolarization that takes place between action potentials is called the *pacemaker potential* or *prepotential*. The steeper the slope of the prepotential, the more rapidly the cell fires. Conversely, as the slope is decreased, the rate of firing slows. This important property of autorhythmic tissues is illustrated in Figure 4.29 and discussed in more detail on page 595. Briefly, however, the unstable membrane potential of autorhythmic cells is caused by a gradual decrease in the permeability of the membrane to potassium ions, which causes a progressive decline in potassium diffusion from the cell following maximal repolarization from the previous action potential. The transmembrane voltage is reduced progressively until the threshold voltage is reached and another action potential is initiated. The concomitant high resting sodium conductance of pacemaker tissues causes a net influx of sodium, and this also contributes to the spontaneous depolarization.

Smooth Muscle

STRUCTURE AND TYPES. Smooth or plain muscle fibers are elongated, spindle-shaped (fusiform) cells whose most obvious histologic property is their lack of the cross-banded pattern observed in striated muscle (Fig. 4.30). Each smooth muscle fiber contains a single nucleus located near the center of the cell. The sarcoplasm is homogeneous and contains many fibrils that run the length of the cell and are presumed to be the contractile elements. The sarcoplasmic reticulum is poorly developed.

Smooth muscle cells exhibit considerable variation in their length in different organs. For example, in the pregnant uterus the fibers can reach a length of 500 μ (0.5 mm), whereas in the human intestine the smooth muscle fibers are about 200 μ (0.2 mm) in length. The smallest smooth muscle cells, found in the walls of blood vessels, may be only 20 μ in length.

Smooth muscle cells have a wide variety of functional roles within the body. Although these elements show a similar morphologic appearance wherever they are found, their responses to pharmacologic and other chemical agents are so diverse

that it is useful to think of smooth muscle cells in the physiologic sense as a number of different cellular species belonging to the same morphologic family. For convenience, however, smooth muscle usually is classified as either visceral or multiunit.

Visceral smooth muscle is found principally as large sheets within the walls of hollow viscera such as the stomach, intestine, bile ducts, ureters, bladder, and uterus. The individual fibers are closely packed in a roughly parallel fashion and form a morphologic and functional syncytium. Hence, low-electric-resistance bridges are present among the individual cells, and spread of local electric excitations can occur readily. Again, like cardiac muscle, the membranes of the individual cells fuse together at their junctions.

Multiunit smooth muscle is composed of individual cellular units arranged more loosely and in a more random pattern than is found in visceral smooth muscle; in addition, intercellular bridges are lacking, so that cell-to-cell impulse conduction usually is absent. In many respects, the contractile properties of multiunit smooth muscle resemble those of skeletal muscle. Generally, nerve impulses are required to initiate the contraction of multiunit smooth muscle, and its contraction is more rapid than that of visceral smooth muscle. For example, the iris of the eye and the blood vessels contain multiunit smooth muscle, which is able to produce delicate, rapid contractions of graded intensity.

PHYSIOLOGIC STIMULI AND MECHANICAL PROPERTIES. Depending upon the location of the muscle tissue in the body, the contraction of smooth muscle is initiated by a variety of stimuli including nerve impulses, local physicochemical changes within the muscle itself, hormones, and other chemical agents. In fact, under physiologic conditions, local stretch or distension of many hollow organs can induce or enhance cellular prepotentials in the muscle fibers, leading directly to increased contractile activity (see also pp. 147, 169). Many specific examples of the various stimuli capable of inducing or inhibiting the activity of smooth muscle in particular situations will be encountered throughout this text; the bioelectric mechanisms that underlie particular stimuli will be considered later (p. 167).

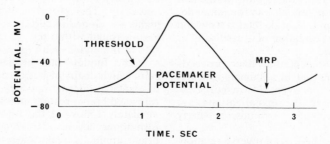

FIG. 4.29. The transmembrane potential of a single cardiac pacemaker cell MRP = maximum repolarization potential. Note that there is no stable rest potential, and that the membrane potential oscillates spontaneously between the MRP and threshold, at which point the cell automatically fires an action potential. See also Figures 4.33 and 4.34.

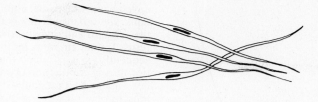

FIG. 4.30.   Smooth muscle fibers showing their fusiform appearance as well as the centrally located nuclei.

The contractile properties of smooth muscle fibers in general differ markedly from those of striated muscle fibers in several important respects. Contraction of smooth muscle is a relatively slow process compared to that of striated muscle (Fig. 4.31). However, smooth muscle is capable of sustaining forceful contractions for long periods of time with a very low concomitant expenditure of energy; similarly, the relaxation of smooth muscle following such a sustained *tonic contraction* is an equally leisurely process (Fig. 4.31).

The sustained tonic contraction of smooth muscle in many instances represents the summation of the effects of individual action potentials, so that, depending on the rate of stimulation, the individual contractions may range from a rhythmic pattern to a complete tetanus.

Spontaneous rhythmic contractions may be superimposed upon the fundamental tonic contraction pattern of smooth muscle, and this property, together with the morphologically irregular orientation of the component fibers, renders it impossible to assign a definite resting length to the smooth muscle elements within any particular organ.

Another mechanical peculiarity of smooth muscle is the variable tension that it exerts at any particular length. Thus, if a strip of visceral smooth muscle is stretched mechanically, at first it exerts an increased tension. However, if the tissue is held at the greater length following stretch, the tension may decline gradually so that the tension ultimately

developed may fall to, or even below, the pre-stretch tension (Fig. 4.32). This phenomenon is called the *stress–relaxation* of smooth muscle. Because of the complete lack of any correlation between the length and tension developed in smooth muscle, no resting length can be defined. Therefore, in some respects smooth muscle behaves similarly to a viscous, amorphous mass (eg, taffy candy) rather than as a highly organized contractile tissue, and this property is called the *plasticity* of smooth muscle.

At the cellular level, the contraction of an individual smooth muscle fiber causes the cell to assume an ellipsoid outline, with many invaginations of the cell membrane at the points where the myofibrils are attached to the membrane itself. In visceral smooth muscle, physical and electric continuity of the individual cell membranes assures transmission of impulses as well as mechanical tension among the individual cells.

In tissues that contain multiunit smooth muscle fibers, on the other hand, neural connections assure coordination among the individual cells, and their mechanical contractions are transmitted to adjacent cells by means of the anastomosing sheaths of reticular connective tissue fibers that invest each muscle fiber.

In marked contrast to skeletal muscle, which can shorten to between 25 and 35 percent of its resting length during a maximal isotonic contraction, smooth muscle is able to contract through

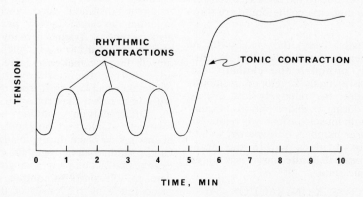

FIG. 4.31.   Rhythmic as well as tonic contractions of smooth muscle. The pattern of contractions may be regular, as shown, or irregular insofar as their force and rhythm are concerned. Tonic contractions of this type of muscle may be quite forceful, sustained for long periods of time, and rhythmic contractions may or may not be superimposed upon the tonic contraction.

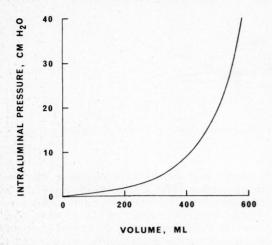

FIG. 4.32. The intraluminal pressure within a hollow viscus such as the stomach or bladder plotted against the volume present in the lumen of the organ. Note that the volume can increase markedly before the pressure increases significantly, a phenomenon known as stress–relaxation of smooth muscle. Similar curves can be obtained by applying external tension to isolated strips of smooth muscle.

much greater distances. This characteristic is of considerable physiologic importance, as it enables certain hollow viscera such as the uterus, stomach, and bladder to alter their lumen diameter from practically zero to many times this width, and then to return to the smaller diameter once again through contraction of the fibers. In this regard, the property of stress–relaxation discussed above is also important, since, during the filling and distension of a hollow viscus such as the bladder, the length of the muscle fibers can increase gradually without developing an increase in tension. Therefore, the pressure exerted upon the contents of the hollow organ — in this case, the bladder — does not increase appreciably until a certain point is reached. The viscus then contracts rapidly and powerfully due to a coordinated and pronounced shortening of the individual smooth muscle fibers, the intraluminal pressure rises due to the increased tension developed, and the contents of the organ are forcibly ejected.

METABOLISM. Qualitatively, the metabolism of smooth muscle fibers is the same as that in other types of muscle, although the capacity for anaerobic glycolysis is highly developed in smooth muscle. Furthermore, the relatively low energy consumption seen during extended periods of tonic contraction in smooth muscle probably is related to the formation of large numbers of low-energy-state actin–myosin cross-bridges within the myofilaments themselves so that little energy is required to maintain the tonic state.

BIOELECTRIC EVENTS, CONTRACTION, AND EXCITATION–CONTRACTION COUPLING. Despite the absence of any microscopic or ultramicroscopic evidence of a highly ordered

molecular arrangement among their contractile proteins, smooth muscle fibers do contain a large quantity of actin and myosin, the same contractile proteins found in muscle fibers generally. Consequently, it is assumed that these elements function during contraction of smooth muscle in a manner similar to that found in skeletal and cardiac muscle.

During the electric excitation of visceral smooth muscle, conduction of impulses among the fibers takes place by a process known as *ephaptic conduction.* That is, the depolarization of one fiber can spread readily between adjacent fibers directly through their low-resistance connections and without the intermediary secretion of any excitatory transmitter substances (p. 195). This process is also involved in the intercellular conduction of impulses within cardiac tissue.

The resting transmembrane potential of smooth muscle fibers is generally low, on the order of about −50 mv. However, regular and irregular fluctuations in the membrane potential of several millivolts often are present, and these are superimposed upon the basic rest potential as discussed further in the next section.

The action potentials recorded from particular smooth muscle fibers are subject to considerable individual variation, as might be anticipated from the various functional types of smooth muscle encountered in the body. The action potentials developed by particular smooth muscle fibers belong to three classes. First, typical rapid spike action potentials such as are seen in skeletal muscle (Fig. 4.20) are found in smooth muscle fibers, for example, in uterine muscle fibers that have been treated with estrogenic hormones. Second, action potentials with a prolonged depolarization plateau like that encountered in cardiac muscle are seen (Fig. 4.28). In this instance, the contractile activity of the smooth muscle persists throughout the duration of the entire action potential. Third, the resting potential of smooth muscle often has small waves, spikes, or ripples, and these electric oscillations may occur at subthreshold voltages or eventually develop sufficient magnitude so that they reach threshold value, in which case an action potential develops. Typical pacemaker potentials also commonly occur in many types of smooth muscle (Figs. 4.29, 4.33), but in visceral smooth muscle the pacemaker loci are found at multiple sites within the tissue. Furthermore, these pacemaker loci can shift from one area to another with time, rather than remain fixed in one area.

Under certain conditions — that is, when a relatively stable rest potential is present — the events taking place during a single action potential can be studied alone together with the mechanical events that accompany a single contraction. Visceral muscle starts to contract approximately 200 msec following the onset of the spike, and this process continues until about 150 msec after the spike is over. Thus, the latent period of smooth muscle is quite long compared to that of skeletal muscle, and the peak of contraction may be reached up to 500 msec following development of the spike. Consequently, the excitation–contraction coupling process in visceral smooth muscle is quite slow compared to that in striated muscle. In skeletal and cardiac muscle fibers, the latent period between depolarization and the onset of contraction is less than 10 msec. Hence, the temporal sequence of events is of much longer duration in smooth muscle than in striated muscle.

The ionic fluxes during excitation of smooth muscle are rather poorly understood. Development of spike potentials presumably is a reflection of the external sodium ion concentration. However, a considerable reduction in the extracellular level of this ion must be effected before the spikes are altered. A potassium ion efflux occurs during contraction of smooth muscle, and the magnitude of this efflux is in direct proportion to the tension developed.

Excitation–contraction coupling in smooth muscle appears to be mediated by calcium ions, whose intracellular concentration increases following depolarization, and the ATP–ADP system provides the energy for the contractile process. The sarcoplasmic reticulum and T-system of smooth muscle both are poorly developed; therefore, the interstitial fluid probably is a major source of the calcium ions necessary to activate the contractile system. There is some recent evidence, however, that a portion of the calcium ions required for the activation of smooth muscle may be obtained from intracellular calcium pools. In particular, the mitochondria and microsomes have been implicated as sources of intracellular calcium for contraction. It also appears that the relative proportions of calcium ions contributed to the overall contractile process by the extracellular fluid and intracellular organelles can vary considerably among smooth muscle fibers derived from different tissues. These biochemical data are also of interest because they substantiate the concept mentioned earlier that various smooth muscle fibers are actually a family of morphologically similar contractile elements which exhibit considerable chemical and functional differences.

Presumably, actual contraction of the smooth muscle fibers takes place by a sliding filament mechanism similar to that described earlier for skeletal muscle (p. 138).

AUTOMATICITY. Like cardiac pacemaker tissue, many smooth muscle fibers exhibit spontaneous rhythmic oscillations of their membrane potential that lead to automaticity of the mechanical contractions when the subthreshold oscillations are of sufficient magnitude to depolarize the cell to its threshold for firing.

As noted earlier, mechanical stretch or distension is an extremely potent stimulus to noninnervated visceral smooth muscle, and is followed by a decline in the resting transmembrane potential to threshold and development of an action potential. Furthermore, stretch also augments the tone of visceral smooth muscle (Fig. 4.33). The inherent rhythmic pattern of automaticity in smooth muscle of the intestine can be modified profoundly not only by the degree of applied stretch but also by chemical agents released in vivo from the terminals of nerves emanating from the sympathetic and parasympathetic branches of the autonomic nervous system (p. 270). As depicted in Figures 4.33 and 4.34, not only stretch but acetylcholine, parasympathetic nerve stimulation, and applied heat enhance the inherent rate of firing of intestinal smooth muscle fibers. These stimuli augment the firing rate by depolarizing the membrane rest potential toward threshold more rapidly. On the other hand, applied epinephrine, sympathetic stimulation, or cooling exert diametrically opposite effects on the firing rate of intestinal smooth muscle, and, as shown in Figure 4.34, may cause hyperpolarization of the membrane away from the threshold potential for firing.

In general, the sympathetic and parasympathetic branches of the autonomic nervous

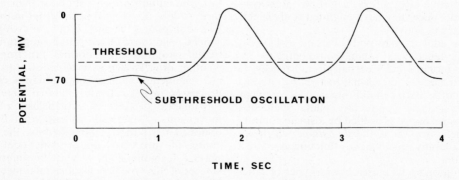

FIG. 4.33. Subthreshold voltage oscillations in the transmembrane potential of a smooth muscle fiber. When the depolarization is of sufficient magnitude to reach threshold, an action potential is generated and the muscle contracts. The actual shape and time course of the action potentials can vary somewhat, depending upon the source of the muscle under study. Mechanical stretch can induce (and/or accelerate) the contractions of smooth muscle, and this effect is produced by depolarization of the membrane to threshold as described above. Cf. Figure 4.34.

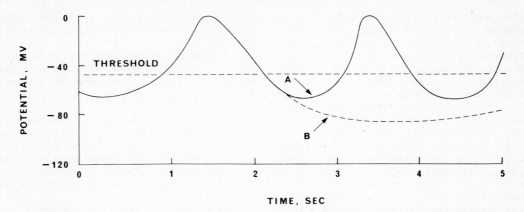

FIG. 4.34. The stimulatory effects of stretch, parasympathetic stimulation, heat, and acetylcholine application on the spontaneous firing rate of intestinal smooth muscle (A). Note that all of these factors accelerate, or steepen, the rate of depolarization of the fiber so that threshold to firing is reached sooner and thus the rate of contraction increases. On the other hand, sympathetic stimulation, cooling, and application of epinephrine hyperpolarize the cell and either decrease the rate of firing or else totally inhibit the generation of action potentials.

Note: The effects of epinephrine or sympathetic stimulation and of parasympathetic stimulation or acetylcholine application upon the pacemaker potentials of the heart are diametrically opposite to those described above for intestinal smooth muscle. The application of heat or cooling exerts similar effects in both tissues.

system exert opposite effects upon the inherent pattern of automaticity in visceral smooth muscle through alterations in the transmembrane potential such as are described above. These nerves do not initiate activity but can modify profoundly the automaticity of the smooth muscle in the viscera, as discussed further in Chapter 7 (p. 406).

THE PHYSIOLOGY OF CONTRACTION IN NONMUSCULAR CELLS AND ORGANELLES

As noted in the introduction to this chapter, during normal embryogenesis most cells are capable of active motility until they reach their ultimate structural and functional destinations within the morphologic pattern of the adult tissue or organ. Furthermore, the formation of an adult pattern by the cells within the tissues of certain organs may necessitate the active contraction of portions of its constituent cells once they have migrated to their permanent sites of activity.

In the adult human body, certain cell types, notably the leucocytes, retain the potential for independent motility under certain circumstances; in addition, a number of specialized epithelial cells contain motile extracellular organelles, or *cilia,* which are actively contractile cellular elements. The spermatozoa are propelled by a single contractile organelle known as the *flagellum,* which appears structurally, and behaves functionally, like a single cilium.

In addition to the more obvious types of cellular motility noted above, there are many instances in which the protoplasm of the cell can flow without any evident external deformation of the cell. This process is called *protoplasmic streaming* or *cyclosis.*

What common morphologic and physiologic denominators may be advanced to account for these highly varied and specialized cellular activities? There is considerable experimental evidence indicating that cellular contractility of the types outlined above is dependent upon the action of two organelles which were introduced in Chapter 3: the *microtubules* (p. 83) and the *microfilaments* (p. 83).

The Roles of Microtubules and Microfilaments in Contractility

The experimental data concerned with the functions of microtubules and microfilaments in cellular contractility have been gained largely through the use of optical and electron microscopy combined with the application of certain drugs to various cells in tissue culture prior to study of their ultrastructure. *Colchicine* is used to interrupt mitosis, since this compound breaks down microtubules in general, and those of the spindle apparatus in particular, so that their normal function is lost. Another drug known as *cytochalasin* disrupts the microfilaments, so that more obvious contractile processes in the cell are inhibited. Interestingly, the inhibitory effect of cytochalasin on cellular movements is reversible, so that, when the drug is removed from the medium bathing the cell, the contractile processes under study continue in a normal fashion. Thus it may be inferred from studies of this type which subcellular mechanism is involved in producing the observed response.

In many instances, the hollow microtubules appear to be relatively passive subcellular structural elements which lend internal support to certain cells or parts of cells, although in the process of mitosis the microtubules of the spindle apparatus themselves appear actively contractile so that they may pull the pairs of chromosomes to opposite ends of the cell during anaphase (p. 109). The microtubules comprising flagella and cilia are also actively contractile structures. In all of these elements that have been studied, the microtubules are composed of a highly ordered array of subunits which consist of a protein known as *tubulin* (p. 83).

Microfilaments, found in many types of cells, are small, threadlike elements that also appear to be actively contractile under certain circumstances. For example, microfilaments are present in a ring surrounding the circular constriction that develops during the telophase stage of cell division. Presumably, contraction of this circular bundle of microfilaments is responsible for the physical division of cells toward the end of mitosis.

There is some evidence which indicates that the microfilaments are composed of a protein similar to actin, at least in the cells of certain lower organisms, and that, as in muscle fibers, calcium ions trigger the contractile process. Microinjection of calcium into normal fertilized frog ova induces the appearance of short microfilaments about the site of the injection as well as a localized contraction of the cell membrane in the same region. In addition, the healing of artificially induced ruptures in the cell membrane requires the presence of calcium ions so that microfilaments aggregate around the wound and contract to bring about apposition of the cut edges of the membrane. The drug cytochalasin reversibly inhibits both of the cellular processes described above. The factors that regulate calcium release, and thus activate the microfilaments normally in vivo, are wholly unknown.

Based upon the foregoing considerations, it appears that microtubules function for the most part as intracellular skeletal elements, whereas microfilaments, as well as the microtubules of some cells, provide actively contractile subcellular organelles whose reversible shortening can induce various types of specialized cellular movements. In certain instances to be cited below, the microtubules and microfilaments of some cells appear to act in a cooperative manner as structural and contractile elements, respectively. Furthermore, it would seem that the proteins which comprise either the tubules or the filaments can assemble themselves into functional macromolecular cellular components in a reversible fashion, depending upon the prevailing conditions within the cell.

Morphologic Alterations in Cells and Cell Patterns

During embryogenesis, all cells are able to undergo considerable changes in their morphologic appearance. These alterations are caused in part by the pressures exerted by the growth and division of adjacent cells in the tissue and in part by a localized contraction of myofilaments within certain cells. Thus, sheets or tubes of epithelial cells can undergo marked changes in their ultimate gross appearance due to the contractile activity of the myofilaments, and the result is a folding or bending of the entire sheet or certain portions of it (Fig. 4.35A). For example, the typical adult pattern of lobules within the lungs and glandular tissues, as well as the folds that develop within the oviduct, are reflections of this type of localized intracellular contractility of the microfilaments. Under experimental conditions, the application of cytochalasin causes cessation of such cellular movement and a regression of the structure to its earlier, simpler pattern, whereas removal of the drug permits resumption of the spontaneous growth pattern.

It is of interest that the injection of estrogenic hormones into the developing female chick embryo

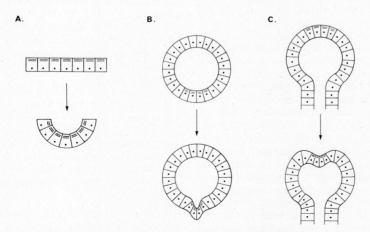

FIG. 4.35. Gross morphologic changes produced in tissues and organs by localized contraction of myofilaments (black lines in individual cells). **A.** Flat sheet of epithelial cells assumes a curved shape when myofilaments shorten. **B.** Oviduct and other tubular organs develop outpocketings when myofilaments shorten. **C.** Salivary glands develop lobules as the result of myofilament shortening. (After Wessels. *Sci Am* 225:76, 1971.)

can induce the premature development of bands of microfilaments across the luminal surface of the epithelial cells of the oviduct, with a concomitant bulging of the wall of the tube as glandular structures develop (Fig. 4.35B). Subsequent treatment with cytochalasin causes a disruption of the microfilament bands and a regression of the protruding glandular tissue to its previous cylindrical form.

Similarly, spontaneous changes in the morphologic features of mammalian salivary glands are brought about by localized activity of myofilaments. In this instance, however, the folding of the tissue results in the formation of deep clefts on the outer surface of the lobules (Fig. 4.35C); this is due to the fact that the microfilaments develop near the external surfaces of the epithelial cells, unlike in the oviduct. Cytochalasin exerts a similar reversible effect upon the development of the salivary gland as upon that of the chick oviduct. It has been observed that recovery from this drug takes place even in the presence of compounds that inhibit the *de novo* synthesis of protein. This fact argues strongly in favor of the reversible depolymerization and repolymerization of preexisting protein sub-units during the process of microfilament formation.

The specific intracellular mechanisms that control formation and activity of the microfilaments during embryogenesis and tissue formation are largely unknown at present; however, endocrine substances are known to play a role in these regulatory processes insofar as the oviduct is concerned.

Ameboid and Related Types of Cellular Motion

The most primitive type of animal locomotion is *ameboid motion,* so called because it resembles the process whereby an ameba glides across a surface by the repetitive extension of projections, or *pseudopodia,* on the leading edge of the cell surface.

In the embryo, practically all cell types are able to move freely by ameboid motion; hence, this process is of major importance to the normal development of the individual. In the adult body as well, a number of cell types can become actively motile under the influence of appropriate stimuli and move independently in relation to their immediate environment. Most commonly, during inflammatory processes the white blood cells, or leucocytes, migrate through the capillary walls by a process called *diapedesis,* and thus become tissue macrophages or microphages. Bacterial invasion and subsequent tissue inflammation can serve as the stimuli to initiate this process. Fibroblasts within the connective tissue, as well as the normally sessile germinal epithelial cells of the skin, also can become independently active, especially following injury. Hence, the migration of fibroblasts and epithelial cells serves to provide elements for the repair of injuries to the tissues.

Normally, leucocytes are spheroid in the bloodstream, but when they contact a solid body, eg, a glass slide, they change shape, emit pseudopodia, and wander about. Cells placed in tissue culture — whether they be endothelial, epithelial, mesenchymal, or other tissue elements — tend to migrate actively away from the original explant following dissolution of the desmosomes that interconnect them. In general, ameboid motion is manifest when the particular cells become attached to a solid substrate.

The rate at which ameboid cells move ranges between 0.5 and 4.6 μ/sec. Neutrophil leucocytes can progress at a maximal rate of around 0.55 μ/sec. This rate may seem quite low; however, if a leucocyte were to move without interruption, it would travel 4.75 cm in 24 hours, a considerable distance in terms of migration through the intercellular spaces of the body.

FACTORS AFFECTING AMEBOID MOTION. In general, ameboid motion is stimulated in quiescent but potentially motile cells by the presence of foreign bodies, injury, or the presence of chemical substances in the tissues. For example, leucocytes can be activated by bacteria or their toxins. This phenomenon is known as *chemotaxis,* and the substance which causes it is called a *chemotactic agent.* Usually, ameboid cells move toward the chemotactic agent — ie, up the chemical gradient of the substance — an effect called positive chemotaxis. Occasionally, however, the ameboid movement is directed away from the source of the chemotactic agent, in which case it is termed negative chemotaxis (see also p. 537).

Another important factor affecting ameboid motion is adhesion of the cell to a solid substrate. Pseudopodia may be extruded by ameboid cells floating in a liquid medium, but progression does not take place. In vivo, connective tissue fibers or the fibrin in blood clots undoubtedly provide the mechanical support necessary for ameboid motion to occur.

The rate of ameboid motion can be altered by modifying the temperature; an increase in temperature such as accompanies a local inflammatory process can accelerate the activity of the leucocytes.

An insufficient oxygen supply can decrease the rate of ameboid motion, but will not cause it to cease entirely.

Calcium ions are necessary for ameboid motion, whereas potassium ions antagonize the stimulatory effects of calcium.

Mechanical stimulation of a pseudopodium can alter the direction of ameboid movement, as discussed below.

MECHANISM OF AMEBOID MOTION. Protoplasm in general has the ability to undergo rapid gel–sol or sol–gel transformations, becoming relatively less or more viscous, respectively. The degree of association or dissociation of protein molecules into microtubules and/or microfilaments is believed to underlie these reversible viscosity transformations; hence, localized contractions of the protoplasm are considered to result from the activity of these organelles.

It has been demonstrated that ATP is responsible for the contractile processes involved in ameboid motion, and a myosinlike protein that exhibits ATPase activity has been isolated from certain ameba. Calcium ions stimulate the ATPase activity of this protein, which is present in the outermost protoplasmic layer, the ectoplasm. Anaerobic glycolysis appears to be the primary energy source for ameboid motion, since inhibitors of glycolysis rapidly abolish movement. The inhibition of aerobic oxidative processes, on the other hand, has little effect upon ameboid motion.

Reversible sol–gel changes caused by the dissociation and reassociation of protein subunits into transient contractile microstructures are believed to be responsible for ameboid movement. As shown in Figure 4.36, protoplasm streams in a regular cyclic pattern throughout the cell. A pseudopodium starts to form at the point where the outer gelated region of the cell becomes a sol, and the adjacent protoplasm flows into this region. As the sol moves forward into the pseudopodium, it flows backward and becomes part of the gel structure. In the temporarily "posterior" region of the cell, the gel now turns into a sol, and the gelated region may contract, thereby forcing the more fluid plasma sol into the pseudopodium where the plasma gel is thinnest. If the tip of an advancing pseudopodium is touched, the plasma gel thickens, and another pseudopodium is formed in another region, thus changing the direction of movement of the cell.

Injection of ATP into an ameboid cell produces contraction and solation of the outer gel layer. If this compound is injected into the "tail," the velocity of protoplasmic streaming is increased, whereas if the injection is into the advancing pseudopodium, the direction of streaming is reversed.

There is no unanimity of opinion as to the active site of the contraction that takes place during ameboid motion, although it is generally conceded that contraction is an important part of this process. Thus, some investigators believe that contraction occurs in the tail region, as noted above, whereas others hold that contractile forces are operative in or near the tip of the pseudopodium (see also discussion in the following section).

DIRECTIONAL MOVEMENTS IN CELLS. Although the sol–gel transformations discussed above undoubtedly take place in all cells undergoing ameboid motion, electron microscopy has revealed in more detail the basic mechanism of this type of motility in certain cells obtained from higher organisms (Fig. 4.37). If a single motile fibroblast or myocardial cell, for example, is studied after culture in a suitable medium, the membrane at the leading edge of the advancing cell is pushed forward, and this edge apparently undulates or flutters up and down. The undulating portion of the membrane attaches to the substrate and then actively contracts, thus pulling the cell forward. The direction in which the cell moves can be altered readily by the activation of different areas of its membrane. Microtubules as well as microfilaments are involved intimately in this process of cellular locomotion. As shown in Figure 4.37, the trailing edge of the cell is supported by a system of microtubules, whereas the contractile leading edge of the cell is powered by a system of microfilaments which pull the cell along the substrate.

Since the proteins that comprise both the microtubules and microfilaments are known to associate and dissociate readily, it is believed that the reversible formation and dissolution of these organelles in a transient fashion enables the cell to undergo marked alterations in its activity as occasion requires.

The contractile mechanism just described also appears to be an important feature in controlling the direction of growth in the elongated processes from nerve cells known as *axons* (p. 157). Experimental work performed on nerve cells growing in vitro has revealed that, at the growing tip of the axon, there is an active contractility of the membrane which is revealed by fluttering or ruffling movements of the tip of the cell process (Fig. 4.38). However, the entire nerve cell, or *neuron,* does not move — only its growing tip. Microfilaments are abundant in the region of growth at the nerve end, whereas an abundance of microtubules is present in parallel axial arrays along the length of the nerve fiber

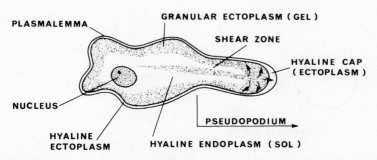

FIG. 4.36. Functional morphology of an ameboid cell. As indicated by the arrow, the direction of movement is toward the right.

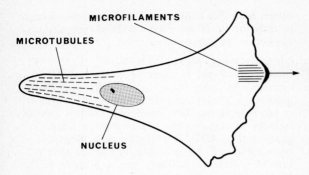

FIG. 4.37. Cellular motility, eg, of a fibroblast, takes place by an undulatory or fluttering movement of the leading edge of the cell (right side of figure). The tip of the leading edge attaches to the substrate, the microfilaments contract, and the cell moves in the direction of the arrow. Microtubules at the opposite end of the cell appear to act as cytoskeletal elements lending rigidity to this region. (Data from Wessels. *Sci Am* 225:76, 1971.)

proper. When growing nerves are treated with colchicine, the microtubule skeleton collapses, the nerve fiber shortens toward the cell body, and organized growth stops. It thus appears that, in nerve cells, movement of the contractile membrane at the growing tip of the axon is controlled by the contraction of localized bundles of myofilaments, whereas the microtubules in this case act merely as cytostructural elements lending rigidity to the entire axon as it develops and thus permitting the moving growth zone to advance.

Possibly the highly ordered directional growth of nerve fibers in vivo is the result of positive chemotactic stimulation, although this is conjecture.

PROTOPLASMIC STREAMING OR CYCLOSIS. Intracellular protoplasmic movements can occur without external deformation of the cell. This type of motility is called *protoplasmic streaming*, or *cyclosis*. Especially prominent in plant cells, cyclosis also can be observed in cells obtained from animal species, particularly when they are studied in tissue culture. As depicted in Figure 4.36, cyclosis can take place in a regular intracellular pattern, in which the gelated ectoplasm is stationary

with respect to the more solated portions of the cytoplasm.

The process of cyclosis, albeit slow, causes some mixing of the cytoplasmic components, including translocation of the smaller free-lying organelles. Cyclosis can serve to distribute intracellular ions and molecules, thereby enhancing concentration gradients across the cell membrane. The complex movements of the chromosomes, centrioles, and other organelles during mitotic division all are caused partially by cyclosis. As noted earlier, cyclosis is an important factor in producing ameboid motion.

In general, factors that decrease cytoplasmic viscosity increase the rate of ameboid motion in general and cyclosis in particular. Hence, suitable alterations in temperature, types of ions as well as their concentrations, and pH changes all can modify the rate of cyclosis. Cyclosis continues in the absence of oxygen; therefore, anaerobic glycolysis is able to provide sufficient energy for this process.

Microtubules and microfilaments both have been postulated as the organelles responsible for developing the propulsive forces during cyclosis. Parallel bundles of microfilaments may be seen at the interface between the moving endoplasm and the stationary ectoplasmic layer, and some investigators believe that such organelles are responsible for producing the contractile energy for cyclosis.

Flagellar and Ciliary Motion

Flagella and cilia share a number of common morphologic and functional attributes; consequently, these motile cellular appendages will be considered together. Flagellar and ciliary motion are caused by minute contractile filaments that are adapted to function solely in liquid media, and both of these cellular elements contract or beat in a rhythmic fashion.

STRUCTURE AND ULTRASTRUCTURE. The following are the principal morphologic distinctions between flagella and cilia that may be observed by means of optical microscopy. These hairlike appendages are called *flagella* if there are only one to several present. In contrast, many *cilia* are found on each cell, and these structures are shorter and thinner than flagella.

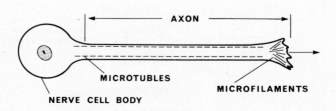

FIG. 4.38. The localization of microtubules and microfilaments within a growing axon. The diameter of the axon has been exaggerated considerably. The direction of growing at the axon tip is indicated by the arrow at the right side of the figure. (Data from Wessels. *Sci Am* 225:76, 1971.)

In the human body, a single flagellum is the propulsive unit of each spermatozoon; ciliated epithelial sheets, on the other hand, cover large areas of the respiratory and genital tracts and are found in other areas as well. Both of these cellular appendages are exceedingly delicate, filamentous processes whose thickness may lie near the limit of resolution of the optical microscope.

At the ultrastructural level, the basic features of the ciliary (or flagellar) apparatus are as follows: (a) The *cilium*, a slender, tapering, cylindrical process, extends upward from the free surface of the cell for a distance of several microns. (b) The *basal body* or *basal granule* is an intracellular organelle similar structurally to the centriole; in fact, the basal bodies arise from the centrioles themselves. (c) In epithelial cells, delicate fibrils called *ciliary rootlets* may arise from the basal granule. These fibrils form a conical bundle whose apex terminates near the nucleus. (d) The basal bodies can give off *horizontal fibrils* which interconnect them laterally in two planes. The morphologic features summarized above are illustrated in Figure 4.39.

The Cilium. All cilia and flagella exhibit an identical basic ultrastructural pattern as shown in Figure 4.39B. Cross sections of these elements, when viewed at a suitable magnification, exhibit a regular pattern of 11 microtubules called the *axoneme*. Two of these elements are centrally located and smaller than the rest. The overall diameter of a cilium is about 2,000 Å. An outer ciliary membrane

surrounds the longitudinal filaments; it is continuous with the plasma membrane of the cell and encloses the tubules, which, in turn, are embedded parallel to the long axis of the cilium within a *ciliary matrix*. The nine outermost tubules are composed of paired substructures, or doublets, each of which has a diameter of approximately 215 Å. The centrally located pair of microtubules are single, as shown in Figure 4.39B. A plane located perpendicular to a line joining two centrally located tubules (dashed line in figure) divides the cilium into symmetrical right and left halves. The plane of ciliary beat is perpendicular to this plane of symmetry. Pairs of projections (arms) are located on one of each of the paired outer tubules, and, when viewed from the base of the cilium, these projections are oriented in a clockwise direction. Filamentous spokes radiate outward from the central sheath investing the two central tubules, and longitudinal microfilaments (seen in cross section as dots) are located at about the midpoint of each spoke.

Throughout the long axis of the cilium, the characteristic 9 + 2 pattern of the microtubules is quite similar, but, at the distal end or tip, the peripheral tubules taper.

The Basal Body. Structurally, the basal body is identical with the centriole, which was described on page 81 and illustrated in Figure 3.20. Similarly to the cilium, the basal body is composed of nine radially disposed groups of tubules, but each group is composed of three rather than two microtubules;

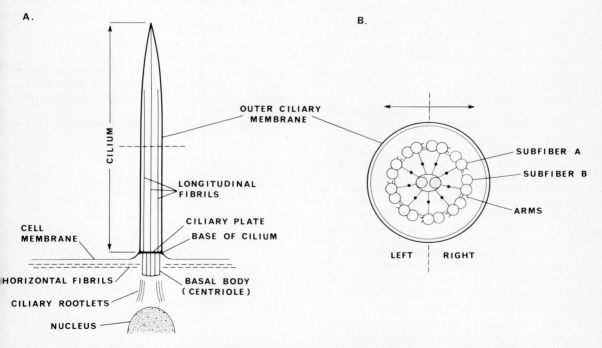

FIG. 4.39. **A.** Structure of a cilium. The structure of a flagellum is similar. **B.** Ultrastructural pattern of the elements as seen in a cross section of a cilium at the level indicated by the dashed line in **A.** The plane of symmetry that divides the cilium into left and right halves is indicated by the vertical dashed line and the direction of beat is indicated by the horizontal arrow. (Data from De Robertis, Nowinski, and Saez. *Cell Biology,* 1970, Saunders.)

thus triplets rather than doublets are the morphologic substructures of the basal body. Furthermore, the basal body has no central tubules as does the cilium, and it is closed at one end by a *terminal plate* where it abuts against the base of the cilium (Fig. 4.39A).

The basal bodies are found in the ectoplasmic layer immediately beneath the free surface of the cell, and they are spaced regularly in parallel rows. Since one cilium originates from each basal body, the cilia exhibit a similar pattern on the cell surface.

The basal body appears to be concerned in some way with the initiation and/or coordination of ciliary beat.

Ciliary Rootlets. Ciliary rootlets originate from the basal bodies in some cells. These structures may be either cross-striated filaments or tubular elements. The physiologic role of ciliary rootlets is unknown.

Horizontal Fibrils. The basal bodies of ciliated epithelial cells give off horizontal fibrils which interconnect these structures in two horizontal planes. Since these fibrils are much more developed in one plane than in the other, this morphologic polarity is believed to underlie coordination of the propagated wave of ciliary contraction, as discussed below.

PHYSIOLOGY OF CILIARY MOTION. In the human body, ciliated epithelial sheets such as are found in the respiratory and genital tracts serve to transport liquid secretions (mucus) and/or solid particulate matter, such as bacteria and dust particles, through the bronchi, and ova through the oviducts. Since all the cilia beat in the same direction, the fluid currents set up are an effective transport mechanism for such materials. Cilia also are located on specialized tissues within certain cavities of the brain, where they serve to circulate the cerebrospinal fluid (p. 561).

A number of epithelial membranes have nonmotile structures that resemble cilia morphologically. These elements are called *stereocilia* in contrast to motile cilia or *kinocilia*. For example, the macula and crista of the inner ear (p. 381) each contain both stereocilia and kinocilia. The epididymis of the male reproductive tract, on the other hand, contains only stereocilia, which apparently assist cellular secretion in some manner.

As noted earlier, the rhythmic flagellar contractions of the tails of the spermatozoa enable these cells to move as independent units; hence, this motility is essential to fertility.

Segments of the rods and cones within the eye have been demonstrated morphologically to be nonmotile ciliary derivatives.

Ciliary Movement. If the beat of cilia upon the surface of an epithelial sheet is observed from above, true rhythmic contraction waves are seen to pass in the same direction along the membrane. That is, the contraction is said to be *metachronic,* and the waves sweep along the epithelial surface

with a rippling motion such as is seen when gusts of wind pass over a field of grain. If, however, ciliary motion is observed from a plane perpendicular to the direction of the motion, then the contraction is said to be *isochronic*. That is, all of the cilia are seen to be in the same phase of contraction simultaneously.

The direction of ciliary beat in any particular epithelial sheet apparently is determined by the cytoplasm of the cells to which the cilia are attached. Therefore, if a small piece of ciliated epithelium is extirpated and then the fragment is reimplanted so that its physical orientation is reversed with respect to the rest of the sheet, the ciliary beat of the fragment continues in a coordinated fashion but in a direction opposite to that of the remainder of the sheet.

Ciliary contractions are quite rapid; for example, the pharyngeal cilia of the frog can beat at a rate of about 15/sec at room temperature. The rate accelerates as the temperature is increased and decreases as the temperature is lowered.

Various specific patterns have been described for the contractions of cilia observed in various organisms and tissues; the most common form of beat observed in mammalian cilia is shown in Figure 4.40. During the effective or forward stroke the cilium is relatively straight and contracts in a single plane, whereas, during the recovery or back stroke, considerable flexion of the cilium is seen as it returns to its earlier position.

Flagella, on the other hand, exhibit an undulant motion in which the contraction wave proceeds back and forth from the point of attachment to the free tip of the appendage.

THE REGULATION OF CILIARY BEAT AND THE MECHANISM OF CILIARY CONTRACTION. Although the actual mechanism of ciliary contraction and its regulation have not yet been worked out in detail, certain basic facts pertaining to these processes are clear.

Action potentials following depolarization of

A. EFFECTIVE STROKE

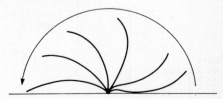

B. RECOVERY STROKE

FIG. 4.40. Pattern of ciliary beat.

the cells are responsible for the excitation of cilia, as is the case for muscular contraction and ameboid motion.

Excitation–contraction coupling in cilia appears to be mediated by the release and/or influx of calcium ions. It has been demonstrated that the increase in the frequency of ciliary beating following mechanical or electric stimulation is proportional to the increase in calcium influx following such stimulation. Furthermore, the intrinsic rate of ciliary beat is proportional to the concentration of extracellular calcium ions. Shifts in the concentration of external calcium ions also can alter the direction of ciliary beat, at least in certain protozoans. Thus, the aforementioned electric events appear to regulate calcium fluxes across the ciliary membrane, and these fluxes, in turn, influence ciliary activity directly.

A protein known as *dynein* has been isolated from the arms of the subfibrils A of the outer tubules (Fig. 4.39B); it is the site of practically all of the ATPase activity in cilia. Another protein whose amino acid composition resembles that of actin derived from skeletal muscle has been isolated from the nine outer tubules of cilia. It has been amply demonstrated that ATP breakdown is necessary for the rhythmic contractile activity of cilia as well as flagella. It is also clear that the mechanism for ciliary contraction is localized within the tubular fibrils of the ciliary shaft proper, rather than an intracellular locus.

Based upon the observations cited above, a *sliding filament model* for ciliary contraction has been proposed. According to this concept, some of the peripheral tubules of the axoneme slide with respect to those of the opposite side of the cilium, and bending results. As noted earlier (p. 153), the microtubules of cilia exhibit a definite morphologic polarity, which is related to the direction of movement of the entire cilium. Thus, the arms located on the A subfibers can be envisaged as interacting with the adjacent microtubular doublet in a selective and coordinated manner, and could cause sequential bridge formation during bending of a cilium in a manner roughly analogous to the repetitive bridge formation that occurs between heavy meromyosin and actin during muscular contraction. The net bending of the cilium would result from the activity of several sliding systems that could develop between adjacent filaments oriented at different angles with respect to each other. The details of this mechanical process and its regulation at the molecular level are, however, completely unknown.

CHAPTER 5

The Physiology of Impulse Conduction in Nerve and Other Tissues

It has been estimated that the human body contains about 14 billion individual nerve cells or *neurons,* and collectively all of these cells constitute the nervous system. The principal function of the nervous system is the rapid communication of coded information in the form of propagated electric signals, or nerve impulses, from one region of the body to another in a selective and integrated fashion. The neurons themselves are highly specialized, both structurally and functionally, for the repetitive generation and conduction of action potentials.

The ability of nervous and other excitable tissues to respond to suitable localized physical and chemical stimuli is called *irritability* (p. 103), whereas the propagation of the local excitation or action potential that develops following a stimulus is termed *conductivity.* Thus, the property of conductivity merely reflects the fundamental capacity of the cell to develop a self-propagating electric signal following the application of an adequate stimulus. The physiologic mechanisms that underlie irritability and conductivity are identical qualitatively in both nerve and muscle cells, although quantitatively the time courses of the electric and other events in muscle are somewhat different from those in nerve. In the discussions to follow, emphasis will be placed on the physiologic mechanisms responsible for conduction in nerve cells and their elongated processes; it must be stressed, however, that these mechanisms are quite similar to those involved in the excitation of skeletal, cardiac, and smooth muscle fibers.

MORPHOLOGIC CONSIDERATIONS

General Organizational Pattern of the Nervous System

The nervous system of higher organisms, including man, customarily is divided into two major regions, the *central nervous system* and the *peripheral nerv-*

ous system. The central nervous system is composed of the brain and spinal cord, whereas the peripheral nervous system is made up of all of the nervous tissue outside of these two organs.

The central nervous system contains the cell bodies of the peripheral nerves in addition to a number of cell types that act as a sort of connective tissue matrix for the neurons and that are termed collectively the *neuroglia* (literally, "nerve glue"). Within the brain and spinal cord, input signals that arise from the external as well as the internal environments are channeled, compared, integrated, and coordinated so that suitable motor responses to these signals are elicited in various effectors (eg, muscles or gland cells).

The peripheral nervous system, on the other hand, consists principally of elongated nerve fibers or cables, each of which is the process or extension of a cell located within the central nervous system or a peripheral ganglion. Functionally, nerves that transmit impulses to the central nervous system from outlying sensory receptors are called *afferent* or *sensory nerves,* whereas nerves that transmit signals from the central nervous system to peripheral effector organs or structures are called *efferent* or *motor nerves.*

The peripheral nervous system also includes a number of bodies known collectively as the *autonomic ganglia.* These structures are composed of nerve cells and their processes which function automatically to regulate the activity of the visceral organs and to control the circulation of the blood. Hence, the autonomic division of the peripheral nervous system acts somewhat independently from central nervous control, but it must be stressed that this physiologic independence is relative rather than absolute in many instances.

The sensory, integrative, and motor functions of neurons outlined above are dependent principally upon the properties of irritability and conductivity. In addition to these characteristics, however, some nerve cells are capable of secreting chemical substances into the bloodstream so that their range of activity is enhanced. Thus, some neurons not only behave as typical conductors for nerve impulses, but also as endocrine systems in that they elaborate hormones.

The individual neurons that in toto comprise the human nervous system are morphologically and trophically discrete units. The functional points of contact within the nervous system through which the individual neurons interact with each other are called *synapses*. These generalizations sometimes are referred to as the *neuron doctrine,* which is merely another way of expressing the cell theory insofar as it applies to the nervous system.

The foregoing brief survey of the nervous system is designed to give the reader a general orientation to this complex topic prior to a discussion of the structure of the neuron and its conduction of impulses. Details of the organization and functions of the entire nervous system will be presented in Chapters 6 to 8.

Structure and Ultrastructure of the Neuron

Of all the cell types encountered within the body, neurons undoubtedly exhibit the broadest range of morphologic diversity. The size and shape of nerve cells, as well as the number and ramifications of their processes, are subject to considerable variation depending upon the region of the nervous system under examination. Presumably these morphologic variations underlie various functional specializations.

In accordance with the neuron doctrine mentioned earlier, the individual neurons comprising the nervous system are morphologically and trophically (nutritionally) independent units which are functionally interconnected at synapses. The cell-to-cell transmission of impulses at synapses takes place in one direction only, and generally such intercellular communication is mediated through the release of chemical substances known as *neurotransmitters* (p. 195).

Neurons in general exhibit the following cytologic characteristics (Fig. 5.1): Each nerve cell has a *cell body* or *soma* that contains the nucleus. The cytoplasm that surrounds the nucleus often is called the *perikaryon* (or *neuroplasm*). Generally the cytoplasm extends into a number of short branching processes known as *dendrites* and one long process called the *axis cylinder* or *axon*. The axon may attain considerable length, and can, in turn, send out a number of branches along its course known as *axon collaterals*. Terminally, the axon may arborize into many fine branches called *terminal buttons* or *telodendria*. These terminations contain microgranules or vesicles in which are stored the neurotransmitter substance of the particular nerve.

THE NUCLEUS. Usually only one nucleus is present in each nerve cell. This large, pale-colored, spherical or ovoid organelle generally is located centrally within the perikaryon. The chromatin is very finely and uniformly dispersed within the nucleus, and a single nucleolus is present.

The nuclear membrane and its pores are quite similar to these elements in other cells as visualized by electron microscopy.

THE PERIKARYON. The cytoplasmic matrix of the neuron (the *neuroplasm* or *perikaryon*) is packed densely with many granular, filamentous, and membranous organelles that are situated in a roughly concentric arrangement about the nucleus. The total mass of the nucleoplasm in the processes of a neuron (axon plus dendrites) may exceed that present in the cell body itself.

Dense networks of *neurofilaments* with a diameter of about 100 Å each are found in the neuroplasm. In the dendrites and axon, these filaments are arranged parallel to the long axis of the processes. Under extremely high magnifications, the individual neurofilaments are seen to be composed of delicate tubular structures *(microtubules)* whose dense walls are about 30 Å in thickness.

When stained with basic dyes, the *chromophilic substance* or *Nissl bodies* are clearly evident under the optical microscope. A major constituent of the Nissl bodies is ribonucleoprotein, and in electron micrographs these elements are

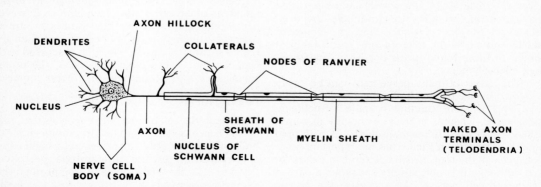

FIG. 5.1. The principal morphologic features of a peripheral motoneuron. The total length of the fiber has been shortened considerably. The Nissl bodies (granular endoplasmic reticulum) are indicated as coarse shading within the soma.

seen to be composed of parallel arrays of the cisternae of a granular endoplasmic reticulum. Ribosomes are attached to the outer surfaces of the membranous cisternae, as in other cell types. Although highly ordered Nissl bodies are present in the dendrites as well as the central mass of the perikaryon, they are lacking in the axon hillock and in the axon itself (Fig. 5.1).

A *Golgi complex* is present in all nerve cells; this relatively coarse, membranous network is found in an arc or a complete circle between the nuclear membrane and the cell membrane.

Small rod-shaped or filamentous *mitochondria* are found in varying numbers in the nerve cell body interspersed among the neurofibrils and Nissl bodies of the perikaryon. These organelles are somewhat smaller in neurons than in nonneural cells and have a diameter between 0.1 and 1.0 μ.

The perikaryon also contains a spherical *centrosome* that encloses a pair of typical *centrioles*. The role of these organelles in adult neurons is unclear, since mature nerve cells are incapable of reproduction in vivo.

The cytoplasmic processes of neurons are their most extraordinary morphologic feature. Practically all nerve cells have two distinct kinds of these processes — dendrite and axon.

The *dendrites* afford the greatest proportion of the functional receptive surface of the nerve cell. Dendrites generally contain Nissl bodies (ie, granular endoplasmic reticulum) as well as mitochondria, and usually several of these extensions of the perikaryon are present on the cell body of each neuron. Microtubular structures about 200 Å in diameter are present within the individual dendrites. The walls of these microtubules are about 60 Å in thickness.

The *axon* or *axis cylinder* differs in several important respects from the dendrites. Each neuron contains only one axon, in contrast to the several dendrites that usually are present. The axon frequently originates from a small elevation of the perikaryon — the *axon hillock* — that lacks Nissl bodies (Fig. 5.1). Nissl bodies are also absent from the axon proper. Generally, the axon is much longer and thinner than the dendrites of the same nerve cell. The cytoplasm within the axon, called the *axoplasm,* contains longitudinally disposed tubules of the endoplasmic reticulum, elongated slender mitochondria, microtubules (neurotubules) similar to those found in the dendrites, and abundant microfilaments.

The axons of many nerve cells can be easily identified by the presence of a *myelin sheath* (Fig. 5.1). This whitish structure is quite refractile in fresh nerve, whereas it blackens readily in neural tissue that is fixed in osmium due to its high lipid content. The myelin sheath has an important functional role in the transmission of nerve impulses (p. 169), although it is not, strictly speaking, a part of the nerve cell per se. At intervals, the myelin sheath of the axon is interrupted at regions called the *nodes of Ranvier* (Fig. 5.1).

A myelin sheath is associated only with axons; consequently, this criterion often provides a useful means for distinguishing dendrites from axons.

During its course, an axon may or may not give off branches or axon collaterals. Generally such collaterals are sent out from the axon at right angles, in contrast to those of dendrites, which tend to branch off at an acute angle from the main dendrite stem.

The principal terminal ramifications of an axon are called the *axon endings* or *telodendria*. Morphologically, the telodendria may branch extensively and form tendrils, networks, or buds that make synaptic contact with the cell body of another neuron or its dendrites. Alternatively, the terminal branches of an axon may contact specialized end organs such as the motor end plates of skeletal muscle (p. 208).

It is essential to interject a note on terminology at this point. The word "axon," as used in the morphologic sense, and as described above, refers to a single elongated process that arises from the cell body of a neuron. The term "axon" also denotes any peripheral nerve fiber that is connected to a sensory receptor (afferent) or to a muscle, gland, or other effector (efferent). In the classic physiologic sense, however, the dendrites are considered to be the receptive (afferent) processes of the neuron and conduct impulses *toward* the cell body, whereas efferent nerve impulses are conducted *away* from the cell body via the axon. This definition is strictly accurate only for spinal motoneurons (Fig. 5.2A). In the discussions to follow, the term *dendrite* or *dendritic zone* will be applied to any neural process or the cell body itself in which the nerve impulse originates, and the *axon* will be defined as that process of the nerve cell which conducts the impulses to the axon telodendria. According to this concept, illustrated in Figure 5.2, the morphologic location of the cell body is irrelevant insofar as the functional definition of an axon is concerned.

THE NERVE CELL MEMBRANE. The perikaryon of each neuron, as well as its process, is invested in a trilaminar unit membrane or plasmalemma which is similar morphologically to that found in other cell types. In addition, however, nerve cells, and especially their processes, are associated intimately with other cell types which give rise to sheaths of various kinds, as discussed in the next section. These coverings are of the utmost significance to the transmission of nerve impulses.

Nerve Fibers

In general, the cell bodies of the neurons that make up the nervous system are located within either the brain, the spinal cord, or peripheral ganglia. Consequently, all of the filamentous peripheral "nerves" in the body actually are protoplasmic extensions or axons of neurons whose cell bodies are grouped together into discrete morphologic and functional units.

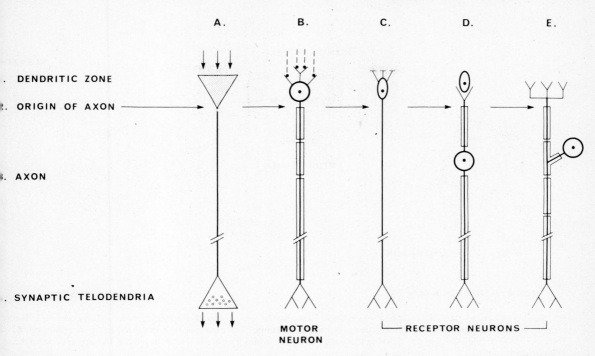

FIG. 5.2. The structure of various types of neuron in functional terms. The dendritic zone (1) is the receptor region of the cell, regardless of the morphologic position of the nerve cell body (indicated in **B, C, D,** and **E** by heavy circles and ellipses) with respect to the axon (2 and 3). The endings of the neurons, or synaptic telodendria, make contact with other neurons or effector (ie, muscular or glandular) cells.

A generalized neuron is shown in **A;** a specific stimulus (vertical arrows) produces a graded electric response in the dendritic zone, all-or-none propagated action potentials in the axon, and the graded release of a neurotransmitter substance (small circles, vertical arrows) at the axon terminals. (After Bloom and Fawcett. *A Textbook of Histology,* 1968, Saunders.)

A *nerve fiber* is made up of an axon together with certain sheaths or coverings. From a developmental standpoint, all of these sheaths, as well as the nerves themselves, are of ectodermal origin. Therefore, the nervous system can be considered to be composed of highly specialized epithelial tissue.

Peripheral nerve fibers in general are enclosed within a sheath composed of Schwann cells. These elements surround the axon from near its origin at a craniospinal nerve root or a peripheral ganglion and accompany the nerve almost to its terminations in the periphery. In addition, the larger peripheral nerve fibers all are enclosed within a *myelin sheath* which is located within the *sheath of Schwann.* Thus two or three discrete membranes may be present around individual nerve fibers depending upon their size: The plasmalemma and the sheath of Schwann are present in all nerves, whereas a myelin sheath is interposed between these two elements in larger axons (Fig. 5.3).

THE SHEATH OF SCHWANN. Sometimes called the *neurilemmal sheath,* the sheath of Schwann is composed of flattened cells that completely invest the axon, or the myelin that surrounds the axon if this structure is present (Fig. 5.3). The Schwann cells are neuroglial cells that have migrated from

the central nervous system as the nerve grows peripherally, and ultimately they form complete neurilemmal sheaths about the nerve fibers. In adult nerves, the nuclei of Schwann cells are flattened and the cytoplasm is much attenuated.

Electron microscopy has revealed that up to 12 axons of individual unmyelinated nerves can occupy recesses within a single Schwann cell. The membrane of the Schwann cell is in close contact with each axon and usually completely surrounding it, thus forming the neurilemmal sheath.

Schwann cells are essential to the sustained life of the axon, and it is believed that they may have some functional role in the nutrition of peripheral nerve fibers. During the regeneration of peripheral nerves following injury, the new axon that grows distally from the nerve cell body follows the conduits provided by Schwann cells (see p. 450).

THE MYELIN SHEATH. Under an optical microscope, the sheath of Schwann and the myelin sheath appear to be distinct entities in stained sections of neural tissue. However, use of the electron microscope has revealed that the myelin actually is a part of the Schwann cell and consists of layers of the membrane of the Schwann cell that are wrapped in a spiral fashion about the axon (Fig. 5.3B). The outermost membrane of the Schwann cell together

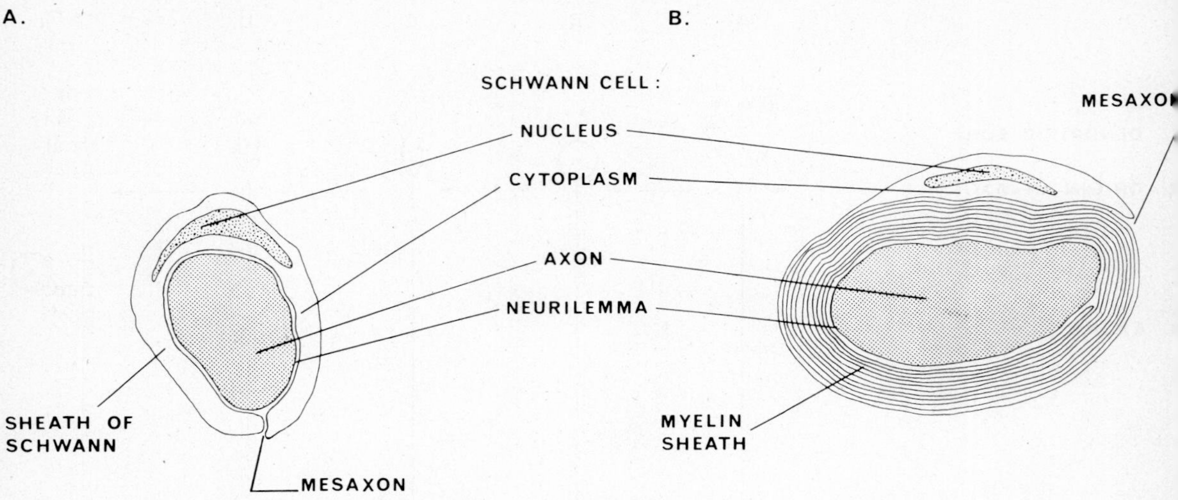

A.

B.

FIG. 5.3. **A.** Cross section of a nonmyelinated axon showing the investment of this structure by a Schwann cell. The mesaxon is the region where the processes of the Schwann cell unite. The axon per se is enclosed in the neurilemma. **B.** Cross section of a myelinated axon, showing the spiral layers of myelin that invest the axon proper and originate from the Schwann cell.

with its closely adherent protein-polysaccharide coating has been resolved by light microscopy as a single layer called the *neurilemma* by classic microscopists. The term "axolemma" once was employed with reference to the inner membrane of the Schwann cell. However, in accordance with current usage, the word "axolemma" is used in this text to denote the plasmalemma of the neuron that invests the axon.

The sheath of Schwann as well as the myelin sheath are punctuated at regularly spaced intervals along the course of the axon; these regions are known as the *nodes of Ranvier* (Fig. 5.1). These nodes mark the points at which the Schwann cells are discontinuous, and the axon is partially exposed in these regions, a fact that is of major physiologic importance to impulse conduction in myelinated nerves (see p. 169).

In longitudinal sections of either fresh or fixed myelinated nerves, the myelin within each segment is interrupted by discontinuities that are at an angle to the long axis of the fiber. These discontinuities are called the *incisures* (or *clefts*) *of Schmidt-Lanterman,* and they represent local regions in which the spiral layers of myelin are separated slightly, although the clefts are bridged by cytoplasm of the Schwann cell.

Within the brain and spinal cord, as well as in peripheral nerves, many fibers are present that have a myelin sheath, although typical Schwann cells are absent in the central nervous system. In brain and spinal cord, a type of glial cell known as the *oligodendrocyte* has a functional role in producing myelination of the fibers. Nodes of Ranvier are

present in the central nervous system, although clefts of Schmidt-Lanterman are not evident.

The myelination of nerves takes place relatively late during fetal development, and this process is not complete until well after parturition has occurred. The various systems of nerve fibers *(tracts)* within the brain and spinal cord are not myelinated simultaneously, but rather at different times during ontogeny.

Nerve Fibers as Components of Peripheral Nerves and Other Neural Structures

Outside of the central nervous system, individual nerve fibers with a thickness ranging between 1 and 30 μ are associated into bundles or *fascicles*. The component fascicles of a grossly visible *peripheral nerve trunk* are united by connective tissue, as shown in Figure 5.4. The outermost layer of connective tissue, the *epineurium,* is composed of longitudinally disposed collagenous fibers and connective tissue cells. Each of the smaller fascicles within a nerve trunk is surrounded by concentric layers of dense connective tissue, the *perineurium.* From the perineurium, delicate, longitudinally oriented collagenous fibers, fibroblasts, and macrophages form a network called the *endoneurium* that surrounds the individual nerve fibers. The exceedingly fine reticular connective tissue fibers that surround each nerve fiber form the *connective tissue sheath of Key and Retzius.* Blood vessels are found principally in the epineurium and perineurium.

There is a good correlation between histologically defined types of nerve fibers and their

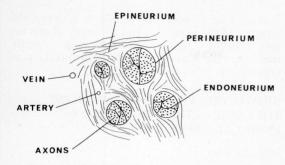

EPINEURIUM

PERINEURIUM

VEIN

ENDONEURIUM

ARTERY

AXONS

FIG. 5.4. A small portion of a nerve trunk in cross section together with the connective tissue sheaths that invest the individual fascicles (four are indicated) and axons. The endoneurium is indicated as partitions within each fascicle; actually this structure ramifies to such an extent that it surrounds every axon with an exceedingly delicate sheath.

physiologic properties. Hence, it is possible to classify nerve fibers according to their diameters. The velocity of impulse conduction as well as the magnitude of the propagated action potential itself vary directly in relation to the diameter of the fiber, as summarized in Table 5.1. The diameters of the individual fibers span a broad and continuous range from the smallest unmyelinated fibers to the largest myelinated fibers, although not all fiber types may be found within any particular peripheral nerve trunk.

Within the brain and spinal cord, ie, the central

nervous system, nerve fibers also are aggregated into anatomic and functional systems. The fibers of the afferent and efferent pathways of the spinal cord, in particular, are clearly defined. Each fiber pathway within the central nervous system has one or more particular functions, some of which are well known, some of which are obscure (see Chapter 7).

Axonal Transport

Basically, peripheral nerve fibers are elongated, highly irritable conductors of electric potentials; but in order to retain their normal morphologic and functional characteristics, the cytoplasmic processes of nerves must be continuous with the perikaryon of the nerve cell body or soma. More than a century ago, Waller demonstrated that section of a peripheral nerve resulted in a typical sequence of degenerative changes in the nerve distal to the cut within a few days. These changes, now known as *Wallerian degeneration,* led to the obvious conclusion that the soma of the neuron is the source of certain nutrients and other materials necessary to maintain the structural and functional integrity of the axon. Thus the nerve cell body exerts a direct nutritional or trophic influence on the processes of the neuron.

It is now clear that the nerve cell body continuously forms new cytoplasm which flows down the axon by a process known as *axoplasmic transport.* Various low- and high-molecular-weight

Table 5.1 TYPES OF MAMMALIAN NERVE FIBERS[a]

FIBER TYPE	FIBER DIAMETER (μ)	CONDUCTION VELOCITY (m/sec)	SPIKE DURATION (msec)	ABSOLUTE REFRACTORY PERIOD (msec)	FUNCTIONS
A (α)	12–22	70–120	0.4–0.5	0.2–1.0	Somatic motor, muscle proprioceptors
A (β)	5–13	30–70	0.4–0.5	0.2–1.0	Touch, kinesthesia (muscle sense), pressure
A (γ)	3–8	15–40	0.4–0.7	0.2–1.0	Motor (to muscle spindles), touch, pressure
A (δ)	1–5	12–30	0.4–1.0	0.2–1.0	Pain, temperature (heat and cold), pressure
B	1–3	3–15	1.2	0.6–1.2	Autonomic pre-ganglionic fibers
C (dorsal root nerves)	0.2–1.2	0.2–2	2	2	Pain, reflexes
C (sympathetic nerves)	0.3–1.3	0.7–2.3	2	2	Postganglionic sympathetics

[a]*Some generalizations on the functional characteristics of nerve fibers are: Motor nerve fibers to the skeletal muscles are thick and heavily myelinated (or medullated). Nerve fibers to visceral smooth muscles are thin and either lightly myelinated or nonmyelinated. The nerves that subserve tactile sensations are medium-sized with a moderate degree of myelination. Nerves that transmit impulses related to pain and taste are thin and have little or no myelin. The sensory fibers of the olfactory nerve always are nonmyelinated. Thus, a reasonably good correlation exists between morphologic and functional types of nerve fibers. (See also Table 5.2.)*

substances have been observed to be transported down the axoplasm from the soma at rates varying between 1 and approximately 400 mm/day. Thus, it appears that both fast and slow axonal transport systems are present. However, the possibility exists that the wide variations observed in the transport rates in different nerves may reflect different physiologic states. In particular, various proteins which cannot be synthesized within the axoplasm itself appear to be transported rapidly. Although the process of simple diffusion may be invoked to explain some aspects of axoplasmic transport, it is clear that a rapid active transport mechanism which requires oxidative phosphorylation and ATP utilization is also present within the axon proper. Furthermore, the application of colchicine, which disrupts the integrity of the axonal microtubular system (p. 148) inhibits the fast transport process. Consequently, these organelles appear to be involved in the transport of proteins from the soma down the axon. In this regard, it is of interest that cyclic adenosine monophosphate (adenosine 3′, 5′-monophosphate, p. 1020) both enhances the movements at the tip of the growing axon (p. 151) and promotes the assembly of microtubules as the axon elongates. Since the protein constituents required for growth of the axon are synthesized in the soma, it may be inferred that axonal growth is accompanied by an enhanced fast transport of proteins, and that the microtubules play an important role in this process.

The axoplasmic transport mechanism shares some common features with those involved in the contraction of muscle. Consequently, it has been postulated that material to be transported down the axon becomes attached to microfilaments within the perikaryon, and the substance-filament complex then slides down the microtubules in a manner analogous to that whereby actin slides along myosin during the contraction of skeletal muscle (p. 138).

In addition to the axoplasmic flow of materials necessary to support growth and to maintain the integrity of the axon, a continual supply of precursors necessary for the elaboration of synaptic transmitters at the nerve endings also must be provided by the same mechanism.

BIOELECTRIC PHENOMENA IN NERVE: THE CONDUCTION OF IMPULSES IN SINGLE FIBERS

Introduction

Nerve (as well as muscle) fibers are highly irritable, elongated, cylindrical conductors enclosed within membranes that insulate them to some extent from the extracellular fluid, which, as an aqueous solution of electrolytes, is a good volume conductor of electricity. The electric conductance of the fiber core depends on the fact that the axoplasm is a thin strand of an electrolytic gel, though much inferior to a metal wire in its ability to conduct electricity. The insulator properties of neural sheaths are not perfect. Thus, a subthreshold electric signal of short duration cannot be propagated for more than a millimeter or so without major distortion and attenuation, for the inherent properties of the surface membrane permit leakage of current and energy is dissipated within the axoplasmic core of the fiber. Due to the unique biologic properties of the axon, however, during the conduction of a nerve impulse the activity of one region of the axon serves to stimulate activity in the adjacent inactive region, so that the impulse is conducted rapidly along the fiber as a moving wave. Furthermore, this process is self-amplifying throughout the length of the axon; therefore, the signal is conducted with an undiminished amplitude from the source of the stimulus to the terminals of the nerve. Briefly, as one region of the fiber develops an action potential, the external surface of the neurilemma becomes negative with respect to the adjacent resting portions of the axon, and currents flow between

Table 5.2 CLASSIFICATION OF SENSORY NEURONS[a]

GROUP	ORIGIN OF FIBERS	FIBER TYPE (As listed in Table 5.1)
Ia	Annulospiral endings in muscle spindles	A (α)
b	Golgi tendon apparatus	A (α)
II	Cutaneous tactile receptors, flower-spray endings of muscle spindles, touch, pressure	A (β)
III	Pain, temperature receptors, crude touch sensations	A (δ)
IV	Pain, crude pressure and touch sensations (unmyelinated fibers)	C (dorsal root nerves)

[a]This numerical classification, sometimes used in sensory physiology, is based upon certain electrophysiologic differences among the particular nerve fibers, eg, the presence or absence of the first afterpotential.

the active and the resting portions of the fiber. This process continues as the impulse progresses, the active regions stimulating the inactive regions in a sequential fashion, and the signal is conducted for long distances. No attenuation of the impulse takes place since each segment of the axon develops a full action potential in response to the local depolarization induced by the active region. That is, a small, localized depolarization of the neurilemma (ie, to the threshold or firing level) is sufficient to cause development of a full-scale action potential of much greater amplitude, and the signal is amplified automatically in each segment of the fiber.

The processes outlined above will be considered in greater detail in the following sections. It should be emphasized that only the conduction of impulses in peripheral nerve fibers will be considered in this chapter. The physicochemical processes to be described also apply to the propagation of action potentials along muscle fibers. The specialized physiologic properties of peripheral receptors as well as those pertaining to synaptic transmission will be dealt with in the next chapter.

The general physiologic characteristics of the nerve fiber or axon may now be summarized. The properties of *irritability* and *conductivity*, mentioned above, enable the axon to respond to appropriate stimuli and to transmit electric impulses from one point to another over widely varying distances within the body. In order to carry out these functions, the nerve fiber must be *morphologically continuous* as well as in a *physiologically suitable condition*. Thus, following passage of a nerve impulse, the fiber remains completely inexcitable for a finite interval of time, called the *absolute refractory period*. When stimulated in its center, an axon can conduct impulses in the normal or *orthodromic* direction as well as in the opposite or *antidromic* direction. (This is in marked contrast to synapses, which can conduct impulses in *one* direction only; p. 198). Normally — that is, in vivo — an *impulse remains within the stimulated neuron* and is propagated along the axon and its branches to the synapses. If a stimulus is at or above threshold, the response of the axon is maximal for the conditions then prevailing; ie, the *nerve fiber obeys the all-or-none law* (p. 121). Lastly, the magnitude or amplitude of the nerve impulse can be altered (or even blocked completely) in a transient fashion by the local action of temperature changes (heat and cold), pressure, a low oxygen concentration (hypoxia), and electric currents, as well as by many chemical agents such as narcotics and anesthetics. In fact, the differential sensitivity of various types of nerve fiber to hypoxia and other agents is of clinical as well as physiologic significance (Table 5.3).

A Note on Methods

TECHNIQUE OF RECORDING ACTION POTENTIALS.
The electric phenomena that occur in nerve fibers during the propagation of impulses

Table 5.3 RELATIVE SENSITIVITY OF MAMMALIAN NERVE FIBERS TO CONDUCTION BLOCK[a]

BLOCKING AGENT	DEGREE OF SENSITIVITY TO BLOCK		
	Greatest	Intermediate	Least
Pressure[b]	A	B	C
Hypoxia	B	A	C
Cocaine, local anesthetics[c]	C	B	A

[a]For classification of fiber types, see Table 5.1.

[b]Mechanical pressure on a nerve can induce a conduction failure in touch, motor, and pressure fibers without affecting the sensation of pain.

[c]Local anesthetics diminish conduction in group C (dorsal root) pain fibers before they have any effect on the group A fibers.

are extremely rapid and transient events which take place during a time span of a few milliseconds. The potential changes that are observed to accompany a nerve impulse are extremely small, being measured in millivolts. Therefore, a cathode-ray oscilloscope coupled to suitable electronic amplification devices is essential to measure the time course as well as the voltage changes that occur during the passage of a nerve impulse. Essentially, the cathode-ray oscilloscope provides an "electronic lever system" in the form of a focused beam of electrons that sweeps horizontally across a screen (the x-axis) at a fixed and predetermined velocity. The magnitude of the vertical deflection of the beam (the y-axis) is a direct measure of the potential under study. This factor, in turn, depends upon the calibration characteristics of the preamplifiers and amplifiers that are included in the circuit. Since, for all practical purposes, an electron beam has no inertia, the time course as well as the potential changes that take place during an action potential or a series of potentials can be determined with considerable accuracy. Usually, a permanent recording of such fast and transient events is obtained directly by photographing the screen of the oscilloscope tube as the experiment progresses. Alternatively, the electric events taking place in the cell or tissue can be recorded on magnetic tape for later playback through the oscilloscope, at which time photographs of significant phenomena can be taken. The diagram of a nerve action potential shown in Figure 5.6 is typical of the recordings made from a single axon by use of the equipment described above and depicted in Figure 5.5.

THE SQUID GIANT AXON. In order to study in detail the bioelectric processes that take place during activity in a single nerve fiber, it is necessary to isolate a single axon and to insert the tip of a microelectrode into that axon so that the effects of stimulation and other experimental procedures can be analyzed (Fig. 5.5). Typically, single mammalian nerve fibers are quite small (less than 30 μ) and many of these elements are bound tightly into compact bundles or nerve trunks by connective tissue. Consequently, the isolation of single mammalian

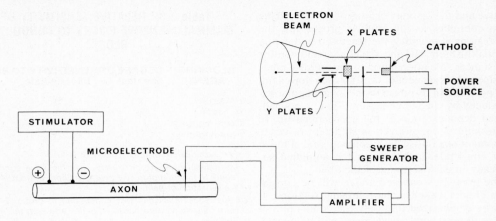

FIG. 5.5.   The equipment used for visualizing action potentials in excitable tissues. The action potentials generally are recorded permanently by photographing the screen of the CRO tube.

nerve fibers for study without causing extensive damage is not feasible. However, a number of invertebrate species contain extremely large nonmyelinated axons which can be isolated and which retain their physiologic properties in vitro for many hours. The largest axons that have been found occur within the mantle of the squid *(Loligo)*. The axons isolated from these marine mollusks range up to 1,000 μ (1 mm) in diameter; thus they provide an ideal biologic model system for the study of the bioelectric properties of nerve cells. There are no major differences between the bioelectric phenomena observed in the squid axon and those that take place in mammalian nerve fibers, so that much of the information on the conduction of impulses has been obtained from use of the squid giant nerve preparation.

By the use of suitable microelectrodes with tip diameters of less than 1 μ, it has also proven feasible to study the electric activity of single mammalian neurons and other cell types in vivo as well as in vitro.

The Resting Potential and Excitation of the Nerve Fiber; The Latent Period and Determination of Conduction Velocity

Like other cells, nerve fibers exhibit a resting transmembrane potential of around -70 to -80 mv, the inside of the cell being negative with respect to the exterior. The genesis of this rest potential, as in muscle and other cells, lies in the unequal distribution of electric charges across the membrane, as discussed in Chapters 3 (p. 98) and 4 (p. 130) and illustrated in Figures 3.35 and 3.36. (pp. 99 and 100).

The ionic basis for the rest potential in nerve fibers may be recapitulated briefly. Sodium ions are transported out of the axon by an active pumping mechanism, whereas potassium is transported into the axoplasm. Because of the electrochemical gradients for these ions developed by the coupled active transport mechanism, sodium ions tend to diffuse into the cell, whereas potassium ions tend to diffuse out of the cell. However, at rest, the membrane is far more permeable to potassium than to sodium. Therefore, the passive efflux of potassium is significantly greater than the passive sodium influx, and because the steady-state neurilemma is relatively impermeable to the anions within the axoplasm, the efflux of potassium is not attended by a concomitant and equal efflux of anions. Consequently, a slight excess of potassium ions builds up in a layer on the external surface of the neurilemma, whereas a similar slight excess of anions lines up on the inner surface of this membrane. The neurilemma is thus polarized, with the outside of the membrane positively charged with respect to the interior (see Fig. 3.35, p. 99).

Nerve fibers have a low threshold for excitation, and chemical, mechanical, and electric stimuli all are effective in producing a response. Generally, electric stimuli are employed under experimental conditions in which the properties of the nerve are under investigation. The characteristics of an effective stimulus were presented in Chapter 4 (p. 120). It was noted in this discussion that, in order to be effective, a stimulus must act for a sufficient length of time *(duration)* as well as reach a sufficient *strength* (Fig. 4.6, p. 121). If, however, the *rate* at which an otherwise adequate electric stimulus reaches threshold of a nerve is too slow, then the fiber may not fire, even though the strength and duration of the stimulus should be adequate to cause excitation. This phenomenon is called *accommodation* and is caused by an adaptation of the fiber to the slowly applied stimulus. Presumably this effect is produced by an alteration in the membrane's threshold to firing.

Following the application of an adequate stimulus, there is an immediate brief deflection of the baseline of the beam on the face of the cathode-ray tube of the oscilloscope. This deflection, known as the *stimulus artifact* (labeled SA in Fig. 5.6), results from leakage of current from the

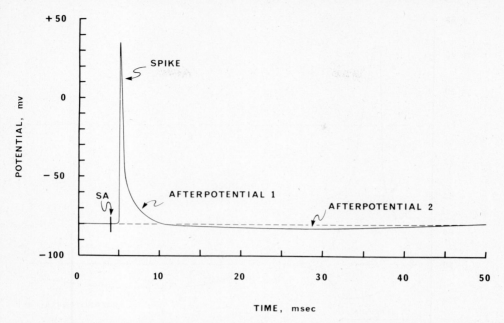

FIG. 5.6. A nerve action potential as recorded photographically from the screen of a cathode ray oscilloscope tube by use of the technique shown in Figure 5.5. SA = stimulus artifact. The rest potential (−80 mv) is indicated by the horizontal dashed line.

stimulating electrodes directly to the recording electrodes. The stimulus artifact is useful, however, as it marks the point in time at which the stimulus was delivered to the fiber.

The interval between the appearance of the stimulus artifact and the appearance of the action potential spike is called the *latent period* (Fig. 5.6). The latent period is manifest by a brief isopotential trace on the oscilloscope screen. The duration of this period depends upon the time required for the impulse to be conducted along the axon from the point of stimulation to the recording electrodes. Therefore, the length of the latent period depends upon the distance between the stimulating and recording electrodes on the axon as well as upon the diameter of the axon itself. The latter factor is an important consideration, since the velocity of conduction is related directly to the diameter of the fiber (see p. 169).

If the duration of the latent period and the distance between the electrodes are known, then the velocity of impulse conduction may be calculated easily for the particular nerve under study. For example, assuming that the distance between the stimulating electrode (the cathode, p. 167) and the recording electrode is 5.0 cm and the latent period is 1.0 msec, then the conduction velocity for that nerve is 5.0 cm/msec or 5.0 meters/sec.

The Action Potential and Related Bioelectric Phenomena

THE ACTION POTENTIAL AND IONIC FLUXES. The onset of an action potential in the nerve

fiber, as in the muscle fiber, is signaled by a progressive depolarization of the membrane. Once the membrane has become depolarized sufficiently, the spontaneous rate of depolarization increases rapidly (at threshold or the firing level), and an all-or-none spike potential develops. As illustrated in Figures 5.6 and 5.7A, the bioelectric components of the nerve action potential are both fast and slow. The rapid transmembrane voltage changes that occur first are called the *spike potential;* these are followed by two much slower afterpotentials, as is the case in the muscle fiber (p. 132). In Figure 5.7, the relationships among the various components of the action potential have been distorted somewhat in order to depict these relationships more clearly.

The duration and magnitude of the two afterpotentials can change without accompanying significant alterations in the rest of the action potential. Hence, it is believed that these slower voltage fluctuations reflect recovery processes that take place in the fiber following bursts of activity. For this reason, the first and second afterpotentials sometimes are not considered to be components of the action potential itself.

The ionic fluxes that occur during development of an action potential were presented in Chapter 4 (p. 135) and may be reviewed briefly at this juncture.

In nerve fibers, as well as in muscle fibers and cells in general, a slight decrease in the resting potential of the cell membrane causes a greater efflux of potassium ions together with an augmented influx of chloride ions. These slight alterations in ionic flux tend to restore the rest potential.

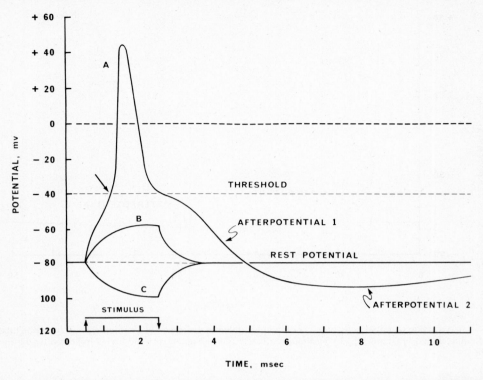

FIG. 5.7. Nerve action potential (A) distorted somewhat to show its individual components. Arrow indicates the firing level. A subthreshold depolarizing current (B), or a hyperpolarizing current (C) produces a transient local excitability change in the axon, as is the case for muscle fibers. See also Fig. 4.21. These local responses, called electrotonic potentials, are discussed in the text.

(See also the discussion of electrotonic potentials and the local response in the next section.) However, in both nerve and muscle cells, if the depolarization proceeds much beyond 7 mv, a unique alteration in the properties of the membrane develops. A voltage-dependent rise in the permeability of the membrane to sodium ions occurs which increases progressively as the membrane potential approaches the threshold to firing; ie, as the transmembrane potential decreases. The electrochemical gradients for sodium ions are inward, and as the membrane potential approaches threshold, the increase in membrane permeability to sodium likewise increases, so that when threshold voltage is reached, the permeability to sodium is so great that further sodium influx takes place in an "explosive" fashion (Fig. 4.23, p. 99). The spike potential develops as a consequence of these cyclic events.

The equilibrium potential for sodium ions in mammalian axons can be calculated using the Nernst equation (p. 99); this value is approximately +66 mv. During the onset of the action potential, the membrane approaches, but does not achieve, this value because the increase in membrane permeability is a rapidly transient phenomenon (Fig. 5.8). Consequently, as shown in the figure, the sodium permeability starts to decline to its resting value even before the spike of the action

potential has fully developed. Furthermore, the reversal of the electric gradient for sodium during the overshoot that develops because of the rapid reversal of the membrane polarity also limits the net sodium influx. Both of these factors — the decrease in sodium permeability of the membrane and the reversal of membrane polarity — act in concert to limit sodium influx and to initiate repolarization of the membrane. A third significant factor stimulating repolarization is the elevation in the permeability of the membrane to potassium ions that accompanies, but lags behind, the increased sodium permeability (Fig. 5.8). The reader will note that the potassium permeability increases more gradually than that for sodium, and also peaks during the declining phase of the action potential. This increase in potassium permeability as repolarization proceeds lowers the barrier to potassium diffusion from the fiber so that a net efflux of potassium from the cell restores the resting membrane potential once more. Concomitantly, the nerve recovers its irritability.

As stressed earlier (p. 98), the *net* quantities of sodium and potassium ions that move during the development of an action potential are but a minute fraction of the total quantities of these ions present. Furthermore, it should be emphasized that these changes occur *only* in the immediate vicinity of the membrane itself (Fig. 3.37, p. 102).

The two afterpotentials may be decreased by

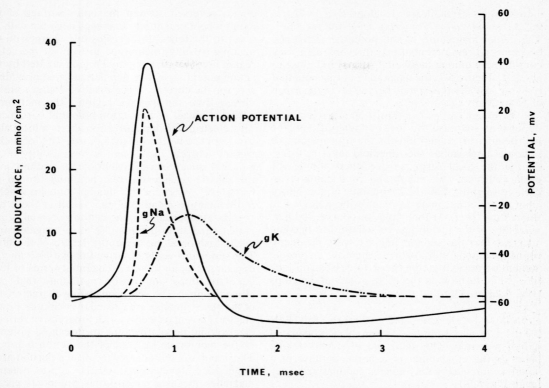

FIG. 5.8. The alterations in permeability of the neurilemma to Na and K ions during an action potential as measured by conductance (g) changes of the membrane.

metabolic inhibitors; thus, as noted earlier, these electric events presumably represent poorly understood metabolic events that are related to a restoration of the normal irritability in nerve following periods of intense activity.

ELECTROTONIC POTENTIALS, THE LOCAL RESPONSE, AND THRESHOLD. Subthreshold stimuli applied to a nerve fiber do not elicit a propagated all-or-none action potential. However, if recording electrodes are placed in close proximity to the stimulating electrodes and subthreshold currents of fixed duration are applied, certain localized effects on the membrane potential can be demonstrated readily.

If the subthreshold currents are applied by the cathode of the stimulating electrodes, a rapid decrease in the membrane potential is observed; this decays exponentially back to the rest potential when the stimulating current is turned off (Fig. 5.7B). If, instead, the subthreshold stimulus is applied under the anode of the stimulating electrodes, the membrane becomes hyperpolarized in a similar transient fashion (Fig. 5.7C). The brief changes in the membrane potential described above are called *electrotonic potentials*. The decrease in membrane potential, or hypopolarization, that occurs upon cathodal stimulation is termed a *catelectrotonic potential*, whereas the hyperpolarization

that occurs upon anodal stimulation is called an *anelectrotonic potential* (see also Fig. 4.22, p. 135).

Catelectrotonic and anelectrotonic potentials are *passive changes* in the membrane potential induced by the removal or addition of electric charge to the membrane by the cathode and anode, respectively. If the intensity of applied current is low, eg, the resting transmembrane potential is displaced by only 20 mv in either a hypopolarizing or hyperpolarizing direction as shown in Figure 5.7B and C, then the magnitude of the voltage change induced in the membrane is directly proportional to the intensity of the stimulus. As the strength of the applied stimulus is increased, this relationship continues to hold true for the anelectrotonic (hyperpolarizing) responses, but *not* for the catelectrotonic (hypopolarizing) responses. The cathodal responses increase to a disproportionately great extent as the intensity of the applied current is increased, and, when the cathodal stimulation is sufficient to cause a depolarization of around 40 mv, ie, at a membrane potential of −40 mv as depicted in Figure 4.7A, the membrane potential commences to fall rapidly and an all-or-none propagated action potential develops. The disproportionately larger effect of progressive increases in the intensity of cathodal stimulation is indicative of the *active role* of the membrane, and is called the *local response*. The point at which an all-or-none action potential suddenly develops (Fig. 5.7A) is the *threshold voltage* or *firing level* for the fiber. The actual magnitude of the potential difference between the rest potential and threshold can vary considerably among individual nerve (and muscle) fibers. In Figure 5.7, the magnitude of this potential is 40 mv; usually it is somewhat lower, in the range of 15 to 20 mv.

As noted earlier (p. 166), there is a voltage-dependent increase in the permeability of the neurilemma to sodium ions as the transmembrane potential approaches the threshold level to firing. During the local response described above, the permeability of the membrane to sodium is increased slightly, but the magnitude of the potassium efflux is adequate to restore the potential to the resting level. At threshold, of course, the permeability of the membrane to sodium has become so great that the sodium influx lowers the membrane potential still further, thereby increasing sodium permeability, and these processes continue in a cyclic fashion as the spike potential develops (Fig. 4.23, p. 136).

CONDUCTION OF THE ACTION POTENTIAL. The nerve impulse is simply a propagated action potential. The sequence of events taking place during conduction of a nerve impulse in a *nonmyelinated* axon may now be summarized. (Refer to Fig. 5.9.)

At rest, the neurilemma is charged or polarized. Positive charges are lined up on the outer surface of the membrane, while negative charges are similarly lined up on the inside. Following the application of an adequate stimulus, an action potential is generated rapidly, and, as the spike develops, the resting polarity of the membrane is abolished. For a very brief interval during the spike, the polarity of the membrane actually becomes reversed so that the outer surface of the membrane momentarily becomes negative with respect to the inner surface. Positive charges on the inactive (resting) membrane in front of the active region flow into the negative region created by the action potential; ie, this depolarized area provides a *sink* for the current. Since positive charges are removed from the inactive regions, the membrane potential ahead of the action potential is reduced. This electrotonic depolarization elicits a local electric response within the membrane. The local electrotonic current increases the permeability of the adjacent steady-state membrane to sodium ions, thereby decreasing the rest potential toward the firing level. When the threshold potential is reached in the inactive membrane as a result of these processes, another all-or-none action potential is generated, and thus the impulse is propagated in a cyclic fashion along the entire length of the fiber. The series of events described above also may be considered as an infinite, sequential series of local depolarizations and action potentials that pass along the fiber. The self-propagating characteristics of the nerve impulse are caused by a circular flow of electric current as well as by a progressive series of electrotonic depolarizations to threshold potential of the membrane in front of the action potential. The impulse is self-amplifying in each portion of the fiber since development of the action potential magnifies the local electrotonic potentials greatly once threshold to firing has been reached.

Since the active region in a nerve is negatively charged with respect to the adjacent steady-state regions, a nerve impulse sometimes is described as a "wave of negativity" that sweeps along the fiber (Fig. 5.9).

The propagation of action potentials along muscle fibers takes place exactly as described for nonmyelinated nerves. The motor end plate regions

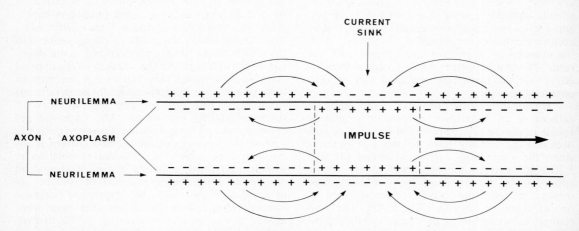

FIG. 5.9. The conduction of an impulse in an unmyelinated nerve. The region of electronegativity (ie, the impulse) provides a sink for current flow as indicated by the curved arrows, and the impulse is propagated toward the right (heavy arrow) as successive regions of the entire membrane become depolarized. The impulse itself is a wave of negativity passing along the fiber.

of skeletal muscle fibers are the most excitable regions of these cells, and provide the physiologic loci for their excitation (p. 210).

The conduction of impulses along *myelinated axons* is based upon a pattern of circular current flow similar to that found in nonmyelinated fibers. There is one important physiologic difference between the conduction in these two types of fiber, however. Myelin is a reasonably good electric insulator; therefore, current flow through the myelinated portions of a nerve is insignificant. Consequently, the depolarization in myelinated nerves leaps from one node of Ranvier to the next, as illustrated in Figure 5.10. The current sink at the active node electrotonically depolarizes the inactive node ahead of the active node to threshold, thereby inducing another action potential in the second node. Thus local depolarizations and propagation of the impulse is an intermittent rather than a continuous process in myelinated nerves, and is called *saltatory conduction* because the impulse jumps in a discontinuous fashion along the fibers. Since only short segments of myelinated nerve fibers depolarize at intervals in order to conduct an impulse, the velocity of conduction is considerably faster in myelinated nerves than in nonmyelinated nerves of comparable diameter.

CHANGES IN THE IRRITABILITY OF NERVE DURING ELECTROTONIC AND ACTION POTENTIALS. The sequences of changes in the irritability of nerve fibers during the presence of electrotonic as well as action potentials are quite similar to those in muscle fibers (Fig. 4.21B, p. 134). The axon undergoes an increase in its threshold to excitation during the passage of anelectrotonic currents, whereas catelectrotonic stimuli lower the threshold as they drive the membrane potential closer to the firing level (p. 134). Similarly, during the local response, the threshold also decreases; however, during the spike potential, the irritability of the axon is at a minimum, and the nerve is said to be refractory to further stimulation. This *refractory period* is divided into two phases. The *absolute refractory period* corresponds to the interval from the onset of firing of an all-or-none action potential and through the spike until repolarization is about one-third completed. During the absolute refractory period, *no* stimulus will elicit another response, regardless of its intensity. The absolute refractory period is followed by the *relative refractory period*. During this interval, a suprathreshold stimulus will elicit another response. The relative refractory period terminates at the first afterpotential, at which point the threshold to excitation once again decreases (Fig. 5.7A). During the second afterpotential, which actually represents a hyperpolarization, the threshold increases once more, and only a supranormal stimulus will excite the tissue. Ultimately, of course, the normal steady-state rest potential of the fiber is reestablished, and, when this occurs, the irritability of the nerve returns to normal.

THE VELOCITY OF IMPULSE CONDUCTION IN RELATION TO FIBER DIAMETER. Despite the fact that the irritability of nerve is inversely related to the diameter of the individual fiber, the *velocity* of impulse conduction in an axon is directly proportional to the diameter of the axon, whether it is myelinated or nonmyelinated. The reason for this is as follows.

In any nerve fiber, the velocity with which the impulse is conducted depends upon the rate at

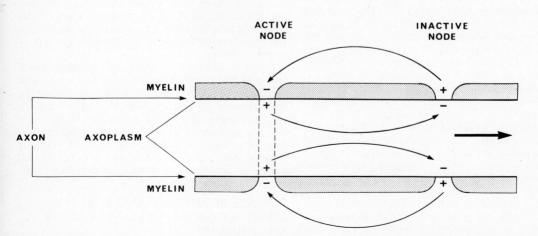

FIG. 5.10. The conduction of an impulse in a myelinated nerve. The active node serves as a current sink; current flows to the next adjacent inactive node as indicated by the curved arrows, the inactive node becomes depolarized in turn, and the impulse is propagated toward the right (heavy arrow). Thus in contrast to unmyelinated nerves, myelinated nerves undergo excitation only at the nodes of Ranvier, and the impulse "jumps" from node to node, a process called saltatory conduction.

which the electric capacity of the membrane ahead of the impulse is discharged beyond the threshold level to firing an impulse. This, in turn, is dependent upon the cable constants of the fiber (which include as major factors the surface capacity of the membrane as well as the electric resistances of the axoplasm and the external medium) and the quantity of electric current that the impulse, or action potential, can generate above the threshold requirement of the resting fiber. Therefore, because the velocity of the impulse depends principally on the electric resistance of the axoplasm longitudinally along the fiber, the velocity of conduction is directly related to the diameter of the fiber, since the greater the diameter, the lower the axoplasmic resistance.

In the mammalian nervous system, space requirements limit the size of the axons, in contrast to the situation found in, for example, the squid. Consequently, the development of myelin sheaths on nerve fibers of a relatively small diameter enables large numbers of conducting elements to be packed into a relatively small volume, while, at the same time, a large quantity of information can be transmitted with extreme rapidity.

FREQUENCY OF IMPULSE DISCHARGE IN NERVE FIBERS. A single nerve impulse is a relatively immutable coded signal, insofar as its amplitude and waveform are concerned. Therefore, not much information can be transmitted in a brief interval of time, and a large number of parallel fibers are necessary to convey detailed information in the form of these coded and stereotyped messages. As noted in the previous section, limitations on the available space within the mammalian nervous system preclude the development of great numbers of excessively large diameter parallel fibers, but the presence of a myelin sheath greatly enhances the velocity of transmission of impulses along single fibers as well as the efficiency with which this process takes place. Therefore, the rate of conduction of impulses is enhanced in myelinated fibers with no great sacrifice in space; and the net result is the transmission of considerably more information per fiber than would be possible otherwise.

Under physiologic conditions in vivo, nerve impulses do not occur singly, as is the case under the experimental conditions set up in the laboratory for elicitation of single action potentials for analysis and study. Rather, nerve fibers in vivo conduct whole series of impulses in succession (ie, in *trains* or *bursts*), and the number of impulses transmitted along a nerve fiber or an entire peripheral nerve trunk per unit of time is called the *frequency of discharge* of the fiber or nerve. Thus, a single axon is able to convey a great deal more information per unit of time by modulating the net frequency or rate at which it discharges and conducts single impulses. The frequency of discharge can range from a few to over 2,000 impulses per second, depending on the properties of the specific fiber involved.

The maximum rate at which any particular axon is able to transmit individual impulses depends strictly upon the absolute refractory period of that fiber (Table 5.1). For example, if the absolute refractory period of a ·C fiber is 2 msec, then that nerve can transmit a maximum of about 500 impulses per second when it is not fatigued.

Almost 70 percent of all fibers in peripheral nerves are type C fibers. Even though these small, unmyelinated fibers conduct impulses at relatively low velocities and frequencies, their relative abundance in nerve trunks permits them to transmit an enormous quantity of information at a great saving in space.

In conclusion, variations in the frequency at which particular axons can transmit series of impulses under different physiologic conditions lend a much greater flexibility to the magnitude of the responses which are evoked in both receptor and effector neurons than would be possible in the absence of this modulation of their activity.

Energy Metabolism of Nerve

A major proportion of the total metabolic energy consumed by nerve is utilized to maintain the charge across the membrane; that is, this energy is required to drive the sodium-potassium pump. The activity of this pump is essential in order to maintain the steady-state resting potential across the neurilemma, as well as to restore the electrochemical gradients for sodium and potassium ions across the membrane following the generation of an impulse. The energy for driving the sodium-potassium pump is obtained directly from the hydrolysis of ATP, and continual aerobic metabolism is critical to the synthesis of this compound in adequate quantities to maintain neural function at normal levels.

During periods of maximal activity, the metabolic rate of nerve can double, as indicated by the increase in oxygen consumption above the resting level. In marked contrast to nerve, the metabolic rate of skeletal muscle can increase up to 100 times above the basal level during periods of intense activity.

Nerve, like muscle, exhibits a resting heat production while it is inactive; an initial heat is evolved above this basal level of heat production during the generation of an action potential; and a recovery heat is associated with the period following activity. The recovery heat measured following the passage of a single impulse along an axon is many times greater than the initial heat that is released during the development of the spike potential itself.

SURFACE POTENTIALS AND THE CONDUCTION OF IMPULSES IN NERVE AND MUSCLE

The bioelectric phenomena discussed so far in this and in preceding chapters have dealt exclusively with the resting and action potentials obtained by recording the voltage changes across the membranes of single cells using the technique illustrated in Figure 5.5.* That is, one recording electrode is inserted into the cytoplasm of the cell while the second remains on the surface, and thus the potential differences across the cell membrane during rest and activity are measured.

In this section, the patterns of bioelectric events as recorded by electrodes placed on the surface of nerve (and muscle) will be presented.

Biphasic and Monophasic Potentials

The technique employed to record surface potentials from single axons (or muscle fibers) in a nonconducting medium such as air is illustrated in Figure 5.11A and described in the footnote.

When the axon is in a steady-state or resting condition, no deflection of the galvanometer needle is seen, since there is no potential difference between the two recording electrodes, and a state of *isopotential* is said to exist along the fiber axis (Fig. 5.11B). When the axon is stimulated (Fig. 5.11C) and an impulse is propagated along the fiber (shaded area in figure), a typical sequence of potential changes develops. As the impulse passes under the recording electrode nearest to the stimulating electrode, this recording electrode becomes negative with respect to the second, more remote recording electrode (Fig. 5.11D) and the potential trace sweeps downward as shown in the graph to the right. As the impulse passes between the two recording electrodes (Fig. 5.11E), a state of isopotential is again achieved momentarily. When the impulse reaches the second recording electrode, the first recording electrode becomes positive with respect to the second, and the trace now shows an upward (or positive) deflection (Fig. 5.11F). Subsequently, the impulse passes away from the second recording electrode, and a state of isopotential develops once more as the fiber repolarizes (Fig. 5.11G).

The entire sequence of negative and positive changes in potential described above is called *biphasic action potential*. The duration of the isopotential interval between the two deflections is proportional to the velocity of impulse conduction within the fiber, as well as the distance between the two recording electrodes.

If the fiber beneath one of the recording electrodes is damaged (as by crushing the axon), the injured region becomes negative with respect to the normal portion of the axon (Fig. 5.11H), and a negative deflection of the galvanometer is observed. This potential difference between the normal and injured regions of the fiber is called the *demarcation potential* and is caused by a *current of injury* that flows between the two regions of the nerve. The magnitude of the demarcation potential is variable, depending upon the extent to which the membrane is damaged as well as the current of

injury flowing between the damaged and normal portions of the fiber, as recorded on the galvanometer, when the external circuit is completed by the electrodes and the galvanometer itself.

If an injured axon in which the demarcation potential is being recorded is stimulated, a transient deflection of the galvanometer to isopotential is seen as the impulse passes beneath the recording electrode situated on the undamaged portion of the nerve (Fig. 5.11I). No deflection in the opposite (positive) direction takes place, since the impulse stops at the injured region. The single deflection that is recorded (trace on right in Figure 5.11I) is called a *monophasic action potential*.

Both biphasic and monophasic action potentials, as well as demarcation potentials, can be obtained exactly as described above from surface electrodes placed upon normal and injured muscle fibers.

The Propagation of Surface Potentials in a Volume Conductor

The body fluids (ie, the blood, lymph, and interstitial fluid) contain high concentrations of electrolytes; consequently, the medium in which the nerves and muscles of the body function is a good conductor of electricity, or, as it is often called, a *volume conductor*. The simple biphasic and monophasic action potentials discussed in the previous section are observed only when a fiber is stimulated in vitro in a nonconducting medium. The alterations in surface potentials that are recorded in a volume conductor basically are like those described above. However, the actual configurations of the waveforms recorded in vivo are somewhat more complex because of the effects of current flow in the extracellular fluids of the body, which act as a volume conductor. Among the factors involved in altering surface potentials recorded in vivo are the position of the electrodes on (or within) the body relative to the direction in which the impulse is being propagated, and the distance between the recording and indifferent electrodes from the active tissue itself.

Generally, when an action potential is recorded in a volume conductor, positive deflections are seen both preceding and following the negative potential, as illustrated in Figure 5.12.

Two important measurements of surface potentials in vivo are used routinely by clinicians for diagnostic purposes as well as for evaluating the efficacy of drugs and other therapeutic measures. These are the *electroencephalogram*, or EEG (p. 420), and the *electrocardiogram*, or EKG (p. 600). In both instances, the action potentials (generated spontaneously by the brain and heart, respectively) are recorded by suitably located surface electrodes. The patterns of these action potentials show typical derangements (due to the presence of local demarcation potentials) when tissue death or injury has occurred due to, for example, a thrombosis or a necrotizing disease process.

---

*The sole exception to this statement is the brief discussion of electromyography presented on page 137.

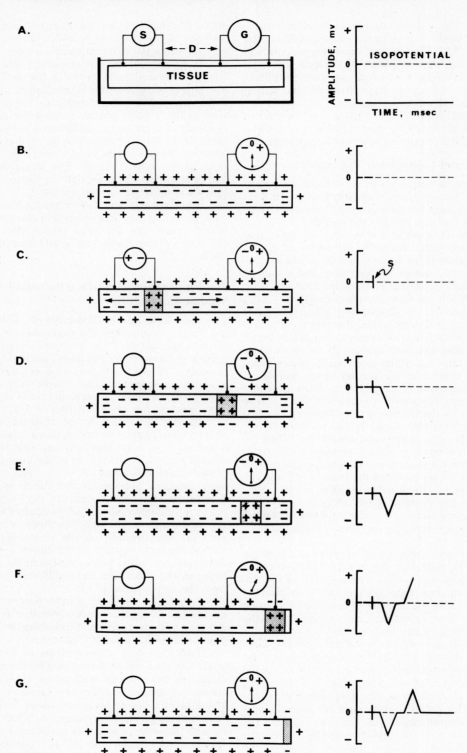

FIG. 5.11. Surface action potentials as recorded directly from irritable tissues. **A.** Technique used to measure directly the surface potentials of a tissue in either air or other nonconducting medium such as mineral oil. S = stimulator; D = distance between stimulating electrodes and recording electrodes; G = galvanometer (eg, strip chart recorder or cathode ray oscilloscope) to measure potential difference between the two recording electrodes. All electric events are graphed on the right side of the figure. **A** through **H.** Genesis of a biphasic action potential; S in part **C.** indicates stimulation of the tissue.

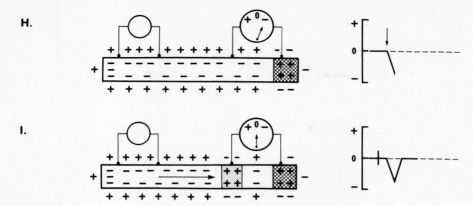

FIG. 5.11 (cont.) **H** and **I.** Genesis of a monophasic surface potential when one electrode is placed upon an injured region of the excitable tissue (nerve or muscle) as indicated by the cross-hatching.

Bioelectric Properties of Mixed Nerves

As discussed earlier in this chapter (p. 160), peripheral nerve trunks generally are composed of large numbers of individual axons of various sizes *(mixed nerves)* that are bound together in a common sheath of connective tissue, the epineurium (Fig. 5.4). It was also stressed that the velocity of impulse conduction along individual nerve fibers is directly proportional to their diameter (p. 169), and that the rate of impulse propagation is far greater in myelinated than in nonmyelinated nerves of the same diameter (p. 169). Futhermore, the thresholds to firing of the individual axons that comprise a mixed nerve may vary considerably, the smaller fibers having lower thresholds to excitation than the fibers with a larger diameter.

THE GRADED RESPONSE. The potential changes that are recorded following the stimulation of a mixed nerve are an algebraic summation of the single all-or-none action potentials developed by each of the axons present in that nerve. Since the thresholds to firing of the single axons, as well as their distances from the stimulating electrodes, can vary considerably, different bioelectric responses can be obtained from a mixed nerve by applying stimuli of different intensities. When subthreshold stimuli are applied to a mixed nerve, none of the fibers is stimulated, and thus no response is observed. When threshold currents are applied, the stimuli excite only those axons which have a low threshold to firing, and only a small potential change is seen. As the intensity of the stimulating current is increased, fibers with higher thresholds to firing are progressively discharged. Ultimately, a *maximal stimulus intensity* is reached, at which point *all* of the axons within the nerve trunk are excited simultaneously. Any further increases in the intensity of the stimulus to supramaximal levels will not evoke a potential of greater amplitude. Thus, in sharp contrast to the all-or-none action

potentials developed by single axons, a mixed nerve can generate a *graded response* depending upon the intensity of the applied stimulus.

Like mixed nerves, the individual fibers within an entire skeletal muscle exhibit slight differences in their thresholds to firing, and there are also significant variations in the distances between the fibers within a muscle and the stimulating electrodes. Consequently, the magnitude of any surface action potential that is recorded in vitro from a whole muscle is proportional to the stimulus intensity between threshold and maximal levels of the applied current. Skeletal muscle as well as nerve thus exhibits a graded electric response, depending upon the intensity of the stimulus employed to elicit the response.

COMPOUND ACTION POTENTIALS. In addition to the graded electric response shown by mixed nerves, the surface action potential recorded from such nerves exhibits a number of peaks; hence it is called a *compound action potential* (Fig. 5.13). Once again, this effect is in marked contrast to the action potentials recorded from a single axon. The complex waveform that is recorded from a mixed nerve trunk is caused by different conduction velocities found in the different groups of fibers within the nerve (Table 5.1). When all of the fibers within a mixed nerve are excited by a maximal stimulus, the impulses generated in the fastest-conducting fibers arrive at the recording electrodes sooner than the impulses propagated along the slower-conducting axons. The greater the distance between the stimulating electrodes and the recording electrodes, the greater the resolution of the peaks generated by the individual groups of fiber types (Fig. 5.13). The number and amplitude of the individual peaks in a compound action potential depend, of course, on the particular nerve under study as well as on whether or not the applied stimulus is maximal or submaximal. It should be emphasized that Figure 5.13 represents a composite diagram; all of the fiber

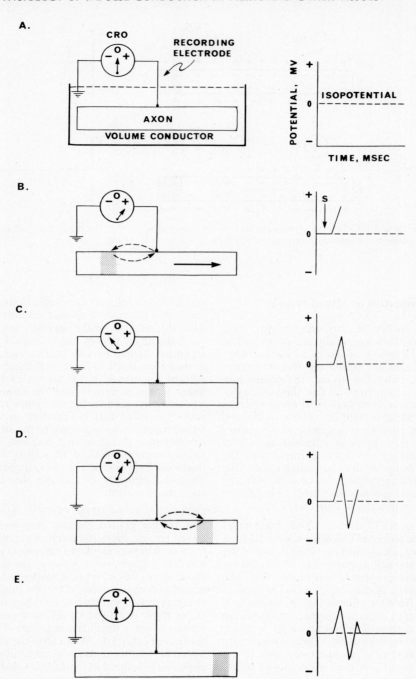

FIG. 5.12. Surface action potential, such as may be seen in an isolated axon placed in a volume conductor (eg, saline solution). **A.** CRO = cathode ray oscilloscope used as a galvanometer; stimulator not shown. Note that the indifferent electrode is placed at some distance from the tissue in the volume conductor as indicated by the symbol for ground. **B.** The impulse (shaded area) is passing from left to right along the axon following stimulation of the tissue (S in graph to right). The current flow, indicated by the curved arrows, causes a positive deflection from isopotential. **C.** As the impulse (negative wave) passes under the recording electrode, there is a downward deflection of the galvanometer. **D.** As the impulse passes to the right and the nerve repolarizes a positive deflection is seen once again. **E.** Ultimately the impulse has passed far enough from the recording electrode so that electrotonic potentials no longer are recorded, and a state of isopotential is achieved once more.

types illustrated are not found in any one particular nerve. The functional aspects of the various fiber types shown in this figure correlate well with the morphologic properties of the same fibers, as summarized in Table 5.1 and discussed on page 161.

In contrast to mixed nerves, whole skeletal muscles do not exhibit compound action potentials because the conduction velocities among the individual fibers do not differ sufficiently to be recorded.

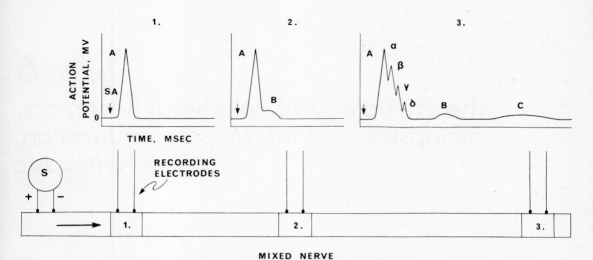

MIXED NERVE

FIG. 5.13. The genesis of a compound action potential following stimulation of a mixed nerve. S = stimulator. Impulse passing toward right and indicated on the bottom of the figure as passing under recording electrodes placed at three successively greater distances from the point of stimulation (numbered boxes). The action potentials as recorded from the three regions of the nerve are indicated by small graphs at the top of the figure. 1. SA = stimulus artifact, indicated by arrow. When the recording electrodes are close to the stimulating electrodes, only one action potential is recorded from the fast-conducting A fibers. 2. When the electrodes are separated more widely along the nerve, two peaks representing the fast-conducting A and slower-conducting B fibers are recorded. 3. When the electrodes are separated still farther, the various components of the A fiber group which conduct at different velocities are resolved (α, β, γ, and δ) and the still slower-conducting B and C fiber groups are seen. It must be stressed that not all of these peaks are recorded from any particular nerve. Rather this conceptual diagram has been designed to illustrate all of the possible fibers whose activity may be recorded from different nerves.

The Physiology of Peripheral Receptors, Synapses, and the Myoneural Junction: The Reflex Arc

Every peripheral nerve fiber, whether it is sensory or motor, eventually ends in a peripheral tissue, where one or more terminal ramifications of the nerve are found. Many nerve fibers branch or arborize as free endings among the nonneural cells of the tissues, whereas others are closely attached to the cells they innervate through morphologically specialized terminations. Sensory nerve endings terminate in *receptors,* and functionally they are dendrites, whereas peripheral nerves that subserve *motor* or *secretory* functions are axons, and their terminals are axon endings (Fig. 5.2, p. 159).

In general, the morphologic pattern at any nerve ending, whether it be sensory or motor, is designed to increase the total area of physical contact between the neural element and the related nonneural cells which it innervates.

All information from the external and internal environments, whether or not this information reaches the conscious level of perception, is conveyed to the central nervous system from various peripheral *sense organs* by means of coded nerve impulses. Within the sense organs themselves are located the *receptor elements* or *cells* that convert or *transduce* various types of energy into propagated action potentials in the sensory nerves. For example, mechanoreceptors are responsive to physical deformation and give rise to sensations of pressure and touch; thermal receptors are sensitive to alterations in the heat content of the environment in their immediate vicinity and give rise to sensations of heat and cold; electromagnetic energy (ie, light waves) are transduced into nerve impulses within the retina of the eye; and chemical energy can be perceived as various tastes or odors. In addition, certain chemoreceptor cells are sensitive to the oxygen and carbon dioxide content of the blood and thus continuously monitor the

physiologic levels of these gases. A more complete listing of the specific types of sensation, or *sensory modalities,* is given in Table 6.1.

The receptor elements of every sense organ within the body are adapted in such a way that they respond to a specific kind of energy at a much lower threshold than do other types of receptor; this energy is called the *adequate stimulus* for the particular receptor involved. This concept sometimes is referred to as the *law of the adequate stimulus.* Although receptors in general are able to respond to various forms of energy other than their adequate stimulus, the threshold to excitation is much greater for these nonspecific types of stimulus. For example, electromagnetic radiations between certain frequencies excite the rods and cones within the retina of the eye quite readily, whereas they do not have any effect whatsoever upon the various cutaneous receptors, including the thermoreceptors. And yet electromagnetic waves of high frequency, such as are used in medical diathermy, can elicit a sensation of warmth in the area which is irradiated (Fig. 1.3, p. 5).

In this chapter, the general mechanisms involved in the functions of certain receptors will be considered first. Next will follow a discussion of the specialized physiologic properties of *synapses,* those regions of the nervous system responsible for functional continuity among the individual nerve elements and integration of incoming sensory impulses to produce an appropriate response in particular *effector organs* or structures (ie, muscular or glandular cells and tissues). The *neuromuscular* or *myoneural junction* in skeletal muscle will receive particular attention, since this specialized region at the termination of the motoneuron upon individual skeletal muscle fibers is of particular physiologic importance.

Table 6.1 A CLASSIFICATION OF SENSORY MODALITIES

SENSORY MODALITY	SENSE ORGAN	CLASSIFICATION
1. Pain 2. Touch 3. Pressure 4. Warmth 5. Cold	Unspecified peripheral nerve endings	Superficial or cutaneous senses (exteroceptors)
6. Stretch of muscle 7. Stretch of muscle 8. Position of joints	Muscle spindles Golgi tendon organs Bare nerve endings	Visceral senses (proprioceptors)
9. Arterial blood pressure	Receptors in walls of carotid sinus, aortic arch	Visceral senses (interoceptors)
10. Venous pressure	Receptors in walls of great veins, atria	
11. Inflation of lung	Endings of vagus nerve in parenchyma of lung	
12. Temperature of blood in head	Hypothalamic cells	
13. Oxygen partial pressure in blood	Aortic and carotid bodies	
14. pH of cerebrospinal fluid	Receptors of medulla oblongata (ventral surface)	
15. Plasma osmotic pressure	Receptors in anterior hypothalamus	
16. Arteriovenous blood glucose difference	Hypothalamic glucostatic cells	
17. Vision 18. Olfaction 19. Audition	Eye: rods and cones Olfactory mucous membrane of nose Ear: cochlea	Special senses (teloceptors)
20. Taste 21. Linear acceleration 22. Rotational acceleration	Tongue: taste buds Ear: utricle Ear: semicircular canals	Special senses

Based upon the foregoing information, the structural and functional components of the *reflex arc,* the pattern that forms the basic physiologic unit of integrated neural activity within the body (Fig. 6.1), will be discussed. The Roman numerals in Figure 6.1 indicate the first four principal divisions of this chapter. The reader will also note in the figure that *any* reflex arc contains a minimum of five principal elements (indicated by the circled Arabic numerals) and that the sequence of topics in the following discussions moves from the periphery toward the central nervous system and back to the periphery once again. Subsequently, the more complex neural patterns underlying the *monosynaptic stretch reflex* as well as certain *polysynaptic reflexes* will be presented.

The final section of this chapter will deal with a number of pertinent clinical states that serve both to review and to summarize certain of the information presented in this as well as the preceding two chapters.

GENERAL MORPHOLOGIC AND PHYSIOLOGIC PROPERTIES OF SOME PERIPHERAL RECEPTORS

Introduction

The simplistic view that the body contains only five major senses (ie, vision, audition, olfaction, gusta-

tion, and touch or pressure) obviously is quite inadequate to cover the much broader range of sensory information that is transmitted to the central nervous system from peripheral sense organs. In Table 6.1 are listed 22 distinct types of sensation, or sensory modalities, and even this list probably is incomplete. The reader will note that some of these sensory modalities are perceived consciously (eg, pain, sensation, vision) whereas others normally are transmitted from receptors to effectors at a wholly unconscious level (eg, the mechanism whereby the oxygen partial pressure of the blood is monitored by the aortic and carotid bodies).

A "sensation" that we "feel" has several distinct properties or characteristics. First, the *quality* of a sensation enables a person to describe that sensation in a subjective fashion. For example, hot or cold, an odor, or a taste are all distinct sensory modalities to the person experiencing them. The quality of a particular sensation may have one to many distinct submodalities; the colors blue, red, and yellow are submodalities of vision, for example, whereas bitter and sour are submodalities of taste. Second, the *intensity* is another basic characteristic of a sensation, since no perceived sensation occurs unless the stimulus achieves a definite subjective threshold level. Third, *localization* of the sensation either as coming from outside of the body or as originating at some particular region within the body is another obvious property of all con-

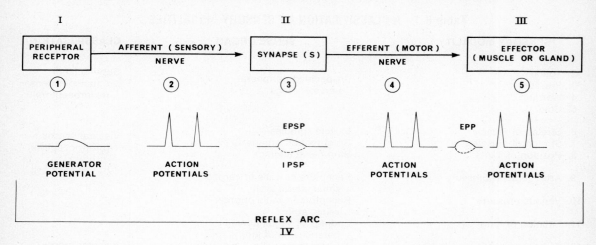

FIG. 6.1. The principal morphologic and functional components of a reflex arc (circled Arabic numerals). The electrophysiologic response of each component of the reflex arc is indicated in the center of the figure immediately below the numbered structure to which it applies. ① The generator potential is a local, graded response. ② Action potentials in the afferent nerve are all-or-none responses. ③ Synaptic potentials are graded, local responses, and may be either excitatory (EPSP, solid curve) or inhibitory (IPSP, dashed curve). ④ Action potentials in the efferent nerve are all-or-none responses. The end plate potential (EPP) at the junction between efferent nerve and effector can be either excitatory (solid curve) or inhibitory (dashed curve). Action potentials in the effector (muscle) are all-or-none responses. The Roman numerals indicate the first four major divisions of Chapter 6.

scious sensory experience. These aspects of sensory physiology will be explored more fully in Chapters 7 and 8; they are introduced at this point, however, to enable the reader to understand more fully the basic concept of sensory modalities before a consideration of the general morphologic and functional aspects of certain peripheral receptors is undertaken.

Classification of Sense Organs

No classification of the sense organs that has been developed to date is entirely satisfactory. The classifications in most common use are given in Table 6.1, and each of these has certain useful features.

According to one scheme, the *superficial* or *cutaneous senses* are those whose receptors are present in the skin and thus respond to stimuli from the external environment in the immediate vicinity of the body surface. The *visceral senses* include sensations that arise within the internal environment from the deeper-lying structures and organs; included in this category are deep pressure and deep pain (not listed in Table 6.1), the *special senses* are vision, olfaction, and taste as well as the senses concerned with equilibrium.

A second classification of sense organs divides the receptors into four major groups:

First, the *exteroceptors* are sense organs that are concerned with the response to various stimuli arising in the immediate external vicinity of the body.

Second, the *proprioceptors* are sensory receptors found within the skeletal muscles and tendons that are concerned intimately with postural and

phasic contractions of the skeletal muscles. These functions must not be confused with those of the sensory receptors in the joints which apprise the individual continually of the position of his body and its parts in space during the waking state. This "position sense" or "muscle sense" is another aspect of proprioception, but it is synthesized at the conscious level from impulses that originate in the cutaneous sensory receptors for touch and pressure, in addition to those impulses that arise in the joints themselves (see also p. 356).

Third, the *interoceptors* are concerned with the response to stimuli that arise within the internal environment.

Fourth, the *teloreceptors* are those receptors that receive sensory information from sources located at a distance from the body (ie, the receptors in the eye and ear).

In addition to the two classifications listed in Table 6.1, other terms sometimes are applied to describe the structures that underlie various sensory modalities. The term *chemoreceptor* is used to denote those receptors in sense organs which are stimulated by appropriate chemical changes in their immediate environment.* Included in this category are the special senses of olfaction and taste, as well as the visceral receptors which monitor alterations

*Sometimes certain deep-lying chemoreceptors also are classed as proprioceptors: for example, the aortic and carotid bodies which continually monitor the partial pressure of oxygen within the blood. The sensory modalities of taste and olfaction are not considered to be proprioceptive sensory modalities.

in the plasma levels of oxygen and carbon dioxide, and osmolality.

The terms *mechanoreceptor* and *pressoreceptor* signify types of receptors that are sensitive to mechanical deformation or pressure changes, respectively. Thus, the proprioceptors within skeletal muscles that are sensitive to tension (ie, the muscle spindles and Golgi tendon organs) may also be classified as mechanoreceptors.

The general term *thermoreceptor* sometimes is used to designate the neural structures that are responsible for detecting alterations in heat energy of the environment.

The term *photoreceptor* is reserved for the receptors in the eye that are sensitive to electromagnetic radiations of wavelengths that fall within the visible spectrum.

The term *nociceptor* is applied to pain receptors, because this sensory modality is elicited by a wide variety of stimuli that have one common property: the ability to be noxious if not actually damaging to the tissues. Such stimuli also can elicit powerful withdrawal reflexes (p. 222).

The terms *somatic* and *visceral* often are used in connection with sensory (as well as motor) phenomena, depending upon the region of the body within which the sensation originates — the more superficial somatic, or the deeper-lying visceral regions.

Sensations of pressure and pain that arise from the cutaneous regions sometimes are considered to be different sensory modalities than the similar sensations of deep pressure and pain that originate within muscles, tendons, and deep fasciae. This distinction is made because qualitatively (ie, subjectively) these sensations are quite different to the person who is experiencing them, a fact that is of considerable clinical importance.

In the present as well as succeeding chapters, the terminology to be used with reference to particular sense organs and sensory modalities will be consistent with the most common usage, and, where applicable, appropriate synonyms for these terms will be included.

Superficial or Cutaneous Receptors

The five sensory modalities that arise from stimulation of the skin or integument (Table 6.1) once were believed to originate in five types of histologically distinct and functionally specialized peripheral end organs. According to this concept, the sense of pain was subserved by bare or naked nerve endings; the touch sensation was elicited by stimulation of Meissner's corpuscles; pressure was the excitatory stimulus for the pacinian corpuscles; a sensation of warmth was evoked by stimulation of Ruffini's end organs; and cold was the effective stimulus for excitation of the Krause's end bulbs. It is now clear that this tidy pattern of morphologic–physiologic specificity is somewhat inaccurate insofar as four of the cutaneous sensory modalities and their receptors are concerned, the possible exception to this state-

ment being the sense of pressure that is evoked by deformation of the pacinian corpuscles (Fig. 6.2). Consequently, there is much controversy over the extent to which histologically discrete receptors underlie particular cutaneous sensations.

It is definitely clear, however, that the distribution of receptors within the skin for the different modalities of cutaneous sensation is localized in small points (ie, it is *punctate*), and that the receptors for these sensory modalities are not distributed uniformly over the entire integumentary surface of the body. Furthermore, there is a definite *functional specificity* insofar as the receptors for pain, touch, pressure, heat, and cold are concerned. Therefore, regardless of the presence or absence of particular histologically defined end organs in specific areas of the skin, there is a *functional differentiation* and *specificity* of the various nerve endings that are sensitive to different types of stimulus and, in turn, give rise to sensations of pain, touch, pressure, warmth, and cold.

FREE NERVE ENDINGS. A free or naked nerve ending in a peripheral tissue is produced by the repeated branchings of an axon into many fine terminals at its end. Near these endings, the axon loses its myelin sheath (if one is present), but its sheath of Schwann remains intact until the ultimate terminations are reached. As explained in Chapter 5 (p. 158) and illustrated in Figure 5.2 (p. 159), the terminal ramifications of a sensory neuron are considered to be dendrites in the functional sense of the word.

The naked terminal branches of a neuron ramify among the cells of the epidermis as well as the dermis, and the ramifications derived from a single neuronal process may anastomose to form a sort of *plexus* or *nerve net*. The nerve nets found at the tips of individual axons may interdigitate or overlap, but they do not form a syncytium, since no protoplasmic interconnections are present among the ultimate branches of the individual axons. Nerve fibers from the superficial cutaneous plexus can ramify over a wide area of skin and end in fine, unmyelinated terminals with beadlike swellings at their tips. The nerve endings associated with hair follicles have a similar interlocking pattern. The branches of one afferent axon may innervate as many as 100 hairs located over an area of several square centimeters of skin, or a single hair follicle may be innervated by the branches of up to 20 single axons.

Only very delicate free nerve endings and nerve networks are found in hairy skin, and these alone must subserve the several sensory modalities that arise from such regions of the body surface.

ENCAPSULATED NERVE ENDINGS. In contrast to hairy skin, smooth or glabrous skin contains both free nerve endings and nerve terminals enclosed within connective tissue capsules of varying complexity. Therefore, from a histologic standpoint, such structures as Meissner's corpuscles, pacinian corpuscles, Ruffini's end organs, and

Krause's end bulbs are well-defined morphologic entities. Whereas it was once thought that the differential sensitivity of cutaneous receptors was conferred upon the nerve terminals by the capsule per se, more recent evidence indicates that the capsules of certain end organs may act as mechanical protective devices against excessively strong stimuli, regardless of the type of stimulus involved.

The histologic structure of the encapsulated nerve endings may be described briefly. In all instances, the nerve endings are surrounded by a connective tissue sheath of variable thickness, depending upon the particular end organ.

Meissner's Corpuscles. Meissner's corpuscles are elongated, generally elliptical structures that are located in the papillae of the skin with their long axes perpendicular to the surface (p. 995). The nerve enters the capsule near the deeper-lying end and spirals upward to terminate in small knobs.

Meissner's corpuscles are found in skin of the tips of the fingers and toes, as well as the palms of the hands and soles of the feet. They also are seen in the skin of the forearm, lips, the mucous membrane at the tip of the tongue, the conjunctiva, and the mammary papillae.

Pacinian Corpuscles. The corpuscles of Vater–Pacini, or pacinian corpuscles, are large (1 to 4 mm in length and 2 mm in diameter) white bodies whose thick connective tissue sheath is organized into concentric layers so that, in cross section, these end organs appear somewhat like microscopic onions cut in half (Fig. 6.2). Each corpuscle is supplied with one or more straight, unmyelinated nerve terminals, the myelin sheath being lost after the nerve enters the connective tissue capsule. The first node of Ranvier is located within the corpuscle, as depicted in Figure 6.2.

Pacinian corpuscles are found in the deeper layers of the skin as well as beneath mucous membranes, in the conjunctiva, heart, pancreas, mesentery, and in other connective tissues. The *genital corpuscles* found in the skin of the external genitalia and the nipples are histologically similar to the pacinian corpuscles.

Because of their relatively large size as well as their ready accessibility in the mesentery of various experimental animals (eg, the cat), pacinian cor-

puscles are ideal for electrophysiologic study. Consequently, much of the information on generator potentials in sensory receptors has been gained by the use of these end organs.

Ruffini's End Organs. The end organs or corpuscles of Ruffini are oval in shape. They are surrounded by a sheath of strong connective tissue within which the nerve fibers ramify into many branches exhibiting numerous irregular swellings or varicosities. The terminations of the nerves within these structures are small, free knobs.

Ruffini's corpuscles are found primarily at the junction of the skin with the subcutaneous tissue and are especially prominent in the fingers, whence they were first described.

Krause's End Bulbs. The terminal or end bulbs of Krause are similar morphologically to the pacinian corpuscles; however, they are smaller and somewhat less complex structurally than the latter.

FUNCTIONAL SPECIFICITY OF CUTANEOUS SENSORY ENDINGS. As described earlier, no encapsulated endings of any kind are found in hairy skin, or in the cornea of the eye, and only delicate free nerve endings and fine networks of nerves about the roots of the hairs are present. Yet hairy skin is sensitive to pain, pressure, touch, warmth, and cold, whereas the cornea is sensitive to pain as well as touch.

On the other hand, smooth or glabrous skin, such as is found on the fingertips and lips, does contain various morphologically specialized end organs as described in the preceding section, and these structures are present in addition to free nerve endings such as are found in hairy skin and the cornea. Nevertheless, the smooth and hairy skin on nearby areas of the body surface show quite similar sensitivities to various stimuli. Thus, it appears that the different modalities of cutaneous sensation are not related to any specific morphologically defined end organs, and that many gradations of sensation are possible (see also p. 177). Furthermore, it has been demonstrated in electrophysiologic studies that individual free nerve endings of C fibers (Table 5.1, p. 161) in the skin can be specifically and keenly sensitive to heat or cold as well as to touch. Therefore, a physiologic specificity rather than a morphologic specificity is present in cutaneous receptors.

Proprioceptors in Skeletal Muscle: Muscle Spindles and Golgi Tendon Organs

A *proprioceptor* can be defined as a sensory receptor that is located deep within the tissues of the body, such as those found in skeletal muscles and their tendons. The proprioceptors of skeletal muscle serve as exquisitely sensitive detectors for alterations in the state of contraction of the particular muscle in which they are located; hence, they are excited by alterations in tension within the muscle itself. Therefore, since mechanical deformation is

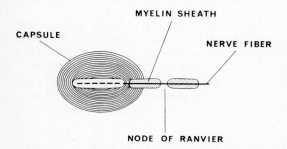

MYELIN SHEATH

CAPSULE

NERVE FIBER

NODE OF RANVIER

FIG. 6.2. Structure of the pacinian corpuscle.

the stimulus for these structures, the proprioceptors of skeletal muscles also can be classified as mechanoreceptors. Some of the receptors in skeletal muscle and tendons are simple bare terminal branches of neurons that are closely applied to the fibers in various patterns, whereas others are morphologically discrete structures. The *neuromuscular spindles,* or simply *muscle spindles,* are encapsulated sensory end organs located among the fibers in the majority of skeletal muscles, whereas the structurally similar *Golgi tendon organs* are found at the junctions between the muscle fibers and their tendons.

In contrast to the cutaneous receptors, there is an excellent correlation between the structure and function of the muscle spindles and the Golgi tendon organs.

THE MUSCLE SPINDLE

Structure. Neuromuscular spindles are found only in the skeletal muscles of higher vertebrates. These end organs are narrow, elongated (0.7 to over 7 mm) structures that generally are located near the junctions of muscles with their tendons. The long axes of the neuromuscular spindles are oriented in parallel with the similar axes of the muscles themselves. Muscle spindles have been demonstrated in all of the skeletal muscles of the body except in the tongue; also, these structures are of scanty occurrence in the ocular muscles.

Histologically, a muscle spindle consists of between two and ten thin skeletal muscle fibers that are enclosed within a dense fibrous connective tissue capsule (Fig. 6.3). The muscle fibers within a spindle are called the *intrafusal fibers,* and characteristically they are quite narrow, with an abundant sarcoplasm and peripherally located nuclei. This structure is in contrast with the *extrafusal fibers,* which make up the contractile machinery of the muscle proper.

In overall appearance, the muscle spindle is thicker in its central region and hence is fusiform

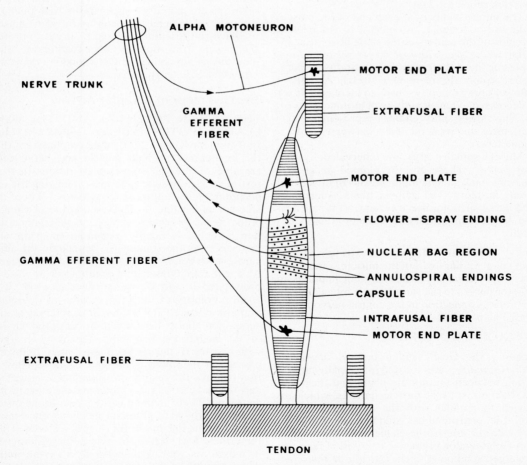

FIG. 6.3. The histology of a muscle spindle and its neural connections. Only one intrafusal fiber is indicated; note that this fiber is arranged in parallel with the extrafusal fiber. The annulospiral endings are continuous with type I(a) sensory nerves, whereas the flower-spray endings are continuous with type II sensory nerves (Table 5.2, p. 162). The gamma efferent fibers innervate the intrafusal fibers, whereas alpha motoneurons innervate the larger extrafusal fibers that comprise the bulk of the muscle.

(ie, spindle shaped). Near the middle of each intrafusal fiber, the typical cross-striations are absent and many nuclei are present. This region is called the *nuclear bag* of the spindle.

Every spindle is innervated by two kinds of sensory nerve fiber. These relatively thick, and hence rapidly transmitting, axons are covered by a thin layer of cytoplasm (derived from Schwann cells) and terminate in either annulospiral or flower-spray endings. The *primary* or *annulospiral endings* are the principal sensory receptors of the muscle spindle. These elements are the endings of afferent neurons whose fibers are about 8 to 12 μ in diameter, and they surround the noncontractile nuclear bag region in the center of each spindle. The intrafusal muscle fibers are attached in parallel with the extrafusal muscle fibers. Therefore, when the entire muscle is stretched, the annulospiral endings also are stretched, and, in response to this mechanical deformation, they discharge afferent impulses of higher frequency. Consequently, the annulospiral endings are the mechanoreceptor for the stretch reflex (p. 218).

The intrafusal fibers at each end of the muscle spindle are contractile, in contrast to the nuclear bag region at the centers of the same fibers. Receptors known as *secondary* or *flower-spray endings* are present near the ends of the intrafusal fibers. These receptors also respond to stretch. Activity in the flower-spray receptors leads to increased flexor and decreased extensor activity of motoneurons (p. 218). The flower-spray endings represent the terminations of smaller nerve fibers between 6 and 9 μ in diameter.

Muscle spindles also are innervated by thin *motor nerves*. These are the *gamma fibers* that arise from small gamma motoneurons in the central nervous system.* The gamma fibers, sometimes called the small motor nerve system, terminate near the ends of the intrafusal fibers in motor end plates (p. 220).

General Function. Mechanical stretch of the annulospiral endings within the muscle spindles increases the frequency of discharge of impulses in the motor nerves to the same muscle by means of a direct reflex connection in the central nervous system (see also p. 219).

THE GOLGI TENDON ORGAN

Structure. The Golgi tendon organs, or *neurotendinal spindles*, are located principally at the junction between the muscle fibers and their tendons. In contrast to the muscle spindles, which are arranged in parallel with the extrafusal muscle fibers, the neurotendinal spindles are found in *series* with the contractile elements of the muscle.

A Golgi tendon organ is composed of fascicles of enlarged tendinous fibers (*intrafusal fascicles*) that are enclosed within a connective tissue capsule. The myelinated sensory nerve fibers that innervate a Golgi tendon organ ramify extensively and terminate in knoblike swellings among the fibers of the tendon (Fig. 6.4). In addition, between 3 and 25 muscle fibers are included within each neurotendinal spindle, as depicted in the figure.

General Function. Since the Golgi tendon organs, unlike the muscle spindles, are in series with the contractile extrafusal muscle fibers, they are stimulated both by passive stretch of the muscle and by its active contraction. Within limits, the force of reflex contraction of a muscle is in direct proportion to the degree to which it is stretched prior to the contraction (p. 218). If the tension within the muscle becomes too great, however, contraction ceases, and the muscle undergoes a sudden relaxation. The Golgi tendon organs are the receptors that are stimulated by the development of an excessive tension within muscle, and they are the sensory elements in the reflex that produces the relaxation under such conditions (p. 221). This *inverse stretch reflex* is an inhibitory mechanism that operates through the alpha motoneurons of the same muscle in which the receptors lie, and it serves to protect the muscle from rupture. Other features of reflex mechanisms involving the Golgi tendon organs will be discussed later (p. 221).

The Physiology of Receptor Activity

THE ACTION POTENTIAL AND THE GENERATOR POTENTIAL: A COMPARISON. In Chapters 4 and 5, it was emphasized that the membranes of both muscle and nerve cells are able to undergo transient alterations in their resting or steady-state potentials following the application of an appropriate stimulus. When the stimulus is capable of depolarizing the cell membrane to its threshold voltage, an all-or-none action potential is rapidly generated (p. 165), and this serves as the stimulus to depolarize the adjacent, inactive area of the membrane. Consequently, a self-amplifying action potential or impulse is propagated rapidly along the fiber (p. 168). The electric polarity of the active region of the membrane is reversed for a very brief interval as the action potential sweeps down the fiber, the outside of the membrane becoming negatively charged with respect to the inside. Immediately following passage of the impulse, the process of repolarization restores the resting transmembrane potential, and the outside of the membrane once again becomes positively charged with respect to the inside as the steady-state potential develops once more and the full irritability of the fiber is restored. Thus the critical first step in the generation of impulses in muscle as well as in nerve fibers is an initial depolarization of the resting membrane.

Under experimental conditions, electricity is the stimulus of choice employed to stimulate excitable tissues in order to study their responses. In the living organism, however, the primary stimuli

---

The large motor neurons of the central nervous system that send axons to the extrafusal fibers of the muscle proper are called alpha motoneurons.

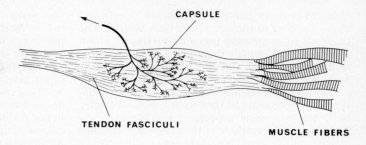

FIG. 6.4. Structure of the Golgi tendon organ. These sensory elements are in series with the muscle fibers, in contrast to the muscle spindles, which are in parallel. (After Gray. *Anatomy of the Human Body,* 28th ed, 1966. Goss CM [ed], Lea and Febiger.)

responsible for eliciting activity in muscles and nerves are *not* electric in nature. Rather, the activity of excitable tissues under physiologic conditions in vivo is evoked by the application of mechanical, chemical, thermal, or electromagnetic (ie, light) energy to the functionally specialized terminals of various afferent nerve fibers, the *receptors*. An adequate stimulus applied to any receptor produces a depolarization of the afferent nerve terminals; the magnitude of this depolarization is, within limits, directly proportional to the intensity of the stimulus (Figs. 6.1, 6.5). If the magnitude of this depolarization within the receptor element is sufficient, action potentials are initiated in the afferent nerve fiber that is continuous with the receptor as the transmembrane voltage of the neurilemma is depolarized to its firing level by the potential generated within the receptor. There-

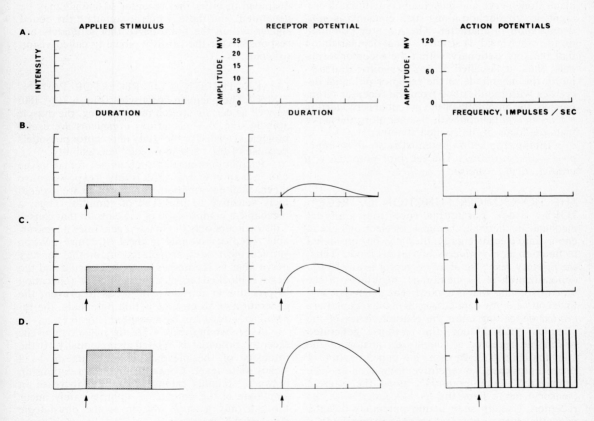

FIG. 6.5. The relationship among stimulus intensity, amplitude of the receptor potential, and frequency of discharge in the nerve that is continuous with the receptor. **A.** In each column, the ordinate and abscissa indicate the units for **B, C,** and **D.** The intensity of the applied stimulus is in arbitrary units (shaded areas, **B, C, D**). **B.** Stimulus applied at vertical arrow produces a receptor potential which is inadequate to evoke an all-or-none response in the nerve. **C.** Stimulus with an intensity double that in **B** evokes a receptor potential that is of sufficient magnitude to cause firing of all-or-none impulses in the nerve continuous with the receptor. **D.** Stimulus of still greater intensity evokes a receptor potential of still greater magnitude, and, as a consequence, the rate of firing of the afferent nerve increases. Note particularly that the receptor potential is coded by amplitude modulation in relation to the stimulus intensity, whereas the frequency of discharge in the afferent nerve is coded by frequency modulation in proportion to the magnitude of the receptor potential.

fore, all-or-none nerve impulses are propagated along the fiber to the central nervous system. The decrease in membrane potential produced within the receptor terminals themselves by the application of an adequate stimulus is called the *receptor potential* or the *generator potential* (Fig. 6.5).

It is clear from the facts presented above that the action potential and the receptor potential differ in several important respects.

First, the magnitude of the action potential has a fixed, all-or-none character; hence, the electric response of nerve (and muscle) is independent of the stimulus intensity, provided that the stimulus exceeds the threshold for excitation of the fiber. The magnitude of the receptor potential, on the other hand, is proportional to the intensity of the applied stimulus. Thus, the receptor potential is a *graded response* in contrast to the all-or-none response of the action potential.

Second, the action potential is conducted along the entire length of the fiber without undergoing a decrease in its amplitude. That is, conduction takes place along nerve and muscle fibers without decrement; ie, conduction along such elements is *nondecremental* in character. The receptor potential, on the other hand, is strictly a local depolarization that remains stationary within the receptor terminals; thus, it is called a *local response*. Furthermore, the magnitude of the receptor potential decreases with the distance from the receptor terminals; ie, the amplitude of the potential is reduced in a *decremental* fashion with distance along the nerve from the locus of the applied stimulus.

Briefly, the action potential is an *all-or-none, propagated response*. The receptor potential is a *graded, local response*.

THE TRANSDUCER FUNCTION OF RECEPTORS.

Since particular receptors convert mechanical, thermal, chemical, or electromagnetic energy into electric energy, they may be considered to function as *transducers* of various kinds. Thus, the primary response of any receptor is an electric depolarization of its membrane, regardless of the type of stimulus that is applied. The magnitude and time course of the depolarization in the receptor are *graded* depending upon the characteristics of the stimulus itself. Hence, the response generated within the receptor is roughly an analog of the stimulus (Fig. 6.5). In turn, the depolarization of the receptor causes a repetitive firing of all-or-none propagated action potentials along the afferent (sensory) nerve fiber that is associated with the receptor. Usually such action potentials develop when the receptor potential reaches an amplitude of about 10 mv. The frequency of discharge in the afferent nerve is directly proportional to the magnitude, or amplitude, of the receptor potential generated by the stimulus in the receptor element. Therefore, when the magnitude of the receptor potential increases above 10 mv, the discharge frequency in the afferent nerve increases accordingly

(see also p. 220). Thus, the rate at which an afferent nerve fiber discharges all-or-none conducted impulses is a direct and coded reflection of the intensity of the stimulus that is acting upon the receptor.

The facts presented above may be restated in physical terms. A receptor converts, or transduces, one type of energy (eg, mechanical or thermal) into an electric signal, whose magnitude is graded, and hence coded, according to the intensity and time course of the stimulus. Therefore, coding of receptor information takes place by *amplitude modulation*. When the local, graded receptor potentials are changed into conducted all-or-none signals in the sensory nerve fiber associated with the receptor element, the intensity of the stimulus is coded by *frequency modulation* (Fig. 6.5).

The concepts presented above apply to all receptors. However, the biologic transduction of energy into receptor potentials sometimes involves additional processes, particularly in the organs that subserve the special senses (Chapter 8). Within the eye, for example, a visual pigment must first be degraded in order that receptor potentials may be generated, and these, in turn, stimulate the coded signals within the optic nerve that ultimately are responsible for the sensory modality called vision (p. 466).

CHARACTERISTICS OF RECEPTOR POTENTIALS.

The pacinian corpuscles (Fig. 6.2; p. 180) as well as certain stretch receptors (eg, the muscle spindle, Fig. 6.3) of various organisms are useful biologic systems for the study of receptor potentials because of their size and ready accessibility.

Pacinian corpuscles such as are found in the cat mesentery are selectively responsive to mechanical deformation, and they also are exceedingly sensitive to this type of stimulus. Thus, a mechanical compression of the capsule that causes a distortion of only 0.2 to 0.5 μ generates a measurable receptor potential in these structures. When surface electrodes are placed upon the sensory nerve fiber as it emerges from the capsule and the nerve is blocked (eg, by compression or application of procaine or tetrodotoxin) so as to prevent the generation of all-or-none action potentials, the receptor potential can be measured (Fig. 6.6).

As shown in Figure 6.7A, the magnitude of the receptor potential is related exponentially to the intensity of the mechanical stimulus up to a steady-state level. Consequently, the relationship between stimulus intensity and the increase in amplitude of the potential is approximately linear through only a small portion of the physiologic range of this receptor.

The rate at which the receptor potential develops with increasing stimulus intensity is depicted in Figure 6.7B.

Another characteristic of the receptor potential generated in the pacinian corpuscle is also illustrated in Figure 6.7B. When the mechanical deformation is of long duration, the receptor potential

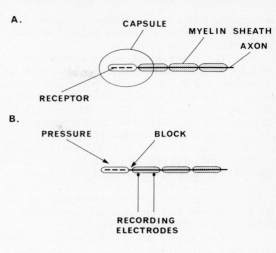

FIG. 6.6. **A.** The receptor portion of the axon terminal within a pacinian corpuscle is indicated by the dashed line. **B.** When the capsule is removed and the first node of Ranvier is blocked (as by locally applied procaine), the receptor potential can be measured independently of any all-or-none electric activity in the afferent fiber when graded mechanical stimuli are applied to the receptor region. Similarly, receptor potentials also can be recorded by mechanical deformation of the capsule following blockade of the afferent nerve. (See also Fig. 6.11.)

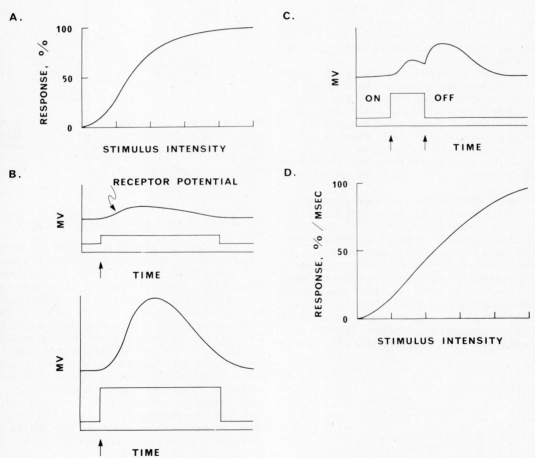

FIG. 6.7. Certain electrophysiologic properties of pacinian corpuscles. **A.** Relationship between the magnitude of a receptor potential and the intensity of the applied stimulus. **B.** The rate of development of the receptor potential, as well as its magnitude, is related directly to the intensity of the stimulus. Note also in both parts of this diagram that the magnitude of the receptor potential decreases with time, even though the stimulus intensity remains constant (cf. Fig. 6.5B, C, D). This phenomenon is called adaptation. **C.** Following application of a mechanical stimulus of short duration to a pacinian corpuscle, two summated receptor potentials can be observed. These are called the on and off response, respectively. **D.** If the intensity of an applied mechanical stimulus is held constant, then the magnitude of the receptor potential depends upon the rate of application of the stimulus.

reaches a peak, then declines to zero. This phenomenon, known as *adaptation*, will be discussed further in the next section.

On the other hand, when a mechanical stimulus of short duration is applied to a pacinian corpuscle, two distinct receptor potentials, known as the *on response* and *off response*, are generated, and can summate (Fig. 6.7C). Because of this characteristic, the receptor potential is not a perfect analog of the stimulus, as noted earlier (p. 184).

Lastly, the response of the pacinian corpuscle depends upon the rate at which the stimulus is applied (Fig. 6.7D). Therefore, if the intensity of the mechanical stimulus is maintained at a constant value, the magnitude of the receptor potential varies directly with the rate at which the mechanical deformation is induced.

Another receptor whose bioelectric properties have been investigated extensively is the muscle spindle of the frog. This structure is similar to, but somewhat less complex than, the muscle spindle of the mammal (p. 180). Mechanical elongation of the spindle by stretch is the adequate stimulus for this type of receptor.

As shown in Figure 6.8, during an applied stretch the receptors within the muscle spindle depolarize to a certain level (phase 1). The potential then declines to a somewhat lower steady-state value that remains approximately constant as long as the stretch is applied (phase 2). When the applied tension is released, the receptor potential reverses its polarity *(off response)* momentarily and then gradually returns to its prestretch resting level (phase 3).

The amplitude of the rapid depolarization that occurs at the onset of stretch (phase 1) is related to the rate at which the muscle is stretched (Fig. 6.9A). The magnitude of the phase 2 steady-state potential that follows varies in accordance with the intensity with which the muscle is stretched and is not related to the rate at which the stretch is applied to the spindle (Fig. 6.9B). The third phase of the spindle receptor potential (the off response) is quite variable in its occurrence, but when it appears at all it is in a hyperpolarizing direction. This fact is in marked contrast to the depolarizing off response

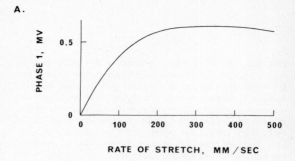

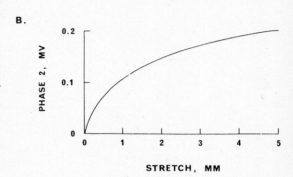

FIG. 6.9. **A.** Relationship between the amplitude of the phase 1 depolarization of the muscle spindle (Fig. 6.8) and the rate at which the muscle is stretched. **B.** Relationship between the phase 2 steady-state potential of the muscle spindle and the extent to which the muscle is stretched.

that is seen in the pacinian corpuscle (Figs. 6.7C, 6.11).

That the unmyelinated nerve terminal is the actual source of the receptor potential has been established by the following types of experiment. First, if the connective tissue capsule is removed from a pacinian corpuscle by microdissection, mechanical stimulation of the bare nerve end still evokes graded receptor potentials whose amplitude depends upon the intensity of the applied stimulus. Second, if the first node of Ranvier is blocked by compression or the application of drugs (eg, procaine), stimulation of either the capsule or the bare nerve end within a pacinian corpuscle still elicits a local, graded receptor potential. Third, if the sensory nerve to a receptor is sectioned and the nerve distal to the cut is allowed to degenerate, the receptor potential is completely abolished.

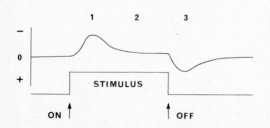

FIG. 6.8. The three phases of electric activity induced by mechanical deformation of the receptors within a muscle spindle. Note that the off response of the spindle is hyperpolarizing, in contrast to the depolarizing off response of the pacinian corpuscle (Figs. 6.7C; 6.11A).

ADAPTATION. As noted earlier, the receptor potentials generated by the pacinian corpuscle and the muscle spindle are not precise analogs of the applied stimulus. If the stimulus is maintained for any length of time (at a constant intensity), the amplitude of the receptor potential decreases spontaneously (Figs. 6.7B, C; 6.8; 6.10). This phenome-

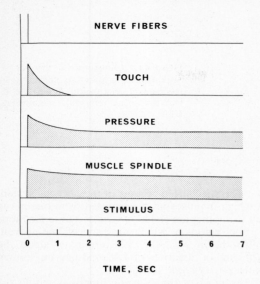

FIG. 6.10. Adaptation. (After Ganong. *Review of Medical Physiology,* 6th ed, 1973, Lange Medical Publications.)

non is called *adaptation.* * In turn, the frequency of impulses in the sensory nerve fiber innervating the receptor also declines as the magnitude of the receptor potential falls.

The rates at which specific receptors undergo adaptation vary considerably. *Rapidly adapting* (or *phasic*) *receptors* include the pacinian corpuscles (Fig. 6.11), the mechanoreceptors around the hair follicles, and the cutaneous receptors for touch and pressure. Phasic receptors signal in an on–off manner. *Slowly adapting* (or *tonic*) *receptors* include the muscle spindles, the Golgi tendon organs, the pressoreceptors in the carotid sinus, the cutaneous receptors for cold and pain, and the mechanoreceptors within the lungs that sense the degree to which these organs are inflated. Tonic receptors continuously signal a prolonged stimulus; eg, the carotid sinus pressoreceptors continually monitor the blood pressure.

The mechanism that underlies adaptation is obscure. Various receptors may even possess different mechanisms for this process. Interestingly, removal of the connective tissue capsule that invests the unmyelinated terminal nerve fiber within the pacinian corpuscle causes this receptor to adapt much more slowly than when the capsule is present; this procedure also abolishes the off re-

*Adaptation also takes place in nerve fibers, so that the frequency of propagated impulses declines with time when a stimulus of constant intensity is applied for prolonged intervals. However, the mechanism that underlies neural adaptation can be wholly independent of adaptation in the particular receptors associated with the nerves. Thus, neural adaptation can involve the ionic mechanisms that underlie generation of the all-or-none spike potential, which are wholly independent of the receptor mechanism.

sponse for reasons that are still unknown (Fig. 6.11A and B).

Actually, the magnitude of the receptor potential at any particular moment is of greater physiologic significance than the rate at which the receptor undergoes adaptation. This is true because the frequency of impulse discharge in the sensory nerve fiber connected to the receptor terminals depends upon the amplitude of the receptor potential at any particular instant, as discussed on page 183.

IONIC FLUXES DURING RECEPTOR EXCITATION. In all receptors that have been investigated, the application of an adequate stimulus depolarizes the nerve terminals. The magnitude of the depolarization is related directly to the intensity of the stimulus, as discussed earlier. There is an inverse relationship between the size of the terminal and the magnitude of the stimulus required to induce a demonstrable receptor potential. For example, a mechanical deformation of only a fraction of a micron can elicit a receptor potential in the nerve terminal of a pacinian corpuscle whose capsule has been removed.†

Since receptor potentials are generated in the exceedingly minute terminals of sensory nerves, their small size precludes the direct recording of transmembrane potentials as the receptor is stimulated. Consequently, what we know of the ionic fluxes that take place during receptor excitation has been deduced from indirect studies.

The steady-state transmembrane potential of receptor terminals, as in other excitable tissues, is generated by electrochemical gradients for Na^+, K^+, and Cl^- that develop exactly as in other cells (p. 102). An adequate stimulus applied to the receptor terminal causes its membrane to become more permeable to sodium, potassium, chloride, and perhaps other ions in a nonselective fashion. Thus, mechanical deformation of a pacinian corpuscle can be envisaged as enlarging the pores in the receptor membrane so that sodium ions can now enter freely, thereby neutralizing the transmembrane potential. A graded response will develop as the stimulus intensity is increased because more pores enlarge as the deformation of the membrane becomes greater. Therefore, the receptor transmembrane potential ultimately will fall to zero.

Since the sodium ion is farthest from its electrochemical equilibrium state in the resting membrane, it might be expected that bathing the receptor in a sodium-free solution would markedly reduce the receptor potential. This fact actually has been demonstrated in unencapsulated terminals of pacinian corpuscles. If the capsule is left on the terminal, however, then lowering the external

†This property of mechanotransduction is found not only in receptor terminals; nerve fibers also can be stimulated to discharge impulses by the application of mechanical compression. However, the deformation required to stimulate a nerve is considerably greater than that required to excite a receptor terminal — on the order of 10 to 15 μ.

A.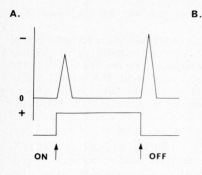

B.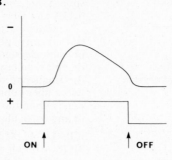

FIG. 6.11. Comparison of the electric responses of a pacinian corpuscle to mechanical deformation. **A.** Capsule intact. **B.** Capsule removed. (See also Fig. 6.6.)

sodium ion concentration has no such effect. Hence, it appears that the capsule itself contains enough sodium for the generation of a receptor potential; since this structure is impermeable to sodium, the minute quantity of this ion that is required for development of the receptor potential remains within the capsule adjacent to the nerve terminal.

Concluding Remarks

In this chapter, only a few simple peripheral receptors have been introduced, and certain physiologic properties of receptors in general have also been presented. The functional roles of the muscle spindle as well as the Golgi tendon organ in reflexes are presented on pages 218 and 221 of this chapter. Additional features of receptors and sensory functions will be treated in Chapter 7, whereas Chapter 8 is reserved for a discussion of the special senses.

In conclusion, the receptor or generator potential appears to be the common physiologic property in most, if not all, receptors. This graded receptor potential, in turn, provides the mechanism whereby the adequate stimulus for each receptor is converted or transduced into coded (ie, frequency-modulated) patterns of impulses in the sensory nerves that are continuous with the receptor terminals. Hence, generator potentials not only are found in muscle spindles and Golgi tendon organs but have also been recorded in such structures as the organ of Corti that subserves the sense of hearing, the organs for olfaction and taste, and many other types of sensory receptor, each of which will be treated later.

MORPHOLOGIC AND PHYSIOLOGIC PROPERTIES OF THE SYNAPSE

Introduction

Fundamentally, the human nervous system is composed of a vast number of individual neurons organized into complex morphologic and functional patterns such that transmission of impulses takes place in one direction only. As noted earlier, according to the *neuron doctrine* (p. 157) each mature nerve cell is an anatomically and trophically independent unit, and no cytoplasmic connections are present between individual neurons.

The specialized regions of the nervous system in which impulses are transmitted from one nerve cell to another are called *synapses*. These junctions are the regions at which the axon of one neuron, called the *presynaptic cell*, terminates upon the soma and/or dendrites of a second neuron, known as the *postsynaptic cell*.

At most of the synaptic junctions within the human body, the transmission of impulses from one nerve cell to the next takes place by the release of a specific chemical mediator, called a *neurotransmitter*, from the terminals of the presynaptic cell. The neurotransmitter substance diffuses across the gap between the presynaptic and postsynaptic membranes to alter the permeability — and thus the electric characteristics — of the postsynaptic membrane. Thus, a *graded response* can be evoked in the postsynaptic membrane, a fact of the utmost importance in producing integrated and modulated activity within the central nervous system. This graded response can be either *excitatory* or *inhibitory*.

Furthermore, synaptic transmission is a *one-way process;* therefore, synapses exhibit the property of *functional polarization*. This characteristic is in marked contrast to nerve fibers, which can transmit impulses with equal facility in either direction.

It is evident from this cursory introduction to synaptic properties that synaptic transmission is not a simple process. The transmission of nerve impulses from the presynaptic to the postsynaptic membrane is *not* analogous to the spark's jumping the gap when a spark plug fires.

At a few synapses, junctional transmission is *electric* in character, and no neurotransmitter is involved. Nevertheless, the spark plug analogy does not apply to these electric synapses either. At certain other synapses, impulse transmission is mediated by both electric and chemical events acting in concert. These so-called *conjoint synapses,* as well as purely electric synapses, are of rare occurrence in the mammalian body. Consequently, the emphasis in this section will be placed upon neurochemical transmission at synapses.

Structure and Ultrastructure of Synapses

GENERAL MORPHOLOGIC CONSIDERATIONS. From a morphologic standpoint, the synapse can be defined as the point of contact between two neurons. This contact may take place between an axon and a dendrite, between an axon and a nerve cell body (soma), or, rarely, between two axons.

There is considerable variation in the number of synapses on different neurons. For example, only a few synapses are found on the granule cells of the cerebellum; a single large motoneuron in the spinal cord, on the other hand, may have over 5,000 individual synapses; and the dendrites of a cerebellar Purkinje cell can have several hundred thousand individual synaptic endings. This anatomic fact underlies the enormous structural and functional complexity of the human central nervous system.

As with the number of synapses found on particular neurons, there is also a tremendous variation in the morphologic appearance of synapses found on different neurons. Each type of nerve cell is characterized by its own morphologic type of synapse. Generally, synapses appear as minute expansions or swellings on the tips of the axon terminals, called the *terminal buttons* (boutons) or *synaptic knobs*. A few examples of synaptic endings of this type are depicted in Figure 6.12.

The terminal ramifications of the axon may also exhibit loose networks or branching patterns that are closely applied to the soma or dendrites of the postsynaptic neuron. For example, the terminal branches of certain presynaptic elements found within the cerebellum and autonomic ganglia form basketlike networks around the soma of the postsynaptic cells.

The area of the membrane on a nerve cell body that is covered by one particular synaptic knob is quite small. However, the total area covered by all of the synaptic knobs that converge upon the soma of a single postsynaptic neuron may reach as high as 40 percent of the entire surface area of the membrane (Fig. 6.13). This anatomic fact has important functional implications (see p. 199).

ULTRASTRUCTURE. Electron micrographs of chemical synapses within the central nervous system indicate that the endings of the axons form varicosities called the *synaptic knobs* or *terminal buttons* (Fig. 6.14). The membrane that completely encloses the presynaptic terminal is separated by a distance of about 200Å from the likewise discrete membrane that invests the perikaryon of the postsynaptic neuron or its dendrite. This space is called the *synaptic cleft*.

Within the synaptic knob are found a number of mitochondria as well as numerous small, round *synaptic vesicles*. The membrane-limited synaptic vesicles range between 200 and 650 Å in diameter and are believed to originate by the process of budding from the ends of the microtubules present in the axon. The synaptic vesicles contain the specific neurotransmitter substance that is elaborated within the presynaptic terminal of the particular axon. Hence, the morphologic appearance of the synaptic vesicles may vary somewhat, depending upon the neurotransmitter that is elaborated by that particular synapse (p. 198).

Bioelectric Properties of Synapses

METHOD OF STUDY. The bioelectric properties of synapses within the spinal cord have been studied intensively using the technique shown in Figure 6.15. A microelectrode is inserted into the soma of a motoneuron located within the ventral horn of the spinal cord, usually of the cat, and the electric events elicited following stimulation of the cell are amplified and recorded.

The microelectrode is inserted by means of a suitable mechanical device (a micromanipulator) into the ventral aspect of the spinal cord; penetration of a single cell is indicated when a stable poten-

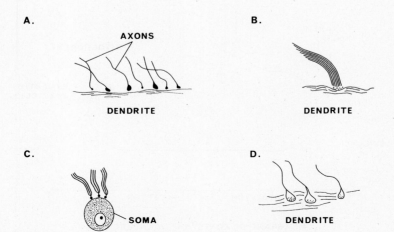

FIG. 6.12. Some examples of different types of synaptic endings. (After Bloom and Fawcett. *A Textbook of Histology,* 9th ed, 1968, Saunders.)

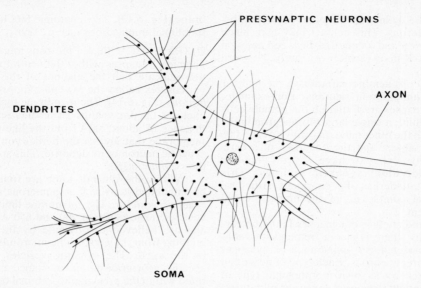

FIG. 6.13.   Synaptic knobs (dots) terminating upon the soma and dendrites of a spinal motoneuron.

tial difference of −70 to −80 mv is observed between the microelectrode within the cell and a suitably placed surface electrode; ie, a typical steady-state transmembrane potential is evident.

The particular cell that has been penetrated by the microelectrode can be identified electrophysiologically by electrical stimulation of the appropriate ventral root as it emerges from the spinal cord (S_2 in Fig. 6.15). The electric activity evoked in the cell can be seen by means of this antidromic stimulation, since the impulse is propagated to the nerve cell body and stops there. Thus, if an action potential is generated in the cell after the application of an antidromic stimulus, the tip of the microelectrode is within a motoneuron and not an interneuron (p. 215). This is true because synaptic conduction is a one-way process; consequently, if the tip of the microelectrode were located in an interneuron rather than a motoneuron, then no all-or-none impulse would be observed following antidromic stimulation. On the other hand, orthodromic stimulation of the dorsal root nerve will induce electric activity in the presynaptic terminals that end upon the soma of the motoneuron under study, and these electric events can be amplified and recorded (S_1, Fig. 6.15).

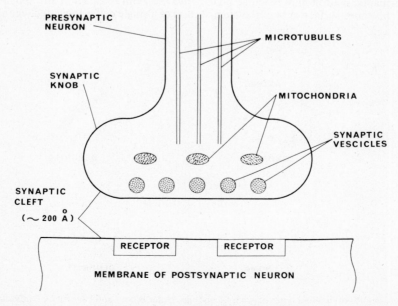

FIG. 6.14.   The ultrastructure of a synapse.

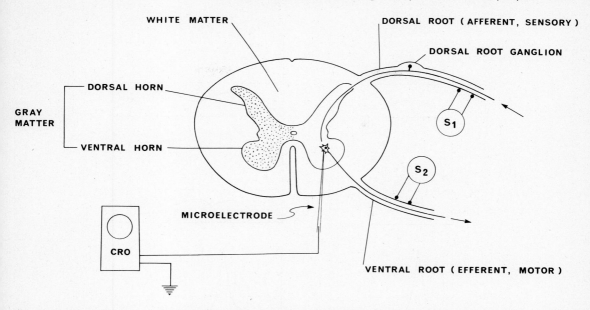

FIG. 6.15. The technique employed for study of the electrophysiologic properties of individual spinal neurons. S_1, S_2 = stimulators; CRO = cathode-ray oscilloscope. The general organizational pattern of the spinal cord in cross section is indicated. Note that the white matter (nerve fiber tracts) surrounds the gray matter (nerve cell bodies, shaded on left). The nerve cell bodies of the afferent neurons lie within the dorsal root ganglion, and the tip of the microelectrode is shown within the cell body of a spinal motoneuron located in the ventral horn. (After Ganong. *Review of Medical Physiology,* 6th ed, 1973, Lange Medical Publications.)

When single stimuli are applied to the dorsal root (sensory, afferent) nerves using the experimental setup described above and illustrated in Figure 6.15, no propagated all-or-none action potentials are evoked in the postsynaptic neuron by the orthodromic impulses caused by these stimuli. Instead, each stimulus produces either a transient, *incomplete depolarization* or else a transient *hyperpolarization* of the postsynaptic cell membrane. These bioelectric phenomena are called the *excitatory postsynaptic potential* and the *inhibitory postsynaptic potential,* respectively.

Presumably, the bioelectric events that accompany synaptic transmission within the brain itself are similar to those recorded from the synapses located within the spinal cord.

EXCITATORY POSTSYNAPTIC POTENTIALS. When a single impulse that has been generated by orthodromic stimulation of a sensory nerve reaches an excitatory synaptic terminal, a slight depolarization of the postsynaptic membrane takes place. The depolarizing response commences approximately 0.5 msec after the afferent impulse enters the spinal cord, peaks between 1 and 1.5 msec later, and then declines exponentially with a time constant of about 4 msec (Fig. 6.16A.1).*

The time constant is defined as the time necessary for the depolarizing response of the postsynaptic membrane to decrease to 1/e (or 1/2.718 = 0.368) of its peak or maximal value.

During the brief interval when this depolarizing potential is present, the excitability of the postsynaptic membrane to additional presynaptic excitatory impulses is enhanced markedly. Consequently, this depolarizing potential is called an *excitatory postsynaptic potential,* or *EPSP.*

The EPSP is caused by a localized depolarization of the postsynaptic cell membrane immediately beneath the active synaptic knob of the presynaptic neuron. The total area of the postsynaptic membrane depolarized is quite small, and thus the current sink generated is too slight to depolarize the entire postsynaptic membrane; ie, not enough current is removed to displace the steady-state voltage of the postsynaptic membrane to its threshold or firing level. However, a single EPSP can *summate* with the depolarizations produced by EPSPs from adjacent synaptic knobs, and, in this manner, the overall depolarization of the postsynaptic membrane may reach threshold, eliciting an all-or-none impulse in the postsynaptic neuron.

Spatial and Temporal Summation in Excitatory Synapses. When more than one excitatory synaptic knob associated with a single postsynaptic neuron is active simultaneously, *spatial summation* (or *multiple-fiber summation*) takes place (Fig. 6.16A). Thus, the activity under one synaptic knob is said to *facilitate* the activity under the other synaptic knobs on the membrane of the same postsynaptic neuron. If a sufficiently large area of the postsynaptic membrane is depolarized at a given

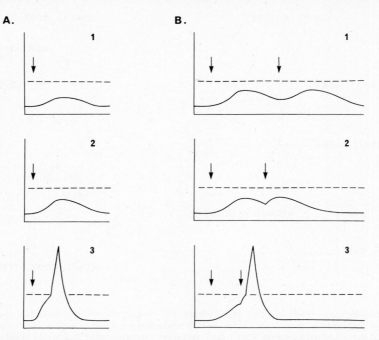

FIG. 6.16.   The electric events recorded from a postsynaptic cell studied by the technique depicted in Figure 6.15. In each of the diagrams, threshold to firing is indicated by a horizontal dashed line, the stimulus is applied at the arrow, and an upward deflection indicates depolarization. Upstroke on ordinate = depolarization; abscissa = time. **A.** Spatial summation: 1 and 2. Subthreshold excitatory postsynaptic potentials of increasing amplitude caused by the progressive firing of more presynaptic terminals per unit time. 3. A sufficient number of presynaptic terminals have discharged in unison so that threshold to firing is reached in the postsynaptic neuron, and an action potential is generated. **B.** Temporal summation: 1 and 2. As the time interval between successive volleys of presynaptic impulses is decreased (arrows), the EPSPs fuse or summate. 3. The impulse frequency in the presynaptic neurons has become sufficiently high that the summated EPSPs depolarize the membrane of the postsynaptic neuron to threshold and an action potential is generated.

instant, the postsynaptic neuron fires an all-or-none propagated impulse as the spatially summated EPSPs collectively reach sufficient amplitude to depolarize the postsynaptic cell membrane to its firing level.

Temporal summation, on the other hand, takes place if the frequency of impulses passing down an individual axon is sufficiently great to generate new EPSPs before the earlier EPSPs have decayed to their steady-state value (Fig. 6.16B). Hence, frequency modulation of afferent impulses underlies temporal summation (see also p. 170).

The EPSP is not an all-or-none response, since its magnitude is proportional to the intensity of the afferent stimulation of the postsynaptic membrane. Only when the EPSP develops sufficient magnitude is an all-or-none action potential generated in the postsynaptic neuron.

Synaptic Delay.   There is a definite time interval between the point at which an impulse reaches the presynaptic nerve terminal and the point at which a response develops in the postsynaptic membrane. This time interval is called the *synaptic delay,* and its duration is about 0.5 msec. Thus, a definite latent period exists during synaptic transmission which is caused by the time required for the neuro-

transmitter substance to be released from the presynaptic terminal, cross the synaptic cleft, and act upon the postsynaptic membrane (see also p. 198).

Due to the synaptic delay, there is a definite proportionality between the total number of synapses in a chain of neurons and the time that it takes for a response to be elicited following stimulation of a receptor. Thus transmission of impulses through a simple reflex arc with one synapse (a *monosynaptic pathway*) is far more rapid than that through a neuron chain comprised of two or more *interneurons* and their synapses (a *polysynaptic pathway*).

Since the minimum time required for transmission across every synapse is at least 0.5 msec, it can be determined readily whether a particular reflex pathway is monosynaptic or polysynaptic simply by measuring the delay in impulse transmission from the dorsal root to the ventral root through the spinal cord.

Ionic Fluxes During the Generation of EPSPs. When the synaptic knob of an excitatory presynaptic neuron is depolarized, the permeability of the postsynaptic membrane lying immediately beneath the knob is increased markedly and in a nonspecific fashion. That is, the permeability to sodium, potassium, and chloride ions increases simultaneously. Therefore, sodium moves down its electrochemical gradient into the postsynaptic neuron, causing an extremely localized depolarization of the postsynaptic membrane beneath the active synaptic knob. But since the area beneath one synaptic knob is so minute, the factors that cause a spontaneous repolarization of the membrane tend to restore the resting or steady-state potential of the postsynaptic membrane (p. 132), and no all-or-none depolarization of the postsynaptic neuron occurs. If, on the other hand, more synaptic knobs are activated, such as through the process of spatial summation, then more sodium ions enter the postsynaptic neuron and, accordingly, the amplitude of the EPSP becomes greater. When the sodium influx becomes sufficiently great, the threshold level to firing is reached and a propagated all-or-none action potential is generated in the postsynaptic neuron.

INHIBITORY POSTSYNAPTIC POTENTIALS. Stimulation of certain presynaptic afferent nerve fibers elicits a hyperpolarization in spinal motoneurons (Fig. 6.17). This effect is in direct contrast to the depolarization produced by stimulating excitatory afferent nerves.

The hyperpolarization caused by stimulation of inhibitory afferent fibers increases the steady-state transmembrane potential of the postsynaptic membrane and thereby reduces the irritability of the motoneuron by driving the membrane potential away from its firing level. Consequently, the hyperpolarizing potential is called an *inhibitory postsynaptic potential*, or *IPSP*.

The IPSP starts to develop between 1.0 and 1.25 msec after the afferent stimulus enters the

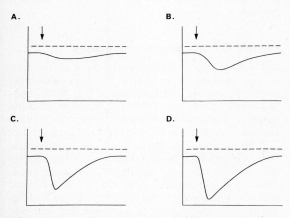

FIG. 6.17. Graded inhibitory postsynaptic potentials recorded by using the technique illustrated in Figure 6.15. Downstroke on ordinate = hyperpolarization of postsynaptic membrane. Threshold to firing is indicated by the horizontal dashed line in each diagram. In **A–D**, the intensity of dorsal root stimulation is increased progressively. Note that the rate of development and the magnitude of the hyperpolarization increase as the intensity of dorsal root stimulation is raised, and spatial summation takes place. IPSPs, like EPSPs, also exhibit temporal summation.

spinal cord, reaches a peak in 1.5 to 2.0 msec, and then decreases exponentially with a time constant of around 3.0 msec.

IPSPs exhibit spatial summation, as shown in Figure 6.17A–D, as the intensity of the stimulus applied to the afferent nerve is increased. Similarly, temporal summation also takes place as the frequency of the afferent nerve stimulation is increased.

Postsynaptic inhibition that is generated through stimulation of inhibitory afferent nerve fibers causing development of an IPSP is called *direct* or *postsynaptic inhibition*.

Ionic Fluxes During the Generation of IPSPs. In contrast to EPSPs, the IPSPs generated in spinal motoneurons appear to be caused by a *selective increase* in the permeability of the postsynaptic membrane to potassium and chloride ions, whereas the permeability to sodium ions remain low. Thus, when an inhibitory synaptic knob becomes activated, there is an increased potassium efflux combined with a similar chloride influx as these ions diffuse down their respective concentration gradients. The net result of these processes is a transfer of negative charge into the cell; thus, the transmembrane potential of the postsynaptic membrane increases.

In contrast to spinal motoneurons, the hyperpolarizing effect on the cardiac pacemaker fibers that is produced by stimulation of the vagus nerve is caused by a selective increase in permeability to potassium ions alone (p. 598).

Neurons That Participate in Postsynaptic Inhibition. When certain dorsal root nerve fibers are stimulated, the impulses pass directly to excitatory synapses on the spinal motoneurons, and EPSPs are generated. There is good experimental evidence to indicate the IPSPs are produced by the terminals of special interneurons that are interposed between the excitatory afferent neuron and the motoneuron, as depicted in Figure 6.18. These interneurons are short and rounded, with thick axons, and are called *Golgi bottle neurons*. Release of an inhibitory neurotransmitter substance from the terminals of the Golgi bottle neuron causes a selective increase in the permeability to potassium and chloride ions in the postsynaptic membrane, as discussed above, so that an IPSP is generated in the motoneuron.

This mechanism is of great physiologic importance because it enables an excitatory input to be transformed within the spinal cord into an inhibitory input, insofar as the activity of the motoneuron is concerned (Fig. 6.18).

ELECTROGENESIS OF THE POSTSYNAPTIC ACTION POTENTIAL. There is a continual barrage of excitatory and inhibitory activity on any postsynaptic neuron, so that the net membrane potential of the soma at any point in time is an algebraic summation of these depolarizing and hyperpolarizing influences. Consequently, the soma of the neuron behaves as an integrator. But when the

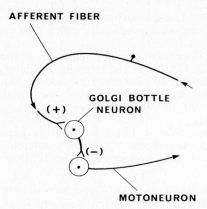

AFFERENT FIBER

GOLGI BOTTLE
(+) NEURON

(−)

MOTONEURON

FIG. 6.18. The conversion of an excitatory input (+) into an inhibitory input (−) at a spinal motoneuron through activation of an inhibitory interneuron (Golgi bottle neuron) which releases inhibitory neurotransmitter at its terminals. The inhibitory neurotransmitter, in turn, directly generates an IPSP in the motoneuron (postsynaptic inhibition).

net depolarization of the membrane is sufficient to reach its threshold to firing, then a conducted all-or-none action potential is generated in the post-synaptic neuron.

The region of the neuron that has the lowest threshold to firing is the unmyelinated *initial segment* of the axon that is located at, as well as immediately distal to, the axon hillock (Fig. 6.19). Thus, electrotonic currents that arise in excitatory or inhibitory synaptic knobs produce depolarizing and hyperpolarizing currents, respectively, within the initial segment of the axon. When the net magnitude of the depolarization caused by the excitatory synaptic knobs depolarizes the cell body sufficiently (eg, around 15 mv), the initial segment of the axon becomes the first region of the neuron to discharge an all-or-none spike potential. The impulse then is propagated in two directions, along the axon proper as well as in a retrograde direction back into the soma of the neuron. Presumably, this transient depolarization of the neurilemma of the soma momentarily obliterates preexisting depolarizing and hyperpolarizing presynaptic influences so that once again the neuron functionally is fully receptive to these opposing electric forces.

ELECTROPHYSIOLOGIC PROPERTIES OF THE DENDRITES. There is considerable experimental evidence which indicates that the dendrites of neurons located within the central nervous system do not conduct impulses like the axons. Rather, they behave functionally as receptors, somewhat like the peripheral receptors discussed earlier in this chapter (p. 182). Consequently, the dendrites appear to function as either sources or sinks for electric currents, which, in turn, can alter the membrane potential electrotonically in the initial segment or other parts of the axon in which action potentials are generated. If the dendrites of a

neuron arborize to any great extent, and if there are many presynaptic terminals ending upon these dendrite branches, then there is a considerable latitude for the interactions between excitatory and inhibitory influences within the neuron as the depolarizing and hyperpolarizing currents ebb and flow.

The electrophysiologic properties of the dendrites are discussed further in relation to their role in the origin of the electroencephalogram (p. 420); however, it should be mentioned here that their participation in this process depends upon their ability to generate local, nonpropagated depolarizing and hyperpolarizing electrotonic potentials in response to excitatory and inhibitory presynaptic influences.

TRANSMISSION AT ELECTRIC AND CONJOINT SYNAPSES. The transmission of impulses at electric synapses does not involve the release of a neurotransmitter substance and takes place as follows. The impulse, upon reaching the presynaptic terminal, directly generates an EPSP in the postsynaptic cell membrane sufficiently great to depolarize the latter to threshold, since there is a low-resistance bridge between the two cells through which current and ions can readily pass. Furthermore, because of this low-resistance bridge, little current loss takes place in the synapse, and the latent period for excitation of the postsynaptic membrane is of very short duration.

In conjoint synapses, the presynaptic and postsynaptic membranes are partly fused and partly separate. Consequently, electric transmission can take place directly through the low-resistance fused regions of the membranes as described above, whereas chemical mediation is necessary to transmit impulses across the cleft between the unfused regions.

Chemical Transmission at Synapses

INTRODUCTION. Immediately prior to the turn of this century, epinephrine (or adrenalin) was isolated from the adrenal gland (Abel, 1898). Subsequently, it was noted that the injection of crude extracts of adrenal tissue into experimental animals caused physiologic responses identical to those elicited by electric stimulation of the sympathetic nerves innervating certain organs and structures (p. 406) — for example, the heart rate accelerated, certain smooth muscles contracted, and the blood pressure increased (Langley, 1901). These findings led another investigator to inject purified epinephrine into experimental animals, and it was found that the purified compound gave identical physiologic responses to those obtained by the injection of crude adrenal extracts or electric stimulation of sympathetic nerves (Elliott, 1904). This researcher then made the brilliant and original suggestion that epinephrine might be liberated from the terminals of sympathetic nerves during their activity, and that this epinephrine, in turn, caused

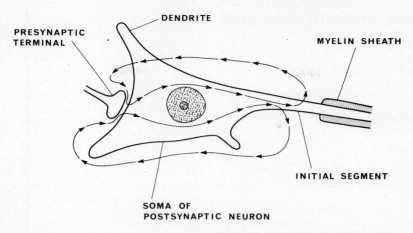

FIG. 6.19.   Electrogenesis of an action potential in a postsynaptic neuron. The electrotonic current flow (indicated by thin lines and arrows) can be excitatory (depolarizing) or inhibitory (hyperpolarizing). The resultant of these opposing influences determines whether or not the depolarization (EPSP) is of sufficient magnitude to generate an action potential in the postsynaptic neuron.

the observed responses in the effector cells, ie, the heart and other muscular tissues. Therefore Elliott postulated the basic concept of neurotransmission by means of chemical mediators.

Some years passed before direct evidence for neurochemical transmission was obtained (Loewi, 1921). This critical experiment was performed upon two isolated, perfused, spontaneously beating frog hearts. The hearts were arranged so that the perfusate from the first heart was pumped through a fine tube and into the second heart. When the vagus nerve to the first heart was stimulated, thereby slowing the rate of its contraction, the rate of beat in the second heart likewise decreased. Therefore, some chemical substance had to have been liberated by the first heart during vagal stimulation and transported in the perfusate to the second heart, where it also exerted its effect. Later, the chemical responsible for this neurotransmitter effect was identified as *acetylcholine*.

If a similar two-heart experiment is performed but the sympathetic or accelerator nerve is stimulated instead of the vagus, then the rate of both hearts accelerates, and the neurotransmitter substance involved has been shown to be *norepinephrine* or *noradrenalin* (von Euler, 1946).

These relatively simple experiments clearly substantiated the earlier concept of neurochemical transmission of nerve impulses.

A *neurotransmitter substance* now can be defined as a chemical which is released from a presynaptic nerve ending following its stimulation and which, upon recognition by a specific receptor in the membrane of the postsynaptic nerve (or other) cell, causes either excitation or inhibition of the physiologic activity in the postsynaptic cell. Neurotransmitters are also called *neurohumoral agents* or *neurohormones*.

It is now clear that many different neurotransmitters are present in the body, and these com-

pounds exert a broad spectrum of physiologic effects upon many vital processes. For example, in addition to their responsibility for synaptic transmission within the nervous system, neurotransmitters also underlie the contraction of skeletal muscle in vivo (p. 210), regulate the heart rate (p. 662), in part regulate the blood pressure through their action upon the diameter of certain blood vessels (p. 667), and stimulate some glands to produce the enzymes that, in turn, synthesize certain hormones (p. 203). In addition, neurotransmitters are the substances which ultimately regulate voluntary bodily movements at the level of the brain and alter both mood and behavioral patterns of the individual.

It is obvious from this brief resumé of neurotransmitter functions that these compounds and their functions in vivo are of enormous physiologic and clinical importance.

NEUROTRANSMITTER SUBSTANCES.   There are four basic criteria that must be met in order for any chemical to be classified as a neurotransmitter. First, nerves must have the requisite enzymes to synthesize the chemical, and the chemical must be present in the nerves. Second, stimulation of the nerves must cause release of the transmitter chemical. Third, following its release, the chemical must react with a specific receptor on the membrane of the postsynaptic neuron or other cell and thereby evoke one or more specific biologic effects. Fourth, there must be mechanisms present locally that are able rapidly to "turn off" the biologic effects of the chemical.

On the basis of these four requirements, two compounds are established firmly as neurotransmitters. These compounds are *acetylcholine* and *norepinephrine* (Tables 6.2 and 6.3). Nerves that contain acetylcholine are called *cholinergic,* whereas nerves that contain norepinephrine are called *adrenergic.*

Table 6.2 NEUROTRANSMITTER SUBSTANCES

A. Proven Neurotransmitters[a]

1. Acetylcholine
2. Norepinephrine (noradrenalin)
3. Epinephrine (adrenalin)
4. Dopamine (dihydroxyphenylethylamine)

B. Putative Neurotransmitters[b]

1. Glycine
2. Aspartic acid
3. Glutamic acid
4. γ-Aminobutyric acid (GABA)
5. Serotonin (5-hydroxytryptamine)
6. Histamine

[a]*See also Table 6.3, Figures 6.20, 6.21, and 6.22.*

[b]*The functions and sites of action of the putative transmitter substances listed in this table will be discussed in context in this and in subsequent chapters.*

There is such strong experimental evidence that *epinephrine* (adrenalin) and *dopamine* are also neurotransmitters in their own right that these compounds have been listed among the proven neurotransmitters in Tables 6.2 and 6.3, despite the fact that these compounds do not quite satisfy all four of the criteria for neurotransmitters given above. Earlier it was thought that dopamine was merely an intermediate compound in the biosynthesis of norepinephrine and epinephrine (Fig. 6.22); however, it is now clear that dopamine is itself a neurotransmitter compound. The structural formulas of the proven neurotransmitters are given in Figure 6.20.

Norepinephrine, epinephrine, and dopamine are known collectively as *catecholamine neurotransmitters* because they have in common a basic chemical structure consisting of a benzene ring upon which are located two hydroxyl groups and an ethylamine side chain (Fig. 6.20).

In addition to the four proven neurotransmitters, a number of other compounds have been implicated as presumed or *putative neurotransmitters* (Table 6.2B) because they fulfill only some of the four requirements listed above. Nevertheless, current experimental evidence strongly indicates that the six compounds listed in the table have definite neurotransmitter functions, and undoubtedly there are other, as yet undiscovered, neurotransmitters present within the body.

Neurotransmitters such as norepinephrine are found in practically all tissues and organs of the body with one noteworthy exception: Nerves are completely absent in the placenta, and hence no neurotransmitter substances are found in this organ.

Biosynthesis of Acetylcholine. The steps in the biosynthesis of acetylcholine are outlined in Figure 6.21. The organic base choline is transported actively into cholinergic neurons, where it reacts with active acetate (acetylcoenzyme A) under the catalytic influence of the enzyme choline acetylase to produce the neurotransmitter *acetylcholine*.

Choline acetylase is such a specific enzyme that its concentration in nerve endings within any portion of the nervous system is presumptive evidence that cholinergic synapses are present in that particular region.

Biosynthesis of the Catecholamine Neurotransmitters. The biosynthetic pathway for dopamine,

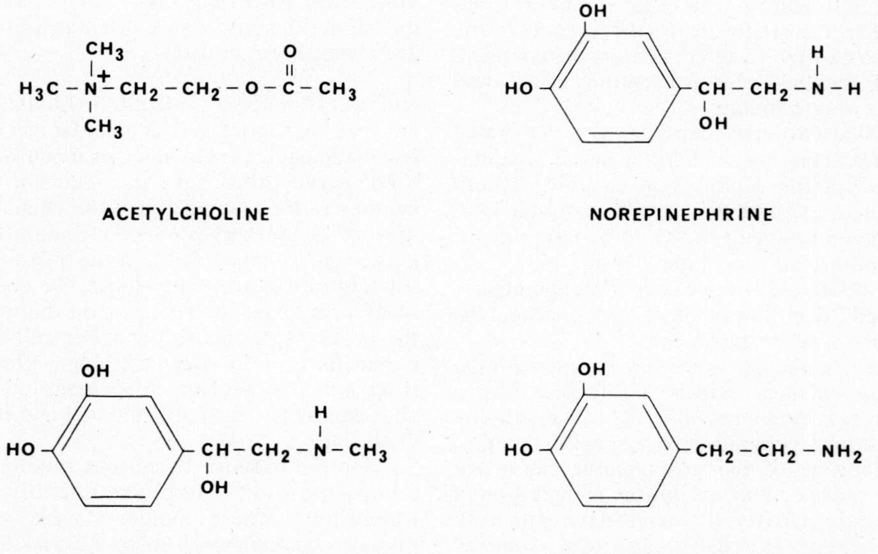

ACETYLCHOLINE

NOREPINEPHRINE

EPINEPHRINE

DOPAMINE

FIG. 6.20. Chemical formulas of the proven neurotransmitters (see also Figs. 6.21, 6.22).

1. $HS-CoA + ATP + Acetate \xrightarrow{\text{Acetylthiokinase}} Acetyl-CoA + H_2O + ADP$

2. $Acetyl-CoA + Choline \xrightarrow{\text{Choline acetylase}} Acetylcholine + HS-CoA$

3. $Acetylcholine + H_2O \xrightarrow{\text{Acetylcholinesterase}} Choline + Acetate$

FIG. 6.21. Outline of the pathways for acetylcholine biosynthesis (1, 2) and degradation (3). HS-CoA = reduced coenzyme A; Acetyl-CoA = acetylcoenzyme A.

norepinephrine, and epinephrine is outlined in Figure 6.22. Note that all three of these catecholamine neurotransmitters are closely related chemically, and that there are only minor differences in their structural formulas, as shown in Figure 6.20.

Three factors involved in regulating the biosynthesis of the catecholamine neurotransmitters are discussed on page 206.

Locus of Neurotransmitter Synthesis. There is good evidence that the enzymes responsible for the biosynthesis of certain neurotransmitters (eg,

Table 6.3   PROVEN NEUROTRANSMITTERS: LOCALIZATION AND GENERAL FUNCTIONS

TRANSMITTER SUBSTANCE	LOCALIZATION	GENERAL FUNCTIONS
Acetylcholine	1. All synapses between preganglionic and postganglionic neurons of autonomic (sympathetic and parasympathetic) nervous system	1. Synaptic transmission
	2. All myoneural junctions	2. Neuromuscular transmission
	3. All postganglionic parasympathetic endings	3. Neuroeffector transmission
	4. Some postganglionic sympathetic endings	4. Neuroeffector transmission
	5. Certain regions of central nervous system	5. Synaptic transmission
Norepinephrine (noradrenalin)	1. Most postganglionic sympathetic endings	1. Neuroeffector transmission
	2. Some peripheral nerve endings	2. Neuroeffector transmission
	3. Central nervous system; eg, dorsal and ventral pathways, cerebrum, cerebellum	3. Synaptic transmission; regulates fine muscular coordination, emotion, alertness
	4. Central nervous system, hypothalamus	4. Synaptic transmission concerned with visceral functions; eg, thermoregulation, hunger, thirst, also reproduction; also modulates behavior and mood
	5. Adrenal medulla	5. Functions as hormone when released into bloodstream
Epinephrine (adrenalin)	1. Tracts in brain stem	1. Synaptic transmission; effects behavior, mood, emotion (?)
	2. Adrenal medulla	2. Functions as hormone when released into bloodstream
Dopamine	1. Sympathetic ganglia	1. Modulates activity of adrenergic neurons
	2. Brain stem: substantia nigra, caudate nucleus	2. Synaptic transmission; functions in nerves that integrate movement, behavior patterns

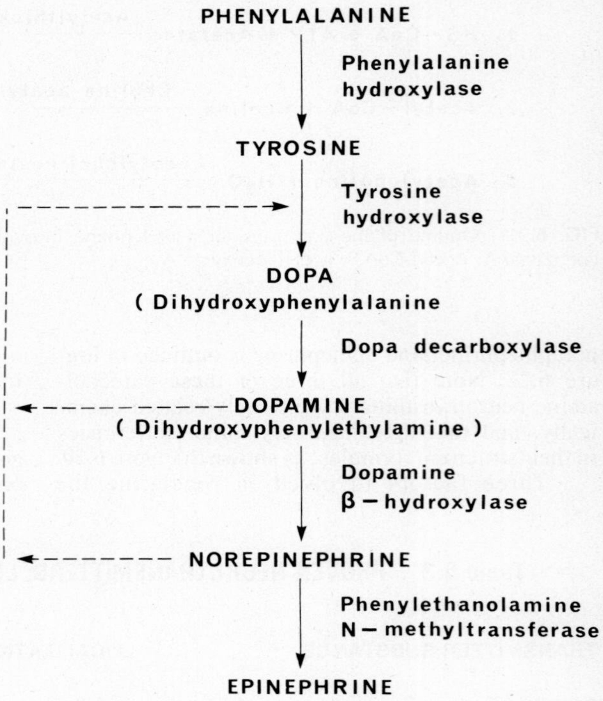

PHENYLALANINE

↓ Phenylalanine
hydroxylase

TYROSINE

↓ Tyrosine
hydroxylase

DOPA
(Dihydroxyphenylalanine

↓ Dopa decarboxylase

DOPAMINE
(Dihydroxyphenylethylamine)

↓ Dopamine
β − hydroxylase

NOREPINEPHRINE

↓ Phenylethanolamine
N − methyltransferase

EPINEPHRINE

FIG. 6.22. Outline of the pathway for catecholamine neurotransmitter biosynthesis. The enzyme responsible for catalyzing each step in the synthesis of dopamine, norepinephrine, and epinephrine are indicated to the right of the arrows. The dashed lines to the left in the figure indicate a negative feedback mechanism whereby the levels of dopamine and norepinephrine control the activity of tyrosine hydroxylase, as discussed in the text. See also Figure 15.46, page 1121.

acetylcholine, norepinephrine) are themselves synthesized within the soma or nerve cell bodies of the presynaptic neurons. The enzymes then are transported down the axons by axoplasmic flow to the presynaptic knobs (p. 150), where the neurotransmitters are synthesized and then packaged, together with some enzymes, in the synaptic vesicles. In this regard, it is interesting to note that colchicine, which blocks the rapid axoplasmic flow mechanism through disruption of the neural microtubular system, causes a depression of synaptic transmission in vivo. Presumably, this defect is caused by a drug-induced reduction in the quantity of enzymes that are available for neurotransmitter biosynthesis within the presynaptic terminals.

Histologically, the acetylcholine-containing vesicles have a clear center, whereas those that contain the catecholamine neurotransmitters have a dense, granular core.

SEQUENCE OF EVENTS DURING SYNAPTIC TRANSMISSION. The dynamic sequence of events that occurs during synaptic transmission is depicted in Figure 6.23. When a nerve impulse arrives at the terminal of a presynaptic neuron, the terminal is depolarized by the action potential so that the permeability of the membrane to sodium and potassium ions is increased. This transient and nonspecific increase in permeability allows calcium ions to diffuse into the presynaptic terminal, thereby enabling the vesicles containing the particular neurotransmitter substance associated with that nerve to fuse with the presynaptic membrane, and the transmitter is released in *packets* or

quanta into the synaptic cleft by the process of *exocytosis* (p. 96). The transmitter then diffuses rapidly across the synaptic cleft, a distance of about 200 Å. Upon reaching the postsynaptic membrane, the transmitter molecules attach to specific receptor sites where an excitatory or inhibitory effect is exerted upon the membrane of the postsynaptic neuron (or effector cell). An excitatory neurotransmitter produces a depolarization of the postsynaptic membrane which is accompanied by development of an EPSP (p. 191), whereas an inhibitory neurotransmitter produces hyperpolarization and development of an IPSP (p. 193).

It must be emphasized that a particular neurotransmitter substance can exert *either* excitatory *or* inhibitory postsynaptic effects depending upon the response of the specific receptors present in the postjunctional membrane. For example, acetylcholine has an excitatory (depolarizing) effect upon the neuromuscular junction (p. 210), whereas the same compound released from the terminals of the vagus nerves within the heart exerts an inhibitory (hyperpolarizing) effect upon the activity of the postjunctional cardiac muscle fibers.

The facts presented above explain a number of important synaptic properties — namely, one-way conduction, synaptic delay, the graded response, temporal summation, and fatigue.

One-Way Conduction. Postsynaptic nerve terminals contain no significant quantities of neurotransmitter substances. Consequently, antidromic stimulation of a synapse through the efferent nerve

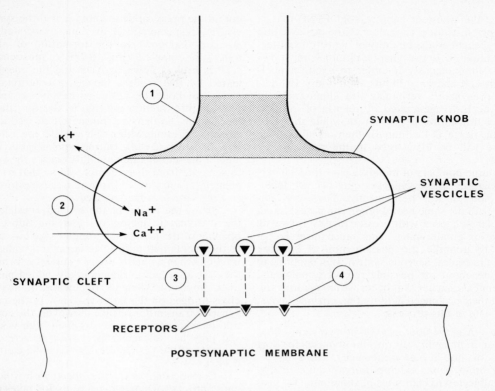

FIG. 6.23. Sequence of events underlying synaptic transmission. ① An impulse depolarizes the entire presynaptic knob. ② As the permeability of the presynaptic membrane to Na⁺ and K⁺ increases during depolarization, Ca⁺⁺ is able to diffuse into the presynaptic terminal. ③ Transmitter substance (black dots) is released from the synaptic vesicles in packets of quanta, and then diffuses across the synaptic cleft (vertical dashed arrows). ④ The neurotransmitter then reacts with specific receptors on the postsynaptic membrane, and an EPSP or IPSP is generated depending upon the nature of the transmitter and receptor.

fiber causes no release of neurotransmitter, and, although the membrane of the postjunctional neuron is depolarized, no transmission of the impulse to the presynaptic membrane can occur, since the transmitter is concentrated within the presynaptic terminals.

This one-way function of the synapse is critical to an orderly, sequential operation of all the elements of the nervous system, and impulses are transmitted forward only when they arrive at presynaptic knobs containing neurotransmitters.

Synaptic Delay.   As mentioned earlier, the time required for an impulse to excite a postsynaptic neuron following presynaptic stimulation is on the order of 0.5 msec (p. 192). This *synaptic delay* represents the sum of the time required for the release and diffusion of the transmitter to the receptor sites on the postsynaptic membrane and the time necessary for activation of the postsynaptic membrane.

The Graded Synaptic Response and Temporal Summation.   The quantity of neurotransmitter released from a presynaptic terminal depends upon the frequency of the nerve impulses propagated to the afferent nerve endings. That is, the transmitter is

not released in an all-or-none fashion; rather, the total quantity of transmitter discharged is related directly to the rate at which impulses arrive at the presynaptic knobs, since each impulse causes release of a certain quantity of transmitter. The amplitude of the postsynaptic potential depends upon the quantity of neurotransmitter released; this, in turn, is a direct function of the frequency of afferent nerve discharge as well as the total number of presynaptic terminals that are active at any given moment. These facts explain the ability of synapses to generate a graded response.

The phenomenon of temporal summation also is explained by the release of more quanta of neurotransmitter per unit of time as the rate of presynaptic impulse discharge increases (p. 191). Therefore, the individual subthreshold postsynaptic potentials can summate in either an excitatory or an inhibitory manner, depending upon the particular transmitter and synapse involved.

Facilitation.   When many presynaptic terminals converging upon a single postsynaptic cell discharge simultaneously but fail to release a sufficient total quantity of neurotransmitter to generate an all-or-none action potential in the postsynaptic neuron, the transient subthreshold EPSP that is developed *facilitates* excitation of the postsynaptic cell by a slight added quantity of neurotransmitter. For example, if 30 presynaptic terminals must discharge simultaneously in order to excite a postsynaptic cell and only 20 fire, then if any 10 additional presynaptic terminals fire simultaneously while the EPSP generated by release of the first burst of transmitter is present (ie, while the postsynaptic nerve is facilitated), then an all-or-none impulse will be discharged in the postsynaptic neuron.

If large numbers of impulses pass through certain synapses, these synapses may become facilitated more or less permanently. Therefore, impulses with the same neural origin can pass through particular synapses with greater facility following such prior conditioning. This mechanism is believed to underlie the phenomenon of *memory* within the central nervous system. It is tempting to speculate about the possible role of fundamental biochemical changes that occur within neurons following their prolonged activity (eg, enzyme levels; see p. 446) in this process.

Fatigue.   A nerve fiber or axon in vivo shows little decrease in its ability to transmit impulses for long periods of time at high frequencies. The synapse, on the other hand, exhibits *fatigue* quite readily when subjected to high rates of stimulation for prolonged intervals. This effect occurs because the rate of transmitter synthesis within the presynaptic terminals cannot keep pace with its rate of utilization; therefore, eventually the transmission of impulses is decreased considerably.

MECHANISM OF NEUROTRANSMITTER RELEASE.   On the basis of research studies, a hypothesis has been formulated to explain the mechanism of neurotransmitter release into the synaptic cleft from the presynaptic terminals.

It has been found that an actomyosinlike protein is present in mammalian brain tissue, and this protein appears to be localized within the presynaptic nerve endings. In order to avoid confusion with the actomyosin of muscular tissue (p. 116), the contractile protein isolated from particulate fractions (or synaptosomes) of brain tissue by differential ultracentrifugation has been named *neurostenin*. This neurostenin in turn is composed of two proteins, *neurin* (analogous to actin), and *stenin* (analogous to myosin).

The proposed model for neurotransmitter release is as follows (Fig. 6.24). There is some evidence that the microtubules of the axon (the neurotubules) may orient the synaptic vesicles in close apposition to the membrane of the presynaptic terminal (Fig. 6.24A). The event that triggers expulsion of the neurotransmitter substance is the sudden, diffusion-driven influx of calcium ions that follows depolarization of the presynaptic membrane and the change in membrane permeability that accompanies this process (Fig. 6.24B). The calcium ions cause the neurin and the stenin present on the presynaptic membrane and the synaptic vesicle, respectively, to become associated in a manner analogous to the formation of an actomyosin complex (Fig. 6.24B). Subsequently, ATP is broken down to ADP and the energy released by this process induces a conformational change in both the vesicular membrane and the presynaptic membrane by a sliding filament mechanism, so that the transmitter as well as any enzymes present within the vesicle are released into the synaptic cleft. During repolarization of the membrane, either calcium efflux causes the vesicle to separate from the receptor site so that its complement of enzyme and transmitter is restored, or else the vesicle is metabolized.

Thus, the fusion of the synaptic vesicle with the presynaptic membrane and the subsequent opening of the membrane are envisaged as similar to the events that take place during muscular contraction when an actin–myosin complex forms and then the thin (actin) filaments are pulled past the thick (myosin) filaments by means of swiveling cross-bridges on the latter (p. 138). In the case of synaptic transmitter release, however, the contraction of neurostenin serves to open both vesicular and presynaptic membranes, with release of the contents of the vesicle into the synaptic cleft (Fig. 6.24C).

Considering the relative speed with which the entire process of synaptic transmission takes place, the exocytotic release of neurotransmitter is indeed a remarkably efficient process.

The mechanism for exocytosis presented above is also applicable to the release of protein-containing secretions elaborated by certain glandular cells. In this way, a mechanism is available for the transport of such large molecules out of the cells in which they are synthesized.

POSTSYNAPTIC RECEPTORS.   Following its release, the neurotransmitter diffuses across the synaptic cleft between the terminal of the presynaptic neuron and the postsynaptic neuron or effector cell. There are specific receptor sites (protein molecules?) located upon the postsynaptic membrane which selectively recognize and react with the neurotransmitter so that the postsynaptic cell is activated in such a way that it is able to carry out its particular function. The response of the postsynaptic cell may be immediate, such as is the case during transmission of an impulse from neuron to neuron or from neuron to skeletal muscle through the genesis of postsynaptic potentials (p. 191). On the other hand, the response may be delayed for minutes or hours, such as in the *de novo* synthesis of enzymes, as discussed below.

Norepinephrine is recognized by two kinds of receptors. These are called the α- and β-adrenergic *receptors* on the basis of their response to various drugs as well as the type of activity that is evoked following their stimulation. Similarly, specific dopamine and acetylcholine receptors are also found in various tissues.

A.

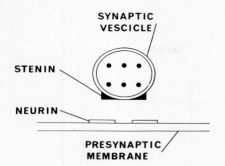

FIG. 6.24. Proposed mechanism for neurotransmitter release. **A.** Synaptic vesicle in proximity to synaptic membrane. Stenin and neurin are contractile proteins located on the membrane of the vesicle and presynaptic membranes, respectively.

B.

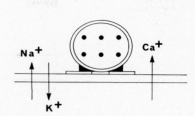

FIG. 6.24. **B.** Depolarization of presynaptic membrane permits sudden influx of Ca^{++} into the presynaptic terminal.

C.

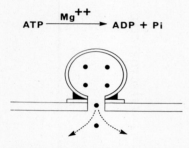

FIG. 6.24 **C.** Ca^{++} permits neurostenin complex to form; contraction of neurostenin opens a gap in the presynaptic and vesicular membranes, thereby releasing transmitter into the synaptic cleft. Note that energy from ATP breakdown is required for this step.

D.

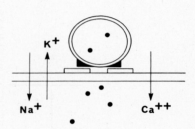

FIG. 6.24. **D.** The events in step B are reversed as the presynaptic membrane repolarizes, thus restoring the steady-state condition, and Ca^{++} diffuses out of the cell.

E.

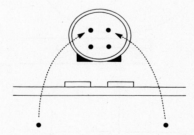

FIG. 6.24. **E.** Following its reaction with postsynaptic receptors, the transmitter is inactivated by uptake into the vesicles. (After Berl, Puszkin, and Nicklas. *Science* 179:441, 1973.)

The β-adrenergic receptors have been studied extensively and will provide an example of long-term receptor activity. β-adrenergic receptors are involved in excitation of effector cells; this response is mediated, in turn, through adenosine 3′, 5′-monophosphate (cyclic AMP, p. 1020) as follows. When a β-adrenergic receptor site is occupied by norepinephrine, the enzyme adenylate cyclase (which is concentrated upon the inner surface of the membrane) becomes activated. Adenylate cyclase

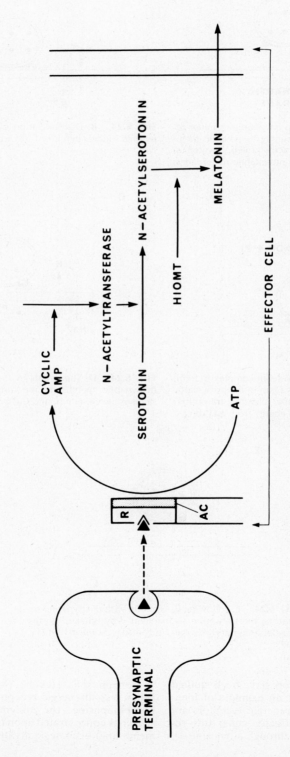

FIG. 6.25. Role of a neurotransmitter (norepinephrine, black triangles) in producing a long-term receptor effect, specifically, the biosynthesis of a hormone (melatonin) in an effector cell (pineal gland). R = receptor; AC = adenylate cyclase (shaded) on inner surface of effector cell membrane; HIOMT = hydroxyindole-O-methyltransferase. (Data from Axelrod. *Sci Am* 230:58; 1974.)

can convert the energy derived from the breakdown of ATP into the synthesis of one or more particular enzymes which then produce a response in the effector cell, eg, the synthesis of a specific hormone (Fig. 6.25).

The general mechanism discussed above is of widespread occurrence in the body, and is of particular importance in the synthesis of, and response to, endocrine substances in general (see also p. 1017).

Cyclic AMP also appears to have an important role in synaptic transmission, as follows (Fig. 6.26). In peripheral ganglia of the cervical sympathetic nervous system, three cell types are present: *preganglionic* or *input neurons* that have their origin within the spinal cord and release acetylcholine at their terminals within the ganglion; *interneurons* that are innervated by the input neurons and that release dopamine from their terminals; *postganglionic* or *output neurons* that send processes to effector tissues and organs situated at a distance from the ganglion and receive synaptic inputs from the preganglionic fibers as well as the interneurons, as depicted in Figure 6.26. Stimulation of the input neurons causes an increase in the concentration of cyclic AMP within the ganglion, a process involving not only dopamine but afferent nerve impulses. Acetylcholine does not participate directly in this process, as shown in Figure 6.26. Apparently, the dopamine that is released from the terminals of the interneuron acts upon specific receptor sites on the postsynaptic membrane to stimulate adenylate cyclase production, and this, in turn, leads to an increase in the level of cyclic AMP within the ganglion. The accumulated cyclic AMP causes an IPSP (hyperpolarization) to develop within the postsynaptic membrane so that the postganglionic neuron becomes less sensitive to excitatory inputs from the preganglionic neuron. Thus, dopamine released from the interneuron appears to modulate the activity of the sympathetic nerves in an inhibitory fashion.

Based upon experimental evidence such as that presented above, the suggestion has been advanced that both short- and long-term memory may be due to the synthesis of specific proteins (enzymes) through the action of adenylate cyclase whose activity has been enhanced by afferent nerve impulses.

INACTIVATION OF NEUROTRANSMITTERS. In order to exert its regulatory function in a precise manner, any neurotransmitter must be inactivated rapidly once it has exerted its effect upon the postsynaptic cell. If this were not the case, the prolonged activity of the transmitter on the postsynaptic receptor would cause a loss of exact control, and normal synaptic function would be much deranged (eg, see strychnine poisoning, p. 232).

There are two mechanisms whereby the inactivation of synaptic transmitters takes place under physiologic conditions. First, the transmitter may be inactivated metabolically through enzymatic conversion into an inactive compound. Second, the transmitter may be inactivated through its uptake into the presynaptic terminals following its release and interaction with the postsynaptic receptors.

Inactivation by Chemical Transformation. In cholinergic nerves, acetylcholine is inactivated rapidly by the enzyme acetylcholinesterase (reaction 3, Fig. 6.21), which degrades this neurotransmitter into choline and acetate.

The catecholamine neurotransmitters are converted into biologically inactive compounds by the action of two enzymes, as outlined in Figure 6.27. Deamination (or oxidation) is catalyzed by *monoamine oxidase (MAO)*, whereas methylation is accomplished by *catechol-O-methyltransferase (COMT)*.

MAO is an important and widely distributed enzyme localized within the mitochondria. It is especially concentrated in brain tissue and adrenergic nerve endings, as well as in the liver and kidneys. MAO deaminates (ie, removes the NH_2 group) from a wide range of compounds, including norepinephrine, epinephrine, and dopamine as well as serotonin, thereby producing physiologically inactive metabolities.

COMT, on the other hand, is not found in adrenergic nerve endings; but, like MAO, it is distributed widely among the body tissues and is found at high concentrations in the cytoplasm of liver and kidney cells. The MAO and COMT present in these two organs are responsible in large measure for the rapid inactivation of *circulating* catecholamines through their transformation into various metabolities (see also p. 1121).

There are thus *two* general pathways for catecholamine metabolism in the body. However, metabolic norepinephrine inactivation at adrenergic nerve endings takes place principally through the deamination of this compound to 3, 4-dihydroxymandelic aldehyde (Fig. 6.27). The latter compound then is oxidized either to 3, 4-dihydroxymandelic acid or 3, 4-dihydroxyphenylglycol, both of which end products are physiologically inactive and can enter the circulation.

Within the past few years, it has become apparent that the inactivation of neurotransmitter substances through their metabolic conversion into physiologically inactive compounds, as in the examples cited above, is a relatively insignificant mechanism.

Inactivation by Neural Uptake. It is now apparent that the principal physiologic mechanism whereby neurotransmitter inactivation takes place is the *neural uptake* by the presynaptic nerve endings of the transmitter previously released (Fig. 6.28). Thus, synaptic activity involves a cyclic release as well as uptake of the transmitter substance by the presynaptic nerves. Once the neurotransmitter has been sequestered within the nerve ending, it is, of course, unable to interact further with the postsynaptic receptors. Therefore, the transmitter is physiologically inactive until it is released again

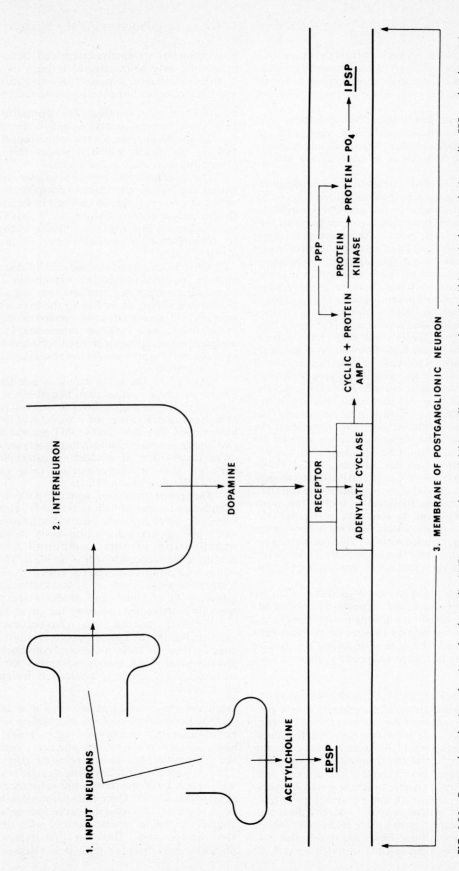

FIG. 6.26. Proposed mechanism of action whereby cyclic AMP acts to produce an inhibitory effect at synapses located within cervical sympathetic ganglia. PPP = phosphoprotein phosphokinase. (Data from Marx: *Science* 178: 1188, 1972; diagram after Greengard, Yale University School of Medicine.)

FIG. 6.27. Outlines of the possible metabolic pathways for catecholamine inactivation. COMT = catechol-O-methyltransferase; MAO = monoamine oxidase. Reactions 1 and 2 both are responsible for the biochemical inactivation of circulating catecholamines, particularly in the liver and kidney. Reaction 2 also takes place locally within adrenergic nerve endings, but is a relatively insignificant mechanism for the inactivation of catecholamine neurotransmitters.

from the presynaptic terminals. Thus, neural uptake of transmitters provides the chief physiologic mechanism whereby neurotransmitters are inactivated following their release from the presynaptic knobs.

REGULATION OF NEUROTRANSMITTER BIOSYNTHESIS. Neurotransmitter substances are not static chemical entities; rather, they are constantly being synthesized, released, recaptured, and metabolically degraded. Furthermore, nerves

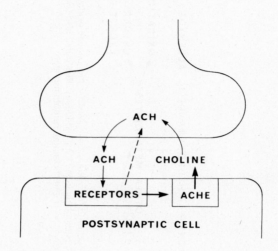

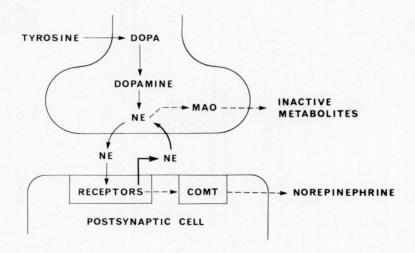

FIG. 6.28. Comparison of neurotransmitter inactivation in cholinergic (**A**) and adrenergic (**B**) neurons. ACH = acetylcholine; ACHE = acetylcholinesterase; DOPA = dihydroxyphenylalanine; MAO = monoamine oxidase; NE = norepinephrine; COMT = catechol-O-methyltransferase. The principal mechanisms for inactivation are indicated by heavy lines, the minor mechanisms by light dashed lines. Acetylcholine inactivation by metabolic transformation is an exceptional case, and most neurotransmitters are inactivated by uptake into the presynaptic terminals following their release.

and synapses themselves are able to undergo considerable variations in their states of activity depending upon the prevailing physiologic conditions. Yet in spite of these dynamic changes, the concentration of neurotransmitters within the tissues stays at remarkably constant levels. There are three general mechanisms whereby the levels of catecholamine neurotransmitters are maintained in vivo.

Feedback Regulation. The feedback regulation of catecholamine biosynthesis depends on the levels of norepinephrine and dopamine within the sympathetic nerve endings. When increased sympathetic nerve activity takes place with a concomitant requirement for more neurotransmitters, tyrosine hydroxylase activity also increases rapidly (Fig. 6.22), and more catecholamine neurotransmitters are synthesized from tyrosine in a given period of time. Conversely, as sympathetic nerve activity diminishes, the concentrations of norepinephrine and dopamine rise, and the increased levels of these compounds exert a negative feedback effect which decreases the activity of tyrosine hydroxylase and thus lowers the rate at which these catecholamines are synthesized (dashed lines in Fig. 6.22).

This fast-acting negative feedback mechanism is brought into play by stress and cold as well as by certain drugs; any one of these factors tends to lower the catecholamine levels within the sympathetic nerve endings through their enhanced activity.

Control Through Neurotransmitter Release. The α-adrenergic receptors are located within the postsynaptic membrane (p. 202). When activated, they inhibit norepinephrine release from the presynaptic endings into the synaptic cleft. Therefore, when the level of norepinephrine rises to a certain level within the synaptic cleft, this neurotransmitter itself directly activates the α-adrenergic receptors so that the release of norepinephrine from the presynaptic terminals is reduced. Conversely, when the norepinephrine concentration within the synaptic cleft falls, the α-adrenergic receptors are turned off and more norepinephrine is released from the presynaptic ending.

The regulatory process described above, like the response of tyrosine hydroxylase to enhanced sympathetic nerve activity, is both fast-acting and a negative feedback mechanism.

Enzyme Synthesis. Following a prolonged discharge of the sympathetic nerves, biosynthesis of certain enzymes that are responsible for producing catecholamines is accelerated. In particular, the concentrations of tyrosin hydroxylase, dopamine β-hydroxylase, and phenylethanolamine N-methyltransferase all increase so that the nerve becomes capable of synthesizing more neurotransmitter (Fig. 6.22).

Factors such as stress and cold, as well as certain psychologic factors, cause a net synthesis* of the three enzymes listed above as a compensatory adaptation within the sympathetic nerves to the enhanced requirement for catecholamine neurotransmitters.

Summary and Conclusions

The specialized bioelectric and biochemical properties of synapses discussed in this section provide the critical underlying physiologic mechanisms whereby integrated and coordinated activity among the neurons present in the human nervous system is accomplished. Thus, such basic synaptic characteristics as the graded response, one-way conduction, synaptic delay, spatial as well as temporal summation, facilitation, and fatigue are of the utmost significance to an understanding of the functional aspects of central nervous processes. Furthermore, the reader will have noted that inhibition can be an active physiologic process and not merely a shutting off of prior activity.

A postsynaptic neuron can be viewed as an *integrator of information* that is received from other, presynaptic neurons. This information is delivered to the postsynaptic neuron in the form of excitatory as well as inhibitory chemical influences that arise in presynaptic terminals. The postsynaptic neuron then sums, or integrates, these influences and responds according to whether or not the information is sufficient to raise the postsynaptic membrane potential to threshold for firing. Thus, the neuron is able to sort incoming information quantitatively and to react accordingly.

Although all neurons exhibit the same fundamental physiologic characteristics in a qualitative sense, individual neurons found in various regions of the central nervous system differ considerably in their quantitative responses. For example, certain neurons have high thresholds to firing, whereas others have low thresholds; some neurons and their synapses fatigue quickly, others more slowly; certain neurons discharge at high frequencies, whereas others fire impulses at much lower rates.

Finally, these general characteristics of neurons, together with their particular synaptic connections within the central nervous system, permit a vast number of regulatory and other functional patterns to be achieved within an individual. The functions of synapses in particular neuronal circuits will be discussed in context throughout the remainder of this text.

The reader should be aware of the distinction between the terms "activity" and "synthesis" as applied to enzymes. The activity of an enzyme may increase without the de novo synthesis of enzyme molecules. For example, release from an inhibition may cause increased activity, as occurs when the concentration of norepinephrine falls and thereby increases the activity of tyrosin hydroxylase. On the other hand, the activity of an enzyme also can increase through the net synthesis of new protein molecules or by other mechanisms. Thus the word "activity" merely indicates the general capacity of an enzyme to perform its functions and no mechanism whereby this effect comes about is implied.

MORPHOLOGIC AND PHYSIOLOGIC PROPERTIES OF THE MYONEURAL JUNCTION AND OTHER NEUROEFFECTOR TERMINALS; EFFECTORS

Introduction

The junction between an efferent (motor) nerve and an effector (either a muscle fiber or a glandular cell) provides the final link in the morphologic and physiologic chain leading to a complete reflex arc (Fig. 6.1). In this section, emphasis will be placed on the structurally and functionally differentiated terminals located between somatic motor nerves and skeletal muscle fibers. These neuromuscular (or myoneural) junctions are known as *motor end plates*. The less specialized terminals of motor nerve endings on cardiac and smooth muscle fibers will also be reviewed.

The more generalized terminals of motor nerves on glandular effectors will be dealt with in context in subsequent chapters, as particular secretory processes and their modification by nerve impulses are encountered.

Structure and Ultrastructure of the Motor End Plate

The histologic appearance of the neuromuscular junction between a motor nerve and its underlying muscle fiber is shown in Figure 6.29A and B. In optical microscopy, a motor end plate appears in cross section as a slightly raised, blisterlike area on the surface of the muscle fiber that typically contains a number of nuclei associated with the Schwann cells (or *teloglial cells*) at the terminal ramifications of the motoneuron (*arborization nuclei*), as well as an accumulation of nuclei of the muscle fiber itself (the *fundamental* or *sole nuclei* of the classic histologist). The relationships and structures described are depicted in Figure 6.29A.

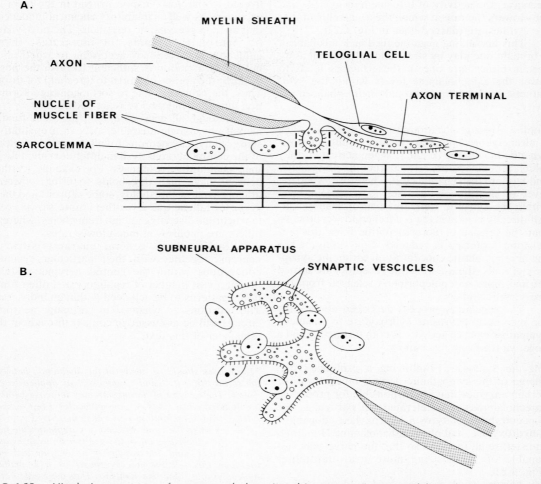

FIG. 6.29. Histologic appearance of a motor end plate viewed in cross section (**A**) and from above (**B**). The small branch of the terminal nerve in A which is enclosed by heavy dashed lines is shown considerably magnified in Figure 6.30. (After Bloom and Fawcett. *A Textbook of Histology,* 9th ed, 1968, Saunders.)

The axon of the motoneuron loses its myelin sheath just outside of the motor end plate, and the nerve terminals then ramify among the aggregated nuclei of the end plate (Fig. 6.29B). The terminal arborizations of the axon are embedded in grooves or depressions located on the surface of the muscle fibers. These recesses are called *synaptic troughs* or *primary synaptic clefts.* Within the primary synaptic clefts at the interface between the surfaces of the axon and the muscle fiber are seen a number of regularly spaced lamellar structures which appear to be related to the underlying sarcoplasm. These specializations are called the *subneural apparatus.*

By means of electron microscopy, the detailed anatomic relationships among the various regions of the motor end plate have been greatly clarified (Fig. 6.30). The terminal Schwann cells cover the outer surface of the axon endings, but *do not* enter the primary synaptic clefts; therefore, in these regions the naked axolemma makes intimate synaptic contact with the underlying sarcolemma. Further-

more, under suitable magnification, the lamellar subneural apparatus of the light microscopist is seen to be a series of *secondary synaptic clefts* that consist of regular invaginations of the sarcolemma itself (Fig. 6.30). The axolemma and sarcolemma within both the primary and secondary synaptic clefts are separated by an extremely thin protein–polysaccharide layer similar to that which envelops the surface of the entire muscle fiber.

Within the nerve terminals of the motor end plate, the axoplasm normally contains a relatively large number of mitochondria. In addition, an abundant supply of small spherical bodies or granules between 400 and 600 Å in diameter are present in this region, and these *synaptic vesicles* appear to be identical with those found in the presynaptic terminals of the central nervous system (p. 189). Presumably, these clear, axoplasmic vesicles contain acetylcholine, the neurotransmitter substance that is responsible for the excitation of skeletal muscle.

Acetylcholinesterase activity is localized

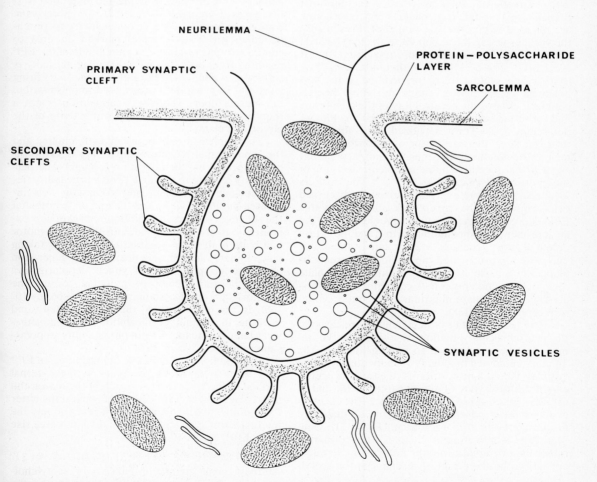

FIG. 6.30. Ultrastructure of a portion of a typical motor end plate, such as is enclosed by dashed lines in Figure 6.29A. The subneural apparatus of the classic microscopist is revealed as secondary synaptic clefts. The shaded ellipses represent mitochondria. (After Bloom and Fawcett. *A Textbook of Histology,* 9th ed, 1968, Saunders.)

within (or immediately adjacent to) the sarcolemma that lines the secondary synaptic clefts, a fact that can be demonstrated readily by the use of appropriate histochemical techniques.

Each motor nerve fiber innervates only one motor end plate. Therefore, no convergence of inputs is present in skeletal muscle.

It is apparent from the facts presented above that the motor end plate represents a highly organized structure in which both axolemma and sarcolemma are modified structurally so as to form an intimate contact with a relatively large area for the rapid one-way transmission of nerve impulses to the muscle fiber. Furthermore, since neuromuscular transmission occurs through release of a neurotransmitter substance (acetylcholine), the transmission of impulses from motor nerves to skeletal muscle occurs by synaptic transmission, exactly as this process takes place in the central nervous system.

Physiology of Neuromuscular Transmission

When an efferent impulse reaches the terminal branches of a motoneuron, packets of acetylcholine are released within the synaptic cleft. Presumably the vesicles within the axoplasm of the motoneuron act as storage sites for the neurotransmitter until it is released. The acetylcholine then diffuses across the synaptic cleft to receptors located in the sarcolemma of the motor end plate; there acetylcholine increases the permeability of the membrane and causes an influx of sodium ions and an efflux of potassium ions. These localized ionic fluxes, in turn, cause a depolarization of the steady-state membrane potential, and thus an *end plate potential (EPP)* is generated in the sarcolemmal region of the motor end plate (Fig. 6.31).* The EPP is localized within the end plate region as discussed below; however, the current sink generated by this EPP depolarizes the adjacent sarcolemma of the muscle fiber to its threshold to firing, and the action potentials thus generated at each side of the motor end plate are propagated rapidly in both directions along the muscle fiber away from the end plate region. The action potentials generated in the muscle fiber then initiate muscular contraction, as described earlier (p. 138).

The total quantity of acetylcholine released at the neuromuscular junction by a single nerve impulse has been estimated to be equivalent to the quantity contained in roughly 75 individual synaptic vesicles. About 2 msec after its release, this acetylcholine diffuses out of the synaptic clefts and is rapidly inactivated, primarily by the acetylcholinesterase present in this region, so that it no longer exerts its excitatory effect (Fig. 6.21). Consequently, inactivation by chemical transformation rather than by uptake into the presynaptic terminal appears to be the principal mechanism for acetyl-

The end plate potential is depolarizing, hence excitatory, and thus it is analogous to the EPSP induced in central synapses by excitatory neurotransmitters.

choline inactivation at the neuromuscular junction, in contrast to most other synapses (p. 203).

THE END PLATE POTENTIAL. The EPP can be studied electrophysiologically only if the all-or-none action potential generated in the muscle fiber by the EPP is blocked (Fig. 6.31). Application of *curare* to the end plate region causes the formation of a strong complex between this drug and the acetylcholine receptors present in the sarcolemma of the motor end plate, as curare prevents formation of a propagated action potential. Consequently, the EPP can be investigated independently of the action potential, provided that the dose of curare employed is sufficiently low to insure that not all of the receptors are blocked so that acetylcholine can generate some EPPs.

Using the technique described above in conjunction with microelectrodes and an oscilloscope for visualizing the electric activity, it has been determined that the EPP developed upon stimulation of an efferent nerve fiber (motoneuron) is a *local response* and that it decreases in magnitude exponentially with distance from the end plate; that is, it decreases *decrementally* (p. 182). Temporal summation can be demonstrated under such experimental conditions. However, in marked contrast to central nervous synapses, the amplitude of the EPP resulting from only *one* nerve impulse normally is sufficient to depolarize the sarcolemma to its firing level and to generate a propagated impulse within the muscle fiber. Therefore, summation is not a requisite to the normal physiologic activity of the motor end plate.

A complex formed between acetylcholine and the sarcolemmal receptors apparently is responsible for producing the alteration in membrane permeability leading to the EPP. The permeability (ie, conductance) to sodium and potassium ions increases *simultaneously,* and the end plate potential thus changes toward an equilibrium level of approximately -10 mv. This bioelectric effect at the motor end plate is in sharp contrast to the events taking place in nerve and muscle fibers during generation of an action potential, in which depolarization drives the transmembrane potential toward the equilibrium value for sodium ions, ie, $+66$ mv (p. 166). In the latter tissues, there is an initial rapid increase in sodium permeability followed somewhat later by the increase in permeability to potassium ions (Fig. 5.8, p. 167).

Acetylcholine is effective in producing an EPP only when it activates receptors at the external surface of the sarcolemmal region of the motor end plate. Hence, microinjection of this transmitter through the muscle fiber so that it reaches only the internal surface of the end plate does not give rise to an EPP.

ACETYLCHOLINE RELEASE DURING REST. At rest, small quanta or packets of acetylcholine are released randomly and spontaneously from the motoneuron terminals. Each of these quanta generates a *miniature end plate potential*

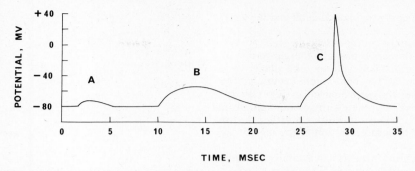

FIG. 6.31. End plate potentials. **A.** Miniature EPP. **B.** Subthreshold EPP. **C.** EPP of sufficient magnitude to induce an action potential in the sarcolemma of the subjacent muscle fiber.

whose magnitude is about 0.5 mv. The quantity of acetylcholine that is released under such conditions is related directly to the calcium ion concentration in the end plate region, and inversely to the magnesium ion concentration.

When a nerve impulse arrives at the nerve terminal, the number of acetylcholine quanta released suddenly increases markedly, and the end plate potential thus produced is several times greater than that required to depolarize the muscle fiber. This effect is sometimes called the *safety factor* of the neuromuscular junction.

MECHANISM OF ACETYLCHOLINE RELEASE. There is good experimental evidence which indicates that a nerve impulse increases the permeability of the nerve terminal to calcium ions, and that acetylcholine is released by a mechanism similar to that encountered in central nervous synapses (p. 200; Fig. 6.24).

FATIGUE. If the rate of stimulation of a motoneuron under experimental conditions is sufficiently high and is continued for a sufficient length of time, the quantity of acetylcholine that is available for release can diminish to such an extent that many impulses are not conducted to the muscle and thus *fatigue of the myoneural junction* occurs. This process is similar to that encountered in neural synapses (p. 200). The frequency of impulse transmission under physiologic conditions, however, is far below that necessary to produce such fatigue, so that this phenomenon rarely occurs in the normal individual.

Neuroeffector Transmission in the Heart

STRUCTURE OF EFFERENT NERVE ENDINGS. Thin, unmyelinated, postganglionic motor nerve fibers arise from intricate plexuses, and their extensively branched but unspecialized terminals form intimate contact with cardiac muscle fibers in the sinoatrial node, the atrioventricular node, and the bundle of His (cholinergic and adrenergic endings), as well as in the ventricular muscle proper (adrenergic endings). No discernible end plates are present, and neurotransmitters are released at the naked portions of the neuron where no sheath of Schwann is present.

The adrenergic neurons may exhibit a few to many varicosities or swellings near their endings that contain dense-core granules; hence, the neurotransmitter liberated by these endings is considered to be norepinephrine. Each varicosity along the terminal ramifications of the neuron is believed to elaborate neurotransmitter, so that one nerve ending together with its many branches can innvervate, and therefore activate, many cardiac effector cells rapidly and simultaneously. Repeated morphologic and functional contacts of this sort between one branch of a terminal axon and several effector cells is called a *synapse en passant*. These facts contrast sharply with the situation found at the motor end plate, in which each nerve fiber innervates only one neuromuscular junction.

Structurally, the cholinergic nerve endings apparently are similar to the adrenergic endings, except that the neurotransmitter (acetylcholine) is contained within synaptic vesicles having a clear center.

NEUROTRANSMISSION IN CARDIAC MUSCLE. As noted earlier, the heart possesses an inherent rhythmicity (p. 143), and the transmembrane potential oscillates continually within the pacemaker cells of the sinoatrial node (Fig. 4.29, p. 144).

Two neurotransmitters are important to the modification of the basic pattern of cardiac rate — namely, norepinephrine (excitatory) and acetylcholine (inhibitory).* Release of norepinephrine from adrenergic nerve endings enhances the spontaneous rate at which the pacemaker cells depolarize to their firing level, thus accelerating the heart rate (Fig. 10.22, p. 599). Conversely, the release of acetylcholine from cholinergic terminals decreases the rate at which these cells depolarize and the heart rate slows (Fig. 10.23, p. 599). If a sufficient quantity of acetylcholine is liberated, the membrane of the pacemaker cells hyperpolarizes to such an extent that

---

*The effects of these compounds upon the force of contraction is considered on page 622.

threshold to firing is not reached at all and the heart stops beating.

The release of norepinephrine and acetylcholine from adrenergic and cholinergic terminals within the heart is believed to take place by a mechanism similar to that which occurs at the synapse and at the neuromuscular junction (p. 200). The transmitter then diffuses outward from the terminals, where it induces either hypopolarization or hyperpolarization of the membrane of the cardiac cells.

Norepinephrine acts as an excitatory neurotransmitter in cardiac muscle because it causes a nonselective increase in the permeability of the membrane to sodium and potassium ions. Consequently, the next flux of these ions in opposite directions across the membrane causes a more rapid depolarization to the firing level of the cell.

Acetylcholine, on the other hand, acts upon the cardiac cell receptors in such a fashion that the permeability to potassium ions alone is increased. Thus, under the influence of this neurotransmitter, the efflux of potassium ions is greater and the membrane becomes hyperpolarized (see also p. 193).

Thus, the excitatory and inhibitory effects of norepinephrine and acetylcholine are the same in the heart as in the neuronal synapses of the central nervous system.

NEUROEFFECTOR TRANSMISSION IN SMOOTH MUSCLE

Structure of Efferent Nerve Endings: Morphologically, the terminals of motor axons in smooth muscle are quite similar to those described above for cardiac muscle, and no specialized nerve endings are present (Fig. 6.32). Near their terminals, the axons ramify extensively, forming a delicate reticulum that generally is in intimate contact with the muscle cells. At points of contact between a nerve fiber and muscle cell, the sheath of Schwann is interrupted and the bare axon touches the fiber. These are the regions where neurotransmitter is believed to be released, after which it is free to diffuse among the adjacent smooth muscle cells.

NEUROTRANSMISSION IN SMOOTH MUSCLE; JUNCTIONAL POTENTIALS.

Both norepinephrine and acetylcholine are the neurotransmitters for smooth muscle. In certain tissues, acetylcholine is excitatory to smooth muscle fibers, whereas in others this compound exerts an inhibitory effect. Similarly, the physiologic effects of norepinephrine can be either excitatory or inhibitory, depending upon the particular smooth muscle involved. In general, however, when acetylcholine is excitatory in a given tissue, then norepinephrine is inhibitory in that tissue, and vice versa. The excitation and inhibition of specific tissues by these two neurotransmitters will be discussed in further detail in relation to the autonomic nervous system (Chapter 7, p. 270).

Apparently the specific receptors present in the membranes of different smooth muscles deter-

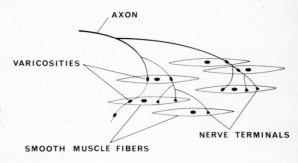

FIG. 6.32. Motor innervation of several smooth muscle fibers indicating varicosities along the course of the terminal branches of a single axon which function as neuromuscular junctions, in addition to the nerve terminals themselves (synapses en passant).

mine whether a particular neurotransmitter is excitatory or inhibitory. For this reason, among others, it was stated earlier that there are many different functional types of smooth muscle, even though morphologically these effector cells appear quite similar (p. 147).

Except for the fact that the process of neurotransmission is relatively slow and of much longer duration, the sequence of events that takes place in smooth muscle is quite similar to that already described for transmission at the motor end plate. When a nerve impulse reaches the terminal end of an excitatory autonomic nerve fiber, the latent period that elapses before any demonstrable change in the membrane potential of the smooth muscle fiber can be detected is around 50 msec. The comple depolarization of the membrane takes about 100 msec, and, if an action potential is not generated in the fiber, then the membrane potential gradually returns to its steady-state level with a time constant of between 200 and 500 msec. The entire sequence of voltage changes just described is called the *junctional potential*. The excitatory junctional potential (EJP) is analogous to both the EPSP of a central neuron and the EPP of a motor end plate.

When the magnitude of the excitatory depolarization in the membrane of a smooth muscle fiber reaches threshold to firing, an all-or-none action potential is generated in precisely the same way as during excitation of a skeletal muscle fiber (p. 132). The normal steady-state transmembrane potential of smooth muscle fibers is about -50 to -60 mv; threshold to firing is reached when depolarization of the membrane becomes approximately -30 to -40 mv.

Inhibition in smooth muscle fibers takes place when the neurotransmitter hyperpolarizes the membrane, thereby driving the rest potential away from threshold to firing and thus producing an *inhibitory junctional potential (IJP)*. This process is essentially the same as that which occurs in central nervous synapses during the generation of an IPSP (p. 193).

The electric responses described above are found in many smooth muscle fibers when a single nerve fiber is stimulated, although the duration of the latent period that is observed before a response is evident varies greatly among the smooth muscles obtained from different tissues. This observation is in accordance with the morphologic and physiologic characteristics of a synapse en passant, as illustrated in Figure 6.32. However, such an experimental finding could also be explained by transmission of the junctional response directly through low-resistance intercellular bridges as well as by the diffusion of the transmitter from one or more sites of release along the nerve to a number of smooth muscle fibers.

Even as miniature EPPs can be recorded in motor end plates at rest (p. 210), so miniature excitatory junction potentials can be seen at times in preparations of smooth muscle. Presumably, these miniature EJPs are caused by the release of single quanta of excitatory transmitter. The miniature EJPs observed in smooth muscle, however, exhibit much variation in both magnitude and duration; these differences can be explained by the unequal distances through which the transmitter diffuses in the tissue, perhaps with a concomitant dilution by the extracellular fluid.

THE REFLEX ARC: MONOSYNAPTIC AND POLYSYNAPTIC PATHWAYS

Introduction

A *reflex* is an automatic, stereotyped reaction, such as movement, that is performed without conscious volition in response to an adequate stimulus delivered to an appropriate receptor. Reflex activity is specific with regard to both the stimulus and the response, so that a particular stimulus elicits a particular response. A *reflex arc* is defined as the entire neural pathway that is involved in a reflex. Thus, a reflex arc is the fundamental anatomic pattern in the human nervous system that underlies the basic functional pattern of integrated neural activity, the reflex.

A reflex arc consists of a minimum of five structurally and functionally continuous and/or contiguous elements, as shown in Figure 6.1. The first element is a sensory end organ or receptor in which an adequate stimulus generates localized, graded receptor potentials that are a rough analog of the intensity of the applied stimulus. These local receptor potentials electrotonically depolarize the membrane of the coterminous afferent nerve fiber, the second element in the reflex arc. All-or-none impulses are propagated in the afferent or sensory nerve fibers to the central nervous system via the dorsal roots of the spinal cord or through appropriate cranial nerves (p. 244). The cell bodies of the afferent spinal nerves lie in the dorsal root ganglia, whereas those of the cranial nerves lie within homologous ganglia present on these nerves themselves.*

Within the spinal cord or other regions of the central nervous system, the afferent neuron synapses with an efferent neuron, the third major element in a simple reflex arc. At the synapse, another graded local electric response is generated which can excite an all-or-none propagated response in the efferent or motoneuron provided that the postsynaptic excitatory state is of sufficient amplitude. The impulse is conducted away from the central nervous system through the ventral roots of the spinal cord in this efferent neuron, the fourth element in the reflex arc, to the effector, which may be either a muscle fiber or a glandular cell.† At the neuroeffector junction, eg, the motor end plate, a local graded response once again is evoked, and the effector cell is stimulated to give its appropriate response. The effector provides the fifth and last element in a simple *monosynaptic reflex arc* as outlined above. In most reflexes found in the body, between one and several hundred interneurons with their synapses are interposed between the afferent and efferent nerves; hence, such reflex pathways are termed *polysynaptic*.

It must be emphasized at this point that in order for *any* reflex act to occur, the entire anatomic sequence of elements outlined above must be intact; if any interruption of the reflex pathway is present, the response will be either altered or abolished completely. Hence, many particular reflexes are of considerable clinical significance as diagnostic aids to the practicing physician (see also p. 227).

The reader will note that there are a minimum of three junctional regions in a reflex arc where a graded response is evoked: the receptor-afferent nerve junction, the synapse between the afferent and efferent nerves, and the neuroeffector junction. A nonpropagated or local graded potential that is proportional in amplitude to the magnitude of the applied stimulus is generated in each one of these three regions, and these graded potentials, in turn, electrotonically depolarize the adjacent nerve or effector membrane so that all-or-none propagated impulses are elicited.

The frequency of afferent impulses is proportional to the intensity of the applied stimulus at the receptor end of the reflex arc; there is also a rough proportionality between stimulus intensity and the frequency of impulses generated in the efferent nerve. However, since the synaptic junction between afferent and efferent neurons within the central nervous system is subject to many converging inputs, the net frequency of impulse generation in efferent nerves is subject to considerable modifica-

---

*A ganglion is defined as a discrete aggregation of nerve cell bodies that usually lies outside of the central nervous system.
†The concept that the dorsal roots of the spinal cord convey afferent impulses centripetally, whereas the ventral roots transmit efferent impulses centrifugally, is called the Bell-Magendie law (see also p. 245).

tion in vivo. Thus, the activities of monosynaptic, and especially polysynaptic, reflex arcs are vulnerable to considerable modification by spatial as well as temporal summation, occlusion, and subliminal fringe effects, as will be discussed later.

The detailed physiologic properties of each region involved in the reflex arc have been presented earlier in this chapter as well as in Chapters 4 and 5. The remainder of this section will be devoted to a discussion of several particular reflexes involving the spinal cord which underlie the control of certain aspects of skeletal muscle function in vivo, as well as to a number of pertinent general factors regarding the central regulation of synaptic functions within the spinal cord that in turn can modify reflex activity.

Some General Characteristics of Reflexes and Their Modification Within the Central Nervous System

THE ADEQUATE STIMULUS. Particular forms of energy trigger specific reflexes; the stimulus responsible for eliciting a reflex is called the *adequate stimulus* for that reflex. The properties of adequate stimuli for various reflexes were discussed on page 176; however, it should be reiterated here that a high degree of specificity usually exists between the stimulus that is applied to a receptor and the response that is evoked.

CENTRAL INHIBITORY MECHANISMS. In order for a reflex to be a smoothly integrated and coordinated response, it is essential that both inhibitory and excitatory mechanisms operate simultaneously in a cooperative manner. Therefore, several physiologic mechanisms within the spinal cord responsible for inhibition at the synapses will be discussed at this juncture.

Direct Versus Indirect Inhibition. The postsynaptic inhibition which develops following release of an inhibitory neurotransmitter and which hyperpolarizes the postsynaptic cell membrane (ie, generates as IPSP) is called *direct inhibition* (p. 193). Direct inhibition is not caused by prior discharge of the postsynaptic nerve. *Indirect inhibition*, on the other hand, can result from such previous activity. For example, indirect inhibition is present in a postsynaptic neuron that has just discharged an impulse because the neuron is in a refractory state. Indirect inhibition can also occur when an inhibitory neurotransmitter is not inactivated immediately following its release and binding to the postsynaptic receptors. The resulting afterhyperpolarization of the postsynaptic membrane thus reduces the excitability of that membrane to depolarizing impulses. In spinal neurons, the magnitude of the afterhyperpolarization may be large and its duration prolonged, especially after extended periods of high-frequency presynaptic activity.

Another type of indirect inhibition can be seen following long periods of presynaptic activity in which the rate of excitatory neurotransmitter release and/or the sensitivity of the postsynaptic membrane to the transmitter is reduced (cf. fatigue, p. 200).

Inhibitory Interneurons and Postsynaptic Inhibition. The central mechanism whereby an excitatory input is transformed into an inhibitory input through the mediation of an inhibitory interneuron is depicted in Figure 6.18 (p. 194). An example of how this mechanism can operate in vivo is given in Figure 6.33. Afferent impulses that arise in muscle spindles in response to stretch can pass directly through one synapse to spinal motoneurons that innervate extrafusal motor units of the same muscle (the *agonist*). These presynaptic impulses generate

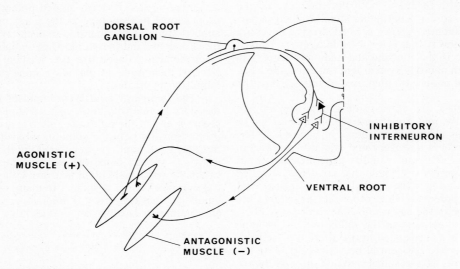

FIG. 6.33. The mechanism underlying the principle of reciprocal innervation of antagonistic muscles through an inhibitory interneuron. The agonistic muscle is excited (+), while the frequency of impulses passing to the antagonistic muscles of the ipsilateral limb is inhibited (−).

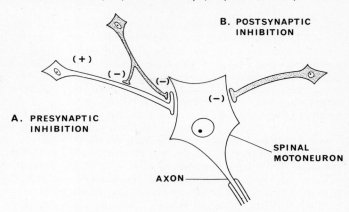

FIG. 6.34. Presynaptic and postsynaptic inhibition. The inhibitory neurons are shaded. **A.** The terminals of an inhibitory neuron (−) can end upon either the axon or the terminals of an excitatory neuron (+). Activity in the inhibitory neuron induces IPSPs in the excitatory neuron, thereby indirectly decreasing the excitability of the motoneuron. Gamma amino butyric acid is the putative neurotransmitter of presynaptic inhibition. **B.** Postsynaptic inhibition is caused by an IPSP generated directly in the membrane of the soma of à motoneuron by an inhibitory neurotransmitter. See also Figure 6.18.

EPSPs in the motorneurons which can summate and thus produce impulses in the postsynaptic motoneurons, thereby stimulating contraction of the same muscle in which the afferent impulses arose.

Simultaneously, IPSPs are generated in interneurons that synapse upon motoneurons which supply the *antagonistic* muscle, so that this muscle relaxes while the agonist contracts. By means of this spinal mechanism, contraction of the agonist is accompanied by relaxation of the antagonist, and a smooth flexion of the joint occurs (cf. reciprocal innervation, p. 221).

Another type of central inhibition that is caused by the activity of an inhibitory interneuron is shown in Figure 6.35 and discussed in relation to the stretch reflex (p. 218).

Presynaptic Inhibition. During presynaptic inhibition, the quantity of neurotransmitter released from a presynaptic excitatory terminal is diminished by the action of a particular neuron that reduces the amplitude of the impulse reaching the presynaptic excitatory knob (Fig. 6.34A). The effect of stimulation of such a *presynaptic inhibitory neuron,* therefore, is to depolarize partially the excitatory terminals that the quantity of excitatory transmitter subsequently released from these endings is diminished. When the presynaptic endings are partially depolarized by the activity of a presynaptic inhibitory neuron, it is true that the membrane potential is driven closer to the threshold to firing and thus excitability increases. More importantly, however, this depolarization reduces the amplitude of the action potential arriving at the presynaptic terminals by an amount equal to the induced depolarization, and this effect proportionally reduces the quantity of excitatory neurotransmitter released by the synaptic knobs.

There is considerable experimental evidence in support of the contention that the mediator released by presynaptic inhibitory neurons is γ-aminobutyric acid (GABA).

Neural Organization of Inhibitory Systems. Both postsynaptic and presynaptic inhibition generally are caused by the stimulation of specific neural circuits that converge upon a particular postsynaptic neuron, as discussed in the previous two sections (Figs. 6.18, 6.33, and 6.34). This effect is sometimes called *afferent inhibition.*

Neurons also may inhibit each other, however, by a negative feedback mechanism, an effect known as *negative feedback inhibition* (Fig. 6.35). For example, spinal motoneurons frequently branch, and the collateral may synapse with an inhibitory interneuron (or *Renshaw cell*) which terminates upon the soma of the same motoneuron as well as upon other motoneurons. Functionally, impulses that arise within the motoneuron excite the inhibitory interneuron, which in turn liberates an inhibitory neurotransmitter. The net effect of this process is that the discharge of impulses by the motoneuron is decreased or stopped completely.

Negative feedback inhibition is also encountered in the limbic system (p. 332) and the cerebral cortex (p. 315), and is caused by recurrent collaterals.

Presynaptic inhibition, as discussed earlier, is also a negative feedback mechanism. In this instance, sensory receptors, eg, in the skin, make functional contacts within the central nervous system so that presynaptic inhibition of their own and adjacent terminals occurs. Furthermore, nerve

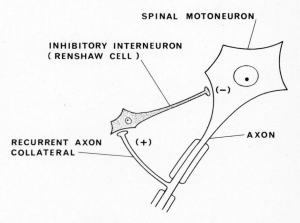

FIG. 6.35. Negative feedback inhibition of a spinal motoneuron through a recurrent axon collateral and an inhibitory interneuron (Renshaw cell).

fibers descending from the medulla oblongata in the pyramidal tracts (p. 267) have their terminations upon the excitatory synaptic knobs of the afferent nerve fibers that originate within the muscle spindles. Since the muscle spindles are the sensory end organs that initiate the stretch reflex (p. 218), inhibitory impulses from the brain stem centers presumably inhibit this reflex to some extent in vivo.

Lastly, a type of inhibition sometimes called *forward-feed inhibition* is seen in the cerebellum. In this region of the brain, excitation of basket cells generates IPSPs in the Purkinje cells; but since both of these cell types are stimulated by the same excitatory inputs, this mechanism presumably limits the time that the excitation persists following any particular burst of afferent impulses.

SUMMATION AND OCCLUSION IN CENTRAL NERVOUS PATHWAYS. In the central nervous system, there is a continuous fluctuation between excitation and inhibition at synaptic junctions, and these opposing influences coordinate as well as modulate reflexes and other neural functions.

A theoretical network of neurons is shown in Figure 6.36. In this conceptual diagram, presynaptic excitatory neurons 1 and 2 *converge* upon postsynaptic neuron A, and presynaptic neuron 2 *diverges* onto postsynaptic neurons A and B. Therefore, excitation of either neuron 1 or neuron 2 will generate an EPSP in postsynaptic neuron A. If, however, neurons 1 and 2 are active simultaneously, then two regions will be depolarized on the soma of postsynaptic neuron A, and these depolarizations can *summate,* giving rise to a conducted impulse in the axonal process of neuron A. Consequently, the EPSP generated in neuron A will have twice the magnitude of the EPSP induced by stimulation of either neuron 1 or neuron 2 alone, so

that the threshold to firing will be reached in postsynaptic neuron A, provided that the summated depolarizations are sufficient to reach the firing level of this neuron. Thus, *spatial facilitation* or *summation* has occurred, because the EPSP generated in postsynaptic neuron A is facilitated by the independent but concomitant activity in each of the two presynaptic neurons.

In the example of spatial facilitation given above, postsynaptic neuron B has not discharged; however, the excitability of this element is increased so that it can fire more readily while the EPSP is present. Therefore, postsynaptic neuron B is said to lie in the *subliminal fringe* of postsynaptic neuron A.

Stated another way, neurons lie in the subliminal fringe if they are not activated directly by a train of afferent impulses, ie, they do not lie in the *discharge zone.* However, neurons in the subliminal fringe do have their excitability increased, depending upon the number of presynaptic terminals ending upon their cell bodies, so that more afferent impulses arriving while these neurons are in this excited state can produce *temporal summation.* The neurons within the discharge zone, of course, have the most presynaptic terminals converging upon them.

Subliminal fringe effects such as described above can exert a considerable effect upon the excitability of neurons both above and below the level of the discharge zone in the spinal cord.

If a dorsal root of the spinal cord is stimulated, the number of afferent neurons that fire (ie, the discharge zone) increases proportionally with the strength of the stimulus up to a maximum as more fibers are activated (p. 344). The magnitude of the subliminal fringe also increases with increasing stimulus intensity, but, as shown in Figure 6.37, the discharge zone and the subliminal fringe never achieve the same magnitude.

Inhibitory impulses exhibit spatial and temporal facilitation (or summation) as well as subliminal fringe responses exactly like those described above for excitatory impulses.

If presynaptic neuron 2 (Fig. 6.36) alone is stimulated at a sufficiently high frequency, postsynaptic neurons A and B will fire because of temporal summation of the EPSPs produced by the individual impulses. Similarly, if presynaptic neuron 3 is stimulated at a sufficiently high rate, postsynaptic neurons B and C also will discharge. Now if neurons 2 and 3 are stimulated at a like high frequency, postsynaptic neurons A, B, and C all will fire. However, the response to a stimulation of presynaptic neurons 2 and 3 *together* is not as great as that which is elicited by the sum of the responses to an *independent* rapid stimulation of presynaptic neurons 2 and 3. This effect is observed because both presynaptic neurons 2 and 3 converge upon postsynaptic neuron B. Consequently, there is a decrease in the net or overall response of the neural circuit on its output side due to the fact that certain presynaptic neurons share a common postsynaptic

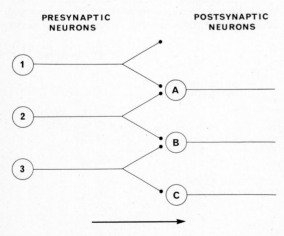

PRESYNAPTIC
NEURONS

POSTSYNAPTIC
NEURONS

FIG. 6.36. Hypothetical nerve circuit to illustrate summation, occlusion, and subliminal fringe effects in the spinal cord. Synapses are indicated by the dots on the presynaptic terminals; the direction of impulse transmission is indicated by the arrow at the bottom of the figure. (See text for explanation.)

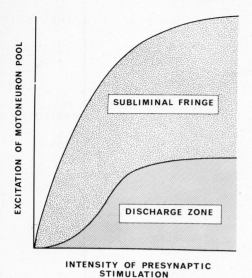

FIG. 6.37. Relationship between the intensity of presynaptic stimulation and the magnitude of the discharge zone and the subliminal fringe. (After Ganong. *Review of Medical Physiology,* 6th ed, 1973, Lange Medical Publications.)

neuron. The phenomenon of a decreased neural output somewhat below that which reasonably could be expected is called *occlusion*.*

Both excitatory and inhibitory subliminal fringe effects as well as the phenomenon of occlusion can exert profound effects upon neural transmission in any specific pathway in the nervous system. Hence, the timing of patterns of impulses within peripheral nerves generally is modified considerably by these mechanisms as the volleys of impulses traverse synapses within the central nervous system.

POSTTETANIC POTENTIATION. The effects of inhibition and facilitation in the central nervous system on neural excitability are of comparatively short duration. Posttetanic potentiation, on the other hand, is another type of facilitation of synaptic activity whose effects on neural excitability are much more protracted.

Posttetanic potentiation occurs in centrally located neurons following intervals of prolonged high-frequency afferent stimulation, and it results in a reduction of the threshold of firing of the neurons to further afferent inputs. That is, the excitability of the neurons is enhanced following such stimulation. Posttetanic potentiation also is seen in autonomic ganglia and at neuromuscular junctions. The potentiation, or heightened excitability, may have a duration of several hours and is

This physiologic usage of the term "occlusion" should not be confused with the contact of the teeth in upper and lower jaws that is an essential component of mastication.

localized strictly within the afferent terminals that have been subjected to the repeated stimulation. Consequently, posttetanic potentiation does not spread to nearby afferent inputs. Inhibitory as well as excitatory inputs exhibit posttetanic potentiation.

There is much experimental evidence in support of the conclusion that posttetanic potentiation is caused by an increase in neurotransmitter release following prolonged afferent stimulation which results in a greater ease of impulse transmission through the specific neural pathway involved (cf. p. 199). Consequently, posttetanic potentiation may be considered to be a rudimentary form of "learning" through repeated use of a given neural circuit (cf., conditioned reflexes, p. 440).

THE FINAL COMMON PATH. As indicated in Figure 6.1, all of the motoneurons that innervate the extrafusal fibers of skeletal muscles form the efferent part of the reflex arc. Thus, all neural excitatory and inhibitory impulses arising in receptors as well as in the brain and spinal cord ultimately must be channeled through the motoneurons to the muscles. The motoneurons, therefore, provide the *final common path* for all such impulses. The cell bodies of the motoneurons that innervate a particular muscle often are called the *motoneuron pool* of the muscle.

A great number of excitatory and inhibitory inputs converge on the cell bodies of the motoneurons in the spinal cord. An average motoneuron can have up to 5,500 individual synaptic knobs terminating upon its surface. Some of the types of input that converge upon a single motoneuron located at one level within the ventral portion of the spinal cord are shown in Figure 6.38. In addition to the multiple inputs shown in this figure, excitatory and inhibitory inputs, usually relayed by interneurons, also converge upon the cell body of any particular motoneuron from different levels of the spinal cord; long, descending (motor) tracts that arise in the brain also synapse upon spinal motoneurons (see Fig. 7.100, p. 366).

It is evident from the facts presented above that the net activity of any given spinal motoneuron is subject to considerable modification from a host of excitatory and inhibitory inputs that continually bombard the nerve cell body.

THE CENTRAL EXCITATORY STATE AND THE CENTRAL INHIBITORY STATE. The physiologic effects of direct and presynaptic inhibition as well as the subliminal fringe excitatory response can be widespread within the spinal cord, but all of these effects generally are evanescent. The spinal cord, however, can also exhibit prolonged alterations in its irritability which are called the *central excitatory state* and the *central inhibitory state*.

A central excitatory state is present when excitatory influences overweigh inhibitory effects; conversely, a central inhibitory state exists when

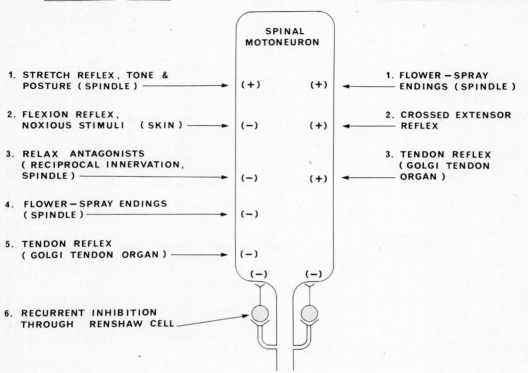

FIG. 6.38. Summary of the major ipsilateral and contralateral inputs to the cell body of a single spinal motoneuron. Excitation = (+); inhibition = (−). (After Ruch and Patton [eds]. *Physiology and Biophysics,* 19th ed, 1965, Saunders.)

inhibitory influences overbalance excitatory effects.

Presumably, these altered states of central irritability are caused by prolonged responses to neurotransmitter substances and/or by the prolonged activity of reverberating circuits (p. 222).

If the central excitatory state is pronounced, then excitatory impulses can spread, or *irradiate,* in an uncontrolled fashion not only to many regions of the spinal cord that regulate somatic functions, but also to regions that control many autonomic activities (see mass reflex, p. 227).

CONCLUDING REMARKS. It is clear from the material presented in the six preceding subdivisions of this section that the fundamental patterns of reflex activity are subject to considerable modification by a wide variety of central nervous influences, despite the fact that all reflexes basically are automatic, stereotyped responses that ultimately are of physiologic significance to the individual. The central mechanisms just presented form the basis for an understanding of many of the integrated functions of the central nervous system in addition to their participation in, and influences upon, somatic reflexes themselves.

In the remainder of this section, a few particular somatic reflexes will be discussed in detail;

many other specific somatic and visceral reflexes will be encountered throughout the remainder of this text.

The Stretch Reflex

The anatomic basis for the *stretch* or *myotatic reflex* is a simple monosynaptic pathway as shown in Figure 6.39. The sensory receptor for the stretch reflex is the muscle spindle (p. 180), and the response elicited is the contraction of the muscle in which the spindle is located. Stretch reflexes are the only monosynaptic reflexes found in the body.

A stretch reflex can be elicited by a sudden deformation of a tendon (as by a sharp tap on the tendon; see p. 227). This rapid mechanical deformation stretches the nuclear bag region of the muscle spindle, thereby distorting the annulospiral endings within the spindle (Fig. 6.3), and receptor potentials are generated in the nerve terminals. The receptor potentials, in turn, induce propagated all-or-none impulses within the class I(a) sensory (afferent) nerve fibers continuous with the annulospiral endings of the spindle (Table 5.2, p. 162). The frequency of impulse discharge in these nerves is directly proportional to the extent to which the endings have been stretched. As shown in Figure 6.39, the afferent impulses pass to the terminals of

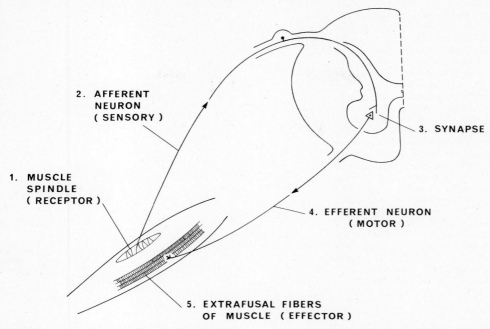

FIG. 6.39. Pathway of the monosynaptic stretch (or myotatic) reflex. The numbering of the individual elements of this reflex arc corresponds to that in Figure 6.1.

the sensory nerve within the spinal cord, synapse directly upon a large, myelinated alpha motoneuron whose cell body lies in the ventral region of the spinal cord, and there generate excitatory post-synaptic potentials. The efferent impulses so produced are then propagated down the motoneuron, and excitatory transmitter (acetylcholine) is released at the motor end plate region. The end plate potentials thus generated depolarize the muscle fibers and trigger the sequence of events leading to muscular contraction.

The net effect of the stretch reflex is to oppose changes in the length of the muscle.

Some further aspects of the stretch reflex and its regulation in vivo now may be considered.

EXCITATION OF THE MUSCLE SPINDLE. The muscle spindle is in parallel with the extrafusal fibers; therefore, if the rate of discharge in the gamma efferent fibers to the spindle remains constant (Fig. 6.3), active contraction of the muscle will reduce the tension applied to the spindle as the entire muscle shortens. Consequently, when a muscle is stretched under such conditions, a sudden volley of afferent impulses is generated within the spindle (Fig. 6.40A), and, as the muscle contracts, reflexly, all discharge in the sensory nerve stops as the external tension applied to the spindle is relieved (Fig. 6.40B). In a like manner, if the extrafusal muscle fibers alone are stimulated directly, then afferent discharge from the spindle is inhibited as the muscle contracts.

Muscle spindles are able to respond to static as well as dynamic changes in muscular activity.

Thus, the receptors in these structures can respond to rapid as well as to continuous stretch. However, the reflex muscular contraction elicited is far stronger when the spindle is stretched rapidly than when it is subjected to a continuous deformation. The more powerful reflex evoked by a rapid stretch, in which the rate of deformation of the spindle is high, is called the *dynamic stretch reflex;* the weaker response produced by a continuous deformation of the spindle is referred to as the *static stretch reflex.*

There is good experimental evidence which indicates that the annulospiral (or primary) nerve endings within the spindle respond to both a continuous stretch and the rate of change of stretch, whereas the flower-spray (or secondary) endings are stimulated mainly by continuous stretch alone.

FEEDBACK REGULATION OF SPINDLE ACTIVITY. The spindle mechanism together with its central reflex connections make up a feedback system that functions to maintain the length of skeletal muscles within definite limits. Hence, stretch of a muscle augments the afferent discharge from the spindle and the muscle shortens reflexly. Conversely, when the muscle shortens, spindle discharge decreases or ceases entirely, so that the muscle relaxes. Thus, an exquisite balance between lengthening and shortening of muscles in vivo is maintained at all times.

In the feedback loop described above and shown in Figure 6.39, the sensitivity of the spindle to the rate of change of stretch tends to offset the effects of the time lag that is an inherent charac-

FIG. 6.40. Electric activity from the annulospiral endings of a muscle spindle (circles to right in each diagram) under various conditions. Note that the spindle is in parallel with the extrafusal fibers, unlike the Golgi tendon organ. **A.** Muscle is stretched; there is some discharge in I(a) fibers from the annulospiral endings. **B.** Muscle is contracted and spindle relaxed; therefore, there is no discharge from the annulospiral endings. **C.** Stretch applied to the muscle spindle causes some gamma efferent discharge (Fig. 6.3) and contraction of the intrafusal fibers, with an attendant discharge in the I(a) sensory fibers through deformation of the nuclear bag region. **D.** Additional tension (or muscular contraction) that is superimposed upon that caused by gamma efferent discharge (**C**) induces a marked acceleration of discharge in the sensory nerve. (Data from Ruch and Patton [eds]. *Physiology and Biophysics,* 19th ed, 1965, Saunders; Ganong. *Review of Medical Physiology,* 6th ed, 1973, Lange Medical Publications.)

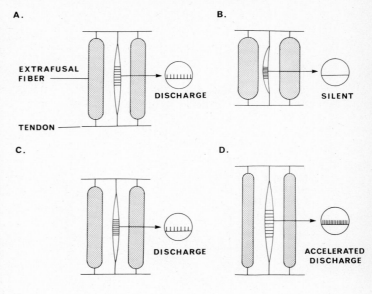

teristic of the neuronal circuit. Nevertheless, this time lag can give rise to some oscillation in the feedback mechanism during voluntary muscular activity; consequently, a *physiologic tremor* with a frequency of about 10 hz is present.

GAMMA EFFERENT DISCHARGE. Direct excitation of the gamma efferent system of the muscle spindle (Fig. 6.3) induces a considerably different pattern of physiologic activity than that which is produced by stimulation of the extrafusal fibers of the muscle. Stimulation of the intrafusal fibers does not produce any demonstrable contraction of the extrafusal fibers of the muscle because these elements are neither powerful enough nor numerous enough to cause shortening of the entire muscle. Increased gamma efferent nerve activity, however, causes a shortening of the contractile ends of the intrafusal fibers so that the nuclear bag region of the spindle is stretched, as shown in Figure 6.40C. The stretch of the nuclear bag region, in turn, stimulates the annulospiral endings to discharge with greater frequency so that volleys of impulses are generated in the class I(a) afferent fibers leading centrally from the spindle. If the entire muscle now is stretched during the enhanced activity of the gamma efferent fibers, additional impulses are generated in the annulospiral endings because of the prior stretch of the nuclear bag region of the spindle. Consequently, the net effect is an increased activity of the I(a) fibers that is superimposed upon the preexisting discharge frequency caused by the prior contraction of the intrafusal fibers that was induced by an enhanced pattern of gamma efferent discharge (Fig. 6.40D). Thus, increased gamma efferent activity increases the sensitivity of the muscle spindles to further stretch. Furthermore, the sensitivity of the muscle spindles themselves is a direct reflection of the frequency of gamma efferent discharge.

REGULATION OF GAMMA EFFERENT DISCHARGE. The activity of the motoneurons comprising the gamma efferent system is controlled to a large extent by descending tracts that arise in the brain stem (the bulboreticular area, p. 305) and that are subject, in turn, to regulation from impulses arising within still higher centers, namely, the cerebellum (p. 307) and the basal ganglia (p. 336) as well as the cerebral cortex itself (p. 324).

Through these central nervous pathways, the irritability of the muscle spindles in different regions of the body is controlled, and thus the thresholds for excitation of various particular stretch reflexes are regulated. In this manner, the constantly varying requirements of postural control are met (p. 367).

In addition to the central nervous factors outlined above, gamma efferent discharge also is increased by anxiety and by stimulation of the skin (particularly by irritating chemical agents). Such increased discharge is also a concomitant of the increased alpha motoneuron activity that initiates muscular contractions. In this instance, the muscle spindles shorten together with the muscle proper so that spindle discharge may persist throughout contraction of the entire muscle. The physiologic role of this simultaneous contraction of extrafusal and intrafusal fibers apparently is related to the coordination and damping of random oscillatory motions during the performance of voluntary movements, since no inhibitory feedback signals originate in the spindles under such circumstances.

In conclusion, reflex muscular contractions are initiated *directly* by excitation of the large alpha motoneurons which receive presynaptic impulses from afferent nerves synapsing directly upon the cell bodies of the motoneurons within the spinal cord. These efferent neurons, in turn, pass directly to neuromuscular junctions on the extrafusal fibers. The gamma efferent system of the intrafusal fibers, on the other hand, reflexly controls muscular contraction *indirectly* through the feedback mechanism described above. Since this feedback mechanism controls a large amount of physical activity through a small energy input, it is sometimes referred to as a *servomechanism.* Any stretch reflex ultimately is stimulated by impulses originating in alpha motoneurons, but the reflex is subject to considerable modification by afferent as well as gamma efferent impulses that originate within or are conducted to the muscle spindles.

REACTION TIME. The time that elapses between the application of an adequate stimulus to a receptor and the appearance of a response is called the *reaction time.* In humans, the reaction time for a simple monosynaptic stretch reflex, eg, the knee jerk, is between 20 to 24 msec.*

It is known from experimental data that the I(a) fibers that originate from the annulospiral endings of the muscle spindles terminate directly upon the cell bodies of alpha motoneurons in the spinal cord whose axons innervate the extrafusal fibers of the same muscle. Since the conduction velocities of impulses within the afferent and efferent nerves are known (Table 5.1, p. 161), and the distance from the muscle to the spinal cord can be measured readily, it is a simple matter to calculate the time consumed by impulse transmission along both afferent and efferent limbs of the reflex arc. When this conduction time is subtracted from the reaction time, the difference between these values is the time required for the impulse to be transmitted through the spinal cord, called the *central delay.* In humans, the central delay for the knee jerk ranges from 0.6 to 0.9 msec; consequently, since the minimum time required for transmission of an impulse through one synapse is 0.5 msec, only one synapse must be traversed as the impulse passes through the spinal cord (cf. p. 192).

The Tendon Reflex

The *tendon reflex,* also called the *inverse stretch reflex,* produces an inhibitory rather than an excitatory response when a muscle is stretched beyond a certain point.

Within limits, the more a muscle is stretched, the stronger the resulting reflex contraction becomes; but when the tension developed within the muscle becomes too great, the contraction abruptly ceases, and relaxation ensues.

The receptor for the tendon reflex is the Golgi tendon organ (Fig. 6.4). Nerve fibers that terminate in the Golgi tendon organ belong to the I(b) group of afferent (sensory) nerve fibers (Table 5.2, p. 162).

The knee jerk is stimulated by a sharp tap upon the patellar tendon which, in turn, stretches the quadriceps femoris muscle.

Excitation of these rapidly conducting fibers results in the generation of IPSPs in the spinal motoneurons innervating the muscle in which the afferent impulses arose (Figs. 6.41). The I(b) afferents from the Golgi tendon organs terminate upon inhibitory interneurons, as shown in the figure; they also form excitatory connections with the antagonists of the muscle (see reciprocal innervation, below).

The Golgi tendon organs, in contrast to the muscle spindles, are in series with the fibers of the muscle and the tendon; thus they are stimulated by passive stretch as well as by active contraction of the muscle. Although these structures have a low threshold to excitation, stimulation of Golgi tendon organs by passive stretch is not particularly significant because the relatively more elastic contractile elements of the muscle absorb most of the stretch (Fig. 6.4, 6.41B). For this reason, a powerful stretch is needed to excite the tendon organs and produce relaxation. On the other hand, the Golgi tendon organs are excited readily by contraction of the muscle (Fig. 6.41C) so that these receptors function as mechanoelectric transducers in a negative feedback system (or servomechanism) which regulates *muscular tension* in a manner comparable to the feedback system of the muscle spindle which regulates the *length* of skeletal muscle.

Physiologically, the tendon reflex appears to have two fundamental attributes. First, the autogenic inhibition produced by this reflex serves to protect a muscle from rupture by inducing relaxation when the contractile force developed within the muscle becomes too great. Second, the tendon reflex provides a mechanism whereby the tension can be maintained at a constant level within the muscle, regardless of the length of the muscle. Many examples of this effect can be adduced; the simple lifting of a light weight by an isotonic contraction of the arm muscles is one illustration of this process.

The Flexor Reflex and the Crossed Extensor Reflex; Reciprocal Innervation of Antagonistic Muscles

POLYSYNAPTIC NEURAL PATHWAYS. As depicted in Figure 6.42, any polysynaptic reflex pathway within the central nervous system can have a complex, branching pattern, and the number of synapses present in each branch can vary greatly. Because of the synaptic delay that is an inherent characteristic of each synapse, excitation of motoneurons is most rapid in those pathways having the fewest synapses in their afferent connections, whereas activity in the pathways with several synapses takes longer to reach and to excite the motoneuron (Fig. 6.42A and B). The net effect of this delayed excitation is that, in a complex polysynaptic pathway, a single stimulus produces a prolonged excitation of the motoneuron, and therefore the response to that stimulus also is prolonged.

A.

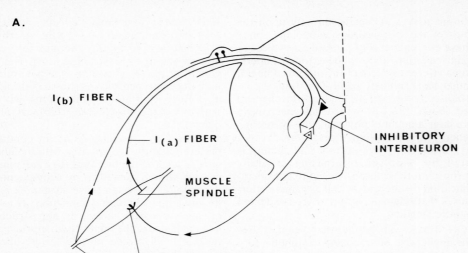

B.

FIG. 6.41.   The tendon reflex. **A.** When a muscle is stretched beyond a certain point, contraction of the muscle is inhibited through excitation of I(b) fibers from the Golgi tendon organs by means of an inhibitory interneuron in the spinal cord, and the muscle relaxes. **B.** The rate of discharge from Golgi tendon organs is low during passive stretch of the muscle. **C.** The frequency of discharge from Golgi tendon organs is high during strong contraction of the muscle. Note that the tendon organs are in series with the contractile elements of the muscle. (Data from Ruch and Patton [eds]. *Physiology and Biophysics,* 19th ed, 1965, Saunders.)

In addition, some of the afferent neurons may branch in such a manner that the impulse is directed backward through part of the circuit, allowing the reexcitation of certain afferent neurons in a cyclic fashion until the signal becomes so weak that postsynaptic excitation does not generate an all-or-none action potential in the motoneuron (Fig. 6.42C). Thus the impulse oscillates or reverberates; this type of neuronal circuit therefore provides another central mechanism whereby the response to a single stimulus is both amplified and prolonged. *Reverberating* or *oscillating* circuits of the type shown in Figure 6.42C are of common occurrence in the spinal cord as well as in the brain. Thus, both multisynaptic pathways and reverberating circuits are important in such regions of the nervous system in extending the period of response to a particular stimulus.

THE FLEXOR OR WITHDRAWAL RE-FLEX.   The flexor or withdrawal reflex is initiated by practically any type of sensory stimulus applied to the skin (Fig. 6.43A). The flexor reflex is elicited strongly by the stimulation of pain endings, such as occurs when the skin of a finger is burned; hence, the withdrawal reflex also is called the *nociceptive reflex*. The withdrawal reflex also is initiated by excitation of the cutaneous touch receptors or by stimulation of the muscles and subcutaneous tissues; however, in these instances the response is less dramatic as well as less prolonged. Therefore, the flexor reflex exhibts a *graded response*, depending upon the type and severity of the stimulus.

The flexor reflex typifies polysynaptic reflexes in general. The response is a contraction of the flexor muscle with a simultaneous inhibition of the extensor muscles in the same limb (ie, the *ipsilat-*

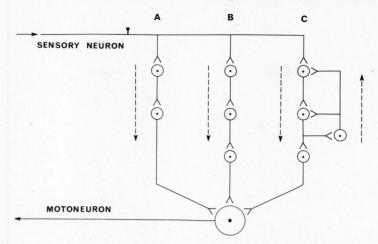

FIG. 6.42. Polysynaptic reflex pathways. **A.** Pathway with one interneuron and three synapses. **B.** Pathway with two interneurons and four synapses. **C.** A reverberating or oscillating circuit. Impulses can pass directly from the sensory neuron through the two interneurons to excite the spinal motoneuron, or else the impulses can feed back through an axon collateral (shown on the second interneuron) and excite a third neuron (right of figure) which, in turn, can reexcite the first and second interneurons in a cyclic fashion. In this way, the activity of the original impulse volley that entered the circuit is prolonged considerably. In actuality, polysynaptic pathways within the brain and spinal cord generally have a considerably greater number of synapses and neurons than are indicated here.

eral side) as shown in Figure 6.43A. Thus, the affected limb is flexed, and hence withdrawn from the stimulus.

THE CROSSED EXTENSOR REFLEX. The application of an extremely strong stimulus to the skin not only evokes flexion of the ipsilateral limb but also causes extension of the opposite or contralateral limb, as illustrated in Figure 6.43B and explained in the legend. Thus, the crossed extensor reflex properly can be considered as a component of the flexor reflex per se.

It is difficult to demonstrate the crossed extensor reflex in an intact subject or animal. However, if the brain is severed experimentally from the spinal cord in an organism such as the cat or dog, moderating impulses from the brain are removed and a so-called *spinal animal* results. When the hind limb of such a spinal animal is pinched, the ipsilateral leg is withdrawn from the stimulus, and the contralateral leg is extended. If the intensity of the stimulus is sufficiently strong, then all four limbs are involved, not merely the hind limbs; the ipsilateral forelimb is extended, whereas the contralateral forelimb is flexed. The spread of excitatory impulses along the spinal cord following intense sensory stimulation is called *irradiation of the stimulus;* the increase in the total number of motor units that are activated by this irradiation is known as *recruitment of motor units*.

AFTERDISCHARGE, REACTION TIME, AND SUMMATION IN THE FLEXOR REFLEX. The flexor and crossed extensor reflexes provide excellent model systems for the investigation of the physiologic properties of polysynaptic reflexes in general. As noted above, the flexor reflex exhibits a graded response, depending upon the intensity of the applied stimulus. Thus, a weak noxious stimulus applied to a limb evokes only a slight withdrawal by a rapid flexion of the affected limb. As progressively stronger stimuli are applied, the extent of the flexion induced becomes greater, may become repetitive, and is more prolonged as the stimulus undergoes irradiation to ever larger areas of the motoneuron pool that supplies the muscles of the affected limb. The prolongation of the response to strong noxious stimuli is caused by repeated firing of the motoneurons and is called *afterdischarge* (Figs. 6.44 and 6.45). The neuronal mechanisms that underlie afterdischarge in polysynaptic neural circuits were discussed at the beginning of this section (p. 218).

As the intensity of a noxious stimulus is increased, the reaction time decreases (p. 221). Both spatial and temporal summation also take place at the synapses involved in polysynaptic circuits (p. 199); and the more intense stimuli evoke impulses of higher frequency in the active neurons. Furthermore, such intense stimuli excite more neuronal pathways than do weaker stimuli. The net effect of these processes is that summation of EPSPs to the firing level takes place more rapidly in the motoneurons, and, accordingly, the reaction time decreases.

PHYSIOLOGIC SIGNIFICANCE OF THE FLEXOR REFLEX. Flexor reflexes can be evoked by mild cutaneous stimulation or by stimulation of muscle spindles. However, powerful flexor responses that withdraw the affected limb from a painful or otherwise noxious stimulus are elicited by stimuli that are harmful, or potentially harmful, to the or-

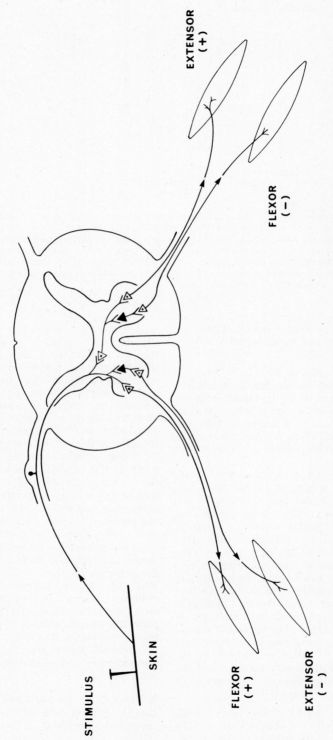

FIG. 6.43. The flexor and cross extensor reflexes. The noxious stimulus is indicated on the left as a nail puncturing the skin, which produces flexion in the ipsilateral limb and extension in the contralateral limb. Excitation = (+); inhibition = (−). Inhibitory neurons within the spinal cord are indicated by solid black triangles. Reverberating circuits (such as indicated in Fig. 6.42C) that prolong the response to this type of stimulus are involved, but not shown in the figure for the sake of clarity (see also Figs. 6.44 and 6.45).

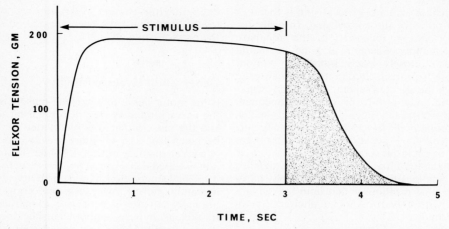

FIG. 6.44. Afterdischarge in a flexor reflex. This response has a sudden onset, as indicated by the rapid development of tension in the flexor muscle recorded on the myogram; fatigue is indicated by a gradual decline in tension as the stimulus persists; and afterdischarge is indicated by the gradual return to a resting tension after cessation of the stimulus (shaded area to right). Afterdischarge is caused by the activity of reverberating circuits (Fig. 6.42C) in the reflex pathway. Compare Figure 6.45. (After Guyton. *Textbook of Medical Physiology,* 4th ed, 1971, Saunders.)

ganism. Such *nociceptive stimuli* result in withdrawal of the affected limb from the noxious stimulus, while the crossed extensor reflex permits continued support of the body, as when one inadvertently steps on a tack with bare feet.

Withdrawal reflexes thus serve as a rapid *protective mechanism* against stimuli which could injure or damage the body. Such responses can be called *prepotent* because they take precedence in the neural pathways of the spinal cord over any other reflex activities that may be taking place when the stimulus is applied.

The phenomenon of afterdischarge, whose duration is related to the intensity of the irritating stimulus, enables the affected limb to be held away from the source of the noxious stimulus for several

seconds, thus allowing sufficient time for activation of additional neural circuits that can move the entire body away from the stimulus itself (Fig. 6.45).

LOCAL SIGN AND FINAL POSITION. The exact spatial position to which a limb is withdrawn following the application of a nociceptive stimulus to the skin depends strictly upon the region on the body surface to which the stimulus is applied. This phenomenon is known as *local sign*. For example, if the medial surface of a limb is stimulated, the response may include abduction as well as flexion, whereas if the lateral surface is stimulated, adduction may accompany the withdrawal response. In any case, the net movement of the affected limb is such as to remove it from the source of the stimulus.

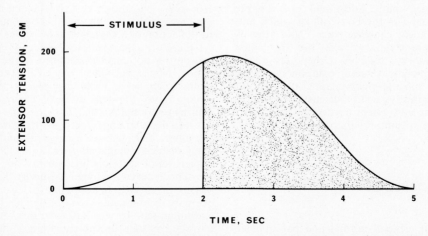

FIG. 6.45. Afterdischarge in a crossed extensor reflex (shaded area). Note that peak tension in this response is not reached until after the stimulus has ended, in contrast to the flexor reflex (Fig. 6.44). (After Guyton. *Textook of Medical Physiology,* 4th ed., 1971, Saunders.)

An excellent example of local sign is found in the scratch reflex of the normal dog. In order for this spinal reflex to be elicited, the adequate stimulus is a number of sequential, punctate, tactile stimuli that are closely spaced in a linear fashion. The response is a precisely directed and energetic scratching of the affected region. Thus, a flea walking along the body surface provides the adequate stimulus for the scratch reflex. However, if the flea jumps from one point to another on the body surface, the two widely separated tactile stimuli are inadequate to elicit the scratch reflex.

The principle of *final position* merely signifies that the original position of the limb is wholly unrelated to the final position of that limb following a flexion or withdrawal response elicited by an adequate stimulus.

FRACTIONATION AND OCCLUSION. A supramaximal stimulation applied to any of the sensory nerves that innervate a limb never evokes as powerful a contraction of the muscles as does direct electric stimulation of the muscles themselves. Therefore, it is apparent that the afferent impulses are *fractionated* within the motoneuron pool of the spinal cord; ie, each impulse is transmitted to only a portion of the entire motoneuron pool that innervates the flexor muscles of that extremity. Furthermore, if all of the afferent nerves to a particular muscle are separated by dissection and then individually stimulated in an artificial fashion, the sum of the single tensions developed in the muscle innervated by these afferent nerves is greater than the tension which is evoked either by direct excitation of the muscle or by stimulation of all of the afferent nerves simultaneously. Thus, it can be seen that different afferent neurons share some of the same motoneurons, and *occlusion* occurs when all of the afferent inputs are stimulated at the same time (Fig. 6.36; p. 216).

ADDITIONAL POLYSYNAPTIC REFLEXES. Besides the flexor reflex and the concomitant crossed extensor reflex, many additional polysynaptic reflexes are found in the body, all of which have similar physiologic characteristics. Thus, the abdominal and cremasteric reflexes are, in a sense, types of withdrawal reflexes. Other polysynaptic reflexes can include visceral components. Many examples of specific polysynaptic reflexes and their neural pathways will be discussed in Chapter 7 as well as in subsequent portions of this text in relation to the specific regulatory functions which they subserve.

CLINICAL CORRELATES

Introduction

In this section, several normal and pathophysiologic states will be described that have a direct bearing upon some clinical aspects of neuromuscular and central nervous activity. The material to be discussed has been selected with the goal of exemplifying certain dysfunctions of normal mechanisms that have been considered in this and the previous two chapters.

Denervation Hypersensitivity

If the motor nerve to a skeletal muscle is severed, the nerve ultimately degenerates distal to the cut, and the muscle that was innervated by the nerve becomes progressively more sensitive to the action of a given quantity of acetylcholine. This phenomenon is called *denervation hypersensitivity* (see also p. 372). Denervation hypersensitivity also is seen in smooth muscle toward the neurotransmitter that normally activates it. However, in marked contrast to skeletal muscle, smooth muscle does not undergo atrophy following denervation. Similarly, denervated glands (except sweat glands) also exhibit hypersensitivity.

There is much experimental data available which indicates that neural elements in general exhibit a like reaction to a loss of their innervation. When higher centers within the central nervous system are damaged or destroyed, the activity of the lower centers that were controlled originally by the higher centers is increased, an effect called the *release phenomenon*. Presumably the release phenomenon is caused by denervation hypersensitivity.

The hypersensitivity that is observed in nerves following their denervation is localized within the elements that were innervated directly by the inactivated neurons; structures further along the pathway are not affected. For example, ventral spinal cord injuries do not cause hypersensitivity to acetycholine within the paralyzed skeletal muscles; rather, this hypersensitivity is localized within the postsynaptic neurons supplied by the destroyed presynaptic neurons.

The primary mechanism responsible for producing denervation hypersensitivity is unclear. However, it is known that, in skeletal muscle, the total area of the sarcolemma that becomes sensitive to the action of acetylcholine increases significantly; therefore, the excitatory effect of this neurotransmitter no longer is confined to the motor end plates, whose net sensitivity to acetylcholine remains unchanged. Following regeneration of the motor nerve, the sensitivity of the muscle fibers returns to normal, so that acetylcholine specifically exerts a depolarizing effect upon the sarcolemma of the motor end plate once again. Denervated postganglionic cholinergic neurons also exhibit a similar expansion of acetylcholine sensitivity.

The denervation hypersensitivity that develops in smooth muscle is believed due, at least in part, to a decrease in monoamine oxidase activity (p. 203). An additional factor contributing to this effect is a diminished uptake of catecholamines into the presynaptic terminals following their release (Fig. 6.28, p. 206). Consequently, higher concentrations of these neurotransmitters are present and thus are able to act upon the postsynaptic receptors.

Clinical Significance of Reflexes

From a neurologic standpoint, various reflexes may be classified as *superficial, deep, visceral,* and *pathologic*. It is worthwhile to reiterate here that a specific reflex is diminished or absent if there is any dysfunction at any point along the entire neurologic pathway for that reflex, from the receptor to the effector. Furthermore, in many instances diminished or absent superficial skin reflexes are of particular diagnostic significance when they are accompanied by excessively strong deep and pathologic reflexes, because this abnormal response pattern is indicative of an upper motor neuron participation in the overall response. Hence, many reflexes and their patterns are of extreme importance to the clinician for diagnostic purposes.

Following are specific examples of particular reflexes belonging to each of the categories listed above.

SUPERFICIAL REFLEXES. Superficial reflexes can be elicited by stimulation of either mucous membranes or the skin.

Examples of mucous membrane reflexes are the *corneal* (or *conjunctival*) *reflex,* in which irritation of the cornea or conjunctiva of the eye causes blinking; the *nasal* (or *sneeze*) *reflex* evoked by irritation of the mucosa of the nasal passages; and the *pharyngeal* (or *gag*) *reflex,* a retching or gagging response to appropriate stimulation of the pharynx.

Also included among the superficial reflexes are the upper and lower abdominal reflexes, which are elicited by stroking the skin on each side of the abdominal wall and which result in a localized contraction of the muscles beneath the area stimulated; the *cremasteric reflex,* which causes elevation of the ipsilateral testicle when the inner aspect of the thigh is stroked; the *gluteal reflex,* which induces contraction of the buttocks when the skin overlying this region is stimulated; and the *plantar reflex,* which causes plantar flexion of the toes in normal adults when the sole of the foot is stroked firmly (cf. Babinski's sign, p. 228). In children, the plantar reflex is often accompanied by retraction of the foot.

DEEP REFLEXES. Included in the category of clinically important deep reflexes are a number of tendon responses, including the *maxillary reflex,* whereby the jaw closes rapidly when the center of the chin is tapped with the mouth slightly open; the *biceps reflex,* which causes flexion of the arm at the elbow when the biceps tendon is tapped; the *triceps reflex,* which causes extension of the arm at the elbow when the triceps tendon is tapped; and the *wrist reflexes,* which cause flexion or extension of the wrist upon percussion of the appropriate tendons.

The *patellar* (or *knee jerk*) *reflex* causes extension of the leg when the patellar tendon is struck, since this stimulus stretches the quadriceps femoris muscle. The absence of this reflex is known as *Westphal's sign*. If the patellar reflex cannot be evoked in the usual manner, however, *Jendrassik's maneuver* is performed; that is, the subject clasps his hands together and then strongly trys to pull his hands apart while the patella is struck. This action tends to facilitate the knee jerk, presumably due to increased gamma efferent discharge initiated by afferent nerve impulses arising in the hands.

A decrease in, or absence of, deep reflexes of the kind summarized above can result from any lesion which disrupts the reflex arc. Thus, receptor damage, peripheral nerve disease, involvement of the dorsal (sensory) columns of the spinal cord, and cerebellar lesions all affect the responses.

Deep reflexes normally are inhibited to some extent by impulses that arise in the higher neural centers; therefore, lesions present within the motor cortex (p. 329) or the pyramidal tracts (p. 267) cause exaggerated responses combined with muscular spasms. Hyperactive reflexes are also seen in strychnine poisoning (p. 232).

VISCERAL REFLEXES. Included among the visceral reflexes are several pupillary reflexes, including the *light reflex,* in which the pupil of the eye constricts rapidly when a beam of light is directed upon the retina; the *consensual light reflex,* which produces constriction of the pupil when a light beam is directed into the opposite eye; the *accommodation reflex,* which causes constriction of the pupils and convergence of both eyes when the subject focuses upon a nearby object; and the *ciliospinal reflex,* which results in dilatation of the pupils when any sensory region is stimulated painfully, as by pinching the skin of the neck. Lastly, the *blink reflex* (of Descartes) causes rapid closure of the eyelids when an object is suddenly and rapidly moved toward the eyes of an unsuspecting subject.

Pressure applied to the eyeball evokes the *oculocardiac reflex*—a slowing of the heart rate (or bradycardia).

The *carotid sinus reflex,* evoked by pressure upon the carotid sinus in the neck, produces bradycardia coupled with a decline in blood pressure.

The bladder and rectal reflexes are physiologic mechanisms that normally control the elimination of urine and feces. Damage to the pelvic autonomic nerves that provide motor impulses to the appropriate sphincters results in incontinence.

The *mass reflex of Riddoch* results in a sudden emptying of the bladder and bowel combined with flexion of the upper and lower extremities, sweating, and rapid fluctuations in blood pressure. This reflex may be evoked in some normal children, but in normal adults these responses are regulated by higher neural centers. The mass reflex can be evoked in normal adults, however, by strong emotional states such as extreme fear. Under pathologic conditions, such as are encountered in chronic paraplegics in whom the spinal cord has been severed completely, the mass reflex can be evoked readily by a relatively mild stimulus applied below the level of the lesion, such as stroking the sole of the foot.

The mass reflex is caused by a loss of normal inhibitory control of higher centers and an irradiation of impulses throughout somatic and visceral branches of the nervous system (p. 218).

The visceral reflexes discussed above will be considered in greater detail in subsequent chapters.

PATHOLOGIC REFLEXES. Included in the category of pathologic reflexes are a number of defense responses, primitive in nature, that are evoked only as a consequence of damage to the upper motoneurons. When the lower motoneurons of the spinal cord are released from the inhibitory influences of the higher centers, particularly the cerebral cortex, this type of reflex activity becomes manifest.

Babinski's sign is an abnormality of the plantar reflex in which the large toe is extended while the small toes fan upward and outward following stimulation of the plantar surface of the foot. Whereas Babinski's sign is an abnormal finding in an adult, it is a perfectly normal response to plantar stimulation in infants before they have learned to walk, a fact which demonstrates the effect of learning or conditioning upon certain automatic responses.

Clonus is another state in which increased gamma efferent discharge is present. The obvious clinical manifestation of clonus is the presence of rhythmic contractions of a skeletal muscle subjected to a rapidly applied stretch maintained for an extended period of time. Thus, *ankle clonus* results in a sustained rapid flexion and extension of the foot that is evoked by dorsiflexing the foot quickly and forcibly while the leg is supported by a hand placed under the popliteal area. *Patellar clonus* is elicited by forcibly depressing the patella while the leg is extended and relaxed. This maneuver results in a rapid up-and-down movement of the patella itself.

The *crossed extensor reflex* (p. 221) is produced by stimulation of the sole of the foot while the subject is supine and both legs are flexed. The stimulation causes extension of the contralateral leg, an observation normally made in spinal animals (p. 223).

The extensor thrust reflex produces an extension of a flexed lower limb when the sole of the same foot is pressed upward, a sign of reduced inhibition from upper center control.

The pathologic reflexes summarized above are but a few examples of the many such diagnostic signs that can be manifest in abnormal states. The reader is referred to a neurology text for a more complete listing of these reflexes.

Variations in the appearance and extent of normal reflexes, as well as the appearance of certain pathologic signs, are of great diagnostic importance to the physician. Furthermore, a detailed knowledge of the specific neural pathways that underlie the particular reflexes is essential to an accurate localization of the lesions produced by various pathologic states.

Muscle Tone and the Lengthening Reaction

The resistance a skeletal muscle affords to an applied stretch is often called *muscle tone* or *tonus*. The tone of a resting muscle in vivo also was defined somewhat earlier as a state of slight, sustained, partial contraction that is both caused and maintained by low-frequency volleys of efferent nerve impulses (p. 129). Consequently, if the motor nerve to a skeletal muscle is severed, the muscle becomes completely relaxed or *flaccid* so that it now offers an extremely slight resistance to stretch. On the other hand, if the frequency of efferent impulses to a skeletal muscle is abnormally high during the resting state, the resulting *hypertonic* or *spastic* muscle is rigid and can provide considerable resistance to applied stretch. Thus, "normal tone" is a poorly defined state that lies somewhere between flaccidity and spasticity. Generally speaking, skeletal muscles are *hypotonic* when the rate of gamma efferent discharge is low, and *hypertonic* when the frequency is high.

In a hypertonic muscle, the following sequence is readily seen. Moderate stretch leads to muscular contraction, which causes a powerful stretch, and this effect, in turn, leads to relaxation of the muscle. For example, passive flexion of an arm at the elbow stretches the triceps muscle, producing a sudden resistance to further flexion as the stretch reflex mechanism is activated within the muscle (p. 218). As the muscle is stretched still further, the tendon reflex is activated (p. 221), the resistance to flexion suddenly fails, and the muscle flexes completely as the extensor contraction is inhibited. This sequence of resistance followed by a sudden relaxation in a limb that is moved passively is called the *lengthening reaction* because it is the manner in which a spastic muscle responds to stretch. In clinical parlance, the lengthening reaction is called the *clasp-knife effect* because it is similar to the closing of the blade of a pocket knife.

Neuromuscular Disorders

DEFINITIONS. The term *myopathy* signifies any disease of a muscle, regardless of its etiology. The general category of neuromuscular disorders includes a number of chronic myopathies that characteristically present a clinical picture of progressive weakness coupled with the gradual atrophy of particular muscle groups. The term *atrophy* generally is applied to muscular dysfunctions that arise from a neural lesion involving either the cell body or the axon of the motoneuron. In contrast to atrophic conditions, *muscular dystrophies* arise from a primary disease process that involves the muscle fibers themselves. Several clinical features that distinguish atrophies and dystrophies are summarized in Table 6.4. A few muscular disease entities and their physiologic manifestations are described on the next page.

Table 6.4 COMPARISON OF MUSCULAR ATROPHIES AND DYSTROPHIES

ATROPHIES	DYSTROPHIES
Usually have onset late in life	Usually have onset in childhood
Affect distal muscle groups	Affect proximal muscle groups
Fasciculations present	Fasciculations not present
Spasticity may be present	Spasticity not present
Not inherited	Inherited

ACUTE ANTERIOR POLIOMYELITIS. Poliomyelitis (Heine–Medin disease) is a disease of the central nervous system that is caused by viral infection. In the acute stage of this disease, a marked and selective destruction of the gray matter of the spinal cord occurs. In particular, the anterior horn (motor) cells of the spinal cord are affected, especially in the lumbar and cervical regions, so that following their destruction the axons arising from the cell bodies of the spinal motoneurons die. This abnormal process effectively denervates the skeletal muscles that were innervated by the nerve fibers originating from the affected nerve cells. The result is a *motor paralysis* of the affected muscles. Assuming that the patient survives the acute phase of the disease, eventually an atrophy of the paralyzed muscles develops due to the lack of normal innervation, as discussed on page 129.

Since the dorsal roots of the spinal cord are relatively unaffected by any of the several strains of virus that cause poliomyelitis, normal sensory functions may be retained by victims of this disease (cf. tabes dorsalis).

McARDLE'S SYNDROME. The hereditary myopathy referred to as *McArdle's syndrome* is characterized by muscular weakness, cramping pain, and contractures (ie, abnormal shortening and hence stiffness) of the skeletal muscles following a bout of moderate physical exercise. Occasionally, myoglobinurea is present.

This syndrome is caused by an absence of the specific enzyme phosphorylase in skeletal muscle; as a result of this deficiency, the victim is unable to convert glycogen to glucose, and thus the muscles are not able to perform glycogenolysis at adequate rates under anaerobic conditions. Therefore, a failure to observe a rise in blood lactic acid during exercise is characteristic of this disease.

The abnormally elevated glycogen levels in the skeletal muscles of patients with McArdle's syndrome has caused this metabolically induced condition to be classed as a glycogen storage disease.

FAMILIAL PERIODIC PARALYSIS. The victim of familial periodic paralysis, another dystrophic condition, is subject to occasional attacks of complete flaccid paralysis of the skeletal muscles that may last from a few minutes to up to several hours, with apparent normalcy of muscular function between attacks. A severe attack of such paralysis may cause death through respiratory failure, al-

though this is uncommon. Decreases in serum potassium and serum phosphate accompany the paralytic attacks; hence, therapy includes the administration of potassium salts. During an attack, the potassium and sodium concentrations within the skeletal muscles are not elevated significantly; however, the muscle loses its normal irritability, and hence it becomes electrically inexcitable. No hyperpolarization of individual muscle fibers is present, as revealed by transmembrane potentials recorded by the use of microelectrodes. Muscle biopsy samples taken during attacks of the disease contain large vacuoles within the endoplasmic reticulum, and the abnormal accumulation of glycogen degradation products found within these vacuoles may account for the simultaneous influx of electrolytes and water into the skeletal muscle fibers, which would preserve the ionic and osmotic balance within the cells.

MYASTHENIA GRAVIS. Myasthenia gravis is a serious neuromuscular disease characterized by profound muscular weakness and fatigability. The disorder affects women more frequently than men and most commonly has its onset between the ages of 20 and 30 years. The primary abnormality is believed to involve faulty neuromuscular transmission at the motor end plates of skeletal muscle. Practically any muscle can be affected, but the disease appears preferentially to involve the muscles innervated by the bulbar nuclei (p. 275), including those of the face, lips, eyes, tongue, throat, and neck.

The primary cause of myasthenia gravis is unknown, although some workers consider this condition to have a metabolic etiology because an abnormally rapid hydrolysis, and hence inactivation, of acetylcholine has been detected in chemical and biologic studies performed upon myasthenic patients.

It has also been suggested that myasthenia gravis is an autoimmune disease, since a number of abnormal antibodies have been found (p. 547), including an antibody against the patient's own skeletal muscle that has been detected in the sera of myasthenic patients.

Recent work also has indicated that the total number of functional postsynaptic acetylcholine receptors present in the skeletal muscles of patients with myasthenia gravis is much lower than that present in normal individuals. Consequently, it is quite possible that multiple etiologic factors are operative in the genesis of myasthenia gravis, at least in some persons.

Clinically, the motor paralysis of myasthenia gravis is accompanied by weakness and rapid fatigue. In particular, speech and swallowing may be markedly affected, especially after these functions have been in use for extended periods. Late in the day, drooping of the eyelids *(ptosis)* may become pronounced. Furthermore, the deep tendon reflexes may exhibit a rapid diminution of intensity when repeated percussion is applied. Of particular importance clinically, paralysis of the respiratory muscles can lead rapidly to death by suffocation during a myasthenic crisis.

Therapy is directed primarily toward restoring function at the neuromuscular junction; therefore, various anticholinesterase drugs are of considerable value, since these compounds prolong the excitatory action of acetylcholine released from the presynaptic endings of the motor end plate.

ELECTROMYOGRAPHY AND NEUROMUSCULAR DISEASE. The topic of electromyography was introduced briefly on page 137. Quite often this technique is useful clinically to distinguish between motoneuron disease and inherent muscular diseases.

When a needle electrode is inserted into a normal muscle in vivo, a rapid volley of surface action potentials is generated by the stimulus of the needle tip itself. Following this transient activity, no additional electric oscillations are observed when voluntary relaxation of the muscle is complete. If the muscle now is contracted voluntarily, action potentials are seen once again, and these progressively increase in frequency as the force of contraction increases to a maximum. The action potentials recorded from the extremities of a normal subject during such a voluntary contraction have an amplitude ranging between 0.2 and 2.0 millivolts, with a total duration of 5 to 10 msec.

Following denervation, and generally after a time lapse of about 3 weeks, abnormalities appear in the electromyogram. *Fibrillation potentials* develop; these small, rapid spikes have an amplitude of less than 0.2 mv, a duration of between 1 and 2 msec, and a frequency of about 2 to 10 per second. The fibrillation potentials are believed to signify the electric activity of individual muscle fibers that become hypersensitive to a given concentration of acetylcholine following destruction of the motor nerve (cf. p. 226). Spontaneous rhythmic discharges of entire motor units, or *fasciculations,* also are seen following denervation. These electric changes are accompanied by large-amplitude (over 4 mv), slow (to 200 msec) *positive sharp waves* — nonpropagated potentials having a positive sign, as indicated by their name.

The locus of a particular disease process within the motoneuron can affect significantly the usual pattern of electric activity that develops following denervation. For example, fasciculations generally are related to disease of the anterior horns of the spinal cord (eg, as in amyotrophic lateral sclerosis) rather than to more peripheral disease processes affecting the motoneurons themselves. Conversely, fibrillatory patterns of electric activity tend to be associated more frequently with peripheral involvement of the nerve than with central activity.

In contrast to the altered patterns of electric activity seen following denervation, inherent dystrophies that affect the muscles themselves yield quite different electromyographic findings. No fibrillations or fasciculations are present, and there is no increase in electric activity upon the insertion of the needle electrode into the muscle. The fundamental effect of inherent myopathies appears to involve the action potentials themselves. Many rapid, low amplitude potentials appear, and polyphasic action potentials characteristically are seen as well. For example, in the hereditary disorder called *myotonia congenita* (Thomsen's disease), there is muscular rigidity, stiffness, and a failure of proper relaxation of the entire muscular system, sometimes accompanied by periodic attacks of generalized muscular spasm (see the next section). Following a contraction, there are bursts of electric activity that resemble, but are not identical to, the electric patterns seen in fibrillation. Thus, in myotonia congenita, an uncoordinated excitation is present *after* a contraction, so that the muscle fails to relax normally. In this regard, it is of interest that a single excitation cannot elicit the myotonic contraction; rather, a series of excitations is necessary to evoke this effect.

MUSCULAR SPASM, CRAMPS, AND MYOEDEMA. A *muscle spasm* can be defined as a repetitive excitation of entire motor units caused by a repetitive impulse discharge of the motoneurons that innervate the muscle. Spasms of this sort are seen in a number of clinical states, including tetanus, hypocalcemic tetany (p. 1072), and during the regeneration of peripheral nerves following injury.

Physiologic tetanus, as defined earlier (p. 125) is a smooth summated contraction of skeletal muscle that is maintained for long intervals and that results from a rapid series of stimuli delivered to and propagated along the sarcolemma of the fibers. The type of muscle determines the frequency at which tetanus is produced, since there is a considerable difference among particular muscles in the rate of stimulation necessary to produce this response.

In contrast to physiologic tetanus, the disease state, also called *tetanus,* is caused by an abnormal firing of motoneurons located within the anterior horns of the spinal cord under the influence of an exotoxin elaborated by the bacillus *Clostridium tetani.* This uncontrolled high-frequency firing leads to muscular spasms of intermittent occurrence, and later a contracture may develop in the affected musculature (cf. strychnine poisoning, p. 232).

Muscle cramps, in contrast to the smooth,

coordinated contractions present in tetanus, are characterized by a rapid but asynchronous firing ring within portions of individual motor units. These uncoordinated excitations tend to spread from portions of an active motor unit to portions of adjacent motor units. Spontaneous twitch contractions of individual fibers may also be present.

Muscle cramps are encountered most frequently in electrolyte disturbances; therefore, it is presumed that the underlying defect is an abnormality in impulse formation and/or conduction that results from an ionic imbalance within presynaptic terminals and possibly also within the sarcolemmal region of the motor end plates.

The terms *idiomuscular contractions* and *myoedema* denote an abnormally prolonged, local contraction of a muscle induced by a direct mechanical stimulation, as through a sharp percussion. This condition is recognized by a small, slowly propagated contractile wave that passes along the muscle following application of the stimulus. No demonstrable alterations in the electromyogram accompany the phenomenon of myoedema, which is seen most commonly in persons suffering from cachexia (ie, the abnormal state characterized by severe malnutrition, emaciation, and debility). The relationship, if any, between myoedema and the muscular atrophy that accompanies cachexia is unclear.

Tabes Dorsalis

Invasion of the central nervous system by the microorganism *Treponema pallidum,* the causative agent of syphilis, can take place either early or late during the progress of this disease. *Tabes dorsalis* or *locomotor ataxia* is a relatively late manifestation of syphilis which usually becomes evident between 15 and 25 years after the initial exposure to *T. pallidum.*

Tabes dorsalis is of interest here because it is the consequence of syphilitic infection of the dorsal (sensory) roots of the spinal cord, with secondary degeneration of the dorsal columns (p. 265). Consequently, the clinical syndrome of tabes includes such sensory abnormalities as shooting pains, paresthesias (eg, morbid cutaneous sensations having no objective cause), and hypesthesias (abnormally diminished cutaneous, especially tactile, sensory modalities). The skeletal musculature is hypotonic due to destruction of the sensory limb of the spindle reflex (p. 218). The deep tendon reflexes also are diminished or absent. For example, the knee jerk or patellar reflex may be lacking, a characteristic finding in tabes known as Westphal's sign.*

Tabes also is characterized by a marked uncoordination of voluntary muscular movements, such as occur during walking, giving rise to a staggering gait. This type of uncoordination, or *locomotor ataxia,* is caused by a destruction of the neural pathways involved in proprioception. Thus, when the patient with tabes is standing with eyes closed and feet held closely together, the body commences to sway, and he may lose balance and fall *(Romberg's sign).* These responses are indicative of dorsal column disease, and in the syphilitic are characteristic of a loss of the proprioceptive sense in the lower extremities.

Pharmacology and the Synapse

Neurotransmission at practically all synapses between nerves in the body is mediated by chemical substances. Similarly, transmission from motoneurons to skeletal muscle and the transmission of autonomic impulses to smooth muscle and glandular cells are also mediated by neurotransmitter agents.

The metabolic systems within the presynaptic regions of contiguous cells, as well as the postsynaptic receptors — both of which underlie neurotransmission — are far more sensitive to the effects of drugs, poisons, and varying respiratory gas concentrations (oxygen and carbon dioxide) than are nerve fibers proper. Thus, polysynaptic pathways are far more vulnerable to the effects of anesthetics and other chemical agents than are monosynaptic channels, a fact of great clinical significance. Consequently, during a state of deep general anesthesia, synaptic transmission can be blocked completely even though axons still are able to conduct impulses when stimulated artificially.

The terminals of all neurons are transducers of one sort or another (p. 184). In particular, presynaptic endings are *electrochemical transducers;* that is, these endings convert the electric energy of nerve impulses into chemical energy. Briefly, there are four steps involved in this transduction process: First, the synthesis and storage of a specific transmitter substance in the presynaptic terminals is necessary. Second, the transmitter must be released into the synaptic cleft. Third, the transmitter has to interact with specific receptors in the postsynaptic membrane. And fourth, the transmitter must be inactivated by uptake into the presynaptic terminal/and or degraded into physiologically inert compounds by specific enzymes that present in high concentrations within or in the immediate vicinity of the nerve endings. From a pharmacologic standpoint, every one of these four steps in synaptic transmission potentially can be inhibited or facilitated by various drugs; thus, patterns of neural function can be altered at will through the use of appropriate chemical agents (see also the next section). Since neurotransmission within the central nervous system and autonomic ganglia as well as at neuroeffector junctions is basically a chemical process, it is possible to regulate selectively not only somatic and visceral motor activity, but also certain higher functions of the brain. For example, emotions, mood, and behavior patterns all can be markedly altered by stimulants or depressants.

*The knee jerk may also be absent in any condition which completely interrupts sensory and/or motor impulse transmission in the femoral nerve.

Thus, *psychotropic drugs* (ie, compounds which affect various chemical processes within the brain itself and thus alter mood and behavior) provide the physician with powerful therapeutic tools. And since all aspects of neural function including mentation itself, reside ultimately in the biologic substrate of chemical transmission of nerve impulses at synapses, the present and future potential for selective manipulation of the various aspects of the individual processes involved in neurotransmission in different regions of the entire nervous system are enormous.

Strychnine Poisoning

A discussion of strychnine poisoning has been included in this section because the physiologic–pharmacologic effects of this compound in vivo provide excellent insight into a number of central nervous functions as well as their alteration by a specific compound.

There are two general mechanisms whereby drugs can stimulate the central nervous system. First, stimulation can be achieved by augmenting synaptic excitation; and second, stimulation can be accomplished by blocking central inhibitory processes. Presynaptic inhibition and postsynaptic inhibition are two well defined mechanisms within the central nervous system (p. 215; p. 214). Strychnine selectively blocks *postsynaptic inhibition,* and therefore this compound is an exceedingly useful pharmacologic tool for the investigation of central nervous functions.

SOURCE, CHEMISTRY, AND USES OF STRYCHNINE. Strychnine is the major alkaloid found in trees and vines of the genus *Strychnos*. For example, the seeds of the tree *Strychnos nuxvomica,* a native of India, contain high concentrations of strychnine as well as another, less potent alkaloid, brucine. The empirical formula of crystalline strychnine is $C_{21}H_{22}N_2O_2$, thus the molecular weight of this compound is relatively high (334.4).

Strychnine is employed as a rat poison and was used to eradicate predators such as wolves during the period of early settlement of the western United States. Clinically, strychnine is of no proven therapeutic value, although it has been used in tonics because of its supposed enhancement of voluntary muscle tone and presumed stimulatory effect upon the gastrointestinal tract. Actually, ingestion of this compound in nontoxic quantities elicits no physiologic effects except for the bitter taste. The primary usefulness of strychnine is as an experimental tool for the "chemical dissection" of central nervous functions.

PHARMACOLOGIC EFFECTS. Strychnine produces excitation in all regions of the central nervous system by selectively blocking postsynaptic inhibition (p. 214). Normally, nerve impulses are channeled very precisely into specific pathways through the action of inhibitory mechanisms. When this physiologic inhibition is blocked by strychnine, all neural activity is intensified, and sensory stimuli produce strongly exaggerated reflex effects. When the concentration of strychnine within the central nervous system rises to a critical level, impulses are able to pass unchecked and in a random fashion throughout the entire nervous system.

Strychnine is an extremely potent convulsion-producing agent. The motor pattern of the convulsions resulting from strychnine poisoning depends upon which muscles at a given joint are the most powerful, since strychnine totally abolishes the inhibitory mechanism that underlies reciprocal innervation of antagonistic muscles (p. 221; Fig. 6.33). In humans, the first effect of strychnine poisoning is rigidity of the facial and neck muscles. Shortly thereafter, reflex excitability becomes greatly increased. Because of this *hyperreflexia,* any sensory stimulus, such as lightly touching the skin, can evoke a violent motor reaction. In the early phase of strychnine poisoning, this reaction is a fully coordinated extensor thrust of the limbs, whereas later a full tetanic convulsion develops — that is, the body may arch or dorsiflex so that only the heels and crown of the head touch the floor. Every skeletal muscle in the entire body is in a maximal tetanic contraction, and respiration stops because the diaphragm as well as the abdominal and thoracic muscles are fully contracted. Repeated periods of violent convulsions alternate with intervals of depression, and both the severity and the frequency of the convulsive episodes are increased by any type of sensory stimulation. Unless prompt therapeutic measures are instituted, the victim of strychnine poisoning dies from an eventual failure of the respiratory center in the medulla oblongata (p. 729) caused by the intermittent hypoxia that develops during the periods of respiratory arrest accompanying the massive convulsions.

During the early phases of strychnine poisoning, the victim is conscious and keenly aware of all external stimuli. The patient also develops a marked apprehension. The muscular contractions are excruciatingly painful.

Therapy for acute strychnine poisoning is directed toward respiratory support as well as prevention of additional convulsions. To control the convulsions, the intravenous injection of a short-acting central nervous depressant such as a barbiturate is the treatment of choice. Usually the hyperreflexia and convulsions induced by strychnine poisoning terminate in about 24 hours.

MECHANISM OF ACTION. As stated earlier, strychnine poisoning involves only postsynaptic inhibition; presynaptic inhibition is not affected by this compound.

There are many known pathways in the spinal cord and brain which mediate postsynaptic inhibition, and it appears that strychnine affects all of them to the same extent. For example, postsynaptic inhibition is an important mechanism for regulating and coordinating the activity of antagonistic muscles, and strychnine abolishes the inhibitory processes involved in this control mechanism (p. 214). Recurrent postsynaptic inhibition in the spinal cord is mediated by Renshaw cells, which are excited by collaterals of motoneuron axons (p. 215; Fig. 6.35). The neurotransmitter that excites the Renshaw cell is acetylcholine. Strychnine blocks recurrent inhibition; however, it does not block the synapse between the Renshaw cell and the collateral of the motoneuron.

Some postsynaptic inhibitory processes within the brain (eg, the cerebellum) are not blocked readily by strychnine, a fact that indicates the presence of strychnine-resistant inhibitory neurotransmitters and/or receptors which differ markedly from the strychnine-sensitive inhibitory regions.

The specific locus of action of strychnine at the synapse itself is unclear. The compound may act at one (or more?) of three possible sites. First, strychnine could alter the receptors on the postsynaptic membrane. Second, the drug could compete with the neurotransmitter itself for receptor sites on the postsynaptic membrane. Third, strychnine could block release of the neurotransmitter from the presynaptic terminals.

CHAPTER 7

The Nervous System

The human nervous system represents the most highly evolved biologic array of neural elements to be found in nature. Within this system, vast numbers of neurons are organized into complex morphologic and functional patterns which are designed specifically for the rapid integration, coordination, and regulation of both voluntary and involuntary bodily processes. In addition, the central nervous system, consisting of the brain and spinal cord, is a highly sophisticated and selective information storage and retrieval as well as processing center for the multitude of sensory impressions that arise constantly in both the external and the internal environments. Consequently the nervous system consists of specialized tissues that form the fast-acting communication system of the body, in contrast to the slower acting endocrine glands which constitute the second major bodily system for the transmission of information. In actual fact, however, the operations of both the nervous and endocrine systems in vivo as regulatory mechanisms for a host of processes are neither independent nor mutually exclusive. Rather, in many instances the functional relationships between neural and hormonal mechanisms are completely inseparable.

The fundamental anatomic and functional unit of the nervous system is the individual neuron, and the synapse provides the physiologic link between the individual neurons in toto. The transmission of information along neuronal processes is by means of coded electric impulses. Synaptic transmission, on the other hand, is mediated by specific chemical substances, or neurotransmitters. The bioelectric properties of neurons as well as the biochemical properties of synapses were presented in extenso in Chapters 5 and 6; these properties and the related material discussed in Chapters 7 and 8 are basic to an understanding of the anatomic and functional attributes of the entire nervous system.

Any comprehension of the physiology of the nervous system must be grounded upon a firm

knowledge of the basic structural patterns of organization among the neural and related morphologic elements that collectively make up this system. Therefore a review of human neuroanatomy will be presented in the first major division of this chapter. This material will serve as a background against which the study of particular functional aspects of the nervous system may be considered individually.

A SURVEY OF HUMAN NEUROANATOMY

Developmental Considerations

THE EMBRYONIC DISC. Approximately 2 weeks after fertilization of the ovum (p. 1190), cellular division has produced a flattened oval plate known as the embryonic disc, and the thickened layers of cells that comprise this structure ultimately become organized into the embryo (Fig. 7.1). The caudal end of the embryonic disc is represented by the body stalk. In the midline, or future longitudinal axis of the embryo, a primitive groove is present, and this lies caudal to a thickened group of cells, the primitive node (Hensen's node). At the bottom of the primitive groove is situated a long mass of cells called the primitive streak. The cells of the primitive streak multiply rapidly, spread laterally beneath the ectoderm and endoderm, and ultimately form the definitive mesoderm of the embryo. The cells of Hensen's node also form mesoderm, but ultimately this mesoderm forms an elongated column in the midline known as the notochord.

The primitive streak is a transient structure. Therefore, as embryogenesis proceeds in a cephalocaudad direction, the primitive streak progressively moves in a caudal direction and ultimately disappears.

THE NEURAL GROOVE AND NEURAL TUBE. During human embryogenesis the nervous system commences to develop during the third

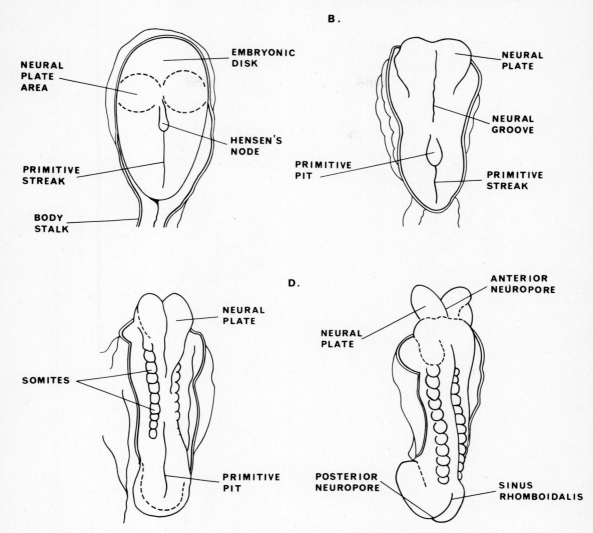

FIG. 7.1. External features seen in the development of the human nervous system from the presomite stage (A) to closure of the neural tube (D). Cephalad end of the embryo is at the top. In D, the neural tube has fused completely except for the anterior and posterior neuropores. Letters in this figure correspond to the same developmental stages shown in cross section in Fig. 7.2. (After Everett. *Functional Neuroanatomy,* 6th ed, 1971, Lea & Febiger; Goss [ed]. *Gray's Anatomy of the Human Body,* 28th ed, 1970, Lea & Febiger.)

week. The superficial ectoderm cephalad to Hensen's node in the embryonic disc undergoes hypertrophy in the midline, thereby forming a structure known as the neural plate (Fig. 7.2A). The neural plate eventually gives rise to all divisions of, and neurons present in, the nervous system of the adult. In addition, the neural plate also gives rise to the neuroglial cells (excepting the mesodermal microglia), the Schwann cells of the peripheral nerves, and the corresponding satellite cells of the peripheral ganglia.

Shortly after its formation, the neural plate invaginates to form a neural groove (Fig. 7.2B) which is bounded laterally on each side by two ridges of tissue known as the neural folds. The neural folds appear most highly developed at the cephalic, or cranial, end of the neural plate.

As the first month of embryogenesis progresses, the neural folds continue to grow extensively so that the free end of each fold becomes apposed with respect to the other (Fig. 7.2C). As the margins of the neural folds touch, the epithelial cells fuse, thereby forming the neural tube (Fig. 7.2D). Subsequently the neural crest develops from the differentiation of cells at the lateral border of each neural fold in the region between the neural plate and the ectoderm of the skin. The latter separates from the neural tube when the neural folds undergo fusion (Fig. 7.2D).

The neural groove closes to form the neural

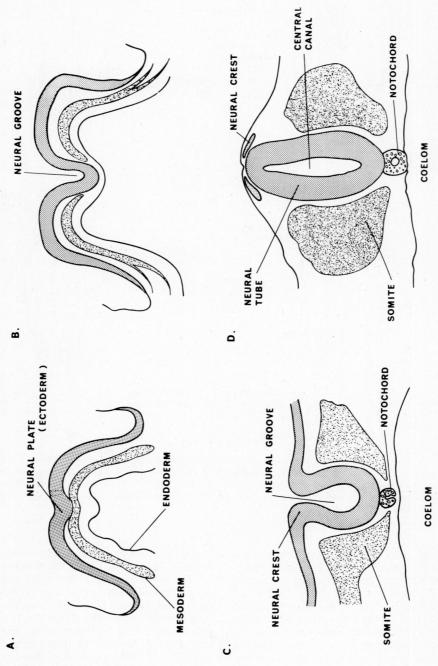

FIG. 7.2. Cross sections of the human embryo at developmental stages corresponding to those shown in Fig. 7.1. (After Everett. *Functional Neuroanatomy*, 6th ed, 1971, Lea & Febiger.)

tube in a definite, but not simultaneous, fashion along the longitudinal axis of the embryo. This fusion takes place first in the future thoracic region, then proceeds in both cranial and caudal directions. Hence the opposite ends of the neural tube remain open for a time, the openings being known as the anterior and posterior neuropores. The anterior neuropore closes when the embryo reaches the 20-somite level of development, whereas the posterior neuropore closes at the 25-somite stage.

Following the complete differentiation and closure of the neural tube, the epithelium of the future nervous system is completely distinct from the superficial ectoderm that invests the embryo, and at this point the neural crests are condensed masses of tissue lying at the dorsolateral borders of the neural tube (Fig. 7.2D).

DEVELOPMENT OF THE SPINAL CORD AND RELATED NEURAL STRUCTURES. Throughout its development, the cranial portion of the neural tube enlarges at a much faster rate than does the caudal portion. Consequently, by the time the neural tube is formed completely, the embryonic brain is delimited clearly, even before closure of the anterior and posterior neuropores has taken place. Shortly after closure of the neural tube, the future brain is evident at the cranial end of the embryo as three hollow, bulblike swellings, as depicted in Figure 7.3. These swellings, known as the primary brain vesicles, are the prosencephalon (forebrain), mesencephalon (midbrain), and rhombencephalon (hindbrain). Caudally, the rhombencephalon continues posteriorly and without distinguishing anatomic landmarks as the myelon or future spinal cord. The development of the spinal cord will be discussed first, as this structure undergoes the

fewest morphologic alterations in reaching adult form. The differentiation of the brain from the three primary vesicles will be discussed in the next section.

At the outset, the myelon is an unsegmented tube that extends from the caudal end of the rhombencephalon to the embryonic tail. When the somatic mesoderm lateral to the neural tube forms individual somites, the myelon becomes constricted into barely discernible segments. Thus a neuromere (or segment of spinal cord) is produced for each corresponding mesodermal somite. In turn, each neuromere gives rise to a pair of spinal nerves that grow peripherally into the mesodermal and ectodermal structures that surround the myelon. In this manner the relationship between particular nerves and peripheral structures (eg, muscular or visceral organs) is developed, a relationship that persists throughout the lifetime of the individual. During subsequent developmental stages, the basic neuromere–somite pattern of innervation can be modified considerably, as through differentiation and torsion of the limb primordia. Nonetheless, the fundamental pattern of neuromere–somite innervation is retained in the adult as a general pattern of muscular innervation in relation to the overlying patches of cutaneous innervation known as dermatomes. The latter are particularly evident as banded areas on the trunk of the body.

The primitive myelon develops into the adult spinal cord as follows. Initially a longitudinal groove develops bilaterally on each side of the luminal surface of the neural tube (Fig. 7.4). This groove, known as the sulcus limitans, bisects the neural tube into dorsal and ventral regions. The dorsal portion of the neural tube now is designated the alar plate, whereas that portion of the neural

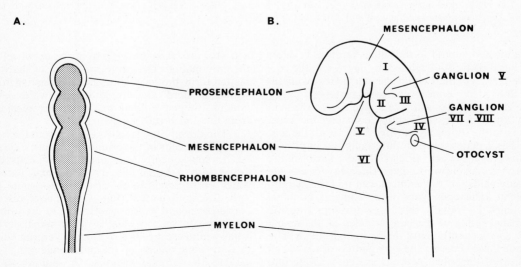

FIG. 7.3. **A.** Longitudinal section of early human brain showing the primary brain vesicles. **B.** Lateral view of human brain showing the primary vesicles. The loci where certain cranial nerves originate are indicated by Roman numerals. (After Everett. *Functional Neuroanatomy,* 6th ed, 1971, Lea & Febiger.)

A.

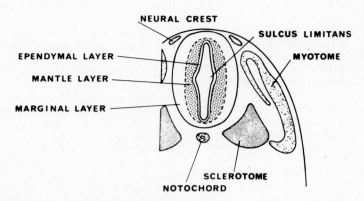

FIG. 7.4. A. Transverse section of an embryo showing components of the neural tube.

B.

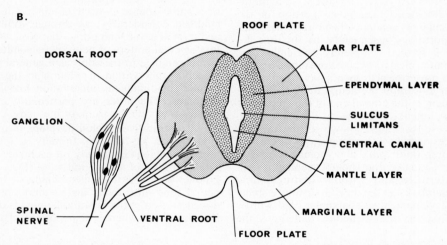

FIG. 7.4. B. Transverse section of the embryonic spinal cord showing the general organization of this structure. (After Everett. *Functional Neuroanatomy,* 6th ed, 1971, Lea & Febiger.)

tube located ventral to the sulcus limitans is called the basal plate. As depicted in Figure 7.4, the roof plate connects the two lateral alar plates, whereas the floor plate connects the basal plates.

The alar plates ultimately give rise to the sensory areas of the adult spinal cord and the basal plates to the motor areas (Fig. 7.20). Both the alar and basal plates undergo differentiation into three distinct layers of cells (Figs. 7.4 and 7.5). From the central canal outward, these three layers are the ependymal layer, the mantle layer, and the marginal layer. The ependymal layer contains germinal elements. Eventually this layer develops into the epithelial lining of the central canal of the spinal cord and the ventricles of the brain in the adult. The mantle layer becomes the gray matter (or nerve cell bodies) of the spinal neurons as cells migrate into this region from the ependymal layer as embryogenesis proceeds. The marginal layer is transformed into the white matter of the spinal cord as

neuronal processes develop from the nerve cells of the gray matter. The nerve processes of the white matter form the ascending and descending tracts within the cord, and in addition the white matter contains the nerve fibers that enter the spinal cord from the dorsal root ganglia.

The dorsal root ganglia develop from the originally diffuse cells of the primitive neural crest (Fig. 7.2C). Subsequently, the neural crest cells condense into a longitudinally segmental pattern, which corresponds to the neuromere–somite pattern described earlier (Fig. 7.4). These discrete condensations of neural crest cells are the primordia of the dorsal root ganglia of the spinal nerves. Some neural crest cells also migrate ventromedially to form components of the autonomic ganglia as well as the adrenal medulla. In addition to the aforementioned roles of neural crest cells, some of these elements also provide neurons to the ganglia of certain cranial nerves; in particular, formation of

A.

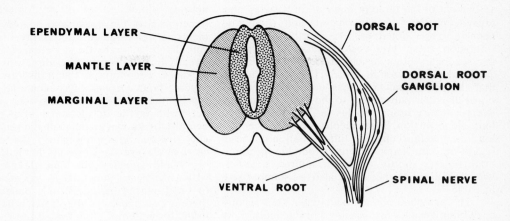

EPENDYMAL LAYER

MANTLE LAYER

MARGINAL LAYER

DORSAL ROOT

DORSAL ROOT GANGLION

VENTRAL ROOT

SPINAL NERVE

B.

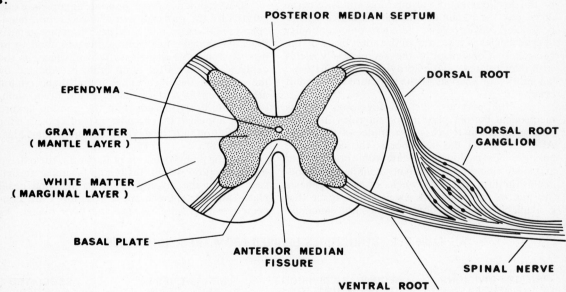

POSTERIOR MEDIAN SEPTUM

EPENDYMA

GRAY MATTER (MANTLE LAYER)

WHITE MATTER (MARGINAL LAYER)

BASAL PLATE

ANTERIOR MEDIAN FISSURE

VENTRAL ROOT

DORSAL ROOT

DORSAL ROOT GANGLION

SPINAL NERVE

FIG. 7.5. Transverse sections of the embryonic neural tube illustrating two stages in the differentiation of various structures. In later stages of development, the ependymal layer persists as a layer of cells that line the central canal of the cord. (After Everett. *Functional Neuroanatomy,* 6th ed, 1971, Lea & Febiger.)

the mesencephalic nucleus of the trigeminal nerve is accomplished by an inward migration of neural crest cells.

An adult spinal nerve develops from cellular elements that are located both inside and outside of the neural tube. Neurons that develop in the spinal ganglia from neural crest cells (pseudounipolar neurons) send a proximal fiber into the alar portion of the developing spinal cord, whereas the distal fiber grows outward and ventrally to form a part of the spinal nerve proper. Consequently the processes of the neurons whose cell bodies are situated within the spinal ganglia provide the bulk of the fibers present within the dorsal (sensory) root of each spinal nerve.

As accumulations of nerve fibers emerge from the basal plate of the neural tube, the ventral (motor) root of the spinal nerve is formed. The nerve fibers that comprise the ventral root are the axons of multipolar motor neurons whose cell bodies lie within the anterior region of the gray matter of the spinal cord.

When the dorsal and ventral roots fuse, the spinal nerve has formed (Fig. 7.5).

The rami communicantes (Fig. 7.10) develop from nerve fibers that emerge from the gray matter of the intermediolateral column of the spinal cord. In addition certain fibers of the rami are derived from the autonomic ganglia and sensory neurons that lie within the spinal ganglia themselves.

DEVELOPMENT OF THE BRAIN. The early development of the three primary brain vesicles (the prosencephalon, mesencephalon, and rhombencephalon) was noted earlier. Shortly after their formation, these three vesicles subdivide into the five secondary brain vesicles as depicted in Figure 7.6B and summarized in Table 7.1. It will be noted that the mesencephalon does not undergo additional division. Subsequent to the formation of the five secondary brain vesicles (ie, the telencephalon, diencephalon, mesencephalon, metencephalon, and myelencephalon), differentiation of these regions into the major structures found in the adult brain follows shortly. These structures, together with their associated cavities, also are summarized in Table 7.1.

The anatomic boundaries between the secondary brain vesicles are obscured in the adult brain by differential patterns of growth, by the formation of various structures on the ventricular walls, and by the development of various large tracts of nerve fibers (eg, the pyramids and the middle cerebellar peduncle). Nevertheless, the embryonic continuity of the vesicles is indicated in the adult brain by the ventricles. Consequently the secondary vesicles with their related cavities provide useful anatomic landmarks for analysis of the structure of the adult brain.

As the nervous system develops, various mechanical factors play important roles in determining the ultimate gross appearance of the brain. After closure of the neuropores, the neural tube enlarges most rapidly in the embryonic cranial area; and since this area is relatively confined, then between the fifth and sixth weeks of embryogenesis the rate of neural growth exceeds the space that is available for expansion of the neural elements.

The mechanical factors involved are changes in neural volume attendant on hypertrophy and hyperplasia of the neural and related cellular elements as well as an elevated intraventricular pressure that is caused by the secretion of fluid from the ependymal cells and the developing choroid plexuses. Therefore, as the cranial end of the embryonic brain increases its volume so much more rapidly than that of the embryonic cranium, the net result is the development of a bend in the neural tube known as the cephalic (or mesencephalic) flexure. At approximately the same time a cervical flexure develops at the junction of brain and spinal cord. The elevated pressure within the neural tube causes the mesencephalon to be displaced gradually in a caudal direction; therefore formation of the cephalic flexure assists in the formation of a third bend in the neural tube known as the pontine flexure. The sequence of events described above is illustrated in Figure 7.7.

Development of the pontine flexure is exaggerated by the rapid growth of the cerebral hemispheres; however in later stages of embryogenesis both the cephalic and pontine flexures tend to be less obvious. Nevertheless, the presence of the cephalic flexure in the adult brain is indicated by the fact that the telencephalon and diencephalon are oriented at a right angle to the brain stem. Furthermore, as the edges of the pontine flexure become apposed, the rhombic lip is in a position for incorporation into the developing cerebellum (Figs. 7.7 and 7.8).

During embryogenesis of the brain, the pattern of organization within this region is similar to that seen within the myelon. That is, alar and basal plates are present and these structures are connected respectively by roof and floor plates. As

Table 7.1 STRUCTURES DERIVED FROM THE NEURAL TUBE

PRIMARY DIVISIONS OF THE NEURAL TUBE	SECONDARY DIVISIONS OF THE NEURAL TUBE	DERIVATIVES IN ADULT BRAIN	ASSOCIATED CAVITIES
Proscencephalon (Forebrain)	Telencephalon	Cerebral hemispheres Corpus striatum Internal capsule	Lateral ventricles and third ventricle (rostral portion)

	Diencephalon	Epithalamus
		Thalamus
		Subthalamus
		Hypothalamus
		  Optic chiasma
		  Hypophysis
		  Tuber cinereum
		  Mammillary bodies
			Most of third ventricle

Mesencephalon (Midbrain)	Mesencephalon	Corpora quadrigemina Crura cerebri	Cerebral aqueduct
Rhombencephalon (Hindbrain)	Metencephalon	Cerebellum Pons	Fourth ventricle
	Myelencephalon	Medulla Oblongata	
Spinal Cord	Spinal cord	Spinal cord	Central canal

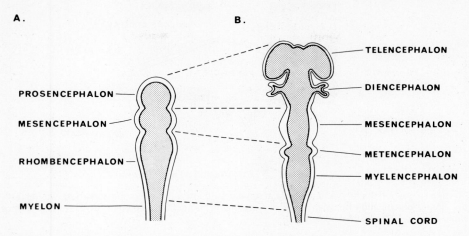

FIG. 7.6. Subdivisions of the embryonic human brain. **A.** Three-vesicle stage. **B.** Five-vesicle stage. See also Table 7.1. (After Arey. *Developmental Anatomy*, 7th ed, 1965, Saunders.)

shown in a cross section of the myelencephalon the sulcus limitans is clearly evident (Fig. 7.8). The very thin roof plate assists in the formation of the choroid plexus, whereas the floor plate becomes an integral part of the median raphe.

The physiologic distinction between the alar (or sensory) and basal (or motor) plates also obtains in the brain; consequently the location of sensory and motor nuclei within the brain stem is similar to that encountered in the spinal cord. However, as the basal plate apparently does not extend cranially beyond the mamillary recess in the diencephalon, this dorsal–sensory, ventral–motor relationship does not hold in an unqualified fashion for the de-

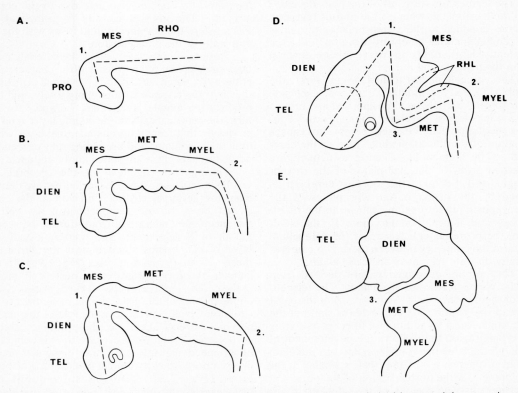

FIG. 7.7. Lateral view of human embryo showing the development of flexures (dashed lines) and the secondary brain vesicles. 1. Cephalic (mesencephalic) flexure. 2. Cervical flexure. 3. Pontine flexure. PRO = prosencephalon; TEL = telencephalon; DIEN = diencephalon; MES = mesencephalon; MET = metencephalon; MYEL = myelencephalon; RHO = rhombencephalon; RHL = rhombic lips. (Data from Everett. *Functional Neuroanatomy*, 6th ed, 1971, Lea & Febiger; Arey. *Developmental Anatomy*, 7th ed, 1965, Saunders.)

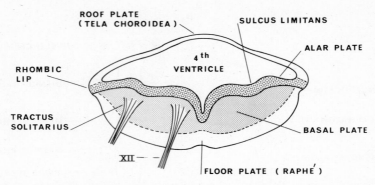

FIG. 7.8.   Transverse section of embryonic medulla at same developmental stage as spinal cord shown in Fig. 7.4, **B.** Note the generally similar organization between these two regions of the central nervous system. (After Everett. *Functional Neuroanatomy*, 6th ed, 1971, Lea & Febiger.)

rivatives of the prosencephalon. Consequently, the neural structures related to the telencephalon and diencephalon presumably originate to a major extent from the alar plate (Table 7.1).

As the myelencephalon develops into the adult medulla oblongata, it undergoes only a slight modification from the basic structure of the spinal cord. Similarly, the sulcus limitans indicates the boundary between the dorsal (sensory) and ventral (motor) regions of the metencephalon.

The tegmentum (or roof) of the pons develops from the basal plate, whereas the fibrous region of the pons is added to the metencephalon per se as the cerebellum and cerebrum develop and concomitantly large tracts of nerve fibers develop within the medulla and pons.

The cerebellum has a dual origin. First, the rhombic lips that are situated laterally on each side of the fourth ventricle become apposed and fuse as the pontine flexure develops (Figs. 7.7 and 7.8). Subsequent growth of the fused rhombic lips causes formation of the flocculonodular lobe of the cerebellum. (The rhombic lips themselves actually are thickened areas of the alar plates of the rhombencephalon.) Second, the corpus cerebelli is formed by the fusion of the alar plates of the metencephalon in the region cephalic to the rhombic lips and flocculonodular lobe. In the adult brain, the posterolateral fissure of the cerebellum indicates the boundary between these two embryologically discrete regions of the cerebellum.

The corpora quadrigemina are derived from the alar plates of the mesencephalon. The tegmentum of the mesencephalon develops from the basal plates of this region. The peduncle (or stalk) of the mesencephalon is an addition of fiber tracts to this embryonic region that follows the development of the large tracts of nerve fibers related to the cerebral hemispheres, a situation that is similar to the development of the fibrous region of the pons.

Three discrete regions of the adult brain are derived from the embryonic diencephalon (Table 7.1). First, the epithalamus develops from the roof plate as well as the upper regions of the alar plates. In addition, the choroid plexus of the third ventricle

also develops from the roof plate. Second, the dorsal thalamus originates from the diencephalic alar plates. Third, the hypothalamus together with its related nuclei and structures develops in the basal region of the diencephalon. It is unclear, however, whether or not the hypothalamic structures have their origin in the diencephalic basal plates.

Lateral evaginations of the alar plates of the telencephalon give rise to the paired cerebral hemispheres (Fig. 7.6). The inner walls of these telencephalic vesicles also give rise to the corpus striatum. The internal capsule is formed at the locus where the corpus striatum fuses with the lateral walls of the diencephalon.

Both the cerebral as well as the cerebellar cortices develop in accordance with a characteristic pattern. As noted earlier, the ependymal cells lining the embryonic spinal cord multiply freely, and thus produce the mantle layer (Figs. 7.5 and 7.9A). In turn, the neuronal cells of the mantle layer send out processes that form a fibrous marginal layer which ultimately contains the ascending and descending nerve tracts of the adult spinal cord (white matter).

During embryogenesis of the cerebral and cerebellar cortices, however, cells that are derived from the germinal elements present in the ependymal layer of the telencephalon as well as metencephalon migrate outward during the second month of development, and thereby establish a layer of cell bodies situated near the surface of these secondary brain vesicles (Fig. 7.9B). In this manner, the gray matter (ie, the nerve cell bodies) of the cerebral and cerebellar cortices is established on the outside of these structures, in contrast to the situation which is present in the rest of the central nervous system. Therefore, the primitive cerebral cortex when examined in cross section is composed of an ependymal layer, a rather dispersed mantle layer, a primordial pyramidal layer, and a marginal layer (Fig. 7.9B). During subsequent embryogenesis, the innermost region of the mantle layer develops into the characteristic and obvious central mass of myelinated nerve fibers (white matter). The six-layered pattern of elements present in the adult cerebral cortex is shown in Figure 7.70.

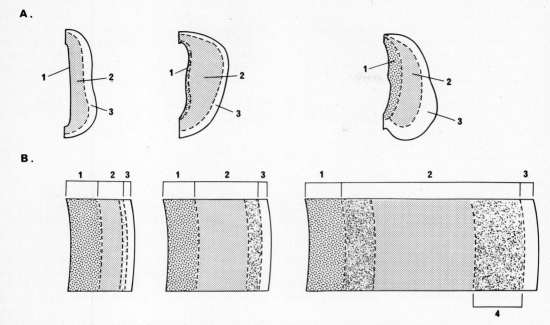

FIG. 7.9. The progressive formation of the spinal cord **(A)** and the cerebral cortex **(B)** in the human embryo. In **A,** the successive layers, from within outward, are: 1. Ependymal layer. 2. Mantle layer. 3. Marginal layer. In **B,** the successive cortical layers (1, 2, 3) are the same as in the cord, however the outermost region of the mantle layer of the cortex differentiates into the pyramidal layer of the embryonic cortex (4). See also Figure 7.70. (After Everett. *Functional Neuroanatomy,* 6th ed, 1971, Lea & Febiger.)

FACTORS THAT INFLUENCE THE GROWTH AND DEVELOPMENT OF THE NERVOUS SYSTEM.

Although the factors that are responsible for governing the developmental patterns within the nervous system are, in general, unknown, certain physicochemical influences can be recognized.

Among the important physical processes that regulate patterns of nerve growth and development is the hydrostatic pressure developed within the neural tube that is caused by an accumulation of fluid secreted by the ependymal cells. As noted earlier (p. 240), this pressure, combined with the differential rates of hypertrophy and hyperplasia of the cells in different regions of the brain, results in profound alterations in the anatomic relationship among the various gross structural elements present in the embryonic and adult brain (p. 246). Consequently, mechanical factors are significant, at least in the development of the central nervous system.

The protoplasm of the tips of growing axons is capable of active ameboid movement (p. 150); hence these neuronal processes are capable of penetrating the intercellular spaces between other cell types. In this regard, the highly organized microfibrillar and microtubular elements of the neurons play an important role in determining this movement (p. 148). Furthermore, ameboid movement of entire neurons (as well as most other embryonic cell types) during embryogenesis is an important factor in determining their ultimate disposition within the nervous system.

A major problem in neuroembryology is: How do neuronal processes become associated with their ultimate target structures or organs during embryogenesis? The adult patterns of innervation throughout the entire body are highly specific; therefore the progression of entire nerve cells (as in the formation of the cerebral cortex and the peripheral ganglia) or their axonal processes (as during the innervation of particular muscles by individual nerve fibers) is a highly selective process. It is clear that chemical factors must play a very decisive role in guiding these developmental patterns. Although the specific biochemical regulatory mechanisms that underlie most of these events in the embryo are unknown, one compound has been isolated and characterized that exerts profound effects upon the generation, growth, and maintenance of specific types of neurons. This compound, known as nerve growth factor (NGF), is a protein that resembles insulin both structurally and functionally.

NGF has been demonstrated to enhance the growth of peripheral sensory neurons in general as well as to stimulate the growth and maintenance of sympathetic neurons in particular. Furthermore, NGF has been demonstrated to affect the regenerative growth of severed axons that arise from centrally located adrenergic neurons, indicating a role of this protein in central nervous functions. Like insulin, NGF appears to act upon specific receptors located on its target cells through stimulation of the biosynthetic and bioenergetic processes of those cells. Hence it has been proposed that sympathetic nerve outgrowth toward particular effector cells in the periphery is through the attractive action of NGF that is bound to the specific receptor sites for NGF upon the target cells. In this regard, it is significant that a quantitative relationship exists between the number of NGF binding sites upon particular cells and the extent of sympathetic innervation of those cells. The nonneural glial cells of the nervous system have been implicated in NGF production, but these elements also appear to elaborate other, as yet unidentified, nerve growth factors.

Interestingly, the functional roles of NGF in vivo do not appear to be confined to neural and related tissues. In fact, NGF has been implicated experimentally as a substance that may be involved in wound healing (p. 547) as well as in carcinogenesis. In this regard, the elaboration of NGF by nonneural glial cells as well as by fibroblasts is particularly significant.

At one time it was believed that neural migration, growth, and organization took place along electric gradients developed between the axons and dendrites of individual neurons. This concept,

known as "neurobiotaxis," remains wholly unsupported by direct experimental evidence. Therefore, it would appear that chemotactic mechanisms involving various as yet unidentified biochemical factors are responsible for the growth and development of the nervous system.

General Anatomic Features of the Adult Nervous System

The nervous system of the adult is divided into two major anatomic regions: the peripheral nervous system and the central nervous system.

THE PERIPHERAL NERVOUS SYSTEM. The peripheral nervous system is composed of the paired spinal and cranial nerves, together with the many ganglia and plexuses that are related to innervation of the visceral organs. All of these structures lie outside of the spinal cord.

Every spinal nerve is connected to the spinal cord by a dorsal (or posterior) root and a ventral (or anterior) root. Fusion of these roots produces the spinal nerve which in turn divides into the dorsal and ventral rami communicantes (or "communicating branches") as depicted in Figure 7.10. Note that

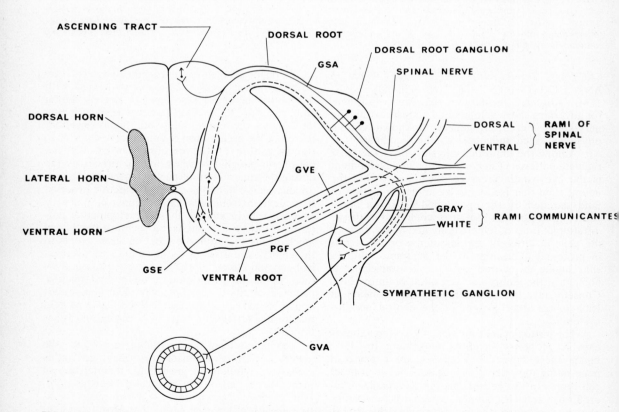

FIG. 7.10. Cross section of the adult spinal cord illustrating the components of a typical spinal nerve and its connections. Nerve cell bodies are indicated by black dots, and synapses by "V" terminations of presynaptic fibers on the nerve cell bodies. The abbreviations used for particular types of fiber are: GSA = general somatic afferent; GSE = general somatic efferent; GVA = general visceral afferent; GVE = general visceral efferent; PGF = postganglionic sympathetic fibers. (Data from Everett. *Functional Neuroanatomy*, 6th ed, 1971, Lea & Febiger; Bloom and Fawcett. *A Textbook of Histology*, 9th ed, 1968, Saunders.)

the dorsal and ventral rami may contain both sensory as well as motor fibers. The rami communicantes connect the spinal nerves to the sympathetic trunk. The white ramus is present only in the thoracic and upper lumbar nerves, whereas the gray ramus is present in all of the spinal nerves.

The dorsal roots of the spinal nerves contain general somatic afferent (or sensory) fibers as well as visceral afferent fibers whose cell bodies lie within the dorsal root ganglion; ie, these fibers originate within the dorsal root ganglion and their processes pass both peripherally as well as centrally into the spinal cord. The peripheral branches of the general somatic afferent nerve fibers innervate somatic receptors via the spinal nerves. The general visceral afferent neurons innervate visceral receptors via the spinal nerves as well as through the several plexuses of the visceral portion of the nervous system. Centrally, the fibers of both the somatic and visceral components of these nerves enter the spinal cord through the dorsal roots, which are exclusively sensory (Fig. 7.10).

The ventral roots of all spinal nerves contain somatic efferent (or motor) fibers which innervate the skeletal muscles. The ventral roots of the spinal nerves located in the thorax, upper lumbar, and middle sacral regions also contain general visceral efferent fibers which terminate in proximity to ganglionic cells of the autonomic nervous system (p. 270) and which provide motor innervation of the visceral organs. Somatic efferent fibers have their origin in neurons whose cell bodies lie within the ventral gray columns of the spinal cord (Fig. 7.10).

The twelve pairs of cranial nerves do not exhibit the same anatomic regularity as do the segmental spinal nerves (Fig. 7.12; Table 7.3). Certain of the cranial nerves are purely sensory, one (the hypoglossal) is entirely motor, and the rest have sensory as well as motor components. The sensory ganglia associated with cranial nerves are similar both morphologically and physiologically to the dorsal root ganglia of the spinal nerves.

From a functional standpoint, there are seven distinct types of nerve fibers that may contribute to the formation of any particular peripheral nerve.

1. *General somatic afferent (GSA) fibers* are present in all of the spinal nerves.* In addition, GSA fibers are present in several cranial nerves. The cell bodies of these fibers are located in the dorsal root ganglia. GSA fibers underlie the transmission of sensory information from exteroceptors (ie, touch, pressure, thermal sensations, pain) and from proprioceptors located in the tendons, muscles, and joint capsules.
2. *Special somatic afferent (SSA) fibers* are present only in the optic and vestibulocochlear nerves. The fibers that comprise the optic nerve transmit the impulses that underlie vision from the retina of the eye to the brain. These neurons have their

cell bodies located within the retina (Chapter 8, p. 468). The vestibulocochlear nerve is made up of axonal processes whose cell bodies are located within the spiral and vestibular ganglia of the inner ear (Chapter 8, p. 506). The dendritic processes of these neurons innervate particular receptors located within the inner ear (p. 506).

3. *General visceral afferent (GVA) fibers* are found in the spinal nerves as well as in certain cranial nerves. GVA fibers innervate receptors located within the viscera of the neck, thorax, abdomen, and pelvic regions. In addition GVA fibers innervate blood vessels and glandular tissues throughout the body. The cell bodies of these neurons lie within the dorsal root ganglia (Fig. 7.10), as well as within the ganglia of certain cranial nerves. GVA neurons form the afferent limbs of visceral reflex arcs, and thus conduct impulses to the conscious level, in particular those impulses giving rise to painful sensations.
4. *Special visceral afferent (SVA) fibers* underlie the sensations of olfaction (smell) and gustation (taste). Consequently SVA fibers are found in the olfactory, glossopharyngeal, vagus, and the intermedius branch of the facial nerve. The cell bodies of the olfactory nerves are located within the olfactory mucous membrane (p. 521), whereas those of the facial nerves are situated within the geniculate ganglia (p. 286). The cell bodies of the glossopharyngeal nerve are located within the inferior ganglion of this nerve, whereas the cell bodies of the vagus are found within the nodose ganglion (ie, the inferior ganglion of the vagus).
5. *General somatic efferent (GSE) fibers* originate from motoneurons whose cell bodies are located within the gray matter of the spinal cord. GSE fibers innervate the skeletal muscles of the body which originate from the mesodermal somites of the embryo. GSE fibers are present in all of the spinal nerves as well as in the oculomotor, trochlear, abducens, and hypoglossal nerves.
6. *General visceral efferent (GVE) fibers* are found in the oculomotor, facial, glossopharyngeal, and vagus nerves; that is, the nerves belonging to the cranial division of the parasympathetic branch of the autonomic nervous system. In addition, GVE fibers are present in all of the thoracic nerves, in the upper two or three lumbar nerves, and in the middle three sacral nerves. GVE fibers take their origin from the cell bodies of neurons located within some nuclei situated within the brain stem as well as within the lateral gray columns of the spinal cord. GVE fibers are distributed to the peripheral ganglia of the autonomic nervous system. The axons of the ganglion cells, upon which the GVE fibers synapse, then innervate cardiac muscle as well as smooth muscle and glandular tissues throughout the entire body.
7. *Special visceral efferent (SVE) fibers* innervate certain skeletal muscles which are derived from the embryonic mesoderm of the visceral (or branchial) arches. Therefore SVE fibers innervate the

*Except for the first cervical nerve, which generally lacks a sensory root.

voluntary muscles of the larynx, pharynx, and soft palate as well as the muscles involved in mastication and facial expression. Consequently SVE fibers are present in the trigeminal, facial, glossopharyngeal, vagus, and spinal accessory nerves. The spinal accessory nerve also innervates the trapezius and sternocleidomastoid muscles, which appear to be derived partly from the embryonic visceral arches.

It should be stressed that no particular nerve contains all seven of the fiber components described above, and that the "special" components are present only in the cranial nerves. The spinal and cranial nerves are summarized, together with their functional components, in Tables 7.2 and 7.3.

THE CENTRAL NERVOUS SYSTEM. The central nervous system consists of the spinal cord and brain. The principal subdivisions of the adult brain include the cerebrum, cerebellum, and brain stem. The brain stem in turn is composed of the diencephalon, mesencephalon, pons, and medulla oblongata (Table 7.1). The gross anatomic relationship among these structures is illustrated in Figures 7.11 and 7.12. A more detailed functional neuroanatomy of the various components of the central nervous system will be presented subsequently.

NEUROGLIA. In addition to an enormous number of neurons, the nervous system also contains an even greater number of interstitial elements or nonneural supportive cells that are known collectively as neuroglial cells or neuroglia (literally, "nerve glue"). The interstitial cells and tissues that are classed as neuroglia include the ependyma, an epithelial membrane which lines the cavities of the ventricles of the brain as well as the central canal of the spinal cord; neuroglial cells together with their processes which form an elaborate meshwork that surrounds the individual neurons within the brain and spinal cord as well as the retina of the eye; and the satellite (or capsular) cells that are found in the peripheral ganglia. The Schwann cells of all peripheral nerves are considered to be peripheral neuroglial elements (p. 159).

Table 7.2 NEUROANATOMIC AND FUNCTIONAL CLASSIFICATION OF THE SPINAL NERVES

REGION IN SPINAL CORD	NERVE FIBER TYPES[a]	GENERAL DISTRIBUTION
Cervical segments C–1 through C–8	GSA, GSE	Back of head, neck (C–1 through C–5) Upper extremity (C–6 through C–8)
Thoracic segments T–1 through T–12	GSA, GSE, GVA, *GVE*	Upper extremity (T–1) Chest, back, abdomen (T–2 through T–12)
Lumbar segments L–1 through L–3	GSA, GSE, GVA, *GVE*	Chest, back, abdomen (L–1 through L–4)
L–4 and L–5	GSA, GSE	Lower extremity (L–3 through L–5)
Sacral segments S–1 and S–2	GSA, GSE, GVA	Lower extremity (S–1 through S–3)
S–3 through S–5	GSA, GSE, GVA, *GVE*	Genitalia, buttocks (S–3 through S–5)
Coccygeal segment Co–1	GSA, GSE	Genitalia, buttocks

[a]*GSA = general somatic afferent (somatic sensory) fibers. Transmit exteroceptive (pain, thermal, touch, pressure) and proprioceptive impulses from sensory endings in skin, body wall, joints, tendons, and muscles to central nervous system.*

GSE = general somatic efferent (somatic motor) fibers. Transmit motor impulses to somatic skeletal muscles that originate in the myotomes. Form bulk of fibers present in ventral roots of spinal nerves.

GVA = general visceral afferent (visceral sensory) fibers. Transmit sensory impulses (usually pain) from visceral structures within the body to the central nervous system.

GVE = general visceral efferent (visceral motor) fibers. Autonomic fibers that transmit motor impulses to smooth as well as cardiac muscle and to glands. This functional category includes fibers of both sympathetic and parasympathetic divisions of the autonomic nervous system (italics in table). The paired sympathetic preganglionic fibers arise from spinal cord segments T–1 through L–3 or L–4 (see also Table 7.3). The preganglionic fibers of the sacral division of the parasympathetic nervous system arise from segments S–2, S–3, and S–4.

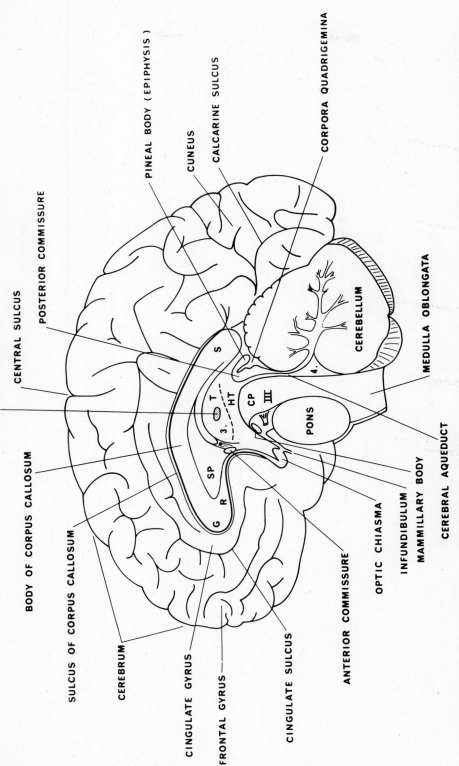

FIG. 7.11. The principal anatomic features of the medial surface of the adult brain. 3. = third ventricle; 4. = fourth ventricle. In addition to the body, the corpus callosum is divided into regions as follows: R = rostrum; G = genu or flexure; S = splenium. The dashed line in the middle of the brain indicates the hypothalamic sulcus which separates the thalamus (T) and the hypothalamus (HT). III = oculomotor nerve. The interventricular foramen is indicated by the black dot located at the tip of the fornix (not labeled) and between the anterior commissure and the third ventricle. (Data from Everett. *Functional Neuroanatomy*, 6th ed, 1971, Lea & Febiger; Goss [ed]. *Gray's Anatomy of the Human Body*, 28th ed, 1970, Lea & Febiger.)

Labels appearing in the figure:

MASSA INTERMEDIA
CENTRAL SULCUS
POSTERIOR COMMISSURE
PINEAL BODY (EPIPHYSIS)
CUNEUS
CALCARINE SULCUS
CORPORA QUADRIGEMINA
BODY OF CORPUS CALLOSUM
SULCUS OF CORPUS CALLOSUM
CEREBRUM
CINGULATE GYRUS
FRONTAL GYRUS
CINGULATE SULCUS
ANTERIOR COMMISSURE
OPTIC CHIASMA
INFUNDIBULUM
MAMMILLARY BODY
CEREBRAL AQUEDUCT
MEDULLA OBLONGATA
CEREBELLUM
PONS

S, T, HT, CP, III, G, R, SP, 3., 4.

Table 7.3 THE CRANIAL NERVES

NUMBER	NAME	NERVE FIBER TYPE(S)[a]	ORIGIN; PRIMARY CELL BODY	PERIPHERAL TERMINATION(S)	PRINCIPAL FUNCTION(S)
I	Olfactory	SVA	Rhinencephalon; olfactory epithelium	Olfactory epithelium	Olfaction
II	Optic	SSA	Diencephalon; retina, ganglionic layer	Bipolar retinal cells to rods and cones	Vision
III	Oculomotor	GSA, GSE, GVE	Superior collicular level; oculomotor nucleus, Edinger-Westphal nucleus	Superior, inferior, medial rectus, and levator palpebrae muscles. Pupillary constrictors and ciliary muscles of eyeball	Proprioceptive impulses; eye movements; accommodation reflex
IV	Trochlear	GSA, GSE	Superior cerebellar peduncle; trochlear nucleus	Superior oblique muscle	Proprioceptive impulses, eye movements
V	Trigeminal	GSA, SVE	Pons; masticator nucleus, semilunar ganglion, mesencephalic nucleus	Face, nose, mouth, jaw	Muscles of mastication; sensory to face, nose, mouth; proprioceptive to tooth sockets and jaw muscles
VI	Abducens	GSA, GSE	Pons; abducens nucleus	Lateral rectus	Eye movements
VII	Facial	GVA, SVA, SVE GVE	Pons; facial nucleus, superior salivatory nucleus, geniculate ganglion	Glands of nose, lacrymal glands, palate; sublingual and submaxillary glands; anterior taste buds	Motor and sensory components to facial region, tongue
VIII	Vestibulo-cochlear:				
	Cochlear division	SSA	Pons; spiral ganglion	Organ of Corti	Audition (hearing)
	Vestibular division	SSA	Pons; vestibular ganglion	Cristae of semicircular canals, maculae of saccule and utricle	Equilibrium

IX	Glosso-pharyngeal	SVA, GVA, SVE, GVE	Medulla; ambiguus nucleus, inferior salivatory nucleus, petrosal ganglion, superior ganglion	Superior constrictor, stylopharyngeus muscles; parotid glands, taste buds (vallate papillae), auditory tube	Motor to pharyngeal region; gustation (taste); motor to parotids; pain, tactile, thermal sensations from posterior tongue, tonsils, and eustachian tubes; regulates blood pressure
X	Vagus	GSA, GVA, SVA, GVE, SVE	Medulla; ambiguus nucleus, dorsal motor nucleus; nodose ganglion; jugular ganglion	Muscles of pharynx and larynx, viscera of thorax and abdomen; pinna of ear	Sensory and motor to thoracic and abdominal viscera, certain skeletal muscles of pharyngeal, laryngeal regions
XI	Spinal accessory	GSA, GVE, SVE	Medulla; accessory nucleus	Sternocleidomastoid and trapezius muscles; portions of laryngeal, pharyngeal muscles; heart (?)	Sensory and motor to muscles of larynx and pharynx; may form components of cardiac branches of vagus
XII	Hypoglossal	GSA, GSE	Medulla; hypoglossal nucleus	Tongue muscles	Sensory and motor to tongue muscles

ªSpecial afferent fibers are found only in the cranial nerves. SVA = special visceral afferent (sensory) fibers that are related only to nerves which subserve olfaction (I) and gustation (VII, IX, X). SSA = special somatic afferent (sensory) fibers that transmit impulses from the special sense organs; ie, the eye (II) and ear (VIII). SVE = special visceral efferent (motor) fibers of cranial nerves which innervate particular skeletal muscles that are derived embryologically from visceral (branchial) arch mesoderm. Some authors prefer the term "branchial motor fibers" to SVE when referring to the nerves that innervate the muscles derived from visceral arch mesoderm (eg, the palatine, pharyngeal, laryngeal, and masticatory muscles as well as those of facial expression that are innervated by branches of cranial nerves V, VII, IX, X, and XI)). GSA = general somatic afferent; GSE = general somatic efferent; GVA = general visceral afferent; GVE = general visceral efferent (see footnote, Table 7.2). The GVE components of cranial nerves III, VII, IX, and X in italics above are the source of the preganglionic fibers of the cranial division of the parasympathetic nervous system.

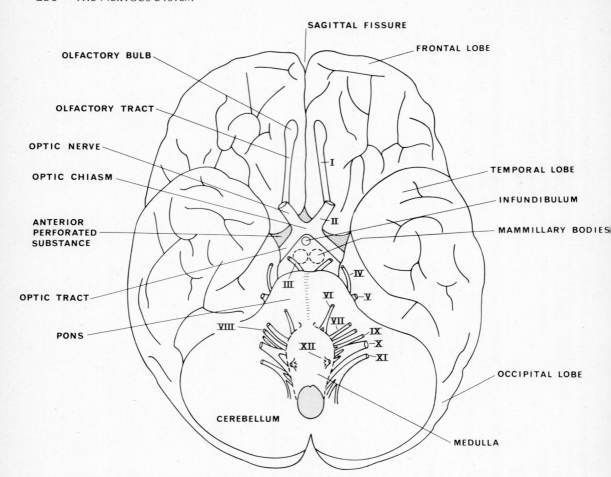

FIG. 7.12. The inferior surface of the human brain showing the points of origin of the cranial nerves (Roman numerals) as listed in Table 7.3. (After Everett. *Functional Neuroanatomy*, 6th ed, 1971, Lea & Febiger.)

The ependyma of the early embryonic neural tube is a simple epithelial layer which lines the cavities of the developing nervous system (Fig. 7.4). In the adult, the embryonic character of the ependyma is retained in some regions, for example, in the choroid plexus (Fig. 7.15). However, the embryonic ependyma undergoes a thickening in most cavities of the developing brain, although the adult ependyma still retains its epithelial character.

Three types of neuroglial (or glial) cells can be distinguished in sections of the central nervous system by the use of appropriate histologic techniques. These glial elements are the astrocytes, oligodendrocytes, and microglia. The first two classes of glial cells, known collectively as the macroglia, are of ectodermal origin, whereas the microglia are derived from embryonic mesodermal cells that migrate into the central nervous system along the branches of the blood vessels. The morphology of the glial cells is shown in Figure 7.13. The reader will note in this illustration that two kinds of astrocyte, protoplasmic and fibrous, are present. Mixed or plasmatofibrous astrocytes also are seen occa-

sionally. Some of the smaller protoplasmic astrocytes form one type of satellite cell when in close association with neurons, although many of the processes of the astrocytes are attached to blood vessels as well as to the pia mater (Fig. 7.16). Similarly, fibrous astrocytes also are related intimately to the blood vessels. In general, however, protoplasmic astrocytes are found in the gray matter of the brain, whereas fibrous astrocytes are found in the white matter.

The oligodendrogliocytes (or oligodendroglia) are found primarily in a close relationship to tracts of nerve fibers, and these glial elements are believed to be responsible for myelinization of the central nervous system. Consequently oligodendrogliocytes are considered homologous to the Schwann cells of the peripheral nervous system. Interestingly, when studied in tissue cultures, oligodendrogliocytes exhibit rhythmic pulsatile movements. The significance of this motor activity in vivo is unknown.

The microglia are of ubiquitous occurrence throughout the central nervous system.

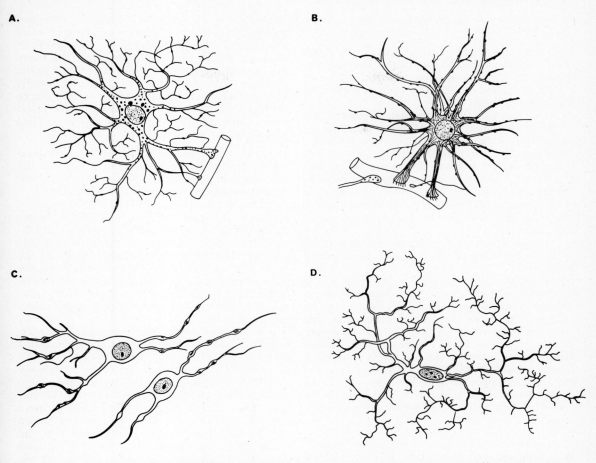

FIG. 7.13. Neuroglial (or interstitial) cells found in the central nervous system. Protoplasmic **(A)** and fibrous **(B)** astrocytes with end feet terminating on capillaries. **C.** Oligodendroglia. **D.** Microglia. (Data from Everett. *Functional Neuroanatomy,* 6th ed, 1971, Lea & Febiger; Bloom and Fawcett. *A Textbook of Histology,* 9th ed, 1968, Saunders.)

Within the adult brain and spinal cord the entire mass of neuroglial tissue provides an exceedingly complex, three-dimensional supporting network of cells and their processes in which the neurons as well as their processes are embedded. This glial framework may be likened roughly to the skeleton of a sponge, within whose interstices the living cells of the organism are situated. However, this analogy is quite limited, because the network of glial cells, like the neurons themselves, does not form a syncytium.

The labyrinthine morphology of the glial tissue serves to encase, and thus to insulate, each neuron from the adjacent neurons, except at their specific points of synaptic contact. The glial tissue serves to isolate central neurons functionally, except at those points where normal impulse transmission takes place. Thus it would appear that the neuroglial cells have an important, albeit indirect, function in the transmission of impulses within the central nervous system because they physically isolate the many pathways that may converge upon any particular neuron, and thus channel the flow of information into specific neural circuits within the brain and spinal cord.

Neuroglial cells also appear to have an important role in the normal metabolic processes of neurons, although the specific details of such participation in neural function are obscure.

Under local or remote pathologic conditions that affect the neurons, the glial cells invariably react, and these elements are the principal loci of origin for tumors in the central nervous system. The microglial cells particularly are able to undergo striking morphologic changes and to exhibit ameboid motion as well as phagocytosis during pathologic states that affect the central nervous system; eg, during infectious disease states.

CONNECTIVE TISSUES OF THE CENTRAL NERVOUS SYSTEM: THE MENINGES. The brain and spinal cord are enclosed in three distinct membranes. From without inward these membranes are the dura mater, the arachnoid, and the pia mater. Collectively these membranes are called the meninges, whereas the arachnoid and pia mater are known conjointly as the leptomeninges.

The Dura Mater. The dura mater is a thick, tough layer of fibrous connective tissue that contains predominantly collagenous fibers together with some elastic fibers, and it is lined on its inner surface by squamous epithelial cells.

The dura surrounding the spinal cord and that of the brain exhibit different relationships to the surrounding bony structures. The lumen of the vertebral canal is lined by periosteum (p. 1064) so that a distinct dural membrane invests the spinal cord (Fig. 7.14). The epidural space that lies between the periosteum and the dura is relatively wide, and it contains connective as well as adipose tissue. The epidural venous plexus also is present within this space.

The dura is attached to the spinal cord laterally by a number of small tooth-shaped (denticulate) ligaments.

The dura mater that invests the brain also is composed of two layers. The outermost layer adheres loosely to the inner surface of the cranium except at the sutures and base of the skull, and serves as the periosteum. Many blood vessels are present, and the dense bundles of collagenous fibers are in contrast to the practically continuous sheetlike layer of more delicate collagenous elements that compose the inner layer. Similarly to the inner surface of the spinal dura, that of the brain also is smooth and covered with a layer of squamous cells.

The Arachnoid. The arachnoid membrane is similar in both spinal cord and brain. This intermediate meningeal structure lacks blood vessels and is smooth on its outer surface. The inner surface of the arachnoid, however, exhibits many delicate branching threads and strands which serve to attach the arachnoid to the pia mater.

The arachnoid membrane crosses over the fissures, or sulci, on the surface of the brain and spinal cord, thereby forming subarachnoid spaces of various dimensions within the sulci.

The Pia Mater. The pia mater is a thin layer of connective tissue which is attached closely to the surface of the spinal cord and brain. The pia is richly vascularized so that the blood vessels of this membrane supply the largest proportion of the blood to the underlying neural tissues. The filamentous strands of the arachnoid membrane are so closely attached to the pia that the leptomeninges often are considered as one membrane, the pia-arachnoid.

The principal extracellular elements that comprise both the pia and arachnoid membranes are bundles of collagenous connective tissue which are invested by networks of delicate elastic tissue. Various cellular elements are present within the pia, including fibroblasts and fixed macrophages. These cell types are particularly abundant near the blood vessels.

Innervation of the Meninges. Both the dura and pia have an abundant innervation. Fine terminal branches of the sympathetic nervous system form elaborate plexuses in the blood vessels, and many free sensory nerve endings also are present in the outermost layer (adventitia) of the blood vessels. The cerebral dura contains many sensory nerve endings in addition to those present in the blood vessels, and the pia of this region also is supplied with abundant neural plexuses.

The Meningeal Spaces. The narrow subdural space, located between the dura mater and the arachnoid, is essentially a serous cavity (p. 565) which contains only a small quantity of fluid.

The subarachnoid space lies between the outer surface of the arachnoid layer and that of the pia and is distinct from the subdural space. The subarachnoid space, in contrast to the subdural space, contains a relatively large volume of fluid. At the peaks of the gyri, or convolutions of the brain, the subarachnoid space is small, but in the sulci it becomes much larger. The subarachnoid space is particularly large throughout the length of the spinal cord (Fig. 7.14).

Within the brain, the subarachnoid space exhibits marked dilatations in certain regions called cisterns (or cisternae). In the cisterns, the arachnoid is separated from the pia by a relatively wide cleft, and the bands of connective tissue that connect these membranes (the trabeculae) are absent. The cisterna magna is the largest of these cisterns, and this space lies at the posterior border of the cerebellum on the dorsal aspect of the medulla oblongata. The fourth ventricle of the brain is in direct communication with this cistern via three openings, the two lateral foramina of Luschka and the medial foramen of Magendie.

THE CHOROID PLEXUSES. As noted earlier, during embryogenesis the neural tube is a hollow structure, and this characteristic is retained in the adult central nervous system. In the spinal cord, the central canal becomes much attenuated or even occluded as development proceeds, whereas in the brain four cavities are present in the adult (Fig. 7.6). The original single cavity of the embryonic neural tube dilates into four distinct cavities, or ventricles, as development proceeds. Thus the adult brain contains two lateral ventricles within the cerebral hemispheres, a third ventricle in the thalamic region, and a fourth ventricle within the pons and medulla oblongata (Table 7.1).

Choroid plexuses are formed in restricted areas within each of the four ventricles, and these structures are responsible for the secretion of a large proportion of the cerebrospinal fluid. Each choroid plexus develops in a region of the brain in which the ependymal layer retains its embryonic properties as a thin aneural epithelium known as the lamina epithelialis. The pia mater, which is closely adherent to the lamina epithelialis, is highly vascular, as well as convoluted; and this pial structure together with the lamina epithelialis is called

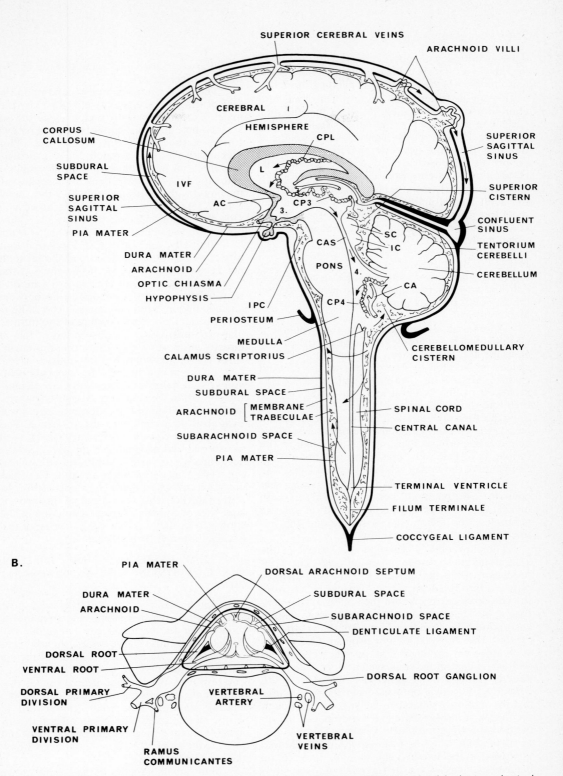

FIG. 7.14. Relationships among the brain, spinal cord, and cerebrospinal fluid. **A.** The right side of the brain and spinal cord (much foreshortened). Arrows indicate the flow of cerebrospinal fluid. L = lateral ventricle; 3 = third ventricle; 4 = fourth ventricle; AC = anterior commissure; IVF = interventricular foramen (of Monroe); L = lateral ventricle; CPL = choroid plexus of lateral ventricle; CP4 = choroid plexus of fourth ventricle; CP3 = choroid plexus of third ventricle; SC, IC = superior and inferior colliculi, respectively; CAS = cerebral aqueduct (of Sylvius); IPC = interpeduncular cistern. The stippled protuberance lying immediately above the superior colliculus (SC) is the epiphysis (or pineal body). **B.** Transverse section of the spinal cord in the spinal canal illustrating the meninges and the manner of exit of the spinal nerves. (See also Fig. 7.15.) (After Goss [ed]. *Gray's Anatomy of the Human Body,* 28th ed, 1970, Lea & Febiger.)

the tela choroidea or choroid plexus. Strictly speaking, however, the specialized pial membrane of the tela choroidea is the true choroid plexus (Fig. 7.15).

From a functional standpoint, the tela choroidea is folded to such an extent that the total surface area exposed to the cerebrospinal fluid is quite large. Similarly, the highly tortuous course of the rich capillary networks within the choroid plexus also exposes a large surface area for the elaboration of cerebrospinal fluid.

The regularly disposed cuboidal cells that form the adult lamina epithelialis contain microvilli on their free surface that are dilated at their tips in contrast to the cylindrical appearance of conventional microvilli (Fig. 7.15).

THE CEREBROSPINAL FLUID (CSF). The entire central nervous system is bathed in cerebrospinal fluid which serves not only to protect it from mechanical injuries but also to assist in its metabolism.

Approximately one-half of the total volume of CSF that fills the ventricles of the brain, the central canal of the spinal cord, and the subarachnoid spaces is secreted by the vessels of the choroid plexuses (Fig. 7.15). The other half of the CSF is produced by the cerebral and pial blood vessels; therefore the composition of the CSF is dependent upon filtration and diffusion of substances from the blood. In addition, facilitated transport as well as active transport mechanisms are operative in the formation of CSF. The major constituents of CSF compared to those of plasma are listed in Table 7.4.

The spaces of the central nervous system that contain CSF apparently communicate freely with one another so that CSF can pass readily among the ventricles, the subarachnoid space, and the extracellular space of the brain per se.

The flow of CSF through the spaces of the central nervous system is as follows. The CSF produced within the substance of the brain passes outward into the subarachnoid spaces. Flow from the choroid plexuses is into the lumens of the ventricles, but these regions do not participate in CSF absorption. The ependymal lining of the ventricles, on the other hand, does not produce CSF; however, some absorption of CSF into the venous system of the central nervous system does take place directly through the ventricular walls. One route whereby CSF flows outward from the ventricular regions into the subarachnoid spaces is through the foramina of Magendie and Luschka located on the dorsal wall of the fourth ventricle (Fig. 7.15).

The CSF which is formed within the choroid plexuses of the lateral ventricles of the cerebral hemispheres passes via the foramina of Monro into the third ventricle. In this region additional CSF is added by the choroid plexus of the third ventricle. The CSF now traverses the aqueduct of Sylvius into the fourth ventricle, where still more fluid is secreted by the choroid plexus of this ventricle. Thence the combined volumes of CSF that were elaborated in each ventricle enter the cerebello-medullary cistern, and then diffuse throughout the subarachnoid spaces.

The large venous sinuses found upon the surface of the brain are heavily enveloped by the dura mater except in certain regions (eg, the sagittal sinus) where the dura mater is pierced by evaginations of the arachnoid membrane. Thus an arachnoid villus is formed by a protrusion of arachnoid mesothelium into a dural venous sinus, and within these regions the CSF is separated from the blood within the sinus only by a thin membrane (Fig. 7.15). Consequently, the arachnoid villi provide the major channels whereby CSF flows directly into the venous circulation of the brain.

Normally, the rate of CSF formation equals its rate of reabsorption, as described above, so that no net accumulation of CSF takes place within the central nervous system. The rate at which CSF is secreted has been determined to lie between 200 and 500 ml per day in human subjects. The rate of CSF production is independent of the intraventricular pressure; however, absorption of this fluid is directly proportional to the intraventricular pressure. Therefore, when the intraventricular pressure falls below about 70 mm Hg, absorption of CSF stops, and large volumes of CSF accumulate rapidly within the central nervous system when the arachnoid villi do not function properly, insofar as their absorptive capacities are concerned. The fluid accumulation caused by this mechanism is known as external hydrocephalus. Similarly, CSF also will accumulate proximally within the ventricles when the foramina of Magendie and Luschka are occluded, or when there is an obstruction to flow within either the foramina of Monro or the aqueduct of Sylvius. The accumulation of fluid resulting from such mechanical damming of the CSF is called internal hydrocephalus.

VASCULAR PATTERNS WITHIN THE CENTRAL NERVOUS SYSTEM; THE BLOOD-BRAIN BARRIER

The Spinal Cord. The anterior and posterior radicular arteries that supply the spinal cord with blood enter the vertebral canal together with the dorsal and ventral nerve roots, thence to form a dense network within the pia mater of the cord. A number of longitudinal arterial channels are present; however, the anterior arterial tract is the most significant, as it produces a large number of small central branches that enter the ventral medial fissure of the cord. These central arteries in turn invade the anterior gray columns laterally and thus provide the chief blood supply to the gray matter of the spinal cord.

Many small branches of the arterial network of the pia mater (the peripheral arteries) enter the white matter of the cord about its entire circumference.

The capillary networks found within the white matter are somewhat diffuse, in contrast with the far more luxuriant capillary networks of the gray matter.

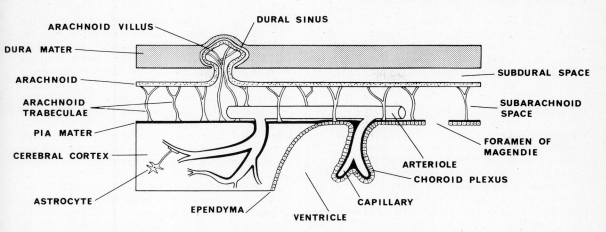

FIG. 7.15. Highly schematic diagram illustrating the relationships among the meninges, their spaces, and the ventricles of the brain. The thickness of the pia mater has been exaggerated considerably. See also Figure 7.16.

The pathways followed by the veins do not parallel those of the arteries. Many small veins exit from the spinal cord about its periphery as well as from the ventral median fissure to form a dense venous plexus located within the pia mater, particularly on the dorsal surface of the cord. Thence blood is drained from this plexus by way of veins that accompany the dorsal and ventral roots of the spinal nerves.

The Brain. The arterial blood supply to the brain is provided chiefly by the paired carotid and vertebral arteries. The vertebral arteries anastomose to form the basilar artery. The circle of Willis is produced by union of the carotids and the basilar artery. Branches of these large vessels enter the pia mater and give off smaller arteries, which in turn enter the substance of the brain, there to form capillary networks. The arteries within the brain are

Table 7.4 CONCENTRATION OF SOLUTES IN HUMAN CEREBROSPINAL FLUID AND PLASMA[a]

SUBSTANCE	CSF[b]	PLASMA
Na^+ (mEq/liter)	147.0	150.0
K^+ (mEq/liter)	2.9	4.6
Ca^{++} (mEq/liter)	2.3	4.7
Mg^{++} (mEq/liter)	2.2	1.6
Cl^- (mEq/liter)	113.0	99.0
HCO_3^- (mEq/liter)	25.1	24.8
Protein (mg %)	20.0	6000.0
Glucose (mg %)	64.0	100.0
Inorganic P (mg %)	3.4	4.7
Urea (mg %)	12.0	15.0
Lactic acid (mg %)	18.0	21.0
pH	7.3	7.4
Pco_2 (mm Hg)	50.2	39.5
Osmolality (mOsm/kg H_2O)	289.0	289.0

[a]*Data from Ganong.* Review of Medical Physiology, *6th ed, 1973, Lange Medical Publications.*
[b]*Normal CSF also contains less than 10 lymphocytes per mm³, but no erythrocytes. The composition of CSF is quite similar to that of the aqueous humor of the eye.*

considered to be end arteries; consequently few anastomotic connections between these vessels are present. Therefore if a major arterial branch is damaged through disease or injury, a physiologically effective collateral circulation is not established.

As is the case within the spinal cord, the densities of the capillary networks found within the white matter are relatively low compared to those present in the gray matter, a fact which some consider to be a reflection of the relative metabolic rates in the two regions.

The central nervous system does not possess any lymphatic drainage channels (p. 553); therefore fluid that is filtered out of the capillaries percolates through the interstitial space of the brain. The vessels that enter the brain from the pia mater are enclosed within perivascular spaces; the latter open on the surface of the brain into the subarachnoid spaces. Thereby the CSF is drained from the substance of the brain outward and toward the meninges.

Venous drainage from the brain takes place chiefly via the deep veins and dural sinuses. In turn, the major proportion of the blood flowing in these vessels enters the internal jugular veins. However, a small volume of venous blood is drained through the pterygoid and ophthalmic venous sinuses through veins leading to the scalp. In addition a small quantity of blood also is drained by the vertebral veins within the spinal canal.

The blood vessels of the cerebral hemispheres exhibit several morphologic peculiarities that are found nowhere else in the body. In the choroid plexuses, there are clefts or interstices between the endothelial cells that comprise the capillary walls. The capillaries within the substance of the brain are similar to those found elsewhere in the body; however, no fenestrations are present. The endothelial cells adhere closely, and a thick basement membrane is present. Furthermore the capillaries within the brain are surrounded by a membrane composed of the end-feet of astrocytes (Fig. 7.16). The end-feet are attached closely to the capillary basement membrane. Thus processes of the astrocytes not only invest, and thereby isolate, central neurons but also these glial elements form membranaceous structures about the cerebral capillaries.

The Blood–Brain and Blood–Cerebrospinal Fluid Barriers. Many years ago it was shown that the injection of certain vital dyes (eg, trypan blue) into adult experimental animals would stain all of the tissues except for large portions of the brain and spinal cord. More recent investigations have demonstrated clearly that only water, oxygen, and carbon dioxide can traverse the membranes of the capillaries within the central nervous system readily, and that these membranes are relatively impermeable to all other substances. This fact is in marked contrast to all other capillaries in the body.

Observations such as those cited above have led to the formulation of the concept that a blood–brain barrier is present in most regions of the central nervous system.

The physiologic mechanisms involved in the exchange of materials across capillary membranes in general will be discussed in Chapter 9 (p. 558); however, it should be mentioned here that there is considerable variation among the transport rates for specific materials across the capillary walls in different organs. Nevertheless, the rate of exchange of many physiologically important compounds and drugs across the cerebral capillaries is so much slower than that of other capillaries of the body that the existence of a functionally specific blood–brain barrier is an inescapable conclusion.

Although the morphologic nature of the blood–brain barrier is a subject of much controversy, it appears that the dense basement membrane of the capillary endothelial cells of the central nervous system and/or the external membrane provided by the end-feet of the astrocytes (Fig. 7.16) have major roles in producing this barrier. Alternatively, some investigators believe that the capillary endothelium of the central nervous system per se is less permeable than in other regions of the body, while other workers feel that morphologically specialized subendothelial regions are the site of the blood–brain barrier. Other students of this problem are of the opinion that not one, but the several structures listed above, form a sequence of morphologic barriers to the rapid transport of substances from the blood to the nervous tissues. Regardless of the ultimate site of the blood–brain barrier, however, it is essential to stress that this barrier is not located at the surfaces of the neurons themselves; rather, the barrier is located between the plasma within the capillaries and the extracellular space of the brain.

The cerebral capillaries are far more permeable at birth than they are in adults; consequently the development of the blood–brain barrier takes place gradually during the first several years of postnatal life.

Since the central neurons are exceedingly sensitive to alterations in their chemical environment, the physiologic role of the blood–brain barrier apparently is to prevent sudden and drastic fluctuations in the composition of the interstitial fluid surrounding these neurons. In this regard, it is important to realize that no substance is excluded completely by the blood–brain barrier from reaching the central neurons. Rather, it is the rate of transport of substances through this region that is of major significance in stabilizing the internal environment of the brain.

A.   **B.**

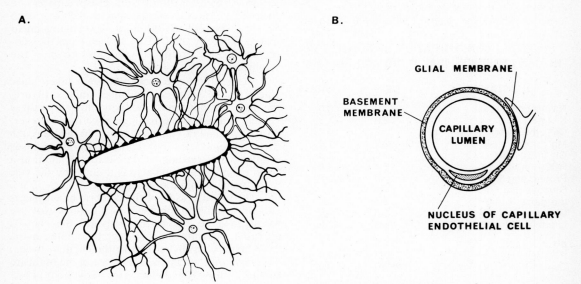

FIG. 7.16.   A. Cerebral capillary surrounded by a membrane composed of the end-feet of fibrous astrocytes. **B.** The termination of a single glial end-foot on a capillary in relation to the basement membrane of the capillary wall. (After Ganong. *Review of Medical Physiology,* 6th ed, 1973, Lange Medical Publications.)

As noted earlier, water, oxygen, and carbon dioxide traverse the blood–brain barrier rapidly. Glucose is transported more slowly, and by a facilitated passive transport mechanism (p. 90). Other materials are transported by diffusion, facilitated passive transport, and active transport (p. 104), cations being transported chiefly by the latter mechanism. The rate at which particular substances are transported by diffusion is proportional to their molecular size as well as their lipid solubility. Thus protein molecules, as well as substances that are bound to protein molecules, cross the blood–brain barrier to only a negligible extent. This fact accounts for the lack of staining of the brain by acidic dyes which become bound to the plasma proteins and thus are rendered unable to traverse the blood–brain barrier.

Once a substance has crossed the blood–brain barrier, transport into the neurons of the brain is accomplished by mechanisms similar to those present in cells generally (p. 85).

There are five small regions of the brain that are located "outside" of the blood–brain barrier. These regions are the pineal gland, the posterior pituitary and the ventral portion of the hypothalamic median eminence in the immediate vicinity of the posterior pituitary, the intercolumnar tubercle, the area postrema, and the supraoptic crest. The reason these areas are not included within the blood–brain barrier is unknown, but speculatively all of these regions have been ascribed chemoreceptor functions of various types.

Clinically, the blood–brain barrier is of considerable importance, as certain drugs (eg, sulfadiazine) enter the brain quite readily, whereas others do not (eg, antibiotics, including chlortetracycline and penicillin). Consequently any rational therapy of diseases of the central nervous system requires that the physician have a thorough knowledge of the permeability of the blood–brain barrier to various drugs. In addition, the blood–brain barrier degenerates at the site of tumor growth or infectious disease processes. Therefore the rate at which substances that enter normal brain tissue very slowly enter diseased brain tissue is accelerated, a fact that is used for diagnostic purposes. For example, albumin labeled with radioactive iodine accumulates rapidly in tumorous, but not in normal, brain tissue.

The blood–cerebrospinal fluid barriers are located at the choroid plexuses. The general principles which underlie the transport of substances into the cerebrospinal fluid are similar to those which govern the transport rate across the blood–brain barrier. That is, molecular size and lipid solubility are of prime importance in controlling the rate of transport from the blood to the cerebrospinal fluid.

The routes whereby metabolites and drugs leave the cerebrospinal fluid are quite different than those whereby they enter this fluid. Thus numerous metabolites as well as drugs can leave the cerebrospinal fluid via the arachnoid villi, and this transport occurs rapidly, regardless of lipid solubility or molecular size. Even large protein molecules, such as plasma albumin, exit rapidly by this route. In fact, the rates at which all substances leave the cerebrospinal fluid are the same. These rates depend in turn upon the rate of bulk flow of the cerebrospinal fluid itself through the arachnoid villi.

Alternatively, certain lipid-soluble metabolites and drugs can leave the brain directly by passive transport across the blood–brain barrier from the extracellular fluid of the brain, a movement that is enhanced by the bulk circulation of the cerebrospinal fluid throughout the brain and into the capillaries.

Terminology

A few terms that are encountered frequently in neuroanatomy are presented and defined in this section. More specific terms will be defined in context in subsequent divisions of this chapter.

Brachium: An arm; a general term used to denote an armlike neural process or structure.

Commissure: A general term used to designate a junction between corresponding anatomic structures, frequently, but not invariably, across the midline of the body.

Fascicle (pl. *fasciculi*): A general term for a small bundle of nerve, muscle, or tendon fibers.

Fissure: A general term for a groove or cleft, particularly a deep fold in the cerebral cortex that includes its entire thickness.

Funiculus (pl. *funiculi*): A general term for cordlike anatomic structures; specifically applied to certain of the columns or tracts within the spinal cord.

Ganglion (pl. *ganglia*): A general term used to denote a group or collection of nerve cell bodies located outside of the central nervous system. Occasionally this term is applied to certain nuclear groups within the brain or spinal cord.

Gyrus (pl. *gyri*): One of the tortuous convolutions (or elevations) on the surface of the brain caused by infolding of the cortex.

Lemniscus (pl. *lemnisci*): A general term for a band or bundle of fibers in the central nervous system.

Nucleus (pl. *nuclei*): A general term which denotes a group of nerve cell bodies located within the central nervous system and directly related to the fibers of a particular nerve. (NB: This usage of the term "nucleus" must not be confused with the cytologic organelle found within individual cells.)

Peduncle: A stalklike or stemlike connecting part.

Plexus: A network. A general term used to denote a network of nerves, lymphatic vessels, or veins.

Ramus (pl. *rami*): A branch. A general term for a smaller structure that branches off from a larger one; eg, the smaller division of a nerve that arises by branching.

Sulcus (pl. *sulci*): A groove or furrow, particularly one of those present on the surface of the brain separating the gyri.

Tract: A bundle of nerve fibers that have a similar origin, course, termination, and function.

In many instances, the compound nouns that make up the names of many fiber tracts within the central nervous system denote the origin and termination of the nerve fibers. Hence these names often afford the reader valuable clues to the function of particular bundles of nerve fibers, especially in the spinal cord. For example, the spinocerebellar tract arises within the cord and sends fibers to the cerebellum; therefore this tract contains ascending or sensory fibers. On the other hand, the corticospinal tract arises within the cerebral cortex, sends fibers to the spinal cord, and therefore is descending or motor in function. Further examples that illustrate this point will be found in Table 7.5.

The Peripheral Nerves and Spinal Cord

The lowest (or first) level for the functional integration of information within the central nervous system takes place within the spinal cord. Within this structure, information encoded in the form of afferent nerve impulses is received from the various sensory receptors located within, and on the surface of, the body. This information then may be acted upon locally within the cord so that appropriate and coordinated somatic or visceral responses to the stimulus take place. More usually, however, the encoded sensory information is relayed to higher brain centers for additional processing and modification; thus more elaborate motor response patterns are evoked.

The general anatomic and functional relationships among peripheral sensory receptors, segmental afferent and efferent neural pathways, and peripheral effectors were presented in Chapter 6 and depicted in Figure 6.1. In the present section, the functional neuroanatomy of the spinal cord in relation to the peripheral nervous system and certain higher brain centers will be considered in somewhat greater detail.

FUNCTIONAL AND ANATOMIC RELATIONSHIPS BETWEEN THE SPINAL CORD, PERIPHERAL NERVES, AND HIGHER NEURAL CENTERS.   Although the spinal cord alone is capable of integrating some sensory information into limited patterns of activity (eg, simple segmental or myotatic reflexes such as were described in Chapter 6), the action of this structure is primarily integrated with the functional activity of the more anterior portions of the brain stem and brain. This integration takes place through ascending fiber pathways which provide information for still higher levels. Similarly, higher brain centers are integrated with, and connected to, lower centers by means of descending fiber pathways, which in turn superimpose their influence upon lower brain and spinal centers. These higher influences subject the lower centers to a modification of their inherent and independent activity by the physiologic mechanisms of facilitation (p. 199) or inhibition (p. 193). Thus the cerebral cortex, which represents the highest level of sensory and motor integration, regulates the activity of the brain stem and spinal cord. In turn, the cerebellum, pons, and medulla acting alone can exercise some control over the activity of the spinal cord, whereas the latter, acting alone and isolated from all higher centers, can maintain only a limited degree of segmental reflex activity.

A discussion of the spinal cord, therefore, is primarily a consideration of its function as a relay system or transmission line for both afferent and efferent impulses; consequently the remainder of this section will be concerned primarily with the fiber tracts in the spinal cord and their central connections that are responsible for mediating the effects described above.

Anatomy of the Spinal Cord.   The spinal cord is an elongated, almost cylindrical mass of neural tissue which lies within the upper portion of the vertebral canal. In adults, the spinal cord is generally between 42 and 45 cm in length, and this structure extends from the first cervical vertebra to the upper border of the second lumbar vertebra. Rostrally the spinal cord is continuous with the medulla oblongata, whereas caudally it terminates in a threadlike structure, the filum terminale which is attached to the first segment of the coccyx.

As mentioned earlier, the spinal cord is invested by three meningeal membranes, the dura mater, arachnoid, and pia mater. The epidural space lies between the dura and the vertebral column, whereas the subdural space lies between the dura and the arachnoid membrane. The subarachnoid space, filled with cerebrospinal fluid, separates the arachnoid from the pia mater. The latter structure invests the cord closely, and sends projections (septa) into the cord proper.

The filum terminale is composed principally of fibrous connective tissue and is continuous with the pia mater caudally.

The 31 pairs of nerves that emanate from the adult spinal cord are arranged segmentally, and are referred to as cervical (C–1 through C–8), thoracic (T–1 through T–12), lumbar (L–1 through L–5), sacral (S–1 through S–5), or coccygeal (C–1), depending upon their points of attachment to the cord (Table 7.2). The individual segments of the cord proper exhibit considerable variations in their length, as indicated by the distance between the points at which successive pairs of spinal nerves emerge. Thus the segments in the cervical and upper lumbar regions are about one-half as long as those in the midthoracic area.

The spinal cord is dilated widely in two regions. In the cervical region the enlargement corresponds to the portion of the cord from which the nerves to the upper extremity originate (the brachial plexus), whereas the lesser enlargement in the lumbar region indicates the point of origin of the nerves that innervate the lower extremity (the lumbosacral plexus).

Table 7.5 PRINCIPAL ASCENDING AND DESCENDING TRACTS OF THE SPINAL CORD

NAME OF TRACT OR BUNDLE[a]	LOCATION IN CORD[b]	MAJOR FUNCTIONS[c]
A. Ascending (sensory) tracts		
1. Fasciculus gracilis (tract of Goll)	Posterior column	Pathway for tactile sensations, position sense, passive movement, spatial discrimination, two-point discrimination, and "vibration"
2. Fasciculus cuneatus (tract of Burdach)	Posterior column	Same as above
3. Fasciculus posterolateralis (dorsolateral fasciculus, tract of Lissaur)	Lateral column	Pathway for pain and thermal sensations
4. Dorsal (posterior) spinocerebellar tract (direct cerebellar tract, tract of Flechsig)	Lateral column	Transmits sensory impulses from proprioceptors of muscles, tendons, and joints to cerebellum, ie, reflex proprioception; also conveys touch, pressure impulses from skin
5. Ventral (anterior) spinocerebellar tract (indirect cerebellar tract, tract of Gowers)	Lateral column	Same as above
6. Dorsal (posterior) spinothalamic tract (anterolateral spinothalamic tract)	Lateral column	Transmits sensory impulses underlying pain and thermal sensations
7. Spinotectal tract	Lateral column	Sensory pathway for visuospinal reflexes (?); nociceptive impulses (?)
8. Ventral (anterior) spinothalamic tract	Anterior column	Sensory pathway for touch and tactile localization
B. Descending (motor) tracts		
1. Lateral corticospinal tract (crossed pyramidal tract)	Lateral column	Transmits motor impulses to skeletal muscles; underlies voluntary movements of body
2. Rubrospinal tract (tract of Monakow)	Lateral column	Transmits motor impulses underlying muscle tone and synergistic contractions of skeletal muscle
3. Dorsal (posterior or lateral) vestibulospinal tract	Lateral column	Transmits motor impulses underlying postural and balance reflexes
4. Olivospinal tract (Helwig's bundle)	Lateral column	Functions unknown; possibly mediates thalamospinal reflexes
5. Reticulospinal tract	Anterior column	Transmits motor impulses underlying muscle tone; controls activity of autonomic nervous system; regulates threshold of peripheral end-organs, eg, sensory pathways and muscle spindles
6. Ventral (anterior) vestibulospinal tract	Anterior column	Transmits motor impulses underlying postural and balance reflexes
7. Tectospinal tract	Anterior column	Transmits motor impulses underlying visuospinal and audiospinal reflexes
8. Ventral (anterolateral) corticospinal tract (direct pyramidal tract)	Anterior column	Transmits motor impulses to skeletal muscles; underlies voluntary movements
C. Ascending plus descending fibers (fasciculi proprii)		
OB Oval bundle (fasciculus septomarginalis)	Dorsal column	
CT Comma tract of Schultze (fasciculus interfascicularis)	Dorsal column	The fasciculi proprii collectively form intrinsic pathways within the spinal cord for the unconscious coordination of spinal reflexes, eg, during voluntary movement and postural adjustments that are initiated by higher centers. Also concerned with integration of reflex activities within the cord, and with coordinating reflex functions at different levels within the cord per se
PGB Posterior ground bundle (posterior intersegmental fasciculus)	Dorsal column	
LGB Lateral ground bundle (lateral intersegmental fasciculus)	Lateral column	
AGB Anterior ground bundle (anterior intersegmental or sulcomarginal fasciculus)	Anterior column	

[a]The numbers and abbreviations used in this column of the table correspond to the tracts and fiber bundles as indicated in Figure 7.20. Alternative names for these structures are given in parentheses. The long, or projection tracts are summarized under headings A and B in the table, whereas the short, association, or intersegmental tracts (collectively known as the fasciculi proprii) are summarized in part C.

[b]The anatomic locations of the principal columns, or funiculi, within the cord are shown in Figure 7.20.

[c]A simple mnemonic device for recalling the general functional and anatomic relationships within the spinal cord, as well as the spinal nerves, is: Sensory Afferent, Motor Efferent = SAME; Dorsal Afferent Ventral Efferent = DAVE.

259

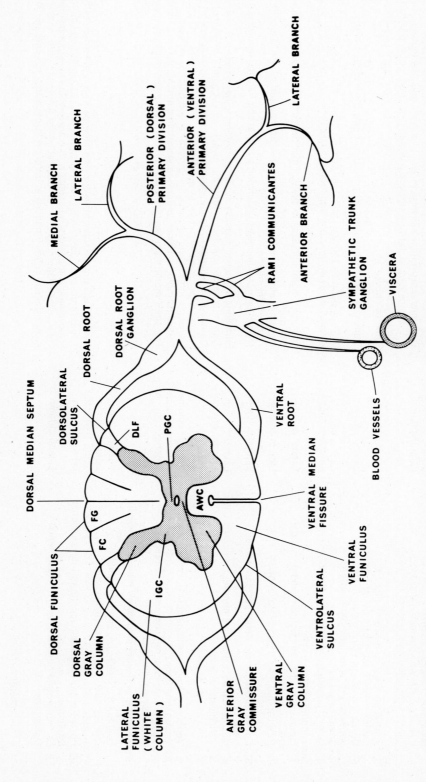

FIG. 7.17. Semischematic illustration showing a transverse section of the spinal cord, the major regions in this structure, and the principal components of a spinal nerve (cf. Fig. 7.10). IGC = intermediolateral gray column; FC = fasciculus cuneatus; FG = fasciculus gracilis; DLF = dorsolateral fasciculus (zone of Lissauer); PGC = posterior gray commissure; AWC = anterior white commissure. (After Chusid. *Correlative Neuroanatomy and Functional Neurology*, 14th ed, 1970, Lange Medical Publications.)

The spinal cord is approximately 25 cm shorter than the vertebral canal. Consequently the lumbar and sacral nerves have extremely long roots which extend from the cord to the intervertebral foramina where the dorsal and ventral roots are located. These elongated roots descend from the cord in a bundle that resembles the tail of a horse; hence this group of nerves collectively is known as the cauda equina.

The general anatomic features of the spinal cord in cross section are illustrated in Figure 7.17 together with the branches of a typical spinal nerve (see also Fig. 7.10). Note that the anterior median fissure and the dorsal (or posterior) median sulcus divide the spinal cord into symmetric left and right halves.

The central canal of the spinal cord is lined by ependymal cells and contains cerebrospinal fluid. This cavity, which extends the entire length of the cord, is continuous anteriorly with the posterior end of the fourth ventricle of the medulla oblongata. The central canal divides the transverse commissure into anterior and posterior gray commissures (Fig. 7.17).

The Gray Matter. The centrally located gray matter of the cord has the general appearance of the outspread wings of a butterfly, and this region is composed of nerve cell bodies, as well as predominantly unmyelinated nerve fibers that are em-

bedded in a framework of neuroglial cells. Although each lateral half of the gray matter commonly is referred to as the dorsal and ventral "horns," the terms dorsal and ventral columns more appropriately describe these regions, as the gray matter extends throughout the entire length of the cord. The large cell bodies of the motor neurons (alpha cells) are located within the ventral columns of the gray matter (Fig. 7.18). These cell bodies have a diameter of about 100 μ, and give rise to axons which exit from the cord by way of the ventral roots. The axon of every motoneuron divides near its peripheral terminations and each branch innervates a skeletal muscle fiber. Thus a single axon together with the several extrafusal muscle fibers which it innervates forms the morphologic and functional entity known as the motor unit (p. 128). The smaller gamma neurons of the ventral gray columns innervate the intrafusal fibers of the muscle spindles and thereby control the threshold of these proprioceptors.

The gray matter of the thoracic and upper lumbar segments of the spinal cord (T–1 through T–12 and L–1 through L–3 or L–4) exhibits a small lateral swelling on each side called the lateral column or lateral ground bundle (Figs. 7.17 and 7.20). The group of nerve cells located within these two regions forms the intermediolateral cell columns. The cell bodies of these neurons in turn give rise to sympathetic preganglionic fibers which exit from

A.

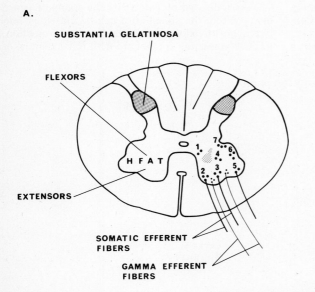

B.

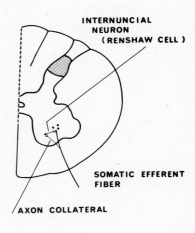

FIG. 7.18. **A.** Cross section of lower cervical spinal cord illustrating the location of the substantia gelatinosa as well as localization of the cell bodies of the alpha and gamma motoneurons. The specific regions of the upper body which are innervated are indicated on the left side; the motoneurons controlling flexor movement lie dorsal to those that control extensor movement in the ventral column. H = hand; F = forearm; A = arm; T = trunk. On the right side of the drawing the motor nuclei are: 1. posteromedial (Clarke's column); 2. anteromedial; 3. anterior; 4. central; 5. anterolateral; 6. posterolateral; 7. retroposterolateral. Gamma efferent fibers arise from very small anterior horn cells, and these axons innervate the small intrafusal fibers of the muscle spindles. **B.** Somatic efferent axons (alpha motoneurons) send collaterals back to the gray matter which may synapse on small medially located inhibitory interneurons. This internuncial neuron pool is indicated in **A** as a small shaded area lying between the posteromedial (1.) and central (4.) neuron pools. (After Chusid. *Correlative Neuroanatomy and Functional Neurology*, 14th ed, 1970, Lange Medical Publications.)

the cord by way of the anterior (ventral) nerve roots.

A clearly delineated group of nerve cells is located at the inner margin of the base of the posterior columns of the gray matter (Clarke's columns). These cells are found almost exclusively within the thoracic region of the cord and form the dorsal nucleus. The dorsal nucleus is the homologue of the nucleus cuneatus of the medulla (p. 265; Fig. 7.21).

The outer region at the base of each posterior column of the gray matter contains an area within which extensions from the gray matter intermesh with filaments of white matter to produce a delicate network known as the reticular formation. This reticular formation is most evident within the cervical region of the spinal cord, and it is continuous with the reticular formation of the medulla oblongata and pons (see p. 302).

Near the apex of the dorsal column is a mass of gelatinous material containing accumulations of small nerve cells that have numerous dendritic processes. This region, called the substantia gelatinosa of Rolando, receives the most delicate myelinated and unmyelinated fibers of the dorsal (posterior) roots (Figs. 7.18 and 7.19).

The White Matter; Tracts of the Spinal Cord

General Considerations. The white matter of the spinal cord completely invests the gray matter, and the three principal anatomic features of this region, the posterior, lateral, and anterior funiculi, are depicted in Figure 7.17. The white matter consists of highly organized bundles (fascicles or tracts) of unmyelinated and myelinated nerve

fibers, although the latter predominate. The nerve fibers of the white matter in turn are embedded within a matrix of neuroglia.

The bundles of nerve fibers, or tracts, located within the white matter of the spinal cord, not only serve to integrate and coordinate sensory and motor functions at any particular level of the cord, but also to interconnect functionally different levels of the cord with each other as well as to various regions within the brain. The principal ascending (sensory) and descending (motor) tracts within the spinal cord are depicted in Figure 7.20A, whereas some of the synaptic connections within the cord among various neurons are shown in part B of the same figure. The functional and certain other properties of the ascending and descending tracts shown in Figure 7.20A are summarized in Table 7.5; note that the numbering of these tracts as given in Figure 7.20A corresponds to the numbering of the same tracts in this table. The general pathways followed by the major ascending (sensory) tracts are shown in Figure 7.21, whereas the major descending (motor) pathways are given in Figure 7.22. Note particularly the levels at which these various sensory and motor pathways cross to opposite sides of the spinal cord or brain. These crossover points, or decussations, are of particular diagnostic significance to the practicing physician in localizing pathologic alterations or lesions within various regions of the central nervous system.

As indicated in Table 7.5, the tracts of nerve fibers within the spinal cord can be classified into two major divisions. First, the long, or projection, tracts connect the spinal cord with other regions of

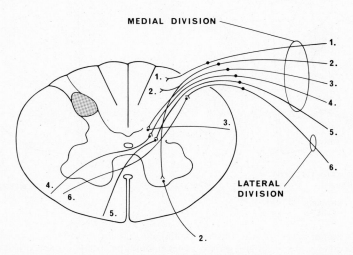

FIG. 7.19. Cross section of spinal cord illustrating the principal loci where dorsal root fibers terminate. Note that the ends of each fiber are numbered. The substantia gelatinosa is shaded on the left side of the figure. 1, 2. Large myelinated fibers which have correspondingly large dorsal root ganglion cells and pass to the dorsal columns. These elements arise from pacinian corpuscles and muscle spindles. 3, 4. Dorsal fibers that terminate on dorsal column cells which give rise to spinothalamic and spinocerebellar tracts. 5. Dorsal fiber which terminates upon a neuron giving rise to the ventral (anterior) spinothalamic tract. 6. Small-caliber neuron subserving pain which terminates on a neuron in the substantia gelatinosa (of Rolando) which in turn gives rise to ascending fiber in the contralateral spinothalamic tract. See also Figure 7.20. (After Chusid. *Correlative Neuroanatomy and Functional Neurology,* 14th ed, 1970, Lange Medical Publications.)

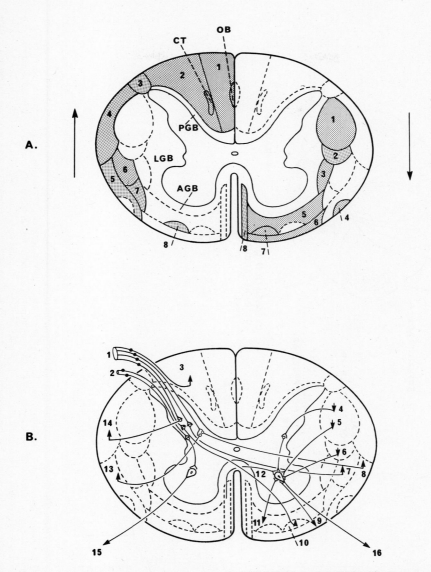

FIG. 7.20. **A.** Principal ascending (left side, arrow pointing upward) and descending (right side, arrow pointing downward) tracts of the spinal cord. Note that the ascending tracts are shaded on the left side of the figure and the descending tracts are shaded on the right. Mixed tracts that contain ascending as well as descending fibers are the oval bundle (OB), the fascicularis interfascicularis or comma tract (CT), the posterior ground bundle (PGB), the lateral ground bundle (LGB), and the anterior ground bundle (AGB). The ascending (sensory) tracts are: 1. fasciculus gracilis; 2. fasciculus cuneatus; 3. fasciculus posterolateralis; 4. dorsal spinocerebellar tract; 5. ventral spinocerebellar tract; 6. dorsal spinothalamic tract; 7. spinotectal tract; 8. ventral spinothalamic tract. The descending (motor) tracts are: 1. lateral corticospinal tract; 2. rubrospinal tract; 3. dorsal vestibulospinal tract; 4. olivospinal tract; 5. reticulospinal tract; 6. ventral vestibulospinal tract; 7. tectospinal tract; 8. anterior corticospinal tract. These tracts, their locations in the cord, and their major functions are listed together with synonymous names in Table 7.5. Note: These fiber bundles do not occupy such neatly delineated areas as indicated in this figure, hence some overlap of fibers, thus of specific functions, can (and does!) occur. **B.** Some of the interconnections among neurons within the spinal cord. 1. Medial division, dorsal root. 2. Lateral division, dorsal root. 3. Dorsal root fibers ascending in posterior columns mediating sensations of touch, spatial discrimination, body position, and movement. 4, 5. Crossed corticospinal fibers connecting indirectly with an anterior horn motoneuron through an internuncial neuron (4) and directly to the same cell (5). 6. Rubrospinal fiber. 7. Sensory fibers entering dorsal spinothalamic tract on contralateral side of cord. 8. Fiber entering contralateral ventral spinocerebellar tract. 9. Vestibulospinal fiber. 10. Fibers concerned with touch and tactile localization entering ventral spinothalamic tract of contralateral side. 11. Reticulospinal fiber. 12. Corticospinal fiber. 13. Sensory fiber entering ipsilateral ventral spinocerebellar tract. 14. Fiber entering ipsilateral dorsal spinocerebellar tract. 15, 16. Alpha motoneurons arising from large anterior horn cells. (After Best and Taylor [eds]. *The Physiological Basis of Medical Practice,* 8th ed, 1966, Williams & Wilkins.)

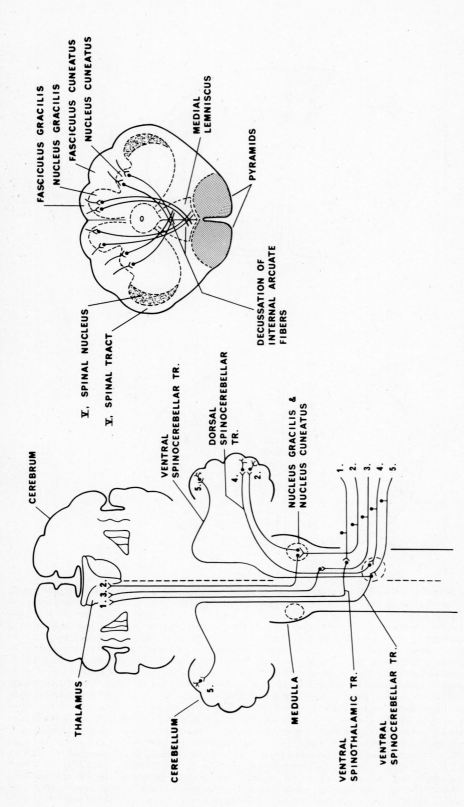

FASCICULUS GRACILIS
NUCLEUS GRACILIS
FASCICULUS CUNEATUS
NUCLEUS CUNEATUS

MEDIAL
LEMNISCUS

PYRAMIDS

DECUSSATION
OF
INTERNAL ARCUATE
FIBERS

V, SPINAL NUCLEUS

V, SPINAL TRACT

VENTRAL
SPINOCEREBELLAR TR.

DORSAL
SPINOCEREBELLAR
TR.

NUCLEUS GRACILIS &
NUCLEUS CUNEATUS

CEREBRUM

THALAMUS

CEREBELLUM

MEDULLA

VENTRAL
SPINOTHALAMIC TR.

VENTRAL
SPINOCEREBELLAR TR.

FIG. 7.21. **A.** The principal long ascending (sensory) tracts (TR) of the cord and their terminations within the brain. The midline is indicated by the vertical dashed line. 1. Pathway for pain and thermal sensations which ascend in dorsal (anterolateral) spinothalamic tract of contralateral side. 2. Pathway for tactile sensations, proprioception, movement, and spatial discrimination. Signals ascend in posterior columns. 3. Pathway for touch and tactile localization. Impulses ascend on ipsilateral side for a variable number of segments in the posterior columns, then undergo decussation and ascend in ventral spinothalamic tract of contralateral side. 4, 5. Ventral and dorsal spinocerebellar tracts (see also Fig. 7.57). Some fibers of dorsal spinocerebellar tract undergo decussation to contralateral side, but are not shown in the figure. **B.** Cross section of the medulla at the level where the internal arcuate fibers undergo decussation. (**A**, After Best, Taylor [eds]. *The Physiological Basis of Medical Practice*, 8th ed,1966, Williams and Wilkins; **B.** After Everett. *Functional Neuroanatomy*, 6th ed, 1971, Lea & Febiger.)

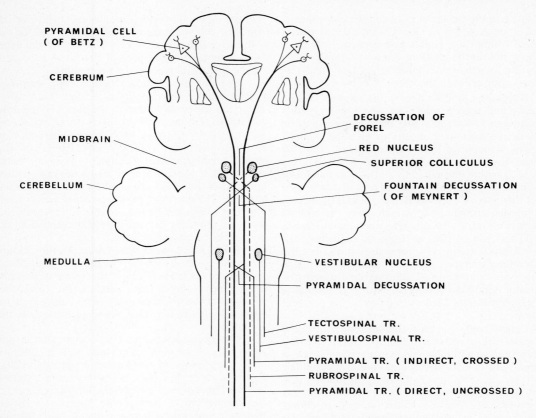

PYRAMIDAL CELL
(OF BETZ)

CEREBRUM

MIDBRAIN

CEREBELLUM

MEDULLA

DECUSSATION OF
FOREL

RED NUCLEUS

SUPERIOR COLLICULUS

FOUNTAIN DECUSSATION
(OF MEYNERT)

VESTIBULAR NUCLEUS

PYRAMIDAL DECUSSATION

TECTOSPINAL TR.

VESTIBULOSPINAL TR.

PYRAMIDAL TR. (INDIRECT, CROSSED)

RUBROSPINAL TR.

PYRAMIDAL TR. (DIRECT, UNCROSSED)

FIG. 7.22. The principal long descending (motor) tracts and their origins within the brain. See also Figures 7.34 and 7.76. (After Best and Taylor [eds]. *The Physiological Basis of Medical Practice,* 8th ed, 1966, Williams & Wilkins.)

the central nervous system. Thus the ascending projection tracts transmit sensory impulses to various higher centers (Table 7.5A; Fig. 7.21). The descending projection tracts, on the other hand, transmit motor impulses to spinal neurons (Table 7.5B; Fig. 7.22). Second, the short, intersegmental, or association tracts originate and terminate entirely within the spinal cord per se (Table 7.5C). Consequently these intrinsic tracts of the cord serve to connect and to integrate functionally (ie, associate) the same as well as different levels of the spinal cord proper. Collectively the short fiber tracts sometimes are called the fasciculi proprii.

The short fiber tracts of the cord have important roles in integrating as well as coordinating various spinal reflexes, as discussed in Chapter 6 (p. 214 et seq.). In this regard, it should be emphasized here that reflexes in vivo are not isolated events, and that any neural activity involving any particular reflex arc always modifies, and is modified by, other portions of the nervous system.

It must be stressed that at different levels of the spinal cord the ascending and descending tracts are not so neatly circumscribed and discrete as they appear in Figure 7.20. Rather, the fibers of the individual tracts exhibit considerable overlap and intermingling with the fibers of adjacent tracts. Furthermore, the sizes of the individual tracts vary

considerably at different levels of the cord and also may occupy slightly different positions relative to each other.

The salient neuroanatomic and functional characteristics of certain major ascending and descending tracts within the cord now may be considered individually.

The Ascending (Sensory) Fiber Tracts

1. *The Posterior White Columns:* The posterior white columns, or fasciculi gracilis and cuneatus, are important components of the dorsal column or funiculus. Anteriorly, the longer ascending fibers from the lower regions of the body are shifted medially by dorsal fibers entering the spinal cord at successively higher levels; thus in the cervical and thoracic regions the posterior funiculus is divided into a lateral fasciculus cuneatus and a medial fasciculus gracilis (Fig. 7.20A; Table 7.5). The fasciculus cuneatus thus contains sensory fibers originating from the upper thoracic and cervical ganglia, ie, the upper trunk, neck, and extremities. The fasciculus gracilis is composed of long ascending fibers arising from dorsal ganglia located in the sacral, lumbar, and lower thoracic regions of the cord. Therefore sensory impulses from the legs and lower extremities are carried in this group of fibers.

The fibers that comprise these two tracts may

terminate in the posterior gray column; however, a number of dorsal root fibers terminate in nuclei located within the medulla, and consequently form the first link in the chain of neurons which constitutes the longest spinal afferent pathway that terminates ultimately in the cerebral cortex.

Ascending dorsal fibers (uncrossed) of the cuneate and gracile tracts are first-order neurons which reach the medulla and there terminate upon cells of the nucleus cuneatus or else the nucleus gracilis, depending upon the particular tract in which they ascended. The second-order neurons originate within the cuneate and gracile nuclei of the medulla (Fig. 7.21). The axons that originate from the cells of these medullary nuclei then form the internal arcuate fibers which cross the midline and pass anteriorly in a bundle, the medial lemniscus. This crossed tract ascends through the pons and midbrain and finally terminates in the ventral posterolateral nucleus of the thalamus, whence third-order relay neurons send axons through the posterior region of the internal capsule to terminate on the appropriate sensory regions of the cerebral cortex.

Functionally speaking, this sensory pathway is the most important one in the cord for conveying highly specific and localized sensory impulses concerned with delicate two-point spatial discrimination (epicritic sensibility) as well as for touch, pressure, and kinesthesia. These sensory fibers also are enormously sensitive to the intensity of stimulation as well as to their temporal relationships. The "vibratory sense" is not a true modality of sensation, but rather is a temporal modulation of the tactile sense.

2. *The Dorsal (Posterior) Spinocerebellar Tract:* This tract consists principally of uncrossed fibers which arise from neurons located in the dorsal nucleus (Clarke's column) of the cord in all of the thoracic and the upper two lumbar segments. The dorsal spinocerebellar tract transmits proprioceptive impulses from the lower extremities and trunk to the cerebellum.

3. *The Ventral (Anterior) Spinocerebellar Tract:* This tract is situated along the lateral periphery of the spinal cord, anterior to the posterior spinocerebellar tract and posterior to the point of exit of the ventral root fibers.

The cells that give rise to the anterior spinocerebellar tract are diffuse and located in the anterior gray matter. The fibers are large and able to conduct impulses at velocities between 70 and 120 m/sec. Nearly all of the fibers cross at the same level of the cord at which they enter.

Two neurons are involved in this pathway to the cerebellum, these being the dorsal root ganglion cells which synapse upon neurons having their cell bodies in the ipsilateral ventral column at the base of the anterior and posterior horns, and the second-order neurons that decussate and ascend directly to the medulla and pons. At the upper pontine level the ventral spinocerebellar tract enters the cerebellum along the dorsal surface of the superior cerebellar peduncle. There a few of the spinocerebellar fibers undergo a second decussation, whereas the majority of the fibers terminate on the ipsilateral side of the cerebellum.

The anterior spinocerebellar tract transmits proprioceptive impulses from the four extremities as well as the trunk to the cerebellum, and the cord cells giving rise to this tract receive both excitatory and inhibitory impulses from ipsilateral and contralateral afferent inputs. The fibers of the spinocerebellar system presumably transmit information related to movement or posture of an entire limb rather than to changes of tension in individual muscles. No proprioceptive sensation arises from impulses conveyed in the anterior spinocerebellar tract, as impulses projected to the cerebellum do not reach the conscious level.

4. *The Cuneocerebellar Tract:* As the column of Clarke is not present above the C–8 level of the spinal cord, afferent fibers of the spinal nerves entering above this level ascend in the fasciculus cuneatus of the ipsilateral side, where they synapse upon cells in the accessory cuneate nucleus in the medulla.

The cuneocerebellar tract originates in the large cells of the cuneate nucleus, which structure is the equivalent of the posterior spinocerebellar tract for the upper limbs. The posterior external arcuate fibers of the cuneocerebellar tract then enter the cerebellum as part of the inferior cerebellar peduncle and terminate in the ipsilateral cerebellar cortex. This tract, in common with the dorsal spinocerebellar tract, transmits impulses from muscle spindles and tendon organs as well as touch and pressure receptors in the skin.

5. *The Dorsal (Posterior) Spinothalamic Tract:* Peripherally, the receptors for pain and thermal sensations send afferent impulses to small as well as to medium-sized dorsal root ganglion cells whose axons enter the spinal cord and synapse with neurons of the dorsal spinothalamic tract, which structure is related closely to the ventral spinothalamic tract. The cells that give rise to the dorsal spinothalamic tract within the cord also appear to arise in several laminae of the cord. The dorsal spinothalamic tract contains an abundance of fibers which pass directly to the ventral posterolateral nucleus of the thalamus like those of the ventral spinothalamic tract.

Dorsal spinothalamic fibers undergo an oblique decussation in the anterior white commissure in the same, or one higher, segment of the cord than their point of origin.

At higher brain-stem levels the dorsal spinothalamic tract also sends abundant collaterals into the tegmentum and reticular formation of the brain.

The dorsal spinothalamic tract is of great clinical importance, since its unilateral section produces a complete loss of pain and thermal sensations on the opposite side of the body extending to one segment below the lesion.

6. *The Ventral (Anterior) Spinothalamic Tract:* The origin of the cells whose processes constitute this tract is obscure, although they appear to originate in several laminae within the spinal cord. It appears well established, however, that the ventral spinothalamic fibers cross in the anterior white commissure, and such decussation takes place gradually across several consecutively ascending spinal segments so that the fibers of this tract ultimately ascend contralaterally in both the anterior and the anterolateral funiculi.

Fibers of the anterior spinothalamic tract ascend uninterrupted by synapses directly to the thalamus where they terminate upon cells of the ventrolateral (posterolateral) nucleus; thence second-order neurons relay impulses directly to the cerebral cortex.

The ventral spinothalamic tract conveys sensory impulses concerned with light touch such as is evoked by stroking glabrous (hairless) skin with a feather. This modality of sensation supplements the deep pressure and discriminatory tactile senses whose afferent impulses are carried in the posterior white columns of the spinal cord.

7. *The Spinotectal Tract:* The cells which give rise to this small tract are unknown. After crossing within the spinal cord, the fibers ascend in an intimate relationship with those of the spinothalamic system, and at the level of the midbrain the spinotectal tract sends fibers medially into the superior colliculus and lateral areas of the central gray substance.

Functionally, the spinotectal tract is believed by some workers to be part of a multisynaptic pathway for transmitting painful (nociceptive) stimuli, whereas others feel that this tract acts in some way to coordinate visuospinal reflexes.

8. *The Dorsal (Posterior) Spinocerebellar Tract:* Certain dorsal root afferent fibers enter the cord and there synapse upon the large cells of Clarke's column (the dorsal nucleus) within the spinal cord. The cells of the dorsal nucleus extend from segments C–8 to L–3 and give rise to the large ascending fibers of the dorsal spinocerebellar tract which ascend uncrossed on the ipsilateral side for the full length of the spinal cord. Within the medulla, the fibers of the dorsal spinocerebellar tract join the posterior cerebellar peduncle, enter the cerebellum, and there terminate in both the anterior and posterior regions of the vermis.

The dorsal spinocerebellar tract is a two-neuron pathway, and the fibers exhibit conduction velocities ranging from 30 to 100 m/sec.

Functionally, sensory impulses arising from muscle spindles and Golgi tendon organs in the lower regions of the body and the legs are transmitted in the dorsal spinocerebellar tract, and these impulses are conducted not only at high rates but also at high frequencies. This tract also requires little spatial summation for discharge. Impulses that are generated by certain exteroceptive stimuli, such as pressure and touch in the skin, also appear to be transmitted by this tract. Thus the posterior spinocerebellar tract conveys impulses from peripheral sensory receptors directly to the cerebellum, these sensory impulses being involved in the delicate coordination of posture and movement of individual muscles in the limbs. There are no higher levels than the cerebellum involved; hence coordination by this tract takes place entirely at a subconscious level.

9. *The Spinoreticular Fibers:* These fibers originate at all levels of the brain stem, presumably from posterior horn cells, and project sensory impulses from the spinal cord to wide areas of the reticular formation of the brain stem after ascending in the anterolateral funiculus. Those fibers which terminate in the medullary reticular formation are largely uncrossed, and they terminate mainly upon cells of the nucleus reticularis gigantocellularis as well as upon parts of the lateral reticular nucleus. The latter nucleus in turn projects fibers to portions of the cerebellum; thus exteroceptive impulses may reach the cerebellum in part through this pathway.

Functionally, the spinoreticular fibers play an important role in the ascending reticular complex, which is involved in the maintenance of consciousness.

10. *Miscellaneous Ascending Fiber Systems:* Spinocortical, spino–olivary, spinovestibular, and spinopontine tracts have been demonstrated experimentally; however, the physiologic significance of these tracts is obscure and largely conjectural at present.

The Descending (Motor) Tracts. The principal descending or motor tracts, their origins, course, terminations, and general functional attributes will be reviewed in this section.

1. *The Pyramidal or Corticospinal Tracts:* These tracts constitute the largest as well as the most important descending nerve fiber system within the human nervous system. They are composed of all fibers which orginate from cells located in the motor region (precentral gyrus) of the cerebral cortex, pass through the pyramid of the medulla, and then enter the spinal cord.

Each of these tracts is composed of over a million individual fibers, 70 percent of which are myelinated; 90 percent of the myelinated fibers are 1μ to 4μ in diameter, whereas most of the remainder are about 5μ to 10μ in diameter. There are also approximately 35,000 large fibers having diameters ranging from 10μ to 22μ in diameter, and these arise mainly from the giant pyramidal cells (of Betz) in the precentral gyrus (Fig. 7.22).

The cortical motor fibers converge in the corona radiata and descend via the internal capsule, crus cerebri, pons, and medulla. At the level of the medulla, the corticospinal tracts emerge to the surface in the pyramid, whereas at the junction of medulla and spinal cord an incomplete decussation of the fibers takes place which gives rise to three tracts, a large lateral or crossed corticospinal tract, an anterior uncrossed corticospinal tract, and a small uncrossed lateral corticospinal tract.

Between 75 and 90 percent of the pyramidal fibers cross in the pyramidal region of the medulla and pass down the cord in the posterior area of the lateral funiculus as the lateral (crossed) corticospinal tract. This tract extends to the most caudal portions of the spinal cord, but becomes progressively smaller as more fibers exit from the gray matter caudally.

The uncrossed ventral or direct pyramidal tract (bundle of Türck) is smaller than the lateral corticospinal tract, and usually it extends only to the upper thoracic region of the spinal cord, although it may occasionally be found still lower. This tract innervates principally the muscles of the upper extremities and neck.

The anterolateral pyramidal tract (of Barnes) is comprised of the remainder of the uncrossed fibers, and this tract passes downward in the more anterior regions within the lateral funiculus of the cord.

The axons of the three corticospinal tracts enter the dorsolateral regions of the spinal gray matter and there terminate on internuncial neurons or directly upon anterior horn motoneurons. These synaptic junctions may take place either before or after the pyramidal fibers cross to the opposite or contralateral side of the cord.

Approximately 55 percent of the total number of pyramidal fibers that are present end in the cervical spinal cord, whereas 20 percent terminate in the thoracic region and 25 percent in the lumbosacral segments. These figures indicate much greater motor control by the pyramidal tracts of the skeletal muscles in the upper extremity as compared to the lower extremity. Destruction of the pyramidal tracts produces motor failure (ie, muscular paralysis) in those regions innervated by these tracts that lie distal to the injury, as well as a loss of muscle tone and local myotatic reflexes in the affected region.

Myelination of the corticospinal tracts commences at about the time of birth, a process which is not complete until around the end of the second year of postnatal life, a fact that is in accordance with the progressively more sophisticated coordination of voluntary movements as the infant develops.

2. *The Rubrospinal Tract:* The rubrospinal tract is a small bundle of fibers arising from the large cells in the posterior region of the red nucleus (nucleus magnocellularis) which is located within the central portion of the mesencephalic tegmentum. The rubrospinal fibers emanate from the medial border of the red nucleus, cross immediately in the ventral tegmental decussation, and then pass downward to the spinal cord where the tract lies anterior to, and partially mingled with, fibers of the lateral corticospinal tract. The rubrospinal tract has not been demonstrated to pass lower than the thoracic region in man.

The rubrospinal tract gives off some collateral fibers (as it passes through the brain stem) to the cerebellum, the facial nucleus, and the medullary lateral reticular nucleus. Uncrossed efferent fibers also arise from the red nucleus which project to portions of the inferior olivary nucleus, the so-called rubrobulbar fibers. Fibers from both the cerebellum and the cerebral cortex also pass to the red nucleus where they synapse on nerve cell bodies as well as the dendrites thereof. Those from the motor regions of the cortex pass ipsilaterally, whereas the efferent fibers from the cerebellum cross the superior cerebellar peduncle. The corticorubral and rubrospinal fibers together constitute a nonpyramidal motor pathway that connects the cortex to various spinal levels.

Rubrospinal fibers do not terminate directly upon anterior horn motoneurons; consequently their impulses are transmitted either to spinal internuncial neurons which then facilitate the activity of alpha motoneurons, or else they affect gamma motoneurons which in turn modify the excitability of the alpha motoneurons indirectly.

Physiologically, the primary function of the rubrospinal tract is believed to be control of the tonus in various groups of flexor muscles.

3. *The Vestibulospinal Tract:* The vestibulospinal tract originates in the lateral vestibular nucleus of the pontine-medullary junction, and it receives fibers from the vestibular division of the vestibulocochlear nerve (cranial nerve VIII) as well as from certain cerebellar regions. The vestibulospinal tract then descends uncrossed throughout the entire length of the spinal cord.

Vestibulospinal fibers are most numerous in the cervical and lumbar spinal segments. The fibers of this tract enter the gray matter where they form axodendritic and axosomatic synapses. Apparently no vestibulospinal fibers terminate directly upon spinal motoneurons, except for a few such terminations in the thoracic region.

Impulses relayed by this important tract originate in the vestibular portion of the ear as well as in specific areas of the cerebellum. Efferent cerebellar fibers from the vermis as well as the fastigial nuclei terminate upon the lateral vestibular nucleus.

Functionally, the vestibulospinal tract exerts a facilitory influence upon reflex activity in the spinal cord as well as upon the mechanisms which control muscle tone. The most important vestibular component, however, is derived from impulses that originate in the inner ear, and which mediate equilibrium as well as posture of the body in a coordinated fashion.

4. *The Reticulospinal Tracts:* The reticulospinal tracts are a system of delicate fibers which have their origin principally in the cells of the medial two-thirds of the reticular formation in the pons and medulla. The reticulospinal tracts descend in the anterior and anterolateral regions of the spinal cord. The fibers originating from the pontine reticular formation traverse the ipsilateral side of the cord almost exclusively, whereas those originating within the medullary reticular formation cross to the opposite side of the cord.

The pontine reticular fibers are the major descending component of the medial longitudinal fasiciculus in the brain stem. Within the spinal cord these fibers descend primarily in the medial portion of the anterior funiculus.

The bilateral medullary reticulospinal fibers are uncrossed for the most part, and are found principally in the anterior portion of the lateral funiculus, although there is no sharp demarcation from the pontine reticulospinal fibers within the spinal cord.

Both pontine and medullary fiber systems apparently traverse the entire length of the spinal cord, and are believed to convey impulses which facilitate motoneurons that innervate the extensor muscles.

Reticulospinal fibers from the medulla terminate in portions of the gray laminae which also receive fibers from corticospinal and rubrospinal tracts. Furthermore, autonomic fibers that arise from the more cephalic regions of the central nervous system also appear to be associated with both of these tracts. These autonomic fibers terminate on visceral motor cells of the gray matter.

Physiologically, the actions of the reticulospinal tract are not altogether clear, although they have been shown to exert certain functional effects under experimental conditions. These effects include inhibition or facilitation of voluntary movement and reflex activity; alterations in muscle tone; modification of respiratory patterns, inspiratory as well as expiratory; pressor and depressor effects on the cardiovascular system; and depressor effects upon the central nervous transmission of sensory impulses. Note that these responses include both somatic as well as visceral (ie, autonomic) components, a fact which correlates with the close anatomic association of reticulospinal and autonomic fibers.

5. *The Tectospinal and Tectobulbar Tracts:* The superior colliculus (which is primarily an optic center) gives rise in its deeper layers to fibers of the tectospinal and tectobulbar tracts. The fibers giving rise to these tracts pass anteromedially, and cross the midline in the dorsal tegmental decussation, descend to medullary levels where the tectospinal fibers enter the medial longitudinal fasciculus, and then pass into the most anterior portion of the anterior funiculus. Most of these motor fibers terminate in the upper four cervical spinal segments, where they synapse on interneurons rather than directly on spinal motoneurons.

The tectospinal fibers are believed to mediate reflex postural movements in response to visual and auditory stimuli.

The tectobulbar fibers pass bilaterally to posterior regions of the mesencephalic reticular formation and to the contralateral pontine and medullary reticular formations.

6. *The Medial Longitudinal Fasciculus:* This composite tract contains descending fibers originating within the medial vestibular nucleus, the reticular formation, the superior colliculus (tectospinal fibers), and the interstitial nucleus of Cajal (interstitiospinal fibers). This tract is well defined only in the upper cervical segments of the spinal cord, although certain of its fibers have been discerned in the sacral region. It is believed that the fibers of the medial longitudinal fasciculus terminate on interneurons located in the medial gray matter of the anterior horn. Fibers of the pontine reticulospinal tract form the largest component of the medial longitudinal fasciculus.

All of the descending tracts described up to this point form links in chains of motoneurons whereby excitatory or inhibitory impulses from higher brain centers ultimately reach the striated muscles of the body via the anterior horn motoneurons.

Autonomic Pathways in the Spinal Cord. Descending visceral (motor) fibers which originate in the important autonomic centers of the hypothalamus, tegmentum of the midbrain, pons, and medulla pass down the spinal cord in the intermediolateral column as well as within the anterior and anterolateral white matter in addition to other diffuse pathways; probably these pathways are interrupted along their course by internuncial or relay neurons. Preganglionic cells for the ultimate innervation of such visceral structures as the heart, smooth muscle, and glandular tissue also descend in the reticulospinal tracts as well as in the fasciculi proprii.

The Fasciculi Proprii. The term "fasciculi proprii" includes a number of shorter fiber systems that lie wholly within the spinal cord, and which are concerned with intrinsic segmental spinal reflexes such as were described in Chapter 6 (p. 213). Such unconscious reflex control of muscular contraction is important during movement and for the maintenance of posture.

Most dorsal root axons involved in simple reflex arcs do not send collaterals directly to anterior horn motoneurons; rather, synaptic connection to motoneurons usually is made indirectly through one or more interposed (internuncial, intercalated) neurons that lie between the afferent and efferent neuronal processes. The internuncial cells then send processes either to motoneurons at the same, higher, and/or lower segmental levels within the spinal cord, thus completing the reflex arc. These ascending and descending intersegmental axonic processes, crossed and uncrossed, which interconnect the same as well as different levels in the spinal cord, also are known as spinospinal columns, and they are found in all of the funiculi of the cord.

The Lower Motoneuron and the Upper Motoneuron. The axons that arise from the anterior horn cells of the spinal cord and which provide motor innervation for all of the skeletal muscles of the body constitute an anatomic and physiologic unit which is called the *final common path* or *lower motoneuron.* In turn, the neurons of the anterior horn can be modified functionally (ie, facilitated or inhibited) by impulses that arise in dorsal root afferent fibers as well as in descending fiber systems, and which converge upon the motoneurons (Fig.

6.38, p. 218). All of the descending fiber systems of the spinal cord as well as certain sensory impulses can influence the functional activity of the lower motoneuron, and collectively these systems sometimes are referred to as the *upper motoneuron*. However, the corticospinal (pyramidal) tracts are of such enormous physiologic and clinical importance that these motor pathways alone sometimes are considered to be the *upper motoneuron*.

The extrapyramidal descending fibers of such pathways as the rubrospinal, vestibulospinal, reticulospinal, and tectospinal tracts regulate the neural mechanisms that affect muscle tone, posture, and reflex activity as well as many automatic synergistic movement patterns of the body. Consequently these extrapyramidal regulatory mechanisms act in a cooperative fashion with the highly sophisticated and precise control of voluntary motor activities that are mediated via the corticospinal fiber systems (p. 267). Therefore for the purposes of this text the term *upper motoneuron* will include all of these convergent physiologic mechanisms, pyramidal as well as extrapyramidal.

THE AUTONOMIC NERVOUS SYSTEM. The material discussed in the previous section dealt principally with the sensorimotor pathways that underlie certain conscious sensations as well as the voluntary and reflex activity of skeletal muscle. The autonomic or vegetative nervous system, on the other hand, functions automatically and primarily below the conscious level. The autonomic division of the peripheral nervous system is distributed to smooth muscle and glands throughout the body as well as to the heart.

The autonomic nervous system exhibits a highly integrated structure and it also functions in a smoothly integrated manner with the rest of the nervous system. From an anatomic standpoint, the autonomic system is divided into the sympathetic nervous system and the parasympathetic nervous system, in accordance with the location of the preganglionic nerve cell bodies. There also are marked functional as well as anatomic differences between the sympathetic and parasympathetic nervous systems.

It is necessary to stress the fact that both of these divisions of the autonomic nervous system are strictly motor (efferent) in function (GVE fibers, Tables 7.2 and 7.3). However, the afferent limbs of certain visceral reflex arcs are composed of fibers which transmit sensory impulses to the central nervous system from peripheral organs and structures, and anatomically these general visceral afferent fibers may accompany the corresponding autonomic nerves (GVA, Tables 7.2 and 7.3).

Functionally, the autonomic nervous system contributes in large measure to the maintenance of the constancy of the internal environment, ie, homeostasis (p. 48).

General Structure of Autonomic Nerves. The autonomic nerves are characterized by a two-neuron chain. The cell body of the primary (pre-ganglionic or presynaptic) neuron is located within the central nervous system. The axon of the primary neuron extends peripherally to synapse upon a secondary (postganglionic or postsynaptic) neuron that is found within one of the many autonomic ganglia that are located outside of the central nervous system. In turn, the axon of the postganglionic fiber terminates within a particular organ or structure.

The total number of postganglionic neurons is considerably greater than that of the preganglionic neurons, the ratio being roughly 30 to 1. Thus one preganglionic neuron can discharge a much larger number of postganglionic cells so that the activity of a relatively large peripheral effector area is controlled by the excitation obtained from the discharge of only a few preganglionic neurons. Consequently the anatomic arrangement between the preganglionic and postganglionic fibers results in a functional amplification of the effector response in the peripheral tissue or end organ.

The Sympathetic Nervous System. The sympathetic, or thoracolumbar, division of the autonomic nervous system arises from preganglionic nerve cell bodies that lie within the intermediolateral cell column of segments T–1 through T–12 plus L–1 through L–3 or L–4 of the spinal cord. The approximate location of the intermediolateral cell column is indicated in Figure 7.20A as the lateral ground bundle (LGB). The sympathetic preganglionic fibers originating in these cells are myelinated for the most part. After these fibers emerge from the cord in the ventral roots, the axons form the white rami communicantes of the thoracic nerves (Fig. 7.23), which in turn send fibers to the paired trunk (paravertebral or vertebral) ganglia of the sympathetic chain which are found laterally on each side of the thoracic and lumbar vertebrae. The general anatomic relationships of the preganglionic and postganglionic sympathetic neurons to their various ganglia and end organs are depicted in Figures 7.24 and 7.26.

The fibers of the white rami may take one of several courses after they enter the trunk ganglia. First, the fibers may synapse directly upon a group of ganglion cells. Second, the fibers may pass up or down the sympathetic trunk and synapse with ganglion cells at a higher or lower level (Fig. 7.23). Third, the fibers may pass directly through the trunk ganglia to synapse within an unpaired collateral (or prevertebral) ganglion (eg, the celiac or inferior mesenteric ganglia, Fig. 7.24).

The nerve branches that arise from the sympathetic trunk may be placed in one of two general categories:

First are the branches that are composed principally of unmyelinated postganglionic fibers, the gray rami communicantes, which pass to all of the spinal nerves (Fig. 7.24).* These gray rami inner-

*Although some general visceral afferent fibers (GVA, Table 7.2) are present within both gray and white rami, these sensory fibers do not belong to the autonomic nervous system.

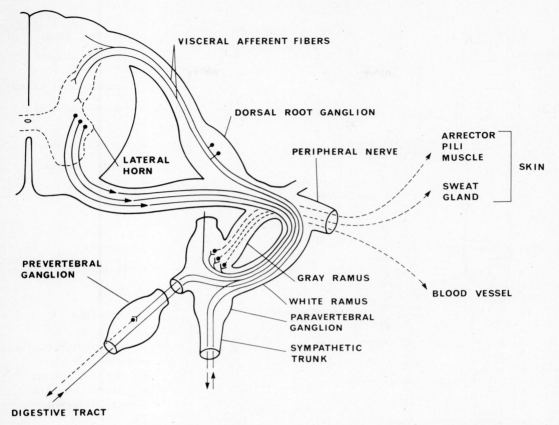

FIG. 7.23. The general pathways involved in sympathetic reflex arcs. Solid lines indicate preganglionic fibers; dashed lines indicate postganglionic fibers. (After Truex and Carpenter. *Human Neuroanatomy,* 6th ed, 1969, Williams & Wilkins.)

vate blood vessels (vasomotor nerves), pilomotor smooth muscles that erect the hairs, and the sweat glands; ie, the somatic areas throughout the body. Branches from the superior cervical ganglion contribute to the formation of plexuses of sympathetic nerves about the internal and external carotid arteries; thence, sympathetic nerves to the head region arise. The three pairs of cervical sympathetic ganglia (the superior, middle, and inferior) give rise to the superior cardiac nerves, which pass to the cardiac plexus located at the base of the heart. From this plexus accelerator fibers are distributed to the myocardium. Sympathetic branches that arise from T–1 through T–5 send vasomotor fibers to the aorta and to the posterior pulmonary plexus. The posterior pulmonary plexus in turn sends fibers to the lungs which dilate the bronchi in these organs.

Second, the chiefly myelinated preganglionic fibers of the white rami communicantes provide the other general category into which the nerve branches of the sympathetic trunk are placed. The splanchnic nerves arise from T–5 through T–12 and then pass to the celiac and superior mesenteric ganglia. Within these ganglia, the preganglionic sympathetic neurons synapse, and the axons of the postganglionic neurons now innervate the abdominal viscera after traversing the celiac plexus (Fig. 7.24). The trunk ganglia in the lumbar region give rise to the lumbar splanchnic nerves, and the preganglionic fibers of these neurons arise within the inferior mesenteric ganglia in addition to the small ganglia located within the hypogastric plexus. Thence postganglionic sympathetic fibers are distributed to the lower abdominal and pelvic viscera (Fig. 7.24).

The Parasympathetic Nervous System. The parasympathetic, or craniosacral, division of the autonomic nervous system has its origin in the cell bodies of preganglionic neurons that are located within the brain stem as well as within the middle segments of the sacral region of the cord (S–2, S–3, and S–4). In contrast to the sympathetic nervous system, the parasympathetic nervous system is distributed exclusively to visceral organs and structures. Again in contrast to the sympathetic nerves, most of the parasympathetic preganglionic fibers pass in an uninterrupted fashion from their central point of origin to the wall of the end organ which they innervate, there to synapse upon the cell bodies of their postganglionic neurons (Fig. 7.25).

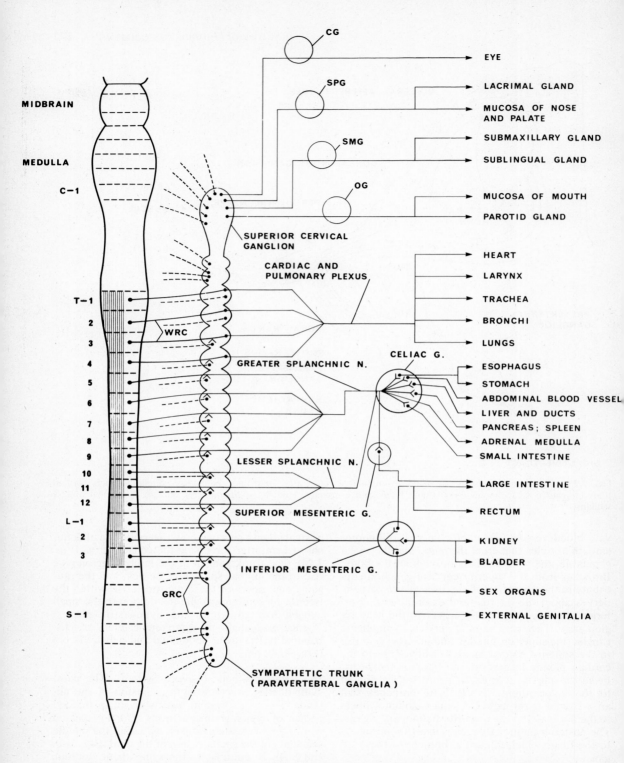

FIG. 7.24. Highly schematic representation of the sympathetic division of the autonomic nervous system. Shaded area in spinal cord at left indicates levels from which sympathetic nerves emerge. The cell bodies of the sympathetic nerves lie in the intermediolateral cell column of the cord. G = ganglion; N = nerve; CG = ciliary ganglion; SPG = sphenopalatine ganglion; SMG = submaxillary ganglion; OG = otic ganglion; WRC = white rami communicantes, indicated by solid lines leaving spinal cord and entering sympathetic trunk; GRC = gray rami communicantes, indicated by dashed lines leaving the sympathetic trunk to become components of all spinal nerves. The collateral (or prevertebral) ganglia include the celiac as well as the superior and inferior mesenteric ganglia. (Data from Best, Taylor [eds]. *The Physiological Basis of Medical Practice,* 8th ed, 1966, Williams & Wilkins; Chusid. *Correlative Neuroanatomy and Functional Neurology,* 14th ed, 1970; Lange Medical Publications; Goodman, Gilman. *The Pharmacological Basis of Therapeutics,* 4th ed, 1970, Macmillan.)

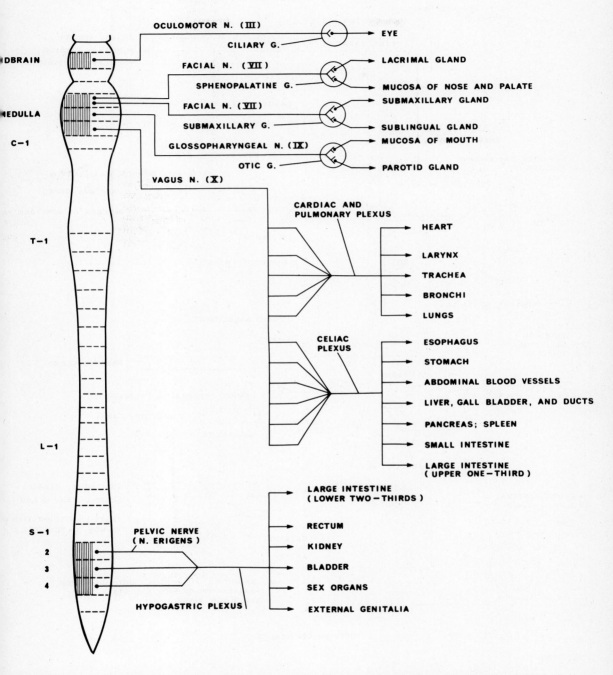

FIG. 7.25. Highly schematic representation of the parasympathetic division of the autonomic nervous system. The shaded areas in the spinal cord at the left indicate the brain and cord levels where the parasympathetic nerves emerge. G = ganglion. Note that parasympathetic preganglionic fibers, in contrast to sympathetics, synapse in the ciliary, sphenopalatine, submaxillary, and otic ganglia. Also, in contrast to the sympathetic system, most preganglionic parasympathetic fibers synapse within the viscera innervated, so that the postganglionic fibers are exceedingly short. (Data from Best and Taylor [eds]. *The Physiological Basis of Medical Practice,* 8th ed, 1966, Williams & Wilkins; Chusid. *Correlative Neuroanatomy and Functional Neurology,* 14th ed, 1970, Lange Medical Publications; Goodman and Gilman. *The Pharmacological Basis of Therapeutics,* 4th ed, 1970, Macmillan.)

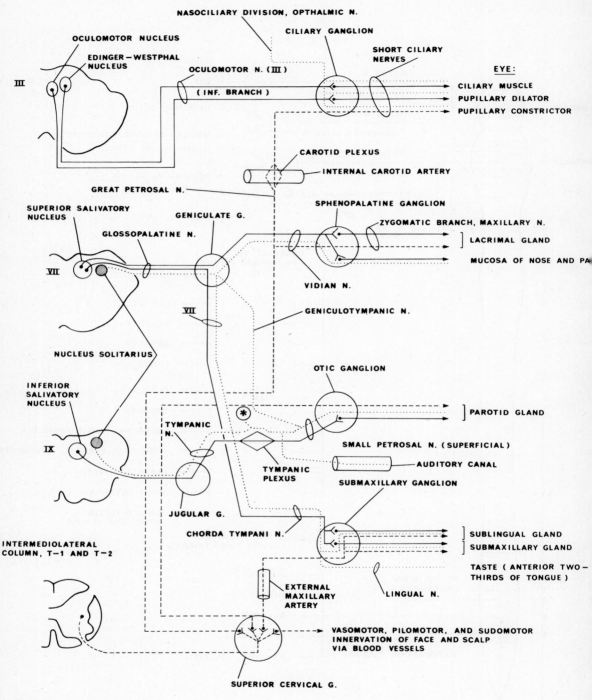

FIG. 7.26. Highly schematic diagram illustrating the autonomic (sympathetic and parasympathetic) innervation of the head. G = ganglion; N = nerve;* = fibers of the trigeminal nerve which course through the otic ganglion to the tensor tympani and palatine muscles. (Data from Chusid. *Correlative Neuroanatomy and Functional Neurology,* 14th ed, 1970, Lange Medical Publications.)

In the intestinal tract, synaptic connection of pre-ganglionic and postganglionic neurons takes place in the plexuses of Meissner and Auerbach (p. 828). The terminal parasympathetic ganglia consist of small accumulations of neurons that are situated close to the structure that is innervated. In particular, this intimate relationship exists in the pelvic organs, eg, the rectum and bladder.

The cranial nerves that contain para-sympathetic preganglionic fibers include the oculomotor (III), facial (VII), glossopharyngeal (IX), and vagus (X), as summarized in Table 7.3 (GVE fibers) and shown in Figures 7.25 and 7.26. The autonomic fibers of these cranial nerves inner-vate particular visceral structures within the head as well as the thoracic and abdominal cavities, whereas the three sacral spinal nerves give rise to the pelvic nerve (nervus erigens) which innervates most of the large intestine, the pelvic viscera, and the genitalia.

The neuroanatomy of the nuclei that give rise to the cranial parasympathetic preganglionic fibers will be discussed further in relation to the struc-tures within the particular regions of the brain stem in which these fibers arise.

Neurotransmission within the Autonomic Nervous System. Acetylcholine (p. 195) is the neuro-transmitter substance responsible for impulse transmission within all autonomic ganglia (sym-pathetic as well as parasympathetic), at all post-ganglionic parasympathetic nerve terminals, and at sympathetic postganglionic nerve terminals in sweat glands (as well as at all motor nerve endings in skeletal muscle). Consequently these structures are referred to as *cholinergic*.

Norepinephrine (or noradrenalin, p. 195) is the chemical mediator at most postganglionic sympathetic nerve terminals (except the sweat glands); hence such terminals are referred to as *adrenergic*.

General Functions of the Autonomic Nervous System. The principal effects obtained by stimula-tion of particular sympathetic and parasympathetic nerves are summarized in Table 7.6. The reader will note that in many instances the stimulation of sympathetic and parasympathetic nerves which innervate the same organ elicit diametrically oppo-site responses.

In vivo, of course, the autonomic nervous sys-tem functions spontaneously, and usually below the conscious level, to provide efferent neural control over a broad range of mechanisms that subserve homeostasis. It must be realized that in the normal body there is an exquisite balance between the ac-tivities of the sympathetic and parasympathetic systems, and that a reflex increase or decrease in the activity of one or both of these systems caused by alterations in the external or internal environ-ments can alter profoundly the net functional state of many structures and organs.

In general, an overall activation of the sym-pathetic system prepares the individual for intense muscular activity such as is required in defense or offense, the so-called fight or flight reaction. Para-sympathetic activity, on the other hand, is con-cerned primarily with those mechanisms respon-sible for maintaining "resting" bodily functions, such as lowered heart rate, promotion of digestive activities, ridding the body of liquid and solid wastes, constriction of the pupil in order to shield the retina from excessive light. Sexual activity, on the other hand, requires a sequential and coordi-nated activation of both sympathetic and parasym-pathetic nerves.

It is important to realize that the functions of the autonomic nervous system are not mutually exclusive of somatic neural activities. Rather, the autonomic nerves participate in a smoothly inte-grated fashion with many somatic reflexes, and con-versely many somatic reflexes may have strong vis-ceral components. Numerous specific examples of visceral and somatovisceral reflexes will be dis-cussed in context throughout the remainder of this book, including the pupillary, respiratory, cardio-vascular, swallowing, vomiting, and genital reflexes.

The Brain Stem

There is some disagreement among various authors as to which structures constitute the brain stem. For the purposes of this text, the following struc-tures will be discussed under this heading: the medulla oblongata, the pons, the mesencephalon (midbrain), and the diencephalon, including the thalamus and its subdivisions, as well as the hypothalamus (Table 7.1). The general anatomic relationships among these structures are shown in Figures 7.11, 7.12, 7.27, 7.28, and 7.29.

All of the cranial nerves are attached to the brain stem (Figs. 7.12 and 7.29), excepting the ol-factory nerve and the spinal portion of the accessory nerve (Table 7.3).

THE MEDULLA OBLONGATA. The medullary region of the brain stem is an extremely important physiologic center for the initiation and coordina-tion of many involuntary (vegetative) reflexes es-sential to the maintenance of life.*

The medulla is superficially continuous with the spinal cord and it extends anteriorly about 2.5 cm to the caudal region of the pons. The central canal of the spinal cord is continuous within the medulla, but this channel expands dorsally at the obex into the large cavity of the fourth ventricle (Fig. 7.27). The roof of the fourth ventricle, the choroid plexus, is made up of a thin sheet of epen-dyma and pia mater (p. 252).

Externally, the pyramids form two longitudinal ridges on the ventral surface of the medulla on each side of the anteromedian fissure. These structures contain the pyramidal (corticospinal) tracts; the de-

---

The medulla oblongata once was called the "bulb," and the term "bulbar" referred to this part of the brain stem; these terms now are obsolete.

Table 7.6 EFFECTS OF AUTONOMIC NERVE STIMULATION[a]

ORGAN OR STRUCTURE	SYMPATHETIC (ADRENERGIC) EFFECTS	PARASYMPATHETIC (CHOLINERGIC) EFFECTS	TYPE OF RECEPTOR[b]
Eye			
Iris: radial muscle	Contraction; dilates pupil (mydriasis)	Contraction; constricts pupil (miosis)	α
Iris: circular muscle			—
Ciliary muscle	Relaxation; lens flattens, accommodates for distant vision (minor effect)	Contraction; lens thickens, accommodates for near vision (strong effect)	β
Smooth muscle: orbit, upper lid	Contraction		—
Heart			
Sino-atrial (SA) node	Increased heart rate (tachycardia)	Decreased heart rate (bradycardia); vagal arrest	β
Atria	Increased contractility, conduction velocity	Decreased contractility, some increase in conduction velocity	β
Atrioventricular (AV) node, conducting system	Increased conduction velocity	Decreased conduction velocity, AV block	—
Ventricles	Increased rates of idiopathic pacemakers, contractility, conduction velocity	Decreased contractility (?)	β
Blood Vessels			
Cutaneous (skin)	Constriction	Dilatation (minor; doubtful physiologic significance)	α
Buccal mucosa	Dilatation		
Coronary	Dilatation, esp. in vivo; constriction (minor effect)[c]	Constriction (minor); dilatation (strong)	α, β
Skeletal muscle	Dilatation[d], constriction[d]	Dilatation[e]	α, β
Cerebral	Constriction (minor effect)	Dilatation (minor; doubtful physiologic significance)	α
Pulmonary	Dilatation, constriction (minor effects)	Dilatation (minor; doubtful physiologic significance)	α, β
Abdominal, pelvic viscera	Constriction (strong); dilatation[d]		α, β
External genitalia	Constriction	Dilatation	—
Salivary glands	Constriction	Dilatation	α
Lung			
Bronchial muscle	Dilatation	Constriction	β
Bronchial glands	Inhibition (?)	Stimulation	—
Stomach			
Stomach wall, motility and tone	Decrease or increase (variable response)	Increase or decrease (variable response)	β
Sphincters	Contraction (generally)	Relaxation (generally)	α
Secretion	Inhibition (?)	Stimulation (marked)	—

Effector organ	Cholinergic impulses response	Adrenergic impulses response	Adrenergic receptor type[b]
Intestine			
Motility and tone	Increase (marked)	Decrease (slight)	α, β
Sphincters	Relaxation (generally)	Contraction (generally)	α
Secretion	Stimulation	Inhibition (?)	—
Gallbladder, Ducts	Contraction	Relaxation	β
Urinary Bladder			
Detrusor	Contraction (generally)	Relaxation (generally)	α
Trigone, sphincter	Relaxation	Contraction	—
Ureter			
Motility and tone	Increase (?)	Increase (generally)	α, β
Uterus			
Nonpregnant		Inhibition[f]	—
Pregnant		Contraction[f]	—
Sex Organs	Erection	Ejaculation	α
Skin			
Pilomotor muscles		Contraction	α
Sweat glands	General secretion (strong response)	Slight local secretion[g]	α
Splenic Capsule		Contraction	—
Adrenal Medulla		Catecholamine secretion (epinephrine, norepinephrine)	—
Pancreas, Acini	Secretion		α
Liver		Glycogenolysis	—
Salivary Glands	Copious, watery secretion	Viscous secretion[h]	—
Lacrimal Glands	Secretion		
Nasopharyngeal Glands	Secretion		

[a] Data from Goodman, Gilman: The Pharmacological Basis of Therapeutics, 4th ed. Macmillan, 1970; Best and Taylor: The Physiological Basis of Medical Practice, 8th ed. Williams & Wilkins, 1966.

[b] These receptor types are based upon the response of various tissues to the action of various sympathomimetic amines, ie, compounds which initiate or mimic sympathetic nerve stimulation. In general, α receptors are most sensitive to epinephrine and least sensitive to isoproterenol; β receptors, on the other hand, are most sensitive to isoproterenol and least sensitive to epinephrine, the effect being inhibitory.

[c] Indirect effects in vivo cause primarily a vasodilatation.

[d] In vivo, released epinephrine produces a β receptor response; ie, vasoconstriction in blood vessels of skeletal muscles and liver. In other abdominal viscera an α response, ie, vasodilatation, is elicited by this catecholamine.

[e] Sympathetic cholinergic nerves induce vasodilatation in skeletal muscle.

[f] Response quite variable, depending on stage of menstrual cycle, and levels of circulating sex hormones (estrogen and progesterone) among other factors. Pregnant and nonpregnant uteri differ in their responses.

[g] For example, palms of hands (so-called adrenergic sweating).

[h] The parotid glands are not innervated by sympathetic nerves.

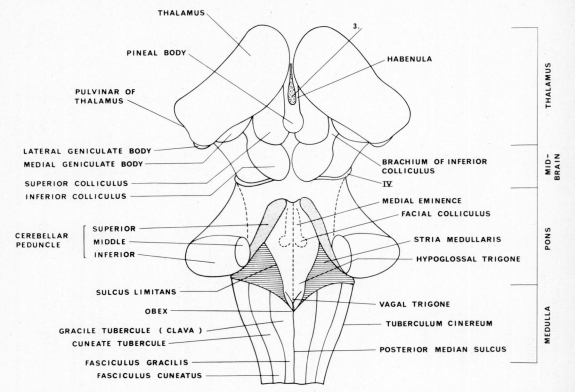

THALAMUS

PINEAL BODY

PULVINAR OF
THALAMUS

LATERAL GENICULATE BODY

MEDIAL GENICULATE BODY

SUPERIOR COLLICULUS

INFERIOR COLLICULUS

CEREBELLAR   SUPERIOR
PEDUNCLE   MIDDLE
             INFERIOR

SULCUS LIMITANS

OBEX

GRACILE TUBERCULE (CLAVA)

CUNEATE TUBERCULE

FASCICULUS GRACILIS

FASCICULUS CUNEATUS

3.

HABENULA

BRACHIUM OF INFERIOR
COLLICULUS

IV

MEDIAL EMINENCE

FACIAL COLLICULUS

STRIA MEDULLARIS

HYPOGLOSSAL TRIGONE

VAGAL TRIGONE

TUBERCULUM CINEREUM

POSTERIOR MEDIAN SULCUS

THALAMUS

MID-
BRAIN

PONS

MEDULLA

FIG. 7.27.   The gross anatomic features of the brain stem from the dorsal aspect. 3 = third ventricle. (Data from Chusid. *Correlative Neuroanatomy and Functional Neurology,* 14th ed, 1970, Lange Medical Publications; Everett. *Functional Neuroanatomy,* 6th ed, 1971, Lea & Febiger; Truex and Carpenter. *Human Neuroanatomy,* 6th ed, 1969, Williams & Wilkins.)

cussation of these fibers terminates the fissure caudally (Figs. 7.28 and 7.29).

The ventrolateral sulcus contains the several roots of the hypoglossal nerve (XII). The dorsolateral sulcus, also continuous with that of the cord, contains roots of the spinal accessory nerve (XI), the vagus (X), and the glossopharyngeal nerve (IX) (Figs. 7.30 and 7.31). Swellings laterally between these sulci represent the olivary body or eminence, and the superior olivary nucleus lies within the medulla in this region of the brain stem.

The dorsal aspect of the medulla presents two low ridges, the fasciculi gracilis and cuneatus. The nuclei within these tracts are marked respectively by small protuberances, the clava and cuneate tubercle (Fig. 7.27). The restiform bodies or inferior cerebellar peduncles form two prominent ridges extending along the sides of the fourth ventricle (Figs. 7.27 and 7.29). In the floor of the fourth ventricle two pairs of swellings are present; their margins taper to a point caudally, and meet in the median sulcus in a region known as the calamus scriptorius. The lateral ridges of this structure are called the vagal trigone; the medial ridges constitute the hypoglossal trigone. The facial colliculus lies rostral to the stria of the fourth ventricle, and the continuation of this ridge anteriorly is the median eminence (Fig. 7.27). A groove on the lateral margins

of these ridges, the sulcus limitans, demarcates the sensory and motor areas of the medulla.

Internally, several of the long nerve fiber tracts from the spinal cord pass directly through the medulla unchanged in position relative to each other; however, others undergo rearrangement within this region of the brain stem.

Within the lower half of the medulla, the central canal is surrounded by a central gray area. This contains the diffuse cells and fibers of the reticular formation, which extends through medulla, pons, and mesencephalon (Fig. 7.50; p. 302). The pyramidal tracts decussate in the most ventral part of the medulla. Dorsally, the fasciculi cuneatus and gracilis terminate in their respective nuclei; their axonic processes then arch downward and cross in the midline in the decussation of the medial lemniscus (Fig. 7.21). Dorsolaterally, the spinal tract and spinal nucleus of the trigeminal nerve (V) are found.

Within the upper half of the medulla is located the inferior olivary nucleus. Many efferent fibers from this region cross and join spinocerebellar fibers in the inferior cerebellar peduncle, whereas other olivocerebellar fibers enter the peduncle without undergoing decussation.

Three symmetric paired areas are found in the dorsal region of the medulla near the floor of the

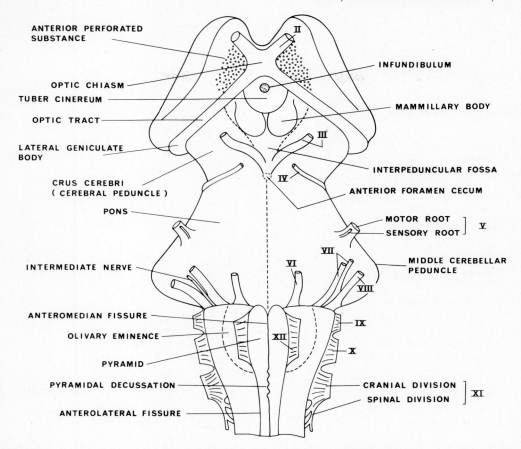

FIG. 7.28. The gross anatomic features of the brain stem from the ventral aspect. (Data from Everett. *Functional Neuroanatomy*, 6th ed, 1971, Lea & Febiger; Truex and Carpenter. *Human Neuroanatomy*, 6th ed, 1969, Williams & Wilkins.)

fourth ventricle. These longitudinal columnar structures consist of the nucleus of the hypoglossal nerve (XII), the dorsal motor nucleus of the vagus (X), and the vestibular nuclei (VIII) which receive afferent fibers from the vestibular branch of the auditory nerve (VIII) (Figs. 7.30 and 7.31).

The nucleus ambiguus is found in the ventrolateral portion of the reticular formation, and its fibers pass into the inferior cerebellar peduncle together with other fibers of the vagus. The solitary tract, an isolated longitudinal fiber bundle, also is present in the reticular formation; it contains afferent fibers of the vagus nerve (X). The adjacent nucleus is called the nucleus of the solitary tract (Fig. 7.30).

The lateral spinothalamic tract lies ventrolateral to the nucleus ambiguus. In this region also are found two large fiber bands which contain the medial longitudinal fasciculus, and the tectospinal and vestibulospinal tracts, the remainder being the medial lemnisci.

Associated with the medulla are four cranial nerves which will be considered individually (see also Table 7.3).

The Hypoglossal Nerve (XII). The hypoglossal

nerve contains some general somatic afferent (GSA) proprioceptive fibers from the lingual muscles. The nerve cell bodies that give rise to these GSA fibers are diffuse. The hypoglossal nerve also is the motor nerve of the tongue muscles. The general somatic efferent (GSE) fibers of the hypoglossal nerve originate in cells of the hypoglossal nucleus (nucleus ambiguus) of the medulla described above (Fig. 7.30).

The Spinal Accessory Nerve (XI). There are two discrete components to the spinal accessory nerve (Figs. 7.30 and 7.32). First, the spinal root contains GSA fibers that terminate in proprioceptors within the neck musculature. These fibers arise from cervical spinal cord segments (C–1 through C–5), ascend through the foramen magnum, and join the second component of XI, the bulbar root, from the medulla. The bulbar fibers run with the spinal fibers for a short distance, then join the vagus nerve and course together with the vagus to terminate with the vagal fibers (Fig. 7.25).

The spinal branch of XI, which consists of *special visceral efferent (SVE) fibers*, descends in the neck and terminates in the sternomastoid and trapezius muscles. Thus, the *spinal* division of this

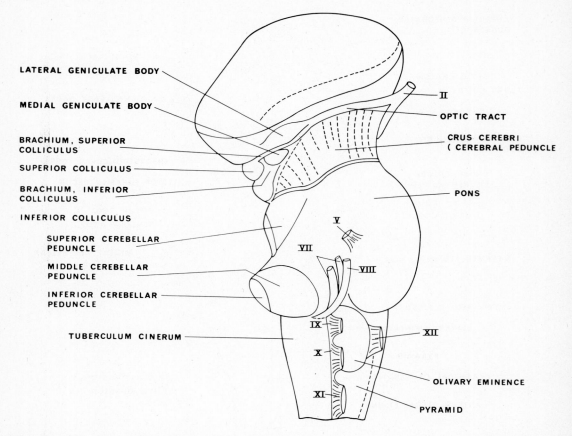

FIG. 7.29. The gross anatomic features of the brain stem from the right lateral aspect. (After Everett. *Functional Neuroanatomy*, 6th ed, 1971, Lea & Febiger.)

nerve actually is a loop of nerve from the brachial plexus and not a true cranial nerve.

The Vagus Nerve (X). The vagus nerve is complex and is made up of both sensory and motor divisions; this nerve also makes important contributions to the cranial division of the parasympathetic nervous system.

The two sensory ganglia of the vagus lie outside of but near the medulla, strung in series on the roots of the vagus nerve like two elongated beads on a string: These are the superior (or jugular) ganglion and the inferior (or nodose) ganglion. GSA fibers from the skin in back of the ear and the posterior wall of the external auditory meatus pass through the auricular branch of the vagus to the superior ganglion.

General visceral afferent (GVA) fibers from the pharynx, larynx, trachea, esophagus, and the thoracic as well as the abdominal viscera conduct impulses centrally to the inferior ganglion of the vagus or else pass on to the nucleus solitarius via the appropriate branches of the vagus. These branches include the pharyngeal, superior and re-

current laryngeal, cardiac, bronchial, and esophageal nerves.

Special visceral afferent (SVA) fibers from diffusely located taste buds in the vicinity of the epiglottis also pass to the inferior (nodose) ganglion of the vagus.

The vagal motor nuclei in the medulla include the dorsal motor nucleus, the salivatory nucleus (also associated with the glossopharyngeal nerve), and the nucleus ambiguus. The latter, found as a column of cells within the reticular formation, contributes bronchiomotor neurons to both the glossopharyngeal as well as vagus nerves, and its axons innervate the voluntary muscles of the soft palate, larynx, and pharynx as SVE preganglionic fibers (Figs. 7.30, 7.31, and 7.32).

General visceral efferent (GVE) preganglionic fibers, which arise from neurons in the dorsal motor nucleus of the vagus, provide motor innervation to thoracic structures and the smooth muscle of the abdominal viscera as far as the level of the splenic flexure of the colon (Fig. 7.25).

The postganglionic fibers of both SVE and GVE fibers originate in the terminal ganglia. It

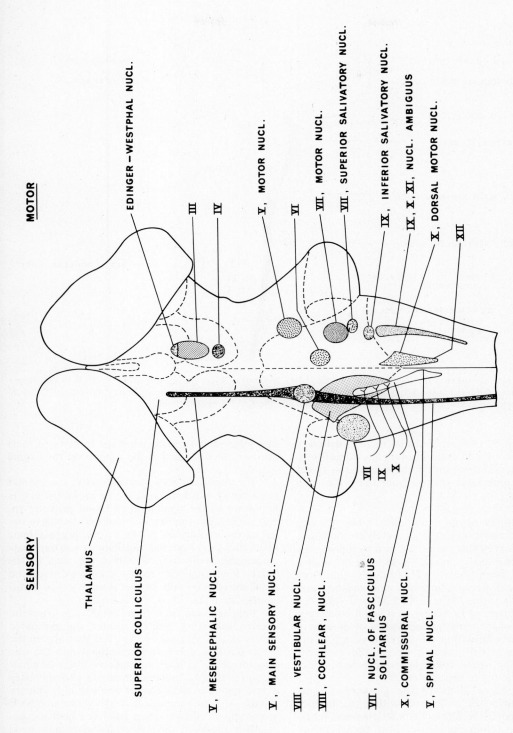

MOTOR

EDINGER – WESTPHAL NUCL.

III

IV

V, MOTOR NUCL.

VI

VII, MOTOR NUCL.

VII, SUPERIOR SALIVATORY NUCL.

IX, INFERIOR SALIVATORY NUCL.

IX, X, XI, NUCL. AMBIGUUS

X, DORSAL MOTOR NUCL.

XII

SENSORY

THALAMUS

SUPERIOR COLLICULUS

V, MESENCEPHALIC NUCL.

V, MAIN SENSORY NUCL.

VIII, VESTIBULAR NUCL.

VIII, COCHLEAR, NUCL.

VII, NUCL. OF FASCICULUS SOLITARIUS

X, COMMISSURAL NUCL.

V, SPINAL NUCL.

VII

IX

X

FIG. 7.30. Dorsal view of brain stem illustrating general locations of the sensory (on left) and motor (on right) nuclei of the cranial nerves. (Data from Chusid. *Correlative Neuroanatomy and Functional Neurology,* 14th ed, 1970, Lange Medical Publications; Goss [ed]. *Gray's Anatomy of the Human Body,* 28th ed, 1970, Lea & Febiger; Carpenter. *Human Neuroanatomy,* 6th ed, 1969, Williams & Wilkins.)

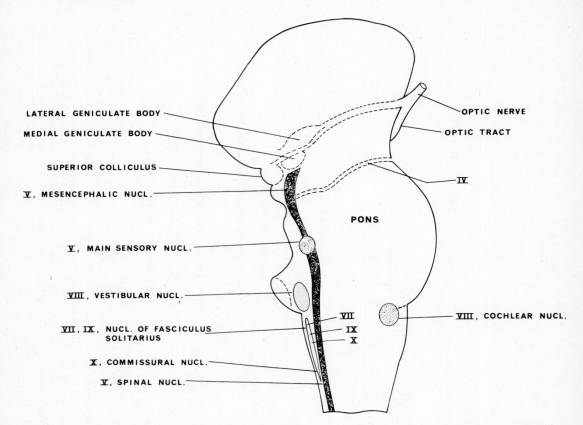

FIG. 7.31.   Lateral view of brain stem showing locations of the sensory nuclei for cranial nerves IV through X. Dorsal aspect of brain stem is to the left in the figure. (Data from Goss [ed]. *Gray's Anatomy of the Human Body,* 28th ed, 1970, Lea & Febiger; Truex and Carpenter. *Human Neuroanatomy,* 6th ed, 1969, Williams & Wilkins.)

should also be remembered that in the medullary region, afferent and efferent branches do not form separate dorsal and ventral roots as in the spinal cord, but enter and emerge from the brain stem together.

The Glossopharyngeal Nerve (IX). The glossopharyngeal nerve is related intimately to the vagus and both nerves share common intramedullary nuclear origins and connections, as well as similar functional components; consequently the glossopharyngeal nerve contains both sensory and motor components (Figs. 7.30, 7.31, and 7.32).

GSA fibers of IX innervate a small cutaneous area back of the ear; the cell bodies of these sensory neurons lie within the superior ganglion and run together with vagal fibers in the auricular branch of the vagus. The axons terminate centrally in the spinal trigeminal nucleus.

GVA nerve fibers of the glossopharyngeal nerve, whose cell bodies lie within the inferior ganglion, enter the fasciculus solitarius and terminate in the solitary nucleus. These fibers convey pain, tactile, and thermal sensations from the mucous membranes of the posterior tongue region, tonsils, and eustachian tubes. More abundant SVA

fibers transmit sensory impulses from taste buds on the posterior third of the tongue via the lingual nerve.

An important SVA branch of IX, the carotid sinus nerve which innervates the carotid sinus, is important in the regulation of blood pressure (p. 672). Centrally, this nerve sends collaterals to the dorsal motor nucleus of the vagus; preganglionic fibers then pass from this vagal nucleus to terminate in ganglion cells located within the atria of the heart. Postganglionic axons then innervate the atrial muscle as well as the sinoatrial and atrioventricular nodes.

Preganglionic GVE fibers of IX arise from the inferior salivatory nucleus, enter the tympanic plexus as the tympanic nerve, and then run in the lesser petrosal nerve to terminate within the otic ganglion (Fig. 7.25). Postganglionic fibers originating within the otic ganglion transmit motor impulses which regulate secretion of the parotid salivary glands. A small component of IX, containing SVE preganglionic fibers, arises from the nucleus ambiguus to innervate the stylopharyngeus muscle via the stylopharyngeal nerve.

The pharyngeal branch of the glossopharyngeal nerve, which originates in the nucleus

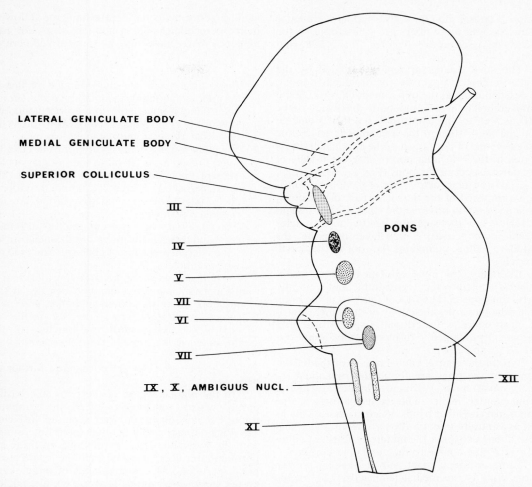

LATERAL GENICULATE BODY

MEDIAL GENICULATE BODY

SUPERIOR COLLICULUS

III

IV

V

VII

VI

VII

IX, X, AMBIGUUS NUCL.

XI

PONS

XII

FIG. 7.32. Lateral view of brain stem showing the motor nuclei of cranial nerves III through XII. Dorsal aspect of brain stem is to the left in the figure. (Data from Goss [ed]. *Gray's Anatomy of the Human Body,* 28th ed, 1970, Lea & Febiger; Truex and Carpenter. *Human Neuroanatomy,* 6th ed, 1969, Williams & Wilkins.)

ambiguus, enters the pharyngeal plexus together with branches of cranial nerves X and V; thence motor fibers are sent to the voluntary pharyngeal muscles that are involved in the various reflexes listed below. Afferent impulses are conveyed in cranial nerves IX and X from both oral and respiratory mucosa, in addition to the proprioceptive impulses that arise from the various striated muscles involved.

Within the medulla oblongata also are found several neuroanatomically nonspecific "centers" which are involved in the functional integration of several complex and physiologically important reflexes. The afferent and efferent nerves which transmit impulses governing these reflexes involve pathways in the "vagal system" (or complex) of nerves; that is, portions of VII, IX, X, and XI. The reflexes listed here, among others, will be discussed fully in relation to the specific organ system (or systems) with which they are associated: (1) cardiac, vasomotor, carotid sinus, and carotid body

reflexes, Chapter 10; (2) the respiratory and cough reflexes, Chapter 11; (3) the swallowing, gag, and vomiting reflexes, Chapter 13.

THE PONS.   The pons (or pons varolii) contains a number of fiber tracts which relay impulses to both lower spinal centers as well as to higher brain centers. The pons also contains the sensory and motor nuclei of cranial nerves V through VIII (Table 7.3).

Anatomically the pons consists of a very large mass of nerve cells, called the pontine nuclei, which is situated rostral to the medulla (Fig. 7.29). The cerebral peduncles are located anteriorly to the pons, whereas the pyramids emerge from its lower (or caudal) margin.

Externally, the ventral surface of the pons exhibits a thick band of transverse fibers which constitute the pons itself. The abducens nerve (VI) exits from the basal sulcus of the pons at its caudal border near the pyramids (Fig. 7.28).

The aforementioned transverse fibers form

compact lateral bundles, the middle cerebellar peduncles, and these fibers attach the pons to the cerebellum lying above it (Figs. 7.27, 7.28, and 7.29).

The small triangular region bounded by the middle cerebellar peduncle (or brachium pontis), the cerebellum, and the upper medulla is called the pontocerebellar angle, and both the facial nerve (VII) as well as the acoustic nerve (VIII) emerge from the brain stem in this region. The trigeminal nerve (V), one of the largest of the cranial nerves, is located in the brachium pontis near the middle of the lateral aspect of the pons (Fig. 7.28).

The triangular rostral floor of the fourth ventricle is composed of the dorsal surface of the pons. Two bands which traverse the sides of this triangular region are the superior cerebellar peduncles (brachia conjunctiva). A thin layer of tissue, the medullary velum, completes the roof of the fourth ventricle.

Internally, the pons exhibits two major subdivisions, dorsally the tegmentum, and ventrally the basilar portion. The roof of this part of the brain stem, which overlies the cavity of the fourth ventricle, has become functionally specialized and anatomically enlarged to form the cerebellum (Figs. 7.11, 7.12, and 7.33).

Within the central caudal region of the basilar portion of the pons are found the pyramidal (or corticospinal) tracts, surrounded by the gray matter of the pontine nuclei. The axons of these nuclei form transverse fiber bands which cross dorsally as well as ventrally to enter the pyramidal tracts, and these fibers also enter the middle cerebellar peduncle where they course to the cerebellar cortex.

In transverse contact with the basilar portion of the pons is the medial lemniscus. The medial longitudinal lemniscus remains near the midline in the floor of the fourth ventricle; this band of fibers contains the tectospinal, the tectobulbar, and the vestibulomesencephalic tracts.

Within the dorsal subdivision of the pons, the tegmentum, the trapezoid body is found as a ventral band of decussating fibers. A small ovoid nucleus, the superior olive, lies dorsolateral to the trapezoid body.

Just dorsal to the superior olive and medial to the nucleus of the spinal root of the trigeminal nerve (V) is found the motor nucleus of the facial nerve (VII). Prior to emerging from the brain stem, the fibers of VII form a loop, the genu of the facial nerve, and along its course this nerve passes just caudal to the nucleus of the abducens nerve (VI) (Figs. 7.30 and 7.32).

Vestibular nuclei are present in the lateral portions of the floor of the fourth ventricle. At this level of the brain stem the individual nuclei are called the lateral and superior vestibular nuclei instead of medial and spinal (inferior), as these nuclei are called in the medulla. The vestibular system of the brain stem, as well as its major projections, is depicted in Figure 7.34.

In sections made at the lower level of the pons

at its junction with the cerebellum, the following paired cerebellar nuclei usually are seen. For convenience these nuclei may be summarized here:

1. *The fastigial nuclei* are situated in the midline on the roof of the fourth ventricle near the vermis. Their afferent fibers arise in the flocculonodular lobe of the cerebellum. Efferent fibers run to the vestibular nuclei in the fastigiobulbar tract.
2. *The globose nuclei* are small groups of cells located immediately lateral to the fastigial nuclei. Afferent fibers course to these nuclei from the paleocerebellum. Efferent fibers from the globose nuclei pass to the red nucleus via the superior cerebellar peduncle.
3. *The emboliform nuclei* are elongated cellular masses found between the globose and dentate nuclei. Their afferent and efferent fibers are similar to the globose nuclei.
4. *The dentate nuclei* are the largest, most lateral of the cerebellar nuclei. Afferent fibers reach this structure from the paleocerebellar and neocerebellar cortex. Efferent fibers run to the red nuclus and the ventrolateral nucleus of the thalamus via the superior cerebellar peduncle.

These cerebellar nuclei are considered further on page 307.

In the middle region of the pons, the basilar portion widens and thickens, while the pyramidal tracts become dispersed into separate fascicles. The longitudinal corticopontine tracts intermingle with the pyramidal fibers in this region; these important motor fibers descend from the frontal, temporal, parietal, and occipital lobes of the cerebral cortex to synapse with cells of the pontine nuclei.

Within the dorsolateral portion of the tegmentum lie two oval nuclei: the superior sensory nucleus of the trigeminal nerve (V) is situated laterally, while the motor (masticator) nucleus of the trigeminal nerve is located medially (Figs. 7.30, 7.31, and 7.32). Small branches of this nerve run dorsally as the mesencephalic root of V, whereas trigeminal fibers emerging from the surface of the pons pass through the middle cerebellar peduncle.

The four cranial nerves associated with the pons now may be discussed individually:

The Vestibulocochlear Nerve (VIII).   The vestibulocochlear nerve contains only special somatic afferent (SSA) fibers; it is composed of two main divisions, the cochlear and auditory.

The cochlear division of VIII arises from sensory cells located within the cochlear duct of the inner ear. The cochlear nerve originates from the bipolar neurons of the spiral ganglion, and the central processes of these cells terminate within the cochlear nuclei of the medulla (Figs. 7.30 and 7.31).

The vestibular branch of VIII forms the vestibular nerve. The peripheral branches of this nerve receive proprioceptive impulses from hair cells located within the maculae of the utricle and saccule

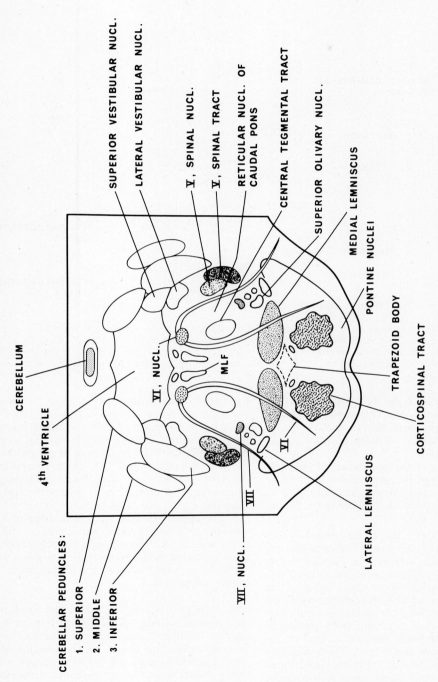

CEREBELLUM

CEREBELLAR PEDUNCLES:

1. SUPERIOR

2. MIDDLE

3. INFERIOR

4th VENTRICLE

SUPERIOR VESTIBULAR NUCL.

LATERAL VESTIBULAR NUCL.

Ⅴ, SPINAL NUCL.

Ⅴ, SPINAL TRACT

RETICULAR NUCL. OF CAUDAL PONS

CENTRAL TEGMENTAL TRACT

SUPERIOR OLIVARY NUCL.

MEDIAL LEMNISCUS

PONTINE NUCLEI

TRAPEZOID BODY

CORTICOSPINAL TRACT

LATERAL LEMNISCUS

Ⅵ, NUCL.

MLF

Ⅵ

Ⅶ

Ⅶ, NUCL.

FIG. 7.33. The pons in cross section at the level of the abducens nucleus. The dorsal region (ie, the tegmentum) contains the reticular formation, ascending and descending tracts, as well as cranial nerve nuclei. The ventral region contains large bundles of corticofugal fibers, the pontine nuclei, and transverse pontine fibers which constitute the middle cerebellar peduncle. MLF = medial longitudinal fasciculus; NUCL = nucleus. The cranial nerves and their nuclei are indicated by Roman numerals. (After Truex and Carpenter. *Human Neuroanatomy*, 6th ed, 1969, Williams & Wilkins.)

285

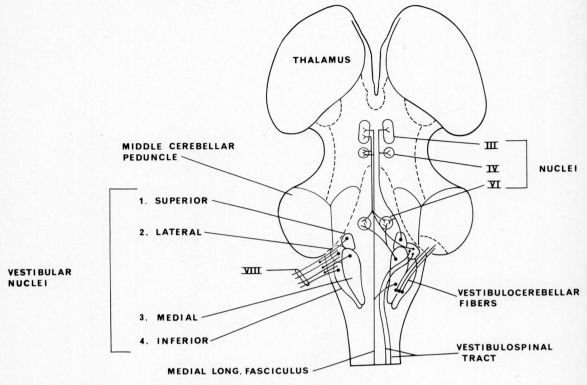

FIG. 7.34. The central connections and some projections of the vestibular nuclei on a dorsal view of the brain stem. (See also Figs. 7.111, 7.112, and 7.113.) (Data from Best, Taylor [eds]. *The Physiological Basis of Medical Practice,* 8th ed, 1966, Williams & Wilkins; Everett. *Functional Neuroanatomy,* 6th ed, 1971, Lea & Febiger; Gatz. *Manter's Essentials of Clinical Neuroanatomy and Neurophysiology,* 4th ed, 1970, F. A. Davis; Truex and Carpenter. *Human Neuroanatomy,* 6th ed, 1969, Williams & Wilkins.)

as well as the cristae of the three semicircular canals of the inner ear (p. 307). The bipolar neurons of the vestibular nerve originate in the vestibular ganglion of the upper medulla–pontine region (Figs. 7.30 and 7.31).

Further details of the neuroanatomic connections and functions of the vestibulocochlear nerve will be presented subsequently in relation to its roles in the maintenance of equilibrium (p. 388) as well as hearing (p. 506).

The Facial Nerve (VII). The facial nerve has four major components: First, general visceral afferent (GVA) fibers which originate in the geniculate ganglion and receive impulses related to deep sensations from the facial region; these impulses are conducted in fibers which are components of the nervus intermedius. Second, special visceral afferent (SVA) fibers pass centrally from peripheral exteroceptors (taste buds) that are located on the anterior two-thirds of the tongue. Their dendrites originate in neurons whose cell bodies are located in the geniculate ganglion, and pass to the tongue via the intermediate and lingual nerves as well as the chorda tympani. The central axons terminate in the nucleus solitarius. Third, general visceral efferent (GVE) fibers of VII or preganglionic parasympathetic fibers, which are components of

the nervus intermedius. These fibers originate in the superior salivatory nucleus and synapse with postganglionic neurons in the submandibular and pterygopalatine ganglia. Fourth, special visceral efferent (SVE) fibers arise from nerve cell bodies in the motor nucleus of VII. These innervate the muscles of the scalp and face, including the platysma, stylohyoid, and posterior digastric muscles.

The special visceral efferent division of VII is considered to be the facial nerve proper, and this nerve subserves all movements concerned with facial expression, and through its motor action on the orbicularis oculi muscles, the facial nerve also closes the eyelids.

The Abducens Nerve (VI). The abducens nerve has general somatic afferent (GSA) fibers which receive impulses from proprioceptive end organs located within the lateral rectus muscles of the eyes. General somatic efferent (GSE) fibers, arising in the abducens nucleus, provide motor innervation to the lateral rectus muscles (Figs. 7.30, 7.31, and 7.32).

The Trigeminal Nerve (V). The trigeminal nerve has two functional components. First, GSA fibers receive impulses from exteroceptors that are located within the skin of the face, orbit, and scalp as well as in the mucous membranes of the mouth and nasal cavity. The neurons of this portion of V arise in the semilunar ganglion (Fig. 7.30). Proprioceptive fibers of the trigeminal nerve receive impulses from neuromuscular spindles located within the muscles involved in mastication; the cell bodies of these fibers are located within the mesencephalic nucleus of the trigeminal nerve.

SVE fibers from the masticator (or motor) nucleus in the lateral tegmentum of the pons enter the mandibular branch and portis minor of the trigeminal nerve to innervate the temporal, masseter, and medial as well as the lateral pterygoid muscles in addition to the tensor veli palatini, tensor tympani, mylohyoid, and the anterior belly of the digastric muscles.

THE MESENCEPHALON OR MIDBRAIN. The midbrain forms a short junction between the pons and diencephalon (Fig. 7.11). Within the midbrain, the small cerebral aqueduct connects the third and fourth ventricles.

Externally, the lower ventral surface of the mesencephalon is made up of two bundles of fibers, the basis pendunculi cerebri or crura cerebri which are separated by the deep interpeduncular space or fossa (Fig. 7.28). At its junction with the cerebral hemisphere above, each peduncle of the crura cerebri is traversed by the optic tract. Caudally, the peduncles run directly into the basilar portion of the pons, and contain the corticospinal, corticopontine, and corticobulbar tracts. These descending (motor) fiber tracts pass from the cerebral cortex to the pons and there form the longitudinal fasciculus of that structure. The oculomotor nerves (III) emerge from each side of the interpeduncular fossa and appear at the surface at the transverse groove between midbrain and pons.

The dorsal surface of the midbrain has four rounded protuberances, the corpora quadrigemina (Fig. 7.27). The pair of rostral elevations are called the superior colliculi, whereas the smaller caudal pair are the inferior colliculi. The trochlear nerves (IV) exit just posterior to the inferior colliculi.

Internally, the midbrain is divided into three regions: the ventrolateral portion or basis pedunculi (crus cerebri); the tegmentum, similar to the pontine tegmentum; and the tectum or roof, which lies just above the cerebral aqueduct and forms the quadrigeminal plate (Fig. 7.35A and B).

Within the lower half of the midbrain, each crus cerebri contains in its central portion the pyramidal tract, on each side of which lie corticopontine fibers. The substantia nigra, a pigmented (brownish) area of gray matter, lies between the peduncles and the tegmentum (Fig. 7.35).

Centrally, the tegmentum contains the decussation of the superior cerebellar peduncle. The medial lemniscus is found more laterally in this region of the brain stem, and its outer border is adjacent to the fibers of the lateral spinothalamic tract. Dorsal

to the spinothalamic fibers, the lateral lemniscus contains the ascending fibers of the special sensory pathway that is involved in hearing. A nucleus of the inferior colliculus lies beneath each of these structures in the tectal region. Fibers that arise within the inferior colliculus, as well as those of the lateral lemniscus, run to the medial geniculate body by way of the brachium of the inferior colliculus (Fig. 7.27). Consequently this brachium is the pathway whereby sensory impulses arising in the cochlea (p. 506) are transmitted from the midbrain to their specific relay nucleus in the thalamus (Fig. 7.36).

Both the rubrospinal and tectospinal tracts originate within the mesencephalon. The cell bodies of each rubrospinal tract lie within the red nucleus of the mesencephalic tegmentum. Crossed fibers from the superior cerebellar peduncle pass into as well as around this nucleus; some of these cerebellar fibers terminate within the red nucleus, whereas others pass rostrally to the thalamus. The tectospinal tract arises from the nucleus of the superior colliculus. Both of these tracts undergo decussation near their points of origin in the midbrain. The rubrospinal tract crosses in the ventral tegmental decussation (decussation of Forel), whereas the tectospinal tract crosses the midline in the dorsal tegmental decussation (fountain decussation of Meynert). The neuroanatomic relationships discussed above are illustrated in Figure 7.22.

The relationships between the midbrain and the two cranial nerves which originate within this region of the brain stem are as follows.

The Trochlear Nerve (IV). The nucleus of the trochlear nerve is situated near the medial longitudinal fasciculus and ventral to the central gray matter in the region of the inferior colliculus (Figs. 7.28, 7.30, 7.32, 7.35A). This cranial nerve contains general somatic afferent (GSA) fibers which innervate the proprioceptive neuromuscular spindles in the superior oblique muscle. General somatic efferent (GSE) fibers arise in the trochlear nucleus and provide motor innervation to the superior oblique muscle of the eye (Table 7.3).

The Oculomotor Nerve (III). The nuclear complex of the oculomotor nerve is situated ventrally in the central gray matter adjacent to the medial longitudinal fasciculus (Figs. 7.30, 7.31, 7.32, and 7.35B). The root fibers of this cranial nerve pass through and around the red nucleus, then converge at their exit within the interpeduncular fossa (Figs. 7.28 and 7.35B).

The oculomotor nerve contains GSA fibers which transmit proprioceptive impulses from the ocular muscles (ie, the superior, middle, and inferior rectus muscles and the inferior oblique muscle). GSE fibers of the oculomotor nerve arise in the oculomotor nucleus (Fig. 7.32) and provide motor innervation for the same four pairs of ocular muscles. The parasympathetic component of III consists of preganglionic general visceral efferent fibers which arise in the accessory oculomotor nu-

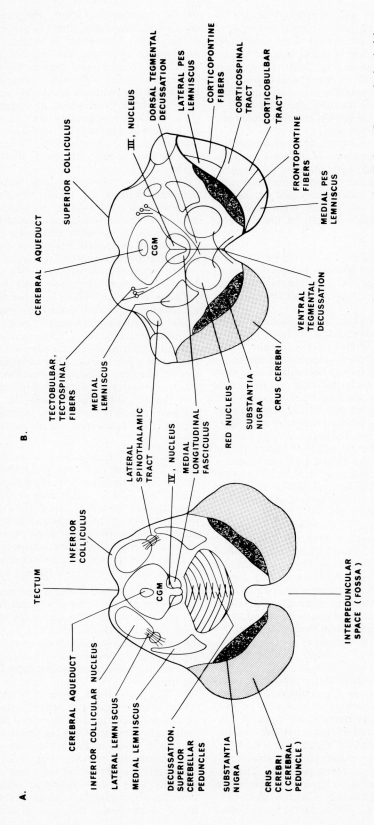

A.

TECTUM

CEREBRAL AQUEDUCT

INFERIOR COLLICULUS

INFERIOR COLLICULAR NUCLEUS

LATERAL LEMNISCUS

MEDIAL LEMNISCUS

DECUSSATION, SUPERIOR CEREBELLAR PEDUNCLES

SUBSTANTIA NIGRA

CRUS CEREBRI (CEREBRAL PEDUNCLE)

CGM

INTERPEDUNCULAR SPACE (FOSSA)

B.

CEREBRAL AQUEDUCT

SUPERIOR COLLICULUS

III, NUCLEUS

DORSAL TEGMENTAL DECUSSATION

LATERAL PES LEMNISCUS

CORTICOPONTINE FIBERS

CORTICOSPINAL TRACT

CORTICOBULBAR TRACT

FRONTOPONTINE FIBERS

MEDIAL PES LEMNISCUS

CGM

TECTOBULBAR, TECTOSPINAL FIBERS

MEDIAL LEMNISCUS

LATERAL SPINOTHALAMIC TRACT

IV, NUCLEUS

MEDIAL LONGITUDINAL FASCICULUS

RED NUCLEUS

SUBSTANTIA NIGRA

CRUS CEREBRI

VENTRAL TEGMENTAL DECUSSATION

FIG. 7.35. **A.** Transverse section of the midbrain at the level of the inferior colliculi. CGM = central gray matter. **B.** Transverse section through the midbrain at the level of the superior colliculi. The particular regions of the cerebral peduncle are indicated within the heavy outline on the right side of this figure. (Data from Everett. *Functional Neuroanatomy,* 6th ed, 1971, Lea & Febiger; Truex and Carpenter. *Human Neuroanatomy,* 6th ed, 1969, Williams & Wilkins.)

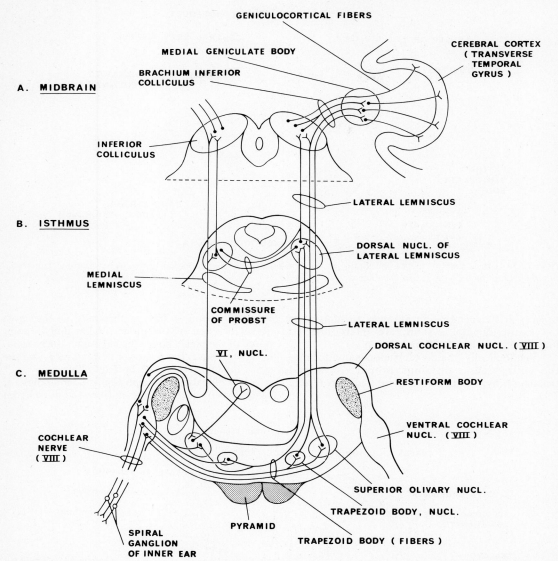

FIG. 7.36. The auditory pathway. (After Truex and Carpenter. *Human Neuroanatomy*, 6th ed, 1969, Williams & Wilkins.)

cleus (Edinger-Westphal nucleus) and which terminate in the ciliary ganglion (Table 7.3; Fig. 7.26).

THE DIENCEPHALON. The diencephalon forms the rostral end of the brain stem, and this structure consists of an oval mass of gray matter that is situated deep within the brain ventral and caudal to the frontal lobes of the cerebral hemispheres (Fig. 7.11). The transitional region between the mesencephalon and diencephalon is shown in Figure 7.37, whereas a cross section of the brain at the level of the diencephalon is presented in Figure 7.38. The fibers of the internal capsule demarcate the diencephalon laterally from the basal ganglia. Ventrally, the diencephalon extends from the optic chiasma to the mamillary bodies, which are in-

cluded in this part of the brain stem (Figs. 7.28 and 7.39). The choroid plexus of the third ventricle forms the roof of the diencephalon, and the two diencephalic halves are separated by the slitlike cavity of this ventricle.

Functionally the diencephalon provides a vital integrative and relay center for impulses passing between the cerebral cortex and the more caudal regions of the brain stem. Consequently, the diencephalon is associated closely with somatic as well as visceral sensory and motor functions. The sensory pathways of the lower centers of the brain stem as well as the reticular arousal system synapse at this level of the brain stem before they transmit (ie, project) sensory impulses onward to the cerebral cortex. Furthermore, portions of the visceral

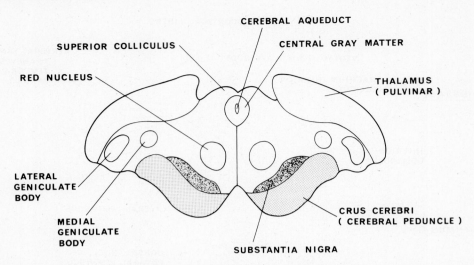

FIG. 7.37. Cross section of the midbrain at the mesencephalic-diencephalic level. (After Everett. *Functional Neuroanatomy,* 6th ed, 1971, Lea & Febiger.)

sensory and visceral motor systems, in addition to the extrapyramidal system, are interconnected functionally within the diencephalon.

Anatomically, the diencephalon is divided into the epithalamus, thalamus proper (or dorsal thalamus), hypothalamus, and subthalamus (Table 7.1).

The Epithalamus. The epithalamus is the most dorsal division of the diencephalon and is composed of the pineal body, the habenula, the habenular commissure, the posterior commissure, and the striae medullaris (Figs. 7.11 and 7.27).

The pineal body (epiphysis) is a small mass of tissue that projects on a stalk from the dorsal roof

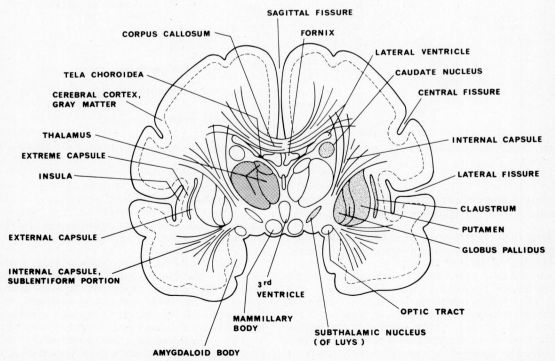

FIG. 7.38. Frontal section through the brain at the level of the diencephalon. The putamen and globus pallidus together make up the corpus striatum. See also Figure 7.80B. (After Everett. *Functional Neuroanatomy,* 6th ed, 1971, Lea & Febiger.)

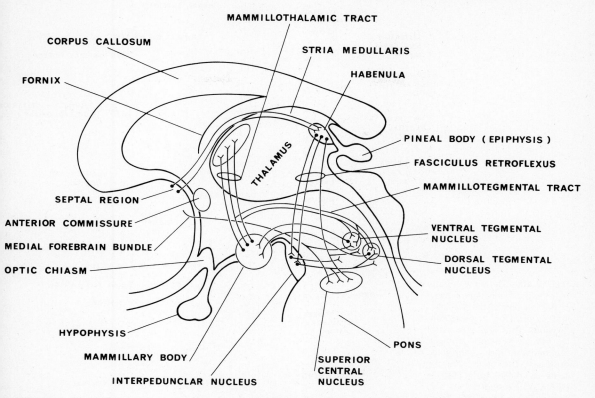

FIG. 7.39.   The limbic pathways that interconnect medial midbrain structures to the telencephalon and diencephalon. (After Truex and Carpenter. *Human Neuroanatomy,* 6th ed, 1969, Williams & Wilkins.)

of the diencephalon in the vicinity of the posterior commissure. The pineal body, or pineal gland, is composed principally of glial cells, although some parenchymal cells, presumably secretory in function, also are present (p. 1172).

The habenular trigone is a small concavity located just forward of each superior colliculus. The trigone region contains the habenular nuclei; these nuclear structures are connected by nerve fibers to the striae medullaris and are interconnected with each other by fibers of the habenular commissure (Fig. 7.39). Afferent fibers to the habenular nuclei originate from the olfactory and septal nuclei as well as in the lateral hypothalamus, and these fibers are carried in the stria medullaris. The habenulopeduncular tract (retroflex tract of Meynert, fasciculus retroflexus) carries efferent fibers that originate within the habenula and terminate in the intrapeduncular nucleus of the midbrain (Fig. 7.39). From here, postsynaptic fibers descend to synapse once again in the nuclei of certain cranial nerves. The habenular nuclei thus appear to be relay stations for olfactory impulses.

The posterior commissure is located ventrally to the base of the pineal body, and this structure consists of a small mass of decussating nerve fibers of the superior colliculi or tectum which traverse the midline on the dorsal side of the cerebral aqueduct (or aqueduct of Sylvius), and which are involved in the visual reflex.

The Thalamus.   The thalamus proper functions as a vital way station for the receipt of sensory impulses which arise within receptors that are located throughout the body (with the exception of olfactory impulses). From the thalamus, this broad range of sensory information is relayed or projected to appropriate areas of the cerebral cortex. In addition, the thalamus integrates and transmits nerve impulses to subcortical areas of the brain and receives afferent impulses that originate within the cerebral cortex, cerebellum, and basal ganglia.

The thalamus is the dorsal protion of the diencephalon and it is composed of two lateral ovoid masses of gray matter that are joined together in the midline by the massa intermedia, a short isthmus of gray matter (Fig. 7.40). Each lateral half of the thalamus is embedded deeply within the corresponding cerebral hemisphere (Fig. 7.11), and is bounded laterally by the internal capsule and medially by the wall of the third ventricle (Fig. 7.38). The roof of the thalamus is formed by the choroid plexus of the third ventricle (Fig. 7.38).

The rostral (or anterior) end of the thalamus is relatively narrow and lies close to the midline (Fig. 7.40). This region delimits the posterior end of the

A.

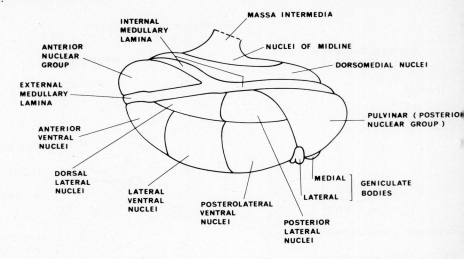

B.

FIG. 7.40. Thalamic nuclei **(A).** Oblique dorsolateral view of the left half of the thalamus and its principal nuclear subdivisions **(B).** Cross sections of the primate thalamus arranged sequentially from anterior to posterior. The nuclei are the same as those designated in part **A** of this figure. (See also Table 7.7.) (**A.** Data from Chusid. *Correlative Neuroanatomy and Functional Neurology,* 14th ed, 1970, Lange Medical Publications; Truex and Carpenter. *Human Neuroanatomy,* 6th ed, 1969, Williams & Wilkins; **B.** After Everett. *Functional Neuroanatomy,* 6th ed, 1971, Lea & Febiger.)

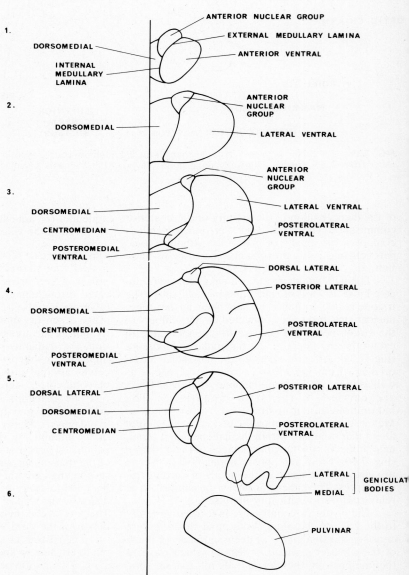

interventricular foramen. The caudal end of the thalamus is somewhat larger, and its medial portion is called the pulvinar. The pulvinar overhangs the superior colliculi. An oval swelling, the lateral geniculate body, is located in this region (Figs. 7.27 and 7.40). Note also in these figures the location of the medial geniculate body. Along the dorsolateral surface of the thalamus runs a groove known as the terminal sulcus which contains a band of fibers, the stria terminalis, which was discussed in the previous section.

The medial surface of the thalamus forms the lateral wall of the third ventricle (Figs. 7.38 and 7.41), and here the aforementioned massa intermedia connects the medial surfaces of each half of the thalamus.

The term thalamic radiation indicates the tracts of nerve fibers that emerge from the lateral surface of the thalamus to enter the internal capsule, thence to terminate in the cerebral cortex. The external medullary lamina, located on the lateral surface of the thalamus adjacent to the internal capsule, is a layer of myelinated fibers. The internal medullary lamina is a vertical partition of white matter which bifurcates anteriorly and thus divides the gray matter of the thalamus into anterior, medial, and lateral portions (Fig. 7.40).

The thalamus contains a large number of individual nuclei whose anatomic and functional relationships are exceedingly complex. Consequently there is some disagreement among various authorities as to the nomenclature, classification (anatomic as well as functional), and subdivisions of the various thalamic nuclei. One scheme of nomenclature assigned to the thalamic nuclei is depicted in Figure 7.40 and summarized in Table 7.7 together with one functional classification (column two).

By definition, a specific thalamocortical nucleus is one which receives anatomically well-defined afferent nerve fibers and which is connected to a clearly delineated area of the cerebral cortex by well-defined efferent thalamic tracts. Included within this category are the principal thalamic sensory relay nuclei (eg, certain of the lateral nuclear group as well as the medial and lateral geniculate bodies).

The nonspecific projection nuclei of the thalamus in contrast to the specific projection nuclei have less well-defined afferent and efferent connections, anatomically speaking. For example, the midline and medial nuclear groups are assigned to the category of nonspecific projection nuclei. The majority of the nonspecific nuclei also are consid-

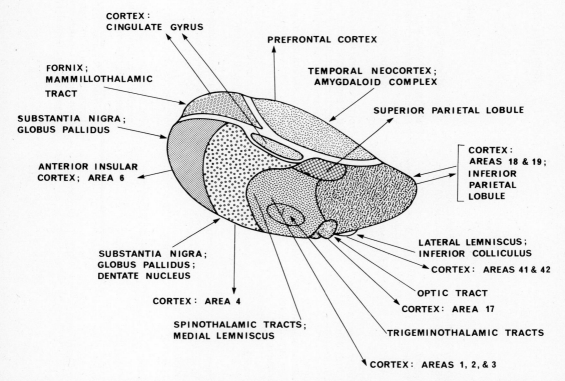

FIG. 7.41. Major thalamic projections to the cerebral cortex. The majority of cortical areas reciprocally project fibers back to the thalamic nuclei whence they receive fibers, however not all of these have been indicated. The shadings used in this figure correspond to those in Figure 7.43, and thus indicate which regions of the thalamus project to which specific cortical areas. See also Figures 7.71 and 7.72 for the location of the cortical areas indicated by Brodmann numbers. See also Table 7.7. After Truex and Carpenter. *Human Neuroanatomy*, 6th ed, 1969, Williams & Wilkins.)

Table 7.7 PRINCIPAL THALAMIC NUCLEI[a]

NEUROANATOMIC CLASSIFICATION	FUNCTIONAL CLASSIFICATION[b]	PRINCIPAL CONNECTIONS		GENERAL FUNCTION(S)
		Afferent Fibers	Efferent Fibers	
1. Anterior Nuclei		From hypothalamus via mammillothalamic tract (of Vicq d'Azyr); higher order olfactory neurons	To cingulate gyrus of cerebral cortex	Part of circuit involved in limbic system (qv); convey olfactory impulses
a) Anteromedial	Nonspecific projection			
b) Anterodorsal	Nonspecific projection			
c) Anteroventral	Nonspecific projection			
2. Midline Nuclei				
a) Cell groups beneath lining of wall, 3rd ventricle	Nonspecific projection	From spinothalamic, trigeminothalamic tracts, medial lemniscus, reticular formation, other thalamic nuclei, hypothalamus		Center for integrating crude visceral and somatic sensations
b) Massa intermedia	Nonspecific projection			
3. Medial Nuclei				
a) Scattered cells in internal medullary lamina (intralaminar nuclei)	Nonspecific projection	From prefrontal cortex, septal areas, basal ganglia, and other thalamic nuclei	To prefrontal cortex	Integrate somatic and visceral sensory impulses before projecting this information to cortex. Association center for synthesis of crude somatic sensations.
b) Dorsomedial	Nonspecific projection	From thalamic nuclei, prefrontal cortex, basal ganglia	To prefrontal cortex	
c) Centromedian	Nonspecific projection	From putamen, caudate nucleus, other thalamic nuclei	To basal ganglia, other thalamic nuclei (?)	Intrathalamic integrating center (?)
4. Lateral Nuclei				
a) Anterior ventral	Nonspecific projection	From globus pallidus via thalamic fasciculus	To corpus striatum; cortex (frontal lobe)	Part of circuit involved in voluntary motor functions
b) Lateral ventral	Specific projection	From cerebellum via superior cerebellar peduncle; globus pallidus via thalamic fasciculus	To cerebral cortex (premotor areas) via posterior limb of internal capsule	
c) Posterolateral ventral	Specific projection	Termination of spinothalamic tracts, medial lemniscus	To sensory areas of cortex (postcentral gyrus) via post. limb of internal capsule	Relay sensory impulses from trunk and limbs[c]
d) Posteromedial ventral	Specific projection	Termination of secondary trigeminal and taste fibers		Relays sensory impulses from face[c]

PRINCIPAL CONNECTIONS

NEUROANATOMIC CLASSIFICATION	FUNCTIONAL CLASSIFICATION[b]	Afferent Fibers	Efferent Fibers	GENERAL FUNCTION(S)
e) Dorsal lateral	Specific projection	From other thalamic nuclei; parietal lobe of cerebral cortex	To cerebral cortex (parietal lobe)	Primary sensory relay nuclei
f) Posterior lateral	Specific projection			
g) Reticular	Nonspecific projection	From entire cerebral cortex, other thalamic nuclei, reticular formation of brain stem	To other thalamic nuclei, tegmentum of midbrain	Functions controversial
5. Posterior Nuclei				
a) Pulvinar	Specific projection	From other thalamic nuclei, cerebral cortex (parietal, temporal, occipital lobes)	To cerebral cortex (parietal, temporal, occipital lobes)	Integrates auditory, visual, somatic impulses (?)
b) Medial geniculate[d]	Specific projection	From brachium of inferior colliculus	To auditory cortex via sublenticular portion of internal capsule (bilateral projection)	Audition
c) Lateral geniculate[d]	Specific projection	From optic tract (cranial nerve II)	To ipsilateral striate cortex via retro-lenticular portion of internal capsule	Vision

[a]Data from Walker: The Primate Thalamus. University of Chicago Press, 1966; Walker: Normal and pathological physiology of the thalamus. In Schaltenbrand, Bailey, (eds): Introduction to Stereotaxis with an Atlas of the Human Brain. Thieme Verlag, Stuttgart, Germany, 1955, pp 291-330; Everett: Functional Neuroanatomy, 6th ed. Lea and Febiger, 1971; Chusid: Correlative Neuroanatomy and Functional Neurology. 14th ed. Lange Medical Publications, 1970.
[b]See text for discussion of the functional classification of the thalamic nuclei.
[c]The posterolateral ventral and posteromedial ventral nuclei of the thalamus are of major physiologic significance as these structures provide the principal thalamic relay system for somesthetic afferent fibers.
[d]The medial and lateral geniculate bodies sometimes are classed together as the metathalamus.

ered to fall within a subcortical classification as they do not project to the cortex or their cortical projection has not been determined (eg, the midline, centromedian, and reticular nuclei).

Alternatively, the thalamic nuclei have been classified as relay nuclei, association nuclei, and nuclei that are connected principally with other subcortical areas (ie, the diffuse conducting or projection system). The relay nuclei are the ventral groups of thalamic cells which transmit impulses to the cerebral cortex, eg, the anterior ventral, the lateral ventral, and the posterolateral nuclei as well as the medial and lateral geniculate bodies. The association nuclei have afferent as well as efferent fiber connections with the cerebral cortex; however, no other specific ascending or descending connections are present. Included among the association nuclei are the anterior nuclear group, as well as the pulvinar, dorsal lateral, and posterior lateral nuclei. The subcortical nuclei of the diffuse conducting system include those nuclei which are located within the internal medullary lamina as well as upon

the extreme lateral or medial aspects of the thalamus; eg, the midline nuclei, the centromedian nucleus, and the reticular nuclei. These regions are termed diffuse projection nuclei because stimulation of any one of the thalamic areas listed above causes a widespread conduction of impulses to the cerebral cortex together with recruitment and alerting of the subject. Consequently these diffuse projection nuclei are considered by some investigators to represent the thalamic portion of the ascending reticular activating system (qv, p. 429).

Some of the major afferent and efferent connections of various thalamic nuclei are depicted in Figures 7.41, 7.42, and 7.43.

The functional attributes of the thalamus will be considered in somewhat greater detail on page 352; however, it should be reiterated here that the thalamus serves as the major relay station for afferent impulses that arise in all sensory end organs throughout the body (excepting the olfactory end organs). Furthermore, the thalamus is the center in which integration of many sensory impulses

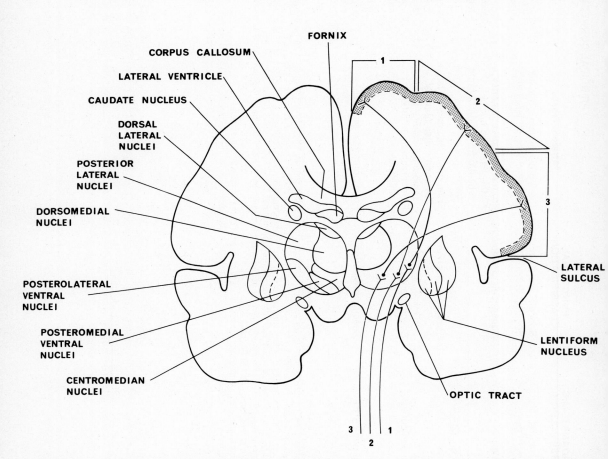

FIG. 7.42.   The sensory tracts and their terminations in the lateral region of the thalamus as well as the fiber projections from these nuclei to the somesthetic cortex via the internal capsule. 1. Sensory fibers from lower extremity project to leg area of cortex. 2. Sensory fibers from upper extremity project to arm area of cortex. 3. Secondary trigeminal area projects to face area of cortex. Note that a spatial orientation of all fiber pathways is maintained at all levels of the nervous system. (See also Fig. 7.44.) (After Everett. *Functional Neuroanatomy,* 6th ed, 1971, Lea & Febiger.)

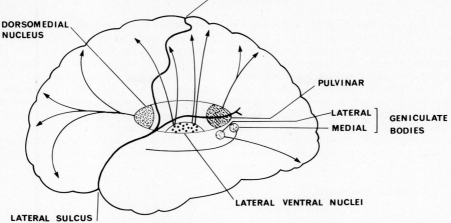

A.

CENTRAL SULCUS

DORSOMEDIAL
NUCLEUS

PULVINAR

LATERAL
MEDIAL
GENICULATE
BODIES

LATERAL SULCUS

LATERAL VENTRAL NUCLEI

FIG. 7.43 A. The principal thalamocortical projections. (After Chusid. *Correlative Neuroanatomy and Functional Neurology,* 14th ed, 1970, Lange Medical Publications.)

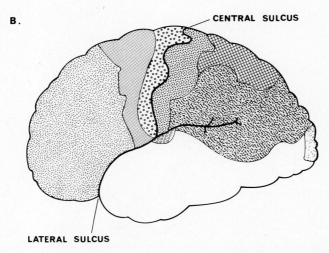

B.

CENTRAL SULCUS

LATERAL SULCUS

FIG. 7.43. B. Lateral surface of the left cerebral hemisphere indicating the cortical projection areas of the thalamic nuclei shown in Figure 7.41. (After Truex and Carpenter. *Human Neuroanatomy,* 6th ed, 1969, Williams & Wilkins.)

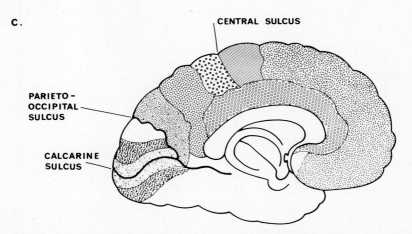

C.

CENTRAL SULCUS

PARIETO-
OCCIPITAL
SULCUS

CALCARINE
SULCUS

FIG. 7.43. C. Same, but viewed from the medial surface of the left hemisphere. The shading code used is identical with that employed in Figure 7.41.

takes place, and this region of the diencephalon appears to be related to the synthesis of visceral and crude somatic sensations which are perceived at the conscious level. In particular, the dorsomedial nuclei are believed to function as an association center for these crude sensations and also to provide the loci where sensations are integrated into subjective feelings that are either pleasant or unpleasant. The thalamus also appears to play a vital role in the integration and coordination of visceral as well as somatic motor responses. The ventrome-

dial portion of the thalamus, including the reticular formation and the intralaminar nuclei, participate together with the reticular formation of the brain stem in the alerting and arousal responses.

Lastly, the lateral geniculate body of the thalamus is a critical relay station for visual impulses that arise within the retina of the eye and which are transmitted via this thalamic nucleus to the visual cortex (Figs. 7.27, 7.28, 7.29, 7.40, and 7.41). The visual pathway is depicted in Figure 7.44 and discussed in greater detail in connection with

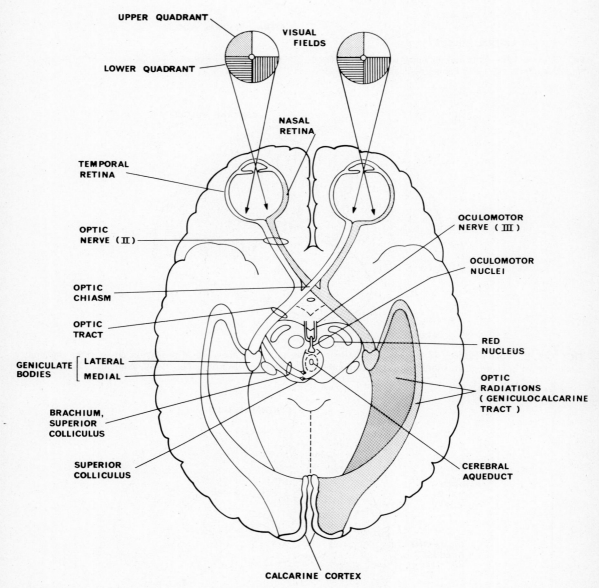

FIG. 7.44. The visual pathway as seen from the ventral aspect of the brain. Note particularly the pathways taken by the fibers of the optic nerves after they undergo partial decussation in the optic chiasm (shaded and unshaded areas). (After Truex and Carpenter. *Human Neuroanatomy*, 6th ed, 1969, Williams & Wilkins. Additional data from Gatz. *Manter's Essentials of Clinical Neuroanatomy and Neurophysiology*, 4th ed, 1970, FA Davis; Everett. *Functional Neuroanatomy*, 6th ed, 1971, Lea & Febiger.)

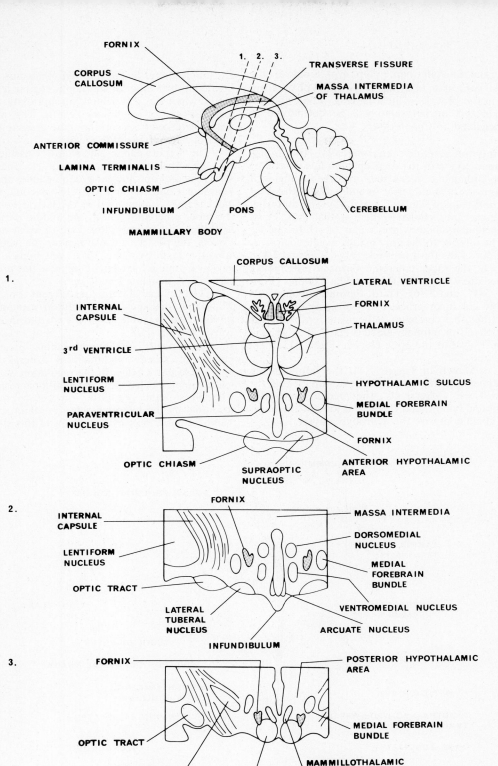

FIG. 7.45. At top, sagittal section through brain stem indicating the relationship of the hypothalamus to adjacent brain structures. Cuts 1, 2, and 3, indicated by dashed lines, are shown in transverse section in the three lower drawings. In all instances, the fornix has been shaded to facilitate comparison of the specific hypothalamic and other structures at each level. (After Everett. *Functional Neuroanatomy,* 6th ed, 1971, Lea & Febiger. Additional data from Truex and Carpenter. *Human Neuroanatomy,* 6th ed, 1969, Williams & Wilkins.)

the visual reflex in Chapter 8 (p. 474). Similarly, the medial geniculate body has an essential physiologic role in the process of hearing (Fig. 7.36), and the role of this nucleus also will be considered in further detail in Chapter 8 (p. 506).

The Hypothalamus. Anatomically, the hypothalamic region of the diencephalon lies ventral to the hypothalamic sulcus and forms the floor as well as portions of the inferior lateral walls of the third ventricle (Figs. 7.11, 7.12, and 7.45). The following structures are included in the hypothalamus. First, the mamillary bodies are a pair of white masses that lie inferior to the gray matter which comprises the floor of the third ventricle (Figs. 7.28 and 7.45). Second, the tuber cinereum forms a prominence immediately rostral to the mamillary bodies (Figs. 7.28, 7.45, and 7.46). Third, the infundibulum is a hollow columnar process that projects downward from the tuber cinereum to the posterior lobe of the hypophysis or pituitary gland (Figs. 7.28, 7.45, and 7.46). Fourth is the optic chiasm (Figs. 7.12, 7.28, 7.44, and 7.45).

As shown in Figure 7.46, the infundibulum continues into the neural or posterior lobe of the hypophysis. Collectively, the median eminence, the infundibulum, and the neural lobe of the pituitary gland comprise the neurohypophysis.

Each lateral half of the hypothalamus can be divided into an anterior supraoptic region, a tuberal region which lies just behind the supraoptic region, and a posterior mamillary region. The preoptic area is situated between the optic chiasm and the anterior commissure (Fig. 7.45).

The hypothalamic nuclei are classified into four principal groups (see Figs. 7.45 and 7.47). First, the anterior nuclear group includes the paraventricular and supraoptic nuclei. Second, the lateral nucleus is composed of the lateral portion of the tuber cinereum. Third, the middle part of the hypothalamus contains the ventromedial nucleus. Fourth, the posterior nuclei include the medial and lateral mamillary nuclei. The cell bodies of the medial mamillary nucleus comprise the protuberance of the mamillary body.

The principal afferent and efferent connections of the hypothalamus to other brain structures are summarized in Table 7.8 and depicted in Figures 7.39, 7.46, and 7.48.

Functionally, the hypothalamus integrates a number of rather straightforward visceral reflexes and it also participates in the regulation of several more complex behavioral and emotional responses. The physiologic regulatory processes that are mediated by the hypothalamus are summarized in Table 7.9, and the particular roles of this structure

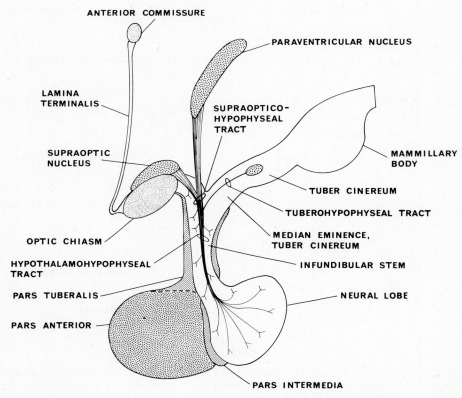

FIG. 7.46. Sagittal section through the hypophysis and hypothalamus illustrating the origins and distribution of the hypothalamohypophyseal tract. (After Everett. *Functional Neuroanatomy,* 6th ed, 1971, Lea & Febiger.)

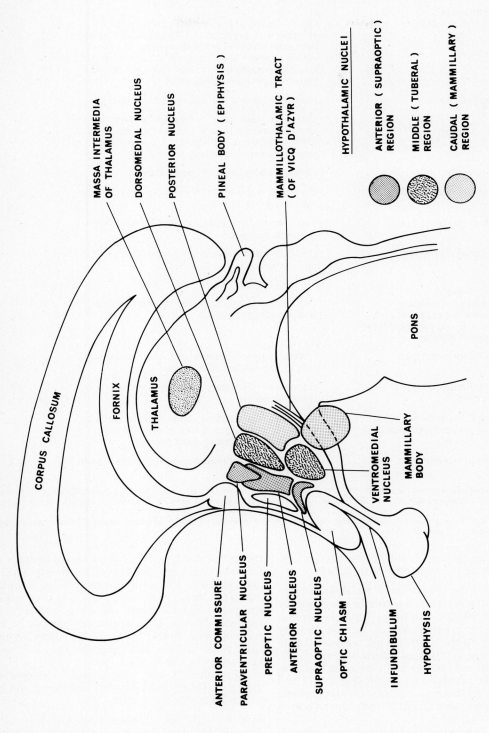

FIG. 7.47. Relationships among the medial hypothalamic nuclei. The shading code used for the three hypothalamic regions is given in the inset to the lower right of the figure. (After Truex and Carpenter. *Human Neuroanatomy*, 6th ed, 1969, Williams & Wilkins. Additional data from Chusid. *Correlative Neuroanatomy and Functional Neurology*, 14th ed, 1970, Lange Medical Publications.)

CORPUS CALLOSUM

MASSA INTERMEDIA OF THALAMUS

DORSOMEDIAL NUCLEUS

POSTERIOR NUCLEUS

PINEAL BODY (EPIPHYSIS)

MAMMILLOTHALAMIC TRACT (OF VICQ D'AZYR)

HYPOTHALAMIC NUCLEI

ANTERIOR (SUPRAOPTIC) REGION

MIDDLE (TUBERAL) REGION

CAUDAL (MAMMILLARY) REGION

FORNIX

THALAMUS

PONS

VENTROMEDIAL NUCLEUS

MAMMILLARY BODY

ANTERIOR COMMISSURE

PARAVENTRICULAR NUCLEUS

PREOPTIC NUCLEUS

ANTERIOR NUCLEUS

SUPRAOPTIC NUCLEUS

OPTIC CHIASM

INFUNDIBULUM

HYPOPHYSIS

Table 7.8 MAJOR AFFERENT AND EFFERENT HYPOTHALAMIC CONNECTIONS[a]

NAME	REMARKS
A. Afferent Pathways[b]	
1. Medial forebrain bundle	Connects septal (or medial) olfactory area of brain with preoptic and hypothalamic areas. Also contains efferent (descending) fibers which connect hypothalamus to midbrain. Several hypothalamic nuclei involved, particularly the ventromedial nucleus.
2. Thalamohypothalamic fibers	From medial and midline thalamic nuclei.
3. Fornix	Connects hippocampus to anterior hypothalamus and mammillary nuclei; possibly others. Efferent fibers may be present. Largest fiber system of hypothalamus.
4. Stria terminalis	Connects amygdala to hypothalamus.
5. Inferior mammillary peduncle	Connects tegmentum of midbrain with hypothalamus.
6. Pallidohypothalamic fibers	From lenticular nucleus to ventromedial hypothalamic nucleus.
7. Retinohypothalamic fibers	Fibers of optic nerve enter hypothalamus from optic chiasm.
B. Efferent Pathways[b]	
1. Hypothalamohypophyseal tract (composed of supraoptico-hypophyseal and paraventriculo-hypophyseal tracts)	Connects supraoptic nuclei to infundibulum and neurohypophysis.
2. Mammillotegmental tract	Connects hypothalamus with reticular formation in tegmentum of midbrain.
3. Mammillothalamic tract (of Vicq d'Azyr)	Connects mammillary nuclei with anterior thalamic nuclei.
4. Periventricular system (includes dorsal fasciculus of Schütz)	Connects hypothalamus with midbrain; efferent fibers to spinal cord. Some afferent fibers from sensory pathways also present.
5. Tuberohypophyseal tract	Connects tuberal region of hypothalamus with neurohypophysis.

[a]*Data from Chusid.* Correlative Neuroanatomy and Functional Neurology, 14th ed. *Lange Medical Publications, 1970; Everett:* Functional Neuroanatomy, 6th ed. *Lea & Febiger, 1971; Ganong.* Review of Medical Physiology, 6th ed. *Lange Medical Publications, 1973.*
[b]The subdivision of hypothalamic connections into afferent and efferent pathways indicates the principal types of fibers contained in each of these pathways.

in each of these processes will be discussed in context throughout the remainder of this book.

The Subthalamus. The ventral thalamus or subthalamus is the brain tissue which lies between the tegmentum of the midbrain and the dorsal thalamus. The hypothalamus is located medially and rostrally to the subthalamus, whereas the internal capsule is situated laterally (Fig. 7.45). Caudally, the red nucleus and substantia nigra extend into the subthalamic region (Figs. 7.35 and 7.37). The subthalamic nucleus (or subthalamic body of Luys) extends caudally to the lateral part of the red nucleus.

The subthalamic nucleus receives afferent fibers from the globus pallidus (p. 336) and thus forms a component of the efferent pathway from the corpus striatum (Fig. 7.49). Located rostrally to the red nucleus are the fields of Forel, whose cells may represent a forward extension of the reticular nuclei (p. 268). The fields of Forel contain fibers of the globus pallidus which are designated H (ventromedial fibers), H_1 (dorsomedial fibers), or H_2 (ventrolateral fibers). The relationships among these fibers and the structures that they interconnect are shown in Figure 7.49.

Functionally the subthalamus is integrated with the extrapyramidal system and hence will be discussed in this context (p. 337).

THE RETICULAR FORMATION OF THE SPINAL CORD AND BRAIN STEM; THE RETICULAR ACTIVATING SYSTEM. The central region of the upper spinal cord and brain stem consists of a

Table 7.9   HYPOTHALAMIC FUNCTIONS[a]

FUNCTION[b]	AFFERENT IMPULSES TO HYPOTHALAMUS	HYPOTHALAMIC AREA RESPONSIBLE FOR INTEGRATION OF FUNCTION
1. Control of Endocrine Secretion		
a) Catecholamines	Emotional stimuli, presumably via limbic system	Dorsomedial and posterior hypothalamus
b) Vasopressin	Osmoreceptors	Supraoptic nuclei
c) Oxytocin	Mechanoreceptors in external genitalia, breast, uterus	Paraventricular nuclei
d) Thyrotropin (thyroid stimulating hormone, TSH)	Thermoreceptors; others (?)	Anterior hypothalamus and anterior median hypothalamic eminence
e) Adrenocorticotropic hormone (ACTH)	Emotional stimuli via limbic system; reticular formation; anterior hypophyseal as well as hypothalamic receptor cells responsive to blood corticoid level; others (?)	Middle portion, median hypothalamic eminence
f) Follicle stimulating hormone (FSH), luteinizing hormone (LH)	Hypothalamic cells responsive to blood estrogen level	Posterior median hypothalamic eminence
g) Prolactin	Tactile receptors in breast	Posterior median hypothalamic eminence (inhibits secretion)
h) Somatotropin (growth hormone, GH)	Receptors unknown	Anterior median hypothalamic eminence
2. Thermoregulation	Cold receptors in skin; thermosensitive hypothalamic cells responsive to blood temperature	Anterior hypothalamus responsive to heat; posterior hypothalamus responsive to cold (thermoregulatory centers)
3. "Appetitive" Functions		
a) Thirst	Osmoreceptors	Lateral superior hypothalamus
b) Hunger	Cells responsive to rate of glucose utilization ("glucostats")	Ventromedial hypothalamic satiety center; ventrolateral hypothalamic hunger center; limbic components also
c) Sexual behavior	Cells responsive to blood estrogen or androgen levels among others	Anterior ventral hypothalamus; in male, pyriform cortex also
4. Defense Reactions		
a) Fear, rage	Special sense organs, neocortex	Diffuse; in hypothalamic and limbic system

[a]*Data from Ganong.* Review of Medical Physiology, 6th ed, *1973, Lange Medical Publications.*
[b]*In addition to the functions listed in this table, there is some experimental evidence which indicates that there may be a "sleep center" and a "waking center" in the hypothalamus (see p. 428).*

poorly defined, diffuse mass of interconnected nerve fibers and cells that are known as the *reticular formation*. The large number of small neurons that comprise this phylogenetically primitive formation are not organized anatomically into discrete tracts or fasciculi. Rather, the neurons that comprise the reticular formation are arranged as complex, intertwining networks containing many nuclei of uncertain function. Consequently these nuclei will not be discussed here.

The general location of the reticular formation is depicted in Figure 7.50. Note that the reticular formation extends in an uninterrupted fashion from the upper spinal cord through the medullary tegmentum and pons to terminate within the thalamic region of the diencephalon. Thus the reticular formation includes the hypothalamus, the subthalamus, and the ventromedial portion of the thalamus (including the intramedial nuclei), as well as the thalamic reticular nucleus.

The reticular system receives many afferent collaterals from the long ascending (sensory) spinal tracts (Fig. 7.20) as well as from the visual, olfactory, auditory, and trigeminal nerve systems. Additional afferent inputs to the reticular formation arise within the cerebellum, basal ganglia, rhinencepha-

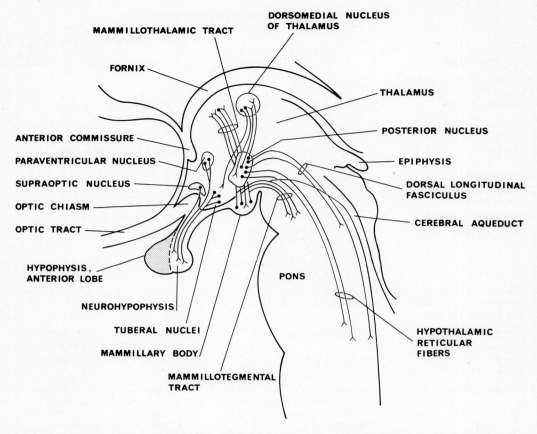

FIG. 7.48. The major efferent hypothalamic pathways. See also Figure 7.39 and Table 7.8. (After Truex and Carpenter. *Human Neuroanatomy,* 6th ed, 1969, Williams & Wilkins.)

lon, and cerebral cortex. Consequently the reticular formation provides a polysynaptic pathway for sensory impulses of all types. In turn, nonspecific efferent fiber pathways from the reticular formation conduct impulses into the higher brain centers, in particular to all regions of the cerebral cortex, as well as caudally into the spinal cord.

Functionally the reticular formation is one of the most important regulatory systems that is located within the central nervous system. The reticular formation of the spinal cord and brain stem provides the neurologic substrate for the so-called *reticular activating system (RAS).* The activity of this system makes possible the alert, conscious (waking) state, whereas a diminution of RAS activity results in the unconscious (or sleeping) state. Thus the RAS exerts a marked influence upon the electric activity of the cerebral cortex. Furthermore, the RAS is concerned intimately with such fundamental processes as maintaining attention; the development of various emotional states and conditioned reflexes; learning; the regulation of skeletal muscle tone and activity; the regulation of visceral functions such as respiration, cardiac rate, vasomotor tone, and gastrointestinal functions; the control of certain endocrine secretions (which in

turn regulate the metabolic rate); the regulation of body temperature; and the modulation of sensory inputs from exteroceptors throughout the body.

The particular functions of the RAS enumerated above will be discussed in context subsequently, but it should be stressed here that this system provides a complex, multisynaptic pathway which receives afferent impulses from all exteroceptors and interoceptors. The large extent to which convergence of these impulses takes place within the RAS in turn tends to abolish any specificity insofar as the particular modalities of sensation are concerned. Consequently the RAS is functionally nonspecific in contrast to the specific neural pathways that are stimulated by only one kind of sensory input (see law of specific nerve energies, p. 120).

From a physiologic standpoint, most of the reticular formation, hence the RAS, is excitatory. In particular, the excitatory regions of the reticular formation are located within the pons, mesencephalon, and diencephalon. The lateral portions of the medullary reticular formation also are excitatory. Collectively these regions are termed the *bulboreticular facilitory area.* On the other hand, a small region of the reticular formation that is lo-

A.

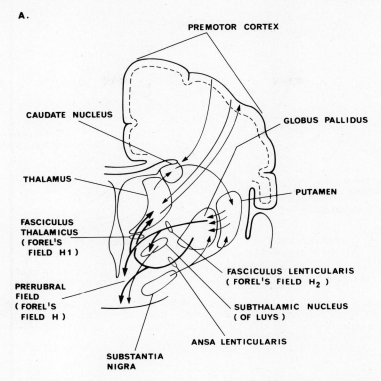

FIG. 7.49. A. Major interconnections of the subthalamus and corpus striatum. (After Gatz. *Manter's Essentials of Clinical Neuroanatomy and Neurophysiology,* 4th ed, 1970, FA Davis.)

B.

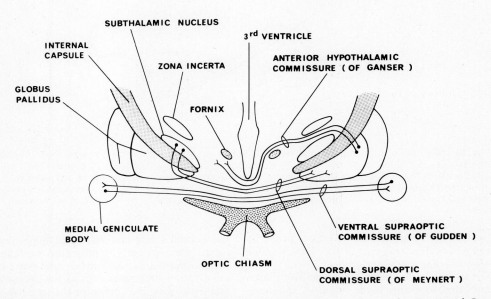

FIG. 7.49. B. The anterior hypothalamic and supraoptic commissures in cross section. (After Truex and Carpenter. *Human Neuroanatomy,* 6th ed, 1969, Williams & Wilkins.)

A.

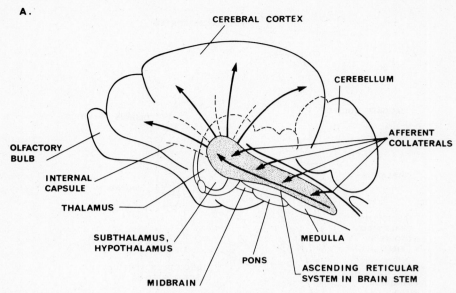

FIG. 7.50. A. Ascending reticular activating system (RAS) of cat brain. Impulses converge on the RAS via afferent collaterals from specific neural pathways, thereby producing excitation of the RAS which in turn evokes nonspecific excitation of the entire cerebral cortex. See also Figure 7.126. (After Best and Taylor [eds]. *The Physiological Basis of Medical Practice,* 8th ed, 1966, Williams & Wilkins.)

B.

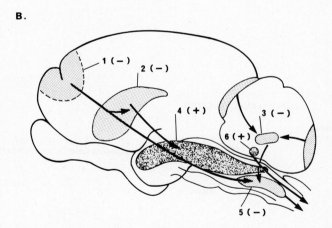

FIG. 7.50. B. Descending suppressor, or inhibitory (−), and facilitory, or excitatory (+), systems (light shading) that act upon the reticular formation of the cat brain. 1. Corticomedullaryreticular pathway. 2. Caudatospinal pathway. 3. Cerebelloreticular pathway. 4. Reticulospinal pathway. 5. Reticulospinal pathway. 6. Vestibulospinal pathway. 1, 2, 3, and 5. exert suppressor effects upon the activity of the reticular formation, whereas 4. and 6. provide facilitatory mechanisms. (After Best and Taylor [eds]. *The Physiological Basis of Medical Practice,* 8th ed, 1966, Williams & Wilkins.)

cated within the ventromedial medulla is primarily inhibitory; therefore this part of the RAS is called the *bulboreticular inhibitory area.*

Normally the bulboreticular facilitory area generates and transmits nerve impulses spontaneously; ie, the neurons are inherently capable of generating repetitive impulses (p. 143). However, in the normal conscious individual inhibitory neural inputs are constantly bombarding the bulboreticular facilitory area, and these inhibitory impulses arise within such structures as the cerebral cortex and basal ganglia. Therefore the intrinsic patterns of activity within the facilitory part of the RAS are reduced considerably.

The Cerebellum

The cerebellum performs a critical, albeit subconscious, role in the control of reflex skeletal muscle tone, in the maintenance of posture and equilibrium, and in the coordination and control of all voluntary muscular activities.

The cerebellum consists of a large and highly organized accumulation of neurons that is located above the medulla and pons and covered posteriorly by the cerebral hemispheres (Figs. 7.11 and 7.12). The cerebellum is attached to the brain stem on each side of the midline by three large bundles of nerve fibers, the superior, middle, and inferior cerebellar peduncles (Figs. 7.27, 7.28, and 7.29).* Thus an abundance of afferent as well as efferent

*The superior cerebellar peduncle also is known as the brachium conjunctivum, the middle peduncle as the brachium pontis, and the inferior peduncle as the restiform body.

connections are present between the cerebellum, brain stem, spinal cord, and cerebral hemispheres. The cerebellum is separated from the overlying cerebral hemispheres by a horizontal sheet of dura mater called the tentorium cerebelli (Fig. 7.14).

The principal gross subdivisions of the cerebellum are the paired lateral cerebellar hemispheres and the median vermis. The principal lobules and fissures that subdivide the cerebellum anatomically are depicted in Figures 7.51 and 7.52.

The flocculonodular lobe, which appears first during embryogenesis, is demarcated by the postnodular and posterolateral fissures from the remainder of the cerebellum which is known as the corpus cerebelli (Fig. 7.52). The flocculonodular lobe, also called the archicerebellum to indicate the early phylogenetic appearance of this region, is connected principally to vestibular structures. The anterior lobe, pyramis, and uvula collectively make up the paleocerebellum. The paleocerebellum is the region within which mainly propriocep-

FIG. 7.51.   Gross anatomy and major divisions of the cerebellum.

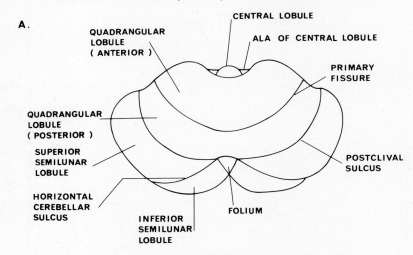

FIG. 7.51.   A. Superior surface.

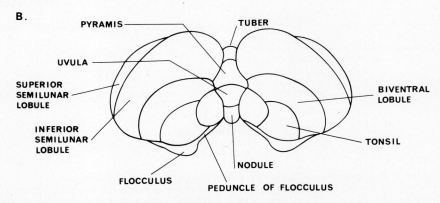

FIG. 7.51.   B. Inferior surface.

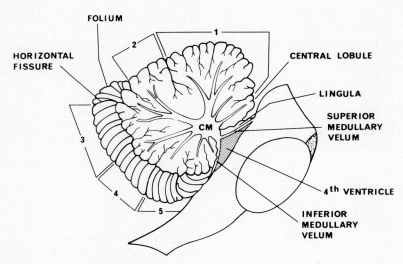

C.

FIG. 7.51. (cont.) **C.** Sagittal section to indicate the lobules of the vermis. 1. Culmen. 2. Declive. 3. Tuber. 4. Pyramis.5. Uvula. CM = central white matter. (After Everett. *Functional Neuroanatomy,* 6th ed, 1971, Lea & Febiger.)

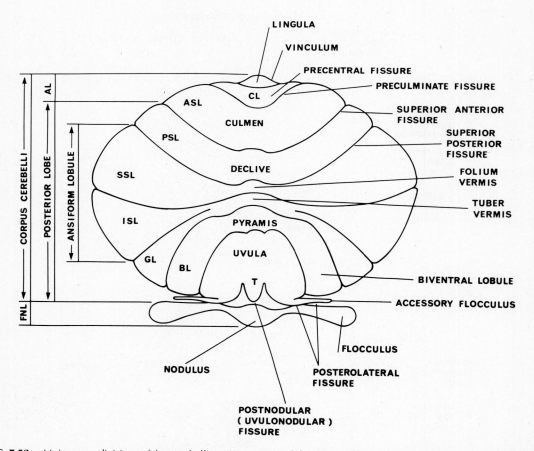

FIG. 7.52. Major gross divisions of the cerebellum. AL = anterior lobe; FNL = flocculonodular lobe; CL = central lobule; ASL = anterior semilunar lobule; PSL = posterior semilunar lobule; SSL = superior semilunar lobule; ISL = inferior semilunar lobule; GL = gracile lobule; BL = biventral lobe; T = tonsil. (After Everett. *Functional Neuroanatomy,* 6th ed, 1971, Lea & Febiger.)

tive and interoceptive impulses are received from the head and body in general. Except for the pyramis and uvula, the posterior lobe is known as the neocerebellum, and this region of the cerebellum developed most recently from a phylogenetic standpoint and concurrently with the cerebral cortex. Therefore extensive neural connections are present between the neocerebellum and cortex via the corticopontile system (Fig. 7.57).

When viewed grossly in cross section, the cerebellum is seen to have an outer gray cortex that surrounds an inner core of white matter within which are embedded four pairs of nuclei. These structures are the dentate, emboliform, globose, and fastigial nuclei (Fig. 7.53). As the emboliform and globose nuclei have similar connections, these structures are known together as the nucleus interpositus. The pathways to and from the individual cerebellar nuclei are summarized in Table 7.10 and depicted in Figures 7.54, 7.55, and 7.56, whereas the principal cerebellar connections are summarized in Figure 7.57.

Upon microscopic examination, the cerebellar cortex is seen to be composed of two layers, an outer molecular layer and an inner granular layer (Fig. 7.58).

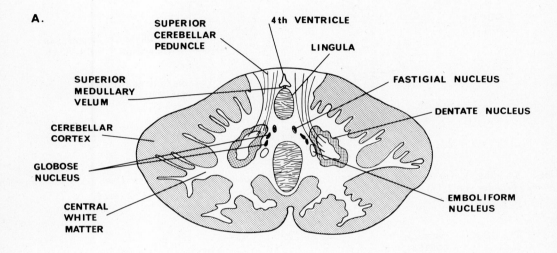

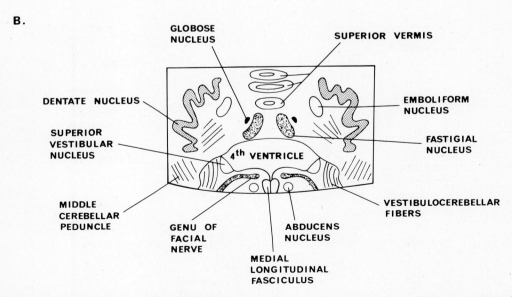

FIG. 7.53. **A.** Cerebellum in horizontal section to illustrate the relationship between the cortical gray matter (shaded around periphery) and the centrally located nuclei within the white matter. **B.** Frontal section through the cerebellum at a level similar to that of the abducens nucleus of the pons (Fig. 7.33) to illustrate the relationships among the cerebellar nuclei. (After Everett. *Functional Neuroanatomy*, 6th ed, 1971, Lea & Febiger.)

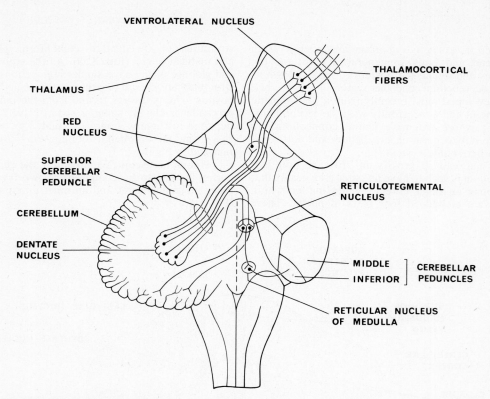

FIG. 7.54. The efferent fibers which project from the dentate nucleus of the cerebellum. (After Truex and Carpenter. *Human Neuroanatomy,* 6th ed, 1969, Williams & Wilkins.)

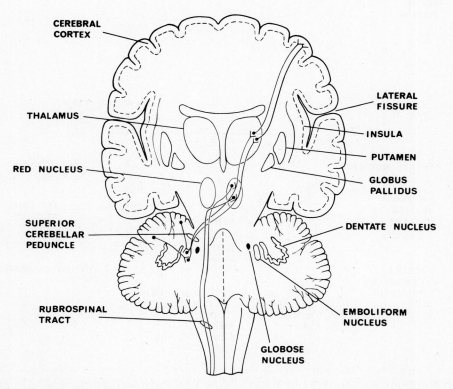

FIG. 7.55. The efferent projections from the emboliform nucleus of the cerebellum to the red nucleus. (After Truex and Carpenter. *Human Neuroanatomy,* 6th ed, 1969, Williams & Wilkins.)

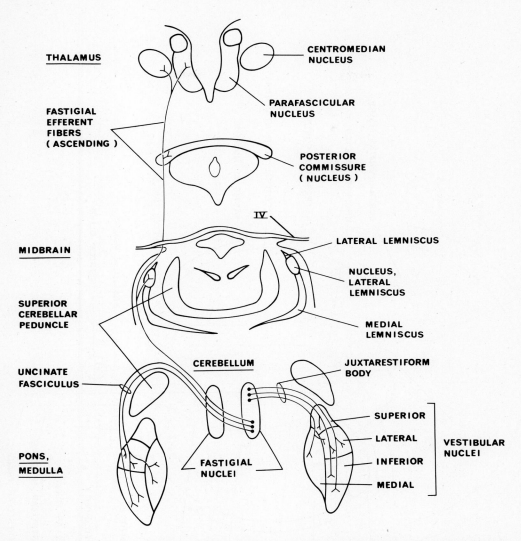

FIG. 7.56. The ascending and descending efferent fiber projections from the fastigial nucleus of the cerebellum. IV = trochlear nerve. (After Truex and Carpenter. *Human Neuroanatomy,* 6th ed, 1969, Williams & Wilkins.)

The outer *molecular layer* contains an abundance of neural processes whose cell bodies are located deeper within the cerebellum; consequently many synapses are present in this region of the cerebellar cortex. Deep within the molecular layer are many so-called basket cells whose axons pass above the Purkinje cells and emit many collaterals, the most obvious being those which form synaptic networks (baskets) about the Purkinje cells. The Purkinje cells in turn are relatively large neurons that are located in a single layer deep within the molecular layer. Functionally, the Purkinje cells are the efferent neurons of the cerebellar cortex, and their axons pass to the cerebellar nuclei. Some of the recurrent axons that arise from these neurons appear to terminate upon other Purkinje cells.

The *granular layer* of the cerebellar cortex contains many small, densely packed cells, the so-called granule cells. Several short, curved dendrites are present on these cells. The axons of the granule cells are longer than the dendrites and they project into the molecular layer, there to undergo bifurcation into layers of parallel fibers. The parallel fibers then synapse upon the spines located on the dendritic processes of the Purkinje cells. The anatomic relationship between the parallel fibers of the granule cells and the dendrites of the Purkinje cells underlies the physiologic properties of spreading (or diffusion) of impulses as well as convergence of impulses within the cerebellar cortex, as one parallel fiber may make synaptic contact with the dendrites of several hundred Purkinje cells.

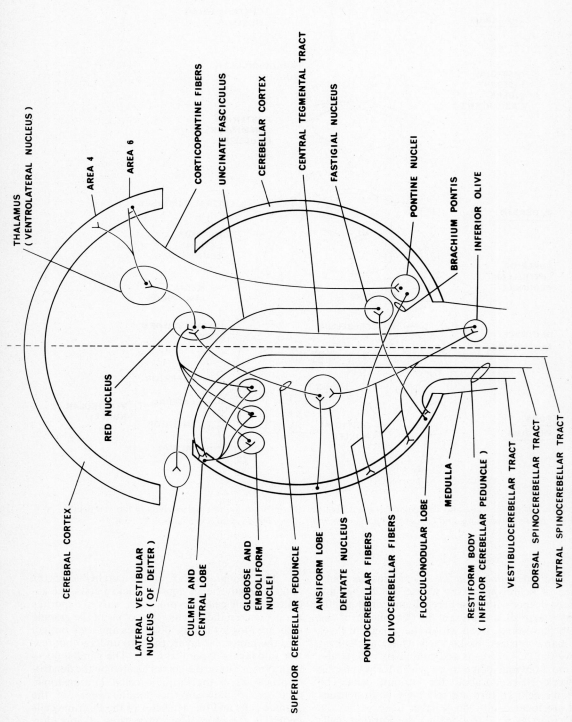

THALAMUS
(VENTROLATERAL NUCLEUS)

AREA 4

AREA 6

CORTICOPONTINE FIBERS

UNCINATE FASCICULUS

CEREBELLAR CORTEX

CENTRAL TEGMENTAL TRACT

FASTIGIAL NUCLEUS

PONTINE NUCLEI

BRACHIUM PONTIS

INFERIOR OLIVE

CEREBRAL CORTEX

RED NUCLEUS

LATERAL VESTIBULAR
NUCLEUS (OF DEITER)

CULMEN AND
CENTRAL LOBE

GLOBOSE AND
EMBOLIFORM
NUCLEI

SUPERIOR CEREBELLAR PEDUNCLE

ANSIFORM LOBE

DENTATE NUCLEUS

PONTOCEREBELLAR FIBERS

OLIVOCEREBELLAR FIBERS

FLOCCULONODULAR LOBE

MEDULLA

RESTIFORM BODY
(INFERIOR CEREBELLAR PEDUNCLE)

VESTIBULOCEREBELLAR TRACT

DORSAL SPINOCEREBELLAR TRACT

VENTRAL SPINOCEREBELLAR TRACT

FIG. 7.57. Highly schematic diagram illustrating the principal afferent and efferent cerebellar connections. Uncinate fasciculus = hook bundle of Russell. (Data from Chusid. *Correlative Neuroanatomy and Functional Neurology,* 14th ed, 1970, Lange Medical Publications; Everett. *Functional Neuroanatomy,* 6th ed, 1971, Lea & Febiger; Gatz. *Manter's Essential of Clinical Neuroanatomy and Neurophysiology,* 4th ed, 1970; F. A. Davis; Truex and Carpenter. *Human Neuroanatomy,* 6th ed, 1969, Williams & Wilkins.)

312

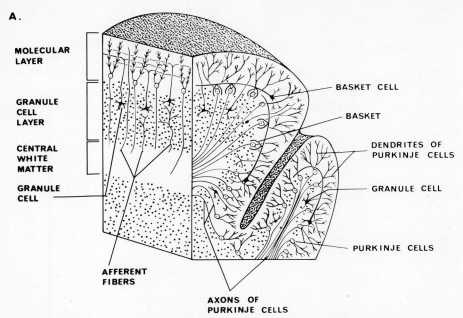

A.

MOLECULAR LAYER

GRANULE CELL LAYER

CENTRAL WHITE MATTER

GRANULE CELL

BASKET CELL

BASKET

DENDRITES OF PURKINJE CELLS

GRANULE CELL

PURKINJE CELLS

AFFERENT FIBERS

AXONS OF PURKINJE CELLS

FIG. 7.58. Microanatomy of the cerebellar cortex. **A.** Longitudinal section of a single cerebellar folium on left and cross sections of two adjacent folia on right illustrating the principal cell layers and synaptic connections. (After Everett. *Functional Neuroanatomy,* 6th ed, 1971, Lea & Febiger.)

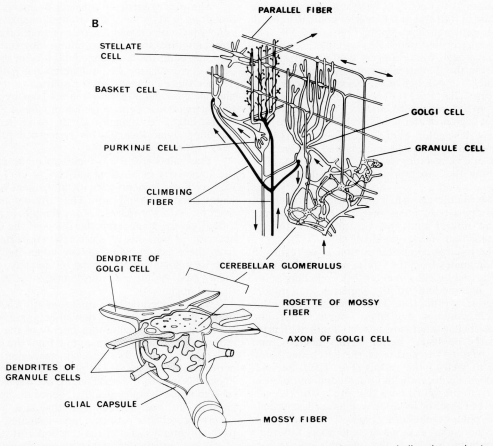

B.

PARALLEL FIBER

STELLATE CELL

BASKET CELL

PURKINJE CELL

CLIMBING FIBER

GOLGI CELL

GRANULE CELL

DENDRITE OF GOLGI CELL

CEREBELLAR GLOMERULUS

ROSETTE OF MOSSY FIBER

AXON OF GOLGI CELL

DENDRITES OF GRANULE CELLS

GLIAL CAPSULE

MOSSY FIBER

FIG. 7.58. **B.** Interconnections among cerebellar neurons. Inset at the bottom shows a cerebellar glomerulus in greater detail. See text for discussion. (After Llinás. The cortex of the cerebellum. *Sci Amer* 232:56, 1975; Inset after Truex and Carpenter. *Human Neuroanatomy,* 6th ed, 1969, Williams & Wilkins.)

Table 7.10   CEREBELLAR NUCLEI AND THEIR CONNECTIONS

NUCLEUS	AFFERENT FIBERS FROM	EFFERENT FIBERS TO
Dentate	Neocerebellar portion of posterior lobe; anterior lobe	Superior cerebellar peduncles, red nucleus, ventrolateral nucleus of thalamus
Emboliform	Paleocerebellum	Red nucleus via superior cerebellar peduncle
Globose	Paleocerebellum	Red nucleus via superior cerebellar peduncle
Fastigial	Flocculonodular lobe	Vestibular nucleus via the uncinate fasciculus (hook bundle of Russell)

Data from Chusid. Correlative Neuroanatomy and Functional Neurology, 14th ed, 1970, Lange Medical Publications; and Everett. Functional Neuroanatomy, 6th ed. 1971, Lea & Febiger.

The cerebellum contains three large bundles of projection fibers, the cerebellar peduncles (Figs. 7.27, 7.28, and 7.29).

The superior cerebellar peduncle (or brachium conjunctivum) passes from the white substance of the upper medial portion of the cerebellar hemisphere and then enters the lateral wall of the fourth ventricle. Thence the major portion of this fiber tract ascends to decussate in its entirety within the midbrain at the level of the inferior colliculi beneath the cerebral aqueduct (Figs. 7.53 and 7.54). The superior cerebellar peduncle contains three principal fiber tracts. First, efferent dentatorubral fibers pass from the dentate nucleus to the contralateral red nucleus and thalamus. Second, the ventral spinocerebellar tract (afferent) enters the cerebellum from the spinal cord, and then terminates in the cortical region of the paleocerebellum. Third, the uncinate fasciculus contains efferent fibers that arise within the fastigial nucleus and terminate in the lateral vestibular nucleus (of Deiter) after traversing the superior cerebellar peduncle (Fig. 7.56).

The middle cerebellar peduncle (or brachium pontis) consists of fibers that originate within the pontine nuclei and run to the contralateral neocerebellum via this peduncle (Fig. 7.57).

The inferior cerebellar peduncle (or restiform body) emanates from the lateral walls of the fourth ventricle and ascends to enter the cerebellum in the region intermediate to the superior and middle cerebellar peduncles. The inferior cerebellar peduncle contains five major tracts. First, the olivocerebellar tract consists of fibers that arise within the opposite inferior olivary nucleus, and which pass to the cortex of the cerebellar hemisphere as well as the vermis (Fig. 7.59). Second, the dorsal spinocerebellar tract contains afferent fibers from the spinal cord that run to the cerebellar cortex (anterior lobe) as well as to the pyramis of the paleocerebellum (Figs. 7.52 and 7.57). Third, the dorsal external arcuate fibers (afferent) pass from the nuclei of the funiculus gracilis and cuneatus of the medulla (Fig. 7.59). Fourth, the ventral external arcuate fibers run from the arcuate

and lateral reticular nuclei of the medulla to the cerebellum via the inferior cerebellar peduncle. Fifth, the vestibulocerebellar tract passes from the vestibular nuclei to the flocculonodular lobe of the cerebellar cortex (Fig. 7.57).

The general functions of the cerebellum may be localized to some extent on the basis of various types of experimental and clinical study.

The phylogenetically ancient archicerebellum functions to keep the body oriented in space, ie, is concerned with the maintenance of equilibrium. Afferent impulses that arise within the labyrinth of the inner ear enter the flocculonodular lobe of the cerebellar cortex via the vestibulocerebellar pathways. Thence these impulses are transmitted to the fastigial nucleus, and efferent impulses pass by way of the uncinate fasciculus (or hook bundle of Russell) to the lateral vestibular nucleus (of Deiter). The flocculonodular lobe thus is concerned primarily with vestibular mechanisms.

The paleocerebellum regulates contraction of the antigravity or postural muscles of the body as well as those concerned with movement and the maintenance of equilibrium. Thus afferent impulses from the extensor muscles, for example, are transmitted via the spinocerebellar tracts of the culmen and central portions of the cerebellar cortex, and from these regions impulses are conducted to the globose and emboliform nuclei. Ultimately, the impulses pass via the superior cerebellar peduncle to the red nucleus.

The neocerebellum functions to check (or brake) discrete voluntary movements, particularly those involving delicate movements of the hands. Afferent impulses are transmitted from the precentral motor cortex to the ansiform region of the neocerebellar cortex via the pontocerebellar tracts. From this part of the cerebellum, the impulses are relayed to the dentate nucleus. Thence the impulses pass to the red nucleus and thalamus via the superior cerebellar peduncle, and ultimately return to the premotor areas of the cerebral cortex.

In conclusion, the cerebellum receives a broad range of sensory information from end organs that are located throughout the body, and this informa-

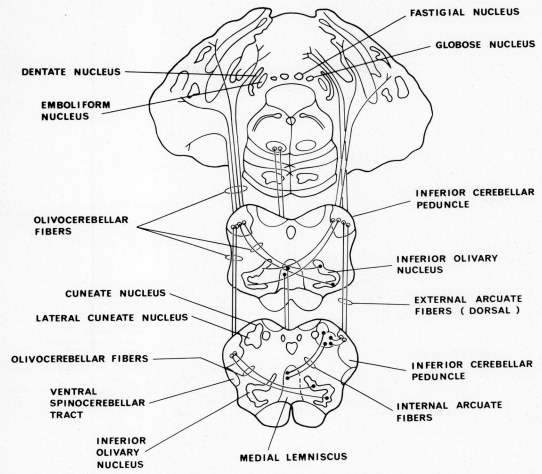

FIG. 7.59. The connections between the lateral nucleus, the olivary nucleus, and the cerebellum. (After Everett. *Functional Neuroanatomy,* 6th ed, 1971, Lea & Febiger.)

tion is coordinated and integrated within the cerebellum itself. Thence, and following appropriate synthesis, the modified (ie, facilitated or inhibited) impulse patterns are projected (or transmitted) from the cerebellum to motor centers that are located within the spinal cord, brain stem, and cerebral cortex. In this manner the activity of the motoneurons that innervate the skeletal muscles of the body is controlled by the cerebellum. As indicated in Figure 7.60, the neural circuits that involve the cerebellum and other central nervous structures, and which underlie the control of complex muscular activities, are excellent examples of feedback loops (or servomechanisms).

The Cerebrum and Related Structures

The cerebrum is by far the largest anatomic region of the human brain, and it contains the highest (or functionally most sophisticated) neural centers to be found in the central nervous system. The cerebrum is so voluminous that it occupies the larger part of the cranial cavity of the skull and covers the cerebellum and brain stem (Figs. 7.11 and 7.12).

Anatomically the cerebrum is divided into two lateral cerebral hemispheres by the longitudinal cerebral fissure, and this sagittal fissure is occupied by a projection of dura mater known as the falx cerebri. The two cerebral hemispheres are joined together by a large mass of white nerve fibers, the corpus callosum. The corpus callosum is the white central commissure which crosses the longitudinal cerebral fissure (Fig. 7.38).

Caudally the cerebral hemispheres are connected to lower brain centers by the continuous nerve fibers of the internal capsule and cerebral peduncle (Figs. 7.12, 7.38, and 7.61).

Deep within each cerebral hemisphere are embedded several large masses of gray matter collectively known as the basal ganglia (Figs. 7.38, 7.49, and 7.80).

The pertinent neuroanatomic features of the cerebral hemispheres and basal ganglia together with a number of their important functional corre-

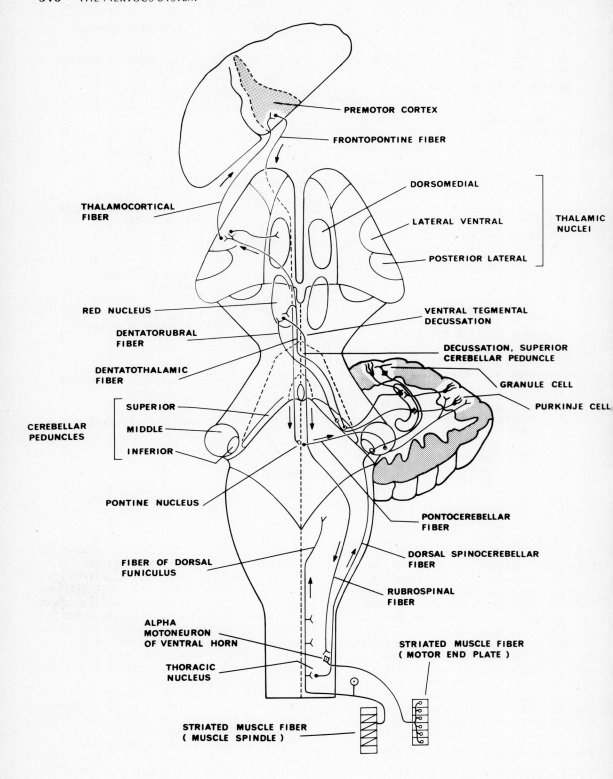

FIG. 7.60. Some of the neural circuits which underlie the coordination of complex motor activity. See also Figure 7.100. (After Everett. *Functional Neuroanatomy,* 6th ed, 1971, Lea & Febiger.)

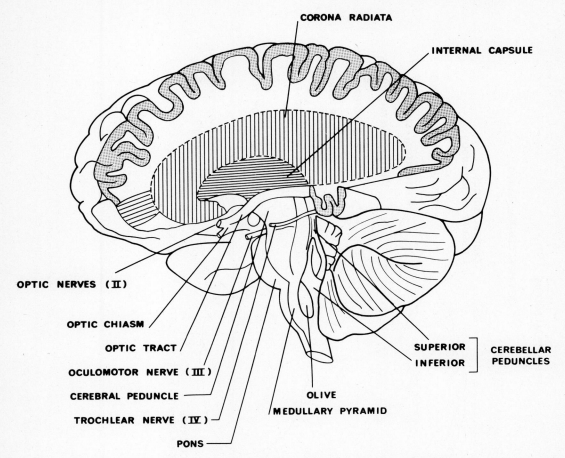

FIG. 7.61. Relationships among the cerebrospinal fibers and other regions of the brain. The stippled area represents the cut gray matter of the cerebral cortex. (After Goss [ed]. *Gray's Anatomy of the Human Body*, 28th ed, 1970, Lea & Febiger.)

lates now may be considered. In this regard, the various anatomic levels within the central nervous system at which certain functions become integrated are summarized in Table 7.11. The reader will note in this table that the cerebral cortex provides the ultimate neural region wherein the so-called higher functions of the human nervous system take place.

THE CEREBRAL HEMISPHERES

Gross Anatomy.   Upon gross inspection, the surfaces of the cerebral hemispheres are seen to be covered with a number of elevated convolutions (or gyri) which are separated by furrows known as fissures (or sulci). The principal gyri and sulci are depicted in Figures 7.62, 7.63, and 7.64.

As indicated in Figures 7.62, 7.64, 7.65, and 7.78, each cerebral hemisphere can be divided into several general regions, the frontal, parietal, occipi-

tal, and temporal lobes, as well as the insula and rhinencephalon.

The *frontal lobe* extends from the most anterior portion of each hemisphere to the central sulcus (or fissure of Rolando) posteriorly, and is delimited inferiorly by the lateral cerebral fissure (or fissure of Sylvius) on the dorsolateral surface (Figs. 7.62 and 7.64). The frontal lobe includes the superior, middle, and inferior frontal gyri, as well as the orbital and the cingulate gyri. The gyrus rectus (or straight gyrus) also is part of the frontal lobe.

The *parietal lobe* includes the region between the central sulcus and the parieto-occipital fissure, and extends laterally to the lateral cerebral fissure (Figs. 7.62 and 7.63). The superior and inferior parietal lobules as well as the supramarginal, angular, and postcentral gyri are the subdivisions of the parietal lobe.

The *occipital lobe* is roughly pyramidal in shape and this region lies behind the parieto-

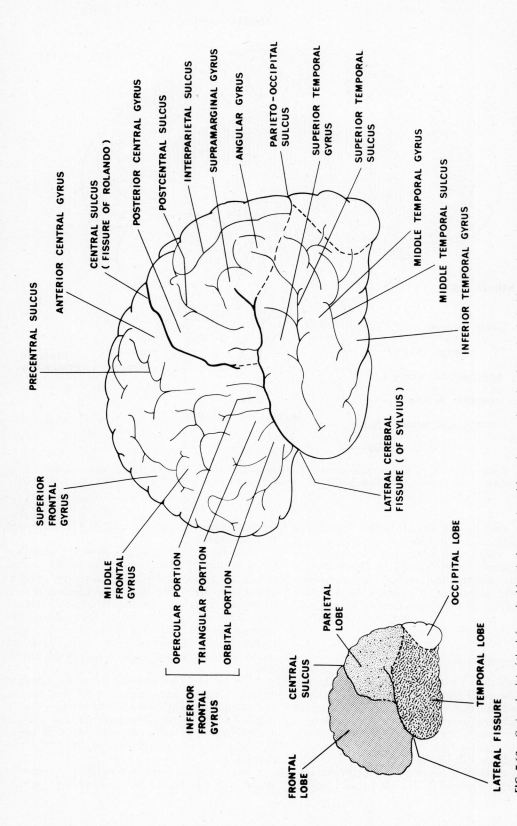

FIG. 7.62. Gyri and sulci of the left cerebral hemisphere as viewed from the superolateral surface. The dashed lines indicate the principal lobes of the cerebral hemisphere as indicated by shaded areas in the small diagram inset at the lower left of the figure. (After Everett. *Functional Neuroanatomy*, 6th ed, 1971, Lea & Febiger. Additional data from Chusid. *Correlative Neuroanatomy and Functional Neurology*, 14th ed, 1970, Lange Medical Publications.)

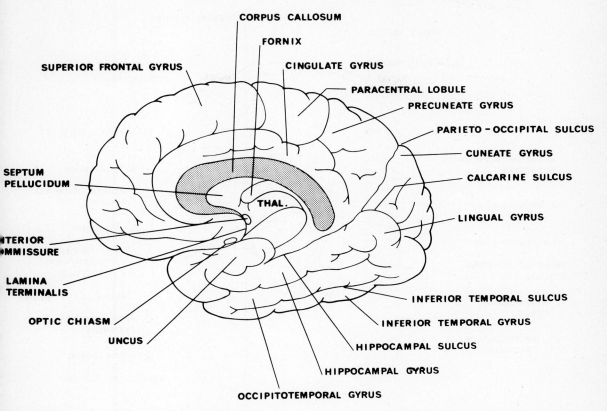

FIG. 7.63. Principal gyri and sulci seen on the medial surface of the left cerebral hemisphere. THAL = thalamus. (After Chusid. *Correlative Neuroanatomy and Functional Neurology,* 14th ed, 1970, Lange Medical Publications. Additional data from Everett. *Functional Neuroanatomy,* 6th ed, 1971, Lea & Febiger.)

occipital fissure (Figs. 7.62, 7.63, and 7.64). The occipital lobe is divided into superior and inferior gyri, and this lobe also contains the cuneus as well as the lingual and fusiform gyri.

The *temporal lobe* is located inferior to the lateral cerebral fissure (of Sylvius), and reaches posteriorly to the level of the parieto–occipital fissure (Figs. 7.62, 7.63, and 7.64). The superior, middle, and inferior temporal gyri as well as the transverse temporal gyrus (Heschl's gyrus) are included in the temporal lobe. In addition, the hippocampal gyrus and its hook-shaped anterior portion (the uncus) also is a part of the temporal lobe.

The *insula* (or island of Reil) is located deep within the lateral fissure (of Sylvius), and can be seen by moving apart the upper and lower edges of this fissure (Figs. 7.38 and 7.65). The insula consists of several short gyri as well as a posterior long gyrus.

The *rhinencephalon* ("nose brain") is a phylogenetically ancient region of the cerebral hemisphere, and it includes those parts of the brain that are involved in the perception of olfactory stimuli, among other functions (see also p. 332). The ovoid olfactory bulb rests upon the ethmoid bone (cribriform plate) where it receives olfactory nerves from the olfactory region of the nasal cavity

(Figs. 7.12 and 7.66). The olfactory tract* divides posteriorly into the lateral olfactory stria which then enters the uncus of the hippocampal gyrus, and the medial olfactory stria which runs medially and to the subcallosal gyrus adjacent to the inferior part of the corpus callosum (Fig. 7.64). The pyriform area of the rhinencephalon includes the anterior portion of the hippocampal gyrus, the uncus, and the lateral olfactory gyrus.

The rhinencephalon also includes the subcallosal and supracallosal gyri, the medial and lateral longitudinal striae which lie along the upper surface of the corpus callosum, and the hippocampus (Fig. 7.78). The hippocampus is composed principally of gray matter and this structure runs the length of the floor of the lateral ventricle where it becomes continuous with the supracallosal gyrus in the splenium region of the corpus callosum (Fig. 7.63). The fornix is a curved fiber tract that emerges from the hippocampal formation. A white layer called the alveus is found upon the ventricular surface of the hippocampus, and this structure contains nerve fibers that lead from the dentate fascia (a thin layer of cortex located upon the upper surface of the

---

Actually the olfactory nerve (cranial nerve I) is not a true nerve; rather it is a fiber tract of the brain.

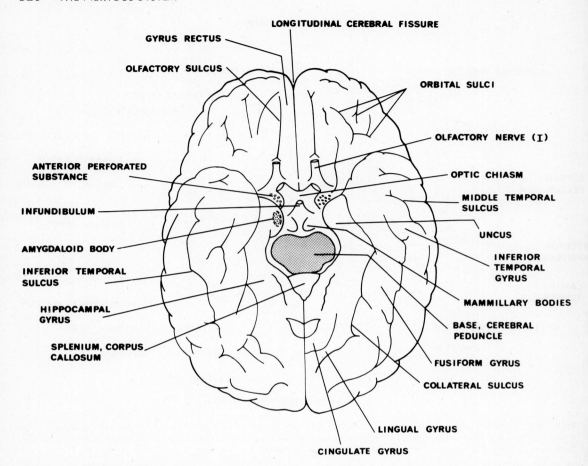

GYRUS RECTUS

OLFACTORY SULCUS

LONGITUDINAL CEREBRAL FISSURE

ORBITAL SULCI

OLFACTORY NERVE (I)

OPTIC CHIASM

MIDDLE TEMPORAL SULCUS

UNCUS

INFERIOR TEMPORAL GYRUS

MAMMILLARY BODIES

BASE, CEREBRAL PEDUNCLE

FUSIFORM GYRUS

COLLATERAL SULCUS

LINGUAL GYRUS

CINGULATE GYRUS

ANTERIOR PERFORATED SUBSTANCE

INFUNDIBULUM

AMYGDALOID BODY

INFERIOR TEMPORAL SULCUS

HIPPOCAMPAL GYRUS

SPLENIUM, CORPUS CALLOSUM

FIG. 7.64. Inferior surface of the cerebrum. The anterior portion of the left temporal lobe has been removed in order to show the location of the amygdaloid body. (After Everett. *Functional Neuroanatomy*, 6th ed, 1971, Lea & Febiger. Additional data from Chusid. *Correlative Neuroanatomy and Functional Neurology*, 14th ed, 1970, Lange Medical Publications.)

hippocampus) and the hippocampus proper. Fibers pass from the alveus to the medial surface of the hippocampus to form the fimbria. The fimbria consist of a flat band of white nerve fibers which rise beneath the splenium of the corpus callosum, bend forward above the thalamus, and thus form the crus of the fornix. The two lateral crura of the fornix are interconnected by the transverse fibers of the hippocampal commissure. Each crus lies adjacent to the underside of the corpus callosum and the two crura unite anteriorly, thereby forming the body of the fornix. The two columns of the fornix in turn arch downward and posteriorly, enter the anterior region of the lateral wall of the third ventricle, and there terminate in the mamillary bodies of the hypothalamus.

The anterior commissure is composed of a band of white nerve fibers that crosses the midline of the cerebrum to interconnect both cerebral hemispheres (Fig. 7.63). The anterior commissure appears to have two distinct regions: The rostral part joins both olfactory bulbs (Fig. 7.67), whereas the more caudal portion interconnects the pyriform areas of each cerebral hemisphere with the other.

The septum pellucidum is a thin-walled partition which is located between the lateral ventricles and which is situated between the fornix and corpus callosum (Fig. 7.63).

The White Matter. The white matter of the cerebral hemispheres is composed of a vast number of myelinated nerve fibers of various sizes as well as neuroglial elements. The core of each cerebral hemisphere includes three types of myelinated nerve fiber. Thus transverse, projection, and association fibers are present.

Transverse Fibers. Transverse (or commissural) fibers provide interconnections between the two cerebral hemispheres. The largest tract of commissural fibers is the corpus callosum. The majority of the fibers present within the corpus callosum arise in various particular regions of one cerebral hemisphere and terminate in similar re-

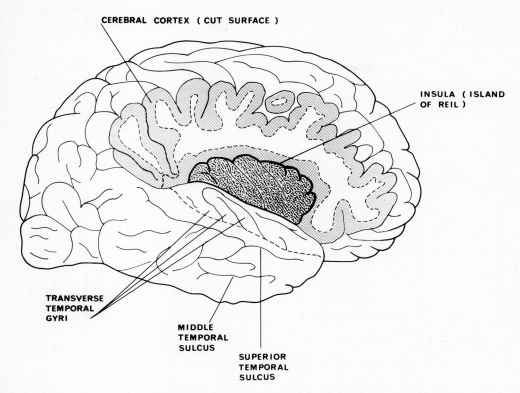

CEREBRAL CORTEX (CUT SURFACE)

INSULA (ISLAND OF REIL)

TRANSVERSE TEMPORAL GYRI

MIDDLE TEMPORAL SULCUS

SUPERIOR TEMPORAL SULCUS

FIG. 7.65. Superior lateral aspect of the right cerebral hemisphere. The lower portions of the frontal and parietal areas have been removed to show the superior surface of the temporal lobe (delimited by sloping dashed line in lower center of figure) and the insula (dark shading circumscribed by heavy black line). (After Everett. *Functional Neuroanatomy*, 6th ed, 1971, Lea & Febiger.)

gions in the opposite hemisphere. The corpus callosum is a thick, bandlike, transverse structure which makes up the roof of the lateral ventricles as well as the roof of the third ventricle (Figs. 7.11, 7.38, and 7.63).

As noted earlier, the anterior commissure interconnects the two olfactory bulbs in its rostral portion (Fig. 7.67); caudally the anterior commissure interconnects the paired pyriform areas of the rhinencephalon.

The hippocampal commissure unites the two hippocampal gyri.

Projection Fibers. Certain projection fibers of the white substance of the cerebral hemisphere traverse the corona radiata to enter the internal capsule (qv, p. 336), and these fibers serve to connect the cerebral cortex with lower regions of the brain as well as the spinal cord (Fig. 7.61).

The principal afferent (or corticipetal) projection fibers include thalamic radiations from the several thalamic nuclei to specific areas of the cerebral cortex, the geniculocalcarine tract which arises in the lateral geniculate body and passes to the calcarine region of the cerebral cortex, and thalamic auditory fibers that arise within the medial geniculate body and radiate to Heschl's gyrus of the auditory cortex (p. 330).

The efferent (or corticofugal) fibers originate within the cerebral cortex, and pass from this region to the thalamus, brain stem, and spinal cord. The corticobulbar and corticospinal tracts, which together comprise the pyramidal motor system of the body, originate from the motor cortex, thence proceed caudally via the internal capsule (p. 336). The corticopontile tracts, which pass from the cerebral cortex to the pons, include the frontopontile tract and the temporopontile tract. The frontopontile tract originates within the frontal lobe of the cerebral cortex and runs to the pontine nuclei. The temporopontile tract originates in the temporal lobe of the cerebral cortex and also terminates in the pontine nuclei. Corticothalamic fibers course from the cerebral cortex to various thalamic nuclei, and the corticorubral tract connects the frontal lobe of the cerebral cortex to the red nucleus located within the midbrain (Fig. 7.22). The fornix sends some projection fibers to the midbrain from their origins within the hippocampus.

Association Fibers. The association fibers of the white matter serve to connect various regions within the same cerebral hemisphere (refer to Figs. 7.68 and 7.69).

Short association (arcuate) fibers connect adjacent gyri. The deeper-lying short association fibers

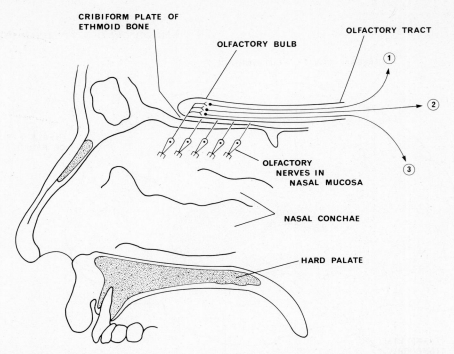

FIG. 7.66. Lateral view of the olfactory nerve. ① Medial olfactory stria to contralateral olfactory bulb via anterior commissure. ② Intermediate olfactory stria to anterior perforated substance and diagonal band. ③ Lateral olfactory stria to nuclei of amygdaloid complex (limbic system) and prepyriform cortex. (See also Figs. 7.12, 7.67, and 7.78.) (After Chusid. *Correlative Neuroanatomy and Functional Neurology,* 14th ed, 1970, Lange Medical Publications. Additional data from Ganong. *Review of Medical Physiology,* 6th ed, 1973, Lange Medical Publications.)

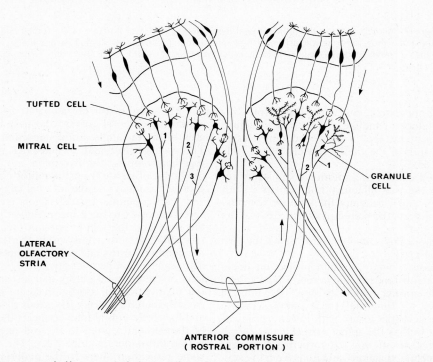

FIG. 7.67. The course of olfactory signals (arrows) from the olfactory mucous membrane to the olfactory cortex via the olfactory bulbs. The bipolar cells of the olfactory mucous membrane project axons into the olfactory bulb where their terminations synapse upon tufted and mitral cells forming structures called olfactory glomeruli. Axons of the mitral cells form the lateral olfactory stria (or tract), whereas the axons of the tufted cells (1, 2, 3) form the medial olfactory stria (or tract) that runs to the contralateral olfactory bulb via the anterior commissure. The granule cells are the elements upon which the fibers of the contralateral tufted cells terminate (3, 2, 1). See also Figure 7.78. (After Everett. *Functional Neuroanatomy,* 6th ed, 1971, Lea & Febiger. Additional data from Ganong. *Review of Medical Physiology,* 6th ed, 1973, Lange Medical Publications.)

A.

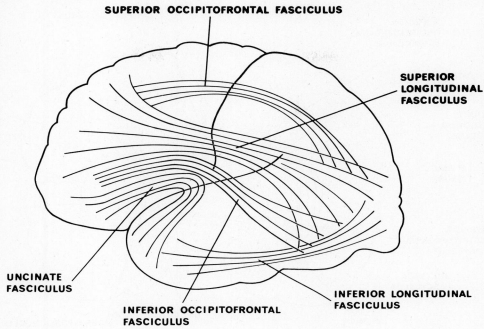

SUPERIOR OCCIPITOFRONTAL FASCICULUS

SUPERIOR
LONGITUDINAL
FASCICULUS

UNCINATE
FASCICULUS

INFERIOR OCCIPITOFRONTAL
FASCICULUS

INFERIOR LONGITUDINAL
FASCICULUS

B.

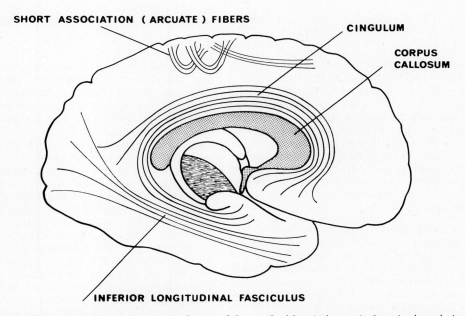

SHORT ASSOCIATION (ARCUATE) FIBERS

CINGULUM

CORPUS
CALLOSUM

INFERIOR LONGITUDINAL FASCICULUS

FIG. 7.68. Principal long and short association pathways of the cerebral hemispheres. **A.** Superior lateral view of left hemisphere. **B.** Medial view. See also Figure 7.69. (After Everett, *Functional Neuroanatomy,* 6th ed, 1971, Lea & Febiger.)

within the cortex are called intracortical fibers, whereas the short association fibers found immediately beneath the cerebral cortex per se are called subcortical fibers.

The long association fibers of the cerebral cortex, in contrast to the short association fibers, serve to connect more distant areas of the cerebral cor-

tex. Six important tracts or bundles of long association fibers are found within each cerebral hemisphere. First, the superior longitudinal fasciculus connects certain regions of the frontal lobe with areas located in the temporal and occipital lobes. Second, the inferior longitudinal fasciculus connects the temporal and occipital lobes. Third, the

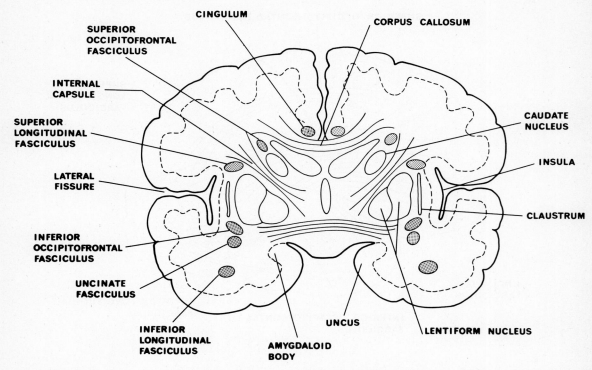

FIG. 7.69. Transverse section of brain illustrating the long association pathways (shaded areas). (After Everett. *Functional Neuroanatomy,* 6th ed, 1971, Lea & Febiger.)

occipitofrontal fasciculus courses dorsally from the frontal lobe, thereby connecting this structure with the occipital and temporal lobes. Fourth, the uncinate fasciculus connects the inferior gyri of the frontal lobe with the anterior temporal lobe. Fifth, the cingulum, which is found within the cingulate gyrus as a white strip, connects the anterior perforated substance of the rhinencephalon and the hippocampal gyrus (Fig. 7.64). Sixth, the principal radiations of the fornix serve to connect the hippocampus to the mamillary body of the hypothalamus (Fig. 7.48).

Histology and Cytoarchitecture of the Cerebral Cortex. The major functional region of the cerebral cortex is composed of a thin superficial layer of neurons and glial elements (gray matter) that ranges between 1.5 and 4.0 mm in thickness, and which has a total surface area of roughly 2,000 cm². And yet this layer, which contains approximately 9 billion individual neurons and covers the surfaces of all of the convolutions of the cerebrum, is responsible for all of the perceptive, integrative, and motor functions of the cerebral cortex.

The cerebral cortex is composed of a great morphologic variety of neural elements; however, these elements are highly organized into a remarkably constant laminar pattern over most cortical regions.

For purposes of discussion, the cerebral cortex frequently is divided into the phylogenetically an-

cient allocortex and more recently evolved isocortex. The allocortex includes those rhinencephalic structures that developed in relation to the sense of olfaction, and includes the hippocampus (or archipallium) and pyriform cortex (or paleopallium). The allocortex does not exhibit the characteristic six-layered organizational pattern of neural elements that is found in the isocortex. The isocortex (also called the neocortex or neopallium) forms by far the largest area of the human cortex, and this region exhibits six distinct histologic layers during some stage of ontogeny (Brodmann). Despite the fact that there is a nonuniform distribution of similar neural elements throughout the isocortex, nevertheless the typical lamination of neurons is present throughout this cortical region.

The six layers of the neocortex can be discerned in Nissl-stained sections of the cortex, and the laminar pattern seen in such preparations is said to represent the cytoarchitecture of the cortex.*

Five principal types of neuron are found within the cerebral cortex:

First, *pyramidal cells* are the most abundant neural components of the cortex, and these elements range in diameter from approximately 10 to 100 μ. Pyramidal cells are classed as small, medium, large, and giant. The dendrite arises from the apex of the pyramidal cell body and runs to-

If the sections of cortex are stained for myelin, the so-called myeloarchitecture is revealed.

ward the cortical surface, there to terminate in many branches. An abundance of smaller dendrites (basal dendrites) emerge from the base and sides of the pyramidal cell bodies and course into the adjoining cortical layers.

The giant pyramidal cells (of Betz), found in the primary motor projection cortex, give rise to axons which control voluntary movements of skeletal muscles on the contralateral side of the body (p. 397).

Second, *satellite* (or *granule*) *cells* are small neural elements that range between 5 and 10 μ in diameter. The satellite cells have numerous branched dendrites which terminate near the cell from which they originate.

Third, *horizontal cells* are relatively small neural elements which have spindle-shaped cell bodies. The processes of the horizontal cells run for long distances in a direction parallel with the outer cortical surface.

Fourth, *cells of Martinotti* are small multipolar neurons that have a rounded or triangular shape, and whose processes terminate exclusively within the cortex. Cells of Martinotti are found in all cortical layers, and their axons pass toward the cortical surface, while their dendrites ramify in all directions.

Fifth, *fusiform cells* are found principally within the deepest layer of the cortex, and their dendrites course into the upper cortical layers. The axons of the fusiform cells enter the white matter, but their termination and distribution are unknown.

The six layers of the neocortex are as follows, from without inward (refer to Fig. 7.70):

I. *Molecular (or Plexiform) Layer:* The molecular layer contains the terminations of dendrites from neurons that lie deeper within the cortex, horizontal cells, and a few cells having short axons (Golgi type II).

II. *External Granular Layer:* The external granular layer is composed of many small granule (or satellite) and pyramidal cells. The dendrites of neurons that lie within the deeper layers pass

A.

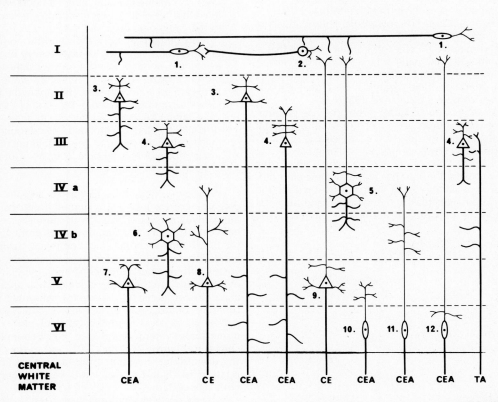

FIG. 7.70. A. Layers of the cerebral cortex (sensory area) presented in highly diagrammatic form. Heavy lines on individual cells represent axons, light lines represent dendrites. The six cortical layers (Roman numerals) are: I, Molecular (plexiform) layer. II, External granular layer. III, Pyramidal layer. IV, Internal granular layer. V, Ganglionic layer. VI, Polymorphic layer. The specific cortical cells found in these layers are: 1. Horizontal cell. 2. Golgi type II cell. 3. Small pyramidal cell. 4. Medium pyramidal cell of layer III. 5. Star pyramidal cell. 6. Stellate cell. 7. Short pyramidal cell. 8. Medium pyramidal cell of layer V. 9. Large pyramidal cell. 10. Short spindle cell. 11. Medium spindle cell. 12. Long spindle cell. Axons found in the central white matter are as follows. CEA = cortical efferent (association) fibers; CE = cortical efferent (projection) fibers; TA = thalamic afferent fibers. (After Everett. *Functional Neuroanatomy,* 6th ed, 1971, Lea & Febiger.)

B.

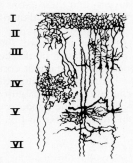

FIG. 7.70. B. Neural interconnections within the cerebral cortex. On the left side are two thalamic afferent fibers, nonspecific and specific respectively. Note particularly the density of the dendritic processes in the different cortical layers. (After Ganong. *Review of Medical Physiology,* 6th ed, 1973, Lange Medical Publications.)

through the rather dense external granular layer, whereas some of the dendrites of cells in the pyramidal layer (layer three) terminate here.

III. *Pyramidal Layer:* The pyramidal layer consists of an outer region that contains mostly medium-sized pyramidal cells and an inner region that is composed principally of larger pyramidal cells. A number of afferent thalamic fibers terminate upon the dendrites of the pyramidal cells located in the inner zone of the pyramidal layer.

IV. *Internal Granular Layer:* The internal granular layer contains an abundance of small pyramidal cells in addition to small satellite cells. Characteristically, this layer has many dense plexuses of fibrils. In the parietal area of the cortex, the plexuses of the internal granular area are composed of afferent fibers that arise within the thalamus, whereas in the temporal and occipital areas the plexuses are formed by afferent fibers

C.

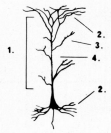

FIG. 7.70. C. Pyramidal cell of the cortex illustrating the sites of presynaptic terminals as follows: 1. Nonspecific afferent fibers of the reticular formation and thalamus. 2. Recurrent collaterals from pyramidal cell axons. 3. Commissural fibers from contralateral hemisphere (mirror image loci). 4. Afferent fibers from specific thalamic sensory projection nuclei. See also Figure 7.127

that arise from neurons located within the medial and lateral geniculate bodies.

V. *Ganglionic Layer:* The ganglionic layer consists of large, medium, and short pyramidal cells. These pyramidal cells are somewhat more sparse, but in general are larger, than those found in layer III. The long dendrites of the large pyramidal cells extend to the molecular layer (I), while their collateral branches and basal dendrites terminate entirely within the ganglionic layer itself. Dendrites of the medium-sized pyramidal cells of layer V terminate in the internal granular layer (IV); thus an abundance of synaptic connections is present between these neurons and thalamic as well as geniculate afferent fibers. Dendrites of the short pyramidal cells terminate entirely within layer V.

VI. *Polymorphic Layer:* The polymorphic layer contains long, medium, and short fusiform (or spindle) cells. Dendrites of the long spindle cells emit collaterals in layer VI, then ascend without yielding additional collaterals to terminate in the molecular layer (I). Dendrites of the medium fusiform cells terminate in the internal granular layer (IV) on the thalamic afferent fibers, whereas the dendrites of the short fusiform cells terminate in the ganglionic layer (V).

In certain areas of the cerebral cortex the presence of myelinated layers of nerve fibers between the cortical layers is responsible for the presence of macroscopically visible white lines. In the striate area of the occipital lobe, the line of Gennari is present on the outer zone of the internal granular layer (IV). Elsewhere in the cortex, a similarly located but thinner line is found, and this is called the external line of Baillarger. At the inner zone of the ganglionic layer (V) is found the internal line of Baillarger.

The cytoarchitecture in various regions of the cerebral cortex has been studied intensively by many investigators in order to attempt a correlation between structure and function in different areas of the brain. One such classification is that developed by Brodmann, who labeled particular areas of the cortex with numbers depending upon the individual cytoarchitectural characteristics of each area. This classification is useful as it gives reference points for the localization of physiologic as well as pathologic processes in the cerebral cortex. The numerical system developed by Brodmann is shown in Figures 7.71 and 7.72, and in the following discussion, the numbering of cortical areas in relation to their functions will be in accordance with this system.

Many cortical functions have been related experimentally to specific brain areas by means of electric and chemical stimulation in addition to ablation of particular cortical areas. Much of this experimental work has been carried out on the brains of higher primates (ie, monkeys and chimpanzees), as well as on human subjects during as well as following neurosurgical procedures. In addi-

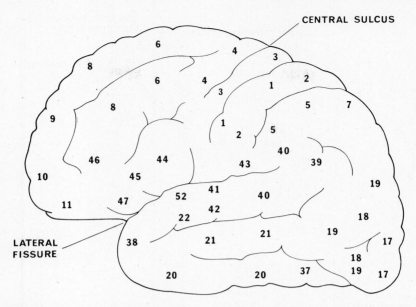

FIG. 7.71. Lateral aspect of the left cerebral hemisphere showing the cortical areas according to Brodmann. Area 8 = frontal eye movement, pupillary change area. Area 6 = premotor area (portion of extrapyramidal system). Area 4 (precentral gyrus) is the primary motor area. Areas 3, 2, 1 (postcentral gyrus) are the primary sensory areas. Areas 5, 6, 7 = secondary sensory association areas. Areas 39, 40 = association areas. Areas 18, 19 = visual association areas. Area 17 = primary visual cortex. Area 41 = primary auditory cortex. Area 42 = associative auditory cortex. Area 44 = motor speech area of Broca. (After Chusid. *Correlative Neuroanatomy and Functional Neurology,* 14th ed, 1970, Lange Medical Publications.)

tion, clinical and postmortem data have added considerably to our knowledge of localized functions in various areas of the brain.

Functional Correlates. The general physiologic properties of the four major cortical regions will be summarized briefly, and this information will be

followed by a more complete discussion of the functions carried out by particular cortical areas (refer to Figs. 7.62, 7.63, 7.64, 7.71, and 7.73).

Area 4 of the frontal lobe is the primary motor area of the cerebral cortex, whereas area 6 forms a part of the extrapyramidal system. Eye movements

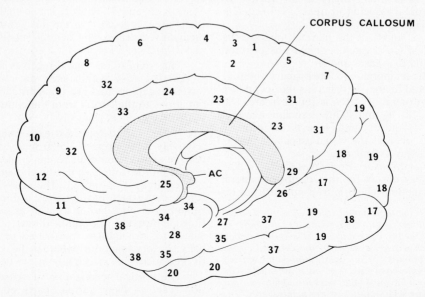

FIG. 7.72. Medial aspect of the left cerebral hemisphere showing the cortical areas according to Brodmann. Functions of specific areas are given in the legend to Figure 7.71. (After Chusid. *Correlative Neuroanatomy and Functional Neurology,* 14th ed, 1970, Lange Medical Publications.)

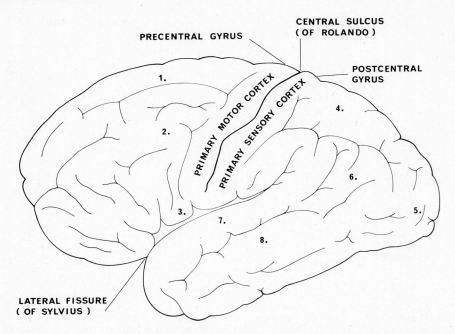

FIG. 7.73. Some effects obtained by electric stimulation of particular areas of the cerebral cortex. Note: The numbers in this figure merely indicate areas stimulated, and are *not* Brodmann areas (Figs. 7.71 and 7.72). Stimuli applied to the primary motor cortex and areas 1., 2., and 3. elicited as follows: 1. Head, eyes, and trunk turn toward the opposite side. Complex synergistic movements of the contralateral arm and leg can be produced. 2. Eyes turn toward contralateral side; no visual aura experienced. 3. Chewing, licking, swallowing movements. 4. Sensory aura in contralateral leg followed by complex synergistic movements. Also, poorly localized and vague but disagreeable sensations. 5. Complex optical hallucinations, including stars, lights, flames, colors. 6. Eyes turn to opposite side of body. 7. Auditory hallucinations, eg, roaring, buzzing. 8. Head, eyes, and trunk rotate to contralateral side coupled to complex synergistic movements of contralateral extremities. The specific sensations and movements evoked by stimulation of the primary sensory and motor cortices are highly localized insofar as regions of the body are concerned, and are depicted in Figures 7.74 and 7.75. (After Chusid. *Correlative Neuroanatomy and Functional Neurology,* 14th ed, 1970, Lange Medical Publications.)

and changes in pupillary diameter are related to area 8. The frontal association areas* are located in areas 9, 10, 11, and 12.

Areas 1, 2, and 3 of the parietal lobe are the primary cortical sensory areas. Sensory association areas are found in areas 5 and 7.

Within the temporal lobe, the primary auditory cortex is located in area 41, whereas the secondary (or associative) auditory area is found in area 42. Areas 20, 21, 22, 38, and 40 are association areas.

Area 17 of the occipital lobe is the primary visual cortex, whereas the visual association areas are found in areas 18 and 19.

The Primary Sensory Projection Cortex. The primary sensory cortex for the receipt of general sensory information which is transmitted from re-

In this context, the term "association" denotes a correlation of neurologic functions involving a high degree of modifiability as well as consciousness. The term "association area" denotes those areas of the cerebral cortex excluding the primary (or principal) motor and sensory areas that are interconnected with each other and with the thalamus by many fibers that pass through the corpus callosum and the white matter of the hemispheres. The association areas are responsible for higher perceptual and mental processes involving memory, reasoning, learning, and so forth.

ceptors located throughout the body is found in areas 1, 2, and 3. This region, called the somesthetic area, is located in the postcentral gyrus immediately dorsal to the central sulcus (Figs. 7.73 and 7.74).

The somesthetic area receives afferent impulses that arise within skin, muscle, joint, and tendon receptors and are transmitted via thalamic radiations to the cortex from the contralateral side of the body.

It has been determined experimentally that a relatively large area of the frontal lobe also can receive sensory impulses (areas 4 and 6), whereas certain motor responses can be elicited by stimulation of areas 1, 2, and 3. Therefore the primary sensorimotor area of the cortex is considered able to subserve both motor as well as sensory functions. However, the region dorsal to the central sulcus (of Rolando) is considered to have principally sensory functions, whereas the region ventral to this sulcus has principally motor functions.

The cortical localization of sensory information received from various parts of the body is indicated in Figures 7.73 and 7.74. Note in these figures the high degree of topographic specificity that is present between the various peripheral re-

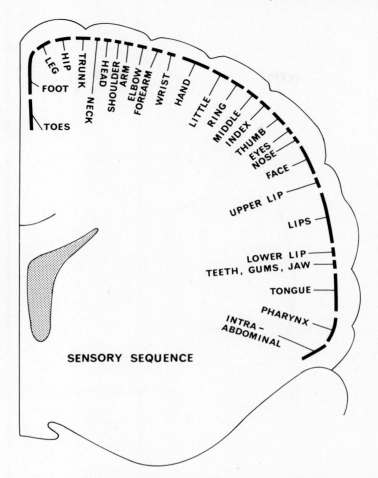

FIG. 7.74. Cross section through the postcentral gyrus of the left cerebral hemisphere (ie, the primary sensory cortex) illustrating the sequence in which sensory phenomena are localized. The sensory sequence in the right hemisphere is an exact mirror image of that shown in this figure. The length of the heavy black bars for each area indicates the approximate brain area that subserves each function in this and the following figure. (After Dawson. *Basic Human Anatomy,* 2nd ed, 1974, Appleton-Century-Crofts.)

gions and the cortical pattern for the expression of these regions.

The Primary Motor Projection Cortex. As indicated in Figures 7.73 and 7.75, the primary motor cortex is found on the anterior wall of the central sulcus as well as in the precentral gyrus immediately adjacent to this sulcus (area 4). This motor area correlates well with the distribution of the giant pyramidal cells of Betz (p. 325); and as noted earlier, these cells control the voluntary contractions of skeletal muscles on the contralateral side of the body. Efferent impulses generated within the Betz cells are transmitted over their axons via the corticobulbar and corticospinal tracts to the nuclei of particular cerebrospinal nerves as depicted in Figure 7.76 (see also Fig. 7.22).

It has been determined experimentally that the topographic organization of the primary sensory areas of the postcentral gyrus bear a mirror-image relationship to those of the primary motor areas of the adjacent precentral gyrus (Figs. 7.74 and 7.75).

Area 4 is the most sensitive region of the cerebral cortex, insofar as motor stimulation is concerned, and it appears that the ganglionic layer (V), which contains the giant pyramidal cells, is responsible for modulating the excitability of the neurons located within this area. Thus facilitation and suppression are phenomena that may be evoked in response to appropriate stimulation of the cortex. Facilitation (or temporal summation) refers to the ability of a neuron to respond by firing an all-or-none impulse in response to a rapidly applied series of subthreshold stimuli (p. 129). Suppression, on the other hand, is an inhibition of the contraction of skeletal muscle following prolonged cortical or subcortical stimulation. Both of these effects, of course, could modify profoundly the contractile state of the somatic musculature in vivo.

Stimulation of area 6, sometimes referred to as the premotor area, evokes contractions of skeletal muscles similar to those elicited by stimulation of area 4. If one extirpates area 4, or sections the

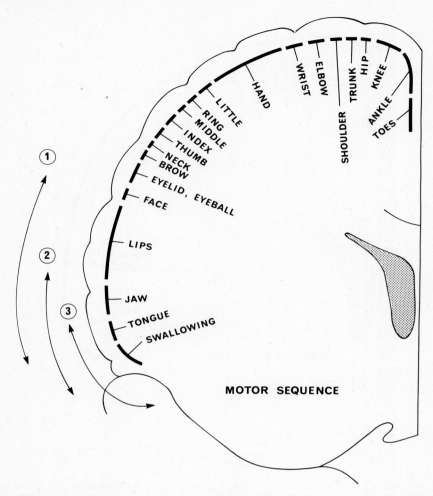

MOTOR SEQUENCE

FIG. 7.75. Cross section through the precentral gyrus of the right cerebral hemisphere (ie, the primary motor cortex) illustrating the sequence in which motor phenomena are localized. The motor sequence in the left hemisphere is an exact mirror image of that shown in this figure. The numbers signify the approximate cortical areas given to complex motor functions as follows. ① Vocalization. ② Salivation. ③ Mastication. (After Dawson. *Basic Human Anatomy,* 2nd ed, 1974, Appleton-Century-Crofts.)

nerve fibers that connect areas 4 and 6, however, stimulation of these regions now produces stereotyped motor responses, including twisting of the body and rotation of the head.

The Primary Visual Cortex. The primary cortical area for the reception of visual impulses is found in the occipital lobe (area 17; the calcarine fissure and nearby parts of the cuneus and lingual gyri). Thus stimulation of this region can produce visual hallucinations including bursts of light, rainbows, or bright stars.

The dorsalmost region of the occipital lobe in primates (including man) is concerned principally with macular vision (p. 471). The more ventral portions of the calcarine fissure, on the other hand, are related to peripheral vision.

The association of visual impulses takes place in areas 18 and 19. Area 18 receives impulses principally from the primary visual cortex (area 17),

whereas area 19 is able to receive impulses from the entire cerebral cortex.

The Primary Auditory Area. The primary cortical area for the receipt of auditory impulses is found in the transverse temporal gyrus (area 41; Heschl's gyrus). This convolution is located within the floor of the lateral cerebral ventricle, and it receives auditory fibers which radiate from the medial geniculate body; the sensory end organ for hearing, in turn, is the cochlea (Fig. 7.36).

Unilateral lesions of the primary auditory area produce only mild deafness, as the perception of auditory impulses from one ear is represented bilaterally on each primary auditory area. This fact is in marked contrast to the unilateral neuroanatomic arrangement of nerve fibers that is found in all other sensory pathways in the brain.

The Olfactory Area. The olfactory area for the receipt of impulses from the olfactory tract is

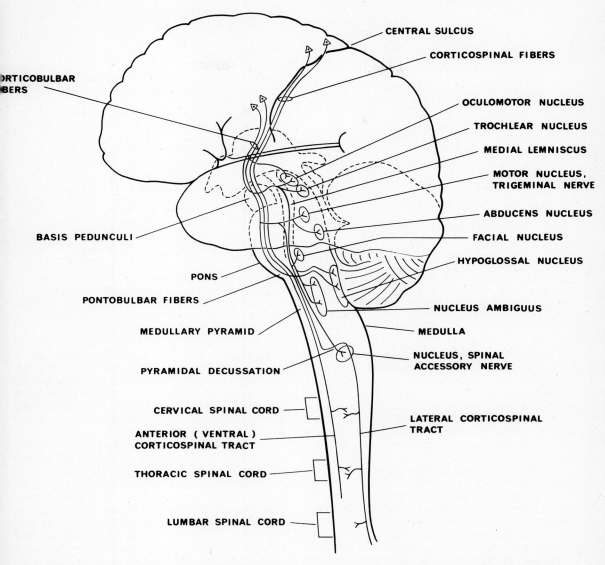

FIG. 7.76. The origin and distribution of the corticospinal (pyramidal) and corticomedullary fibers. See also Figure 7.22. (After Everett. *Functional Neuroanatomy,* 6th ed, 1971, Lea & Febiger.)

located within the uncus as well as the nearby regions of the hippocampal gyrus of the temporal lobe.

Primary Association Areas of the Cerebral Cortex. The association areas of the sensory and motor regions of the cerebral cortex (Fig. 7.77) are connected by a number of tracts within the white matter which contain association fibers (p. 321; Figs. 7.68 and 7.69). The association areas are of major importance for the maintenance of the higher mental processes in man, and these will be considered in more detail in relation to specific topics that fall under this heading.

Briefly, however, the frontal lobe anterior to area 4 is concerned with higher intellectual and psychic functions, eg, thought processes. This or-

bitofrontal area, which includes areas 9, 10, 11, and 12, receives afferent nerves from the dorsomedial thalamic nuclei which in turn receive hypothalamic connections.

When stimulated electrically, the posterior region of the orbital surface (area 47) as well as the anterior region of the insula evoke profound autonomic responses, as well as alteration of blood pressure and respiratory inhibition. When the anterior cingulate area is stimulated (area 24) inhibition of skeletal muscle tone as well as autonomic responses are elicited. Furthermore, extirpation of this area in lower primates (male monkeys) reduces their aggressive behavior somewhat so that the animals become more tractable and less fearful.

The primary association areas located within

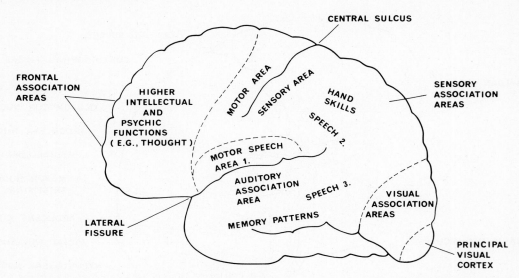

FIG. 7.77. Some general functional areas of the left cerebral cortex. The crude localizations shown here are those as would be found primarily in the categorical hemisphere of a right-handed person as discussed in the text. (Cf. Figs. 7.71 and 7.73.) (Data from Best and Taylor [eds]. *The Physiological Basis of Medical Practice,* 8th ed, 1966, Williams & Wilkins; Chusid. *Correlative Neuroanatomy and Functional Neurology,* 14th ed, 1970, Lange Medical Publications; Guyton. *Textbook of Medical Physiology,* 4th ed, 1971, Saunders.)

the parietal lobe (areas 5 and 7) are necessary for the integration of cutaneous sensations with other perceptual cues. Thus a subject can recognize various common objects placed in his hand when his eyes are closed. This function is called the stereognostic sense. The parietal lobe also contains secondary sensory and motor association areas that are broadly connected to the primary motor and sensory areas.

Complex memory patterns are retained within the temporal lobes of the cortex.

The aphasias further illustrate the importance of the cortical association areas to the higher mental processes. An aphasia can be defined as a defect or loss of the power of expression, whether by speech, writing, or signs (ie, the use of symbols or language). The term aphasia also denotes a loss of comprehension of spoken or written language.* Various clinically recognized aphasias can result from lesions in the association areas of the brain. In right-handed individuals, aphasias are produced by lesions in the contralateral (or the dominant) hemisphere, whereas the opposite situation obtains for left-handed individuals. For example, motor aphasia results from damage to the motor speech area of Broca (area 44). The patient can move both tongue and lips but is incapable of producing the coordinated movements necessary for coherent speech. Agraphia, or the inability to write words, often accompanies motor aphasia. Lesions within the posterior region of the superior temporal gyrus

This type of aphasia is somewhat analogous to hearing an unfamiliar foreign language spoken, or seeing that language in written form. The normal individual would hear or see the words, but they would be completely uncomprehensible.

cause sensory aphasia (area 39). The victim cannot understand the meaning of spoken words, although hearing may be unimpaired. Damage to the angular gyrus (areas 39 and 40) may result in word blindness; that is, the patient is unable to understand the meaning of written words although vision is unimpaired.

In conclusion, it must be emphasized that the association areas and their functions are closely integrated, and that many of the higher sensorimotor functions of the brain do not take place in circumscribed regions. Rather, higher mental processes such as thought or memory involve the participation of large cortical association areas that function in a cooperative fashion and are connected by association fibers. It is, of course, impossible to localize anatomically any particular mental process.

The Limbic System (Rhinencephalon). The term rhinencephalon was applied earlier in this text to designate those phylogenetically ancient structures of the cerebral hemisphere that are concerned with the perception of olfactory stimuli (p. 319). However, since the rhinencephalic portion of the brain has been demonstrated unequivocally to possess a number of important functional roles in a variety of physiologic processes in addition to olfaction, the term limbic system generally is employed preferentially to denote the rhinencephalic structures of the cerebral hemisphere. In fact, the terms rhinencephalon, limbic lobe, and limbic system often are used interchangeably. Henceforth the term limbic system will be used in this text to denote collectively the following structurally and functionally interconnected neural elements: the limbic lobe and hippocampus together

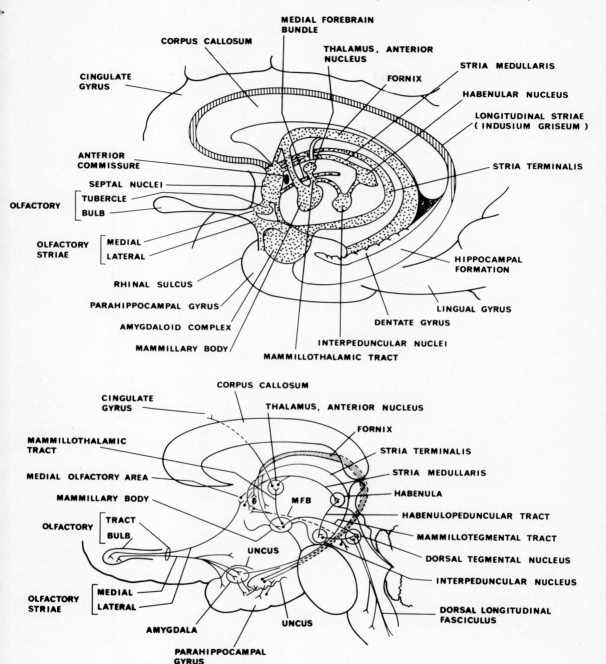

FIG. 7.78. **A.** The relationships among the component structure of the limbic system (rhinencephalon) as found in a medial section of the right cerebral hemisphere. **B.** Similar view of the limbic system illustrating the interconnections among the component structures. MFB = medial forebrain bundle. (See also Figures 7.66 and 7.67 for details of peripheral elements of the olfactory nerve and Figure 7.79 for details of the limbic cortex.) (**A.** After Truex and Carpenter. *Human Neuroanatomy,* 6th ed, 1969, Williams & Wilkins. **B.** After Everett. *Functional Neuroanatomy,* 6th ed, 1971, Lea & Febiger.)

with their related subcortical neural connections; the amygdala; the septal nuclei; the hypothalamus; the anterior thalamic nuclei; portions of the basal ganglia (qv, p. 336). The epithalamus sometimes is considered part of the limbic system. The relation-

ships among these structures are depicted in Figure 7.78.

The general anatomic relationship between the cortical part of the limbic system and the entire cerebral hemisphere is shown in Figure 7.79. Each

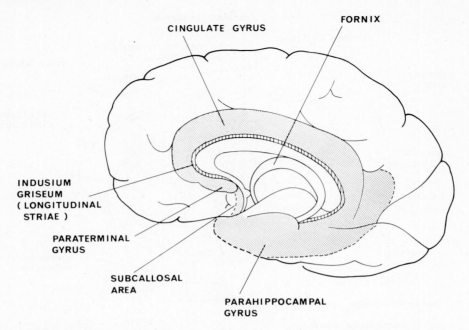

FIG. 7.79.   Medial surface of right cerebral hemisphere showing the limbic cortex (shaded area). (After Everett. *Functional Neuroanatomy*, 6th ed, 1971, Lea & Febiger.)

cortical lobe of the limbic system is composed of a band of primitive (allocortical) tissue that surrounds the hilum of the cerebral hemisphere. Histologically, this allocortical tissue does not have the characteristic six-layered cytoarchitectural pattern found in the neocortex.

The principal afferent and efferent connections of the limbic system are given in Figure 7.78. The fornix connects the hippocampus to the mamillary bodies, and these structures in turn are connected to the anterior thalamic nuclei via the mamillothalamic tract (of Vicq d'Azyr; p. 291). The anterior nuclei of the thalamus then project fibers to the cingulate gyrus, and from this cortical area neural connections pass to the hippocampus, thereby completing a closed neural circuit or loop, sometimes referred to as the Papez circuit after the worker who described it originally.

One peculiarity of the limbic system is that it has only a few connections with the neocortex. A small number of nerve fibers pass from the frontal lobe to the nearby limbic structures, and possibly some indirect functional connections are present via the thalamus.

The functions of the limbic system now may be summarized briefly, and this information has been gained not only from stimulation and ablation experiments performed on animals, but from clinical studies as well.

In man, the role of the limbic system in olfaction is relatively minor (p. 332). In general, the limbic system is concerned with a wide variety of autonomic somatosensory and somatomotor responses. Specifically, the activity of the limbic system is concerned initimately with such complex

phenomena as emotions of rage and fear as well as their genesis; the control of various biologic rhythms; sexual behavior; motivation; and together with the hypothalamus, drinking and feeding behavior. These manifestations of limbic function will be dealt with more fully in later sections of this chapter.

Electric stimulation of various components of the limbic system can evoke a wide variety of responses, including the inhibition of respiratory movements, shivering, and swallowing, as well as voluntary movements that are initiated by discharge of impulses from the premotor area of the cerebral cortex. Alternatively, such stimulation can elicit chewing, swallowing, facial-vocal movements, licking, tonic movements of the extremities and trunk, changes in pupillary diameter, piloerection, and salivation, as well as reflex micturition and defecation. The inherent electric activity of practically the entire cerebral cortex also can be modified by stimulation of various limbic structures.

Certain components of the limbic system, such as the anterior limbic region as well as the posterior orbital surface, appear to exert an inhibitory role upon those mechanisms in the brain stem that are related to the expression of emotions such as fear and anger. Lesions in these structures produce hyperactivity.

Physiologically, the activity of the neocortex can modify emotional behavior and the opposite also is true, despite the lack of abundant neural pathways between the limbic system and other cortical areas. Nevertheless, one of the basic characteristics of emotional behavior is that it cannot be

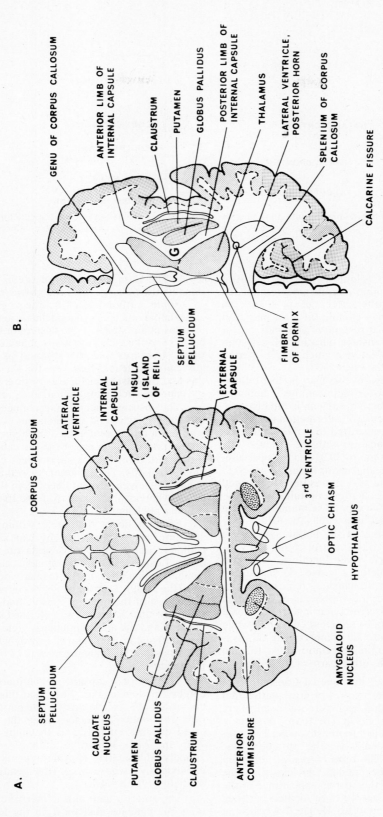

FIG. 7.80. **A.** Transverse section through the cerebral hemispheres at the level of the anterior commissure. **B.** Horizontal section through the right cerebral hemisphere showing the relationships among the anterior limb, genu (or flexure, G), and the posterior limb of the internal capsule. (**A.** After Chusid. *Correlative Neuroanatomy and Functional Neurology,* 14th ed, 1970, Lange Medical Publications. **B.** After Gatz. *Manter's Essentials of Clinical Neuroanatomy and Neurophysiology,* 4th ed, 1970, FA Davis.)

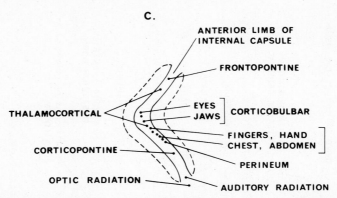

FIG. 7.80. C. Detail of the internal capsule illustrating the general spatial orientation of certain ascending and descending tracts. See also Figures 7.82 and 7.83. (After Gatz. *Manter's Essentials of Clinical Neuroanatomy and Neurophysiology,* 4th ed, 1970, FA Davis.)

evoked or terminated voluntarily. In this regard, it is of interest that limbic circuits once stimulated exhibit a prolonged after-discharge (p. 223). Consequently this fact may explain (at least partially) the fact that emotional responses generally are of long, rather than of transient, duration. Hence such responses tend to persist for a much longer interval than do the stimuli that elicit them.

Memory for recent events may be lost following bilateral lesions of the hippocampus in humans, indicating that the limbic system is involved in some way in the process of information retention.

It has been demonstrated that the limbic system has major relationships with the reticular formation of the brain stem (p. 302), and thus presumably has a role in the alerting or arousal process.

Lastly, the limbic system has been implicated in the hypothalamic regulation of pituitary activity, and possibly this limbic control is manifest via the hypothalamic neurosecretory cells (p. 1030).

Since the functions of the limbic system are related so intimately to the control of a broad variety of autonomic and other functions, the term "visceral brain" sometimes is used to denote the limbic structures. In conclusion, it is unclear at present which structural components of the limbic system should be assigned particular functional roles, a fact which can be appreciated readily when one considers the anatomic complexity of this system as well as its intimate relationships to, and interconnections with, other parts of the brain.

The Basal Ganglia and the Internal Capsule. The basal ganglia are large masses of gray matter that are located deep within each cerebral hemisphere (Figs. 7.38, 7.49, and 7.80). The term "basal ganglia" usually is applied to those nuclear masses called the caudate nucleus, the putamen, and the globus pallidus. The term *corpus striatum* (or *striatum*) denotes the caudate nucleus plus the putamen, because of the striped appearance of the bundles of white matter that are located between these two nuclear structures. Hence this region of the brain sometimes is referred to as the striatal

level. The term *lenticular* (or *lentiform*) *nucleus* denotes the putamen plus the globus pallidus.

Some authors also include the claustrum, amygdaloid nucleus, red nucleus, and substantia nigra as components of the system of basal ganglia because of the elaborate interconnections between these structures and the striatum as well as the lenticular nucleus. However, for the purposes of this text the basal ganglia are considered to be only the caudate nucleus, putamen, and globus pallidus.

The principal neural interconnections among the basal ganglia and the thalamus as well as the substantia nigra are depicted in Figure 7.81. Many afferent nerve fibers pass from the caudate nucleus to the putamen; thence short fibers course to the globus pallidus. The putamen as well as the globus pallidus receives afferent fibers from the substantia nigra, whereas fibers pass from the thalamus to the caudate nucleus.

Efferent fibers that arise within the corpus striatum leave this nucleus by way of the globus pallidus. Some of these efferents run through the internal capsule and form a bundle, called the fasciculus lenticularis, on the medial side of the capsule (Fig. 7.49). Additional fibers course to the mesial region of the internal capsule and form a loop called the ansa lenticularis. Both the fasciculus lenticularis and the ansa lenticularis yield terminals to the subthalamic nucleus (of Luys), whereas other fibers pass to the thalamus by way of the fasciculus thalamicus.

The internal capsule is continuous with the cerebral peduncle (Fig. 7.61), and this structure is composed of a wide band of white matter that separates the lenticular nucleus from the medially located caudate nucleus and thalamus (Figs. 7.38, 7.80, and 7.82). When the brain is sectioned horizontally, the internal capsule appears as a widely expanded V, whose apex (the genu) is directed medially (Fig. 7.80B).

The anterior limb of the internal capsule separates the caudate nucleus from the lenticular nucleus, and this region contains thalamocortical as

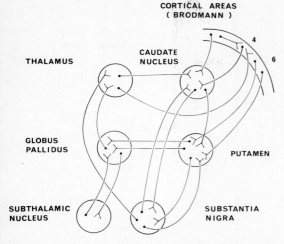

CORTICAL AREAS
(BRODMANN)

THALAMUS

CAUDATE
NUCLEUS

4

6

GLOBUS
PALLIDUS

PUTAMEN

SUBTHALAMIC
NUCLEUS

SUBSTANTIA
NIGRA

FIG. 7.81. Highly schematic diagram illustrating the interconnections among the basal ganglia and the cerebral cortex. (After Ganong. *Review of Medical Physiology,* 6th ed, 1973, Lange Medical Publications.)

well as corticothalamic fibers which interconnect the lateral thalamic nucleus with the frontal lobe of the cerebral cortex. The anterior limb of the internal capsule also contains frontopontine (or frontopontile) tracts that descend from the frontal lobe to nuclei within the pons, as well as fibers that run from the caudate nucleus to the putamen (Fig. 7.83).

The posterior limb of the internal capsule is located between the thalamus and the lenticular nucleus. This region sometimes is divided into lenticulothalamic, retrolenticular, and sublenticular regions (Fig. 7.83). The anterior portion of the lenticulothalamic region contains the corticobulbar and corticospinal tracts which provide motor innervation to the arm and leg. Corticorubral fibers from the frontal lobe of the cortex course together with the fibers of the corticospinal tract, to terminate within the red nucleus. The retrolenticular portion of the internal capsule contains fibers from the lateral thalamic nucleus to the postcentral gyrus of the cortex. The sublenticular region of the internal capsule lies beneath the lenticular nucleus and contains parietotemporopontine fibers from the temporal and parietal lobes of the cerebral cortex that pass to pontine nuclei. In addition, the sublenticular region contains auditory radiations from the medial geniculate body that pass to the transverse temporal gyrus (Heschl's gyrus), as well as optic fibers that course from the lateral geniculate body to the calcarine cortex.

It should be reiterated here that all of the afferent pathways to the cerebral cortex (excepting the olfactory pathway) have synaptic junctions (or relay stations) within the various thalamic nuclei, and that in every instance the axon of the final sensory neuron passes to the cortex via the internal capsule, which in turn is continuous with the cerebral peduncle.

The exact functions of the basal ganglia are a mystery, although experimental as well as clinical data support the contention that these nuclei have an important role in the control of bodily movements.

The caudate and lenticular nuclei, in combination with the fibers of the internal capsule which lie between these structures, comprise an important structural unit of the extrapyramidal system, the corpus striatum (qv, next section). From the corpus striatum efferent fibers course to the globus pallidus, and this unit also receives afferent fibers from the frontal lobe (areas 4 and 6) as well as the thalamus and hypothalamus (Fig. 7.81).

An important efferent route from the globus pallidus passes to the cerebral and brain-stem nuclei by way of the ansa lenticularis (Fig. 7.49).

When the basal ganglia are stimulated electrically, skeletal muscle tone may be inhibited, and movements that originate within the cortex either may be evoked or inhibited, although such results usually are negative.

The globus pallidus and the lateral thalamic nuclear groups apparently are regions of the brain within which many efferent (motor) pathways converge. Consequently these nuclei would seem to act as regulatory as well as integrating centers for motor activities, in addition to functioning as relay centers for afferent impulses to the cerebral cortex.

It has been demonstrated experimentally that certain neurons within the basal ganglia discharge impulses only during slow and smooth eye movements, whereas they do not fire during rapid and jerky (or saccadic) eye movements such as are found when one reads a printed page. These facts tend to corroborate the hypothesis that the basal ganglia are involved in some way with the control of movement. Furthermore certain clinical states involving the basal ganglia also tend to support this idea (see Clinical Correlates, p. 456).

The Extrapyramidal System. The extrapyramidal system is a physiologic (ie, functional) rather than an anatomic unit whose normal activity depends upon the presence of an intact corticospinal (or pyramidal) system. Until fairly recently, the extrapyramidal system generally was considered to be made up of all of the motor systems of the central nervous system with the exception of the pyramidal tract (p. 267). As more anatomic and physiologic data were obtained, however, it became apparent that a clear distinction between the pyramidal and extrapyramidal systems was not present. Hence the extrapyramidal system has come to be regarded as a functional rather than as an anatomic entity. This system is composed of extrapyramidal areas of the cerebral cortex, the thalamic nuclei that are connected to the corpus striatum, the corpus striatum per se, the subthalamus, and the rubral as well as the reticular systems. The extrapyramidal system reaches the segmental levels of neural distribution via rather devious routes wherein the neural chains are interrupted by synapses located within the basal

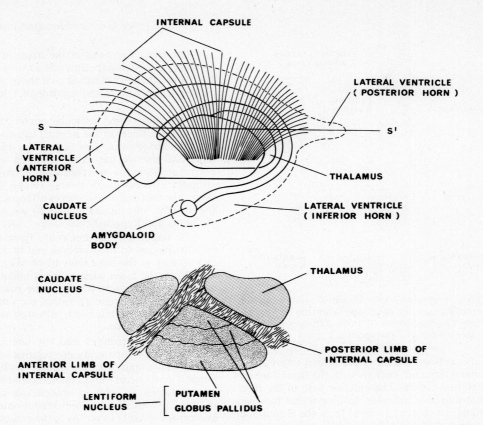

FIG. 7.82. The basal ganglia and adjacent structures as seen in a lateral view. A horizontal cross section of the basal ganglia through the plane S—S' is shown at the bottom of the figure. See also Figures 7.80 and 7.83. (After Chusid. *Correlative Neuroanatomy and Functional Neurology,* 14th ed, 1970, Lange Medical Publications. Additional data from Everett. *Functional Neuroanatomy,* 6th ed, 1971, Lea & Febiger.)

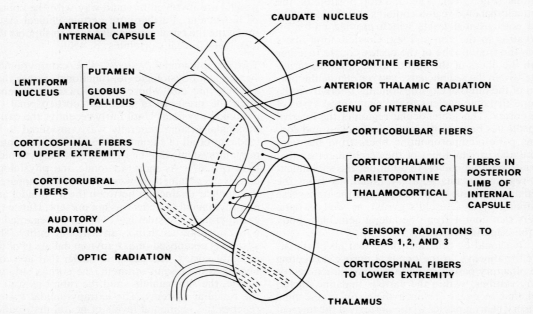

FIG. 7.83. Horizontal section illustrating the regions of the internal capsule and its fiber systems in relation to adjacent structures. See also Figures 7.80 and 7.82. (After Everett. *Functional Neuroanatomy,* 6th ed, 1971, Lea & Febiger.)

338

ganglia and subcortical ganglia and in the reticular formation. This fact contrasts markedly to the more direct neural pathways traversed by the·elements of the pyramidal system.

The extrapyramidal system may be considered as having three levels of functional integration: first, at the cortical level; second, at the level of the basal ganglia (striatal level); and third, at the midbrain or tegmental level.

The bulboreticular facilitory and inhibitory areas of the brain stem (pp. 304-5) are connected by afferent fibers to certain areas of the cerebral cortex, the corpus striatum (ie, the caudate and lenticular nuclei plus the intervening portion of the internal capsule), and the anterior part of the cerebellum.

Functionally the extrapyramidal system operates to coordinate associated movements and alterations in posture, as well as to integrate the functions of the autonomic nervous system.

Damage to, or lesions present within, any of the functional levels of the extrapyramidal system may conceal, or abolish completely, voluntary movements of the body, and involuntary movements now appear (see p. 456).

Concluding Remarks

The foregoing brief review of the anatomic and physiologic properties of the nervous system will serve as a background against which the specific functions of this system can be presented in somewhat greater detail. The student is urged to bear in mind the various anatomic as well as functional levels of organization when reading the material to follow, as in this way a coherent picture of the many neural activities that occur in vivo will be obtained more readily (see Table 7.11).

NEUROPHYSIOLOGY

The general characteristics and particular functions of the nervous system to be considered in this section have been divided arbitrarily into five major categories for the purposes of discussion. First, the general physiologic properties of nerves and neuronal circuits will be reviewed. Second, the sensory functions of the nervous system will be presented, and this material will be followed, third, by a consideration of voluntary and involuntary motor functions and their regulation. Fourth, the so-called higher functions of the cerebral cortex will be examined together with certain behavioral patterns, including emotional responses. Fifth, the overall metabolic properties of neural tissues in general, and the brain in particular, will be summarized. It is important to realize that these categories are not mutually exclusive, and that some overlap among the topics to be presented in these five general areas is unavoidable. Nevertheless, the subdivision of the material as outlined

Table 7.11   LEVELS OF INTEGRATION OF NEURAL FUNCTIONS[a]

LEVEL OF INTEGRATION IN CENTRAL NERVOUS SYSTEM	FUNCTION
Spinal cord	Spinal reflexes
Medulla oblongata	Respiratory, cardiovascular, and other reflexes
Medulla	Antigravity (postural) reflexes
Cerebellum[b]	Coordinate rapid (saccadic, ballistic) movements
Midbrain	Righting reflexes
Midbrain, thalamus	Locomotor reflexes
Hypothalamus, limbic system	Emotional and instinctual responses; visceral reflexes
Basal ganglia[b]	Coordinate slow (ramp) movements
Cerebral cortex[b]	Memory, conscious perception, initiative, learning, volitional motor activities; other "higher functions," including conditioned reflexes

[a]*The term "integration" as used in this context signifies a combination (or coordination) of various functions so that they operate together in a cooperative fashion to accomplish a common "physiologic goal."*
[b]*See discussion of cerebellum, basal ganglia, and cerebral cortex in relation to somatic motor functions (p. 396 et seq.).*

above will give the reader a basis from which to synthesize for himself a coherent idea of the complex and closely integrated functions of the human nervous system.

General Functional Properties of Nerves and Neuronal Circuits

The physiologic characteristics of neurons and some simple neuronal circuits were presented in extenso in Chapters 5 and 6; however, certain of these properties will be reviewed briefly as such material is basic to a clear understanding of the functions of more complex neural circuits (refer to Fig. 6.1, p. 178).

At the endings of sensory nerves, particular stimuli are converted into a local, graded electric response whose intensity is a rough analog of the stimulus intensity (p. 184). In turn, this local response at the nerve ending induces a series of propagated all-or-none impulses in the afferent nerve fibers whose frequency is a reflection of the intensity of the applied stimulus (Fig. 6.5, p. 183). When the impulses arrive at a synaptic junction, a local graded electric response is evoked in the membrane of the postsynaptic cell, and this response can be either excitatory or inhibitory, depending upon the particular neuron and neurotransmitter involved (pp. 191, 193). Impulse transmission in the efferent nerve is, once again, an all-or-none process, but at the terminals of the efferent nerve a local graded electric response occurs in the postsynaptic cell membrane, whether it be an effector element (ie, muscle or gland cell) or another neuron (p. 210). It is essential to realize that the graded excitatory or inhibitory bioelectric properties of synaptic junctions give rise to a high degree of operational flexibility within the complex neural circuits found in the central nervous system.

Ordinarily peripheral nerves are composed of bundles of individual nerve fibers, and the velocity of impulse transmission is directly proportional to the diameter of each particular fiber (p. 169). The same principle also holds true within the various fiber tracts of the central nervous system.

When discussing impulse transmission within the nervous system, the term *signal* often is used. A signal can be defined as a specific pattern of individual nerve impulses within a particular nerve or fiber tract. In turn, either the amplitude or the frequency of the signal can be modified, as will be discussed subsequently.

Trains of individual nerve impulses (signals) are the means whereby data or *information* is transmitted from point to point within the nervous system. Although the impulses conducted in all nerve fibers are the same from a bioelectric standpoint, their sources and ultimate destinations within the brain and other regions of the nervous system determine the type of information that is transmitted. Thus fibers of the optic nerve convey visual information, whereas the auditory nerve transmits information related to hearing. Similarly,

the pyramidal tracts conduct information in the form of nerve impulses that underlie the voluntary contractions of skeletal muscles; ie, these tracts convey motor information.

The remainder of this section will be concerned with the physiologic aspects of signal transmission in peripheral nerves, nerve tracts, and in neuronal pools. In addition, the various mechanisms whereby signals are integrated, or processed, within particular neuronal circuits will be discussed.

SIGNAL TRANSMISSION IN PERIPHERAL NERVES AND NERVE TRACTS; AMPLITUDE AND FREQUENCY MODULATION. It is a rare event when a signal is transmitted in a single nerve fiber under physiologic conditions. Rather, signals in vivo are conducted in peripheral nerves or centrally located nerve tracts which in turn are composed of large numbers of parallel nerve fibers of various sizes (p. 173). As noted earlier, the intensity or strength of the signals that are transmitted within the nervous system is subject to considerable variation through amplitude modulation and frequency modulation.

Amplitude Modulation of Signals: Spatial Summation. Regardless of the fact that each nerve impulse transmitted by a single nerve fiber invariably is all-or-none in character (p. 168), a peripheral nerve or centrally located fiber tract can give rise to signals of graded intensity merely by virtue of the fact that stimuli of progressively greater intensity cause proportionally more individual fibers within the nerve bundle to discharge, and this effect in turn leads to a modulated signal of greater amplitude or intensity. This phenomenon, known as *spatial* (or *multiple fiber*) *summation* is illustrated in Figure 7.84. In this figure note also that the nerve terminals overlap to a great extent. Therefore as the intensity of the stimulus is increased, progressively more fibers discharge in the regions peripheral to the area immediately beneath the point at which the stimulus is applied, and this area is called the *receptive field* of the nerve.

The increase in the number of individual nerve fibers that discharge as the intensity of the stimulus is increased also is called *recruitment* of additional fibers (see also p. 129).

Frequency Modulation of Signals: Temporal Summation and Signal Averaging. A second method whereby signals of variable intensity are generated within peripheral nerves or fiber tracts is through alterations in the frequency at which impulses are transmitted by the individual fibers. As depicted in the curves given in Figure 7.85, the individual impulses are averaged so that a smooth, graded signal is developed, and the strength or intensity of the signal is proportional to the frequency of impulse transmission. For example, if 100 nerve fibers innervate a skeletal muscle, and each fiber transmits one impulse per second, a weak but smooth contraction of the muscle is produced because the indi-

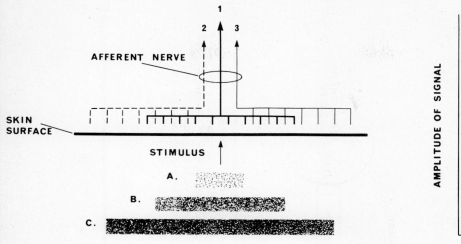

FIG. 7.84. Spatial (or multiple fiber) summation. The net effect of this process is modulation of the amplitude of the signal. A mild stimulus applied at one point (arrow) would stimulate only fiber 1 in the afferent nerve and give rise to a signal of amplitude or intensity A (light shaded bars). A moderate stimulus would excite fibers 1 and 2, and concomitantly give rise to a signal of amplitude B (intermediate shaded bars). A strong stimulus would excite fibers 1, 2, and 3 simultaneously and elicit a signal of amplitude C (dark shaded bars). The receptive fields of each fiber are depicted as the horizontal ramifications of each fiber of the afferent nerve; but it must be remembered that these fields in vivo are roughly circular *not* linear patterns.

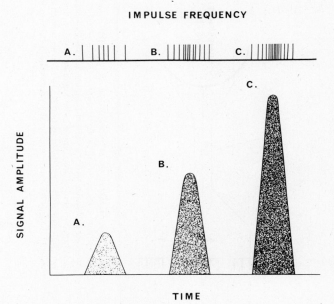

FIG. 7.85. Temporal summation of nerve impulses, an effect that also is called frequency modulation. At A, the impulse frequency is low; consequently the average signal also is low as shown in the graph. In B, as the impulse frequency increases, so does the net amplitude of the signal, and in C when the impulse frequency is quite high so too is the amplitude of the signal that is transmitted along the nerve. The phenomenon of temporal summation sometimes also is called signal averaging, and this effect can take place either in a single fiber or in an entire nerve (or tract). Each impulse in any particular fiber is, however, an all-or-none response, the net gain in signal amplitude merely being caused by more impulses being transmitted per unit time.

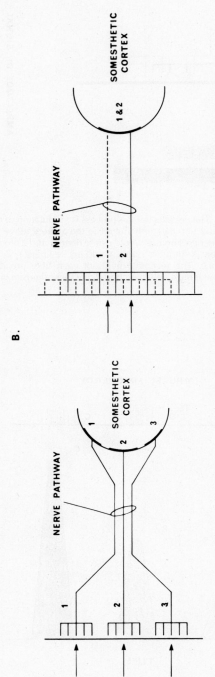

FIG. 7.86. A. Three discrete peripheral nerves upon stimulation (arrows) transmit three distinct signals to three discrete areas of the somesthetic cortex so that the three stimuli are perceived individually. The afferent signals are transmitted to the brain in separate, spatially oriented pathways, and the peripheral receptive fields of each afferent neuron, as well as its central sensory receptive area, do not overlap. **B.** When two stimuli are applied close together so that the overlapping receptive fields of two neurons (1, 2) are excited simultaneously, the two signals will be perceived as one stimulus even though each is transmitted in a spatially oriented pathway, because the central sensory receptive areas for each peripheral receptive field overlap in the somesthetic cortex as well as in the periphery. See also Figure 7.99.

342

vidual impulses that in toto make up the signal are spaced out in time. If, however, each fiber in the nerve fires at a rate of 50 impulses per second, the signal is considerably stronger, but the muscle still undergoes a smooth, coordinated, tetanic contraction (see also p. 126).

The process whereby an increase in the frequency of neural discharge causes an increase in the average strength of the signal is known as *temporal summation*, and this type of summation takes place by frequency modulation as described above. In turn, the average signal intensity at any point in time merely reflects the total number of impulses being transmitted per second by the nerve or fiber tract, and this frequency in turn modulates the signal intensity.

Signal averaging, such as described above, also takes place at synaptic junctions within the central nervous system. Since each postsynaptic neuron is bombarded with hundreds to thousands of single impulses per second, the postsynaptic membrane continually must summate and average all of the excitatory and inhibitory inputs to give fairly smooth alterations in the excitatory condition of the postsynaptic neuron (see also p. 207).

Briefly, volleys of nerve impulses are merely series of individual electric pulses rather than a single, regular signal. Consequently all of the terminals within the nervous system (ie, the postsynaptic membranes) are able to average the incoming impulses in such a manner that smooth electric signals of graded intensity are produced. Similarly, impulses are averaged at neuroeffector junctions. In both instances, a coordinated response is achieved in the postsynaptic cells.

SPATIAL ORIENTATION OF SIGNALS WITHIN THE NERVOUS SYSTEM. The specific neuroanatomic organization of nerve fibers into afferent pathways within the peripheral nerves, spinal cord, and brain enables one to perceive and to localize different types of sensory information with variable degrees of precision. Similarly, an equally specific neuroanatomic organization of efferent pathways from higher brain centers to the periphery enables one to move at will particular groups of skeletal muscles, and similarly the autonomic nervous system is able to control many visceral functions at unconscious levels. Therefore signal transmission in vivo takes place through specific neural pathways that are *spatially oriented* throughout the entire nervous system.

A simple example to illustrate the concept of spatial orientation of signals within the nervous system is shown in Figure 7.86A. If pressure is applied to the skin at three discrete and separate points (eg, on the forearm), the sensory nerve lying just beneath each of the points of stimulation is excited most strongly, whereas the adjacent nerve fibers are stimulated less strongly. Consequently, the signals that arise within the afferent fibers immediately beneath each point of stimulation are transmitted ultimately to specific areas of the sensory cortex via

discrete, spatially oriented neural pathways, and there perceived as pressure applied to three localized areas of the skin. If, however, the stimuli are applied close together on the arm so that the receptive fields of the three cutaneous areas that are stimulated overlap to any great extent, then the three separate stimuli will be perceived as a single, more generalized sensation of pressure (Fig. 7.86B). Nevertheless the spatial pattern of orientation of the signals still is maintained throughout the entire pathway from periphery to cortex (see also p. 425).

SIGNAL TRANSMISSION AND PROCESSING IN POOLS OF NEURONS. The entire central nervous system can be regarded both anatomically and functionally as being composed of a multitude of individual, but interconnected, neuronal pools which vary considerably in size and complexity. Although each neuronal pool is organized to perform certain characteristic functions, nevertheless all of these structures share certain common physiologic properties insofar as the transmission and processing of signals within the central nervous system are concerned.

Specifically, the dorsal (sensory) and ventral (motor) gray matter of the spinal cord may be regarded conceptually as forming two neuronal pools. Similarly the many specific nuclei and ganglia located within the brain stem, cerebellum, and cerebrum also may be regarded as forming discrete neuronal pools. Actually the entire cerebral cortex per se may be regarded as one enormous neuronal pool. In each instance, a neuronal pool has afferent and efferent connections with other pools, and in many cases shorter association fibers interconnect various regions within the same neuronal pool.

As depicted in Figures 6.36 (p. 216) and 7.87, a typical neuronal pool has a large number of input as well as output fibers, only a few of which are shown in the illustrations. Note that the processes of each incoming fiber ramify extensively so that these terminal processes make extensive synaptic contact with the dendrites or cell bodies of many postsynaptic neurons within the pool.* The region of a neuronal pool that is innervated by the terminals of a single input neuron is called the *excitatory* (or *stimulatory*) *field* of that neuron. Note also that the largest number of input terminals form synaptic junctions primarily near the center of the excitatory field, and that relatively fewer fibers innervate neurons lying at the periphery of this region.

Summation, Inhibition, and Convergence. Under physiologic conditions, excitation of a single presynaptic terminal rarely causes discharge of impulses in the postsynaptic neuron (p. 191). Rather, a large number of presynaptic terminals must transmit impulses to the same postsynaptic neuron

---

Actually each input fiber usually divides several hundred times, and accordingly the terminal fibers innervate a similarly large number of postsynaptic elements within a given pool of neurons.

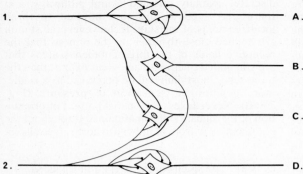

1. ————— A.

2. ————— B. **FIG. 7.87.** A simple neuronal pool. 1, 2 = terminals of input (presynaptic) neurons. A, B, C, D = output (postsynaptic) neurons. See text for discussion of the functional properties of such pools.

C.

D.

either simultaneously or in a brief interval of time in order for the membrane potential of the postsynaptic neuron to reach its threshold and to discharge. On the other hand, subthreshold stimuli temporarily render the postsynaptic neuron more excitable by inducing a partial depolarization of the membrane, and the neuron now is said to be *facilitated*. While the postsynaptic neuron is in the facilitated state, additional impulses can induce discharge by either of two basic mechanisms. The first is by *spatial facilitation* (or *spatial summation*). As shown in Figure 7.87, if four impulses are required for the discharge of neuron B, and neuron 2 discharges while neuron B is facilitated through prior subthreshold excitation by impulses from neuron 1, then the impulses which have *converged* on neuron B from neurons 1 and 2 are sufficient to cause the excitation of neuron B. The second is by *temporal summation*. In this instance, impulses arriving at the three terminals on neuron C from input neuron 1 are sufficient to facilitate neuron C, but not adequate to cause discharge of neuron C. If, however, three additional excitatory impulses are conducted to the terminals of neuron 1 on postsynaptic neuron C while the latter is facilitated, then the membrane potential of neuron C now is depolarized to threshold, and this neural element now discharges through temporal summation.

Alternatively, if neurons 1 and 2 fire simultaneously, then the excitatory impulses that reach all of the terminals of these neurons may be sufficient to discharge the postsynaptic neurons in the center of the discharge zone (A and D) of neurons 1 and 2, as well as to cause either discharge or facilitation of postsynaptic neurons B and C. And assuming that only facilitation has occurred in postsynaptic neurons B and C, then either spatial or temporal summation can occur in these postsynaptic neurons provided that additional impulses arrive at the presynaptic terminals while these neurons are facilitated.

Thus the signal output of a neuronal pool is regulated both by threshold and by subthreshold input signals (see also p. 191).

The concept of a discharge zone and a facilitated zone in a pool of neurons is illustrated in Figure 7.88 (see also Fig. 6.37, p. 217). As noted above, the terminals of a single input neuron are most apt to discharge those postsynaptic neurons which lie near the center of the excitatory field for that particular input neuron (shaded area in Fig. 7.88). Consequently, the facilitated postsynaptic neurons tend to lie at the periphery of the discharge zone as relatively fewer postsynaptic terminals of the input neuron reach this area, and thus these outlying terminals tend to facilitate rather than to discharge the more peripherally located postsynaptic neurons.

Facilitation as well as inhibition of a neuronal pool can take place through input fibers that converge* upon the postsynaptic output fibers from a number of sources. For example, the number of excitatory as well as inhibitory inputs to a single spinal motoneuron is quite large, as shown in Figure 6.38 (p. 218). Whether or not a particular input neuron is excitatory (facilitory) or inhibitory depends, of course, upon the particular neurotransmitter that is liberated at the presynaptic terminals of that neuron (p. 195).

As depicted in Figure 7.89, the basic neuronal circuit for excitation or inhibition is quite simple, and whether the postsynaptic neuron is excited to discharge or not depends upon the net summated effect of the incoming excitatory as well as inhibitory influences upon the postsynaptic membrane potential at any particular moment in time (p. 191). Some particular examples of inhibitory inputs to neuronal pools are given in Figures 6.34 (p. 215), 6.35 (p. 215), and 7.90. In the latter figure, a basic neuronal circuit is illustrated in which an excitatory input is converted into excitatory as well as inhibitory outputs through the inclusion of an inhibitory interneuron (cf. Fig. 6.34).

Briefly, the convergence of excitatory (facilitory) as well as inhibitory input signals within pools of neurons is responsible for determining the patterns of the resulting output signals that are transmitted to other neuronal pools within the nervous system or to effector structures and or-

*In this context, the term "convergence" denotes control of a single output neuron by two or more input nerve fibers.

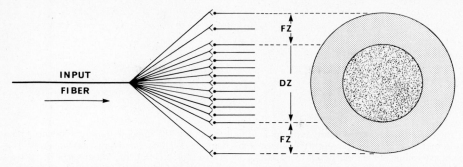

FIG. 7.88. Discharge and facilitated zones of a neuronal pool. The input fiber to the pool ramifies as shown to the left. When the signal amplitude passing along the input fibers (arrow) is sufficient, a number of postsynaptic fibers in the center of the postsynaptic field are depolarized to threshold, and fire all-or-none impulses; this region is called the discharge zone (DZ) of the input fibers. A number of postsynaptic fibers located at the periphery of the input fiber terminals are merely facilitated by the input signal hence do not fire all-or-none impulses. Consequently this area is called the facilitated zone (FZ). The relative areas of the discharge and facilitated zones are indicated by the relative areas of the shaded concentric circles to the right of this figure.

gans. Thus convergence permits the summation and correlation of information derived from a variety of sources within particular neuronal pools so that the ultimate response is an integrated effect of this processed information. Many specific examples of the effects of convergence of signals will be encountered in subsequent discussions (eg, in the control of voluntary muscular activity, p. 396).

Divergence. The term "divergence" signifies that excitation of a single input fiber to a neuron pool stimulates many output fibers of the same pool. The two principal types of divergence are depicted in Figure 7.91.

Divergence in the Same Tract. When divergence takes place within the same nerve tract, the incoming signal is amplified because it spreads to a progressively larger number of fibers in traversing successive neuronal pools. For example, divergence in the same tract is typical of the cortico-

spinal (or pyramidal) pathway which underlies the voluntary control of skeletal muscles. Thus when a single giant pyramidal cell of the motor cortex is stimulated experimentally, a single efferent impulse is transmitted into the spinal cord. However, this impulse can stimulate a number of internuncial neurons, each one of which can stimulate several hundred motoneurons in the anterior horn of the cord. In turn, each motoneuron can stimulate up to several hundred individual muscle fibers. The net result of this sequence of events is that the original stimulus has diverged (or has been amplified) many thousands of times (Fig. 7.91).

Divergence Into Multiple Tracts. As shown in Figure 7.91B, input signals to a neuronal pool may diverge so that the output signals go in different tracts, ie, diverge, as they leave the pool. Thus divergence into multiple tracts permits the same information to be transmitted simultaneously to

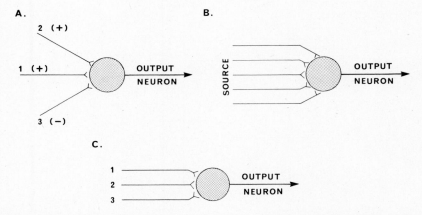

FIG. 7.89. The principle of convergence in neuronal pools. In all examples impulses are transmitted from left to right. **A.** Neuron 1 is the primary source of impulses for excitation (+) of the output neuron. Additional neurons (2, 3) also may converge upon the output neuron, and these secondary neurons may be excitatory (+) or inhibitory (−) so that the net result is either facilitation or inhibition of the output neuron. **B.** Multiple fibers from a single source converge upon a single output neuron. All of these inputs in turn can be either excitatory or inhibitory. **C.** Fibers from three different sources (1, 2, 3) can converge upon a single output neuron, and these multiple inputs can be excitatory and/or inhibitory as discussed for **A.**

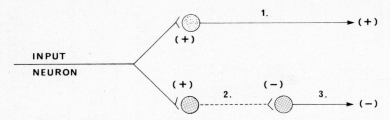

FIG. 7.90.   The conversion of an excitatory input to an inhibitory output. The upper terminals of the input neuron liberate an excitatory neurotransmitter, consequently the output of this neuron is excitatory and postsynaptic neuron 1. is stimulated (+). If, however, a terminal of the input neuron synapses upon, and excites, an inhibitory interneuron, 2. (such as a Renshaw cell in the cord) whose neurotransmitter is inhibitory (−), then the activity of the postsynaptic neuron contiguous with the inhibitory interneuron also will be inhibited (3.).

quite different regions within the nervous system. For example, sensory signals that are transmitted in the dorsal funiculi of the spinal cord pass both to the cerebellum and via the thalamus directly to the cerebral cortex (Fig. 7.21). Similarly, within the thalamus all sensory information (excepting olfactory signals) is relayed to various thalamic nuclei as well as to specific cortical areas.

SIGNAL TRANSMISSION THROUGH NEURONAL POOLS IN SERIES.   Under physiologic conditions, signals are transmitted to various regions of the nervous system through a series of neuronal pools that are located in a definite sequence with respect to each other. For example, sensory information that arises within cutaneous exteroceptors is transmitted through first-order afferent neurons

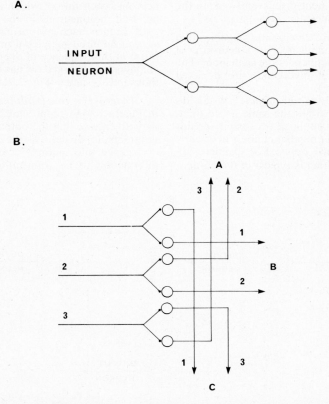

FIG. 7.91.   The principle of divergence. Small dots represent presynaptic terminals whereas open circles represent postsynaptic neurons. **A.** Divergence in a single pathway results in a net amplification of the output signal. **B.** Divergence of fibers (1, 2, 3) in a single tract into three separate output pathways (A, B, C) results in a simultaneous transmission of the same signals to three separate areas of the central nervous system. Note: The principle of divergence applies only to nerve tracts, and not to single nerve fibers and their ramifications. (Data from Guyton. *Textbook of Medical Physiology,* 4th ed, 1971, Saunders.)

that enter the spinal cord through the dorsal root. Thence second-order neurons that arise within the cord proper or the gracile and cuneate nuclei of the medulla transmit the signals to the thalamus. Within the thalamus, third-order neurons arise which relay the signals to the cortex. This pathway is illustrated in Figure 7.92.

Since the adjacent sensory end organs within the skin overlap to a great extent, and since the terminal nerve fibers of each successive neuronal pool within the central nervous system also diverge greatly, one might think that signals from discrete areas of the skin might become "blurred" or mixed because of the peripheral overlap as well as the central divergence of the nerve fibers which would give rise to facilitation of the signals within the series of neuronal pools. Yet the afferent signals from the skin are perceived consciously in a reasonably faithful and discretely localized pattern for the following reasons. First, the anatomic organization of the fibers within the peripheral nerves and central nerve tracts tends to preserve the spatial orientation of the signals within the nervous system and the neuronal pools thereof. Second, the cutaneous nerve fiber having the most sensory end organs in the particular region stimulated (ie, the center of the receptive field; p. 340) will be activated to the greatest extent whereas adjacent fibers will be stimulated less strongly. Third, when the afferent signal reaches each neuronal pool, the largest number of active presynaptic terminals will lie at the center of the excitatory or stimulatory field; therefore only the spatially related post-synaptic neuron nearest the principal input (or pre-synaptic) neuron will be stimulated strongly, so that the signal does not tend to diverge greatly in each neuronal pool of the series.

Nevertheless, if the discrete incoming signal passed through a sequence of neuronal pools that were highly facilitated, the signal could diverge to such an extent in each pool that the ultimate cortical perception of the signal would be quite diffuse or distorted. Actually, the extent to which facilitation or inhibition takes place in each neuronal pool can fluctuate spontaneously; therefore convergence of the signal (with sharpening of the perceptual "image") predominates at certain times, whereas divergence is present at others (with a concomitant diffusion of the perceptual "image"). The continual flux between greater and lesser degrees of facilitation appears to be regulated by fibers emanating from particular areas of the cortex that control the extent to which incoming impulses are facilitated in specific neuronal pools. Thus a feedback mechanism that operates through centrifugal nerve fibers damps the extent to which each neuronal pool is facilitated or inhibited, and thereby incoming signals are "focused" within particular neural pathways in each successive pool of neurons. The result of this physiologic mechanism is the discrete, functionally localized transmission of signals between individual neuronal pools within the central nervous system, as shown in Figure 7.92.

In addition to corticofugal fibers that regulate

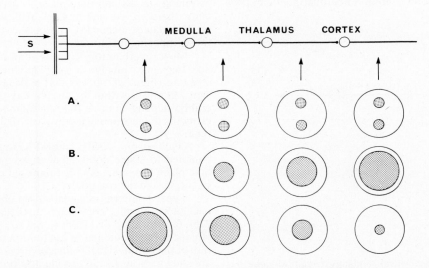

FIG. 7.92. The organization and transmission of signals from a peripheral receptor in the skin to the somesthetic cortex. S = stimuli applied to receptors (horizontal arrows at left). **A.** Discrete point-to-point transmission of two stimuli from the periphery to the cortex when all successive neuron pools in the pathway are normally excitable. **B.** Divergence of a signal evoked by one stimulus in successive neuronal pools owing to facilitation in these pools. This facilitation is indicated by successively larger shaded areas from left to right. Because of divergence, the initial stimulus is perceived over a diffuse area of skin rather than as a discrete point. **C.** Convergence of a signal evoked by one stimulus in successive neuronal pools owing to inhibition in these pools. This inhibition is indicated by successively smaller shaded areas from left to right. Convergence results in a neural "focusing" of the signals so that a diffuse stimulus applied to a relatively large cutaneous area is perceived as coming from a much smaller and localized area. See also Figure 7.99. (Data from Guyton. *Textbook of Medical Physiology,* 4th ed, 1971, Saunders.)

the extent to which neuronal pools are facilitated, *lateral inhibitory circuits* are present, at least in some areas of the nervous system, which also tend to sharpen or "focus" the output signals from pools of neurons by inhibiting the neurons lying toward the periphery of the principal excitatory field (Fig. 7.93). Lateral inhibitory circuits appear to be present in many sensory pathways including those mediating somesthetic sensations and vision.

AFTER-DISCHARGE. Up to this juncture, only mechanisms whereby signals are transmitted rapidly through neuronal pools have been discussed in this chapter. Frequently, however, an input signal to a neuronal pool can induce a protracted signal output that is called *after-discharge,* and this effect can persist for long intervals after the input signal has terminated. There are three fundamental mechanisms which underlie after-discharge.

Synaptic After-Discharge. The discharge of presynaptic nerve terminals upon the soma or dendrites of a postsynaptic neuron can generate an excitatory postsynaptic potential (EPSP) (p. 191) that outlasts the presynaptic discharge by many milliseconds. For example, an input signal to a spinal motoneuron may have a duration of 1 msec; however, excitation of the postsynaptic neuron may last for up to 15 msec, and this type of after-discharge persists for as long as an EPSP is present in the postsynaptic element.

The Parallel Circuit. As shown in Figure 7.94, the ramification of an input neuron into several parallel chains of neurons within a neuronal pool can pro-

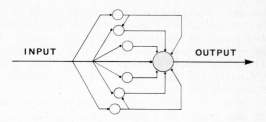

FIG. 7.94. A simple parallel after discharge circuit which operates primarily on the principle of synaptic delay.

duce after-discharge. In such a parallel circuit, a single input signal follows several pathways which ultimately converge upon a single output neuron, and since the delay in impulse transmission at each synapse is about 0.5 msec, then a series of temporally spaced impulses converge sequentially upon the postsynaptic neuron and thereby produce after-discharge.

The Reverberatory (or Oscillatory) Circuit. The characteristics of a simple reverberatory, or oscillatory, circuit were introduced in Chapter 6 (p. 222) and depicted in Figures 6.42, 6.44, and 6.45. Some additional reverberatory circuits are illustrated in Figure 7.95 in a sequence of increasing complexity. In each instance, incoming signals are fed back through collateral branches or fibers so that reexcitation of the system occurs in a cyclic fashion thereby prolonging greatly a very brief input signal. As shown in Figure 7.95, facilitation or inhibition of incoming signals can take place in oscillatory circuits; therefore the output signal will be protracted or attenuated, depending upon the resultant of these opposing influences.

Most reverberatory circuits are fairly complex and thus may involve a large number of neural elements (Fig. 7.95). In a circuit of this type, the net signal intensity within the circuit may vary considerably, depending upon the total number of neural elements that are active at any particular moment in time.

Oscillation within neural circuits is not confined to single pools of neurons, but also can take place through a circuit composed of a number of successive neuron pools.

Typical patterns of signal output obtained from oscillatory circuits are shown in Figures 6.44 (p. 225), 6.45 (p. 225), and 7.96. In the latter figure, the effects of inhibition and of facilitation on the duration of the output signal from a reverberating circuit are shown. Note also in this illustration that following the brief input stimulus (eg, 1 msec), the intensity of the output signal peaks rapidly, declines gradually to a certain level, and then stops completely. Since other regions of the brain can influence any particular neuronal pool through facilitory and inhibitory influences, an extremely precise control of the duration of after-discharge is obtained in vivo. For example, the pattern of signals recorded from motoneurons to flexor muscles

A.

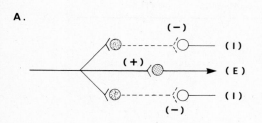

B.   **C.**

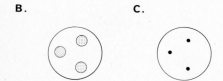

FIG. 7.93. A. A lateral inhibitory circuit whereby the effects of excitation of one input neuron of a neuronal pool (E) produces inhibition of adjacent fibers (I) through activation of inhibitory interneurons (coarse shading, dashed lines). **B.** Normal output pattern of a neuron pool not subject to lateral inhibition. **C.** Output pattern of same neuron pool as in B subjected to lateral inhibition. Note that the three stimulated points are sharpened considerably, hence perception of the stimuli would be more acute and better localized. (After Guyton. *Textbook of Medical Physiology,* 4th ed, 1971, Saunders.)

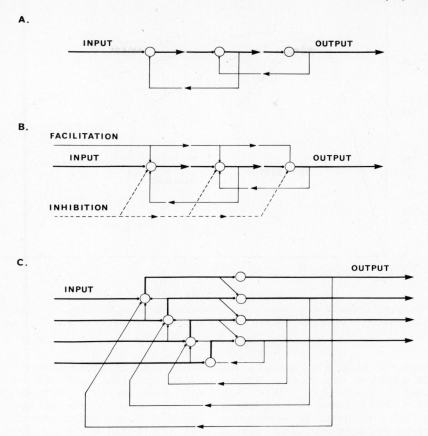

FIG. 7.95. Three types of oscillatory (or reverberatory) circuits. In all examples the principal input and output pathways are indicated by heavy lines. Only a few collaterals are shown in each example. The arrows indicate the direction of impulse propagation. **A.** Simple reverberatory circuit consisting of only three neurons and two collaterals which feed back impulses to two of the neurons thereby inducing oscillation. **B.** More complex oscillatory system involving facilitatory as well as inhibitory neurons which can alter considerably the duration of the spontaneous activity. Thus facilitation would increase the tendency toward spontaneous oscillation, whereas inhibition would decrease this effect or else terminate the oscillation suddenly. See Figure 7.96. **C.** Oscillatory circuit containing many parallel input as well as output fibers. At each synapse, the terminal fibers of the presynaptic fiber can diffuse widely. The net output signal of a circuit such as this is quite flexible, depending upon the total number of parallel nerve fibers that are active at any particular moment. (Data from Guyton. *Textbook of Medical Physiology,* 4th ed, 1971, Saunders.)

following stimulation of a pain fiber is quite similar to that shown in Figure 7.96. (The flexor reflex was discussed on page 221.)

Synaptic fatigue (p. 200) is a major factor which determines the duration of after-discharge in any reverberatory circuit. Thus rapid fatigue would decrease the period of the after-discharge, whereas a slower onset of synaptic fatigue would prolong the after-discharge.

Obviously if any particular oscillatory circuit contains more neurons and more collateral fibers than another, the duration of the output signal will be longer in the former than in the latter. Consequently there is a direct relationship between the total number of neural elements that participate in any given oscillatory circuit and the period of the after-discharge from that circuit.

The after-discharge of signals from different oscillatory circuits may range from a few milli-seconds to several hours. In fact, the phenomenon of consciousness (or wakefulness) possibly is related to the oscillatory circuits located in the basal region of the brain. Hence signals that arise within reverberating circuits located in the reticular activating system are believed to induce sustained reverberations within cortical circuits that are related to the waking state (see also p. 427).

REPETITIVE SIGNAL DISCHARGE FROM NEURONAL POOLS. Certain pools of neurons are able to generate output signals in a continuous fashion even in the absence of input signals.

Similarly to other excitable tissues (eg, cardiac pacemaker cells, p. 143) neurons can discharge continuously and in a repetitive manner provided that their membrane potentials decrease spontaneously below their threshold level to firing (p. 143). Apparently the membrane potential of certain

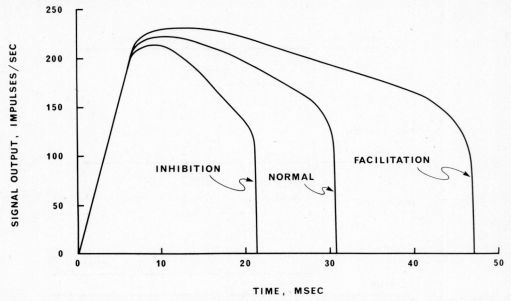

FIG. 7.96. The signal output of an oscillatory (or reverberatory) circuit similar to that illustrated in Figure 7.95B, following a single input stimulus delivered at zero time. (After Guyton. *Textbook of Medical Physiology,* 4th ed, 1971, Saunders.)

neurons is able to discharge in this manner so that such neurons continually emit trains of impulses whose inherent frequency can be accelerated by facilitory inputs or decelerated by inhibitory inputs. In the latter situation, the spontaneous, rhythmic impulse discharge even can be stopped completely, provided that the intensity of the inhibitory inputs is sufficiently great. Thus input signals can modify profoundly the intrinsic pattern of spontaneous neural activity, even though such activity resides within the autorhythmic neurons themselves.

Another mechanism whereby continuous signals could be emitted from a neuronal pool would be through the operation of an oscillatory circuit which did not exhibit fatigue readily. Similarly to the situation which exists for autoexcitable neurons discussed above, facilitory input signals would increase the output signals, and inhibitory inputs would attenuate or suppress completely the output signals. Note again that in this situation the two kinds of input do not induce, but merely alter, the fundamental pattern of activity within the oscillatory circuit, ie, the input signals regulate the output signals. This type of neural control is encountered within the autonomic nervous system, and thereby heart rate, vascular tone, motility and tone of the gastrointestinal tract, and the extent to which the iris is constricted all are regulated.

RHYTHMIC OUTPUT OF SIGNALS FROM NEURONAL POOLS. As noted in the previous section, certain neural circuits are able to generate rhythmic outputs. For example, the rhythmic signals which are responsible for respiration have their sources within the reticular matter of the pons and

medulla (p. 729), and these repetitive stimuli persist throughout the lifetime of the individual. On the other hand, rhythmic movements such as are involved in walking require that input signals be received by the appropriate neuronal pools in order that output signals be transmitted to the appropriate groups of leg muscles.

Facilitory or inhibitory input signals can modulate the output of spontaneous as well as induced rhythmic activities in a manner quite similar to that observed in situations where a continuous signal output is found. For example, and as illustrated in Figure 7.97, the rhythmic signals that pass down the phrenic nerve to induce contraction of the diaphragm during respiration are enhanced considerably by a reduced arterial oxygen partial pressure (or Po_2) so that the amplitude of the rhythmic signals increases considerably. In other words, a relative oxygen deficiency in the arterial blood within the carotid body causes these chemoreceptor structures to transmit facilitory input signals to the respiratory center, thereby increasing the strength of the signals passing down the phrenic nerve to the diaphragm (see also p. 737).

Sensory Functions of the Nervous System

The sensory end organs located throughout the body provide the ultimate sources for all information received by the central nervous system. Stated another way, each particular type of sensory receptor generally has its lowest threshold to a specific kind of energy flux that may arise within either the internal or the external environment.* The receptor

*Receptors thus exhibit the property of "differential sensitivity" to various types of stimuli.

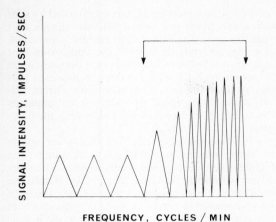

SIGNAL INTENSITY, IMPULSES/SEC

FREQUENCY, CYCLES/MIN

FIG. 7.97. The rhythmic oscillatory output of the respiratory center. Note that the signal intensity (or amplitude) in the phrenic nerve to the diaphragm, as well as the signal frequency in the same nerve, increase significantly upon stimulation of the carotid body. Such an effect can be induced by reducing the arterial oxygen partial pressure within the carotid body per se. (After Guyton. *Textbook of Medical Physiology*, 4th ed, 1971, Saunders.)

then converts (transduces) this energy flux into signals, or patterns of nerve impulses, which then are transmitted via appropriate neural pathways to their corresponding neuronal pools located within the spinal cord and brain. Finally, and assuming that the signals reach higher brain levels, the sensory information is so integrated that one usually can recognize and localize the type as well as the source of the stimulus.

The most fundamental subconscious motor activities of the body (eg, monosynaptic reflexes, p. 213) originate in sense organs, as do certain far more complex voluntary and involuntary responses. For example, such basic and generalized sensations as hunger, thirst, fatigue, pain, nausea, and sexual feelings not only involve a number of intricate neural pathways, but also can initiate complex behavioral patterns which lead to their amelioration.

Actually, a single "pure" sensation may be considered an abstraction as in the adult body various individual sensations are so modified by past experience and by comparison with one another that they are transformed into *perceptions*. For example, a rapidly repeated mechanical stimulus applied to the skin evokes the sensation which is perceived as vibration. The distinction between a perception and a sensation is important, as it often affords the clinician valuable clues to malfunctions at different levels of the nervous system.

In physiologic terms, a *sensation* has three fundamental characteristics. First, the *quality* is that subjective difference between individual sensations which enables a person to name the many

sensory modalities, eg, cold, warmth, touch, pain, and so forth. The visual sense has many submodalities which are so different that different terms are applied to each; eg, the colors red, yellow, and blue. Similarly, sweet, salty, sour, and bitter are the basic submodalities applied to the sensation of taste.

Second, the *intensity* is another fundamental attribute of sensation. Thus a stimulus must achieve a certain definite strength in order to exceed the subjective threshold for its perception.

Third, a sensation appears to arise within a certain region of the body or the external environment. Thus sensations are localized, ie, they originate from a particular *locus*, and appear to be projected to their source by the activity of the brain. Exactly how the brain accomplishes this feat is, in general, unknown (see also p. 364).

In addition to quality, intensity, and locus which are characteristic properties of each sensation, the comparison of two or more sensory stimuli applied concomitantly or in succession gives rise to a *discrimination* or *judgment* concerning the stimuli. Thus perception of the size, shape, texture, position, and temperature of various objects requires that a subjective interpretation or discrimination among the stimuli be achieved. The ability to discriminate among various objects, their spatial relationships, and so forth often assumes great importance in neurologic diagnosis.

Lastly, sensory information produces a subjective response, in addition to those of quality, intensity, and locus, that may be either "pleasant" or "unpleasant," and this characteristic of a sensation (or of a perception) is called *affect*. For example, mild alcoholic intoxication can produce a pleasant affect or euphoria, whereas the hallucinations that may accompany delirium tremens following a prolonged bout of heavy drinking produce quite the opposite type of affect.

Sensations may be classified broadly into four general categories which depend largely upon the anatomic features of the sensory end organs and neural pathways involved (see Table 6.1, p. 177): (1) Superficial or cutaneous sensations are transmitted via the cutaneous branches of spinal as well as certain cranial nerves and include touch, pressure, and thermal sensations (warmth and cold), as well as pain. (2) Deep sensations are conveyed by those branches of the spinal nerves that innervate the skeletal muscles, and certain cranial nerves also participate in the transmission of deep sensations. Included in this category are muscle, tendon, and joint sensibility as well as the kinesthetic sense in which changes in spatial position of the limbs are perceived. Deep pressure and deep pain also may be included here. (3) Visceral sensations are conducted by impulses transmitted along the general visceral afferent nerve fibers that accompany autonomic nerves. Visceral pain as well as certain organic sensations (eg, hunger, thirst, sexual sensation, nausea) also may be classified among the visceral sensations. (4) The cranial nerves underlie the functions of the special senses, and this category includes vision, audition, taste, olfaction, and the vestibular functions of the inner ear which are related to posture, balance, and movement.

The cutaneous, deep, and visceral sensations outlined above often are referred to as somatic sensations, in contrast to the special senses.

The foregoing classification will serve as a broad working outline for a consideration of the several types of sensation listed above.

CUTANEOUS SENSATIONS

General Anatomic Considerations. The end organs which underlie the sensations of touch, pressure, warmth, cold, and pain were discussed in Chapter 6 (p. 178), whereas the functional types of nerve that transmit impulses from these peripheral receptors to the central nervous system are summarized in Table 5.1 (p. 171). The neuroanatomic basis for the cutaneous sensations was presented earlier in this chapter (eg, Figs. 7.20, 7.21, 7.31, 7.40, 7.41, 7.42, 7.43, 7.61, and 7.74).

By way of summary, the afferent nerve fibers that arise from peripheral sensory nerve endings terminate upon interneurons located at many levels of the spinal cord and there make polysynaptic connections with motoneurons that are involved in particular reflex arcs. In addition, however, the afferent fibers also synapse upon the cell bodies of neurons of the long ascending (or projection) tracts which relay impulses to the cerebral cortex, as depicted in Figures 7.20 and 7.21. The dorsal root fibers become separated into specific and relatively discrete tracts that subserve particular sensory modalities once they have entered the cord. Thus fibers transmitting signals related to delicate touch and pressure (as well as proprioception) ascend in the dorsal columns (ie, the fasciculi gracilis and cuneatus) to the medulla oblongata where these fibers synapse within the nuclei gracilis and cuneatus. Additional fibers that subserve the sensations of touch and pressure synapse upon neurons located within the dorsal gray column of the cord, and the axons of these neurons then cross the midline to ascend principally in the ventral spinothalamic tract.

The fibers which transmit signals underlying the sensations of pain, warmth, and cold are located somewhat more laterally, and these fibers synapse in the dorsal gray matter of the cord. The second-order neurons which arise from these synaptic junctions then cross the midline and ascend in the dorsal (or anterolateral) spinothalamic tract (Table 7.5).

Collateral branches from the afferent fibers which enter the dorsal columns course to the substantia gelatinosa, a pale-staining region located upon each dorsal gray column (Fig. 7.18). Functionally it appears that these afferent collaterals to the substantia gelatinosa act to modify the neural inputs to other cutaneous sensory systems, including that which subserves pain. Consequently the dorsal gray column behaves physiologically as a

"gate" within which the signals transmitted by sensory nerve fibers are converted into signals that enter the ascending tracts, and current experimental evidence tends to support the conclusion that signal transmission through this gate depends upon the source as well as the pattern of impulses that impinge upon the substantia gelatinosa. Parenthetically it should be noted also that efferent impulses descending from the brain also can modify the functional activity of this gate.

The second-order neurons that arise within the nuclei gracilis and cuneatus decussate in the midline and ascend in the medial lemniscus to terminate within the thalamus (Fig. 7.21). Within the brain stem, these neurons are joined by fibers of the spinothalamic tracts as well as sensory fibers that originate within certain cranial nerves, in particular fibers of the trigeminal nerve. Signals that underlie the sensations of pain, warmth, and cold in the head region are relayed via the spinal nucleus of the trigeminal nerve, whereas signals related to pressure, touch, and proprioception are transmitted primarily by way of the principal sensory and mesencephalic nuclei of the same nerve (Fig. 7.30).

The thalamus and its specific ascending sensory nuclei were discussed earlier (p. 291), and it was emphasized that this diencephalic structure with its associated nuclei forms the major relay station through which all sensory information, regardless of its source, must funnel before it is transmitted to the cerebral cortex.* Thus the spinothalamic tracts, the fibers of the medial lemnisci, and the related specific ascending sensory systems all terminate in particular sensory relay nuclei located within the thalamus (Fig. 7.40). Within these thalamic nuclei, the ascending sensory systems synapse upon neurons whose axons pass via the internal capsule to the postcentral gyrus of the cerebral cortex as the thalamic radiation (Figs. 7.41 and 7.43).

Within the thalamus, as in other regions of the specific sensory systems, there is a high degree of spatial orientation of the fibers so that various parts of the body are represented in this structure both accurately and distinctly (p. 426).

Within the postcentral gyrus of the cerebral cortex the fibers of the thalamic radiation are so organized that the various regions of the body are represented in a definite sequence (Fig. 7.74). Furthermore, as indicated by the relative sizes of the cortical areas in this figure, there is a directly proportional relationship between the cortical representation for afferent signals from any particular region of the body and the number of cutaneous sensory receptors that are located within that region.

Experimental studies that have been performed upon the primary sensory area of the cortex indicate that an extremely discrete point-for-

---

The major exception to this statement being olfactory signals.

point localization of peripheral regions of the body is present. Thus electric stimulation of specific areas of the postcentral gyrus gives rise to sensations that are projected to the corresponding part of the body by the subject. Generally the sensations evoked by such stimulation are those of tingling, numbness, or a feeling of movement. However, by using extremely fine electrodes, reasonably "pure" sensations of touch, warmth, cold, and pain can be evoked.

The neural elements located within the postcentral gyrus apparently are organized in columns similarly to those found in the visual cortex (p. 430). Therefore when a particular sensory signal is received, all of the cortical cells that are located within that specific column are activated.

The superior wall of the sylvian fissure contains a secondary sensory receiving area; however, the representation of the various regions of the body is neither as complete nor as detailed in this area as it is in the postcentral gyrus. The functions of this secondary sensory receiving area are unclear, although it has been implicated in the perception of pain and perhaps in the coordination of some complex motor activities.

Impulses that arise within cutaneous receptors also may be transmitted to the primary motor area, the precentral gyrus.

General Physiologic Considerations. The general physiologic mechanisms which underlie the action of cutaneous receptors were presented in Chapter 6 (p. 182). Briefly, every sensory end organ is specialized so that it most readily converts (or transduces) a particular type of energy into a graded and localized receptor potential. In turn the receptor potential excites volleys of all-or-none impulses in the sensory fiber that is related to the receptor. The signals thus produced in the periphery are transmitted to the central nervous system along specific neural pathways, and within the central nervous system the signals for each modality of sensation still follow discrete pathways to localized areas of the brain and there produce excitation. Consequently each modality of sensation is transmitted along a particular neural pathway, and the sensation perceived by the individual as well as the region of the body to which it is localized depend upon the area of the brain that has been stimulated by the afferent signals.

The intensity of any sensory modality that is perceived can be varied by altering the frequency at which impulses are transmitted along the afferent nerve or by changes in the total number of receptors that are stimulated either simultaneously or successively. Furthermore, central excitatory (or inhibitory) states, among other factors, can modify the ultimate sensory perception appreciably in accordance with the physiologic properties of neuronal circuits discussed earlier in this chapter (p. 340 et seq.). It is important to realize, however, that an increase in the intensity of the stimulus that is applied to any given sense organ

exerts little, if any, effect upon the quality of the sensation that is evoked by stimulation of that receptor.

Only the cutaneous senses display *epicritic sensibility*. That is, the receptors for touch, pressure, warmth, cold, and pain are arranged spatially within the skin so that an accurate localization of these sensations can be made. This punctate representation of the cutaneous senses depends upon the numerical distribution of the several types of receptor involved, and thus varies widely from region to region on the body surface. If one carefully "maps" the skin by stimulating with a delicate hair, the sensation of touch is elicited only from skin areas that overlie touch receptors. In a like manner pain, pressure, and thermal sensations can be evoked only by appropriate stimulation of the skin immediately above the points where the receptors for these sensory modalities are located.

Touch, Pressure, and Vibration. The touch or tactile sense, which is one of the four basic cutaneous sensory modalities, is known to be subserved in man by several functional types of mechanoreceptor. One type of mechanoreceptor adapts rapidly to a continuously applied stimulus of uniform intensity, whereas the second type adapts far more slowly (see adaptation, p. 186).

The rapidly adapting cutaneous tactile receptor discharges for only 0.2 second following a sustained touch applied to the skin or while a hair shaft is bent.* These receptors have a low threshold and are attached to myelinated nerve fibers that range between 2 and 8 μ in diameter in hairy skin and between 6 and 11 μ from smooth (glabrous) skin. The receptive fields (Fig. 7.86) of the rapidly adapting touch receptors are large and range between 1.5 and 5.0 cm² in area.

The more slowly adapting tactile receptors discharge for many seconds following an initial rapid adaptation. The frequency of afferent neural discharge from these receptors is proportional to the pressure that is applied and may achieve rates of over 300 impulses per second. The receptive field is between 1 and 2 mm in diameter in both hairy and glabrous skin.

Human skin also contains a physiologically nonspecific mechanoreceptor which responds to cooling as well as to pressure. This receptor also is unique because it discharges spontaneously at the low frequency of one impulse per second. The functional significance of this dual pressure–temperature sensitivity is unknown, although it is believed that these cutaneous receptors give rise to a combined touch-pressure sensation when stimulated mechanically.

Pacinian corpuscles (p. 180) are rapidly adapting mechanoreceptors that are located subcutaneously as well as within deeper-lying tissues rather

---

*In this regard hairs tend to act as levers, thus amplifying the mechanical deformation of the mechanoreceptors that surround their roots.

than in the skin per se. However, these receptors are quite sensitive to mechanical deformation, and because they adapt so rapidly (in a fraction of a second) they are able to detect rapid movements of the tissues, such as are encountered during the application of vibratory stimuli.

Free nerve endings are of ubiquitous occurrence within the skin and many other tissues, and these elements also can detect touch and pressure as well as painful and thermal stimuli. For example, the cornea of the eye contains only free nerve endings, and yet this structure is quite sensitive to light touch and pressure as well as to painful stimuli.

The sensation known as vibration (or pallesthesia) is elicited when a vibrating tuning fork is placed upon the skin. A "buzzing" sensation or "thrill" is felt at the point of application of the tuning fork. Actually, the vibratory sense is not a distinct sensory modality. Rather it involves a rapidly repeated sequence of mechanical stimuli applied to the touch-pressure receptors discussed above, and in addition the subcutaneous pacinian corpuscles are particularly sensitive to this type of intermittent stimulation.

Vibratory sensibility has been called "osseous sensation" because the application of a tuning fork over a bone intensifies the response. Actually, this concept of an "osseous sensation" is erroneous, as placing the tuning fork above a bone merely amplifies the stimulus mechanically. Actually the fingertips have the lowest threshold to vibratory stimuli. In all instances, vibratory sensibility has a punctate representation on the skin which corresponds to the pressure-sensitive areas.

The end organs that are sensitive to touch-pressure stimuli are not distributed uniformly over the body surface. Rather they are most abundant in the skin of the lips and fingers and most scanty in the skin of the trunk. Touch receptors are particularly numerous around hair follicles as well as in the subcutaneous tissue beneath glabrous skin.

Warmth and Cold. There are two temperature senses, one for warmth and one for cold. If one regards normal skin temperature as physiologic zero, the warmth receptors fire more rapidly as the temperature rises above this point, whereas the cold receptors fire more rapidly when the temperature declines below it.

It is important to realize that the adequate stimulus for both the warmth and cold receptors is heat, as "cold" is merely a relative decrease in heat energy. Temperature is, of course, merely an arbitrary scale whereby the heat energy is measured conveniently, but temperature per se has no energy dimension.

Like the touch-pressure sense, thermal receptors are distributed over the body surface in a punctate fashion, and when one maps the skin carefully, small areas are found which respond only to the application of warmth whereas other areas respond only to cold. The skin areas lying between the heat-sensitive and cold-sensitive areas are insensible to either type of stimulus.

The cold-sensitive areas far outnumber those which are sensitive to warmth. For example on the skin of the forearm, about 14 cold-sensitive areas are present per square centimeter, whereas only one or two heat-sensitive regions are present in the same area.

There is no correlation between the physiologically evident presence of heat and cold receptors and the existence of two morphologically discrete end organs that subserve these sensory modalities. The early concept that Krause's end bulbs (p. 180) were the receptors for cold and Ruffini's end organs (p. 180) were the receptors for warmth has been disproved. Experimental evidence indicates that certain small myelinated nerve fibers belonging to the A (δ) or type III group of sensory fibers (Tables 5.1 and 5.2, pp. 161, 162) transmit impulses related to thermal sensations to the postcentral gyrus of the cerebral cortex via the dorsal (anterolateral) spinothalamic tracts and thalamic radiation.

Actually the thermoreceptors are located within the subepithelial tissues; consequently it is the temperature of the subcutaneous tissue that underlies the type of sensation evoked by stimulation of these receptors. For example, standing on a cool tile floor elicits a greater sensation of cold in the soles of the feet than does standing on a bath mat at the same temperature, because the tile floor conducts heat more rapidly from the subcutaneous tissues of the feet than does the bath mat. This fact also demonstrates another basic physiologic property of receptors in general, namely that the rate of change of stimulus intensity often is an important determinant of receptor excitation.

Both warmth and cold receptors adapt at relatively fast rates and are able to discharge at steady temperatures. Warmth receptors fire at temperatures ranging between 20 C and 45 C; however, their maximal discharge frequency occurs between 37.5 C and 40 C. Cold receptors, on the other hand, fire in the range between 10 C and 41 C, whereas their maximal discharge rate takes place between 15 C and 20 C. If the temperature rises above 45 C the warmth receptors stop firing, and incongruously, the cold receptors also commence firing simultaneously with the pain receptors. Consequently it would appear that the sensation of extreme heat is evoked by the concomitant discharge of cold and pain receptors, and not the warmth receptors.

In addition to the physiologically specific warmth and cold receptors discussed above, there is a group of thermal receptors that respond not only to a decline in temperature but also exhibit pronounced mechanoreceptive properties as discussed in the previous section. The physiologic role of these mechanoreceptors, which fire spontaneously at a neutral skin temperature, and in the absence of any obvious mechanical deformation, is unknown.

Pain. The sensation of pain may be considered as a protective mechanism, a physiologic warning to the individual that all is not well, and thus in a sense all forms of pain may be considered as pathologic. Clinically the different types of pain and the region of the body from which the pain emanates often are of considerable diagnostic importance to the physician.

The sensory end organs which subserve pain are ubiquitous; hence they are found in practically every tissue of the body. Three types of pain are recognized, depending upon their source on or within the body: first, cutaneous (or superficial) pain that is localized upon the body surface; second, deep pain that arises from muscles, tendons, joints, and fascia; third, visceral pain which is evoked by appropriate stimulation of the visceral organs. Cutaneous and deep pain together often are referred to as somatic pain.

Various terms have been applied to describe the character of the different types of pain which may be experienced. Thus cutaneous pain often is described as pricking, sharp, or burning, and may be either localized or diffuse. For example pricking the finger with a needle elicits a sharp, localized (epicritic) pain, whereas burning a large area of the skin evokes the familiar burning sensation. Deep and visceral pain in contrast to cutaneous pain usually are poorly localized (ie, protopathic), and are designated by such terms as aching, dull, nauseous, throbbing, and so forth. The perception of pain has, of course, an enormous subjective component (an unpleasant affect) which usually leads the individual to take immediate action to alleviate the situation causing the pain. Therefore painful sensations are protective in contrast to the other modalities of sensation which are informative (or gnostic).

The receptors for pain are unique in that they are not specialized physiologically; therefore, several types of energy are adequate stimuli to evoke painful sensations. Thus electrical, mechanical, and thermal as well as chemical stimuli of many types all are able to elicit painful sensations if the intensity of the stimulation is sufficiently great. The one common physiologic denominator which underlies these various types of stimuli is that all are capable, or potentially capable, of producing tissue damage or destruction. Thus pain receptors are functionally specific naked nerve endings that are able to respond to a broad range of energy types; however, painful sensations are not produced by overstimulation of other receptors.

The receptors for this broad range of painful, or nociceptive, stimuli are naked nerve endings, and the impulses arising in these nerve terminals are transmitted to the central nervous system by two systems of nerve fibers. One fiber system is composed of small myelinated type A (δ) fibers that conduct impulses at rates between 12 and 30 m/sec, whereas the second fiber system is made up of unmyelinated C fibers which conduct impulses at rates that range between 0.2 and 2.0 m/sec (Table

5.1, p. 161).* The processes of both of these fiber systems terminate upon neurons of the dorsal or anterolateral spinothalamic tract; thence pain signals ascend to the posterolateral and posteromedial thalamic nuclei. From these thalamic nuclei relay neurons transmit the signals to the postcentral gyri of the cerebral cortex (Table 7.5, Fig. 7.23).

The fact that there are two functionally specific neural pathways which transmit pain signals underlies the physiologic observation that there are two kinds of pain. Thus a noxious stimulus rapidly evokes a sharp and localized sensation of pain which is followed by a dull, diffuse, aching sensation having an unpleasant affect. These two kinds of pain are called fast (epicritic) pain and slow (protopathic) pain, or first and second pain.

The time lag between the perception of fast and slow pain depends upon the distance from the brain at which the painful stimulus is applied. This fact, together with other evidence, indicates that fast pain is transmitted by way of type A (δ) nerve fibers, whereas slow pain is transmitted via type C fibers.

Sensory stimuli of many types are perceived in the absence of the cerebral cortex, and this fact is particularly true for the perception of pain. The primary receiving area for sensory information, the postcentral gyrus, appears to be concerned with the precise, discriminative, and meaningful interpretation of painful sensations; however, the mere perception of such sensations takes place at lower brain levels, particularly within the thalamus.

Painful stimuli generally evoke powerful withdrawal and behavioral reactions which remove the body, or the affected part of the body, from the locus of the noxious stimulation. Pain also is unique among the senses in that it contains a strong and inherent emotional component. Sensory information that is transmitted to the brain from the special senses may be neutral or it may elicit a secondary emotional response that may have a pleasant or an unpleasant affect which depends to a great extent upon past experience. Painful sensations, however, give rise to an intrinsic and unpleasant affect, which appears to be caused by the neural connections of pain fibers within the thalamus (eg, see thalamic syndrome, p. 376).

The view has been advanced that the ultimate stimulus for pain is chemically mediated, and that the diverse types of stimuli that evoke this sensation all are able to liberate a chemical within the tissues which in turn stimulates the pain receptors. For example, kinins are polypeptides which can be released from proteins by the action of proteolytic enzymes following the application of noxious

---

The C fibers are not associated exclusively with the transmission of pain signals, but also have been implicated in the conduction of impulses related to the sensations of touch, warmth, and cold.

stimuli, and these substances are able to cause intense pain when injected (see also page 648 for a more thorough discussion of kinins). The pain caused by injection of kinins is of short duration, a fact that would correlate with their rapid degradation in vivo. Histamine has been implicated as another chemical mediator responsible for eliciting painful sensations. However, this compound may have an indirect action in producing pain, as there is some experimental evidence which indicates that histamine injected in vivo produces a liberation of kinins at the site of injection.

Although cutaneous, deep, and visceral pain share certain common features (ie, receptor type, effective stimuli, neural pathways, and mechanism for initiation), nevertheless there are certain important differences among the painful sensations that arise in these three general regions of the body. Consequently, certain particular features of deep and visceral pain will be contrasted with superficial pain in the appropriate sections to follow.

Itch and Tickle. Itching sensations can, under certain abnormal circumstances, approach the discomfort caused by pain. Whenever the sensation of cutaneous pain is lost, so is the sensation of itching. Thus when painful sensations are blocked by anesthesia, neurologic lesions, or surgical operations which interrupt the spinothalamic tracts, itching sensations also are lost. Furthermore, the temporal patterns of both burning pain and itch sensations are similar. For these reasons, it appears that a C-fiber system similar to that which transmits pain signals also conveys impulses related to the sensation of itch, and that the endings of these fibers are extremely sensitive to light tactile and other stimuli.

Itch has a punctate representation on the skin surface, although the distributions of areas that are sensitive to itch and to pain are different. For example, itching occurs only in the skin, eyes, and certain mucous membranes, but not in the deeply-lying tissues and viscera. In addition, low-frequency stimulation of pain fibers produces pain, but not itch, whereas intense high-frequency stimulation of itch-sensitive areas of the skin may produce more intense itching but not pain. Thus the receptors to itch and pain stimuli would appear to differ, even though afferent impulses for both sensations are transmitted in C-fiber systems.

A very mild, repetitive, low-frequency tactile stimulation applied to the skin produces tickle, particularly when the stimulating agent (eg, an insect) moves along the skin surface. Insofar as affect is concerned, tickle may be pleasant, if not too intense; itch usually is irritating, and pain is unpleasant.

Histamine as well as kinins have been implicated in itch. Thus it is of interest that the seed pods of a woody tropical vine known as cowage (*Mucuna pruritum*) are covered with barbed hairs that cause intense itching. In turn it has been demonstrated that these hairs contain a proteolytic enzyme which presumably liberates kinins in the skin which then stimulate the terminals of the C-fiber system that subserves itch.

DEEP SENSATIONS. If the skin overlying a particular region of the body, say an arm, is anesthetized completely so that the strictly cutaneous sensations are lost, three deep sensibilities persist. These are the proprioceptive or muscle sense, the deep pressure sense, and the deep pain sense. In contrast to the cutaneous senses discussed in the previous section, perception of the deep sensations is more diffuse, hence more poorly localized, than are the sensations arising in the cutaneous region of the body.

Proprioception. The proprioceptive or muscle sense* is a largely ignored and usually unconscious sense that provides an individual with knowledge of the position and physiologic state of the muscles, joints, limbs, and other parts of the body.

The sensory end organs that subserve the proprioceptive sense are free nerve endings, the muscle spindles (p. 181), the Golgi tendon organs (p. 182), the pacinian corpuscles (p. 180), and the spray endings in the tendons. These muscle and tendon receptors continuously monitor sensory information that is related to the state of the skeletal muscles throughout the body: first, active contraction; second, passive stretch, ie, the length of particular muscles and muscle fibers; and third, the tension developed within individual muscles whether this is caused by active contraction or by passive stretch.

Proprioceptive signals are transmitted up the spinal cord to the brain in the dorsal columns (Table 7.5, Fig. 7.20), and a large proportion of this input passes directly to the cerebellum. However, some proprioceptive signals are transmitted via the medial lemnisci and thalamic radiations to the cerebral cortex. In addition, there is some experimental evidence which indicates that proprioceptive impulses reach conscious levels in the anterolateral columns of the spinal cord. Impulses that are generated within the muscle spindles as well as the Golgi tendon organs apparently do not reach conscious levels. Therefore a conscious perception of the spatial position of the various parts of the body at any particular moment is dependent upon impulses that arise within receptors which are located in as well as around the joints (ie, the joint capsules). The end organs that participate in the conscious aspects of proprioception therefore include the spray endings within the tendons, which adapt very slowly and are similar histologically to the Golgi tendon organs, as well as the pacinian corpuscles found within the ligaments and

---

Synonyms for these terms include "muscle, tendon, and joint sensibility," "kinesthesia," and "position sense" and "the appreciation of passive movement."

synovia. The latter structures are, of course, most sensitive to movements and changes in limb position because of their rapid rate of adaptation. In addition to the impulses which arise from the spray endings and pacinian corpuscles, signals that arise concomitantly in touch as well as pressure receptors contribute to the synthesis of a conscious pattern which enables the individual to have a continual knowledge of the position of his body in space. The contributions of the inner ear to spatial orientation are considered on page 381.

Deep Pressure.   The sensation of deep pressure is elicited by exerting a firm mechanical compression over muscles and tendons. Deep pressure, in contrast to the more superficial cutaneous sensations, is relatively diffuse and thus is not clearly localized. The deep pressure sense persists after cutaneous anesthesia or section of the cutaneous nerves and is associated with mechanical stimulation of certain of the receptors which subserve the proprioceptive sense and which were discussed in the previous section.

Deep Pain.   The sensation of deep pain, as well as of deep pressure, can be evoked by firm pressure over both muscles and tendons as both of these structures are exceedingly sensitive to noxious stimuli. The signals which underlie deep pain originate in muscles, tendons, and joints and in general are transmitted via the mixed nerves that innervate the muscles. Deep pain has a dull or aching quality which appears to originate beneath the skin; however, localization is poor as the sensation usually undergoes radiation, a phenomenon that is discussed in the next section in relation to visceral pain.

In contrast with cutaneous or superficial pain, deep pain is not only poorly localized, but also is accompanied by definite autonomic responses including nausea and alterations in heart rate and blood pressure as well as sweating.

The deep structures have a differential sensitivity to noxious stimuli. Thus in order of ascending threshold to the injection of irritant chemicals (eg, hypertonic saline solution) the periosteum exhibits the greatest sensitivity to painful stimuli. The ligaments, fibrous capsules of the joints, tendons, fascia, and the bodies of the muscles show a progressive decrease in their sensitivity to such stimuli in the order listed. Within these structures there is a definite histologic correlation between the density of the delicate fiber networks of free nerve endings (similar to those present in the skin) and the pain sensitivity of each particular structure. Thus a positive correlation exists between sensitivity to deep pain and the density of the peripheral neural elements that underlie this sensation.

Insofar as the adequate stimuli that can evoke deep pain are concerned, intense mechanical compression, trauma, or infection can produce a sensitivity so great that even a slight touch or move-

ment can elicit an excruciating sensation of deep pain (eg, during acute attacks of gout). Rhythmic muscular contractions such as are performed during isotonic exercise (eg, walking), and even sustained contractions that are interrupted frequently, usually produce no pain unless the muscles contract without an adequate blood supply, ie, become ischemic. The deep pain that develops following muscular ischemia is called intermittent claudication if it occurs in the leg, whereas it is termed angina if it develops in the myocardium of the heart.

If, however, a muscle contracts rhythmically while the blood supply is occluded, pain develops rapidly. The pain is not caused by the tension developed within the muscle during its contraction, as it persists even during the intermittent periods of relaxation, and until an adequate blood flow is restored, at which point the pain diminishes rapidly. Alternatively, if the exercise is stopped after pain has developed but while the blood flow remains occluded, the pain continues until such time as the flow is reestablished, and then it subsides rapidly once again. Experimentally determined facts such as these have led to the postulate of a chemical substance which is responsible for causing the pain that develops in ischemic muscles. Presumably, during the ischemic period, the chemical is produced metabolically within the tissues and the concentration rises locally until it is sufficient to stimulate the pain endings. Following restoration of an adequate blood flow the substance is metabolized and/or rapidly washed out of the ischemic tissues. Originally Lewis suggested that this chemical mediator of deep pain was a substance he called the "P factor." Subsequent research has failed to reveal the nature of this compound; however, kinins as well as potassium ion have been implicated as possible agents responsible for producing deep pain.

Prolonged contractions of skeletal muscle that may induce painful sensations (eg, "cramps") can originate either in higher brain centers or through the activity of somatic or visceral reflexes. The sustained contractions in turn lead to tissue ischemia which exacerbates the painful sensations. The process whereby stimuli that evoke deep pain in muscles can produce prolonged spasms can be envisaged as a cyclic mechanism as follows: Noxious stimulus → afferent impulses in deep pain fibers → facilitation of central interneuron pools (?) → sustained reflex contraction of muscle with attendant ischemia → additional afferent pain signals → continued reflex muscular contraction, and so forth. The question mark in this cycle indicates that no definitive experimental proof has been obtained of a prolonged facilitory effect; however, it is clear that such treatments as massage or the injection of local anesthetics into the affected region can afford the victim relief and apparently this is caused by breaking the cycle. Note also that this cycle is a positive feedback mechanism and hence is self-amplifying.

VISCERAL SENSATIONS AND REFERRED PAIN. As noted earlier, the cutaneous, deep, and visceral sensations collectively are termed the somatic senses. On the other hand, the autonomic nervous system is, by definition, composed entirely of efferent or motor pathways. Nevertheless, both sympathetic and parasympathetic nerves are accompanied by general visceral afferent nerve fibers which have major physiologic roles in reflex arcs that subserve homeostatic mechanisms throughout the body, albeit most of these operate at unconscious levels. For example, a number of specialized receptors are located in the viscera including baroreceptors, chemoreceptors, and osmoreceptors which detect and monitor changes within the internal environment (Table 6.1, p. 177). The afferent nerves that arise from these receptors form reflex arcs with various autonomic nerves, and thus such receptors serve to regulate, usually at unconscious levels, the functions of the particular organ systems to which they are related. The physiology of these receptors will be discussed in context together with the organs or organ systems whose activity they regulate. In addition, afferent fibers that accompany sympathetic and parasympathetic nerves can mediate the signals interpreted as visceral pain.

In the viscera, the receptors for the various sensory modalities that are perceived at conscious levels are quite similar to those found in the skin, although considerable differences in the distribution of these end organs are present in the two regions. The viscera contain only a few thermal and pressure receptors, but no proprioceptive end organs are present. Pain receptors are present within the viscera, and although these endings are far less abundant than in somatic structures, appropriate stimuli can induce severe visceral pain.* Although pacinian corpuscles are abundant in the mesentery, their function in this region is unknown.

General visceral afferent fibers transmit impulses from the various types of receptor located within visceral structures and organs to the central nervous system via sympathetic as well as parasympathetic pathways as discussed earlier (pp. 269, 270), and the cell bodies of these neurons are located within the dorsal root ganglia as well as in the ganglia of certain cranial nerves (Table 6.1, p. 177). Thus visceral afferent fibers are found in the facial (VII), glossopharyngeal (IX), and vagus (X) nerves (Table 7.3), as well as in the thoracic, upper lumbar, and sacral dorsal roots of the spinal nerves (Table 7.2). The trigeminal nerve (V) also may contain some general visceral afferent fibers that originate within the eye. The splanchnic nerves contain both type A (β) and A (δ) afferent fibers; however, only type A (δ) fibers are found in the vagus and pelvic nerves.

Interestingly, the intestinal tract can be manipulated, cut, and even cauterized (burned) without discomfort through an abdominal incision made under local anesthesia.

In addition to the general visceral afferent fibers that accompany the autonomic nerves, visceral pain signals also are conveyed along somatic afferent fibers that innervate the body wall in the thoracic and abdominal cavities as well as the diaphragm, biliary tract, some parts of the pericardium, and the roots of the mesentery.

The signals that convey visceral sensations to conscious levels are transmitted along the same pathways as those for somatic sensations, viz, by way of the spinothalamic tracts and thence by thalamic radiations to the postcentral gyri of the cerebral cortex. Within these primary cortical receptive areas the visceral sensory areas intermingle with those for somatic sensations.

Visceral sensations are very poorly localized (ie, protopathic), because of the paucity of receptors in the visceral organs, and such sensations usually consist of vague feelings of discomfort, pressure, or fullness, if not actual pain.

Visceral pain has a distinctly unpleasant effect: often it is associated with nausea; autonomic manifestations such as vomiting, sweating, and alterations in heart rate as well as blood pressure are its common accompaniments. Furthermore, visceral pain frequently radiates, or is referred, to other regions of the body than the locus from which it originates.

Visceral Pain. As practically anyone can testify from personal experience, visceral pain can be excruciating despite the anatomic scarcity of pain receptors within the viscera. The pain receptors that are located within the walls of hollow viscera are particularly susceptible to excitation by mechanical distension of the organs within which they lie. For example a large accumulation of gas within the intestinal tract can produce quite painful symptoms. This situation can be reproduced experimentally by inflating within the gastrointestinal tract a balloon that has been connected to a tube and then swallowed. The pain evoked by this maneuver fluctuates in severity as the gut contracts and relaxes upon the balloon. A similar intestinal colic is seen in patients having an intestinal obstruction when the dilated portion of the intestine lying just above the obstruction contracts upon the blocked region thereby evoking intermittent bouts of pain.

Thus the adequate stimuli for visceral afferents include spasms or strong contractions, particularly if ischemia is present; rapid distension against a resistance; chemical irritants; and mechanical factors, in particular when the organ is hyperemic, eg, the stomach. Thus the increased blood flow that accompanies the inflammation of infectuous processes may render the affected visceral structure(s) exquisitely sensitive to other types of stimuli, such as mechanical displacement. Normally, of course, the rhythmic mechanical activities of such contractile visceral organs as the heart, stomach, and intestines do not discharge pain fibers. However, the normal functions of these organs can become quite painful when accompanied by an inadequate blood flow.

Similarly to deep somatic pain, visceral pain can initiate reflex contraction of skeletal muscle that is located in the vicinity of the irritable locus (p. 357). The reflex muscular contraction usually is seen in the abdominal wall, and it is most pronounced when the inflammatory process involves the peritoneal lining of the abdomen; hence somatic as well as visceral afferent pain fibers are involved.

Classically, the clinical signs of an inflammatory process within an abdominal viscus (eg, the appendix) are pain and tenderness accompanied by such autonomic changes as hypotension and sweating in addition to spasms of the overlying abdominal muscles. The tenderness is caused by the enhanced sensitivity of the pain receptors located within the viscus, the autonomic symptoms result from stimulation of visceral reflexes, and the spasms of the abdominal muscles are caused by their reflex contraction, presumably due to the operation of a vicious cycle, as outlined earlier (p. 357).

Referred Pain. Quite frequently deep as well as visceral pain is interpreted as arising in a somatic structure or region of the body other than that in which the stimulus causing the pain is situated. This phenomenon is called *referred pain,* as the pain is inaccurately localized or referred to another structure than its site of origin. Cutaneous sensations, in contrast to deep and visceral pain, are never referred.

Occasionally visceral pain may be localized as well as referred. In such a situation the pain appears to spread or *radiate* from the affected structure or organ to a remote locus.

An understanding of referred pain and a knowledge of the usual sites to which pain is referred, or projected, from each of the viscera are of enormous importance clinically. A few specific examples of pain referral may be cited. Cardiac pain often is referred to the inner aspect of the left arm, whereas central diaphragmatic irritation (eg, that caused by a hiatal hernia) can result in pain in the tip of the shoulder. Pain caused by distension of the ureter is referred to the testicle. Pain originating within a maxillary sinus may be referred to the adjacent teeth. Thus general somatic afferent as well as general visceral afferent fibers can become involved in referred pain. It must be realized that these are only a few of the many examples of referred pain that are encountered in clinical medicine, and that the sites of referral are not absolute. Thus, cardiac pain also may be referred to the abdomen, shoulder, right arm, or the neck. Furthermore, referred pain has a strong psychologic as well as physiologic component, so that a patient may refer pain to a site that has been traumatized previously, eg, to the region of an old

surgical wound. This phenomenon is called *habit reference,* and is learned on the basis of previous experience.

Generally when pain is referred the faulty projection of the sensation is to a structure or organ which developed from the same embryonic segment (or dermatome) as the structure or organ in which the pain has its origin. This concept is known as the *dermatomal rule.* For example, the heart and arm share the same segmental origin within the embryo so that the nerve supply to each of these structures also shares a common origin; and despite a relative movement between the positions of these structures as ontogeny proceeds, their innervation is carried along as they move apart during growth and development. Therefore in the adult the same pattern of sensory innervation with relation to the level of the central nervous system is retained by each structure as this pattern was established in the early embryo. Similarly, the diaphragm migrates during embryogenesis to its adult location from its site of origin in the neck region, and the phrenic nerve is carried along with it as development proceeds. Lastly, the testicle also migrates in conjunction with its nerve supply from the urogenital fold within the abdomen; the kidney and ureter also take their origin from the same primitive urogenital ridge, and thus the adult kidney, ureter, and testicle share common central sources of innervation in the adult.

The dermatomal rule provides a neuroanatomic basis for an understanding of referred pain, but what physiologic mechanisms are involved? Two theories have been advanced to explain the phenomenon of referred pain. The first of these theories explains referred pain largely by the convergence of impulses (p. 686), whereas the second is based principally upon the effects of facilitation (p. 344).

Convergence and Referred Pain. As described above, the afferent nerves which arise from visceral structures enter the central nervous system at the same level as do the afferent nerves from the somatic structures to which that pain is referred. It is a fact of neuroanatomy that there are considerably greater numbers of sensory fibers in peripheral nerves than there are ascending fibers in the dorsal spinothalamic tract. Consequently there is doubtless a high degree of convergence between peripheral sensory fibers and the cell bodies of spinothalamic neurons. The convergence theory of referred pain states that both somatic and visceral afferents converge upon the same spinothalamic neurons, as indicated in Figure 7.98A. Somatic pain is far more common than visceral pain; therefore the brain has learned to interpret the afferent signals which ascend in this pathway as coming from a pain stimulus that arises in a specific somatic region. Thus when the same afferent pathway is stimulated by signals that originate in visceral afferent nerves, the signal that reaches the sensory cortex is identical, and is misinterpreted as having arisen within the somatic area, ie, projection of the sensation is faulty (see also the law of projection, p. 365).

In this regard, it was remarked earlier that past sensory experience has an important role in referred pain, and that the pain may be referred to a somatic region or structure, however remote from the actual locus of the pain stimulus it may be, that previously had been the site of a painful trauma. Observations of this type lend support to the conclusion that central learning processes underlie the interpretation of convergent signals, to some extent at least.

Facilitation and Referred Pain. The second theory concerning the origin of referred pain presumes that afferent impulses from visceral structures produce subliminal fringe effects (p. 216) that lower the excitatory threshold of spinothalamic neurons which receive afferent fibers from somatic areas (Fig. 7.98B). Therefore any slight activity in the pathways transmitting pain impulses from somatic regions, and which normally would become extinguished (or occluded, p. 217) within the spinal cord, is facilitated and thus reaches conscious levels.

If convergence were the sole explanation for the phenomenon of referred pain, then local anesthesia of the somatic reference area should not abolish the pain. On the other hand, if subliminal fringe facilitory effects were the primary physiologic mechanism underlying referred pain, then anesthesia of the reference area should abolish the pain. Actually, however, variable effects result from local anesthetics applied to the reference area. Thus in cases of severe pain such anesthesia generally does not affect the perception of referred pain, whereas if the pain is minor it may be abolished completely. Therefore it would appear that convergence as well as facilitation each has a role in the development of referred pain, and that under particular conditions the functional emphasis may shift from a predominance of one of these mechanisms to the other.

Central Inhibition and Pain. There is much evidence, experimental as well as clinical, which indicates that signal transmission in the neural pathways that mediate superficial as well as deep pain can be blocked centrally by inhibition or occlusion (p. 216). Thus stimulation of the splanchnic nerves can evoke bioelectric potentials in the postcentral gyrus; however, no such response is elicited if the stimulation is preceded (or accompanied by) the stimulation of somatic afferent nerves. In addition, severe traumatic injuries incurred having periods of intense excitement (eg, as during a battle) may not be perceived as painful until much later. Lastly, touching an area that is emitting painful sensations tends to reduce the intensity of the pain.

The substantia gelatinosa presumably is one region wherein the inhibition of pain signals takes place (p. 262). Collateral branches emitted by the touch fibers that enter the dorsal column also enter the substantia gelatinosa, and it has been suggested that the impulses which are transmitted in these collaterals (or in the interneurons upon which the collaterals terminate) inhibit the transmission of pain signals from the dorsal root fibers to the neurons of the spinothalamic tract. Presynaptic inhibition (p. 215) appears to be involved in this process, and thus chronic stimulation of the dorsal columns has been employed as a technique for relieving intractable pain. The relief from pain that is obtained by use of this procedure could be due either to antidromic conduction of impulses to the functional gate located within the substantia gelatinosa (p. 262) or to the orthodromic conduction of impulses to another, but functionally similar, gate located within the brain stem.

In this regard it is interesting that chronic pain sometimes can be relieved by inserting acupuncture needles at various points on the body surface. According to the gate control theory of pain, the afferent pain signals are subject to a selective inhibition within the substantia gelatinosa as well as within the posterior columns of the spinal cord by other sensory impulses. Thus it is of interest to consider the possibility that the afferent signals evoked by acupuncture needles may block selectively the mechanism whereby pain impulses are permitted to cross this gate, and thus no conscious perception of the painful stimulus is evoked. Similarly, a central inhibitory state as well as an increase in threshold of the pain centers in the brain could be induced by afferent impulses from the skin that are stimulated by acupuncture needles; thus conscious reception of pain stimuli would be inhibited.

Localized deep or visceral pain such as is present in muscular spasms, following a sprain, in rheumatic and arthritic conditions, or the referred pain in the skin above a visceral inflammation often can be ameliorated by the application of counterirritants. Thus local application of heat (eg, by poultices, heat lamps, hot water bottles, or rubefacient liniments) to the affected skin areas presumably is efficacious because of the production of an inhibitory state in central pathways that transmit pain.

Similarly, the relief of pain that is claimed for the use of acupuncture needles could be explained on the same basis, the needles providing the counterirritant stimuli; but as pain has a strong psychologic rather than physiologic component in certain instances the possibility of suggestion in such therapy cannot be overlooked. Interestingly, it has been reported that the clinical application of acupuncture for the relief of pain is most successful in those conditions which are physiologically reversible and in which there is no anatomic damage to the tissues involved.

THE SPECIAL SENSES. The special senses include vision, audition, those sensory functions of the inner ear which are concerned with the orien-

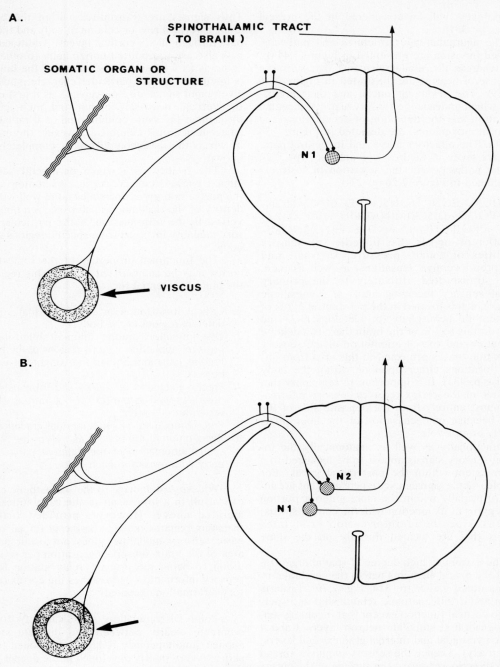

A.

SPINOTHALAMIC TRACT
(TO BRAIN)

SOMATIC ORGAN OR
STRUCTURE

N 1

VISCUS

B.

N 2

N 1

FIG. 7.98.   The two theories of referred pain N1, N2 = spinothalamic neurons. Heavy arrow pointing to hollow viscus in each diagram signifies a painful stimulus, eg, mechanical distension. **A.** Convergence theory. **B.** Facilitation theory. See text for explanation of figure, and law of projection (p. 365). (After Ganong. *Review of Medical Physiology,* 6th ed, 1973, Lange Medical Publications.)

tation of the body in space (ie, equilibrium) as well as the detection of acceleratory and rotational movements, and the chemical senses of olfaction and gustation (Table 6.1, p. 177). These special senses, with the exception of the equilibrial func-

tions of the inner ear, will be discussed in Chapter 8 (p. 360) together with the adequate stimuli necessary to excite each receptor type. The anatomy and functions of the inner ear, insofar as these topics pertain to equilibrium and acceleratory reflexes, will be considered in the present chapter (p. 381).

The neuroanatomic structures and pathways involved in vision are presented in Figure 7.44 (p. 298), whereas the similar elements that subserve the sense of hearing are illustrated in Figure 7.36 (p. 289). The vestibular nuclei and their projections, which underlie the sensory aspects of spatial orientation and the detection of acceleratory and rotational movements, are depicted in Figure 7.34 (p. 286). The olfactory nerve and its central pathways are given in Figures 7.66 and 7.67 (p. 322), and the pathways for the sensation of taste are diagrammed in Figure 7.26 (p. 274).

CORTICAL FUNCTIONS AND SENSATION: TWO-POINT DISCRIMINATION, LOCALIZATION, PROJECTION, AND STEREOGNOSIS. The afferent signals that underlie the sensory modalities of touch, pressure, thermal, and some proprioceptive sensations, as well as pain, all are transmitted ultimately to the primary cortical sensory receiving area, or somesthetic cortex, which is located in the postcentral gyrus of each cerebral hemisphere (p. 328; Figs. 7.73 and 7.74).* In this region of the brain there is a definite localization and spatial orientation of the sensory information that is projected to this area from different locations either upon or within the body (see also p. 435). It is important to remember that each side of the cortex receives sensory information almost entirely from the opposite side of the body, because of decussation of the fiber tracts at lower brain levels.

Functionally as well as anatomically the incoming sensory signals appear to be localized to a certain extent within the postcentral gyrus. For example, tactile signals excite neurons that are located principally within the more anterior portion of this gyrus at its junction with the central sulcus, whereas kinesthetic information stimulates neurons that are located dorsally on the same gyrus.

The extremely high degree of spatial organization of the neural elements within the cerebral cortex was noted earlier (p. 324; Fig. 7.70). Little is known about this anatomic relationship as it pertains to cortical function except that incoming sensory signals first stimulate neuron layers 3 and 4 (ie, the pyramidal and internal granular cell layers, respectively). Thence the sensory impulses spread in two directions: toward the surface of the cortex and inward to the deeper-lying layers of neurons.

*Since the functions of the secondary cortical receiving area for sensations are so poorly understood they will not be considered here, except to state that afferent signals converge upon this region of the brain from both the dorsal column and spinothalamic systems. Furthermore artificial stimulation of this area sometimes elicits complex bodily movements, a fact which implies that the secondary cortical sensory area may have a role in the sensory control of some motor functions, in addition to the perception of pain.

Thus signals are transmitted within the cortex along a vertical row or column of cells and thereby pass through all six cortical layers. Additional signals also are transmitted horizontally to other cell columns as well as to other areas of the brain. It is assumed that the particular characteristics of a given input signal are "perceived" as it first enters the vertical neuronal column and then spreads throughout the same column, and as it radiates to adjacent neuronal columns; however, the manner in which this is accomplished is completely unknown.

The facility with which peripheral sensory stimuli evoke localized cortical excitation within the postcentral gyrus correlates quite well with the density of the thalamocortical projection fibers involved in the transmission of the particular sensory modality involved from specific regions of the body.

The functional properties of the somesthetic cortex may be summarized briefly. This region of the brain is essential for:

1. Critical localization of sensations that arise in different regions of the body.
2. Discrimination among different intensities of pressure applied to various regions of the body.
3. Precise judgment among the weights of various objects.
4. Precise judgment of shapes and forms (stereognosis), as well as the textures of various objects which are not seen.
5. Precise judgment of small thermal gradations.
6. Recognition of the precise as well as the relative spatial orientation of the different parts of the body with respect to each other.

Widespread damage to the somesthetic cortex can result in a loss of all of the capabilities enumerated above. However, provided that the thalamus remains intact, anesthesia for all of the basic sensory modalities does not occur as this area of the brain subserves sensation per se in addition to being the primary relay station for all sensory information that reaches the cortex (olfactory information excepted).

Two-Point Discrimination. The capacity to discriminate clearly between two discrete stimuli applied simultaneously is highly developed in the skin and very poorly developed in the deeper-lying tissues and organs of the body. Thus if two compass points touch the skin simultaneously, the minimum distance the points must be separated in order to be perceived as two distinct stimuli is called the two-point threshold. On the fingertips, the ability to perceive two points is retained, even though the points of the compass are only 1 to 3 mm apart. The ability to discriminate between two independent stimuli depends upon the sensation of touch plus the cortical ability to recognize the two stimuli as separate entities rather than as one. The neural mechanisms for the perception of one- and two-point stimuli are illustrated in Figure 7.99A and B, and the effects of overlap of individual peripheral receptive fields on this response are shown in Figure 7.86B.

The magnitude of the two-point threshold for tactile sensations, determined as described above, varies considerably on various parts of the body; thus points stimulated on the back must be separated by roughly 70 mm before they can be discerned as two individual stimuli.

The magnitude of the two-point threshold is approximately that of a single receptor field, at least on the hands; but because of the complex intermingling of fibers between adjacent receptive fields, the neural basis for the phenomenon of two-point discrimination probably is far more complex than indicated by this discussion.

Two-point thresholds also can be determined as described above, by using appropriate pairs of stimuli, for the other cutaneous senses, ie, warmth and cold as well as pain.

The ability to discriminate critically among cutaneous stimuli lies in the anatomic as well as functional accuracy with which afferent signals are transmitted from the cutaneous receptors to the cortex via the dorsal columns of the spinal cord, ie, the fasciculi gracilis and cuneatus (Table 7.5). Furthermore these fast-conducting tracts also are able to convey rapidly changing patterns of impulses as well as signals having a broad spectrum of frequencies. Consequently, the dorsal columns are well suited for the rapid transmission of epicritic cutaneous sensibilities that have a broad range of modalities and intensities.

These functional capacities of the dorsal column system are in marked contrast to the spinothalamic system which conveys sensory impulses related to pain, warmth, cold, crude touch, and crude pressure as well as sexual sensations. In fact, there are four major differences between the spinothalamic and dorsal column systems.

1. The conduction velocities for impulses in the spinothalamic tracts are only 20 to 50 percent of those found in the dorsal column system.

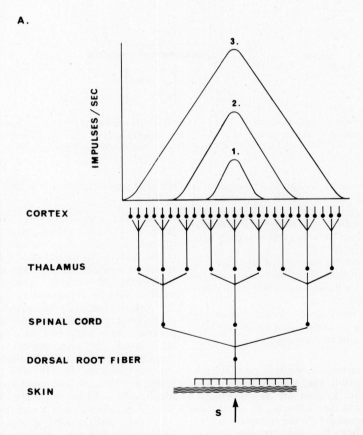

FIG. 7.99. The transmission of signals through multiple pathways to the somesthetic cortex. **A.** A single punctate stimulus (S) applied to the skin will excite progressively greater numbers of impulses in the cortex depending upon the intensity of the stimulus. 1. weak stimulus; 2. moderate stimulus; 3. strong stimulus. The subjective perception of the stimulus intensity thus will depend upon the cortical area stimulated as indicated by the areas beneath the three graphs at the top of the figure. (After Guyton. *Textbook of Medical Physiology,* 4th ed, 1971, Saunders.)

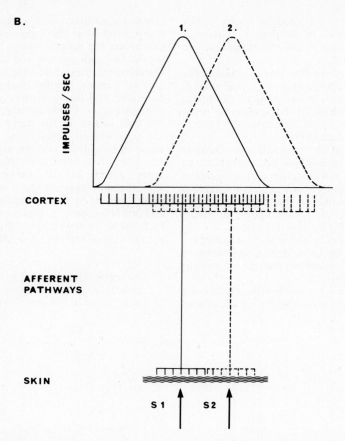

B.

FIG. 7.99. (cont.) B. Two punctate stimuli (S1, S2) of a given intensity applied to the skin are transmitted independently to the cortex and may (or may not) be perceived as two points depending upon the separation of the cortical areas that are excited. See also Figures 7.86 and 7.92. (After Guyton. *Textbook of Medical Physiology*, 4th ed, 1971, Saunders.)

2. The spatial orientation of signals transmitted in the spinothalamic tracts is far less precise than for those transmitted in the dorsal column system. This is particularly true in the dorsal spinothalamic tract which conveys pain and thermal signals (Table 7.5).
3. The graduations in the intensity of signals transmitted by the spinothalamic tracts also are far less precise than for those transmitted via the dorsal column system.
4. The capacity for transmission of high-frequency, repetitive impulses is nonexistent for all practical purposes in the spinothalamic system.

The facts summarized above indicate that from a functional standpoint the signals transmitted by the spinothalamic system are far less accurately localized than those conducted by the dorsal column system. Consequently the receptors which relay afferent signals to the spinothalamic system do not display nearly as low a two-point threshold to stimulation as those of the dorsal column system; therefore any two-point discrimination for

signals transmitted by these fiber tracts is rough at best.

Localization and Projection. All somatic sensations, in addition to their properties of quality and intensity, have a localization upon the surface of the body. This aspect of somatic sensation also is called *topognosis*. Damage to central afferent pathways can derange seriously the ability to localize stimuli; consequently functional testing of this sensory property is important clinically. For example, if the subject is stimulated by light touch, pressure, or a warm or cool object when the eyes are closed, the discrepancy between the point stimulated and the point indicated by the subject can be measured readily, and this gives the examining physician an indication of the error of localization. Even normally, this error varies widely on different regions of the body surface.

Localization and projection are related phenomena; however, there is a difference between these two aspects of sensory experience. Projection is concerned primarily with the general external or internal region from which a sensation appears to come, whereas localization is concerned primarily with the more precise location, either external or internal, in which the sensation appears to arise. Ultimately, of course, all sensory processes take place in the brain, but there is no conscious awareness of these central events. Rather, the various sensory experiences are projected to the external environment or to a peripheral structure or organ in the body, and this projection to the locus from which the stimulus arises is a learned response. Thus light appears to come from various external sources, either directly or indirectly by reflection. Similarly, sounds are projected to their sources. Other stimuli such as thirst, hunger, pain, nausea, muscle sense, labyrinthine sensations including acceleration, deceleration, as well as rotational movements and sexual feelings all are projected to the internal environment. Thermal sensations, on the other hand, may be projected either to the skin or to the immediate surroundings depending upon the temperature.

One important aspect of the projection of sensation may be stated as follows: Stimulation of any sensory pathway at any point that lies central to the sense organ itself evokes a sensation that is projected to the periphery and not to the point where the stimulus was applied. This statement, sometimes called the *law of projection,* is of particular importance clinically, and many examples of this law could be given. For instance, the phenomenon of referred pain (p. 359) illustrates this fundamental principle. The "phantom limb" of amputee patients provides another example of the law of projection. Thus an amputee frequently will experience complex, often quite painful, sensations and even perceive movements in the amputated limb that are projected in such a way that it feels as though the limb were still present.

The physiologic mechanisms that underlie the phenomenon of projection are for the most part unknown, although the high degree of spatial orientation of the neural elements within the nervous system (p. 343) doubtless plays a major role in this process.

Stereognosis. Normal persons can identify various common objects (eg, keys, different coins) quite readily merely by handling them and without seeing them, a sensory ability known as *stereognosis.* Not only are relatively intact tactile, pressure, and sometimes thermal senses necessary to accomplish stereognosis, but a cortical component also is involved. Thus faulty stereognosis provides an early indication of cortical damage, in particular to the primary somesthetic area in the postcentral gyrus. Occasionally, however, a cortical lesion responsible for impaired stereognosis may be found in the parietal lobe posterior to the postcentral gyrus, and this functional defect is present even though the perceptions of touch and pressure stimuli are normal. The loss of the ability to judge

the shapes or forms of various objects is termed *astereognosis.*

Somatic Motor Functions and Their Regulation

Ultimately the physiologic activity of any skeletal muscle in vivo depends upon the frequency and pattern of impulse discharge from motoneurons whose cell bodies are located within the anterior columns of the spinal cord as well as in the motor roots of certain cranial nerves. The efferent processes of these motoneurons provide the final common pathways to the somatic muscles throughout the body (p. 217). In turn, the signal output of all somatic motoneurons is regulated by an enormous number of excitatory (or facilitory) as well as inhibitory input pathways which converge upon each motoneuron from various neuronal pools located in the central nervous system; and thereby these input signals to the somatic motoneurons control the physiologic activity of individual muscles.

A few of the possible inputs to a single spinal motoneuron are depicted in Figure 6.38 (p. 218). However it must be stressed that in addition to the localized, segmental inputs shown in this illustration, many other afferent signals that originate in various brain centers can impinge upon the motoneurons of the final common path. The principal neural structures whose integrated multiple inputs to the spinal motoneurons underlie postural regulation as well as control of voluntary and involuntary somatic muscular activities are shown in Figure 7.100.

In vivo, the contraction of somatic muscles requires that efferent signals be transmitted to the neuromuscular junctions (motor end plates, p. 208), and that muscular contractions be abnormal, or even absent, if there is a failure of impulse generation and/or conduction at any point in any of the neural circuits leading to the terminations of the motoneuron upon the muscle fibers (p. 227). Furthermore, central inhibitory processes which decrease the signal traffic to particular skeletal muscles are responsible for reducing or "turning off" contractile activity in vivo, and thus the balance between excitatory and inhibitory inputs to the centrally located motoneurons, whose processes constitute the final common pathways, determines the net contractile state of individual skeletal muscles at any particular moment in time. Consequently any discussion of movement, posture, and equilibrium must consider the central neural pools which are responsible for transmitting integrated signal inputs to the motoneurons, thereby controlling their output.

The convergent signal inputs to the motoneurons that arise in the structures indicated in Figure 7.100 underlie three principal somatic motor functions, no one of which is mutually exclusive of the other two: First, the inputs adjust the

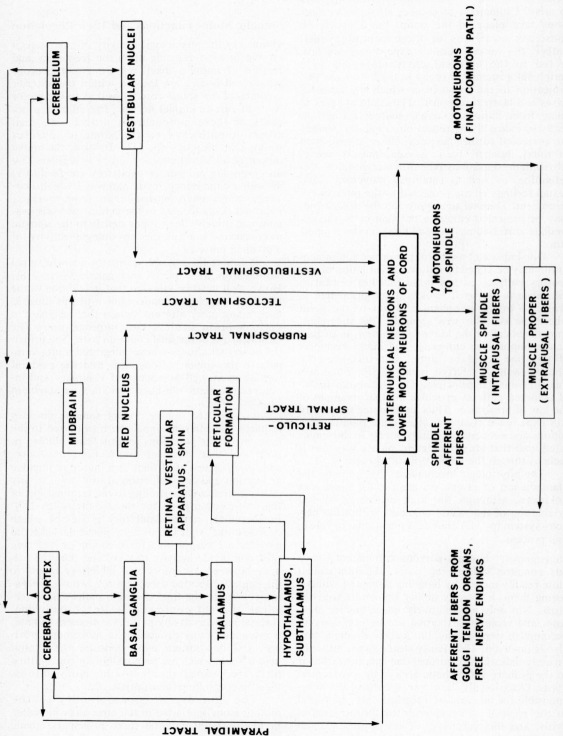

FIG. 7.100. The principal neuroanatomic structures and pathways that underlie the control, integration, and coordination of somatic muscular contractions in vivo. The direction of signal transmission in particular pathways is indicated by arrows. See Figure 7.101 for details of the extrapyramidal pathways.

posture of the entire body through the selective activation of certain muscles or groups of muscles. Therefore a steady physiologic background is continually provided against which specific voluntary movements can take place. Second, the input signals from certain higher brain centers can initiate highly coordinated (or skilled) voluntary muscular contractions that lead to specific movements. Third, the inputs to the motoneurons are so patterned physiologically that the activity of particular muscles, or groups of muscles, is rendered smooth and accurate (or precise) as well as graded (insofar as the force of contraction is concerned). The functional patterns of signal inputs to the somatic motoneurons thus are organized both qualitatively (ie, the net inputs are either excitatory or inhibitory) and quantitatively (insofar as signal frequency and amplitude are concerned). The physiologic resultant of these processes leads to smooth, graded, and coordinated contractions of agonistic muscles, with a simultaneous, equally smooth, graded, and coordinated relaxation of the antagonistic muscles.

The arbitrary functional trichotomy outlined above will serve as a general outline within which the various aspects of somatic motor activities in vivo can be considered in a sequence of ascending complexity.

REFLEXES, POSTURE, AND MOVEMENT. A number of simple reflex arcs were analyzed in Chapter 6 (p. 213 et seq.), and in this section the participation of these and other somatic reflexes in the control of posture* and certain movements will be discussed. The levels within the central nervous system at which a few postural and locomotor reflexes are integrated have been given in Table 7.11, and the major reflexes involved in the maintenance of posture itself are summarized in Table 7.12. Since the functional roles of these reflexes in vivo will be stressed henceforth, the reader will note occasional marked differences between certain aspects of reflexes as elicited by artificial stimuli applied under experimental conditions for the analysis of synaptic functions and the same reflex when evoked in vivo by natural stimuli. Thus it is essential to realize that any reflex activity as it occurs in the intact, unanesthetized subject always is initiated in some receptor end organ and that the afferent signals that excite the motoneurons usually are asynchronous because the receptors not only discharge out of phase with each other, but also fire in a repetitive manner. These facts are in distinct contrast to the reflex activity that is elicited by artificial stimulation of, for example, a peripheral nerve trunk or dorsal root in which many different functional types of afferent fiber may be present and thus excited simultaneously and in single bursts de-

*For the purposes of this text, the term "posture" signifies the relative arrangement or orientation of the various parts of the body in space.

pending on the frequency of stimulation. Furthermore, the anesthetic employed may so alter the responses to artificial stimulation that the actual biologic significance of the reflex to the organism is largely obscured. Experimental procedures such as those just mentioned are, of course, not only useful but valid means for obtaining quantitative data of value in analyzing, for example, the temporal course of synaptic processes; however, such data are of no great value in understanding reflexes as they take place in the unanesthetized person or animal subjected to natural environmental stimuli.

Muscle Tone and the Stretch Reflex. Although it is impossible to distinguish postural movements from voluntary movements in a clear-cut fashion, nevertheless it is relatively easy to define a number of reflexes that underlie posture and postural adjustments (Table 7.12). These reflexes serve two major functions: First, they maintain the human body in an upright position with respect to the force of gravity; and second, these reflexes furnish continuously the motor adjustments that are essential for the maintenance of a postural background upon which voluntary movements are superimposed.

The reflex motor adjustments which are responsible for the maintenance of posture are of two general kinds. The sustained tonic contraction of certain skeletal muscles maintains the upright posture of the body for long periods of time, and certain *static reflexes* are responsible for this effect. The continual, dynamic adjustments of the basic postural tone which provide the physiologic background for voluntary movements are carried out by short-acting rapid or *phasic reflexes*. Both static and phasic postural reflexes are integrated at various levels of the central nervous system from the spinal cord to the cerebral cortex and both types of reflex are mediated principally through the extrapyramidal pathways (Fig. 7.101).

Normal skeletal muscles studied in vivo may appear to be fully relaxed; nevertheless these structures are in a partially contracted state at all times, and this effect is known as *muscle tone* (or *tonus*). Resting muscles exhibit a certain resilience to palpation; and furthermore when such resting muscles are stretched passively as by flexing a joint, a resistance is encountered which is totally unrelated to the performance of any conscious activity by the subject. Thus the reduced contractile activity at rest as well as the reflex contraction upon mechanical stretch are the chief physiologic (as well as clinical) manifestations of muscle tonus.

Section of, for example, either dorsal (sensory) or ventral (motor) roots to the spinal cord immediately abolishes muscle tone, indicating that the maintenance and regulation of this contractile state is caused by reflex activity and is *not* a property of skeletal muscle per se. The subject of muscle tone was discussed in relation to muscular contraction in vivo (Chapter 4, p. 129 et seq.), but has

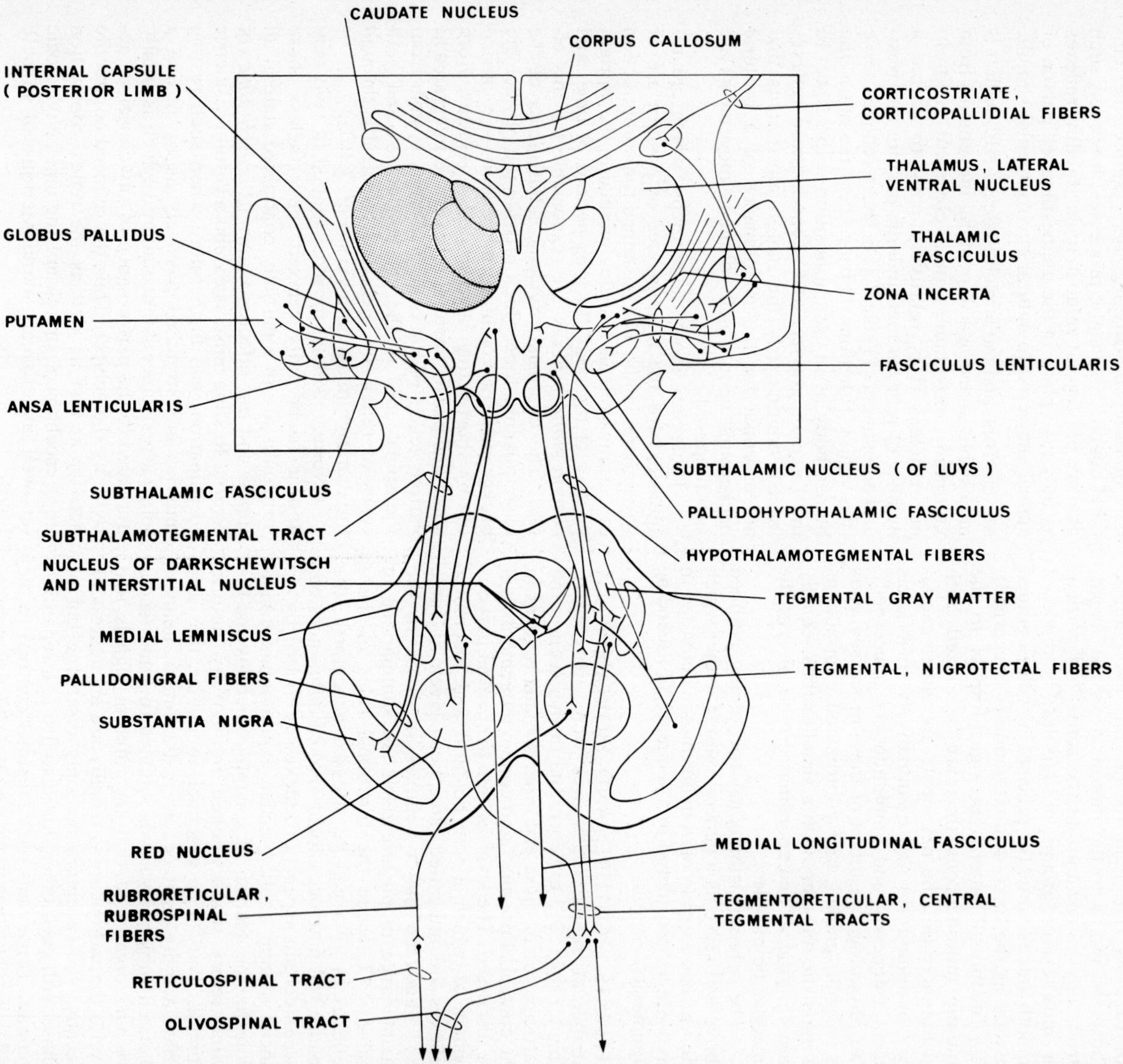

FIG 7.101. The relationships among the nuclei, and the ... (After Everett: Functional Neuroanatomy, 6th ed, 1971, Lea & Febiger.)

CORPUS CALLOSUM

CAUDATE NUCLEUS

INTERNAL CAPSULE (POSTERIOR LIMB)

GLOBUS PALLIDUS

PUTAMEN

ANSA LENTICULARIS

SUBTHALAMIC FASCICULUS

SUBTHALAMOTEGMENTAL TRACT

NUCLEUS OF DARKSCHEWITSCH AND INTERSTITIAL NUCLEUS

MEDIAL LEMNISCUS

PALLIDONIGRAL FIBERS

SUBSTANTIA NIGRA

RED NUCLEUS

RUBRORETICULAR, RUBROSPINAL FIBERS

RETICULOSPINAL TRACT

OLIVOSPINAL TRACT

CORTICOSTRIATE, CORTICOPALLIDIAL FIBERS

THALAMUS, LATERAL VENTRAL NUCLEUS

THALAMIC FASCICULUS

ZONA INCERTA

FASCICULUS LENTICULARIS

SUBTHALAMIC NUCLEUS (OF LUYS)

PALLIDOHYPOTHALAMIC FASCICULUS

HYPOTHALAMOTEGMENTAL FIBERS

TEGMENTAL GRAY MATTER

TEGMENTAL, NIGROTECTAL FIBERS

MEDIAL LONGITUDINAL FASCICULUS

TEGMENTORETICULAR, CENTRAL TEGMENTAL TRACTS

been reviewed briefly here as the *stretch* (or *myotatic*) *reflex* (p. 218) and is the fundamental neural mechanism for maintaining tone in skeletal muscles. In addition to the role of this reflex in keeping muscles in a state of slight partial contraction at all times, the stretch reflex is able to increase the contractile state of certain muscle groups under physiologic conditions, and thus provide a background of postural muscle tone upon which voluntary movements can be superimposed. In other words, the basic reflex underlying an upright posture is the stretch reflex, which is present to varying degrees in all skeletal muscles. Thus muscle tone is responsible for the maintenance of the posture that is characteristic of any particular species.

The sensory receptor for initiation and maintenance of the stretch reflex is the muscle spindle, whose morphology and physiology were discussed in extenso in Chapter 6 (p. 180; p. 181). Briefly, the muscle spindle functions in the stretch reflex as follows. Mechanical deformation of a skeletal muscle by stretch causes a like deformation of the receptor terminals (annulospiral endings) within the spindles located within that muscle (Fig. 6.3, p. 181). As the intrafusal muscle fibers of the spindle lie parallel to the extrafusal fibers of the muscle proper, the excitation of the afferent endings within the spindle is roughly proportional to the degree of stretch that is imposed upon the entire muscle. The afferent fibers from the spindles enter the spinal cord via the dorsal root, and there make synaptic contact directly with motoneurons of the ventral columns (Figs. 6.39 and 6.40; p. 219 and p. 220). Thus the pathway for the stretch reflex is unique in that it is monosynaptic, and thus no interneurons are present as in the majority of reflex arcs.

Functionally, the stretch reflex produces a local, contractile response of the muscle within which the excited spindles lie, and the force of this contraction is proportional, within limits, to the applied force which stretches the muscle. The contractile response of the muscle to stretch ceases immediately when the force applied to the muscle is removed, ie, the myotatic reflex exhibits no after-discharge (p. 348). Furthermore the latent period of the stretch reflex is short (less than 20 msec).

It is important to realize that the higher motor centers of the central nervous system transmit signals to skeletal muscles by two routes. The first of these routes is via impulses that converge upon the large anterior horn cells or alpha (α) motoneurons (p. 182). The second route is through impulses that converge from higher centers upon the smaller gamma (γ) motoneurons (p. 182). Excitation of the gamma efferent motoneurons produces contraction of the intrafusal muscle fibers that are located within the neuromuscular spindles (p. 220), and the increased tone of the intrafusal fibers of the spindle that results from such excitation renders these structures more sensitive to applied stretch.

Consequently the reflex tone of particular muscles can be adapted physiologically to different situations by means of this feedback mechanism. It appears that the very slight alteration in tonus of the intrafusal fibers that is induced by different rates of discharge of the gamma efferent system is quite adequate as a stimulus to activate the alpha motoneurons which in turn control the reflex contractile state of the entire muscle. The feedback regulation of spindle activity and gamma efferent discharge were discussed in Chapter 6 (p. 219 et seq.).

The reticular formation (p. 302) has been demonstrated to regulate discharge of the gamma efferent system. Thus stimulation of the facilitory region of the reticular substance, which is located in the pons and mesencephalon, enhances discharge of the gamma efferent fibers of the muscle spindles, and thereby the stretch reflex is enhanced. Conversely, stimulation of the medullary inhibitory area of the reticular formation (through the reticulospinal tract) results in an inhibition of gamma efferent discharge and a diminution of the myotatic reflex.

Since postural reflexes are responsible primarily for maintaining the body in an upright position, the stretch reflexes are most readily elicited in the extensor or antigravity muscles (see also discussion of decerebrate rigidity below). The stretch reflex which is, after all, the physiologic mechanism that underlies most postural reflexes, is turned off immediately when the force applied to the muscle that induces spindle deformation is removed. In addition the stretch reflex also is inhibited in vivo by four other mechanisms: (1) painful (or nociceptive) stimuli applied to an extremity rapidly and powerfully inhibit the stretch reflex, giving rise to flexion of the affected part (p. 218); (2) stimulation of afferent cutaneous nerves will inhibit the stretch reflex; (3) stretch applied to the antagonistic flexor muscles will inhibit the stretch-induced contraction of the agonists (see reciprocal inhibition, p. 221); (4) the application of excessively powerful stretch to a contracted muscle (or to its tendon) also will evoke inhibition of the stretch reflex and give rise to the so-called lengthening reaction (p. 228). The four inhibitory mechanisms just outlined serve as physiologic devices which tend to protect the muscle, or the organism, from injury.

As noted above, the antigravity muscles are the primary loci within which postural tonus develops in most species. In humans, the principal antigravity muscles are the elevators of the jaw (the masseters), the retractors of the neck, the extensors of the back, the supraspinatus, the extensors of the knee and ankle (the soleus, gastrocnemius, and vastocrureus), and possibly the ventral abdominal wall muscles. If these muscles relax completely, as during coma, collapse of the body ensues. Conversely, in the normal, alert individual the tonus developed by operation of the stretch reflex in the muscles listed above is primarily responsible for the upright static posture of the body.

An extremely important factor in the control of posture is the alteration in the threshold of the myotatic reflexes whose integrative centers are located within the spinal cord. Thus changes in the excitability of the alpha motoneurons in the cord directly affect these responses, whereas alterations in the discharge frequency of the gamma efferent (or small motor nerve) system indirectly affect the postural reflexes. And even though the basic mechanism whereby the tone in skeletal muscles is controlled resides in the stretch reflex, which is integrated within the spinal cord, the degree of tonus these muscles develop in turn is mediated by the motoneurons and thus is affected markedly by signals that descend from higher neural centers and which alter the excitability of these neurons. For example, afferent signals that arise in receptors located within the neck muscles

or labyrinth of the inner ear can alter markedly the background tonic state of the postural muscles which is established by the stretch reflex of the cord. Likewise extrapyramidal pathways that originate from centers located within the cerebrum, midbrain, and cerebellum transmit excitatory, facilitory, or inhibitory impulses which converge upon the motoneurons of the final common path and thereby alter the general tonic state of the voluntary muscles, bring about delicate adjustments in the tonic state of particular muscles, and maintain the distribution of tone among groups of agonistic as well as antagonistic muscles. The tone of particular groups of muscles also may be altered via the spinal centers through bombardment of these centers by impulses that originate in quite unrelated muscle groups or in cutaneous receptors. For example, the positive supporting reaction (Table 7.12) can be influenced by afferent signals that originate within the neck muscles.

Table 7.12 MAJOR POSTURAL REFLEXES[a]

REFLEX	STIMULUS	RECEPTOR	RESPONSE	INTEGRATIVE CENTER
Stretch reflexes	Stretch	Muscle spindles	Contraction	Spinal cord and medulla
Positive supporting reaction	Contact with palm or sole	Proprioceptors in distal flexors	Foot and leg extended to support body	Spinal cord
Negative supporting reaction	Stretch	Proprioceptors in extensors	Release of positive supporting reaction	Spinal cord
Labyrinthine reflexes (tonic)	Gravity	Otolithic organs of inner ear	Extensor contraction (rigidity)	Medulla
Neck reflexes (tonic)	Turning head: 1. To side 2. Up 3. Down	Proprioceptors in neck	Alteration in pattern of extensor contraction: 1. Extension of limbs on same side to which head is turned 2. Legs flex 3. Arms flex	Medulla
Righting reflexes (labyrinthine)	Gravity	Otolithic organs of inner ear	Head maintained in level position	Midbrain
Righting reflexes (neck)	Stretch of neck muscles	Muscle spindles	Righting of thorax, shoulders, pelvis	Midbrain
Righting reflexes (head on body)	Pressure on side of body	Exteroceptors	Righting of head	Midbrain
Righting reflexes (body on body)	Pressure on side of body	Exteroceptors	Righting of body when head sideways	Midbrain
Righting reflexes (optical)	Visual cues	Eyes	Righting of head	Cerebral cortex
Placing reactions	Visual, exteroceptive, and proprioceptive stimuli	Eyes, exteroceptors, proprioceptors	Feet placed on substrate so that body is supported	Cerebral cortex
Hopping reactions	Lateral movement while standing	Stretch receptors in muscles	Hops, such that limbs remain in position to support body	Cerebral cortex

[a]*Data from Ganong.* Review of Medical Physiology, 6th ed, *1973, Lange Medical Publications.*

An important functional characteristic of the tonic contraction of skeletal muscles, such as underlies the postural reflexes, is the low energy expenditure that accompanies the tonus. Thereby posture is maintained for extended periods of time with no (or very little) evident fatigue. For example, this effect is manifest clearly in normal humans by the maintained closure of the jaw as well as during standing or sitting and in decerebrate rigidity (qv, below). Similarly, in the frog or toad during amplexus, the clasping reflex may persist for many hours without any evident diminution of intensity or development of fatigue.

The increased metabolic rate above the resting level that accompanies prolonged tonic contractions is somewhat less than that which develops when movements are performed (ie, when external work is done). The physiologic mechanism whereby this economy in energy expenditure is effected is as follows. During tonic contractions different groups of muscle fibers within particular muscles contract asynchronously with respect to each other. Therefore, at any particular moment only a small proportion of the total number of fibers present in the entire muscle are active, and when these active fibers relax other groups of inactive fibers contract. Consequently the groups of fibers alternate between rest and activity, and thus tonic contraction of the entire muscle can be maintained for long periods of time without the development of fatigue (see also p. 128 et seq.).

Transection of the Spinal Cord. In the normal human or animal all of the individual somatic motor responses are components of an overall pattern of motor activity, and the various anatomic levels at which certain motor functions are integrated have been summarized in Tables 7.11 and 7.12. At higher neuroanatomic levels the neural patterns and their connections assume an ever-increasing complexity, and thus underlie progressively more complex motor activities. However, when the neural axis is transected completely, then the physiologic activities which are integrated by regions of the nervous system located below the section are deprived of, and thus released from, any form of functional control by the higher brain centers. Thus the functions of the nervous system that are integrated by the centers that are located below the transected region often appear to be exaggerated, and hyperactivity of reflexes may be seen. Under certain circumstances, this release phenomenon may be caused by the removal of inhibitory regulation that originates within the higher neural centers.

The physiologic responses of animals and humans following complete transection of the spinal cord will serve to illustrate the integration of a number of reflexes at the spinal level. In this regard there is a direct relationship between the degree of cortical development in a given species of animal and the functional roles of the cortex in that species, a phenomenon that is called *encephalization*. As a consequence of this effect,

cortical ablation produces far more serious functional derangements in primates than it does in lower vertebrates, eg, fish, frogs, chickens, mice, cats, or dogs. In addition, the effects of encephalization may not only be responsible for quantitative neurologic differences among various species but qualitative differences as well. The latter fact is, of course, extremely important insofar as extrapolating data obtained from experiments performed on subprimate vertebrate species to humans is concerned. The comparative effects of spinal transection in various species will illustrate further the concept of encephalization.

Transection of the spinal cord causes an immediate flaccid paralysis of the somatic musculature below the level of the section. This complete loss of muscle tone in the affected region is accompanied by a total depression of the stretch reflex and other extensor reflexes. Tendon reflexes such as the knee jerk (p. 227) also are absent, the blood pressure falls, and vascular as well as visceral reflexes cannot be evoked. This condition, known as *spinal shock,* also is accompanied by a slight (-2 to -6 mv) hyperpolarization of the spinal motoneurons. Gradually the depression of reflex activities that accompanies spinal shock wears off, and the duration of this shock state is directly proportional to the degree of encephalization of the motor functions in the particular species of animal. Thus in frogs, the duration of spinal shock is quite brief, lasting for only a few minutes. In rabbits, tendon reflexes can be elicited a few minutes after spinal cord transection, whereas in the cat and dog such reflexes reappear within an hour or two. In monkeys, on the other hand, spinal shock lasts for several days, whereas in humans spinal shock usually persists for at least two weeks following spinal-cord transection.

During recovery from spinal shock, the blood pressure gradually returns to normal levels, and vascular as well as visceral reflexes also become functional once again.

The time required for restoration of full reflex activity in either humans or animals following cord transection is prolonged considerably if complications develop. Thus, for example, malnutrition or infections greatly retard the recovery from spinal shock, although the mechanisms that underlie this effect are unknown.

The mechanism responsible for the production of spinal shock is unclear in detail. However, it is believed that the loss of continuous excitatory (facilitory) signals which are generated in the vestibular nuclei as well as the reticular formation and are transmitted via descending pathways to reinforce the activity of spinal reflex centers seem, in part, to be responsible for the etiology of spinal shock.*

Although the shock state could be induced by the inhibitory effect of the local injury attendant

*The shock state in general is considered in Chapter 16, p. 1237 et seq.

upon transection of the cord, this factor appears to be eliminated by one important experimental observation. After an animal has recovered from spinal shock, if a second transection of the cord is performed inferior (or caudally) to the original transection, no second shock state develops following the second transection. This fact substantiates the conclusion that the state of spinal shock is induced, in part at least, by a cessation of the normal excitatory stimuli which act upon spinal centers to condition their activity.

The threshold for the spinal reflexes declines steadily once these responses reappear following an interval of spinal shock, and ultimately reflex hyperexcitability is established. This effect has been explained on the basis of denervation hypersensitivity. That is, the neurotransmitters released by the intact spinal excitatory terminals exert a greater-than-normal stimulatory effect upon the interneurons and motoneurons following removal of higher control. Alternatively, some evidence is available to substantiate the conclusion that collaterals grow from intact presynaptic afferent neurons, and thus a greater total number of excitatory terminals develop upon the spinal interneurons and motoneurons.

In humans, the earliest reflex to make its appearance following a period of spinal shock often is a negligible contraction of the adductors and flexors of the leg following application of a nociceptive stimulus. However, tendon reflexes (such as the knee jerk) may appear first in some patients.

As time passes and the spinal reflexes become progressively more excitable, even slight nociceptive stimuli not only may cause prolonged withdrawal of the stimulated extremity from the stimulus, but also induce pronounced and complex flexion and extension movements in the other three extremities. The flexion movements in particular take place for long periods of time, and contractures may develop in the affected flexors.

The stretch reflexes also become hyperactive with the passage of time following spinal shock, and so do the more complex responses which are based upon this type of reflex. Thus if one exerts pressure with a finger upon the sole of the foot of a chronic spinal animal (ie, an animal whose cord has been transected) the limb generally extends and "tracks" the finger as the experimenter moves his hand. This positive supporting reaction (also called the magnet reaction, or extensor thrust reflex, Table 7.12) involves not only tactile receptors but also proprioceptive elements, and by means of this reflex the limb becomes rigid and thereby is able to support the weight of the body against the force of gravity. The cessation of the positive supporting reaction also is an active effect, at least partially. Consequently the negative supporting reaction (or extensor flexion reflex) is elicited by stretch of the extensor muscles.

A number of autonomic reflexes are integrated in the spinal cord, and following the period of spinal shock can be elicited readily (see also p. 227). Thus reflex contractions of the distended rectum or bladder occur following spinal cord transection although the latter organ is emptied incompletely. At rest, the blood pressure may remain between normal limits; however, the highly accurate feedback control normally provided by the baroreceptors is lacking (p. 672). Therefore undamped oscillations in blood pressure are common accompaniments of spinal cord transection. Similarly, blanching of the skin, caused by unrestrained vasoconstriction, and intervals of sweating also take place.

The chronic spinal animal or patient also exhibits other reflex activities which are integrated within the spinal cord. Genital massage performed on male spinal animals or humans can evoke erection and ejaculation indicating that these reflexes are integrated at the spinal cord level. However, such reflexes are merely a small part of a highly complex and coordinated sexual pattern that is integrated at many central nervous levels in addition to the cord in normal individuals.

In chronic spinal animals or humans afferent signals often are able to spread (or irradiate) without much restraint from one reflex center to another within the cord. Therefore even low-intensity noxious stimuli applied to the skin may irradiate the autonomic centers and elicit contractions of the bladder and rectum, oscillations in blood pressure, sweating, and pallor. These effects collectively are known as the *mass reflex*.

Decerebrate Rigidity. Decerebrate rigidity provides an experimental situation in which certain mechanisms involved in the higher control of neural functions related to posture and equilibrium may be studied.

When the brain stem of an experimental animal such as a dog or cat is transected at any level between the superior colliculi and the vestibular nuclei (eg, at the superior border of the pons), a sustained contraction of the antigravity muscles develops immediately and without an intervening period of spinal shock. The limbs and body of the animal assume a typical attitude known as *decerebrate rigidity* following the operative procedure.

Both the extensor and flexor muscles are involved; however, in the dog and cat the spastic contractions are most evident in the extensor muscles.* The animal exhibits a characteristic appearance in which the limbs are extended rigidly, the head is retracted, the jaws are closed, the tail is

---

Frogs, which normally squat, exhibit primarily a flexor rigidity following decerebration. The extremities are flexed and held closely against the body. Sloths also show flexor rigidity following decerebration, as these arboreal animals habitually hang upside down from the branches of trees, this usual posture being maintained by the flexors of arms and legs. Decerebrate birds, on the other hand, exhibit wing flexion with leg extension; thus such animals assume a posture that is typical of their resting postion after decerebration.

elevated, and the back is straight and stiff (Fig. 7.102A). If the animal is placed upon its feet, the limbs are able to support the body weight and the position assumed is an exaggeration ("a caricature") of the normal standing posture of the animal. All of the stretch reflexes, including the knee jerk, are hyperactive. The righting reflexes (p. 376) are abolished; however, the tonic neck and labyrinthine reflexes still are present (p. 379 et seq.).

The occurrence of true decerebrate rigidity in man is rare. Decorticate rigidity occurs more commonly and this condition may develop following the appearance of lesions of the cerebral cortex in which a large proportion of the brain stem remains intact. The position of the limbs with respect to the rest of the body in decerebrate and decorticate humans is depicted in Figure 7.103.

When the lesions are present in the midbrain, the decerebrate rigidity which develops in humans is quite similar in pattern to that which develops in dogs and cats following experimental decerebration (Fig. 7.103A). That is, extensor tone predominates in all four extremities with pronation of the forearms. The skeletal muscles exhibit the lengthening and shortening reactions discussed below, and tonic neck as well as labyrinthine reflexes

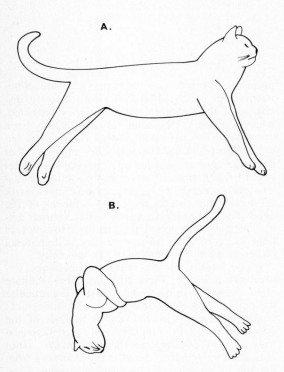

FIG. 7.102. A. Decerebrate cat with intact labyrinths suspended to show extensor rigidity. **B.** Decerebrate, labyrinthectomized cat. Following labyrinthectomy, the head drops forward which causes flexion of the forelimbs with concomitant extension of the hindlimbs. (After Best and Taylor [eds]. *The Physiological Basis of Medical Practice,* 8th ed, 1966, Williams & Wilkins.)

can be evoked. Thus when the head is flexed upon the neck, the arms flex at the elbows whereas the legs remain extended and rigid. Rotation of the head to either side causes the extremities on that side to extend, whereas the extremities on the opposite side of the body flex (or relax). Conditions which induce decerebrate rigidity in man are, in general, incompatible with life.

Decorticate humans, in contrast to decerebrate individuals, usually exhibit flexion of the arms at the elbows, wrists, and fingers (Fig. 7.103B, C, and D). The legs generally show extension. Frequently this condition is seen in patients having a degenerative lesion that is located within the internal capsule.

If one tries to flex passively an extended limb of a decerebrate (or for that matter chronic spinal) preparation, considerable resistance is offered by the limb. But if the force applied to the limb is increased sufficiently the stretch reflex (which, of course, underlies the resistance) suddenly is inhibited and the limb collapses. This phenomenon is called the *clasp-knife effect,* because of its similarity to the sudden closure of the blade of a pocket knife.

Once the extensor tone of the limb has decreased because of central inhibition of the extensor tone, the limb now may be flexed readily and it will remain in any position in which it is placed. The relaxation of the extensor muscles which underlies the sudden flexion of the limb is termed the *lengthening reaction,* whereas the contraction of the extensors which permits the flexed limb to remain in any position in which it is placed is called the *shortening reaction.*

The two reactions just described lend the muscles a certain plasticity. The lengthening response is believed to be caused by inhibitory impulses that arise in both the muscle as well as the tendon when the force applied to the muscle exceeds a certain threshold. The sudden relaxation (lengthening reaction) of the muscle appears to be a protective mechanism which prevents rupture of either the extensor muscle or its tendon when an excessively strong mechanical stimulus is presented.

When the extensor muscle of a knee has been inhibited by an applied force and the knee flexes, the extensor muscle of the contralateral leg contracts strongly, ie, is stimulated reflexly. This phenomenon is called Philippson's reflex; this reflex affords yet another example of the principle of reciprocal innervation of antagonistic muscles (p. 221).

There is much experimental evidence which indicates that the mechanism underlying decerebrate rigidity consists of the release of a lower brain center (or centers) from the inhibitory control of a higher brain center, and thereby the normal stretch reflexes are facilitated considerably in a nonspecific fashion. This facilitation in turn is caused by the operation of two factors. First, there is an enhanced overall excitation of the

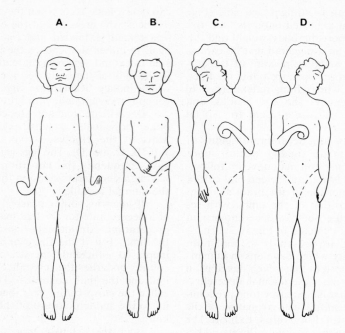

FIG. 7.103. **A.** Decerebrate rigidity in man. Note position of arms and hands, as well as extension of head and feet. **B, C,** and **D.** Decorticate rigidity. In **B,** the victim is lying supine with head facing forward. In **C** and **D** the head is turned to the right and left respectively causing alterations in the position of the hands and arms through activation of tonic neck reflexes. (After Ganong. *Review of Medical Physiology,* 6th ed, 1973, Lange Medical Publications.)

motoneuron pool below the lesion or transection, and second, a net increase in the rate of discharge of the neurons of the gamma efferent system takes place. The areas of the cat brain in which electric stimulation produces facilitation or inhibition of stretch reflexes are shown in Figure 7.50A (p. 306), and with the exception of the vestibular nuclei (Fig. 7.34), these areas function by enhancing or diminishing the sensitivity of the muscle spindles.

The facilitory area that is located in the reticular formation of the brain stem (p. 305) discharges spontaneously, and presumably this effect is caused by afferent inputs in a manner analogous to discharge of the reticular activating system (Fig. 7.50A). In contrast to the facilitory area, however, the area in the brain stem that inhibits gamma efferent discharge does not fire spontaneously. Rather, fibers that converge upon this region from the cerebral cortex as well as the cerebellum cause discharge in the inhibitory area, as depicted in Figure 7.50A. The inhibitory area located within the basal ganglia may act either through descending neural pathways as indicated in the same illustration, or through excitation of the cortical inhibitory center.

Facilitory and inhibitory impulses descend from the reticular areas in the lateral funiculus of the spinal cord. Therefore when the brain stem is cut through at the superior border of the pons, two-thirds of the inhibitory areas that send impulses to the reticular inhibitory area are eliminated (Fig. 7.50A, 1 and 2). Consequently discharge of the facilitory area continues unabated, whereas firing of the inhibitory center has been decreased. The net result is that the balance between the facilitory and inhibitory impulses that converge upon the gamma efferent neurons swings toward facilitation. Thereby gamma efferent discharge is increased and a generalized hyperactivity of the stretch reflex develops. Removal of the cerebellar inhibitory area from a decerebrate animal intensifies the preexisting rigidity. However, in humans, cerebellar damage results in hypotonia rather than an exacerbation of the spasticity.

The vestibulospinal pathways enhance decerebrate rigidity as these structures also facilitate stretch reflexes. In marked contrast to the reticular pathways, however, the vestibulospinal tracts course principally in the ventral funiculi of the cord, and the rigidity that is caused by an increased frequency of discharge in these pathways is not abolished by severing the afferent pathways from the spastic muscles (ie, by deafferentation). Consequently the rigidity that is mediated via the vestibulospinal facilitory center and the vestibulospinal tracts is caused by a direct effect upon the spinal motoneurons, rather than an indirect effect which is mediated through the gamma efferent motoneurons. If the latter system were involved, then deafferentation of the spastic muscles would immediately result in their relaxation.

The Tendon Reflex, Flexion Reflex, and Crossed Extensor Reflex. The *tendon reflex** is a spinal reflex which is evoked readily by suddenly tapping the tendon of practically any skeletal muscle in the body (p. 218). Many tendon reflexes which can be evoked in specific muscles are of great neurologic importance, insofar as diagnostic purposes are concerned.

The tendon reflex is characterized by a short latent period and absence of any after-discharge, and the reaction is a rapid, single contraction of the muscle in response to a sudden stretch. Actually the monosynaptic stretch (or myotatic) reflex discussed in the previous section underlies the tendon reflex, and the sudden contraction of the muscle in response to a single stretch merely reflects an abnormal stimulus. Thus when one taps either the tendon or the belly of a skeletal muscle, the muscle is stretched for only a very brief period of time. The stretch receptors within the muscle are stimulated in a synchronous manner; and as the afferent pathway from the muscle is composed largely of nerve fibers having similar diameters (hence similar conduction velocities), the afferent signals reach the spinal cord almost simultaneously. The motoneurons in turn also respond almost synchronously to the afferent signal; the efferent discharge also is practically synchronous and thus the muscle contracts briefly and in a manner similar to the twitch which follows excitation of a muscle through its motor nerve by a single electric stimulus.

Under natural conditions, stretch generally is imposed upon the muscles by the force of gravity, and sudden brief-acting mechanical deformation of the muscles only takes place under physiologic conditions, eg, when one touches the ground again following a jump. The sustained force of gravity, on the other hand, causes a highly asynchronous afferent discharge because this long-term deformation induces repetitive firing in large numbers of receptors whose thresholds may be quite different. Furthermore, various degrees of stretch may be imposed upon different receptors, even within the same muscle; hence the frequencies of afferent impulse discharge can vary considerably. The net result of the operation of both of these factors is that the motoneurons normally are excited in an asynchronous manner which corresponds to the asynchronous pattern of afferent discharge, and therefore a smooth tonic contraction of the muscle results, and this contraction is maintained for long periods of time. An upright posture is maintained continually and automatically against the force of gravity; and as discussed earlier, the stretch reflex is the fundamental mechanism responsible for the maintenance of this posture. The tendon reflex is merely a small functional part of the entire pattern of the stretch reflex, and a part which generally is evoked by artificial means.

The *flexion reflex* is another spinal reflex which is integrated wholly within the cord (p. 222 et seq.); hence this type of response can be elicited readily in the spinal animal by such noxious stimuli as pinching or burning an extremity. The response to such stimuli is protective, and results in flexion and withdrawal of the affected (or ipsilateral) limb from the noxious stimulus. Since the afferent fibers that transmit impulses from the flexor muscles make reciprocal synaptic connections in the cord with the extensor muscles in the same limb (Fig. 6.43.), the flexor muscles contract smoothly while simultaneously the extensor muscles of the same limb are inhibited. This physiologic effect is called the *reciprocal innervation of antagonistic muscles* (p. 221).

The reflex withdrawal of a limb subjected to a noxious stimulus is related directly to the contraction of the extensors and relaxation of the flexors in the contralateral limb, and thereby as the stimulated limb is withdrawn from the painful stimulus, the opposite limb extends automatically in order to support the weight of the body which suddenly is shifted to it when the ipsilateral limb flexes. The contralateral excitation of the extensor muscles with concomitant flexor inhibition is called the *crossed extensor reflex*. Actually the crossed extensor reflex is merely a component of the flexion reflex which is activated by noxious stimuli, and not a separate reflex.

The flexor reflex is noteworthy because of its broad receptive field as well as its action on large groups of muscles. In general, a noxious stimulus applied to any cutaneous region on the distal part of a limb induces contraction of all of the flexor muscles present in the limb. This contractile response is not rigidly stereotyped, however, because the relative intensity of the contractions in different muscles varies depending upon the particular locus at which the stimulus is applied. Furthermore, the actual spatial pattern of limb movement as well as the ultimate spatial position of the limb can vary considerably. Thus the relationship between the pattern of reflex movement and the site at which the initial stimulus was applied is termed the *local sign* (p. 225), and because of this effect the withdrawal of the limb is physiologically effective regardless of the area which is injured.

The functional characteristics of the flexion reflex are typical of polysynaptic (or multisynaptic) reflex pathways in general. For example, the large receptive field, which includes deep as well as cutaneous receptors, is typical of a multisynaptic reflex arc. Similarly, the selective manner in which excitatory signals are funneled to the ipsilateral flexor muscles whereas the ipsilateral extensor muscles are inhibited simultaneously is another characteristic of multisynaptic reflex arcs. A third characteristic of the flexion reflex is *after-discharge;* that is, the excitation of the muscle greatly outlasts the duration of the stimulus (p.

The inverse stretch reflex (p. 221) also is referred to as the tendon reflex, but it is inhibitory rather than excitatory. It is apparent that usage of the term "tendon reflex" is ambiguous; hence it must be specified whether the response is excitatory or inhibitory following stretch of a muscle. The terms "stretch reflex" or "myotatic reflex" more accurately describe the sudden contraction that follows a tap on a tendon (p. 218).

223). This phenomenon is, of course, to be expected in polysynaptic circuits within which a cyclic transmission of impulses prolongs excitation of the motoneurons through circuits of interneurons for long periods after the initial afferent signals have ended (eg, see Figs. 6.44, p. 225, and 6.45, p. 225, Figs. 7.94, 7.95, and 7.96).

The Extensor Thrust Reflex and Rhythmic Stepping Movements.   In chronic spinal animals the application of gentle pressure to the soles of the feet evokes contraction of the extensor muscles so that the limb becomes rigid. This response is called the *extensor thrust reflex* (see also p. 228).

The efferent nerves which mediate the extensor thrust reflex form part of the plantar nerves; however, electric stimulation of these nerves merely evokes ipsilateral flexion and extension of the contralateral limb. These observations illustrate the point that was made earlier (p. 375) concerning the abnormality of reflex patterns that can be initiated when artificial stimulation is applied to abnormal loci in order to excite reflex activity. Actually, the extensor thrust reflex probably is involved in standing and walking under physiologic conditions.

If light touch is applied to the hind footpad of a chronic spinal dog, rhythmic stepping or walking movements sometimes can be elicited. If, however, the animal is held in an upright position by a sling placed around the body so that the hip extensor muscles are stretched, rhythmic stepping movements of the hindlimbs can be stimulated with far greater ease. Thus the normal walking movements that can be evoked in such a spinal animal presumably are initiated by an alternate rhythmic excitation of the touch- and pressure-sensitive receptors on the soles of the feet and by a similar excitation of the stretch-sensitive afferent terminals in the spindles of the hip muscles. However, this view of the spinal mechanism which underlies rhythmic stepping movements is rather simplistic, as indicated by the following experiments.

First, if one denervates surgically all of the tactile receptors from the legs of an animal such as the cat, normal walking movements are unimpaired.

Second, rhythmic stepping movements of alternate limbs can be induced by stimulating the afferent fibers which originate from each hindlimb. Thus if one excites only one afferent peroneal nerve, the ipsilateral extensor muscles relax, whereas the contralateral extensor muscles contract. If, however, stimuli of equal intensity are applied simultaneously to both afferent nerves then rhythmic and alternate contraction and relaxation of flexors and extensors is obtained, and the rhythmic movements persist for as long as the dual stimuli are applied. In this experiment the afferent inputs to each side of the spinal cord not only are equal, but also are continuous rather than intermittent.

Third, rhythmic alternating contractions of similar muscles in opposite legs can be evoked as described just above even after the limbs are totally deafferented.

Based upon experiments such as these, it is an inescapable conclusion that the physiologic mechanism which underlies reflex stepping movements must reside within the cord per se. Furthermore, this mechanism is not solely dependent for its operation upon alternate afferent inputs from cutaneous receptors in the soles of the feet and the muscle spindles of the hip muscles.

Static Reflexes

The Righting Reflexes.   The static reflexes that control the spatial orientation of the head relative to the trunk of the body and also regulate the motor adjustments of the eyes and limbs in relation to the position of the head are stimulated by proprioceptive impulses that arise in receptors located within the neck muscles, the vestibular apparatus of the inner ear (semicircular canals, utricle), the retina of the eye (optical righting reflexes), and the limb muscles and body wall. As indicated in Table 7.12, these complex reflexes are integrated within the medulla and midbrain. In contrast to the general postural and other reflexes discussed up to this juncture, the righting reflexes (to be considered in this section) as well as the supporting reactions and the tonic labyrinthine and neck reflexes (to be discussed in the two succeeding sections) are concerned with the posture of the limbs and the trunk as this posture is affected by movements of other extremities as well as by movements of the head and neck. Consequently these static reflexes have a number of important functions. For example, static reflexes assist in maintaining the upright spatial orientation of the body; they function in restoring the body to an upright posture following movements; and such reflexes also participate in the postural adaptations that accompany such movements as walking, running, jumping, and so forth.

Two kinds of experimental preparation have been employed extensively for the investigation of static as well as phasic reflexes,* and the general functional characteristics of each of these preparations will be discussed briefly before the individual postural reflexes are considered in some detail.

1. *The Midbrain Preparation:* If the brain stem of a cat or dog is sectioned through the neural axis at the rostral level of the brain stem (ie, through the cerebral peduncles and at the level of the superior colliculi, Figs. 7.27 and 7.28), a so-

---

*A phasic reflex *is defined as an active (dynamic) and coordinated movement which occurs in response to an adequate stimulus. A* static reflex, *on the other hand, is defined as a reflex which is concerned primarily with the maintenance and adjustment of posture, as well as those movements that are concerned with righting of the body. Hence the righting reflexes have a dynamic or active component even though classed as "static" reflexes.*

called *midbrain preparation* is obtained. (A similar preparation is called a *thalamus animal,* if the cerebral hemispheres are extirpated whereas the optic thalamic nuclei are left intact.)

Such preparations retain their righting reflexes, and can regulate their body temperature through segmental reflexes that control vasoconstriction as well as execute highly coordinated reflex acts. In contrast to decerebrate preparations, the patterns of muscular tone seen in midbrain animals are similar to those found in normal animals. However, extensor tone becomes pronounced when the animal lies quietly upon its back or is held away from the substrate. In the decerebrate preparation the rigidity is pronounced because there are no phasic postural reflexes present which modify the static postural reflexes.

Chronic midbrain animals are able to rise from a lying position, stand, walk, and right themselves. When performing active movements, midbrain animals show no sign of decerebrate rigidity, which is, after all, a static manifestation of excessive tone in the postural muscles.

Animals whose midbrains are intact, but severed from higher centers, also exhibit pupillary reflexes in response to light of various intensities (p. 475), but only when the optic nerves are undamaged. Nystagmus, the reflex eye movements that are evoked by rotational acceleration (p. 393), also is present in such preparations.

2. *The Decorticate Preparation:* A second type of preparation which has been employed extensively for studies of various reflexes involves the surgical removal of each cerebral cortex. Such a procedure, called *bilateral decortication,* causes few motor defects in lower vertebrates (eg, cats and dogs) because the basal ganglia are the major centers responsible for the coordination of motor activities in such species. Primates, however, generally are unable to walk after bilateral decortication and they exhibit marked alterations in the muscular tone of the extremities (see encephalization, p. 371). For example, after decortication a monkey usually assumes a typical position in which it lies on its side, and the limbs beneath the body are extended whereas the uppermost limbs are flexed. The limbs upon the upper side of the body show a marked *grasping* or *holding reflex* as indicated by flexion of the fingers. This general pattern of motor activity is reversed completely when the animal is turned over and placed upon the opposite side of the body.

The righting reflexes as well as the tonic neck and labyrinthine reflexes are clearly evident following decortication, although those righting reflexes subserved by the optical pathways are, of course, lost completely.

In general, decorticate animals have all of the reflexes that are present in midbrain animals, and these reflexes usually appear shortly after the operative procedure has been performed. This fact is in distinct contrast to midbrain animals in which a period of up to three weeks must elapse before such responses become evident. Thus it would appear that neural centers which are located in the region between the rostral end of the midbrain and the cerebral cortex tend to facilitate those reflexes whose integrative centers lie in the brain stem per se.

Decorticate animals can be kept alive more readily than midbrain animals, principally because the hypothalamus is present. Therefore the visceral functions, including thermoregulation, which are controlled by this region of the brain remain within normal limits (see p. 407).

Some muscular rigidity is seen in the decorticate animal, and this defect is caused by loss of the cortical region which inhibits gamma efferent discharge by way of the reticular formation as depicted in Figure 7.50. Similarly to the rigidity that appears following transection of the central nervous system at any point above the rostral end of the midbrain, the rigidity that develops in decorticate animals disappears when phasic postural reflexes are evoked, and thus is seen only when the animal is in the resting state.

The most obvious deficiency seen in decorticate animals is their inability to function in the present with respect to experiences that were learned in the past; ie, learned reactions are lost completely as, after all, a major cortical function, in primates especially, is information storage. Although decorticate animals can develop conditioned reflexes (ie, learn) under the influence of special types of training, no learning or conditioning takes place in such animals when they are maintained in an ordinary laboratory situation.

Further reference to midbrain and decorticate preparations as well as to decerebrate preparations will be made in context as particular static and kinetic reflexes are discussed in this and the following sections. In this regard it is important to emphasize that many reflexes are "unmasked," hence can be demonstrated far more readily, in animals lacking higher neural controls as in many instances these response patterns become exaggerated.

It is a fundamental and typical characteristic of animals in general that the head is oriented in space with respect to the force of gravity, and furthermore the normal relationship of the body position to that of the head also is characteristic. For example, if a normal cat is held upside down by the feet and then released, the body turns rapidly as the animal falls so that the normal postural attitude is assumed and the animal lands upon its feet in the head-up position. There are five distinct types of reflex which underlie such complex righting behavior (Table 7.12).

1. *Labyrinthine righting reflexes* that act upon the neck muscles can be demonstrated either in a midbrain preparation or in a blindfolded normal animal. If the animal is suspended in the head-down position by holding the pelvis, the head flexes upward until it reaches its normal position with respect to the force of gravity. The head now

is maintained in this position by the action of the labyrinthine righting reflexes which induce contraction of the neck muscles. If the body of the suspended animal now is turned or rotated into different positions, the head moves so that it remains in the same relative position in space; ie, compensatory movements keep the head oriented upright in space. If, however, the utricle (p. 381) or labyrinths are destroyed the head now hangs limply as though the animal were dead.

Since the labyrinthine proprioceptors are responsible for producing the reflex tone in the neck extensor muscles, if one destroys the labyrinths of a decerebrate animal (eg, a cat), the head no longer is held erect, but now drops into a fully flexed position.

The rigidity seen in the forelimbs of the decerebrate preparation is sustained by proprioceptive reflexes which are elicited by stretch of the extended neck muscles in addition to proprioceptive impulses that arise in labyrinthine structures. If labyrinthectomy is performed on a decerebrate animal, the flexed position of the head which now is assumed stimulates the firing of proprioceptive impulses in the spindles of neck muscles, and these impulses lead to a decline in the extensor tone of the forelimbs, which now assume a strongly flexed position on the chest of the animal (Fig. 7.102B). If at this point the neck muscles of the labyrinthectomized-decerebrate preparation are deafferented, movements of the head now cause no alterations in the extensor tone of any of the four limbs.

2. *The neck-righting reflex which acts upon the body* also can be demonstrated either in a midbrain preparation or in a blindfolded normal animal. Thus when the animal is placed upon its side the head is raised into the normal upright position by the action of the labyrinthine reflexes discussed above. When the neck muscles contract reflexly the head is rotated, and this movement in turn stretches the neck muscles so that proprioceptive signals are transmitted to a motor center located within the upper cervical segments of the spinal cord, and thus the muscles which rotate the body are stimulated. The thorax rotates first and this movement is followed by rotation of the pelvis so that the body is brought into the upright position in relation to the head.

3. *The body-righting reflex acting upon the head* is somewhat similar to the neck-righting reflex that acts upon the body as described above. If a labyrinthectomized animal is laid upon its side so that the cutaneous exteroceptors are stimulated by pressure on one side of the body only, the head is raised to the upright position but only *after* the body has undergone rotation. This righting reflex is caused by an unequal stimulation applied to both sides of the body, because if one lays a board upon the upper surface of the animal, and this board has an identical weight to that of the subject, then no righting reflex is elicited, hence no compensatory movement of the head is seen.

4. *The body-righting reflex that acts upon the body* can be demonstrated in either a decorticate or normal animal which is placed upon its side. If the head is held firmly upon the surface so that the effects of the cervical, and labyrinthine, reflexes as well as the body-righting reflex that acts upon the head are eliminated, nonetheless the body of the animal tends to assume an upright position. In this instance, the hindquarters tend to rotate to the upright position, and presumably the stimulus for this reflex is initiated by an unequal pressure exerted upon the skeletal muscles.

5. *The optical righting reflexes* are elicited by visual stimuli that excite the retina of the eye, and in contrast to all of the other righting reflexes discussed above, an intact cerebral cortex is an essential component of the optical reflexes. Thus the midbrain or the decorticate preparation lacks these reflexes, as the center for visual impressions lies within the occipital lobes of the cerebral cortex. Such optical righting reflexes have an important function in regulating the orientation of the head in space in species such as primates, dogs, and cats. For example, if a labyrinthectomized animal is suspended in the air in order to obviate the effects of the three types of body-righting reflexes discussed above, the animal is able to orient the head in space. If, however, the animal is blindfolded, then this ability is lost.

The various types of righting reflexes can be demonstrated readily in the normal human infant. For example, a baby lying prone will lift its head into an almost vertical position, and when the child is blindfolded and held in midair by the pelvis, movement into different positions stimulates movement of the head into the normal upright position.

The Supporting Reactions. The simultaneous reflex contraction of both extensors and flexors together with other antagonistic muscles locks or fixes the joints, and thereby the limbs become rigid structures which are capable of supporting the weight of the body against the force of gravity. The extensor rigidity seen in the decerebrate animal exemplifies this normal process in an abnormal fashion.

Under natural conditions, the muscular contractions which lead to rigidity of the limbs are called the *positive supporting reaction* or *extensor thrust reflex* (Table 7.12), and this response is evoked by excitation of a number of elements which, under physiologic conditions, act together in a smoothly coordinated fashion.

First, afferent impulses are discharges from the flexor muscles located in the distal segments of the limbs, eg, the ankles. Thus the pressure of the soles of the feet upon the ground is the adequate stimulus for evoking simultaneous contractions of both the flexors and extensors of the knees, and thereby the legs are extended stiffly.

Second, stretch (myotatic) reflexes are initiated in the plantar flexors (of the ankles and toes), and thereby any tendency for the development of an abnormally great extensor contraction of the muscles acting upon these joints is minimized. (In contrast to this physiologic situation, the decerebrate animal exhibits such an exaggerated extensor tone.) Similarly, any tendency in the direction of overextension at the knee (or the elbow) also is minimized by means of the reflex which is initiated by stretch of the flexor muscles attached to these joints. Furthermore, if the knee (or the elbow) tends to bend when supporting the weight of the body, a stretch reflex is elicited from the extensor muscles and this effect precludes any decrease in the supporting function of the limb. In this manner, the tone of extensors and flexors is regulated in a smooth and reciprocal manner so that the weight of the body is supported fully at all times.

Third, afferent impulses which are discharged from the pressoreceptors located within the subcutaneous layers in the soles of the feet reinforce the afferent impulses of proprioceptive origin. Such exteroceptive impulses from the soles obviously are present only when the individual is standing.

The magnet reaction (p. 372) illustrates clearly how such an external stimulus as pressure applied to the soles of the feet causes reinforcement of the proprioceptive stimuli. Thus in a decerebrate (or decerebellate) preparation, a gentle pressure applied by a finger to the pad of a foot causes the limb to extend and to track the movements of the finger. The experimenter perceives the movement as though the limb were attached to, and pulled out by, the movement of his finger; hence the name "magnet reaction."

In contradistinction to the positive supporting reaction, the *negative supporting reaction* underlies relaxation of the limb muscles and thereby the joints are unlocked so that the limbs can be flexed and then may assume a new position. The negative supporting reaction is initiated by raising the sole of the foot off the substrate and by plantar flexion of the toes and ankle. In this manner the exteroceptive and proprioceptive (ie, stretch) stimuli to the plantar flexors are removed. However, in addition to this passive removal of exteroceptive and proprioceptive stimuli, the plantar flexion and the subsequent stretching of the dorsiflexor muscles of the ankles and toes cause relaxation of the knee extensors and concomitantly contraction of the flexors takes place. All of these mechanisms act together to unlock the rigid pillarlike structure of the leg.

The rhythmic alternation of the positive and negative supporting reactions is a fundamental mechanism whereby walking movements take place (see also p. 376).

The Tonic Labyrinthine and Neck Reflexes. The statotonic reflexes that arise in the labyrinth and neck primarily affect either the tone of skeletal muscles of the limbs and body or else the tone of the eye muscles; consequently each of these groups of tonic labyrinthine and neck reflexes may be considered individually.

1. *The tonic labyrinthine and neck reflexes that act upon the limbs* alter the tonus of the skeletal muscles in such a manner that the various parts of the body maintain a suitable spatial position in relation to any particular position of the head in space. Since the righting reflexes are minimized in the decerebrate animal, this type of preparation is particularly suitable for demonstrating the influence of statotonic labyrinthine and neck reflexes upon the limbs and body.

The proprioceptive endings that initiate each of these general groups of postural reflexes are located in two regions. First, the proprioceptors found within the labyrinth of the inner ear, ie, the utricle (p. 381), give rise to those labyrinthine reflexes which are activated by alterations in the position of the head in space. Second, the tonic neck reflexes are stimulated by receptors located within the neck muscles. These receptors are activated when the position of the head is changed in relation to that of the body.

It is important to realize that the labyrinthine reflexes tend to increase (or decrease) the extensor muscle tone in all limbs simultaneously; that is, the tone in all four limbs changes in the same direction. In contrast to this concerted effect of the labyrinthine reflexes, the neck reflexes generally exert opposite influences upon the tone in the forelimbs and hindlimbs.

Therefore in an animal such as a cat which is standing upon all four legs, the effect of the *tonic labyrinthine reflexes* is to increase the extensor tone in all of the limbs concomitantly when the head is rigidly extended. Conversely, the extensor tone in the four limbs is decreased when the head is flexed strongly, and this effect also is due to the action of the labyrinthine neck reflexes.

The activity of the *tonic neck reflexes* alone can be seen most clearly in the decerebrate animal following labyrinthectomy. Thus in such a preparation a downward flexion of the head toward the sternum evokes flexion of the forelimbs with a concomitant extension of the hindlimbs. In contradistinction to the effects of such ventriflexion, extension (or dorsiflexion) of the neck produces an opposite reflex effect, so that the hindlimbs now are extended whereas the forelimbs are flexed.

The cooperative effects of the labyrinthine and neck reflexes now may be reviewed as observed in the decerebrate preparation with intact labyrinths. When the neck of such an animal is forcibly ventriflexed, the activity of the tonic neck reflexes tends to reinforce that of the tonic labyrinthine reflexes on the forelimbs but to antagonize this effect on the hindlimbs. The general response to this ventriflexion maneuver is to induce relaxation of the forelimbs with a concomitant marked increase in the extensor tone of the hindlimbs.

If, on the other hand, the neck is dorsiflexed, the tonic neck reflexes enhance the effect of the tonic labyrinthine reflexes insofar as the extensor tone of the forelimbs is concerned, whereas the extensor tone in the hindlimbs is decreased somewhat. Thus when dorsiflexion of the head is performed, the effect of the tonic neck reflexes upon the extensor tone of the hindlimbs once again is predominant. Therefore extension of the forelimbs continues; however, the hindlimbs tend to undergo relaxation.

If the head of an animal is rotated in the direction of the frontal plane of the skull, an increased extensor tone develops in forelimbs and hindlimbs on the side of the body toward which the head is rotated. Simultaneously the extensor tone is decreased in the limbs on the opposite side of the body. When the head is turned laterally toward one shoulder, a similar response is observed.

The physiologic significance of the several tonic labyrinthine reflexes that act upon the limbs can be appreciated most clearly when one observes an intact animal such as a cat. For example, when the animal changes the horizontal direction in which it is moving, as in turning a corner, the limbs on the side to which it turns become more rigid in order to support the body weight.

When a cat looks upward, as at a canary in a suspended cage, the forelimbs extend whereas the hindlimbs flex. The back becomes tilted upward in such a manner that the head and eyes are in a more suitable position for observation of the bird; and in addition the orientation of the body is such that the animal can leap from this position readily. Conversely, when a cat looks downward, as into a plate of milk or a mouse hole, the forelimbs flex and the hindlimbs extend, thereby tilting the back downward, a posture which likewise is distinctly advantageous to the animal.

2. *The tonic labyrinthine and neck reflexes that act upon the eyes* produce alterations in the tone of the ocular muscles that are analogous to those which affect the posture of the body as a whole.

Any change in the position of the head with respect to the body when the neck is immobilized is followed by compensatory eye movements. Similarly, any change in the position of the head with respect to the position of the body following labyrinthectomy also is accompanied by compensatory eye movements. Observations of this type indicate clearly that both labyrinthine as well as neck reflexes underlie tonic alterations; hence the compensatory movements of the ocular muscles.

If the head is turned downward, the eyes rotate upward and remain in this position for as long as the head is kept in the ventriflexed position. This compensatory rotation of the eyes is caused by a reflex increase in tone of the inferior oblique and superior rectus muscles, whereas a concomitant decrease in tone is induced in the superior oblique and inferior rectus muscles. If, on the other hand, the head is turned upward, a compensatory downward movement of the eyes takes place, the tonus changes in the individual ocular muscles being in the opposite direction to those described above for ventriflexion of the head.

When the head is turned to one side or the other, the internal and external rectus muscles of each eye contract in a coordinated manner so that each eye is rotated inward or outward relative to

the new position of the head. Thus both eyes rotate in a direction that is opposite to the direction in which the head is moved, and thereby the eyes exhibit a tendency to maintain the same position in space that they had prior to rotation of the head. Consequently, the visual field of each eye also tends to remain the same as it was before the movement of the head.

It is important to realize that the actual physical movements of the eyes which were discussed above are carried out by statokinetic reflexes that are initiated in the semicircular canals of the inner ear (p. 391). This mechanism is quite different from the statotonic reflex which originates in the utricle, and which maintains the position of the eyes while the head remains in the altered position following its movement.

The Placing and Hopping Reactions.   Placing reactions are those movements which place the feet upon a surface with considerable accuracy (Table 7.12). A number of such reactions have been described in the normal cat.

First, a cat is held in the air with legs hanging down and the chin elevated so that it cannot see below or in front. Contact of the backs of the forelegs with the edge of a table is followed immediately by a rapid movement of these limbs which brings the paws, pads down, upon the edge of the table with considerable accuracy.

Second, the forelegs of a cat are held down and the chin is brought in contact with the edge of a table. Release of the forepaws is followed immediately by their being raised and placed upon the table on either side of the chin. Generally this placing movement is followed by extension of the limbs and the animal rises to a standing position.

Third, the forelegs or hindlegs of a cat which is sitting or standing near the edge of a table are pushed over the edge. The legs are lifted immediately and placed back in the positions which they occupied before being displaced.

The three placing reactions described above are initiated by excitation of exteroceptors located upon the body surface, as well as by impulses that arise in muscle and tendon proprioceptors.

Fourth, abduction performed upon a limb of a standing cat results in the immediate return of the limb to its earlier position. This reflex is initiated entirely by afferent signals that arise in muscle and tendon proprioceptors.

Fifth, a cat is blindfolded and held in the air with its forelegs hanging free. The head is moved toward an object such as a wall. When the vibrissae (whiskers) contact the obstacle, the forelegs are raised immediately and the paws are placed upon the surface of the object with considerable precision. This reaction is initiated by touch receptors located around the roots of the vibrissae. A similar placing reaction can be evoked by touching any part of a foot to the surface while the cat is held as described above.

Sixth, a cat is lowered toward a table top

which it can see. The limbs are extended while the animal is being lowered in order to support the body immediately upon contact with the surface. This placing reaction is controlled by visual reflexes.

The *hopping reactions* are rapid limb movements which serve to keep the limbs in position to support the weight of the body when the animal is pushed horizontally. For example, a cat is supported in midair so that only one limb is supporting the body. The animal then is pushed in various directions, but the limb which supports the body hops rapidly in the direction in which the body is moved so that the foot remains directly underneath the shoulder (or hip). Similarly, a sudden push that throws a person or animal off balance is accompanied by hopping reactions until the balance is regained. Presumably stretch (myotatic) reflexes underlie the hopping reactions.

Both placing and hopping reactions are controlled largely through the primary sensory and motor areas of the cerebral cortex. Consequently these reactions are seriously impaired in decorticate preparations. This cortical regulation is manifest strictly on the reactions of the opposite side of the body. That is, the movement patterns evoked on one side of the body are regulated entirely by the contralateral cerebral cortex.

Kinetic or Acceleratory Reflexes. The labyrinthine reflexes are classified as *kinetic* because movements of the head evoke such responses. Actually, however, the adequate stimulus for the labyrinthine reflexes discussed in the previous section is not head movement per se, but rather the labyrinthine receptors are sensitive to acceleratory (or deceleratory) movements in either a linear or an angular (rotatory) direction that is above a certain rate; ie, these receptors detect changes in the rate of movement. It is more accurate to designate the labyrinthine reflexes as *acceleratory reflexes,* but with full recognition of the fact that the term "acceleratory" in this context signifies both increasing (positive) as well as decreasing (negative) rates of change of head position.

The anatomy of the inner ear will be reviewed first, as the structure of the paired labyrinths and their specialized sensory receptors is fundamental to an understanding of the functions of these end organs in the initiation and modification of both the static and acceleratory reflexes which underlie posture and equilibrium.

Functional Anatomy of the Labyrinth. The labyrinthine receptors, which are innervated by the vestibular branch of cranial nerve VIII, consist of five individual regions of highly developed and specific neuroepithelium; three of these areas are termed cristae (or crests) and two are termed maculae (or spots). The cristae and maculae in turn are located at particular sites within a complex system of interjoined fluid-filled ducts and sacs called the *membranous labyrinth.* The membranous labyrinth is composed of the three semicircular canals, the *utricle,* and the *saccule.* The membranous labyrinth lies within the bony labyrinth of the skull which is merely a system of tunnels or passages that are located within the petrous region of the temporal bone, and which closely surround the membranous labyrinth. The bony labyrinth is depicted in relation to the membranous labyrinth in Figure 7.104, whereas the latter structure together with the innervation of the vestibular receptors is shown in Figure 7.105.

The *membranous cochlear duct* is located within the bony cochlea of the bony labyrinth (Fig. 7.104), and this structure contains the receptor that is responsible for the sense of hearing, the *organ of Corti* (Fig. 7.106). However, by convention, a distinction is made between the auditory and the nonauditory or vestibular labyrinth. The present discussion is concerned primarily with the vestibular labyrinth (or simply the labyrinth) whereas the cochlea and its functions will be discussed in Chapter 8 (p. 506).

The membranous labyrinth is filled completely with a clear fluid called the endolymph (or otic fluid). The endolymph is similar to intracellular fluid in its ionic composition; consequently there is a much higher concentration of potassium than of sodium (cf. Table 3.3, p. 54 and Table 8.3, p. 513). The space between the membranous labyrinth and the surrounding walls of the bony labyrinth is filled with perilymph (or periotic fluid), and this liquid has roughly the same composition as cerebrospinal and other interstitial fluids (Table 8.3, p. 513). Additional details concerning the physiology of the labyrinthine fluids are given in Chapter 8 (p. 511).

There are three semicircular canals associated with each labyrinth (Fig. 7.104). Each *semicircular canal* terminates in a dilated end called the *ampulla,* within which is located a specific sensory end organ known as the *crista ampullaris.* The semicircular canals lie in three planes that are found at roughly right angles to each other. Therefore the canals are called the *horizontal* (also called lateral or external), the *superior* (or anterior), and the *posterior* (or vertical). The gross relationships among these structures are shown in Figure 7.104. The close relationship between the ampullae of the horizontal and superior canals at the anterior end of the labyrinth is clearly evident in this figure. Both of these ampullae are continuous with the utricle and empty into this region near the macula (Fig. 7.106). The ampulla of the posterior canal, on the other hand, opens into the opposite end of the utricle in proximity to the narrow end of the same canal. The narrow ends of the superior and posterior semicircular canals, both of which lie in the vertical plane, fuse and thus form a single structure known as the *common crus* (or *crus commune*), and both the superior and posterior canals are in direct communication with the utricle through this structure.

The horizontal semicircular canals that are found on each side of the head lie in the same

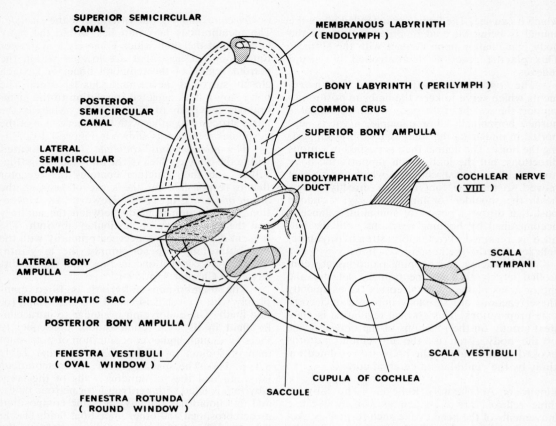

SUPERIOR SEMICIRCULAR
CANAL

MEMBRANOUS LABYRINTH
(ENDOLYMPH)

BONY LABYRINTH (PERILYMPH)

POSTERIOR
SEMICIRCULAR
CANAL

COMMON CRUS

SUPERIOR BONY AMPULLA

LATERAL
SEMICIRCULAR
CANAL

UTRICLE

ENDOLYMPHATIC
DUCT

COCHLEAR NERVE
(VIII)

LATERAL BONY
AMPULLA

SCALA
TYMPANI

ENDOLYMPHATIC SAC

POSTERIOR BONY AMPULLA

FENESTRA VESTIBULI
(OVAL WINDOW)

SCALA VESTIBULI

CUPULA OF COCHLEA

FENESTRA ROTUNDA
(ROUND WINDOW)

SACCULE

FIG. 7.104. Human bony labyrinth as seen on right side. The position of the membranous labyrinth within the bony labyrinth is indicated by dashed lines, and the endolymphatic sac (which actually lies in the plane parallel to and behind the utricle and lateral bony ampulla) is indicated by shading. Both the superior semicircular canal and the cochlea have been opened to show their internal structure. See also Figures 7.105 and 7.106. (After Best and Taylor [eds]. *The Physiological Basis of Medical Practice,* 8th ed, 1966, Williams & Wilkins.)

plane, their circular portion being directed outward as well as downward. Thus when the head is in an erect position, the horizontal canals actually are inclined backward and downward at an angle of approximately 30°.

The superior and posterior semicircular canals, both of which lie in the vertical plane, each make an angle of approximately 45° with the sagittal and frontal planes of the skull. Consequently the superior canal of one ear is almost in the same plane as the posterior canal of the opposite ear. Furthermore, the superior or posterior canal of each ear is at right angles to the comparable canal on the other side of the head, and also these canals are at right angles to the two other canals of the ear on the same side of the head. These spatial arrangements are shown in Figure 7.107.

The receptor organ of each semicircular canal, the *crista ampullaris,* is located transversely within the ampulla (crosshatched areas in ampullae, Fig. 7.106). Each crista is composed of a ridge of neuroepithelial cells that has a layer of secretory epithelial cells at its base called the planum

semilunatum (Fig. 7.108). The neuroepithelial elements of the crista ampullaris consist of two kinds of mechanosensory hair cells that are referred to as types I and II (Fig. 7.109). The type I hair cells are round-bottomed, flask-shaped, and surrounded by a chalicelike afferent nerve terminal. The type II hair cells are cylindrical, and have many small afferent as well as efferent nerve endings upon their bases. Each type of hair cell has between 40 and 60 stereocilia upon the free surface, whose rootlets are firmly implanted within a terminal web (or cuticular plate) that is located at the apex of each cell. In addition to the stereocilia, a single kinocilium is found upon the surface of each hair cell, and this kinocilium resembles morphologically the motile cilia such as are found, for example, in the trachea (p. 706).

The stereocilia of the hair cells are lined up in neat rows and exhibit a gradation with respect to their length. Thus the longest stereocilia (40 to 75 μ) are found in immediate proximity to the kinocilium (Figs. 7.109 and 7.110). *Supporting cells* lie between the hair cells, and the nuclei of

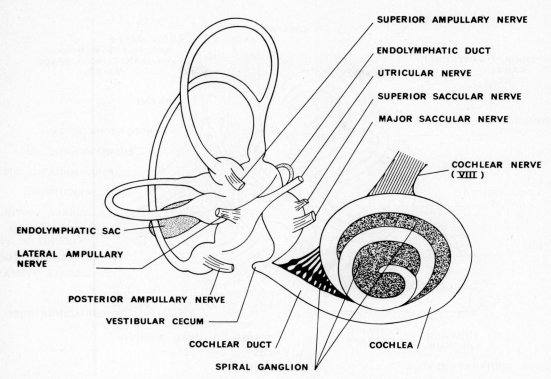

SUPERIOR AMPULLARY NERVE

ENDOLYMPHATIC DUCT

UTRICULAR NERVE

SUPERIOR SACCULAR NERVE

MAJOR SACCULAR NERVE

COCHLEAR NERVE (VIII)

ENDOLYMPHATIC SAC

LATERAL AMPULLARY NERVE

POSTERIOR AMPULLARY NERVE

VESTIBULAR CECUM

COCHLEAR DUCT

SPIRAL GANGLION

COCHLEA

FIG. 7.105. Human membranous labyrinth as seen on right side. The innervation of the vestibular end organs and the organ of Corti are indicated. (After Best and Taylor [eds]. *The Physiological Basis of Medical Practice,* 8th ed, 1966, Williams & Wilkins.)

these elements are located near the basement membrane whereas microvilli are found upon their free (or apical) surfaces (Fig. 7.109). A peculiar gelatinous structure known as the *cupula* extends from each crista to the roof of each ampulla (Fig. 7.108). Within the cupula the sensory elements of the hair cells (the cilia) are enclosed in exceedingly delicate, parallel tubules. The cupula has been shown to be the critical moving part that is involved in the stimulation of the hair cells. The cupula is composed principally of mucopolysaccharides which apparently are secreted by the supporting cells of the crista.

The utricle and saccule, sometimes called the *otolith organs,** are interconnected by way of the *utriculosaccular duct* (Fig. 7.106), whereas the saccule is connected to the cochlear duct via the *ductus reuniens* (R in Figure 7.106). The endolymphatic duct arises from the utriculosaccular duct, and this structure passes by way of the osseous vestibular aqueduct to end in the closed endolymphatic sac that rests upon the intracranial surface of the petrous bone, as shown in Figure 7.106.

The *maculae* of the utricle and the saccule are composed of neuroepithelium, which is similar his-

The term "otolith" literally means "ear stone."

tologically to that found in the cristae. Therefore type I and type II hair cells are present in these structures, together with supporting cells (Fig. 7.110). The maculae have no cupula, however. Instead each macula is covered by a gelatinous *otolithic membrane* within which are embedded large numbers of *otokonia*. The otokonia are small crystals of calcium carbonate (calcite) and have a relatively high specific gravity (ie, around 2.9). Consequently these elements add significantly to the physical mass of the otolithic membrane into which the sensory elements of the hair cells (ie, the cilia) project.

The macula of the utricle is situated upon the anterior and inferior walls of the utricular sac. The posterior region of this structure is inclined backward at an angle of about 30° so that it lies parallel with the horizontal semicircular canal. The anterior third of the utricular macula is inclined sharply upward like the tip of a snowshoe. The maculae on each side of the head lie in roughly the same horizontal plane.

The macula of the saccule is located upon the medial wall of the saccule in an approximately vertical anteroposterior plane.

The principal neural connections between the vestibular apparatus and the central nervous sys-

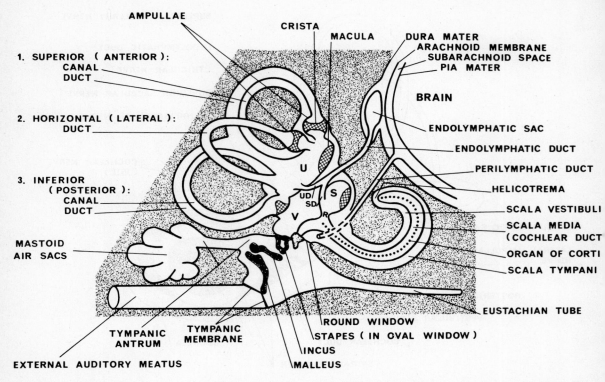

FIG. 7.106. The relationships among the various structures and regions of the inner ear. U = utricle; S = saccule; UD = utricular duct; SD = saccular duct; V = vestibule; R = ductus reuniens. (After Best and Taylor [eds]. *The Physiological Basis of Medical Practice,* 8th ed, 1966, Williams & Wilkins.)

tem and the relationships of these elements to various effectors should be reviewed briefly at this juncture (see Fig. 7.34, p. 286).

Afferent impulses that arise within the proprioceptors located within the five neuroepithelial areas of the nonauditory labyrinth are transmitted to the medulla via the fibers of the vestibular division of cranial nerve VIII (Table 7.3, p. 248). The cell bodies of the bipolar vestibular neurons are found within Scarpa's ganglion that lies within the internal auditory meatus. The dendrites (afferent processes) of these vestibular neurons in turn innervate the sensory end organs of the semicircular canals, utricle, and saccule by means of synaptic junctions directly upon the type I and type II hair cells. The nerve fibers that terminate as cuplike (chalice or calyx) endings about the type I hair cells have a somewhat greater diameter than those of the type II cells.

The majority of the axons of the vestibular

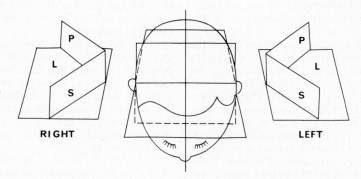

FIG. 7.107. Spatial relationships among the planes of the head and those occupied by the semicircular canals. S = superior (or anterior) semicircular canal; P = posterior semicircular canal; L = lateral semicircular canal. Note that the plane of the superior (or anterior) canal on one side lies parallel to that of the contralateral posterior canal. (After Best and Taylor [eds]. *The Physiological Basis of Medical Practice,* 8th ed, 1966, Williams & Wilkins.)

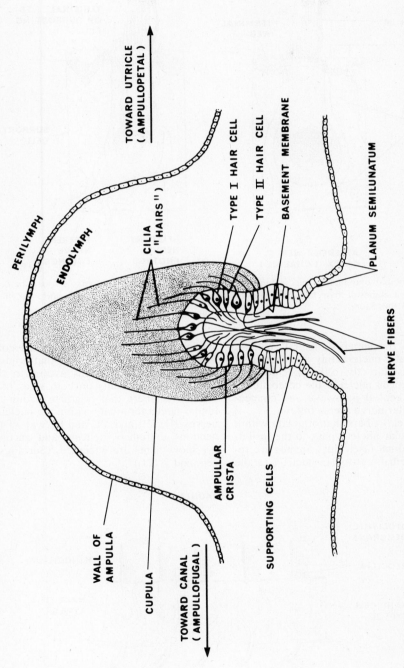

FIG. 7.108. The general anatomic features of an ampullar crista. The gelatinous mass of the cupula is shaded. (After Best and Taylor [eds]. *The Physiological Basis of Medical Practice*, 8th ed, 1966, Williams & Wilkins.)

A.   **B.**

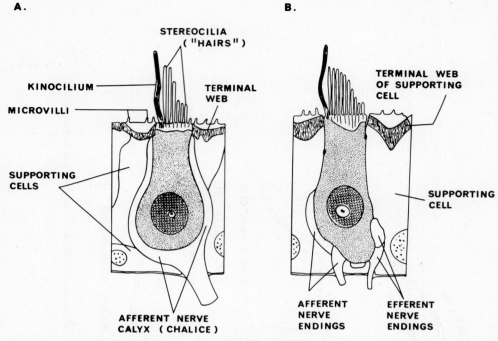

FIG. 7.109. The principal ultramicroscopic features of the type I **(A)** and **(B)** vestibular sensory (hair) cells. Note particularly the different shapes of these cells as well as their specific innervation. (After Bloom and Fawcett. *A Textbook of Histology,* 9th ed, 1968, Saunders.)

neurons course to the superior, lateral, medial, and inferior vestibular nuclei. Only a few afferent vestibular fibers pass directly to the flocculonodular lobe and the fastigial nuclei of the cerebellum.

The principal central pathways and connections of the vestibular nerve are as follows:

1. Ascending relay fibers that originate within the superior vestibular nucleus pass to the ipsilateral medial longitudinal fasciculus. Signals are relayed from the medial vestibular nucleus to the

medial longitudinal fasciculus of the contralateral side of the brain stem. Fibers that originate in the inferior vestibular nucleus ascend in the medial longitudinal fasciculus on both sides of the brain stem. The fibers that originate within these vestibular nuclei terminate within the nuclei of cranial nerves III, IV, and VI, and by way of these connections extremely delicate and accurate reflex eye movements are executed. (See Figs. 7.34, p. 286, and 7.111.)

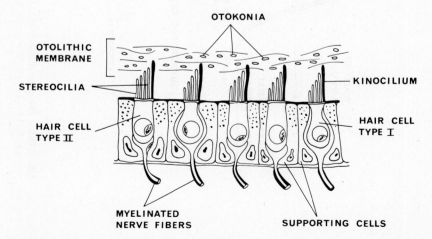

FIG. 7.110. Certain histologic features of the utricular neuroepithelium. (After Best and Taylor [eds]. *The Physiological Basis of Medical Practice,* 8th ed, 1966, Williams & Wilkins.)

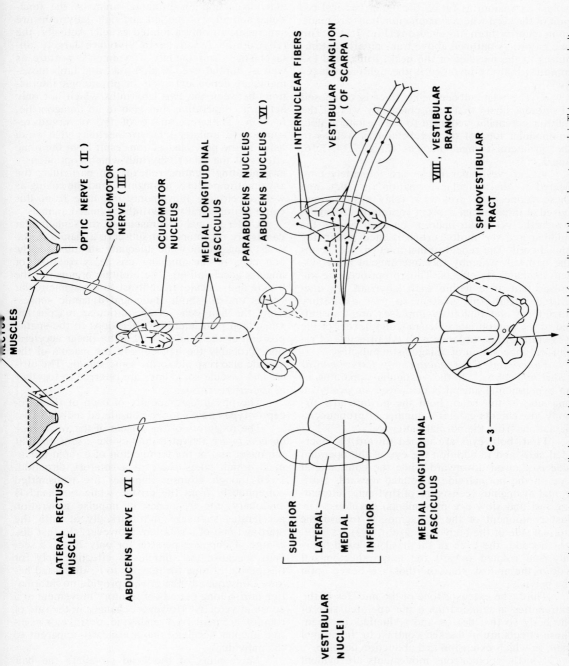

FIG. 7.111. Relationships among the vestibular nuclei, the medial longitudinal fasciculi, and the oculomotor nuclei. This system underlies movement of the eyes in the horizontal plane. (After Best and Taylor [eds]. *The Physiological Basis of Medical Practice*, 8th ed, 1966, Williams & Wilkins.)

2. Fibers that descend from the lateral vestibular nucleus (of Deiter) constitute the vestibulospinal tract (p. 268), whereas those fibers which originate within the inferior vestibular nucleus descend in the medial longitudinal fasciculus (spinal extension) as far as the upper cervical region of the cord where synaptic junctions are made upon anterior horn motoneurons (Fig. 7.112). The two pathways outlined above transmit efferent impulses to the muscles of the neck, trunk, and extremities that are involved in the righting reflexes (p. 376).

3. The vestibulocerebellar tract is composed of afferent fibers that pass chiefly by way of the inferior cerebellar peduncle to the ipsilateral flocculonodular lobe of the cerebellum as well as to the ipsilateral fastigial nucleus (Figs. 7.34, 7.56, and 7.113).

4. The vestibular nuclei are intimately connected to the reticular formation (p. 302), and these connections provide a relay system composed of internuncial neurons (or interneurons) between the vestibular neurons on each side of the midline and the motoneurons located within the spinal cord. The reticulospinal tract (p. 268) forms the pathway whereby this internuncial relay system mediates its effects. Thus proprioceptive impulses that arise within each labyrinth are integrated within this system so that excitatory (facilitory) and inhibitory impulses are transmitted to the appropriate extensors and flexors of the extremities in accordance with the principle of reciprocal innervation of antagonistic muscles.

Functional Consequences of Experimental Labyrinthectomy. If the vestibular apparatus of an experimental animal is destroyed surgically on one side of the head, both the posture and the body movements exhibit a number of pronounced deviations from the normal patterns.

First, both eyes are turned toward the operated side, and in addition the eye of the operated side is skewed downward while the contralateral eye on the normal side is rotated upward. Horizontal nystagmus (consisting of rhythmic, alternating fast and slow eye movements) develops. The fast component of the nystagmus is toward the normal side of the body (see also p. 393).

Second, the back of the head is flexed laterally and rotated toward the labyrinthectomized side of the animal while the thorax is flexed upon the pelvis.

Third, the extensor tone of the muscles of the extremities is reduced upon the operated side of the body so that flexion and adduction are seen; these effects are in marked contrast to the normal side, in which extension and abduction develop.

Fourth, spontaneous movements are evident in an animal following unilateral labyrinthectomy, and all of these movements take place in the direction of the operated side. For example, the subject may roll (or rotate) about a horizontal axis. Circling also may be evident, in which the animal rotates its body about a vertical axis. In addition, the head may move from side to side (head nystagmus), or the body may tend to fall toward the operated side.

Normal Labyrinthine Functions. In normal individuals and experimental animals, the functional activities of the left and right labyrinths are synergistic to only a limited extent. Actually, the labyrinthine functions in vivo are largely antagonistic. Consequently a symmetric posture as well as normal eye, head, trunk, and limb movements are dependent upon the physiologic interactions between the two labyrinths which not only act in a cooperative, but also in an antagonistic, fashion. Therefore many of the observed responses to unilateral labyrinthectomy that were listed in the previous section result from the tonic effects of the intact labyrinth which continuously is generating righting reflexes. In particular, the asymmetric posture, nystagmus, and the rolling as well as circling movements originate from this cause in unilaterally labyrinthectomized animals.

Under normal circumstances, the semicircular canals are receptors that subserve dynamic sensations; consequently the adequate stimulus for the hair cells of the ampullar cristae is rotation (or angular acceleration). The otolithic organs (ie, the utricle and saccule) in contrast to the semicircular canals underlie both static and dynamic senses. Thus the hair cells of the utricular macula are stimulated principally by alterations in the spatial position of the head as well as by linear acceleration. Probably the hair cells of the macula of the saccule also respond to the same stimuli. The utricle and saccule sometimes are referred to together as *gravity receptors*.

The physiologic activity of each of these receptor types now may be considered individually.

The receptors of the cristae of the *semicircular canals* are activated (and evoke a sensation) at the onset and at the termination of a rapid movement which takes place in a rotatory direction. Even though afferent impulses are transmitted continuously from the cristae while the head is stationary, the frequency of impulse generation accelerates suddenly and markedly at both the start and end of a rotation. However, the fast discharge of impulses persists for only 20 to 25 seconds, because the stimulus is effective only for this period of time for reasons to be discussed below. Consequently the cristae provide no information during long periods of rotatory movement at a constant velocity. However, changes in the rate of rotation are perceived readily; ie, positive or negative angular accelerations are clearly apparent to the individual.

Movements of the head stimulate the hair cells of the cristae indirectly, and this stimulation is mediated via movements of the endolymph found within the semicircular canals. In turn the movements of the endolymph are induced directly by head movements, and the specific canal under-

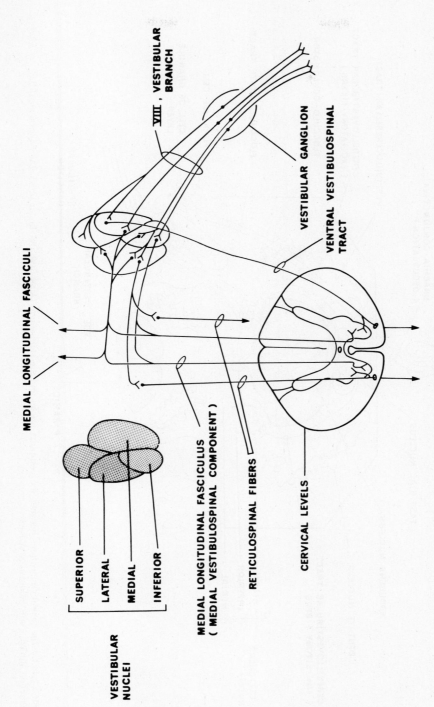

MEDIAL LONGITUDINAL FASCICULI

VIII , VESTIBULAR BRANCH

VESTIBULAR GANGLION

VENTRAL VESTIBULOSPINAL TRACT

VESTIBULAR NUCLEI

SUPERIOR

LATERAL

MEDIAL

INFERIOR

MEDIAL LONGITUDINAL FASCICULUS (MEDIAL VESTIBULOSPINAL COMPONENT)

RETICULOSPINAL FIBERS

CERVICAL LEVELS

FIG. 7.112. The neural connections between the vestibular nuclei and the ventral columns of the spinal cord. (After Best and Taylor [eds]. *The Physiological Basis of Medical Practice*, 8th ed, 1966, Williams & Wilkins.)

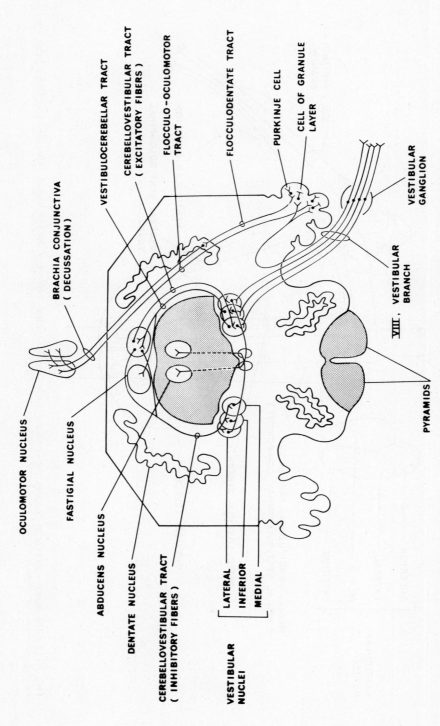

OCULOMOTOR NUCLEUS

BRACHIA CONJUNCTIVA
(DECUSSATION)

VESTIBULOCEREBELLAR TRACT

CEREBELLOVESTIBULAR TRACT
(EXCITATORY FIBERS)

FLOCCULO - OCULOMOTOR
TRACT

FLOCCULODENTATE TRACT

PURKINJE CELL

CELL OF GRANULE
LAYER

VESTIBULAR
GANGLION

VIII , VESTIBULAR
BRANCH

PYRAMIDS

VESTIBULAR
NUCLEI

LATERAL
INFERIOR
MEDIAL

CEREBELLOVESTIBULAR TRACT
(INHIBITORY FIBERS)

DENTATE NUCLEUS

ABDUCENS NUCLEUS

FASTIGIAL NUCLEUS

FIG. 7.113. The neural connections between the vestibular nuclei and the cerebellum. (After Best and Taylor [eds]. *The Physiological Basis of Medical Practice*, 8th ed, 1966, Williams & Wilkins.)

going stimulation at any particular moment in time depends entirely upon the plane in which the rotation acts.

In human beings, the semicircular canals are fine capillary tubes having a diameter that ranges between 100 and 200 μ, and the membranous labyrinth is a closed and fluid-filled system that is supported externally by perilymph, connective tissue, and the rigid walls of the bony labyrinth of the skull. The endolymphatic movements that take place during rotation of the head act as a "fluid drive" stimulus for excitation of the hair cells of the cristae as follows. For example, as the head turns rapidly from a resting position toward the left (ie, in a counterclockwise direction) about a vertical axis so that the horizontal canals are stimulated, at first the endolymph does not move with the walls of the canal due to the inertia of the fluid. This delay in the fluid movement is similar to a flow or displacement of the endolymph toward the right (ie, in a clockwise direction) in both horizontal canals. However, because of the position of each ampulla with respect to the other, such a flow of the endolymph takes place toward the ampulla in the left canal (the ampullopetal direction), whereas the endolymph flows away from the ampulla in the right canal (the ampullofugal direction). Therefore the cupula of each crista of the horizontal canals is deflected toward the utricle in the left ampulla, but away from the utricle in the right ampulla (Fig. 7.108). The bending of the cupula results in a mechanical deformation of the hairs embedded in the base of this gelatinous structure, and the bending of the hairs provides the adequate stimulus for excitation of the neural elements of the cristae. However, the stimulus is most effective when the hair cells of the horizontal canals are deflected toward the utricle rather than away from this structure. Therefore immediately following the onset of a counterclockwise rotation about a vertical axis, the hair cells of the left horizontal canal are stimulated more effectively than are those of the right canal. This pattern of excitation is, of course, reversed when the head is rotated in a clockwise direction; that is, toward the right.

By way of summary, a counterclockwise rotatory movement of the head about a vertical axis stimulates the left horizontal canal whereas a clockwise rotation stimulates the right canal. In other words, the receptor elements of the cristae of the horizontal semicircular canals are stimulated only when a flow of endolymph toward the ampulla (ampullopetal flow) takes place.

In contradistinction to the horizontal canals, the receptor elements located within the anterior and posterior (ie, the vertical) semicircular canals are stimulated only when the displacement of the endolymph takes place in a direction away from the ampulla (that is, in an ampullofugal direction). This functional difference between the excitation of the horizontal and vertical semicircular canals is known as *Ewald's law*, and is caused by the mor-

phologic orientation and polarization of the cilia upon the hair cells of the cristae. Thus the groups of cilia located upon the apical surface of each hair cell exhibit a particular spatial arrangement in that the kinocilium always is located at the periphery of the bundle of cilia (Figs. 7.110 and 7.114). In the cristae of the horizontal canals, the kinocilium always is found on that side of the ciliary bundle which faces the utricle. In the cristae of the vertical canals, on the other hand, the kinocilium is located on the side of the ciliary bundle away from the utricle. Based upon this morphologic arrangement, it has been postulated that flexion of the hairs toward the kinocilium induces depolarization of the receptor cell membrane, and thus accelerates the rate of discharge in the afferent nerve fibers which innervate the hair cells. Conversely, displacement of the hairs in the opposite direction

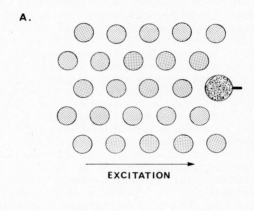

EXCITATION

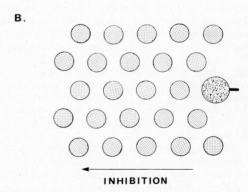

INHIBITION

FIG. 7.114. The ciliary bundles of a vestibular hair cell cut in a plane lying parallel with the apical surface of the cell. The large groups of light-shaded circles represents the pattern of stereocilia. The larger single coarse-shaded circle to the right having a small projection represents the kinocilium and its basal foot. Note the morphologic asymmetry of the ciliary pattern. **A.** When the deflection of the cilia is toward the kinocilium (ie, to the right as indicated by the arrow) excitation of the nerve terminals connected to the hair cell results. **B.** When the deflection of the cilia is away from the kinocilium (ie, to the left as indicated by the arrow) inhibition of the nerve terminals connected to the hair cell results. Thus a functional as well as morphologic polarization is present in the vestibular hair cells.

results in a hyperpolarization of the membrane, and a reduced frequency of impulse discharge in the afferent fibers results.

Any semicircular canal yields a maximal response when rotated in the same plane which that canal occupies (this principle sometimes is called *Flourens' law*). Therefore no excitation is evoked in a horizontal canal when it is rotated in a plane located at right angles to the canal. However, the vertical canals can give some response when they are rotated in planes other than that which elicits a maximal excitation of these receptors; ie, in the diagonal plane. Either a rapid inclination of the head laterally toward a shoulder (ie, rotation about an anteroposterior axis) or a similar inclination of the head backward or forward (ie, rotation about a transverse axis) stimulates the vertical semicircular canals.

As noted earlier the vertical semicircular canals are located diagonally within the head and the anterior canal on one side occupies the same plane as the posterior canal of the opposite side (Fig. 7.107). Therefore a diagonal rotatory movement in the forward direction will excite an anterior canal to a maximal extent, whereas the activity within the opposite posterior canal will be inhibited. A similar rotation in the backward direction will act in a diametrically opposite fashion; that is, a posterior canal will be excited maximally while the contralateral anterior canal will be inhibited.

If a rotational stimulus persists for more than a few seconds, the inertia of the endolymph is surmounted, and the cupula returns to its resting position. Hence no deflection of the hair cells is present and the rate of impulse generation in the afferent fibers from the canal decreases sharply. When the movement ceases, however, the momentum of the endolymph produces a sudden deflection of the cupula in an opposite direction to that which took place at the onset of the movement. These facts explain why no sensation of movement is felt during a prolonged rotation at a constant velocity, and also why a sensation of turning about some axis in an opposite direction is perceived as deceleration occurs or when the rotation stops suddenly.

The foregoing discussion clearly indicates that the semicircular canals are well adapted to indicate rapid head movements (or rotations) in one plane or another. However, these receptors are suited very poorly to signal continuous rotation at a constant velocity, a fact which is consistent with the rare occurrence of such movements during a normal lifetime. Thus continuous rotatory movements tend to bewilder the individual rather than to enable him to adapt his posture successfully. Furthermore, when such rotatory movements terminate the person undergoes the illusory perception that rotation in the opposite direction is taking place, an effect which is caused by the pendulumlike swing of the cupula in the direction opposite to that which took place at the start of the rotation. A trained figure skater or ballet dancer can minimize

such effects of rapid rotation at a constant velocity by performing an alternate series of accelerations and decelerations of the head during each rotation by fixing the eyes on one point for as long as possible and then rapidly turning the head 180° in the direction of the rotation.

In contrast to the semicircular canals, the *otolithic organs* or *gravity receptors* (ie, the utricle and saccule) are sensitive to both static as well as to dynamic stimuli. The physiologic role of the utricle as the sensory receptor underlying the static reflexes has been discussed earlier (p. 379). In addition to the force of gravity, the otolithic organs also are excited by linear acceleration so that forward, backward, up or down, and lateral movements all are effective stimuli to these structures. In addition, slow inclination of the head on its anteroposterior or transverse axes also stimulates the gravity receptors, and a rapid deceleration from a linear movement as well as a sudden alteration in the velocity of such a movement also serve as effective stimuli to the gravity receptors. The utricular receptors also are excited by centrifugal force as well as by angular (rotatory) acceleration.

The adequate stimulus for excitation of the hair cells of the *utricle* appears to be a mechanical deformation of the cilia caused by the pull of the otoliths. As indicated in Figure 7.110, the hair cells of the gravity receptors and their cilia exhibit a regular pattern of morphologic polarization; however, the physiologic importance of this spatial arrangement is not understood. Presumably, however, the individual signals which arise within each utricle are constantly undergoing a process of evaluation within the medullary vestibular nuclei so that the attitude of the head is adapted continuously in order to reduce the differences between the individual signal inputs to a minimum.

The utricular receptors discharge spontaneously and at a very low frequency when the head is maintained in a level position. However, when the head is tilted laterally the firing rate increases tremendously in the afferent utricular nerve fibers on the side toward which the head is inclined. Concomitantly the resting discharge of impulses on the contralateral side of the head stops completely. The signal frequency that develops when the head is tilted laterally from the vertical position is proportional to the angle at which the head is inclined. When the head is maintained in the inclined position, the signal frequency decreases gradually because the gravity receptors undergo a slow adaptation to a constant stimulus (p. 186). When the head is tilted about a transverse axis (so that the nose is either elevated or depressed) the utricular receptors also increase their firing rates; however, quantitatively the response so evoked is far less than that induced by a lateral inclination of the head.

Unlike the utricular responses enumerated above, the role of the *saccule* is not wholly clear. Despite the fact that the saccule resembles other vestibular structures morphologically, and thus would appear to be an integral component of the vestibular mechanism, some investigators believe the saccule to be a part of the auditory mechanism, at least in certain lower vertebrates (eg, frogs).

Principal Physiologic Responses to Stimulation of the Semicircular Canals. Excitation of the semicircular canals can be achieved by any of four types of applied stimuli. First, rotation of the head around a vertical, transverse, or anteroposterior axis (ie, angular acceleration): As this is the most important type of stimulus to the semicircular canals from a physiologic standpoint, the effects of rotatory stimuli will be considered in greatest detail in this section. Second, caloric (or thermal) stimulation of the semicircular canals is accomplished by douching (or irrigating) the external ear canal with hot or cold water. Third, electric (or galvanic) stimulation of the semicircular canals is carried out by passing a direct current through the labyrinth. Fourth, mechanical stimuli that cause displacement of the cupula also excite the semicircular canals effectively. The latter two kinds of stimulus are primarily of experimental rather than of clinical interest; hence they will not be considered further.

Rotation: When a human or animal is subjected to rotation, the attendant stimulation of the semicircular canals causes nystagmus, vertigo, certain alterations in the tone and coordination of the neck and limb muscles, and a number of autonomic reactions.

Nystagmus: If the head of a human or animal subject is rotated forcibly and suddenly about a vertical axis the eyes tend to move or drift in the opposite direction to that in which the head is turned and then to return to the original forward position with a rapid twitchlike (or jerky) movement. And if the rotation is continued in the same plane, then a rhythmic, jerky, back-and-forth movement of the eyes occurs, an effect that is called *nystagmus*.

Nystagmus occurs both while the rotation is taking place *(per-rotational nystagmus)* as well as after the rotation has ended *(post-rotational nystagmus)*.

When the eyes are kept open during rotation, the resulting nystagmus has two components. First, the movement of visual images across the retina produces an *optokinetic nystagmus* which persists for only as long as the rotation continues. This component of the nystagmus can be eliminated merely by closing the eyes. Second, the stimulation of the semicircular canals that accompanies rotation produces *vestibular nystagmus.* When the optokinetic component of rotational nystagmus is removed by closing or covering the eyes so that the apparent motion of the surroundings is no longer visible to the subject, then only the vestibular component is present, and this can be recorded by suitably placed electrodes. Vestibular per-rotational nystagmus only persists for about 20 sec; in other words, this component of nystagmus

only is manifest while the cupula is displaced as the consequence of the angular acceleration (p. 383).

The post-rotational vestibular nystagmus which develops in man can be seen readily by rotating a subject in a revolving (Bárány) chair with the eyes closed at a rate of about 30 revolutions per minute (eg, 10 complete revolutions in 20 seconds). The head is tilted forward at an angle of about 30° from the vertical position in order to stimulate the horizontal canals to a maximal extent, as this position places these structures almost precisely in the plane of the rotation (p. 382). The rotation is stopped abruptly, and following this sudden angular deceleration the subject opens his eyes, at which point the vestibular nystagmus is evident as a rapid horizontal back-and-forth movement of the eyes. The nystagmus is composed of a relatively slow drift of the eyes toward one side followed by a rapid, jerky return of the eyes to the normal (ie, forward) position. The post-rotatory vestibular nystagmus persists for approximately 20 seconds in a normal individual, and the rapid phase of the eye movements is in a direction that is opposite to that in which the rotation took place. Conventionally, therefore, *the direction of the rapid eye movements is considered to be the direction of the nystagmus,* regardless of the stimulus which caused the nystagmus in the first place. Therefore a left per-rotational horizontal nystagmus is induced by rotation to the left, while at the termination of such rotation, the nystagmus reverses its pattern and thus becomes a right horizontal post-rotational nystagmus.

The optokinetic and vestibular components of nystagmus are in the same direction (ie, they are synergistic) when the eyes are held open during the rotation. However, as deceleration takes place under such conditions, these two components take place in opposite directions (ie, they are antagonistic) because the optokinetic component does not undergo reversal during an angular deceleration.

Nystagmus can assume two other forms in addition to the horizontal type described above. These are *vertical nystagmus,* in which the eyes oscillate in an up-and-down direction, and *rotatory nystagmus,* in which the eyes swivel in a circular pattern about an anterioposterior axis. Both of these kinds of nystagmus are evoked by the concomitant excitation of both vertical semicircular canals (ie, the anterior and posterior canals). Vertical nystagmus is elicited when the head is tilted at an angle of approximately 90° from the vertical and placed upon one shoulder or the other. Thus the sagittal plane of the head is in the direction of the plane of the rotation. Rotatory nystagmus, on the other hand, is produced when the frontal plane of the head is inclined forward at an angle equal to, or greater than, 90°. Rotatory nystagmus also is induced when the head is tilted backward at an angle between 30° and 60°. Both vertical and rotational nystagmus are evident when the subject brings his head upright following a period of rotation.

The slow component of vestibular nystagmus originates within the semicircular canals, whence afferent impulses are transmitted to the vestibular nuclei. From these centers efferent impulses are transmitted via the medial longitudinal fasciculus to the motor nuclei that control the ocular muscles. The rapid component of vestibular nystagmus depends entirely upon central neural mechanisms that lie within the brain stem between (and including) the oculomotor and vestibular nuclei. The specific pathways that mediate the rapid component of nystagmus are unclear, however.

Optokinetic nystagmus originates within the retinas of the eyes, and the afferent signals that underlie this type of eye movement are transmitted via the optic nerve and tract as well as the superior colliculus to the medial longitudinal fasciculus and oculomotor nuclei whence efferent impulses to the eye muscles originate.

Vertigo: The term "vertigo" denotes an illusion of movement, a sensation that the external environment is revolving or whirling about the subject *(objective vertigo),* or that the subject himself is revolving in space *(subjective vertigo).* The term "dizziness" is *not* synonymous with vertigo. Dizziness is a more general term than vertigo. Dizziness signifies a disturbed relationship to space or a disequilibrium; a sensation of unsteadiness with a feeling of movement within the head (or giddiness).

In either vertigo or dizziness, the equilibrium may be so deranged that the individual staggers and may even fall. For example, true vertigo can be induced in a human subject by rotation of the body in a Bárány chair with the head inclined forward at a 30° angle. During deceleration, the subject will experience the illusion (ie, false sense) that he is spinning in the opposite direction to the rotation that was applied in the horizontal plane. Thus a sensation of counterrotation is experienced.

Vertigo as well as dizziness are caused by many other factors than rotation of the body and are common symptoms of many neurologic as well as somatic and visceral diseases (eg, vertigo often accompanies cardiovascular, renal, and gastrointestinal diseases as well as many toxic states). Furthermore, vertigo or dizziness frequently accompany alcoholic intoxication, motion sickness (seasickness, carsickness, airsickness), and so forth.

It was noted earlier that the labyrinths can exert a considerable effect upon the ocular muscles. Conversely, vertigo or dizziness can result from abnormal activity of the eye muscles themselves upon labyrinthine functions. Consequently vertigo and such attendant symptoms as headache and even nausea are common effects of eye strain. Similar effects also can be produced by observing the landscape from a moving vehicle, and there is a strong visual component in the etiology of seasickness. For example, merely watching the gyra-

tions of a ship's mast often is sufficient to induce an attack of mal de mer in many individuals.

Regardless of the clinical or other condition to which vertigo may be related, in all cases this symptom is caused directly by stimulation of the semicircular canals and/or the central neural connections of these receptors.

Alterations in Muscular Tone and Coordination: When the horizontal semicircular canals are stimulated in experimental animals by rotation about a vertical plane, the angular motion induces tonus changes as well as bilateral movements. Both of these responses to such a rotation tend to compensate for this movement and permit the animal to maintain its balance. Thus if a frog is rotated to the right, the tonus of the neck musculature is enhanced upon the left side and diminished upon the right side of the body. Concomitantly, the head turns toward the left. The limbs, in particular the right hindlimb, undergo extension, whereas those on the left side of the body undergo flexion. When the rotatory motion is terminated the animal generally moves toward the left.

Similar alterations in tonus and compensatory movements can be observed following stimulation of the vertical semicircular canals induced by a rapid inclination of the animal. For example, excitation of the left anterior vertical semicircular canal by a sudden inclination of the body toward the left and in a forward diagonal direction induces contraction of the left forelimb. On the other hand, stimulation of the right anterior vertical semicircular canal caused by similarly inclining the body forward and to the right elicits a like movement of the right forelimb. Tilting the animal backward either to the left or to the right stimulates muscular contraction in the left hindleg or right hindleg, respectively.

Normal human subjects experience no difficulty in putting their finger upon a particular spot with their eyes closed once they have placed the same finger on the identical spot with the eyes open. Following rotation, the subject will be able to perform the same maneuver as long as the eyes are open; however, when the eyes are closed, the subject cannot place his finger upon the same spot. Hence the finger *past-points* or deviates from the mark on one side or other of (or above or below) the target spot following rotation and when the eyes are closed. The actual direction in which the error of finger placement takes place is related to the direction of the rotation and to the spatial position of the head during the rotation; that is, upon the particular semicircular canal(s) that was stimulated. The phenomenon of past-pointing is a voluntary motor response, and it is *not* a reflex. The error in placement of the finger upon the target spot following rotation is the result of an unconscious motor compensation being made in the opposite direction to the rotation in response to the inaccurate sensory information that is elicited by the rotation. Past-pointing and the vertigo which accompanies this effect of rotation occur in opposite directions. The quantitative error in past-pointing following a period of rotation is referred to as the past-pointing test of Bárány.

Several other post-rotational muscular responses can be observed in human subjects. For example, when the body is subjected to a rotation so that the horizontal canals are excited, after cessation of the motion the head turns in the direction of the rotation when the eyes are kept closed. Alternatively, when the rotation is performed in such a manner that one or another of the vertical semicircular canals is excited, and the head is brought to the normal (upright) position following the rotation, the body leans, or actually falls, to one side or the other, or backward or forward, depending upon the position of the head during the rotation (ie, depending upon which canals were stimulated).* Essentially, the falling response to excitation of the vertical canals represents a past-pointing reaction of the whole body. Consequently the direction in which the body falls (or leans) is in the opposite direction to the vertigo, and the subject labors under the illusion that he is leaning to one side. Thus he compensates for this wholly inaccurate and inappropriate sensory information by leaning in the opposite direction in a compensatory fashion, and if walking is attempted the subject falls down.

Autonomic Responses: Stimulation of the semicircular canals in human subjects frequently is accompanied by a number of autonomic symptoms, including pallor, sweating, nausea, and vomiting. A slight decrease in heart rate (or bradycardia) may be observed together with a concomitant diminution of blood pressure.

During rotation, the pupils of the eyes undergo constriction whereas pupillary dilatation can be observed when the rotatory motion is terminated.

In general, drugs that are employed to combat motion sickness (eg, Dramamine, Bonamine) exert their effects upon the autonomic nervous system, and not directly upon the semicircular canals or vestibular mechanisms. On the other hand, optokinetic and vestibular nystagmus both can be inhibited by the administration of barbiturates in small doses.

Caloric stimulation: A second technique whereby the semicircular canals may be excited under physiologic conditions is by syringing one ear at a time with either warm or cold water (thermal stimulation.)† The physiologic responses to this type of stimulus are quite similar to those observed following rotation in a Bárány chair; however, each labyrinth can be excited individually, a fact that is of distinct advantage to the examining physician. The head is tilted backward at an angle of approximately 60° and thus the horizontal canals are in a vertical position, the ampullae being upward. Lavage of the external ear canal with the test fluid induces a significantly greater alteration in the temperature of the endolymph in the portion of the canal which is located nearest to the external auditory meatus than in the portion of the canal which lies more deeply within the head. The convection currents that are induced in the endolymphatic fluid by the douche produce a deflection of the cupula, and horizontal nystagmus as well as vertigo develop. The alteration in temperature of the endolymph within the canal lags behind the lavage of the external meatus by several seconds, and the actual direction taken by the convection currents in the endolymph is determined by the relative temperature of the irrigating fluid. The convection currents themselves are produced merely by alterations in the specific gravity of the endolymph that are induced by heating or cooling.‡ Therefore irrigation of the ear canal with a cold fluid produces convection currents which flow away from the ampulla (ampullofugal flow), whereas lavage with a hot douche fluid produces the opposite effect (ampullopetal flow). Thermal excitation of the vertical semicircular canals can be accomplished by irrigation of the ear canal with the head in the upright position, and a rotatory nystagmus develops following such a maneuver.

In normal individuals, the nystagmus that is elicited following caloric stimulation of the semicircular canals persists for about one to two minutes. On the other hand, in patients who have been treated for prolonged intervals with the antibiotic streptomycin (eg, for tuberculosis), the degeneration of the neuroepithelial elements that is induced by the ototoxic effects of this drug on the vestibular apparatus may be such as to eliminate completely, or to attenuate severely, this normal response to thermal stimulation of the semicircular canals.

A note of caution is pertinent. Following excitation of the vertical canals, the subject is wholly unable to control his fall when the head is elevated to an upright position. And since the fall may be rapid and quite violent, the observer must be ready to catch the subject in order to prevent injury.

†*Under clinical conditions, ice water frequently is employed as the thermal stimulus.*
‡*The interesting postulate has been advanced that the nystagmus which develops in an otherwise normal subject following the ingestion of a copious volume of an alcoholic beverage may be due to alterations in the specific gravity of the endolymph that are caused by the alcohol per se. In turn, a localized decrease in endolymphatic specific gravity that is produced by alcohol could induce convection currents within the endolymph of the semicircular canals in a manner similar to those caused by thermal stimuli as described in the text, and thereby nystagmus would result.*

HIGHER BRAIN CENTERS AND THE INITIATION, REGULATION, AND COORDINATION OF MOVEMENT. Up to this juncture, only the automatic (reflex) mechanisms which underlie posture and certain basic movements have been discussed, together with certain of the central mechanisms (in particular the vestibular components) that control these reactions. However, in addition to these stereotyped patterns of motor response, all higher organisms, especially man, are capable of initiating and controlling at will numerous motor functions that vary considerably in their complexity. Thus voluntary muscular activities of various types can be learned and thereby superimposed upon the fundamental and innate patterns of reflex motor functions. The present discussion will be concerned with an analysis of the principal neural elements and physiologic mechanisms that are involved in the initiation, control, and coordination (or integration) of movements that involve the skeletal muscles so that a synthesis of the entire process of voluntary muscular activity in vivo can be attempted.

In the past, a large proportion of the experimental efforts designed to obtain data which would lead toward a better understanding of the higher functions of the brain were directed toward an analysis of sensory inputs and their processing in neuronal pools from the peripheral receptor up to the cortical centers. Recently, however, it has been suggested that an even deeper understanding of the enormously complex functions of the human brain can be achieved through analysis of the motor outputs of the central nervous system rather than through an exclusive preoccupation with study of the sensory inputs. When viewed in this context, the effector or motor activities of the brain assume a primary rather than a secondary functional role in the successful and continuous adaptation of the organism. Consequently the so-called higher intellectual functions of the human brain including rational thought, creative ability (imagination), abstract perception, and other mental activities can be viewed as means to an evolutionary goal rather than as the goal per se. Such mental capabilities thus enable one to refine his innate motor activities to an extremely high degree of sophistication, and the greatly enhanced motor skills that result therefore are of direct benefit to the individual in performing an ongoing biologic adaptation to the demands of an existence of ever-increasing complexity.

The study of voluntary motor functions in vivo is fraught with considerable difficulty, and only within the past few years have suitable techniques been developed whereby such studies can be carried out in unanesthetized animals. In particular, only recently have suitable techniques been developed whereby the electric activity of single neurons located within specific areas of the brain can be recorded simultaneously with the contractions of particular muscles in unanes-thetized subjects (principally monkeys). This technical advance, coupled with suitable conditioning (or training) schedules prior to the experiment, has proven to be of inestimable value to the gradual elucidation of central mechanisms that underlie the initiation, regulation, and coordination of voluntary somatic movements. Obviously the subject must be conscious and thus able to participate actively in the experiment if the volitional aspect of muscular control and coordination is to be studied effectively. This fact is in marked contrast to the investigation of sensory and certain reflex phenomena which are amenable to electrophysiologic study in anesthetized subjects. For example, one may investigate readily the physiology of the visual receptors by the use of light flashes as adequate stimuli and microelectrodes implanted within various regions of the nervous system as recording tools, anesthetized animals being quite suitable for such studies. On the other hand, the study of voluntary eye movements in response to particular environmental stimuli cannot be conducted under similar experimental conditions because the subject must be able to perceive and give conscious attention to suitable controlled visual stimuli. Furthermore the subject also must be able to execute voluntary and coordinated eye movements in response to such stimuli.

The two major technical advances that were necessary in order to fulfill the requirements for study of voluntary motor activities in unanesthetized primates can be outlined briefly. First, suitable conditioning (p. 442) methods were developed whereby the animals could be trained to execute specific movements on cue. These learned movements were such that they could be observed readily, modified systematically, and recorded accurately. Second, and of critical importance to the investigation of voluntary movements, was the design and construction of a miniaturized apparatus containing a microelectrode which can be clamped permanently to the skull of an experimental animal without undue discomfort. This apparatus is constructed in such a manner that the microelectrode can be implanted within specific regions of the brain, and also, once implanted, the electrode cannot be dislodged readily by normal head movements. Both of these requirements having been met, it now is possible to investigate the bioelectric activity of specific regions of the brain in vivo when the experimental animal executes learned movements and to relate temporally these movements to the activity of single neurons that are located within particular areas of the brain. The cerebral cortex, cerebellum, and basal ganglia are of particular importance in the regulation of voluntary muscular activities, so that data obtained by the use of the techniques mentioned above (in addition to information derived from classic stimulation, ablation, pathologic, and pharmacologic studies) will be presented in the following sections.

Review of Neural and Other Elements that Underlie Somatic Motor Functions. It should be emphasized and reiterated here that the contraction of every skeletal muscle fiber in vivo requires that efferent signals be transmitted to the neuromuscular junctions (motor end plates) of the muscle fibers (p. 208), and that muscular contractions are either abnormal or totally absent if there is a failure of adequate impulse generation and/or transmission at any point in the entire neural pathway leading to the terminations of the motoneurons upon the muscle fibers (p. 227). Furthermore, central inhibitory processes which decrease the signal traffic to particular skeletal muscles are responsible for reducing, or "turning off," contractile activity in vivo; and thus the balance between excitatory and inhibitory inputs to the centrally located motoneurons whose processes constitute the final common pathway determines the net contractile state of the individual skeletal muscles at any particular moment in time (p. 269).

The principal neural and related elements that underlie and regulate all of the somatic motor activities of the body are depicted in Figure 7.100. The general physiologic characteristics of the cerebral cortex, the pyramidal and extrapyramidal systems, the cerebellum, and the basal ganglia now may be reviewed.

It has been known for over a century that artificial stimulation of certain localized areas of the cerebral cortex would evoke discrete, albeit twitchlike, contractions of particular muscles located upon the opposite side of the body to the site of application of the stimulus (Figs. 7.73 and 7.75). Hence the primary motor cortex, located within the precentral gyrus of each cerebral hemisphere, long has been considered to be the highest level of the brain responsible for the control and integration of voluntary muscular activity. In the light of recent investigations, however, the premotor area of the cerebral cortex does not appear to occupy such an exalted position, and now it would appear that the motor cortex actually occupies a physiologically "low" position insofar as its direct relationship to the excitation of the skeletal muscles in vivo is concerned. Thus the precentral gyrus of the cortex does not appear to be involved in the initiation of voluntary muscular activities; rather it appears that the primary functional role of this cortical area is related to the highly precise (or accurate) control of voluntary muscular contractions.

The nerve fibers of the pyramidal system, which includes the lateral and ventral corticospinal tracts (Table 7.5; Figs. 7.20 and 7.22), originates from the precentral gyrus of the cortex as well as in related cortical areas. Thence the fibers descend via the internal capsule to the medullary pyramids (p. 269; Figs. 7.28 and 7.60). From the pyramids about 10 percent of the pyramidal fibers terminate directly upon spinal motoneurons in humans. The remaining 90 percent of the pyramidal fibers terminate

upon interneurons located within the cord, and through these intermediate structures synaptic connections are made upon spinal motoneurons.

The extrapyramidal motor system is composed of the nonpyramidal efferent fiber tracts which are summarized in Table 7.5, B and C, and depicted in Figures 7.22 and 7.101. It is important to realize that these nonpyramidal fiber tracts project from cortical levels to the motoneurons whose cell bodies lie within the brain stem by way of a series of neurons. Thus the extrapyramidal efferent pathways which originate in the cortex send fibers to the basal ganglia (Figs. 7.100 and 7.101). Thence one or more neurons pass to the tegmental and/or reticular nuclei that are situated in the brain stem.

The physiologic role of the extrapyramidal system in relation to the pyramidal motor system in humans is not at all clear. In particular, the specific role(s) of the individual basal ganglia in the integration of coordinated muscular activities is obscure. However, recent experimental work, taken in conjunction with pathologic studies performed at autopsy upon victims of certain neurologic diseases, tends to support the general conclusion that the basal ganglia act selectively in vivo to coordinate slow (or ramp) movements.

In this regard, it is pertinent that a pyramidal system is found *only* in mammals, and that in birds and lower vertebrates the basal ganglia provide the most highly developed neural centers that are present for the coordination of motor functions. The cerebral cortex in such organisms is, of course, quite rudimentary. Thus the movements performed by such phylogenetically less advanced animals as birds, amphibians, reptiles, and fishes tend to be largely automatic or stereotyped, relatively gross (crude), and repetitive in character. Furthermore, highly precise (accurate) movements are lacking in such species, and this is exactly the situation which is found in the movement patterns of human infants before the adult structural and functional relationships have become established in the pyramidal tracts; ie, the descending cortical fibers that are associated with these pathways. Interestingly, if the adult human pyramidal system is destroyed or damaged severely by disease or injury, the ability to perform a considerable number of voluntary motor functions is retained or becomes restored gradually. Presumably these compensatory effects are brought about through the operation of the intact extrapyramidal pathways.

In addition to the cerebral cortex and basal ganglia, the third major area of the brain which is responsible for the integration and coordination of somatic motor functions is the cerebellum. Because of its relatively "simple" structural and functional organization at the cellular level (Fig. 7.58), the physiology of the cerebellar cortex is understood far better than is any other region of the brain. In general, the cerebellum functions as a major regulatory center for the integration of complex patterns of voluntary movement. The cerebellum also appears to be concerned largely with the coordination of rapid (ballistic) movements, but it is *not* involved with the initiation of those movements. The latter is a complex process which may involve a few (or many) cortical areas whose outputs apparently converge upon the precentral gyrus (ie, the primary motor cortex that is located within the precentral gyrus) as well as the secondary cortical motor areas.

General Physiology of the Cerebellar Cortex: Cytoarchitectural and Functional Correlates. The cerebellar cortex is one region of the mammalian brain in which a fairly detailed correlation can be made between cytoarchitectural features and functional properties. Therefore a summary of the structural and general physiologic characteristics of the cerebellar cortex will serve as a model to illustrate the operation of a relatively simple pattern of specific neural elements within the brain, one which is related intimately to the regulation of voluntary muscular contractions in vivo.

The gross anatomy of the cerebellum was presented earlier (p. 307 et seq.) and illustrated in Figures 7.51 through 7.57. Like the cerebral cortex, the outermost layer of the cerebellar cortex consists of a layer of gray matter; that is, a highly organized layer of nerve cell bodies. The microscopic anatomy of the cerebellar cortex, which is of particular relevance to the following discussion, is illustrated in Figure 7.58 (p. 313). Furthermore, the basic neural circuit to be described below is found in *all* vertebrates, regardless of their degree of phylogenetic advancement.

The cerebellar cortex contains only seven morphologically distinct types of nerve element which are arranged in a highly repetitive pattern throughout this structure. The *climbing fibers* and the *mossy fibers* conduct impulses to the cerebellar cortex (ie, they form afferent systems), whereas the axons of the *Purkinje cells* provide the sole means whereby signal outputs are transmitted away from the cerebellar cortex (ie, they provide the efferent system). The four other neural elements of the cortex are the *Golgi cells*, the *basket cells*, the *granule cells*, and the *stellate cells*. All four of these interneurons are found solely within the cerebellar cortex; hence they serve to interconnect the three other cell types.

The Purkinje cells are the most complex type of neuron found within the nervous system. Each of these elements consists of a large, highly branched system of dendrites that transmit impulses to a bulb-shaped soma that in turn emits a long thin axonal process. The dendrites of an average Purkinje cell may form up to 100,000 individual synapses with afferent nerve terminals. This number of synapses far exceeds those present upon any other type of neuron in the central nervous system.

The Purkinje cells are the key neurons of the cerebellar cortex. These cells are located throughout the cortical region within which their cell bodies form a continuous layer at the junction between the molecular layer and the granule cell layer (Fig. 7.58A). The numerous dendrites which extend outward toward the free edge of the cortex from the cell bodies of the Purkinje cells comprise a region called the *molecular layer*. The axons of the Purkinje cells that penetrate deeper into the cortex form part of the *granule cell layer*. Actually, the axons of the Purkinje cells course entirely through the cortex to make synaptic junctions with neurons whose cell bodies are located within the cerebellar nuclei themselves (Figs. 7.57 and 7.115). The cerebellar nuclei also are innervated by collaterals of the climbing and mossy fiber systems; therefore all the signals passing to the cerebellar cortex also are transmitted simultaneously to the cerebellar nuclei (p. 309). Within these nuclei, the afferent signals are integrated with those efferent signals which have originated within the cortex. The resultant efferent signals then are transmitted to other areas of the brain as well as to the spinal cord, and thence to the skeletal musculature of the body.

The elaborate dendritic patterns formed by the Purkinje cells have a particular functional significance. All of the many individual ramifications of the dendrites that arise from a single Purkinje cell lie in one plane, like the single page of a book, and furthermore the planes of all of the Purkinje cell dendrites found in a localized cerebellar region also are parallel with respect to each other. Therefore the patterns formed by the dendrites of adjacent cells are stacked in orderly ranks, like the pages of a closed book. In fact, the term *folium* (Latin "leaf"), which is applied to the single lamella that comprises the cerebellar cortex, merely reflects this cytoarchitectural pattern.

Although the processes of the Purkinje cells constitute the sole ouput (or efferent) system of the cerebellar cortex, it must be realized that this region of the brain does not function merely as a relay system for neural information that arises elsewhere in the body, but rather it forms a vital integrative center for many somatic motor activities. In turn, cerebellar functions are wholly dependent upon the morphologic interconnections among the seven neural elements that constitute the cortex as well as with other regions of the nervous system. Consequently the functional morphology of the individual neural elements that comprise the cerebellar cortex must be reviewed briefly, as this material will provide a background for a more detailed consideration of cerebellar function at the cellular level.

As noted above, the climbing fibers and the mossy fibers constitute the two principal afferent systems that underlie signal transmission to the Purkinje cells. However, these two fiber systems differ considerably in both their morphologic and

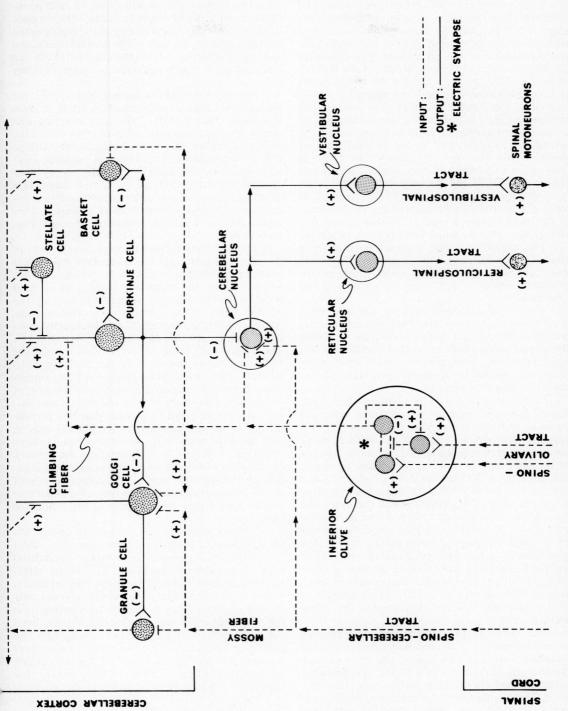

FIG. 7.115. The functional interrelationships among the cerebellar neurons and other regions of the central nervous system. (+) = excitatory synapse; (−) = inhibitory synapse. (After Llinás. The cortex of the cerebellum. *Sci Am* 232:56, 1975.)

399

functional relationships to the Purkinje cells. Nevertheless both of these systems are present throughout the entire cerebellar cortex.

Each *climbing fiber* exhibits a one-to-one morphologic relationship with an individual Purkinje cell. The climbing fibers themselves represent the axons of neurons whose cell bodies lie entirely outside of the cerebellum. For example, climbing fibers can originate in cells of the inferior olive (Fig. 7.57, p. 312), and the branching axons of these cells innervate both the cerebellar nuclei as well as the cerebellar cortex. The relationship between a climbing fiber and a single Purkinje cell is established during the early development of the cerebellum in utero, and as the network of dendrites of the Purkinje cell grows, the climbing fiber follows to form a pattern which corresponds closely to that of the Purkinje cell dendrites. In fact, the term "climbing fiber" was applied to these axons originally because of the resemblance of this neural growth process to that in which a vine grows upon the trunk and branches of a tree.

The actual synaptic contact between the climbing fiber axons and the dendrites of the Purkinje cells takes place only at points where groups of minute spines project from the surface of the dendrites.

Up to 300 individual synaptic contacts have been counted between one Purkinje cell and its climbing fiber, a fact which correlates well with the experimental observation that stimulation of a single climbing fiber elicits a prolonged, high-frequency discharge of action potentials in the corresponding Purkinje cell.

In marked contrast to the climbing fiber which makes multiple synaptic contacts with only one Purkinje cell, the *mossy fiber* makes only a few synaptic junctions with each of the many Purkinje cells with which it is associated. Again in contrast to the climbing fiber, the mossy fibers terminate indirectly upon the Purkinje cells by way of interneurons (the *granule cells*) which are found just beneath the layer of Purkinje cells in the cerebellar cortex. Thus the granule cells function to increase enormously the number of Purkinje cells that can be excited by one afferent mossy fiber. The granule cells are extremely abundant. In fact, it has been calculated that the total number of these elements within the cerebellar cortex of the human may be 10 times greater than the entire number of cells previously believed to comprise the entire brain!

The axons of the granule cells are directed toward the surface of the cerebellar cortex into the molecular layer. In this region, the axon of each granule cell bifurcates, each of the two branches being directed horizontally and in opposite directions. The orientation of the horizontal branches of the granule cells not only is parallel with each other, but also is perpendicular with respect to the plane occupied by the dendrites of the Purkinje cells (Fig. 7.58B, p. 313).

The parallel fibers of the granule cells make synaptic contact with the Purkinje cells by way of spines that are found upon the terminal ends of the dendrites of the Purkinje cells. Usually each parallel fiber contacts any single Purkinje cell only once or twice. However, the majority of the inputs to the Purkinje cells are by way of the parallel fibers, and as noted earlier, up to 100,000 parallel fibers can make synaptic junctions upon a single Purkinje cell.

Mossy fibers are excitatory and so are the granule cells that are stimulated by these afferent processes (Fig. 7.115). Thus both climbing and mossy fiber systems produce excitation within the cerebellar cortex. However, owing to the extreme differences between the number of synaptic junctions made by the climbing and the mossy fiber systems upon the Purkinje cells, stimulation of a climbing fiber produces an extremely localized excitation, whereas stimulation of a mossy fiber yields a broadly diffuse excitation of many Purkinje cells.

The interneurons that form an integral part of the neuronal pattern of the cerebellum have short axons. Both the *basket cells* and *stellate cells* are located within the molecular layer, and each of these elements receives signals by way of the parallel fibers of the granule cells. Hence these neurons exert their functional effects directly upon the Purkinje cells as depicted in Figure 7.115. The basket cells make synaptic contact upon the soma and lower dendrites of the Purkinje cells. whereas the stellate cells form synaptic connections principally upon the dendrites proper of these cells. The axons of both the stellate and basket cells are oriented in a plane that lies perpendicular to the parallel fibers of the granule cells as well as perpendicular to the axis of the Purkinje cells (Fig. 7.58, p. 313). Consequently the network of elements within the molecular layer of the cerebellum consists of three fundamental types of neural process, each of which rests in a plane that lies at right angles with respect to the other two processes.

Within the granule cell layer of the cortex, the third type of cerebellar interneuron, the *Golgi cells,* forms components of a highly specialized type of synaptic connection. The Golgi interneuron receives impulses from the parallel fibers; however, the dendrites of the Golgi cells not only receive excitatory signals that are transmitted from the parallel fibers of the granule cells, but also from the mossy fibers (Fig. 7.115). Consequently the cerebellar Golgi cells are components of a particular synaptic arrangement called the *cerebellar glomerulus* (Fig. 7.58B).* The cerebellar glomerulus, which is located within the granular

*The word "glomerulus" denotes a tuft or cluster, and in anatomic nomenclature this general term signifies such a structure; eg, a cluster of nerve fibers or blood vessels (see also p. 768).

layer of the cortex, thus consists of a complex of three different synaptic inputs. These inputs are derived from mossy fibers that are enveloped by dendrites of granule fibers, and these microstructures in turn are surrounded by the axons of Golgi cells.

The three kinds of cerebellar interneuron described above have been shown experimentally to have inhibitory properties. Thus the basket, stellate, and Golgi cells all exhibit this functional property as determined by electrophysiologic studies, although the biochemical transmitters elaborated by the particular terminals have yet to be elucidated. In this regard, note that impulse transmission across the synaptic junctions of the spino-olivary tract within the inferior olive are electric rather than mediated by a neurochemical substance (Fig. 7.115). That is, presynaptic depolarization directly produces a corresponding excitation (depolarization, EPSP) or an inhibition (hyperpolarization, IPSP) of the postsynaptic membrane.

From a physiologic standpoint, excitation of the cerebellar cortex can, in principle, be evoked by stimulation of a single climbing fiber which elicits a strong response from a single Purkinje cell. However, the response that is produced by stimulation of a mossy fiber is far less precise, owing to the more diffuse synaptic connections of these axons.

The entire sequence of events that takes place during excitation of the neurons of the cerebellar cortex has been postulated as follows, insofar as a single neuronal circuit is concerned (refer to Fig. 7.115). Afferent stimulation of one climbing fiber stimulates the production of a strong volley of impulses from one Purkinje cell. Afferent inputs from a very small group of mossy fibers, on the other hand, excites a large number of Purkinje cells, as well as all three kinds of inhibitory interneuron, but this effect takes place indirectly through excitation of the processes of the three types of interneuron, ie, the stellate, basket, and Golgi cells. The sign (ie, excitation or inhibition), magnitude, and duration of the response to the particular input by these neurons are restricted by the particular neurotransmitters involved, as well as their quantal release (p. 198). The excitation of the Purkinje cells that is caused by the parallel fibers of the granule cells becomes inhibited rapidly by the outputs of both the stellate and basket cells. However, the axons of the stellate and basket cells are oriented at right angles to the parallel fibers. Consequently the inhibitory effect is not restricted merely to the excited Purkinje cells. Rather, the Purkinje cells that are located adjacent to each side of the primary locus of the stimulated Purkinje cells also are inhibited markedly. The net physiologic effect of this process is that the excitatory vs. inhibitory response is focused sharply between the stimulated and inhibited groups of cells.

Experimental evidence has been obtained that the Purkinje cell per se exerts an inhibitory function.* Consequently, the entire signal output of the cerebellar cortex merely represents a highly selective and integrated inhibition of the neurons that in toto constitute the entire mass of the cerebellum. In addition, it has been demonstrated that of all the neuronal elements present solely within the cerebellar cortex, only the granule cells and their parallel fibers are excitatory (or facilitory). Consequently the stellate, basket, Purkinje, and Golgi cells all exert inhibitory functions as shown in Figure 7.115. Thus discharge of the centrally located nerve cells that give rise to the afferent climbing and mossy fibers also rapidly stimulates cerebellar nuclei, and thereby the activity of the cerebral and spinal nerve systems that are innervated by these neurons also is modified. However, this excitatory activity of the cerebral and spinal neurons is "turned off" quite rapidly by the efferent signals that originate within the cerebellar cortex itself; that is, within the negative feedback loops formed by the basket, stellate, and Golgi cells.

When the projections of specific afferent input fibers from particular regions of the body to the cerebellar cortex are determined, it is found that each input terminates at a specific locus upon the cortex. This morphologic arrangement is quite similar to that encountered in the cerebral cortex, and it indicates that a high degree of spatial orientation of the afferent nerve fibers (p. 307), hence afferent signals, also is present in the cerebellar cortex (Fig. 7.116). It also has been determined that the afferent neurons from particular regions of the body are oriented in bands which lie parallel to the midline, and that large areas of the cortex are given to representation of these regions.

As noted earlier, a major source of the climbing fibers has been demonstrated to lie within the inferior olive (Fig. 7.57, p. 312). In addition, it has been found experimentally that the afferent fibers which emanate from this medullary region are projected to the cerebellar cortex in bands that run in an anteroposterior direction. Furthermore, a pharmacologic excitation of the inferior olive causes muscular tremors. Consequently it would appear that the inferior olive, together with the afferent climbing nerve fibers that this structure sends to the cerebellar nuclei and Purkinje cells of the cortex, is a component of an efferent motor system that is involved in the muscular contractions which are associated with movement. This concept is substantiated still further by the finding that when "maps" of the proprioceptors from various regions of the body are placed upon the longitudinal bands which are related to the climbing fiber system of the cerebellar cortex the resulting patterns of climbing fiber systems tend to overlap

---

In this regard the Purkinje cell of the cerebellum is similar functionally to the Renshaw cell of the spinal cord (p. 215).

A.   B.

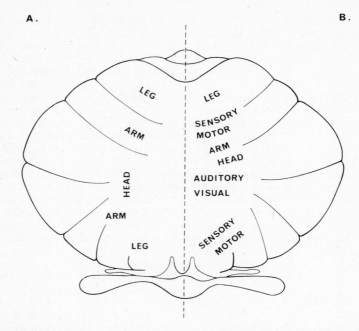

FIG. 7.116. General spatial localization of cerebellar functions in the monkey brain. **A.** Tactile areas. **B.** Cortico-cerebellar projections. Note the overlap of sensorimotor functions with the tactile functions **(A).** In addition, the flocculonodular lobe has been shown specifically to control eye position with respect to the orientation of body and head. Thus one can fix the eyes on one point while the body is moving. The vermis apparently regulates the rapid (saccadic) eye movements that are essential to visual tracking of moving objects. (After Truex and Carpenter. *Human Neuroanatomy*, 6th ed, 1969, Williams & Wilkins. Additional data from Llanás. The cortex of the cerebellum. *Sci Am* 232:56, 1975.)

those of the afferent projections for several regions of the body. Therefore it has been postulated that the climbing fiber system is directly responsible for the regulation of rapid (ballistic) and synchronous contractions of muscle groups. Presumably this regulatory effect includes the muscles of more than one extremity at a time. Inhibition of the cerebellar nuclei via the activity of the Purkinje cells must follow immediately after excitation of the cerebellar nuclei so that a very brief, albeit strong, efferent signal is generated by these structures.

The in vivo functions of the mossy fiber system are less clear than are those of the climbing fiber system, although it has been postulated that activation of the mossy fibers may be involved in the excitation of large cerebellar cortical areas rather than individual cells, as well as in the modulation and detection of specific patterns of incoming signals. In this regard, the Golgi cells also may have a role in ascertaining the types of information that are transmitted to the cerebellar cortex.

It has been demonstrated unequivocally, however, that the mossy fiber system is concerned intimately with the exceedingly precise and delicate movements which are involved in visual coordination. The flocculonodular region of the cerebellum has been demonstrated to regulate the orientation of the eyes with respect to the spatial position of the body. Thus the eyes can remain

fixed upon one point even though the body is moving. The cerebellar vermis, on the other hand, appears to be involved in the generation of the rapid (saccadic) eye movements that are essential to visual tracking. Such eye movements are, for example, a critical part of reading printed material.

Interestingly, it has been demonstrated that excitation of Purkinje cells by the mossy fiber system occurs some 25 milliseconds prior to the onset of an actual eye movement. Thus it may be inferred that the cerebellar control of voluntary movements which is mediated by the mossy fiber system can correct errors *before* any movement actually begins. Consequently the cerebellum appears to have an inhibitory function as noted earlier.

Certain general features of cerebellar function now may be summarized. The cerebellum clearly represents an important regulatory area of the brain which is essential to the organization of complex voluntary movements, but the cerebellum does not initiate such movements. In fact, movements can take place even when the cerebellum is extirpated. However, the cerebellum is essential for the normal integration and coordination of afferent signals from many different sources so that the resulting movements are smoothly regulated and efficient.

The cerebellum also has been shown capable of coordinating such complex movements as walking in experimental animals despite the fact that all sensory (proprioceptive and forebrain) inputs to this structure were eliminated. However, this capacity was lost irrevocably following ablation of the cerebellum. Further recent experimental work suggests that still another function of the cerebellum appears to be concerned with a constant evaluation of the internal conditions or state of the central nervous system per se. Thus the ventral spinocerebellar tract can transmit information that is unrelated to such commonly listed sensory modalities as touch-pressure and proprioception (Table 7.5, p. 259). On the contrary, this tract may convey information that is related solely to the functional state of the inhibitory interneurons that are located within the spinal cord.

In conclusion, the cerebellum is concerned not only with the coordination and integration of many stereotyped motor activities as might be anticipated from the extremely repetitive pattern of its cytoarchitecture, but it appears that this region of the brain also exhibits a certain flexibility insofar as its adapting certain of these motor functions to rapidly changing environmental or individual requirements.

Summary of the Neural Mechanisms That Underlie Voluntary Movements. The principal neural elements reponsible for the integration and regulation of somatic muscular contractions are depicted schematically in Figure 7.100. All of the individual neural systems and their components which are illustrated in this figure, and which underlie the coordinated activity of the skeletal muscles under physiologic conditions, have been discussed in extenso earlier in this book. Consequently the material to be presented in this section will be confined primarily to a general consideration of the physiologic roles of the cerebrum, the cerebellum, and the basal ganglia in the control of voluntary skeletal muscle contractions in vivo (see also motivation, p. 414). In addition, certain aspects of the functional activity of the muscle spindle will be presented in relation to the functions of this structure in regulating the tension exerted by the skeletal muscles in normal individuals.

It has been realized for slightly over a century that certain areas of the cerebral cortex regulate somatic movements. Thus Jackson postulated before 1870 that the human brain contained specific sensory as well as motor areas which were concerned with the integration of sensations and movements. His conclusion was based upon the uncoordinated, convulsive movements seen in patients suffering from epilepsy, as well as upon the absence of normal motor functions in other patients following a stroke. The postmortem location of blood clots or ruptured cerebral vessels in specific areas of the brain provided Jackson with concrete evidence which could be correlated with the premortem neurologic findings in such patients, and he concluded that the area of the cere-

bral cortex most particularly involved in the regulation of movement was located in the region of the brain whose blood flow is derived from the middle cerebral artery. Direct experimental evidence in support of this contention was provided in 1870, when it was found that electric stimulation of one cerebral hemisphere would produce muscular contractions on the contralateral side of the body in the dog (Fritsch and Hitzig). Shortly thereafter the area of the cerebral cortex that had been proposed as the motor area by Jackson was actually proven to have this function (see also Jacksonian epilepsy, p. 460).

The convulsive, jerky, or spastic contractions that occur in patients suffering from epilepsy, or that are evoked artificially by electric stimulation of the motor cortex, are, however, totally unlike the smoothly coordinated and purposeful movements of groups of antagonistic muscles that can be initiated voluntarily. And as noted earlier (p. 396), two fundamental advances in methodology had to be perfected before further significant experiments could be performed on the mechanisms involved in the control of voluntary muscular activity. Briefly, suitable conditioning techniques had to be developed for experimental animals (primates), and also a method had to be made available whereby the electric activity of individual cells located within specific regions of the brains of the conscious and trained subjects could be recorded. The data obtained by use of these major technologic advances, taken in conjunction with evidence obtained from the application of classic neurophysiologic, neurologic, and neuropathologic techniques, have yielded valuable new insights into the functions of three major areas of the brain in the control of voluntary movement.

The motor cortex of the cerebral hemispheres, the cerebellum, and the basal ganglia have long been known to be interconnected anatomically as well as functionally (see Fig. 7.100). Thus destruction of particular areas of the primary motor cortex (p. 329; Fig. 7.75) can produce a total paralysis of specific skeletal muscles. On the other hand, damage to the cerebellum or to the basal ganglia induces abnormal movements, but does not suppress movement completely in marked contrast to cortical injury. Thus cerebellar disease or injury can be manifest by the appearance of a severe muscular tremor that becomes most evident during voluntary activity and is at a minimum when the muscles are relaxed. On the other hand a patient suffering from Parkinson's disease, a neurologic condition that results from malfunction of (or injury to) the basal ganglia, exhibits, among other symptoms, hypertonicity of the skeletal muscles, tremor, and a retarded ability to initiate certain voluntary movements (see also discussion of this condition on p. 456). The motor defect that is present in a person suffering from Parkinson's disease is related principally to the functional manner in which particular skeletal muscles are used rather than to the anatomic groups of muscles that are involved in specific movements. For example a rapid, or ballistic, movement may be accomplished readily by a patient with Parkinson's disease, whereas the initiation of a slow movement in the identical muscles may be carried out only with great difficulty.

Experimental and clinical data such as those just presented clearly indicate that the functions of the cerebral cortex, cerebellum, and basal ganglia are closely related. The normal manner and sequence in which each of these neuronal pools is activated in vivo thus is of major significance in an understanding of voluntary motor functions in vivo. Experiments have been performed in monkeys whereby the electric discharges that originate within single neurons located in all three of these major regions of the brain have been recorded together with the contraction of the specific muscles involved in particular movements. Hence the temporal sequence of neuronal excitation prior to the performance of a voluntary movement is of considerable importance in determining which elements form components of the neuronal pools and circuits that underlie voluntary movements.

Using the techniques described briefly on page 396, it has been demonstrated unequivocally that the neurons of the primary motor cortex (ie, the postcentral gyrus) discharge impulses several milliseconds *before* the subject executes simple movements that involve contraction of the hand muscles. Similar experiments, in which the impulses originating within the primary somesthetic or sensory cortex (ie, the precentral gyrus) were recorded, indicated that these cells fired *after* the muscular contractions had been initiated, a fact which supports the general conclusion that these sensory elements are not components of the neuronal circuit that initiates the primary contraction of the muscles. Thus it appears that the cells of the primary somesthetic cortex may regulate movement through feedback processes, once movement has commenced (cf. p. 328).

The cerebellum also receives many afferent sensory inputs, in particular from proprioceptive end organs throughout the body as well as from the vestibular apparatus of the inner ear. By means of experiments such as those performed upon single cells of the cerebral cortex, it also has been shown that the cerebellar cells fire impulses *before* the onset of muscular contraction. This finding has proved of considerable interest, as the cerebellum generally has been considered to regulate voluntary movement by feedback processes that originate within the muscles themselves, processes which were thought to be initiated *after* the onset of contraction.

Lastly it has been demonstrated that neurons located within the basal ganglia also discharge impulses *before* the onset of muscular contractions.

Since not only the cortex but also the cerebellum and basal ganglia fire before the onset of voluntary muscular contractions, the physiologic interrelationship among these neuronal pools must undergo a reevaluation. From the anatomic standpoint, large tracts of nerve fibers pass directly to the cerebellum and the basal ganglia from the entire cerebral cortex (Fig. 7.100). Conversely, both the cerebellum and the basal ganglia send large fiber tracts back to the cortex by way of the thalamic nuclei. Consequently the cerebellum and basal ganglia can receive visual, vestibular, auditory, and somatic sensory information either directly from the periphery or indirectly by way of the cerebral cortex. These structures then integrate this afferent information, and then transmit the modified signal patterns back to the cortex. Consequently it appears that the cerebellum and basal ganglia are the "functionally higher" centers insofar as the integration of voluntary motor activity is concerned, and that the motor region of each cerebral cortex actually is related in a much more direct fashion both anatomically and functionally to the motoneurons of the spinal cord than is either the cerebellum or the basal ganglia.

Further studies have indicated that the patterns of efferent impulses from the cerebral cortex appear to regulate to some extent both the length (ie, displacement) and the tension (ie, force) developed in the muscles during their voluntary contractions. For example, under certain experimental conditions the signals developed in the neurons of the motor cortex have been correlated with the force and rate of the contractions rather than to the degree of shortening (ie, displacement) produced by the contraction. Thus if a person lifts different weights at the same rate and over a similar distance, the cortical signal patterns to the muscles involved will be considerably different when a heavier rather than a lighter weight is moved, a fact that is reflected by the activation of more (or less) individual motor units (p. 128) in the particular muscles involved.

It also has been determined experimentally that the motor cortex normally is involved in the regulation of body movements regardless of whether the movement is of a reflex nature (ie, innate) or whether it is a learned response. In both of these instances, the discharge of the neurons within the cerebral cortex which is related to control of the movement is *not* related to the conditions under which the movement is performed, but rather the pattern of impulse discharge is related to the actual work performed by the muscles.

On the other hand, in areas of the cerebral cortex that lie *outside* of the primary motor cortex, the discharge of impulses occasionally is related more specifically to the psychic background (ie, intentions or circumstances) against which the voluntary movement takes place than to the muscular activity itself. This statement holds true regardless of whether the movement is learned or reflex, voluntary or involuntary, rapid or slow. It has been demonstrated in the eye that rapid, jerky (saccadic) movements (such as occur during reading) and smooth following or tracking movements (such as maintain a fixed image of a moving object upon the retina) are subserved by two sets of neurons that are located within the frontal region of the cortex, and this statement holds true despite the fact that contractions of the same ocular muscles underlie both types of movement. Thus the functional activity of neurons located in the area of the frontal cortex that regulates eye movements is related to the type of movement rather than to the muscular contractions themselves. In marked contrast to this situation, within the primary motor cortex the same neurons have been found to control the contractions of particular skeletal muscles, regardless of the circumstances or psychic background in which the movement takes place.

It has been postulated that neurons of the cerebellum and the basal ganglia are excited selectively depending upon the particular kind of movement that is involved. The principal physiologic role of the cerebellum is to "plan" and to initiate rapid (saccadic or ballistic) movements. The basal ganglia, on the other hand, are concerned with the initiation of slow (or ramp) movements. The cerebral cortex, on the other hand, is involved in the more precise integration of both fast as well as slow movements. The general clinical and experimental evidence in support of this hypothesis is quite good.

The precise manner in which the neurons of the basal ganglia as well as the cerebellum function in the regulation of slow and fast movements is unclear at present. However, it is evident that cells of the basal ganglia tend to become activated preferentially during slow movements, whereas those of the cerebellum are excited in like manner during the initiation of rapid movements. Furthermore, and in view of the extremely fast on–off switching characteristic that is exhibited by the output side of the individual neuronal circuits of the cerebellum (p. 398), it would appear that this structure behaves physiologically as an enormously complex and highly organized on–off relay device which is suited admirably to the rapid and continuous regulation of ballistic movements of particular skeletal muscles. The cerebellum operates in this fashion not only during the initiation of a movement but also while the movement is in progress. The continuous input of impulses from peripheral receptors to the cerebellum, as well as through the reciprocal connections between the cerebral cortex and the cerebellum, also provides information so that the degree of excitation or inhibition of specific voluntary muscles is controlled rapidly, precisely, and continuously during their activity.

Possibly the basal ganglia perform in a like manner to the cerebellum as biologic on–off switching devices during the initiation and regulation of slow movements, although this statement is entirely conjectural at present.

The physiologic relationship among voluntary movement, the muscle spindles, and the stretch reflex now will be considered. The anatomy and physiology of the muscle spindles have been discussed elsewhere (pp. 181, 219), and likewise the stretch reflex also has been treated in detail (pp. 218, 367). Although some workers still contend that a separate "muscle sense" is present, and that the afferent impulses responsible for producing this "sensation" arise from the nerve terminals located within the muscle spindles (Fig. 6.3, p. 181) throughout the body, there is abundant experimental evidence which supports the conclusion that normally the impulses arising within these end organs exert their functions solely as essential elements in the servomechanism that regulates muscular contraction in a manner that is *not* perceived consciously. This statement does not imply that cutaneous as well as deep pressure and pain sensations that arise within the muscles in response to appropriate stimuli are not perceived acutely (pp. 178, 356); and similarly, proprioceptive impulses that originate within receptors that are located within the tendons and joint capsules also give rise to signals that inform one of the position of his limbs and body in space.

Experimental work performed by many investigators during the past few years has demonstrated conclusively in the baboon and cat, for example, that the afferent impulses which arise within the muscle spindles are transmitted to the postcentral gyrus of the cerebral cortex, ie, to the primary sensory or somesthetic area. But it appears that these animals, as well as humans, are totally unaware of the signals arising from these peripheral end organs, a fact that is in marked contrast to impulses that arise in, for example, the visual, auditory, vestibular, and cutaneous receptors.

As noted earlier (p. 220), the intrafusal muscle fibers of the spindle are contractile at each end, but not appreciably so in their centers at which points the annulospiral (or primary sensory) endings are located. Thus when the intrafusal fibers contract, the annulospinal endings are deformed mechanically, and afferent impulses are generated in the nerve fibers that are continuous with these primary endings. When contraction of the intrafusal fibers is followed by a quantitatively similar contraction of the extrafusal fibers that comprise the principal mass of the muscle, then the tension will be removed from the sensory region of the spindle and the afferent nerve will become electrically silent (see Fig. 6.40, p. 220). Actually the annulospiral endings of the muscle spindle are exceedingly sensitive to *differences* in length between the intrafusal and extrafusal fibers, so that when these elements are out of alignment with respect to each other the primary endings discharge afferent impulses. That is, the muscle spindle is exquisitely sensitive to any misalignment between the intrafusal and extrafusal fibers and thus discharges afferent impulses when the rates of contraction of these two elements is dissimilar. Consequently, the spindle fires when contraction of the extrafusal fibers does not parallel quantitatively the contraction of the intrafusal fibers and vice versa. That is, the spindle is sensitive to transient differences in length between the intrafusal and extrafusal elements. And since the efferent innervation of the intra and extrafusal fibers is for the most part separate, then the intrafusal fibers of the spindles can be excited independently of the extrafusal fibers, as there is no particular physiologic reason why these two elements must act in concert.

The most clearly established function of the afferent impulses that arise in the spindles is their participation in the excitation of the stretch reflex (p. 218), and as noted earlier, a major physiologic role of this response is in the automatic (or reflex) maintenance of posture as well as muscular tone in general under different circumstances; ie, under various conditions of load (p. 219).

What, however, is the physiologic role of the muscle spindle in the control of voluntary movement? When the spindles contract at a particular rate, then the annulospiral (primary) endings within these structures are stimulated, and thus generate afferent impulses, only if the rate of contraction (or rate of shortening) of the extrafusal fibers does not parallel that of the intrafusal fibers; ie, a misalignment of the two sets of fibers develops during an active contraction. In this manner, the stretch reflex is able to compensate automatically for alterations in load that may develop during active movement (= active contraction), as a more rapid shortening of the intrafusal fibers of the spindles actually would excite contraction of the extrafusal fibers (ie, the principal mass of the muscle) by means of the stretch reflex. Therefore if the load on a particular muscle is increased suddenly during a voluntary contraction, the rate of efferent impulse generation would be enhanced, and conversely if the load is decreased, then the rate of efferent impulse generation would decline proportionately so that the contraction would be inhibited. Direct experimental evidence has been obtained in human subjects that the extremely fast reflex compensation described above actually does take place. It will be apparent to the reader that in this physiologic context the stretch reflex functions as a servomechanism whose regulatory activity follows, rather than precedes, the initial mechanical activity. Consequently an individual not only can position his limbs voluntarily with extreme accuracy because of the precision with which this automatic servomechanism operates in regulating the contraction of specific muscles, but also a person can alter voluntarily the rate of change of limb position within certain limits.

All of the motor effects of the servomechanism which operates through the stretch reflex as described above take place at the subconscious level. Presumably, since electric excitation of the somesthetic cortex *follows* rather than precedes voluntary movements (p. 404), the afferent impulses from the spindles which are transmitted to the cortex (as well as to the cerebellum and/or basal ganglia?) may serve to regulate quantita-

tively the net efferent output to the muscles from the cortical (and other?) motor centers. Thereby the direct as well as indirect excitation, facilitation, and inhibition of the motoneurons that constitute the final common pathway to specific voluntary muscles would be controlled with an exceedingly high degree of precision.

Autonomic Functions and their Regulation

The autonomic nervous system continuously provides motor signals which control the functions of the visceral organs, glands, and blood vessels in such a manner that a physiologically constant internal environment is maintained throughout the lifetime of the individual. In turn this homeostasis is initiated through a vast array of sensory inputs which impinge upon the preganglionic neurons of the autonomic nerves so that the signal outputs of these neurons are stimulated to a continuous state of activity which, however, exhibits marked quantitative fluctuations with time. These fluctuations in turn depend upon the particular internal and external conditions at any given moment.

The neuroanatomy of the sympathetic and parasympathetic divisions of the autonomic nervous system was presented in detail on page 270 et seq., and illustrated in Figures 7.10, 7.17, 7.24, 7.25, and 7.26. The general functions of the autonomic nervous system are revealed to some extent by artificial stimulation of particular sympathetic and parasympathetic nerves, and these data are summarized in Tables 7.6 and 7.11. However, the in vivo roles of many autonomic reflexes, as well as certain bodily responses which have important autonomic components, are best discussed in detail in context with the particular organ systems with which these reflexes and responses are related. Consequently the following presentation will be limited for the most part to a survey of a number of important autonomic and related effects as they occur in vivo, and thus this discussion will provide the reader with a summary of, and an introduction to, the more detailed considerations of individual autonomic reflexes and related automatic responses which follow throughout the remainder of this text.

Like the somatic nervous system, the autonomic nervous system exhibits a hierarchy of levels within the central nervous system at which visceral functions are integrated. Thus many simple autonomic functions are integrated wholly within the spinal cord, whereas particular higher brain centers are essential for the integration of the more complex autonomic responses. For example, the spinal animal (p. 223) exhibits a normal blood pressure at rest, and coordinated reflex contractions of the full rectum or bladder can take place in a relatively normal manner. On the other hand, the more complex visceral reflexes which regulate the pupillary responses to light (p. 475) as well as those underlying pupillary accommodation for distance (p. 476) are integrated within the midbrain.

The medulla oblongata (p. 275) and the hypothalamus (p. 300; Table 7.8) are two particularly important regions of the central nervous system that are linked inseparably to the integration of many complex autonomic responses as well as to several complex physiologic processes that have autonomic components. The hypothalamus, in addition to its major role in the regulation of a number of autonomic functions (Table 7.9), also operates together with the limbic system as a physiologic unit, a unit which provides the neurologic substrate for emotional responses as well as innate (or instinctual) behavior.

VISCERAL RESPONSES AND THE MEDULLA OBLONGATA. The term *vital centers* is applied to those regions located within the medulla which are responsible for regulating the tone of the blood vessels (ie, vasomotor control), the heart rate, the force of myocardial contractions, and the process of respiration.* Thus damage or injury to the cardiovascular and/or respiratory centers generally is fatal.

The afferent fibers that send inputs to the vital centers of the medulla usually originate in highly specialized mechano- and chemoreceptors. Thus, for example, the receptors of the aortic and carotid sinuses (p. 672) as well as the aortic and carotid bodies transmit afferent signals which result in extremely delicate, precise, and graded motor adjustments insofar as heart rate and caliber of the blood vessels are concerned. Furthermore it appears that certain specialized receptor elements may be located within the medulla per se. The cardiac and vasomotor reflexes are discussed in Chapter 10 (pp. 668, 666) and the respiratory reflex in Chapter 11 (p. 729).

The medullary region of the brain stem also contains the neural centers which integrate a number of autonomic reflexes in addition to those outlined above. Coughing (p. 738), sneezing (p. 739), swallowing (p. 836), gagging, and vomiting (p. 902) all are complex responses that are integrated within the medulla oblongata. It is important to note that some of these reflexes have voluntary as well as autonomic components. For example, the swallowing reflex is elicited by the voluntary propulsion of a mass of food or liquid toward the back of the pharynx, at which point this mechanical stimulation induces the fully automatic swallowing reflex proper (see also Chapter 13, p. 837).

VISCERAL RESPONSES AND THE HYPOTHALAMUS. The principal hypothalamic functions listed in Table 7.9 range from the integration of

relatively straightforward visceral reflexes to extremely complex behavioral and emotional responses. Nonetheless, in all of these processes the hypothalamus serves as the major physiologic integrative center, and in every case a clear-cut stimulus–response is evident. Stimulation of the hypothalamus under experimental conditions will evoke certain autonomic responses to be discussed below; however it also appears that many normal autonomic functions which are elicited in vivo by hypothalamic excitation are components of more complex physiologic reactions, and that the hypothalamus initiates such autonomic responses secondarily to the onset of certain emotional states, eg, fear and rage.

Experimental stimulation of the anterior superior hypothalamic region (as well as many other areas of the hypothalamus) may induce contraction of the urinary bladder, and for this reason a so-called parasympathetic center was postulated to lie within the anterior hypothalamus. However, hypothalamic excitation produces few other clear-cut parasympathetic reactions so that the experimental evidence in favor of the existence of a hypothalamic "parasympathetic center" remains unconvincing.

On the other hand, artificial excitation of other hypothalamic areas, particularly the lateral regions, evokes pupillary dilatation, hypertension, and erection of the hairs (piloerection) among other characteristic adrenergic responses. In vivo, however, this pattern of diffuse adrenergic response does not result from afferent regulatory signals that arise within the viscera per se; rather such a response is evoked by emotional stimuli, particularly fear and rage. Such adrenergic reactions also are touched off by appropriate thermal stimulation (ie, an extremely cold environment), and thus form part of the overall physiologic mechanism that is concerned with thermoregulation.

Dilatation of the blood vessels in skeletal muscle has been observed following stimulation of the dorsal region of the hypothalamus in the midline. This experimental finding, taken in conjunction with other evidence, tends to support the conclusion that the hypothalamus may be a component in the sympathetic vasodilator system that has its origin in the cerebral cortex, and such a cholinergic motor system has been implicated as the physiologic mechanism responsible for dilatation of the blood vessels within the skeletal muscles that takes place immediately following the onset of exercise (p. 1177).

When the dorsomedial and posterior hypothalamic areas are stimulated, an increased secretion of catecholamines (ie, epinephrine and norepinephrine) takes place in the adrenal medulla (p. 1118). Such an enhanced adrenal medullary secretion is a basic physiologic characteristic that accompanies the emotions of fear and rage, and this response in vivo also may be related to excitation of the sympathetic vasodilator system.

---

*The term "center," as used in this physiologic context, is defined as a collection of neurons that is concerned with the performance of a particular function; a given functional center may or may not correspond exactly to a particular and well-defined neuroanatomic structure or nucleus (or to a collection of two or more nuclei). Similarly, the word "area" may be used as a general term to describe either a specific, limited surface or a particular functional region.

The physiologic mechanisms whereby the hypothalamus integrates the secretion of a number of endocrine substances (or hormones) are summarized in Table 7.9 (p. 303). The mechanism whereby secretion of the catecholamines (p. 1123), vasopressin (p. 1043), oxytocin (p. 1046), thyrotropin (p. 1033), adrenocorticotropic hormone (p. 1035), follicle-stimulating hormone (p. 1039), prolactin (p. 1039), and somatotropin (p. 1030) is regulated by the hypothalamus will be discussed in context with each specific hormone. Similarly, the role of the hypothalamus in thermoregulation is presented in detail in Chapter 14 (p. 1000).

As indicated in Table 7.9, the hypothalamus also has been implicated as a factor that is involved in the physiologic states of wakefulness and sleep. As an examination of these functional states is wholly inseparable from a consideration of the reticular activating system and other neural elements, the possible role of the hypothalamus in wakefulness and sleep will be presented as an integral part of the discussion of these functional states (p. 428). Similarly, as the physiologic interactions between the hypothalamus and the limbic system are so intimate, the roles of these neuronal pools in certain behavioral states will be discussed on page 418.

The remainder of this discussion will be concerned with a consideration of the three so-called appetitive functions with which the hypothalamus is concerned directly. These functions, as listed in Table 7.9, are thirst, hunger, and sexual behavior. Actually these fundamental drives evoke patterns of activity which involve many organ systems in addition to the hypothalamus, systems whose activation leads to complex behavioral responses on the part of the individual, which in turn lead to amelioration or satiation of the basic drive. Actually, therefore, the sensation of thirst merely provides a physiologic stimulus whereby a net deficit of water in the body is corrected and thereby homeostatic balance is restored through the ingestion of fluid. Similarly, hunger leads to food-seeking behavior and the consumption of food so that bodily energy stores are maintained.

Thirst and the Regulation of Water Balance. "Thirst" may be defined as a sensation which is referred to the mouth and pharyngeal region and is associated with a craving for liquids. Usually this sensation is associated with a desire for water. Thus a localized dryness of the oral mucosa per se is related to the sensation of thirst; therefore a certain relief from this sensation can be achieved by sucking a damp cloth or an ice cube. Experimental animals (rats) with induced lesions in the areas of the hypothalamus that control drinking behavior will ingest small quantities of water when their pharyngeal mucosa becomes dry, although a generalized tissue dehydration per se no longer is an effective stimulus to induce drinking.

The central neural mechanism that regulates water intake is located within the hypothalamus.

As illustrated in Figure 7.46 (p. 300), a large tract of unmyelinated nerve fibers has its origin in neurons whose cell bodies are located within the supraoptic and paraventricular nuclei of the hypothalamus. This hypothalamohypophyseal tract courses through the median eminence and stalk of the pituitary, then terminates in the neural lobe of the hypophysis. The hypothalamohypophyseal tract has a critical physiologic role in the production and release of the hormones of the posterior pituitary. In humans, *antidiuretic hormone* (sometimes also called *vasopressin;* see also p. 89) is the most important endocrine substance that is elaborated by the posterior pituitary (see also pp. 1026, 1043). Antidiuretic hormone stimulates the reabsorption of water into the bloodstream from the fluid that has been filtered into the tubules of the kidneys (p. 796); consequently this hormone is a mjaor factor which regulates the quantity of water which is lost from the body to the urine. Thus the volume of urine excreted per unit time is, in general, inversely proportional to the quantity of antidiuretic hormone carried in the bloodstream to the renal tubules. In turn the release of antidiuretic hormone under physiologic conditions depends upon the presence of a morphologically and functionally intact hypothalamohypophyseal tract. Pathologic or experimental damage to this system, or to the neural lobe of the pituitary itself, produces a condition known as *diabetes insipidus*. In this condition (not to be confused with diabetes mellitus, p. 798), enormous volumes (up to 20 liters per day in humans) of a dilute urine are excreted, a condition known as *polyuria*. In turn this tremendous loss of water results in an abnormally great thirst which leads to the ingestion of large volumes of water, a condition known as *polydipsia*.

In experimental animals it has been demonstrated that electric stimulation of the hypothalamohypophyseal tract by means of electrodes chronically implanted within this system would induce the secretion of enough antidiuretic hormone to inhibit the diuresis produced by the prior ingestion of a large volume of water.

In contrast to the majority of glandular tissues, such as the adrenal medulla (p. 1118) and salivary glands (p. 846), in which nerve impulses induce secretory activity, the neural lobe of the pituitary is composed principally of nerve endings that are derived from the hypothalamohypophyseal tract. A few connective tissue elements also are present. These nerve terminals do not end upon secretory cells; rather they end in close proximity to the blood vessels of the neural lobe. Furthermore, not only the terminations of the nerve fibers, but also the fibers themselves, as well as the cell bodies of these neurons which lie within the supraoptic and paraventricular nuclei, all contain a high concentration of membrane-limited secretory vesicles or granules. Briefly, it appears on the basis of data obtained from histologic and physiologic studies that the entire hypothalamohypophyseal system is composed of *neurosecretory cells* that synthesize, transport, and release antidiuretic hormone. These hypothalamic neurosecretory cells also exhibit bioelectric properties which in all respects are similar to those found in more conventional neurons; and although it is not yet entirely clear what the precise role of this electric activity is insofar as the release of antidiuretic hormone in vivo is concerned, nevertheless it may be suggestive that electric stimulation of the hypothalamohypophyseal system causes an enhanced secretion of antidiuretic hormone.

In vivo, the excretion of water by the kidneys is dependent upon the size of the entire body pool of water. The ingestion of large volumes of water produces a relative dilution of the body fluids and a rapid decrease in the volume of water that is reabsorbed from the renal tubules with a concomitantly increased rate of urine flow or a diuresis (p. 798). Conversely, a prolonged decrease in water intake results in an increased rate of water reabsorption from the renal tubules which results in the excretion of a correspondingly small volume of more concentrated urine per unit time. Homeostatic regulation of the body water content is achieved by exceedingly precise changes in the release of antidiuretic hormone, and in turn these changes in the rate of secretion of this substance are mediated through minute alterations in the osmotic pressure of the blood. This fact can be demonstrated by the experimental injection of hypertonic and hypotonic solutions into the internal carotid artery. Intracarotid injection of hypertonic saline stimulates the release of antidiuretic hormone and a decreased urine output in relation to the quantity of liquid ingested (ie, an *oliguria*). On the other hand, dilution of the carotid arterial blood causes a decrease in antidiuretic hormone secretion and a diuresis ensues. Based upon experimental data such as these, it has been postulated that there are *osmoreceptors* present somewhere within the brain, and presumably these receptors lie within the supraoptic and paraventricular nuclei of the hypothalamus (Fig. 7.47). Hence these osmoreceptor elements constitute a physiologic mechanism whereby the homeostatic regulation of antidiuretic hormone release, and consequently water excretion, is controlled automatically. In addition to the osmoreceptor mechanism for controlling the secretion of antidiuretic hormone that was described above, it also appears that the total volume of body water per se also may have a role in the physiologic control of antidiuretic hormone secretion. Hence the existence of volume receptors as well as osmoreceptors has been postulated (see discussion on p. 804). In view of the critical importance that water plays in all bodily processes, it would, de facto, be rather surprising if the quantitative regulation of the body water store were subserved by merely one homeostatic mechanism.

Presumably satiety occurs when the degree of hydration is such that there is no osmotic or volume imbalance acting upon either osmo- or volume receptors, and thereby the sensation of thirst is alleviated.

In addition to playing a vital role in the control of water elimination from the body, the hypothalamus also is responsible for regulating water intake. Thus experimental lesions which are placed in suitable hypothalamic areas can decrease, or even abolish completely, water intake in experimental animals. Such alterations in drinking patterns sometimes can be obtained without changing feeding behavior. Conversely, in the dog and goat, electric stimulation of the lateral superior hypothalamic region (ie, between the columns of the fornix and the mamillothalamic tract; Fig. 7.45) by means of chronically implanted electrodes results in a tremendous increase in water intake. In fact, the polydipsia resulting from such an experimental manipulation may result in an increase in body weight of up to nearly 50 percent above the normal weight for the particular animal. Furthermore, the injection of minute quantities of hypertonic saline into the same areas of the hypothalamus as those subjected to electric stimulation also causes polydipsia. This finding implies that the neurons responsible for producing this effect are sensitive to alterations in osmotic pressure like those of the hypothalamohypophyseal system.

Hunger and the Regulation of Food Intake. "Hunger," or "appetite" may be defined as the normal desire or craving for food. The sensation of hunger usually is referred in a general way to the lower chest or epigastric region. It should be mentioned here that the intermittent gnawing hunger pangs that may be experienced correlate very poorly with the actual patterns of gastric contractions which occur during the fasting state. Consequently the simplistic notion that special "hunger contractions" give rise to the subjective sensation of hunger is invalid, and it has been demonstrated conclusively in human subjects that no particular type of gastric contraction is related in a specific manner to hunger pangs. There is, incidentally, no qualitative difference between the contractions of the stomach which are observed during periods of digestion and during periods of fasting.

As is the case with thirst, the hypothalamus is related initimately to hunger and to the regulation of food intake. It has been shown experimentally that a *hunger* or *feeding center* is present in the ventrolateral hypothalamus, and that a *satiety center* is present in the ventromedial hypothalamus (Fig. 7.45). Therefore under normal circumstances these two centers interact in such a reciprocal manner as to regulate the desire for food. Artificial stimulation of the feeding center elicits eating behavior in conscious animals, whereas destruction of this center causes a loss of appetitie, or *anorexia,* so severe that the animal starves to death. Stimulation of the ventromedial satiety center, on the other hand, causes the animal to stop eating, but lesions induced experimentally in the same region induce a pattern of gross overeating, or *hyperphagia,* such that the animal becomes extremely obese, a condition known as *hypothalamic obesity* (see also p. 1215). When the feeding center is destroyed experimentally in rats which have bilateral lesions already present in the satiety center, anorexia results. This experiment indicates that the satiety center operates under physiologic conditions by exerting an inhibitory effect upon the feeding center. In other words, the surprisingly precise equilibrium between food intake and energy expenditure under normal circumstances that underlies the maintenance of a constant adult body weight in turn apparently results from an equally precise balance between the reciprocal activity of the feeding and satiety centers of the hypothalamus. According to this concept, the ventrolateral hunger center is dominant, and thus impulses from the satiety center presumably inhibit the desire to continue eating.

An important consideration is the source of the afferent inputs which provide information to the feeding and satiety centers concerning the status of the energy stores of the body (Fig. 7.47, Table 7.9). Neither radical gastrectomy nor denervation of the stomach and intestinal tract prevent the development of the hypothalamic hyperphagia that results in gross overeating and the onset of obesity following the induction of small bilateral lesions in the ventromedial nucleus whose activity underlies satiety. Therefore it is unlikely that increased afferent inputs from the gastrointestinal tract underlie the drive to increase food intake to pathologically high levels. There is, however, good experimental evidence which suggests that the activity of the feeding and satiety centers are controlled by the rate of glucose metabolism of the cells located within each of the centers; hence these cells have been called *glucostats.* A low blood glucose level, or hypoglycemia, is presumed to stimulate the feeding center and to inhibit the satiety center, whereas an elevated blood sugar level, or hyperglycemia, has the opposite effect. In support of these contentions it has been determined experimentally that hypoglycemia caused a reduction in the electric activity of the ventromedial satiety center, while that of the ventrolateral feeding center was enhanced somewhat. Hyperglycemia again produced reciprocal alterations in the electric cavity of both of the centers, but in diametrically opposite directions to the changes induced by hypoglycemia.

Stated another way, it has been suggested that when the arteriovenous difference in blood glucose concentration in the feeding center is low, the rate of glucose utilization within the glucostatic cells also is low; consequently the activity of the feeding center is uninhibited and the individual "feels hungry," the sensation being referred to the epigastric region. On the other hand, when the rate of glucose utilization is high, the feeding center is inhibited by the enhanced activity of the glucostatic cells of the satiety center and the individual no longer feels hungry.

Both of these statements of the *glucostatic hypothesis* for the regulation of appetite are essentially similar, and the experimental data currently available strongly supports the conclusion that the glucostatic mechanisms within the hypothalamic feeding and satiety centers are major physiologic regulatory systems insofar as feeding behavior is concerned. This hypothesis also explains the markedly increased appetite, or *polyphagia,* which is present in diabetes mellitus (p. 1077), a disease which is caused by a deficiency in the secretion of the pancreatic hormone insulin (p. 1078). The blood sugar level in diabetics is abnormally high, whereas the actual metabolism of glucose by most of the cells of the body is abnormally low. The lack of insulin prevents the normal transport of sugar into, and utilization by, the cells of the body. Neural tissues in general, and the brain in particular, do not require insulin to utilize glucose effectively. The ventral hypothalamic cells apparently are an exception to this general rule. Consequently the rate of glucose utilization by the hypothalamic cells is related directly to the quantity of circulating insulin like other cells of the body so that a deficiency of this hormone results in a decrease in glucose utilization, the activity of the feeding center is unrestrained, and polyphagia develops.

The rate of uptake and utilization of glucose by the cells of the ventromedial satiety center normally is quite high, as demonstrated by studies in which radioactive glucose was used as a tracer substance. If gold thioglucose, which is toxic to and produces clear-cut lesions in the ventromedial nuclei, is administered, the animals (mice) develop hypothalamic hyperphagia and subsequent obesity. The gold thioglucose performs a "chemical ablation" of the cells of the satiety center in the ventromedial nucleus so that the activity of the feeding center is manifest without inhibition.

Experimental stimulation of the mamillary and premamillary regions as well as damage to the amygdaloid nuclei of the limbic system have been demonstrated to produce enhanced feeding behavior. These neural areas also appear to be concerned in some way with the regulation of appetite, although the normal physiologic roles of these areas is obscure.

In contrast to the afferent inputs which regulate the activity of the feeding and satiety centers, not much information is available concerning the efferent pathways from these hypothalamic centers. However, it is apparent that both of these regions must give rise to tracts which innervate the motor nuclei of the brain and spinal cord which are involved in the extremely complex motor behavior involved in acquiring and ingesting food.

A number of factors are related to the control of food intake in addition to the primary regulatory mechanisms that are located in the hypothalamus.

For example, a warm environment depresses whereas a cold environment enhances the appetite, and both of these responses are related directly to the net metabolic rate, hence, overall energy consumption by the body. Similarly the quantity of physical work (or exercise) performed is an important factor in regulating appetite through the quantity of energy consumed. In humans, cultural and environmental factors are extremely important determinants of food intake, as are prior conditioning experiences related to the sight, odor, and taste of food. At times the physical availability of food alone is the major factor in determining intake.

The net effect of the operation of the several physiologic and other mechanisms which govern hunger is that in normal adult animals and humans the caloric intake balances the energy expenditure of the body over long periods of time so that the body weight is maintained within close tolerances (see also p. 409.).

Sexual Behavior and the Hypothalamus. Like thirst and hunger, the basic sexual drive results in the development of extremely complex behavioral patterns for its amelioration, and behind this activity lies a number of equally complex physiologic mechanisms. Thus many endocrine and neural factors operating together in a smoothly coordinated fashion are indispensable to normal mating behavior. The following discussion will emphasize the role of neural processes in mating with particular stress being placed upon the role of the hypothalamus in these processes. For a thorough discussion of the role of specific endocrine glands and their secretions in the reproductive physiology of both male and female, the reader is referred to Chapters 15 (p. 1016) and 16 (p. 1076).

Sexual behavior is integrated at all levels of the nervous system. Hence penile erection and ejaculation can be evoked in humans and animals following complete transection of the spinal cord (p. 452), although such sexual activity is fragmentary in that it represents only a small and isolated part of an entire behavior pattern that is present normally. It is impossible to evoke complete and normal patterns of sexual behavior in either paraplegics or in spinal animals, despite the fact that certain fundamental sexual reflexes are integrated within the cord. When the brainstem and greater part of the midbrain are intact, stimulation of the vagina will elicit some of the postural adaptations necessary for mating, but again this behavior is only a small part of the entire pattern necessary for successful copulation.

Thus copulation per se merely consists of a number of reflexes that are integrated in spinal and lower brain centers. However, the behavioral patterns that attend this activity, the sexual drive itself which leads to copulation, and the highly coordinated sequence of physiologic mechanisms in both male and female that lead ultimately to pregnancy are controlled in large measure by the hypothalamus acting in conjunction with the limbic system (p. 332; Fig. 7.78).

There is an inverse relationship between the phylogenetic status of a particular species of animal and the direct importance of hormones in the initiation of mating behavior. The hormones involved are testosterone in the male (p. 1137) and ovarian hormones, ie, estrogens (p. 1158) and progesterone (p. 1160), in the female. Thus in rats mating behavior is dependent strictly upon the availability of a sexually receptive female (ie, a female in estrus or heat) to a sexually mature male, and in both sexes of this species mating behavior is determined entirely by sex hormones. Copulation in rats which have had no prior sexual contact or experience proceeds quite normally at the first attempt; learning does not appear to play a role in mating in this species. Such behavior is innate (or instinctual) and is triggered solely by the presence of certain chemical substances.

On the other hand, conditioning or learning plays an extremely important role in the development of mating behavior of primates, including man. In humans, sexual functions are closely related to the high degree of encephalization present in this species and hence are conditioned to a large extent by social as well as by psychic influences.

The hypothalamus is essential to ovulation and to the development of estrus, or heat, in lower mammals.* Estrus in turn leads to mating behavior in the majority of mammals, and this effect as well as ovulation is mediated by way of hypothalamo-hypophyseal pathways. Therefore if suitably placed lesions are made in the anterior hypothalamus, the animal fails to come into estrus and mate, whereas ovulation per se may be normal. These facts indicate that these biologic mechanisms are regulated independently by separate hypothalamic regions. Furthermore the injection of estrogenic (female) hormones into such operated experimental animals does not cause the reappearance of estrus, a fact which suggests strongly that the hormones normally responsible for the production of estrus exert their functions upon a particular hypothalamic region, and that this region also is crucial to the development of an entire somatic behavior pattern which results in mating.

When lesions are introduced in the hypothalamus between the ventromedial nucleus and the mamillary body, ovarian atrophy and a cessation of estrus ensue. Presumably these effects are caused by injury resulting from the operative procedure upon the release of pituitary gonadotropins (p. 1037). Conversely, artificial stimulation of the ventromedial nucleus also can induce ovulation.

*The term "estrus" signifies the restricted period of sexual receptivity which occurs in a cyclic fashion in female mammals (other than primates). Periods of estrus are marked by an intense sexual drive. The sexual cycle of the human female as well as a few other primates of reproductive age is called the menstrual cycle (p. 115).

The sexual behavior of male animals also is affected by the hypothalamus and other regions of the brain. Artificial stimulation of the tuberal region causes ejaculation in rats, whereas stimulation of the medial forebrain bundle and adjacent hypothalamic areas (Fig. 7.45) induces penile erection coupled with marked emotional displays in monkeys. In castrated rats, which have a greatly diminished, or completely absent, sex drive, the implantation of pellets of testosterone into the hypothalamus restores the normal pattern of sexual behavior. In normal rats, suitably located anterior hypothalamic lesions will abolish completely all interest in sexual matters, whereas lesions situated in the mamillary region may cause enhanced sexual activity in male rats.

Generally complete extirpation of the neocortex will inhibit sexual activity in male experimental animals, whereas subtotal cortical ablation can produce a partial inhibition of such activity. The extent to which sexual behavior is decreased by such a procedure is not related to the degree to which motor defects are induced concomitantly by the surgical manipulation. The diminution of sexual behavior following subtotal cortical ablation is most evident when the frontal lobes are involved.

In marked contrast to the decrease in sexual behavior noted above which follows cortical injury, when bilateral lesions are placed in the pyriform cortex of the limbic system overlying the basal amygdalar nucleus (Fig. 7.117) the animal usually develops a pronounced hypersexuality. Thus cats and monkeys subjected to this operative procedure not only copulate with adult females of their own species, but also attempt copulation with immature females, other males, animals of other species, and even with inanimate objects. Such in-

discriminate and misdirected sexual behavior clearly lies beyond the normal range for the particular species involved. This excessive sexual activity was found to decrease somewhat following castration but was not itself induced by a greater secretion of testosterone following the initial operative procedure upon the pyriform cortex.

In most mammals the sexual activity of the male is relatively continuous,* whereas that of the female occurs in a cyclic manner called the *estrous cycle* which is dependent upon oscillations in the blood estrogen level. During the periods of heat, or estrus, the increase in the circulating estrogen level which occurs at this point in the estrous cycle causes an alteration in the behavior of the female such that the company as well as the sexual advances of the male no longer are rejected. Rather, during estrus the female actively seeks out the male and attempts to mate. The females of certain species, particularly the rabbit and ferret, come into estrus and remain so until pregnancy intervenes. In such species, ovulation per se is due to a neuroendocrine reflex which operates through the hypothalamus. Direct stimulation of the genitalia and other sensory areas at the time of copulation excites the hypothalamus; in turn the signals from this region of the brain are transmitted directly to the pituitary and cause the release of gonadotropic hormones which then are carried by the bloodstream to the ovaries where they induce rupture of the ovarian follicles and liberation of the ova. In many other species, including hu-

*Notable exceptions to this statement are found in the male rhinoceros and elephant, which species enter a state of rut at intervals that coincide with peaks of greatest sexual excitation.

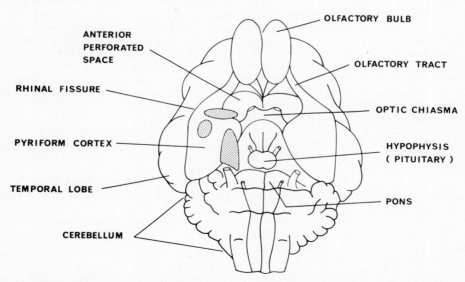

FIG. 7.117. Ventral surface of cat brain showing areas (shaded) where lesions placed in the pyriform cortex over the amygdaloid nucleus cause hypersexuality. (After Ruch and Patton [eds]. *Physiology and Biophysics,* 19th ed, 1966, Saunders.)

mans, the ova are liberated spontaneously and in a cyclic fashion. In lower mammals this release of the female germ cells coincides with estrus. In humans, this event takes place at the approximate midpoint of the menstrual or sexual cycle as discussed on page 1145 et seq.

In female experimental animals, extirpation of the neocortex and/or the limbic cortex completely eliminates the male-seeking behavior by the female during estrus. However, the other aspects of this physiologic process remain unchanged, including the slight rise in body temperature that occurs. Induced lesions of the pyriform and amygdaloid regions do not produce hypersexuality in the female as they do in the male, although discrete lesions placed in the anterior hypothalamus do abolish the behavioral effects of estrus but without altering the normal pituitary–ovarian cycle of hormonal changes.

The implantation of tiny pellets of estrogen into the anterior hypothalamus will induce estrus in ovarectomized rats; however implantation of the hormone into other parts of the brain as well as into regions that are located outside of the brain produces no such response. Consequently it would appear that some hypothalamic structure is responsive to the circulating estrogen level and is activated by a high circulating estrogen level so that the sexual behavior typical of the estrous state is induced.

Additional physiologic effects of testosterone and estrogens are discussed in Chapter 15 (eg, pp. 1139, 1162).

VOLUNTARY REGULATION OF "INVOLUNTARY" FUNCTIONS. Generally the functional activities of the autonomic nervous system are carried out at wholly subconscious levels and furthermore these activities cannot be regulated ordinarily through a conscious effort of the will. Although several investigators have studied the physiologic effects of meditation (yoga and Zen Buddhism) on a number of readily measured functional parameters, the results often were conflicting, perhaps because of the differences in technique employed by the subjects to induce the meditative state itself. Recently, however, several controlled investigations have been carried out upon normal human subjects all trained in the same technique of meditation. During this state, called "transcendental meditation," the subject achieves an extremely relaxed physical condition while remaining fully conscious and alert. In general, the specific functional alterations observed in subjects during transcendental meditation were consistent with several of those alterations which were observed earlier in yogas and practitioners of Zen meditation. These data are summarized in this section, as it now is clear that a number of supposedly involuntary processes are indeed subject to modification through the application of conscious mental effort on the part of the subject. The potential clinical significance of some of these findings will be indicated in context. In the studies to be cited, each subject served as his own control.

During periods of transcendental meditation, the rates of both oxygen consumption and carbon dioxide production declined sharply, but the ratio of carbon dioxide excretion to oxygen consumption (volume/volume) remained unchanged for all practical purposes when compared to the nonmeditation control values. The respiratory rate decreased slightly as did the minute volume of air respired (ie, the volume of air breathed per minute measured in liters; p. 726). These facts suggest strongly that the overall metabolic rate of the subject declines during periods of meditation. Furthermore, the rate of decrease in oxygen consumption that is seen during meditation is quite different from that observed during sleep (Fig. 7.118).

The mean arterial blood pressure remained at fairly low and constant values during periods of meditation. However, it is interesting to note that this blood pressure was achieved during the quiet interval which preceded meditation, and that the blood pressure did not change appreciably following the meditation. The heart rate was observed to

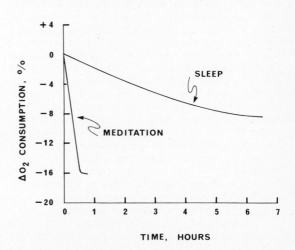

FIG. 7.118. Comparison of the effects of transcendental meditation and sleep on the rate of oxygen consumption. Hypnosis per se causes no alteration in the rate of oxygen consumption (see also text). (Data from Wallace and Benson. The physiology of meditation. *Sci Am* 226:84, 1972.)

decline very slightly during meditation, on an average of three contractions per minute.

The partial pressures of oxygen and carbon dioxide in the arterial blood (P_{O_2}, p. 745, P_{CO_2}, p. 750) remained practically unaltered during meditation, but the blood pH declined very slightly, a fact that was indicative of negligible metabolic acidosis (p. 763).

The electric resistance of the skin increases sharply (up to four times the control value) during meditation. Supposedly this factor is related to the "degree of relaxation"; that is, the greater the resistance, the more at ease the subject is.

The blood lactic acid concentration is a direct indicator of anerobic metabolism (p. 1187). During transcendental meditation, the blood lactate concentration was found to decline at a rate which was up to four times as rapid as that measured in the nonmeditating subject who was merely resting in a supine position, and the lactate concentration remained low even at the end of the postmeditation interval. The explanation for this effect appears to lie in the fact that during meditation blood flow through the skeletal muscles increases significantly, as indicated by flow measurements in the forearm. This factor would increase the rate of delivery of oxygen to the muscles during meditation, and thus anaerobic production of lactic acid would be minimized as the principal source of this compound in vivo is within the skeletal muscles.

As noted above, the blood pressure and heart rate remain at essentially constant levels during meditation; therefore under such circumstances alterations in blood flow are affected principally by changes in the caliber of the vessels (p. 666). Consequently it appears that during meditation the vascular bed in the skeletal muscles undergoes vasodilatation, and it has been suggested that the cholinergic sympathetic vasodilator nerves (p. 667) may be responsible for this effect. It also has been suggested that patients having anxiety neuroses, as well as patients suffering from essential (p. 695) or renal (p. 813) hypertension might benefit markedly from transcendental meditation and the concomitant sharp reduction in blood lactate that accompanies this state. It is known, of course, that such neurotic patients tend to develop a marked hyperlacticacidemia when under stress, and it has been demonstrated experimentally that injections of lactic acid into such patients (or even normal persons) can precipitate anxiety attacks. Furthermore, both classes of hypertensive noted above exhibit higher-than-normal blood lactate levels in the resting state than do nonhypertensive subjects. It is tempting to speculate upon the possible physiologic role of the low blood lactate concentration found in meditating subjects insofar as the etiology of this state is concerned. Furthermore, the potential clinical value of meditation to the hypertensive or anxiety neurotic is such as to warrant further investigation in this area.

In addition to the physiologic and biochemical alterations discussed above, electroencephalographic recordings obtained from meditating subjects revealed that a pronounced increase in the alpha-wave activity of the cortex takes place in all subjects, and in some individuals a concomitant development of theta-wave activity was observed in the frontal area of the brain. The general topic of electroencephalography is discussed on page 420, and a detailed comparison between the electroencephalographic findings during sleep and transcendental meditation is presented on page 433. Suffice it to say at this point that from a physiologic standpoint the state induced by transcendental meditation in no way resembles that evoked by sleep, hypnosis, autosuggestive techniques, or operant conditioning.

The overall functional pattern that is evoked voluntarily by meditation in trained subjects appears to resemble an integrated reflex that includes a number of autonomic responses that are coordinated by the central nervous system, and in contrast to the sympathetic "fight or flight" reaction that involves activation of the entire sympathetic nervous system, it would appear that transcendental meditation exerts quite the opposite effects with the net result that the individual is extremely relaxed, though in full possession of his mental faculties.

Other studies have demonstrated clearly that suitable conditioning techniques can be employed to train a person or an experimental animal to control certain individual visceral functions at will. These studies, which contrast somewhat with the more widespread physiologic responses evoked by transcendental meditation, will be discussed in context together with the phenomenon of learning (p. 441).

Neurophysiology of Behavior: Motivation and Emotion

The term "behavior" can be defined as any activity which is performed by a particular organism, especially activity which can be observed externally and measured in quantitative terms regardless of whether the activity is induced by internal or external stimulation. Thus in broadest terms, behavior is anything or everything that an organism does in response to various stimuli.

Behavioral responses in toto are not the result of the activity of mutually exclusive regions of the nervous system. Rather, behavior is a function of the entire nervous system acting as a coordinated unit. This statement also should be construed to mean, however, that certain types of behavior can be (and are) dominated by the activity of certain regions of the nervous system that may act in concert with other organ systems of the body.

In this section the physiologic mechanisms which underlie motivation and emotion will be discussed. The problems encountered in studies of motivation tend to overlap those encountered in studies on emotion. The reader must realize that the topics presented under each of the following two subheadings have been separated arbitrarily. Thus the material concerned with intracranial autostimulation has been considered under the heading of motivation, although it could have been considered just as readily under emotion.

MOTIVATION. The term "motivation" can be defined as any of the forces which regulate the behavior of an organism such that this behavior is directed toward the satisfaction of certain basic needs or the achievement of certain goals. Motivation can be considered to be made up of two aspects, one of which guides the *direction* in which the motivation is channeled and is chiefly a product of learning (p. 439), whereas the other is concerned with the intensity or *drive* which underlies the behavior. In this section, emphasis will be placed upon the physiologic mechanisms that underlie the drive aspect of motivation.

From an experimental standpoint, the drive component of motivation can be studied quantitatively by observing the measurable alteration in one aspect of behavior after altering the intensity of some stimulus. The stimulus which is varied can be either external or internal with respect to the organism. For example, food or water deprivation can be varied quantitatively to give variable internal stimuli, whereas alterations in the intensity of applied sound, light, or electric shocks all will provide similarly variable external stimuli. In turn, the effects induced by the application of such interoceptive or exteroceptive stimuli can be estimated by measuring any of several behavioral changes, including, for example, gross motor activity, alterations in autonomic responses, latent period of a learned response, choice made between two or more different stimuli, and the intensity (or force) of a response to a given stimulus. It should be borne in mind, however, that the relationship between motivation and learning is exceedingly complex, so that even under the best of experimental conditions the quantitative measurements of motivation per se as derived from studies on either human or animal subjects are only refined approximations, and not absolute certainties.

When the observed drive (or intensity) of a particular behavior pattern changes, it is tacitly inferred that the activity of the central nervous system has changed, although direct proof of this contention may be lacking. Nevertheless, several well-known classes of stimuli can alter markedly the behavioral patterns of humans as well as experimental animals; these factors and their physiologic roles in behavior may be summarized briefly.

Afferent Input Stimuli. The afferent inputs that influence motivated behavior can be divided into exteroceptive and interoceptive stimuli.

Exteroceptive Stimuli. Generally speaking, any sensory input can alter the drive aspect of motivated behavior, provided that the magnitude (or intensity) of the stimulus is sufficiently great. For example, extremes of temperature, very loud sounds, and strong illumination all can motivate avoidance behavior under certain circumstances. It appears that those sensory inputs having a marked affect upon the subject either in a pleasant or unpleasant (painful) direction exert the most pronounced motivational effect.

In humans, it also has been found that sensory deprivation can be as deleterious to the subject as an abnormally strong sensory input. Thus a pronounced deterioration of higher mental faculties including the ability to reason and memory, followed a prolonged (6-month) interval of isolation in one subject. Furthermore, hallucinations (ie, false sensory impressions) commonly develop in subjects undergoing extended periods of sensory deprivation, and the ability to perform even simple motor tasks also declines appreciably. Most important to the present discussion, however, is the fact that isolation can produce an extreme state of apathy or lassitude in the subject.

Interoceptive Stimuli. Sensory inputs to the central nervous system that arise within the body per se, in particular the viscera, are an important source of afferent signals that underlie motivated behavior. Thus alterations in the caliber of blood vessels and mechanical displacement or stretching of internal organs may evoke afferent impulses that in turn stimulate extremely complex behavioral patterns to alleviate the condition. The sensation of pain (p. 355) is the major example of a visceral sensation that induces such behavior patterns.

Humoral Stimuli. The chemical agents which are included under the heading of humoral substances consist of a variety of hormones, metabolites, nutrients, and drugs, all of which are carried by the bloodstream to the blood–brain barrier (p. 256). After crossing this barrier, these humoral agents can exert their effects directly upon particular groups of nerve cells within the brain and thereby alter motivated behavior accordingly (see also p. 231). In addition, the temperature as well as the osmotic pressure of the blood per se also can exert direct influences upon certain behavioral patterns.

Hormones. Two examples of the direct effects of endocrine substances upon particular areas of the central nervous system that are involved in behavior may be cited. First, it has been shown experimentally that sex hormones can lower the response threshold of the reticular formation to electric stimulation. Second, epinephrine has been demonstrated to facilitate the transmission of impulses in the ascending reticular activating system.

Metabolites and Nutrients. Certain metabolites have been considered to have a role in the onset of fatigue, which in turn motivates one toward entering the sleep phase of the diurnal wakefulness–sleep cycle. Although lactic acid has been implicated in this motivational process, the relationship between the two has not been established unequivocally.

During the course of normal metabolism, the concentration of some constituents of the blood decreases whereas the concentration of other blood constituents may increase. In turn, these alterations in the concentration of blood nutrients may appear across the blood–brain barrier so that an altered function of the neurons themselves develops. For example, a relative deficiency of glucose in the brain that is caused by the presence of an excessively high insulin level may bring about marked behavioral changes that can range from symptoms resembling those of alcoholic intoxication to outright convulsions (p. 232).

Thermal Stimuli. The temperature of the blood itself provides an extremely potent physical stimulus to the activation of complex motivated behavioral patterns that are involved in the maintenance of body temperature. The general topic of mammalian thermoregulation has been considered in Chapter 14 (p. 994 et seq.); however it should be mentioned here that the maintenace of body temperature within very narrow limits is exceedingly important to survival. Mammals (as well as birds) have evolved a very complex and precise thermoregulatory system that involves certain automatic components which underlie the bodily mechanisms that balance heat production against heat loss. In addition to the automatic thermoregulatory mechanisms of the body, motivated behavioral responses also are invoked by thermal stresses in either direction, and these behavioral patterns tend to balance discrepancies between heat load and heat loss in the body. For example, an animal may seek shelter or curl up in a ball (thereby diminishing the surface exposed to heat loss) in order to conserve body heat in a cold environment. A human being, on the other hand, may alter the amount of clothing being worn or adjust the thermostat in a building in order to achieve the same goal.

In addition to the motivated behavior patterns discussed above that are potentially mediated by alterations in blood temperature per se, it should be mentioned that any marked alteration in blood or body temperature undoubtedly will influence behavior directly, merely because of the effects of temperature changes upon the rate of all chemical reactions in the body, including those taking place in brain and peripheral nervous tissues.

Osmotic Pressure. It was noted earlier that the injection of a hypertonic saline solution into the carotid arteries will induce highly motivated drinking behavior (p. 409). Consequently it would appear that alterations in the tonicity of the plasma itself can affect the irritability of certain nerve cells in such a manner as to influence markedly certain behavior patterns concerned with thirst.

Inherent Neural Stimuli: Circadian Rhythms. Practically all plants and animals exhibit regular cyclic variations in many of their functions. The duration of particular functional cycles can vary markedly (eg, the menstrual cycle lasts for about 28 days), however among the most evident are those whose duration is approximately 24 hours in length, and these cycles are called *diurnal* or *circadian rhythms*. In man as well as in lower mammals the daily oscillations in body temperature, sodium and potassium excretion, white blood cell (leucocyte) count, adrenocortical function, urine volume, and wakefulness and sleep are the best-studied cycles, although many other diurnal fluctuations in particular functions are known.

As discussed above, afferent inputs and humoral influences are the most readily detectable means whereby the activity of the central nervous system; hence behavioral intensity, is altered, however many changes in behavior cannot be ascribed to these sources. Certain of the circadian rhythms enumerated above thus have been ascribed to the operation of a so-called biologic clock mechanism. Thus rhythmic processes taking place either in the environment external to the organism, or within the organism itself but outside of its central nervous system, then affect either the afferent or humoral mechanisms which in turn alter the activity of the central nervous system and thereby alter behavior indirectly and in a cyclic fashion. On the other hand, some rhythmic physiologic alterations in the activities of various organisms indicate that the central nervous system per se exhibits cyclic fluctuations in its activity that are independent of extraneural cyclic events. It would appear that the central nervous system has inherent mechanisms which can operate in a cyclic ("clocklike") manner, and thus modify behavior even though an external stimulus is absent.

In certain lower organisms (insects, fish, birds) the morphologic patterns of organization within the nervous system are so arranged that appropriate stimuli can evoke stereotyped, often complex and highly motivated, but unlearned, behavioral patterns called *instincts*. In mammals, however, it is difficult to ascertain the actual role of instincts in behavior, because such innate behavior as may be present is dominated by the far greater capacity for learning that is present. Hence instinctual behavior is governed to a great extent by learned behavior patterns so that the two merge and become, de facto, indistinguishable.

From a neuroanatomic standpoint, it appears that the mechanisms which control some of the diurnal rhythms are located within the limbic system (p. 332; Figs. 7.78 and 7.79). For example, damage to the fornix alters the adrenocortical rhythm, whereas more generalized lesions of the limbic system provoke abnormal sleep–wakefulness and body temperature cycles without interfering with thermoregulation per se.

Affect and Motivated Behavior: Neural Centers Which Underlie Approach and Avoidance Responses. Within the past 20 years, it has been determined experimentally that a number of hypothalamic and closely related structures are concerned intimately with the affective component of sensory experience (see also p. 355). Thus if an animal is confined to a cage having a lever that can be pressed, eventually the animal depresses this lever by accident. Now if electrodes are implanted in the brain of the animal, and the lever is arranged as a switch which can be manipulated by the animal to deliver an electric stimulus to the brain each time the lever is pressed, the affective properties of the stimulus can be deduced from the behavior of the animal. Thus when the electrode tip is located in certain regions of the brain (eg, the tegmentum, posteromedial hypothalamus, and septal nuclei) the animal will stimulate itself repeatedly for long periods of time, even to the point of exhaustion, whereas if the electrode tip is situated in other brain areas (eg, the posterolateral hypothalamus, dorsal midbrain, entorhinal cortex) the stimulus causes the animal to avoid pressing the bar at all following the initial excitation.

The areas of the brain in which implantation of the electrodes causes repetitive voluntary stimulation are far larger than those from which an avoidance reaction can be evoked. However each of these regions is distinct from the other, even though the two areas may be closely situated within the brain.

Experiments such as those described above have given rise to the concept that the hypothalamus and related structures contain two general affective sensory regions which variously are termed "reward" and "punishment," "approach" and "avoidance," or "pleasure" and "pain" centers, depending upon the positive or negative behavior patterns that are elicited by autostimulation of particular brain areas.

It is clearly impossible for an investigator to ascertain or to sense precisely what an experimental animal feels or perceives when the reward (or approach) system is stimulated electrically through its own efforts, but clearly the evoked sensation is such that repetition of the stimulation is in order, and this repetitive behavior has been interpreted as arising from a pleasurable affective response. Alternatively, however, it is just as reasonable to suppose that the presence of the electrode tip in the so-called reward center actually provides an extremely strong irritation of some type and the animal quickly learns that rapidly repeated lever pressing will ameliorate this situation for as long as the mechanical behavior continues (death of the animal may intervene, but satiation, insofar as lever pressing is concerned has not been observed).* In this view, the repeated lever pressing

*Verbal reports given by patients (ie, schizophrenics and epileptics) in whose brains electrodes were implanted chronically indicate that the sensations evoked by autostimulation generally were "pleasurable," and caused "relief of tension." Interestingly, certain subjects who exhibited the highest lever-pressing rates could not describe why they performed this activity at all. Conversely, if the electrodes were placed in areas of the brain in which autostimulation normally was avoided, stimulation of these areas evoked sensations ranging from a slight uneasiness or fear to outright terror.

by an experimental animal is somewhat analogous to the behavior of a confirmed heroin addict who seeks the drug no longer as a pleasurable and temporary escape from reality, but rather to avoid the intrinsic misery of withdrawal symptoms.

Similarly it is impossible for an investigator to appreciate what actually is felt by an experimental animal when the avoidance system of the brain is stimulated.

Regardless of the interpretation given to the experiments on stimulation of the approach system of the brain, it is clear that from a physiologic standpoint much normal human, as well as animal, behavior is motivated by primary stimuli which give rise to a pleasant affect as well as by other stimuli which give rise to an unpleasant, or even painful, affect.

EMOTION. The term "emotion" can be defined as a physiologic departure from a condition of homeostasis during which a state of strong mental excitement (or depression) is present. The respiratory, cardiovascular, neuromuscular, endocrine, gastrointestinal, cutaneous, and other physiologic changes which are accompaniments of emotion prepare the individual for overt behavioral responses or acts which may or may not be carried out. Thus normal emotion is a temporary functional imbalance that is characterized by profound autonomic (visceral) and somatic as well as psychic alterations; and as noted earlier (p. 336), emotion cannot be initiated or terminated voluntarily, the latter owing to the prolonged after-discharge of the limbic system whose functions are linked inseparably with those of the hypothalamus to the physiologic manifestations of the emotional state.

Emotions involve several distinct parameters, including cognition, conation, and affect as well as the specific physical changes summarized above. *Cognition* signifies an operation of the mind whereby one becomes aware of objects of thought or perception; this term also signifies an awareness of a sensation, and frequently its cause (see also p. 462). *Conation* indicates that component of emotion (and other mental states) in which a positive and conscious tendency to act is involved; often conation is referred to in the vernacular as "willpower." *Affect* is, of course, the subjective component of any sensation. In emotional states the affective component is either "pleasant" or "unpleasant," but never "neutral."

The common words employed to describe or indicate various emotions clearly indicate their affective (or subjective) content, insofar as human experience is concerned; ie, learning plays an important role in the expression of emotion. For example, the terms love, desire, joy, elation, surprise, anger (or rage), fear, gloom, hate, anxiety, and disgust all may have somewhat different connotations, but in general all of these words signify either a pleasant or an unpleasant affect when used in a particular context.

In brief, therefore, emotion is a state of mind

as well as a behavioral response, a state which is accompanied by a motivation either toward or away from an object or event.

The following discussion is limited primarily to the neurophysiologic aspects of emotion as studied by neuroanatomic, neurosurgical, electrophysiologic, and observational techniques. Electroencephalography (p. 420) is a particularly useful technique for detecting objective signs of excitement and alerting in experimental animals as well as human subjects.

Functional Manifestations of Emotion. When studied under controlled laboratory conditions in human subjects, emotional responses have been found to have parasympathetic as well as sympathetic components. That is, both divisions of the autonomic nervous system can participate simultaneously in the altered physical signs present during emotional states, and these signs may include erection of the hairs, tachycardia, hypertension, pupillary dilatation, inhibition or stimulation of the gastrointestinal tract and urinary bladder, and so forth.

Emotions can be accompanied by marked excitement or depression. In the former instance the mental processes are accelerated, whereas in the latter they become dull, retarded, or sluggish.

Neuroanatomic Basis of Emotion: The Hypothalamus and Limbic System. The vascular, glandular, and visceral alterations that take place during emotional states are regulated by the autonomic nervous system (p. 270), which in turn is controlled in large measure by the hypothalamus (p. 300). This structure also is the neural center which underlies the genesis of certain emotional states including rage as well as elaborate patterns of sexual behavior, but many other structures are involved in these responses in addition to the hypothalamus.

When small lesions are induced within the ventromedial nuclei of the hypothalamic region, the experimental animal develops a state of rage* and makes accurately directed and ferocious attacks upon anyone in the vicinity. When artificial stimulation by means of electrodes chronically implanted within the hypothalamus is employed, it is possible to localize rather accurately the hypothalamic regions that are responsible for evoking two general behavioral states. First, a pattern similar to rage (or fight) behavior can be evoked in which the animal (eg, cat) growls, hisses, extends his claws, flattens the ears, arches the back, erects the hairs (piloerection) develops pupillary dilatation, and exhibits other sympathetic responses. Second, a pattern similar to fear (or flight) can also be evoked, and this response includes pupillary dilatation, glancing back and forth, turning the head back and forth laterally as

*This response sometimes is called "sham rage."

though seeking an avenue of escape, and ultimately flight. These experimentally evoked behavioral patterns are complete and they involve highly integrated somatic and visceral reactions of the entire body, not merely isolated parts. In brief, the responses elicited by stimulation of appropriate hypothalamic areas resemble those exhibited normally upon presentation of a suitable stimulus to a cat, eg, a dog.

It appears that the fight (or defense) reaction is localized within the dorsal portion of the central thalamus which surrounds the descending part of the fornix; this "center" then extends rostrally into the preoptic and ventral septal area as well as caudally into the posterior hypothalamus within which region the tissue dilates into the central gray matter of the midbrain (Fig. 7.47). Consequently the part of the brain from which the fight or defense reaction can be elicited consists of a narrow (1 to 1.5 mm) band or lamina of tissue which traverses the entire hypothalamus. According to certain workers, the flight reactions are evoked by stimulation of the external portion of this lamina, whereas fright (or flight) reactions are elicited by stimulation of the central part of the lamina. Other investigators feel that the neural mechanisms which subserve the flight reaction merely are located somewhat rostrally to those which underlie the escape reaction.

Similar fight or flight reactions can be evoked by stimulation of the amygdalar portion of the limbic system (Fig. 7.78).

Despite the lack of unanimous agreement as to the precise areas of the brain wherein the rage and fear responses are localized, nevertheless it is apparent that an area which starts in the telencephalon, traverses the hypothalamus, and extends into the midbrain is the neural region wherein aggressive motor behavior as well as escape reactions, together with their autonomic components, are integrated. Furthermore, it has been demonstrated that excitation of other hypothalamic areas than those discussed above, as well as the cerebral cortex and hippocampus, does not evoke emotional responses. However, it has been shown that a fiber system which involves the ventromedial hypothalamic nuclei apparently is involved as a check upon emotional activity, since destruction of these regions results in unrestrained aggressive behavior.

The other neural system which plays an intimate role in the genesis of emotional states is the limbic system (p. 332; Figs. 7.78 and 7.79). It must not be inferred that the limbic system operates independently of the hypothalamus merely because the functions of these two parts of the brain in the development of emotional responses have perforce been discussed separately. Quite the contrary; these two parts of the brain are inseparable from a functional standpoint.

The olfactory function of the limbic system is discussed in relation to the sense of smell in Chapter 8 (p. 522). It is clear, however, that only a small portion of this system is activated by olfactory stimuli, and the functions of the nonolfactory structures of the limbic system are related intimately to emotional states.

The major anatomic components of the limbic system may be reviewed briefly (see also p. 332; Figs. 7.78 and 7.79). The principal structures that comprise the limbic system are two rings of limbic cortex and a number of related subcortical nuclei, including the amygdala, the septal nuclei, and the anterior thalamic nuclei.

Experiments performed by stimulation or ablation of specific regions within the limbic system indicate that this part of the brain not only is concerned with olfaction, but also with feeding behavior (ie, sniffing, salivation, chewing, licking, swallowing, retching, and gagging). Acting in concert with the hypothalamus, the limbic system also is involved in the control of certain autonomic functions, biologic rhythms (p. 416), sexual behavior (p. 411), and in motivated behavior (p. 414), as well as in the emotions of rage and fear.

Stimulation of the limbic system causes a number of autonomic responses, in particular alterations in heart rate, blood pressure, and respiratory rate. Since these responses are evoked by stimulation of many limbic structures, there do not appear to be any localized centers for the integration of these responses. Consequently it would appear that the autonomic effects which are mediated by the limbic system are parts of more complex behavioral patterns including emotional responses. Stimulation of the amygdaloid nuclei, on the other hand, evokes the complex motor responses noted above that are related directly to feeding behavior (eg, chewing, licking, and so forth). Lesions in the amygdala occasionally cause some hyperphagia.

The defense responses evoked by stimulation of the amygdala are like those produced by stimulation of the hypothalamus and midbrain. However, these reactions are produced with greater facility when the amygdala is stimulated directly, but it is clear that this structure exerts its effects indirectly and through excitation of hypothalamus and midbrain, although the specific pathway by which these effects are transmitted is unclear.

When the amygdala is stimulated simultaneously with a hypothalamic region known to produce an aggressive behavior pattern, sometimes the effects of the dual stimulation are facilitory. More frequently, however, the aggressive behavior is moderated by such dual stimulation, and these two responses have been obtained by stimulation of two entirely discrete areas within the amygdala. These data support the contention that the amygdala is located at a higher neural level than the hypothalamus insofar as the integration of emotional behavior is concerned.

When the septal region of the limbic system is extirpated, in experimental animals such as rats,

the responses to a situation which normally would produce extreme anxiety or fear in the animals are decreased considerably. On the other hand, such operated animals become quite vicious, and will attack each other or the experimenter readily.

Following bilateral ablation of the anterior temporal lobes in monkeys, portions of the limbic system, in particular the amygdala, generally are damaged. Following such a procedure, the animals suffer a number of profound behavioral and other disturbances including psychic blindness (ie, visual agnosia), and a marked compulsive exploratory behavior develops, together with loss of normal aggressive tendencies. The animals concomitantly become quite docile, passive, and unresponsive, and their dietary habits become deranged. That is, a herbivore may be converted into a carnivore. Collectively these postoperative symptoms are known as the Klüver–Bucy syndrome.

In contrast to the loss of aggressive behavior which accompanies the Klüver–Bucy syndrome, other kinds of emotional response, in particular sexual activity, become increased markedly. The experimental subjects (monkeys) became hypersexual, and the normal sex drive of the male often becomes channeled into clearly abnormal directions (see p. 412 and Fig. 7.117).

Some years ago it was proposed that the limbic cortex and related structures together with certain thalamic and hypothalamic regions formed a neuroanatomic unit that functionally was responsible for the genesis of emotions (Papez circuit, p. 334). It has been demonstrated that the emotional changes which are observed to take place in the Klüver–Bucy syndrome are caused by loss of the pyriform cortex and the subjacent amygdalar nuclei. The docility which also develops following amygdalectomy is specific, but such a placid animal sometimes can be made fierce and subject to bouts of rage by placing lesions in the ventromedial hypothalamic nucleus. On the other hand, amygdalectomized animals, which have become docile following the operation, occasionally can be rendered savage by subsequently placing lesions in the septal region.

The hippocampus has anatomically well-defined neural connections to the hypothalamus by way of the fornix, and with the anterior cingulate gyrus by way of the mamillary bodies and the anterior thalamic nuclei. Nevertheless, the significance of these pathways in the genesis and integration of emotions has not been established clearly. Furthermore it has been proven experimentally that the hippocampal formation does not underlie any demonstrably obvious type of emotional behavior.

It has been proposed that the Papez circuit (p. 334) mediates the cognitive as well as other subjective features of emotion, and the hippocampal formation in particular seems to have been implicated to some extent in the subjective aspects of emotion. Beyond these rather nebulous statements, nothing more edifying on the relationships between the neuroanatomic and neurophysiologic aspects of emotion can be presented with certainty at present.

Drugs and Behavior Modifications.   For basic scientific as well as clinical reasons, considerable interest has developed in the action of a number of drugs that alter human behavioral patterns. Pharmacologic agents that produce such effects on behavior include *tranquilizers,* such as reserpine, which decrease anxiety; *psychotomimetics,* such as mescaline or lysergic acid diethylamide (LSD), which produce hallucinations and other psychotic symptoms; and *psychic energizers,* such as amphetamines, which are antidepressant agents that increase drive and interest as well as serve to elevate the mood. The action of many psychoactive drugs appears to be through modification of synaptic transmission in various regions of the brain, consequently the discovery of such chemical agents has led to considerable research on the biochemistry and functional characteristics of various neurotransmitter agents (see discussion of neurotransmitters in Chapter 6, p. 195 et seq.).

Electrophysiologic Properties of the Cerebral Cortex: Electroencephalography

All living cells exhibit characteristic bioelectric properties, as emphasized repeatedly throughout earlier chapters of this book. In particular, the tissue of the cerebral cortex consists of a vast, highly organized array of neuronal and glial elements that is characterized by a continuous rhythmic oscillation of electric potential (the *spontaneous rhythm*) in addition to a number of larger-voltage responses that occur locally and as the result of excitation of specific receptors (the *evoked potentials*).

It has been known for about a century that spontaneous, rhythmic variations in voltage can be recorded from the brain of an unanesthetized animal. However, systematic analysis of these "brain waves" resulted from the pioneering efforts of Berger, who introduced the term *electroencephalogram (EEG)* to denote the recording made of the alterations in electric potential that can be obtained from various parts of the brain by means of electrodes placed upon the skull of an intact human or animal subject. The EEG also can be obtained by placing the electrodes directly upon the pial surface of the cerebral cortex after the skull has been opened, in which case the recording sometimes is called an *electrocorticogram.*

An EEG recording may be *bipolar;* in this instance, the potential difference or voltage generated by the brain is measured between two active electrodes placed upon the skull or cortex. Alternatively, a *monopolar* recording of the EEG indicates the changes in potential at a single active lead, which are compared with a "stable" or "indifferent" electrode that is located on the body at some distance from the cortical electrode. In practice, establishing a valid reference electrode for

recording monopolar EEGs is extremely difficult, consequently bipolar recording of EEGs is the method of choice. The general technique used for recording EEGs from human subjects is illustrated in Figure 7.119. Although only two leads are shown, in practice many more leads actually are employed, and the records obtained may be subjected to computer analysis.

EEGs yield information that is of interest to the clinician as an index to the functions of the brain in health and disease. In addition, EEGs provide objective data concerning certain functions of the brain during various functional or behavioral activities, for example during the waking and sleeping states. The spontaneous wave patterns encountered in the EEGs obtained from normal humans now will be discussed together with the evoked cortical potentials that can be recorded following stimulation of particular sensory receptors. The latter technique also has proven to be of major importance in mapping the pathways from specific peripheral sensory receptors to their final terminations within particular regions of the cerebral cortex.

SPONTANEOUS CORTICAL RHYTHMS.   When suitable leads are placed upon the head of a normal human subject, extremely small and irregular potentials can be recorded from the frontal, parietal, temporal, and occipital regions (Figs. 7.119 and 7.120). The actual voltages recorded are in the range of microvolts (μv), and the maximum peak-to-peak voltage recorded from normal human subjects is about 50 μv.

Characteristic differences are present between EEGs obtained from different regions of the brain, the specific pattern that is recorded being dependent upon the physiologic conditions prevailing at the time of recording as well as upon the technique used for making the EEG (ie, a bipolar or unipolar arrangement of the electrodes).

The normal EEG is composed of a large number of waves having irregular amplitudes as well as frequencies. However, during any particular physiologic state one frequency usually dominates, and this property is far more significant than is the amplitude of the individual waves in the interpretation of EEGs. Thus the frequency of the dominant EEG wave in any recording is the basis upon which EEG patterns are named, and these frequencies are denoted by Greek letters as follows.

If the predominant wave frequency of an EEG lies between 8 and 14 per second, an *alpha rhythm* is present. This pattern characteristically is recorded with electrodes which are placed upon the scalp of an adult human sitting with the eyes closed and the mind at rest. The amplitude of the alpha waves thus recorded is about 50 μv. A *beta rhythm* is present when the frequency of the dominant rhythm is 14 to 60 waves per second. A *theta rhythm* is indicated by a frequency of 4 to 8 cycles per second, whereas a *delta rhythm* is man-

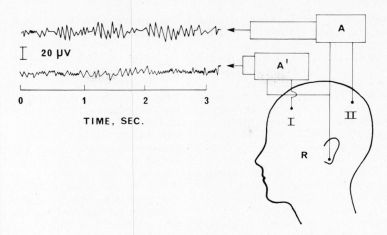

FIG. 7.119. Technique used for recording electroencephalograms (EEGs). A, A' = amplifiers; I = frontal lead; II = occipital lead; R = reference lead. Note the low amplitude, fast wave activity recorded between I and R compared to the higher amplitude, slower activity recorded between lead II and R. (After Ruch and Patton [eds]. *Physiology and Biophysics,* 19th ed, 1966, Saunders.)

ifest by fewer than 4 waves per second. A number of other minor rhythms are recognized in addition to those defined above, but in all instances the amplitude of the waves is related inversely to the characteristic frequency of the particular wave pattern. Furthermore, it is a convention of electroencephalographic terminology that high-frequency waves (of short duration) are called *fast activity,* whereas low-frequency waves (of long duration) are termed *slow activity.*

FACTORS THAT MODIFY SPONTANEOUS CORTICAL RHYTHMS. The pattern of frequency of the EEG is quite labile, and thus is modified by a host of factors. In humans the frequency of the dominant rhythm varies with age. At birth the principal rhythms are 0.5 to 2.0 waves per second and 20 to 50 per second, the latter resembling a fast adult beta pattern. During childhood, theta rhythms are most evident. The delta rhythm of infancy as well as the theta rhythm of

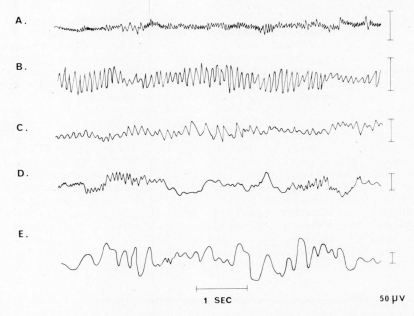

FIG. 7.120. Tracings of electroencephalographic patterns recorded from human subject during various states. **A.** During excitement, ie, the attentive, alert state. The activity is of low amplitude, fast, generally irregular, and "desynchronized" in that no consistent rhythmic wave patterns are present. **B.** Subject relaxed with eyes closed. Note regular alpha rhythm. **C.** Subject drowsy. **D.** Subject asleep. Note two groups of regular waves with a frequency of about 14/sec, the so-called sleep spindles. **E.** Deep sleep, irregular slow waves of high amplitude. (After Ruch and Patton [eds]. *Physiology and Biophysics,* 19th ed, 1966, Saunders.)

childhood are seen most clearly when recorded over the occipital and temporal areas of the brain. Gradually, however, these characteristic EEG patterns of immaturity change to the alpha rhythm of the adult so that between the ages of 14 and 19 years the adult EEG pattern is developed.

The particular EEG pattern that becomes established during maturation is a specific characteristic of the individual (like the fingerprints), and this functional property remains unaltered, assuming that no injury or disease of the brain occurs subsequently.

The frequencies of EEG's that are recorded from the frontal and parietal areas generally are higher than those recorded from the occipital region.

The alpha frequency of the adult EEG pattern can be decreased markedly by a reduction in the blood sugar level, a low body temperature, a high arterial partial pressure of carbon dioxide, or a decline in the blood concentration of adrenal glucocorticoid hormones. In general, therefore, those factors which reduce metabolism simultaneously tend to decrease the rate of alpha discharge in the brain. Conversely, a high blood glucose level, an elevated body temperature, a low arterial Pco_2, or an increase in the circulating adrenal glucocorticoid level all accelerate the alpha wave frequency (ie, the fast-activity component of the EEG increases). Hyperventilation (p. 735), or forced respiration, sometimes is employed clinically in order to reveal latent abnormalities in the EEG, as this procedure markedly lowers the arterial Pco_2 in the brain, and thus enhances fast activity in the EEG.

The alpha rhythm of the adult cortex normally predominates during the waking state, and this pattern is recorded when the subject is relaxed, the mind is wandering, and the eyes are closed. If, however, the eyes are opened so that a strong visual stimulus is presented suddenly and the attention is thereby fixed, the alpha waves (Fig. 7.120B) are replaced by a high-frequency, irregular rhythm having a low voltage, but which has no dominant rhythmic pattern. This phenomenon, which is called *alpha blocking,* also is caused by any type of sensory excitation or mental concentration, and is illustrated in Figure 7.121. Alpha blocking also is called *desynchronization,* because it indicates that the coordinated, or synchronized, activity of the neural elements which underlie production of the alpha wave pattern is interrupted by the sensory stimulus. Since desynchronization is induced by sensory excitation, and is correlated with the

aroused (or alert) state, this phenomenon also is called the *arousal* or *alerting* response (see also p. 428). Once the sensory stimulus no longer engages the subject's attention, the slower alpha rhythm reappears, as shown to the right in Figure 7.121.

Another factor that exerts a considerable influence upon the dominant pattern of the EEG is the state of arousal of the individual; and as shown in Figure 7.120, sleep per se exerts a profound effect upon the EEG pattern. Sleep will be discussed subsequently in relation to the waking state and to the participation of the reticular activating system in these processes (p. 428 et seq.). It should be mentioned at this juncture, however, that there is an interesting direct relationship between the predominant frequency of the EEG and the evident state of arousal of the subject.

PHYSIOLOGIC MECHANISMS THAT UNDERLIE THE ELECTRIC ACTIVITY OF THE CORTEX. The alpha and slow-wave sleep patterns of the EEG exhibit rhythmic waves; it would appear that neural elements are firing in a regular and repetitive manner to produce these recorded electric oscillations. Actually, there is considerable evidence which indicates that the EEG depends upon the electric characteristics of the cortical cells themselves. In particular, the outermost two cell layers of this structure (ie, the molecular or plexiform layer and the external granular layer, Fig. 7.70A), have been shown to be responsible for generating the largest portion of the primary evoked response of the cortex, to be discussed below (p. 424).

The appearance of rhythmic waves in the EEG suggests that neural elements are firing in a synchronous rather than in a random fashion, and that such activity could be explained, at least partially, by simultaneous and coordinated oscillatory or reverberatory activity in a large number of closed neuronal circuits (p. 348). Alternatively, certain neurons could exhibit autorhythmicity (p. 143), and thus drive adjacent groups of neurons to rhythmic activity.

The best evidence that is available at this time indicates that the rhythmic changes recorded in the cortical EEG apparently are caused by current flow in the oscillating dipoles that are produced upon the dendrites of the cortical cells and their cell bodies by fluctuating afferent inputs. The superficial layers of the cortical gray matter contain relatively few cells; however the abundant dendrites in this region form an exceedingly dense mass of interdigitating fibers (Fig. 7.70B). Since

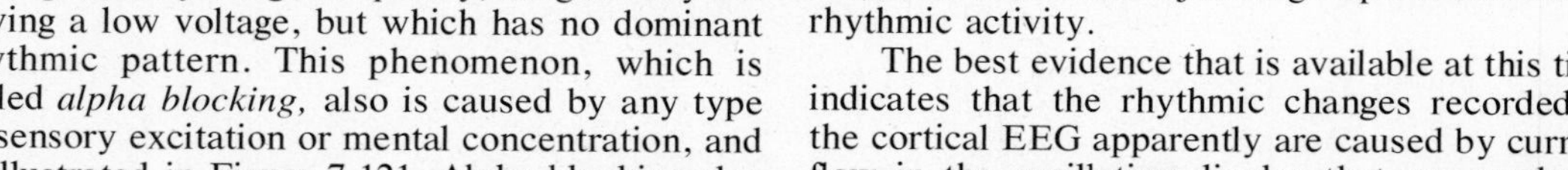

S

FIG. 7.121. Desynchronization of the EEG in the rabbit induced by an olfactory stimulus (S). Note the sudden appearance of irregular, high frequency, low amplitude waves upon application of the stimulus. (After Ganong. *Review of Medical Physiology,* 6th ed, 1973, Lange Medical Publications.)

dendrites usually do not transmit all-or-none action potentials (p. 163), local (or nonpropagated) hyperpolarizing and hypopolarizing potentials are produced on the dendritic processes by excitatory (facilitory) or inhibitory endings on the dendrites. In turn, activity in these presynaptic endings produces a localized current flow into (and out of) the postsynaptic nerve cell body as well as its dendritic processes. Consequently the functional interrelationship between the nerve cell body and its dendrites becomes an ever-changing dipole, and the current flow in this dipole would induce detectable oscillations in a volume conductor (eg, the interstitial fluid) as depicted in Figure 7.122. If the net sum of the dendritic activity is positive with respect to the cell body, the cell is hyperpolarized and consequently less excitable. Conversely, when the net sum of the electric activity on the dendrites is negative relative to the cell body then the cell is hypopolarized, hence nearer to threshold, and thus hyperexcitable. In other words, the net dendritic potentials fluctuate rhythmically and thereby the excitability of the entire nerve cell membrane will be influenced by the electrotonic spread of such current fluctuations. Therefore the electric characteristics, hence the ir-ritability, of a large number of nerve cells in a particular area of the cortex can change simultaneously, provided that no afferent (ie, desynchronizing) inputs are received which would disturb this regular process; and when the electric activity of large groups of nerve cells fluctuates in synchrony, rhythmic electric waves are detectable in the EEG.

There are two factors which lead to the synchronous electric activity of large groups of cortical dendrites such that rhythmic waves appear in the EEG. First, afferent impulses transmitted from the thalamus and other subcortical areas can induce synchronous cortical activity. A shallow, circular cut made in the cortex does not desynchronize the alpha rhythm, provided that the circulation is undamaged; therefore a lateral spread of impulses in the cortex is not essential to synchrony. However, the rhythmic wave pattern becomes much deranged if the deep-lying connections beneath the cortical area within the circular cut are severed. The effects of stimulation of particular thalamic areas upon the cortical potentials are considered on page 426.

The second mechanism whereby wave patterns are synchronized in the cortex is illustrated

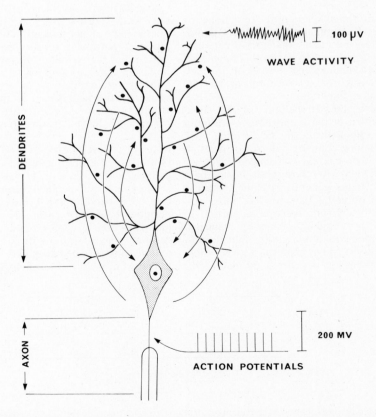

FIG. 7.122. The electric activity as recorded from the dendrites compared to that recorded from an axon. The ebb and flow of current on the active presynaptic terminations on the dendrites (black dots) causes potential fluctuations that are recorded as wave activity. In contrast to these strictly localized potential alterations, all-or-none action potentials are transmitted along the axon. (After Ganong. *Review of Medical Physiology,* 6th ed, 1973, Lange Medical Publications.)

in Figure 7.123, and this type of synchrony takes place between adjacent and parallel dendritic processes. When two (or more) nerve fibers are in lateral contact with each other in a volume conductor, and one fiber is stimulated, the electric properties of the second fiber are modified appreciably even though the second fiber may not discharge an all-or-none impulse. The excitability of the second fiber is altered by passage of the impulse down the first fiber as follows. Flow of current into the depolarized region on the active fiber (Fig. 7.123B) hypopolarizes the adjacent region of the second fiber, but hyperpolarizes the second fiber at two other points, and thus areas which are more positively charged with respect to other regions of the nerve develop at loci where inhibitory (ie, hyperpolarizing) nerve terminals are active. This purely electric transmission of impulses, or of electric effects, from the membrane of one nerve fiber to that of another fiber in lateral contact with it is called *ephaptic conduction,* and such physical junctions are called *ephapses.* This physical and functional arrangement is in marked contrast to conventional synapses (p. 188), and it will be apparent to the reader that since no transmitter substance is involved, ephaptic conduction is entirely an electric effect which can take place in either direction.

Since the outermost layers of the cerebral cortex are composed essentially of extremely dense networks of neural processes that are oriented in similar planes with respect to each other, then the generation of current sinks and sources in adjacent fibers tends to synchronize current flow by the process of ephaptic conduction.

The cortex of the cerebellum (p. 309; Fig. 7.58) as well as that of the hippocampus are both quite similar morphologically to the cerebral cortex, insofar as having complex patterns of parallel subpial dendritic processes overlying a layer of neurons is concerned. Therefore, both the cerebellum and hippocampus generate rhythmic oscillations in surface potential that are like those recorded in EEGs made from the cerebral cortex.

In contrast to the synchronizing mechanisms discussed above, it was noted earlier that desynchrony of the alpha wave pattern occurs when specific sensory inputs to the cortex are increased,

and the rhythmic pattern of the EEG is altered so that an irregular low-voltage activity develops. Stimulation of the specific sensory nuclei and pathways up to, but not above, the level of the midbrain causes desynchronization. However, stimulation of the specific sensory relay nuclei of the thalamus (Table 7.7) or the primary cortical sensory areas does not produce desynchronization. In contrast to these responses, high rates of stimulation applied to the reticular formation in the tegmentum of the midbrain as well as to the nonspecific thalamic projection nuclei produces desynchronization of the EEG and will awaken a sleeping animal (see also p. 422). When massive lesions are placed in the midbrain such as to interrupt the medial lemnisci in addition to other specific ascending sensory pathways, desynchronization still occurs following appropriate sensory stimulation. However, lesions induced in the tegmentum of the midbrain which interrupt the reticular activating system, but without disturbing the specific sensory systems, are accompanied by a synchronized EEG pattern which is not modified by sensory stimulation. Experimental animals which have the midbrain lesions described above are conscious, whereas animals having tegmental lesions remain comatose for extended intervals. Consequently it would seem that the afferent inputs to the cortex which are responsible for producing the desynchronization which follows stimulation of particular sensory endorgans are transmitted up the specific sensory systems to the midbrain. Thence the signals enter the reticular activating system by way of collateral fibers; subsequently the signals pass to the thalamus and ultimately reach the cerebral cortex via the nonspecific thalamic projection system.

EVOKED CORTICAL POTENTIALS. The cerebral cortex not only manifests continuous electric activity as described in the previous sections, but also a number of specific types of electric response can be evoked in the cortex by stimulation of peripheral sensory receptors or loci along the pathways that ascend from the receptors to the cerebral cortex. Careful study of such *evoked potentials* has revealed that two fiber systems functionally (as well as morphologically) connect the

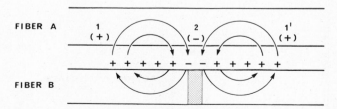

FIG. 7.123.   The electric properties of two adjacent nerve fibers. Current flows into the depolarized (stippled) area of active fiber B from the surrounding membrane as an impulse passes along the nerve. At points 1 and 1' on the membrane of the inactive fiber A, positive charges build up (+), thus the membrane becomes slightly hyperpolarized in these two regions. At point 2 of fiber A, on the other hand, positive charges are "removed" so that the membrane undergoes a slight depolarization at this point.

receptor endorgans to the cortex. The first of these fiber systems transmits impulses from the sense organ directly to specific cortical areas via a chain of three or four interneurons, and this system exhibits an extremely high degree of spatial orientation of its component fibers; hence the signals transmitted by this system are oriented similarly (p. 343), so that a high degree of sensory localization results. The second fiber system branches from the direct pathway outlined above at the levels of the medulla and midbrain, and thus afferent signals are transmitted to the cortex indirectly via the reticular formation of the brain stem and diencephalon. The terminations of this second fiber system within the cortex thus are quite diffuse, in contrast to those of the direct afferent pathway. Another component of the diffuse fiber system diverges at the thalamic level.

The functional characteristics of each of these two afferent fiber systems are understood chiefly from the potentials which they produce in the cerebral cortex as well as by the manner in which these systems induce or alter spontaneously rhythmic cortical activities.

From a technical standpoint, the electric events which occur following excitation of a sense organ can be recorded by an exploring electrode placed upon the cortex, an indifferent electrode being located some distance away upon or within the body. The exploring electrode can be placed either intra- or extracellularly; however, in the latter instance the net electric activity of a much greater number of neurons may be sampled than with an intracellular electrode (cf., microelectrode studies of individual cerebellar neurons, p. 401).

Following excitation of a particular sensory endorgan, two general types of potential are generated and can be recorded from the cerebral cortex. First, a sudden potential change called the unit spike can develop, and this activity persists for only one or two milliseconds. Second, the other type of potential develops relatively slowly, is of long duration, and is called the primary evoked response (or primary evoked potential).

The *unit spike* (Fig. 7.124A) is related to the electric field in the immediate proximity of a single active cortical neuron, hence is recorded by microelectrodes whose tip diameter is relatively small (ie, on the order of 2 μ) in order to prevent damage to the cell (or group of cells) under study. Recordings of unit spike activity have provided important data concerning the specificity of individual

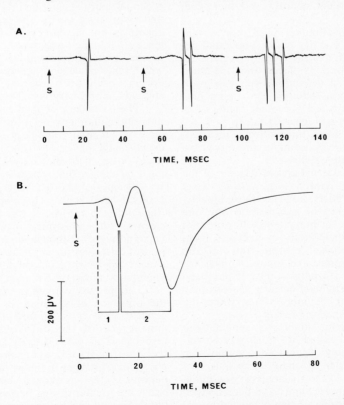

FIG. 7.124. **A.** Unit spikes recorded extracellularly from a single neuron of the postcentral gyrus (monkey) at different times following delivery of a single stimulus (S). Note that the cortical neuron can fire one or more times in response to a single stimulus. The latency of response also is subject to spontaneous variation. **B.** Evoked electric response recorded from the cat sensory cortex following contralateral sciatic nerve stimulation (S). 1 = primary evoked response; 2 = secondary response. (**A.** Data from Ruch and Patton [eds]. *Physiology and Biophysics,* 19th ed, 1966, Saunders. **B.** Data from Ganong. *Review of Medical Physiology,* 6th ed, 1973, Lange Medical Publications.)

neurons and their pathways insofar as receptive fields are concerned (p. 340).

Primary evoked potentials are detected by placing a large electrode on the cortical surface; hence the recordings reflect the summated activity of a large number of neurons which are located beneath the electrode. Subsequent excitation of a sense organ, a peripheral sensory nerve, or a thalamic relay nucleus produces a diphasic potential having a magnitude up to 1 mv, the potential being localized specifically in the sensory receptive areas of the cortex (Fig. 7.124B). In fact, the ultimate cortical locus where the primary evoked potential can be recorded from stimulation of a particular sense organ is so specific that the evoked potential has been used to map the specific primary sensory areas in the postcentral gyrus of the cortex (eg, Fig. 7.74). This technique has been used to map the projections of the cutaneous, retinal, and cochlear receptors upon the cortical sensory areas. Such a neurophysiologic procedure is valid for determining the spatial orientation of peripheral as well as central endings because the electric activity in the direct, rapid-conducting sensory pathways not only is synchronous, but also because these pathways have an exquisite spatial orientation of their component neurons and their fiber processes.

As indicated in Figure 7.124, the actual configuration of the primary evoked potential depends upon the position of the recording electrode. Thus when this electrode is placed upon the cortical surface, the initial deflection usually is positive, and this phase is followed by a more prolonged negative phase. On the other hand, when the exploring electrode is inserted "deep" (ie, 0.2 to 0.3 mm) into the cortex, this pattern is inverted; ie, the initial deflection is negative, and this phase is followed by a positive deflection.

Apparently the external granular layers of the cortex (Fig. 7.70) are related to the genesis of the evoked cortical potentials, however these bioelectric phenomena cannot be ascribed to any one neural source. The primary evoked potential commences as electric activity ascends from below into the terminal ramifications of the thalamocortical afferents, and then continues for an interval of 20 to 30 milliseconds. It has been demonstrated that individual cortical neurons may discharge spikes one or more times while the primary evoked potential is present.

If the experimental animal is anesthetized lightly, the primary evoked potential may be followed by a short volley of low-frequency positive waves (ie, about 10 per second.) These *repetitive waves* are produced in the thalamus by the initial afferent volley of impulses and have been considered to be an after-discharge that is caused by oscillatory activity between the cortex and the thalamus (p. 348).

If, on the other hand, the animal is anesthetized deeply, the specific primary evoked potential frequently is succeeded by a second positive–negative potential change, as indicated to the right in Figure 7.124B. This *secondary evoked discharge* (or *response*), in contrast to the highly localized primary evoked response discussed above, is quite diffuse and it appears simultaneously over most of the cortex and other parts of the brain after a rather consistent latency of between 30 and 80 milliseconds following the initial stimulus.

The diffuse secondary response is not altered by a circular cut through the gray matter which isolates a "button" of tissue containing the locus of the primary evoked potential from all lateral connections. This fact, together with the consistent latent period that elapses before the appearance of the secondary evoked response, argues in favor of the conclusion that this electric activity must arise from subcortical regions. Apparently the diffuse or nonspecific thalamic projection system (Table 7.7; p. 293) is involved in the genesis of the secondary evoked discharge.

Furthermore, within the primary sensory receptive areas of the cortex (p. 327) the neurons which fire during the course of the primary evoked discharge do not fire while the secondary discharge is taking place. Consequently a different group of neurons must be involved during the secondary process. Lastly, the secondary evoked discharge is not similar everywhere throughout the cortex. Thus the secondary activity that is elicited in the association areas is different from that which takes place simultaneously within the visual cortex, both of which were evoked by one stimulus, ie, primary visual stimulation.

The secondary evoked response is related intimately to another bioelectric phenomenon. When an experimental animal is anesthetized rather deeply, and obvious sensory inputs are absent, trains or bursts of almost synchronous waves having a frequency that ranges between 8 and 12 per second can be recorded from large areas of each cerebral hemisphere. These spontaneously occurring *cortical burst waves*, like the alpha waves and sleep spindles illustrated in Figure 7.70, tend to wax and wane. Although such spontaneous bursts of cortical waves are not true evoked potentials, a similar bioelectric phenomenon can be evoked by stimulation of the reticular system of the thalamus.

The primary evoked potential is unaltered by the presence of spontaneous bursts of cortical activity; however such activity does prevent the appearance of the diffuse secondary evoked response. This burst activity apparently is indicative, but not a cause, of the spontaneous physiologic changes in the irritability of the central nervous system, as evoked cortical responses generally are greater during periods of burst activity, but depressed during the intervals between bursts.

Spontaneous cortical bursts are produced through activation of a diffuse (nonspecific) thalamic projection system, in contrast to the repetitive waves which are generated in the specific thalamic projection nuclei.

The secondary evoked potential also is related to another bioelectric phenomenon called the *recruiting response*.* Thus a series of electric stimuli applied to the intralaminar thalamic nuclei causes the appearance of a train of negative-then-positive diphasic potential waves to appear in both cerebral hemispheres following a lag period of 15 to 60 milliseconds. If the frequency of the applied stimulation lies between 5 and 15 per second, then the amplitude of the evoked potentials gradually increases at first (Fig. 7.125A). Subsequently, the trains of evoked potentials rhythmically increase and decrease in amplitude, their frequency ranging between 8 and 12 per second. The configuration and frequency of the potentials evoked by the recruiting response are like those of the spontaneous cortical bursts. Furthermore, the spatial distribution of the recruiting response in the cortex is identical with that observed when spontaneous burst activity occurs.

The functional classification of the thalamic nuclei (p. 293; Table 7.7) has been developed as the consequence of studies done on the cortical potentials which are evoked following stimulation of the thalamic nuclei. Thus extremely localized stimulation of the relay and association nuclei produces a similarly localized focal activity in the cerebral cortex following a latent period of 1 to 5 milliseconds; hence these structures are called *specific thalamic nuclei*. Excitation of the other

The recruiting response of the cerebral cortex must not be confused with recruitment. In the latter situation, a reflex response increases gradually to a maximum when a prolonged stimulus of constant intensity is applied (cf. p. 340).

thalamic nuclei produces diffuse, bilateral, cortical potentials which are indicative of a recruiting response; these structures are called the *nonspecific thalamic nuclei*. The latter nuclei discharge the cortical nuclei which in turn generate the continuous electric activity of the brain that is recorded as the EEG.

Wakefulness and Sleep

The normal diurnal wakefulness–sleep cycle is one of the most obvious circadian rhythms that can be experienced as well as observed, and the mental and physical benefits which accrue to the individual from alternating periods of wakefulness and sleep are equally evident. From a strict physiologic standpoint, however, the specific benefits that are derived from intermittent periods of sleep are far less clear. It is plain that sleep directly affects the performance of the entire nervous system, as prolonged bouts of sleeplessness (pervigilium) result in a progressive deterioration of one's mental and behavioral characteristics. Apathy, listlessness, and mental dullness or irritability may be the accompaniments of extended periods of wakefulness. In turn, the irritability may yield to outright psychotic manifestations, provided that the lack of sleep is extended for a sufficiently long interval. In fact, such neurologic symptoms can appear following deprivation of one component of normal sleep, namely the rapid eye movement (REM) phase of this state as discussed on page 432.

The particular functional effects of sleep upon the various somatic and visceral structures of the body are even more nebulous than those seen to

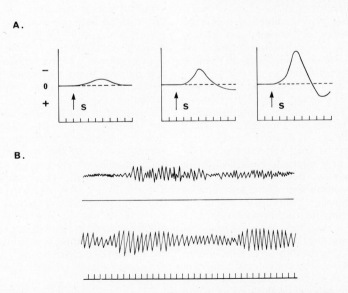

FIG. 7.125. **A.** Recruiting response to three successive stimuli of constant intensity which were delivered singly to the intralaminar (nonspecific) region of the thalamus. The polarity is indicated on the left. Note that the amplitude of the evoked response recorded from the cortex increases progressively. **B.** Electric responses of cortex to rhythmic electric stimulation of the intralaminar portion of the thalamus. Upper trace, "spontaneous" burst of activity; lower trace, waxing and waning response. (Data from Ruch and Patton [eds]. *Physiology and Biophysics,* 19th ed, 1966, Saunders.)

occur in the nervous system, and there is even some evidence which indicates that no impairment of such functions occurs even though these structures are not subjected to a regular wakefulness–sleep cycle.

During periods of wakefulness, the nervous system exerts certain general influences upon somatic and visceral functions. Thus tone of the skeletal muscles is increased and net sympathetic discharge likewise is augmented during the waking state. Conversely, during periods of sleep, impulse transmission to the somatic musculature may fall so low that tonus of the skeletal muscles becomes practically nil. Concomitantly, the discharge of the sympathetic nervous system also decreases markedly, whereas parasympathetic nervous activity may (or may not) be enhanced. A number of additional physiologic changes accompany the altered patterns of neural activity seen during the sleep state. For example the body temperature declines somewhat leading to a fall in the metabolic rate to basal levels (qv, p. 989). The heart rate declines and the vessels of the cutaneous vascular bed dilate; consequently the arterial blood pressure falls. The excretory function of the kidneys declines, whereas the activity of the gastrointestinal tract may increase somewhat depending upon whether or not the functional state of the parasympathetic nervous system is augmented.

For the purpose of the following discussion, the waking, aroused, or alert state can be defined as that condition in which a person or animal is capable of responding to various sensory stimuli and of having subjective experiences. Conversely, normal or physiologic sleep can be considered to be a reversible state of mental as well as physical rest during which volition and consciousness are partially or completely absent. Sleep also may be defined as a reversible behavioral state during which a characteristic and immobile posture is assumed, and the individual is relatively insensitive to external or internal stimuli.*

It must not be inferred that wakefulness and sleep are sharply distinct and mutually exclusive phenomena. Rather there is a smooth and graded continuum from keen alertness through drowsiness to light sleep and thence to deep sleep. Furthermore, depending upon internal and external conditions, an individual may fluctuate back and forth, both readily and rapidly, between any of these states. Therefore, wakefulness and sleep actually consist of many degrees of intensity, and these may shade either perceptibly or imperceptibly into each other.

The neuroanatomic and neurophysiologic substrates upon which the waking and the sleeping states are based now will be considered, and this material will be followed by a discussion of the particular physiologic aspects of the various sleep states.

ROLES OF THE HYPOTHALAMUS AND THALAMUS IN WAKEFULNESS AND SLEEP. When lesions are placed in the posterior hypothalamus of experimental animals a state of prolonged and comatose sleep ensues. Similarly, artificial stimulation of the dorsal hypothalamus of conscious animals induces sleep. Experimental observations such as these have led to the postulate that "wakefulness" and "sleep" centers are present in the hypothalamus. However, more recent experimental studies performed upon the reticular activating system (RAS, p. 302) as well as the nonspecific thalamic projection nuclei have indicated that the role of the hypothalamus, if any, in producing or regulating sleep is neither direct nor specific. Thus it is important to mention that the posterior hypothalamic lesions that produce a sleeplike comatose state in experimental animals also damage RAS fibers as they course to the thalamus and cortex. Consequently, the effects of such posterior hypothalamic lesions doubtless can be explained on the basis of destruction of a part of the RAS that is involved in producing wakefulness, rather than due to destruction of a presumed hypothalamic center that performs this function.

Similarly, the interpretation of experiments in which stimulation of dorsal hypothalamic areas produces sleep in conscious animals also can be criticized. Thus the dorsal hypothalamic areas which appear to induce sleep when stimulated lie extremely close to the nonspecific thalamic projection nuclei. Sleep is induced when the frequency of stimulation is at a rate of about 8 per second.† However, this frequency of stimulation is the same as that required to produce sleep when applied to the nonspecific thalamic projection nuclei, the orbital surface of the frontal lobe, and certain regions of the brain stem. Consequently it would appear that the effects of hypothalamic excitation are caused by stimulation of a nonspecific and diffuse neural system whose stimulation produces sleep, and thus the hypothalamus is not involved directly in the mechanism responsible for inducing this state.

Insofar as the thalamic component of a possible "sleep center" is concerned, it has been shown that electric stimulation of the lateral and ventral portions of the massa intermedia by chronically implanted electrodes will induce sleep in conscious animals. In contrast to the effects of hypothalamic stimulation described above, thalamic stimulation produces an organized behavioral pattern leading to sleep. A definite sequence of sleep-related activities is evoked by thalamic stimulation, including yawning, seeking an appropriate sleeping site, lying down, closing the eyes, somatic relaxation, and ultimately sleep.

ROLE OF THE RETICULAR ACTIVATING SYSTEM (RAS) IN WAKEFULNESS AND SLEEP. Although the actual source of an indi-

---

*In contrast to physiologic sleep, coma is a state of unconsciousness from which the patient or animal cannot be aroused.

†Stimulation at a higher frequency than 8 per second elicits an arousal response.

vidual's capacity to perform intellectual as well as volitional motor functions, to learn, and to perceive as well as to localize particular sensations lies within the cerebral cortex, the cortex itself cannot perform these functions unless it is "turned on" or "awake" so that a conscious or alert state is present. The principal neuroanatomic structure which is critical to production of a conscious state in the cortex of a sleeping person or animal is the *reticular formation* (p. 304). This phylogenetically ancient core of the medulla oblongata and midbrain also has a number of other functions, including centers which participate in the regulation of respiration, cardiac rate, blood pressure, somatic movements, and the secretion of certain endocrine glands. The reticular formation also has been shown to participate in the modulation or control of sensory inputs to the brain.

About 25 years ago it was discovered that stimulation of the reticular formation of a sleeping or drowsy cat by means of a chronically implanted electrode would awaken the animal as peacefully as though it awakened naturally. This experiment provided direct evidence that excitation of the reticular formation was responsible for producing

arousal or alerting of the cortex. Subsequently, the reticular formation of the brain stem and midbrain was named the reticular activating system (RAS) because of its ability to induce the waking state when stimulated.

As noted earlier (p. 303), all of the major sensory pathways of the body emit enormous numbers of collaterals which enter the reticular formation. Thereby somatic, visceral, auditory, visual, and other sensory signals from practically all regions of the body not only are transmitted directly to the cortex via the appropriate thalamic nuclei,* but also are shunted into the reticular formation (Figs. 7.50 and 7.126). Thus various sensory inputs can excite the reticular formation so that this structure in turn transmits signals to the cortex producing arousal (or wakening) of the cortex. The aroused or awakened cortex then is able to interpret directly the sensory signals which are being received.

The function of the RAS in producing arousal is nonspecific in that it responds identically to any sensory input as shown in Figure 7.126. Hence the

Olfactory signals excepted.

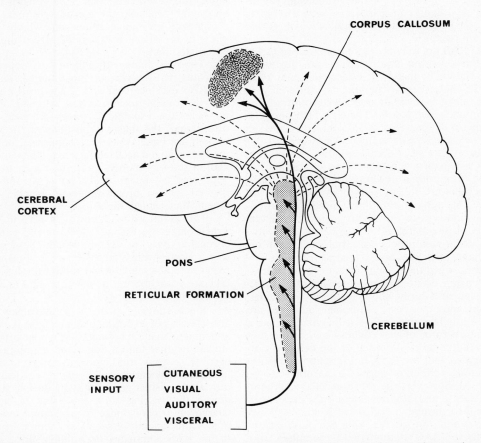

FIG. 7.126. The reticular formation of the human brain. Various sensory inputs send nonspecific impulses into the reticular formation via collaterals as well as to specific sensory areas of the cerebral cortex (shaded). In turn, the reticular formation sends nonspecific impulses throughout the cortex (curved dashed lines) to "awaken" the entire brain. See also Figure 7.50. (After French. The reticular formation. In Teyler et al. *Altered States of Awareness,* 1972, Freeman.)

response of the RAS to any sensory stimulus is to arouse or alert the brain and not to relay any particular information concerning the input signal per se. The signals that are relayed from the RAS are not focused upon any one particular cortical area as are those transmitted by the specific sensory systems. Rather the cortical inputs from the RAS are diffuse.

The RAS apparently is able to learn to discriminate among various sensory inputs. Thus a stimulus which would awaken one individual instantly is completely ineffective when applied to another person. For example, a person who has lived for a long period near a railroad depot would sleep undisturbed throughout the inevitable racket caused by the operations of the depot, whereas a newcomer to the area would be disturbed and awakened constantly by the barrage of auditory (and perhaps olfactory) stimuli. On the other hand, the individual who was conditioned to living near the sounds of the depot might awaken instantly at the relatively slight noise produced by a door opening in his home in the middle of the night, and this selectivity to different stimulus intensities also appears to reside within the RAS.

The RAS also plays a role in focusing the attention selectively on particular sensory inputs during the waking state, a function which is discussed on page 434.

The RAS not only is responsible for producing cortical arousal, but also it is necessary for maintaining the individual in a conscious state once aroused. Therefore if a person has suffered permanent injury to, or pathologic destruction of, the RAS, an irreversible coma ensues which only can terminate in death (see also Clinical Correlates, p. 461. When the RAS is uninjured, however, it can maintain a state of wakefulness (but not of consciousness) even when the cortex is absent. For example, in a normal newborn infant whose cortical functions are nil, periods of wakefulness are present throughout the day. Similarly, anencephalic monsters (ie, infants born with a congenital absence of the cerebral hemispheres) may live several years with proper care, but such creatures never have any true understanding of, or communication with, their environment. Nevertheless, in both instances cited above intervals of wakefulness are present during which feeding, swallowing, digestion, and other basic physiologic activities are carried out. The conclusion is inevitable that a rudimentary form of wakefulness can be present even in the complete absence of the cerebral cortex provided that the RAS can function adequately. For long periods of normal wakefulness to be maintained, however, the presence of the cerebral cortex is essential, and the alert state appears to be sustained through a reciprocal feedback of signals between the cortex and the RAS. Consequently the RAS is stimulated by impulses from some areas of the cortex as well as by impulses that originate in the sensory receptors that are located throughout the body. This fact can be de-

monstrated, for example, by electrically stimulating the orbital surface of the frontal lobe or the superior temporal gyrus in monkeys. This localized cortical stimulation will arouse a sleeping monkey and cause augmented reticular activity in such an animal. However, if the stimulation is applied to these cortical areas while the monkey is awake, the animal may immediately cease whatever prior activity in which it was engaged, and look about intently, if not to say questioningly, for want of a better term to describe the evoked response. Consequently it appears that the cerebral cortex, acting in concert with the RAS via diffuse systems of corticofugal fibers, has an important role in focusing attention during the waking state as well as providing a pathway whereby intracortical processes can initiate and sustain arousal of the individual. For example, direct fiber pathways pass into the RAS from nearly all areas of the cortex. However, the sensorimotor cortex (ie, the precentral and postcentral gyri), the frontal cortex, the cingulate gyrus, the hippocampus and other limbic structures, as well as the basal ganglia give rise to especially prominent corticofugal pathways that course into the reticular formation. Furthermore, such corticoreticular fiber systems conceivably might underlie the alerting responses which accompany emotional states and which can take place in the absence of any clearly demonstrable external stimulus.

There is some experimental evidence which indicates that the various functions carried out by the RAS are localized and it has been postulated that the caudal portion of the reticular formation underlies the wakefulness–sleep patterns. On the other hand, the thalamic portions of the reticular system are believed to be responsible for regulating the localized arousal patterns which constantly shift or scan from one area of the cortex to another with time as the normally active person is confronted with a broad range of specific sensory stimuli that change rapidly.

Impulses that ascend in the RAS desynchronize preexisting synchronized patterns of cortical activity (p. 422); however, the mechanism that underlies this effect is obscure. Furthermore, experimental stimulation of the RAS inhibits spontaneous cortical burst activity (p. 426) and the recruiting response (p. 427).

Since stimulation of the RAS causes arousal and alerting of experimental animals, it might be anticipated that such excitation would produce an elevated rather than a depressed irritability of the cortical neurons. It is known, however, that certain neurons located in the sensory cortical projection areas discharge at random during sleep but during the waking state the same neurons discharge only in response to particular sensory inputs. Consequently it seems as though the activity of the RAS eliminates the random firing that occurs during sleep by inhibiting cortical activity to some extent, and thereby the affected neurons are unhampered by extraneous impulses so that they now are able to respond solely to specific sensory inputs. In other words, the input signal is tuned more sharply even though the overall sensitivity of the entire system to excitation (ie, the net irritability) is decreased somewhat.

Certain pharmacologic agents initiate (or mimic) the physiologic action of the sympathetic nervous system on the RAS. Thus the catecholamines (epinephrine and norepinephrine) as well as amphetamine all cause behavioral alerting as well as electroencephalographic arousal patterns. This effect is brought about through a reduction in the threshold to firing of the reticular neurons located within the brain stem. Under physiologically stressful conditions, the adrenal medulla normally secretes quantities of catecholamines directly into the bloodstream (p. 1123). One major effect of the generalized adrenergic discharge that occurs during emergencies is an intensification of the alert state that in turn is essential to an effectual response to the emergency state per se.

The reinforcement of the alerting response to the secretion of catecholamines during periods of physiologic stress is mediated indirectly. That is, the epinephrine and norepinephrine liberated from the adrenal medulla do not act directly upon the excitability of the reticular neurons. Rather, these agents increase the blood pressure (ie, they exert a pressor effect) and it has been demonstrated that an increase in blood pressure alone, from whatever cause, will concomitantly increase the irritability of the neurons that comprise the reticular formation; that is, these neurons apparently possess a mechanoreceptor function.

In conclusion, what physiologic mechanisms appear to underlie the normal wakefulness–sleep cycle?

It is clear that awakening or arousal from the sleeping state is produced when sensory (or afferent) input signals to the cerebral cortex via the RAS reach a sufficient intensity to "turn on" or "activate" the cortex in a general fashion. The particular sensory receptors that provide these input signals are irrelevant, as the cortical arousal produced by the RAS is nonspecific and crude. It is far from clear, however, how these sensory inputs to the RAS become quantitatively adequate to trigger normal awakening following a night's sleep. One clue is available which perhaps may be relevant insofar as this process is concerned. A number of investigators have demonstrated experimentally the presence of as yet unidentified biochemical factors in cerebrospinal fluid as well as blood plasma of well-rested animals which can induce wakefulness when injected or transfused into recipient animals. Consequently it is tempting to speculate upon the possibility that under physiologic conditions one (or more) specific humoral substances can reach a sufficiently high level in the body fluids that the irritability of the RAS increases markedly. Thus any minor sensory input to the RAS would be adequate to produce cortical arousal.

Alternatively it could be suggested that normal arousal following a period of sleep is triggered by the depressed activity of a sleep center (or system) whose function when fully active is to inhibit the activity of the RAS and/or the cortex either directly or via certain humoral agents. Reciprocal activity between a "wakefulness system" (the RAS) and a "sleeping system" (several structures, particularly the thalamus?) would result in the normal wakefulness–sleep cycle.

In any event, once the waking state of the cortex is achieved, the alert or aroused condition normally is maintained for a number of hours by a continuous bombardment of the RAS by sensory inputs that arise in somatic as well as visceral structures, and in turn these inputs, in particular those from the muscle spindles,* serve to maintain cortical arousal via the RAS. In addition, the cerebral cortex, once aroused, transmits impulses back to the RAS, and these signals automatically tend to enhance RAS activity. In other words, a positive feedback loop is established functionally between the cortex and the RAS during periods of wakefulness. Lastly, any sympathetic neural activity that is present in the waking state would tend, a priori, to enhance the activity of the RAS, although the quantitative significance of this factor under physiologic conditions is unknown.

As the waking state becomes prolonged, the familiar symptoms of "drowsiness" or "fatigue" develop. Undoubtedly many factors contribute to these symptoms, including not only marked physical activity which induces neuromuscular fatigue, but also monotony and boredom. The latter two factors, of course, would contribute to drowsiness by decreasing sensory inputs to the RAS, hence the degree of cortical arousal would decline automatically. In addition, it has been shown experimentally that humoral agents are present in the plasma of fatigued subjects which can produce sleep in rested animals. Presumably such chemical agents have a role in the physiologic onset of sleep through their depressant action on certain areas of the nervous system (eg, the RAS and/or cortex), and possibly their stimulant action on a "sleep center" or "system" of which thalamic areas probably are a component. According to this view, therefore, normal arousal following an adequate period of sleep could be due to the elimination and/or metabolic inactivation of the humoral agent (or agents) responsible for producing sleep, so that the depressed excitability of the RAS neurons caused by the humoral agent returned to "normal" or "physiologically active" levels. The cir-

---

Physical activity, which sharply increases afferent signals to the central nervous system from the skeletal muscles, is a well-known factor in holding sleep in abeyance. On the other hand complete muscular relaxation, which markedly reduces such afferent signal traffic, can induce sleep even though the subject is not fatigued.

cumstantial evidence in support of this contention will be presented in the next section.

In this regard, it is of interest that the ability of general anesthetics to produce the unconscious state depends (at least partially) upon their ability to inhibit or depress conduction within the RAS (see also Clinical Correlates, p. 461).

Alternatively, the subjective feelings of drowsiness or fatigue which lead to sleep on a more or less regular daily basis also could be induced, in part at least, by a depletion of neurotransmitter substances in the neuronal pools which comprise the arousal system. This view is all the more attractive since the RAS is composed of enormous numbers of polysynaptic neural circuits; therefore depletion of transmitter substances conceivably would occur more readily here than in more direct pathways having fewer synapses. In any event, and regardless of the specific neural and biochemical mechanisms that may be involved, when the signal output of the RAS to the cerebral cortex decreases below a certain level, sleep ensues. There also is a strong implication that excitation of certain thalamic structures may participate actively in the behavioral patterns that lead to the sleeping state.

PHYSIOLOGIC CORRELATES OF SLEEP STATES. The EEG has been used extensively to gain some objective insight into the neurophysiologic alterations that occur during sleep. Some of the changes in the EEG that are recorded from normal human and animal subjects can be related directly to the depth of sleep. In addition to EEGs, eye movements, body temperature, reaction time, stimulus intensity for arousal, sequence in which various sleep patterns appear, electromyographs (p. 231), heart as well as respiratory rates, and the fluctuations in blood levels of certain hormones all have been studied and correlated with one another in normal subjects during periods of natural sleep.

It has been recognized since ancient times that there are two clearly different levels of sleep. However, only within the past half century has use of the electroencephalograph enabled investigators to demonstrate conclusively that there are fundamentally two sleep patterns that alternate with each other throughout a period of normal repose. Thus, a period of light sleep accompanied by slow brain waves of large amplitude characteristically is followed by a period of deep sleep in which the bioelectric activity of the cortex is accelerated markedly. Each of these two distinct sleep states may now be considered individually.

Light Sleep. Immediately following the onset of the sleep state in a relaxed and drowsy human or animal subject,* the alpha rhythm of the EEG is replaced by a series of low-frequency waves hav-

ing a larger amplitude than those seen during the alert state (Fig. 7.120). The subject can be awakened readily from this state of *light sleep*. As the period of light sleep progresses, groups of alphalike synchronized waves called *spindles* appear in the EEG, and some muscular tone still is present. Occasional gross body movements occur, but any eye movements that take place during light sleep are slow and of rare occurrence. The respiratory cycle is regular and deep, the heart rate is rhythmic and slow compared to the waking state, and the blood pressure also is somewhat below waking levels. In the male, penile erections do not occur during light sleep. Such mentation as occurs is repetitive and thoughtlike.

Light sleep also is called *synchronized sleep, slow sleep,* or *spindle sleep,* because of the characteristic EEG patterns that may be recorded. Light sleep also is called *non-rapid eye movement* (or *non-REM*) sleep because of the absence of conjugate eye movements during this state.

Deep Sleep. The light sleep state described above gives way after variable intervals throughout the entire sleep period to an entire spectrum of physiologic changes that mark the onset of periods of *deep sleep*. The high-amplitude slow waves seen in the EEG during light sleep generally are replaced in deep sleep by low-voltage, high-frequency, irregular (desychronized) waves that are far more characteristic of the alert than the sleeping state, although regular theta waves (p. 420) occur consistently at the hippocampal level. Hence deep sleep is also called *desynchronized sleep*. Despite this high level of cortical activity during deep sleep, arousal from this state is difficult, and complete muscular atony also is present. These observable facts present a distinct physiologic paradox; hence deep sleep also is called *paradoxical sleep*.

During periods of deep sleep, rapid lateral or back-and-forth (occasionally up-and-down) eye movements take place in a conjugate fashion, and these eye movements have been demonstrated to be associated with periods of dreaming. Consequently deep sleep is also called *rapid eye movement* (REM) *sleep* or *dreaming sleep*.

REM sleep is accompanied by involuntary twitches of the somatic musculature at irregular intervals, rhythmic contractions of the stapedius and tensor tympani muscles of the middle ear, a shallow respiration of variable rate, and a rapid heart rate. The blood pressure also may vary somewhat during REM sleep. Penile erections occur frequently in males during REM sleep, and in those individuals who exhibit bruxism (spasmodic or rhythmic grinding of the teeth not related to mastication), this condition is manifest during intervals of deep sleep.

Periods of light sleep alternate with periods of deep sleep throughout the night, REM sleep intervals superseding light sleep about every 60 to 90 minutes. Typically, periods of REM sleep become

*The cat is an extremely useful animal for investigating sleep phenomena, as this species normally sleeps for about two-thirds of the time.

longer toward morning, and in young adults approximately 20 percent of an average night's sleep is spent in the state of paradoxical sleep. In fact, studies in humans have indicated that there is an inverse relationship between the percentage of the total sleep period spent in REM sleep and age. Thus, infants a few days old spend approximately 50 percent of their sleep time in REM sleep, despite the fact that dreaming is unlikely because of their limited cortical experience. At the other end of the scale, persons in their seventh decade spend only 15 percent of their total sleep time in REM sleep. During the intervening years, between infancy and old age, proportionate declines in REM sleep also are seen with age.*

Possible Mechanisms of Light and Deep Sleep. If humans or animals are awakened every time an interval of REM sleep commences so that a net deficit of REM sleep is produced, the subjects may (or may not!) become irritable and "anxious." Subsequently, when allowed to sleep undisturbed, REM-deficient human and animal subjects spend a considerably greater proportion of their total sleep time in REM sleep than otherwise would be the case. These studies, among others, have led some investigators to the conclusion that a deficiency of paradoxical sleep causes a buildup of some biochemical factor, possibly an amine, which normally would be eliminated during periods of REM sleep. Although the present evidence for this statement is not wholly conclusive, nevertheless it is convincing (cf. p. 431).

Insofar as the neuroanatomic–biochemical system involved in the production of light sleep is concerned (in the cat, at least) it has been demonstrated that the nuclei of the median raphe of the brain stem are the chief source of serotonin (p. 648) in the brain. Subtotal destruction of these nuclei induces a state of relative sleeplessness (ie, the total percent of each day spent in sleep declines appreciably); consequently it would appear that this neurochemical system may play a role in the cyclic inactivation of the RAS leading to light sleep, as the decrease in net serotonin concentration caused by subtotal brain stem injury causes an increased wakefulness.

On the other hand, the neurochemical elements responsible for causing REM or paradoxical sleep appear to reside in the dorsal pons, specifically in an area known as the *locus caeruleus*. If the brain stem of a cat is sectioned at the upper level of the pons, a rhythmic cyclic variation between the states of wakefulness and paradoxical sleep develops. The latter state can be detected readily, of course, by rapid eye movements as well as atony of the skeletal muscles, in particular those of the neck. If, however, the brain stem is sectioned at the caudal level of the pons and im-

mediately rostral to the medulla, paradoxical sleep no longer is seen. Therefore, the area of the brain stem responsible for inducing paradoxical sleep lies in the middle of the pons, and further investigations have indicated that this region is the locus caeruleus, whose cells contain an abundance of *norepinephrine* (p. 195). It appears that this neurochemical system plays a role in the etiology of REM (or paradoxical) sleep like that of serotonin in the genesis of light sleep.

Insofar as the mechanisms related to the development of atony in the skeletal muscles during paradoxical sleep are concerned, it appears that the source of this inhibition lies principally within the spinal cord.

The physiologic roles of REM sleep and dreaming are quite obscure, and from a phylogenetic standpoint this type of sleep occurs only in mammals and birds. The latter animals only exhibit very brief intervals of REM sleep (ie, up to 15 seconds); consequently only 0.5 percent of the total sleeping time is occupied with paradoxical sleep. Mammals, on the other hand, spend a considerably greater proportion of their total sleeping time in REM, up to 20 to 30 percent in higher mammals such as apes and man.

In conclusion, it is evident that the "sleeping state" actually represents two distinct and individual physiologic conditions which alternate normally and "spontaneously" throughout the normal period of repose. Furthermore, it is evident that each of these two states of sleep represents a highly complex physiologic state, and neither is accomplished simply by "turning off" the cortex via decreasing afferent inputs from the RAS, as other neurochemical systems undoubtedly are active participants in the process.

A COMPARISON OF PHYSIOLOGIC ALTERATIONS DURING SLEEP AND TRANSCENDENTAL MEDITATION. Certain physiologic changes that are observed during transcendental meditation were discussed earlier (p. 413), and at that time it was noted that these changes have little in common with those observed during normal sleep. For example, oxygen consumption declines rapidly within a few minutes after a subject commences a period of transcendental meditation, hypnosis causes no appreciable change in this parameter, and during sleep several hours must elapse before oxygen consumption decreases significantly (Fig. 7.118).

The blood concentration of carbon dioxide rises significantly during sleep, a fact which indicates a decrease in respiration. Concomitantly there is a negligible decrease in the pH of the blood (ie, it becomes more acid), and this factor can be related to the decrease in pulmonary ventilation that accompanies sleep, and not to metabolic changes like those which accompany transcendental meditation.

The increase in skin resistance which accompanies sleep takes place at a much slower rate, and

It should also be mentioned here that the total sleep requirement also declines as aging takes place. Thus an infant may sleep 16 hours per day, whereas a septuagenarian may sleep only 6 or 7 hours.

is of a significantly lower magnitude, than that which is measured during transcendental meditation.

The EEG patterns characteristic of the sleeping state are composed primarily of low-frequency waves (roughly 12 to 14 cycles per second) which have a relatively high amplitude (ie, voltage), together with a heterogeneous collection of lower-amplitude waves having a number of frequencies. This typical sleep pattern is not seen during transcendental meditation.

In marked contrast to either the sleeping state or to periods of transcendental meditation, during hypnosis the EEG assumes the pattern that typically is found in the mental state which has been suggested to the subject. Similarly, the alterations in heart rate, blood pressure, respiration, and skin resistance all follow patterns like those of the suggested state in a hypnotized subject.

THE MODULATION OF SENSORY INPUTS: ATTENTION. If any higher organism, particularly a mammal such as man, were forced to give equal attention to, as well as act upon, all of the sensory stimuli to which it is exposed constantly in the course of everyday life, the total mass of this sensory information converted into signals within the nervous system would be such as to overwhelm the circuits responsible for producing orderly and effective response patterns by the organism. As such effective responses are critical to survival not only of the individual but also of the species, it would appear a priori that certain brain mechanisms must have evolved which function to protect the individual from sensory overload and to direct his attention selectively to those particular stimuli which require action most urgently. In the case of the human species, interest as well as mere biologic urgency plays a considerable role in the priority given by the individual to particular stimuli. Consequently attention and the neurophysiologic mechanisms which underlie and direct this process are an important aspect of the waking state. The term "attention" per se means a selective awareness of a part or an aspect of the environment, as well as a selective responsiveness to one type of stimulus. Obviously, therefore, experience (ie, learning) plays a critical role in determining which factors warrant attention by any particular species.

The remainder of this section will present a brief summary of several experimental findings which indicate that the brain contains specific mechanisms which are able to regulate the generation of sensory impulses, as well as to control their transmission in specific sensory systems.

Stimulation of the reticular formation in the medulla inhibits transmission of sensory impulses at the first synapse in the lemniscal and spinothalamic systems as well as in the trigeminal nuclei (see also discussion of reticular inhibitory and facilitory systems, page 304; Fig. 7.50).

Sensory transmission in the cochlear nucleus of the auditory system (p. 284) can also be inhibited by excitation of the cochlear nerve, although some investigators feel that this effect is caused in part by contraction of the stapedius and tensor tympani muscles (p. 511). Nevertheless, the cochlear nerve contains efferent fibers which pass to (and terminate within) the cochlea, and whose stimulation directly inhibits the generation of impulses within the organ of Corti (p. 507). Similarly, the optic nerve also contains efferent fibers, and when these fibers are stimulated, retinal activity is either increased or decreased.

The foregoing observations clearly indicate that the central nervous system contains neural mechanisms which are able to enhance or diminish the afferent signal input traffic to the brain, and these mechanisms may operate through their effects upon the specific sensory pathways or upon the sense organs themselves. Some of these mechanisms which regulate the sensory input traffic volume are found in the reticular formation of the brain stem and diencephalon, whereas others are not. For example, stimulation of the sensory cortex can produce inhibition at presynaptic terminals which are located within the cuneate and gracile nuclei. Similarly, other inhibitory mechanisms are located in the dorsal horns of the spinal cord (p. 464) and the specific thalamic nuclei as well as in the cerebral cortex. All of these mechanisms share one common function, and that is to modulate (or regulate) the volume of sensory inputs to the central nervous system.

Another effect upon sensory input which is altered by stimulation of the brain can be mentioned. When physiologic excitation of central nervous structures causes a generalized release of epinephrine, the threshold to firing of the cutaneous receptors is decreased considerably. This response serves to increase markedly the excitability of the peripheral receptors in emergency states during which an overall adrenergic response occurs, and signal output from the receptors is enhanced greatly.

When a human or animal subject focuses his attention upon a particular object, the bioelectric activity of some of the mechanisms which modulate sensory input can be evoked and recorded. For example, electrodes can be implanted chronically within the cochlear nucleus of a cat to record the electric activity of this structure. Following recovery, each time a sharp noise (eg, a click) is produced, a spike potential can be recorded from the cochlear nucleus. However, when the animal's attention is focused upon a particular stimulus such as a mouse or the odor of fish, no potential can be evoked by the clicking sound. However, when either the mouse or fish odor stimulus is removed, a cochlear potential is generated once again by the sound. It is apparent that when the attention is confined to a particular and narrow area or object in the external environment, at least part of this attention-focusing effect is caused functionally by a decrease in afferent input signals generated in the peripheral receptors themselves. Hence afferent signal transmission to the cerebral cortex may be reduced in the periphery, rather than centrally, during periods of focused attention.

The regulatory effect of central nervous structures upon the sensitivity of the muscle spindles provides another example of central regulation of sensory input (p. 369). Hence the net activity of the gamma efferent neurons that pass to the spindles is controlled not only by facilitory but also inhibitory effects through descending tracts which are located in the lateral funiculi of the spinal cord; and as discussed earlier (p. 306), stimulation of one portion of the reticular formation of the brain stem facilitates gamma efferent discharge whereas stimulation of other portions of the reticular formation inhibits this process (Fig. 7.50). In turn, the net gamma efferent discharge at any particular moment regulates the tension of the intrafusal fibers, hence the sensitivity of the spindles per se (p. 220). Similar to the reticular formation, the cerebellum (p. 307), the basal ganglia (p. 336), and the motor areas of the cerebral cortex (p. 329) all contain inhibitory centers which modulate sensory imputs in relation to the control of somatic motor activities.

Higher Cortical Functions; Learning and Memory

The higher mental faculties of the human species, collectively termed the "mind" or "intellect," have their principal neuroanatomic origin within the cortical neurons of the cerebral hemispheres. The term "mind" can be defined as those functions of the brain whereby an individual becomes aware of his surroundings and of their arrangement in space as well as in time, and by which he experiences (or perceives) feelings, emotions, and desires. Additional functional attributes of the mind enable the individual to remember, reason, attend to (or concentrate upon) either general or specific matters (stimuli), either of external or internal origin, to decide (or judge), and to associate ideas. In short, the mind enables one to feel, to think, to reason, and to synthesize ideas. In turn, an "idea" or "thought" can be defined, to paraphrase Darwin, as an abstract secretion of the neural elements of the brain.

It is patently ridiculous to assign the source of the mind to any transcendental, supersensible (ie, metaphysical), or vitalistic origin which lies beyond the limits of human experience. For example, in support of this statement, one merely has to reflect for a moment upon the complete lack of mental capacities in a congenital anencephalic monster (p. 430) or upon the derangement of mental (as well as physical) abilities which can follow necrosis of neural tissue in certain regions of the brain subsequent to a cerebrovascular accident in a previously normal individual. The studies to be discussed below, which were performed upon "split-brain" humans and animals further substantiate the concept that the "mind" is a product of some billions of neurons which are linked together both morphologically and functionally in fantastically complex patterns. Although these patterns and their workings are at present obscure to a large extent, the mind is nevertheless amenable to ultimate understanding, but only insofar as this understanding is grounded upon sound and rational physicochemical principles.

There are five basic techniques whereby some of the phenomena associated with higher mental processes as well as learning and memory can be studied in human and animal subjects. First, clinical observations made in humans can be correlated with the locus as well as the magnitude of brain damage seen at autopsy. Second, the information obtained from clinical studies can be extended by stimulation of specific cortical regions during neurosurgical operations that are carried out under local anesthesia. Third, subcortical structures may be stimulated by means of chronically implanted electrodes in patients suffering from, for example, Parkinson's disease or epilepsy. Fourth, the study of conditioned reflexes in humans and animals has shed some light upon the processes of learning and memory (qv, p. 439). Fifth, the alterations in psychic functions that develop subsequent to the performance of specific surgical procedures can be studied. Of particular interest to the present discussion are observations made upon patients following complete section of the corpus callosum, so that each cerebral hemisphere functions as an independent unit.

GENERAL PSYCHIC FUNCTIONS OF THE CEREBRAL HEMISPHERES (NEOCORTEX). The properties of learning and memory are not functions which reside within any circumscribed region of the brain in either humans or animals (p. 436); however, the centers which underlie the so-called higher functions of the nervous system, in particular those related to language (ie, speech and reading), appear to be localized to some extent in the neocortex, an area of the brain which achieves its highest level of development in the human species (p. 324).

From the standpoint of comparative anatomy, humans exhibit the greatest ratio between brain weight and body weight of any animal, and the most obvious gross anatomic feature of the human brain is the enormous and disproportionately great development of the three principal association areas of this structure. The *frontal association area* lies rostral to the motor cortex; the *temporal association area* lies between the superior temporal gyrus and the limbic cortex; and the *parieto–occipital association area* is found between the somesthetic cortex and the visual cortex (see, for example, Figs, 7.71, 7.72, 7.73, and 7.77).

To recapitulate briefly, these association areas are a portion of the six-layered mantle of neocortical gray matter which invests the surfaces (cortices) of the cerebral hemispheres. In turn, the slight histologic differences which Brodmann used to classify various cortical areas (Figs. 7.71 and 7.22) usually, but not invariably, can be correlated with functional differences.

The interconnections among the individual neurons of the neocortex (Fig. 7.70) form an exceedingly complex network. Thus the descending axons of the larger elements of the pyramidal cell layer emit collateral fibers which feed back via interneurons to the dendrites of the cells from which they originate. Consequently, highly complex oscillatory (or reverberatory) mechanisms can develop (p. 348). Such recurrent collateral fibers can also be related functionally to adjacent neurons of the same layer, whereas some of these fibers can also terminate upon inhibitory interneurons which then end upon the original neuron so that negative feedback loops are formed (p. 49).

On the other hand, the large and complex dendritic processes of the neurons lying deep within the neocortex can receive ascending nonspecific thalamic projection fibers, reticular afferent fibers, and intracortical association fibers which can terminate within all layers of the neocortex (Fig. 7.70). In addition, the deep-lying neocortical cells can receive specific thalamic afferent fibers that terminate in layer IV of the cortex (ie, the internal granular layer). From a functional standpoint, the role of the intracortical association fibers is unclear, as these elements can be sectioned without producing any obvious effects. From a physiologic standpoint, use of the term "association area," although convenient, is not entirely valid, as these areas of the neocortex must have far more complex and subtle functions than a mere interlinking of different regions of the brain.

It has been clearly established from an anatomic as well as a functional standpoint that the cerebral hemispheres of normal humans are asymmetric. For example, in one study it was demonstrated that a greater anatomic development in the left auditory area was present in 65 percent of the brains examined whereas 11 percent of these organs exhibited a more pronounced development in the same region of the right cerebral hemisphere. The remaining 24 percent of the brains studied in this series had an equal development of the auditory area in each hemisphere. Other investigations have established the fact that human language functions, including the analysis of spoken words, are localized to a somewhat greater extent in one cerebral hemisphere than in the other as discussed below, whereas the analysis of music occurs in the contralateral hemisphere.

The hemisphere which is concerned with the understanding of symbols (such as words) and the classification (or categorization) of ideas often has been referred to as the *dominant hemisphere* and its opposite member as the *nondominant hemisphere;* however now it is apparent that the second hemisphere is just as specialized in other functional areas than is the dominant hemisphere. Perception of temporal and spatial relationships

either in the real or in the abstract sense takes place in the nondominant hemisphere. Hence the nondominant hemisphere is responsible for performing such mental functions as the identification of objects through their shape (see stereognosis, p. 365), and the recognition of faces as well as themes in music. Because of these facts, the concept of a "dominant" and "nondominant" cerebral hemisphere appears to be outmoded and less accurate than a concept in which each hemisphere is considered to be specialized functionally in a mutually complementary manner to its contralateral partner. One cerebral hemisphere is specialized to subserve the functions involved in language and is called the *categorical hemisphere,** whereas the other is specialized for the perception of temporal and spatial relationships and is called the *representational hemisphere.*

Damage to the categorical hemisphere causes *aphasias,* whereas extensive lesions in the representational hemisphere do not produce such a defect, because the language functions are localized roughly within the neocortex of the categorical hemisphere. An understanding of printed as well as spoken words and the expression of ideas in writing and in speech may be defective when an aphasia is present, even though vision, hearing, and motor functions may remain normal (see also p. 462 and p. 463 for a further discussion of the aphasias).

When lesions are present in the representational rather than in the categorical hemisphere, *astereognosis* results so that the individual is unable to recognize common objects merely by handling them. The term "agnosia" signifies a general inability to recognize particular objects through the use of a specific sensory modality, even though the functions of that sensory modality are normal. Parietal lobe damage usually is associated with agnosias.

The complementary specialization of hemispheric functions is related to left- and right-handedness. Thus the left hemisphere is the categorical hemisphere in right-handed individuals. In left-handed persons, on the other hand, the right cerebral hemisphere is the categorical hemisphere in only about 30 percent of such individuals. In the remaining 70 percent of left-handed persons, the left hemisphere is the categorical hemisphere.

Damage produced in either the categorical or the representational hemisphere in an adult induces the characteristic defects associated with these regions, and these defects persist for long periods of time. Children who must undergo hemispherectomy (eg, for brain tumors) usually regain the functions subserved by the missing hemisphere quite rapidly, and regardless of which hemisphere was extirpated.

**This use of the term "categorical" denotes "explicit" or "specific". Therefore none of the various philosophic meanings assigned to the word "categorical" are meant or implied in this context.*

When the neocortex of the *frontal lobes* is dissected out of a primate, at first the animal exhibits apathy. This interval is followed by the development of an abnormally great motor activity (hyperactivity) so that the animal walks or paces constantly. The overall "intelligence" appears to be unaltered in general, and the responses to tests which require a prompt reaction to an environmental stimulus are unimpaired, whereas those responses which require the use of information that was learned earlier are abnormal. In human patients, frontal lobectomy causes defects in the recognition of events in their normally occurring temporal sequence. A lobectomized human may be unable to recall how long ago he saw a specific picture used as a stimulus. Right frontal lobectomy causes a major defect in the ability to recognize visual stimuli, whereas following left frontal lobectomy the principal defect is seen in tests that involve words as stimuli.

A mental patient can suffer from incapacitating tensions caused by delusions, phobias, and compulsions as well as by real (or imaginary) failure to perform adequately. If the white matter of the frontal lobes is severed from the rest of the brain by an operation called a *frontal* (or *prefrontal*) *lobotomy*, the tension is blocked despite the fact that the delusions, phobias, and other symptoms still are present. In other words, the patient becomes indifferent to the symptoms. Similarly an indifference to extreme and intractable pain develops after frontal lobotomy. From a practical standpoint, however, frontal lobotomy can be followed by a number of additional psychic and behavioral alterations which can be quite dissimilar among different patients depending upon the preoperative experience, personality, and environment, the quantity and site of the brain damage produced by the surgery, and so forth. For example, a given patient may become irresponsible, highly profane, vacillating, stubborn, indifferent to others and to the usual social amenities following lobotomy. Nowadays, tranquilizers and other psychotherapeutic drugs rather than lobotomy are the methods of choice for the treatment of mental disease. Nevertheless the clinical studies that have been performed on lobotomized patients have yielded some insight regarding the general functions of the frontal regions of the neocortex in humans.

Similarly, bilateral ablation of the temporal lobes in monkeys resulting in development of the Klüver–Bucy syndrome (p. 419) has yielded some information concerning this part of the brain. The effects of this experimental operation were discussed earlier in relation to the neurophysiology of behavior (p. 414), but may be reviewed briefly. Monkeys become docile, hyperphagic, and the males become hypersexual following bilateral temporal lobectomy. All of these effects can be ascribed to removal of components of the limbic system. The operated monkeys also exhibit visual agnosia as well as an abnormally great increase in oral activity as manifest by a continual and repeated mouthing, licking, and biting of all movable objects in their environment. It has been postulated that this behavior pattern could be induced by an inability to identify objects and/or to a memory loss caused by extirpation of the hippocampus. Such Klüver–Bucy animals also are distracted readily, every stimulus, minor or not, being explored regardless of its relevancy.

In humans, bilateral hippocampal lesions produce an impairment of memory for recent events. Some patients who have suffered bilateral injury to the amygdaloid nuclei and pyriform cortex may develop hypersexuality. In other words, certain aspects of the Klüver–Bucy syndrome are seen in humans suffering from lesions in particular neuroanatomic loci; however, the abnormalities which are observed following temporal lobe lesions in particular, as well as those caused by neocortical lesions in general, do not lend themselves well to the synthesis of any coherent and overall hypothesis of intellectual functions at this time.

On the other hand, a number of extremely important observations concerning motor as well as psychic functions of the individual cerebral hemispheres have been made in animals as well as humans who have undergone a total surgical section of the corpus callosum (p. 320), the enormous transverse band of white fibers which functionally interconnects the two cerebral hemispheres (eg, see Fig. 7.11, 7.63 and 7.69).

About 25 years ago it was found that cutting through the corpus callosum as well as the optic chiasm of the cat resulted in a preparation in which each cerebral hemisphere not only functioned independently of the other, but also whose visual input from each eye went only to the ipsilateral hemisphere (Fig. 7.44). When the animal was faced with a problem to solve, and only one eye was exposed, the visual information could pass only to the ipsilateral cortex. The animal could learn to perform a certain action, or solve a problem, based upon the information supplied solely to one eye. However, when the "informed" eye was covered, and the other "uninformed" eye was exposed to the same problem, the animal was unable to recognize the situation as the same problem, and necessarily was forced to learn the solution over from the beginning by use of the other half of the brain. This experiment indicates conclusively that the mammalian brain is indeed a dual organ, and this fact has been confirmed repeatedly by more recent studies.

To date a number of human patients have undergone section of the corpus callosum for therapeutic reasons, ie, for the treatment of uncontrollable epilepsy. Since this operation did not appear to impair the mental faculties of experimental animals, it was hoped that the procedure would confine the epileptic seizures to one of the hemispheres. Actually, callosal section eliminated practically all epileptic attacks, even those occurring unilaterally. Thus it would appear, at least superficially, as though the corpus callosum served to facilitate the epileptic attacks, presumably via some type of feedback mechanism.

The remainder of this discussion will present the results of a number of studies performed upon "split-brain" human and animal subjects.

Callosal section causes no apparent change in the personality, general intelligence, or temperament of the patient, although some alterations in daily behavior are evident. For example, the patients tended to favor the right side of the body insofar as their response to sensory stimuli was concerned. The right side usually is controlled, of course, by the left hemisphere (categorical hemisphere). For an extended period of time following the surgery, the patients rarely performed spontaneous movements with the left side of the body, and likewise tended to ignore (or to be oblivious of) sensory stimuli applied to the left side.

Tests were employed to examine the responses to visual stimulation presented in such a way that both left and right halves of the visual field were scanned individually, and as lights were flashed in a horizontal sequence the patients were requested to describe what they saw (see Fig. 7.44, p. 298). Every patient stated that lights were flashed in the right half of the visual field, but when lights were flashed only in the left half of the visual field, the patients reported that no lights were seen. The patients were not blind in the left field, however, because when they were asked to point to, not describe verbally, the flashing lights, their responses were almost entirely accurate. It was demonstrated that the right cerebral hemisphere is almost the same as the left insofar as visual perception is concerned. But since the speech centers of the brain are localized in the left (the categorical) hemisphere in most individuals, then it was not surprising that the patients were unable to state visual perception in the right hemisphere verbally, although perception of the stimulus per se was unimpaired as indicated by the ability to respond correctly when manual signals were used.

Stereognostic tests in which the patients were requested to describe and name various objects merely by touch resulted in findings that were similar to those obtained by visual stimulation. That is, if an object was held in the right hand, sensory signals were transmitted to the left hemisphere and the subjects were able not only to name, but to describe, the test object verbally. But when the object was placed in the left hand, whence sensory information is transmitted principally to the right hemisphere, the subjects could not describe the object verbally, although nonverbal matching tests whereby the test object was compared to a similar object at random revealed that stereognosis was unimpaired. Interestingly, it was also found that each cerebral hemisphere receives some sensory input from the ipsilateral side of the body as well as from the contralateral side. The latter input, of course, is the principal source of

sensory information normally, and the ipsilateral sensory input to each hemisphere was found to provide relatively crude information such as where the stimulus is located upon the body surface as well as whether or not a stimulus is present at all. However, the ipsilateral input cannot ordinarily relay signals that convey information about the qualitative aspects of the stimulus or test object.

When tests of motor control were applied to patients following callosal section, it was found that the left cerebral hemisphere exerted normal control over the right hand. However, the left hemisphere exerted less precise control over the left hand. The same results were obtained insofar as the right hemisphere was concerned. That is, the right hemisphere exerts full control over movements of the left, but not the right, hand. When the two individual hemispheres were in conflict insofar as initiating completely different movements in the same hand were concerned, the hemisphere on the side contralateral to the hand usually became dominant and the movement was carried out in accordance with signals emanating from the hemisphere on the opposite side of the body.

The results obtained on the intellectual capacities of split-brain human patients now may be summarized. A number of visual as well as tactile tests were employed in these studies, and these tests were administered in such a way that the test information was transmitted to only one hemisphere at a time. It was found that when visual or tactile stimuli were presented only to the categorical left hemisphere, the subjects were able to describe the stimulus (eg, a picture) accurately and normally not only orally, but also in writing. Conversely, when similar stimuli were directed to the right hemisphere, no accurate oral or written responses were evoked, and only random guesses as to the nature of the stimulus were offered, assuming that the subject made any attempt to verbalize a response at all. When nonverbal identification of visual and tactile stimuli was substituted for verbal answers, it became clearly evident that the right hemisphere possessed a considerable ability to perform accurately. But even when the subject held an unseen object in his left hand, he still was unable to describe it verbally, a fact which suggests that the left hemisphere was isolated completely from the right hemisphere, insofar as perception and knowledge are concerned. Additional tests did reveal that the right hemisphere had a certain capacity to comprehend language, although this ability was limited to mere recognition of common words by association of these words with the object they described and vice versa. But verbalization of the names of the objects was impossible when only the right hemisphere was involved.

Auditory inputs to one ear are transmitted equally to both sides of the brain. Tests were devised to investigate the ability of the right hemisphere alone to answer questions put to the subject orally. This was accomplished by having the subject use his left hand to retrieve various common objects from a bag hidden from his sight. Even abstract statements concerning the nature of the objects were followed by correct retrievals, but even when the correct test object was being held in the left hand, the subject was unable to name or to describe the object verbally. Hence it was clear that tactile information was being transmitted exclusively to the right hemisphere.

The capacity to perform specific feats with language differs considerably between the left and right hemispheres among individual patients who have undergone callosal section. However, it has been demonstrated quite clearly that the right (or representational) hemisphere was greatly inferior to the left (or categorical) hemisphere in the particular subjects which have been studied. Interestingly, the right hemisphere may respond accurately in tests to a concrete noun, whereas verbs are understood rather poorly. The grammar function is also poorly developed in the right hemisphere as indicated, for example, by the inability of this hemisphere to form plurals.

A number of different neurologic observations point to the conclusion that up to the age of about four years, a child has an approximately equal capacity and skill for language functions in each cerebral hemisphere. Beyond this age, however, a functional separation develops between the hemispheres insofar as proficiency with language is concerned. This separation of functions ultimately results in the presence of a categorical (usually the left) and a representational (usually the right) hemisphere in the adult, as discussed earlier (p. 436). The reason for this functional dichotomy between the hemispheres with advancing years is completely unknown.

Although language functions are highly developed in the categorical hemisphere, regardless of which side of the brain it happens to be on in any particular individual, the representational hemisphere does have some extremely important and specialized functions (see also p. 462). For example, it has been demonstrated in patients following callosal section that the left hand (ie, the right hemisphere) was able to perform quite accurately mechanical operations such as piling blocks or making drawings that involved spatial, or three-dimensional, information provided by a visual pattern. Conversely, the right hand under instructions from the left (categorical) hemisphere was unable to perform the same operations. Insofar as emotional expression is concerned, it appears that each hemisphere functions to an equal extent in the generation of an emotional response.

Collectively the studies performed upon human patients as well as animals following section of the corpus callosum indicate that two functionally independent brains are present within one individual, and that each brain is capable of performing highly sophisticated mental functions independently of the other. Furthermore, there is abundant evidence to substantiate the conclusion that each cerebral hemisphere of the normal adult is specialized to perform certain functions well and others poorly, and that these functions tend to supplement each other. Interestingly, there is direct evidence which indicates that following callosal section monkeys are able to handle twice as much visual information as they could prior to the surgery. This fact raises an interesting question regarding the possibility that a split-brain individual could learn twice as much information per unit time, or learn to perform twice the number of manual tasks in the same time span. One thing is evident, however, from all of the studies cited above. Section of the corpus callosum so that each cerebral hemisphere can function independently of the other produces an individual having "two independent spheres of consciousness within the same cranium." Therefore the psychic properties of consciousness, awareness, and perception are divisible in the literal sense of the word!

In conclusion, what may be said regarding the normal function of the corpus callosum? Some experimental evidence indicates that signals related to complex visual patterns are transmitted from one hemisphere to the other across the corpus callosum. It also appears that nerve impulses related to learning and memory are encoded, and these complex signals then are transmitted across the corpus callosum so that both hemispheres presumably are able to learn (and to recall?) the information simultaneously. However, the foregoing statements must be regarded as tentative rather than conclusive, and far more detailed factual information must be obtained before the physiologic role of the corpus callosum can be stated accurately.

NEUROPHYSIOLOGY OF LEARNING. The ability to learn is a fundamental characteristic of man as well as of many other organisms, and as consequences of the learning process, habit patterns, skills, and behavior are modified by experience. The capacity for learning (as well as that for memory) lies wholly within the nervous system and may be considered to reside in a sort of "functional plasticity" of this complex system whereby physicochemical modifications can occur as the result of experience. Although the mechanisms which underlie neural plasticity remain obscure for the most part, it is quite clear that the activities of the nervous system can be formed (or molded) by experience in such a way that responses to various stimuli can be altered considerably. Furthermore, as a result of this plasticity the nervous system is also able to store and to retrieve information through its memory functions (which actually are inseparable from the learning process), and can even recover functionally in some instances even though irreversible neural damage has been incurred in certain areas of the brain (see p. 461).

One particularly enigmatic feature that is

shared in common by the memory as well as the learning process is the temporal scale upon which these events take place. Thus the majority of neurophysiologic processes have a duration lasting for a few milliseconds or perhaps a few seconds. Learned information may (or may not!) take considerably longer to "store," and once committed to memory the information may be retained for many years. These generalizations imply strongly that the fundamental mechanisms which underlie the learning process are quantitatively as well as qualitatively different from those underlying simple reflexes.

The remainder of this section will deal principally with experimental studies concerned with learning, and in particular with certain overall behavioral modifications which are brought about by experience. The discussion of memory in the next section will consider several modifications that have been demonstrated to take place in the nervous system at the ultramicroscopic and molecular levels following a period of learning.

Some General Aspects of Learning. Frequently, and erroneously, it is assumed that the learning process is an exclusive function of the cerebral hemispheres. However, primitive forms of "learning" can be demonstrated readily to take place in many species including, for example, insects and worms. The octopus, a particularly interesting marine mollusc, has a demonstrably remarkable capacity for discrimination learning. In fact, this species can even exhibit a rudimentary form of intelligence, provided that one is content to define this term as a "comprehension" of, as well as an ability to respond appropriately to, peculiar environmental situations.

Furthermore, "learning" phenomena have been demonstrated to occur in mammals at subcortical as well as spinal levels in the central nervous system. For example, posttetanic potentiation (p. 217), which is the facilitation observed to take place in a synaptic pathway after repeated stimulation, is an excellent example of this kind of subcortical learning phenomenon.

The pigeon provides another example of a submammalian organism which is able to learn quite rapidly a number of highly complex and sophisticated visual discriminations.

Despite the facts summarized above, the more highly advanced learning processes that take place in man require the participation of the cerebral cortex even though the brain stem is also involved. The structural changes that have been observed to take place subsequent to a period of learning are discussed on page 445.

Learning Processes. Fundamentally two principal types of learning process are recognized, depending upon whether or not the subject plays a passive or an active role. During *classic* (or *Pavlovian*) *conditioning*, as well as during *habituation*, the subject plays a relatively involuntary or passive role; hence this type of learning is

considered to represent a relatively "primitive," "simple," or "generalized," learning process. On the other hand, during *operant conditioning* (also called *instrumental conditioning* or *trial-and-error learning*) the subject exerts a considerable degree of voluntary control over the learning process so that this type of conditioning is considered to represent learning at a much "higher" level than either classic conditioning or habituation.

The principal features of each of these learning processes now may be discussed individually, but it should be stressed that each of these processes is extremely important to the individual organism insofar as net learned information is concerned.

Classic Conditioning. Classic conditioning results in the formation of learned responses called *conditioned reflexes.* A conditioned reflex in turn is an automatic (or reflex) response to a stimulus which previously did not evoke the response. This effect is produced by repeatedly pairing the stimulus which normally does not produce the response (the *conditioned stimulus,* CS) with a second stimulus which normally produces a specific and innate response (the *unconditioned stimulus,* US). When the unconditioned and conditioned stimuli are applied in the correct temporal sequence and for a sufficient number of times, then ultimately presentation of the conditioned stimulus alone will evoke the response which originally could be elicited only by the unconditioned stimulus.

Among the classic experiments performed by Pavlov, salivation was induced in the dog in response to placing meat in the mouth (the unconditioned stimulus). If a bell was rung immediately before the meat was placed in the mouth, and this process was repeated a number of times, then eventually ringing the bell (ie, the conditioned stimulus) alone was sufficient to evoke salivation.

In order for conditioning of this type to take place at all, the conditioned stimulus *must* precede the unconditioned stimulus. If the conditioned stimulus is applied after the unconditioned stimulus no conditioned reflex develops. The conditioned response follows the conditioned stimulus by a time interval which is practically identical to that which separated the conditioned stimulus and the unconditioned stimulus during the training period. This delay between stimulus and response may range from a second or so up to 1.5 minutes, and when a considerable time such as this elapses between the application of the conditioned stimulus and the development of the conditioned response, the latter is called a *delayed conditioned reflex.*

When first developed, a conditioned reflex can be elicited not only by the conditioned stimulus but also by similar stimuli, eg, by ringing bells having slightly different tones to evoke salivation in the example cited above. If only one particular conditioned stimulus (ie, a bell having one specific tone) is reinforced by repetition, whereas the similar stimuli are not repeated, then an animal can learn to discriminate between the different stimuli with amazing accuracy, and this process is called *discriminative conditioning*. The phenomenon whereby the response to stimuli other than the conditioned stimulus is eliminated exemplifies the effect of internal inhibition, as discussed below. Discriminative conditioning is an exceedingly useful tool for studying color vision and pitch discrimination, among other sensory modalities, in various subhuman species.

Once a conditioned reflex becomes established, if the conditioned stimulus is applied a sufficient number of times in the absence of the unconditioned stimulus, eventually the conditioned response disappears. The process whereby this effect takes place is called *internal inhibition* (or *extinction*). On the other hand, if the animal is distracted or disturbed by an external stimulus which is presented immediately following the conditioned stimulus, the conditioned reflex may not take place, and this effect is called *external inhibition*. If, however, the conditioned reflex is *reinforced* occasionally by pairing the unconditioned and conditioned stimuli, the conditioned response will remain unaltered for extremely long periods of time.

An enormous number of specific somatic and visceral effects can be evoked as conditioned responses following application of suitable conditioning stimuli to humans as well as to animals. Insofar as the visceral responses are concerned, it has been demonstrated clearly that these supposedly automatic functions can be altered readily by appropriate discriminative conditioning regimes in a manner quite similar to those used to produce somatic conditioned reflexes; ie, a person can learn to alter his visceral functions in response to appropriate stimuli. For example, heart rate, blood pressure, intestinal contractions, rate of urine formation, and the caliber of the blood vessels all can be increased or decreased by the use of appropriate discriminative conditioning techniques (cf, p. 431). In this regard, it must be noted that conditioned reflexes are particularly difficult to form if the unconditioned stimulus evokes a purely motor response. Furthermore, the signals evoked by excitation of the sensory endings in the muscle spindles cannot be used as afferent inputs to form a conditioned reflex,* whereas afferent signals from the skin, eyes, or ears readily participate in such a function. On the other hand a conditioned response can be formed with relative facility, provided that the unconditioned stimulus is coupled to either a pleasant or unpleasant affect. Thus stimulation of the "pleasure" or reward system in the brain (p. 417) provides strong *positive reinforcement* for such learned visceral responses as were

outlined above, whereas stimulation of the "pain" or avoidance system (p. 417) of the brain (as well as a painful cutaneous stimulation) provides an equally strong *negative reinforcement* for such conditioned responses.

One fact is obvious concerning the neurophysiologic mechanism that underlies learning by classic conditioning techniques. That is, during this process a new functional connection (or connections) must be formed within the nervous system. For example, in the classic study by Pavlov concerned with salivation in response to ringing a bell it is obvious that a functional pathway is developed during the conditioning period between the auditory pathway and the autonomic centers in the brain that control salivation.

Although decortication can depress the formation of many conditioned reflexes, it is also known that the new functional connections present after a conditioned reflex has become established may or may not involve the cortex. For example, if a complex sensory pathway (such as is involved in vision or hearing) is necessary in order to transmit the conditioned stimulus, then the cortical sensory area that underlies perception of the particular sensory modality involved also must be intact, otherwise no conditioned response can develop. The remainder of the cortex, however, is unnecessary so that conditioned reflexes which do not involve discrimination and which are formed in response to simple sensory stimulation can develop when the entire cortex is absent. Under circumstances such as these, of course, subcortical structures obviously provide the neural substrate within which new functional pathways and connections are produced during conditioning.

Habituation. When a human or animal is exposed to a new (or novel) sensory stimulus, the subject becomes alert, and actively pays keen attention to the stimulus. The EEG exhibits a diffuse arousal (ie, desynchronized) pattern (p. 420), and evoked secondary electric responses become clearly evident in many parts of the brain. The behavioral response described above has been called the "What is it?" reflex, but more commonly the response is known as the *orienting reflex* (Pavlov).

When the stimulus is "neutral," that is, it evokes neither a pleasant nor an unpleasant affect, repetition produces progressively less and less response in the subject. Ultimately the arousal changes in the EEG are not observed following habituation, and the subject learns to ignore the stimulus completely. The gradual diminution in the behavioral response which is produced by the repeated application of a stimulus without reinforcement is known as *habituation*.

Habituation generally develops after the stimulus has been applied only a few times and it occurs in many species; therefore it has been used experimentally as a model of plasticity in the nervous system.

If a stimulus used to produce habituation in an

This fact is also considered by some investigators as specific evidence that proprioceptive information from the spindles is received and acted upon within the central nervous system at wholly subconscious levels.

animal now is paired with a second stimulus which elicits arousal (as evidenced by the EEG pattern), typical classic conditioning takes place. Thus after a few applications of the paired stimuli, the neutral stimulus to which the subject exhibited habituation now produces arousal (ie, desynchronization of the EEG) when administered alone. This type of conditioned response to a neutral stimulus is called *electrocortical conditioning* (or the *alpha block conditioned reflex*). If such electrocortical conditioning is not reinforced, extinction takes place rapidly; and concomitantly with this effect, the EEG develops quite regular, high-amplitude waves in the cortical area which was associated with the formation of the reflex in the first place. Experimental data such as these suggest that extinction (or unlearning) is similar to learning in that an active neural process is involved. Therefore extinction is not merely a passive subsidence of a previously active mechanism in the nervous system.

The functional connections produced during electrocortical conditioning appear to be formed at (or beneath) the thalamic level of the brain. Thus such conditioning is not altered by severing the lateral connections of the sensory areas of the cortex; however, this response does not develop following placement of lesions in the nonspecific thalamic projection nuclei.

The general arousal response to any particular stimulus can evolve into a behavioral reaction in which the attention is focused sharply and selectively (p. 429; p. 430); and when this state is present then extraneous sensory inputs are inhibited (p. 434). At a behavioral level, habituation and conditioning to the arousal value of various stimuli are common and everyday human experiences. For example, one can become habituated very rapidly to commercial advertisements which are presented visually and/or orally, and unless such stimuli produce a particular affect they will be ignored totally. On the other hand, a person becomes so conditioned to the sound of his own name that a specific arousal response is evoked immediately when that particular name is spoken. The adaptive value of such habituation and arousal responses are obvious.

Operant Conditioning. During classic (or Pavlovian) conditioning the subject has a passive role in the learning process so that the paired stimuli that elicit this type of learning always are presented in a stereotyped temporal pattern irrespective of the behavioral response that is evoked. In marked contrast to classic conditioning, operant conditioning results in the animal being taught (ie, learning) to make a response voluntarily, and this voluntary response in turn determines whether or not the conditioning is *reinforced*. In other words, the subject must perform a task or "operate on the environment," hence the name "operant conditioning" (Skinner). Affect is an important aspect of operant conditioning, hence the unconditioned stimulus may be either a pleasant or an unpleasant event (eg, food or an electric shock), whereas the conditioned stimulus is a signal (eg, a light flash or a sound) which alerts the subject to perform the task. Motor responses which are conditioned in such a way that they enable an animal to avoid a noxious stimulus are called *conditioned avoidance* (or *escape*) *reflexes*. On the other hand, the subject may be taught to perform a task following which a reward is given (eg, food), and this reward tends to reinforce the learned behavioral response.

In practice, operant conditioning involves an apparatus wherein an animal, such as a monkey or rat, can perform specific motor responses which then can be observed and/or recorded readily and quantitatively. In other words, the apparatus is an instrument upon which the subject performs a task, hence this type of learning also is called *instrumental learning* or *instrumental conditioning*.

A large number of problem boxes, discrimination apparatuses, and mazes have been constructed with which various aspects of instrumental conditioning can be investigated in many species. Only one such apparatus will be discussed here, the so-called Skinner box. In this type of apparatus, which can take a great variety of specific forms depending upon the species and problem under study, an animal such as a rat can be trained to press a lever each time a light flashes and thereby receive a food pellet. This exceedingly simple type of stimulus–response pattern clearly illustrates the principal feature of instrumental learning, namely that the animal must participate (or respond) actively in order to be reinforced. In practice, instrumental responses can be *shaped* or *patterned* according to *schedules of reinforcement* so that the animal might receive a food pellet only after each third press on the lever or after a particular time interval. In such a way the ultimate behavioral response that the investigator wishes to achieve is not produced at random but by gradual and deliberate shaping by successive reinforcements.

Animals can be taught, for example, to displace distinctively shaped or colored blocks, as well as to respond only to specific tones, light intensities, and colors following the administration of specific cues. In this way studies on cortical learning as well as localization and specificity of sensory functions can be carried out on normal as well as on experimentally altered animals with considerable accuracy following periods of so-called *discrimination learning*. For example, the conditioned avoidance response, mentioned above, often is used in practical pharmacologic tests of psychoactive drugs, such as the tranquilizers, which modify behavior. For example, an animal can learn to avoid an electric shock delivered through a grid in the cage floor by pressing a bar, and this behavior can be recorded against time. Following administration of a specific dose of a particular drug, the quantitative alteration in the bar pressing over a period of time can be taken as an index of the effects of the drug on that particular behavioral response in the animal under study.

However extrapolation of such data to human behavior may (or may not!) be valid.

In conclusion to this discussion, the term *reinforcement* now may be defined as it applies both to classic and to instrumental learning. During classic conditioning, reinforcement merely refers to the pairing of the conditioned and unconditioned stimuli in time, and as noted earlier, the former *always* must precede the latter in order for conditioning to occur at all. In instrumental learning, on the other hand, the term reinforcement signifies the pairing of a *response* on the part of the subject to some specific consequence of the original stimulus, for example, food, or the avoidance of a shock.

Interhemispheric Transfer of Learning. Classic as well as instrumental conditioning techniques have played vital roles in establishing to some extent the functional roles of the brain structures which underlie the interhemispheric transfer of learned information, and as noted earlier (p. 439), the corpus callosum plays a key role in this process. Two particularly relevant studies on this area of the brain were performed about 50 years ago in Pavlov's laboratory. If a normal dog is conditioned by the classic technique to salivate in response to tactile stimulation of one particular locus on the body surface, stimulation of the same point on the opposite side of the body evokes the same response. If, however, the corpus callosum was sectioned prior to the conditioning, the salivatory response was limited to cutaneous stimulation on only one side of the body. If the corpus callosum was sectioned prior to conditioning, then individual conditioned responses could be produced in each hemisphere. More recently it has been demonstrated unequivocally that not only the corpus callosum, but also the optic chiasm as well as the anterior and posterior commissures all play essential roles in the transfer and sharing of visual as well as other types of sensory information between the two cerebral hemispheres. It has been demonstrated experimentally in monkeys that signals transmitted via the corpus callosum are able to trigger memory traces that are stored in the contralateral hemisphere, but this structure does not underlie the transfer of *memory traces,* or *engrams,* between one hemisphere and the other. On the other hand, engrams actually can be transmitted from one hemisphere to the other by encoded signals that are carried in the anterior commissure. Stated in the language of the computer scientist, the corpus callosum only can command a memory "read-out," whereas the anterior commissure can induce interhemispheric memory storage or "write-in."

MEMORY AND RELATED MECHANISMS

Types of Memory. Even as learning causes alterations in response that are based upon prior experience, so memory enables an organism to store learned information for retrieval at a later time.

Hence memory is that mental faculty whereby particular sensations, impressions, and ideas are stored for recall.

Fundamentally it appears that three interlocking mechanisms are operative in causing memories, because there are basically three types of memory, each of which in turn is classified by its temporal characteristics. First there is *short-term memory* which underlies the immediate recall of events that occur on a moment-to-moment basis. This type of memory enables an individual to process hundreds of sensory impressions for a second or so or until the most important of these impressions can be acted upon; the overwhelming majority of these impressions, however, rapidly become extinct (or are forgotten). Second, *intermediate memory* of sensations and events lasts somewhat longer, and recall is on the order of from minutes to hours. This type of memory, for example, enables one to remember an address until the destination is reached. Third, *long-term memory* represents the highest level of information storage. This type of memory consists of sensory and other impressions that are gathered throughout a lifetime; and, once acquired, such memories are indelible. Learning at this memory level usually requires several hours to become complete, but once the information is stored, or consolidated, it remains stable for the lifetime of the individual.

Long-term memory is remarkably persistent, even when severe brain damage is present, and this fact contrasts sharply with the severe impairment or total loss of memory for recent events that may occur in persons suffering from various neurologic diseases or lesions.

Neuroanatomic Substrates of Memory. What can be said of the specific areas of the brain which are involved in the storage of memories? As is the case with the learning process itself, no one specific area of the central nervous system can be invoked as the primary area for all types of memory storage. However, the temporal lobe appears to be intimately concerned with storage of experiences that occurred in the distant past, at least in patients suffering from epilepsy in this region of the brain. Studies have been performed on such patients in that various regions of the temporal lobe, exposed by local anesthesia, were stimulated artificially. The result was that exceedingly detailed and complete patterns of visual and/or auditory memories, which frequently could not be recalled voluntarily, were evoked by the stimulation of particular loci on the temporal cortex. The memories produced by such temporal lobe stimulation are "replayed" accurately with a time sequence similar to that of the original event, but only for as long as the stimulation is continued. Frequently the patient felt as though he were an "outsider" viewing the memory replay as a detached observer. The evoked memories were complete, even to subjective impressions of the environmental situation surrounding the recalled events; however it is quite interesting that no reports of olfactory sensations were reported (or recorded) in connection with the recollections evoked by electric stimulation of the temporal cortex.

The facts described above differ markedly from the specific visual, auditory, and other sensory responses evoked by similar artificial stimulation of particular cortical sensory areas, again in conscious patients. For example, direct stimulation of the visual cortex (Fig. 7.71) evokes bright luminous objects (eg, stars or streaks) or dark forms. On the other hand, a veritable kaleidoscope of colors may be perceived. These evoked visual images may either move or remain stationary, but in no way do they resemble anything seen in the environment despite the fact that they exhibit form and color.

Direct stimulation of the auditory area (Fig. 7.71) can elicit ringing, buzzing, or knocking sounds; but no words or auditory memories having regular patterns are evoked.

Stimulation of the somesthetic cortex can produce a sensation of numbness, tingling, or a sense of actual movement. Perception of an unpleasant odor may follow stimulation of the cortical olfactory area, and a strong taste similarly may follow excitation of the gustatory area.

It is apparent from these facts that excitation of particular cortical sensory receptive areas follows generalized and innate (or inherited) patterns, whereas responses evoked by stimulation of the memory areas of the temporal lobe are quite different, and thus represent as well as reflect the specific experiences of that particular individual. In other words, the memories evoked by temporal lobe stimulation are a "hallucination of recollection."

The memory traces (or engrams) themselves do not appear to be stored within the temporal lobes. Rather it appears that excitation of the temporal lobes merely fires or "cues" an associative circuit thereby releasing impressions that are stored elsewhere in the brain so that an earlier impression or perception reaches the conscious level.

Occasionally a patient may alter his interpretation of his immediate environment following stimulation of different areas of the temporal lobes other than those which are involved in producing memories. Thus application of the stimulus may cause the patient to feel that an event which is currently in progress has occurred previously, or he may feel strange in familiar surroundings or vice versa. Such an untimely sense of familiarity with new events (or surroundings) is called the *déjà vu phenomenon.* Occasionally this occurs in normal individuals, although more frequently such a phenomenon is seen in an epileptic as an *aura,* the sensation which occurs just before a seizure develops.

In an experimental animal, the retrieval of a learned response is nullified completely if the brain is subjected to an electric shock (or the entire animal is chilled sharply) within five minutes of the training session. If, however, the animal is subjected to such treatment several hours after the training period, no failure of learning occurs, as indicated by the ability of the subject to perform adequately. Experiments such as these indicate that a definite period of time is necessary in order that long-term memory engrams be *consolidated* or *encoded* in the brain, and that the memory trace is labile (or unstable) for a period of time after exposure to the stimulus which produces the learning.

The hippocampus and its interconnections are known to be concerned intimately with the consolidation of learned information. For example, when humans suffer bilateral destruction of the ventral hippocampus, or when the same region of the brain is damaged in experimental animals, the memory for recent events becomes seriously deranged. Human patients who have suffered such hippocampal destruction still possess normal long-term memory, and such persons are able to cope adequately with everyday activities, but only so long as they pay close attention (ie, concentrate) on what they are doing at all times. If such patients are distracted from any task for even a very short period of time, however, all recollection of what was being done is lost, together with any plans that were present for undertaking new activities. Consequently, patients suffering from bilateral ventral hippocampal lesions are unable to form new intermediate and long-term memories in a normal manner.

Further evidence for an important role of the hippocampus in the formation of memories is obtained from human patients who have electrodes chronically implanted in this region of the brain. Thus stimulation of the hippocampus through these electrodes so as to cause disruption of the normal pattern of repetitive bioelectric activity in the hippocampal neurons produces a loss of memory for events that took place in the recent past. Furthermore, certain drugs which alter or impair the recall of recent events also can induce abnormal electric discharge patterns in the hippocampus.

Sometimes alcoholics suffering from brain damage concomitantly exhibit a pronounced loss of memory for recent events, and it appears that there is a good correlation between this derangement and the consistent finding of pathologic alterations in the mammillary bodies of such patients. Therefore it would seem that these structures play some role, albeit obscure at present, in the genesis of memory storage, ie, engram formation.

Physicochemical Mechanisms Involved in Learning and Memory. Any general hypotheses that are advanced to explain the learning and memory functions of the nervous system must be able to account satisfactorily for the mechanisms that underlie and characterize short-term, intermediate, and long-term (or stable) memory.

The process of learning as well as that of memory may involve merely the functional creation of new circuits within the nervous system in response to experience (ie, appropriate stimuli), whereas the basic and enormously complex neuronal patterns that underlie the morphologic aspects of these functions of the brain are developed highly and almost completely during prenatal life so that by the time the individual is born such patterns are clearly established. In turn, this complex neuroanatomic circuitry is determined during embryonic life by the genetic constitution of the particular organism, and furthermore it appears that the pathways formed in the brain during this stage develop in part because of specific chemical interactions between the membranes of particular neurons; ie, *cell recognition,* a common phenomenon in biologic systems, plays an important role in this process. In addition, the individual neurons, which in toto comprise the nervous system, do not undergo mitosis in postembryonic life, a fact which undoubtedly confers a certain measure of long-term anatomic stability on functional circuits, once these circuits have become established through learning processes. Furthermore, such long-term neuroanatomic stability doubtless plays some role in the stability, hence the retention, of long-term memory engrams.

An important distinction exists between the *formation of an engram* during the learning process and the *storage of an engram* as a part of the past experience of the individual. Thus it has been suggested that the acquisition and long-term storage of memories occurs in two stages. According to this view, during the learning process an immediate, unstable, and short-term memory trace (or engram) is formed and stored temporarily. Subsequently, this short-term memory engram may be "consolidated" into a long-term (or permanent) memory engram, and presumably the short-term engram serves as a "template" which facilitates this consolidation process.

The physical mechanisms which have been postulated to underlie the three types of memory belong to two general categories. In the first, a bioelectric process is involved, whereas in the second, morphologic changes in the neurons, synapses, and/or biochemical changes within the neurons themselves are deemed responsible for producing the memory engram.

Mechanisms Involved in Short-Term and Intermediate Memory. It has been postulated that the memory traces for recent events are produced by the activation of reverberatory (or oscillatory) circuits within the brain (qv, p. 348). Information is stored as memory engrams by closed circuits of neurons whose components reactivate each other in a cyclic fashion.

As noted earlier, electrically induced convulsions, deep anesthesia, and certain neuropathologic situations tend to abolish memory for recent events (p. 444), whereas such treatment does not interfere with long-term memory patterns. These facts would tend to indicate that reverberatory mechanisms are not involved in all learning and memory processes. However, there is some evidence that reverberatory mechanisms are responsible, at least in part, for the temporary storage of new information prior to the development of a permanent memory engram.

A generalized kind of learning can be said to have taken place whenever the nervous system functions in a different manner following a period of activity. The phenomenon of posttetanic potentiation (p. 217) can be cited as merely one example of this type of learning. Thus when the presynaptic limb of a monosynaptic reflex arc is stimulated for a prolonged interval at a high frequency, a hyperexcitability of that reflex develops which may persist for several hours. It has been suggested that this phenomenon might provide a useful biologic model in which to study the mechanisms of synaptic plasticity in general and the structural basis of learning in particular. Similarly, habituation has also been suggested as a model for such investigations (p. 441).

Mechanisms Involved in Long-Term Memory. Although the precise nature of the long-term memory engram remains an enigma, the fact that such memories are quite resistant to physical injury of the brain (eg, as through concussion) as well as to electric shock (p. 461) would suggest that permanent memory engrams develop as the consequence of actual morphologic and/or biochemical changes within the brain. Furthermore, the time interval required for the development of a permanent memory engram (or consolidation of a temporary memory trace) would tend to substantiate the hypothesis that physicochemical alterations are involved in the process of long-term storage of information as memories.

It has been demonstrated beyond doubt and in many species that the central nervous system can be modified structurally, chemically, and functionally by experience, and that the stimuli producing these modifications are effective regardless of whether they arise within the external or internal environments.

At the gross level, it was observed nearly two centuries ago that animals subjected to a period of intensive training exhibited visibly more complex cerebellar convolutions than did untrained siblings of the same species. Subsequently it has been demonstrated clearly that the total mass of the cortex is related to the learning process. For example, rats which were trained to perform various activities, or which were exposed to visually stimulating or physically complex environments, had somewhat thicker and heavier cerebral cortices than did litter mates maintained under isolated or monotonous conditions.

Careful study of electron micrographs made on sections of brains taken from rats which were subjected to a stimulating environment has revealed that the total number of spines on the dendrites of cortical pyramidal cells is considerably greater than the number of dendritic spines found on similar neurons in the brains of rats maintained in an "impoverished" environment (Fig. 7.127A).

The cell bodies of cortical neurons from rats exposed to a stimulating environment were larger than comparable cells of rats maintained in a monotonous environment. Similarly the total number of glial elements present in the cortex increases significantly when sensory experience is enriched.

Furthermore, it has been shown that the cross-sectional area of the synaptic junctions in the occipital cortex of rats exposed to a stimulating environment were about 50 percent larger than those present in the cortices of rats from a monotonous environment (Fig. 7.127B). The rats which had been deprived of a rich sensory experience, however, exhibited a quantitatively smaller number of synapses per unit area of brain tissue. Still other experiments have confirmed that there is an increased area of synaptic contact in rats that have had a wide prior sensory experience, and that an increase in the area of synaptic contact is related inversely to the total number of synapses per unit mass of tissue. These morphologic observations tend to support the contention that learning and memory are due, at least in part, to a distinctive pattern of addition as well as deletion of synaptic contacts between individual neurons con-

comitantly with reciprocally related variations in the total area of each synaptic junction. Furthermore, it appears that these processes may take place simultaneously, with the net result that the functional contacts between individual neurons in the brain are subject to considerable variation in number depending upon prior experience. It must be reiterated that the total number of neurons present in the nervous system does not change regardless of experience; however, studies such as were cited above clearly indicate that the cerebral neurons are capable of adapting morphologically to changes in their functional state, and that the selective morphologic alterations which take place at the synaptic junctions during the learning and storage of memories may represent an encoding of information at the morphologic level.

The visible morphologic changes noted above clearly reflect the fact that biochemical alterations must accompany the learning and storage of new information. What, then, can be said of the specific biochemical changes that have been found in the brains of animals (rats) following their maintenance in a highly stimulating environment, when these values are compared to the levels of similar components derived from the brains of suitable control animals? The occipital cortex of rats that were exposed to an enriched environment contains quantitatively higher levels of total protein, and in particular the levels of the enzymes, acetylcholinesterase, cholinesterase, and hexokinase are elevated. The quantity of deoxyribonucleic acid (DNA) per milligram of tissue was found to decrease significantly, whereas the level of ribonu-

A.   **B.**

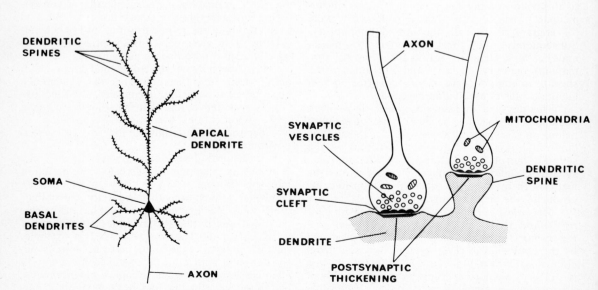

FIG. 7.127. **A.** Pyramidal neuron of cerebral cortex illustrating pattern of dendritic spines. **B.** Synaptic junctions directly between an axon and a dendrite (left) and between an axon and a dendritic spine (right). See also Figure 7.70. (After Rosenzweig, Bennett and Diamond. Brain changes in response to experience. *Sci Am* 226:22, 1972.)

cleic acid (RNA) decreased to an insignificant extent. Therefore the ratio of RNA to DNA increased significantly in rats subjected to a stimulating environment.

The facts summarized briefly above leave no room for doubt that learning and storage of new information can produce definite morphologic and biochemical changes in the brain, although the mechanisms that underlie these processes remain completely obscure, as does the mechanism underlying the recall or association of stored information, sometimes in response to specific stimuli or cues.

Insofar as the form in which learned information may be stored in the brain is concerned, a number of investigators have claimed that several specific and different biochemical compounds are synthesized in the brain in response to the stimulus of experience, and these compounds in turn are responsible for all of the enormously complex phenomena related to learning and memory. Thus peptides, proteins, glycoproteins, and RNA all have been implicated as "memory molecules" by various workers. It also has been claimed that learned information can be transferred from trained to naive animals by injection of suitable extracts prepared from the brains of the former. The evidence for these statements at present is interesting and at best circumstantial, but hardly compelling.

There is no doubt that polypeptide, protein, glycoprotein, and ribonucleic acid molecules each has the potential to store an enormous amount of coded information. It is also true that RNA serves as the template for synthesis of specific protein molecules, and that neurons synthesize more protein than any other cell type in the body. Furthermore it can be demonstrated clearly in animals that chemical agents which block protein synthesis also inhibit the consolidation of long-term memory engrams, but do not block memory for short-term learned responses. It is clear that protein synthesis (hence RNA activity) must provide at least a part of the mechanism which results in the formation of long-term memory engrams. However, such studies afford no insight whatsoever as to how such a chemical mechanism operates in detail, since one is studying a general physiologic process rather than a specific memory. Nevertheless the clues yielded by such biochemical studies provide data that ultimately will be of assistance in resolving the problems of memory functions, insofar as chemical effects on synaptic transmission (ie, neurotransmitter release) and other aspects of this phenomenon are concerned.

From a neurophysiologic standpoint, it is clear that the phenomena of learning and long-term memory do not involve merely the formation of new functional circuits or pathways which are localized in one portion of the brain or another, and which discharge on cue when appropriate stimuli are administered. On the contrary, there is considerable evidence which indicates that memory traces are redundant, and thus may be stored in multiple loci in the brain. It is equally clear that the presence of cortico–cortical circuits is essential to normal learning and memory processes. One fact is evident, however. Learning causes the functional synchronization of large numbers of neurons that may be located in widespread cortical areas and in turn this process involves excitation of some of these neurons together with a concomitant inhibition of others, so that specific patterns of neural activity are evoked in the brain, and these patterns of neural discharge have been postulated to give rise to specific memories.

At present, it is evident that the factual information concerning the mechanisms which underlie learning and memory are fragmentary at best. Ultimately, however, these phenomena will have to be explained in terms of biochemical as well as biophysical mechanisms, because neither of these two processes operates in a mutually exclusive fashion in any single living cell, let alone in the entire nervous system.

Some years ago synaptic transmission was thought to be a chemical process by some investigators, and a wholly bioelectric process by others. Subsequently it was learned that bioelectric and biochemical mechanisms, acting in a coordinated fashion, both were responsible for normal synaptic transmission. It would be most surprising if learning, memory, and the recall of stored information did not involve a similar functional cooperation between biochemical and bioelectric processes in widespread areas of the central nervous system.

Some Metabolic Characteristics of the Nervous System

Under normal conditions and at "rest" the nervous system, and in particular the human brain, has an inordinately high requirement for energy which is supplied primarily by the aerobic oxidation of glucose (p. 137). Consequently a continuous and uninterrupted supply of glucose and oxygen are critical requirements for survival of the nervous system, in particular the brain, throughout the lifetime of the individual. Although the total mass of the brain amounts to only some 2 percent of the total body weight of an adult, this organ alone has been estimated to utilize between 25 and 45 percent of the total oxygen consumed by the body under basal conditions, ie, during complete mental and physical rest. Similarly, the brain utilizes about 65 percent of the total circulating glucose supply, the remaining third of this nutrient being utilized to a major extent by the skeletal muscles. The human brain alone requires approximately 100 to 145 grams of glucose per day, a quantity that yields some 400 to 600 calories.

As demonstrated and emphasized about 60 years ago by Barcroft, "There is no instance in which it can be proved that an organ increases its activity, under physiological conditions, without also increasing its demand for oxygen." The nervous system is no exception to this general principle. However the proportional increase in energy consumption above the basal level during intense mental activity by the brain is relatively low, amounting to only 3 or 4 percent above the resting level. Skeletal muscles, on the other hand, can increase their energy consumption up to several hundred percent above their basal level of energy consumption during the performance of intense physical work.

Nerve cells in the adult body do not perform any obvious mechanical work nor do they undergo mitosis. Furthermore, the brain is exceedingly vulnerable to even a transient interruption of its glucose and oxygen supply so that irreparable damage can result from a deficient supply of these blood constituents of even a few minutes' duration. Therefore, what are the chief processes that account for the exceedingly high energy requirements of the nervous system?

First of all, the metabolic energy derived from the oxidation of glucose must be made available in such a form as can be utilized readily for a number of specific processes by the individual neurons. The synthesis of high-energy phosphate compounds, in particular adenosine triphosphate (ATP), lies at the center of all energy utilization by all neural (and other) cells of the body. Thus ATP synthesis basically accounts for the extremely high net energy consumption by the nervous system, as an uninterrupted and quantitatively adequate supply of this compound is critical to all aspects of neural function. In turn, the ATP made available to each neuron is channeled into three major systems, each of which requires considerable energy for its maintenance.

First, a large proportion of the total energy expended by neural tissue is required for the continuous maintenance of the active transport mechanisms at the cell membranes which underlie neuronal irritability. Thus operation of the sodium-potassium pump, which maintains and restores electrochemical gradients in all of the neural elements of the body, requires a large and continuous source of energy (as ATP) in order to function normally (p. 101).

Second, neural tissue is extremely active metabolically, hence a large fraction of the total energy consumption by the nervous system is consumed in specific biosynthetic reactions that in turn utilize ATP directly as an energy source. For example, the synthesis of adequate quantities of neurotransmitters (eg, acetylcholine, epinephrine, norepinephrine, and serotonin) requires large amounts of energy. Furthermore, the maintenance, repair, and specific morphologic alterations that all neural elements undergo during the course of their continued existence are dependent completely upon biosynthetic reactions. In particular, the rate of protein synthesis is extremely high in the nervous system and the elaboration of proteins requires the expenditure of a considerable quantity of energy, which once again is utilized in the form of ATP (p. 104).

Third, the factual evidence that is currently available clearly indicates that not only do neurons manufacture a host of specific biochemical substances, but also that the neurons contain mechanisms to transport certain of these compounds intracellularly, sometimes for considerable distances and at relatively fast rates (eg, as from the soma of a spinal motoneuron to the terminations of the axonal processes in the foot). Transport mechanisms of this type require energy expenditure for their maintenance as well as their continued operation (p. 198).

All tissues of the body, including neural tissues, contain three interconnected metabolic pathways whereby energy is stored in the form of high-energy phosphate bonds, principally in the form of ATP (p. 104). These metabolic pathways are, first, the Embden–Myerhof mechanism which underlies anaerobic glycolysis; second, the Krebs tricarboxylic acid cycle which is an aerobic pathway; and third, the electron transport system, which also is an aerobic mechanism. Each of these pathways is presented and discussed in detail in Chapter 14 (p. 931 et seq.). Briefly however, the participation of each of these biochemical mechanisms in the synthesis of ATP can be summarized as follows.

The glycolytic pathway underlies the metabolic degradation of glucose to pyruvate (under aerobic conditions) or to lactate (under anaerobic conditions), but only a relatively small quantity of ATP is synthesized by this latter mechanism (p. 936).

The tricarboxylic acid cycle is an aerobic mechanism which oxidizes the pyruvic acid produced by the glycolytic pathway into carbon dioxide and water (p. 945); hence this cytoplasmic mechanism requires the presence of an adequate supply of molecular oxygen in order to operate at all. This oxygen in turn unites with hydrogen ions (ie, protons) that are derived from pyruvate as well as certain of the intermediate compounds of the tricarboxylic acid cycle by way of the cytochrome and flavoprotein enzymes that comprise the respiratory chain (p. 931). The latter system, located within the mitochondria, is composed of a series of catalysts which are responsible for the stepwise aerobic transport of reducing equivalents (ie, electrons and hydrogen) as well as their eventual reaction with molecular oxygen to form water (p. 933).

The total number of high-energy phosphate bonds that are synthesized during the conjoint operation of the tricarboxylic acid cycle and the respiratory chain is considerably greater than the number synthesized during anaerobic glycolysis alone. However, the simultaneous and continuous operation of all three of these biochemical mechanisms in the presence of an adequate supply of glucose and oxygen assures the nervous system of an uninterrupted supply of ATP that is quantitatively sufficient to meet all of its energy requirements.

CLINICAL CORRELATES

Derangement of neural functions can range from abnormal neuromuscular transmission to defects in "higher" cortical functions including learning, memory, and affective behavior. Sensory and/or motor defects encompass an entire gamut that extends from mild and transient, through incapacitating and chronic, to rapidly lethal.

From a clinical standpoint, the definition (or diagnosis) of the specific problem at hand is fundamental to the rational therapy or management of any disease entity, and this fact is exemplified with great clarity by many of the nervous disorders which produce specific *symptom complexes,* or *syndromes*. In turn, fundamental neuroanatomic and neurophysiologic considerations are critical to such an evaluation. The classification of neural (and other) disorders according to their etiology not only affords one basis for the diagnosis and subsequent treatment of the individual patient, but also can yield an occasional insight toward the better understanding of the normal functions of particular structures or regions of the nervous system. Thus pathophysiologic events may have their primary origin in congenital, vascular, traumatic, infectious, neoplastic, metabolic, toxic, degenerative, and idiopathic as well as iatrogenic factors. The principal causative agent (or agents) just listed may, in turn, induce one or more secondary adaptive or maladaptive physiologic responses.

The specific topics that were selected for discussion in the remainder of this chapter will serve to illustrate the several points which were brought out in this introduction, as well as to review certain aspects of neuroanatomy and neurophysiology which were presented earlier. The first four principal subdivisions of this section deal with some general as well as specific dysfunctions that involve peripheral nerves, spinal cord, brain stem, and cerebellum. The remainder of this section will be concerned with discussions of a number of particular clinical situations, primarily from the standpoint of the pathophysiology involved.

Lesions and Regeneration of Peripheral Nerves

When a mixed peripheral nerve is severely injured or severed, a complete loss of all modalities of sensation, including proprioception, as well as paralysis of the skeletal muscles that are innervated by that nerve, take place distal to the lesion. The affected muscles are completely flaccid, absolutely no reflex activity can be evoked from the denervated region, and eventually total atrophy of the muscles develops.* A comprehensive knowl-

---

A number of somatic reflexes of clinical importance were presented in Chapter 6 (p. 227 et seq.).

edge of the gross anatomy underlying the ramifications and distribution of peripheral nerves is essential to the accurate diagnosis of lesions in specific nerves.

It is important to realize, however, that not all paralytic conditions are total so that some innervation may remain and the muscle may be weakened but not completely paralyzed, the reflexes can be decreased but not eliminated entirely, and only subtotal atrophy develops. For example, *multiple neuritis* involves an inflammation or partial destruction of peripheral nerves, the lesions may be widespread, and occur on both sides of the body, being most evident in the distal regions of the extremities. Clinically, multiple neuritis is characterized by muscular weakness and atrophy together with diffuse, poorly circumscribed cutaneous regions which exhibit marked sensory changes. On the other hand, the neuritis may be localized (mononeuritis) so that it affects only one nerve trunk or a small group of nerves.

Etiologic factors that underlie these neuropathies include chronic intoxication (alcohol, benzene, sulfonamides, carbon disulfide); specific infections (syphilis, meningitis, diphtheria, pneumonia, mumps, tuberculosis); metabolic causes (gout, diabetes mellitus, pregnancy, rheumatism); and nutritional factors (vitamin deficiencies, beriberi, cachectic states).

Following section of a motor nerve to a skeletal muscle, as the nerve degenerates the muscle becomes exceedingly sensitive to the pharmacologic effects of acetylcholine, a phenomenon called *denervation hypersensitivity* (p. 226). In contrast to skeletal muscle, smooth muscle does not undergo atrophy following denervation but it too develops an acute sensitivity to the chemical transmitter that normally stimulated it while innervated.

When any nerve fiber is cut in two or otherwise damaged irreversibly, the portion of the fiber that has been detached from the soma degenerates, and as this process takes place the fiber loses its myelin sheath. In addition to causing irreversible damage to the part of the fiber distal to the cut, severing the axon from the nerve cell body exerts a deleterious effect upon the nerve cell body per se. Thus the Nissl bodies (p. 157) undergo *chromatolysis* during the several weeks following the injury. During this process, the cytoplasmic ribonucleic acid granules lose their normal staining properties so that they appear to "dissolve" in the cytoplasm. Some of the neurons die and degenerate following chromatolysis, however others recover as RNA synthesis is restored. This *retrograde degeneration* that may accompany chromatolysis provides an important experimental technique for determining the cells from which specific fibers originate in the central nervous system. Thus serial sections made several weeks after placing a discrete lesion can be stained for Nissl substance, and the specific regions which show chromatolysis are indicative of the loci which contain the nerve cell bodies of the damaged fibers.

Following the complete section of a peripheral nerve, the fibers central to the cut send branches outward. These branches are outgrowths from the axon near its cut tip, and as many as 50 may develop from a single axon. Concomitantly the neurilemmal cells of Schwann at the proximal end of the severed nerve proliferate rapidly and migrate so as to bridge the gap between the cut ends of the nerve. The neurofibrils within the proximal end of the nerve divide longitudinally and commence to sprout from the end of the nerve. Some of these regenerating axon tips successfully cross the gap between the severed ends of the nerve and each may enter one ultramicroscopic neurilemmal tubule leading to the peripheral terminals of the nerve, the growth rate ranging between one and four millimeters per day. The entrance of a single regenerating axon into a neurilemmal microtubule that is connected to a sensory or motor terminal apparently is governed strictly by chance, the tubules being empty because of degeneration of the axon. Thus if the severed endings of a peripheral nerve are carefully apposed and sutured, function sometimes can be restored. If, however, the gap between the severed nerve endings is about 3 millimeters wide, the fibers growing outward from the central stump tend to form a dense meshwork which results in a bulbous swelling at the tip of the proximal end of the cut nerve called a *neuroma*. Once this has taken place, functional regeneration is unlikely to occur. Furthermore, if sensory fibers are present in the neuroma, it can be exquisitely sensitive to pressure. Hence, neuroma formation can be a serious complication following limb amputation.

Fibers of the spinal cord and brain do not regenerate to any significant extent following their destruction.

Lesions of the Spinal Cord

DORSAL COLUMNS AND POSTERIOR ROOTS. *Tabes dorsalis,* a type of neurosyphilis, will serve as an example of a disease which causes bilateral lesions in the dorsal roots of the cord as well as in the posterior funiculi. As might be anticipated from the area of the cord that is affected preferentially, *paresthesias* (abnormal or perverted sensations), including shooting and burning pains, pricking sensations, and formication* may be present. The loss of the proprioceptive pathways in the dorsal roots and posterior columns of the cord results in a pronounced failure of muscular coordination or *ataxia*. Consequently, the patient walks with his legs apart, the head is inclined toward the ground, and the eyes are fixed in the same direction. The legs are thrown abnormally high and the feet are slapped down. Vision is essential to

maintenance of posture, and the pattern of *locomotor ataxia* that develops in tabes victims is reminiscent of that seen in chronic alcoholics.

The paresthesias and bouts of severe pain that accompany the early stages of tabes dorsalis are caused by irritation of the dorsal roots and posterior columns. As degeneration of the cord continues, a diminished pain sensitivity develops, especially in the testicles and Achilles' tendons (the latter symptom, known as *Abadie's sign,* is presumptive evidence for tabes). The ankle jerk and patellar reflex disappear, whereas the vibratory sense diminishes as the disease progresses. The Romberg sign is positive; that is, the body sways uncontrollably or the patient falls down when standing with the feet close together and the eyes closed.

ANTERIOR COLUMNS. *Acute poliomyelitis* (Heine–Medin disease or infantile paralysis) is produced by a virus which preferentially destroys the nerve cell bodies of the anterior columns of the spinal cord, particularly in the cervical and lumbar enlargements of this structure (p. 258). The specific muscles that are paralyzed depend upon the precise distribution of the motoneurons which are damaged irreversibly, and the affected muscles are flaccid whereas their reflexes generally are absent. Several weeks after the infected motoneurons have been destroyed, muscular atrophy develops. This effect may be extensive so that eventually the muscular tissue may be replaced entirely by connective and adipose tissue, and severe physical deformities develop.

CENTRAL GRAY MATTER. *Syringomyelia* is a condition of unknown origin (ie, it is idiopathic)† in which softening and cavitation take place around the central canal of the spinal cord and brain stem. The affected area of the cord is surrounded by an area of gliosis (ie, the glial cells exhibit an enormous proliferation, particularly the astrocytes).

Clinically syringomyelia is characterized by muscular atrophy, weakness, assorted sensory defects, evidence of injury to the long tracts, and trophic abnormalities.

Commonly the lesion is found in the cervical enlargement of the spinal cord such that the fibers of the lateral spinothalamic tracts are interrupted as they decussate just ventral to the central canal (Fig. 7.21). Consequently, and as the lateral spinothalamic tracts transmit signals related to pain and thermal sensations from dermatomes that are located bilaterally, the results of syringomyelia can include a loss of pain and thermal sensibility having a segmental distribution in the upper extremities on both sides of the body (cf. p. 452). However, the spinothalamic tracts from the lower

---

*A sensation like that produced by small insects crawling over the skin.

†*Some investigators consider this disease to be caused by an inadequate fusion of the neural tube (p. 235) and the persistence of embryonic tissue that gives rise to the excessive glial elements.*

regions of the body are unaffected, therefore no such impairment is present in the lower extremities. Muscular weakness as well as alterations due to involvement of the sympathetic fibers that are located centrally in the cord characteristically accompany syringomyelia.

When the lesion is unilateral, thermal sensations are lost and analgesia develops in the skin that is innervated by, and on the ipsilateral side of, the diseased segments of the cord themselves. The sensations of touch and proprioception are unaffected until later in the progress of the disease. Such a condition, in which one modality of cutaneous sensation is lost whereas others are preserved, at least temporarily, is called *sensory dissociation.*

If the disease process extends into the medulla oblongata *(syringobulbia)* then signs that the nuclei of the cranial nerves are affected may appear, and when the spinal nucleus of the trigeminal nerve is involved, a characteristic loss of sensation develops over the facial region. The skeletal muscles which are supplied by the ventral column cells of the diseased segments become atrophic and functionally weak, such effects frequently being manifest first in the small muscles of the hand.

If the white matter is not involved in the disease process, then the sensory and motor defects produced by syringomyelia are localized and distributed segmentally. On the other hand, if the descending motor pathways are involved in the disease process, a spastic paralysis will develop caudally to the lesion, whereas pressure exerted upon the dorsal columns and spinothalamic tracts will result in a loss of sensation caudal to the lesion. Consequently the actual extent of the sensory loss, its type, distribution, and bilateral representation will vary from patient to patient depending upon the extent to which particular tracts are involved in the disease process.

Since thermal and pain sensations are absent in some areas, the patient with syringomyelia readily can suffer local minor injuries (such as a cut or burn) which go unperceived and thus may develop into serious disorders.

Since the sympathetic nervous centers in the lateral columns of the cord may be involved in the lesion produced by syringomyelia, vasomotor malfunctions and heavy sweating (or no sweating at all) may result. The abnormalities of vasomotor function in turn can lead to trophic disturbances in the periphery due to circulatory inadequacy, including gangrene and ulcers.

ANTERIOR HORNS AND PYRAMIDAL TRACTS. *Amyotrophic lateral sclerosis* is a fatal condition of unknown etiology which typically involves degeneration of the motoneurons in the anterior gray columns of the spinal cord in addition to bilateral destruction of the pyramidal tracts. Sensory alterations generally do not accompany this disease. Some muscles exhibit weakness and

atrophy, whereas others are spastic and hyperreflexic, the specific pattern of neuromuscular defects noted in any particular patient being a reflection of the specific pattern of neural damage in the spinal cord. In its classic form, amyotrophic lateral sclerosis first is manifest by skeletal muscle weakness, atrophy, and fasciculation of the muscles of the hands and arms. Subsequently, spastic paralysis of the legs develops, and the motor nuclei of the cranial nerves may deteriorate as well.

POSTERIOR AND LATERAL FUNICULI. *Subacute combined degeneration* of the spinal cord involves the white matter of this structure, and practically always is associated with pernicious anemia although it may be related occasionally to other types of anemia or nutritional disturbances. The disease process involves a degeneration of the myelin sheaths of the posterior funiculi and the pyramidal tracts of the cord, although the gray matter usually is not affected. Subsequent to demyelination, the axons are destroyed and the neural elements are replaced ultimately by glial tissue. These pathologic changes are seen most commonly in the pyramidal (corticospinal) and cerebellar tracts as well as in the dorsal columns.

From a clinical standpoint, the major features of subacute combined degeneration of the cord are muscular weakness and spasticity; a deranged sense of position as well as of passive movement of the limbs, defects which in turn underlie ataxia and a positive Romberg's sign; the ability to perceive vibration, as well as the ability to carry out spatial discriminations are lost or impaired to varying degrees.

The paresthesias seen in this condition may include tingling, pricking, and burning sensations which often appear prior to the loss of particular sensory capacities and frequently are related to a peripheral neuropathy that is associated with the primary disease process. Frequently the tendon reflexes are abnormally active since descending motor pathways are involved, but such responses may be lost completely as the disease progresses, provided that the peripheral neuropathy becomes exacerbated sufficiently. A positive Babinski sign (or reflex) is seen bilaterally, when the outer margin of the sole of the foot is stroked from the heel toward the little toe. Normally, such a maneuver elicits flexion of all toes in an adult. However, if the corticospinal fibers are interrupted at any point along their course between the upper region of the precentral gyrus (Fig. 7.60) and the muscles of the big toe, the latter structure will dorsiflex, whereas the other toes may (or may not) fan out laterally.*

There is no consistent parallel between the extent of the nervous defects exhibited by the patient with subacute combined degeneration of the cord and the degree of anemia that is present. Furthermore, the degeneration is not caused by the anemia per se, despite the fact that the two dis-

---

A positive Babinski sign is of normal occurrence in infants before they have learned to walk.

eases are associated. This is demonstrated clearly by the clinical observation that the neurologic deterioration proceeds unabated despite correction of the anemia by administration of folic acid. Conversely, progress of the neurologic degeneration can be arrested by the administration of cyanocobalamin (vitamin B_{12}, p. 883).

TRANSECTION OF THE SPINAL CORD. A sudden and total disruption of the integrity of the spinal cord at one level can result either from a traumatic cause (eg, a stab or gunshot wound or a fracture and spinal dislocation) or from the operation of an acute inflammatory process (eg, transverse myelitis). In either event, the cord is transected functionally so that an immediate and total loss of all voluntary control over the skeletal muscles caudal to the lesion ensues together with a complete loss of all sensation in the same region of the body.

The term *paraplegia* is applied to the total paralysis of both lower extremities resulting from any lesion of the spinal cord, whereas if the cord is severed completely in the lower cervical region all four extremities are paralyzed, a condition known as *quadriplegia*. If, on the other hand, the cord is transected in the upper cervical region, death rapidly ensues as the diaphragm and other respiratory muscles no longer are controlled by impulses from the respiratory center (qv, p. 729).

The condition of spinal shock which develops immediately following complete interruption of the spinal cord, as well as the progressive return of certain reflexes after cord transection, are discussed on page 371 et seq.

HEMISECTION OF THE SPINAL CORD. When the spinal cord is hemisected laterally, as by a "clean" knife or bullet wound, a condition known as the *Brown–Sequard syndrome* develops (refer to Figs. 7.20, 7.21, and 7.22). The functional defects seen in this syndrome as well as the damaged structures responsible for causing them can be summarized as follows:

Injury to the fiber tracts that are located in the *same side of the cord* as the lesion produces an ipsilateral motor paralysis (*hemiplegia*) below the lesion, because of pyramidal tract injury. Hence damage to this tract results in the clinical signs of spasticity, some hyperreflexia, loss of superficial reflexes, and the development of a positive Babinski sign.

Concomitantly, damage to the posterior funiculus on the same side of the cord as the injury results in a loss of proprioceptive ability and a loss of the ability to perceive vibrations, as well as a loss of the capacity to perform tactile (touch) discriminations; all of these defects are apparent caudal to the injury on the same side of the body as the lesion. Ataxia cannot be demonstrated clearly because of the muscular paralysis.

Injury to the fiber tracts which are located on the *opposite side of the cord* from the lesion results in a loss of thermal and pain sensations on the contralateral side of the body because of damage to the dorsal (anterolateral) spinothalamic tract. This defect is seen to commence one or two segments below the lesion, because of the fact that the fibers of this tract ascend in the same side of the cord for a segment or two before undergoing decussation to the opposite side (p. 266).

Damage to the ventral (anterior) spinothalamic tract causes only a slight alteration (or no change whatsoever) in the sense of simple touch.

Only rarely is a lesion located so precisely within the cord as to produce all of the classic signs of the Brown–Sequard syndrome described above. More frequently, the lesions encountered in clinical practice are limited to one portion of the cord, and thus produce an incomplete Brown–Sequard syndrome whose specific signs and symptoms reflect the exact location and extent of the lesion in the patient under examination.

LOCAL INJURY TO SPECIFIC CORD SEGMENTS AND NERVE ROOTS. In addition to the clinical states which are caused by destroying signal transmission in the long ascending and descending tracts of the cord, other clinical symptoms can result from injury to specific dorsal and ventral roots at the same level as the lesion. Such symptoms, when present, are of assistance to the physician in localizing the particular segments of the spinal cord that are involved in the case under evaluation.

Thus excitation of fibers in the dorsal root area on the same side of the lesion can induce paresthesias or radicular pain which is localized over the particular segments (or dermatomes) that are innervated by the specific nerve whose root is affected. Similarly, total destruction of dorsal root causes anesthesia over the specific dermatome that is supplied by the affected dorsal root. Conversely, destruction of a specific ventral root causes flaccid paralysis of only the specific skeletal muscles whose innervation is affected, and the sensory innervation of that area is unaffected.

THROMBOSIS. The anterior spinal artery courses in the ventral median sulcus (p. 254), and terminal branches of this vessel supply the ventral columns, the dorsal spinothalamic tracts, and the corticospinal (pyramidal) tracts. The posterior columns and dorsal funiculi of the cord receive their blood supply from a pair of dorsal spinal arteries which are independent of the anterior spinal artery. The effects of a clot or thrombosis therefore depend upon which vessel, hence which region of the cord, is affected primarily. For example, a clot, or thrombus, which forms in the anterior spinal artery in the cervical region causes flaccid paralysis, fibrillation, and eventually atrophy of the muscles that are innervated by the motoneurons whose cell bodies are located at the level of the cord where the lesion is located, and which are killed and degenerate following the thrombosis. Pyramidal tract involvement results in violent, involuntary, and sudden bilateral muscular contractions below the level of the thrombus, and this *spastic paraplegia* can be exceedingly painful. Concomitant damage to the dorsal spinothalamic tracts sometimes may result in a loss of thermal as well as painful sensations below the lesion. The onset of the symptoms caused by thrombosis of the anterior spinal artery not only is sudden, but also may be accompanied by severe pain.

TUMORS LOCATED WITHIN THE SPINAL CORD. Neoplasms are classified depending upon their location within the spinal canal. Thus extradural tumors are located outside of the dura mater (Fig. 7.14) whereas intradural tumors are located within the dura mater. In turn, intradural neoplasms are located either outside of the spinal cord (extramedullary tumors) or within the spinal cord (intramedullary tumors). There is a definite correlation between the location of a tumor within the spinal canal and its pathologic nature. For example, an extramedullary tumor usually is benign, but can progressively exert pressure on the cord as it enlarges. Generally the nerve roots are compressed first as the tumor grows, and this mechanical factor is responsible for the pain that is manifest over the dermatomes which are innervated by the affected roots. Subsequently, and as the tracts within the cord become involved, a complete or partial Brown-Sequard syndrome is exhibited.

The temporal sequence in which particular symptoms appear often can provide an indication as to where the tumor is located. Thus a loss of thermal and pain sensations on the right side of the body followed by spastic paralysis on the left side suggests that the neoplasm has originated around the ventrolateral portion of the cord on the left side. Similarly, a loss of the normal proprioceptive sense on the right side with a subsequent development of spastic paralysis on the left side implies that the tumor is exerting pressure on the dorsomedial portion of the spinal cord on the left side.

Specific neurologic symptoms that may develop subsequent to compression of the spinal cord include sensory loss, disturbances in the operation of the bladder and rectal sphincters, and paresthesias that become exacerbated during physical exercise, coughing, or straining. Localized skeletal motor weakness, paralysis, and muscular atrophy can occur if the ventral (motor) roots or columns of the cord are affected by the tumor.

Nerve fibers can be damaged to various degrees, so that in some instances pressure that is exerted on a nerve or tract may inhibit transmission of impulses, but not cause irreversible destruction of the fibers. Thus the pressure exerted by tumors, blood clots, the edema and swelling caused by traumatic injuries, and/or by displaced intervertebral discs all may produce symptoms that are indicative of spinal cord dysfunction, and yet may prove to be completely reversible following appropriate therapy. The ultimate prognosis, of course, depends upon the extent or severity of the compression as well as the length of time the abnormal pressure has been exerted upon the neural elements themselves, and the same factors obtain regardless of whether one is dealing with the spinal cord or the brain per se.

Lesions of the Brain Stem

The complex structures of the brain stem are located in such a relatively small volume of neural tissue that a restricted lesion generally affects a number of these structures; consequently a bewildering array of clinical signs can result. Hence the correct interpretation of these signs rests on a clear understanding of the neuroanatomy and neurophysiology of the nuclei and fiber tracts (p. 275). Only three of the many possible syndromes which can result from lesions in specific areas of the brain stem will be described. However, these symptom complexes will serve to illustrate the potential effects of thrombosis, hemorrhage, tumor development, or degeneration in particular areas of this part of the central nervous system.

BASAL PORTION OF MEDULLA. A number of cranial nerves run near the lateral aspect of the pyramidal tract before they exit from the brain stem (eg, see Figs. 7.28 and 7.29). A lesion which includes the nerve as well as the pyramidal tract at one point causes paralysis of the nerve on the same side as the lesion and hemiplegia on the opposite side. For example, a lesion that involves the left pyramid as well as the left hypoglossal nerve produces a right hemiplegia together with a motor paralysis of the muscles in the left half of the tongue. Paralysis of the arm and leg is on the contralateral side to the lesion because the pyramidal tract undergoes decussation to the right side caudal to the site of the lesion, as illustrated in Figure 7.22. Symptoms of spasticity, including increased tonus of the skeletal muscles, hyperreflexia, pathologic reflexes, and loss of superficial reflexes, are present. The tongue deviates to the left when it is protruded from the mouth, because of the paralysis of the muscles on the left side of this structure, and eventually these paralyzed muscles will undergo total atrophy.

Another important sign of hyperreflexia that frequently is associated with spastic states in general is *clonus,* and this neurologic sign may be present when a lesion develops in the basal medullary region (as well as in other parts of the central nervous system). Clonus is manifest when the normal (or physiologic) asynchrony of discharge from the motoneurons involved in a stretch reflex is lost (p. 218), and thereby a series of jerky, irregular contractions takes place in a repetitive manner, and these contractions may be superimposed upon a single tonic contraction (Fig. 7.128). Thus in a hyperreflexive patient, clonus may be initiated by applying a sustained stretch to a muscle, and then tapping on the tendon of that muscle. The tendon reflex which results from this stimulus is followed by a series of irregular twitches which continue for a prolonged interval, provided that the muscle is maintained in the stretched condition.

Clonus is explained by the fact that the stretch receptors are arranged in parallel within the muscle spindle (p. 182) so that a sharp rap on the tendon initiates a synchronous volley of afferent impulses which in turn suddenly discharge the alpha motoneurons to the muscle, also in a synchronous manner. The sudden jerklike contraction which develops then releases the spindles from the tension caused by the maintained stretch, therefore they stop firing and the afferent stimulus to the motoneurons is terminated. As a consequence, the muscle now undergoes relaxation so that the spindles once again are placed under tension which leads to the initiation of another synchronous afferent signal discharge which again fires the alpha motoneurons, and the cycle is repeated.

Clonus that persists for any length of time always is symptomatic of hyperreflexia, which in turn indicates damage to the central nervous system, although normal individuals sometimes may exhibit small clonic oscillations of relatively short duration. In normal individuals, of course, gamma efferent discharge to the spindles is modulated by inhibitory (as well as facilitory) impulses from higher neural centers which automatically and continually regulate the net spindle sensitivity. But when lesions are present in higher motor pathways, then the normal balance between excitatory and inhibitory impulses that impinge upon the neurons of the gamma efferent pathway to the spindles is altered in favor of excitation, so that the spindles (and the myotatic reflexes these structures subserve) become hyperexcitable and the stage is set for clonus to be manifest when an appropriate stimulus is applied.

Ankle clonus affords a specific example of this type of response. This clonic reflex can be evoked in a suitable patient by a sudden, sharp dorsiflexion of the foot, stretch of the gastrocnemius muscle being maintained by placing the hand upon the sole, the opposite hand being held in the popliteal space to support the leg. A rapid, sustained, and repetitive plantar flexion and extension at the ankle indicates a positive response.

In conclusion, it must be stressed that a clonic response is a *neurologic sign* of a lesion that is located somewhere in the motor pathway of the central nervous system, and that it is *not* a sign of any specific disease.

UPPER PORTION OF MEDULLA. The presence of a small discrete lesion in the lateral reticular formation of the upper medulla may affect simultaneously the dorsal spinothalamic tract as well as the nucleus ambiguus. If the lesion is present, for example, on the left side, there is a loss of thermal and pain sensations on the right side of the body with the exception of the facial region. These sensory defects occur on the contralateral side of the body because the fibers of the dorsal spinothalamic tract undergo decussation near the point of origin (Fig. 7.21). When the left nucleus ambiguus is destroyed, the voluntary muscles that are innervated by the left glossopharyngeal and vagus nerves are paralyzed. The left side of the soft palate is unable to contract so that difficulty in swallowing becomes apparent. Furthermore, when the patient attempts to speak, the uvula and palate are pulled toward the normally innervated right side of the pharynx, and loss of function of the left vocal cord causes the voice to become hoarse during phonation.

When a lesion of somewhat greater size is

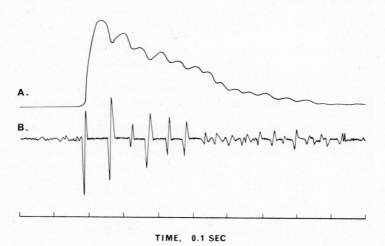

TIME, 0.1 SEC

FIG. 7.128. Clonus. **A.** Mechanical record of contractions. **B.** Electric record from same muscle (quadricepis). A slight prior stretch of the muscle evoked a tonic reflex discharge of impulses to the muscle as shown by the asynchronous waves to the left in **B.** Subsequently a tap applied to the tendon of the muscle elicited a sudden sharp contraction (sudden rise in curve **A**), followed by a characteristic clonic response as indicated in both the mechanical and electric records. (After Ruch and Patton [eds]. *Physiology and Biophysics,* 19th ed, 1966, Saunders.)

present in the same region of the left upper medulla, the tractus solitarius as well as the medial lemniscus may be affected. Thus in addition to the defects noted above, destruction of the solitary tract causes anesthesia of the pharyngeal mucosa on the left side of the oral cavity as well as a concomitant loss of the sensation of taste on the left side of the tongue. Functional interruption of the fibers of the medial lemniscus produces a loss of the proprioceptive and tactile discriminative senses on the right side of the body, in addition to anesthesia for temperature and pain senses on the same side of the body which were incurred following development of a smaller lesion.

PONTOCEREBELLAR ANGLE. When a neoplasm develops from the Schwann cells that form the sheath of the vestibulocochlear nerve as this structure exits from the brain stem, the lateral portion of the caudal region of the pons gradually becomes compressed adjacent to the pontocerebellar angle (eg, Fig. 7.29). The symptoms, which include deafness, a deficiency of normal labyrinthine responses, and occasionally horizontal nystagmus, are progressive and resemble those which follow damage to the vestibulocochlear nerve per se (see also Meniere's syndrome, p. 463). Subsequently the cerebellar peduncles undergo compression as the tumor enlarges so that *cerebellar asynergia* (or *dyssynergia*) becomes evident on the ipsilateral side of the body to the lesion (see next section). Injury to the spinal tract as well as to the trigeminal nucleus eliminates the corneal reflex and produces a reduced thermal and pain sensibility of the facial region on the same side as the lesion. When fibers of the facial nerve (VII) also are involved, a peripheral facial paralysis is also present on the homolateral side of the body as the lesion.

Lesions of the Cerebellum

Characteristically, lesions of the cerebellum or of its pathways are distinguished by *asynergies (asynergias, dyssnergias),* that is, a failure in the normal, smoothly coordinated and integrated action among groups of muscles. The general neuronanatomic centers which underlie the regulation of synergistic contractions at the cerebellar level may be summarized briefly (see also p. 307 et seq.). The cerebellar vermis (Fig. 7.116) controls the contractions of the bilaterally innervated muscles of the head, neck, and trunk. The left and right cerebellar hemispheres regulate antagonistic (ie, flexor and extensor) contractions on the same (or homolateral) side of the body. This fact is in marked contrast to the cerebral hemispheres, which are concerned with functions located on the contralateral side of the body. At the spinal cord level the neural centers are essential to the normal bilateral control of muscular activities in the extremities, eg, walking (p. 376). The physiologic mechanism which ultimately is responsible for all forms of synergy regardless of the neural center

involved is, of course, reciprocal innervation of antagonistic muscles (p. 221).

The cerebellum is no different from any other organ or structure in that it is susceptible to hereditary (or congenital) defects, as well as to neoplastic, traumatic, inflammatory, vascular, toxic, degenerative, functional, and metabolic derangement. The signs and symptoms of cerebellar dysfunction, as well as of the afferent and efferent pathways to and from this structure, are similar regardless of the etiologic factor (or factors) involved. Furthermore, all of the signs related to cerebellar disorders involve motor functions.

Several types of *ataxia,* or failure of normal muscular coordination, characteristically are associated with cerebellar disease.

First, ataxic disturbances of *gait* and *posture* may be marked. Thus damage to the midline portion of the cerebellum causes difficulty in maintaining a standing, or upright, posture. The loss of equilibrium that is present is caused by a defect in the synergistic action of the muscles and not by a functional defect in the pathway that mediates conscious proprioception (ie, the dorsal spinocerebellar tract, Table 7.5). Closing the eyes does not exacerbate this form of ataxia, and the gait is staggering and may resemble that seen in acute alcoholic intoxication. If the lesion is localized in one cerebellar hemisphere, there is a tendency of the body to fall toward the same side as that in which the lesion is located, and if the victim attempts to walk a straight line with the eyes closed, he may swerve or reel erratically toward the side of the damaged hemisphere. Vertigo (dizziness), unsteadiness, and irregularity of movement thus are typical signs of cerebellar ataxia. While walking, the legs are held wide apart, appear loose, and the limb taking the forward step starts to move gradually; but suddenly it is thrown forward erratically and the step ends with the foot thrust to the floor in a stamping movement.

Second, ataxia due to cerebellar damage can be manifest as *decomposition of movement,* sometimes called the ''by-the-numbers'' phenomenon. That is, a physical action which requires the smoothly coordinated movement of several joints becomes uncoordinated so that the movement takes place in several discrete phases rather than as a smooth continuum. For example, in bringing a hand with food to the mouth, the joints of the wrist, elbow, and shoulder may be moved individually in a jerky fashion rather than as a single, smooth, and synchronized movement. A coarse *intention tremor* may be induced by the movement, and this tremor will disappear once the movement is completed (see also below).

Third, cerebellar damage can result in *dysmetria*. The patient is unable to estimate the range of voluntary movement, so that when reaching a hand toward an object the hand either stops before the object is reached or else it overshoots the goal. For example, when the patient is requested to point a finger directly at one of the examining physician's fingers he may point consistently to one side, a phenomenon that is called past-pointing. Similarly the subject may touch his cheek with a finger when asked to touch his nose.

Fourth, *adiadochokinesia* (or *dysdiadochokinesia*) is an inability to stop one movement and immediately start another movement in the opposite direction; ie, to perform rapidly alternating movements. For example rapid pronation and supination of the hands, tapping of the fingers, or flexion and extension of a hand cannot be carried out by a patient who exhibits this manifestation of cerebellar ataxia.

Fifth, *scanning speech* is caused by asynergy of the muscles used in phonation. The spacing of the sounds is irregular, and pauses, as for punctuation, syllables, emphasis, and so forth are made in the wrong places. Another type of speech defect which may be caused by cerebellar lesions is *dysarthria;* in this instance the speech is slurred and explosive.

In addition to the five general manifestations of ataxia listed above that may be related to cerebellar dysfunction, several other defects can be related to lesions in this region of the central nervous system. Cerebellar damage can result in *hypotonia,* an effect which is manifest by the decreased muscle tone that is evident upon palpation. A *pendular knee jerk* reflex sometimes is observed in conjunction with, and as a consequence of the hypotonia. When this sign is present, the leg swings freely back and forth several times following stimulation of the patellar tendon.

The skeletal muscles which are affected by cerebellar lesions are weaker and undergo fatigue far more readily than do normal muscles, this condition being known as *asthenia.*

When involuntary movements (or tremors) are associated with cerebellar disease, the tremors usually become evident during purposeful (voluntary) movements, but are absent or decreased when the muscles are at rest. Such movements, called *intention tremors,* are generally caused by lesions which affect the efferent pathways that are located within the superior cerebellar peduncle. The gross movements associated with intention tremors are coarse and irregular (ie, arrhythmic).

Lastly, cerebellar lesions frequently are accompanied by nystagmus (p. 393), and such eye movements are induced by the stimulation of vestibular fibers located within the cerebellum by the lesion. Alternatively the nystagmus may be caused by mechanical compression of the vestibular nuclei of the brain stem which are located ventral to the cerebellum (Fig. 7.111).

In conclusion, it is important to reiterate that the left cerebellar hemisphere controls the left side of the body, and the right cerebellar hemisphere controls the right side. Consequently any signs of cerebellar dysfunction on one side of the body reflect a cerebellar defect that is located on the ipsilateral (or homolateral) side of the cerebellum.

This fact is in sharp contradistinction to the situation which is present with lesions of the cerebral hemispheres, and which always cause defects upon the contralateral side of the body.

Disorders of the Basal Ganglia

Two general types of defective movements are seen in humans suffering from diseases of the basal ganglia (p. 336 et seq.). *Hyperkinetic movements* are seen in athetosis, ballism, and chorea; in these conditions excessive as well as abnormal patterns of movement are present. *Hypokinetic* as well as hyperkinetic movements are seen in Parkinson's syndrome (or disease).

ATHETOSIS, BALLISMUS, CHOREA. There are many similar features among the three dyskinesias (abnormal movements) known clinically as athetosis, ballismus, and chorea. All of these conditions in common exhibit movements that appear to be voluntary, but which are not, and thus are completely beyond the control of the patient. The movements themselves are pronounced and can even become quite violent; however no characteristic tonus changes accompany the movements, as is the case in parkinsonism (qv, next section).

In *athetosis,* the limbs writhe continuously, slowly, and in a sinuous manner. These involuntary movements are particularly evident in the hands and may appear as though these extremities are grasping or performing avoidance reactions. Antagonistic muscles may undergo simultaneous contraction. Damage to the lenticular nucleus (p. 336) and the corpus striatum (p. 336) both have been implicated as sources of the defects which cause the abnormal movements of athetosis.

In *ballismus,* the involuntary movements are quite violent, the limbs being flung wildly about by contractions of the proximal muscles of the extremities. That is, the movements are rapid, hence ballistic. When the movements are restricted to only one side of the body, as frequently happens, the condition is called *hemiballismus.* Ballismus appears when the subthalamic nuclei (of Luys, p. 336) are damaged as by a thrombus or hemorrhage. Thus hemiballismus may appear with exceeding rapidity following a hemorrhage in the contralateral subthalamic nucleus.

In *chorea* a considerable variety of rapid and jerky but well-coordinated movements occur continuously. As is the case with the other dyskinesias, the movements are involuntary and are useless insofar as being of some purpose to the patient is concerned.

PARKINSON'S SYNDROME (PARALYSIS AGITANS). This syndrome, originally described by James Parkinson about one and one half centuries ago, has a gradual and insidious onset, progresses slowly, and generally develops in persons in the fifth or sixth decade of life. A number of etiologic factors are known to produce Parkinson's disease, including cerebral arteriosclerosis, trauma to the head, neurosyphilis, cerebrovascular accidents, carbon monoxide or manganese poisoning, or following an attack of endemic encephalitis or influenza. Parkinsonism also can develop as a serious complication of therapy with phenothiazine tranquilizers (iatrogenic etiology), or in many instances the cause simply is unknown (idiopathic etiology).

The principal clinical findings of well-developed parkinsonism are the hypokinetic sign, *akinesia* (or *poverty of movement*), as well as the hyperkinetic features of *tremor* and *rigidity*. The absence of voluntary motor activity can be quite pronounced, as the patient experiences considerable difficulty in initiating such movements. There also is a marked decrease in normal *associated movements* that take place unconsciously, eg, swinging the arms while walking, and the spectrum of facial expressions related to the emotional aspects of thought, speech, and language, as well as the normal gestures and motor manifestations of mild mannerisms that are common to everyone.

The rigidity or hyperkinesia of parkinsonism is not spastic, and the myotatic reflexes are normal, although Jendrassick's maneuver (ie, clenching the hands firmly together) to enhance a weak tendon jerk reflex does not alter such responses in Parkinson's disease.

Passive flexion of an extremity has been likened to bending a lead pipe as the rigid muscles are stiff although somewhat plastic. Occasionally a limb will "give" and then "catch" again in a series of jerks as it is being moved. Hence the terms *lead-pipe rigidity* and *cogwheel rigidity* sometimes are applied to the muscles of a patient having Parkinson's disease.

A clasp-knife effect can be seen when a spastic muscle suddenly loses resistance as it is stretched passively (p. 228). In Parkinson's disease, however, no such response occurs during passive flexion of the rigid limbs.

The intermittent tremor of Parkinson's disease occurs through regular alternating contractions of antagonistic muscles at a frequency that ranges between two and six contractions per second. In sharp contrast to the intention tremor that is associated with cerebellar lesions (p. 456), the tremor of Parkinson's disease is most pronounced while the patient is at rest and tends to disappear or be reduced as a movement takes place. Tremors often are of a type called "pill-rolling" and thus involve the thumb, index finger, or wrist. Occasionally a to-and-fro tremor of the head is present. Fatigue and emotional states tend to exacerbate the tremor. The facial expression on the other hand is characteristically fixed and masklike.

The rigidity of the skeletal muscles may be such that the patient has difficulty in arising from a chair, turning around while standing, or turning over in bed. Even normally simple actions such as washing and dressing become exceedingly difficult without assistance, and voluntary movements are carried out sluggishly and with great effort. Charac-teristically, the posture is stooped, and while walking a *festination gait* is seen. That is, the victim of paralysis agitans leans forward and takes shuffling steps that are quite short, commence slowly, and become rapid as they progress. Frequent rapid accelerations are seen so that the patient "appears to be chasing his center of gravity."

Parkinson's disease cannot be reproduced readily in experimental animals. In humans, however, this condition clearly is produced by diffuse brain damage. In particular, the basal ganglia are involved, although the specific neurophysiologic and neuroanatomic mechanisms that underlie the tremor, rigidity, and poverty of movement that are the principal features of this disease remain obscure. In humans, the parkinsonian syndrome sometimes can be alleviated surgically by additional destruction of the very nuclei whose malfunction underlies the abnormal condition in the first place, an interesting paradox! Thus extirpation of the motor cortex, section of the crura cerebri (cerebral peduncles), posterolateral spinal cordotomy, destruction of the medial globus pallidus, section of the ansa lenticularis, and lesions induced in the ventrolateral thalamic nuclei all have been claimed to diminish the rigidity and tremor of parkinsonism. If nothing else, these facts indicate that an exceedingly diverse and complex pattern of neural elements is involved in the etiology of Parkinson's disease.

During recent years, considerable interest has focused upon a biochemical mechanism for the etiology of Parkinson's disease as well as for the therapy of this disorder. In particular, parkinsonism has been related to an abnormal metabolism of certain amines in the basal ganglia, and perhaps the hypothalamus as well.

Under normal circumstances, the concentration of dopamine in certain parts of the brain is high whereas that of norepinephrine is low. Dopamine (dihydroxyphenylethylamine), is, of course, the normal precursor of norepinephrine (p. 196; Fig. 22), and this compound has been implicated as a neurotransmitter that is involved in motor functions. Under normal conditions the basal ganglia of humans, especially the caudate nucleus and putamen, contain high dopamine concentrations whereas the norepinephrine concentration in these structures is quite low. In Parkinson's disease the dopamine level of the caudate nucleus and putamen falls to about 50 percent of the normal level. Concomitantly the hypothalamic norepinephrine level also falls somewhat, but not to such a great extent. In addition, several pharmacologic agents that induce undesirable side effects in patients that resemble certain features of Parkinson's syndrome are known to alter the normal metabolism of dopamine in the brain. All of these facts provide some evidence, albeit circumstantial, that the normal metabolism of certain amines in the basal ganglia is deranged in Parkinson's disease, and that the two states may be related causally.

In support of this contention, it has been found clinically that L-dopa (levo-dihydroxy-phenylethylamine), which is the immediate biochemical precursor of dopamine in vivo (Fig. 6.22, p. 196), is extremely useful in the treatment of Parkinson's disease when administered in a sufficiently large dose. L-dopa, unlike dopamine per se, is able to cross the blood-brain barrier in quantities sufficient to produce a demonstrable increase in brain dopamine concentration.

Epilepsy

Epilepsy is a transient state, a syndrome, and not a particular disease entity. The epileptic state is characterized by abrupt and transitory changes in the normal functions of the brain so that somatic sensory, somatic motor, autonomic, and/or psychic symptoms which may (or may not) be accompanied by alterations in consciousness are manifest. Concomitant changes in the brain waves accompany epileptic attacks, hence profound alterations in the EEG can be recorded during such periods.

From an etiologic standpoint, epilepsy frequently is caused by abnormally functioning brain tissue that results from trauma, infections, or unknown agents (ie, idiopathic epilepsy). The latter tends to occur in families, hence congenital or genetic factors may be suspected in such instances. The most common factors related to the development of epilepsy in children are birth trauma which may be related to anoxia, head injuries, cerebrovascular accidents, infections of the brain with attendant inflammatory processes, and congenital malformations of the brain.

It should be mentioned that seizures which are identical with those seen during episodes of idiopathic epilepsy may take place in patients suffering from organic brain disease (eg, intracranial infection, cerebral vascular accidents, tumors). Similar epileptiform attacks also may be precipitated by metabolic defects, such as during hypoglycemia, uremia, and hypocalcemia (see also p. 1072).

Epileptic seizures may be precipitated in some persons who are susceptible to such attacks by physical events such as auditory, tactile, and visual stimuli, whereas in other patients the seizures occur typically during sleep. Additional factors may participate indirectly in the genesis of epileptic seizures, including a lack of food and sleep, emotional tension, or the ingestion of alcoholic beverages.

The pattern of epileptic seizures in any given patient tends to be stereotyped, although there is a wide disparity in the patterns of attacks that are seen among different patients. For example, epileptic seizures may take place in groups over a period of hours or of days. If the seizure arises in the motor cortex, the patient tends to exhibit a temporary paresis known as *Todd's paralysis* following the attack.

PATHOPHYSIOLOGY. Considerable research has been conducted on possible biochemical and physiologic mechanisms that could be invoked to explain the pathophysiology of epilepsy, but the fundamental causes that underlie seizures remain obscure. It is clear, however, that seizures are more likely to appear when organic brain damage is present than in persons with a normal central nervous system. Furthermore, histologic examination of the brains from patients who had idiopathic epilepsy usually fails to reveal any specific alterations in the tissue that can be correlated with the functional defect.

Possibly a seizure could be initiated by hyperexcitability of the neurons in one (or more) reverberatory (or oscillatory) circuits, such as might be found in the reticular activating system, any specific lesion being a sufficient stimulus to induce such hyperexcitability. Alternatively, specific neural pathways could become facilitated, hence hyperexcitable, so that synchronous discharges from reverberatory circuits could produce firing in these specific pathways that would lead to the particular signs and symptoms observed.

Presumably, an epileptic seizure could be terminated physiologically by active fatigue of the synapses involved in the circuits responsible for the seizure in the first place, although the sustained attacks seen in status epilepticus (qv, p. 460) might seem to cast some doubt upon this idea. Alternatively, active inhibition through negative feedback circuits might play some role in breaking the cycle of events initially leading to the seizure.

Our current state of ignorance in this area is clearly defined by the vague and speculative nature of the foregoing remarks concerning the initiation and termination of an epileptic attack. The possible role of the corpus callosum in the etiology of epilepsy was presented earlier (p. 437).

As noted earlier (p. 447), the overall metabolic activity of the brain varies to a certain extent depending upon its state of functional activity. When there is a generalized acceleration of neuromuscular activity, such as occurs during convulsive states, it is not surprising that the net metabolic activity of this organ increases. Nevertheless, there is no significant alteration of the metabolic activity in specific regions of the brain that can be correlated with particular functional states.

The level of acetylcholine as well as the activity of cholinesterase have been shown to be approximately proportional to the size and density of the neurons in all regions and cortical layers of the normal brain. On the other hand, cortical foci which discharge abnormal signals resulting in epileptiform seizures in experimental animals have been shown to have an increased cholinesterase activity as well as alterations in the bound acetylcholine level during induced epileptic seizures. Such findings have been considered to be evidence that a deranged acetylcholine metabolism underlies the brain abnormalities present in epileptics.

There also is considerable experimental evidence which suggests that mechanisms other than an abnormal acetylcholine metabolism may be involved in the production of epileptic seizures. Thus an abnormal distribution of potassium has been found in the brains of experimentally convulsed animals and the onset of experimental seizures may be heralded by an increased rate of nitrogen metabolism as indicated by an elevated level of ammonia synthesis.

From an experimental and clinical standpoint, metabolic poisons, vitamin deficiencies (eg, of pyridoxine), defects in the metabolism of certain amino acids (particularly glutamic acid), metals (eg, aluminum), infectious agents, and even particular dietary components all have been shown capable of producing epileptiform seizures. Despite this broad spectrum of biologic models, however, the underlying mechanism(s) responsible for causing epilepsy remains unknown.

CLINICAL PATTERNS IN EPILEPTIC STATES. A number of systems are used in the classification of epileptic states, and these are based primarily upon etiologic factors (when known), clinical symptoms, and anatomic localization of the underlying lesion, as well as the age of the patient when the symptoms became manifest.

Several types of seizure will be discussed below in accordance with the clinical findings present in each.

Petit Mal Seizures. Epileptic seizures may occur in which the patient exhibits a transient, even momentary, loss of consciousness so brief that neither the patient nor his intimates are aware that an attack has occurred, and convulsive movements or falling of the body do not occur. Thus, the so-called triad of petit mal seizures includes myoclonic jerks, akinetic seizures, and a very transient loss of consciousness, termed "brief absences."

The *myoclonic jerks* can be manifest in either limbs or muscles and such rapid, involuntary contractions can be seen in either whole muscles, portions of muscles, or in whole groups of muscles that are either widespread or restricted to one region of the body. Such myoclonic contractions can be synchronous or asynchronous, and may occur without a loss of, or any evident change in, consciousness. Myoclonic jerks are more frequent in the morning and while falling asleep, and even normal persons may exhibit such sudden contractions while drowsy or during periods of light sleep.

An *akinetic attack,* or seizure, is characterized by an abrupt loss of postural tone such that the patient slumps over before he is able to catch himself. Alternatively, the victim may slump so that the knees or body just touch the ground before recovery is achieved.

During petit mal attacks, the EEG typically shows "doublets" that appear with a frequency of about three per second. That is, each wave consists of a spike coupled to a rounded wave as depicted in Figure 7.129.

Grand Mal Seizures. An attack of grand mal epilepsy often is signalled by an *aura* which warns the patient that a seizure is imminent. Generally the aura is specific for each particular patient and it may consist of a somatic or visceral sensation (eg, numbness or nausea), a visual image, an odor, or a flash of memory (see also the déjà vu phenomenon, p. 444).

Following the aura, loss of consciousness takes place and the patient drops to the ground or floor. In falling, vocalization as well as physical injury can occur. Convulsions generally ensue, the patient lying rigid for a minute or so. During this phase, the somatic muscles are in a state of sustained tonic contraction. This *tonic phase* subsequently gives way to a *clonic stage,* during which massive, rhythmic, synchronous contractions of the skeletal muscles take place throughout

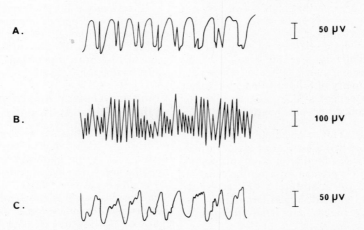

A.   ⌐ 50 μV

B.   ⌐ 100 μV

C.   ⌐ 50 μV

FIG. 7.129. Tracings of EEGs recorded during: **A.** petit mal; **B.** grand mal; and **C.** psychomotor seizures. (After Guyton. *Textbook of Medical Physiology,* 4th ed, 1971, Saunders.)

the body. During these convulsions, normal control over certain viscera such as the bladder and bowels often is lost so that urination and defecation occur. Rarely, fractures of the bones attend the violent convulsions; however, self-inflicted tongue injuries are common as the patient's jaw musculature contracts spasmodically.

Following the convulsive phase, a period of stuporous sleep ensues which may last for a variable interval, but generally ranges between one and four hours' duration. The patient has no clear memory for events during this postconvulsive interval, even if he is awake, and upon full recovery from the seizure the somatic musculature may be exceedingly painful.

During grand mal seizures, the EEG exhibits a fast activity pattern during the tonic phase (Fig. 7.129B), but at the time each clonic jerk takes place slow waves are seen, and during the recovery phases the slow waves may persist.

Interestingly, hemodialysis of the blood as part of the therapy for chronic renal failure has been demonstrated to precipitate grand mal epileptic seizures. Furthermore it has been shown that the conduction velocity of peripheral nerves in such patients is reduced. The precise mechanism whereby grand mal seizures are precipitated by dialysis is unclear, although it would seem that such episodes are related to the rapid alterations in blood biochemistry that are induced by dialysis and that if the dialysis procedure is prolonged the incidence of such abnormalities is reduced, as the changes in blood pH and urea concentration are induced more gradually.

Jacksonian Epilepsy. Epileptic seizures that are caused by a focal irritation of a specific region of the primary motor area of the cerebral cortex may be restricted to a specific peripheral region on the contralateral side of the body (p. 267). The patient may remain conscious during the attack. The seizure also may ramify over the rest of the adjacent motor cortex rather than remaining localized in one specific focal area so that adjacent peripheral regions ultimately may become involved in the convulsive motor responses.

Jacksonian epileptic seizures generally are associated with relatively specific organic brain damage such as is associated with scar tissue following traumatic injury or following a hemorrhage. Alternatively, an epileptic focus can develop as the result of a localized neoplastic growth.

Status Epilepticus. This exceedingly dangerous situation develops when a number of major (grand mal) seizures follow each other in rapid succession so that the interval between any two seizures is relatively brief, or completely absent. The patient becomes exhausted, hyperthermia may develop, and death frequently intervenes during such attacks.

Psychomotor Seizures. Included under this heading are essentially all types of epileptic seizure that do not comply with the classic definitions of petit mal, grand mal, and focal jacksonian seizures.

Patterned, stereotyped movements, seemingly purposeful movements, incoherent speech, emotional lability, rotation of the head and eyes, lip smacking, writhing and twisting movements of the limbs, dimming of consciousness, and amnesia are common accompaniments of psychomotor seizures. EEG changes frequently associated with this form of epilepsy include slow round waves (Fig. 7.129C), spikes, sharp waves and/or combinations of these patterns, usually recorded from foci on the temporal lobes, and these abnormalities may become much more evident during periods of light sleep in patients who are subject to this type of seizure.

Febrile Convulsions. In very young children, fever and convulsions frequently are related. Furthermore, the first seizure manifest by an epileptic child usually is a febrile convulsion. In fact, it has been claimed that febrile convulsions are much more frequent among children whose parents have a history of epilepsy.

The relationship between fever and convulsions has been attributed to a number of possible causes. First, the fever and the convulsions both are caused by an infectious microorganism and/or its toxic metabolic byproducts. Second, the immature brain may react to high fever and/or an infectious agent by discharging random impulses that result in a convulsion. Third, excessive fluid intake (ie, hydration) as well as any drugs employed to combat an infection may precipitate convulsions. Fourth, the fever per se develops secondarily from the liberation of excess heat energy from the muscles as they convulse. Fifth, during convulsive seizures, the hypothalamic thermoregulatory centers become "reset" at a higher level so that fever develops. Sixth, an abnormal reaction may be induced in the brain by an infection such that a local (or generalized) irritable focus develops and convulsions ensue.

Frequently children develop psychomotor seizures following one or more episodes of febrile convulsions, although the prognosis varies considerably. Nonfebrile convulsions develop subsequent to febrile convulsions in most of the patients having a history of the latter.

Coma and General Anesthesia: Dysfunctions of the Reticular Activating System

The reticular activating system (RAS) occupies a particularly exposed neuroanatomic site at the rostral end of the midbrain and in the posterior hypothalamus because it is displaced medially by other fiber systems (Fig. 7.126). Consequently small lesions, such as may be induced by a tumor, hemorrhage, or other cause will produce a coma that may last for an extended period of time. Experimentally it has been shown that the trauma induced by a concussion will inhibit normal bioelectric activity in the RAS.

One major effect produced by general anesthetics such as ether, halothane, nitrous oxide, or barbituric acid derivatives is to produce a loss of consciousness. It has been demonstrated clearly that such chemical agents depress conduction in the ascending RAS as well as in the midbrain reticular formation. The excitatory influence of impulses ascending to the cortex from the RAS is inhibited markedly and consciousness is lost (see also p. 428).

The physiologic effects of anesthetics on the RAS may reflect merely a general inhibition of synaptic conduction in all areas of the central nervous system rather than a selective (or specific) effect on the synapses of the RAS. In any event, the specific sensory pathways contain only two to four synapses, whereas the reticular pathways can involve hundreds of synapses. And at every synapse, the excitatory postsynaptic potential (EPSP, p. 191) must reach a certain magnitude before that cell will fire. Since the neural pathways through the RAS contain so many synapses, on this basis alone they are far more vulnerable to inhibition by any agent, chemical or otherwise, which exerts a depressant effect on synaptic conduction.

Narcolepsy and Catalepsy

The clinical syndrome characterized by the intermittent occurrence of periods of uncontrollable and usually inappropriate sleep is termed *narcolepsy*. The hypothalamus has been implicated in some cases of this disorder.

To summarize briefly, hypothalamic lesions such as may result from influenza, encephalitis, or neoplasms can result in abnormalities of fat, carbohydrate, or water metabolism; emotional manifestations, eg, bouts of crying or laughing; sympathetic or parasympathetic effects; derangement of normal sexual functions; and disorders of the normal sleep pattern (see also p. 407).

Although narcolepsy can be attributed to hypothalamic lesions in many instances, frequently there is no overt damage to this region of the brain, hence the term *idiopathic narcolepsy* is applied in such cases. Nevertheless, a person suffering from attacks of idiopathic narcolepsy may exhibit such hypothalamus-related signs as obesity, polyuria, or abnormal sexual functions, which suggest a physiologic if not an evident pathologic impairment of the normal functions of the hypothalamus. The role of the RAS in the etiology of narcolepsy is obscure, but this matter would bear close experimental scrutiny as it is extremely doubtful that the narcoleptic state would develop independently of the system which normally is related so inseparably to the conscious and unconscious states.

Clinically, the duration of individual narcoleptic attacks can range from a few seconds to several hours, and a number of such episodes can occur with or without warning at generally inappropriate times throughout the day, such as while in conversation, attending a lecture, or driving an automobile. The attacks resemble normal sleep, and the diurnal nocturnal sleep pattern of a narcoleptic patient exhibits no marked peculiarities.

During a narcoleptic attack, the patient may exhibit a sudden loss of muscle tone, hence weakness, so that he sinks to the ground, particularly when the narcoleptic attack is triggered by a sudden emotional stimulus such as mirth, anger, fear, or surprise. This transient hypotonia, known as *catalepsy,* frequently is associated with narcolepsy; and although narcolepsy may be present without catalepsy, the reverse situation is exceedingly rare.

The simultaneous occurrence of narcolepsy and catalepsy suggests that similar pathophysiologic mechanisms may be operative in the two states. However, the hypotonus observed during a cataleptic seizure is not explicable on the basis of any presently known physiologic mechanism.

The interesting suggestion has been advanced that narcoleptic and cataleptic attacks are relics of primitive behavioral defense mechanisms which are unmasked in humans by disease. Thus certain lower organisms, such as the opossum, may enter a state similar to narcolepsy when seriously endangered.

Retrograde Amnesia

The human brain has an amazing capacity to recover functionally from irreversible damage to its structure, and it is a frequent clinical observation that a major physical or other derangement which appears immediately after traumatic or vascular injuries to the central nervous system may subsequently and spontaneously improve to such an extent that the original defects may disappear completely. Nonetheless, and despite the functional improvement, the structural damage remains. Consequently the mechanism which underlies this effect is of major importance to a better understanding of the learning (p. 439) and memory (p. 443) phenomena.

In this regard it is of interest that both experimentally as well as clinically it can be demonstrated readily that a loss of recent memory frequently occurs following a brain concussion or electroshock therapy. That is, *retrograde amnesia* develops for those events which immediately preceded the physiologic insult. In humans, the period for which recollection is totally lost may extend backward in time for days, weeks, months, or even years, although long-term memory is completely unaffected, a fact which testifies eloquently for the stability of such memory engrams (p. 445).

In experimental animals, memory for learned responses, as evidenced by behavioral studies, is lost if the subject is anesthetized or subjected to an electric shock within a few minutes after a training session. On the other hand if the physiologic insult is postponed for several hours, no such memory loss occurs. Studies such as these suggest, but do not prove conclusively, that the consolidation of recently acquired information so that stable memory engrams develop sometimes may require a fairly long interval. In humans the period for which retrograde amnesia is present may (or may not!) reflect the time required for the entire consolidation process to become complete; however this remark is largely conjectural.

Aphasia, Apraxia, and Agnosia

When the memory for learned response is lost because of damage to the cerebral cortex, this derangement of the intellectual functions, which are localized approximately in the neocortex per se, can take any of three forms; viz., aphasia, apraxia, or agnosia.

APHASIA. The term "aphasia" denotes a defect or loss of the power to express oneself in speech, writing, or gestures (ie, by the use of symbols) or of the loss of the ability to comprehend spoken or written language due to lesions in the brain centers which are responsible for these functions. An aphasia thus is a disturbance of the senory and/or motor areas of the brain that is caused by brain damage, regardless of the specific pathogenesis. Aphasias are *not* caused by defects in the sense organs themselves (ie, in vision or audition), in the muscles which are essential to speech, or mental defects.

A number of classifications have been devised in an attempt to correlate specific brain lesions with particular functional abnormalities. As one consequence of these efforts, the terminology relating to these defects in the ability to understand and to use language is utterly confusing!

In most general terms, aphasias can be divided into two broad categories: *sensory (receptive) aphasias* and *motor (expressive) aphasias.* Each of these principal categories in turn can be subdivided further. *Word deafness* is an inability to comprehend words that are spoken. *Word blindness* is an inability to comprehend written words. *Agraphia* is the inability to express ideas by writing them down. The inability to express ideas in speech commonly is termed *motor aphasia.* The category of motor aphasia can be subdivided into *fluent aphasia,* in which normal speech occurs, and this speech even may be quite rapid although key words are missing, and *nonfluent aphasia,* in which speech is slow, labored, and the victim has difficulty in selecting any words to express himself. In fact, nonfluent aphasia in its extreme form may result in a vocabulary that contains only a few words in toto. Frequently patients with nonfluent aphasia speak only a few *automatic words,* eg, name the months of the year, or the words being used at the time of the lesion developed. A patient with nonfluent aphasia for reasons unknown may utter only obscenities in an automatic and repetitive fashion.

As discussed earlier (p. 436), the cerebral hemispheres exhibit a complementary functional dominance, so that in any particular individual only one of these structures is concerned primarily with speech, the *categorical hemisphere.* One conception of the general areas of the brain in which lesions produce particular aphasias is illustrated in Figure 7.130.

APRAXIA. The term "apraxia" signifies the inability to perform a highly skilled or complex movement when requested to do so. Apraxia is not caused by muscular paralysis (or paresis), ataxia, alterations in the sensory receptors, or by a defective comprehension of the problem (ie, confusion). Thus in the classic description of apraxia, the patient was unable to stick out his tongue when so requested, although he was able to perform normally complex semiautomatic movements of this structure such as were involved in chewing and swallowing *(motor apraxia).*

Ideational apraxia signifies that the patient is unable to formulate or to remember the concept necessary to the performance of a particular motor act. *Ideomotor apraxia* denotes a situation in which the patient is unable to carry out a specific act correctly even though older and stereotyped motor acts may be performed spontaneously as well as repetitiously.

The apraxias usually are accompanied by specific lesions of the brain. Thus ideational apraxia often is a manifestation of certain diffuse brain lesions such as cerebral arteriosclerosis, whereas motor apraxia usually is related to a lesion that is located in the precentral gyrus. Ideomotor apraxia, on the other hand, usually is related to lesions that are located within the supramarginal gyrus of the categorical hemisphere.

AGNOSIA. The term "agnosia" denotes an inability to recognize common objects (see also p. 436). That is, the patient is unable to interpret sensory stimuli correctly.

Five general types of agnosia sometimes are recognized. First, in *astereognosis* (or *tactual agnosia*) objects cannot be recognized by handling them (p. 365). Second, in *auditory agnosia* (or *psychic deafness*) the patient is unable to recognize objects from aural cues. This type of agnosia is similar in some respects to the aphasia called word blindness. Third, in *visual agnosia* (or *psychic blindness*), the patient neither can perceive the meaning of objects or colors that are seen, nor can he appreciate the significance of visual space. Fourth, in *autotopagnosia* (also called *body-image agnosia* or *finger agnosia*) there is a failure to recognize the different parts of the body and their relationships, to discriminate between left and right, or to perceive the relationship between the body and objects in the immediate environment. Fifth, in *anosognosia,* the patient is unaware of disease or may deny the existence of disease per se.

Like the apraxias, the agnosias generally are associated with lesions in particular areas of the brain. For example, anosognosia may be present

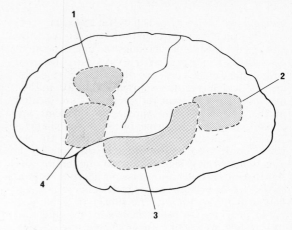

FIG. 7.130. Lateral aspect of the left cerebral hemisphere illustrating one concept of the regions of the brain responsible for language functions (shaded areas). 1. Damage affects ability to write. 2. Damage causes word blindness. 3. Damage causes word deafness. 4. Damage produces a loss of articulate speech. According to this concept, therefore, particular aphasias are related to lesions in specific areas of the brain. (After Ruch and Patton [eds]. *Physiology and Biophysics,* 19th ed, 1966, Saunders.)

when a lesion is evident in the supramarginal gyrus of the parietal lobe, whereas autotopagnosia may be associated with lesions that are located in the posteroinferior region of the same lobe.

Meniere's Syndrome (Paroxysmal Labyrinthine Vertigo)

Meniere's syndrome typically is manifest by intermittent bouts of severe vertigo, or dizziness, that are coupled with deafness and tinnitus. That is, an auditory nerve stimulation is present such that the patient hears "sounds" such as ringing, buzzing, or roaring, or clicking.

This syndrome frequently develops in men between the fourth and sixth decades of life for unknown reasons, although an abnormal accumulation of serous fluid in the cochlear duct is suspected (Fig. 7.106). In turn this so-called endolymphatic hydrops, when maximal, could exert a mechanical pressure upon the vestibulocochlear nerve such that this structure would be excited abnormally, and this would result in the attacks of severe vertigo which is the principal symptom of this disease. A similar explanation could be offered for the tinnitus.

The vertigo may be so severe that the patient falls violently to the ground as though thrown down, and consciousness may be lost briefly during such an attack (see p. 394 et seq. for a discussion of normal vestibular functions). Surrounding objects may appear to spin around the victim, and autonomic responses such as nausea, vomiting, and heavy sweating are frequent accompaniments. Any given attack may persist for a few minutes or last for up to several hours, and the frequency with which the attacks occur can vary considerably in any individual. Headache, deafness caused by defective neural conduction, and tinnitus tend to occur as well as to persist between the attacks of vertigo.

The hearing loss tends to be progressive and in the overwhelming majority of cases this loss tends to be found in one ear. Nystagmus (p. 393) may be seen during attacks of vertigo, and the

Bárány or caloric tests may reveal an abnormal labyrinthine function (p. 395). The patient exhibits a hypersensitivity to loud sounds.

Meniere's syndrome can follow head injuries or infections of the middle ear; however in many instances this condition appears in the absence of any demonstrable damage to either the ear or nervous system.

Intractable Pain and Percutaneous Cervical Cordotomy

Intractable pain and percutaneous cordotomy have been selected for discussion as the concluding topics in this chapter as they not only exemplify the exceedingly complex nature of a seemingly straightforward neural phenomenon (viz, the sensation of pain) but also illustrate the pragmatic value of a knowledge of neuroanatomy and neurophysiology to the management of an exceedingly difficult clinical problem.

NEUROANATOMIC AND NEUROPHYSIOLOGIC CONSIDERATIONS. As noted earlier, pain in any form may be considered to represent the operation of an abnormal process, a biologic warning system (p. 355). Viewed from a strictly physiologic standpoint, the mechanism underlying the sensation of pain consists merely of nerve impulses transmitted from the periphery, particularly in the so-called pain pathways of the nervous system. The dorsal (or anterolateral) spinothalamic system is especially significant in relaying such impulses from specific peripheral regions of the body to the thalamus and thence to the cortex where the signals are perceived as painful sensations (p. 351). This view is, of course, entirely too simplistic, because pain is an exceedingly complex sensation, as the interpretation of the painful stimulus by the mind is considerably different than the mere biologic reception and transmission of information pertaining to the stimulus which is coded as nerve impulses having a certain frequency.

Although a number of hypotheses have been advanced to explain the phenomenon of pain, the *gate control theory* affords a useful and reasonable explanation of this sensation (p. 360). To recapitulate briefly, this theory includes the interdependent operation of three neural systems located within the spinal cord and that are activated by signals which are transmitted centrally following stimulation of pain receptors that are located in the periphery. The three cord systems are composed of, first, neurons of the substantia gelatinosa, located in the dorsal horn of the spinal cord; second, the fibers of the dorsal column which relay signals toward the brain; and third, the first-order transmission neurons located in the dorsal column, and which convey pain signals in the dorsal horn. The substantia gelatinosa per se provides the neural matrix for the gate control system which in turn modulates the afferent impulse patterns before they excite or facilitate the transmission neurons. The afferent impulse patterns in the dorsal column in turn function as a regulatory mechanism which can initiate selective processes in the brain that can alter the modulating characteristics of the gate control system proper. And lastly, the transmission cells are able to initiate neural mechanisms which make up the functional system that underlies the perception and response to pain. Thus pain phenomena, according to the gate control theory, are caused by the functional interplay among these three systems.

Impulses that underlie painful sensations are transmitted to the brain, once they have traversed the gate in the cord, via synapses that are located in the dorsal column. Thence neurons decussate in the ventral commissure of the cord to pass rostrally in the ascending anterolateral tracts of the cord. Most of these ascending fibers are located within the spinothalamic tract, although five discrete pathways have been implicated in the projection of pain sensations to the cerebral cortex via the thalamic projection nuclei. Thus two pathways involved in the transmission of painful sensations project to the somatosensory thalamic nuclei, thence to the related cortical regions via the dorsal column-lemniscal and dorsolateral tracts. The dorsal spinothalamic tract per se actually has been shown to contain the other three functional tracts that convey impulses related to pain. One of these tracts contains few synapses and is composed of fibers which are highly oriented in space so that this tract conveys signals related to the perception of rapidly perceived, highly localized pain. The remaining two spinothalamic subdivisions contain a greater number of synapses than does the fast-acting tract, and these fibers terminate in the nonspecific intralaminar nuclei of the thalamus. Consequently this system transmits the signals related to slowly conducted diffuse pain. This portion of the thalamic system typically exhibits much convergence, has connections with nuclei that receive visual as well as auditory stimuli, and thus it underlies the transmission of signals that are more concerned with the perception of pain (ie, the affective component of pain) rather than with the physiologic mechanisms that are involved in the central reception of pain signals. Consequently impulses that are related to pain inputs are not simply localized to specific neuroanatomic pathways and neurophysiologic circuits. Rather, the inputs are able to spread to neural systems that are involved with highly subjective aspects of the perception of pain in addition to the mere sensory perception of painful stimuli. Furthermore, the functional activity of the input pathways that subserve the phenomena related to the perception of pain can be altered by a number of descending polysynaptic influences which can operate in a reciprocal fashion, and which have their origin in the somatosensory and related cortical areas as well as in the reticular formation.

MANAGEMENT OF INTRACTABLE PAIN. Quite frequently in clinical practice it is impossible to effect a cure, or even an amelioration, of the underlying disease process so that not only objective signs but also subjective symptoms remain. Advanced malignant disease of various organs provides an example to illustrate this point. Alternatively, it may prove feasible to eradicate a disease process, and yet the patient becomes totally incapacitated because of the presence of chronic, intractable pain; that is, the pain is of long duration and is quite refractory to cure, control, or even symptomatic relief.

The physician can select from a wide range of methods for the relief of pain depending upon the specific case. For example, the use of various drugs, the intrathecal injection of phenol, alcohol or saline; thalamotomy; transcutaneous electric stimulation of the spinal cord; and surgical or percutaneous cordotomy all have a place in the management of pain.* Each of these therapeutic procedures has advantages and disadvantages insofar as the physician and patient are concerned, and certain of these techniques are employed only as a last resort as a means to alleviate the misery and suffering that can accompany the advanced states of malignant diseases.

The technique of percutaneous cervical cordotomy was selected for discussion here as it not only illustrates certain basic physiologic principles related to pain phenomena, but also as the clinical results obtained by use of this technique are such as to warrant consideration.

Briefly, the technique of percutaneous cordotomy involves the selective destruction of the dorsal (anterolateral) spinothalamic tract by means of a radiofrequency current that is passed through a needle electrode which is inserted through the skin and implanted in the spinal cord between C–1 and C–2. This current produces a permanent, discrete, and highly localized lesion by electrocoagulation of the tissue whose size depends upon the total time the current is passed. The current is

The clinical role, if any, of acupuncture (p. 360) in the rational management of pain must await further objective study and evaluation.

given in several increments of a few seconds each, the reactions of the patient being tested between times. Placement of the tip of the needle electrode is accomplished by means of a stereotactic micromanipulator and visualized by either lateral and anterior roentgenograms or by viewing the position of the needle on a unidirectional cineradiographic apparatus. The spinal cord per se is outlined by the injection of air following removal of a similar volume of cerebrospinal fluid. Since percutaneous cervical cordotomy is carried out under a local anesthetic on the side of the cord *opposite* to that on which it is desired to produce the permanent analgesia, the patient is awake, alert, cooperative, and thus able to report his sensations between individual applications of current. If somatic motor weakness develops, the procedure is stopped or postponed. Between application of successive bursts of radiofrequency current, the level of analgesia on the contralateral side of the body is checked verbally with the patient as well as by pricking the skin with a pin. The process is repeated until the necessary level of analgesia is reached in the affected area.

The principal effect of unilateral percutaneous cordotomy is contralateral pain loss, this effect being coupled with a similar loss of thermal sensations (ie, heat and cold) on the contralateral side of the body to the induced lesion. Since the total quantity of radiofrequency current passed determines the magnitude of the lesion, it is possible to produce lesions of graded size so that analgesia can be obtained selectively in the pain pathways from S–5 to C–5 levels. Unilateral cordotomy produces no loss of the touch sensation and usually no motor, respiratory, or bladder problems develop. Equally important to a severely ill patient is the fact that percutaneous cordotomy involves little stress, such as would be involved in a standard surgical cordotomy.

Ideally the patient who undergoes percutaneous cordotomy has unilateral pain but little alteration in lung function. Thus a patient with carcinoma of the lung and little respiratory deficit makes a good candidate for this procedure, and few complications are likely to arise.

Bilateral cordotomy, on the other hand, can give rise to complications, particularly of micturition and respiration. Thus catheterization of the urinary bladder becomes necessary, although some recovery of function may occur spontaneously as time passes. Respiratory complications can become serious following bilateral cordotomy, particularly when the phrenic nerves are damaged on both sides. This effect will occur if analgesia is necessary at high levels. These complications may be avoided by performing a percutaneous cordotomy on the contralateral side of the cord to the side of the body which is experiencing pain to a higher (ie, more cephalad) level. Then an entirely different method for alleviating the pain on the other side of the body can be applied. For instance, a left percutaneous cordotomy for pain on the right side could be followed by a left intrathecal block (eg, with alcohol) for residual pain on the left side. Alternatively, respiratory problems can be minimized by performing a second cordotomy at a spinal level that is below the outflow of the phrenic nerves so that the integrity of diaphragmatic innervation is retained.

In conclusion, the technique of percutaneous cervical cordotomy is a relatively simple, efficient, and low-risk technique which entails minimal stress for the patient, but which can afford selective relief from severe intractable pain that arises in certain regions of the body.

CHAPTER 8

The Special Senses

The physiologist is not concerned with philosophic concepts of human experience in relation to sensory phenomena and the conscious perception thereof. Rather, it is relevant that he examine and evaluate the functional performance of the sense organs in terms of their capacities and limitations, as well as in terms of their functional patterns in time and space as described by physicochemical alterations in response to specific changes in environmental energy levels (ie, stimuli). When viewed from this standpoint, human beings as well as experimental animals *respond* to particular physical or chemical stimuli rather than *feel* them. In turn, the responses made by any higher organism to a particular environmental change consist of a definite sequence of five events. First, there are physicochemical changes in functionally specific receptor elements. Second, these lead to alterations in the firing pattern of the afferent neural processes connected to these receptor cells. Third, the afferent neural signals cause excitation of neurons located within the central nervous system. Fourth, efferent neurons are excited. Fifth, the efferent neural discharge pattern induces alterations in the physiochemical properties of specific effector elements so that a motor response is evoked (ie, glandular secretion or muscular contraction). This sequence restates briefly the general components as well as the functional attributes of the reflex arc, discussed in detail previously (Chap. 6, p. 213). In this chapter, emphasis will be placed upon the demonstrable effects of specific stimuli upon the receptor elements that underlie the special senses of vision, hearing, gustation (or taste), and olfaction (or smell). The vestibular functions of the inner ear, which may be considered as the fifth special sense, were discussed earlier (p. 388 et seq.).

THE EYE AND VISUAL PROCESSES

The eye is the peripheral organ of vision wherein light of certain wavelengths initiates a physiologic process that is manifest in the subjective sensation of vision.

In all vertebrates, the eyes are complex organs that structurally resemble a camera obscura. Each eye is essentially a spherical sac filled with transparent media having opaque walls and a transparent front to admit light (the cornea). The eye also contains a lens for focusing an image on the photoreceptive cells of the retina as well as a variable aperture, the iris, to control the amount of light permitted to impinge upon the retina. From a functional standpoint, however, the conception of the eye as a camera is wholly inadequate. For example, the retina contains not only light-sensitive cells, but other elements that refine and process the impulses generated in the photoreceptors before they are transmitted to the brain. Furthermore, the diameter of the pupil is not only controlled by the ambient light intensity, but also by excitation of specific autonomic nerves (eg, sympathetic stimulation causes pupillary dilatation). Thus, not only light, but also emotions, can produce transient alterations in pupillary diameter.*

General Anatomy of the Eye

The human eyeball is roughly spherical, and about 24 mm in diameter (Fig. 8.1). At its anterior end, or pole, lies the transparent cornea (about 11 mm in diameter) consisting of five layers of epithelial cells as shown in Figure 8.2. From the front backward, these epithelial layers are: (1) the corneal epithelium; (2) the anterior elastic lamina (of Bowman); (3) the substantia propria (or stroma); (4) the posterior elastic lamina (of Descemet); and (5) an endothelial cell layer. Behind the cornea lies the anterior chamber, filled with the aqueous

---

An excellent example of this involuntary effect is the momentary pupillary dilatation that can be observed in a card player who suddenly perceives that he has been dealt an exceedingly good hand.

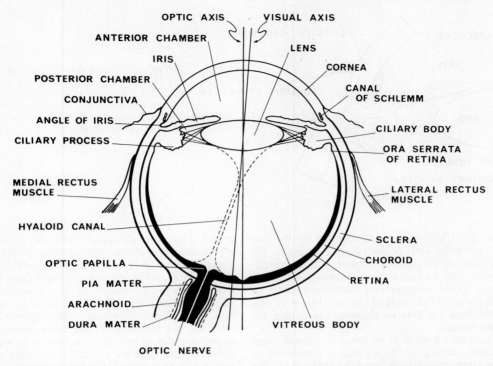

OPTIC AXIS   VISUAL AXIS

ANTERIOR CHAMBER

IRIS

LENS

CORNEA

POSTERIOR CHAMBER

CANAL OF SCHLEMM

CONJUNCTIVA

ANGLE OF IRIS

CILIARY BODY

CILIARY PROCESS

ORA SERRATA OF RETINA

MEDIAL RECTUS MUSCLE

LATERAL RECTUS MUSCLE

HYALOID CANAL

SCLERA

OPTIC PAPILLA

CHOROID

PIA MATER

RETINA

ARACHNOID

DURA MATER

VITREOUS BODY

OPTIC NERVE

FIG. 8.1. The gross structures and axes of the human eye as seen in a horizontal section viewed from above. The cuplike retinal depression through which the optic axis passes is the fovea (Fig. 8.12), a region of the retina wherein only cones are present, hence the point at which maximal visual acuity is achieved in daylight. The spheroid curvature of both cornea and eyeball have been exaggerated deliberately to emphasize the fact that the entire eye consists of portions of two spherical structures superimposed upon each other. (Data from Goss [ed]. *Gray's Anatomy of the Human Body,* 28th ed, 1970, Lea & Febiger; Bloom and Fawcett. *A Textbook of Histology,* 9th ed, 1968, Saunders.)

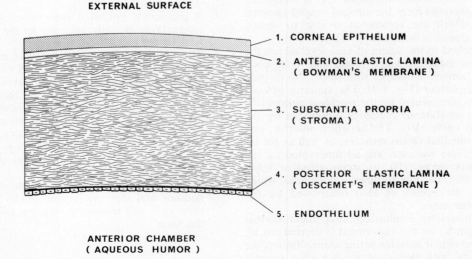

EXTERNAL SURFACE

1. CORNEAL EPITHELIUM

2. ANTERIOR ELASTIC LAMINA (BOWMAN'S MEMBRANE)

3. SUBSTANTIA PROPRIA (STROMA)

4. POSTERIOR ELASTIC LAMINA (DESCEMET'S MEMBRANE)

5. ENDOTHELIUM

ANTERIOR CHAMBER (AQUEOUS HUMOR)

FIG. 8.2. The layers of the human cornea as seen in cross-section. The cornea is somewhat thicker than the sclera, being about 0.9 mm thick in the center and 1.1 mm peripherally. (Data from Bloom and Fawcett. *A Textbook of Histology,* 9th ed, 1968, Saunders.)

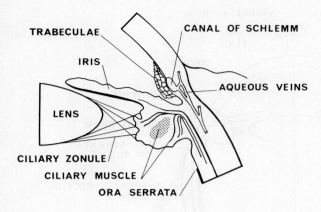

TRABECULAE

CANAL OF SCHLEMM

IRIS

AQUEOUS VEINS

LENS

CILIARY ZONULE

CILIARY MUSCLE

ORA SERRATA

FIG. 8.3. The principal anatomic features of the junction between the sclera and the cornea. (Data from Goss [ed]. *Gray's Anatomy of the Human Body,* 28th ed, 1970, Lea & Febiger.)

humor (qv, Chap. 9, p. 563) secreted by the columnar epithelial cells of the ciliary process on the ciliary body in the posterior chamber. The crystalline lens, held in place by suspensory ligaments attached to the circular and meridional fibers of the ciliary body, lies behind the iris (Fig. 8.3). Behind the lens lies a large chamber containing the vitreous humor.

The entire eyeball is invested in a tough outer protective coating (or tunic) that is composed of fibrous connective tissue, the sclera (Fig. 8.1). The transparent cornea forms the anterior portion of the sclera. Within the sclera lies another, richly vascular, middle layer consisting of the choroid (or uvea). The aforementioned ciliary body and iris are subdivisions of the choroid layer. The third, innermost layer of the eyeball is the complex nervous tunic or retina, which contains the photosensitive elements of the eye, the rods and cones.

The exposed part of the eyeball is covered by a delicate mucous membrane, the conjunctiva, which is reflected onto the inner surfaces of the eyelids. The conjunctival membrane is moistened by fluid secreted from the almond-shaped lacrimal glands, which are somewhat similar to serous salivary glands in structure (p. 819).

Attached to the sclera of each eyeball are the six ocular muscles, which rotate and move each eye in a conjugate manner with respect to its contralateral partner (Fig. 8.4). The superior oblique muscle is inneravated by the trochlear nerve (IV); the external (lateral) rectus is innervated by the abducens nerve (VI). The superior, inferior, and internal (medial) rectus muscles, as well as the inferior oblique muscles, are all innervated by the oculomotor nerve (III), as summarized in Table 7.3 (p. 248). The superior levator palpebrae, which elevates the eyelid, is also innervated by the oculomotor nerve.

It should be emphasized that under normal circumstances no eye movement is carried out by one of the ocular muscles acting alone. Rather, the muscles of both eyes contract in such a manner that the two eyes generally move in unison and both turn in the same direction at once (except during convergence for near vision). This effect is called *conjugate deviation,* and the physiologic mechanism underlying this process is the reciprocal innervation of antagonistic muscles (p. 221). For example, when the eye is rotated outward, the external rectus contracts, as do the superior and inferior oblique muscles, while their antagonists, viz. the inferior, external, and superior rectus muscles, are inhibited. It has been demonstrated experimentally in primates (monkeys) as well as in cats that the superior colliculus (p. 269) plays a major role as a neural integrative center for coordinating tracking movements of the eyes and head when following moving objects.*

The Retina

This complex structure is actually an extension of the central nervous system. In cross-section, the retina is composed of ten histologically discrete layers, as illustrated in Figure 8.5. In summary, these layers are as follows:

1. A *pigmented epithelial layer* consisting of a single layer of cells adjacent to the choroid tunic of the eyeball
2. The rods and cones *(bacillary layer),* which are actually the processes of rod and cone cells whose cell bodies are located in layer 4; these elements are the photoreceptors of the eye.
3. The *external (outer) limiting membrane*
4. The *outer nuclear layer,* which contains the cell bodies and nuclei of the rod and cone cells
5. The *outer plexiform (synaptic) layer,* which forms the connections between rod and cone cells and the bipolar neurons
6. The *inner nuclear layer,* which contains bipolar neurons and their nuclei; the nuclei of amacrine cells and horizontal cells are also within this layer.

*The superior colliculus, in addition to its major role in coordinating eye and head movements, also appears to be involved in the orientation of the head and eyes in response to auditory and somatic sensory stimuli.

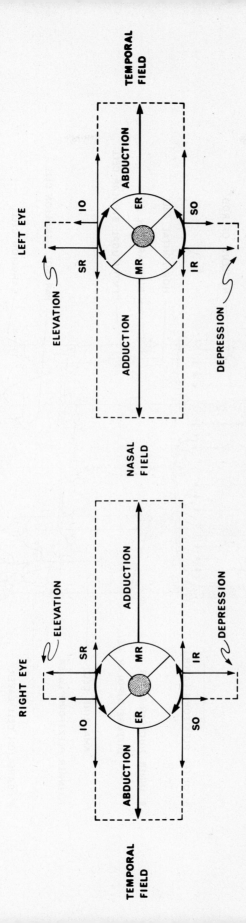

FIG. 8.4. Schematic diagram of both eyes viewed from the front showing the movements induced by contraction of the individual ocular muscles. The abbreviations and principal actions of each ocular muscle are: I.O. = inferior oblique (elevation); E.R. = external, or lateral, rectus (abduction); M.R. = medial, or internal, rectus (adduction); S.R. = superior rectus (elevation); S.O. = superior oblique (depression); I.R. = inferior rectus (depression). (Data from Best and Taylor [eds]. *The Physiological Basis of Medical Practice*, 8th ed, 1966, Williams & Wilkins.)

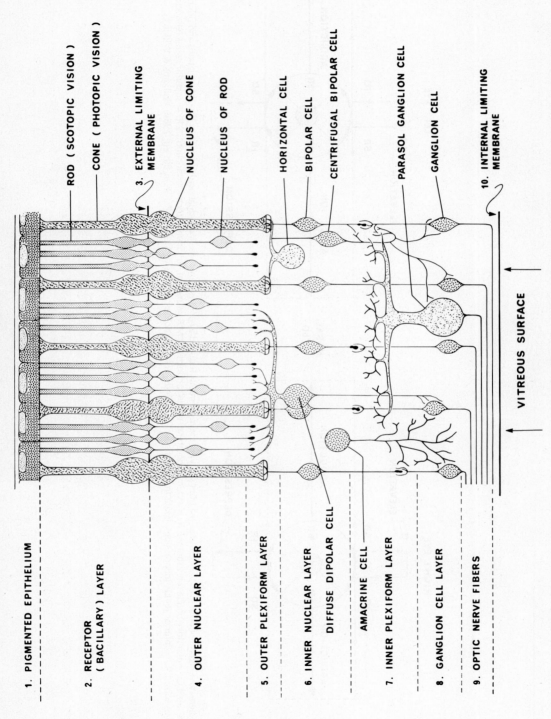

CHOROID SURFACE

- ROD (SCOPTIC VISION)
- CONE (PHOTOPIC VISION)
- 3. EXTERNAL LIMITING MEMBRANE
- NUCLEUS OF CONE
- NUCLEUS OF ROD
- HORIZONTAL CELL
- BIPOLAR CELL
- CENTRIFUGAL BIPOLAR CELL
- PARASOL GANGLION CELL
- GANGLION CELL
- 10. INTERNAL LIMITING MEMBRANE

VITREOUS SURFACE

1. PIGMENTED EPITHELIUM

2. RECEPTOR (BACILLARY) LAYER

4. OUTER NUCLEAR LAYER

5. OUTER PLEXIFORM LAYER

6. INNER NUCLEAR LAYER
   - DIFFUSE DIPOLAR CELL
   - AMACRINE CELL

7. INNER PLEXIFORM LAYER

8. GANGLION CELL LAYER

9. OPTIC NERVE FIBERS

FIG. 8.5. The principal layers of the retina and their major interconnections as seen in cross-section. The vertical arrows at the bottom of the figure indicate the direction of the light path and the numbers indicate the specific retinal layers from the choroid outward. (After Everett. *Functional*

470

7. The *inner plexiform* (synaptic) *layer,* in which connections are made between the bipolar cells and the ganglion cells
8. *Ganglion cell layer*
9. *Optic nerve fiber layer*
10. The *internal* (or inner) *limiting membrane*

Normally, the retina is completely transparent, but in order to stimulate the rods and cones, light must first pass through all the retinal elements, including certain blood vessels, as shown by the vertical arrows at the bottom of Figure 8.5. Furthermore, the only region in the body where blood vessels can be examined directly under physiologic conditions is in the retina. This fact has considerable importance in the early detection and diagnosis of certain vascular diseases which may be manifest first as alterations in the normal appearance of the larger superficial vessels within the fundus of the eye. The ophthalmoscope (Fig. 8.6) thus is an indispensable diagnostic instrument to the practicing physician (see also Clinical Correlates, p. 501).

The greater portion of the retina is nourished by blood from vessels located within the choroid layer of the eyeball (Fig. 8.1). Other vessels, however (eg, arterioles and venules), also enter and leave the retina together with the optic nerve at the optic papilla or "blind spot," whence these vessels radiate outward to cover the entire surface of the retina that lies in contact with the vitreous humor. (For demonstration of the blind spot see Fig. 8.7.)

In one specialized region of the retina called the *optic disc* (or *macula lutea*) there is a complete absence of rods so that in the central depression within this region (the *fovea centralis*) only cones are found. The fovea is the concavity in the retina through which the optic axis of the eye passes (Fig. 8.1), and in this region the cones are relatively small, abundant, and densely packed together. In addition, the synaptic relationship of the cones in the fovea to the bipolar neurons is on a one-to-one basis; therefore, in the foveal region, visual acuity is sharpest provided that the light intensity is sufficient (see also p. 456, and Fig. 8.12). Consequently, certain ocular reflexes automatically tend to focus objects being viewed upon the foveal region of each eye. At night or in dim light, of course, one must look deliberately either above or to one side of an object in order to visualize it, as the cones become totally ineffective under such conditions; hence, the image must be focused upon a more peripheral region of the retina that contains rods.

Optical Properties of the Eye

Prior to a discussion of the physiology of vision, certain elementary optical principles concerned with image formation by a lens must be reviewed; these properties are summarized in Figure 8.8 and are considered individually below:

REFRACTION OF LIGHT. In vacuo or in air, light travels at a velocity of about 3×10^5 km/sec. In solids, this velocity is much reduced. The ratio of the velocity of light in air to its velocity in another transparent medium, which can be either a liquid or a solid, defines the *refractive index* of that medium. (The refractive index of air itself is 1.00.) Thus, the refraction, or bending, of light rays depends upon the ratio of the refractive indices of

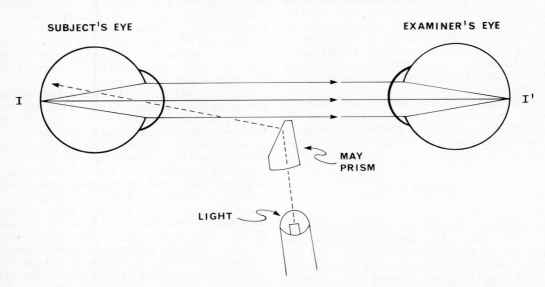

FIG. 8.6. Light paths during ophthalmoscopic examination. A narrow beam of light (dashed line) is directed at an angle into the subject's eye by the May prism of the ophthalmoscope, thereby illuminating the retina of the subject's eye. The retinal vessels at I form an image (I') upon the examiner's retina. If necessary, appropriate lenses are interposed in the light path between the subject's eye and the examiner's eye in order that the image I' be sharply focused. (After Ruch and Patton [eds]. *Physiology and Biophysics,* 19th ed, 1966, Saunders.)

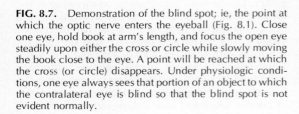

FIG. 8.7. Demonstration of the blind spot; ie, the point at which the optic nerve enters the eyeball (Fig. 8.1). Close one eye, hold book at arm's length, and focus the open eye steadily upon either the cross or circle while slowly moving the book close to the eye. A point will be reached at which the cross (or circle) disappears. Under physiologic conditions, one eye always sees that portion of an object to which the contralateral eye is blind so that the blind spot is not evident normally.

the two transparent media, and also upon the angle at which the beam of light strikes the interface between these two media (Fig. 8.8A).

CONVEX LENSES. The convergence of parallel light rays from a distant source (ie, beyond about 7 meters) by convex lenses of focal lengths f_1, f_2, and f_3 is illustrated in Figure 8.8B, C, and D. Image formation by a similar lens is illustrated in part E of the same figure. Note that light passing through the center, or optical axis, of any lens is not refracted and that the image behind the lens is always inverted vertically as well as reversed horizontally.

The identical situation obtains in image formation on the retina of the eye except that the focal length of the vertebrate lens is not fixed, and that the brain interprets the inverted, reversed image as "normal"; that is, upright and in correct horizontal register. Hence, light rays, either parallel or diverging from a nearby point source, can be focused at the same point behind the lens, provided that the curvature of the lens can change, or accommodate for different distances. The lens of the human eye can perform such a feat, so that normal image formation upon the retina is quite sharp for objects located from relatively close to the eye out to infinity.

The relationship between the focal length of a lens, f, the distance between a point source of light from the lens, d, and the distance the image is focused behind the lens, d^1, is given by the expression:

$$1/f = 1/d + 1/d^1$$

REFRACTIVE POWER OF LENSES AND THE EYE. The more a lens bends (or refracts) light rays, the greater its refractive power, and this property of lenses is measured in *diopters*. In a convex lens, 1 diopter is equal to 1 meter divided by the focal length (f) of that lens. A lens with a refractive power of +1 diopter can focus parallel light rays at a point 1 meter behind the lens.

As pertains to the human eye, the approximate individual refractive indices may be summarized as follows: Air = 1.0; cornea = 1.38; aqueous humor = 1.33; lens = 1.40; and vitreous humor = 1.34. Note that the greatest difference between any two of these refractive indices in the eye is found at the air–cornea interface.

The emmetropic (or normal) human eye has a total refractive power of about +59 diopters. Of this total, roughly +41 diopters are accounted for by three physical factors: first, the large difference between the refractive indices of the air (1.0) and the cornea (1.38) mentioned above; second, the marked convexity of the outer corneal surface*; and, third, the distance of the cornea from the retina.

Interestingly, the crystalline lens of the eye in vivo contributes only about +18 diopters to the total refractive power of the eye. This is caused by the relatively slight differences in refractive index between the lens and the aqueous humor on the one hand, and the lens and the vitreous humor on the other. (An isolated crystalline lens suspended in air, however, has a refractive power of +150 diopters!)

In simple optical calculations involving the eye, the so-called "reduced eye" is often used; that is, the influences of all the individual refractive surfaces are summed algebraically so that a single lens is considered to be present with its center 15 mm in front of the retina, and this reduced eye has a total of +59 diopters refractive power when accommodated for distant vision (Fig. 8.8F).

ACCOMMODATION OF THE LENS. As stated above, the lens of the human eye can change its refractive power significantly, a process called *accommodation*. The range involved is from 18 to a maximum of 32 diopters (in young children), and during accommodation, the actual shape of the lens as seen in cross-section changes from slightly convex to very convex.

The physiologic mechanism underlying the process of accommodation to distant and near vision is as follows. Normally, the lens in situ is held in a somewhat flattened shape by tension exerted upon its periphery by the radial suspensory ligaments of the ciliary zonule (Fig. 8.3). These ligaments pull the edges of the lens toward the ciliary body and insert upon the meridional and circular smooth muscle fibers located within this structure. When the muscles of the ciliary body contract, they pull the insertions of the ligaments forward, thereby releasing a certain amount of tension on the lens, which then automatically tends to assume a greater convexity. (A completely isolated lens will assume a spherical form because of the high

The concave inner surface of the cornea in contact with the aqueous humor produces a −5 diopter "neutralization" of the other refractive surfaces of the eye.

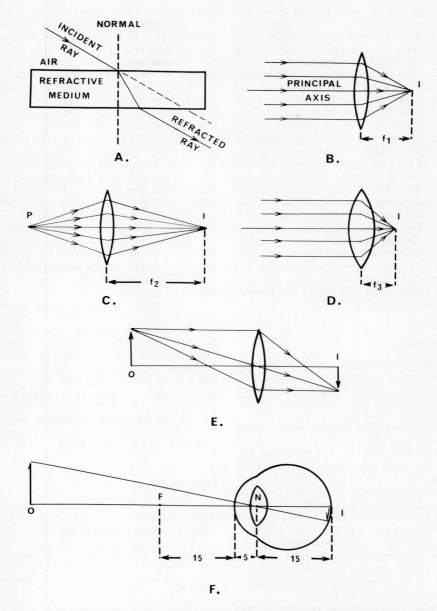

FIG. 8.8. Some general optic principles. **A.** Refraction toward the normal plane when light enters a refractive medium of greater density than air and away from the normal plane as the light enters the less dense medium once again. **B.** Convergence of parallel light rays by a convex lens of focal length f_1 to form an image at I. **C.** Diverging light rays from a point source (P) are focused into an image (I) by a convex lens having a focal length f_2. **D.** Parallel light rays are focused close behind a strongly convex lens of short focal length, f_3. **E.** Image formation by a convex lens. O = object; I = image. Note that the image is inverted (as well as transposed horizontally). **F.** Image formation (I) of an object (O) by the human eye. F = focal length; N = nodal point. This diagram also illustrates the so-called "reduced eye"; hence the numbers indicate the dimensions in mm.

degree of elasticity of the fibers which cover its surface, as well as the viscous nature of the protein within its core.) Thus, accommodation to near vision is a passive physical response on the part of the lens as a consequence of a sphincterlike action

of the muscles within the ciliary body. (See also p. 494.)

The change in the curvature of the lens during accommodation for near vision can be demonstrated readily by observing the Purkinje images in a normal human subject when the eye is relaxed, or accommodated for distant vision (Fig. 8.9A), and when it is focused upon a nearby object (Fig. 8.9B). In a darkened room, a candle is held to one side of the head of a subject and three images will be seen by the observer after a little practice. The images are: (1) a bright upright reflection of the candle from the convex surface of the cornea; (2) a second larger and fainter image of the candle reflected from the convex anterior surface of the crystalline lens; and (3) a much smaller inverted but brighter image reflected from the posterior surface of the lens which is acting as a concave mirror. Now, if the subject focuses his eye upon a nearby object (Fig. 8.9B), the first and third (1 and 3) images do not change appreciably; however, the second image (2) moves significantly closer to the first and also becomes somewhat smaller. This result signifies that during the process of accommodation the anterior surface of the lens assumes a greater convexity. The refractive power of this structure is increased and the more divergent rays from a nearby object are focused sharply upon the retina (Fig. 8.8C).*

The Visual Pathway

Vision takes place when light rays impinge upon the eyeball, are refracted by the cornea and lens, and form an inverted reversed image upon the retina, as described in the previous section. Within the brain, the entire visual pathway is spatially organized to a high degree so that the signals transmitted from the periphery are faithfully reproduced as patterns in specific cortical areas. Despite the fact that each optic tract (ie, that portion of the optic pathways which lies behind the optic chiasm) conveys information from the temporal part of the ipsilateral retina as well as from the nasal portion of the contralateral retina, the perceived image is seen as upright and in correct horizontal register. This apparent distortion of signal position is compensated for by central mechanisms. For example, the prefrontal motor areas as well as the parietal somesthetic regions are similarly inverted and transposed.

The general neuroanatomic features of the visual pathway are shown in Figures 7.44 (p. 298) and 8.37. The photosensitive rods and cones in the retina (layers 2 and 4, Fig. 8.5) convert (or transduce) light energy into nerve impulses; these structures are the first-order neurons in the visual pathway, and they synapse with the second-order bipolar neurons that are located in the outer plexiform layer (layer 5). These second-order neurons in turn synapse with third-order neurons in the inner plexiform layer (layer 7), whose cell bodies (found in layer 8) send axonal processes

to the layer of fibers that ultimately converge to form the optic nerve (layer 9). The optic nerve (cranial nerve II) then exits from the eyeball at the optic papilla (Fig. 8.1). The optic nerve is not a true cranial nerve, but rather it represents an evagination of fiber tracts from the diencephalon.

Fibers from the macula lutea, where visual acuity is the sharpest, enter the temporal side of the optic papilla (disc). After passing through the sclera, fibers of the optic nerve pass directly to the optic chiasm located just in front of the pituitary gland in the sella turcica of the sphenoid bone of the skull. The partial decussation of optic nerve fibers in the chiasm is diagrammed in Figure 7.44. Fibers from the temporal halves of each retina converge at the chiasm but leave it without crossing, whereas fibers originating in the nasal halves of each retina cross to the opposite side.

Behind the chiasm, the optic fibers diverge and pass as optic tracts to the left and right lateral geniculate bodies of the thalamus, as well as the pretectal area rostral to the superior colliculus (see below).

Cells within the lateral geniculate bodies send fibers (fourth-order neurons?) which form the geniculocalcarine tract (or optic radiation) to the occipital cortex of each cerebral hemisphere. The cortical region that receives the optic radiation surrounds the calcarine fissure located on the medial side of the occipital lobe. This cortical visual area (Brodmann's area 17 or striate area) is adjacent to areas 18 and 19, which are important cortical regions for visual perception and for certain visual reflexes, eg, visual fixation (see Figs. 7.70, 7.71, and 7.73; pp. 325, 327, and 328).

Optic Reflexes

In addition to the process of vision, there are a number of additional reflexes that underlie certain motor activities related to the eyes.

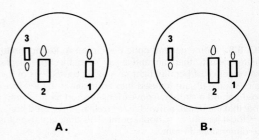

FIG. 8.9. The effect of accommodation for near vision upon the Purkinje images. **A.** Eye at rest; that is, accommodated for distant vision. **B.** Eye accommodated for a nearby object. 1 = image of candle reflected from the air–cornea interface. 2 = image reflected from the aqueous humor–lens interface. 3 = image reflected from the lens–vitreous body interface. See text for additional explanation. (After Ruch and Patton [eds]. *Physiology and Biophysics*, 19th ed, 1966, Saunders.)

The movement of the Purkinje images caused by accommodation of the lens for near vision should not be confused with the Purkinje shift, which is discussed on p. 489.

THE DIRECT AND INDIRECT LIGHT REFLEXES. When light is flashed into only one eye, the pupil of that eye constricts; this response is called the direct light reflex. Simultaneously, the pupil of the other eye also constricts, but to a lesser extent, this latter effect being called the indirect (or consensual) light reflex.

The sensory receptors for these reflexes are the retinal rods and cones. The afferent pathway is the same as that of the visual pathway, described previously, as far as the level of the thalamus. Instead of entering the geniculate bodies, however, fibers pass to the pretectal area situated immediately rostral to the superior colliculus. Thence, internuncial neurons arise, and these neurons send axons to the accessory nucleus, a rostral subdivision of the oculomotor nucleus (Fig. 7.30). Cells in this accessory oculomotor nucleus (Figs. 7.28 and 7.32; Table 7.3) in turn give rise to efferent fibers whose axons exit from the midbrain in the oculomotor nerve (III) and terminate in the ciliary ganglion. From there, parasympathetic postganglionic visceral efferent fibers enter the eyeball to innervate the circular smooth muscles that form the sphincter of the iris (Fig. 7.26, p. 274). The passive dilatation of the pupil that takes place in dim light is caused by a reduction in the tonic flow of parasympathetic impulses to these circular smooth muscles. Conversely, in stronger light, the signal flow to the circular smooth muscles of the iris via the parasympathetic nerves increases markedly; pupillary constriction takes place. The light reflex thus can be construed as a protective mechanism that prevents excessively high levels of illumination from damaging the retina, as the degree of pupillary constriction is, within limits, directly proportional to the light impinging upon the eye.

Although, strictly speaking, the sympathetic innervation of the iris is not part of the neural pathway underlying the light reflex, this topic will be treated here as part of the overall control mechanism for regulating pupillary aperture. The radial smooth muscle fibers of the iris are innervated by sympathetic nerve fibers originating in the intermediolateral horn cells of the first thoracic segment of the spinal cord which pass to the superior cervical ganglion, and then to the eyeball along progressively smaller branches of several arteries (see Fig. 7.26, p. 274). Contraction of these radial fibers dilates the pupil; thus, the sympathetic nerves act in an antagonistic fashion to the circular muscle fibers and parasympathetic stimuli.

Active dilatation of the pupils occurs involuntarily when an individual is suddenly startled or when a "pleasurable" object is viewed. For an example of this effect, see footnote, page 466.

THE VISUAL FIXATION REFLEX. When the head and eyes are turned voluntarily toward an object that attracts attention, these "rough tuning" mechanisms bring the image into the same approximate position on each retina. As noted earlier (p. 468), conjugate deviations cause the eyes to move in such a manner that the visual axes (Fig. 8.1) of both eyes are maintained parallel with respect to each other or else these axes converge toward a common point during accommodation for near vision, as discussed in the following section. In the latter instance, the medial (internal) rectus of one eye acts together with the lateral (external) rectus of the contralateral eye. The fovea of each eye (the point on each retina having keenest vision under conditions of adequate illumination) is automatically focused upon the object. Under normal circumstances, one cannot voluntarily cause the visual axes to diverge beyond parallel. If this were to happen, double vision *(diplopia)* would result. The *fixation reflex* is essential to produce identical correlation or "fine tuning" of the two visual fields so that diplopia does not occur.

The afferent pathway of the fixation reflex is identical with the visual pathway from the retina to the visual cortex. The precise course of the efferent pathway is not clear; however, it is known that it commences in the occipital lobe of the cortex and passes to the pretectal region of the superior colliculus (see also p. 287). Axons, whose origins are in nerve cell bodies located within the pretectum, pass via the medial longitudinal fasciculus to the motor nuclei of the oculomotor (III), trochlear (IV), and facial (VI) nerves. When the images received by the visual cortex from the left and right retinas are not aligned correctly, afferent impulses are initiated and transmitted through the occipitotectal tracts and then by way of the appropriate cranial nerves to the ocular muscles in order to bring the eyes into register for correct fixation.

For example, this *optokinetic reflex* is demonstrated by reading a line of print. The eyes move rapidly in a series of unconscious, jerky *(saccadic)* movements with periods of fixation in between to allow for visualization and perception of the individual words. Reading speed can thus be accelerated by special training which has been designed to increase the number of letters visualized per "glance" and/or to decrease the time spent on each fixation or pause.*

NEAR-POINT REFLEXES. When viewing distant objects (eg, beyond about 7 meters), the eyes and their visual axes are parallel with respect to each other. When viewing an object at progressively shorter distances, however, the medial rectus muscles contract more strongly so that the images in each eye focus on the identical region of each retina. This *convergence reflex* is necessary in order that the images fuse in the cortex; otherwise, diplopia would result.

A second reflex involved in near vision is accommodation of the lens to increase its refractive power by changes in tension of the ciliary muscles,

---

These facts underlie the success of so-called "speed-reading" programs.

as discussed earlier in this chapter (p. 472). Since the process of accommodation necessitates muscular effort, fatigue of the ciliary muscle can take place; in fact, the ciliary muscles of the eyes are among the most active in the body. The capacity of the lens to accommodate and thus focus on nearby objects is quite limited, however. The light rays reflected from an object located very close to the eye cannot be focused on the retina at all. The minimum distance at which an object can be seen clearly is called the *near point* of vision, and this distance increases gradually and progressively throughout the lifetime of the individual. For example, the average near point of a child 8 years old is about 8.5 cm, whereas at 20 years of age the near point has increased to around 10.5 cm. By the age of 60 years, the near point has receded to around 83 cm from the eye.

The increase in the near point is caused chiefly by a physiologic decrease in the elasticity in the lens with advancing age. Consequently, the curvature of this structure can change to a much more limited extent in older persons; ie, accommodation decreases markedly. Generally, the near point in normal individuals has increased to such an extent between the ages of 40 and 45 that glasses with convex lenses are necessary to read or perform close work. This gradual hardening of the crystalline lens is called *presbyopia*.

The pupillary constriction reflex (p. 475) also plays a role in adaptation of the eyes and close vision. Optically, a smaller pupillary diameter results in an increased depth of field, in a fashion similar to the action of the diaphragm in a photographic camera. Thus, constriction of the pupils takes place when the eyes are focused on a nearby object, and this response takes place independently of the light intensity.

The tripartite response of accommodation, convergence, and pupillary constriction is known as the *near response*. This effect can be evoked voluntarily by deliberately focusing the eyes upon a nearby object.

The center for the convergence reflex within the brain is considered to be in cells of *Perlia's nucleus* in the oculomotor region, whence motor fibers pass to the medial rectus muscles. Additional fibers from this nucleus probably also pass to the accessory oculomotor nucleus, and induce the associated pupillary constriction and accommodation reflexes.

PROTECTIVE REFLEXES OF THE EYE. The blink (wink) reflex is elicited involuntarily when an object suddenly appears in front of the eyes. Afferent retinal impulses pass directly to the tectum of the midbrain, and then are transmitted along the tectobulbar tracts to the facial nerve nuclei. The facial nerves then activate the orbicularis oculi muscles to close the eyelids.

If the visual stimulus is extremely strong, more extensive reflex activity is involved. In this instance, the tectal region sends impulses over the

tectospinal tract as well as to the tectobulbar tract, and in addition to winking, there is a generalized "startle" or "jump" response of the entire body, the arms being thrown over the face.

The pupillary reflex protects the eyes from excessively high levels of illumination, in addition to regulating the diameter of the iris during accommodation for near vision.

In addition to the protective reflexes noted above, the eyes are protected physically from injury by their location within the bony walls of the orbit. Furthermore, dessication of the cornea is prevented by tears that are secreted from the lacrimal glands. Each lacrimal gland is situated within the upper part of the orbit, and the tears flow across the surface of each eye to empty into the nasal cavity via the lacrimal duct. The blink reflex which closes the eyelids serves to distribute the tears over the corneal surface as well as to protect the eyes from physical injury.

General Physiologic Mechanisms Related to Vision

The formation of a reasonably faithful facsimile, or physical image, of an external object upon the retina of the eye by the crystalline lens, as discussed in previous sections of this chapter, is merely the first step in the visual process, albeit a step that is essential to normal vision. The retinal image must next be encoded as a pattern of nerve impulses which the cerebral cortex can utilize effectively in order to synthesize an approximately accurate facsimile of objects and events in the external environment, a perception that includes detail, sharpness of contour, and color.

There are two kinds of photoreceptors in the retina of the eye that underlie the photochemical aspects of vision, and each type of light-sensitive cell is specialized to perform quite different functions, although these two cellular elements are associated intimately within the retina. The light-sensitive cells involved in vision are the elongated rods and cones whose nuclei are situated in the outer nuclear layer of the retina and whose photosensitive cellular processes pass through the outer limiting membrane to abut upon the pigmented epithelial layer (Fig. 8.5).

The rods are specialized for twilight and night vision (also called *scotopic vision*); hence, the threshold to excitation of these elements is quite low (Table 8.1). When the eye is exposed to dim light (or total darkness), a chemical process takes place within the rods which renders them far more sensitive to light. This chemical process, called *dark adaptation,* is accompanied by pupillary dilatation so that more light is admitted to the eye (see also p. 487). As a result, the human retina is able to utilize exceedingly low levels of illumination. The dark-adapted eye is unable to record sharp boundaries, small details, or colors, however. In this regard, the rods are able to respond to light of all colors (ie, all wavelengths of the visible spec-

Table 8.1 LIGHT INTENSITY AND VISION[a]

BRIGHTNESS[b] (LUMINANCE, MILLILAMBERTS)	PERCEPTION OF INTENSITY (APPARENT BRIGHTNESS[c])	VISUAL RESPONSE
10^{-7}	Visual threshold following dark adaptation	
10^{-6}		
10^{-5}	White surface illuminated by moonless sky at night	Scotopic (twilight and dark) vision
10^{-4}		
10^{-3}	White surface illuminated by moonlit sky at night	Transition between scotopic and photopic vision
10^{-2}		
10^{-1}	Newsprint read with difficulty	
1		
10^1	Comfortable reading	
10^2	Perform most delicate visual tasks	
10^3	Luminance of white paper in direct sunlight	Photopic (daylight, color) vision
10^4		
10^5		
10^6	Filament of incandescent lamp	
10^7		
10^8	Carbon arc	
10^9	Sun	Irreversible retinal damage
10^{10}	Atom bomb, to 3 msec after detonation	

[a]*Data from Best and Taylor (eds).* The Physiological Basis of Medical Practice, *1966, Williams & Wilkins.*
[b]*The term* brightness *(or luminance) is restricted to the intensity of the physical stimulus.*
[c]*The term* apparent brightness *(also called luminosity or brilliance) refers to the subjective visual response to a stimulus having a given brightness.*

trum) except for deep red. But no colors are perceived following excitation of the rods, regardless of the wavelength of the light that excites them. Despite this fact, the activity of the rods in dim light makes the difference between total blindness and the ability to achieve some visual orientation, even though objects at night are perceived as irregular crude shapes and dark masses.

In contrast to the rods, the cones of the retina form a system of receptors specialized to function in daylight when a high level of illumination is present (Table 8.1). Under such circumstances, objects are seen clearly, considerable detail is evident, and many shades of color are present in the visual image. In daylight, the pupil constricts so that visual images are sharpened; therefore visual acuity is at a maximum in daylight.

The human eye is sensitive to an extremely narrow band of wavelengths on the electromagnetic spectrum. These wavelengths lie between 723 and 397 mμ, and thus range between long infrared waves and short ultraviolet rays (Figs. 1.3, p. 5; 8.10, and 8.11; Table 8.2).

As might be anticipated from the unequal morphologic distribution of rods and cones within the retina as discussed below, the response of different areas of this structure to various wavelengths of light in the visible spectrum also differs. Stated another way, the eye is quite selective in its response under different conditions.

As shown in Figure 8.12, the fovea in cross section appears as a shallow bowl or depression in the retina. The total width of the fovea (or ''macula lutea'') is about 1.5 mm, and this region contains only cones. In the center of the fovea, the layers of the retina to zone 5 (Figs. 8.5 and 8.12), as well as the retinal blood vessels, are pushed aside so that a very shallow depression is present in the exact center of the fovea (the fovea centralis). In the floor at the center of the fovea are found the longest, smallest, and most densely packed cones; hence, visual acuity in daylight (p. 484) is greatest in this region of the retina. Furthermore, the lack of neural elements and blood vessels over the photosensitive elements in this region of the retina means that light waves do not

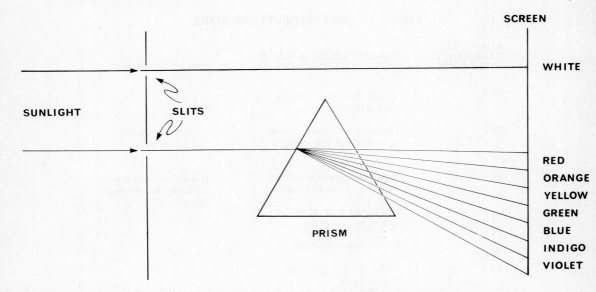

FIG. 8.10.   The dispersion of white light (sunlight) by a prism to produce the visible spectrum. If all of the colors produced by dispersion through the prism are added together once again by passing them through a second prism, a beam of white light is produced. The specific wavelengths of the individual colors (hues) of the visible spectrum are listed in Table 8.2 and depicted in Figure 8.11.

have to travel through these more superficial layers in order to excite the foveal cones, another mechanism that contributes materially to sharpening the visual acuity in this region.

The area of the fovea that is completely free of rods is approximately 0.5 mm in diameter and may contain as many as 30,000 cones having an average diameter of 1.5 μ; hence, only photopic vision is possible following excitation of this retinal area by bright light. This foveal area is also exceedingly sensitive to the perception of specific colors (p. 485). In the peripheral regions of the retina, the cones are much larger, ranging between 5 and 8 μ in diameter. In contrast to the cones, the rods are functional only in dim light or darkness, and these elements range in size from between 2 and 4 μ near the center of the retina to 4 or 5 μ near the periphery.

When the rods and cones are compared experimentally, insofar as the brightness or luminosity of light necessary to produce maximum excitation of these elements is concerned, it can be shown that the wavelength necessary for maximum photopic (cone) vision is about 510 mμ (5,100 Å) and for scotopic (rod) vision is around 550 mμ (or 5,100 Å), as shown by the curves in Figure 8.13. Thus, the eye exhibits two distinct patterns of response when excited, with different wavelengths of light each of which has, however, an identical energy level.

It has been calculated that the absolute threshold for visual excitation (ie, the minimum quantity of light that can be seen) under favorable conditions appears to lie close to the theoretical minimum. That is, the photoreceptors of the eye are able to detect one quantum of light, and only 54 quanta of light impinging upon the cornea can be perceived. Reflection and absorption by the ocular media within the eyeball have been allowed for, as well as the spatial distribution of the rods within the retina, and thus it has been determined that one quantum of energy is sufficient to excite a rod. Since one quantum of energy will degrade one molecule of visual purple in accordance with the photochemical equivalence law (of Einstein), it appears that during the course of evolution the vertebrate eye has achieved the theoretical maximum limit of sensitivity. The threshold for excitation of a cone in the fovea, on the other hand, appears to lie between five and seven quanta.

It is important to realize that vision ultimately is synthesized from a pattern, or mosaic, of punctate retinal areas that have been excited by the visual stimulus. The nerve impulses arising in these areas are encoded in some way and sent to various cortical areas via specific pathways, but the ultimate destinations of these signals in the brain do not comprise a simple reflection of the same pattern that arose peripherally in the retinas. For example, in Figure 7.44 (p. 298), note the pathways of the fibers of the two optic tracts following the partial decussation of the optic nerve fibers (see also Fig. 8.37). The cerebral mechanisms whereby a visual image is synthesized within the cortex and then appropriately projected to the external environment are at present unknown.

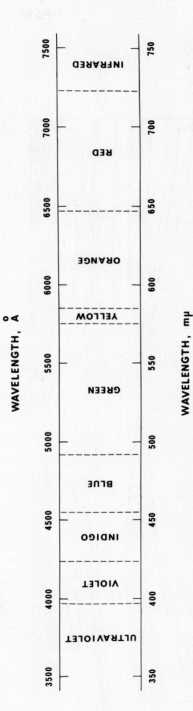

FIG. 8.11. The visible spectrum. The energy of light waves is directly proportional to their wavelength; consequently waves at the red end of the spectrum have a higher energy than those toward the blue end. For this reason, red waves are refracted least following passage through a prism and violet waves are refracted most (Fig. 8.10). See also Figure 8.32.

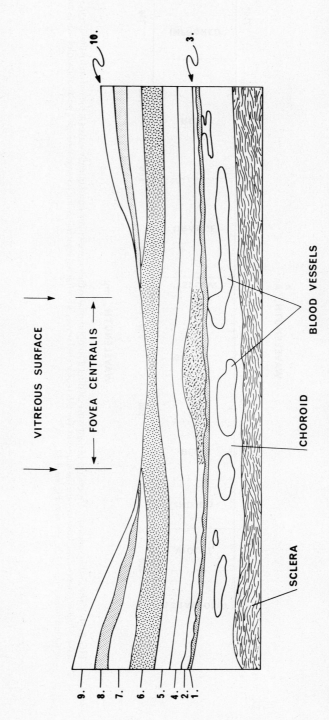

FIG. 8.12. The fovea of the retina in cross-section. Note the sharp decrease in thickness in the foveal region, at which point maximal visual acuity is present. The vertical downward arrows at the top of the figure indicate the direction of the light path and the numbering of the specific retinal layers corresponds to those given in Figure 8.1. See also Figure 8.5. (Data from Bloom and Fawcett. *A Textbook of Histology*, 9th ed, 1968, Saunders.)

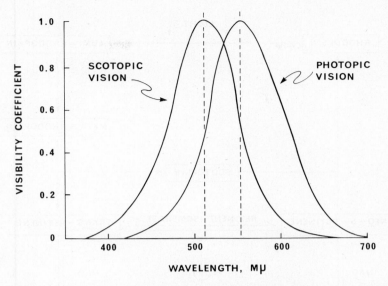

FIG. 8.13. Curves illustrating the *relative* visibility (luminosity) perceived by the rods (scotopic vision) when stimulated by dim light compared to that of the cones (photopic vision) when stimulated by bright light. The visibility coefficient is an indication of the relative sensitivities of the rods and cones, and each curve was placed upon a similar scale by arbitrarily adjusting the maximum visibility coefficient in each instance to equal one. The peak of the visibility curve for scotopic vision thus falls at about 510 mμ, whereas that for photopic vision at 550 mμ. On an *absolute* intensity scale, however, the scotopic visibility curve would lie much below that for photopic vision as the cones are far less light sensitive than the rods. (Data from Ruch and Patton [eds]. *Physiology and Biophysics,* 19th ed, 1966, Saunders.)

Photochemical and Bioelectric Processes in Scotopic Vision

The ultimate mechanism for the detection of light energy, whether by the rods or the cones, is through the degradation of photosensitive chemical compounds, a process which in turn excites the nerve endings within the retina. Thus, the rods and cones of the eye are photoelectric transducers.

The photochemical mechanisms whereby the rods respond to light energy are well understood and will be discussed in this section. The principles that underlie the photochemistry of the cones are quite similar and will be discussed subsequently.

THE RHODOPSIN–RETINENE CYCLE. Around the end of the last century, it was observed that frog retinas were pinkish in color when dissected out in dim light, but these structures rapidly bleached to a pale yellow color when exposed to daylight (Kühne). This simple experiment proved fundamental to understanding the photochemical basis of vision. The photosensitive pink form of the retinal pigment was named *rhodopsin* (or *visual purple*) and subsequent observations led to the conclusion that this material was concentrated in the outermost segments of the rods (Fig. 8.5).

As rhodopsin is bleached by light, this photochemical alteration must underlie scotopic vision by the rods, provided that light waves having different wavelengths, but of the same intensity (ie, having the same energy), would bleach rhodopsin

at the same rate. This effect has, in fact, been demonstrated conclusively (Fig. 8.13); therefore rhodopsin is the pigment that underlies scotopic vision by the rods.

The rhodopsin found within the rods of the human eye has now been isolated and characterized. Present in the rods at a concentration of around 40 percent, rhodopsin in turn is composed of a carotenoid pigment called *retinene* in combination with a specific protein called *scotopsin*. (The general term *opsin* signifies the specific proteins associated with the rods or the cones; ie, scotopsin and photopsin, or with specific pigment moieties, eg, rhodopsin.) The retinene in rhodopsin is a form of 11 *cis*-retinene known as *neo-b-retinene,* an important fact because only in this form can the retinene combine with scotopsin to form the photosensitive pigment rhodopsin. *Cis–trans* isomerism was discussed on p. 30.

As depicted in Figure 8.14, the absorption of light energy by rhodopsin causes this compound to decompose as the *cis*-form of retinene is converted into an all-*trans*-form. The *trans*-form of retinene has a chemical composition (ie, constitution) identical with the *cis*-form; the shape of the molecule (ie, the conformation), however, is now such that it no longer fits the reactive loci on the scotopsin molecule. The *trans*-retinene molecule tends to move away from the scotopsin, forming, as the immediate product, a loose scotopsin : *trans*-retinene complex known as *lumirhodopsin.*

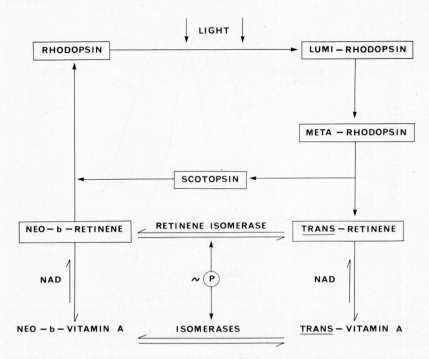

FIG. 8.14. The chemical transformations of the rhodopsin cycle as found in the rods. The symbol ~ Ⓟ indicates that metabolic energy is required for the reaction to proceed and the double arrows signify reversible reactions. (After Guyton. *Textbook of Medical Physiology,* 4th ed, 1971, Saunders.)

Lumirhodopsin, however, is exceedingly unstable; consequently, it splits practically instantaneously into meta-rhodopsin. The latter compound is also unstable, so that it too is degraded into scotopsin plus all-*trans*-retinene, a process that requires several seconds. Excitation of the rods takes place during the breakdown of rhodopsin to scotopsin plus all-*trans*-retinene; hence, the signals are generated in afferent neural pathways which ultimately are interpreted as vision.

The restoration of rhodopsin takes place as indicated in Figure 8.14 so that the rhodopsin–retinene visual cycle is completed. The all-*trans*-retinene first is converted into neo-b-retinene, the *cis*-conformation of the pigment, a reaction that is catalyzed by the enzyme *retinene isomerase*. The requirement for metabolic energy to accomplish this process is signified in Figure 8.14 by the symbol ~ Ⓟ. Following the generation of *neo-b*-retinene, this compound recombines spontaneously with scotopsin to form rhodopsin, as this reaction is exergonic. The rhodopsin thus formed is stable until light once again causes the photolysis of rhodopsin.

Insofar as the mechanism of rod excitation following the degradation of rhodopsin is concerned, it is clear that some ionized free radicals are formed momentarily when the all-*trans*-retinene and scotopsin molecules separate. In turn, these free radicals produce rod excitation. Presumably, this effect is caused through the interaction of the free radicals with the resting bioelectric potential of the rod membrane per se (see also discussion on p. 182).

RECEPTOR POTENTIALS DURING ROD EXCITATION. Four types of bioelectric potential can be obtained from the retina, depending upon the size and location of the electrode used to record such activity: (1) A steady potential can be recorded between the front and back of the eyeball (the *corneoretinal potential*). The amplitude of this potential ranges between 30 and 60 mv, the choroidal side being negative. When the steady corneoretinal potential is recorded from the tissues in the immediate vicinity of the eyes, it can be used to detect eye movements (as during REM sleep, p. 432). (2) Electric responses can be obtained from single bipolar retinal cells. (3) Action potentials can be obtained from single ganglion cells and optic nerve fibers using suitably placed microelectrodes. (4) The sequence of compound and rapid (phasic) potential changes that can be recorded extracellularly following excitation of the retina by light is called the *electroretinogram (ERG,* Fig. 8.15).

Meticulous analysis of the wave components of the ERG has been accomplished by the use of drugs as well as microelectrodes inserted into the retina at different specific levels. The negative *rod receptor potential* is abrupt in onset, and this potential is manifest initially as the *A* wave; how-

A.

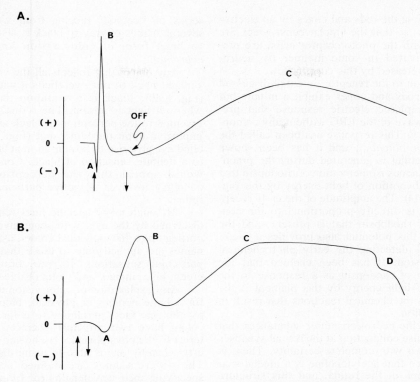

B.

FIG. 8.15. Electroretinograms illustrating the wave forms obtained by artificial excitation of the retina by light. In each figure the stimulus was applied at the upward-directed arrow and terminated at the downward-directed arrow. **A.** Electroretinogram recorded from the vitreous body illustrating the three typical deflections (A, B, C) as well as a positive off-response. **B.** In this instance a negative off-response is indicated at D. See text for additional discussion. (**A.** After Ganong. *Review of Medical Physiology,* 6th ed, 1973, Lange Medical Publications. **B.** After Ruch and Patton [eds]. *Physiology and Biophysics,* 19th ed, 1966, Saunders.)

ever, the slow decline of this receptor potential causes the remnant negativity, which is recorded from the ERG of a retina that contains principally rods. The electropositive *B* wave is produced in the inner nuclear layer (layer 6, Fig. 8.5). The decay of a direct current component that is also produced in the inner nuclear layer gives rise to the off response that is a characteristic finding in rod ERGs. The *C* wave is generated in the pigmented epithelium (layer 1, Fig. 8.5).

If the ERG is recorded from an area of the eye containing only cones (ie, the fovea), the compound pattern of bioelectric potential changes is similar to that seen in Fig. 8.15A, save that the receptor potential of the cones decays more rapidly than that of the rods. Since this decline in potential takes place prior to the direct current component, which has an opposite sign, a small positive deflection is seen following termination of the stimulus that is called the *D* wave, as the currents sum algebraically (Fig. 8.15B).

The receptor potentials of both rods and cones are unique in comparison with all other types of receptor potential (p. 184); that is, following generation of receptor potentials in these retinal elements, the outer as well as the inner por-

tions of the rods and cones are positive with respect to the terminals of the axons. Current flows in an extracellular pathway from the outer and inner segments of the rods and cones toward the axon terminals, a route that is diametrically opposite to that followed by the current flow in other receptors, and this current flow serves as a generator potential to produce nerve impulses (cf. p. 187).

In addition, it has been demonstrated by use of microelectrodes inserted into single rods and cones that the membranes of both the outer (layer 2, Fig. 8.5) and inner (layer 4) segments of these photosensitive elements undergo hyperpolarization while the generator potential is present. In darkness, small photoreceptor currents flow continuously, and these currents have been found to maintain the entire rod or cone cell in a partially depolarized condition. The presence of light, however, selectively decreases the sodium permeability in the outer segments of both the rods and cones, and this response effectively reduces the magnitude of the depolarizing current that flows in the dark. Therefore, both outer and inner segments of the rods and cones become hyperpolarized. This hyperpolarization apparently is conducted along the rods and cones by an electrotonic process, so that the bipolar cells, which are contiguous with the photoreceptor cells, are presumably activated in some manner by neurotransmitters released by this current flow.

In addition to the receptor potential discussed above, both rods and cones exhibit a minute but measurable biphasic electric response that precedes the A wave of the ERG without any demonstrable latency. This response has been called the *early receptor potential,* and it has been shown that this potential is generated during the practically instantaneous isomerization of rhodopsin that follows the absorption of light energy by this pigment (Fig. 8.14). The amplitude of the early receptor potential is directly proportional to the concentration of rhodopsin that is present, and the recording of this potential also provides a useful technique for detecting the early photochemical events in vision. It has been claimed that the isomerization of rhodopsin as a response to the absorption of light energy by this pigment is the initial step in the chemical reactions that result in visual excitation.

The specific neurotransmitter substances that mediate impulse conduction at the retinal synapses are not known with complete certainty. There is some evidence that acetylcholine is a mediator at some junctions in the retina, and this structure also contains fairly high concentrations of dopamine, 5-hydroxytryptamine, gamma-aminobutyric acid (GABA), and substance P (a polypeptide).

Photopic Vision

SOME PROPERTIES OF COLOR. The sensations of specific colors (or *hues*) that one perceives following excitation of the retina with particular wavelengths of the visible spectrum (Fig. 8.11, Table 8.2) make up the *chromatic series.* The extraspectral color purple is also a member of the chromatic series. Incongruous as it may seem, the

Table 8.2 THE VISIBLE SPECTRUM

COLOR	RANGE OF WAVELENGTH	
	mμ	Å
Red[a]	723–647	7,230–6,470
Orange	647–585	6,470–5,850
Yellow	585–575	5,850–5,750
Green[a]	575–492	5,750–4,920
Blue[a]	492–455	4,920–4,550
Indigo	459–424	4,550–4,240
Violet	424–397	4,240–3,970

Note: The energy of any electromagnetic wave is directly proportional to its wavelength and frequency, or v = nλ (p. 6). Consequently, when white light is dispersed by passing it through a prism, red light is refracted the least and violet light is refracted the most, the other colors lying in between these two extremes (Fig. 8.10).

[a]*Red, green, and blue-violet are the primary additive colors.*

series of "colors" ranging from white through several shades of gray to black is also considered to be a form of color vision known as the *achromatic series.*

Any object that reflects all the visible rays of sunlight back to our eye elicits a white sensation (Fig. 8.10). Black is a sensation that is evoked when all light is absent. The retina, however, is essential to the perception of black as light waves impinging upon the blind spot (Fig. 8.7) or on a blind eye give rise to no sensation at all rather than to a definite sensation of black. Consequently, it would appear that the perception of black definitely involves an active participation by the retina.

Although many specific hues (colors) can be discerned by the eye, with some investigators reporting up to 160 individual colors, usually specific names are applied only to those that give rise to particular sensations, ie, red, orange, yellow, green, blue, indigo, and violet (Table 8.2).

Any explanation of color vision must account for a large number of physical, biophysical, and psychologic facts pertaining to color. In general, colors have two principal characteristics of interest to the physiologist: (1) The *hue* is that property whereby one is able to name distinct colors. The different hues are defined accurately by specifying their wavelengths (or frequencies). No light waves are colored inherently; rather, the retina and brain give rise to the perception of different colors (or hues). (2) The degree of *saturation* of a color is that property which defines its chromatic purity, ie, the freedom from dilution by white. A light (or pigment) which contains only a small white component is considered to be highly saturated, whereas if much white is present, the diluted color is said to be unsaturated (ie, a pale or pastel shade).

The *additive primary colors* are red, green, and blue, because mixing (or fusing) these colors in appropriate proportions will give rise to the sensation of any color in the visible spectrum, as well as white and the extraspectral color purple (Fig. 8.16). Every primary color has a *complementary color;* that is, when the primary color is mixed with its complement, white results. For example, as indicated in Figure 8.16, green + magenta yields white. Colors that lie close together on the visible spectrum on fusion give rise to an intermediate color. For example, red and yellow give orange when mixed together. On the other hand, colors that lie farther apart on the spectrum than the distance between the complementaries give rise to various shades of purple.

The colors of objects in the external environment are determined largely by the subtractive rather than the additive process (refer to Fig. 8.16). If one holds a red and green filter together in front of an arc light, the eye perceives black, as the red filter absorbs all the spectral rays save for those lying in the red, and the green filter absorbs all the spectral rays save for those in the green

A.

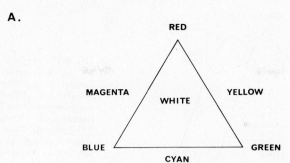

B.

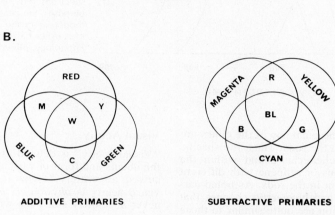

ADDITIVE PRIMARIES   SUBTRACTIVE PRIMARIES

FIG. 8.16. A. Color triangle. The additive primary colors are located at the corners and the subtractive primaries are located on the sides of the triangle. The arrangement is such that the sum of any two additive primaries (corners) gives the subtractive primary lying between them whereas the sum of the three additive primaries yields white (center of triangle). Colors opposite each other on the triangle are complementary, ie, when added to each other they yield white. For example, magenta + green = white. **B.** Additive and subtractive color mixing are discussed in the text. The abbreviations are: M = magenta; Y = yellow; C = cyan; W = white; G = green; B = blue; R = red; Bl = black. (Data from White. *Modern College Physics,* 3rd ed, 1956, Van Nostrand.)

region. No light waves are transmitted through the pair of filters and the eye perceives a black sensation. Colors resulting from the absorption of light waves are known as *subtractive colors,* and the colors of objects in the external environment are principally subtractive colors. A red bowl appears red because it absorbs all other colors of light and reflects only red. The bowl per se has no color; light generates the color that is interpreted as such by the retina and the brain.

COLOR VISION. A number of theories have been advanced to explain color vision, none of which is entirely satisfactory. Certain aspects of the classic *Young–Helmholtz trichromat theory* of photopic vision have the merit of experimental confirmation, and will be discussed here. Basically, the trichromat theory assumes that the retina contains three types of cone cell each of which gives rise to a particular color sensation, ie, red, green, and violet. This theory also maintains that each of these types of cone contains a different photochemical substance that undergoes photolysis when acted upon by light having suitable wavelengths; hence, specific nerve fibers and cortical terminals are involved in the perception of diffe-

rent hues. Furthermore, when all three cone types are stimulated equally, a sensation of white results, because, after all, each type of cone responds to one of the three primary colors. Last, the trichromat theory postulates that the perception of any color is determined by the relative frequency of impulses (or patterns of signals) that are transmitted to the visual cortex by way of each of the three nerve fiber systems that are connected to each of the three cone systems.

Much experimental evidence has been gathered which supports the conclusion that the human retina contains three distinct types of cone and that the functional interaction among these elements determines the colors that are perceived. As shown in Figure 8.17, three different photosensitive pigments have been found in different cones, each of which exhibits a peak light absorption at a particular wavelength. Furthermore, the photochemicals in the cones are almost identical chemically with those involved in scotoptic vision by the rods, the major difference between them being in the opsins, or protein moieties, of the visual pigments. In the cones, these *photopsins,* as the proteins are called, have been shown to differ chemically from scotopsin. In other respects, the

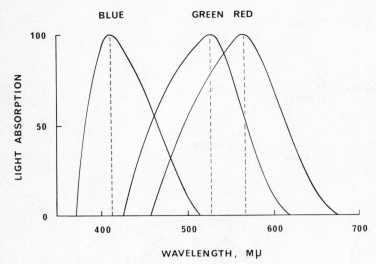

FIG. 8.17. Light absorption by the blue-, green-, and red-sensitive cones of the human retina plotted as the percent of the maximum absorption against the wavelengths of the light. The peak absorbances lie at 430 (blue), 535 (green), and 575 (red) mμ. Note that the maximal absorption of light by the "red" cone actually lies in the orange color band of the spectrum (Fig. 8.11). For comparative purposes, the peak of light absorption for rhodopsin in the rods lies at about 505 mμ. (After Guyton. *Textbook of Medical Physiology*, 4th ed, 1971, Saunders.)

retinene moieties of the pigments in the cones are identical with those in the rods. Thus, the specific color-sensitive photochemicals found within the cones are combinations of retinene with different opsins than are present in the rods. As noted earlier, the biochemical and bioelectric changes that occur in photopic vision are quite similar to those discussed in relation to scotopic vision.

The trichromat theory of color vision is in good agreement with the law of specific nerve energies (p. 120), since each photosensitive pigment merely serves to stimulate a nerve fiber, and the color that is perceived ultimately is dependent upon the termination of the fiber in the brain.

Experimental evidence does not support the view that there are three separate and independent neural pathways from the retina to the brain, each of which subserves impulse transmission from a particular type of cone. In this respect, the trichromat theory appears to be in serious error. Actually, it appears that the retina per se contains functional mechanisms that are able to encode the information pertaining to colors into specific on and off responses in single fibers of the optic nerve. Since color vision is not completely developed in the fovea, and as convergence is present to a high degree, it is evident that the retina functions to refine and process the raw data provided by excitation of the photosensitive retinal elements before it is transmitted to the brain. Thus, the visual receptors outside of the fovea converge upon the ganglion cells (Fig. 7.89, p. 345), and some of the synaptic connections appear to be excitatory, whereas others are inhibitory (Fig. 7.90, p. 346; see also divergence, Fig. 7.91, p. 346). Of particular significance to photopic (color) vision is the demonstrated fact that the outputs of cones, having different spectral sensitivities, converge upon the same cell in the lateral geniculate body, such inputs being mutually inhibitory.

Visual Acuity

"Visual acuity" is defined as the extent to which the details and form of various objects can be perceived accurately.* One factor that underlies visual acuity is *physiologic nystagmus*. Eyes are never absolutely still even when gazing fixedly at one point or object. Rather, the eyes continuously move in a jerky fashion that may include horizontal, vertical, and/or rotatory movements. These small ocular excursions are functionally important, despite the fact that the individual retinal photoreceptors do not adapt rapidly to constant illumination because of the extremely high rate at which the visual pigments are degraded and resynthesized. Physiologic nystagmus is important because the neurons connected to the photoreceptors *do* adapt rapidly, hence a minute, but continual, shift in the position of the retinal image is essential in order to prevent the object being viewed from disappearing rapidly. This fact can be demonstrated by use of an optical system so arranged to track the eye movements that the retinal image always falls upon the same point on the retina itself.

Visual acuity is also affected by such factors as the optical mechanisms that form images within the eye; retinal parameters, including the level of photochemicals in the rods and cones (p. 481); the wavelength and intensity of the visual stimulus; the simultaneous (or successive) contrast between the visual stimulus and the background; the temporal exposure to the visual stimulus; and the state of adaptation of the eye.

From a clinical standpoint, visual acuity generally is defined in terms of the minimum distance that two lines can be separated and still be perceived individually. This *minimum separable* dis-

*Visual threshold, in contrast to visual acuity, is the minimum quantity of light that can be perceived (p. 478).

tance is frequently measured in practice by use of Snellen letter charts, which are read from a distance of 6 meters (20 feet). The subject recites aloud the smallest line of type that he can perceive, and the results are stated in fractional terms. That is, the numerator of the fraction is 20, this number representing the distance (in feet) that the chart is placed from the subject when it is read. The denominator of the fraction is the maximum distance from the chart that can be read; hence, normal vision is indicated as 20/20.

The size of the letters in the Snellen charts is so designed that the vertical type size in the smallest line that a normal person can read at a distance of 6 meters subtends a visual angle of 5 minutes (Fig. 8.18). Each line in these letters subtends 1 minute of arc; consequently, the minimum separable in a normal person equals a visual angle of 1 minute, and this corresponds to an excitation of two points in the retina (fovea) that are about 2 μ apart, the average diameter of the cones in this region being 1.5 μ.

Retinal Sensitivity

The retina has an amazingly great capacity to regulate its sensitivity automatically to the intensity as well as the wavelength of light that enters the eye. For example, in contrast to peripheral cutaneous receptors that are sensitive to a range in stimulus intensity between roughly a few hundredfold to a few thousandfold from threshold to maximal excitation, the eye can exhibit a similar range of sensitivity up to one millionfold between the limits of maximal dark and maximal light adaptation (Table 8.1).

Several adaptive mechanisms and properties of the eye under different conditions will be discussed in this section, including information coding by the retina, dark and light adaptation, critical fusion frequency, afterimages, and the Purkinje shift. All these topics share the common feature of involving various aspects of retinal sensitivity under different conditions.

INFORMATION CODING. As in other receptors, the afferent nerve fibers of the eye display a graded increase in discharge frequency as the intensity of the stimulus is increased. Again, as in most other receptors, the retina has been demonstrated to exhibit a logarithmic relationship between the stimulus intensity and the discharge frequency in the steady-state adapted afferent nerve fibers; that is, the afferent discharge frequency is directly proportional to the log of the stimulus intensity within certain limits of the latter. It has been demonstrated in the eye of the horseshoe crab (*Limulus* sp.) that the amplitude of the static generator potential is directly proportional to the log of the stimulus intensity, whereas the discharge frequency in the optic nerve is related in a linear fashion to the amplitude of the generator potential. The logarithmic function enters the visual process at the photoelectric transduction stage of the visual process. This relationship between the discharge frequency of the afferent fibers and the intensity of the light that excites the photoreceptors of the eye enables these cells to provide continuously graded as well as frequency-coded information to the brain. This information in turn signals the type of stimulus (ie, the color) as well as its intensity between threshold and the intensity at which a maximal generator potential is developed by the photoreceptors.

The general principles summarized above doubtless are too simplistic insofar as information coding in the vertebrate retina is concerned, because of the exceedingly complex multisynaptic connections that are present within this structure. Nevertheless, such facts provide a clue as to the mechanisms whereby the specific retinal elements may encode particular information for transmission to the visual cortex and other cerebral areas, such information ultimately being perceived as visual experience.

LIGHT AND DARK ADAPTATION. The sensitivity of the rods is roughly proportional to the antilogarithm of the concentration of rhodopsin, and presumably a similar relationship obtains for the pigments in the cones. Therefore, an exceedingly small decrease in rhodopsin concentration causes an enormous decrease in the sensitivity of the rods. For example, when the concentration of rhodopsin decreases about 0.5 percent, the rod sensitivity concomitantly declines about 3,000 times. It is apparent that the sensitivity of the retinal photosensitive elements to light is dependent

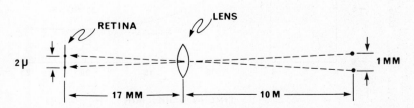

FIG. 8.18. Calculation of the approximate maximal visual acuity for two point sources of light by means of a direct proportion; ie, 1 mm/10 m = X/17 mm; X = 1.7 μ, Rounded off this value = 2.0 μ. The centers of two foveal cones are located about 2 μ apart on the retina; consequently the maximal visual acuity is slightly greater than the diameter of a single foveal cone (about 1.5 μ). (After Guyton. *Textbook of Medical Physiology,* 4th ed, 1971, Saunders.)

to a major extent upon the concentrations of the photochemicals present in the rod and cone cells. Furthermore, the sensitivity of the rods and cones to light can be altered greatly by exceedingly small changes in the concentration of the photosensitive chemicals, a mechanism that serves to adapt the sensitivity of the retina automatically to varying light intensities.

If one remains in bright light for a period of time, most of the photochemicals in the rods as well as the cones are converted into retinene and the opsins (Fig. 8.14); hence, the visual threshold rises gradually. A great deal of the retinene in the rods and cones is transformed into vitamin A. As the result of these biochemical responses to light, the sensitivity of the eye is reduced considerably, a phenomenon called *light adaptation*. Thus, when one passes from a dimly lit building into bright sunlight, the eyes feel uncomfortable for several minutes until the mechanisms described above have increased the visual threshold sufficiently, a process that is complete in about 5 minutes.

Actually, however, light adaptation is merely a loss of *dark adaptation*. When one remains in a dimly lit or dark environment for a period of time, practically all the retinene and opsins in the rods and cones are transformed into photosensitive pigments. Simultaneously, large quantities of vitamin A are changed into retinene, and this compound in turn is transformed into still more photosensitive chemicals, the final concentrations of the

latter being determined by the quantity of opsins that are available in the rods and cones. As a result of these chemical changes, the photoreceptor cells gradually become exceedingly sensitive to minute amounts of light (p. 500).

The time-course of dark adaptation is illustrated in Figure 8.19. The initial decline in the visual threshold is small in magnitude, although it takes place rapidly. Dark adaptation of the cones is responsible, because this early effect is noted only in the fovea which, of course, contains only cones (Fig. 8.19A). Dark adaptation of the rods in the extrafoveal regions of the retina gives rise to the second component of dark adaptation, an effect that develops far more slowly (Fig. 8.19B). The net result of dark adaptation increases retinal sensitivity enormously.

It was noted earlier that the rods are essentially insensitive to light at the deep red end of the spectrum. The practical importance of this fact is that radiologists, sailors, aircraft pilots, and others who must have maximal visual sensitivity in dim light can wear dark red goggles in bright light, and thus avoid the necessity of spending 20 or so minutes in darkness to achieve dark adaptation.

The key role of vitamin A in dark adaptation is discussed in relation to night blindness (p. 500).

AFTERIMAGES. When one gazes fixedly at a particular black and white object, the brightly illuminated parts of the visual scene induce light adapta-

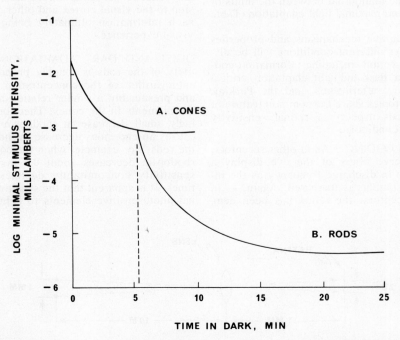

FIG. 8.19. Curves illustrating the time-course of dark adaptation obtained by plotting the visual threshold against the time spent in darkness. A. Adaptation of the cones is essentially complete in about five minutes (vertical dashed line). B. Adaptation of the rods is complete in 20 to 25 minutes. (Data from Ruch and Patton [eds]. *Physiology and Biophysics,* 19th ed, 1966, Saunders.)

tion in the retina, and the dimly lit parts of the scene similarly induce dark adaptation in the corresponding parts of the retinal image. Thus, retinal areas that have become excited by light have an increased visual threshold (ie, decreased sensitivity) whereas the retinal areas that were exposed to a lower illumination concomitantly have developed a decreased visual threshold (ie, increased sensitivity). If the person transfers his gaze from the object and looks at a white surface, the same visual image is perceived; however, the previously light areas appear dark, and vice versa. This reversal of the light and dark areas is called the *negative afterimage,* a phenomenon that follows dark and light adaptation in specific retinal areas. The negative afterimage persists for as long as any dark or light adaptation of the retinal elements is present. Similarly, if one stares fixedly at a colored object (eg, green), the corresponding visual pigment in the appropriate green-sensitive cones undergoes photolysis. Subsequently, when the same cones are subjected to white light, the red- and blue-sensitive photochemicals have been acted upon to a lesser extent. These elements now respond to the white light to a greater extent, and the afterimage takes on a red-violet (ie, purple) hue.

THE PURKINJE SHIFT. Another phenomenon related to dark adaptation, hence to retinal sensitivity, is the Purkinje shift; that is, during the process of dark adaptation when the ambient illumination is too dim (ie, the light intensity is too low) to excite the cone mechanism, the environment will be perceived only in terms of light and shade, and of course no colors will be evident. The subject will be unable to read fine print, and spectroscopic evaluation reveals that the eye has become most sensitive to light in the blue-green region of the spectrum (Fig. 8.11). Colors in this part of the spectrum appear particularly bright, whereas the eye becomes totally blind to the longer wavelengths found at the red end of the spectrum. This alteration in the relative sensitivity of the eye to different wavelengths, and which occurs at lower light intensities, is called the *Purkinje shift.* It accounts for the fact that red flowers appear black at twilight, whereas blue flowers appear gray or white. Similarly, the Purkinje shift is responsible for the particularly vivid green appearance of foliage or grass on a heavily overcast day.

CRITICAL FLICKER FUSION FREQUENCY AND THE PERCEPTION OF MOTION. The retina is able to fuse individual flashes of light so that they are perceived as coming from a continuous source, provided that the rate at which the flashes are emitted is sufficiently high. A single instantaneous flash of light will excite the retinal photoreceptors for 100 to 200 milliseconds, and because the single excitation persists, rapidly successive flashes become fused so that the individual flashes are seen as a continuous steady beam of light. The rate at which a flickering light is perceived as a steady beam of light is called the *critical flicker fusion frequency.*

As illustrated in Figure 8.20, the critical frequency at which fusion of single flickers takes place is related directly to the light intensity, hence the retinal sensitivity. At a low level of illumination, flicker fusion can occur even though the rate is as low as 5 or 6 flashes per second, whereas if the intensity is high, the critical fusion frequency rises up to 60 flashes per second. These differences result, at least partly, from the fact that the rods (which function primarily in dim light) are unable to detect as rapid shifts in light intensity as are the cones, which normally operate at higher light intensities.

From a practical standpoint, the effect of critical flicker fusion frequency is employed in both motion pictures and television. Images are projected on the screen at a rate of 24 frames per second for motion pictures, and at a rate of 60 frames per second for television. In either event, the fusion of single images in rapid succession gives rise to the illusion of continuous motion.

It is now pertinent to inquire: How does the eye perceive the motion of objects in the external environment? Here again the analogy with a camera fails completely, because the eye has no shutter, whereas either a still or motion picture camera does contain such an element, and thus is able to capture static patterns. The continuous, rapid scanning of the electron beam across the face of a cathode-ray tube in a televison set also provides a function like that of a camera shutter. The eye operates without a shutter, however, and in an incredibly precise manner can detect many details in a complex scene such as is encountered at a busy city intersection; that is, people and vehicles, all moving at different velocities and in different directions, can be perceived with amazing clarity. Similarly, the eye can determine, with considerable accuracy, the position of small, rapidly moving objects in both space and time. Examples of this effect are to be found in handball, baseball, tennis, and in many other sports and activities.

The fundamental difference between the eye and the camera is that the eye has evolved as a receptor system for detecting motion. Thus, continuous alterations in the visual pattern of the stimulus are essential to normal ocular function, and as noted earlier, when the image of an object is held at constant intensity on one point of the retina, the image rapidly fades because of adaptation of the neural elements of the visual pathway (p. 474). Consequently, the vertebrate retina is the receptor which functions to analyze continuously alterations in light flux over long periods of time, and if there is no change in the patterns of light impinging upon the photosensitive retinal elements, no neural response which ultimately culminates in visual perception can take place. The extremely rapid and continuous response of the retinal photoreceptor cells to minute fluctuations in light flux results in the clear perception of motion

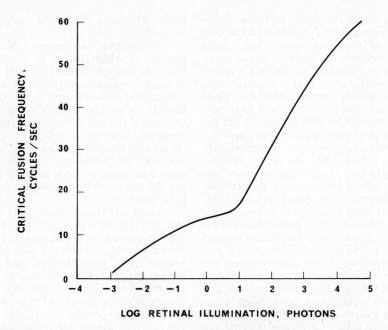

FIG. 8.20. Graph illustrating the relationship between the critical fusion frequency and the intensity of the illumination. Note that the light intensity and critical fusion frequency are related directly to each other. (Data from Ruch and Patton [eds]. *Physiology and Biophysics,* 19th ed, 1966, Saunders.)

in the external environment and sharp perceptual images are obtained without blurring. Indeed, it is difficult to conceive of an unmoving organism in a wholly static environment. Even when one sits in a quiet room the eyes are normally shifting their position constantly in either a conscious or an unconscious manner (physiologic nystagmus, p. 393), so that the flow of visual information to the retinas likewise is shifting to new optic patterns that in turn give rise to ever-changing visual impressions.

Binocular Vision

In human subjects and in the majority of primates, as well as in cats and birds of prey, the visual field of each eye covers an arc of about 170 degrees and the visual field of each eye overlaps that of the other eye to a considerable extent (Fig. 8.21). This overlap of the visual fields results in *binocular vision,* which in turn provides the animal with an important mechanism whereby objects in the external environment can be located with extreme accuracy.* This visual capability of localizing objects is called *stereopsis* (or *solid vision*). Stereopsis in turn depends upon the fusion within the brain of

two slightly different (or disparate) views of the same object.

From a neuroanatomic standpoint, the extent to which binocular vision is present in a given organism depends in part upon the number of uncrossed fibers that pass from one retina to the ipsilateral visual cortex. As the extent to which overlap of the individual visual fields of the eyes increases, so too does the number of ipsilateral fibers. In man, the overlap is quite large (Fig. 8.21), so that about 50 percent of the fibers that comprise the optic nerve (and optic tract) do not undergo decussation within the optic chiasm (Fig. 8.22; see also Figs. 7.44, p. 298, and 8.37, p. 504).

As indicated in Figure 8.22, within the visual cortex individual neurons can receive simultaneous input signals from specific points on both retinas, and as each neuron of the visual cortex is receiving excitatory signals from both eyes it is "seeing" in two directions at once. This neural arrangement within the cortex enables one not only to perceive visual direction with respect to the head, but also to estimate the distance of an object by means of *binocular parallax,* as illustrated in Figure 8.23. Hence, the underlying cause of stereopsis is the horizontal disparity between the two retinal images that are transmitted to the visual cortex from the retina of each eye. In turn, the brain has mechanisms whereby it is able to select specific portions of each retinal image that are similar in that the portions selected each represent, and are images of, the same characteristic features of the object being viewed in space.

In contrast to organisms exhibiting binocular vision, many species have their eyes located upon opposite sides of the head so that little (if any) overlap of the visual fields is present, and each eye scans a different scene. This effect, called panoramic vision, *is present in such organisms as fishes, certain birds (eg, the pigeon), and lizards (eg, the chameleon).*

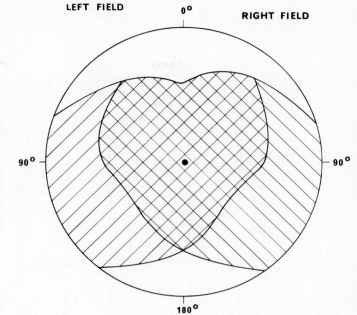

LEFT FIELD   0°   RIGHT FIELD

90°   90°

180°

FIG. 8.21. The monocular visual field for each eye is indicated by the irregular outlines that enclose the areas indicated by diagonal lines. The binocular visual field is indicated by the irregular area in the center of the figure where the monocular visual field of each eye overlaps that of the contralateral eye, when both eyes are focused upon the same object (black dot). Note that each visual field is not circular, as the medial portion is cut off by the nose and the superior portion is cut off by the orbital roof. (After Ganong. *Review of Medical Physiology,* 6th ed, 1973, Lange Medical Publications.)

The three-dimensional spatial orientation of the neural elements from the retina to the visual cortex is exceedingly precise. In fact, the degree of neuroanatomic segregation of nerve fibers into spatially oriented layers within the lateral geniculate body is directly related to the degree of binocular vision present in the particular organism under examination. Therefore, man, who possesses a considerable degree of binocular overlap, exhibits a very distinct segregation of neural inputs in the lateral geniculate body. Furthermore, this anatomic compartmentalization of nerve fibers is enhanced still further by a functional segregation in that inhibitory connections are present among the corresponding neural elements of neighboring layers of the lateral geniculate body.

The cortical neurons underlying vision, in marked contrast to those found in the retina or lateral geniculate body, require extremely precise stimuli for their excitation. Experimental studies have shown that each cortical element requires that a visual stimulus having clearly defined spatial characteristics be admitted to the retina in order to excite the cortical neuron. Neurons that exhibit similar spatial requirements for orientation of the stimulus on the retina are concentrated together on the cortical surface and run in columns inward toward the white matter (see also p. 343). Electrophysiologic studies have revealed that within each column, certain neurons have a reasonably direct input from the retina to the cortex *(simple neurons),* whereas other neural elements, the so-called *complex neurons,* presumably signal the output of the entire neural column. The latter cells function electrically as though they are activated by the net input derived from a considerable

number of simple neurons (Fig. 8.22, neurons 5 and 6).

It has proven feasible to map specific regions in the visual field of each eye wherein a single visual stimulus will cause excitation of a specific neuron in the visual cortex, and each such region is called the *response field* for the stimulus. The response fields of each eye have been found to lie in slightly different regions of each retina; a spatial disparity is present (Fig. 8.23B). This effect gives rise to very slightly different patterns of central neural excitation, and thus the perception of this retinal disparity by the individual is interpreted as giving the object a solidity or position in space, an effect which neither eye alone can accomplish.

Additional visual cues important in the localization of objects in space as well as in the determination of their relative size or distance from the observer are use of the angle subtended by objects of known size; shadows and colors; the effort required for accommodation; and the effect of motion parallax, that is, those movements in which the velocity and/or direction of near and far objects differs perceptibly. Thus, monocular as well as binocular cues both are significant for the perception of depth. At any distance from the observer, say over 30 meters, binocular vision diminishes to the vanishing point as the angle of retinal disparity becomes so small as to become imperceptible, and objects are seen as though they were two-dimensional rather than three-dimensional. This fact can be confirmed readily by viewing stereoscopic photographs with an appropriate slide viewer. Nearby objects that lay about 1 meter from the camera when photographed exhibit a strong three-dimensional appearance, whereas

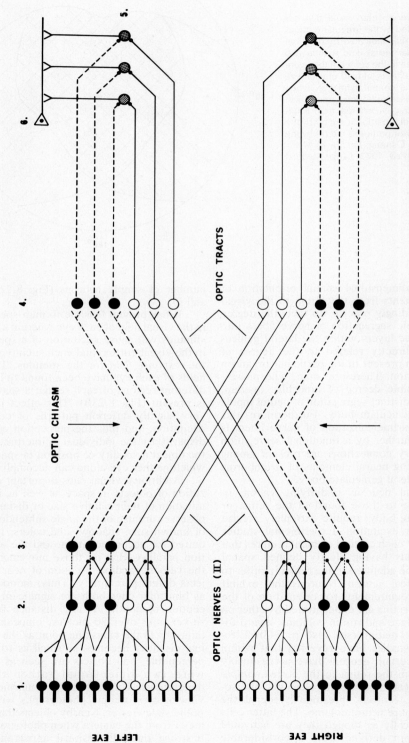

FIG. 8.22. Highly schematic representation of the neural elements and pathways involved in binocular vision. The neural pathway from each eye to the primary visual cortex consists of six elements, three of which are found in the retina. 1. Photoreceptor cells (rods and cones). 2. Bipolar retinal cells. 3. Ganglion cells of retina. 4. Lateral geniculate body. 5. Simple cortical cell. 6. Complex cortical cell. Note that half of the nerve fibers emanating from the retinal ganglion cells undergo decussation in the optic chiasm whereas the other half course to the cells of the ipsilateral visual (calcarine) cortex (5, 6). See also Figure 7.44, page 298. (After Pettigrew. *Sci Am* 227:84, 1972.)

492

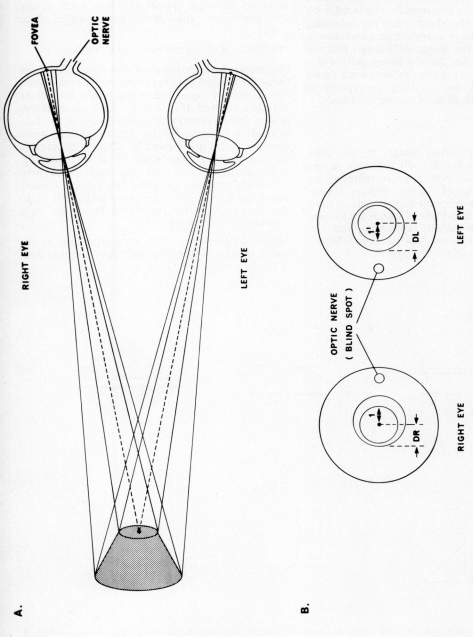

FIG. 8.23. **A.** When a person views a three dimensional object there is a disparity between the two retinal images that is called *binocular parallax*. In the figure, each eye is focused upon the same point (dot) on a truncated cone (such as a lampshade, bucket, or cork) and the binocular parallax is indicated by the relative positions of the image of the object upon the retina of each eye as indicated by the solid lines. **B.** Diagrams of the fundus of each eye showing the fovea as a dot in the center. The smaller circles representing the retinal images of the bottom of the truncated cone shown in part **A** of the figure both are equidistant from the fovea (ie, 1 = 1′), and these images are formed upon precisely similar points on the retina, hence have zero disparity. However, the larger circles which represent the retinal images of the upper rim of the cone are displaced horizontally in relation to the smaller circles and the fovea of each eye. Consequently the distance DR does not equal the distance DL; hence these images lie on disparate points on the retina. From a quantitative standpoint, the amount of the retinal disparity is the difference between DR and DL, and this value generally is expressed as an angle. (After Pettigrew. *Sci Am* 227:84, 1972.)

those objects that lay in the middle distance (roughly around 10 meters from the camera) have a much less pronounced three-dimensional appearance. On the other hand, objects that lay about 30 or more meters from the camera when photographed appear to be completely "flat" or two-dimensional when viewed stereoscopically. In this example, each lens of the stereoscopic camera merely photographs a horizontally displaced view of the same scene. Therefore, when the individual pictures are mounted in a single frame and seen in a suitable viewer (the distance between the two frames being corrected for the interpupillary distance of the observer), each eye sees each frame independently and the resulting perception of nearby objects is one of their having stereopsis.

Optical Illusions

The interpretation of visual images by the brain may be incorrect insofar as one's judgment of size and shape, as well as perception of relationships between, and distance of, various objects, is concerned. Such errors are called *optical deceptions* or *optical illusions,* hundreds of which are known. Several examples of various types of optical illusions are given in Figures 8.24 through 8.28 and explained in the legends to the figures.

Clinical Correlates

Visual defects can be caused by physical, biochemical, physiologic, and genetic factors, and such defects in turn can range from mild and readily corrected dysfunctions to total blindness. The few examples of particular visual and related defects discussed in the remainder of this section will serve to illustrate specifically how each of the etiologic factors cited above can produce visual errors.

"PHYSIOLOGIC ABNORMALITIES" OF VISION

Presbyopia. The normal or *emmetropic eye* can focus the parallel light rays reflected from a distant object sharply upon the retina when the ciliary muscle is relaxed. (Fig. 8.29A). Such a normal eye is also able to produce accommodation of the lens through contraction of the ciliary muscle when nearby objects are viewed (p. 472). With advancing age, however, the elasticity of the lens (Fig. 8.30) decreases, so that the near point progressively increases (p. 476). This gradual hardening of the lens is called *presbyopia* (p. 494), a defect that is corrected readily by lenses having a suitable curvature (Fig. 8.33).

Spherical Aberration. The crystalline lens of the eye is unable to focus the light rays entering near

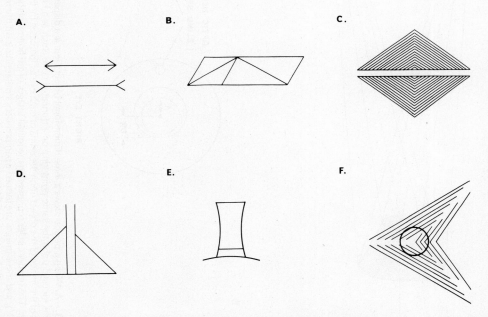

FIG. 8.24. Optical illusions with lines and angles. **A.** The Müller-Lyer illusion. Both horizontal lines are of equal length. **B.** The diagonal lines within each parallelogram are the same length. **C.** The two horizontal lines are parallel as well as straight. **D.** The lower right-hand line if extended will intersect the left-hand line where it joins the vertical. **E.** The brim of the hat is the same length as the height. **F.** The perfect circle appears to be distorted. (Data from White. *Classical and Modern Physics,* 1940, Van Nostrand.)

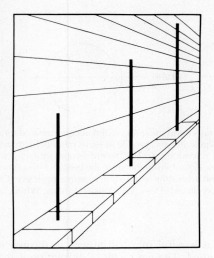

FIG. 8.25. An optical illusion in perspective; that is, the picture appears to have depth (three dimensions) when in reality it is flat (ie, only two dimensions are present). (Data from White. *Classical and Modern Physics,* 1940, Van Nostrand.)

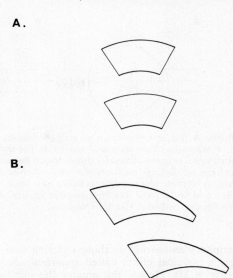

FIG. 8. 27. Optical illusions of area. In both **A** and **B** the similar figures have an equal area, although the slanting lines at the ends cause the lower figure in each example to appear larger than the one above. (Data from White. *Classical and Modern Physics,* 1940, Van Nostrand.)

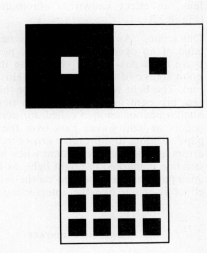

FIG. 8.28. Optical illusions illustrating the phenomenon of irradiation (or brightness contrast). **A.** The large and small black and white squares are of equal size. When an image of these squares is formed upon the retina, the rods and cones that lie in the retina immediately beyond the white edges of the small square are stimulated by the active retinal elements in their immediate vicinity. Consequently the small white square appears to be larger than the corresponding small black square. **B.** Gray spots are seen at the intersections of the white line, an effect that is produced by a similar mechanism to that described for part **A** of the figure. (After White. *Classical and Modern Physics,* 1940, Van Nostrand.)

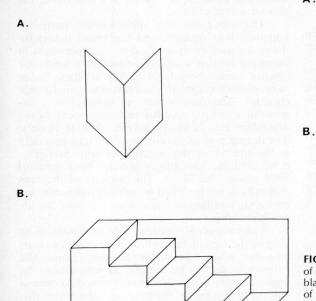

FIG. 8.26. Optical illusions to illustrate fluctuation of the attention. **A.** The folded sheet of paper appears to open either toward or away from the reader. **B.** The flight of steps is seen either from above looking down or from below looking up. (Data from White. *Classical and Modern Physics,* 1940, Van Nostrand.)

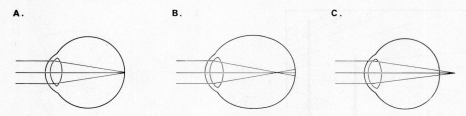

FIG. 8.29.   A. The emmetropic (or normal) eye focuses parallel light rays upon the retina when accommodated for distant vision; ie, when the ciliary muscle is relaxed. **B.** The myopic (or nearsighted) eye is unable to focus parallel light rays upon the retina when accommodated for distant vision because the eyeball is too long. **C.** The hypermetropic (or farsighted) eye cannot focus parallel light rays upon the retina when accommodated for distant vision because the eyeball is too short. Note that light rays come to a focus in front of the retina in the myopic eye **(B)** and behind the retina in the hypermetropic eye **(C).** (Data from Best and Taylor [eds]. *The Physiological Basis of Medical Practice,* 8th ed, 1966, Williams & Wilkins; White. *Modern College Physics,* 3rd ed, 1956, Van Nostrand.)

its periphery as sharply as those entering near its center, a physical effect called *spherical aberration* (Fig. 8.31). Hence, the greater the pupillary diameter (as in dim light) the greater the spherical aberration, and the less sharp the retinal image becomes.

Chromatic Aberration.   The refractive power of the crystalline lens is different for different hues of light, an effect that is quite similar to that of a prism (Fig. 8.10). Consequently, different colors are focused at slightly different points behind the lens, an effect known as *chromatic aberration* (Fig. 8.32).

Diffraction.   As light waves pass near the sharp edge of an opaque object, every point on every wave front passing the object can itself act as a point source of new light waves (Huygens' principle). The light waves passing near the opaque object are bent and thus travel in new directions. Such a phenomenon is called *diffraction,* and this occurs as light waves pass over the edge of the pupil, leading to *diffractive errors* in vision. Such errors are particularly evident when pupillary constriction is marked in bright light, as then interference patterns are produced on the retina. Such an effect is much like the intersecting waves seen

when a handful of small stones is thrown into a quiet pond. The net result of diffraction is that the sharpness of the visual image is related directly to the pupillary diameter.

Cataract.   As the crystalline lens undergoes aging, alterations in the color of the lens may accompany the loss of plasticity and elasticity resulting from the physiologic sclerosis noted earlier. The normally transparent crystalline lens may become tinted with amber, reddish, or brownish discolorations, and these in turn result in a filtering of the shorter waves of light.

The term *cataract* is applied to any partial or complete lens opacity, and the visual defect induced by such an opacity can range from slight to complete loss of vision. Commonly, a lenticular opacity cannot be related to any etiologic factor save advancing age; this condition is called *senile cataract.* The development of cataract is a degenerative and not an inflammatory process, as the crystalline lens is totally devoid of blood vessels. The deeper part of the cortex of the lens generally is the site where the opacity develops first (Fig. 8.30), and the lens swells because fluid accumulates among the fibers. The nucleus of the lens generally is not involved in cataract formation, except that swelling of the cortex narrows the anterior chamber.

The cataract is considered to be *immature* as long as the superficial cortical layers remain clear; however, when these layers become opaque, the cataract is said to be *mature,* at which time the water content of the lens has declined to normal values. Following the mature stage, the cortex disintegrates, and drying and shrinkage of the entire lens ultimately take place.

The fundamental alteration in cataract development is a progressive coagulation of the proteins that comprise the lens itself, an effect that has been related in part to the long-term action of ultraviolet light upon the lens, as well as the thermal damage caused by infrared waves in some instances (eg, in glassblowers' cataract). The lens normally protects the retina of the eye from ul-

ANTERIOR SURFACE

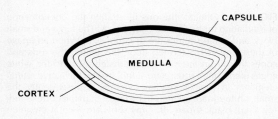

POSTERIOR SURFACE

FIG. 8.30.   The general structural features of the human lens. (After Best and Taylor [eds]. *The Physiological Basis of Medical Practice,* 8th ed, 1966, Williams & Wilkins.)

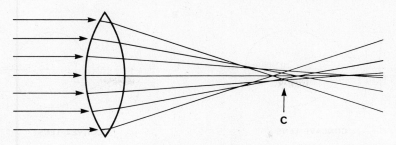

FIG. 8.31. In spherical aberration, the radius of curvature, hence the refractive power, of a lens is not perfect. Consequently light rays entering near the periphery are focused farther away from the lens than those entering closer to the center. C = circle of least confusion. (Data from Best and Taylor [eds]. *The Physiological Basis of Medical Practice,* 8th ed, 1966, Williams & Wilkins; White. *Modern College Physics,* 3rd ed, 1956, Van Nostrand.)

traviolet radiation damage by absorbing electromagnetic waves having lengths between 350 and 450 mμ (Fig. 8.11). The absorption of these rays, however, can induce physical alterations in the proteins of the lens itself, and these lead to coagulation. The lenticular proteins are denatured (p. 56) by light and/or heat waves in such a way that they subsequently become susceptible to clotting or coagulation into an opaque flocculent mass. The latter process occurs only in the presence of certain mineral salts, notably calcium, as well as magnesium, and the coagulation process is intensified in the presence of certain organic compounds such as dextrose and acetone. The latter compounds are present in elevated concentrations in the blood of diabetic patients; the high incidence of cataract in such patients appears to be related causally to such factors. It is not surprising that lenses containing cataracts differ biochemically in several respects from normal lenses. Specifically, the concentrations of insoluble albuminlike proteins increases markedly, whereas the concentration of soluble lenticular proteins as well as glutathione decreases. The oxygen uptake decreases markedly. The calcium concentration of such abnormal lenses increases greatly, whereas the sodium as well as magnesium content increases moderately. On the other hand, the potassium concentration declines far below normal levels as cataracts develop.

Once a cataract has developed to the point that light transmission is severely impaired and vision thus obscured, the entire lens is removed surgically. Subsequent to this operation, the eye loses a large proportion of its total refractive power (p. 472); a strong convex lens must be placed in front of the eye in order to correct the refractive defect in vision that has been created (Fig. 8.33).

REFRACTIVE ERRORS. Spherical aberration, chromatic aberration, and diffraction are intrinsic optical defects present to some extent in any optical system, whereas presbyopia normally accompanies the aging process. As noted earlier, the normal eye refracts parallel rays of light (ie, those arising from infinity or a distance greater than 6 meters) onto the retina without any accommodation; that is, the emmetropic (normal) eye (Fig. 8.29A) has a static refraction equal to zero. On the other hand, if the far point of an eye lies at some point other than infinity, it is said to be *ametropic*. Myopia and hypermetropia are the two forms of ametropia.

Myopia. The myopic (or nearsighted) eye is defective in that the parallel light rays emanating from a source at infinite distance come to a focus in front of the retina. The rays then cross; hence, a diffusion circle (or blurred image) is formed upon the retina (Fig. 8.29B). Generally, the principal defect in myopia is an eyeball that is too long, and

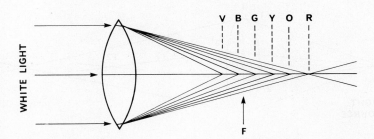

FIG. 8.32. In chromatic aberration, the different wavelengths of light are focused at various distances behind the lens depending upon their particular energy levels (cf. Figs. 8.10 and 8.11). V = violet; B = blue; G = green; Y = yellow; O = orange; R = red. (Data from White. *Modern College Physics,* 3rd ed, 1956, Van Nostrand.)

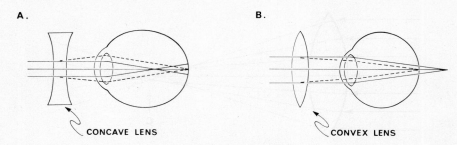

FIG. 8.33. **A.** Myopia is corrected by the use of concave lenses which diverge the light rays before they enter the eye (dashed lines); the refracted rays are able to focus sharply upon the retina. **B.** Hypermetropia is corrected by the use of convex lenses which converge the light rays before they enter the eye (dashed lines) so that the refracted rays now are able to focus sharply upon the retina. Convex lenses also are used to correct the defective accommodation that develops in presbyopia as well as following cataract extraction. (After Guyton. *Textbook of Medical Physiology,* 4th ed, 1971, Saunders.)

only rarely is the refractive power of the lens too great.

In myopia, the far point is quite close to the eye as the light rays must diverge from their source in order to form a sharp image upon the retina. The far point may lie only a few centimeters from the eye if the myopia is extreme. Accommodation for the far point is relaxed in the myopic eye like that found in the emmetropic eye, since increasing the refractive power of the lens merely would cause an increased blurring of the image.

Myopia is corrected readily by means of concave lenses that diverge the parallel light rays before they enter the eye, as illustrated in Figure 8.33A.

Hypermetropia.   The hypermetropic (or "farsighted") eyeball, in contradistinction to that present in myopia, is too short for the refracting power of the lens; the image is focused behind the retina when parallel light rays enter the eye (Fig. 8.29C). The unfocused light rays passing through the retina form diffusion circles, hence a blurred image.

Also in contrast to the myopic individual, the person with hypermetropia must accommodate when looking at distant objects so that the parallel rays are focused upon the retina.

As illustrated in Figure 8.33B, hypermetropia is corrected by use of a convex lens that causes the parallel rays of light from a distant source to converge before they enter the eye.

Astigmatism.   Astigmatism is a defect in refraction such that the optical system of the eye does not focus the rays of light into a cone about all meridians (or planes). Rather, the rays passing through different meridians are focused at different points, because of different curvatures of the optical system in different meridians about the optic axis of the system (Fig. 8.34). A meridian of the optical system having a greater curvature will have a greater refractive power; the image will be formed in front of the retina. The converse of this statement also is true.

Actually, a slight astigmatism is present in all optical systems, including the human eye. When the visual acuity is reduced seriously, astigmatism is considered abnormal. The meridian in which astigmatism is present can be detected clinically by use of charts such as that illustrated in Figure 8.35. The patient looks at the center and is requested to tell which group(s) of lines is blackest and sharpest, and which is lightest and most blurred.

Astigmatism is most commonly a result of irregularities in the curvature of the cornea, and less frequently of such abnormalities of the crystalline lens. It was noted earlier that the major refractive power of the eyeball is obtained at the air-cornea interface (p. 472); therefore, relatively small defects in the regular curvature of the cornea can produce marked astigmatic defects.

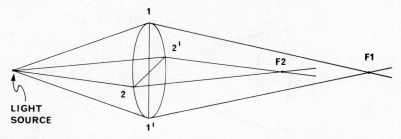

Fig. 8.34.   The optical mechanism of astigmatism. Light rays from a single point source are focused at two places behind the lens (F1 and F2) as the radius of curvature (hence the refractive index) in the vertical plane (1–1') is less than the refractive index in the horizontal plane (2–2'). Astigmatism is corrected by the use of cylindrical lenses, as discussed in the text. (Data from Guyton. *Textbook of Medical Physiology,* 4th ed, 1971, Saunders.)

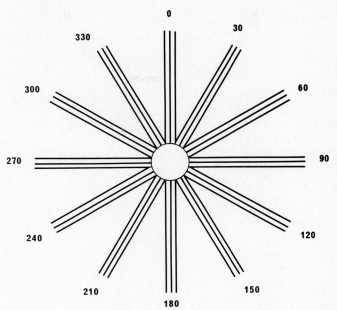

FIG. 8.35. Line chart for detecting astigmatism as discussed in text.

Astigmatism is corrected by cylindric lenses that are convex in the meridian of the cornea (or crystalline lens), which has the smaller curvature. For example, if the curvature is greater in the horizontal plane of the eye, the cylindrical lens (which refracts in only one plane) that is employed is convex only in the vertical meridian.

ARGYLL–ROBERTSON PUPIL. A particular ocular defect known as the Argyll–Robertson pupil frequently develops in patients having advanced neurosyphilis, and in whom central neural degeneration has occurred (eg, tabes dorsalis, p. 231). In this condition, both the direct and indirect (consensual) light reflexes (p. 475) are absent; however, the accommodation and convergence reflexes (pp. 472, 475) are still present and may even be abnormally hyperactive. In addition, the affected pupil is generally smaller than normal; ie, myosis is present, and the pupil does not dilate following the application of painful stimuli, or dilate completely following the topical application of mydriatic drugs (such as atropine) to the eyeball. Patients exhibiting the Argyll–Robertson pupil have this defect quite independently of any visual defects that may also be present. The Argyll–Robertson pupil may be unilateral, although most commonly it is of bilateral occurrence.

Although the Argyll–Robertson pupil is seen most commonly in, and is an important diagnostic sign of, neurosyphilis, this abnormality may also be encountered occasionally when central nervous system damage has been produced by, for example, encephalitis or chronic alcoholism.

The precise neuroanatomic site of the lesion that gives rise to the Argyll–Robertson pupil is unknown, although the posterior commissure synapses within the oculomotor nucleus, and pathologic changes within the iris per se have been implicated in causing this defect.

DIPLOPIA AND STRABISMUS. Normally, when one looks at an object, each eye forms a separate image of the object on each retina. In turn, the signals generated within each retina are transmitted to the brain and there fused in such a manner that the two separate retinal images are perceived subjectively as one conscious image. This characteristic of binocular vision may be considered to reside in the stimulation of corresponding retinal points in each eye. If, however, the ocular muscles act in an unequal fashion so that the two retinal images of the same object are not focused upon the corresponding retinal points in each eye, the separate images are not fused and perceived consciously as one image. Two images of the object are "seen" or perceived, and this false projection of the visual field is called *diplopia* or *double vision*.

Thus, if the ocular muscles are weak or paralyzed, or if an unequal action develops between the ocular muscles of the two eyes, diplopia results. For example, transient diplopia can develop in acute alcoholic intoxication because of a partial paralysis of the ocular muscles.

Strabismus is an involuntary deviation of the eyes such that the visual axes assume a position in relation to each other that is quite abnormal with respect to the physiologic condition. Thus, the eye movements in strabismus are not coordinated. The eyes may assume any of three possible basic spatial relationships to each other, as well as combinations of these relationships, depending upon the specific ocular muscles involved (Fig. 8.36).

Some investigators consider the possibility that strabismus is caused by an abnormal fixation

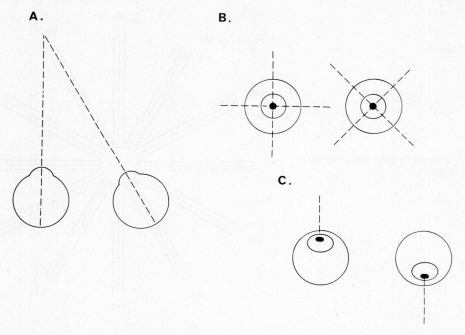

FIG. 8.36.   The three fundamental types of strabismus. **A.** Horizontal strabismus. **B.** Torsional strabismus. **C.** Vertical strabismus. In each instance, the dashed lines indicate the relative positions of the eyes. (After Guyton. *Textbook of Medical Physiology,* 4th ed, 1971, Saunders.)

of the neural mechanisms that underlie normal fusion of both retinal images. When a child attempts to focus both eyes upon the same object, one eye is able to achieve this focus correctly, whereas the contralateral eye does not. Alternatively, both eyes may focus correctly upon objects, but successively rather than simultaneously. The neural patterns that underlie normal conjugate eye movements become abnormally fixed or "set," and thus the eye movements do not fuse properly, and the individual has developed strabismus.

If the abnormal position of the eyes present in strabismus is congenital or has been experienced for some time, diplopia usually is not present. The reason for this lies in the fact that the so-called "false" retinal image formed in the weaker (or squinting) eye induces the development of an area in the peripheral retina that functionally behaves as does the fovea in the stronger (or sound) contralateral eye. This "pseudo fovia" or "false macula" in the weak eye assumes the function of the fovea in the normal eye, the result being that the two images undergo fusion within the brain so that essentially normal binocular vision results. If, however, a strabismus (or squint) of this type which has been present for a long time is corrected surgically so that the visual axes are now aligned parallel with respect to each other, diplopia ensues because the anatomic fovea of the operated eye does not have a functional correspondence with that of the contralateral normal eye.

NIGHT BLINDNESS AND VITAMIN A DEFICIENCY. Since vitamin A is an essential constituent for the synthesis of retinene (Fig. 8.14), a lack of this nutrient can produce visual abnormalities if the deficiency is sufficiently prolonged and severe (see also p. 885). Night blindness, or *nyctalopia,* is one of the earliest visual manifestations of vitamin A deficiency, and this abnormality can develop through either an inadequate intake or impaired absorption of the fat-soluble vitamin A. Similarly, malfunction of the cones can also develop as a consequence of a lack of vitamin A, and if this condition is chronic, ultimately an irreversible degeneration of the photoreceptor cells can take place. Subsequently, the other neural elements of the retina undergo a similar pathologic change.

Normally, the vitamin A utilized by the eye for the synthesis of retinene is stored in the pigmented epithelial layer of the retina (Fig. 8.5). Furthermore, the liver contains large stores of this substance, which are readily available when needed, so that for nyctalopia to develop a severe vitamin A deficiency must be present for weeks, or even months, before visual abnormalities appear. Furthermore, when vitamin A is administered to a deficient individual, the visual abnormalities are reversed with extreme rapidity provided that irreversible retinal degeneration has not yet occurred.

The retina, in common with other neural tissues, also requires other vitamins in order to perform its normal functions, the B complex being particularly important (p. 878). Nicotinamide is especially significant, as this compound provides an indispensable constituent of nicotinamide adenine dinucleotide (NAD, p. 933), a coenzyme essential to the interconversion of vitamin A and retinene in the rhodopsin cycle (p. 481).

RETINOPATHIES. As noted earlier (p. 471), the fundus of the eye is the only area in the body where a vascular bed can be examined readily under physiologic conditions. Direct ophthalmoscopic examination of the eye sometimes can reveal to the clinician pathologic vascular changes long before more overt symptoms of particular diseases are manifest. Alternatively, a fundic examination can often help to reveal the progress of a disease, and this information is quite useful in determining the prognosis or effects of therapeutic measures in a given patient.

A few examples of the retinopathies observed in cardiovascular and hematologic disease states may be presented.

Chronic hypertension (p. 695) is attended by a gradual narrowing and increasing tortuosity of the retinal arteries, and eventually yellowish exudates may appear in the retina. Severe hypertension is related to a localized narrowing of the arteries, hemorrhages, white spots (cytoid or cell-like bodies), and papilledema (ie, edema associated with swelling of the optic papilla).

A sudden, or acute, rise in the blood pressure far above normal levels, such as is seen during a *hypertensive crisis* or during *eclampsia,* causes the retinal arteries to narrow markedly. The entire retina, including the papilla, assumes a pale, milky appearance. Fluid accumulating in subretinal regions may then induce retinal detachment with an attending total loss of central as well as some peripheral vision. In contrast to the retinal appearance during chronic hypertension, an *acute hypertensive episode* may occur with the appearance of only a few retinal hemorrhages and exudates or even none of these at all, whereas the associated papilledema is less red than that seen in other hypertensive states. The retinal reactions discussed above may be entirely reversible.

In *congestive heart failure* (p. 1244), the retinal veins alone are engorged; in cyanotic heart disease, not only are the veins engorged, but the entire vascular system of the retina assumes a darkish appearance.

Ophthalmoscopic examination of the fundus in patients having *atherosclerosis* reveals that the retinal vessels have an irregular diameter, are narrowed, and appear yellow in color. The end arteries practically disappear, and in older patients the extreme vasoconstriction underlying this effect results in a loss of peripheral vision. Papilledema, hemorrhages, and exudates are not caused by atherosclerosis per se. Rather, these signs develop secondarily to the hypertension that accompanies the atherosclerosis. Occlusion of the central retinal artery, emboli, and ischemic optic neuropathies are the underlying processes whereby artherosclerosis causes visual losses.

Retinopathies accompany blood (hematologic) disorders, and although such signs are not diagnostic for any particular disease, several of these abnormalities may be seen consistently in any severe *anemia*. Round as well as flame-shaped hemorrhages, exudates, Roth's spots,* papilledema, and pallor all may be seen in the retina of a patient suffering from an anemia or from a bleeding disease. *Polycythemia* (p. 569), on the other hand, is associated with dilated dark veins and arterial as well as venous thrombosis, all of which may be seen by retinal examination.

SOME COMMON EYE DISEASES

Rapid Loss of Vision. When vision is lost suddenly, rapid diagnosis is critical because such a symptom is frequently a manifestation of a central nervous or vascular disease that may require prompt attention.

Vascular conditions that affect the retinal vessels include diabetes, occlusion of the peripheral veins, hemorrhagic disorders, and inflammation of the veins. All these conditions can lead to a loss of vision of varying severity through hemorrhage into the vitreous body (Fig. 8.1). In fact, such *vitreous hemorrhage* may be the first sign of a retinal perforation that may ultimately result in retinal detachment.

Subretinal hemorrhage is generally seen near the macula, and this in turn causes a loss of central vision. A previous history of histoplasmosis is frequently associated with this condition, particularly in young and middle-aged patients. In older persons, however, senile macular degeneration generally causes a gradual visual loss, although visual loss can take place suddenly through a subretinal hemorrhage.

If the central retinal artery becomes occluded, the visual defect may be so severe that the patient is unable to perceive (or can barely perceive) light. Commonly, atherosclerosis, which causes narrowing of the retinal vessels, is the cause of *central retinal artery occlusion,* although emboli (such as may arise in heart or lungs), polycythemia, and arteritis all can produce this abnormality.

Occlusion of the central retinal vein can produce a severe but subtotal visual loss. The fundic veins are greatly dilated and have a tortuous course. Once again, atherosclerosis is the most frequent etiologic factor. If a sclerotic artery exerts pressure upon the central vein either within the retina or behind the cribiform plate, *central retinal vein occlusion* results.

Retinal detachment may take place either rapidly or slowly. Most patients perceive a gradual loss of peripheral vision and/or spots floating before the eyes before the visual loss ensues. Follow-

*Localized white retinal areas each surrounded by a hemorrhagic area.

ing pupillary dilatation, the detached retina generally can be seen readily during ophthalmoscopic examination.

Frequently the *optic nerve* itself is involved in abnormal processes leading to profound visual loss. For example, *ischemic papillopathy* is the term applied to a disease process involving the nerve head in the retina, or the optic nerve immediately behind this structure. The optic papilla appears white, pale, and irregular because of edema, and the retinal veins may exhibit a slight distention. One or more small hemorrhages usually are present near the edematous locus; however, blindness is the most evident sign of this condition. *Optic nerve infarction,* which occurs more posteriorly in the optic nerve than ischemic papillopathy, exhibits symptoms that are identical with the latter; however, in the early stages there are no visible signs in the optic disc. Atrophy of the optic nerve develops in both of these conditions. Once again, atherosclerosis is the most frequent cause of optic nerve infarction, although temporal (also called cranial or giant cell) arteritis is most important clinically. The latter vascular condition is manifested by severe temporal headache for periods of up to 2 weeks before the onset of any visual or other neurologic signs. Appropriate therapy can be administered to prevent irreversible damage to the optic nerve, ie, injection of corticosteroids (p. 1112).

Glaucoma. The term *glaucoma* signifies a marked elevation of the intraocular pressure. Normally, the pressure of the fluids within the eye exerts a total force of 10 to 22 mm Hg, a measurement that can be made readily, albeit indirectly, by use of an instrument called a *tonometer.* Following application of a topical anesthetic, the cup-shaped piston on the tonometer is pressed gently against the cornea. The minimum pressure required to deform the eyeball slightly is considered to be the intraocular pressure, and this value is read directly from the calibrated scale on the instrument. Parenthetically, it should be noted that this simple test, like an ophthalmoscopic examination of the retina, should form a routine part of any complete physical examination, as only in this way can the primary glaucomas be detected in their early stages when these conditions are most amenable to therapeutic intervention.

There are two kinds of *primary glaucoma,* the acute congestive and chronic simple forms. In addition, there are many processes whereby an increase in the intraocular pressure can develop secondary to another pathologic mechanism *(secondary glaucomas).* In either event, the abnormally high intraocular pressure ultimately destroys the retinal elements in an irreversible fashion so that permanent blindness develops.

Acute Congestive Glaucoma. In this condition, the peripheral iris is forced against the peripheral recess of the anterior chamber (Figs. 8.1 and 8.3); hence, acute congestive glaucoma is also called *closed-angle glaucoma.* The effect of this dislocation is to obstruct the outflow of aqueous humor into its normal channel, the canal of Schlemm. Symptoms associated with this condition may include short periods of foggy, blurred, or halo vision which occur in darkness or are related to episodes of emotional stress. The eyeball feels hard when finger pressure is exerted on it, and the intraocular pressure ranges between 30 and 90 mm Hg as determined by tonometry.

Therapy must be instituted early in the course of the disease, and this is carried out by the administration of certain drugs that constrict the pupil (eg, eserine or pilocarpine) as well as other compounds that inhibit the action of the enzyme carbonic anhydrase (q.v., p. 757) such as acetazolamide, ethoxzolamide, or dichlorphenamide. The second phase of treatment involves surgical intervention in the form of an iridectomy. This procedure creates a channel through which aqueous humor can flow from the space posterior to the iris into the anterior chamber.

Chronic Simple Glaucoma. The intraocular pressure elevations caused by chronic simple (or open-angle) glaucoma produce irreversible loss of peripheral, and ultimately central vision. The symptoms at first may be somewhat nebulous, eg, a vague ocular aching or discomfort or fatigue while reading. Subsequently, unilateral or bilateral loss of peripheral vision may be the first major symptom, whereas a loss of visual acuity in the foveal areas is manifest somewhat later.

As the early symptoms of chronic simple glaucoma are so insidious, the tonometric procedure for evaluating pressure is exceedingly valuable, and pressure elevations that range between 25 and 50 mm Hg above normal are commonly found in this condition.

The early detection of glaucoma before irreparable retinal damage has occurred is critical to effective therapy. As with acute congestive glaucoma, the initial treatment of chronic simple glaucoma is by means of miotic drugs to constrict the pupil as well as with carbonic anhydrase inhibitors. If the pressure remains uncontrolled and at high levels, surgical intervention must be employed in order to provide a channel for the absorption of aqueous humor through a subconjunctival route.

EFFECTS OF LESIONS IN THE CENTRAL VISUAL PATHWAYS. The neuroanatomic pathways involved in vision were presented in Figure 7.44 (p. 298). Lesions at various sites along these pathways can be pinpointed with considerable accuracy by the specific effects they produce in the visual fields, as shown in Figure 8.37.

The visual fields can be plotted in detail by use of an instrument known as a *perimeter.* In practice, the visual fields can be screened for gross abnormalities by a technique known as the *confrontation method:* The patient is asked to close one eye and then fix the gaze of the other eye straight ahead on one point. An object such as a pencil or a finger is then introduced from a point outside of the normal visual periphery and slowly moved toward the center of the visual axis. The point at which the object is perceived first is noted and the process is repeated in several meridians about the visual axis. A rough estimate of the visual field for that eye can then be made. The process is repeated for the second eye.

Complete visual loss, or blindness, is described with respect to the visual fields involved rather than with reference to the affected retinal area. When atrophy of the optic nerve is present, both visual fields may be contracted, or a centrally located region where visual loss is present (*scotoma*) may be found. Restricted visual fields without any evident organic lesions are found in some hysteric (psychoneurotic) patients (Fig. 8.37, 1). In such persons, visual perceptions are similar to those obtained when looking through a pair of gun barrels ("tubular" or "tunnel" vision). Tubular vision may also be caused by retrobulbar or optic neuritis.

Nerve fibers from the nasal half of each retina undergo decussation in the optic chiasm; consequently, the fibers that make up the optic tracts consist of neural processes arising from the temporal half of one retina and the nasal half of the contralateral retina (Fig., 8.22). Stated another way, each optic tract underlies one-half of the visual field. Light rays enter the eye through the pupil, and these rays also travel in straight lines. The nasal half of each retina functions to perceive objects that are located in the temporal visual field, and the temporal half of each retina underlies the perception of objects located in the nasal field. A lesion that completely interrupts one optic nerve causes blindness in that eye (Fig. 8.37, 2), whereas a lesion that involves only one optic tract causes a visual loss in one half of the visual field (Fig. 8.37, 5). Such a defect is called a *homonymous hemianopia*. The defect is a half-blindness that affects the same side of both visual fields. On the other hand, a lesion that involves only the lateral region of one optic nerve produces a unilateral nasal hemianopia (Fig. 8.37, 4).

Lesions that affect the optic chiasm produce destruction of the optic nerve fibers that arise from the nasal portion of each retina, resulting in a *heteronymous hemianopia* (Fig. 8.37, 3). The defect in this instance is a half-blindness that affects the opposite side of each visual field. The fibers of the optic chiasm can be destroyed by tumors of the pituitary that enlarge to such an extent that they expand out of the sella turcica to encroach upon the chiasm.

Other specific defects in the visual fields are classed as left or right, bitemporal or binasal, as indicated in Figure 8.37.

The optic nerve fibers that originate from the upper quadrants of the retinas, and that serve vision in the lower halves of the visual fields, synapse in the medial region of the lateral geniculate body. On the other hand, the nerve fibers that course from the lower quadrants of the retinas end in the lateral regions of the lateral geniculate bodies. The geniculocalcarine fibers originating in the medial portion of each lateral geniculate nucleus in turn end upon the superior lip of the calcarine fissure, whereas those fibers originating in the lateral regions of the geniculate nucleus terminate upon the inferior lip of the calcarine fissure (Fig. 8.38). In addition, fibers from the lateral geniculate nucleus that underlie macular vision become separated from those that underlie peripheral vision. The macular fibers then terminate on the calcarine fissure, as shown in Figure 8.38. As a consequence of this neuroanatomic localization, lesions of the occipital lobe may produce specific visual defects in particular quadrants of the visual fields, as indicated in Figure 8.37, 6 and 7. As shown in this illustration, either the upper or the lower quadrant may be involved.

Commonly, an effect known as *macular sparing* can accompany occipital lesions, ie, the macular vision remains intact, whereas peripheral vision is lost. This effect is caused by the fact that the representation of the macula on the visual cortex is anatomically distinct from that of the peripheral fields, and is also much larger than that of the peripheral fields (Fig. 8.38). As a consequence, large areas of the visual cortex must be destroyed before peripheral and macular vision both are involved.

Bilateral destruction of the visual cortex in man causes practically total blindness. The primary visual receiving area (Brodmann's area 17) also has a role in visual discrimination, whereas visual association areas 18 and 19 appear to be involved with depth perception, visual orientation of the body in space, and the transmission of visual information from the visual cortex to other brain areas (see Figs. 7.71, 7.72, 7.73, and 7.77).

The nerve fibers underlying the direct light reflex (p. 475), wherein ipsilateral pupillary constriction follows a light that is shone into one eye leave the optic tracts ventral to the geniculate bodies and enter the pretectal region. Blindness in conjunction with an intact pupillary light reflex is usually caused by a lesion that lies dorsal to the optic tracts (Fig. 7.44).

COLOR BLINDNESS

Mechanisms and Classification. The most commonly used classification of color blindness is based upon the Young–Helmholtz trichomat theory of vision. If one or more particular types of cone are absent from the retina, or if such an element functions poorly, the person is unable to discern certain colors. As shown in Figure 8.17, there are three primary photosensitive pigments in the cones, each of which responds maximally to a particular range of wavelengths in the visible spectrum. If red-sensitive cones are missing from the retina, light in the range of about 525 to 625 nm can excite

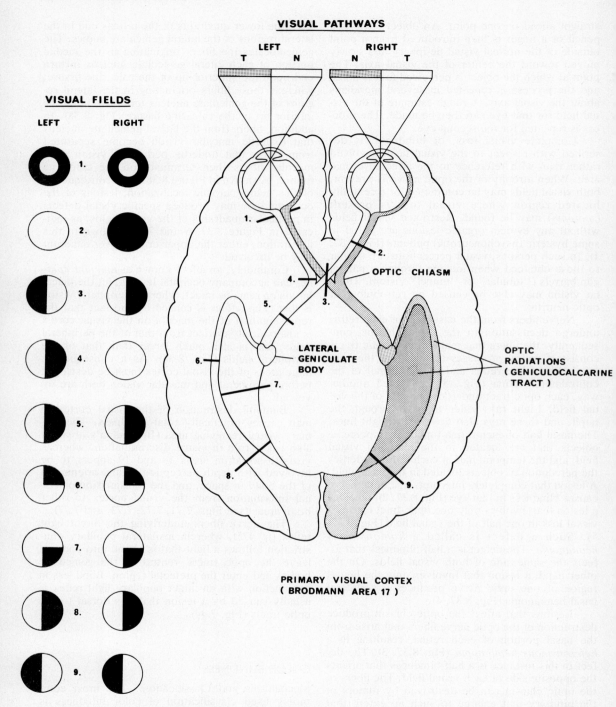

VISUAL PATHWAYS

VISUAL FIELDS

FIG. 8.37. Defects in the visual fields (black areas) produced by lesions in specific regions of the visual pathways (numbered black bars). 1. Circumferential blindness or "tubular vision." 2. Total blindness of right eye. 3. Bitemporal hemianopia. 4. Left nasal hemianopia. 5. Right homonymous hemianopia. 6. Right homonymous superior quadrantanopia. 7. Right homonymous inferior quadrantanopia. 8. Right homonymous hemianopia (defect is similar to 5, but the lesion is located closer to primary visual cortex). 9. Left homonymous hemianopia. A combination of lesions located as in 8. and 9. would result in total blindness in both eyes. See also Figure 7.44, p. 298. (Data from Chusid. *Correlative Neuroanatomy and Functional Neurology,* 14th ed, 1970, Lange Medical Publications; Gatz. *Manter's Essentials of Clinical Neuroanatomy and Neurophysiology,* 9th ed, 1970, Davis.)

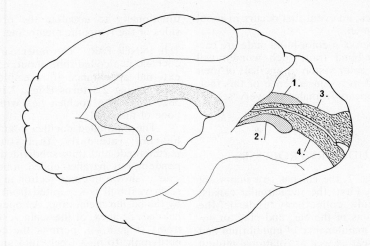

FIG. 8.38.   Projections of the retina on the primary visual cortex of the right cerebral hemisphere (medial aspect). 1. Upper peripheral retinal quadrant. 2. Lower peripheral retinal quadrant. 3. Upper macular quadrant. 4. Lower macular quadrant. Note that the total area for cortical representation of the macula is considerably larger than that of the entire peripheral retina. (After Ganong. *Review of Medical Physiology,* 6th ed, 1973, Lange Medical Publications.)

only green-sensitive cones. The ratio of excitation of the cones cannot change as the hue changes from green to red, and as the result of this effect, all colors appear identical to the individual.

If, on the other hand, green-sensitive cones are absent, colors in the range from green to red can excite only red-sensitive cones and once again the individual can perceive only one color. When a person lacks either red- or green-sensitive cones he is classified as "red-green color blind," an effect that is denoted by the suffix -*anopia*. If, however, a person has one (or more) type of anomalous cone, albeit that cone still is able to respond normally, but to a limited extent, the person is said to have a "color weakness" rather than color blindness; this effect is denoted by the suffix -*anomaly*.

The prefixes prot- (red), deuter- (green), and trit- (blue) are frequently employed to denote the particular defects in specific cone systems. Thus, persons having normal color vision as well as those having protanomaly, deuteranomaly, and tritanomaly are all called *trichromats,* as all three cone systems are present and functional, although one of these systems may be weak. Persons having only two cone systems are called *dichromats,* and such individual may have protanopia, deuteranopia, or tritanopia. Lastly, *monochromats* have only one cone system, and such persons are limited in visual experience to black and white as well as various shades of gray.

"Blue weakness" (or tritanomaly) results from a functional defect in the blue-sensitive cones, and occasionally such elements may be lacking completely in the retina *(tritanopia).* As indicated in Figure 8.17, the blue-sensitive cones are excited maximally in a spectral range that lies at much shorter wavelengths than either green- or red-sensitive elements. If the blue receptors are either defective or absent, the individual perceives a far greater proportion of green, yellow, orange, and red than of blue. Both tritanomaly and tritanopia are quite rare, however.

Etiology of Color Blindness.   Some cases of color blindness arise as complications of various ocular diseases; however, by far the majority of such defects are inherited. A total of about eight percent of all males in the human population are subject to some form of abnormal color vision, whereas only 0.4 percent of females are similarly afflicted. Deuteranomaly is most frequently encountered, followed by deuteranopia, protanopia, and protanomaly.

These abnormalities are inherited as recessive, sex-linked characteristics; that is, they are caused by a mutant gene in the X chromosome, a defect that prevents a normal development and/or function of the three photosensitive retinal elements. Since the defect in color vision is inherited as a recessive trait, color blindness will not be manifest provided that another X chromosome contains the genes essential to development of the three types of color-sensitive cones.

Human males have only one X chromosome; consequently, all three color genes that underlie the development of all three types of color photoreceptor must be present in the X chromosome in order that trichromat vision develop normally. Human females, on the other hand, have two X chromosomes, one of which is inherited from each parent; consequently, since color blindness is a recessive trait, females will exhibit a defect in color vision only when both of the X chromosomes contain abnormal genes, an event that occurs in about one out of 250 women.

Female children of a color-blind male are carriers of the color-blind trait; such women will transmit defective color vision to one-half of their male children, and as a consequence of this fact, color blindness appears in males of every second generation.

THE EAR AND AUDITORY PROCESSES

As noted in Chapter 7, the ear has two important receptor functions. First, the semicircular canals, utricle, and saccule collectively underlie the labyrinthine functions of the ear, and play an important role in the maintenance of equilibrium and the detection of linear as well as rotational motion of the head and body. The labyrinthine functions of the ear were discussed earlier (p. 388 et seq.). Second, the ear functions as the auditory apparatus subserving the sensory modality of hearing. The auditory structures of the ear perform a function generally similar to that of the labyrinthine structures in that the organ of hearing also functions as a mechanoelectric transducer. Insofar as hearing is concerned, however, sound waves rather than acceleratory and deceleratory forces are translated into nerve impulses that ultimately are perceived as sound when transmitted to appropriate cortical areas.

General Anatomy of the Ear

The auditory apparatus, shown in Figure 7.106 (p. 384), is divided into three principal regions: the external, the middle, and the inner ear. The external and middle portions of the ear are concerned primarily with the accurate transmission of sound waves from the environment to the inner ear. The inner ear contains the *organ of Corti,* which is the sensory receptor responsible for the conversion of sound energy into nerve impulses for transmission to the auditory regions of the brain.

THE EXTERNAL EAR. This region of the auditory apparatus consists of the external ear (or auricle) and the external auditory meatus (or auditory canal); the external ear is limited within the head by the external surface of the tympanic membrane (or eardrum).

THE MIDDLE EAR. Attached to the inner surface of the cone-shaped tympanic membrane is the outermost of the three bony ossicles, the malleus (hammer), which in turn is articulated with the middle ossicle, the incus (anvil); the third osseous structure of the middle ear is the stapes (stirrup), which is connected to the outer side of the oval window (or fenestra ovale) of the cochlea (Fig. 7.106).

The air-filled cavity of the middle ear is connected to the nasopharynx, hence the external environment, by the eustachian tube, which serves functionally to equalize the pressures on both sides of the tympanic membrane.

THE INNER EAR. The inner ear, also called the cochlea, is a coiled tube about 3.5 cm long, and in external appearance it resembles a spiral snail shell having 2.5 turns (Figs. 7.104 and 7.105; p. 382, 383). The cochlea lies within the temporal bone of the skull.

The fluid-filled cochlea contains the organ of hearing; internally, this structure is compartmented, and thus subdivided into three independent but parallel coiled tubes lying side by side as shown in cross-section in Figure 8.39. The scala vestibuli is separated from the scala media by the basilar membrane. An opening at the terminal end, or apex, of the scala vestibuli, the helicotrema (Fig. 8.44), permits the cochlear fluid, or perilymph, to pass freely into and from the scala tympani, which is sealed where it abuts the middle ear by the flexible round window situation adjacent to the oval window (see Figs. 7.104, 7.105, and 7.106).

The scala vestibuli is separated from the scala media (or cochlear duct) by the vestibular membrane of Reissner. The cochlear duct contains endolymph, and it is bordered by the organ of Corti, which contains the sensory epithelium or hair cells of the auditory apparatus itself, as depicted in Figure 7.40.

The Auditory Pathway

Potentials that originate in the sensory hair cells of the organ of Corti are transmitted to, and initiate impulses within, the dendritic processes of bipolar cells of the spiral ganglion located within the bony modiolus or central pillar of the cochlea (Figs. 8.39 and 8.40). These neurons form the cochlear division of cranial nerve VIII (Table 7.3, p. 248). Axons of these nerve cells enter the brain stem as the cochlear nerve at the junction of the medulla and pons (Fig. 7.28, p. 279). Fibers of the cochlear nerve then pass to the anterior and posterior auditory nuclei of the brain stem. Within these cochlear nuclei, synaptic connection is made with second-order neurons, whose afferent fibers run to the lateral lemniscus and ascend to the midbrain and thalamus. These neuroanatomic relationships are depicted in Figure 7.36 (p. 289). Note that many of these ascending auditory fibers undergo decussation to the opposite side of the brain stem; however, many ascend directly in the lemniscus on the ipsilateral side of the brain. These facts underlie the bilateral representation of the sensation of hearing in the cerebral cortex.

There are several specialized groups of cells that lie along the auditory fiber path as it passes toward the midbrain. These are the nucleus of the trapezoid body, the superior olive, and the nucleus of the lateral lemniscus. Each of these regions receives auditory fibers, as indicated in Figure 7.36.

A few fibers of the lateral lemniscus (or central acoustic tract) pass directly to the medial

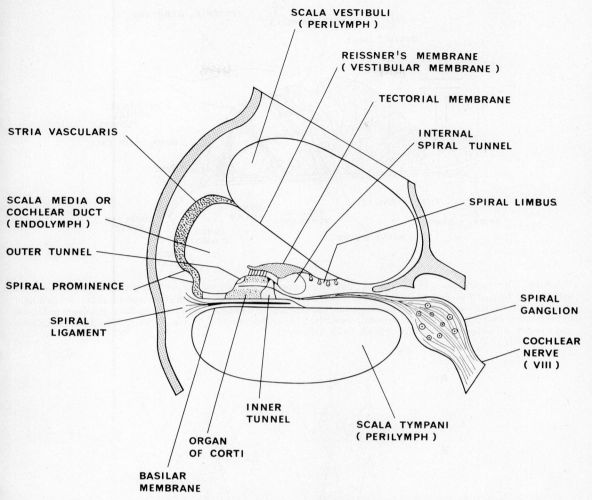

SCALA VESTIBULI
(PERILYMPH)

REISSNER'S MEMBRANE
(VESTIBULAR MEMBRANE)

TECTORIAL MEMBRANE

STRIA VASCULARIS

INTERNAL
SPIRAL TUNNEL

SCALA MEDIA OR
COCHLEAR DUCT
(ENDOLYMPH)

SPIRAL LIMBUS

OUTER TUNNEL

SPIRAL PROMINENCE

SPIRAL GANGLION

SPIRAL
LIGAMENT

COCHLEAR
NERVE
(VIII)

INNER
TUNNEL

SCALA TYMPANI
(PERILYMPH)

ORGAN
OF CORTI

BASILAR
MEMBRANE

FIG. 8.39. The relationships among the cochlear structures in cross-section. (Data from Bloom and Fawcett. *A Textbook of Histology,* 9th ed, 1968, Saunders.)

geniculate body; the remainder terminate in the nucleus of the inferior colliculus. Thence, axons proceed to the medial geniculate body via the brachium of the inferior colliculus (Fig. 7.36).

The final specialized relay stations of the auditory pathway are the medial geniculate bodies, which are located in the thalamus. Thence, afferent fibers contained in the auditory radiations connect the medial geniculate bodies with the anterior transverse temporal gyrus (of Heschl) in the cerebral cortex (Fig. 7.65, p. 321). This region (Brodmann's area 41, Figs. 7.71, 7.72, and 7.73; pp. 327, and 328) is the primary area for reception of auditory stimuli. Auditory signals are perceived in this cortical region; however, their intelligent recognition and correct interpretation depend upon their transmission to other cortical areas via association pathways.

Efferent (descending) nerve fibers also have been demonstrated in all portions of the auditory pathway, and these systems are believed to act as feedback loops. For example, the olivocochlear bundle that passes from the superior olive to the organ of Corti probably serves to inhibit impulses that arise from the latter structure, and presumably such mechanisms serve to protect the brain from excessively loud sounds (p. 511).

Mechanisms of Hearing

SOME PHYSICAL CHARACTERISTICS OF SOUND. "Sound" can be defined as the sensation perceived when alternating or cyclic phases of compression and rarefaction of the molecules in the external environment impinge upon the tympanic membrane. Since sound is produced by the longitudinal vibration of molecules in the external environment, the frequency and amplitude of these changes in pressure can be plotted against time, the result being a series of waves (Fig. 8.41).

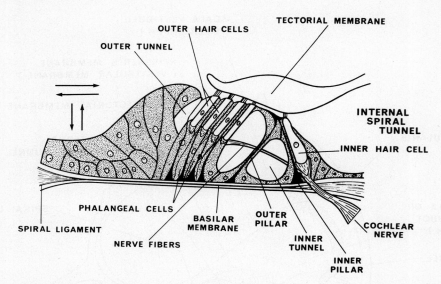

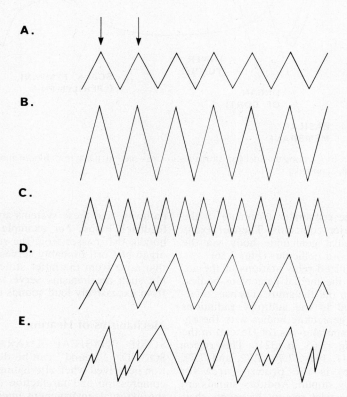

FIG. 8.40.   The microanatomy of the organ of Corti. Movements of the basilar membrane (horizontal and vertical arrows at left) induced by sound waves impart a shearing effect to the hairs of the hair cells which in turn are imbedded in a rigid structure, the tectorial membrane. This bending of the hairs excites the hair cells, and thereby auditory signals are generated in the terminal fibers of the cochlear nerve that invest the bases of the hair cells. (cf. Fig. 7.109, p. 386.) (Data from Bloom and Fawcett. *A Textbook of Histology,* 9th ed, 1968, Saunders.)

FIG. 8.41.   Some properties of sound waves. In all examples, the amplitude (vertical deflection) of the waves equals the pressure change whereas the horizontal axis indicates time. **A.** Record of a pure tone. One complete cycle is indicated by the distance between the two vertical arrows. **B.** The amplitude of this tone is twice that in **A,** hence the sound is louder. **C.** The tone has the same amplitude as in **A,** but the frequency is twice as great; thus the pitch is higher. **D.** Complex wave forms repeated regularly are the basis of musical sounds. **E.** Irregular wave forms having no rhythmic pattern are perceived as "noise." (Data from Ganong. *Review of Medical Physiology,* 6th ed, 1973, Lange Medical Publications.)

These waves are usually called *sound waves,* and the velocity of transmission of such movements in general is directly proportional to the density of the medium, and it is faster in solids and liquids than in air. The velocity of sound transmission (in air) is also slightly greater with increasing temperature or at high altitudes, the latter effect being caused principally by the higher daytime temperatures encountered far above the surface of the earth. The velocity of sound at sea level at 20 C is about 344 meters per second, whereas at the same altitude and at 0 C it decreases to 331 meters per second. In contrast to air, water transmits sound at a velocity of approximately 1,430 meters per second.

Sound transmission from source to receiver requires the presence of a physical medium such as air; consequently, no sound is transmitted in a vacuum.

In general, the *loudness* of a sound is related to the *amplitude* of the wave that produces the sound, whereas the *pitch* is related principally to the *frequency* (or cycles per second) of the waves that produce the sound (Fig. 8.41). Therefore, as the amplitude of a sound increases, so does the loudness of that sound, and the greater the frequency the higher the pitch becomes.

As depicted in Figure 8.41D, when a sound wave has a repetitive or periodic, though complex, pattern the sound is perceived as musical. On the other hand, nonrepetitive (aperiodic) vibratory waves give rise to a subjective sensation of "noise." The majority of sounds that are interpreted as "musical" are composed of a wave having a primary frequency that determines the pitch combined with a number of harmonic vibrations *(overtones);* the latter give the sound its *timbre* (or *quality*). Thus, variations in timbre permit the identification of the sound produced by different musical instruments despite the fact that all are playing notes having the same pitch.

In practice, the amplitude (loudness) of sound waves is measured on a relative scale calibrated in *decibels.* The *bel* in turn is defined as the logarithm of the ratio of the intensities of any sound to that of a standard sound. Thus, 1 bel = log (intensity of sound/intensity of a standard sound). Since the intensity of a sound is directly proportional to the square of the sound pressure, 1 bel = 2 log (pressure of sound/pressure of a standard sound). A decibel is, of course, 0.1 bel.

The standard reference level for sound that has been adopted for practical use by the Acoustical Society of America is equal to 0 decibel at a pressure of 2×10^{-4} dynes/cm². This value was chosen because it represents the auditory threshold for hearing in most human subjects. In Figure 8.42, several common sounds are listed together with their approximate decibel levels; but when examining this figure it is important to realize that the decibel scale is logarithmic and not linear. A level of zero decibels corresponds to the arbitrary standard for the intensity of the auditory

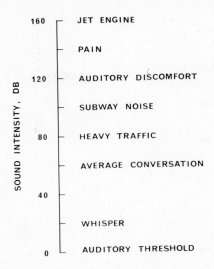

FIG. 8.42. Decibel (DB) scale for commonly-experienced sounds. The auditory threshold (0 DB) is defined as a pressure of 2.04×10^{-4} dyne/cm² (or 10^{-16} watt/cm²), and represents the auditory threshold for the average human ear. Therefore O decibels does not signify the complete absence of sound, but rather an arbitrary standard of sound intensity. (Data from Ganong. *Review of Medical Physiology,* 6th ed, 1973, Lange Medical Publications.)

threshold (ie, 2×10^{-4} dyne/cm²) and *not* a total absence of sound. The intensity of sounds that lie between threshold and those that are potentially destructive to the organ of Corti range over a 10^{12} (or trillion)-fold difference. Furthermore, it has been determined experimentally that a displacement of the tympanic membrane smaller than the diameter of a hydrogen atom (ie, 10^{-8} cm) is effective in producing an auditory sensation.*

The human ear can detect sounds lying between approximately 20 and 20,000 hertz (1 Hz = 1 cycle per sec, p. 6). Other species, particularly the dog, rat, cat, bat, and porpoise, can detect much higher frequences. For example, dogs and rats can perceive sounds that range up to 40,000 cycles per second; cats to 45,000; bats to about 100,000; and porpoises up to around 130,000.† The latter two organisms, of course, employ this capability for navigation by "radar" and "sonar," respectively.

As depicted in Figure 8.43, the auditory threshold of the human ear varies with the pitch of the sound, and the greatest sensitivity is found between 1,000 and 4,000 Hz. This sensitivity (or pitch discrimination) decreases sharply at higher and lower frequencies, however.

The pitch of the voice of an average adult male is about 120 Hz, whereas that of the average

---

*How relatively gross pressure changes of, and vascular movements within, the tympanic membrane fail to produce random or rhythmic sound is unknown.
†Some cetaceans can generate sounds having frequencies ranging up to 150,000 Hz.

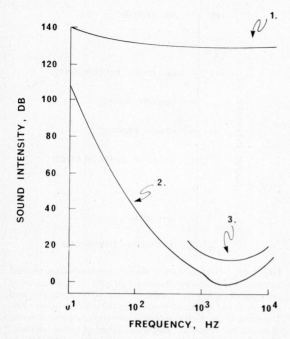

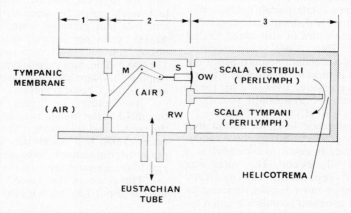

FIG. 8.43. Audibility curves for the human ear. DB = decibel. 1. Threshold for physical sensation in ear. 2. Threshold for hearing as determined under ideal conditions. 3. Curve for threshold of hearing obtained by audiometry as discussed in text. When sounds reach an intensity level in the vicinity of about 140 DB they are perceived as painful or other types of physical sensation in addition to being heard (curve 1). (Data from Ruch and Patton [eds]. *Physiology and Biophysics,* 19th ed, 1966, Saunders; Ganong. *Review of Medical Physiology,* 6th ed, 1973, Lange Medical Publications.)

female is about 250 Hz. Most persons can distinguish roughly 2,000 pitches, a far lower number than is perceptible to a musician.

GENERAL OUTLINE OF THE AUDITORY PROCESS. The ear transforms the pressure changes that are the essential feature of sound waves into physical movements of the tympanic membrane (Fig. 8.44). These movements of the tympanic membrane in turn are translated into movements of the auditory ossicles so that the footplate of the stapes induces movements of the oval window, and these movements are transmitted as waves to the endocochlear fluids. The pressure waves induced in the cochlear fluids then cause movements of the basilar membrane of the organ of Corti, which is located within the cochlear duct (Fig. 8.39), and thus a shearing or twisting motion is induced in the processes of the hair cells whose distal tips are embedded in the tectorial membrane (Fig. 8.40). The mechanical deformation of the hairs is trandsuced by the hair cells into nerve impulses within the terminal fibers of the cochlear branch of the vestibulocochlear nerve (VIII), and when these signals are transmitted to appropriate cortical areas, as discussed earlier, they are perceived as sound.

EXTERNAL SOUND TRANSMISSION. The structrues of both the external and the middle portions of the auditory apparatus serve to transmit sound waves to the cochlea (or inner ear) with very little distortion.

Sound waves impinging upon the external surface of the tympanic membrane induce in-and-out oscillations of this structure which are faithfully transmitted to the ossicles of the middle ear. The tympanic membrane thus functions as a *resonator,* as it is able to duplicate with surprising accuracy the vibration produced at the sound source. The tympanic membrane also stops vibrating almost instantly when the sound ends; therefore, this structure is said to be *critically damped* for all practical purposes.

The malleus is tightly connected to the incus

FIG. 8.44. The physical transmission of sound waves from the external environment to the inner ear. The pathway followed by the sound waves is indicated by the arrows. 1. External ear; 2. Middle ear (M = malleus; I = incus; S = stapes); 3. Inner ear. OW = oval window; RW = round window. The scala media and organ of Corti are not indicated in the inner ear. The double vertical arrows at the level of the middle ear (2) indicate that the eustachian tube is the route whereby the air pressure within the middle ear is equalized with that of the external environment. See also Figure 7.106 (p. 384).

by ligaments, and the latter structure in turn is articulated with the stapes (Fig. 8.44). The footplate of the stapes is attached to the oval window of the cochlea. The malleus and incus usually move as a unit or lever system whose fulcrum lies at the edge of the tympanic membrane, whereas the stapes is articulated with the incus in such a fashion that the back-and-forth rocking movement of the footplate is imparted directly as an in-and-out motion to the oval window of the cochlea by the mechanical pressure of the sound waves.

The *tensor tympani* is a tiny muscle that is innervated by efferent fibers of the mandibular nerve, a branch of the trigeminal nerve (V). It is attached to the handle of the malleus, and through its contraction exerts tension so that the tympanic membrane is stretched at all times. Therefore, sound waves striking any part of the typanum are transmitted to the inner ear.

The ossicles of the middle ear do not amplify the actual movements of the stapes and the membrane of the oval window. Rather, the amplitude of the deflections is diminished about 25 percent below that generated at the handle of the malleus. The ossicles, however, are organized to form a single lever system (Fig. 8.44); therefore, the force exerted upon the oval window by movements of the stapes is approximately 1.3 times greater than that produced by sound waves impinging upon the tympanum. Furthermore, the average surface area of the tympanic membrane is 55 mm², and that of the stapes in contact with the oval window about 3.2 mm², a ratio of about 17:1. The mechanical advantage between tympanum and oval window is 1.3 × 17 or 22-fold; consequently, the pressure applied to the cochlear fluid is approximately 22 times greater than is exerted by the pressure of a sound wave acting directly upon the tympanic membrane. This so-called *impedance matching* in the ear is important, because more dense substances (such as liquids) require far greater forces to achieve the same wave motion that is induced in less dense media (such as air) by the action of relatively weak forces. As a consequence of these physical facts, the matching of sound waves in the air and in the cochlear fluid is about 50 to 75 percent accurate between the range of about 300 to 3,000 cycles per second.

The *resonant frequency* of a vibrating system depends upon two components, the inertia as well as the elasticity, and this physical property can be defined as that frequency at which the greatest oscillatory response is obtained with the smallest energy input. Hence, at its resonant frequency any system can vibrate most readily. The ossicles and suspensory ligaments of the ear have a broadly tuned natural resonant frequency that ranges between about 700 and 1,400 cycles per second. This natural resonant frequency, however, is damped by the viscosity of the ligaments and muscles attached to the ossicles; therefore, excessive resonance is prevented, and sound waves in the vicinity of 1,200 cycles per second can be transmitted

through the external and middle regions of the ear with greatest facility.

The external auditory meatus also functions as an air column resonator, having a natural resonant frequency of about 3,000 cycles per second. If this resonant effect is combined with that of the system of ossicles, the transmission of sound from the external environment to the cochlea in the human ear is most efficient between frequencies of 600 and 6,000 cycles per second, and transmission is less effective both above and below these limits.

When sound waves are conducted to the fluid of the inner ear by way of the tympanic membrane and the auditory ossicles, the process is called *ossicular conduction*. Sound waves can also induce oscillations in the membrane that closes the round window, a process called *air conduction*. Air conduction is of no importance to normal hearing, and ossicular conduction ordinarily is the physiologic route for the transmission of sound from the environment to the inner ear. A third type of sound conduction is present in the ear, viz., *bone conduction*. Since the cochlea is embedded in the temporal bone of the skull, vibrations of a tuning fork or other vibrating objects placed directly on the skull produce considerable bone conduction. Exceedingly loud sounds can also be transmitted to the inner ear via this route.

The *tympanic,* or *sound attenuation reflex* is important primarily in preventing damage to the cochlea that might be caused by the physical impact of the extremely loud sounds that also serve to initiate this response.* Thus, when the small muscles located in the inner ear (ie, the tensor tympani and the stapedius) contract simultaneously, the handle (manubrium) of the malleus is moved inward, whereas the footplate of the stapes is drawn outward, and thereby sound transmission through the middle ear is decreased considerably as the simultaneous contraction of these muscles causes the normally flexible system of auditory ossicles to become quite rigid. Sound transmission is thereby decreased appreciably, especially at lower frequencies. The tympanic reflex, which has a latent period of only 10 msec, also permits the ear to adapt automatically to sound of quite different intensities such as are encountered frequently during the course of everyday living.

COCHLEAR SOUND TRANSMISSION. As indicated in Figures 7.106, 8.39, and 8.40, the inner ear consists of three fluid-filled tubes, two of which (the scala vestibuli and the scala tympani) are con-

---

*It has been demonstrated experimentally that stimulation of the efferent fibers of the olivocochlear bundle also decreases the response of the cochlear nerve to auditory stimulation. Presumably, this effect provides another protective mechanism against overstimulation of the cochlea in vivo by extremely intense sounds. (See also p. 344 for a general discussion of peripheral inhibition of afferent inputs.)

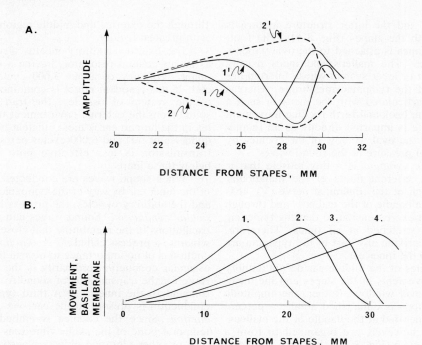

FIG. 8.45. Traveling waves in the cochlea. **A.** Curves 1 and 1′ represent a traveling wave at two successive moments in time. Curves 2 and 2′ show the net amplitude of the waves at any particular instant. The vertical axis indicates the relative amplitude of the waves. **B.** Physical movement (or displacement) of the basilar membrane induced by four particular sound frequencies in relation to the point of maximum excitation of the organ of Corti as measured from the stapes. 1. 1,600 Hz; 2. 800 Hz; 3. 400 Hz; 4. 50 Hz. Note that the higher frequencies are most effective near the base of the cochlea whereas the lower frequencies are more effective near the apex. These facts provide the basis for the place theory of cochlear function. (After Ganong. *Review of Medical Physiology,* 6th ed, 1973, Lange Medical Publications.)

nected via a passage at their apex, the helicotrema (Fig. 7.45). Pressure exerted on a confined fluid is transmitted equally throughout that fluid (Pascal's law, p. 4). Since the membranous structures of the cochlea are confined within rigid bony cavities in the skull, a gradual pressure that is exerted slowly on the oval window by the stapes causes perilymph to move from the scala vestibuli through the helicotrema into the scala tympani and the round window bulges outward into the middle ear cavity, as depicted in Figure 8.44. If, however, the stapes vibrates rapidly, no *net* movement of fluid can occur in the brief interval between the successive oscillations in pressure; instead, the fluid wave is transmitted to the basilar membrane and causes this structure to rock back and forth with the vibrations as indicated by the arrows in Figure 8.40. The generation of such *traveling waves* in the cochlea is depicted in Figure 8.45A.

There are about 20,000 *basilar fibers* that extend outward from the central pillar, or *modiolus,* of the cochlea. Like harp strings, these fibers exhibit a graded range in length from about 0.04 mm at the cochlear base to 0.5 mm at the helicotrema. An important physiologic consequence of this anatomic fact is that the shorter basil fibers that are located near the cochlear base are stimu-

lated more readily by high-frequency vibrations than are the longer fibers at the helicotrema, which are stimulated to move more readily by low-frequency vibrations (Fig. 8.45B).

A second factor involved in the graded differential response of the cochlea to sound waves of different frequencies from base to helicotrema lies in the degree of fluid loading imposed upon the basilar membrane. In turn, this fluid loading produces an unequal inertia throughout the length of the organ of Corti. This effect of inertia is lowest near the oval window and greatest at the helicotrema. Thus the so-called *volume elasticity** of the basilar fibers is 100 times smaller at the base of the cochlea than at the helicotrema, a fact that accounts for a range of sensitivity to different resonant frequencies that amounts to around seven octaves.

The pattern of traveling waves of several frequencies is shown in Figure 8.45B. Note that a sound wave of any particular frequency has a low amplitude as it enters the cochlea, becomes progressively stronger, and then attains a maximum intensity when it reaches the area having the same maximum resonant frequency as itself within the

---

**The product of basilar fiber length times the loading factor.*

basilar membrane. At this point, the energy of the wave is dissipated in imparting movement to that particular area of the basilar membrane, and thus the wave travels no further along the cochlea. Consequently, the primary method whereby discrimination among sounds of different frequencies is achieved by the inner ear lies in the specific *pattern of stimulation* in the fibers of the cochlear nerve within specific regions of the basilar membrane; that is, the distance from the oval window to the point of maximum excitation of the cochlear nerve is related inversely to the frequency of the vibrations that initiate the wave. Thus, high-frequency sounds generate waves that reach their peaks near the base of the cochlea, whereas low-frequency sounds induce waves that reach their maxima near the apex of the cochlea (Fig. 8.45B).

The walls of the scala vestibuli are rigid; however, Reissner's membrane is quite flexible, and similarly the basement membrane is also flexible. Consequently, the peaks of waves generated in the perilymph of the scala vestibuli depress both Reissner's membrane as well as the basilar membrane toward the scala tympani, and such displacements of fluid as those occurring within the perilymph of the scala tympani as neutralized by expansion of the round window into the air of the middle ear. Hence, sound waves produce a deformation of the basilar membrane, and the region of the basilar membrane where the distortion achieves a maximum is determined by the pitch (frequency) of the sound wave. This statement is known as the *place theory of cochlear function.*

PHYSIOLOGY OF THE ORGAN OF CORTI. The organ of Corti lies on the basilar membrane as shown in Figure 8.40. The sensory hair cells within the organ of Corti that lie upon, and are stimulated by, movement of the basilar membrane, are of two types. First, there is a single row of *internal hair cells,* 12 μ in diameter; the entire organ of Corti contains about 3,500 of these elements. Second, there are three or four rows of *external hair cells,* which average 8 μ in diameter; about 20,000 of these cells are present in toto. The bases of both of these types of hair cells are entwined closely with a network of endings of the cochlear branch of cranial nerve VIII, fibers of which pass to specific loci within the *spiral ganglion of Corti,* which is located within the modiolus of the cochlea (Fig. 7.105, p. 383). Axons from the spiral ganglion in turn send fibers to the cochlear nerve, and from there to the central nervous system, as described earlier in this chapter (p. 506).

The superficial portions of the hair cells are held firmly within the *reticular lamina,* which is composed of surface extension of the pillar, while the distal ends of the hairs at the free surface are embedded within the tectorial membrane, as shown in Figure 8.40. The reticular lamina is rigidly attached to the basilar fibers via the *pillars* (or *rods*) *of Corti,* so that up-and-down movements of the basilar membrane rock the tips of the hair cells back and forth, thus inducing torsion or

bending in the hair cells. This mechanical deflection in turn induces excitation in the fibers of the cochlear nerve which invest the bases of the hair cells.

DEVELOPMENT OF RECEPTOR POTENTIALS IN THE HAIR CELLS. Changes of electric potential are produced in the hair cells by the alternate to-and-fro bending of the hairs as the basilar membrane is rocked by sound waves. How this receptor potential develops in the hair cells by mechanical deformation and then stimulates the nerve endings that envelop these structures is unknown. There may be direct electrical excitation of the neuronal endings, and/or possibly a chemical transmitter is involved in the transduction process (cf., p. 188).

In any event, it is clear that when the basilar fibers bend upward, that is, toward the scala vestibuli, the membrane potential of the hair cell is decreased; ie, the cell depolarizes. The opposite situation takes place when the basilar membrane moves down; ie, a hyperpolarization of the membrane of the hair cell is induced. Afferent impulses are generated in the fibers of the cochlear nerve when the hair cells depolarize; however, impulse generation ceases immediately when the hair cells hyperpolarize.

THE ENDOCOCHLEAR POTENTIAL AND THE COCHLEAR MICROPHONIC. In addition to the intermittent generator potentials developed within the individual hair cells following their deformation, the scala media (or cochlear duct) and tectorial membrane as a whole also have a steady potential which is some 80 mv positive to the perilymph, even when the ear is not stimulated by sound waves (Fig. 8.46). Energy for the maintenance of this *endocochlear potential* is supplied by a sodium pump located within the stria vascularis. This potential of +80 mv, measured with respect to the perilymph within the scala vestibuli, is not dependent upon the ionic composition of the endolymph per se (Table 8.3), but it is highly sensitive to a reduced oxygen supply. Static displacement of the basilar membrane can alter the endocochlear potential up or down by several milli-

Table 8.3 APPROXIMATE COMPOSITION OF ENDOLYMPH, PERILYMPH, AND CEREBROSPINAL FLUID[a]

SUBSTANCE	ENDOLYMPH	PERILYMPH	CSF
K+	145	5	4
Na+	16	150	152
Cl-	107	122	122
Protein	15	50	21

Note: Ionic constituents are expressed in mEq/liter; protein in mg/100 ml.

[a]*Data from Ganong.* Review of Medical Physiology. *6th ed, 1973, Lange Medical Publications.*

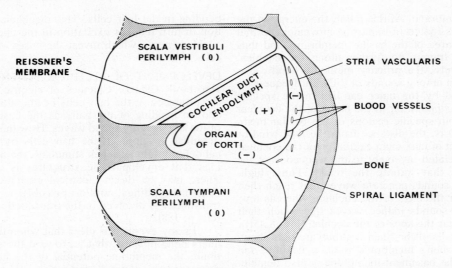

FIG. 8.46.   Cross-section of the cochlea indicating the localization of the positive endocochlear potential (within the heavy line) in relation to the perilymph (O). The negative signs within the organ of Corti and stria vascularis denote the fact that the intracellular potentials of these elements are negative with respect to the perilymph. (After Ganong. *Review of Medical Physiology*, 6th ed, 1973, Lange Medical Publications.)

volts. When the −70-mv transmembrane potential of the hair cells with respect to the perilymph is added to the +80-mv potential of this fluid, there is a *net* potential difference between the upper margin of the hair cells and the endolymph amounting to 150 mv. This high electric potential is believed to increase the sensitivity of these cells to mechanical deformation of their hairs.

When a sinusoid or a complex sound wave (eg, music) is applied to the tympanic membrane, a potential is developed in the vicinity of the cochlea which can be recorded, and this potential closely reflects or "mirrors" the wave form used as the stimulus. This potential is called the *cochlear microphonic*,* and it is caused by stimulation of the outer hair cells. The cochlear microphonic is not a series of action potentials in the auditory nerve, as discussed below, although the microphonic can be recorded from this source as the waves spread decrementally along the auditory nerve (unlike true action potentials) as well as outward from the cochlea. Furthermore, the cochlear microphonic appears without any apparent threshold and with an insignificant latency. Presumably, a piezoelectric effect in the hair cells is involved in the generation of the cochlear microphonic, as this response is proportional to the magnitude of the displacement of the basilar membrane.

Detailed study and analysis of the cochlear microphonic recorded from various regions of the

Since the cochlear microphonic so accurately reproduces the frequency and amplitude of the sound being played, it is possible to reproduce a song played to an experimental animal merely by feeding the cochlear microphonic directly into an audio amplifier.

inner ear have provided direct evidence for the standing wave theory of cochlear transmission discussed above, and also these data substantiate the place theory of tonal localization in the cochlea, which was also considered earlier.

ACTION POTENTIALS IN THE AUDITORY NERVE.   Both mechanically and electrically, the cochlea behaves as an audio-frequency (or acoustic) analyzer, the analysis being encoded in all-or-none impulses that are generated in the individual fibers of the auditory nerve.

Although the physiologic mechanism whereby depolarization of the membranes of the hair cells excites impulse generation in the individual terminals of the auditory nerve is unknown, it appears that the cochlear microphonic, together with the summated negative (ie, depolarizing) potentials of the hair cells, functions cooperatively to produce a generator potential. Presumably, the endocochlear potential serves to maintain the receptor elements in a highly excitable state by continuously generating a charge separation between the hair cells and the hairs themselves.

As discussed earlier, the principal factor that determines the subjective pitch of a sound is the site (or place) in the organ of Corti that is stimulated maximally by the sound waves. Hence, the traveling wave induced by a given tone causes a maximal deflection of the basement membrane in one region of the organ of Corti; therefore, maximal receptor excitation is induced in the same region. The linear distance between the maximally stimulated region and the oval window in turn is related inversely to the frequency of the sound waves; therefore, sounds with a high pitch (frequency) produce a maximal stimulation at the base of the cochlea, whereas sounds with a low pitch induce maximal excitation at the apex of the cochlea. The neural pathways leading from specific regions of the organ of Corti to the brain also exhibit a high degree of spatial orientation.

An additional factor appears to be involved in the perception of different pitches when the frequencies of the sounds lie below 2,000 Hz. This factor is the actual pattern of the signals in the auditory nerve which is generated by the sound. Thus, at low frequencies, each nerve fiber responds by generating one action potential per sound wave; the ratio between the frequency of the sound waves and the frequency of the nerve impulses generated thereby is 1:1, a phenomenon known as the *volley effect*. As the frequency of the sound waves increases, however, the enhanced impulse frequency in the auditory nerve is perceived mainly as an increase in the intensity (or loudness) of the sound rather than as an increase in the pitch. In turn, there are three general mechanisms whereby the frequency of the impulses generated within the auditory nerve can be increased in direct proportion to the intensity of the sound: (1) As the intensity or amplitude of a sound increases, the amplitude to which the basilar membrane is displaced also increases. (2) As the intensity of the sound increases, spatial summation takes place within the organ of Corti as progressively greater numbers of hair cells on the borders of the primary oscillating region of this structure are stimulated. (3) Some hair cells are excited only when the sound intensity reaches a sufficiently high level and as the excursions of the basilar membrane likewise reach a certain amplitude.

Changes in the intensity of a sound are detected by the ear in proportion to the cube root of the sound intensity.* This fact contrasts sharply with the mere perception of sound intensity (or loudness) which, as noted above, is directly related to the frequency of the sound waves above a certain level. Thus, the ear can detect changes in the intensity of sound from softest to loudest that encompass a one trillionfold range (or a 10^{12}-fold range of intensities). This extraordinarily huge range is compressed into a ten thousandfold (10^4) range, however, through the operation of the power law noted above. The bel system for measuring sound intensities empirically was discussed on p. 509 (Fig. 8.42).

CENTRAL NEURAL MECHANISMS AND AUDITION. The neuroanatomic features of the auditory pathway were presented in detail earlier (p. 284 et seq.). From a functional standpoint, the responses of single second-order neurons in the auditory pathway to stimulation by sound waves are similar to those recorded from individual auditory nerve fibers; however, one important difference is

present between the two types of neuron. A sharper decline in the electric activity is seen in the second-order (medullary) neurons than is present in the first-order neurons of the auditory nerve as the frequency of the sound increases. As the intensity of the sound is increased, however, the spectrum of frequencies to which a response occurs becomes broader in both the first- and second-order neurons. Presumably, the greater functional specificity of the second-order medullary neurons of the auditory pathway is produced by an inhibitory mechanism (presently unknown) that is located within the brain stem, and which "turns off" the activity in the second-order neurons.

The crossed and uncrossed neural pathways that ascend to the auditory cortex from the cochlea possess an extremely high degree of spatial organization (p. 343). In human subjects, sounds having a low pitch are perceived in the anterolateral portion of the auditory cortex, whereas sounds having a high pitch are represented posteromedially in the same general region of the brain (Figs. 7.71, 7.72, and 7.73).

Individual neurons in the auditory cortex respond not only to a definite pitch, but also to such factors as the direction from which an auditory stimulus originated, as well as to the beginning, duration, and frequency of the stimulus. There is good experimental evidence to indicate that the primary auditory cortex is not responsible for the perception of sounds, at least in some species. Consequently, the primary auditory cortex in man may be involved principally in the recognition of patterns of sounds, with the analysis of sound characteristics, and with the localization in space of particular sounds.

Insofar as sound localization is concerned, two factors appear to be important, depending upon the frequency of the sound. At frequencies below 1,000 Hz, the temporal difference between the arrival of sound waves emanating from the same source in the two ears (ie, the phase difference) is of major importance in localizing the external source of a sound. On the other hand, at frequencies above 1,000 Hz, the difference in the intensity of the sounds impinging simultaneously upon the two ears seems to be the major factor in localizing an external sound. In this instance, of course, the loudest sound perceived would be that in the ear lying closer to the source of the sound.

As already noted, a large number of neurons in each side of the auditory cortex receive inputs from one cochlea, and thus are able to respond either to a maximal or to a minimal extent when sounds are out of temporal phase when they arrive at each ear (Fig. 7.36, p. 289). Hence, in one ear, the response of the neurons on both sides of the cortex which receive inputs from that ear is delayed by a fixed time interval relative to the arrival time of the sound at the contralateral ear. The fixed time interval can vary, of course, among individual neurons — a factor that lends considerable flexibil-

---

*The so-called **power-law** *states that alterations in the intensity of any stimulus are interpreted by, and in terms of, some power function of the stimulus intensity.*

ity to the accurate perception of sounds that arise in different external regions.

In addition to the neural factors enumerated above, the sounds that originate from in front of a person differ somewhat in quality from those that originate behind the individual, and it appears that these differences are caused by the forward tilt of the external ears on each side of the head, a purely physical factor that assists in sound localization.

As might be anticipated, lesions of the auditory cortex can completely derange the ability to localize sounds, an effect that can be produced experimentally in many species of animal. Pathologic lesions in the human auditory cortex produce a similar effect in man.

Clinical Correlates

DEAFNESS. Loss of the sense of hearing can range from a minor inability to perceive certain particular sound frequencies to complete deafness. From a clinical standpoint, deafness may be caused by an impairment of sound transmission in the external or in the middle ear, an effect that is called *conduction deafness*. Alternatively, deafness may be caused by damage to the organ of Corti or to the specific neural pathways that subserve hearing, and this type of defect is called *nerve deafness*.

Conduction deafness can be produced by a large accumulation of ear wax (cerumen) in the external auditory meatus or by foreign bodies, either of which can obstruct sound transmission in this part of the auditory system. Similarly, destruction of the auditory ossicles or massive scarring and thickening of the tympanic membrane caused by repeated infections in the middle ear *(otitis media* or *tympanitis)* will also produce a marked derangement of normal hearing. If the attachment of the footplate of the stapes to the oval window is too rigid, sound transmission will also be impaired. Hearing aids, which amplify sounds before transmitting them to the cochlea via bone conduction (p. 511), are of value primarily in compensating for a hearing loss of the conduction type, as such a device cannot substitute for a damaged cochlea or auditory nerve.

Nerve deafness can be caused by a number of factors including heredity (in many cases, deaf–mutism is inherited as a simple autosomal recessive), toxic degeneration of the auditory nerve induced by streptomycin, Meniere's disease (p. 463), medullary hemorrhages, and by tumors that involve the auditory nerve or cerebellopontine angle. Furthermore, a loss of the ability to perceive specific tones can be produced by exposure to extremely loud sounds for prolonged intervals. Such "occupational deafness" with attendant cochlear and/or nerve damage can develop in persons such as riveters; workers using pneumatic chipping tools in shipyards or boiler factories, where the steel amplifies the sounds; rock musicians; jet aircraft mechanics; jackhammer operators; and so forth. Properly designed ear plugs afford some protection to persons engaged in these occupations, as such devices effectively screen out sounds that otherwise would enter the external auditory meatus.*

Conduction deafness and nerve deafness can be distinguished quite readily by a number of simple tests that employ a tuning fork. For example, in order to test an ear for bone conduction, a vibrating tuning fork is placed in front of the ear. When the subject informs the observer that he no longer can hear the vibrations, the butt of the still-vibrating fork is placed immediately against the mastoid process. If bone conduction is better than air conduction, the subject will hear the sound made by the tuning fork once again, and conduction deafness is considered to be present. On the other hand, if the subject cannot hear the sound of the weakly vibrating tuning fork when the butt is applied to the mastoid process, bone conduction probably has decreased as much as air conduction. Consequently, any deafness that is present in the ear presumably is caused by cochlear or neural damage rather than by a conduction defect, and nerve deafness is assumed to be present.

An objective measure of auditory acuity whereby the clinician can obtain an accurate picture of the hearing defects present in a given patient is by use of an instrument called an *audiometer*. This device consists of one earphone connected to an electronic oscillator that is able to generate pure tones that range from low to high frequencies. The instrument is so calibrated that the zero intensity level of the sound emitted at each frequency is loud enough just to be heard by a normal ear. The volume control on the audiometer, however, is calibrated in such a manner that the intensity of each tone can be altered to increase the loudness either above or below zero for every tone. Thus, if the volume of a particular tone must be increased 20 decibels above normal before the subject is able to perceive it, the subject is considered to have a hearing loss of 20 decibels for that specific tone.

Generally, when one carries out a hearing test with an audiometer, about 10 tones that cover the entire auditory spectrum are tested individually with each ear. The hearing loss is determined for each tone; then it is plotted on a chart, and the resulting audiogram depicts the hearing loss throughout the entire auditory spectrum of that individual.

The audiometer also contains an electronic vibrator, in addition to an earphone, and this de-

---

In this regard it should be mentioned that ear plugs should never *be worn while diving, even to very shallow depths, as any pressure differential that is developed between the water and the external and/or middle ear could force the plugs through the tympanic membrane and thereby completely destroy the integrity of the auditory ossicles. Conduction deafness would result.*

vice is used for testing bone conduction from the mastoid process to the cochlea.

THE CHEMICAL SENSES

Gustation

The sensation of gustation, or taste, arises from impulses that are generated within receptor cells of specialized end-organs, the taste buds, which are located in the mouth and found primarily on the tongue. There are at least four basic sensations of taste: sweet, sour, salty, and bitter. The taste buds may be considered as chemoelectric transducers which respond in varying degress to stimulation by various specific chemical substances in solution that come into direct contact with the taste buds. Presumably, the extremely broad range of flavors (or tastes) that can be perceived by an average person represent a subtle "blending" of the four basic taste modalities, much as the eye can perceive an enormous range of colors even though only three specific types of photopic visual receptor are present in the retina. Furthermore, it is possible, although not proven experimentally, that there are many less obvious subclasses of the four sensations of taste.

LINGUAL PAPILLAE AND MORPHOLOGY OF THE TASTE BUDS. The anterior dorsal surface of the tongue is covered by numerous small protuberances called *papillae*. The papillae are of three principal types as illustrated in Figure 8.47. First, the *filiform papillae* are arranged in diverging rows to the right and left of the midline, and they lie parallel to the V-shaped gustatory region. Second, the *fungiform papillae* are scattered individually among the filiform papillae, and these structures are especially numerous near the tip of the tongue.

Third, the large *vallate* (or *circumvallate*) *papillae* are situated along the arms of the V-shaped gustatory region, toward the back of the tongue. Only ten to twelve of these structures are present on the human tongue. In addition to these three gross types of lingual papillae, a few paired *foliate papillae* are found in a rudimentary condition upon the posteriolateral surfaces of the tongue. In adult humans, there are in toto about 10,000 taste buds, a number that tends to decrease beyond middle age.

The taste buds are all similar histologically, regardless of the specific primary taste sensation that is elicited most readily from each region of the tongue (Fig. 8.48). Taste buds are found in large numbers upon the circumvallate papillae, with moderate frequency on the fungiform papillae all over the surface of the tongue, and in moderate numbers in the foliate papillae.

Scattered taste buds are also located on the soft palate, the glossopalatine arch, on the posterior wall of the pharynx, and on the posterior surface of the epiglottis.

Microscopically, each taste bud appears as a pale ovoid body, about 72 μ long; it is imbedded in the surrounding epithelium (Fig. 8.49). Each taste bud opens upon the epithelial surface through a small outer *taste pore*. Two cell types are present, the spindle-shaped *supporting cells* and the *neuroepithelial taste cells*. The latter are distributed among the supporting cells, and there are between 4 and 20 of these elements per taste bud. The neuroepithelial cells are rod-shaped and slender with a central nucleus and have a short taste hair (3

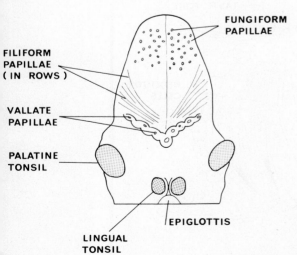

FIG. 8.47. Dorsum of tongue showing the general arrangement of the papillae upon which the taste buds are located. (After Bloom and Fawcett. *A Textbook of Histology,* 9th ed, 1968, Saunders.)

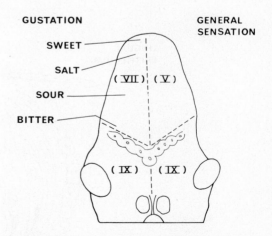

FIG. 8.48. Sensory innervation of the various regions of the tongue. The general areas of this structure within which the primary taste sensations can be elicited most readily also are indicated. V = trigeminal nerve; VII = facial nerve; IX = glossopharyngeal nerve. Afferent fibers of these cranial nerves which subserve both gustation as well as general sensations from the tongue course to the specific central nuclei within the brain in the same nerves (cf. Fig. 8.50). (After Ganong. *Review of Medical Physiology,* 6th ed, 1973, Lange Medical Publications.)

μ long by 0.2 μ in diameter) on their free end which projects through the lumen of the outer taste pore into the oral cavity. The taste hairs are believed to be the actual receptor surfaces for the initiation of gustatory sensations. The base of each taste cell is closely invested in a network of terminal nerve fibers which are stimulated by the taste cells.

Interestingly, the taste buds degenerate completely when denervated; should neurons subsequently grow to the epithelial surface of the tongue, new taste buds develop from unspecialized epithelial cells in the vicinity of the nerve terminals. These facts again illustrate the important trophic influence of innervation on certain structures (cf. the effects of denervation of skeletal muscle, p. 130).

NEUROANATOMY OF GUSTATION. Afferent impulses from the taste buds are received and transmitted to the brain via three separate neural pathways, each of which serves a different general area of the mouth and tongue (Table 7.3, p. 248; Figs. 8.48 and 8.50): (1) Impulses from the anterior two-thirds of the tongue pass first into the trigeminal nerve (V), from there via the chorda tympani to the facial nerve (VII), and then into the tractus solitarius within the brain stem (p. 275). (2) Gustatory sensations from the circumvallate papillae as well as the posterior regions of the mouth are transmitted via afferent fibers of the glossopharyngeal nerve (IX) to the tractus solitarius. (3) Some impulses subserving taste that arise from the base of the tongue (as well as the pharynx) are transmitted to the tractus solitarius by way of the vagus nerve (X).

All nerve fibers that subserve the sensation of taste converge on the nucleus of the tractus solitarius, where they synapse; second-order neurons then pass to a small region within the thalamus that is located medially to the terminations of the dorsal column-medial lemniscal system (Table 7.7, p. 294). Third-order neurons that originate in the thalamus then transmit gustatory impulses to the cerebral cortex, probably to the parietal opercular-insular area. This region lies at the lateral margin of the postcentral gyrus within the sylvian fissure (Fig. 7.74, p. 329).

Efferent impulses that arise in the tractus solitarius and pass directly to the superior and inferior salivatory nuclei provide motor impulses for the secretion of saliva by the parotid and submaxillary glands (Fig. 7.26, p. 274). The regulation of this process is considered in detail in Chapter 13 (p. 846).

PHYSIOLOGY OF GUSTATION

Chemical Stimuli and the Primary Gustatory Sensations. The effective stimuli that are required to elicit each of the primary gustatory sensations can be considered individually. The general localization of these sensations upon the tongue is shown in Figure 8.48.

Sweet. No single group of chemical substances evokes the sensation of sweetness. A large number of organic compounds belonging to many diverse chemical classes produce this sensation. Thus glycols, alcohols, sugars, amino acids, sul-

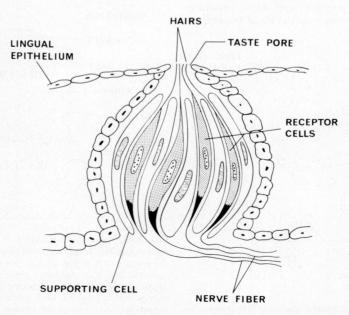

FIG. 8.49. Diagram of a taste bud in cross-section. Morphologically all taste buds are similar regardless of their differing functional sensitivity in different regions of the tongue (Fig. 8.48). (After Ganong. *Review of Medical Physiology,* 6th ed, 1973, Lange Medical Publications.)

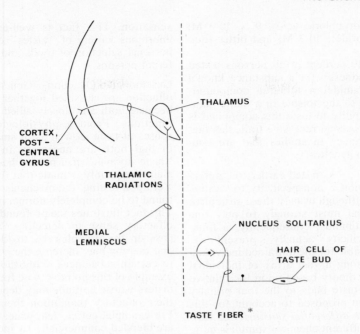

CORTEX,
POST –
CENTRAL
GYRUS

THALAMUS

THALAMIC
RADIATIONS

MEDIAL
LEMNISCUS

NUCLEUS SOLITARIUS

HAIR CELL OF
TASTE BUD

TASTE FIBER *

FIG. 8.50. The general neural pathway subserving the sensation of taste. *Taste fibers pass to the cortex via branches of the facial (VII) and glossopharyngeal (IX) nerves as indicated in Figure 8.48. The midline is indicated by the vertical dashed line. (After Ganong. *Review of Medical Physiology*, 6th ed, 1973, Lange Medical Publications.)

fonic acids, ketones, aldehydes, halogenated acids, amides, and esters all "taste" sweetish in varying degrees. Although proteins in general elicit no gustatory sensation, several specific polypeptides have recently been demonstrated to elicit a sweet taste, whereas others have been shown to produce bitter or sour sensations.

Among inorganic compounds, only certain salts of lead and beryllium taste sweet; these compounds are, incidentally, highly toxic. It is interesting to note that saccharin, which is totally unrelated chemically to common table sugar (sucrose), exerts a sweetening effect that is approximately 600 times greater than sucrose.

Sour. Acids are responsible for producing a sour taste, and the intensity of the sensation evoked is roughly proportional to the logarithm of the hydrogen ion concentration. The hydrogen ion (H^+) itself rather than the anion excites the receptor elements.

Salty. The anions of ionized salts are mainly responsible for producing a salty taste. The "quality" of the salty taste varies considerably, because different inorganic salts excite fundamental taste sensations other than saltiness alone. The cationic portions of such compounds may also contribute to the overall sensation evoked. In general, halogens are particularly effective in eliciting a salty taste, although some organic compounds also have a salty taste.

Bitter. The chemical substances that evoke a bitter sensation are primarily organic compounds belonging to two major chemical classes: alkaloids

and long chain organic substances. Some inorganic compounds also taste bitter. Thus, compounds such as quinine, strychnine, morphine, nicotine, urea, caffeine, and colchicine, as well as calcium, ammonium, and magnesium salts, all taste bitter despite the fact that several of these substances have little in common from a strict chemical standpoint.

Localization of Taste Sensations. Sweet substances usually are perceived and localized most clearly at the tip of the tongue, sour at the sides, and bitter at the back (Fig. 8.48). The last sensation, if sufficiently intense, usually causes rejection of the substance from the mouth. Doubtless this response exerts a protective function for the individual, as very low concentrations of exceedingly toxic alkaloids found in certain plants evoke an extremely unpleasant bitter sensation. For example, strychnine can be detected at a concentration of 1.6×10^{-6} moles/liter.

Threshold for Taste. The threshold for taste varies considerably among the four primary taste sensations, but in all instances both the threshold as well as intensity discrimination in gustation is a crude approximation at best. Consequently, around a 30 percent change in the concentration of any substance being tasted is necessary before any perceptible change in stimulus intensity can be detected.

The approximate threshold concentrations of certain "standard compounds" used as stimuli for taste tests may be summarized: Sweet (sucrose),

10^{-2} M; sour (hydrochloric acid), 9×10^{-4} M; salty (sodium chloride), 10^{-2} M; and bitter (quinine), 8×10^{-6} M.

About 15 to 39 percent of all persons tested exhibit "taste blindness" for a substance known as phenylthiocarbamide, a thiourea compound, when it is applied to the tongue in a dilute solution. Since the inability to taste this compound is inherited as an autosomal recessive trait, this fact is of much importance in studies that are concerned with human genetics.

Taste Discrimination. As noted earlier, the different taste buds exhibit a nonspecificity to various chemical stimuli, although usually these structures individually respond most strongly to only one type of primary stimulus; consequently, a functional "pseudospecificity" actually is present

Since it is possible for the untrained individual to discriminate among a vast array of individual taste sensations, a theory based upon the relative excitability of the taste buds to the four primary stimulants has been proposed to account for this fact. As the *relative differential sensitivity* of various taste buds to different chemical stimuli is quite variable, it has been proposed that some (unspecified) area of the brain can detect the ratios of excitation of different taste buds located on different regions of the tongue; hence, the signals thereby elicited are interpreted within the brain as various flavors. The nebulosity of this statement clearly indicates our current ignorance in this area of physiology.

Receptor Potentials of Taste Buds. In common with all other sensory receptors, appropriate chemical stimulation of a neuroepithelial cell within a taste bud results in the partial depolarization of that cell from its rest potential toward some threshold value. This decrease in negativity within the cell, or generator (receptor) potential, in turn is proportional to the logarithm of the concentration of the stimulant, as discussed on p. 515.

The specific mechanism whereby the receptor cells subserving gustation are depolarized by chemical substances is unknown, as is the mechanism whereby the generator potential induces impulses in the terminal nerve fibers. Possibly a neurotransmitter substance depolarizes the terminal nerve fibers surrounding the neuroepithelial cells of the taste buds. It is known, however, that application of an effective stimulant to a taste bud produces a burst of high-frequency afferent impulses within a few milliseconds. Several seconds later, however, the rate of impulse formation declines to a much lower steady signal level; ie, adaptation takes place rapidly in the neuroepithelial elements of the taste buds.

Affective Properties of Taste. The affective, or subjective, attributes of taste are "pleasant" and "unpleasant," although prior conditioning of the individual contributes markedly to what a person considers to be a pleasant or an unpleasant taste sensation. This fact is well-exemplified by the enormous variety of "tastes" encountered in the personal selection of foods and beverages by different persons.

Gustation and Olfaction. Generally, gustation and olfaction are classified together as visceral senses because both are closely allied to gastrointestinal functions. From a physiologic standpoint, the senses of taste as well as of smell are inseparable in that both are necessary in order that foods "taste normal." If a person has a cold, the complaint frequently is made that food is "tasteless." Yet upon testing, the mechanisms of gustation are found to be completely normal in such individuals. This fact illustrates the profound importance of the olfactory sense in detecting "tastes," as odors pass up the nasopharynx to the olfactory region and thereby may increase the gustatory sensitivity to certain substances many thousandfold. An apt example of this point is to be found in professional tasters whose aptitude may depend far more upon their olfactory than upon their gustatory sense. For example, coffee, tea, wine, and whiskey often are blended commercially by individuals who possess an abnormally keen development of their olfactory as well as gustatory senses. Conditioning (or training) of these senses also would be another indispensable factor in such individuals.

Olfaction

The olfactory receptors, like the taste receptors, are chemoreceptors that are excited by molecules in solution in the fluids of the nasal mucosa. From an anatomic standpoint, however, these senses exhibit a considerable number of individual differences. The olfactory receptors are *teloreceptors* (or distance receptors, p. 520), in contrast to the gustatory receptors. Furthermore, the olfactory receptors are unique in that there is no thalamic relay of impulses to the cortex from the receptors of the olfactory membrane. Lastly, there is no specific neocortical projection area for olfactory signals.

Of all the senses, olfaction is the least well understood from an objective standpoint. The anatomic location of the olfactory membrane itself renders study difficult, and since olfaction is largely a subjective phenomenon, it is quite difficult to study it, with any degree of accuracy, in lower organisms. Furthermore, in humans, the olfactory sense is primitive when compared with that of certain other animals. Thus, in many subhuman primates and other mammals, the sense of smell is of critical importance in two essential phases of survival, viz., in obtaining food and in reproduction. For example, wild carnivorous species sometimes must depend largely upon their olfactory sense to track their prey as well as to detect females that are in estrus (p. 1152). In addition, many species of animal (eg, certain female moths) are known to elaborate specific sex attractants or *pheromones*, which can be detected over the relatively enormous distance of several miles by members of the opposite sex. In this regard, it is of interest that certain compounds, believed to be pheromones, have been isolated from the vaginal secretions of normal sexually mature human females; however, the role of such compounds (if any) in eliciting reproductive behavior in the human species is quite inferior to conditioned olfactory, visual, and auditory cues. Hence, "synthetic pheromones" (ie, perfumes) are far more effective as sexual attractants in the human species as discussed on p. 523.

MORPHOLOGY OF THE OLFACTORY MEMBRANE. The sensory receptors for olfaction lie in the olfactory membrane, which is situated in the superior part of each nostril (Figs. 7.66 and 67; p.

322). Medially, this structure is reflected downward over the surface of the nasal septum, while laterally it covers the superior turbinate as well as the upper surface of each middle turbinate. The total surface area of the olfactory epithelium is about 5 cm², and in general this structure is composed of pseudostratified columnar epithelium.

The *olfactory cells* are the receptors for the sense of smell (Fig. 8.51). These elements are actually bipolar neurons derived embryologically from the central nervous system (Fig. 7.67). There are between 10 and 20 million olfactory cells scattered among the supporting (or sustentacular) epithelial cells within the olfactory epithelium. On the mucosal surface of each olfactory cell are large numbers of minute olfactory hairs (or cilia), each

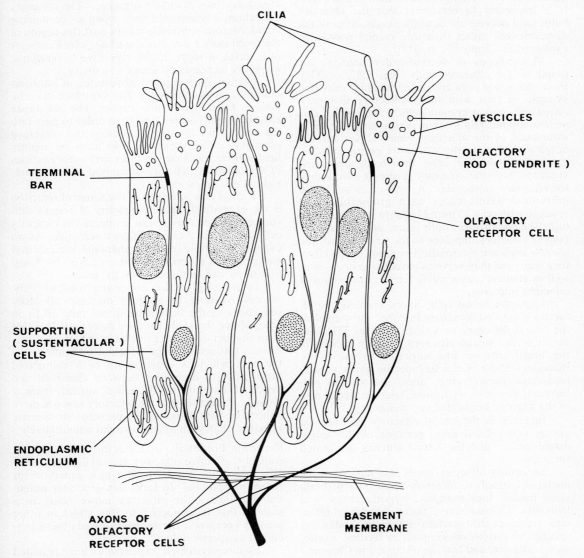

FIG. 8.51. The structural elements of the olfactory mucosa. (After Ruch and Patton [eds]. *Physiology and Biophysics,* 19th ed, 1966, Saunders.)

of which is about 3 μ long by 0.3 μ in diameter. These cilia project into the mucous coating which lines the surface of the nasal cavity, and these structures are believed to respond to appropriate chemical stimuli which in turn excite the olfactory cells. Interspersed among the olfactory cells are Bowman's glands, which secrete a layer of mucus onto the surface of the olfactory membrane.

NEUROANATOMY OF OLFACTION. As indicated above, the olfactory cells themselves are bipolar neurons; their thin central processes converge, form filaments, and run through the fenestra of the cribiform plate as the olfactory nerves (I), as depicted in Figure 7.66 (p. 322). These nerves pierce the olfactory bulbs and there synapse with neurons in the rostral portion of each of the bulbs. It is important to remember that the olfactory bulbs (and nerves) are actually evaginations of the diencephalon, rather than true cranial nerves, as explained in Chapter 7, (p. 319).

Three types of second-order neurons are found in the olfactory bulb (Fig. 7.67, p. 322). First, the *mitral cells* are large triangular cells that synapse in turn with many olfactory filaments, tufted cells and/or granule cells. Functionally, this neuroanatomic relationship presumably results in summation of the olfactory stimulus. Second, the *tufted cells* are smaller and more peripherally located than the mitral cells; their bushy-appearing synapses with the olfactory filaments are termed the *olfactory glomerulus*. A total of about 25,000 individual axons enter each glomerulus and synapse with only 25 mitral cells (in addition to the tufted cells). In turn, both mitral and tufted cells convey olfactory impulses to the brain. Third, the *granule cells* are the smallest neurons of the olfactory bulb, and they serve as associative neurons as well as elements responsible for the summation of olfactory impulses.

From the mitral cells, axons traverse the olfactory tract and terminate in either the medial or the lateral olfactory area (Fig. 7.78, p. 333). The nuclei of the medial olfactory area are located in the brain anterior and superior to the hypothalamus. These nuclei include the gyrus subcallosus, the paraolfactory area, the septum pellucidum, the olfactory trigone, and the medial part of the anterior perforated substance.

Included in the lateral olfactory area are the uncus, prepyriform area, portions of the amygdaloid nuclei, and the lateral anterior perforated substance.

Secondary olfactory tracts also pass from the nuclei of both of the olfactory areas into the brain stem nuclei, hippocampus, hypothalamus, and thalamus. Consequently, these secondary olfactory tracts control certain automatic bodily responses to various odors, such as feeding, excitement, pleasure, and fear, as discussed in Chapter 7 (p. 406).

The lateral olfactory area also sends secondary nerve tracts to the prefrontal and temporal regions of the cerebral cortex. In these regions, the more complex olfactory responses are perceived and integrated. These responses include the association of olfactory with visual, somatic, tactile, and other sensations.

PHYSIOLOGIC ASPECTS OF OLFACTION

Primary Olfactory Sensations. There is at present no wholly adequate or even satisfactory system for the classification of basic (or primary) olfactory sensations such as exists for taste. One such system groups the primary odors as ethereal, putrid, pungent, pepperminty, camphoraceous, musky, and floral. This list obviously is more eloquent than informative!

Thresholds for Olfactory Excitation. The olfactory epithelium is stimulated only when air containing an odoriferous substance moves past this region of the respiratory tract. Thus, sniffing, which consists of a series of short, rapid, repetitive inspirations, markedly increases olfactory sensitivity.

Certain general physical features of odoriferous substances are known. The substance must be volatile to at least some extent. The substance must be slightly water soluble in order to pass into the air:mucus interface covering the olfactory epithelium. The substance also must be slightly lipid soluble, because the hairs and outer portions of the olfactory cells are composed of lipid materials as discussed earlier.

A chief characteristic of the sense of olfaction is the exceedingly minute quantity of certain substances that are required to stimulate the olfactory cells and evoke a clear olfactory sensation. As an extreme example of this phenomenon, methyl mercaptan at a concentration of 2.5×10^{-11} mg/ liter of air can be detected by the nose.

Interestingly, the concentrations of substances required to produce maximal olfactory sensations in the human being are only 10 to 50 times above threshold for their detection. This fact is in distinct contrast to the ear, which can perceive a trillionfold range of sound intensities, and the eye, which is sensitive up to a millionfold range of light intensities between threshold and maximal excitation. It would appear from a biologic standpoint that the olfactory sense is more concerned with detecting the presence or absence of odors, than with assessing them quantitatively.

Receptor Potentials in Olfactory Cells. Presumably, effective olfactory stimuli depolarize the olfactory cells, thereby producing a generator (or receptor) potential. In turn, this generator potential depolarizes the olfactory nerve fibers in a manner that is analogous to this effect in other sensory receptors, the resulting signals being perceived as specific odors.

Electrophysiologic experiments have revealed that the mitral cells of the olfactory bulb exhibit continuous and spontaneous rhythmic activity (p.

349), and that stimulation of the olfactory cells generates trains of impulses which are superimposed upon this rhythmic background pattern of impulse formation. In other words, impulses from the olfactory epithelium modulate the inherent pattern of electric activity of the mitral cells.

Adaptation. About 50 percent adaptation takes place in the first few seconds after stimulation of the olfactory receptor cells; thereafter, adaptation proceeds very slowly. Since an odor is not perceived for more than a minute or so during continuous exposure to an odoriferous environment, it would appear that there is a pronounced central nervous system component involved in this rapid adaptation. Such a central mechanism for adaptation has also been postulated for the taste receptors.

Selectivity of the Olfactory Receptors. Two theories have been advanced to explain the selective response of the olfactory cells to various stimulants.

Physical Theory. The physical theory presumes that there are physically different receptor sites located on the membranes of the olfactory cilia which permit specific stimulants to become adsorbed selectively onto different olfactory cells, thereby stimulating them selectively to produce generator potentials. This idea is substantiated to some extent by the observation that chemical substances with markedly different chemical properties, yet with similar molecular shapes, smell the same.

Chemical Theory. The chemical theory assumes that receptor chemicals located within the cilia of the olfactory cells react with various specific olfactory stimulants. Thus, the type of receptor chemical present in any given receptor cell determines the type of substance that will stimulate that cell. Presumably, the olfactory cell membrane permeability is increased by the stimulant–receptor chemical reaction, and receptor potentials are thereby elicited.

Quite possibly, the actual mechanism that underlies the olfactory process may involve both of these postulated mechanisms to some extent. Undoubtedly, there are additional (and as yet undiscovered) factors involved in the differential sensitivity to various stimuli that is exhibited by olfactory cells.

Affective Properties of Olfactory Sensations. Subjective characteristics of various odors, as with tastes, may be pleasant, or unpleasant. As mentioned earlier, both of these senses, either separately or acting in conjunction, are of considerable importance to the rejection or selection of food, as well as in certain other far more complex attributes of human behavior. As an example of the latter, by association with previous experience, the odor of a specific perfume alone may suffice to elicit both autonomic and somatic activity patterns of the most far-reaching sort in the suitably conditioned adult male.

Anosmia. Destruction of one olfactory bulb produces a unilateral loss of the olfactory sense, a deficit called *anosmia*. Colds accompanied by a severe inflammation of the nasal mucosa *(acute catarrhal rhinitis)* can produce a reversible bilateral anosmia.

CHAPTER 9

The Body Fluids and Compartments

The indispensable role of water as the vital matrix in which all of the life processes take place was emphasized in Chapter 2 (p. 45), and the physical as well as chemical properties of water as related to such processes were discussed in Chapter 1 (p. 16).

In the present chapter a number of specialized body fluids and the so-called compartments in which they are found will be examined, emphasis being placed upon the blood and lymph.

The term "compartment" is convenient in a descriptive sense when applied to the different regions of the body within which particular fluids are located normally; however, this term also is misleading as it implies a fairly rigid physical isolation of substances as well as the concepts pertaining thereto. For this reason it must be stressed at the outset that the several fluid compartments found within the human body do not form an inflexible and sharply defined series of discrete chambers or "containers" for various solutions. Rather, these compartments, their fluids, and the particular functions that each subserves form a number of fluid systems which may be interrelated to a greater or lesser extent in a complex and continuously dynamic fashion. Furthermore water, which provides the common physiologic denominator in all of these regions of the body, is freely interchangeable among all of these compartments (Fig. 2.1, p. 46).

It also is important to emphasize that not only the volume but also the composition of the body fluids normally are regulated by homeostatic mechanisms so that extremely close tolerances are maintained at all times between these parameters. The specific physiologic mechanisms concerned with the regulation of volumes and constituents of the various fluid compartments, in particular the blood, will be discussed in context in this and in subsequent chapters.

BODY FLUID COMPARTMENTS AND THEIR QUANTITATIVE MEASUREMENT

Water constitutes between 45 and 75 percent of the total mass of the human body, and thus it is the most abundant single bodily constituent. The approximate distribution of water among the tissues is given in Table 9.1. Normally, and despite the wide range in the total body water cited above, the quantity of water that is present in the body of any particular individual in caloric balance is remarkably constant. The water content, however, is related inversely to the total quantity of body fat present. Thus a lean person has a relatively higher body water content than an obese individual. Women tend to have a lower total quantity of body water (about 52 percent of the total body weight) than do men (about 63 percent of the total body weight) because of the significantly greater subcutaneous adipose tissue deposits in females. The largest proportion of water to body weight is found in infants.

Basically, there are three major fluid compartments in the human body (Fig. 9.1). First, the extracellular fluid compartment includes the blood plasma (about 4.5 percent of the total body weight), the interstitial fluid (about 16 percent of the total body weight), and the lymph (about 2 percent of the total body weight). The interstitial (or tissue) fluid forms the actual internal environment of the body as it bathes all of the cells and tissues, provides nutrients, and removes wastes from the immediate vicinity of these structures.

Second, the intracellular fluid compartment is comprised of the sum of all the intracellular fluids within all of the cells of the body. This compartment is neither a homogeneous nor a continuous phase, and from a chemical standpoint the various cell types vary as greatly in their composition as they do morphologically. Therefore, any designa-

Table 9.1   DISTRIBUTION OF BODY WATER IN THE TISSUES OF A 70 KG MAN[a]

TISSUE	APPROXIMATE PERCENT OF BODY WEIGHT	LITERS WATER PER 70 KG	WATER IN TISSUE, PERCENT
Skin	18	9.07	72
Muscle	41.7	22.10	75.6
Skeleton	15.9	2.45	22
Brain	2.0	1.05	74.8
Liver	2.3	1.03	68.3
Heart	0.5	0.28	79.2
Lungs	0.7	0.39	79.0
Kidneys	0.4	0.25	82.7
Spleen	0.2	0.10	75.8
Blood	7.0-8.0	4.65	83.0
Intestine	1.8	0.94	74.5
Adipose Tissue	±10.0	0.70	10

[a]*Data from Pitts.* Physiology of the Kidney and Body Fluids, *2nd ed, 1968, Year Book Medical Publishers.*

tion of a specific "intracellular fluid compartment" is more convenient than accurate. Between 30 and 40 percent of the total body weight is intracellular water.

Third, the transcellular fluid represents a specialized part of the extracellular fluid, and this compartment consists of the intraocular, cerebrospinal, pleural, pericardial, peritoneal, and synovial fluids. The endolymph and perilymph of the inner ear (Table 8.3, p. 513), as well as the digestive fluids (ie, the salivary, gastric, hepatic, and intestinal secretions), are also included in the transcellular fluid "compartment." However, with reference to the digestive fluids, it must be remembered that topologically speaking anything within the lumen of the open-ended gastrointestinal tract is actually outside of the body (Chapter 13, p. 815).

The discontinuous transcellular fluid fractions have in common one particular feature: They are isolated from the blood plasma not only by the

capillary endothelium, but also by an essentially continuous additional layer of epithelial cells.

The transcellular water forms only a minor portion of the entire body mass, the digestive secretions amounting to only 1 to 3 percent of the total body weight.

A brief comment on the quantitative measurement of the volume of the body fluid compartments in human subjects is in order. Theoretically, if one had a substance which could be injected into the body, and which would only mix with the water in a specific compartment, then the total volume of that compartment could be calculated by use of a direct proportion from the concentration of that substance in a sample of the fluid withdrawn from the compartment after complete mixing and dilution had taken place. However, such a calculation requires that several assumptions be made: (1) Complete mixing of the substance must take place. (2) The substance must not diffuse rapidly into a compartment other than the one whose volume is being measured. (3) The substance must be detectable quantitatively and in low concentrations with reasonable accuracy. (4) The material used as a tracer must be nontoxic. (5) The injected substance cannot be excreted rapidly or metabolized by the body to any significant extent. (6) The substance must not be adsorbed onto the cells or other constituents (eg, proteins) of the compartment being measured. (7) The injected substance itself must not alter the distribution of the fluid; for example, by changing the permeability of the capillary walls.

The *dilution principle* described above may be applied not only to the measurement of the plasma volume, but also to the measurement of the extracellular fluid compartment. In practice, the plasma volume of the vascular bed can be determined with reasonable accuracy by use of a dye known as Evans blue (T-1824). A known quantity of dye is injected into the patient, and after a suitable equilibration period (about 10 minutes), a blood sample is withdrawn and the dye concentration is determined colorimetrically on a specific

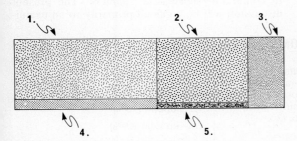

FIG. 9.1. Water distribution in the several body compartments. 1. Intracellular water (30–40 percent). 2. Interstitial water (16 percent). 3. Plasma (4.5 percent). 4. Transcellular water (1–3 percent). 5. Lymph (1–3 percent). Collectively the water in compartments 2, 3, 4, and 5 comprises the extracellular fluid of the body. Although the total volume of water in each of the body fluid compartments normally is maintained at a relatively constant value, the dynamic flux (or exchange) of water among these compartments is quite high. (Data from Pitts. *Physiology of the Kidney and Body Fluids,* 2nd ed, 1968, Year Book Medical Publishers.)

volume of plasma. For example, after the injection of 200 mg of Evans blue into an 80-kg subject, the final plasma concentration of the dye was found colorimetrically to be 0.055 mg/ml after mixing with the circulatory system. Consequently, the plasma volume is determined by the expression:

$$\text{Volume of distribution} = \frac{\text{Quantity of dye administered}}{\text{Concentration of dye in sample}}$$

$$\text{Plasma volume} = \frac{200 \text{ mg}}{0.055 \text{ mg/ml}} = 3,600 \text{ ml}$$

And 3.6 liters are equal to 4.5 percent of the subject's total body weight. Knowing the percentage of red blood cells that are present per unit volume of whole blood (the hematocrit*). The total blood volume can be calculated readily from the total plasma volume.

Plasma albumin labeled with radioactive iodine (^{131}I) or red blood cells tagged with isotopic iron or chromium (^{59}Fe or ^{51}Cr) can also be used clinically to determine the total plasma or blood volume, the radioactivity being assayed quantitatively in a sample of plasma or whole blood by means of a scintillation counter.

In order to measure total body water, water labeled with deuterium oxide (D_2O or "heavy water") can be used to label all of the H_2O in the body. However, the final calculation must be corrected for the loss of water by excretion in urine, feces, sweat, and through respiration during the equilibration period following injection of the labeled water.

The total volume of the interstitial water can be obtained indirectly by subtracting the total plasma water from the total extracellular water, whereas the intracellular water can be obtained by subtracting the extracellular water from the total body water. The values for both the interstitial water and intracellular water are only rough ap-

*The hematocrit is obtained by centrifuging a sample of whole blood in a calibrated tube.

proximations; but since the total plasma volume, measured as described above, is of the greatest significance to the physician, measurement of the volumes of the other compartments ordinarily is of more academic than practical interest.

BLOOD: A LIQUID TISSUE

As indicated in Table 9.1, the total blood volume is about 8 percent of the total body weight. This complex body fluid actually is a liquid tissue, and fundamentally blood consists of several types of *formed elements* (or *cells*) that are suspended in an aqueous fluid matrix (the *plasma*) in which are dissolved a complex spectrum of organic molecules of various sizes as well as various inorganic ions (Table 9.2). The biochemical fabric of the blood undergoes rapid and continuous changes, and yet the overall quantitative patterns of individual constituents normally remain surprisingly constant, as does the total volume of the blood, despite the extremely broad fluctuations in the intake and elimination of water.

Physiologically the blood serves as the major bulk transportation and distribution system for materials within the body. The diversity of the functions performed by the blood is indicated by the following summary: transport of respiratory gases to and from the tissues; transport of nutrients, waste products, and electrolytes; the regulation of pH; thermoregulation; regulation of the water content of the tissues; defense of the body against invasion by foreign bodies and microorganisms. An especially important physiologic integrative role of the blood is to serve as a vehicle in which endocrine substances of the most diverse sorts are carried to regions far removed from their sites of origin, either to specific target organs or to the body tissues in general. Further details of these and other individual processes will be found in context in subsequent chapters. The present discussion will be confined primarily to an exami-

Table 9.2   MAJOR COMPONENTS OF HUMAN BLOOD[a]

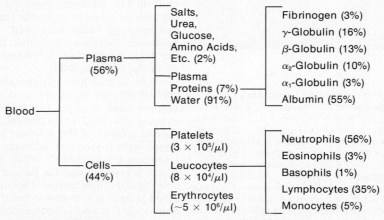

[a]Data from Bishop (ed). Overview of Blood, 1974. Blood Information Service.

nation of some of the basic properties of blood per se as an essential body fluid together with certain important functions of this vital and all-pervading tissue.

Anatomic and Chemical Morphology of Blood

Whole blood consists of a highly complex fluid (the plasma) in which are suspended several types of cells or formed elements (Table 9.2). Tables 9.3 and 9.4 summarize some of the important constituents of normal blood as well as certain physical and chemical properties of this tissue. This outline is not intended to be complete; rather it is designed to serve the reader as a general basis for the present as well as subsequent discussions involving the blood and its several diverse physiologic roles within the body.

A.

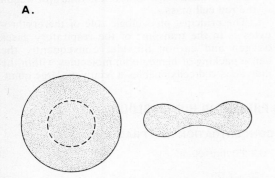

B.

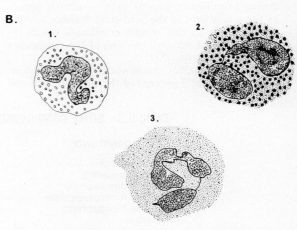

FIG. 9.2. Formed elements of the blood, not drawn to same scale. **A.** Erythrocyte, top and side views to illustrate biconcave shape.

FIG. 9.2. **B.** Granulocytes. 1. Basophilic leukocyte. 2. Eosinophilic leukocyte. 3. Neutrophilic leukocyte.

C.

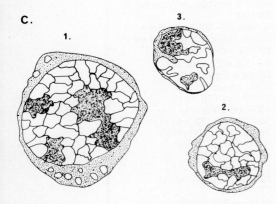

D.

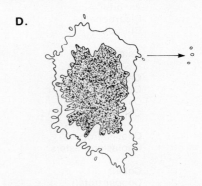

FIG. 9.2. **D.** Megakaryocyte and platelets.

FIG. 9.2. **C.** Lymphocytes. 1. Large. 2. Medium. 3. Small. Age of the lymphocytes is related inversely to their size.

E.

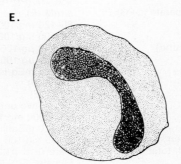

FIG. 9.2. **E.** Monocytes. (Data from Bloom and Fawcett. *A Textbook of Histology*, 9th ed, 1968, Saunders.)

ERYTHROCYTES. Mature, circulating, human red blood cells (erythrocytes or corpuscles) are biconcave discs having a mean diameter of ~ 8 μ and a thickness of approximately 2.2 μ at the periphery, and 1.0 μ at the center (Fig. 9.2). In dried films the diameter of these cells is about 0.5 μ smaller because of the shrinkage which occurs during the drying process.

Because of osmotic changes, the red blood cells normally are slightly larger (by about 0.5 μ during exercise) in venous than in arterial blood, as there is a shift in the acid–base balance toward the acid side in the venous capillaries which results in a slight osmotic swelling of the corpuscles in these vessels (p. 751).

The *stroma*, or framework, of the red cell comprises about 5 percent of the total wet red cell mass. This stroma is composed primarily of lipids, and these lipids are combined to an extent of about 50 percent with an albuminoid protein. Certain antigenic properties are associated with this lipoprotein component of the erythrocytes; for example, the A, B, and Rh antigens are present in this lipoprotein complex, which is called *elenin*.

The respiratory functions of the erythrocyte are carried out by the red pigment *hemoglobin* which is present in the erythrocytes at a concentration of around 32 percent of the total wet weight of the red cell mass.

The principal physiologic role of the erythrocytes is in the transport of the respiratory gases oxygen and carbon dioxide; consequently the dense packing of hemoglobin molecules within the individual red cells enables a relatively large quan-

Table 9.3 SOME COMPOUNDS PRESENT IN HUMAN BLOOD[a]

COMPOUND[b]	CONCENTRATION AND FRACTION[c]
Acetoacetic acid + acetone ("ketone bodies")	0.2-2.0 mg/100 ml (S)
Adenosine diphosphate	100 μM (E)
Adenosine monophosphate	13 μM (E)
Adenosine triphosphate	1 μM (E)
Aldosterone	3–10 mg/100 ml (P)
Alpha amino acid nitrogen	3.0–5.5 mg/100 ml (P)
Amino acid (total)	300 mg/100 ml (P); 120 mg/100 ml (E)
Ascorbic acid	0.4–1.5 mg/100 ml (Fasting B)
Bicarbonate	25 mM
Bilirubin	Direct: 0.1–0.3 mg/100 ml (S) Indirect: 0.2–1.2 mg/100 ml (S)
Calcium	8.5–10.5 mg/100 ml; 4.3–5.3 mEq/liter (S)
Carbon dioxide	26–28 mEq/liter (S)
Carotenoids	0.08–0.4 μg/ml (S)
Ceruloplasmin	27–37 mg/100 ml (S)
Chloride	100–106 mEq/liter (S) 102 mM (P); 78 mM (E)
Cholesterol	150–280 mg/100 ml (S)
Cholesterol esters	$\cong$ 60–75% of total cholesterol (S)
Cobalt	17 μM (E)
Copper (total)	100–200 μg/100 ml (S) 17 μM (E)
Cortisol (17-hydroxycorticoids)	5–18 μg/100 ml (P)
Creatine	40 μM (P); 600 μM (E)
Creatinine	0.7–1.5 mg/100 ml (S) 450 μM (P); 160 μM (E)
2,3-Diphosphoglycerate (DPG)	4 mM (E)
Glucose (Folin)	80–120 mg/100 ml (Fasting, B)
Glucose (true)	70–100 mg/100 ml (B)
Iodine (protein bound)	3.5–8.0 μg/100 ml (S)
Iron	50–150 μg/100 ml (S)
Lactic acid	0.6–1.8 mEq/liter (B) 1 mM (P)
Lead	3 μM (E)
Lipase	Below two units/ml (S)
Lipids, total	450–1000 mg/100 ml (S)
Magnesium	1.5–2.0 mEq/liter; 1–2 mg/100 ml (S)
Nonprotein nitrogen	15–35 mg/100 ml (S)
Phosphatase, acid, total[d]	Male: 0.13–0.63 sigma units/ml (S) Female: 0.01–0.56 sigma units/ml (S)

Table 9.3 (Cont'd)

COMPOUND[b]	CONCENTRATION AND FRACTION[c]
Phosphatase, alkaline[d]	2.0–4.5 Bodansky units/ml (S)
Phospholipids	145–225 mg/100 ml (S)
Phosphorus, inorganic	3.0–4.5 mg/100 ml (adult, S)
	0.9–1.5 mM/liter (adult, S)
Potassium	4 mM (P); 111 mM (E)
Protein	
Total	6.0–8.0 gm/100 ml (S)
Albumin	3.5–4.5 gm/100 ml (S)
Globulin	2.0–3.0 gm/100 ml (S)
Pyruvic acid	0–0.11 mEq/liter (P)
Riboflavin	500 nanomolar (E)
Sodium	136–145 mEq/liter; 310–340 mg/ 100 ml (S)
	140 mM (P); 6 mM (E)
Sulfate	0.5–1.5 mEq/liter (S)
Transaminase (serum glutamic oxaloacetic transaminase, SGOT)[d]	10–40 units/ml (S)
Urea nitrogen (blood urea nitrogen, BUN)	8–25 mg/100 ml (B)
Uric Acid	3.0–7.0 mg/100 ml (S)
Vitamin A	0.15–0.6 μg/ml (S)

[a]*Data from Bishop (ed).* Overview of Blood, *1974. Blood Information Service;* Ganong: *Review of Medical Physiology, 6th ed, 1973,* Lange Medical Publications.

[b]*The substances and the values listed for them in this table are intended to give an indication of the biochemical complexity of blood, and are not to be construed as absolute. Individual blood levels of specific blood components can vary among normal individuals or in the same individual at different times. Furthermore, the normal values for specific blood constituents can differ considerably among different laboratories, depending upon the technique used for their determination. Consequently the data presented in this table are illustrative, but of no particular diagnostic significance.*

[c]*The abbreviations used for the several blood fractions are: (B) = whole blood; (P) = plasma; (S) = serum; (E) = erythrocytes.*

[d]*The activities of over 50 specific enzymes have been detected in the various blood fractions, only three of which are listed in this table.*

tity of gas to be transported by each cell. Furthermore, since the red cells are biconcave discs, each of which has an average surface area of approximately 120 μ^2 and a volume of 85 μ^3, this shape presents a much greater surface/volume ratio (1.4/1) than would be possible if the cells were, for example, spherical (Fig. 9.2). Thus, a maximal surface area is exposed per unit mass of hemoglobin for gas exchange. The significant physiologic consequence of this fact is that a facilitation of oxygen and carbon dioxide exchange in the lungs as well as the tissues is present merely because of the biconcave shape of the red cells.

Erythrocytes, although nonnucleated when mature, do carry on a number of important metabolic and respiratory functions, although on a limited scale. Thus the enzymes of the glycolytic system (p. 137) are present within the red cells, as well as all of the blood glutathione. Other organic as well as inorganic salts are also present in these cells. Potassium, for example, is found in very high concentration within the erythrocytes. Carbonic anhydrase, the essential enzyme that is responsible for the rapid conversion of molecular carbon dioxide gas and water into carbonic acid

(and vice versa) is also found in elevated concentrations in the erythrocytes (p. 757).

Normally the erythrocytes circulating within the body are extremely flexible and elastic structures which can be deformed readily upon passage through the capillaries or other small vessels, only to resume their shape immediately when the external forces that were exerted upon them are removed.

Erythropoiesis. Various tissues of the body are involved successively in the production of red blood cells, or *erythropoiesis,* depending upon the age of the individual. During the first several weeks of embryonic life, erythrocytes are produced by the yolk sac, whereas this function is taken over by the liver primarily, as well as to some extent by the spleen and lymph nodes, throughout the middle trimester of gestation. Later, and following birth, the red cells are formed selectively by the bone marrow as indicated in Table 9.5. Prior to adolescence, the marrow of all bones contributes erythrocytes; however, beyond this stage, the erythropoietic function is usurped gradually by the marrow of the membranous bones, in particular the sternum, pelvis, vertebrae, and ribs. At the same time, the marrow of the long bones becomes fatty so that erythropoiesis no longer occurs at these sites in the adult.

Even in the adult, however, extreme and prolonged stimuli, such as hematologic disease, can induce erythropoiesis once again in the long bones, liver, spleen, and even the lymph nodes. Therefore these tissues retain a latent potentiality for erythropoiesis throughout the lifetime of the individual.

The gradual morphologic differentiation of

Table 9.4 SOME AVERAGE NORMAL VALUES FOR HUMAN BLOOD[a,b]

DETERMINATION	VALUE	
pH	7.35–7.45	
Osmolality	285–295 mOsm/l (S)	
P_{O_2}	100 mm Hg (arterial B)	
	40 mm Hg (mixed venous B)	
P_{CO_2}	40 mm Hg (arterial B)	
	46 mm Hg (mixed venous B)	

	ADULT	
	Male	Female
Total erythrocyte count	$4.5–5.5 \times 10^6$ cells/μl	$3.8–4.8 \times 10^6$ cells/μl
Hemoglobin	14–18 gm/100 ml (B)	12–16 gm/100 ml (B)
Hematocrit (HCT)	40–54%	37–47%
Mean corpuscular volume (MCV) $= \dfrac{Hct \times 10}{RBC\ (10^6/\mu l)}$	84–92 fl	84–92 fl
Mean corpuscular hemoglobin (MCH) $= \dfrac{Hb \times 10}{RBC\ (10^6/\mu l)}$	28–32 pg	28–32 pg
Mean corpuscular hemoglobin concentration (MCHC) $= \dfrac{Hb \times 10^6}{Hct}$	32–34 gm/100 ml	32–34 gm/100 ml
Mean erythrocyte diameter = mean diameter of 500 cells in blood smear	7.5 μ	7.5 μ
Red cell mass as fraction of body weight	27–33 ml/kg	24–30 ml/kg
Plasma volume	4.0–5.0 liters	
Total blood volume	60–80 ml/kg body weight	
Reticulocytes	0.2–2% of total (E)	
Leucocytes (white blood cells, WBC), total	$4–11 \times 10^3/\mu l$	
Neutrophils	40–75% of total WBC	
Lymphocytes	20–45% of total WBC	
Monocytes	2–10% of total WBC	
Eosinophils	1–6% of total WBC	
Basophils	1% of total WBC	
Platelets	$2–4 \times 10^5/\mu l$	
Sedimentation rate (erythrocyte sedimentation rate, ESR)	0–6 mm hr	
Bleeding time	To about 11 minutes	
Coagulation time	5–11 minutes	
Prothrombin time	10–14 sec	
Clot retraction	30–60 minutes	
Osmotic fragility	0.5% NaCl: hemolysis starts	
	0.35% NaCl: hemolysis complete	

[a]*Data from Bishop (ed).* Overview of Blood, *1974. Blood Information Service; Ganong.* Review of Medical Physiology, *6th ed, 1973, Lange Medical Publications.*

[b]*The general comments and abbreviations given in the footnotes to Table 9.3 also apply here.*

primitive, generalized cells which are found at random throughout the bone marrow into mature red blood cells may be summarized in a stepwise fashion (Table 9.5): (1) primordial stem cell; (2) hemocytoblast; (3) basophil erythroblast (hemoglobin synthesis commences in these cells); (4) polychromatophil erythroblast (the specific staining characteristics of these cells depend upon the particular mixture of hemoglobin and basophilic material that is present); (5) normoblast (the nucleus shrinks and more hemoglobin is synthesized at this stage); (6) reticulocyte (the hemoglobin now has risen to a 34 percent concentration and the nucleus undergoes autolysis and resorption); (7) erythrocyte (the reticulum of the nucleus is resorbed completely and the cells enter the vascular circulation of the bone marrow by the process of *diapedesis,* or direct passage through the endothelial walls of the capillary membranes).

During stages 1 through 5 the various cells continue to divide mitotically, so that their total numbers increase enormously.

Less than 0.5 to 1.0 percent reticulocytes are found in normal circulating blood so that an elevated reticulocyte count is indicative clinically of an increased activity of the bone marrow. Similarly, normal peripheral blood contains no normoblasts, erythroblasts, or hemocytoblasts, although all of these elements, as well as reticulocytes, are abundantly present in the bone marrow.

Physiologic Regulation of Erythropoiesis. The total red blood cell mass in the body remains constant within close limits so that a normal circulation of the blood is not hindered due to an excessively high blood viscosity. Nevertheless a sufficient quantity of erythrocytes is present at all times to ensure adequate tissue oxygenation.

Normally the average life of a circulating red blood cell is around 120 days, as determined by labeling red cells with isotopes such as ^{59}Fe, ^{51}Cr, or ^{15}N, and by subsequently following the rate of disappearance (or turnover rate) of such tagged cells at intervals following their injection into a normal subject. It has been estimated that approximately 3×10^6 red cells are destroyed by the body each second. Furthermore, in the normal adult it has been calculated from the rate of bile pigment formation that about 6.0 gm of hemoglobin are degraded and replaced daily. Observations such as these clearly indicate that there must be an equal rate of synthesis both of erythrocytes and of hemoglobin in order that the total red cell mass of the body remain within the limits that are essential for maintenance of a homeostatic equilibrium insofar as these blood elements are concerned.

Two general mechanisms are known to stimulate red blood cell production by the bone marrow. First, hypoxia (or a decreased oxygen partial pressure) is a major physiologic stimulant to erythropoiesis. Hypoxia in turn can be induced by long-term exposure to high altitudes (above 3,000 meters) or by excessive destruction of the red blood cells within the body due to various abnormal processes. Each of these factors serves as a powerful stimulant to erythropoietic mechanisms within the bone marrow. Second, a glycoprotein has been shown to be present at elevated levels in the blood during hypoxia. Several tissues, but chiefly the kidney, appear to be responsible for the synthesis of this endocrine factor which variously is called erythropoietin, hemopoietin, or erythropoietic stimulating factor.

An interesting negative feedback mechanism exists with regard to the rate of erythrocyte production and its regulation; this mechanism is summarized in Figure 9.3. Tissue hypoxia, acting upon the kidney, induces an increased rate of erythropoietin synthesis. This endocrine substance in turn is carried by the blood to the bone marrow where it stimulates (in about 3 to 4 days) both the rate of maturation of the red cells as well as the total number of mitotic cells present at all stages of erythropoiesis. The latter effect occurs primarily at the stem cell level; hence the rate of conversion of these primordial cells into hemocytoblasts is accelerated markedly. As the blood level of circulating erythrocytes increases, the hypoxia within the renal tissue is alleviated (negative feedback), and thus erythropoietin synthesis decreases. The decreased erythropoietin level in turn reduces the rate of erythropoiesis, and thereby the concentration of circulating erythrocytes is stabilized. This exquisitely delicate mechanism ensures that sufficient red blood cells

Table 9.5 STAGES IN THE DEVELOPMENT OF ERYTHROCYTES[a]

1. Primitive stem cell → committed stem cell
2. Hemocytoblast (proerythroblast)
3. Basophil erythroblast (early erythroblast)
4. Polychromatophil erythroblast (intermediate erythroblast)
5. Normoblast (late erythroblast)
6. Reticulocyte
7. Mature erythrocyte

[a]*Data from Bloom and Fawcett:* A Textbook of Histology, *9th ed, 1968, Saunders; Bishop (ed).* Overview of Blood, *1974. Blood Information Service.*

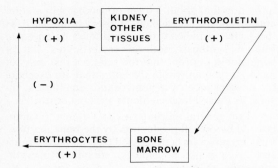

FIG. 9.3. The negative feedback mechanism that regulates the rate of erythropoiesis via a relative hypoxia. (Data from Guyton. *Textbook of Medical Physiology,* 4th ed, 1971, Saunders.)

are present for adequate tissue oxygenation but without increasing their total mass so greatly as to interfere with normal circulation by increasing the viscosity of the blood excessively. In conclusion, it is evident that hypoxia does not act directly upon erythropoiesis; rather hypoxia exerts its effect indirectly through the negative feedback mechanism just described and illustrated in Figure 9.3.

Hemoglobin Synthesis and Iron Metabolism. The synthesis of hemoglobin in red cells commences in the erythroblasts and continues through the normoblast stage. In fact the young erythrocytes released from the marrow into the bloodstream continue to synthesize hemoglobin for several days, albeit at very low rates, even as the nuclear membrane is undergoing final degeneration and resorption.

The steps in the biochemical synthesis of *heme,* the respiratory pigment moiety of the hemoglobin molecule, are outlined in Figure 9.4. Interestingly, the complex heme molecule is synthesized primarily from two simple organic compounds, acetic acid and the amino acid glycine.

A glance at Figure 9.4 will reveal the literally central position of one iron atom in each heme molecule. This metal not only is essential to the formation of normal hemoglobin, but also other important compounds in the human body such as the red pigment that is found in muscle (myoglobin) and various specific intracellular enzymes such as the cytochromes, peroxidase, cytochrome oxidase, and catalase.

The essential features of iron absorption and metabolism are summarized in Figure 9.5. The total quantity of this metallic element in the human body amounts to about 4 gm, most of which is found in hemoglobin (about 65 percent). Approximately 1 percent of the total iron is found in the various cellular oxidative enzymes listed above, and 4 percent is present as myoglobin. The blood plasma protein transferrin contains some 0.1 percent of the total iron while 15 to 30 percent of the body iron pool is stored as ferritin, chiefly in the liver.

The normal human male excretes about 0.6 mg of iron per day in the feces. Hemorrhage increases iron loss; consequently menstruation results in an average iron loss that amounts to 1.5 mg per day in women.* These minimal losses are made up for by a diet that contains adequate quantities of available iron (p. 893).

Iron is absorbed by an active transport mechanism principally from the duodenum and upper region of the small intestine. Once the body pool for this element is "saturated," no more (or very little) is absorbed; rather, the element is recycled within the body. Thus, as indicated in Figure 9.5, the total quantity of iron in the body is closely regulated by a negative feedback mechanism operating through apoferritin, transferrin, and the intestinal mucosal cells themselves. Thus excess dietary iron is not absorbed; rather it is excreted in the feces (see also p. 893 et seq).

Additional Factors Essential for Erythrocyte Maturation. A number of specific substances in addition to iron are essential to the normal maturation of red blood cells.

The Intrinsic Factor and Vitamin B_{12} (Cyanocobalamin). Normally the gastric secretions contain a mucopolypeptide (or mucopolysaccharide?) which is called the *intrinsic factor.* This substance enables dietary vitamin B_{12} to be absorbed from the gastrointestinal tract as follows. Intrinsic factor and vitamin B_{12} bind together chemically so that the vitamin B_{12} is not digested by the intestinal enzymes. Once absorbed through the intestinal mucosa, cyanocobalamin is released in free form and then stored in the liver (up to 1 mg), to be released as needed at the rate of about 1 μg per day in order to promote normal maturation of the erythrocytes.

Vitamin B_{12} is an essential factor in erythrocyte maturation as it is necessary for the synthesis of certain deoxyribonucleic acid (DNA) molecules. (Cyanocobalamin is discussed further on page 883.)

Folic Acid (Pteroylglutamic Acid). Folic acid (p. 882) promotes red cell maturation through a mechanism which is similar to that for vitamin B_{12}; ie, through the formation of DNA. However, the specific nucleotide syntheses promoted by each of these compounds are different.

Pyridoxine (Vitamin B_6). Pyridoxine (p. 880) appears to play some role in the maturation of red blood cells in humans; however, the specific biochemical mechanism in which this compound is involved has not yet been elucidated.

Trace Elements. Minute quantities of cobalt as well as of copper ions are essential to the normal rate of erythrocyte maturation. Apparently these elements are necessary in certain synthetic

*Actually many sexually mature females (ie, those exhibiting regular menstrual cycles) may have barely minimal, if not slightly inadequate, iron stores despite the fact that overt signs of anemia are absent (p. 567).

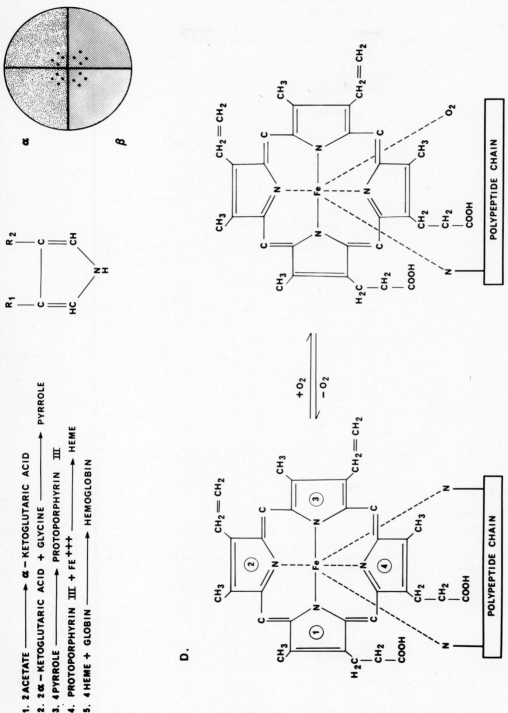

A.

1. 2 ACETATE ⟶ α – KETOGLUTARIC ACID
2. 2 α – KETOGLUTARIC ACID + GLYCINE ⟶ PYRROLE
3. 4 PYRROLE ⟶ PROTOPORPHYRIN III
4. PROTOPORPHYRIN III + FE^{+++} ⟶ HEME
5. 4 HEME + GLOBIN ⟶ HEMOGLOBIN

FIG. 9.4. A. General summary of the major steps involved in the synthesis of hemoglobin. **B.** Structure of the pyrrole ring. R_1 and R_2 = different radicals. **C.** The entire hemoglobin molecule is a globular structure that is composed of two similar α polypeptide (globin) chains and two similar β polypeptide (globin) chains, each one of which contains a single molecule of heme. The black dots signify the four pyrrole rings in each polypeptide subunit. **D.** Structure of deoxygenated and oxygenated heme illustrating the central position of the iron atom in the molecule as well as the readily reversible linkage of oxygen to the iron. (**A, B, D.** Data from Guyton. *Textbook of Medical Physiology*, 4th ed, 1971, Saunders. **C.** Data from Ganong. *Review of Medical Physiology*, 6th ed, 1973, Lange Medical Publications.)

533

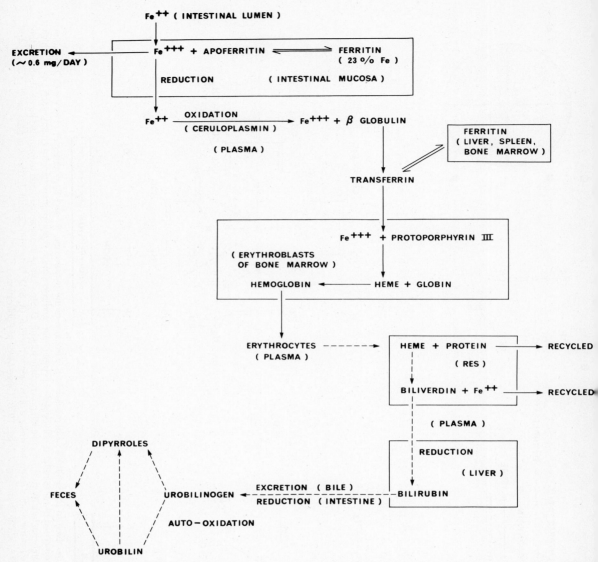

FIG. 9.5. General summary of iron transport and metabolism. RES = reticuloendothelial system. An extremely small quantity of iron as well as bile pigments are excreted daily in the urine (not shown in figure) as well as in the feces. (Data from Harper. *Review of Physiological Chemistry,* 13th ed, 1971, Lange Medical Publications; Ganong. *Review of Medical Physiology,* 6th ed, 1973, Lange Medical Publications; Guyton. *Textbook of Medical Physiology,* 4th ed, 1971, Saunders.)

reactions which are involved in the production of hemoglobin.

Hemolysis and Crenation of Erythrocytes. Under normal circumstances, hemoglobin cannot diffuse out of red cells either in plasma or in certain artificial aqueous solutions which have an osmotic pressure equal to that present within the cells themselves *(isotonic solutions).* Thus a sodium chloride solution containing 0.9 gm salt/100 ml

This biologic usage of the term "normal" must not be confused with the use of the same term in reference to a chemical concentration of a solute (p. 13).

water is said to be *isotonic* or *normal** with respect to the red cells, as when erythrocytes are placed in such a medium no net osmotic movement of water takes place. If, however, red cells are placed in diluted (or *hypotonic*) saline (for example, a 0.5 percent salt solution), or in distilled water, then water moves into the cells down its osmotic gradient causing them to swell at first, and then to rupture (Fig. 9.6A), thereby liberating hemoglobin into the solution. This process is called *hemolysis,* and it has occurred because of the unrestricted osmotic movement of water into the cells (p. 23). The empty red cell membranes found after hemolysis are called "ghosts."

A.

B.

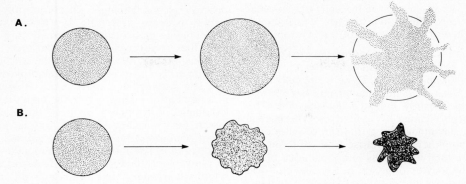

FIG. 9.6. A. Hemolysis of erythrocytes takes place in a hypotonic solution; the membrane ruptures and releases hemoglobin. **B.** Crenation or shrinking of erythrocytes takes place in a hypertonic solution with an attendant wrinkling of the membrane.

Conversely, placing red cells in a more concentrated (ie, *hypertonic*) solution, for example 1.5 percent saline, will cause osmotic movement of water out of the red cells with an attendant shrinking and wrinkling of the membrane, a process that is called *crenation* (Fig. 9.6B).

If, however, erythrocytes are added to an isotonic solution of a substance such as urea, to which the erythrocyte membrane is freely permeable, then osmotic hemolysis will also result. Hemolysis occurs because the urea diffuses down its own concentration gradient (ie, into the red cells) and in so doing water also is carried into the cells causing them to rupture in a manner similar to that which takes place in hypotonic saline solutions.

Hemolysis also can be produced by exposing erythrocytes to chemical agents in vitro such as lipid solvents or detergents, both of which disrupt the normal integrity of the red cell membrane and thus release the hemoglobin.

The extent to which an isotonic saline solution can be diluted before hemolysis commences is a practical measure of the fragility of the erythrocytes. Clinically the osmotic fragility of erythrocytes is tested by placing a sample of blood into each of a series of saline solutions having graded and progressively decreasing salt contents, starting with a 0.9 percent sodium chloride solution. Generally hemolysis of normal erythrocytes starts at a concentration of around 0.5 percent NaCl and is complete at a concentration of 0.35 percent NaCl. Hemolysis is visible macroscopically when the release of hemoglobin from the ruptured cells causes the specimen to turn a clear reddish color, whereas if hemolysis has not occurred then the suspension of red cells exhibits turbidity when a beam of light is passed through it.

Hemolysis, whether induced by chemical or mechanical means, assumes major importance in certain clinical situations since the free hemoglobin liberated in the bloodstream in addition to the erythrocyte ghosts and stroma of these cells can exert a number of deleterious physiologic effects.

For example, the hemoglobin which is liberated by hemolysis of red cells during an acute malarial attack imposes a severe excretory burden on the kidneys. When the plasma level of hemoglobin exceeds about 0.15 gm/100 ml (150 mg percent) then it is excreted in the urine, which fluid then turns a dark red, brown, or even black. The hemoglobinuria in this situation has given rise to the term "blackwater fever," a common appellation for malaria.

LEUKOCYTES (WHITE BLOOD CELLS). Leukocytes are classified morphologically on the basis of their staining characteristics and specific microscopic appearance as either *granulocytes* or *agranulocytes* (Fig. 9.2, Table 9.2). Leukocytes in general are larger than erythrocytes and range in size from 8 to 15 μ in diameter, the size depending upon the particular type of cell. Red corpuscles outnumber the white cells by about 600 to 1 in a typical blood smear.

Granulocytes. These cells are placed in three groups according to the specific staining characteristics of their cytoplasmic granules (Fig. 9.2 and Table 9.6).

Polymorphonuclear eosinophils. These cells usually possess a bilobed nucleus and abundant coarse oval granules in the cytoplasm which stain bright red with eosin (an acid dye). Eosinophils are not numerous; usually only 2 to 4 percent are present in a normal differential white cell count.

Polymorphonuclear basophils. The numerous coarse basophilic granules in the cytoplasm of these cells stain deep blue with basic dyes, such as methylene blue. In fact, the granules may be so abundant as to obscure the multilobed nucleus. Usually basophils are present in a differential count to an extent of only 0.15 percent.

Polymorphonuclear neutrophils. These are the most abundant (about 60 percent) of the leucocytes. The fine cytoplasmic granules of these large (10 to 14 μ) cells stain violet with neutral dyes, ie, mixtures of acidic and basic dyes. The nuclei of the neutrophils exhibit a variable number of lobes, depending upon the age of the cell, the so-called *Arneth stages*. Thus, five distinct stages can be distinguished in the life cycle of the neutrophils.

Agranulocytes. These nongranular leukocytes generally are divided into three classes:

Small Lymphocytes. These white cells range between 7 and 10 μ in diameter and have a relatively large, slightly indented nucleus which becomes deeply stained by basic dyes. The nucleus in turn is surrounded by a thin rim of paler cytoplasm. Lymphocytes comprise from 20 to 30 percent of the total white cells in normal adult blood. Small lymphocytes originate in lymphatic as well as lymphoid tissue, hence are found in abundance in the spleen and lymph nodes. Consequently, they are the most common cells found in the lymph.

Large Lymphocytes. These cells are similar in general appearance to the small lymphocytes; however, as indicated by their name, they are somewhat larger, ranging between 10 and 14 μ in diameter. Found occasionally in the blood of young children, large lymphocytes in the adult are encountered primarily in lymphoid tissue where they are considered to represent a juvenile form of the small lymphocytes.

Monocytes. These agranular leucocytes are large (from 10 to 18 μ in diameter), and generally possess an indented, kidney-shaped nucleus which is surrounded by a relatively more abundant cytoplasm than the lymphocytes. Monocytes constitute about 5 to 7 percent of the white cells, and are currently believed to represent a transitional stage in the formation of polymorphonuclear leukocytes. They should perhaps be considered together with the "polymorphs" as part of the circulating elements of the reticuloendothelial system (p. 555).

The granulocytes listed above are believed to originate from the same common primordial "stem cells" as do the erythrocytes within the myeloid tissue of the bone marrow (Table 9.6). Early in the

process of morphologic differentiation from a common stem cell, the typical basophilic, eosinophilic, or neutrophilic granules are formed. Thus these cell types may be distinguished readily from each other, as well as from the erythroblasts, at quite early stages in their metamorphosis.

The lymphocytes and monocytes are formed in a number of organs within the body, notably in lymphatic or lymphoid tissues.* Specifically the lymph nodes, spleen, thymus, tonsils, and lymphoid regions of the gut (p. 826) are sites for the production of lymphocytes and monocytes.

In general, the formation of leukocytes by the lymphatic system and bone marrow requires amino acids and vitamins that are similar in both type and quantity to those required for the production of erythrocytes. The white blood cells are especially sensitive to a deficiency of folic acid so that in extreme debilitating diseases leukocyte formation may be seriously depressed, even though more of these elements usually are needed in such conditions to combat infectious processes.

The life span of the various types of white blood cells varies considerably once these elements have entered the circulation. The three kinds of granulocyte each has an average life in the blood of about 12 hours, whereas during an active infectious process, the survival time may be curtailed to two hours. The lymphocytes enter the bloodstream continuously via the lymphatic drainage system (qv, p. 553); however, they cycle back rapidly from the blood capillaries through the interstitial fluid, thence back to the lymphatic system to enter the lymph nodes once again. Therefore a given cell may remain in the bloodstream for

---

The term "lymphatic" denotes a specific relationship to a lymphatic organ (ie, a lymph node) or to a lymphatic vessel (qv, p. 554). The term "lymphoid" refers to tissues that histologically resemble those of the lymphatic system in certain respects, eg, the splenic pulp.

Table 9.6   STAGES IN THE DEVELOPMENT OF LEUKOCYTES[a]

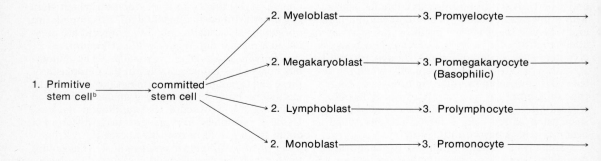

[a]*Data from Bishop (ed).* Overview of Blood. *1974, Blood Information Service.*
[b]*All of the formed elements of the blood, including the erythrocytes, are believed to originate from the same primitive stem cell.*

only a few hours at a time, even though the total life span of lymphocytes in general ranges between 100 and 300 days. This long survival time may be altered markedly, of course, depending upon the tissue needs for these cells. The *monocytes* pass freely between the blood and the interstitial space; their life span is unknown.

The primary function of the leukocytes is to provide a rapid and specific defense mechanism for the body against invasion by infectious microorganisms. The white blood cells, in particular the neutrophils and monocytes, are capable of rapid, independent motility. This property of ameboid motion (p. 150), coupled with their avid proclivity for the ingestion of foreign particulate matter (*phagocytosis*, p. 94), enables the white cells to pass rapidly in large numbers out of the capillaries by a process known as *diapedesis* into a region that is, for example, inflamed by the presence of bacteria. The leukocytes then are free to wander among the tissues and to engulf and thereby destroy large numbers of microorganisms.

Chemical substances within the tissues induce movement of the leukocytes either toward or away from the source of the substance, ie, either up or down the concentration gradient of the substance. These processes respectively are known as *positive* and *negative chemotaxis*. Thus, in a situation where a positive chemotactic gradient of a substance exists, the leukocytes extend pseudopodia toward the direction of the highest concentration of the substance and thus progress actively toward the source of the substance. For example, certain bacterial toxins elicit a positive chemotactic response (see also p. 550).

Selectivity on the part of the phagocytic leukocytes is essential; otherwise normal bodily components would be destroyed by these cells. Three factors play a role in this selectivity: (1) *Roughness* of the surface of a particle increases the possibility of phagocytosis. (2) The *surface electric charge* of particulate matter is important. Electropositive surfaces (such as may be found on dead cells or foreign particles) are more apt to be phagocytized than the negatively charged cellular components of the body. (3) Foreign particulate material may first combine with a circulating globular protein called an *opsonin* in the body. This mechanism promotes phagocytosis, as it permits the leukocytes to adhere closely to the foreign matter, a process which assists phagocytic ingestion. Furthermore, certain specific protein-free foreign substances known as *haptens* must combine with a carrier protein in the body in order that they be recognized as a foreign substance and thus produce an antigenic effect (see also p. 547).

Once ingested by a leukocyte, digestion of the foreign particle commences, provided that the foreign matter is amenable to chemical degradation.* Phagocytic cells in general are rich in intracellular proteolytic enzymes which can digest foreign proteins. Lipases which can digest the lipid component of certain bacterial cell walls (eg, tuberculosis bacilli) also are abundant within the macrophages. In addition, certain bactericidal substances are found in phagocytic cells. For example, polymorphonuclear neutrophilic leukocytes contain phagocytin as well as lysozyme, both important bactericides. Another mechanism of particular importance for the destruction of foreign microorganisms is the presence of peroxidase enzymes in the leukocytes. The activity of a peroxidase in the presence of hydrogen peroxide (H_2O_2) is toxic to bacteria, as the enzyme can oxidize directly certain organic constituents of these microorganisms, thereby killing them.

*Inert materials, such as carbon particles, are not degraded but rather are "eliminated" from the interstitial spaces by storage within phagocytic cells. When the phagocyte dies, a similar cell merely engulfs the released particle once again.

Table 9.6 (Cont'd)

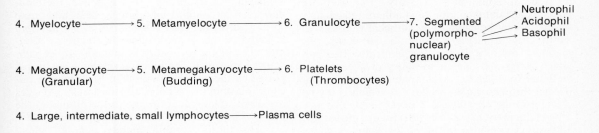

THROMBOCYTES (PLATELETS). The platelets are derived from the cytoplasm of certain giant polymorphonuclear cells (up to 40 μ in diameter) called *megakaryocytes* which are found in the myeloid red marrow of adults, and in the liver and spleen and other blood-forming tissues of embryos (Table 9.6). In turn, the megakaryocytes arise from hemocytoblasts whose nuclei hypertrophy, become horseshoe-shaped, then constricted into several lobulations following an increase of the total cytoplasmic mass. Consequently, the megakaryocyte nucleus alone undergoes five or six mitotic divisions, while concomitantly the cytoplasmic reticulum becomes more complex. Groups of vesicles surrounded by *platelet demarcation membranes* ultimately appear within the cytoplasm. These small partitioned-off cytoplasmic units then are shed from the parent megakaryocyte as platelets. The platelets themselves are found in the circulating blood as small nonnucleated cytoplasmic fragments about 2.5 μ in diameter and which contain the aforementioned vesicles (Fig. 9.2). Two kinds of granule or vesicle are present in the circulating platelets. Some of these granules, found centrally within the platelets as clumps or in chains, stain with azure blue or neutral red. A second type of platelet granule, which is scattered singly and at random throughout the platelets, stains with Janus green.

The life of circulating platelets has been estimated to range between 8 and 11 days. Functionally, the platelets have an important part in the mechanism of blood clotting (p. 542), and in addition these blood elements appear to have other important physiologic roles. For example, thrombocytes are quite "sticky," hence they tend to agglutinate together in masses and to form deposits upon roughened surfaces as well as upon foreign bodies such as bacteria. This response tends to enhance the bodily defenses against infection, as the leukocytes tend to phagocytize microorganisms more readily when the latter have been "pretreated" as noted above. Thus the platelets would tend to alter the surface properties of the microorganisms so that phagocytosis would be accomplished more readily. Furthermore platelets accumulate upon, and adhere tenaciously to, defects in the capillary endothelial wall as well as to cuts in larger vessels. In so doing, the platelets help to fasten mechanically the clot which forms subsequent to injury of a vessel, as well as to liberate substances necessary for normal thromboplastin production which is a vital element in the coagulation process itself. Platelets also liberate a vasoconstrictor material, 5-hydroxytryptamine or *serotonin*, which reduces the lumen size of injured vessels, thereby still further facilitating the complex process of hemostasis (p. 541). The platelets also contain materials which exhibit strongly antigenic properties as discussed on page 547.

PLASMA. The complexity of this fluid matrix of the blood is indicated by the listings of some of its principal constituents in Tables 9.2, 9.3, and 9.4. Normally, the plasma levels of every one of these components are regulated by homeostatic mechanisms to within rather narrow limits in an individual, so that the concentrations found in the blood do not show wide fluctuations.

The homeostatic mechanisms whereby other specific plasma constituents are regulated will be presented in context in subsequent chapters. Thus, the critically important maintenance of blood pH will be discussed in relation to respiratory physiology (Chapter 11) as well as under renal physiology (Chapter 12). Salt and water metabolism and their regulation are also linked intimately to renal function, as well as to certain endocrine substances (Chapter 15). Other specific metabolites and substances found in the plasma will be considered where appropriate in relation to digestion and metabolism (Chapters 13 and 14). Therefore, in this section emphasis will be placed upon the several blood proteins by way of introduction to their important functions in the blood clotting mechanism, the existence of blood types, and immune reactions, each of which will be discussed below (Fig. 9.7A).

Whole plasma contains three types of protein, *albumin, globulins,* and *fibrinogen.* These proteins may be separated readily by the process of electrophoresis (Table 9.2, Fig. 9.7B). That is, the various plasma proteins migrate at different rates in an electric field depending upon their individual molecular size and electric charge (see ampholytes and the isoelectric point, p. 23), as well as on the pH of the buffer which is used for the separation. In addition, the nature of the substrate upon which the separation is carried out (eg, on sheets of filter paper, starch blocks, or cellulose acetate films) can be chosen in order to effect the cleanest separation of the different plasma proteins.

It is important to remember that electrophoretic separation of the plasma proteins does not yield chemically pure molecular species. Hence all of the fractions contain some carbohydrate and lipid material which presumably is conjugated to the protein. For example, the albumin fraction contains bilirubin (p. 853), and about 75 percent of the total protein-bound plasma lipid is found in the α- and β-globulin fractions. Albumin also binds reversibly with many other substances in which combined form they are transported by the plasma (Table 9.7). For example calcium, phosphorus, iodine, certain drugs, and dyes (eg, Evans blue) all are found in the plasma in the protein-bound form, at least to a certain extent.

The term "serum" merely denotes whole plasma from which the fibrinogen has been removed by the process of clotting as discussed on page 545.

Sources of the Plasma Proteins. The hepatic parenchymal cells are the primary site for synthesis of plasma albumin to the extent of several

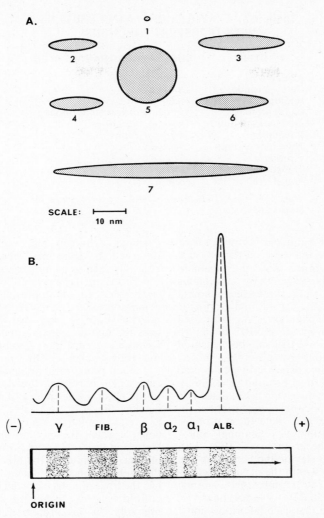

FIG. 9.7. **A.** Relative sizes and shapes of several protein molecules in relation to glucose (1). 2. Albumin (69,000); 3. α_1 lipoprotein (200,000); 4. β_1 globulin (156,000); 5. β_1 lipoprotein (1,300,000); 6. γ globulin (90,000); fibrinogen (400,000). The numbers in parentheses indicate the molecular weights of the proteins in daltons. **B.** Electrophoretic separation of the plasma proteins when a band of whole plasma is placed on the origin. The arrow indicates the direction of migration of the proteins toward the anode (+) and the upper curve gives the relative quantities of the proteins in the individual fractions. γ = gamma globulin; Fib. = fibrinogen; $\beta, \alpha_2, \alpha_1$ = globulins; Alb. = albumin. (**A.** Data from Harper. *Review of Physiological Chemistry,* 13th ed, 1971, Lange Medical Publications. **B.** Data from Best and Taylor [eds]. *The Physiological Basis of Medical Practice,* 8th ed, 1966, Williams & Wilkins.)

grams daily, although other cells (eg, the Kupffer cells) also are thought to contribute to the manufacture of this protein. Fibrinogen is produced exclusively in the liver. The various globulin fractions, including the α_1, α_2, β, and γ components, probably are synthesized by the reticuloendothelial cells of the body as a whole (p. 555), although about 80 percent of the total globulin synthesis has been shown to occur in the hepatic Kupffer cells. The lymphocytes and plasma cells, on the other hand, appear to be the main source of the γ globulin fraction which is concerned with the immune mechanisms of the body; ie, the γ globulins are synthesized in extrahepatic tissues.

Functions of the Plasma Proteins. The principal physiologic roles played by the plasma proteins may be summarized:

First, fibrinogen, which is similar structurally in the configuration of its molecule to such fibrous proteins as myosin and collagen, is indispensable for the coagulation (or clotting) of the blood (p. 545). Hence fibrinogen is an essential hemostatic element.

Second, the plasma proteins lend viscosity to the blood, thereby exerting an important influence upon the normal blood pressure as discussed in the next chapter (p. 625).

Third, the plasma proteins are important in the regulation of blood acid–base balance (Chapters 11 and 12; p. 756 and p. 804).

Table 9.7 EXAMPLES OF PLASMA PROTEINS WITH TRANSPORT FUNCTIONS[a]

PROTEIN	SUBSTANCES TRANSPORTED
Albumin	Pigments, ions
Ceruloplasmin	Copper
Haptoglobin	Hemoglobin
β-Lipoprotein	Lipids, cholesterol, hormones
α-2-Macroglobulin	Plasmin, hormones
Prealbumin	Thyroxin, retinol
Thyroxin-binding-protein	Thyroxin
Transcortin	Cortisol
Transferrin	Iron

[a]*Data from Bishop (ed).* Overview of Blood, *1974. Blood Information Service.*

Fourth, both the globulin and fibrinogen fractions exert an important influence on the physical stability of the blood. Since the cells remain in suspension in whole blood only by continuous mixing within the circulation, these elements tend to settle out when blood is drawn, and clotting is prevented by an anticoagulant. The so-called *erythrocyte sedimentation rate* (ESR), an important clinical diagnostic tool, thus can be measured in millimeters per hour by placing a blood sample mixed with an anticoagulant in a calibrated tube that is held in a vertical position. The rate at which settling of the erythrocytes takes place is a direct function of the plasma concentrations of fibrinogen and globulin; and since these proteins tend to facilitate agglutination (clumping) and/or rouleau formation of the erythrocytes (Fig. 9.8), clinical situations which elevate the fibrinogen and globulin concentrations within the plasma accelerate the sedimentation rate. Approximate normal values of the ESR are listed in Table 9.4; however, it is important to realize that each laboratory must establish its own range of normal values for the ESR. In pathologic situations, the ESR generally is increased; very rarely is it decreased. For example, extremely high sedimentation rates are observed in septicemia, whereas a moderate increase is seen in pulmonary tuberculosis.

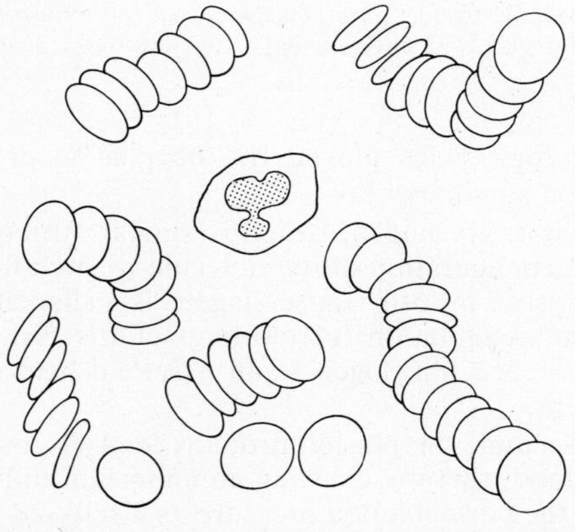

Fifth, the three types of plasma protein collectively exert an important influence upon the total osmotic pressure of the blood (p. 646), since their molecules are so large that they do not normally pass through the capillary membrane. This colloid osmotic pressure thus amounts to some 25 to 30 mm Hg (see also p. 558).

Each protein fraction exerts an osmotic pressure which is related directly to its concentration in the plasma, and inversely related to its molecular size (Fig. 9.7A and Table 9.2). Since fibrinogen is a relatively large molecule and is present in low concentration, this protein exerts a negligible effect upon the total colloid osmotic pressure of the plasma. Albumin, on the other hand, is present in the highest concentration in the plasma, but also has the lowest molecular weight. Consequently the colloid osmotic pressure of plasma depends primarily on this protein. Since albumin has an osmotic activity which is some 2.4 times that of the globulin fraction, then 80 percent of the total plasma osmotic pressure is caused by this one protein alone.

Sixth, antibodies, or immune substances, which react with microbial antigens are found in the γ-globulin fraction of the plasma proteins, and it is this fraction which is used after purification for clinical immunization against infectious hepatitis,* poliomyelitis, tetanus, measles, and other diseases. Other antibodies such as the isoagglutinins (p. 550) also are present in the β- as well as the γ-globulin fractions (p. 538).

Seventh, hormones, enzymes, and clotting factors all are found in the α- and β-globulin fractions of the plasma proteins which also serve as carrier molecules for the intravascular transport of many substances as indicated by a number of particular examples which are cited in Table 9.7. Similarly, certain plasma proteins also function to bind metals such as iron for transport in the bloodstream.

FIG. 9.8. Rouleaux formation by erythrocytes. Note that the cells become stacked upon each other like plates. A leukocyte is indicated at the center of the figure. (After Best and Taylor [eds]. *The Physiological Basis of Medical Practice,* 8th ed, 1966, Williams & Wilkins.)

Vaccines currently are available for both infectious (or type A) as well as serum (or type B) viral hepatitis.

Eighth, the plasma proteins also form a general pool which functions as a protein reserve from which the body can draw temporarily during times of fasting or periods of inadequate protein intake (p. 875).

It is important to realize that the concentrations of the various protein fractions in plasma can change independently of one another, and this effect can occur either with or without a concomitant alteration in the quantity of total protein. Thus in a number of pathologic states, the albumin and globulin levels may show a reciprocal relationship. That is, the plasma albumin/globulin ratio may change markedly, although the total concentration of the plasma proteins remains constant.

Hemostasis

The several physiologic mechanisms whereby the body is protected against a massive loss of blood that could follow a traumatic injury may be discussed in the general sequence of their temporal occurrence. Three mechanisms are involved in physiologic hemostasis; viz, vasoconstriction, platelet agglutination, and coagulation of the blood itself (Fig. 9.9).

VASOCONSTRICTION. Immediately following rupture of a blood vessel, the wall of the vessel contracts, thereby reducing the size of the lumen and decreasing the blood flow through that vessel. This *localized vascular spasm* is initiated by the operation of three mechanisms.

First, afferent nerve impulses, presumably transmitted from the injured region along pain fibers, enter the spinal cord whence efferent sympathetic impulses travel to the damaged vessels and their immediate vicinity causing reflex contraction and narrowing of the vessels. This immediate *neurogenic spasm* (ie, spasm induced by nerves) lasts for only a short time, however, then a second mechanism of vasoconstriction ensues. In toto, the neurogenic vascular spasm persists from less than one to several minutes. Second, direct traumatization of a vessel causes a localized *myogenic vascular spasm* to occur. This sustained (20- to 30-minute) effect results from the direct mechanical stimulation of the contractile elements themselves which are located within the vessel wall; hence the term "myogenic" is used to describe this process. Third, vasoconstrictor substances (eg, serotonin) may be liberated from the damaged tissues or blood elements (eg, the thrombocytes) which in turn cause contraction of the vessel walls through a chemical mechanism. The type of initial trauma is important in determining both the extent as well as the duration of myogenic vasoconstriction; thus cleanly severed vessels bleed more freely than do vessels which have been ruptured by a crushing or tearing injury. The net effect of vasoconstriction is to decrease the blood flow, thereby allowing time for the two other hemostatic mechanisms to become activated.

HEMOSTASIS AND THE PLATELETS. The second event following injury to a vessel and the attendant vasospasm involves the platelets (thrombocytes). These elements normally are round or oval discs; however, upon contact with a wettable surface (such as damaged capillary endothelium) the platelets now undergo a dramatic change in their physicochemical characteristics. First of all, the

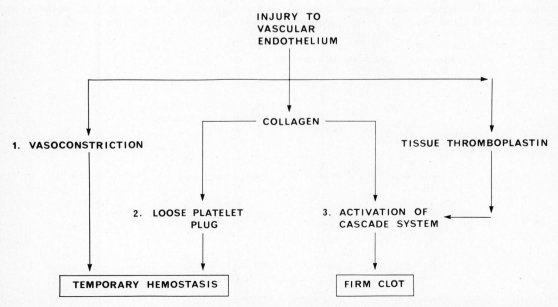

FIG. 9.9. Summary of the three physiologic mechanisms involved in hemostasis (1, 2, 3). The numbers in the figure correspond to the similar headings in the text under which each of these mechanisms is discussed. (Data from Ganong. *Review of Medical Physiology*, 6th ed, 1973, Lange Medical Publications.)

platelets become irregularly shaped with numerous projections. Second, and more important, however, the platelets become sticky so that they now become attached readily to the damaged endothelium of the vessel as well as to each other. Because of this adhesion, multiple layers of platelets are formed until a small loose plug forms in the injured region of the vessel. This plug of thrombocytes alone may be adequate to effect hemostasis, provided that the rupture in the vessel wall is sufficiently minute. Usually the platelet mass does not interfere with circulation through the vessel even though it is adequate to effect complete hemostasis.

Platelet breakdown also releases serotonin which in turn acts directly upon the contractile elements of the damaged vessels to enhance vasoconstriction as noted above.

BLOOD CLOT FORMATION OR COAGULATION. The final mechanism which is involved in hemostasis is evoked rapidly (15 to 20 seconds) or slowly (several minutes), depending upon the severity of the initial trauma. Three to six minutes after the rupture of a vessel, regardless of cause, the damaged region becomes filled with a gelatinous clot, and after 30 to 60 minutes the clot retracts, a process which is known as *syneresis*, and which closes the vessel still further.

Details of the complex mechanisms involved in blood clotting remain obscure despite the enormous amount of research effort which has been expended in this important area of physiology. The present discussion will be limited to a consideration of certain basic aspects of this phenomenon which appear to be reasonably well established.

Two general groups of substances have been found in blood and tissues which influence coagulation. Some of these materials, called *procoagu-lants,* expedite clotting, whereas others, termed *anticoagulants,* inhibit this process. In the aggregate, some 30 of these substances, or factors, have been described. One point appears clear. Blood will or will not coagulate, depending upon an extremely delicate balance between the net effects of the procoagulants and anticoagulants within the body.

It is also well established that in general the mechanism of blood clotting occurs in three sequential steps as indicated in Figure 9.10. First, a *prothrombin activator* is formed as the consequence of damage to the blood itself or of injury to a vessel. Second, the prothrombin activator in turn catalyzes the transformation of *prothrombin* into *thrombin.* Third, the thrombin, functioning enzymatically, converts fibrinogen into *fibrin threads* which entrap red cells, platelets, and plasma, thus forming the actual clot.

These three steps, which are summarized in Figure 9.10, now may be examined in somewhat greater detail. The actual chemical synthesis of prothrombin activator, which initiates clotting, is poorly understood. There are, however, two systems whereby the coagulation of blood can be initiated (Table 9.8, Fig. 9.11). These systems consist of an *extrinsic mechanism* which is external to the blood itself, and an *intrinsic mechanism* that is present within the blood.

The extrinsic system is responsible principally for the initiation of clotting within the body when blood vessels and tissues are damaged. Such trauma liberates a substance from the damaged tissue which is called "tissue thromboplastin." Actually tissue thromboplastin consists of a complex mixture of lipoproteins having one or more phospholipid (eg, cephalin) moieties. Tissue thromboplastin in turn must react with several blood factors (V, VII, and X), as well as calcium,

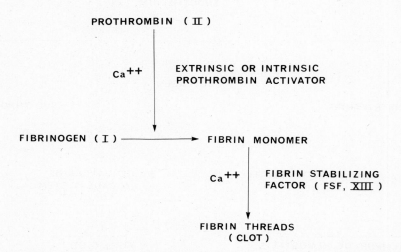

FIG. 9.10. General summary of the blood clotting mechanism. The Roman numerals in this figure correspond to those of the clotting factors listed in Table 9.8 (cf. Fig. 9.11). (Data from Guyton. *Textbook of Medical Physiology,* 4th ed, 1971, Saunders.)

Table 9.8   FACTORS INVOLVED IN THE COAGULATION OF BLOOD[a]

NUMBER OF FACTOR	NAME(S)	REMARKS
I	Fibrinogen	Average concentration in plasma 3000 mg/liter. MW = 341,000 daltons. Composed of three polypeptide chains (glycoprotein and globulin). Thrombin cleaves two pairs of α- and β-chains to yield insoluble fibrin (3% of molecule).
II	Prothrombin	Trace is present in plasma. MW = 68,000 daltons. One glycoprotein chain present. When activated, converted to thrombin, MW $\cong$ 39,000.
III	Thromboplastin, thrombokinase, platelet factor III	Lipoprotein
IV	Calcium ion (Ca^{++})	—
V	Proaccelerin, labile factor, prothrombin accelerator, accelerator globulin	Globulin
VI	(None designated)	—
VII	Stable factor, serum prothrombin conversion accelerator (SPCA), proconvertin, prothrombin conversion factor (PCF)	β-Globulin
VIII	Antihemophilic factor (AHF), antihemophilic globulin (AHG), antihemophilic factor A	β_2-Globulin
IX	Christmas factor, plasma thromboplastin component (PTC), antihemophilic factor B	β-Globulin
X	Stuart factor, Stuart-Prower factor	α-Globulin
XI	Plasma thromboplastin antecedent (PTA), antihemophilic factor C	β_2-Globulin
XII	Hageman factor, glass factor	α-Globulin
XIII	Fibrin-stabilizing factor (FSF), L–L factor, fibrinase, fibrin serum factor, Laki–Lorand factor	FSF is a — SH (sulfhydril) enzyme which, in presence of Ca^{++}, reacts with fibrin to cross link fibrin molecules into long polymeric threads.

[a]Data from *Mountcastle (ed). Medical Physiology, 12th ed, Vol I, 1968, Mosby; Bishop (ed). Overview of Blood, 1974. Blood Information Service.*

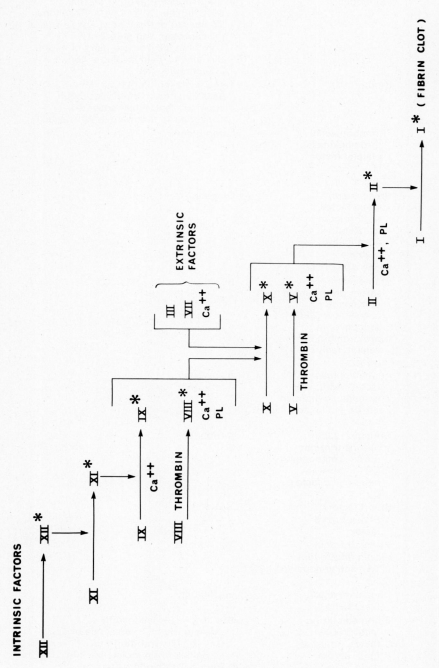

FIG. 9.11. Diagram summarizing the cascade system of blood clotting. The Roman numerals denote the specific clotting factors listed in Table 9.8, whereas the asterisks signify activation of these factors (cf. Fig. 9.10). (Data from Bishop [ed]. *Overview of Blood*, 1974. Blood Information Service, Buffalo, NY; Mountcastle [ed]. *Medical Physiology*, 12th ed, Vol. I, 1968, The CV Mosby Co.)

before it becomes "prothrombin activator" that is capable of converting prothrombin into thrombin.

The intrinsic system for initiating the clotting process resides wholly within the blood itself. The platelets have a critical role in this intrinsic mechanism (blood drawn very carefully into a clean, nonwettable container clots very slowly in contrast to blood "contaminated" by fluids released by damaged tissues). As indicated earlier, platelets become extremely sticky when in contact with a damaged vessel. Hence many of these elements disintegrate rapidly under such conditions, thereby liberating into the blood *platelet factor* (III) which contains a phospholipid. Similarly to tissue thromboplastin, this platelet factor must also react with several protein blood factors including V, VIII, IX, X, XI, and XII (Table 9.8) before it becomes a prothrombin activator which can initiate clotting. However, in contrast to tissue thromboplastin, the intrinsic platelet factor III acts far more slowly than does the tissue thromboplastin of the extrinsic mechanism. Thus factor III may require up to an hour to become fully effective whereas tissue thromboplastin may exert its maximal effect in only a few minutes. Exposure of blood to *collagen* also initiates the intrinsic clotting mechanism.

Both intrinsic as well as extrinsic systems are important for initiating the clotting process in vivo, but principally the latter. The so-called *cascade system,* which is illustrated in Figure 9.11, converts the loose platelet plug into a dense clot and this firm plug effectively seals the ruptured vessel in most instances (Fig. 9.9).

Certain specific blood proteins involved in the clotting process now can be discussed in somewhat greater detail (see Table 9.8).

First, *prothrombin* is an unstable α_2-globulin having a molecular weight of about 69,000. This component of the plasma proteins is found normally at a concentration of about 15 mg percent in the plasma, and it is split readily into two smaller compounds. One of the hydrolytic products is thrombin which has a molecular weight of about 34,000, almost precisely half that of prothrombin.

Second, *fibrinogen* has a high molecular weight (340,000) and it occurs as a constituent of the plasma proteins at concentrations ranging between 100 and 700 mg percent. Because of this high molecular weight, fibrinogen normally is not found in the lymphatic and interstitial fluids. However, when capillary permeability is increased markedly in abnormal situations, fibrinogen is present in these fluids, and clotting of interstitial fluid and/or lymph now can occur (p. 553).

Thrombin is a proteolytic enzyme which acts on fibrinogen to split off two low-molecular-weight polypeptides per molecule of fibrinogen, thereby forming one molecule of *activated fibrin* or *fibrin monomer* (Fig. 9.9). Huge numbers of these monomeric molecules of fibrin now can undergo rapid polymerization, and thus form extremely long fibrin threads, which in turn form the framework or reticulum of the clot. Calcium ions

in addition to another factor called the *fibrin stabilizing factor* (XIII) augment the bonding between the fibrin monomers as well as the polymeric fibrin chains themselves, thereby strengthening and stabilizing the fibrin threads.

The clot itself ultimately consists of a three-dimensional network of fibrin threads in which are entrapped blood cells, plasma, and platelets. The fibrin threads adhere tenaciously to the damaged blood vessel and thus anchor the clot in place.

Shortly after the jellylike clot is formed, it commences to shrink or contract, a process that is known as *syneresis*. The clear plasma expressed from the clot during syneresis now lacks essentially all of the fibrinogen and blood factors responsible for clotting, and this fluid now is called *serum*. Syneresis causes the edges of the damaged vessels to become more closely apposed, or pulled together, and this process constitutes the final event in hemostasis.

Platelets are required in large numbers for normal clot retraction to occur. These structures adhere in huge numbers to, and thus help bond together, the individual fibrin threads. Hence in some unknown way the platelets facilitate the shortening of elements within the fibrin meshwork, enabling syneresis to occur.

It has been known for centuries that cadaver blood may reliquefy and not clot again; such blood has been employed clinically in the past for transfusion since no anticoagulants were needed. The mechanism which is responsible for producing liquefaction of clotted blood is called the *fibrinolytic system*. This process is summarized in a simplified manner in Figure 9.12, and basically the mechanism involves the enzymatic hydrolysis of the fibrin clot by means of a fibrinolysin called *plasmin*. Note that the fibrinolytic system itself is activated by activated factor XII. Consequently when the clotting mechanism is initiated so too is the mechanism for the eventual dissolution of the clot.

PREVENTION OF CLOTTING AND ANTICOAGULANTS. Normally, intravascular clotting appears to be inhibited, first by the extreme smoothness of the endothelium of the entire vascular bed in contact with the blood, and second by a negatively charged monomolecular layer of protein which is adsorbed upon the endothelial surface of the vessels and which actively repels the platelets and intrinsic blood factors that are responsible for initiating the clotting process. These two factors effectively prevent *contact activation* of the platelets and clotting factors within the blood so that the blood remains liquid in vivo.

Within the blood itself, the important anticoagulants are those which prevent the activation of thrombin. The two most significant of these antithrombin agents are the fibrin threads themselves which form during coagulation, and an α-globulin fraction of the plasma proteins called

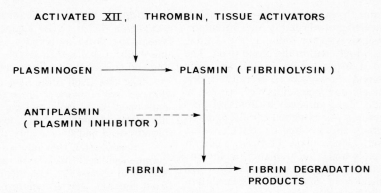

FIG. 9.12. Outline of the fibrinolytic system for the dissolution of blood clots. (Data from Bishop [ed]. *Overview of Blood,* 1974. Blood Information Service, Buffalo, NY; Ganong. *Review of Medical Physiology,* 6th ed, 1973, Lange Medical Publications.)

antithrombin.* During clot formation, around 90 percent of the thrombin is removed locally from the blood near the clot by adsorption upon the fibrin threads; consequently spread of the clotting process is limited. The remaining thrombin is inactivated by combination with plasma antithrombin in about 10 to 20 minutes, so that the clot remains localized at the injured site and does not spread throughout the entire cardiovascular system with disastrous consequences. The fibrinolytic system also tends to minimize excessive clot formation (Fig. 9.12).

Heparin, a powerful anticoagulant, is produced and secreted under physiologic conditions by the pericapillary basophilic *mast cells,* and perhaps also by the circulating basophilic cells of the blood. From a chemical standpoint, heparin is a sulfated mucopolysaccharide which is composed of repeating units of sulfated glucosamine and glucuronic acid. The mechanisms whereby the anticoagulant action of this compound is manifest may be summarized briefly. First, heparin prevents the activation of factor IX, the plasma thromboplastin component (PTC). Second, acting in conjunction with an albumin cofactor, heparin inhibits the action of thrombin on fibrinogen, thereby preventing the formation of fibrin threads. Third, heparin assists in the deactivation of thrombin by accelerating the rate at which thrombin reacts with antithrombin. Fourth, heparin augments the quantity of thrombin which is adsorbed upon any fibrin clots that may be present thereby effectively sequestering the thrombin.

Since heparin can inactivate the blood coagulatory mechanisms both in vitro as well as physiologically in vivo, this compound is of con-

Actually several antithrombin components are present. Interestingly, a so-called transamidase system *may be responsible for clot formation in vivo without the participation of thrombin at all. Thus in arteriosclerosis, pregnancy, or in women taking the contraceptive pill, the transamidase system can convert fibrinogen molecules into a reticular network having peptide bonds but without the activation of any prothrombin.*

siderable clinical importance, for example, in preventing clot formation during surgical procedures.

Clotting can also be prevented by removing free calcium ions from whole blood. The reader will note in Figure 9.11 that this metallic ion is a critical requirement in several stages of the coagulation mechanism. Therefore whole blood treated with oxalate or citrate ions will not clot, as the calcium has been removed chemically.

In conclusion to this section, the student should remember that the two plasma proteins essential to normal clot formation, prothrombin and fibrinogen, are both synthesized primarily by the liver. Diseases of this organ can either delay or completely inhibit normal blood coagulation. Furthermore, prothrombin formation cannot take place in the liver without an adequate supply of vitamin K (p. 889). In the absence of this vitamin a prolonged clotting time or a bleeding tendency may result. In this regard, it is of considerable clinical importance to distinguish between clotting time, bleeding time, prothrombin time, and thromboplastin generation time, as all of these measurements provide the physician with a quantitative assessment of various aspects of the clotting mechanism.

Clotting time signifies an in vitro test in which a sample of blood is collected and placed in a chemically clean glass tube treated with silicone to prevent wetting and which is tilted once about each 30 seconds until coagulation occurs, an effect which usually is manifest in five to eight minutes.

Bleeding time signifies the time that is required for clot formation in vivo after an earlobe or fingertip has been pierced. The bleeding time normally ranges between three and six minutes, but is rather crude as well as inaccurate because the size of the instrument used to make the wound as well as the hyperemia present in the pierced region either individually or together can affect the bleeding time significantly.

Prothrombin time gives an indication of the total quantity of prothrombin that is present in the blood. A blood sample is treated with oxalate immediately following withdrawal so that no prothrombin can be converted into thrombin. Subsequently a large excess of calcium ion as well as tissue thromboplastin (tissue extract) is mixed rapidly with the oxalate-treated blood sample. The time required for clotting to occur gives an indication of the quantity of prothrombin present. Normally, clotting occurs in roughly 10 seconds; however, as in all other clinical laboratory tests, the precise method employed can influence the results considerably. Consequently, it is essential that standards be instituted and adhered to closely by each particular laboratory.

The *thromboplastin generation time* provides a sensitive test for the detection of defects in the formation of intrinsic thromboplastin (also called prothrombinase or intrinsic prothrombin-converting principle), and therefore of the factors involved in the production of this substance. A sample of plasma from the patient, his serum, calcium chloride, and platelets (or a platelet lipid substitute) all are incubated together simultaneously. At intervals, the clotting time of a normal citrated plasma sample is determined following the addition of aliquots of the incubation mixture.

Inflammation

The process of inflammation occurs as a physiologic response of the body to tissue injury, regardless of whether the injury is caused by chemicals, a burn, trauma, or bacterial invasion. The chemical *histamine*, among other substances, is liberated by the damaged tissue. The damage in turn causes an increase in local blood flow by dilatation of the vessels in conjunction with an elevated capillary permeability to fluid (p. 576) as well as to the plasma proteins; thus a *local edema* results. Local clotting of the fluid exudate from both blood and lymphatic capillaries results as a consequence of fibrinogen leakage; hence circulation in the injured region stagnates. In this manner the inflamed region essentially becomes walled off from the rest of the body, thereby effectively decreasing the rate of spread of toxins or bacteria. Phagocytic cells, especially neutrophils, are attracted to the inflamed region in large numbers by positive chemotactic agents, and these cells now adhere to the damaged capillary walls at the locus of inflammation, a process known as *margination*. Subsequently these cells move into the damaged tissue by diapedesis and then proceed to phagocytize the cellular debris and any microorganisms that may be present.

Wound Healing

After a blood clot has formed, fibroblasts commence to migrate into the injured region within a few hours so that ultimately a fibrous connective tissue scar develops from these cellular elements within one or two weeks at the site of the original trauma. The fibrinolytic mechanism discussed earlier and shown in Figure 9.12 is activated automatically during the clotting process, and this system effectively dissolves the clot gradually as wound healing progresses. One of the plasma proteins is a euglobulin called *plasminogen* (or *profibrinolysin*). During blood coagulation, a quantity of this substance becomes incorporated into the clot and then becomes activated by the presence of previously activated factor XII (Table 9.8). Some time later the plasminogen within the clot is transformed into a powerful proteolytic or digestive enzyme similar to trypsin of the pancreas. This enzyme, called *plasmin* (or *fibrinolysin*) then proceeds to digest, or lyse, the fibrin within the clot as well as any other proteinaceous blood substances that are present in the vicinity, such as fibrinogen and prothrombin, as well as activated factors V, VIII, and XII. Therefore, plasmin reduces the ability of the blood to coagulate in the immediate neighborhood of the clot, and the clot itself is dissolved and rendered fluid.

Some Basic Immunologic Concepts

GENERAL TYPES OF IMMUNITY. Certain basic terms and concepts related to the complex field of immunology must be introduced prior to a discussion of the major blood groups and blood typing.

The human body is said to have a certain degree of natural immunity or resistance to infection by microorganisms or their toxins. Consequently the blood contains a large number of substances which act directly on microorganisms or their toxins to destroy or inactivate them. Examples of these general substances are (1) *interferon,* a large protein which inactivates many viruses; (2) *properdin,* another large protein which reacts directly with gram-negative bacteria to destroy them; (3) certain *basic polypeptides* which similarly react with gram-positive bacteria; (4) *lysozyme,* a mucolytic polysaccharide which dissolves certain bacteria. These materials, among many others, confer on the human body a considerable degree of natural immunity against invasion by microorganisms.

In addition, however, the human body also possesses the ability to develop an *acquired immunity* in response to the specific stimuli provided by invading microorganisms, viruses, and their toxins as well as to foreign tissues and other substances.

Acquired immunity to an appropriate foreign organism or material can be developed by the body in two ways. First, the stimulus of a suitable foreign agent, or *antigen,* results in the development of specific proteins to that substance, which proteins then form part of the circulating gamma globulin (immunoglobulin) fraction of the plasma proteins. This type of response is called *humoral immunity,* and it develops slowly, taking weeks or even months to appear following antigenic invasion. Second, in *lymphocytic* (or *cellular*) *immunity* the antigenic substance reacts with the lymphocytes in such a manner that these cells become sensitized specifically to the particular antigen; later the sensitized lymphocytes can attach to the antigenic agent and destroy it.

Antigenic agents in general must have a molecular weight which is greater than 10^4 daltons in order to evoke an immune response, and they also must possess repeating active prosthetic groups or radicals on the surface of their molecule. Hence proteins and the large polysaccharides are usually antigenic. For example, since almost all bacterial toxins (as well as the organisms themselves) are composed of proteins, large polysaccharides, or mucopolysaccharides, these substances are all strongly antigenic.

Many nonprotein compounds which have a molecular weight below 10^4 daltons can act as incomplete antigens, or *haptens.* Such materials alone are incapable of stimulating antibody production. However, when a hapten is coupled to a suitable carrier protein, then, and only then, is it capable of stimulating an immune response, ie, antibody production. In humoral immunity, the specificity of the antibody so developed is directed primarily toward the hapten. In cell-mediated immunity, on the other hand, the antibody is directed toward both the hapten and the carrier protein. Examples of haptens that may induce this type of immune response are certain drugs, exfoliated dermal cells and dandruff, chemicals which may be present in drugs, and some industrial chemicals.

The gamma globulin antibodies formed in response to the stimulus of a given antigen are, in general, capable of reacting solely and specifically with that one antigen and no other. The antigen–antibody reaction may at times involve a very weak link between the two substances, probably via hydrogen bonds.

CELL-MEDIATED AND HUMORAL IMMUNITY. The cell-mediated and humoral immune responses are of such great physiologic and clinical importance that they must be considered in somewhat greater detail. The cell-mediated immune response is active against invasion of the body by viruses and fungi; this system also is responsible for the initiation of foreign tissue and tumor rejection. The humoral immune system, on the other hand, is active against bacterial infections as well as reinfection by virus. There are two functionally distinct mechanisms present in the body, one of which is responsible for mediating cellular immunity, the other humoral immunity. At times, both of these mechanisms can act in a cooperative fashion, although in general each functions independently of the other. Not surprisingly, there are two functionally distinct cell populations in lymphoid and other body tissues (especially the blood) and each of these is responsible for mediating either cellular or humoral immunity. During their developmental stages, both of the two kinds of lymphoid cells have identical morphologic appearances, and they cannot be distinguished on this

basis alone. Furthermore, each class of lymphocyte originates from the same primitive hemopoietic stem cells that give rise to all of the other formed elements of the blood (Tables 9.5 and 9.6).

The two kinds of circulating lymphocyte are called *thymic lymphocytes* and *bursal lymphocytes* or *T-cells* and *B-cells,* respectively, depending upon their developmental fate (Fig. 9.13). As indicated in this figure, the ultimate site of origin of all immunologically active cells in the mammal is the embryonic yolk sac; however, the stem cells that originate in this structure subsequently pass through liver and spleen before moving on to their eventual habitat in the bone marrow. Note that the thymic lymphocytes (T-cells) are formed by the passage of stem cells through the thymus whereas the bursal lymphocytes (B-cells) must pass through certain other tissues in the mammal in order to be converted into bursal lymphocytes.*

Lymphoid cells found within the thymus are called *thymocytes* in order to distinguish them from the T-cells themselves, which are merely circulating lymphocytes that have a thymic origin.

As indicated in Figure 9.13, when a T-cell is activated by a specific antigen, it responds by the production of molecules of a specific substance called a *lymphokine,* and this material now assists in the inactivation of the antigenic material which initiated the synthesis of the lymphokine in the first place. This mechanism, in brief, summarizes the function of the T-cells in cell-mediated immunity. On the other hand, a thymus-independent B-cell may be converted into a circulating plasma cell through the cooperative action of activated T-cells, and the plasma cell then can elaborate a specific humoral antibody against the specific antigen. Alternatively an antigen may convert a B-lymphocyte in the peripheral circulation directly into a circulating plasma cell which also is capable of secreting the appropriate antibody against that antigen.

In brief, therefore, cellular immunity is mediated by the T-cells, whereas humoral immunity is mediated by a second class of cells derived from B-lymphocytes, the plasma cells. Under certain circumstances, a cooperative interaction between both of these distinct functional types of lymphocyte occurs in the presence of some antigens, as indicated by the double arrow in Figure 9.13.

Within each broad class of lymphocyte there is a vast number of individual and functionally specific cells; hence it is a remarkable fact that each T- or B-cell can recognize one, and only one,

In the bird, nonspecific stem cells are converted into B-cells exclusively in the bursa of Fabricius, *a mass of lymphoid tissue located near the cloaca. Mammals lack such a discrete lymphoid structure; nevertheless stem cells are converted into typical bursal cells, and it appears that several tissues may be involved in this process, including possibly the appendix and other intestinal lymphoid tissues. The liver and spleen also may participate in B-cell formation in the mammal.*

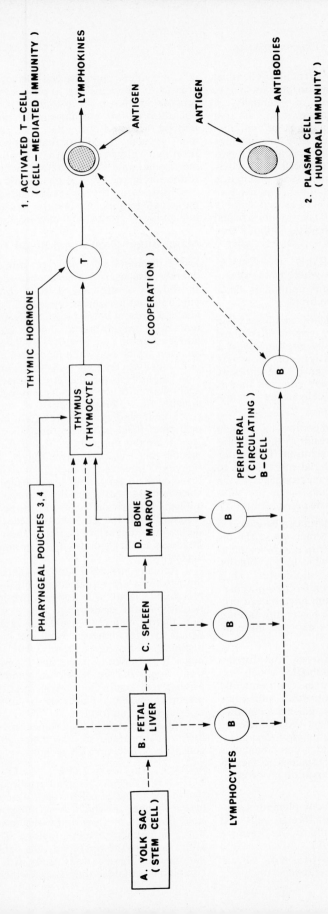

FIG. 9.13. The development and differentiation of T- and B-lymphocytes. The embryonic yolk sac (A) is the ultimate source of all stem cells which are the precursors of erythrocytes (Table 9.5) and leukocytes (Table 9.6), as well as lymphocytes. In the embryo (dashed lines), stem cells from the yolk sac pass in succession to the liver (B), the spleen (C), and ultimately the bone marrow (D). From each of these organs, the stem cells pass either to the thymus (thymocytes) where they are converted into, and released as, T-cells which underlie cell-mediated immunity, or the stem cells pass to various other structures (see text) where they are converted into B-cells which underlie humoral immunity. Note that the thymus per se is derived from the third and fourth pharyngeal pouches of the embryo, and that this structure can secrete a hormone which later may affect the fate of the circulating T-cells. Mutual cooperation between circulating T- and B-cells is indicated by the slanted dashed line connecting these elements. (After Cooper and Lawton. *Sci Am* 231:58, 1974.)

antigenic substance by the particular chemical groupings which are present on the molecule. Since there are literally millions of potentially antigenic materials present in nature, it appears that there are an equally large number of immunologically active cells which are able to recognize each of them. Once an antigen has activated a particular type of T- or B-cell, that cell now undergoes rapid mitotic division until a large population (or *clone*) of the cell which is alone responsive to that specific antigen is present. In fact, a suitably activated B-cell (or plasma cell*) can synthesize and secrete approximately 2,000 identical antibody molecules per second until it dies several days later.

MECHANISMS OF ANTIGEN INACTIVATION. The five mechanisms whereby the inactivation of antigens by humoral antibodies takes place now may be summarized briefly.

Neutralization. A toxic antigenic substance upon gaining entrance into the body may be neutralized by combination with its antibody, presumably by a "masking," hence inactivation, of the toxic prosthetic groups on the antigen by the antibody molecules. The antigen–antibody complex then is phagocytized and digested by cells of the reticuloendothelial system (p. 555).

Precipitation. Suitable juxtaposition of antigen and antibody molecules may result in their combination and precipitation from solution. The resulting particulate matter is rendered more liable to phagocytosis and digestion by macrophages of the reticuloendothelial system than when the molecules were present in a more dispersed form.

Agglutination. Dispersed bacteria can be clumped or agglutinated by antibodies which cause the organisms to attach to each other in enormous numbers in a manner that is similar to the binding together of antigen–antibody molecules. Thus agglutination of bacteria serves a twofold purpose. First, following agglutination the bacteria are prevented from invading the host tissue any farther, and second, the organisms are exposed more directly to phagocytosis.

Lysis. Antigenic bacteria, viruses, and foreign cells also are destroyed by antibodies which first attach to their outer membranes. Subsequently *complement,* a complex lipoprotein substance which normally is present in the plasma, combines with the antigen; the complement then penetrates and ruptures the membrane of the microorganism, a process called lysis. Note that complement is inactive until an antibody first attaches to the antigenic cell, following which activation of the complement takes place.

Opsonization. Antibodies and complement sometimes may become attached to foreign cells and yet may be unable to destroy them. The presence of complement upon the cell surface markedly increases its susceptibility to phagocytosis, and this process is known as *opsonization.* Through the operation of this mechanism the phagocytic rate may be increased up to 100 times above that normally present. It appears that complement can intermingle physically with the phagocytic cell membrane, and thereby greatly facilitate ingestion of the foreign cell by the phagocytic cell.

Following this brief introduction to immunologic terminology and mechanisms, the major human blood groups now may be considered.

Major Blood Groups and Blood Typing

Early attempts at blood transfusion were frequently disastrous before it was recognized that the bloods of different persons contain different antigenic and immune properties. Thus the cellular antigens present in the blood of one individual may react with the plasma antibodies of the blood of another person resulting in a so-called *transfusion reaction* which may take place either immediately or at indefinite time intervals following the transfusion. Hemolysis of the erythrocytes as well as agglutination of the red cells may occur in the recipient, the clinical severity of the transfusion reaction depending upon the extent to which either or both of these processes take place.

A person normally does not form antibodies against his own cellular antigens *(autoimmune reaction);* however, if such an individual is transfused with incompatible blood from another person, then antibodies will develop against *all* of the antigens not originally present in the blood of the recipient. It is indeed fortuitous that a number of the major antigens are common among various individuals so that bloods can be grouped and typed accordingly.

Actually, around 30 antigens capable of inducing reactions are found commonly in the various blood cells, and these antigens are localized particularly in the erythrocyte membranes. In addition, there are more than 100 other known antigens which have less potency insofar as inducing transfusion reactions are concerned, but which assume major importance in tissue and organ transplantation. Some of the lesser antigenic groups are primarily of academic interest in studying inheritance, establishing parentage or racial interrelationships, and so forth.

Of major clinical importance insofar as transfusion is concerned are two major groups of antigens, the A–B–O and Rh groups.

A–B–O BLOOD TYPES AND GROUPS: AGGLUTINOGENS. Human erythrocytes contain two major antigens, called types A and B. Because of the manner in which these antigens are inherited, however, a given individual may have one, both, or neither of these factors. The A and B antigens make the red blood cells liable to agglutination when mixed (transfused) with other blood containing suitable antibodies, which are called

---

*A plasma cell is a descendant of a lymphocyte, but it is different functionally from the parent lymphocyte.

agglutinins, in the plasma. Consequently, the type A and type B antigens are called *agglutinogens*.* Therefore it is on the basis of these agglutinogens that bloods are typed into groups for the purpose of transfusion.

There are four major A–B–O groups into which donor and recipient bloods are classified, depending upon the specific agglutinogens which are present in the red cells. These groups are summarized together with their genotypes in Table 9.9.

Inheritance of Agglutinogens. The allelomorphic genes which determine the agglutinogens in any individual are determined by the pattern of these genes in the parents, and these genes can be any of three types, A, B, or O. Thus in one parent the genotype, or combination of two allelomorphic genes, can be OO, OA, OB, AA, BB, or AB. Every person has one of these six patterns of different genotypes, and there are a total of 21 possible genetic combinations between the male and female. An example will clarify this. If an AA father mates with an OB mother, there are only two possible genotypes in the offspring, and these are OA and AB. If, however, an OA father mates with an AB mother, there are now four possible combinations in the offspring, OA, OB, AA, and AB.

There is no particular dominance among the O, A, or B allelomorphic genes. The type O gene, however, is essentially without function, so that there is no type O agglutinogen produced in the red cells. Types A and B genes, on the other hand, do induce the development of powerful cellular agglutinogens. Consequently, if any genes on the parental chromosomes are type A and/or B, then one or both of these agglutinogens will be present in the erythrocytes of the offspring.

AGGLUTININS. If type A agglutinogens are absent from the erythrocytes of an individual, then the plasma of that person develops circulating antibodies which are known as anti-A agglutinins (Table 9.9). A similar situation obtains for an individual lacking type B agglutinogens, but in this

*Agglutinogens also are referred to as group-specific substances.

instance the plasma contains anti-B agglutinins. On the other hand, the plasma of type O persons contains both anti-A as well as anti-B agglutinins, whereas that of type AB individuals has no agglutinins whatsoever.

The agglutinins, which are not present in the plasma at birth, reach a maximum concentration (or *titer*) in children at about 10 years of age. Subsequently the agglutinin titer declines gradually throughout the rest of the person's life. These antibodies are globulins which are produced by the plasma cells which form the other humoral antibodies. What is the antigenic stimulus to produce these specific circulating agglutinins, since these proteins neither develop, nor are they found, in the presence of the appropriate A and B agglutinogens? It is believed that suitable antigenic substances to stimulate agglutinin formation are ingested or acquired via bacterial infection, although this is hardly a complete and satisfactory answer to this complex question.

TRANSFUSION REACTIONS. If improperly matched bloods, such as red cells containing type A agglutinogens and blood containing anti-A agglutinins, are transfused into a recipient or mixed together in vitro, the red cells clump together (or agglutinate) into coarse, macroscopically visible masses. This type of antigen–antibody reaction results from the bivalent nature of the agglutinin. Thus each molecule of agglutinin can attach to two erythrocytes causing the latter to adhere to one another. The resulting clumps then are transported through the circulatory system to small vessels which they occlude. Assuming that the degree of this reaction is not too severe and that the patient survives, the reticuloendothelial system phagocytizes the agglutinated cell masses within a few days, thereby releasing hemoglobin into the bloodstream.

Parenthetically, it should be noted here that the sudden occlusion of blood vessels by agglutinated blood cells is termed *thromboembolism*. The term *thrombus* in turn signifies an aggregation of blood factors (especially fibrin and platelets) with entrapment of cellular elements which may cause vascular obstruction (or *occlusion*) at the site of

Table 9.9 MAJOR HUMAN BLOOD GROUPS AND THEIR RELATIVE OCCURRENCE[a]

BLOOD GROUPS	GENOTYPE[b]	AGGLUTINOGENS (ANTIGENS)	AGGLUTININS (ANTIBODIES)	PERCENT OCCURRENCE[c]
O	OO	None	Anti-A + Anti-B	47
A	OA or AA	A	Anti-B	41
B	OB or BB	B	Anti-A	9
AB	AB	A and B	None	3

[a]Data from Guyton. Textbook of Medical Physiology, 4th ed, 1971, Saunders.

[b]During meiosis (maturation or reduction division) of the sex cells (sperm or ova), the genotypes are divided so that after this process is complete each haploid chromosome contains only one allelomorphic gene.

[c]The relative frequencies of occurrence of the various blood groups are for a caucasian population and are approximate only.

origin. "Thrombosis" denotes ·the formation or presence of a thrombus.* In contrast to the term thrombus, an *embolus* is defined as a clot or a foreign mass that is transported by the blood from its site of origin (or point of entrance into the body) to a smaller vessel which becomes occluded thereby. Thus *embolism* signifies a sudden blockage of an artery by a clot or foreign material (eg, air, bacteria) which has been transported to its final point of arrest by the current of blood.

Occasionally antibodies called *anti-A* and *anti-B hemolysins* develop in the plasma in addition to the agglutinins. These hemolysins, when present, do not cause agglutination of the red cells, but rather they induce hemolysis (p. 534) by their reaction with red cells and complement, the mechanism being similar to the lysis of microorganisms by complement as described earlier (p. 550).

Usually in a transfusion reaction erythrocyte agglutination is the principal event seen as hemolysins are less commonly present in plasma than are agglutinins. Ultimately, however, some lysis of red blood cells takes place even though agglutination was the principal response, as the reticuloendothelial macrophages ultimately destroy the clumped cells, thereby liberating hemoglobin into the bloodstream.

BLOOD TYPING. This determination usually is performed in routine clinical practice by diluting a small quantity of the unknown blood type with normal (isotonic) saline in order to obtain a red cell suspension. A small amount of this suspension then is mixed, usually on a microscopic slide, with drops of known anti-A and anti-B typing sera which are prepared from the blood of individuals having extremely high titers of these antibodies. A few minutes after mixing the unknown blood and the typing serum, the droplets are examined either grossly or microscopically for signs of agglutination of the cells. If clumping has occurred, then an immune reaction has taken place between the cells and that particular serum. The results of this procedure are summarized in Table 9.10.

In a clinical emergency requiring immediate transfusion, it is not necessary to know the actual blood types of the donor and the recipient. In such a situation, the bloods can be tested for compatibility by *cross-matching*. That is, a suspension of donor red cells is mixed with a sample of defibrinated recipient serum. Conversely, a sample of the donor serum also is mixed with a sample of red cells obtained from the recipient. If no agglutination is observed, the two bloods probably are sufficiently compatible to proceed safely with the transfusion.

The cells of individuals having type O blood contain neither type A nor type B agglutinogens (Table 9.9). Small volumes of type O blood can be

Table 9.10 BLOOD TYPING

| ERYTHROCYTES (AGGLUTINOGENS) | SERUM[a] | |
	Anti-A Agglutinin	Anti-B Agglutinin
O	−	−
A	+	−
B	−	+
AB	+	+

[a]A minus sign indicates no reaction; ie, the red cells remain suspended whereas a plus sign indicates agglutination (clumping) of the cells.

transfused into practically any recipient without the development of a severe and immediate agglutination of the recipient's red cells. Persons having type O blood sometimes are called *universal donors*. If, however, large volumes of type O blood are transfused into an incompatible recipient, either an immediate or a delayed agglutination of the recipient's erythrocytes can occur because the agglutinins which were infused will not be diluted sufficiently to preclude a reaction from taking place.

On the other hand, persons having type AB blood sometimes are called *universal recipients* because their plasma contains neither anti-A nor anti-B agglutinins. Small volumes of all other blood types can be infused into such persons without a deleterious response. If, however, a large volume of incompatible blood is transfused into a type AB recipient, the agglutinins in the donor blood could accumulate to such an extent that the recipient's cells would agglutinate. Usually, however, it is very rare during transfusion for the recipient's cells to become agglutinated to any major extent, since the donor blood plasma is diluted immediately by the recipient plasma; hence the titer of infused agglutinins is decreased too rapidly to cause serious agglutination of the recipient's erythrocytes. However, the donor blood does not significantly dilute the recipient's plasma agglutinins; consequently agglutination of the donor's cells can occur readily if the bloods are not compatible. For these reasons, sterile plasma which has been lyophilized (ie, dried while frozen under high vacuum) and which contains no erythrocytes is used extensively for transfusion in certain clinical situations as no agglutinogens are present (see transfusion in various types of shock, p. 1243 et seq).

Rh FACTORS AND BLOOD TYPES. Another important group of blood factors is found in the so-called Rh system.† Unlike the A–B–O agglutinins which develop spontaneously, the Rh agglutinins generally are formed as the consequence of exposure to a specific antigen, but before sufficient antibodies have developed to the antigen to cause

---

Frequently a distinction is made clinically between thrombosis and the normal activation of the clotting mechanism.

†*The designation Rh is derived from the first two letters of the name rhesus, a species of monkey in whose blood this factor was discovered originally.*

any significant transfusion problems. There are about eight types of Rh agglutinogens, each of which is called an Rh factor. Four of these agglutinogens are *Rh negative* and four are *Rh positive.* The mechanism underlying the inheritance of these Rh factors is not completely clear, although the gene for each is known to be a dominant mendelian trait. Therefore if a chromosome contains a particular one of these allelomorphic genes, then the corresponding Rh factor is present in the blood of that individual.

About 85 percent of Caucasians are Rh positive, whereas 15 percent are Rh negative. If the erythrocytes from an Rh positive individual are injected into an Rh-negative person, then anti-Rh agglutinins develop slowly, the maximum titer being reached some two to four months later. Repeated exposure to the Rh antigen results in a markedly elevated titer of anti-Rh agglutinins (or "sensitization") in that individual.

Anti-Rh agglutinins behave similarly to the anti-A and anti-B plasma agglutinins discussed above. The Rh agglutinins also attach to the Rh-positive red cells causing them to agglutinate, and hemolysis occurs secondarily to phagocytosis of the clumped red cells by the reticuloendothelial cells.

Blood is typed for the various Rh factors using specific anti-Rh sera as described above for the A–B–O agglutinogens.

One particularly important condition that can result from the operation of the Rh factors outlined above is erythroblastosis fetalis; this disease will be considered under Clinical Correlates (p. 573).

THE LYMPH AND TISSUE FLUIDS

During its passage through the capillaries of the vascular system, the blood normally loses a certain quantity of water and protein as well as salts and other plasma constituents to the interstitial spaces, thereby forming the *interstitial fluid,* and as noted earlier, this fluid is the internal environment within which the cells and tissues of the body live and by which they are nourished (Fig. 9.1).

The *lymph* essentially is interstitial fluid which enters the lymphatic vessels. Hence the composition of lymph is far more variable than that of the blood for a number of reasons (Table 9.11). For example, the specific region of the body in which it originates, the nutritional state of the individual, and the degree of bodily activity all influence not only the composition but also the rates of formation and flow of this important body fluid. Following a discussion of the morphologic aspects of the human lymphatic system, the several factors affecting lymph composition and flow will be discussed.

General Functions of the Lymphatic System

Fundamentally, the lymphatic system provides an alternative and indirect route whereby interstitial fluid and its various constituents can be returned

Table 9.11 COMPOSITION OF LYMPH COMPARED WITH THAT OF PLASMA[a, b]

SUBSTANCE (SPECIES[c])	LYMPH[d]	PLASMA
Na (M)	127.0 mEq/liter (T)	127.0 mEq/liter
K (M)	4.7 mEq/liter (T)	5.0 mEq/liter
Ca (M)	4.2 mEq/liter (T)	5.0 mEq/liter
Cl (M)	98 mEq/liter (T)	96.0 mEq/liter
Glucose (M)	124 mg % (T)	123.0 mg %
Urea (D)	23.5 mg % (C)	21.7 mg %
Creatinine (D)	140.0 mg % (C)	137.0 mg %
Amino acids (D)	4.8 mg % (C)	4.9 mg %
Phosphorus (D)		
Total	11.8 mg % (C)	22.0 mg %
Inorganic	5.9 mg % (C)	5.6 mg %
Protein (D)		
Total	4.89 gm % (T)	7.04 gm %
Albumin	2.34 gm % (T)	3.15 gm %
Globulin	2.36 gm % (T)	3.33 gm %

[a]*Data from Best and Taylor (eds). The Physiological Basis of Medical Practice, 8th ed, 1966, Williams & Wilkins; Ruch, Patton (eds). Physiology and Biophysics, 19th ed, 1966, Saunders.*
[b]*The data included in this table are approximate only, as the composition of the lymph can vary considerably depending upon many factors including the tissue and its state of metabolic activity. Thus lymph contains many enzymes, hormones, and metabolic by-products of the cells in addition to the substances listed. On passage through the lymph nodes, lymph also acquires a large cellular population, principally lymphocytes. Thus thoracic duct lymph may contain between 8×10^4 and 12×10^4 of these elements per mm^3, whereas peripheral lymph normally contains only a few hundred cells per mm^3.*
[c]*M = man; D = dog.*
[d]*T = thoracic duct lymph; C = cervical (peripheral) lymph.*

to the blood. Unlike the closed circulatory system, in which blood passes in a cyclic fashion throughout a complete circuit (Fig. 10.15, p. 588), the lymphatics provide a one-way system of drainage channels from the cells and tissues that lead back to the blood vascular system (Fig. 9.14). Thereby fluids and substances which cannot be removed from the interstitial regions via direct reabsorption into the blood capillaries are removed continually from the interstitial compartment and at rates which are commensurate with their formation.

All of the tissues of the body have a lymphatic drainage system. Exceptions to this general statement are the superficial regions of the skin, the central nervous system (see also p. 681), the deeper-lying regions of the peripheral nerves, the bones, and the endomysium of the skeletal muscles. Actually, however, even these tissues possess microscopic interstitial spaces through which fluid can percolate; therefore eventually the fluid from these regions also reaches a plexus of lymphatic vessels.

Approximately 90 percent of the water which is filtered from the arterial end of a vascular capillary is reabsorbed by the time the blood reaches the venous end of the vessel as discussed on page 558. The remaining 10 percent of the filtered water, now called interstitial fluid, enters the terminal

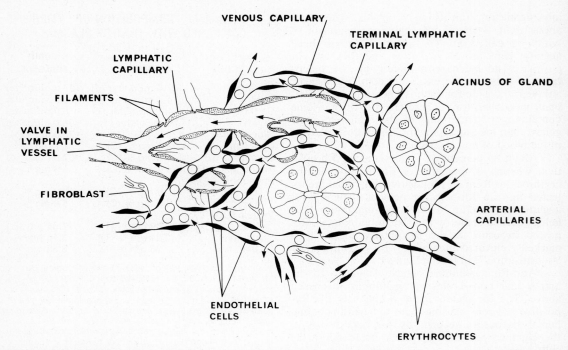

FIG. 9.14.   The histologic relationships between a capillary network and the terminal lymphatics in an exocrine gland. The arrows signify the direction of flow of blood and lymph into and out of the interstitial fluid space. (Data from Guyton. *Textbook of Medical Physiology,* 4th ed, 1971, Saunders; Ruch and Patton [eds]. *Physiology and Biophysics,* 19th ed, 1966, Saunders.)

lymphatic capillaries and as lymph is returned ultimately to the blood.

The fine structure of a single blind-ending terminal lymphatic capillary is shown in Figure 9.14. Note that the individual endothelial cells which form the walls of these vessels are contiguous but not continuous, as is the case with the endothelium of the vascular capillaries (Figs. 9.14 and 9.17). The endothelial cells of the lymphatics merely overlap somewhat at their edges, thereby lending a high degree of permeability to the lymphatic capillaries. In fact, actual gaps may be present between the cells, and the basement membrane is imperfect or absent completely. Substances of high molecular weight, such as proteins, can pass almost unimpeded from the interstitial fluid into the lumen of the lymphatic vessels through the pores, a situation which is quite unlike that found in blood capillaries.

Note also in Figure 9.14 the inward-facing "flap-valve" disposition of the endothelial cells which permits only a one-way flow of interstitial fluid. The *anchor filaments* attached to the exceedingly delicate endothelial cells also tend to lend structural support to the vessels. As a consequence of this overlapping cellular arrangement, interstitial fluid, once it has entered a terminal lymphatic capillary, cannot escape readily. This is true because an increased pressure within the vessel closes the pores in the vessel wall by causing mechanical apposition of the edges of adjacent en-

dothelial cells. In addition, the lymphatic channels having a diameter larger than 100 to 200 μ are invested with progressively heavier layers of elastic and collagenous connective tissue fibers, as well as smooth muscle bundles. The larger lymphatics are equipped with delicate paired valves which are oriented in the direction of lymph flow, and these structures arise as endothelial folds from the innermost layers of the walls of the lymph vessels. Immediately above the paired flap valves the lymphatic vessel may be provided with an abundant array of smooth muscle fibers whose contraction would assist in the propulsion of the lymph.

The minute individual lymphatic capillaries throughout the body branch and anastomose freely to form elaborate plexuses. The ever-larger lymphatic vessels ultimately join either the *thoracic duct* or the *right lymphatic duct.* The thoracic duct drains lymph from practically all of the lower body regions and this lymphatic vessel empties into the venous system at the junction of the left subclavian and internal jugular veins. The thoracic duct also receives lymph from the left arm as well as the left thoracic and head regions prior to emptying into the venous circulation. The right lymphatic duct receives lymphatic drainage principally from the right side of the head and neck, right arm, and portions of the right thorax; this vessel empties into the veins at the junction of the right internal jugular and subclavian veins.

Small quantities of lymph also can enter the venous system of the lower body directly in the inguinal region, and possibly at loci in the abdomen as well.

All of the larger lymphatic vessels are abundantly provided with valves so that the flow of lymph, albeit sluggish, is assured of a unidirectional course toward the venous circulation.

Lymph Nodes and the Reticuloendothelial System

Situated along the larger lymphatic vessels at intervals are found highly organized flattened and rounded structures known as the *lymph nodes* (Fig. 9.15). These nodes range between 1 and 25 mm in diameter, and are especially prominent during certain infections in the cervical, axillary, and inguinal regions. Lymph nodes essentially are filters composed of masses of lymphatic tissue (or nodules) which are contained within a capsule that is composed of connective tissue penetrated by afferent and efferent lymphatic vessels.

As noted above, the lymphatic vessels, unlike the vascular capillaries, are freely permeable to large molecules, and even to foreign particulate matter such as microorganisms. Consequently all such foreign matter upon entry into the interstitial compartment of the body must traverse the lymphatic system prior to entering the general circulation.

The lymphatic tissue within the lymph nodes comprises an essential portion of the so-called *reticuloendothelial system* (RES) of the body. The reticuloendothelial system itself is composed of various cells lining many vascular channels in addition to the lymph nodes, and these cells are capable of actively phagocytizing foreign bodies such as bacteria and viruses in addition to forming antibodies against these infectious agents as discussed earlier. Phagocytic cells of the reticuloendothelial system are found in the bone marrow, spleen, liver (Kupffer cells), and lymph nodes among many other tissues. All of these phagocytic cells are closely related and described by the general term *reticulum* (or *mesenchymal*) cells.

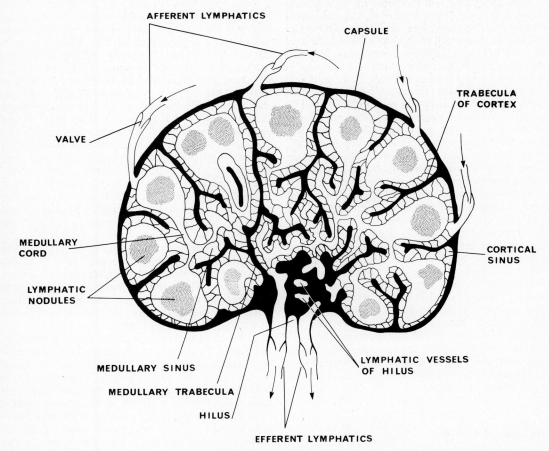

FIG. 9.15. A cross section of a lymph node showing the relationship between the afferent and efferent vessels and disposition of the valves. The pattern of lymph flow is indicated by the arrows. The trabeculae of the cortex have their origin in the capsule (heavy black lines) and these structures divide the cortex into ampullae. Note that the trabeculae of the medulla are continuous with those of the cortex. Blood vessels are not indicated. (After Bloom and Fawcett. *A Textbook of Histology,* 9th ed, 1968, Saunders.)

The stem or reticulum cell is a primitive element regardless of where it is found; hence it retains the latent potentiality of undergoing differentiation into many quite specialized cell types as depicted in Figure 9.16. Thus, reticulum cells in bone marrow (stem cells) can form either hemocytoblasts, which in turn form erythrocytes, or else may evolve into myeloblasts for the production of leukocytes (Tables 9.5 and 9.6). The reticulum cells also can form lymphocytes in the lymph nodes (Fig. 9.13). The actively ameboid tissue *histiocytes* (or *clasmatocytes*) which wander constantly through the tissues as macrophages are related closely to the reticulum cells. These macrophages ingest and digest foreign particulate matter, microorganisms, and dead or necrotic body cells themselves. Histiocytes in traumatized tissues also can turn into fibroblasts and thereafter form collagen fibers to assist in wound healing. Primitive reticulum (or stem) cells also can develop into specialized *plasma cells*, which are essential to the formation of humoral antibodies (p. 548).

Monocytes and lymphocytes, as well as histiocytes, all can become actively motile macrophages under appropriate conditions, and these elements are important in inflammation (p. 547) as well as in infections. Such elements thus function as components of the reticuloendothelial system to remove bacteria as well as necrotic body cells.

Consequently the principal bodily defenses against infection or invasion by foreign materials depend both upon a reservoir of primordial stem (reticulum) cells and upon specialized cells developed therefrom which can either engulf and digest the foreign agent, or else, over longer periods of time, form specific immunoglobulins against the invading agent or substance.

A specific and dramatic example of the efficacy of the lymph nodes as mechanical filters for bacteria may be cited. Experimentally, the iliac and popliteal lymph nodes of dogs have been perfused for over an hour with solutions of virulent streptococci (2.5×10^7 colonies/ml), and at the end of this interval the thoracic duct lymph was found to be sterile! Viruses are not removed by nodal filtration, however, and it should also be remembered that the lymph nodes themselves, once badly infected, can serve as focal sources for generalized infection of the bloodstream, although normally bacteria are destroyed within the lymph nodes after their removal from the lymph. Similarly the lymphatic channels can serve as ducts for the metastasis of cancer.

Formation of Lymph

Normally the volume of lymph is equal to the difference between the volume of fluid which leaves the vascular capillaries at their arterial end and the volume which is reabsorbed once again at the venous end. Thus lymph formation balances the quantity of vascular fluid which is exchanged in any tissue or organ of the body, and in man roughly two to four liters of lymph per day leave the peripheral tissues, and this fluid is continually being returned to the vascular circulation. Lymph formation represents only a very small fraction of the total flux of water across the capillaries. The latter has been estimated in man to be about 70 percent of the total blood volume flowing through a given capillary bed per minute. However, normal lymph flow is essential to the maintenance of normal tissue volume, because if this excurrent lymph flow is blocked, swelling of the tissue rapidly ensues and edema develops (p. 579).

The most important functions of normal lymphatic drainage of the body are to remove excess proteins and fluid from the interstitial spaces. There is no other channel than the lymphatics whereby proteins can be returned to the circulatory system, so that the continual removal of the protein from the interstitial space is critical to the life of the individual; and if this process is interfered with in any major way, normal capillary dynamics become so deranged that continued survival is impossible (p. 558).

As stated earlier in this section, lymph is interstitial fluid which has entered the lymphatic vessels; the composition of this fluid reflects in nearly identical fashion that of the interstitial fluid in that part of the body whence it was derived.

The interstitial fluid protein concentration is low, on the average being about 1.8 gm/100 ml; this value also is the average protein concentration in the lymph which is derived from most of the peripheral tissues. Lymph produced in the liver, however, may have a protein concentration as high as 6 percent, and intestinal lymph may contain 3 to 4 gm percent protein. Since about one-half of the total lymph formed in the body is derived from the liver and intestines, the thoracic duct lymph contains roughly between 3 and 5 gm percent protein in contrast to about 7 gm percent in plasma (Table 9.11).

The important role of the lymphatics in the absorption of nutrients, especially fats, from the gastrointestinal tract, will be examined in Chapter 13 (p. 866). It should be mentioned in passing, however, that after a fatty meal the thoracic duct lymph may contain a concentration of fat that can be as high as 1 or 2 percent.

The fluid that is filtered out through the arterial end of the peripheral capillaries under the driving force of a hydrostatic pressure head developed by the heart (p. 617) normally contains around 0.15 percent protein, although the interstitial fluid itself usually contains about 10 times more protein than this. The final concentration of about 1.5 percent protein in the lymph results largely from the reabsorption of water at the venous end of the capillary, a process which concentrates the protein, whereas the concomitant reabsorption of protein back into the vascular capillary is negligible. The net result of the operation of these factors is that a marked concentration of the interstitial fluid (hence lymph) proteins develops.

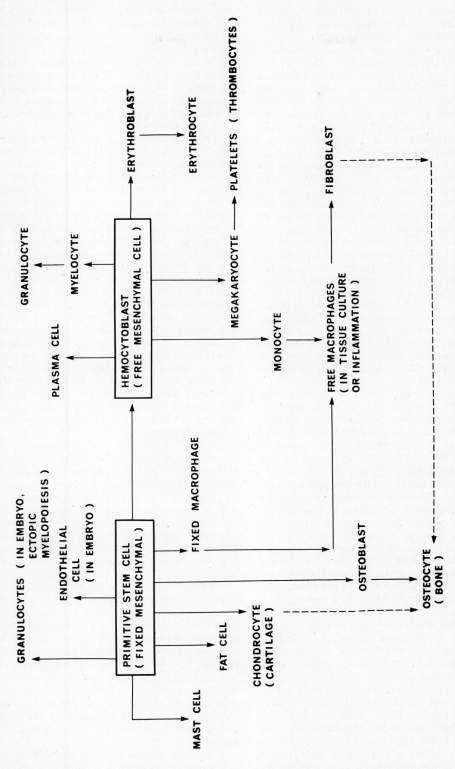

FIG. 9.16. The overall relationships among the primitive fixed stem (or reticular or mesenchymal) cell and the various cells of the blood and connective tissues. The dashed lines indicate potential transformations which can occur under special circumstances only (eg, neoplastic growth). Not all steps in the development of each cell type are indicated (cf. Tables 9.5 and 9.6; Fig. 9.13). (Data from Bloom and Fawcett. *A Textbook of Histology,* 9th ed, 1968, Saunders.)

Capillary Dynamics and Interstitial Fluid

Certain functional aspects of capillary dynamics now must be considered in relation to the formation of interstitial fluid, hence the lymph.

The vascular capillaries are the actual functional portion of the cardiovascular system, as it is here that the actual exchange of materials between the blood and the interstitial fluid takes place and vice versa. The true capillaries are the smallest vessels found in the blood vascular system, and these vessels have a diameter that ranges between 7 and 9 μ. The capillaries are composed of a single layer of endothelial cells covered by a thin basement membrane so that the entire thickness of the entire capillary wall is about 0.5 μ (Fig. 9.17). The ultrastructure of the capillary wall is such that minute pores between adjacent endothelial cells are present which have a diameter between 80 and 90 Å. Although these submicroscopic pores represent only 0.001 percent of the total surface area of the capillaries, they are of considerable functional importance as through these regions water and dissolved substances pass from the blood to the interstitial fluid and vice versa.

Diffusion (p. 85) is the major process whereby water and other substances traverse the capillary membrane, and this process can be exceedingly rapid. For example, the rate of diffusion for water across the capillary wall is about 40 times greater than the linear flow rate of plasma through the capillary.

The volume transfer of water between the capillaries and interstitial fluid across the endothelial interface is regulated by four major factors, and these factors in turn determine whether fluid moves out of, or into, the plasma. These important parameters involved in capillary dynamics may be considered individually (refer to Fig. 9.18).

First, the functional mean intracapillary pressure, which has been measured experimentally both directly and indirectly, appears to be approximately 17 mm Hg. Although the intracapillary pressure fluctuates widely under various physiologic conditions, it is generally conceded that at the arterial end of the capillary the average hydrostatic pressure is around 32 mm Hg, whereas at the venous end it has fallen to about 15 mm Hg. This hydrostatic pressure tends to force fluid out of the capillaries.

Second, the interstitial fluid colloid osmotic pressure also tends to cause movement of fluid out of the capillaries, and this osmotic flow of water results from the movement of a normally small quantity of plasma proteins through the larger pores in the capillary membrane, thereby increasing the net osmotic pressure of the interstitial fluid (average 4.5 mm Hg; range 0–10 mm Hg). The plasma proteins found in the interstitial fluid are chiefly albumin and globulin, which are able to penetrate the capillary wall far more readily than the much larger fibrinogen molecules. Parenthetically, however, it should be noted that lymph clots when withdrawn from the body, and presumably the mechanism involved in this process is similar to that which is present in plasma (Figs. 9.10 and 9.11). Therefore some fibrinogen must be able to escape from the lumen of the vascular capillaries into the extracellular fluid.

As the total volume of interstitial fluid in the human body is around 12 liters, the total quantity of protein in the interstitial fluid compartment is almost equal to that present in the plasma, despite the lower net concentration of protein that is found in the interstitial fluid.

Third, the interstitial fluid (or tissue) pressure tends to move fluid inward across the vascular capillary membrane. This pressure, like the capillary pressure, has been found extremely difficult to measure directly because of the tiny spaces involved (on the order of 1 μ). However, indirect measurement of this factor by several techniques has indicated that this pressure normally is around −7 mm Hg; that is, the interstitial fluid pressure actually is 7 mm Hg below atmospheric (ambient) pressure. Most investigators, however, consider the hydrostatic pressure of the tissues to range from 0 to +10 mm Hg. Hence this positive tissue pressure would tend to move fluid inward across the capillary membrane from the interstitial space, as indicated in Figure 9.18. A negative tissue pressure would, on the other hand, tend to steepen the outward hydrostatic pressure gradient between the capillary lumen and the interstitial fluid. This process must be subjected to further experimental evaluation before a definitive statement can be made insofar as any physiologic role can be ascribed to the negative interstitial fluid pressure in capillary dynamics.

Fourth, the plasma colloid osmotic pressure (or oncotic pressure) also tends to move fluid into the capillaries. This osmotic force is developed by the plasma proteins which are found in solution in the plasma at a total concentration of about 7 gm percent, compared to around 1.8 gm percent in the interstitial fluid. Most of the osmotic pressure developed in the plasma is due to the presence of these proteins which behave as a colloidal suspension even though they are in true solution within the plasma.

Furthermore, since the capillary membrane is impermeable to the vast majority of the plasma protein molecules that are present, these molecules increase the total colloid osmotic pressure of the plasma some 50 percent, because of the operation of the Donnan equilibrium as discussed on page 25. As the direct consequence of the Donnan effect, and since the plasma proteins are negatively charged nondiffusible ions (anions), a large number of cations, principally sodium, remain in the capillary plasma thereby increasing the total number of osmotically active particles in the plasma.

The total colloid osmotic pressure of normal human plasma is around 25 mm Hg; and of this total, about 17 mm Hg are produced by the dis-

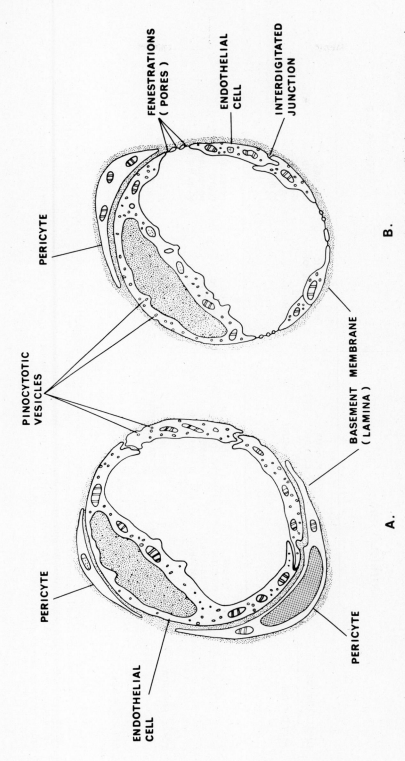

PINOCYTOTIC
VESICLES

PERICYTE

FENESTRATIONS
(PORES)

ENDOTHELIAL
CELL

INTERDIGITATED
JUNCTION

PERICYTE

BASEMENT MEMBRANE
(LAMINA)

PERICYTE

ENDOTHELIAL
CELL

A.

B.

FIG. 9.17. The ultramicroscopic relationships among the principal elements that comprise a continuous (or "muscular") capillary having an uninterrupted endothelium (A), and a fenestrated capillary in which the endothelium varies in thickness, the thinnest areas having pores that are closed by extremely thin diaphragmatic membranes (B). The average wall thickness of both of these commonly-encountered types of capillary measured through the cytoplasmic portion of their component endothelial cells is about 0.5μ. (After Bloom and Fawcett. *A Textbook of Histology*, 9th ed, 1968, Saunders.)

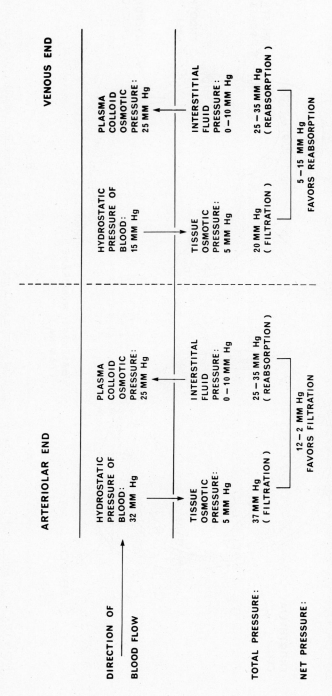

FIG. 9.18. Normal capillary dynamics in man according to the Starling hypothesis. The integrated mean capillary hydrostatic pressure is 17 mm Hg (Note: this mean pressure is *not* the arithmetic mean pressure; see discussion on page 635). It is important to realize that there is a smooth and gradual transition in the magnitude of the hydrostatic pressures of the blood between the arteriolar and venous ends of the capillary and not a sudden and abrupt change as indicated between the left and right halves of the figure (vertical dashed line). See also Figure 9.20. (Data from Ruch and Patton [eds]. *Physiology and Biophysics,* 19th ed, 1966, Saunders.)

solved protein, and 8 mm Hg by the cations held in the plasma by the electric charge of the proteins through the participation of the Donnan equilibrium.

The relative fluid balance between capillary plasma and interstitial fluid results from an extremely delicate interplay among the four factors considered above, as well as the capillary permeability to specific compounds.

Physiologic Mechanisms Regulating Lymph Flow

As mentioned earlier, only 10 percent of the total plasma fluid which is filtered from the capillaries finds its way back to the blood via the interstitial fluid and lymph. In humans, the total lymph flow into the venous circulation from all sources has been estimated to be about 120 ml/hr, a volume that amounts to only 0.001 percent of the calculated rate at which water diffuses across the capillaries in both directions (p. 645), and only 10 percent of the rate of water filtration from the arterial ends of the vascular capillaries to the interstitial spaces in the entire body. Therefore the net lymph flow represents an extremely small fraction of the total water exchange between the plasma and interstitial fluid.

There are two principal mechanisms whereby the physiologic regulation of the rate of lymph flow is achieved, the interstitial fluid pressure and mechanical compression of the lymphatic vessels per se.

INTERSTITIAL FLUID PRESSURE. As stated above, the mean interstitial fluid pressure at rest normally is subatmospheric, being around -7 mm Hg. Some authors cite normal mean values for this factor as ranging between 0 and $+10$ mm above atmospheric for tissue pressures, and these values are taken as an indication of the interstitial fluid pressure. Regardless of the actual numerical value of the interstitial fluid pressure with respect to the outside of the body, however, an elevation of this factor will increase the flow of interstitial fluid into the terminal lymphatics, thereby directly increasing the rate of lymph flow. Anything which tends to increase the interstitial fluid pressure will augment the rate of lymph flow. Several factors which elevate lymph flow are, first, an elevated intracapillary pressure or an increased capillary permeability; second, a reduced plasma osmotic pressure; and third, an increased interstitial fluid osmotic pressure. The latter factor would operate through increasing the net flux of water into the interstitial fluid compartment.

It is important to realize that the values cited above for interstitial pressure are mean levels only, and that under physiologic conditions these values can (and do) fluctuate markedly and rapidly both above and below the average values which are recorded under experimental conditions in anesthetized animals.

THE "LYMPHATIC PUMP." Since not only the larger lymphatic capillaries but also the entire system of lymphatic vessels are supplied liberally with valves, then any mechanical pressure or massage, regardless of its cause, exerted upon a tissue or lymphatic vessel will cause movement of lymph in one direction only. This "lymphatic pump" ensures the slow but steady one-way progression of the lymph back into the blood vascular system. The rate of lymph flow thus may be augmented between 3 and 14 times above the resting level during exercise, because of the active muscular contractions, as well as by passive movements of the body, arterial pulsations, or external manipulation (compression) of the tissues.

Contraction of the smooth muscle fibers in the lymphatic vessels per se also would assist in the propulsion of lymph, although man lacks rhythmically pulsatile lymph hearts such as are found in the frog, bat, rat, and guinea pig.

Any process which impedes normal lymph flow or which increases the rate of interstitial fluid formation without a concomitant increase in the rate of lymph flow leads to a net accumulation of interstitial fluid or *edema*. The several mechanisms of edema formation are considered on page 576.

MISCELLANEOUS BODY FLUIDS

In this section a number of specialized extracellular body fluids will be discussed. These transcellular fluids, although present in the human body in rather small total individual volumes (Fig. 9.1), all serve a number of important and specific physiologic roles. The reader should note the intimate interrelationship that exists among the various transcellular fluids, the capillary blood, the interstitial fluid, and the lymphatic drainage. Although the individual transcellular fluids may have a composition that differs considerably from that of plasma, nevertheless water provides the continuum among these various fluids, and ultimately all are derived from blood constituents.

Cerebrospinal Fluid

The total volume occupied by the brain and spinal cord is approximately 1,650 ml; approximately 140 ml of this total volume is occupied by the cerebrospinal fluid (CSF). A principal function of this fluid is to act as a cushion for the brain (p. 254), and as both brain and cerebrospinal fluid have essentially the same specific gravity, the brain literally floats within the skull.

Cerebrospinal fluid is found within the ventricles of the brain and in the subarachnoid space (between the arachnoid membrane and pia mater) which surrounds both the brain and spinal cord as well as in the cisterns, which are merely dilatations of the subarachnoid space in certain regions (refer to Figs. 7.14, p. 253, and 7.15, p. 255). The most important of these dilatations are the *cisterna magna,* which is located beneath and behind the cerebellum, the *cisterna pontis*, found on the ventral aspect of the pons, and the *cisterna basalis*, which also contains the arterial loop known as the circle of Willis.

FORMATION AND CIRCULATION OF CEREBRO-SPINAL FLUID.　The principal site of formation of the cerebrospinal fluid is in the *choroid plexuses* which are located within the ventricles of the brain (Fig. 7.15, p. 235). These choroid plexuses are found in the temporal horns of the lateral ventricles, in the posterior region of the third ventricle, and in the roof of the fourth ventricle. These highly vascular tufted structures are composed of many small pouches which project into the ventricles of the brain, and are composed of pia mater, capillaries, and the ependymal cells which line the ventricles. The choroid plexus has an afferent arterial and efferent venous blood supply. The pia and ependymal cells form a richly vascular double-layered membrane termed the *tela choroidea*. The cerebrospinal fluid is secreted into the ventricles and its solutes are derived from blood passing through the capillaries of the tela. Some additional cerebrospinal fluid probably is formed by the ependymal cells lining the ventricles and spinal cord, although the total volume of CSF derived from this source probably is quite low.

The cerebrospinal fluid is being formed continually and reabsorbed at a rate of approximately 750 ml per day, and the net pressure of the CSF is regulated by a balance between these two factors. An average value for the cerebrospinal fluid pressure in a normal supine individual is 130 mm water (about 10 mm Hg), although considerable variations above and below this value may be encountered in practice.

The circulation of the cerebrospinal fluid takes the following pathway (see also Fig. 7.14, p. 253; discussion on p. 254). From the lateral ventricles, the fluid passes through the foramina of Monro into the third ventricle; together with the additional fluid formed in the latter chamber, the cerebrospinal fluid then traverses the aqueduct of Sylvius into the fourth ventricle where more fluid is produced. The combined CSF fluid volumes thence pass through the two lateral foramina of Luschka as well as the antral medial foramen of Magendie into the cisterna magna.* Thence the cerebrospinal fluid passes upward to the arachnoid villi of the cerebrum through the subarachnoid spaces and the small tentorial opening of the mesencephalon, whence it is reabsorbed.

ABSORPTION OF CEREBROSPINAL FLUID.　The main route of absorption of the excess cerebrospinal fluid produced each day is through the arachnoid villi. These structures project from the subarachnoid spaces into the venous sinuses of the brain, and the trabeculae of the arachnoid membrane bulge through the venous walls, and thus

A useful mnemonic device for these foramina is Luschka = lateral, Magendie = medial.

provide highly permeable regions which allow free passage of cerebrospinal fluid, including protein molecules and particulate matter smaller than 1 μ in diameter, into the blood. The interstitial fluid which is formed in the brain is absorbed partly by the arachnoid villi; hence together with the perivascular spaces these villi perform an important lymphatic function in the central nervous system as this structure lacks any true lymphatic drainage as mentioned earlier.

The perivascular spaces themselves are formed as the superficial vessels of the brain penetrate inward, thereby carrying a layer of pia mater with them as they course away from the subarachnoid space (Fig. 7.15, p. 255). The open space which is located between the pia and the vessel is termed the *perivascular space,* and these regions persist along the course of the vessels down to the level of the arterioles and venules. There is no pia mater present about the capillaries of the brain. The perivascular spaces thus provide the main channels for the transport of interstitial fluid formed by the brain to the arachnoid villi, and the subarachnoid space capillaries from which this fluid is absorbed.

COMPOSITION OF CEREBROSPINAL FLUID.　Certain major constituents of the cerebrospinal fluid were listed in Table 7.4 (p. 255) and compared with plasma. Note that the composition of cerebrospinal fluid differs in certain important features from that of plasma. For example, the sodium content is about 7 percent higher, whereas glucose is 30 percent lower and the potassium content also is lower, being around 40 percent less than that found in plasma. Note also the considerably lower concentrations of protein and glucose in cerebrospinal fluid compared to plasma. Obviously the cerebrospinal fluid is not a simple plasma filtrate; rather it is a secretion which is produced actively by the choroid plexus. Furthermore, judging from the quantitative differences between CSF and plasma for specific solutes, the choroid also is quite selective in its functions.

It is believed that the cuboidal choroid epithelial cells secrete sodium ions into the cerebrospinal fluid by an active transport mechanism (p. 92). This active process in turn builds up a net positive charge; hence chloride and other anions move passively together with the sodium into the cerebrospinal fluid. This ionic surplus increases the osmotic pressure of the fluid to about 160 mm H_2O within the ventricles of the brain; therefore large volumes of water, containing other permeable solutes, move across the choroid membrane and into the ventricles. The glucose concentration remains low, as this substance is not as freely diffusible as water (p. 90). Potassium ions, presumably, are transported actively out of the cerebrospinal fluid; hence their concentration is maintained at low levels.

Interestingly, the total concentration of osmotically active constituents within cerebrospinal fluid is 9 milliosmols above the plasma, and this value corresponds to an osmotic pressure of 160 mm H_2O as noted above, a fact which provides the basis for the osmotic theory of cerebrospinal fluid formation presented above.

CEREBROSPINAL FLUID PRESSURE. The pressure of cerebrospinal fluid within the central nervous system depends upon two reciprocal factors: the relative rates of secretion and absorption of the CSF. Normally these factors are in an exceedingly delicate balance so that the total volume, hence the pressure, remains essentially constant when measured under identical conditions in any particular individual (cf p. 254). Any imbalance between these two factors can lead to severe defects, some of which are presented on page 574 et seq.

BLOOD–BRAIN AND BLOOD–CEREBROSPINAL FLUID BARRIERS. The anatomic and functional characteristics of the blood–brain barrier were discussed in Chapter 7 (p. 256) but may be reviewed briefly at this point.

The membranes between the blood and cerebrospinal fluid as well as the blood and the brain per se are not freely permeable to all substances. These so-called *blood–brain* and *blood–cerebrospinal fluid barriers* are present in the choroid plexus and in the vascular tree of all other regions of the brain parenchyma, save the hypothalamic region, which is particularly sensitive to variations in the concentrations of extracellular fluid components. Functionally, both of these barriers are freely permeable to water, oxygen, and carbon dioxide; however, they are only slightly permeable to electrolytes (eg, sodium, potassium, and chloride ions), and practically impermeable to certain drugs that are used therapeutically.

The blood–cerebrospinal fluid barrier is caused by the high functional impermeability of the choroid epithelial cells to certain substances, as the passage of such materials into the cerebrospinal fluid requires that the substance first must pass through both the capillary endothelium and subsequently the epithelial cells of the choroid plexus (Fig. 7.15, p. 255).

Perilymph and Endolymph

The major constituents of spinal fluid, perilymph and endolymph, are compared in Table 8.3 (p. 513).

PERIOTIC FLUID OR PERILYMPH. The perilymph fills the periotic labyrinth of the inner ear, which consists of the vestibule, semicircular canals, scala vestibuli, and scala tympani (Fig. 7.106, p. 384). The perilymph also is in direct communication with the cerebrospinal fluid in the subarachnoid spaces which surround the brain via the *perilymphatic duct.*

The perilymphatic fluid is considered to be an ultrafiltrate of blood. The capillaries of the spiral ligament have been implicated both in the formation as well as in the reabsorption of perilymph.

OTIC FLUID OR ENDOLYMPH. The endolymphatic fluid is found within the otic labyrinth, ie, in the cochlear duct (scala media), utricle, saccule, and within the three semicircular canals. The endolymph probably is secreted by cells of the stria vascularis (Fig. 8.39, p. 507), as well as by secretory cells which are associated with the neuroepithelial structures of the vestibule. The endolymphatic sac, which lies between the dura and periosteum of the interior surface of the petrous bone, has been implicated in the reabsorption of endolymph (Fig. 7.106, p. 384).

Despite the minute volume of endolymph which is available for sampling and analysis, it has been established definitely that the ionic composition of this fluid is quite different from either perilymph or cerebrospinal fluid as indicated in Table 8.3 (p. 513). Endolymph contains very high concentrations of potassium ions, and thus resembles intracellular fluids in this regard (see Table 3.3, p. 54; Table 4.1, p. 131). Perilymph, on the other hand, is similar to other interstitial fluids insofar as its ionic composition is concerned.

The fluid within the organ of Corti (tentatively named *cortilymph*) must have a composition at least resembling perilymph, as the high potassium ion concentration of the endolymph would completely preclude impulse conduction within the unmyelinated fibers of the hair cells. Possibly the ionic constitution of the endolymph plays a role in the genesis of the endocochlear potential which was described on page 513, although this statement is conjecture which at present is unsubstantiated by any direct experimental evidence.

Intraocular Fluids: Aqueous Humor and Vitreous Humor

The intraocular fluids of the eye perform the important function of distending the eyeball mechanically; hence the intraocular fluids maintain the retina and cornea in an exceedingly precise alignment with respect to the lens and other elements of the eye.

THE AQUEOUS HUMOR. This clear, freely flowing liquid portion of the intraocular fluid is found to the front, as well as at the sides, of the lens, as shown in Figure 8.1, p. 467.

Formation, Composition, and Properties of Aqueous Humor. In the human eye, aqueous humor is formed at an average rate of about 0.1 ml^3 per minute, and nearly all of this fluid is secreted by the ciliary processes. These structures are folds having a total area of about 6 cm² in each eye, and they extend from the ciliary body into the posterior chamber. This latter region is circumferentially located around the eyeball, behind the iris, and just behind the ligaments of the lens (Fig. 8.3, p. 468).

The exceedingly transparent aqueous humor has a composition resembling that of an ultrafiltrate of plasma; hence the protein concentration is quite low, ranging between 10 and 20 mg per 100 ml in humans. Other significant differences between the plasma and the aqueous humor are that the latter fluid has a lower glucose and higher lactic acid content. Furthermore, the ascorbic acid (vitamin C) concentration in aqueous humor is between 15 and 20 times higher than that of plasma. The pH of the aqueous fluid is between 7.1 and 7.3, the specific gravity is 1.002 to 1.004, and the refractive index is 1.33.

The aqueous humor presumably is derived from the blood within the capillaries of the ciliary processes by a mechanism similar to that already described for the formation of the cerebrospinal fluid (p. 562). That is, active intraocular transport of sodium results in the osmotic movement of water and additional solutes into the eyeball. In partial support of this concept for the formation of aqueous humor, it has been determined that the osmolar activity of this fluid is 3 to 6 milliosmols higher than that of the plasma. This difference is far more than sufficient to produce a water movement across the membrane which is quantitatively adequate for the normal production of aqueous humor. Furthermore, studies conducted directly upon the epithelial membrane of the ciliary processes have revealed that sodium is transported actively by this tissue in quantities which again are quantitatively adequate to account for the intraocular osmotic activity that is essential to the operation of an osmotic mechanism for the production of aqueous humor. Other substances, including amino acids, also are transported actively and thereby secreted into the aqueous humor of the eye.

The capillaries of the choroid and the retina function primarily to maintain the integrity of the latter structure; however, minute quantities of fluid from these vessels also can pass into the aqueous humor of the vitreous body.

The capillaries within the iris are concerned primarily with the nutrition of this structure. However, because of its strategic position in the middle of the aqueous humor (Fig. 8.1), approximately 40 percent of certain ions in the aqueous humor can diffuse rapidly both into and out of the iris. Phagocytic properties are well developed in the surface epithelial cells of the iris; consequently particulate matter is eliminated readily from the aqueous humor by this mechanism, and this in turn helps to maintain the crystal clarity of this fluid.

Circulation and Absorption of Aqueous Humor. Following its secretion by the ciliary processes, the aqueous humor flows between the ligaments supporting the lens, through the pupil, and into the anterior chamber of the eye (Figs. 8.1 and 8.3). Thence the aqueous humor passes into the angle between the lens and the iris, the so-called iridocorneal angle, through the trabeculae, and into the canal of Schlemm (Fig. 8.3, p. 468). The latter structure actually is a thin-walled vein running circumferentially around the eye. This vessel is unusual in its extremely high degree of permeability (even small particulate material and proteins can pass readily through its wall), and also for the fact that it is usually filled with aqueous humor instead of blood, despite its being a true vein in the anatomic sense. In addition, the small veins which drain the canal of Schlemm also are usually filled with aqueous humor; hence these vessels are called *aqueous veins*. As one consequence of this unusual venous arrangement, a lymphatic drainage system for the eye is unnecessary because of the direct removal of proteins together with water and other substances by their direct passage into a vein (the canal of Schlemm). This situation is unique, as only in the eye do proteins pass directly into the venous circulation.

VITREOUS HUMOR. Sometimes also called the *vitreous body* (or *corpus vitreum*) because of its gel-like consistency, this portion of the intraocular fluid owes its viscosity to a dense fibrillar network of transparent proteinaceous fibers. The aqueous humor which fills the eyeball behind the lens is freely, but very slowly, interchangeable within this matrix. Thus very leisurely diffusion is the process whereby the vitreous humor and the substances contained therein are exchanged, a fact that is in marked contrast to the active flow of liquid within the aqueous humor proper which fills the anterior part of the eyeball.

INTRAOCULAR PRESSURE. Normally the average intraocular pressure when measured directly with a manometer connected to a needle inserted within the anterior chamber ranges from 10 to 22 mm Hg (average about 16 mm Hg). Indirectly, the intraocular pressure can be measured by means of a tonometer, a device which is placed directly upon the surface of the anesthetized cornea (see also p. 502). A pressure sufficient to deform the cornea is applied by a piston within this instrument, which is calibrated in such a way that the applied force can be related directly to the intraocular pressure.

Although not entirely understood in detail, the mechanism for regulating the intraocular pressure at a constant level throughout the lifetime of an individual is remarkably precise. It is believed that the principal regulatory mechanism for the intraocular pressure resides in the outflow resistance against the flow of aqueous humor into the canal of Schlemm. Thus an augmented intraocular pressure presumably lowers the outflow resistance by distending the trabeculae leading to the canal of Schlemm, thereby increasing the outflow of aqueous humor until the pressure becomes stabilized at normal levels once again.

Obviously alterations in the rate of secretion of the aqueous humor at a constant outflow resistance, as well as variations in the resistance to outflow of the aqueous veins themselves, also could affect the intraocular pressure markedly. The effects of such alterations and the development of glaucoma were discussed in Chapter 8 (p. 502).

Fluids Within the "Potential" Body Spaces

The transcellular fluids included within this broad classification are in the *pleural, pericardial, peritoneal,* and *synovial cavities* (or both the joints and the bursae) as well as the *tunica vaginalis* surrounding the testes. Normally, these regions of the body are "dry" (actually, moist), and contain at most only a few milliliters of fluid which serves to lubricate the surfaces of the various organs and structures which move in contact with each other. In certain pathologic circumstances, however, these regions can swell to contain enormous volumes of fluid, hence the term "potential" spaces is applied to such regions of the body.

Since the membranes surrounding each of the several potential spaces listed above are freely permeable to water as well as to most solutes, they offer little resistance to the passage of fluid and electrolytes in either direction between the space and the surrounding interstitial fluid.

Protein collects in the various potential spaces, as it does in the interstitial fluids, by leakage out of the capillaries. However, each of these spaces is related either directly or indirectly to terminal lymphatic vessels as discussed below, so that normally only very small quantities of free liquid are present.

The lymphatic pump, as discussed earlier, operates almost continuously to move fluid into or toward lymphatic drainage vessels from the potential spaces. Therefore various movements of the body are associated in general with maintenance of the potential spaces, as each of these movements may exert a mechanical compression which forces (or "milks") the fluid into or toward lymphatic vessels. Protein and other osmotically active substances thus are removed continuously, and the spaces normally remain "dry" save for a thin film of liquid on the surface of the various structures which are present in any particular space.

Any pathologic changes in capillary dynamics as discussed on page 576, and which cause extracellular edema, also can produce a greatly increased fluid volume and higher pressures within the potential spaces. Consequently, a decreased plasma colloid osmotic pressure, an increased colloid osmotic pressure of the fluid within the space, an increased capillary pressure or permeability, or an obstruction of the lymphatics which drain a potential space all can produce an increased fluid volume within that space. The fluid which collects following the operation of any one of these mechanisms is known as a *transudate* if it is characterized by a low content of cells and/or by a low concentration of materials derived from cells, eg, protein. Thus transudates in general are characterized by a low viscosity, ie, a high fluidity. An *exudate,* on the other hand, typically contains high concentrations of protein, cells, and the solid material derived from cells such as is seen following the operation of an inflammatory process. Consequently the viscosity of an exudate is much higher than that of a transudate. Furthermore a transudate usually (but not invariably!) is sterile, whereas exudates usually are not, since they frequently are produced by the action of microorganisms such as bacteria. Excessive fluid in the peritoneal cavity is referred to specifically as *ascites,* whereas a generalized edema or interstitial fluid accumulation is known as *anasarca.*

The several potential spaces and certain unique features which are related to each of them now may be considered.

THE PLEURAL CAVITY.   The mechanism whereby diffusion of fluid into and out of this space takes place is exactly the same as that which obtains in the tissues, save that a very permeable serous membrane, called the parietal and visceral pleura (depending upon its location) is situated between the capillaries of the thoracic structures and viscera of the pleural cavity itself. The mediastinal and lateral surfaces of the parietal pleura contain numerous lymphatics; therefore during expiration the intrapleural pressure rises thereby forcing fluid into, as well as along, the lymphatics (p. 561).

Within the pulmonary capillaries, the blood pressure is relatively low, about 5 to 10 mm Hg (Chapter 10, p. 564). The plasma proteins and electrolytes, however, exert a net colloid osmotic pressure of some 25 mm Hg; thus a continuous pressure exists at the surface of the visceral pleura of some 20 mm Hg, and this pressure tends to withdraw fluid from the pleural cavities into the pulmonary capillaries. As a consequence, the net pressure of the film of fluid within the intrapleural space averages -10 to -12 mm Hg with respect to atmospheric. Since this continuous negative pressure is greater than the elastic (or "collapse") force of the lungs (-4 mm Hg), these structures remain permanently expanded as discussed on page 716. The thin film of fluid that lines the pleural cavity is essential, however, to the lubrication of the lungs during normal respiratory excursions.

THE PERICARDIAL CAVITY.   This potential space surrounding the heart has similar fluid dynamics to those found within the pleural cavity. The intrapericardial hydrostatic pressure also is negative, and like that of the pleural cavity, the pressure within the pericardial cavity also fluctuates rhythmically due to contractions of the heart as well as the respiratory motions (Chapter 10, p. 617). The small quantity of pericardial fluid functions like that present in the pleural cavity.

THE PERITONEAL CAVITY.   The fluid dynamics of the peritoneal cavity are similar to those found in other potential spaces of the body. Hence fluid is filtered through both parietal and visceral peritoneal membranes, and this serous fluid also is reabsorbed through them. However, the peritoneal cavity is more subject to the development of a fluid excess than are the other potential spaces for two reasons. First, whenever the pressure within the hepatic sinusoids increases above 5 to 10 mm Hg large quantities of fluid containing a high concentration of protein exude through the surface of the liver directly into the peritoneal cavity. The result is increased osmotic pressure within the peritoneal fluid that results in an osmotic movement of water into the peritoneal cavity; thus the volume of fluid increases until equilibrium is attained. Second, the capillary pressure within the visceral peritoneum is much higher than in other regions of the body because of the vascular resistance to blood flow to the liver that is present within the portal veins (p. 689). Any increase in the intrahepatic resistance is reflected in a retrograde increase in the vascular pressure within the visceral capillaries, and this pressure increase can lead directly to an increased filtration of fluids from the visceral peritoneal capillaries directly into the abdominal cavity.

Infection is a very common cause of increased fluid volume (or *ascites*) in the peritoneal cavity; the exuded cells and debris may obstruct the lymphatics, thereby leading to an increased protein concentration and a resultant increased osmotic pressure in the ascites fluid with an attendant failure of fluid reabsorption. Elevated central venous pressures which may be induced by portal vein occlusion, hepatic cirrhosis, carcinoma, or cardiac failure all can lead to ascites fluid accumulation within the abdominal cavity.

Lymphatic drainage from this potential space is effected principally through large vessels located throughout the peritoneal cavity and especially on the inferior surface of the diaphragm. The latter structure, of course, moves rhythmically during respiration thereby pumping large quantities of lymph into the thoracic duct with each movement.

THE SYNOVIAL CAVITIES. The synovial cavities include the bursae of the skeletal muscles as well as the joint cavities and the tendon sheaths. These tissue spaces contain large quantities of mucopolysaccharides of unknown origin which add viscosity to the synovial fluid (or synovia), thus augmenting its lubricant function. In fact, the viscosity of the synovia resembles that of raw egg white.

The synovial fluid is a typical interstitial fluid, save for the polysaccharides referred to above, and the fibrous tissue membranes lining the several synovial cavities normally offer only slight resistance to reabsorption, hence drainage, of this fluid from these spaces.

Intracellular Fluids

The approximate compositions of certain intracellular fluids have been summarized in Tables 3.3 (p. 54) and 4.1 (p. 131). Because of their special significance to the bioelectric aspects of excitation, conduction, and contraction, the intracellular ionic and other constituents of muscle and nerve have received more research attention than the intracellular components of many other cell types. The intracellular composition of erythrocytes also has been investigated in some detail because it is a simple matter to prepare large quantities of these

elements for analysis, and this fact is reflected by some of the data included in Tables 9.3 and 9.4.

It is, of course, impossible to make any accurate generalizations about the intracellular fluids of different cell types because extremely wide qualitative as well as quantitative differences are found normally in the cell populations that make up the various body tissues. Furthermore, even the same cell type may exhibit marked differences in its composition depending upon such factors as age, state of functional activity, and so forth. In addition to the tables referred to above, specific data concerning the intracellular components and biochemical structure of particular cell types will be cited in context throughout the remainder of this book.

CLINICAL CORRELATES

Anemias

GENERAL CONSIDERATIONS AND CLASSIFICATION. "Anemia" can be defined as a clinical situation in which a subnormal concentration of circulating erythrocytes or of hemoglobin is present. This condition in turn can result from a decreased manufacture of red cells, a decreased hemoglobin synthesis by the precursor cells which give rise to the erythrocytes, an increased loss of red cells from the circulation, or an increased rate of destruction of circulating red cells. Furthermore, anemia can be produced by the simultaneous or successive operation of a combination of these factors. Under normal circumstances, the rate of erythrocyte production is in an exceedingly delicate balance with the rate of erythrocyte destruction as summarized in Figure 9.19; consequently any factor causing an imbalance between these processes such that red cell or hemoglobin manufacture are impaired can lead to an anemia. Actually, anemia is not of itself a disease entity; rather it should be considered as a symptom of some underlying malfunction.

Under normal circumstances when mature erythrocytes enter the circulation from the bone marrow of the adult, these elements remain functionally viable for about 100 to 120 days. Despite the fact that the red cells lack a nucleus, and thus are unable to reproduce or to synthesize quantities of protein, they nevertheless contain enzyme systems within their cytoplasm which are capable of metabolizing glucose and other substrates, although at progressively decreasing rates. As an erythrocyte ages, it gradually becomes more fragile, presumably because its metabolic machinery is unable to cope with the increased necessity for replacing essential cellular constituents such as enzymes at rates which are commensurate with prolonged survival. The plasmalemma of the erythrocyte becomes more fragile and eventually ruptures due to mechanical factors. The hemoglobin and cell fragments thus released are phagocytized by reticuloendothelial elements in the spleen and

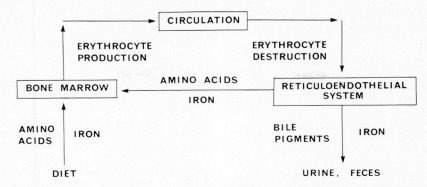

FIG. 9.19. The balance between red cell production and destruction in a normal person such that the total circulating red cell mass remains constant. Erythrocyte production (1×10^{10} red cells and 0.3 gm hemoglobin produced/hr) = erythrocyte destruction (1×10^{10} red cells and 0.3 gm hemoglobin destroyed/hr). A net quantity of about 3×10^{13} erythrocytes and 900 gm hemoglobin are present in the circulation at all times. The iron absorbed per day from exogenous (dietary) sources is equal to the quantity of this element which is lost in the urine and stools during the same period. (See also Figure 9.5.) When the rate of erythrocyte destruction exceeds the rate of erythrocyte production and/or when the rate of hemoglobin degradation exceeds that of synthesis, anemia develops. (Data from Ganong. *Review of Medical Physiology,* 6th ed, 1973, Lange Medical Publications.)

elsewhere. As indicated in Figures 9.4 and 9.5, the iron and heme may be recycled or else the latter may be converted into the bile pigment bilirubin.

Anemias can develop through defects in any stage in the normal life cycle of the red cell as outlined above. In general the anemias can be classified in broadest terms as, first, anemias which are associated with increased blood loss or destruction, and second, anemias which are caused by defective blood formation.

A few examples of each of these general types of anemia will be cited following a brief survey of the major clinical signs and symptoms which can be found in anemia and which may (or may not) be present in any particular patient.

The skin and/or mucous membranes exhibit pallor, and bruising occurs readily. Iron-deficiency anemia may be accompanied by fissures of the lips and brittle nails. Punctate hemorrhages *(petechiae)* may be present in the skin in the anemia which accompanies scurvy, whereas glossitis and atrophy of the papillae of the tongue can accompany vitamin B_{12} deficiency.

The cardiovascular and respiratory systems exhibit a number of typical alterations in anemia including an accelerated heart rate *(tachycardia),* palpitations, angina, claudication, cardiac failure, difficulty in breathing *(dyspnea),* and an accelerated respiratory rate *(tachypnea).* Several of these disorders can be related directly to the reduced oxygen-carrying capacity of the blood produced by the anemia.

Persons having chronic anemia, but who are healthy otherwise, do not exhibit dyspnea following physical exertion until their blood hemoglobin falls to around 7.5 gm percent (from a normal level of about 15 or 16 gm percent). Physical weakness is a significant clinical sign of anemia when the hemoglobin concentration is about 6.0 gm percent,

and dyspnea at rest develops when the circulating hemoglobin level has fallen to approximately 3 gm percent. Cardiac failure develops when the hemoglobin level reaches about 2 gm percent.

Neural symptoms of anemia can include fatigue, headache, irritability, dizziness, and tinnitus. Outright delirium may accompany vitamin B_{12} deficiency, whereas sickling disorders or anemias caused by lead poisoning may be responsible for convulsive episodes.

Gastrointestinal disorders such as loss of appetite, nausea, diarrhea, or constipation may be present in vitamin B_{12} deficiency. Hemolytic anemias may be accompanied by jaundice.

Miscellaneous signs and symptoms which are found in various types of anemia may include abdominal pain, arthritis, or gout.

ANEMIAS RESULTING FROM BLOOD LOSS OR INCREASED BLOOD DESTRUCTION

Posthemorrhagic Anemias. Anemias that develop following blood loss may be acute, such as are seen after a rapid hemorrhage (eg, following a blood donation), or chronic, as the consequence of a bleeding peptic ulcer or uterine bleeding.

Subsequent to a rapid hemorrhage, the volume of fluid lost from the circulation is replaced with water and electrolytes in a few hours, and the plasma itself is replaced in several days; a low concentration of erythrocytes is still present, however. Assuming that a second hemorrhage does not occur, the red cell concentration is restored to normal levels within three to four weeks in persons having a nutritionally adequate diet.

In chronic blood loss, on the other hand, the absorption of iron from the intestines (p. 871) may be inadequate to form hemoglobin at a rate that is sufficient to equal the rate at which hemoglobin is being lost. The erythrocytes formed may have a low hemoglobin concentration, giving rise to a *hypochromic microcytic anemia* which is typical of iron deficiency.

Hemolytic Anemias. The hemolytic anemias are the consequence of increased red blood cell destruction produced by a variety of etiologic factors. Usually, however, in hemolytic anemias the erythrocyte fragility is far greater than normal so that the cells rupture easily (ie, undergo hemolysis) upon passage through the capillaries. As a consequence of this fact, and even though the rate of erythrocyte production is completely normal, the life span of the red cell is so short that a serious anemia ensues due to the rapidity with which mature erythrocytes are destroyed.

Many of these abnormalities in red cell fragility are inherited, although this statement is not true in all types of hemolytic anemia. The defect may be due to abnormalities of the red cells themselves or to external causes acting upon the erythrocytes.

Abnormal Structure of the Red Cells. In *familial hemolytic anemia* (also known as *hereditary spherocytosis*) the erythrocytes are spherical rather than biconcave discs, and the cells are of an abnormally small size (ie, microcytic). Since these spheroid erythrocytes cannot be compressed without rupture, they hemolyze readily upon passage through the capillaries. Furthermore, the spherical shape renders the red cells far more likely to be phagocytized than is the situation with normal erythrocytes.

Thalassemia (Cooley's anemia, or Mediterranean anemia) is another inherited type of anemia in which the erythrocytes are also small and have fragile membranes; hence these elements too are readily susceptible to rupture upon passage through the capillaries.

Extracorpuscular Causes of Hemolytic Anemias. Infectious agents such as malarial parasites, viruses, bacterial septicemia (eg, that caused by *Clostridium welchii*), and nonprotozoan parasites all can produce hemolytic anemias.

Hemolytic anemias that are produced by chemical agents may be dose-dependent or else due to the development of a hypersensitivity to such agents. Included in this general group are lead, coal-tar derivatives, quinine, and sulfonamides (see discussion on mechanism of this effect, p. 550).

Physical agents such as heat also can produce hemolysis, as do some vegetable and animal toxins. Fava beans and certain snake venoms (eg, of pit vipers or crotalids) belong to this category of hemolytic agents.

Isoagglutinins (anti-Rh factors) can produce hemolytic disease of the newborn or erythroblastosis fetalis (p. 573).

Inherited hemolytic anemias include sickle-cell anemia, which is present in about 40 percent of some West African Black populations, and around 10 percent of the entire Black American population. An abnormal type of hemoglobin (type S*) is present in the erythrocytes of these individuals so that when the erythrocytes are exposed to low partial pressures of oxygen, the hemoglobin precipitates into elongated crystals within the cells, ie, its solubility decreases. In turn, the cells become elongated by the abnormal crystals into a characteristic curved or sickle shape. The precipitated hemoglobin also damages the cell membrane leading to extreme fragility of the erythrocytes; hence a serious hemolytic anemia develops.

A dangerous positive feedback cycle may be initiated in persons having this condition, wherein a lowered tissue P_{O_2} induces sickling; the sickling and consequent hemolysis in turn impede the circulation so that the P_{O_2} progressively falls still further. Once this pathologic cycle is initiated it progresses rapidly, leading to a profound decrease in the total circulating red cell mass, and death of the patient may ensue within a few hours following the onset of such a hemolytic crisis.

Persons who are heterozygous for S hemoglobin have a sickle-cell trait, but only rarely develop a severe anemia. Persons who inherit the sickle-cell trait in a homozygous fashion, however, tend to develop the disease in its most severe form.

ANEMIAS DUE TO IMPAIRED BLOOD FORMATION

Nutritional Anemias. Included in this category is the hypochromic microcytic anemia produced by a dietary iron deficiency, by malabsorption of this element from the intestine, or by chronic hemorrhage. Insufficient dietary iron is usually the cause of hypochromic anemia in infants and adolescents, during pregnancy, and in menstruating females. In all of these situations the iron requirement of the body is increased, but not met; hence the synthesis of heme is impaired (Figs. 9.4 and 9.5). In adult males, on the other hand, hemorrhage is the usual cause of hypochromic microcytic anemia.

A second type of nutritional anemia is addisonian pernicious anemia. In this condition there is a dietary deficiency of, or a failure in the absorption and/or utilization of, the specific antianemic factor cyanocobalamin (vitamin B_{12}, p. 883). This deficiency in turn results in a type of hyperchromic macrocytic anemia, ie, the cells are strongly colored and abnormally large.

The anemia associated with infestation by a large intestinal parasite, the tapeworm *Diphyllobothrium* (sp.), belongs to this category of nutritional anemias. The organism competes with the host for vitamin B_{12}, thereby producing a relative deficiency of this compound in the gut of the individual harboring the parasite, and as a consequence erythrocyte maturation in the bone marrow is impaired seriously.

Valine is present instead of glutamic acid on the number six position of the β-polypeptide chain of S hemoglobin.

In this regard it is important to stress that vitamin B$_{12}$, intrinsic factor produced by the stomach mucosa, and folic acid all are essential to the normal production and maturation of erythrocytes in the bone marrow. An inadequate supply or loss of any one of these factors can result in an inadequate rate of proliferation of bone marrow elements to form erythrocytes in sufficient numbers to meet the bodily requirements; hence a pernicious anemia develops. Furthermore, such red cells as are formed under conditions that favor such a maturation failure are usually larger than normal (macrocytic) and have abnormal shapes, and their membranes are abnormally fragile. Such erythrocytes hemolyze readily, leaving the patient with a severe deficit of red cells.

Anemias Due to Interference with Bone Marrow Function. Idiopathic aplastic anemia results from a failure of the bone marrow to produce formed elements for the blood; however, any obvious etiologic factors are absent in this condition. Aplastic anemia, on the other hand, may result from total destruction or suppression of the function of the bone marrow such as may be caused by gamma radiation, toxic agents including industrial chemicals, or drugs used in therapy. The following causative agents which produce suppression of bone marrow activity leading to anemias of varying degrees of severity may be listed: X-rays, gold salts, benzene, arsphenamine, radium, and occasionally, bacterial (including syphilitic) toxins. The red bone marrow may be replaced by fatty tissue so that serious impairment of hemopoiesis results. In addition, renal disease, malignant tumors, and the replacement of bone marrow by fibrous tissue (myelofibrosis) all can produce anemias of various types and of various degrees of clinical severity.

POLYCYTHEMIA. A net increase in the total circulating red cell mass of the body is called polycythemia. Under physiologic conditions, a polycythemia may develop following a prolonged stay at high altitudes (p.1226). Under such conditions, the relative lack of oxygen stimulates the erythropoietic organs to produce more red cells, and the result of this process is such that the circulating red blood cell count may range between 6 and 8 × 10^6 red cells per cubic millimeter instead of the usual values (Table 9.4).

Polycythemia vera (or *erythremia*), in contrast to physiologic polycythemia, is a chronic and self-perpetuating disorder that is caused by a tumorous abnormality of the erythropoietic organs such that the red cell count may rise to as high as 11 × 10^6 cells per cubic millimeter, while the hematocrit may rise up to 80 percent. In polycythemia vera the production of leukocytes and platelets also becomes abnormally high, although this response is not as evident as is the enormous increase in the total number of red cells.

During the course of polycythemia the total blood volume may double in some patients, and as a consequence the entire cardiovascular system becomes markedly engorged. Furthermore, as the viscosity of the blood may increase to as much as five times above normal, many capillaries become plugged with red cells.

In patients with polycythemia the enormous increase in blood viscosity causes venous return to the heart (p. 671) to decrease, whereas the increase in blood volume tends to offset this decrease to some extent; consequently the cardiac output (p. 657) remains at about normal values.

Because of the large total blood volume, the circulation time for the blood throughout the body also rises appreciably and may be increased up to twice the normal value. Thus in a normal person the time required for the circulation of a single red cell from the left ventricle throughout the systemic and pulmonary circuits and back to the left ventricle may be around 60 seconds. In polycythemia vera, however, this circulation time may be doubled to 120 seconds. In other words, the velocity of blood flow in any particular vessel is sharply reduced in polycythemia; that is, the blood flow becomes sluggish.

The mean arterial blood pressure is approximately normal in about two-thirds of all patients with polycythemia, whereas in the remaining third it is elevated. The physiologic compensatory mechanisms present in the body can maintain a normal systemic arterial blood pressure in most individuals suffering from polycythemia despite the tremendous increase in blood viscosity. (The interrelationships among the several factors which normally are responsible for the maintenance of blood pressure are discussed in Chapter 10, page 624 et seq.)

Since normal skin color in Caucasians depends in large measure upon the quantity of blood present in the subcutaneous veins, in polycythemia the sluggish blood flow combined with the massive vascular engorgement tends to give the skin (and mucous membranes) a bluish or dusky cast. In addition, the reduced rate of blood flow also permits a greater quantity of reduced hemoglobin to form in the cutaneous capillaries before the blood enters the superficial venous plexus. Since reduced hemoglobin is bluish compared to oxygenated hemoglobin which is red, the person with polycythemia vera commonly exhibits *cyanosis*.

Disorders of Blood Coagulation

As described earlier in this chapter, the blood clotting mechanism necessarily was presented as though the process occurred in a series of individual and sequential stages (Figs. 9.10 and 9.11). Actually, however, once the coagulation process has been initiated it proceeds as a smooth continuum. Hence the normal coagulation mechanism has been likened to a system of linked electronic multipliers in which each protein factor involved in coagulation behaves first as a substrate and then as an enzyme in a definite series of chemical reactions in order to amplify the final output (or end product) of the system which is the clot per se.

Once clotting was initiated, it would proceed unchecked until all of the fibrinogen contained in the entire vascular bed underwent proteolysis were it not for a number of circulating blood factors which inhibit diffuse (or disseminated) clotting in vivo as well as others which dissolve the formed clot. Many of these anticlotting factors are poorly understood, but it is clear that the fibrinolytic system is activated by some of the same factors which are part of the clotting mechanism (Fig. 9.12).

Normally there is a large excess of the factors responsible for coagulation in the blood; therefore the concentration of any one factor must fall well below 50 percent of its physiologic level before a tendency toward hemorrhage occurs. On the other hand, the normal fluidity of the blood is maintained in vivo by an exquisite balance between anticoagulant and procoagulant factors so that intravascular clotting does not occur.

From a clinical standpoint, a hemorrhagic disorder can be produced by defects in the blood vessels, platelets, or any of the proteins involved in the clotting mechanism itself (Table 9.8). Consequently, the accurate diagnosis of any particular bleeding disease not only is dependent upon a precise medical history but also upon accurate and comprehensive laboratory data.

Patients who exhibit a true coagulation defect have a predisposition to hemorrhage (or *hemorrhagic diathesis*) because of a deficiency of a plasma protein factor that is essential for normal blood clotting. Furthermore, there is in general a direct relationship between the magnitude of the deficiency and the bleeding tendency. Bleeding disorders can be either inherited or acquired.

INHERITED BLEEDING DISORDERS. A number of bleeding disorders are known to be inherited. In this regard it is essential to reiterate the basic fact that each gene controls the synthesis of a single protein (p. 62), and as the clotting factors are specific proteins, then a defective gene for any clotting factor will result in a deficiency of that factor.* Several examples of bleeding conditions and the deficient plasma factor underlying them may be listed, all of which deficiencies lead to an excessive bleeding tendency. First, deficiency of Factor XI is inherited as an autosomal dominant, and can lead to abnormal bleeding following trauma. Second, Christmas disease is inherited as an X-chromosome-linked recessive, and the important symptoms of this disease are summarized in the so-called *hemophilia constellation.*† Third, a

deficiency of Factor VII is inherited as an autosomal recessive and the hemophilia constellation also is present in this condition. Fourth and fifth, deficiencies in ability to synthesize Factors V or X both are inherited as autosomal dominants. Once again the hemophilia constellation is evident in both of these conditions save that hemarthroses are rarely seen in patients having a deficiency of Factor X. Sixth, a defective ability to synthesize Factor VIII, or antihemophiliac globulin (AHG), is inherited as an X-linked recessive, and this deficiency gives rise to classic hemophilia. This condition will be discussed in somewhat greater detail as a typical example of bleeding disorders in general. Hemophilia, of course, gives evidence of all of the signs of the hemophilia constellation.

Classic Hemophilia. Classic hemophilia, caused by a relative lack of Factor VIII, will serve first of all to emphasize a point made earlier; viz, the severity of any bleeding disease is related directly to the relative deficiency of the particular factor involved. Therefore a patient who has between 15 and 25 percent of the normal Factor VIII concentration in his blood generally has a much lower tendency to bleed excessively than does a patient having a 1 percent of normal level of this factor.

Although rare, classic hemophilia is the most commonly encountered coagulation defect. Although it has been considered for some time that a relative deficiency of Factor VIII alone was responsible for the abnormalities seen in classic hemophilia, recent work has indicated that hemophiliacs do synthesize a plasma protein whose structure is similar to that of Factor VIII, albeit this protein is unable to function in the clotting mechanism as does Factor VIII (Fig. 9.10).

The locus or loci of Factor VIII synthesis within the body are unclear, although this protein is known to have a biologic half-life in the circulation that ranges between 6 and 12 hours in normal individuals. However, when a bleeding hemophiliac is infused with Factor VIII, then the half-life of this protein may be shortened considerably.

Hemophilia is a disease whose symptoms persist throughout the lifetime of the individual. In fact, the excessive bleeding tendency often is seen first during circumcision of the infant. As growth proceeds, hemarthroses often take place leading to a fusion and immobility of the affected joints (*ankylosis*). Either spontaneous or posttrauma hemorrhage into soft tissues occurs frequently, and hematuria may persist for weeks, even months. Intracranial bleeding is rare; however, when an epidural or subdural hematoma does develop it is a major clinical problem.

Over long periods of time, hemophilia has been reported to alternate between prolonged asymptomatic intervals which are followed by periods during which massive hemorrhages follow each other in rapid succession. The clinical signs of hemophilia tend to become ameliorated with

See Table 9.8 for the synonymous names of the clotting factors given by Roman numerals in the text.
†*The hemophilia constellation includes bleeding after minor trauma, bruising, hemarthroses (ie, extravasation of blood into a joint or its synovial cavity), and hematuria (blood in urine).*

advancing age. The reasons for this effect are not entirely clear.

Hemophilia is diagnosed with greatest accuracy by determining the Factor VIII level in the plasma directly through quantitative assay.

ACQUIRED BLEEDING DISORDERS. In general, inherited defects in coagulation nearly always involve a lack of one clotting factor, whereas acquired bleeding disorders (except for thrombocytopenia) always involve deficiencies in several factors. Commonly encountered bleeding disorders include those which develop secondary to anticoagulant therapy, those which are associated with severe liver disease, and those which develop secondary to a deficiency of vitamin K (qv, p. 889). The latter condition will be discussed somewhat more extensively, as it illustrates several aspects of the defective physiologic mechanisms encountered in acquired bleeding disorders.

Vitamin K Deficiency and Clotting Defects. The name vitamin K is applied to a number of naphthoquinone derivatives which are necessary for the synthesis of Factors II (prothrombin), VII, IX, and X by the liver. The vitamin-K-active substances are fat soluble; consequently the presence of bile salts is essential to their absorption from the gastrointestinal tract. Since vitamin K is synthesized readily by the intestinal flora, the quantity of vitamin K in the diet presumably is not a major limiting factor in its availability as an essential factor in human nutrition.

A bleeding tendency can develop secondary to a vitamin K deficiency, as in such a situation the synthesis of a number of clotting factors is impaired directly. Any of four principal factors may be involved in the etiology of this defect.

First, if a person is maintained on a poor diet together with the prolonged therapeutic administration of antibiotics that inhibit the intestinal flora, vitamin K deficiency can be produced with attendant defects in the clotting mechanism. Oral vitamin K administration rapidly corrects this defect. Interestingly, hemorrhagic disease of the newborn is caused by a lack of vitamin K, which is believed to result from inadequate intestinal flora.

Second, if fat absorption is impaired chronically, whether by a biliary obstruction or by a primary bowel disease such as the enteritis of ulcerative colitis, then a vitamin K deficiency can result. Parenteral vitamin K therapy rapidly corrects the defect in coagulation.

Third, if the liver parenchyma is the primary site of a major disease process, the synthesis of the vitamin-K-dependent blood Factors II, VII, IX, and X will be curtailed. In addition, the synthesis of Factor V as well as Factor I (fibrinogen) also may be impaired in severe hepatic disease.

Fourth, therapeutic administration of anticoagulants such as bishydroxycoumarin (or its analogs) results in a decreased synthesis of the vitamin-K-dependent clotting factors. Such anticoagulants function, of course, by inhibiting the normal action of vitamin K, and therefore are of much use in clinical medicine for the prevention as well as treatment of intravascular clotting. If these compounds are administered with too great an enthusiasm, an iatrogenic hemorrhagic diathesis can be induced.

Heparin, a sulfated mucopolysaccharide, is another powerful anticoagulant that is often employed for the prevention as well as treatment of intravascular thrombosis. This compound inhibits the action of activated Factor II (thrombin) and also blocks the early stages of coagulation, although its detailed mechanism of action is not clear.

ABNORMALITIES OF THE PLATELETS. Abnormal bleeding may be present in association with qualitative platelet defects or else an increased platelet count (thrombocytosis). However, the most frequent cause of a hemorrhagic diathesis that involves the platelets is a reduction in the total number of these elements in the circulating blood, or thrombocytopenia.

Thrombocytopenia. A net decrease in the concentration of circulating platelets can be caused by a decreased production, an increased destruction, or an abnormally great utilization of these elements. In all instances, however, a hemorrhagic diathesis is present. There is some experimental evidence which indicates that the normal manufacture of platelets in the bone marrow is controlled by a circulating endocrine substance called *thrombopoietin* (cf erythropoietin, p. 531). However, the nature, locus of synthesis, and mechanism of action of thrombopoietin are unknown. The role of this substance, if any, in the development of thrombocytopenia is likewise obscure.

A large number of primary and secondary bone marrow defects can induce thrombocytopenia. For example, bone marrow disorders such as idiopathic aplasia, leukemia, various deficiency states (eg, vitamin B_{12} or folic acid), drugs, chemicals, and ionizing radiation all can produce thrombocytopenia and an associated hemorrhagic diathesis by inhibiting platelet development in the bone marrow.

Similarly, a number of abnormal conditions can induce thrombocytopenia by accelerating the rate of destruction of the platelets. For example, certain disorders of the connective tissue system of the body such as systemic lupus erythematosus, congestive splenomegaly, or thrombotic and idiopathic thrombocytopenic purpura all can produce an increased rate of platelet destruction and clinical thrombocytopenia. Similarly certain bacterial or protozoan infections as well as drugs such as quinine or quinidine also can produce a thrombocytopenia. Based on the foregoing brief and incomplete summary, it can be appreciated that the etiology of platelet deficiencies can be quite varied, and in general when normal platelets are transfused into patients having thrombocytopenia, the rate of destruction of the transfused platelets is accelerated for reasons that are usually unknown.

Idiopathic (or Immune) Thrombocytopenic Purpura (ITP). ITP will serve as an example of a chronic thrombocytopenia which is produced by a known etiologic agent, specifically a circulating factor in the plasma which accelerates the rate of platelet destruction. Thus when plasma from a patient having ITP is transfused into a normal individual, the recipient develops thrombocytopenia following the infusion. If a woman has ITP, frequently her offspring exhibit thrombocytopenia at birth, and normal platelets when infused into such patients exhibit an accelerated rate of destruction as noted earlier.

The humoral factor responsible for ITP is adsorbed by the platelets and will affect the survival of autologous as well as homologous platelets. In addition this substance is found in a specific gamma globulin fraction of the plasma; hence it is presumed to be an incomplete antibody of a type called an *isoantibody*. That is, the patient appears to have manufactured an antibody against his own platelets which accelerates their destruction in vivo.*

Clinically ITP is characterized by a tendency to bruise easily following minor injuries, and large hemorrhages called *ecchymoses* (in contrast to punctate hemorrhages or *petechiae*), which are visible through the epidermis, are typical signs of ITP. The general term *purpura,* in fact, denotes a number of disorders all of which are characterized by the presence of these quite visible purplish or brownish red skin discolorations. In addition, patients with ITP exhibit a prolonged bleeding time after suffering even minor injuries so that hemorrhage from the nose, gastrointestinal tract, and/or bladder is commonly seen. A splenic enlargement (or splenomegaly) that is evident upon palpation rarely is seen in ITP, and anemia also is rare unless the hemorrhage has been extensive.

Petechiae in patients with ITP frequently are seen in skin areas which lie beneath relatively restrictive clothing or underclothing.

Miscellaneous Thrombocytopenias. Another type of thrombocytopenia, which also is caused by an increased rate of platelet destruction, is seen in some patients as a response to drug therapy, particularly following quinine and quinidine administration. This so-called *drug-induced immune thrombocytopenia* is caused by the formation of an antibody to the drug by the patient's immune system. In turn, the immune complex is adsorbed upon the platelets which are passive participants in the process. Following adsorption on the platelets, the immune complex activates complement (p. 550) and then the platelets are destroyed prematurely either in the circulation or by the reticuloendothelial system. (This mechanism is similar to that which is operative in certain drug-induced hemolytic anemias, p. 568.)

The onset of drug-induced thrombocytopenia, in contrast to ITP, is abrupt and severe following administration of the drug, but the signs disappear rapidly when the drug is withheld.

Patients having enlarged spleens (or splenomegaly), regardless of the cause, may exhibit thrombocytopenia. The platelets of such hypersplenic individuals do not exhibit a decreased survival time. Rather the thrombocytopenia is simply a reflection of a physical redistribution of the total circulating number of platelets due to an enlarged (ie, congested) vascular pool within the spleen.

Chronic alcoholics frequently exhibit a moderate thrombocytopenia following a period of heavy drinking. The cause of this thrombocytopenia is unknown, and the platelet counts tend to increase rapidly to normal or even rise above normal levels when the alcohol consumption is stopped.

A transient and generally mild isoimmune thrombocytopenia occasionally develops in adult patients following several transfusions, or in rare instances is seen in newborn infants. This type of thrombocytopenia is caused by the synthesis of antibodies which are produced as the result of stimulation of the humoral immune system by minor antigens found in the platelets. In the transfused patients, antibody production is stimulated by the foreign platelet antigens, however these antibodies also accelerate the rate of destruction of the patient's own platelets.

BLEEDING AND ASPIRIN. Aspirin (acetylsalicylic acid) is both a common and frequently used drug. Therefore it is important to realize that this compound is known to extend the bleeding time, to inhibit the normal aggregation of platelets, and to inhibit the normal release of platelet clotting factors following their aggregation. Such abnormalities can be demonstrated in the blood of persons who ingest only one 300-mg aspirin tablet per day. Hemorrhage of clinical severity is rare in normal individuals; however, in a patient with a hemorrhagic diathesis there is a strong possibility that aspirin could intensify this bleeding tendency. (See page 901 for a discussion of the mechanism of aspirin absorption.)

Such so-called autoimmune diseases are not uncommon, as antibodies developed by the immune system may react with host cells in vivo producing a clinical disorder. Hence the immune response to a viral or a bacterial infection may result in the development of a pathologic situation rather than being of benefit to the individual; eg, lymphocytic choriomeningitis and acute streptococcal glomerulonephritis are but two examples of the many known autoimmune diseases.

Maternofetal Incompatibility: Erythroblastosis Fetalis

Erythroblastosis fetalis is typically a disease of the newborn infant which is characterized by a gradual agglutination and then a phagocytosis of the erythrocytes. In the majority of such patients the mother is Rh negative, whereas the father is Rh positive (p. 552). The infant in turn has inherited the Rh-positive characteristic from the father, whereas the mother has developed anti-Rh agglutinins against the positive Rh antigen which then have been transferred into the fetus across the placental barrier (p. 1195) to cause agglutination and destruction of the neonatal erythrocytes.

During a first pregnancy an Rh-negative woman generally does not develop a sufficiently high titer of anti-Rh agglutinins to harm the infant. However, when the Rh-negative woman has a second Rh-positive child she may have become so sensitized by the first pregnancy that anti-Rh agglutinins develop with extreme rapidity during her second pregnancy. Thus about 3 percent of the second babies which are born to Rh-positive fathers and Rh-negative mothers exhibit some clinical signs of erythroblastosis, whereas 10 percent of the third infants born to such parents exhibit the disease. The incidence of erythroblastosis rises far more rapidly with the third and successive pregnancies.

Rh-negative women can develop anti-Rh agglutinins only when the fetus is Rh positive. Since about 55 percent of Rh-positive males are heterozygous, only about 25 percent of their offspring are Rh negative. Following the birth of a baby having erythroblastosis it is uncertain whether or not future children will inherit the disease as they could be Rh negative and thus not susceptible. Furthermore the anti-Rh response becomes attenuated with an increasing time interval between successive pregnancies. Longer rather than shorter intervals between pregnancies would appear to be more favorable to Rh-positive male and Rh-negative female parents.

The effects of the maternal anti-Rh antibodies on the fetus are brought about by a gradual prenatal diffusion of these plasma factors through the placental membrane into the fetal bloodstream or by direct bleeding into the fetal circulation late in pregnancy or during parturition through placental damage. In the fetal blood the maternal anti-Rh antibodies cause a gradual agglutination of the fetal red cells, and these cell masses tend to occlude the smaller fetal blood vessels. Eventually the agglutinated cells hemolyze, thereby releasing free hemoglobin into the blood. The reticuloendothelial system now converts this hemoglobin into bilirubin (Fig. 9.4), and when the concentration of this compound in the plasma reaches a sufficiently high level a yellowing of the skin (jaundice) takes place.

From a clinical standpoint, the newborn infant with erythroblastosis fetalis is not only jaundiced but quite anemic. Since the maternal anti-Rh agglutinins present in the infant's circulation are not inactivated for a month or two postpartum, then progressively more erythrocytes are destroyed. The circulating hemoglobin concentration in a baby suffering from erythroblastosis fetalis continues to decline during this period and if the level of this pigment falls much below about 7 gm percent the infant frequently dies (cf p. 765).

In response to the relative anoxia produced by the anemia, the hemopoietic tissues of the erythroblastotic infant manufacture erythrocytes at an exceedingly high rate in order to replace those destroyed by hemolysis. The liver and spleen become much enlarged and then commence to manufacture erythrocytes once again. Hence the latent potentiality of these organs to produce erythrocytes is manifest, a function which occurs normally only during the middle of gestation (p. 529).

Since erythropoiesis is taking place at such a high rate in erythroblastosis, it is not surprising that many immature formed elements, particularly erythroblasts (Fig. 9.5), are released into the circulation, a fact which has given rise to the name of the disease itself.

Generally the anemia alone is responsible for the death of infants with erythroblastosis fetalis. However, other degenerative abnormalities frequently are encountered in this disease, particularly defects which result in brain damage. Thus children who survive the anemia may suffer from permanent mental derangement or destruction of the motor areas of the brain. In large measure, this central neural damage is caused by the precipitation of bilirubin in the nerve cells *(kernicterus)*, and this effect in turn causes destruction of the cells.

The principal therapeutic technique employed for erythroblastosis fetalis is to replace the newborn infant's blood as completely as possible with Rh-negative blood. Thus several hundred milliliters of Rh-negative blood are infused slowly (over a period of hours) while the Rh-positive blood of the infant is withdrawn simultaneously. In this manner the titer of anti-Rh agglutinins is reduced to innocuous levels so that no further erythrocyte destruction of any consequence takes place and ordinarily no further transfusions are necessary. Following this procedure, the transfused Rh-negative red cells are replaced gradually with erythrocytes produced by the infant, and this replacement takes place over a period of roughly two months. During the same interval, however, the anti-Rh agglutinins have been completely inactivated.

Histocompatibility

Within recent years the transplantation of entire organs has assumed importance as a therapeutic technique. As a result, the study of histocompatibility between donor and recipient tissues by means of tissue typing has become a subject having much practical interest. Blood transfusion is, after all, a kind of tissue transplantation, and it was stressed earlier that this fluid is in fact a liquid tissue (p. 526). In general terms, the principles of histocompatibility which underlie the transplantation of cutaneous, cardiac, renal, hepatic, and other tissues are identical with those discussed earlier for the transfusion of blood groups (p. 550). That is, the blood groups found in the erythrocytes and the antigenic constituents of the cell walls of all other tissues are quite similar. In fact, A–B–O blood group antigens themselves provide a strong obstacle to histocompatibility between the tissues of different individuals. Consequently, and as is the case in blood transfusion, tissue used for transplantation to a type AB patient ("universal recipient," p. 552) can be taken from an A, B, AB, or type O donor, whereas tissue which is transplanted into a type O patient ("universal donor," p. 552) should be obtained from a type O donor.

There is another important histocompatibility locus in man besides the A–B–O locus, and it is called HL-A because it is examined by the use of antisera which type human leukocytes.* In actual fact, however, the single HL-A locus appears to be a region having not one but two major loci (called LA and Four), and each of these loci is found in close relationship to the other on the same chromosome. Furthermore, each of these loci has a large number of alleles.

The antisera employed for identifying the different types of HL-A individuals are obtained from patients who have had multiple transfusions or from women who have had several children, ie, who are multiparous. The toxic effects of the antiserum on the leukocytes are studied microscopically, and such cytotoxic effects are indicative of a positive reaction, even as agglutination of red cells is indicative of a positive reaction when typing blood (p. 552).

The likelihood of histocompatibility is far greater between siblings, particularly between identical twins, than between a parent and his or her offspring because of the genetic pattern whereby the HL-A alleles are inherited. Among unrelated individuals the proportion that is compatible is quite small, whereas the proportion of compatible siblings is always at least 25 percent.

The HL-A system does not test for still other (and weaker) histocompatibility loci. Thus siblings who exhibit apparently ideal A–B–O and HL-A match, according to the most accurate tissue typing techniques that are available, eventually will reject a renal or cutaneous graft. The use of drugs or sera which suppress the immune system must

be employed, a procedure which perforce renders the patient quite susceptible to infection.

Abnormalities of Cerebrospinal Fluid Pressure

As noted earlier, the normal pressure of the cerebrospinal fluid is maintained by an exceedingly delicate balance between the rate of formation and the rate of absorption of this fluid (p. 563). Any factor or agent which alters this balance likewise can change the pressure of the CSF greatly. For example, a brain tumor which blocks the normal reabsorptive pathway for cerebrospinal fluid (eg, in the tentorial opening), can produce tremendous increases in CSF pressure which may range up to 500 mm of water (37 mm Hg). Infection and intracranial hemorrhage also can induce marked elevations in the cerebrospinal fluid pressure. Normally, free cells are almost completely absent from this fluid, and the large quantities of these elements which appear during infection and in conjunction with an inflammatory swelling can block reabsorption of CSF by the arachnoid villi almost totally, thereby inducing massive increases in cerebrospinal fluid pressure.

Hydrocephalus, or excessive water within the cranium, also can augment the cerebrospinal fluid pressure greatly, and this condition may assume one of two clinical forms. Noncommunicating hydrocephalus, such as is caused by a block in the aqueduct of Sylvius, causes massive dilatation of the two lateral ventricles as well as the third ventricle of the brain, with a concomitant increase of cerebrospinal fluid pressure and a flattening of the brain against the skull. In newborn infants the entire head may swell to enormous dimensions because of such a defect. Permanent psychomotor damage may be extensive. Communicating hydrocephalus generally is caused by excessive development of the choroid plexus in newborn infants. Therefore in this condition the rate of formation of cerebrospinal fluid, consequently its pressure, are increased greatly, although the functional reabsorptive capacity as well as the anatomic channels for reabsorption are normal. Here, too, severe brain damage may result from the mechanical deformation of the neural tissue together with marked enlargement of the head.

Blockage of the central canal of the spinal cord (as by a blood clot or tumor) results in a marked fall in the pressure of the cerebrospinal fluid below the obstruction, as the cerebrospinal fluid normally must pass upward to the brain in order to be reabsorbed.

Anatomically, the dura mater of the brain follows the optic nerve to the eye as a connective tissue sheath which terminates at the sclera, If the pressure of the cerebrospinal fluid is elevated markedly, this pressure is transmitted directly to the sheath of the optic nerve, thence to the retinal artery and vein which enter this sheath just behind the eyeball to pass into the retina of

---

The A signifies that this was the first human leukocyte locus to be specified.

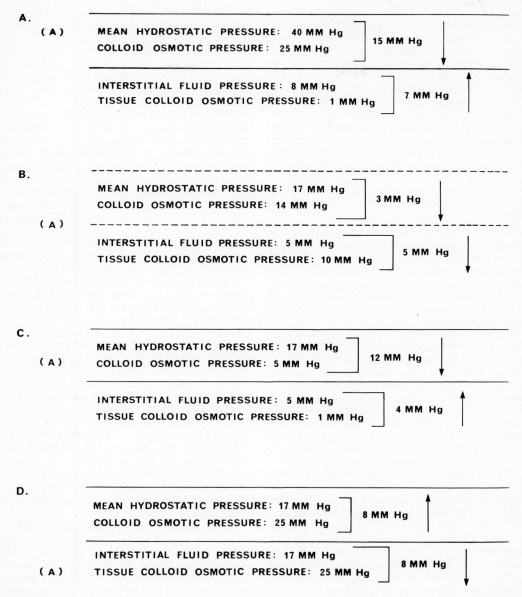

FIG. 9.20. The four possible mechanisms for the genesis of edema. In all instances, blood flow is from left to right in the capillary (horizontal parallel lines). In each example, the principal abnormality (A) is indicated. The vertical arrows to the right signify the direction of the net pressures for filtration (downward) or reabsorption of fluid (upward). **A.** An elevated mean hydrostatic pressure results in an abnormally high outward (filtration) pressure of 8 mm Hg at the venous end of the capillary, hence an elevated rate of interstitial fluid formation and edema. **B.** An elevated capillary permeability (dashed lines) results in an excessive loss of plasma protein from the capillary to the interstitial fluid. Hence a concomitant decrease in the plasma colloid osmotic pressure as well as a net increase in the tissue colloid osmotic pressure develops so that a greater-than-normal loss of fluid from capillary to interstitial fluid occurs (net outward pressure = 8 mm Hg) and edema develops. **C.** Hypoproteinemia results in a marked decrease in the colloid osmotic pressure of the plasma; hence an excessive net loss of fluid to the interstitial space occurs along an 8 mm Hg gradient, resulting in edema formation. **D.** Blockage of the lymphatic drainage channels results in a net accumulation of protein in the interstitial fluid and a concomitant increase in the tissue colloid osmotic pressure. Although the filtration and reabsorption pressures are about equal, edema develops nevertheless, because any excess fluid filtered out of the capillaries above that reabsorbed cannot be removed by the lymphatics; hence a massive edema ensues. Furthermore, any increase in the concentration of protein in the interstitial fluid would result in an osmotic removal of still more water from the plasma, thereby worsening the edema. The numerical values presented in this figure are indicative, but not necessarily absolute. For the normal interrelationships among the four dynamic factors which normally are responsible for regulating the volume of interstitial fluid (thus preventing edema), see Figure 9.18.

the eyeball together with the optic nerve. Therefore an elevated pressure in the optic sheath hinders the outflow of blood in the retinal veins, thereby inducing increased capillary pressure throughout the eyeball with an attendant retinal edema (p. 501). The tissues of the optic disc are especially sensitive to an increased capillary pressure; the edema developed within this region of the eye is much greater than elsewhere in the retina, and thus the degree of papilledema of the optic disc can be observed directly by an ophthalmoscope (p. 471). This observation in turn is indicative of the alteration in cerebrospinal fluid pressure.

Edema

The rate of formation of interstitial fluid by the filtration of water and solutes from the plasma must be balanced precisely by the rates of lymph formation and lymph flow in order that a net accumulation of interstitial fluid, or *edema,* not develop. Therefore under normal circumstances the fluid which is filtered from the capillaries in excess of that reabsorbed by the same vessels is prevented from accumulating in the interstitial spaces by lymph formation at a rate which balances that of production of interstitial fluid (Fig. 9.18). If, however, there is a disturbance in any one (or more) of the physiologic regulatory mechanisms so that excess interstitial fluid accumulates within the tissues, then the condition of edema develops as the consequence of the derangement of the normal interchange and distribution of fluid. Edema per se is not a particular disease; rather this condition is a sign of an underlying disorder.

Edema may be generalized throughout the body or else localized in certain regions, depending upon the mechanism involved in its production (Fig. 9.20). The four specific types of edema and some of their potential causes are given here, as they illustrate clearly the effects produced by a derangement of the normal mechanisms of fluid balance among the blood, interstitial fluid, and lymph. The distinction among these mechanisms is important clinically as therapy is directed toward ameliorating the causative factor.

First, an elevated capillary pressure induces edema as more fluid is filtered out of than is returned to the capillary (Fig. 9.20A). Hence excess fluid collects in the interstitial spaces until the interstitial fluid pressure increases sufficiently to balance the abnormally high intracapillary pressure.

Pathologic factors which increase the venous pressure produce this type of edema. For example, cardiac failure (principally of the right side of the heart, p. 1248) tends to produce an increased systemic venous pressure, and this results in a serious and generalized "cardiac edema" because the increased venous pressure is transmitted in a retrograde fashion to the capillaries. Venous thrombosis, on the other hand, causes a localized edema peripheral to the obstruction due to the increased intracapillary pressure in the vessels that normally are drained by the occluded vein.

Second, an increased capillary permeability causes excessive loss of plasma proteins to the interstitial fluid with an attendant rise in the colloid osmotic pressure of this extravascular compartment, whereas the osmotic pressure of the plasma concomitantly decreases as the plasma proteins are lost. The resulting osmotic imbalance causes the interstitial fluid pressure as well as volume to tend to increase, thereby producing edema (Fig. 9.20B).

Burns, insect bites, drugs, and bacterial toxins which interfere with the normal capillary permeability all may cause this type of edema.

Third, a decreased net plasma protein concentration, if of sufficient magnitude, can reduce the normal colloid osmotic pressure of the plasma to such an extent that edema develops (Fig. 9.20C). In this situation, more fluid leaves the capillary than can be reabsorbed and progressive edema develops as the excess fluid accumulates in the interstitial space. Not only are severe burns attended by a loss of protein at the site of the injury due to an increased capillary permeability which causes edema as described above, but also this localized protein loss is complicated by an overall reduction in the circulating plasma protein concentration. If this net protein loss from the plasma is severe enough, then a generalized edema develops. Severe malnutrition with an attendant decrease in the total plasma protein concentration also can produce this type of edema (p. 875).

Fourth, lymphatic obstruction produces edema directly because the protein normally drained from the tissue spaces via these channels is retained in the interstitial space, increasing the interstitial fluid colloid osmotic pressure markedly. As a consequence, excess fluid tends to collect in the tissues producing a massive edema. For example, infestation with filarial worms (nematodes) can produce lymphatic occlusion to such an extent that a limb or the scrotum may become enormous; the resulting severe edema is referred to specifically as *elephantiasis.* Lymphatic obstruction and edema also may occur following radical surgery for cancer in which malignant lymph nodes are deliberately extirpated and their vessels tied off in order to prevent spread *(metastasis)* of the tumor.

CHAPTER **10**

Cardiovascular Physiology

The basic function subserved by the human cardiovascular system is to maintain the complex physical and chemical structure of the aqueous internal environment of the body, the interstitial fluid. Both the composition and the volume of this compartment must be kept within homeostatic limits that are compatible with prolonged survival of the cells, tissues, and organs, both individually and collectively, as integral components of the entire organism. The cardiovascular system may be considered to be the principal bulk transport system within the body for water and innumerable other materials. Among the major substances carried by the blood, both to and from the tissues, are the respiratory gases (Chap. 11), the waste products of tissue metabolism (Chap. 12), and essential nutrients and metabolites (Chaps. 13 and 14). Furthermore, since the circulation of blood within the cardiovascular system underlies the volume transport of water within the body, this function is also critical to the regulation of body temperature (Chap. 14) as well as to the maintenance of the pH of the body fluids in general (Chaps. 11 and 12).

In addition, the cardiovascular system, through its transport functions, underlies many vital integrative, adaptive, and bodily protective mechanisms. In contrast to the manner in which these functions are carried out by the nervous system, the integrative and other cardiovascular mechanisms, although similar, in general operate at much slower rates. For example, specific endocrine substances (hormones, Chap. 15) as well as antibodies (Chap. 9) are transported in the bloodstream from their sites of production either to specific target organs or to generalized regions of the body, where their effects ultimately are manifest after a time lapse that can be measured in seconds, hours, or even days. The effects of neural activity, on the other hand, are usually evident within milliseconds after the initial stimulus.

The heart and blood vessels, which comprise the human cardiovascular system, must be considered as a dynamic organ complex which constantly is able to react and to adapt automatically and rapidly to internal as well as to external demands imposed upon it by the constantly changing requirements of the individual. Fundamentally, the internal environment of the body is maintained in a relatively constant state, one that is compatible with life solely through the continuous exchange of materials and energy, and this goal is achieved only by the cyclic action through the activity of the heart in maintaining the circulation of the blood within a closed system of vessels throughout the lifetime of the individual.

The topics to be presented in this chapter will be concerned primarily with the basic structural and functional aspects of the cardiovascular system. In addition, the several mechanisms whereby a considerable degree of flexibility of response to various situations may be imposed upon the basic physiologic pattern of activity of the various elements of this system will also be examined. The functional role and response of the cardiovascular system as one major participant in the overall bodily adaptations to a number of physiologic as well as pathophysiologic states will be examined in detail in Chapter 16 (p. 1176).

At the outset, it should be emphasized that the quantitative values cited in the text, tables, and figures of this chapter are illustrative, but not absolute. Discrepancies among particular data presented by different authors often are a result, in part, of the study techniques employed in conducting similar experiments in different laboratories (eg, different anesthetics); such discrepancies also most particularly can be a result of the considerable biologic differences that may exist among individuals of the same or different species of animal (eg, dogs or cats) that have been used for experimental study, as well as among different human subjects.

FUNCTIONAL ANATOMY OF THE CARDIOVASCULAR SYSTEM

Essentially, the cardiovascular system is composed of two parallel muscular pumps that are situated within a single organ, the heart, and this organ, by its coordinated rhythmic contractions, imparts both kinetic and potential energy to a mass of blood contained in a closed system of conduits or blood vessels, consisting of the arteries, capillaries, and veins (Table 10.1). Unidirectional flow of the blood within the circulatory system is achieved anatomically by strategically located valves as well as functionally by suitable pressure gradients from arterial to venous circuits throughout the body.

THE HEART

GROSS MORPHOLOGY. The adult human heart is essentially a complex pulsatile blood vessel which has a large lumen and extremely muscular walls. This vessel has its embryologic origin partly in the fusion of simple paired endothelial tubes of mesodermal origin (Fig. 10.1). These embryonic vessels in turn form the lining of the adult heart, the *endocardium* (Fig. 10.2). Cells of the embryonic mesoderm external to the endothelial primordia in turn give rise to the *myocardium*, or heart muscle proper, as well as the *epicardium*, the outermost serous layer of the heart. The complex form and structural relationships found within the adult heart are direct consequences of the differential growth of the embryonic tubular vessels resulting in their flexion as well as a concomitant reversal of their primitive cephalocaudal positions within the body (Figs. 10.3 and 10.4). Thus, in the adult, the sinus venosus region of the primitive embryonic heart lies cephalad (anterior) to its original site of development during early intrauterine life. In addition, the development of partitions or septa between the two atria and the two ventricles, as well as the formation of certain valves, further compartments the heart into the four discrete chambers found in the adult form of this organ.*

The adult human heart is a blunt cone-shaped organ enclosed in a dense fibrous sac called the *pericardium* (Figs. 10.5, 10.6, and 10.7). It is located within the mediastinal space of the thoracic cavity between the lungs, and lies anterior to the trachea and the thoracic aorta. The heart rests obliquely upon the superior surface of the diaphragm, so that about two-thirds of the organ is located to the left of the midline of the body. In size, the heart weighs approximately 320 gm in the adult male and 260 gm in the adult female; its dimensions are about 12 cm in length, 9 cm in width, and 6 cm in its anteroposterior diameter.

The two thin-walled upper chambers, the right and left atria, are separated by the interatrial septum, and serve principally to convey blood to the right and left ventricles, respectively; the latter structures are relatively thick-walled, and are divided by the interventricular septum. The two thick-walled ventricles perform the major pumping functions of the heart.

The right atrium receives venous blood from the superior and inferior venae cava; the blood then passes to the right ventricle, and hence passes via the pulmonary artery to the lungs in the so-called *lesser circulation*. Oxygenated blood from the lungs then enters the left atrium via the four (two left and two right) pulmonary veins, from whence it passes to the heavy-walled left ventricle. It is next pumped into the aorta and from there to the *greater*, or *systemic, circulation* for distribution throughout the entire body.

For a comparison of the patterns of blood flow in the maternal and fetal circulations, see Figure 16.19 (p. 1207); for a diagram of blood flow within the fetus itself, see Figure 16.20 (p. 1208).

Table 10.1   APPROXIMATE BLOOD DISTRIBUTION AND PRESSURES IN AN ADULT MALE[a]

REGION	(ml)	VOLUME[b] (percent)		MEAN INTERNAL PRESSURE (mm Hg)
Heart, diastole	360	7		—
Pulmonary Circulation				
Arteries	130	3		15
Capillaries	110	2	440	10
Veins	200	4		5
Systemic Circulation				
Aorta, large arteries	300	6		100
Arterioles	400	8		60
Capillaries	300	6	4,200	30
Venules	2,300	46		20
Veins	900	18		10

[a]*Data from Mountcastle (ed). Medical Physiology, 12th ed, Vol. 1, 1968, CV Mosby; Ruch and Patton (eds). Physiology and Biophysics, 19th ed, 1966, Saunders.*
[b]*Data estimated for a 75-kg individual having a body surface area of 1.85 m², age 40 years, total blood volume 5.0 liters.*

A.

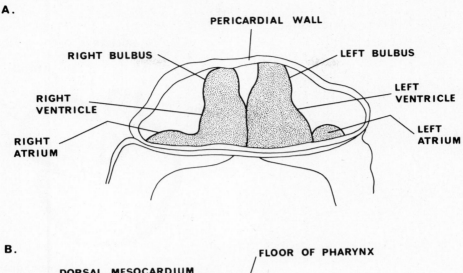

B.

FIG. 10.1. Simple tubular stage in the development of the human heart. **A.** Pericardial cavity opened on ventral side to show paired cardiac tubes. **B.** Sagittal section illustrating same features. Both diagrams are of the six-somite stage. (After Arey. *Developmental Anatomy,* 1943, Saunders.)

The arrangement of the valves within the heart is critical to the maintenance of an effective circulation (Fig. 10.8). There are no valves present at the junctions of the superior and inferior venae cava with the right atrium; however, the opening between the right atrium and the right ventricle is guarded by the tricuspid valve, which opens readily in a direction toward the ventricle only. The opening between the right ventricle and the pulmonary artery similarly has a three-cusped valve, the *pulmonary semilunar valve,* whose flaps open toward the pulmonary artery. Valves are also absent at the point where the pulmonary veins enter the left atrium, but the *mitral* (or *bicuspid*) *valve,* which has two flaps (cusps), is present at the junction between the left atrium and left ventricle. A three-cusped structure, the *aortic semilunar valve,* is located at the junction of the aorta with the left ventricle (Fig. 10.9). Similarly, to the right side of the heart, the cusps of the mitral and aortic

semilunar valves are so directed as to offer little resistance to the flow of blood from left atrium to the aorta.

The tough connective tissue cusps (or flaps) of the several cardiac valves are attached to fibrous structures, the *chordae tendineae,* which in turn are anchored to the endocardial wall by the *papillary muscles,* as illustrated in Figure 10.9. The tendineae and papillary muscles thus prevent eversion of the valves during ventricular contraction *(systole).* These structures also help to appose and thus seal the free margins of the valve cusps during contraction of the heart; consequently, they serve to prevent regurgitation (backflow) of blood. During ventricular filling *(diastole),* however, the blood can pass readily from the atria to the ventricles without significant resistance, as the cusps of these valves are quite pliant, and open readily when pressure is applied to them in the appropriate direction.

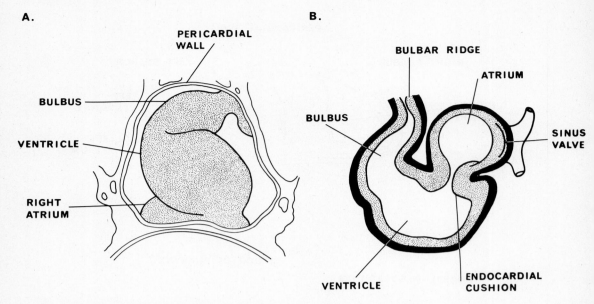

FIG. 10.2.   Human heart in early stage of flexion. **A.** Ventral view, 11-somite stage. **B.** Sagittal section at same stage. (After Arey. *Developmental Anatomy,* 1943, Saunders.)

The right and left *coronary arteries,* which supply the heart muscle or *myocardium,* arise from the aorta immediately beyond the origin of this structure from the left ventricle (Fig. 10.9). The several coronary arteries, after branching to supply the myocardium with blood, form a number of *cardiac veins* (Fig. 10.6), which empty into the *coronary sinus* (Fig. 10.8); this vessel in turn empties into the right atrium between the atrioventricular aperture and the opening of the inferior vena cava. The orifice between the coronary sinus

and right atrium is guarded by the *valve of the coronary sinus (valve of Thebesius).* This semilunar valve is incompetent; hence, regurgitation of some blood from atrium to sinus can occur normally.

MICROSCOPIC ANATOMY OF THE ATRIAL AND VENTRICULAR MUSCULATURE.   The contractile myocardium of the human heart consists of densely packed striated muscle fibers (Figs. 10.10 and 10.11), which differ in three important re-

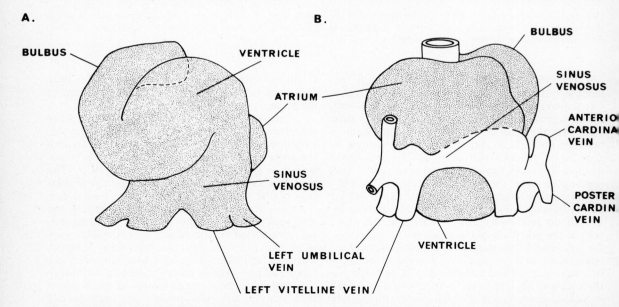

FIG. 10.3.   Human heart in advanced stage of flexion. **A.** Ventral view, 16-somite stage. **B.** Dorsal view, 22-somite stage. (After Arey. *Developmental Anatomy,* 1943, Saunders.)

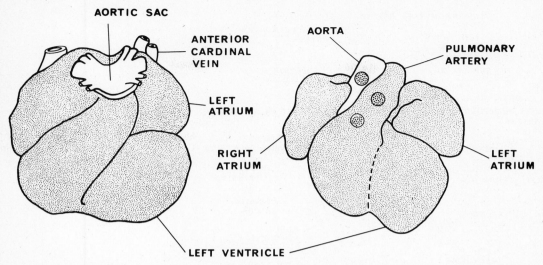

FIG. 10.4. Morphology of ventral aspect of human heart at 5-mm stage **(A)** and at 11-mm stage **(B).** (After Arey. *Developmental Anatomy,* 1943, Saunders.)

spects from skeletal muscle (Figs. 4.1, p. 115; 4.2, p. 117). These major differences are summarized below.

1. Cardiac muscle fibers are not simple cylindrical units; rather, they bifurcate and anastomose freely with adjacent fibers, thereby forming an

intricate three-dimensional network (Fig. 10.10). This morphologic arrangement has important functional implications.
2. The myocardial fibers do not form an anatomic syncytium (ie, one huge multinucleate cell), as was believed for many years. Rather, the fibers are made up of individual cellular units that are

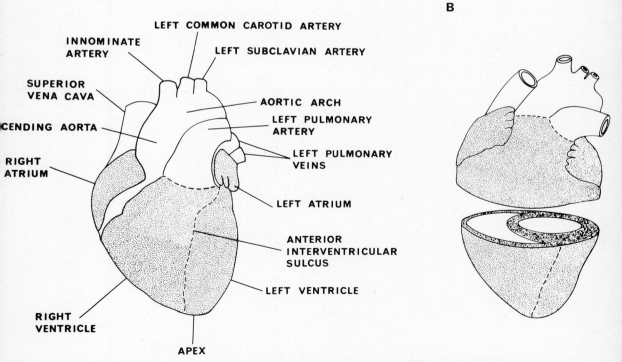

FIG. 10.5. **A.** Adult heart, ventral aspect. **B.** Transverse section through ventricles to illustrate the relative thickness of the ventricular myocardium of right and left sides. Note also that right ventricle partially envelops left ventricle. (**A.** After Goss [ed]. *Gray's Anatomy of the Human Body,* 28th ed, 1970, Lea & Febiger. **B.** After Guyton. *Textbook of Medical Physiology,* 4th ed, 1971, Saunders.)

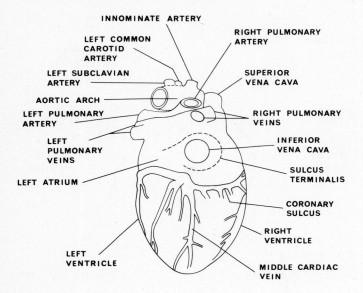

FIG. 10.6.   Adult heart, inferior aspect. (After Goss [ed]. *Gray's Anatomy of the Human Body,* 28th ed, 1970, Lea & Febiger.)

joined together end to end by specializations of the cell surfaces called *intercalated discs;* these structures, which run transversely across the individual fibers, have two important physiologic properties: (1) They provide low-resistance bridges or pathways for the rapid conduction of electric impulses within the myocardium (p. 600). Thus, the entire atrial and ventricular musculatures form two separate and distinct functional, but not anatomic, syncytia. (2) In addition, the intercalated discs, by maintaining a firm union between adjacent myocardial cellular units, serve to transmit myofibrillar tension from one cellular unit to the next, and thus link the entire myocardial structure into a single contractile unit.

3. The elongated nuclei of the myocardial cellular units generally are located deep within the fiber rather than just beneath the sarcolemma, as is the case with skeletal muscle (Fig. 10.10).

The cytology of cardiac muscle is quite similar in certain respects to that of skeletal muscle (Fig. 10.11). The pattern of the myofibrillar cross-striations and the nomenclature of the bands as A, I, M, H, and Z are identical in both types of muscle (Fig. 4.1, p. 115). The thin sarcolemma of cardiac muscle is also quite similar to that of skeletal muscle; however, the mitochondria are far more numerous and the sarcoplasm is relatively more abundant in the former. The contractile material in heart muscle is divided into fascicles by the rows of

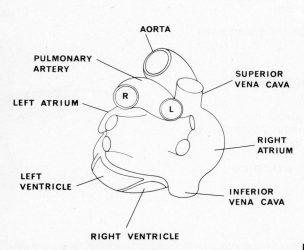

FIG. 10.7.   Base of heart, dorsal (posterior) aspect. (After Goss [ed]. *Gray's Anatomy of the Human Body,* 28th ed, 1970, Lea & Febiger.)

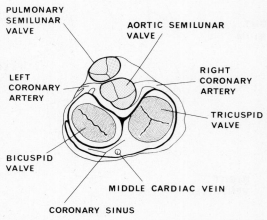

FIG. 10.8.   Valves of the heart and related structures as seen from above following the removal of both atria and arterial trunks. (After Goss [ed]. *Gray's Anatomy of the Human Body,* 28th ed, 1970, Lea & Febiger.)

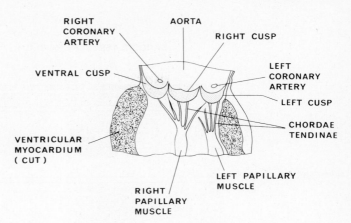

FIG. 10.9. Root of aorta cut open and edges spread in order to show cusps of aortic semilunar valve, points of origin of left and right coronary arteries, and papillary muscles. (After Goss [ed]. *Gray's Anatomy of the Human Body,* 28th ed, 1970, Lea & Febiger.)

mitochondria within the interfibrillar sarcoplasm; and around the centrally located nuclei, the myofibrils diverge. Thus, a fusiform-shaped region of sarcoplasm is delineated about each nucleus (Fig. 10.10). A small Golgi apparatus is located near one pole of each nucleus, and glycogen droplets are quite abundant in the sarcoplasm, far more so than in skeletal muscle. Fat droplets are also found in the conical sarcoplasmic regions at the poles of each nucleus.

The *intercalated discs* appear in suitably stained histologic preparations of cardiac muscle as heavy transverse lines which occur at fairly regular intervals and are found only in the region of the I bands. The transverse regions of the inter-

calated discs may show a stepwise appearance, rather than traverse the individual fiber completely, as the apposing ends of these structures interdigitate freely with each other by means of projections, ridges, and papilla-shaped prominences. The individual cell membranes of the adjacent intercalated discs of two individual cardiac cells thus follow an irregular parallel course separated by a cleft some 200 Å in width.

Electron microscopy has revealed a pattern of myofibrillar structure within cardiac muscle quite similar to that found in skeletal muscle. A regular alternating arrangement of thick *(myosin)* and thin *(actomyosin)* filaments is observed at extremely high magnification, such as is shown in Figure 4.3,

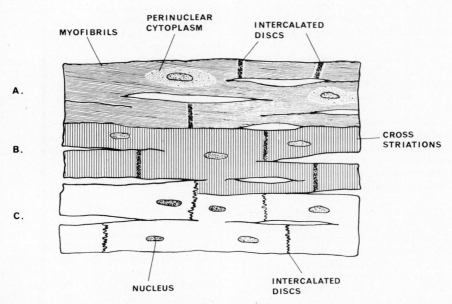

FIG. 10.10. Several features of cardiac muscle. Note the repeated branchings and anastomoses among individual fibers. **A.** Myofibrils are depicted by fine lines. Note that perinuclear cytoplasmic areas have no myofibrillar elements. **B.** Vertical lines represent typical cross striations as in skeletal muscle (Fig. 4.1, p. 115). **C.** The intercalated discs under extremely high magnification are seen as interlocking, or interdigitating, membranous processes which delimit each cardiac muscle fiber.

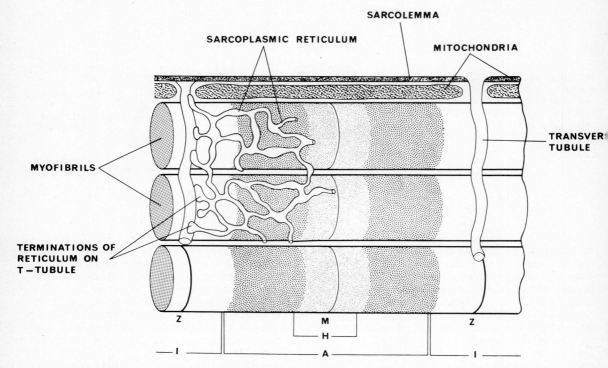

FIG. 10.11. The ultrastructure of mammalian cardiac muscle. Note that no terminal cisternae, hence triads, are present as in skeletal muscle (Fig. 4.2, p. 117). (Data from Bloom and Fawcett. *A Textbook of Histology,* 9th ed, 1968, Saunders.)

p. 117. Hence, the contractile machinery of cardiac muscle is identical with that found in skeletal muscle.

The *transverse tubular system* (or *T-system*) of cardiac muscle is relatively much larger than that of skeletal muscle (Fig. 10.11; cf Fig. 4.2, p. 117). In human heart muscle, however, these transverse sarcolemmal invaginations are found at the level of the Z lines instead of at the A–I junctions. Consequently, as the sarcomeres of cardiac muscle have a resting length of about 2.2 μ, no point in a cardiac muscle cell is more than 2 or 3 μ from the extracellular fluid space, either within or at the surface of the fiber. This anatomic fact has important implications insofar as rapid excitation–contraction coupling is concerned, as well as for the similarly rapid exchange of materials between the interior of the cell and the interstitial fluid.

The *sarcoplasmic reticulum* of cardiac muscle fibers is relatively more simple than that of skeletal muscle, and it consists of a plexiform network of tubules which occupy clefts among the myofilaments. Within this reticulum, there are no continuous transverse elements comparable to the cisternae of skeletal muscle; hence, no well-defined triads are present.

SPECIALIZED TISSUES OF THE HEART. The principal function of the atrial and ventricular myocardial cells discussed above is to contract, hence to pump the blood throughout the pulmonary and systemic vascular circuits. There is, however, within the heart a distinct system of specialized muscle cells whose primary function is not to contract, but rather to generate the rhythmic bioelectric stimuli that are necessary to ensure the repetitive contractions of the myocardium proper. There is also present a system of specialized tissues whose primary function is to conduct the rhythmic impulses in such a fashion as to ensure a smoothly coordinated sequence of activation of auricles and ventricles so that the heart as a whole behaves as an efficient pump. The elements of the impulse-generating and -conducting systems are illustrated in Figure 10.12 and will be discussed individually.

The Sinoatrial Node (Keith-Flack Node). The sinoatrial (SA) node is small, being about 3 mm wide by 10 mm in length in the human heart, and it is located under the epicardium within the posterior wall of the right atrium at the junction of the superior vena cava with the atrium. The individual cells comprising the sinoatrial node are fusiform in shape, distinctly smaller than typical myocardial cells, have a single nucleus, and are arranged in a network embedded within a fairly dense connective tissue framework (Fig. 10.13). Lacking intercalated discs, sinoatrial nodal cells possess abundant sarcoplasm and relatively few, although typical, myofibrillae having the usual cross-striations. This latter characteristic enables these cardiac elements

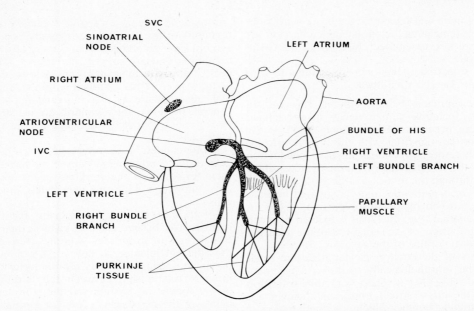

FIG. 10.12. The specialized conducting tissue system of the heart in relation to the sinoatrial node. SVC = superior vena cava; IVC = inferior vena cava. (After Ruch and Patton [eds]. *Physiology and Biophysics,* 19th ed, 1966, Saunders.)

to be distinguished from the fibroblasts with which they are closely associated and which they markedly resemble.

The sinoatrial nodal cells are capable of spontaneously generating rhythmic impulses at a rate that is faster than any other tissue in the heart; this node is referred to as the *cardiac pacemaker* (p. 595).*

All adult cardiac tissues, specialized or otherwise, retain a latent potentiality to contract spontaneously, albeit at markedly different rates.

Since the sinoatrial nodal cells, which actually are specialized cardiac muscle fibers, are themselves capable of generating rhythmic impulses independently of any external influences, the origin of the human heartbeat is said to be *myogenic* in character; ie, the beat originates in muscular tissue.

The sinoatrial node is also richly innervated by terminal axons of both the sympathetic and parasympathetic (vagal) branches of the autonomic nervous system; these nerves serve to modify the inherent rate of discharge of impulses by the

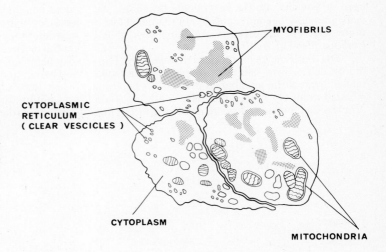

FIG. 10.13. A cross section of three pacemaker cells in the sinoatrial node. Note the abundant cytoplasm, scanty myofibrils, and numerous mitochondria. (After Jensen. *Intrinsic Cardiac Rate Regulation,* 1971, Appleton-Century-Crofts.)

sinoatrial node and thus alter the basic heart rate. Peripheral ganglia of the parasympathetic nervous system are also associated closely with the tissue of this node.

The sinoatrial nodal fibers are continuous with typical atrial myocardial fibers.

The Atrioventricular Node (Node of Tawara). In mammals, including human beings, no specialized conducting tissue has been demonstrated so far that links the sinoatrial with the atrioventricular node. Therefore, functional impulse conduction between the two regions of the heart takes place directly through the myocardial fibers of the atria.

The *atrioventricular node* is situated beneath the endocardium in the lower portion of the interatrial septum (Fig. 10.12); it is located between the point of origin of the septal leaf of the tricuspid valve and the orifice of the coronary sinus.

Fibers of the atrioventricular node are small and similar morphologically to those of the sinoatrial node (Fig. 10.13). There is gradual transition between the atrioventricular node and the common atrioventricular bundle, which is a continuation of this portion of the conducting system.

The Common Atrioventricular Bundle (Bundle of His). The bundle of His originates from the anterior part of the atrioventricular node with which it is continuous, and enters the interventricular septum within which it divides into the right and left bundle branches; these branches in turn ramify to supply both the right and left ventricle, respectively. Beneath the endocardium of each of these ventricular chambers, the bundle branches divide considerably, thereby forming extensive systems of delicate fibers which ultimately terminate in close contact with myocardial fibers of each ventricle.

The cells of the common atrioventricular bundle (of His) are small and similar in appearance to those of the atrioventricular node, but as they proceed downward into the two bundle branches, the cells gradually become considerably larger, have a quite different appearance, and are now called *Purkinje fibers* (Fig. 10.14).

The Purkinje cells of the left and right bundle branches have one or two spherical or ovoid nuclei located in an abundant clear sarcoplasm situated centrally within the cells. Both glycogen droplets and mitochondria are abundant within this sarcoplasm; the scanty and typically striated myofibrillae are located peripherally around the sarcoplasm, as indicated in Figure 10.14. The myofibrils that are present in Purkinje fibers are present singly or in small groups; and unlike these contractile elements as found within true myocardial cells, they are oriented in a random pattern. In general, however, the myofibrils are usually found to lie roughly parallel to the long axis of the Purkinje cells.

Intercalated discs rarely are present in the cells of the conducting system of the heart, and at their terminations the Purkinje fibers apparently are continuous with the ventricular myocardial cells.

Purkinje cells are noteworthy for their irregular, often bizarre shapes, as well as their numerous irregular processes which often envelop the adjacent myocardial fibers. As a direct functional consequence of this morphologic fact, the intimate relationship between the Purkinje cells of the conducting system and the contractile elements of the myocardium assures a rapid activation of the latter.

Similarly to the sinoatrial node, the atrioventricular node and common bundle are innervated heavily by unmyelinated terminals of both sympathetic and parasympathetic (vagus) nerves. Numerous ganglion cells of the latter are also found in these tissues, especially within the extensive neural plexuses located at the base of the heart. Furthermore, both the atrial and ventricular myocardial fibers themselves are innervated heavily by unmyelinated nerve fibers. No specialized nerve endings are found either on the cells of the conducting system or on the myocardial cells. This fact is in distinct contrast with the situation that is present in skeletal muscle. In the heart, the thin axonic processes merely terminate as bare endings close to the cell surfaces. The axoplasm within the actual nerve terminations, however, does contain numerous minute vesicles that are identical in appearance with those present in the membranes of neuromuscular junctions and synapses within the central nervous system. It is clear that these cardiac nerves play a definite role in the regulation of heart rate, and it is assumed that the vesicles represent the excitatory or inhibitory neuromuscular transmitters that are responsible for modifying the inherent cardiac activity in vivo.

A.

MYOFIBRILS

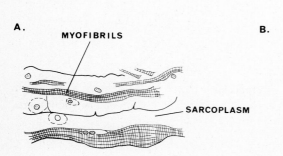

SARCOPLASM

B.

FIG. 10.14 Purkinje fibers of the atrioventricular node. (Data from Bloom and Fawcett. *A Textbook of Histology,* 9th ed, 1968, Saunders.)

THE BLOOD VESSELS

Blood pumped from the left ventricle of the heart into the systemic circulation is distributed to the various organs and tissues throughout the entire body by an elaborate branching system of closed tubes, the blood vessels. The individual components of this circulatory system are illustrated in Figure 10.15. The vessels of the circulatory system have quite dissimilar morphologic, hence physiologic, properties. The major types of vessel will be considered individually, although actually they form a sequential continuum insofar as blood flow is concerned from the aorta to the smaller arteries and arterioles, then to the capillaries, thence to the venules and veins, and lastly to the venae cavae and back to the right atrium of the heart. A like sequence of vessel types is found in the lesser, or pulmonary, circulation, from the right ventricle to the left atrium.

Specific variations in this basic anatomic sequence of vessels are encountered in several regions of the body, and these will be emphasized where appropriate.

ARTERIES AND ARTERIOLES. The basic morphologic arrangement of the principal structures that comprise the walls of all arteries, regardless of their caliber, is as follows (Fig. 10.16). First, the innermost layer is known as the *tunica intima,* and its elements are disposed predominantly in a longitudinal direction. Second, the next major layer is called the *tunica media.* This coat provides the thickest portion of the arterial wall, and its elements are arranged in a circumferential fashion. Third, the outermost coat is known as the *tunica adventitia,* and the majority of the structural elements that are located within this layer are oriented parallel with the long axis of the vessel. Both the fibrous and cellular components of the adventitial region of the arterial wall are confluent progressively with the connective tissue that envelops every blood vessel.

The internal elastic membrane (or *elastica interna*) is situated between the tunica intima and the tunica media, whereas the external elastic membrane *(elastica externa)* is found between the tunica media and the tunica adventitia.

Depending upon their relative diameter and distribution of their components, as well as their distance from the heart, several particular types of artery may be recognized.

Elastic Arteries. The large elastic or conducting arteries include the aorta, innominate, subclavian, common carotid, and pulmonary. These vessels contain within their walls prominent series of fenestrated bands and sheets composed of a scleroprotein, *elastin,* which infiltrate primarily the intima and media, giving the larger of these vessels a distinct yellowish appearance in cross section when they are examined in the fresh condition. The elaborate fenestrated membranes tend, in fact,

to obscure the delineation of the tunica intima from the tunica media.

The endothelial cells lining the large arterial vessels tend to be polygonal in outline rather than being elongated parallel to the axis of the vessel.

The wall of a large elastic vessel, ie, the aorta, is thick. In proportion to the cross-sectional area of the lumen, however, the aortic wall is relatively thinner than the wall of the smaller muscular arteries (Table 10.2). The walls of some large thick-walled arteries themselves contain their own arterial and venous blood supply. These vessels in turn are called *vasa vasorum,* or "vessels of the vessels."

Muscular Arteries. The muscular arteries are blood vessels that are of an intermediate caliber between the large elastic arteries and the arterioles, and these vessels are the most numerous arterial structures to be found in the body. Consequently, the muscular arteries form the major distributing arterial system.

Beneath the endothelium that lines the muscular arteries, a well-defined elastica interna is located (Fig. 10.16). The tunica media of these vessels is composed almost entirely of smooth muscle fibers arranged in concentric layers and embedded in an abundant interstitial matrix that is composed of glycoprotein together with bundles of collagenous connective tissue fibers. Elastic connective tissue fibers are also present within the tunica media of muscular arteries, and these structural elements form loose networks that surround the individual layers of smooth muscle cells. A well-developed elastica externa generally is found in the distributing arteries, and on the outer surface of this membranous structure are located numerous fascicles of unmyelinated axons which contain mitochondria and synaptic vesicles. Usually the terminal axonic processes of these nerves do not traverse the elastica externa, but stimulation of such a nerve results in the diffusion of transmitter substance through the external elastic layer such that depolarization and excitation of the peripheral smooth muscle cells of the tunica media may result. A concomitant cell-to-cell transmission of the impulse thus evoked may then occur so that ultimately the entire mass of smooth muscle cells that comprises the tunica media may contract.

The tunica adventitia of the muscular arteries may be even thicker at times than is the media. The adventitia of these vessels is composed of bundles of collagen, elastin fibers, and fibroblasts that are oriented primarily in a longitudinal direction, insofar as the long axis of the vessel is concerned. These arterial components are continuous with the surrounding loose connective tissue, and thus their movements permit the vessel to change its diameter readily and without restraint, although the longitudinal disposition of the adventitial components prevents excessive retraction, should the vessel be cut.

Because of the profound importance of

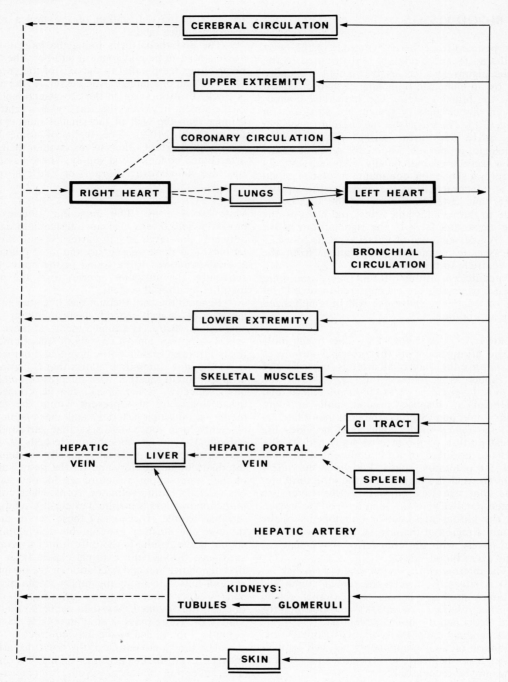

FIG. 10.15. The relationships in the human circulatory system. Arteries are depicted by solid lines, and veins by dashed lines. Note that series resistances in the cardiovascular system branch into parallel resistances at the capillary level (Figs. 10.19 and 10.52). The circulation to special regions of the body are indicated by a line under the box representing the organ or system. (The renal circulation is considered in Chap. 12, p. 772.)

smooth muscle fibers to the dynamic regulation of the caliber of arterial blood vessels, and thus to hemodynamic patterns, the properties of this type of contractile cell now must be reviewed briefly (see also p. 144 et seq). Unlike both skeletal and

cardiac muscle fibers, smooth muscle cells derive their name from the fact that their myofibrils lack cross-striations (Fig. 4.30, p. 145).

Smooth muscle fibers are thin spindle-shaped structures, each containing a single elongated nu-

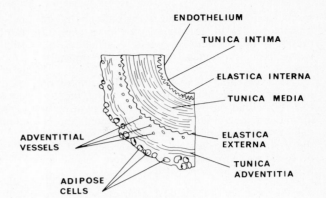

FIG. 10.16. Cross section of a typical artery to illustrate the several layers of tissue that are present in such vessels. (After Bloom and Fawcett. *A Textbook of Histology,* 9th ed, 1968, Saunders.)

cleus. These cells are quite variable in length, depending upon the tissue or organ in which they are found. For example, in the walls of small arteries, these structures may be as short as 20 μ, and where such fibers occur in sheets or layers, as in the blood vessels, the centrally placed nuclei of the individual fibers are offset (or staggered) so that only cross sections cut through the center of any particular cell reveal a nucleus.

The homogeneous-appearing sarcoplasm of smooth muscle fibers can be shown, by special techniques, to contain numerous, densely packed longitudinal myofibrils which are doubly refractile (birefringent) under the polarizing microscope, although no trace of alternating bands such as are characteristic of striated muscle fibers is evident. These thin myofibrils, which presumably are the principal if not sole contractile elements of smooth muscle cells, occupy the major portion of the sarcoplasm. A few mitochondria, a Golgi apparatus, and a few tubular structures representing a granular endoplasmic reticulum are also present in smooth muscle cells.

The myofilaments found in smooth muscle apparently are only of the thin type, unlike the situation which obtains in striated muscle, in which both thick and thin filaments have been demonstrated clearly. The proteins actin and myosin both have been isolated from smooth muscle as well as from cardiac and skeletal muscle, however,

and these fibrous elements doubtless have a key functional role in the contraction of smooth muscle as well as in the other two kinds of contractile cell.

Smooth muscle is usually considered morphologically to represent a single type of cellular element; however, the widely diverse physiologic and pharmacologic properties of these contractile elements in various tissues and organs tend to substantiate the view that functionally, at least, there are numerous kinds or populations of "smooth muscles." The various specific properties of the smooth muscle, which is found in different regions of the body, will be dealt with as appropriate. Certain general functional properties of these cells which are relevant to the cardiovascular system may be summarized here, however. First, the contraction of smooth muscle is far slower than that of striated muscle, but a forceful contraction can be maintained for prolonged time intervals. Second, excitation of smooth muscle can be induced by nerve impulses, endocrine substances, or merely by mechanical stretch of the muscle fibers (p. 147). Third, a series of rhythmic contractions, or a single sustained contraction, which can be induced by the stimulus of mechanical distention, are of particular importance in certain visceral organs, notably the gastrointestinal tract (p. 835) and the urinary tract and bladder (p. 812), as well as in certain blood vessels (p. 666).

Vascular smooth muscle, such as is encoun-

Table 10.2 PROPERTIES OF HUMAN BLOOD VESSELS[a]

BLOOD VESSELS	LUMEN DIAMETER	AVERAGE WALL THICKNESS	TOTAL CROSS-SECTIONAL AREA (estimated; cm², all vessels)
Aorta	2.5 cm	2 mm	4.5
Distributing artery	0.4 cm	1 mm	20
Arteriole	30 μ	20 μ	400
Capillary	6 μ	1 μ	4,500
Venule	20 μ	2 μ	4,000
Vein	0.5 cm	0.5 mm	40
Vena cava	3 cm	1.5 mm	18

[a]*Data from Ganong.* Review of Medical Physiology, *6th ed, 1973, Lange Medical Publications.*

tered in the arteries and arterioles, behaves physiologically in a manner similar to skeletal muscle in that its activity generally is induced by nerve impulses, hence, neuromuscular transmitters.

It is essential to realize that the caliber, or diameter, of the muscular arteries (and the arterioles) within the vascular bed depends directly upon the degree of contraction of the circumferentially oriented smooth muscle fibers located within the walls of these vessels, as this basic fact has profound and far-reaching implications insofar as hemodynamics are concerned (p. 637).

The Arterioles. The smallest arteries, which are 0.3 mm or less in diameter, are called arterioles. The tunica intima of these vessels consists merely of a layer of endothelium, and this inner coat is surrounded by a tunica media comprising from one to several layers of smooth muscle cells, depending upon the size of the vessel. In the larger arterioles, an internal elastic membrane is present between intimal and medial coats.

The smooth muscle cells located in the tunica media of the arteriolar walls range between 15 and 20 μ in length, and these fibers are arranged in a circular fashion about the lumen when the vessel is viewed in cross section. The smooth muscle cells thus are oriented transversely with respect to the long axis of the vessel when seen from a surface view, and this situation is similar to that which was described above in relation to the larger muscular arteries. Furthermore, the smooth muscle cells are of considerable functional importance in the regulation of the caliber of the arterioles, a situation again like that encountered in the muscular arteries.

The tunica adventitia of arterioles is similar in thickness to that of the tunica media, and this adventitial layer is composed of loose connective tissue in which are found longitudinally disposed collagenous and elastic fibers in addition to fibroblasts. The adventitial layer merges with the surrounding connective tissue in a manner reminiscent of the situation encountered in the larger arteries. An external elastic membrane is absent in arterioles.

Special Arteries. Within the arteries of the lower limbs, the tunica media is far more highly developed than is the case in similar vessels within the arms, a morphologic fact that is related to the greater hydrostatic pressure encountered in the vessels of the legs.

Arteries that are located within the skull have a thin wall and a well-developed internal elastic membrane, as these vessels are protected by the skull from external mechanical influences. In the cerebral vessels, elastic connective tissue fibers are almost completely lacking in the intima, media, and adventitia.

In conclusion to the foregoing discussion concerned with the morphology of the arterial portion of the circulatory system, the anatomic transition between one type of artery and another may be rather abrupt and obvious, or else progressive and subtle so that an intermediate zone which has certain characteristic elements of each type of artery is present (eg, in the so-called metarterioles). Of particular significance here is the fact that the endothelial lining of the *entire* cardiovascular system presents an exceedingly smooth and continuous uninterrupted channel to the flow of blood.

Physiologic Correlates with Arterial Structure. The contractions of the heart are intermittent, or pulsatile; blood is ejected from this organ in spurts rather than as a smooth continuous stream. If the walls of the arteries and arterioles were rigid (or inflexible), the flow of blood through the capillaries likewise would be intermittent. The large arteries near the heart normally are quite elastic; consequently, during each contraction (*systole*) of the ventricles, only a fraction of the total work performed during the contraction is expended in moving the column of blood along the peripheral vessels. The remainder of the work performed is expended in dilating the elastic walls of the larger arteries, and the potential energy thus accumulated in these vessels during systole is expended as elastic recoil of the vessel during the subsequent relaxation (*diastole*) of the ventricles. This rhythmic alternate expansion and elastic recoil of the larger arteries acts as an ancillary pumping mechanism to the heart. The net result of the operation of this mechanism is that close to the heart blood flow is intermittent (Fig. 10.50), whereas in the peripheral arteries and arterioles a continuous smooth flow is present despite the fact that the heart is an intermittent pump.

As indicated above, contraction and relaxation of the smooth muscle fibers present within the arterial and arteriolar walls can influence markedly the flow and distribution of blood to various organs and tissues, whereas differences in the arterial caliber in turn have a direct effect upon peripheral vascular resistance, hence, blood pressure. Contraction and dilatation of muscular arteries in turn are regulated directly by vasoconstrictor and vasodilator nerves of the autonomic nervous system which terminate in the walls of the vessels at the outer limit of the tunica media (Table 7.6, p. 276).

Lastly, the smooth muscle fibers present in the arterial wall are normally in a sustained state of partial contraction which is called *tone*. The vascular tone is, of course, subject to great modification under physiologic conditions, depending upon, for example, existing hemodynamic factors, activity of the individual, neural influences, and hormonal effects.

All the factors regarding arteriolar functions summarized above will be considered individually and in greater detail in subsequent sections of this chapter.

THE CAPILLARIES. The structure of the capillaries, which are continuous with the smallest arterioles, was introduced in Chapter 9 (p. 558). Together with the smallest venules (qv, below), the capillaries are the physiologically important vessels; hence, most of the actual exchange of water, respiratory gases, and other substances between blood and interstitial fluid takes place through the capillary walls (Figs. 9.14, 9.17, and 9.18).

Elongated endothelial cells with tapering ends are the principal elements comprising the capillary walls, and these cells are aligned with their long axes roughly parallel to the similar axes of the vessels. In the smallest capillaries, a single endothelial cell may extend around the entire vessel, whereas in the larger capillaries two or three cells may be required to encompass the lumen.

The diameter of the capillaries ranges within narrow limits that are related approximately to the diameter of the erythrocytes. In general, therefore, the caliber of these vessels is between 6 and 8 μ. It is important to realize, however, that the capillaries are quite active, and, depending upon the functional state of the tissue or organ, the capillary lumen may increase or decrease markedly in diameter, thereby increasing or decreasing the blood flow to that tissue.

The capillaries are associated closely with fixed macrophages and mesenchymal cells, and occasionally nerve fibers are present. Specialized Rouget cells with their extensive processes closely applied to the capillary walls once were thought, because of their contractile properties, to behave similarly to smooth muscle cells and to constrict the capillaries (the so-called precapillary sphincters were thought to operate by such a mechanism). It appears, however, that the capillary endothelial cells themselves are capable of contraction, so that the participation of these "pericytes" in regulating capillary diameter and blood flow is unnecessary and of doubtful significance in vivo.

There are no intercellular clefts between the endothelial cells that comprise the capillary walls, as once was believed. There are, however, two principal morphologic types of capillary as revealed by electron microscopy. These are continuous (or muscle) and fenestrated capillaries.

Continuous Capillaries. These structures are found in skeletal, cardiac, and smooth muscle, as well as in other tissues (Fig. 9.17A; p. 559), and they have an unbroken endothelium which is between 0.2 and 0.5 μ thick. The cells are covered on the outside and supported by a continuous thin *basal lamina* or *basement membrane* which is composed of a collagenlike material.

The nuclei of the endothelial cells in this type of capillary are ovoid in outline and centrally located within the cells, whereas the ends of the cells taper gradually and appose directly with the membranes of adjacent endothelial cells. These junctions between the cells may be either smooth or there may be interdigitations. In the junctional regions between adjacent endothelial cells, the separate and discrete membranes of each of the pair of cells are only a few hundred angstroms apart, and a small fold of one or both cells may protrude into the lumen of the vessel. In this regard, the capillaries of brain and retina exhibit an important peculiarity of structure. In these capillaries, the endothelial cell boundaries have actual tight junctions *(zonulae occludentes),* in which regions the adjacent membranes of the endothelial cells actually fuse, thereby obliterating the intercellular cleft entirely.

Fenestrated Capillaries. These structures are found in the renal glomeruli (Chap. 12, p. 778), in the intestinal glands (Chap. 13, p. 827), and in the endocrine glands (Chap. 15, p. 1024). Fenestrated capillaries are defined by the presence of very thin regions of the endothelium (around 500 Å in thickness), which are perforated by extremely thin circular *pores* or *fenestrae* ("windows") having a diameter of about 900 Å. These pores are not open "holes" in the capillary endothelium in most instances, as actually they are closed by an extremely thin diaphragm, which is enclosed on the outside by the basement lamina of the capillary (Fig. 9.17B; p. 559). The fenestrations may vary in size or number, depending upon their location, and they may be quite regular in their occurrence. The total surface area of the fenestrations relative to that of the entire capillary is quite small.

Within the glomeruli of the kidney, the fenestrations seen in the capillaries are true pores; hence, these openings are not closed by membranous diaphragms as indicated above. Furthermore, in these renal vessels, the basal lamina is about 3 times thicker than in other fenestrated capillaries.

In addition to the two kinds of true capillary discussed above, there is a third important type of terminal vascular channel in the human body. This type of channel is called the *sinusoid,* a large, usually irregular-shaped vascular space that is lined by connective tissue, the latter being continuous with the vascular wall (when such is present) and with the parenchyma of the organ itself. Sinusoids are found in the liver, spleen, bone marrow, and some endocrine glands, such as, for example, the pituitary and adrenal glands. In turn, sinusoids may be classified as discontinuous or fenestrated.

Discontinuous sinusoids such as are found in the liver contain phagocytic cells (Kupffer's cells) of the reticuloendothelial system. Typical lining endothelial cells are found in some regions of these tortuous "vessels," whereas they are absent in other portions of the same channel. A basal lamina, if present, occurs intermittently; the term *discontinuous* is applied to these minute vascular channels.

Fenestrated sinusoids such as are found in the adrenal and pituitary glands lack phagocytic cells and have no physical gaps or spaces between the cells. A continuous basal lamina is present, and the thin endothelium contains pores that are closed by exceedingly thin diaphragmatic membranes, as described for fenestrated capillaries.

It is evident from the brief discussion presented above on the capillaries and sinusoids that there is a considerable morphologic variation and specialization of these vessels, and that in turn important functional differences exist among these structures in various regions of the body. The entire capillary bed of the human body presents a vast surface area for the interchange of materials between blood and tissues, an area that has been estimated to be around 6,300 m². This would, in its total extent, comprise a strip of endothelium roughly 9 km in length by approximately 30 cm in width, and which is so thin that it would produce a cylinder about the diameter of a lead pencil if rolled tightly.

The exchange of substances (primarily, water and the pressure relationships involved in this process) across the capillary walls was introduced on page 558 (Fig. 9.18). The important topic of capillary permeability will be discussed further in context in relation to the exchange of specific materials across the endothelial walls of particular capillaries, a process which occurs both rapidly and with an insignificant expenditure of energy. It is sufficient to mention here that in general capillaries appear to behave in many respects as though they were inert membranes containing submicroscopic pores which are around 30 Å in diameter and which were freely permeable to water and crystalloid materials. The capillaries are relatively impermeable to the passage of large molecules such as plasma proteins, however.

THE VEINS AND VENULES.   The conduits that return blood from the capillary beds throughout the tissues and organs of the body to the heart are the venules and veins (*dashed lines,* Fig. 10.15). These vessels become larger, fewer, and progressively thicker-walled as they converge, and the venous channels usually course together with a corresponding artery from any particular organ or tissue. Veins are more plentiful than arteries, their walls are thinner, and their diameters are larger. The total volume capacity for blood on the venous side of the circulatory system is considerably greater than that on the arterial side; hence one frequently hears the term *venous reservoir* used in connection with this portion of the systemic circulation (Table 10.1).

The thin-walled veins are much less elastic than are the arteries as well as more supple; hence the veins tend to collapse readily. Generally small (venules), medium, and large caliber veins are distinguished, although individual veins may exhibit considerable morphologic variations in various regions of the body.

In general, veins have three layers similar to those found in arteries. A tunica intima, tunica media, and tunica adventitia sometimes may be discerned, although in many veins these coats are not clearly defined (Fig. 10.17). The media, in particular, may be poorly developed. Muscular and elastic properties are poorly developed in veins; however, a greater proportion of connective tissue is present in these vessels than in arteries.

Venules.   These small veins are formed when several capillaries anastomose giving rise to a single vessel about 20 μ in diameter. The venule is composed of a single endothelial layer invested with a thin jacket of collagenous fibers and fibroblasts. In venules about 45 μ in diameter, scattered cells that appear to resemble partly differentiated smooth muscle fibers make their appearance between the connective tissue and endothelial layers, and these elements become progressively more numerous as the caliber of the vessel increases. Therefore, small veins that have a diameter around 200 μ have a layer of typical fusiform smooth muscle cells present in their walls.

In the larger venules, elastic connective tissue fiber networks appear; the intima in these vessels consists only of a layer of endothelium, the media consists of one to several scattered layers of smooth muscle fibers, and the adventitia is composed of collagenous fibers and fibroblasts. Most of the fibrous components of the larger venules are oriented parallel with the long axis of the vessel.

Venules, especially the smaller ones, apparently have an active role in the exchange of mate-

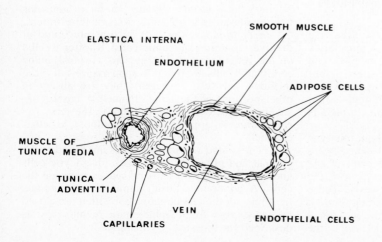

FIG. 10.17.   Cross section of a small artery with its accompanying vein for comparative purposes. (After Bloom and Fawcett. *A Textbook of Histology,* 9th ed, 1968, Saunders.)

rials with the interstitial space, and these vessels also have an important function in the vascular changes related to inflammation (p. 547). Consequently, small veins having a caliber of around 50 μ are particularly susceptible to permeability alterations induced by certain chemical agents, notably serotonin and histamine. In this regard, the venules may be even more sensitive to the influence of such compounds than are the capillaries themselves.

Medium-Caliber Veins. These vessels range in diameter from 2 to 9 mm, and include, for example, the deeper-lying as well as the cutaneous veins of the extremities located distal to the popliteal and brachial veins.

The tunica intima of medium-caliber veins consists of irregular, polygonally shaped endothelial cells; occasionally, elastic fibers are found in this layer, and externally this region may be circumscribed by a network of elastic connective tissue fibers. The tunica media in medium-sized veins is considerably thinner than in comparable arteries, and this layer consists principally of longitudinally oriented collagenous fibers and fibroblasts that separate the circumferentially disposed smooth muscle cells. The tunica adventitia, which is generally thicker than the media, is composed of connective tissue having longitudinal collagenous fiber bundles as well as networks of elastic fibers. Some longitudinal smooth muscle fibers may also be found in the adventitial layer, especially in proximity to the media.

Large-Caliber Veins. The azygos, renal, superior mesenteric, splenic, external iliac, portal, and the superior as well as the inferior vena cava all are vessels belonging to the category of large-caliber veins.

The tunica intima in the large veins has the same structure as in the medium-sized veins; however, the connective tissue layer may be as great as 70 μ in thickness. The tunica media is poorly developed and, if present at all, resembles that of the medium-caliber veins. The tunica adventitia comprises the greater part of the wall of large veins, and this layer is composed chiefly of loose connective tissue having thick elastic fibers as well as numerous axially oriented collagenous fibers. Adjacent to the tunica media (if it is present), or adjacent to the intima in the absence of the media, the adventitial layer contains histologically obvious longitudinal coats of smooth muscle together with complex networks of elastic fibers.

Valves are present in the larger veins, and these structures are usually found in pairs that are situated opposite to each other. Valves also are present within many veins of medium caliber, especially those found in the extremities. The valves are composed of connective tissues and appear as semilunar outpocketings of the internal surface of the vessel wall whose free ends are oriented in the direction of the blood flow (Fig. 10.18). The mechanical activity of the valves thus prevents

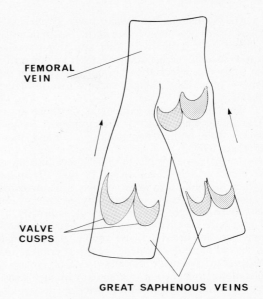

FEMORAL VEIN

VALVE CUSPS

GREAT SAPHENOUS VEINS

FIG. 10.18. Section of a vein laid open in order to show the valves. Arrows indicate the direction of blood flow. (After Goss [ed]. *Gray's Anatomy of the Human Body*, 28th ed, 1970, Lea & Febiger.)

movement of blood away from the heart. In this regard, it should be stressed at this point that the venous (as well as the cardiac) valves, like those in the lymphatic vessels, are exceedingly pliant structures that open or close passively depending upon the relative pressure exerted by the blood or lymph on the two sides of the individual cusps or flaps. Normally, these valves close tightly (ie, their flaps become apposed snugly together) when the "upstream" pressure exceeds the "downstream" pressure in any particular vessel or chamber of the heart.* No backward leakage, or *regurgitation,* of blood or lymph can take place, and thus a one-way passage of the fluid is assured.

In general, a terminal arteriole gives rise to a capillary network, which vessels in turn anastomose to form a venule. The transitional regions among these several types of vessel usually are gradual. If, however, blood passes from one set of capillaries directly into another large vessel, thence to a second set of capillaries prior to returning to the systemic circulation, the anatomic arrangement is called a *portal system.* An example of a portal system is found within the kidney. In this case, an *afferent arteriole* forms a capillary tuft known as a *glomerulus.* The glomerular capillaries then anastomose once again to form an *efferent* arteriole, which in turn forms a second capillary network surrounding the renal tubule (Fig. 12.5, p. 774).

The *hepatic portal vein* is another example of

Eversion is prevented and apposition of the cusps of the tricuspid and bicuspid valves is assisted by contraction of the papillary muscles during ventricular systole.

a portal system in the human body. This vessel arises from capillary networks in the abdominal viscera; within the liver, this structure ramifies into a second, series-connected set of channels, the sinusoids, which then anastomose to form the hepatic vein. This arrangement is shown in Figure 10.15.

The functional significance of these morphologic specializations of the circulatory system will be discussed in Chapter 12 (p. 772) and Chapter 13 (p. 827).

There is one further terminal specialization of the blood vessels found within the body. In many regions, terminal arterioles not only are connected to venules indirectly via capillary networks, but also these vessels may be connected directly by *arteriovenous anastomoses.* Such direct connections between arterioles and venules usually arise as heavily muscular branches of the arterioles, and these arterioles pass directly to small venules. Vasomotor innervation is abundant in such anastomotic vessels; therefore, the arteriovenous anastomoses constrict sharply when their motor nerves are stimulated, and when such arteriovenous anastomoses are constricted, the blood which earlier had flowed through them now becomes shunted to the capillary bed distal to the anastomosis. On the other hand, when the anastomotic connection is open, blood now passes directly to the venule, thereby bypassing the capillary bed. Consequently, arteriovenous anastomoses are of considerable importance in the regulation of blood supply in a number of tissues under various physiologic conditions.

A specialized and highly organized type of arteriovenous anastomosis is the *glomus,* a structure which is found in the nailbeds, ears, and the pads of the toes and fingers. The glomus is considered to be of importance in the regulation of temperature and conservation of body heat (Chap. 14, p. 994).

In conclusion to this section, the anatomic relationships between an arteriole, capillary bed, and venule are illustrated in Figure 10.19, together with an arteriovenous anastomosis, whereas some general characteristics of the blood vessels discussed above are summarized in Table 10.2.

CARDIAC PHYSIOLOGY

The sole function of the heart is to take a mass of blood returning to it via the systemic and pulmonary veins and to impart both kinetic and potential energy to that mass of blood by the intermittent, coordinated, and sequential contractions of the atria and ventricles, then to eject the blood into the pulmonary and systemic vascular circuits. The mechanisms within the heart underlying this rhythmic pumping function provide the basis of the material in this section. Before entering upon a detailed consideration of the individual aspects of cardiac physiology, however, it is necessary to clarify a basic, though vitally important, principle with regard to the overall action of the heart. This is the *excitation–conduction–contraction concept*

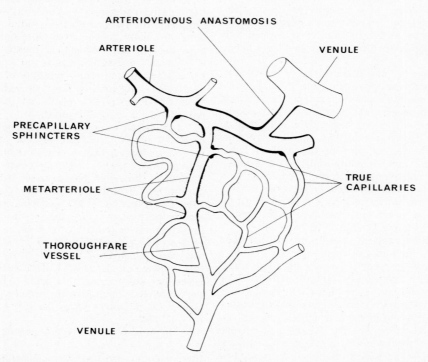

FIG. 10.19. A small area of a capillary bed. The vascular walls become thicker in regions where smooth muscle fibers are present. (After Ganong. *Review of Medical Physiology,* 6th ed, 1973, Lange Medical Publications.)

of cardiac activity, which may be explained briefly as the normal sequential chain of events that take place during a single cardiac cycle. *Excitation* of the heart takes place by means of rhythmic electric impulses that are generated spontaneously within the sinoatrial nodal tissue. These locally generated impulses then are *conducted* directly throughout the atrial myocardium and thence to the ventricular musculature via the specialized conducting tissues of the atrioventricular node, common bundle, and left and right bundle branches (Fig. 10.12). These transmitted impulses in turn stimulate contraction of the myocardial tissue of the heart in a rhythmic fashion.

In the normal heart, this series of events is inseparably linked; however, it is important to remember that under abnormal conditions this sequence may be interrupted or altered, resulting in the most bizarre clinical manifestations. For example, should the rhythmic, conducted electric impulses become uncoupled from the contractile mechanism of the ventricular myocardium, mechanical cardiac arrest will occur, although the electric pattern of excitation and conduction throughout the heart, as evidenced by the electrocardiogram, may appear to be normal. Of course, under such circumstances no blood is pumped, the blood pressure rapidly falls to zero, and death ensues unless immediate and drastic remedial measures are instituted.*

The physiologic mechanisms involved in each of the three aspects of cardiac function described above will now be considered individually.

Cardiac Excitation: Pacemaker Function of the Sinoatrial Node

The specialized muscular tissue of the sinoatrial node is capable of producing, spontaneously and rhythmically, the small electric impulses which are responsible for intiating the chain of events that result in the heartbeat. The impulses arising within this node dominate the activity of the entire heart simply because their inherent rate of generation by the sinoatrial cells is considerably faster (around 70 to 80 impulses/minute) than that of the other autorhythmic cardiac tissues, including the atrial musculature proper, the atrioventricular nodal cells (about 50 impulses/minute), or the Purkinje fibers (about 20 to 40 impulses/minute). Thus, an impulse that originates within the sinoatrial node is conducted through the atrial myocardium, and induces discharge in the atrioventricular node, bundle of His, and bundle branches before any of these tissues can reach its own particular threshold for spontaneous discharge. After discharge, the sinoatrial nodal cells complete their recovery cycle first so that once again the sinoatrial node is able to fire before any of the other autorhythmic cells.

Such an uncoupling effect may be demonstrated by chilling the heart several degrees below normal body temperature, a technique frequently employed during cardiac surgery when the patient's blood is being circulated and oxygenated by a suitable mechanical device.

and the process repeats in a cyclic fashion. The sequence of activation of the various regions of the mammalian heart, together with typical transmembrane potentials recorded from different regions, is illustrated in Figure 10.20.

The mechanism of repetitive firing in living cells in general was introduced on page 143. Cardiac pacemaker cells, in common with all other spontaneously active tissues, have no stable rest potential. Following their discharge during the previous electric cycle, the transmembrane potential of the cells is regenerated during the repolarization phase to a certain voltage called the maximum repolarization potential *(MRP)*. Immediately after this potential has been reached, the cell begins to depolarize slowly and spontaneously; ie, the pacemaker cell leaks both voltage and current until a finite voltage level, the threshold, has been reached, at which point an all-or-none action potential develops. Then the cell immediately commences to repolarize once more. This sequence of events is illustrated in Figure 10.21.

The quantitative difference (in millivolts) between the maximum repolarization potential and the threshold potential to firing is termed the *prepotential*, or more aptly, the *pacemaker potential*. This prepotential results from a relatively slow depolarization from the maximal value.

The configuration of the cardiac pacemaker potential and the myocardial action potential differs from those developed in other tissues in several important respects (Fig. 10.20). First, the time course of the electric cycle is much slower in comparison with, for example, a rhythmically discharging neural sensory receptor or nerve (eg, Fig. 5.6, p. 165). Second, the maximum repolarization potential developed by cardiac pacemaker cells is far lower (-55 to -60 mv) than is the rest potential found in other body cells, for example, skeletal muscle fibers, which can have a rest potential in the neighborhood of -80 to -90 mv. Third, the overshoot of the action potential usually is negligible in cardiac cells compared to, for example, nerve, in which the overshoot may reach $+20$ mv or more above the isopotential level.

It is obvious from the facts just presented that the membrane of cardiac pacemaker cells normally is unstable insofar as its permeability to ions is concerned, a fact that is in marked contrast to the situation present in skeletal muscle and nerve (cf. pp. 130, 164, and 210). This specialized functional property of such cells to "leak" voltage and current, and which results in their spontaneous depolarization, has been attributed by many investigators to an extremely high "resting" sodium permeability or conductance. The low maximum repolarization potential itself is also considered to be a result of this high conductance for sodium ion. This membrane property in turn results in a gradual decrease from the maximum repolarization potential to a finite threshold voltage, at which point an action potential is generated and then

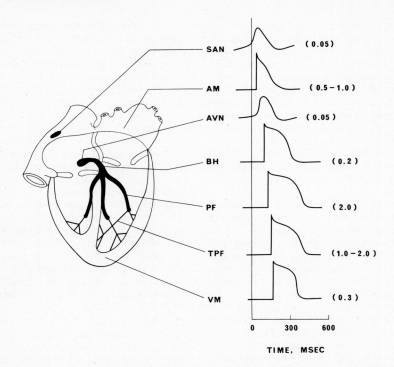

FIG. 10.20. Normal sequence of events in cardiac excitation and conduction. The configuration of the action potentials in each region is shown and the numbers in parentheses to the right of each action potential signify the approximate conduction velocity in meters per second through that particular tissue. Note particularly the extremely low conduction velocity in the atrioventricular node, a fact which permits complete atrial excitation to take place before ventricular excitation is initiated (cf. Fig. 10.24). SAN = sinoatrial node; AM = atrial musculature; AVN = atrioventricular node; BH = bundle of His; PF = Purkinje fiber; TPF = terminal Purkinje fibers; VM = ventricular musculature. (After Jensen. *Intrinsic Cardiac Rate Regulation,* 1971, Appleton-Century-Crofts; Ruch and Patton [eds]. *Physiology and Biophysics,* 19th ed, 1966, Saunders.)

propagated in all directions throughout the atrial muscle, much as the ripples proceed outward from the point where a stone has been thrown into a pond. Repolarization involves a reduced permeability (conductance) to sodium ion with a concomitant marked increase in the permeability (conductance) to potassium, so that the transmembrane voltage tends to approach the potassium equilibrium potential. As noted above, however, the membrane permeability to sodium ion is somewhat higher in pacemaker than in other excitable tissues. Therefore, as the permeability to potassium decreases progressively during the repolarization phase of the electric cycle, the transmembrane potential also begins to decrease from near the potassium equilibrium potential toward a lower steady-state value, because of the high sodium permeability. This steady-state value is so low, however, that the membrane potential reaches threshold first and another all-or-none action potential (impulse) is generated (see also p. 143).

There is, however, equally strong evidence that the development of the diastolic depolarization in cardiac pacemaker cells may not depend entirely upon a high "resting" sodium conductance, but that another ion, possibly calcium, is implicated

in this phenomenon. Regardless of the ultimate ionic mechanism (or mechanisms) which is demonstrated to underlie the genesis of the prepotentials in cardiac pacemaker cells, it is evident that the action potentials developed by these cells do depend in large measure upon changes in sodium and potassium conductance which are similar to those described on page 132 and illustrated in Figure 4.24 (p. 136) for nerve and skeletal muscle. Furthermore, a coupled sodium–potassium pump is operative in cardiac pacemaker fibers as in other excitable cells (p. 92).

The rate at which the normal human heart beats is quite subject to change, and may vary considerably depending upon many factors both extrinsic and intrinsic to the pacemaker cells. These factors impinge simultaneously upon the pacemaker and modify its inherent spontaneous frequency of activity. Thus, the heart rate may vary from around 70 contractions per minute during rest to maximal levels approaching 200 contractions per minute during extreme muscular exertion or sexual excitement (see Chap. 16, p. 1178 and p. 1182).

Within the sinoatrial node proper, only a small group of cells discharge impulses most rapidly at

A.

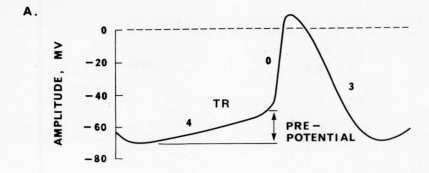

B.

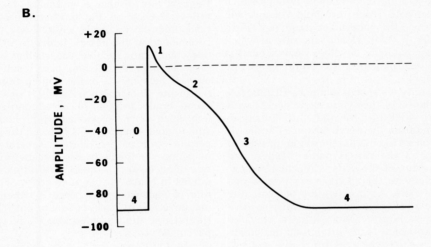

C.

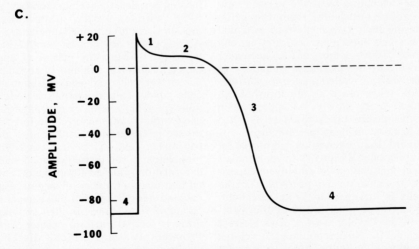

FIG. 10.21. **A.** Pacemaker potential as recorded from the sinoatrial node. **B.** Action potential recorded from an atrial fiber. **C.** Action potential recorded from ventricular fiber. The numbers signify the various phases of the action potential. Note that the pacemaker potential has no stable rest potential (phase 4); rather, these cells spontaneously depolarize to a threshold (TR) level, at which point an action potential develops. Note also that the pacemaker potential lacks a spike (phase 1) as well as a plateau (phase 2). See also Figure 10.22. (After Hoffman and Cranefield. *Electrophysiology of the Heart,* 1960, The Blakiston Division, McGraw-Hill.)

any given moment, and thus these cells are the primary functional pacemaker for a variable period of time. Subsequently, other groups of nodal cells assume dominance insofar as their firing rate is concerned. The specific anatomic locus of pacemaker activity within the sinoatrial node shifts from time to time under physiologic conditions, and when the electric impulses that drive the entire heart originate anywhere within the sinoatrial node, a *sinus rhythm* is said to be present.

What, then, is the mechanism whereby the pacemaker cells alter their inherent rate of discharge, and thus influence the overall heart rate? Any factor that changes the rates of ionic flux across the pacemaker cell membrane, thereby altering the rate (or slope) of the diastolic depolarization phase of the electric cycle through which the pacemaker cells reach threshold voltage, will directly influence the pacemaker frequency (Figs. 10.22 and 4.29, p. 144). Hence, the rate at which the pacemaker potential develops ($\Delta mv/\Delta time$) is the principal determinant of the overall heart rate (Fig. 10.22A).

Consequently, as shown in Figure 10.22B, following stimulation of the sympathetic nerves with an attendant liberation of the catecholamine norepinephrine from the nerve terminals in the vicinity of the pacemaker cells, the slope, or rate, of depolarization of the cell is much steeper, and threshold to firing of the cells is reached sooner, thereby accelerating the spontaneous rate of discharge of the cells. An acceleration of heart rate above its previous frequency is called a *positive chronotropic response*. Clinically, an elevated heart rate is known as *tachycardia,* and this condition is said to exist when the heart rate exceeds 100 beats per minute.

Conversely, upon stimulation of the vagus, liberation of acetylcholine from the nerve terminals drives excess potassium from the cells, thereby increasing their transmembrane potential (ie, the cells become hyperpolarized) and thereby reducing the rate of development of the pacemaker potential. Slowing of the discharge frequency of the pacemaker, which is called a *negative chronotropic response,* takes place as the threshold to firing is reached more slowly, as indicated in Figure 10.22C. Clinically, bradycardia exists when the heart rate is below 60 beats per minute.

In the event that excessive quantities of acetylcholine are applied directly to the pacemaker or strong vagal stimulation is employed experimentally, the transmembrane potential of the pacemaker cells is driven away from threshold, and the attendant hyperpolarization of the membrane results in the development of a stable rest potential (Fig. 4.34). No prepotential develops; thus, the membrane potential does not reach threshold to firing, and the heart stops beating. During prolonged vagal stimulation and the resulting cardiac arrest, however, the heart may recommence beating spontaneously once again, as the enzyme cholinesterase gradually inactivates

the acetylcholine, thereby permitting the pacemaker cells to discharge to threshold voltage once again, and thus fire spontaneously. This phenomenon is known as escape of the heart from vagal inhibition, or, more simply, *vagal escape.* These facts are illustrated in Figure. 4.34, p. 148.

The heart rate in human subjects normally is governed to a large extent by neural effects which are mediated through chemical transmitter agents, and the reflexes involved in cardioregulation will be discussed on page 668. Other humoral agents that influence heart rate, either directly or indirectly, will be considered in context as appropriate.

Although the temperature of the pacemaker cells has a profound and direct influence on their frequency of discharge, normally this factor has little influence on the heart rate, as man is a homeothermic organism whose body temperature is regulated within narrow limits (p. 994). Elevated body temperatures, however, such as occur during fever or hyperthyroidism, may result directly in marked tachycardia in the patient through warming of the sinoatrial nodal cells. Conversely, profound bradycardia is observed during hypothermia, such as is occasioned by prolonged immersion in cold water or as is induced deliberately for certain surgical procedures. Several electrolyte imbalances in the body also can alter markedly the rate of pacemaker discharge, resulting in wide variations in heart rate, as ultimately, of course, the rate of pacemaker discharge is dependent upon, and governed by, the concentrations of ions in the interstitial fluid immediately adjacent to the pacemaker cells.

In addition to the factors enumerated above, which affect their frequency, pacemaker cells are influenced directly by mechanical stimuli, in a fashion similar to certain sensory receptors (p. 659). Consequently, pacemaker cells are pressure-sensitive, mechanoelectric transducers. An important distinction should be mentioned, however, between the effects of applying a sustained mechanical stimulus of constant intensity to pacemaker cells and sensory endings. The latter cells rapidly exhibit adaptation under such conditions (p. 186), whereas pacemaker cells do not, at least within a reasonable time interval. Thus, the inherent frequency of impulse discharge by pacemaker cells can be increased within limits in proportion to the degree of mechanical stretch or deformation imposed upon them. This fundamental and inherent ability of sinoatrial nodal cells to respond directly to such mechanical influences by an appropriate change in their spontaneous rate of firing is termed *intrinsic cardiac rate regulation.* Presumably, mechanical distention of the pacemaker fibers increases their surface area and thereby augments their permeability to the various ionic species involved in the genesis of the pacemaker potential so that the observed prepotential develops faster, and accelerated activity of the cells ensues as described above. Conversely,

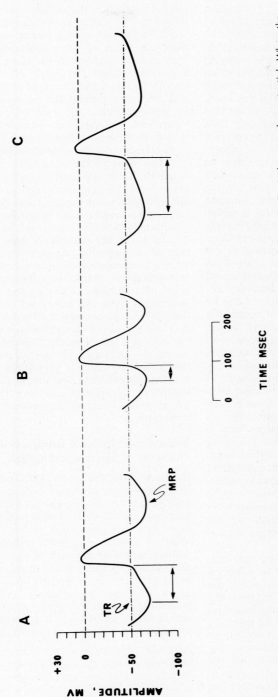

FIG. 10.22. Alterations in the pacemaker potential which result in a faster or slower rate of firing. **A.** Typical pacemaker potential. When the transmembrane voltage decreases from a maximum repolarization potential (MRP = −70 mv) to threshold (TR = −50 mv), the cell fires spontaneously. **B.** Cardioacceleration, such as accompanies sympathetic stimulation, is caused by an increased rate of depolarization to threshold for firing; hence, the cell reaches threshold and thus fires sooner. **C.** Cardioinhibition, such as accompanies vagal stimulation, is caused by a decreased rate of depolarization to threshold for firing; hence, the cell takes longer to depolarize to threshold and thus fires later. See also Figure 4.34, p. 148. (After Jensen. *Intrinsic Cardiac Rate Regulation,* 1971, Appleton-Century-Crofts.)

removal of the excess load that produced mechanical deformation of the cells results in a diminution of the frequency to the lower, predeformation value.

Although this intrinsic pacemaker mechanism probably exerts little influence in the regulation of normal heart rate in the intact mammal, the latent importance of a second aspect of this same intrinsic cardiac rate regulatory mechanism is manifest in certain clinical situations, in which the induction of abnormal loci of pacemaker function can take place through mechanical stimulation of the heart. It was emphasized earlier that all the tissues that comprise the entire human heart, including both the specialized conducting tissues and the contractile myocardial elements themselves, are potentially capable of generating spontaneous impulses under appropriate pathologic or experimental conditions. The mechanism is similar to that whereby the sinoatrial node performs this pacemaker function normally. Thus, any factor that enables a region of the heart to develop prepotentials rapidly can become an abnormal, or ectopic, pacemaker, which then can compete with the normal, or entopic, pacemaker of the sinoatrial node, an effect that may result in the development of pronounced disturbances of the normal rhythm, ie, *arrhythmias*. Ectopic functional pacemakers can be induced, for example, in both atrial and ventricular musculature by the mechanical irritation of the endocardium that is attendant upon the procedure of diagnostic cardiac catheterization. In this instance, the mechanical irritation of the endocardium by the catheter tip produces a repetitive discharge of impulses within the myocardium in a random fashion, resulting in arrhythmias ranging from premature systoles to fibrillation. These iatrogenic arrhythmias usually revert to a normal pattern of sinus activity after various intervals following withdrawal of the catheter from the right atrium. This type of response is an excellent example of another aspect of the intrinsic cardiac rate regulatory mechanism, viz, spontaneous activity can be induced by prolonged mechanical stimulation in cardiac tissue which normally is driven by impulses generated in the sinoatrial nodal pacemaker.

Consequently, the intrinsic cardiac rate regulatory mechanism can, first, control to a limited extent the normal spontaneous rate at which the sinoatrial pacemaker cells fire, depending upon the degree to which these cells are stretched mechanically. Second, mechanical stimulation of normally quiescent myocardial tissue is able, under appropriate circumstances, to induce rhythmic prepotentials, hence ectopic foci of pacemaker activity, in any region of the heart.

By way of summary, the specialized cells of the cardiac pacemaker generate spontaneous repetitive electric impulses (action potentials) as a direct consequence of rhythmic fluctuation of the transmembrane voltage to a threshold value. Therefore, the membranes of sinoatrial nodal cells

possess an inherent electric instability that enables them to generate pacemaker potentials, which in turn activate the adjacent myocardial cells of the atrium. The frequency with which the pacemaker potentials (hence the action potentials) are produced is the direct consequence of the rate at which the prepotentials develop; or, stated another way, the slope of the prepotential curve determines the rate at which the sinoatrial nodal cells fire. The steeper the slope of the prepotential curve, the faster the threshold to firing is reached, and the more rapidly the pacemaker fires. Conversely, reducing the rate of prepotential development prolongs the time necessary for the transmembrane potential to reach threshold; therefore, a slowing of pacemaker activity ensues.

The slope of the pacemaker potential curve, hence the heart rate itself, is regulated normally by nerve impulses that cause the liberation of chemical transmitter substances from the nerve terminals (norepinephrine and acetylcholine), and these compounds directly accelerate or decelerate the rate at which the pacemaker fires by altering the rate of prepotential development. Furthermore, such additional factors as the temperature, ionic concentrations, endocrine substances, and the degree of mechanical distention to which the pacemaker cells are subjected may all influence the net rate of pacemaker firing by their direct action on the above-described mechanism. In this regard, the effect of mechanical deformation of the pacemaker cells upon the normal mammalian heart rate in vivo probably is very slight, although there is abundant evidence that the intrinsic cardiac rate regulatory mechanism can participate directly in the genesis of ectopic pacemaker activity, hence arrhythmias.*

Intracardiac Impulse Conduction and the Elements of Electrocardiography

In this section, the second principal aspect of cardiac physiology introduced on page 594 will be discussed, in particular the normal conduction (or transmission) of electric impulses throughout the heart from the pacemaker region. The conducted impulses in turn result in the orderly, coordinated, and sequential contraction of the atria and ventricles. Following this presentation, certain basic aspects of electrocardiography will be introduced. Since this technique deals exclusively with measuring and recording the electric phenomena within the heart in vivo, it is of fundamental importance to the practicing physician.

IMPULSE TRANSMISSION. It was stated earlier in this chapter that the human heart is composed of two large functional syncytia joined anatomically

---

*The intrinsic cardiac rate regulatory mechanism has been demonstrated unequivocally to play an important physiologic role in the regulation of heart rate in a large number of submammalian vertebrate as well as invertebrate species.

Table 10.3 SEQUENCE OF EVENTS IN CARDIAC EXCITATION[a]

REGION	APPROXIMATE TIME FOR EXCITATION (msec)	CONDUCTION VELOCITY (m/sec)
Sinoatrial node	0	0.05
Atrioventricular node	40	0.1
Left atrial appendage	100	1.0
Bundle of His	130 to 160	2[b]
Right ventricle, anterior surface	190	0.3[c]
Ventricular apex, surface	220	0.3[c]
Posterior basal area	260	0.3[c]

[a]*Data from Ganong.* Review of Medical Physiology, *6th ed, 1973, Lange Medical Publications;* Guyton. Textbook of Medical Physiology, *4th ed, 1971, Saunders; Ruch and Patton (eds).* Physiology and Biophysics, *19th ed, 1966, Saunders.*
[b]*Rate is about the same for left and right bundle branches.*
[c]*Rate of conduction in Purkinje system per se is about 1.0 m/sec. Values listed in table are for ventricular myocardium per se.*

through the atrioventricular node, the individual cellular elements that compose these syncytia being connected by low-resistance bridges, the intercalated discs (p. 583). The rhythmic localized depolarizations that originate in the sinoatrial node spread radially and directly throughout the atrial myocardium. As noted earlier, there is convincing anatomic evidence that no specialized conducting system is present within the atria. The action potentials generated within the pacemaker region of the sinoatrial node depolarize the contiguous atrial myocardial cells directly, and thus the impulses are transmitted outward in a wavelike fashion from the excitatory locus (Fig. 10.23). Stated another way, the current produced by the nodal cells flows longitudinally and thereby lowers the resting transmembrane potential of adjacent atrial cells to their threshold for firing. These areas in turn depolarize, and thus propagated action potentials spread radially over the entire atrial myocardium. The velocity of impulse conduction in specific regions of the heart (in meters per second) is given to the right in Figure 10.20. The temporal sequence in which the different regions of the heart are activated is given in Table 10.3.

The velocity of conduction of the impulses within the sinoatrial node, although difficult to measure, is quite low, and this value has been estimated to be around 0.05 meter per second. Spread of the impulse through the atrial myocardium proper gradually becomes somewhat faster than transmission within the sinoatrial node; therefore, impulse propagation outside of the node ranges between 0.5 and 1.0 meter per second in the atrial muscle proper.

As the propagated impulse wave traverses it, the upper portion of the atrioventricular node is excited in passing while the impulse continues to sweep across the atria so that ultimately the tip of the left auricular appendage is excited. This effect is similar to "crossing a T," as shown in Figure 10.24. There is a significant temporal delay in spread of the excitation within the atrioventricular nodal region, a fact which is of considerable importance to the normal and sequential activation of the heart. As the impulse approaches the upper (or atrial) border of the atrioventricular node, the velocity of conduction becomes sharply reduced from that encountered in the atrial musculature proper. Thus, by the time the impulse reaches the center of this node, it is traveling at about the same rate as in the pacemaker cells of the sinoatrial node, or 0.05 meter per second. Correlated with this decrease in the conduction velocity, there is a concomitant decrease in the rate of development, or upstroke, of the action potentials within the cells of the central nodal region (Fig. 10.20C), which apparently is linked to a *decremental conduction* of the impulse as indicated in Figure 10.25A. In decremental conduction, the actual magnitude of the voltage diminishes from one end of the impulse path to the other, an effect that contrasts sharply with nondecremental conduction such as is found in nerve and muscle fibers (Fig. 10.25B). Both decremental conduction and the decreased rate of action potential development in the atrioventricular node contribute directly to a reduced conduction velocity in this region of the heart. The propagation of the impulse through the central portion of the atrioventricular node, in consequence, accounts for about 30 msec of the total atrioventricular nodal delay. The latter amounts to between 80 and 120 msec from the time of excitation of the upper region of the node to exit of the impulse at the bundle of His. Presumably, this slowing of impulse transmission is occasioned to some extent by small fiber size as well as by extensive branching of the fibers, which physically lengthens the overall pathway which the impulse must traverse.

After the central region of the atrioventricular node has been passed, the impulse velocity once

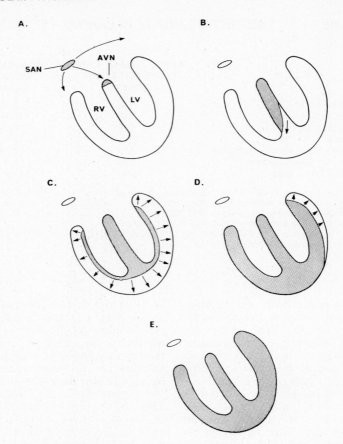

FIG. 10.23. Normal spread of excitation throughout the heart is indicated by shading. **A.** Sinoatrial node (SAN) depolarizes at zero time and atrial excitation is complete in about 50 msec. **B.** Following an 80- to 120-msec delay in the atrioventricular node (AVN), the impulse activates the interventricular septum from left to right; thence, it passes to the subendocardial surface of both ventricles **(C).** Note that the right ventrical (RV) is excited more rapidly than the left ventricle (LV) because it is thinner-walled. However, the endocardial surface of both ventricles is excited almost simultaneously, ie, in about 50 msec after the impulse enters the common bundle. Note also that the excitation pathway is from endocardium to epicardium (arrows), which is exactly the reverse of the pathway taken during repolarization (cf. Fig. 10.30). **D** and **E** show events late in, and at the end of, ventricular depolarization, respectively. (Data from Ganong. *Review of Medical Physiology,* 6th ed, 1973, Lange Medical Publications; Ruch and Patton [eds]. *Physiology and Biophysics,* 19th ed, 1966, Saunders; Best and Taylor [eds]. *The Physiological Basis of Medical Practice,* 8th ed, 1966, Williams & Wilkins.)

again increases to around 1.0 to 1.5 meters per second and the configuration of the action potential here is identical with that found in the bundle of His (Fig. 10.20D).

It thus is apparent that the atrioventricular nodal region has three functionally, if not morphologically, specialized and discrete regions, a conclusion based upon the physiologic characteristics of impulse transmission in different regions of the node. In this regard, no detailed histologic studies have been forthcoming on the atrioventricular node insofar as correlating possible anatomic specialization of the several regions of this structure with their functional properties.

Once an impulse enters the bundle of His and the left and right bundle branches, there is a marked increase in the conduction velocity to around 3 to 4 meters per second. This increase may be attributed to the much greater diameter of

the fibers in this region (Purkinje cells, for example, are 50 to 70 μ in diameter), as well as to their infrequent branching, which results in a more direct pathway for impulse propagation. Action potentials recorded from Purkinje cells characteristically exhibit a rapid upstroke, high amplitude, and a prolonged duration (Fig. 10.20E).

Upon entering the ventricular myocardium, the impulse slows down once again to around 0.4 to 1.0 meter per second. This effect is caused principally by the smaller fiber size (15 to 20 μ diameter), as well as by the frequent occurrence of branching by the muscle cells themselves. As illustrated in Figure 10.23, the spread of excitation within the thick-walled ventricles is from the endocardial toward the epicardial surface. The last region of the heart to be activated is the posterior basal area of the left ventricle.

In summary, therefore, the sequence of con-

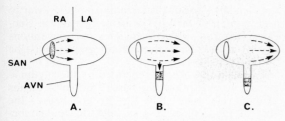

RA LA

SAN

AVN

A. **B.** **C.**

FIG. 10.24. Atrioventricular nodal delay. **A.** An impulse generated in the sinoatrial node (SAN) of the right atrium passes outward toward the left atrium (LA). As the impulse passes the atrioventricular node (AVN), this tissue is excited, **B.**, but because of the low velocity of nodal conduction (about 0.05 m/sec, Fig. 10.20), the impulse excites the left atrial appendage before ventricular excitation commences,**C.**

duction of impulses within the heart that initiates the coordinated rhythmic contractions of this organ is summarized in Table 10.3 and illustrated in Figure 10.23.

Functionally, the atrioventricular nodal delay is of extreme significance, as it permits complete excitation (hence contraction) of both the right and left atria prior to ventricular excitation (and contraction). Hence, the blood within the atria is ejected prior to the onset of ventricular systole. Furthermore, the relatively fast conduction velocity, which is an inherent property of the bundle of His, the left and right bundle branches, and the Purkinje tissues (Fig. 10.20), permits an almost simultaneous as well as smoothly coordinated excitation of both left and right ventricles, so that the ensuing contraction of both of these chambers is also smoothly coordinated and almost simultaneous. If this were not the case, the heart would be an inefficient pump indeed, as certain excited portions of the ventricles would be undergoing contraction, whereas regions not yet activated would merely bulge passively as the consequence of a localized increase in the intraventricular pressure.

The property of decremental conduction within the atrioventricular node proper (Fig. 10.25), in addition to contributing physiologically

to normal atrioventricular nodal delay, has another aspect of clinical importance. If, for any reason, either the amplitude or the rate of increase of the amplitude of the impulse after traversing the atrioventricular node proper has become attenuated to such an extent that the impulse fails to depolarize, hence excite the cells of the bundle of His to threshold, *heart block* of varying degrees develops. If no impulses can pass through the atrioventricular region from the atria, a condition of total (complete, or third-stage) heart block is present, and the ventricles will be paced slowly by impulses originating primarily within the bundle of His. The net result of total block is that the atria will continue to respond normally to activation by the sinoatrial node (say at a frequency of 80 beats per minute), but now the ventricular rate will be considerably slower, as these chambers are being driven by an ectopic pacemaker in the bundle of His, which may have an inherent discharge frequency as low as 40 to 50 per minute (see p. 696). An incomplete block also may develop, and in such an event an occasional impulse does manage to pass through the blocked atrioventricular nodal region from the atria. Clinically, the degree of block now is described by the arithmetic ratio of atrial to ventricular impulses. For example, various degrees of partial block may be observed, in which these ratios may be 2:1, 3:1, and so forth. Lastly, impairment of atrioventricular conduction may be such that only the conduction of impulses from atria to ventricles is delayed beyond the normal value of 0.2 second.

BASIC PRINCIPLES OF ELECTROCARDIOGRAPHY. The principal technique that is readily available to the physician for assessing (and evaluating) the functional integrity of the *excitation* and *conduction* mechanisms within the heart of a patient is called *electrocardiography*. This technique, it should be emphasized, can detect anomalies of the contractile machinery if, and only if, these mechanical anomalies result in, or from, disruption of the normal excitation–conduction process described above. In other words, the elec-

FIG. 10.25. **A.** When conduction with decrement is present, the amplitude (or intensity) of the signal decreases with distance from its source, an effect that is believed to take place in the atrioventricular node. **B.** Conduction without decrement in which there is no alteration in signal amplitude with distance from its source. This is the type of all-or-none conduction seen normally in nerve and muscle.

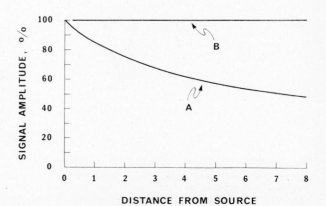

trocardiogram (EKG)* tells nothing directly about the mechanical properties of the heart, that is, the ability of this organ to contract and perform work. Rather, the EKG reflects solely the *electric performance* of the heart at the moment during which the recording was made, and that performance alone! It is, however, true that many clinically observed electric abnormalities give rise to (or are the consequence of) mechanical defects; however, the student should recognize clearly these limitations of the EKG when taken alone as a diagnostic tool. Notwithstanding this precautionary note, the vital and undisputed importance of electrocardiography as a clinical diagnostic procedure is great.

The remainder of this section will be devoted to a presentation of certain fundamental concepts related to electrocardiography, especially as these pertain to the normal excitation of, and conduction of, impulses within the human heart.

Introductory Remarks. The human body fluids are good conductors of electricity because of their ionic constituents. The body as a whole forms a good volume conductor for electric potentials that are generated by the individual cells within the heart. A recording from the body surface of the changes in the *algebraically summed potentials* (or voltages) that are generated simultaneously at any given moment by the entire mass of individual myocardial cells that make up the heart when plotted against time is called an *electrocardiogram* or *EKG*.

This abbreviation is derived from the German term Elektrokardiograph, and will be used preferentially in this book in order to avoid confusion with the abbreviation for electrocorticogram, or ECG (p. 420).

Conventions Used in Electrocardiography. Prior to a discussion of the physiologic origin of the electrocardiogram, several techniques and conventions employed in clinical electrocardiography must be introduced.

The small voltages generated by the rhythmic electric activity of the heart are recorded by means of metal electrodes that may be attached to the skin of the subject; a good electric contact between skin and electrodes is assured by a film of abrasive conducting paste that is rubbed on the skin at the point of contact with the electrode. Extraneous electric signals ("noise") from muscle contractions and other sources are minimized by having the patient in a supine and relaxed condition while the EKG is being recorded.

The impulses generated by the heart and recorded through the surface electrodes are amplified electronically to an extent that is sufficient to drive a galvanometer attached to a recording pen, which writes directly by use of a heat stylus or ink upon a moving strip of paper (Fig. 10.26A). By convention, a 1.0-cm vertical deflection of the pen above the zero (or baseline) position on the chart represents +1.0 mv. The horizontal speed of the chart is 25 mm (2.5 cm)/sec, so that a 1.0-mm horizontal deflection on the chart is equal to 40 msec. These conventions are depicted in Figure 10.26B.

The actual configuration of the recorded potentials will depend upon the exact placement of the leads upon the surface of the body; therefore, certain standard locations of the electrodes are now used routinely in clinical electrocardiography. The limbs may be considered merely as conductors, and since current flows only in the body

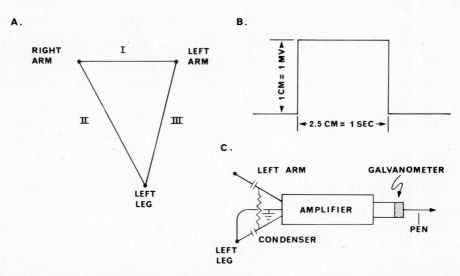

FIG. 10.26. **A.** Conventional positions of standard bipolar limb leads used in electrocardiography. **B.** Conventions for standarization and paper speed. Thus 1 mv = 1 cm on the vertical scale, and 1 mm on the horizontal scale = 0.04 sec, the net horizontal speed of the chart paper being 2.5 cm/sec. **C.** Equipment used for electrocardiography (cf. Fig. 10.33). (After Ruch and Patton [eds]. *Physiology and Biophysics,* 19th ed, 1966, Saunders.)

fluids, records obtained from any area on the arms and legs are the same as those which would be obtained by placing the electrodes at the points of attachment of the limbs to the body. The extremities are used as convenient points of attachment for the electrodes. The right leg generally is used as a ground terminal. The active, or exploring, electrode (or lead) may be either bipolar or unipolar.

Bipolar Leads. In bipolar recording of electrocardiograms, two active leads are used, so that the potential difference that exists between two limbs at any particular moment is recorded. The *standard limb leads* are designated by the Roman numerals I, II, and III and depicted in Figure 10.26C.

When the convention originally established by Einthoven is used, lead I connects the left arm with the right arm, so that when the left arm is relatively positive (or the right arm negative) an upward deflection of the galvanometer pen is produced in the recorder.

Lead II connects the right arm with the left leg, and lead III connects the left arm with the left leg. In leads II and III, an upward (positive) deflection of the galvanometer is produced by the appropriate lower limb being positive (or the upper limb negative) with respect to the arm with which it is connected. These leads are shown in Figure 10.26A and were established by Einthoven so that the principal deflection of the pen in each lead would be upright in a normal subject.

Unipolar or V Leads. When unipolar recordings are made, the active electrode is connected to an indifferent electrode which has zero potential Fig. 10.26A

At the present time, 12 standard leads are used routinely in clinical electrocardiography, as described below and illustrated in Figures 10.26A and 10.28. Thus, in addition to the three standard bipolar leads (I, II, and III), three unipolar limb leads which are designated VR (right arm), VL (left arm), and VF (left foot) are recorded together with six precordial (chest or fourth) leads which are denoted V_1 through V_6 (Fig. 10.28). By convention, when the precordial (or exploring) unipolar electrode is relatively positive, an upright wave is inscribed on the chart of the recorder, whereas a downward deflection is traced when the same electrode is relatively negative. In toto, six limb and six chest leads are recorded in standard clinical electrocardiography.

Augmented limb leads, as indicated by the small letter "a" (aVR, aVL, aVF), are in general use at present. The augmented recordings are made between one limb and the other two limbs as illustrated in Figure 10.27B. This electrode arrangement has the advantage that the amplitude of the recorded potentials is amplified some 50 percent but without any change in the configuration of the recorded waveforms. The reason for this amplification is given on page 612.

In addition to the leads described above, the electric activity of the atria may be studied in greater detail by means of a unipolar electrode placed within a catheter and then swallowed by the patient. Esophageal leads are denoted by the letter "E," followed by the distance in centimeters from the catheter tip to the teeth; eg, E30 would signify a record made with an esophageal lead whose tip lies 30 cm inferior to the teeth.

Further conventions relevant to electrocardiography will be introduced as appropriate in subsequent sections.

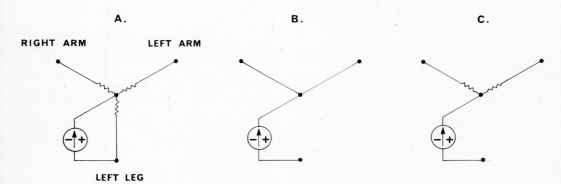

FIG. 10.27. Types of connection used in electrocardiographic recording by means of unipolar leads. **A.** All extremities are connected through a network of resistors. The network in turn is connected to the negative terminal of the amplifier. The exploring (positive) electrode is connected to each extremity in order to record unipolar limb leads or else placed upon the precordial positions illustrated in Figure 10.28 (Wilson). **B.** A modified unipolar limb lead recording circuit. The potential of one electrode on a limb is connected to the positive terminal of the amplifier, and this voltage is recorded against that of the other two electrodes which are connected without resistors (Goldberger). **C.** Method of unipolar limb recording in which the potential recorded from one extremity is compared with the potentials at two other points which are connected through resistors. The circuit depicted in **B** is in general use for unipolar limb lead recording, whereas the circuit shown in **A** is in use for recording unipolar precordial leads (Wilson). (After Ruch and Patton [eds]. *Physiology and Biophysics,* 19th ed, 1966, Saunders.)

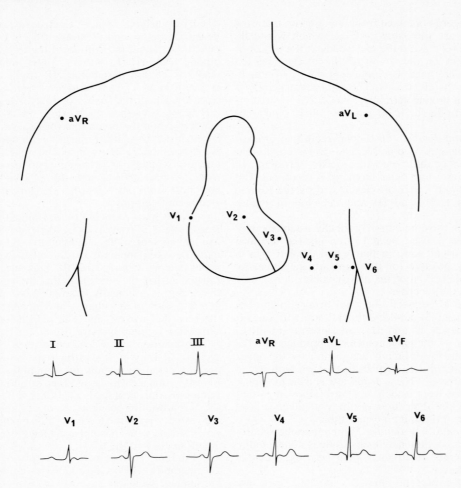

FIG. 10.28. Positions of the standard unipolar chest (or precordial) leads which are used in electrocardiography together with typical electrocardiographic recordings obtained from all twelve standard leads. (After Ganong. *Review of Medical Physiology,* 6th ed, 1973, Lange Medical Publications.)

THE CONCEPT OF ELECTRIC DIPOLES; THE ELECTROCARDIOGRAM. As noted earlier (p. 604), the EKG is merely an amplified recording which represents the algebraically summed individual action potentials developed by the entire heart at any moment and measured at a distance from their source. The potentials thus recorded are essentially *surface potentials* in contrast to *transmembrane potentials* recorded from a single fiber by means of a microelectrode inserted within that fiber (pp. 90, 130, 163; Figs. 5.5, p. 164; 5.6, p. 165; 5.11, p. 172; 5.12, p. 174).

An understanding of the complex biphasic waveforms generated by the heart and recorded by an EKG is best obtained by consideration of the simple biphasic voltage changes that are generated upon stimulation of a muscle. As illustrated in the figures in Chapter 5, the passage of an impulse along an excitable tissue results in transient deflections of the galvanometer whose sign (positive or negative) depends upon the instantaneous location of the wave front of the action potential relative to the position of the electrodes on the tissue, or, in the case of the heart, upon the body surface. Therefore, it is both convenient and accurate to think of the action potential developed by the heart as a *wave of negativity* sweeping outward from the point of stimulation both over the surface and through the entire mass of the tissue, as indicated in Figure 10.29. Similarly, repolarization may be visualized as a *wave of positivity* sweeping over the depolarized tissue from the point of stimulation, as depicted in Figure 10.30. The net result is a recorded waveform similar to that shown in Figure 10.31. Note in the latter figure that the recorded waves that represent the depolarization and repolarization processes contain equal areas which are shaded in the illustration.

Since the sequential activation of the heart by the conducted excitatory wave results in certain areas being negatively charged relative to other regions at the same instant, the entire heart itself

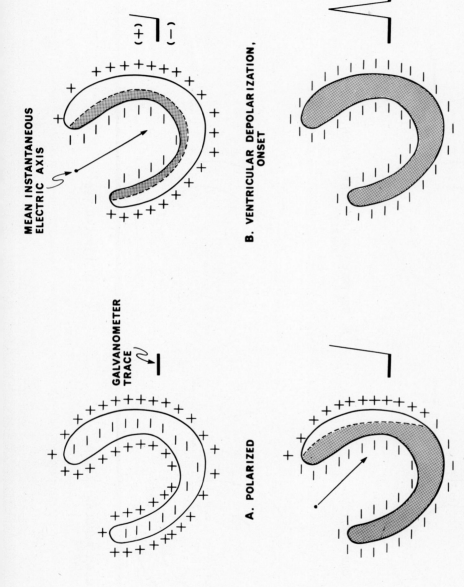

MEAN INSTANTANEOUS
ELECTRIC AXIS

GALVANOMETER
TRACE

A. POLARIZED

B. VENTRICULAR DEPOLARIZATION,
ONSET

C. VENTRICULAR DEPOLARIZATION,
LATE

D. DEPOLARIZED

FIG. 10.29. **A.** Depolarization of the heart and deflections of the galvanometer produced thereby. **B.** The mean instantaneous electric axis of the heart (*arrows*) is a vector which shifts its direction, depending upon the stage of ventricular depolarization. See also Figure 5.11, p. 172 and 5.12, p. 174. (After Burch and Winsor. *A Primer of Electrocardiography,* 5th ed, 1966, Lea & Febiger.)

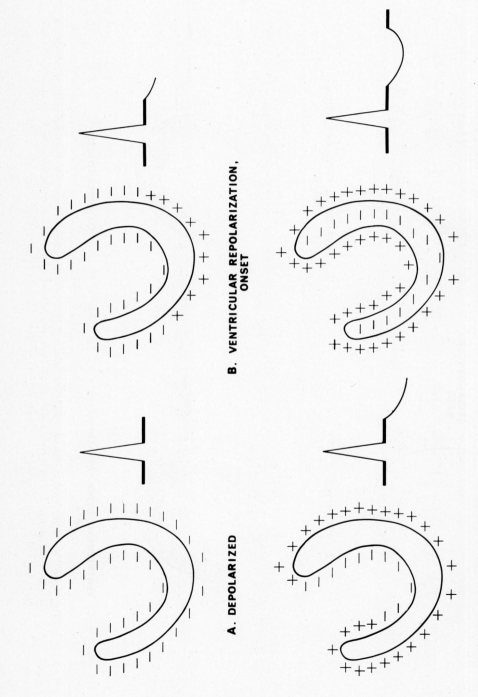

A. DEPOLARIZED

B. VENTRICULAR REPOLARIZATION, ONSET

C. VENTRICULAR REPOLARIZATION, LATE

D. REPOLARIZED

FIG. 10.30. Repolarization of the heart and deflections of the galvanometer produced thereby. (After Burch and Winsor. *A Primer of Electrocardiography,* 5th ed, 1966, Lea & Febiger.)

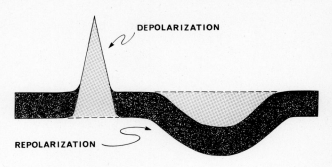

FIG. 10.31. Depolarization as well as repolarization of the heart result in the production of two waves which cover identical areas, hence the net quantity of electric charge involved also is equal during each of these processes. See also Figures 5.11, p. 172, and 5.12, p. 174.

often is considered to be an *electric dipole*. As a dipole is merely an electrically charged area that has an opposite polarity to another area a short distance away (p. 16), recordings of the *direction* and *amplitude* of the instantaneously occurring series of dipoles during a single cardiac cycle may be considered to be the elements that constitute an EKG (Fig. 10.32).

A typical normal EKG is shown in Figure 10.33, together with the various individual waves and intervals. The average temporal values and interpretations thereof for a normal EKG in a young adult are listed in Table 10.4.

It must be reiterated that the recorded EKG represents the algebraic sum (or resultant) of all of the single cardiac action potentials as the excitation wave (which consists of depolarization, overshoot, and repolarization) spreads over adjacent cells of the heart. Therefore, as indicated in Figures 10.32 and 10.33, positive and negative monophasic waves are summated, giving a final recorded waveform which has a modified *amplitude* and *configuration* as shown in these illustrations.

The principal factor governing the relative magnitude of the various component waves of the entire electrocardiographic complex is the relative *mass of tissue* involved in the generation of these potentials in each particular region of the heart. The P wave is small, as it represents the sum of the atrial fiber depolarization process and the total atrial mass is relatively small. The QRS complex represents ventricular depolarization, and as the total mass of these two chambers makes up the principal bulk of the heart, the amplitude of these potentials is considerably greater than that of the P wave. The T wave represents ventricular repolarization. Occasionally, a small upright U wave is encountered following the T wave; this is believed to be related to the slow repolarization of the papillary muscles. The magnitude and configuration of the individual waves will, of course, vary, depending upon the actual location of the electrodes upon the body surface as shown in Figure 10.28.

Note that the amplitudes of all the electrocardiographic potentials in millivolts are quite small when compared with those of transmembrane potentials obtained from single fibers. This difference is caused by attenuation of the signal in proportion to the distance from the heart at which the EKG is recorded; that is, the body is far from being a "perfect" volume conductor for electricity.

Following atrial depolarization, the repolarization wave of these chambers, the *atrial T wave*, follows essentially the same pathway through the atria as did the depolarization wave. Any EKG changes accompanying repolarization of the atria are minute (less than 1 mm in amplitude), however, or are obscured completely on the EKG by the ventricular QRS complex, which is of considerably greater magnitude. When it is observed, and then only rarely, the deflection of the atrial T

Table 10.4 NORMAL ELECTROCARDIOGRAPHIC PARAMETERS[a]

	NORMAL DURATION (range, sec)	REMARKS
PR interval[b]	0.16 (0.12 to 0.20)[c]	Atrial depolarization, impulse conducted through AV node
QRS duration	0.08 (to 0.10)	Ventricular depolarization
QT interval	0.40 (to 0.43)	Ventricular depolarization and ventricular repolarization
ST interval (QT — QRS)	0.32	Ventricular repolarization

[a]*Data from Ganong.* Review of Medical Physiology, *6th ed, 1973, Lange Medical Publications.*
[b]*Measured from beginning of P wave to onset of QRS complex (see Fig. 10.33).*
[c]*The duration of this interval decreases with an increase in heart rate.*

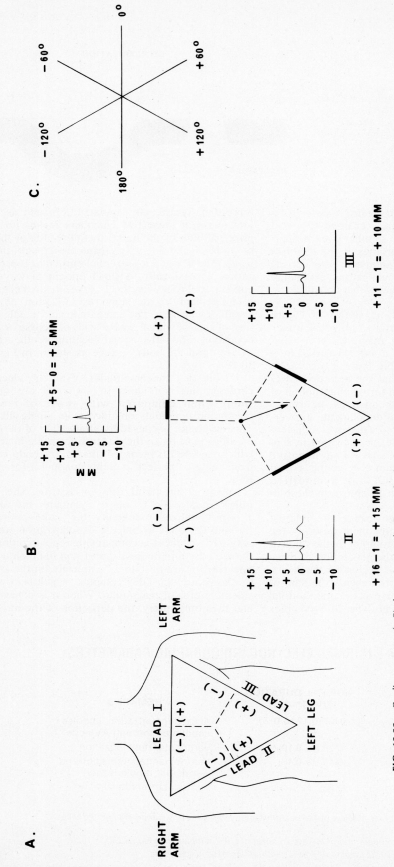

FIG. 10.32. Cardiac vectors. **A.** Einthoven's triangle. Perpendicular lines drawn toward the center from the midpoint of each side of this equilateral triangle intersect at the center of electric activity (*dot*). **B.** Calculation of the mean QRS vector. In each of the three standard leads (I, II, III), the distance which is equal to the height of the R wave less the height of the greatest negative deflection in the QRS complex is measured from the side of the triangle which represents that lead. These quantities now are plotted on the Einthoven triangle as indicated by the heavy bars, and perpendicular lines are drawn from the ends of the individual vectors. An arrow now is drawn from the electric center of the triangle through the point at which the perpendiculars from the three individual vectors of the three standard leads intersect, and this arrow represents both the magnitude and direction of the mean vector of the QRS complex. **C.** The direction of the mean QRS vector, determined as in **B**, is given by these reference axes. (After Ganong. *Review of Medical Physiology*, 6th ed, 1973, Lange Medical Publications.)

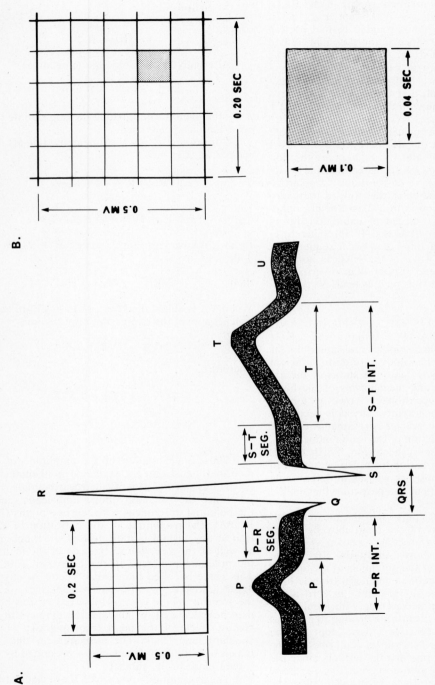

FIG. 10.33. Definitions of the standard segments (SEG.) and intervals (INT.) of the EKG. **A.** The coordinates of the standard paper used in electrocardiography are shown to scale at the upper left. **B.** Enlarged segment of electrocardiographic chart paper illustrating the dimensions employed conventionally. Shaded area in upper chart (= 1 mm²) is enlarged below. The J point is located at the point of the junction between the QRS complex and the S–T segment. The J point is important, as it signifies the exact time at which the depolarization wave has completed its passage through the ventricles of the heart. The J point also provides a zero potential reference for the entire electrocardiographic complex. (After Burch and Winsor. *A Primer of Electrocardiography,* 5th ed, 1966, Lea & Febiger.)

wave is opposite in sign, ie, negative, to the P wave which preceded it, as shown in Figure 10.31.

Subsequent to ventricular depolarization, the algebraic sum of the electric alterations that underlie ventricular repolarization appear in the EKG, first as the isopotential S–T segment (Fig. 10.33). This interval corresponds roughly to that portion of the single cardiac fiber action potential termed the phase 2 plateau (Fig. 10.34). The isoelectric S–T segment in turn gives rise to the period of rapid repolarization called phase 3 in the single-fiber action potential, and this phase corresponds rather closely in time with the appearance of the T wave on the surface EKG. As shown in Figure 10.23, the ventricular epicardial fibers are the last to depolarize during the electric cycle of the heart (see also Table 10.3). But these fibers are the first to repolarize, as they do so faster than the endocardial fibers. The wave of ventricular repolarization spreads from the epicardial toward the endocardial surface, which is in a diametrically opposite direction to the spread of excitation during depolarization (Figs. 10.23 and 10.35). Therefore, as a consequence of this paradoxical sequence of reactivation, the T wave of the EKG is normally upright in certain leads rather than inverted (eg, standard leads I and II as well as leads V_1 through V_6; Fig. 10.28).

THE ELECTRIC AXIS OF THE HEART AND VECTORCARDIOGRAPHY. Usually the EKGs recorded from the three bipolar and nine unipolar lead positions described above are quite sufficient for the physician to evaluate the status of the excitatory and conducting mechanisms of the heart in a given patient. Occasionally, however, a more detailed electrocardiographic analysis is necessary in order that a more precise assessment be made of the extent and location of the lesion induced by, for example, a coronary occlusion. Furthermore, such additional information may also be helpful in establishing a prognosis as well as in determining the response of a patient to a particular therapeutic regimen.

For such clinical purposes, the mean instantaneous electric axis of the heart may be determined by means of a procedure known as *vectorcardiography*. As shown in Figure 10.32, the algebraic sum of all of the potentials at the apex of an electrically conducting equilateral triangle having a current source at its center is zero at all times. This fact is in accordance with Kirchhoff's second law of electric currents flowing in a closed circuit. Such a triangle (or closed loop) can be constructed with the heart at its center by placing electrodes on both arms and the left leg, since the heart rests within a volume conductor, the body (Einthoven's triangle). In such a triangle, then, the algebraic sum of any electric complex in lead I and the same complex recorded simultaneously in lead III is equal to that complex recorded in lead II at the identical instant. This statement is *Einthoven's law of electrocardiography*.

Stated mathematically, and in keeping with Kirchhoff's second law of circuits, Einthoven's law becomes:

$$I + III = II$$

Alternatively, this expression becomes

$$VR + VL + VF = 0$$

when the symbols designating the appropriate unipolar limb leads are used.

Consequently, in the augmented leads described on page 605:

$$aVR = VR \frac{(VL + VF)}{2} \tag{1}$$

and

$$2aVR = 2VR - (VL + VF) \tag{2}$$

From Einthoven's triangle, since VR + VL + VF = 0,

then

$$VR = -(VL + VF) \tag{3}$$

When VR is substituted for −(VL + VF) in equation (2) above, the expression now becomes

$$2aVR = 2VR + VR$$

Therefore,

$$aVR = 3/2VR = 1.5VR \tag{4}$$

Thus, augmentation of the amplitude of the recorded signal can be explained mathematically on the basis of this law of electrocardiography.

Since the standard bipolar limb leads record the potential differences between two points (Fig. 10.26A), the magnitude of the deflection in each lead at any instant is indicative of the *magnitude* as well as the *direction* in the axis of the lead for the potential produced in the heart (Fig. 10.32). This value is called the cardiac vector or electric axis.* Using the Einthoven triangle (Fig. 10.32), the *electric vector* (or *electric axis*) of the heart may be calculated for any specific moment in the electric cycle of the heart from any two of the standard limb leads. The measurement thus obtained is in the two dimensions represented by the frontal plane of the subject.

The assumption upon which these calculations are based (ie, that the lead positions form an equilateral triangle with the heart in its exact

---

*A vector being defined mathematically as a line segment denoting both the magnitude and direction of some finite parameter, in this instance, electric potential.

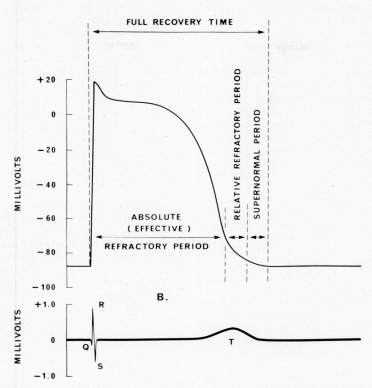

FIG. 10.34. Temporal correlation between a transmembrane potential **(A)** and the surface EKG **(B)** of the ventricle. (After Hecht. *Ann NY Acad Sci* 65:700, 1957.)

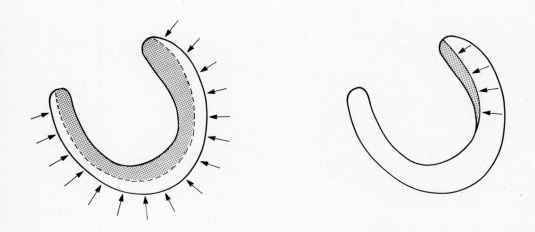

A. VENTRICULAR REPOLARIZATION, EARLY

B. VENTRICULAR REPOLARIZATION LATE

FIG. 10.35. Ventricular repolarization. Note that repolarization proceeds from the epicardial toward the endocardial surface; ie, this process follows a reverse pathway to that taken by the impulse during ventricular depolarization (cf. Fig. 10.23).

center) is not entirely correct in detail, but is of sufficient accuracy that clinically useful quantitative approximations are obtained.

The mean QRS vector, which is called the electric axis of the heart, often is plotted by using the average deflection of the QRS complex in the three standard limb leads, as illustrated in Figure 10.32. Note that this is a *mean vector* for the entire electric cycle of the heart in contrast to an *instantaneous vector*. In the latter, the QRS deflections are measured by mathematical integration of the electric complexes. Usually, however, in routine clinical practice the algebraic differences between the positive and negative peaks in the QRS complex are measured in millimeters and plotted as shown in Figure 10.32.

The normal mean QRS vector, or electric axis of the heart, is considered to lie between 0° and + 100° on the coordinate system illustrated in Figure 10.32. In infants under 6 months of age, however, the axis is toward the right (+130°), whereas in babies between 1 and 5 years of age, the axis moves toward the left, the average being around +50°. At puberty, the average axis is around +65°, and in the adult it shifts to the left once again, the average value being around +58°. All these changes reflect alterations in the relative position of the heart (particularly the ventricles) within the chest, as growth and development proceed.

Clinically, right axis deviation (above + 100°) in an adult may be found in certain disease states, including right ventricular hypertrophy or dilatation, mitral stenosis, pulmonary hypertension, large interatrial septal defects, and acute right ventricular failure. Left axis deviation, on the other hand, is found in left ventricular dilatation, left ventricular hypertrophy, and left bundle branch block.

In contrast to the mean electric axis of the heart, determined as described above for the QRS complex, the peaks of the arrows, which represent all the instantaneous cardiac vectors, can be plotted during the cardiac cycle. If the tops of all these arrows are connected as shown in Figure 10.36, the line connecting all of them forms a series of three loops, one for the P wave, one representing the QRS complex, and one representing the T wave. The loops can be "drawn" electronically and then projected onto the screen of a cathode-ray oscilloscope. Such loops are termed *vectorcardiograms*, and they can be produced in the frontal, horizontal, and sagittal planes by using the appropriate limb, precordial, and esophageal leads.

Certain electrocardiographic abnormalities are described and illustrated in the final section of this chapter (p. 696, et seq).

Mechanical Activity of the Heart: Cardiac Contractility

The final link in the chain of events following the rhythmic excitation of the heart by the sinus pacemaker, and the coordinated, sequential conduction of the impulses throughout the atria and ventricles, is the attendant mechanical contraction or systole in response to these impulses.

In this section, certain properties of cardiac muscle will be discussed as an introduction to the events which take place during an entire cycle of cardiac activity.

SOME GENERAL FUNCTIONAL PROPERTIES OF CARDIAC MUSCLE. Unlike skeletal muscle, cardiac muscle is rhythmically contractile as well as involuntary. In this regard, cardiac muscle shares a common feature with visceral (smooth) muscle, which also carries out its functions, under ordinary circumstances, totally removed from conscious volition.*

It has been emphasized that the heart is composed of two functional syncytia that are interconnected physiologically through the atrioventricular node. Therefore, because of this property, the atrial syncytium and the ventricular syncytium both obey the all-or-none law (p. 121). Hence, an adequate stimulus applied to any single atrial myocardial fiber may result in the propagation of the action potential throughout the entire atrial mass, and a similar response takes place when a ventricular fiber is stimulated. Normally, of course, atrial impulses are conducted downward through the atrioventricular node *(orthodromic conduction)* to stimulate the ventricles as depicted in Figure 10.20.

From an electrophysiologic standpoint, cardiac muscle has certain peculiarities that are of importance to the regulation of its contractile properties. As shown in Figures 10.20 and 10.21, the action potential duration of cardiac muscle fibers is considerably longer than that of either skeletal muscle or nerve (Figs. 5.6, p. 165; and 5.7, p. 166). The resting membrane potential of a myocardial fiber is around −80 to −85 mv, and approximately –90 mv in a Purkinje fiber, in comparison with the somewhat lower potential of a typical pacemaker cell in the sinoatrial node (p. 595). The rate of spike development (about 2 msec) and the overshoot, as well as their genesis, are quite similar in cardiac muscle fibers to those found in nerve and skeletal muscle. As shown in Figure 10.21, however, the duration of the action potential is considerably longer in myocardial fibers than in other excitable cells, and this may be as long as 150 to 300 msec. This long action potential is caused specifically by the plateau (phase 2) which follows the spike as well as by a much slower rate of repolarization of the myocardial cell back to its rest potential. Although the detailed mechanisms that underlie this prolongation of the repolarization phase in cardiac muscle are unknown, the important physiologic consequence of

---

By means of certain training procedures, however, an individual or animal can learn to exert a certain degree of voluntary control over some visceral functions, including heart rate (p. 413).

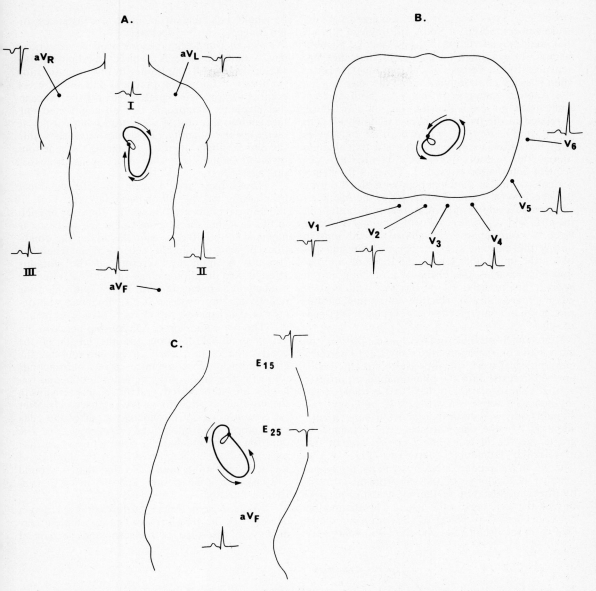

FIG. 10.36. Vectorcardiographic loops in a normal subject. The small inner loops are formed by the P wave, the large outer loops by the QRS complex. **A.** Frontal plane. **B.** Horizontal plane. **C.** Sagittal plane. E_{15} and E_{25} represent the two standard esophageal leads. See also Figure 10.32. (After Ganong. *Review of Medical Physiology,* 6th ed, 1973, Lange Medical Publications.)

this fact is that both the *absolute* and the *relative refractory periods* in cardiac muscle are quite long, as indicated in Figure 10.21. Therefore, as the direct result of this basic fact, cardiac muscle normally cannot be thrown into tetanus, as an excessive rate of stimulation will produce merely an occasional contraction, when, and only when, the muscle has recovered sufficient irritability to respond to another stimulus. Artificially applied electric stimuli can produce repetitive contractions of cardiac muscle before the end of the plateau phase, but under physiologic conditions such a response is not seen. For this reason, the "abso-

lute" refractory period in the heart sometimes is referred to as the *functional refractory period*. In atrial muscle, for example, the functional refractory period is around 150 msec, and the relative refractory period an additional 30 msec. The functional significance of the long refractory period is obvious, as tetanus of the heart would prove rapidly fatal.

In cardiac muscle the proteins that make up the contractile machinery, as well as the contractile process per se, are quite similar to, if not identical with, those found in skeletal muscle, and which were described in detail in Chapter 4, page

116 et seq. Thus, the contractile elements of the heart are composed of actin and myosin strands that undergo shortening by the sliding filament mechanism described on page 138 when they are activated by the release of calcium ions attendant upon passage of an impulse along the fiber.

Another common property of both skeletal and cardiac muscle is that the force of contraction is related to the initial fiber (or sarcomere) length (Fig. 10.37); that is, when cardiac muscle is stretched prior to its contraction within definite limits, the tension developed during the subsequent contraction is greater. Thus, in the heart, a greater filling of the chambers with blood during the diastolic (or relaxation) phase of the mechanical cardiac cycle results in a greater distention of the atria and ventricles, and subsequently a greater force of contraction is developed by the muscle during the ensuing systole. This relationship between the initial fiber length and force of contraction is known as the *Frank–Starling law of the heart*. The implications of this law will be discussed more fully in relation to normal cardioregulation in a later section of this chapter (p. 658, et seq.).

The inherent (intrinsic) mechanical property of muscle to contract more forcefully (ie, to perform more physical work) in response to stretch should not be confused with the phenomenon of intrinsic cardiac rate regulation (p. 659) that is found in the rhythmically active sinoatrial cells, although the stimulus (physical deformation) is the same in both situations. In the one instance, stretch induces a greater degree of *mechanical contraction* of the myofibrillar elements themselves (Fig. 10.37), whereas in the second, stretch augments the inherent rate of discharge of the sinoatrial nodal cells by direct modification of the permeability properties of the cell membrane itself. The distinction between these two effects of stretch on *force* and *rate* is an important one, and the two events may be wholly independent or, in certain instances, interrelated to various degrees.

Another important general property of cardiac muscle is that there is a direct relationship between the overall amplitude (in mv) and duration (in msec) of the action potential and the force of contraction (ie, work performed) by the myocardial elements. It appears that an augmented voltage and/or duration of the action potential per se increases the quantity of calcium released within the muscle fibers, thereby enhancing the reactions involved in the contractile process, as discussed on page 139.

These facts would seem to belie the basic proposition of the all-or-none law (p. 121). It should be remembered, however, that it is implicit in this law that the response of a cell or tissue is maximal to an effective stimulus under the conditions existing within the cell or organ when the stimulus is applied. Therefore, if the stimulus strength and duration change so that they alter the fundamental chemical and/or physical mechanisms underlying the response itself, obviously the degree of the response will change to some extent.

There is also a direct relationship between the action potential duration and the length of the contraction of the heart. In cardiac muscle, the contraction phase commences a few milliseconds following the onset of the action potential and following the latent period, and the muscle remains in the contracted state for a few milliseconds after the action potential terminates; ie, after the cells are completely repolarized. Consequently, the duration of the mechanical contraction of cardiac muscle is determined principally by the duration of the action potential itself, a value that is around 250 msec in atrial muscle, and 300 msec in ventricular muscle.

An increase in heart rate thus takes place by a decrease in the time involved in a complete cardiac cycle; therefore, both electric and mechanical

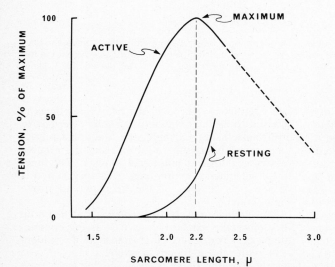

FIG. 10.37. The effect of stretch upon the force of contraction of cardiac muscle. The smaller resting curve represents the passive tension developed with stretch, whereas the large (active) curve represents the tension developed during contraction. Note that when the sarcomere length exceeds 2.2 μ, the tension developed falls off rapidly, an effect such as occurs in cardiac failure. See also Figure 4.13, p. 125.

events are shortened when cardioacceleration takes place. During acceleration of the heart, however, the action potential duration as well as the time the heart is in mechanical systole do not decrease as much as does the time that is required for an entire cycle. A disproportionate increase in the ratio between the contraction time and the relaxation time is found during tachycardia (Table 10.5). At a resting heart rate of around 75 beats per minute the contraction period or systole occupies approximately 40 percent of the total cycle, whereas at a rate of about 200 beats per minute, this value becomes around 65 percent. As a consequence of this fact, at the lower heart rate, 60 percent of the time involved in each cardiac cycle is available for diastolic filling, whereas at the elevated rate, the time available for this process has decreased to about 35 percent of the entire time for each cardiac cycle. This fact is of great importance physiologically as well as clinically, since during severe tachycardia the heart does not remain in diastole for a sufficiently long interval to permit complete filling of the chambers with blood prior to the onset of the next contraction.

If the heart is stimulated by a second impulse immediately following a contraction, ie, before the myocardium has recovered fully, the force of the second, or premature, contraction generally is lower than that of the first. Presumably, the reason for the second contraction being weaker lies in the fact that the concentrations of the chemical substances involved in the contractile process are not restored fully. In addition, diastolic filling (hence distention of the myocardium) is incomplete, and the operation of these two factors results in the diminished force of the second, or premature contraction. On the other hand, as the interval between successive contractions of the heart increases, the force of the contractions becomes stronger.

Bouts of *premature ventricular contractions* such as were described above may be observed in otherwise normal individuals, and they may occur during periods of excessive fatigue or mental strain. They also may be induced in some individuals by intemperate cigarette and/or coffee consumption.

An integrated synthesis of the entire sequence of events that occur during a normal cardiac cycle in the human heart now may be presented.

Electric and Mechanical Correlates of the Cardiac Cycle: The Heart as a Pump

The coordinated rhythmic activity of the heart results in a pulsatile ejection of blood into the pulmonary artery and aorta. In performing this function, sufficient kinetic and potential energy is imparted to the mass of blood expelled at each stroke, or beat, of the heart to propel the blood through the capillaries. In other words, the heart performs an enormous quantity of mechanical work, in the physical sense of the term, throughout the lifetime of an individual.

Obviously, in the normal, nonfailing heart, the volume of blood ejected by both the right and left ventricles must be equal over a period of time, although this is not necessarily true on an individual stroke-to-stroke basis.

The entire cardiac cycle is composed of a period of relaxation called *diastole,* during which filling of the chambers takes place. This interval is followed by a contraction phase termed *systole,* during which the blood is expelled from the heart under pressure. Since blood is an aqueous liquid, hence incompressible, the pressure developed during the systolic contraction of the heart muscle is transmitted undiminished to the blood within its chambers according to Pascal's first law (p. 624).

The electric and mechanical events that take place during a complete cardiac cycle are summarized in Figure 10.38, together with the sounds that are developed by the heart during its activity. These aspects of cardiac function will be discussed individually following a brief introductory consideration of the role of the pericardium and the valves in the normal activity of the heart.

THE PERICARDIUM. This fibrous double-walled sac which surrounds the entire heart contains several milliliters of serous fluid (p. 565), and it provides a smooth, lubricated surface for the movements of the heart. The potential volume of the pericardial cavity is greater than the volume of the normal heart in diastole. If the heart suddenly be-

Table 10.5 RELATIONSHIP AMONG HEART RATE, ACTION POTENTIAL DURATION, AND RELATED CYCLIC PHENOMENA[a]

DURATION OF ACTION POTENTIAL	HEART RATE[b]	
	75 beats/minute	200 beats/minute
Entire cardiac cycle	0.80	0.3
Systole	0.27	0.16
Action potential duration	0.25	0.15
Absolute refractory period, duration	0.20	0.13
Relative refractory period, duration	0.05	0.02
Diastole, duration	0.53	0.14

[a]*Data from Ganong.* Review of Medical Physiology, *6th ed, 1973, Lange Medical Publications.*

[b]*All values listed in table are in seconds.*

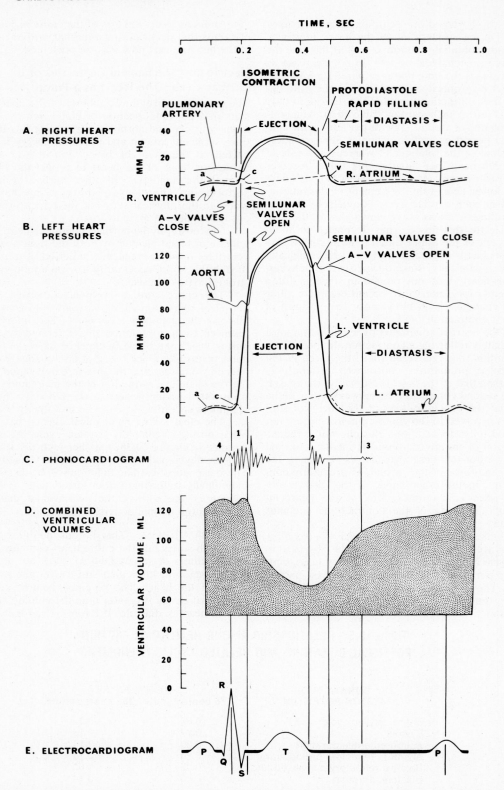

FIG. 10.38. Sequence of mechanical, audible, and electric events during one complete cardiac cycle. (Data from Best and Taylor [eds]. *The Physiological Basis of Medical Practice,* 8th ed, 1966, Williams & Wilkins; Guyton. *Textbook of Medical Physiology,* 4th ed, 1971, Saunders.)

comes excessively dilated, the intrapericardial pressure rises markedly even with a small increase in the overall cardiac volume; therefore, diastolic expansion of the organ is restricted (especially the right ventricle), and this effect in turn protects the left ventricle from overload and the pulmonary circulation from congestion, to some extent. This type of adaptation takes place in acute situations. If the heart enlarges gradually, on the other hand, eg, during a gradually developing congestive failure (p. 695), the pericardium also enlarges by stretching so that it accommodates to the increased volume of the heart.

The congenital absence of a pericardium, however, does not impair normal cardiac function in either human subjects or animals.

CARDIAC VALVES. The valves of the heart assure a continuous one-way flow of blood in the heart (Fig. 10.8). The principal factor that determines the opening and closing of the valves is the relative pressure of the blood upon their opposite surfaces; ie, their movement is passive in response to fluctuations in the blood pressure. Valve closure also is influenced to some extent by eddy currents set up by movements of the blood in the various chambers and vessels, which currents also facilitate apposition of the pocketlike cusps or leaflets of the valves (Fig. 10.39A).

The papillary muscles and chordae tendineae do not exert any marked influence on the closure of the atrioventricular (tricuspid and mitral) valves; rather, these structures prevent eversion of the valves into the atria during ventricular systole, and the contraction of the papillary muscles at the onset of systole merely serves to tighten the tendons of the chordae, hence to appose the valve leaflets somewhat (Fig. 10.39B).

It is important to realize that during atrial systole the cusps of the mitral and tricuspid valves are not forced against the ventricular wall, but in fact the cusps of the valves lie in a position that is halfway open as the consequence of the two opposing currents of blood. The pressure of blood flowing from the atria keeps the valves open, whereas currents of blood pressing upon their ventricular surface tend to close them. The overall position of the valves is delicately poised between

the open and closed positions. At the end of atrial systole, the pressure of the blood flowing into the ventricle decreases and the valve leaflets become apposed. Full valve closure is effected by the sudden marked increase in the intraventricular pressure at the onset of ventricular systole (Fig. 7.38).

Normally, blood does not leak backward (ie, *regurgitate*) from one chamber or vessel to another at any time during the cardiac cycle when the valves are closed. In the event that regurgitation of blood does occur during a phase of the cardiac cycle when the valves normally are closed, the audible sound that is heard with a stethoscope is termed a *murmur* (p. 626). Abnormalities of the valve leaflets, such as leaks and murmurs, may be congenital or they may be produced by a disease process, eg, rheumatic fever, which roughens the cusps of individual valves so that they do not appose properly. Such a condition is often referred to as incompetence of the valves or simply valvular incompetence.

In a normal individual, the valves are said to be *patent* when they open fully and no obstruction is present to the outward passage of blood from any chamber of the heart.

In the following discussion, which is concerned with the various phases of the normal cardiac cycle, the reader is urged to note particularly the sequential role of the various cardiac valves, as the action of these structures provides convenient "physiologic landmarks" on the curves in Figure 10.38 with which to relate the various pressure and volume relationships within the atria and ventricles.

ROLE OF THE ATRIA IN THE CARDIAC CYCLE. The atria and the arterial trunks are attached to the superior aspect of the four valve rings of dense connective tissue that comprise the fibrous skeleton of the heart. The thin-walled atrial musculature is arranged in two bands that radiate from the sulcus terminalis, one of which is common to both atria and encircles them, the other being placed at right angles to the first and separate in each atrium.

The net volume capacity of the two atria is somewhat greater than that of the two ventricles; therefore, sufficient blood normally is present in

A.

B.

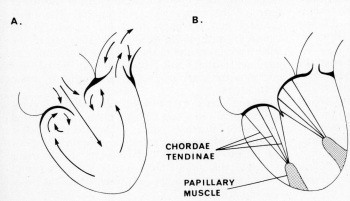

CHORDAE TENDINAE

PAPILLARY MUSCLE

FIG. 10.39. The mechanisms involved in closure of the cardiac valves. **A.** Arrows indicate the general pathways taken by eddy currents, and partial closure of the valves resulting from these currents. **B.** Anatomic relationships between chordae tendinae, papillary muscles, and valve cusps. (After Best and Taylor [eds]. *The Physiological Basis of Medical Practice,* 8th ed, 1966, Williams & Wilkins.)

the atria during diastole to assure complete filling of the ventricular chambers.

The atria are filled normally by a continuous inflow of blood from the great veins, ie, the superior and inferior venae cavae and the pulmonary vessels. About 70 percent of this incoming blood enters the ventricles directly prior to atrial contraction, which event produces an additional 30 percent ventricular filling. Atrial systole thus increases the ventricular pumping efficiency by about 40 to 50 percent. The earlier view that the atria function dynamically as auxiliary pumps to fill the ventricles must be discarded in the light of recent experimental evidence. In this regard, since the normal heart possesses a latent capacity for pumping blood which is some 300 to 400 percent above the bodily requirements, a functional inability of the atria to pump blood into the ventricles does not cause any cardiac insufficiency at rest. During exercise, however, the loss or impairment of this atrial function may become evident, and symptoms of an inadequate cardiac output appear.

In Figure 10.38, an electrocardiographic record is shown in order that the mechanical events of the cardiac cycle can be compared and related directly to the electric events that initiate them. Thus, the P wave immediately precedes the onset of atrial systole; the conducted wave of excitation spreads outward from the sinoatrial pacemaker region across right and left atria, as described earlier, resulting in a smooth and sequential activation of these chambers.

When pressure traces are recorded directly from the right and left atria, each of the recordings shows three distinct peaks, called the a, c, and v atrial pressure waves (see appropriate records, Fig. 10.38).

First, the *a wave* is produced by atrial systole. In the right atrium, this peak rises to a maximum at around 4 to 6 mm Hg, whereas in the left atrium the pressure rises to between 6 and 9 mm Hg.

Second, the *c wave,* which is seen at the onset of ventricular systole, is caused by three factors: (1) the reflux of some blood into the atria that takes place when ventricular systole commences and as the atrioventricular valves are closing; (2) a bulging of the atrioventricular valves toward the atria caused by the increased intraventricular pressure developed during contraction of the ventricles; (3) traction exerted on the atrial muscle per se, which is caused by the ventricular muscle during contraction of the latter.

Third, the *v wave,* which originates near the end of ventricular systole, is caused by the gradual accumulation of blood in the atria while the atrioventricular valves are closed. Following ventricular systole, these valves open and the blood flows rapidly into the ventricle, so that the v wave disappears rapidly near the end of ventricular diastole.

The *volume flow* of blood into the atria closely parallels the pressure curves, and also shows three phases, as described above.

One particularly important fact should be noted, and it is shown in Figure 10.38; that is, the atrial pressure curves lie *above* the corresponding ventricular pressure curves until such time as the ventricular pressures rise above the corresponding atrial pressures, at which point the atrioventricular valves close, and the atrial pressures now fall *below* the ventricular pressures. Stated another way, during ventricular filling, the atrium and ventricle on each side of the heart form a single, continuous functional chamber, and the blood flows down the pressure gradient which at that moment exists between the upper and lower regions of this single chamber.

Before proceeding to a discussion of the events that take place during ventricular filling and contraction, it is necessary to point out that although the dynamic events which take place on both sides of the heart during the cardiac cycle are, in general, similar, there is a definite asynchrony between, as well as certain differences in the duration of, various portions of the cardiac cycle on the left and right sides of the heart. The onset of right atrial contraction precedes that of the left atrium, whereas the onset of left ventricular systole precedes that of the right ventricle. This statement is true even though right ventricular ejection of blood commences prior to, and is slightly later than, left ventricular ejection.

ROLE OF THE VENTRICLES IN THE CARDIAC CYCLE. The ventricular chambers and the atrioventricular valves are attached to the inferior surface of the fibrous skeleton of the heart. Functionally, the ventricular myocardium itself is made up of large muscle bundles that are arranged into two principal groups. These groups are the *spiral muscles* and the *deep constrictor muscles.*

The *superficial spiral muscles* originate from the tricuspid and mitral rings, and they cover the surface of both ventricles to a depth of about 1 mm, coursing diagonally along the surface of the heart to converge at the ventricular tip, or apex, where they constitute the full thickness of the ventricular wall. At the apex, these muscles penetrate to the endocardial surface and there form the thin inner layer of spiral muscle as well as the lower third of the interventricular septum. The inner layer of muscle spirals obliquely upward in the ventricles at an angle of approximately 90° to the superficial layer and in the opposite direction. The papillary muscles are formed during the upward course of the inner layer of spiral muscles by some of its fibers.

Sandwiched between the thin exterior and interior spiral muscles lie the circularly oriented *constrictor muscles,* which constitute the basilar two-thirds of the interventricular septum as well as the lateral wall of the left ventricle. In the right ventricle, this middle layer is thin, so that the wall of this chamber is composed principally of the inner and outer layers.

The physiologic consequences of these anatomic facts are important. Since the superficial and inner spiral muscle fibers are oriented in opposite directions, their contraction develops tension which is antagonistic; therefore, the net result of their simultaneous activity is a longitudinal shortening of the ventricular cavities. In the right ventricle, where little circular muscle is present, this type of contraction is dominant so that ejection of blood from the right ventricle is accomplished predominantly by a shortening of this chamber from base to apex resulting from the combined action of the spiral muscles. In the left ventricle, on the other hand, contraction of the large mass of circularly arranged fibers results primarily in a reduction of ventricular transverse diameter; hence, the circumference of the chamber is decreased markedly, but with little shortening of the ventricle from base to apex. As a consequence of this alteration in circumference, the lateral wall of the left ventricle exhibits considerable movement when compared to the right ventricle during systole. The diameter change accounts for most of the power developed as well as the marked volume change seen during left ventricular ejection, because the volume of a cylinder decreases as the square of the radius, and the left ventricular chamber is essentially a cylinder having a conelike apex, an anatomic fact that contrasts to the roughly triangular cross-sectional area of the right ventricle (Fig. 10.5B).*

With the techniques available at the present time, it is impossible to assess quantitatively and accurately the degree of myocardial fiber shortening which is associated with ventricular ejection. Indeed, for that matter, what is the actual "resting length" of a given myocardial fiber in vivo? The general relationships between the pressure developed by the contraction of the heart and the wall tension developed at a given diastolic filling volume as indicated by the radius of curvature of the chambers (Laplace's law) will be discussed on page 627, as the interrelationships among these three parameters provide the currently most useful way of evaluating cardiac performance for comparative purposes.

In the thin lateral wall of the right ventricle, presumably all the fibers shorten to a similar extent during systole, whereas in the thick-walled left ventricle, the myocardial cells nearest the endocardium necessarily must shorten somewhat more than do the more superficial fibers in order to eject a specific volume of blood.

It has been established clearly, however, that the heart does not empty completely during ventricular systole, so that a certain residual volume of blood is still present in the ventricles at the end of the ejection phase of the cardiac cycle (Fig. 10.38). Thus, both the right and left ventricles expel during systole only a fraction of the total mass of blood that entered their chambers during diastole. The volume ejected during a single systole of the left ventricle is called the *stroke volume* of the heart, and in a human subject at rest the stroke volume is approximately 60 ml. This value amounts to about 55 to 65 percent of the total volume of blood contained within that ventricle during diastole, or around 130 ml.† About 35 to 45 percent of the total diastolic blood volume remains in the left ventricle of the heart at the end of the ejection phase of the cardiac cycle.

Within about 40 to 50 msec after the onset of the Q wave of the EKG, ventricular systole commences (Fig. 10.38). When the pressures developed within the right and left ventricles exceed their respective atrial pressures, the tricuspid and mitral valves close; thereupon the intraventricular pressure curves rise sharply. During the 20 to 30 milliseconds while the ventricles are developing sufficient wall tension, hence intraventricular pressure, to force open the pulmonary and aortic semilunar valves, ventricular contraction is taking place although no blood is being ejected. Since the ventricular myocardial fibers do not shorten, this interval is called the *period of isometric* or *isovolumetric contraction;* ie, the tension developed by the ventricular myocardium is increasing but no physical shortening of the fibers is taking place.

When the right intraventricular pressure rises above approximately 8 to 10 mm Hg, and that of the left ventricle exceeds about 80 mm Hg, the ventricular pressures push the pulmonary and aortic semilunar valves open; now blood immediately flows into both the pulmonary artery and aorta from the right and left ventricles, respectively.

Approximately 50 percent of the total volume of blood ejected during a single cycle is emptied from the heart during the earliest quarter of the entire systolic interval; the remaining 50 percent of the ejected blood is emptied during the ensuing half of the systolic interval. This part of the systolic interval is called the *isotonic period of ejection;* ie, the muscle is shortening actively during expulsion of the blood (Fig. 10.38).

Note that the rate of ejection of blood is rapid at first, as indicated on both right and left ventricular pressure curves in Figure 10.38, and then the pressures level off as the peaks are reached. Subsequently, the intraventricular pressures decline rapidly at the end of systole and as ventricular diastole commences.

Ventricular diastole is composed of four phases: (1) protodiastole, (2) isometric relaxation, (3) a period of rapid inflow, and (4) a period of reduced inflow or diastasis. The first effect of ventricular relaxation is a drop in the intraventricular

---

*The traditional concept of ventricular ejection being accomplished by a shortening of the chambers which is accompanied by a rotation, hence a "wringing" effect, no longer is tenable.

†From a clinical standpoint, a normal "ejection fraction" (ie, stroke volume/end diastolic volume) lies between 0.55 and 0.65.

pressure and the closure of the pulmonary and aortic semilunar valves, which closures are marked by a sharp notch or incisura which is most evident on the aortic pressure curve. The interval between the onset of diastole and closure of the aortic semilunar valve is called the *protodiastolic phase,* and it lasts for about 20 msec. The end of the protodiastolic interval (ie, closure of the aortic valve) coincides with the second heart sound; consequently, this sound usually is considered to be indicative of the end of diastole and the onset of systole. As the ventricles relax, the elevated blood pressures that were developed during systole in the pulmonary artery and aorta tend to force blood back toward the ventricles, and when these arterial pressures exceed the respective intraventricular pressures, the semilunar valves close once again, as indicated in Figure 10.38. Then, for the next 30 to 60 msec, the ventricles continue to relax; however, no significant change in volume takes place during this period, and all the valves remain closed. This interval is called the *isometric* or *isovolumetric relaxation phase,* and it continues until the atrial pressures once again exceed those in the ventricles so that the atrioventricular valves open and thereby the *rapid inflow phase* commences, an interval that lasts for around 60 milliseconds. The inflow of blood slows markedly during the last phase of diastole, which is called the *reduced inflow phase* or *diastasis.* The duration of the reduced inflow phase is around 230 milliseconds. Then a new cardiac cycle commences.

At the bottom of Figure 10.38, a representation is given of the alterations in the volume of both ventricles simultaneously. Note that in general an inverse relationship is present between the atrial and ventricular pressures and the volume of the total ventricular mass.

Throughout the entire cardiac cycle, the aortic pressure consistently is higher than the pressure developed in the pulmonary artery. Thus, the peak systolic pressure in the aorta is around 120 mm Hg normally, whereas the comparable systolic pressure in the pulmonary artery of a normal adult is only 25 mm Hg, or approximately one-fifth that in the aorta. The comparable relative end diastolic pressures are 80 mm Hg in the aorta and around 10 mm Hg in the pulmonary artery. Hence, the *pulse pressure* (defined as the systolic pressure minus the diastolic pressure in mm Hg) in the aorta is 120 − 80 = 40 mm Hg, whereas the pulse pressure in the pulmonary artery is 25 − 10 = 15 mm Hg.

As shown in Figure 10.38, the end diastolic pressures in both ventricular cavities are rather low, with a small pressure gradient always present between the left ventricle (0 to 5 mm Hg mean diastolic, and 5 to 12 mm Hg end diastolic) and the right ventricle (0 to 3 mm Hg mean diastolic, 0 to 5 mm Hg end diastolic). The end diastolic atrial pressures also amount to several mm Hg, and the left atrial pressure (2 to 12 mm Hg) again is consistently higher than that found in the right atrium (0 to 5 mm Hg).

HEART SOUNDS. As the direct consequence of the mechanical activity of the heart and valves, as well as the attendant sudden acceleration and deceleration of the masses of blood passing through its chambers, which result in the production of eddy currents (p. 619), several distinctly audible vibrations are produced within the heart during certain phases of the cardiac cycle. The individual heart sounds are shown in Figure 10.38 in temporal relationship to the other events of the cardiac cycle. The sound waves produced by the heart during a single cycle exhibit a wide frequency range. Although sophisticated electronic visualization and recording of the heart sounds is possible, from a practical standpoint the technique of direct *auscultation* with a stethoscope gives the most information to the trained ear of an examining physician. This statement holds true despite the fact that the stethoscope neither amplifies sounds accurately nor transmits them faithfully to the ear.

Direct auscultation reveals two distinct heart sounds; occasionally, a third sound may be heard, whereas a fourth sound is revealed by graphic recording. The first two sounds take place primarily during ventricular systole, the latter two during ventricular diastole. The individual sounds now may be discussed, both as to their characteristics and their etiology.

The *first heart sound* commences with the rapid rise in intraventricular tension during the isometric phase of ventricular systole, and it reaches a maximum at the onset of ventricular ejection. This sound is of relatively long duration, low in pitch, and soft in quality.

There are three major factors involved in the production of the first heart sound: (1) A valvular component is produced by closure of the atrioventricular valves combined with the tension developed within the chordae tendineae and papillary muscles as the intraventricular pressure rises. (2) A vascular component is caused by the sudden turbulent surge of blood from the ventricles as well as the shock wave which is transmitted to the aortic and pulmonary arterial walls (see streamlined and turbulent flow, p. 626). (3) Contraction of the ventricular muscle per se induces vibrations of the myocardial mass.

When recorded electronically, the first heart sound is composed principally of a series of vibrations having in total a duration that ranges between 90 and 160 milliseconds and a frequency that ranges between 33 and 110 cycles/sec. These vibrations are of low amplitude at the outset, rise to a maximum at the end of the isometric period, and then diminish with a time course similar to that of their development.

The first sound may be heard most clearly and at maximum intensity when a stethoscope bell is placed over the fifth left intercostal space. At this location, the mitral component of the sound predominates; listening with the stethoscope bell placed over the inferior tip of the sternum is of use in determining abnormal sounds produced by the tricuspid valve.

The *second heart sound* coincides in time with the dip in the incisura of the aortic pressure curve, and this sound is shorter in duration, sharper, and of a higher pitch than the first sound. Verbally described as "lub" and "dup," separated by a brief pause, the first two heart sounds demarcate the onset and termination of ventricular systole. The interval between the beginning of each of these sounds, recorded phonocardiographically, gives an accurate determination of the length of ventricular systole in man.

The second heart sound is caused by vibrations that are induced in the blood columns and arterial walls when the semilunar valves develop tension rapidly following their sudden closure.

The second heart sound lasts for about 100 msec, and it commences approximately 90 msec following the peak of the electrocardiographic T wave. The frequency of this sound averages around 50 cycles/sec.

The second sound normally is "split" into two components when the stethoscope bell is placed in the pulmonary area, ie, along the upper left sternal margin. Phonocardiographic recordings indicate that the first component of the second sound is synchronous with the dicrotic notch of the carotid arterial pressure pulse wave; hence, this sound represents closure of the aortic valve. The second component of the split second sound is produced by closure of the pulmonary arterial valve.

The relative delay between the closure of the aortic and pulmonary valves is augmented during inspiration when the increased venous return of blood to the right heart prolongs the ejection time of the right ventricle (see p. 652). During expiration, on the other hand, the situation is reversed; ie, the slightly greater mass of blood which was ejected from the right ventricle to the pulmonary circuit now reaches the left side of the heart and this factor now prolongs left ventricular ejection slightly. Thus, pulmonary and aortic valve closures now become closer together in time. Inspiration widens the split normally present in the second heart sound to around 50 milliseconds, whereas this time interval narrows to around 20 milliseconds, or even disappears, during expiration.

The *third heart sound* sometimes may be heard in normal hearts. This sound follows the second sound by around 80 milliseconds, and its duration is about 40 milliseconds. Commonly heard in young adults, the third sound may be heard best at the apex of the heart.

The third heart sound can be induced to appear, or to become intensified, by manipulations which increase blood flow into the atria, such as is accomplished by exercise or by assuming a supine position.

It is believed that the third sound is caused by vibrations of the ventricular walls that are induced by the surge of blood that fills them toward the end of the rapid diastolic filling phase of the cardiac cycle.

The *fourth heart sound* generally cannot be heard in normal individuals. The phonocardiogram, however, sometimes reveals several small presystolic vibrations, usually having two components, as shown in Figure 10.38. The first component is associated with the atrial a wave; hence it is caused by atrial systole. The second component of the fourth sound occurs after the peak of atrial systole has been reached, and this component can be obtained by intracardiac phonocardiographic recording more often from the ventricles than the atria. It is believed that this second component of the fourth sound is produced by vibrations induced directly in the ventricular wall by the jet of blood expelled from the atria.

The accurate interpretation of both normal and pathologic auscultatory findings, as with EKGs, is a skill requiring considerable practice and experience to develop. It is recommended that the student not only practice auscultation by listening to various subjects directly with a stethoscope, but also obtain a phonographic recording such as are commercially available in order to gain some basic familiarity with both normal and abnormal heart sounds.

SUMMARY. The material presented in this section has been a review of the pertinent facts concerning the physiologic mechanisms whereby the heart performs its closely timed and smoothly coordinated function of pumping the blood throughout the pulmonary and systemic circuits. Briefly, the heart possesses the fundamental properties of rhythmic excitability, conductivity, and contractility. The myocardium also possesses the property of mechanical distensibility in response to appropriate filling pressures.

Following discussions of the basic elements of hydrostatic and hydrodynamic principles and their relationship to systemic and pulmonary hemodynamics, the physiologic mechanisms that are present in the human body, and that regulate normal cardiovascular activity in vivo, will be considered.

FUNDAMENTAL HYDROSTATIC AND HYDRODYNAMIC PRINCIPLES AS APPLIED TO HEMODYNAMICS

A knowledge of the basic physical laws and principles that underlie the behavior of fluids at rest (hydrostatics) and in motion (hydrodynamics or hydraulics) is necessary to an understanding of the normal events that take place continuously within the human cardiovascular system. Such an understanding is, of course, critical to an appreciation and evaluation of the many pathologic changes in cardiovascular physiology that may be encountered in clinical practice. The physical principles governing fluids at rest and in motion will be discussed prior to a consideration of their specific roles in cardiovascular physiology.

A fluid is defined as a substance that responds to even very slight shearing forces by changes in its shape (deformation); ie, fluids offer no permanent resistance to changes in their shape. Unlike gases, which also are fluids, aqueous liquids (such as blood) are completely incompressible for all practical purposes.

In contrast with solids, which strongly resist alterations in both their shape and volume by virtue of the property of elasticity, all fluids, hence liquids, exhibit a property known as viscosity, a property that may be considered as an internal resistance to flow (or to deformation) when the fluid is acted upon by an external shearing force.

Laws of Fluids at Rest: Hydrostatics

Pascal formulated the three laws of fluids at rest. These laws deal with fluid pressure, which is defined as a *force* exerted per unit area (expressed in dynes/cm², or as the height of a column of liquid, eg, mm Hg or cm H$_2$O), which is exerted by that fluid.

The three laws of Pascal are:

1. Pressure within a confined fluid is transmitted equally in all directions throughout the mass of the fluid.
2. The pressures at points that lie in the same horizontal plane within a fluid are equal. This law holds within the mass of the fluid, but not at its free surface, where atmospheric pressure is exerted upon the fluid.
3. In a fluid at rest, the pressure increases in direct proportion to the depth below the free surface. This third law may be stated mathematically as: $P = \rho gh$. In this expression, the sumbol P is the pressure in dynes/cm² when ρ is the density of the fluid expressed in gm/ml, g is the force produced by the acceleration of gravity (980 cm/sec/sec), and h is the depth in cm beneath the surface of the fluid.

This relationship is of practical significance other than its having an important role in cardiovascular physiology, as it forms the basis for use of U-tube and reservoir manometers in the determination of absolute pressures. In these instruments, the net difference induced by the pressure in the levels of two columns of mercury, saline, or water is measured directly, usually in mm or cm of the liquid. Thus, use of the equation presented above gives the absolute pressure of a mercury column 10 mm high in dynes/cm², using the density for mercury of 13.6. Thus, $\rho gh = 13.6 \times 980 \times 1.0 = 13,328$ dynes/cm².

Similarly, when physiologic saline ($\rho = 1.04$) or water ($\rho = 1.00$) is used, the appropriate absolute values for pressures determined by using these substances as manometer fluids can be calculated. A useful factor is that used for converting manometer readings from mm Hg to cm saline (or water). Thus, by dividing ρ Hg by ρ saline, 13.6/1.04 = 13.1 is obtained. Therefore, 1 mm Hg pressure = 13.1 mm (or 1.31 cm) of physiologic saline. Similarly, a pressure of 1 mm Hg = 1.36 cm of water ($\rho = 1.00$). And since the density of blood is equal to about 1.055, a pressure of 1 mm Hg is equal to a pressure of 1.29 cm of blood.

Laws of Fluids in Motion: Hydrodynamics or Hydraulics

The following basic laws and principles governing the physical parameters involved in hydraulics (fluids in motion) apply in the strictest sense only to newtonian fluids (p. 627) that are moving in a streamline fashion within a rigid (nondistensible) system. Despite the fact that the cardiovascular system is highly elastic, however, this in no way detracts from the fundamental importance of these laws and principles in understanding and interpreting hemodynamic patterns within the human body. Such modification of these laws as is perforce necessary for their application to cardiovascular physiology will be emphasized in the discussions to follow.

RELATIONSHIPS AMONG PRESSURE, FLOW, AND RESISTANCE: THE POISEUILLE–HAGEN EQUATION. The fundamental law expressing the relationship between flow, F, the pressure, P, and the resistance, R, is given by the relationship $F = \Delta P/R$; hence, $\Delta P = RF$, and $R = \Delta P/F$. This expression is *Poiseuille's law*, and it defines the fact that the volume of liquid flowing through a system per unit time (F) is directly proportional to the pressures at both ends of the system (ΔP or $P_1 - P_2$), and inversely related to the resistance to flow within the system. Hence, Poiseuille's law is analogous to Ohm's law for electric currents; that is, $E = I/R$, as discussed on page 8.

Expressed mathematically in somewhat greater detail, this basic law becomes the Poiseuille–Hagen equation, which can be written as follows:

$$F = (\Delta P) \left(\frac{\pi}{8}\right)\left(\frac{1}{\eta}\right)\left(\frac{r^4}{L}\right) = (\Delta P)\frac{(\pi r^4)}{(8L)(\eta)} \quad (1)$$

$$\Delta P = P_1 - P_2 = \frac{8\eta LF}{\pi r^4} \quad (2)$$

$$R = \frac{(8\eta L)}{(\pi r^4)} \quad (3)$$

In these equations, P represents the pressure in the system, whereas the expression ΔP (or $P_1 - P_2$) represents the driving force of pressure gradient that is present, ie, the difference in the two pressures per unit length of a tube or between the two ends of a tube having a length L. It is important to realize that the pressure factor (ΔP) represents the potential energy that is present within the system. The symbol η represents the coefficient of viscosity for the fluid; F is the flow rate; and r is the radius of the tube.

In the Poiseuille–Hagen equation, if ΔP is expressed in dynes/cm², η is in poises, and if the dimensions of the tube (the radius, r, and its length, L) are in cm, the flow rate, F, will be in ml/sec.

Note that the resistance to flow in the Poiseuille–Hagen equation is dependent upon two factors. First, the viscosity of the fluid, η, and second, the geometry of the tube itself, which in turn depends upon the radius and the length of the tube.

RELATIONSHIPS AMONG VISCOSITY, VELOCITY, AND PRESSURE. The term *viscosity* may be defined as "a lack of slipperiness between adjacent layers of fluid." This definition, coined by Newton, still is valid as well as conceptually useful. Hence, if one considers fluid moving within a tube under the force of a driving pressure, as shown in Figure 10.40, the velocities of adjacent layers (or laminae) of fluid assume a parabolic distribution along the axis of the tube at a given ΔP. The fluid layer adjacent to the wall of the tube has zero velocity, whereas adjacent laminae have a progressively greater velocity, until a maximum velocity is reached along the central axis of the tube as illustrated in Figure 10.40.

It is necessary to realize that the resistance in a fluid system is not based upon "friction" between the wall of the tube (or blood vessel) and the fluid moving within it. Rather, the resistance depends largely upon the internal cohesiveness (or viscosity) of the fluid, and it is this "lack of slipperiness" which causes the fluid to resist shear by the external driving force exerted upon it.

The velocity of flow at any given radius from the central axis of the tube may be obtained from the expression:

$$Vr = Vm\ (1 - r^2/R^2)$$

In this equation, Vr = the velocity at any radius (r) from the central axis of the tube; Vm = the maximum velocity along the central axis of the tube; and R = the radius of the tube. Thus the velocity of the fluid lamina at the fluid:wall interface is 0, since $r = R$ at this point.

The *maximum velocity* in terms of the coefficient of viscosity and the pressure gradient can be expressed as:

$$Vm = \frac{(\Delta P)(R^2)}{4\eta}$$

In this expression, Vm = the maximum velocity along the axis of the tube; ΔP = the pressure gradient, ie, the pressure drop per unit length of tube, or $(P_1 - P_2)$; R = the radius of the tube; and η = the coefficient of viscosity of the fluid.

THE RELATIONSHIP BETWEEN PRESSURE AND VELOCITY OF FLOW: BERNOULLI'S PRINCIPLE. Even as the pressure in a moving fluid system is a measure of the potential energy of that system, so the velocity with which the fluid moves is a measure of the kinetic energy within that system. Thus, the kinetic energy of a fluid may be calculated readily from the equation:

$$KE = 0.5\ (\rho)\ (V^2)$$

In this expression, KE = kinetic energy; ρ = the density of the fluid; and V = the velocity. Hence if the density of the fluid is expressed in CGS units, ie, in gm/ml, and the velocity is stated in cm/sec, then the kinetic energy is given in ergs/ml and this is the equivalent of dynes/cm². Consequently, the value for kinetic energy has the same physical dimensions as those for pressure. In a moving fluid, of course, the kinetic energy is derived from the potential energy, that is, the pressure per se.

Bernoulli's principle states that in a fluid system, the lateral pressure exerted at any point in the system is inversely proportional to the velocity of flow at that point in the system. Stated another way, the lateral pressure exerted by a moving fluid within a system is directly proportional to the cross-sectional area of the system. The velocity of flow in a cylindrical tube is, of course, related inversely to the square of the radius; thus, in such a tube the smaller the bore the greater the flow rate at a given driving pressure and viscosity. It also is obvious that a specific volume of fluid must pass two points on the tube in the same time interval, regardless of the bore size of the tube at these points. These facts are illustrated in Figure 10.41.

If, as shown in Figure 10.42A, the pressure within a fluid system is measured by a suitable manometer or other device attached to the end of the tube, all the potential and kinetic energy of the fluid within the tube will be recorded as an *end pressure*. If, however, the measuring device is located so as not to interrupt the flow, as shown in Figure 10.41B, the *lateral* or *side pressure* will be recorded, and this value will be lower than the end pressure by an amount that is equivalent to the kinetic energy of flow. This is true because according to Bernoulli's principle the total energy, ie,

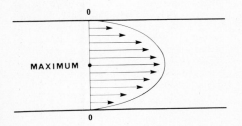

FIG. 10.40. Streamline flow. Note that flow reaches a maximum along the central axis of the tube and is zero at the wall. The several arrows denote the fact that the flow is laminar; that is, the fluid is made up of an infinite number of layers (laminae) which slide past each other without mixing.

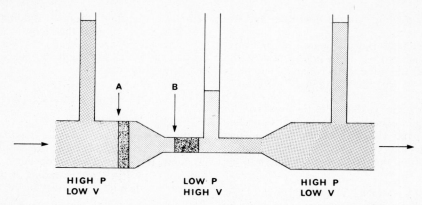

FIG. 10.41. Bernoulli's principle expresses the relationship between tube diameter, pressure (P), and the velocity of flow (V) of a moving fluid. When the tube has a large diameter, the fluid pressure is high, but the flow velocity is low. Conversely, when the tube diameter is small the pressure is low, but the velocity is high. Stated another way, the volume of fluid (dark shading) that passes point A per unit time must equal the volume of fluid passing point B during the same time interval.

potential energy or pressure plus the kinetic energy, must remain constant.

STREAMLINE VERSUS TURBULENT FLOW. If the velocity at which a fluid is forced through a system of rigid tubes is increased progressively, the flow is linear with the driving pressure up to a certain critical velocity, and the adjacent laminae in the fluid slide past each other smoothly without intermixing. This phenomenon is known as *streamline flow* (Fig. 10.40).

At flow rates above the critical velocity, however, the flow no longer is streamlined; that is, eddy currents and vortices develop within the

fluid, adjacent laminae intermix freely, and *turbulent flow* now exists (Fig. 10.43).

Thus, if one plots the flow rate against the driving pressure, as shown in Figure 10.43, at and above the critical velocity, there is a break or inflection in the curve, and turbulent flow ensues; below the inflection point, streamline flow is present and the flow obeys Poiseuille's law.

When streamline flow is present, the pressure energy is used principally to overcome the viscosity of the fluid, whereas in turbulent flow this potential energy now is used primarily to generate the kinetic energy of the eddies. A new law deter-

A.

B.

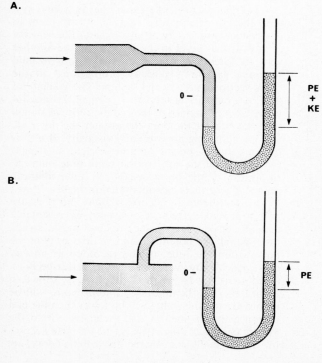

FIG. 10.42. **A.** Measurement of end pressure. When the tube is occluded completely, the pressure measured on the manometer is the sum of the potential energy (PE) and the kinetic energy (KE), the latter being converted into pressure energy. **B.** Measurement of side pressure. Only the potential energy is measured, as the kinetic energy of flow through the open tube does not affect the height of the manometer fluid. Side pressures are shown in Figure 10.41. (Data from Ruch and Patton [eds]. *Physiology and Biophysics,* 19th ed, 1966, Saunders.)

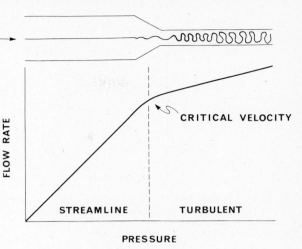

FIG. 10.43. At the critical velocity, streamline flow changes into turbulent flow, and eddy currents develop together with audible sounds. The Poiseuille relationship holds for streamline flow, that is, at flow rates below the critical velocity, whereas the Reynolds equation describes the situation that exists at flow rates above the critical velocity. (After Ruch and Patton [eds]. *Physiology and Biophysics,* 19th ed, 1966, Saunders.)

mines how the fluid behaves at flow rates above the critical velocity. According to Poiseuille's law, the resistance to flow depends principally on the viscosity. However, since the kinetic energy = 0.5 (ρV^2), the resistance to flow at or above the critical velocity depends primarily upon the density (ρ) of the fluid, according to the relationship determined by Reynolds. Thus,

$$V_c = K\eta/\rho R$$

In this equation, V_c = the critical velocity in cm/sec; η = viscosity of the fluid in poises; ρ = density in gm/ml; R = radius of the tube in centimeters; K = a constant (Reynolds number).

The constant, K, is approximately 1,000 for many fluids, including blood, if it is assumed that a long straight tube having a uniform diameter is used for the determination of the critical velocity. The Reynolds number becomes considerably smaller in tubes that are curved or in which there is a narrowing.

One further point with regard to streamline and turbulent flow should be emphasized here. Streamline flow is silent, whereas turbulent flow produces audible sounds, and this fact underlies the genesis of most of the sounds in the cardiovascular system, including heart murmurs. Thus, for example, when the lumen of an artery becomes partially occluded, the velocity of blood flow through the occluded region accelerates in accordance with Bernoulli's principle; and if the velocity of the flow remains below the critical velocity of the blood, no audible sounds are heard, as the flow is streamlined (silent). If, however, the velocity of blood flow through the occluded region exceeds the critical velocity for blood, turbulent flow develops and audible sounds can be heard in the occluded vessel (see Korotkoff's sounds, p. 632).

NEWTONIAN AND NON-NEWTONIAN FLUIDS. As discussed above, all the physical properties of fluids have been considered as though the fluids were newtonian, and a newtonian fluid (liquid) is defined as one whose viscosity remains constant when the velocity of the flow changes. A non-newtonian fluid, on the other hand, exhibits what is called *anomalous viscosity;* that is, the viscosity of a non-newtonian fluid changes according to the velocity with which it is moving through the system.

In the normal (physiologic) range of blood flow rates that are encountered in the human cardiovascular system, blood behaves as though it were a newtonian fluid as long as it is moving. There is a considerable difference, however, in the viscosity of blood in normal motion compared to the viscosity at rest or at very low flow rates. In these latter two situations, blood exhibits anomalous, or non-newtonian, properties.

LAPLACE'S LAW. Laplace's law is of considerable importance to cardiovascular physiology, as it relates the three physical parameters of pressure, tension, and radius in the heart and blood vessels (as well as in hollow viscera generally). For example, the equilibrium which is present among the three factors acting upon the distensible blood vessels can be analyzed by means of Laplace's law so that the peculiar shape of flow–pressure curves in the cardiovascular system as well as the critical closing pressure of small blood vessels can be explained (Fig. 10.44; see also p. 642). Mathematically, this law is expressed:

$$P = T/(1/R_1 + 1/R_2)$$

In this equation, P = the distending pressure within a hollow system, called the *transmural pressure.* The transmural pressure is measured by the *difference in pressures* that exists between the outside and inside of the hollow viscus at equilibrium, and this pressure is defined in terms of dynes/cm², that is, the pressure on one side of the wall minus the pressure on the other side. T = the tension developed in the wall of the elastic hollow

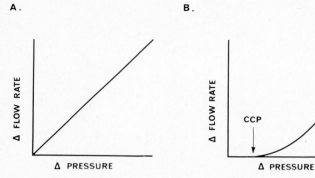

FIG. 10.44. **A.** The changes in flow rate in a rigid tube are directly proportional to the changes in pressure. **B.** In blood vessels, the relationship between flow and pressure are nonlinear. Furthermore, at some finite critical closing pressure (CCP) that is above zero, blood vessels, particularly capillaries, tend to collapse, so that the flow rate becomes zero. (After Ganong. *Review of Medical Physiology*, 6th ed, 1973, Lange Medical Publications.)

system in dynes/cm; and R_1 and R_2 = the two principal radii of curvature of the system at any point on the surface of the wall measured in cm. These parameters are illustrated in Figure 10.45.

Consequently, Laplace's law states that the distending pressure in a hollow elastic system *at equilibrium* is equal to the wall tension divided by the two principal radii of curvature of that system.

If the equation above is related to a spherical object, such as a bubble, $R_1 = R_2$; therefore, the expression becomes:

$$P = 2T/R$$

In an elastic cylinder such as a rubber tube or blood vessel, however, one radius is infinite; the Laplace relationship becomes:

$$P = T/R$$

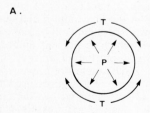

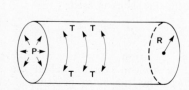

FIG. 10.45. Forces present in the wall of a cylindrical tube or blood vessel at equilibrium. T = wall tension of vessel in dynes/cm vessel length; P = pressure in dynes/cm²; and R = radius of vessel in cm. See discussion of Laplace's law in text. (After Ruch and Patton [eds]. *Physiology and Biophysics*, 19th ed, 1966, Saunders.)

An example will serve to illustrate the Laplace relationship. Consider two tires that are round in cross section at sea level, and that are inflated to the same extent, namely, 4.13×10^6 dynes/cm² (this value is equivalent to a pressure of 60 pounds/inch²). Atmospheric pressure at sea level is equal to 1.01×10^6 dynes/cm². Therefore, the transmural pressure across the wall of each tire is the same; that is, $4.13 - 1.01 = 3.12 \times 10^6$ dynes/cm². Now assume that one of the tires is for a bicycle and has a radius of 1.5 cm, whereas the other fits an automobile and its radius is 10.0 cm.

Since $P = T/R$, $T = (P)(R)$, so the wall tension in the bicycle tire is equal to $(3.12 \times 10^6) \times (1.5) = 4.68 \times 10^6$ dynes/cm; similarly, the wall tension in the auto tire is found to be 31.2×10^6 dynes/cm. Thus, the wall tension in the larger auto tire is approximately 6.7 times greater than that in the bike tire simply because of the difference in the radii between the two tires (10/1.5 = 6.7). In other words, the smaller the radius of the tire (or heart or blood vessel) the lower the wall tension necessary to balance the distending pressure becomes (Fig. 10.45).

The situation that obtains in blood vessels is identical to that presented above. For example, in the human aorta, which has a radius of approximately 1.3 cm and a mean internal pressure of 1.3×10^5 dynes/cm², the wall tension is 170,000 dynes/cm. In the vena cava, on the other hand, the vessel radius is about 1.6 cm, and the mean internal pressure is 1.3×10^4 dynes/cm². In this large vein, the wall tension is only 21,000 dynes/cm. In marked contrast to both of these large vessels, however, a capillary with a radius of only 4 μ and a mean internal pressure of 4×10^4 dynes/cm² has a wall tension of merely 16 dynes/cm, a fact which explains why capillaries, although exceedingly delicate structures, are relatively difficult to rupture under normal circumstances. In other words, the chief reason for their strength lies in their extremely small radius.

Laplace's law also explains why a dilated,

hence failing, heart is working against a considerable mechanical disadvantage. Thus, when the radius of a particular chamber of the heart increases, the wall tension developed by the myocardium of that chamber during systole also must increase considerably in order to develop the same intraluminal pressure as a heart chamber that has a normal radius. The dilated heart must perform considerably more work than the normal heart in order to achieve the same physiologic performance level.

Similarly, Laplace's law applies to the relationship between wall tension, radius, and pressure as found in the bladder, stomach, intestine, lungs, uterus, and ureters.

HEMODYNAMICS IN THE SYSTEMIC CIRCULATION

The human cardiovascular system forms a continuous circuit for the movement and distribution of blood throughout the entire body. Consequently, and since the blood vessels comprising this system have a finite capacity, the volume of blood pumped by the heart during a given interval of time must equal the flow through the various subdivisions of the circulatory system during the same period. It also follows from this basic concept that if a quantity of blood is shunted from one region of the body to another, for example by the process of vasoconstriction, the vessels in another region of the body must dilate and thus increase their capacity by a volume that is equal to that of the displaced blood in order that the total flow remain the same. This statement assumes, of course, that the total blood volume remains constant, and that there is no loss from the circulation as by hemorrhage.

As shown in Figure 10.15, the two major subdivisions of the human cardiovascular system are the pulmonary and the systemic circuits. It follows from the concept presented above that the total blood flow through each of these series-connected but morphologically independent subdivisions of the circulatory system must be identical over a given time interval in any individual. This important concept is axiomatic to normal cardiovascular physiology.

It is important to remember also that the physical laws and principles introduced in the previous section hold strictly true in the mathematical sense only for the behavior of perfect fluids in rigid tubes. The blood itself is not by any means a "perfect" fluid, and furthermore it is composed of two phases, the plasma and formed elements. Therefore, since blood is both a complex and heterogeneous mixture of liquid as well as solid phases, it is not surprising that this fluid exhibits certain anomalous properties in its physical behavior. Furthermore, the various types of blood vessel that make up the cardiovascular system itself are readily distensible as well as contractile to various degrees. The arteries, capillaries, and veins form a

highly elastic and dynamic system of conduits rather than an inert system of rigid tubes. The foregoing statements, however, in no way detract from the fundamental importance of the basic laws of hydrostatics and hydrodynamics. These laws are essential to an understanding and interpretation of the complex patterns of hemodynamic activity within the human body in a general, if not a quantitatively precise, way. The topics to be discussed thus will be related to the specific physical laws presented earlier insofar as possible.

In this section, the relationships between pressure, flow, resistance, and other factors within the human systemic cardiovascular system will be dealt with at various morphologic and functional levels within the vascular tree, starting at the aorta. Certain specialized functional attributes of the pulmonary circulation will be introduced in a later section of this chapter (p. 654) and their role in respiratory function will be integrated with the material in Chapter 11.

Pressure-Volume Relationships within the Vascular System: Distensibility of Blood Vessels

Since the blood vessels are not a system of rigid tubes, but are elastic conduits, they possess the property of *distensibility*. This property varies quantitatively to different degrees in the various types of blood vessel; however, distensibility in all instances can be expressed as the proportional increase in volume for each mm Hg increment of pressure change within the lumen of the vessel.

The walls of arteries are relatively stronger than are those of veins; the latter structures average an eightfold greater capacity for distensibility than do the arteries. A given increase in pressure within a vein will augment the volume capacity to a far greater extent than the same pressure will increase the volume capacity of an artery having a similar diameter to that of the vein.

In studying hemodynamics, it is far more useful to know the total volume of blood that can be stored in a specific region of the circulation than it is to know the distensibility of single vessels; the concept of overall distensibility is used in cardiovascular physiology, and the physical term *compliance* (or *capacitance*) is applied to this concept. The compliance of a system is measured by the increase in volume that is caused by a specific increase in the pressure. Expressed mathematically, this concept can be defined by the expression:

$$\text{Compliance} = \frac{+\Delta V}{+\Delta P}$$

In this equation, $+\Delta V$ signifies the increment of volume change in a system with a given increase in the pressure within that system, $+\Delta P$.

In dealing with various regions of the circulation, or the heart itself, it is convenient to express the relationship between pressure and volume as a pressure–volume curve, as shown in Figure 10.46. This illustration indicates that the compliance of the vena cava is considerably greater quantitatively than that of the aorta. This fact is of considerable physiologic importance, as it exemplifies the fact that normally a very small incremental change in pressure in the veins in general will suffice to increase the total volume capacity of these vessels enormously; hence, the veins form important blood reservoirs, and this functional property results directly from the ability of the veins to alter their net volume capacity readily with only slight alterations in their internal pressure.

One additional concept must be introduced in context with the concept of vascular compliance per se. This is the phenomenon of *delayed compliance*. Delayed compliance signifies that a vessel in which the pressure has been increased by elevating the volume gradually loses this pressure with time; ie, the vessel dilates progressively because of the inherent property of stress–relaxation of the smooth muscle in the vascular wall (Fig. 10.47). This property of stress–relaxation is characteristic of all smooth muscle, not merely that present in the blood vessels; however, in the vascular bed delayed compliance is an important adaptive mechanism to a drastically increased total blood volume, such as is present following a massive transfusion. In this regard, it should be emphasized that delayed compliance occurs to a far greater extent in the veins than in the arteries. Conversely, the vascular bed can adapt itself gradually to a decreased total circulating blood volume, such as occurs following a large hemorrhage, so that the tonus of the vascular smooth muscles gradually increases and thereby compensates for the decreased total circulating blood volume.

Measurement of Blood Pressure

EFFECTS OF BODY POSITION ON BLOOD PRESSURE. Even though the blood is in continuous motion, it is obvious that Pascal's laws determine to some extent the normal hydrostatic pressures that can be recorded from various regions of the body in a subject who assumes different bodily positions.

Because of the influence of the force of gravity upon the mass of blood within the circulatory system, the effect of this gravity factor upon the observed blood pressure is manifest quite clearly when the subject is standing in a vertical position (Fig. 10.48). For this reason, the mean arterial blood pressure measured in the ankle of a standing individual is around 180 mm Hg. Hence, the observed pressure value recorded in the ankle is higher by the factor ρgh (around 80 mm Hg) than is the mean blood pressure within the aorta at heart level (about 100 mm Hg). Similarly, in arteries that are located above the heart, the blood pressure will be lower than that recorded at heart level by an amount that is equal to the same factor (ρgh). The mean arterial blood pressure in the head is approximately 60 mm Hg.

If, on the other hand, the subject is lying on his back in a horizontal position, the mean arterial blood pressures in all regions of the body are approximately equivalent with respect to the blood pressure recorded at heart level. Similar gravitational effects are exerted on venous pressures, and will be considered on page 651.

Therefore, from a clinical standpoint, when arterial blood pressures are measured in the arm or leg of a patient who is not lying down, these values must be corrected to heart level for comparison with standard values that were obtained at heart level. The factor ρgh is added if the measurement is made above heart level, and subtracted if the pressure is recorded from below heart level. This correction factor, in mm Hg, can be approximated by dividing the difference in centimeters between heart level and the level at which the measurement is taken by 13. This factor in turn is obtained by dividing the density of mercury, 13.6, by the density of blood, 1.055.

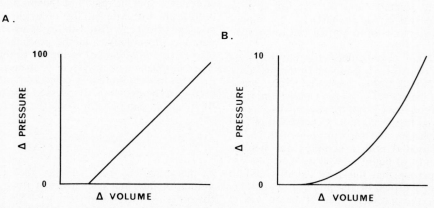

FIG. 10.46. The pressure-volume relationships in the aorta (**A**) and vena cava (**B**). The volume scales are identical for both curves, but the pressure scale for the aorta is 10 times that employed for the vena cava. (After Ganong. *Review of Medical Physiology,* 6th ed, 1973, Lange Medical Publications.)

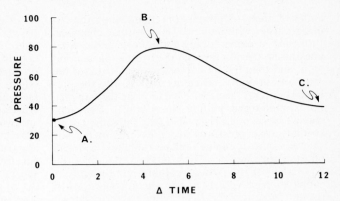

FIG. 10.47. Delayed compliance. When a hollow organ is distended, ie, as when it is filled with fluid (A), the pressure rises to some maximum value with time (B). As the volume is maintained, however, the wall tension gradually decreases; hence, the pressure falls with time (C). Pressure and time scales in arbitrary units.

Clinically, the measurement of blood pressure is usually performed on the arm of the patient while he is seated in the upright position. Therefore, this level is sufficiently close to that of the heart so that no correction is required.

METHODS FOR DETERMINING BLOOD PRESSURE. The blood pressure can be determined either *directly* or *indirectly*. If a hollow tube, or cannula, is inserted into a blood vessel and then connected to either a water or a mercury manometer, the blood pressure can be measured directly. Alternatively, the cannula can be tied into the vessel and then connected to a calibrated strain gauge which in turn is connected to an amplifier, whose output causes deflections of a pen that writes upon the moving strip of paper of an oscillograph.

If the vessel is tied off beyond the point of insertion of the cannula, an end pressure is recorded as depicted in Figure 10.42A. In this situation, flow in the vessel is interrupted completely. Therefore, all the kinetic energy of flow is converted into pressure energy and is recorded as such. On the other hand, if a T tube is placed within the vessel so that the pressure is measured in the side arm of the tube (Fig. 10.42B), the lateral pressure within the vessel, which now is recorded, is lower than the end pressure by a quantity that equals the kinetic energy of flow, if it is assumed that the pressure drop in the vessel segment caused by resistance of the vessel is insignificant. This statement is valid because, in a blood vessel or other tube, the total energy of a moving fluid (which energy consists of the sum of the pressure energy and the kinetic energy of flow) is a constant, and this is merely another way of stating Bernoulli's principle (p. 625).

It is important to realize that the fall in pressure that can be observed in any particular section of the arterial system is caused by the operation of two factors: (1) the resistance to flow offered by the vessel, and (2) the conversion of the potential energy of the blood into kinetic energy. The fall in blood pressure that is occasioned by the blood overcoming the resistance of the vessel is lost irrevocably as heat; however, the fall in blood pressure that is caused by the conversion of potential energy into kinetic energy as the vessel narrows is a reversible process, because the potential energy increases as the vessel widens once again, in accordance with Bernoulli's principle (Fig. 10.41).

Clinically, direct cannulation of vessels for the determination of various pressures and obtaining samples is used for diagnostic purposes in the procedure known as *cardiac catheterization*. Cardiac catheters usually are closed at the tip, but a small opening is present at the side near the end of the

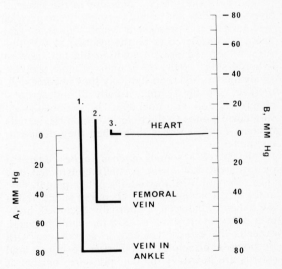

FIG. 10.48. The effects of gravity (ρgh) on intravascular pressures. Scale on left **(A)** indicates the increment in venous pressure that would be present at each level as a result of the effects of gravity. The heavy black bars on the left signify the height to which a column of blood would rise in a manometer tube connected to a vessel 1. in the ankle, 2. the femoral vein in groin, or 3. the right atrial level, the subject being in a standing position. Corresponding pressures with the subject supine are 10 mm Hg (1), 7.5 mm Hg (2), and 5 mm Hg (3). Scale on right **(B)** gives the increment or decrement in mean arterial pressure at each level, the mean arterial pressure in all large arteries at heart level being about 100 mm Hg. (Data from Ganong. *Review of Medical Physiology,* 6th ed, 1973, Lange Medical Publications.)

catheter; thus, they sample predominantly the lateral pressure within the vessel or cardiac chamber. Intracardiac catheters are introduced under local anesthesia, usually through the brachial or femoral vein, and then the tip can be manipulated into the right atrium and ventricle as well as into the pulmonary artery. The exact location of the catheter tip is monitored with a fluoroscope. The subclavian or carotid artery also may be catheterized so that aortic pressures and pressures in the left side of the heart can be obtained similarly to those on the right side.

Other than cardiac catheterization, direct cannulation of vessels is not a technique used in routine practice for determining arterial blood pressure on patients, so that this value must be determined *indirectly*. Arterial blood pressure is measured indirectly by the *auscultatory technique,* with an instrument called a sphygmomanometer. The inflatable cuff of the sphygmomanometer is placed about the upper arm, and the bell of a stethoscope is placed over the brachial artery at the elbow. The cuff now is inflated well above the anticipated systolic blood pressure in the brachial artery.* Since the artery is occluded completely, no sound is heard. As the pressure within the cuff is lowered slowly, however, a tapping sound is heard concomitantly with each heartbeat at the point where the systolic pressure in the brachial artery is slightly above cuff pressure at the peak of systole. The cuff pressure at which sounds are first heard is considered to be the systolic blood pressure. When the pressure within the cuff now is lowered still further, the sounds become progressively louder, and then suddenly they become dull, muffled, and finally they usually disappear completely (Korotkoff's sounds). The diastolic pressure corresponds to the point at which the sounds become muffled rather than when they disappear completely.

Korotkoff's sounds appear to be generated by turbulent flow in the partially constricted brachial artery (p. 626, Fig. 10.43). When patent, flow in the artery is streamlined, therefore silent. When the lumen of the artery is partially occluded, however, the velocity of blood flow through this region exceeds the critical velocity in accordance with Bernoulli's principle, turbulence develops, and sounds are heard. If the cuff pressure remains at any point above the diastolic arterial pressure, blood flow will be interrupted for at least a portion of the diastolic interval; hence, the sounds have a sharp and abrupt (ie, staccato) quality. At the point where the pressure within the cuff is immediately above diastolic pressure the artery still is slightly constricted; hence, turbulent flow is continuous, the sounds now having a muffled character.

According to Pascal's first law (p. 624), the pressure developed within the cuff of the sphygmonanometer is transmitted through skin, fascia, and muscle to the brachial artery, thereby occluding this vessel as the pressure is raised.

Certain standard procedures must be adhered to when determining blood pressures by the auscultatory method. First, and as noted above, the cuff of the sphygmomanometer should be placed on the arm at heart level in order to avoid the effects of gravity upon the column of blood. Second, false high readings will be obtained on the arms of extremely obese individuals when standard width cuffs are used. Use of a wider cuff obviates this difficulty. Third, if the cuff is inflated upon the arm and left for too long, reflex vasoconstriction may be produced by the discomfort, and false high readings will be obtained. Fourth, the blood pressure should be measured in both arms at the first visit of the patient, because large differences between the blood pressure in each arm (or leg) are indicative of a vascular obstruction.

The systolic pressure alone can be determined readily by a technique known as *palpation.* In practice, the sphygmomanometer cuff is inflated so as to occlude blood flow through the brachial artery, and the radial artery is palpated as the cuff pressure is reduced gradually. Systolic pressure is taken at the point at which a radial pulse is first palpable. Systolic pressures obtained by palpation generally are several millimeters of mercury below those obtained by auscultation, because of the difficulty encountered in detecting the first beat precisely.

When the auscultatory method is used for the measurement of blood pressure, palpation of the radial pulse until it disappears will ensure that false low readings will not be made, as Korotkoff's sounds may disappear completely in some patients at pressures between systolic and diastolic. This phenomenon is called the *auscultatory gap.*

NORMAL BLOOD PRESSURE. By convention, arterial blood pressure data are recorded as systolic pressure/diastolic pressure, it being understood the numbers signify that the units employed are mm Hg. In normal young adults, the blood pressure in the brachial artery, when determined at rest in the sitting or lying position, is about 120/70 mm Hg. The arterial pressure is equal to the cardiac output multiplied by the peripheral resistance, as discussed in the next section. Any condition that alters either one or both of these factors will affect blood pressure directly. Any factor (eg, excitement) that increases the cardiac output tends to increase primarily the systolic pressure, whereas if the peripheral resistance is elevated, the diastolic pressure tends to increase.

According to some studies, normal females of a given age group tend to have slightly higher systolic and diastolic blood pressures than do normal males who belong to the same age group, and the blood pressures of both females and males tend to increase progressively with age. These facts are illustrated in Figure 10.49. Other studies, however, have indicated that the situation is diametrically opposite, in that males tend to have higher blood

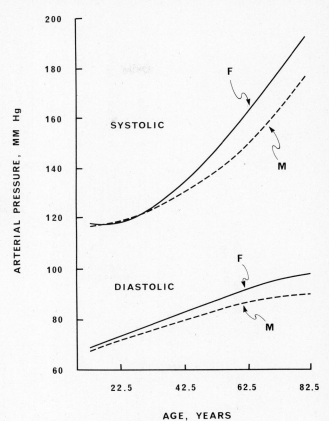

FIG. 10.49. Mean systolic and diastolic arterial pressures for females (F) and males (M) plotted against age. (Data from Harvey, Johns, Owens, and Ross [eds]. *The Principles and Practice of Medicine,* 18th ed, 1972, Appleton-Century-Crofts.)

pressures up to around the age of 50 years, at which point no significant differences are apparent between males and females, but in both sexes the values again tend to increase in proportion to age.

Arterial Hemodynamics

The important relationships between pressure, flow, resistance, volume, and relative cross-sectional area of the systemic circulation are given in Figure 10.50. Reference should be made to this illustration throughout the discussions of arterial and arteriolar as well as capillary and venous hemodynamics, as it contains in graphic form all the pertinent concepts that are discussed in the text. The approximate resting blood flow to various organs and tissues is summarized in Table 10.6.

During each systolic phase of the cardiac cycle, the left ventricle ejects a quantity of blood into the aorta which is known as the *stroke volume*. In turn, the *cardiac output* is equal to the product of *heart rate* and *stroke volume* in ml/min; ie, cardiac output is a *flow rate*. In turn, blood flow per se is merely the quantity (volume) of blood that passes a given point in a stated time interval. Since the left ventricular stroke volume in an adult at rest is about 60 ml, at a heart rate of 80/minute the cardiac output would be 60 × 80 = 4,800 ml/minute or

4.5 liters/minute. The clinical measurement of cardiac output will be discussed on page 640.

The mass of blood that is ejected into the aorta during each stroke (contraction) of the heart has a maximum systolic pressure (or total fluid energy) imparted to it which is about 1 mm Hg below that found in the left ventricle (Fig. 10.38), or around 119 mm Hg. The total energy imparted to this mass of blood by the heart as it enters the aorta is largely potential energy, which in turn dilates the proximal, highly elastic region of the aorta as well as its large branches, which lie close to the heart. Some kinetic energy also is imparted simultaneously to the column of blood as it moves from the left ventricle into the aorta.

During the diastolic runoff phase following closure of the aortic semilunar valves, the potential energy that was stored during cardiac systole in the aortic wall now is transmitted gradually by elastic recoil of the vessel to the mass of blood in its lumen so that the blood now receives sufficient kinetic (or flow) energy to move down the vessel, attended by the concomitant pressure wave which also is transmitted down the aorta and its branches. These facts are illustrated in Figures 10.38 and 10.50. In this regard, it should be mentioned that the velocity of transmission of the pulse wave (ie, the pressure wave) down the aorta

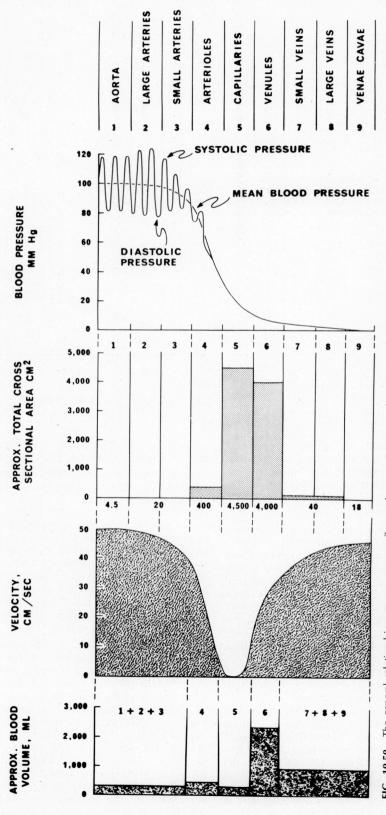

FIG. 10.50. The general relationships among pressure, flow, volume, resistance, and cross-sectional area in the systemic circulation. (Data from Burton. *Physiology and Biophysics of the Circulation,* 1965, Year Book Medical Publishers; Ganong. *Review of Medical Physiology,* 6th ed, 1973, Lange Medical Publications; Guyton. *Textbook of Medical Physiology,* 4th ed, 1971, Saunders.)

Table 10.6 RELATIVE BLOOD FLOW TO VARIOUS ORGANS AND TISSUES[a]

ORGAN OR TISSUE	FLOW[b] (ml per minute)	TOTAL CARDIAC OUTPUT (percent)
Heart	150	3
Bronchial region	150	3
Skeletal muscle (inactive)[c]	750	15
Brain	700	14
Kidneys[d]	1,100	22
Skin[c]	300	6
Bone	250	5
Liver	1,350	27
Portal circulation	1,050	21
Arterial circulation	300	6
Adrenal glands	25	0.5
Thyroid gland	50	1
All other tissues	175	3.5
Total	5,000	100

[a]*Data from Guyton. Textbook of Medical Physiology, 4th ed, 1971, Saunders.*

[b]*Values compiled for an individual under basal conditions.*

[c]*The volume flow through skeletal muscle can increase up to twentyfold during strenuous exercise. Blood flow in the skin also can show extremely wide fluctuations, depending upon the need to conserve or liberate heat.*

[d]*Note that the perfusion of the kidneys alone accounts for over one-fifth of the total cardiac output, a fact that is related directly to the function of the kidneys.*

is about 15 times faster than the rate of blood flow in this vessel, and in more distal arteries the pulse wave may travel up to 100 times faster than the rate of blood flow.

It is an interesting and seemingly paradoxical fact that the pulse pressure actually increases in magnitude as the arterial blood pressure is recorded in the aorta at successively greater distances from the heart, as shown in Figure 10.50. It would thus appear that the movement of this blood is contradicting a basic physical law by moving up a pressure gradient in flowing toward the capillaries, since the arithmetic mean blood pressure does increase slightly with distance from the heart. In fact, however, the *integral mean blood pressure* decreases in accordance with distance from the heart, so that the blood moves down the pressure gradient that exists between the aorta and the capillaries, as shown in Figure 10.50. The integral mean pressure at various levels in the arterial tree is obtained by integrating mathematically the entire area under the arterial pressure curve, as shown in Figure 10.50 and explained in Figure 10.51.

The physical principle whereby the blood is driven through the circulatory system is given by the Bernoulli equation. The Bernoulli equation in turn is based upon the fundamental law of conservation of energy, that is:

$$E = (PV) + (\rho gh) + (0.5mV^2)$$

In this relationship, E = the total energy in a liquid flowing in a streamline pattern; PV = the pressure energy (or potential energy) of the system; ρgh = the potential energy factor, in which ρ = the density of the fluid in grams per milliliter, g = the acceleration of gravity or 980 cm/sec^2, and h = the height of the fluid column in cm;* and 0.5 mV^2 = the kinetic energy factor resulting from the velocity of flow, m being the mass of blood in gm.

In accordance with this equation, the total energy, E, within a moving fluid system (that has no resistance) is a constant; hence, as the potential energy increases, the kinetic energy decreases proportionally, and vice versa. Therefore, the actual driving force for the circulation of blood within the cardiovascular system is the effective pressure gradient that exists between an upstream point and a downstream point, expressed as the *total energy gradient*, that is, the difference between the total energy (E) that is present at the two points in the circulation as defined by the Bernoulli equation. From a practical standpoint, this energy gradient generally is expressed as the difference in the mean pressures (ΔP) between two points in the cardiovascular system.

It may be appreciated readily, from a consideration of this mathematic relationship, that the sum of the potential energy (mean pressure) and

*Thus, if a mercury column having a density (ρ) of 13.6 is used for measurement, the pressure corresponding to 1 mm Hg is equal to 13,6 × 980 × 0.1 = 1,333 dynes/ cm^2. Similarly, the pressure corresponding to 1.0 cm of saline = 1.04 × 980 × 1.0 = 1,019 dynes/cm^2.

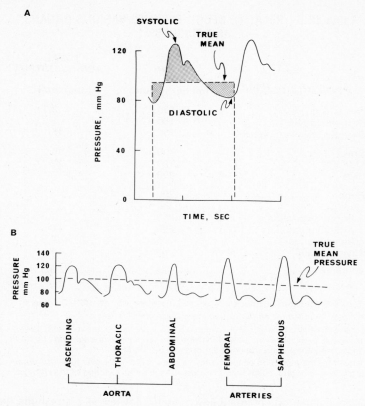

FIG. 10.51. A. The integral (area or true) mean blood pressure. This value is equal to the level of the horizontal dashed line (the shaded areas above and below this point are equal); hence, the integral pressure is not the arithmetic mean of the systolic and diastolic pressures. **B.** Note that this mean value decreases from the aorta toward the peripheral arteries despite the fact that the pulse pressure increases. It is this slight decrease in the true mean pressure that causes the blood to move toward the arterioles. A mercury manometer is a better instrument for measuring the true mean arterial pressure than is a highly sensitive electronic transducer, because of the inertia of the mercury which damps the rapid phasic and systolic and diastolic oscillations in blood pressure. (Data from Burton. *Physiology and Biophysics of the Circulation,* 1965, Year Book Medical Publications; Mountcastle [ed]. *Medical Physiology,* 12th ed, Vol. I, 1968, The Mosby Co.)

the kinetic energy (flow rate) at any given point in the cardiovascular system is constant; therefore, when the velocity of flow is greater, the potential energy is lower at that point, and vice versa. The total driving energy for the blood is, of course, opposed by the resistance offered to flow by the walls of the vessels themselves. Hence the Bernoulli equation given above is not quantitatively applicable to the cardiovascular system.

The *mean velocity* of blood flow in the aorta is reported to range between 30 and 50 cm/sec; this value falls to around 10 cm/sec in the smaller arteries, as shown in Figure 10.50. As these vessels branch so that the total cross-sectional area at any particular level of the circulatory system is greater, the velocity decreases in accordance with Bernoulli's principle. The continuity equation, which is derived from this principle, states that the velocity of flow times the cross-sectional area of a vessel is constant at all points; therefore, $(V_1) \times (A_1) = (V_2) \times (A_2)$.

Since the aortic flow is phasic, namely, it changes rapidly during various phases of the car-

diac cycle, the velocity of flow in this vessel actually ranges from around 120 cm/sec during systole to a transient reversal of direction near the left ventricle as the aortic semilunar valves close during the onset of diastole (Fig. 10.51). In more distal regions of the aorta, as well as in the more peripheral arteries, the systolic velocity also is greater than that in diastole. Flow away from the heart, however, is continuous rather than pulsatile, because the vessels that have been stretched during systole exhibit elastic recoil during diastole. Thus, at the arteriolar level of the arterial tree, there is a gradual and progressive damping of the intermittent oscillations of the systolic and diastolic phases (Fig. 10.50), so that a smooth and continuous blood flow through the capillaries results, although sometimes a minute pulse wave may be transmitted to the capillaries and even to the venules. Insofar as overall systemic hemodynamics are concerned, this fact is insignificant.

It is important to remember the general concept that wherever the total cross-sectional area of the vascular tree changes (Fig. 10.50, Table 10.2), there is a concomitant conversion of kinetic energy to pressure energy or vice versa, depending upon whether the total area increases or decreases. The total energy at any level in the cardiovascular system, of course, remains constant, a fact that is a direct consequence of the Bernoulli principle, and although this interconversion of energy is of some importance in the systemic arterial circulation, it is of far greater significance in the pulmonary circulation and for venous filling of the heart. As shown in Table 10.1, the proportion of the total blood volume within the arterial portion of the circulation at any time amounts to around 11 percent, the relative distribution being roughly 2 percent in the aorta, 8 percent in the arteries, and 1 percent in the arterioles. The arteries and arterioles sometimes are referred to as *resistance vessels,* whereas the veins, which act as blood reservoirs and contain about 64 percent of the total circulating blood volume, are called *capacitance vessels* (Table 10.1).

At the level of the aorta and its large branches, there is little resistance to flow of blood because the diameter, hence the radius, of the vessels is relatively large (Table 10.2). At the level of the smaller arteries, and principally the arterioles, there is a sharp increase in the peripheral resistance owing to the smaller radius of these vessels, and a concomitant drop in the mean arterial pressure takes place as shown in Figure 10.50. Since the blood flow is directly proportional to the difference in pressure at both ends of a vessel and inversely related to the resistance according to Poiseuille's law, small changes in arteriolar caliber can effect tremendous changes in the blood flow, either in specific regions or throughout the entire body.

As noted earlier, the smooth muscle coat within the arteriolar wall normally is in a state of slight partial contraction, or tonus; thus, one speaks of "vascular tone" in the cardiovascular system. The arteriolar smooth muscle also is capable of contracting to such an extent that the lumen of the vessels in a specific region of the vascular tree may be obliterated completely, thereby shunting the blood to other regions of the body. For example, a generalized splanchnic vasoconstriction can result in the shunting of a large volume of blood to the skeletal muscles, such as when an individual is performing strenuous physical exercise. Conversely, the arterioles are also capable of dilatation to a considerable degree. Thus, active vasoconstriction or vasodilatation by the arterioles is of considerable importance in the regulation of blood flow and pressure, both locally and throughout the entire body.

Since the blood pressure at any specific point within the circulatory system is equal to the product of the flow rate* times the resistance at that point, or $P = F \times R,$ in accordance with Poiseuille's law (p. 624), an increase in the total peripheral resistance results in a proportional increase in systemic arterial blood pressure, if it is assumed that the flow remains constant. Similarly, since $F = P/R,$ an increase in the total peripheral resistance will result in a marked reduction in blood flow, provided the mean arterial pressure remains constant. Last, since $R = P/F,$ any increase in blood pressure at a constant flow rate or a decrease in the flow rate at a constant pressure will result in an increased resistance to flow.

It should be noted that the resistance factor (R) cannot be measured directly, and can only be calculated by use of Poiseuille's law or the Poiseuille–Hagen equation. Thus, if a pressure differential (ΔP) of 1 mm Hg exists between two points in a vessel and the flow is 1 ml/sec the resistance is said to be one peripheral resistance unit (PRU). This expression is wholly arbitrary, but at times it is useful for comparative purposes. In CGS units, on the other hand, resistance to flow is expressed as dyne sec/cm⁵, and can be calculated from the equation:

$$R = \frac{\rho g h \times \Delta P}{F}$$

In this expression $\rho g h = 13{,}330$ dynes/cm² (p. 624), ΔP is stated in mm Hg, and the flow, $F,$ in ml/sec.

According to the Poiseuille–Hagen equation, the resistance factor, $R,$ is directly proportional to the viscosity and length of the tube, but inversely related to the fourth power of the radius:

$$R = \frac{8\eta L}{\pi r^4}$$

Normally, both the viscosity of the blood and the length of the vessels remain constant. Because of active constriction and dilatation of the arterioles, however, the radius of these vessels can change markedly, and since the resistance to flow varies inversely as the fourth power of the radius, the blood flow will be increased 16-fold when the arteriolar radius is doubled. Conversely, decreasing the radius by half will decrease the flow 16-fold. On the other hand, doubling the pressure at a constant vessel radius merely will double the blood flow. Consequently, the alterations in arteriolar caliber are of vital importance in the regulation of blood flow within the body.

The *total peripheral resistance* is defined as the resistance to blood flow that is produced by the entire systemic vasculature. Note also that in the systemic circulation the vessels that produce these resistances may be either in series or in parallel (Figs. 10.15 and 10.52). Thus, at the arteriolar level, the overall degree of constriction of these vessels, or vasomotor tone, assumes a role of the utmost importance in determining the total peripheral resistance in the entire systemic circulation of the body. Thus, in essential hypertension, a very slight degree of arteriolar vasoconstriction can produce an exceedingly great elevation in the

---

The symbol Q also is used to indicate the volume flow of blood per unit time, particularly in respiratory physiology.

systemic blood pressure as the resistance is inversely related to the fourth power of the radius of the vessels. The blood pressure rises markedly, an effect that requires the heart to perform considerably more work in order to maintain an adequate output (cf p. 695).

The contribution of viscosity to the peripheral resistance is a small factor in the normal individual. Plasma has a viscosity about 1.8 times that of water; however, whole blood has a viscosity ranging from 2 to 15 times greater than that of water, and this increase in viscosity is brought about primarily by the erythrocytes. As the hematocrit increases (p. 527), so does the viscosity of the blood, eg, as in polycythemia (p. 569), Conversely, when the hematocrit is low, as in anemia, the viscosity of the blood decreases in proportion to the erythrocyte deficiency (Fig. 10.53).

Temperature has a direct effect upon the viscosity of the blood, so that marked cooling of the extremities or hypothermia induced for surgery may increase the blood viscosity to such an extent that a significantly reduced blood flow may ensue at a given pressure. Normally, however, the relationship between temperature and blood viscosity is of minor importance.

As discussed on page 627, blood exhibits certain anomalous properties unlike a true newtonian fluid whose viscosity is independent of flow (shear) rate or tubing size. Therefore, particularly at low velocities, the pressure–flow characteristics of blood are anomalous. Hence, in very small tubes or blood vessels (such as arterioles), the apparent (or relative) viscosity* is decreased, a phenomenon known as the Fahraeus–Lindqvist effect. Blood thus behaves more like water when flowing in smaller vessels than in larger ones, an effect that is

The relative viscosity is defined as the ratio of the flow of water to that of a fluid such as blood under identical conditions of temperature, tube size, and pressure gradient.

similar to a reduced hematocrit, as shown in Figure 10.53. In other words, the viscosity of blood flowing in small tubes or vessels approaches that of plasma. This phenomenon, also known as the *sigma effect,* is caused by the erythrocytes flowing through vessels that have a diameter only slightly greater than that of the cells themselves. If one calculates the blood viscosity, using the Poiseuille equation, but by summing the layers of fluid as though they were as thick as are the erythrocytes themselves (instead of assuming the laminae to have infinite thinness as in the classic approach to this problem), the equation more accurately predicts the viscosity changes in blood under different physiologic conditions.

Such a mathematical treatment explains only partially the anomalous viscosity of blood flowing in small vessels. Another aspect of this problem and a further elucidation of the Fahraeus–Lindqvist effect is to be found in the *axial accumulation* and *axial streaming* of the erythrocytes in small vessels, ie, those that have a lumen diameter of less than 500 μ. As shown in the velocity profile of a blood vessel (Fig. 10.40), the flow rate is most rapid at the center or axis of the tube. A layer of slower moving low-viscosity plasma at the vessel wall thus exerts a greater pressure upon the erythrocytes than does the plasma in the faster moving stream along the central axis of the vessel in accordance with the Bernoulli principle. The erythrocytes tend to become oriented in the center of the vessel, and thus flow (or stream) preferentially along the central axis with its lower relative pressure. Normally, the viscosity of blood is constant at physiologic flow velocities, so that the Fahraeus–Lindqvist effect and axial streaming of the erythrocytes are maximal at velocities lower than the usual flow velocities encountered in the circulatory system. Furthermore, the Fahraeus–Lindqvist effect is offset to a great extent by the fact that at very low flow velocities such as normally exist within the smallest arterioles and capillaries, the viscosity of the blood increases sharply,

A.

B.

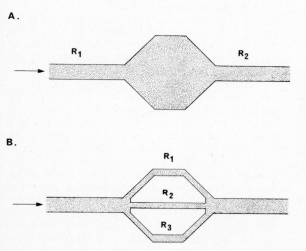

FIG. 10.52.   A. Series resistances in the circulatory system. The total resistance in this case is given by the sum of the individual resistances, ie, $R_t = R_1 + R_2 + - - - R_n$. **B.** Parallel resistances in the circulatory system. The total resistance in this instance is given by the sum of the reciprocals of the individual resistances, ie, $R_t = 1/R_1 + 1/R_2 + 1/R_3 + - - - 1/R_n$. Note similarity to the method whereby resistances are calculated in electric circuits (p. 8).

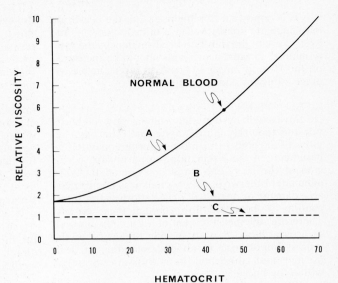

FIG. 10.53. Relationship between blood viscosity and hematocrit (A), compared to the viscosity of water (C). The viscosity of normal blood having a hematocrit of 45 percent is indicated by a dot in the center of curve A. Curve B = plasma. (After Guyton. *Textbook of Medical Physiology,* 4th ed, 1971, Saunders.)

simply because of its relatively slow flow rate (Fig. 10.50). The viscosity thus can increase by as much as 10 times from this physical effect alone, an effect that presumably is caused by some adhesion of the erythrocytes to each other (rouleaux formation, p. 540) as well as to the vessel walls when the blood is moving at low velocities.

In the normal person, the decreased viscosity caused by the Fahraeus–Lindqvist effect and the increased viscosity caused by reduced velocity of flow in small vessels are inconsequential. When pathologically elevated hematocrit levels are encountered, however (eg, in polycythemia), or in shock states when the blood pressure and flow are reduced already (p. 1237), the augmented viscosity of the blood caused by the reduced flow rate can exacerbate the clinical problems already existing in the patient by further reducing blood flow to, hence oxygenation of, the tissues.

Because of the axial accumulation of erythrocytes in small vessels (eg, arterioles), the plasma close to the wall of these vessels tends to have a lower hematocrit than does the blood flowing in the center of the vessel. Smaller branches leaving at right angles to the axis of the relatively larger vessel tend to receive a disproportionately large quantity of blood from near the wall of the larger vessel, blood which has a lower hematocrit than the axial blood. This phenomenon is called *plasma skimming,* and presumably it is responsible for the fact that the hematocrit of capillary blood averages 25 percent lower than that of blood in the rest of the body.

By way of summary of the material presented above, it is essential to reiterate that the three inseparably interrelated factors of pressure, flow, and resistance form the basis for any understanding of hemodynamics. Thus, the *blood flow* through a vessel, or at any level of the entire circulatory system for that matter, is governed by two factors: (1) the *pressure difference* between the two ends of the vessel and (2) the *resistance* that hinders the flow. Thus, $F = \Delta P/R$.* Note that it is the pressure difference, ΔP, that causes the flow, and not the absolute pressure within the vessel (if the absolute pressure were, for example, 100 mm Hg throughout the entire segment of vessel, no flow could occur, as no pressure gradient exists). The other two algebraic forms of the flow equation are of equal importance and should be understood thoroughly:

$$\Delta P = F \times R \text{ and } R = \Delta P/F$$

In any clinical evaluation, not only the blood pressure but also the heart rate is recorded as a routine measurement. Sometimes it also is necessary to know the cardiac output. The determination of heart rate and the measurement of blood flow will now be considered.

HEART RATE. The heart rate is obtained readily by counting the pulses in the radial artery in the wrist for a specific interval (usually one minute), and is expressed in beats per minute.

The heart rate also may be obtained easily from electrocardiographic tracings. The interval between two beats is the reciprocal of the heart rate, so that the elapsed time in seconds between two identical and successive waves divided into one gives the heart rate. Therefore, if the time interval recorded between two R waves (these are used, as they are prominent in all leads) is 1 sec, the heart rate is obviously 60/minute; and if the interval between two similar waves on an electrocardiogram, as measured by the distance between the vertical grid lines on the chart (25 mm = 1 sec) is 0.75 sec, the heart rate is 60/0.75 × 80 beats/ minute.

*Note the similarity between this relationship and Ohm's law, ie, $I = E/R$ (p. 8).

It is important to remember that the heart rate may fluctuate rapidly because of many factors, and even the act of respiration itself influences the rate. A number of similar intervals should be measured on the EKG in order to arrive at a reasonable average value for a given patient.

MEASUREMENT OF BLOOD FLOW. In the cardiovascular system, the volume flow of blood per unit time through the aorta and other vital regions of the body is the most important quantitative parameter in the circulation. In particular, perfusion of the heart and brain must at all times be adequate; otherwise, unconsciousness and irreparable damage or death will ensue shortly after interruption of the blood flow through these organs. Therefore, the measurement of aortic flow, as well as arterial blood flow to specific organs, is of the utmost importance to an understanding of both normal and abnormal hemodynamic patterns within the body.

The majority of the techniques available for the direct determination of arterial blood flow in various regions of the body in humans are applicable only in special situations (eg, during surgery) or during experimental procedures on animals. Since the anesthetic, the surgical manipulations with their attendant trauma, respiratory, and other factors, all can influence profoundly the measurement of blood flow, the quantitative values obtained by use of such procedures must be considered as approximations only. Several of the instruments employed for the direct measurement of blood flow will be mentioned; however, emphasis will be placed upon the two techniques that are employed for measuring cardiac output (hence aortic flow) in humans, both of which are clinically significant.

Several types of flowmeters are available for measuring the mean blood flow through an artery. These instruments include the mechanical stromuhr, the rotameter, the bubble flowmeter, the thermostromuhr, and the turbinometer. Considerable basic information on arterial flow has been derived from the use of these instruments in experimental animals, usually dogs, despite a number of obvious disadvantages that are inherent in the use of such instruments, including the requisite for anesthesia, the injection of an anticoagulant, the insertion of the instrument between the severed ends of the vessel in which flow is being measured, and their inherently slow response time. The last factor is important because arterial blood flow is highly phasic (ie, it changes rapidly, depending upon the phase of the cardiac cycle); hence, rapid fluctuations in flow will not be recorded. These devices cannot be used satisfactorily to measure venous flow because of the resistance drop across the two ends of the instrument. In an artery, on the other hand, because of the much greater pressure, this factor is negligible.

In addition to the instruments listed above, a number of flowmeters have been developed that are exquisitely sensitive to phasic changes in blood flow. Included in this category are differential pressure, ultrasonic, and electromagnetic flowmeters. The latter two kinds of instruments have the distinct advantage that they can be used to monitor venous and arterial blood flow in intact, unanesthetized animals following their surgical implantation.* Furthermore, as the sonar and electromagnetic sensing devices are contained within a cuff that may be placed around the intact vessel under study, no anesthetic is required save for that necessary during the implantation procedure. In fact, the electromagnetic flowmeter has been used successfully on human subjects during surgical operations to monitor blood flow to specific organs.

The two procedures in routine use clinically for measuring cardiac output in patients are the direct Fick method and the indicator dilution method.

The Direct Fick Method. According to the Fick principle, the quantity of a substance that is taken up by the body as a whole (or by a specific organ) per unit time is equal to the arterial level of that substance minus the venous level (or the arteriovenous difference). The arterial blood must, of course, be the only source of the substance being measured.

The Fick principle thus may be used to determine cardiac output, hence aortic blood flow, provided the oxygen consumption (in ml/minute), the arterial oxygen concentration (in ml O_2/liter of blood), and the mixed venous oxygen concentration (also in ml O_2/liter) are known. Thus, left ventricular output in liters/min equals the oxygen consumption divided by the difference between arterial and venous oxygen concentrations, so that if the oxygen consumption measured by a spirometer in a patient is 275 ml/minute, and by analysis the arterial blood oxygen concentration is 190 ml/liter and the mixed venous blood oxygen concentration is found to be 140 ml/liter the cardiac output is:

$$\text{C.O.} = \frac{275}{190 - 140} = \frac{275}{50} = 5.5 \text{ liters/minute}$$

The basic equation for the Fick method is:

$$F = \frac{Q_0}{A_0 - V_0}$$

*In one particularly interesting study, the blood flow of free-ranging giraffes was monitored successfully by radiotelemetry of the signals generated by an electromagnetic flowmeter. These animals, incidentally, are of particular significance because of the chronic "hypertension" normally present in their carotid arteries. These vessels, of course, must supply blood to the brain, which in turn is located several meters above the level of the heart in an adult specimen.

In this equation, F = the mean blood flow per unit time Q_0 = the oxygen consumed per minute; and $A_o - V_o$ = the arteriovenous concentration difference of oxygen per volume of blood. Therefore, when one knows the cardiac output as well as the heart rate (measured during the same interval as was the oxygen consumption), it is a simple matter to determine the average stroke volume. For example, if the cardiac output, as determined by the Fick technique, was 5,500 ml/minute and the heart rate was 85 beats/minute, the average stroke volume was about 65 ml.

The arterial blood oxygen concentration is of course constant throughout the body, so that a sample of blood can be obtained with a hypodermic needle (under anaerobic conditions) from any artery. There is a considerable difference in the venous oxygen content of the blood returning to the heart in veins coming from various organs and regions of the body, however. In order to obtain a truly representative mixed venous blood sample, cardiac catheterization is employed (p. 631), and the sample is taken anaerobically from the pulmonary artery. Right atrial, and even right ventricular, blood samples may not be completely mixed; hence, they are not considered to represent thoroughly mixed venous blood. Recall that in streamline flow, the laminae within the blood remain separate, so that mixing occurs only when turbulent flow is present, as discussed on page 626.

Obviously, the application of the Fick principle to humans will be in serious error if the subject is not in a "steady state," ie, if gas is being liberated or stored, such as occurs during a shift in the balance between aerobic and anaerobic metabolism, an effect that bears no relationship to blood flow per se.

Originally, the Fick principle was used indirectly in conjunction with rebreathing a foreign gas, such as N_2O or acetylene. The foreign gas was assumed to be in equilibrium with venous blood after several respiratory cycles; however, this indirect Fick method is not especially reliable because of inherent errors in this and the other assumptions upon which it is based. The indirect method will be considered in no further detail; however, it was mentioned because it represents an attempt to avoid the necessity of taking a mixed venous blood sample directly from the pulmonary artery and thus to avoid cardiac catheterization to obtain the sample.

The Indicator Dilution Method.   The indicator dilution principle is based upon the fact that a quantity of a substance, S, which is mixed uniformly throughout a volume of blood, V, has a concentration, C. Thus $S = (V) \times (C)$. By rearranging this equation so that $V = S/C$, the dilution method can be used to estimate quantitatively the various body fluid compartments (Chap. 9) by using suitable materials as indicators that equilibrate within the entire compartment under investigation. This relationship, which is related to the Fick principle, can also be used to measure the flow rate, because blood flow is equal to the volume of fluid passing a given point per unit time, or: $Q = V/t$. Thus, by substituting $S/C = V$ in the flow equation, one gets:

$$Q = \frac{S/C}{t}$$

In the Stewart–Hamilton single-injection method for the determination of cardiac output, a known quantity of an indicator such as Evans blue, ^{131}I-labeled serum albumin, or, more usually, indocyanine green is injected rapidly, the best results being obtained by injection of the material into the pulmonary artery. Usually, however, the dye is injected intravenously. The substance is mixed with, and is diluted by, the blood in passing through the lungs and left ventricle. The arterial blood concentration of the dye then is measured, and the cardiac output is calculated. As shown in Figure 10.54, when indocyanine green dye is used as an example of an indicator substance, when the arterial blood is pumped through a suitable photoelectric densitometer after injection of a known quantity of the dye into the venous circulation, a curve can be recorded. This curve relates the concentration of dye (in mg/ml) against time. The concentrations of dye in the blood at various fractions of a minute then are plotted semilogarithmically against the time in seconds, and the curve then is extrapolated to zero (first circulation of the dye), as shown in the illustration, since the dye will recirculate. Because of dye recirculation, the curve will not return to the baseline. The total area under the curve thus represents the concentration per unit time of dye for one passage through the left ventricle. The concentration/interval now can be estimated either by a planimeter or by adding up (ie, integrating) the areas found under individual 1-sec portions of the curve. The sum of these areas divided by the number of samples taken during 1 minute (60 sec), gives the concentration per minute of dye. Since the quantity of dye, S, injected initially is known, the cardiac output may be calculated from the expression:

$$Q = \frac{S}{\Sigma_0^\infty \, Cdt}$$

In this equation, Q = flow rate in milliliters/minute; S = the quantity of indicator injected; C = instantaneous concentration of indicator in the arterial blood.

Thus, if 5.0 mg of indocyanine green dye are injected into a subject at rest, and the flow in 40 sec averages 1.5 mg dye/liter:

$$Q \,(40 \text{ sec}) = \frac{5 \text{ mg}}{1.5 \text{ mg/liter}} = 3.3 \text{ liter in 40 sec}$$

Hence the cardiac output/minute is:

$$\text{C.O.} = Q \,(60 \text{ sec}) = 3.3 \times \frac{60}{40} = 5.0 \text{ liters/minute}$$

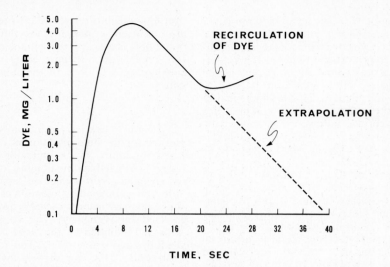

FIG. 10.54. Indicator dilution technique for the determination of cardiac output. At zero time *(arrow)*, a known quantity of dye is injected rapidly into a vein; the dye curve is obtained on serial arterial blood samples. Cardiac output equals the total quantity of indicator injected divided by its average concentration in arterial blood after a single passage through the heart. Hence, the curve must be extrapolated to zero at the point where an upswing (signifying recirculation) becomes evident. (After Ganong. *Review of Medical Physiology,* 6th ed, 1973, Lange Medical Publications.)

With an elevated cardiac output, the circulation time required for a complete curve to develop will be considerably less, as the same quantity of dye will be pumped by the heart in a much shorter interval of time.

It should be remembered that the cardiac output, as determined by either the direct Fick method or the indicator dilution technique, yields mean values which, when carefully performed, are accurate within about ±10 percent. There are several sources of error inherent in each method, and it should be emphasized that the values obtained by use of these techniques represent only the average values of cardiac output obtained over a period of time ranging from less than 1 to several minutes, and that the stroke volume may change considerably from beat to beat, even in a subject at rest, owing to variations in respiratory activity with attendant fluctuations in the intrathoracic pressure and neural tone, among other factors.

For practical reasons, the indicator dilution technique has broader clinical application than does the direct Fick method.

Capillary Hemodynamics

At any given moment, only 6 percent of the total circulating blood volume is found within the entire capillary bed. Yet in a physiologic sense, this relatively small volume of blood is the most important in the body because only in the capillaries does the vital and continual exchange of materials between the blood and interstitial fluid take place to any great extent.

Certain features of capillary hemodynamics and fluid exchange were introduced in Chapter 9, page 558, and in Figures 9.17, 9.18, and 9.20 (pp. 559, 560, and 575, respectively), in relation to the physiology of the body fluids. In this section, some of this material will be reviewed briefly and integrated with other functionally related components of the cardiovascular system, principally the arterioles and metarterioles.

Morphologically, as illustrated in Figure 10.19, the terminal arterioles give rise to transitional regions between the arteries and the true capillaries, which are known as metarterioles. Therefore, the parallel networks of true capillaries arise within each tissue and organ of the body from the metarterioles. Surrounding each capillary at its junction with the metarteriole is a single circumferentially arranged smooth muscle fiber known as the *precapillary sphincter.*

The terminal arterioles, which are only a few mm in length, have diameters ranging between 8 and 50 μ, and these vessels branch repeatedly, each arteriole giving rise to between 10 and 100 individual capillaries. The capillaries in turn are between 5 and 8 μ in diameter. The human body has roughly 10 billion capillaries which have, in the aggregate, more than 100 m^2 of surface area for the exchange of materials between the blood and interstitial fluid.

CAPILLARY FRAGILITY AND LAPLACE'S LAW. The capillary walls are extremely thin, around 1 μ; consequently, they are extremely fragile. Since the diameter of individual capillaries also is very small, however, their relative strength under physiologic conditions can be understood readily in terms of Laplace's law (p. 627). As the wall tension developed in an elastic tube is directly proportional to the transmural pressure times the diameter, in accordance with this law, the wall tension developed within the capillaries is exceedingly low. Therefore, capillaries are able to resist normal intraluminal blood pressure, since they have an approximate wall tension of only 15 dynes/cm, whereas in comparison, a large vessel such as the aorta can develop a wall tension of up to 200,000 dynes/cm.

CRITICAL CLOSING PRESSURE. An important physiologic effect of a markedly reduced blood pressure is to be found in the *critical closing pressure* that is associated with small blood vessels. In a system composed of rigid tubes, the relationship between pressure and flow of a uniform fluid is linear, as depicted in Figure 10.44A. As shown in Figure 10.44B, this relationship is nonlinear in blood vessels in vivo because of their property of distensibility. When the pressure falls in a small blood vessel, a level is reached ultimately at which there is no blood flow through the vessel, despite the fact that the pressure within the vessel is not zero (CCP, Fig. 10.44B). The critical closing pressure is partly a result of the necessity for a certain minimal pressure to be present in order that the erythrocytes be forced through the vessel, especially through capillaries having diameters smaller than erythrocytes (about 8 μ), and it is partly a result of the fact that the surrounding tissues and interstitial fluids themselves exert an external pressure directly upon the vessel. Even though this external pressure is quite small, when it exceeds the intravascular pressure, the vessel collapses. The intravascular pressure at which flow ceases defines the critical closing pressure. In inactive tissues many capillaries are collapsed because their intraluminal pressure is below the critical closing pressure because of constriction of metarterioles and precapillary sphincters.

CAPILLARY BLOOD FLOW AND VASOMOTION. Capillary blood flow is intermittent rather than continuous and steady; that is, the blood flows in more or less rhythmic surges through the capillary bed of any particular organ rather than in a regular stream. Underlying this pulsatile flow through the capillaries is the regular contraction and relaxation of the metarterioles and precapillary sphincters at an approximate rate of from 2 to 12 times per minute. This phenomenon, known as *vasomotion*, is not related to the regular pulsatile changes in arterial pressure that are attendant upon the systolic and diastolic arterial pressures developed by the heart. Rather, the underlying mechanism involved in vasomotion is the alternate constriction and relaxation of the circular smooth muscle fibers of the metarterioles and precapillary sphincters themselves. This smooth muscle in turn is stimulated to contract when the oxygen supply, and also presumably the metabolite concentration within a given tissue, is adequate, and when the blood pressure within a given capillary falls below the critical closing pressure, flow through that vessel ceases. As hypoxia develops in the surrounding tissue following the arrest of blood flow through a

particular capillary, the reduced oxygen concentration, and perhaps also the increased levels of tissue metabolites, exert a depressing effect upon the smooth muscle fibers which then relax producing vasodilatation. Thus, flow through the capillary is restored once again.

As tissue activity is increased (eg, in skeletal muscle during exercise), the frequency of this contraction–relaxation cycle is accelerated so that capillary blood flow is augmented markedly. The principal mechanism for the regulation of blood flow at the tissue and cellular level is the degree of constriction or dilatation of the metarterioles as well as the degree of tone exhibited by the precapillary sphincter muscle fibers. As pointed out above, the oxygen concentration of the tissue appears to be the primary stimulus governing this regulatory mechanism, although sympathetic nervous influences and various chemical agents may have some role in this process. The principal effects of the latter two factors, of course, are manifest most strongly in the regulation of flow at the level of the arterioles.

In any capillary bed, however, a certain proportion of these minute vessels do not exhibit closure under any circumstances. These thoroughfares or main channels always remain open and therefore convey blood continuously, whereas it is the so-called "nutrient," or "true" capillaries which exhibit vasomotion as described above. It is through the walls of the nutrient capillaries that the principal bulk transport of materials into and out of the interstitial space takes place.

Sometimes the changes in caliber observed in the capillaries merely reflect the passive alterations in diameter which occur in these vessels secondarily to changes in their transmural pressure.

Regardless of the fact that blood flow is intermittent in many capillaries, there is such an enormous number of these vessels present in most tissues* that their function on an organ level becomes averaged. One may speak of the net or average rates of blood flow, capillary pressure, and transfer of substances between blood and interstitial fluid (and vice versa). These concepts hold, regardless of the intermittent behavior exhibited by individual capillaries, so that in the remainder of this discussion as well as throughout the rest of this book, the average values for these aspects of capillary function will be considered.

RELATIONSHIPS AMONG CAPILLARY PRESSURE, RESISTANCE, AND VELOCITY. As shown in Figure 10.50, the pulsatile arterial pressure waves become progressively damped as they reach the arterioles; that is, the oscillations between systolic and diastolic pressures become less extreme and, consequently, the pulse pressure decreases. This damping of the pressure waves is a result of the combined effects of a greater vascular resistance and an increasing distensibility of the smaller ves-

Cartilage, bone, and teeth being notable exceptions to this statement.

sels. By the time the blood reaches the arterial end of the capillaries, there is a fluctuation of only 5 mm Hg (or less) in the blood pressure, and at the venous end there usually is none.*

The decrease in arterial pressure within any given portion of the cardiovascular system is directly proportional to the resistance, and according to Poiseuille's law, $\Delta P = (R) \times (F)$. The mean blood pressure falls very little within the aorta and larger arteries (100 mm Hg mean pressure). Furthermore, the pressure declines only 5 to 7 percent (to around 93 to 95 mm Hg) by the time the blood reaches the arteries as small as 3 mm in diameter, because the resistance in these larger vessels is quite low (Fig. 10.50). The resistance in the smaller arteries then increases rapidly, so that the mean blood pressure at the beginning of the arterioles now is only about 85 mm Hg. The resistance encountered by the blood within the arterioles is the greatest in any region in the systemic circulation. In fact, these vessels account for about half of the total resistance of the entire systemic circuit. Consequently, the blood pressure decreases about 55 mm Hg within the arterioles, and thus blood enters the capillary bed at a mean pressure that can range between 25 and 32 mm Hg. In turn, this value has fallen to about 10 mm Hg by the time the blood has traversed the capillary bed and has reached the venules. These facts illustrate the point that the capillary resistance is approximately 40 percent of that encountered within the arterioles.

Between the venules and the right atrium, the blood pressure falls from 10 mm Hg to zero.

As shown in Table 10.2, the total cross-sectional area of the capillaries is around 100 times greater than that of the aorta, or about 4,500 cm². The direct consequence of this fact is that capillary flow in an individual at rest is relatively slow, averaging between 0.07 and 0.3 mm/sec in accordance with Bernoulli's principle. The comparable flow velocity within the aorta thus is 1,000 times greater, or roughly 30 cm/sec.

Since capillaries are, on an average, only 0.3 to 1.0 mm in length, the transit time for blood from the arterial to the venous end of a typical capillary is only 1 to 3 seconds. When this brief interval is considered, it is interesting to imagine how rapidly the diffusion of gases, water, and metabolites must take place, not only independently but also simultaneously, and in certain special instances in an interdependent fashion as well.

EXCHANGE OF MATERIALS ACROSS THE CAPILLARY WALL.   The structure of capillaries found in various regions of the body was discussed earlier. It is necessary to preface this section on the exchange of materials across the capillary membrane with a brief review of certain morphologic details that are relevant to a consideration of transcapillary movement of various substances.

The individual endothelial cells comprising the capillary wall rarely abut in an edge-to-edge fashion (Fig. 9.17, p. 559). Rather, the intercellular junctions between the endothelial cells consist principally of irregular overlapping regions about 0.5 to 1.0 μ in extent, and the gap between the individual cells is around 90 to 150 Å. This space is generally filled with a mucopolysaccharide, and particulate matter (including leucocytes and lipid droplets) can pass readily through these intercellular gaps. This fact indicates the presence of *functional pores* which have been reported by various authors to range between 40 and 90 Å in diameter, even though openings, as such, may not always be visible by electron microscopy. These junctional regions between adjacent capillary endothelial cells doubtless also function to a certain extent in the transport of fluid and electrolytes across the capillary wall, although the basement membrane itself is considered by some investigators to constitute the "true" functional membrane of the entire capillary surface.

There are two mechanisms whereby transcapillary exchange of materials takes place. These mechanisms are *diffusion* and *filtration*. A third mechanism of doubtful significance in vivo is called *cytopempsis* (vesicular transport).

Capillary Diffusion and Permeability.   The most important mechanism whereby water and dissolved substances traverse the capillary wall from the plasma and the interstitial fluid is by the physical process of *diffusion* (see also Chap. 1, p. 23). The process of diffusion in turn is caused by the random motion of molecules that is induced by thermal agitation.

In mathematical terms, the rate of passive transport or diffusion of any substance in solution can be defined by the relationship:

$$\frac{dQ}{dt} = D_s A_s \frac{dc}{dx}$$

In this expression, the rate of diffusion of a specific quantity of a substance (Q) per unit time (t) having a diffusion coefficient (D_s) through a particular cross-sectional area (A_s) is directly proportional to the concentration gradient of the substance (dc/dx). Note particularly in this equation that the concentration gradient, dc/dx, is an extremely important factor in determining the rate of transport of specific materials by diffusion; hence, this factor is the major determinant of the rate of transport of substances across the capillary wall. Capillary permeability thus is determined by a combination of the diffusion rate, dQ/dt, as well as by the properties of the membrane itself with respect to specific substances.

Capillary permeability also can be defined as the extent to which a given substance can traverse the membrane in either direction. Since the capillary wall is essentially a passive barrier, the princi-

At times, some pulse pressure may persist into the ends of the venules adjacent to the capillaries, but this is a negligible factor insofar as the overall hemodynamic pattern of the individual is concerned.

pal factors which determine the transport of substances across this wall are their molecular size and shape as well as their lipid solubility. Thus, the membrane exhibits the property of *selective permeability* toward different materials (Table 10.7).

Substances having a molecular weight ranging up to about 5,000 daltons appear to cross the capillary membrane readily. An important exception to this general statement is plasma albumin, which has a molecular weight of around 70,000. This protein "leaks" from the capillary circulation at a rate of around 0.1 percent of the total circulating plasma protein per minute, as determined by radioactive tracer studies, so that in 24 hours the total quantity of albumin filtered from the blood is equal to the entire mass of circulating plasma protein (see also p. 526).

Water, which has a molecular weight of 18, is exchanged across the capillary membranes of the entire body in both directions at a fantastically high rate, and this value amounts to around 65 percent of the total blood volume per minute! Stated another way, water molecules can diffuse across the capillary membrane at a rate which is 40 times greater than the rate at which the blood plasma traverses the capillary in a linear direction. Since the net rates of diffusion of water out of and into the capillary are so nearly equivalent, the overall plasma volume remains almost precisely constant as the blood traverses the capillary from the arterial to the venous end.

The gases oxygen and nitrogen, as well as the noble gases in the atmosphere, are soluble in water to a limited extent, but very soluble in lipid. Because of their high oil/water solubility ratio (or *partition coefficient,* p. 88), these substances can move readily across the capillary membrane, as this structure essentially is a bimolecular lipid layer. Carbon dioxide, on the other hand, is relatively more soluble in water than in lipid; therefore, its partition coefficient is much lower. The rates of transcapillary passage of electrolytes such as Na^+ and Cl^- as well as crystalloids such as glucose and urea are in general intermediate between those for water and gases (Table 10.7).

It is important to understand that there is a considerable difference between the total surface area of a capillary and the net effective area that is available on the vessel surface through which a substance can actually pass. The latter, obviously, is a fraction of the former, so that the total area available for diffusion of a particular substance may be quite limited. Since the capillary barrier may be considered as a tube composed of a bimolecular layer of lipids, having small aqueous regions or "functional pores," it is not surprising that for all practical purposes substances that have a high partition coefficient can utilize essentially the entire surface area of the capillary for their transport, whereas other materials that have a low partition coefficient are transported only through the restricted areas formed by the aqueous "pores." This fact is indicated by the symbol A_s in the diffusion equation given above.

The physiologically important consequence of these facts is that such important substances as oxygen and carbon dioxide can utilize essentially the entire surface area of the capillaries; their rate of transport is far greater than that of materials that are limited to the aqueous pore regions. Various anesthetic gases and ethanol are further examples of highly lipid-soluble materials that also are transported rapidly through the capillary wall.

It should be remembered that even the slight physiologic concentration differences found across the capillary wall between the plasma and interstitial fluid are sufficient to account for a rapid and large net transport of materials down these gradients by the process of diffusion alone. The continuous movement of the body fluids per se naturally prevents a static condition of equilibrium from being reached, and this movement also assures the constant maintenance of these essential gradients.

Since the process of diffusion is critically dependent upon the concentration gradient as well as upon the total effective surface area that is available to a specific substance for transport, precapillary vasodilatation will increase the effective surface area by stretching the vessel wall as well as by increasing to some extent the permeability of the wall itself. Both of these factors will enhance

Table 10.7   RELATIVE PERMEABILITY OF THE MAMMALIAN CAPILLARY TO CERTAIN LIPID-INSOLUBLE MOLECULES[a]

SUBSTANCE	MOLECULAR WEIGHT (daltons)	DIFFUSION COEFFICIENT (cm² per sec)	MOLECULAR RADIUS (cm)	PERMEABILITY (cm³/sec/100 gm)
Water	18	3.4×10^{-5}	1.5×10^{-8}	3.7
NaCl	58	2.0×10^{-5}	2.3×10^{-8}	1.8
Urea	60	1.95×10^{-5}	2.6×10^{-8}	1.8
Glucose	180	0.91×10^{-5}	3.7×10^{-8}	0.64
Inulin	5,500	0.21×10^{-5}	12×10^{-8}	0.036
Serum albumen	69,000	0.85×10^{-5}	36×10^{-8}	0.000

[a]*Data from Selkurt (ed). Physiology, 2nd ed, 1966, Little, Brown and Co.*

the rate of exchange of materials across the capillary wall.

In conclusion to this discussion, reference should be made to the considerable morphologic differences that are found in capillaries in various specific regions of the body (p. 590). Since the development of fenestration and the basement membrane itself exhibit considerable differences among capillaries in various organs and tissues, it is hardly surprising that normal capillary permeability shows considerable variations in different regions of the body. For example, intestinal and liver capillaries are highly permeable to protein, whereas these vessels in skeletal muscle, heart, lung, and skin are relatively impermeable to proteinaceous materials.

Capillary Filtration.   The second principal method whereby transcapillary exchange of materials takes place is by capillary filtration. By this mechanism, the bulk transport of water, electrolytes, and crystalloids is accomplished. The filtration process is a hemodynamic event, and is dependent upon the hydrostatic pressure of the blood within the capillary at any point in the vessel. The filtration rate thus depends upon the filtration pressure, which in turn is equal to the hydrostatic pressure at any point in the capillary minus the hydrostatic pressure of the interstitial fluid at that point (see Fig. 9.18, p. 560).

The filtration pressure is opposed by the oncotic pressure within the capillary lumen, which results from the osmotic activity of the plasma proteins. An inward-directed osmotic gradient exists between the interstitial fluid and the lumen of the vessel across the capillary wall, which amounts to around 25 mm Hg.

Depending upon the technique employed for its measurement, the interstitial fluid pressure has been reported to range between −7 mm Hg and 0 mm Hg, or even between 0 and +10 mm Hg (cf p. 561). In any event, the interstitial fluid hydrostatic pressure is very low, perhaps even slightly below atmospheric pressure in certain instances.

The dynamic events that underlie capillary filtration were presented in detail in Chapter 9 (p. 558). It should be reiterated here that the volume of water filtered from the plasma almost, but not quite, equals that reabsorbed by the same capillary along its length. There is a very slight net gain in interstitial fluid volume, and this in turn is maintained in homeostatic balance by the lymphatic drainage of this bodily compartment, as discussed on page 553 et seq.

Cytopempsis and Pinocytosis.   A third mechanism of rather doubtful significance for the transport of large particles, and perhaps protein molecules, may be present within the endothelial cells of the capillary walls themselves. In contrast to the far more important processes of diffusion and filtration, *cytopempsis,* or *vesicular transport,* as it sometimes is called, involves the production of minute vesicles which form around the molecules at the luminal border of the capillary by invaginations of the plasma membrane. The material then is engulfed, and transported through the cytoplasm within the vesicle to be discharged on the interstitial surface of the capillary (Fig. 9.17).

Cytopempsis should not be confused with the phenomenon of *pinocytosis* (p. 94), whereby water or plasma is ingested and transported as discrete, membrane-limited vesicles through the cytoplasm of the cell to be discharged, together with any solutes present, on the opposite surface.

Although the mechanism of pinocytosis possibly may transport extremely small quantities of water, salts, and crystalloids across the capillary wall, the total amount of such low-molecular-weight substances moved by this mechanism is quantitatively negligible. In fact, these materials are actually transported at rates in excess of 10^7 times faster by diffusion and filtration than could be accounted for by pinocytosis alone.

CAPILLARY REACTIONS TO MECHANICAL AND HUMORAL INFLUENCES.   The endothelial cells of the capillary walls have the property of irritability, and thus have an inherent, although strictly limited, ability to change their shape. As mentioned earlier, the ability of capillaries as a whole to exhibit independent contractility is doubtful, and it is clear that these vessels, in the mammal at least, lack innervation as well as contractile precapillary cells; that is, the Rouget cells do not perform a vasoconstrictor function as once was believed.

Control of flow through individual capillaries thus is regulated primarily by the one or more smooth muscle fibers of the precapillary sphincter (Fig. 10.19), and at the tissue or organ level by the richly innervated and muscular terminal arteriolar elements. Blood flow may, for example, be shunted from the skeletal muscles to the splanchnic region and vice versa. Hence, constriction of the arterioles causes a diminution of capillary pressure and sometimes collapse of these vessels (p. 643), whereas conversely, arteriolar vasodilatation causes an increased intracapillary pressure. Similarly, an elevated venous pressure increases capillary pressure and also has the important concomitant effect of reducing the reabsorption of fluid, because the hydrostatic (filtration) pressure will approach the oncotic pressure within the capillary, thereby reducing the net reabsorptive pressure (Fig. 9.20A, p. 575).

Mechanical Stimulation and the Triple Response.   Certain gross observations made on the skin and its underlying vascular bed give some insight into the capillary reactions in normal and pathologic situations. If the skin of the forearm is stroked gently with a blunt instrument, a pale line appears several seconds after the stimulus has been applied. The pallor increases gradually for about a minute and then fades slowly during the ensuing several minutes. This *white reaction* is caused by a constriction of the capillaries which is induced directly by the mechanical stimulus.

On the other hand, if the skin is stroked firmly rather than gently, the triple response appears. First, a *red line* becomes evident after a latent period of only a few seconds. The extent and duration of this response depend upon the intensity of the stimulus, and the red line is caused by an active dilatation of the capillaries which is brought about by the release of histamine or other vasodilator substances attendant upon an injury sustained by the tissues surrounding the vessels. This red line effect also is independent of blood flow to, and innervation of, the vessels themselves. Second, a *red flare* develops in about half a minute (or perhaps longer) after the stimulus has been applied. This vasodilatation is caused by a local phenomenon known as the *"axon reflex."* That is, afferent impulses travel up pain nerve fibers in the skin in the normal centripetal (or *orthodromic*) direction until they reach local, peripheral branches of the same nerve; then they pass down the branches again *(antidromic conduction)* to the arterioles, where a vasodilator substance of unknown composition is released, and this substance in turn dilates the same vessels for up to several centimeters on each side of the line along which the stimulus was applied (Fig. 10.55). Hence, this diffuse arteriolar dilatation gives rise to the flushed appearance of the skin, which is called the flare. Obviously, the neural mechanism that produces the red flare is not a true reflex, as no central neural connections are involved (p. 177). Third, in individuals having particularly sensitive skin, a localized edema called the *wheal* develops subsequent to the formation of the red flare. Release of histamine, and probably other vasodilator substances, from the injured tissues causes such an intense local arteriolar vasodilatation with a concomitant constriction of the venules that a plasma-rich exudate escapes from the capillaries (whose permeability also increases) into the interstitial region. The resulting local edema is known as the wheal.

Other examples of this type of localized triple vascular response in the skin are to be found in mosquito or spider bites, as well as bee and scorpion stings. In such instances, the noxious organism injects a number of complex toxic humoral (chemical) substances which then act directly upon the blood vessels at the site of the attack, giving rise to any (or all) of the various phases of the triple response described above.

Humoral Stimulation. Mechanical stimulation usually produces its effects on the capillaries in an indirect fashion, primarily by releasing various chemicals which in turn produce the observed effects. Histamine (Fig. 10.56) is a very powerful dilator of the smaller vessels, and this compound exerts its effect directly upon the contractile cellular elements of the vascular endothelium, and independently of any innervation that may be present. This compound is released locally at the site of tissue injury, presumably by the mast cells, which also contain an abundance of heparin.*

In large (ie, experimental) doses, histamine causes the systemic blood pressure to fall, owing to the generalized arteriolar and capillary dilatation, which in turn produces a concomitant fall in the peripheral resistance. The inherent permeability of these vessels also increases; consequently, an attendant loss of water, plasma proteins, and even erythrocytes from the capillaries ensues.

Carbon dioxide also is a capillary vasodilator, and this compound appears to exert its effect on these vessels independently of the alterations in pH that it produces (p. 759). Lactic acid, produced during exercise (p. 137), is another physiologic humoral factor that has capillary vasodilator properties.

The prostaglandins (p. 1173) are a functionally and chemically diverse group of compounds which are derived from, and related to, certain fatty acids and phospholipids. These substances form another group of vasoactive substances which presumably have a vital (but at present poorly understood) role in normal and abnormal vascular function. The role of these compounds is not confined to the capillaries, but also includes other widely dissimilar regions of the cardiovascular system as well as the entire body itself.

The catecholamines epinephrine and norepinephrine are vasoconstrictors of the arterioles and capillaries of the skin and mucosa, and thereby these compounds produce blanching of the skin and give rise to the common expression, "turn white with fright." These compounds are

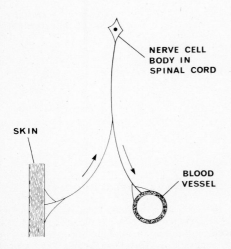

NERVE CELL
BODY IN
SPINAL CORD

SKIN

BLOOD
VESSEL

FIG. 10.55.   Pathway of the "axon reflex."

*There also are high concentrations of histamine in the pituitary gland as well as in nearby median eminence of the hypothalamus. In this region, the histamine is not synthesized by mast cells.

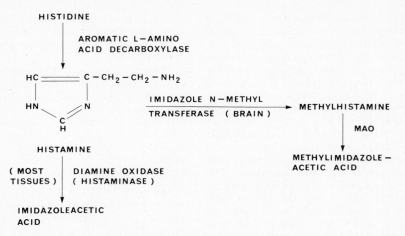

FIG. 10.56. Outline of the pathways involved in the synthesis and degradation of histamine. MAO = monoamine oxidase. (Data from Ganong. *Review of Medical Physiology,* 6th ed, 1973, Lange Medical Publications.)

not confined to producing vasoconstriction, however, as they also dilate the vessels within the muscles and other organs (Table 7.6, p. 276).

There also are a number of substances that influence the permeability of the capillaries and other vessels directly. Normally, of course, there is a considerable range of permeabilities to various substances encountered in the capillaries and other vessels that are located within various specific organs and tissues in the body, as noted on page 645. It also should be mentioned, however, that the normal range of permeability to specific substances varies little in the identical vessels of similar tissues.

Histamine, as mentioned above, increases capillary permeability. This compound, which is a normal constituent of most tissues as well as certain specific cells, is probably bound (or sequestered) by attachment to extracellular and intracellular proteins, so that in its storage form it is inactive physiologically. Histamine is liberated in free form (or activated) by appropriate chemical and physical stimuli. Among such chemical stimuli atropine, morphine, and curare may be mentioned.

Serotonin (5-hydroxytryptamine, Fig. 10.57) is another compound that exerts profound effects upon capillary permeability. This substance, found normally in the platelets of the blood as well as in mast cells, gives a response identical to that of histamine when injected intradermally. Thus, it is a powerful vasodilator.

Certain miscellaneous polypeptides, known as kinins, exert a marked influence on capillary permeability (Fig. 10.58). Leucotoxin, which can be isolated from exudates of inflamed tissues, promotes a localized migration of leucocytes from the capillaries into the interstitial space. Other kinins, produced by a partial digestion of plasma proteins, also augment capillary permeability.

A number of nonspecific nucleotides and nucleosides which are formed during the degradation of nucleoproteins also can produce alterations in capillary permeability. Although these substances have a probable role during inflammatory processes, their function in normal capillary activities is wholly unknown.

Certain enzymes, known generically as spreading factors, also contribute to an increased capillary permeability, since their normal sub-

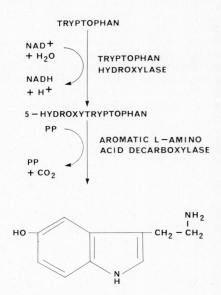

SEROTONIN (5 – HYDROXYTRYPTAMINE)

FIG. 10.57. The biosynthetic pathway for serotonin. NAD^+ = nicotinamide adenine dinucleotide; NADH = reduced nicotinamide adenine dinucleotide; PP = pyridoxal phosphate. (Data from Ganong. *Review of Medical Physiology,* 6th ed, 1973, Lange Medical Publications.)

A.

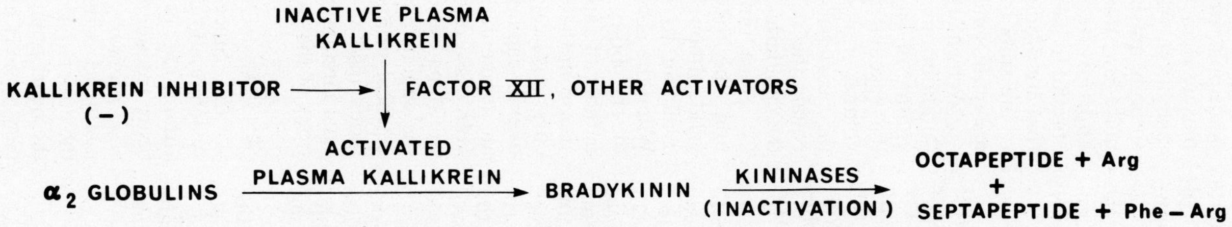

INACTIVE PLASMA
KALLIKREIN

KALLIKREIN INHIBITOR ────→ FACTOR XII , OTHER ACTIVATORS
(−)

ACTIVATED

α₂ GLOBULINS ── PLASMA KALLIKREIN ──→ BRADYKININ ──KININASES──→ OCTAPEPTIDE + Arg
                                                    (INACTIVATION)   +
                                                                          SEPTAPEPTIDE + Phe − Arg

B.

Arg − Pro − Pro − Gly − Phe − Ser − Pro − Phe − Arg

FIG. 10.58. A. Proteolytic formation and degradation of bradykinin. (−) = inhibition. **B.** Amino acid sequence in bradykinin. (See page 970 for abbreviations of amino acids.) (Data from Ganong, *Review of Medical Physiology*, 6th ed, 1973, Lange Medical Publications.)

strates are chondroitin sulfate and hyaluronic acid, both of which substances are components of the matrix which forms the ground substance between cells. Hyaluronidase, for example, by its enzymatic action depolymerizes the mucopolysaccharides of the ground substance, thereby lowering the viscosity of this material so that other substances can diffuse more readily through intercellular junctions. By this mechanism the injection of hyaluronidase produces a transient edema following its intravenous injection.

The relationship between capillaries and the development of various types of edema was considered earlier (p. 576, et seq; Fig. 9.20, p. 575).

Venous Hemodynamics

The venous portion of the systemic circulation is not merely a low-pressure system of converging passive conduits which return blood from the capillaries to the right atrium of the heart. In actuality, the venous circulation also provides an important blood reservoir, which has a large and variable capacity, and which continuously supplies the right side of the heart with relatively deoxygenated blood.

RELATIONSHIPS BETWEEN PRESSURE, RESISTANCE, AND FLOW IN THE VEINS. The mean pressure in the venules of a relaxed human subject *lying supine* is around 10 mm Hg (13.0 cm H_2O), and this pressure decreases progressively to about 0.04 mm Hg (0.5 cm H_2O*) within the great veins at their junction with the right atrium (Fig. 10.50). This pressure head, or vis a tergo, which drives the blood toward the heart, results from the residual arterial pressure at the venous end of the capillaries, and it can be modified profoundly by a number of factors, each of which will be discussed below.

Also as illustrated in Figure 10.50, the velocity of blood flow increases in the venous circulation in accordance with the Bernoulli principle, because of the convergence of these vessels at successively higher levels. This convergence, in turn, progressively and effectively decreases the total cross-sectional area on the veins at levels closer to the heart (Table 10.2).

Since the veins provide low-resistance channels to blood flow when they are distended, the net pressure drop from the venules to the venae cava is relatively small, amounting to only a few cm of water.

As mentioned above, the veins function as a large and variable-capacity blood reservoir which operates at low pressure because of the ease with which these vessels can be distended (p. 592). Normally, the veins are not completely distended with blood, so that they are partially collapsed, hence, flattened or ovoid in cross-sectional outline in a supine individual. And since the walls of these

vessels are thin, elastic, and pliant, they expand readily with very small increments of internal pressure (p. 627). A vein that is empty, or nearly so, can readily increase its volume capacity sixfold with an intraluminal pressure rise of only 1 mm Hg. Beyond this pressure level, however, the increase in vessel capacity decreases sharply with additional pressure increments, as the smooth muscle elements in the vessel walls oppose further distention. Although the forces developed by the smooth muscle present in the venous walls are quite low, these muscular elements can respond to appropriate neural, chemical, and hormonal stimuli to some extent. Any resistance changes that are produced in the veins through constriction of the smooth muscle fibers present in their walls are quite small, however.

When the cardiovascular system is in "equilibrium" (eg, in a resting individual) the volume of blood that is present in each region of the vascular tree is approximately constant. The volume of blood that is returned to the right side of the heart per unit time by the venous circulation equals the volume of blood that is received from the capillaries during the same interval of time.

As a broad concept regarding venous function, the following generalization appears to be valid, although no direct experimental evidence is available to confirm this statement at present. When the net volume capacity of the capillary bed of the body increases because an increased number of capillaries are active, a concomitant decrease in the capacity of the venous reservoir occurs passively. Conversely, a decrease in the overall capillary volume capacity would result in an augmented capacity within the venous reservoir. As a consequence of these general interrelationships between venous and capillary networks, the venous circulation would appear to have a reciprocal relationship with the arteries and capillaries insofar as the regional distribution of the total circulating blood volume within various portions of the cardiovascular system is concerned.

CENTRAL VENOUS AND RIGHT ATRIAL PRESSURES. Since blood from all the veins in the systemic and coronary vessels flows into the right atrium, the pressure within this chamber of the heart frequently is referred to as the *central venous pressure*. The right atrial pressure in the normal heart is the resultant of two major factors: (1) the effective energy with which the left ventricle of the heart pumps the blood into systemic and coronary vessels and (2) the tendency of the blood to pass from the peripheral vessels toward the heart down a continuous pressure gradient and the various factors which act on this property (Fig. 10.50).

Right atrial pressure normally is about 0 mm Hg, with respect to the atmospheric pressure surrounding the body, and a marked and prolonged elevation of the central venous pressure is indicative of a serious pathologic situation, such as is present in right heart failure (p. 1248), when the

---

*For all practical purposes, the mean venous pressure at the right atrium is 0 mm Hg.

myocardium is unable to pump effectively the blood which is returned to it by the venous circulation. On the other hand, right atrial pressure may fall to as low as −5 mm Hg below atmospheric. This negative pressure is similar to that encountered normally within the pericardial and intrapleural spaces surrounding the heart, and such low pressures are encountered when the venous return is inadequate (eg, during severe hemorrhage), or when the heart is pumping (and thus ejecting) the blood returned to it with extreme efficiency.

The right atrial pressure fluctuates continuously throughout the cardiac cycle, as shown in Figure 10.38, thereby giving rise to the three pressure waves, which are reflected backward to some extent in the great veins. This retrograde phenomenon is the origin of the *central venous pulse,* which sometimes can be observed in the jugular vein in the neck of a recumbent individual.* Thus, the central venous pulse waves are caused by the rhythmic activity of the right atrium and ventricle and are coupled to the attendant cyclic movement of the blood within the great veins.

FACTORS REGULATING VENOUS RETURN TO THE HEART. There are several interrelated factors that determine venous blood pressure, hence the return (volume flow) of blood, to the right heart. Although at present no techniques are available for the quantitative assessment of the contribution of each of these factors to the overall venous return under different physiologic conditions, it is clear that each of these factors plays a definite, although variable, role in maintaining a continuous and adequate blood supply to the right heart.

Vis a Tergo. The vis a tergo is a driving force that is derived ultimately from left ventricular contraction, and it represents the residual pressure that is present within the venules after the blood has traversed the arteries, arterioles, and capillary bed. The pressure within the venules only amounts to around 10 mm Hg; however, when the blood reaches the right atrium this pressure energy has been dissipated almost completely in overcoming the slight resistance offered by the veins. Thus, the mean right atrial pressure is approximately 0 mm Hg.

It was stated earlier that the veins offer little resistance to blood flow when they are distended, ie, in a person lying in the supine position. Actually, however, when an individual is sitting or standing, many of the large veins in the upper portions of the body are collapsed and thus offer a considerably greater resistance to blood flow toward the heart. Within the chest, however, the veins remain distended at all times because of the negative intrathoracic pressure within this region of the body (p. 716).

The Siphon and Hydrostatic Effects. In an uninterrupted column of fluid, such as is present in the venous circulation, the flow rate from one point to another depends upon the pressure difference (ΔP) between the two points, and not upon the levels of the vessel connecting the two points.

In the highly elastic cardiovascular system, when an individual rises from a supine position to a sitting or standing posture, a quantity of blood tends to distend the veins in the regions of the body below the heart (Fig. 10.48), thereby temporarily decreasing venous return to the right atrium. Once the dependent veins are filled under the increased pressure that develops peripherally, an adequate venous return to the heart is restored. As one arises from the supine position, a negative pressure develops within the veins of the head region above the heart level, and if the cardiac output is inadequate because of a decreased venous return to the right heart, the cerebral blood flow falls, and either dizziness or *syncope* (fainting) ensues. The operation of the so-called muscle pump is a critical factor in preventing this hypotensive effect.

Once a steady-state has been achieved insofar as venous return to the heart is concerned in a standing individual, and the dependent blood vessels have been distended by an adequate blood volume, the siphon effect is in operation so that the blood in effect does not have to be "pumped uphill" from the lower extremities; ie, the fluid column is continuous from the left ventricle to the right ventricle, and since both of these chambers are essentially on the same level, little pressure energy is required in moving blood from the foot to the right atrium.

Because of the effects of gravity acting on the blood, the pressure within the veins increases some 1 mm Hg (1.3 cm H_2O) for each 1.36-cm distance below the level of the heart (Fig. 10.48). Since this pressure results from the force of gravity acting upon the mass of blood, the pressure that is developed is known as the hydrostatic pressure and the magnitude of this factor in turn is dependent on the factor ρgh as discussed on page 624.

Within the human cardiovascular system, the hydrostatic pressure factor is especially pronounced in the feet of an individual who is standing quietly in an erect position. A static pressure of about +80 mm Hg develops in the veins of the feet of an average person, compared to a pressure of 0 mm Hg within the right atrium of the same person, merely because of the vertical distance from feet to heart (Fig. 10.48). The actual height of the individual would, of course, cause some variation in the quantitative value of the hydrostatic pressure in the feet.

Within the veins of the arms, the intravascular pressure at the level of the top rib is approximately +6 mm Hg, because the subclavian vein is com-

---

The central venous pulse should not be confused with the minute residual pulsations that sometimes occur in the venules and are the result of transmission of arteriolar pulse waves through the capillary bed.

pressed as it passes over this rib. If the pressure within the veins of the hand is 30 mm Hg because of a relatively low position of the arm, the total hydrostatic pressure within the veins of the hand is $(+6) + (+30) = +36$ mm Hg.

The large veins within the neck are almost completely collapsed when an individual assumes the upright position because of the atmospheric pressure exerted upon the skin of the neck. Therefore, along the entire length of these veins the net intravenous pressure is zero.

Within the skull, which is a rigid structure, the veins do not collapse. Negative pressures can develop within the dural sinuses when a person is in a standing position. For example, within the sagittal sinus, the venous pressure can be calculated to be around -10 mm Hg because of the hydrostatic difference that is present between the top and the base of the skull. If the sagittal sinus is opened during surgery, air may be aspirated into the bloodstream, and serious aeroembolism, with death of the patient, may ensue.

The venous pressure within the legs can be altered directly through changes in the intraabdominal pressure. Normally, the pressure within the peritoneal cavity is around $+2$ mm Hg. This intraabdominal pressure can be elevated greatly, for example, during pregnancy or by the presence of large abdominal tumors. In such situations, the pressure within the femoral veins now must exceed that within the abdominal cavity before the intraabdominal veins can open and blood can return from the lower extremities to the right heart.

It is important to realize that the hydrostatic factor also is operative on the blood within the arteries; therefore, if the mean arterial blood pressure is measured within one foot of a standing individual, this pressure will be higher by around $+80$ mm Hg than the mean blood pressure measured at heart level (Fig. 10.48).

The large veins within the thoracic cavity normally remain distended with blood at all times, since the intrapleural pressures are always negative with respect to atmospheric pressure. During respiration (inspiration and expiration), however, the fluctuating intrapleural pressures developed by the movements of the chest wall directly produce rhythmic fluctuations in venous return to the right side of the heart, as described below.

The Skeletal Muscle Pump.   An extremely important mechanism for assuring a continuous and adequate venous return to the right side of the heart is the so-called *venous* or *muscle pump*. Because of the hydrostatic factor (ρgh), the venous pressure in the feet of a person standing quietly would be around $+80$ mm Hg. The valves within the veins are disposed so that blood can flow readily only toward the heart, however; thus, when the skeletal muscles contract, the veins within the contracting muscles are compressed so that a quantity of blood is "milked" out of these veins and thus that blood is propelled toward the right

heart. Hence, the blood contained within the veins of active muscles tends to move away from the areas of compression, and because of the valves this milking effect of the muscles promotes venous return. Consequently, the pressure within the veins of the feet in an adult who is walking only amounts to around 25 mm Hg.

If, however, a person stands quietly so that the venous pump becomes inoperative, the venous pressures within the lower legs can rise to the full hydrostatic value of 80 mm Hg in a few seconds. Under such circumstances, the intracapillary blood pressure is elevated sharply and fluid leaks into the interstitial spaces, thereby producing a local edema or swelling of the feet (cf p. 576). Furthermore, if the venous return becomes so inadequate that cardiac output falls markedly, thereby decreasing the cerebral blood supply, syncope ensues when sufficient fluid and blood have pooled within the lower extremities.*

Varicose veins result ultimately from pathologic abnormalities of the veins such that they no longer function properly. Pregnancy and prolonged standing, for example, stretch the peripheral veins, thereby increasing their cross-sectional area; however, the valves do not increase in size simultaneously with the increased area of the veins. Venous return eventually becomes inadequate and stretches the veins still further so that backward flow can take place through the distorted valves, and ultimately the valves are destroyed with the consequence that the superficial veins bulge and become knotted or varicose. Edema develops in the legs of such individuals even after short intervals of standing; however, there is some venous return to the right heart as the operation of the skeletal muscle pump causes the blood to flow along the path of least resistance, which is toward the heart.

Effects of Respiration upon Venous Return: The Thoracic Pump.   The central venous pressure and blood flow in the great veins both are markedly affected by natural and artificial respiration. During inspiration, blood flow increases in both the superior and inferior venae cavae, and this effect in turn is transmitted to the peripheral veins. Conversely, the blood flow decreases during expiration (Fig. 10.59). The mechanisms that underlie these effects are as follows. Central venous pressure at the entrance to the right atrium is very low but finite, because of the residual vis a tergo from the arterial circulation. This pressure amounts to 0.04 mm Hg (0.5 cm H_2O). During the inspiratory phase of the respiratory cycle, the intrathoracic pressure falls to around -6 mm Hg (8.1 cm H_2O) with respect to atmospheric pressure, whereas during expiration, the pressure within the thorax rises to

*Venous return, is, of course, a prime determinant of cardiac output, and thus a normal (nonfailing) heart is able to pump out all the blood returned to it over a period of time.

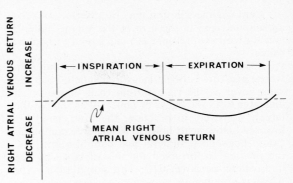

FIG. 10.59. The effect of respiration upon venous return to the right atrium.

around -2.5 mm Hg (3.4 cm H_2O). Therefore, during inspiration, the transmural pressure of the great veins in the thoracic cavity is the algebraic sum of the intravenous pressure and the intrathoracic pressure, or $(-6$ mm Hg$) - (+0.5$ mm Hg$) = -5.5$ mm Hg. Similarly, during expiration the pressures are: $(-2.5$ mm Hg$) - (+0.5$ mm Hg$) = -2.0$ mm Hg. These negative transmural venous pressures tend to expand the intrathoracic structures and also permit blood to flow readily into the thorax, and this occurs to a relatively greater extent during inspiration than during expiration.

Furthermore, during inspiration, the diaphragm descends into the abdominal cavity somewhat, thereby increasing the intraabdominal pressure. This factor tends to steepen the pressure gradient that develops during inspiration between the intrathoracic and intraabdominal vessels, and thus also facilitates venous return to the heart.

The effect of inspiration described above is exerted principally upon the thin-walled intrathoracic veins, to a lesser extent on the atria, and to a negligible degree on the relatively rigid, thick-walled ventricles and arteries.

Venous return to the heart is decreased markedly when the so-called *Valsalva maneuver* is performed. If a forced expiration is carried out with a closed glottis, for example, as during labor (childbirth) or when straining at stool, the increased intrathoracic pressure that is developed compresses the veins in this region, thereby diminishing venous return considerably.

The converse of the Valsalva effect, called *Müller's experiment,* is a forced inspiration against a closed glottis. In this situation, the highly negative intrathoracic pressure that is developed exerts a "suction effect," so that the pressure within a peripheral vein may fall by as much as 50 cm H_2O.

Total Circulating Blood Volume. If the total circulating mass of blood is increased (as by transfusion), the venous return may be augmented somewhat.* The transfusion of a quantity of fluid into

In human subjects, the spleen is not an important blood reservoir as it is in some mammalian species, such as the dog.

the cardiovascular system does not necessarily assure that such an effect will be observed, however, as the additional volume of fluid may serve merely to distend any partially collapsed vessels within specific regions of the venous portion of the systemic circulation. Therefore, a net increase in the central venous pressure may not be observed.

In this regard, it is important to remember that alterations in venous pressure and flow within one region of the venous circulation do not necessarily reflect similar changes in other regions of the body. In fact, there may be a considerable and wholly independent change in central venous pressure without a similar concomitant change in peripheral venous pressures elsewhere in the body. It is very difficult to predict with any degree of accuracy, from the effects of altering venous pressure in one region of the body, what is happening in another region.

By way of summary, the net venous return to the right side of the heart in a normal individual is the resultant of the interplay of the several factors discussed above, and the force of right ventricular contraction in turn is balanced delicately with the inflow of blood from the great veins to the right atrium; thus, no net accumulation of blood takes place in this chamber. If, however, the heart fails, blood gradually accumulates within the veins, and the back pressure caused by the dammed-up blood causes a rise in venous pressure once these vessels are distended to their full capacity. Stated another way, the nonfailing heart is able, over a period of time (rather than on a systole-to-systole basis), to pump out all the blood that is returned to it by the veins; therefore, no stasis or accumulation of blood occurs anywhere within the cardiovascular system.

THE MEASUREMENT OF VENOUS PRESSURE. The *central venous pressure* can be measured only by direct cannulation of the great veins within the thorax, as for example, during diagnostic intracardiac catheterization of the right heart (p. 632).

Clinically, however, the peripheral venous pressure gives a reasonably good indication of central venous pressure in most instances; therefore, a needle connected to a manometer containing sterile saline solution may be inserted into the antecubital vein in the arm of the supine patient, and the venous pressure is read directly in cm of saline.

From a theoretical point of view, the vein whose pressure is being measured should be at the same level as the right atrium (half of the chest diameter from the back, or about 10 cm above the table); however, in practice, the arm usually is lowered below heart level to assure free communication between peripheral and central veins, and the reading is corrected for this distance by the hydrostatic factor (p. 624). The quantitative value which is obtained in cm saline can be converted to mm Hg by dividing by the density of mercury, or 13.6.

Peripheral venous pressure increases with the distance from the right atrium, hence, the point of measurement; the average corrected pressure in the antecubital vein is about 7.0 mm Hg compared with 4.5 mm Hg within the central veins.

The central venous pressure may also be estimated readily with some accuracy in the supine patient whose head is elevated slightly above the heart by noting the point above heart level at which the external jugular vein collapses. The vertical distance in cm above the right atrium at which this collapse occurs (ie, the point at which the pressure within the vein falls to zero) is an indication of the venous pressure. Similarly, one can also note the distance above heart level at which the superficial veins on the back of the hands collapse.

The central venous pressure is decreased in shock and during negative pressure artificial respiration, but increased during straining, positive pressure respiration, failure of the right heart, and increased total circulating blood volume. Thus, in advanced congestive failure of the right heart, the pressure within the antecubital vein may rise to levels above 20 mm Hg.

HEMODYNAMICS OF THE PULMONARY CIRCULATION

The pulmonary circuit differs anatomically as well as functionally from the systemic circulation in several important respects. Therefore, certain features of the pulmonary circulation will be contrasted with those in the systemic circulation. This material will also serve as an introduction to the physiologic mechanisms involved in respiration to be presented in Chapter 11.

Anatomic Considerations

The pulmonary vascular bed has three principal morphologic differences when compared to the systemic circulation: (1) The walls of the pulmonary artery and its principal branches are approximately one-third the thickness of the aortic wall. (2) The small pulmonary arteries, in contrast to the arterioles of the systemic circulation, are com-

posed principally of endothelium and have very little smooth muscle in their walls. (3) The pulmonary capillaries are relatively large and have numerous anastomoses. Therefore, each alveolus in the lungs is surrounded by a capillary network.

Functional Considerations

The six principal functional differences between the pulmonary and systemic circulations are as follows: (1) The pulmonary circuit is confined entirely within the thoracic cavity, and thus is subjected to continuously fluctuating, but always negative, intrathoracic pressures. (2) The pulmonary circulation, which occupies a unique anatomic location between the left and right sides of the heart, is the only vascular circuit in the body to receive the entire cardiac output. (3) The perfusion pressure within the pulmonary circulation amounts to approximately 13 percent of that of the systemic circuit. Consequently, the pulmonary circulation is a low-pressure (as well as low-resistance) system.* (4) The average transit time for blood passing through the pulmonary circulation is equal to that of the systemic circuit; however, the linear distance traversed by the blood in the pulmonary vessels is considerably less, because of the huge surface area of the capillary bed within the lungs. (5) Pulmonary blood flow is controlled primarily by the systole-to-systole pressure differential between the right ventricle (ie, the pulmonary arterial pressure) and the left atrium. Upon this pressure difference are superimposed variations in flow which are induced by the rhythmically fluctuating variations of intrathoracic pressure during respiration. (6) The pulmonary arterial system is unique in that it carries reduced hemoglobin. Similarly, the pulmonary veins also are unique in that they contain oxygenated blood. Furthermore, because of the difference in gas pressures between the alveoli of the lungs and the capillaries, together with the tremendous surface area of these vessels, instantaneous transfer of large volumes of gases takes place between the alveoli of the lungs and blood within the pulmonary capillaries, as discussed in Chapter 11.

Relationships Between Pressure, Volume, Flow, and Resistance in the Pulmonary Circulation

The entire pulmonary vascular bed should be viewed conceptually as a highly elastic (distensible) low-pressure circuit. Pulmonary arterial blood pressure normally is around 25 mm Hg systolic, 10 mm Hg diastolic, and the mean pressure is approximately 15 mm Hg. Left atrial diastolic pressure has a mean value of 5 mm Hg; therefore, the pressure gradient within the pulmonary circulation

*This statement does not hold true for the fetus or newborn infant.

amounts to only 10 mm Hg, compared to an overall gradient within the systemic circulation of about 100 mm Hg (Fig. 10.60).

The volume output of blood from the right ventricle per minute must, of course, equal that of the left ventricle (approximately 5.0 liters/minute at rest). The total volume of blood in the pulmonary bed at any moment amounts to about 0.5 liter or roughly 9 percent of the total blood volume (Table 10.1), and of this total volume present in the pulmonary vessels, less than 100 ml is in the capillaries of the lungs. As indicated in Figure 10.38, blood flow within the pulmonary circuit is highly phasic. In the main pulmonary artery, flow occurs almost entirely during right ventricular systole, and the diastolic flow is nearly zero. There is little or no diastolic runoff into the pulmonary capillaries; blood flow in these vessels is pulsatile and apparently intermittent, as at the end of diastole the pressures within the pulmonary vascular bed reach equilibrium for a brief interval.

The mean velocity of blood flow within the root of the pulmonary artery is similar to that in the aorta, or roughly 40 cm per second. This flow rate declines rapidly within the pulmonary vascular bed, and then rises again in the pulmonary veins. Hence, the velocity of the blood in the lung vessels is such that it takes an erythrocyte approximately 0.75 sec to traverse a pulmonary capillary during rest and less than 0.30 sec during exercise.

According to Poiseuille's law, the resistance to flow in the pulmonary circulation should be proportional to the pressure difference divided by the flow (ie, $R = \Delta P/F$). The resistance to flow under various physiologic conditions thus should provide information concerning the mechanisms which regulate pulmonary flow. The relationship between pressure and flow within the pulmonary circuit is extremely difficult to measure, however. Furthermore, because of the high degree of distensibility in this vascular circuit, the pulmonary circulation is considerably different quantitatively from the systemic circulation. Only rough approximations for pulmonary vascular resistance are available under various physiologic conditions.

It is clear at present from experimental studies, however, that the pulmonary vessels can dilate readily and passively as a consequence of increased blood flow, thereby decreasing the resistance factor considerably. It also is clear that the pulmonary vessels in general (cf. p. 712). Furtherresistance to blood flow than do comparable systemic vessels, because of the large caliber of the pulmonary vessels in general (cf p. 712). Furthermore, a much lower vascular tone is developed in the pulmonary arterioles because of the absence of a highly developed smooth muscle coat in the walls of these vessels such as is found in systemic arterioles. If the blood supply to one lung is interrupted by occlusion of a major branch of the pulmonary artery, the flow through the other lung is doubled with only a very slight increase in pulmonary arterial pressure.

Because of the relatively large caliber of all vessels within the pulmonary vascular bed, pressure changes within the pulmonary veins are readily transmitted backward (ie, in a retrograde direction) to the arterial side of the pulmonary circuit. Increased total circulating blood volume can augment lung volume significantly. This fact has considerable clinical importance. For example, in left heart failure the augmented intraventricular pres-

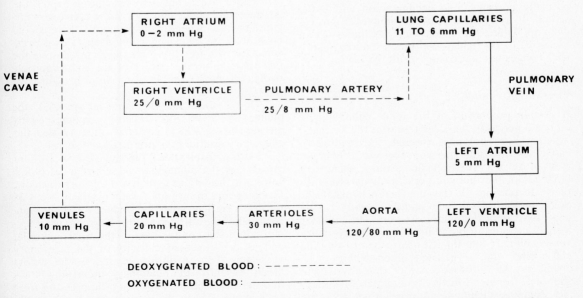

FIG. 10.60. Intravascular pressures in the pulmonary and systemic circuits. (Data from Comroe. *Physiology of Respiration,* 1965, Year Book Medical Publications.)

sure in the left heart is reflected backward into the pulmonary vascular bed, thereby inducing massive congestion of blood, and edema develops readily within the lungs.

The pulmonary vessels are innervated liberally with sympathetic vasoconstrictor fibers; therefore, if the cervical sympathetic ganglia are stimulated experimentally, there is an attendant decrease in pulmonary blood flow amounting to as much as 30 percent. The net capacity of the pulmonary blood reservoir can be decreased significantly by operation of this neural mechanism.

Hypoxia increases pulmonary arterial pressure, but this pressure is affected to an insignificant extent by certain stimuli which markedly influence the systemic arterial pressure. For example, during strenuous exercise, the pulmonary blood flow increases to the same extent as does the systemic blood flow up to sixfold or sevenfold; however, the pressure within the pulmonary artery increases to an insignificant extent because of the ease with which the pulmonary vessels can dilate to accommodate the increased blood volume.

It is not at all clear, however, the extent to which the vasoconstrictor and other mechanisms participate in the regulation of normal pulmonary blood flow in vivo under various physiologic states.

Pulmonary Capillary Pressure

The mean hydrostatic pressure within the pulmonary capillaries averages around 10 mm Hg, whereas the plasma oncotic pressure is about 25 mm Hg. Consequently, there is a pressure gradient of approximately 15 mm Hg directed toward the lumen of the pulmonary capillaries and which normally prevents fluid accumulation within the alveoli. If hydrostatic pressures above 25 mm Hg develop within the pulmonary capillaries as in "left-sided" or "backward" heart failure, pulmonary congestion and edema result (p. 1250).

Physiologic Shunts

A certain proportion of the total blood volume (about 2 percent), which is found in the systemic arteries, does not pass through the pulmonary capillaries, and thus has not received its complement of oxygen. The bronchial arteries, which are branches of the thoracic aorta, provide blood directly to portions of the lung parenchyma itself. Some of this blood then returns directly to the heart by way of the pulmonary veins. In addition, a small quantity of blood flows from the coronary vessels directly into the left heart, thereby diluting still further the oxygenated blood within these chambers. As a consequence of this physiologic shunting, the blood in the systemic arteries actually contains slightly less oxygen than does the blood which has been equilibrated with air in the alveoli of the lungs.

The Pulmonary System as a Blood Reservoir

The pulmonary veins, in a similar manner to the systemic veins, provide an important reservoir of blood because of the ease with which they can be distended. The total blood volume within the pulmonary circuit can increase up to 40 percent when a person lies down, whereas this blood is diverted into the systemic circulation when the standing position is assumed. This redistribution of blood in the supine (recumbent) position is the cause of the decrease in vital capacity that occurs when a person is lying down (p. 726). Similarly, this pulmonary engorgement is also responsible for producing the difficulty in breathing that is experienced in the supine position (orthopnea) in persons suffering from congestive heart failure.

Cyclic Respiratory Effects upon Pulmonary and Systemic Blood Pressures and Flows

As described on page 653, during inspiration there is an augmented influx of blood into the right atrium and ventricle. Paradoxically, the pulmonary arterial pressure falls during this phase of the respiratory cycle but the pulmonary arterial flow increases during inspiration. The fall in pulmonary arterial pressure during inspiration is caused by the increased circumference of the pulmonary vessels, which enlarges their capacity during this phase of the respiratory cycle, and this effect in turn is produced by traction of the lung tissue surrounding these vessels as they expand. This effect more than compensates for the absence of a change in the pressure gradient, as well as for the increased volume of blood entering the pulmonary circulation during inspiration; hence, the pressure within the pulmonary artery falls.

These effects are reversed during expiration. As the total volume capacity of the pulmonary circuit decreases when the lungs empty, there is concomitantly an increased pulmonary arterial pressure, and this pressure increase occurs despite the fact that the stroke output of the right ventricle decreases upon expiration, as described on page 652.

During the Valsalva maneuver (p. 653), the pulmonary vessels are forcefully compressed by the lung tissue which surrounds them so that the pulmonary arterial pressure increases sharply.

During inspiration, the augmented volume capacity of the pulmonary vessels reduces blood flow into the left atrium in a transitory fashion. This response in turn lowers the systolic output of the left ventricle, and thus induces a slight transient fall in the aortic pressure. Subsequent to several additional contractions of the right ventricle, and as the increased capacity of the pulmonary vessels is reduced during expiration, the excess blood present in the pulmonary circuit now flows into the left ventricle and the aortic pressure rises once more. Thus, the aortic pressure normally waxes and wanes according to the phase of the respiratory cycle. The fluctuations seen on systemic arterial blood pressure tracings recorded directly (eg, by catheterization of the aorta or of another large systemic artery close to the heart) are caused by the physiologic mechanism just described. If the respiratory motions are recorded simultaneously with the systemic blood pressure, it will be seen that the blood pressure commences to fall at the onset of inspiration and reaches its minimum in the latter half of this respiratory phase. Subsequently, the blood pressure commences to increase, and it reaches a maximum level toward the end of expiration.

MECHANISMS OF CARDIOVASCULAR REGULATION

There are a number of important control mechanisms within the body for the continuous and precise regulation and integration of the functional processes of the cardiovascular system (Fig. 10.61). The heart and blood vessels themselves possess a considerable degree of autonomy insofar as their functions and regulation thereof are concerned, but superimposed upon these more or less localized controls are more generalized nervous and humoral control systems which serve to coordinate and integrate still further the relatively independent local mechanisms. The cardiovascular regulatory mechanisms can be classified into two broad areas: first, those mechanisms that are inherent (or intrinsic) to the heart and blood vessels themselves, and second, the more generalized systemic control mechanisms, whose influence is superimposed upon the inherent mechanisms to various degrees under different circumstances. This

second category includes the complex of neural and hormonal regulatory systems.

The delicate and coordinated interplay among all these intrinsic as well as extrinsic functional systems thus maintains the homeostatic level of cardiovascular function in accordance with bodily requirements under a wide variety of circumstances.

The majority of cardiovascular adaptations within the body are met by alterations in the total blood flow throughout the circulatory system rather than by changing individual patterns of flow in such a way as to redistribute the blood in accordance with different functional requirements imposed by specific organs or systems. The total blood flow in turn is a reflection of the cardiac output. This latter term denotes the quantity (or volume) of blood which is ejected per minute by either the right or the left ventricle alone and not to their combined volumes. Normally, the flow through the pulmonic and systemic circuits is both sequential and equal since these circuits are in series; the output of each ventricle is also identical over a period of time. This statement is accurate save for the volume of blood that perfuses the bronchial arteries directly, and thus drains directly into the left ventricle via the pulmonary veins.

The average stroke volume can be determined readily by dividing the mean cardiac output by the heart rate. Since the stroke volume represents the quantity of blood ejected during each systole (ie, it is the difference between the end diastolic and end systolic blood volumes within either the right or left ventricle), it is helpful to consider the regulation of stroke volume in terms of those factors that

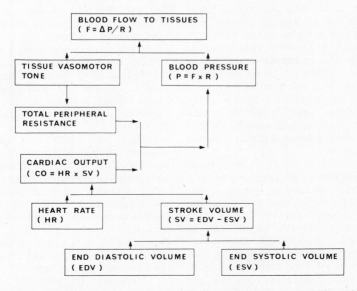

FIG. 10.61. Summary of the dynamic events within the cardiovascular system. As indicated, there are four major determinants of systemic arterial blood pressure and cardiac output. 1. Total peripheral vascular resistance as determined by vasomotor tone of the vessels, principally the arterioles. 2. Heart rate. 3. End diastolic ventricular volume. 4. End systolic ventricular volume. (After Selkurt [ed]. *Physiology,* 2nd ed, 1966, Little, Brown, and Co.)

alter this parameter directly, as the stroke volume in turn underlies the cardiac output over a period of time.

The factors that influence the stroke volume are: (1) the heart rate, which is a direct function of sinoatrial pacemaker activity; (2) the end diastolic volume, which in turn is determined by the venous filling pressure, the distensibility (or resistance to filling) of the ventricular walls, and the atrial mechanism, which fills the ventricles with a variable quantity of blood immediately before each ventricular systole; and (3) the end systolic volume, which is a direct reflection of the extent to which the ventricular myocardium contracts during the systolic phase of the cardiac cycle.

Before proceeding into a discussion of the specific intrinsic and extrinsic mechanisms concerned with the regulation of these factors, hence cardiac output, it is important to mention that under certain circumstances there is a definite shunting of blood into certain regions of the body in response to functional requirements. For example, during physical exercise, blood flow increases in skeletal muscle (p. 693); during periods of heat loss (or retention), blood flow increases (or decreases) in cutaneous regions; and during digestion, mesenteric blood flow increases greatly. It should be reiterated, however, that alterations in blood flow occur principally through changes in the output of the heart. The intrinsic and extrinsic factors that regulate this important aspect of cardiac function now will be discussed individually.

Inherent (Intrinsic) Cardioregulatory Mechanisms

At rest, the cardiac output is around 5.0 liters per minute in an adult male, whereas during heavy exercise this volume can increase up to roughly 20 liters per minute, although higher values have been reported. The heart possesses two inherent mechanisms whereby this enormous increase in its output can take place. These are (1) an increase in the rate of its contractions per minute, which is a direct reflection of the frequency at which the sinoatrial node is firing, and (2) the force with which the heart contracts at each systole, thereby ejecting a greater stroke volume at higher pressure during each contraction.

In vivo under different states of physiologic activity, both of these factors are operative to varying degrees, and the net cardiac output at any time is the resultant between them. It is important to mention at the outset, however, that in some individuals an increased cardiac output may be achieved principally by an increased force of contraction, the concomitant rate changes being relatively negligible (as in a highly trained athlete, p. 1188), whereas in other individuals both rate and force changes exert a more balanced influence to achieve the same result. Furthermore, under different circumstances in the same individual, both of these two factors may be operative to varying degrees to bring about an increased cardiac output. No strict rule can be enunciated whereby one or the other of these inherent cardiac mechanisms can be predicted to assume primary importance in producing an adequate cardiac output under all physiologic conditions. Therefore, only the general principles pertinent to an understanding of these adaptive mechanisms within the heart will be presented.

WORK PERFORMED BY THE HEART AND LAPLACE'S LAW. The sudden impact of the isometric phase of ventricular systole upon the blood may be likened to a bat striking a baseball. The energy that is transmitted directly to the ball by the bat imparts momentum to the ball in addition to potential as well as kinetic energy. During the follow-through with the bat, which is analogous to the isotonic phase of systole, the ball is moving at a constantly accelerating velocity, even though the greatest portion of the energy has been transferred to it already. Likewise, the mass of blood ejected during systole continues to accelerate and move into the aorta during the isotonic phase of ventricular contraction, although the blood already has received the major proportion of its total energy.

Cardiac work may be defined as the total quantity of energy imparted to the blood during its systolic ejection from the ventricles, and a large proportion of the cardiac work is performed in imparting potential (or pressure) energy to the mass of blood received under low pressure from the veins. A minor proportion of the total energy imparted to the blood during systole is used in accelerating the blood to a velocity sufficient for its ejection through the pulmonary and aortic valves, and this fraction of the total work (normally around 2 to 4 percent) of the heart represents the kinetic energy of flow.

Right ventricular work is equal to the stroke volume times the mean ejection pressure minus the right atrial pressure. A similar relationship is used to express left ventricular work, and if the stroke volume is stated in ml, while the pressure is expressed in dynes per cm^2, the work performed is in ergs.

Right ventricular work amounts to roughly one-seventh that performed by the left ventricle, because of the difference in the systolic pressures developed by the two sides of the heart.

The kinetic energy delivered by either ventricle is equal to 0.5 times the total mass of blood ejected during systole multiplied by the square of the ejection velocity; that is, $KE = 0.5 \ (mV^2)$. When the mass is stated in gm of blood ejected, and the velocity is given in cm per sec, once again the work output is in ergs.

Normally, the potential energy imparted to the blood represents 96 to 98 percent of the total energy expended by the heart during systole, and the remainder, kinetic energy, is necessary to produce the rapid acceleration of the blood during the first quarter of systole as it is ejected from the ventricle. In pathologic situations, however, such as aortic stenosis (ie, narrowing of the aortic lumen), when the blood flows at a considerable velocity through the narrowed valvular orifice, up to 50 percent of the total work performed by the heart may be required merely to produce the kinetic energy of flow necessary to eject the blood from the left ventricle.

As in skeletal muscle, the chemical energy utilized by heart muscle during its contraction is derived primarily from the aerobic metabolism of glucose and fatty acids (Chap. 14, p. 945), and to a lesser extent from other substrates. The relationship between the energy expended by the heart and the work performed by this organ is roughly proportional to the degree of tension developed by the heart during contraction multiplied by the time that tension is maintained.

Since the energy expended during muscular contraction is converted largely into heat, and only a small fraction is utilized for the contractile process itself, the efficiency of cardiac contractions may be expressed as the ratio of useful work performed to the total energy expenditure. Thus, the efficiency of a normal heart under an average load amounts to only 5 to 10 percent, whereas during maximum work output the efficiency may increase to as high as 15 to 20 percent in an average individual. In a trained athlete, however, the efficiency may rise to as much as 25 percent during exercise.

Cardiac work can be correlated readily with Laplace's law (p. 627). The pressure developed within the ventricles depends not only upon the tension developed by the ventricular myocardium during contraction, but also upon the geometric changes in the size and shape of the heart during the systolic phase of the cardiac cycle. The load imposed upon the heart is dependent principally upon the tension developed by the musculature, as well as upon the time that tension is maintained, and both of these factors in turn regulate the oxygen requirement of the myocardium. Thus, the external load imposed upon the heart by pumping the blood actually is a relatively minor factor insofar as the total cardiac work is concerned.

In accordance with Laplace's law, the pressure developed by the heart is directly proportional to the tension developed by the ventricular muscle during systole, and inversely related to the principal radii of curvature of the chambers. If one ventricle of the heart is dilated, as in failure, it functions at a distinct mechanical disadvantage, as noted earlier (p. 629). For example, if the diameter of a ventricle is doubled, the radii of curvature are likewise doubled, so that the tension developed within the wall of that chamber must be twice as great in order to produce the same systolic pressure, and thus the total wall tension that is developed within the muscle is increased fourfold. Consequently, any increase in cardiac size directly increases the work load on the heart.

Furthermore, an accelerated heart rate will increase the tension–time factor involved in the contractility of cardiac muscle, as the time involved during each systole is not shortened appreciably during a bout of tachycardia. Rather, acceleration takes place principally at the expense of diastole, thereby also placing an increased load on the heart, as diastole is the only phase of the cardiac cycle during which the organ can "rest."

Finally, an increased blood pressure itself imposes a greatly augmented load on the heart, since, in accordance with Laplace's law, the tension developed by the myocardium is proportional to the pressure. As the systemic (and/or pulmonary) arterial pressure rises, the ventricular tension developed at each systole also must increase accordingly in order to maintain the pressure at the same level.

Laplace's law must be considered in the effective management of patients having certain forms of cardiac disease, for a strong emotion, such as anger, which raises both heart rate and blood pressure to high levels, can impose a far greater load (strain) upon the heart than will light exercise.*

INTRINSIC CARDIAC RATE REGULATION. The inherent property of the pacemaker cells of the sinoatrial node to respond to changes in their state of physical deformation by an appropriate alteration in their spontaneous firing rate is termed intrinsic cardiac rate regulation. This basic property of autorhythmic tissues in general, and cardiac pacemaker cells in particular, was introduced on page 598. Briefly, an increased mechanical deformation of pacemaker cells increases their firing rate, whereas removing the added tension decreases the rate once again. This rate effect is evident only between certain limits, both of frequency and applied tension, so that if the pacemaker cells are firing initially at an elevated rate, mechanical deformation of the tissue will not further increase the rate of firing. In fact, added stretch actually may decrease the spontaneous pacemaker rate under such circumstances.

In the normal human heart this mechanism, although definitely present within the sinoatrial nodal cells, doubtless plays a minor role in regulation of heart rate; therefore, cardiac rate regulation in vivo is governed principally by the extrinsic neural and humoral factors, to be discussed later in this section, and whose direct influence upon the pacemaker is both rapid and profound.

Nonetheless, the latent ability of myocardial cells in general to respond to sustained mechanical stimuli by becoming rhythmically functional ectopic (abnormal) pacemakers undoubtedly reflects

---

For example, the great British physician and surgeon, John Hunter, who was subject to repeated and severe attacks of angina pectoris, stated that his "life was in the hands of any rascal who chose to annoy and tease him." Ultimately, in fact, he succumbed to such an attack which was induced by violent disagreement with his colleagues at a meeting in St. George's Hospital.

the inherent property of cardiac tissues in general to assume pacemaker properties (ie, to exhibit spontaneous diastolic depolarization) under abnormal conditions. For example, during intracardiac catheterization, atrial as well as ventricular arrhythmias are of common occurrence, whereas the pattern generally reverts to a sinus rhythm once again as the catheter tip is withdrawn from the heart.

INTRINSIC CARDIAC MECHANISMS FOR THE REGULATION OF STROKE VOLUME. It was stated earlier in this chapter (p. 594) that the three events responsible for the normal rhythmic activity of the heart take place in a sequential and smoothly coordinated order. Thus, the rhythmic process of cardiac excitation by the pacemaker tissue of the sinoatrial node leads to conduction of these impulses throughout the myocardium, and in turn the conducted impulses produce stimulation of the contractile machinery of the heart in a rhythmic fashion.

It should be emphasized at this point that even as the sinoatrial nodal cells of the pacemaker contain an intrinsic pressure-sensitive mechanism for governing their rate, so also the contractile elements of the myocardium contain intrinsic mechanisms that regulate the force with which they contract. Used in this context the word *intrinsic* merely signifies that the property under discussion, whether *rate* or *force,* is an inherent (or innate) property of the tissues themselves. The physiologic mechanisms involved are, of course, two distinct and independent entities, although functionally they may or may not operate simultaneously to a certain extent.

Venous return to the heart is one of the principal determinants of cardiac output. It is axiomatic in cardiac physiology that the normal (or nonfailing) heart, acting within physiologic limits, pumps out all the blood that is returned to it, so that no excess blood pools or stagnates within the veins. As venous return to the heart changes, depending upon the peripheral blood flow through specific and localized regions of the body, so too the myocardium adapts automatically to the changing volumes of blood returned to it by the veins. Thus, the venous input may vary from as low as 3 liters per minute during sleep to as high as 20 or more liters per minute during exercise. The heart, therefore, can adapt readily to the minute-to-minute or even moment-to-moment alterations in venous return. And totally unlike a mechanical pump whose stroke output is limited strictly by the volume displaced by a piston, and whose minute output is determined solely by the rate at which the piston moves, the output of the heart is governed not only by its rate, but also by the stroke volume. The latter factor, in turn, is strictly dependent upon venous return.

This inherent property of the heart to adapt to augmented (or diminished) diastolic filling by appropriate alterations in myocardial contractility is known commonly as the Frank–Starling "law" of the heart (Fig. 10.62).* Thus, one mechanism whereby the heart adapts to an increased diastolic filling volume is that when the myocardial fibers are stretched to a greater extent, the ensuing contraction is more forceful, and the extra blood thus is pumped automatically into the arteries. The property of contracting more forcefully when the muscle is stretched within physiologic limits (ie; when the initial fiber length is greater, Fig. 10.37) is not unique to cardiac muscle. Skeletal muscle also exhibits the same property as described in Chapter 4 (p. 124). Presumably, this effect is caused by the actin and myosin filaments within the muscle fibers reaching a mechanically and chemically optimum position relative to each other when stretched, and thus maximal shortening takes place upon appropriate stimulation. Furthermore, stretch prior to contraction increases the proportion of the energy that is expended by the heart (as well as skeletal muscle) in performing useful work. The effect of stretching on the contractile force developed by the heart, as described above, sometimes also is referred to as *heterometric autoregulation.*

There is an additional, though relatively minor, factor that increases the efficacy of cardiac contraction when the volume of this organ is increased. This mechanism takes about half a minute to become fully operative rather than immediately on a stroke-to-stroke basis. An increase in cardiac metabolism per se develops gradually when the heart is distended by an elevated filling with blood. Under these conditions, the heart reaches a new and higher metabolic steady state for as long as the increased filling, hence stretch, is maintained. Since no changes in fiber length are involved in this process, this effect is called *homeometric autoregulation.*

The ability of an isolated heart to alter its stroke volume independently of neural and endocrine factors is the result of an intrinsic property of the contractile elements of the myocardium. The tension produced by contraction of cardiac muscle gradually becomes greater with added load up to a maximum point (around 200 percent of the unstretched length), after which the tension developed by the muscle decreases. When an isolated strip of cardiac muscle is preloaded (p. 125), there is a *contractile tension* superimposed upon the resting tension at each load. The contractile tension developed at each load is equal to the difference between the total tension developed during contraction and the resting tension, and the total tension increases as the muscle is stretched by added increments of load up to a maximum point,

---

Actually, however, certain basic observations pertaining to this principle as well as its formulation were reported by Sewall and Donaldson working in collaboration a number of years previous to the independent work of Frank and Starling.

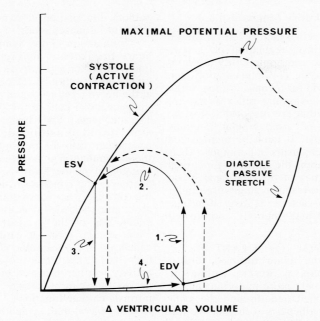

FIG. 10.62. Intracardiac volume-pressure relationships. 1. Isovolumetric contraction phase. 2. Ejection. 3. Isovolumetric relaxation phase. 4. Filling. Normal mechanical cycle is indicated by the lines connecting these four sequential events. Systole, which commences at the end diastolic volume (EDV), first produces the isovolumetric intraventricular pressure change shown, and terminates at the end systolic volume (ESV). Isovolumetric relaxation then ensues and the cycle repeats. An increase in the EDV *(dashed lines)* produces an increase in stroke work, since myocardial contractility is increased in accordance with the Frank–Starling effect (heterometric autoregulation). See also Figures 10.63 and 10.64. (After Selkurt [ed]. *Physiology,* 2nd ed, 1966, Little, Brown, and Co.)

beyond which the contractile tension will diminish as the load is increased still further (Fig. 10.37).

In the classic dog heart–lung preparation of Starling,* the aortic resistance (or impedance) against which the ventricles pump blood can be controlled artificially; therefore, aortic pressure can be regulated, since $\Delta P = F \times R$. Venous inflow to the right ventricle also can be changed at will by adjusting the height of a reservoir above the right atrium. Combined left and right ventricular volumes, venous pressure (at the right atrium), left ventricular outflow resistance, and aortic blood pressure all can be monitored continuously in such a preparation, and the effects of manipulating these various parameters studied.

Using this experimental procedure, the following important facts concerning the heart have been demonstrated. First, if the aortic pressure is raised by compression of the artificial resistance, both the systolic and diastolic volumes of the combined ventricles are increased, but no change in the stroke volume takes place. The increased energy released by the ventricles contracting against an increased peripheral resistance (or impedance) is provided by an increase in the degree of diastolic distention, and the stroke volume remains the same, despite an increased outflow resistance.

Second, when the arterial pressure is regulated so that it remains at a constant level, and the venous inflow is augmented by increasing the height of the reservoir attached to the right atrium, the cardiac output increases, an effect that is caused by an increased diastolic filling of the cardiac chambers. There also is an attendant, marked increase in the stroke volume when the venous return is increased.

To summarize these points, the isolated heart responds to an increase in either a pressure load or a volume load by an increased distention of the ventricular walls, and in turn this distention causes an increase in the length of the individual muscle fibers. Thus, the force of contraction (ie, mechanical work performed) during each contraction is proportional to the initial fiber length (ie, the extent of the preload). Therefore, cardiac output can be increased substantially with no concomitant increase in heart rate, as shown in Figure 10.62. An augmented stroke volume combined with a tachycardia would, of course, increase the cardiac output to an even greater extent.

In the light of more recent experimental evidence, it appears that this concept of cardioregulation is too simplistic, and extrinsic mechanisms that modify the intrinsic myocardial contractile processes are of major importance in vivo. Furthermore, in the normal individual, the extent to which the heart is free to dilate is limited by the anatomic restrictions imposed by the pericardium.

*In this preparation, the heart is denervated; hence, it is not subject to extrinsic neural influences.

Extrinsic Cardioregulatory Mechanisms

In this section, the principal factors relevant to an understanding of extrinsic regulation of cardiac function insofar as rate and contractility are concerned will be discussed. This presentation also will serve as background for a consideration of the several important cardiovascular reflexes that are operative in vivo and that are responsible for the homeostatic regulation of cardiovascular functions in the normal individual. It must be emphasized that the extrinsic mechanisms superimpose their regulatory functions upon the intrinsic functional and regulatory properties of the heart and blood vessels. Thereby cardiovascular adaptation under various states of bodily activity is assured to take place in an integrated and smoothly coordinated fashion.

EXTRINSIC CONTROL OF HEART RATE. Superimposed upon the minor intrinsic cardiac rate regulatory mechanism present within the pacemaker cells are vastly more important neural influences which are mediated through the sympathetic and parasympathetic branches of the autonomic nervous system. The distribution of these nerves to the heart is shown in Figure 10.67. The sinoatrial and atrioventricular nodes are richly innervated by both sympathetic and parasympathetic nerve endings. Insofar as the myocardium itself is concerned, the atria are innervated to a limited extent by both types of autonomic nerve; however, the ventricular musculature is innervated principally by sympathetic fibers, whereas sym-

pathetic innervation is somewhat more prevalent in the atrial musculature.

Stimulation of the sympathetic nerves to the heart produces a marked cardioacceleration (tachycardia or a positive chronotropic response) by the direct action of the norepinephrine liberated at the nerve terminals upon the rate of discharge of the pacemaker cells of the sinoatrial node (p. 598). The mechanism underlying this rate response is a more rapid diastolic depolarization of the pacemaker cells to threshold for firing. Sympathetic stimulation also increases the excitability (or irritability) of the heart as a whole, and thereby decreases the conduction time for impulses passing through the atrioventricular node (ie, the atrioventricular delay is shortened). In addition, stimulation of the sympathetic nerves markedly increases the force of cardiac contractility (also called a *positive inotropic* effect) as shown in Figure 10.63. This important aspect of sympathetic activity will be discussed below. The sympathetic nerves thus can be considered to augment all the intrinsic cardiac functions.

In general, parasympathetic stimulation (ie, of the vagus nerves) exerts the opposite effects to those listed for the sympathetic nerves. The right vagus is distributed predominantly to the sinoatrial node and the left vagus innervates principally the atrioventricular node; stimulation of the former decreases the heart rate (ie, it produces a bradycardia or a negative chronotropic response), whereas left vagal stimulation may induce primarily ventricular slowing subsequent to the development of heart block.

Vagal stimulation liberates the neurotransmitter substance acetylcholine at the nerve endings, and this compound in turn causes a decreased rate of diastolic depolarization to threshold for firing of the pacemaker cells (p. 598), so that a reduced heart rate ensues. If the vagal stimulation is sufficiently intense, these cells hyperpolarize to a steady-state voltage and then cease firing entirely so that quiescence of the heart ensues.

Vagal stimulation also increases the atrioventricular conduction time significantly, an effect that is brought about by reducing the excitability of the junctional tissues. If this response is sufficiently pronounced, atrioventricular block occurs, and the atria and ventricles no longer are functionally connected (p. 601). The atria now beat independently of the ventricles, and the latter chambers also assume a slower rate of contraction, as discussed on page 696. Vagal stimulation reduces cardiac contractility only to a negligible extent (Fig. 10.63).

The participation of the cardiac accelerator (sympathetic) and depressor (vagus) nerves in the complex of cardiovascular reflexes that are active in vivo will be considered subsequently.

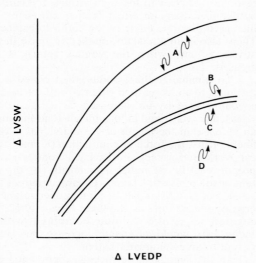

FIG. 10.63. Left ventricular stroke work (LVSW), obtained by multiplying stroke volume times mean systolic blood pressure and left ventricular end diastolic pressure (LVEDP). A. Effect of sympathetic stimulation on circulating catecholamines. B. Basal level. C. Vagal stimulation. D. Cardiac failure. (After Selkurt [ed]. *Physiology*, 2nd ed, 1966, Little, Brown, and Co.)

EXTRINSIC CONTROL OF MYOCARDIAL CONTRACTILITY. As stated in the previous section, sympathetic stimulation with the attendant libera-

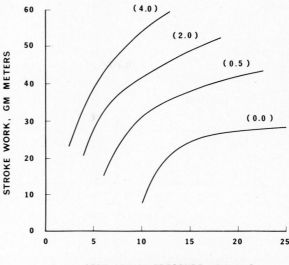

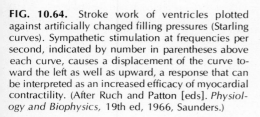

FIG. 10.64. Stroke work of ventricles plotted against artificially changed filling pressures (Starling curves). Sympathetic stimulation at frequencies per second, indicated by number in parentheses above each curve, causes a displacement of the curve toward the left as well as upward, a response that can be interpreted as an increased efficacy of myocardial contractility. (After Ruch and Patton [eds]. *Physiology and Biophysics,* 19th ed, 1966, Saunders.)

tion of norepinephrine from the nerve terminals exerts a profound direct influence upon the contractile machinery of the heart muscle itself (Fig. 10.63). Length–tension curves such as these are readily obtained experimentally by using isolated myocardial strips or papillary muscles, in heart–lung preparations, or in the exposed hearts of otherwise intact experimental animals. Studies of cardiovascular adaptation performed in intact human or animal subjects, however, indicate that the ability of the heart to adapt in vivo under various conditions may not be entirely a result of the regulatory effects of the Frank–Starling (heterometric) autoregulatory mechanism. Thus, other mechanisms that alter the capacity of the ventricular myocardium to contract may be even more important under certain circumstances, and it appears that the stroke volume in the intact mammal is regulated only partially by the length of the cardiac muscle fibers and the aortic pressure. Sympathetic stimulation not only increases heart rate, but it also causes the myocardial fibers to develop a greater force of contraction at a given fiber length (Fig. 10.64). As a consequence of this effect, when the strength of ventricular contraction increases at a specific fiber length, more of the blood that is present within the ventricles is ejected during systole; therefore, the end systolic volume decreases. Hence, the cardioregulatory functions of the adrenergic neurotransmitter liberated at the sympathetic nerve endings are twofold, as this substance (norepinephrine) produces both *positive chronotropic* as well as *positive inotropic effects*. In addition, sympathetic stimulation also can be shown to accelerate the rate at which the myocardial fibers shorten.

It obviously is impossible to ascertain the actual fiber length within the three-dimensional network of myocardial fibers that are present in the intact heart under different conditions; several types of ventricular function curves are used to assess the contractile properties of the heart experimentally. For example, as shown in Figure 10.64, the stroke work in gram-meters may be plotted against the atrial filling pressure. The family of curves illustrated in this figure is called the stroke work–output curve; it not only illustrates the fact that as atrial pressure increases the stroke work also increases until a maximum point is reached, but this relationship also clearly indicates the effects of sympathetic stimulation upon the force of cardiac contraction. Another type of ventricular function curve relates the atrial filling pressure in mm Hg to the cardiac output in liters per minute.

Actually, however, despite an enormous amount of research work, there is no unanimity of opinion as to what mechanisms are the most important in regulating myocardial contractility in vivo. Furthermore, until techniques are developed whereby this important aspect of cardiac physiology can be studied in intact, unanesthetized subjects, under various physiologic conditions, no valid quantitative evaluation of the effects of extrinsic control mechanisms upon cardiac contractility will be forthcoming.

Intrinsic Vascular Regulatory Mechanisms

The vascular system has a number of intrinsic mechanisms that are responsible for regulating its own functions independently of external control. Even as the heart possesses certain intrinsic mechanisms that enable it to regulate its frequency and force of contraction to a certain extent, so also the blood vascular system has a number of intrinsic mechanisms for the regulation of blood flow which are independent of nervous and other extrinsic regulatory factors. In fact, local blood flow

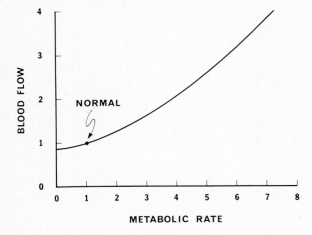

FIG. 10.65. The relationship between tissue blood flow (X normal on ordinate) and the metabolic rate (X normal on abscissa). (After Guyton. *Textbook of Medical Physiology*, 4th ed, 1971, Saunders.)

within various organs and tissues is regulated normally to a great extent solely by the presence of such intrinsic adaptive mechanisms within the vascular bed.

It is obvious that if all the vessels within the circulatory system of the body were to dilate to full capacity simultaneously, the total peripheral resistance would drop to a negligible value; cardiac output and blood pressure would approach zero. If the heart and vascular system are to function properly, the majority of the blood vessels must be at least partially constricted at all times. In order that this homeostatic goal be achieved, vascular constriction in certain regions of the body must be very delicately balanced against vascular dilatation in other regions in order that the cardiac output is adequate to meet tissue needs under all circumstances.

It must be stressed at the outset of this discussion that any consideration of blood flow is primarily a discussion of vasoconstriction and vasodilatation and the specific factors that produce these alterations in vascular caliber ($F = \Delta P/R$).

LOCAL REGULATION OF BLOOD FLOW IN ACCORDANCE WITH TISSUE REQUIREMENTS. As shown in Table 10.6, there are considerable differences among the individual blood flows to various organs and tissues within the body. The blood flow through the brain, liver, and kidneys is relatively enormous, despite the fact that these organs represent only a small fraction of the total body mass.* Skeletal muscle, on the other hand, which has the greatest total mass of any structure in the body (around 34 to 40 percent of the total body weight), has a blood flow at rest that amounts to only 15 percent of the total cardiac output; however, the arteriovenous oxygen concentration difference is quite high in muscular tissue.

The differences in blood flow among various specific organs are reflected only partially by their

The carotid body has the highest flow rate per mass of tissue of any organ in the body (p. 674).

relative metabolic activities. In brain and liver, although the metabolic rates are moderate, the blood flow is disproportionately greater than their individual metabolic requirements would indicate. The extremely high blood flow to the kidneys (approximately 20 percent of the entire cardiac output) is a reflection of the principal function of these organs, which is to excrete urine, rather than an indication of their metabolic rate. Resting skeletal muscle, on the other hand, has a low metabolic rate, and this is directly indicative of a low blood flow rate. However, the metabolic rate of skeletal muscles can increase as much as 50-fold during strenuous exercise, with a concomitant increase in blood flow up to 20 times above the resting value. Similarly, the blood flow to the skin also is quite variable, and this fact is of considerable importance to the homeostatic regulation of body temperature (p. 688).

Despite considerable variations among the total blood flows to individual organs, which in turn reflect different and unique functional characteristics of these structures, there is an important and fundamental mechanism for the local regulation of blood flow. This is the response of the vascular bed within specific tissues or organs to the metabolic requirements of that tissue or organ itself under various states of physiologic activity. As shown in Figure 10.65, there is a rough proportionality between the metabolic rate and blood flow, and this fact is true of any tissue or organ. An increase in the metabolic rate to a level eight times above the resting value augments the blood flow roughly fourfold. It is important to realize that the increase in blood flow to a tissue lags behind the increase in the metabolic rate of that tissue, so that initially the metabolic rate increases more rapidly than does the flow rate.

According to Poiseuille's law, the blood flow through the vascular bed of an organ is directly proportional to the pressure and inversely related to the vascular resistance. A small increase in the caliber of the vessels results in a pronounced fall in the resistance to flow, since the resistance factor varies as the fourth power of the radius (p. 624). Thus, a marked increase in flow occurs with a very small change in vessel diameter, when it is assumed that the blood pressure remains constant. Any local factors that underlie the degree of constriction or dilatation of the vessels within a given vascular bed are of the utmost importance to the intrinsic regulation of local blood flow. The principal factors that directly affect vascular caliber locally now may be considered individually.

Tissue Oxygen Requirements and Local Blood Flow. Tissues in general require additional oxygen when their metabolic activity increases (as during exercise, p. 447), or when the blood saturation of oxygen falls (as at high altitudes, p. 1223). Under these conditions, regional blood flow increases markedly. Thus, the tissue oxygen levels as well as the tissue oxygen requirements influence the degree of vasodilatation of the small blood vessels directly. There is in the majority of (but not all!) tissues an inverse relationship between oxygen concentration and flow, and decreased oxygen levels thus produce a direct arteriolar and metarteriolar dilatation. The smaller the vessel, the greater is the direct effect of lowered oxygen concentration in reducing tone in that vessel. The smallest arterioles, metarterioles, and precapillary sphincters are extremely sensitive to reduced local oxygen concentrations. In fact, cyclic vasomotion, the rhythmic contractions of the precapillary sphincters (p. 642), is accelerated by lowered tissue (or blood) oxygen concentrations; vascular channels that are located downstream from these regions remain open for proportionately longer intervals when there is less oxygen in the immediate vicinity.

Presumably, the smooth muscle cells, which are abundantly present in the smallest vessels and precapillary sphincters, act as chemoreceptors, and thus are able to regulate their contractile tension, hence the caliber of the vessel or sphincter, in accordance with the local availability of oxygen.

This local intrinsic mechanism for the regulation of blood flow represents a simple negative feedback control system. The tissue oxygen requirements thus are exquisitely balanced with local blood flow, and the smooth muscle cells of the vascular bed themselves can act autonomously as the sensory elements to regulate their own degree of constriction or dilatation in accordance with local conditions.

Tissue Metabolites and Local Blood Flow. In addition to tissue oxygen requirements, a number of tissue metabolites and neurohumoral agents have been implicated in the control of local blood flow. Because of the extremely low concentrations of these substances and their complex interrelationships insofar as the regulation of vascular caliber is concerned, any quantitative assessment of their actual role in local blood flow regulation in vivo is a matter of conjecture at present.

Vasodilator substances, which can act on the smallest vessels to augment blood flow locally, include such compounds as carbon dioxide, lactic acid, histamine, potassium and hydrogen ions, as well as adenosine and phosphorylated adenosine compounds such as adenosine triphosphate (ATP). Other vasoactive compounds include norepinephrine, acetylcholine, and 5-hydroxytryptamine (serotonin). Presumably, many other nonspecific and unidentified substances also exert a similar effect.

During active secretion in the sweat glands, in the exocrine portion of the pancreas, and in the salivary glands, a specific peptide called bradykinin is formed by hydrolysis (Fig. 10.57). This substance has pronounced local vasodilatory properties, and possibly it may be involved in the increase in local blood flow that accompanies activity of these secretory tissues. In addition, a number of other chemically related vasoactive polypeptides are formed within various tissues, and collectively these peptides are known as kinins. Although the actual physiologic role of the kinins is obscure at present, it has been postulated that they may play a role in the regulation of local blood flow in tissues other than glands, such as the skin, where they have been implicated in the process of thermoregulation (p. 688).

Kinins are similar in their actions to histamine in that they produce relaxation of vascular smooth muscle with a concomitant increase in capillary permeability. In contrast to this specific effect on blood vessels, kinins also produce contraction in visceral smooth muscle.

A physiologically widespread group of compounds, called prostaglandins, also should be mentioned here (see also p. 1173). Certain of these substances appear directly concerned with the regulation of vascular tone, hence, blood flow. One of these compounds also has the ability to alter the normal elasticity of erythrocytes so that these cells become unable to undergo sufficient deformation to pass readily into the smaller capillaries, and thereby the cells are "shunted" into larger vascular channels. Although at present it is impossible to evaluate the role of such an effect on the regulation of blood flow in vivo, it should be borne in mind that such an effect has been demonstrated to exist, and that eventually this process may prove to be an important mechanism for controlling blood flow locally under certain circumstances.

It is important to realize that within the terminal vascular bed of any particular organ, some of the vessels may respond strongly to one agent but not to others, so that metabolic influences may predominate in metarterioles and precapillary sphincters, whereas in the arterioles, extrinsic neural factors in general exert a considerably greater control over vascular caliber. Thus, profound differences in the type of control of vessel caliber may exist within the same vascular bed.

LOCAL REGULATION OF BLOOD FLOW IN ACCORDANCE WITH VARIATIONS IN ARTERIAL PRESSURE. The smooth muscle cells of the smallest blood vessels have the property of inherent myogenic automaticity similar to that exhibited by the sinoatrial nodal cells of the cardiac pacemaker. Hence, these muscular elements exhibit the independent rhythmic contractions that underlie vasomotion (p. 147), and this inherent property is wholly independent of extrinsic nervous, physical, or chemical influences. It must be stressed, however, that extrinsic physical, neural, and chemical factors can (and do) greatly modify the intrinsic pattern of the contractions. Thus, the smooth muscle cells present at the terminal branches of the vascular tree can maintain autonomously a certain degree of partial contraction or tone, and this tone can be modified greatly by the superimposed effects of neural and other extrinsic factors, one mechanical factor being variations in the intravascular pressure itself.

It would be expected that an increase in arteriolar pressure would cause a proportionate increase in blood flow according to Poiseuille's law ($F = \Delta P/R$), but in certain vascular beds of the body, for example those in skeletal muscle and particularly those present in the kidney, the flow remains relatively constant despite wide fluctuations in the arterial pressure. This fact is shown in Figure 10.66. As the mean arterial pressure rises acutely between about 80 mm Hg and 175 mm Hg, the local blood flow stays relatively constant. The inherent ability of a vascular bed to maintain a constant flow rate despite wide fluctuations in arterial pressure is termed *autoregulation* (or *self-regulation*) *of flow*. Figure 10.66 also illustrates the

fact that the autoregulatory mechanism is more effective over a much wider range of arterial pressures when this pressure is raised very gradually (ie, over a period of several weeks) rather than suddenly.

The mechanism underlying the phenomenon of autoregulation lies in the ability of smooth muscle to respond to mechanical stretch (induced by an increased intraluminal pressure) by a delicately graded contraction, thereby decreasing the diameter of the vessels and increasing the vascular resistance. The net flow remains constant. Humoral factors also have been implicated as being operative in producing autoregulation of vascular caliber. Such chemical influences on vascular diameter perhaps are more effective in long-term autoregulation, whereas a rapid rise in blood pressure might be more apt to produce vasoconstriction by the direct, rapid mechanical stimulation of the vascular smooth muscle.

Extrinsic Vascular Regulatory Mechanisms

Superimposed upon the intrinsic local control mechanisms for the regulation of blood flow are two general mechanisms of considerable importance for the integrated control of vascular tone, hence blood flow, within the body as a whole. These are neural and humoral (or chemical) regulatory mechanisms.

VASOMOTOR NERVES. Similarly to the heart, arterioles are in general richly innervated by two

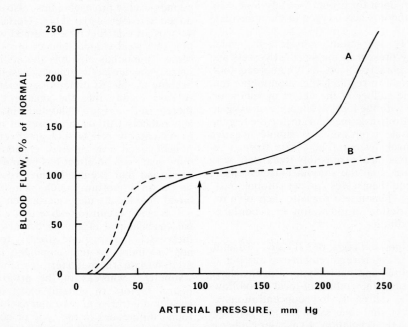

FIG. 10.66. Effect of increasing the arterial pressure upon the blood flow through a muscle. Vertical *arrow* indicates normal blood flow at a mean arterial blood pressure of 100 mm Hg. A. Solid curve shows effect on flow if the blood pressure is raised suddenly, ie, over a few minutes. B. Dashed curve indicates the effect on blood flow if the pressure is raised gradually, ie, over a period of weeks. (After Guyton. *Textbook of Medical Physiology*, 4th ed, 1971, Saunders.)

types of nerve fibers. Those nerve fibers that cause contraction of the smooth muscle coat upon stimulation are termed *vasoconstrictor nerves,* whereas those nerve fibers which produce inhibition and relaxation of the arteriolar smooth musculature are termed *vasodilator nerves.* The general term *vasomotor nerves* refers to both vasoconstrictor and vasodilator nerves.

Sympathetic Vasoconstrictor Nerves.   These nerves belong to the thoracolumbar, or sympathetic, division of the autonomic nervous system. The sympathetic vasoconstrictor fibers arise from nerve cells located within the lateral horns of the gray matter of the spinal cord. In man, they extend from the first thoracic segment (T-1) to the second or third lumbar segment (L-2 or L-3) of the cord (p. 270). Despite their relatively limited site of origin in the spinal cord, all the arterioles of the body are supplied with terminal ramifications of the sympathetic vasoconstrictor fibers. For additional details concerning the sympathetic innervation of various organs and regions of the body, see Figures 7.24 and 7.26 (p. 272 and p. 274), respectively) as well as Table 7.6 (p. 276).

In toto, these sympathetic vasoconstrictor nerves form a powerful mechanism throughout the body for the regulation of vascular tone in the arteriolar and smaller vessels. Under normal physiologic conditions, these nerves discharge at a frequency of around one or two impulses per second, and this rate is sufficient for the maintenance of normal vascular tone. Under conditions of maximum physiologic excitation, this rate can increase to around ten impulses per second. It is important to remember that despite the presence of vasodilator nerve fibers on the vessels located in several regions of the body, a reduction in the frequency of impulses transmitted along a vasoconstrictor nerve generally results in a passive relaxation of the smooth muscle within the walls of the vessels innervated by this nerve; dilatation of the vessel follows. This general fact is of fundamental significance to an understanding of the neural regulation of blood flow. Norepinephrine is the chemical transmitter released upon stimulation of the sympathetic vasoconstrictor fibers, and this humoral agent acts directly upon the vascular smooth muscle. Contraction of the arteriolar smooth muscle follows, which reduces the vascular caliber so that in turn the peripheral resistance is greatly increased. The sympathetic vasoconstrictor fibers regulate primarily the diameter of the resistance vessels of the systemic circulation, in particular the arterioles and metarterioles. The diameter of the capillaries, on the other hand, appears to be regulated principally by intrinsic local factors.

Sympathetic nerve fibers also have a profound influence upon the regulation of heart rate and cardiac contractility, as discussed on page 662. In addition, sympathetic vasoconstriction of the capacitance vessels, principally the veins and venules, can alter markedly venous return to the heart; and as a consequence of this effect, the cardiac output can be altered profoundly. Therefore, very small changes in the caliber of the veins can produce marked changes in the total volume of blood that is contained within these vessels, and in this manner the overall capacity of the venous reservoir can be decreased and thus a greater volume of blood is shunted into the arterial side of the systemic circulation. For example, if the stroke output of the ventricles is equivalent to around one or two percent of the total blood volume contained within the systemic veins, a generalized venoconstriction of only two percent will decrease the overall venous capacity for blood to such an extent that cardiac diastolic filling will double. In contrast to this effect in the venous circulation, a proportionately small vasoconstriction of the resistance vessels (arterioles) would augment the total peripheral resistance to only a negligible extent.

Vasodilator Nerves.   The nerves that convey impulses which mediate relaxation of vascular muscle tone, or vasodilatation, belong to both divisions of the autonomic nervous system, the sympathetic and parasympathetic. Furthermore, antidromic vasodilator impulses are transmitted in nerves belonging to the sensory or posterior branches of the spinal nerves.

Sympathetic Vasodilator Fibers.   The terminal branches of some fibers that belong anatomically to the thoracolumbar (sympathetic) nervous system are classed functionally as efferent vasodilator fibers. These nerve endings are exlcusively cholinergic; hence, the neurotransmitter released is believed to be acetylcholine. It must be emphasized that efferent dilator sympathetic nerve fibers are distributed solely to skeletal muscle, and normally the more powerful sympathetic vasoconstrictor effects overshadow the influences of the vasodilator nerves. During fainting, however, peripheral blood flow, such as can be measured in the forearm, increases greatly, suggesting the presence of vasodilator activity. Similarly, sympathetic vasodilator fibers may have a physiologic role in augmenting blood flow to the skeletal muscles at the onset of a bout of exercise (p. 693).

Parasympathetic Vasodilator Fibers.   Both the cranial and sacral divisions of the parasympathetic nerves contain vasodilator fibers. These nerves reach peripheral vessels via the vagus, the glossopharyngeal, and the chorda tympani nerves, as well as via the sacral outflow of the pelvic nerve. The parasympathetic vasodilator nerves innervate circumscribed areas of the body, for example, the cerebral vessels, salivary glands, tongue, external genitalia, and probably the bladder, as well as the rectum. These structures all receive parasympathetic vasodilator fibers.

Although these parasympathetic nerves are not tonically active as are the sympathetic vasoconstrictor fibers, their distribution to specific regions of the body, each of which has certain specialized functions, suggests that their role in vascular adaptation is localized within, and related to, the physiology of the specific organ or tissue that they innervate. Despite the fact that the action of the parasympathetic vasodilator fibers is quite refractory to inhibition by atropine (which normally blocks parasympathetic nerves), these parasympathetic vasodilator nerves are considered to be cholinergic.

Antidromic Vasodilator Impulses. According to the Bell–Magendie law (p. 213), dorsal root fibers transmit impulses only in a centripetal direction, ie, toward the spinal cord. An exception to this is found when vasodilator impulses are transmitted centrifugally along sensory nerves in the periphery to produce vasodilatation. Such impulses are called antidromic, which means "to run against." If the stimulus arises peripherally, ie, in the skin, the impulses travel up to a bifurcation in the nerve and then down another branch of that nerve to a blood vessel, causing vasodilatation. This is the so-called axon reflex, which was described on page 647, and is illustrated in Figure 10.55. Vasodilatation also can occur when the stimulus arises within a dorsal root ganglion and is transmitted peripherally in the normal, or orthodromic, direction. The axon reflex is of significance to the regulation of blood flow only in tissues that possess an abundant supply of pain fibers. For example, the mucous membranes and the skin contain many pain fibers; hence, the axon reflex may assume some importance as a local regulatory mechanism in these regions.

The neurohumoral transmitter of antidromic vasodilator impulses is nonspecific. Thus, histamine (or a histaminelike substance), adenosine triphosphate, and acetylcholine all have been implicated as humoral agents responsible for producing antidromic vasodilatation. In the skin, at least, it is clear that any stimulus that damages the tissues can produce this type of vascular response. For example, burns, frostbite, and foreign agents such as bacteria all increase local blood flow by the antidromic mechanism. Thus, the bodily defense and repair mechanisms in the superficial tissues are aided by the increased blood flow (see also inflammation, p. 547).

REGULATION OF VASCULAR CALIBER BY THE ADRENAL MEDULLA. The role of circulating catecholamines (epinephrine and norepinephrine), which are secreted into the bloodstream by the adrenal medulla, is negligible insofar as the normal regulation of vascular tone is concerned (Chap. 15, p. 1124), since the influence of the sympathetic vasoconstrictor fibers completely dominates vascular tone under ordinary circumstances. The major exceptions to this general statement are found in the heart and in skeletal muscle, in which the augmented secretion of physiologic quantities of epinephrine by the adrenal medulla can produce maximal vasodilatation of the vascular beds. This effect is seen during excitement, in strong emotional states such as fear, and during strenuous exercise such as in a competitive athletic contest. In all these situations, adrenal medullary secretion is greatly enhanced by hypothalamic stimulation. In summary, therefore, local autonomous control and direct motor innervation of the resistance blood vessels are the two factors that are primarily responsible for the regulation of vascular tone, hence, blood flow. The secretions of the adrenal medulla, on the other hand, assume importance only under special circumstances.

In the hepatic parenchyma, which lacks any direct sympathetic innervation, epinephrine liberated by the adrenal medulla plays an important direct role in the metabolic processes of the hepatic cells, as discussed on page 689, and this compound also may possibly have some part in the regulation of vascular tone in this organ.

Neural and Chemical Regulation of Cardiovascular Functions in Vivo

In order to function effectively in maintaining a homeostatic equilibrium in the body at levels that are compatible with life, the entire cardiovascular system must operate as a coordinated and integrated functional unit at all times. Since the individual regions within this system must work together in a harmonious fashion, a number of extremely sophisticated neural and chemical regulatory mechanisms are present, and the operation of these mechanisms assures an adequate flow of blood to and from the tissues under widely diverse activity states.

The most important of these controls are mediated by reflexes involving several regions of the central nervous system. In a number of instances, the chemical composition of the blood itself provides the essential stimulus to cardiovascular regulation, and in general terms, these regulatory mechanisms fundamentally consist of a number of delicately poised negative feedback controls for regulation of the functional activities of the heart and blood vessels.

Only those cardiovascular regulatory mechanisms that are considered to play an important role in the normal human body will be considered to any extent. It should be realized, however, that other cardiovascular regulatory systems are known although their relative physiologic significance is either minor or obscure.

Figure 10.67 shows the principal neural pathways involved in the various cardiovascular reflex mechanisms (see also Figs. 7.24, p. 272; and 7.25 p. 273). Reference should be made to these figures throughout the following discussions.

CENTRAL NERVOUS CONTROL OF CARDIOVASCULAR FUNCTIONS: GENERAL CONSIDERATIONS. Certain aspects of the extrinsic modification superimposed by specific types of nerve fiber upon the inherent and local regulatory mechanisms that are present within the heart and

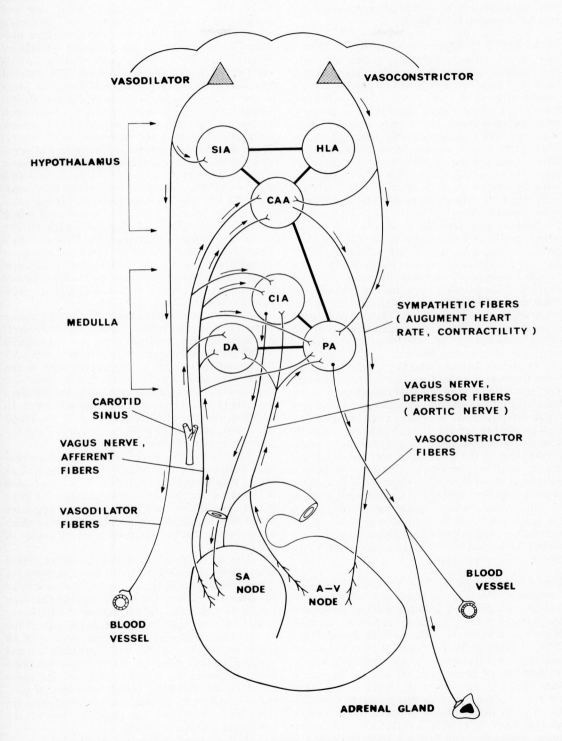

FIG. 10.67. Cardiovascular reflex systems. SIA = sympathetic inhibitory area; HLA = heat loss area; CAA = cardioaccelerator area; CIA = cardioinhibitory area; DA = depressor area; PA = pressor area. (After Best and Taylor [eds]. *The Physiological Basis of Medical Practice*, 8th ed, 1966, Williams & Wilkins.)

blood vessels have been presented already. The several neural regions involved in the regulation of these functions by the central nervous system will now be discussed individually.

Spinal Cord. Preganglionic sympathetic neurons may, under certain conditions, exhibit spontaneous rhythmic activity that is independent of their excitation by afferent impulses or central nervous excitation (p. 350). This autonomous rhythmicity (or efferent neural "tone") may result from markedly altered oxygen and carbon dioxide partial pressures, which in turn directly affect the excitability of the spinal neurons. More importantly, afferent sensory impulses from various sources are able to elicit spinal reflexes that affect vasomotor tone directly (Fig. 7.23, p. 271). The efferent pathway of this reflex is formed by the sympathetic vasoconstrictor fibers.

Stimuli such as cold or mechanical trauma (which produces pain), when applied to the skin, invoke segmentally distributed vasoconstriction, not only superficially, but also within the intestinal vessels. Conversely, heating the skin produces moderate vasodilatation, presumably because of a reflex diminution of constrictor tone in the appropriate segmental efferent fibers. Thus vasodilatation in most instances takes place passively and as the direct consequence of inhibition of constrictor tone in the appropriate fibers.

In human subjects who have a chronic transverse lesion at elevated levels in the spinal cord (p. 452), the distention of a hollow viscus such as the bladder results in an enormous increase in the systemic arterial blood pressure, even to levels as high as 300/140 mm Hg. In such patients, it can be demonstrated that there is a pronounced vasoconstriction in the extremities (eg, the hand) together with a concomitant reduction in the peripheral blood flow. This response is caused by sensory impulses arising in the distended bladder or other viscus which spread or radiate within the spinal cord and thus produce generalized peripheral vasoconstriction throughout the body by a diffuse activation of the segmental spinal vasomotor reflex mechanism.

Medulla Oblongata. The sympathetic vasoconstrictor effects noted above normally are controlled principally by a rather diffuse vasomotor area located in the reticular formation on the floor of the fourth ventricle of the medulla oblongata (p. 275). Laterally in this portion of the brain stem, an area is present that causes vasoconstriction and an acceleration of heart rate, or tachycardia, when stimulated (the pressor area). Medial to this pressor region is another area that produces vasodilatation upon stimulation; it is called the depressor area. It must be emphasized that this vasomotor center within the medulla is not made up of clearly defined and specific neuroanatomic nuclei. Rather, the vasomotor center consists of a large and diffuse functional area that regulates overall vaso-

constriction and vasodilatation of the resistance vessels (principally the arterioles) throughout the body. This center also controls heart rate and myocardial contractility. Furthermore, it should also be emphasized that the depressor area produces vasodilatation of the resistance vessels by inhibition of the vasoconstrictor neurons, which are located within the spinal cord. The detailed interactions between pressor and depressor areas within the medulla are unknown, although doubtless there is some mutual (reciprocal?) interaction between them. It is clear, however, that pressor (or excitatory) fibers and depressor (or inhibitory) fibers descend in anatomically discrete and separate regions of the spinal cord. These independent fibers then probably converge on the same spinal neurons. Thus, through their activity they modulate the inherent discharge frequency of the vasoconstrictor neurons within the spinal cord which in turn form the final common path to the peripherally located resistance blood vessels. As a consequence, the degree of vasoconstriction or vasodilatation within the blood vessels at any moment is determined principally by modifications of the impulse frequency in the sympathetic vasoconstrictor fibers.

The action of the medullary vasomotor center in turn is governed by still higher control centers, and these centers consist of areas that are situated within the hypothalamus and cerebral cortex. Afferent impulses from widespread regions of the body may converge directly upon both the hypothalamic and cortical centers, as do impulses that arise within other central neural structures. In addition, the chemical composition of the blood itself also regulates directly the activity of the vasomotor center to some extent.

The medullary vasomotor center also regulates both cardiac rate and contractility as well as the relative peripheral resistance; thus, this region of the brain stem controls cardiac output to a large extent. In addition, the vasomotor center regulates blood flow to specific organs in response to afferent impulses that are received from all tissues of the body.

The neural elements of the vasomotor center also exhibit a spontaneous rhythmic activity (or automaticity) that is similar to the rhythmicity of the sinoatrial node of the heart. Impulse discharge continues even after the elimination of all afferent neural impulse input from sources outside of the center itself. It can be shown that removal of all higher nervous control (ie, from hypothalamus and cerebral cortex) does not affect the blood pressure. Sectioning the spinal cord below the vasomotor center, however, results in a loss of normal vascular tone, so that arterial blood pressure falls because of passive vasodilatation of the peripheral resistance vessels.

In this regard, when the spinal cord is transected as by an injury, the blood pressure decreases at first and then very gradually increases, because the musculature within the denervated vessels eventually develops an inherent peripheral tone that results in a slight vasoconstriction, and an increase in blood pressure follows.

Hypothalamus. The hypothalamic region of the brain contains areas that have both excitatory and inhibitory influences upon the degree of sympathetic tone that is transmitted to peripheral vessels, and this effect is mediated indirectly via the influence of these areas upon the medullary vasomotor center. In addition to these small poorly defined hypothalamic areas, there is also present an important heat loss center in the anterior region of the hypothalamus. This region plays an important role in adjusting blood pressure by regulating the discharge frequency to the vasoconstrictor sympathetic nerves in the cutaneous region. Local cooling of, or electric stimuli applied to, the heat loss center produces vasoconstriction, and a rise in blood pressure follows. Conversely, local warming of this hypothalamic region produces a vasodilatation, and a concomitant fall in blood pressure in the skin takes place.* The arteriovenous shunts (or anastomoses), as well as the arterioles and metarterioles, are the most important vascular elements that are involved in heat loss from the skin (Chap. 14, p. 1003).

Cerebral Cortex. There is good evidence that the motor and premotor cortices play important roles in causing vasoconstriction of cutaneous, renal, and splanchnic blood vessels, whereas a concomitant vasodilatation is produced in skeletal muscle. The overall effect of cortical stimulation is a pronounced rise in the systemic arterial blood pressure. These cortical centers are believed to have an important regulatory influence on blood pressure in response to such stimuli as anxiety or pain.

It is worthwhile to reiterate here that measurements of systemic blood pressure alone do not necessarily give a reliable indication of the state of overall vasoconstriction or vasodilatation within the body. Thus, constriction of the resistance vessels in one region of the circulatory system may be offset completely by a concomitant dilatation in other regions of the body, so that the mean arterial pressure remains about the same. It is imperative to measure flow through specific regions under different conditions in order to arrive at a reasonable conclusion insofar as relative vasoconstriction or vasodilatation are concerned.

Central Control of Sympathetic Vasodilator Fibers

The cholinergic sympathetic fibers that mediate vasodilatation were discussed on page 667. Their central pathway is anatomically and functionally separate from the medullary vasomotor center that regulates sympathetic vasoconstrictor activity throughout the body.

The specialized vasodilator efferent fibers have their origin in the motor region of the cerebral cortex, and they pass to, and synapse within, the hypothalamus. From the hypothalamus, efferent fibers now pass through the ventrolateral region of the medulla, and from there on to the lower spinal motor neurons. These sympathetic vasodilator fibers are presumed to respond to cerebral excitation, either in conjunction with, or in anticipation of, skeletal muscle activity (p. 1178).

Vascular Reflexes Elicited by Somatic and Visceral Nerve Stimulation

Although these somatic and visceral reflexes are, in general, of relatively minor significance in overall circulatory regulation under ordinary circumstances, at times they can assume importance. Stimulation of a peripheral nerve, such as the sciatic or a sensory cranial nerve, may evoke either an increase or decrease in the systemic arterial blood pressure. The net response in turn depends upon two factors, the strength and the type of the stimulus. The reflex arc involves: (1) the afferent fibers of the stimulated nerve; (2) the vasomotor centers within the central nervous system; and (3) the efferent vascular nerves, which can be either constrictor or dilator.

A pressor reflex requires a much stronger stimulus to be evoked than does a depressor response. A highly painful stimulus would evoke a pressor response. In either situation, however, the overall magnitude of the response varies with the total number of afferent fibers that are stimulated.

It is important to reiterate that not all pressor or depressor impulses must reach the medullary or higher brain centers in order to be effective, as the spinal cord itself contains neural centers that can mediate complete reflex arcs. Cortical and other higher centers undoubtedly influence the pressor response to a painful stimulus, as does the adrenal secretion of catecholamines.

The depressor response can be evoked in normal persons merely by distention of certain hollow viscera, such as the bladder,† since there is abundant afferent innervation in this organ which is quite sensitive to excitation by stretch. In addition, mild stimulation of such regions as the peritoneum, mesentery, abdominal viscera, anus, vagina, and spermatic cord also can elicit a depressor reflex.

Last, such reflex vascular responses as were discussed above may be elicited either with or without attendant alterations in the systemic arterial blood pressure. In normal persons, acute bladder distention can induce blushing, ie, peripheral

---

*Vasoconstriction would, of course, tend to reduce heat loss from the skin, whereas vasodilatation would have the opposite effect.

†This is in distinct contrast to the situation which obtains following cord transection, in which bladder distention results in development of an acute hypertensive state (p. 670).

vasodilatation of the facial region, but with an attendant venoconstriction; hence, the blood pressure may remain constant. Such a mechanism doubtless is operative, for example, in those brides whose wedding schedule has been too closely planned to allow for normal physiologic exigencies.

CARDIOVASCULAR REFLEXES: PRESSORECEPTOR MECHANISMS.   The basic mechanism underlying the regulation of systemic arterial blood pressure in vivo consists of a number of reflexes that are evoked by mechanical stimulation of pressure-sensitive receptors located within the adventitial layer of the walls of certain large vessels. The activity of these pressoreceptors, or baroreceptors as they are often called, is controlled by the arterial blood pressure itself. A continuous monitoring effect on both heart rate and the tone of the systemic resistance vessels is achieved. The sensory receptors that initiate these important cardiovascular reflexes are localized primarily in the aortic arch and carotid sinus regions of the arterial system. The particular reflexes that are elicited by the mechanoelectric transducer effect that is caused by distention of these vascular areas will be discussed individually.

The Depressor Reflex.   The branch of the vagus nerve (X), which is known as the cardiac depressor or aortic nerve (of de Cyon and Ludwig), is wholly afferent and depressor in function. The peripheral terminals of this nerve provide the receptors of the depressor reflex, and these terminals are distributed principally to the wall of the aortic arch and upper portion of the thoracic aorta. There is also some innervation of both ventricles by branches of this nerve, and possibly the coronary and pulmonary vessels receive some terminal fibers of the aortic nerve.

When the depressor nerve is sectioned, stimulation of the peripheral (cardiac) end produces no effect, as only afferent impulses are transmitted in this structure. Excitation of the central end, however, produces a marked slowing of heart rate (bradycardia), as well as a pronounced fall in the systemic arterial blood pressure, which is called the depressor effect. These responses are produced by two factors. First, the bradycardia and negative inotropic effect result from stimulation of the efferent vagal fibers which constitute the motor pathways of the depressor reflex arc and which originate within the medullary vasomotor centers (Fig. 10.67). Normally, both left and right vagal efferent pathways are involved in this reflex in vivo. The slightly reduced contractility of the heart, or negative inotropic response, probably is a result of the direct effects of acetylcholine on the myocardium (Fig. 10.63). Second, vasodilatation of certain peripheral resistance vessels of the systemic circulation is the principal factor responsible for the fall in blood pressure elicited by the de-

pressor reflex, and the vasomotor pathways discussed earlier (p. 666) form the efferent portion of this reflex arc. Inhibition of the medullary pressor area of the vasomotor center results in a passive dilatation of the resistance vessels, primarily in the splanchnic region but also in the skeletal muscles and skin to some extent, and as a consequence the blood pressure declines.

The bradycardia elicited by the depressor reflex has only a slight effect upon the fall in blood pressure observed following depressor nerve stimulation, whereas vasodilatation and the concomitant decrease in the total peripheral resistance are primarily responsible for this effect.

The depressor reflex can be evoked artificially by mechanical or electrical stimulation of the aortic wall; however, under physiologic conditions, the rhythmically fluctuating intraaortic pressure itself is the most important stimulus. The timing of impulse transmission along the aortic nerve is synchronous with the heartbeats, so that the physiologic stimulus that regulates the activity of this reflex is caused by the rhythmic pulsatile dilatation of the aortic wall. The frequency of impulse transmission in this nerve, on the other hand, is directly proportional to the intraaortic pressure. Consequently, a rise in the systemic arterial blood pressure stimulates an increased activity of the depressor reflex, causing the blood pressure to become lower, whereas a fall in blood pressure induces the diametrically opposite effect. Hence, the blood pressure and heart rate at rest are regulated closely by the operation of the depressor reflex. This important reflex also provides an excellent example of negative feedback control within the cardiovascular system, as the mean arterial blood pressure is maintained within narrow limits by the operation of this mechanism.

The Carotid Sinus Reflex.   A second important mechanism that is involved in the normal regulation of heart rate and systemic arterial blood pressure is the carotid sinus reflex.

At the bifurcation of each common carotid artery into its internal and external branches is found a small swelling known as the carotid sinus. This structure contains mechanoreceptors which are the sensory elements of the carotid sinus reflex. These sensory elements in turn consist of masses of pale-staining, epithelioid cells which lie among the connective tissue fibers that are present within the adventitia of the sinus wall. The receptors are richly supplied with terminal fibers of the sinus nerve, a branch of the glossopharyngeal nerve (IX). Afferent impulses are thus transmitted centrally from the carotid sinus regions on each side of the body to the medullary vasomotor centers (Fig. 10.67). The efferent portion of the reflex arc is provided by the vagus nerve, whereas the sympathetic vasoconstrictor and vasodilator fibers once again form the efferent link to the systemic resistance vessels in the periphery.

External pressure applied directly to the

carotid bifurcation in a human subject so as to augment the pressure within the carotid sinus usually produces a pronounced bradycardia as well as vasodilatation and a marked decline in blood pressure or hypotension. Similar pressure, when it is applied to the common carotid artery below the sinus (so as to induce a fall in pressure within the sinus), generally induces tachycardia, vasoconstriction, and an attendant rise in the systemic arterial blood pressure. Epinephrine also may be liberated. Consequently, the carotid sinus mechanism can mediate both depressor as well as pressor effects.

Studies of impulse frequency within the sinus nerve indicate that impulses are discharged throughout the entire cardiac cycle. When the pressure within the carotid artery is quite high, however, the impulse frequency normally increases during systole and a silent period is recorded during diastole as the arterial pressures fluctuate in accordance with the phases of the cardiac cycle.

Any generalized and sustained increase in the systemic arterial blood pressure increases the frequency of impulse generation in the sinus nerve. In addition, a greater number of sensory receptors are excited within the carotid sinus itself, so that although these elements exhibit a gradual adaptation in the face of a constant and sustained stimulus (ie, a prolonged mechanical distention of the arterial wall), there is little decline in the frequency of impulse formation during cardiac systole under physiologic conditions.

The carotid sinus reflex affords another excellent example of the operation of a feedback control mechanism in the cardiovascular system.

Together, the aortic and sinus nerves behave as buffer nerves for the physiologic maintenance and regulation of the systemic arterial blood pressure. These nerves, together with the reflexes that they elicit, also constitute essential mechanisms for assuring that an adequate circulation is maintained in the brain at all times. Furthermore, the increased diastolic blood pressure and the elevated heart rate that are observed when a subject rises from the recumbent to a sitting or standing position also appear to be mediated via the reflexes subserved by these nerves. The aortic and carotid reflexes also play an important role in counteracting the effects of gravity upon the cardiovascular system, which was discussed earlier (p. 630).

It must be emphasized that the reciprocal as well as mutual interactions between the aortic and carotid pressoreceptor reflexes in response to various arterial pressures are far more complex than was outlined above. It also should be mentioned that the specific mechanisms that are involved in baroreceptor reactions to various arterial pressures are understood only partially. Not only the blood pressure per se, but also the inherent elasticity of the pressoreceptor elements themselves (or their resistance to deformation by varying the intravascular pressure) may play an extremely important part in the regulation of blood pressure.

Some Cardiovascular Reactions in the Normal Individual. In vivo, the level of the blood pressure at any particular moment may be considered to be the resultant of all afferent impulses, both excitatory and inhibitory, which converge upon the vasomotor centers. Ordinarily, impulses arising in the aortic arch and carotid sinus dominate the cardiovascular regulatory pattern. Impulses from the skin, skeletal muscles, viscera, and the higher central nervous areas, discussed earlier, also may exert a significant influence upon this basic pattern of cardiovascular activity. For example, the occasional influence of higher neural centers upon specific peripheral vascular responses is evidenced by such psychomotor reactions as pallor, blushing, erection of the sexual organs, and certain kinds of fainting (syncope). Even a slight excitation of the psychic centers can exert a marked influence upon certain vascular reactions. Thus, the psychogalvanic response, in which changes in the electric resistance of the skin occur, is underlaid by an altered cutaneous blood flow. Measurement of this factor and its involuntary alterations is one parameter recorded in the polygraph (or "lie detector") test.

There is also an important, and generally reciprocal, vascular relationship between the skeletal muscles and the cutaneous as well as splanchnic vessels. Thus, epinephrine dilates the vascular bed within the muscles, while simultaneously it produces constriction of the vessels located within the skin and splanchnic regions. Consequently, a greater proportion of the total circulating blood volume is redistributed to the muscles during extreme activity while relatively less blood now goes to the skin and visceral regions. Similarly, cooling the body, as by immersion in cold water, induces cutaneous vasoconstriction, whereas the vessels within the skeletal muscles again dilate.

Effects of Pressoreceptor Reflexes upon the Heart. It is important to realize that the influence of the aortic and sinus reflexes is not confined solely to their effects upon the peripheral resistance, hence, blood pressure. Thus, any sympathetic vasomotor stimulation that is mediated via these reflexes also directly alters not only the rate but the contractility (ie, the force of contraction) of the heart itself.

There is also some experimental evidence indicating that the stroke work performed by the heart can be increased enormously by carotid sinus stimulation, and this response takes place independently of the heart rate, whereas the peripheral resistance was found to increase only slightly in the same study. Experiments such as this strongly suggest that the pressor effects of carotid sinus pressoreceptor stimulation are primarily and directly upon the heart, and only secondarily a result of an augmented peripheral resistance that is caused by peripheral vasoconstriction. In this regard, pressoreceptor reflexes are well known to induce venoconstriction under certain conditions, and this factor alone would augment cardiac output markedly by increasing the venous return to the right heart.

CARDIOVASCULAR REFLEXES: CHEMORECEPTOR MECHANISMS. In addition to the pressure-sensitive mechanoreceptor terminals of the aortic and carotid nerves, which functionally integrate the arterial blood pressure with activity in the vasomotor center, there are also present in the aortic and carotid bodies epithelioid cells that are richly endowed with nerve endings. These elements are chemoreceptors which continuously monitor the chemical composition of the arterial blood. Histologically, the aortic and carotid bodies are identical, and these tiny structures consist of irregular conglomerations of pale-staining cells that are embedded within the connective tissue of the adventitia and are applied closely to the endothelial cells lining the sinuses. The carotid bodies are richly vascular structures and have an abundant network of sinusoidal capillaries, a fact that accounts for a relatively enormous normal blood flow.

Terminal filaments of the depressor nerve innervate the left and right aortic bodies, whereas the carotid bodies are similarly innervated by the carotid sinus nerve, as depicted in Figure 10.68. As with the pressoreceptor reflexes, afferent impulses are conducted from the aortic bodies, together with the aortic branches of the vagus nerves, while fibers of the sinus branch of the glossopharyngeal nerve transmit impulses from the carotid body to the vasomotor center in the medulla.

The left and right carotid bodies appear to have the highest known blood flow per unit mass of tissue of any structure in the body. This has been estimated to be around 2,400 ml blood per 100 gm carotid tissue per minute, compared to a flow to the left ventricle of about 90 ml per 100 gm myocardial tissue per minute.

Both the aortic and carotid chemoreceptors continuously monitor the arterial blood. The receptor cells within these structures are extremely sensitive to the partial pressure of oxygen (Po_2) and carbon dioxide (Pco_2), as well as to a lesser degree to the relative hydrogen ion concentration as indicated by blood pH. These chemoreceptors are stimulated by a relative lack of oxygen or anoxia, by elevated carbon dioxide levels or hypercapnia, and to some extent by reduced pH levels or acidosis.

The cellular mechanisms whereby the

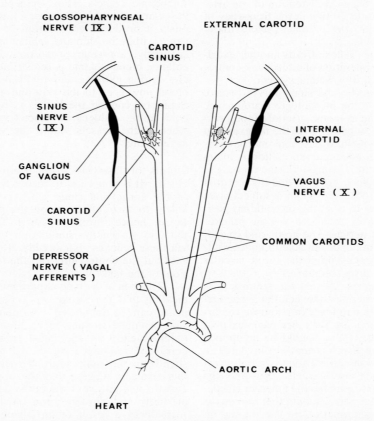

FIG. 10.68. Innervation of the carotid sinus and aortic arch. (After Best and Taylor [eds]. *The Physiological Basis of Medical Practice,* 8th ed, 1966, Williams & Wilkins.)

chemoreceptor cells of the aortic and carotid bodies sense alterations in the chemical composition of the blood are completely unknown. Electrophysiologically, however, it can be demonstrated that these chemoreceptor structures are activated when the partial pressure of CO_2 in the blood rises above approximately 30 mm Hg, and when the Po_2 falls below around 90 mm Hg. As the Pco_2 increases and the Po_2 decreases, a generalized chemoreceptor discharge occurs. The chemoreceptor response to anoxia is far greater than that to hypercapnia; ie, the aortic and carotid bodies are more sensitive to oxygen lack than they are to elevated blood CO_2 levels.

At rest, it is doubtful that the chemoreceptors play any significant role in cardiovascular regulation. Acute hypoxia, however, which can be produced experimentally by inhalation of gas mixtures having a low oxygen and an elevated nitrogen concentration, induces systemic arterial hypertension concomitant with vasoconstriction in the limbs and intestine. Such a response does not occur if the chemoreceptors are inactivated. The elevated blood pressure apparently is not mediated solely by vasoconstriction of the resistance vessels, although direct and simultaneous studies on cardiac output and the capacitance vessels (veins) have not been conducted.

It has been established definitely that the chemoreceptors contribute to maintenance of an effective circulation of blood following a decrease in the total circulating blood volume, eg, after a massive hemorrhage. Experimentally, for example, if blood withdrawal proceeds until the mean arterial pressure is lowered to around 40 mm Hg, intense ''spontaneous'' firing of the chemoreceptors ensues with a concomitant rise in blood pressure. This firing can be inhibited by increasing the partial pressure of oxygen in the blood, whereas section of the sinus nerves produces a further decline in blood pressure. As the baroreceptors appear to be completely inoperative at blood pressure levels below 40 mm Hg, this vascular response can be caused only by vasoconstriction induced by the chemoreceptor reflex. Perhaps the veins as well as the arterial resistance vessels are involved in this pressor response; however, this fact has not been proven experimentally.

In general, both carbon dioxide and oxygen are considered to have dual effects on the cardioregulatory mechanisms; that is, in addition to a direct peripheral effect upon the aortic and sinus chemoreceptor cells, these compounds also exert a direct central effect upon the medullary vasomotor areas.

Carbon dioxide doubtless has a relatively minor effect upon the overall circulation that is mediated via the aortic and carotid chemoreceptor reflexes. This is overshadowed, however, by the direct central effect of this gas upon the neurons of the medullary vasomotor center. An increased blood concentration of carbon dioxide stimulates peripheral vasoconstriction indirectly via a direct excitation of the vasomotor center, whereas in the presence of a reduced blood carbon dioxide level, vasodilatation occurs. The latter response is the result of a reduced tonic discharge to the sympathetic vasoconstrictor fibers from these central areas. Carbon dioxide also has a direct and opposite effect upon the peripheral vessels themselves, so that an increased blood carbon dioxide level causes dilatation, whereas a decreased blood carbon dioxide produces vasoconstriction.

The situation insofar as oxygen concentration is concerned is similar to that seen with different levels of carbon dioxide. Presumably, the central effects of a lowered blood oxygen concentration are mediated to some extent through the aortic and carotid bodies; therefore, reflex peripheral vasoconstriction occurs. In contrast to this central effect, however, locally within the peripheral vessels a hypoxic condition within the blood leads to vasodilatation.

These general remarks are based upon experimental studies, performed largely in animals. It should be realized, however, that it is almost impossible to make a clear physiologic distinction among the results of various experiments as to which factors predominate in normal cardiovascular regulation in vivo, because of the enormously complex interrelationship of Po_2, Pco_2, pH, and metabolic effects in various tissues and organs, both locally and in the body as a whole. For these reasons, it is extremely difficult to extrapolate such data to the intact animal or human. Insofar as the normal physiologic variations in the blood gases and other factors within the human body are concerned in the regulation of the cardiovascular system, it is clear that the chemoreceptor reflexes as well as the direct central effects of Po_2, Pco_2, and pH upon these reflexes are very important. But it is impossible at present to assess the relative physiologic significance of such factors individually.

CARDIOVASCULAR REFLEXES: THE RIGHT HEART. The carotid and aortic baroreceptor and chemoreceptor reflexes have been the most thoroughly studied of all of the cardiovascular reflex mechanisms. Their role in the control of normal circulatory patterns within the body is the best understood despite considerable gaps in our knowledge concerning their functions in vivo.

Laboratory experiments also have demonstrated the presence of other vascular reflex mechanisms that can be elicited by appropriate stimulation of certain regions of the chambers within the heart itself as well as the intrathoracic vessels. It is important to realize that the massive surgical trauma that is necessary to perform such experiments, in addition to the necessity for anesthesia and artificial respiration, may perforce modify normal responses profoundly. Despite these and other technical factors, it is clear that a number of other important reflex cardiovascular regulatory mechanisms are present in the body.

Although a quantitative evaluation of the physiologic role of such mechanisms in the body cannot be attempted at present, nonetheless, certain of these cardiopulmonary mechanisms will be described. Eventually, they even may prove to be of equal, or even greater, significance in the regulation of normal physiologic processes than the presently better-known reflexes. In fact, some of these mechanisms may prove ultimately to have an even more important position in the etiology of clinical dysfunctions within the cardiovascular system than is their role in normal individuals.

It has been demonstrated conclusively that within the atrial walls, as well as in the great veins, there are functional pressure-sensitive nerve endings, ie, mechanoreceptors. These pressoreceptors respond to intraatrial pressure alterations by firing, the afferent impulses being transmitted centripetally by fibers that form components of the vagus nerves. During atrial systole, type A nerve fibers are stimulated, whereas during the passive dilatation of the atria in late diastole, type B nerve fibers fire.

A right heart reflex involving these neural elements as the sensory (afferent) link can be demonstrated by perfusing the right atrium artificially at elevated pressures. Such a procedure induces bradycardia and systemic hypotension, provided the vagi are intact. Bilateral vagotomy abolishes this response. Topical application of atropine to the heart in order to block vagal action also abolishes the bradycardia resulting from the augmented right atrial perfusion pressures, but not the systemic hypotension that develops concomitantly with right atrial distention. Hence, reflex peripheral vasodilatation takes place simultaneously with right atrial distention. The receptors mediating this response are located within the atria, but not the ventricle.

A second reflex that involves the right heart can be elicited experimentally by partial occlusion of the pulmonary artery. When this manipulation is performed, pulmonary arterial blood flow increases. The appearance of this reflex necessitates the presence of intact sympathetic innervation to the right heart, but it has been shown that the vagi and the medullary vasomotor centers are not involved. This reflex probably is caused by a localized reflex sympathetic stimulation of the right ventricle, with a concomitant augmentation of myocardial contractility; hence, the stroke volume of this chamber is increased.

Another right atrial reflex was proposed many years ago by Bainbridge on the basis of the effects on the heart which were produced by the intravenous injection of blood or saline into experimental animals (dogs). Infusion of such fluids caused tachycardia which persisted for as long as the right atrial pressure remained elevated, and this response was abolished only following bilateral vagotomy. During the ensuing years, a considerable number of workers adduced much conflicting evidence concerning the presence or absence of a "Bainbridge reflex," which automatically adapts heart rate to right atrial filling pressure via alterations in impulse frequency along the vagus and sympathetic nerves. (The right atrial receptors presumed to underlie this process never were clearly defined.)

It is now apparent that the tachycardia observed upon the infusion of fluids into the right atrium reflects the direct effect of augmented distention of the sinoatrial cells upon their spontaneous discharge frequency, ie, the intrinsic cardiac rate regulatory mechanism discussed on page 659, as a tachycardia can be induced readily by right atrial distention of the denervated hearts of many species.

In this regard, it is important to mention that bilateral vagotomy in the dog produces a large increase in the spontaneous frequency of the heartbeat, as normal vagal tone in this species acts as a "dragging brake" upon the inherent discharge frequency of the sinoatrial cells in this species (in human subjects and other mammals, this physiologic effect of vagal tone is far less pronounced). Vagotomy would abolish this normal inhibitory vagal effect, so that increased right atrial distention produced in the dog by intravenous infusion in the face of preexisting control tachycardia no longer would exert any additional effects upon the rate. This fact undoubtedly underlies many of the conflicting data that have been published concerning the Bainbridge reflex.

Conversely, if the spontaneous heart rate in vagotomized animals is reduced by stimulation of the distal ends of the vagus nerves, or by the injection of acetylcholine, an increased right atrial pressure once again will induce tachycardia by stimulation of the intrinsic rate regulatory mechanism within the sinoatrial tissue. Furthermore, this inherent rate effect of distention upon the sinoatrial pacemaker tissue can be demonstrated by increasing the venous return, hence right atrial pressures, in human subjects as well as in experimental animals after heart transplant surgery. In such situations, of course, all extrinsic neural influences are completely eliminated. Obviously, under such conditions, no reflex arc can be present, and the response observed by Bainbridge is not a true reflex in any sense of the term. Rather, this response merely reflects an intrinsic pressure sensitivity on the part of the sinoatrial nodal cells themselves.

CARDIOVASCULAR REFLEXES: THE LEFT HEART

Left Atrial Receptors. There is some experimental evidence that pressure-sensitive neural elements are located within the left atrium, perhaps also in the pulmonary veins, and these elements may be involved as sensory receptors in a reflex that regulates blood volume. Hence, congestion of the pulmonary vascular bed, ie, elevated intravascular pressures produced by various experimental techniques, sometimes causes diuresis, according to some workers. Thus, it has been suggested that intracardiac and/or arterial baroreceptors regulate aldosterone secretion by the adrenal cortex (p. 1099) and thus control the blood volume indirectly by the effect of this hormone on urine formation. The factual details of this mechanism remain obscure (see Chap. 12, p. 801).

Left Ventricular Receptors. Under appropriate experimental conditions in which the aortic pressure may be held constant, it can be demonstrated that an increased left ventricular pressure produces a reflex bradycardia, vasodilatation of peripheral blood vessels, and a concomitant systemic hypotension. Cervical vagotomy or pharmacologic blockade of these nerves abolishes this reflex.

Since increasing the left atrial pressure alone fails to elicit this response, presumably the receptors responsible for these effects are located within the left ventricular myocardium and perhaps in the aortic arch as well. The systemic vascular resistance can be shown to be reduced significantly during experimental elevation of the left ventricular systolic pressure, or else during the simultaneous elevation of both the left ventricular diastolic pressure and the mean left atrial pressure. Significantly, an increase in left intraventricular pressure was found to produce a decreased venous return following a large experimentally induced increase in the systemic blood volume. This effect has been shown to be caused by a marked peripheral vascular dilatation, which in turn arises reflexly from mechanical stimulation (stretch) of the left heart and/or aortic root.

These observations suggest that left ventricular pressoreceptor stimulation evokes a reflex inhibition of venous tone, although the physiologic significance of this reflex in vivo is equivocal, since relatively large (ie, unphysiologic) ventricular pressure increases are necessary to demonstrate this systemic flow response experimentally.

The Coronary Chemoreflex. This effect, also known as the Bezold-Jarisch reflex, can be elicited upon injection of veratrine into the left circumflex artery of the heart. Such a procedure results in bradycardia as well as a fall in blood pressure, and these two effects can be shown to be independent of one another. The hypotension is caused by reflex systemic vasodilatation. Both of these responses can be eliminated by bilateral vagotomy. On the other hand, injection of veratrine into the right coronary artery or into the coronary branches that supply the atria does not elicit this reflex, so that there appear to be receptors located specifically within the left ventricle that are supplied with blood by the left circumflex artery.

At present, this reflex appears to be an interesting "physiologic curiosity" which can be evoked only under artificial conditions.

CARDIOVASCULAR REFLEXES: THE PULMONARY CIRCULATION. It is now recognized that the pulmonary vessels possess tone, and that this tone can be altered by extrinsic vasomotor nerves which both constrict and dilate these vessels actively; that is, vasoconstrictor as well as vasodilator fibers are present in the pulmonary circulation, and thus pulmonary vascular resistance can be changed directly by neural influences. Evidence has also been obtained which demonstrates that carotid body stimulation by perfusion of this structure with venous blood causes a fall in intrapulmonary blood pressure. This chemoreceptor response to a low oxygen partial pressure causes a decrease in pulmonary vascular resistance, and this response can be abolished by severing the carotid sinus nerve through bilateral cervical vagotomy, or by the injection of atropine. The specific vascular region within the pulmonary bed which is involved in this response is unknown.

In addition to these effects, two general classes of pulmonary reflex have been demonstrated experimentally.

The Pulmonary Depressor Reflex. In cross-perfusion experiments that are performed with two dogs, blood can be pumped directly from the right atrium of the donor animal into the pulmonary artery of a recipient. The outflow from the pulmonary vein of the recipient now is returned to the jugular vein of the donor, thereby bypassing the left atrium of the recipient animal.

In such an experimental preparation, a pronounced increase in the perfusion pressure in the pulmonary artery of the recipient animal may result in bradycardia as well as systemic hypotension in the donor animal. These effects are eliminated by sectioning the vagi in the donor. Furthermore, the injection of certain drugs into the pulmonary artery also may induce a decrease in the heart rate and systemic arterial blood pressure, effects which are once again abolished by vagotomy.

Experimental observations such as these suggest the presence of a depressor chemoreflex within the pulmonary circulation, and it is felt by some investigators that the sensory receptors responsible for these effects lie within the pulmonary veins.

Reflexes Elicited from the Smaller Pulmonary Vessels. A vast amount of research effort has been expended in order to ascertain whether or not reflexes arise within the arterioles, capillaries, and venules of the pulmonary circuit, but the results of such studies are wholly equivocal at present.

It is known that sympathetic stimulation (ie, of the stellate ganglia) produces largely vasoconstrictor effects in the pulmonary vessels, and also it may reduce blood flow through the lungs by as much as 30 percent. There is no compelling evidence, however, that pulmonary vasoconstriction can be evoked under physiologic conditions by excitation of pulmonary vascular receptors.

Summary

Based upon the foregoing account of the various intrinsic and extrinsic mechanisms that regulate and integrate the activity of the cardiovascular system, certain general statements can be made concerning the normal overall functional activity of the various components of these mechanisms in the human body.

1.   The heart and blood vessels exhibit a considerable degree of autonomy, insofar as regulation of their own activity is concerned. The heart possesses intrinsic mechanisms that regulate both the rate and the force of its beat, whereas the various elements that constitute the vascular system itself all possess to a certain degree the ability to respond to various local stimuli, such as degree of distention, by an appropriate regulation of their caliber in accordance with local bodily requirements for blood flow.
2.   The inherent regulatory mechanisms within both the heart and blood vessels are in turn subject to the influence of a number of extrinsic regulatory mechanisms that are both neural and humoral. The neural mechanisms for the regulation of heart rate and/or contractility operate through both parasympathetic and sympathetic divisions of the autonomic nervous system, whereas the humoral agents may be either local or general insofar as their sources and physiologic effects are concerned.

BLOOD FLOW THROUGH SPECIAL BODY REGIONS

In certain regions of the body, blood flow and its regulation must be considered independently of other organs and tissues. Included in the category of specialized vascular regions of the body are the coronary circulation to the heart, the cerebral circulation, the splanchnic circulation, and the circulation to the skin. Certain other specialized vascular regions include the pulmonary circuit (p. 659); renal blood flow (Chap. 12, p. 772); uterine and placental blood flow (Chap. 15, p. 1195); the fetal and neonatal circulation (Chap. 16, p. 1207); and circulation in the erectile tissues (Chap. 15, p. 1136).

The Coronary Circulation

The importance of the coronary circulation under both normal and pathologic circumstances cannot be overemphasized. Approximately one-third of all deaths in the United States occur each year through some form of coronary involvement, and practically all elderly persons exhibit, to various degrees, some manifestation of coronary vascular disease.

MORPHOLOGIC CONSIDERATIONS.   The two principal coronary arteries providing the vascular supply to the myocardium arise from sinuses, which are located behind the cusps of the aortic semilunar valve at the root of the aorta (Fig. 10.9). The openings of these vessels are patent throughout the cardiac cycle, as eddy currents (Fig. 10.39A) deflect the cusps of the valve away from the orifices of the coronary arteries during systolic ejection of blood from the left ventricle.

The left coronary artery is distributed primarily to the left ventricle, whereas the right coronary artery supplies principally the right ventricle, and in some instances a portion of the left ventricle as well. In around 80 percent of normal individuals, flow through the right coronary artery is somewhat greater than through the left, whereas in the remaining 20 percent, left coronary flow is greater.

Venous blood leaves the myocardium through two main channels. The left ventricle is drained principally via a system of superficial veins which empty into the coronary sinus (Fig. 10.8). The small anterior cardiac veins drain independently and directly into the right atrium. These vessels handle approximately 70 percent of the total coronary flow and they receive blood mainly from the left ventricle. The rest of the heart is drained by a deep system of thebesian veins which empty directly into all chambers of this organ. The deep system of vessels also contains arteriosinusoidal vessels, which again empty directly into the chambers. A few arterioluminal vessels are present, and these also form direct connections between the coronary arterioles and the atrial and ventricular chambers.

In man, coronary arterioles normally are somewhat less than 40 μ in diameter; however, in coronary arterial disease, there is evidence that these vessels both enlarge and become more abundant per unit mass of tissue.

NORMAL CORONARY BLOOD FLOW AND PRESSURE GRADIENTS.   In an average human being at rest, the coronary blood flow is around 225 to 250 ml per minute. This flow rate amounts to approximately 0.8 ml per gm of tissue, or 5 percent of the total cardiac output.

During heavy exercise, cardiac output can increase by as much as sixfold and the work output 10-fold, whereas the coronary blood flow is elevated a disproportionate fourfold to fivefold under such conditions. An increased efficiency of cardiac contractility during severe exertion apparently compensates for this relatively lower increase in blood flow as the total energy output by the heart increases.

In a manner similar to skeletal muscle, the myocardium compresses the blood vessels within its structure during contraction, and during systole the pressure within the left ventricle is only slightly less than that within the aorta (Fig. 10.69A). Blood flow through the left ventricular coronary vessels during various phases of the cardiac cycle is highly phasic in character, as illustrated in Figure 10.69A. During systole of the heart, compression of the coronary vessels in the superficial portion of the left ventricle is nil, and some flow is present in the subendothelial portion of the myocardium throughout the cardiac cycle, despite the compression of the myocardium. In dia-

A. LEFT HEART

B. RIGHT HEART

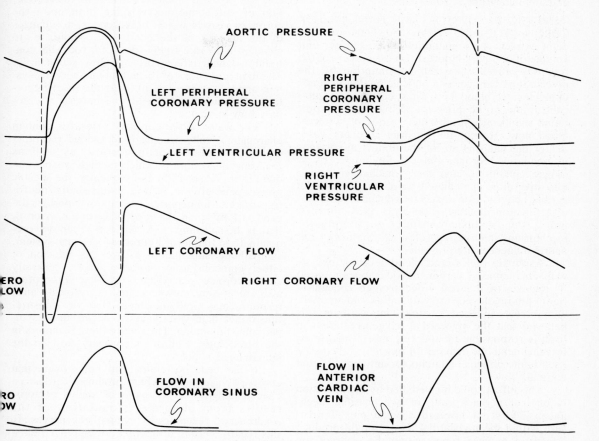

AORTIC PRESSURE

LEFT PERIPHERAL CORONARY PRESSURE

RIGHT PERIPHERAL CORONARY PRESSURE

LEFT VENTRICULAR PRESSURE

RIGHT VENTRICULAR PRESSURE

LEFT CORONARY FLOW

RIGHT CORONARY FLOW

ERO LOW

FLOW IN CORONARY SINUS

FLOW IN ANTERIOR CARDIAC VEIN

RO OW

FIG. 10.69. Curves indicating phasic changes in left and right coronary pressures and flows in relation to ventricular and aortic pressures. (After Best and Taylor [eds]. *The Physiological Basis of Medical Practice,* 8th ed, 1966, Williams & Wilkins.)

stole, when the myocardium relaxes completely, blood flow no longer is impeded; therefore, a rapid blood flow occurs through the left ventricular capillaries throughout this phase of the cardiac cycle.

In the right ventricle, however, the pressure differences between this chamber and the aorta are sufficiently great during both systole and diastole that blood flow is not markedly altered during any phase of the cardiac cycle (Fig. 10.69B).

The diastolic phase of the cardiac cycle is shortened appreciably during tachycardia; coronary blood flow is diminished during periods when the heart rate is accelerated markedly.

Since blood flow ceases momentarily in the subendocardial portion of the left ventricular myocardium during systole (Fig. 10.69A), this region of the heart is the most common locus of myocardial infarction as well as of ischemia. In patients having a stenosis (or narrowing) of the aortic valves, for example, left ventricular coronary blood flow is decreased sharply, because in such individuals the pressure in the left ventricle must be significantly higher than in the aorta in order to eject any blood at all. The coronary vessels therefore undergo severe compression during systole, and this compression in turn can lead to the development of serious myocardial ischemia of a progressive nature.

Coronary blood flow also is impaired when the aortic diastolic pressure is low. During congestive heart failure, when the venous pressure is elevated markedly, coronary flow also is reduced because the effective coronary perfusion pressure is lowered because of the elevated pressure in the right heart.

For technical reasons, accurate measurement of coronary blood flow in human subjects is difficult. An intracardiac catheter can be used to sample blood from the coronary sinus and the Kety method (p. 683) can be applied to the heart, if it is assumed that the N_2O (nitrous oxide) content of the blood in the coronary veins is representative of the total myocardial efflux. The figures given for coronary blood flow at the beginning of this section were obtained by using the Kety method for determining coronary flow.

INTRINSIC REGULATORY MECHANISMS FOR CORONARY BLOOD FLOW.

Since the myocardium extracts large quantities of oxygen per unit volume of coronary blood at rest (65 percent of the total O_2 present), oxygen consumption by the myocardium can increase only as the consequence of an elevated blood flow. Thus, coronary vascular caliber and aortic pressure changes are of profound significance in altering the rate of coronary blood flow. Hence, the intrinsic coronary vasodilator mechanisms are of the utmost importance in regulating coronary blood flow under physiologic conditions, and these mechanisms come into operation immediately upon an increase in cardiac activity. Conversely, a reduced level of cardiac function diminishes coronary flow. Furthermore, the intrinsic coronary autoregulatory mechanisms operate regardless of whether or not neural influences are present.

Asphyxia (a lowered blood level of oxygen with an increased carbon dioxide concentration) or hypoxia (a reduced blood oxygen concentration) both invoke coronary dilatation and an attendant increase in blood flow to the myocardium. It is believed that the myocardial oxygen requirement itself is responsible for this response. Thus, in effect the metabolic demand for oxygen by the tissue itself is the principal stimulus for autoregulation of coronary flow.

This effect could be mediated by the operation of one, or perhaps both, of two possible mechanisms. First, a lowered oxygen partial pressure within the myocardial tissues could produce directly a reduced tonus within the walls of the coronary vessels themselves, leading to their vasodilatation. Second, a relative hypoxia (anoxia) could stimulate the liberation of vasodilator compounds locally within the tissues of the heart. For example, adenosine compounds have been implicated in this effect, and other chemical agents such as prostaglandins or kinins also may be demonstrated eventually to be involved in this process.

Regardless of the other mechanism(s) ultimately shown to be involved in coronary vasodilation in vivo, it is clear that hypoxia per se causes such a response, and thus blood flow to the myocardium is regulated directly in accordance with the metabolic requirement for exygen by the cardiac muscle fibers themselves.

Hypercapnia and a reduced pH, ie, increased blood acidity, both have been shown experimentally to alter coronary blood flow, but to a negligible extent.

EXTRINSIC REGULATORY MECHANISMS FOR CORONARY BLOOD FLOW.

The extrinsic autonomic nerves to the heart can influence coronary blood flow either directly or indirectly. A direct action of sympathetic or parasympathetic stimulation would result from the liberation of the transmitter substances norepinephrine or acetylcholine and the effects of these compounds upon the musculature of the coronary vessels. The indirect action of autonomic nerve stimulation upon the coronary vessels would be secondary to changes in the activity of the heart as a whole, ie, to changes in the rate and/or force of myocardial contraction which affect coronary flow secondarily. The indirect actions of the autonomic nervous system appear, at present, to exert the most important influences upon the normal pattern of coronary flow in vivo.

The walls of the coronary arterioles contain pharmacologically specific α-adrenergic receptors which mediate vasoconstriction, as well as β-adrenergic receptors which mediate vasodilatation (Table 7.6, p. 276). Consequently, the specific physiologic effects of the sympathetic neurotransmitters, epinephrine and norepinephrine, depend strictly upon the type of receptor, α or β, that is being stimulated. Since β receptors predominate in the normal coronary vascular system, sympathetic stimulation, as well as infusion of either epinephrine or norepinephrine, generally causes coronary vasodilatation. Some α receptors also are present, however, and occasionally an α response may predominate so that vasoconstriction rather than vasodilatation occurs upon sympathetic stimulation. The role of these receptors in the physiologic control of coronary flow in the human heart is not known.

It has been postulated, but not proven, that coronary vasospasm, which produces cardiac ischemia and the severe pains of angina pectoris, results from abnormal stimulation of the α-adrenergic receptors by sympathetic nerves. In this regard, certain drugs that are employed clinically to relieve angina by dilating the coronary vessels may possibly act upon these α receptors. Accordingly, such compounds as amyl nitrite, nitroglycerin, aminophylline, and papaverine are used clinically to relieve anginal pains by dilating the coronary vessels, thus alleviating the concomitant myocardial ischemia.*

There is one situation in which a clear-cut neural control of coronary flow is evident. In a normal individual, when the systemic blood pressure falls markedly, such as after a massive hemorrhage, the reflex sympathetic discharge that is stimulated by this event produces coronary vasodilatation and an increased blood flow to the heart ensues. This response takes place while the splanchnic, cutaneous, and renal vessels become constricted at the same time, and flow through these regions is reduced sharply. Consequently, the circulation to the heart, like that to the brain,

---

Alternatively, nitroglycerin could act in angina by reducing cardiac work, thereby lowering the oxygen requirement of the heart so that in turn a lower coronary flow would be adequate and anginal pains would be relieved. Nitroglycerin also might act by dilating collateral coronary channels, thus allowing more blood to be shunted to the ischemic region.

is maintained at the expense of flow to less immediately vital organs. This preferential redistribution of flow to heart and brain is of great importance in physiologic emergencies.

The Cerebral Circulation

The brain, like the heart, continuously requires an adequate blood supply. Interruption of blood flow to this region of the nervous system for even a few seconds produces loss of consciousness, whereas a marked reduction or cessation of flow for even a few minutes can produce either death or irreversible brain damage.

MORPHOLOGIC CONSIDERATIONS. The entire arterial blood supply to the brain in man, save for a small amount which flows in the anterior spinal artery to the medulla, is carried by four vessels, ie, the two internal carotid arteries and the two vertebral arteries. The vertebral arteries anastomose to form the basilar artery of the hindbrain. The circle of Willis in turn is formed by the union of the basilar artery and the internal carotids, and from this structure the six large arterial vessels that supply the cerebral cortex take their origin (Fig. 10.70).

In man, a relatively small proportion of the total blood flow to the brain is carried by the vertebral arteries so that the internal carotids convey most of the blood to this structure. Injection of various tracer substances into one carotid artery results in a distribution of the material principally within the cerebral hemisphere on the same side of the body as that on which the injection was made.

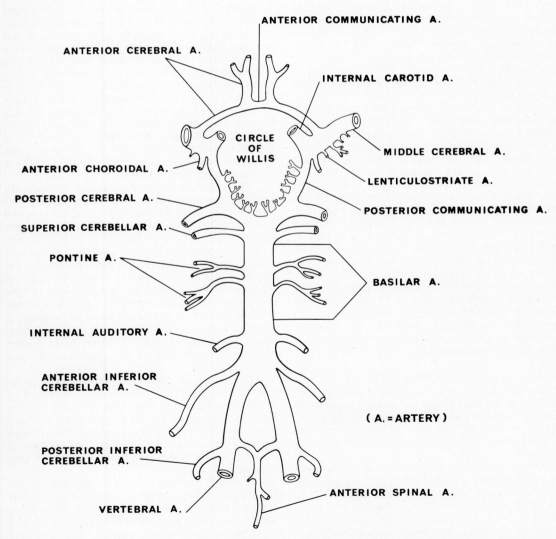

FIG. 10.70. The circle of Willis and the major arteries supplying the brain. (After Chusid. *Correlative Neuroanatomy and Functional Neurology*, 14th ed, 1970, Lange Medical Publications.)

This response probably results partially from the fact that normal intravascular flow is streamlined, but more importantly because the normal pressures within both carotids are equal; hence, no cross-perfusion occurs. Even though the pressures within the anastomotic channels of the circle of Willis do become unequal, the flow is still channeled left side to left side, and right side to right side. For example, partial occlusion of one internal carotid artery, especially in older patients, may induce serious symptoms of cerebral ischemia, indicating an insufficient blood flow to certain portions of the brain on the ipsilateral side as the occlusion. In man there are some precapillary anastomoses between the cerebral arterioles; however, blood flow through such channels is wholly inadequate to sustain an adequate circulation and to prevent infarction following the occlusion of a cerebral artery.

Blood drains from the brain principally by way of the deep veins and dural sinuses (p. 255), and in man these channels empty primarily into the jugular veins. A small quantity of venous blood also drains into emissary veins which run to the scalp, as well as into the vertebral veins in the spinal canal.

The cerebral vessels themselves exhibit several unique features. First, within the choroid plexuses (p. 252) there are physical gaps present between the endothelial cells that comprise the capillary walls, although the epithelial cells that comprise the choroid proper interdigitate to form a very dense three-dimensional network. Second, the capillaries within the substance of the brain itself are similar to these vessels as found within other parts of the body; however, no fenestrations are seen in the endothelium such as are present in certain other organs (Fig. 9.17, p. 559). A thick basement membrane is present in the cerebral capillaries, and the endothelial cells that comprise these vessels are tightly bound together. Third, the capillaries of the brain are surrounded by a membrane composed of the end-feet of neuroglial cells known as astrocytes (Fig. 7.13, p. 251). The numerous end-feet of the multiple processes on these cells are attached closely to the capillary basement membrane, and thus invest the capillary walls almost completely (Fig. 7.16, p. 256).

THE BLOOD–BRAIN BARRIER. The concept of a blood–brain barrier was introduced on page 256. Because of the physiologic and clinical importance of this functionally specialized region of the brain, certain features of this part of the cerebral circulatory system will be reviewed in the following discussion.

When certain materials such as acidic dyes (eg, trypan blue) are injected into living animals, the neural tissue does not become stained; thus, the existence of a blood–brain barrier was postulated in order to account for this phenomenon. Subsequent research has demonstrated clearly that only water, oxygen, and carbon dioxide can traverse the cerebral capillaries readily to enter the neural tissue. All other substances are exchanged across this membrane slowly, and although the rate of transcapillary exchange of different materials varies widely among various organs in the body, this process in the cerebral capillaries is so different quantitatively when compared to other regions that the concept of a specific blood–brain barrier is valid. Insofar as permeability is concerned, it is the rate of transfer of various materials across the blood–brain barrier that is of the utmost importance, since all substances can traverse this region and enter the brain, at least to some extent and given sufficient time.

As noted, water, oxygen, and carbon dioxide are able to cross the blood–brain barrier both readily and with extreme rapidity. Glucose also is transported with reasonable facility from the blood into the neural tissue. Sodium, potassium, magnesium, chloride, bicarbonate, and phosphate ions in plasma require between 3 and 30 times longer to come into equilibrium with cerebrospinal fluid than they do to equilibrate with the interstitial fluid in other regions of the body. Urea also penetrates into the brain and cerebrospinal fluid very slowly compared with other tissues. Only minute traces of catecholamines and proteins are able to cross the blood–brain barrier. Normally, a pH difference exists between the blood and brain, and this difference amounts to around 0.1 unit. In general, substances having a high lipid solubility (or a higher partition coefficient, p. 88) tend to cross the blood–brain barrier more rapidly than do those having a lower lipid solubility.

The blood–brain barrier develops during early years of postnatal life so that the cerebral capillaries are far more permeable in newborn children than in adults. Thus, in severely jaundiced infants (eg, in erythroblastosis fetalis, p. 573), bile pigments can penetrate directly into the central nervous system. In combination with the relative asphyxia which also is present in such infants, the pigment accumulation in the brain can lead to serious damage to the basal ganglia, and thus a severe form of icterus neonatorum known as kernicterus (p. 573). In adults, severe jaundice does not affect neural function, because the blood–brain barrier prevents the entrance of bile pigments into the brain.

Certain areas within the brain appear to lie "outside" of the blood–brain barrier, as judged by the results of staining experiments. These regions are (1) the pineal body, (2) the posterior pituitary together with the ventral area of the median hypothalamic eminence, (3) the intercolumnar tubercle, (4) the supraoptic crest, and (5) the area postrema.

Although the physiologic importance of the greater permeability in these regions of the brain is unclear, it has been postulated that these areas may be various types of chemoreceptors, hence must be in more direct contact with systemic blood. The area postrema, for example, can induce vomiting when stimulated by appropriate chemical changes in the plasma. There also is some experimental evidence indicating that vasopressin (antidiuretic hormone) secretion is regulated by osmoreceptors, which are located outside of the blood–brain barrier (Chap. 12, p. 796; Chap. 15, p. 1044). Furthermore, the median hypothalamic eminence may be acted upon by circulating cortisol and sex hormones, so that anterior pituitary gonadotropin secretion and adrenocorticotropic hormone (ACTH) production are regulated by this mechanism. The pineal body, supraoptic crest, and intercolumnar crest remain enigmatic, insofar as any specific chemoreceptor functions are concerned.

The highly selective permeability characteristics of the blood–brain barrier doubtless operate to maintain within very precise qualitative and quantitative limits the composition of the internal environment in which the cerebral neurons live and perform their manifold functions. These neural elements are extraordinarily sensitive to even minute changes in the ionic composition of their environment; their normal functions require an exceedingly delicate homeostatic balance among the various factors that act upon them. Hence, the blood–brain barrier furnishes an additional protective mechanism to the various other homeostatic mechanisms found in the body for the regulation of the quantitative as well as qualitative consistency of the interstitial fluid compartment.

Clinically, an understanding of the blood–brain barrier is of the utmost importance because the physician must understand the permeability of drugs through this region in order to treat central nervous diseases effectively. For example, sulfadiazine and erythromycin can penetrate the blood–brain barrier with facility, whereas the antibiotics chlortetracycline and penicillin cross this region only to a very limited extent.

Another aspect of clinical importance is the fact that the blood–brain barrier breaks down in regions of the brain that are infected, irradiated, or at the locus of tumors. In the last instance, for example, such growths can be localized accurately by the injection of radioactive iodine-labeled albumin, which becomes concentrated in the tumor but not in normal tissue where the barrier is intact. Thus, the radioactivity stands out as discrete shadows on x-ray films and delineates the area involved in the pathologic process.

CEREBROSPINAL FLUID. Cerebrospinal fluid was considered in Chapter 9, page 561 et seq, in relation to the body fluids and compartments (see also p. 254).

NORMAL CEREBRAL BLOOD FLOW. In an adult human, the normal blood flow through the brain is around 55 ml per 100 gm of tissue per min; the extreme range is between 40 and 65 ml/100 gm tissue/min, or about 750 ml of blood per min for the entire brain. In children, the value is around 105 ml/100 gm of tissue/minute, but at puberty this value declines to adult levels. This flow rate is approximately 15 percent of the entire resting cardiac output, and even under extremely different states of activity (eg, between sleep and strenuous exercise) the cerebral blood flow does not vary appreciably from the resting level. The control systems for regulating cerebral blood flow thus are attuned closely, and exceptions to this rule of a constant cerebral blood flow occur only in situations of severe anoxia or extreme carbon dioxide excess.

The normal regional blood flow through various portions of the brain varies widely. Evidence obtained on lower animals indicates that the mean flow is highest in the inferior colliculus, averaging 1.8 ml/gm tissue/minute. Flow in cerebral white matter is only around 0.2 ml/gm tissue/minute, whereas cerebral gray matter has an approximately sixfold greater flow, being around 1.2 ml/gm tissue /minute. The white matter of the spinal cord exhibits the lowest flow rate, 0.15 ml/gm tissue/minute. Other specific regions of the brain have blood flows intermediate to those cited.

ESTIMATION OF CEREBRAL BLOOD FLOW. In accordance with the Fick principle (p. 640), the blood flow through any organ can be measured by determining the quantity of a substance which is removed from the blood by an organ per unit time (the rate of extraction, Q_s) divided by the arteriovenous concentration difference of that substance, or expressed mathematically:

$$F = \frac{Q_s}{A_s - V_s}$$

In the Kety or nitrous oxide (N_2O) method for the determination of cerebral blood flow, the subject inhales small (ie, subanesthetic) quantities of nitrous oxide, and as this gas is taken up by the brain an equilibrium between blood and brain levels of N_2O is achieved in about 10 to 14 minutes. At equilibrium, the N_2O concentration in cerebral venous blood is equal to that in the brain, since the blood–brain partition coefficient for this gas is 1.0. Consequently, the concentration of N_2O in cerebral venous blood divided by the mean arteriovenous N_2O difference during the equilibration interval is equal to the mean cerebral blood flow *(CBF)* per unit mass of brain tissue.

Cerebral blood flow can be calculated readily, if it is assumed that one knows the total quantity of N_2O taken up by the brain as well as the integral mean arteriovenous difference of blood N_2O concentration. Alternatively, the mean arteriovenous blood difference for N_2O can be estimated from a plot of the arterial and venous blood concentrations at different times during the equilibration period, as shown in Figure 10.71. By use of a previously calculated proportionality factor, the total quantity of N_2O absorbed per 100 gm brain tissue can be obtained by multiplying the arterial blood concentration of N_2O at the 10-minute level times this factor.

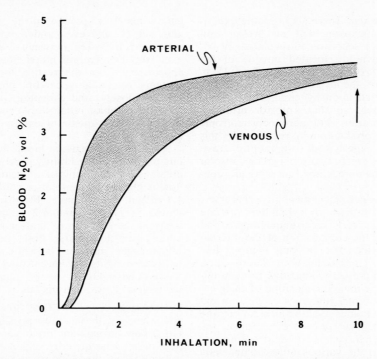

FIG. 10.71. Plot of uptake of N_2O by the brain based upon the arteriovenous blood differences in N_2O concentration for use in determining cerebral (or coronary) blood flow by the Kety method. Saturation of tissue with gas is complete in about 10 minutes (vertical arrow). (After Guyton. *Textbook of Medical Physiology,* 4th ed, 1971, Saunders.)

By use of the Fick equation, the cerebral blood flow values can be calculated in ml/100 gm of brain/minute and this figure can be converted into an approximate value for total cerebral blood flow *(TCBF)* by multiplying by 1,400 gm, which is the average weight of the adult human brain.

In human subjects, when the Kety method is used to determine cerebral blood flow, serial arterial blood samples can be obtained from any artery, as the systemic arterial concentration of all substances is the same throughout the body. Cerebral venous blood is obtained by intravenous puncture of the jugular bulb with a needle, and this dilatation of the internal jugular vein lies at the point of exit of the vein from the skull. The subject then inhales N_2O for 10 minutes, during which period blood samples are drawn sequentially at intervals. The N_2O concentration in the samples is determined and the mean arteriovenous N_2O difference is calculated.

The Kety method also can be used to determine the mean coronary blood flow (p. 678). In this procedure, mixed venous blood samples from the heart are obtained by means of an intracardiac catheter whose tip lies within the coronary sinus in the right atrium, and a total heart mass of 280 gm is substituted for the 1,400-gm weight used for calculations involving the brain. Otherwise, the procedure and calculations are identical to those used for determining cerebral blood flow.

The Kety method for the determination of

cerebral blood flow has several limitations. First, it is essential to obtain mixed cerebral venous blood in order that cerebral blood flow be determined with any degree of accuracy. In humans, sampling the jugular venous blood apparently satisfies this requirement in the majority of cases. Second, since cerebral blood flow is determined over a 10-minute period, the value for flow rate thus obtained gives only an average flow rate during this interval, and it tells nothing about regional differences in blood flow through different parts of the brain. Furthermore, no indication of variations in the regional blood flow is given by the Kety method. This fact is important, as vasoconstriction (or vascular occlusion) in one region of the brain could be offset by a compensatory vasodilatation in other regions so that the calculated value for total cerebral blood flow would be unchanged and thus would appear normal. Third, ischemia in one region of the brain obviously cannot be detected with the Kety method, as this technique determines only the total blood flow per unit mass of brain tissue, and as an ischemic (or occluded) region would take up little (or no) N_2O, an inaccurate value would be obtained.

INTRINSIC REGULATION OF THE CEREBRAL CIRCULATION. In contrast to other organs in which blood flow increases markedly when they become active, cerebral blood flow is regulated so that it remains relatively constant despite widely fluctuating physiologic conditions. Even vigorous mental activity does not alter the total cerebral blood flow appreciably. The several factors known to affect total cerebral blood flow may be considered individually.

Carbon Dioxide and Cerebral Blood Flow. The level of carbon dioxide in the cerebral tissues is of major importance in regulating the caliber of the cerebral vessels. Hence, an increased arterial P_{CO_2} is a powerful vasodilator of the cerebral vessels, whereas a reduced P_{CO_2} produces cerebral vasoconstriction (Fig. 10.72).

During hyperventilation, when an abnormally large quantity of carbon dioxide is excreted from the blood via the lungs (p. 737), the cerebral symptoms that develop (eg, dizziness) are caused, in large measure, by the lowered arterial P_{CO_2} level, because the attendant cerebral vasoconstriction induces a transient ischemia, hence a relative oxygen deficiency in the brain. Thus, irritability of the cerebral neurons is decreased markedly by elevated carbon dioxide levels. Therefore, when the P_{CO_2} rises, the cerebral vessels dilate, and the increased blood flow that results as a consequence of this vasodilatation rapidly removes the excess carbon dioxide from the brain in addition to providing more oxygen at the same time. Therefore the P_{CO_2} has a dual role in the homeostasis of cerebral blood flow.

Oxygen and Cerebral Blood Flow. In a manner similar to the vascular beds of the heart and skeletal muscle, cerebral blood flow increases markedly when the P_{O_2} of the blood (or in the brain tissue itself) falls. The resulting vasodilatation tends to restore oxygenation of the tissues toward normal.

It should be noted that there is an exceedingly delicate interplay between oxygen and carbon dioxide levels in the brain, insofar as the state of vasoconstriction, hence, blood flow, is concerned. The cerebral vessels are far less sensitive to oxygen deficiency as a vasodilator stimulus than they are to elevated carbon dioxide levels; however, the reciprocal regulation of the levels of these two gases provides the principal stimulus to the intrinsic regulation of cerebral vascular caliber, hence, blood flow in the brain. Thus, asphyxia provides a very powerful stimulus to an enormous increase in cerebral blood flow, because of the vasodilatation produced by the combined stimuli of hypoxia and hypercapnia.

Arterial Pressure and Cerebral Blood Flow: Autoregulation of Vascular Caliber. As shown in Figure 10.73, the cerebral vessels are able to adapt automatically their pressure–flow relationship over an extremely broad range of mean arterial pressures (cf Fig. 10.66). Thus, the cerebral vascular bed is able to autoregulate the caliber of its vessels within close limits so that blood flow remains essentially constant in the face of wide fluctuations of arterial pressure.

Intracranial Pressure and Cerebral Blood Flow. Since the brain, spinal cord, and cerebral ves-

FIG. 10.72. Relationship between cerebral blood flow and the arterial P_{CO_2} level. Note that cerebral blood flow is normal (100 percent) at an arterial P_{CO_2} level corresponding to 40 mm Hg. (After Guyton. *Textbook of Medical Physiology*, 4th ed, 1971, Saunders.)

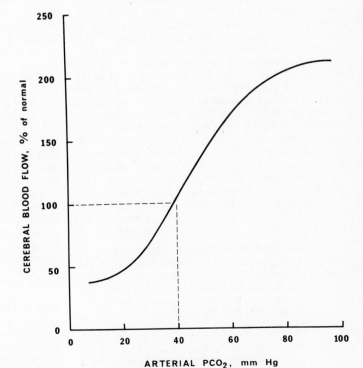

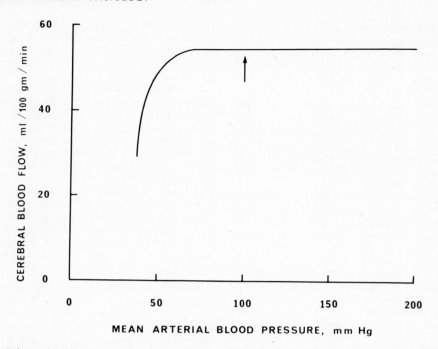

FIG. 10.73. Relationship between cerebral blood flow and mean arterial blood pressure. Note that the blood pressure must fall considerably below normal (about 100 mm Hg, *arrow*) before cerebral hypotension develops, whereas marked hypertension has no effect on flow through the brain. See also Figure 10.66. (After Guyton. *Textbook of Medical Physiology,* 4th ed, 1971, Saunders.)

sels of an adult human are encased within a rigid bony vault, and since the brain tissue, blood, and cerebrospinal fluid are incompressible for all practical purposes, the total volume of all these elements must remain approximately constant at all times in health as well as in disease. This statement is called the *Monro–Kellie doctrine.** The significance of this hypothesis lies in the fact that the cerebral vessels are compressed when the intracranial pressure rises, and this pressure can increase through alterations in either the arterial or venous pressures. Therefore, an elevated venous pressure decreases the cerebral blood flow by inducing compression of the cerebral vessels as well as by reducing the effective perfusion pressure (ie, by lowering the arteriovenous pressure gradient), and thereby this mechanism helps to compensate for altered arterial pressures at head level, and it also exerts a protective effect against rupture of the cerebral vessels while straining at stool or during parturition (Valsalva maneuver, p. 653).

pH and Cerebral Blood Flow. Alterations in the relative concentration of hydrogen ion within the blood or tissues of the brain produce negligible changes in the cerebral blood flow. Therefore, this factor appears to be inconsequential in normal regulation of blood flow to the brain.

*Exceptions to this generalization are found in the cranium of a young child prior to fusion of the sutures as well as in an adult with a cranial defect.

EXTRINSIC REGULATION OF THE CEREBRAL CIRCULATION. The cerebral vessels are innervated both by adrenergic vasoconstrictor fibers and by parasympathetic vasodilator fibers. The physiologic role of these nerves in the normal regulation of cerebral vascular caliber in man, hence blood flow, is extremely doubtful, as it appears at present as though no vasomotor influences are involved in this process, normal blood flow being achieved for the most part by the intrinsic mechanisms described above.

Should the intracranial pressure become elevated rapidly to about 35 mm Hg, a marked reduction in cerebral blood flow occurs, and this reduced blood flow in turn causes a relative cerebral ischemia. The ischemia stimulates the medullary vasomotor center so that the systemic arterial blood pressure increases through peripheral vasoconstriction (Cushing's reflex). Concomitantly, the cardioinhibitory center is stimulated, and a decreased respiratory rate as well as a bradycardia are produced. The increased systemic arterial blood pressure tends to maintain cerebral blood flow, and this increase in pressure is related directly to the rise in intracranial pressure over a wide range. If the intracranial pressure rises to a point where it exceeds the arterial blood pressure, cerebral blood flow stops completely.

It is uncommon for the activity of the neurons of the brain to increase to such an extent that an overall increase in the metabolic rate occurs, but if such a generalized increase in neuronal activity does occur (as during massive convulsions), the blood flow can increase up to 50 percent above normal. Conversely, some anesthetics may reduce cerebral blood flow up to 40 percent through their generalized depressant effect upon the metabolic processes in the brain.

An interesting extrinsic modification of blood flow to the brain occurs when light is flashed suddenly into the eyes of experimental animals (Fig. 10.74). In response to this stimulus, the regional blood flow is increased within the occipital cortex, superior colliculi, and lateral geniculate bodies, although the total blood flow through the brain remains essentially unchanged.

INTEGRATED CONTROL OF CEREBRAL BLOOD FLOW. The several mechanisms discussed above operate in a coordinated fashion at all times to maintain an adequate flow of blood through the brain within closely controlled limits. Most important in this respect are the intrinsic mechanisms that are present within the cerebral vascular bed itself; that is, the direct response of the vessels to alterations in carbon dioxide and oxygen levels, as well as the inherent vascular autoregulatory mechanism for maintaining a constant flow despite wide fluctuations in the systemic arterial blood pressure.

The Cutaneous Circulation

Blood flow through the skin plays two major roles. First, in the nutrition of the skin tissues themselves, and second, in the critical process of controlling body temperature or thermoregulation (p. 687). The skin is also an important organ insofar as protection of the body against invasion by foreign agents or microorganisms is concerned; the cutaneous inflammatory response to such stimuli was discussed on page 547. Local vascular reactions within the skin to other appropriate stimuli, the white reaction (p. 647), the triple response (p. 647), and the axon "reflex" (p. 646) also have been discussed.

In this section, the more general cutaneous vascular reactions and their control will be considered.

MORPHOLOGY OF THE CUTANEOUS VASCULAR BED. In addition to the usual arrangement of arterioles, capillaries, and venules which subserve a nutritive function in the dermal region, there also is a highly developed subdermal venous plexus found subcutaneously throughout the body. In certain regions of the body, arteriovenous anastomoses also are found (Fig. 10.19). These blood shunts directly connect the arteries with the subdermal venous plexus within the volar surfaces of the hands and feet, lips, fingers, toes, and ear-lobes.

The specialized arteriovenous anastomoses have thick muscular walls which are heavily innervated by sympathetic vasoconstrictor nerve fibers, which in turn exert their stimulatory effect on the vascular smooth muscle by secretion of the neurotransmitter norepinephrine. During maximum neural activity, the anastomoses can constrict to such an extent that blood flow into the venous plexuses is practically nil. When maximally dilated, on the other hand, a very rapid flow of warm blood from deeper regions of the body takes place between the arterioles and the venous plexus. Hence, the cutaneous arteriovenous shunts function as "adjustable radiators" in controlling heat loss from the body surface under different physiologic conditions.

The widespread capillary and venous plexus networks found beneath the skin of the entire body surface also function as major blood reservoirs. When severe stress is imposed upon the cardiovascular system, for example by hemorrhage, strenuous exercise, or anxiety, the vessels that comprise the venous plexuses beneath the skin of the entire body can constrict strongly under the influence of sympathetic stimulation. This widespread effect in turn augments the total circulating blood volume by an estimated 5 to 10 percent, and thereby the subcutaneous venous plexus can act as a blood reservoir of significant magnitude under certain conditions. In shock, however, elevating the patient's body temperature (eg, by the excessive use of blankets) to too great an extent can offset this beneficial effect of peripheral venoconstriction by causing peripheral and cutaneous vasodilatation through the agency of warm blood

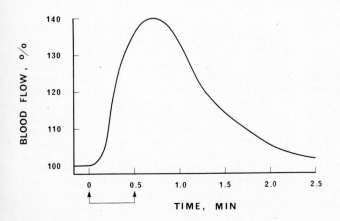

FIG. 10.74. Blood flow increases in the occipital region of the brain when a light is flashed into the eyes of an experimental animal for a few seconds, as indicated by the distance between the vertical arrows. (After Guyton. *Textbook of Medical Physiology*, 4th ed, 1971, Saunders.)

inhibiting the hypothalamic vasoconstrictor center (p. 300), and the shock state will be worsened (p. 300).

The skin also affords a region where certain vascular reactions as well as blood changes can be observed directly, and this fact is of much importance for the clinical diagnosis of certain diseases.

CUTANEOUS BLOOD FLOW. The blood flow through the skin may range from around 1 ml/100 gm of skin/minute to around 150 ml; thus, the flow through this region is among the most labile of any in the body. In turn, this wide range of blood flow rates reflects any alterations in the metabolic rate of the entire body, as well as the environmental temperature. The small volume of blood flow required for the nutrition of the skin itself has an almost negligible role in thermoregulation. At ordinary skin temperatures, the volume flow required for thermoregulation alone is about 10-fold greater than that necessary to maintain the normal integrity and functions of the skin tissues themselves. But in situations of extreme cold, vasoconstriction may so curtail skin blood flow, especially in the extremities, that even this minimal function is impaired with serious consequences to the individual.

In an average adult, blood flow to the skin of the entire body amounts to around 0.25 liter per m² of body surface area, or a total of approximately 300 ml per minute (Table 10.6). During maximum vasodilatation of the cutaneous vascular bed, such as occurs during extremely warm environmental conditions, the blood flow can increase around sevenfold, so that the total flow now is approximately 2.1 liters per minute. This factor alone can account for a significant proportion of the total cardiac output under such conditions. Consequently, patients with cardiac failure can become seriously decompensated (ie, unable to adapt) because of the added stress imposed upon their hearts by hot weather.

EXTRINSIC REGULATION OF CUTANEOUS BLOOD FLOW. In contradistinction to most organs of the body, local (or intrinsic) mechanisms for normal regulation of blood flow to the skin play an insignificant and limited role in this tissue, and extrinsic neural mechanisms are of the utmost importance in controlling cutaneous flow, hence in thermal adaptation of the body as a whole to various environmental temperatures.

In the vicinity of the preoptic region of the anterior hypothalamus is located a center that regulates body temperature by controlling the caliber of the cutaneous vessels (p. 300). Therefore, warming this region of the diencephalon induces a generalized vasodilatation throughout the entire cutaneous vascular bed and sweating, whereas cooling exerts the opposite effects. These responses are mediated principally by an overall inhibition of sympathetic vasoconstrictor tone (Fig. 10.67), and the resultant passive dilatation of the superficial vessels and the specialized cutaneous arteriovenous anastomoses results in a massive increase in cutaneous blood flow to the subdermal venous plexuses with an attendant increase in heat loss. In regions of the body that lack such anastomoses, including the trunk, arms, and legs, arteriolar dilatation through inhibition of the hypothalamic constrictor influence increases flow to the venous plexuses markedly, and thus more warmed blood is exposed to heat loss at the skin:environment interface through large areas of the body surface. Otherwise normal individuals who have undergone amputation of both legs well above the knee suffer severely impaired thermoregulatory ability in hot weather as the total surface area of these two extremities is considerable.

The participation of sympathetic vasodilator fibers in the active control of cutaneous blood flow is debatable. Some investigators believe that the secondary increase in blood flow to the skin above that which is caused by passive dilatation of the vessels, and which occurs only when sweating takes place, is the result of active sympathetic vasodilator nerve fibers whose effect is mediated directly by acetylcholine and superimposed upon the passive dilatation that is mediated by inhibition of constrictor tone. Other workers are of the opinion that acetylcholine mediates this secondary active vasodilatation by liberation of the polypeptide bradykinin through hydrolysis of globulin in the interstitial fluids. Bradykinin per se is, of course, a powerful vasodilator substance (p. 648). It must be pointed out, however, that experimental inhibition of the bradykinin mechanism does not prevent the augmented cutaneous blood flow that occurs during sweating, so that both the mechanism and the actual physiologic role of this phenomenon of a second increase in blood flow during sweating are quite unclear at present.

INTRINSIC REGULATION OF CUTANEOUS BLOOD FLOW. Under physiologic conditions, regulation of blood flow through the skin by means of local control of the vascular caliber is of minor importance. Nevertheless, there are two situations in which intrinsic mechanisms may exert an important influence. These situations are reactive hyperemia and the direct effects of cold upon the peripheral circulation.

Reactive Hyperemia. If the blood flow to a region of the skin is decreased markedly for approximately half an hour, as by sitting in one position, or by means of tight clothing or a partially inflated sphygmomanometer cuff, the subsequent release of the pressure will result in a pronounced reddening of the skin in the previously ischemic region. This postocclusion flush of the skin is called reactive hyperemia, and it results from a marked but temporary increase in local blood flow which is mediated by the effects of hypoxia on the blood vessel caliber (principally the arterioles). A chemical vasodilator substance released from the anoxic tissues also may play a role in the etiology of reactive hyperemia.

Reactive hyperemia can also occur in many organs other than the skin (eg, the coronary vessels) in response to mechanical occlusion and temporary ischemia, but this reponse normally is visible only in the cutaneous region.

Effects of Cold upon the Peripheral Circulation. Down to a temperature of around 15C, the cutaneous vessels constrict, and the vasoconstriction is caused principally by an increased sensitivity to nerve stimulation at lower temperatures as well as by reflex vasoconstriction that is mediated through spinal neurons. Below a skin temperature of around 15C, however, the cutaneous vessels commence to dilate owing to the direct effect of cold upon the vessels themselves; therefore, maximal vasodilatation is achieved at around 0C. Presumably, the lowered temperatures directly paralyze the vascular contractile mechanism and/or block transmission of vasoconstrictor nerve impulses to the vessels. In any event, the increased local blood flow provides a protective mechanism, as it helps to prevent freezing of exposed regions of the body, particularly the hands, face, and ears. This physiologic effect also is the cause of the ruddy complexion seen in extremely cold weather.

SKIN COLOR AND TEMPERATURE. Since the relative skin color in Caucasians is a result principally of the color of the blood within the capillaries and venules, rapid blood flow when the skin is warm (or flushed) results in a pinkish or reddish appearance. Conversely, when the skin is cold, the blood flows much more slowly, and most of the oxygen is extracted before the blood leaves the capillaries; hence, the skin takes on a darker, more bluish hue.

In situations when cutaneous vasoconstriction is pronounced (eg, as in extreme fright), the blood within the skin is diverted into other, deeper-lying regions of the cardiovascular system; the individual assumes an ashen, whitish pallor, which in turn is the color of the subcutaneous connective tissue per se.

Photographs of various regions of the skin in patients suffering from peripheral vascular disease sometimes are helpful diagnostically. Called *thermograms,* such photographs are made with infrared film, which is extremely sensitive to heat. Areas with an abundant blood flow are warm, and as a consequence are more greatly exposed on the photographic negative; they appear as white or pale regions in the positive prints. Ischemic (relatively cool) regions, on the other hand, do not expose the film to as great an extent; they form pale areas on the negative and dark areas in the print. Simultaneous photography of both hands, feet, or legs sometimes can reveal the peripheral vascular manifestations of such conditions as Raynaud's disease, which is characterized by severe peripheral vasoconstriction upon exposure to cold, usually in the extremities, followed by vascular dilatation and reactive hyperemia; peripheral

arteriosclerosis with impaired circulation; and Buerger's disease, also called thromboangiitis obliterans, a condition which is caused by a vasoconstriction that results from inflammation of the sheaths containing the arteries, veins, and nerves in peripheral regions of the body. In such patients, merely the act of walking causes severe vasoconstriction of the vessels in the calf, and extreme ischemic pain results. This symptom, called intermittent claudication, is akin to the vasospasm and ischemia seen in angina pectoris as well as in Raynaud's disease.

The Splanchnic Circulation

As illustrated in Figure 10.15, blood from the intestines, as well as the pancreas and spleen, drains through the portal vein to the liver, and thence via the hepatic veins to the inferior vena cava. These vessels comprise the portal circulatory system. Collectively, the arterial blood supply to the liver plus the portal circulatory system are known as the splanchnic circulation, and the total flow through the splanchnic circulation comprises a large proportion of the total cardiac output (Table 10.6).

HEPATIC BLOOD FLOW AND ITS REGULATION. The total hepatic blood flow in an adult at rest is approximately 1,500 ml per minute, or around 27 percent of the total cardiac output. Of this total flow, about two-thirds (1,000 ml/minute) is received from the portal vein, whereas the remaining third (500 ml/minute) arrives via the hepatic artery.

Within the liver, the portal blood flows through the hepatic sinusoids, which are in intimate contact with the cords or plates of hepatic parenchymal cells; thence, the blood enters the central veins of the hepatic lobules before passing on to the vena cava. This peculiar morphologic arrangement is illustrated in Figure 10.75.

The hepatic artery supplies the structural components of the liver and the walls of the bile ducts; this blood then drains into the hepatic sinusoids, where it mixes with the portal blood. The hepatic arterial contribution to the liver is of the utmost importance, as loss of this flow can produce necrosis of the liver tissue with rapidly fatal consequences.

Since the liver affords a significant resistance to the flow of blood from the portal system to the vena cava, the blood pressure is relatively high in the portal vein, amounting to around 10 mm Hg. Pressure within the hepatic vein is about 5 mm Hg. As a consequence of this elevated portal venous pressure, the pressures within the portal venules and capillaries also are correspondingly high; there is a greater tendency for the pressure within these vessels to become abnormal than in any other region of the body (p. 565).

The mean hepatic arterial pressure is around 90 mm Hg. And since the hydrostatic pressure within the sinusoids is quite low relative to the

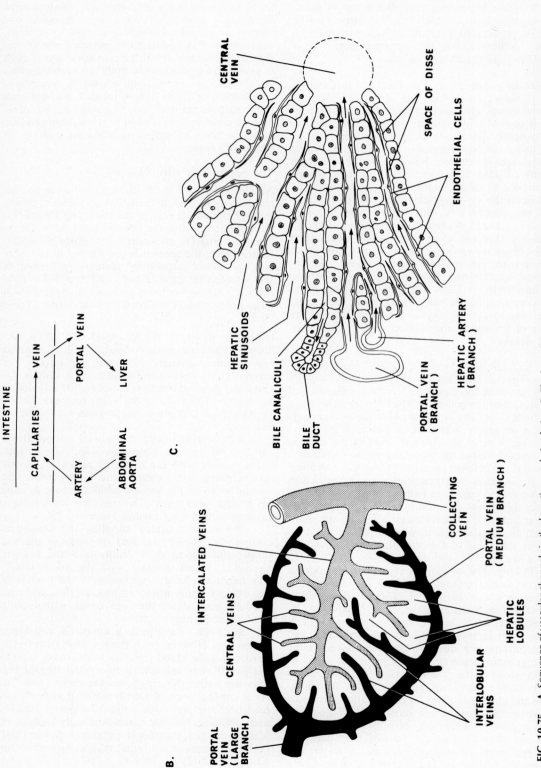

A.

INTESTINE

CAPILLARIES → VEIN

ARTERY → PORTAL VEIN → LIVER

ABDOMINAL AORTA

C.

CENTRAL VEIN

SPACE OF DISSE

ENDOTHELIAL CELLS

HEPATIC SINUSOIDS

BILE CANALICULI

BILE DUCT

PORTAL VEIN (BRANCH)

HEPATIC ARTERY (BRANCH)

B.

INTERCALATED VEINS

CENTRAL VEINS

COLLECTING VEIN

PORTAL VEIN (MEDIUM BRANCH)

HEPATIC LOBULES

INTERLOBULAR VEINS

PORTAL VEIN (LARGE BRANCH)

FIG. 10.75. A. Sequence of vascular channels in the hepatic portal circulation. **B.** The branches of the portal vein (dark shading) are separated from the branches of the portal vein (light shading) by hepatic lobules. **C.** The morphology of a portion of a hepatic lobule. The centripetal blood flow within the sinusoids and the centrifugal flow of bile within a canaliculus are indicated by arrows. (**B.** and **C.** After Bloom and Fawcett. *A Textbook of Histology*, 9th ed, 1968, Saunders.)

690

hepatic arterial pressure, there must be a very steep fall in pressure along the course of the hepatic arterioles themselves.

The branches of the portal vein that are located in the liver contain smooth muscle fibers within their walls that are innervated by adrenergic vasoconstrictor nerve fibers. These motor nerves pass to the liver via the ventral roots of spinal nerves T_3 to T_{11} as well as by way of the splanchnic nerves. The hepatic sympathetic plexus innervates the hepatic artery, and is entirely vasoconstrictor in function. In fact, there are no vasodilator fibers to the liver.

The liver as a whole is able to dilate and constrict passively in response to altered vascular pressures, so that this organ is capable of storing large quantities of blood. Consequently, the entire liver forms another important venous reservoir in the body. Normally, the total blood volume contained within the liver, including the hepatic veins and sinuses, is around 650 ml (or about 13 percent of the total circulating blood volume).

Circulation through the peripheral regions of the liver ordinarily is rather sluggish; therefore active perfusion of the hepatic parenchyma is present only through certain regions. If the central systemic venous pressure rises, however (as during right heart congestive failure), the branches of the portal vein dilate passively so that the total hepatic blood volume and liver size increase significantly. Conversely, when there is a widespread adrenergic discharge such as occurs in response to a precipitous fall in the systemic arterial blood pressure, the branches of the hepatic portal vein located within the liver constrict so that the portal vein pressure itself rises. A more active hepatic blood flow ensues, although much of the liver parenchyma is bypassed. Most of the blood within the liver now enters the systemic circulation. A concomitant vasoconstriction of the hepatic artery also tends to divert blood flow from the liver. Mesenteric arteriolar constriction can also reduce portal blood flow under such circumstances.

PERMEABILITY OF THE HEPATIC SINUSOIDS. These vascular channels within the hepatic parenchyma are lined with an endothelium similar to that found in the capillaries; however, the permeability of this endothelium is so great that even protein molecules can traverse this boundary with relative ease (Fig. 10.75). Since the hydrostatic pressure within the sinusoids is relatively low (around 5 to 7 mm Hg), most of this protein diffuses back into the plasma, although hepatic lymph contains the highest concentration of protein of any lymph in the body. In fact, hepatic lymph flow amounts to approximately 50 to 66 percent of the total lymph produced by the body.

Kupffer cells, which are phagocytic macrophages lining the sinusoids of the liver, are very efficient in removing foreign particulate matter from the portal blood. Bacteria, which normally are absorbed, together with nutrients from the intestinal tract, are ingested by the phagocytic action of these cells in the amazingly brief contact interval of about 10 milliseconds. Thus, Kupffer cells provide an efficient system for the removal of foreign particulate matter from the intestinal tract before the blood reaches the systemic circulation.

REGULATION OF INTESTINAL BLOOD FLOW. Approximately 80 percent of the total portal blood flow comes from the stomach and intestines; the remaining 20 percent comes from the spleen and pancreas. The regulation of blood flow within the gastrointestinal tract itself appears to be carried out principally through the operation of intrinsic mechanisms, as is the situation in most organs. Blood flow to the musculature of the stomach and intestines is controlled independently of blood flow to the mucosa and submucosa, where absorption and glandular activity take place (qv, p. 827). Thus, when contractions of the gastrointestinal tract increase, so too does blood flow to the muscular layers. On the other hand, when the rate of secretion is elevated, blood flow to the mucosal and submucosal layers likewise increases selectively.

Although the particular mechanisms that underlie these responses are not wholly understood, it is clear that a decreased oxygen supply produces vasodilatation of the gastrointestinal vascular bed, whereas an adequate oxygen supply produces vasoconstriction. Thus, autoregulation of flow would result as a direct consequence of the local metabolic rate, in similarity to other organs.

Stimulation of the gastrointestinal glands also may liberate an enzyme locally which frees the vasodilator peptide bradykinin from a plasma protein (p. 648). In turn, this vasodilator substance could produce a local vasodilatation, thereby augmenting blood flow locally. There is some experimental evidence, however, suggesting that this may not be a valid mechanism for the physiologic control of blood flow in the gastrointestinal tract.

The sympathetic nerves have an important vasoconstrictor influence on the entire gastrointestinal tract. Stimulation of the vascular musculature of this region via these nerves can reduce blood flow to almost zero. Normal blood flow to the gastrointestinal region can be "turned off" for all practical purposes, and thereby this blood is diverted to other parts of the body when needed. For example, the skeletal muscles and heart receive a significantly greater total volume flow of blood because of the operation of this mechanism during strenuous exercise.

Gastrointestinal distention per se also can change the splanchnic blood flow markedly. Such a mechanical effect thus can stretch the circumferential vessels that pass around the gut, and this effect in turn results in a partial occlusion of these vessels, but most especially the veins, and this elevated outflow resistance raises the intracapillary pressure. This response, if sufficiently pronounced, can result in a very rapid loss of fluid and protein into the lumen of the gastrointestinal tract in addition to causing serious edema of the wall of the gut. A large quantity of plasma may be lost from the circulation in a brief period, and as a consequence pronounced distention of the gut is a common etiologic factor in causing shock through plasma loss.*

THE SPLEEN.   The spleen is unnecessary for survival in man; however, several hundred ml of blood can be stored in this organ. Unlike the situation that obtains in lower animals such as cats and dogs, the human splenic capsule is nonmuscular and thus incapable of active contraction. Dilatation of the vascular bed within this organ, however, permits a passive storage of the volume of

Anything present in the lumen of the gut is literally outside the body (p. 525).

blood mentioned above, whereas a constriction of the splenic vessels themselves, which in turn is mediated by sympathetic stimulation, can extrude a large proportion of this stored blood into the general circulation by reducing the total volume capacity of these vessels. In man, however, the physiologic operation of this system is not clearly understood, and the spleen generally is not considered to function as a true blood reservoir.

Blood is sequestered in the spleen in two regions: the red pulp and the venous sinuses (Fig. 10.76). In the pulp, the capillaries are quite permeable, so that blood, including erythrocytes, is extruded from the capillaries into this region, whence it passes slowly into the venous sinuses. The latter structures in addition receive blood directly from small vessels, so that when the spleen dilates passively these regions become engorged.

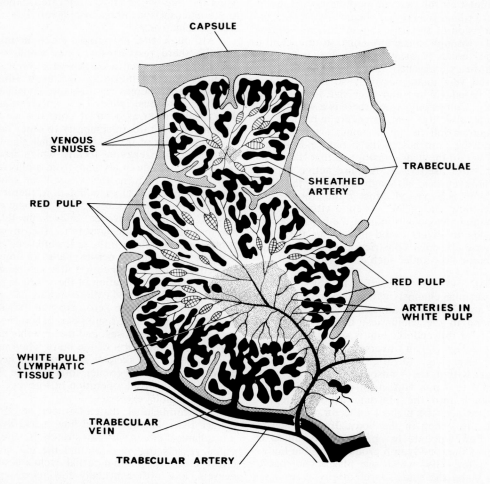

FIG. 10.76.   The general anatomy of the spleen. The white pulp consists of lymphatic tissue, and the red pulp of sequestered erythrocytes plus some leukocytes. (After Bloom and Fawcett. *A Textbook of Histology,* 9th ed, 1968, Saunders.)

The storage of erythrocytes within the red pulp of the spleen results in a slight concentration of the erythrocytes, so that the systemic hematocrit is reduced slightly as the spleen dilates. Expulsion of the blood stored within the spleen during vasoconstriction can thus increase the total circulating erythrocyte mass to a certain extent; in severe stress situations a greater total quantity of oxygen per unit volume of blood may be carried. This functional response is clearly evident in lower organisms, at least; however, in man the physiologic details of this process remain obscure, but it is clear that the spleen is unnecessary to prolonged survival.

The spleen also functions as a hematopoietic organ in the fetus (p. 529), and it also provides an important site of formation for phagocytic reticuloendothelial cells (monocytes and lymphocytes). These elements are found in the white pulp of the spleen. Blood percolating through the white pulp of the spleen thus is subjected to removal of foreign particulate bodies such as bacteria in a manner similar to that which takes place in the lymph nodes.

Blood Flow within the Skeletal Muscles

In a resting individual, blood flow through skeletal muscle averages around 6 ml/100 gm of tissue/minute. During strenuous exercise, this value can increase up to 20-fold, so that the flow now becomes approximately 100 ml/100 gm of muscle/minute. When a muscle is contracting rhythmically in an isotonic pattern, the flow rises and falls sharply as the vessels are compressed during contraction, thereby expressing the blood (see also venous pump, p. 652). During an isometric contraction, blood flow through a muscle can be reduced to almost zero (p. 1180).

During intervals of rest, only a small proportion of the total number of capillaries in any particular skeletal muscle are open, perhaps 10 percent of the total number that are present (see critical closing pressure, p. 643). During severe exercise, all the capillaries become functional, so that this factor alone accounts for the major proportion of the increased blood flow. Furthermore, another important physiologic effect results when the capillary bed opens completely. The total distance that oxygen and other nutrients must diffuse from the capillaries in order to reach the muscle fibers is reduced considerably, and likewise the distance that carbon dioxide and other waste products must travel to be eliminated also is reduced.

Blood flow through skeletal muscle is regulated by intrinsic as well as extrinsic mechanisms.

INTRINSIC MECHANISMS FOR REGULATION OF BLOOD FLOW IN SKELETAL MUSCLE. Local factors generated within skeletal muscles during their activity and which act directly upon the arterioles to produce vasodilatation are primarily responsible for the enormous increase in blood flow that occurs in contracting skeletal muscle. Furthermore,

three intrinsic factors appear to operate simultaneously upon the vessels to produce this vasodilator effect.

Oxygen Concentration. The level of oxygen present within the muscular tissues themselves undoubtedly is one of the most important local factors regulating directly the degree of constriction or dilatation of the vessels. Since muscle utilizes oxygen quite rapidly during aerobic contractile activity, the concentration of this gas becomes reduced rapidly in the interstitial fluids of active muscle, and vasodilatation follows, either because the arteriolar smooth muscle cannot remain contracted in a relative hypoxia, or because vasodilator compounds are liberated. Regardless of the ultimate mechanism, however, it is clear that hypoxia per se is a powerful stimulus for vasodilatation in skeletal muscle.

Vasodilator Substances. Although the quantitative role of vasodilator materials cannot be assessed with any degree of certainty at present, several compounds, ions, and metabolites are known to produce vasodilatation in skeletal muscle during activity. Adenosine (or one of its compounds, eg, adenosine triphosphate) has been implicated in this effect. Similarly, potassium ions, carbon dioxide, lactic acid, and acetylcholine all are liberated within, and each is capable of producing dilatation of, the vascular bed within skeletal muscle during periods of activity.

Arterial Blood Pressure. During bouts of exercise, the mean arterial blood pressure rises, and this factor alone can elevate blood flow markedly in skeletal muscle. Thus, a slight increase in this factor (eg, +20 mm Hg) can distend the arterioles significantly, thereby increasing their volume capacity; consequently, the blood flow also is increased considerably. Also, since $F = \Delta P/R$, and the pressure is increasing while the resistance is decreasing because of vasodilatation, the simultaneous operation of both of these factors augments blood flow tremendously. The elevated arterial pressure will tend to raise many inactive capillaries above their critical closing pressure, so that flow through this portion of the vascular bed of the muscles increases greatly.

EXTRINSIC MECHANISMS FOR REGULATION OF BLOOD FLOW IN SKELETAL MUSCLE. Maximal dilatation of the vascular bed within skeletal muscle can occur through the operation of one or more of the intrinsic local autoregulatory mechanisms described above. Extrinsic neural mechanisms, although present, play a relatively minor role in the physiologic regulation of blood flow through these tissues, except as noted below.

Sympathetic Vasoconstrictor Fibers. The neurotransmitter substance liberated by the sympathetic vasoconstrictor nerves is norepinephrine. Maximal stimulation of the sympathetics to skeletal muscle, however, results in only a 75 percent reduction of blood flow, in contrast to most other regions of the body, where such stimulation can reduce blood flow to practically zero. Nevertheless, this response can be of physiologic significance under conditions of severe stress, such as shock, in which a marked reduction of flow through the muscles would help to alleviate the overall cardiovascular distress.

During vigorous exercise, the adrenal medulla (p. 1118) liberates significant quantities of the catecholamines epinephrine and norepinephrine into the systemic circulation. The circulating norepinephrine then can act upon the α receptors of the vessels within the muscles to produce vasoconstriction in a manner similar to that of the norepinephrine that is liberated from the sympathetic terminals by nerve stimulation. The epinephrine released from the adrenals, on the other hand, has a vasodilator effect in many instances, and this response is believed to be a result of the action of this compound upon the β receptors in the vascular walls (Table 7.6, p. 276).

Sympathetic Vasodilator Fibers. The sympathetic vasodilator nerve fibers of the skeletal muscles liberate acetylcholine upon stimulation, which compound in turn produces vasodilatation of the vascular bed. Maximum stimulation of the sympathetic vasodilator nerves can increase blood flow up to 400 percent under experimental conditions.

The sympathetic vasodilators are stimulated via a pathway which originates from the cerebral cortex, and which is related closely to the cortical areas for control of voluntary motor activity. The fibers then pass downward via the hypothalamus and brain stem to the spinal cord, thence to the periphery (Fig. 10.67).

This neural dilator mechanism is believed to initiate vasodilatation in the skeletal muscles at the onset of voluntary activity, and before the local vascular regulatory mechanisms have had time to dilate the vessels (cf p. 1178). Thus, for example, this mechanism would come into operation in a track athlete awaiting the starting gun during the "anticipatory phase" before a race. Thereby, cortical stimulation dilates the vascular bed of the skeletal muscle and blood flow will increase markedly. This mechanism would provide an important initial adaptation of the muscles to increased activity. It must be remembered, however, that in general this neural effect is secondary to the intrinsic regulatory mechanisms present within the vascular bed of skeletal muscle, and that these intrinsic mechanisms come into full operation immediately after the onset of activity and supersede the neural vasodilator effects. Furthermore, maximal vasodilatation in muscular tissue also can (and does) take place without any extrinsic neural influences.

When blood flow through the skeletal muscles throughout the body increases, there is a concomitant rise in the cardiac output, which in turn supplies the increased demand for blood by the muscles. The principal factors involved in this process may be summarized here, whereas the overall physiologic adaptations to exercise will be discussed further in Chapter 16, page 1177. The three principal factors in the adaptation of cardiac output to exercise are: (1) vasodilatation within the contracting muscles, which causes an increased flow of blood into the vena cava, thereby elevating venous return, hence right atrial pressure, and directly increasing the cardiac output; (2) contraction of the muscles, causing a rhythmic compression of the vessels within their structure, which greatly augments venous return to the heart (in fact up to 8 percent of the total blood volume can be expressed from the muscles and thus returned to the heart and lungs by the operation of this mechanism); (3) sympathetic stimulation via excitation of the medullary centers, which increases heart rate and contractility directly. In addition, such stimulation induces vasoconstriction of certain blood reservoirs such as are found in the cutaneous and splanchnic regions, thereby diverting an increased blood volume to the working skeletal muscles.

All these mechanisms operate in a smoothly coordinated fashion in order to carry out an integrated and effective physiologic transition from rest to full muscular activity.

CARDIAC METABOLISM

Several quantitative differences between the energy metabolism of cardiac and skeletal muscle are worthy of summary.

Actively contracting skeletal muscle obtains its energy principally from the anaerobic degradation of glucose or glycogen to lactic acid (p. 936; Fig. 14.25). Resting skeletal muscle, on the other hand, derives a large proportion of its energy from the aerobic oxidation of acetoacetate (a ketone body) and fatty acids (p. 963; Fig. 14.35), so that under conditions of limited activity, glucose utilization in skeletal muscle is negligible. Similarly, smooth muscle derives its contractile energy chiefly from the metabolism of fatty acids and acetoacetic acid, whereas glucose is utilized to only a slight extent.

Cardiac muscle is rich in the pigment myoglobin, the enzymes of the tricarboxylic acid cycle (or Krebs' cycle, p. 946; Fig. 14.29, and the electron transport system (p. 931; Fig. 14.20). Consequently, heart muscle utilizes principally aerobic reactions in order to obtain the continuous supply of adenosine triphosphate that is critical to its sustained contractions. For this reason, the heart, unlike skeletal muscle, has an insignificant capacity to contract anaerobically; an adequate oxygen supply to the myocardium is an essential factor to continued survival.

Cardiac muscle removes only very small quantities of glucose from the coronary blood. On the other hand, the myocardium in a resting individual obtains most of its ATP through the oxidation of fatty acids (p. 963). The oxidation of acetoacetic and lactic acids also provides a small proportion of the total ATP required during rest. When the metabolic rate is accelerated, however, as during the performance of physical work, the myocardium actively removes lactic acid from the coronary blood and utilizes it directly as a substrate for the production of energy. The blood level of lactic acid increases during vigorous exercise because it is produced at a highly accelerated rate in the contracting skeletal muscles.

In general, insulin administration causes the glycogen concentration of skeletal muscle (but usually not the liver) to increase somewhat. Cardiac muscle, however, is insensitive to this effect of insulin so that the quantity of glycogen in this tissue actually may decrease from the level seen in the skeletal muscles.

CLINICAL CORRELATES

A number of clinically significant aspects of cardiovascular physiology were mentioned in context in the present chapter. Discussions of the pathophysiology of shock and of congestive heart failure were deferred until Chapter 16 (p. 1237 and p. 1244, respectively), so that these conditions could be discussed as examples of the cardiovascular and respiratory adaptations of the entire body under abnormal stress brought on by particular pathologic situations.

In this section, an important and fundamental distinction between cardiac adaptation and performance in the normal and diseased heart first will be made. This discussion will be followed by a brief examination of the effects of anemia upon the cardiovascular system. Last, several representative cardiac abnormalities and their electrocardiographic manifestations will be considered.

Cardiac Reserve in the Normal and Failing Heart

It has been stressed repeatedly throughout this chapter that the normal heart is able to pump out all the blood that is returned to it over a period of time by the venous system, both at rest, when this input is low, and during strenuous exercise, when the quantity of blood returning to the right atrium is increased greatly. Basically, therefore, the heart not only is able to adapt to an increased *volume load,* but also it is capable of adapting to various *pressure loads,* and this statement describes, in briefest terms, the normal adaptive functions of this organ.

There are, of course, many etiologic factors that can induce cardiac failure, but fundamentally the failing heart is unable to adapt to an increased volume and/or pressure load as can a normal heart, so that eventually cardiac output declines progressively until a point is reached that is incompatible with continued survival.

Essential hypertension will serve to illustrate the long-term effects of increased pressure and volume loads upon the left ventricle. The origin of essential hypertension is unclear, but the elevated systolic and diastolic blood pressures which are the sole manifestations of this process in its earlier stages are caused by a generalized systemic vasoconstriction, and since $F = \Delta P/R$, the left heart must perform considerably more work to maintain an adequate flow rate than is normal in overcoming the elevated diastolic resistance, especially over long periods of time. This is true because a relatively large proportion of the total cardiac work performed during systole must be used solely to elevate the blood pressure within the left ventricle above the aortic diastolic pressure in order that the ejection phase of the cardiac cycle can take place. Thus, the total isometric work imposed upon the left ventricular myocardium in hypertension is considerably greater than in a heart working against normal, and thus lower, end diastolic arterial pressures. Over long periods of time, a heart which must operate against such a physiologic disadvantage may compensate by dilatation and hypertrophy; however, beyond certain limits, failure ultimately ensues. The failure in turn develops progressively in the heart because the dilated ventricle has to develop a higher wall tension during systole merely to overcome the abnormally high diastolic blood pressure in the aorta in accordance with Laplace's law (p. 267). The failing heart now must cope not only with an abnormal pressure load, but also with an abnormal volume load, and this a failing heart cannot do!

Cardiovascular Effects of Anemia

As discussed on page 569, blood viscosity is related directly and principally to the concentration of circulating erythrocytes. In severe anemia the viscosity of the blood may fall to 50 percent of its normal value. According to the Poiseuille relationship, blood flow is related inversely to the peripheral resistance; venous return to the heart, hence the cardiac output, may be increased greatly in anemias, thereby imposing a severe volume load on the heart, as discussed in the previous section.

An additional factor that increases the cardiac output (hence cardiac work) in anemia is the hypoxia itself. This relative oxygen deficiency, present in anemia, in turn causes the blood vessels within the tissue to dilate (p. 665), thereby lowering the peripheral resistance still further. And this factor, when added to the decreased blood viscosity inherent in anemia, also tends to elevate the venous return, and thus the cardiac output is elevated still further. Therefore, one major effect of severe anemia is the imposition of a significantly increased work load on the heart.

The elevated cardiac output in turn tends to compensate for many of the symptoms of anemia, even though the quantity of oxygen carried per unit volume of blood may be as low as 25 percent of normal. If the blood flow to the tissues of the body is sufficiently high, normal or near-normal quantities of oxygen can be delivered to the tissues at rest or even during times of mild physical activity. An individual having severe anemia is always on the borderline of serious hypoxia (p. 764); if the seriously anemic person is subjected to strong emotion, or if such a person exercises strenuously, the already maximal cardiac output cannot increase sufficiently to meet the bodily oxygen requirements, as the heart has no remaining cardiac reserve. Acute cardiac failure may develop.

Some Electrocardiographic Manifestations of Cardiac Abnormalities

As stated on page 604, electrocardiograms are useful diagnostically only when the abnormality present in the heart affects the electric properties of this organ. Mechanical defects are detected by use of the electrocardiographic technique only when such defects in turn cause an abnormality of impulse generation or propagation through the myocardium.

In this section, a few representative cardiac arrhythmias and other abnormalities will be discussed briefly as examples of typical defects that can be detected and diagnosed electrocardiographically. The components of a normal EKG were defined in Figure 10.33, and the normal electrocardiographic waveforms obtained from the twelve standard leads are depicted in Figure 10.28.

All regions of the myocardium and conduction system of the heart (Fig. 10.12) retain the latent potentiality of developing pacemaker activity; under abnormal conditions, any part of the heart can assume a pacemaker function. This effect occurs when the sinoatrial node is depressed, when the atrioventricular node is blocked, or when an ectopic focus of irritability in either atria or ventricles commences to discharge. Such defects can be seen readily in the EKG. Not all irregularities of cardiac rhythm are of pathologic significance, however.

ALTERED SINUS RHYTHMS. Clinically, tachycardia is considered to be present when the heart rate exceeds 100 contractions per minute. As illustrated in Figure 10.77, a marked acceleration of the rate at which the sinoatrial node generates impulses results in tachycardia. Such accelerated heart rate can result from an increased body temperature, autonomic nerve stimulation, or toxic effects of various agents directly upon the myocardium.

Bradycardia is the opposite of tachycardia (Fig. 10.78); the sinoatrial node is generating impulses at a slower-than-normal rate, about 70 per minute in a normal young adult. Any cardiovascular reflex that elicits vagal stimulation can produce bradycardia. Furthermore, trained athletes, in contrast to untrained individuals, achieve the major proportion of their cardiac output by means of a greatly increased stroke volume; such athletes frequently exhibit marked physiologic sinus bradycardia.

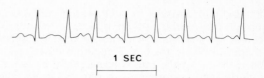

FIG. 10.77. Sinus tachycardia, recorded with lead I. (After Guyton. *Textbook of Medical Physiology,* 4th ed, 1971, Saunders.)

ATRIOVENTRICULAR BLOCK. Occasionally, a patient may be encountered in whom the impulses generated by the sinoatrial node are blocked before they excite the atrial muscle; hence, the P waves disappear from the EKG, and a new but much slower rhythm is initiated by impulses that generally arise in the atrioventricular node (Fig. 10.79). The resulting QRS complex and T waves are normal in appearance.

When incomplete block of impulse conduction between atria and ventricles is present, the P–R interval is prolonged beyond the normal value of 0.16 second when the heart rate is normal. If, however, the P–R interval is longer than about 0.20 second, first-degree incomplete heart block is said to exist (Fig. 10.80A), and when the P–R interval reaches 0.30 second or longer, second-degree heart block is present (Fig. 10.80B). When complete heart block develops, as shown in Figure 10.81, the P waves of the EKG become completely dissociated from the QRS complexes and T waves.

In some patients having atrioventricular block, this condition occurs at sporadic intervals, normal atrioventricular impulse conduction being seen between the episodes of complete atrioventricular block. Thus Adams-Stokes disease (or syndrome) is a condition that is characterized by sudden attacks in which consciousness is lost owing to a reduced cerebral blood flow as the direct consequence of the episodic development of complete heart block. Convulsions may or may not accompany the block. The implantation of electronic pacemakers in such patients to accelerate the ventricular rate, hence to augment the cardiac output, is of benefit.

PREMATURE BEATS. Premature beats, or extrasystoles, are contractions that occur before the next successive contraction normally would be expected. Generally, premature beats are caused by the activity of ectopic foci of irritability within the heart such as might be caused by local ischemia; toxic effects of such agents as drugs, nicotine, or caffeine on the myocardium or conducting system; or the mechanical stimulus of the tip of an intracardiac catheter (p. 632).

As shown in Figure 10.82, an atrial extrasystole has occurred. The electrocardiographic waves of the extrasystole are essentially normal, but note the slightly prolonged interval between the end of

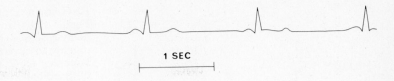

FIG. 10.78. Sinus bradycardia, recorded with lead III. (After Guyton. *Textbook of Medical Physiology*, 4th ed, 1971, Saunders.)

1 SEC

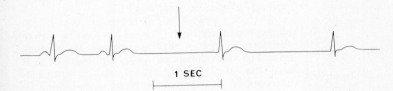

FIG. 10.79. Block of sinoatrial node, commencing at arrow, with development of an atrioventricular nodal rhythm, recorded with lead III. (After Guyton. *Textbook of Medical Physiology*, 4th ed, 1971, Saunders.)

1 SEC

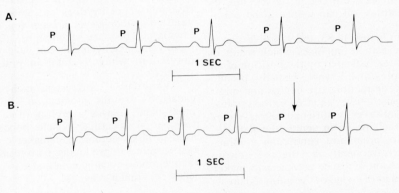

A.

B.

FIG. 10.80. **A.** Prolonged P–R interval, recorded with lead II. **B.** Partial atrioventricular block, recorded with lead V$_3$. Note absence of QRS complex at arrow. (After Guyton. *Textbook of Medical Physiology*, 4th ed, 1971, Saunders.)

FIG. 10.81. Complete atrioventricular heart block recorded with lead II. (After Guyton. *Textbook of Medical Physiology*, 4th ed, 1971, Saunders.)

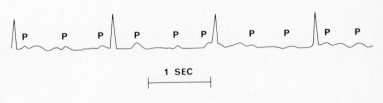

1 SEC

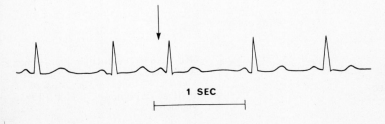

1 SEC

FIG. 10.82. Atrial premature beat (*arrow*) recorded with lead I. (After Guyton. *Textbook of Medical Physiology*, 4th ed, 1971, Saunders.)

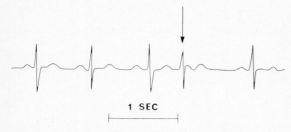

FIG. 10.83. Atrioventricular nodal premature beat *(arrow)*, recorded with lead III. (After Guyton. *Textbook of Medical Physiology,* 4th ed, 1971, Saunders.)

the extrasystole and the next successive electrocardiographic complex.

Atrial premature beats are of frequent normal occurrence in average persons as well as trained athletes, whose hearts are extremely strong. Nevertheless, such abnormal conditions as a prolonged loss of sleep, emotional overactivity, or overindulgence in tobacco, coffee, or alcohol, as well as certain drugs, all can induce atrial extrasystoles.

As shown in Figure 10.83, the ventricle also is able to generate ectopic impulses which result in premature beats, as indicated by the position of an extra QRS complex and T wave. Usually, premature ventricular contractions are benign and originate from the same causes as atrial extrasystoles. In many cases, however, ventricular extrasystoles may also have their origin in ventricular pathology. For example, premature ventricular contractions may be seen frequently in patients following a coronary occlusion, because ectopic pacemaker loci are produced at the edge of the infarcted area.

PAROXYSMAL TACHYCARDIA. In some individuals, bouts of atrial tachycardia can occur without warning, and these in turn give rise to an

EKG having a pattern similar to that illustrated in Figure 10.84. In this patient, the heart rate suddenly rose from a rate of about 95 contractions per minute to around 150 contractions per minute. Note that the P wave is inverted before each of the QRS complexes during the tachycardia, and furthermore that the P wave also is superimposed to some extent upon the preceding T wave; hence, the ectopic pacemaker focus was located in the atria. Similarly, paroxysmal tachycardia can result from an irritable focus that is located in the atrioventricular node or bundle of His.

Atrial as well as atrioventricular nodal paroxysmal tachycardia can be induced by any of the several factors that are known to produce atrial and ventricular extrasystoles; however, as is also the situation with the latter arrhythmias, episodes of paroxysmal tachycardia can occur in otherwise healthy individuals.

In contrast to atrial paroxysmal tachycardia, ventricular paroxysmal tachycardia is a serious matter for two reasons (Fig. 10.85): (1) This arrhythmia generally does not appear unless major ventricular damage is present (eg, infarction following a coronary occlusion). (2) Ventricular paroxysmal tachycardia can lead to ventricular fibrillation, which is fatal in practically all instances.

Ventricular paroxysmal tachycardia can be caused by digitalis intoxication. Quinidine, on the other hand, causes a prolongation of the refractory period of cardiac muscle, hence frequently is used to block ectopic pacemaker regions that are responsible for producing paroxysmal ventricular tachycardia. Procaine amide also is employed clinically for similar reasons.

ATRIAL AND VENTRICULAR ARRHYTHMIAS. At times, the atria or the ventricles contract in an uncoordinated manner and with extreme rapidity.

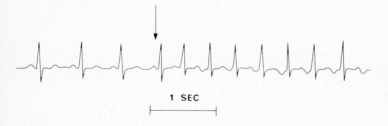

FIG. 10.84. Atrial paroxysmal tachycardia, onset at arrow, recorded with lead I. (After Guyton. *Textbook of Medical Physiology,* 4th ed, 1971, Saunders.)

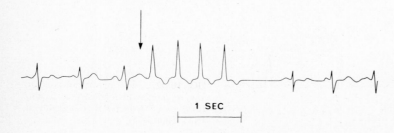

FIG. 10.85. Ventricular paroxysmal tachycardia, onset at arrow, recorded with lead III. (After Guyton. *Textbook of Medical Physiology,* 4th ed, 1971, Saunders.)

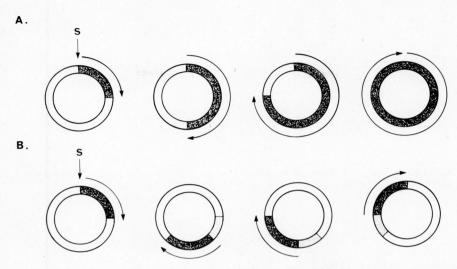

FIG. 10.86. **A.** In the normal heart, application of a stimulus (S) such that the impulse can pass in one direction only (ie, clockwise in figure) results in extinction of that impulse when it travels in a complete circle and reaches its starting point, as the tissue still is refractory to reexcitation (solid black). **B.** Circus movement. Any of the three possible mechanisms discussed in the text result in the ring of tissue being fully excitable once again at the point where the stimulus was applied (S) when the impulse has traveled in a complete circle. Hence, the tissue can be reexcited, and the impulse passes around and around the tissue. (The black region in each diagram again signifies the absolute refractory period, the stippled region the relative refractory period.)

When the atrial contractions range between 200 and 300 per minute, the condition generally is called *flutter*, whereas the wholly uncoordinated contractions that occur at rates above 300 per minute are called *fibrillation*. Fibrillation can occur in either atria or ventricles.

Flutter and fibrillation are explained by either of two theories. In the first theory, many ectopic foci are believed to discharge continuously and simultaneously throughout the atrial (or ventricular) myocardium so that the resulting contractions are random, hence wholly uncoordinated. The cardiac output of the affected chambers falls appreciably (in flutter) or becomes nil (in fibrillation).

The second theory that is invoked to explain flutter and fibrillation is called *circus movement*, because a single impulse travels in an endlessly repetitive circular pattern throughout the myocardium, as depicted and explained in Figure 10.86. The fundamental premise involved in the circus movement theory of flutter and fibrillation rests upon the concept that as the impulse travels in a circle, the timing of the impulse is such that as it reaches its starting point, the tissue no longer is refractory. The tissue can be reexcited once again (Fig. 10.86B) in contrast to the condition that obtains in normal tissue, in which a refractory condition is present at the starting point so that the im-

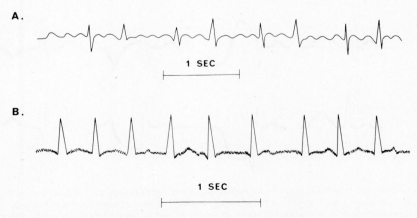

FIG. 10.87. **A.** Atrial flutter, recorded with lead II. **B.** Atrial fibrillation, recorded with lead I. (After Guyton. *Textbook of Medical Physiology,* 4th ed, 1971, Saunders.)

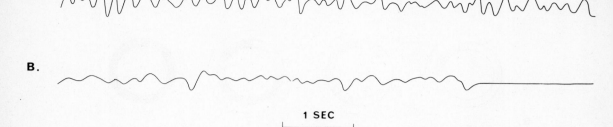

1 SEC

FIG. 10.88 **A.** and **B.** Ventricular fibrillation terminating in asystole and death of the patient, recorded with lead V₄. (After Ganong. *Review of Medical Physiology,* 6th ed, 1973, Lange Medical Publications.)

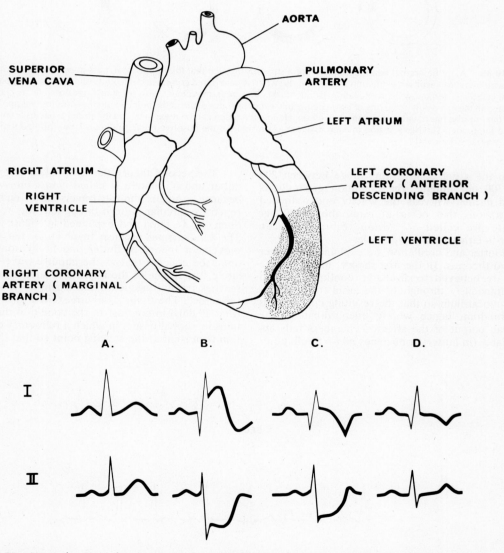

FIG. 10.89. Upper drawing. Infarct in anterior descending branch of left coronary artery (shaded area surrounding solid black vessel). Lower drawing. Lead I and lead II EKGs. **A.** Normal EKG for comparative purposes. **B.** Acute phase of myocardial infarction, immediately following occlusion of vessel shown in upper part of figure. **C.** Subacute phase. **D.** Chronic phase some time following the infarction. (After Burch and Winsor. *A Primer of Electrocardiography,* 5th ed, 1966, Lea & Febiger.)

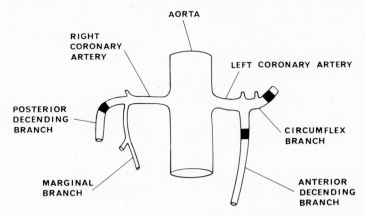

FIG. 10.90. The coronary arterial system and the three common sites of coronary artery occlusion (black areas), each of which gives rise to typical electrocardiographic abnormalities as depicted in Figure 10.89. (After Burch and Winsor. *A Primer of Electrocardiography,* 5th ed, 1966, Lea & Febiger.)

pulse becomes extinct when its starting point is reached once again (Fig. 10.86A).

Three factors would favor the circus movement of impulses in abnormal states: (1) the length of the circular pathway; if this path were long, a condition met in dilated hearts, circus movement would occur; (2) a decreased rate of impulse conduction, such as occurs from block of the Purkinje system, which would favor circus movement; (3) if the refractory period of the muscle became appreciably shortened, such as in response to drugs (eg, epinephrine) or following massive vagal discharge, this response also would predispose the myocardium to a circus movement type of impulse conduction.

There is, in fact, evidence that both ectopic foci as well as circus movement are involved to some extent in the etiology of flutters as well as fibrillation.

Atrial Fibrillation. The electrocardiographic patterns seen in atrial flutter as well as atrial fibrillation are shown in Figure 10.87. Note that in flutter the rate of atrial excitation as given by the rapidity of development of the P waves is around 300 impulses per minute, whereas the ventricular rate as indicated by the QRS complexes is around 125 per minute. Note also that only a certain and irregular percentage of the atrial impulses are able to excite the ventricle, owing to the long refractory period of the atrioventricular nodal tissue.

In atrial fibrillation, no distinct P waves are developed and the ventricular rhythm is quite irregular because the intervals between the time of arrival of impulses at the atrioventricular node likewise are irregular. Furthermore, the refractory period of the atrioventricular node determines the irritability of this tissue, governing whether or not a given atrial impulse will cause ventricular excitation.

Ventricular Fibrillation. In ventricular fibrillation, no typical and regular electrocardiographic pattern is evident in any lead, and only small random oscillations of the voltage are seen (Fig. 10.88). The ventricular myocardium contracts in a wholly uncoordinated, random fashion. If one holds a fibrillating heart in the hand, the sensation evoked is described aptly by the expression "holding a bag of worms."

Following the onset of ventricular fibrillation, the cardiac output, hence blood pressure, rapidly falls to zero. Unless immediate and drastic therapeutic measures (eg, cardiac massage and electric defibrillation) are instituted, rapid death is the inevitable consequence of ventricular fibrillation.

THE ELECTROCARDIOGRAPH AND MYOCARDIAL INFARCTION. As depicted in Figure 10.89, a myocardial infarct will produce typical electrocardiographic alterations, which in turn can be subjected to vector analysis (p. 612) in order to localize the particular region of the heart that has undergone infarction. In the particular example given, the infarct is located in the anterior apical region of the heart. Two other common sites for coronary occlusion in addition to the anterior descending branch of the left coronary artery are shown in Figure 10.90. Blockage of either of these vessels also gives rise to typical electrocardiographic abnormalities such as are depicted in Figure 10.89 for blockage of the anterior descending branch of the left coronary artery.

In conclusion, it must be stressed that the electrocardiographic alterations presented in this section are illustrative only, and represent but a small sampling of the multitude of such alterations that can be encountered in clinical practice.

CHAPTER 11

Respiratory Physiology

The general concept of man as an "aquatic animal," whose life processes are carried out wholly within an aqueous medium of closely controlled composition and volume, was emphasized in Chapter 2. It also was mentioned there that gas exchange with the environment takes place at an air:fluid interface within the lungs, and that, prior to their effective transport or utilization by the individual, the respiratory gases must pass into aqueous solution. The lungs and other respiratory structures in turn are protected from dessication by their anatomic location within the thoracic cavity, as well as by the exceedingly thin layer of liquid that normally coats the surfaces of the various regions of this organ system.

Since man is an exceedingly complex terrestrial organism having a substantial total mass and volume, the physical process of diffusion alone is inadequate to provide sufficient molecular oxygen to the billions of individual cells that comprise his body, or to remove from them the carbon dioxide produced during the normal metabolic processes that attend living. Consequently, two intimately related physiologic systems are essential to provide a continuous and adequate supply of oxygen to the most remote cells of the body, and to remove carbon dioxide from them. These are the respiratory system, which essentially is a gas exchange system, and the cardiovascular system (Chap. 10, p. 577), which is designed for the bulk transport of gases (and other nutrients) both to and from the individual cells through circulation of the blood. These two systems must function in an integrated fashion at all times. Even in a healthy individual, the so-called matching of gas volume to pulmonary capillary blood flow is never equal; ie, this matching is imperfect because the inspired air is not distributed equally to all the capillaries in all the alveoli. Ventilation of the lungs is uneven in distribution. This fact assumes major importance in cardiopulmonary disease, in which the normally imperfect matching of air distribution within the

lungs to pulmonary blood flow may become grossly exaggerated.

Fundamentally, the respiratory system exposes large volumes of water-saturated air to large volumes of blood in order to effect rapid gas transfer between the blood and the environment. The cardiovascular system in turn transports and distributes this gas to the capillaries where exchange once again can occur rapidly. Basically, of course, the physiologic goal is the exchange of respiratory gases between the cells and their immediate environment, the interstitial fluid. Thus, the respiratory system functions as an air pump, whereas the heart functions as a blood pump; and the blood vessels themselves comprise a distributing system for the blood and the gases carried by this liquid tissue.

The term *respiration* denotes four distinct but interrelated processes. First, and in the most general sense, respiration means the absorption of oxygen and elimination of carbon dioxide by the entire body. This process occurs exclusively within the lungs, and gas exchange at this level sometimes is referred to as *external respiration*. The term *respiration* is also used to indicate the mechanical act of breathing. Second, respiration means the utilization of oxygen and excretion of carbon dioxide by the individual cells, tissues, and organs of the body and their immediate environment, generally, the interstitial fluid. This process may also be referred to as *internal respiration*. Third, the transport of oxygen and carbon dioxide within the body is occasionally referred to as the *respiratory function* of the blood. Fourth, the term *respiration* (or more aptly, *tissue respiration*) signifies the roles of oxygen and carbon dioxide in specific metabolic reactions and processes at the molecular level.

In summary, the reader should be aware that the term *respiration* includes several processes: the exchange of oxygen and carbon dioxide between the atmosphere and the blood within the lung capillaries; the transport of these gases by the blood; the exchange of oxygen and carbon dioxide with the interstitial fluid and the cells of the body; and the participation of oxygen in the metabolic processes of the cells at the molecular (ie, metabolic) level.

An additional critical function of respiration is in the maintenance of blood pH, so that the neural as well as chemical mechanisms involved in respiratory regulation during various physiologic states are topics which form an important part of any discussion of mammalian respiration. The material to be presented in this chapter will be concerned with external respiration, gas transport in the blood, and tissue gas exchange following a brief summary of the anatomic features pertinent to an understanding of the physiology of the respiratory system. Respiration at the metabolic level will be dealt with extensively in Chapter 14 (p. 907).

PHYSIOLOGIC ANATOMY OF THE RESPIRATORY SYSTEM AND RELATED STRUCTURES

There are two principal functional divisions of the respiratory tract: the *conducting portion*, which connects the exterior of the body with the *respiratory portion*, where the actual exchange of gases between air and blood occurs.

The conducting airways and related accessory structures of the respiratory tract include the following elements listed in sequence of their occurrence from the exterior to the lungs: the nasal cavity and paranasal sinuses; the mouth, nasopharynx, and pharynx; the larynx; the trachea; the left and right primary bronchi; the secondary bronchi, or bronchia; the bronchioles; and the terminal bronchioles (Fig. 11.1). The respiratory structures of the lung, which are continuous with the terminal bronchioles of the con-

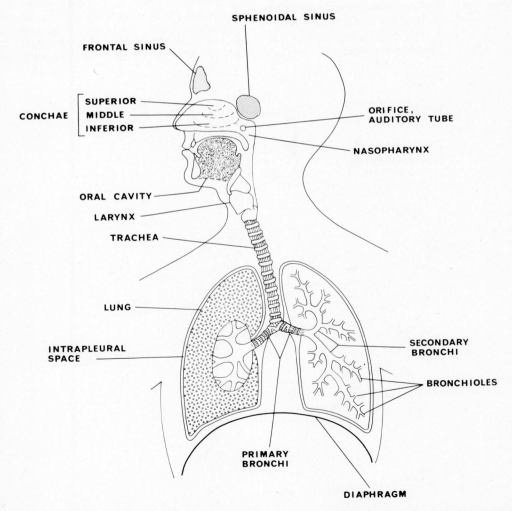

FIG. 11.1. Gross anatomy of the respiratory system. (After Dawson. *Basic Human Anatomy,* 2nd ed, 1974, Appleton-Century-Crofts.)

ducting portion, include the respiratory bronchioles, alveolar ducts, alveolar sacs, and alveoli (Figs. 11.2 and 11.8).

In addition to these two major divisions of the respiratory system proper, there are a number of specific skeletal muscles involved in the mechanical aspects of pulmonary ventilation, and which function as essential components of the respiratory pumping mechanism.

The anatomic structures involved in each of these categories now may be considered sequentially.

Conducting Airways of the Respiratory Tract

This division of the respiratory system consists of a series of hollow conduits for air which connect the exterior of the body with the respiratory structures where the actual gas exchange takes place. Physiologically, these passageways are "inert" tubes whose importance lies in their ability to transport and to exchange large volumes of air between lungs and environment in a rapid manner.

NASAL CAVITY AND ACCESSORY SINUSES. The nose is a hollow structure composed of cartilage,

bone, connective tissue, and muscles, and which is lined by the integument through the anterior nares as far as the vestibule. Stratified squamous epithelium lines the surfaces of this region. The hairs found in the nose are presumed to assist in the exclusion of particulate matter in the inspired air by acting as coarse filters.

The remainder of the nasal cavity is lined with pseudostratified ciliated epithelium which secretes mucus. The olfactory sensory areas are provided with a highly specialized type of ciliated neuroepithelium (the olfactory epithelium) whose morphology and function was discussed in Chapter 8 (p. 520).

The ciliated epithelium of the nose is similar to that found in the larynx and trachea. Goblet cells are distributed abundantly among the ciliated cells, and a basement membrane separates the epithelial layer from the deeper lying connective tissue lamina which contains the mucous glands whose ducts empty upon the epithelial surface. The secretions from these glands are of importance in preventing dessication of the nasal cavity.

Underneath the epithelium of the lower nasal conchae (or turbinates) are found abundant venous plexuses that serve to warm the air passing

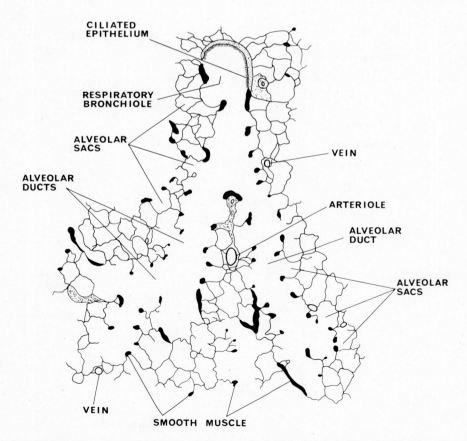

FIG. 11.2.   A section through a respiratory bronchiole and two alveolar ducts of the lung. (After Bloom and Fawcett. *A Textbook of Histology,* 9th ed, 1968, Saunders.)

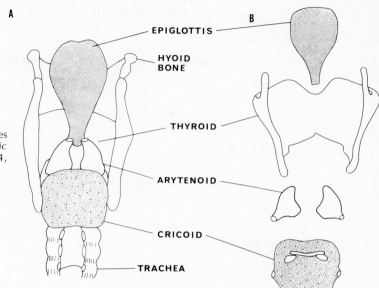

FIG. 11.3. Skeleton **(A)** and cartilages **(B)** of the larynx. (After Dawson. *Basic Human Anatomy*, 2nd ed, 1974, Appleton-Century-Crofts.)

EPIGLOTTIS

HYOID BONE

THYROID

ARYTENOID

CRICOID

TRACHEA

through the nose. The tissue supplied with the venous plexuses is capable of substantial engorgement; however, this nasal tissue differs from true erectile tissue (p. 1134) in that septa provided with smooth muscle are absent.

The accessory nasal sinuses (frontal, ethmoid, sphenoid, and maxillary) are found as cavities within the respective bones of the skull, and these sinuses are connected directly with the nasal cavity. The ciliated epithelium that lines the four pairs of sinuses has fewer glands than does the nasal cavity. The beat pattern of the cilia is toward the nasal cavity, so that the mucous sheet moves in this direction and toward the exterior.

LARYNX.   Inspired air moves from the nasal cavity by way of the nasopharynx and the pharynx to the larynx (Figs. 11.1 and 11.3).

The nasal portion of the pharynx is lined with ciliated columnar epithelium, whereas the oral region is covered with stratified squamous epithelium, continuous with that in the mouth above and the esophagus below. The pharynx has a dual physiologic role in that it serves both as the passage for air to the lungs and as a conduit for ingested material from the mouth to the esophagus (p. 821).

The larynx not only serves as a portion of the airway to the lungs but also as the organ of phonation. This roughly tubular, elongated structure has an irregular shape, being somewhat wider in its superior aspect, where it is continuous with the pharynx, and somewhat narrower below, where it is continuous with the trachea. The larynx is composed of nine cartilages, three of which are paired and three unpaired; these structures are summarized in Table 11.1. In addition, the larynx contains connective tissue as well as related extrinsic and intrinsic striated muscles, and is lined with a mucous membrane.

The extrinsic laryngeal muscles connect the larynx to adjacent structures of the neck and assist in swallowing (see deglutition, p. 836). The five intrinsic muscles of the larynx connect the laryngeal cartilages and by their contraction alter the shape of the cavity within the larynx; thereby they have an important role in phonation (Table 11.1).

The epiglottis is leaf-shaped and projects upward posterior to the root of the tongue (Fig. 11.4). The free end of this structure is located above the opening of the larynx; during the act of swallowing the laryngeal orifice is closed by elevation of the larynx so that it becomes approximated to the end of the epiglottis; thus, solids or liquids are prevented from entering the trachea.

Within the larynx proper are located two pairs of shelflike folds which project inward from the lateral walls. The superior folds, called the ventricular folds or false vocal cords, are stationary. Immediately beneath the ventricular folds lie the true vocal folds or true vocal cords, which are capable of movement. The vocal cords each contain one inferior thyroarytenoid ligament medially, and each of these ligaments is bordered on its lateral surface by the thyroarytenoid muscle. When these muscles contract, thereby approximating the thyroid and arytenoid cartilages, the vocal cords relax. The space between the true vocal cords is known as the glottis (or rima glottidis).

Sound, but not speech, is produced by vibrations induced in the vocal cords as expired air is forced between them through the glottal opening. The intrinsic laryngeal muscles primarily determine the width of the opening of the glottis, as well as the tension of the vocal cords themselves. Pitch of the voice depends upon the relative length of the vocal cords as well as their tension. The anteroposterior dimension of the space between

Table 11.1 CARTILAGES AND MUSCLES OF THE LARYNX[a]

UNPAIRED CARTILAGES

Thyroid	Largest of the laryngeal cartilages; the midline laryngeal prominence formed by this structure is known as the "Adam's Apple"
Cricoid	Only complete cartilaginous ring in respiratory tract
Epiglottis	Leaf-shaped; projects upward at base of tongue; guards opening of larynx

PAIRED CARTILAGES

Arytenoid	The vocal ligaments are attached to these cartilages, and their movements control the tension of these ligaments
Corniculate	Sometimes fused with arytenoids
Cuneiform	

EXTRINSIC MUSCLES

Omohyoid Sternohyoid Sternothyroid Thyrohyoid Stylopharyngeus Palatopharyngeus	Support the larynx and suspend it from adjacent bones
Constrictors, inferior Constrictors, superior	Act as sphincters

INTRINSIC MUSCLES

Cricothyroideus	Lifts anterior part of cricoid cartilage, thereby elongating and exerting tension on vocal cords
Posterior cricoarytenoideus	Produces lateral rotation of the arytenoid cartilages, thereby separating the vocal cords and opening the glottis
Lateral cricoarytenoideus	Produces medial rotation of the arytenoid cartilages, thereby approximating the vocal cords and closing the glottis
Arytenoideus	Approximates the arytenoid cartilages and closes the glottis
Thyroarytenoideus	Tilts the arytenoid cartilages toward the thyroid cartilage; thus shortens and relaxes the vocal cords

[a]*Data from Dawson. Basic Human Anatomy, 2nd ed, 1974, Appleton-Century-Crofts; Goss (ed). Gray's Anatomy of the Human Body, 28th ed, 1970, Lea & Febiger.*

the vocal cords is around 23 mm in men and 18 mm in women; therefore, women generally have higher pitched voices than do men. Short, tight cords produce higher notes, whereas longer, more relaxed cords produce lower tones. The timbre, or quality, of the voice in turn depends upon a number of factors, including the size and shape of the resonance chambers: the ventricules, the vestibule of the larynx, the pharynx, mouth, nasal cavities, and the accessory or paranasal sinuses. Consequently, the shape of the glottis is capable of considerable variation during several phases of respiration as well as when different sounds are produced, eg, as in talking and singing.

The anterior surface of the epiglottis and the medial surfaces of the vocal cords are covered by stratified squamous epithelium. In adults, ciliated epithelium generally begins at the base of the epiglottis, and it continues down into the larynx, trachea, and bronchi. These cilia are about 3.5 to 5.0 μ in length, and their pattern of beat is toward the mouth. Foreign bodies, such as dust particles and bacteria, as well as mucus, thus are transported toward the exterior surface of the body.

TRACHEA. The trachea is the first segment of the respiratory tract that is concerned solely with the function of respiration. The trachea is both flexible and thin-walled (Fig. 11.5). In the adult, this tubular structure is approximately 11 cm in length. The trachea is continuous with the larynx above, and it terminates within the thoracic cavity after passing

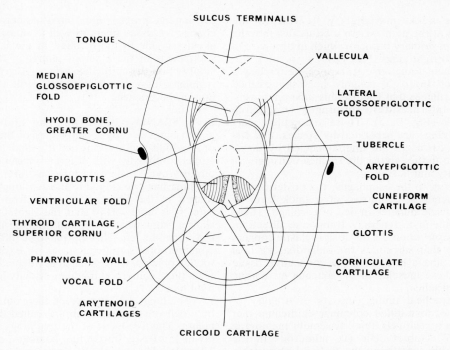

FIG. 11.4. Epiglottis and vocal cords as seen from above. (After Dawson. *Basic Human Anatomy,* 2nd ed, 1974, Appleton-Century-Crofts.)

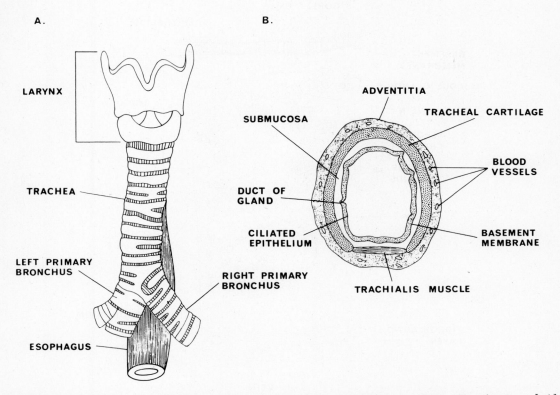

FIG. 11.5. Gross anatomy of the trachea as seen from the ventral aspect **(A)** and in cross-section **(B).** (After Goss [ed]. *Gray's Anatomy of the Human Body,* 28th ed, 1970, Lea & Febiger; Dawson. *Basic Human Anatomy,* 2nd ed, 1974, Appleton-Century-Crofts.)

down the neck immediately ventral to the esophagus at the point where it bifurcates into the two main or primary bronchi, which in turn pass to the left and right lungs (Fig. 11.1).

A principal anatomic feature of the trachea is a series of 16 to 20 roughly horseshoe-shaped hyaline cartilages that enclose this tube on its ventral and lateral sides, forming a supporting framework (Fig. 11.5). These incomplete cartilaginous rings are separated by spaces, and the individual cartilages are interconnected by dense fibroelastic tissue, giving the trachea considerable extensibility and flexibility.

Adjacent to the esophagus, the posterior wall of the trachea is composed of a thick layer of smooth muscle bundles in lieu of cartilage. These muscles run in a generally transverse direction, and they insert upon the dense fibroelastic connective tissue that surrounds the individual tracheal cartilages, and a layer of loose connective tissue attaches these muscles to the mucous membrane lining the trachea.

The tracheal lining consists of a layer of ciliated pseudostratified columnar epithelium that rests upon a relatively thick basement membrane (Fig. 11.6). Goblet cells are numerous in the ciliated epithelium lining the trachea, and within

the lamina propria are found numerous elastic connective tissue fibers as well as numerous small glands similar to those found in the larynx. The short ducts of these serous glands of the trachea empty upon the epithelial surface. Aggregations of lymphatic tissue are also found in the lamina propria of the trachea.

LUNGS AND BRONCHI. The highly elastic paired lungs occupy a major part of the total volume within the thoracic cavity, and their shape varies continually with different phases of the respiratory cycle. The right lung has three lobes, the left lung has two (Fig. 11.7). Each lung receives a branch of the primary bronchus from the same side. It is noteworthy that the right bronchus is shorter, somewhat wide, and oriented more vertically than the left. This morphologic difference has considerable significance, as it explains why foreign objects, when inhaled accidentally (eg, buttons, pins, coins, and the like), usually lodge in the right lung.

The lungs are invested on their outer surfaces by a delicate serous membrane called the visceral pleura. The five lobes of the two lungs in turn are divided grossly into large numbers of lobules. These lobules are roughly pyramid-shaped masses

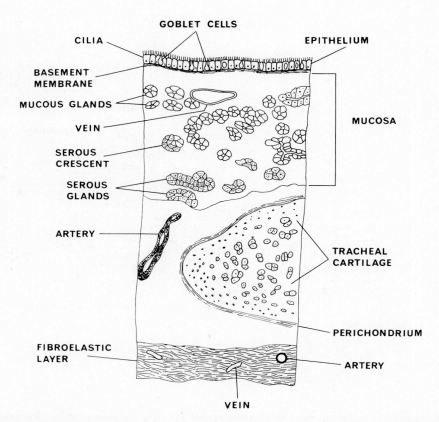

FIG. 11.6. The histology of the trachea in cross-section. (After Bloom and Fawcett. *A Textbook of Histology*, 9th ed, 1968, Saunders.)

A. LUNGS, COSTAL SURFACE

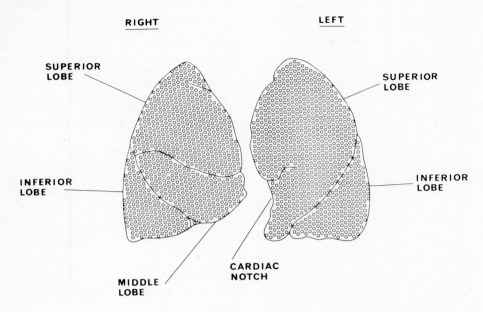

B. LUNGS, MEDIASTINAL SURFACE

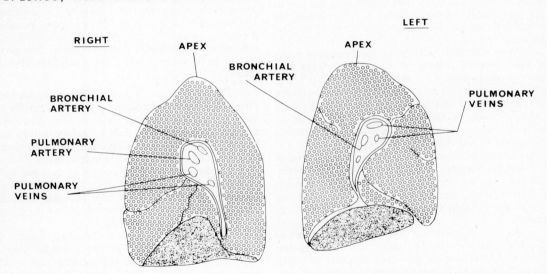

FIG. 11.7. Gross anatomy of the lungs. (After Dawson. *Basic Human Anatomy,* 2nd ed, 1974, Appleton-Century-Crofts.)

of pulmonary tissue whose apexes are directed toward the hilus (or the root) of the lungs, and whose bases are oriented toward the surface. The primary lobules comprise the functional units of the lung.

The two main bronchi, which arise from bifurcation of the trachea, enter the lungs at the hilus on the left and right sides, and then course downward and outward to branch into two, smaller secondary bronchi on the left side and three on

the right (Fig. 11.1). These secondary bronchi in turn give rise to innumerable smaller branches called bronchioles, and these bronchioles in turn give rise to an estimated 50 to 80 terminal bronchioles within each lobule.

Prior to entering the lungs, the extra-pulmonary bronchi are quite similar in structure to the trachea; however, once the lungs are penetrated, the cartilaginous rings disappear, to be replaced by cartilaginous plates of irregular shape that entirely surround the bronchi. The intrapulmonary bronchi are cylindrical in shape, and not flattened on one side, such as is the case in the trachea and extrapulmonary bronchi. When the plates of cartilage become more unevenly distributed around the bronchi and bronchioles, the layer of smooth muscle completely surrounds these structures. This smooth muscle layer is of considerable importance in governing the caliber of the bronchioles, hence, airway resistance, as discussed on page 722. The cartilage disappears completely when the bronchioles become approximately 1 mm in diameter.

The mucous membrane that lines the bronchi is continuous with that of the trachea, and is composed of the same type of ciliated epithelium. Adjacent to the mucosa is a layer of smooth muscle fibers arranged in fascicles or bundles. Unlike the situation found in blood vessels (p. 587) and the gut (p. 817), these smooth muscle cells do not form a closed ring; rather, they form an irregular interdigitating network. The openings within this network of smooth muscle fibers become progressively larger as the size of the bronchioles decreases. Abundant elastic connective tissue fibers form an intimate relationship with the smooth muscle cells. These two structural components have an important role in the normal changes in lung volume that accompany the physical movements of respiration. A rich network of blood vessels is also present in this myoelastic layer.

The outermost portion of the bronchial walls is composed of a layer of connective tissue that also contains numerous elastic fibers. This bronchiolar coating also invests the cartilaginous plates of the bronchi, and is continuous with the connective tissue of the surrounding lung tissue as well as that connective tissue which accompanies the larger blood vessels into the lung.

Glands, both serous and mucous, are found in the various ramifications of the bronchi and bronchioles until the cartilage disappears. The glandular structures generally lie beneath the muscular layer, and their ducts empty upon the mucosal surface.

As the diameter of the bronchi and bronchioles progressively decreases, the layers within their walls become thinner, less distinct, and may fuse. The smooth muscle lamina, however, remains distinct up to the end of the respiratory bronchioles, and may even continue into the alveolar ducts themselves.

Respiratory Structures of the Lung

As mentioned earlier, the functional unit of the lung within which the process of external respiration takes place is the primary lobule. This unit consists of a respiratory bronchiole, alveolar duct, alveolar sac, and alveoli, together with all the related blood vessels and lymphatics, as well as associated nerves and connective tissue (Fig. 11.2).

In a gross section of lung tissue, the respiratory portion is seen as a spongy network of large spaces separated by thin partitions or septa. This network is punctuated by larger bronchi and blood vessels of various sizes which course through the lung parenchyma. Microscopically, thick sections of pulmonary tissue appear as an irregular honeycomb in which the alveolar sacs and polyhedral-shaped alveoli are seen, and this structure in turn is laced with bronchioles and alveolar ducts. The alveolar sacs and alveoli themselves open on the alveolar ducts.*

RESPIRATORY BRONCHIOLES. The respiratory bronchioles are short, tubular structures that are continuous with the ends of the terminal bronchioles. Respiratory bronchioles have an initial diameter of around 0.5 mm, and for a short distance they are lined with ciliated columnar epithelium (Fig. 11.8). Goblet cells and cartilage are absent. A short distance along their course, the cells lose their cilia and become cuboid. The walls of respiratory bronchioles consist principally of collagenous connective tissue, and smooth muscle bundles are found as a network; sparse elastic fibers are also present.

On the opposite side of the bronchiole along which runs a branch of the pulmonary artery are found several alveoli that appear as outpocketings. These alveoli are the first respiratory structures encountered in the lung, hence their presence gives rise to the term *respiratory bronchiole*. The respiratory bronchioles in turn branch, thus giving rise to between two and eleven radiating alveolar ducts.

ALVEOLAR DUCTS. The alveolar ducts are long, thin-walled tubular structures which generally assume a tortuous course and in turn give off a number of branches, which branches in turn may branch again. Numerous alveolar sacs and alveoli form pouches which provide outpocketings on the alveolar ducts. The blind-ending alveoli are polyhedral sacs that open only upon the surface facing the alveolar ducts. Since the alveolar sacs are densely packed together, their openings form most of the wall of the alveolar ducts.

Between the openings of the alveolar sacs, the wall of the alveolar duct is supported by smooth muscle cells as well as collagenous and elastic fibers.

ALVEOLAR SACS AND ALVEOLI. The alveolar sacs, which branch from the alveolar ducts, contain between two and four (or even more) single alveoli; individual alveoli also can arise directly from the alveolar ducts. The alveoli themselves are extremely thin-walled, polyhedral sacs, and the side of these structures which faces the alveolar duct is always absent. Thus, air may pass

*Some histologists consider the area bounded on one side by the end of an alveolar duct and on the other sides by the openings of the alveolar sacs to represent a distinct anatomic entity called the atrium (Fig. 11.8).

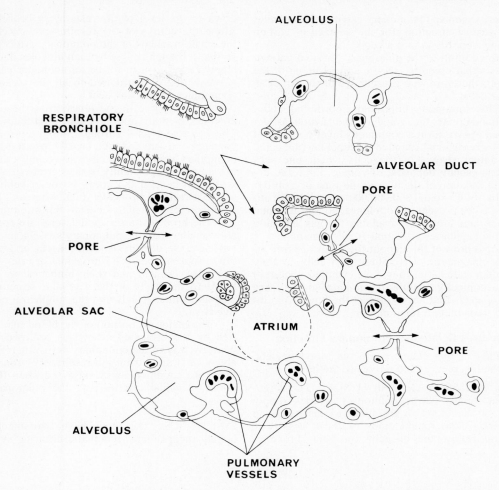

FIG. 11.8.   The respiratory structures of the lung. The atria are indicated by circles, and these spaces are limited upon one side by the openings of the alveolar sacs and on the other by the termination of the alveolar duct. See also Figure 11.2. (After Bloom and Fawcett. *A Textbook of Histology,* 9th ed, 1968, Saunders.)

freely between the ducts, sacs, and cavities within the alveoli.

The most obvious structural feature of the alveoli is the dense, freely anastomosing, capillary network within their walls. Also present are networks of branching reticular connective tissue as well as scattered elastic fibers. These connective tissue elements form the delicate framework that supports the capillaries. The pulmonary capillaries tend to protrude into the alveolar lumen, thereby presenting a larger proportion of their surface to the alveolar air.

Gaps known as alveolar pores are present in the septal walls separating adjacent alveoli. These minute openings have a diameter of about 7 to 9 μ. Although their functional significance is not entirely clear, it is possible that they provide a limited collateral air circulation during secondary bronchial obstruction, and thus hinder atelectasis (an incomplete expansion or collapse of the lungs).

During pulmonary diseases such as pneumonia, the alveolar pores also could provide direct pathways for the spread of bacteria between adjacent alveoli.

The openings of the alveolar sacs are surrounded entirely by a ring of collagenous and elastic connective tissue fibers, and these structural elements pass between adjacent sacs and lend structural support to the wall of the alveolar duct.

The alveolar lining is of considerable interest and physiologic importance. Only after the advent of the electron microscope was it possible to demonstrate unequivocally that there is a thin, continuous, cellular covering of the alveoli. The air–blood barrier or alveolocapillary membrane consists of three distinct layers: (1) the alveolar epithelium, (2) an interstitial space in which a basement lamina is present, and (3) the capillary endothelium itself (Fig. 11.9). These three structures form the morphologic interface between alveolar oxygen and pulmonary capillary blood, and, in a reverse direction, for the passage of carbon dioxide from blood into alveoli.

In addition to the pulmonary (or small) alveolar cells that comprise the innermost lining of the alveoli, there are also present septal (or large) alveolar cells.

The squamous pulmonary epithelial cells, which comprise the lining of the alveoli as described above, form a continuous layer punctuated only by occasional rounded or cuboid septal cells. The actual shapes of the cells which one sees microscopically depends, of course, upon the degree of inflation of the lung at the time of fixation.

In total, some 300,000,000 alveoli are estimated to be present in both lungs, so that the total surface area normally available for gas diffusion into the capillaries is around 70 m² in man.

Also present throughout the lungs are free alveolar phagocytic cells. These dust cells, as they are often called after they have engulfed (ingested) microscopically visible particulate matter, are macrophages that act to protect the lungs from inhaled particulate foreign matter.

Structures Related to Pulmonary Function

Included in this category are the blood vessels, lymphatics, pleura, and nerves of the lungs.

BLOOD VESSELS OF THE LUNGS. The large-caliber, highly elastic pulmonary arteries supply most of the blood to the lungs as discussed on page 654. The ramifications of these vessels in general follow the bronchi and their branches down to the level of the respiratory bronchioles. Along the course of the respiratory bronchioles, the small branches of the pulmonary artery divide so that a branch passes to each alveolar duct; this blood vessel in turn forms an abundant capillary network that is distributed to all the alveoli that communicate with this duct.

Venules arise in the lungs from the pleural, alveolar, septal, and alveolar duct capillaries. These vessels pass through the intersegmental connective tissue and then anastomose to form the pulmonary veins. The pulmonary venules thus follow a path independently of their arteriolar supply, so that in general the artery is found behind and above its related bronchial duct, whereas the vein is in front of as well as beneath this structure.

The bronchial arteries and veins are considerably smaller than the pulmonary vessels, and these vessels arise directly from the aorta or from the intercostal arteries. The bronchial arteries course along the bronchi and their branches are distributed to the walls of these tubular structures as well as to their glands and the subpleural connective tissue.

The major proportion of the blood that is carried by the bronchial arteries is returned to the heart by way of the pulmonary veins; therefore, in alveoli that arise from the respiratory bronchioles there are capillary anastomoses between the terminal bronchial arteries and the pulmonary arteries.

LYMPHATICS. The lungs have two major, and largely independent, lymphatic drainage systems.

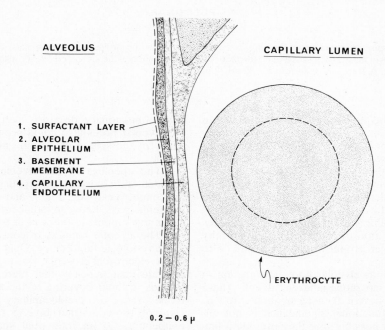

ALVEOLUS   CAPILLARY LUMEN

1. SURFACTANT LAYER
2. ALVEOLAR EPITHELIUM
3. BASEMENT MEMBRANE
4. CAPILLARY ENDOTHELIUM

ERYTHROCYTE

0.2 – 0.6 μ

FIG. 11.9. The pulmonary blood–air barrier at high magnification. (Data from Bloom and Fawcett. *A Textbook of Histology*, 9th ed, 1968, Saunders; Comroe. *Physiology of Respiration*, 1965, Year Book Medical Publishers; Guyton. *Textbook of Medical Physiology*, 4th ed, 1971, Saunders.)

The first of these systems is found in the pulmonary tissue, the second in the pleura itself. Both systems drain into lymph nodes located at the hilus of the lungs.

Pulmonary Lymphatics. The pulmonary lymphatics include those branches that drain the bronchi, pulmonary artery, and pulmonary vein.

The bronchial lymphatics terminate in the alveolar ducts; there are no lymphatics beyond these structures. The branches of the pulmonary lymphatic vessels anastomose with the lymphatic plexuses that are found surrounding the pulmonary artery and vein. Several lymphatic trunks drain the pulmonary artery. The lymphatics associated with the pulmonary vein begin as radicles located in the alveolar ducts and the pleura. All the pulmonary lymphatics drain into the hilus of each lung. From the lymph nodes found in the hilar regions, lymphatic trunks anastomose, thereby forming the right lymphatic duct which in turn provides the principal drainage system of both right and left lungs.

Lymphatics of the Pleura. The lymphatics of the pleura form a dense network of vessels that has both small and large meshes. The larger meshes, which are composed of the larger lymphatic vessels, delineate the lobules of the lungs. These lymphatics possess many valves in contrast to the pulmonary lymphatics, so that lymphatic flow is directed toward the hilus rather than toward the pulmonary tissue itself. The pleural lymphatics then anastomose, and the larger vessels thus formed then drain into the lymph nodes at the hilus of the lung.

PLEURA. As mentioned earlier, the surfaces of the lungs are invested closely with a thin serous membrane, the visceral pleura. This membrane consists of a layer of collagenous and elastic connective tissue fibers that contain fibroblasts as well as macrophages. The surface of this membrane is covered with a layer of mesothelial cells, similarly to the peritoneum. The morphologically similar layer of tissue that lines the outer wall of the thoracic cavity is called the parietal pleura.

These two extremely smooth and moist pleural layers are quite important physiologically, as the lungs are able to glide freely along the surface of the thoracic cavity without rub or friction during the continuous volume changes of these organs which normally accompany respiratory movements.

NERVES. Branches of the vagus (X) and branches of the inferior cervical and first thoracic sympathetic ganglia form the pulmonary nerve plexuses located at the root of the lungs.

Parasympathetic bronchial constrictor fibers are derived from the vagus, whereas bronchodilator fibers arise from the aforementioned sympathetic ganglia. Physiologically, these nerves are important in regulating the caliber of the smaller branches of the pulmonary airways which are abundantly supplied with smooth muscle fibers, as indicated earlier in this chapter (see also Table 7.6, p. 276).

The pulmonary blood vessels are also supplied with parasympathetic and sympathetic nerve fibers; however, the physiologic role of these nerves is not clear at present (p. 677). The sympathetic nerves also innervate, and can constrict, the bronchial arteries.

Muscles Involved in Respiration

The skeletal muscles that are responsible for producing the rhythmic respiratory movements necessary to ventilate the lungs are summarized in Table 11.2, together with certain general remarks concerning their function.

Summary

The respiratory tract is a highly complex and anatomically sophisticated distributing system for inspired air. Starting with paired nasal tubes (and/or the mouth), air passes through a continuous single tube, composed of the larynx and trachea, until the bifurcation into the primary bronchi is reached. The right and left main bronchi in turn each subdivide repeatedly to a total of around 20 subdivisions, so that in the aggregate around 10^6 respiratory bronchioles and alveolar ducts are present in the human lungs. Each of these terminal portions of the respiratory tract is provided with numerous blind-ending saccular structures, the alveoli, of which there are a total of around 3×10^8 present in both lungs. The diameter of the alveoli varies between 75 and 300 μ, and they may be close to the hilus, or as far as 20 to 30 cm from this region.

Within the lungs, the successive subdivisions of the respiratory tree ultimately result in the generation of an enormous total surface area for gas exchange in the alveoli, around 70 m², or about 40 times the total surface area of the body itself. Yet, the thickness of the barrier between air and blood itself is less than 0.1 μ! The millions of short thin-walled capillaries which invest the alveoli also expose the blood to a huge total surface area, estimated by various investigators to lie between 70 and 140 m² in man. The exchange of gases between the external and internal environments is a highly rapid and efficient process, a transfer process that is accomplished solely by the physical process of diffusion (p. 85).

Another remarkable feature of the respiratory system lies in the extremely wide range of its capacity for the volume (or bulk) transport of air into and out of the lungs. At rest, the adult human oxygen requirement is such that only 250 ml O_2/min need to be transferred from the environment to the blood. Thus, it is important to realize that during rest only about 5 percent of the total pulmonary surface available for gas exchange is in actual use. Therefore, a large margin of physiologic reserve is available to the individual. During

Table 11.2　RESPIRATORY MUSCLES[a]

MUSCLES	FUNCTION
Inspiration	
Diaphragm	During quiet breathing, pulmonary ventilation is accomplished almost entirely by contraction of this muscle
External intercostals	Pull ribs forward and upward
Sternocleidomastoids	Raise sternum
Scalenes	Raise first two ribs
Anterior serrati and scapular elevators	Raise ribs
Spinal erectus	
Expiration	
Internal intercostals	Pull ribs backward and downward
Abdominal recti	Compress abdominal viscera upward on diaphragm; pull downward on ribs
Posterior inferior serrati	Lower ribs

[a] *Data from Dawson.* Basic Human Anatomy, *2nd ed, 1974, Appleton-Century-Crofts; Goss (ed).* Gray's Anatomy of the Human Body, *28th ed, 1970, Lea & Febiger.*

periods of extreme exertion, however, the oxygen requirement may become as high as 5,500 ml O_2/minute, or more than 20 times the resting level. Similarly, the right ventricle of the heart may propel only 4 liters of blood per minute through the pulmonary capillaries at rest, whereas this volume flow may increase to over 20 liters/minute during periods of maximal exercise.

It is important to remember that the lungs frequently are the locus of inflammatory diseases which do not leave an impaired function after recovery has taken place. On the other hand, certain infections (such as tuberculosis) destroy large masses of pulmonary tissue, and healing is always accompanied by the formation of connective tissue scars. There is no evidence to indicate that regeneration of pulmonary tissue can occur, and the normal pulmonary reserve mentioned above becomes critically important to the patient where some tissue has been destroyed. In this regard, the lungs exhibit a characteristic similar to that found in the central nervous system, in which neurons are never replaced once they have been destroyed (p. 106).

MECHANICS OF PULMONARY VENTILATION

It is worthwhile at this point to contrast certain features of the blood pump to the lungs with the air pump that ventilates these organs. The symbols and abbreviations used in pulmonary physiology are summarized in Table 11.3, and the composition of dry air is given in Table 11.4.

Blood is propelled through the lungs by the rhythmic contractions of a muscular pump, the right ventricle. Unidirectional flow of the blood is assured by valves. The tricuspid valve prevents regurgitation of blood into the right atrium from the corresponding ventricle during its systole, whereas the pulmonary valve assures that there is no backflow of blood into the right ventricle during

Table 11.3　SYMBOLS AND ABBREVIATIONS USED IN PULMONARY PHYSIOLOGY[a]

FOR GASES

1. Primary Symbols for General Variables
   V = gas volume
   $\dot{V}$ = gas volume/unit time
   P = gas pressure (or partial pressure)
   F = fractional concentration in dry gas phase

   f = respiratory frequency (breaths/unit time)
   D = diffusing capacity
   T = transfer factor
   R = respiratory exchange ratio (= CO_2 output/O_2 uptake)

FOR BLOOD

1. Primary Symbols for General Variables
   Q = volume of blood
   $\dot{Q}$ = volume flow of blood/unit time
   C = concentration of gas in blood
   S = percent saturation of Hb with O_2 or CO
2. Secondary Symbols Which Modify Primary Symbols
   a = arterial blood
   v = venous blood
   c = capillary blood
   p = plasma
   b = blood in general
   s = steady-state; shunt blood
   ac = alveolar component
   e = effective

Table 11.3 (Cont'd)

2. Secondary Symbols Which
   Modify Primary Symbols
   - I = inspired gas
   - E = expired gas
   - A = alveolar gas
   - T = tidal gas
   - D = dead space gas
   - M = membrane
   - L = lung

STPD = 0°C, 760 mm Hg, dry (standard temperature and pressure, dry)

BTPS = body temperature and pressure, saturated with water vapor

ATPS = ambient temperature and pressure, saturated with water vapor

A dash (—) placed above any symbol denotes a *mean* value; eg, $\overline{P}_{AO_2}$ = mean alveolar oxygen

A dot ( ) placed above any symbol denotes a *time derivative*; eg, $\dot{Q}c$ = blood flow through pulmonary capillaries/minute

The chemical symbols O_2, CO_2, N_2, H_2O, CO, N_2O, and so forth also are used as subscripts to modify the general and secondary symbols listed in table; eg, P_{cCO} = partial pressure of carbon monoxide in pulmonary capillary blood.

[a]*Data from Best and Taylor (eds).* The Physiological Basis of Medical Practice, *8th ed, 1966, Williams & Wilkins; Ruch and Patton (eds).* Physiology and Biophysics, *19th ed, 1966, W. B. Saunders.*

the diastolic phase (Fig. 10.8, p. 582). The blood then passes through a system of low-pressure conduits, the pulmonary arteries, to a system of vessels where gases are exchanged, ie, the pulmonary capillaries. Thence, the blood is returned to a second pumping chamber, the left ventricle, via a system of collecting vessels, the pulmonary veins, for distribution throughout the body (Fig. 10.15, p. 588 and 10.60, p. 655).

The air pump, on the other hand, has no valves, and air is moved back and forth through the same passageways rather than in a complete circuit. These passageways conduct and distribute fresh air to the alveoli, and they collect alveolar gas as well. Since essentially no gas is exchanged in the conducting tubes, this region is called the physiologic dead space (p. 726).

Table 11.4 COMPOSITION OF DRY ATMOSPHERIC AIR AT SEA LEVEL[a]

GAS	MOL FRACTION (percent)
Nitrogen (N_2)	78.09
Oxygen (O_2)	20.95
Carbon Dioxide (CO_2)	0.033
Argon (A)	0.93
Neon (Ne)	1.8×10^{-3}
Helium (He)	5.24×10^{-4}
Krypton (Kr)	1.0×10^{-4}
Hydrogen (H_2)	5.0×10^{-5}
Xenon (Xe)[b]	8.0×10^{-6}
Ozone (Oz)[b]	1.0×10^{-6}
Radon (Rn)[b]	6.0×10^{-18}

[a]*Data from Weast (ed).* Handbook of Chemistry and Physics, *54th ed, 1973, Cleveland, CRC Press.*

[b]*Ozone and radon are particularly susceptible to variation at sea level and above.*

Unlike the heart, which is a positive-pressure blood pump, the air pump is a *negative*, or more correctly, a *subatmospheric pressure pump.** This subatmospheric pressure pump functions by enlarging the thorax actively during the inspiratory phase of the respiratory cycle, so that the pressure within the alveoli falls below atmospheric pressure; thus, air at atmospheric pressure flows into the lungs down the gradient created by thoracic enlargement. Inspiration is thereby analogous to using a soda straw. When one "sucks" on the straw, a partial vacuum is created within the air column above the surface of the liquid, and the relatively higher external atmospheric pressure forces fluid up the straw and into the mouth.

In contrast to inspiration, expiration is largely a passive process. Normally, the elastic recoil of the lungs and structures of the thoracic wall reverse the pressure gradient slightly so that air now flows to the exterior of the body. The thoracic muscles of expiration are also capable of strong contractions during expiration when there is a requirement for large volumes of air, as during exercise, or when an obstruction to expiratory flow is present. These facts, together with the various pressures involved in respiration, now will be considered in greater detail.

Mechanisms of Inspiration and Expiration

The rhythmic increase and decrease in lung volume during respiration is accomplished in two ways. First, the alternate contraction and relaxa-

*The term negative *is incorrect when used in this context. A negative pressure is below zero mm Hg. The pressures developed during inspiration, as well as the intrapleural pressure itself, actually are only a few mm Hg below atmospheric pressure (760 mm Hg at sea level); thus, the term* subatmospheric *will be used to indicate such pressures.*

A. EXPIRATION

B. INSPIRATION

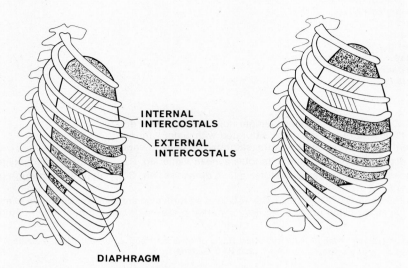

INTERNAL
INTERCOSTALS

EXTERNAL
INTERCOSTALS

DIAPHRAGM

FIG. 11.10. Mechanics of breathing. **A.** During expiration, the ribs are depressed and the diaphragm is relaxed so that it is highly convex when viewed from below. **B.** During inspiration, the ribs are rotated upwards and outwards and the diaphragm contracts, thereby lengthening the thoracic cavity. (After Ganong. *Review of Medical Physiology,* 6th ed, 1973, Lange Medical Publications; Guyton. *Textbook of Medical Physiology,* 4th ed, 1971, Saunders.)

tion of the diaphragm alternately lengthens and shortens the thoracic cavity. The attendant intrathoracic pressure alterations cause movement of air into and out of the lungs. Second, the elevation and depression of the rib cage increase and decrease the anteroposterior diameter of the chest, producing concomitant pressure changes. The skeletal muscles involved in these processes are summarized in Table 11.2, together with a brief description of their individual functions; however, normal quiet breathing is accomplished principally by diaphragmatic contraction alone.

Those muscles that elevate the rib cage are classified as inspiratory muscles, whereas those that depress this structure are expiratory muscles. At the end of expiration, the ribs project forward and downward from the spinal column (Fig. 11.10A); when the internal intercostal muscles contract, the ribs now extend forward in a more horizontal plane and some outward rotation also is achieved, so that the anteroposterior chest diameter is increased about 20 percent (Fig. 11.10B). During a period of maximum respiratory effort, the increase in thoracic diameter alone can account for over half of the increase in lung volume.

Immobilization of the chest wall does not impair pulmonary ventilation in a normal individual at rest, as the diaphragm alone is capable of moving adequate volumes of air into the lungs.

The contraction of the diaphragm (and the accessory respiratory muscles) thus can be considered to function in a manner analogous to a piston in a cylinder (eg, a hypodermic syringe). Pulling on the plunger creates a partial vacuum; therefore, air moves into the cylinder. Pushing on the plunger expels the air. Similarly increasing the intrathoracic volume inflates the lungs and vice versa (Figs. 11.11 and 11.12).

Elastic Recoil of the Lungs and Intrapleural Pressure

As mentioned earlier, the lungs and thoracic wall are highly elastic structures. The normal *elastic recoil* of the lungs themselves may be attributed to two factors. First, the thin film of fluid that lines the alveoli exerts surface tension (see also surfactant, p. 721), and this physical effect tends to collapse these delicate structures. Second, the elastic connective tissue that is present within the lung parenchyma normally is stretched to some extent and thus it produces tension when the lungs are distended within the thoracic cavity (see also compliance of lungs, p. 719). These two factors produce elastic recoil of the lungs, and thus these organs tend to pull away from the thoracic wall (Fig. 11.11). This elastic recoil in turn creates an intrapleural pressure within the thoracic cavity that is slightly below the corresponding atmospheric pressure. At the end of expiration, the normal intrapleural pressure is −3 to −4 mm Hg, measured with respect to the ambient atmospheric pressure (Fig. 11.12).

The lungs remain expanded within the thoracic cavity at all times because the intrapleural pressure is subatmospheric. Hence, the normal tendency of the lungs to exhibit elastic recoil is counterbalanced by the tendency of the chest wall to recoil in the opposite direction. Consequently, the lungs remain expanded and fill the thoracic

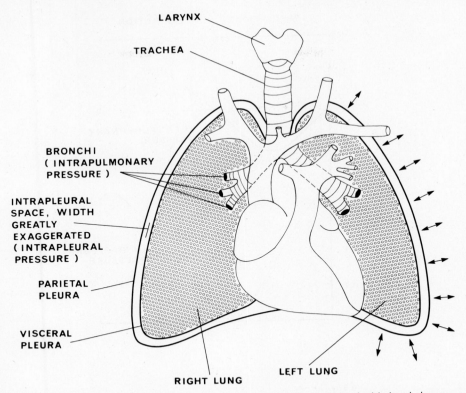

LARYNX

TRACHEA

BRONCHI
(INTRAPULMONARY
PRESSURE)

INTRAPLEURAL
SPACE, WIDTH
GREATLY
EXAGGERATED
(INTRAPLEURAL
PRESSURE)

PARIETAL
PLEURA

VISCERAL
PLEURA

RIGHT LUNG

LEFT LUNG

FIG. 11.11.   The major intrathoracic structures and sites for measuring pressures. The double-headed arrows on the right side of the figure signify the elastic recoil of the lungs away from the thoracic wall, which force is countered exactly by the cohesion of the visceral pleura to the parietal pleura by surface tension as well as by the negative intrapleural pressure per se. See Figure 11.12 for intrapulmonary and intrapleural pressures during inspiration and expiration. (After Ganong. *Review of Medical Physiology,* 6th ed, 1973, Lange Medical Publications.)

cavity completely during all phases of the respiratory cycle.

During inspiration and expiration, as the intrathoracic volume changes (Fig. 11.11), the visceral pleura covering the lungs glides along the parietal pleura of the thoracic cavity, the two surfaces being connected by a thin film of serous fluid. This process is analogous to two sheets of glass separated by a thin film of oil or water. Even as the glass sheets are able to slide readily past each other when a small shearing force is applied in a plane parallel to their long axes, so the lungs glide smoothly along the parietal pleura during inspiration and expiration. Once again, in analogy to the sheets of glass, the film of liquid causes the two pleural surfaces to cohere closely to each other at all times. Thus, it is extremely difficult to separate either the glass sheets or the pleural membranes by a force applied at right angles to their surfaces.

As the lung tissues are stretched simultaneously with their increase in volume during inspiration, the elastic recoil also increases. A still lower intrapleural pressure is produced. This reaches a maximum value of around −6 mm Hg at the end of inspiration during quiet breathing. (These facts for quiet respiration are illustrated in Figure 11.12.) Furthermore, forced inspiratory efforts can in-

crease the intrapleural pressure to values as low as −30 mm Hg. Such an effect occurs, for example, during extreme physical exertion.

If the chest wall is opened, or if air is injected into the pleural cavity so that the intrapleural and atmospheric pressures are equalized, the lungs collapse because the force of elastic recoil is no longer counteracted by a subatmospheric intrapleural pressure. This condition is known as pneumothorax.

The intrapleural pressure in human subjects can be measured with reasonable accuracy by means of an inflatable balloon attached to a catheter and swallowed so that it lies within the intrathoracic portion of the esophagus adjacent to the trachea; the other end of the catheter is attached to a water manometer and the pressures can be measured directly in cm H_2O. Because of the peculiar anatomic relationship between esophagus and trachea within the thorax (Fig. 11.5), the pressure alterations within the balloon reflect closely the changes in intrapleural pressure during respiration. Thus the intraesophageal pressure represents the pressure difference across the lung wall, since normally the esophagus is relaxed (except during the act of swallowing) and therefore it exerts no appreciable pressure of its own upon the balloon.

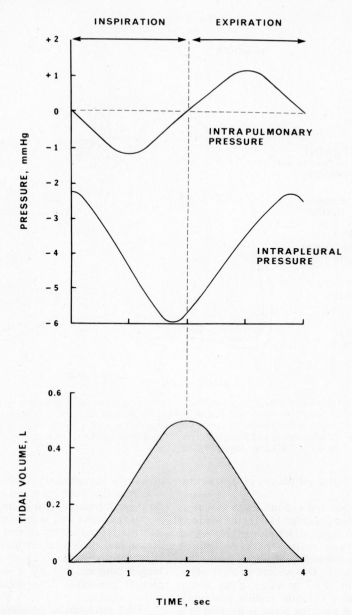

FIG. 11.12. Alterations in intrathoracic (or intrapleural) and intrapulmonary pressures during one complete respiratory cycle. (After Ganong. *Review of Medical Physiology*, 6th ed, 1973, Lange Medical Publications.)

Intrapulmonary Pressures During Respiration

The normal pressure fluctuations that occur during quiet inspiration and expiration are shown in Figure 11.12. As noted above, inspiration is an active process, so that the intrapulmonary pressure drops about 1 mm Hg below atmospheric pressure at the peak of inspiration, and the elastic recoil of the lungs and chest wall pulls the lungs back to the expiratory position, until the recoil pressures of lungs and chest wall just balance.

In Figure 11.12, note especially that the intrapulmonary pressure (ie, the pressure within the air passages and alveoli) fluctuates slightly below and above ambient atmospheric pressure during inspiration and expiration, respectively; in contrast to this situation, the intrapleural pressure always remains subatmospheric, and becomes more so at the peak of inspiration, as discussed in the previous section.

During forced expiration, the maximum intrapulmonary pressure developed may rise to around 80 mm Hg above atmospheric pressure, whereas the maximum intrapulmonary pressure achieved during a forced expiration may be as high as 120 mm Hg (cf. p. 653).

The rhythmic oscillations of the intrapulmonary pressure discussed above, and illustrated in Figure 11.12, are responsible for the movement of air into and out of the lungs, and thus these oscillations underlie pulmonary ventilation itself.

Compliance of the Lungs and Thorax

Compliance means the volume change produced by a unit change of pressure; this concept was introduced in relation to cardiovascular physiology on page 629, and it is equally useful in pulmonary physiology as a convenient way of measuring the distensibility (or elasticity) of the lungs and thoracic structures.

Elasticity is a physical property of matter that enables it to return to its resting shape after deformation by an external force. According to *Hooke's law,* the deformation of an elastic body, such as a spring, is directly proportional to the applied force, up to the elastic limit of the spring (Fig. 11.13). Similarly, certain tissues of the lungs and thorax exhibit elastic properties, and when a force is applied to these structures, in this case pressure, the resulting *volume* change is proportional to the applied force (pressure) within limits. When the volume is increased, the pressure also changes accordingly. When the force is removed, the elastic properties of the tissues restore the

original volume. This concept is illustrated in Figure 11.14.

In pulmonary physiology, the term *compliance* is used to measure static pressure-volume relationships; that is, compliance = $V/\Delta P$, and the units employed for expressing ΔP are usually liters/cm H_2O. In contrast with compliance, the term *resistance* is employed with reference to the measurement of pressure-flow relationships involving motion, eg, in electricity, or in respiratory and cardiovascular physiology; hence, the term *resistance* is applicable only to dynamic measurements.

It is a relatively simple matter to measure the compliance of the lungs alone in an open-chest anesthetized animal or when the lungs have been isolated from the body. Under such experimental conditions, it is found that the lung compliance alone is about 0.22 liter/cm H_2O, ie, the lungs expand 220 ml for each 1.0 cm H_2O pressure increment. In vivo, however, the lung compliance is reduced to around 0.13 liter/cm H_2O because of the necessity for expanding both the lungs and the thoracic cage; the muscles, bones, tendons, and other structures of the thoracic wall all have inertia as well as inherent viscoelastic properties which must be overcome during inspiration, so that the volume change of the lungs plus thoracic structures (ie, the total pulmonary compliance) is much lower for a given pressure increase than for the lungs alone.

Hence, work must be performed by the respiratory muscles to expand the thoracic cavity and by the lungs in order to breathe. The pressure relationships within the lung and thorax in vivo are shown in Figure 11.15.

In man, the pulmonary (or lung) compliance can be determined by measuring the pressure developed within an intraesophageal balloon during respiration (p. 717). The pressure is recorded at

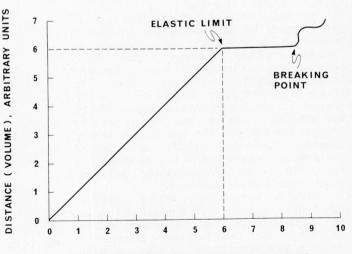

FIG. 11.13. Hooke's law. That is, in any elastic structure the increase in length or volume of the structure is directly proportional to the increase in applied force or pressure until the elastic limit is reached. Throughout their physiologic range, this law applies to normal lungs. (Data from Comroe. *Physiology of Respiration,* 1965, Year Book of Medical Publishers; White. *Modern College Physics,* 3rd ed, 1956, Van Nostrand.)

DISTANCE (VOLUME), ARBITRARY UNITS

ELASTIC LIMIT

BREAKING POINT

FORCE (PRESSURE), ARBITRARY UNITS

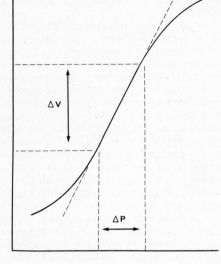

FIG. 11.14. The pressure–volume relationships in a hollow elastic structure, for example, the thorax or lung. Thus a given change in pressure (ΔP) causes a given change in volume (ΔV) and thus compliance, C, is equal to $\Delta V/\Delta P$. Note that the relationship between ΔV and ΔP is linear for a considerable distance (slanted dashed line) in accordance with Hooke's law (Fig. 11.13). The slope of this line also defines the compliance. (After Selkurt [ed]. *Physiology*, 2nd ed, 1966, Little, Brown.)

the end of expiration and again while holding the breath after inspiring a known volume of air. This process is repeated several times following inhalation of successively greater volumes of air, and the changes in the intraesophageal pressure then are plotted against the change in lung volume. The slope of the curve gives the total pulmonary compliance in liters/cm H_2O (Fig. 11.15).

The elastic recoil of the lung cannot be measured directly by determining the intrapleural pressure at one lung volume alone, because the technique employed (an intraesophageal balloon) records changes in intrapleural pressure far more accurately than it does absolute pressures. The change in pressure that produces a change in volume (or vice versa) is relatively accurate, although this technique necessitates that measurements be made at two volumes. The measurement of lung compliance during slow respiratory movements by the technique described above is illustrated in Figure 11.15B, the measurement during rapid respiration in part C of the same figure.

Since the compliance is related directly to lung volume, in order to be a meaningful indicator of elastic recoil, both compliance and the lung volume itself must be known. Therefore, in order to ascertain whether or not lung tissue has normal elasticity, both the compliance and the lung volume at the time the compliance is measured must be known; and thus compliance per

lung volume is called the *specific compliance*. This value normally is around 0.6 liter/cm H_2O.

A decreased specific compliance indicates that a loss of elasticity and increased rigidity of the pulmonary tissues are present, such as occur in pleural or interstitial fibrosis. In these pathologic situations, the fibrous and collagenous tissues have different length-tension relationships than do the normally abundant elastic fibers which they replace; hence, the specific compliance is decreased. Pulmonary compliance may also decrease in conditions where the lung tissue becomes edematous, such as in pneumonia, or where there are obstructed airways. In emphysema, on the other hand, the specific compliance may increase if there is damage to, or diminution of, the elastic fibers but without a concomitant replacement by fibrous or collagenous connective tissue.

If the compliance of both lungs and thorax together (ie, the total pulmonary compliance) is considered, one must also be aware that any condition that reduces the ability of the thoracic cage to expand, and thus hinders lung ventilation, will reduce the overall compliance. Fibrotic pleurisy, a broken rib, or congenital abnormalities that restrict movements of the chest (eg, scoliosis or kyphosis), and paralyzed respiratory muscles all can produce a defective pulmonary ventilation and thereby lead to various degrees of functional impairment.

Surface Tension and Recoil of the Lungs

An important factor involved in lung recoil is to be found in the physical force developed by surface tension within the alveoli. This physical property of air:liquid interfaces was discussed on page 17.

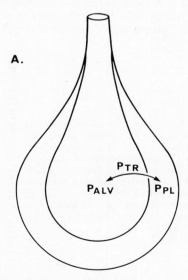

FIG. 11.15. A. Intrathoracic and intrapulmonary pressure relationships. The intrathoracic or intrapleural pressure (P_{PL}) is different from the alveolar or intrapulmonary pressure (P_{ALV}). The arithmetic difference between these two pressures is the transmural pressure of the lung (P_{TR}).

B.

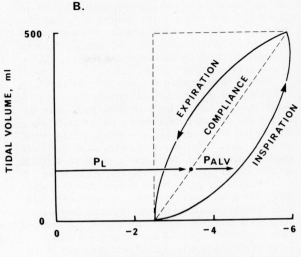

FIG. 11.15. (cont'd)   B. Plot of the pressure against volume for the lung and thorax. The compliance is obtained by measuring the tidal volume and the intrapleural pressure (when air flow is zero, as discussed in the text) over a range of different tidal volumes. Loops are obtained by measuring the pressure and volume continuously during slow breathing.

Within the delicate alveoli, which are lined with a thin aqueous film, the property of surface tension at the interface with the air tends to cause these structures to assume the least surface area, ie, to become spherical like minute bubbles located at the ends of equally minute capillary tubes. According to Laplace's law, the wall tension developed in a sphere is proportional to the transmural pressure as well as the radius of the sphere (p. 627). The surface-tension recoil developed by the lung can be calculated, assuming that the alveolar liquid film develops the same surface-tension force as that of plasma, that is, 50 dynes/cm. Using this assumption, the surface-tension recoil of the lung would be 20,000 dynes/cm², and this force would have a pressure that is around 20 cm H_2O (1 cm H_2O = 980 dynes). This estimated value is reasonably close to that obtained experimentally in fully inflated lungs. But in lungs that are only partially inflated, this figure is far too high, by a five- to tenfold factor, so that at smaller lung volumes the surface tension is only around 5 to 10 dynes/cm rather than 50 dynes/cm.

Ingenious experiments have demonstrated the presence of a layer of detergentlike molecules that covers the surface of the alveolar liquid film. This surfactant appears to be a complex phospholipid (such as dipalmitoyl lecithin), and may act either alone or in combination with a protein. Surfactant appears to be synthesized by the alveolar cells themselves for release into the lumen of the alveoli. Native surfactant reduces the surface tension, like a detergent, by reducing the cohesive (or attractive) force between the water molecules at

C.

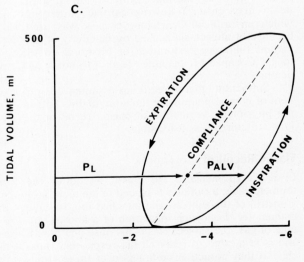

FIG. 11.15. (cont'd)   C. Pressure–volume measurements obtained during a rapid respiratory cycle. The alveolar pressure and transmural pressure of the lung also are shown in both **B** and **C** at one point during inspiration. (After Selkurt [ed]. *Physiology,* 2nd ed, 1966, Little, Brown.)

the air:liquid interface. Interestingly, a film of liquid covered with a layer of lung surfactant develops a far higher surface tension when stretched than when compressed, ie, the surface tension changes inversely according to the degree of compression or expansion of the film.

These facts explain why the fully inflated lung exhibits a far greater surface tension than the partially inflated lung. The surfactant molecules decrease the surface tension to a greater extent when the alveoli are smaller; ie, the film is compressed, so that the surfactant molecules are relatively more concentrated, and thus can exert a much greater effect upon the overall surface tension.

What is the physiologic significance of surfactant? First, an alveolus with a relatively low surface tension requires a low air pressure to maintain its radius and volume. A lower surface tension thus diminishes the total muscular work necessary to ventilate the lungs. Second, the presence of surfactant within the alveoli tends to stabilize these structures. Although the alveoli have different diameters, all are inflated at the same pressure throughout the lung. According to Laplace's law, smaller alveoli would have a greater wall tension, and tend to collapse at a given pressure, whereas larger alveoli would tend to overinflate at the same pressure because of their inherently lower wall tension. Third, the tendency toward elastic recoil in the lungs which is caused by surface tension alone, is greatest when the lungs are inflated (at the end of inspiration), and this tendency diminishes as the surface tension decreases (toward the end of expiration). Fourth, if surfactant is not present, as the alveoli become smaller during expiration, whereas if their wall tension remains the same, they collapse. For example, if surfactant is not present in the lungs of a newborn baby, the lungs cannot inflate properly (p. 1209) and death ensues from the atelectasis. This condition is known as *hyaline membrane disease,* or the *respiratory distress syndrome.* Similarly, in an adult, death may ensue following cardiac surgery in which an artificial heart–lung bypass was used, because of atelectasis caused by inactivity of surfactant following restoration of breathing to the lungs. Collapse of the lungs leads to anoxia and death.

Surfactant, therefore, is an essential component of normal pulmonary function, and its role in the physiologic regulation of elastic recoil in the lung structures is considerable.

Elastic Recoil of the Thorax

In order to measure the elastic recoil of the thoracic cage, a catheter is placed in the mouth (or a nostril) of the subject, and attached to a suitable manometer. Following inhalation of a known volume of gas, the subject closes his nose and mouth and is requested to relax all his respiratory muscles so that no active contraction of these structures is present. The intraoral (or intranasal) pressure gives the alveolar pressure. From these data, total compliance can be obtained graphically, as described earlier (Fig. 11.15).

The intrapleural pressure is determined simultaneously with the total compliance by means of an intraesophageal catheter, as described on page 717, so that the pulmonary compliance also is obtained. This process is repeated several times following inspiration of different known volumes of air, and the thoracic cage compliance is determined from the relationship:

1/Total compliance = 1/Pulmonary compliance + 1/Thoracic cage compliance

The approximate normal compliance values for a young adult male are: Total compliance (lungs and thoracic cage) 0.1 liter/cm H_2O. Pulmonary compliance 0.2 liter/cm H_2O. Thoracic cage compliance 0.2 liter/cm H_2O.

Resistance to Air Flow and the Work of Breathing

During quiet respiration in a normal individual, most of the work that is performed by the respiratory muscles is used to overcome the elastic recoil of the lungs and thorax, and very little energy is required to move the air through the respiratory passageways.

The same physical laws and factors that govern the flow of a liquid (fluid) in closed tubes determine the resistance to air flow in the respiratory tract. Poiseuille's relationship for laminar flow ($Q = \Delta P/R$) describes the resistance to air flow within the respiratory tree, but only to a limited extent because of a number of complicating factors. First, large regions of the respiratory tract, unlike most of the circulatory system, are not composed of a series of smooth cylinders. Rather, the respiratory conduits are irregularly-shaped tubes, whose length and diameter are constantly fluctuating. Second, some of these conduits are arranged in parallel (the nostrils), and these in turn are followed by other tubes which are arranged in series (the nasopharynx, larynx, and trachea). Subsequently, the system divides and subdivides repeatedly into parallel tubular systems, including the bronchi, bronchioles, terminal bronchioles, respiratory bronchioles, and alveolar ducts. Air flow into and out of the lungs is not always laminar, but becomes turbulent in certain regions, and eddies may form because of the considerable velocity that develops in some parts of the respiratory tract (Fig. 10.43, p. 627). Third, there are also constrictions in the airways, and these in turn produce orifice flow so that the gas must accelerate under such conditions in accordance with Torricelli's theorem. This relationship states that the rate of discharge of a fluid through an orifice is proportional to the cross-sectional area of the orifice multiplied by the linear velocity of the fluid. Fourth, the density of the inhaled gas itself may change, such as when an individual is breathing special oxygen–helium mixtures or is at a high altitude, where the atmospheric pressure is lower (p. 1223), or is diving, while breathing compressed air (p. 1231).

Hence, a number of complex factors are involved in air flow within the respiratory tract, and these in turn all are involved in production of the overall airway resistance within the respiratory system.

Resistance, as stated on page 637, is a dynamic property which cannot be measured directly, but must be obtained by calculation. By analogy to an electric circuit (p. 8), the total resistance of a number of elements in parallel, ie, below the tracheal bifurcation, is equal to the sum of the reciprocals of the individual resistances:

$$1/R \text{ total} = 1/R_1 + 1/R_2 + 1/R_3 + \ldots 1/R_X$$

Within the respiratory tree, the total resistance would thus become enormous at the level of the alveolar ducts were it not for three important factors. First, as each of the airways divides, each of the branches is only a trifle smaller than the branch from which it originated. Second, at each point where branching occurs, the flow decreases by half. Third, the respiratory tubes become shorter as they divide and become narrower.

These three factors act in combination to reduce effectively the total resistance of the airways throughout the successive branchings of the respiratory tract. The overall airway resistance in normal adults has been estimated to range between 0.05 and 1.5 cm H_2O/liter/second. During quiet breathing, around 50 percent of the total airway resistance is encountered in the nasal passages; breathing through the mouth effectively reduces this resistance. A small part of the overall resistance is present in the glottis, while the remainder is found in the branching tubes in the remainder of the respiratory tree.

Clinically, the airway resistance may increase because of the operation of one or more of the following conditions: constriction of the smooth muscle in the bronchioles; edema and swelling of the bronchiolar tissues; inflammation or mucous congestion; plugging of the lumen of the airways by edema fluid, mucus, exudate, or foreign matter; cohesion of the mucosal surfaces by surface tension; compression of the bronchioles; fibrosis; and collapse of the bronchioles because of loss of alveolar elastic fibers, which exert traction on the bronchiolar tubes. Thus, such clinical conditions as asthma or obstructive emphysema result in a greatly increased airway resistance.

Little change in airway resistance accompanies quiet respiration, because traction on the airways by surrounding tissues causes them to become wider as well as longer during inspiration, since resistance is related inversely to the radius of a tube (p. 624). In fact, maximum inspiration actually decreases airway resistance to a demonstrable extent. At the end of a maximum expiration, however, the airway resistance increases considerably because of the artificially elevated intrapleural pressure that compresses and/or collapses respiratory bronchioles and alveolar ducts. Maximum air flow rates, as determined by a device called a spirometer, are used clinically to determine whether or not obstructive pulmonary disease is present in a patient. This technique is illustrated in Figure 11.16.

From a physiologic standpoint, airway resistance is governed principally by the extent of dilatation or constriction of the bronchioles and alveolar ducts (p. 710). Parasympathetic cholinergic impulses constrict the circularly oriented smooth muscle fibers of the airways, thereby markedly increasing their resistance to air flow in accordance with Poiseuille's law. Sympathetic adrenergic impulses, on the other hand, relax the bronchial smooth muscle fibers producing dilatation and a concomitant lowered resistance to air flow. The reflex mechanisms which underlie these neural effects on airway resistance are discussed on page 733.

Certain drugs that modify airway resistance appreciably may be mentioned. These compounds may act indirectly, through stimulation of autonomic ganglia, by liberating neurotransmitter substances, or by stimulation or inhibition of receptor sites within the tissues themselves. Other compounds act directly upon the cells in a manner wholly unrelated to nerves or receptor sites. For example, isopropylarterenol, epinephrine, and norepinephrine produce bronchodilation by stimulating the receptor sites associated with terminals of the sympathetic postganglionic fibers. Anticholinesterase agents as well as acetylcholine act at parasympathetic receptor sites, and thus produce bronchoconstriction. The anticholinesterase agents function by blocking the action of cholinesterase; therefore, acetylcholine is not hydrolyzed and its concentration rises within the tissues to a more active level. Histamine produces strong constriction of the alveolar duct sphincters, and thus closes the alveolar ducts.

In a physical sense, work is defined as a force times a distance. In the respiratory tract, this relationship becomes the summated effect of pressure times the instantaneous volume of air moved so that work = $\int PVd$ in which P is the pressure, V is the volume, and d is the distance.

As noted earlier, during quiet respiration, expiration normally is a passive event. If, however, the airway resistance suddenly increases, such as occurs during the sudden bronchospasm of an asthmatic attack, the patient must perform considerable physical work during expiration in order to overcome the additional resistance in the respiratory tract by increasing the pressure differential between the lungs and the exterior of the body. This effort can become exhausting if prolonged. Forced expiration is necessary under such circumstances in order to empty the lungs during a sufficiently brief interval (normally around 3 seconds) prior to the next inspiration.

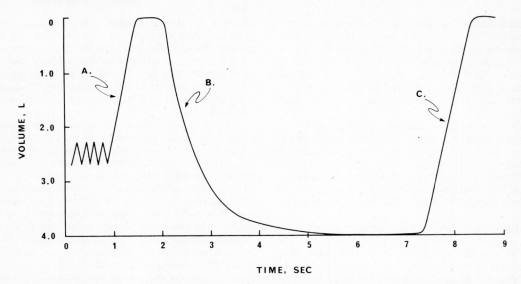

FIG. 11.16. Maximal inspiratory and expiratory flow rates as obtained by a spirometer. A subject having a vital capacity of 4 liters is requested to inspire to his maximum ability (A) and then to expire both as rapidly and fully as possible (B). Then, from full expiration, the subject is asked to inspire both fully and rapidly once again (C). The volumes of air are recorded on a chart moving at constant speed, so that the air flow rates during inspiration and expiration can be obtained readily, for example by the slopes of the curves. (After Comroe. *Physiology of Respiration*, 1965, Year Book Medical Publishers.)

It is important to remember that airway resistance increases directly with an increased velocity of flow. Therefore, patients having obstructive pulmonary diseases, such as asthma, bronchitis, emphysema, pulmonary fibrosis, and lung tumors, tend to breathe slowly, and the work performed during respiration may increase appreciably, particularly during periods of increased overall bodily activity. It is important to remember that the greater the rate of volume flow of air, the greater the pressure gradient becomes that is necessary to overcome the resistance of the airways.

Pulmonary Volumes and Capacities

The possible alterations in lung volume under different physiologic conditions are best defined and explained by reference to Figure 11.17, in which the four *pulmonary volumes* and the four *pulmonary capacities* are shown on the left and right sides of the figure, respectively.

The specific lung volumes and capacities to be mentioned in the following discussion represent average values for normal young adult males. It is important to remember that all pulmonary volumes and capacities are between 20 to 25 percent lower in females. There is also a direct relationship between body size (as well as athletic capacity) and the magnitudes of these volumes. With advancing age, however, all these values tend to become lower.

PULMONARY VOLUMES. The sum of the four pulmonary volumes represents the maximum volume to which the lungs can be expanded in vivo.

The resting expiratory level can be taken as a convenient base line from which the four pulmo-

nary volumes are determined (Fig. 11.17). During quiet respiration, the elastic recoil of the lungs and thoracic cage causes the lungs to collapse passively to a relaxed (but not completely collapsed!) condition, and the volume of air then present within these organs is known as the resting expiratory level.

The tidal volume (TV) is the total volume of air inhaled and exhaled during quiet breathing. This volume usually is around 400 to 500 ml, but it can be considerably greater in large (or athletic) individuals. This fraction of the total pulmonary volume is so named because of its resemblance to oceanic tides, which move back and forth along the same path.

The *inspiratory reserve volume (IRV)* is the volume of air that can be inhaled above the normal tidal volume, usually about 3,000 ml.

The *expiratory reserve volume (ERV)* is that volume of air which can be expelled by forceful expiration following a normal tidal expiration. This volume is usually around 1,100 ml.

The *residual volume (RV)* is the quantity of air remaining in the lungs following a forceful expiration. This volume is approximately 1,200 ml.

PULMONARY CAPACITIES. For convenience in discussing certain events pertaining to respiratory physiology, it is often useful to consider simultaneously two or more of the volumes described above. Such combinations of respiratory volumes are known as pulmonary capacities.

The *Functional Residual Capacity (FRC)* is equal to the expiratory reserve volume plus the residual volume, ie, *(ERV + RV)*. This is the volume of air remaining in the lungs at the resting expiratory level as described above, and following a normal tidal expiration. The *FRC* volume is around 2,300 ml.

The *Inspiratory Capacity (IC)* is equal to the tidal volume plus the inspiratory reserve volume (ie, *TV + IRV*). This is the total quantity of air an individual can inspire starting at the normal resting expiratory level, followed by a maximal inhalation. This volume is about 3,500 ml.

The *Vital Capacity (VC)* is the sum of the inspiratory reserve volume, the tidal volume, and the expiratory reserve volume (ie, *IRV + TV + ERV*). The vital capacity thus represents the maximum volume of air that an individual can exhale during a maximal forced expiration follow-

ing a maximal forced inspiration. This volume is approximately 4,600 ml.

The *Total Lung Capacity (TLC)* represents the sum of the four individual lung volumes, that is, the inspiratory reserve volume, the tidal volume, the expiratory reserve volume, and the residual volume (ie, *IRV + TV + ERV + RV*). Therefore, the total lung capacity is the total volume of gas that is contained in the lungs at the end of a forced inspiration, around 5,800 ml.

GENERAL PHYSIOLOGIC SIGNIFICANCE OF THE PULMONARY VOLUMES AND CAPACITIES. As noted earlier, the volume of air contained within

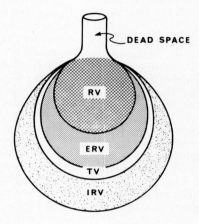

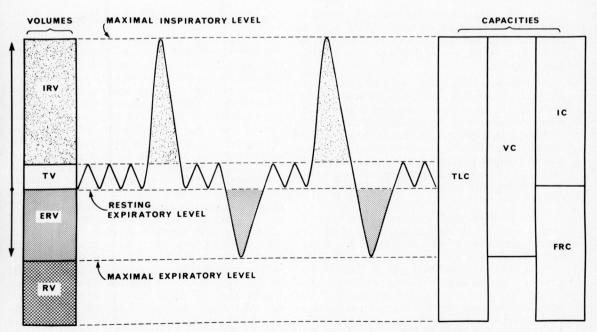

FIG. 11.17. The lung volumes and capacities. The upper diagram illustrates the four principal lung volumes. RV = residual volume; ERV = expiratory reserve volume; TV = tidal volume; IRV = inspiratory reserve volume. The same shading is used in the lower part of the figure to indicate the same volumes. The four lung capacities are shown to the right of the figure. TLC = total lung capacity; VC = vital capacity; IC = inspiratory capacity; FRC = functional residual capacity. See text for additional discussion. (After Comroe. *Physiology of Respiration,* 1965, Year Book Medical Publishers.)

the lungs depends upon the sex of the individual as well as body size and general configuration.

It is equally important to note that alterations in body position can markedly affect the pulmonary volumes and capacities. Most of these parameters decrease when a person lies down, first, because the abdominal contents then can exert pressure upon the diaphragm, and second, because the augmented pulmonary blood volume in the recumbent position decreases the available space for pulmonary air (p. 656). These facts are of importance clinically, because a patient suffering from marginal cardiac and/or pulmonary disease may experience exacerbation of the symptoms while lying in the supine position in which the vital capacity is decreased, whereas the symptoms are ameliorated while sitting up.

The pulmonary compliance (p. 713) and the strength of the respiratory muscles themselves are also significant factors insofar as the vital capacity of an individual is concerned. Hence, if pulmonary compliance is decreased by any factor that prevents normal expansion of the lungs, the vital capacity also is impaired. For example, tuberculosis, pulmonary carcinoma, emphysema, chronic bronchitis or asthma, and fibrotic pleurisy all can reduce pulmonary compliance and thus reduce vital capacity. Consequently, measurement of this parameter by means of a spirometer is an important clinical aid to the physician in assessing the progress of various pulmonary diseases (Fig. 11.16).

If the respiratory muscles are partially paralyzed, as, for example, following poliomyelitis or spinal cord injuries, the vital capacity may be reduced to 500 ml, a level that is barely capable of sustaining life.

Pulmonary congestion, such as results from left ventricular failure, decreases the vital capacity markedly because lung compliance also is diminished because of the presence of excess fluid within the lung tissues. The measurement of vital capacity, therefore, is a useful clinical determination in patients having left ventricular failure, as such data indicate the progress of the disease, and may also help to evaluate the degree of pulmonary edema that is present.

The residual volume is that fraction of the total quantity of air that cannot be expelled from the lungs even by forceful expiration. Physiologically, the residual volume is very important, because it provides the alveoli with air between respiratory cycles so that the blood is oxygenated and carbon dioxide is eliminated continuously. If this were not the case, marked fluctuations in these blood gases would ensue between breaths. The residual volume of air in the lungs may be considered as exerting a "damping effect" on the oscillations of blood gas concentrations which otherwise would become far more pronounced than they actually are.

The Minute Respiratory Volume

The total volume of fresh air moved into the respiratory tract per minute is known as the minute respiratory volume or total ventilation, and this volume is equal to the product of the tidal volume and the respiratory rate. Therefore, if the tidal volume is 500 ml and the respiratory rate is 12/minute, the minute respiratory volume is $500 \times 12 = 6,000$ ml/minute (or 6 liters/minute). This figure is average, and considerable variation in the minute volume above as well as below 6 liters/minute may be encountered. It is obvious that a markedly elevated tidal volume would be compensated for by a reduced respiratory rate and vice versa.

The respiratory rate may rise to as high as 50/min, and the tidal volume can increase to equal the vital capacity; however, at extremely rapid respiratory rates, the tidal volume cannot be maintained for any length of time at levels much above one-half of the vital capacity.

The maximum breathing capacity is the total volume of air that can be moved into and out of the respiratory tract per minute during forced voluntary respiratory movements. This volume can reach as high as 170 liters/minute for short intervals. The same individual, however, can sustain a minute volume of over 100 liters/minute during heavy exercise for prolonged intervals. The respiratory system has an enormous reserve, so that the minute respiratory volume can increase up to 25-fold for a short period, and up to 20-fold for longer periods in normal individuals.

Anatomic and Physiologic Dead Space

As noted earlier in this chapter, the air that enters the respiratory tract is "inert" (insofar as respiratory gas exchange is concerned) until it has passed through the conducting airway (the nasal passages, mouth, pharynx, larynx, trachea, bronchi, and bronchioles). The total internal volume of these passageways is known as the *anatomic dead space*. This volume is difficult to determine in the laboratory, but tables have been prepared for estimating this value on the basis of age, sex, and tidal volumes as well as lung volumes. A useful rule of thumb is that the anatomic dead space in an adult is roughly equal in (ml) to the "ideal weight" of that individual in pounds (see Table 16.3, p. 1217).

The anatomic dead space is subject to change, depending upon the body position of the subject. In a healthy young man lying supine, the approximate value for the anatomic dead space is about 115 ml, and this value increases to around 156 ml in the semirecumbent position. During maximal expiration, the value falls to around 110 ml, whereas during maximal inspiration the anatomic dead space increases to about 230 ml, because of the increased total volume of the conducting airways as the pulmonary structures dilate, hence increase in overall volume capacity. As noted earlier, the values for the anatomic dead space are proportionally lower in women.

In normal persons, the anatomic dead space is equivalent to the physiologic dead space, because all terminal ducts and alveoli are capable of functioning in a normal lung. If, however, some of the respiratory ducts and alveoli are nonfunctional because of disease, the volume of these tissues must be added to the anatomic dead space to give the physiologic dead space, a term sometimes used to distinguish between the two situations. In other instances, the ratio of pulmonary blood flow to ventilation in some alveoli may become so low that these alveoli must be included as part of the functional dead space. In certain diseases, the physiologic dead space may become ten-fold greater than the anatomic dead space, and thus the value for this factor may become as high as 1 or 2 liters.

Pneumonectomy, tracheotomy, and asthma and its attendant bronchoconstriction all decrease the anatomic and physiologic dead space. The anatomic dead space is increased in bronchiectasis, in which the air ducts are larger, or in emphysema, in which the total lung volume, including the air spaces, is enlarged.

The anatomic dead space can be increased by breathing through a tube, the increment of total dead space being equal to the volume of the tube itself. If the volume of the physiologic dead space plus the volume of the tube exceed the vital capacity of the individual, he will suffocate rapidly because he merely is moving the same air back and forth in the tube and no fresh air is being received by the lungs.

Alveolar Ventilation

The ventilation of the alveoli themselves is the critical factor in the entire process of pulmonary ventilation. Total ventilation (or minute respiratory volume) is the quantity of air that enters the respiratory tract per minute, whereas alveolar ventilation represents the volume of fresh air that enters the alveoli at each breath or per minute. The alveolar ventilation is always a fraction of the total ventilation; ie, it is less than the total ventilation, and the extent to which it is less depends upon the anatomic dead space, the tidal volume, and the frequency of respiration.

The rate of alveolar ventilation is dependent upon the volume of fresh air that enters the alveoli per minute. Therefore, the alveolar ventilation in milliliters per minute is equal to the respiratory rate times the tidal volume minus the volume of the anatomic dead space. Therefore, if it is assumed that there is a tidal volume of 500 ml, a respiratory rate of 12/minute, and an anatomic dead space of 150 ml, the alveolar ventilation = $(12 \times 500) - 150 = 5,850$ ml/minute.

If the tidal volume were to fall so low that it equaled the dead space volume, no alveolar ventilation would occur, regardless of the respiratory rate. Conversely, if the tidal volume were to increase and become several liters, the effect of the dead space volume upon alveolar ventilation would become relatively insignificant. Thus, if the tidal volume becomes 5 liters, such as occurs during severe exercise, and the dead space volume is 200 ml, the alveolar ventilation per minute is 96 percent of the total pulmonary ventilation rate (or minute respiratory volume).

Since alveolar ventilation is a major factor in determining the oxygen and carbon dioxide concentrations within the alveoli, any discussion of gas exchange must take into consideration the alveolar ventilation. It should be reiterated emphatically at this point that the anatomic (or physiologic) dead space, tidal volume, and respiratory frequency (hence, the total pulmonary ventilation), are of significance only insofar as they influence alveolar ventilation itself!

Physiology of the Respiratory Passages

In addition to their principal function as conduits for air to and from the lungs, the upper respiratory passages have three other important functional attributes. These structures act as air filters, heat exchangers, and humidifiers. Therefore, collectively the upper respiratory passages serve to condition the inspired air before it reaches the exceedingly delicate respiratory membranes within the lungs.

FILTRATION. Large bits of particulate matter in the inspired air may be removed by the hairs within the nostrils. Far more important for removal of inspired particulate matter, however, is the turbulence created within the air column as it passes through the nasal cavity. The air passing through this region impinges upon the nasal septum, turbinates, and pharyngeal wall so that eddy currents are produced, and thus particles are trapped in the mucous coating which lines the walls of the nasal cavity. In addition, both the upper and lower portions of the respiratory tract are lined with ciliated epithelial membranes and the beat pattern of these cilia is such that they constantly propel the mucous sheet toward the pharynx together with the particulate matter trapped within it, so that it is either swallowed or expectorated. This mechanism for cleaning the inspired air is so efficient that practically no particles larger than about 6 μ in diameter enter the lungs via the nose. Most of the dust in inspired air is removed by the nasal and other upper respiratory passages. Particles having a size less than 0.5 μ in diameter generally remain suspended in the alveolar air, and are thus exhaled, such as, for example, the solid matter in cigarette smoke.

Particulate matter that does become entrapped within the alveoli is engulfed by phagocytic macrophages ("dust cells," p. 712). An excess of such foreign material, however, can produce pulmonary fibrosis leading to permanent lung damage. An example of this condition is to be found among coal miners who develop a type of disease known as "black lung," or anthracosis, from the long-term inhalation of carbon dust. Other types of pneumoconiosis are encountered among workers who inhale beryllium, talc, silica, asbestos, or other dusty material. Obviously, under such conditions, the normal air filtration mechanism is inadequate to cope with the quantity of material that is inhaled.

HEATING.   Another process that is involved in air-conditioning by the upper respiratory tract is the warming of the air as it passes over the extensive surfaces within the nasal cavity, including the septum and turbinates. Normally, the temperature of the inspired air rises to within a few degrees of body temperature by the time it reaches the lower trachea. In extremely cold (ie, subzero) environments, however, precautions must be taken to prewarm the air and/or to avoid inhaling too rapidly. Under such circumstances, the heat exchange may be insufficient, so that the lungs and other pulmonary structures can actually freeze, with grave consequences.

HUMIDIFICATION.   A very important physiologic role of the upper respiratory tract in conditioning the inspired air is that of adding moisture. As in the case of warming, humidification is about 97 percent complete, so that the air is nearly saturated with water vapor as well as close to body temperature by the time the lower trachea is reached. This function is extremely important, as in patients breathing cool air through a tube following tracheotomy, the drying effect on the lungs can lead to a condition that renders them liable to a severe pulmonary infection.

The extremely important physiologic consequence of the partial pressure developed by the water vapor in the alveolar air will be considered on page 741.

Artificial Respiration

In conclusion to this section on the mechanics of pulmonary ventilation, a few remarks on artificial respiration are germane.

MANUAL TECHNIQUES.   In general, manual techniques for performing artificial respiration depend upon a rhythmic mechanical compression of the thorax, so that elastic recoil fills the lungs, perhaps assisted by lifting part of the body. At best, a maximum tidal volume of only 500 to 700 ml can be maintained for long periods using such methods. Two manual techniques will be described:

The back-pressure arm-lift (Holger Nielson) method now is considered among the better manual methods for artificial respiration. The patient, lying face down with arms folded above the head, is faced by the operator, who is kneeling. The operator performs two movements in sequence, which are repeated rhythmically 12 to 15 times a minute. First, the thorax is compressed firmly for several seconds at the base of the scapulae. Second, the operator's hands then are transferred to

the upper arms of the patient, and the arms are lifted. Thus, expiration and inspiration are induced; this cycle is repeated for as long as is necessary. A free airway must be maintained at all times through mouth and/or nose to the trachea. The tongue must not be allowed to obstruct the laryngeal opening; vomitus or mucus must be removed.

Mouth-to-mouth artificial respiration is performed by the operator taking a deep breath and then rapidly exhaling into the mouth of the patient. This process is repeated rhythmically as described above, and, of course, the airway must be clear. Physiologically, the expired air contains sufficient oxygen to sustain life, and the carbon dioxide present actually may assist in stimulating the respiratory center of the victim (p. 735).

MECHANICAL TECHNIQUES.   There are many types of artificial respirators available to supply oxygen or air. These devices have a safety valve that prevents excessive positive (ie, above atmospheric) pressures being applied to the lungs, an effect that is capable of inflicting severe pulmonary damage on the patient. Rhythmic positive, and sometimes subatmospheric, pressures are applied through a mask that covers the face. The safety valves on modern resuscitators prevent the development of a positive pressure greater than 14 mm Hg above atmospheric, and they also prevent the negative pressure from falling lower than 9 mm Hg below the ambient atmospheric pressure. These pressure limits give excellent ventilation of normal lungs. It must be remembered, however, that when air is forced under positive pressure into the lungs, the intrathoracic pressure also rises, and this effect may occur to such an extent that it may exceed the venous pressure in the venae cavae. Hence, venous return to the right heart may be decreased seriously. Moreover, cardiac output can be reduced to such an extent by positive-pressure respiration that death of the patient may occur. Therefore, excessively high (or low) intrathoracic pressures (above 20 to 30 mm Hg above atmospheric pressure) must be avoided when using positive-pressure resuscitators.

REGULATION OF RESPIRATION

Throughout the lifetime of an individual, the pulmonary and cardiovascular systems must act continuously and in a coordinated and integrated fashion in order to provide all the bodily tissues with an adequate supply of blood and oxygen. Similarly, carbon dioxide in excess of that needed by the body must be eliminated continuously. Neither of these systems alone can accomplish these tasks, and to meet the ever-changing requirements of the cells, both the air pump and the blood pump must be regulated with an exquisite degree of precision so that bodily requirements are met at all times and with a minimum expenditure of energy. Additionally, the quantities of gases and blood supplied to (and by) these pumps must be balanced throughout all regions of the lungs.

The regulation of respiration involves many physiologic activities other than merely providing air for gas exchange with the blood. Included among these activities, which involve respiration in varying degrees, the following examples may be cited: coughing, yawning, laughing, speaking, singing, sneezing, sobbing, sighing, sucking, sniffing, straining, snoring, hiccuping, and vomiting. Some of these complex acts are voluntary while others are involuntary, but all of them require the participation of respiratory, and at times nonrespiratory, muscles in an exact and precisely coordinated sequence. In addition, these respiratory alterations may be attended by concomitant, equally complex, changes in the pattern of cardiovascular activity. Some of these special acts that involve and modify respiration are discussed later in this chapter (p. 738).

There are many individual factors involved in respiratory regulation and its modification under various circumstances. Therefore, it should be emphasized that normally several of these regulatory factors are in operation simultaneously either in a cooperative and synchronous fashion to augment respiration, or one factor may act in opposition to another. Furthermore, any one factor that is capable of altering the breathing pattern can activate other respiratory and/or circulatory patterns so that secondary factors now come into play. These facts must be borne in mind throughout the following discussion, as the single factors that influence respiration when acting alone will be considered first. Subsequently, a presentation of the simultaneous interplay of a number of regulatory factors involved in the respiratory process will be undertaken.

It is important at this point to define certain terms used in connection with the physiology of normal and abnormal respiratory patterns. *Eupnea* means normal quiet breathing, in which inspiration is active and expiration is passive. *Tachypnea* (or *polypnea*) signifies an accelerated frequency of breathing. *Hyperpnea* denotes an abnormal increase in the rate and depth of respiration. *Apnea* is a cessation of respiration in the expiratory position. *Apneusis* is a cessation of respiration in the inspiratory position. *Apneustic breathing* consists of apneusis which occasionally is terminated by expiration. This type of respiration may be rhythmic. *Gasping* consists of a regular (rhythmic) or irregular (arrhythmic) spasmodic inspiratory effort that is terminated suddenly. *Hyperventilation* is an increased alveolar ventilation relative to the metabolic rate, so that the alveolar Pco_2 falls below 40 mm Hg (p. 759). This term is also used in connection with abnormally prolonged deep breathing. *Hypoventilation* signifies a decreased alveolar ventilation relative to the metabolic rate so that the alveolar Pco_2 rises above 40 mm Hg. This term is also used in connection with a decrease of the air in the lungs below the normal

quantity. *Biot's respiration* consists of a series of uniformly deep gasps, followed by apnea.

Periodic respiration (or *Cheyne–Stokes respiration,* p. 739) is marked by a sequence of respiratory cycles in which the tidal volume gradually increases, followed by a sequence of respiratory cycles having a gradually decreasing tidal volume.

Additional terminology employed by physiologists and clinicians in relation to specific respiratory processes will be introduced and defined in context.

The neural mechanisms for respiratory regulation and control will now be presented, and this material will be followed by a consideration of the primary chemical stimuli and other influences that evoke appropriate physiologic changes in the basic respiratory pattern.

Neural Regulation of Respiration and the Respiratory Centers

Unlike the heart, whose spontaneous rhythm normally originates with impulses generated within the cells of the sinoatrial node (p. 595), the skeletal muscles that underlie respiratory activity possess no inherent rhythmicity. Consequently, severing the motor nerve to the diaphragm (the phrenic nerve) as well as the nerves that innervate the intercostal muscles results in total paralysis of the respiratory mechanism. It is evident that the central nervous system must contain the neural centers that are responsible for initiating and coordinating the respiratory muscles, and thus provide the stimuli critical to the effectual rhythmic operation of the air pump. Extensive investigation of this problem for the past century and a half, with a variety of experimental techniques, has revealed the existence of three important and interconnected areas or "centers" that are located within the medulla and pons and that are concerned with normal respiratory activity. Collectively these areas comprise a *respiratory center.* The word *center,* as used in this context, means a localized region or area located within the brain or spinal cord which is capable of eliciting appropriate physiologic activity upon stimulation or other experimental manipulation. Such usage of the term *center* is not synonymous with a discrete, relatively well-delineated, anatomic nucleus as defined in Chapter 7 (p. 257); rather, it signifies a more diffuse area having demonstrable functional characteristics.

The three major parts of the respiratory center are the medullary center, the apneustic center, and the pneumotaxic center. Figure 11.18 illustrates the location of these centers in the cat brain.

THE MEDULLARY RESPIRATORY CENTER. This bilateral area is located medially within the reticular formation of the caudal medulla, and below the inferior part of the fourth ventricle.

This region of the brain stem possesses inherent rhythmicity, so that repetitive motor impulses

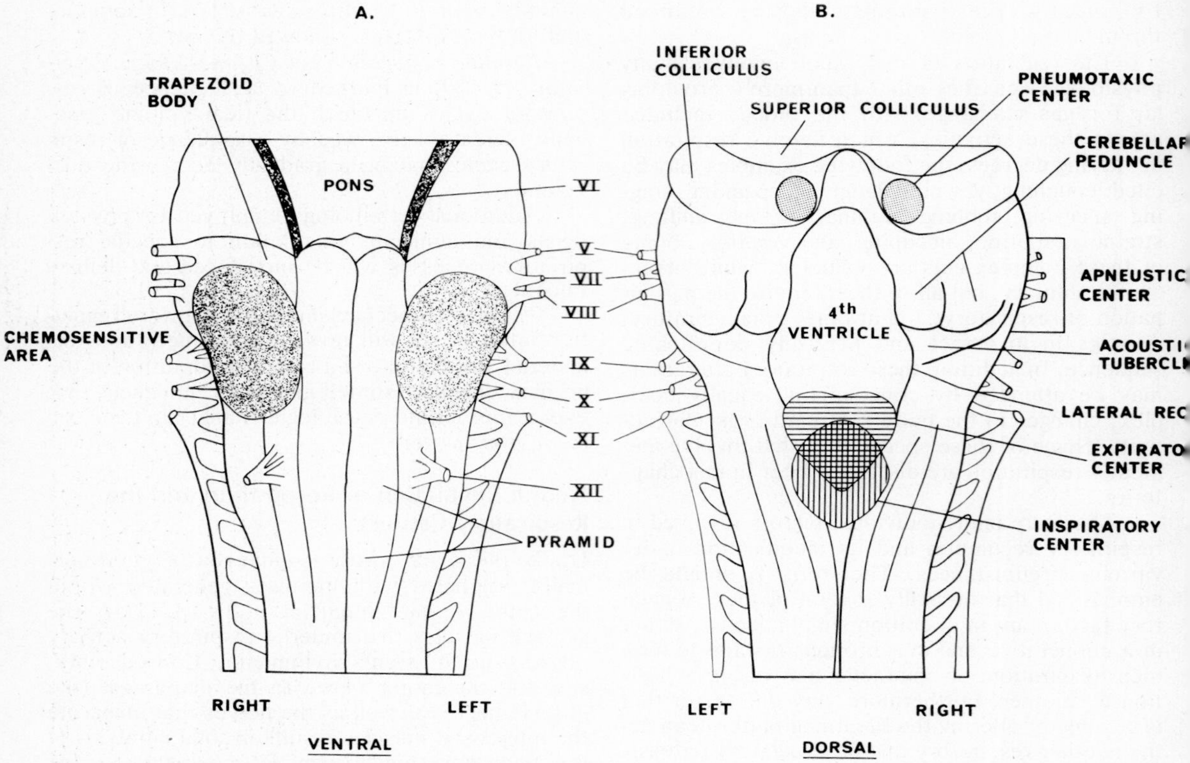

FIG. 11.18. Respiratory centers in the pons and medulla of the cat brain. **A.** Ventral surface of brain stem. **B.** Dorsal surface of brain stem. Roman numerals signify cranial nerves. (After Comroe. *Physiology of Respiration,* 1965, Year Book Medical Publishers.)

are generated spontaneously, even though all afferent impulse traffic to this center is abolished.

The medullary respiratory center contains all the neurons that are essential to maintenance of the fundamental coordinated sequence of inspiration and expiration. However, the spontaneous rhythm of breathing becomes somewhat irregular when the medullary center is isolated from the other respiratory centers, as shown in Figure 11.19C.

Functionally, the medullary center can be divided into inspiratory and expiratory regions. The neurons of the respiratory region overlap the neurons of the expiratory region to some extent. Electric stimulation, as well as recordings of spontaneous activity, reveals that these two portions of the center act in an antagonistic fashion with respect to each other.

The inspiratory neurons normally exhibit spontaneous bursts of activity between 12 and 15 times per minute. Thus, if all the cranial nerves are severed and the brain stem is transected superior to the pons, regular breathing is continuous, ie, all the various elements of the respiratory center are present. If, however, the brain stem now is transected through the inferior border of the pons, the inspiratory neurons discharge continuously, so that apneusis results.

In contrast to the rhythmically active inspiratory region, the expiratory neurons of the medullary center require stimulation in order to become activated. Apparently, the inspiratory and expiratory neurons are linked closely, probably through two oscillating feedback circuits (Fig. 11.20), so that inspiratory activity inhibits expiratory activity, and vice versa, in order to produce smoothly coordinated respiratory movements. Normal respiration at rest is accomplished solely by active contraction of the inspiratory muscles, and expiration, being passive, simply consists of inhibition of the muscles of inspiration. Thus, a rhythmic and basic respiratory pattern is established.

It was once thought that the medullary respiratory center itself was the area of the brain that responded to an elevated arterial blood Pco_2 to evoke the classic CO_2 hyperpnea. According to current evidence, however, it appears that there are specialized chemoreceptors located upon the ventrolateral surfaces of the superior portion of the medulla (Fig. 11.18A), and apparently these receptors respond to blood hydrogen ion concentration, rather than directly to the Pco_2 of the arterial blood. These sensory elements, known as central hydrogen ion receptors, are responsive to alterations in the Pco_2 because this factor changes the hydrogen ion concentration, hence pH, in their immediate vicinity.

The medullary respiratory center receives in-

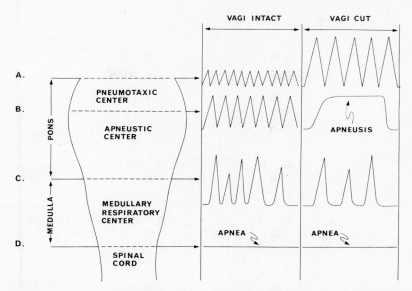

VAGI INTACT VAGI CUT

A.

PNEUMOTAXIC
CENTER

B.

APNEUSTIC
CENTER

APNEUSIS

PONS

C.

MEDULLARY
RESPIRATORY
CENTER

APNEA

APNEA

MEDULLA

D.

SPINAL
CORD

FIG. 11.19. Respiratory patterns following complete transection of the brain stem at four levels, A, B, C, and D. (After Comroe. *Physiology of Respiration,* 1965, Year Book Medical Publishers; Ganong. *Review of Medical Physiology,* 6th ed, 1973, Lange Medical Publications.)

formation from numerous receptors, both directly and indirectly (Fig. 11.21), and this region is connected with the higher respiratory centers in the pons as well as with areas of the hypothalamus and the reticular activating system. Possibly, all impulses that stimulate the respiratory muscles synapse within the medullary respiratory center. There is some evidence, however, that the cerebral cortex may have pathways that are connected directly to the respiratory muscles without any synaptic junctions in the medulla or pons.

THE APNEUSTIC CENTER. This portion of the respiratory center, as well as the pneumotaxic area, is located bilaterally within the reticular substance of the pons (Fig. 11.18). If both the pneumotaxic center and the left and right vagi are severed, a sustained apneusis results, as shown in Figure 11.19B, because of a continuous discharge of the inspiratory center of the medulla.Therefore, the normal physiologic role of the apneustic center appears to be that of a group of tonically discharging neurons located within the caudal portion of the pons, which in turn drives the inspiratory neurons of the medullary center. The apneustic center is not necessary for maintaining the basic respiratory rhythm, but it does appear to have the basic role in respiration of facilitating inspiration.

In vivo, the normal discharge from the apneustic center to the inspiratory center appears to be inhibited intermittently by impulses arising within, or passing through, the pneumotaxic center. Also, afferent impulses related to the degree of lung inflation, and transmitted by the vagi, impinge directly upon the apneustic center.

The activity of the apneustic center also is modulated by the arterial Pco_2 so that the mag-

nitude of inspiration is increased by an elevation of this blood factor. Stimulation of certain sensory nerves also augments the activity of the apneustic center directly.

THE PNEUMOTAXIC CENTER. This bilateral area, located within the reticular substance of the rostral pons, functionally is intimately related to the apneustic and medullary centers (Fig. 11.18). Physiologically, the pneumotaxic center does not appear to exhibit spontaneous rhythmic activity. Rather, it receives information from extrinsic sources, possibly through a negative feedback loop originating within the medullary inspiratory center (Fig. 11.20B). Thus, during inspiration, afferent impulses arise in the inspiratory center of the medulla, and these impulses are transmitted to the pneumotaxic center, which in turn generates impulses that pass to, and stimulate, the expiratory center, as well as concomitantly producing inhibition of the inspiratory center.

Consequently, the function of the pneumotaxic center is somewhat similar to that of the inflation reflex (p. 734). The neurons in this region of the pons thus appear to facilitate rhythmic respiration, although they do not augment inspiration, as does the apneustic center. Furthermore, the pneumotaxic center is not necessary to mediate the respiratory responses to vagal reflexes, somatic pain, decreased arterial Po_2, or increased arterial Pco_2.

SUMMARY OF THE RESPIRATORY CENTERS AND THEIR FUNCTIONS. The medulla oblongata contains a region that is concerned with the initiation of breathing — the respiratory center — and this may be considered as having neurons that mediate

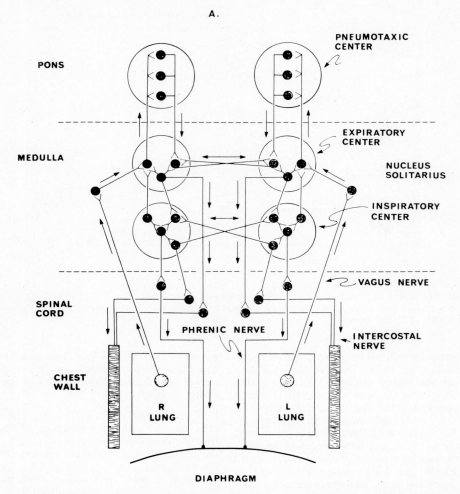

FIG. 11.20.   A. Interconnections among the neural elements that comprise the respiratory centers.

both inspiration and expiration. Closely related physiologically to this center are two additional regions located within the pons, the apneustic and pneumotaxic centers. The former appears to facilitate inspiration, the latter expiration.

The medullary inspiratory center has the property of spontaneously generating rhythmic bursts of impulses that stimulate inspiration by stimulating contraction of the appropriate respiratory muscles (Table 11.2). The process of inspiration presumably is facilitated by impulses arising from the apneustic center, so that inspiration (at rest) lasts for around 2 seconds. In turn, the pneumotaxic center exerts an inhibitory influence upon the inspiratory neurons (during rest) while concomitantly exciting the expiratory neurons during periods of increased activity.

Because of the functional interrelationships among the three centers, a negative feedback mechanism between medulla and pons has been postulated. This hypothetical feedback circuit operates to provide intermittent discharge of the inspiratory neurons. According to this conception,

afferent impulses from the medullary inspiratory center are relayed by collateral axons to the pneumotaxic center. This center in turn generates impulses that pass back to the medulla, thereby inhibiting the inspiratory center; they also excite the expiratory center neurons during periods of increased activity (but not at rest). This inhibition lasts for about 3 seconds at rest because of the phenomenon of afterdischarge (p. 348). In turn, this prolonged afterdischarge results from the length of the multisynaptic pathway involved in transmission of the impulses from the pneumotaxic center to the medulla (Fig. 11.20A). The onset of the next inspiration occurs at the end of the period of afterdischarge so that the entire respiratory cycle during quiet breathing lasts for approximately 5 seconds.

It is important to emphasize that the vagus nerves mediate not one, but several respiratory reflexes. A considerable number of diverse types of sensory endings are found within the lungs, thorax, heart, and great vessels, as well as in the abdominal viscera. The majority of these endings

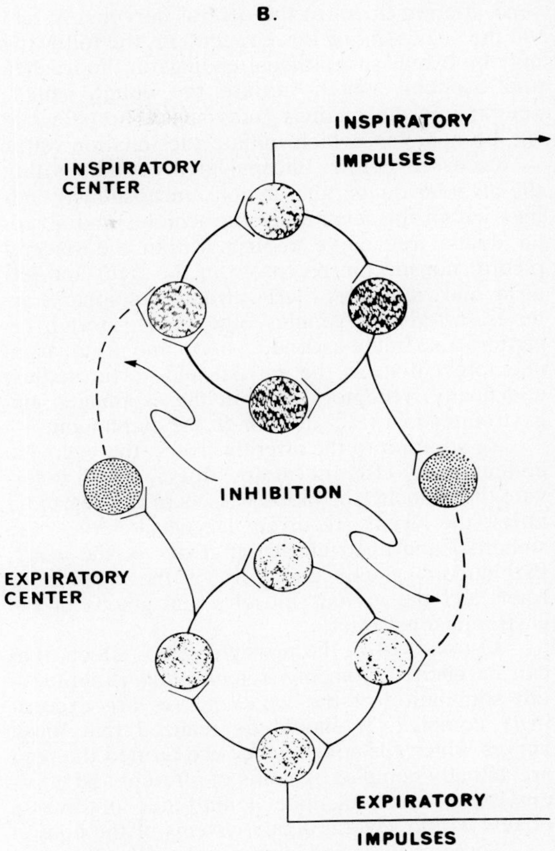

B.

FIG. 11.20. (cont'd) **B.** Hypothetical feedback mechanism for explaining rhythmic activity of the respiratory center. (**A.** After Best and Taylor [eds]. *The Physiological Basis of Medical Practice,* 8th ed, 1966, Williams & Wilkins. **B.** After Guyton. *Textbook of Medical Physiology,* 4th ed, 1971, Saunders.)

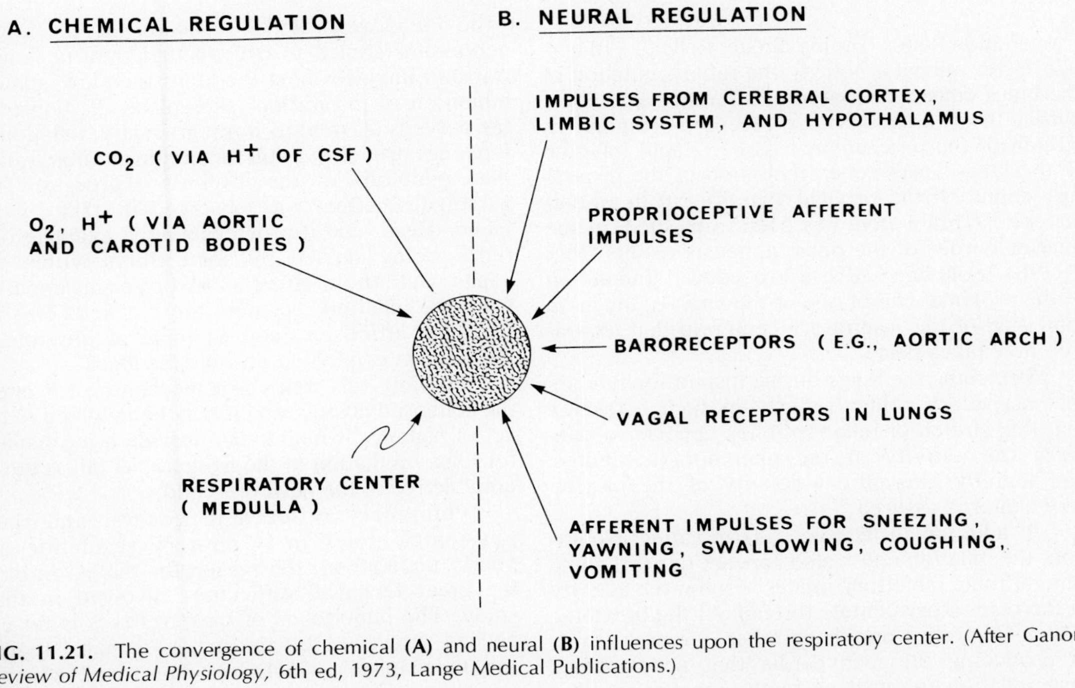

INSPIRATORY
IMPULSES

INSPIRATORY
CENTER

INHIBITION

EXPIRATORY
CENTER

EXPIRATORY
IMPULSES

A. CHEMICAL REGULATION

B. NEURAL REGULATION

IMPULSES FROM CEREBRAL CORTEX, LIMBIC SYSTEM, AND HYPOTHALAMUS

CO_2 (VIA H^+ OF CSF)

O_2, H^+ (VIA AORTIC AND CAROTID BODIES)

PROPRIOCEPTIVE AFFERENT IMPULSES

BARORECEPTORS (E.G., AORTIC ARCH)

VAGAL RECEPTORS IN LUNGS

RESPIRATORY CENTER (MEDULLA)

AFFERENT IMPULSES FOR SNEEZING, YAWNING, SWALLOWING, COUGHING, VOMITING

FIG. 11.21. The convergence of chemical (**A**) and neural (**B**) influences upon the respiratory center. (After Ganong. *Review of Medical Physiology,* 6th ed, 1973, Lange Medical Publications.)

send afferent fibers to the central nervous system via the vagi. Among these receptors, the following may be listed: subepithelial endings in the trachea and bronchi, which initiate the cough reflex; smooth muscle endings (not muscle spindles) in the bronchioles, which initiate the inflation reflex as discussed below; uncapsulated endings within the alveolar ducts, and alveoli; uncapsulated endings within the respiratory bronchioles and alveolar ducts; free nerve endings within the visceral pleura; nonspecific receptors in the right and left atria and ventricles, left atrium, pulmonary arteries, arterioles, venules, and veins; stretch receptors also in the ascending aorta and aortic arch; chemoreceptors in the cartoid and aortic bodies; and finally, receptors, also in the esophagus and gastrointestinal tract inferior to the diaphragm.

In addition to the afferent fibers, the vagi also contain many efferent (motor) fibers, which innervate the bronchi, the blood vessels in the heart and lungs, the larynx (recurrent laryngeal nerve), the sinoatrial and atrioventricular nodes of the heart, the mucous glands and cilia of the respiratory tract, and the various muscles and glands of the gastrointestinal tract.

Consequently, the possible reflex effects that can be elicited from physiologic (and experimental) stimulation of the vagus nerves are exceedingly complex. It should be realized that those nerves which for convenience are termed the *vagi* are actually complex systems of afferent and efferent fibers which mediate a multitude of diverse effects within several organ systems of the body.

Three of the more important vagal reflexes involving the lungs and respiratory center deserve mention.

The Inflation Reflex (Hering–Breuer Reflex). In certain lower animals, notably the rabbit, inflation of the lungs causes discharge of afferent vagal fibers within these organs whose receptors respond to stretch of the parenchymal tissue. Rapid inflation of the lung causes reflex inhibition of the inspiratory center. If the cervical vagi are cut in an animal whose brain stem has been transected, at the inferior border of the pons, apneusis results (Fig. 11.19B). Following such a procedure, stimulation of the proximal end of one of the vagi results in an inhibition of the inspiratory center so that expiration now takes place.

Stretching the lungs during inspiration inhibits the inspiratory center reflexly in such a fashion that this stretch or inflation reflex appears to reinforce the activity of the pneumotaxic center, and thereby intermittent activity of the inspiratory center is assured.

In addition to its role in terminating inspiration, the inflation reflex also appears to exert some general tonic inhibitory influence upon the activity of the respiratory centers throughout the breathing cycle. This reflex may have other effects, such as in producing the tachycardia that accompanies lung inflation, as well as in the mechanism that dilates the bronchioles during inspiration and constricts them during expiration.

In human beings, as opposed to lower mammalian species, the inflation reflex is weak unless the lungs are distended to a considerable extent. This reflex could be considered as a protective mechanism that prevents overinflation (hence overdistention) of the lungs. In newborn babies, on the other hand, this reflex is quite well developed, but it decreases in efficacy during the first several days of postnatal life.

It is interesting to note in passing that the inflation reflex, which was elucidated by Hering and Breuer over a century ago (1868), was one of the earliest examples studied of a negative feedback mechanism being involved in a physiologic process.

The Deflation Reflex. Hering and Breuer also elucidated another respiratory reflex known today as the deflation reflex. When the lungs are deflated, an increased force and rate of respiration occur. The details of this mechanism are obscure at present, however. Presumably, the rapidly adapting receptors for this reflex are located within the smallest air ducts and alveoli and the afferent impulses are transmitted within the finest branches of the vagi.

Although not yet demonstrated in man, this reflex may possibly have a role in the increased ventilation observed when the lungs are abnormally deflated (eg, in pneumothorax, or in the deep inspiratory movements that are necessary to prevent atelectasis). This statement is, however, conjectural.

"The Paradoxical Reflex" (Head). If the vagus nerves are cooled to 0 C so that they no longer transmit impulses, and the lungs then are inflated, inhibition of inspiration takes place. If, however, the nerve is allowed to warm gradually, inflation of the lungs produces a still further inspiration, rather than inhibition of the inspiratory process. This paradoxical effect was observed in 1889 by Sir Henry Head, and it provides a good example of a reflex being carried by some fibers within the vagus while those afferents that transmit impulses from the inflation receptors are blocked. This effect also affords a good example of physiologic mechanism employing positive feedback.

Possibly this reflex is a mechanism for opening collapsed alveoli, and it may be involved in the act of sighing. It might also provide a mechanism for lung ventilation in the neonate, as this reflex is quite active in the newborn child.

Pulmonary reflexes in general and their physiologic roles in respiratory regulation are poorly understood; the reason for this is, in part, the great technical difficulties involved in their study. The importance of these reflexes is far exceeded, however, by the major chemical regulatory mechanisms, which exert a profound influence upon the respiratory center.

Chemical (Humoral) Regulation of the Respiratory Centers

When the metabolism of the body is increased, as for example during periods of strenuous exercise, there is an increased requirement of the tissues for oxygen as well as a need to eliminate the excess carbon dioxide produced. It is not surprising that the respiratory center is exquisitely sensitive to the partial pressures of these two gases (Po_2 and Pco_2), as well as to the hydrogen ion (H^+) concentration of the blood. Therefore, activity in this part of the brain stem adapts the rate of alveolar ventilation to the bodily demands in such a manner that the alveolar ventilation rate is adjusted almost precisely to the need for oxygen, and thus blood Po_2 and Pco_2 are barely altered in the normal individual, even in periods of extreme respiratory stress.

Therefore, when one speaks of the chemical (or humoral) regulation of respiratory activity, reference is made principally to the effects of changes in oxygen, carbon dioxide, and/or hydrogen ion levels in the body fluids on the control of respiration.

The effects of alterations in carbon dioxide and hydrogen ion concentration are exerted for the most part directly upon the respiratory center per se, whereas oxygen acts principally upon certain peripheral chemoreceptors which in turn affect the respiratory center indirectly.

EFFECTS OF CARBON DIOXIDE AND HYDROGEN ION UPON THE RESPIRATORY CENTER. An elevation of the arterial blood Pco_2 or H^+ concentration (ie, a fall in pH, p. 755) leads to an increase in the activity of the respiratory center so that hyperventilation ensues. Respiration is stimulated by an increase in either of these factors so that alveolar ventilation may increase up to sevenfold above the basal state. Although Pco_2 and pH both can alter alveolar ventilation markedly under normal physiologic conditions, the arterial Pco_2 level is by far the more important factor involved in normal respiratory regulation. Conversely, if arterial blood Pco_2 and pH fall, activity of the respiratory center becomes inhibited so that hypoventilation results.

The small receptor areas for detecting changes in blood carbon dioxide and hydrogen ion concentration appear to be located on the ventrolateral surfaces of the brain stem (Fig. 11.18A) and are not found within the medullary respiratory center proper. These medullary chemosensitive regions are especially sensitive to pH alterations of the cerebrospinal fluid and/or possibly interstitial fluid of the brain itself. It is mainly the Pco_2 in blood, however, that regulates the pH of the cerebrospinal fluid for the following reasons. Hydrogen ions do not diffuse rapidly from blood into the cerebrospinal fluid, whereas carbon dioxide does diffuse rapidly. The carbon dioxide that enters the cerebrospinal fluid compartment in solution immediately combines with water to form carbonic acid, which in turn dissociates, giving rise to hydrogen ions, and these hydrogen ions in turn act directly upon the chemosensitive receptors which then stimulate the medullary respiratory center proper.

EFFECT OF OXYGEN UPON RESPIRATORY CONTROL: THE AORTIC ARCH AND CAROTID BODY CHEMORECEPTORS. Located near the carotid bifurcation in the neck on each side of the body are located the two small carotid bodies, while two or more tiny aortic bodies are found near the arch of the aorta (see Chap. 10, p. 674). As illustrated in Figure 11.10, these structures are connected to the medullary respiratory center via afferent branches of the carotid sinus, glossopharyngeal, and vagus nerves. The structure of each aortic and carotid body is similar, and it consists of groups of chemoreceptor cells as well as the supporting or sustentacular cells which invest the chemoreceptor elements, and which appear to isolate the chemoreceptor cells from the sinusoidal blood vessels with which each glomus is richly supplied. Unmyelinated nerve fibers are in intimate contact with the cell membranes of the chemoreceptor cells, and these fibers too are held in place by the sustentacular cells.

The normal rate of blood flow through the aortic and carotid bodies is enormous, considering their size. For example, each carotid body has a mass of around 2.0 mg; and a blood flow of 0.04 ml/minute. This flow rate in turn is the equivalent of 2,000 ml/100 gm tissue per minute! This compares with a blood flow of about 55 ml per minute for the brain and 420 ml for the kidney. Thus, blood flow through the aortic and carotid bodies is by far the highest for any tissue in the body, and the oxygen requirement of the cells in these structures can be met solely by dissolved oxygen rather than oxygen that is bound to hemoglobin. These receptors are not stimulated by carbon monoxide poisoning (p. 749) or by the hypoxia of severe anemia (p. 767), because the dissolved oxygen is unaffected, whereas the combined oxygen is reduced greatly under such conditions.

Ordinarily, the frequency of impulse transmission is relatively slow in the afferent nerves from the aortic and carotid bodies; hence, it ranges between 7 and 12 impulses per second; however, this rate increases markedly when the arterial blood Po_2 decreases or if the blood Pco_2 increases, as shown in Figure 11.22. As the impulse frequency is increased, the respiratory center is stimulated and the minute respiratory volume also increases proportionally.

The large blood flow through the aortic and carotid chemoreceptors, combined with their ability to respond to changes in blood Po_2 or Pco_2 by appropriate alterations in impulse transmission, renders them important "fine tuning" homeostatic mechanisms for the regulation of respiration. These chemoreceptors become strongly stimulated when the arterial Po_2 falls to between 30 and 60 mm Hg. This fact is especially important as this also is the range in which the arterial hemoglobin saturation is decreased sharply. Under such circumstances, the medullary respiratory center is stimulated strongly to produce increased alveolar ventilation by the aortic and carotid chemoreceptor mechanism. Normally, however, as long as the arterial Po_2 remains above 100 mm Hg, there is little effect on alveolar ventilation caused by this factor alone, ie, a reduced Po_2. Stated another way, any fall in arterial Po_2 must be well below the normal (physiologic) range in order that the chemoreceptors be stimulated by the relative oxygen deficiency. Hence, the bodily detection mechanisms for oxygen deficiency are relatively insensitive when compared to the exquisite delicacy with which minute alterations in carbon dioxide and hydrogen ion levels are sensed. The respiratory minute volume is proportional to the metabolic rate, but the physiologic link between ventilation and metabolism is through carbon dioxide and the hydrogen ion concentration, rather than through the oxygen level of the blood. Therefore, the aortic and carotid chemoreceptors are stimulated principally by an elevated arterial Pco_2 or an increased H^+ concentration, or secondarily by a marked decrease in the Po_2. This peripheral chemoreceptor activity in turn stimulates the respiratory center to alter the minute respiratory volume, hence alveolar ventilation, in an appropriate direction so that the blood gas and hydrogen ion concentrations are maintained within close tolerances.

It should also be reiterated here that an increase in the arterial Pco_2 or H^+ can strongly influence the respiratory center directly as well as indirectly through the aortic and carotid chemoreceptor mechanisms. Oxygen deficiency (or hypoxia), on the other hand, can affect ventilation

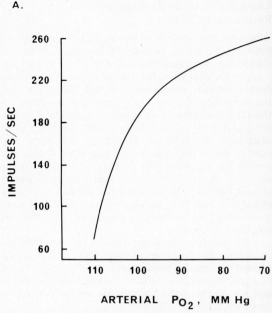

A.

FIG. 11.22. **A.** Relationship between frequency of impulse generation in the carotid sinus nerve and arterial Pco_2.

to a significant extent only through an indirect action on the peripheral chemoreceptors, as a severe hypoxia tends to depress rather than to stimulate the respiratory center. This peripheral effect is observed only when the arterial Po_2 is reduced greatly.

PULMONARY VENTILATION AND CARBON DIOXIDE. If tissue metabolism increases so that there is an increased production of carbon dioxide,

B.

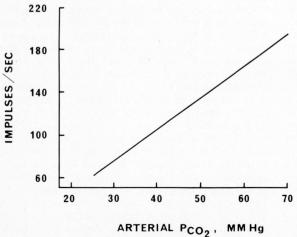

FIG. 11.22. **B.** Relationship between frequency of impulse formation in the carotid sinus nerve and arterial Po_2. (After Ganong. *Review of Medical Physiology,* 6th ed, 1973, Lange Medical Publications.)

ventilation is stimulated by the mechanisms described above so that the rate of carbon dioxide elimination by the lungs is increased until the arterial blood P_{CO_2} falls to a normal level once again. The stimulus to an increased ventilatory activity is eliminated, and the respiratory rate now falls to resting levels. A delicate balance exists within the body between the rate of carbon dioxide production and elimination, and the operation of the feedback mechanism described above serves to regulate the level of carbon dioxide production and the respiratory rate.

If a gas mixture containing an elevated level of carbon dioxide is inhaled, hyperventilation results. As the alveolar P_{CO_2} rises, this increases the arterial P_{CO_2} level; when blood containing a higher P_{CO_2} reaches the respiratory center, this serves as the stimulus to trigger an increased minute respiratory volume, so that the alveolar P_{CO_2} level now falls toward normal as pulmonary ventilation increases. The alveolar P_{CO_2} does not reach normal levels, however, so that a new steady-state level is reached at which the alveolar P_{CO_2} is somewhat elevated, and the state of hyperventilation persists for as long as excess carbon dioxide is being inhaled. These facts explain the discrepancy between a relatively high P_{CO_2} in the inspired gas mixture, and the relatively low alveolar P_{CO_2} observed during hyperventilation induced by breathing a gas with a high CO_2 content.

There is, between limits, essentially a linear relationship between the alveolar P_{CO_2} and the minute respiratory volume (Fig. 11.23). When these limits are exceeded so that the alveolar P_{CO_2} rises excessively (eg, when the CO_2 in the inspired gas rises above 7 percent), CO_2 elimination from the body is impaired and carbon dioxide commences to accumulate within the body, a condition known as hypercapnia. The hypercapnia in turn causes depression of the central nervous system, including the respiratory center itself, and eventually a type of coma known as carbon dioxide narcosis ensues.

These facts also explain why a small quantity of carbon dioxide (usually 5 percent) is included in oxygen mixtures (95 percent) that are used for inhalation therapy, as this gas at low levels serves to stimulate the respiratory center and to cause hyperventilation, thereby also increasing the alveolar P_{O_2}.

On the other hand, during periods of voluntary forced ventilation (hyperventilation) while breathing normal atmospheric air, the alveolar P_{CO_2} decreases markedly. Therefore, a state of hypocapnia ultimately results, as an excess of blood carbon dioxide is eliminated through the lungs. A transient inhibition of the respiratory center ensues, and pulmonary ventilation decreases or stops entirely until the blood carbon dioxide level once again reaches physiologic levels. Skin divers take advantage of this fact by hyperventilating mildly prior to a dive in order to be able to stay under water longer. The hyperventilation does not increase the total quantity of oxygen in the body appreciably; rather, it reduces the respiratory drive temporarily by lowering the blood CO_2 level somewhat, and thereby the activity of the respiratory center is decreased.

PULMONARY VENTILATION AND OXYGEN DEFICIENCY (HYPOXIA). If the oxygen content of the inspired air is lowered, there is some concomitant increase in the minute respiratory volume; however, as long as the P_{O_2} of the inspired air exceeds 60 mm Hg, this effect is slight, although some changes in respiratory patterns may be observed at P_{O_2} levels below 100 mm Hg. Oxygen partial pressures below 100 mm Hg do stimulate marked impulse discharge of the aortic and carotid chemoreceptors (Fig. 11.22A); however, this mechanism tends to be counteracted by two factors which come into operation in the normal

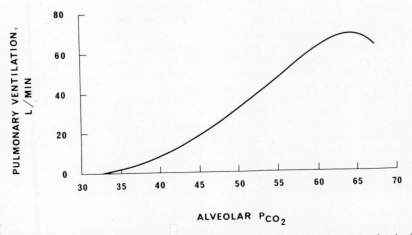

FIG. 11.23. Relationship between alveolar P_{CO_2} and pulmonary ventilation. (After Guyton. *Textbook of Medical Physiology*, 4th ed, 1971, Saunders.)

individual. First, the increase in ventilation that takes place tends to lower alveolar P_{CO_2}, and thus to inhibit respiration. Second, reduced hemoglobin is a weaker acid than is oxygenated hemoglobin (p. 751). There is a slight decrease in the free hydrogen ion concentration of arterial blood when the arterial P_{O_2} level falls and the hemoglobin becomes less saturated with oxygen. This decrease in hydrogen ion concentration in turn tends to inhibit respiration. Below an ambient P_{O_2} of 60 mm Hg, however, the stimulatory effects of hypoxia on respiration are clearly manifest, as they become strong enough to override the two mechanisms described above.

If one holds the alveolar P_{O_2} level constant, experimentally, while altering the amounts of carbon dioxide that are inhaled, it can be shown that the respiratory sensitivity to increased arterial carbon dioxide levels also is increased in the presence of hypoxia.

THE BREAKING POINT. A subject can inhibit his respiration voluntarily for some time, but ultimately such voluntary control cannot be maintained and the respiratory reflexes cause breathing to commence once again. The point at which respiration can no longer be inhibited is called the breaking point.

The breaking point in turn is the result of a rise in arterial P_{CO_2} as well as of a concomitant fall in P_{O_2}. Hyperventilation with ambient air, as described for skin divers (p. 737), can prolong to some extent the interval for which the breath can be held voluntarily. Similarly, prior breathing of 100 percent oxygen has the same effect in prolonging the time of breath-holding before the breaking point has been reached.

PULMONARY VENTILATION AND SHIFTS IN ACID–BASE BALANCE. The effects of respiration on acid–base balance will be discussed more fully later in this chapter (p. 759); however, certain aspects of this problem must be introduced here in relation to the regulation of pulmonary ventilation.

In the condition of metabolic acidosis (p. 763) such as occurs in diabetes mellitus resulting from the accumulation of fixed acid (ketone bodies) in the blood, there is a marked respiratory stimulation known as Kussmaul breathing (p. 1077). The hyperventilation attendant upon this condition lowers alveolar P_{CO_2}, thereby producing a compensatory decrease in the blood hydrogen ion concentration. On the other hand, a metabolic alkalosis such as is seen in patients following an extended period of vomiting with an attendant loss of hydrochloric acid from the body, results in a depression of respiration so that arterial P_{CO_2} rises, and thus the hydrogen ion concentration increases toward normal; ie, some compensation takes place.

If an increase in pulmonary ventilation (hyperventilation) occurs which is not secondary to an increase in arterial hydrogen ion concentration, the fall in P_{CO_2} reduces the hydrogen ion concentration below normal, and this condition is known as a respiratory alkalosis. Conversely, a decrease in ventilation (hypoventilation) which is not secondary to a fall in the plasma hydrogen ion concentration will produce a condition known as a respiratory acidosis.

Nonchemical Regulation of Respiration

In addition to the several reflexes mentioned earlier, the process of respiration undoubtedly is influenced markedly by afferent impulses from the cerebral cortex to the medullary respiratory center. This is true since inspiration and expiration can be performed under voluntary control, although normally this act is unconscious. Since emotional stimuli and pain also affect respiration markedly, there are probably also afferent pathways from the hypothalamus and limbic system to the respiratory center, although none of the pathways that mediate these effects have been identified.

Both active as well as passive movements of the joints can stimulate respiration, and proprioceptive impulses from these peripheral regions probably help to augment pulmonary ventilation during exercise. Hence, afferent pathways from proprioceptors located in joints, tendons, and muscles presumably mediate this effect by their convergence upon the respiratory center (Fig. 11.21).

A number of other specialized physiologic acts that involve respiration and its modification now will be considered.

Special Acts Involving Respiration

In this section, brief mention will be made of some respiratory adjustments that occur under certain circumstances. The important physiologic changes that occur during expiration against a closed glottis (straining or Valsalva's maneuver, p. 653), or during inspiration with a closed glottis (sucking, Müller's maneuver, p. 653), are circulatory rather than respiratory. Phonation, ie, the production of sounds, was discussed on page 705.

THE COUGH REFLEX. The cough reflex is an important protective mechanism in the human, as this response functions to clear the upper airways of foreign matter or mucus and to force it into the throat.

The sensory receptors for the cough reflex are located principally in the bronchi and trachea, and these elements are exquisitely sensitive to irritation, whether by chemical (eg, ammonia vapor) or mechanical (eg, mucus) stimuli. Afferent impulses for the cough reflex are conveyed from the respiratory passageways (including the terminal bronchioles and alveoli) to the medulla via vagal afferents. Once the impulses arrive in this region of the brain stem, a series of wholly automatic events are set in action. First, several liters of air are inspired; second, the vocal cords appose closely and the epiglottis closes; third, the abdominal muscles, external intercostals, and other respiratory muscles contract vigorously, thereby sharply elevating the intrathoracic pressure to over 100 mm Hg; fourth, the vocal cords and epiglottis open suddenly so that the air, which is under considerable pressure within the lungs, is expelled forcefully and at an enormous velocity, amounting to several hundred kilometers per hour. An important factor in producing this rate of air expulsion during a cough is the extremely high intrathoracic pressure which collapses the noncartilaginous part of the trachea, as well as the bronchi and smaller airways. Hence, the air actually rushes through narrow slits rather than cylindroid tubes during coughing. The net effect of the airstream moving at high velocity is to dislodge foreign matter from the airways.

THE SNEEZE REFLEX. This reflex is quite similar to coughing save that the nasal passages are involved, rather than the lower respiratory airways. Chemical or mechanical stimuli applied to the nasal mucosa initiate the generation of impulses which are transmitted in afferent fibers of the trigeminal nerve (V) to the medulla, where the reflex is triggered. The series of events that follows is quite similar to those involved in coughing, except that the uvula is depressed. Large quantities of air once again pass at high velocity through the nose as well as the mouth, and this rapidly moving airstream helps to clear the nasal passages of foreign matter.

SWALLOWING. The act of swallowing or deglutition inhibits respiration in expiration, preventing the accidental inhalation of liquid or food. Deglutition is discussed in relation to gastrointestinal physiology (p. 836).

HICCUPING. The phenomenon of hiccuping is caused by an intermittent spasmodic contraction of the diaphragm, often accompanied by attendant contractions of other respiratory muscles. Periods of hiccuping occur in normal individuals, and this condition also may accompany a variety of clinical states, such as heart, liver, renal, esophageal, gastric, abdominal, and brain disease. There is no satisfactory explanation of the cause of hiccuping, although there is no paucity of theories concerning its origin.

Hiccuping, if severe, frequent, and prolonged can endanger life, as it may prevent sleeping, eating, and drinking. Various treatments used for hiccup include swallowing, which inhibits the medullary respiratory center; holding the breath; drugs; or phrenic nerve block. Rebreathing from a bag to elevate the carbon dioxide level, or inhalation of 95 percent oxygen, 5 percent carbon dioxide mixtures actually may stimulate hiccuping.

SNORING. Snoring is merely loud breathing through the mouth which is produced by moving air causing vibrations of the relaxed soft palate and edge of the posterior pillars when an individual is asleep. The social importance of this natural phenomenon is best judged by the considerable space devoted to its discussion and "cure" by the writers of popular columns.

Cheyne–Stokes Respiration

The pattern of breathing in an individual having Cheyne–Stokes respiration is illustrated in Figure 11.24A. As shown in this figure, the tidal volume increases then decreases in a regular series of cycles that are repeated over and over. This type of respiratory pattern may occur in normal individuals (eg, in a hypoxic environment such as at high altitude), but more often it is seen in patients suffering from neurologic and/or circulatory disorders. In some patients, neurologic factors, such as central nervous system damage, cause hyperventilation even though the arterial P_{CO_2} is subnormal (ie, below 40 mm Hg). Thus, arterial P_{CO_2} is not responsible for the respiratory drive; neither is anoxia the cause, as elevating the arterial P_{O_2} to normal levels does not alter the breathing pattern. Such patients may exhibit no demonstrable heart disease; therefore, a neurologic disorder appears to be responsible for the observed respiratory abnormality.

In other patients, who suffer from heart disease which impairs the circulation time considerably, there is a definite lag between the time well-oxygenated blood leaves the pulmonary capillaries and reaches the chemoreceptors that sample the arterial P_{O_2} and P_{CO_2}. Because of this time lag, the chemoreceptors are behind, temporally speaking, so that the feedback impulses that are necessary to effect the adaptation of ventilation with arterial blood gas levels on a moment-to-moment basis are not achieved. Therefore, wide fluctuations in the minute-to-minute respiratory drive result in alternate over-and-under compensation, and the successively increasing and decreasing tidal volumes of Cheyne–Stokes respiration ensue. The chemoreceptor feedback mechanism tends to oscillate continuously; ie, this mechanism is not damped properly, because of prolongation of the circulatory time and the attendant fluctuations in the P_{O_2} and P_{CO_2} levels (Fig. 11.24B).

It should also be mentioned that many patients having congestive heart failure with a concomitant prolonged lung-chemoreceptor circulation time do not exhibit Cheyne–Stokes respiration.

Possibly some undetected neurologic disorder precipitates the arrhythmia known as Cheyne–Stokes respiration, whereas the circulatory abnormality merely exacerbates or sustains it.

PULMONARY GAS EXCHANGE, GAS TRANSPORT BY THE BLOOD, AND TISSUE GAS EXCHANGE

In this section, the physiologic mechanisms whereby alveolar oxygen is transported to the cells of the body will be presented, together with a dis-

A.

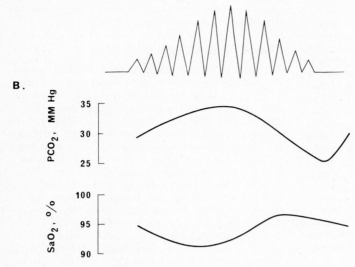

B.

FIG. 11.24. **A.** Pattern of Cheyne–Stokes respiration. Note cyclic alterations in tidal volume. **B.** Arterial P_{CO_2} and arterial oxygen saturation curves correlated with respiratory cycles. (After Comroe. *Physiology of Respiration,* 1965, Year Book Medical Publishers.)

cussion of the interrelated process of carbon dioxide excretion. The student is urged to review and to become thoroughly familiar with the fundamental process of diffusion, the gas laws, and the concept of pH before proceeding with the material that follows. For reference, these important concepts have been summarized in Chapter 1, page 10 et seq., and will be reiterated in context as deemed necessary.

Pulmonary Gas Exchange

The sole factor involved in pulmonary gas exchange is the wholly passive physical process of diffusion.

Within the alveoli, oxygen must diffuse rapidly across the morphologic air–blood barrier described on page 711, and illustrated in Figure 11.9, in order to reach the blood; conversely, carbon dioxide must pass simultaneously in the reverse direction. The key to understanding the movement of oxygen and carbon dioxide in the lungs as well as within the entire body rests on the elementary fact that there is a net downhill pressure gradient for oxygen between the alveoli and the tissues, whereas the pressure gradient for carbon dioxide is from tissues to alveoli. This concept is illustrated in Figure 11.25. The rate of diffusion in turn is dependent upon the concentration gradient (or pressure difference), the distance traveled, the cross-sectional area for diffusion, and the density of the gases involved. Diffusion coefficients for the several important respiratory gases are: oxygen, 1.0; carbon dioxide, 20.3; and nitrogen, 0.53. These comparative figures give the rate of diffusion of a gas through a given area for a given distance, and are dependent upon the solubility and molecular weight of the specific gas. Oxygen (molecular weight = 32) diffuses somewhat more rapidly than does carbon dioxide

(molecular weight = 44) within the alveolar gas phase.

The alveoli are quite small, having diameters of approximately 100 μ, so that the maximum distance gas molecules must move to enter the capillary blood also is quite small; hence, the process of gas diffusion within the alveoli is quite fast, being nearly complete in about 2 msec.

The pulmonary gas exchange of oxygen and carbon dioxide takes place at an interface between gas and liquid; therefore, the solubility of the gas in the liquid is important in maintaining the concentration gradient. In turn, the solubility of a given quantity of gas in a liquid is directly proportional to the partial pressure of the gas at a given temperature (Henry's law, p. 11), if it is assumed that the gas does not combine chemically with the liquid.

Carbon dioxide is 24 times more soluble in water than is oxygen at a given partial pressure and temperature; the diffusion rate of carbon dioxide within a liquid phase is high compared to that for the rate of movement of oxygen in the same medium. Therefore, carbon dioxide diffuses more rapidly within (and into) a fluid than does oxygen, even though carbon dioxide diffuses more slowly within the alveolar gas phase per se, as described above. This effect results in a steep gradient for carbon dioxide between pulmonary capillary blood and the alveolar air. Diffusion of carbon dioxide from the pulmonary capillaries into the alveoli never becomes a clinical problem except when a patient has an impaired diffusion to such an extent that he must subsist on 100 percent oxygen.

Since the gases important to respiration are quite soluble in lipids, they are also quite soluble in cell membranes, as these structures are composed of lipids to a great extent. The respiratory gases are able to diffuse through these structures very rapidly and with little hindrance. The principal limitation imposed upon the movement of gas molecules in tissues is their rate of movement through the tissue fluids themselves. Therefore, the rate of diffusion of gases through tissues, as well as across the alveolar membrane itself, is practically the same as that for diffusion through water, as indicated by the diffusion coefficients listed above.

COMPOSITION OF ALVEOLAR AIR. Alveolar air does not have the same composition as atmospheric air, as shown in Table 11.5. Note especially that inspired air becomes saturated with water vapor (humidified) as it enters the respiratory tract, as discussed on page 728. Water exerts a vapor pressure of 47 mm Hg at 37 C, or normal body temperature. This vapor pressure is dependent solely upon the water temperature, as shown in Table 11.6. Table 11.5 shows the alterations in the concentration (hence, the partial pressure) of carbon dioxide in alveolar and expired air compared to atmospheric air.

The addition of water vapor to the inspired air lowers the partial pressure of oxygen from 159 mm Hg in the inspired air to around 104 mm Hg in the alveolar air; this factor essentially dilutes the concentration of the individual gases within the lungs. Expired air has a slightly higher oxygen content (or concentration of oxygen) than does alveolar air, because the exhaled alveolar air is mixed with dead space air during expiration. Since the dead space air has had no oxygen removed, it will raise the oxygen concentration of the alveolar air when the two fractions mix.

The oxygen concentration, hence the Po_2, within the alveolar air at any given moment, depends upon the alveolar ventilation rate (p. 727) as well as the rate at which oxygen is absorbed into the pulmonary capillary blood. The interrelationship between these two variables is shown in Figure 11.26; it is assumed here that the individual is absorbing and utilizing oxygen at a rate of 250 ml/min. If the level of oxygen utilization increases, as for example during exercise, the rate of pulmo-

nary ventilation must increase proportionally in order to maintain the alveolar Po_2 at the normal level of 104 mm Hg.

The alveolar carbon dioxide concentration, on the other hand, is dependent upon the rate of metabolic production of this gas which is being excreted continually into the alveoli, where it is removed by the process of ventilation. Consequently, the Pco_2 within the alveoli depends, first, upon the rate of elimination of carbon dioxide from the pulmonary capillary blood; and second, upon the rate at which it is eliminated from the lungs by ventilation. As shown in Figure 11.28, the rate of carbon dioxide excretion normally is around 200 ml per minute, so that with a resting alveolar ventilation rate of around 4 liters per minute, the alveolar Pco_2 is 40 mm Hg. Furthermore, the alveolar Pco_2 increases in direct proportion to the carbon dioxide excretion rate, and it decreases as the alveolar ventilation rate increases; ie, as the alveolar ventilation rate increases the alveolar Pco_2 falls (Fig. 11.27).

GAS DIFFUSION AND THE RESPIRATORY MEMBRANE. The four factors that influence the diffusion of gases across the respiratory membrane may now be summarized.

Surface Area. The total respiratory surface area of the lungs is made up of the sum of the individual alveolar membrane surface areas and amounts to around 70 m^2 in the normal individual. Of this total area, around 20 percent (14 m^2) is in active use at rest. Any disease or pathologic condition that reduces the total surface area for pulmonary gas diffusion may not impair respiratory functions at rest, but can cause pulmonary insufficiency during even mild exertion. Pneumonectomy, pulmonary edema, emphysema, or pulmonary fibrosis all decrease the total surface area of the lungs that is available for gas exchange.

Thickness. Normally, the respiratory membrane, which presents a barrier for gas diffusion between alveoli and pulmonary capillary blood, is only about 0.20 μ in thickness; hence, the rate of gas

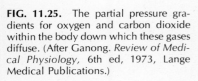

FIG. 11.25. The partial pressure gradients for oxygen and carbon dioxide within the body down which these gases diffuse. (After Ganong. *Review of Medical Physiology*, 6th ed, 1973, Lange Medical Publications.)

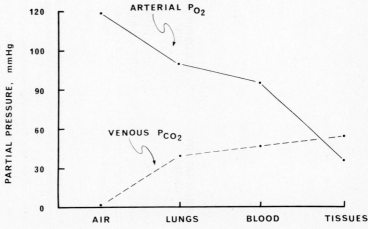

Table 11.5 PARTIAL PRESSURES OF RESPIRATORY GASES AT SEA LEVEL [a]

RESPIRATORY GASES	ATMOSPHERIC AIR[b]	HUMIDIFIED (TRACHEAL) AIR[c]	ALVEOLAR AIR	EXPIRED AIR	ARTERIAL BLOOD	MIXED VENOUS BLOOD
Oxygen (P_{O_2}):	159.0 (20.84)	149.3 (19.67)	104.0 (13.6)	120.0 (15.7)	100 (13.2)	40 (5.7)
Nitrogen (P_{N_2}):	597.0 (78.62)	563.4 (74.09)	569.0 (74.9)	566.0 (74.5)	573 (75.4)	573 (81.2)
Carbon Dioxide (P_{CO_2}):	0.3 (0.04)	0.3 (0.04)	40.0 (5.3)	27.0 (3.6)	40 (5.2)	45 (6.4)
Water Vapor (P_{H_2O}):	3.7 (0.50)	47.0 (6.20)	47.0 (6.2)	47.0 (6.2)	47 (6.2)	47 (6.7)
Total	760 (100.0)	760 (100.0)	760 (100.0)	760 (100.0)	760 (100.0)	705 (100.0)

Note: The partial pressures (P) are given in mm Hg followed (in parentheses) by the percentage concentrations at 38 C. Note that the partial pressure of each gas is directly proportional to its concentration in the mixture, and that the total pressure is equal to the sum of the individual partial pressures (Henry's law). Thus, the partial pressure of a gas in a mixture = (Total Pressure) × (Percent Concentration of Gas)/100. The values listed are for a normal adult male, resting, at sea level, barometric pressure = 1 atmosphere = 760 mm Hg.

[a] Data from Comroe. Physiology of Respiration, 1965, Year Book Medical Publishers.

[b] See Table 11.4 for more detailed composition of dry air.

[c] Note that inspired air becomes saturated with water vapor (P_{H_2O} at 37 C = 47 mm Hg, Table 11.6) as it passes through the nose, mouth, and pharynx. The total gas pressure in the trachea must equal 760 mm Hg. Therefore, at sea level 760 − 47 = 713 mm Hg total air pressure is available for respiration, and the P_{O_2} of moist inspired air is 21.9 percent of 713 mm Hg or 149 mm Hg, not 159 mm Hg as would be the case if the air passing to the lungs were completely dry. At the altitude of Denver, Colorado, where the altitude is 1,610 meters, the mean barometric pressure is about 640 mm Hg. Consequently, 640 − 47 = 593 mm Hg is the total air pressure in the trachea; thus, the P_{O_2} at this altitude is about 144 mm Hg. At higher altitudes, the air pressure decreases in accordance with the relationship given in Figure 16.25 (p. 1225).

Table 11.6 RELATIONSHIP BETWEEN WATER VAPOR PRESSURE AND TEMPERATURE[a]

TEMPERATURE (°C)	P_{H_2O} (mm Hg)
0	4.6
5	6.5
10	9.1
15	12.6
20	17.5
25	23.8
30	31.8
35	42.2
37[b]	47.0 (46.6)
40	54.9
50	92.0
100	760

[a]*Data from Comroe. Physiology of Respiration, 1965, Year Book Medical Publishers.*

[b]*The inspired tracheal and alveolar air are saturated with water vapor having a P_{H_2O} of 47 mm Hg. This value must be taken into account when calculating the P_{O_2} of inspired air, as indicated in the footnote to Table 11.5.*

diffusion across this morphologic interface is essentially the same as through a layer of water having the same thickness. Edema fluid, if present in the pulmonary interstitial spaces and/or alveoli, can increase the magnitude of the physical barrier to gas diffusion. Similarly, pulmonary fibrosis can also increase the physical distance the gas must diffuse from the alveoli to the capillary blood. Since the rate of diffusion is inversely proportional to the distance the gas must travel, any factor that increases the membrane thickness several times above normal can hinder pulmonary gas exchange appreciably.

Pressure Differences. The net transfer of a gas from the alveoli into the blood (or vice versa) is dependent to a considerable extent upon the partial pressures of the gas on each side of the respiratory membrane. Hence, oxygen moves into the bloodstream, whereas carbon dioxide moves into the alveoli from the blood in accordance with their individual differences in partial pressures (Fig. 11.25).

Diffusion Coefficient. This factor is wholly physical, and it depends primarily upon the molecular weight of the gas and its solubility in the membrane. As explained above, the diffusion rates of the respiratory gases are almost identical within the respiratory membrane as they are in water; however, because of differences in their individual diffusion coefficients (p. 740), carbon dioxide dif-

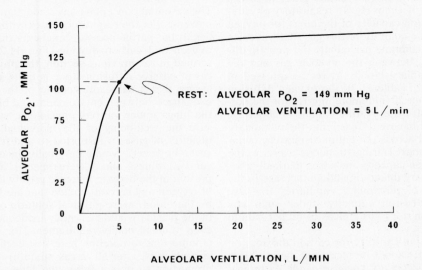

FIG. 11.26. Relationship between the rate of alveolar ventilation, the rate of oxygen absorption, and the alveolar P_{O_2}. (After Guyton. *Textbook of Medical Physiology*, 4th ed, 1971, Saunders.)

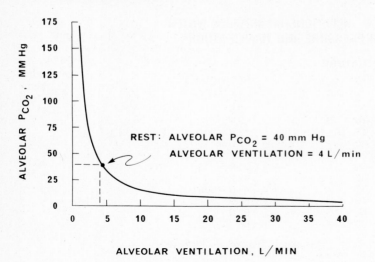

FIG. 11.27. Relationship between the rate of alveolar ventilation, P_{CO_2} excretion, and alveolar P_{CO_2}. (After Guyton. *Textbook of Medical Physiology*, 4th ed, 1971, Saunders.)

fuses around 20 times more rapidly than oxygen in an aqueous liquid.

DIFFUSION CAPACITY. The movement of oxygen into the blood as well as the transfer of carbon dioxide into the alveoli is strictly in accordance with the process of passive physical diffusion. There is no evidence for the existence of any active transport mechanism being involved in this process in the lungs. Thus, for example, if the P_{O_2} of alveolar air is 104 mm Hg and the P_{O_2} of mixed venous blood in the pulmonary capillaries is 40 mm Hg, oxygen moves very rapidly into the blood down this pressure gradient. The oxygen dissolved in the plasma then enters the erythrocytes, where it combines with hemoglobin, as discussed in detail in the next section.

The diffusion capacity for a gas is defined as the volume of a gas that diffuses through a membrane per minute down a pressure gradient across the membrane of 1 mm Hg. In respiratory physiology, the diffusion capacity of the lungs for oxygen is defined as the quantity of this gas that traverses the alveolar membrane per minute per mm Hg difference in P_{O_2} between the alveolar gas and the pulmonary capillary blood. At rest, expressed in terms of *STPD* (Table 11.3), the oxygen diffusion capacity is about 20 ml/minute/mm Hg for the normal young adult male. A total volume of about 250 ml oxygen per minute diffuses into the pulmonary capillary blood across the entire respiratory membrane of the lungs. During strenuous exercise, the oxygen diffusion capacity increases markedly, because of capillary dilatation and an increase in the total number of pulmonary capillaries that are open. The diffusion capacity under such circumstances can rise to as high as 65 ml per minute per mm Hg.

The diffusion capacity for carbon dioxide, on the other hand, cannot be measured directly with any degree of accuracy because of the extremely rapid transfer rate of this gas from blood to alveoli. Since the P_{CO_2} of pulmonary capillary blood is 46 mm Hg, however, and that of the alveoli is 40 mm Hg (Table 11.5), carbon dioxide diffuses rapidly into the alveoli down this gradient (Fig. 11.25).

Since the P_{CO_2} of blood leaving the lungs is 40 mm Hg, carbon dioxide must be able to diffuse out of the blood with extreme rapidity. Furthermore, since the diffusion capacity has been shown experimentally to be related directly to the diffusion coefficient, the diffusion capacity for carbon dioxide must exceed that for oxygen by at least 20-fold.

Carbon dioxide thus passes through all biologic membranes with extreme facility, so that carbon dioxide retention is rarely a clinical problem whereas damage to the respiratory membrane can impair oxygen uptake seriously.

Blood Gas Transport Between Lungs and Tissues

The essential concept to any understanding of the movement of the respiratory gases within the body lies in the partial-pressure gradients that exist for oxygen and carbon dioxide (Fig. 11.25). As described in the previous section, the physical process of passive diffusion alone is quite adequate to assure that a sufficient volume of these gases is exchanged between environment and blood within the lungs under various physiologic states to assure the well-being of the individual. The total quantity of gases transported to and from the tissues, however, would be totally inadequate for survival unless there were special mechanisms present in the body to increase greatly the volumes of oxygen and carbon dioxide that are transported by the blood. Mere physical solution of the gases in the plasma would be wholly inadequate to meet even minimal bodily requirements for survival. It is important to reiterate here that each gas in the air (which is, after all, a gas mixture) dissolves in proportion to its solubility coefficient and partial pressure (Henry's law, p. 11), and each of the gases of air while in solution exerts a partial pressure as though it were in the free state. Thus, one speaks of P_{O_2}, P_{CO_2}, P_{N_2}, and so forth with reference to the partial pressure of each gas in simple solution as though it were in the free state, and once again the partial pressue of each gas in solution is in proportion to the concentration of the gas within the solution.

The necessity for increasing bulk transport of respiratory gases by the blood is met by a series of rapidly reversible chemical reactions that greatly enhance the total carrying capacity of the blood for oxygen and carbon dioxide. Oxygen dissolved in the blood reacts chemically and reversibly with the protein hemoglobin in the erythrocytes, so that the total carrying capacity of blood for this gas is increased 70-fold above simple solution alone (see Fig. 9.4, p. 533). Carbon dioxide dissolved in the blood, on the other hand, enters into a series of reversible chemical reactions that enhance the carrying capacity of blood for this gas 17-fold. It should be noted in passing that the chemical combination of these gases effectively removes them from solution, thereby effectively permitting more gas to pass into the aqueous phase of blood by steepening the concentration gradient so that more of the gas is dissolved and can react chemically. The gas content of normal blood is summarized in Table 11.5

The remainder of this section will deal specifically with those chemical reactions and processes that are involved in oxygen and carbon dioxide transport from lungs to tissues and vice versa.

OXYGEN TRANSPORT. The blood flow to a tissue determines the quantity of oxygen delivered to that tissue, and in turn, blood flow is determined by the cardiac output and the state of the vascular bed, ie, degree of constriction of the vessels.

Variations in the quantity of dissolved oxygen exert little influence upon the quantity of oxygen delivered to the tissues. Rather, blood flow, the concentration of hemoglobin (usually expressed in grams percent), and the oxygen affinity of hemoglobin (usually expressed as percent saturation or in volumes percent; that is, ml oxygen per 100 ml of blood) are the important factors that determine the oxygen transport to the tissues. The latter two factors will be discussed here.

Hemoglobin. This red pigment that is found within the erythrocytes (p. 528), and that transports oxygen, is a globular protein having a molecular weight of 64,450 daltons. It is comprised of four subunits, each containing a heme moiety that is chemically conjugated to a polypeptide. A derivative of porphyrin, each heme moiety contains an iron atom chemically bound in the reduced or ferrous state (Fe^{++}), as shown in Figure 9.4 (p. 533). Collectively, the four polypeptides comprise the globin part of the hemoglobin molecule in a normal adult human; two of these subunits contain one type of polypeptide, and two are of another type. These polypeptides are termed α-chains

(having 140 amino acid residues) and β-chains (having 146 amino acid residues). Individuals possessing equal quantities of α- and β-chains in their hemoglobin are said to have hemoglobin A, although other types of hemoglobin, such as A_2 also may be present, depending upon the specific chemical nature of the amino acids that constitute the polypeptides.

Deoxygenated hemoglobin (or deoxyhemoglobin, abbreviated Hb), binds reversibly to oxygen, forming oxyhemoglobin, as discussed below. When blood is exposed either in vivo or in vitro to oxidizing agents or drugs, the ferrous iron (Fe^{++}) in the heme molecule is oxidized to ferric iron (Fe^{+++}) producing a dark-colored compound called methemoglobin.

Hemoglobin is especially useful as an oxygen carrier because of the chemical reactions that take place between the two compounds. Since one hemoglobin molecule contains four subunits, each containing a complex of porphyrin with one ferrous iron atom (a heme moiety), each heme subunit can combine reversibly with one atom of molecular oxygen in accordance with the following scheme:

$$Hb_4 + O_2 \rightleftharpoons Hb_4O_2 \ (25 \text{ percent saturated})$$
$$Hb_4 + O_2 \rightleftharpoons Hb_4O_4 \ (50 \text{ percent saturated})$$
$$Hb_4 + O_2 \rightleftharpoons Hb_4O_6 \ (75 \text{ percent saturated})$$
$$Hb_4 + O_2 \rightleftharpoons Hb_4O_8 \ (100 \text{ percent saturated})$$

These reactions are termed oxygenation (not oxidation!), since the iron remains in the ferrous (Fe^{++}) state. Sometimes the above reactions are abbreviated $Hb + O_2 \rightleftharpoons HbO_2$, although it is important to remember that each complete molecule of hemoglobin can bind four molecules of oxygen.

The reactions outlined above are extremely rapid, so that the equilibrium can shift to the right in less than 10 msec in the lungs. The process of deoxygenation of oxygenated hemoglobin also is extremely rapid within the capillaries of the body tissues.

A curve that relates the percent oxygen saturation of hemoglobin to the partial pressure of oxygen (P_{O_2}) is called an oxygen dissociation curve. Such a curve is shown in Figure 11.28. Note the characteristic sigmoid configuration. Successive oxygenation of the four heme moieties in a hemoglobin molecule progressively increases the affinity for oxygen by the remaining heme moieties; hence, the affinity of the hemoglobin for the fourth oxygen molecule is much greater than for the first. This progressive shift in affinity is the cause of the sigmoid shape of the oxygenation curve. As hemoglobin binds oxygen, the β polypeptide chains move closer together, and this shift apparently is the cause of the change in affinity of hemoglobin for oxygen.

In blood that is equilibrated with 100 percent oxygen ($P_{O_2} = 760$ mm Hg), the hemoglobin becomes 100 percent saturated, and each gram of hemoglobin now contains 1.34 ml of oxygen. The

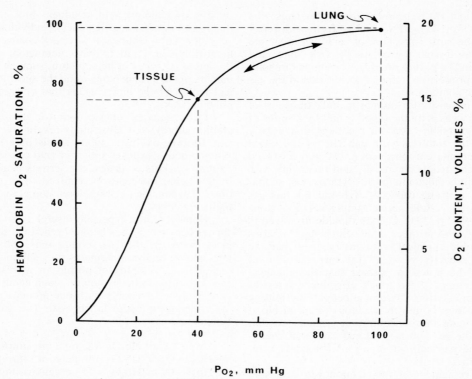

FIG. 11.28. Hemoglobin dissociation curve for oxygen, pH 7.40, 38 C. Note the reversible shift in the upper portion of the curve between the lung and tissue (arrow) where the hemoglobin receives and gives up oxygen respectively. (After Ganong. *Review of Medical Physiology,* 6th ed, 1973, Lange Medical Publications.)

hemoglobin concentration in normal blood of men is around 16 gm per 100 ml, and that in women is around 14 gm per 100 ml. Since the average concentration of hemoglobin is 15 gm per 100 ml, 100 ml of blood contains $15 \times 1.34 = 20.1$ ml of oxygen which is chemically bound to hemoglobin when fully (100 percent) saturated. The quantity of dissolved oxygen in the blood, on the other hand, is in linear proportion to the Po_2; therefore, 0.003 ml of this gas dissolves per 100 ml of blood per mm Hg Po_2.

At the venous end of the pulmonary capillaries in vivo, the hemoglobin in the blood is approximately 97.5 percent saturated with oxygen and the Po_2 is about 97 mm Hg. In systemic arterial blood, however, the hemoglobin is only 97 percent saturated. This slight decrease in blood saturation is caused by dilution of pulmonary venous blood by blood which passes through the physiologic shunt (p. 656) and thus bypasses the lungs. The normal ventilation/perfusion ratio is indicated in Figure 11.29, together with the effects of the venous shunt in determining the ultimate Po_2 of arterial blood.

Systemic arterial blood thus contains an approximate total volume of 19.8 ml of oxygen per 100 ml, and of this total volume, 19.5 ml of oxygen are bound to hemoglobin and only 0.3 ml is dissolved in solution.

At rest, the hemoglobin of venous blood is only 75 percent saturated, so that the total oxygen content now is around 15.2 ml oxygen per 100 ml of blood. Consequently, at rest the tissues extract approximately 4.6 ml of oxygen per 100 ml of blood which perfuses them. Of this total, about 4.4 ml represent the oxygen released from combination with hemoglobin, and the remainder comes from oxygen carried in solution in the plasma. In this manner, the normal resting volume of 250 ml of oxygen is transported from lungs to tissues by the blood.

The ability (or affinity) of hemoglobin to bind oxygen reversibly, as measured by the oxygen dissociation curve, is modified by three important factors.

First, as shown in Figure 11.30B, a fall in pH toward the acidic end of the scale causes the entire oxygen dissociation curve to shift to the right so that a higher Po_2 is required for hemoglobin to bind a particular volume of oxygen at the higher pH level. In other words, at a given Po_2, the oxygen saturation of hemoglobin decreases with increasing pH, as shown in the figure. Conversely, a rise in pH toward the alkaline end of the scale causes the dissociation curve to shift toward the left. As indicated in Figure 11.30A, an increased Pco_2 has the same effect as a decreased pH on the oxygen dissociation curve, because the increased

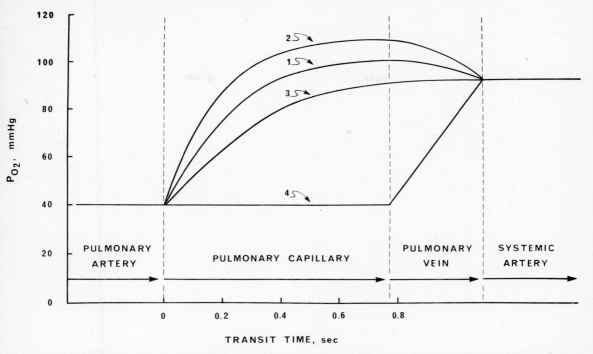

FIG. 11.29. The effect of alveoli having high (1), normal (2) and low (3) ventilation/perfusion ratios upon the ultimate P_{O_2} of the arterial blood. 4 = the venous shunt. See also Figure 11.36. Note also the transit time for blood flowing through a pulmonary capillary. (After Selkurt [ed]. *Physiology,* 2nd ed, 1966, Little, Brown.)

P_{CO_2} also lowers the pH. (The blood buffers and the respiratory influence on blood pH will be considered more fully on page 755 as well as in Chapter 12, p. 805).

The decreased oxygen affinity of hemoglobin, which is observed when the blood pH falls, is known as the *Bohr effect*. This phenomenon is related intimately to the fact that reduced (deoxygenated) hemoglobin (Hb) is more active in binding H^+ than is oxyhemoglobin (HbO_2). Since the pH of blood falls as its carbon dioxide content increases, as the P_{CO_2} rises (in the tissues), the oxygen dissociation curve shifts to the right and oxygen binding decreases. By far the greatest proportion of the oxygen that is carried by the hemoglobin is unloaded in the tissue capillaries as the direct consequence of the fall in P_{O_2} in this region. An additional small quantity of oxygen, as well as the large volume, is released from the hemoglobin because of the rise in P_{CO_2} in the tissues. This causes a concomitant slight shift in the oxygen dissociation curve to the right due to the pH effect.

Second, an increased temperature shifts the oxygen dissociation curve in a direction similar to that caused by a fall in pH, ie, to the right. Conversely, a shift to the left is observed at temperatures below 38 C. In normal individuals, the effects of such temperature fluctuations on the oxygen dissociation curve are slight (Fig. 11.30C).

Third, a compound called 2,3-diphosphoglycerate (2,3-DPG) also has an important effect on the oxygen dissociation curve of hemoglo-bin. This compound, which is present within the erythrocytes at a high concentration, thus has been shown to exert an effect upon the affinity of hemoglobin for oxygen. The synthetic pathway for 2,3-DPG is shown in Figure 11.31, and the precursor for this compound is 1,3-diphosphoglycerate, a glycolytic product of the Embden-Meyerhof pathway (p. 936).

A highly polar anion, 2,3-DPG binds tightly to the β-chains of deoxyhemoglobin (Hb), whereas it does not bind to oxyhemoglobin (HbO_2). One mole of 2,3-DPG thus reacts with one mole of HbO_2 according to the reaction:

$$HbO_2 + 2,3\text{-}DPG \rightleftharpoons Hb\text{---}2,3\text{-}DPG + \uparrow O_2$$

An increased concentration of 2,3-DPG shifts the equilibrium position of this reaction to the right so that more oxygen is liberated from hemoglobin.

The factors that affect the level of 2,3-DPG in the erythrocytes are not fully known at present, although the concentration of this compound decreases at lower (more acid) pH levels, and rises once again with an elevated pH. Thyroid hormones and exercise that is prolonged beyond an hour both increase 2,3-DPG concentration in the erythrocytes so that the affinity for oxygen by hemoglobin is lower; ie, the oxygen dissociation curve shifts to the right. In addition to the increased 2,3-DPG concentration that develops during exercise, concomitantly the body temperature is raised somewhat and the blood pH becomes

A. PCO_2

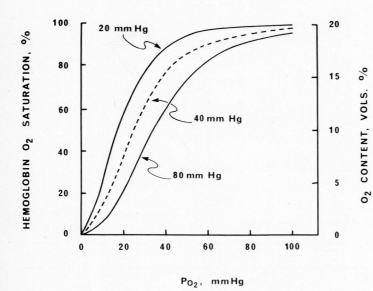

FIG. 11.30.   **A.** Effect of altering the PCO_2 on the hemoglobin dissociation curve. Note that a low PCO_2 shifts the curve to the left, whereas a high PCO_2 shifts the curve to the right.

lower through increased production and accumulation of carbon dioxide and metabolic products such as lactic acid. These additional factors that come into play during physical exertion promote a shift in the oxygen dissociation curve to the right; hence, they facilitate unloading of oxygen in the active tissues.

Oxygen release from each unit of hemoglobin flowing through muscular tissues during exercise also is facilitated by a much lower tissue PO_2 level than is present during rest, and this effect alone tends to steepen the blood-tissue oxygen concentration gradient. The last factor involved in the process of oxygen release in the tissues is the steep slope of the oxygen dissociation curve at low PO_2 values. Large quantities of oxygen are liberated from each unit of hemoglobin per unit fall in PO_2 level (Fig. 11.25).

A hypoxic environment, such as is encountered at high altitudes (p. 1223), also produces a

B. pH

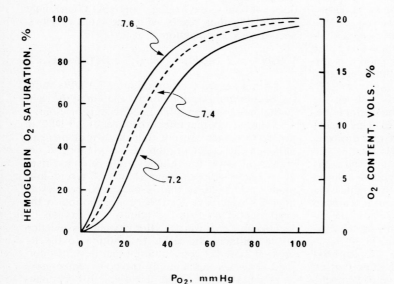

FIG. 11.30.   **B.** Effect of pH on the hemoglobin dissociation curve. Note that the effect of low (acidic) pH is similar to an elevated PCO_2.

C. TEMPERATURE

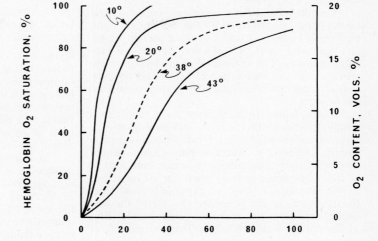

FIG. 11.30. (cont'd) C. Effect of temperature changes on the hemoglobin dissociation curve. (After Ganong, *Review of Medical Physiology,* 6th ed, 1973, Lange Medical Publications.)

significant increase in 2,3-DPG concentration secondary to an increase in blood pH, so that the dissociation curve for hemoglobin once again shifts toward the right, thereby augmenting the quantity of oxygen that is available to the tissues.

Fetal hemoglobin (known as hemoglobin F) has a greater affinity for oxygen than does adult hemoglobin (hemoglobin A), and this enables the fetus to extract more oxygen from the maternal blood flowing through the placenta (p. 1206). Hemoglobin F has a greater oxygen affinity than adult hemoglobin A because the β-chains are replaced by α-polypeptide chains in the fetus, and the latter have a much poorer ability to bind 2,3-DPG than do the β-chains found in adult hemoglobin. Consequently, hemoglobin F has a greater ability to take up oxygen.

Certain types of hemoglobin found clinically in adults exhibit an abnormally high affinity for oxygen, so that the resulting tissue hypoxia is a sufficient stimulus to augmented erythrocyte production and polycythemia (an elevated erythrocyte count, p. 569) develops.

Whole blood that has been stored in a blood bank gradually loses its normal compliment of 2,3-DPG; therefore, tissue oxygen release within the transfused recipient of such blood is lowered, and the benefits of such transfusion into an already hypoxic individual may be minimal. The rate of loss of 2,3-DPG in stored whole blood can be reduced by using a citrate–phosphate–dextrose solution as a preservative.

Affinity of Hemoglobin for Carbon Monoxide. Although carbon monoxide (CO) can hardly be con-

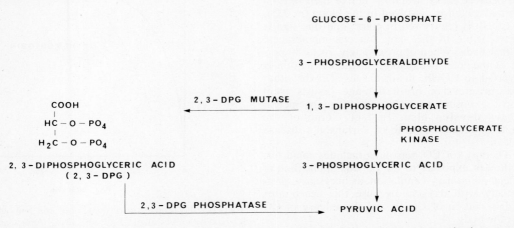

FIG. 11.31. Biosynthesis and metabolism of 2,3-DPG. (After Ganong. *Review of Medical Physiology,* 6th ed, 1973, Lange Medical Publications.)

sidered a "normal" constituent of the respiratory gases, its clinical importance as a poison, as well as its interesting physiologic attributes, renders its discussion worthwhile at this point.*

Carbon monoxide, a colorless, odorless, and tasteless gas, is produced by the incomplete combustion of carbon. It was known in ancient times, and was used by the Greeks and Romans to execute criminals. The emissions of modern gasoline engines contain over 6 percent CO, and this gas also is produced abundantly by poorly ventilated gas heaters.

CO is highly toxic to man, as it reacts in a relatively irreversible fashion with hemoglobin to form carboxyhemoglobin (or carbonmonoxyhemoglobin, COHb), so that the hemoglobin cannot take up oxygen. The affinity of hemoglobin for CO is over 200 times greater than for oxygen, but COHb liberates CO at an extremely slow rate. Furthermore, in the presence of COHb, the dissociation curve for the remaining HbO_2 (Fig. 11.28) shifts toward the left, so that the quantity of oxygen released in the tissues is decreased, an effect that tends to worsen the hypoxia produced by the CO itself.

Carbon monoxide poisoning can be listed as a form of anemic hypoxia, since there is a decreased quantity of hemoglobin available to carry oxygen, even though the total concentration of hemoglobin circulating in the blood is unaffected. Since CO has such a great affinity for hemoglobin, there is a gradual formation of HbCO when the alveolar Pco is greater than 0.4 mm Hg. The total quantity of HbCO formed ultimately depends not only upon the concentration of CO in the inspired air but also upon the duration of the exposure.

Carbon monoxide poisoning produces symptoms that are similar to those found in any form of hypoxia (p. 764), but headache and nausea are especially prominent. There is little respiratory stimulation because the Po_2 of dissolved oxygen remains at normal levels in the arterial blood. The aortic and carotid chemoreceptors are not stimulated because these structures derive their oxygen from that which is dissolved in the blood (p. 735).

COHb is cherry-red in color, and this sign is clearly visible in the skin, mucous membranes, and nail beds. When approximately 70 to 80 percent of the total hemoglobin in the circulation is converted to COHb, death ensues. On the other hand, progressive brain damage results from chronic exposure to CO, and attendant mental changes develop also. In fact, some of the symptoms resemble those of Parkinson's disease (p. 456).

Therapy for CO poisoning includes rapid termination of exposure to the gas and application of artificial pulmonary ventilation, as necessary, preferentially with oxygen, because this accelerates the dissociation of COHb and the release of CO from the body.

CO causes more peacetime deaths than all other gases combined.

Carbon monoxide, although directly toxic to the intracellular cytochrome enzyme systems (p. 931), plays no role in clinical poisoning by this gas, since the quantity of CO required to inhibit the cytochromes is 1,000 times greater than the lethal dose. Death intervenes long before the cytochromes are affected.

Myoglobin. In the skeletal muscles, especially in those muscles that are specialized for sustained contraction, a respiratory pigment other than hemoglobin is present, and this pigment is called myoglobin. Myoglobin is similar to hemoglobin in that it also contains iron; however, it binds only 1 mole of oxygen, compared to 4 moles for hemoglobin (p. 745).

In configuration, the dissociation curve of myoglobin is a rectangular hyperbola, rather than sigmoid, as shown by the solid curve in Figure 11.32. Since the dissociation curve for myoglobin lies to the left of the hemoglobin dissociation curve (dashed curve), this muscle pigment can take up oxygen from blood, and it releases oxygen only at very low Po_2 levels. Since the Po_2 level in heavily exercising muscle falls nearly to zero, however, and since the blood supply is cut off during active contraction of the muscle, myoglobin may provide some oxygen under these extreme conditions. But the normal contribution of myoglobin to the total oxygen consumption of muscle is negligible in man. †

CARBON DIOXIDE TRANSPORT. Carbon dioxide gas is produced continuously throughout life in widely varying amounts as an end product of tissue metabolism (Chap. 14, p. 945).

†*Myoglobin is an important oxygen source in such diving mammals as seals, sea lions, whales, porpoises, and walruses.*

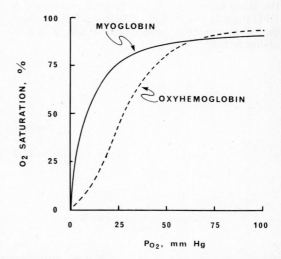

FIG. 11.32. Comparison of myoglobin and hemoglobin dissociation curves at pH 7.40 and 38 C. (After Ganong. *Review of Medical Physiology*, 6th ed, 1973, Lange Medical Publications.)

The solubility of carbon dioxide in aqueous solution is greater by a factor of 20-fold than is that of oxygen; therefore, far more carbon dioxide is present within the blood in simple solution than oxygen under identical conditions of temperature and pressure. Carbon dioxide that diffuses into the erythrocytes from the tissues rapidly combines with water to form carbonic acid, a reaction that is catalyzed at very high rates by the enzyme carbonic anhydrase:

$$CO_2 + H_2O \xrightleftharpoons[\text{anhydrase}]{\text{carbonic}} H_2CO_3$$

The carbonic acid in turn now dissociates into a hydrogen ion and a bicarbonate ion according to the following expression:

$$H_2CO_3 \rightleftharpoons H^+ + HCO_3^-$$

The hydrogen ions thus produced by the dissociation of carbonic acid are buffered as discussed in the next division of this chapter, principally by hemoglobin. The bicarbonate ions formed in the erythrocytes as described above then diffuse into the plasma.

As the oxygen saturation of hemoglobin decreases in the blood traversing the capillaries, the buffering capacity of this respiratory pigment increases because deoxyhemoglobin binds more hydrogen ion than does oxyhemoglobin.

A small fraction (about 20 percent) of the total carbon dioxide that is carried to the lungs is present in the blood as carbamino compounds that are formed by the reaction between carbon dioxide and the amino groups of proteins in both the erythrocytes and in the plasma. The amino groups of hemoglobin, for example, can react with carbon dioxide as follows:

$$R{-}N\overset{\displaystyle H}{\underset{\displaystyle H}{\Big\langle}} + CO_2 \rightleftharpoons R{-}N\overset{\displaystyle H}{\underset{\displaystyle COOH}{\Big\langle}}$$

Hemoglobin itself produces by far the greatest total quantity of carbamino compounds, although a similar reaction takes place in the plasma directly between carbon dioxide and the amino groups of the several plasma proteins.

When the blood P_{CO_2} level is above 10 mm Hg, the total quantity of carbamino-Hb that is produced remains essentially constant, because of the fact that the presence of more carbon dioxide also increases the quantity of hydrogen ion formed. Therefore, the $R{-}NH_2$ groups of the protein are blocked by the formation of $R{-}NH_3^+$ groups because of this availability of greater quantities of hydrogen ion that react with, and thus limit, the availability of, $R{-}NH_2$ groups for carbamino compound formation.

As the direct physiologic consequence of deoxyhemoglobin being able to form carbamino compounds far more readily than oxyhemoglobin, carbon dioxide transport in venous blood is achieved more readily than in arterial blood.

As stated above, the total quantity of carbon dioxide that is carried to the lungs in the carbamino form amounts to around 20 percent of the total; however, only a small quantity of free carbon dioxide is present in aqueous solution in venous plasma, since the quantity of this gas that is hydrated directly is very small because of the absence of the enzyme carbonic anhydrase outside of the erythrocytes.

The electric and osmotic changes caused by the reactions described above now may be considered.

The elevation of bicarbonate ion (HCO_3^-) concentration that takes place within the erythrocytes as they pass through the capillaries is much greater than it is within the plasma. Bicarbonate ion diffuses into the plasma from the cells because of the concentration gradient produced across the erythrocyte membrane. In this manner, approximately 70 percent of the bicarbonate ultimately reaches the plasma after being formed in the erythrocytes. Since the protein anions normally cannot cross the cell membrane, and since the electrochemical pump for sodium and potassium ions prevents free diffusion of these cations, electrochemical neutrality is maintained by the diffusion of a chloride ion into the erythrocyte for each bicarbonate ion which leaves. This phenomenon is known as the *chloride shift*, or the *chloride–bicarbonate shift*, and, as a consequence, the chloride content of the erythrocytes in venous blood is considerably greater than that in arterial blood. The chloride shift is illustrated in Figure 11.33, together with certain other pertinent reactions discussed above that pertain to carbon dioxide and oxygen transport.

The rapidity with which the chloride shift takes place is noteworthy; it is practically complete in 1 msec. (The transit time for an erythrocyte through a pulmonary capillary is given in Fig. 11.29).

Since every protein molecule within an erythrocyte has an abundance of negative charges, whereas there is only one charge on each bicarbonate and chloride ion, the total number of osmotically active particles within the erythrocytes increases as bicarbonate ions accumulate, and the hydrogen ions produced by the dissociation of carbonic acid are buffered. The osmotic activity within the erythrocytes increases, and water is taken up to restore osmotic equilibrium. Therefore, the erythrocytes increase in size so that the hematocrit of venous blood normally is around 3 percent greater than that of arterial blood. When the chloride moves out of the erythrocytes in the lungs, the erythrocytes shrink once again.

The overall transport of carbon dioxide in erythrocytes and plasma is summarized in Figure

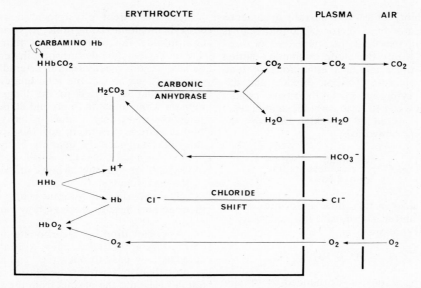

FIG. 11.33. The chloride shift and related reactions in the blood during oxygen uptake and carbon dioxide elimination in the lung. (After Best and Taylor [eds]. *The Physiological Basis of Medical Practice,* 8th ed, 1966, Williams & Wilkins.)

11.34; the relative extent to which each of the various mechanisms already discussed is involved is also indicated in this figure. In arterial blood, a total of approximately 50 ml of carbon dioxide is transported per 100 ml blood. Of this quantity, 3 ml is in solution, 3 ml is in carbamino compounds, and 44 ml is present in the plasma as bicarbonate ion. Within the capillaries of the tissues, around 5 ml of carbon dioxide is added per 100 ml of blood so that the total CO_2 concentration of venous blood is around 55 volumes percent. Of this increment, about 0.5 ml remains in solution in the plasma water, 1.0 ml forms carbamino compounds, and 3.5 ml forms bicarbonate ion. Consequently, blood pH falls to 7.36 in the veins from a level of 7.40 in the arteries.

Within the lungs, the processes described above undergo reversal, so that the excess 5 ml of carbon dioxide is excreted into the alveoli by diffusion. At rest, around 200 ml of carbon dioxide per minute is transported from tissues to lungs and eliminated. In 24 hours, this quantity of carbon dioxide is equivalent chemically to 12,500 mEq of hydrogen ion.

During exercise, of course, significantly greater quantities of carbon dioxide are produced in the tissues, transported to the lungs, and eliminated from the body.

Capillary-Tissue Gas Exchange

Within the capillary beds of the tissues throughout the body, the mechanism for respiratory gas exchange is basically that found in the lungs, except that the process is reversed. Thus, simple diffusion alone is responsible for movement of oxygen out of the capillaries to the interstitial fluid, whereas carbon dioxide moves from cells to capillary blood in the opposite direction, each gas simply passing down its own pressure gradient (Fig. 11.25).

OXYGEN DIFFUSION FROM CAPILLARIES TO CELLS. The Po_2 within the interstitial fluid adjacent to a capillary is low and quite variable; the interstitial fluid Po_2 is around 40 mm Hg, compared to a Po_2 within the capillary that is approximately 95 mm Hg at its arterial end. Consequently, a pressure gradient for the diffusion of oxygen out of the capillary of 55 mm Hg is present at the arterial end of the vessel. As the blood passes through the capillary, the oxygen diffuses out very rapidly; thus, the capillary Po_2 soon falls to that of the surrounding interstitial fluid, or about 40 mm Hg. Venous capillary blood has a Po_2 that is near that of interstitial fluid. The net distances involved in this transfer process are extremely short in general, and the total surface area for gas exchange is extremely large. This mechanism is quite efficient, so that capillary gas exchange is practically instantaneous.

When the blood flow to a tissue is increased (or decreased), considerably greater (or lesser) total quantities of oxygen are carried to the tissues per unit time. The maximal limit to which the tissue Po_2 may conceivably rise is determined solely by the arterial blood Po_2, or 95 mm Hg, as illustrated in Figure 11.35. If tissue metabolism increases markedly, eg, as in an exercising muscle, far more oxygen is utilized. The interstitial fluid Po_2 falls, a response which causes vasodilatation and an increased blood flow (p. 647). Conversely,

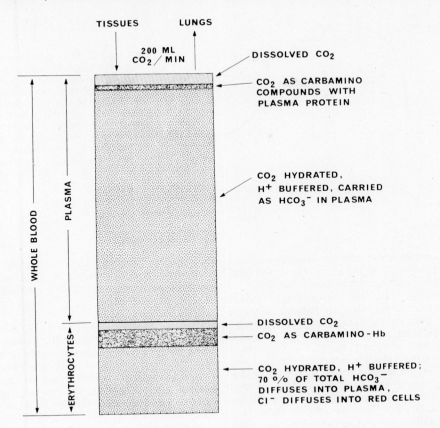

FIG. 11.34. Summary of carbon dioxide transport in the blood. (After Ganong. *Review of Medical Physiology*, 6th ed, 1973, Lange Medical Publications.)

the tissue P_{O_2} rises toward that of the arterial blood when the metabolic rate decreases after the cessation of activity.

The factors involved in oxygen transport and unloading by the blood were discussed earlier, and the changes in P_{O_2} throughout the circulatory system are summarized in Figure 11.36. Since around 97 percent of the total oxygen transported by the blood is found in combination with hemoglobin, any factor that produces a decrease in the hemo-

FIG. 11.35. The relationship between oxygen consumption, blood flow, and interstitial fluid P_{O_2}. (After Guyton. *Textbook of Medical Physiology*, 4th ed, 1971, Saunders.)

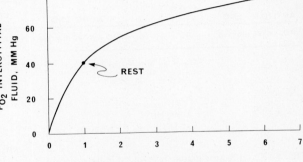

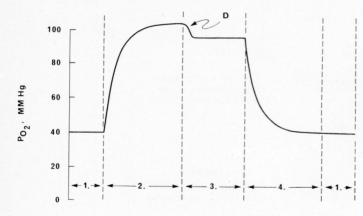

FIG. 11.36. The approximate changes in P_{O_2} of the blood throughout the circulatory system. Note the slight drop in arterial P_{O_2} because of dilution (D) of oxygenated blood from the pulmonary capillaries by deoxygenated blood from the pulmonary shunt. 1. Mixed venous blood; 2. Blood in pulmonary capillaries; 3. Systemic arterial blood; 4. Blood in systemic capillaries. (See also Fig. 11.29.) (After Guyton. *Textbook of Medical Physiology*, 4th ed, 1971, Saunders.)

globin concentration (or impairs its carrying capacity for oxygen) will influence the interstitial fluid P_{O_2} similarly to a reduced blood flow.

In summary, then, the tissue P_{O_2} is determined by an exceedingly delicate balance between two factors: first, by the rate of oxygen transport to the tissues, and second, by the rate of tissue utilization of that oxygen (Fig. 11.37). Oxygen is being used by the cells continuously, so that the intracellular P_{O_2} is always lower than that of the surrounding interstitial fluid regardless of the metabolic state or other process; ie, when the interstitial fluid P_{O_2} falls for any reason, there is a concomitant fall in the intracellular P_{O_2}, and vice versa. As mentioned earlier, oxygen diffuses with extreme rapidity through cell membranes. As a consequence of this fact, the intracellular P_{O_2} is almost the same as that of the interstitial fluid.

In some instances, the distances from capillary to cell are relatively great, so that the intracellular P_{O_2} may be as low as zero; however, this value can range up to 40 mm Hg. The average value for the intracellular P_{O_2} found in experimental animals (dogs) is approximately 5 mm Hg. As the metabolic machinery of the cell can be supported at an intracellular P_{O_2} ranging between 1 and 5 mm Hg, this value is quite sufficient for normal functional integrity to be maintained.

CARBON DIOXIDE DIFFUSION FROM CELLS TO CAPILLARIES. Within the cells, carbon dioxide is being produced continuously as the result of various metabolic processes (Chap. 14, pp. 944, 955); the intracellular P_{CO_2} tends to increase. This tendency is balanced by the extreme rapidity with which carbon dioxide diffuses into the interstitial fluid and from there into the capillary blood. Since carbon dioxide can diffuse some 20 times faster than oxygen, the intracellular fluid P_{CO_2} is around 46 mm Hg, whereas adjacent to the capillaries this value is 45 mm Hg, giving rise to a 1 mm Hg extravascular pressure gradient. These facts are shown in Figure 11.38. Blood entering a capillary bed has an arterial P_{CO_2} of approximately 40 mm Hg. As this blood traverses the capillaries, carbon dioxide is taken up from the interstitial fluid so that the venous blood P_{CO_2} is around 45 mm Hg, or approximately the same as that found in the ambient interstitial fluid. Since the diffusion

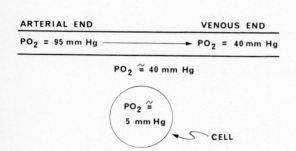

FIG. 11.37. Pressure gradients involved in the diffusion of oxygen from a capillary to a cell. (Data from Guyton. *Textbook of Medical Physiology*, 4th ed, 1971, Saunders.)

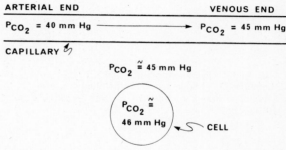

FIG. 11.38. Pressure gradients involved in the diffusion of carbon dioxide from a cell to a capillary. (Data from Guyton. *Textbook of Medical Physiology*, 4th ed, 1971, Saunders.)

coefficient of carbon dioxide is high the P_{CO_2} of venous blood comes to within a fraction of a mm Hg of an actual equilibrium with the P_{CO_2} value of the interstitial fluid.

Alterations in blood flow and tissue metabolism affect the tissue P_{CO_2} in precisely the opposite manner in which they influence the tissue P_{O_2} as described above (Fig. 11.39). The lowest limit to which the interstitial fluid P_{CO_2} can fall is the same as that of the arterial blood as it enters the tissue capillaries, or around 40 mm Hg under normal conditions. An elevation of the P_{CO_2} value above this minimum level thus depends upon blood flow and metabolic rate. A decreased blood flow alone causes an elevated venous P_{CO_2}, whereas a greatly increased blood flow can lower the tissue P_{CO_2} to almost 40 mm Hg, its lower limit in arterial blood. Augmented metabolic rates, especially in the face of low blood flow rates, can elevate the tissue P_{CO_2} to very high levels; conversely, reduced metabolic rates lower tissue P_{CO_2} levels proportionately.

Within the lungs, the P_{CO_2} of the deoxygenated blood (flowing in the pulmonary artery) is around 45 mm Hg, while the alveolar P_{CO_2} is about 40 mm Hg (Table 11.5). This 5 mm Hg gradient alone is quite adequate to effect rapid transfer of carbon dioxide from blood to alveoli, especially in view of the high diffusion coefficient for this gas. The pulmonary transfer of carbon dioxide thus is essentially complete in only 400 msec (0.4 sec), and since the transit time for blood through a pulmonary capillary is about 800 msec (Fig. 11.29), there is more than enough time for carbon dioxide exchange to be completed.

BLOOD BUFFERS AND RESPIRATORY REGULATION OF BLOOD ACID–BASE BALANCE

The system of blood buffers collectively provide a critical homeostatic mechanism for the maintenance and regulation of blood and interstitial fluid pH within the body. It is essential for survival that the aqueous internal environment in which the cells and tissues live and function be protected at all times from wide fluctuations either above or below a tolerable pH level. Furthermore, it is of the utmost importance that the clinician have a clear understanding of the concept of pH and the blood buffers, as well as how the respiratory processes affect these mechanisms in the human body in both normal and certain abnormal situations. The equally important renal contribution to the regulation and maintenance of blood pH will be discussed in Chapter 12, page 804.

The concepts of pH, pK, and buffers as well as the Henderson-Hasselbalch equation were developed in Chapter 1, page 20. The reader should familiarize himself thoroughly with this basic material before proceeding further into this discussion, although some of the earlier material will be reviewed cursorily.

In the following sections, a consideration of the blood buffers and their normal physiologic roles will be followed by an examination of the important respiratory contributions to the regulation of blood pH.

The Blood Buffers

It was noted throughout the earlier sections of this chapter that pH, or hydrogen ion concentration,

FIG. 11.39. Relationships among blood flow, metabolic rate, and tissue P_{CO_2}. (After Guyton. *Textbook of Medical Physiology*, 4th ed, 1971, Saunders.)

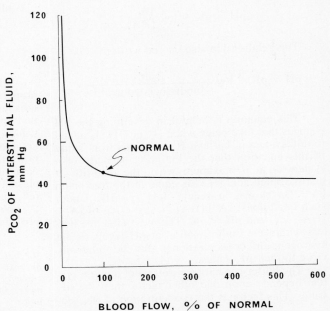

exerts an important influence upon many facets of respiratory control and regulation.

The normal hydrogen ion concentration of arterial blood is maintained close to a pH of 7.40, whereas the pH in venous blood is about 7.36, for reasons described on page 752. In a normal individual, the physiologic range of blood pH is between 6.8 and 7.8, and fluctuations above and below these extremes can have serious consequences if not rapidly compensated.

A buffer, by definition, is an aqueous solution of any acid (or base) and its salt which can resist a change in pH upon the addition of more acid (or base). The two substances involved in a system that produces this effect sometimes are called a buffer pair. In the system: $HA \rightleftharpoons H^+ + A^-$, in which HA is the undissociated acid, A^- is the anion (HA and A^- are the buffer pair in this example), and H^+ is hydrogen ion. Addition of a stronger acid to the system will cause the equilibrium to shift to the left; thus, a certain proportion of the hydrogen ions form more undissociated salt so that the net increase in hydrogen ion, hence decrease in pH, is less than it would be otherwise. A like situation also holds true for bases.

According to the law of mass action, the mathematical product of the concentrations of the chemical products in a reaction divided by the concentration of the reactant is equal to a constant. Therefore,

$$K = \frac{[H^+][A^-]}{[HA]}$$

When this equation is solved for $[H^+]$, and put in pH notation (ie, the pH is the negative log of the hydrogen ion concentration), the Henderson–Hasselbalch equation is obtained (see also p. 22):

$$pH = pK + \log \frac{[A^-]}{[HA]}$$

This relationship also can be written:

$$pH = pK + \log \frac{[\text{salt}]}{[\text{undissociated acid}]}$$

This equation is helpful in describing the pH changes that take place when H^+ or OH^- is added to any system of buffers. It is obvious from study of this equation that the buffering capacity of a system is at a maximum when the quantity of undissociated acid [HA] is equal to the free anion $[A^-]$, as shown in Figure 1.11 (p. 21). Under such conditions, $[A^-]/[HA] = 1$; the log of $[A^-]/[HA] = O$ and pH = pK. The most effective buffers within the body function at pH levels (ie, at hydrogen ion concentrations) that lie relatively close to their pK values.

The buffer capacity (or power) of a given system depends upon one other factor than the pK value. This factor is the concentration of the sub-

stances that are involved in the buffer pair. Thus, the capacity of a buffer is related to the pK, and it also is directly proportional to the concentrations of the buffer substances.

Although the important blood buffers are described individually for the sake of convenience, it must be emphasized that within the body fluids, these buffers function as an integrated system (p. 757, isohydric shift) whose physiologic role is to maintain continuously the pH of the blood within narrow tolerances.

Principal Blood Buffer Systems

Three major types of compound are involved in the important buffer systems of the body. These are: first, the systems of proteins within the plasma and erythrocytes; second, the carbonic acid–bicarbonate system; and third, the phosphate buffers.

PROTEIN BUFFER SYSTEMS. Within the blood, the plasma proteins are especially important as buffers. This is true because both their free amino and carboxyl groups can dissociate:

$$R—NH_3^+ \rightleftharpoons R—NH_2 + H^+$$

And

$$R—COOH \rightleftharpoons R—COO^- + H^+$$

Written in the form of the Henderson–Hasselbalch equation, these relationships become, respectively:

$$pH = pK_{R-NH_3^+} + \log \frac{[R-NH_2]}{[R-NH_3^+]}$$

And

$$pH = pK_{R-COOH} + \log \frac{[R-NH_2]}{[R-COOH]}$$

Since proteins exhibit both acidic and basic properties (ie, they are ampholytes, or amphoteric substances, p. 23), they can operate as buffers that oppose each other. Furthermore, since the pK values of these acidic and basic protein systems are close to pH 7.4, they form by far the most powerful as well as quantitatively the most abundant buffering system of the body.* Therefore, the plasma proteins themselves make a highly significant contribution to the regulation of blood pH.

Within the erythrocytes, hemoglobin also provides another important protein buffer system

*In fact, the intracellular proteins alone can account for approximately 75 percent of the total chemical buffering capacity of the intracellular fluids, and this effect in turn assists in the regulation of the pH of all the extracellular fluids.

for the blood. This pigment is not only present in large quantities, but also the imidazole groups of the histidine residues are abundant in hemoglobin (33 per mole). Therefore, hemoglobin has a six-fold greater total buffering capacity than do the plasma proteins. The free amino and carboxyl groups of the hemoglobin molecules themselves contribute little to the overall buffering capacity of this compound, so that the histidine residues alone are of prime importance in producing this buffering effect.

Another fact of importance with regard to the action of hemoglobin as a buffer is unique to this compound. The imidazole groups in deoxyhemoglobin dissociate to a lesser extent than do those of oxyhemoglobin. Consequently, deoxyhemoglobin is a weaker acid, thus a better buffer, than is oxyhemoglobin.

THE CARBONIC ACID–BICARBONATE BUFFER SYSTEM.

Within the body, the activity of the carbonic acid–bicarbonate buffer system is not particularly strong because the pK of this system is 6.1, relative to an average blood pH of 7.4. Approximately 20 times more of the bicarbonate buffer is in the form of bicarbonate ion than is present as dissolved carbon dioxide (or carbonic acid).

$$CO_2 + H_2O \rightleftharpoons H_2CO_3 \rightleftharpoons H^+ + HCO_3^-$$

Therefore $[HCO_3^-]/[H_2CO_3] = 20/1$.

Nevertheless, this system does provide one of the most important buffers in the blood because the carbonic acid (H_2CO_3) level in the plasma is in equilibrium with dissolved carbon dioxide, and the ratio between these compounds in turn is regulated by respiration. If hydrogen ion is added to blood, the bicarbonate ion decreases, since more carbonic acid is formed; ie, the equilibrium in the reaction given above shifts to the left. The overall blood carbonic acid concentration would rise, except that the carbonic acid is converted into water and carbon dioxide, and the latter is eliminated by the lungs: $H^+ + HCO_3^- \rightleftharpoons H_2CO_3 \rightleftharpoons H_2O + \uparrow CO_2$.

If enough hydrogen ion were added to the blood to decrease the plasma bicarbonate ion concentration by one-half, the blood pH would fall from 7.4 to 6.0. Any additional carbonic acid formed by an increased metabolic rate, however, is eliminated by the lungs. In addition, the increased blood hydrogen ion level also stimulates respiration (p. 737), so that a fall in the alveolar P_{CO_2} results, thereby removing still more carbonic acid from the lungs by increasing the diffusion gradient for CO_2 between the pulmonary capillary blood and the alveoli.

In the absence of the enzyme carbonic anhydrase, the reaction $CO_2 + H_2O \rightleftharpoons H_2CO_3$ is quite slow in either direction, and as noted earlier, plasma contains none of this enzyme. Carbonic anhydrase, however is found in abundance within the erythrocytes.* Carbonic anhydrase is a protein having a molecular weight of 30,000 daltons, and it contains one atom of zinc in each molecule.

THE PHOSPHATE BUFFER SYSTEM.

This relatively minor buffer is composed of the buffer pair $H_2PO_4^- \rightleftharpoons HPO_4^= + H^+$. Because the pK of this buffer pair is 6.8, it is relatively close to that of blood which is around pH 7.4. Despite the fact that this buffer is operative on a reasonably good portion of its titration curve, however, the concentration of phosphates in plasma is only one-sixth that of the carbonic acid–bicarbonate system. As a consequence, the total buffering capacity of the phosphate system is quite limited; ie, it is also one-sixth that of the bicarbonate system.

THE ISOHYDRIC SHIFT (OR PRINCIPLE).

The foregoing discussion considered each of the blood buffer systems as a separate entity rather than as a unit of an interrelated system whose primary function is to regulate closely the pH of the body fluids. The common chemical denominator among these individual buffer systems is the hydrogen ion (H^+). Whenever the concentration of the hydrogen ion changes, regardless of cause, this effect in turn causes a shift in the balance of all the individual buffer systems at the same time so that the pH is maintained at a constant value. This effect is called the *isohydric principle,* and as a consequence of this principle, the buffer systems exert a mutual buffering action on each other.

The isohydric principle can be stated simply in algebraic form by using three buffer pairs, $[HA_1]/[A_1^-]$, $[HA_2]/[A_2^-]$, and $[HA_3]/[A_3^-]$:

$$H^+ = K_1 \frac{[HA_1]}{[A_1^-]} = K_2 \frac{[HA_2]}{[A_2^-]} = K_3 \frac{[HA_3]}{[A_3^-]}$$

Restated in terms of pH this expression becomes:

$$pH = pK_1 + \log \frac{[A_1^-]}{[HA_1]} = pK_2 + \log \frac{[A_2^-]}{[HA_2]}$$
$$= pK_3 + \log \frac{[A_3^-]}{[HA_3]}$$

It is apparent from inspection of this relationship that any alteration in the hydrogen ion concentration will cause a corresponding shift in the [salt]/[undissociated acid] ratios of the several buffer pairs so that the pH will remain constant, and this is exactly how the isohydric principle operates within the body fluids to maintain blood pH within very narrow limits.

Carbonic anhydrase also is found in high concentrations within renal tubular cells (p. 805), as well as in the gastric parietal cells that elaborate hydrochloric acid (p. 847).

BUFFERING OF THE BODY FLUIDS IN VIVO. The blood buffers are not by any means the only buffer systems in the body. As noted on page 756, the intracellular proteins themselves exert a powerful buffering effect and they, in turn, affect the extracellular fluid pH indirectly. In fact, only 15 to 20 percent of the acid load of the blood is handled by the blood buffers in metabolic acidosis, while an additional 20 to 25 percent is buffered by the carbonic acid–bicarbonate buffer pair of the interstitial fluid. The balance of the excess acid load in turn is buffered by the intracellular proteins as well as intracellular organic phosphate compounds. When the intracellular proteins and organic phosphates bind hydrogen ion, however, they liberate sodium and potassium ions, which then diffuse into the interstitial space, so that during periods of metabolic acidosis the level of these ions rises in the extracellular fluids. In metabolic alkalosis, on the other hand, approximately 30 to 35 percent of the excess hydroxyl ion is buffered within the cells. During respiratory acidosis and alkalosis, almost all the buffering takes place at the intracellular level.

Respiratory Regulation of Blood Acid–Base Balance

The three major buffers of the blood discussed earlier were the plasma proteins, hemoglobin, and the carbonic acid–bicarbonate system. The latter buffer system is unique because carbonic acid in the reaction $CO_2 + H_2O \rightleftharpoons H_2CO_3 \rightleftharpoons H^+ + HCO_3^-$, is converted into carbon dioxide and water, and the carbon dioxide is excreted by the lungs. Any increase in the carbon dioxide concentration within the body fluids reduces their pH toward the acidic side. Conversely, any decrease in the carbon dioxide concentration increases the pH toward the alkaline side. By means of this mechanism, therefore, the respiratory system is capable of altering the pH in either direction.

The equilibrium within this buffer system is described by the Henderson–Hasselbalch relationship:

$$pH = pK_{H_2CO_3} + \log \frac{[HCO_3^-]}{[H_2CO_3]}$$

If the symbol $[H_2CO_3]$ in this equation represents both the concentration of carbonic acid plus the dissolved carbon dioxide, the pK is 6.1, and the bicarbonate/carbonic acid ratio is 20/1 (or 20). In turn, the quantity of dissolved carbon dioxide and carbonic acid that is present in the blood is proportional to the Pco_2. The plasma bicarbonate in turn is equal to the total plasma carbon dioxide minus the dissolved carbon dioxide, the carbonic acid, and the carbamino form of carbon dioxide.

A clinically useful form of the Henderson–Hasselbalch equation has been obtained by deriving certain constants experimentally, and it is written as follows:

$$pH = 6.10 + \log \frac{[\text{Total } CO_2] \, aP_{CO_2}}{aP_{CO_2}}$$

The bicarbonate ion concentration cannot be measured directly, but the arterial blood pH, total carbon dioxide, and Pco_2 can be, so that in this expression if the total carbon dioxide ([total CO_2]) of true plasma is expressed in mM/liter, and the Pco_2 is given in mm Hg, a = 0.0314. In clinical practice, however, nomograms usually are employed to plot the acid–base characteristics of arterial blood.

Since carbon dioxide is being produced in the body as a product of intermediary metabolism, this gas is continually diffusing from the cells into the interstitial fluid and blood. Thence, it is transported to the lungs, excreted by diffusion into the alveoli, and eliminated therefrom by pulmonary ventilation. The removal of a given volume of carbon dioxide from the lungs, however, is not an instantaneous process, but rather one that requires several minutes for completion, as there is a gradual and progressive dilution (or "washout") of carbon dioxide from the lungs by that carbon dioxide that was eliminated more recently.

If the metabolic rate increases, the rate of carbon dioxide formation and its blood level also increase, and conversely, if the metabolic rate falls, then so too does the blood level of carbon dioxide fall. If the rate of pulmonary ventilation increases, however, the rate of carbon dioxide elimination increases, and the converse of this statement also is true. Therefore, by way of summary, the extracellular fluid level of carbon dioxide at any given moment is determined and controlled by an exceedingly delicate balance between the metabolic rate and the rate of pulmonary excretion of this gas. Thus, the relationship between the rate of alveolar ventilation and the carbon dioxide concentration in the body fluids is as follows, if it is assumed that the metabolic rate stays constant: $CO_2 \propto 1/\text{alveolar ventilation}$. Since the blood pH decreases with an increase in carbon dioxide concentration, alterations in alveolar ventilation also change the blood hydrogen ion concentration. These alterations in hydrogen ion concentration in turn may result in respiratory acidosis and respiratory alkalosis, topics which will be discussed below.

At this point it is important to reiterate that not only does the respiratory rate (hence alveolar ventilation) affect the blood hydrogen ion concentration, but also that the hydrogen ion concentration per se in turn directly affects the alveolar ventilation rate by its direct action on the medullary respiratory center (p. 729). There is, consequently, an overall reciprocal effect between blood pH (hence blood hydrogen ion concentration) and alveolar ventilation, such that if the blood pH falls (ie, a relative acidosis develops), the alveolar ventilation rises. The converse of this statement also is true. If the blood pH declines from 7.4 toward 7.0, pulmonary ventilation may be increased up to 5 times above the normal resting value, whereas an increase in pH above 7.4 may reduce pulmonary ventilation to one-half the normal value. The foregoing statements are illustrated in Figure 11.40.

Since the respiratory center can respond directly to alterations in blood pH (ie, changes in blood hydrogen ion concentration), and the alveolar ventilation rate in turn can regulate hydrogen ion concentration, the respiratory system is governed by a typical feedback regulatory mechanism. This feedback mechanism operates as follows. When the blood hydrogen ion concentration falls below normal levels, alveolar ventilation is stimulated by the direct action of the excess hydrogen ions upon the medullary respiratory center. The concentration of carbon dioxide within the blood (and other extracellular fluids) thus is reduced, so that the hydrogen ion concentration falls toward a normal level, thereby eliminating the stimulus to an increased ventilatory rate. Conversely, if the blood pH rises, the respiratory center is inhibited, and carbon dioxide (hence hydrogen ion) levels gradually rise in the extracellular fluids until a normal hydrogen ion level is reached once again. If, however, the blood pH falls or rises markedly, because of some factor extraneous to the respiratory tract, such as occurs in metabolic acidosis or alkalosis, this respiratory feedback mechanism alone cannot return the hydrogen ion concentration to normal values.

Based on the foregoing discussion, one may conclude that the respiratory system itself acts as a physiologic "buffer system," in contrast to the chemical buffer systems discussed earlier. In fact, many such physiologic buffer systems (otherwise known as feedback mechanisms) are present in the body for the automatic regulation of numerous functions within homeostatic tolerances.

Respiratory Acidosis and Alkalosis

Any disturbance of alveolar ventilation that prevents carbon dioxide excretion from proceeding at a rate that is sufficient to satisy the existing bodily needs is termed *respiratory acidosis* (Fig. 11.41A). In this situation, the balance between carbon dioxide production and elimination is faulty, so that the common denominator in respiratory acidosis, of whatever cause, is some impairment of alveolar ventilation.

Respiratory alkalosis, on the other hand, may be defined as alveolar ventilation in excess of that required by the body for the normal excretion of carbon dioxide (see hyperventilation, p. 735), so that the arterial Pco_2 falls to inappropriately low levels (Fig. 11.41A). As is the situation with res-

piratory acidosis, the balance between carbon dioxide production and elimination is impaired; however, in respiratory alkalosis there is an excessive excretion of carbon dioxide, rather than an excessive retention of this gas, as in metabolic acidosis.

A physiologically increased alveolar ventilation rate (eg, such as occurs during exercise) does not fall into the category of hyperventilation, and respiratory alkalosis normally does not occur under such circumstances.

Examination of the Henderson–Hasselbalch equation for the carbonic acid–bicarbonate system reveals that the principal changes in arterial Pco_2 that occur with respiratory derangements are alterations in the ratio $[HCO_3^-]/[H_2CO_3] = 20/1$. Therefore, a reduced alveolar ventilation raises the arterial Pco_2 above 40 mm Hg, and thereby decreases this ratio by increasing the carbonic acid concentration. Since carbonic acid is also in equilibrium with bicarbonate ion, the plasma bicarbonate rises. This results in a shift in the ratio of bicarbonate to carbonic acid, so that a new equilibrium is attained, but at a lower pH, and a respiratory acidosis ensues. This effect is illustrated in Figure 11.41B, in which the pH is plotted against the plasma bicarbonate ion concentration. In a similar fashion, a decline in the arterial Pco_2 below 40 mm Hg evokes a respiratory alkalosis, as also shown in Figure 11.4B.

The changes in pH shown in this illustration take place independently of the operation of any compensatory mechanism; hence, they are termed *uncompensated respiratory acidosis (URAc)*, or *uncompensated respiratory alkalosis (URAlk)*. In either case, compensatory changes are produced in the kidneys, so that the acidosis or alkalosis of the blood is adjusted toward a normal pH. The renal compensatory mechanisms involved will be discussed in Chapter 12, p. 808.

ETIOLOGIC FACTORS RELATED TO RESPIRATORY ACIDOSIS. As stated earlier, an impairment of normal alveolar ventilation is the underlying feature in respiratory acidosis. The rate and depth of respiration that govern alveolar ventilation in turn are regulated by the respiratory centers in the medulla and pons. Consequently, depression of these respiratory centers, regardless of cause, can result in hypoventilation. Hypoxia such as can be caused by anemia, a greatly elevated carbon dioxide concentration, diseases of the central nervous system (eg, trauma, increased cerebrospinal fluid pressure), anesthetics, narcotics, and sedatives all can produce impaired alveolar ventilation by their action on the respiratory centers, and a respiratory acidosis can develop.

Even in the presence of a physiologically normal respiratory center, alveolar ventilation can be seriously reduced if the transmission of impulses to any or all the respiratory muscles is blocked. For example, spinal cord lesions, myas-

A.

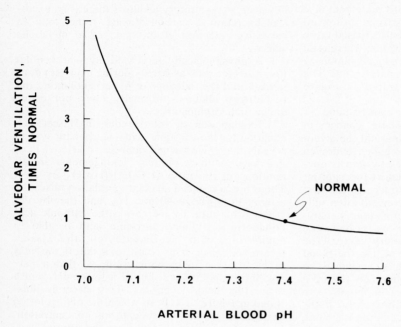

FIG. 11.40.   A. Effect of change in arterial blood pH on alveolar ventilation rate. The dot on the curve signifies a normal ventilation rate (1 on ordinate) at pH 7.40.

B.

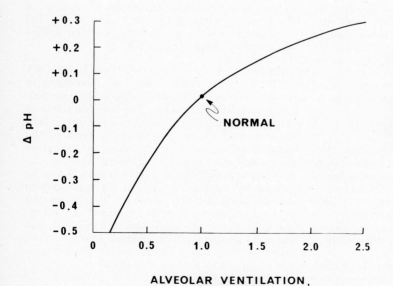

FIG. 11.40.   B. Effect of changes in the rate of alveolar ventilation on the pH of the body fluids. (After Guyton. *Textbook of Medical Physiology*, 4th ed, 1971, Saunders.)

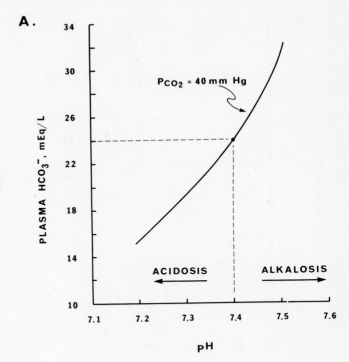

FIG. 11.41.   A. Effect of alterations in plasma bicarbonate upon the pH of the blood at a normal arterial P_{CO_2} of 40 mm Hg.

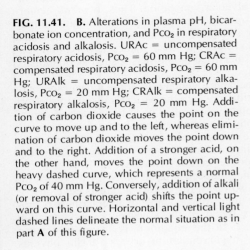

FIG. 11.41.   B. Alterations in plasma pH, bicarbonate ion concentration, and P_{CO_2} in respiratory acidosis and alkalosis. URAc = uncompensated respiratory acidosis, P_{CO_2} = 60 mm Hg; CRAc = compensated respiratory acidosis, P_{CO_2} = 60 mm Hg; URAlk = uncompensated respiratory alkalosis, P_{CO_2} = 20 mm Hg; CRAlk = compensated respiratory alkalosis, P_{CO_2} = 20 mm Hg. Addition of carbon dioxide causes the point on the curve to move up and to the left, whereas elimination of carbon dioxide moves the point down and to the right. Addition of a stronger acid, on the other hand, moves the point down on the heavy dashed curve, which represents a normal P_{CO_2} of 40 mm Hg. Conversely, addition of alkali (or removal of stronger acid) shifts the point upward on this curve. Horizontal and vertical light dashed lines delineate the normal situation as in part **A** of this figure.

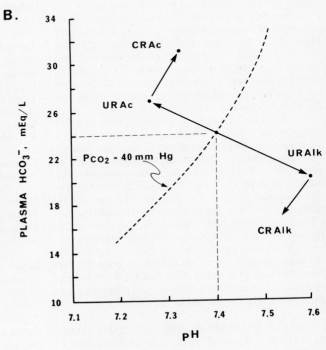

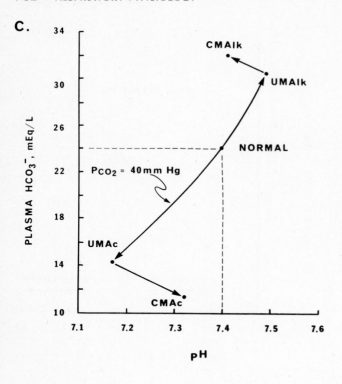

C.

FIG. 11.41. (cont'd) **C.** Alterations in plasma pH, bicarbonate ion concentration, and P_{CO_2} in metabolic acidosis and alkalosis. UMAc = uncompensated metabolic acidosis, P_{CO_2} = 40 mm Hg; CMAc = compensated metabolic acidosis, P_{CO_2} = 21 mm Hg; UMAlk = uncompensated metabolic alkalosis, P_{CO_2} = 40 mm Hg; CMAlk = compensated metabolic alkalosis, P_{CO_2} = 48 mm Hg. See also Chapter 12, p. 808. (After Ganong. *Review of Medical Physiology*, 6th ed, 1973, Lange Medical Publications.)

thenia gravis (p. 229), and peripheral neuritis, in addition to such drugs as curare, succinylcholine, and nerve gases as well as toxins (eg, botulin) all inhibit alveolar ventilation by this mechanism.

Alveolar hypoventilation also can occur in the presence of various diseases that result in impairment of the function of the respiratory muscles. Included in this category are severe obesity, congenital or acquired thoracic abnormalities, and arthritis. Pulmonary movement also can be hindered by intrathoracic effusions (eg, blood or pleural fluid) and pneumothorax.

Pulmonary disease in general forms the last category of clinical situations in which alveolar ventilation is markedly impaired. Chronic congestive heart failure, atelectasis, and tumors are all restrictive diseases in which distensibility (ie, compliance) of the lungs is impaired. Obstructive diseases that impair alveolar ventilation include asthma, chronic bronchitis, and emphysema. Obstruction of an upper airway may also occur, as by a tumor, foreign body, or tracheal stenosis.

Often alveolar hypoventilation is caused by the simultaneous operation of several etiologic factors, such as in a congestive heart failure patient with chronic pulmonary disease who receives drugs therapeutically for his pulmonary edema.

ETIOLOGIC FACTORS RELATED TO RESPIRATORY ALKALOSIS. Alveolar hyperventilation results in a lower-than-normal arterial P_{CO_2}, so that a respiratory alkalosis develops consequent to the elimination of excess quantities of carbon dioxide.

Hyperventilation leading to a respiratory alkalosis may be induced by drugs such as salicylates, 2,4-dinitrophenol, and paraldehyde. Hormones such as progesterone (p. 1160) and epinephrine augment respiration, probably by a direct stimulant action on the central nervous system. Lesions of the central nervous system also can increase alveolar ventilation. Thus, cerebral hemorrhage, trauma, encephalitis, and meningitis all may lead to a respiratory alkalosis. Fever, shock, and hyperthyroidism all may induce hyperventilation, as can interstitial lung diseases such as pulmonary fibrosis.

Mechanical ventilation of the lungs, especially ventilation in which the patient's respiratory effort is completely controlled as to both rate and depth, frequently can result in alveolar hyperventilation.

The chemoreceptor mechanism, although relatively insensitive, also may be activated under physiologic circumstances, such as by the anoxemia (reduced arterial blood P_{O_2}) of high altitudes, so that a respiratory alkalosis develops.

An interesting example of alveolar hyperventilation is found in the so-called hyperventilation syndrome. In this condition, recurrent or chronic alveolar hyperventilation occurs for no obvious reason in patients who are tense and anxious. Such patients often complain of numbness and tingling in the extremities, light-headedness, and spasms of flexion in the wrists. Electroencephalographic (EEG) abnormalities also are seen. The symptoms, at least in part, develop concomitantly with the EEG changes and sometimes are related quantitatively to the rate of drop as well as to the extent in fall of the arterial Pco_2 level.

Metabolic acidosis (see below) normally results in the development of an increased alveolar ventilation and a respiratory alkalosis.

Metabolic Acidosis and Alkalosis

Metabolic acidosis is a disturbance in the extracellular fluid pH resulting from the addition to the blood of noncarbonic acids, which are stronger than the buffer acids. The blood pH becomes lower, causing an uncompensated metabolic acidosis, as shown in Figure 11.41C (UMAc). An excellent example of metabolic acidosis is found in diabetic acidosis (p. 1077), in which large quantities of the nonvolatile keto-acids acetoacetic acid and β-hydroxybutyric acid accumulate in the blood of the patient.

In metabolic acidosis, the addition of fixed acids to the blood, such as in the example given above, results in buffering of the hydrogen ions while the blood levels of hemoglobin, protein, and bicarbonate ions fall. The carbonic acid thus formed is converted into water and carbon dioxide, and the latter is excreted rapidly by the lungs. If a sufficiently large quantity of fixed acid were added to the blood, however, enough to decrease the plasma bicarbonate ion by 50 percent, and the carbonic acid did not form carbon dioxide that was excreted rapidly, the pH would fall to around 6.0, and death would ensue rapidly. But the carbonic acid level *is* regulated so that the blood pH only falls to approximately 7.1. This is the situation that is found in uncompensated metabolic acidosis. In addition to this mechanism, the excess hydrogen ion concentration in the plasma also stimulates respiration, thereby eliminating still more carbonic acid so that the blood level of this compound is actually reduced, and this respiratory compensation tends to elevate the pH still further (CMAc, Fig. 11.41C). Then renal compensatory mechanisms excrete the excess hydrogen ion so that the buffer anion levels return to normal (p. 804).

Metabolic alkalosis, on the other hand, is an abnormal condition in which there is a rise in the blood pH (or a decrease in its hydrogen ion concentration), if it is assumed that no secondary (or compensatory) changes occur (Fig. 11.41C; UMAlk). In this situation, the plasma bicarbonate concentration rises. The respiratory compensation for a metabolic alkalosis is a fall in alveolar ventilation in response to the elevated pH, so that the arterial Pco_2 level is increased (CMAlk). This mechanism operates to lower the blood pH toward normal, while concomitantly elevating the plasma bicarbonate ion still further. The extent to which the respiratory compensation takes place in metabolic alkalosis is limited by the aortic and carotid chemoreceptors which stimulate the respiratory center only if the arterial Po_2 level falls significantly (p. 736). The renal compensation of metabolic alkalosis will be discussed on page 809.

ETIOLOGIC FACTORS RELATED TO METABOLIC ACIDOSIS. Clinically, any factors that increase the production and/or retention of noncarbonic acids can induce metabolic acidosis. The keto-acids found in diabetic acidosis were mentioned earlier. Since acetoacetic acid has a pK of 3.58, and β-hydroxybutyric acid has a pK of 4.39, and these compounds are ionized almost completely in the extracellular fluids, it may be appreciated readily how "flooding" the body (and its buffer systems) with a large quantity of fixed acids can induce a metabolic acidosis. To a quantitatively lesser extent, these keto-acids also are found at elevated levels in the ketosis that accompanies starvation (p. 1212).

Severe hypoxemia (ie, a lowered arterial blood Pco_2) results in excess lactic acid production, and a metabolic acidosis may result. Drugs and chemicals such as paraldehyde, ethylene glycol (antifreeze), methanol, and salicylates, also can cause severe metabolic acidosis if ingested in sufficient quantities.

Severe diarrhea and intestinal fistulas can cause an enormous loss of bicarbonate ion along with the bowel contents, and therefore can be an etiologic factor in producing metabolic acidosis.

In the uremia that accompanies most chronic forms of kidney disease, metabolic acidosis is common, since renal excretion of hydrogen ion as well as ammonia is faulty (p. 804).

ETIOLOGIC FACTORS RELATED TO METABOLIC ALKALOSIS. An increased loss of fixed acid from the body is a relatively common cause of metabolic alkalosis, although a decreased metabolic production of such acid is insignificant clinically as a causative factor in this condition. Thus, severe and prolonged vomiting and external gastric fistulas cause extensive loss of hydrochloric acid from the body, as well as depletion of the extracellular fluid volume, leading to metabolic alkalosis.

Diuretic therapy with drugs such as mercurials or thiazides also can induce metabolic alkalosis. Adrenal steroid therapy or, occasionally, excess adrenal hormone production, which in turn causes excessive sodium and bicarbonate ion to be retained by the kidney, can cause a metabolic alkalosis. Imbalances in potassium ion metabolism, which results in a chronic depletion of this substance, also may produce a metabolic alkalosis. Conjugate bases, such as bicarbonate of soda, when ingested to excess, or when bicarbonate ion is reabsorbed too extensively from the urine by the kidney, can also lead to a metabolic alkalosis.

In patients suffering from chronic pulmonary insufficiency and respiratory acidosis, therapy with a respirator may be sufficient to precipitate an episode of metabolic alkalosis, provided an atendant bicarbonate diuresis is not achieved.

It is evident from the brief listing of some of the etiologic factors that can induce metabolic acidosis and metabolic alkalosis that renal rather than respiratory mechanisms are intimately concerned with these pathophysiologic conditions. Therefore, further discussion of these topics will be deferred until after a consideration of the physiologic mechanisms that are involved in the excretion of various substances by the kidneys (Chap. 12, p. 783).

HYPOXIA

In conclusion to this chapter, a number of situations that result in a reduced or insufficient oxygen supply to the tissues, or *hypoxia*, will be summarized.*

There are three principal types of oxygen deficiency, depending upon the etiologic factor involved. Two or more of these types of hypoxia may be present in an individual simultaneously, and each kind also may be present either as an acute or chronic condition.

Arterial Hypoxia

The condition of arterial hypoxia is characterized by a subnormal oxygen partial pressure in the arterial blood, ie, the arterial P_{O_2} is below 100 mm Hg. Both the oxygen-carrying capacity and the blood flow rate either are normal or elevated.

Four general factors may contribute to a condition of arterial hypoxia, and each of these factors in turn may be caused by a number of additional contributing factors:

1. The inspired air contains a low P_{O_2}:
   (a) Inhalation of an artificial gas mixture containing a lower P_{O_2} than is found in atmospheric air.
   (b) High altitude (p. 1223).
   (c) Rebreathing air in a closed space. In this situation, the arterial P_{CO_2} also is elevated, together with the reduced P_{O_2}.
2. Decreased pulmonary ventilation can occur in a variety of circumstances:
   (a.) Pneumothorax, whether spontaneous or artificially induced, (eg, to collapse a lung during convalescence from tuberculosis). The arterial P_{CO_2} is not affected in this situation.
   (b) Obstruction of an airway, as by a tumor. The arterial P_{CO_2} is elevated.
   (c) Paralysis or weakness of the respiratory muscles with a concomitant deficiency in thoracic movements during respiration. The arterial P_{CO_2} also is elevated.
   (d) Inhibition or depression of the respiratory centers by drugs or other agents. The arterial P_{CO_2} also is elevated.
3. Insufficient Oxygenation in Abnormal Lungs:

   (a) Impaired diffusion of gases across the alveolar air-blood barrier, such as occur in interstitial fibrosis. The arterial blood P_{CO_2} is not affected in this situation.
   (b) Inadequate matching of air in the alveoli with the pulmonary capillary blood. The arterial P_{CO_2} is elevated.
   (c) Constriction of the bronchioles, as in asthma. The arteriolar P_{CO_2} is elevated.
   (d) The alveoli are filled with fluid, as in pneumonia, pulmonary edema, hemorrhage, and in drowning. The arterial P_{CO_2} is elevated.
4. Arteriovenous shunts: Arteriovenous shunts are found in a wide variety of cardiac and/or vascular abnormalities, eg, as inherited defects. Such defects permit the mixing of oxygenated and unoxygenated blood so that a net hypoxia is present.

Anemic Hypoxia

Anemic hypoxia is characterized by a reduced oxygen-carrying capacity of the blood. The arterial P_{O_2} as well as the blood flow rate is normal or even may be elevated (p. 566).

1. Anemias of all types, in which there is a subnormal hemoglobin concentration in the blood which in turn results in a lowered oxygen-carrying capacity and hypoxia.
2. Hemoglobin which is so altered chemically that it cannot combine normally with oxygen; ie, the hemoglobin has been converted into methemoglobin after poisoning with acetanilid, ferricyanides, nitrites, chlorates, or other compounds.
3. Hemoglobin which has combined with a gas other than oxygen, for example, carboxyhemoglobin formed by union with carbon monoxide.

Hypokinetic Hypoxia

Sometimes called "stagnant anoxia," the condition of hypokinetic hypoxia is manifest by a reduced blood flow rate, and it can be either general or localized in the body.

1. General hypokinetic hypoxia is found in such clinical situations as shock (p. 1237), hemorrhage (p. 1242), or congestive heart failure (p. 1244).
2. Localized hypokinetic hypoxia occurs in embolism, thrombosis, or regional vasoconstriction.

Asphyxia and Cyanosis

The term asphyxia (or more commonly suffocation) refers to situations in which a hypoxia (characterized by a lowered arterial P_{O_2}) is combined with a hypercapnia† (ie, an elevated arterial,

---

The term anoxia (meaning, literally, without oxygen) often is used synonymously with hypoxia; however, the latter term is preferable, as it implies oxygen deficiency or lack, rather than the total absence of oxygen.

†The term hypocapnia refers to a reduced arterial P_{CO_2}.

hence also tissue, P_{CO_2}). In a number of the conditions outlined above in which an arterial hypoxia was present, hypercapnia also was present; drowning may be cited as an example of this situation. In some cases, however, such as are encountered at high altitudes, although an arterial hypoxia is present, hypercapnia is not.

The physiologic responses to asphyxia are caused by a reduced arterial P_{O_2} as well as by an elevated P_{CO_2}, the latter being the more important stimulus. In the outlines presented above, a certain degree of asphyxia is present in a number of hypoxic conditions merely because of an impaired excretion of carbon dioxide which is a concomitant of the oxygen deficiency.

Dyspnea (labored or difficult breathing), as well as respiratory stimulation, is a prominent clinical feature in asphyxia. The elevated P_{CO_2} is more important than the lowered P_{O_2} in producing the respiratory stimulation.

Cyanosis may be present in arterial or hypokinetic hypoxias, but not in histotoxic* or anemic anoxias. Thus, cyanosis may appear if asphyxia is sufficiently advanced. Cyanosis is defined as a diffuse bluish or dusky appearance of the skin and mucous membranes which is caused by an excess of reduced hemoglobin in the blood of the superficial capillaries and subcutaneous venous plexuses. The blue color of the skin depends solely upon the absolute quantity of reduced hemoglobin in the capillary blood, and it does not depend upon the relative proportions of reduced hemoglobin and oxyhemoglobin.

Normal blood contains approximately 15 gm hemoglobin per 100 ml. It has been found experimentally that when about 5 gm of reduced hemoglobin per 100 ml is present, cyanosis will appear. When the hemoglobin is fully saturated with oxygen, it contains about 20 volumes percent of this gas, and 0.75 gm hemoglobin combine with 1.0 ml oxygen. Consequently, 5 gm hemoglobin combines with approximately 6.7 ml of oxygen, and 5 gm of reduced hemoglobin is present when the blood contains 13.3 volumes percent of oxygen. As a general rule, cyanosis appears clinically when the capillary blood of the patient reaches a level of oxygen unsaturation that ranges between 6 and 7 volumes percent.

A further symptom of asphyxia is found in the elevated systemic arterial blood pressure that results from stimulation of the medullary vasoconstrictor center by the elevated P_{CO_2}. This situation does not hold, however, in the terminal stages of this condition, so that vasodilatation and a fall in blood pressure develop in the terminal stages of asphyxia.

Chronic (ie, long-term) physiologic adaptations to varying degrees of hypoxia or asphyxia include an elevated erythrocyte count, hematocrit, hemoglobin concentration, and oxygen-carrying capacity of the blood. These responses all result from the reduced P_{O_2}. In addition, there is an increased renal excretion of acid (p. 804) that compensates for the rise in pH which is caused by carbon dioxide retention in asphyxic states. The long-term adaptive mechanisms to chronic hypoxia will be discussed further in Chapter 16, p. 1226, under the heading of high-altitude physiology.

Hypoxia and Blood Oxygen-Carrying Capacity

The blood carrying capacity for oxygen in three particular types of hypoxia now may be discussed and compared.

ARTERIAL HYPOXIA. In arterial hypoxia, both arterial P_{O_2} as well as the blood concentration of oxygen (in volumes percent) are reduced. The blood has a normal (or even elevated) oxygen carrying capacity of 20 volumes percent or greater; however, the P_{O_2} to which it is exposed in the alveoli is too low to effect normal saturation of the hemoglobin, a value that normally is around 97 percent (Fig. 11.28). It is important to realize, however, that this partially saturated blood can unload as much oxygen per unit volume in the capillaries as it could if it were fully saturated, because the tissues are operating at a subnormal oxygen partial pressure. This normal unloading occurs because the pressure gradient for oxygen diffusion from the capillaries to the tissues is still the same even though the arterial blood is not fully saturated; ie, the volume of oxygen carried per unit volume of blood is lower in arterial hypoxia.

For example, 20 volumes percent of oxygen normally is present in arterial blood and 15 volumes percent is present in mixed venous blood. If arterial hypoxia develops because of a reduced P_{O_2} in the atmospheric air, the arterial P_{O_2} may fall to 15 volumes percent. But the quantity of oxygen that is unloaded in the tissues by the capillary blood still would be 5 volumes percent, and thus the mixed venous blood now would contain only 10 volumes percent oxygen. In this instance, the hemoglobin would reach 50 percent unsaturation in the venous blood, compared with 25 percent in the normal situation. In this regard, it is important to reiterate here that the concentration of oxygen in a gas mixture is proportional to its individual partial pressure in that mixture, and vice versa, in accordance with Henry's law (p. 11).

ANEMIC HYPOXIA. In anemic hypoxia, the arterial P_{O_2} is normal, so that the hemoglobin saturation is close to 100 percent; however, the quantity of oxygen transported by the blood is reduced in proportion to the reduction in the concentration of normal hemoglobin in the blood. For example, if anemic blood carries 10 volumes percent oxygen, rather than the normal 20 volumes percent, and it gives up 5 volumes percent of this

*A histotoxic anoxia is caused by metabolic poisons which inhibit respiration at the tissue (metabolic) level. Thus, cyanide produces histotoxic anoxia.

oxygen in the capillaries, 50 percent desaturation of the hemoglobin has been produced, instead of the normal 25 percent desaturation, and the mixed venous blood now will contain only 5 volumes percent oxygen.

HYPOKINETIC HYPOXIA. It is characteristic of hypokinetic hypoxia that both the Po_2 and the concentration of oxygen (in volumes percent) in the arterial blood are normal. The rate of blood flow is reduced, however, far below normal; ie, it is "stagnant" or "hypokinetic." But the tissues continue to use oxygen at a normal rate; the tissue Po_2 falls, thereby causing more oxygen to unload per unit volume of blood that is perfusing the tissues. If the blood flow becomes extremely low while the tissues themselves are metabolically active, practically all the oxygen delivered by the blood may be extracted; the tissue Po_2 in such a situation may approach zero, and irreparable damage thus may ensue if the situation is allowed to exist for any length of time.

CHAPTER 12

Renal Physiology

The cardinal role of the kidneys is to maintain the constancy of the aqueous internal environment within which the cells of the body both live and carry out their individual activities. In contrast with the lungs, whose sole function is to maintain the composition of the blood insofar as oxygen and carbon dioxide are concerned, the kidneys must perform a variety of functions in regulating not only the volume, but also the composition of the extracellular fluid compartment. Thus, the kidneys perform a regulatory as well as an excretory function.

One of the most important renal functions is in the maintenance of water balance within the body, as, after all, this compound is the major constituent of the blood and tissues. In addition, the kidneys must regulate individually and with exquisite precision the bodily levels of a number of individual inorganic ions (electrolytes) as well as the levels of a multitude of organic compounds. For example, sodium, potassium, calcium, magnesium, chloride, sulfate, phosphate, and bicarbonate, as well as glucose and urea levels, all are regulated by the kidneys by virtue of their excretory function. In addition to their excretory function, the kidneys also have a vital role in the regulation of blood acid–base balance through the excretion of acid radicals. These concepts are summarized most aptly in a quotation from H. W. Smith: "It is no exaggeration to say that the composition of the blood is determined not by what the mouth ingests but by what the kidneys keep; they are the master chemists of our internal environment, which, so to speak, they synthesize in reverse."

The view that the kidneys merely "produce urine" is entirely too simplistic, because our terrestrial mode of life itself, both physical and mental, is wholly dependent upon the constancy of our internal environment; and this constancy in turn is achieved throughout the life span of an individual by the functions of the kidneys.

Fundamentally, the kidneys perform their ex-cretory function by producing an ultrafiltrate of blood in which waste products and excess materials are eliminated from the body, while simultaneously cells and plasma proteins are retained. The kidneys must also selectively reabsorb essential substances from this ultrafiltrate of blood in quantitatively adequate amounts to prevent their irrevocable loss from the body. Toward achieving these functional goals, a variety of homeostatic mechanisms operate through the kidneys.

In addition to the paired glandular kidneys, the human urinary system also consists of a pair of tubular excretory ducts, the ureters; a storage organ for urine, the bladder; and a single conduit through which urine is voided to the exterior of the body, the urethra.

The intrinsic mechanisms whereby the kidney performs its basic functions as well as the extrinsic controls by which these functions are regulated will form the bulk of the material to be presented in this chapter.

FUNCTIONAL ANATOMY OF THE KIDNEY AND RELATED URINARY STRUCTURES

General Morphology

The principal gross anatomic features of the kidney are shown in Figure 12.1. Human kidneys are paired glandular organs located retroperitoneally on the posterior wall of the abdominal cavity, and they are situated laterally to the vertebral column. These structures are bean-shaped and approximately 10 to 12 cm long, 5 to 6 cm wide, and 3 to 4 cm thick. The medial border of each kidney contains a deep concavity, the hilus, from which a large excretory duct, the ureter, emerges and passes downward to the urinary bladder. The bladder is located directly behind the pubic bone within the pelvis.

A thin capsule of dense collagenous connective tissue encloses each kidney, and the glandular

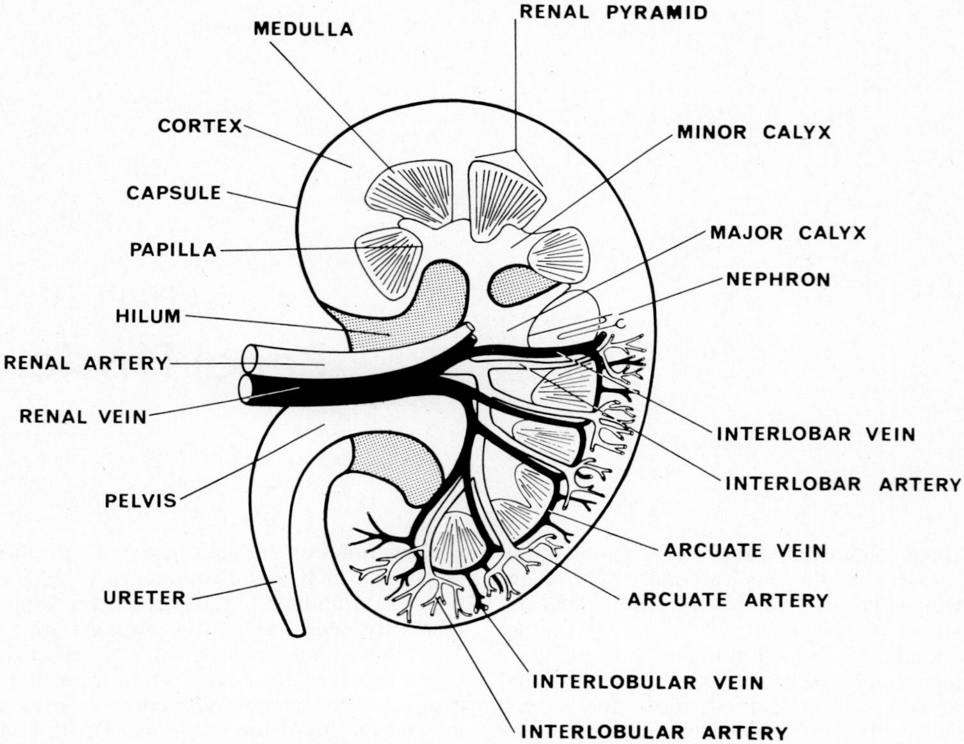

FIG. 12.1. Gross anatomy of the kidney in vertical section. (After Dawson. *Basic Human Anatomy,* 2nd ed, 1974, Appleton-Century-Crofts; Goss [ed]. *Gray's Anatomy of the Human Body,* 28th ed, 1970, Lea & Febiger.)

tissue surrounds a large cavity, the renal sinus, which contains the renal pelvis, a funnel-shaped dilatation of the upper end of the ureter. The renal pelvis in turn subdivides into primary and secondary branches, the major and minor calyces (singular, calyx). The renal artery and vein as well as the renal nerves also pass into the kidney via the renal sinus.

Gross visual inspection of a hemisected kidney reveals a reddish-brown cortex surrounding a lighter-colored grayish medulla. The medulla consists of between 8 and 18 conical renal pyramids; each of the apexes of the pyramids (or papillae) projects into the lumen of a calyx, whereas the bases of the pyramids are directed toward the cortex. The boundaries of each pyramid are delineated by masses of darker cortical tissue which extend inward to form the renal columns (of Bertin). A renal lobe is defined as a renal pyramid together with the cortical tissue that invests it.

The substance of each pyramid grossly exhibits radially striated brown lines that converge toward the apex of the papilla. The straight portions of the uriniferous tubules (see below), together with the blood vessels running parallel with them, make up these striations. At the tip of each renal papilla is the area cribrosa, which is fenestrated, having between 10 and 25 tiny openings. In

this region, the terminal portions of the uriniferous tubules empty into a minor calyx.

Uriniferous Tubules

The tubules comprising the greater portion of the kidney have two major parts: (1) the *nephron,* which corresponds to the secretory portions of other glands, and (2) the *collecting duct* or *tubule,* which transports urine to the renal pelvis (Fig. 12.2).

THE NEPHRON. The structural and functional units of the kidney are the nephrons. Each kidney contains about one million individual nephrons, which are arranged morphologically and function in parallel. Each of these secretory structures can form urine, so that a consideration of renal function in general can be based upon a discussion of the functions of a single nephron.

Upon microscopic examination, the nephron is seen to be composed of six morphologically discrete regions, each of which occupies a distinct location in cortex or medulla. The general anatomy of a nephron is illustrated in Figure 12.2 and the six general regions are so numbered in the figure. The sequential relationship of the morphologic regions within each nephron are as follows: (1) The glomerulus; a tuft of blood vessels is

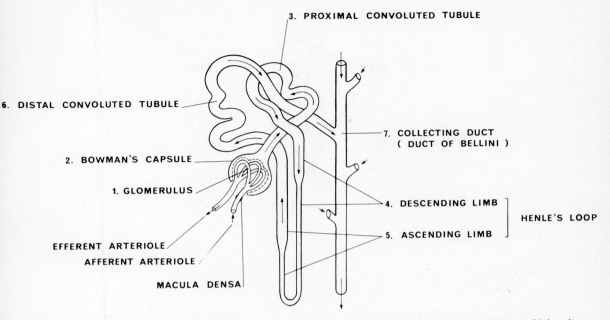

FIG. 12.2. Morphology of the nephron. (After Smith. *Lectures on the Kidney,* 1943, Lawrence, Kansas, University Extension Division, University of Kansas.)

located within, (2), Bowman's capsule. (3) Proximal convoluted tubule. (4) Thick and thin segments of the descending limb of Henle's loop. (5) Thin and thick segments of the ascending limb of Henle's loop. (6) The distal convoluted tubule. This final portion of the nephron in turn is continuous with a collecting duct (7).

The proximal and distal convoluted tubules both are located in the renal cortex near the renal corpuscle. The part of the nephron which is located between the convoluted segments forms Henle's loop as it passes from the cortex into the medulla and back into the cortex (Fig. 12.2). The individual portions of a nephron now may be discussed individually.

Renal Corpuscle. The renal (or Malpighian) corpuscle consists of a double-walled cup of squamous epithelium (Bowman's capsule) which surrounds a capillary tuft consisting of some 50 blood vessels arranged in parallel, the glomerulus. The outer wall of Bowman's capsule is called the parietal layer whereas the visceral layer of this structure closely invests the glomerulus. The narrow space between the two layers of the corpuscle is called Bowman's space.

Blood enters the vascular pole of Bowman's capsule through an afferent arteriole, a branch of the renal artery, which passes through the glomerular capillaries, and exits via an efferent arteriole also situated at the vascular pole.*

This vascular arrangement of two arteries in series with a capillary bed situated between them is unique to the kidney.

The ultrafiltrate of plasma, known as the glomerular filtrate, is produced within the renal corpuscles by blood flowing through the glomeruli; however, the renal blood supply is linked so inseparably to renal function that it will be discussed in detail in the next section.

In the adult kidney, the cells of the visceral layer of Bowman's capsule become so modified that they scarcely resemble epithelial cells, especially when studied in detail by electron microscopy. These cells, called podocytes, send radiating processes which terminate in end-feet (or pedicles), which in turn invest the glomerular capillaries in a complex morphologic arrangement (Fig. 12.3). The adjacent pedicles are separated by slits about 250 Å in width called filtration slits or slit pores. Experiments have shown that smaller protein molecules (eg, below a molecular weight of around 50,000) pass through these slit pores readily, but the openings present a barrier to larger molecules, giving rise to the interpretation that these structures are responsible for the differential permeability to proteins of different molecular sizes and shapes on the part of the glomerular capillaries. Furthermore, this morphologic arrangement of filtration slits situated between the foot processes of the podocytes produces a relatively large surface area for filtration from the glomerular vessels.

The endothelial cells of the glomerular capillaries are quite attenuated and have abundant circular fenestrations, or pores, between 500 and 1,000 Å in diameter. Presumably, these pores function to enhance the permeability of the glomerular capillaries.

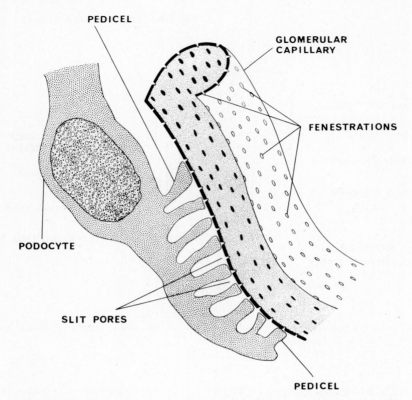

FIG. 12.3. Ultrastructure of a glomerular capillary loop. (After Bloom and Fawcett. *A Textbook of Histology,* 9th ed, 1968, Saunders.)

Proximal Convoluted Tubule. The proximal convoluted tubule is composed of a layer of cuboidal epithelium, which in turn has a brush border that is composed of microvilli. This tubule is continuous with the parietal squamous epithelium of Bowman's capsule at the urinary pole of the capsule. The proximal convoluted tubule is about 60 μ in diameter and approximately 14 mm in length. Its course is quite tortuous, and this part of the nephron comprises the principal bulk of the renal cortex itself.

The proximal convoluted tubule is continuous with the thick portion of the descending limb of Henle's loop (pars recta).

Henle's Loop. Henle's loop consists of the pars recta, the descending and ascending thin limbs, and the ascending thick limb, which is merely the straight region of the distal tubule.

There is an abrupt transition between the straight portion of the proximal renal tubule and the descending thin limb of Henle's loop. Hence, the epithelium suddenly changes from cuboidal to squamous cells, the latter having a thickness between 0.5 and 2 μ. The brush border ends abruptly and the cells of the thin segment appear similar to capillary endothelial cells. The lengths of the limbs in Henle's loop vary from 4.5 mm to over 10 mm.

The thick ascending limb of Henle's loop (or straight portion of the distal tubule) averages about 9 mm in length, and it is composed of cuboidal epithelial cells which lack a brush border. This segment of the uriniferous tubule enters the renal cortex and then passes to the renal corpuscle of its own nephron, where it attaches to the vascular pole of the capsule, primarily to the juxtaglomerular cells of the afferent arteriole. The juxtaglomerular cells in turn are located among the smooth muscle cells in the wall of the afferent arteriole just before they enter the capsule. These cells contain abundant cytoplasmic granules, and are considered to be the source of renin, a potent systemic vasoconstrictor (or pressor) substance.

The epithelium of the straight portion of the distal tubule which lies in contact with the afferent arteriole is thickened into an elliptical, disc-shaped area, measuring about 40 by 70 μ in man and called the macula densa (Fig. 12.2). The macular region lies opposite the afferent arteriolar juxtaglomerular cells, and this morphologic arrangement suggests a functional interaction between the two regions. This supposition is substantiated by the experimental finding that the rate of secretion by the juxtaglomerular cells is related functionally to the enzymatic activity of the cells of the macula densa. Therefore, both of these regions are referred to collectively as the *juxtaglomerular apparatus.* This structure is of considerable interest as it has been implicated in the production of renin, a proteolytic enzyme which after its release acts upon a plasma α_2-globulin, which in turn ultimately becomes a pressor substance called angiotensin. Thus, the juxtaglomerular apparatus of the kidney functions as an endocrine organ, as discussed in Chapter 15 (p. 1169). Furthermore, the kidney also produces another endocrine substance (from an unknown site) which is known as erythropoietin, and this compound stimulates hematopoiesis, as discussed on page 531. Thus, the kidney acts not only as an excretory organ, but also as an endocrine organ.

Distal Convoluted Tubule. The distal convoluted tubule of the nephron is highly tortuous, and generally it forms a loop directed toward the surface of the renal cortex above its own renal corpuscle. In length, the proximal convoluted tubule is around 5 mm, whereas it is between 20 and 50 μ in diameter. No brush border is present, and the cuboidal epithelium is lower than in the proximal tubule.

Collecting Tubules. The connections between the nephrons and the collecting tubules are located in the renal cortex along the medullary rays. The distal tubules are continuous with the collecting tubules, several of which in turn converge on long straight tubules known as the papillary ducts (of Bellini). The papillary ducts, which have a diameter of 100 to 200 μ, open upon the area cribrosa at the base of each papilla.

Histologically, the collecting tubules and papillary ducts are composed of epithelial cells which range progressively from cuboidal to tall columnar elements, and morphologically the cells in the collecting ducts are quite different from the cells of the excretory portion of the nephron.

The total length of a nephron is estimated to range from 3.0 to 3.8 cm, whereas that of the collecting tubules is approximately 2.0 cm.

Excretory Passages for Urine

The conduits, which serve to transport urine from the excretory tissue of the nephron to the outside of the body, consist of the calyces, renal pelvis, ureters, bladder, and urethra. All these structures possess a well-developed smooth muscle layer in their walls; the spontaneous contractions of this muscular coat propel the urine from the calyces within the renal pelvis toward the bladder.

CALYCES, PELVIS, URETERS, AND BLADDER. All these structures have in common a similar morphologic pattern; however, the walls become progressively thicker in the sequence from the calyces to the bladder. A diagram illustrating the principal morphologic features of the ureter in cross section will exemplify the accessory urinary structures (Fig. 12.4).

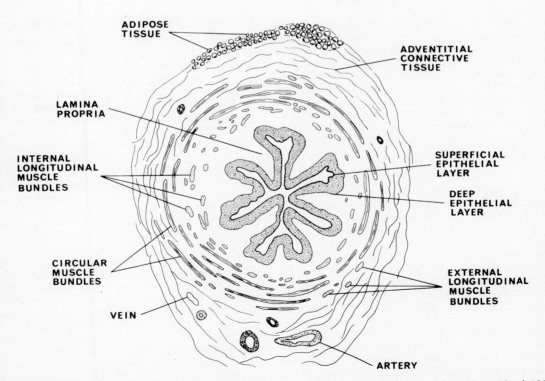

FIG. 12.4. Cross-section of ureter in the contracted state. (After Bloom and Fawcett. *A Textbook of Histology,* 9th ed, 1968, Saunders.)

The inner surfaces of calyces, renal pelvis, ureters, and bladder are all lined with a mucous membrane of transitional epithelium which normally is impermeable to the urinary constituents. No submucosa is present, so that the lamina propria attaches directly to the smooth muscle layer. The latter is covered by adventitial connective tissue. No true glands are present in these accessory urinary structures. The smooth muscle fibers in the walls of the urinary passageways are loosely oriented so that there is an inner longitudinal layer and an outer circular layer. At the lower end of the ureters, a third outermost longitudinal layer of smooth muscle fibers is present, and this is especially well developed in the bladder. In contrast to the digestive tract (qv, p. 817), the layers of muscle within the walls of the urinary passages are poorly defined, so that the anastomosing fibers are present as a loose meshwork, separated by abundant connective tissue and elastic networks.

The ureters enter the bladder wall at an oblique angle; consequently, the pressure developed by the urine within the bladder is sufficient to keep their orifices closed and prevent retrograde flow. A fold of bladder mucous membrane also occludes the openings and facilitates this process.

The urinary bladder possesses a strong and heavy muscular coat whose layers are not sharply defined. In the vicinity of the trigone, smooth muscle fibers in dense bundles are oriented in a circular pattern around the internal urethral orifice, thereby forming the internal sphincter of the bladder.

The accessory urinary structures are provided with abundant blood vessels, lymphatics, and nerve plexuses. In particular, the bladder is provided with a sympathetic nerve plexus (plexus vesicalis) which is located in the adventitial layer. This neural apparatus is formed partly by branches of the hypogastric nerves, and partly by the pelvic nerves which originate from the sacral outflow. The mucous membrane is provided with abundant free endings of sensory (afferent) nerve fibers.

URETHRA

Male. The male urethra serves two functions: (1) it is the terminal portion of the urinary tract and (2) it conveys semen to the exterior of the body (p. 1131). In an adult male, the urethra is between 18 and 20 cm in length and it is divided into three segments: (1) The *pars prostatica* is the short proximal urethral segment which is surrounded by the prostate. The posterior wall in this region forms an elevation known as the colliculus seminalis; in the midline of this structure, the utriculus prostaticus opens. Situated on either side of the colliculus seminalis are the two openings of the ejaculatory ducts as well as the abundant openings of the prostatic gland per se. (2) The *pars membranacea* is another short (18 mm) segment of the urethra which extends distally from the lower end of the prostate to the bulk of the corpus spongiosum penis. (3) The *pars spongiosa* (or *pars cavernosa*) is approximately 15 cm long and it passes through the corpus spongiosum of the penis.

Transitional epithelium is found in the prostatic portion of the urethra, whereas the pars membranacea and spongiosa both are lined by stratified or pseudostratified epithelia. Occasional mucous (goblet) cells are found among the epithelial cells.

Female. The adult female urethra is about 25 to 30 mm in length, and its mucous membrane is composed of stratified squamous epithelium with occasional areas of pseudostratified epithelium. Clear mucous cells are present.

The mucous membrane is surrounded by an abundant coat of smooth muscle fibers whose inner layer exhibits a longitudinal pattern, the outer a circular arrangement. In its distal portion, a sphincter of striated muscle is present which reinforces the action of the smooth muscle insofar as the voluntary retention of urine is concerned.

RENAL CIRCULATION AND BLOOD FLOW

A clear understanding of renal function is impossible without a basic knowledge of the morphologic arrangement of the vascular bed of the kidney as well as the functional pattern of blood flow through this organ. This section will deal with these two subjects prior to a discussion of the mechanisms involved in the various renal functions.

Morphologic Aspects of Renal Blood Supply

The gross anatomic features of the blood supply to the kidney are shown in Figure 12.1. The renal arteries, which are large branches of the abdominal aorta, enter the hilus of each kidney and there divide into two major groups of branches which pass toward the dorsal and ventral regions of the kidneys. Within the adipose tissue surrounding the pelvis, these branches divide into smaller interlobar arteries which enter the kidney and run toward the periphery in the columns between the lobes (or pyramids) of the kidney. At the base of the pyramids, the interlobar arteries bend over to run parallel with the surface of the organ at the junction between cortex and medulla. These vessels now are called arcuate arteries. At regular intervals, still smaller interlobular arteries arise from the arcuates, and course in a radial direction toward the surface of the organ. In turn, the interlobular arteries give rise to the afferent arterioles (Fig. 12.2), which then form the tufts of glomerular capillaries. The glomeruli then anastomose, forming the unique efferent arterioles.

In the outermost region of the cortex, the efferent arterioles have a small diameter and they branch repeatedly, forming the intertubular (or peritubular) capillary network of the cortex, as shown in Figure 12.5A. On the other hand, the efferent arterioles of deeper-lying glomeruli (juxtamedullary glomeruli) are of large caliber and pass into the medulla where they branch into bundles of thin-walled vessels somewhat larger than ordinary capillaries. These vessels are called vasa recta (or "straight vessels"), as depicted in Figure 12.5B. Efferent vessels of both the juxtamedullary glomeruli as well as the vasa recta provide branches to an intertubular (or peritubular) capillary network, which is located in the renal medulla. The important physiologic consequences of the peritubular capillary network of the cortical nephrons as well as the vasa recta of the juxtamedullary nephrons will be discussed subsequently.

The individual groups of vasa recta undergo sharp turns at various levels within the medulla so that they loop back toward the cortex from different levels, running parallel with, and close to, the vessels from which they originate. The vessels that comprise the descending arterial limb of these vascular bundles form extensive networks, or retes, and are somewhat smaller than the returning vessels which comprise the venous portion of the network. The arterial (descending) portion of the rete is composed of a continuous endothelium, whereas the venous (ascending) portion of the vascular bundle has a very thin fenestrated endothelium.

The morphologic arrangement whereby afferent and efferent limbs of the retes are in very close proximity to the tubules of the nephrons, combined with their large net surface area, promotes the rapid and efficient countercurrent exchange of diffusible substances between the vascular bed and the renal tubules and vice versa.

Within the outer layers of the renal cortex, the capillaries drain toward the surface by radial branches of the superficial cortical veins. These vessels in turn join the stellate veins on the kidney surface. The stellate veins then anastomose with interlobular veins which drain into the arcuate veins, and the latter accompany the arcuate arteries (Fig. 12.1).

Within the deeper parts of the renal cortex, the efferent venous vessels of the retes empty into the radially situated deep cortical veins; these abundant vessels course parallel with an equivalent number of interlobular arteries. Blood in the deep cortical veins drains inward to the arcuate veins, thence to the interlobar veins. The interlobar veins ultimately converge in the hilus forming the renal vein, which empties directly into the inferior vena cava.

The important distinctions in the pattern of blood flow between cortical and juxtamedullary nephrons which were noted above are summarized below:

Comparison of Blood Flow Patterns Through Cortical and Juxtaglomerular Nephrons (see Fig. 12.5):

1. *Cortical Nephron:* Interlobar artery, arcuate artery, interlobular artery, afferent arteriole, glomerulus, efferent arteriole, peritubular capillary network, venule, interlobular vein, arcuate vein, interlobar vein

2. *Juxtaglomerular Nephron:* Interlobar artery, arcuate artery, interlobular artery, afferent arteriole, glomerulus, efferent arteriole:

Path 1: peritubular capillary network ⎱ venule, arcuate vein,
Path 2: vasa recta ⎰ interlobar vein

The total surface area of the renal capillaries in man is about 12 m². This area is approximately equal to the entire surface area of the renal tubules, which also is around 12 m². The total volume of blood within the renal capillaries at any given moment is between 30 and 40 ml.

Functional Aspects of Renal Blood Flow

The total blood flow through both kidneys of a resting adult weighing 70 kg amounts to between 1.1 and 1.3 liters per minute. This figure is approximately 22 to 25 percent of the entire cardiac output (Table 10.6, p. 635). The percentage of the total cardiac output which passes through the kidneys per minute is known as the renal fraction. Among normal individuals, the value for total renal blood flow may vary by as much as ±0.5 liter, so that the renal fraction may range between 10 and 30 percent of the total cardiac output.

One may surmise, from the complex patterns of vascular elements within the various regions of the kidney already discussed, that the regional blood flow similarly may exhibit considerable differences. This is indeed the case, as animal (unanesthetized dog) experiments have clearly indicated. For example, when the blood flow rate through the renal cortex was about 470 ml per 100 gm of tissue per minute (ie, approximately 76 percent of the total renal blood flow), the outer medullary flow at the same time was about 130 ml per 100 gm of tissue per minute (21.5 percent) and the inner medullary flow was only 15 ml per 100 gm of tissue per minute (2.5 percent). The physiologic significance of these profound differences in blood flow will become apparent in the discussion of countercurrent mechanisms in the kidney.

The figures cited above indicate a normally rapid cortical flow; however, intense sympathetic stimulation can reduce this value to nearly zero through vasoconstriction.

MEASUREMENT OF RENAL BLOOD FLOW. In experimental situations, renal blood flow through one or both kidneys may be measured directly by means of flow meters of various types (eg, Doppler or electromagnetic). In the human subject, however, the Fick principle is more useful for technical reasons (p. 640). In applying the Fick principle to the kidneys, instead of measuring the amount of a substance taken up per unit time by a tissue and dividing this value by the arteriovenous concentration difference for the substance, the renal plasma flow is determined by measuring the quantity of a substance which is excreted from the plasma per unit time into the urine, and this value is divided by the arteriovenous blood difference in

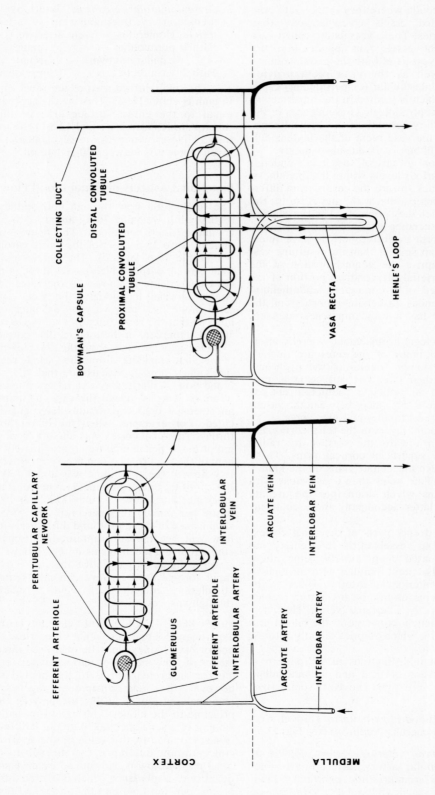

FIG. 12.5. The vascular supply to a cortical nephron (**A**) and a juxtaglomerular nephron (**B**).

the concentration of the substance determined at the same time.

This application of the Fick principle to the measurement of renal plasma flow is valid provided that the quantity (or concentration) of the material used as an indicator does not change within the erythrocytes as they traverse the vascular bed of the kidney. Furthermore, the indicator substance must be readily measurable in arterial and venous renal plasma. It must not be metabolized, stored, or synthesized by the kidney, nor can it of itself produce alterations in renal blood flow. Obviously, the substance also must be nontoxic. The renal blood flow in turn is obtained from the renal plasma flow by dividing by 1 minus the hematocrit reading.

Two substances are used commonly in the determination of renal plasma flow in human subjects. These compounds are para-aminohippuric acid (PAH) and iodopyracet (Diodrast). Both of these materials have a high extraction ratio; that is, the arterial concentration minus the renal venous concentration divided by the arterial concentration, ie, the A–V difference/arterial concentration. This value can be expressed as a percent. In low concentrations, the extraction ratio is over 90 percent for PAH, for example, since it is excreted both by glomerular filtration (p. 783), as well as by tubular secretion. Since over 90 percent of the plasma PAH is cleared (or removed) from the arterial blood during one passage of the blood through the kidney, it is common practice to calculate the effective renal plasma flow *(ERPF)* by dividing the urinary PAH level by the plasma PAH level. The PAH level in the renal venous plasma is thus ignored, leading to a small error. (The term *ERPF* itself signifies that the renal venous PAH concentration was not taken into account.) Expressed mathematically this relationship becomes:

$$ERPF = C_{PAH} = \frac{U_{PAH} - V}{P_{PAH}}$$

In this equation, *ERPF* = effective renal plasma flow in ml/minute; C_{PAH} = clearance of para-amino-hippuric acid in ml/minute; U_{PAH} = urinary PAH concentration in mg/ml; V = urinary flow in ml/minute, P_{PAH} = plasma level of PAH. In the above expression, it should be noted that *ERPF* is equal to clearance, an important concept in renal physiology which will be discussed further on page 783. For the present, however, it should be stressed that clearance is a flow rate, and *not* a quantity.

In man, the *ERPF* averages around 625 ml per minute. The actual renal plasma flow can be obtained from the effective renal plasma flow by dividing the *ERPF* by the extraction ratio for the indicator used. In the case of PAH, the extraction ratio is 0.9 so that the actual renal plasma flow is 625/0.9 = 700 ml per minute.

RENAL HEMODYNAMICS. The total renal blood flow to both kidneys of around 1,100 ml per minute can be considered to perfuse the nephrons in a normal individual. Some investigators, however, believe that a small fraction of this total volume (about 3 percent) passes through arteriovenous shunts (Trueta shunts), which afford a direct connection between the afferent arterioles and the venules, thereby bypassing the glomeruli and the capillary networks surrounding the tubules.

As noted earlier, blood flow through the various regions within the kidney is quite variable, the medulla receiving a much lower proportion of the entire renal blood flow than does the cortex. Consequently, the vasa recta only receive about 2 percent of the renal blood flow; therefore, circulation in this part of the kidney is quite slow in comparison with that of the cortex.

As described earlier in this chapter, the vascular supply to the nephron is unique. It is composed of two distinct capillary networks (or plexuses), the glomerulus and the peritubular capillaries, which are separated by the efferent arteriole.

The two principal regions that afford considerable resistance to blood flow through the renal vascular bed are the afferent and efferent arterioles (Fig. 12.2). Because of the high outflow resistance on the efferent arteriole, the hydrostatic pressure developed within the glomerular capillaries is quite high. This is an extremely difficult pressure to measure directly because of the size and location of the vessels involved, so that it is variously estimated to range between 50 and 90 mm Hg, compared to a mean systemic arterial blood pressure of 100 mm Hg in the afferent arteriole. A value of around 70 mm Hg for glomerular capillary pressure appears to be a reasonable figure; but regardless of its absolute magnitude, the important fact is that the glomerular capillary pressure is quite high. As blood passes from the efferent arteriole into the peritubular capillary bed (the low-pressure region of the system), the pressure falls still further to a level of around 15 mm Hg, as measured directly by micropuncture techniques. These pressure relationships in the vascular bed of the nephron are shown in Figure 12.6.

Each kidney is encased tightly within a tough fibrous sheath of connective tissue, the capsule, which normally imposes a strict limitation on increment volume changes by these organs, such as are occasioned by altered blood flow. The total tissue pressure (or intrarenal pressure) of the kidney has been estimated at about 10 mm Hg above atmospheric; this elevated pressure is similar to that found within the proximal tubule. The renal interstitial fluid pressure, on the other hand, has been estimated to be approximately equivalent to atmospheric pressure.

Should the total renal volume rise, because of edema or other cause, the capsule limits the degree of possible expansion of the entire kidney, so that the intrarenal pressure increases significantly. This augmented tissue pressure in turn decreases the glomerular filtration rate, and can lead to a suppression of urinary formation known as anuria.

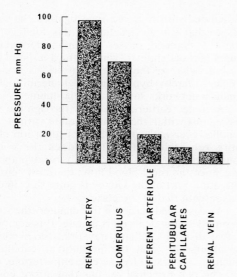

FIG. 12.6. Hydrostatic pressures at various points in the nephron. (Data from Ganong. *Review of Medical Physiology,* 6th ed, 1973, Lange Medical Publications; Guyton. *Textbook of Medical Physiology,* 4th ed, 1971, Saunders.)

REGULATION OF RENAL BLOOD FLOW. The intrinsic neural, chemical, and miscellaneous factors that appear to influence normal renal blood flow in vivo will now be considered in a sequence of their probable relative significance.

Autoregulation of Renal Blood Flow. The concept of an intrinsic autoregulatory mechanism within various organs of the body which maintains a relatively constant blood flow in the face of widely varying arterial perfusion pressures was presented on page 666. In the kidney, the autoregulatory mechanism appears to be quite highly developed. As shown in Figure 12.7, the arterial perfusion pressures to a kidney can be varied considerably (eg, in dogs, between about 90 to over 200 mm

Hg), yet the blood flow in ml/gm per gram of tissue per minute remains constant. In accordance with Poiseuille's law, the flow is directly proportional to the pressure, but inversely related to the resistance $(Q = \Delta P/R)$, so that as the pressure increases, the vascular resistance also must increase simultaneously in order that the flow remain constant over the pressure range indicated above. An increased resistance indicates that vasoconstriction has taken place, all other factors remaining constant. Experimental evidence has implicated the afferent arterioles as the principal regions affected. Consequently, glomerular and peritubular capillary pressures fall, and the net flow remains constant.

The detailed mechanisms involved in the phenomenon of renal autoregulation of blood flow remain obscure. It is clear, however, that a completely denervated kidney can autoregulate its blood flow normally; hence, neural components do not appear to play a significant role in this effect. The increased pressure alone is known to act as a direct mechanical stimulus on the vascular smooth muscle, thereby producing vasoconstriction. In addition to the clearly demonstrated direct mechanical effect of stretch in producing vasoconstriction, chemical feedback mechanisms operating through the renin mechanism of the juxtaglomerular apparatus (p. 1169), as well as the osmotic and/or the sodium ion concentration mechanisms, have been postulated to have a role in the autoregulation of renal blood flow. Quite possibly these additional mechanisms play a secondary role to mechanical autoregulation and thus maintain renal blood flow under special circumstances. This statement is, however, entirely conjectural.

In summary, an inherent (or intrinsic) mechanism within the kidney plays the most significant role in maintenance of a constant renal blood flow (20 to 25 percent of the cardiac output),

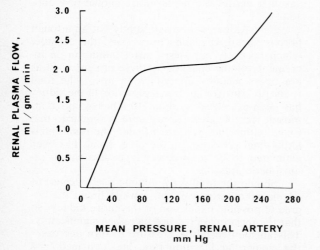

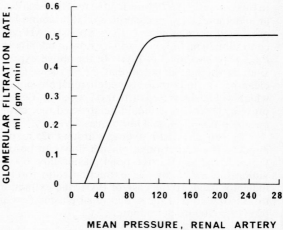

FIG. 12.7. Renal autoregulation of blood flow. (After Ganong. *Review of Medical Physiology,* 6th ed, 1973, Lange Medical Publications.)

despite wide fluctuations in systemic arterial pressure. Direct mechanical stimulation of the vascular smooth muscle, primarily of the afferent arteriole, results in vasoconstriction and an increased resistance. This factor possibly is aided by secondary chemical influences which operate through localized feedback mechanisms within the individual nephrons.

Neural Regulation of Renal Blood Flow. The kidneys are innervated by the renal nerves, which enter the hilus together with the blood vessels. Many sympathetic efferent fibers are present in these nerves, as well as afferent fibers, which may transmit impulses related to pain. It has been demonstrated that some of the afferent fibers from the kidney apparently have a baroreceptor function. Other functions of these visceral afferents are unknown.

The sympathetic fibers to the kidney are predominantly vasoconstrictor, and these fibers are distributed primarily to the afferent and efferent arterioles, although nerve terminals of obscure function also are found on tubular and juxtaglomerular cells.

Stimulation of the anterior region of the temporal lobe of the cerebral cortex, the medullary vasomotor center, and portions of the brain stem, as well as the renal nerves themselves, all produce renal vasoconstriction. In man, there appears to be no fundamental tonic discharge in the renal nerves. When the systemic arterial blood pressure falls, however, the generalized vasoconstrictor response induced by a reduced discharge in the pressoreceptor endings in the aortic arch and carotid bodies also includes renal vasoconstriction.

Hypoxia induces renal vasoconstriction; however, this effect only takes place when the arterial oxygen concentration falls below 50 percent of normal. This response is mediated by the aortic and carotid chemoreceptors, which in turn stimulate the medullary vasomotor center.

Experimental evidence suggests that when the renal nerves are stimulated, salt excretion is diminished by a mechanism that is independent of the vasoconstriction of the afferent arterioles produced by stimulation of these nerves; thus, the effect is mediated by a change in glomerular filtration itself. This effect could be direct (ie, on the tubular cells themselves) or else secondary to alterations in renal medullary blood flow. Another possibility is that augmented neural activity could stimulate the juxtaglomerular cells, so that vasoconstriction is the consequence of increased renin secretion.

It must be reemphasized that the kidney possesses a considerable degree of autonomy insofar as regulation of its own blood flow is concerned, and that renal denervation in man is not followed by any impairment of normal renal function.

Chemical and Miscellaneous Factors That Affect Renal Blood Flow. A number of chemical as well as physiologic stimuli can produce vaso-constriction or vasodilatation of the vascular bed within the kidney.

Renal vasoconstriction is produced by the catecholamines epinephrine and norepinephrine. These compounds in low doses act primarily upon the afferent arterioles, hence the glomerular pressure; the filtration rate is maintained, despite the fact that total renal blood flow is decreased. Larger doses of these amines reduce the glomerular filtration rate. Epinephrine also causes vasoconstriction of the renal veins, thereby increasing the total intrarenal pressure. Barbiturates and ether, as well as certain other anesthetics, also produce renal vasoconstriction.

Renal vasodilatation is mediated not by nerves but by a number of other agents. For example, a high-protein diet, bacterial pyrogens, and certain drugs all increase renal blood flow. As an example of the latter, hydralazine (Apresoline, a sympatholytic compound used in the treatment of hypertension) is unique, as it lowers systemic blood pressure through its direct vasodilator effect upon the systemic vessels while at the same time it increases renal blood blow.

RENAL PHYSIOLOGY

During the production of urine, the kidneys do not synthesize new material to any appreciable extent. Rather, they eliminate as waste excess water and electrolytes in addition to metabolic by-products, particularly nitrogenous compounds, which are produced elsewhere in the body. Foreign materials are treated similarly, since they are also transported to the kidneys in solution in the blood. These are the excretory functions of the kidney, and they are carried out in conjunction with the equally important conservative functions of this organ, whereby water, electrolytes, and other materials are retained within the body in quantities that are sufficient to maintain a delicate and critical homeostatic balance among the individual constituents of the blood. In addition, the kidneys play a vital role in the regulation of blood pH, both normally and under certain abnormal conditions. These topics will be discussed individually in the sections to follow. It must be emphasized that a variety of physical as well as biologic mechanisms are employed by the kidney, either singly or in combination, in order to achieve the several physiologic goals outlined above. These mechanisms are: filtration, passive diffusion (or passive transport, including osmosis), active secretion (or active transport) and selective reabsorption, countercurrent exchange, and pinocytosis.

The general functions of the several specific regions in the nephron will be considered first. This material will be followed by a discussion of how the quantitative measurements of these functions are obtained clinically. The mechanisms whereby specific materials are handled by the normal kidney as well as the role of this organ in the regulation of blood acid–base balance will be presented last.

General Functions of the Nephron

The basic physiologic role of the nephron is to remove or clear certain substances from the blood plasma as it passes through the specialized vascular bed surrounding the nephron. In particular, excess water, metabolic end-products (eg, urea, uric acid, creatinine, sulfates, and phenols), as well as excess electrolytes (eg, sodium, potassium, and chloride) must be removed from the body.

The basic mechanisms whereby these processes are accomplished are as follows: (1) A large proportion of the total fluid of the plasma passing through the glomerulus is filtered through the glomerular membrane and the visceral layer of the glomerular epithelium into the capsular space (Bowman's space) of the nephron. (2) As this ultrafiltrate of plasma* passes through the tubules, the unnecessary materials are not reabsorbed, whereas substances required by the body, especially water, glucose, and many electrolytes, are reabsorbed selectively into the plasma of the peritubular capillaries. Thus, urine essentially represents an aqueous solution of various materials that were rejected by the tubules of the nephron.

The magnitude of these processes can be appreciated from the following example. Approximately 20 percent of the total volume of plasma fluid passing through the glomeruli of both kidneys is filtered. Therefore, the glomerular filtration rate (GFR) in an average man amounts to some 125 ml per minute; and this in turn is equivalent to 7.5 liters per hour, or 180 liters per day! Since the total urine output amounts to only 1 liter per day on the average, more than 99 percent of the glomerular filtrate is reabsorbed by the tubules. And at the filtration rate cited above (125 ml per minute), the kidneys filter in 1 day a quantity of liquid which is equivalent to 4 times the total body water, 15 times the entire extracellular fluid volume, and 60 times the entire plasma volume! Filtration rates of women average about 10 percent lower than those of men.

Certain individual physiologic attributes of different specific regions of the nephron now may be discussed individually.

GLOMERULUS. Three principal factors govern filtration across the glomerular membrane. These factors are permeability, hydrostatic pressure, and osmotic pressure.

Permeability. The glomerular membrane acts as though it contained pores that are approximately 75 to 100 Å in diameter, and thus these vessels exhibit a permeability that is around 100 times greater than that of the capillaries in skeletal muscle. Hence, compounds having a molecular weight up to around 70,000 daltons (depending upon the

*A plasma ultrafiltrate contains all of the elements of plasma fluid except protein molecules or cells; ie, it contains salts, sugars, amino acids, glucose, and so forth in concentrations that are similar to those of plasma (see Table 9.3, p. 529).

shape of the molecules) are found in the glomerular filtrate. Some low-molecular-weight proteins, polypeptides, electrolytes, and nonelectrolytes can pass readily into the glomerular filtrate. Globulins, which generally have molecular weights above 90,000 daltons, are excluded from the glomerular filtrate.

There is evidence that some plasma albumin (molecular weight 69,000 daltons) is filtered through the glomeruli to a limited extent, but this protein normally is reabsorbed in the proximal convoluted tubule so that none appears in the urine.

When extensive hemolysis occurs in the body (as, for example, in hemolytic anemia) so that hemoglobin (molecular weight 64,500) is present in the plasma at a high concentration, this protein appears in the glomerular filtrate and also may be found in the urine. The "blackwater fever" of malaria provides an excellent example of the passage of hemoglobin from blood to urine following extensive intravascular hemolysis.

Since the concentration of low-molecular-weight proteins and polypeptides in plasma normally is quite low, only minute traces of such compounds are found in the glomerular filtrate.

In many renal diseases, glomerular capillary permeability is increased markedly so that quantities of albumin and other proteins now are found in the urine and proteinuria results.

Hydrostatic and Osmotic Pressures within the Glomeruli. In addition to permeability, the pressure gradients developed across the glomerular membrane are of considerable importance in governing filtration in the glomeruli. As emphasized previously, the intraglomerular hydrostatic pressure is considerably higher (around 70 mm Hg) than in more conventional capillaries. This is because of the short and direct connection of the afferent arterioles to the interlobular arteries as well as the relatively high resistance imposed by the efferent arterioles. The hydrostatic pressure within the glomerular capillaries is opposed by an intracapsular hydrostatic pressure of around 15 mm Hg as well as by an oncotic pressure developed by the plasma proteins of around 25 mm Hg. The hydrostatic or filtration pressure within the glomeruli thus is opposed by an osmotic pressure of 25 mm in addition to a hydrostatic pressure of 15 mm Hg in Bowman's capsule. Therefore, the net pressure opposing filtration is 40 mm Hg, and the net filtration pressure within the glomeruli is 70 $-40 = 30$ mm Hg when the mean systemic arterial pressure is normal (100 mm Hg).

Since plasma proteins normally cannot diffuse across the barrier formed by the glomerular membrane in significant quantities, a Donnan effect develops on the monovalent diffusible ions (p. 25). Hence, these ions become redistributed so that the concentration of anions becomes approximately 5 percent greater in the glomerular filtrate than in the glomerular plasma, whereas the concentration of monovalent cations becomes about 5 percent lower in the filtrate than in the plasma. From a general physiologic standpoint, this effect is negligible so that the composition of the glomerular filtrate is quite similar to that of the plasma except for the presence of proteins and cells.

The total glomerular filtration rate at any time is the resultant of the three factors discussed above, because in man, unlike in some lower organisms (eg, frogs), all the nephrons in both kidneys are functional continuously rather than in an intermittent fashion.

Alterations in the glomerular filtration rate may be produced by a number of factors. Note especially in Figure 12.7 that over the normal range of blood pressures within which renal blood flow is autoregulated (90 to around 200 mm Hg), the GFR is also autoregulated. Hence, changes in renal vascular resistance also tend to stabilize the GFR. If, however, the mean systemic arterial pressure falls below 90 mm Hg, there also is a concomitant sudden fall in the GFR, which is accentuated by a baroreceptor-mediated reflex constriction of the renal vascular bed.

If the efferent arterioles are constricted to a greater extent than are the afferent arterioles, the GFR increases as the consequence of an increased glomerular hydrostatic (filtration) pressure. The opposite situation also obtains, as the GFR falls when the efferent arterioles dilate. It should be noted that in both of these instances the tubular blood flow is reduced.

Additional factors that can alter the glomerular filtration rate in a predictable fashion can be summarized: (1) changes in systemic arterial blood pressure; (2) alterations in the hydrostatic pressure within Bowman's capsule (as by ureteral obstruction or renal edema); (3) changes in the oncotic pressure of the plasma proteins; (4) increased permeability of the glomeruli (such as is caused by various diseases); and (5) a decrease in the total area of the glomeruli available for filtration (as by total or subtotal nephrectomy, or by diseases that destroy the glomeruli with or without an attendant destruction of the tubules). All these factors can, and do, alter the GFR.

PROXIMAL CONVOLUTED TUBULE. A number of substances are removed from the fluid within this portion of the nephron by an active transport mechanism. The operation of such a mechanism in turn requires an expenditure of energy in order to move these solutes against their concentration gradients and back into the plasma. Since the fluid within this segment of the nephron is isosmotic with respect to plasma, however, water moves passively out of the tubule and along the osmotic gradients that develop when the solutes are actively transported out of the proximal tubule. Thus, isotonicity of tubular fluid with respect to the plasma is maintained at all times.

It may be calculated that by the time the fluid has reached the end of the proximal convoluted tubule 75 percent of the solutes in the glomerular filtrate and a like quantity of the filtered water have been removed from the filtrate and returned to the plasma.

The reabsorption of such plasma proteins as are present in the glomerular filtrate takes place in the proximal convoluted tubule. The mechanism whereby this occurs is believed to be pinocytosis, a process in which the hydrated albumin molecules are actively ingested by the tubular cells and transported in vesicles across the membrane (p. 94).

Within the cells of the renal tubules, in common with other cells which possess active transport mechanisms, there is a definite maximal rate or transport maximum *(Tm)* at which each system can transport a specific solute. The quantity of a given solute that is transported is proportional to its concentration up to a maximum level, at which point the mechanism becomes saturated. An increase in the concentration of the solute beyond the *Tm* results in no significant increase in the quantity of solute transported. In the kidney, when the *Tm* is exceeded, the substance (eg, glucose) appears in the urine although it is not found there as a normal constituent. On the other hand, certain transport systems have a *Tm* that is almost impossible to saturate under physiologic conditions.

HENLE'S LOOP AND THE COUNTERCURRENT EXCHANGE MECHANISM. Henle's loop of the juxtamedullary nephrons runs deeply into the pyramids of the renal medulla before merging with the distal convoluted tubules of the cortex and the collecting ducts (Fig. 12.5). There is a gradual increase in the osmolarity of the interstitial fluid of the medullary pyramids, so that the osmolarity of the interstitial fluid which surrounds the tips of the papillae at the apexes of the pyramids is 3 to 4 times that of plasma. The ascending limb of Henle's loop is impermeable to water. Sodium is removed from the tubular fluid in this segment of the nephron by an active transport mechanism, however. The fluid in the descending limb of Henle's loop becomes progressively more hypertonic to the plasma as water moves down its osmotic gradient toward the hypertonic interstitial fluid in Henle's loop. In the ascending limb, on the other hand, the fluid becomes progressively more dilute as sodium is pumped out while the water remains behind, as noted above. By the time the tubular fluid has reached the top of Henle's loop, it has become hypotonic with respect to plasma, because of the active transport of sodium, but not of water, out of the tubular lumen.

Therefore, when the tubular fluid has passed through the entire Henle's loop, there has been a net decrease in the total volume of fluid amounting to approximately 5 percent. Since about 75 percent of the total volume of the glomerular filtrate was removed in the proximal tubule, a total of around 80 percent of the original volume of fluid filtered through the glomeruli has been returned to the plasma by the time the fluid has reached the end of Henle's loop.

The formation of a dilute (hypotonic) urine thus is a relatively simple process. Since many solutes that are contained in the tubular fluid are absorbed by active transport mechanisms, and since water is absorbed only by passive diffusion (osmosis), dilution of the urine is achieved by absorption of solutes while the water remains behind, as the ascending limb of Henle's loop is impermeable to water. The formation of a concentrated (hypertonic) urine, on the other hand, is a more complex process which involves the reabsorption of water from the porous collecting ducts, a process which is regulated closely by the secretion of antidiuretic hormone from the posterior lobe of the pituitary gland (p. 1043).

The elements of a countercurrent exchange system are as follows: (1) The inflow to a countercurrent system must run parallel to the outflow for some distance. (2) The conduits for both inflow and outflow in a countercurrent system must be in close approximation to each other. (3) The flow within the two loops of the system must be in opposite directions, ie, must run counter to each other. A simple example of the operation of a countercurrent system is shown in Figure 12.8, in which a heat exchanger, such as would be found in a boiler or water heater, is depicted and explained.

Within the kidney, the anatomic and flow relationships in the two limbs of Henle's loop fulfill the requirements of a countercurrent exchanger for materials rather than for heat, although the underlying principles of operation are similar in both cases. Likewise, the configuration and flow pattern within the vasa recta also fulfill the requirements for a countercurrent exchanger. Consequently, the kidney possesses not one, but two functionally re-

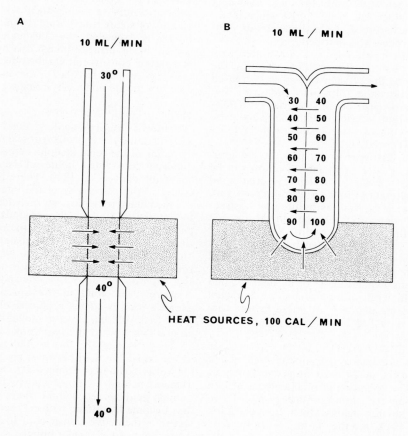

FIG. 12.8. The operation of a thermal countercurrent exchange system. **A.** Heater surrounds a pipe and raises the temperature of the water flowing through the pipe 10 C. **B.** Pipe is bent into a tight U shape, the inflow being adjacent to the outflow. The same heater is applied to the base of the U, and the temperature of the water still is raised 10 C; however, the incoming water is warmed by the outflow. Hence a temperature gradient is established throughout the length of the pipe such that the temperature of the water at the bend is raised from 90 C to 100 C. Incidentally, countercurrent heat exchange takes place in the webs of the feet of aquatic and marine birds (eg, ducks, geese, gulls), as well as in the flippers of marine mammals (eg, seals, sea lions, whales) exactly as described for the artificial heat exchanger. By use of this mechanism, core body heat is preserved, and also the extremities are prevented from freezing in extremely cold weather. (After Pitts. *Physiology of the Kidney and Body Fluids,* 2nd ed, 1968, Year Book Medical Publishers.)

lated and interdependent countercurrent systems which operate in times of water deprivation to produce a concentrated (hypertonic) urine. Figure 12.5B illustrates the morphologic situation as found in the juxtamedullary nephrons of the kidney.

As stated earlier in this section, there is a progressive increase in the osmolality of the interstitial fluid within the pyramids of the renal medulla from the base toward the apex at the papillae. There is also a similar progressive increase in the osmolality of the tubular fluid within the descending limb of Henle's loop to the bend in this structure, which lies deep within the medulla. Active transport of sodium from the ascending limb of Henle's loop produces the osmolal gradient found within the interstitial fluid of the medulla.* The active absorption of sodium ions from the tubular fluid causes an imbalance of electric charges; therefore, chloride and other anions also move across the walls of the tubules into the interstitial fluid to maintain electric neutrality, and this passive movement of anions, together with the active transport of sodium, thus further increases the osmotic pressure of the interstitial fluid in the vicinity of the ascending limb of Henle's loop.

This osmotic gradient develops within the medulla, but not in the renal cortex for two reasons: first, medullary blood flow is sluggish compared to that within the cortex; removal of accumulated sodium chloride from the interstitial fluid is a relatively slow process. Second, the principal reason for the presence of the high medullary concentration of osmotically active substances is because of the countercurrent mechanisms that operate within Henle's loop as well as in the vasa recta. These two mechanisms can be discussed individually.

The Countercurrent Mechanism in Henle's Loop. As shown in Figure 12.9, the descending limb of Henle's loop is quite permeable to sodium (and to chloride), whereas the ascending limb has an active transport mechanism to absorb sodium from the tubular fluid and thus add it to the interstitial fluid. Chloride also moves out passively in the ascending limb. Therefore, each time a sodium ion is transported actively out of the ascending limb, the ion rapidly diffuses into the descending limb. This ionic movement now increases the sodium (and chloride) concentration progressively and in a cyclic fashion as the tubular fluid flows downward toward the tip of Henle's loop. Additional sodium ion is continuously being added to the system by the glomerular filtrate; therefore, this sodium also is transported continuously out of the ascending limb and recycled back to the descending limb. This cyclic process, plus the continual addition of more sodium (and chloride) to the tubular urine as more glomerular filtrate is

Some sodium also may be absorbed by an active transport mechanism in the descending limb, but this is conjectural.

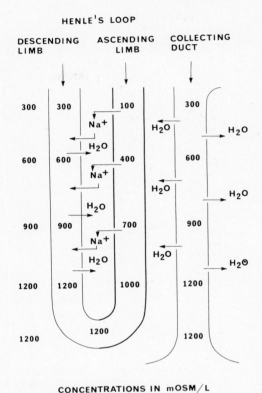

FIG. 12.9. Countercurrent multiplication of concentration within Henle's loop. Hypertonic urine is produced by the kidney through the operation of this mechanism. (After Pitts. *Physiology of the Kidney and Body Fluids,* 2nd ed, 1968, Year Book Medical Publishers.)

produced, results in a large increase in the intramedullary concentration of sodium chloride at the tip of the loop, and this mechanism as it operates within Henle's loop is known as a countercurrent multiplier.

The Countercurrent Mechanism in the Vasa Recta. There also is a countercurrent flow of blood within the vasa recta, as shown in Figure 12.10. This factor also contributes to the extremely high medullary concentration of sodium chloride in the sense that it prevents washout of most of the salt. Blood flowing down in the descending branches of the vasa recta receives sodium chloride from the interstitial fluid by the process of passive diffusion. This in turn causes the osmolar concentration of the blood to increase progressively to a maximum concentration at the tips of those vessels which lie in the deepest region of the medulla. As the blood subsequently flows upward toward the cortex, most of the excess sodium chloride gained by the blood during its downward passage is lost to the interstitial fluid once again. By the time the blood leaves the medulla, its osmolar concentration is only slightly higher than when it entered the descending branch of the vasa recta. Therefore,

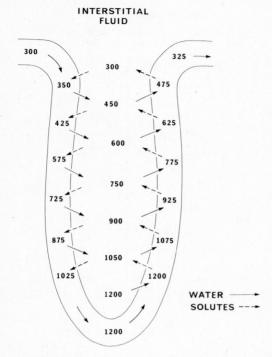

INTERSTITIAL
FLUID

300

325 →

300

350   475

450

425   625

600

575   775

750

725   925

900

875   1075

1050

1025   1200

1200

WATER ———→
SOLUTES --→

1200

FIG. 12.10.   The countercurrent exchange mechanism in the vasa recta. The operation of this mechanism results in the production of an extremely steep osmolar gradient between the cortex and medulla of the kidney. (After Guyton. *Textbook of Medical Physiology,* 4th ed, 1971, Saunders; Pitts. *Physiology of the Kidney and Body Fluids,* 2nd ed, 1968, Year Book Medical Publishers.)

blood flowing away from the vasa recta removes only a trace of sodium chloride from the interstitial fluid of the renal medulla. The type of countercurrent mechanism found in the vasa recta is called a countercurrent exchanger in contrast to the countercurrent multiplier system found within Henle's loop.

DISTAL CONVOLUTED TUBULE AND COLLECTING DUCTS.   The posterior pituitary gland produces a substance known as vasopressin or antidiuretic hormone (ADH), whose level is under the control of a hypothalamic–neurohypophyseal mechanism, as discussed in detail in Chapter 15 (p. 1044). Antidiuretic hormone (or vasopressin) acts directly upon the epithelial cells of the distal convoluted tubules and collecting ducts to regulate the permeability of the cell membranes to water. When antidiuretic hormone is present in any quantity, the permeability of these membranes is increased; therefore, passive diffusion of water and other substances out of the distal tubules and collecting ducts is enhanced, and the remaining urine becomes more concentrated as a result. The dilute fluid leaving Henle's loop thus loses water rapidly by osmosis in the distal convoluted tubule so that its osmolality reaches equilibrium with the cortical

interstitial fluid. This more concentrated tubular fluid then flows downward once again in the collecting duct to the medulla, where it is exposed to the hyperosmolar interstitial fluid of this region (Fig. 12.5A and B). In the medulla, large additional quantities of water are absorbed osmotically from the fluid within the collecting ducts; the concentration of the urine now approaches that of the medullary interstitial fluid. Therefore, the maximum concentration of the urine that can be attained by the kidney is about 4 times the osmolality of the glomerular filtrate or of the plasma.

By way of summary, therefore, the antidiuretic hormone of the posterior pituitary acts directly upon the epithelial cells of the distal tubules and collecting ducts to increase the reabsorption of water from these regions and thus to assist in the production of a more concentrated urine. This hormonal effect is enhanced by the anatomic location of the collecting ducts in the renal medulla, as well as by the extremely high osmolar concentration of solutes which is present in the medullary interstitial fluid. In turn, the high solute concentration within the interstitial fluid compartment is produced and maintained by the countercurrent mechanisms within Henle's loop and the vasa recta. Thus, a steep osmotic gradient is provided for tubular fluid to diffuse rapidly outward when the "pore size" of the membrane is increased under the influence of antidiuretic hormone. Conversely, when this substance is present in diminished quantities, the pores "shrink," so that less water is reabsorbed from the distal convoluted tubule and collecting ducts; a more dilute (hypotonic) urine of greater volume is produced as the consequence.

Clinically, it is far easier to measure the specific gravity of urine with a hydrometer than to determine its osmolality, although both techniques are in current use. This measurement gives the physician an idea of the concentrations of solutes that are present in the specimen. An ultrafiltrate of plasma has a specific gravity of 1.010, whereas that of maximally concentrated urine is approximately 1.035. It is important to remember, in interpreting alterations in specific gravity of a solution, that both the kind as well as the concentration of the solute are contributory factors.

SUMMARY OF COUNTERCURRENT MECHANISMS.   It should be emphasized here that the countercurrent exchange mechanisms are wholly passive, and the principal factor that is involved in the formation of a gradient of increasing osmolality between bases and apexes of the pyramids is the active transport of sodium, since the countercurrent multiplication mechanism depends upon the concentration of sodium in the interstitial fluid of the medulla. In the ascending limb of Henle's loop, sodium is pumped actively out of the tubular fluid so that the interstitial fluid becomes hyperosmotic. Water leaves the tubular fluid all along the descending limb, whereas sodium and urea diffuse into the fluid of the descending limb, which fluid thus becomes progressively more concentrated (Fig. 12.9). The maximum concentration of solutes is found at the tip of Henle's loop. The ascending limb is impermeable to water regardless of whether or not vasopressin is present; therefore, as sodium is pumped out of the tubule by the active transport mechanism of the tubular cells, the fluid becomes hypo-osmolar (hypotonic) by the time it reaches the end of the limb and enters the distal convoluted tubule.

If the sodium and urea in the interstitial fluid of the medullary pyramids were removed rapidly by the circulation, the osmotic gradient in this region would be eliminated. Partly because circulation in the vasa recta is sluggish, but principally because of the countercurrent exchange function performed by these vessels (Fig. 12.10), these solutes remain within the interstitial fluid. Sodium diffuses out of the vessels which transport blood toward the cortex, whereas sodium diffuses into those vessels which transport blood down into the pyramid. Water diffuses out of the descending vessels and into the ascending vessels. Solutes are recycled within the medulla, whereas water tends to bypass this region, and thereby the hypertonicity is maintained.

The process of countercurrent exchange, although wholly passive as noted above, is dependent upon a two-way movement of water and solutes across the permeable walls of the vasa recta. The osmotic gradient along the pyramids, however, could not be maintained by the vasa recta without the simultaneous countercurrent multiplication of solute concentration by Henle's loop, Thus, the two countercurrent systems act in concert.

Antidiuretic hormone in turn acts to maintain the countercurrent mechanism by producing renal vasoconstriction (the pressor effect of this hormone), thereby reducing medullary blood flow. ADH also may play another role in this process by stimulating the sodium transport mechanism within the ascending limbs of Henle's loop.

Specific Renal Functions

Basically, the formation of urine involves three renal functions: (1) glomerular filtration, (2) tubular reabsorption, and (3) tubular secretion. Each of these topics, together with pertinent examples of specific materials handled by the various mechanisms discussed earlier, will be considered next. A comparison of the average composition of glomerular filtrate and urine is presented in Table 12.1.

CONCEPT OF RENAL CLEARANCE AND THE DETERMINATION OF GLOMERULAR FILTRATION RATE (GFR). Renal clearance is defined as the volume of plasma (in ml) from which a given substance is cleared (or removed) by the kidneys in 1 minute. Thus, clearance is a flow rate, usually expressed in milliliters per minute. It should be noted that clearance of substances other than inulin are

calculated flow rates rather than values which have been measured in absolute terms. The renal clearances of a number of substances are compared in Figure 12.11.

In order to obtain valid measurements of the glomerular filtration rate in man, the substance used for determination of this renal function must have certain properties. (1) The substance must be nontoxic and inert metabolically; that is, it must not bind to the plasma proteins or be stored in the kidney. (2) The material must be filtered freely through the glomeruli only. It must not be absorbed or secreted either actively or passively by the glomeruli or tubules. (3) The substance must not exert any effect on any renal function or on renal blood flow. (4) Accurate quantitative determination of the substance must be possible in both blood and urine. The polysaccharide inulin is the only compound known which fills these requirements satisfactorily, so that it can be used to give an accurate measurement of GFR, as illustrated in Figure 12.12. Inulin is a polymer of fructose having a molecular weight of 5,200 daltons, and it is found in artichokes and dahlia tubers.

Glomerular filtration rate is determined clinically by injecting a large quantity of inulin to load the body fluids, and this is followed by infusion of more inulin to keep the arterial plasma level constant. When equilibrium has been reached with the body fluids, a closely timed urine specimen is collected. A blood sample is obtained for plasma determination of inulin halfway through the urine collection. The plasma and urine concentrations of inulin are determined, and the clearance is calculated as described below.

By the definition of clearance, the quantity of inulin found in a specific volume of urine that is excreted per unit time must have come from the volume of plasma which contained this quantity of inulin. If inulin is designated by the subscript *In,* the GFR equals the concentration of inulin in the urine, U_{In}, times the urine flow per unit time, V, divided by the arterial plasma level of inulin, or P_{In}. Thus, the plasma clearance of inulin is:

$$C_{In} = \frac{U_{In}V}{P_{In}}$$

Note particularly that the units for clearance are in milliliters per minute.

Since inulin is not metabolized appreciably in the body, a venous blood sample can be used without introducing serious error. Thus if $U_{In} = 28$ mg/ml, $V = 1.2$ ml/minute, and $P_{In} = 0.27$ mg/ml, then:

$$C_{In} = \frac{28 \times 1.2}{0.27} = 125 \text{ ml/minute}$$

This value, which was cited earlier, is the normal GFR for an average-sized man (Fig. 12.12). The clearance of inulin is equal to the rate of glomerular filtration, because it measures the

Table 12.1 COMPOSITION OF GLOMERULAR FILTRATE COMPARED TO THAT OF URINE[a]

Substance	GLOMERULAR FILTRATE (125 ml/minute) Quantity Filtered/minute	Concentration	URINE (1 ml/minute) Quantity/minute	Concentration	Plasma Clearance/minute
Electrolytes					
Na^+	17.7 mEq	142 mEq/liter	0.13 mEq	128 mEq/liter	0.9
K^+	0.6 mEq	5 mEq/liter	0.06 mEq	60 mEq/liter	12
Ca^{++}	0.5 mEq	4 mEq/liter	0.005 mEq	4.8 mEq/liter	1.2
Mg^{++}	0.4 mEq	3 mEq/liter	0.02 mEq	15 mEq/liter	5.0
Cl^-	12.9 mEq	103 mEq/liter	0.013 mEq	134 mEq/liter	1.3
HCO_3^-	3.5 mEq	28 mEq/liter	0.014 mEq	14 mEq/liter	0.5
$H_2PO_4^-$ } $HPO_4^=$ }	0.3 mEq	2 mEq/liter	0.05 mEq	50 mEq/liter	25
$SO_4^=$	0.1 mEq	0.7 mEq/liter	0.033 mEq	33 mEq/liter	47
Organic Compounds					
Glucose	125 mg	100 mg percent	0 mg	0 mg percent	0
Urea	33 mg	26 mg percent	18.2 mg	1820 mg percent	70
Uric acid	3.8 mg	3 mg percent	0.42 mg	42 mg percent	14
Creatinine	1.4 mg	1.1 mg percent	1.96 mg	196 mg percent	140
Inulin	——	——	——	——	125
Iodopyracet					560
PAH					585

aData from Guyton. Textbook of Medical Physiology, 4th ed, 1971, Saunders.

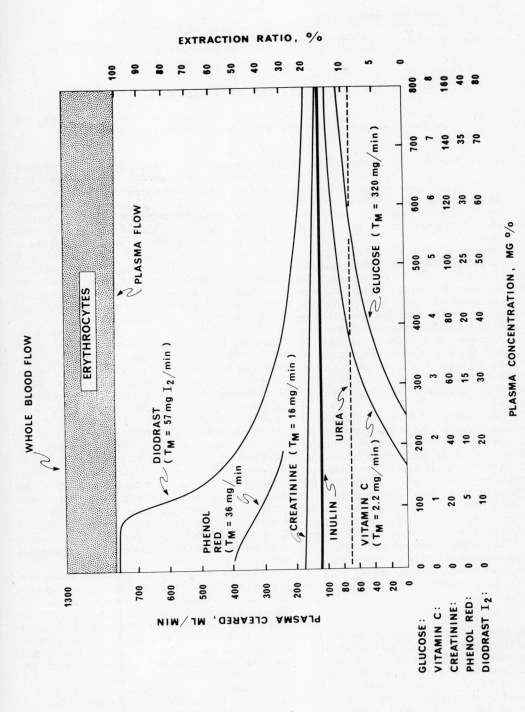

FIG. 12.11. The excretion of several compounds by the human kidney. If the clearance of any compound is below that of inulin, tubular reabsorption has taken place. If the clearance of any compound is greater than that of inulin, tubular secretion has taken place. (After Smith. *Lectures on the Kidney,* 1943, Lawrence, Kansas, University Extension Division, University of Kansas.)

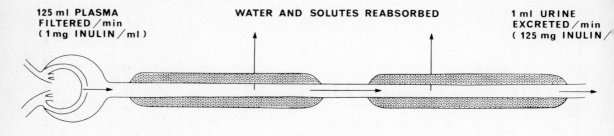

WATER AND SOLUTES REABSORBED

1 ml URINE
EXCRETED/min
(125 mg INULIN/

NO INULIN SECRETED OR REABSORBED

FIG. 12.12. The principles underlying the determination of glomerular filtration rate by inulin clearance. (After Pitts. *Physiology of the Kidney and Body Fluids,* 2nd ed, 1968, Year Book Medical Publishers.)

volume of plasma filtered by the glomeruli in 1 minute.

The clearance of any substance other than inulin is a virtual volume, not a real volume, because no single milliliter of blood has all of the substance removed from it during one transit through the renal circulation.

Note in Figure 12.11 that the curve for inulin clearance is a straight line. This means that the clearance of inulin is strictly dependent upon the glomerular filtration rate, and that inulin clearance does not depend upon the plasma level of this substance or upon the rate of urine formation. Furthermore, inulin is neither excreted nor absorbed by the tubules; therefore, it provides a base line for comparison with the way in which other materials are handled by the kidney. Note, however, in Figure 12.13 that the rate of excretion of inulin $(U_{In}V)$ is a linear function of the plasma level of this compound.

The filtration fraction is defined as the ratio of the glomerular filtration rate to the renal plasma flow. The technique used for the determination of renal plasma flow, using the Fick principle and para-aminohippuric acid (PAH) as the indicator, was discussed on page 774. Normally, in man, the value of the filtration fraction varies between 0.16 and 0.20. A fall in the systemic arterial blood pressure produces a smaller decrease in the glomerular filtration rate than in the renal plasma flow; consequently, the filtration fraction increases. This situation is encountered during the early course of congestive heart failure, and the altered filtration fraction may have significance in the etiology of this pathologic condition.

TUBULAR REABSORPTION. Since glomerular filtration is the first step in the formation of urine, and since many constituents of the glomerular filtrate are present in lower concentrations, or completely absent, in the urine, reabsorption of some or all of these materials must have occurred. Inspection of Table 12.1, which compares the composition of the glomerular filtrate with that of urine, illustrates this point. The importance of tubular reabsorption also is emphasized by the fol-

lowing example. A 70-kg man contains approximately 40 liters of water, about 3.5 liters of which is plasma. At a glomerular filtration rate of 125 ml per minute, a total 180 liters of water is filtered from the glomeruli each 24 hours, so that an enormous quantity of water must be reabsorbed and recycled. Furthermore, 180 liters of glomerular filtrate contains 0.5 kg of glucose, 0.25 kg of bicarbonate, and 0.1 kg of amino acids. Yet, the quantities of these substances which actually are excreted in the urine during a 24-hour period are but negligible fractions of the total amounts which were filtered during the same time period.

The quantity of any substance which is filtered by the glomeruli equals the product of the GFR (or the inulin clearance, C_{In}) and the plasma level of the substance, P, or $(C_{In}) \times (P)$. During the process of tubular absorption, some or all of the substance may be removed from the glomerular filtrate. Conversely, during tubular secretion, more of the substance can be added to the glomerular filtrate. Therefore, the quantity of the substance excreted in the urine is equal to the quantity that has been filtered plus the net quantity transported (T_x) in either direction by the tubules. Thus, when there is net tubular reabsorption, the value for T_x is negative, whereas when net tubular secretion is present, the value of T_x is positive. In Figure 12.11, the compounds that show a negative T_x lie below the curve for inulin clearance, whereas those showing a positive T_x lie above this curve.

Stated another way, the clearance of any substance is equal to UV/P, so that if no net absorption (or excretion) of the substance occurs, its clearance equals the GFR. If there is net tubular absorption, however, the clearance of the substance is below the GFR and the T_x is negative. On the other hand, if the clearance is greater than the GFR, net tubular secretion has occurred, and the T_x now is positive (Fig. 12.11).

In situations where the net concentrations of substances in the glomerular filtrate are altered, either by tubular reabsorption or by secretion, the clearance values now are virtual rather than actual volumes. A virtual volume provides only an index of renal function, for comparison with an actual

A.

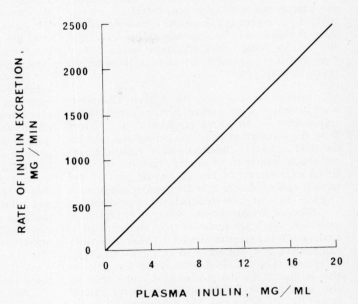

FIG. 12.13. A. Relationship between the rate of inulin excretion in the urine and plasma inulin level.

B.

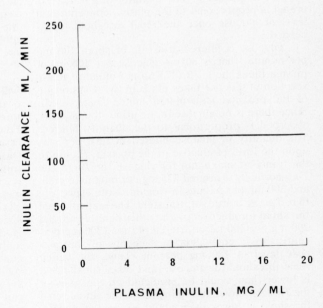

FIG. 12.13. B. Relationship between inulin clearance and plasma inulin concentration. (After Pitts. *Physiology of the Kidney and Body Fluids,* 2nd ed, 1968, Year Book Medical Publishers.)

(absolute) volume, such as is obtained from inulin clearance.

As noted earlier, the renal tubules handle particular materials by a variety of mechanisms, including a number of active transport systems, as well as by passive transport (diffusion, or, in the case of water, osmosis).

One property of active transport systems must be considered before entering upon a discussion of

the tubular absorption of various specific materials. Similarly to active transport systems in other tissues of the body, the renal active transport mechanisms each exhibit a maximal rate at which they can operate, ie, they can become saturated (Fig. 3.30, p. 91). This maximal rate is known as the transport maximum (or *Tm*) for any particular solute. The quantity of the solute that is transported per unit time is therefore proportional to its concentration up to the *Tm* for that solute. But at higher concentrations than its *Tm,* the transport mechanism becomes saturated; therefore, no significant increment in the quantity transported occurs. It also should be emphasized, however, that the *Tm* for some active transport systems is so high that it is impossible to saturate them for all practical purposes.

Substances Reabsorbed by Active Transport Mechanisms. A substance is considered to be transported by an active mechanism if it moves from the tubular lumen to the peritubular interstitial fluid against a chemical and/or electric gradient. Such transport, of course, requires the continual expenditure of energy on the part of the tubular epithelial cells in order to move substances "uphill" against such electrochemical gradients.

Glucose Reabsorption. Glucose is typical of materials which are removed from the urine by a transport-maximum-limited (*Tm*-limited) mechanism, and will serve to exemplify this type of system (Fig. 12.11). Normally, all the glucose that is filtered by the glomeruli is reabsorbed so that none appears in the urine except for minute traces (ie, a few mg per 24-hour period). If, however, the plasma glucose level rises above a critical value known as the renal threshold, glucose appears in the urine. The quantity of glucose excreted is proportional to the plasma concentration level of glucose once the renal threshold is exceeded.

Glucose is filtered at a rate of about 100 mg per minute, that is 80 mg glucose per 100 ml of plasma times the GFR (125 ml per minute). Reabsorption of glucose takes place in the first portion of the proximal convoluted tubule (Fig. 12.15). Since there is no glucose in the urine, the quantity absorbed is proportional to the quantity filtered. The quantity of glucose filtered in turn depends upon the level of glucose in the plasma, P_G, up to the transport maximum for glucose, or Tm_G. The Tm_G normally is around 375 mg per minute in men, and 300 mg per minute in women. Note especially that *Tm* is a rate of transfer. The actual renal threshold for glucose, on the other hand, is around 200 mg per 100 ml of arterial plasma (200 mg percent); this is equivalent to a venous plasma glucose level of 180 mg percent. Thus, the actual renal threshold for glucose (and other substances) is lower than would be predicted on the basis of the *Tm*. As shown in Figure 12.14, the actual renal threshold is lower than would be predicted theoretically, if it is assumed that the *Tm* for all the tubules are identical, and that all the glucose is removed from each tubule. In man, the data obtained empirically deviate from the theoretical curve as shown in the figure, and this deviation is called splay. This splay can be attributed to the fact that all the tubules do not have an identical filtration rate (Tm_G). In addition, some glucose is not reabsorbed when the quantity filtered is below the Tm_G, because the reactions involved in the

glucose transport mechanism are not entirely reversible. Consequently, the rounding (or splay) of the curve is an indication of the extent to which the carrier molecules in the transport mechanism bind glucose. The more rounded the curve, the less the carrier binds to the substrate, which in this case is glucose.

The details of the glucose transport mechanism still are obscure. It is clear, however, that if a given mechanism transports more than one substance, that substance having the greatest affinity for the carrier is absorbed from the tubular fluid to the exclusion of other substances; ie, competition for the carrier occurs. This is especially true if the substance having the greatest affinity for the carrier also is present in the highest concentration. For example, glucose, fructose, galactose, and xylose all are absorbed by the same transport mechanism. Glucose, however, is absorbed preferentially from the tubular urine because it has the highest affinity for the carrier of all these sugars. The glycoside phlorhizin blocks the reabsorption of all these carbohydrates, so that a type of diabetes ensues.

Reabsorption of Sodium and Potassium. The majority of the active transport systems that reabsorb specific solutes are located in the proximal convoluted tubules (Figs. 12.15 and 12.16). The principal exception to this general statement is the active transport mechanism for the reabsorption of sodium ion, which is located in all portions of the nephron except in the descending limb of Henle's loop, which is, however, highly permeable to this ion, as noted earlier.

The absorption of sodium and potassium ions will be considered in some detail, as the handling of these ions by the kidney exemplifies certain features of cellular transport mechanisms in general. Furthermore, as the active transport of sodium is essential to the establishment of electric, concentration, and osmotic gradients for the passive transport of chloride, urea, and water, respectively, this topic warrants more than cursory treatment.

Sodium and potassium ions are absorbed by a special type of active mechanism known as gradient–time-limited transport. Unlike the *Tm*-limited mechanisms discussed above, which are confined to the transport of so many milligrams of solute per minute, the transport system for sodium ion appears to be limited by the gradient that can be established across the tubular wall in the time that the tubular fluid is in contact with the epithelium. The absorption of chloride, although wholly passive, also is a gradient–time-limited process, because it follows the electrochemical gradient that is established by the active transport of sodium ions out of the tubular lumen.

The reabsorption of ions and water from the proximal convoluted tubule is isosmotic. Since the principal ions — sodium, chloride, and bicarbonate — account for around 95 percent of the total osmotic activity of the plasma and the glomerular

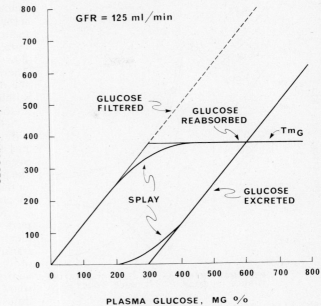

FIG. 12.14. Relationship between the rate of glucose excretion and the plasma level. The curved lines (splay) give a more accurate picture of the actual reabsorption and excretion of glucose in vivo than do the theoretically ideal curves which break sharply. (After Pitts. *Physiology of the Kidney and Body Fluids,* 2nd ed, 1968, Year Book Medical Publishers.)

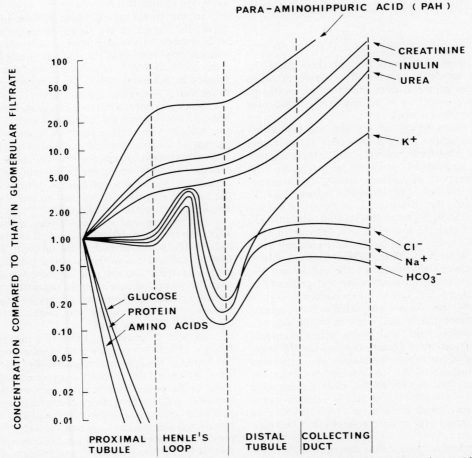

FIG. 12.15. Alterations in the concentration of various substances in different regions of the nephron. (After Guyton. *Textbook of Medical Physiology,* 4th ed, 1971, Saunders.)

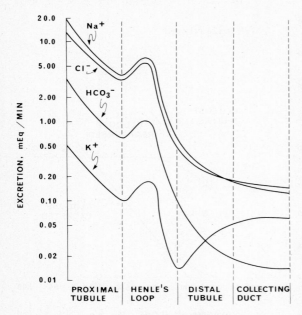

FIG. 12.16. Alterations in the concentration of four major electrolytes in different regions of the nephron. (After Guyton. *Textbook of Medical Physiology*, 4th ed, 1971, Saunders.)

filtrate, these ions must be absorbed from the proximal tubule at osmotically equivalent rates. Hence, the proximal convoluted tubule performs no osmotic work, as it neither concentrates nor dilutes the tubular urine.

The reabsorption of sodium and potassium ions is an active process, whereas chloride and water are absorbed passively along their electrochemical and osmotic gradients, respectively. The active transport of sodium across the proximal tubular epithelium is illustrated in Figure 12.17. In this system, the potassium concentration within the cell is high, whereas the sodium and chloride concentrations are low. The cell interior is electronegative with respect to the lumen as well as to the peritubular fluid. Thus, sodium ions are able to diffuse between the lumen of the tubule and the cell interior along an electrochemical gradient. In order to keep the intracellular concentration of sodium low, this ion is transported actively across the cell membrane at the junction of the cell and the peritubular fluid, and so an active transport system for sodium is present only within the outer membrane of the tubular epithelial cells. Continual active transport of the sodium outward into the peritubular fluid results in the maintenance of a gradient to this ion; therefore, as sodium ion is pumped out, more diffuses in from the luminal fluid, perhaps by a carrier-mediated passive mechanism.

The principal factor that determines the electric potential across the peritubular membrane is the diffusion of potassium and chloride. The potential across the luminal cell membrane is not determined by the ratio of the transmembrane concentration gradients. Rather, it is generated by the current times resistance fall that occurs as current flows from the lumen of the tubule into the cell. Therefore, the electromotive force in this circuit is the diffusion potential; perhaps the sodium pump within the peritubular membrane also contributes to this effect.

Chloride ions are reabsorbed passively (by diffusion) through the luminal border of the proximal tubules, although the details of this process are unclear. The reabsorption of chloride through the peritubular membrane also is passive, although in this case the passive diffusion takes place down an electrochemical potential gradient.

Since the net reabsorption of potassium takes place across the luminal cell membrane against an electrochemical potential difference, an active transport mechanism for this ion must be present at this boundary. Net movement of potassium across the peritubular membrane from the surrounding fluid to the cytoplasm occurs by passive diffusion, although a coupled sodium–potassium exchange mechanism also may be present, as shown in Figure 12.17.

In the distal tubular cell, shown in Figure 12.18, the lumen of the tubule is electronegative with respect to the peritubular fluid. The transmembrane potentials of the epithelial cells vary between −70 to −90 mv; consequently, the net difference in potential across the luminal cell membrane varies between −20 to −40 mv.

Since the intracellular sodium concentration is low and the cell interior is negative with respect to the lumen, net reabsorption of sodium occurs against an electrochemical gradient. Therefore, an active transport mechanism is involved in this process. In addition, active potassium uptake occurs at the peritubular cell border. Whether or not the sodium extrusion is coupled to the potassium uptake in this region is unknown.

Adrenal mineralocorticoids (eg, aldosterone) affect the transport of sodium as well as potassium in both the proximal and distal tubules. The pattern of reabsorption in the proximal tubules, however, is largely stereotyped, as noted earlier, so that the more delicate specific regulation of individual urinary constituents occurs in the distal nephron, which includes the distal convoluted tubule and collecting duct.

By way of summary, there is a potential difference of −70 mv between the tubular lumen and the inside of the proximal tubular cells (Fig. 12.17). This potential difference is approximately the same as the potential difference recorded between the cells and the peritubular interstitial fluid. The potential difference between the distal tubular lumen and the tubular cells, on the other hand, is only 30 mv more positive than the interior of the cells (Fig. 12.18). Therefore, a net potential difference of 40 mv is present between the lumen of the distal tubule and the peritubular interstitial fluid,

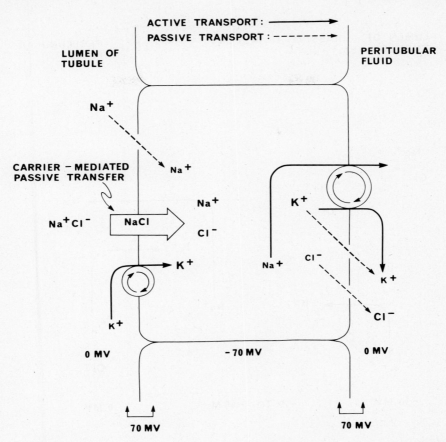

FIG. 12.17. Ion transport in an epithelial cell of the proximal convoluted tubule of a nephron. Dashed lines = passive diffusion down electrochemical gradients; solid lines = active transport systems; large arrow = carrier-mediated passive transport system for NaCL. (After Pitts. *Physiology of the Kidney and Body Fluids,* 2nd ed, 1968, Year Book Medical Publishers.)

the tubular lumen being negative with respect to the extracellular fluid. Across the collecting ducts, a net difference in potential of around 20 mv has been recorded.

Active Reabsorption of Other Substances. In addition to glucose, sodium, and potassium, other substances actively reabsorbed by the tubules include bicarbonate, organic anions (eg, citrate, malate, α-ketoglutarate, and lactate), as well as phosphate, amino acids, creatine, sulfate, uric acid, and ascorbic acid (vitamin C). Ketone bodies (acetoacetic acid and β-hydroxybutyric acid) also are absorbed by active transport mechanisms. The *Tm* values of these substances range from very low to immeasurably high. A number of specific and individual transport systems also are present for different amino acids, and some transport mechanisms for different substances may involve a step that is common to each. Some mechanisms may transport in either direction, so that a given substance may be reabsorbed or secreted by the same transport system, depending upon the prevailing conditions. Thus, hydrogen ion may be exchanged for an equivalent

quantity of sodium ion, which is absorbed by active transport.

The various active transport mechanisms in the kidney, similarly to the active transport mechanisms in other organs of the body, can be inhibited either competitively or noncompetitively. The practical significance of this statement is found in two examples: (1) The action of a number of diuretics used clinically depends upon their ability to block certain transport systems selectively. (2) In the treatment of gout, the active reabsorptive mechanism for uric acid is inhibited by drugs such as phenylbutazone (Butazolidin) and probenecid (Benemid).

Substances Reabsorbed by Passive Transport. Passive transport is simply the physical diffusion of materials down a concentration gradient, and no energy is expended directly upon the substances so transported to effect their transfer. Energy must be expended, however, to establish the concentration gradients along which diffusion takes place. Thus, active reabsorption of sodium is the fundamental process that establishes the concentration,

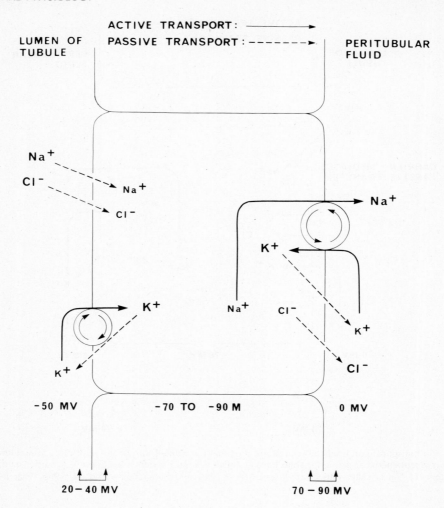

FIG. 12.18. Ion transport in an epithelial cell of the distal convoluted tubule. Other notations as in Figure 12.17. (After Pitts. *Physiology of the Kidney and Body Fluids,* 2nd ed, 1968, Year Book Medical Publishers.)

electric, and osmotic gradients that are necessary for the absorption of urea, chloride, and water, respectively, from the glomerular filtrate.

Diffusion of water and solutes from the renal tubules is quite a rapid process owing to the extremely small diameter of the different regions of the tubules, as well as the concentration gradients developed by the active transport of sodium.

Passive Transport of Urea. Urea exhibits a clearance of around 75 ml per minute in a normal man at urine flows above 2 ml per minute (a condition of polyuria). Since the glomerular filtration rate is 125 ml per minute, the urea present in 50 ml of filtrate must be reabsorbed each minute (Figs. 12.11 and 12.15). The expression $1 - (C_u/C_{In})$ enables the fraction of the filtered urea that is reabsorbed to be calculated. Therefore, $1 - 75/125 = 1 - 0.6 = 0.4$, or 40 percent of the filtered urea, is reabsorbed when a polyuria exists in the subject.

As shown in Figure 12.19, when the urea-to-inulin clearance ratio is plotted against urine flow,

it is found that urea clearance decreases sharply at a flow rate below around 2 ml per minute.

Since the urea clearance is consistently below the actual glomerular filtration rate (Fig. 12.11), and the rate of absorption of this compound depends upon the rate of urine flow, obviously urea clearance cannot be used to obtain accurate figures for glomerular filtration rate, and the clearance of this substance is a virtual rather than an absolute value.

The passive absorption of urea takes place down the concentration gradient that is established by the active transport of sodium ion, and therefore is typical of substances that are reabsorbed by the mechanism of passive diffusion.*

In contrast to mammals, marine elasmobranch fishes (eg, sharks, skates, and rays) absorb urea from the tubular urine by an active transport mechanism in such large quantities that their blood is maintained hyperosmotic to sea water.

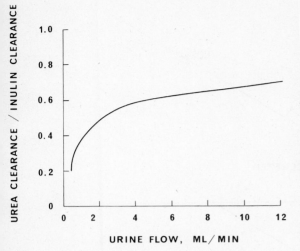

FIG. 12.19. Relationship between the clearance ratio urea/inulin and the rate of urine flow. (After Pitts. *Physiology of the Kidney and Body Fluids,* 2nd ed, 1968, Year Book Medical Publishers.)

TUBULAR SECRETION. In order to study the tubular reabsorption or secretion of any substance, it is essential to know the rate at which that substance is filtered by the glomeruli, as well as the rate at which it is excreted in the urine. In many respects, tubular secretion is similar to tubular reabsorption. The principal difference between the two processes lies in the direction in which the materials are transported, ie, in the orientation of the transport system. Tubular secretion thus takes place from the peritubular interstitial fluid, whereas tubular reabsorption takes place in the reverse direction.

Similarly to active reabsorption, active secretion also requires transport of material against an electrochemical gradient; therefore, a continuous expenditure of energy is essential to this process. There are three types of tubular secretory mechanisms that are analogous to the three absorptive mechanisms discussed in the previous section: (1) active secretory mechanisms that exhibit a *Tm*-limited transport capacity; (2) active secretory mechanisms that exhibit a gradient–time-limited transport capacity; and (3) passive transport mechanisms, in which substances diffuse down electric and/or concentration (electrochemical) gradients. Each of these mechanisms now will be discussed in relation to the types of materials secreted.

Tm-Limited Mechanisms. There are three known *Tm*-limited active transport mechanisms in the kidney which secrete various groups of compounds, both natural (ie, normally occurring or produced within the body) and artificial (ie, dyes, drugs, and other chemicals).

Tm-Limited Transport Mechanism for the Secretion of Organic Acids and Miscellaneous

Compounds. Substances actively secreted by this *Tm*-limited transport mechanism include hippuric acid, phenol red, and other sulfonphthalein dyes, para-aminohippuric acid (PAH), chlorothiazide, penicillin, and other compounds (Fig. 12.11). Included in the latter category are certain urologic contrast agents containing iodine, eg, iodopyracet, Shiodan, and Topax. Creatinine also may be secreted by this mechanism.

The secretion of PAH typifies this type of active tubular secretory mechanism. As shown in Figure 12.20, the quantity of PAH filtered by the glomeruli is a linear function of the plasma level, but the secretion of this organic acid increases only until the point at which the tubular maximum, or Tm_{PAH}, has been attained. Thus, when the plasma level of PAH is low, the clearance of this compound is high, but as the plasma level of PAH rises above the Tm_{PAH}, the clearance of PAH falls progressively so that eventually it approaches the inulin clearance (Figs. 12.11 and 12.20). This effect results from the fact that the quantity of PAH secreted actively becomes a progressively smaller fraction of the total quantity secreted as the plasma level increases.

Another compound that presumably is secreted by this *Tm*-limited active transport mechanism is creatinine. The clearance curve for creatinine is shown in Figure 12.11. The creatinine-to-inulin ratio in normal man is around 1.4 at low plasma levels of this metabolite, which indicates that some tubular secretion occurs. The *Tm* for creatinine is quite low (around 16 mg per minute); therefore, as the plasma level of this substance increases, the clearance ratio decreases, as shown in Figure 12.11. Tubular secretion of creatinine is depressed rapidly to the inulin clearance level by large doses of iodopyracet, which compete for the same loci in the creatinine transport mechanism.

In addition to the compounds listed earlier that are transferred by this *Tm*-limited active transport mechanism, ethereal sulfates, steroids, and other glucuronides, as well as 5-hydroxyindoleacetic acid (the chief metabolite of serotonin), also are secreted actively by the same system. All these compounds are weak anions, and since active competition exists among them for secretion, a single transport mechanism is implicated in their transfer.

The active transport mechanism discussed above is limited exclusively to secretion within the proximal convoluted tubule.

Transport Mechanism for Strong Organic Bases. The substances secreted by this *Tm*-limited active transport mechanism once again include both naturally occurring and synthetic compounds; however, all these substances share the common property of being strong organic bases. Naturally occurring compounds secreted by this active mechanism include, for example, guanidine, methylguanidine, histamine, choline, thiamine, and piperidine. Synthetic compounds, which are

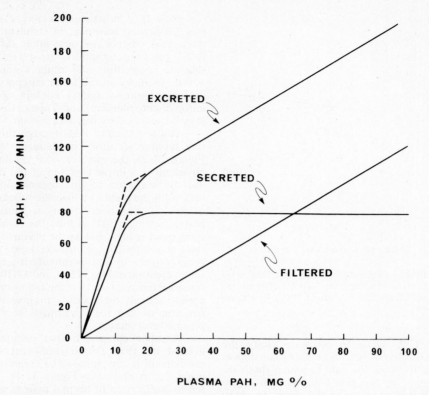

FIG. 12.20. Relationship between the rates of PAH filtration, secretion, and excretion in man as influenced by the plasma PAH concentration itself. (After Pitts. *Physiology of the Kidney and Body Fluids,* 2nd ed, 1968, Year Book Medical Publishers.)

handled similarly, include hexamethonium, tetraethylammonium, tetramethylammonium, and tolazoline (Priscoline).

The dynamic processes that are involved in the *Tm*-limited active transport system for organic bases are quite similar to those involved in the active transport system which secretes organic acids. Competition among individual compounds for secretion also is exhibited by the organic base active transport system, and it, too, is located in the proximal convoluted tubule.

Transport Mechanism for Ethylenediaminetetraacetic acid (EDTA). The *Tm*-limited transport mechanism for EDTA, a synthetic chemical, appears to be different from the transport systems which secrete acids and bases. The presence of a distinct system which handles EDTA alone implies the possible existence of other renal transport mechanisms; however, this mechanism is, at present, more of academic than clinical interest.

Gradient–Time-Limited Active Secretory Mechanisms. Hydrogen ions are transported from the tubular cells into the tubular fluid throughout the entire length of the nephron against an electrochemical gradient; therefore, transport of this ion is an active process.

The different portions of the nephron are specialized insofar as their ability to transport hy-

drogen is concerned. Hence, the proximal convoluted tubule can transport large quantities of hydrogen ion against a small gradient; a difference of only 0.5 pH unit is the maximum that can be established in this region of the nephron between the blood and the tubular fluid. The collecting duct, on the other hand, can transport only small quantities of hydrogen ion. In this region, however, the hydrogen ions are transported against a steep gradient so that a pH difference of 3.0 pH units (a 30-fold gradient) can be established between blood and urine in this part of the nephron. In severe acidosis, the kidney can excrete urine which has a maximum acidity of pH 4.4. The maximum gradient of hydrogen ion that can be developed by the cells of the collecting duct is around 1000 to 1.

When the intracellular concentration of hydrogen ions is high, large quantities of hydrogen are secreted by the tubules. This situation also prevails when the transtubular gradient between blood and urine is low because of the presence in the urine of quantities of buffer compounds which have an appropriate pK. On the other hand, when the intracellular hydrogen ion concentration is low, relatively little hydrogen ion is secreted.

There is a reciprocal relationship between the secretion of hydrogen and potassium ions. Potassium ions are excreted by a passive transport mechanism down a transtubular electric potential gradient, as discussed in the next section. It is sufficient to mention here that when blood potassium is elevated above normal (a condition known as hyperkalemia), the intracellular potassium concentration of the tubules is high, whereas the hydrogen ion concentration is low. Consequently, only small quantities of hydrogen ion are secreted actively, while a considerable quantity of potassium is secreted passively along the transtubular potential gradient.

In the opposite situation, hypokalemia (ie, a reduced blood potassium), there is a relatively low concentration of potassium ions within the tubular cells, so that the quantity of hydrogen ions is above normal. Therefore, an abundant secretion of hydrogen ions into the urine occurs, while potassium ions are retained. This topic will be dealt with further in conjunction with the renal regulation of acid–base balance.

Passive Secretion by the Renal Tubules. Similarly to passive tubular reabsorption, passive secretion from the tubules involves no energy expenditure; rather, it is movement of materials down electrochemical gradients. Energy is required, however, for the development of the electrochemical gradients down which the transfer takes place.

Diffusion-Trapping of Weak Acids and Bases. The passive secretion of certain weak acids (eg, salicylic acid and phenobarbital) and bases (eg, ammonia, procaine, quinine, and quinacrine) occurs by a process which has been termed diffusion-trapping. The clearance of these substances from the plasma in turn depends upon urinary pH. Since the pK values of these compounds range between 6.8 and 9.4, alterations in urinary pH between 8 and 5 can change the plasma clearance from quite low values (indicating a marked reabsorption) to extremely high values (indicating tubular secretion).

For example, in plasma which has a pH of 7.4, the basic substances are present as equilibrium mixtures of the charged cation and the unionized free base. Since the lipid solubility of the free base in the plasma membrane is far greater than is that of the cationic form, the free base penetrates the membrane much more easily than does the ionized form. In urine which has an acidic pH, for example, 5.0, the weak base also is present as an equilibrium mixture of both ionized and un-ionized forms. The equilibrium is shifted in favor of the cationic form, however. As the free base diffuses from the plasma (pH 7.4) into the urine (pH 5.0), hydrogen ion binds with it and forms the cationic form, which is far less able to penetrate the membrane and thus be reabsorbed. By this mechanism, the free base is secreted by passive diffusion into the tubular lumen and there converted by urinary hydrogen ion into a cationic form to which the tubular membrane is far less permeable. The free base is trapped within the urine by conversion into an ionic species. Moreover, this chemical process assures the maintenance of a concentration gradient for the base between the blood and the urine so long as the pH of the latter remains low, ie, acidic. If the urinary pH becomes elevated, eg, to around 8, the diffusion gradient is reversed and the weak base is reabsorbed passively by diffusion into the tubules and thence to the peritubular fluid.

A similar process occurs in the case of passive secretion by diffusion-trapping of weak acids, although the only materials that are known to be handled by a diffusion-trapping secretory mechanism are salicylic acid and phenobarbital.

The secretion of the weakly basic substance ammonia will be considered on page 807, in relation to its role in renal acid–base regulation.

Passive Secretion of Potassium. Normally, potassium is cleared from the plasma to an extent of only 20 percent of that filtered by the glomeruli. Therefore, the major portion of the filtered potassium is reabsorbed, principally by an active transport mechanism shown to be located primarily within the proximal convoluted tubules, where the bulk of potassium reabsorption takes place.

Most of the potassium found in the urine, however, is added to the tubular fluid by the process of passive secretion as the urine moves through the distal convoluted tubules and collecting ducts. The net quantity of potassium that is secreted into the urine in these parts of the distal nephron is determined principally by the transtubular potential as well as the intracellular potassium concentration. If the bodily stores of this electrolyte are depleted sufficiently, active reabsorption in not only the proximal but also the distal nephron is stimulated so that only small amounts of potassium appear in the urine, and thereby the body stores are conserved.

REGULATION OF EXTRACELLULAR FLUID VOLUME AND COMPOSITION

The vital roles played by the kidneys in relation to other organ systems in the overall physiologic regulation of water, electrolyte, and acid–base balance in the body will now be considered. Certain physiologic mechanisms whereby the functions of the kidneys themselves are controlled also will be considered in relation to these topics.

It has been emphasized repeatedly that the cells of the body can live and function properly only so long as the volume and composition of the body fluids remain within fairly narrow limits. Therefore, a basic understanding of the physiologic mechanisms whereby normal homeostasis of the body fluids is attained is of the utmost importance to the clinician.

The kidneys participate directly in the regulation of four individual properties of the extracellular fluids (ie, the blood and interstitial fluid) of the body: (1) They control the total volume of the extracellular fluids; that is, they maintain and regulate the overall water balance within the body. (2) The concentration of specific ions is regulated individually by the kidneys. In some instances, the ratios of certain ions to each other also are regulated, either directly or indirectly. (3) As a direct consequence of the first and second processes, the kidneys control the osmolality of the extracellular fluids. (4) The kidneys, in cooperation with the respiratory system, regulate the hydrogen ion concentration, ie, the pH of the extracellular fluids, and thereby play an important physiologic role in the regulation of acid–base balance in the body.

All four of these functions are dynamic processes which are active simultaneously to a greater or lesser extent throughout the lifetime of an individual. In order to achieve clarity of presentation, each of these four topics must be dealt with individually, and it is important to remember that they are interdependent to a certain extent rather than mutually exclusive entities.

Water Balance and the Regulation of Total Body Water Volume

A number of specific physiologic processes relating to the renal mechanisms that are concerned with water loss and retention have been discussed earlier in this chapter. Only general concepts relating to the overall control of extracellular fluid volume in the body will be discussed in this section.

WATER BALANCE. The net quantity of water in the body depends upon a net balance between the total daily intake versus the total loss from the body during the same interval. This relationship is shown in Figure 12.21.

Water Intake. In an average adult, the total daily water intake amounts to approximately 2.4 liters. The major proportion of this volume, about 66 percent (1.46 liters), is obtained by oral intake of water (and other beverages); the remainder of this total volume is obtained from ingested food as well as from metabolic water (around 150 to 250 ml per day). The latter is derived from the oxidation of hydrogen in food, as discussed in Chapter 14 (p. 933).

Water Loss. The routes of water loss from the body are shown in Figure 12.21. An individual in water balance loses approximately 2,400 ml per day, a volume equivalent to that gained. About 1,000 ml is lost in the urine, 100 ml in the sweat, 200 ml in the feces, and a negligible quantity is lost in tears and expectorated saliva. The remaining 1,100 ml, however, is lost by evaporation through the lungs (300 to 500 ml per day) as well as by direct diffusion through the skin as sensible and insensible perspiration. The latter may amount to between 300 and 600 ml per day; however, in patients suffering from extensive burns, this volume may increase to as much as 5 liters per day.

As noted in Chapter 11 (p. 728), the inspired air quickly becomes saturated with water vapor to a partial pressure of 47 mm Hg prior to expiration. Water loss via the lungs usually amounts to several hundred milliliters per day, and this volume can increase considerably in very cold weather because of the reduced partial pressure of water vapor in the atmosphere. Conversely, water loss via the respiratory tract may decrease in warm weather because the atmospheric (ambient) vapor pressure of water can increase in proportion to the temperature. Therefore, the relative humidity of the atmosphere at a given temperature plays a role, though a minor one, in water loss from the respiratory tract.

During periods of extremely hot and dry weather, water loss increases markedly from heavy sweating, particularly if a person is performing heavy manual work. Total water loss thus can exceed 4 liters per hour under such conditions and serious depletion of body stores of water (and salt) can ensue.

Exercise per se increases water loss, first, by increasing the respiratory rate, and second, by increasing the total heat production of the body, so that sweating is more pronounced.

PHYSIOLOGIC REGULATION OF WATER BALANCE BY THE KIDNEYS. There is an important physiologic mechanism which controls the renal excretion of water, and thereby regulates the total quantity of water that is retained by the body under various circumstances. This mechanism is called the osmoreceptor–antidiuretic hormone system. The specific physiologic role of antidiuretic hormone in the regulation of cellular permeability to water within the distal nephron (ie, the distal convoluted tubule and collecting duct) was described on page 782. In the following discussion, the role of this hormone in the overall mechanism involved in water conservation by the body will be considered.

Located within the supraoptic nuclei of the anterior hypothalamus outside of the blood–brain barrier are certain specialized neural cells which are extremely sensitive to alterations in the concentration of the extracellular fluids. Since these cells respond specifically to changes in the osmolality of the fluid surrounding them, they are termed *osmoreceptors*. If the osmolality of the extracellular fluid increases (ie, becomes relatively more hypertonic), the cells shrink as water moves from them by osmosis, and their rate of impulse discharge increases. Conversely, if the intracellular fluid volume decreases (ie, becomes more hypotonic), these cells swell as water moves into them, again by osmosis, and their discharge frequency now decreases.

The impulses arising within the osmoreceptor cells of the supraoptic nucleus then are transmitted through the pituitary stalk into the posterior pituitary gland where, depending upon their frequency, they stimulate or depress the secretion of antidiuretic hormone (ADH), as shown in Figure 12.22. This hormonal substance is transported in the bloodstream to its locus of action in the distal nephron. Consequently, a sophisticated feedback mechanism controls the rate of water absorption

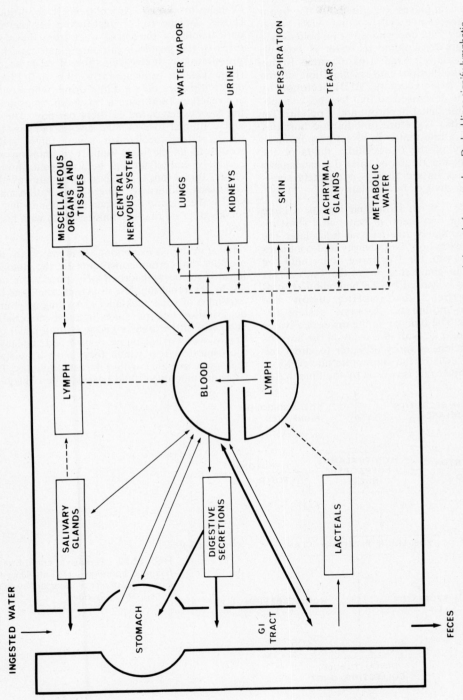

FIG. 12.21. The various aspects of water balance in the body. Heavy arrows denote principal channels for water loss. Dashed lines signify lymphatic drainage from interstitial fluid compartments of various organs and tissues. Solid lines signify exchange of water among blood, tissues, and organs. The metabolic water produced daily by certain intracellular chemical reactions represents only a very small fraction of the total body water. Note also that, from a topologic standpoint, fluid which enters the gut now is outside of the body.

by the distal nephron. An elevated ADH level increases the permeability of the epithelial cells to water; more water is reabsorbed, a relative hemodilution ensues, and this in turn decreases the frequency of the impulses generated by the hypothalamic osmoreceptor cells so that ADH secretion diminishes. The opposite situation holds when a relatively greater volume of water is present, such as after taking a large drink of water. In this situation, hemodilution causes inhibition of the hypothalamic osmoreceptors, ADH secretion by the pituitary is decreased so that less water is absorbed from the distal nephron, and a polyuria (or diuresis) results. Therefore, antidiuretic hormone is secreted at a rate that is in direct proportion to the osmolality of the extracellular fluids of the body, in particular the blood, and the mechanism which underlies its secretion is exquisitely sensitive to minute changes in osmolality.

Water Diuresis. If an individual ingests a large volume (one or two liters) of water in a short interval, approximately one-half an hour later the rate of urine excretion rises sharply and continues at an elevated rate for 1 or more hours until the excess fluid is eliminated from the body. This water diuresis is caused by the hemodilution produced by the liquid acting upon the osmoreceptor–antidiuretic hormone system described above. The time lag in the onset of a water diuresis is in part a result of a delay in the absorption of a sufficient quantity of water to affect the osmoreceptor cells of the supraoptic nuclei. Most of the delay, however, is caused by the time that is necessary for the ADH already present within the distal nephron to become inactivated so that an increased volume of water can be eliminated.

Diabetes Insipidus. Any pathologic condition that causes damage to, or destruction of, the supraoptic nuclei or the nerve tract from these structures to the posterior pituitary gland causes suppression or complete cessation of ADH secretion (Fig. 12.23). Subsequently, the individual so afflicted passes only a dilute urine, whose volume can reach extraordinary levels, sometimes amounting to between 5 and 15 liters per day. The rapid and continual loss of fluid causes the onset of a compensatory mechanism, severe thirst, and this results in an elevated liquid consumption which generally can offset the hemoconcentration that ensues for as long as the person remains awake or conscious. Otherwise, severe dehydration may result if the victim is unable to maintain his fluid intake at a rate that is commensurate with his fluid loss.

Obligatory Urine Excretion. It is important to remember that certain constituents of the glomerular filtrate can be reabsorbed only poorly, or not at all, by the renal tubules, urea and creatinine being examples of such substances. As long as glomerular filtration occurs, some of these materials remain within the tubules and prevent a small quantity of water from being reabsorbed, because of the osmotic effect which they exert by virtue of their mere presence within the renal tubules. Thus, a certain amount of fluid passes, albeit sluggishly,

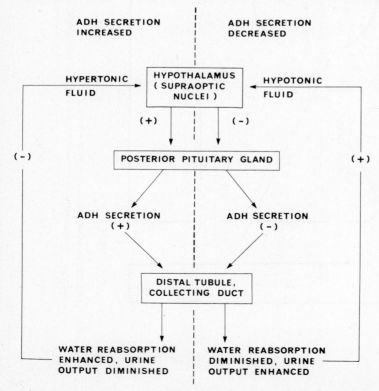

FIG. 12.22. Feedback control of antidiuretic hormone secretion. (Data from Guyton. *Textbook of Medical Physiology,* 4th ed, 1971, Saunders.)

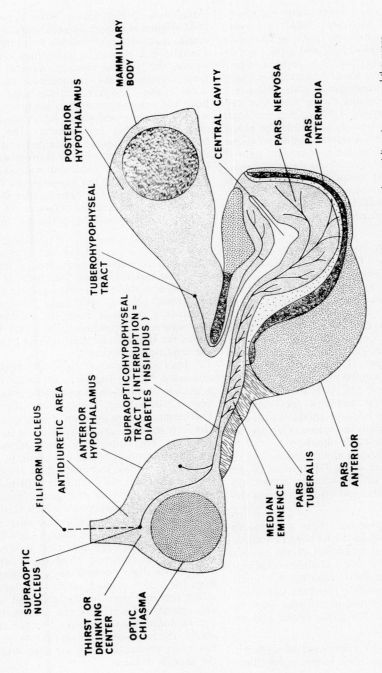

FIG. 12.23. Midsagittal section through hypothalamus (cat) illustrating the thirst (or drinking) center, the antidiuretic area, and the supra-opticohypophyseal tract (SOH). Injury to this tract induces diabetes insipidus. (After Ganong. *Review of Medical Physiology*, 6th ed, 1973, Lange Medical Publications; Chusid. *Correlative Neuroanatomy and Functional Neurology*, 14th ed, 1970, Lange Medical Publications.)

SUPRAOPTIC NUCLEUS

THIRST OR DRINKING CENTER

OPTIC CHIASMA

FILIFORM NUCLEUS

ANTIDIURETIC AREA

ANTERIOR HYPOTHALAMUS

SUPRAOPTICOHYPOPHYSEAL TRACT (INTERRUPTION = DIABETES INSIPIDUS)

POSTERIOR HYPOTHALAMUS

MAMMILLARY BODY

TUBEROHYPOPHYSEAL TRACT

CENTRAL CAVITY

PARS NERVOSA

PARS INTERMEDIA

MEDIAN EMINENCE

PARS TUBERALIS

PARS ANTERIOR

through the tubules at all times. The minimum quantity of urine that is required to rid the body of its normal body metabolic wastes is known as the obligatory urine excretion. The volume of obligate urine output amounts to around 400 ml per day.

Ingestion of Hypertonic Electrolyte Solutions. Since the kidneys of man are unable to excrete solutes at a total concentration that is greater than about 5 times that of plasma (1,400 mOsm per liter), the continued ingestion of hypertonic saline solutions (eg, sea water, about 3 percent total salt) leads to a serious derangement of water balance.* The human kidney cannot concentrate the salt load for excretion beyond 1,400 mOsm per liter; therefore the excess electrolytes within the blood following ingestion of hypertonic saline solutions are neither eliminated by the kidneys nor diluted by the other body fluids. A severe water and electrolyte imbalance and intracellular dehydration, among other disturbances, ultimately ensue, with fatal consequences if sufficiently prolonged. Intense continual thirst also attends the ingestion of hypertonic saline solutions.

REGULATION OF WATER BALANCE BY THE THIRST MECHANISM. The principal factor regulating water intake is the thirst mechanism. The sensation of "thirst" may be defined as a conscious desire for water which is regulated by physiologic necessity.

Located within the hypothalamus, and situated laterally and caudally to the supraoptic nuclei, are two regions known collectively as the thirst or drinking center. This area is shown in Figure 12.23, and it is known to overlap the region that stimulates the release of ADH from the pituitary. Consequently, this anatomic fact explains why an increased water intake associated with certain stimuli that are discussed below often are related to increased renal water retention.

The thirst mechanism is stimulated by any of four major physiologic causes: (1) dryness of the buccal (oral) mucosa; (2) low cardiac output; (3) extracellular dehydration; and (4) intracellular dehydration.

Dryness of the Oral Mucosa. This stimulus to thirst probably is related to the localized extracellular and intracellular dehydration of the tissues within the mouth. This is not a complete explanation of the origin of the thirst sensation, however. For example, atropine, which prevents salivation, resulting in a dryness of the mouth that leads to drinking, does not alter the extracellular and intracellular fluid volumes.

Low Cardiac Output. A hemorrhage of sufficient magnitude to lower the cardiac output significantly

*In marine birds, such as seagulls, a special salt gland located above the beak enables these animals to secrete by active transport an extremely concentrated salt solution so that the body fluids remain isotonic, as any sodium chloride in excess of bodily requirements is eliminated selectively.

(p. 1242) will induce a sensation of acute thirst. A similar effect also is observed in acute cardiac failure, in which the cardiac output is inadequate, even though the extracellular fluid volume may be normal.

Extracellular Dehydration. When the extracellular fluid volume is reduced significantly, regardless of cause, a sensation of intense thirst develops. For example if a person is maintained on a low salt intake, the osmolar concentration as well as the total volume of the extracellular fluids declines considerably. The person drinks large quantities of fluid, but unless sodium chloride is available in the water (eg, while performing heavy manual labor in an extremely hot environment), the normal extracellular fluid volume is not restored, so that an elevated water consumption is prolonged until salt is provided. When sufficient sodium chloride is provided, however, the sensation of thirst disappears promptly. A low extracellular fluid volume, and not an insufficiency of water, is responsible for initiating the sensation of thirst in this instance.

Intracellular Dehydration. When the extracellular fluid volume is normal but the intracellular fluid volume is reduced, a sensation of intense thirst also is stimulated. Thus, ingestion of hypertonic electrolyte solutions or the injection of a similar hypertonic solution will cause thirst by osmotic withdrawal of water from the intracellular fluid compartment. Drinking plain water rapidly restores the intracellular fluid volume to normal, and the sensation disappears.

Interestingly, when the gastrointestinal tract is distended by the ingestion of water per se, the sensation of thirst disappears before the water has been absorbed, ie, immediately after drinking. Furthermore, mere inflation of a balloon within the stomach similarly can alleviate thirst for around 15 minutes. This inhibitory mechanism operating through distention of the gut tends to inhibit the tendency to drink too much fluid prior to its absorption into the body. The act of drinking and the attendant fullness of the gastrointestinal tract stimulate sensory receptors which transmit afferent impulses to the hypothalamic thirst center and inhibit it. The other three factors known to initiate thirst, and which were discussed above, however, all share one common physiologic denominator. Thus intracellular dehydration ultimately results from the action of each of these factors, and this effect possibly may be the principal, if not the only, stimulus for eliciting the sensation of thirst.

In addition to the several factors mentioned above, the renin–angiotensin system (p. 1169) also has been implicated as a possible mechanism in the genesis of thirst.

Regulation of Extracellular Fluid Ion Concentrations

From a physiologic standpoint, the most significant cations in the extracellular fluid are sodium, potassium, calcium, and magnesium, whereas the important anions are bicarbonate, chloride, and phosphate. Two anions of lesser significance are sulfate and nitrate.

The precise regulation of the cations is especially important, because significant alterations in the concentrations of these substances can produce a severe derangement of normal function. For example, increased (or decreased) plasma sodium and/or potassium levels can induce profound derangements in the bioelectric properties of nerve and muscle cell membranes, leading to an impairment of their normal function. Changes in the level of calcium ions directly alter the permeability of cell membranes in general. Thus, a high concentration of calcium ion decreases cellular permeability, whereas a low concentration increases it, with concomitant changes in normal cellular activity. Although the precise function of magnesium is unknown, elevated levels of this ion depress both the nervous system and muscular contraction.

The strict regulation of anion concentrations, on the other hand, is not so critical as is the need for cation regulation. The ratio of chloride to bicarbonate is of great importance, however. This topic will be dealt with on page 802 of this chapter. Phosphate ion is essential because, first of all, this material combines with calcium to form the salts that comprise bone (p. 1064), and second, phosphate per se is essential for the maintenance of intracellular chemical reactions. The level of phosphate ion must be regulated within reasonable limits in order that normal bone formation and replacement be achieved and also so that normal cellular function can be maintained at all times.

REGULATION OF SODIUM ION CONCENTRATION. Since sodium forms approximately 90 percent of the total extracellular cations, this substance is the most important single cation that must be regulated. The mechanism responsible for regulating sodium concentration in the extracellular fluid is found in the kidney and adrenal cortex.

As shown in Figures 12.15 and 12.16, sodium is absorbed from various parts of the nephron. This reabsorptive mechanism for sodium is extremely adaptable in its function under different circumstances. Thus, on a limited sodium intake, practically all the filtered sodium is reabsorbed, so that less than 1 gm per day is excreted in the urine. Alternatively, if the sodium intake is high, as much as 30 gm per day of sodium can be excreted by normal kidneys.

Sodium absorption by the renal tubules is regulated primarily by the blood concentration of aldosterone, an adrenocortical hormone. Although the adrenal cortex secretes several mineralocorticoids which influence salt excretion (p. 1096), aldosterone has by far the greatest potency of any of them (by a factor of 30 times). Therefore, aldosterone accounts for about 95 percent of the total sodium reabsorption which is caused by all of the adrenal mineralocorticoids. The rate of sodium ab-

sorption from the tubular fluid is accelerated by aldosterone in all parts of the tubules; however, this effect is most important in the ascending limb of Henle's loop as well as in the distal convoluted tubules and collecting ducts. When aldosterone is absent, almost no sodium is reabsorbed for about 3 days,* whereas an excess of this hormone causes almost complete reabsorption of this cation. Urinary sodium concentration can fluctuate over an extremely wide range, depending upon the level of aldosterone in the blood.

Aldosterone secretion by the cortices of the adrenal glands in turn can be regulated by one (or more) of four factors acting singly or in combination.

Low Sodium Concentration. A reduced extracellular fluid sodium concentration per se provides an important direct stimulus to the adrenal cortex to secrete aldosterone. A feedback mechanism controlling the blood level of this hormone is present. A reduced sodium level stimulates aldosterone secretion. The aldosterone in turn, acting directly upon the resorptive mechanism within the renal tubules, causes more sodium to be removed from the tubular fluid so that the blood level of sodium rises. The elevated blood sodium level then serves to inhibit the release of more aldosterone by the adrenal cortices, and the system becomes stabilized within narrow limits. Conversely, elevated blood sodium levels inhibit aldosterone secretion, and urinary output of sodium increases as tubular reabsorption of this ion is reduced.

Elevated Potassium Concentration. Aldosterone secretion is directly related to blood potassium concentration. Only very slight changes in the concentration of this cation are necessary to double the output of this adrenal hormone from the adrenal glands.

Physical Stress. Aldosterone secretion is increased to a slight extent by trauma (eg, burns, a broken bone) or by extreme physical exertion. This type of stimulus acts indirectly upon the adrenals via the pituitary gland, which liberates increased quantities of adrenocorticotropin (adrenocorticotropic hormone, ACTH). Pituitary corticotropin in turn stimulates the adrenals to secrete larger amounts of another general class of adrenal hormones known as glucocorticoids (p. 1097), and aldosterone secretion also is increased to a very small extent by the action of adrenocorticotropin.

Reduced Cardiac Output and Blood Volume. A decreased cardiac output or a reduced blood volume increases the rate of aldosterone secretion, so that the increased sodium retention increases the total extracellular fluid volume secondarily to the sodium retention, and the increased blood volume in turn increases cardiac output toward normal

*After 3 days, the kidney begins to reabsorb sodium normally and in the absence of aldosterone. This phenomenon is called "aldosterone escape."

levels so that adequate tissue blood flow is restored.

The detailed mechanism or mechanisms whereby a lowered cardiac output and/or reduced blood volume stimulates aldosterone release from the adrenal cortex remain obscure. This statement also is true concerning the effect of the other three stimuli for aldosterone release discussed above, although several theories have been advanced to account for the effects observed.

REGULATION OF POTASSIUM ION CONCENTRATION. Although potassium usually is considered to be primarily an intracellular cation, the extracellular fluid concentration of this cation nonetheless is of considerable importance to normal physiologic activity. Therefore, the renal mechanism whereby the blood potassium level is regulated is also of considerable importance. The two principal mechanisms whereby the extracellular potassium level is regulated are aldosterone feedback and the extracellular fluid potassium level itself.

The Aldosterone Feedback Mechanism. This feedback system functions in a precisely opposite direction to that described for sodium regulation. When sodium is reabsorbed from the renal tubules, large quantities of cations are transported from the tubular to the peritubular fluid, resulting in the development of a strong electronegative potential (-40 to -120 mv) within the lumen of the tubule. Potassium, being a cation, diffuses passively down this electric gradient (which was generated by active sodium transport) into the tubular fluid from the tubular cells. Thence it is excreted in the urine. Potassium ions are exchanged by a passive transport mechanism for sodium ions which are simultaneously and actively transported out of the tubular fluid, hence reabsorbed. This exchange is enhanced markedly by increased aldosterone blood levels, and depressed by a lowered aldosterone concentration. The feedback system is completed by the blood potassium level itself. An elevated blood potassium level enhances the rate of aldosterone secretion, as discussed above, so that more sodium is reabsorbed. Concomitantly, however, more potassium is excreted. In this manner, the blood level of potassium is lowered toward normal once again. Consequently, the overall effect of the aldosterone feedback mechanism is to regulate the sodium-to-potassium ratio in the extracellular fluids.

Extracellular Fluid Potassium Level. The second mechanism that is involved in the regulation of potassium level of the extracellular fluids is the potassium concentration within these fluids themselves. Thus, an elevated blood potassium ion concentration results in an increased secretion of this cation by passive transport into the electronegative tubular fluid from the peritubular fluid and epithelial cells of the distal nephron, since a steeper diffusion gradient is present under such circumstances. The aldosterone feedback system operating cooperatively with the passive transport system together produce an exceedingly precise control of extracellular fluid potassium concentration, so that it remains within 10 percent of an average value of 4.5 mEq per liter.

REGULATION OF MAGNESIUM ION CONCENTRATION. Magnesium ion levels in the extracellular fluids apparently are regulated by a passive diffusion mechanism which is similar to that described above for potassium, although little is known of this process in detail. It is clear, however, that increased magnesium levels in the blood promote decreased renal absorption of this cation; the converse of this situation also holds true. Magnesium, in common with potassium, is principally an intracellular rather than an extracellular cation.

REGULATION OF CALCIUM ION CONCENTRATION. The blood level of this important extracellular fluid constituent is regulated primarily by the hormone of the parathyroid glands, which will be discussed in detail in Chapter 15 (p. 1062). The renal tubules also appear to regulate the blood calcium ion level directly to some extent. An increased concentration of calcium in the extracellular fluids results in a markedly elevated urinary excretion of this cation. Part of the renal regulation of the calcium level in the extracellular fluids also may reside in the parathyroid mechanism, although the details of the overall mechanism that controls calcium excretion and absorption by the kidneys are unknown.

REGULATION OF EXTRACELLULAR FLUID ANION CONCENTRATIONS. The total concentration of anions in the extracellular fluid is regulated indirectly (or secondarily) as a consequence of the primary (or direct) regulation of the cations discussed above. The reason for this is as follows. When a cation is transported actively (ie, absorbed) from the lumen of the renal tubule, an electronegative state is generated within the tubular lumen, so that anions, and particularly chloride ion, diffuse passively out of the tubule in order to maintain electric neutrality within the tubular lumen. Consequently, the identical mechanisms that are involved directly in the reabsorption of cations indirectly produce reabsorption of anions. For example, aldosterone stimulates active reabsorption of sodium, and thus indirectly stimulates the reabsorption of anions, mainly chloride, since this is the principal anion found in the glomerular filtrate (approximately 75 percent of the total anions present is Cl^-).

The chloride-to-bicarbonate ratio is far more significant than the absolute concentrations of each of these ions. This important topic will be discussed subsequently in relation to the renal regulation of acid–base balance.

Phosphate ion exhibits a definite transport maximum, so that when the concentration of this anion in the glomerular filtrate is exceeded, the excess phosphate now is excreted in the urine. Below the tubular maximum, all the phosphate is reabsorbed. Since the normal intake of phosphate (as in milk) generally exceeds this tubular maximum (around 0.1 millimol of phosphate per minute), and the plasma level is above threshold (approximately 0.8 millimol per liter), the excess phosphate above threshold passes into the urine. Conversely, if phosphate intake is insufficient to exceed these values, phosphate is reabsorbed completely from the glomerular filtrate and thus is retained by the body.

The role of the parathyroid glands in regulating phosphate metabolism will be discussed on page 1062, as this anion is closely related to calcium from a metabolic standpoint.

Minor anions such as sulfates, nitrates, urates, lactates, and amino acids all exhibit individual transport maxima. Therefore, these substances are regulated by an "overflow process" similar to that described above for phosphate ion, and so when they are present within the extracellular fluids at concentrations that are below their individual thresholds, they are reabsorbed completely from the glomerular filtrate.

Regulation of Extracellular Fluid and Blood Volumes

In this section, a general summary of the regulatory mechanisms that are concerned with the overall regulation of extracellular fluid and blood volumes will be presented. Certain individual aspects of these topics have been discussed earlier, so that cross-reference to them will be made as appropriate. In this way, functional processes covered independently in the earlier material can be viewed in perspective as part of an integrated physiologic mechanism that functions to maintain overall fluid balance within the body.

Fundamentally, three distinct physiologic mechanisms regulate extracellular fluid and blood volume: (1) an inherent (or intrinsic) mechanical system that consists of the processes involved in capillary dynamics and fluid exchange therein, as well as hemodynamic effects that are exerted directly upon the excretion of fluid within the kidneys; (2) endocrine factors that influence fluid volume, most particularly those hormones regulating the excretion of extracellular fluid by the kidneys; and (3) certain reflexes that may influence the ability of the kidneys to excrete fluid.

REGULATION OF INTERSTITIAL FLUID VOLUME. The extracellular fluid compartment of the body fluids consists of the interstitial fluid and the plasma (p. 524). Interstitial fluid normally does not accumulate within the tissues in excess because of the action of two mechanisms: (1) the osmotic reabsorption of water by the capillaries at their venous ends (p. 558), and (2) the continual drainage of excess interstitial fluid from the tissue spaces by the lymphatics (p. 556). Normally, these two mechanisms remove excess interstitial fluid almost as rapidly as it is formed; therefore, only minute volumes of interstitial fluid are present at any time among the cells of the body. The interstitial fluid volume in turn is regulated by the quantities of mucopolysaccharides in the interstitial spaces. Therefore, young children and babies, whose bodies contain much higher concentrations of mucopolysaccharides than older persons, have a higher proportion of interstitial fluid than do older persons; ie, younger persons are more juicy. In myxedema, a consequence of hypothyroidism (p. 1057), the patient has an abnormally elevated synthesis of polysaccharides so that an excess of interstitial fluid is present, giving the patient a bloated or "doughy" appearance.

It is apparent, therefore, that under normal circumstances the volume of interstitial fluid is regulated principally by the tissues themselves, rather than by the circulatory system. Abnormal fluid retention within the body such as occurs in cardiovascular or renal disease as well as in imbalance of the normal capillary hemodynamics causes excess quantities of fluid to be retained and edema ensues (p. 576).

REGULATION OF BLOOD VOLUME. Any discussion pertaining to the control of blood volume is inseparable from a consideration of certain cardiovascular factors, since the two systems function as an integrated unit in regulating the volume of this body fluid compartment.

As emphasized in Chapter 10, the intrinsic controls of the circulation play a major role in the regulation of normal cardiovascular function. The three principal factors involved may be reviewed briefly: (1) The intrinsic property of the heart to pump blood which is returned to it by the great veins into the arteries of the systemic circulation. The quantity of blood which is pumped out in turn depends upon venous return and the automatic alterations in the contractility of the myocardium in accordance with the Frank–Starling mechanism (p. 660). (2) There is a long-term intrinsic autoregulatory mechanism which governs pressure-flow relationships within individual organs, particularly the kidney (p. 776). (3) The arterial pressure exerts a direct influence upon the excretion of fluid by the kidneys into the urine. These mechanisms are relatively slow; consequently, more rapidly acting neural and humoral (endocrine) controls superimpose their effects upon the more slowly acting intrinsic controls.

The elements of the feedback system controlling the total plasma volume, hence the blood volume, are as follows: When the extracellular fluid volume becomes elevated above normal limits, approximately one-third of the excess fluid remains in the plasma so that the blood volume increases. This increased blood volume in turn increases venous return to the heart, thereby increasing the cardiac output. The elevated cardiac output results in an increased systemic arterial blood pressure, because the increased flow directly raises the pressure in accordance with Poiseuille's law, and also because the increased blood flow through the vascular beds of individual tissues augments the total peripheral resistance, and thus secondarily elevates the blood pressure by an additional increment. The elevated arterial pressure in turn stimulates the kidneys to secrete the excess fluid volume, so that the total plasma volume returns toward normal. As this process continues, the stimulus (ie, elevated blood volume) to increased fluid excretion by the kidneys is removed so that the blood pressure falls once again to within normal limits.

The osmoreceptor–antidiuretic hormone (ADH) mechanism, discussed earlier, also plays a significant role in the process of regulating blood (hence interstitial fluid) volume in addition to the purely mechanical role of the intrinsic regulatory mechanism described above. Furthermore, the aldosterone system of the adrenal cortex for regulating electrolytes within the extracellular fluids also plays a significant role in the retention of salts as well as water by the kidneys. The relative importance of these endocrine mechanisms now may be considered in perspective, and in relation to the inherent mechanical regulatory mechanism.

Excess ADH can change the total amount of water in the body fluids by approximately 5 percent in either direction before the altered blood volume is opposed by the intrinsic mechanism.

The aldosterone mechanism, once it has achieved a maximal increase in total blood volume of about 10 percent above normal (and the blood pressure has risen concomitantly about 20 mm Hg), also is opposed by the intrinsic mechanical system. Consequently, an increased fluid excretion ensues which balances the ability of aldosterone to cause further reabsorption and retention of electrolytes and water. The ADH system is thus far more significant for the regulation of osmolality of the body fluids than it is for extracellular fluid volume regulation.

REFLEXES INVOLVED IN THE REGULATION OF EXTRACELLULAR FLUID VOLUME. Practically all the cardiovascular reflexes play some role in the control of blood volume. For example, the baroreceptor reflex (p. 672), when stimulated by a reduced systemic arterial blood pressure, induces reflex afferent arteriolar vasoconstriction in the kidney. This results in a decreased glomerular filtration rate, so that fluid is retained by the body, and this effect tends to restore the arterial pressure toward normal.

There also is some experimental evidence which indicates that certain reflexes originating in the great veins and atria may have a more specific and direct function in the regulation of blood volume. Within the walls of these structures, especially in the left atrium, are located mechanoreceptors called volume receptors. Thus, distention of these structures, as by an increased venous return and pressure (ie, by an elevated blood volume), is believed by some investigators to stimulate a reflex causing increased renal blood flow and urinary excretion. Afferent nerve impulses transmitted to the hypothalamus from these receptors are supposed to reduce the secretion of antidiuretic hormone, so that the kidneys excrete more water, and the plasma volume is reduced toward normal. Equally compelling experimental evidence, however, indicates that the participation of a reflex mechanism for the regulation of total blood volume in vivo is of minor significance, if it is assumed that such a mechanism is present at all. Consequently, the status of neural mechanisms in the regulation of total blood volume is unclear at present. Hence, the ADH, aldosterone, and intrinsic mechanical system collectively appear to have the major roles in the regulation of total blood volume.

The Role of the Kidneys in the Regulation of Hydrogen Ion Concentration

The kidneys serve not only to regulate the volume and composition of the extracellular fluids, but also provide an excretory function to rid the body of excess or unwanted materials. In addition, these organs perform an equally important function in the regulation of acid–base balance within the body. In this section, the mechanisms whereby hydrogen ions in urine and extracellular fluids are regulated by the renal tubules will be discussed in relation to other substances which are intimately related to this process. This presentation will be followed by a summary of the role of the kidneys in metabolic acidosis and alkalosis.

In order to protect the body fluids against excessive fluctuations in hydrogen ion concentration, and thus to prevent acidosis or alkalosis, three specialized control and regulatory systems are present in the body to maintain homeostatic equilibrium of the pH within normal limits: (1) All the body fluids are provided with acid–base buffer systems which combine immediately with excess acid or alkali and thereby prevent marked changes in hydrogen ion concentration, ie, pH (p. 755). (2) Appreciable changes in hydrogen ion concentration are followed by appropriate stimulation (or inhibition) of the respiratory center, so that pulmonary ventilation is altered. This effect in turn leads to an altered rate of carbon dioxide elimination, and thus the hydrogen ion concentration tends to return toward normal (Fig. 11.41B, p. 761). (3) Most pertinent to this discussion, if the hydrogen ion concentration is altered from normal levels, the kidneys compensate for this change by excretion of either an acid or an alkaline urine.

Thus, these three mechanisms operate in an integrated, cooperative fashion in order to effect restoration of extracellular fluid pH to normal values. Temporally speaking, the acid–base buffer systems act within a fraction of a second when excess hydrogen ion is present. The respiratory system requires several minutes to come into full operation when the concentration of hydrogen ion is altered suddenly. In contrast to these two rapidly acting systems, the renal mechanisms, which are the most powerful of the three acid–base regulatory mechanisms, require from several hours to several days to restore the hydrogen ion concentration to normal levels.

REGULATION OF HYDROGEN ION CONCENTRATION BY THE KIDNEYS. In accordance with the Henderson–Hasselbalch equation, the kidneys regulate the hydrogen ion concentration primarily by increasing or decreasing the bicarbonate ion concentration within the body fluids. Therefore, this relationship can be written:

$$pH = pK + \log \frac{[HCO_3^-]}{[CO_2 + H_2CO_3]} \qquad (1)$$

$$pH = P_{KH_2CO_3} + \log \frac{[HCO_3^-]}{[H_2CO_3]} \qquad (2)$$

The HCO_3^- concentration in the numerator is expressed in millimoles rather than in molarity, as is usually the case when concentrations are indicated by brackets. Since the Pco_2 can be determined readily in the laboratory, the denominator in the log expression in equation (1) can be replaced by the Pco_2 multiplied by the factor 0.03, if it is assumed that the temperature is 38 C, and the ionic strength is 0.15. At this temperature and ionic strength, the pK of the entire carbonic acid pool is 6.1. Therefore, a useful working equation for physiologic studies is:

$$pH = 6.1 + \log \frac{[HCO_3^-]}{0.03 \; Pco_2} \qquad (3)$$

The important facts to be derived from equation (3) are as follows. An increase in the concentration of dissolved carbon dioxide (Pco_2) shifts the acid–base balance toward the acid side (or lowers the pH). Conversely, an elevated bicarbonate ion concentration shifts the acid–base balance toward the alkaline side (or raises the pH).

Thus, as pointed out earlier, the concentration of carbon dioxide (Pco_2) in the body fluids can be altered markedly by increasing or decreasing pulmonary ventilation, and thereby the respiratory system can regulate the pH of the body fluids to some extent. The kidneys, however, can alter directly the concentration of bicarbonate within the body fluids, so that the pH is increased (toward the alkaline side) or decreased (toward the acid side). These two principal hydrogen ion regulatory systems function by changing the ratio between the two components of the bicarbonate buffer system shown in equation (3) above.

In order that the kidney may regulate hydrogen ion concentration by increasing or decreasing the bicarbonate ion concentration of the body, several interrelated chemical reactions take place within the renal tubules. These reactions include: tubular secretion of hydrogen ions; sodium ion reabsorption; urinary bicarbonate ion excretion; and ammonia secretion by the tubular epithelium.

Secretion of Hydrogen Ions by the Renal Tubules. The mechanisms whereby the proximal and distal tubules, as well as the collecting ducts of the nephrons, secrete hydrogen ions into the tubular fluid are summarized in Figure 12.24.

Renal hydrogen ion secretion commences within the tubular cells with carbon dioxide, which combines with water to form carbonic acid, a reaction catalyzed by the enzyme carbonic anhydrase (p. 757). The carbonic acid then dissociates into hydrogen and bicarbonate ions; the hydrogen ion is secreted through the cell membrane into the tubular lumen. As shown in Figure 12.24, hydrogen ion secretion is coupled intimately to sodium ion reabsorption by the tubular cells. Consequently, a sodium ion is reabsorbed for each hydrogen ion secreted.

Although the exact mechanism whereby hydrogen ions are secreted by the tubules is obscure at present, it is clear that the collecting ducts can concentrate hydrogen ions in the tubular urine to a maximum of about 1,000-fold over their concentration in the peritubular interstitial fluid. Thus, a minimum urinary pH of 4.5 (the limiting pH) can be achieved by the activity of the collecting tubules. In contrast to this, the proximal tubules secrete over 80 percent of the total hydrogen ion that is eliminated in the urine; however, their ability to concentrate hydrogen ion is only threefold, so that the pH can only be reduced about 0.5 unit in this region of the nephron. The minimum pH that the tubular fluid can attain in this region of the nephron is about 6.9 (ie, 0.5 pH unit below the peritubular fluid pH of 7.4). Between these extremes, the distal tubules have an intermediate capacity for secretion and concentration of hydrogen ions.

The series of chemical reactions which is involved in the secretion of hydrogen ions commences with carbon dioxide. Consequently, the rate of these reactions is governed directly by the concentration of carbon dioxide (ie, the Pco_2) in the extracellular fluid in accordance with the law of mass action. Any factor that increases the extracellular fluid Pco_2 accelerates the secretion of hydrogen ion and a more acid urine results. For example, an increased metabolic rate, as during exercise, or a decreased pulmonary ventilatory rate, both can augment the rate of hydrogen ion secretion. Conversely, factors that reduce carbon dioxide levels concomitantly decrease the rate of hydrogen ion secretion, ie, inactivity or increased pulmonary ventilation.

Relationship between Hydrogen Ion Secretion and Bicarbonate Ion Reabsorption. During the process of glomerular filtration, bicarbonate ion passes continuously into the glomerular filtrate. These

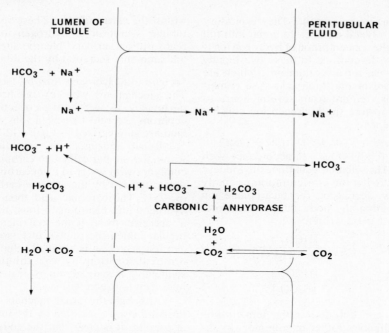

FIG. 12.24. The chemical mechanism whereby titratable acid is formed in the epithelial cells of the renal tubules. (Data from Pitts. *Physiology of the Kidney and Body Fluids,* 2nd ed, 1968, Year Book Medical Publishers; Ganong. *Review of Medical Physiology,* 6th ed, 1973, Lange Medical Publications.)

bicarbonate ions are accompanied by cations, principally sodium. As the sodium and bicarbonate ions enter the tubular fluid, large quantities of the sodium ions are reabsorbed, chiefly in exchange for hydrogen ions, as noted above. Simultaneously, the hydrogen ions react chemically with the bicarbonate ions, as shown in Figure 12.25, to form carbonic acid. This compound in turn dissociates into carbon dioxide and water. The carbon dioxide thus produced now diffuses almost quantitatively through the epithelial cells back into the peritubular fluid, whereas the water passes into the urine. If a sufficient number of hydrogen ions are present within the tubular fluid, the bicarbonate ions are almost entirely removed from the tubular fluid, so that essentially none pass into the urine.

As shown in Figure 12.25, the formation of each hydrogen ion within the tubular cells is accompanied by the simultaneous formation of a bicarbonate ion by the dissociation of carbonic acid. The bicarbonate ion thus formed in association with a sodium ion reabsorbed from the tubule then diffuse together into the extracellular fluid surrounding the tubule. Electrochemical neutrality is maintained. The net result of this series of reactions is that a mechanism is present in the renal tubule for the reabsorption of bicarbonate ions from the tubular fluid, and this mechanism operates principally in the cells of the proximal convoluted tubule.

A certain paradox exists in this mechanism for the reabsorption of bicarbonate ions, however, because the bicarbonate ions which are reabsorbed

from the tubular fluid are synthesized de novo within the tubular epithelial cells, and thus are not the identical bicarbonate ions which were present originally in the tubular fluid, as close inspection of Figure 12.25 will indicate.

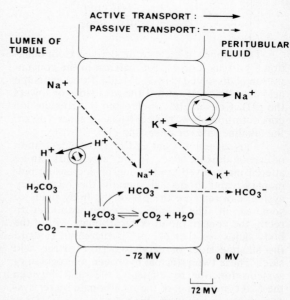

FIG. 12.25. The reabsorption of bicarbonate ion in an epithelial cell of the proximal convoluted tubule. Conventions as given in legend to Figure 12.17. (After Pitts. *Physiology of the Kidney and Body Fluids,* 2nd ed, 1968, Year Book Medical Publishers.)

TRANSPORT OF HYDROGEN IONS FROM TUBULES TO URINE.

There are two different mechanisms whereby excess hydrogen ions secreted into the tubular fluid are transported in the urine. In both of these mechanisms, however, the hydrogen ions react chemically with other substances in the tubular fluid for transport into the urine.

Reaction with Buffers in the Tubular Fluid. In addition to the bicarbonate buffer discussed above, excess hydrogen ions can combine with phosphate buffer for transport in the tubular urine. Quantitatively, disodium phosphate (Na_2HPO_4) is about 4 times more abundant in the glomerular filtrate than is sodium acid phosphate (NaH_2PO_4). Surplus hydrogen ions, upon entering the tubules, can react with disodium phosphate, as illustrated in Figure 12.26. This reaction forms monosodium phosphate, which is excreted in the urine. In this reaction, one sodium ion is released, as shown in Figure 12.26, and this ion is reabsorbed from the tubules instead of the hydrogen ion. As described earlier, this sodium ion combines with the bicarbonate ion which is synthesized de novo during the secretion of hydrogen ions (Figs. 12.24 and 12.25). The consequence of these reactions is a net increase in the extracellular fluid sodium bicarbonate concentration, and this is the chemical mechanism whereby the kidney alleviates a condition of acidosis in the body fluids.

Ammonia Secretion. The second mechanism for the transport of hydrogen ions from tubular fluid to the urine is by their reaction with ammonia. The principal reactions whereby ammonia is synthesized are summarized in Figure 12.27, together with the mechanisms within the renal tubules whereby ammonium ions (NH_4^+), formed by the reaction of ammonia (NH_3) with hydrogen ions, are excreted.

The epithelial cells of the proximal and distal tubules all synthesize ammonia continuously, and this compound diffuses into the tubules where ammonium ion is formed by reaction with hydrogen ions for excretion in association with chloride and other tubular anions. The result of these reactions, as summarized in Figure 12.27, is a net increase in the sodium bicarbonate concentration in the peritubular interstitial fluid, as described above for the reactions involving tubular phosphate buffers.

The ammonia mechanism for transport of hydrogen ions is particularly important physiologically, because the principal anion in the tubular fluid is chloride, and only small quantities of hydrogen ion could be transported by direct reaction with chloride for the following reason. Hydrochloric acid is a strong acid so that if much of this substance were present within the tubular fluid, the pH would decline rapidly below 4.5, and thereby additional hydrogen ion secretion would be inhibited. When hydrogen ions and ammonia

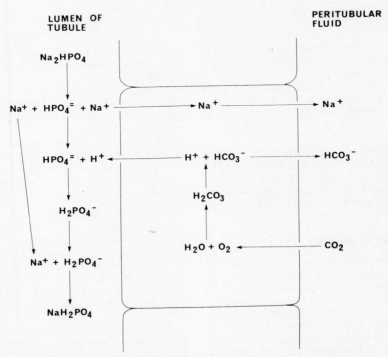

FIG. 12.26. Chemical mechanism in the renal tubules whereby hydrogen ion is buffered by phosphate. (After Ganong. *Review of Medical Physiology*, 6th ed, 1973, Lange Medical Publications.)

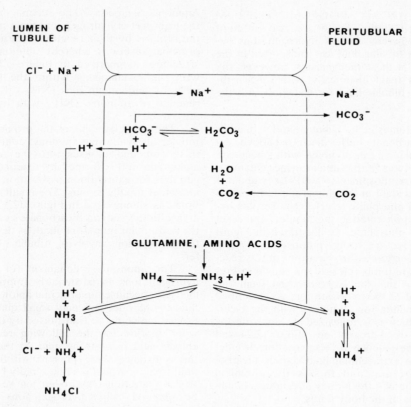

FIG. 12.27. Ammonia synthesis by the renal tubular epithelial cells and neutralization of hydrogen ion by reaction of hydrogen with the ammonia in the tubular lumen. (After Guyton. *Textbook of Medical Physiology,* 4th ed, 1971, Saunders.)

combine, the resulting ammonium chloride is a neutral salt; therefore, no significant fall in pH occurs within the tubule.

As indicated in Figure 12.27, about 60 percent of the ammonia synthesized in the renal tubules is derived from glutamine, whereas the remaining 40 percent originates from asparagine and certain amino acids.

The ammonia-secreting mechanism is highly adaptable, as indicated by the fact that if the tubular fluids remain highly acidic for long periods of time, the rate of ammonia formation rises steadily, ultimately by as much as tenfold after several days.

Renal Acid–Base Regulation

The respiratory mechanisms involved in acid–base regulation were presented in Chapter 11 (p. 755), and certain features of respiratory as well as metabolic acidosis and alkalosis were presented in Figure 11.41 (p. 761) in order that the physiologic mechanisms which underlie these abnormalities could be compared readily. The renal mechanisms for the secretion of hydrogen ions and the reabsorption of bicarbonate ions have been discussed above. The homeostatic mechanisms whereby the

kidneys regulate an abnormal body fluid pH toward normal values now will be examined.

By convention, the terms *metabolic acidosis* and *metabolic alkalosis* refer to all situations in which the normal acid–base balance is deranged, except those conditions that are produced by an excess or deficiency of carbon dioxide in the body fluids. Dissolved carbon dioxide is termed a *respiratory acid* (even though it is produced metabolically!), whereas all other acids are called *metabolic* or *fixed acids,* whether they are produced by faulty metabolism or ingested.

METABOLIC ACIDOSIS. The condition of metabolic acidosis can result from an excess production of, or an accumulation of, excess acids in the body (eg, as in diabetes mellitus or uremia); IV (parenteral) administration or ingestion of fixed acids; or loss of alkali from the body fluids (as in severe diarrhea or prolonged vomiting of intestinal contents which contain large quantities of bicarbonate).

All these clinical conditions have a common physiologic denominator; specifically, they produce an increased hydrogen ion concentration in the extracellular fluids. The proportion of carbon dioxide to bicarbonate ions, equation (3) p. 805, increases in the body fluids and the pH falls. In compensation for this effect, an increase in hydrogen ion secretion takes place. The rate of hydrogen ion secretion becomes elevated far above the rate of filtration of bicarbonate ion, so that the secreted acid cations no longer have sufficient bicarbonate with which to react. The excess acid, however, combines with the buffers in the tubular fluid as described earlier (Figs. 12.26 and 12.27), and the products of these reactions are excreted in the urine.

As illustrated in Figure 12.25, two other events take place simultaneously whenever a hydrogen ion is secreted: (1) a sodium ion is absorbed from the tubular fluid into the epithelial cell, and (2) a bicarbonate ion is synthesized within the epithelial cell itself. The sodium and bicarbonate ions then diffuse together from the inside of the epithelial cell into the interstitial fluid surrounding the tubule, so that there is a net increase of sodium bicarbonate in the extracellular fluid. This effect in turn increases the ratio of bicarbonate to acid in the peritubular interstitial fluid in accordance with the Henderson–Hasselbalch equation, and the pH of this fluid now rises. In this manner, renal compensation for metabolic acidosis takes place, as summarized in Figure 11.41 (p. 761). The buffering mechanisms involved in metabolic acidosis are shown in Figure 12.28.

Physiologic Effects of Metabolic Acidosis. Depression of the central nervous system is the principal physiologic effect of a metabolic acidosis so severe that the compensatory mechanisms are unable to cope with the excess hydrogen ion. A patient suffering from this condition loses his orientation and then becomes comatose when the blood pH falls below 7.0; persons dying of uremic or diabetic acidosis generally are in a state of coma when death ensues.

Since the increased hydrogen ion concentration of acidosis stimulates the respiratory center, the increased rate and depth of respiration are useful diagnostic signs of metabolic acidosis. In contrast to this situation, pulmonary ventilation generally is depressed in respiratory alkalosis.

METABOLIC ALKALOSIS. Metabolic alkalosis is far less common clinically than is metabolic acidosis. Metabolic alkalosis usually is seen following the ingestion of excessive amounts of alkaline drugs or chemicals, eg, sodium bicarbonate. Such compounds are employed in the treatment of peptic ulcer or gastritis. Prolonged vomiting of the gastric contents, with an attendant loss of hydrochloric acid, also can lead to metabolic alkalosis through retention of excess bicarbonate in the extracellular fluids.*

Metabolic alkalosis has an etiology which is directly opposite to that which causes metabolic acidosis: that is, the ratio of bicarbonate ion to

*Note that if lower gastrointestinal tract contents are lost, a metabolic acidosis ensues.

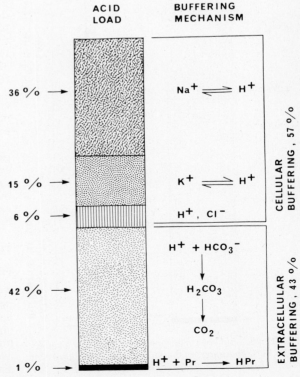

FIG. 12.28. Buffering mechanisms in metabolic acidosis. (After Pitts. *Physiology of the Kidney and Body Fluids,* 2nd ed, 1968, Year Book Medical Publishers.)

dissolved carbon dioxide increases so that the pH likewise increases above 7.4 in accordance with the Henderson–Hasselbalch equation. This effect serves as the stimulus for the renal tubules to increase the ratio of bicarbonate ions filtered to the hydrogen ions secreted. Consequently, the exquisite balance in the bicarbonate to dissolved carbon dioxide ratio is upset, and increased quantities of bicarbonate ions enter the tubular fluid. As no bicarbonate ions can be reabsorbed without involving a hydrogen ion, as shown in Figure 12.25, all the excess bicarbonate ions which are present in metabolic alkalosis are carried into the urine in association with sodium and other cations. The net result of this process is to remove excess sodium bicarbonate from the extracellular fluid, and thereby renal compensation for the metabolic alkalosis occurs. The buffering mechanisms involved in metabolic alkalosis are shown in Figure 12.29.

Since a net loss of bicarbonate ions from the extracellular fluid decreases the ratio of bicarbonate to carbon dioxide back toward normal in accordance with the Henderson–Hasselbalch equation, the pH of the extracellular fluids is reduced toward normal levels.

Physiologic Effects of Metabolic Alkalosis.   In contrast to a metabolic acidosis, metabolic alkalosis

produces hyperexcitability of the central nervous system. Both the central nervous system and the peripheral nerves are affected; however, the latter are more susceptible, so that they are affected first. Spontaneous firing of impulses occurs, resulting in muscular tetany, which usually commences first in the arms. In severe cases, respiratory arrest may occur, with death following tetanic spasms of the respiratory muscles. Usually a metabolic alkalosis results in extreme nervousness or convulsions, especially in patients who are subject to epileptic seizures.

Once again, in marked contrast to metabolic acidosis, a state of metabolic alkalosis depresses pulmonary ventilation. In both types of disorder, however, respiratory compensation alone can only restore the blood pH about halfway to a normal level. Thus, the renal compensatory mechanisms are critical in effecting a complete homeostatic restoration of normal extracellular fluid pH to around 7.4.

In this regard, the maximum total capacity of both kidneys to eliminate acid or alkali is around 500 millimoles per day. If this total capacity is exceeded and larger quantities of acid or alkali enter the body fluids than can be dealt with by the kidneys, a severe metabolic acidosis or alkalosis ensues.

URINARY pH.   Under normal conditions, hydrogen ions are secreted by the kidneys at a rate of 3.50 millimoles per minute, whereas bicarbonate is filtered at a rate of 3.49 millimoles per minute. Since practically all the bicarbonate ions are removed from the tubular fluid by the mechanism described earlier (Fig. 12.25), a small excess of hydrogen ions remains in the tubular fluid and reacts with other substances (buffers, ammonia) for excretion in the urine. As shown in Figure 12.30, the rate of gain or loss of bicarbonate ions is related closely to extracellular fluid pH.

The compensatory mechanisms just discussed can alter the urinary pH from a physiologic minimum of 4.5 (during metabolic acidosis when excess hydrogen ion is being excreted) up to 8.0 when excess alkali is being excreted. Since between 50 and 100 millimoles more of acid are synthesized each day by the body than of alkali, a small quantity of this acid is lost continuously in the urine of normal individuals. Even when the extracellular fluid pH is at a normal value of 7.4, the normal urinary pH is acid, usually around 6.0.

FUNCTIONS OF THE URINARY BLADDER

Filling of the Bladder

As noted earlier, the walls of the ureters contain three layers of smooth muscle. Urine is transported from the renal pelvis to the bladder by unidirectional waves of rhythmic peristaltic contractions of the ureters which occur at a rate of one to five per minute. The urine enters the bladder in spurts which are synchronous with the peristaltic

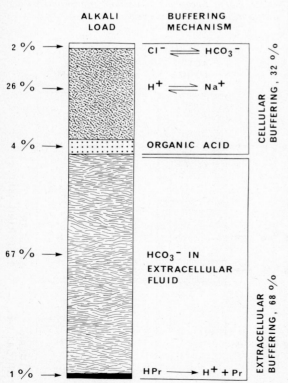

FIG. 12.29.   Buffering mechanisms in metabolic alkalosis. (After Pitts. *Physiology of the Kidney and Body Fluids,* 2nd ed, 1968, Year Book Medical Publishers.)

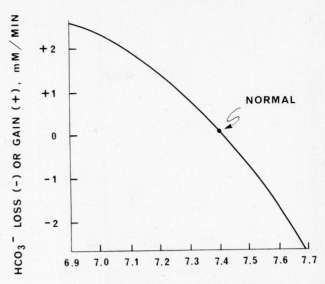

FIG. 12.30. Relationship between the pH of the extracellular fluid and the rate-of-change of bicarbonate ion concentration in the body fluids. (+) = gain in bicarbonate ion; (−) = loss of bicarbonate ion. (After Guyton. *Textbook of Medical Physiology,* 4th ed, 1971, Saunders.)

waves. There are no ureteral sphincters; however, as these structures enter the bladder at an oblique angle, the ureters tend to remain closed except during the passage of a contraction wave. Thereby urine does not reflux from the bladder back up the ureters toward the kidneys.

Emptying of the Bladder (Micturition)

Similarly to the ureters, the wall of the bladder also is composed of three layers of smooth muscle. The outermost layer, known as the detrusor muscle, is responsible primarily for emptying the bladder during the act of urination (or micturition). The internal urethral sphincter is composed of extensions of the detrusor muscle on each side of the urethra. The external urethral sphincter is composed of striated muscle and is located farther along the urethra from the bladder than is the internal sphincter. The innervation of the bladder is summarized in Table 12.2. The act of micturition itself is a spinal reflex which may be facilitated or inhibited by higher nervous centers, in a manner similar to the act of defecation. As the urine enters the bladder, the intravesicular pressure does not rise appreciably at first, ie, the smooth muscle fibers adapt to the increased stretch without a significant change in tonus. In other words, the smooth muscle of the bladder thus exhibits the property of plasticity so that when it is stretched the initial tension is not maintained; therefore, only small pressure increments develop as the bladder fills. As the micturition reflex is initiated, however, a sudden steep rise in pressure occurs. These facts can be recorded by a cystometrogram (Fig. 12.31), in which a pressure versus volume curve is plotted from data obtained by a urethral catheter after injection of known volumes of fluid into the bladder.

The impulse to void is perceived first at a bladder volume of around 150 ml; a pronounced sense of discomfort or fullness occurs at a volume of about 400 ml.

Figure 12.31 shows that the relatively flat portion of the curve, obtained as the bladder is filled with successive increments of fluid, is a reflection of Laplace's law. According to this law, the pressure is proportional to the wall tension of a hollow elastic organ but inversely related to the radius of the viscus; therefore, the wall tension of the bladder increases as the organ fills. In the bladder, the radius also increases as filling takes place, so that until this organ is markedly distended, the pressure increase exerted on the luminal contents is small.

During micturition, the external urethral muscles as well as the perineal muscles relax, the detrusor muscle contracts, and urine is expelled through the urethra by the sudden pressure increase in the bladder (Fig. 12.31C). The internal urethral sphincter appears to have no part in micturition; its only role appears to be in the male during ejaculation, when it prevents the reflux of semen into the bladder.

It is not clear at present how the mechanism of voluntary micturition is initiated; however, the ability to maintain the external urethral sphincter in a contracted state is definitely a learned response.

During the act of urination, one of the first events which occurs is the relaxation of the muscles of the pelvic floor so that the ensuing downward pull exerted upon the detrusor muscle can stimulate its contraction. The external urethral sphincter and perineal muscles can be contracted voluntarily, thereby preventing or inhibiting urine flow once it has begun. After micturition, the female urethra empties by gravity, whereas the

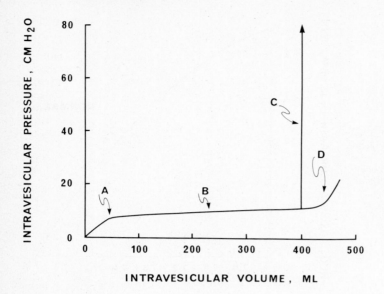

FIG. 12.31. Cystometrogram of normal human urinary bladder. A. Slight pressure increment develops when bladder starts to fill. B. Plasticity of smooth muscle of bladder wall enables bladder to fill markedly with very small pressure increment (see text). C. A high pressure develops rapidly when micturition occurs. D. Pressure–volume relationships that would have developed had the subject not micturated (C). (After Ganong. *Review of Medical Physiology,* 6th ed, 1973, Lange Medical Publications.)

male urethra is emptied by rhythmic contractions of the bulbocavernosus muscle.

Bladder contraction normally is under the control of a reflex, although the smooth muscle fibers in the wall of this viscus do possess intrinsic contractile properties. Stretch receptors within the wall of the bladder stimulate reflex contraction at a lower threshold than that at which the inherent contractile response is initiated. Sensory fibers running within the pelvic nerves (Table 12.2) form the afferent limb of the voiding reflex; the efferent parasympathetic fibers to the bladder also course within these nerves. This reflex is integrated within the sacral region of the spinal cord.

In an adult, reflex bladder contraction is stimulated normally when the volume of urine in the bladder reaches between 300 and 400 ml. Sym-

pathetic nerves evidently have no role in micturition.

Facilitory and inhibitory centers for the micturition reflex are located within the brain stem. The former is located in the pontine region and posterior hypothalamus, whereas the latter is situated in the brain stem. The superior frontal gyrus of the cerebral cortex has been implicated in the voluntary control of micturition; however, other cortical areas also are involved, as indicated by stimulation experiments.

Three types of neurologic lesion can induce bladder dysfunction: (1) Interruption of afferent nerves from the bladder, for example, diseases of the dorsal roots, such as the tabes dorsalis of advanced syphilis, abolish all reflex contractions of the bladder. (2) Complete denervation of the blad-

Table 12.2 INNERVATION OF THE URINARY BLADDER AND URETHRA[a]

STRUCTURE INNERVATED	PERIPHERAL NERVES	FUNCTION
Afferent Fibers		
Dome of bladder	Hypogastric nerves	Pain
Detrusor muscle	Pelvic nerves	Pain and afferent impulses for micturition (voiding reflex)
Urethra	Pudendal nerves	Pain
Efferent Fibers		
Somatic, S_3–S_4	Pudendal nerves	Stimulate external sphincter
Parasympathetic, S_2–S_4	Pelvic nerves	Stimulate contraction of detrusor muscle, internal sphincter during ejaculation in male
Sympathetic; lower thoracic, upper lumbar	Superior hypogastric plexus and hypogastric nerves	Unclear. Innervate bladder in general as well as internal sphincter

[a]*Data from Guyton.* Textbook of Medical Physiology, *4th ed, 1971, Saunders.*

der, in which both afferent and efferent pathways are destroyed. (3) Immediately following spinal cord transection such as may be caused by traumatic injury, all facilitory and inhibitory pathways descending from the brain are eliminated. In this situation, the bladder overfills, and urine leaks through the sphincters producing a situation known as overflow incontinence. Once the condition of spinal shock has passed, the voiding reflex returns; however, all voluntary control over micturition is permanently lost, eg, as in paraplegics.

RENAL OXYGEN CONSUMPTION

In man, the rate of oxygen consumption by both kidneys is about 20 ml per minute. Since renal blood flow is quite high in proportion to the total mass of both kidneys (ie, 1,100 ml per minute per 300 gm renal tissue in man), it is not surprising that this high flow rate is more a reflection of the work performed by the kidneys in maintaining the composition of the extracellular fluid than it is a reflection of the metabolic requirements of these organs themselves. Furthermore, because of this high flow rate, the arteriovenous oxygen difference in the kidney also is low, being only 17 ml per 100 ml blood. By way of comparison, the systemic blood of the body as a whole has an arteriovenous oxygen difference that ranges between 4 and 6 ml per 100 ml blood. Nevertheless, the kidneys, which comprise less than 0.5 percent of the entire body weight, utilize about 8 percent of the total oxygen consumption at rest, ie, 20 ml per minute.

The single renal function correlated most clearly with total oxygen consumption is the active transport of sodium. The oxygen consumption of the cortical region is roughly 9 ml per 100 gm renal tissue per minute, while that of the medullary region amounts to only 0.4 ml per 100 gm tissue per minute.

CLINICAL CORRELATES

Although many types of renal disease are known, several abnormalities are shared in common by a number of these pathologic states. Commonly, the urine contains protein, erythrocytes, leucocytes, and casts (or small masses of precipitated protein) in various renal diseases. Additionally, in renal disease, the kidney sometimes loses its ability to produce a concentrated or dilute urine, and uremia, proteinuria, and abnormal sodium ion retention all may be associated with abnormal renal function. Renal diseases such as glomerulonephritis, pyelonephritis, and polycystic disease all can result in the secondary development of systemic hypertension, ie, a marked and chronic elevation of the systemic arterial blood pressure. The role of renin, an endocrine product elaborated by the kidneys and having a marked pressor effect, will be considered in Chapter 15 (p. 1169).

A few specific abnormalities of renal function now may be discussed individually.

Water Intoxication

The maximal urine output that can be achieved during water diuresis is about 16 ml per minute when the kidneys are excreting an average osmotic load. If the rate of water intake exceeds this excretory value of 16 ml per minute for a prolonged interval, distention of the cells of the body owing to excess water uptake from the hypotonic extracellular fluid takes place, and symptoms of water intoxication become manifest. Osmotic distention of central neurons within the brain results in the development of convulsions and coma. Eventually, death intervenes unless the excess water load is eliminated before irreparable neural damage ensues.

Water intoxication also can develop following the therapeutic administration of antidiuretic hormone, provided that water intake is not decreased concomitantly. In addition, water intoxication can occur from a release of excessive quantities of endogenous ADH as a consequence of stimuli such as surgical trauma.

Diuretics

Sometimes it is desirable to increase the total urinary output, whereas at other times it is necessary to increase the output of specific urinary constituents; consequently, various diuretics are in use clinically. The mechanism of action of a few of these substances will be reviewed briefly, as such material will provide a useful summary of certain factors affecting urinary volume as well as electrolyte excretion.

Water and ethyl alcohol produce a diuresis by inhibiting antidiuretic hormone secretion via the hypothalamic osmoreceptor mechanism.

Ingestion of a large quantity of an osmotically active substance such as a sugar will produce an osmotic diuresis in which the urine solute concentration approaches that of the plasma.

Xanthine compounds (eg, caffeine) presumably exert their weak diuretic effect via an increase in glomerular filtration rate combined with a decreased tubular reabsorption of sodium ion.

Acidifying salts such as NH_4Cl and $CaCl_2$ also produce diuresis. Thus, ingestion of NH_4Cl results in a dissociation of the compound into ammonium and chloride ions. The NH_4^+ dissociates into ammonia (NH_3) and H^+. The ammonia then is converted into urea; therefore, ingesting an acid salt such as ammonium chloride is similar to ingesting HCl. The hydrogen ion is buffered, whereas the chloride ion is filtered through the glomeruli together with sodium ion, an effect which maintains electric neutrality of the system. Water and sodium ion thus are lost in the urine insofar as sodium ion is not replaced by hydrogen ion from the renal tubules.

Organic mercury salts (eg, mercaptomerin or meralluride) induce diuresis by inhibiting sodium reabsorption in the nephron. These compounds also inhibit potassium secretion.

Carbonic anhydrase inhibitors (eg, acetazolamide) cause diuresis by reducing hydrogen ion secretion through decreasing the supply of carbonic acid. Consequently, sodium ion secretion is enhanced because hydrogen ion secretion is decreased. Bicarbonate ion reabsorption also is inhibited. Since hydrogen and potassium ions compete not only with each other, but also with sodium ion, the reduced hydrogen ion secretion which is induced by carbonic anhydrase inhibitors serves to enhance the elimination of potassium ion.

Thiazide compounds (eg, chlorothiazide) produce diuresis by their inhibitory action on the reabsorption of sodium ion in the distal part of Henle's loop as well as in the proximal region of the distal tubule.

Some additional examples of diuretic substances and their mechanisms of action are summarized in Table 16.15 (p. 1255).

Proteinuria

In a number of renal diseases, the permeability of the glomerular capillaries can increase to such an extent that large quantities of plasma protein are lost in the urine, a condition known as proteinuria. Since by far the largest proportion of this protein is albumin, the condition frequently is called albuminuria. In particular, when nephrosis is present, the quantity of protein that is lost in the urine may exceed the synthetic capacity of the liver to replace the lost plasma protein. The resulting hypoproteinemia may achieve such a magnitude that the plasma volume declines, whereas the plasma oncotic pressure likewise decreases to such an extent that a generalized edema develops.

Uremia

When the metabolic products derived from protein catabolism are not excreted properly so that these compounds accumulate in the blood, a syndrome called uremia becomes manifest. The symptoms of uremia are loss of appetite (anorexia), lethargy, nausea, vomiting, confusion and other signs of mental deterioration, convulsions, and coma. Anemia always is associated with chronic uremia because erythropoiesis is inhibited in such patients.

The severity of the uremia can be evaluated by measuring the quantitative increase above normal in the blood urea nitrogen, nonprotein nitrogen, and creatinine levels. It appears, however, that the accumulation of urea and creatinine in the blood is not responsible for the symptoms of uremia. Rather, other compounds such as phenolic substances and organic acids have been implicated in producing the symptoms noted above.

Regardless of their exact nature, the toxic agents which produce the symptoms of uremia can be removed from the blood by periodic dialysis against a solution having an appropriate composition in a machine known as an "artificial kidney."

Loss of Renal Capacity to Concentrate and Dilute the Urine

In many types of renal disease, the volume of urine that is eliminated becomes progressively greater and the urine becomes more dilute as the kidneys gradually lose their capacity to form a more concentrated urine. Although the kidney may be able to produce a dilute urine, in long-standing renal disease the kidney may only be able to excrete urine having the same osmolality as that of plasma. This fact means that the kidney now has lost the ability to produce either concentration or dilution of the urine. These defects in turn are a result in part of a defect in the countercurrent mechanisms discussed earlier in this chapter (p. 779). The principal defect, however, is a loss in the total number of functioning nephrons. Thus, surgical removal of one kidney results in a loss of 50 percent of the total number of functional nephrons, an effect that also can be produced by renal disease. But the total quantity of electrolytes and other materials excreted by such patients is not halved; consequently, each of the functional nephrons that remain must eliminate through filtration and secretion more osmotically active materials. An osmotic diuresis appears to be involved, as the osmolality of the urine becomes closer to that of the plasma. If, however, most of the nephrons are destroyed, a severely curtailed urine flow (or oliguria) results, and in extreme cases, a total cessation of urine flow (or anuria) may ensue. In the latter instance, the symptoms of uremia inevitably develop.

CHAPTER 13

Gastrointestinal Physiology and Nutrition

Green plants, as well as certain bacteria and protozoa, are autotrophic organisms; that is, the cells of these species are able to utilize carbon dioxide (or carbonates) as their sole source of carbon, and simple inorganic nitrogenous compounds (such as nitrates or ammonia) as precursors for the metabolic synthesis of a vast array of complex organic compounds. Man, on the other hand, is a heterotrophic organism, and as a result he must obtain the various substances required for his overall nutrition from sources outside of his body. Ultimately, therefore, man is wholly dependent upon autotrophic organisms for an exogenous source of the many carbon- and nitrogen-containing compounds required to sustain his life. In addition to the necessity for sufficient quantities and types of carbohydrates, proteins, fats, and vitamins to be present in the diet, man also requires an adequate supply of inorganic compounds (electrolytes) as well as water in order that overall nutritional balance be maintained.

In general terms, the subject of gastrointestinal physiology deals with three relatively distinct, but closely interrelated, processes. First, ingested water and various foodstuffs must be transported in bulk quantities throughout the alimentary canal, and solid waste matter (ie, undigested material and cellular debris) eliminated from the body. Second, the process of digestion transforms many nutrients into forms that may be utilized by the body, as most ingested foodstuffs cannot be absorbed or used directly. Digestion is partly a mechanical, but primarily a chemical process, which in turn requires a number of special enzymes to catalyze the breakdown of various complex foodstuff molecules into simpler forms (Table 13.1). Consequently, the secretion of various juices containing specific digestive enzymes and other substances forms an important subject in the general topic of gastrointestinal physiology. The cells and accessory glands that elaborate these secretions are also important in this regard. Third, once the digestive processes have converted the more complex ingested substances into simpler chemical and/or physical forms, the various individual nutrients are absorbed (or assimilated) from the lumen of the gut. The absorbed materials now may be utilized readily by the cells of the body, either directly or after further metabolic modification in various organs, principally in the liver.

In summary, the topic of gastrointestinal physiology deals with the structures and processes that underlie *transport, digestion,* and *absorption* of various nutrients, as well as the physiologic regulation of these functions. Each of these broad areas of gastrointestinal physiology will be considered individually in the following sections, and these discussions will be followed by a brief review of the general nutritional requirements of man for specific substances.

It is worth mentioning at the outset that materials within the lumen of the gastrointestinal tract are outside of the body in a topologic sense, and so remain until they have traversed the membrane of mucosal cells lining the gut to enter the substance of the body proper. This seemingly abstruse fact is of much clinical importance, however, as gastric pumping combined with lavage is thus quite effective in removing toxic materials from the stomach before they are absorbed into the cells of the body. This technique is employed with good results in certain types of poisoning. Furthermore, severe nutritional deficiencies can develop in persons on a wholly adequate diet if the absorption of essential nutrients is faulty for any reason (eg, chronic diarrhea).

FUNCTIONAL ANATOMY OF THE GASTROINTESTINAL TRACT AND RELATED STRUCTURES

General Features

The digestive system of man is a greatly elongated, sinuous tube that commences with the lips and terminates at the anus. The morphologically and

Table 13.1 SUMMARY OF THE MAJOR DIGESTIVE ENZYMES[a]

SOURCE	ENZYME (ZYMOGEN)[b]	ACTIVATOR[c]	SUBSTRATE(S)	FUNCTION/PRODUCTS
Salivary glands	Ptyalin (salivary α-amylase)	———	Starch	Hydrolyzes α-1, 4-glucosidic bonds of starch (Fig. 13.21) producing maltose, maltotriose, and α-limit dextrins (Fig. 13.20).
Gastric glands	Pepsins (pepsinogens) Gastricsin	HCl	Proteins, polypeptides	Produce scission of peptide bonds (Fig. 13.25) adjacent to aromatic amino acids (Fig. 13.22).
Pancreas, exocrine	Trypsin (trypsinogen)	Enterokinase	Proteins, polypeptides	Produces scission of peptide bonds adjacent to basic amino acids.
	Chymotrypsins (chymotrypsinogens)	Trypsin	Proteins, polypeptides	Produces scission of peptide bonds adjacent to aromatic amino acids.
	Carboxypeptidase A (procarboxypeptidase A)	Trypsin	Proteins, polypeptides	Produces scission of amino acids with terminal carboxy group and having aromatic or branched aliphatic side chains.
	Carboxypeptidase B (procarboxypeptidase B)	Trypsin	Proteins, polypeptides	Produces scission of amino acids with terminal carboxy group and having basic side chains.
	Pancreatic lipase	Emulsifying agents	Triglycerides	Diglycerides, monoglycerides, and fatty acids (Fig. 13.23).
	Pancreatic α-amylase	Chloride ion	Starch	Same as Ptyalin
	Elastase (proelastase)	Trypsin	Elastin, other proteins	Produces scission of bonds which are adjacent to neutral amino acids.
	Ribonuclease	———	RNA	Nucleotides (Table 13.9).
	Deoxyribonuclease	———	DNA	Nucleotides (Table 13.9).
	Phospholipase A (prophospholipase A)	Trypsin	Lecithin	Lysolecithin
Intestinal mucosa	Enterokinase	———	Trypsinogen	Trypsin
	Aminopeptidases	———	Polypeptides	Produces scission of N-terminal amino acid from peptide.
	Dipeptidases	———	Dipeptides	Two amino acids.
	Maltase	———	Maltose, maltotriose	Glucose
	Lactase	———	Lactose	Glucose plus galactose.
	Sucrase (= invertase)	———	Sucrose	Fructose plus glucose.
	Isomaltase	———	α-limit dextrins	Glucose
	Nuclease, related enzymes	———	Nucleic acids	Pentoses; purine and pyrimidine bases.
	Intestinal lipase	———	Monoglycerides	Glycerol, fatty acids.

[a]*Data from Ganong.* Review of Medical Physiology, *6th ed, 1973, Lange Medical Publications; Kassell and Kay.* Science *180: 1022, 1973; White, Handler, and Smith.* Principles of Biochemistry, *4th ed, 1968, The Blakiston Division, McGraw-Hill.*

[b]*There is some experimental evidence which indicates that the precursors (or zymogens) of pepsin, trypsin, chymotrypsin, the carboxypeptidases, and elastase do possess enzymatic activities of varying degrees, at least in vitro. Just what significance this fact has in vivo is unclear.*

[c]*It appears that certain proenzymes can undergo autoactivation in vitro, ie, the activation process is initiated by the zymogen per se, and then the activation process itself is accelerated by the activated enzyme. Again the role of this process in vivo is unclear.*

functionally discrete, but sequentially connected, portions of the digestive tract are the mouth, fauces and pharynx, esophagus, stomach, small intestine, large intestine, and rectum (Fig. 13.1). The epithelial membrane lining all of these regions of the alimentary canal is not keratinized but moistened continuously by the secretions of goblet cells and mucus-secreting glands which lubricate the surface. This epithelial layer exhibits various specializations in different regions of the digestive system, depending upon whether it is involved in secretion and/or absorption, or whether it is designed for protection. These morphologic specializations of the epithelium will be dealt with as appropriate in the sections to follow.

In addition to the specific regions of the gastrointestinal tract proper there are a number of accessory glands associated with this system whose secretions are of the utmost importance to the normal processes of digestion. In this category are the salivary glands, the liver, and the exocrine pancreas. These organs will be discussed in context with the appropriate region of the digestive system where the secretions of these glands exert their particular functions.

The epithelial structures of the digestive tract are derived from endoderm during the course of embryonic development, and the muscular and connective tissue layers of the gastrointestinal tract develop from the visceral mesoderm. As shown in Figure 13.2 the adult gastrointestinal tract is comprised of five distinct layers. These are, from within outwards, the mucosa, the mus-

cularis mucosae, the submucosa, the muscularis externa and the adventitia (or serosa). The mucous membrane is comprised of a layer of epithelium underlain by a connective tissue lamina propria. The muscularis externa surrounding the mucous membrane is composed of smooth muscle fibers. In many regions of the digestive tract the mucous membrane is delineated from the connective tissue submucosa by a thin layer of smooth muscle, the muscularis mucosae. In regions of the alimentary canal in which a definite muscularis mucosae is absent, the lamina propria merges gradually into the submucosa.

During embryonic development, the epithelial membrane lining the digestive tract produces numerous outgrowths which vastly increase the surface area of the gut. In addition, many invaginations of this membrane form, and these epithelium-lined glandular or cryptlike structures may secrete either digestive enzymes or mucus to lubricate the surface membrane. Certain of these embryonic glands, eg, the pancreas and liver, do not remain within the epithelium where they originated, but become separate organs that remain anatomically connected to the gut in the adult by ducts. This adult connection marks their point of origin in the embryonic epithelial membrane.

The wall of the digestive tract in the oral cavity, esophagus, and rectum is covered by a layer of connective tissue which in turn provides attachment of these structures to adjacent tissues. The stomach and intestines, on the other hand, are covered externally with a serous membrane and are

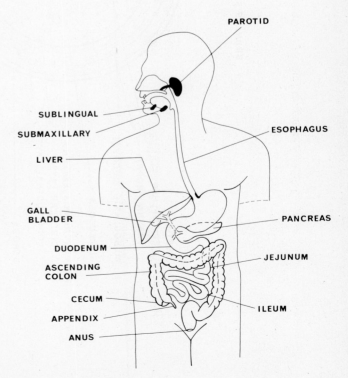

FIG. 13.1. Gross anatomy of the gastrointestinal tract. (After Dawson. *Basic Human Anatomy,* 2nd ed, 1974, Appleton-Century-Crofts; Guyton. *Textbook of Medical Physiology,* 4th ed, 1971, Saunders.)

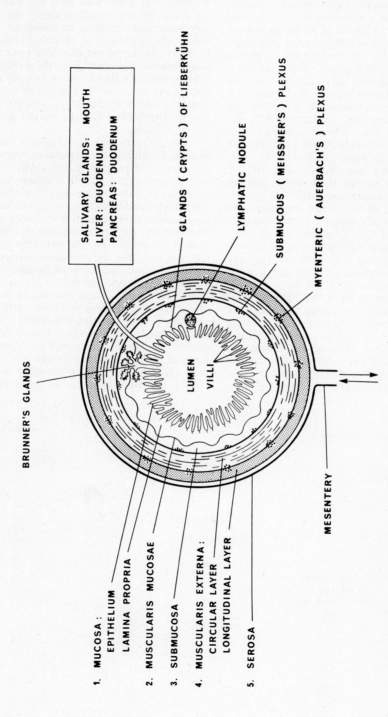

SALIVARY GLANDS: MOUTH
LIVER: DUODENUM
PANCREAS: DUODENUM

BRUNNER'S GLANDS

GLANDS (CRYPTS) OF LIEBERKÜHN

LYMPHATIC NODULE

SUBMUCOUS (MEISSNER'S) PLEXUS

MYENTERIC (AUERBACH'S) PLEXUS

LUMEN
VILLI

MESENTERY

1. MUCOSA:
   EPITHELIUM
   LAMINA PROPRIA

2. MUSCULARIS MUCOSAE

3. SUBMUCOSA

4. MUSCULARIS EXTERNA:
   CIRCULAR LAYER
   LONGITUDINAL LAYER

5. SEROSA

FIG. 13.2. The principal morphologic features of the digestive tract. The principal layers are numbered. Brunner's glands are found only in the duodenum, and no villi are present in the colon. (After Bloom and Fawcett. *A Textbook of Histology*, 9th ed, 1968, Saunders.)

suspended by the mesenteries within the peritoneal cavity so that they can move freely within this space.

The vascular supply to the gastrointestinal tract is abundant. In addition to supplying oxygen and nutrients, the same blood vessels also transport a large proportion of the absorbed digestive products away from the mucous membrane (Fig. 10.15, p. 588). The remainder of the absorbed products are conveyed within the intestinal lymphatics.

An abundant autonomic innervation is present within the wall of the gastrointestinal tract, consisting of ganglia and neural plexuses. These nerves are concerned intimately with the regulation of gastrointestinal motility as well as with certain secretory activities in the digestive tract.

It should be emphasized that the processes of transport, digestion, and absorption of various substances by the gastrointestinal tract represent an orderly and stepwise sequence of events, and the functional state of each region of this system directly modulates the functional activity of the following regions.

The Oral Cavity

The first subdivision of the gastrointestinal tract, the mouth, is bounded anteriorly by the lips and laterally by the cheeks. The roof of the mouth is comprised of the hard and soft palates. The floor of the mouth consists of the tongue, certain muscles, and the alveolar arch of the mandible. The latter structures are covered by a mucous membrane that is reflected from the surface of the tongue at its root to the inner surface of the cheeks, and covers the underlying muscles and alveolar arch of the mandible.

The oral mucous membrane is composed of stratified squamous epithelium similar to the skin; however, this epithelial layer normally does not undergo complete cornification in the oral cavity. The oral mucous membrane is extremely sensitive to various stimuli because of the presence of abundant sensory endings which are branches of the trigeminal nerve (V).

The mouth contains the tongue, gingiva (or gums), the teeth and the terminal ends of the salivary gland ducts.

TONGUE. The tongue, composed primarily of striated muscles and covered by mucous membrane, is essential to articulate speech and it also plays an important role in the mastication of food as well as in the act of deglutition (or swallowing). The tongue also contains the taste buds that were considered among the special senses (Chap. 8, p. 517).

GINGIVA. The gingiva (or gums) are composed of dense fibrous connective tissue covered by a mucous membrane. The gingival tissue surrounds the necks of the teeth and is reflected into the tooth sockets (or dental alveoli) where it merges with the periostium lining these openings in the maxillary and mandibular bones.

TEETH. The teeth have an important role in the mechanical disintegration of food (or mastication) prior to the act of deglutition. The teeth grind and/or tear the food into small pieces, thereby greatly increasing the surface area for subsequent digestion by various enzymes. During the chewing process, food particles are also mixed with the salivary secretions, and thereby thoroughly moistened.

Humans develop two sets of teeth during their life span. During childhood, 20 deciduous (primary or milk) teeth are present. By the time an individual has reached the age of 18 to 21 years, the deciduous teeth usually have been replaced by the permanent teeth. There are 32 of these structures in the adult which are equally divided between the upper and lower jaws; ie, there are 16 teeth in each jaw. The teeth are classified according to their general function. The four incisors in each jaw are designed for shearing or cutting. The two cuspids in each jaw are designed for holding and tearing. The four premolars and six molars in each jaw are designed for grinding.

The structure of a typical tooth is depicted in Figure 13.3.

SALIVARY GLANDS. There are three pairs of salivary glands associated with the oral cavity. These structures are compound tubuloalveolar glands whose secretions are conveyed by individual ducts to the oral cavity, where the combined secretions form the saliva.

Gross and Microscopic Anatomy of the Salivary Glands. The histology of a salivary gland is shown in Figure 13.4. The three types of salivary gland may be considered individually.

First, the parotid glands are the largest of the salivary glands, and these structures lie on the lateral aspects of the face in front of (and below) the external ear. The parotids are enclosed in a connective tissue capsule, and their terminal ducts (Stensen's ducts) empty into the vestibule of the mouth opposite the second upper molars.

The parotids elaborate a watery secretion containing salts and the salivary enzyme ptyalin (Table 13.1); hence the glandular tissue is composed almost exclusively of cuboidal serous cells whose homogenous cytoplasm contains an abundance of refractile secretory granules that are located between the nucleus and the free surface of the cells bordering the lumen of the glandular alveoli. The shape of the nucleus reflects the functional state of the gland. Therefore, in an actively secreting gland whose cells contain numerous cytoplasmic granules the nucleus is spherical, small, and it is located in the basal region of the cell.

The vascular supply to the parotid glands is derived from branches of the internal carotid ar-

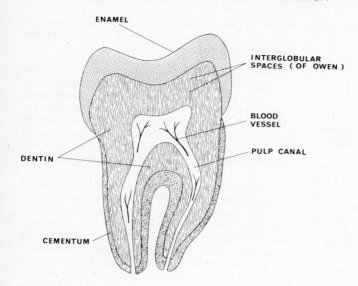

FIG. 13.3. Morphology of a tooth in vertical section. (After Dawson. *Basic Human Anatomy*, 1966, Appleton-Century-Crofts.)

tery, and drainage is via tributary vessels to the external jugular vein.

Second, the encapsulated submaxillary (or submandibular) glands are roughly the size of a walnut and they are located beneath the mandibular angle in front of the lower portion of the parotids. The ducts of the submaxillary glands (ducts of Wharton) pass forward along the sides of the root of the tongue, and open into the floor of the mouth upon each side of the frenulum of the tongue.

The submandibulars are mixed glands. That is, they contain both serous and mucin-secreting mucous cells, although the former predominate. The mucous cells contain pale droplets of mucigen in their cytoplasm, a substance that is the precursor of the viscous mucin.

The blood supply to the submandibular glands is via branches of the lingual and external maxillary arteries. Venous drainage is through the lingual and deep facial veins.

Third, the sublingual glands lack a distinct connective tissue capsule, and are the smallest of the salivary glands. These structures lie beneath the sublingual mucous membrane covering the floor of the mouth, and each is drained by 8 to 20 small ducts which empty into the oral cavity.

The sublinguals are mixed glands that contain both serous and mucous cells. Mucous cells are more abundant than in the submandibular glands, and in general serous cells are a minor component of the overall cell population of the sublingual glands.

The vascular supply to the sublingual glands is derived from the submental branch of the facial artery as well as the sublingual branch of the lingual artery. Tributaries of the submental and sublingual veins drain blood from these glands.

The salivary glands are all innervated by sympathetic as well as parasympathetic nerves (Table 7.6, p. 276; Fig. 7.26, p. 274). The sympathetic nerves are vasoconstrictors; hence they form plexuses on the blood vessels of the glands. The parasympathetic fibers, on the other hand, induce secretion as well as vasodilatation. Afferent fibers from the glands are also present, and these fibers participate in the salivatory reflex when stimulated directly.

Saliva. In addition to the combined secretions from the three pairs of salivary glands, saliva also contains the secretions of the small buccal, labial, and lingual glands of the mouth. The collective secretions of all of these glands amounts to a vol-

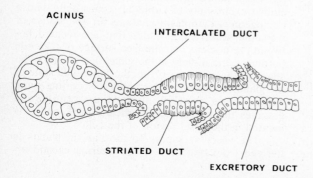

FIG. 13.4. Terminal portion of a submandibular salivary gland together with its duct. (After Bloom and Fawcett. *A Textbook of Histology*, 9th ed, 1968, Saunders.)

ume that ranges between 1,000 and 1,500 ml per day.

Saliva is composed principally of water (99.5 percent), but it also contains enzymes, mucin, salts, and desquamated epithelial cells. It is important to realize that saliva is a complex mixture of individual secretions; consequently the composition of this fluid can vary considerably depending upon the extent to which the various individual glands participate in its formation. Furthermore, the secretion of even one gland also can vary markedly, depending upon the type of stimulus producing the secretion.

Fauces and Pharynx

The fauces is a narrow opening (or passageway) covered by a mucous membrane which marks the junction between the mouth and the oral pharynx. This circumscribed region is limited above by the soft palate, at the sides by the anterior pillars (glossopalatine arches) and posterior pillars (pharyngopalatine arches) of the fauces, and below by the dorsal aspect of the tongue. Two almond-shaped masses of lymphoid tissue, the palatine tonsils are located one on each side of the fauces, between the anterior and posterior pillars. The palatine tonsils are part of a group of lymphoid organs (Waldeyer's tonsillar ring) which afford protection to the respiratory and digestive openings of the body. These lymphoid structures (including the palatine tonsils, pharyngeal tonsils or adenoids, the tubal, and the lingual tonsils) all contain abundant phagocytic cells and lymphocytes which remove foreign particulate matter and microorganisms prior to their entry into the body via the digestive or respiratory systems.

The oral pharynx is continuous with the fauces at its posterior margin, the nasal pharynx above, and the laryngeal pharynx below, hence the oral pharynx is a common passageway for liquid, food, and air, as noted on page 836. The pharynx is located between the soft palate and hyoid bone, and between the hyoid and inferior margin of the cricoid cartilage this passageway becomes the laryngeal pharynx. The laryngeal pharynx is continuous with the esophagus at the level of the cricoid cartilage, and the anterior wall of this structure contains the laryngeal aperture.

The pharyngeal muscles will be discussed in relation to the act of swallowing. The entire pharyngeal region is lined by an epithelial membrane containing abundant mucous cells.

Esophagus

The esophagus is the first portion of the digestive tube proper, and this structure is designed to transport food and fluids rapidly from the pharynx to the stomach. The digestive tube proper serves only the digestive system, and it is composed of the esophagus and stomach, as well as the small, and large intestines. The overall length of this structure is about 9 meters.

GROSS ANATOMY. The esophagus is an anterio-posteriorly flattened muscular tube approximately 25 cm in length, and it starts at about the level of the cricoid cartilage where it forms a continuation of the pharynx. The esophagus is continuous with the stomach, and it enters the abdominal cavity through the esophageal opening in the diaphragm. The esophagus is located vertically in the body immediately posterior to the trachea; the abdominal portion of this structure is short, being approximately 2 cm in length.

HISTOLOGY. The esophageal wall contains all of the five layers which are characteristic of the digestive tube generally, ie, mucosa (epithelial layer and lamina propria), muscularis mucosa, submucosa, muscularis externa, and adventitia (Fig. 13.2).

The mucous membrane is about 500 to 800 μ in thickness, and it is composed of stratified squamous epithelium that is similar to that of the pharynx with which it is continuous. At the cardia of the stomach there is a sudden transition of the epithelium from stratified squamous to simple columnar; consequently gross inspection of the gastroesophageal junction reveals a clearly delineated boundary between the two structures. The esophagus is white and the gastric mucosa is pink.

In the vicinity of the cricoid cartilage, the muscularis mucosae of the esophagus replaces the elastic layer of the pharynx. This mucosal tissue consists of elastic fiber networks and longitudinally oriented smooth muscle fibers. The esophageal muscularis mucosae near the stomach is around 300 μ in thickness.

The submucous layer of the esophagus is composed of collagenous and elastic connective tissue fibers, and accumulations of lymphocytes are present around the glands. The submucous layer and the muscularis mucosae together exhibit many longitudinal folds so that the lumen of the esophagus is irregular in cross section. However, as a mass of food (or bolus) is swallowed, the folds expand and become smooth.

The muscularis externa is between 0.5 and 2 mm thick, and in the upper third of the esophagus both the outer and inner layers of this tissue are composed exclusively of striated muscle. In the middle third, smooth muscle fibers appear, and in the lower third of the esophagus the muscularis externa is composed entirely of smooth muscle. The muscle fibers within the layers of the muscularis externa of the esophagus are not regularly disposed, ie, in circular and longitudinal patterns. Rather, spiral and oblique bundles of fibers are encountered frequently. A loose tunica adventitia composed of connective tissue surrounds the esophagus and connects it with adjacent structures.

The esophageal glands are irregularly distributed, small, compound structures that have elaborately branched tubuloalveolar secretory portions containing only mucous cells. These glands are of two types, and are termed esophageal glands proper and esophageal cardiac glands. The secretions of both types of esophageal gland serve to lubricate the bolus of food during its passage from pharynx to stomach.

The esophagus is innervated by parasympathetic fibers of the vagus nerve as well as by sympathetic nerves, and the parasympathetic fibers form plexuses within the esophageal wall. The myenteric plexus (of Auerbach) is found between the two layers of the muscularis externa; the submucous plexus (of Meissner) is situated within the submucosa itself (Fig. 13.2).

Stomach

The stomach is the most widely dilated portion of the digestive tract, and it functions both to store as well as to digest food. Within the stomach, solid food ultimately is converted into a semifluid mass by the contractions of its muscular wall combined with the mixture of the food with the glandular secretions of the gastric mucous membrane.

Although food in the upper region of the stomach may remain solid for relatively long periods, in the lower part of this organ the food becomes transformed into a pulpy fluid mass known as the chyme, and the chyme then is ejected into the small intestine in small quantities once a proper consistency has been achieved. Thus the action of the stomach on ingested food is partly mechanical and partly chemical.

GROSS ANATOMY. In an adult, the stomach is approximately 25 cm in length and 10 cm wide, and it has a capacity of roughly one liter. This organ extends from the distal end of the esophagus above to the first portion of the small intestine (the duodenum) below (Fig. 13.1).

In shape, the stomach generally is described as having the shape of a "J," although in fact the shape of the organ varies considerably under different circumstances. Several regions within the stomach are described, as shown in Figure 13.5. The principal mass of the stomach lies to the left of the midline in the upper left abdominal quadrant, and it is most firmly attached to the body at the cardia, where it joins the esophagus. The remainder of the organ hangs relatively free within the abdominal cavity. Consequently, the stomach is capable of varying its position considerably, especially following drinking, eating, and changes in body position.

The esophageal entrance to the stomach is guarded by the cardiac sphincter; the pyloric sphincter is located at the junction of the pylorus with the duodenum.

When empty, the stomach assumes a contracted state, so that its lumen is not much larger than that of the intestine, but the organ can dilate markedly with little change in intraluminal pressure because of the plasticity of its smooth muscle fibers.

HISTOLOGY. The stomach has the five layers typical of the entire digestive tract (Fig. 13.2). The mucous membrane is smooth in the distended stomach; however, when contracted it exhibits many grossly evident longitudinal folds (or rugae). Still smaller invaginations of the epithelial surface give rise to the gastric pits or foveolae gastricae (Fig. 13.6). The gastric pits and the surface epithelium of the stomach are lined by a tall columnar epithelium which commences suddenly at the cardia below the esophageal epithelium. In the pylorus, the intestinal epithelium replaces the gastric columnar epithelium.

The gastric epithelial cells secrete a mucus that acts as a protective and lubricating coating for

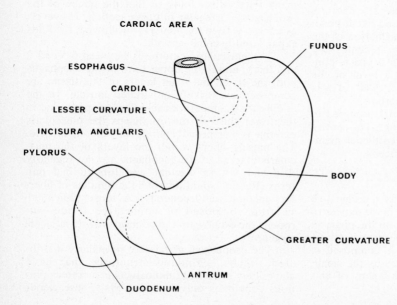

FIG. 13.5. Gross anatomy of the various regions of the human stomach. (After Dawson. *Basic Human Anatomy*, 2nd ed, 1974, Appleton-Century-Crofts.)

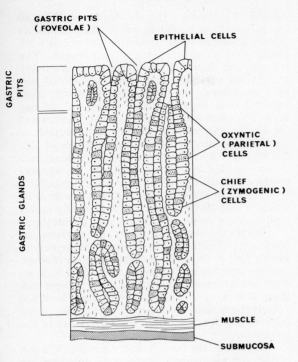

GASTRIC PITS (FOVEOLAE)

EPITHELIAL CELLS

GASTRIC PITS

GASTRIC GLANDS

OXYNTIC (PARIETAL) CELLS

CHIEF (ZYMOGENIC) CELLS

MUSCLE

SUBMUCOSA

FIG. 13.6. Histology of the gastric mucosa showing the glandular elements. (After Bloom and Fawcett. *A Textbook of Histology,* 9th ed, 1968, Saunders; Davenport. *Sci Am* 226:86, 1972.)

the surface of the stomach. These cells exhibit a rapid turnover rate, as they are continuously desquamated into the lumen of the stomach to be replaced by new cells every three days. Regeneration of the surface epithelial layer takes place only in the deeper regions of the foveolae as well as in the necks of the gastric glands. The new epithelial cells produced by mitosis push upward to replace those desquamated at the surface.

In cross section, the abundant gastric glands open into the bottoms of the gastric pits, and three general regions are described in the stomach depending upon the morphologic appearance of these glands. Cardiac glands are found in the vicinity of the cardia in a ringlike area between 5 and 30 mm in width. The glands of the fundus or gastric glands proper (oxyntic glands) are located in the fundus and proximal two-thirds of the stomach. Finally, the pyloric glands are found in the pyloric region of the stomach. These regions are not sharply delineated, however; therefore some overlap of these glandular types occurs in the marginal zone between any two regions.

The gastric glands are all simple, branched, tubular structures, and they are the major source of gastric juice. The gastric glands are vertically oriented with respect to the epithelial surface, and they are packed tightly together in the mucosal layers through which they extend completely (0.3 to 1.5 mm). The bases of these structures almost

reach the level of the muscularis mucosae (Fig. 13.6).

Four cell types are found in various proportions within the gastric glands.

Chief (or Zymogenic) Cells. The chief cells are located as a simple layer upon the inner surface of the basement lamina of the gastric glands, and they are found adjacent to the lumen in the lower portion of the gland tubule. In fresh preparations made after a period of fasting, the cytoplasm of the chief cells is filled with large, refractile granules which are believed to be pepsinogen, the precursor of the proteolytic digestive enzyme, pepsin (Table 13.1). The nucleus of the chief cells is spherical, and these cells are cuboidal or low columnar. Microvilli are present upon their free surfaces.

The fine structure of the chief cells is quite similar to that of other cell types which secrete proteinaceous enzymes, such as the pancreatic acinar cells. Thus gastric chief cells contain a well-developed Golgi apparatus as well as an extensive granular endoplasmic reticulum.

The chief cells also have been implicated in the formation of gastric intrinsic factor (or gastric antipernicious anemia factor) which is indispensible for the absorption of vitamin B_{12} (p. 872).

In nursing babies, gastric juice also contains an enzyme called rennin (not to be confused with renin from the kidney, p. 1169). Rennin induces coagulation of milk, facilitating the digestion of this foodstuff in infants. The chief cells of the gastric glands are believed to be the source of this enzyme.

Parietal (or Oxyntic) Cells. The parietal cells are located singly among the chief cells. Parietal cells are large and have a spherical or pyramidal appearance. A single spherical nucleus is present generally, and the cytoplasm of parietal cells stains deeply with acidic dyes. No secretory granules are present in these elements.

The most characteristic morphologic feature of the parietal cells is the secretory canaliculus, which is present as a loose network surrounding the nucleus and which is continuous with invaginations of the surface membrane. The secretory canaliculus is lined by elongated microvilli. Electron microscopy also reveals an abundance of densely packed mitochondria in the cytoplasm of the parietal cells.

There is excellent evidence from several lines of investigation that the parietal cells of the gastric glands secrete the hydrochloric acid of the gastric juice.

Mucous Neck Cells. The irregularly shaped mucous neck cells are located between the parietal cells in the neck of the gastric glands, and are of relatively infrequent occurrence. The basal nuclei usually are flattened, and the cytoplasm of unstained mucous neck cells contains pale, transparent granules.

The mucous neck cells, as well as the surface

epithelium of the stomach, secrete mucus which forms a protective layer upon the surface, thus presumably inhibiting autodigestion of the stomach during life. Following death, however, the gastric mucosa disintegrates rapidly because of the actions of the various digestive secretions.

Argentaffin (Enterochromaffin) Cells. Argentaffin cells are found singly between the basement lamina and the zymogenic cells in the gastric glands, and are similar to the argentaffin cells found in the intestine. Their shape is rounded or flattened, and the cytoplasm contains numerous small granules which stain heavily with silver or chromium salts. Argentaffin cells are responsible for the synthesis and storage of serotonin (5-hydroxytryptamine).

Within the stomach, the lamina propria forms a delicate meshwork of reticular and collagenous fibers between the gastric glands.

The muscularis mucosae of the stomach is composed of two layers of smooth muscle, an inner circular, and an outer longitudinal layer. In addition, smooth muscle fibers are oriented from the inner circular layers towards the luminal surface of the organ, and these fibers are located between the glands. Hence contraction of these muscle fibers compresses the mucous membrane, and presumably assists in emptying the glands of their secretory products.

The large blood and lymphatic vessels of the stomach are located within the submucous layer, which in turn is composed principally of dense connective tissue containing some fat cells, abundant mast cells, lymphoid cells, and eosinophils.

Three layers of smooth muscle comprise the muscularis externa: an inner oblique layer, a middle circular layer, and an outer, chiefly longitudinal, layer. The longitudinal esophageal fibers continue in the outermost layer of the stomach, whereas in the vicinity of the pylorus, the middle circular fibers form a thick sphincter which assists in the control of gastric emptying, a process which is due principally to the precisely coordinated contraction of all of the muscular coats.

The muscular wall of the stomach is able to adapt to marked alterations in the volume of the contents without concomitantly large alterations in the pressure in its lumen, a response that is quite similar to that of the bladder as it fills with urine (Fig. 12.31, p. 812).

The serous membrane investing the stomach is a thin connective tissue layer surrounding the muscularis externa, and its outer surface is covered with mesothelium.

Small Intestine

The part of the digestive tract lying between the stomach and large intestine is the small intestine, and this structure in turn is divided into three segments, the duodenum, the jejunum and the ileum.

GROSS ANATOMY. The duodenum is 25 to 30 cm in length; it is mainly retroperitoneal and is closely adherent to the posterior abdominal wall; consequently no mesentery is present. The jejunum and ileum, on the other hand, are freely mobile and attached to the dorsal abdominal wall by a mesentery. The three segments of the small intestine total about 7 meters in length, and this organ has three major functions: first, to transport the chyme onward from the stomach; second, to continue digestion of the chyme by means of special digestive juices elaborated by intrinsic and accessory glands; and third, to absorb the nutrients produced by the digestion of various foodstuffs.

The small intestine is largest in diameter at its junction with the stomach, and it gradually tapers until reaching its junction with the large intestine.

Within the abdominal cavity, the jejunum and ileum are supported by the fanshaped mesentery (a double fold of the peritoneum) in a series of loops that are attached to the dorsal abdominal wall, an anatomic arrangement that tends to prevent tangling or kinking of the small intestine. The blood vessels, lymphatics, and nerves which pass to the intestine are located within the two layers of the mesentery.

In general, the basic morphologic organization of the various regions of the small intestine is quite similar, so that the following general description will apply equally to the duodenum, jejunum, and ileum. Specific morphologic differences among these regions will be noted in context.

The small intestine exhibits two important structural modifications which greatly enlarge the total surface area that is present for the absorption of nutrients, but without increasing the total length of the digestive tube. These modifications are the grossly visible plicae circulares (valves of Kerckring) and the microscopic intestinal villi (Fig. 13.7).

The plicae circulares are permanent ridgelike folds which extend into the lumen of the intestine, and are formed by the mucosa, muscularis mucosae, and submucosal layers. These structures are absent in the first few centimeters of the duodenum; they reach their greatest height in the proximal jejunum, thence gradually diminish in size throughout the course of the ileum so that they are absent once again from the middle of this structure onwards.

The plicae not only increase the absorptive area of the intestine, but also mix chyme and digestive juices and slow down the rate of transport of the chyme so that a more thorough absorption of nutrients can take place.

The intestinal villi are minute flattened (in the duodenum) or fingerlike (in the ileum) projections of the mucous membrane that range between 0.5 and 1.5 mm in length, and which cover the entire surface of the intestinal mucosa. The tips of many of the intestinal villi are bifurcated, particularly in young babies.

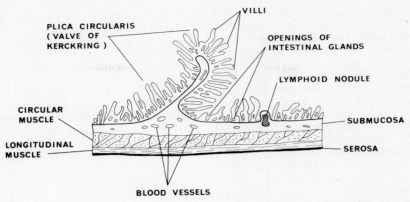

FIG. 13.7.   Portion of small intestine showing plica circularis and lymphoid nodules. (After Bloom and Fawcett. *A Textbook of Histology,* 9th ed, 1968, Saunders; Goss [ed]. *Gray's Anatomy of the Human Body,* 28th ed, 1970, Lea & Febiger.)

HISTOLOGY.   Located at the bases of the villi are vast numbers of openings of the intestinal glands (or crypts of Lieberkühn), as shown in Figure 13.8. These simple tubular structures are about 300 to 450 μ in length, and traverse the mucous membrane in a direction that is vertical with its surface, their bases almost reaching the muscularis mucosae.

A simple columnar epithelium covers the free surface of the intestinal mucous membrane, and three morphologically distinct cell types are present. These cells are the absorptive cells, goblet cells, and argentaffin cells.

The absorptive cells are columnar epithelium. They are about 20 to 25 μ in length, and have a prismatic shape. At the free surface a striated border composed of microvilli is present. Mitochondria are abundant in the apical cytoplasm, be-

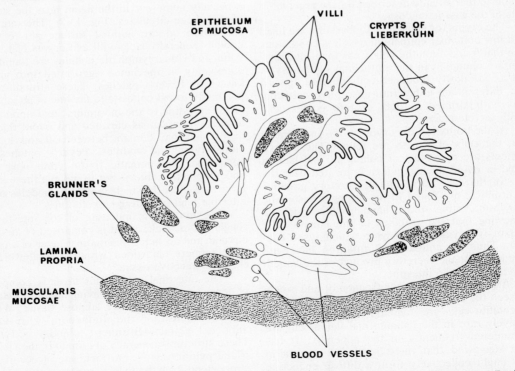

FIG. 13.8.   Cross-section of small intestine showing villi and crypts of Lieberkühn. (After Bloom and Fawcett. *A Textbook of Histology,* 9th ed, 1968, Saunders.)

neath a clear cytoplasmic layer known as the terminal web, which in turn is located immediately below the striated luminal border. The bases of the absorptive cells rest upon a thin basement membrane (or basal lamina) which in turn is attached to the lamina propria. The nuclei of the absorptive cells are elongated, and an elaborate Golgi apparatus is located at the apical end of these structures.

The epithelial cells are produced from undifferentiated cells which line the crypts, and subsequently move upward along the sides of the villi as mitotic division proceeds within the crypts. Desquamation occurs at the tips of the villi, and as also is the situation in the stomach, the rate of replacement of the intestinal epithelial cells is quite rapid.

Goblet cells are found scattered in a random fashion among the absorptive cells. The goblet cells themselves are unicellular glands, hence their apical cytoplasm is filled with mucigen droplets. The goblet cell nucleus is flattened, and a well-developed Golgi complex is located between the nucleus and the mucigen droplets found in the apical cytoplasm just below the free luminal surface of the cell.

Argentaffin cells are the third type of cell found in the intestinal epithelium. These elements are encountered occasionally at the bases of the villi, but they occur more commonly in the crypts of Lieberkühn. The morphology of these cells is similar to that already described for the argentaffin cells of the gastric mucosa.

The crypts (or glands) of Lieberkühn are lined by a low columnar epithelium which is continuous with that found upon the villi. Cells in mitotic division are abundant in the epithelium of the crypt, and as the newly produced cells migrate upward they differentiate either into absorptive epithelial cells with striated borders or into goblet cells which secrete mucus.

At the bottom of the crypts in the small intestine, the cells of Paneth are found. These large elements are pyramidal or columnar, and have a round or ovoid nucleus which is located near the base of the cell. Large apical secretory granules (or droplets) are found in the strongly basophilic cytoplasm. Animal studies have provided evidence suggesting that the granules of the Paneth cells contain lysozyme, an enzyme which lyses bacteria. Thus an antibacterial function has been postulated for the secretory product of the Paneth cells in the human intestine.

Argentaffin cells occur frequently and singly between the cells lining the crypts of Lieberkühn. Argentaffin cells are common in the duodenum, relatively rare in the jejunum and ileum, but are encountered frequently in the appendix. It has been suggested that these scattered argentaffin cells might collectively form a diffuse endocrine system known as the enterochromaffin system. There is strong experimental evidence that the argentaffin cells elaborate serotonin (5-hydroxytryptamine), which in turn is a potent stimulant to smooth muscle contraction. In this regard, it is of interest that argentaffinomas, or carcinomatous tumors of the argentaffin cells, frequently are accompanied by physiologic symptoms of excessive serotonin liberation. Although the physiologic role of the argentaffin cells of the intestinal tract is poorly understood, it is possible that under certain circumstances release of serotonin from these elements may stimulate peristalic action of the gastrointestinal tract.

The lamina propria of the intestinal mucous membrane is composed of a reticulum (or network) of argyrophilic fibers that contains smooth muscle fibers which arise from the muscularis mucosae. In this reticulum also are located fixed cells having pale oval nuclei, and these cells may become macrophages under appropriate circumstances. The smooth muscle strands of the reticulum are most evident within the central part of the villi, where they are oriented parallel to the long axis of these structures. In addition, smooth muscle fibers also are located around the terminal branches of the lymphatic plexus of the villi which in turn is known as the central lacteal. Delicate argyrophilic networks also course from the muscularis mucosae along the blood vessels and also invest the crypts of Lieberkühn.

Abundant, isolated, small (about 0.5 to 3 mm in diameter) lymphatic nodules are also found in the lamina propria. Such lymphatic structures are especially numerous in the ileum near the surface of the plicae circulares (Fig. 13.7). The intestinal epithelium on the luminal surface above these nodules generally lacks villi and crypts of Lieberkühn. Individual lymphatic nodules are found predominantly in the ileum aggregated into groups known as Peyer's patches. These structures are invariably found opposite to the line of attachment of the intestine to the mesentery, and upon gross inspection appear as elevated and ovoid thickenings which may contain between 30 and 40 individual lymphatic nodules. Peyer's patches are composed of lymphatic tissue containing lymphocytopoietic centers, but in old age these structures as well as the single lymphatic nodules of the intestine tend to undergo involution.

The muscularis mucosae of the intestine is roughly 40 μ thick, and it is composed of an inner circular and an outer longitudinal layer of smooth muscle fibers embedded within a framework of elastic connective tissue.

The intestinal submucous layer is composed of dense connective tissue having numerous elastic fibers together with some adipose tissue. Within the submucosa of the duodenum, a layer of duodenal glands (Brunner's glands) is found. These structures generally are found in the vicinity of the pyloric region, but their site of occurrence may extend from the pyloric region of the stomach into the upper jejunum. The secretory portions of the tubular Brunner's glands are highly branched and coiled; their ducts open into a crypt of Lieberkühn after traversing the muscularis mucosae. The alkaline (pH 8.2 to 9.3) secretion of the duodenal glands is clear and viscous, and it is believed to exert a protective function upon the duodenal mucosa against the damaging effects of the highly acid gastric juice by virtue of the alkalinity, mucus, and perhaps by the bicarbonate content which would serve to neutralize the highly acid gastric contents as they enter the upper small intestine.

The muscularis externa of the small intestine is composed of two highly developed layers of smooth muscle (inner circular, outer longitudal) between which is located the sympathetic myenteric nerve plexus of Auerbach. Lastly, the serosal layer of the intestine is composed of a layer of mesothelial cells.

The Large Intestine (Colon)

GROSS ANATOMY. The colon extends from the ileocecal junction to the anus (Fig. 13.1). Approximately 1.5 meters long, the large intestine originates in the lower right side of the abdominal cavity in a blind ending sac known as the cecum. The appendix is a blind ending evagination of the cecum which is about 9 cm in length. The large intestine ascends from the cecum to the liver along the posterolateral wall of the abdominal cavity (ascending colon), crosses to the left side (transverse colon), and then turns sharply downward (descending colon), again along the posterolateral wall of the abdominal cavity. During its downward course it forms an S-shaped curve (sigmoid colon), and in the midline of the pelvic area it turns downward through the pelvis (rectum) to terminate at the anus (Fig. 13.16).

HISTOLOGY. The mucosa of the large intestine is smooth, in contrast to the mucosa of the small intestine. Villi are not found beyond the ileocecal valve, and no folds are present in the large intestine save in the rectum. A simple columnar epithelium having a thin striated border lines the large intestine, and the glands of Lieberkühn are straight tubules somewhat longer than those in the small intestine, their length being about 0.5 mm (0.7 mm in the rectum). Goblet cells are more numerous in the large than in the small intestine. Occasionally argentaffin cells are found at the bottom of the crypts together with proliferating epithelial cells. Generally, no cells of Paneth are present in the colon.

The lamina propria of the large intestine is quite similar morphologically to that of the small intestine, and diffuse lymphatic nodules are present which extend into the submucous layer. A well-developed muscularis mucosae is present in the colon, and this layer is composed of longitudinal and circular smooth muscle bundles. The submucosa is typical of that of the rest of the intestinal tube; however the outer longitudinal layer of the muscularis externa of the large intestine is condensed into three thick, longitudinal bands known as the taenia coli. The taenia coli disappear in the rectum, so that the muscularis externa once again becomes a continuous layer surrounding the intestinal tube. The serous coating of the large intestine contains the epiploic appendages (or appendices epiploicae); these pendulous structures consist of adipose tissue, and hang from the outer surface of the colonic wall near the taenia coli.

In the anal region, the rectal columns of Morgagni are formed by longitudinal folds of the mucous membrane. The crypts of Lieberkühn disappear in the vicinity of the rectal columns, and about two cm above the anal orifice, stratified squamous epithelium appears. At the level of the external anal sphincter, the superficial layer is similar in structure to that of the skin and sebaceous and circumanal glands are present. In this region, a plexus of large veins is present in the lamina propria. These vessels are known as hemorrhoids when abnormally dilated and varicose.

Blood Vessels of the Digestive Tract

Several gastric arteries are distributed to the ventral and dorsal surfaces of the stomach. The intestinal arteries reach both the small and the large intestine through the mesentery and then divide into large branches that enter the muscularis externa. Subsequently, these vessels penetrate the submucous layer where they form an extensive submucous plexus. In both the stomach and the colon, some branches of the submucous plexus provide capillaries which supply the muscularis mucosae, whereas other branches form capillary networks surrounding the glands of the mucosa.

The capillary networks surrounding the glands form veins of rather large diameter, and these vessels in turn form a glandular venous plexus which is located between the muscularis mucosae and the bottoms of the glands. Branches of this venous plexus then pass into the submucosa and form a submucous venous plexus*; then large veins pass together with the arteries through the muscularis externa to enter the serous membrane.

The submucous arterial plexus of the small intestine gives off some branches which course upon the inner surface of the muscularis mucosae and then give rise to capillary networks which invest the individual crypts of Lieberkühn in a manner similar to the vascular pattern seen in the stomach. In contrast to the stomach, however, other arteries of the submucous plexus also send branches to the villi, so that each of these structures receives one or more of these vessels. These small vessels enter each villus at its base and then ramify into a dense capillary network just beneath the epithelium. Close to the tips of the villi one or two venules arise from the capillary network, and these vessels pass downward to join the glandular venous plexus in the mucosa. From there, branches of the glandular venous plexus pass to

In the stomach, the veins of the submucous plexus have valves.

(and anastomose with) the vessels of the submucous venous plexus. These intestinal veins have no valves, but valves are present in the veins as they pass through the muscularis externa. Valves also are lacking in the mesenteric collecting veins.

Of major physiologic importance is the fact that ultimately nearly all of the blood that drains the gastrointestinal tract passes into the hepatic portal vein, then through the liver, as discussed on page 689.

Lymphatic Drainage of the Digestive Tract

In the wall of both the stomach and the colon, the lymphatics originate as an elaborate network of large capillaries in the mucous membrane surrounding the glands. These lymphatic structures are deeper lying than the blood capillaries. The lymphatic capillaries now pass to the inner aspect of the mucous membrane and there form a plexus of delicate vessels from which branches traverse the muscularis mucosae. The lymphatic vessels then form a second plexus in the submucosa. Valves are present in the submucous lymphatic plexus, and larger lymphatic vessels course through the muscularis externa from this network. Within the muscularis externa there is located still a third lymphatic plexus; branches of this latter system anastomose with the branches of the submucous plexus, and the larger lymphatic vessels thus formed now follow the blood vessels into the retroperitoneal tissues.

In contrast to the stomach and colon, the terminal lymphatic vessels of the small intestine, called central lacteals, are blind-ending structures terminating near the tip of each villus (Fig. 13.9). The wider duodenal villi may have two or more interconnected central lacteals. The walls of these lymphatic structures are composed of thin endothelial cells, and when they are distended the caliber of the lacteals is much greater than that of the blood capillaries. At the bases of the villi, the central lacteals anastomose with the aforementioned lymphatic plexus of the mucosa which surrounds the glands. The remainder of the lymphatic drainage from the various layers of the small intestine is quite similar to that described above for the stomach and the colon, except that an additional dense plexus of lymphatic vessels is present in the muscularis externa between the circular and longitudinal layers of smooth muscle. Extensive lymphatic capillaries also are present about the single and aggregated lymphatic nodules of the small intestine.

The lymphatics of the small intestine are essential to the absorption of fat, as discussed on page 866.

Innervation of the Digestive Tract

The general pattern of innervation of the gastrointestinal tract appears similar throughout this structure, and it consists of an extrinsic and an intrinsic nerve supply. The extrinsic nerves are

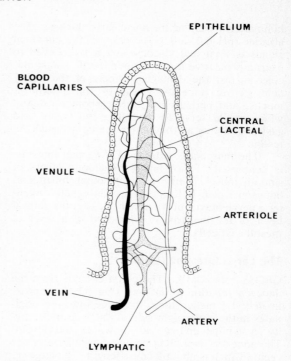

FIG. 13.9. Functional anatomy of a single villus in longitudinal section. (Data from Bloom and Fawcett. *A Textbook of Histology,* 9th ed, 1968, Saunders; Goss [ed]. *Gray's Anatomy of the Human Body,* 28th ed, 1970, Lea & Febiger; Guyton. *Textbook of Medical Physiology,* 4th ed, 1971, Saunders.)

preganglionic parasympathetic (vagus) fibers as well as postganglionic sympathetic fibers from the celiac plexus.

The intrinsic nerve supply is derived from nerve cells and their processes which are located within the wall of the intestine itself (Fig. 13.2). Between the circular and longitudinal layers of the muscularis externa are found groups of nerve cells and bundles of nerve fibers. This extensive system of neural elements is known as the myenteric plexus of Auerbach. Similarly, in the submucosa the neural elements form the submucous plexus of Meissner. Both of these plexuses comprise the intrinsic neural mechanism of the intestinal wall. The cell bodies of the ganglia are interconnected by unmyelinated extrinsic and intrinsic nerve fibers, and the majority of these nerve fibers are processes of the intrinsic neurons, although some extrinsic vagal and a few sympathetic fibers also are present. Sympathetic fibers also innervate the intestinal blood vessels. The role of these nerves in the regulation of gastrointestinal motility is discussed on page 842.

Accessory Glands and Structures of the Digestive Tract: Liver, Gallbladder, and Exocrine Pancreas

In addition to the salivary glands, there are two other important glands associated with the digestive tract. These organs are the liver and the exocrine pancreas.

LIVER

General Considerations. This vitally important visceral organ is the largest gland in the body, weighing approximately 1,500 gm in an adult. The liver functions both as an endocrine gland as well as an exocrine gland. The endocrine functions of the liver include the synthesis and secretion of a large number of compounds directly into the bloodstream, whereas the exocrine function includes the production and secretion of bile through a system of ducts into the duodenum. In addition to its important digestive function, the bile also serves as an important medium for the excretion of certain detoxified compounds and waste materials into the intestine for their eventual elimination from the body.

A clear understanding of the interrelationships between hepatic morphology and function depends in turn upon a knowledge of the blood supply of the liver and its relationship to the general circulation (see also p. 689 and Fig. 10.75, p. 690). A large volume of venous blood is received by the liver from the intestinal tract and spleen via the portal vein. A smaller quantity of arterial blood reaches the liver by way of the hepatic artery. The hepatic veins drain the blood from the liver directly into the inferior vena cava close to the heart. Consequently, the hepatic circulation lies between the venous drainage from the intestinal tract and the systemic circulation so that nearly all of the substances absorbed from the intestinal tract pass through the liver via the portal blood before reaching the systemic circulation. The major exceptions to this general statement are the fats or lipids. These compounds are transported principally in the chyle by way of the intestinal lymphatics directly to the thoracic duct, whence they enter the systemic circulation.

Gross Anatomy. The liver rests immediately beneath the diaphragm in the upper right quadrant of the abdominal cavity, and this organ appears as a large wedge whose shape has been altered by the pressure of adjacent structures. The base of the wedge lies against the right wall of the abdominal cavity, whereas the apex extends toward the left and terminates just to the left of the midline (Fig. 13.10).

The liver is composed of four lobes. These are the large left and right lobes, together with the smaller caudate and quadrate lobes. The superior (or diaphragmatic) surface of the liver is dome-shaped and fits closely against the abdominal surface of the diaphragm. The inferior surface of the liver is roughly concave, and exhibits many gross irregularities because of the pressure of adjacent organs, in addition to the inherent lobular divisions.

The liver is attached to the lower surface of the diaphragm and other abdominal organs by several double folds of peritoneum known as ligaments. These are the falciform, coronary, triangu-

lar, and round ligaments as well as portions of the lesser omentum.

The porta is a deep, narrow fissure lying transversely across the liver which is especially prominent on its posterior surface, and this groove separates the caudate and quadrate lobes. The porta is the region in which the portal vein, hepatic artery, lymphatics, bile ducts, and nerves gain access to the hepatic parenchyma.

Histology. The liver may be considered as a compound tubular gland whose surface is covered with a thin, but dense, fibrous connective tissue layer known as Glisson's capsule. The capsular tissue spreads out from the porta and then follows the ramifications of the blood vessels as they enter the substance of the liver. Ultimately, the connective tissue of Glisson's capsule provides a supporting framework or stroma for the hepatic cells, and it divides the liver into small hexagonal morphologic units known as lobules (Fig. 13.11).

The hepatic parenchyma is composed of epithelial cells which are arranged in layers (or laminae) to form a continuous, three-dimensional network. The laminae which comprise this network are arranged in a radial fashion about a central vein which is located at the center of each lobule (Fig. 13.11). These central vessels represent the terminal branches of the hepatic veins. In cross section through a lobule, the cords of hepatic cells, which are oriented radially about the central vein, are seen to lie between a similarly oriented system of vascular channels, the hepatic sinusoids, which in turn are related intimately to the surfaces of the hepatic epithelial cells. The sinusoids are interconnected through fenestrations, so that they form a continuous vascular network which is exposed to an enormous surface area of hepatic parenchymal cells.

Around the periphery of the polygonal lobule, a portal canal is located at each corner. This structure consists of connective tissue stroma in which are situated small branches of the hepatic portal vein and the hepatic artery, in addition to a small branch of a bile duct (or ductule). Between the hepatic cells within the lobule are located the terminal bile ductules, and it should be noted that the flow of bile is toward the ductule within the portal canal, ie, it is centrifugal. Thus, unlike the majority of glands whose parenchymal tissue surrounds the ducts which drain them, the hepatic lobule is oriented about the central vein rather than the excretory ducts.

Blood flow through the hepatic lobule, in contrast to that of the bile, is from the periphery toward the central vein, ie, it is centripital. Portal venous blood together with hepatic arterial blood enters the lobule peripherally, percolates through the sinusoid, and then enters the central vein. These facts are illustrated in Figure 10.75 (p. 690). This morphologic arrangement of the hepatic lobule in relation to its blood flow gives rise to the important physiologic consequence that the parenchymal cells nearest the periphery receive portal

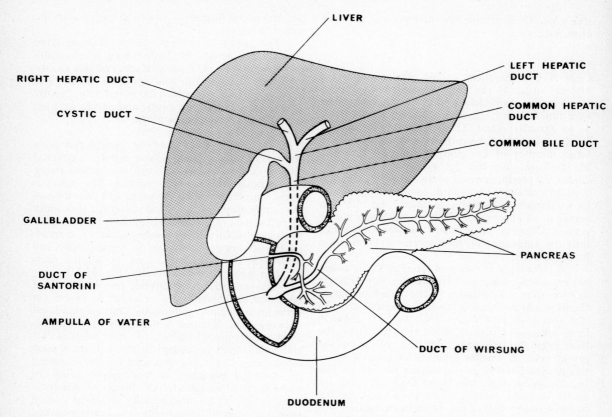

FIG. 13.10. Gross anatomy of the liver showing its relationship to the gallbladder, pancreas, and duodenum, as well as the interconnections among these structures. (After Ganong. *Review of Medical Physiology,* 6th ed, 1973, Lange Medical Publications.)

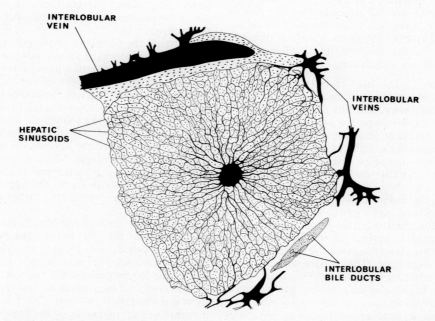

FIG. 13.11. General histologic features of the hepatic lobule (cf. Fig. 10.75, p. 690). (After Bloom and Fawcett. *A Textbook of Histology,* 9th ed, 1968, Saunders.)

and arterial blood first, whereas the cells located about the central vein receive blood last. This may result in an unequal distribution of substances within the hepatic cells (eg, glycogen or lipid) and has given rise to the concept of concentric, graded zones of activity in the cells about the central vein of the hepatic lobule. Those cells lying nearest the center of the lobule thus would exhibit the least activity, whereas those cells in the middle region would be intermediate, and those lying nearest the periphery therefore would have the greatest metabolic activity. Blood flow through the lobule is such that there would be a concentration gradient of oxygen as well as other substances within the blood of the sinusoid, the concentrations being highest at the periphery of the lobule. This concept also has important implications in the etiology of certain pathologic situations involving the liver.

The use of electron microscopy has resolved certain questions concerning the fine structure of the hepatic sinusoids and the elements that line these irregular channels. It appears that two cell types are present in the sinusoids. These are, first, typical endothelial cells, which have a small, dark staining nucleus, and second, the larger stellate cells of Kupffer. The latter cells are fixed macrophages, and they have a large, oval nucleus with an obvious nucleolus. The Kupffer cells have irregular processes which partially surround or even cross the sinusoids. The property of phagocytosis is highly developed in the Kupffer cells, as they readily and rapidly ingest foreign particulate matter from the blood circulating through the sinusoids. There also is some evidence that the epithelial cells can metamorphose into stellate phagocytic cells of Kupffer under appropriate conditions.

Another important question concerning the ultrastructure of the sinusoid has been resolved by electron microscopy. In certain areas, the cells lining the sinusoids are in contact and may overlap at their junctions as in typical capillaries; however, in other regions of the sinusoid, the cell borders are separated by openings ranging between 0.1 and 0.5 μ across. The lining of the sinusoid is discontinuous, a rare situation in the circulatory system of mammals, including man. Plasma (but not erythrocytes) is able to enter freely the perisinusoidal space of Disse, a space which once was considered to be an artifact. The space of Disse lies immediately beneath the lining of the sinusoid, and there the plasma comes into rapid and direct contact with the underlying hepatic parenchymal cells, as shown in Figure 13.12. Thus exchange of materials between the blood within the sinusoid and the hepatic cells is accomplished with extreme rapidity and efficiency.

Terminal lymphatic vessels have not been demonstrated within the lobular tissue; however, there is some speculation that the plasma within the space of Disse may contribute to the formation of lymph by the liver.

The hepatic parenchymal cells are polyhedral, having at least six surfaces. These surfaces may be exposed to one of three situations, as depicted in Figure 13.12. First, the surface may abut on a perisinusoidal space; second, it may be adjacent to a bile canaliculus; third, it may be exposed to an adjacent hepatic cell. Microvilli are present on the cell surfaces that are in contact with the perisinusoidal spaces, and these submicroscopic structures greatly increase the net surface area of the parenchymal cells which is exposed to the plasma that percolates out of the sinusoids into the space of Disse.

Nuclei of the hepatic parenchymal cells are large and spherical. One or two of these structures may be present in each cell, but mitotic figures are rare in the normal liver. Following injury, however, when regeneration is taking place mitoses are abundant, as the liver has an enormous capacity for self-repair. The mitochondria of the hepatic cell are abundant and primarily filamentous.

The cytoplasm of the hepatic cell exhibits marked variations in its appearance that are dependent partially upon the functional condition of the cell. For example, following a meal the fat and glycogen content of the hepatic cells increases considerably and in turn these materials give rise to an abundance of cytoplasmic vacuoles, droplets, and particles readily seen by either light or electron microscopy.

In the livers of well-nourished subjects, basophilic bodies are evident in the cytoplasm as abundant masses or threads of deep-staining basophilic material. These nucleic acid-rich structures are the loci of protein synthesis within the hepatic cell, and not storage sites for protein, as protein storage does not occur to any considerable extent within the normal liver. Electron microscopy reveals both a smooth and a rough endoplasmic reticulum, and abundant ribosomes are associated with the cisternae. The latter structures are the same as the basophilic bodies visualized with conventional light microscopy (Fig. 13.12).

The compact Golgi network usually is situated near a bile canaliculus, and since several such canaliculi are present in association with each cell, there are generally several of these organelles in each hepatic cell (Fig. 13.12). Presumably, the Golgi complex is related to the secretion of bile constituents. The bile canaliculi are minute, extracellular, anastomosing channels that run between adjacent liver cells throughout the hepatic parenchyma. Ultimately these canaliculi form a continuous network throughout the liver. The walls of these structures are merely localized specializations of the cell surfaces of adjacent hepatic cells. Apparently no permanent intracellular bile canaliculi are present. The terminal branches of the bile ducts (ductules or cholangioles) enter the lobule together with the terminal branches of the portal vein between two lobules (Fig. 10.75, p. 690).

The finest branches of the lymphatics can be seen together with those of the portal vein, but as noted above, no lymphatic vessels have been dem-

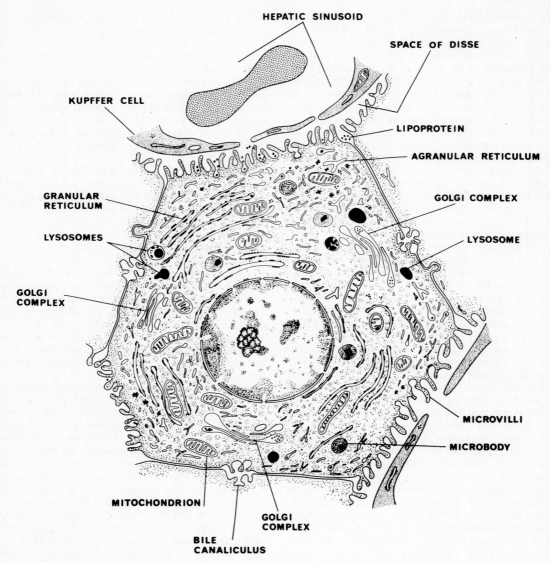

FIG. 13.12. Ultrastructure of a single hepatic cell in relation to a hepatic sinusoid, Kupffer cell, space of Disse, and bile canaliculi. (After Bloom and Fawcett. *A Textbook of Histology,* 9th ed, 1968, Saunders.)

onstrated to be present within the hepatic lobule itself. Presumably, the plasma within the space of Disse traverses the outside of the lobule, exchanges material with the parenchymal cells, and then moves into the interstitial space surrounding the portal vein, bile duct, and the lymphatic vessels. There it becomes the tissue fluid which is drained as lymph by the lymphatic vessels.

Outline of Hepatic Functions. The general functions of the liver may be summarized at this point, leaving certain details of these processes for subsequent consideration.

Practically all of the biochemical reactions involved in intermediary metabolism (Chap. 14) can occur in the liver. Because of this fact, as well as the key position occupied by this organ within the circulatory system, the liver performs many critical roles in the handling of various substances that are absorbed from the gastrointestinal tract in addition to altering them chemically into forms suitable for use by other specialized body tissues.

The liver is essential to the maintenance of a normal blood glucose concentration and this process is one of the most vital functions performed by this organ. The liver also has an important part in the transport of blood lipids, as well as in the maintenance of constant levels of these compounds in the bloodstream. Furthermore, hepatic lymph contains large quantities of plasma protein which thus enter the systemic vascular system indirectly.

The plasma proteins, with the exception of the gamma globulins, are synthesized by the liver, and these synthetic reactions take place at an extremely rapid rate.

The most significant excretory function of the liver is found in the formation of bile acids and bile pigments by this organ.

The liver is also responsible to a great extent for the metabolic detoxification of many compounds and drugs. In fact, the liver shows remarkable adaptive properties, so that the prolonged administration of certain drugs (eg, barbiturates) stimulates the enzymatic machinery that is responsible for their degradation into biologically inactive (or less active) compounds.

The liver also plays an important role in the metabolism of certain hormones, as discussed in Chapter 14.

In summary, then, the liver is a "metabolic factory" that simultaneously performs a multitude of important biochemical and other functions. Normally these functions are not interdependent; consequently it is important to realize that one aspect of liver function can be deranged under abnormal circumstances, whereas other hepatic functions may remain normal (or nearly so). The normal liver has not only a great functional reserve, but also a tremendous capacity for regeneration when damaged. Normal, or nearly normal, hepatic function tests performed clinically do not necessarily indicate that this organ is completely free of damage. Furthermore, rapid death follows massive damage to, or extirpation of, the liver.

BILE DUCTS AND GALLBLADDER

Bile Ducts. The hepatic cells secrete bile constituents into the bile canaliculi (Fig. 13.12), and these structures are continuous with the canals of Hering which in turn communicate with the interlobular bile ducts. The smallest branches of the bile ducts are composed of cuboidal epithelium, and have diameters ranging between 15 and 20 μ. The small branches of the portal vein are surrounded by an anastomotic network of interlobular bile ducts. As the porta of the liver is approached, the bile ducts become progressively larger, and the cells of their epithelium become taller. At the level of the transverse fossa, the principal bile ducts arising from each lobe anastomose, giving rise to the hepatic duct. The cystic duct of the gallbladder joins the hepatic duct, and these combined ducts pass to the duodenum as the common bile duct or ductus choledochus (Fig. 13.10).

The extrahepatic bile ducts are composed of tall columnar epithelium, and the mucosa exhibits a much-folded appearance. Scattered bundles of smooth muscle fibers appear in the common bile duct, and the longitudinally as well as obliquely oriented muscle fibers gradually form a more distinct layer close to the duodenum. This muscular layer acts as a sphincter to control the flow of bile as described below.

Gallbladder. The gallbladder is a pear-shaped hollow organ which lies in a fossa that is located on the inferior surface of the liver near its anterior margin. The gallbladder is attached closely to the posterior surface of the liver, in a region devoid of the peritoneum which covers the remainder of the free surface of the liver.

The gallbladder measures approximately 10 by 4 cm in an adult human, and it has an approximate capacity of 1 or 2 ml of bile per kilogram of body weight. This viscus can hold roughly 40 ml or more of fluid when completely filled. The organ is divided into a blind-ending fundus, a body, and a neck. The latter structure continues as the cystic duct, mentioned above and illustrated in Figure 13.10.

The wall of the gallbladder is composed of an inner mucous layer covered with a single layer of tall columnar cells; a lamina propria which contains simple tubuloalveolar glands made up of cuboidal epithelium; a layer of smooth muscle; and a perimuscular layer of relatively dense collagenous connective tissue having some elastic fibers. This perimuscular connective tissue layer completely invests the organ, and in turn is covered by the serosa that surrounds most of the gallbladder.

The cystic artery supplies the gallbladder with blood, and the gallbladder is drained principally by veins that empty into hepatic capillaries. To a lesser extent, blood from the gallbladder also drains into the cystic branch of the portal vein.

The gallbladder contains a rich lymphatic supply within the lamina propria and connective tissue layers. These lymphatic vessels drain ultimately into the cisterna chyli.

Splanchnic sympathetic nerves and vagal fibers innervate the gallbladder. Possibly both excitatory as well as inhibitory nerves are present in each of these divisions of the autonomic nervous system, although experimental evidence as to their individual functions is contradictory. Clinically the sensory endings to the gallbladder are of greater significance, as overdistension or spasms of the biliary tract may induce reflex alterations in normal gastrointestinal motility and even inhibit respiration.

Functionally the gallbladder serves as a bile reservoir, and this fluid apparently is secreted continuously, although probably at various rates. Following ingestion of fat or meat the gallbladder contracts forcibly and thus expels its contents into the duodenum. The hormone cholecystokinin which is elaborated by the mucosa of the small intestine also stimulates contraction of the gallbladder under physiologic conditions.

The gallbladder normally exerts a considerable ability to concentrate the bile as it is produced by the liver, and this function is carried out by removal of water from the hepatic bile. However, no major secretory function of the gallbladder per se has been demonstrated unequivocally, although the glandular epithelial cells of this organ apparently can secrete some mucus.

The choledochoduodenal junction is that region of the duodenal wall penetrated by the ductus choledochus as well as the pancreatic duct. The sphincter of Oddi in this region is composed of the musculus proprius which is common to both the bile as well as to the pancreatic ducts, and contraction of this structure results in shunting of the bile against the secretory pressure of the liver, so that it backs up into the gallbladder, where it is stored during fasting (Fig. 13.10). Following a meal, the sphincter relaxes and the gallbladder contracts, so that bile is ejected through the ductus choledochus into the duodenum. Cholecystokinin also causes relaxation of the sphincter of Oddi, in contrast to its stimulatory action on contraction of the smooth muscle of the gallbladder itself.

Exocrine Pancreas. The pancreas is the second largest gland that is associated with the digestive tract, the liver being the largest. The pancreas is approximately 20 cm in length, and averages 100 gm in weight. Two distinct glandular tissues are found in the pancreas. These are an endocrine portion, the islets of Langerhans (discussed on page 1075), and an exocrine portion whose function is to secrete a digestive juice containing a number of important enzymes (Table 13.1) through a series of ducts into the duodenum.

Gross Anatomy. The pancreas lies retroperitoneally posterior to the stomach at about the level of the second lumbar vertebra (Fig. 13.10). The head of this gland is directed toward the right, and is firmly attached to the central part of the duodenal loop in which it lies. The body and tail run transversely toward the left across the posterior wall of the abdominal cavity and terminate in proximity to the spleen. A thin layer of loose connective tissue surrounds the pancreas, but does not form a definite capsule. The gland is lobulated, and the outlines of the larger lobules are visible macroscopically.

Histology. Septa pass from the connective tissue which invests the pancreas into the substance of the gland, and excretory ducts, blood vessels, nerves, and lymphatics run in these septa. These connective tissue partitions also divide the pancreas into lobules; therefore this organ is a compound acinous gland.

The acini that produce the digestive juice (the external secretion) are either short tubules or rounded structures, and are composed of a single layer of pyramidal epithelial cells having a basally located spherical nucleus which surrounds a centrally located lumen whose diameter varies according to the functional state of the gland. The epithelial cells rest upon a basal layer, and are supported by reticular connective tissue fibers. Delicate secretory capillaries are located between the individual acinar cells.

There is a good general correlation between the functional state of the exocrine pancreatic cells and their histologic appearance (Fig. 13.13). When active, the apical region of the fresh cell is filled with refractile, spherical, secretion droplets, which are also called zymogen granules. The basal region of the cell appears optically homogenous or faintly striated because of the presence of abundant filamentous mitochondria.

An extensive tubular granular endoplasmic reticulum is found in the basal half of the pancreatic acinar cells together with parallel systems of cisternae, and free ribosomes are encountered frequently in the cytoplasmic matrix. These morphologic features are common to all cells which elaborate proteinaeous secretions. The supranuclear Golgi apparatus is comprised of parallel cisternae arranged in layers, together with minute vesicles and low-density vacuoles of secretory material.

The digestive enzymes of the pancreas are believed to be synthesized within the basal cytoplasm of the acinar cells, and then they are concentrated within the lumen of the tubules of the sarcoplasmic reticulum. Then the synthesized secretory products pass into the Golgi apparatus where they are "packaged" in the vesicular elements of this organelle, and then condensed into zymogen droplets for secretion by coalescence with the plasmalemma at the apex of the acinar cell (see exocytosis, p. 96 and Fig. 3.33, p. 96).

Pancreatic Duct System. Within each acinus, the lumen is continuous with that of a small duct surrounded by so-called centroacinar cells. The centroacinar cells are pale staining in conventional histologic sections. The terminal ducts that are surrounded by centroacinar cells then drain into the intralobular (or intercalated) ducts which in turn are lined by a low cuboidal epithelium whose cells resemble centroacinar cells. The secretory products then pass into the larger interlobular ducts that are lined by columnar epithelium. Occasional argentaffin and goblet cells are also found in the lining of the interlobular ducts.

The interlobular pancreatic ducts ultimately anastomose with the two main pancreatic ducts. The duct of Wirsung is the larger of the two; it commences in the tail of the organ and courses throughout the pancreas while receiving numerous branches, becoming larger as it approaches the duodenum (Fig. 13.10). At the head of the pancreas, the duct of Wirsung courses parallel with the ductus choledochus, and it may share the same opening into the duodenum with the ductus, or else the duct of Wirsung may discharge its contents independently in the ampulla of Vater at the choledochoduodenal junction. The accessory pancreatic duct of Santorini lies cranial to the duct of Wirsung and it is approximately 5 cm in length. The flow from both excurrent pancreatic ducts is regulated by the sphincter of Oddi.

Pancreatic Blood Supply, Lymphatic Drainage, and Innervation. The arterial blood supply to the pancreas is derived from branches of the celiac and superior mesenteric arteries. The branches of the celiac are the pancreaticoduodenal

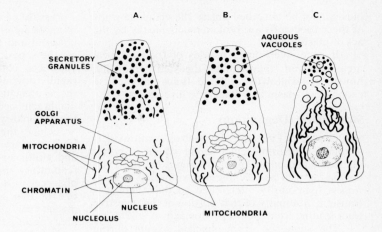

FIG. 13.13. Morphologic changes seen in a secretory cell during rest (A), activity (B), and after prolonged secretory activity (C). (After Bloom and Fawcett. *A Textbook of Histology,* 9th ed, 1968, Saunders.)

and splenic arteries, whereas the inferior pancreaticoduodenal artery is a ramification of the superior mesenteric artery. In addition, some blood reaches the pancreas via branches of the hepatic artery. Veins running parallel with the arteries are present in the pancreas, and these vessels drain blood into the hepatic portal vein or into the splenic vein.

The lymphatic drainage of the pancreas is primarily into the celiac nodes, although the morphologic details of the terminal lymphatic structures within the pancreas are unclear.

Innervation of the pancreas is principally by unmyelinated nerve fibers derived from the celiac plexus and terminating about the acini. Sympathetic ganglion cells are abundant within the interlobular connective tissue. Myelinated fibers of the vagi also are present, and it is believed that the terminations of these nerves (filled with typical synaptic vescicles) are responsible to some extent for regulating pancreatic secretion.

The islets of Langerhans which comprise the endocrine tissue of the pancreas are scattered throughout the acinar tissue. These elements are discussed in relation to the hormones in Chapter 15 (p. 1075).

GASTROINTESTINAL MOTILITY AND ITS REGULATION

Fundamentally, ingested materials are transported through the digestive tract by the development of a pressure gradient from mouth to anus.* In turn, the creation of pressure gradients within the various regions of the gastrointestinal tract depends upon the operation of three general factors: first, the intrinsic contractile properties of the smooth muscle of the gut; second, visceral reflexes that

Development of pressure gradients by the gastrointestinal tract is a far more intermittent process than it is within the cardiovascular system. Furthermore, the pressures developed within the gut usually are far lower than those produced by the heart.

may initiate as well as modify intrinsic gastrointestinal movements; and third, the action of several gastrointestinal hormones secreted by certain regions of the mucosa, transported via the circulation, which then regulate the activity of the stomach, intestines, and gallbladder.

In addition to the propulsion of food, the motions of the various structures of the digestive tract also serve to mix ingested material with the several digestive secretions, to soften it, and to churn it in such a fashion that absorption of nutrients can occur more readily.

Chewing (Mastication)

Solid food taken into the mouth is reduced by the act of mastication into smaller particles to facilitate swallowing, and it is also mixed with saliva which moistens and lubricates the food mass. In addition, the digestion of carbohydrate commences in the mouth by the action of ptyalin in the saliva.

The teeth can exert considerable force during the process of mastication, up to 25 kg by the incisors and 90 kg by the molars.

Mastication is in part a voluntary act, and in part a chewing reflex initiated by the tactile stimulus of a food mass (bolus) against the teeth, gums, and anterior portion of the hard palate. This sensory stimulus inhibits the muscles of the jaw that are involved in mastication or jaw closure. These muscles, the masseters, are innervated by the mandibular division of the trigeminal nerve, and their inhibition causes the jaw to open. This effect in turn initiates a stretch reflex which leads to contraction of the masseters. As the jaw closes, the bolus of food is compressed against the oral sensory areas once more, contraction of the masseters is inhibited, the cycle repeats in a rhythmic fashion, and chewing thus occurs.

Since the digestive enzymes act solely at the surface of food particles, the rate of digestion is related directly to the extent to which food is masticated. Furthermore, as cellulose cannot be digested by humans, the nutrients (eg, starch) which are contained within most fruits and raw vegetables cannot be absorbed unless the cellulose walls of these foodstuffs are broken mechanically by the act of mastication.

If excessively large masses of food material are swallowed, they will be digested eventually; however, swallowing a large bolus may cause esophageal spasm and epigastric discomfort.

Swallowing (Deglutition)

MECHANICAL EVENTS. The act of swallowing is a complex neuromuscular event which is partly voluntary and partly reflex in origin. During swallowing (as well as belching and vomiting) respiration is temporarily interrupted so that there is a transient continuity of that portion of the digestive tract leading from mouth to stomach.

The sequence of mechanical events during swallowing is as follows. The voluntary stage of swallowing is initiated when a bolus of food is collected upon the tongue, and this bolus is forced backward into the pharynx by the upward movement of the tongue. The lips are closed, thereby sealing off the anterior portion of the mouth. Henceforth, the act becomes reflex (automatic), and cannot be inhibited voluntarily. When the bolus reaches the back of the mouth, the pharyngeal stage of swallowing is initiated by stimulation of receptor areas in the pharynx. Afferent impulses from these receptors are transmitted to the brain stem, and a sequence of reflex events now is initiated. First, the soft palate becomes elevated and this structure closes the nasopharynx, thereby preventing food from refluxing into the nasal cavities. This event takes place as the bolus of food passes backward over the tongue. Second, the palatopharyngeal muscles also contract. This action, combined with the contraction of the soft palate, effectively prevents any loss of pressure from mouth to nasal cavity. Third, the bolus, which lies in a groove on the midline of the tongue, passes backward into the oral pharynx by the action of the tongue moving backward while exerting pressure against the hyoid bone, an action that may be likened to that of a piston. Fourth, respiration is inhibited briefly, the vocal cords are approximated, and the larynx is raised so that the epiglottis hangs over the glottis as the bolus passes by. Thus food cannot pass into the trachea. It is important to realize that vocal cord approximation is necessary to prevent food from entering the trachea; ingested material does not pass into the trachea if the epiglottis is removed surgically, but damage to the vocal cords, their muscles, or the innervation thereof can result in strangulation during the act of swallowing. Fifth, the bolus flows over and around the epiglottis, and the hypopharyngeal sphincter* about the esophageal entrance relaxes. The bolus now flows through the lower pharynx into the upper esophagus, driven by

a pressure of about 10 mm Hg which is generated by movements of the tongue. Sixth, as the larynx is raised and the hypopharyngeal sphincter relaxes, the superior constrictor muscle of the pharynx contracts, resulting in a peristaltic wave passing rapidly downward over the pharyngeal muscles to the esophagus (Fig. 13.14A). Relaxation of the diaphragm occurs while the bolus passes through the hypopharynx.

Once the bolus has passed the clavicular level, the larynx descends, the respiratory passage and glottis open, the tongue moves forward and respiratory movements are resumed. The entire pharyngeal stage of swallowing occupies around one or two seconds. To recapitulate briefly the mechanical events which take place during the pharyngeal stage of swallowing, the trachea is sealed off, the esophagus opens, and a rapid wave of peristaltic contraction which originates in the pharynx forces the bolus into the upper esophagus. The pressure within the pharynx rises to a maximum of about 100 mm Hg as the bolus passes through it.

The pharyngeal stage of swallowing is followed immediately by the esophageal stage. The esophagus acts merely to conduct food from the pharynx to the stomach; its mechanical activity is designed to carry out only this one function.

Once a bolus of food has entered the upper esophagus, the peristaltic wave which originated in the pharynx spreads into, and continues in, the esophageal musculature as a primary peristaltic wave. The maximal esophageal pressure developed is around 30 to 120 mm Hg just behind the bolus. This movement sweeps the food mass downward into the stomach in around four to eight seconds in a person in the vertical position. In a recumbent individual, however, food takes about five to ten seconds to reach the stomach because of the lack of a downward gravitational pull upon the esophageal contents.

If a bolus of food does not enter the stomach completely under the influence of a primary peristaltic wave which has originated in the pharynx, then one or more secondary peristaltic waves are initiated within the esophageal musculature, and these waves continue to be generated until the esophageal contents are emptied completely into the stomach.

Approximately 5 cm above the cardia of the stomach, the circular muscle of the esophagus is slightly hypertrophied, and it forms a physiologic (or functional) gastroesophageal sphincter. This region, however, is quite similar morphologically to the rest of the esophagus except for the hypertrophy.

The gastroesophageal sphincter is in a state of tonic contraction under normal circumstances. This fact is in contrast to most of the esophagus which is fully relaxed unless food is being swallowed. As a peristaltic wave passes down the esophagus, however, the gastroesophageal sphincter relaxes ahead of the peristaltic wave so that the

---

This structure consists of the cricopharyngeal muscles, which usually are in a state of tonic contraction.

A. PERISTALSIS

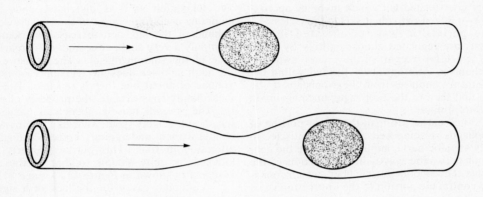

B. SEGMENTATION

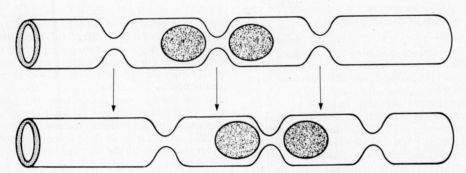

FIG. 13.14. A. Peristaltic waves propel a mass of chyme (shaded) along the intestinal tract for variable distances. **B.** Segmentation contractions "chop" the chyme into small masses and thereby mix it intimately with the digestive juices as well as expose it to the intestinal wall for nutrient absorption.

bolus is transported into the stomach after a delay of only a second or so. This effect is termed receptive relaxation of the sphincter. If the gastroesophageal sphincter does not relax properly, the clinical condition known as achalasia is present and difficulty in swallowing is experienced.

The principal function of the gastroesophageal sphincter is to inhibit any reflux of the gastric contents back into the esophagus. The hydrochloric acid and proteolytic enzymes secreted by the stomach are highly corrosive to the esophageal epithelium, except for the portion of the esophagus immediately adjacent to the cardia of the stomach.

Under normal circumstances, the pressure exerted by the gastroesophageal sphincter is about 5 mm Hg higher than that pressure present within the stomach during quiet respiration, and in turn the intragastric pressure under these circumstances is about 10 mm Hg higher than is the intrathoracic pressure.

Man, in common with other nonruminant animals, does not exhibit reverse peristalsis of the esophagus.

SWALLOWING REFLEX. The act of swallowing can be induced voluntarily or initiated reflexly by stimulation of a number of regions in the mouth and pharynx, such as the pillars of the fauces, or the sides of the posterior wall of the hypopharynx. The glossopharyngeal nerve and superior laryngeal branch of the vagus form the afferent pathways to the swallowing center which is located within the medulla oblongata. Following activation of this nerve center, the entire act of swallowing is carried out automatically by a sequential efferent neural discharge through the nuclei of several cranial nerves (V, VII, X, XI, and XII) as well as the nucleus ambiguus, the retrofacial nucleus (of Jacobsohn), and motoneurons arising in the upper cervical segments of the spinal cord (C–1 through C–3).

Discharge of these nuclei to the appropriate muscles takes place in accordance with a definite and sequential pattern. Therefore swallowing is a smoothly coordinated act which involves approximately two dozen individual skeletal muscles.

The peristaltic waves generated in the esophagus are regulated almost entirely by vagal reflexes which form a part of the general swallowing mechanism. Afferent fibers located within the vagus transmit impulses from the esophagus to the medulla and then to the esophagus once again via vagal efferent fibers.

As noted earlier, in man the upper portion of the esophagus is composed of striated muscle, so that only somatic nerve impulses underlie the generation of peristaltic waves in this portion of the esophagus. The smooth muscle fibers which constitute the contractile portion of the lower esophagus are also under extrinsic control by the vagus nerve. However, secondary peristaltic waves can be initiated by a bolus of food forced into this region if the vagal pathways to the esophagus are interrupted, and these waves now are initiated by the intrinsic myentric plexus of Auerbach (Fig. 13.2). Normal relaxation of the gastroesophageal sphincter also appears to be regulated by the myenteric plexus of the esophagus.

AEROPHAGIA AND ERUCTATION. Inevitably during the processes of eating and drinking some air is swallowed (aerophagia), and this ingested air may be absorbed, passed into the intestinal tract, or eliminated by regurgitation (eructation). The increased intragastric pressure induced by a large gas volume causes the bubble, which normally is situated in the fundus of the stomach, to move beneath the cardia. During swallowing, as the gastroesophageal sphincter relaxes, flow of the gas from the stomach into the esophagus ensues, the pressures within these regions becoming equal momentarily. Belching occurs as the gas progresses through the pharynx and oral cavity at a sufficient velocity and with adequate force to induce audible vibrations of the structures in these regions.

Reflux of some gastric contents during eructation or other situations may induce a burning sensation in the epigastric region, an effect commonly known as "heartburn."

Gastric Motility

The stomach performs three general mechanical functions. First, food is stored in bulk quantities within this organ until such time as it can be processed most efficiently by the intestinal tract; second, food is mixed with the gastric secretions which convert it into a pasty, semifluid mixture (chyme); and third, the chyme is emptied from the stomach at a rate consistent with the most effective rates of digestion and absorption by the small intestine.

STORAGE OF FOOD. As food enters the stomach from the esophagus, it passes in ever

widening concentric masses from the cardia toward the wall of the fundus and body, and the gastric musculature undergoes progressive receptive relaxation as it is distended, thus accommodating itself to the mass of food swallowed. The volume of the stomach increases markedly, but with only a very slight concomitant increase in the tone of the smooth muscle. The pressure within the adult stomach does not rise appreciably until a volume of about one liter is reached. The gastric musculature thus exhibits the property of plasticity. The smooth muscle fibers can change their length appreciably (somewhat like pulling on taffy candy) without undergoing a correspondingly great alteration in tone. This plastic property of the stomach is quite similar to that of the urinary bladder, as shown in Figure 12.31, page 812.

A second reason for the lack of a significant pressure increment as the stomach volume increases is found in the law of Laplace. As the gastric volume increases, so does the radius of curvature of the stomach walls; and thus the wall tension decreases as the radius of curvature increases.

MIXING AND PROPULSION OF FOOD. The gastric glands secrete the digestive fluids, and these secretions would only affect the food directly in contact with the gastric lining were it not for weak tonus waves (or mixing waves) which pass at intervals over the organ. These shallow contractions commence at a pacemaker region that is situated near the cardia, and sweep over the fundus and body of the stomach toward the pylorus at a rate of about three per minute, thereby gradually moving the outer layer of food mixed with digestive secretions, toward the antrum (Fig. 13.5). In the antral region, the peristaltic contraction waves become much stronger about an hour after a meal; thus mixing is more complete, and the food mass gradually assumes a still greater degree of fluidity for the following reasons. Within the antrum, strong peristaltic movements directed toward the pylorus knead and churn the food intimately with the digestive secretions, and also a large quantity of the food mass is squirted backward toward the body of the stomach through the ring of constriction, whereas only a few milliliters pass onward through the pyloric sphincter at each central contraction. Thus the food is reduced to a relatively homogenous semifluid of pasty consistency known as chyme.

In a manner similar to the much weaker tonus waves discussed above, the strong peristaltic waves of the antrum occur at a rate of approximately three per minute and these waves are also initiated by the pacemaker region in the cardia. These antral waves become especially prominent in the vicinity of the incisura angularis (Fig. 13.5); then they spread progressively over the antrum, pylorus, and into the first few millimeters of the duodenum. As the stomach gradually empties, the strong peristaltic waves commence higher up on the body of the stomach above the incisura and propel any remaining stored food toward the antrum where it mixes with the chyme already present in this region.

The intragastric pressure exerted upon the chyme during strong contractions of the antrum may be as great as 70 cm H_2O.

In summary, the empty stomach is relaxed and small, and the weak contractions that sweep continuously over its muscle from the pacemaker region in the cardia exert little influence upon the intragastric pressure. During hunger, strong contractions initiated at the cardia also pass over the stomach at a rate of approximately three per minute; however, recent studies indicate that such "hunger contractions" with attendant augmentation of the intragastric pressure play little, if any, role in the sensation of hunger which is experienced by a fasting individual.

Following a meal, the stomach dilates passively with little pressure increment in order to accommodate its volume to that of the ingested material (storage function); it concomitantly secretes a juice that commences the processing of the foodstuffs into chyme (digestive function); and then gradually the stomach propels the chyme in measured quantities into the intestinal tract (transport function), where further digestion and absorption of nutrients take place. Thus the stomach performs its several physicochemical digestive functions as a smoothly coordinated unit.

GASTRIC EMPTYING. Gastric emptying essentially is the consequence of strong peristaltic waves that sweep chyme from the stomach into the duodenum, and the propulsive force generated by the peristaltic contractions of the stomach normally is counteracted to some extent by resistance within the pyloric region. The pylorus normally remains in a state of slight tonic contraction, and a pressure gradient of approximately 4 cm of water is present between the stomach and the duodenum. This gradient is sufficient to permit only water and highly fluid chyme to pass selectively into the duodenum. When the antral peristaltic waves increase in intensity the contractions become especially strong at the incisura angularis; then these contractions sweep down the antrum toward the duodenum at an inherent frequency of about three per minute. As each wave progresses downward, the smooth muscles of the pyloric sphincter and the proximal duodenum are inhibited and relax (receptive relaxation), and several milliliters of chyme then are ejected from the pylorus into the duodenum. This mechanical action of the stomach sometimes is referred to as the "pyloric pump."

REGULATION OF GASTRIC MOTILITY AND EMPTYING. The rate of gastric emptying in general is regulated by one major factor, and that is the type of food ingested. Therefore a carbohydrate-rich meal passes from the stomach within a few hours, protein-rich food takes somewhat longer, and gastric emptying is slowest after a meal which contains a high fat content. In turn, underlying the regulation of gastric motility and emptying, are two basic kinds of physiologic mechanisms. These mechanisms are neural and humoral (or chemical). Hence psychic (emotional) factors can play a major role in the rate at which the stomach contracts and empties following a meal.

The Enterogastric Reflex. This inhibitory reflex is a physiologically important neural mechanism underlying the rate of gastric peristalsis; hence it governs to a large extent the rate at which the stomach empties. The enterogastric reflex is mediated via two principal motor pathways. First, vagal afferent fibers course to the medulla from the stomach and in turn these afferents inhibit the activity in vagal efferent pathways to the stomach. Second, there is reason to believe that a portion of the enterogastric reflex is mediated by way of the celiac ganglion and sympathetic pathways to the stomach. Thus it appears that both divisions of the autonomic nervous system participate in the enterogastric reflex.

Generally speaking, sympathetic stimulation of the appropriate nerves to the stomach inhibits the force of peristalsis, and parasympathetic stimulation augments the intensity of gastric contractions. In neither case is the inherent rate of contractions altered. Reciprocal activity between these two motor systems is present, so that when one is enhanced, the other decreases and vice versa. The inhibitory effect of the enterogastric reflex upon gastric motility depends upon net activity of the sympathetic branch of the autonomic nervous system at a given moment in relation to the relative activity in the parasympathetic nerves at the same time.

The chief stimuli that elicit the enterogastric reflex and thereby cause inhibition of the stomach musculature are duodenal distension by an excessive volume of chyme; irritants within the intestine; excessive acidity of the chyme (ie, a pH below 4), the products of protein digestion; and the relative hypertonicity or hypotonicity of the chyme itself. All of these stimuli cause a physiologic decrease in the rate of gastric emptying by stimulating the enterogastric reflex, and thus a regulatory balance is achieved between the rate at which the stomach empties and the ability of the intestinal tract to process the chyme.

The tonicity of the chyme is an especially important factor in gastric emptying, as stimulation of the enterogastric reflex prevents an excessive and too rapid flow of hypertonic or hypotonic fluids into the intestine. The general electrolyte balance of the body is not altered significantly during absorption of the intestinal contents, as the net water flux into (or out of) the small intestine during digestion is such as to produce an isotonic chyme.

Humoral Factors. In addition to the enterogastric reflex, certain hormonal factors also influence the rate of gastric motility. Protein, fats, and acid in the chyme stimulate the liberation of secretin (p. 849) and cholecystokinin (CCK, p. 849) from the duodenal mucosa. These substances in turn pass to the stomach via the circulatory system and there inhibit gastric motility as well as the secretion of gastric juice. Another hormone, enterogastrone, has also been implicated in the regulation of gastric motility. Enterogastrone is liberated from the duodenal mucosa into the bloodstream under the stimulus of fat in the duodenum; then it passes to the stomach, where it inhibits gastric motility, and thus prolongs the emptying of fats into the intestinal tract, an effect which permits a longer time for their intestinal digestion as well as assimilation.

Psychic Influences. Emotional factors also exert a marked influence upon gastric motility and secretion, although these effects are largely unpredictable. Excitement may accelerate gastric emptying where fear or depression may inhibit it. Direct observations of the gastric mucosa in human subjects having gastric fistulas have revealed that anger or hostility induced hyperemia and turgor of the stomach. Painful injury may result in total inhibition of gastric motility and emptying for up to 24 hours or even longer.

The Pylorus. Under physiologic conditions, contraction of the pyloric region of the stomach per se plays only a minor role in the regulation of gastric emptying.

Motility and Transport in the Small Intestine

The movements of the small intestine may be divided for convenience into several categories. These movements in turn either stir the chyme so that it is mixed intimately with the digestive secretions, or else propel the chyme toward the large intestine.

Mixing of the intestinal contents is accomplished primarily by tonus changes, pendular movements, and especially by segmentation contractions. In addition, the villi themselves are quite motile and the movements of these structures serve to enhance the absorption of nutrients by the epithelial cells in addition to helping transport lymph within the lacteals. In addition to the four general types of intestinal movement summarized above, peristaltic contractions sweeping unidirectionally from the oral towards the anal end of the gut propel the chyme through the small intestine toward the colon.

TONUS CHANGES. Tonus movements consist of localized changes in the rate and/or degree of contraction of the smooth muscle of the gut, and such changes are found in all regions of the digestive tract. Tonus changes are not related to the volume or type of the intestinal contents, and these contractile movements appear to reflect the inherent contractile ability of the intestinal smooth muscle fibers.

PENDULAR MOVEMENTS. Pendular movements are localized to a few centimeters of the intestine, and they occur with a frequency of about 11 per minute. Pendular movements are weak contractions that pass in an oral as well as anal direction within short segments of the gut, and presumably they assist in the thorough mixing of the chyme since their propulsive force is slight. The net effect of these waves is a to-and-fro motion, or back-and-forth oscillation, of the intestinal contents; hence the term pendular movement is applied to this type of intestinal activity.

Similarly to the tonus changes, pendular movements also are caused by the inherent contractile properties of the smooth muscle cells rather than by extrinsic or intrinsic nerves.

SEGMENTATION CONTRACTIONS. By far the most significant factor in mixing the chyme is the segmentation contractions exhibited by the small intestine. These ringlike constrictions usually occur at fairly regular intervals along the intestine, as shown in Figure 13.14B. The contracted regions then relax, and constrictions now appear in the relaxed regions between the original constrictions. Thus the chyme is kneaded thoroughly with the digestive juices, by the alternate contraction and relaxation of the gut. Of equal physiologic importance, the chyme also is completely exposed by this mixing action to the absorptive surface of the intestinal mucosa.

Experimental studies on dogs and humans have shown that the fundamental electric rhythm underlying the phenomenon of segmentation in the small intestine originates near the entrance of the bile duct into the duodenum, and conduction of the impulses from this region takes place at a rate of approximately 20 cm per second. The basic rhythm for segmentation in man is fairly constant at around 10 per minute in the duodenum and is not affected by extrinsic nerves, fasting, or feeding. A contractile gradient may possibly exist in man along the intestine, so that the rate of appearance of segmentation contractions diminishes progressively with distance from the duodenum. Such an activity gradient has been shown to be present in the dog intestine.

Unlike the frequency with which segmentation occurs, the force of these ringlike contractions varies considerably. During fasting, segmentation contractions are absent or very weak; however, they become quite strong following a meal, and vagal stimulation also augments the force, but not the rate of these movements.

Experimental work also has shown that an intact myenteric nerve plexus, but not extrinsic innervation, is essential for normal segmentation to occur in the gut.

The chyme gradually is rotated in a counterclockwise direction as it passes down the gut, since the longitudinal smooth muscle actually is arranged in a similar long counterclockwise spiral in the wall of the intestine.

MOVEMENTS OF THE VILLI. As described earlier, the villi possess smooth muscle fibers arranged in parallel with the long axes of the villi, and these fibers originate in the muscularis mucosae. Contraction of these muscle fibers causes the villi to shorten or to bend, and thereby the transport of lymph from the central lacteals into the lymphatic system presumably is facilitated. Alternate contraction and relaxation of the villi is especially prominent after a meal, and the movements of these structures appear to be regulated by local reflexes within the intrinsic neural plexuses of the intestine which in turn are stimulated by the presence of chyme. In addition, a hormone known as villikinin is believed to be secreted by the mucosa, and this substance presumably also stimulates the motility of the villi.

PERISTALTIC CONTRACTIONS. The major propulsive movements which transport chyme through the intestinal tract are the peristaltic waves (Fig. 13.14A). In order that sufficient time be permitted for adequate digestion and absorption to occur within the small intestine, the peristaltic contractions of this organ are a series of weak propulsive movements. Thus chyme is transported down the intestine by a sequence of peristaltic waves each of which only moves a short distance, then becomes extinguished, and is succeeded by another wave which commences just caudally. Therefore the net effect is a leisurely stepwise progression of the chyme toward the ileocecal sphincter. Peristaltic waves normally exhibit great variation in the rate at which they travel as well as in their intensity. Peristaltic waves can originate at any point in the small intestine, and they pass toward the anus at a rate of one or two cm per second. Normally such waves become extinct after travelling for only a few centimeters.

Chyme passes through the small intestine at an average rate of one cm per minute, and thus normally between three and ten hours are required for chyme to pass from the pylorus of the stomach to the ileocecal sphincter at the entrance of the large intestine.

Following a meal the motility of the small intestine increases sharply, although the actual rate of propulsion of the chyme is reduced. The increased peristaltic activity under such circumstances is caused by a gastroenteric reflex, and this reflex is initiated by physical distension of the stomach. The impulses then are conducted down the intestinal wall via the myenteric plexus of Auerbach, and result in an increased overall excitability of the small intestine insofar as both motility and secretion are concerned.

In addition to the general gastroenteric reflex that is located wholly within the neural elements of the intestinal tract, local peristaltic waves also can be initiated experimentally as well as physiologically by distension of small, isolated segments of gut. A deep circular constriction (the peristaltic wave) forms orally from the point of stimulation, and this constriction then passes slowly and smoothly toward the ileocecal sphincter. This localized response to stretch of the intestine is known as the myenteric reflex. It thus may be appreciated readily that under physiologic circumstances distension of the intestine by chyme forms the basis of an important local physiologic stimulus for the propulsion of material through the small intestine.

Extrinsic innervation is not critical to normal peristalsis. A segment of gut can be removed and then resutured into the body and peristalsis continues as before and without interruption through the reimplanted segment. If, however, the segment of gut is reversed before suturing it back into the body so that the oral end of the segment is joined to the anal end of the cut intestine, peristaltic waves stop (ie, are blocked) at the reversed segment. Within the reversed segment itself peristaltic waves now pass in an oral direction; that is, in what was the normal (anal) direction before the segment was reversed. There is, consequently, a pronounced physiologic gradient of activity along the entire length of the small intestine from the oral to the anal end. Hence the oral end exhibits the greatest irritability, and the anal end the least. The mechanism underlying this functional polarity along the small intestine is unknown.

Peristaltic waves do not occur locally if the intrinsic myenteric plexus of Auerbach is absent, damaged, or anesthetized. The receptors involved in the myenteric reflex apparently lie within the intestinal mucosa itself.

One process of the bipolar ganglion cells that are located within the submucous plexus of Meissner enters the mucosa whereas the other passes to the myenteric plexus. The myenteric reflex whereby peristaltic contractions are initiated locally in the gut may be summarized as follows. Mechanical stretch of the intestinal mucosa stimulates the bipolar cells of the submucous plexus and the impulses then are transmitted to, and activate, the neurons of the myenteric plexus. Impulses generated in the neurons that comprise the myenteric plexus in turn induce contraction of the smooth muscle fibers that are innervated by these secondary neurons. Serotonin and a material called substance P both have been implicated as having roles as neurotransmitter substances in this localized reflex arc within the gut.

In conclusion, it should be pointed out that peristaltic waves of various types are of importance to transport of materials not only in the small intestine, but also in the stomach, esophagus, colon, ureters, and fallopian tubes. In other words, peristalsis is a general functional mechanism that is present in several hollow organs and structures of the body.

As noted above, peristaltic contractions may show considerable variation in both their intensity and in the distance that they travel along the gut. In humans, very strong, rapid peristaltic waves that sweep chyme through long segments of the small intestine are considered abnormal. Such a wave is termed a peristaltic rush, and it may be induced by purgatives, such as castor oil.

CONTRACTILITY OF THE MUSCULARIS MUCOSAE. Like the smooth muscle coats of the muscularis externa, the muscularis mucosae of the small intestine also is stimulated to contract principally by local reflexes (Fig. 13.2). Contraction of this muscular layer induces the formation of folds in the mucosa itself. These folds have a variable length. In addition, fibers from this muscular layer enter the villi and by their contraction cause these structures to shorten and/or bend as noted earlier.

The folding produced in the intestinal mucosa by contraction of the muscularis mucosae serves to increase the absorptive area of the epithelium which is exposed to the chyme, and thus the rate of absorption of nutrients from the intestinal lumen also is increased.

AUTONOMIC INNERVATION AND INTESTINAL MOTILITY It is evident from the discussions presented above that intestinal motility is largely independent of extrinsic nerves, and that the small intestine functions normally as a coordinated, relatively autonomous, unit without such extrinsic regulation. Total sympathectomy does not affect intestinal motility and vagotomy only decreases intestinal motility in a transitory fashion; therefore shortly after the latter procedure is performed, a normal intestinal motility is regained. Perhaps the vagus has a role in the overall coordination of the activity within various regions of the intestine, but if so, such an action is not critical to the normal functions of the intestinal tract.

Topical application of acetylcholine or vagal stimulation can induce a transient stimulation of the movements of the villi, whereas sympathetic stimulation is inhibitory to the motility of these structures.

The Ileocecal Valve and Sphincter

Peristaltic waves developed in the small intestine gradually propel the chyme from a given meal to the ileocecal sphincter, where passage of the residue now is blocked until the next meal is ingested. This results in a prolongation of the time which the chyme remains in the ileum, and thus intestinal absorption is enhanced still further. Following the next meal, peristalsis in the ileum is stimulated by a gastroileal reflex, and this increased activity now expels the residual chyme into the cecum through the ileocecal sphincter (Fig. 13.15).

Another significant physiologic action of the ileocecal sphincter is to prevent the reflux (or regurgitation) of chyme, now termed feces, from the colon back into the ileum. Anatomically the lips of the ileocecal valve extend into the lumen of

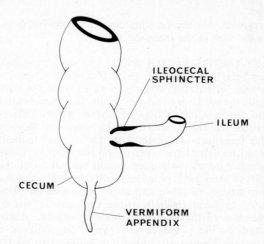

FIG. 13.15. General morphology of the ileocecal valve. Emptying of ileum is facilitated by a more liquid intestinal contents, as well as by distension of this structure and chemical irritation, both of which stimulate peristalsis and relax the sphincter. On the other hand, distension or chemical irritation of the cecum and/or proximal colon both inhibit peristalsis of the ileum and stimulate contraction of the ileocecal sphincter. (After Guyton. *Textbook of Medical Physiology*, 4th ed, 1971, Saunders.)

the cecum (Fig. 13.15), and this structure can resist a reverse pressure of up to 60 cm of water. The ileocecal sphincter itself consists of a thickened muscularis externa within the ileum for several centimeters before the ileocecal valve is reached as shown in Figure 13.15, and normally this muscle is in a state of mild tonic contraction. Emptying of the ileal contents into the colon is delayed for several hours as noted above, except following a meal that causes an enhanced peristalsis within the ileum. Roughly 5 ml of chyme enters the colon under the influence of each enhanced peristaltic wave; approximately 300 to 500 ml of isotonic chyme enters the colon each day.

The extent to which the ileocecal sphincter is contracted largely determines the rate at which chyme enters the cecum, and this tonus in turn is controlled largely by reflexes which arise in the cecum itself. Distension of the cecum is the primary stimulus for augmenting the contractile state of the sphincter and thus delaying the passage of chyme into the colon. Furthermore, local irritation of the cecum (such as in appendicitis) also stimulates contraction of the ileocecal sphincter and may even result in a spasm of this structure to such an extent that emptying of the ileum is inhibited completely. The myenteric plexus of Auerbach presumably is responsible for these local reflex effects, although the vagus may play some role in the regulation of activity of the ileocecal valve.

A number of reflexes which may arise in visceral structures other than the gastrointestinal tract (eg, the peritoneum or kidneys) also can induce spasm of the ileocecal sphincter. Therefore passage of the intestinal contents into the colon is delayed or even blocked completely. Sympathetic stimulation also increases the tonic contraction of the ileocecal valve.

Motility and Transport in the Large Intestine

Basically the colon (Fig. 13.16) has two major functions: first, the absorption of water and electrolytes (including sodium and chloride) from the 300 to 500 ml of isotonic chyme which enters it each day, and second, the storage of the waste products of intestinal digestion (feces) until such a time as they may be eliminated from the body by the act of defecation. In keeping with the nature of these functions, the movements of the large intestine are quite sluggish, and as is the case with the small intestine, these movements either mix or propel the colonic contents.

The total mass of feces eliminated per day is roughly 250 gm, of which about 70 percent is water; consequently the net water absorption per day from the lumen of the large intestine is around 400 ml. Since the small intestine normally absorbs about 8,000 to 9,000 ml of water per day, it is hard to reconcile this well established fact with the often stated concept that the colon is an important water conservation organ.

Experiments performed upon normal human subjects using isotopically labeled water have indicated that there are two opposing water fluxes in the colon, as in other segments of the intestinal tract. The water flux from lumen to blood exceeds that from blood to lumen by around 1.5 ml per minute; it has been calculated that the entire human colon potentially could absorb about 2.5 liters of water per 24 hours.

Net sodium and chloride absorption occur in the colon whereas potassium and bicarbonate are secreted; consequently the osmolality of the fluid remains unchanged as it passes through this organ.

Disease states can alter markedly the fluxes of water and other constituents in the colonic fluids. It is important to remember that the principal source of the water and electrolytes lost from the body in diarrhea is the small intestine. The large volumes of water passed into the large intestine literally swamp the colonic reabsorptive mechanisms. Normally the quantity of sodium and potassium lost from the body in the feces is an insignificant fraction of the total body pool of these electrolytes. However, in severe and prolonged diarrhea, serious electrolyte imbalances occur which can be fatal. For example, a cholera patient may lose up to 12 liters of water per day through the large intestine as well as 1,600 mEq of sodium ion in addition to 240 mEq of potassium ion.

Organic substances which enter the colon include desquamated cells, mucus, and enzymatic secretions of the proximal regions of the gut. Dietary carbohydrate, protein, and fat have been digested and absorbed by the time the food residue reaches the colon; hence only negligible quantities of these substances are present in the feces. Cellulose, of course, is for the most part undigested, and this carbohydrate provides bulk in the stools.

Ingested bacteria are killed principally in the stomach by the hydrochloric acid; only a few of these microorganisms are found in the stomach and small intestine. The population of bacteria rises in the ileum and reaches a maximum in the colon. Consequently, the small quantity of protein found normally in the stool consists largely of protein derived from intestinal microorganisms (bacteria, yeasts, and fungi).

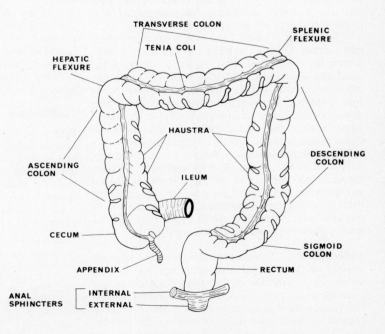

FIG. 13.16. Gross anatomy of the human colon. (After Ganong. *Review of Medical Physiology*, 6th ed, 1973, Lange Medical Publications.)

MIXING IN THE LARGE INTESTINE. In common with the small intestine, the colon exhibits transverse ringlike segmentation movements which may be so strong as to obliterate the lumen in the constricted region almost completely. Concomitantly with the segmentation of a region of the colon about 2.5 cm in width, the longitudinal muscle fibers of the three taenia coli contract, and the combined contraction of the circular and longitudinal muscles causes the intervening relaxed portions of the colon to bulge outward in outpocketings known as haustrations. Occasionally the haustral contractions pass slowly toward the anus, and thus they propel the fecal matter slowly toward the rectum. However, the chief effect of these large intestinal contractions is to roll and knead the feces so thoroughly that approximately 80 percent of the water in the chyme has been absorbed by the time the feces have passed into the rectum. Tonus changes similar to those in the small intestine are also found in the colon, and these facilitate to some extent mixing of the fecal matter.

MASS MOVEMENTS IN THE COLON. Unlike the small intestine, regular peristaltic waves do not occur in the colon to any great extent. Instead, mass movements occur in the large intestine two or three times a day, and these extremely fast and strong peristaltic rushes propel the intestinal contents toward the rectum. Similarly to other peristaltic waves, the mass movements are initiated by local distension of the large intestine, require an intact myenteric plexus, and are wholly independent of extrinsic innervation. The mass movements propel the feces for some distance along the colon, but they are not perceived normally by the individual, and are not related to defecation. Mass movements can originate in any region of the large intestine, but they are of most frequent occurrence in the transverse or descending colon.

Following a meal, the duodenocolic reflex may initiate hyperexcitability of, and mass movements in, the colon. This reflex is initiated by duodenal distension, and it is transmitted to the colon principally via the myenteric plexus; therefore extrinsic denervation does not affect the duodenocolic reflex to any great extent. A somewhat weaker gastrocolic reflex is also present. Quite possibly the autonomic nerves may serve to reinforce these intrinsic reflexes, as extrinsic denervation of the intestinal tract serves to weaken the mass movements in the large intestine. Mass movements in the colon may also be initiated by local irritation within the colon, as, for example, in patients with ulcerative colitis, who may exhibit such contractions almost continuously.

DEFECATION. Tonic constriction of the smooth muscle of the internal anal sphincter, as well as voluntary contraction of the striated muscle of the external anal sphincter retains fecal matter within the rectum until defecation takes place. Normally the act of defecation is controlled by the defecation reflex, which involves the following sequence of events. Distension of the rectum by fecal matter initiates afferent nerve impulses which radiate through the myenteric plexus, and thereby initiate peristaltic contractions in the descending and sigmoid colon. These waves in turn propel the feces toward the anus, and the internal anal sphincter is relaxed by receptive relaxation. If the external anal sphincter (which is innervated by the pudendal nerve) is relaxed voluntarily, defecation ensues. Distension of the rectum also results in conscious perception of the desire to defecate.

The defecation reflex just described is quite weak; thus another reflex is involved, and this augments markedly the evacuation of the rectum. This second reflex involves the sacral segments of the spinal cord. Stimulation of the afferent nerve fibers within the rectum results in the transmission of nerve impulses to the spinal cord, and then via the nervi erigentes motor impulses are transmitted to the descending colon, sigmoid flexure, and internal anal sphincter. These signals markedly reinforce the peristaltic waves in the large intestine from the splenic flexure to the anus, and concomitantly they inhibit the tonic contraction of the internal anal sphincter.

The explusion of feces also is facilitated by means of straining (Valsalva maneuver, p. 653), which markedly increases the intraabdominal pressure. Thus a deep breath is taken and the abdominal muscles are contracted against a closed glottis; concomitantly the external anal sphincter is relaxed voluntarily. This additional effect may be initiated voluntarily or else as a result of the afferent signals entering the spinal cord as the consequence of rectal distension.

In newborn babies or in persons having a transected spinal cord, the lower colon empties automatically because of the presence of the defecation reflex and the absence of any voluntary control over the external anal sphincter. In normal persons, however, voluntary constriction of this sphincter causes the defecation reflex to be inhibited within a few minutes, so that the urge to defecate does not return for some time, usually several hours.

Gallbladder

Bile is secreted continually by the liver, but in normal individuals this secretion is shunted into, and stored within, the gallbladder because the sphincter of Oddi remains in a state of tonic contraction most of the time. However, under the specific stimulus of fat in the duodenum, the intestinal mucosa liberates a hormone known as cholecystokinin–pancreozymin.* This substance is

*A hormone is defined as a chemical substance produced in a glandular organ or tissue, and secreted directly into the bloodstream (or lymph). It is then carried to a target tissue or organ for regulation of the functional activity of that tissue or organ; hence endocrine structures are termed ductless glands. This is in marked contrast to exocrine glands whose secretions are released through tubes or ducts onto an internal or external body surface.

transported in the bloodstream to the gallbladder causing it to contract rhythmically, and thus to eject the bile. Simultaneously, as peristaltic waves sweep along the duodenum, the wave of relaxation which precedes each peristaltic contraction also relaxes the sphincter of Oddi. The gallbladder ejects bile into the duodenum in small spurts because of its rhythmic contractions combined with an intermittent relaxation of the sphincter of Oddi; these two factors regulate emptying of the gallbladder. Another factor which may cause some relaxation of the sphincter of Oddi is the presence of food in the mouth.

Cholecystokinin and other gastrointestinal hormones are released by the presence of food in the stomach and duodenum; however, fat is the most potent stimulant for cholecystokinin secretion, hence contraction of the gallbladder. Incidentally, substances which cause contraction of the gallbladder are known as cholagogues, and substances which stimulate bile secretion from the liver are called choleretics. Other substances that exhibit cholagogue activity include acid, magnesium sulfate, and the products of protein digestion.

Local anesthesia of the duodenal mucosa with procaine or cocaine prevents the release of cholecystokinin; therefore presumably the release of this hormone is controlled by a local reflex mechanism.

Transit Time in the Intestinal Tract

Following administration of a test meal to a human subject, the first portion of the food arrives at the cecum in approximately four hours. After about nine hours have elapsed, all of the undigested residue of the meal has entered the colon.

The first part of the meal reaches the hepatic flexure of the colon in six hours, whereas nine hours elapse before the splenic flexure is attained, and it is twelve hours before the material arrives in the pelvic colon. Afterwards transport to the rectum is considerably slower, and around one quarter of the meal still may be in this region after three days. Markers ingested with the test meal (eg, dyes, colored beads) have shown that total recovery of such tracer material takes longer than a week, an indication of the normally sluggish rate of progression of fecal residue through the large intestine. The remains of one meal catch up, and mingle with, the residue of previous meals in the large intestine. Approximately 20 percent of the entire colonic contents pass through this segment of the gut and are eliminated each day. A slow rate of passage of the contents through the colon also permits the intestinal flora to digest certain compounds (such as cellulose) that are not attacked by the usual digestive enzymes. The net contribution of the large intestine to the overall energy balance of man is negligible. This organ does not actively absorb amino acids and glucose. However, trace nutrients are synthesized by the intestinal flora, eg, folic acid, biotin, riboflavin and nicotinic acid, and these substances may be absorbed by passive transport from the lumen of the colon in nutritionally significant quantities.

SECRETIONS OF THE GASTROINTESTINAL TRACT

The secretions produced by the many glands associated with the gastrointestinal tract have two principal functions (Table 13.2): first, to provide enzymes essential for the digestion of various foodstuffs, and second, to elaborate mucus which serves to lubricate and protect all regions of the alimentary canal from mechanical as well as chemical injury. Glands that secrete enzymes are present in all regions of the digestive tract from the mouth to the distal ileum, whereas mucous glands are present from mouth to anus.

In the present section, emphasis will be placed upon the various glands which elaborate the digestive secretions, the general functions of these secretions, and the physiologic mechanisms which regulate the activity of these glands.

It is important to stress at the outset that the majority of the digestive secretions are elaborated in response to the stimulus of food within the gastrointestinal tract, and that there is a delicate balance between the quantity produced and the quantity that is necessary to assure adequate digestion. In addition, the individual components of a given digestive fluid also may vary markedly, depending upon the kinds of foodstuff which are present in the alimentary tract. Specific examples of these general statements are cited in the discussions to follow.

Salivary Secretion

SALIVA. The morphology of the salivary glands was presented on page 819. Man secretes a total of between 1,000 and 1,200 ml of saliva per day, and this liquid has an average pH ranging between 6.0 and 7.0.

Saliva has a number of significant functions. It moistens the oral cavity and thus facilitates swallowing; it acts as a solvent for the molecules responsible for stimulating the taste buds; it lubricates the lips and tongue, thereby assisting in speech; and it also flushes particulate matter from the mouth and teeth, thus assisting in keeping the mouth and teeth clean. An antibacterial effect also may be produced by the saliva.

Since the pH of saliva is around neutrality, this secretion is saturated with calcium and as a consequence the teeth do not loose this cation to the fluids in the mouth, a situation which may develop if the salivary pH becomes more acid.

Saliva contains a digestive enzyme known as ptyalin (or salivary α-amylase, Table 13.2) which plays a minor part in the digestion of starch. This enzyme is most active at the pH of the oral cavity.

Saliva is produced not only by the three pairs of salivary glands (the parotids, submaxillaries, and sublinguals), but the abundant small buccal glands also contributed to the total volume of this fluid by their secretion of a mucus (mucin).

Table 13.2 DAILY SECRETION OF DIGESTIVE FLUIDS[a]

FLUID	VOLUME	pH
	ml/day	
Saliva	1,000–1,200	6.0–7.0
Stomach (gastric secretion)	2,000–3,000	1.0–3.5
Pancreas (exocrine secretion)	1,200–2,000	8.0–8.3
Bile secretion	600–700	7.8
Brunner's gland secretion	50	8.0–8.9
Succus entericus (intestinal juice)	2,000–3,000	7.8–8.0
Large intestine	60	7.5–8.0

[a]Data from Guyton. Textbook of Medical Physiology, 4th ed, 1971, Saunders.

Two specific kinds of secretion are present in saliva. First, there is a serous (or watery) secretion containing ptyalin. This type of secretion is elaborated exclusively by the parotid glands and to some extent by the submaxillary glands. Second, a mucous secretion is the chief product of the sublingual and buccal glands. The submaxillary glands also secrete some mucus; the latter are mixed glands as they contain serous as well as mucous cells.

REGULATION OF SALIVARY SECRETIONS. Secretion by the salivary glands is under direct neural (reflex) control. The parotid glands are innervated by the glossopharyngeal nerve, whereas the submaxillary and sublingual glands are innervated by the facial nerve. Parasympathetic nerve stimulation causes the secretion of a copious volume of a watery saliva from the parotids, and this saliva has a relatively low concentration of organic material. A marked vasodilatation takes place in the gland following parasympathetic stimulation, and this vasodilatation occurs concomitantly with secretion. The vascular effect may be due to a local release of the nonpeptide bradykinin when the salivary glands are secreting actively.* Cholinergic agents, such as atropine, block salivary secretion.

In contrast to parasympathetic stimulation, stimulation of the sympathetic nerves in man induces the secretion of a small volume of saliva from the submaxillary glands, and this saliva is rich in organic compounds. Sympathetic stimulation has no effect on secretion by the parotids.

Under physiologic conditions, salivation is induced reflexly by the thought, sight, or odor of food as well as by the presence of food in the mouth, stomach, and upper intestines, as discussed in Chapter 7 (p. 440) in relation to conditioned reflexes. Salivation is especially pronounced when irritating foods are present in the gastrointestinal tract, or when one is nauseated.

The reflex neural regulation of salivation is mediated via impulses to the submaxillary and sub-

lingual glands via the superior salivatory nuclei in the pons–medullary junction, whereas the inferior portions of the same nuclei control the activity of the parotid gland. These pathways are shown in Figure 7.26 (p. 274).

As noted above, salivation can be stimulated (or inhibited) by impulses reaching the salivatory nuclei from higher centers in the central nervous system, such as the cerebral cortex, anterior hypothalamus, and amygdala. Hence conditioned reflexes play an important role in the normal regulation of salivary secretion, as the sight, odor, or thought of food can stimulate this process.

Esophageal Secretion

The secretions of the esophageal glands are mucoid only, and these secretions function solely for lubrication of food during the process of swallowing. As noted earlier, mucous glands are located in the gastrointestinal tract from mouth to anus; thus the general properties and functions of mucoid secretions throughout the alimentary canal may be summarized conveniently at this point.

Mucus is composed of water, electrolytes, and a mixture of several types of mucopolysaccharide, so that this material is quite viscous and slimy. Although the chemical nature of mucus may vary somewhat in various regions of the gastrointestinal tract, all types of this secretion share in common a number of physicochemical properties regardless of the site of origin. First, mucus is capable of adhering tightly to food particles, and of spreading over their surfaces as a thin film. In addition, mucus coats the epithelial lining of the intestinal tract, thereby providing a slippery surface (ie, it has a low resistance to shear) which lubricates the gut, and also serves to protect the epithelium mechanically from the corrosive action of the digestive fluids. Second, from a chemical standpoint, mucus is quite resistant to digestion by the gastrointestinal enzymes, giving additional protection to the epithelial lining of the gut. Third, the mucopolysaccharides present in mucus are, like all proteins, amphoteric substances capable of buffering small quantities of acids and alkalis, thus affording still further protection to the gastrointestinal epithelium from the action of the digestive fluids.

The mucus secreted by the compound glands in the vicinity of the gastroesophageal junction exerts a physical as well as chemical protective influence against gastric juices which may reflux into the lower esophagus, and a lack of this protective secretion may result in ulceration of the lower esophagus.

Additional specific functions of mucus in the other regions of the gastrointestinal tract will be noted in context.

Gastric Secretion

GASTRIC JUICE. The cells of the gastric glands secrete a total volume ranging between two and three liters of gastric juice per day. This digestive

*Bradykinin also is released in the exocrine pancreas and sweat glands when these organs are active.

fluid contains a number of substances, as summarized in Table 13.3. In addition, gastric mucous cells and glands secrete a thick alkaline mucus which forms a thin coating on the stomach wall; thus it is of great importance in protecting the epithelial lining of the stomach from inflammation and/or autodigestion (autolysis).

Functions of the Gastric Glands. The tubular gastric (or fundic) glands secrete digestive juices, and are composed of three morphologic cell types, as noted earlier. Of particular significance to the present discussion are the chief (or peptic) cells which secrete pepsinogen, and the parietal (or oxyntic) cells which secrete hydrochloric acid. The latter are especially prominent in the glands of the fundus.

Pepsinogen Secretion. The proteolytic enzyme pepsin, which degrades ingested native proteins into polypeptides, is secreted by the chief cells of the stomach in an inactive form, known as pepsinogen. Pepsinogen in turn is contained in the zymogen granules found in the chief cells of the gastric glands. When pepsinogen is secreted into the gastric lumen, it comes into contact with hydrochloric acid and pepsin, which has been formed earlier. Cleavage of the pepsinogen molecule now takes place so that more active pepsin is produced. During this process, the pepsinogen molecule, which has a molecular weight of 40,400 daltons, undergoes scission to pepsin, which has a molecular weight of 32,700 daltons. The enzyme pepsin is active enzymatically only in a highly acid medium (optimum pH 2.0), and it is inactivated above a pH of 5.0. Consequently, the secretion of hydrochloric acid is essential to protein digestion in the stomach by pepsin.

Actually gastric juice contains several types of pepsin, as noted in Table 13.3. For example, pepsin I has an optimum activity at pH 3.0. On the other hand, the proteolytic activity of pepsin II, which is the most common type of pepsin in the stomach, is optimal around pH 2.0.

Hydrochloric Acid Secretion. The parietal cells of the gastric glands secrete free hydrochloric acid into the lumen of the stomach by a mechanism presumably similar to that presented in Figure 13.17. These cells are capable of performing the osmotic work necessary to concentrate hydrogen ions to a level over 4×10^6 times greater than they are found in arterial blood; hence the resulting solution contains 150 millimols of HCl per liter and has a pH as low as 0.87, a concentration which is equivalent to a 0.17 N HCl solution! In order to concentrate the hydrogen ions to such an extent, an expenditure of over 1,500 calories of energy per liter of gastric juice is required.

Analysis of samples of relatively pure specimens of gastric juice secreted by the parietal cells indicates that their secretion is essentially isotonic with blood. This corresponds to a solution containing 150 mEq each of H^+ and of Cl^- per liter. For comparative purposes, plasma contains 10^{-5} mEq of H^+ and about 100 mEq of Cl^- per liter.

Table 13.3[a] PRINCIPAL CONSTITUENTS OF NORMAL GASTRIC JUICE[b]

Water

Cations: H^+, Na^+, K^+, Mg^{++}

Anions: Cl^-, $HPO_4^=$, $SO_4^=$

pH: 1.0

Pepsins I–III: Act primarily on tryptophan, phenylalanine, tyrosine, methionine, and leucine residues bearing a carbonyl group.

Gastric Lipase: This enzyme actually is a tributyrase, therefore it acts principally upon tributyrin (a constituent of butterfat) rather than on fats in general.

Gastric Amylase: Has an insignificant role in starch digestion.

Gastric Urease: This enzyme comes from bacterial contamination of the gastric contents, rather than from the gastric mucosa per se, and it forms ammonia from urea within the food mass or gastric juice.

Intrinsic factor

Gastricsin

Mucus

[a]*Data from Ganong.* Review of Medical Physiology, *6th ed, 1973, Lange Medical Publications; White, Handler, and Smith.* Principles of Biochemistry, *4th ed, 1968, The Blakiston Division, McGraw-Hill.*
[b]*In the fasting state.*
See also Table 13.1.

Although the cellular mechanisms whereby hydrogen ion secretion by the parietal cells takes place against such an enormous concentration gradient are not understood in detail, it is clear that the enzyme carbonic anhydrase has a key role in this process. The principal source of the secreted hydrogen ion likewise is still unclear; however this ion may be produced by the ionization of water. Additional hydrogen ion also may come from substrates (eg, glucose) via the flavoprotein–cytochrome enzyme system (Chap. 14, p. 931). For each hydrogen ion secreted, a hydroxyl ion is produced within the cell (Fig. 13.17), and this hydroxyl ion then is neutralized by another hydrogen ion generated by the dissociation of a molecule of carbonic acid. The bicarbonate ion liberated by this process diffuses out of the parietal cell and into the bloodstream, whereas the carbonic acid concentration is maintained by the intracellular hydration of carbon dioxide under the catalytic influence of carbonic anhydrase, as this enzyme is present in a high concentration within the gastric mucosa. As physiologic consequences of these reactions, blood leaving the stomach is more alkaline and has an elevated bicarbonate content; arterial blood coming to the stomach has a greater carbon dioxide content, hence a lower pH than the venous blood.

Aerobic glycolysis provides the energy for the active transport of hydrogen ion across the luminal membrane of the parietal cell, and two mols of

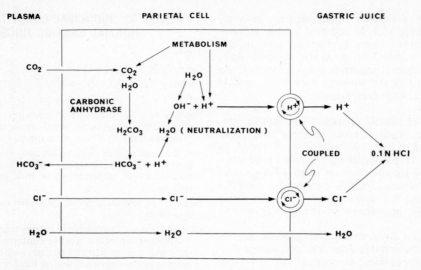

FIG. 13.17. Hypothetical mechanisms underlying gastric acid secretion, as well as intracellular neutralization mechanisms operative during this process. (Data from Davenport. *Physiology of the Digestive Tract,* 2nd ed, 1966, Year Book Medical Publishers; Ganong. *Review of Medical Physiology,* 6th ed, 1973, Lange Medical Publications; Guyton. *Textbook of Medical Physiology,* 4th ed, 1971, Saunders.)

hydrogen ion are secreted per mol of oxygen utilized. Chloride ion is also secreted actively by the parietal cells against an electrochemical gradient, as the luminal membrane of the gastric mucosa is around 60 millivolts negative with respect to the serosal membrane. Generally the active secretion of hydrogen ion is coupled to that of chloride ion, although if hydrogen secretion stops then a minimal secretion of chloride still takes place.

Following a meal, the rate of secretion of hydrochloric acid by the stomach is elevated markedly, and this secretion may be sufficient to raise the pH of the systemic blood slightly. Concomitantly the urine also becomes more alkaline, and this effect is called the postprandial alkaline tide.*

Histamine is a powerful stimulant to gastric acid secretion, and this effect is mediated by cyclic adenosine monophosphate. Since the gastric mucosa normally has a high concentration of histamine, liberation of this compound has been implicated as the chemical mediator in the stimulation of acid secretion by other factors. However, chemical agents in addition to histamine also appear to have a role in acid secretion by the stomach (eg, gastrin).

GASTRIC SECRETION OF MUCUS. The cardiac and pyloric glands are similar morphologically to the gastric glands. However, unlike the latter structures, chief and parietal cells are scarce, and the glands found in the cardiac and pyloric regions of the stomach consist almost entirely of mucous

cells identical with the mucous neck cells of the gastric glands themselves (Fig. 13.6). All of the mucous cells secrete a thin mucus that affords physicochemical protection to the gastric mucosa.

Between the openings of the gastric glands on the surface of the gastric mucosa is found an almost unbroken layer of mucous cells that continuously secrete an abundance of a thick, alkaline mucus which covers the luminal surface of the stomach to a thickness of about one millimeter. This coating not only provides a low friction surface for the passage of food, but it also provides an important protective layer for the gastric mucosa. Even minor irritation of the stomach wall stimulates these superficial mucosal cells to produce large quantities of viscid mucus, and when this secretory process is abnormal for any reason, peptic ulcers can develop.

REGULATION OF GASTRIC SECRETION. Gastric secretion as well as gastric motility is controlled by neural and hormonal mechanisms, unlike secretion by the salivary glands, which is regulated exclusively by neural factors.

Neural Mechanisms. Nervous regulation of gastric secretion is mediated via fibers of the vagus (parasympathetic) nerve. Impulses which result in gastric secretion arise in the dorsal motor nuclei of the vagi located in the medulla; the impulses then are transmitted by way of the vagus nerves to the myenteric plexus of the stomach, and thereby to the gastric glands. The net result of parasympathetic stimulation is the secretion of large quantities of pepsin and acid by the gastric glands. Vagal stimuli also induce a greatly augmented secretion of mucus by the cardiac and pyloric glands as well as by the mucous neck cells of the gastric glands.

The postprandial alkaline tide has been invoked as an explanation of the marked lassitude one experiences following ingestion of a heavy meal.

Vagal stimulation also causes the antral region of the stomach to secrete the hormone gastrin. Consequently, the vagus nerve exerts a direct effect upon the secretion of gastric juice by the gastric glands, and indirectly increases their secretion by the liberation of gastrin, as discussed below.

Humoral Mechanisms. A number of hormones have been discovered whose function is to regulate certain aspects of gastrointestinal function, and three of these substances have been isolated in pure form and two of them have been synthesized. Since the functions of the several gastrointestinal hormones overlap somewhat, these substances will be discussed together in this section.

Gastrin. Gastrin is a heptadecapeptide (ie, it contains 17 amino acid residues), and is synthesized and secreted by certain epithelial cells which are located in the antral region of the gastric mucosa in response to several types of stimulus. Thus gastrin is secreted following vagal stimulation as well as following local mechanical distension of the antrum by food. In addition, gastrin secretion is also stimulated by certain secretagogues (ie, substances which enhance secretion), including low concentrations of alcohol, digestive products of protein, or certain substances in specific foods. Conversely, gastrin secretion is inhibited by overdistension of the antral mucosa as well as by the presence of acid. The latter inhibitory effect takes place through the operation of a local feedback mechanism to be described below.

When gastrin is absorbed into the bloodstream, it is transported to the gastric glands where it stimulates principally the oxyntic (or parietal) cells, and to a lesser extent the chief (or zymogenic) cells. Thus acid secretion by the parietal cells may increase up to eight times under the influence of gastrin, whereas the enzyme-rich secretion of the chief cells may increase only three times under the same circumstances. Consequently, gastrin stimulation results in the production of a very acid gastric juice, but a juice which also contains an elevated concentration of digestive enzymes.

The maximal increase in the rate of gastric secretion produced by gastrin is somewhat less quantitatively than that produced by vagal stimulation; however gastrin produces a more prolonged effect (ie, the duration lasts for several hours) than does vagal stimulation. Furthermore, vagal stimulation acting together with gastrin produces an enhanced rate of gastric secretion that is greater than the sum of each factor acting alone. In other words, the overall effect of both stimuli acting together is synergistic. It thus appears that both the neural and endocrine mechanisms act in concert under physiologic conditions to cause sufficient production of gastric juice to digest an entire meal.

Histamine also stimulates gastric secretion in a manner that is practically identical to that of gastrin; however, this amino acid derivative has a potency which amounts to only 80 percent of that of gastrin.

Gastric acid secretion apparently is regulated by a feedback mechanism, as noted earlier. Therefore, as the pH of the gastric fluids decreases below 2.0, the increased acidity itself completely blocks gastrin secretion and thus the stomach is protected from autodigestion by the highly acid juices that would be produced in unlimited quantities unless an inhibitory mechanism were present. Furthermore, this feedback mechanism also helps to maintain the optimum pH level that is essential to the normal function of the peptic digestive enzymes themselves.

Cholecystokinin–Pancreozymin. It was believed at one time that a gastrointestinal hormone known as cholecystokinin stimulated rhythmic contractions in the gallbladder, as described on page 844, whereas a distinctly different hormone, called pancreozymin, stimulated secretion of pancreatic juice having a high concentration of enzymes. It is now evident that one hormone which exhibits both of these regulatory activities is secreted by the mucosa of the proximal small intestine. This hormone is called cholecystokinin–pancreozymin, and it is released when there are products of protein digestion or fats in the lumen of the duodenum.

Thirty-three amino acid residues are present in porcine cholecystokinin–pancreozymin, and the five terminal residues of this polypeptide are identical with those found in gastrin. In fact, chemical removal of the sulfate ($—SO_3H$) group on the tyrosine residue near the C-terminal of cholecystokinin–pancreozymin converts the biologic activity of this hormone into that of gastrin.

Secretin. The polypeptide hormone called secretin contains 33 amino acid residues; this substance is also secreted by the mucosa in the upper region of the small intestine. Secretin is released under the stimulus of acid and/or the products of protein digestion, and it in turn induces the secretion of a pancreatic juice which has a high concentration of bicarbonate ions.

Glucagonlike Immunoreactive Factor. The mucosae of the stomach and small intestine contain a substance known as glucagonlike immunoreactive factor (or intestinal glucagon) which is secreted from these regions in response to the stimulus of ingested sugar. This hormonal substance, in common with pancreatic glucagon (p. 1091), can stimulate the release of insulin from the β cells of the islets of Langerhans (p. 1076), although chemically intestinal glucagon probably differs from the pancreatic hormone.

Interactions Among the Gastrointestinal Hormones. The principal (ie, most obvious) functions of the individual gastrointestinal hormones were summarized in the preceding sections. It has become evident that the functions of these hormones overlap significantly. Thus gastrin and cholecystokinin–pancreozymin both stimulate: (1) gastric motility; (2) acid and pepsin secretion by the stomach; (3) secretion of enzymes and bicarbonate by the pancreas; (4) bicarbonate secretion in the bile by the liver; (5) gallbladder contraction;

(6) intestinal motility; (7) secretion of insulin by the endocrine pancreas; (8) secretion by Brunner's glands of the duodenum.

Secretin, on the other hand, inhibits acid secretion by the stomach and it also inhibits both gastric and intestinal motility. Secretin, however, also stimulates the other tissues and organs listed above which are also stimulated by gastrin and cholecystokinin–pancreozymin. That is, secretin enhances pancreatic enzyme and bicarbonate secretion, bicarbonate excretion in the bile, gallbladder contraction, insulin secretion, and secretion by Brunner's glands.

Pancreatic glucagon has practically all of the actions demonstrated for secretin, but it does not stimulate pepsin secretion. The three principal gastrointestinal hormones are known to exert a multiplicity of actions. In addition to their numerous independent functions, the several gastrointestinal hormones can also interact with each other in a complex fashion, as summarized in Table 13.4.

Other Hormonal Factors Which Influence Gastric Secretion. In addition to the neural and humoral factors discussed above, the internal secretions of the adrenal cortex and medulla also may have some role in the regulation of gastric secretion. Thus adrenocorticotropic hormone secreted by the adenohypophysis stimulates an increase in the circulating level of adrenal glucocorticoids, which hormones in turn may cause an increased gastric acid and pepsin secretion, at least under experimental conditions.

Different types of stress can influence gastric secretion in various unpredictable ways, presumably because of release of catecholamines from the adrenal medulla. Epinephrine and norepinephrine themselves inhibit gastric secretion, and these hormones are secreted in quantity during periods of stress (eg, during hypoglycemia).

On the other hand, the glucocorticoid hormones of the adrenal cortex can decrease the quantity as well as change the composition of the mucus which is secreted by the gastric mucosa. Since this mucus is critical to the protection of the gastric mucosa against the corrosive actions of hydrochloric acid and pepsin, glucocorticoid administration to a patient during long-term therapy may induce ulcers by causing a breakdown in the mucous barrier so that pepsin and acid can erode the gastric epithelium.

Physiologic Regulation of Gastric Secretion

Normal gastric secretion usually is divided arbitrarily into three distinct phases (cephalic, gastric, and intestinal) for the purposes of discussion. It should be realized that these phases are mutually interdependent and show considerable overlap in vivo.

CEPHALIC PHASE. Gastric secretion is stimulated reflexly and directly by the presence of food in the mouth, and the vagus nerves transmit the impulses responsible for this effect. In the cephalic phase of gastric secretion, however, the sight, odor, or even the thought of food can also stimulate elevated gastric secretion via reflexes which have been conditioned from early childhood. Such conditioned reflexes have their neural origin within the cerebral cortex or within the appetite centers of the hypothalamus and amygdala (ie, within the diencephalon and limbic system), and the impulses

Table 13.4 SOME ACTIONS AND INTERACTIONS OF THE GASTROINTESTINAL HORMONES[a]

HORMONE(S)	HYDROGEN ION SECRETION BY STOMACH	EFFECT ON PANCREATIC SECRETION OF	
		BICARBONATE	ENZYMES
1. *Individual Actions*			
Gastrin	Strong stimulation	Weak stimulation	Strong stimulation
Cholecystokinin–pancreozymin	Weak stimulation	Weak stimulation	Strong stimulation
Secretin	Inhibition	Strong stimulation	Weak stimulation
2. *Combined Actions*			
Gastrin + cholecystokinin–pancreozymin	Competitive inhibition	Stimulation is additive[b]	Stimulation is additive[b]
Gastrin + Secretin	Non-competitive inhibition	Synergism[c]	Synergism[c]
Cholecystokinin–pancreozymin + secretin	Non-competitive inhibition	Synergism[c]	Synergism[c]

[a]*Data from Ganong.* Review of Medical Physiology, *6th ed, 1973, Lange Medical Publications.*
[b]*The effects of each hormone are proportional to its concentration up to the maximal effect which is produced by each hormone acting alone.*
[c]*The stimulatory effects of both hormones acting together (in maximal doses) is greater than the maximal stimulatory effect produced by either hormone acting by itself. This phenomenon also is called potentiation.*

arising in these neural centers are transmitted to efferent vagal pathways and thus to the stomach by way of the dorsal motor nuclei of the vagi which in turn are located within the medulla.

As noted earlier, hypoglycemia produces a stimulation of gastric secretion, presumably via the effect of the low blood glucose level acting directly upon the hypothalamus and perhaps also via the secretion of adrenocorticotropic hormone.

The cephalic phase of gastric secretion accounts for between 10 and 20 percent of the total gastric secretion which normally is produced following the ingestion of a meal.

Psychic influences have a marked effect on gastric secretion as well as on gastric motility so that hypersecretion by the gastric mucosa is observed when the subject is angry or tense. Thus strong emotional stimuli can elicit excessive gastric secretion at physiologically inappropriate times, and this response is believed to be a major factor in the development of peptic ulcers. Conversely, fear or depression can decrease gastric blood flow and markedly inhibit gastric secretion.

GASTRIC PHASE. Food which has entered the stomach stimulates gastrin secretion directly, and this hormone in turn is responsible primarily for the prolonged moderate elevation in the rate of gastric secretion that takes place while food is present in the stomach.

In addition to the gastrin mechanism, food within the stomach also causes local reflexes in the intrinsic plexuses of this organ as well as more generalized vasovagal reflexes. The impulses that subserve the latter are transmitted to the brain stem and then back to the gastric glands via the vagi. Therefore, both of these parasympathetic neural mechanisms increase the rate of gastric secretion which already is elevated by the gastrin mechanism.

The gastric phase of secretion amounts to roughly 65 percent of the total daily secretion of gastric juice, and thus it accounts for most of the total daily volume of between two and three liters of gastric secretions produced by the stomach (Table 13.2).

INTESTINAL PHASE. After the stomach has nearly emptied following a meal, the presence of chyme within the small intestine also stimulates the additional secretion of small quantities of gastric juice. Experimental evidence suggests that this effect is caused by the products of protein digestion within the stomach (eg, polypeptides). Furthermore, the duodenal mucosa possibly may secrete a gastrinlike hormone in response to this stimulus; however, this has not been proven unequivocally. Extrinsic denervation of the small intestine does not inhibit the intestinal phase of gastric secretion.

In contrast to the effect of protein digestive products upon gastric secretion, the presence of acid, carbohydrates, and fat in the duodenum inhibit acid and pepsin secretion by the stomach as well as gastric motility via an enterogastric reflex. Quite possibly these inhibitory effects also are due, in part, to the effects of secretin and cholecystokinin–pancreozymin.

Approximately 5 percent of the total gastric secretion per day is produced during the intestinal phase.

Pancreatic Secretion

PANCREATIC JUICE. The exocrine pancreas secretes between 1,200 and 2,000 ml per day of an important digestive fluid which is rich in bicarbonate and a number of enzymes (Table 13.5). The pH of pancreatic juice is about 8.0. This alkalinity, together with the neutrality or slight alkalinity of the bile and the intestinal juices themselves (the succus entericus), acts to neutralize the acidity of the gastric chyme as it enters the duodenum. The pH of the duodenal chyme is raised to between 6.0 and 7.0; therefore, when the chyme reaches the jejunum it is approximately neutral. Consequently, the intestinal contents almost never exhibit an acidic reaction.

Pancreatic juice contains a number of potent enzymes for the digestion of proteins, carbohydrates, fats, and other compounds (Table 13.5). The proteolytic enzymes secreted by the pancreas include trypsin and two chymotrypsins; these enzymes cleave whole and partially digested proteins. Carboxypeptidase is another pancreatic enzyme, and it attacks peptide chains at their ends, thereby liberating the terminal amino acid with its free carboxyl group. In addition, a ribonuclease

Table 13.5 PRINCIPAL CONSTITUENTS OF NORMAL HUMAN PANCREATIC JUICE[a]

Water
Cations: Na^+, K^+, Ca^{++}, Mg^{++}, pH = 8.0 − 8.3
Anions: HCO_3^-, Cl^-, $SO_4^=$, $HPO_4^=$
Albumin, globulin
Trypsinogen → Trypsin
    (Acts primarily on arginine, lysine residues bearing a carbonyl group)
Chymotrypsinogens → Chymotrypsins
    (Acts primarily on tryptophan, phenylalanine, and tyrosine residues bearing a carbonyl group)
Procarboxypeptidase A → Carboxypeptidase A
    (Acts primarily on C-terminal bond of tyrosine, tryptophan, and phenylalanine)
Procarboxypeptidase B → Carboxypeptidase B
    (Acts primarily on arginine and lysine residues bearing a carbonyl group)
Prophospholipase A → Phospholipase A
Ribonuclease
Deoxyribonuclease
Elastase
Pancreatic lipase
Pancreatic α-amylase

[a]*Data from Ganong.* Review of Medical Physiology, *6th ed, 1973, Lange Medical Publications; White, Handler, and Smith.* Principles of Biochemistry, *4th ed, 1960, The Blakiston Division, McGraw-Hill.*

and deoxyribonuclease are present in pancreatic juice, and these enzymes split ribonucleic acid and deoxyribonucleic acid, respectively. Pancreatic α-amylase hydrolyzes starches, glycogen, and many other carbohydrates into disaccharides. However, this enzyme does not hydrolyze cellulose, an important high molecular weight polysaccharide found in plant material. Pancreatic lipase hydrolyses neutral fats into glycerol and fatty acids.

The proteolytic and hydrolytic enzymes of the pancreas are synthesized in, as well as secreted by, the acinar tissue in an inactive form, and they become enzymatically active only after they are secreted into the gut. Thus trypsin is secreted as a precursor (or proenzyme) which is called trypsinogen, and this substance is activated by an enzyme known as enterokinase which in turn is secreted by the intestinal mucosa in response to the presence of chyme in the gut. Trypsinogen also is activated to a slight extent by trypsin, which already has been formed in the gut. The chymotrypsinogens are activated by trypsin, and procarboxypeptidase appears to be activated by enterokinase. Prophospholipase A is converted into active phospholipase A by the action of trypsin.

If the pancreatic enzymes were secreted in their active forms rather than as inactive precursors, the pancreas itself would be digested rapidly by its own products. The pancreatic acinar cells responsible for secretion of the proteolytic enzymes also secrete another polypeptide known as trypsin inhibitor. This material is found intracellularly within the cytoplasm surrounding the zymogen granules. Thus trypsin inhibitor prevents the activation of trypsinogen, not only within the secretory cells, but also in the lumens of the acini as well as in the pancreatic ducts themselves. Since trypsin activates the other proteolytic enzymes of the pancreas, trypsin inhibitor also inhibits the premature activation of the other proteolytic enzymes within this organ. If, however, this mechanism fails for one reason or another, a usually fatal acute pancreatitis quickly develops.

Regulation of Pancreatic Secretion

In a manner similar to gastric secretion, pancreatic secretion is regulated by neural as well as humoral mechanisms. Hormonal factors, however, are the major determinants of pancreatic secretory activity, in contrast to the stomach, in which nervous activity predominates in the regulation of secretion.

HORMONAL REGULATION. As chyme enters the small intestine, a voluminous secretion of pancreatic juice ensues, principally in response to the hormone secretin.* This hormone is a small polypeptide that is found in the mucosa of the upper small intestine in an inactive form, prosecretin.

Secretin was the first hormone identified.

Hydrochloric acid stimulates the release and activation of secretin, which then is absorbed into the bloodstream. The presence of chyme in the small intestine also stimulates secretin release to some extent, but hydrochloric acid is by far the most important factor in this process.

Secretin carried in the bloodstream stimulates the pancreas to secrete a large volume of watery fluid which has an extremely high bicarbonate content (up to 145 mEq/liter), but which has very low enzyme and chloride levels.

The secretin mechanism is quite important physiologically, as this hormone is released in significant quantities when the pH of the upper intestinal mucosa falls below about 4.0. The resulting copious flow of alkaline pancreatic juice which quickly ensues then neutralizes the hydrochloric acid in the duodenum according to the reaction $HCl + NaHCO_3 \rightarrow NaCl + H_2CO_3$. Thus a neutral solution of sodium chloride remains in the duodenum, whereas the carbonic acid rapidly dissociates into water and carbon dioxide. The latter compound then is absorbed into the body fluids.

The rapid neutralization of the acid gastric contents by the alkaline pancreatic juice in the duodenum immediately inhibits the proteolytic activity of pepsin, since the optimum pH for the activity of this enzyme is about 2.0. Consequently, this is an important protective mechanism for the duodenal mucosa, as this region of the gastrointestinal tract cannot long withstand the erosive action of gastric juice. Failure of any aspect of this biochemical mechanism results in the development of duodenal ulcers.

Another important function of the abundant alkaline pancreatic secretion produced by secretin is to provide a pH suitable for the maximal activity of the various pancreatic enzymes. All of these digestive substances function most effectively in a neutral or slightly alkaline environment, and the secretion of a pancreatic juice having a pH of around 8.0 assures that this requirement is met. The secretion of a watery pancreatic juice under the stimulus of secretin has been termed hydreletic stimulation.

In addition to secretin, another hormone, called pancreozymin, is liberated from the upper intestinal mucosa by the presence of chyme in the gut. Proteoses and peptones produced by the partial gastric digestion of proteins are especially potent stimulants to the release of pancreozymin. Pancreozymin is also transported in the circulation to the pancreas, but unlike secretin it stimulates the secretion of a pancreatic juice having a high concentration of the digestive enzymes summarized earlier. This process has been called ecbolic secretion. The effect of pancreozymin thus is quite similar to that produced by vagal stimulation.

During active secretion by the exocrine pancreas the vasodilator polypeptide bradykinin is released, and this humoral agent is deemed responsible by some investigators for the increased blood flow observed during the secretory process.

Hepatic Secretion of Bile

BILE. The secretion of bile is but one of the numerous functions of the liver, some of which are summarized in Table 13.6.

As described earlier, bile is secreted continuously by the hepatic parenchymal cells through a system of ducts into the bile duct whence it passes into the duodenum when the sphincter of Oddi relaxes (Figs. 13.10, 13.11, 10.75, p. 690).

Bile itself is a complex fluid containing a number of components, which are listed in Table 13.7. Bile contains no digestive enzymes, but it is of importance in digestion because of the bile salts it contains. These compounds in turn perform the important task of emulsifying fats within the intestine, thereby increasing enormously the total surface area of these substances which is exposed to the action of the pancreatic and intestinal lipases, as will be discussed subsequently. An adult human liver secretes between 600 and 700 ml of bile per day, and this fluid generally is slightly on the alkaline side of neutrality.

Bile salts are sodium and potassium salts of the bile acids (Fig. 13.18) which are conjugated to glycine or taurine, resulting in the formation of glychocholic and taurocholic acids (Fig. 13.19). Recent evidence has indicated that deoxycholic acid apparently is formed exclusively within the intestine by the action of bacteria upon cholic acid. Exclusion of bile from the intestine results in a loss of up to 25 percent of the ingested fat in the feces.

The bile pigments biliverdin and bilirubin form glucuronides and these degradation products of

Table 13.7[a]   COMPOSITION OF HUMAN BILE[b]

COMPONENT	APPROXIMATE PERCENT OF TOTAL
Water[c]	97.0
Bile salts	0.7
Bile pigments	0.2
Fatty acids	0.15
Lecithin	0.1
Fat	0.1
Cholesterol	0.06
Inorganic salts	0.7
Alkaline phosphatase	Trace

[a]*Data from Ganong.* Review of Medical Physiology, *6th ed, 1973, Lange Medical Publications.*
[b]*Sample taken from the hepatic duct.*
[c]*The bile is concentrated during storage within the gallbladder by the absorption of water, so that gall bladder bile contains around 89 percent water in contrast to about 97 percent in the hepatic duct bile.*

hemoglobin are responsible for the golden yellow color of bile (Fig. 9.5, p. 534). Substances that stimulate bile secretion by the liver are known as choleretic agents, and the bile salts themselves exhibit this property when given orally.

A number of compounds are excreted in the bile including cholesterol, alkaline phosphatase, adrenocortical and other steroid hormones, and certain drugs. These substances as well as the bile salts themselves may be reabsorbed from the gut in various amounts via the enterohepatic circulation, and thus recycled. This mechanism is of importance, for example, in preventing an excessive and irrevocable loss of the bile salts in the feces.

REGULATION OF BILE SECRETION. The rate at which bile secretion may be altered in the body is accomplished by neural, hormonal, and choleretic agents as well as by alterations in hepatic blood flow per se.

Neural. Vagal stimulation can increase the rate of hepatic bile secretion markedly.

Hormonal. The hormone secretin can stimulate an increased rate of bile secretion up to 80 percent above basal levels, while concomitantly accelerating the rate of hydreletic secretion by the pancreas. It should be emphasized, however, that this hormone does not increase the rate of bile acid synthesis.

Choleretic Agents. Bile salts per se can stimulate bile secretion, and the bile salts which are secreted into the intestine are largely reabsorbed via the enterohepatic circulation rather than excreted in the feces. Consequently, these reabsorbed bile salts are reutilized over and over, hence they provide an important physiologic stimulus for the maintenance of a normal bile flow by the liver.

Hepatic Blood Flow. Within definite limits the rate of bile flow is directly proportional to the rate

Table 13.6   SUMMARY OF MAJOR HEPATIC FUNCTIONS[a]

1. Maintains blood glucose level; stores glucose as glycogen.
2. Manufactures and secretes bile.
3. Synthesizes ketone bodies.
4. Performs biochemical reduction and conjugation of adrenal and gonadal steroid hormones.
5. Synthesizes and secretes certain plasma proteins (eg, albumen).
6. Synthesizes urea.
7. Performs a multitude of detoxification reactions involving drugs and toxins.
8. Stores certain vitamins (eg, vitamins A and D).
9. Exerts a protective function against invasion of the body by microorganisms and other foreign bodies.
10. Plays a role in the regulation of visceral and systemic blood flow.
11. Performs a host of biochemical transformations that are involved in the intermediary metabolism of carbohydrates, fats, and proteins.

[a]*This list includes only the principal general functions of the liver, and does not attempt to list the innumerable individual metabolic reactions which are carried out by this organ, some of which are considered in detail in Chapter 14 and other sections of this book.*

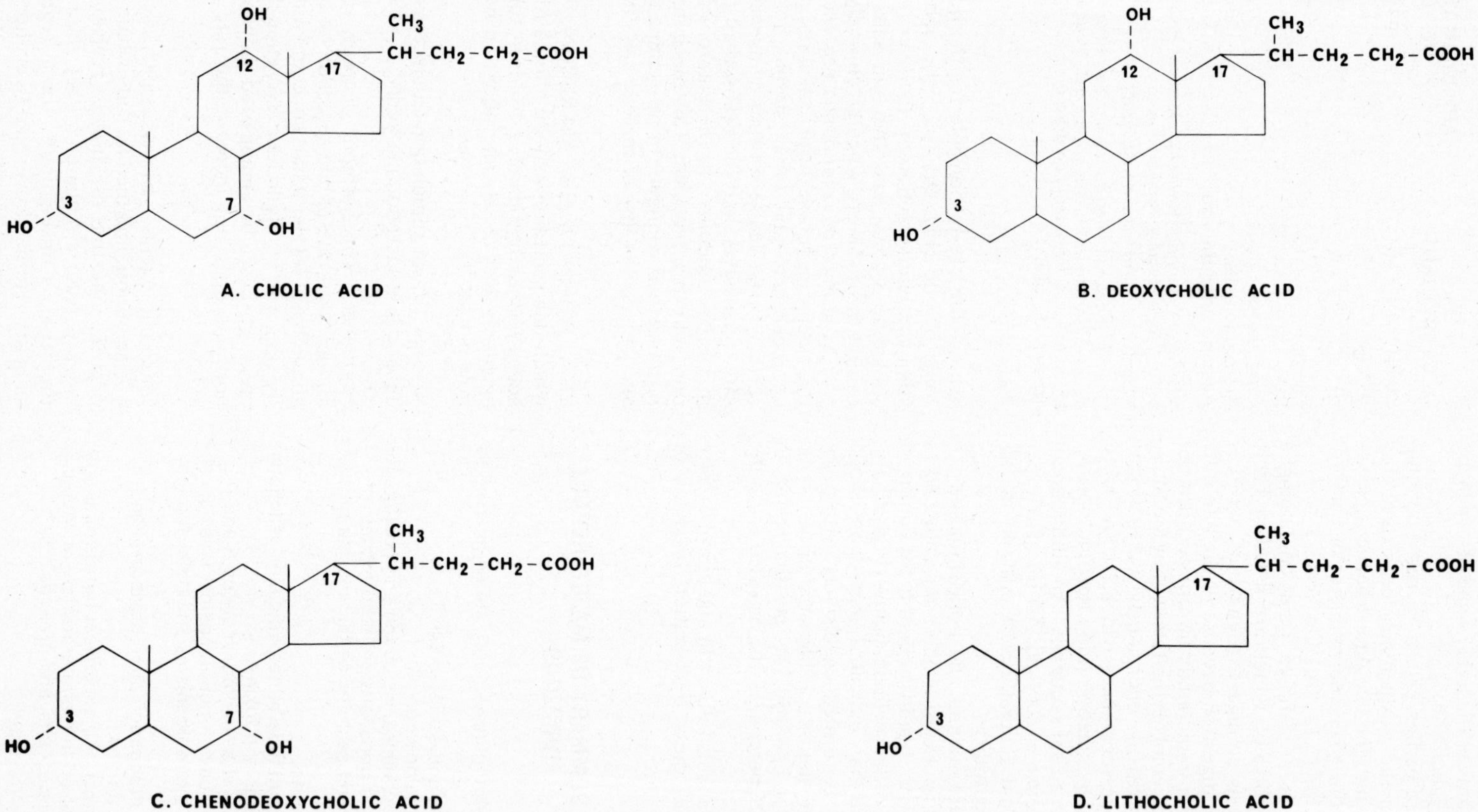

FIG. 13.18. Structural formulas of the four bile acids. (After White, Handler, and Smith. *Principles of Biochemistry*, 4th ed, 1968, The Blakiston Division, McGraw-Hill.)

$$C_{23}H_{26}(OH)_3 \overset{\overset{\displaystyle O}{\displaystyle \|}}{C} - \overset{\overset{\displaystyle H}{}}{N} - CH_2 - COOH$$

A. GLYCOCHOLIC ACID (CHOLYLGLYCINE)

$$C_{23}H_{26}(OH)_3 \overset{\overset{\displaystyle O}{\displaystyle \|}}{C} - \overset{\overset{\displaystyle H}{}}{N} - CH_2 - CH_2 - SO_3H$$

B. TAUROCHOLIC ACID (CHOLYLTAURINE)

FIG. 13.19. In bile, the acids depicted in Figure 13.18 are combined via an amide linkage to the amino acids glycine and taurine to give glycocholic (**A**) and taurocholic (**B**) acids respectively. (After White, Handler, and Smith. *Principles of Biochemistry*, 4th ed, 1968, The Blakiston Division, McGraw-Hill.)

of hepatic blood flow. This fact can be demonstrated readily in an isolated, perfused liver, and doubtless this mechanism participates to some extent in the regulation of bile flow in vivo.

Small Intestinal Secretions

INTESTINAL DIGESTIVE JUICES. The epithelial cells of the crypts of Lieberkühn in the small intestine secrete between two and three liters per day of a digestive fluid which sometimes is called the succus entericus. The crypts are located on the entire surface of the intestinal mucosa, with the exception of the duodenal region where the Brunner's glands are located. The secretions of the intestinal glands are practically pure extracellular (interstitial) fluid having a pH in the range between 6.5 and 7.5. The intestinal secretions are reabsorbed rapidly by the villi; hence there is a continuous circulation of fluids from the crypts to the villi; this fluid provides an aqueous medium for the absorption of materials from the lumen of the small intestine.

In the absence of cellular debris, the succus entericus contains almost no enzymes save for enterokinase, which activates trypsin, and a small quantity of an amylase. The mucosal epithelial cells, on the other hand, contain large quantities of digestive enzymes which probably digest the food still further during the process of absorption. In this regard, the rapid turnover rate of the intestinal epithelial cells may be of some significance to digestion, as the continual breakdown of these cellular units would liberate the many enzymes contained within their cytoplasm in a free state within the intestinal lumen. Since the average life of an intestinal cell is only 48 to 72 hours, this continual replacement of the epithelium also rapidly heals any injury the intestinal mucosa may sustain.

The enzymes which are found in the small intestine are as follows: (1) A number of peptidases are present. These substances are proteolytic enzymes which produce cleavage of polypeptides into their constituent amino acids. (2) A small quantity of intestinal amylase is present in the gut, and this enzyme converts polysaccharides into disaccharides. (3) Four enzymes are present in the intestinal fluids which split disaccharides into

monosaccharides. These include a lactase, a sucrase, a maltase, and an isomaltase. (4) An intestinal lipase also is present, and this enzyme degrades neutral fats into fatty acids and glycerol.

REGULATION OF SECRETION IN THE SMALL INTESTINE. Secretion of the succus entericus by the small intestine is regulated principally by neural, and perhaps also by hormonal, factors.

Neural Regulation. The chief mechanism for the regulation of small intestinal secretion is via local reflexes involving the intrinsic neural plexuses of the gut. Therefore, a stimulus of prime importance to the secretion of large quantities of intestinal juice from the crypts of Lieberkühn is mechanical distension. Tactile stimuli and irritating agents also can produce a marked increase in the volume of intestinal secretion. The presence of chyme within the small intestine is a major regulatory factor for secretion by the small intestine; therefore, the quantity of fluid that is secreted under physiologic conditions is related directly to the amount of food present within the intestinal lumen.

Stimulation of the extrinsic parasympathetic nerves to the small intestine can double or even treble the basal secretion. However, compared to the effect of distension, the extrinsic nerves are a relatively minor factor in the overall regulation of intestinal secretion in vivo.

HORMONAL FACTORS. There is some experimental evidence that one or more hormones (which collectively are called enterocrinin) are liberated from the intestinal mucosa by the presence of chyme, and this substance (or substances) may play a role in intestinal secretion. However, this endocrine mechanism does not appear to be of any great significance in the physiologic regulation of intestinal secretion.

SECRETION OF MUCUS. Brunner's glands, which line the first several centimeters of the duodenum (ie, usually between the pylorus and papilla of Vater), are especially important in the protection of this region of the small intestine from the erosive action of gastric juice. In response to such stimuli as acid, vagal stimulation, direct tactile stimuli, and perhaps gastrointestinal hormones, these glandular structures immediately secrete large quantities of a viscid mucus.

Secretion from the Brunner's glands is inhibited sharply by sympathetic stimulation; consequently, the duodenal bulb is unprotected by mucus during emotional states involving pronounced sympathetic discharge, eg, anger. Chronic emotional states may contribute significantly to the development of peptic ulcer in this part of the gastrointestinal tract.

Goblet cells, which are abundantly scattered over the entire surface of the intestinal mucosa, also secrete a protective mucus in response to the presence of chyme or irritants. Presumably the secretion of these cells, as well as the goblet cells within the necks of the intestinal glands, is regulated principally by local reflexes involving the intrinsic neural plexuses of the gut.

In view of the facts reviewed above, it is clear that small intestinal secretion in vivo is regulated primarily by local reflex mechanisms, and as a consequence the small intestine exhibits a high degree of autonomy in the performance of its normal secretory functions.

Large Intestinal Secretions

MUCUS. The large intestine is provided with enormous numbers of goblet cells both in the glands (crypts) as well as on the mucosal surface. These cells secrete quantities of a viscous mucus having pH of about 8.0, and this is the only major secretion of the large intestine. In this region of the gastrointestinal tract the mucus serves not only to protect and lubricate the intestinal wall, but also to bind the fecal material together. In addition the mucus serves to protect the colon from acids formed by the enormous amount of bacterial activity that takes place within the fecal matter itself.

The principal regulatory mechanism for mucus secretion in the large intestine appears to be mediated via local mechanical stimulation or irritation of the epithelial cells, and the enhanced secretion of mucus which results from such stimulation probably is caused by the activation of local intrinsic reflexes in the colonic wall.

In addition to this regulatory mechanism, stimulation of the nervi erigentes, which provide parasympathetic innervation to the distal segment of the colon, induces a copious secretion of mucus by the large intestine.

WATER AND ELECTROLYTES. Irritation of the large intestinal mucosa, eg, during bacterial or other infections, results in the secretion of large quantities of water and electrolytes in addition to mucus. This water and electrolyte secretion serves not only to dilute the irritant, but the colonic distension also stimulates rapid movement of the watery feces to the anus, causing diarrhea. Water and electrolyte loss from the patient may result in dehydration of the body tissues, and a severe electrolyte imbalance may develop which can have rapidly fatal consequences, especially in infants. Similarly, death may result in an adult if the diarrhea is sufficiently prolonged and severe (eg, as in cholera). The electrolytes lost during diarrhea contain large quantities of potassium; therefore a hypokalemic paralysis also may ensue following an extended bout of diarrhea.

DIGESTION AND ABSORPTION

The three major classes of foodstuffs utilized by the human body as sources of energy, as well as for other metabolic processes, are carbohydrates, fats, and proteins. In general, these substances cannot be absorbed directly from the gastro-

intestinal tract in their natural (or native) form as they are ingested. Consequently, the process of digestion is essential to the orderly breakdown of these complex materials by the action of various enzymes (Table 13.1) into chemically simpler compounds which then can be absorbed or assimilated into the body fluids (blood or lymph) via the intestinal epithelial cells and thus serve as nutrients. In addition to the three major foodstuffs, adequate human nutrition also requires the absorption of certain ions and vitamins from the ingested food, as well as an adequate quantity of water.

The processes whereby the digestion and absorption of the major foodstuffs take place in the body form the basis of the present section, together with a discussion of the mechanisms involved in the absorption of major electrolytes, vitamins, and water.

In general, the physical mechanisms that underlie the absorption of various substances into the blood or lymph from the lumen of the gastrointestinal tract include passive transport (or diffusion), active transport, solvent drag, and pinocytosis. The application of these processes to the transport of specific nutrients will be discussed in context in the sections which follow.

The total surface area of the mucosa of the human intestine available for the absorption of nutrients has been estimated to be around 75,000 cm² whereas that of the colon is only 2,500 cm². The discrepancy between these two figures is due chiefly to the fact that the large intestine lacks villi.

The microvilli that compose the brush border of the epithelial cells of the small intestine themselves increase the total surface area for nutrient absorption about 600-fold. Stated another way, the total absorptive area of the human small intestine is 4,500 m².*

Capillary blood flow to the small intestine is about 1 liter per minute at rest, ie, under basal conditions. During the hyperemia that accompanies the digestion of food, the flow increases markedly, whereas blood flow to the gut decreases sharply during periods of exercise. Total lymph flow through the left thoracic duct averages between 100 and 200 ml per hour.

Only a few processes regulate intestinal absorption, as the gut absorbs nutrients in a rather indiscriminate fashion. Hence body composition is regulated by appetite, thirst, and the availability of specific nutrients.

The absorptive potentiality of the small intestine greatly exceeds ordinary physiologic requirements. About 50 percent of this organ can be removed without deleterious effects insofar as adequate nutrient absorption is concerned. However, if conditions are such that digestion and/or absorption becomes inadequate (eg, in various intestinal diseases), then the defects in nutrient or

*This value is equivalent to an area of about 1.1 acres.

water absorption may become sufficiently great to endanger life.

Hydrolysis and Digestion

Complex carbohydrates, fats, and proteins result from the condensation of appropriate chemical subunits into larger compounds according to the general reaction R_1—OH + R_2—H → R_1—R_2 + H_2O. Hydrolysis, on the other hand, is the reverse of this reaction, ie, R_1—R_2 + H_2O → R_1—OH + R_2—H. Note that during the condensation of the two radicals, R_1 and R_2, a molecule of water is produced, whereas during hydrolysis one molecule of water is utilized.

Fundamentally, the digestion of all three major foodstuffs by the enzymes of the gastro-intestinal tract merely involves the hydrolysis of complex compounds into less complex subunits that are more readily absorbed by the intestinal mucosa. The only difference in the digestion of these three types of food lies in the specific enzymes required for catalyzing the hydrolysis of each group of substances.

Carbohydrates

The principal carbohydrates in the human diet are polysaccharides and disaccharides, and these compounds in turn are formed by the chemical condensation of two or more monosaccharides (usually hexoses or six-carbon sugars), as described in the previous section. The principal carbohydrates found in the usual human diet are the disaccharides sucrose (cane sugar), and lactose (milk sugar), as well as various starches. The latter substances are high molecular weight polysaccharides (ie, polymers of glucose) found in most foodstuffs, but especially in grains. Starches provide the single most important source of carbohydrate for human nutrition. Minor sources of carbohydrate in the diet include such compounds as glycogen, pyruvic acid, lactic acid, dextrins, pectins, and alcohol, as well as the monosaccharides glucose and fructose.

Although cellulose is a major carbohydrate constituent of plant tissue, and thus ingested by man in large quantities, there are no enzymes in the human digestive tract that are capable of hydrolysing this material.* Consequently, cellulose is not a food; rather, it contributes to the total mass (or bulk) of the feces.

It should be mentioned that the starch present in most foods in their native (or natural) state is contained or "packaged" within small granules that in turn are invested with a cellulose membrane. Therefore, unless this membrane is mechanically disintegrated or destroyed by cooking, the starch remains unavailable to the action of the digestive enzymes of the human gastro-intestinal tract.

DIGESTION. The biochemical mechanisms involved in carbohydrate digestion are summarized in Figure 13.20.

During the process of mastication, food is mixed with saliva, and this secretion contains the enzyme ptyalin (or α-amylase) whose chief source is the parotid glands. Ptyalin acts most effectively at an optimal pH of 6.7, and it catalyzes the hydrolysis of starch into two disaccharides, maltose and isomaltose. However, as food remains in the mouth for only a brief interval during mastication, then only an insignificant fraction of the total starch present in the food is hydrolyzed in the mouth before deglutition takes place. Within the stomach, however, ptyalin may continue to act for up to an hour within the center of the food mass before the fundic contents are mixed with the acid gastric secretions. Once the pH of the food in the stomach declines below approximately 4.0, the activity of ptyalin is inhibited completely. Before this inhibition takes place, up to 40 percent of the ingested starches will have been converted into maltose by the action of ptyalin.

Starches and disaccharides also are hydrolysed within the stomach by the hydrochloric acid secreted there, but only to a negligible extent.

Within the small intestine, an α-amylase secreted by the pancreas rapidly hydrolyses any undigested starches into maltose and isomaltose. This pancreatic enzyme is similar to, but somewhat more potent than, the salivary α-amylase. Therefore, by the time the chyme has reached the end of the jejunum, practically all of the starch has been converted into maltose and isomaltose in addition to some maltotriose and α-limit dextrins. The latter substances are starch breakdown products which generally are composed of eight glucose subunits.

The brush border of the small intestinal epithelial cells (principally of the jejunum) also contains four enzymes in varying concentrations. These enzymes are sucrase (or invertase), maltase, isomaltase, and lactase, and these enzymes hydrolyze the disaccharides sucrose, maltose, isomaltose, and lactose, respectively, into their constituent monosaccharides as they contact the intestinal epithelium (Figs. 3.11, 3.12, and 3.13). Maltase also catalyzes the breakdown of maltotriose and α-limit dextrins into glucose. The monosaccharides then are absorbed rapidly into the portal blood of the villi, as described below. The ultimate digestive products of ingested carbohydrates absorbed into the bloodstream are therefore monosaccharides.

An average diet contains more starches than either sucrose or lactose; therefore about 80 percent of the final digestive products of carbohydrate are glucose, whereas fructose and galactose each constitute approximately 10 percent of the total carbohydrate breakdown products.

Even termites and cows require particular symbiotic protozoans in their intestinal tracts in order to digest cellulose.

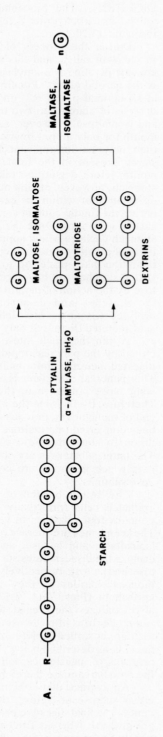

FIG. 13.20. General summary of carbohydrate digestion. **A.** Starch ultimately is degraded to single glucose molecules (Ⓖ). **B.** Sucrose is degraded to one glucose and one fructose (Ⓕ) molecule. **C.** Lactose is degraded to one glucose and one galactose (ⒼⒶ) molecule. (After Guyton. *Textbook of Medical Physiology*, 4th ed, 1971, Saunders.)

It is important to mention that lactase activity usually is low or absent in most adult humans and in children beyond the age of two to four years. In fact, some newborn infants may lack this enzyme completely, so that lactose (or milk sugar) is tolerated poorly when ingested. Human milk contains about 7.5 gm percent lactose, and cow's milk has 4.5 gm percent of this carbohydrate. A deficiency or complete absence of the specific β-galactosidase, known as lactase, results in the development of a watery diarrhea because of the osmotic effects of this sugar within the large intestine. In addition, fermentation of this sugar by the colonic flora gives rise to organic acids (eg, lactic acid) and large volumes of carbon dioxide which in turn can lead to flatulence, belching, and abdominal cramps. Lactose intolerant individuals, therefore, can suffer symptoms ranging from mild discomfort to malnutrition or even death in extreme cases, for example, infants who are wholly unable to digest quantities of this carbohydrate. These facts assume special significance because powdered milk is an economical, convenient, and highly nutritious foodstuff that is supplied to many impoverished areas of the world.

From a biochemical standpoint, starches are high molecular weight branched polymers of glucose. The commonest starch in the human diet is amylopectin, in which the glucose molecules form long chains in an α-1, 4-linkage, and the branches are formed by α-1, 6-linkages (Fig. 13.21). Starch is digested by salivary α-amylase (ptyalin), and pancreatic α-amylase in the intestine. Both of these enzymes hydrolyse α-1, 4-glucosidic bonds, but leave the α-1, 6-glucosidic bonds, terminal α-1, 4-linkages, and α-1, 4-linkages next to branches untouched. Consequently, α-amylase digestion yields the disaccharides maltose and isomaltose, and the trisaccharide maltotriose, and α-limit dextrins. The latter substances are branched polymers which have an average of eight glucose molecules.

Disaccharides are hydrolyzed within the brush border (ie, the microvilli) of the epithelial cells of the intestinal mucosa as they are being absorbed, a process called intracellular digestion (Table 13.8). Free disaccharidases are present in very low concentrations within the intestinal lumen itself, and such disaccharidases as are present within the intestinal lumen doubtless reflect the breakdown products of desquamated epithelial cells. Only monosaccharides are found in portal blood as the consequence of absorption. Since digestion of the disaccharides maltose, isomaltose, sucrose, and lactose is intimately and inseparably related to absorption of these sugars, these two processes will subsequently be considered at greater length.

ABSORPTION.   As discussed in the previous section, the principal form of dietary carbohydrate for

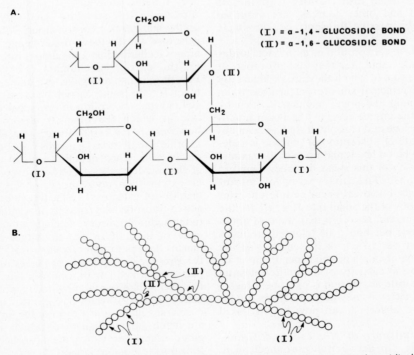

FIG. 13.21.   A. Branch point found in amylopectin (starch) and glycogen. (I) = α-1,4-glucosidic bond; (II) = α-1, 6-glucosidic bond. **B.** Amylopectin, showing branching structure. Each small circle represents a glucose molecule whose chemical linkages are shown in part **A** of this figure. (After White, Handler, and Smith. *Principles of Biochemistry,* 4th ed, 1968, The Blakiston Division, McGraw-Hill.)

Table 13.8   MAJOR HYDROLYTIC ENZYMES FOR DISACCHARIDES[a]

ENZYME	SUBSTRATE(S)	PRINCIPAL SITE OF ACTION
Lactase	Lactose, including both α- and β- glucosides and galactosides	Duodenum, upper jejunum
Isomaltase	Isomaltose	Jejunum, upper ileum
Maltases	Maltose	Jejunum, upper ileum
Sucrase (invertase)	Sucrose, maltose	Jejunum, ileum

[a]Data from Davenport, Physiology of the Digestive Tract, 2nd ed, 1966, Year Book Medical Publishers.

humans is plant starch, which in turn is composed of glucose molecules which are condensed into high molecular weight polymers having both straight and branched chains. During the process of digestion, free glucose molecules are liberated. Glucose, derived from the hydrolysis of starch, is the most important single sugar absorbed from the digestive tract from a quantitative standpoint.

Among the common sugars, only glucose, as well as galactose, which is derived from the hydrolysis of lactose, are absorbed from the intestinal lumen by an active transport mechanism. All other monosaccharides, whether present as such in the diet or derived from disaccharides (eg, fructose from sucrose) and other sources, are absorbed from the gut by passive transport, ie, by simple diffusion.

In general, the absorption of monosaccharides by man is quite rapid; ie, they are absorbed at about the same rate at which they are produced by digestion. The principal sites for sugar absorption are the duodenum and upper jejunum, so that by the time that the chyme reaches the ileum practically all of the ingested carbohydrate has been absorbed. The rate of absorption of glucose is independent of the blood level, and no overall maximal rate of absorption of this sugar has been found in normal humans. The absorption of monosaccharides does, however, occur at different rates in various regions of the small intestine. Hence the entire duodenum can absorb a maximum of about 20 gm of glucose per hour, the jejunum around 6 grams per hour per 15 cm of its length, and the ileum approximately 4 gm per hour per 15 cm of its length. In vivo the capacity of the entire intestine for monosaccharide absorption is extremely high.

The two mechanisms whereby certain monosaccharides are absorbed can now be discussed individually.

Active Transport of Glucose and Galactose. Basically, an active transport mechanism imparts energy to the substance being transported, so that the substance is moved against a concentration and/or electric gradient (p. 92). The intestinal mechanism by which glucose and galactose

are absorbed appears to be quite similar to that which is present in the renal tubules whereby glucose is reabsorbed from the glomerular filtrate.

The six characteristics of the intestinal active transport mechanism for the absorption of glucose and galactose may be summarized briefly. First, metabolic inhibitors, such as dinitrophenol, cyanides, iodoacetic acid, and phlorhizin inhibit active transport of glucose and galactose; oxidative metabolic energy is mandatory for this absorptive process. Second, the intestinal active transport mechanism is specific for two sugars alone, ie, glucose and galactose. The sugars being transported must have at least six carbon atoms, a D-pyranose ring structure, and a hydroxyl group on the number two carbon atom. Both glucose and galactose satisfy these requirements. Third, each monosaccharide has a maximum rate of transport, ie, the carrier mechanism can be saturated. Fourth, glucose and galactose compete with each other for the same carrier mechanism within the cell, so that absorption of one sugar decreases the rate at which the other is absorbed. Glucose, however, has a relatively greater affinity (about twofold) than does galactose for the transport mechanism. Consequently, a given concentration of glucose depresses galactose transport to a greater extent than does the presence of a similar quantity of galactose. Fifth, the glucose–galactose active transport carrier system is intimately linked to the sodium transport mechanism. Thus the presence of sodium in the intestinal contents enhances the absorption of these sugars. Conversely, the presence of glucose enhances sodium transport, and if sodium transport is blocked, glucose absorption likewise is inhibited. Sixth, as in other active transport systems, the rates at which glucose and galactose are absorbed from the intestine are much greater than those which are found for monosaccharides which are absorbed by passive transport (diffusion) alone.

Although the details of the carrier system involved in the glucose–galactose absorption mechanism have yet to be elucidated in detail, the general aspects of this process can be visualized, at least in broad outline. Glucose and galactose, liberated by the hydrolytic processes of digestion, are found in high concentration within the brush borders of the intestinal epithelial cells. In this region, the sugars combine with a carrier which transports them through the cytoplasmic matrix of the mucosal cells to the basal membrane, where they are discharged into the interstitial fluid; then they are absorbed by the capillaries of the portal system.

Glucose uptake by intestinal epithelial cells is not stimulated by insulin, a fact that is in sharp contrast to the effect of this hormone on glucose uptake by skeletal muscle cells. Once glucose has entered the mucosal cells of the intestine, roughly 10 percent of this compound is oxidized by aerobic metabolism and the lactate thus produced is transported ultimately to the portal capillaries of the liver. The remainder of the absorbed sugar is also transported as glucose directly to the portal system of the liver.

Passive Transport of Fructose and Other Monosaccharides. As noted earlier, the hydrolytic enzymes which digest disaccharides normally are present in very low concentrations within the intestinal lumen. Therefore, disaccharides are not split into monosaccharides as they are liberated from starch within the gut or as they are ingested (eg, sucrose). Rather, disaccharides undergo hydrolysis within the brush border of the intestinal epithelial cells as they are absorbed by diffusion from the luminal contents. Since monosaccharides are liberated within the brush border in high concentration, they are free to diffuse through the cell toward the basal region rather than back into the intestinal lumen; ie, they move down their own concentration gradients within the cytoplasm of the cell.

The hydrolysis of sucrose yields one molecule each of glucose and of fructose. The free glucose now combines with, and is transported by, the intracellular carrier system discussed in the previous section. Fructose, on the other hand, accumulates within the epithelial cells, from where it ultimately diffuses into the portal system. A fraction of the ingested fructose is also converted into lactic acid and transported to the liver in this form.

Fructose can be absorbed directly from the intestinal lumen by a process of "facilitated diffusion" when this sugar is administered as the free monosaccharide. Thus the rate of absorption of fructose is much slower than that of glucose or galactose, which are transported by an active mechanism, and somewhat faster than that of other sugars (such as mannose), which are absorbed solely by passive transport. It should be reiterated that under normal circumstances, disaccharides are absorbed and transported as such in only negligible quantities.

Protein

In order to remain in nitrogen balance, adults require a daily intake of protein, and growing children necessarily have a proportionally greater need for this type of nutrient, as discussed on page 874. Major sources of protein in the human diet include meat, dairy products, and vegetable products. The native proteins derived from these sources are composed in turn of long chains of amino acids that are bonded together by peptide linkages (p. 55). Each protein has certain characteristic physical, chemical, and biologic properties which are conferred upon it by the types of amino acid residues that are present in the molecule, as well as by the particular sequence in which the amino acids are arranged in the molecule. During the process of digestion, all ingested native protein, save for an insignificant fraction, is hydrolyzed into its constituent amino acids, which then are absorbed. During this digestive process, ingested native protein also loses all of its biologic specificity which once conferred upon it antigenic properties.

Between 10 and 30 gm of protein is secreted per day in the digestive fluids and about 25 gm of protein is derived from desquamated cells of the gastrointestinal tract itself. The protein from these sources is added to the ingested protein, digested, and the amino acids derived therefrom are also absorbed from the intestinal lumen. Consequently, only about 10 percent of the total quantity of protein in the gut is lost from the body each day in the feces.

DIGESTION. Protein digestion, or proteolysis, commences in the stomach, where the pepsins cleave certain peptide linkages of the ingested proteins (Fig. 13.22, Table 13.1). Specifically, these enzymes split the amino acid residues bearing the carbonyl group of tryptophan, phenylalanine, tyrosine, methionine, and leucine as well as a number of acidic amino acids. The net result of this hydrolytic process is a complex mixture of proteoses, peptones, and polypeptides having quite diverse molecular weights. Since hydrochloric acid activates the pepsinogens, and the low pH of the stomach is necessary for the full proteolytic activity of the pepsins, when the gastric contents reach the duodenum peptic activity ceases completely upon contact with the alkaline pancreatic juice. Gastric digestion of protein is unnecessary, however, since individuals having achlorhydria (a condition characterized by failure of the gastric mucosa to secrete hydrochloric acid and pepsin), remain in nitrogen balance upon a diet containing the usual native proteins. Consequently, intestinal proteolysis is adequate to produce sufficient free amino acids in such individuals.

In normal persons, the extent to which gastric digestion of protein takes place exhibits wide variations, depending upon several factors: first, the secretion of gastric juice and the factors which influence this process; second, the physical state (ie, particle size) of the proteinaceous foods which are ingested; third, mixing in the body and antrum

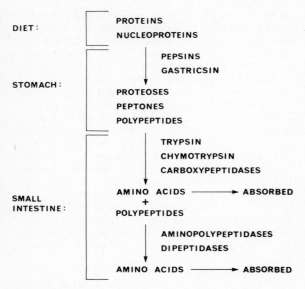

DIET:
PROTEINS
NUCLEOPROTEINS

STOMACH:
PEPSINS
GASTRICSIN

PROTEOSES
PEPTONES
POLYPEPTIDES

SMALL
INTESTINE:
TRYPSIN
CHYMOTRYPSIN
CARBOXYPEPTIDASES

AMINO ACIDS ⟶ ABSORBED
+
POLYPEPTIDES

AMINOPOLYPEPTIDASES
DIPEPTIDASES

AMINO ACIDS ⟶ ABSORBED

FIG. 13.22. Summary of protein digestion. See also Figure 13.25. (After Guyton. *Textbook of Medical Physiology,* 4th ed, 1971, Saunders.)

of the stomach; fourth, the rate at which the stomach empties. Thus the extent of gastric protein digestion may range from only one percent or so of the total protein ingested during a meal, up to 15 percent of the total. Consequently, any ingested protein is transported to the duodenum in the chyme as a complex mixture of undigested muscle fibers, a solution of native protein, and a complex mixture of the products of gastric protein digestion (proteoses, peptones, polypeptides). Only a very small quantity of free amino acids is liberated by gastric protein digestion. Therefore, most protein digestion takes place in the intestine.

The proteolytic enzymes trypsin and chymotrypsin that are secreted in the pancreatic juice rapidly degrade the mixture of protein and proteinaceous materials received in the chyme from the stomach into small polypeptides and dipeptides. The specific amino acid residues forming the loci within the polypeptide chains which are hydrolysed by the pancreatic enzymes are summarized in Table 13.1. The elastase which is listed in this table hydrolyses the fibrous proteins that are components of the connective tissue and ligaments of meat. A collagenase also is present in pancreatic juice, and this enzyme aids in the digestion of collagen. The pancreatic carboxypeptidases, together with the intestinal aminopeptidases and dipeptidases, then cleave the small polypeptide and dipeptide fragments into their constituent free amino acids; thus some free amino acids are released within the intestinal lumen. However, the aminopeptidases and dipeptidases are concentrated on the surface of, or just within, the brush border (microvilli) of the intestinal epithelial cells, like certain disaccharidases. There-

fore, large quantities of free amino acids are released locally in this region of the intestinal epithelium.

In addition to the various proteolytic enzymes discussed above, the surface of the intestinal epithelial cell as well as its cytoplasmic region are also the sites for the hydrolysis of a number of other compounds, including nucleic acids derived from the digestion of ingested nucleoproteins as depicted in Table 13.9. Thus purine and pyrimidine nucleotides are hydrolyzed on the cell surface into nucleosides and inorganic phosphate.

Phosphatases are also found upon the surfaces of the intestinal cells (but not in the intestinal lumen), and these enzymes hydrolyze such compounds as glucose-1-phosphate, phenyl phosphate, adenylic acid, hexose diphosphate, β-glycerophosphate, fructose-6-phosphate, and adenosine triphosphate among many others.

As with the gastric digestion of protein, the rate of intestinal digestion and absorption of proteins can vary markedly, and these processes in turn are governed by the rate at which protein is transported into the intestine from the stomach. For example, 50 percent of a test meal consisting of lean meat is transported into the duodenum within one hour following ingestion, and about 80 percent by the time three hours have elapsed. Once it reaches the duodenum, proteinaceous food is digested rapidly (in a matter of an hour or less), so that digestion of 5 percent protein (again in a test meal) is roughly 25 to 80 percent complete by the time the food reaches the end of the duodenum. Approximately 80 percent of the exogenous food protein together with endogenous protein has been digested and absorbed by the time the residue enters the proximal ileum; therefore only about 10 percent protein is found in the stools, and this protein does not reflect exogenous protein. Rather it is endogenous protein, such as is derived from mucoid secretions, desquamated intestinal epithelial cells, and microorganisms.

In addition to the exogenous and endogenous protein sources discussed above, the capillaries of the normal human gastrointestinal tract apparently are more permeable to protein than are, for example, the capillaries of muscle. Consequently, the interstitial fluid of the digestive tract has a relatively high concentration of protein that comes from the plasma, and a small quantity of this protein "leaks" into the lumen of the gut, where it is digested together with the protein from all other sources. The actual quantity of protein that is lost from the plasma in this fashion is unknown, but in normal persons it is an insignificant quantity. However, in certain clinical states the rate of plasma protein loss via the intestinal route may exceed the rate of replacement by hepatic synthesis; therefore hypoalbuminemia develops.

ABSORPTION

Native Protein. Normal adults can absorb only minute amounts of undigested (native) protein, if

Table 13.9   SUMMARY OF NUCLEOPROTEIN DIGESTION AND ABSORPTION[a]

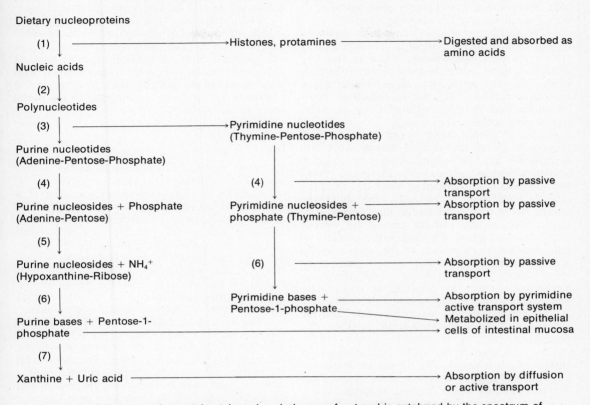

(1) Hydrolysis of dietary nucleoproteins takes place in lumen of gut and is catalyzed by the spectrum of proteolytic enzymes secreted by the stomach and pancreas (Table 13.1).
(2) Hydrolysis occurs in lumen of gut, catalyzed by pancreatic and intestinal mucosal enzymes.
(3) Hydrolysis is carried out on cell surfaces under the influence of intestinal mucosal phosphodiesterases.
(4) Hydrolysis occurs rapidly on and/or within the intestinal mucosal cells and is catalyzed by nucleotidases of the intestinal mucosa.
(5) Rapid deamination of purine nucleosides is catalyzed by adenosine deaminase of the intestinal mucosa.
(6) Phosphorolysis takes place slowly under the influence of several phosphorylases present in the intestinal mucosa.
(7) Oxidation of purine bases takes place through catalysis of the reaction by xanthine oxidase.

[a] *Data from Davenport.* Physiology of the Digestive Tract, *2nd ed, 1966, Year Book Medical Publishers.*

any, because the intestinal mucosa is impermeable to the large molecules. It has been demonstrated that infants can absorb sufficient albumen to become sensitized immunologically to this protein. The human fetus does not synthesize antibodies (ie, the plasma lacks gamma globulins); hence the passive immunity conferred upon the child at birth is due to the placental transfer of maternal antibodies. In addition, the colostrum secreted by the human mammary glands immediately following parturition and prior to the onset of lactation contains globulin antibodies, and these substances are ingested and absorbed intact by the suckling neonatal infant. The antibodies are protected from digestion by the pancreatic enzymes of the child by a state of neonatal achlorhydria (ie, an absence of gastric hydrochloric acid) combined with a

trypsin inhibitor which also is present in the colostrum.

Pinocytosis is the physiologic mechanism underlying the absorption of native proteins from the lumen of the neonatal gut. However, beyond 36 hours postpartum, the intestine loses its capacity to absorb intact native protein. The reasons underlying this phenomenon are unknown. The globulins which are absorbed during the first day and a half following birth are transported in the intestinal lymph (not in the portal blood) to the systemic plasma, where they confer passive immunity upon the neonate.

Amino Acids.   Following ingestion of a protein meal, there is a rapid transient increase in the amino nitrogen concentration of portal blood which reflects the fact that amino acids are absorbed into the body as rapidly as they are liberated from protein. Consequently, free amino acids do not accumulate in the intestinal lumen.

The absorption of all L-amino acids produced by digestion of native protein is accomplished by active transport mechanisms, which in turn are dependent upon oxidative metabolism in the cells of the intestinal epithelium. Three distinct active transport systems for amino acids have been characterized experimentally using in vitro biochemical techniques.

Neutral Amino Acids. The system which transports all of the neutral amino acids (eg, methionine, glycine, phenylalanine, tryptophan) exhibits some optical specificity. Thus D-methionine is transported at about one-third the rate of L-methionine, and D-methionine inhibits the absorption of the L-isomer of this amino acid. Competition is also present among the various amino acids transported by this single carrier system. Another characteristic of the neutral amino acid transport system is that the acid transported must contain a free carboxyl group; therefore, amino acids which are esterified or reduced (to an alcohol) are not transported by this mechanism. A free α-hydrogen also must be present, and the side chain of the transported amino acid must be neutral.

Basic Amino Acids. The active transport system which is responsible for the absorption of L-lysine, L-arginine, L-cystine and DL-ornithine operates at a rate which is about 90 percent lower than that for the neutral amino acids.

Proline, Hydroxyproline, and Other Compounds. The third mechanism in the intestinal mucosa transports L-proline, hydroxyproline, betaine, dimethylglycine, and sarcosine. Both proline and hydroxyproline exhibit a far greater affinity for this carrier system than they do for the mechanism which transports neutral amino acids. Consequently, these amino acids always are transported selectively by this third transport system.

Active transport of amino acids by any of the three carrier systems outlined above is inhibited when sodium transport is blocked. The relationship between these two processes, as in the case of carbohydrate absorption, is obscure, as are the details of the actual carrier mechanisms themselves.

Practically no free amino acids are found in the intestinal contents during protein digestion. This observation reflects the fact that the rate of amino acid absorption is far greater than is the rate of release of amino acids during proteolysis. Therefore, the rate of amino acid absorption is determined and limited, not by the rate of absorption itself, but rather by the rate of protein degradation.

Amino acids tend to accumulate within the epithelial cells of the intestinal mucosa as they are absorbed actively, since they enter the mucosal surface at a much faster rate than they are discharged through the basal surface of the cells. The latter process presumably occurs by passive transport (diffusion) of the amino acids into the portal capillaries through the intervening interstitial fluid.

Most amino acids are not metabolized to any significant extent after absorption, the two notable exceptions to this general statement being aspartic and glutamic acids. A small fraction of each of these compounds undergoes transamination with pyruvic acid in the epithelial cells, thereby forming alanine which in turn is transported to the portal blood by diffusion.

Fat

The most common fats (or lipids) in the human diet are neutral fats (also called tryglycerides or triacylglycerols). These compounds consist of a molecule of an alcohol (glycerol) which is condensed with three fatty acid molecules, as shown in Figure 3.14 (p. 68). The fatty acid radicals are coupled to the glycerol by low-energy ester linkages, so that an acyl radical is formed. Neutral fats also may consist of one or two fatty acids linked to the glycerol, and the terms monoglyceride (monoacylglycerol) or diglyceride (diacylglycerol) are applied to such compounds.

The fatty acids found in natural (ie, dietary) fats of both plant and animal origin in turn confer upon the fat its physical and chemical properties, for example, the melting point. The fatty acids themselves are straight chain (aliphatic) hydrocarbons, and may be saturated (ie, no double bonds are present in the molecule) or unsaturated (one or more double bonds are present). A few commonly occurring fatty acids are listed in Table 3.7 together with their structural formulae. If a neutral fat molecule contains two or more different fatty acids, then it is said to be a mixed triglyceride. The fatty acids in most dietary lipids consist almost exclusively of saturated palmitic (C_{16}) and stearic (C_{18}) acids as well as unsaturated oleic (C_{18}, one double bond) and linoleic (C_{18}, two double bonds) acids. Milk fat forms a notable exception to this general statement, as it is the only common food which contributes fatty acids of shorter chain lengths (C_4 to C_{14}) to the diet in any quantity.

Phospholipids are also found to a limited extent in any mixed diet. These compounds consist of a linkage between one glycerol alcohol radical and the phosphoric ester of an organic base generally choline or inositol), as shown in Figure 3.15 (p. 69).

Cholesterol esters, which contain a fatty acid linked to a sterol nucleus, are also considered to be fats. Cholesterol, since it exhibits certain of the chemical and physical properties of fats in general and is metabolized in a similar fashion, is also included as a dietary fat.

DIGESTION. Fat digestion in the stomach is negligible; consequently, most fat is transported to the duodenum as it is ingested. The presence of fat in the duodenum retards gastric emptying so that the rate at which fat enters the intestine is regulated by this factor. Gastric emptying of a high fat meal (ie, which contains around 50 gm fat) is quite slow, being spaced out over an interval of about 4 to 6 hours. Thus it appears that the maximal ability of the intestine to absorb fat may be at a rate of approximately 10 grams per hour. The normal small intestine absorbs all dietary fat, so that fecal fat represents only that which is derived from the intestinal mucosa plus any fat that is synthesized by bacterial action in the colon.

Fats are insoluble in water and do not mix with chyme. Furthermore, most ingested natural or artificial emulsions (eg, as in mayonnaise or homogenized milk) are broken down in the stomach.

The most important enzyme for the intestinal degradation of fats is pancreatic lipase. In turn, prior to attack by this enzyme, the fat is emulsified by such agents as bile salts and acids, protein, fatty acids, monoglycerides, cholesterol, lecithin (a phospholipid), and its derivative, lysolecithin. These emulsifying agents act as detergents, and thus physically disperse the fat into minute droplets (0.5 to 1.0 μ in diameter) which are stable within the pH range of the intestine, which ranges between 6.0 and 8.5 (see also discussion of micelle formation, Chap. 1, p. 15). Emulsification serves to increase the total surface area of the fat particles enormously, as the surface area is inversely proportional to the droplet diameter, and thus to facilitate digestion of triglycerides by the water-soluble pancreatic lipase which can act only at the surface of the droplets.

The most stable emulsion of fats is formed in the intestine, under the influence of bile salts, glycerides, and fatty acids. The latter two groups of substances are produced by the action of pancreatic lipase acting upon a less stable emulsion which is produced initially by the action of bile salts and/or lysolecithin. Consequently, the digestive products of this pancreatic enzyme themselves tend to stabilize the fat emulsion which ultimately forms.

In addition to their participating directly in the process of emulsification, the bile salts also combine with free fatty acids (which are liberated during hydrolysis of triglycerides by pancreatic lipase), as well as with most of the monoglycerides and some cholesterol. The result of this process is the formation of negatively charged, spherical micelles which are polymolecular aggregates ranging between 30 and 100 Å (3 to 10 mμ) in diameter (Fig. 1.8, p. 15). Thus the fatty acids and monoglycerides liberated from the emulsion droplets by pancreatic lipase are rendered water soluble, and in this manner the free fatty acids and monoglycerides can be absorbed readily by the epithelial cells while the bile salts are liberated once again and passed down the intestine for eventual reabsorption in the ileum.

The epithelial cells of the intestinal mucosa also contain a small amount of an intestinal (or enteric) lipase, and this intracellular enzyme together with pancreatic lipase hydrolyzes neutral fat into free fatty acids and glycerol. Some monoglycerides also are produced by the action of this enzyme.

Digestion is accomplished quite rapidly in the duodenum once the fat enters this region of the small intestine. Pancreatic lipase, acting on the emulsion droplets which were produced as described above thus can liberate up to 80 percent of the total fatty acids present in a triglyceride by the time the fat has reached the middle of the duodenum.

ABSORPTION.   Fat absorption commences within the duodenum so that about 50 percent of the total fat contained in a test meal (eg, 30 gm) is digested and absorbed by the time the residue has reached the end of the duodenum. Approximately 95 percent of the total fat in a test meal has been absorbed by the time 50 to 100 cm of the ileum have been traversed.

The reader should note particularly that neutral fat absorption involves both hydrolysis and re-synthesis of triglycerides. As discussed above, hydrolysis of fat occurs principally within the intestinal lumen under the influence of pancreatic lipase, but this process also proceeds within the epithelial cells of the mucosa under the influence of the enteric lipase.

Pancreatic lipase only degrades emulsified fats (triglycerides) into free fatty acids and 2-monoglycerides, but since the ester bond energy of these compounds is low, then some of the fatty acid which is present in the 2-position of the glycerol can migrate spontaneously to the 1-position, and thus be rendered susceptible to hydrolysis by this enzyme. Hydrolysis of 2-monoglycerides is accomplished intracellulary by the intestinal lipase after their passive absorption from the micelles which were formed within the lumen of the gut, This second lipase is localized within the microsomes of the epithelial cells, and it hydrolyses only short-chain fatty acids (eg, the eight carbon chain of octanoic acid) in contrast to pancreatic lipase which attacks long-chain fatty acids (eg, palmitic acid).

As the digestion (hydrolysis) of fat proceeds, the monoglycerides and free fatty acids that are released form highly stable micelles with bile salts in the intestinal lumen, as discussed in the previous section. Since these micelles are fat (as well as water) soluble, they can now penetrate the lipid phase of the cell membranes readily, and no special absorptive mechanism is present. However, the cells at the apices of the villi absorb the fat from the micellar solution more readily than do those at the sides. The size of the micelles is of the utmost importance to their absorption, as they must pass between the microvilli of the brush border in order to be absorbed. Micelles larger than about 650 Å (0.065 μ) are not absorbed, and furthermore, no micelles are absorbed through the tips of the microvilli that form the brush border of the intestinal epithelial cells.

Therefore, on the basis of currently available evidence, it appears that only those micelles containing long-chain free fatty acids and monoglycerides are absorbed, whereas larger micelles containing long-chain diglycerides and triglycerides are not.

The bile salts present in the micelles are separated in some unknown fashion from the fatty acids and monoglycerides during the absorptive process. Subsequently the bile salts are absorbed into the blood after they have passed to the ileum.

During active fat absorption, the apical cells of the villi contain more fat than do those at the sides. Microscopically visible droplets of lipid are not apparent within the cells, except beneath the terminal web in the endoplasmic reticulum, and fat droplets are localized in the cells only in the supranuclear region. The microsomes of the reticulum contain the enzymes which resynthesize the triglycerides. Experimental evidence indicates that the rate of fat absorption is equal to the rate of triglyceride resynthesis within the mucosal cells; therefore, there is no net accumulation of free fatty acids, glycerol, or dissolved fat during absorption.

As fat absorption proceeds, droplets of resynthesized triglycerides gradually fill the endoplasmic reticulum, pass to the supranuclear region of the cell, and are coated with a layer of phospholipid. Then these droplets pass out of the cells at the sides and become concentrated above the basement membrane in the basal region of the epithelium. The fat droplets then enter the central lacteal of the villi in large quantities (Fig. 13.9), and thus enter the principal lymphatic drainage channels of the gut as minute (0.1 to 3.5 μ diameter) droplets known as chylomicrons. Only negligible quantities of fat enter the portal capillaries directly.

The chylomicrons are coated with phospholipid. These droplets are composed of, by weight, roughly 5 percent phospholipid in addition to 90 percent triglyceride, 1 percent cholesterol, 4 percent free fatty acids, and a small quantity of protein. The phospholipid found in intestinal lymph is derived almost exclusively from de novo synthesis within the mucosal cells of the intestine, and dietary phospholipids apparently play no role in this process.

Most of the water-soluble constituents of dietary fats, including some glycerol and short-chain fatty acids, enter the portal capillaries by passive transport (diffusion). Presumably these substances enter the portal vessels via the fenestrations in the basement membrane of these vessels, whereas the chylomicrons are too large to pass through these channels. These fatty droplets, on the other hand, are able to enter the lacteals readily because of the open channels between the lumen of these vessels and the interstitial space. Furthermore, pinocytotic vesicles also may be responsible in part for transport of chylomicrons into the lumen of the lacteals.

Within the lymph of terminal lacteals, the chylomicrons are forced onward by active contractions of the smooth muscles of the villi, as well as by contractions of the intestinal smooth muscle as a whole. Chylomicrons do not return to the lymph once they have reached the systemic circulation. During active fat absorption, the rate of lymph flow in the thoracic duct increases markedly, and this situation may persist for up to 12 or more hours following ingestion of a high-fat meal.

Certain general features of the intracellular resynthesis of triglycerides from absorbed fatty acids and glycerol as well as other metabolic reactions involving lipids within the intestinal mucosa now may be summarized. After their absorption, 1-monoglycerides are hydrolyzed by intestinal (enteric) lipase into fatty acids and glycerol; the same intracellular lipase also may degrade some 2-monoglycerides which were not hydrolyzed by pancreatic lipase in the intestinal lumen prior to their absorption. Only long-chain fatty acids as well as unhydrolyzed 2-monoglycerides are resynthesized into triglycerides and phospholipids, and these syntheses take place before the fat leaves the cell for transport into the lymph.

As noted earlier, the enzyme system which is responsible for the resynthesis of triglycerides is localized in the microsomes of the endoplasmic reticulum. During resynthesis, the fatty acids react with adenosine triphosphate and coenzyme A to form fatty acyl~CoA. This reaction in turn is catalyzed by a kinase which requires magnesium ion for its action. The fatty acyl~CoA then reacts with either monoglycerides to yield diglycerides, or with diglycerides to yield triglycerides. Fatty acyl~CoA also may react with L-α-glycerophosphate, and thus form lysophosphatidic acid, which in turn can react with an intermediate of cytidine and thereby becomes a phospholipid. Alternatively, the lysophosphatidic acid may lose its phosphate to form a diglycerate. This intracellular esterification process is stimulated by certain conjugated (but not unconjugated) bile salts, including glycocholate and taurocholate.

During resynthesis, no fatty acids having less than 8 (octanoic acid) or even 10 (decanoic acid) carbon atoms in their chains are incorporated into the triglycerides. In fact, the quantities of lauric (C_{12}) and myristic (C_{14}) acids which are found in resynthesized triglycerides are disproportionately low when compared to their fractional occurrence in naturally occurring ingested triglycerides. Consequently, only fatty acids having chain lengths that are greater than 14 carbon atoms are the principal compounds involved in the intracellular resynthesis of triglycerides; the intracellular triglyceride biosynthetic mechanism is quite selective insofar as the fatty acids used for this process is concerned.

The glycerol released by hydrolysis of triglycerides in the intestinal lumen is metabolized by one of three separate pathways. First, it may pass unchanged to the liver where it is utilized for the synthesis of glycogen. Second, the glycerol may be oxidized into carbon dioxide within the intestinal mucosal cells. Third, the glycerol may be utilized directly for the intracellular resynthesis of triglycerides as described above. Some of the glycerol which is utilized for triglyceride resynthesis also may be derived from glucose, the biochemical precursor for this reaction being L-α-glycerophosphate (p. 941).

The phospholipids of the intestinal mucosa exhibit a rapid turnover rate during fat absorption; however, their net concentration remains constant throughout the absorptive process. Fatty acids are incorporated into triglycerides and phospholipids at different rates, but fatty acids do not require incorporation into phospholipids in order to be absorbed.

The role of phospholipids in forming the covering membrane with which chylomicrons are surrounded was mentioned earlier. Chylomicrons form a very stable emulsion as the consequence of this phospholipid coating; however enzymes which digest the phospholipid (eg, lecithinase) readily "break" such emulsions.

The overall processes of fat digestion and absorption are summarized in Figure 13.23.

Cholesterol is of ubiquitous occurrence in the human diet, principally as the free alcohol, and this compound absorbed by adults at the approximate rate of 10 mg per kilogram body weight per day. Within the intestine, cholesterol is also derived from two additional sources: first, from the bile (between one and two gm per day); and second, from secretions as well as desquamated intestinal epithelial cells (roughly 0.3 gm per day).

Within the intestine, cholesterol mixes freely with fats, bile salts, and a cholesterol esterase which is present in pancreatic juice. As the result of the action of this esterase, practically all of the cholesterol esters which are present in the gut are hydrolyzed, and the free cholesterol is incorporated into micelles for absorption along with other lipid constituents. Those esters of cholesterol that are refractory to hydrolysis are not absorbed.

Different sterols are incorporated into micelles with different degrees of facility; however, bile salts are essential to micelle formation, hence cholesterol absorption. Concomitant fat absorption is not essential for cholesterol absorption; however, the simultaneous absorption of neutral fats markedly stimulates cholesterol absorption. In this regard, unsaturated fats are more efficacious than are saturated fats insofar as facilitation of cholesterol absorption is concerned. In all, between one and three grams of cholesterol is transported from the intestine to the lymph per day.

Cholesterol is esterified with fatty acids to some extent within the intestinal mucosal cells, although by far the largest fraction of this compound is present in these cells as the free alcohol. About 66 percent of the cholesterol in the chylomicrons is present as the esterified form, and the fatty acids for this esterification come from the intracellular pool of fatty acids as well as ingested fats. Linoleic acid is the fatty acid most commonly employed for cholesterol esterification.

In general, the absorption of dietary sterols other than cholesterol can be predicted on the basis of their lipid solubility. Thus the greater the solubility of the sterol in lipid, the greater its total absorption as well as rate of absorption. This fact is of some clincial importance, as certain essential fat-soluble substances and vitamins require the presence of bile for their absorption. Carotene, a precursor of vitamin A, as well as vitamin K itself, both require the presence of bile in the gut in order to be absorbed. The absorption of these substances as well as of vitamins D and E is stimulated by the presence of a small quantity of fat. Presumably this effect is due to the absorption of the vitamins within micelles, and the formation of these microstructures is enhanced by concomitant fat absorption.

Such clinically important hormonal sterol derivatives as corticosterone, cortisone, cortisol, and testosterone are transported via the portal circulation after their absorption from the gut, and they do not enter the lymphatic circulation.

Other than carbohydrates, proteins, and fats, the nutrients essential to humans (ie, water, electrolytes, and water-soluble vitamins) do not require prior digestion in order to be absorbed. The physiologic mechanisms whereby these substances are absorbed following ingestion will now be considered individually.

Water Absorption

Five to 10 liters of water enter the small intestine of the normal adult human each day. This water is derived from exogenous food and drink as well as from endogenous salivary, gastric, biliary, pancreatic, and intestinal secretions. The absorption of water in the stomach is negligible, and since of this ten liters only 0.5 liter enters the colon, the small intestine must absorb water at a minimum rate of about 200 to 400 ml per hour. The upper limit to water absorption, assuming that any limit is present, is imposed by the maximal capacity of the body to handle a water load (p. 795). Maximal water excretion when the body is in a steady state amounts to a rate of about one percent of the body weight per hour; eg, 600 ml per hour urine is produced by a 60 kg individual who is drinking quantities of water at intervals sufficient to produce a maximum diuresis.

If one liter of water is drunk at one time, absorption of this water is complete in less than one hour. Consequently the maximal rate of water absorption must exceed this value.

The principal site for water absorption following a meal is the upper portion of the small intestine, although water is absorbed throughout the

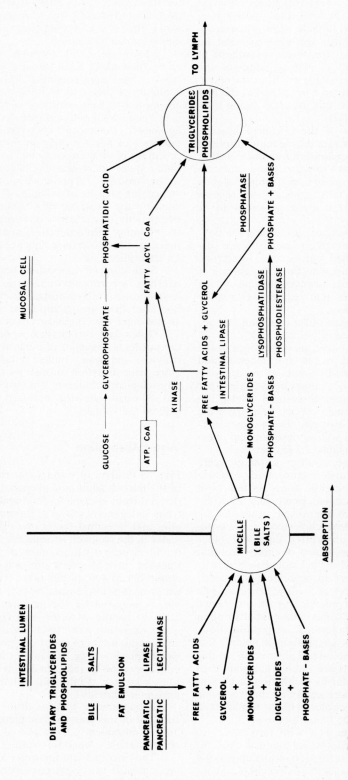

FIG. 13.23. Summary of fat digestion and absorption. See also Figure 13.26. (Data from Davenport. *Physiology of the Digestive Tract*, 2nd ed, 1966, Year Book Medical Publishers; Guyton. *Textbook of Medical Physiology*, 4th ed, 1971, Saunders.)

length of this structure as well as to a limited extent by the colon.

The important physiologic aspect of water absorption in the duodenum is that the net movement (or flux) of water is in the direction which maintains isotonicity of the intestinal contents with the blood. The large absorption of water in this region of the gut in turn reflects the absorption of osmotically active substances (principally glucose and amino acids) from the duodenum. In addition, a lowered osmotic pressure develops to some extent in the duodenum following neutralization of gastric acid, a process which results in the production of water. Water can move with facility either into or out of the intestinal lumen, depending upon the total osmotic activity of the intestinal contents at any particular time.

The net flow of water across the intestinal wall is the result of two large unidirectional fluxes (Fig. 13.24). First, one flux is directed from the intestinal lumen to the interstitial fluid and then to the blood or lymph. The second flux is directed in the opposite direction, that is from the interstitial fluid into the intestinal lumen. Experimental studies have indicated that the rates of these fluxes in both directions are quite high; however, the flux of water from the intestinal lumen to the blood is approximately 10 times faster than is the net absorption of this substance. Since the net absorption of water reflects the difference between the two large but oppositely directed fluxes, a small change in either one or the other flux can produce a large change in the net flow of water in one direction or the other. Thus when one liter of water containing deuterium (D_2O) as a tracer is ingested, 50 percent of the ingested deuterium appears in the blood in about 3 minutes, whereas it takes about half an hour for net absorption of 500

ml (again 50 percent) of the ingested water. The rapid appearance of the deuterium in the blood is indicative of an extremely rapid intestine-to-blood water flux which is some 10 times greater than the net rate of water absorption.

The actual pathways whereby water traverses the intestinal mucosa are unknown, but this tissue behaves as though small pores were present in the epithelial cell membranes. These postulated pores have been calculated to have a diameter of 7.2 nm and to occupy around 0.1 percent of the total surface area of the membrane. These pores are impermeable to solutes having a molecular weight greater than 200. Consequently when water is absorbed via the membrane pores, low molecular weight compounds, such as xylose, may be transported passively along with the water. This phenomenon is known as solvent drag.

Two general hypotheses have been advanced to explain the rapid movement of water across the intestinal mucosa. The first theory proposes a specific mechanism within the cells which rapidly transports vast numbers of individual water molecules, and pinocytotic vacuoles have been invoked as the bulk carriers of water together with certain solutes (eg, sodium and chloride) into and through the epithelial cells. The second hypothesis considers that water is transported passively as a secondary effect which results from the active absorption of solutes via specific active transport mechanisms. For example, active sodium ion absorption results in an osmotic gradient between the intestinal lumen and the cytoplasm of the cell; therefore, water moves passively along this gradient and into the cell. Perhaps one or the other absorptive mechanism predominates, depending upon the particular circumstances which are present at any moment in time.

The net absorption of water is affected profoundly by the osmotic pressure and electrolyte composition of the intestinal contents. Experimental studies in dogs have demonstrated that when hypotonic solutions were placed within the intestinal lumen, the flux of water from blood to intestinal lumen was high, ie, around 1 ml per minute, but the flux from the lumen to the blood was much higher, that is, around 2 ml per minute. Therefore, net absorption of water was quite rapid when the intestinal contents were hypotonic. On the other hand, when an approximately isotonic solution was present in the gut, the flux from blood to lumen remained unchanged. Under these circumstances, the flux from the lumen to the blood was decreased; therefore this flux was only slightly greater than that in the opposite direction. The net absorption of water was quite slow. Finally, the presence of a hypertonic solution within the intestine caused only a slight increase in the flux of water from blood to lumen; however, the flux in the opposite direction is diminished significantly. The net flow reversed and the volume of water within the lumen of the gut increased, thereby diluting the hypertonic solution.

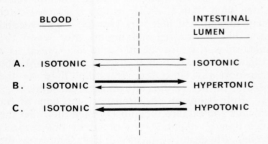

FIG. 13.24. The fluxes of water between the blood and the intestinal lumen in relation to the net transport of water. A. The blood–lumen and lumen–blood fluxes are equal and oppositely directed; therefore no net transport of water occurs. B. Hypertonic solution in gut. The blood–lumen flux of water (heavy arrow) now exceeds the lumen–blood flux (light arrow); therefore a net movement of water into the intestinal lumen takes place until the intestinal contents are isotonic once again. C. Hypotonic solution in gut. The lumen–blood flux of water (heavy arrow) now exceeds the blood–lumen flux (light arrow); therefore a net movement of water into the blood takes place until the intestinal contents become isotonic once again.

If an osmotically active solute which cannot be absorbed is present within the intestinal lumen, then water must be added to dilute it to isotonicity. Thus Epsom salts ($MgSO_4 \cdot H_2O$) are an effective saline cathartic, as both magnesium and sulfate ions are absorbed poorly, and, as sufficient water must be present to assure isotonicity, the total volume of liquid in the gut increases considerably, accelerating intestinal motility through distension.

Based upon the foregoing remarks, it is evident that the net movement of water from (or into) the intestinal lumen is quite variable, and this movement adapts readily to the osmotic situation that prevails within the gut at any particular time. Thus the osmolality of chyme entering the duodenum is adjusted to isotonicity quite rapidly by the addition or absorption of water, and thus the intestinal contents remain isotonic throughout their passage through the remainder of the digestive tract.

Absorption of Major Cations

SODIUM. Experimental studies using isotopic sodium (^{24}Na) as a tracer have shown that in the dog small intestine there are two oppositely directed and rapid fluxes for this ion (as for water), and that net absorption of sodium from the lumen of the gut reflects a slight difference in magnitude between these two fluxes. The sodium fluxes in the duodenum are large, but equal, so that there is no net absorption of sodium from this region of the gastrointestinal tract when isotonic sodium chloride is present. The rate of net sodium absorption is quite high in the jejunum, lower in the ileum, and still lower in the colon.

Sodium flux from the blood to the intestinal lumen is independent of the concentration of this electrolyte in the lumen for all practical purposes; however, the flux from the lumen to the interstitial fluid, hence the blood, increases in direct proportion to the sodium concentration in the intestinal lumen.

Sodium can be absorbed by the intestinal mucosa against an electrochemical gradient. Therefore, an active transport mechanism (sodium pump) is present within the epithelial cells of the mucosa, and metabolic energy is essential to this active transport process. The jejunum exhibits a high rate of aerobic metabolism; sodium transport ceases when glucose is absent. The aerobic metabolic rate is considerably lower in the ileum, but in this region of the intestine sodium transport can be maintained by oxidation of pyruvate or citrate when glucose is lacking. Anoxia or metabolic poisons (such as dinitrophenol) completely block intestinal sodium transport, as is the case in all other tissues which possess mechanisms for the active transport of this ion.

POTASSIUM. As is the case for sodium ion, there are also two opposing fluxes across the intestinal wall for potassium ion. However the rate of transfer of potassium ion is only a small fraction (5 percent) of the flux rate of sodium ion.

The blood–lumen and lumen–blood fluxes for potassium are equal in the duodenum; no net transport of potassium occurs in this region of the intestinal tract. Small potassium fluxes are also encountered in the jejunum and ileum. Within these regions of the digestive tract there is net absorption of both water and sodium, and as these substances are absorbed, potassium also is absorbed from the intestinal contents. The absorption of potassium in the jejunum and ileum apparently occurs by passive transport (diffusion) down its own electrochemical gradient into the mucosal cells as the consequence of water absorption. No active transport mechanism for this cation has yet been identified in the gut.

CALCIUM. Calcium is absorbed throughout the entire length of the small intestine, although the absorption of this cation takes place most rapidly in the duodenum. The rate of net absorption of calcium ion is directly proportional to the concentration of calcium within the intestinal lumen up to approximately 10 mM. The rate of calcium absorption is depressed by the presence of a sufficient concentration of magnesium ion in the intestinal contents (above approximately 4 mM).

A specific active transport mechanism is responsible for the absorption of calcium, and in turn this system requires energy for its operation, which in turn is derived from high energy phosphates that are produced by aerobic metabolism.

The net transport of calcium involves two distinct steps. First, calcium is absorbed rapidly from the intestinal contents. The rate of this absorptive process increases in direct proportion to the concentration of calcium which is present in the intestinal lumen (in the range from 0 to 1 mM). Second, absorbed calcium then is discharged into the interstitial fluid, but at a rate which is much slower than that at which it was absorbed. The second step has a limiting maximum rate; when the mucosal surface of the intestine is exposed to a calcium concentration in excess of 0.4 mM, the transport rate into the interstitial fluid on the serosal surface of the cells remains constant.

Both steps in the transport mechanism for calcium ion require vitamin D in the diet, and the ingestion of vitamin D (or a precursor of vitamin D called dihydrotachysterol) rapidly stimulates the rate of intestinal absorption of calcium, especially from the duodenum. This stimulatory effect on calcium absorption is most pronounced in vitamin D-deficient (rachitic) subjects.

Calcium must be in the ionized state (Ca^{++}) in order to be absorbed at all, as the transport mechanism is specific only for the ionized form of this metal. Consequently, agents which precipitate or chelate calcium ion within the intestinal tract effectively prevent calcium absorption. Thus oxalate, glucuronate, or citrate effectively prevent the absorption of calcium from the intestinal tract.

MAGNESIUM. Magnesium ion absorption is relatively constant throughout the small intestine, but none of this ion is absorbed from the colon. Magnesium apparently is absorbed passively by diffusion from the intestinal contents along with the movement of water.

IRON. On an average diet, an adult human ingests roughly 10 mg of iron per day and of this total quantity an adult male absorbs on an average between 0.5 and 1.0 mg. A sexually mature female, on the other hand, absorbs 1.0 to 1.5 mg of iron per day in order to compensate for that lost due to menstrual hemorrhage. The loss of iron from the body through desquamated cells is equal to the rate of iron absorption in the absence of hemorrhage, so that the total quantity of iron in the body pool of this electrolyte remains constant at approximately 4.0 grams.

The total amount of iron absorbed from the gut increases with the quantity ingested. However, the fraction of the total amount that is ingested decreases as the ingested quantity increases. Since the quantity of iron which is absorbed by normal individuals is not regulated by the amount needed by the body, excess absorption of iron above that which is lost eventually leads to the deposition of an iron-containing pigment in the tissues, a condition known as siderosis.

An excessive iron intake actually depresses absorption of this electrolyte in normal persons, whereas in patients with true iron deficiency the rate of absorption is between two and five times greater than in a normal subject, assuming that both are given the same dose of this cation. Following hemorrhage, the rate of iron absorption increases temporarily, but this effect is not manifest until about three days after bleeding.

Iron contained in vegetables as well as in the organic compounds of food (eg, myoglobin of muscle) is absorbed poorly. However, iron salts are absorbed more readily, and ferrous iron (Fe^{++}) is absorbed to a far greater extent than is ferric iron (Fe^{+++}). Large quantities of ascorbic acid (vitamin C) when added to the iron facilitate its absorption. The ascorbic acid acts by keeping the iron in the reduced (ferrous) state. Acid also facilitates the absorption of iron; therefore more iron is absorbed from the gastric contents as they enter the duodenum than from neutral solutions. Thus the effects of ascorbic and hydrochloric acids upon iron absorption tend to be synergistic.

The principal physiologic sites for iron absorption are the duodenum and upper jejunum, so that little, if any, is absorbed from lower regions of the digestive tract.

As is the case with calcium ion, iron absorption involves two steps. The first step is rapid and the second is much slower. Iron first enters the mucosal cells and then it passes into the blood via the interstitial fluid. Iron tends to accumulate within the intestinal epithelial cells, where it is present in two (or perhaps more) forms. One form of iron is distributed throughout the cytoplasm of these cells, whereas the other form is present as iron hydroxide micelles within ferritin molecules. The transport of iron from the mucosal epithelium of the intestine is regulated by the need for iron in body tissues and processes in general (Fig. 9.5, p. 534). Thus during recovery from hemorrhage or during pregnancy a large quantity of the iron stored in the intestinal mucosal cells is released and is transported in the plasma in combination with the globulin transferrin. But when the bodily iron requirement is low, only small amounts enter the plasma from the mucosa.

Under physiologic circumstances, the absorption of iron ceases because of an unknown alteration in its state in the lumen of the intestine which prevents its further absorption. Furthermore, iron absorption from the gut also ceases when the mucosal cells of the intestine become "saturated" with parenterally administered iron. Iron from the body pool can also enter the mucosal cells and thus regulate the intestinal absorption of ingested iron.

As noted earlier, the mucosal cells of the intestine have a high turnover rate and therefore a short life-span which amounts to about 48 to 72 hours. Most of the iron contained in these epithelial cells is lost to the feces as these cells are desquamated. Thus loss of iron from the intestinal mucosa provides a significant route for loss of this substance from the body.

Absorption of Major Anions

CHLORIDE. According to the law of electric neutrality, a solution must contain equal numbers of anions and cations. Hence the active transport of sodium ions out of the intestinal lumen during their absorption requires that an equivalent number of anions be absorbed simultaneously.* Therefore, intestinal absorption of chloride (as well as nitrate, iodide, bromide, and other anions) results predominantly from the active transport of sodium ions, and the anion moves passively out of the lumen of the intestine along the electric gradient that is created by the active transport of sodium.

Not only chloride, but also the other anions that are absorbed by diffusion down an electric gradient are absorbed at about the same rate as that of sodium. The intestine does not discriminate among these substances.

There is, however, some experimental evidence indicating that the intestinal epithelium is capable of absorbing certain anions independently of cation absorption. The ileum is able to absorb chloride more rapidly than sodium so that a large concentration gradient for chloride develops across the epithelium of the ileum. The jejunum also exhibits this property, but to a lesser extent. The

---

*Alternatively, an equivalent number of cations could be transported into the intestinal lumen.

secretion of bicarbonate into the intestinal lumen maintains electric neutrality during this active process.

The chloride concentration within the intestinal lumen per se does not affect the net rate of chloride absorption.

BICARBONATE. Bicarbonate ion is absorbed rapidly by the jejunum, presumably because the mucosa of this region secretes some acid. The hydrogen ion secreted reacts with the bicarbonate of the chyme, thus giving rise to a markedly elevated P_{CO_2} (average 100 mm Hg); the average pH of the chyme is around 6.5.

The ileum, on the other hand, actively secretes bicarbonate; and, as noted above, during chloride absorption bicarbonate ion gradually replaces the chloride ion in the lumen, so that the pH of the intestinal contents rises. If the activity of carbonic anhydrase is inhibited by drugs (eg, acetazolamide), both the secretion and reabsorption of bicarbonate are blocked.

Since bicarbonate is secreted by the mucosa of the ileum, the pH of the ileal contents of humans lies between 7 and 8. Also, proportionately more bicarbonate and less chloride is found in the intestinal contents as the chyme approaches the ileocecal valve.

PHOSPHATE. All regions of the small intestine absorb phosphate ion. Presumably this anion is absorbed by passive transport from the intestinal contents. Simultaneous absorption of glucose has no influence upon phosphate absorption, and phosphorylation of sugar is not a requisite to absorption of phosphate.

Absorption of the Vitamin B Complex

The water-soluble vitamins which are included in the vitamin B complex include thiamine (and thiamine phosphate), nicotinic acid (and nicotinamide), folic acid, flavine mononucleotide, biotin, and pantothenic acid. All of these compounds have relatively low molecular weights; their molecules are small enough to be absorbed directly from the intestinal contents by passive transport in quantities which are sufficient for human nutrition.

Vitamin B_{12} (or cyanocobalamin, molecular weight 1,357), on the other hand, is the largest water-soluble nutrient molecule that is essential to humans and that is absorbed on its native form by the intestinal mucosa. This B vitamin requires a special mechanism for absorption which is understood only in general outline. Only one microgram per day is the adult requirement for this vitamin, but this minute quantity is critical for the normal maturation of erythrocytes (p. 532).*

*Lest the reader assume that one microgram of cyanocobalamin is an insignificant quantity, one may calculate readily from the molecular weight of this compound and Avogadro's number (ie, 6.02×10^{23} molecules per mole) that one microgram of this vitamin contains approximately 4.45×10^{14} individual molecules!

Despite the fact that vitamin B_{12} is present in the diet as one (or more) coenzymes rather than as the free vitamin, the absorption of all forms of this substance is accomplished in the same fashion. Within the stomach, vitamin B_{12} forms a strong complex with a heat-liable mucoprotein known as intrinsic factor. In man, this protein has a molecular weight greater than 53,000 and one milligram of intrinsic factor combines with 25 micrograms of vitamin B_{12}. The intrinsic factor is secreted by the oxyntic gland area, possibly by the chief cells. The intrinsic factor–vitamin B_{12} complex then is taken up by the mucosal cells, probably by pinocytosis. Once within the epithelial cells, cyanocobalamin is liberated from its complex with the protein, and then the vitamin is transported into the blood. The mechanism whereby vitamin B_{12} reaches the bloodstream is unknown, but it has been postulated that special carrier substances are responsible for the release, transport, and utilization of this substance.

The cobalt in vitamin B_{12} can be labeled with ^{60}Co in order to trace its rate of absorption and elimination. Ingestion of the radioactive vitamin reveals that it is absorbed into the blood following a lag period of about 3 hours, and it reaches a peak blood concentration at around 8 hours. Subsequently, the radioactivity disappears from the blood in roughly 10 days.

Atrophy of the gastric mucosa first results in a failure to secrete hydrochloric acid, followed by a pepsinogen deficiency, and finally no intrinsic factor is elaborated. This condition is responsible for the failure to absorb enough vitamin B_{12} to permit normal erythrocyte maturation, so that pernicious anemia develops. Parenteral administration of this vitamin circumvents the absorption defect, and thus prevents the development of the macrocytic anemia.

A patient suffering from pernicious anemia and who lacks intrinsic factor must ingest between 1 and 10 mg of cyanocobalamin per day in order to absorb a nutritionally adequate quantity of the vitamin. This daily level is 1,000 to 10,000 times greater than that necessary for prevention of pernicious anemia in a normal individual.

THE ELEMENTS OF HUMAN NUTRITION

It was stated in the introduction to this chapter that man is a heterotrophic organism, and that as such he must obtain a variety of organic compounds in his diet which were synthesized ultimately by autotrophic organisms. In addition to organic compounds, man also requires an exogenous supply of certain minerals, and, of course, an adequate supply of water.

The term *nutrition* may be defined in the broadest sense as the sum of the processes whereby an animal (or a plant, for that matter) takes in and utilizes food substances. In animals such as man, the entire process of nutrition involves the ingestion, digestion, absorption (or assimilation), and utilization of nutrient substances. The material presented so far in this chapter has dealt with three aspects of human nutrition. The chemical processes that underlie the physiologic utilization of specific food substances (or metabolism) will form the subject of the next chapter. In the present section, the specific organic and inorganic nutrients required by the human body will be outlined.

For the purpose of this discussion, a food (or more precisely, a nutrient) may be defined as an ingested substance which supports growth, maintains body functions, is used for the repair or replacement of tissues and/or is used as an energy source. Within the scope of this definition, water, vitamins, and inorganic electrolytes all are essential nutrients (or ''foods''), although none of these substances yields energy.

In order to remain in an adequate or physiologically normal nutritional state, the human diet should contain sufficient total calories to provide energy, an adequate quantity of protein to maintain nitrogen balance, and sufficient amounts of vitamins, minerals, and water to supply the specific needs for these substances.

It must be borne in mind that the individual need for specific nutrients can vary widely, depending upon such factors as sex, age, general bodily configuration, environmental temperature, as well as the extent and type of physical activity in which the person is engaged. Such physiologic states in adult women as menstruation, pregnancy, and lactation also impose increased demands upon the body for some essential nutrients, and thus necessitate compensatory dietary adaptations.

The science of nutrition deals not only with these specific nutritional problems in normal individuals but is also concerned intimately with such problems as those related to the pathologic effects of malnutrition and deficiency states, the special dietary requirements of surgical patients (including parenteral nutrition), geriatric nutritional requirements, allergic conditions arising from foods, and the specialized diets which may become necessary in specific disorders. In the latter category may be included such widely diverse clinical states as congenital metabolic diseases, anemias, neurologic and psychiatric disorders, skin diseases, gout, osteoarthritis, diabetes mellitus, diseases of the gastrointestinal tract, including malabsorption, hepatic, biliary, cardiovascular, renal, and urinary tract dysfunctions.

It is obvious from this cursory summary that a proper application of current nutritional knowledge is of the utmost clinical importance.

Caloric Requirements

The calorie is a measure of the heat energy which is generated by the complete combustion (or oxidation) of a substance (p. 982). The caloric content of foods generally is determined by measuring the heat produced by burning a known quantity of a foodstuff in pure oxygen within a bomb calori-

meter (Fig. 14.54, p. 983). From such data, tables may be compiled which give the total caloric content of various foods. It should be borne in mind, however, that the energy which is contained in indigestible fibers, particularly cellulose in fruits and vegetables, contributes to the total caloric value of a foodstuff as determined by bomb calorimetry. But as stated earlier such substances are not digested and absorbed by the human body; consequently the actual calories which are obtained by the body from eating a portion of a specific food that contains a high percentage of such indigestible materials will be somewhat lower than the empirically determined caloric value. However, the caloric values of various foods as listed in nutritional tables are sufficiently accurate for the practical purpose of determining the dietary intake of energy-producing foods.

The complete oxidation of a pure carbohydrate yields 4.1 kilocalories of energy per gram; protein also yields 4.1 kilocalories per gram; but fat yields 9.3 kilocalories per gram and ethyl alcohol yields about 7 Cal per gram. Therefore fats are the most concentrated source of energy in the diet.

Over a period of time, the average total caloric intake per day must equal the total energy expenditure of the body (in the form of heat production and work performed) in order that body weight be maintained. If the caloric intake is inadequate, the body stores of protein and fat are broken down to supply energy (ie, they are catabolized), and weight loss ensues. On the other hand, if the caloric intake exceeds the energy expenditures of the body for any length of time, obesity develops as the ingested calories above those required to supply energy are stored as fat (p. 1216). The total daily caloric requirements of the diet should be met by a balanced combination of carbohydrate, protein, and fat, as discussed briefly in the following section.

The approximate total daily caloric requirements for males and females of different ages are listed in Tabel 13.10, and a few examples of the energy required in order to perform various types of physical work are given in Table 14.16. Energy balance of the body will be discussed in more detail in Chapter 14.

Carbohydrate

Although adequate nutrition is quite possible in the absence of carbohydrates in the diet, these substances generally comprise the major energy source for humans. The proportion of carbohydrate in the diet in turn varies widely, and is governed largely by economic and ecologic factors. For example, in tropical regions as much as 80 percent of the diet may consist of starches and sugars, but in the polar regions only 20 percent of the total dietary calories may be derived from carbohydrates, the balance of the caloric requirements being met by animal protein and fat.

Agricultural practices and economic status

Table 13.10[a]   MEAN STANDARD VALUES FOR THE BASAL METABOLIC RATE[b]

AGE, YEARS	MALES	FEMALES
6	53.00	50.62
8	51.78	47.00
10	48.50	45.90
12	46.75	44.28
14	46.35	41.45
16	45.72	38.85
18	43.25	36.74
20	41.43	36.18
25	40.24	35.70
30	39.34	35.70
35	38.63	35.70
40	38.00	35.70
45	37.37	34.94
50	36.73	33.96
55	36.10	33.18
60	35.48	32.61
65	34.80	32.30

[a]Data from Boothby, Berkson, and Dunn. Am J Physiology 116:468, 1936.

[b]The values listed are in Cal/m²/hr. Note the decline in BMR with age in both sexes which is a direct reflection of a reduced total caloric requirement. Note also that females consistently exhibit a lower BMR than males. See also Table 14.16, p. 994.

also markedly affect the proportion of carbohydrate which is consumed by a given population, so that there is an inverse relationship between the level of income and carbohydrate intake among different countries and among different population groups within the same country. A rough rule of thumb advanced by nutritionists in the United States is that carbohydrates should supply approximately 50 percent of the total energy requirement.

Vegetables and vegetable products contribute the bulk of the carbohydrate to the diet, and such common foods as potatoes, rice, legumes (beans and peas), cereals, corn, cane and beet sugar, and flour products, all are rich in carbohydrates. Many fruits contain an abundance of sugar; hence these foods too provide rich carbohydrate sources.

Protein

Proteins form indispensable components of every living cell in the body, and they are also found in most extracellular fluids. In general, proteins are very large molecules whose size, structure, and composition, as well as other physical and chemical properties vary over an extremely wide range. Individual natural proteins are composed of about 20 amino acids which are held together by peptide linkages (Fig. 13.25). In turn the number, type, and sequence of the individual amino acids in a given protein molecule confer upon it the several properties listed above. The number of individual amino acid residues found in particular proteins may range from several hundred to many thousands, and because of their molecular size, proteins cannot ordinarily traverse semipermeable membranes. Because of this property, proteins normally are digested into their constituent amino acids by the various enzymes which are present in the gastrointestinal tract and absorbed as individual amino acids. The metabolism of these nitrogenous compounds will be surveyed in the next chapter; however from a nutritional standpoint, certain amino acids are deemed essential and must be obtained from exogenous sources because man has no metabolic capacity to synthesize these compounds. Other amino acids can be synthesized from metabolic precursors which are available in the human body if these amino acids are not supplied in the diet. Such compounds are called nonessential amino acids.

Of the naturally occurring amino acids, ten are essential for growth and maintenance (nitrogen balance) in white rats. These essential amino acids are arginine, histidine, phenylalanine, methionine, leucine, isoleucine, valine, lysine, threonine, and tryptophan.*

The nonessential amino acids include alanine, aspartic acid, cystine, glutamic acid, glycine, hydroxyproline, proline, serine, and tyrosine.

The human requirement for essential amino acids appears to be quite similar to, if not identical with, that of the white rat. The ten essential amino acids listed above are deemed necessary to normal growth and maintenance in the human species.

It also should be stressed that the terms *essential* and *nonessential* refer only to dietary re-

*Histidine may be an essential amino acid for normal growth of children, whereas arginine is required for normal maintenance of nitrogen balance in adults. Both of these compounds are included among the essential amino acids.

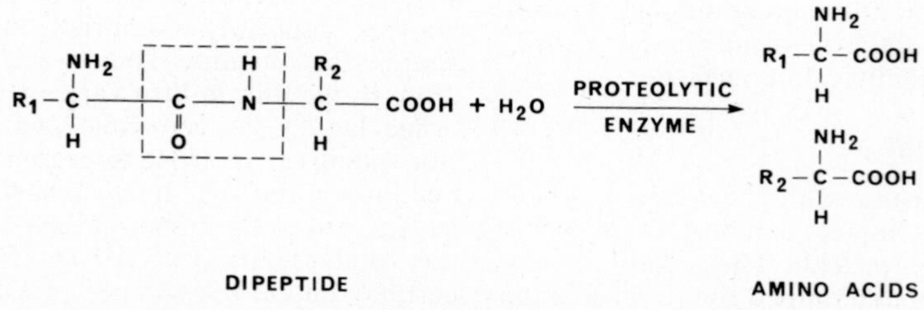

FIG. 13.25. A peptide bond (in a box at left) in a dipeptide and the proteolytic cleavage of the dipeptide into two individual amino acids. R_1, R_2 = individual radicals. (Cf. Fig. 13.22.)

quirements, and have no significance insofar as their importance in intermediary metabolism is concerned.

It has been demonstrated in humans that if the exogenous (ie, dietary) supply of total amino acids is marginal, then simple carbon and other compounds (eg, urea, glutamic acid, ammonium salts) spare the nonessential amino acids by virtue of the fact that these substances are able to provide nitrogen for growth and maintenance. Under conditions of severe nitrogen deprivation, nonessential amino acids can be synthesized from carbon precursors that are present in the metabolic pool of the body, using nitrogen which is present in urea and ammonium salts.

From a nutritional standpoint, food proteins may be classified according to their ability to supply the body with essential and other amino acids that are necessary for the synthesis of tissues, hormones, and enzymes. The term *biologic value* is used to indicate the amount (percentage) of absorbed nitrogen which is retained by the organism for use in the synthetic processes listed above. A given dietary protein may have a high or low biologic value, depending upon whether or not it supplies an adequate quantity of the essential (as well as nonessential) amino acids. Digestibility is another factor which is directly related to the quality of a protein, as adult humans can absorb only free amino acids that are liberated by complete hydrolysis of the native protein molecule. In this regard the method of cooking is of the utmost importance, as it may enhance the digestibility of protein. Thus boiling in water increases the digestibility of wheat protein, presumably by altering the structure of the gluten molecules. On the other hand, overcooking, especially by dry heat or during frying, may destroy certain essential amino acids (in addition to other nutrients) such as lysine, and thus render the nutritive value of the protein less than optimal.

The quality of various proteins can be summarized into two general categories. Grade I proteins are nutritionally complete, and are found in such foods as milk, eggs, meat, and fish. The animal proteins in these foods contain amino acids in roughly those proportions which are required for protein synthesis in the tissues of the human body. Grade II proteins are nutritionally incomplete, and contain different proportions of the amino acids than are found in grade I proteins. Some grade II proteins have either low concentrations of, or lack entirely, one or more of the essential amino acids. Most plant proteins fall into the grade II category, and thus include the proteins in wheat, rice, corn, beans, and nuts. A carefully selected mixture of plant foods if ingested in sufficient quantity may provide adequate nutrition insofar as the essential amino acids are concerned, but the inclusion of animal protein in the diet is most desirable.

The minimal adult requirement for grade I (or complete) protein which is necessary to supply the ten essential amino acids is one gram of protein per day per kilogram of body weight for both men and women living in a temperate climate, provided that the total caloric intake in addition to the protein intake is adequate to meet the individual energy requirements.

As in the case of carbohydrate, there is a correlation between the intake of animal (complete) protein and the economic level of the population, as foods containing this type of protein are more valuable nutritionally, but much more expensive. Animal protein, if possible, should constitute at least one-third of all dietary protein. Foods included in this category include milk, milk products (eg, cheese), meats, fish and other sea foods, poultry, and eggs. Plant proteins are readily available in cereals such as oats, wheat, rice, corn, barley, and rye as well as in the seeds of legumes (beans, peas, chick peas). In highly technologic societies such as the United States, the refinement of wheat flour and other grain products is carried to such an extent that many common cereal foods are rendered nutritionally marginal at best, whereas the whole grains from which the refined products were prepared are far more nourishing. Thus, for example, most of the common heavily advertised "breakfast foods" on the American market contain only minimal quantities of essential nutrients (eg, vitamins) other than starch. Skim milk powder, on the other hand, is a relatively inexpensive source of complete, high quality protein, provided that a lactose intolerance is not present (p. 859). Likewise, soybeans and flour prepared therefrom contain proteins of good nutritional quality.

Two states of malnutrition will serve to exemplify the effects of long-term ingestion of a biologically inadequate protein and/or insufficient calories.

The major protein in corn (zein) is deficient in both tryptophan and lysine. A syndrome known as kwashiorkor develops in persons who must depend exclusively upon corn as their major dietary staple. Kwashiorkor in turn is a severe protein deficiency state characterized by retarded growth, alterations in cutaneous and hair pigmentation, edema, and pathologic hepatic changes including fatty infiltration, necrosis, and fibrosis. Mental apathy, pancreatic atrophy, gastrointestinal disorders, anemia, hypoalbuminemia, and dermatoses also develop in kwashiorkor. The dermatoses of kwashiorkor are pellagroid, in that patches of skin desquamate, leaving pink raw areas.

Marasmus is another malnutrition syndrome which involves both a low protein and a low caloric intake, and develops chiefly during the first year of life. Marasmus, which now is considered to be related to kwashiorkor, is characterized by a retardation of growth and a progressive wasting of subcutaneous fat and skeletal muscle. The appetite and mental alertness usually are maintained, in contrast to kwashiorkor. If sufficiently prolonged, the symptoms of kwashiorkor or marasmus may be irreversible, despite substitution of a nutritionally adequate regime for the deficient diet. Specific vitamin deficiencies can of course complicate either kwashiorkor or marasmus.

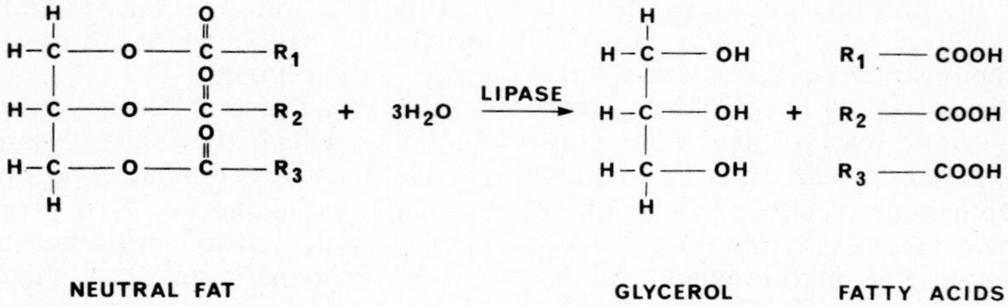

FIG. 13.26. The structure of a neutral fat (or triglyceride) and its hydrolysis into glycerol plus three fatty acids. R_1, R_2, R_3 = aliphatic chains of the three individual fatty acid moieties. (Cf. Fig. 13.23.)

Fat

Fat is an essential component of living tissues, and among other functions, it provides the major reserve energy source of the body. Lipids and fatty acids therefore form a necessary part of the diet.

Upon oxidation each gram of fat provides more than twice the caloric energy of an equivalent quantity of carbohydrate or protein; consequently the body lipids provide a concentrated fuel reserve in times of caloric deficit. Thus fat is the only source of energy which the body is able to store in quantity, and when needed it is mobilized readily from the body depots which are located subcutaneously, in the peritoneal cavity, (peritoneal and omental fat depots), around the kidneys and ovaries, and among the skeletal muscles. The depot fats are principally triglycerides (Fig. 13.26). Surplus dietary carbohydrates and proteins, as well as fats themselves, which are eaten in excess of caloric and other nutritional requirements of the body are converted into, and stored as, depot lipids. In fact, there appears to be no upper limit to the quantity of fat storage which is possible in the human body.

Certain fatty acids are essential nutrients, and as there is apparently no de novo synthesis of these compounds in the body, they must be derived from dietary sources. Included among the essential fatty acids are the unsaturated linoleic and archidonic acids. The role of these compounds in the biosynthesis of prostaglandins will be discussed on page 1173, and normal skin structure of babies also depends upon the presence of these fatty acids in the diet. A typical eczematous dermatitis develops in infants deprived of arachidonic and/or linoleic acid, and the severity of the dermatitis is inversely related to the plasma level of the fatty acids. The infantile dermatitis responds specifically to the administration of arachidonic and linoleic acids; however, similar lesions do not occur in adults deprived of these compounds.

Other unsaturated fatty acids which are of biologic importance include oleic and linolenic acids. The important dietary saturated fatty acids include acetic, butyric, palmitic, and stearic acids.

The principal sources of dietary fats include vegetable oils (such as cottonseed, corn, olive, soy,

peanut, and sesame), margarine, butter, milk, cheese, lard, meats, poultry, eggs, fish, and nuts. In natural fats, the triglycerides usually are mixed. If the mixture is solid at room temperature it is called a fat, if liquid it is called an oil.* The melting point of a fat depends, of course, upon its constituent fatty acids. Hard fats contain principally saturated fatty acids, while a high proportion of unsaturated fatty acids is present in liquid oils. The term polyunsaturated signifies that two or more double bonds are present in the fatty acid molecule. Treatment of an unsaturated fatty acid with hydrogen atoms in the presence of a suitable catalyst changes the double bonds into single bonds. This process, known as hydrogenation, is used commercially to convert liquid vegetable oils into solid fats, as in the process of making margarine.

Cholesterol is found throughout the human body, but most especially in bile, nervous tissue, and blood, and it is present as the free alcohol or esterified with fatty acids. Cholesterol is the precursor of the bile salts, as well as the adrenal cortical and sex hormones. In the skin, cholesterol (and certain closely related substances) is converted into active vitamin D under the influence of sunlight or ultraviolet radiation. Cholesterol also facilitates the absorption of free fatty acids from the intestine, as well as their transportation in the blood by the formation of cholesterol esters which are more soluble and emulsify more readily than do the free fatty acids.

Only foods of animal origin supply cholesterol in the diet; however, this compound is synthesized continuously by the human liver and many other organs (including the arteries) so that an exogenous supply is not critical from a nutritional standpoint. The rate of cholesterol synthesis in the body is related inversely to the presence of this compound in the diet, and the synthesis and breakdown of this cholesterol occurs simultaneously and continually; the cholesterol level in various body tissues remains fairly constant.

*The general term lipid signifies a fatty substance of biologic origin; hence petroleum derivatives are not considered to be lipids.

There are three types of phospholipid which are of great biologic significance. These are the lecithins, cephalins, and sphyngomyelins. Lecithins and cephalins are composed of glycerol, two fatty acids, and a nitrogenous base such as choline. Sphyngomyelins on the other hand contain glycerol, one fatty acid moiety, phosphoric acid, and two nitrogenous bases.

The phospholipids appear to be constituents of all living cells, and they are essential to the structure of the cell wall as well as the membrane which surrounds the mitochondria. Phospholipids also have an important role as intermediary metabolites in the transport and utilization of fatty acids, and they also provide an important reserve of fatty acids in a form which is easily metabolized.

Dietary fat is essential to human nutrition, despite the fact that it has been implicated in the genesis of atherosclerosis. In addition to the general and specific functions of fat in the body as a compact, metabolically labile form of high energy stored fuel, and as an essential constituent of the cells and tissues themselves, several other attributes of dietary fats should be mentioned. Hence there is a close reciprocal relationship between the metabolism of fat and of carbohydrate, although the latter is the substrate which is oxidized preferentially by the body. Fat delays gastric emptying and also decreases intestinal motility. These effects delay the onset of hunger, and consequently fat in meals has a high satiety value. Fat also lends palatability to a meal, especially by its addition of flavor and aroma to various foods. An important metabolic property of dietary fat is that it spares protein as an energy source; ie, fat is oxidized preferentially to protein. Dietary fat is also necessary in order to assure a supply of essential fatty acids and to promote absorption of the fat soluble vitamins.

There is as yet no unanimity of opinion among various responsible agencies as to what constitutes an optimal level of fat in the diet or what mixture of essential fatty acids appears to be most favorable to normal human nutrition.

As is the case with carbohydrate and protein, the proportion of fat in the diet of a given population appears to be regulated to a large extent by economic factors. In underdeveloped, overpopulated areas of the world, fat may contribute as little as 10 percent or less to the total caloric intake whereas in the United States and Canada this figure may approximate 40 percent. Among individuals, the intake of fats in such foods as meat, eggs, and dairy products tends to increase in direct relation to income and thus the ability to purchase such foods in greater quantities.

At present, in the United States at least, some emphasis is being placed upon the intake of a greater proportion of unsaturated over saturated fatty acids in the diet, and a daily intake of fats has been recommended which contributes around 25 percent of the total calories ingested.

Vitamins

The vitamins are an important group of organic compounds which are essential for the normal metabolism of other nutrients, hence for the maintenance of normal physiologic processes in the body. Since vitamins are not synthesized by humans, they must be obtained from exogenous sources, ie, in the diet and/or through the action of the intestinal flora in certain instances. The individual vitamins exhibit considerable differences among each other insofar as their chemical composition and ultimate functions are concerned.

Vitamins are found in different quantities in various foods and although most foods contain several vitamins, no single food contains all of them in adequate quantities to satisfy the requirements of the body. A few examples of typical foods which contain reasonable quantities of the vitamin under discussion are mentioned in the following sections.

Vitamins are necessary in relatively minute quantities (ie, they are micronutrients) in contrast to carbohydrates, proteins, fats, certain electrolytes, and water, which are required in much larger quantities (macronutrients). Most vitamins function as catalysts, and so are components of enzyme systems which accelerate critical metabolic reactions. Only small quantities of vitamins are necessary in the diet, as any given enzyme system is degraded to a negligible extent while performing its function. On the other hand, the macronutrients, their metabolites, and end products are involved directly in these chemical reactions. Much greater quantities of these substances are utilized in the various metabolic processes of the body.

All of the recognized vitamins have been isolated in pure, crystalline form and characterized chemically. These major achievements, together with an understanding of the roles of the vitamins in human metabolic reactions, were given impetus because of the classic deficiency diseases (eg, rickets, scurvy, beriberi, and pellagra) which have been known since ancient times, and which develop when these substances are absent, or present in restricted quantities, in the diet. The human requirement for vitamins is expressed either in terms of milligrams of the pure compound or in terms of units. The latter are also based upon definite quantities of the pure compounds.

A useful method for classifying the vitamins into two groups is based upon their solubility in either water or oil. This system is also helpful because the vitamins included in each group also possess certain common physiologic properties. Thus most of the water-soluble vitamins (Table 13.11A) have proven to be components of essential enzyme systems; they are not stored in the body in significant quantities; they are unrelated to dietary lipids, and are absorbed independently of such substances; they are excreted in small amounts in the urine; and a sustained dietary in-

Table 13.11[a]   VITAMINS ESSENTIAL FOR HUMAN NUTRITION[b]

1. *Water Soluble*
   a. Thiamine hydrochloride (vitamin B_1, aneurin): $C_{12}H_{17}ClN_4OS \cdot HCl$; MW = 337.
   b. Riboflavin (vitamin B_2): $C_{17}H_{20}N_4O_6$; MW = 376.
   c. Niacin (nicotinic acid): $C_6H_5NO_2$; MW = 123.
      Niacinamide (nicotinamide): $C_6H_6N_2O$; MW = 122.
   d. Pyridoxine hydrochloride (vitamin B_6): $C_8H_{11}NO_3 \cdot HCl$; MW = 205.
   e. Biotin: $C_{10}H_{16}N_2O_3S$; MW = 244.
   f. Pantothenic acid: $C_9H_{17}NO_5$; MW = 219.
   g. Folic acid (folacin): $C_{19}H_{19}N_7O_6$; MW = 441.
   h. Cyanocobalamin (vitamin B_{12}): $C_{63}H_{88}CoN_{14}O_{14}P$; MW = 1,355.
   i. Ascorbic acid (vitamin C): $C_6H_8O_6$; MW = 176.

2. *Fat Soluble*
   a. Vitamin A (retinol): $C_{20}H_{30}O$; MW = 286.
   b. Vitamin D; $C_{28}H_{44}OC_{28}H_{44}O$; MW = 793; calciferol (vitamin D_2): $C_{28}H_{44}O$; MW = 396.
   c. α-Tocopherol (vitamin E): $C_{29}H_{50}O_2$; MW = 430.
   d. Menadione (vitamin K): $C_{11}H_8O_2$; MW = 172.

[a]*Data from Stecher (ed). The Merck Index, Rahway, NJ, Merck & Co, Inc, 1968.*
[b]*This list includes the vitamins currently recognized as being necessary for adequate nutrition of humans. Other organic accessory factors may be required in trace amounts; however at this time a definite requirement has not been established. Synonyms for the vitamins listed are in parentheses and the vitamins are listed in this table together with their empirical formulas in the same sequence in which they are discussed in the text. MW = molecular weight.*

take is necessary to avoid dysfunctions arising from a decrease in their concentration in the body. The fat-soluble vitamins (Table 13.11B) on the other hand are associated with the lipids of foods and they are absorbed together with dietary fats. Clinical states which interfere with fat absorption also derange the absorption of the fat-soluble vitamins. These substances are also stored in the body in moderate concentrations, so that a daily supply is not necessary, and the fat-soluble vitamins are not excreted in the urine.

Only those 13 vitamins generally considered to be essential to human nutrition will be discussed.

WATER-SOLUBLE VITAMINS

Thiamine (Vitamin B_1). Thiamine (Fig. 13.27) is water soluble and reasonably stable in dry heat, but sensitive to oxidation at an alkaline pH. The molecular weight is 337.

Cocarboxylase, also called thiamine pyrophosphate (TPP) or diphosphothiamine (DPT), is the pyrophosphoric ester of thiamine. This com-

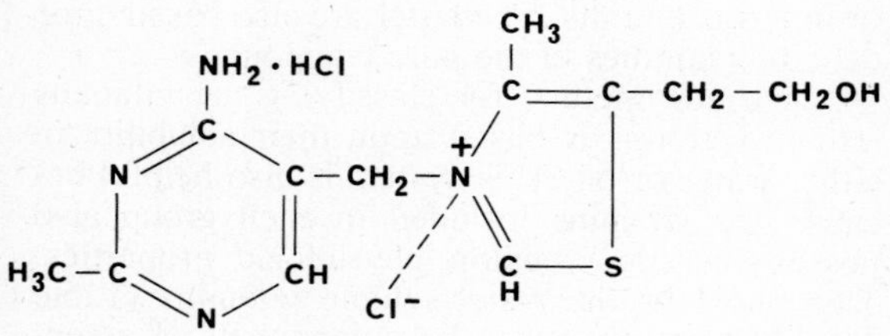

pound is a vital coenzyme in decarboxylation reactions which are catalyzed by specific carboxylases. Thus cocarboxylase participates in nearly all of the oxidative decarboxylation reactions which lead to the formation of carbon dioxide in the body (Fig. 14.25, p. 938). Cocarboxylase is of major importance in the oxidative decarboxylation of pyruvic acid, so that if cocarboxylase is lacking a critical step in carbohydrate metabolism does not occur, and the result is a net accumulation of pyruvic and lactic acids.

Cocarboxylase is also a coenzyme in the direct oxidative pathway of glucose metabolism, and it acts in the transketalase reaction.

The majority of the decarboxylating enzymes which have been studied extensively to date consist of cocarboxylase and magnesium linked to a protein which is specific for the substrate involved. All nucleated cells can form appropriate carboxylase enzyme systems, provided that thiamine is present in adequate quantities.

Thiamine deficiency produces a biochemical lesion which is caused by deranged pyruvate metabolism and an accumulation of metabolic intermediates, as described above. Clinically this deficiency is manifest as any of several forms of beriberi. The symptoms of this disease may include peripheral neuritis (polyneuritis), muscular atrophy, and edema fluid accumulates within the body cavities. These symptoms are ameliorated dramatically when thiamine treatment is instituted.

Thiamine occurs in many foods; however, few of these contain the vitamin in appreciable concentrations. Relatively rich sources of thiamine are lean and organ meats (liver, kidney, heart), pork, eggs, whole (or enriched) cereals, berries, nuts, legumes, and green leafy vegetables. Since thiamine is water soluble, it may be leached out of

THIAMINE HYDROCHLORIDE

FIG. 13.27.   Structural formula of vitamin B_1 (thiamin).

food during cooking, and/or destroyed by heat oxidation, especially at elevated pH levels.

Thiamine also is synthesized by the intestinal flora, and in persons who are maintained on diets that are marginal in the vitamin, this can be an important supplementary source.

The human requirement for thiamine is related directly to the energy expenditure of the body through carbohydrate metabolism. In turn, the thiamine requirement is related directly to the proportion of carbohydrate in the diet. Beriberi is endemic in areas where the overall food intake contains a very high percentage (up to 80) of dietary calories in the form of carbohydrate. The total requirement for thiamine is also increased in persons having an elevated net metabolic rate (hence energy expenditure), such as is encountered during heavy exercise, pregnancy, and lactation, as well as in febrile states and hyperthyroidism.

The Food and Nutrition Board has recommended a daily intake of 0.4 mg of thiamine per 1,000 dietary calories for adults. With an adequate diet containing a variety of foods, the thiamine intake will be adjusted roughly to the total caloric intake; however, on diets containing less than 2,000 total calories per day, a minimum daily intake of 0.8 mg of this vitamin is recommended.

Since the body does not store thiamine to any great extent, the entire body store of the vitamin is sufficient for only a few weeks.

Riboflavin. When dissolved in water, riboflavin (Fig. 13.28) exhibits a yellow-green fluorescence. This vitamin has a molecular weight of 376, and is reasonably stable to heat, but is sensitive both to light and to ultraviolet radiation.

When riboflavin is present in the form of riboflavin mononucleotide (riboflavin-5-phosphate), or as more complex flavin adenine dinucleotides in combination with specific proteins, this vitamin plays a role of great importance as a catalyst in a number of cellular oxidations (Fig. 14.22, p. 934). Mammalian tissues have a variety of distinct flavoprotein enzyme systems, each of which is comprised of a riboflavin-containing prosthetic group (or coenzyme) combined with a substrate-specific protein molecule (or apoenzyme).

Deficiency experiments performed on human subjects do not give rise to any specific riboflavin deficiency syndrome. However, certain nonspecific clinical symptoms which may (or may not) be encountered when the dietary intake of riboflavin is restricted can include inflammation and fissures at the angles of the mouth with desquamation and encrustation (angular stomatitis). Ocular lesions may include an abnormal corneal vascularization by capillary proliferation, visual fatigue, and photophobia. Inflammation of the tongue (glossitis) and seborrheic dermatitis may also be seen in pronounced riboflavin deficiency. Since riboflavin is synthesized by the intestinal flora, an acute deficiency may not develop in man because of the small quantities of the vitamin which are made available from this exogenous source.

Appreciable quantities of dietary riboflavin are obtained from such organ meats as heart, liver, and kidney, as well as from milk, cheese, meats in general, eggs, whole grains, legumes, and green leafy vegetables. Since riboflavin is only slightly soluble in water and is relatively heat stable (in the absence of light), major losses of this vitamin usually do not occur during cooking.

As is the case with thiamine, a relationship between caloric intake and riboflavin requirement has been established, so that the Food and Nutrition Board has recommended an adult daily intake of 0.6 mg of riboflavin per 1,000 calories ingested, with a proportionally greater intake for persons performing heavy physical labor. The requirements for infants, children, and adolescents, as well as in women during pregnancy and lactation are not so clear-cut as are those for adults.

Niacin (Nicotinic Acid). Niacin, or nicotinic acid, (Fig. 13.29) is soluble in hot water and relatively stable in the presence of heat, as well as in solutions of dilute acids and alkalis. Niacinamide (or nicotinic acid amide) is far more soluble in water than is niacin; however both of these compounds have a similar biologic function, as described below.

Nicotinic acid as such is present in plant tissues; in animal tissues the vitamin is present as the physiologically active amide form. In man, ingested niacin is converted readily into

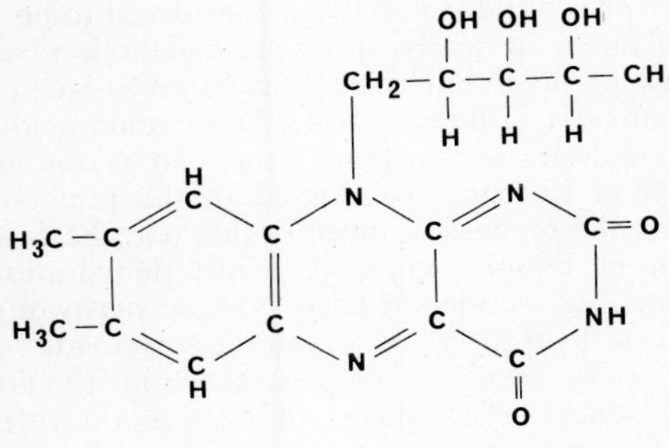

RIBOFLAVIN

FIG. 13.28. Structural formula of Vitamin B_2 (riboflavin).

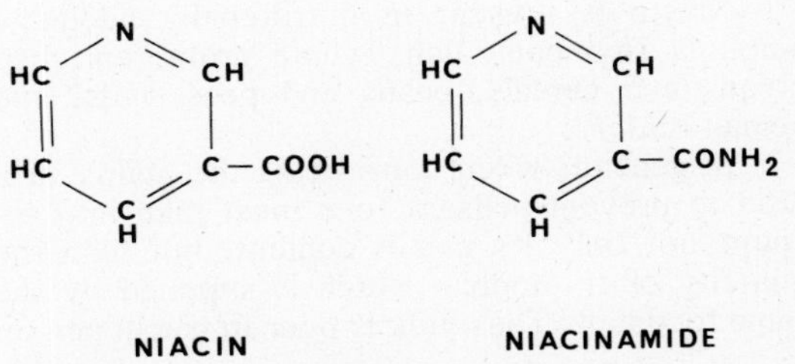

NIACIN   **NIACINAMIDE**

FIG. 13.29. Structural formulas of niacin (nicotinic acid) and niacinamide (nicotinic acid amide).

niacinamide. Niacinamide in turn is an essential constituent of coenzyme I (nicotinamide adenine dinucleotide, NAD; Fig. 14.21, p. 933), and of coenzyme II (nicotinamide adenine dinucleotide phosphate, NADP). Coenzymes I and II are the prosthetic (ie, functional) groups in a number of essential intracellular oxidation–reduction enzyme systems, and these enzyme systems in turn are critical to the metabolism of all major nutrients. Thus, for example, coenzyme I when coupled to a specific apoenzyme (ie, protein) forms the dehydrogenase which is necessary for the conversion of lactic acid to pyruvic acid. Similarly, coenzyme I when coupled to other appropriate apoenzymes, is responsible for the metabolic conversion of alcohol to acetaldehyde and β-hydroxybutyric acid to acetaldehyde. When coupled to specific apoenzymes, coenzyme II also forms essential dehydrogenases, which in turn are responsible for reactions involving hexose phosphate, citric acid, and many other substrates.

Niacin (or niacinamide) deficiency results in development of the clinical picture of pellagra. The precise relationship between the organic symptoms of this condition and the underlying metabolic dysfunctions are unclear. The principal symptoms of pellagra include mental changes (ie, dementia), dermatitis, diarrhea,* and stomatitis, the tongue becoming smooth, red, and painful. Administration of niacin in adequate amounts results in a rapid reversal of these symptoms.

In large (ie, therapeutic) doses niacin (but not niacinamide) exerts the pharmacologic effect of inducing mild vasodilatation. This response includes a transient hypotension with dizziness, flushing of the face, and an elevated skin temperature. The amide is utilized in pharmaceutical preparations of this vitamin to avoid these undesirable pharmacologic effects.

Niacin is synthesized in mammalian tissues from the amino acid tryptophan, but this conversion cannot occur in the absence of pyridoxine (vitamin B_6). Since it takes about 60 mg of tryptophan to serve as the precursor for one mg of niacin, pellagra is endemic in areas which depend principally upon corn as a major dietary staple. This food is deficient in niacin per se, and the protein in corn (zein) also lacks sufficient tryptophan for the endogenous synthesis of niacin by the body, a deficiency which could complicate the nutritional defects in kwashiorkor.

Niacin is present in nutritionally adequate amounts in meats, fish, whole grain, enriched breads and cereals, beans and peas, nuts, and peanut butter.

In general, when considering the ability of a food to prevent pellagra, one must take into account not only its niacin content, but also the quantity of tryptophan which is supplied by the same foodstuff. Thus milk is poor in niacin per se,

but is quite active in preventing the development of pellagra, because it is a rich source of tryptophan.

Since the tryptophan in proteinaceous foods can be converted into niacin (assuming the presence of an adequate supply of pyridoxine in the diet), it is not possible to define in simple terms the human requirement for this vitamin. Thus the current (1936) recommendation of the Food and Nutrition Board has expressed the human requirement for niacin in terms of niacin equivalents, taking into account the fact that 60 mg of tryptophan can serve as the precursor for 1 mg of niacin. Thus 6.6 mg of niacin equivalents per 1,000 calories per day has been recommended for infants, children, and adults. During pregnancy, an additional intake of 3 mg of niacin equivalents per day is recommended (in keeping with the increased caloric intake), and this value increases to 7 mg per day during lactation. As with the other vitamins, these recommendations include a generous safety margin above the minimal dosage that is required to prevent the appearance of deficiency symptoms.

Pyridoxine (Vitamin B_6). Pyridoxine is water soluble and reasonably heat stable, but it is unstable in vitro insofar as oxidation and ultraviolet radiation are concerned. Three naturally occurring and related compounds have potential vitamin B_6 activity. These compounds are pyridoxine, molecular weight 169; pyridoxamine, molecular weight 168, and pyridoxal, molecular weight 167 (Fig. 13.30). Both pyridoxine and pyridoxamine are converted into pyridoxal during normal metabolic processes, and the latter compound is the biologically active form of vitamin B_6.

Pyridoxal has vital functional roles in the normal metabolism of amino acids, carbohydrates, and fats, as well as in certain other processes. Pyridoxal-5-phosphate is the coenzyme in a large number of enzyme systems which are critical to the normal metabolism of amino acids, for example, in transaminase, decarboxylase, and desulfurase reactions. In fact, the number and variety of the reactions involving this vitamin as coenzyme is so large that pyridoxal may be considered to be a critical factor in nearly all of the metabolic reactions which involve the interconversion as well as the nonoxidative degradation of the amino acids. These include, for instance, the conversion of tryptophan to niacin (discussed in the previous section); the process of racemization (eg, the conversion of D- to L-glutamic acid); dehydration reactions (eg, serine→H_2O + NH_3 + pyruvate); and desulfhydration (eg, cysteine→pyruvate + NH_3 + H_2S). Hence all of these reactions require the presence of pyridoxal in a catalytic role. Pyridoxine also appears to be necessary for the active transport of certain amino acids, such as tyrosine and methionine, during their absorption from the intestine.

In fat metabolism, pyridoxine appears to be

The triad dementia, dermatitis, and diarrhea sometimes is referred to as the "three D's" of pellagra.

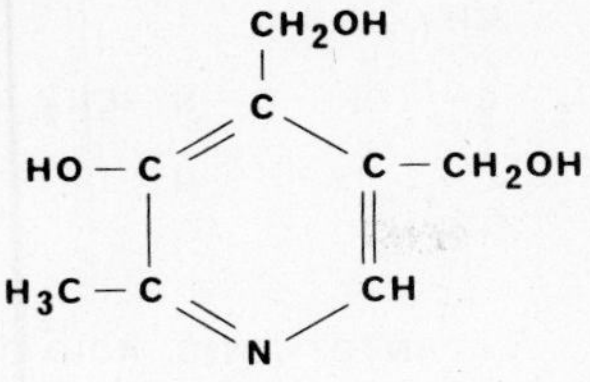

FIG. 13.30. Structural formulas of pyridoxine (vitamin B_6), pyridoxal, and pyridoxamine.

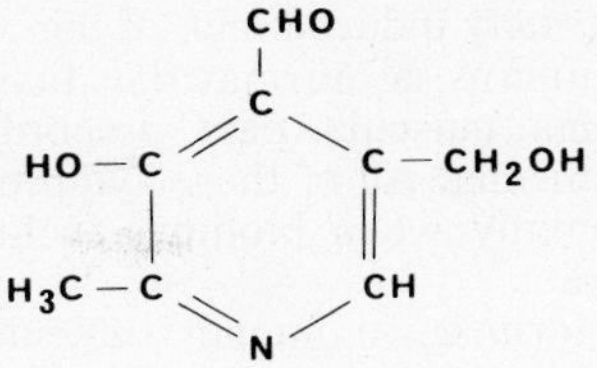

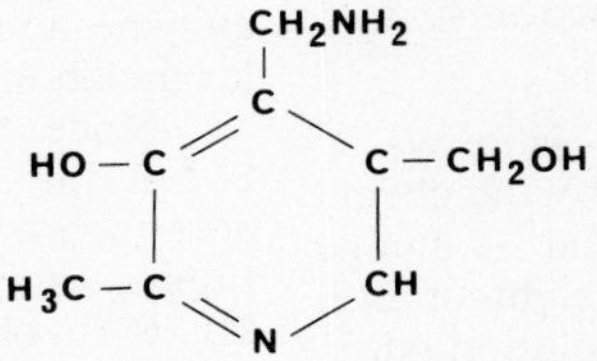

specifically involved in the synthesis of the highly unsaturated arachidonic acid from linoleic and linoleic acids.

Pyridoxine also has a key role in carbohydrate metabolism, as this vitamin is a major component of the enzyme phosphorylase which catalyzes the hydrolysis of glycogen into glucose-1-phosphate.

In humans, pyridoxine deficiency causes a sebhorreic dermatitis around the eyes, nose, mouth, and behind the ears. The fissures which appear about the lips and angles of the mouth (cheilosis) as well as the glossitis and stomatitis (a generalized inflammation of the oral mucosa) are not distinguishable from the similar lesions which are found in riboflavin and niacin deficiency. Certain patients present symptoms of a peripheral sensory neuropathy and a subsequent motor impairment. The administration of even small doses of pyridoxine (ie, 5 mg per day) stimulates the rapid reversion of these clinical symptoms to normal.

Pyridoxine is found in reasonable quantities in such foods as meats, whole grain cereals, corn, soybeans, peanuts, and many vegetables. Difficulties in the analytical techniques currently available for determining the pyridoxine content of foods precludes a complete listing.

No official standards have been established as yet for the requirement of this vitamin in man; however, this requirement has been estimated to range between one and two milligrams per day for adults, and this quantity of pyridoxine is provided readily by an ordinary mixed diet. Some individuals appear to have an inherent (hereditary?) need for greater quantities of pyridoxine (between 5 and 10 mg per day) than do others, and a similarly increased need for this vitamin also becomes apparent during pregnancy.

Biotin. Biotin is slightly soluble in cold water, but quite soluble in hot water, and is heat stable although readily oxidized. Biotin (Fig. 13.31) combines with a glycoprotein found in raw (but not cooked) egg white called avidin to form a stable complex which is not hydrolyzed by the digestive enzymes. Hence in this special circumstance biotin cannot be absorbed from the intestine so that it is nutritionally unavailable.

In similarity with the other vitamins of the B complex, biotin presumably functions as a coenzyme, and it is believed to act in the carboxylation and decarboxylation reactions of oxaloacetate, malate, aspartate, and succinate. This vitamin also has been implicated as a cofactor in the biosynthesis of aspartate, citrulline, and unsaturated fatty acids as well in pyruvate oxidation (Chap. 14).

There is no distinct disease known in man caused by an underlying biotin deficiency, al-

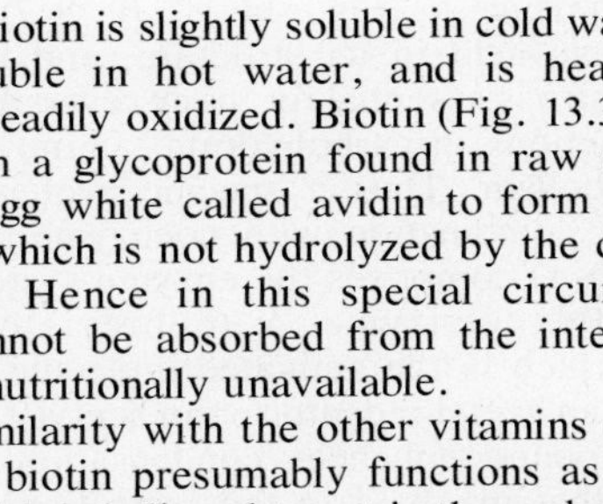

BIOTIN

FIG. 13.31. Structural formula of biotin.

though experimentally induced lack of this vitamin produces in humans a dermatitis, lassitude, anorexia, insomnia, muscular pain, precordial distress, and mild anemia. All of these symptoms are ameliorated promptly when biotin is included in the diet once again.

Studies performed on human subjects have indicated that biotin is synthesized in sufficient quantities by the intestinal flora to supply the nutritional requirements of humans. An exogenous source of this vitamin is unnecessary.

Biotin is found in high concentrations in meats, egg yolks, peanuts, mushrooms, and chocolate.

Pantothenic Acid. Pantothenic acid is a water soluble compound which is produced by the chemical combination of pantoic acid and β-alanine (Fig. 13.32), and has a molecular weight of 219. Pantothenic acid is stable in neutral solution but is unstable in acid or alkali. The hydroxy analog of pantothenic acid, panthenol, is a viscous oil, and it exhibits the same biologic activity as the vitamin itself. The calcium salt of pantothenic acid is a crystalline solid, and this compound is the most extensively used form of the vitamin.

Coenzyme A (Fig. 14.28, p. 947) requires pantothenic acid in its structure, and this coenzyme in turn is essential for many reversible acetylation reactions in carbohydrate, amino acid, and fat metabolism. Thus a specific protein (apoenzyme) in combination with coenzyme A (abbreviated Co-A) comprises the enzyme system.

Acetyl coenzyme A (abbreviated acetyl Co-A), which is the activated molecule, may act either as an acetyl receptor or an acetyl donor, and this compound enhances condensation reactions such as that involved in the formation of citrate from oxaloacetate in the Krebs cycle (Fig. 14.29, p. 948). Acetyl Co-A also functions as the recipient of acetyl radicals generated during the metabolic oxidation of pyruvate, citrate, and fatty acids; these radicals then are transferred to other compounds.

Acetyl Co-A also conjugates with acyl (R—COO$^-$) groups in addition to acetyl radicals; hence it is active in the synthesis of fatty acids, cholesterol, and other sterols.

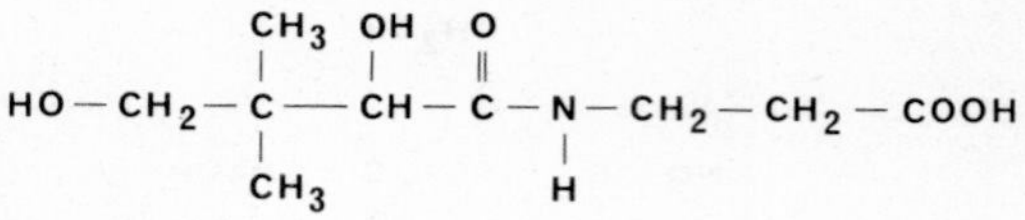

PANTOTHENIC ACID

FIG. 13.32. Structural formula of pantothenic acid.

Pantothenic acid is a vital constituent of coenzyme A; the importance of this vitamin in cellular metabolism is evident.

Since there is an intimate relationship between the tissue level of pantothenic acid and adrenocortical function, pantothenic acid (functioning as a constituent of coenzyme A) presumably is involved in the activity of this endocrine gland.

No clinical syndrome which is caused by a deficiency of pantothenic acid in the diet is known, despite the importance of this vitamin to many facets of intermediary metabolism. The daily requirement for this substance is satisfied even by poor diets. Pantothenic acid (the name of this compound means "dervied from everywhere") is ubiquitous in its occurrence in both plant and animal tissues. Organ meats, egg yolk, peanuts, whole grains, cereal brans, and vegetables all are good sources of this vitamin, whereas meats, milk, and fruits also contain reasonable although somewhat lower quantities. It appears that ordinary cooking procedures do not lead to an appreciable loss of pantothenic acid.

Folic Acid (Pteroylglutamic Acid, Folacin). Folic acid is water soluble and heat labile in acidic solutions. The molecular weight of this compound is 441.

In natural products, folic acid (Fig. 13.33) occurs together with the closely related folinic acid and both form conjugates with glutamic acid. These conjugates are hydrolyzed during digestion; therefore free folic and folinic acids are released. Although both of these compounds are biologically active, folinic acid is the more potent of the two.

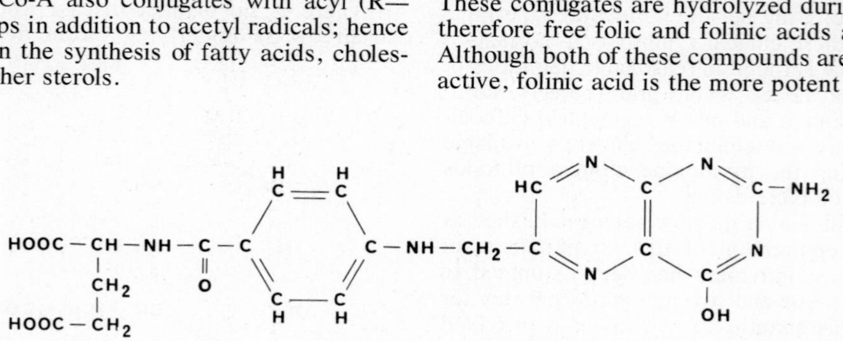

FOLIC ACID

FIG. 13.33. Structural formula of folic (pteroylglutamic) acid.

Within the body, folic acid is converted into folic acid by reduction and the addition of a formyl group, a process that is enhanced by the presence of ascorbic acid (vitamin C). Presumably other metabolic reactions then convert folinic (and folic) acid into an active coenzyme, although the metabolic activities of folic and folinic acids in vivo have not been established precisely. Experimental evidence favors the view that a folinic acid coenzyme functions as a catalyst in the transfer of 1-carbon units, even as pantothenic acid functions in the metabolism of 2-carbon units. When folic or folinic acid is present as a coenzyme linked to various substrate-specific proteins (apoenzymes), the resulting enzymes are essential to the catalysis of such reactions as the conversion of glycine to serine; the methylation of homocysteine to form methionine; the production of creatine from guanidoacetic acid; and the synthesis of ribose nucleic acids, thymine desoxyriboside, as well as the synthesis of porphyrin compounds, including hemoglobin.

Clinically, the most significant role of folic acid resides in its ability to facilitate the synthesis of nucleoproteins, as these compounds are critical to the maturation of the erythrocytes. Therefore, when an insufficient quantity of nucleoproteins is present, hematopoiesis is arrested at the megaloblast stage of red cell maturation in the bone marrow. The result is a characteristic and abnormal cytologic pattern in the blood which incudes macrocytic anemia, leukopenia, thrombocytopenia, and multilobed neutrophils. Thus in humans a deficiency of folic acid in the diet results in a macrocytic anemia which is similar to pernicious anemia, except that the deranged central nervous functions which accompany pernicious anemia are absent. Additional nonspecific symptoms of folic acid deficiency may include gastrointestinal malabsorption, diarrhea, gastrointestinal lesions, and glossitis. Folic acid therapy induces reversion of the anemia to a normal blood picture, as well as amelioration of the other symptoms. Folic acid also is useful in the therapy of the malabsorption syndrome (sprue) as well as in certain other types of anemia, eg, the megaloblastic anemias of infancy and pregnancy.

Considerable amounts of folic acid are present in liver, kidney, leafy vegetables, lima beans, nuts, and whole grain cereals. However, cooking and storage of the food may cause up to half of this vitamin to be lost. The intestinal flora also synthesize folic acid, although the quantity derived from this source is unknown.

The daily human requirement for folic acid is not known precisely; however, it appears that a dietary intake of between 0.5 and 1.0 mg per day should be adequate to meet the normal physiologic requirements for this vitamin. This statement is based upon the fact that anemias that have been caused by a deficiency of folic acid respond to therapeutic doses of the vitamin administered at levels ranging between 5 and 15 mg per day.

Furthermore, a dose of only 200 μg per day of folic acid was found to be the minimum quantity of the vitamin which was effective therapeutically in cases of infantile megaloblastic anemia.

Cyanocobalamin (Vitamin B_{12}). Cyanocobalamin (Fig. 13.34) is unique among the vitamins in two respects. It is the only vitamin that contains an essential mineral element (cobalt), and in addition it is the only compound known that contains cobalt which is also essential to life. Vitamin B_{12} is water soluble, and the compound is unstable in acids, alkalies, and light. The molecular weight of cyanocobalamin is approximately 1,450.

Cyanocobalamin is believed to be involved in the metabolic transfer of single carbon intermediates, principally methyl groups. For example, such transmethylation reactions are involved in the synthesis of choline from methionine, of glycine from serine, and of methionine from homocysteine. Cyanocobalamin is also involved in purine metabolism, as well as in the synthesis of pyrimadine bases. This vitamin affects in some manner the metabolism of folic acid, and it may be necessary for production of the folinic acid coenzyme moiety. By virtue of its role in the metabolism of purines and pyrimidines, cyanocobalamin is essential to the synthesis of nucleic acids as well as nucleoproteins, and thus it is essential to the normal production of erythrocytes by the bone marrow. Lastly, vitamin B_{12} is essential to the normal metabolism of neural tissue, although the exact manner in which it functions in this regard is unknown.

The absorption of cyanocobalamin from the gastrointestinal tract, and the role of intrinsic factor in this process, were discussed earlier (p. 872). It has been postulated that intrinsic factor either liberates cyanocobalamin from a strong protein–vitamin complex (in which state it is present in foods) or else it participates in the transport of the vitamin across the intestinal mucosa during absorption. Regardless of the mechanism, however, intrinsic factor is known to be essential for the normal absorption of cyanocobalamin, and in patients suffering from pernicious anemia, the fundamental defect is in the gastric mucosa (atrophy) so that no intrinsic factor is secreted, cyanocobalamin no longer is absorbed from foods, and pernicious anemia develops. In addition to the classic Addisonian blood picture found in this disease, a progressive neurologic degeneration occurs, with symptoms ranging from functional impairment of the peripheral sensory receptors and peripheral neuritis to sclerosis of the posterior and lateral columns of the spinal cord. This disease ultimately terminates in death unless parenteral cyanocobalamin therapy is instituted and sustained, as the fundamental gastric malfunction remains uncorrected. The abnormal blood picture in pernicious anemia is reverted to normal by folic acid therapy. The deranged neural function is not ameliorated by therapy with folic acid. Treatment

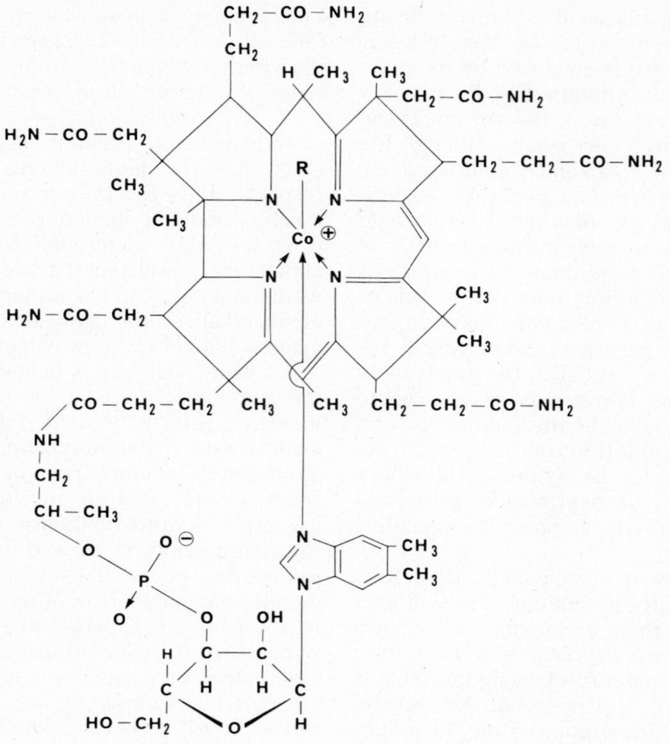

COBALAMIN

FIG. 13.34. Structural formula of cobalamin. Several compounds having the prosthetic group about the cobalt atom which is shown in this figure exhibit "vitamin B_{12}" activity. When R is a cyanide group (CN^-), the compound is known as cyanocobalamin, commonly referred to as vitamin B_{12}.

of pernicious anemia with folic acid alone is hazardous because of the continuing neurologic deterioration which takes place in the absence of vitamin B_{12}.

Cyanocobalamin is found primarily in foods of animal origin; consequently, liver and kidney are excellent foodstuffs insofar as their content of this vitamin are concerned. Muscle meats and fish contain vitamin B_{12} in moderate quantities, whereas milk, vegetables, and cereals are relatively poor sources.

The daily human requirement for cyanocobalamin has not been established definitely; however, as 1 μg of crystalline vitamin per day is sufficient for maintenance of a patient with pernicious anemia, this level may correspond approximately to the normal absorption (and requirement) of this vitamin from the gastrointestinal tract of normal adults.

Ascorbic Acid (Vitamin C). Ascorbic acid is water soluble and sensitive to oxidation, but relatively stable in acid solution. The molecular weight of this vitamin is 176.

As shown in Figure 13.35, ascorbic acid is readily oxidized in a reversible fashion to the less stable, but still biologically active, dehydroascorbic acid. If dehydroascorbic acid is oxidized still further, then the loss of antiscorbutic activity (ie, ability to prevent scurvy) is complete and irreversible. Heat, and alkaline pH, iron and copper salts, and oxidative enzymes, as well as exposure to light and air all enhance the irreversible oxidation of ascorbic acid and reduce the potency of this vitamin in foods.

Ascorbic acid is necessary to the normal metabolism of amino acids, especially to the oxidation of tyrosine and phenylalanine, and this vitamin may function as a coenzyme in these metabolic processes. Ascorbic acid is critical for the occurrence of many hydroxylation reactions, and it also assists in the conversion of folic acid into folinic acid.

Ascorbic acid is believed to act as a carrier in one or more intracellular hydrogen transfer systems, and thereby it would regulate oxidation–reduction potentials within cells. However, this postulate must await experimental verification. Ascorbic acid (in large quantities) also increases the intestinal absorption of iron in normal as well as in iron-deficient persons, an effect that is due,

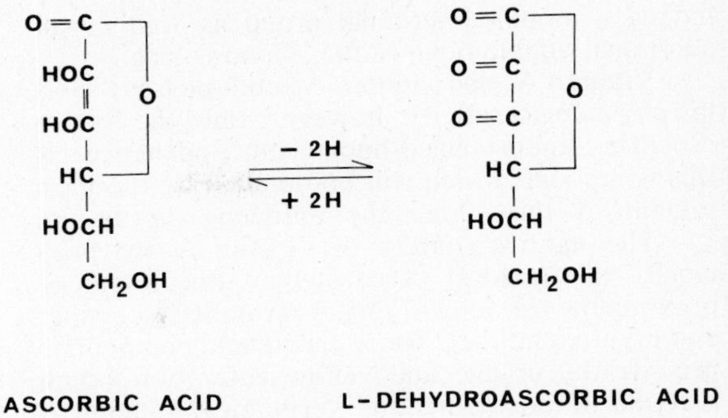

FIG. 13.35. Structural formulas of ascorbic acid (vitamin C) and dehydroascorbic acid.

presumably, to the chemical reducing property of the vitamin.

Vitamin C is essential to the production of the intercellular matrices in collagenous and fibrous connective tissue as well as for the normal synthesis of mucoproteins and collagen in such structures as cartilage, bone, teeth, and skin. Consequently, vitamin C is essential to the normal formation of bones and teeth through its effects upon the synthesis of chondroid and osteoid. Ascorbic acid is also necessary for the production of callus during the physiologic repair of fractures, as well as in the healing of burns and wounds in which the formation of a collagenous intercellular matrix is the first stage in the regenerative process of the damaged tissues. Ascorbic acid is also necessary for the maintenance of a normal physiologic permeability of the capillary walls.

Ascorbic acid is absorbed from the gut by simple diffusion, and as the rate of this process depends upon the concentration gradient across the mucosa, intestinal absorption of vitamin C is regulated directly by the quantity of the vitamin ingested. High concentrations of ascorbic acid normally are found in the pituitary gland, adrenals, corpus luteum, and thymus, whereas this vitamin is present at lower levels in such tissues as the brain and viscera. Such a saturation of certain bodily tissues and organs with ascorbic acid results in stores of the vitamin which are sufficient to prevent deficiency symptoms for several months on an inadequate intake of this nutrient.

Man is one of the few animal species which must derive his vitamin C from exogenous sources. Severe and prolonged deficiency of ascorbic acid in the diet results in the classic symptoms of scurvy, in which state the symptoms related to a weakened intercellular substance are most obvious. The patient with scurvy exhibits swollen, tender joints; prolonged wound healing; friable and spongy gums; loose teeth; and an overall weakness. There are abundant diffuse hemorrhages, and these hemorrhages appear most frequently under the skin and mucous membranes as well as near the bones and joints. Infantile scurvy is the form of this disease seen most frequently in the United States. The response of the patient to vitamin C therapy is quite rapid.

Among the foods in which vitamin C is found in quantity are citrus fruits (and their juices), raw or slightly cooked vegetables including peppers, cauliflower, kale, turnip greens, potatoes, tomatoes, and cabbage.

It is important to realize that the ultimate concentration of ascorbic acid in fruits and vegetables that reach the consumer is dependent upon the conditions under which they were grown, stored, and prepared for consumption. For example, the total quantity of sunlight available during the ripening of tomatoes governs their final vitamin C concentration to a large extent. The ascorbic acid content of fruits and vegetables is reduced sharply by leaching of the vitamin during washing, oxidation through standing at room temperature, and in the cooking (or canning) process itself. A splendid procedure for ensuring maximal loss of vitamin C from prepared foods is to store them at lukewarm to warm temperatures on steam tables until served, a technique commonly utilized in cafeterias.

There is no unanimity of opinion in the recommendations for an adult daily intake of vitamin C, although the minimum quantity of this vitamin that is necessary to prevent scurvy in adults is on the order of 10 mg per day. Currently the Food and Nutrition Board suggests a daily allowance of 30 mg per day for infants, 75 mg per day for adults, and still larger quantities for pregnant and lactating women. Actually, a daily allowance of vitamin C ranging between 50 and 100 mg does not actually saturate the tissues, although some tissue retention and excretion of the vitamin in the urine do take place at these levels.

FAT-SOLUBLE VITAMINS

Vitamin A. Chemically pure vitamin A is a viscous fat soluble alcohol. This vitamin actually is found in two closely related chemical forms. Vitamin A_1 is that chemical form of this compound which commonly is denoted by the term *vitamin A*, and it is found in mammals and marine fishes (Fig. 13.36). Vitamin A_1 also is relatively heat stable, although it is readily inactivated by oxidation. Vitamin A_2, on the other hand, is found in fresh water fish, and it differs chemically from vitamin A_1 in having a modified terminal group as well as an additional double bond in the β-ionone ring.

Vitamin A_1 and vitamin A_2 both possess similar physiologic activity; however, only the former is of prime importance in human nutrition; hence it is this compound which will be denoted by the term vitamin A throughout the following discussion.

The natural sources of vitamin A are principally in the form of carotenoid precursors or provitamins (ie, α-, β-, and γ-carotene, lycopene, and cryptoxanthine). These carotenoid compounds provide the orange and yellow coloration which is found in most fruits and vegetables. Vitamin A is synthesized in the body of man from these carotenoids following their hydrolysis. Thus one or two molecules of the vitamin are produced, depending upon the number of β-ionone rings which are present in the precursor. For example, β-carotene contains two β-ionone rings in its molecule; hence it yields two molecules of vitamin A upon hydrolysis. Therefore β-carotene is the most important provitamin A from a quantitative standpoint. This metabolic conversion to vitamin A occurs principally in the cells of the intestinal mucosa, and it also may occur in the liver, although to a more limited extent.

Vitamin A has three major physiologic roles in the body: first, in the growth and maintenance of epithelial tissues; second, in vision; and third, in the growth and development of bones and teeth.

A supply of vitamin A is critical to the normal morphologic and functional integrity of the epithelial tissues, as well as to the normal growth of epithelial cells in general. Vitamin A deficiency causes atrophy of the epithelial cell layer, and this defect in turn is followed by a marked proliferation of the underlying basal cell layer, and the latter eventually undergoes metaplasia so that a keratinized stratified epithelial layer develops in vitamin A deficiency. The normal secretory functions of mucous membranes are disturbed in vitamin A deficiency to such an extent that eventually secretion may cease entirely. Such deranged secretory function may be observed in the mouth, respiratory and urinary tracts, prostate, and seminal vesicles, in the female genital tract, and in the eyes and the glands associated with these sensory organs. In epithelial tissues whose primary function is protective (eg, the skin) vitamin A deficiency merely induces excessive keratinization.

Vitamin A is also necessary for normal vision, as it is found in the retina as an essential constituent of the visual pigment rhodopsin (Fig. 8.14, p. 482). As discussed in Chapter 8 (p. 481), rhodopsin (or visual purple) is converted in a stepwise fashion into visual white upon exposure to light. Although this reaction is reversible, and some visual purple ultimately is resynthesized in this manner, the regeneration of the pigment is incomplete quantitatively. Therefore, a continuous supply of vitamin A is essential for the complete resynthesis of a normal complement of rhodopsin in the eye. Vitamin A deficiency causes impaired vision in dim light. This failure of normal dark adaptation (or night blindness) is reversed quickly by the inclusion of suitable quantities of vitamin A or its precursors in the diet.

Lastly, vitamin A is essential to the normal growth and development of the bones and teeth. In these functions, the vitamin appears to maintain a normal, orderly, functional relationship between the activity of osteoblasts and osteoclasts as discussed on page 1067.

Vitamin A is found only in animal foods, and this compound is provided in abundance by a diet that is rich in liver, eggs, butterfat, or fish liver oils. The carotenoids which are precursors of vitamin A, on the other hand, are abundant in such foods as dark green leafy vegetables (eg, spinach), deep yellow vegetables (eg, carrots), tomatoes and tomato products, liver sausage, butter, fortified margarine, and cheese which has been manufactured from whole milk. Thus a large proportion of the vitamin A which is required for normal bodily activities is synthesized in vivo from precursors that were ingested as part of a normal diet.

The presence of bile in the intestinal tract (as well as overall conditions which favor the absorption of fat), is essential to the absorption of vitamin A as well as carotenoid pigments. Biliary and exocrine pancreatic dysfunctions, chronic diarrhea, absorption defects including sprue and celiac disease are all conditions in which malabsorption of vitamin A and the carotenoids is found.

Normally the liver stores over 90 percent of the total body pool of vitamin A, and the concen-

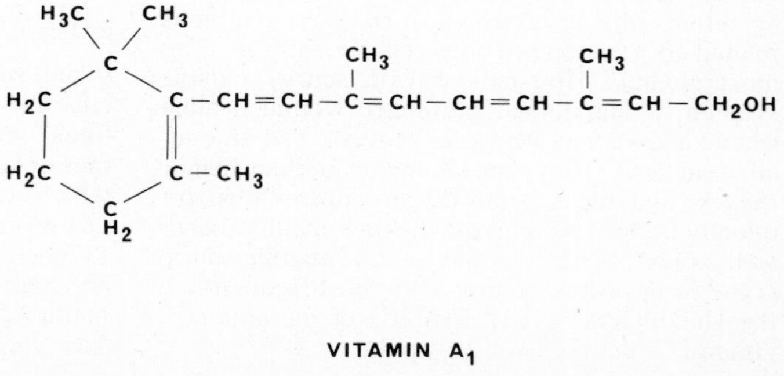

FIG. 13.36. Structural formula of vitamin A_1.

VITAMIN A_1

tration of the vitamin present in this organ increases gradually from infancy to advancing age.

An abnormally high chronic intake of carotene can produce a carotinemia with an attendant yellowing of the skin. A chonically excessive intake of vitamin A per se can induce a toxic syndrome called hypervitaminosis A. Typical symptoms of hypervitaminosis A include nausea, headache, loss of hair, abnormally high bone fragility, deep bone pain, hepatomegaly and splenomegaly (abnormal enlargement of liver and spleen, respectively), drying and peeling of the skin, and a generalized pruritus (intense itching). The blood alkaline phosphatase is also elevated. Hypervitaminosis A usually is caused by dosing children with concentrated vitamin A preparations, an overzealous mother usually being responsible. This condition is spontaneously reversible when the abnormally high vitamin A intake is ended.

Vitamin A and carotenoid pigments are not extracted from foods by water during cooking to any appreciable extent; however, the former compound is quite sensitive to oxidation, especially when high temperatures are present. Cooking and canning procedures which exclude oxygen tend to minimize losses of vitamin A.

The International Unit (IU) of vitamin A as defined by the World Health Organization is the biologic activity of 0.3 μg of crystalline vitamin A alcohol (or 0.344 μg of vitamin A acetate). Similarly, the IU of provitamin A is defined as the biologic activity of 0.6 μg of chemically pure β-carotene. It has been shown that ingested carotene on the average has about one half of the biologic activity of vitamin A. For practical purposes 2,000 units of provitamin A in a food is equivalent to 1,000 units of vitamin A.

The Food and Nutrition Board has recommended a daily intake of 5,000 IU vitamin A for adults. This allowance is somewhat smaller for infants (1,500 IU per day) and children (4,500 IU), and larger for pregnant (6,000 IU) and lactating (8,000 IU) women.

The minimum vitamin A requirement for humans has been determined to be approximately 6 μg per kilo body weight per day.

Vitamin D.   Although there are a number of compounds which are able to prevent or to cure rickets, only two of these are of practical significance. These compounds are vitamins D_2 and D_3. Vitamin D_2, also known as calciferol (Fig. 13.37), is manufactured synthetically from the plant sterol ergosterol under the influence of ultraviolet radiation (activated ergosterol). Crystalline vitamin D_2 in turn is soluble in fat and fat solvents, and is stable to heat, oxidation, acids, and alkalis. Vitamin D_3 is quite similar to vitamin D_2, and is produced by sunlight or ultraviolet radiation of the skin from 7-dehydrocholesterol. Vitamin D_2 does not occur naturally, in contrast to vitamin D_3. However, both compounds exhibit the same biologic potency in humans.

Vitamin D augments the absorption of dietary calcium and phosphate from the gastrointestinal tract and also specifically reduces urinary phosphate excretion by stimulating reabsorption of this anion from the renal tubules. Vitamin D also maintains physiologically optimum calcium and phosphorus levels in the plasma, and by virtue of this action facilitates normal mineralization of bones during growth in children, or repair of bones following their fracture in adults. In addition, vitamin D probably acts specifically at the locus of mineral deposition during the development of bony structures, and there is some evidence that vitamin D is also necessary for the conversion of organic to inorganic phosphorus in bone.

Vitamin D deficiency results in the failure of the intestine to absorb adequate quantities of calcium and phosphorus, together with a concomitant elevated loss of these minerals in both urine and feces. These defects are accompanied by a decline in serum calcium and phosphate levels (principally the latter), so that calcium now must be mobilized from bone in order to maintain adequate blood

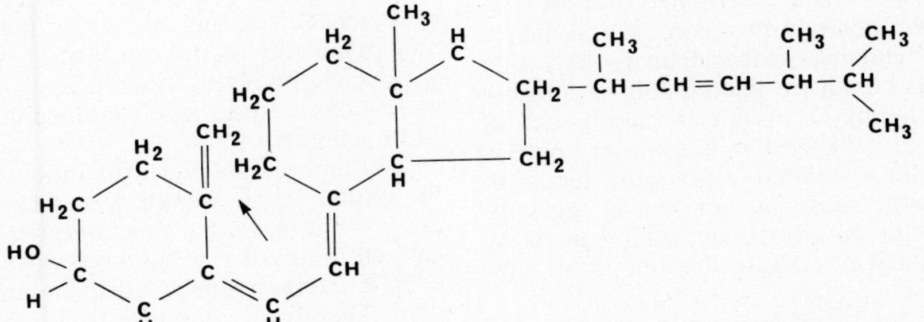

VITAMIN D_2

FIG. 13.37.   Structural formula of vitamin D_2.

levels of this cation. Osteomalacia results from this demineralization of bone in adults. In children, on the other hand, vitamin D deficiency causes the condition known as rickets. Since the bones are growing actively at the epiphyses in young children, the cartilagenous matrix forms normally; however, calcium and phosphate are not deposited in vitamin D deficiency and thus normal mineralization of the cartilage model does not take place. Since the cartilage cells do not degenerate, the capillaries which normally invade the region of growth as the cartilage model disintegrates cannot advance. The result is the development of a so-called rachitic metaphysis which appears clinically as a swelling, or a beadlike appearance at the ends of the ribs as well as a broadening of the ends of the long bones. Delayed closure of the anterior fontanelle occurs in infants, and the skull remains soft, because of the retarded or inadequate mineralization.

The defects summarized above coupled with the normal mechanical stresses imposed on the body by the force of gravity in turn give rise to skeletal malformations, including curvature of the spine, pelvic and thoracic deformations, and bowed legs. Furthermore, the teeth of rachitic children erupt late, are poorly formed, and decay readily.

In common with the other fat-soluble vitamins, vitamin D is absorbed together with the dietary lipids, and is associated with dietary fats. Thus conditions which interfere with fat absorption reduce the absorption of vitamin D. Once absorbed, vitamin D is stored principally in the liver, but also to some extent in the skin as well as in the brain, bones, and spleen. These reserves presumably account for the infrequent occurrence of vitamin D deficiency in adults. In rapidly growing infants such stores are limited; therefore, rickets is more commonly seen in clinical practice than is osteomalacia.

Certain animal foods contain high concentrations of natural vitamin D. These include marine fishes, especially those that have high levels of body oils such as sardines, herring, and salmon. Fish (eg, cod) liver oils are extremely rich in vitamin D, but their potency in regard to this factor renders them medicines rather than foods.* Egg yolks as well as beef liver also contain reasonable amounts of vitamin D. Artificial enrichment of many foods (eg, milk and milk powder, butter, margarine) with irradiated ergosterol tends to minimize the wide seasonal variation in the vitamin D content of natural foods, which in many instances is related directly to the amount of sunlight available.

A chronic intake of excessive quantities of vitamin D results in a toxicity syndrome similar to the effects of hyperparathyroidism. Characteristic signs of this hypervitaminosis D syndrome include loss of appetite (anorexia), headache, drowsiness, vomiting, and diarrhea. Polyuria and polydipsia may also be seen. Elevated serum calcium and phosphate levels are present, and abnormal calcium deposition occurs in the heart and major blood vessels, as well as in the renal tubules, among other soft tissues. These effects are reversible, provided that the excessive intake of the vitamin is stopped in time. As with hypervitaminosis A, overzealous maternal administration of concentrated vitamin D preparations is a frequent cause of hypervitaminosis D.

The standard for vitamin D is the International Unit (IU), and this represents the biologic activity of 0.025 microgram of vitamin D_3. Therefore 1 mg of vitamin D_3 is equivalent to 40,000 IU of vitamin D.

The Food and Nutrition Board of the National Research Council recommends a daily intake of 400 IU of vitamin D for infants, children, adolescents, and pregnant as well as lactating women. The daily requirement for other adults is unspecified.

Vitamin E. Four closely related compounds exhibit vitamin E activity. These compounds are α-, β-, γ-, and δ-tocopherol. These oils are soluble in fat solvents, stable to heat and acids, but unstable to alkalis, ultraviolet radiation, and oxidation. All four of the tocopherols exhibit strongly antioxidant properties, so that they prevent oxidation (hence rancidity) of the natural oils with which they normally are associated.

Alpha-tocopherol (Fig. 13.38) is by far the most potent vitamin E compound from a physiologic standpoint. The role of vitamin E in human nutrition (if any!) is not at all clear at present, although a deficiency of this vitamin produces characteristic physiologic changes in certain animals. For example, vitamin E deficiency results in degeneration of the germinal epithelium and loss of sperm motility in the male rat, and sterility ensues. Fetal death and fetal resorption also occur in pregnant female rats as consequences of vitamin E deficiency. However, there is no conclusive proof of any need for this accessory nutrient factor in human nutrition at present, and no deficiency state specifically involving vitamin E has been described in humans despite much undocumented publicity to the contrary.†

Vitamin E is a powerful antioxidant, inhibiting the oxidation of vitamin A and fatty acids which are present in natural fats. Because of this chemical action in vitro, it has been postulated that vitamin E exerts a similar effect on animal tissues in vivo. Therefore, vitamin E is supposed to prevent oxidation of the unsaturated lipid constituents of the mitochondria in all cells. In addition, it is be-

---

*Polar bear liver contains such enormous concentrations of vitamins A and D that it is highly toxic to man when ingested!

†For example, vitamin E has no proven role in affecting any aspect of human sexual physiology, a myth derived by extrapolation from the known physiologic effects of this compound on the fertility of male rats.

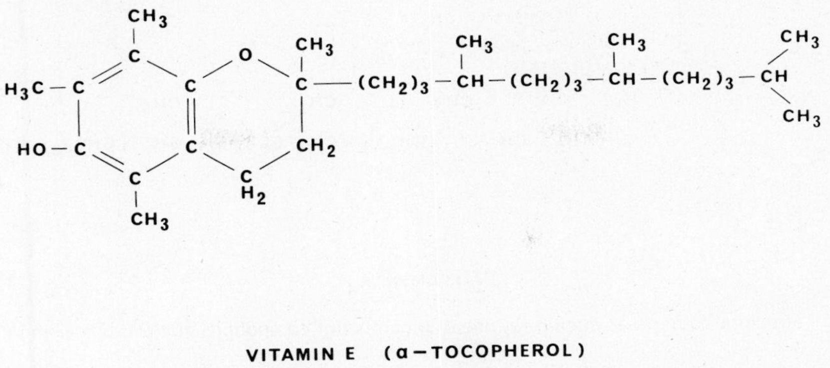

VITAMIN E (α−TOCOPHEROL)

FIG. 13.38. Structural formula of one compound exhibiting vitamin E activity, α-tocopherol.

lieved by some workers that the lipid of the erythrocyte wall similarly is protected from oxidation by vitamin E, especially when a high proportion of fatty acids is present in this structure because of an elevated dietary intake of unsaturated fatty acids. These ideas are of some interest, especially since vitamin E can induce a favorable hematologic response in undernourished infants with macrocytic anemia and an increased in vitro red cell fragility when this vitamin is given in conjunction with folic acid and cyanocobalamin. Thus tocopherols may play a role in nucleic acid metabolism in close association with the latter two vitamins; however, at present, our factual knowledge in this area is wholly unsatisfactory.

Vitamin E is ingested by man in all mixed diets, and is abosrbed from the gut together with lipids and the other fat-soluble vitamins. Such foods as liver, eggs, and wheat-germ oil are quite rich sources of vitamin E.

Vitamin K. A large number of naphthoquinone compounds are known which exhibit vitamin K activity. Menadione (vitamin K_3, or 2-methyl-1, 4-naphthoquinone) is both the simplest chemically, as well as biologically the most potent of these compounds (Fig. 13.39). This synthetic compound is fat soluble and heat stable, but is labile to alkali, strong acids, oxidation, and light. The natural naphthoquinones having vitamin K activity are related chemically to menadione.

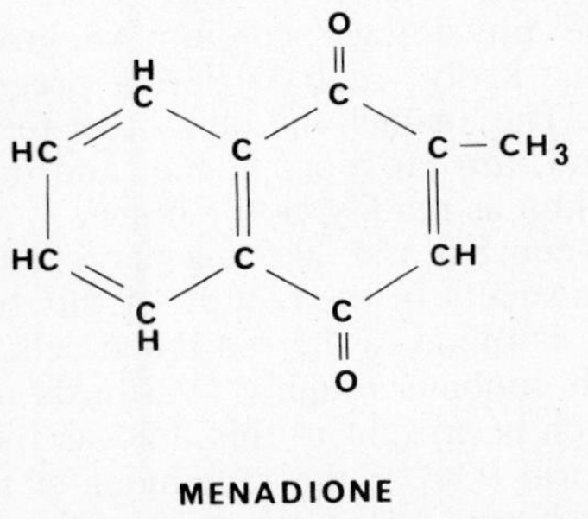

Vitamin K is derived from two principal natural sources. Vitamin K_1, or 2-methyl-3-phytyl-1, 4-naphthoquinone, is ingested with foods of plant origin, especially leafy vegetables (Fig. 13.40). On the other hand, vitamin K_2 (2-methyl-3-difarnesyl-1,4-naphthoquinone) is synthesized by the flora within the intestinal tract.

Vitamin K catalyzes the synthesis of prothrombin by the liver, and thus it is essential for the maintenance of a normal blood clotting mechanism in man, although the vitamin does not form a part of the prothrombin molecule per se. In vitamin K deficiency there is a prolonged bleeding time because of impaired prothrombin synthesis, and perhaps the production of other clotting factors as well. Vitamin K also may participate in electron transport, and thus a role of this substance in oxidative phosphorylation has been postulated.

In common with other fat-soluble vitamins, conditions which derange normal intestinal fat absorption also impair the absorption of vitamin K, so that a hypoprothrombinemia and prolonged clotting time may ensue. Since a large quantity of vitamin K is produced by the intestinal flora and then absorbed from the intestinal lumen, chronic sterilization of the gut with antibiotics and sulfa drugs can induce hypoprothrombinemia. This is an important consideration in patients being readied for surgery. Neonates have a low blood prothrombin level for several days following delivery and until an adequate intestinal flora develops.

No human deficiency of vitamin K due to a dietary lack of this substance has been established. Rather, the synthesis and absorption of vitamin K_2 from the intestine (rather than dietary sources) appears to be the major source of vitamin K for humans, although this compound is widely distributed in natural sources. For example, pork liver and green leafy vegetables (such as spinach) are extremely rich in vitamin K. Tubers, fruit, and grains also contain the compound, although in considerably lower concentrations. Eggs and milk contain still lower amounts of vitamin K.

MENADIONE

FIG. 13.39. Structural formula of menadione, a synthetic compound that exhibits vitamin K activity.

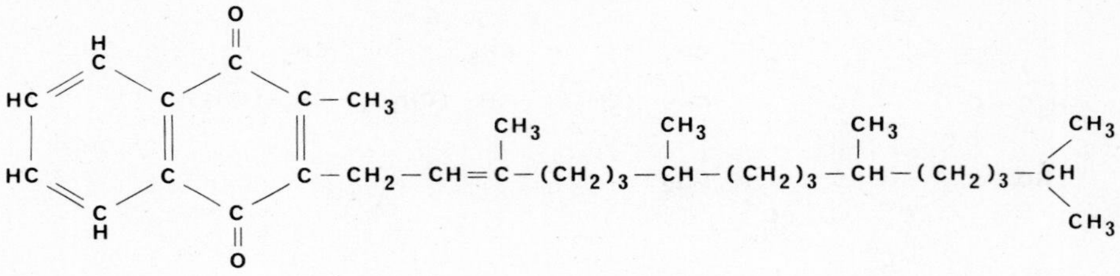

VITAMIN K₁

FIG. 13.40. Structural formula of vitamin K_1, one of a number of compounds that exhibits vitamin K activity.

Under normal conditions, no specific dietary requirements for vitamin K have been established, and this is reasonable when one considers how this factor is supplied to the body. However, if a possibility of neonatal hemorrhage exists (as in premature birth or erythroblastosis), then between 2 and 5 mg of synthetic vitamin K_3 (menadione) may be given to the mother during labor, or a dose of between 1 and 2 mg of this compound may be administered to the postpartum infant to preclude development of hypoprothrombinemia.

Minerals

A general survey of the some two dozen chemical elements which currently are deemed essential to the structural and functional integrity of living tissues was presented in Chapter 3 (p. 52; Table 3.2, p. 53). The physiologic roles of many of the inorganic nutrients or mineral elements (ie, electrolytes) have been considered in relation to specific topics throughout this book. In this section, those inorganic nutrients which have demonstrable functions in the human body will be reviewed. In addition, several additional mineral elements shown to be necessary for normal growth, development, and function in lower animals, as well as their potential significance in humans will also be presented. The 14 elements essential to normal functions of the human body are: calcium, phosphorus, sodium, potassium, magnesium, chlorine, sulfur, iron, copper, cobalt, iodine, manganese, zinc, and fluorine. In addition, molybdenum, selenium, chromium, tin, vanadium, and silicon also have been demonstrated to be necessary in minute traces for the normal growth and development of rats, and quite possibly these substances also are necessary to the normal function of human cells and tissues. All of these elements must be derived from the diet, but the three elements most likely to be in deficient supply in the human diet are calcium, iron, and iodine.

The critically important role of water, not only as a structural component of living matter, but also as the milieu in which all life processes take place has been stressed earlier in this book (Chap. 2). A brief review of this substance as a component of the diet will be presented at the end of this section.

At the outset, it is important to stress the fact that the mineral elements in general do not function as independent entities within the body, even as there are complex nutritional and metabolic interrelationships among carbohydrates, proteins, and fats. Rather, certain minerals may exhibit mutually synergistic or antagonistic physiologic and metabolic interrelationships, and that in the normal person these properties are balanced within exceedingly delicate homeostatic limits. Specific examples of these general facts will be cited in the text which follows.

CALCIUM. Both calcium and phosphorus serve as the principal structural elements of the skeleton in the form of the complex mineral apatite. This mineral salt is composed of calcium phosphate and calcium carbonate which are present in definite proportions, and arranged in a specific crystal lattice. Thus approximately 99 percent of the total body calcium is present in bones (and teeth), whereas the remainder is present in the body fluids and soft tissues. Bone must not be considered to be merely an inert structural and supporting framework for the body. Actually, the skeleton serves as a store of calcium which is in dynamic equilibrium with that of body and tissue fluids so that the rate of exchange (or flux) of this element is, in fact, more rapid than the deposition rate for new bone. Calcium is mobilized most rapidly from the trabecular regions of the bones, especially when the physiologic calcium requirement increases markedly, such as during pregnancy and lactation. The enamel and dentine of teeth, on the other hand, are far more stable, and do not provide calcium as readily as the bones.

Between 8.5 and 10.5 mg percent calcium is present normally in the blood. Of this total quantity, none is found in the red blood cells, whereas the serum contains roughly two-thirds of the calcium which is present in this fluid as the soluble, ionized form (Ca^{++}); the remainder of the serum calcium is bound to the plasma proteins.

In turn, the blood calcium functions in maintaining the normal irritability of nerve tissue, and it is also essential to the blood clotting mechanism. Therefore, reduced blood calcium levels such as are encountered in hypoparathyroidism increase the irritability of neural tissue to such an extent that tetany and/or convulsions may ensue. Elevated calcium levels, on the other hand, depress the normal irritability of nerve. An appropriate blood level of calcium is also essential to normal cardiac contractility. In this regard, it is important to reiterate that the relative concentrations of calcium, sodium, potassium, and magnesium are of the utmost importance in the maintenance of normal irritability in nerve as well as muscle. Furthermore, a magnesium deficiency may induce a concomitant widespread calcium deposition in the soft tissues of the body in addition to producing defects in the bony structures.

It is axiomatic in physiology that the body conserves and utilizes a substance more efficiently when there is a relative deficiency of that substance, and this general rule applies to calcium as well as to a host of other substances. Thus absorption of calcium from the small intestine is enhanced as the dietary supply of this mineral diminishes. Conversely, under normal circumstances, calcium absorption increases to some extent, but not proportionately, when the dietary intake is increased markedly. Other factors are also involved in calcium absorption. Absorption of this mineral is aided by the presence of vitamin D and the normal secretion of hydrochloric acid by the stomach as well as by a relatively low pH in the intestinal tract. The presence of lactose in the gut also enhances dietary calcium absorption, although the precise mechanism which is responsible for this effect is unclear.

Dietary calcium may be lost through the feces by formation of insoluble compounds or complexes with other dietary constituents in the gut, although such losses generally are minor. For example, spinach and rhubarb contain relatively high concentrations of oxalic acid, and the insoluble calcium oxalate which is produced by the reaction of these substances interferes with normal calcium absorption. Phytic acid (found in cereal seeds and bran) also forms insoluble calcium salts with free calcium in the intestine, and again such a process may interfere with normal absorption of the mineral. Poor digestion or excessive quantities of fat in the intestine also render calcium unavailable for absorption through the formation of insoluble soaps. In general, however, the ingestion of a reasonable quantity of calcium-rich foods in the diet tends to compensate for any losses through the mechanisms described above.

Calcium is excreted by the kidney when the blood level exceeds about 7 mg percent and the renal threshold to calcium ion excretion apparently is regulated to a certain extent by parathyroid hormone. Calcium is also lost from the body to a significant extent via the feces. Some of this loss is derived from dietary calcium by mechanisms such as those described above, but a high percentage of the total fecal calcium represents loss from the body store itself. This fraction of the total fecal calcium may be derived from the gastrointestinal secretions as well as from desquamated cells of the intestinal tract.

Heavy sweating provides another route for loss of small quantities of calcium from the body. On the order of one to 20 milligrams per hour thus may be lost through the skin.

In children, a vitamin D deficiency generally is responsible for rickets. This disease may also be caused by an inadequate intake of calcium and phosphorus, or by a profound imbalance in the ratio between calcium and phosphorus intake.

Adults may develop osteomalacia as the consequence of calcium (and vitamin D) deficiency. This condition is rare in the United States. Osteomalacia should not be confused with osteoporosis, a relatively common condition found in older subjects whose bones tend to become demineralized.

The estimated calcium requirements for humans vary widely depending upon such factors as the age of the individual as well as the source of the recommendation itself. The Food and Nutrition Board of the National Research Council recommends an intake of 0.8 grams per day of calcium for infants; 1.3 to 1.4 grams per day for adolescents during rapid pubertal growth; and 0.8 grams per day for adults to maintain their body stores of the mineral. Pregnancy and lactation greatly enhance the need for calcium by females, so that an intake of 1.3 grams per day is recommended during such periods of increased physiologic requirements for this mineral.

Milk and dairy products, egg yolk, shellfish, canned sardines and salmon (with bones), and green vegetables (such as kale and broccoli) are all good dietary sources of calcium.

PHOSPHORUS. Phosphorus has the distinction of participating in more functions in the body than does any other mineral element. Approximately 80 percent of the total body phosphorus is combined with calcium in the form of the mineral apatite, and in this form it adds both strength and rigidity to bones and teeth. The remaining 20 percent of the body phosphorus pool is found in the extracellular fluids as well as in various chemical forms within every cell of the body. In fact, a complete review of phosphorus metabolism would entail a discussion of nearly all of the metabolic processes of the body because of the ubiquitous occurrence of this element. A few general examples of the many biologic roles of phosphorus may be cited here. Phosphorylated compounds are essential to carbohydrate, protein, and fat metabolism, as well as to the development of energy by contractile (muscular) tissues. Phosphorus is also essential to normal blood chemistry, metabolism of neural tissues, skeletal growth, tooth development, and fatty acid transport. Phosphate also has an important role as a component of numerous enzyme systems, and it is essential to energy storage and transfer in the form of phosphorylated compounds, eg, adenosine triphosphates and diphosphates.

Of the some 35 to 45 mg percent of phosphorus which normally is found in blood, between 3 to 4.5 mg is present as inorganic phosphate, and in this chemical form it is able to participate readily in many chemical reactions. Under normal circumstances, an inverse relationship exists between the serum calcium and the serum inorganic phosphate.

Approximately 70 percent of the total phosphorus that is ingested in foods is absorbed normally. Presumably intestinal phosphatases liberate simple phosphate compounds from the chyme prior to their absorption. As is the situation with calcium, phosphate absorption is facilitated by an acid medium, although magnesium, iron, and aluminum, when present in quantity, interfere with phosphorus absorption through formation of insoluble phosphate compounds. Normally, however, any loss of phosphate via the synthesis of such compounds is insignificant.

The diet should supply roughly equal amounts of both calcium and phosphorus, so that if the diet presents marked deviations in the calcium/phosphorus ratio then the absorption of both minerals becomes defective, and a calcium deficiency may ensue.

Phosphorus absorption is enhanced markedly by the presence of vitamin D, and quite possibly this stimulatory effect is secondary to the augmented calcium absorption which is produced by this vitamin.

Fecal phosphorus represents both the unabsorbed element which was ingested with the food as well as phosphorus which has been secreted into the intestinal tract. Urinary phosphorus is primarily inorganic phosphate, and the quantity of phosphorus lost from the body per day via this route is dependent upon the quantity absorbed from the gut, among other factors. During periods of starvation, catabolism of the body tissues liberates considerable phosphorus which then is excreted in the urine. Conversely, a person whose dietary intake of phosphorus is constant, but who is metabolizing carbohydrate extensively (as during intense physical work) exhibits a decreased urinary phosphate excretion.

A phosphorus deficiency syndrome has not been encountered in man, as the average diet contains a sufficiency of this element. As a general rule of thumb, if the calcium intake is adequate, then the phosphorus intake should be approximately equal to that of calcium. Ordinarily those foods which satisfy the calcium and protein requirements of the body also satisfy the phosphorus requirements.

The Food and Nutrition Board recommends an adult phosphorus intake which is about 1.5 times that of calcium. On the other hand, infants as well as pregnant and lactating women require approximately the same quantities of both of these minerals in the diet.

Such foods as meat, poultry, fish, and eggs are excellent dietary sources of phosphorus. Milk, cheese, legumes, and nuts also contain adequate quantities of this mineral. Cereal grains contain much of their phosphorus as phytic acid; consequently the availability of the mineral from such food sources is of doubtful significance from an overall nutritional standpoint.

SODIUM.   Sodium and potassium both are necessary for the normal physiologic processes of the body; however, dietary deficiencies in either of these minerals are rare.

Nearly all of the sodium of the body is found in the extracellular fluids, and this mineral comprises about 93 percent of the total cations in the blood. Sodium ion is essential to such physiologic processes as acid–base regulation, and is important to the regulation of osmotic pressure in the extracellular fluids. Usually about 90 percent of the ingested sodium is excreted in the urine either as sodium chloride or as sodium phosphate. During periods of intense sweating, however, loss of sodium in the perspiration is the major route whereby this mineral is excreted. If sweating is prolonged, an acute sodium depletion of the body may ensue, leading to vascular collapse, headache, weakness, and cramps of the skeletal muscles. Furthermore, chronic vomiting or diarrhea and adrenocortical insufficiency can all produce a need for additional sodium. Increased dietary or parenterally administered sodium chloride may be indicated to ameliorate the effects of the relative salt deficiency produced by any of the conditions summarized above. Conversely, a salt-restricted diet may be indicated in certain forms of cardiac and renal disease wherein it is desired to reduce the total volume of the extracellular fluid compartment.

Daily salt intake (as sodium chloride) may range between 2 and 20 gm. According to the Food and Nutrition Board, an intake of 5 gm of sodium chloride per day is liberal, but if the water intake is over 4 liters per day, then 1 gm of salt should be taken per liter of water in excess of this volume.

POTASSIUM.   In contradistinction to sodium, potassium is the major intracellular cation of the body. Potassium exerts a profound influence upon the irritability of nerve, as well as upon the contractility of skeletal, cardiac, and smooth muscle.

Potassium deficiency can result from deranged renal function or renal disease, chronic diarrhea, and diabetic acidosis. Hypokalemia (a reduced blood potassium level) may also be caused iatrogenically during the maintenance of the body fluid volume for prolonged intervals by intravenous infusion (ie, parenteral administration) of an isotonic glucose and sodium chloride solution which contains no potassium.

The symptoms of potassium deficiency include weakness of the skeletal muscles, increased nervous irritability, psychic disorientation, and cardiac arrhythmias.

Abnormally high blood potassium levels are frankly toxic and result in cardiac arrhythmias, so that in severe hyperkalemia (in excess of 10 mM per liter), the irregularities may progress to cardiac arrest. Hyperkalemia cannot be induced in normal persons through dietary administration of potassium. Therefore, such a condition results from renal disease or the oliguria (ie, a markedly reduced excretion of urine) which accompanies dehydration and shock. In both of these situations, there is failure of the kidneys to excrete potassium normally.

The normal dietary requirement for potassium is approximately the same as that for sodium, and most common proteinaceous and vegetable foods contain adequate quantities of potassium to satisfy the daily requirement for this mineral.

MAGNESIUM. Magnesium is a normal constituent of soft tissues as well as bone. The normal physiologic irritability of nervous tissue as well as contractility of skeletal and cardiac muscle is dependent upon an appropriate balance between magnesium and calcium ions. Magnesium is also essential as an activator in many enzyme systems which are responsible for phosphate transfer, including, for example, the myokinase, and creatine kinase systems. Magnesium ion is also required to activate pyruvate oxidase and pyruvic carboxylase.

A diet which contains between 250 and 300 mg per day of magnesium suffices to maintain normal adults, although the actual requirement for this mineral has not been established.

A deficiency produced by dietary magnesium lack has not been characterized in man. However, a magnesium deficiency syndrome which resembles hypocalcemic tetany and is characterized by muscle tremor, choreiform movements, and even delirium and convulsions, may develop secondarily to certain abnormal conditions. Included in this category are the toxemia of pregnancy, severe renal disease, chronic loss of gastrointestinal secretions, a voluminous diuresis induced by drugs, and the protracted intravenous infusion of fluids lacking in magnesium. Parenteral administration of magnesium reverses these symptoms rapidly.

Such plant foods as nuts, whole wheat, flour, brown rice and legumes are good sources of magnesium, whereas meats, poultry, and fish are comparatively poor sources of this mineral.

IRON. Iron is an important constituent of hemoglobin and myoglobin molecules, as well as of the cytochromes and other enzyme systems. This mineral is essential to the normal bulk transport of oxygen as well as in those respiratory processes at the cellular level (ie, electron transport, p. 931), from which energy is derived.

As noted earlier, ferrous iron (Fe^{++}) is absorbed from the gut far more readily than is ferric iron (Fe^{+++}), and presumably the latter is reduced to the ferrous state as it is released from foodstuffs by the gastrointestinal digestive processes. An acid medium, such as is provided by normal gastric secretion, is far more favorable to iron absorption than is an alkaline medium. Similarly, the role of ascorbic acid, and the presence of sulfhydryl compounds (substances which contain an —SH radical) also promote iron absorption, and this effect presumably is due to the ability of these substances to provide a reducing environment for ferric iron. The efficiency with which dietary iron is absorbed is also enhanced in persons living at high altitudes.

An abundance of phosphates in the gut inhibits iron absorption, because the iron compounds which are formed by the reaction of these two substances are insoluble, and the iron phosphates thus formed are eliminated in the feces.

Under normal circumstances, approximately 10 percent of the total iron ingested is absorbed; however this quantity increases markedly (up to around 25 percent) when there is a deficiency of this mineral in the body. Consequently, the principal factor which determines iron absorption from the gut is the bodily requirement for this substance.

At present, the mucosal blockage theory generally is accepted as the explanation of how iron is absorbed from the gut only as needed so that the body does not store large quantities of this mineral (see Fig. 9.5, p. 534). Apparently iron enters the mucosal cells of the intestinal tract and combines with a specific protein, called apoferritin; thereby an iron–protein complex called ferritin is formed, and this complex in turn yields its iron to the blood whenever iron is required by other tissues of the body. Apoferritin in the mucosal cells may complex additional iron as it is absorbed but only until the mucosal cells are saturated with ferritin, at which point further iron absorption ceases. Thus when ferritin releases its iron to the blood, as for hemoglobin synthesis or other bodily requirements, the apoferritin produced thereby now is free to combine with more iron, and the inhibition to absorption is relieved temporarily. The operation of this delicately poised homeostatic mechanism thus depends upon the bodily requirements for this mineral, and gross stimuli, such as the hypoxia encountered at high altitudes or an acute hemorrhage, result in an imbalance within the body so that iron absorption now is enhanced markedly.

Iron is present in the ferric state in plasma where it is bound to a specific globulin in a complex known as siderophilin. This iron–protein complex thus transports iron and a large percentage of the mineral is utilized in the bone marrow for the synthesis of hemoglobin (Fig. 9.4, p. 533). Minute quantities of copper (see next section) also are required for the incorporation of iron into hemoglobin as well as into the cytochrome enzymes.

The total body pool of iron in a normal adult averages approximately 4 gm, of which about 1 gm is stored, chiefly in the liver and spleen. This reserve iron is stored as an intracellular protein complex, either as ferritin or as hemosiderin. The iron stored in these forms is mobilized readily when need arises; therefore, an anemia develops only after these bodily iron reservoirs are depleted markedly.

The approximate life-span of an erythrocyte is around 120 days, and when the cell degenerates the iron is retained and reutilized for synthesis of new hemoglobin. Despite the highly economical recycling of iron within the body, small quantities of the mineral are lost daily via such pathways as desquamated mucosal and epithelial cells, sweat, hair growth, and excretion in the urine, as well as to a limited extent, by excretion in the bile and feces. Thus adult males lose a total of approximately 1 mg of iron per day, whereas in females throughout the interval from puberty to menopause, the average daily loss is somewhat greater (about two milligrams per day), because of periodic menstrual blood loss. Furthermore, pregnant women provide a total of approximately 300 to 500 mg of iron to the fetus during the entire nine months gestation period; the dietary requirement for this mineral remains consistently higher than normal throughout pregnancy.

Additional factors which can augment iron loss from the body significantly are occult (or hidden) intestinal bleeding, hemorrhagic injuries, surgery, and blood donations.

Iron deficiency is manifest clinically as hypochromic anemia, as discussed in Chapter 9. Briefly, this condition is characterized by a subnormal total quantity of circulating hemoglobin, regardless of whether or not the erythrocyte count is normal or reduced. The blood oxygen carrying capacity is lowered and the typical clinical manifestations of this condition are similar to those of other anemias, and include dyspnea during exertion, weakness, pallor of the skin and mucous membranes, chronic fatigue, and headache.

Iron deficiency anemia may result from an inadequate supply of iron in the diet and thus have a nutritional basis, or through inadequate absorption of the element. Several of the conditions described above also may contribute to the overall loss of iron from the body through an increased rate of blood loss. In addition, intestinal parasites, hemorrhoids which bleed chronically, peptic ulcers, and frequent nose bleeds all predispose the individual to the development of an iron deficiency anemia. Furthermore, if the diet of such persons is minimal, or actually deficient, in iron, then the person is particularly susceptible to the development of an iron deficiency anemia. As milk is a notably poor source of iron, infants are particularly prone to develop an iron deficiency anemia unless their diet is supplemented with this mineral from other sources.

It is essential to realize that the mere availability of an adequate dietary iron supply is not the sole factor involved in preventing iron deficiency, hence a low hemoglobin content in the blood, and furthermore, that the overall nutritional pattern is of major importance in preventing anemia. There-fore in addition to iron, the availability of nutritionally adequate proteins and water-soluble vitamins of the B-group, as well as the total caloric intake, are all significant nutritional factors in preventing anemia.

As noted above, the adult male loses approximately one milligram of iron per day, whereas adult, sexually mature females lose on the average about twice this quantity of iron. Since only 10 percent of the dietary iron is absorbed, the estimated daily intake of this mineral is in the range of 10 to 20 mg per day for normal adults. The Food and Nutrition Board recommends between 8 and 12 mg of iron per day for children, 10 to 15 mg for adolescents and adults, and 20 mg for pregnant or lactating women. It should be reiterated that these values are estimated requirements only, and the daily intake should be increased to compensate for obvious blood loss or in conditions that are unfavorable to normal iron absorption.

A number of foods are good sources of iron, and these foods are listed in an order of decreasing quality insofar as their iron content is concerned: liver, heart, kidney, lean meats, shellfish, egg yolk, legumes (including dried beans), dried fruits, nuts, green leafy vegetables, whole-grain and enriched cereals, cereal products, and milk. Since milk is an extremely poor source of iron, the inclusion of enriched cereals, meat, egg yolk, and vegetables in diets for infants is justified at a reasonably early age.

COPPER.   The enzyme tyrosinase (or polyphenol oxidase) requires copper as an essential constituent of the molecule; this enzyme is necessary for the synthesis of the pigment melanin in the body. Copper is also an essential constituent of the enzyme butyryl coenzyme A dehydrogenase. Quite possibly copper also is associated with other oxidation–reduction enzyme systems, and it is believed that this mineral also plays a role in enhancing the absorption of iron from the gut. Furthermore, copper apparently is important to the synthetic mechanism whereby iron is incorporated into the hemoglobin molecule; however, the precise mechanism whereby this hematopoietic function of copper is carried out is unknown.

Absorption of copper takes place in the upper segments of the small intestine. In the blood, approximately 50 percent of the total copper is found within the erythrocytes, whereas the remainder is found in the plasma, where it is found as a copper: α-globulin compound known as ceruloplasmin at a level between 25 and 30 mg percent, eight copper atoms being present per molecule of ceruloplasmin. Formation of this compound is essential to the normal absorption and regulation of copper stores within the body, as once the plasma ceruloplasmin is "saturated" with copper, no more of the mineral is absorbed. Copper is stored primarily in the liver, and it is excreted chiefly in the bile.

Hepatolenticular degeneration (Wilson's disease) is a rare disorder of copper metabolism which tends to occur in certain families for reasons unknown (ie, it is a familial disease). In patients having this disease normal ceruloplasmin formation is defective so that between 0 to 5 mg percent of this protein is found in the plasma. As a consequence, there is an unlimited and unregulated copper absorption, and the mineral is retained mainly in the brain and liver of such individuals. Thus a gradual degeneration of the lenticular nucleus takes place, generally with attendant neurologic symptoms, and concomitantly a derangement of hepatic functions which ultimately leads to cirrhosis occurs. Death ensues unless therapy is instituted, chiefly in the form of a diet low in copper, combined with chelating agents which prevent excessive copper absorption from the intestinal tract.

This disease has been cited here as a specific example of how a biochemical lesion can result in a defect in the homeostatic regulation of an absorptive mechanism for a specific mineral. Note that there is an overall similarity between ceruloplasmin formation as the homeostatic regulatory agent for copper absorption, and the formation of ferritin from apoferritin within the intestinal mucosa, as described earlier.

A copper deficiency syndrome has not been defined in adult humans, although rarely infants with hypochromic anemia are benefited to a greater extent from the therapeutic administration of copper simultaneously with iron.

The adult human requirement for copper is 2 mg per day. The average diet contains much more than this level, so that copper balance is maintained readily.

COBALT. Cobalt is an essential constituent of the cyanocobalamin (vitamin B_{12}) molecule (Fig. 13.34), and this is probably the only function of the mineral in the body.

Although cobalt is absorbed readily, little is retained in the body (human tissues are unable to synthesize vitamin B_{12}), as the mineral is excreted quite freely in the urine. A cobalt deficiency has not been described in humans, and the only known requirement for this mineral is for that contained in cyanocobalamin, which, of course, must be obtained from exogenous sources.

ZINC. Zinc is an essential component of several enzymes which catalyze a number of critical metabolic reactions. Thus carbonic anhydrase contains zinc as an indispensable constituent of the molecule, and this enzyme is found in erythrocytes as well as in gastric and renal tubular cells. In the erythrocytes, carbonic anhydrase accelerates carbon dioxide exchange in the lungs at a rate which is compatible with life, whereas in the stomach the enzyme participates in the manufacture of hydrochloric acid (Fig. 13.17). In the kidney, carbonic anhydrase is essential to the regulation of acid–base balance.

Zinc is also a constituent of the digestive enzyme carboxypeptidase, and it may also be present in certain dehydrogenases. Zinc is not a normal component of insulin and is unnecessary to the physiologic action of this hormone. However, when insulin is complexed artificially with zinc, the rate of absorption of the hormone following its subcutaneous injection is prolonged, so that less frequent injections of the hormone are necessary.

Zinc is found in most foods, and no human deficiency state involving this mineral has been described, nor has a daily human requirement for zinc been defined.

IODINE. Iodine is a critical element for the synthesis and composition of thyroid hormone (thyroxine); thus it is an essential nutrient for humans. This is the only known physiologic role of iodine.

One major factor which regulates thyroxine production is the availability of iodine in the diet, and a lack of this element leads to the development of simple (or endemic) goiter. The details of thyroxine synthesis and the specific role played by iodine in the individual reactions leading to thyroxine production by the thyroid gland are presented in detail in Chapter 15 (p. 1050 et seq).

The adult dietary requirement for iodine (as iodide, I^-) ranges between 0.15 and 0.30 mg per day. During puberty as well as pregnancy the iodine requirement is increased, in the latter instance because the fetus can obtain its iodine only from the maternal supply of the mineral. Lactation is another physiologic situation which greatly increases the human iodine requirement, as quantities of iodide are secreted in the milk. Therefore, the iodide requirement of a nursing mother is practically doubled when milk flow is maximal.

If food is grown on iodine-poor soil, the iodine content of the crops may be inadequate to satisfy the small but essential human requirements for iodine, and endemic goiter is prevalent in such regions, especially if they are geographically remote from the ocean. For example, simple iodine deficiency goiter occurs in such regions of the world as the Great Lakes Basin, Central America, the Pyrenees, the Alps, and the interior of China.

Sea foods and vegetables which are grown on iodine-rich soils are the best natural sources of iodine, although the distribution of such foods may be haphazard at best. Consequently, the use of iodized salt has proven to be the most practical method for ensuring an adequate dietary intake of iodine on a wide scale. Commercial iodized salt in the United States contains 0.01 percent potassium iodide, so that if the average daily salt intake of an adult is 5 gm, the concomitant daily intake of iodide from this source amounts to 0.50 mg. This quantity of iodine is roughly twice the daily requirement, providing a sufficient quantity of this element for a reserve supply. Iodates are used instead of iodide in certain areas (eg, Central America) as they are more stable in less highly refined salt.

Since the principal source of iodine, in the United States at least, is iodized salt, it is important to remember that when a patient is placed on a salt-restricted or salt-free diet, the usual iodine intake is curtailed sharply. This fact is of particular importance in patients whose condition requires a drastic reduction in salt intake, for example in certain types of cardiovascular and renal disease, or in pregnant women who exhibit a tendency to develop toxemia. Supplementation of the diet with iodine may be indicated in such patients.

Another important mineral element is supplied automatically in adequate quantities for normal human nutrition by the ingestion of common salt (sodium chloride) in the diet, regardless of whether or not it is iodized. This is the element chlorine, which is present as chloride. This substance is an important anion component of the extracellular fluids, and it also is essential to hydrochloric acid secretion by the stomach, to mention two of its physiologic roles. However, no deficiency of this substance is physically or chemically possible on a diet that is even barely compatible with mere survival.

MANGANESE. Arginase, an enzyme that is essential to the formation of urea, requires manganese as a component of its molecular structure. This mineral also serves as an activator for several of the enzymes which participate in the citric acid cycle, in particular the decarboxylation reactions which involve the dicarboxylic and tricarboxylic acids. Thus manganese plays several important roles in the body, but is required in only minute quantities. Therefore, it is termed an essential micronutrient.

Manganese is found commonly in foods of both animal and plant origin. No requirement for this mineral has been established for man, and no deficiency syndrome is known in humans.

MOLYBDENUM. Molybdenum is a component of the molecular structure of two enzymes, aldehyde oxidase and xanthine oxidase; consequently this mineral is considered to be an essential micronutrient. In these enzymes, the molybdenum presumably is involved in the linkage of the flavin nucleotide (the coenzyme or prosthetic group) to the substrate-specific protein (the apoenzyme). Molybdenum also is believed to have a catalytic role in the oxidation of fatty acids.

No characteristic deficiency syndrome for molybdenum has been described for humans.

FLUORINE. The element fluorine, as fluoride ion, is present normally in bones and teeth, and an adequate, though small, intake of fluorine has been shown to be important in the production of maximal protection against dental caries both in children and in adults. This effect appears to be the only role of fluorine in the body; hence fluorine deficiency results in a decreased resistance to dental caries, but no other lesion has been demonstrated in either hard or soft tissues of man or other animals when fluorine intake is restricted.

In nature fluorine is of wide but uneven distribution. The fluorine content of water depends upon the chemical composition of the geologic structures (soils and rocks) through which the water percolates prior to consumption. Many foods (especially sea foods and tea) also contain quantities of fluorine.

The absorption of fluorine from the gut depends principally upon the solubility of the compounds involved, but there is little information available on the metabolism of this element within the body.

Approximately 0.2 parts per million (ppm) of fluorine are found in normal human blood, and 0.1 ppm in saliva. Ingested fluorine is excreted almost totally in the urine, up to an intake of approximately 3 mg per day, and minute quantities also are eliminated in the sweat and feces.

In areas where the natural water supply to a community (such as in certain localities in Colorado and the Texas panhandle) contains a high concentration of fluorine (two to six ppm), a brownish mottling of the tooth enamel develops owing to an excessive intake of the element. On the other hand, if the natural water supply contains about 1 ppm fluorine, there is a significant decrease in the occurrence of dental caries in comparison with the national average. The mechanism whereby fluorine reduces the incidence of tooth decay is unknown, but the mineral apparently is incorporated into the structure of the tooth enamel during development of the teeth, so that solubility of the enamel in the acids that are produced by bacteria in the mouth is reduced. It has also been postulated that fluoride acts as a bactericide, although this possibility is less likely.

Fluoridation of public water supplies is carried on extensively in the United States at present in an attempt to reduce the occurrence of dental caries among young persons. Generally 1 ppm fluorine is added to the water. Toxic symptoms or other untoward side effects have not been observed after fluoridation of water at this low concentration. However, the overall incidence of dental caries is reduced significantly in children and adolescents by such a program in individuals up to 16 years of age. This fact has been well documented by evidence derived from well-controlled experimental studies.

An ordinary diet provides between 0.2 and 0.3 mg of fluorine per day, and if the water supply contains 1 ppm fluorine, then an adult will ingest roughly 1.5 mg per day total fluorine and the intake will be somewhat less in younger persons. This quantity of fluorine is quite sufficient to reduce the incidence of tooth decay, particularly in persons below 16 years of age.

It must be emphasized that normal tooth development requires not only the presence of fluorine, but an overall diet adequate in all other essential nutrients, including total calories, proteins, vitamins, and other minerals. In particular, calcium, phosphorus, and vitamins A, D, and C are essential to this process.

A further, even more fundamental, role of fluorine in biologic processes is suggested by experimental findings to be discussed in the next section.

EXPERIMENTAL STUDIES ON THE REQUIRE-MENTS FOR TRACE ELEMENTS. The technical difficulties involved in demonstrating the biologic need for a trace element increase enormously as the total quantity of that element in the body is reduced. Furthermore, as the element may be needed in minute quantities to exert a maximal effect, the use of ultrapure reagents and diets as well as an elaborately and rigidly controlled experimental environment all are critical factors in studies that are designed to ascertain whether or not an animal requires a given mineral for normal growth, reproduction, or other bodily processes.

Since practically all of the chemical elements listed in the periodic table may be found in living cells, the question immediately arises as to which elements are essential to normal functions and which are merely present fortuitously.

Carefully performed experiments, which were carried out under rigidly controlled conditions, have demonstrated beyond reasonable doubt in lower animals (principally rats), that extremely minute quantities of selenium, chromium, tin, vanadium, and silicon are needed by living tissues in addition to the other minerals discussed above.

Selenium deficiency will induce hepatic necrosis and muscular dystrophy in rats; however, death may be prevented by inclusion of 0.1 ppm of selenium in the diet. Selenium deficiency can also render the rat erythrocytes sensitive to damage through oxidation by metabolically synthesized hydrogen peroxide. Evidence has been obtained that normal glutathione peroxidase activity in the erythrocyte is present only when selenium is available. This enzyme normally functions to catalyze the oxidation of glutathione using hydrogen peroxide; thus the latter compound is prevented from oxidizing the reduced (ferrous) iron in hemoglobin to the ferric state. Apparently selenium is an essential component of the glutathione peroxidase molecule. Further studies with selenium performed on chicks definitely indicate that this element is essential to life, at least in this species.

Similarly a deficiency state has been demonstrated for chromium in rats, and this state is characterized by a failure of growth, corneal lesions, an abbreviated life-span, and a defect in sugar metabolism. Recent evidence indicates that chromium ion acts synergistically with insulin in some presently unknown manner, and some cases of human diabetes perhaps may result from a defective chromium metabolism. The potential importance of experimentally derived facts such as this one to clinical medicine is obvious.

The normal growth rate of rats is inhibited by the absence of tin in the diet, and a similar defect is found in rats deprived of vanadium.

In addition to the effects of fluorine discussed earlier, recent studies in rats have indicated that minute quantities of the element (0.5 ppm) are essential to the normal growth of these animals. This fact suggests a far more basic role for fluorine than had been suspected heretofore.

It is quite probable that trace elements such as those described in this section function in general as integral components of specific enzymes or as cofactors which promote or enhance the activity of certain enzymes.

It may be assumed with reasonable certainty that the list of essential trace elements will be expanded in the future as experimental studies in this area continue. However, it is doubtful that human requirements for the minerals present in such minute quantities will be established, nor will deficiencies in these substances be encountered clinically, as any reasonable diet contains such elements in more than adequate quantities to sustain normal functions.

WATER. The essential nature of this compound to the life processes has been emphasized repeatedly throughout this textbook. As an inorganic nutrient, water is second only to oxygen insofar as its physiologic importance is concerned, and a continual daily supply of this substance in adequate quantity is critical to survival.

Water is a unique compound for several reasons, including its chemical and physical properties. Water serves as the major constituent of living systems, including the human body; it is the medium of chemical and physical transport within the body; it serves as the milieu in which all of the metabolic reactions in the body take place; and it is essential to the process of thermoregulation as well as to the elimination of metabolic wastes from the body. Water may be considered an indispensable nutrient, and it is hardly a cliché to reiterate here that life and its multitude of physical and chemical processes could not exist in its presently known abundance of forms in the absence of this compound.

A daily total intake of water amounting to approximately 2.5 liters per day is recommended for normal adults by the Food and Nutrition Board. The ingestion of one milliliter of water per calorie of food is a good general rule for water intake in persons subject to moderate temperature and exercise (see Fig. 12.21, p. 797). In actual fact, however, the water intake may fluctuate markedly either above or below this arbitrary standard, although normally water intake both as fluids per se and in foods is regulated unconsciously by the thirst mechanism in keeping with the physiologic requirements of the body. This statement obviously does not apply to infants or to extremely ill persons. For example, voluminous, prolonged diarrhea or chronic vomiting can result in rapid dehydration and even death due to cardiovascular collapse. Infants are particularly susceptible to death from such causes, as their homeostatic mechanisms for the regulation of water balance are not fully developed. Water deprivation is also common in unconscious or comatose patients so that an adequate fluid supply is a prime requisite in the care of such individuals. Such fluid must, of course, be administered either parenterally or via a gastric tube, as an unconscious patient is unable to ingest fluid normally, and an attempt to force such an individual to swallow can result in suffocation.

Summary and Conclusions

The basic elements required for adequate normal human nutrition have been reviewed briefly. Essential components of the normal diet include foods which contain the three basic nutrients, carbohydrates, proteins, and fats; these compounds, in addition to their specific functions in the body collectively, must also satisfy the total energy (or caloric) requirements of the particular individual, taking into account such factors as age, sex, environment, and work performance.

In addition to these three bulk nutrients, some twelve additional organic substances, the vitamins, are essential to normal growth and well-being. All of these compounds must be derived from the diet, as they are not synthesized in the body and they are needed in extremely small or trace quantities to ensure normal health. Included in this category are the water-soluble vitamins (thiamine, riboflavin, niacin, pyridoxine, and cyanocobalamin, as well as folic, pantothenic, and ascorbic acids), and in addition, three fat soluble vitamins (A, D, and K) are also necessary. The role of vitamin E in human nutrition remains obscure; however, as the tocopherols are practically ubiquitous in the various foodstuffs used by humans, it appears that a deficiency state involving this vitamin is unlikely to appear in normal, otherwise well-nourished persons.

One further comment with regard to the vitamins (which applies to certain other nutrients as well) is in order. The reader should be aware of the important distinction between the quantity of a vitamin required by a person in order to maintain normal functions, and that quantity required for the therapeutic correction of a clinical deficiency of that vitamin.

In addition to the organic nutrients, an adequate diet must include a number of important inorganic (or mineral) elements in the diet in various quantities. In particular, calcium, phosphorus, sodium, potassium, magnesium, iron, copper, cobalt, zinc, manganese, iodine, and chloride are essential nutrients, and these substances are required in daily amounts ranging from gram to microgram quantities. Other elements which may be needed in trace amounts include fluorine, molybdenum, selenium, chromium, tin, and vanadium, although such minerals, if actually required by humans, are needed only in exceedingly minute quantities, and doubtless all of these micronutrients are present in sufficient amounts in any adequate diet.

Other than the normal nutritional requirements of the human body, a specific dietary regimen is of the utmost importance in the management of many clinical conditions, such as were summarized on page 873. The exact details of foods to be included in, or omitted from, such therapeutic diets are to be found in textbooks on nutrition.

CLINICAL CORRELATES

Included in this section are a few representative malfunctions of the gastrointestinal tract and related organs, which are significant because they exemplify clinical situations that can arise when the normal (or physiologic) mechanisms in this complex system are deranged. The principal clinical symptoms that may be evidence of underlying disease of the gastrointestinal tract include dysphagia, heartburn, abdominal pain, hemorrhage, nausea and vomiting, diarrhea, and constipation.

Dysphagia

Difficulty in swallowing, or dysphagia, is an important symptom of alimentary tract disorder indicating a malfunction of the normal reflex swallowing mechanism. Dysphagia is experienced only during the passage of a bolus while swallowing, and it may or may not be related to a sensation of pain.

Dysphagia may be produced by any of a number of underlying causes, which may include abnormalities of the sensory or motor pathways to the esophageal musculature (eg, pharyngitis or bulbar poliomyelitis); involvement of the neuromuscular junction (eg, myasthenia gravis); involvement of the esophageal striated muscle per se (eg, muscular dystrophy); stenosis, or narrowing, of the esophageal lumen (eg, when hypertrophy of the thyroid gland is present, or by a benign or malignant tumor); or a failure of the normal peristaltic reflex mechanism (eg, achalasia).

PARALYSIS OF THE SWALLOWING REFLEX. If the normal swallowing mechanism is paralyzed either completely or partially, the malfunctions may include the following signs: first, complete failure of the swallowing act; second, failure of the cricoid esophageal sphincter to remain closed during normal respiration so that large volumes of air are aspirated into the esophagus; third, food may reflux into the nose during swallowing because the soft palate and uvula do not close the posterior nares properly; and fourth, food can enter the trachea as well as the esophagus, because of a failure of the glottis to close.

For example, under deep surgical anesthesia, total paralysis of the swallowing mechanism occurs. A quantity of chyme may be vomited into the pharynx, and as the normal swallowing reflex is inhibited, the chyme is not swallowed, but rather is aspirated into the trachea, so that the patient may choke to death.

ACHALASIA. If the last few centimeters of the esophagus above the stomach fail to relax properly during the normal swallowing reflex, a condition known as achalasia exists, and passage of food into the stomach is delayed or inhibited. This condition results from an inherent defect or a disease process in the myenteric plexus of the lower esophagus. In either event, the property of receptive relaxation of the lower esophagus is abnormal, and the tone of the esophagus near the cardia remains abnormally high during swallowing. Eventually, dilatation of the esophagus proper occurs above the physiologic block because of the presence of food masses and a condition known as megaesophagus develops. In achalasia, it is essential to reduce the tone of the cardiac sphincter (preferably by dilatation), so that the combined forces of pharyngeal contraction and gravity empty the esophagus in a manner approximating the normal.

ESOPHAGEAL STENOSIS. Commonly dysphagia is caused by a narrowing of the esophageal lumen by neoplastic or scar tissue, and the stricture is too narrow to permit a bolus of food, especially when large and solid, to pass readily.

Esophageal carcinoma, principally a disease of the elderly, is manifest first by intermittent dysphagia. Usually, however, this symptom appears only when the neoplasm has advanced to such a stage that surgery is rarely curative. Esophageal carcinoma is about five times more prevalent in men than in women.

Benign inflammatory stricture of the esophagus is produced most commonly by peptic or reflux esophagitis, next most commonly by ingestion of corrosive substances (eg, lye), and least commonly by ulceration of the columnar epithelium of the lower esophagus by reflux of the gastric juice.

Heartburn

Although often minor, heartburn may be indicative of a more serious underlying condition. Basically, heartburn, or more accurately dyspepsia, is caused by reflux of the acid gastric juice into the esophagus (gastroesophageal reflux).* Typically this condition results in a burning sensation in the subxyphoid region, which may extend upward into the chest, and even into the neck, shoulders, and ulnar surfaces of the arms when severe. Thus differentiation from angina pectoris or peptic ulcer may be difficult.

Heartburn may appear after a meal, and it is

There is one reported case of "true heartburn" in which a depressed individual attempted suicide by thrusting a red hot poker into his thoracic cavity through the fifth interspace 8 cm to the left of the midline, thereby incurring an 8-cm² burn on the left ventricle.

exaggerated by bending over or lying down. Overindulgence in alcoholic beverages contributes markedly to this symptom. Drinking bland liquids, standing up, and most especially ingestion of antacids tends to ameliorate the symptom.

Heartburn may be encountered as a symptom in clinical states such as hiatal hernia or esophagitis. The latter is a serious chronic inflammation of the esophagus which can allow a prolonged and severe gastroesophageal fluid reflux. Patients having a weak cardiac sphincter may develop esophagitis. Dysphagia may or may not accompany heartburn.

Gastritis

Inflammation of the gastric mucosa, or gastritis, can result from the presence of irritating foods within the stomach, bacterial action, or the corrosive effects of the gastric secretions themselves. Alcohol, for example, can cause gastritis accompanied by heartburn, and voluminous salivary secretion also may be present because of salivary reflexes which are initiated within the irritated gastric mucosa.

Gastric Atrophy

Roughly one-third of the patients suffering from chronic gastritis regardless of cause eventually develop atrophy of the gastric mucosa, so that hydrochloric acid secretion ceases (anacidity) or is reduced (hypoacidity), because of a failure of the gastric glands. Pepsin secretion may also be reduced or completely absent in gastric atrophy; however, the absence of hydrochloric acid automatically prevents any pepsin that may be secreted from exerting its normal hydrolytic function; therefore gastric digestion is decreased markedly or ceases completely.

From a clinical standpoint, the hydrochloric acid secreted by the gastric mucosa may be divided into two fractions, free and combined acid. The free acid is determined first on a sample of gastric juice by titrating to a pH of 3.5 with one indicator (dimethylamino-azobenzene); the same secretion then is titrated to pH 8.5 with a second indicator (phenolphthalein), and this second determination gives the combined acid. Normally, there is little free acid in the stomach when food is present and the combined acid forms the largest fraction of the total gastric acid, whereas if the stomach is empty, then the secretion normally contains a much higher percentage of free acid.

Gastric anacidity, however, does not necessarily decrease the overall digestion of food, as the proteolytic and other enzymes which are secreted into the intestinal tract by the pancreas and which also are present within the mucosal cells themselves are quite capable of digesting the bulk of the dietary protein and other nutrients, particularly if the food was well chewed before swallowing.

A more significant feature of gastric atrophy than the failure of gastric digestion per se is the lack of secretion of intrinsic factor by the parietal cells. The relationship between this defect and the development of pernicious anemia due to a concomitant deficiency of cyanocobalamin (vitamin B_{12}) was discussed on page 568.

Ulcer Formation

An ulcer may be defined as a local defect or excavation on the surface of an organ which is caused by sloughing of necrotic inflammatory tissue. A peptic ulcer is an ulceration which is found on the surface of the stomach or duodenum, whereas the terms "gastric ulcer" and "duodenal ulcer" refer more specifically to the loci where these defects are found.

In general, two major interrelated factors contribute to the formation of ulcers in humans. First, any agent that blocks the normal protective mechanisms of the stomach against the erosive action of gastric juice can induce ulceration. Second, any factor that increases the rate of production of gastric juice also can produce peptic ulcers.

Under normal circumstances, the epithelial cells of the gastric mucosa are fused together at their apices, forming tight junctions, and these junctions provide a physical barrier against the back-diffusion of hydrogen ions into the mucosa from the lumen of the stomach, as there is a steep concentration gradient in the lumen to serosal direction. In addition, the surface epithelial cells of the gastric mucosa (in common with the rest of the digestive tract) exhibit an extremely high rate of desquamation and replacement so that any actual damage to the gastric mucosa is repaired quickly, within a matter of days or even hours. Finally, the mucous cells of the normal stomach secrete a protective coating which is composed of an alkaline mucus. This material normally coats the lining of the stomach with a layer about one mm thick which neutralizes the acid. Consequently, the alkaline mucus provides chemical as well as physical protection to the gastric mucosa against invasion and destruction by the gastric juice, providing the normal stomach with an effective physicochemical barrier against autodigestion and the development of ulcers.

Current evidence favors the view that ulcer formation in the stomach and related structures is caused principally by the erosive action of hydrochloric acid, in particular the free hydrogen ions, once the normal protective barrier is breached. There is no compelling evidence to indicate that pepsin itself is involved in the process of peptic ulcer formation.

Since cell membranes in general, and the membranes of the gastric epithelial cells in particular, consist in part of a bimolecular lipid layer, any detergentlike substance which could emulsify this layer in effect would destroy the protective mucus barrier of the mucosal layer. In the stomach this would permit invasion of the mucosa

by hydrochloric acid which in turn would lead to ulcer formation. The bile contains two effective types of detergent, bile salts as well as lysolecithin. Both of these substances in actual fact can produce gastric ulcers, and many gastroenterologists now are of the opinion that bile, regurgitated into the stomach, is responsible for ulcer production in many patients by virtue of the detergent action of this secretion upon the lipid components of the gastric epithelial cells.

The sequence of events in ulcer formation by this mechanism therefore may be envisaged somewhat as follows. First, reflux of bile causes a destruction of the protective coating on the mucosal surface. Second, hydrogen ions penetrate the damaged mucosa, and stimulate the nerves located within the stomach wall, thereby producing strong gastric contractions which apparently accounts for the spasms of excruciating pain that are felt by ulcer patients. Concomitantly, sodium ions diffuse rapidly into the lumen of the stomach. Third, the acid also stimulates the release as well as the rate of synthesis of histamine. The histamine which is released in turn stimulates increased acid secretion, and this compound also dilates the precapillary sphincters as well as the capillary bed of the mucosa. Thus blood flow and blood pressure within the vascular bed of the mucosa are elevated. Since histamine also augments the permeability of capillaries, fluid and proteins are extruded from these vessels leading to a local edema. Fourth, the further rapid diffusion of acid into the mucosa also is responsible for destruction of the capillary walls and the attendant severe hemorrhaging of gastric ulcers. Such bleeding is most pronounced in cases when high back diffusion of acid is combined with strong gastric contractions.

The second protective mechanism, namely the rapid regrowth of the epithelial surface, assumes considerable importance here. Thus ulcer formation by this process may be envisaged as a type of positive feedback mechanism, which if unchecked, can lead to most serious consequences for the patient.

In humans, the development of peptic ulcer in some patients appears to be related to such neural factors as severe anxiety and stress which in turn stimulate the secretion of excessive quantities of gastric juice. Thus, for example, there is a high correlation between the occupation of an individual and the incidence of ulcers. Since emotional factors can produce abnormally great stimulation of the dorsal motor nucleus of the vagus nerve via impulses transmitted from the cerebrum, the result is an excessive rate of flow of gastric juice, especially in the periods between meals and during sleep. This excessive secretion, especially of hydrochloric acid, can lead to ulcer formation by the mechanism described above, but most especially in individuals whose normal protective barrier is weakened by gastrointestinal irritation or other factors. For example, alcohol or drugs such as aspirin can damage the mucosal barrier severely in susceptible persons. Alcohol is one of the few substances that are absorbed readily through the gastric mucosa, as this compound is soluble both in lipid and in water. Although alcohol per se is insufficient to damage the normal barrier of the gastric mucosa, in combination with other substances, particularly aspirin, this compound can be exceedingly harmful.

Aspirin (acetylsalicylic acid) alone can destroy the mucosal barrier of the stomach, hence can cause ulceration and gastric hemorrhage. The severity of this reaction depends upon individual susceptibility, and it also may vary in the same individual at different times, for reasons unknown.

Salicylates (such as aspirin) ionize in accordance with the pH of the medium. Thus in a neutral solution, the carboxyl group ionizes, and the ionized compound is relatively insoluble in lipid (Fig. 13.41A). In an acid medium such as in the stomach, however, the carboxyl group is unionized, and in this state the compound is readily soluble in fat; hence it can penetrate the lipid portion of the cell membrane readily (Fig. 13.41B), and thus aspirin can diffuse into the mucosa rapidly (Fig. 13.41C). Once within the mucosal cells, the compound ionizes instantly, and thus it is unable to enter the gastric lumen once again so that it now diffuses into the interstitial fluid (Fig. 13.41D). Hence a steep concentration gradient develops for diffusion of the aspirin into the mucosa, and in turn this gradient depends upon the acidity of the stomach.

Aspirin destroys the normal mucosal barrier which prevents entrance of acid into the mucosa, and it also damages the mucosa by inducing hemorrhage. These effects are more pronounced in

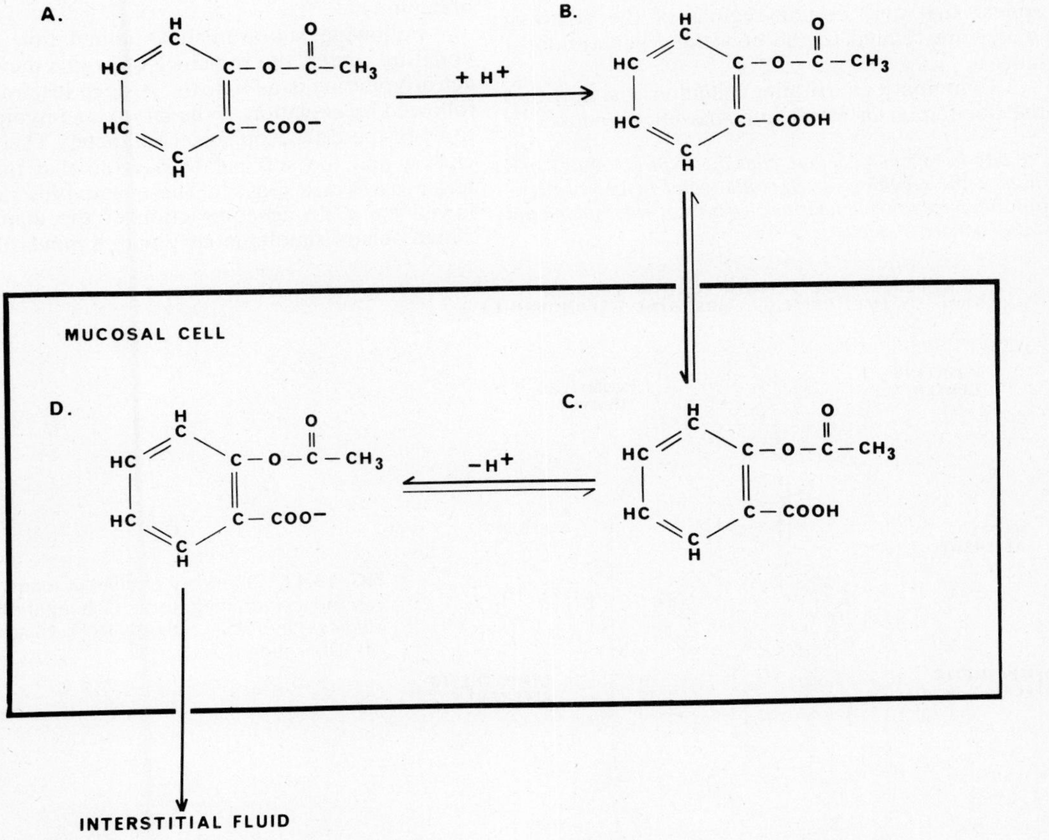

FIG. 13.41. The several steps in the absorption of aspirin (acetylsalicylic acid) by the gastric mucosal cells. See text for explanation.

the presence of alcohol, therefore a combination of these two compounds can exert a highly deleterious effect upon the normal protective mechanisms of the stomach.*

Clinically, peptic ulcers are seen most commonly in regions which are least protected by the mucus barrier, such as at the cardia, along the lesser curvature, especially at the pyloric junction, and in the duodenum.

Nausea and Vomiting

The sensation of nausea generally, but not always, is a prodrome (ie, premonitory symptom) of vomiting. This conscious sensation reflects the subconscious stimulation of a medullary area which is related to, or is part of, the vomiting center discussed below.

Nausea can be produced by irritation of the gastrointestinal tract proper, by impulses which arise in the brain and are related to motion sickness, or by impulses which arise in the cerebral cortex. Examples of the latter type of stimulus include psychic factors such as noxious odors or extremely disturbing environmental factors such as a severe accident. At times vomiting can occur without the prodrome of nausea. Thus it would appear that only certain regions of the vomiting center are related to the conscious perception of nausea.

Commonly an irritating stimulus arising within the duodenum or small intestine elicits nausea; a

It has been claimed that chewing aspirin tablets and diluting the resulting powder with water before swallowing minimizes the deleterious effects of this compound on the gastric mucosa.

forceful contraction of the small intestine follows concomitantly with relaxation of the stomach, so that the intestinal contents regurgitate into the stomach as a prelude to the act of vomiting.

Vomiting is a mechanism whereby the upper gastrointestinal tract empties its contents in response to severe irritation, hyperexcitability, or excessive distension. The stomach and duodenum are the regions of the gut which provide the most potent stimuli to this reflex, although vomiting can occur through appropriate stimulation of any region of the digestive tract.

Afferent impulses for vomiting are transmitted via sensory nerves which are located in both the vagus and sympathetic nerves to the vomiting center which is located in the medulla (Fig. 13.42). This center is found at about the level of the dorsal motor nucleus of the vagus. The vomiting act per se is induced by efferent impulses which are transmitted from the medullary vomiting center to the upper gastrointestinal tract through branches of certain cranial nerves, including the trigeminal (V), facial (VII), glossopharyngeal (IX), vagus (X), and hypoglossal (XII). Additional motor pathways involved in the vomiting reflex are by way of the spinal nerves to the abdominal muscles and diaphragm.

Following suprathreshold stimulation of the vomiting center, the sequence of events during the act of vomiting is as follows. A deep inspiration is followed by elevation of the larynx and hyoid bone to open the cricoesophageal sphincter. The glottis closes, and the soft palate rises so that the posterior nares are closed. These activities are followed by a strong contraction of the diaphragm which occurs simultaneously with a powerful con-

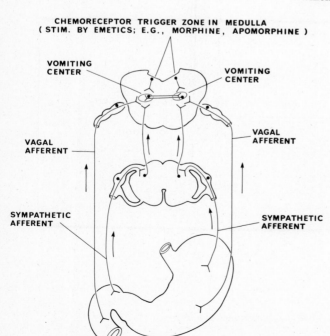

CHEMORECEPTOR TRIGGER ZONE IN MEDULLA
(STIM. BY EMETICS; E.G., MORPHINE, APOMORPHINE)

VOMITING CENTER

VOMITING CENTER

VAGAL AFFERENT

VAGAL AFFERENT

SYMPATHETIC AFFERENT

SYMPATHETIC AFFERENT

FIG. 13.42. Summary of afferent inputs which can induce vomiting. (After Ganong. *Review of Medical Physiology*, 6th ed, 1973, Lange Medical Publications.)

traction of the abdominal muscles. Consequently, the stomach is compressed strongly so that the intragastric pressure rises sharply and suddenly. Lastly, the gastroesophageal sphincter opens and the gastric contents are expelled vigorously through the esophagus.

In addition to impulses which arise within the gut proper, vomiting can also be elicited by stimulation of a chemoreceptor area (or zone) which is located bilaterally in the floor of the fourth ventricle of the brain and superior to the area postrema. Significantly, the direct stimulation of this area by certain drugs (eg, morphine, apomorphine, and certain digitalis compounds) can trigger this zone and initiate vomiting. If the chemoreceptor zone is destroyed, vomiting induced by this type of stimulus is inhibited, although appropriate stimuli applied to the gastrointestinal tract proper still can elicit vomiting.

Motion also can induce vomiting, because the mechanical stimulus of the receptors within the labyrinth of the ear causes impulses to be transmitted either directly, or indirectly via the vestibular nuclei, to the cerebellum. Then impulses pass via the uvula and nodule of the cerebellum to the chemoreceptor area, and vomiting is then triggered by excitation of this region of the brain stem.

The Dumping Syndrome

It is essential that hypertonic solutions are diluted to an isotonic state before they enter the jejunum. However, if a hypertonic solution suddenly enters the jejunum, a number of symptoms develop rapidly, including a sense of epigastric distension and pain, nausea, pallor, sweating, dizziness, and ultimately fainting. This collection of symptoms is called the dumping syndrome, and is caused by the sudden arrival of hypertonic fluid in the jejunum. This stimulus in turn produces a massive and generalized vasodilatation in the skeletal muscles throughout the body as well as a moderate vasodilatation in other regions, causing the blood pressure to fall dramatically; fainting ensues.

The dumping syndrome may appear in normal persons following, for example, the ingestion of exceedingly large amounts of carbohydrates, or following a normal meal in patients who have undergone partial gastrectomy.

Recovery from the dumping syndrome is spontaneous in about two hours following its onset.

Dysfunctions of the Small Intestine:

GALLSTONES. Probably the most commonly encountered clinical dysfunction of the small intestine is secondary to the presence of stones in the gallbladder or bile ducts, although the functional capacity of this segment of the gut usually is not impaired unless the flow of bile into the intestine is blocked by such concretions. If the bile flow is blocked, however, the absence of this secretion prevents normal fat digestion and absorption, so that the feces contain large amounts of fats (steatorrhea).

PANCREATIC FAILURE. Since food is digested principally in the upper one-third of the small intestine, it is rare when malnutrition develops due to failure of this process. In fact, a large segment of the small intestine can be removed surgically without the appearance of serious malnutrition in the patient. If, however, the normal secretion of pancreatic juice is inhibited for any reason then serious digestive and absorptive defects ensue because of the absence of such enzymes as trypsin, chymotrypsin, pancreatic lipase and amylase, and carboxypeptidase.

A deficiency of pancreatic secretion is encountered in such disease states as pancreatitis, following blockage of the ampulla of Vater by a gallstone, or following the surgical extirpation of the head of the pancreas because a malignancy has developed in this region.

In the absence of the appropriate pancreatic enzymes, or when the quantity of these enzymes is diminished appreciably, a large and significant proportion of the carbohydrates, proteins, and fats which are ingested are neither digested nor absorbed. Consequently, voluminous feces containing a high proportion of fat are eliminated.

An inflammation of the pancreas is termed pancreatitis, and this condition can be manifest in either an acute or a chronic form. Most commonly, acute pancreatitis develops following blockage of the ampulla of Vater by a gallstone. Thus bile as well as pancreatic juice is prevented from reaching the intestine. More importantly, however, the damming up of the pancreatic juice results in an accumulation of the precursors of the proteolytic enzymes, especially trypsinogen, within the ducts and tissues of the pancreas itself. Ultimately the trypsin inhibitor is overwhelmed by the rising concentration of trypsinogen, and a small quantity of active trypsin is generated. Once this has taken place, the trypsin in turn activates ever larger quantities of trypsinogen, chymotrypsinogen, and other enzymes in a snowball-rolling-downhill fashion so that an uncontrolled, rapidly progressive, activation of all of the proteolytic enzymes that are contained within the pancreas and its ducts takes place. Autolysis of the pancreas then rapidly ensues, and this autodigestion may be of sufficient magnitude that the ability of the pancreas to elaborate proteolytic and other enzymes may be partially or even completely destroyed, assuming that the patient survives the acute attack. If the active proteolytic enzymes erode through the surface of the pancreas and enter the abdominal cavity, then a chemical peritonitis develops, and death generally intervenes rapidly.

Following the acute phase the pancreatitis subsides, but the partial or total destruction of the pancreatic tissue results in a permanent malfunction of this organ, insofar as its exocrine secretion is concerned. Usually pancreatitis does not affect the islets of Langerhans markedly, so that the secretion of insulin is unimpaired despite an inadequate secretion of pancreatic juice into the intestine.

Chronic pancreatitis, in contrast to the acute form of the disease develops much more slowly. However, in general the causative mechanism is similar to that involved in acute pancreatitis.

Pancreatitis also may be caused by other conditions which are conducive to inflammation of the ampulla of Vater so that blockage or stenosis of the lumen results; for example, as by the chronic ingestion of excessive quantities of alcohol.

MALABSORPTION OF NUTRIENTS: SPRUE. Under certain circumstances, foods may be digested thoroughly, but they are not absorbed properly from the small intestine. The several individual conditions which cause a decreased absorption of nutrients are designated collectively by the term sprue. In young children, the term commonly applied to this condition is celiac disease.

The cause of idiopathic sprue in adults or of celiac disease in children is unknown. However, in many patients suffering from these conditions a marked improvement in the nutritional state can be obtained by eliminating foods containing the protein gliadin (a gluten) from the diet. Gliadin is found naturally, for example, in wheat and rye flour. Apparently gliadin is not digested completely by some individuals, and certain polypeptides (containing glutamine) which are produced by the partial degradation of the gliadin molecule actually are toxic to the intestinal mucosa. Consequently, the epithelial cells which are produced within the crypts of Lieberkühn die before migrating to the tips of the villi. Thus the latter structures eventually become attenuated or even may disappear completely, and thereby the total absorptive surface of the gut is reduced enormously.

Tropical sprue is a different type of malabsorptive process, and this condition frequently responds to therapy with antibacterial agents, although no specific causative organism or agent has been identified. Apparently an inflammation of the intestinal mucosa is present in tropical sprue, so that an impaired fat absorption is the principal defect in the early stages of this disease. Since the dietary fat appears in the feces largely in the form of soaps (ie, as salts of fatty acids), rather than as triglycerides, the defect in tropical sprue apparently lies in the absorptive mechanism rather than in the digestive process itself. At this stage, sprue often is termed idiopathic steatorrhea, ie, fatty feces of unknown cause.

Regardless of their ultimate causative mechanisms, the conglomerate of diseases which are known collectively as "sprue" can result in pronounced malnutrition in severe cases. Since the absorption of practically all of the essential nutrients, including carbohydrates, proteins, fats, and certain vitamins is defective, severe wasting of the body tissues develops, despite the ingestion of an adequate diet.

Certain particular defects in sprue may be listed. Decalcification of the bones (osteomalacia) develops due to a general calcium deficiency; blood clotting is prolonged because of insufficient absorption of vitamin K, and a pernicious (macrocytic) anemia may develop because cyanocobalamin (vitamin B_{12}) and folic acid are not absorbed properly. This constellation of signs is also called the malabsorption syndrome.

ILEITIS AND APPENDICITIS. If a portion of the ileum becomes inflamed, the condition is termed regional ileitis. Although the cause of the condition is unknown, it could result from an active infection, as is usually the situation in appendicitis. In either of these conditions, the patient experiences a sensation of midabdominal visceral pain and severe inhibition of normal gastrointestinal motility occurs. The symptoms of acute intestinal obstruction are elicited, especially spasms of abdominal pain and vomiting.

INTESTINAL OBSTRUCTION. Any segment of the digestive tract may become obstructed, and the abnormalities which then arise depend upon the region of the gut in which the blockage has occurred. Intestinal obstruction may result, for example, from tumors, inflammatory disease, hernias, intestinal spasm, paralysis of normal motility in a particular segment of the gut, or from ulceration or peritoneal adhesions which lead to a fibrotic constriction.

Pyloric obstruction may lead to a fibrotic constriction as the consequence of a peptic ulcer. Chronic vomiting of the gastric contents ensues, and this in turn lowers the overall nutritional state of the body. Furthermore, pyloric constriction with the attendant vomiting produces a net loss of hydrogen ions from the body, and varying grades of alkalosis develop (p. 809).

When the intestine develops a block distal to the stomach, the intestinal, pancreatic, and biliary secretions are regurgitated backward. The vomitus contains these secretions together with those of the stomach itself. In this type of obstruction, the patient loses significant quantities of water and electrolytes so that severe dehydration ensues. However, if the loss of acids and bases are about equal then little alteration in the blood acid–base balance is observed. On the other hand, if the block occurs near the distal end of the ileum, then more base than acid may be vomited; therefore an acidosis can develop (p. 808).

It is important to realize that the small intestine dilates markedly proximal to any obstruction, and large quantities of water, electrolytes, and even plasma proteins then are lost from the body in this region. The blood–lumen flux of water and electrolytes is normal, but the flux in the opposite direction is reduced by distension, and therefore a net accumulation of these substances within the intestinal lumen occurs. The intestinal wall also becomes edematous, and this factor combined with the distension reduces blood flow and may result in necrosis and rupture of the gut. Any appreciable loss of water, electrolytes, and plasma proteins results in a marked diminution of the plasma volume and hemoconcentration (up to 35 percent); cardiovascular collapse or shock frequently develops. Renal insufficiency or failure also may accompany the cardiovascular collapse. A positive feedback mechanism operates in a cyclic fashion to worsen the situation as follows. More water loss to the intestine in turn leads to further intestinal distension, and this promotes a still greater loss of water, electrolytes, and plasma protein, with an attendant progressive diminution of plasma volume. If left unchecked, this vicious cycle can have fatal consequences for the patient.

Lastly, when the obstruction occurs distally in the large intestine, feces accumulate progressively in the colon; severe constipation develops at first, but without excessive vomiting. Ultimately, when the large intestine becomes completely filled with fecal material, severe vomiting develops, and a backward reflux of the colonic contents through the ileocecal valve may take place up to the stomach in some patients so that the vomitus contains fecal material. Abnormal reverse peristalsis may play a role in this process, although such a mechanism has not been proven. Unless corrected a prolonged obstruction of the colon can result in rupture of the organ. Dehydration and cardiovascular collapse also develop as consequences of the intense chronic vomiting. Intestinal obstruction also results in a marked loss of appetite (anorexia), as well as severe abdominal pain (intestinal colic).

Dysfunctions of the Large Intestine

DIARRHEA. Excessively rapid passage of the intestinal contents through the colon is termed diarrhea. However, as noted earlier (p. 843), the principal source of water and electrolytes in the stools in diarrhea is the small intestine rather than the colon.

Diarrhea is caused most frequently by two conditions: first, excessive stimulation of colonic motility via the parasympathetic nervous system (neurogenic diarrhea) and second, by an infection of the gastrointestinal tract (gastroenteritis). The osmotic diarrhea which is produced by chemical irritants or cathartics (such as magnesium sulfate) was discussed on page 870.

Neurogenic Diarrhea. Excessive nervous tension or periods of psychic stress (eg, such as may develop when facing an important examination) can induce hypermotility of the colon combined with an abnormally great secretion of mucus in this region of the gut. These effects result directly from overstimulation of the colon by the parasympathetic nervous system, and may lead to diarrhea. In addition, chyme, water, and electrolytes also may be propelled more rapidly through the small intestine during such periods, and for the same underlying reason; the total loss of water and electrolytes may be considerable and the diarrhea is exacerbated by the contribution from the small intestine. Alleviation of the causative factor responsible for the tension will cause reversion to a normal colonic action; however drugs such as paregoric sometimes are helpful in the acute stages of diarrhea.

In the condition known as ulcerative colitis large areas of the colonic mucosa develop ulcers, and this condition, in common with that present in neurogenic diarrhea, often is associated with states of psychic tension. However, the actual cause of the ulceration itself is unknown, although the colonic mucosa exhibits a chronic inflammation. The motility of the large intestine is increased so markedly in patients with ulcerative colitis that mass movements in this region may occur almost continuously, rather than for a few minutes per day as in normal individuals. At the same time, colonic secretion is enhanced, so that episodes of diarrhea occur frequently in patients suffering from ulcerative colitis.

Ulcerative colitis, if sufficiently advanced, may require radical surgical intervention in order to effect a cure. However, in many instances a simple removal of the factors which produce the nervous tension alone may be sufficient to ameliorate the condition.

A chronic diarrhea ranging from mild to severe, depending upon the patient and the severity of the disease, also may result from bacterial infections or infestation with parasites, such as *Entameba histolytica*. The dysentery resulting from amebiasis also may be accompanied by ulceration of the colonic mucosa, as the ameba responsible for producing the disease secretes a proteolytic enzyme which hydrolyzes the colonic mucosa.

CONSTIPATION. Constipation, the physiologic opposite of diarrhea, signifies the slow passage of feces through the large intestine. In fact, chronic constipation aptly has been termed "a peculiar disorder of the bowel and the mind." From a physiologic standpoint, however, retarded movement of fecal matter permits ample time for water absorption and the fecal matter which accumulates in the descending colon becomes hard and dry.

Early conditioning undoubtedly plays a significant role in the development of normal bowel habits, but if the defecation reflexes become attenuated after long periods of time through conscious inhibition, then constipation may develop. On the other hand, gastrocolic and duodenocolic reflexes give rise to a mass movement in the colon as well as the urge to defecate, and usually such colonic movements occur in the morning, generally following breakfast. If this stimulus to defecation is not ignored, then regularity of bowel habits becomes established and development of constipation is rendered unlikely. In this regard, the routine inclusion of an adequate quantity of undigestible cellulose in the diet ("roughage") to give bulk to the feces is of considerable importance to the maintenance of normal bowel motility.

From a clinical standpoint, constipation may result from contractile spasm of the sigmoid colon or from a loss of tone of this organ (atony). Even under normal circumstances the contractile activity is weak in the large intestine; therefore factors which inhibit even slightly the normal propulsive mechanism in the sigmoid or descending colon result in constipation.

In a clinical state known as megacolon (or Hirshsprung's disease), constipation is severe enough so that evacuation of the bowels takes place only once a week, or even less frequently. The enormous quantity of fecal matter which accumulates in the large intestine causes marked dilatation of this organ, hence the name applied to this condition.

Megacolon is caused most frequently by a congenital absence of the myenteric plexus in a portion of the sigmoid. Consequently, this defective region is relatively nonmotile, and block occurs in the defective segment. Thus feces accumulate proximal to the affected region giving rise to distension and megacolon, whereas distal to the block the sigmoid tends to shrink.

In common with other types of intestinal block discussed above, the neurogenic block leading to megacolon may be partial rather than total; all degrees of severity of this condition may be encountered clinically in this type of intestinal dysfunction.

There is no concrete evidence that toxic products are produced or that they accumulate within the colon during mild constipation, and which give rise to the symptoms of this disorder through their absorption into the bloodstream. The phrase "intestinal intoxication" is a gross misnomer, and the condition implied by this term is a complete fallacy insofar as any factual basis is concerned.

CHAPTER 14

Intermediary Metabolism, Energy Balance, Thermoregulation

In the broadest sense, the term *metabolism* can be defined as the sum total of all of the individual, highly integrated, sequential chemical reactions that are associated with the various life processes. The phrase *intermediary metabolism* denotes specific chemical reactions as well as the orderly and sequential patterns followed by these reactions taking place continuously within cells, tissues, and the organism as a whole. Since living cells constitute an open system from the standpoint of thermodynamics, a constant supply of energy is required in order to maintain the structural and functional integrity of these units of living matter. Many of the chemical reactions that take place in living systems, particularly those concerned with the degradation of various substances, yield energy, whereas other biochemical reactions, especially those involved in the synthesis of larger from smaller molecules, require energy in order to take place.

The chemical reactions and processes whereby the human body obtains and utilizes energy from the degradation of various energy-producing compounds (nutrients) are of particular relevance to the study of physiology, and this aspect of intermediary metabolism will receive major emphasis in this chapter.

The potential chemical energy of the basic foodstuffs (ie, carbohydrates, fats, and proteins) is converted metabolically into forms of energy that can be utilized directly by the organism. Ultimately a large fraction of this energy is dissipated principally as heat, and this aspect of energy metabolism will be considered in the second major section of this chapter. The homeostatic mechanisms whereby the human body normally regulates its temperature within narrow limits (ie, thermoregulation) will be discussed in the third section.

The overall process of tissue formation or synthesis in the body often is referred to as anabolism, and that of tissue breakdown is called catabolism. In actual fact, these extremes represent opposing, but reversible, chemical reactions whereby small molecules are combined into larger ones during anabolism, and larger molecules are degraded into smaller ones during catabolism. The synthesis (anabolism) and degradation (catabolism) of protoplasm deals largely with protein metabolism, because proteins generally form the major constituent of living matter, although water, nucleic acids, various lipids and carbohydrates as well as certain inorganic constituents are also involved. Hence there is a constant exchange (or turnover) of organic as well as inorganic substances in the body.

From a general metabolic standpoint, when the overall anabolic processes in the body exceed the catabolic processes, growth of the individual takes place; therefore growth occurs throughout the normal period of immaturity. During the period of maturity the anabolic and catabolic processes are in balance so that no major alteration in tissue mass normally occurs. During the process of aging, the tissue mass decreases as the catabolic rate gradually comes to exceed the anabolic rate. The latter phenomenon may also be observed in an exaggerated form during acute starvation when the nutrient molecules required for synthetic reactions are unavailable to the metabolic machinery of the body, as well as during certain pathologic states such as in the malabsorption syndrome, or in diabetes.

Of major importance to the study of physiology are the chemical reactions that provide energy in a usable form for the performance of the multitude of cellular and bodily functions. In brief, the specific metabolic reactions that release energy for the performance of mechanical, electric, chemical, thermal, and osmotic work are essential to all aspects of life itself, and will be stressed in the following presentation.

In the broadest sense of the word, the study of metabolism involves not only the chemistry of tissue synthesis and degradation, but also the chemistry of the formation of various specific compounds essential to the operation and regulation of the metabolic machinery of the cells, as well as those reactions designed primarily for the generation of energy. It must be emphasized that the general patterns of the chemical reactions involved in these various primary biochemical processes are not mutually exclusive; rather they generally overlap and are integrated. For example, compounds produced during primary energy-producing reactions may also be utilized in synthetic reactions. Further specific examples of this important fact will be encountered frequently in the material that follows.

The practical consequences of various metabolic derangements are of great importance to the practicing physician, and cannot be overemphasized. Therefore, a clear understanding of the normal metabolic pathways involved in energy production and utilization within the body, as well as certain other aspects of metabolism, is essential. Finally, a large part of the science of pharmacology as well as the direct application of this science to clinical medicine involves the use of vitamins, hormones, and drugs that affect the metabolic processes either in human tissues or the metabolism of various organisms that may have invaded the body.

The reader should be thoroughly familiar with the general types of organic compounds and their chemical reactions, which are summarized in Chapter 1, before undertaking a study of the following material.

INTERMEDIARY METABOLISM

It is remarkable that metabolic reactions take place in the body quite rapidly under mild conditions of temperature and pH, and that very few unnecessary side reactions occur. Furthermore, many of the chemical reactions occurring readily in living systems are difficult, if not impossible, to duplicate under laboratory conditions, even under drastic extremes of temperature, pressure, acidity, and concentrations of reactants.

The reason underlying this seeming incongruity rests upon the fact that most metabolic reactions are catalyzed in vivo by specific enzymes. The study of intermediary metabolism is inseparable from a discussion of the enzymes involved in specific reactions and of the properties and mode of operation of the enzymes in general. In fact, it is a truism that the study of enzymes and the reactions they catalyze is the study of life itself. The topic of intermediary metabolism will be introduced by a brief survey of certain important physicochemical properties of enzymes. Following this survey, the topics of bioenergetics and oxidation–reduction reactions will be reviewed. This material will serve as a background for a presentation of the basic features of intermediary metabolism, with emphasis on those biochemical processes whereby the cells of the body derive the energy necessary to carry out their manifold functions. Viewed collectively, the enzymes form the largest and most complex group of proteins known.

Enzymes

DEFINITIONS AND GENERAL PROPERTIES. Enzymes are proteins that markedly accelerate the rate of chemical reactions in living systems. They may be considered as organic (or biologic) catalysts, in whose absence most of the metabolic reactions in the body would take place far too slowly for compatibility with life. Examples of a number of particular enzymes have been given in context throughout this book, eg, the digestive enzymes were discussed in the previous chapter.

Enzymes may exhibit considerable specificity insofar as the biochemical reactions they accelerate are concerned. This is in sharp contrast to the effects of inorganic catalysts, which generally function in a nonspecific manner. The specificity of enzymes will be considered further in the next section.

Because the metabolic processes of the body involve an enormous number of individual chemical reactions, a correspondingly large number of enzymes is also involved. In this regard, it is interesting to note that living cells, whether of plant or animal origin, share many common biochemical pathways; and underlying these orderly, sequential patterns of chemical reactions, the various specific enzymes and cofactors responsible for catalyzing these processes are also quite similar.

The enzymes characterized thus far may be divided into two broad categories. The first are *simple protein enzymes,* which consist merely of specific proteins, as the name implies. In this instance the enzymatic activity depends upon the structure of the protein molecule alone. The second are *complex enzymes,* which contain two parts. In addition to a specific protein moiety known as the *apoenzyme,* a nonprotein (ie, small molecule) organic cofactor known as a *coenzyme** must be bound to the protein (apoenzyme) in order for the enzyme to exhibit catalytic properties (Table 14.1). Hence, enzymes of this type may be considered to be conjugated proteins. The biologically active combination of apoenzyme with the cofactor or coenzyme is called a *holoenzyme.* Dissociation of these two essential components re-

---

The term prosthetic group *sometimes was used to denote a coenzyme attached strongly to an apoenzyme by covalent bonds; however this term is now obsolete.*

Table 14.1 COFACTORS REQUIRED BY CERTAIN ENZYMES[a]

COENZYME[b]	SUBSTANCE TRANSFERRED
Nicotinamide adenine dinucleotide (NAD)	Hydrogen atoms (electrons)
Nicotinamide adenine dinucleotide phosphate (NADP)	Hydrogen atoms (electrons)
Flavin mononucleotide	Hydrogen atoms (electrons)
Flavin adenine dinucleotide (FAD)	Hydrogen atoms (electrons)
Coenzyme Q	Hydrogen atoms (electrons)
Thiamin pyrophosphate	Aldehydes
Coenzyme A	Acyl groups
Cobamide coenzymes	Alkyl groups
Pyridoxal phosphate	Amino groups

METALLIC COFACTORS[c]	ENZYME(S)
Zn^{++}	Alcohol dehydrogenase, carbonic anhydrase, carboxypeptidase
Mg^{++}	Phosphotransferases, phosophohydrolases
Fe^{++} or Fe^{+++}	Peroxidase, catalase, cytochromes, ferredoxin
Mn^{++}	Arginase, phospho-transferases
Cu^{++}	Cytochrome oxidase, tyrosinase
K^+	Pyruvate phosphokinase (also requires Mg^{++})
Na^+	Adenosine triphosphatase (ATPase) of plasma membrane (also requires K^+ and Mg^{++})

[a]*Data from Lehninger.* Biochemistry, *1970, Worth Publishers.*

[b]*Coenzymes generally function as intermediary electron carriers, or else they carry the specific atoms or functional groups that are transferred in the net reaction.*

[c]*Generally, metallic cofactors may function as a bridging group; thus they serve to bind enzyme and substrate together through formation of a coordination complex. Alternatively, the coenzyme may function as the catalytic group per se.*

sults in complete loss of enzymatic activity, and neither component alone comprises an enzyme.

Some enzymes require merely a metallic ion as a coenzyme (or cofactor) rather than a complex organic molecule. Nevertheless, dissociation of the apoenzyme–coenzyme complex still results in complete loss of biologic activity. Enzymes that are coenzyme-dependent bind their cofactors with varying affinities. In most instances, the cofactor may be removed simply by dialysis, but in some cases the coenzyme is bound firmly to the apoenzyme by covalent bonds.

In general, the compound or metabolite that is acted upon by any enzyme is called the substrate.

CLASSIFICATION OF ENZYMES. In order to achieve a rational classification of enzymes based upon the types of reaction that are catalyzed as well as the mechanisms of the reactions themselves, the International Union of Biochemistry has adopted the system of nomenclature outlined in Table 14.2. Note that all known enzymes are grouped under six major headings, and under each of these headings are grouped a variable number of subheadings, each of which indicates the types of reaction catalyzed by specific enzymes.

In accordance with this system of nomenclature, the name of an enzyme consists of two parts: first, the name of the substrate or substrates acted upon by that enzyme, and second, the type of reaction catalyzed, ending in the suffix "-ase." For example, under heading 1.1 in the table, the classic name of the group of enzymes responsible for catalyzing the oxidation of alcohol was known by the general designation of dehydrogenase. Under the present system of nomenclature, this name has been succeeded by the designation alcohol:NAD oxidoreductase, which in turn catalyses the following general type of reaction:

Alcohol + $NAD^+ \rightarrow$ aldehyde (or ketone) + NADH + H^+.

In many instances, however, the classic names of enzymes have been retained. Thus under heading 3.4 in Table 14.2 several of the hydrolases which are responsible for catalyzing the hydrolysis of peptide bonds are still referred to by their clas-

Table 14.2[a] CLASSIFICATION OF ENZYMES[b]

OXIDOREDUCTASES: Catalyze the oxidation–reduction reactions between two substrates, S_1 and S_2. S_1 reduced + S_2 oxidized = S_1 oxidized + S_2 reduced. This class includes enzymes previously known as dehydrogenases or oxidases.

Groups acted upon:

1.1	CH—OH (alcohol)
1.2	C=O (carbonyl)
1.3	CH=CH (ethylene)
1.4	CH—NH$_2$ (amino)
1.5	CH—NH (imino)
1.6	NADH, NADPH

TRANSFERASES: Catalyze the transfer of functional groups, G (other than hydrogen), between two substrates, S_1 and S_2. Thus S_1—G + S_2 = S_2—G + S_1

Groups acted upon:

2.1	One-carbon groups
2.2	Aldehydic or ketonic groups
2.3	Acyl groups
2.4	Glycosyl groups
2.7	Phosphate groups
2.8	Sulfur-containing groups

HYDROLASES: Catalyze the hydrolysis of various bonds in a number of compounds, including the following types of compound or bonds:

3.1	Esters
3.2	Glycosidic bonds
3.4	Peptide bonds
3.5	Other C—N bonds
3.6	Acid anhydrides

LYSASES: Catalyze the removal of groups from substrates by reactions other than hydrolysis so that double bonds result, according to the general reaction:

$$\begin{array}{cc} X & Y \\ | & | \\ C{-}C \end{array} \rightarrow X{-}Y + C{=}C$$

Bonds formed by lysases include:

4.1	C=C
4.2	C=O
4.3	C=N
4.4	C=S
4.5	C—halide (eg, Cl)

ISOMERASES: Included here are all enzymes that catalyze the interconversion (isomerization) of optical, geometric, or positional isomers:

5.1	Racemases and epimerases
5.2	*Cis-trans* isomerases
5.3	Enzymes that catalyze the interconversion of aldoses and ketoses

LIGASES: Enzymes that catalyze the formation of bonds between two compounds coupled to the cleavage (scission) of a pyrophosphate bond in ATP or a similar compound. Bonds formed by ligases include:

6.1	C—O
6.2	C—S
6.3	C—N
6.4	C—C

[a]*Data from Harper*. Review of Physiological Chemistry, *13th ed, 1971, Lange Medical Publications; Lehninger*. Biochemistry, *Worth Publishers. For an alternative system of enzyme classification, see West, Todd, Mason, and VanBruggen*. Textbook of Biochemistry, *4th ed, 1966, Macmillan*.

[b]*This classification, based upon reaction types and mechanisms, was prepared by the International Enzyme Commission of the International Union of Biochemistry*.

sic names; eg, rennin, cathepsin, pepsin, and chymotrypsin.

Because of the current flux in the state of enzyme nomenclature, the specific enzymes referred to throughout this chapter will be designated by the name in most common usage. Where deemed necessary in the interest of clarity, the alternative name for the enzyme will be included in parentheses.

DISTRIBUTION OF ENZYMES. As discussed in Chapter 3, the electron microscope has elucidated the morphology of a number of cellular organelles with a degree of clarity impossible with the light microscope. This visual technique, coupled with biochemical studies performed upon subcellular particulates and other cellular components that have been isolated by differential (or fractional) ultracentrifugation, has yielded many profound new insights into the highly complex structural and functional organization of enzymes and enzyme systems within the cell. Numerous intracellular metabolic activities have been localized and associated with specific subcellular particulates and actual organelles. The earlier concept that regarded a cell merely as a simple "bag of enzymes" has had to be discarded in favor of a concept that recognizes the high degree of molecular as well as morphologic organization present within the cell. According to this modern view, the enzymes, substrates, and cofactors within the cell are normally present for the most part in a spatially arranged and compartmentalized fashion, and such a "molecular anatomy" is in strict keeping with the known facts regarding the orderly, stepwise progression of metabolic events within the cell. Therefore, it is important that the student of physiology have an overall understanding of the intracellular distribution and localization of enzymes.

In addition to intracellular enzymes, the human body also requires the presence of a number of extracellular enzymes in order to function normally. Two obvious examples of the activity of extracellular enzymes may be cited here. Both the process of blood clotting and the normal

Table 14.3 IMPORTANT EXTRACELLULAR ENZYMES

NAME OF ENZYME[a]

1. Amylase, salivary (ptyalin)
2. Pepsins
3. Gastricsin
4. Trypsin
5. Chymotrypsin
6. Amylase, pancreatic
7. Lipase, pancreatic
8. Enterokinase
9. Carboxypeptidases
10. Plasmin
11. Alkaline phosphatase
12. Acid phosphatase
13. Thrombokinase
14. Renin
15. Angiotensinase

[a]*Note that the largest number of extracellular enzymes is associated with the gastrointestinal tract and the process of digestion (Chapter 13).*

digestion of foods are catalyzed by specific enzymes as discussed in Chapters 9 and 13, respectively. Many other examples of extracellular enzymes are to be found throughout this book. The major extracellular enzymes are summarized in Table 14.3.

Lastly, and of great significance to the practicing physician, it has become evident during recent years that in many pathologic states certain enzymes normally found within cells may appear in abnormal quantities in the blood plasma or serum. Alternatively the concentration or activity of certain extracellular enzymes may be elevated or depressed, with concomitant derangement of normal function. Therefore, the normal and abnormal distribution patterns of certain intracellular and extracellular enzymes are of extreme importance from a clinical standpoint.

The remainder of this section will be devoted to a summary of the normal intracellular localization of enzymes insofar as is known currently, and this discussion will be cross-indexed to the pertinent morphologic features of the cell that were discussed in more detail in Chapter 3.

Intracellular Localization of Enzymes

Plasma Membrane. The plasma membrane provides the interface between the cell interior and its immediate environment, and this structure carries out a number of vital enzyme-mediated functions. Most importantly, the selective permeability of the plasma membrane controls the internal chemical environment of the cell with extreme precision, a regulatory function critical to normal biochemical function (p. 85).

Various enzyme systems are required for active transport of substances into and out of the cell across the plasma membrane, and these enzymes are localized within the protein layers of this membrane (Fig. 3.22, p. 75). Included in this category

is the specific adenosine triphosphatase enzyme system that appears to underlie the active transport of such ions as sodium and potassium (p. 92). As stressed earlier, such transport requires the continuous expenditure of energy which in turn is derived from the breakdown of ATP via enzymatic catalysis. The transport of other low molecular weight substances (eg, sugars), may also be mediated by active transport systems, or take place by passive transport alone.

The various transport systems of the cell membrane, hence regulation of the internal metabolic processes of the cell itself, are regulated in turn by other chemical substances, such as hormones (Chap. 15). Presumably hormones act through their effects on specific or general enzymatic processes, although these regulatory influences have not been defined with certainty.

Cytoplasmic Matrix. All of the enzymes concerned with the metabolic breakdown of simple sugars to pyruvic acid, a process known as glycolysis, are found in the soluble (ie, nonparticulate) fraction of the cytoplasm; hence these enzymes presumably are associated with the cytoplasmic matrix (p. 74). The pathway of glycolysis and the individual enzymes involved in this process are shown in Figure 14.25.

Mitochondria. Among the largest organelles to be found in the cytoplasm are the mitochondria (Fig. 3.25, p. 78). These structures have been aptly named the "power plants" of the cell, as a portion of the energy released from the oxidation of nutrients is trapped by the formation of high-energy chemical bonds of adenosine triphosphate within these structures, as will be described later in this chapter.

As depicted in Figure 3.25 and Figure 14.1, the mitochondrion is bounded by an outer membrane. A second inner membrane is folded into cristae, so that the interior of the mitochondrion is divided physically into minute compartments. Each of the two unit membranes of the mitochondrion in turn consists of alternate layers of lipid and protein molecules, as depicted in Figure 3.21 (p. 73). The complex of enzymes concerned with cellular respiration is believed to be localized in a precise sequence on the protein layer, whereas the enzymes concerned with oxidative phosphorylation are thought to be similarly organized within or upon the lipid layers of the mitochrondrial membrane. The enzymes that catalyze biochemical transformations of the citric acid cycle are found within the fluid matrix of the mitochondrion.

Lysosomes. The lysosomes contain digestive enzymes (hydrolases) which break down various large molecules such as proteins fats, and nucleic acids into smaller units — molecules that in turn can be metabolized readily by the several enzyme systems within the mitochondria (Fig. 3.27, p. 82).

Although lysosomes do not appear to have a definite internal structure, they are surrounded by

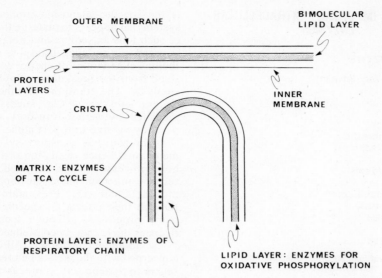

FIG. 14.1. Highly schematic diagram illustrating a portion of a mitochondrion. See also Figures 3.24, 3.25, and 14.20. (Data from Harper. *Review of Physiological Chemistry*, 13th ed, 1971, Lange Medical Publications.)

a lipoprotein unit membrane that normally isolates their hydrolytic enzymes from the cytoplasm of the cell. Therefore, the enzymes within the lysosome normally are unable to act upon cytoplasmic substrates. If, however, the lysosomal unit membrane ruptures or becomes damaged, then these enzymes are released and lysis (or digestion of the cell) rapidly follows.

Endoplasmic Reticulum. The anastomosing network of cytoplasmic canaliculi that comprise the endoplasmic reticulum may contain numerous granules or ribosomes (Fig. 3.26, p. 80). The ribosomes in turn are rich in ribonucleic acid (RNA), and thus form the metabolic loci for protein synthesis within the cell. A granular endoplasmic reticulum is most highly developed in those cells most actively engaged in the de novo synthesis and secretion of protein. For example, hepatic and pancreatic cells both exhibit a well-developed endoplasmic reticulum.

Extracellular Enzymes. For convenient reference, the major extracellular enzymes normally found in the body fluids have been summarized in Table 14.3.

ENZYME SPECIFICITY. Enzymes generally catalyse only one, or at most, only a few chemical reactions. This specificity of action is one of the most important single properties of these organic catalysts, as a host of metabolic processes within the cell may thus be regulated with exquisite precision, depending upon minute changes in the catalytic efficiency of particular enzymes. Such enzymatic control in turn is essential to the normal functioning of living organisms at all levels of their organization.

Despite their specificity, the majority of enzymes are also capable of catalyzing the same general type of reaction (eg, oxidation–reduction, or phosphate transfer), when two or more substrates are closely related from a structural standpoint. In fact, reactions with alternate substrates take place quite often in living systems, provided that the alternate compounds are present in sufficiently high concentration, and that the affinity between the substrate and a given enzyme is sufficiently high.

In general terms enzymes can exhibit optical specificity and group specificity.

Optical Specificity. On the whole, enzymes exhibit absolute optical specificity, insofar as at least a part of the substrate molecule is concerned. The only important exception to this general rule is the group of epimerases (or racemases, Table 14.2) that interconvert optical isomers. For example, the enzymes involved in the glycolytic pathway (Fig. 14.25), as well as the direct oxidative pathway (Fig. 14.29), for carbohydrate metabolism only catalyze the interconversion of L-phosphorylated sugars, and do not affect the D-isomers of the same compounds. Likewise, an overwhelming majority of the enzymes found in mammalian tissue act exclusively on L-isomers of the amino acids. A notable exception to this generalization is D-amino acid oxidase, which is found in mammalian liver and kidney.

In many instances, substrates appear to form three bonds with an enzyme molecule, as illustrated in Figure 14.2, and such attachment of an enzyme to its substrate can induce asymmetry in an otherwise symmetric molecule. Actually, since enzyme molecules have a three-dimensional structure, the representation of this attachment depicted in Figure 14.3 probably reflects the actual situation more accurately.

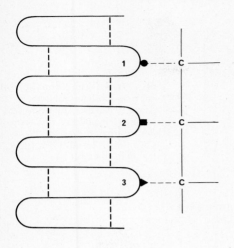

ENZYME   **SUBSTRATE**

FIG. 14.2. Three active sites (1, 2, 3) in an enzyme–substrate complex.

Group Specificity. A second major characteristic of enzymes is their specific affinity for certain chemical groups within substrate molecules. For example, trypsin acts specifically upon peptide bonds, alcohol dehydrogenase acts only upon hydroxyl (—OH) groups, and esterases act solely on ester linkages. A large number of chemically related substrates may be altered by a given enzyme, provided that each substrate contains the appropriate chemical groups that are acted upon by that particular enzyme.

Other enzymes may exhibit a more narrowly defined type of group specificity than that described above. For example, chymotrypsin selectively hydrolyzes those peptide bonds wherein the carboxyl group involved is formed from certain aromatic amino acids, that is, tryptophan, tyrosine, or phenylalanine.

Other examples of group specificity will be found in the discussions of individual enzymatic process.

THE ACTIVE SITE AND THE MECHANISM OF ENZYME ACTION. It is clear from the brief resumé of enzyme specificity presented above that a specific region of the enzyme molecule functions as an active or catalytic site where the substrate is acted upon. Originally, the concept of a "template" or "lock and key" model was developed to visualize this process, as depicted in Figure 14.4. However, this hypothetical model assumes that the active site is rigid, a situation which simply does not exist, so that the template concept has largely been superceded by the induced-fit model of the enzyme substrate complex; this view is depicted in Figure 14.5. In this conception, the active site is flexible rather than rigid; hence the substrate causes an actual alteration in the conformation of the enzyme.* Rotation or unfolding of a protein about its bonds constitutes a change in conformation.

In accordance with the induced fit model, conformational changes in the enzyme protein induced by the substrate align certain amino acid groups spatially so that substrate binding and/or catalysis may take place.

The experimental evidence upon which the substrate-induced conformational change hypothesis rests is quite sound. However, it is not particularly clear which amino acid residues may actually constitute the active site (or sites) on the enzyme molecule, even though the complete primary structure of the protein (ie, the amino acid sequence) may be known in complete detail. It is clear, nonetheless, that even though some individual amino acid residues may be quite remote from one another on the primary enzyme molecule, they may approach each other quite closely during the substrate-induced conformation change.

Regardless of the actual molecular nature of the active site and the way in which it interacts with the substrate, the concept of an enzyme–

The term conformation as used here means a change in the mean (or average) positions of the atomic nuclei with respect to each other, not changes in covalent (or other) bonds.

FIG. 14.3. Three active sites (1, 2, 3) in an enzyme–substrate complex. In contrast to Figure 14.2, the active sites in this figure are represented in three rather than two dimensions.

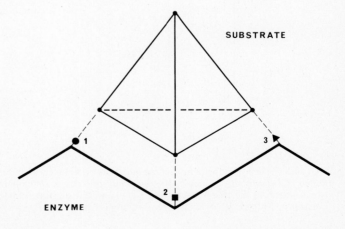

SUBSTRATE

ENZYME

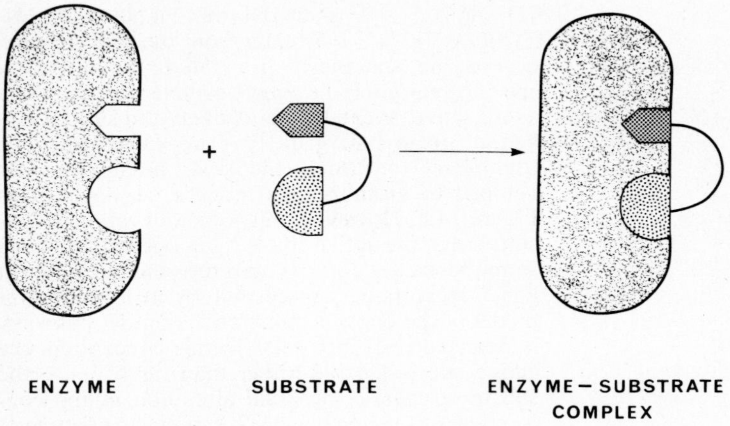

FIG. 14.4. The formation of an enzyme–substrate complex in accordance with the "lock-and-key," or template, hypothesis of Fisher. (Data from Harper. *Review of Physiological Chemistry,* 13th ed, 1971, Lange Medical Publications.)

ENZYME   SUBSTRATE   ENZYME – SUBSTRATE
COMPLEX

substrate complex is basic to an understanding of the prevalent theory of enzyme action advanced by Michaelis and Menten. This theory postulates the formation of an enzyme–substrate complex as the first step in the action of any enzyme. Next, this enzyme–substrate complex liberates the enzyme and the catalytic products in the second step. These statements are summarized in the following relationships:

Step 1: Enzyme + Substrate $\rightleftharpoons$ Enzyme–Substrate Complex
Step 2: Enzyme–Substrate Complex $\rightleftharpoons$ Enzyme + Products

Experimental support for this fundamental two-phase hypothesis of enzyme action is quite strong; however, it is now clear that in many instances several more intermediate stages are involved. The mechanism of enzyme action may be indicated somewhat more accurately as a series of reactions, ie, $A \rightleftharpoons B \rightleftharpoons C \rightleftharpoons D \rightleftharpoons E$.

MAJOR FACTORS THAT MODIFY THE RATE OF ENZYME ACTIVITY. The efficiency of an enzyme can be expressed in terms of the number of moles of substrate transformed per mole of purified enzyme per unit time. It has been found that enzymes may catalyze the transformation of up to 10,000 to 1,000,000 moles of substrate per minute per mole of purified enzyme! Also, depending upon the specific circumstances, enzyme-

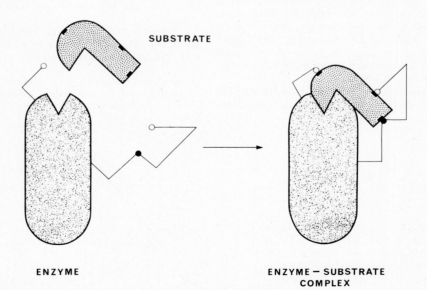

SUBSTRATE

ENZYME   ENZYME – SUBSTRATE
COMPLEX

FIG. 14.5. The formation of an enzyme–substrate complex by an induced fit of the conformation of the enzyme molecule to that of the substrate molecule. The open and solid circles on the enzyme molecule represent active loci on the protein which attach to the binding sites (black bars) on the substrate molecule. (Data from Harper. *Review of Physiological Chemistry,* 13th ed, 1971, Lange Medical Publications.)

catalyzed reactions in living tissues are quite frequently reversible.

Another important general feature of catalyzed reactions is the fact that the quantity of catalyst has no stoichiometric relationship with the quantity of material that is transformed.

The remainder of this section will be concerned with a brief summary of the nine principal factors that affect the catalytic activity of enzymes.

Substrate Concentration.   Only under certain conditions does the velocity of an enzyme-catalyzed reaction parallel the substrate concentration. Under completely controlled experimental conditions, and when the substrate concentration is low, such a correlation generally can be shown. This relationship is shown in Figure 14.6. In addition, if the substrate concentration is too high then the rate of the reaction may actually be depressed as shown to the right of the arrow in Figure 14.6.

Note also in this figure that the rate of enzyme activity does not increase beyond a certain substrate concentration; ie, the curve reaches a plateau, and in keeping with the Michaelis–Menten concept, the enzyme becomes completely saturated with substrate. Presumably this plateau effect is caused by the substrate molecule combining with the enzyme, and after a given time interval the reaction products are liberated. Following a second brief time interval another substrate molecule combines with the enzyme, and the process is repeated. Thus the total reaction time for a single substrate molecule is the sum of the times for its attachment to, and liberation of the product from, the enzyme molecule. If the substrate is present in a sufficiently high concentration, a minimum time may be required for attachment to the enzyme, and thus at or above this concentration the enzyme is acting catalytically at its maximum velocity. At some optimal substrate concentration the rate of enzyme activity reaches a maximum.

The practical significance of experimentally determined facts such as those outlined above lie in their application in the clinical laboratory as well as in the research laboratory for the determination of enzyme activities. If the substrate concentration employed in a given enzyme assay is sufficiently high, the variable of substrate concentration is eliminated and the calculations involved in determining the enzyme concentration are much simplified.

Enzyme Concentration.   If the substrate concentration is held at a constant level, then the velocity of an enzyme-catalyzed reaction is directly proportional to the concentration of the enzyme. This fact is shown in Figure 14.7.

Effect of Reaction Products.   The linear relationship between the concentration of a purified en-

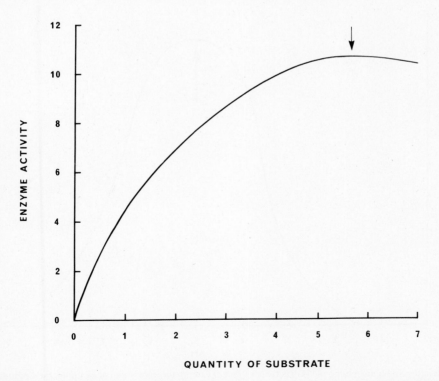

FIG. 14.6. Relationship between the concentration of substrate and enzyme activity. Note that beyond an optimum substrate concentration (arrow) the activity of the enzyme commences to decrease. (After West, Todd, Mason, and Van Bruggen. *Textbook of Biochemistry*, 4th ed, 1966, Macmillan.)

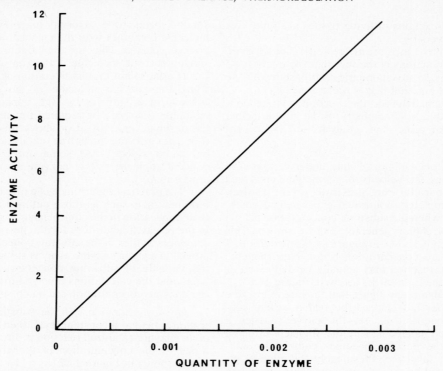

FIG. 14.7. Relationship between the concentration of an enzyme and enzyme activity. (After West, Todd, Mason, and Van Bruggen. *Textbook of Biochemistry,* 4th ed, 1966, Macmillan.)

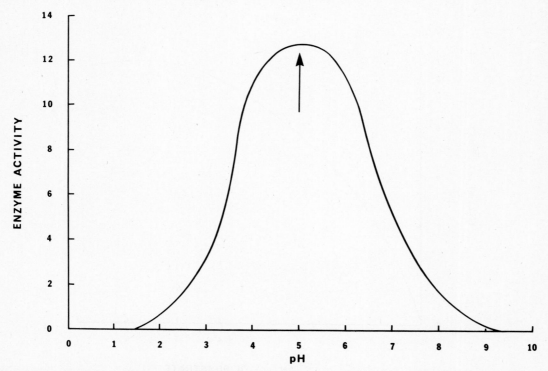

FIG. 14.8. Relationship between the pH and activity of an enzyme. In this hypothetical example, the optimum pH (arrow) is at 5.0. In actuality, the pH optima for various enzymes may lie well above or below this pH, and the curves may be skewed. (Data from Harper. *Review of Physiological Chemistry,* 13th ed, 1971, Lange Medical Publications; West, Todd, Mason, and Van Bruggen. *Textbook of Biochemistry,* 4th ed, 1966, Macmillan.)

zyme and its effect upon reaction rate, discussed above (Fig. 14.7), can be upset by the addition of the reaction products to the system. The net result of this effect is that the presence of reaction products in sufficiently high concentrations can markedly inhibit the usual activity of the enzyme.

pH. Every enzyme has a pH optimum at which it functions at a maximum velocity; the catalytic activity of an enzyme is regulated to a great extent by the pH of the system in which it is operating and the pH optimum may be altered greatly by changes in such factors as temperature, time, substrate concentration, and so forth. Most importantly, however, even slight alterations on either side of the optimum pH of an enzyme can result in major changes in reaction rates. In fact, local changes in pH undoubtedly shift the equilibrium point between hydrolysis and synthesis for some enzymes in vivo.

In general, the effect of altering pH on the rate of enzyme activity in vitro as depicted on a graph is a bell-shaped curve (Fig. 14.8).

The mechanism(s) whereby pH changes alter the rate of enzyme activity are poorly understood in detail; however, such factors as enzyme–substrate complexing, and the ionization of the protein moieties involved in this process doubtless participate to some extent in producing this effect.

Temperature. Chemical reactions in general proceed at a faster rate as the temperature rises, regardless of whether or not catalysis is involved.

The velocity of reactions catalyzed by enzymes generally increases up to around 50 C. Beyond this temperature, most enzymes undergo heat inactivation as the proteins become denatured. In the normal human body enzymes are, of course, maintained at about 37 C, and the effects of extreme fluctuations in temperature are of more academic than of practical interest insofar as the velocity of enzyme-catalyzed reactions is concerned.

Temperature optima for reaction velocities have been determined empirically under specific conditions for a number of enzymes. The temperature coefficient, or Q_{10}, is defined by the ratio of the reaction velocities at two specific temperatures, ie, $T_2^\circ + 10^\circ/T_1^\circ$ (or T_2°/T_1° where $T_2 = T_1 + 10^\circ$). The majority of enzymes from mammalian and other sources have been found to exhibit a Q_{10} of 3.0 or lower.

Time. It has been demonstrated that the optimum temperature for the velocity of enzyme-catalyzed activity exhibits a time-dependence as well as a pH dependence. When this additional factor is taken into account then the optimum temperature for the activity of a number of enzymes from warm-blooded organisms is around 37 C only if the time is measured in hours. If the time is measured in minutes, then the temperature optimum is far lower, whereas if the time is measured in days, then the temperature optimum may rise above 70 C. Thus a time factor is an important consideration

in strictly defining the conditions that regulate the velocity of enzyme activity.

Oxidation State of the Enzyme. Mild oxidation (as by aeration) can reversibly inactivate a group of enzymes generally referred to as the sulfhydryl enzymes. Conversely, enzymes of this type are reactivated by certain reducing agents. Included in this category are such enzymes as urease, succinic dehydrogenase, and the intracellular catheptic enzymes.

Enzymes belonging to this category are activated by natural reducing agents such as glutathione and cysteine. Experimental work has indicated that an actual reduction of disulfide linkages takes place in the molecule, so that sulfhydryl groups are formed, and this reaction produces activation of the enzyme. This process may be visualized according to the following scheme:

$$2 \text{ Enzyme}\!-\!\text{SH} \underset{\text{reduced}}{\overset{\text{oxidized}}{\rightleftharpoons}}$$

$$\text{Enzyme}\!-\!\text{S}\!-\!\text{S}\!-\!\text{Enzyme} + 2 \text{ H}^+$$

Reduced form: active
Oxidized form: inactive

Under appropriate conditions, this reaction is completely reversible, and doubtless in living tissues various compounds play essential roles in maintaining certain key enzymes in their proper oxidation states at all times.

Activators. Many metallic ions and molecules act as cofactors and thus serve to activate certain apoenzymes; a number of essential coenzymes are summarized in Table 14.1.

The activation of the proenzyme pepsinogen by hydrogen ion (H^+), and the significance of this effect on intragastric protein digestion (hydrolysis) was discussed on page 847, and as outlined above. glutathione and cysteine are important activators of the sulfhydryl complex of enzymes.

Certain enzymes are themselves capable of activating other enzymes (or even proenzymes) with important physiologic consequences. For example, enterokinase can activate trypsinogen and produce active trypsin (p. 852).

The inactive precursors of certain enzymes are termed zymogens. Trypsinogen, chymotrypsinogen, pepsinogen, and the procarboxypeptidases are all zymogens. In the activation of zymogen molecules, a specific and unique peptide bond is hydrolysed; therefore this fundamental change in the primary structure of the protein involved in each instance presumably results in the formation of a catalytically active site or locus on the molecule. Certain experimental data obtained on the structural changes that occur locally during the activation of the chymotrypsinogen molecule support this contention. Alternatively, such an activation mechanism could expose a previously shielded active site on the proenzyme molecule, thereby making it available for binding to substrate molecules.

Inhibition. The practical significance of enzyme inhibition in the practice of medicine cannot be overemphasized, since the pharmacologic action of many drugs depends to a great extent on the inhibition of enzymes. For example, the action of antibiotics and sulfonamides depends upon their ability to inhibit certain microbial enzyme systems so that the organisms cannot metabolize properly, and thus are unable to grow or to reproduce. However, it must be stated that in most instances the specific manner in which the activity of enzymes is affected by drugs is unknown. Nonetheless, experimentation with various enzyme inhibitors has yielded many new drugs of great practical significance in clinical medicine.

In any discussion of enzyme inhibition, a clear distinction must be made at the outset between denaturation of an enzyme that causes loss of its activity, and true competitive or noncompetitive inhibition. Since enzymes are proteins, they can be denatured or precipitated by physical as well as by chemical agents, resulting in a complete loss of their catalytic (ie, biologic) activity. Such denaturation, however, is not considered true inhibition in the biochemical sense of the term. For example, heat or trichloroacetic acid will coagulate enzymes, and thereby alter their structure so radi-

cally that they can no longer exert a catalytic function (Fig. 14.9).

Competitive inhibition means that there is a competition between the inhibitor and substrate molecules for the enzyme in the formation of enzyme–substrate or enzyme–inhibitor molecules. The rate of inhibition in any specific reaction is regulated first by the relative affinities of the substrate and the inhibitor molecules for the same binding site on the enzyme molecule, and second, by the relative concentrations of the substances involved in the reaction. As enzymes are not absolutely specific, molecules similar chemically to normal substrate molecules may react (ie, bind) with active sites on the enzyme, thereby preventing such molecules from participating in catalysis of reactions involving normal substrate molecules. Characteristically the enzyme–inhibitor complex formed during competitive inhibition can be reversed by increasing the concentration of the usual substrate acted upon by the enzyme. That is, competitive inhibition is a reversible process.

Noncompetitive inhibition, in marked contrast to competitive inhibition, is an irreversible process; hence, the enzyme–inhibitor complex does not dissociate upon increasing the concentration of normal substrate. Noncompetitive inhibition may occur in two ways. First, the reaction of the inhibitor with the enzyme may occur at the locus on the enzyme molecule which normally is occupied

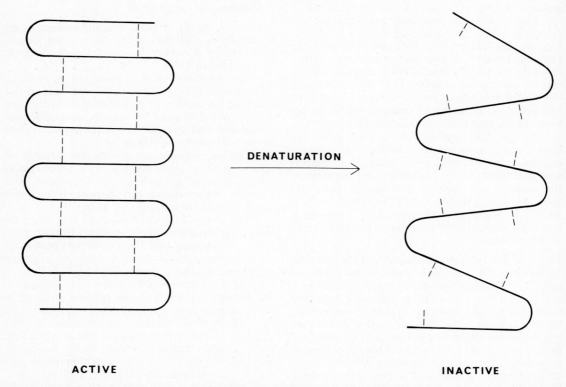

DENATURATION

ACTIVE

INACTIVE

FIG. 14.9. The process of denaturation of an enzyme which thereby causes its inactivation. (Data from Harper. *Review of Physiological Chemistry,* 13th ed, 1971, Lange Medical Publications.)

by substrate molecules. Or, second, the inhibitor may react at a locus in the enzyme molecule other than the active site, and thereby the enzyme is prevented from effectively catalyzing its normal reaction. Regardless of which mechanism is involved, inhibition takes place to an extent that is governed by the affinity of the inhibitor for the enzyme as well as by its concentration. Also, as noted above, increasing substrate concentration does not reverse noncompetitive inhibition.

The feedback inhibition of enzymes is an important regulatory mechanism in living cells. In particular, the first enzyme in a metabolic pathway is inhibited by the end-product of that reaction so that a decrease in the formation of a product takes place when that product already is present in an adequate concentration within the cell. Systems involving the synthesis of certain pyrimidines provide an example of this type of naturally occurring inhibitory mechanism.

Interestingly, experimental evidence favors the view that an enzyme is capable of being inhibited by the end-product of its synthesis only when that enzyme is in a specific conformation, and the inhibition itself takes place by means of a reversible physical deformation of the enzyme molecule induced by the product molecule. Binding of the inhibitor takes place at a locus other than the active site for binding of the normal substrate, thereby decreasing the affinity of the enzyme molecule for the normal substrate. Thus, feedback inhibition of an enzyme system is not simple competition between substrate and inhibitor molecules for the active site. Rather, the inhibitory effect upon the enzyme just described has been called *allosteric transition,* and the changes involved are termed *allosteric reactions.*

ENZYME SYNTHESIS AND ENZYME INDUCTION. Little information is available at present concerning the mechanisms that govern enzyme synthesis by mammalian tissues, despite the fundamental importance of this process. Some information in this area has been derived from experiments (largely performed on microorganisms) using compounds known as inducers (or inductors). An inducer may be defined as a substrate molecule, or a chemically related compound that is capable of stimulating the formation of the enzyme that acts upon it. Hence the substrate induction of enzymes is an important tool in studying experimentally the production of these biocatalysts.

A question of fundamental importance in the subject of enzyme induction is whether or not there is a net de novo synthesis of enzyme molecules that results directly in an increased activity. Alternatively, there may be a decreased rate of breakdown of preexisting enzyme molecules and/or an activation of more enzyme molecules that are already present in the system. The latter effect could be achieved either by removal of an enzyme inhibitor, or by a more efficient apoenzyme–coenzyme interaction.

Although it is often impossible to ascribe a single definite cause to an induced increase in the activity of a specific enzyme, it has nonetheless been demonstrated in certain instances that this increase results from a de novo synthesis of enzyme protein molecules, hence a net increase in the concentration of the enzyme. Usually, however, one speaks of an increase in enzyme activity that is induced by a substrate, thereby neatly avoiding any definite implication of a specific biochemical mechanism in the process.

Bioenergetics

The cells and tissues of all living organisms are open systems that continuously exchange matter and energy with their environment; such systems are in a dynamic steady-state rather than in a state of true thermodynamic equilibrium. The continuous orderly procession of enzyme-catalyzed chemical reactions involved in these exchange processes in turn constitute the subject of metabolism. The general topic of bioenergetics (sometimes called biochemical thermodynamics) deals with those reactions that provide chemical energy in forms utilizable by living systems. The maintenance of normal body temperature, the mechanical work performed by muscles and cilia, the production of nerve impulses in all neural tissues, the active transport of substances against their concentration gradients, and a host of biosynthetic reactions, all require the transformation of energy. The energy required by living systems to carry out these critical functions is derived from the controlled oxidation of foodstuffs and made available as chemical energy released by the breakdown of certain high-energy compounds. In turn, the individual reactions that release energy are catalyzed and regulated by specific enzyme systems.

Ultimately, of course, all of the energy utilized by living organisms is derived from the electromagnetic radiation of the sun, and this radiant energy in turn is generated by nuclear reactions, which proceed at a temperature of 6,000 K. In particular, the fusion of hydrogen nuclei to form helium with the release of a quantum of energy as gamma radiation takes place according to the equation:

$$4_1H^1 \rightarrow H_e{}^4 + 2_1e^\circ + h\nu.$$

The energy released by this process is represented by $h\nu$ in the equation, where h equals Planck's constant and ν is the wavelength of the emitted gamma radiation. Following a number of intricate reactions whereby the emitted gamma radiation is absorbed by electrons (e) most of the gamma radiation subsequently is released from the sun as photons (or quanta) of light energy (Table 1.3, p. 5). This light energy in turn is absorbed by the chlorophyll of green plants on earth, and is converted into the chemical energy of reduced carbon compounds, such as glucose, by the complex process of photosynthesis. Consequently, solar energy is the ultimate source of all biologic energy upon the earth, and the stepwise oxidation of reduced carbon compounds derived from green plants by all living systems, including the human body, provides the energy necessary for the performance of thermal, mechanical, chemical, electric, and osmotic work upon which life itself depends.

It is imperative to stress at this juncture that the useful energy released in biologic systems for the performance of work of any type by the oxidation of a nutrient represents only a small fraction of the total energy that is obtained from the complete combustion of the same nutrient. In other words, during the oxidation of a given foodstuff molecule, much of the energy is lost as heat and only a small percentage of the total energy released by the reaction is converted into forms of chemical energy that can be utilized directly by cells to perform biologic work of various kinds, and this fraction is called the free energy of the system.

The remainder of this section will be concerned with a brief discussion of certain fundamental concepts pertaining to energy and energy exchange that are basic to a clear understanding of energy metabolism in the body.

ENERGY AND WORK. From the standpoint of physics, energy may be defined as the capacity to perform work (see also p. 3). Both energy and work share the same physical dimensions of mass, length (or distance), and time, according to the relationship ML^2/T^2.

All forms of energy potentially are interconvertible into each other, but whenever such conversion takes place, the quantity of energy produced must equal precisely the quantity of the other form of energy which has been transformed. Stated another way, energy can neither be created nor destroyed, but merely transformed (or converted) from one form into another. This statement is known as the first law of thermodynamics or the law of the conservation of energy, and it applies equally to both living systems and to inanimate systems. In accordance with this law, therefore, the total energy of the entire universe is constant. For the more immediate purposes of this discussion, the energetics of particular systems will be considered, a system being defined as the particular assemblage of matter under study.

The second law of thermodynamics states that all processes, whether chemical or physical, take place in such a direction that the entropy of the system and its surroundings (including the entropy of the entire universe) increases to the maximum possible for the particular conditions of temperature, concentration, and pressure that are involved in the process. Stated another way, when the process reaches equilibrium, then the entropy is at a maximum. For the time being, entropy may be defined as the degree of randomness or disorder within a system.

Some important examples of several kinds of energy may be mentioned. Mechanical systems (such as muscles) exhibit potential and kinetic energy; chemical energy underlies all bodily metabolic processes; electric energy is of the utmost importance in the nervous system; and radiant (electromagnetic) energy is the ultimate source of energy for all biologic processes upon the earth, as well as being essential to the sense of vision per se.

In turn, energy, regardless of its form, is the product of an *intensity factor* and a *capacity factor*. Specifically, chemical energy represents the product of the chemical potential (intensity) and the number of moles that are transformed by the reaction (capacity). The intensity factor determines the force driving the reaction and the direction in which the energy flows, whereas the capacity factor determines how far the process goes; ie, the net quantity of a substance that undergoes reaction.

Only the free energy, ΔF, which is released when a system undergoes change is capable of performing work, and in chemical systems, a reaction is spontaneous only if free energy is given off. If free energy is yielded (ie, lost) during a reaction, then the free energy of the entire system declines by a finite capacity, and this is indicated by a negative sign $(-\Delta F)$. Conversely, if the free energy of the system increases during an energy change in that system, the increase is indicated by a positive sign $(+\Delta F)$. In the system $A + B \rightleftharpoons C + D$, if the system yields 5,000 Cal, then this fact is denoted by the symbol $-\Delta F$. Conversely, if the system gains 5,000 Cal, then this net gain in energy is signified by the symbol $+\Delta F$.

In order for a chemical reaction to proceed spontaneously, the free energy must decrease, and the ΔF must be negative in sign $(-\Delta F)$, and the extent to which the reaction occurs is dependent upon the magnitude of the decrease in free energy $(-\Delta F)$ in the reaction. Therefore, the decrease in free energy during a reaction is a direct measure of the driving force or chemical potential energy for the reaction to take place.

As the cell is the fundamental structural as well as functional unit of living organisms, the topic of biologic energy transformation, or bioenergetics, may best be considered first at this level of bodily organization, prior to a discussion of overall energy metabolism in the body.

The reader will note that the discussion presented above circumvents deliberately and entirely any of the philosophic concepts as to what actually constitutes "energy" per se. Rather, the physical measurement of, and the useful work performed by, the controlled release of energy in living systems is of the utmost importance to an understanding of the topic of bioenergetics; hence it will be stressed in the following material.

CHEMICAL ENERGY. Since chemical energy plays a key role in all of the metabolic processes within the body, a clear understanding of the general concepts regarding the release and transfer of this type of energy is imperative.

It was stated earlier that all of the chemical energy required by living organisms is obtained ultimately from the absorption of light energy by the chlorophyll within the cells of green plants. This absorbed solar energy then is converted into chemical energy; and this chemical energy in turn is utilized for the reduction of atmospheric carbon dioxide into glucose. During this complex process, the plant generates oxygen which then is released into the atmosphere as an excretory byproduct of photosynthesis. The net equation for the synthesis of glucose and the concomitant production of oxygen by the process of photosynthesis may be written:

$$6\ CO_2 + 6\ H_2O + nh\nu \rightarrow C_6H_{12}O_6 + \uparrow 6\ O_2.$$

The light quanta which provide the energy for this reaction to proceed are indicated by the factor $nh\nu$.

Thermodynamically, therefore, the biosynthesis of glucose from carbon dioxide and water is an endergonic process that requires an energy input in the form of light quanta; ie, energy is necessary for the reaction to proceed to the right. There is also a large increase in the free energy of the system during the reduction of carbon dioxide to glucose, whereas a concomitant decrease in the entropy or randomness of the system also takes place (see next section). During the formation of glucose by the process of photosynthesis, the change in energy involved has been determined to be +673,000 cal/mole (ΔH). Conversely, when one mole of glucose is oxidized completely (ie, burned) to carbon dioxide and water, this stored chemical energy is released, −673,000 cal/mole now are released as heat energy, and the reaction now is termed exergonic or energy yielding. Note that the quantities of energy involved during the synthesis and the oxidation of one mole of glucose are identical, in accordance with the law of conservation of energy.

Similarly, other organic compounds generate different quantities of heat energy when they are completely oxidized (or burned) regardless of whether this oxidation takes place in vitro (as in a bomb calorimeter), or in living tissues. Only a small fraction of the total energy released during oxidation is available as free energy which is available to perform useful work, the remainder appearing as heat.

The quantity of energy released during the oxidation of any specific organic compound (sometimes called the heat of combustion for that compound when this is measured in vitro), depends upon the composition of the individual substance and its specific chemical structure, as well as the total energy that is found in the combustion products themselves. The total quantity of energy released by any organic compound (ΔH), therefore, is the difference between the energy content of the substance prior to oxidation, and the total energy content of the oxidation products derived therefrom. This statement holds true regardless of the specific oxidative pathway involved, provided that the end-products of the reaction are identical. The complete oxidation of one mole of glucose to carbon dioxide and water in a bomb calorimeter or in a hepatic cell yields 673,000 calories. In the liver cell, however, the glucose is oxidized in a definite series of orderly, sequential reactions and the energy is liberated gradually in small increments, each increment corresponding to an appropriate stage in the oxidative process. In sharp contrast to this gradual, stepwise biologic oxidative process, the oxidation of a mole of glucose to carbon dioxide and water in vitro proceeds in an almost explosive fashion. Yet in both situations, in vivo as well as in vitro, the total quantity of energy yielded is identical ($\Delta H = -673,000$ cal/mole). Thus, the heats of combustion for certain compounds as determined under in vitro conditions in a bomb calorimeter provide a most useful index of the ability of these substances to supply energy in living systems.

As noted earlier, endothermic reactions absorb heat as they proceed, whereas exothermic reactions evolve heat. Chemical reactions in general are associated with energy changes, and ultimately these energy changes are manifest by measurable heat changes. Usually only chemical reactions that liberate energy (ie, exothermic reactions) proceed spontaneously, since energy flows "downhill" from a higher to a lower level. From a practical standpoint, however, a catalyst may be required to make an exothermic reaction proceed. In the cells of the body, most of the chemical reactions involved in bioenergetic processes are, of course, exothermic, and the catalysts that are required for these reactions to take place at rates which are compatible with life are specific cellular enzyme systems. Endothermic reactions, on the other hand, take place only when energy is supplied to the system, and this energy is usually provided to the system in the form of heat. The heat for endergonic biochemical reactions most often is supplied by an exergonic reaction that is coupled to the endergonic reaction; consequently, there is a net gain in the free energy of the latter, and many biosynthetic reactions in living systems take place in accordance with this general principle.

FREE ENERGY, ENTHALPY, AND ENTROPY. In physical as well as in chemical systems, the capacity to perform useful work depends upon the expenditure of energy, and a certain quantity of this energy may be stored for future use. The energy of a tightly wound watch spring is stored as potential energy, and gradually this energy is converted, or transduced, through a system of gears and levers in order to move the hands of the watch. In other words, the potential mechanical energy stored in the spring is transformed spontaneously into kinetic energy. Similarly, in a chemical system, potential chemical energy may be stored in the form of covalent bonds for release and utilization at some future time. Some examples of spontaneous chemical changes are the combustion (oxidation) of such substances as gasoline or glucose into carbon dioxide and water, the hydrolysis of a protein into its component amino acids, and the chemical combination of oxygen and hydrogen to form water. Such chemical changes are spontaneous, but only if they are accompanied by a decrease in chemical potential energy. Within living systems, in particular the cells of the human body, chemical potential energy is stored in various compounds such as glucose, glycogen, and lipids, for later controlled release and utilization in various physiologic processes. Life would be impossible if biochemical reactions occurred rapidly; the controlled and gradual liberation of stored chemical potential energy from various compounds by accelerating the rate at which spontaneous chemical reactions takes place is one of the major and critical functions of enzymes. Thus, it does not necessarily follow from the fact that a reaction is spontaneous that it can also occur at a rate sufficient for compatibility with life. Activation of the reaction is necessary.

In any physical or chemical system, the two laws of thermodynamics must be obeyed. It was mentioned earlier that during the oxidation (combustion) of any compound free energy is liberated, and only this free energy is capable of doing work. The term *enthalpy* is defined as the heat energy released or consumed in a system operating at a constant pressure. It follows that if the pressure within a system is held constant, then the free energy is equal to the enthalpy energy. The term *entropy*, on the other hand, defines the energy unavailable to perform useful work in a system, and entropy is associated with an increased disorder or randomness within that system as free energy is liberated during the particular process involved. These thermodynamic concepts are illustrated in Figure 14.10.

A simple physical example illustrating the concepts of free energy and entropy are shown in Figure 14.10A. If two identical copper blocks having different temperatures are closely apposed, then heat will flow from the hot block on the left (dark shading) into the cold one on the right (light shading) until some intermediate temperature is reached when the system is at equilibrium, and the temperature throughout both of the blocks is the same, ie, isothermal conditions prevail. The total energy in the system has remained constant, but the spontaneous flow of heat from left to right causes a reduction in the overall free energy with a concomitant increase in the entropy or randomness of the entire system. This reaction is irreversible; ie, the heat never flows back spontaneously into the block on the left once equilibrium is

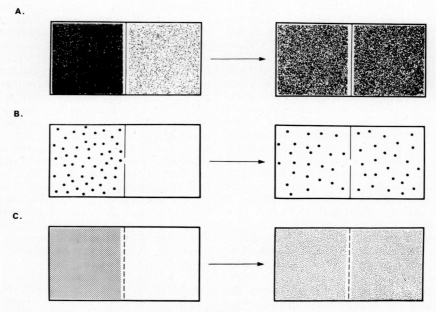

FIG. 14.10.   An increase in entropy (or randomness) in three physical systems. **A.** Heat flows from a relatively warmer (dark shading) to a relatively cooler (light shading) body until the temperatures of both are equal (intermediate shading). **B.** Gas molecules flow through an orifice from a region of high pressure to a region of low pressure. **C.** Solute molecules diffuse from a region of high concentration through a semipermeable membrane to a region of low concentration. Such flows of energy or materials *never* reverse themselves spontaneously. (After Lehninger, 1965, *Bioenergetics*, Benjamin.)

reached, thereby increasing the free energy and decreasing the entropy of the system. Stated another way, at equilibrium the energy within this system is in a randomized state and thus is unavailable to perform useful work. Similarly, gas molecules pass from a region of high pressure to one of low pressure (Fig. 14.10B) and solute molecules diffuse from a region of high concentration to one of low concentration (Fig. 14.10C). In these instances too, once a state of thermodynamic equilibrium is achieved, the total energy in the system is randomized, and thus the entropy is at a maximum.

In Figure 14.11 the free energy and entropy changes are shown for a hypothetical chemical compound, A, undergoing spontaneous oxidation to another compound, B. Note that during this transition, the total energy of the system remains constant, the free energy declines, and the entropy increases proportionally (diagonal arrow). As in the physical example of the copper blocks presented above, this chemical reaction is irreversible, and when compound A, which is in a more highly ordered state, undergoes oxidation to compound B, a less ordered (ie, more randomized)

state is achieved as the free energy stored in compound A is liberated.

In the examples cited above, the first law of thermodynamics is obeyed insofar as the conservation of energy is concerned, but the operation of this law gives no clue as to the direction of the reaction. According to the second law, however, the reaction proceeds in such a direction that there is a tendency of the total energy within the system to approach equilibrium with a concomitant decrease in the free energy of the entire system, while at the same time a proportionate increase in the disorder or randomness of the system occurs, the latter process being denoted by the term entropy. Thus energy may be said to flow "downhill" from regions or compounds having a higher energy state (less random, more ordered, low entropy) to regions or compounds having a lower energy state (more random, less ordered, high entropy). Therefore, at equilibrium the free energy of the system is at a minimum, and the entropy of the system is at a maximum (Fig. 14.11). In order to reverse such a spontaneous process, energy must be supplied to the system in order to achieve an increase in free energy and decrease the entropy,

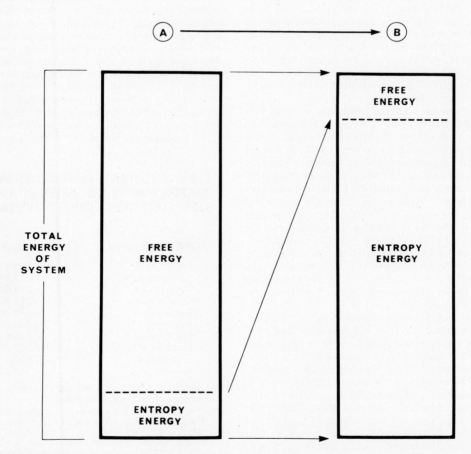

FIG. 14.11. The decrease in free energy and the increase in entropy energy (diagonal arrow) when a high energy compound, Ⓐ, is transformed into a low-energy compound, Ⓑ.

as in the synthesis of a glucose molecule from its precursor substances, carbon dioxide and water. It is essential to recognize the fact that no work can be obtained from a chemical reaction at thermodynamic equilibrium, and, furthermore, the free energy change in any chemical reaction is highest when the activities of both the reactant and its products are farthest removed from the equilibrium state! In addition, under such conditions the most work (or energy) is obtained from a system if there is a decrease in the free energy ($-\Delta F$), as discussed below.

It should be reiterated at this point that the laws of thermodynamics are not concerned with the actual pathways whereby energy exchange takes place, but rather with the initial and final energy states of the system under scrutiny, whether it be of a physical or chemical nature.

Mathematically, the foregoing general statements may be represented by the expression

$$\Delta F = \Delta H - T \Delta S.$$

In this equation, ΔF represents the incremental change in the free energy of a system that is available to perform useful work (sometimes also denoted by the symbol ΔG); ΔH is the total heat energy of the system in calories; T is the absolute temperature ($0°C = 273°$ absolute); and ΔS denotes the entropy change of the system in calories per mole degree. As stated earlier, at constant pressure the free energy change is equal to the enthalpy; therefore ΔF is equivalent to the enthalpy under such conditions.

Spontaneous chemical reactions liberate free energy and thus are said to be exergonic, whereas reactions that are not spontaneous require a gain in free energy in order to proceed, and are termed endergonic. Spontaneous (exothermic) reactions liberate heat to their surroundings as they proceed, and this heat loss is denoted by a minus sign preceeding their Δ value (ie, $-\Delta F$). The molar enthalpy of glucose is $-673,000$ cal/mole. On the other hand, the molar enthalpy for the dissociation of acetic acid ($CH_3—COOH + H_2O \rightarrow CH_3COO^- + H_3O^+$) is $+1,150$ cal/mole. Hence the heat that is required for an endothermic reaction to take place is denoted by a plus sign (ie, $+\Delta F$), and there is an increase in the free energy of the system as the reaction proceeds.

By way of summary, free energy may be considered as that fraction or component of the total energy of a system which is capable of performing work under *isothermal* conditions. Thus, free energy declines during an irreversible process, whereas the entropy of the system increases simultaneously. During all spontaneous and irreversible physical and chemical processes, there is a decline in free energy until an equilibrium is attained, at which point the free energy of the system is at a minimum and the entropy is at a maximum. These conditions define the state of thermodynamic equilibrium in a closed system. Free energy is use-

ful energy, and entropy represents degraded energy. Quantitatively, the magnitudes of the ΔH values presented in Table 14.4 represent the driving force that enables these compounds to undergo spontaneous chemical degradation, and thus to liberate free energy ($-\Delta F$). The physicochemical mechanisms wereby enzymes catalyze this energy release will be discussed subsequently.

The thermodynamic equation $\Delta F = \Delta H - T\Delta S$ applies equally to energy conversion within both inanimate and animate systems. For example, the operation of a heat engine (whether steam or internal combustion) depends upon the transduction of the kinetic energy of an expanding gas into mechanical work. The actual amount of work that is performed by such an engine depends, in part, upon the difference in temperature between the gas within the cylinder before it expands (free energy at a maximum, entropy at a minimum), and the temperature of the exhaust gas after it has expanded (free energy at a minimum, entropy at a maximum). During expansion, the gas yields a portion of its total energy in moving the piston. The important point to stress is that a temperature differential is essential to the performance of useful work by any heat engine. On the other hand, living systems in general, and the cells of the human body in particular, perform the biochemical transduction of energy under essentially isothermal conditions, ie, at approximately constant temperature, since there is a net balance between heat production and heat loss in the cell and in the body as a whole. Therefore, no temperature differential exists among different regions within a cell or among different cells within a tissue. How, therefore, is work accomplished (or energy transduced) by the biochemical reactions within a cell under

Table 14.4[a]   HEATS OF COMBUSTION TO CARBON DIOXIDE AND WATER OF A FEW REPRESENTATIVE ORGANIC COMPOUNDS[b]

PURE COMPOUND	ΔH cal/mole
Acetic acid	$-209,400$
Acetaldehyde	$-279,000$
Pyruvic acid	$-279,000$
Ethyl alcohol	$-327,600$
D-glucose	$-673,000$
Sucrose	$-1,349,600$
Lactic acid	$-326,000$
Palmitic acid	$-2,380,000$
Tripalmitin	$-7,510,000$
Glycine	$-234,000$

[a]*Data from West, Todd, Mason, and Van Bruggen. Textbook of Biochemistry, 4th ed, 1966, Macmillan.*

[b]*When the combustion is carried out at constant pressure, the heat change is known as the enthalpy change and it is negative in sign ($-$) when the heat is lost to the environment, such as in a bomb calorimeter. Divide ΔH in cal/mole by the molecular weight $\times$ 1,000 to get the caloric value of the substance in Cal/gm. The factor 1,000 is to convert small calories (cal) to kilocalories (Cal).*

such a condition of equal temperature? The answer to this important question lies partly in the fact that living cells are open systems and thus exchange matter with their environment. This situation is in marked contrast to closed systems, which attain a true thermodynamic equilibrium. Although the biochemical constituents of living cells may appear to be present in constant amounts, actually they are not present in a true thermodynamic equilibrium. Rather, all substances and compounds within living cells are present in a dynamic steady state, so that the rate of formation of a particular constituent is equal to the rate of degradation or elimination of the same constituent. From the standpoint of energetics, a major function of the living cell is to maintain its biochemical reactions in a state that is far removed from a thermodynamic equilibrium in order that free energy $(-\Delta F)$ can be provided by certain other chemical reactions in order to support the basic life processes. This continuous free energy supply is accomplished through the intake of nutrients and the oxidative metabolism thereof. The excretion of waste products generated by this energy metabolism in turn is an essential feature of the total metabolic pattern, and one that utilizes a significant proportion of the free energy derived from the oxidative metabolic processes themselves.

From the standpoint of energy exchange, the metabolic reactions within the body whereby nutrients (foods) are oxidized in order to provide energy to maintain still other biochemical reactions in states that are far removed from equilibrium is essential so that these other reactions may proceed spontaneously and thus yield energy. Throughout the lifetime of an organism, the metabolic processes continually decrease the entropy of the living system as a whole; when all of the reactions within an organism reach a thermodynamic equilibrium, then that organism is dead! Stated another way, it is oftentimes the task of the physician to prevent the patient from prematurely achieving this thermodynamic equilibrial state, so that his goal is to minimize those processes that lead to an irreversible and permanent increase in the entropy of the bodily processes. Ultimately, however, since living systems exist in a finite period of time, as well as space, and their physicochemical reactions proceed at a finite rate, such systems become irreversible. This concept of a living system once again stands in sharp contrast to a purely thermodynamic (ie, theoretically ideal) physicochemical system in which time per se is an inconsequential factor.

Biologic systems are isothermal; consequently, the heat energy liberated by the oxidative metabolic processes cannot be converted directly into mechanical, chemical, or electric energy as described earlier. Therefore, transduction of free energy derived from the oxidation of foodstuffs within the body into useful work is accomplished by a chemical linkage or chemical coupling of energy-yielding oxidative reactions to such bodily processes as syntheses, active transport of materials, physical work via muscular contraction, and bioelectric phenomena including nerve conduction. The synthesis of certain high-energy phosphate compounds (in particular adenosine triphosphate [ATP]) within the cell is the means whereby a portion of the total free energy that is liberated by certain oxidative reactions is trapped or harnessed for immediate and/or subsequent use by these energy-dependent vital processes within the body. Prior to a discussion of actual coupling reactions within the cell and the paramount role of ATP in trapping free energy for a multitude of biochemical and physiologic processes, it is essential to discuss one further aspect of enzyme activity: specifically, how do enzymes function as catalysts from a standpoint of the energetics that are involved in oxidative reactions?

ENZYMES AND THE ENERGY OF ACTIVATION. A discussion of the energy of activation in regard to enzyme function was deferred until after the concepts of free energy and entropy were developed so that the reader would have a better understanding of the basic principles underlying the fundamental mechanism whereby enzymes control the rates of chemical reactions in living cells.

For a chemical reaction to occur spontaneously, there must be a decrease in free energy $(-\Delta F)$, and the magnitude of this free energy change is the chemical potential (or driving force) for the reaction. Nevertheless, merely because a decrease in free energy $(-\Delta F)$ does occur in a given reaction, it does not necessarily follow that a reaction automatically must take place. This "chemical reluctance" to react results directly from the fact that a certain quantity of energy of activation is required for the molecules involved in a given process to react at all, as illustrated in Figure 14.12.

In accordance with the kinetic, or collision, theory of chemical reactions, in order for a chemical reaction to take place, the molecules involved in that reaction must collide with sufficient energy to overcome the energy barrier for reaction, as illustrated and explained in Figure 14.13. At ordinary temperatures all molecules are in constant motion, a fact attested to by the phenomenon of diffusion. Furthermore, an increase in the temperature of a system, hence its heat energy, causes a proportional increase in the rate of diffusion because of the increased thermal motion of the molecules involved. Conversely, a decrease in the temperature of a diffusing system reduces the kinetic energy of the molecules, and likewise lowers the rate of this process.

In the absence of enzymes, most of the chemical reactions encountered in living cells take place at extremely low rates because of the mild temperatures that are involved. Nevertheless, even under such a mild temperature state the molecules undergo constant thermal agitation, and collisions be-

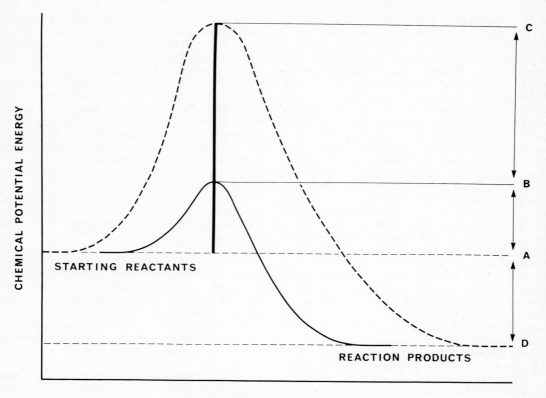

CHEMICAL POTENTIAL ENERGY

STARTING REACTANTS

REACTION PRODUCTS

C

B

A

D

KINETIC ENERGY OF REACTION

FIG. 14.12. The energy of activation that is required to induce a reaction (heavy vertical line) in the absence (A–C) and in the presence (A–B) of a catalyst, eg, an enzyme. In either event, the net chemical energy released by the reaction (A–D) is the same whether it is catalyzed or not. See also Figure 14.13. (After West, Todd, Mason, and Van Bruggen. *Textbook of Biochemistry,* 4th ed, 1966, Macmillan.)

tween molecules occur frequently. However, the molecules do not react rapidly under such conditions because most of them do not have enough kinetic energy to surmount the energy barrier that is present to hinder their reaction. Consequently, the spontaneous reaction rates are extremely slow. If, however, the kinetic energy of the molecules is increased markedly by raising the temperature to a sufficient extent, then the reactions will proceed spontaneously at much faster rates.

In living cells, the presence of enzymes allows the same reactions to occur rapidly, but under the mild temperature conditions which prevail in such systems. The mechanism whereby catalysis is accomplished by an enzyme is depicted in Figure 14.13. Stated briefly, an enzyme decreases the energy of activation to such an extent that the reaction can now proceed rapidly at a far lower temperature than would be possible in its absence. For example, the hydrolysis of sucrose into glucose and fructose when catalyzed by hydrogen ion requires an energy of activation of approximately 25,500 calories. If the same reaction is catalyzed by the specific enzyme invertase, then the energy of activation is reduced to 9,000 calories.

As noted earlier, there is considerable experimental evidence indicating that enzymes function by the formation of an enzyme–substrate complex of some sort. This complex then decomposes to yield the products of the reaction plus the free enzyme once again. This statement may be summarized as follows:

enyzme + substrate $\rightleftharpoons$
enzyme–substrate complex $\rightleftharpoons$
enzyme + products.

Note that the reaction between the enzyme and its substrate is indicated as being potentially reversible, and that the enzyme–substrate complex may be interpreted as an intermediate stage in the reaction, regardless of the direction in which the reaction is proceeding.

To reiterate briefly, from the standpoint of thermodynamics the energy yielded or absorbed during a chemical reaction depends only upon the initial and final energy states of the reactants and products involved. These energy transformations are wholly independent of the actual pathway or mechanism whereby the energy is exchanged, so

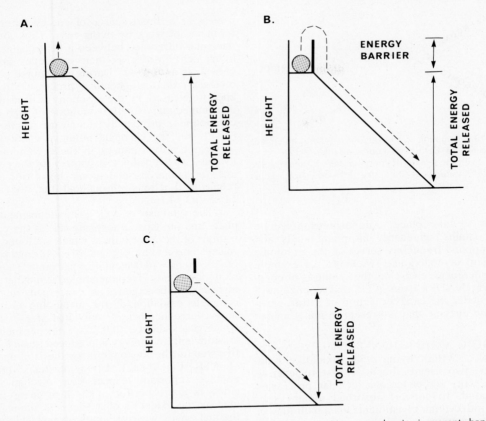

FIG. 14.13. Representation of the energy barrier for chemical reactions. **A.** No energy barrier is present; hence a very slight push will cause the ball to roll down the inclined plane, ie, the reaction proceeds spontaneously with little energy input. **B.** An energy barrier (vertical bar) is present; hence an increment of energy equivalent to that barrier must be added to the system before the potential energy can be released. **C.** An enzyme effectively reduces or "tunnels through" the energy barrier so that the energy of activation for the reaction is reduced considerably. See also Figure 14.12. (Data from Harper. *Review of Physiological Chemistry,* 13th ed, 1971, Lange Medical Publications.)

that only the overall energy changes involved in the reaction are of importance. Thus, enzymes neither alter the initial or final state of a reaction, nor do they affect the net energy exchanges involved in any given chemical process. Rather, enzymes allow a reaction to take place at a lower energy of activation than would be possible otherwise, and the reactions in both directions are accelerated markedly. The net direction of an enzyme–catalyzed reaction depends, of course, on such additional factors as the law of mass action (p. 19), substrate concentrations, and so forth. In fact, merely by altering the substrate–product concentration ratio it is possible to change the free energy (ΔF) relationships of a reaction so that the net direction in which the reaction takes place is reversed. Thus, enzymes affect only the energy barrier against a reaction. Enzymes do not in any way affect the ΔF of the reactants themselves, nor do they alter the initial and final states of those reactants. The ΔF is determined only by the potential chemical energy of the reactants that participate in a given process.

From the standpoint of bioenergetics, enzyme–catalyzed reactions may now be placed into three general categories, according to the type of process involved. The first are the exergonic reactions, whereby the system loses free energy ($-\Delta F$). Exergonic reactions go to completion for all practical purposes. A few of the specific enzymes involved in reactions of this type are urease, lipase, and catalase. The second are reactions that involve little change in free energy ($\Delta F \cong 0$). Such reactions reach a chemical equilibrium, ie, the substrate and product molecules of the reactants are present in constant amounts. For example, the reaction between glycogen and inorganic phosphate to form glucose-1-phosphate is reversible, and at equilibrium, approximately 77 percent glycogen and 23 percent glucose-1-phosphate are present. The third are the endergonic reactions ($+\Delta F$), which require an input of free energy in order to proceed. This type of reaction generally is linked or coupled directly to an exergonic reaction (Fig. 14.14). The exergonic reaction supplies the energy necessary for the endergonic

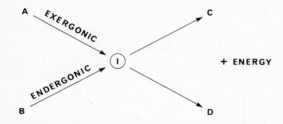

FIG. 14.14. Coupling of an exergonic (A–D) to an endergonic reaction (B–C) through a common intermediate compound, I, with the release of energy as heat. (After Harper. *Review of Physiological Chemistry,* 13th ed, 1971, Lange Medical Publications.)

reaction to take place. Adenosinetriphosphate (ATP) contains high-energy phosphate bonds, and this compound frequently serves as the exergonic component of endergonic reactions. An example of an endergonic reaction is the synthesis of coenzyme II (NADP) from ATP and coenzyme I (NAD) under the catalytic action of a transphosphorylase enzyme that is present in many animal tissues.

BIOLOGIC OXIDATION–REDUCTION REACTIONS. Within living cells, oxidative reactions have two primary functions: first, to generate usable energy for endergonic cellular processes; and second, to convert absorbed dietary substances into cellular constituents via a multitude of synthetic reactions.

When a mole of glucose is oxidized completely to carbon dioxide and water in vitro under standard conditions, the free energy yielded by this process $(-\Delta F)$ amounts to $-673,000$ calories/mol.* However, the thermodynamics of this reaction give no clues as to the mechanisms whereby the cell can utilize the free energy that is derived from such an oxidative process for the performance of useful chemical, electric, or mechanical work. Furthermore, the rapid oxidation of even a small fraction of a mole of glucose within a cell would generate sufficient heat energy to derange cellular functions completely and permanently.

It was stressed earlier that the oxidation of fuel in a gasoline or steam engine causes the expansion of a gas or water vapor in such a manner that useful work is performed, and that machines of this type can operate only if a temperature differential is present in the system. Living cells, however, operate at an essentially constant temperature so that under the isothermal conditions encountered in biologic systems, heat energy per se cannot be used, even theoretically, to perform work directly. The heat energy released by the oxidation of carbohydrates, fats, and proteins is

useless as a direct source of energy for the performance of work by living cells. Thus, one fundamental difference between a heat engine and a living cell, insofar as their energetics are concerned, is as follows. During the course of many biologic oxidative chemical processes, the free energy released by the reactions is used directly and immediately for the synthesis of high energy compounds, principally adenosine triphosphate (ATP), and the chemical potential energy stored (or temporarily trapped) in these compounds is available to the cellular machinery for the performance of numerous endergonic tasks (ie, energy-requiring functions) throughout the body (Figs. 14.15 and 14.16).

The synthesis of ATP from adenosine diphosphate and inorganic phosphate (Pi) at the concentrations of these substances which are found in living systems and at 37 C and pH 7.4, requires a ΔF of $+8,000$ to $+10,000$ calories per mole. Hence the oxidation of any given metabolite is most efficient, insofar as bioenergetics is concerned, when the complete oxidation of the metabolite is divided into a large number of individual steps, each of which yields about $+10,000$ calories per mole; that is, sufficient free energy is produced at each oxidative step for the concomitant synthesis of one mole of ATP (Fig. 14.17). Living systems differ in another important respect from mechanical engines, in that a portion of the heat energy generated by the oxidation of nutrients is eliminated as thermal waste, whereas simultaneously a relatively large fraction of the free energy derived from the oxidative processes is harnessed (or trapped) directly in another form of chemical potential energy. Therefore, the metabolic machinery of living cells operates with an incomplete loss of heat unlike an engine. Moreover, free energy may be transferred repeatedly among various compounds under isothermal conditions, unless the chemical energy is lost from the cellular system as heat energy. Regardless of the number of individual steps involved in such energy transformations, the net energy exchange in the cell is identical with that measured when the heat is produced by com-

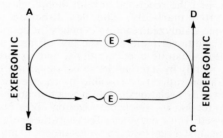

FIG. 14.15. Transfer of energy from an exergonic reaction (A–B) to an endergonic reaction (C–D) through formation of a high-energy intermediate compound, ~ Ⓔ. (After Harper. *Review of Physiological Chemistry,* 13th ed, 1971, Lange Medical Publications.)

*Some authors list the free energy value for glucose as $-686,000$ calories/mol, a discrepancy which needs reconciliation.

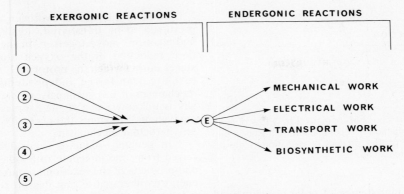

FIG. 14.16. Transfer of energy in biologic systems takes place through formation of a common intermediate,~ Ⓔ, by various exergonic reactions (① through ⑤), and that intermediate can serve as an energy source for various endergonic reactions. Adenosinetriphosphate (ATP) is the high-energy intermediate for most biologic processes. (After Harper. *Review of Physiological Chemistry,* 13th ed, 1971, Lange Medical Publications.)

bustion of the same compounds in a bomb calorimeter.

Each transformation of energy in a cell is relatively inefficient, because each reaction that takes place involves some loss of energy as heat with a concomitant gain in entropy by the entire system (ie, the body). Paradoxically, however, this relative inefficiency of the metabolic transformations within the body lends direction to the biochemical processes involved. In a series of oxidative stages $A \rightleftharpoons B \rightleftharpoons C \rightleftharpoons D$, if the change in free energy for each of the steps, ΔF, were zero (ie, the equilibrium constant for each reaction = 1.0), then A would be converted at equilibrium into a mixture having equal concentrations of A, B, C, and D. But if the ΔF for each step in this reaction sequence was both negative and large, especially that for $C \rightleftharpoons D$, then A would be converted almost quantitatively into D.

Thus, a principal use of the free energy derived from biologic oxidations is to maintain the body in a state far removed from thermodynamic

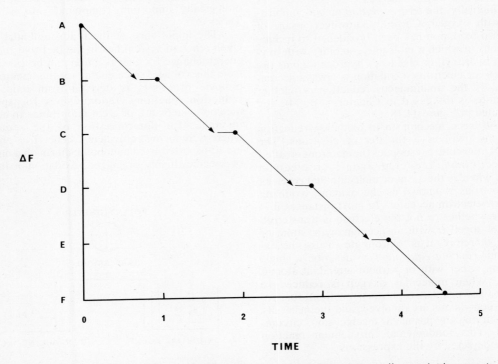

FIG. 14.17. In biologic systems, energy is released from high-energy compounds in small, stepwise increments. That is, total free energy of the compound (ΔF) is not released all at once (A–F). Rather the energy is released in small quantities by a number of reactions, A–B, B–C, and so forth.

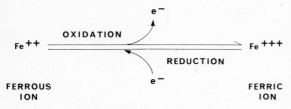

FIG. 14.18. Oxidation is accompanied by a loss of electrons, whereas reduction is accompanied by a gain of one (or more) electrons. The process of oxidation is always accompanied by reduction.

equilibrium. A high degree of order is maintained within the components of living cells and throughout the lifetime of the individual entropy is kept at a minimum.

The remainder of this section will describe the general principles involved in energy exchange through biologic oxidation–reduction reactions as well as the mechanisms whereby the (free) energy liberated by such oxidative processes is trapped as high-energy phosphate bonds which then can be utilized readily to provide energy for cellular functions.

Biologic Oxidation–Reduction Reactions and the Terminal Respiratory Chain. All of the work performed by living systems depends upon a continuous supply of free energy that is supplied by certain oxidation–reduction reactions found in the metabolic pathways of the cell.

Originally, the term *oxidation* meant combination with oxygen. Currently, however, usage of the term oxidation has been broadened to include reactions in which a molecule combines with oxygen or hydroxyl, or else loses hydrogen atoms (ie, protons) or electrons. Oxidation is always accompanied by the simultaneous reduction of another molecule, as discussed in Chapter 1 (p. 14), and illustrated in Figure 14.18.

The basic mechanism in oxidation–reduction reactions involves a transfer of electrons. The atom or molecule that supplies the electrons is called the reductant, reducing agent, or electron donor, whereas the atom or molecule that receives the electrons is known as the oxidant, oxidizing agent or electron acceptor. In most instances it is unclear whether or not the electron is transferred alone or together with an accompanying atom or group; therefore, it is appropriate to use the expression *reducing equivalent* to describe a single electron, either with or without attendant atomic nuclei. When molecular oxygen is reduced to water (a key cellular respiratory reaction), four reducing equivalents are involved, and the necessity for specifying any particular mechanism is circumvented. This fundamental intracellular reaction may be depicted by the expressions:

$$O_2 + 4H \cdot \rightarrow 2H_2O \qquad (1)$$

$$O_2 + 4H^+ + 4e^- \rightarrow 2H_2O \qquad (2)$$

In the first equation, hydrogen atoms (H·) are presumed to be transferred, so that the protons and electrons move together. In the second equation, reduction of the oxygen atoms occurs in stages during which the protons (H$^+$) and electrons (e$^-$) are transferred independently. Since the mechanism of this seemingly simple process actually is unknown, use of the expression "reducing equivalents" obviates the necessity for specifying any particular mechanism.

In general, reducing equivalents are transferred from one molecule to another because the end-products of the reaction are more stable. Free energy is liberated during this process, and, provided that intermediate or transition states are not involved in the reaction which in turn require large quantities of energy for activation (Fig. 14.12), the reaction will proceed to equilibrium. The position of the equilibrium is determined by the "pressure" of the reducing equivalent from the electron donor toward the electron acceptor, and since electrons are involved, this equilibrial state can be determined empirically by the electric potential for the reaction (E), which can be measured under standard conditions of temperature, pH, and concentration of reactants (Fig. 14.19). The relative power of chemical substances to act as oxidants or reductants is expressed in volts or millivolts under specified conditions, and tables have been prepared to describe the relative electron pressures of various specific reactions. Some examples of typical oxidation–reduction (or redox) potentials of biologically important compounds are listed in Table 14.5.

The importance of the redox potential for a given reaction system lies in the fact that the free energy change for that reaction can be calculated from this voltage; since energy for the operation of biologic processes is derived from oxidation–reduction reactions, redox values for specific metabolic steps are of great importance in certain instances for determination of the sequence whereby reducing equivalents, hence free energy, are transported from metabolites through a number of specific, highly organized, oxidation–reduction

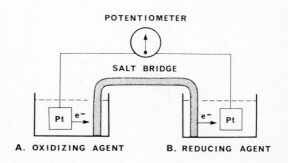

FIG. 14.19. Oxidation–reduction cell for determining redox potentials in vitro. (After West, Todd, Mason, and Van Bruggen. *Textbook of Biochemistry,* 4th ed, 1966, Macmillan.)

Table 14.5 SOME REDOX POTENTIALS IN MAMMALIAN OXIDATION SYSTEMS[a]

SYSTEM	E'_0, VOLTS[b]
Oxygen/water	+0.82
Cytochrome a; Fe^{+++}/Fe^{++}	+0.29
Cytochrome c; Fe^{+++}/Fe^{++}	+0.22
Ubiquinone; oxidized/reduced	+0.10
Cytochrome b; Fe^{+++}/Fe^{++}	+0.08
Fumarate/succinate	+0.03
Flavoprotein; oxidized/reduced	−0.12
Oxaloacetate/malate	−0.17
Pyruvate/lactate	−0.27
Acetoacetate/β-hydroxybutyrate	−0.19
NAD^+/NADH	−0.32
H^+/H_2	−0.42
Succinate/α-ketoglutarate	−0.67

[a]*Data from Harper. Review of Physiological Chemistry, 13th ed, 1971, Lange Medical Publications; West, Todd, Mason, and VanBruggen. Textbook of Biochemistry, 4th ed, 1966, The Macmillan Co.*

[b]*In oxidation/reduction reactions which involve a change in free energy (ΔF), the change in free energy is directly proportional to the tendency of the reactants to donate or to accept electrons, a fact which is expressed quantitatively as an oxidation-reduction (or redox) potential. By convention, the redox potential of a system (E_0) is compared against the the potential of a hydrogen electrode, and at pH 0.0 this value is equal to 0.0 volts. In biologic systems, however, the redox potential (E'_0) is expressed at pH 7.0, and at this pH the potential of the hydrogen electrode is −0.42 volts. The oxidizing agent in each system listed in this table has a greater capacity for taking up and holding electrons than do all systems lying below it, and less capacity than all systems lying above it. Furthermore, the reducing component of each system has a greater capacity to yield electrons than does the reducing component of any of the systems lying above it. Therefore, for example, the oxygen/water system oxidizes all systems lying below it in the table, whereas all systems below the oxygen/water system reduce it. And as the differences between E'_0 values of individual oxidation/reduction systems increases, then so too do the oxidizing and reducing powers between the two systems.*

carrier stages until the terminal oxidation step is reached. The terminal oxidase is an enzyme that catalyzes reduction of most of the molecular oxygen utilized by the cell during its respiratory processes.*

The enzyme system responsible for the transport of reducing equivalents from metabolites to molecular oxygen contains many individual components that are highly organized into a so-called terminal respiratory chain of enzymes. This multienzyme system breaks down oxidation–reduction processes in the cell into a series of small steps, each of which involves a small potential (hence free energy) difference, so that efficient recovery of the free energy released at each step is obtained. If this functionally coordinated system of enzymes were lacking in the cell, reducing equivalents would pass in a single large step from

*The term "respiration" as used in this chapter denotes cellular (or metabolic) respiration and not exchange of oxygen and carbon dioxide, as discussed in Chapter 11.

metabolite to molecular oxygen, and the efficient recovery of free energy would be impossible.

The first step in electron transport within the cell consists of dehydrogenation (or a loss of protons [H^+] from metabolites, and these reactions are catalyzed by substrate-specific dehydrogenases. The many individual dehydrogenases all share the same property of transferring reducing equivalents into a terminal carrier system, and this multienzyme system is shared in common by the majority of the dehydrogenases.

The basic principles of biologic oxidation–reduction (ie, energy transfer) reactions may now be summarized.

First, the enzymes that catalyze biologic oxidation–reduction reactions are dehydrogenases, carriers of reducing equivalents, and oxidases. Specific examples of these enzymes will be encountered in the text and figures that follow.

Second, the activation of specific hydrogen atoms by specific dehydrogenases is the first process involved in the transfer of reducing equivalents in the terminal respiratory chain.

Third, the transfer of hydrogen atoms (protons, reducing equivalents) from metabolites to carriers of reducing equivalents is the second process in the terminal respiratory chain.

Fourth, the terminal respiratory chain, discussed below, consists of certain of the enzymes summarized above that are organized into a highly specific pattern for the transfer of reducing equivalents.

Fifth, the combination of reducing equivalents from the carrier enzymes with molecular oxygen is the final stage in terminal oxidation.

The Respiratory Chain. Within the mitochondria (Fig. 14.1), most of the useful energy that is released by the oxidation of nutrients is harnessed (or trapped) in the form of the high-energy compound ATP. All of the utilizable energy that is derived from the oxidation of fatty acids and amino acids, as well as practically all of the energy yielded by the oxidation of carbohydrates, is liberated within the mitochondria. In order to achieve this functional goal, the mitochondria contain the highly organized and functionally integrated series of enzymes known as the *respiratory chain*. The enzymes of this system are concerned with the orderly transport of reducing equivalents in the form of hydrogen ions (or protons) and electrons to their final terminal reaction with molecular oxygen to form water.

The terminal respiratory chain within the mitochondrion is depicted in Figure 14.20 in a sequential fashion of increasing redox potential from left to right. Reducing equivalents (hydrogen and electrons) thus flow through the chain in a stepwise fashion from the relatively more electronegative toward the more electropositive components of the system. The release of free energy, of course, parallels the change in redox potential at each step and as discussed in the footnote to the

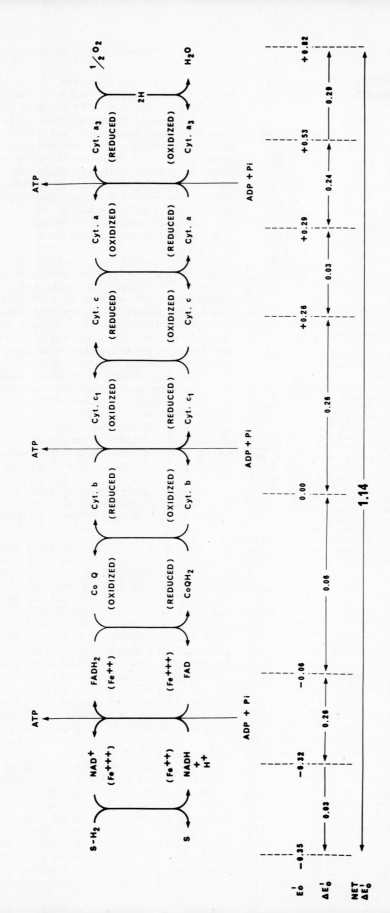

FIG. 14.20. The respiratory chain of enzymes. Three moles of ATP are formed through oxidative phosphorylation of ADP. The energy flow is indicated by the redox potentials for individual reactions (E_0') at the bottom of the figure, as explained in the text. See also Figures 14.25 and 14.29. (Data from Harper. *Review of Physiological Chemistry,* 13th ed, 1971, Lange Medical Publications; West, Todd, Mason, and Van Bruggen. *Textbook of Biochemistry,* 4th ed, 1966, Macmillan; White, Handler, and Smith. *Principles of Biochemistry,* 4th ed, 1968, The Blakiston Division, McGraw-Hill.)

figure. Furthermore, the redox potentials for the individual reactions give a clue as to the sequential position of each enzyme within this system.

The overall flow of reducing equivalents, hence energy, in the terminal respiratory chain may be summarized briefly as follows:

substrate → nicotinamide dehydrogenases → flavoprotein enzymes→ coenzyme Q→ cytochromes → cytochromes → molecular oxygen.

The structure of nicotinamide adenine dinucleotide, or NAD (sometimes referred to as diphosphopyridine nucleotide, or DPN, in earlier texts), is shown in Figure 14.21. This compound is the essential coenzyme (prosthetic group) of a group of dehydrogenases. Similarly, the coenzyme flavin adenine dinucleotide, or FAD, is shown in Figure 14.22, and that of the recently-discovered coenzyme Q is depicted in Figure 14.23.

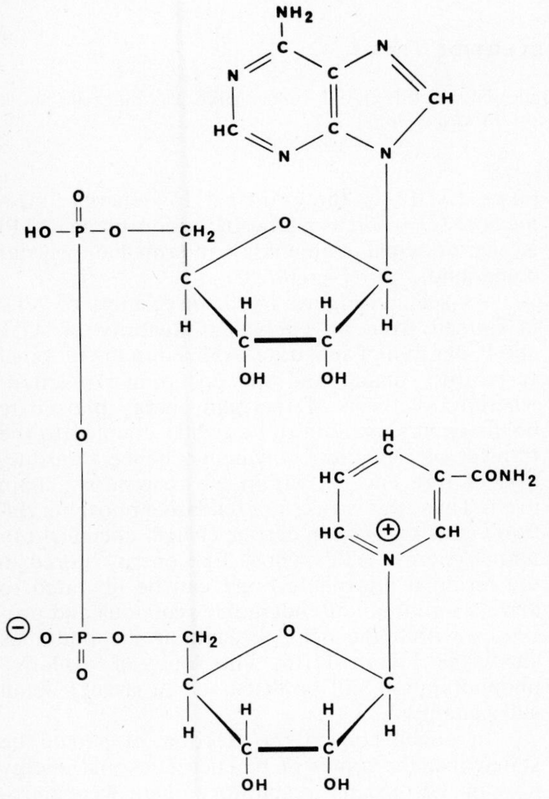

NICOTINAMIDE ADENINE DINUCLEOTIDE (NAD⁺)

FIG. 14.21. Structural formula of the coenzyme nicotinamide adenine dinucleotide, a common electron acceptor in biologic systems. (After White, Handler, and Smith. *Principles of Biochemistry,* 4th ed, 1968, The Blakiston Division, McGraw-Hill.)

As shown in Figure 14.20, the overall change of potential through the entire terminal respiratory chain is 1.14 volts, and since 0.15 volts represents a change in free energy of 7 Cal, then the overall change in free energy $(-\Delta F)$ of this carrier system is approximately 5.26 Cal. Theoretically, therefore, about 7 mols of ATP could be synthesized by this terminal respiratory process per mol of substrate oxidized. In actuality, only 3 mols of ATP are generated as shown in Figure 14.20. The remainder of the free energy released during terminal oxidation is lost as heat. The water produced by the reduction of molecular oxygen in the last step of terminal oxidation becomes part of the water pool of the mitochondrion.

It is evident from inspection of Figure 14.20 that reducing equivalents are passed from one carrier enzyme to the next in the terminal respiratory chain in a highly ordered and sequential fashion. Furthermore, the five cytochrome enzymes are found in mitochondria in almost constant molar proportions to one another (1:1). This fact, together with the finding that many components of the terminal respiratory chain are found in close association with the mitochondrial membranes, has suggested the conclusion that the carrier enzymes themselves have a precise spatial orientation in the membranes of these cellular organelles.

CAPTURE OF FREE ENERGY DURING OXIDATION: OXIDATIVE PHOSPHORYLATION. In living systems, an exergonic chemical process that involves a loss in free energy $(-\Delta F)$ may be used to drive a coupled endergonic chemical process in such a manner that some of the free energy released during the exergonic reaction is captured, the remainder being lost from the system as heat. For example, two metabolites, A and B, may react to form a common intermediate, I, before they proceed to yield metabolites C and D, as shown in Figure 14.14. As indicated in this diagram, when substance A releases energy in being transformed to substance D, the energy released during this process is used to synthesize a high-energy intermediate, I, which in turn supplies energy for the conversion of substance B to substance C. In this scheme it is mandatory that some of the free energy released during the conversion of A to D be liberated as heat in order to drive the reaction to the right, and furthermore, the intermediate must be common to both reactants A and B; ie, the reactions are coupled together through the intermediate I.

Another mechanism whereby an exergonic reaction may be coupled to an endergonic reaction is by the synthesis of another compound having a high chemical potential energy. This additional compound, Ⓔ, is incorporated into the endergonic reaction, so that free energy is transferred indirectly from the exergonic pathway via the intermediate compound, designated~ Ⓔ, to the endergonic pathway. Such a process is depicted in Figure 14.15. In this conceptual diagram, Ⓔ rep-

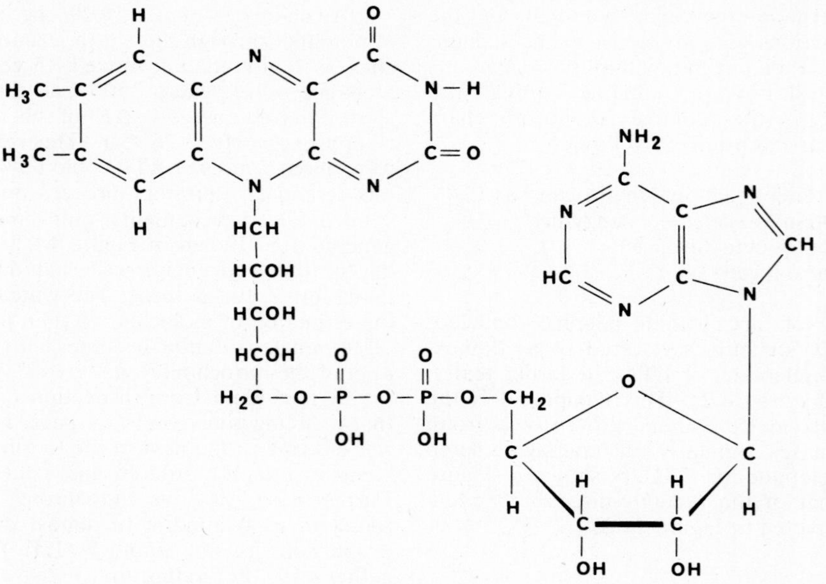

FLAVIN ADENINE DINUCLEOTIDE (FAD)

FIG. 14.22. Structural formula of the coenzyme flavin adenine dinucleotide (FAD). (After White, Handler, and Smith. *Principles of Biochemistry,* 4th ed, 1968, The Blakiston Division, McGraw-Hill.)

resents the compound of low chemical potential energy and ~Ⓔ represents the compound having a high chemical potential energy. The biologic advantage of this mechanism over the one described above and depicted in Figure 14.14, lies in the fact that unlike the common intermediate, I, Ⓔ does not necessarily have to be related chemically or structurally to A, B, C, or D. Thus, Ⓔ can act to transduce energy from a multitude of exergonic reactions to an equally great number of endergonic reactions or cellular processes as shown in Figure 14.16. In living cells, of course, adenosine diphos-

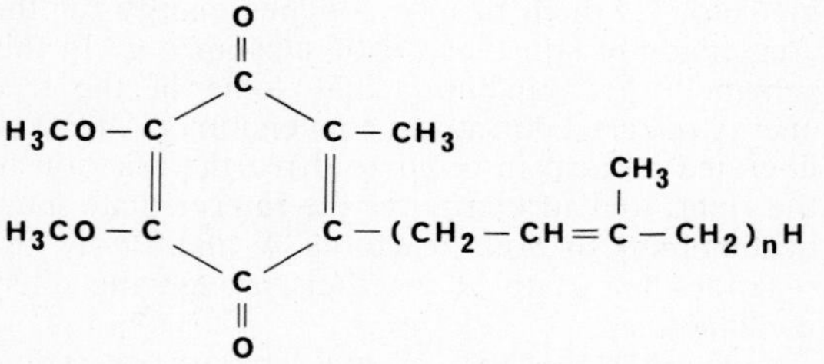

COENZYME Q (UBIQUINONE)

FIG. 14.23. Structural formula of coenzyme Q. (After White, Handler, and Smith. *Principles of Biochemistry,* 4th ed, 1968, The Blakiston Division, McGraw-Hill.)

phate (ADP) is the principal low-energy intermediate, Ⓔ, whereas adenosine triphosphate (ATP) is the principal high-energy intermediate carrier compound, ~Ⓔ (Fig. 14.24).

As seen in Figure 14.20, three mols of ATP is formed from an equivalent quantity of ADP and P_i per mol of substrate oxidized in the terminal respiratory chain, and the endergonic reactions whereby synthesis of the high energy phosphate bonds occurs are said to be tightly coupled to the transfer of reducing equivalents, hence transduction of free energy, within the respiratory chain itself. Thus, this process of oxidative phosphorylation yields a common carrier of high chemical potential energy, ATP, whose free energy stored in the terminal phosphate bond can be liberated to drive a multitude of endergonic reactions and processes within the cells throughout the body, as shown in Figure 14.16. The topic of oxidative phophorylation will be discussed in greater detail subsequently.

In conclusion to this section, it should be stated that the topics of bioenergetics and energy flow in the terminal respiratory chain were introduced prior to a discussion of intermediary metabolism so that the reader may have some idea of energy transduction and utilization in living systems. It is essential that the concepts presented above be understood clearly prior to a consideration of the specific metabolic reactions that are involved in the stepwise oxidation of nutrients.

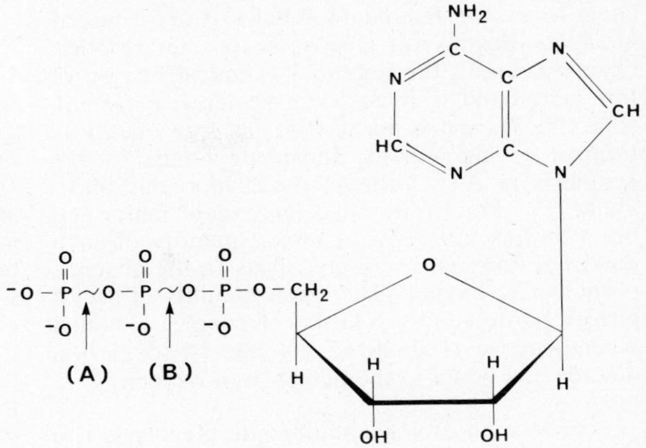

FIG. 14.24. Structural formula of the high-energy compound adenosine triphosphate (ATP). Hydrolysis of the terminal phosphate bond (A) releases 7,400 cal/mole (7.4 Cal/mole) at pH 7.0, whereas hydrolysis of the inner phosphate bond, (B), releases 7,500 cal/mole (7.5 Cal/mole). (Data from White, Handler, and Smith. *Principles of Biochemistry*, 4th ed, 1968, The Blakiston Division, McGraw-Hill.)

Survey of Energy Metabolism

In this section, the principal aspects of the specific metabolic reactions whereby carbohydrates, fats, and proteins are metabolized in the cells of the body to yield usable energy will be presented. In addition, a brief discussion will be included of certain of the more important interconversions and biosynthetic reactions that take place among these basic nutrients.

CARBOHYDRATE METABOLISM. As noted in Chapter 13, only a few carbohydrates form a major proportion of the total calories ingested by the average human being, and normally over half of the total energy requirements of the body are met by the oxidation of these compounds. In particular, carbohydrate metabolism in the animal body deals primarily with the metabolism of the hexose sugar, glucose, and of substances that are related chemically to this compound.

Remarkably few specific dietary carbohydrates are utilized by cells to provide energy. Polysaccharides such as starches, amylase, and pectin supply most of the dietary carbohydrate, and the disaccharide, sucrose, is usually present in the diet in variable quantities. The disaccharide, lactose (or milk sugar), may assume special importance in the nutrition of infants, provided that a lactose intolerance is not present (p. 859). Monosaccharides are present in the adult diet only in negligible quantities. Ultimately, however, the chief monosaccharides produced by the digestive processes are the metabolically interconvertible monosaccharides glucose, fructose, and galactose. From a quantitative standpoint, fructose is of major importance only if the sucrose intake is quite high, and similarly galactose is of major quantitative importance only if lactose forms the principal dietary carbohydrate source. Normally dietary fructose and galactose are converted into glucose within the liver.

Pentose sugars are a minor dietary carbohydrate source, and D-ribose as well as D-2-deoxyribose are synthesized within the body.

From the standpoint of bioenergetics, there are three metabolic pathways for carbohydrate that are of major significance in the animal body.

The first is *glycolysis,* or the Embden–Myerhof pathway, whereby glucose is oxidized under anaerobic conditions to pyruvate and lactate.

The second is the *tricarboxylic acid cycle,* also known as the citric acid cycle or the Krebs cycle. This is the ultimate common metabolic pathway whereby carbohydrates, fats, and proteins are metabolized to carbon dioxide, and eventually to water via the terminal respiratory chain. This series of reactions takes place under aerobic conditions, in sharp contrast to the anaerobic glycolytic pathway.

The third, the *hexose monophosphate shunt* (HMS), provides an alternative metabolic route for the complete oxidation of glucose to carbon dioxide and water.

These three pathways for carbohydrate metabolism will receive major emphasis in the discussions to follow. From a physiologic standpoint, however, the process of glycogenesis (ie, the synthesis of glycogen from glucose), as well as the process of glycogenolysis (ie, the breakdown of glycogen, principally to glucose) must also be considered, as they are of the utmost physiologic importance. Lastly, the de novo formation of glucose or glycogen from noncarbohydrate sources, a process called gluconeogenesis, is another aspect of carbohydrate metabolism that will be discussed in context.

It is apparent from this brief summary that glucose occupies the key position in carbohydrate metabolism; however, it should be mentioned at this point that the oxidative metabolism of carbohydrates is linked inseparably to that of fats and proteins.

Glycolysis. In mammalian cells, the principal metabolic destiny for glucose is to enter reaction chains whereby the potential chemical energy of this compound is released in a stepwise fashion (Fig. 14.17), and a portion of this free energy is trapped as high-energy phosphate bonds by the synthesis of ATP from ADP and inorganic phosphate (P_i). Practically all cells (except nerve and brain tissue) can derive a minor quantity of such energy from glucose by glycolysis in the absence of molecular oxygen. The major quantity of useful energy harnessed as ATP is, of course, obtained when glucose is oxidized completely to carbon dioxide and water in the presence of oxygen.

From a physiologic standpoint, glycolysis is a most useful process. For example, in a rapidly contracting muscle ATP is made available under relatively anaerobic conditions so that the muscle can continue to perform work despite a relative oxygen deficiency (see also p. 137). Furthermore, part of the glycolytic series of biochemical reactions must take place in order to provide the intermediates for the tricarboxylic acid cycle to function at all.

Reactions of Glycolysis. The net (or overall) reaction for glycolysis is:

$$glucose + 2\ ADP + 2\ P_i \rightarrow$$
$$2\ lactic\ acid + 2\ ATP.$$

Note that there is a net synthesis of two molecules of ATP for each molecule of glucose degraded, and that two molecules of lactic acid are also produced by this reaction.

The individual reactions involved in glycolysis are summarized in Figure 14.25, and indicated by heavy arrows. The enzymes that catalyze these reactions are summarized in Table 14.6, together with certain of their important properties. The circled numbers for the individual glycolytic reactions that are discussed below are identical with those given in Figure 14.25 and Table 14.6. Note that glycolysis is a linear rather than a cyclic pathway for the oxidation of glucose, and that the enzymes responsible for the reactions involved in glycolysis are found in the cytoplasm, rather than in intracellular organelles, as indicated to the left in Figure 14.25.

Although carbon atoms derived from glucose ultimately may appear in carbon dioxide, in other hexoses, pentoses, lipids, amino acids, purines and pyrimidines, as well as in many other compounds, glucose has one major fate in the mammalian cell, and that fate is oxidation to carbon dioxide and water through phosphorylation to glucose-6-phosphate:

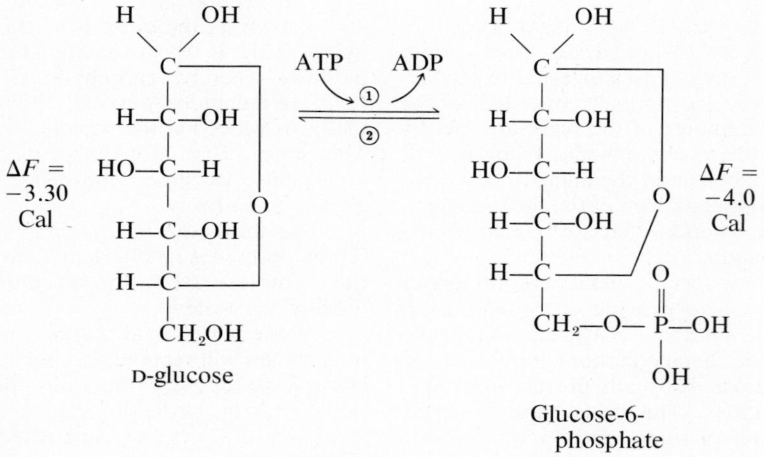

D-glucose

Glucose-6-phosphate

$\Delta F = -3.30$ Cal

$\Delta F = -4.0$ Cal

Actually, carbohydrate metabolism may be considered to start with glucose-6-phosphate, as it is this ester rather than glucose itself that may enter any of several metabolic pathways. Because of the large $-\Delta F$ of the hexokinase reaction [-4.0 Cal, (①)], this process is practically irreversible under intracellular conditions.

The reverse of glucose phosphorylation, the hydrolysis of glucose-6-phosphate, (②), can take place only in liver, intestine, and kidney (but not in skeletal muscle) and thereby free glucose is released once more to the blood for transport by the circulatory system to other cells of the body. In liver, the enzyme D-glucose-6-phosphatase catalyzes this reaction; hence, this enzyme is of extreme physiologic importance to the homeostatic regulation of blood sugar level in vivo.

Glucose-6-phosphate may also be converted into glucose-1-phosphate as the first stage in nucleoside synthesis, or else it may be oxidized to 6-phosphogluconic acid as the initial step in the 6-phosphogluconate pathway (Table 14.9).

The most important fate of glucose-6-phosphate in the body is its conversion into fructose-6-phosphate:

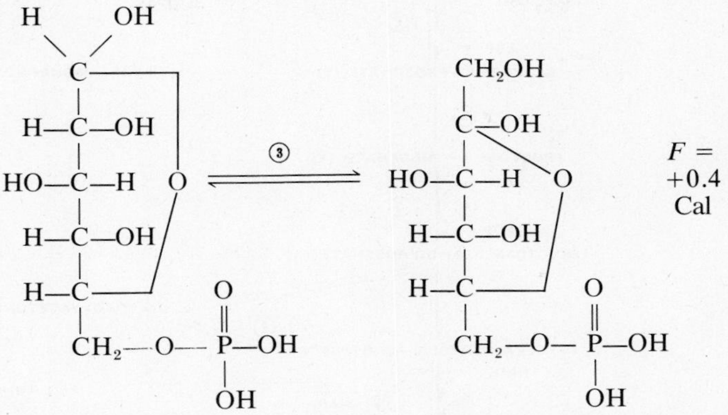

Glucose-6-phosphate

Fructose-6-phosphate

$$F = +0.4 \text{ Cal}$$

Reaction ③ proceeds readily in either direction.

In the next glycolytic reaction, fructose-6-phosphate is converted into fructose-1,6-diphosphate under the specific catalytic influence of the enzyme phosphofructokinase; hence, the next step in glycolysis is called the phosphofructokinase reaction:

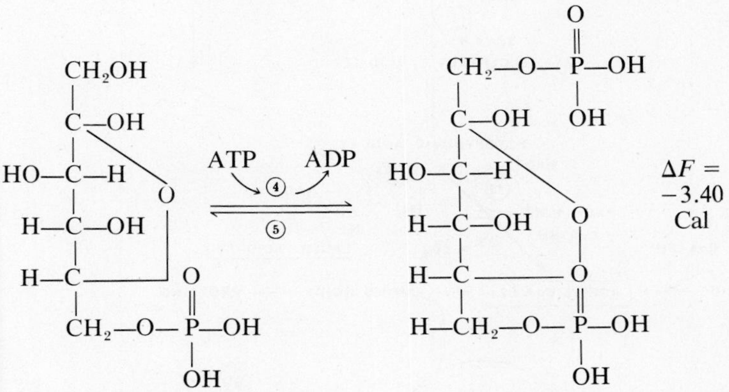

Fructose-6-phosphate

Fructose-1,6-diphosphate

$$\Delta F = -3.40 \text{ Cal}$$

This reaction provides a major regulatory step in the glycolytic pathway. That is, the activity of phosphofructokinase is altered by the existing concentration of various modifiers (or inducers) within the cell. Thus, when the intracellular concentration of ATP as well as citrate, fatty acids, or ketone bodies is high, the activity of phosphofructokinase is decreased. Conversely, the activity of this enzyme is stimulated by high ADP or AMP concentrations, as well as by insulin, inorganic phosphate (P_i), fructose-6-phosphate, and fructose-1, 6-diphosphate.

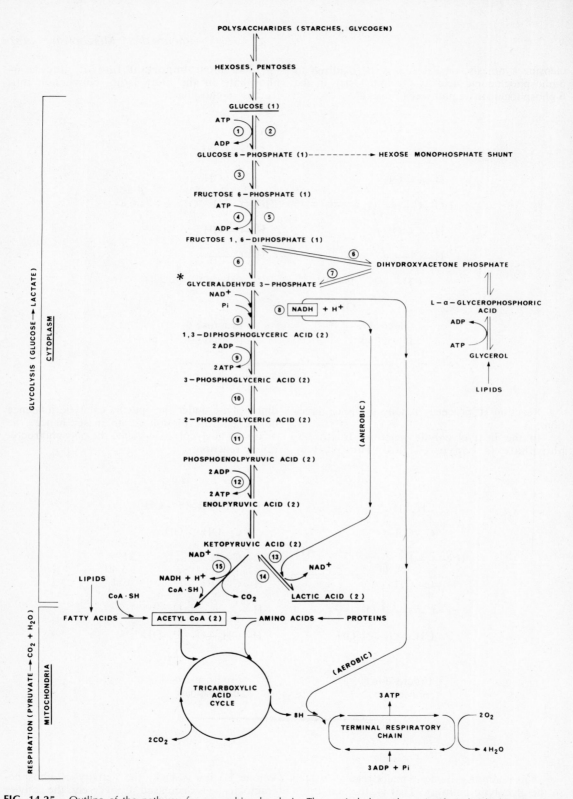

FIG. 14.25. Outline of the pathway for anaerobic glycolysis. The encircled numbers signify individual reactions as discussed in the text, and the enzymes that catalyze these reactions are summarized in Table 14.6. The numbers in parentheses following the names of individual compounds signify the number of moles formed from one mole of glucose. Note that following the glyceraldehyde-3-phosphate step (*) two moles of each compound are formed. The relationship of the aerobic respiratory pathway (tricarboxylic acid cycle, Fig. 14.29) and the terminal respiratory chain (Fig. 14.20) to the glycolytic pathway is indicated at the bottom of the figure. (Data from Harper. *Review of Physiological Chemistry*, 13th ed, 1971, Lange Medical Publications; Lehninger. *Biochemistry,* 1971, Worth Publishers; West, Todd, Mason, and Van Bruggen. *Textbook of Biochemistry,* 4th ed, 1966, Macmillan.)

Table 14.6 SUMMARY OF ENZYMES THAT PARTICIPATE IN GLYCOLYSIS[a]

REACTION NUMBER FIGURE 14.25	ENZYME (SYNONYM)[b]	COENZYME OR ACTIVATORS	REMARKS
①	Hexokinase (glucokinase)	Mg^{++}	Hexokinases are relatively non-specific enzymes that are of widespread occurrence in living cells. Liver hexokinase catalyzes the phosphorylation of glucose, fructose, and mannose. Gluco-kinases catalyze the phosphorylation of D-glucose only.
②	Glucose 6-phosphatase	Mg^{++}	Found in microsomes of hepatic cells; catalyzes dephosphorylation of glucose-6-phosphate, thereby releasing free glucose to blood-stream.
③	Phosphoglucose isomerase (phospho-hexose isomerase, phosphoglucoisomerase, hexose phosphate isomerase)	—	This enzyme is specific for glucose 6-phosphate and fructose 6-phosphate.
④	Phosphofructo-kinase	Mg^{++}, ATP	This allosteric (or regulatory) enzyme is inhibited by high ATP or citrate concentrations within the cell, and stimulated by low ATP concentrations; ie, by high ADP or AMP concentrations.
⑤	Diphosphofructose phosphatase	Mg^{++}, ATP	Activity increased by high ATP concentrations, and decreased by high ADP or AMP concentrations, in contrast to phosphofructokinase.
⑥	Aldolase (muscle aldolase, liver aldolase)	—	Muscle or liver aldolase requires no coenzymes or activators, but each of these is a distinct enzyme.
⑦	Triose phosphate isomerase (phospho-triose isomerase)	—	———
⑧	Glyceraldehyde 3-phosphate dehydro-genase (phospho-glyceraldehyde dehydrogenase)	NAD^+	Contains sulfhydryl groups (SH) that are essential to enzymatic activity. NAD^+ specifically re-quired as oxidant.
⑨	3-Phosphoglyceric acid kinase (phospho-glycerate kinase)	Mg^{++}	———
⑩	Phosphoglycero-mutase	Mg^{++}	2,3-diphosphoglyceric acid is required as the intermediate in this reaction.
⑪	Enolase	Mg^{++}, Mn^{++}	———
⑫	Pyruvic acid kinase (pyruvate kinase; pyruvate phosphokinase)	Mg^{++}, K^+	———
⑬	Lactic acid dehydro-genase (lactate dehydrogenase)	NAD^+	———
⑮	Pyruvic oxidase system (pyruvate dehydrogenase system)	CoA, NAD^+, Lipoic acid, Mg^{++}, FAD, and Thiamine pyrophosphate (TPP)	This oxidase system consists of a "team" of four closely-knit enzymes and 6 cofactors.

[a]Data from Lehninger. Biochemistry, 1970, Worth Publishers.
[b]Synonyms for the names of certain enzymes are given in parentheses.

Since the synthesis of fructose-1,6-diphosphate takes place with a free-energy change of about -3.4 Cal/mol, this reaction is essentially irreversible. The hydrolysis of fructose diphosphate is mediated by another enzyme, diphosphofructose phosphatase (⑤). These two enzymes operate in a "cooperative" fashion within the cell, because the phosphatase acts effectively only at high ATP concentrations, and is inhibited by AMP, in marked contrast to the effects of these modifiers upon the activity of phosphofructokinase.

The sixth step in glycolysis involves the cleavage of fructose-1,6-diphosphate between C–3 and C–4 to yield one molecule each of D-glyceraldehyde 3-phosphate and dihydroxyacetone phosphate under the catalytic influence of aldolase:

Fructose-1,6-diphosphate   Glyceraldehyde-3-phosphate   Dihydroxy-acetone phosphate

This so-called aldolase reaction is reversible; however, the normal intracellular concentrations of glyceraldehyde-3-phosphate and dihydroxyacetone phosphate are quite low; therefore, reaction toward the right is favored.

Only glyceraldehyde-3-phosphate can enter the glycolytic pathway directly. Thus, dihydroxyacetone phophate is converted (reversibly) into glyceraldehyde-3-phosphate by the enzyme triose phosphate isomerase:

Dihydroxy-acetone phosphate   Glyceraldehyde-3-phosphate

Hence, in the glycolytic pathway, each molecule of hexose now has been degraded into two 3-carbon molecules, as indicated by the numbers in parentheses to the right of the names of the individual compounds in Figure 14.25.

The scission of fructose-1,6-diphosphate into the two molecules of triose phosphate (⑥) and the interconversion of dihydroxyacetone phosphate into glyceraldehyde-3-phosphate (⑦) completes the first phase of glycolysis. In this first phase, one glucose molecule has undergone phosphorylation and cleavage in preparation for the second phase of glycolysis, which consists of oxidoreductions and phosphorylation steps during which ATP is synthesized.

Note that since one molecule of glucose forms two molecules of glyceraldehyde-3-phosphate, both halves of each glucose molecule now traverse the same metabolic pathway.

The oxidation of glyceraldehyde-3-phosphate to 1,3-diphosphoglyceric acid under the catalytic influence of glyceraldehyde-3-phosphate dehydrogenase is the eighth step in glycolysis:

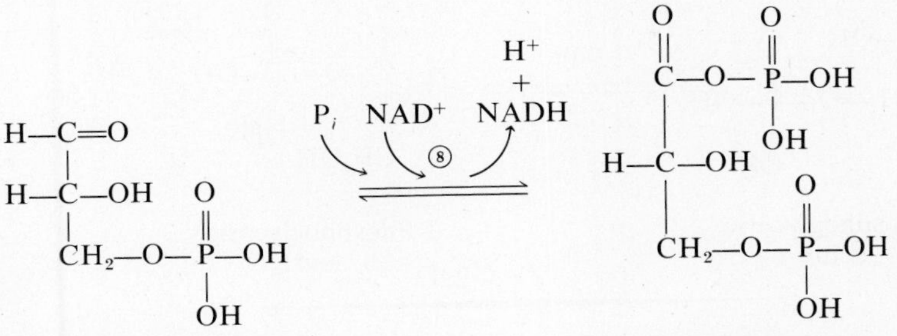

Glyceraldehyde-
3-phosphate
(Low-energy
phosphate)

1,3-Diphospho-
glyceric acid
(High-energy
phosphate)

$\Delta F = +1.5$ Cal

In this important glycolytic reaction, the aldehyde group of glyceraldehyde-3-phosphate is oxidized to a carboxyl group, and the energy is conserved as a high-energy phosphorylated product, namely 1,3-diphosphoglyceric acid. The coenzyme that acts as oxidizing agent in this process, nicotinamide adenine dinucleotide (NAD⁺, Fig. 14.21), accepts electrons from the aldehyde group of the glyceraldehyde-3-phosphate and ultimately transports these electrons to pyruvate later in the glycolytic sequence (⑬). The hydrogen atom that is removed from the substrate during the reduction of NAD⁺ (oxidized form) to NADH (reduced form) is lost to the immediate environment as a hydrogen ion (H⁺).

Note that under aerobic conditions, the reducing equivalents which are derived from reaction ⑧

pass directly from NADH to the terminal respiratory chain as shown in Figure 14.25.

Actually, reaction ⑧ is two distinct but coupled processes, one of which is highly endergonic, and the other highly exergonic. The important fact, however, is that the net reaction (as written above) conserves the energy that is derived from the oxidation of the aldehyde as a high-energy phosphate group in 1,3-diphosphoglyceric acid.

The 1,3-diphosphoglyceric acid (an acid anhydride) which is formed by the reaction described above now reacts with ADP so that the 1-phosphate group of 1,3-diphosphoglyceric acid is transferred to the ADP molecule, thereby forming ATP:

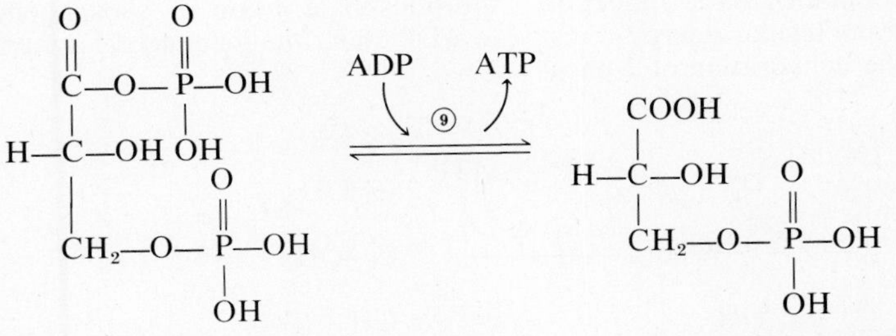

1,3-Diphosphoglyceric
acid

3-Phosphoglyceric
acid

$\Delta F = -4.50$ Cal

Since reaction ⑨ is strongly exergonic, it tends to draw reaction ⑧ to completion; hence the equilibrium is far to the right for both reactions ⑧ and ⑨, and the energy derived from the oxidation of an aldehyde to a carboxyl group has been stored as the phosphate bond energy of ATP. This energy storage is temporary pending utilization of the ATP for other processes (eg, syntheses, muscular contraction, and so forth).

Phosphoglyceromutase now catalyses the conversion of 3-phosphoglyceric acid into 2-phosphoglyceric acid:

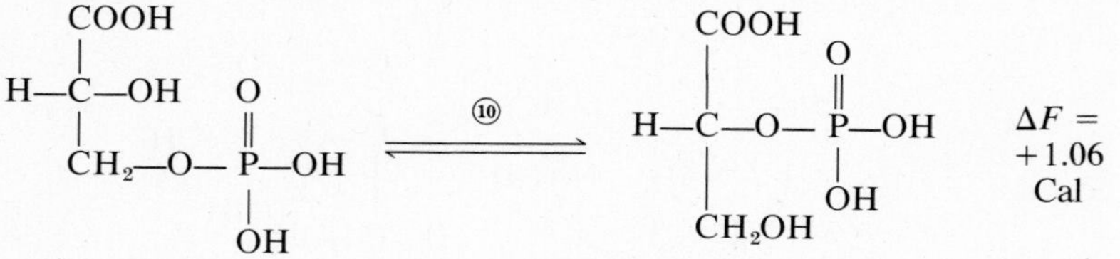

$$
\begin{array}{ccc}
\text{COOH} & & \text{COOH} \\
| & & | \\
\text{H—C—OH} \quad \text{O} & \xrightarrow{\;⑩\;} & \text{H—C—O—P—OH} \\
| \quad\quad \| & & | \quad\quad\quad\; \| \\
\text{CH}_2\text{—O—P—OH} & & \text{CH}_2\text{OH} \quad \text{OH}
\end{array}
$$

$\Delta F = +1.06$ Cal

3-Phosphoglyceric acid → 2-Phosphoglyceric acid

Reaction ⑩ is quite reversible because of the small free-energy change that is involved.

In reaction ⑪, 2-phosphoglyceric acid now undergoes dehydration to phosphoenolpyruvic acid under the catalytic influence of enolase:

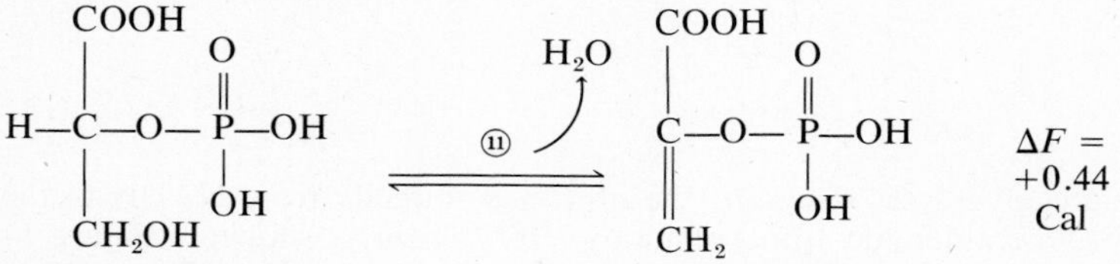

$$
\begin{array}{ccc}
\text{COOH} \quad \text{O} & \xrightarrow[\;⑪\;]{H_2O} & \text{COOH} \quad \text{O} \\
\text{H—C—O—P—OH} & & \text{C—O—P—OH} \\
\text{CH}_2\text{OH} \quad \text{OH} & & \text{CH}_2 \quad \text{OH}
\end{array}
$$

$\Delta F = +0.44$ Cal

2-Phosphoglyceric acid (Low-energy phosphate) → Phosphoenolpyruvic acid (High-energy phosphate)

Note that this is the second reaction of the glycolytic chain in which a high-energy phosphate bond is generated. Specifically, this reaction is an example of an intramolecular oxidoreduction, since the removal of a water molecule from 2-phosphoglyceric acid produces a greater oxidation of C–2, with a concomitant greater reduction of C–3; thus, a large intramolecular energy redistribution occurs during the dehydration of 2-phosphoglyceric acid.

A divalent cation, either Mg^{++} or Mn^{++}, is required specifically as a cofactor for the enolase reaction (⑪), and the metallic ion complexes with the enzyme before the substrate is bound.

In reaction ⑫, the enol form of pyruvic acid is formed. This compound is produced by transfer of the phosphate group of phosphoenolpyruvic acid to ADP under the influence of pyruvic acid kinase:

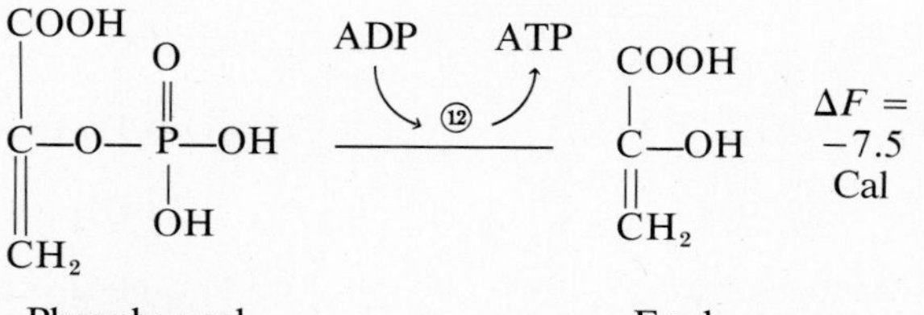

$$
\begin{array}{ccc}
\text{COOH} \quad \text{O} & \xrightarrow[\;⑫\;]{ADP \quad ATP} & \text{COOH} \\
\text{C—O—P—OH} & & \text{C—OH} \\
\text{CH}_2 \quad\quad \text{OH} & & \text{CH}_2
\end{array}
$$

$\Delta F = -7.5$ Cal

Phosphoenolpyruvic acid → Enol-pyruvic acid

The enzyme pyruvic acid kinase requires Mg^{++} or Mn^{++} to act, and complexing must take place before the substrate can be bound. K^+ is also required for this reaction, and this cation functions as the physiologic activator of the enzyme–substrate complex.

Reaction ⑫ is quite exergonic; hence it is irreversible under prevailing intracellular conditions.

Enolpyruvic acid in turn undergoes spontaneous tautomerization to the keto form with which it is in equilibrium:

$$
\begin{array}{ccc}
\text{COOH} & & \text{COOH} \\
| & \text{(spontaneous)} & | \\
\text{C—OH} & \rightleftharpoons & \text{C}{=}\text{O} \\
\| & & | \\
\text{CH}_2 & & \text{CH}_3
\end{array}
$$

Pyruvic acid   Pyruvic acid
(Enol form)   (Keto form)

The final step in the sequence of glycolytic reactions involves the reduction of pyruvic acid to lactic acid under anerobic conditions, a step that is catalyzed by the enzyme lactic acid dehydrogenase:

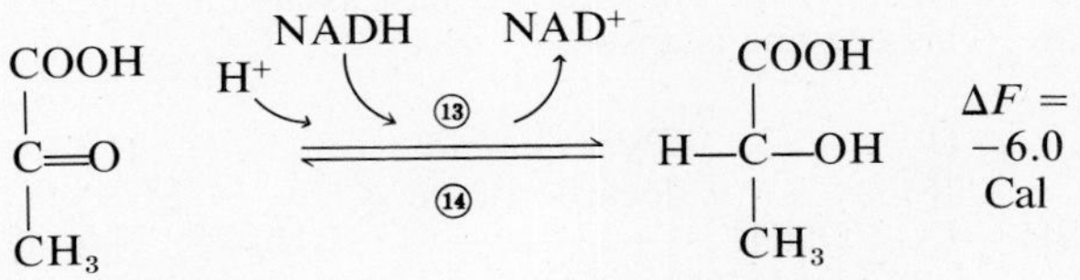

$$
\begin{array}{ccc}
\text{COOH} & \xrightarrow[\text{⑭}]{\text{H}^+ \;\; \text{NADH} \;\; \text{NAD}^+ \;\; \text{⑬}} & \text{COOH} \\
| & & | \\
\text{C}{=}\text{O} & & \text{H—C—OH} \\
| & & | \\
\text{CH}_3 & & \text{CH}_3
\end{array}
\qquad
\begin{array}{l}
\Delta F = \\
-6.0 \\
\text{Cal}
\end{array}
$$

Pyruvic   Lactic
acid   acid

The electrons required for this anaerobic reduction are carried by NADH, and derived from the oxidation of glyceraldehyde-3-phosphate to 3-phosphoglyceric acid (⑧). NAD is present within the cell in minute amounts compared to that of carbohydrate. Therefore, if the NAD^+ that was reduced to NADH during the oxidation of glyceraldehyde-3-phosphate were not reoxidized by electron transfer to pyruvate, then anaerobic glycolysis would stop as soon as all of the available NAD^+ was reduced to NADH. This effect, of course, is prevented by the action of lactic acid dehydrogenase, which enables NADH to be reoxidized to NAD^+ once again.

Lactic acid, the product of anaerobic glycolysis, is a metabolic dead end. Once formed, lactate can be metabolized further only by reconversion into pyruvic acid by the reversal of reaction ⑬, ie, reaction ⑭, and the pyruvic acid then is degraded through the pathways of pyruvate metabolism discussed in the next section.

In sharp contrast to the intermediate phophorylated compounds produced during glycolysis, lactate and pyruvate can diffuse readily out of the cells in which they are produced (principally muscle). Normally pyruvate is converted rapidly into lactate under anaerobic conditions in vivo, and the lactate thus formed then passes rapidly into the bloodstream, whence it is removed by the liver. Ultimately the lactate is reconverted in part back into glucose and glycogen. Thus, lactic acid escapes from cells as a metabolic waste product during periods of intense muscular activity when the cardiovascular system is unable to supply an adequate quantity of oxygen to the tissues, so that relatively anaerobic conditions prevail. During recovery, when sufficient oxygen is available once again, the excess lactic acid produced during exercise is reconverted into glucose and glycogen. This process, known as the lactic acid cycle (or Cori cycle, after its discoverer), is depicted in Figure 14.26, and will be discussed further in Chapter 16 in connection with the physiology of exercise (p. 1187).

Glycolysis and Alcoholic Fermentation. It is noteworthy that the sequence of chemical reactions described above, and that result in the formation of pyruvate from glucose under anaerobic conditions in mammalian tissues, is identical with the process that takes place in many microorganisms, especially yeast. However, in sharp contrast to animal tissues, the fate of pyruvic acid in

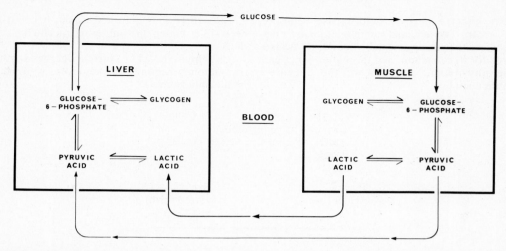

FIG. 14.26. The lactic acid (or Cori) cycle. See text for discussion. (After Harper. *Review of Physiological Chemistry*, 13th ed, 1971, Lange Medical Publications.)

yeast is an irreversible decarboxylation to acetaldehyde, a reaction that is catalyzed by the enzyme pyruvic acid decarboxylase. This enzyme is absent in animal tissues; similarly, lactic acid dehydrogenase is absent in yeast cells.

$$CH_3-\overset{\overset{\displaystyle O}{\|}}{C}-COOH \xrightarrow[Mg^{++}]{\text{Pyruvic acid}\atop\text{decarboxylase}} CH_3-\overset{\nearrow H}{\underset{\searrow O}{C}} + CO_2$$

Pyruvic acid   Acetaldehyde

The acetaldehyde produced in this fashion thus replaces pyruvic acid as the oxidant for the NADH resulting from the oxidation of glyceraldehyde-3-phosphate (reaction ⑧, Fig. 14.25). In the next step, acetaldehyde is reduced by NADH, this reaction being catalyzed by another enzyme which also is lacking in animal cells, alcohol dehydrogenase:

$$CH_3-\overset{\nearrow H}{\underset{\searrow O}{C}} \xrightarrow[\text{dehydrogenase}]{\overset{\text{NADH}\quad\text{NAD}^+}{\diagdown\quad\diagup}\atop\text{Alcohol}} CH_3-CH_2OH + H^+$$

Ethyl alcohol

The two anaerobic reactions found in yeast and summarized above, together with the glycolytic pathway glucose → pyruvate (Fig. 14.25), collectively are termed *alcoholic fermentation*. In fact, the anaerobic glycolysis of glucose to lactate as found in animal cells is often referred to as a fermentation process.

Some General Aspects of Glycolysis, Energetics and Efficiency. The reader will have noted that no molecular oxygen was utilized anywhere in the entire glycolytic sequence, 1 glucose → 2 lactate (Fig. 14.25). Furthermore, no net oxidation or reduction took place in this process, even though two individual steps in the glycolytic sequence are oxidoreduction reactions involving NAD$^+$ and NADH (viz., reactions ⑧ and ⑬).

The approximate net yield of usable energy derived from anaerobic glycolysis may now be calculated. One molecule of ATP is utilized at each kinase step. Thus, one molecule of ATP is required for the phosphorylation of one molecule of glucose (reaction ①), and one molecule of ATP also is required for the phosphorylation of each molecule of fructose-6-phosphate (reaction ④). When 1,3-diphosphoglyceric acid is converted into 3-phosphoglyceric acid (reaction ⑨), and when phosphoenolpyruvic acid is transformed into enol-pyruvic acid, two molecules of ATP are generated for each of these reactions. Thus, there is a net gain of two moles of ATP for each mole of glucose metabolized to lactate under anaerobic conditions (Table 14.8).

Although the actual free energy change (ΔF) for glycolysis is unknown under physiologic conditions, the overall efficiency of this process may be determined roughly. In vitro, under standard conditions, the degradation of glucose to lactate yields 5.6 Cal/mole of free energy. Since the production of two moles of ATP from ADP and P_i requires approximately 1.6 Cal of free energy under standard experimental conditions, it follows that the theoretical efficiency for the glycolytic process is $1.6/5.6 \times 100 = 29$ percent. Therefore, about one-third of the total chemical potential energy of glucose is stored as chemical energy in high-energy phosphate bonds of ATP during the process of anaerobic glycolysis alone.

Control of Glycolysis. The inherent regulatory mechanisms that control the rate of anaerobic glycolysis in mammalian cells are poorly understood. It is evident, however, that the overall rate of the glycolytic process must be determined by substrate availability, ATP utilization, and the concentrations as well as the activities of the various enzymes involved. For example, certain experimental studies (performed on yeast cells) have demonstrated definite rhythmic or oscillatory pattern insofar as NADH concentration is concerned. This oscillatory behavior with time results from a continuous alteration in the rate of the phospho-fructokinase reaction as substrate, product, and modifiers change their individual concentrations.

Aerobic Metabolism of Pyruvic Acid. Pyruvic acid, in a manner similar to glucose-6-phosphate, plays a key role in metabolism, and as shown in Figure 14.27, pyruvate may participate in a variety of metabolic reactions. However, the one reaction of major importance in most mammalian cells involving this intermediate is the oxidation of pyruvate to acetyl coenzyme A (abbreviated acetyl CoA, or CoA-SH; see Fig. 14.28) under aerobic conditions, and the subsequent oxidation of acetyl CoA to carbon dioxide and water in the tricarboxylic acid cycle (Fig. 14.29).

The high-energy compound acetyl CoA is formed in the presence of an abundant supply of oxygen from carbohydrates, lipids, and amino acids. Therefore, this coenzyme occupies a vital and central role in metabolism; however, for the purposes of the present discussion the oxidation of pyruvate will be stressed.

The conversion of pyruvate to acetyl CoA takes place according to the following net reaction:

$$
\begin{array}{l}
\mathrm{CH_3} \\
| \\
\mathrm{C{=}O} + \mathrm{CoA{-}SH} + \mathrm{NAD^+} \\
| \\
\mathrm{COOH}
\end{array}
\xrightarrow{\;\;⑮\;\;}
\begin{array}{l}
\mathrm{H^+} \qquad \mathrm{CH_3} \\
\qquad\quad | \\
\qquad\quad \mathrm{C{=}O} + \mathrm{NADH} \\
\mathrm{CO_2} \quad | \\
\qquad\quad \mathrm{S{-}CoA}
\end{array}
\qquad
\begin{array}{l}
\Delta F = \\
-8.0 \\
\mathrm{Cal}
\end{array}
$$

Pyruvic Acid   Acetyl CoA

The mechanism of this multistep process actually is one of the most complex to be found in metabolism. It is catalyzed by an enzyme complex known as pyruvic oxidase, and the operation of this system involves six cofactors (Table 14.6, ⑮). Essentially, however, pyruvate first is oxidized to acetic acid (CH_3—COOH), which in turn is converted into the acetyl derivative of coenzyme A (CoA—S—CO—CH_3) as depicted above, and this latter compound now reacts directly with oxaloacetic acid to form citric acid (Fig. 14.29).

The conversion of pyruvate to acetyl CoA is an irreversible process for all practical purposes because of the extremely large ΔF that is involved (approximately −8.0 Cal).

The reaction of acetyl CoA (sometimes called active acetyl) with oxaloacetic acid to form citric acid is the first step in the tricarboxylic acid cycle.

Tricarboxylic Acid Cycle. The process of glycolysis releases only a small fraction of the total chemical potential energy present in the glucose molecule. This fact may be summarized by the following two expressions:

  1. Glycolysis:
    Glucose → 2 Lactic acid; $\Delta F = -5.6$ Cal
  2. Respiration:
  Glucose + $6O_2 \rightarrow 6CO_2 + 6H_2O$; $\Delta F = -673.0$ Cal

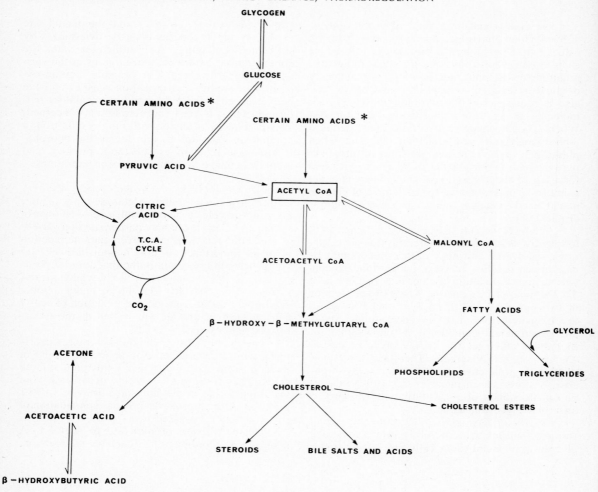

FIG. 14.27. Outline of the metabolism of pyruvate. Note the key position occupied by acetyl CoA in this scheme.* For the fates of particular amino acids see Table 14.11, and Figure 14.52. (Data from West, Todd, Mason, and Van Bruggen. *Textbook of Biochemistry*, 4th ed, 1966, Macmillan; White, Handler, and Smith. *Principles of Biochemistry*, 4th ed, 1968, The Blakiston Division, McGraw-Hill.)

The end products of anaerobic glycolysis (reaction 1) specifically produce two molecules of lactic acid which leave the muscle cell, as they cannot be metabolized any further. Yet these lactic acid molecules contain almost the same potential chemical energy as does the original glucose molecule, because lactate contains carbon atoms that are in the same oxidation state as in glucose; ie, the lactic acid molecule is highly organized, thus containing much potential free energy. The entropy of the molecule is low.* On the other hand, carbon dioxide and water are the end-products of tissue respiration (reaction 2), and the carbon dioxide molecule contains much less free energy as its atoms are completely oxidized (ie, the entropy of CO_2 is high). Furthermore, during respiration the quantity of energy derived from an individual reaction is far greater when the electrons are transferred from an intermediate com-

pound to an electron acceptor such as oxygen. This effect is in marked contrast to anaerobic glycolysis, in which the electrons are transferred to another type of electron acceptor, namely pyruvate.

The major final pathway whereby all foodstuffs (carbohydrates, lipids, proteins) are oxidized in the body to carbon dioxide and water is known as the tricarboxylic acid cycle, the citric acid cycle, or the Kreb's cycle. This overall oxidative process is termed cellular respiration, or simply respiration. Furthermore, once again in marked contrast to the glycolytic process, the enzyme systems responsible for catalyzing the aerobic oxidation of pyruvate to carbon dioxide and water are localized within the mitochondria, rather than found in the cytoplasm (Fig. 14.25, left side).

Reactions of the Tricarboxylic Acid (TCA) Cycle. The overall plan of the TCA cycle is shown in Figure 14.29, and the enzymes that catalyze the individual reactions involved are summarized in Table 14.7. The numbering of

Specifically the methyl carbon of lactic acid is more reduced than is the carboxyl carbon of this compound.

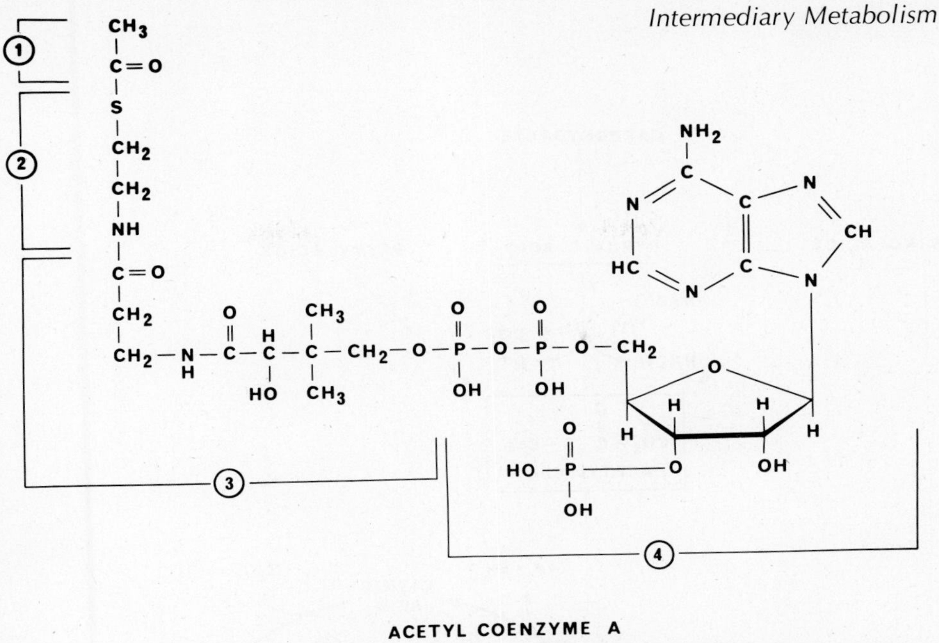

ACETYL COENZYME A

FIG. 14.28. Structural formula of acetyl CoA. ①= acetyl group; ②= β-mercaptoethylamine; ③= pantothenic acid; ④ = adenosine 3'-phosphate-5'pyrophosphate. The formation of one mole of acetyl CoA is equivalent, in terms of free energy harnessed, to the formation of one mole of ATP from ADP + P_i. (After White, Handler, and Smith. *Principles of Biochemistry,* 4th ed, 1968, The Blakiston Division, McGraw-Hill.)

specific reactions, as well as other conventions, are similar to those employed in the earlier discussion of the glycolytic pathway.

Fundamentally, the TCA cycle functions only to provide electrons to the enzymes of the terminal respiratory chain so that oxygen ultimately is reduced to water, while ATP is generated by the several coupled oxidative phosphorylation reactions. This process was discussed in detail earlier and is shown in Figure 14.20.

The net reaction of the TCA cycle is as follows for one revolution:

$$CH_3-COOH + 2H_2O \rightarrow 2CO_2 + 8H^+ + 8e + \text{energy}.$$

Thus the overall effect of one complete turn of the TCA cycle is the oxidation of one molecule of acetic acid (as acetyl CoA), and the production of two molecules of carbon dioxide, and, ultimately, four molecules of water. The reader should note

that ATP, P_i, and molecular oxygen do not participate directly in the TCA cycle; rather, these substances are involved only when the reducing equivalents generated by the oxidation of acetate are passed along to the terminal respiratory chain. Furthermore, each turn of the TCA cycle, when linked to the respiratory chain generates 12 molecules of ATP from ADP by the oxidation of the eight H atoms that are produced by the components of the cycle.

At physiologic pH, the acids involved in the citric acid cycle are present as anions, eg, oxaloacetate, citrate, and so forth. However, for the sake of clarity, these compounds are referred to and depicted in the text and figures as their un-ionized forms.

The first reaction of the TCA cycle is a condensation between one molecule of active acetate (acetyl CoA) and a molecule of oxaloacetic acid to form one molecule of citric acid, a reaction catalyzed by citrate synthetase:

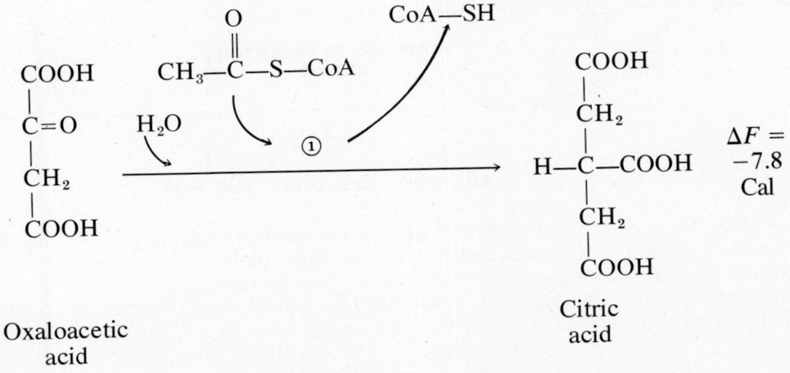

Oxaloacetic acid

Citric acid

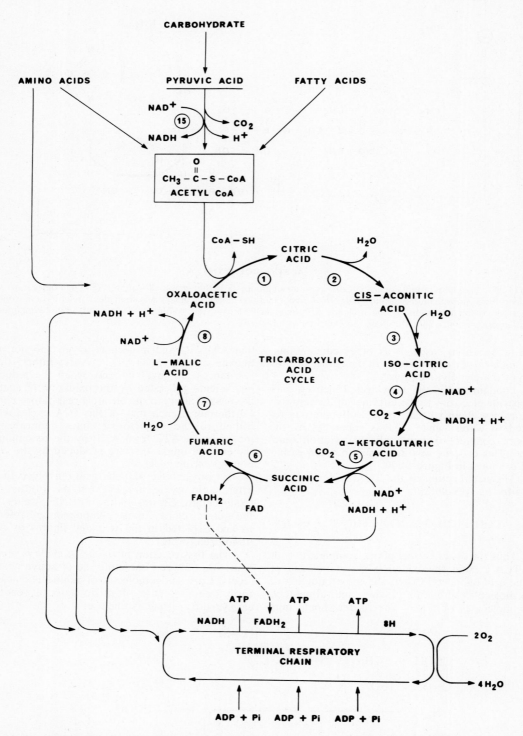

FIG. 14.29. Outline of the tricarboxylic acid cycle in relation to the terminal respiratory chain. The encircled numbers signify individual reactions as discussed in the text, and the enzymes that catalyze these reactions are summarized in Table 14.7. See also Figures 14.20 and 14.24. (Data from Harper. *Review of Physiological Chemistry,* 13th ed, 1971, Lange Medical Publications; Lehninger. *Biochemistry,* 1971, Worth Publishers; West, Todd, Mason, and Van Bruggen. *Textbook of Biochemistry,* 4th ed, 1966, Macmillan.)

Table 14.7 SUMMARY OF ENZYMES THAT PARTICIPATE IN THE TRICARBOXYLIC ACID CYCLE[a]

REACTION NUMBER; FIG. 14.29	ENZYME	COFACTORS AND ACTIVATORS	REMARKS
①	Citrate synthetase	Oxaloacetate	Favors formation of citrate. Inhibited by physiologic levels of ATP (= negative modifier of rate).
②	Aconitase	Fe^{++} Cysteine	The ferrous ion forms a stable chelate complex with citrate, and this is essential for the enzyme to act upon this substrate.
③	Aconitase	Fe^{++} Cysteine	
④	Isocitric acid dehydrogenase	Mg^{++} or Mn^{++} ADP	An allosteric enzyme; this enzyme generally is the most important rate-limiting step in the TCA cycle.
⑤	α-Ketoglutarate dehydrogenase complex (includes succinic acid thiokinase, a nucleoside diphosphokinase, and α-ketoglutarate oxidase)	CoA, thiamine pyrophosphate (TPP), lipoic acid, NAD^+	This complex of three enzymes is similar to the pyruvic acid dehydrogenase system. Oxidation at substrate level by this kinase yields 2∼Ⓟ .
⑥	Succinic acid dehydrogenase	——	Electrons passed to FAD rather than NAD as in all other reactions of TCA cycle.
⑦	Fumarase	——	Reaction specific for transformation of fumaric acid.
⑧	Malic acid dehydrogenase	NAD^+	The action of this enzyme regenerates oxaloacetic acid; thereby the TCA cycle recommences.

[a]*Data from Lehninger.* Biochemistry, *1970, Worth Publishers.*

The extremely high change in free energy $(-\Delta F)$ for this reaction strongly favors the formation of citrate under physiologic conditions.

It is quite significant biologically that the enzyme involved in this reaction, citrate synthetase, is inhibited by physiologically elevated levels of ATP. Therefore, ATP normally serves as a negative modifier for the synthesis of citric acid from acetyl CoA. When the cellular need for oxidizing acetyl CoA to generate ATP is low, then the metabolism of acetyl CoA is diverted into other pathways, such as the formation of fatty acids and/or acetoacetic acid.

Aconitase catalyses the reversible transformation of citric acid into *cis*-aconitic acid, thence to isocitric acid according to the equilibria:

| Citric acid (90 percent) | *cis*-Aconitic acid (4 percent) | Isocitric acid (67 percent) |

The figures in parentheses below the intermediates in these reactions indicate the approximate equilibrium mixture of the components by percentages at pH 7.4 and 25 C. Note that a reversible addition of a water molecule takes place. The compounds involved are highly stereospecific, insofar as their structure is concerned in relation to that of the aconitase molecule. That is, the transformation of citric acid into *cis*-aconitic acid results from an asymmetric dehydration, and isocitric acid is formed as the consequence of a stereospecific hydration. A single enzyme, aconitase, appears to be responsible for catalyzing both of these processes.

Although the equilibria for reactions ② and ③ favor the formation of citric acid, under physiologic conditions the reaction goes strongly to the right because isocitric acid is oxidized rapidly to α-keto-glutaric acid under the catalytic influence of mitochondrial NAD^+-linked isocitric dehydrogenase:

$$
\begin{array}{ll}
\text{COOH} \\
| \\
\text{CH}_2 \\
| \\
\text{H—C—COOH} \\
| \\
\text{HO—C—H} \\
| \\
\text{COOH} \\
\text{Isocitric} \\
\text{acid}
\end{array}
\qquad
\begin{array}{l}
NAD^+ \qquad NADH + H^+ \\
\qquad\qquad CO_2 \\
\xrightarrow{\quad④\quad} \\
\end{array}
\qquad
\begin{array}{ll}
\text{COOH} \\
| \\
\text{CH}_2 \\
| \\
\text{CH}_2 \qquad \Delta F = \\
| \qquad\qquad -5 \\
\text{C=O} \qquad\quad \text{Cal} \\
| \\
\text{COOH} \\
\alpha\text{-Ketoglutaric} \\
\text{acid}
\end{array}
$$

Isocitric dehydrogenase is an extremely important rate-limiting enzyme which possesses an allosteric site which may react with small molecules and thus be activated. Hence isocitric dehydrogenase requires Mg^{++} or Mn^{++} for its activation. Furthermore, this enzyme is linked to NAD^+, and it also specifically requires ADP, in addition to a divalent ion, as an activator. Therefore, if the intramitochondrial level of ADP rises, the activity of the enzyme increases also, so that the rate of isocitrate oxidation, hence the rate of acetate oxidation by the entire TCA cycle increases. This effect in turn causes a net increase in the rate of electron transport, hence coupled oxidative phosphorylation of ADP to ATP in the terminal respiratory chain, and the overall result is a more rapid net synthesis of ATP. Conversely, if the ATP concentration (or the NADH level) rises, this decreases the ADP concentration, thereby reducing the activity of isocitrate dehydrogenase once again. By means of this mechanism the rate of the entire TCA cycle is regulated with exquisite precision by the effects of ADP and NAD^+ levels upon the catalytic activity of isocitrate dehydrogenase.

Although two additional isocitric dehydrogenases are known to be present in the cytoplasm of mammalian cells, apparently only one such enzyme is present within the mitochondria and this intracellular isocitric dehydrogenase is responsible for catalysis of the isocitric acid → α-ketoglutaric acid conversion in the TCA cycle. Oxalosuccinic acid (not shown) is not a proven intermediate in this process within the mitochondria.

The oxidation of α-ketoglutaric acid to succinic acid is a complex, multienzyme-catalyzed process that involves the formation of succinyl CoA as one intermediate. These reactions may be outlined as follows:

$$
\begin{array}{ll}
\text{COOH} \\
| \\
\text{CH}_2 \\
| \\
\text{CH}_2 \\
| \\
\text{C=O} \\
\alpha\text{-Keto-} \\
\text{glutaric} \\
\text{acid}
\end{array}
\quad
\begin{array}{l}
\text{CoA—SH} \quad H^+ + \\
\quad NAD^+ \quad\; NADH \\
\xrightarrow{\quad⑤a\quad} \\
\qquad\qquad CO_2
\end{array}
\quad
\begin{array}{ll}
\text{O=C—S—CoA} \\
| \\
\text{CH}_2 \\
| \\
\text{CH}_2 \\
| \\
\text{COOH} \\
\text{Succinyl} \\
\text{CoA}
\end{array}
\quad
\begin{array}{l}
\text{GDP} \qquad \text{GTP+CoA} \\
\xrightarrow{\quad⑤b\quad} \\
\text{Pi}
\end{array}
\quad
\begin{array}{ll}
\text{COOH} \\
| \\
\text{CH}_2 \\
| \\
\text{CH}_2 \qquad \Delta F = \\
| \qquad\qquad -8 \\
\text{COOH} \qquad\; \text{Cal} \\
\text{Succinic} \\
\text{acid}
\end{array}
$$

GDP = Guanosine Diphosphate; GTP = Guanosine Triphosphate

These reactions are catalyzed by a multienzyme complex known as the α-ketoglutarate dehydrogenase complex; this group of functionally integrated enzymes is analogous to the pyruvic acid dehydrogenase system.

The guanosine triphosphate which is produced by reaction ⑤ᵇ may phosphorylate ADP to ATP directly (GTP + ADP → GDP + ATP). Guanosine triphosphate may also be used directly for many other diverse metabolic processes, eg, the activation of fatty acids. Thiamine pyrophosphate (TPP, derived from vitamin B₁) also is an essential cofactor for the oxidation of α-ketoglutaric acid to succinic acid. The intermediate in this initial step in the oxidative process is α-hydroxy-γ-carboxypropyl thiamine pyrophosphate (not shown above), which is formed by the reaction of α-ketoglutaric acid with TPP with the attendant elimination of a carbon dioxide molecule.

The oxidation of succinic acid to fumaric acid is the sixth step in the TCA cycle:

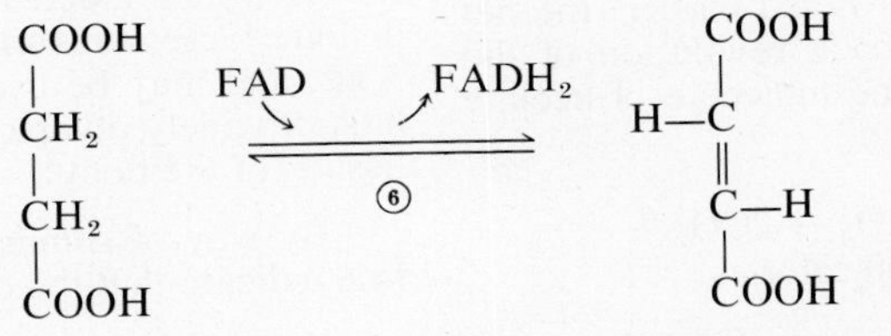

Succinic acid → Fumaric acid (*trans*-form)

This dehydrogenation, catalyzed by succinic acid dehydrogenase, is the only oxidative reaction in the TCA cycle in which NAD (ie, any pyridine nucleotide) is not involved; thus reducing equivalents enter the terminal respiratory chain at the FAD level (Fig. 14.29), rather than at the NAD level, and two, rather than three molecules of ATP are generated by this reaction as shown to the right in the figure.

Fumaric acid now is hydrated reversibly to form L-malic acid, this reaction being catalyzed by the highly stereospecific enzyme fumarase:

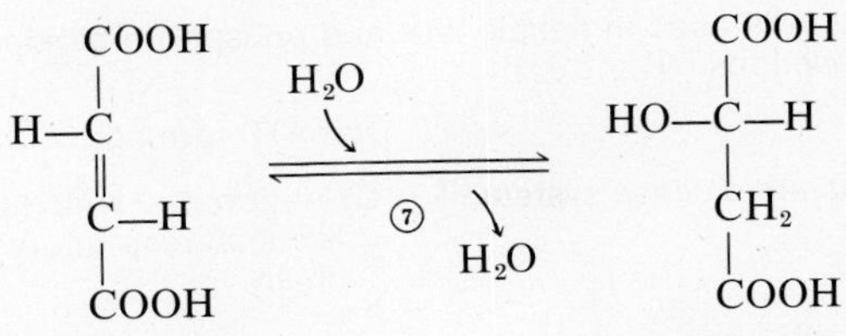

Fumaric acid → L-malic acid

In the eighth and final step of the TCA cycle, malic acid is oxidized to yield oxaloacetic acid, in a reaction catalyzed by malic acid dehydrogenase; this reaction requires NAD^+:

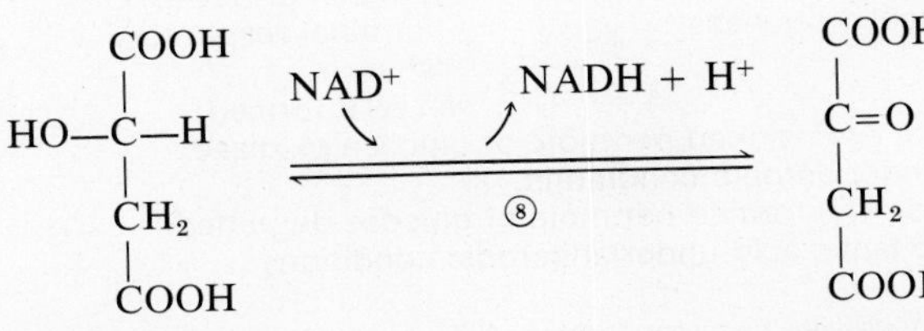

L-Malic acid → Oxaloacetic acid

With this reaction the TCA cycle is completed. The oxaloacetic acid regenerated by the dehydrogenation of L-malic acid now is available for condensation with acetyl CoA and the process can be repeated.

It is interesting to observe, that theoretically at least, one molecule of oxaloacetic acid may be reused in the TCA cycle an infinite number of times (see also anaplerosis, p. 953).

General Aspects of the TCA Cycle; Net Energy Yield and Approximate Efficiency. The TCA cycle is a biochemical mechanism whereby acetate, α-ketoglutaric acid, fumaric acid, and malic acid are oxidized. As noted earlier, the net result of this process for each revolution of the cycle is the oxidation of one molecule of acetate with the liberation of energy:

$$CH_3\text{—}COOH + 2O_2 \rightarrow 2CO_2 + 2H_2O + 209,000 \text{ calories.}$$

As shown in Table 14.8, each revolution of the TCA cycle generates 12 molecules of ATP. Stated another way, the oxidation of 1 mole of acetate yields 12 moles of ATP, which is equivalent to approximately 8.4 Cal. Since the oxidation of 1 mole of acetic acid in vitro under standard conditions yields 209 Cal, then the efficiency of the TCA cycle in producing useful free energy that is stored as high-energy phosphate bonds of ATP is $(8.4/209) \times 100 \cong 40$ percent.

The oxidation of metabolites in a gradual, stepwise fashion by the TCA cycle combined with the terminal electron transport chain releases definite increments of energy which are stored as ATP, and may be used as "common coin" for a host of widely diverse cellular activities within the tissues of the body.

By way of summary, the data given in Table 14.8 indicate that the complete oxidation of 1 mole

Table 14.8 YIELD OF HIGH-ENERGY BONDS BY THE CATABOLISM OF GLUCOSE[a]

PATHWAY	REACTION NUMBER AND ENZYME INVOLVED	MODE OF ~℗ GENERATION	NUMBER OF ~℗ GENERATED PER MOL OF GLUCOSE
A. Glycolysis (Figure 14.25)	⑧ Glyceraldehyde 3-phosphate dehydrogenase	Oxidation of 2 NADH by terminal respiratory chain	6
	⑨ 3-Phosphoglyceric acid kinase	Oxidation at substrate level	2
	Pyruvic acid kinase	Oxidation at substrate level	2
	⑫ Less ATP used in hexokinase and phosphofructokinase reactions (① and ④)		10 −2
		Net ATP formed:	8
B. Tricarboxylic acid cycle (Figure 14.29)	⑮ Pyruvate oxidase system	Oxidation of 2 NADH by terminal respiratory chain	6
	④ Isocitric acid dehydrogenase	Oxidation of 2 NADH by terminal respiratory chain	6
	⑤ₐ α-Ketoglutaric acid dehydrogenase	Oxidation of 2 NADH by terminal respiratory chain	6
	⑤ᵦ Succinic acid thiokinase	Oxidation at substrate level	2
	⑥ Succinic acid dehydrogenase	Oxidation of 2 $FADH_2$ by terminal respiratory chain	4
	⑧ Malic acid dehydrogenase	Oxidation of 2 NADH by terminal respiratory chain	6
		Net ATP formed:	30
	Total ATP formed per mole of glucose oxidized under aerobic conditions:		38
	Total ATP formed per mole of glucose degraded to lactic acid under anaerobic conditions:		2

[a]*Data from Harper.* Review of Physiological Chemistry, *13th ed, 1971, Lange Medical Publications; White, Handler, and Smith.* Principles of Biochemistry, *4th ed, 1968, The Blakiston Division, McGraw-Hill; West, Todd, Mason, and VanBruggen.* Textbook of Biochemistry, *4th ed, 1966, Macmillan.*

of glucose would give a total theoretical yield of 38 moles of ATP:

$$C_6H_{12}O_6 + 38 \text{ ADP} + 38 \text{ P}_i \rightarrow$$
$$6 \text{ CO}_2 + 6 \text{ H}_2O + 38 \text{ ATP}$$

If one assumes that around +8,000 cal/mole ΔF is required for the formation of each high-energy phosphate bond of ATP, and that the glucose molecule contains 673,000 cal/mole of total potential chemical energy, then $38 \times 8,000/673,000 \times 100 = 45$ percent of the total energy from a mole of glucose is conserved (or harnessed) as useful phosphate bond energy. This figure also indicates the efficiency of the entire oxidative process.

Respiration and Oxidative Phosphorylation. It is evident from the material presented thus far that a major physiologic goal of the TCA cycle (Fig. 14.29) combined with the terminal electron transport chain is to conserve free energy which is liberated by the stepwise oxidation of carbohydrates (and other nutrients) as high-energy phosphate bonds of ATP.* These tightly coupled oxidative phosphorylation reactions are known to occur within the mitochondria, in contrast to the process of glycolysis that takes place within the cytoplasm. The underlying mechanism(s) whereby the fundamental transformation of free energy takes place among various compounds remains unknown despite the large research effort which has been directed toward solution of this problem.† Nevertheless, it has been amply demonstrated under controlled experimental conditions that the oxidation of one molecule of NADH takes place with a net formation of three molecules of ATP. An identical quantity of ATP also is formed per atom of oxygen consumed; therefore the P/O ratio is 3/1 or 3. Thus, regardless of the substrate involved, it is evident that ATP is synthesized concomitantly with the reoxidation of NADH which in turn is produced by oxidation of the substrate. Furthermore, with the sole exception of the oxidation of α-glutaric acid (reaction ⑤b, p. 950), the oxidation of a substrate only serves to provide reducing equivalents to the terminal electron transport chain (Fig. 14.20).

A P/O ratio of 3 should be regarded as minimal, since under certain restricted experimental conditions a P/O ratio as high as 6 has been found per mole of NADH oxidized. Hence, theoretically at least, the process of oxidation of acetate by the TCA cycle could yield six moles of ATP instead of three, with an attendant efficiency of 50 percent (assuming that the $\Delta F = +8,000$ cal/mole ATP generated), or even higher.

Control of Cellular Respiration. It is implicit in the concept of oxidative phosphorylation that if

these intramitochondrial reactions are tightly coupled to respiration (ie, both processes are inseparable), then respiration can take place over such phosphorylation pathways only at a rate which is commensurate with the availability of ADP and P_i. In actual fact, it has been proven unequivocally that the rate of cellular respiration is an inverse function of the ATP/(ADP + P_i) ratio.

It may be concluded that the tight coupling of oxidation to phosphorylation, as present in normal mitochondria, is a fundamental mechanism whereby the rate of oxidation of nutrients (whether carbohydrate, lipid, or protein) is regulated by the cellular requirements for utilizable energy. As a corollary to this statement, the utilization of ATP as an energy source to drive the many cellular processes automatically increases the quantity of ADP and P_i that are available to react in the coupling processes, and thus allow respiration to proceed in an uninterrupted fashion.

Another important factor is involved in respiratory control at the cellular level. An enzyme known as adenylic acid kinase is widely distributed in various tissues, and this enzyme catalyzes the reaction:

$$2 \text{ ADP} \underset{②}{\overset{①}{\rightleftharpoons}} \text{ATP} + \text{AMP}$$

The physiologic consequence of this process is to increase the utilization of the energy derived from ATP when this compound is being used at a rate faster than it is being synthesized (reaction ①), ie, when the supply of ATP becomes limited within the cell. In the reverse direction, ②, this reaction is the first step in the synthesis of ATP from adenosine monophosphate (AMP, Fig. 14.30). AMP is produced in many synthetic processes, eg, during the activation of amino or fatty acids. AMP is also of considerable physiologic importance as this compound (rather than ADP) acts either as a positive or a negative modifier that controls the rates of a number of enzymatic processes. For example, the phosphofructokinase (④, Fig. 14.25), diphosphofructose phosphatase (⑤, Fig. 14.25), and isocitric acid dehydrogenase (④, Fig. 14.29) reactions are all regulated by the intracellular AMP concentration.

Anaplerosis. At each revolution of the TCA cycle, a molecule of oxaloacetic acid is regenerated to initiate the next turn of the cycle (Fig. 14.29) and were it not for several additional metabolic fates for oxaloacetate, this process could continue indefinitely. However, oxaloacetic acid (in addition to α-ketoglutaric acid and succinyl CoA) can be diverted to metabolic pathways other than the TCA cycle. For example, oxaloacetic acid may be decarboxylated to pyruvic acid and carbon dioxide, a reaction that occurs both spontaneously as well as by enzymatic catalysis. A net loss of oxaloacetate by this process inevitably would decrease the rate at which the TCA cycle

---

Another physiologic role of the TCA cycle is in the formation of α-ketoglutaric acid for glutamine synthesis (Fig. 14.49).

†*Furthermore, it is doubtful that this process will be clarified in detail until such time as the actual nature of free energy itself is understood!*

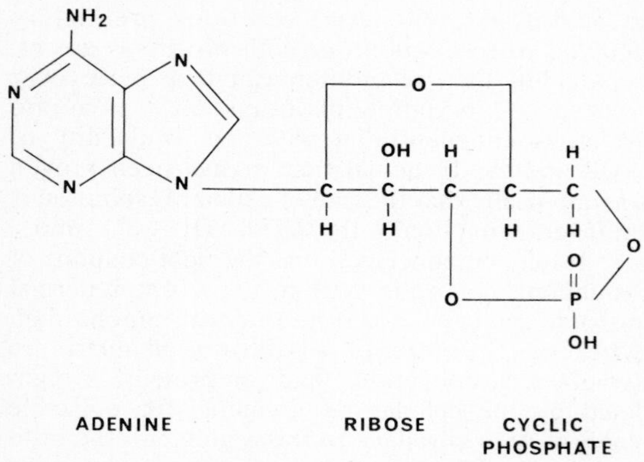

ADENINE　　　**RIBOSE**　　**CYCLIC PHOSPHATE**

FIG. 14.30. Structural formula of cyclic 3′, 5′-adenosine monophosphate. (After Harper. *Review of Physiological Chemistry*, 13th ed, 1971, Lange Medical Publications.)

operates, were it not for other reactions which compensate for this loss. Thus oxaloacetic acid may arise by transamination of glutamic acid or this amino acid also may be metabolized to α-ketoglutaric acid. The oxaloacetate and α-ketoglutarate arising from these reactions then may enter the TCA cycle. However, in normal liver as well as in growing cells of various types, both of these compounds are transaminated readily back to the respective amino acids, so that an alternative mechanism must be available to supply these essential TCA cycle intermediates.

Several reactions take place within cells that replenish the supply of keto acid through the fixation of carbon dioxide with pyruvic acid, a process known as anaplerosis, are present within cells. Three of these important reactions shown in order of decreasing quantitative physiologic importance are:

$$\text{Pyruvic acid} + CO_2 + ATP \xrightarrow{Mg^{++}} \quad (1)$$
$$\text{oxaloacetic acid} + ADP + P_i$$

This reaction is catalyzed by pyruvic acid carboxylase, and has a specific requirement for acetyl CoA. Although the acetyl CoA does not actually participate in the carbon dioxide fixation reaction, it serves as a positive modifier insofar as the activity of the pyruvic acid carboxylase is concerned.

$$\text{Pyruvic acid} + CO_2 + NADH + H^+ \underset{}{\overset{Mn^{++}}{\rightleftarrows}} \quad (2)$$
$$\text{L-malic acid} + NAD^+$$

This reaction has a similar mechanism to that catalyzed by the isocitric acid dehydrogenase complex, and it is catalyzed by the so-called malic enzyme. Since malic acid is a normal TCA cycle intermediate, and since this compound is oxidized readily into oxaloacetic acid by NAD^+ in the presence of malic acid dehydrogenase, this reaction also provides a ready source of oxaloacetate for the uninterrupted operation of the TCA cycle.

$$\text{Phosphoenolypyruvic acid} + CO_2 + \quad (3)$$
$$\text{Inosine diphosphate} \underset{}{\overset{Mg^{++}}{\rightleftarrows}}$$
$$\text{Oxaloacetic acid} + \text{Inosine triphosphate}$$

The enzyme responsible for catalysis of this reaction is phosphoenolpyruvic acid carboxykinase. This enzyme is quite widely distributed in animal tissues (eg, liver and muscle), and it is activated significantly by acetyl CoA. Actually, it appears that this third anaplerotic reaction is of doubtful quantitative importance as a source of oxaloacetate in vivo. Rather, the reverse process, ie, the formation of phosphoenolpyruvic acid by this reaction, is of major significance in the process of gluconeogenesis.

The Phosphogluconate Oxidative Pathway (Hexose Monophosphate Shunt). The glycolytic pathway (Fig. 14.25; Table 14.6) is but one of two major series of reactions whereby glucose is metabolized in mammalian cells. The second important mechanism for this process, which like the glycolytic mechanism also is found in the cytoplasm, is known as the phosphogluconate oxidative pathway or the hexose monophosphate shunt (abbreviated HMP shunt). This direct pathway for glucose metabolism involves a complex series of reactions that are summarized in Table 14.9. The HMP shunt involves several intermediate compounds and enzymes that are found in the glycolytic pathway in addition to several unique intermediates and transformations.

The phosphogluconate pathway is a biochemical mechanism whereby glucose is oxidized completely as well as independently of the tricarboxylic acid cycle. It is also an important mechanism for the production of $NADPH_2$ which is essential to the synthesis of fatty acids, as well as the action of a number of hyroxylases. Furthermore, this pathway acts as a source of D-ribose for nucleic acid synthesis as well as a source of 4- and 7-carbon sugars.

Examination of the individual reactions involved in the phosphogluconate pathway (Table 14.9) reveals that no ATP is required once glucose has been phosphorylated to glucose-6-phosphate. For the operation of this pathway, it will be noted that four of the enzymes which participate in glycolysis are necessary. These enzymes are: first, phosphoglucose isomerase, to convert fructose-6-phosphate to glucose-6-phosphate (③, Fig. 14.25); second, diphosphofructose phosphatase, to cause the hydrolysis of fructose-1,6-diphosphate to fructose-6-phosphate (⑤, Fig. 14.25); third, aldolase, to catalyze the formation of fructose-1,6-diphosphate from the triose phosphates (⑥, Fig. 14.25); and fourth, triose phosphate isomerase to catalyze the interconversion of glyceraldehyde-3-phosphate and dihydroxyacetone phosphate (⑦, Fig. 14.25). These four enzymes normally partici-

Table 14.9 SUMMARY OF REACTIONS OF THE PHOSPHOGLUCONATE OXIDATION PATHWAY (HEXOSE MONOPHOSPHATE SHUNT)[a]

NET REACTION (6) Glucose-6-phosphate + (12) NADP $\longrightarrow$ (5) Glucose-6-phosphate + (6) CO_2 + (12) $NADPH_2$ + (12) H^+ + P_i[b]

① (6) Glucose-6-phosphate + (6) NADP $\xrightarrow{\text{Glucose-6-phosphate dehydrogenase}}$ (6) 6-phosphogluconate + (6) $NADPH_2$ + (6) H^+

② (6) 6-Phosphogluconate + (6) NADP $\xrightarrow{\text{Phosophogluconic acid dehydrogenase}}$ (6) Ribulose-5-phosphate + (6) $NADPH_2$ + (6) H^+ + (6) CO_2

③ a (2) Ribulose-5-phosphate $\xrightarrow{\text{Pentose epimerase}}$ (2) Xylulose-5-phosphate

  b (2) Ribulose-5-phosphate $\xrightarrow{\text{Pentose isomerase}}$ (2) Ribose-5-phosphate

  c (2) Xylulose-5-phosphate + (2) Ribose-5-phosphate $\xrightarrow{\text{Transketolase}}$ (2) Sedoheptulose 7-phosphate + (2) Glyceraldehyde-3-phosphate

④ (2) Sedoheptulose-7-phosphate + 2-glyceraldehyde-3-phosphate $\xrightarrow{\text{Transaldolase}}$ (2) Erythrose 4-phosphate + (2) Fructose-6-phosphate

⑤ a (2) Ribulose-5-phosphate $\xrightarrow{\text{Pentose epimerase}}$ (2) Xylulose-5-phosphate

  b (2) Xylulose-5-phosphate + 2 erythrose 4-phosphate $\xrightarrow{\text{Transketolase}}$ (2) Glyceraldehyde 3-phosphate + (2) Fructose-6-phosphate

⑥ Glyceraldehyde-3-phosphate $\xrightarrow{\text{Triose isomerase}}$ Dihydroxyacetone phosphate

⑦ Dihydroxyacetone phosphate + glyceraldehyde 3-phosphate $\xrightarrow{\text{Aldolase}}$ Fructose-1,6-diphosphate

⑧ Fructose-1,6-diphosphate $\xrightarrow{\text{Phosphatase}}$ Fructose-6-phosphate + P_i

⑨ (5) Fructose-6-phosphate $\xrightarrow{\text{Hexose isomerase}}$ (5) Glucose-6-phosphate

[a]*Data from White, Handler, Smith. Principles of Biochemistry, 4th ed, 1968, The Blakiston Division, McGraw-Hill.*
[b]*The number of moles of each substrate and intermediate involved in these reactions is placed within parentheses ( ) to the left of each reactant. See text for further discussion of the reactions presented in this table.*

pate in the physiologic synthesis of hexose from pyruvate or triose moieties by reversal of the usual glycolytic sequence.

General Aspects of the Phosphogluconate Pathway. By reference to Table 14.9, one may visualize a metabolic pathway for glucose-6-phosphate whereby the hexose enters continuously and carbon dioxide leaves as the sole carbon byproduct of the reaction sequence. The individual balanced equations presented in this table describe the overall conversion of 6 moles of glucose-6-phosphate to 5 moles of glucose-6-phosphate with the evolution of 6 moles of carbon dioxide. This is indicated by the equation for the net reaction describing this entire sequence which is placed at the top of the table. The reader should note that the only reaction in this series in which carbon dioxide is evolved is through the oxidation of 6-phosphogluconic acid (reaction ②).

It is important to realize that the phosphogluconate pathway is strictly an oxidizing mechanism accomplished solely by reactions ① and ② (Table 14.9), in which NADP is reduced to $NADPH_2$ and CO_2 is evolved. The oxidation of 6 moles of hexose phosphate by these two reactions

causes 12 pairs of electrons (ie, 24 electrons in all) to be transferred to NADP. This is the quantity of electrons which is required for the complete oxidation of 1 mole of glucose to carbon dioxide. Following reaction ②, 6 moles of ribulose 5-phosphate remain, and if these are utilized for the synthesis of nucleotides, then they only have to undergo isomerization to ribose-5-phosphate. On the other hand, if the total number of carbon atoms present in the 6 moles of pentose were reorganized to produce 5 moles of hexose, then the net consequence of this effect would be to oxidize completely 1 of the 6 moles of hexose phosphate originally present. Such a rearrangement of the carbon atoms into hexose commences in reactions ③ and ④, leaving a balance of 2 moles of pentose and 2 of tetrose. The 2 reactions summarized under ⑤ employ the 2 moles of pentose to generate 2 additional moles of pentose and 2 of triose (⑤b).

Reactions ⑥ and ⑧ now transform the 2 triose moieties into one hexose molecule. Note that in each series of reactions, fructose-6-phosphate is the final hexose generated (reactions ⑤a and ⑤b; reaction ⑧), and this compound ultimately undergoes isomerization (reaction ⑨) to glucose-6-phosphate which now may reenter this metabolic sequence at step ①. Thereby the process continues in a cyclic manner.

Physiologic Significance of the Phosphogluconate Oxidative Mechanism. It is important to realize that in mammalian striated muscle, no direct oxidation of glucose occurs via the phosphogluconate pathway. Rather, the catabolism of glucose takes place in this tissue exclusively via glycolysis and the TCA cycle. In liver, however, a significant fraction (over 30 percent) of the total carbon dioxide arising from glucose oxidation is derived by oxidation of this carbohydrate in the phosphogluconate pathway. In certain tissues, namely the mammary gland, the adrenal cortex, leukocytes, adipose tissue, and testis an even higher percentage of glucose catabolism takes place by the oxidation of phosphogluconate.

Certain physiologic advantages accrue to the organism from this direct oxidative pathway for glucose, in addition to the fact that the HMP shunt forms pentoses for nucleotide synthesis. First, no ATP is required other than for the initial formation of glucose-6-phosphate. Second, the operation of the phosphogluconate mechanism is wholly independent of the availability of the 4-carbon dicarboxylic acids of the TCA cycle. Third, the phosphogluconate pathway utilizes NADP exclusively as the electron acceptor.

It is uncertain at present whether or not ATP can be synthesized efficiently by the intramitochondrial oxidation of $NADPH_2$ that has been formed within the cytoplasm; however, if this did take place, then 3 moles of ATP would be generated per mole of $NADPH_2$ formed, and this in turn would yield 36 moles of ATP for the complete oxidation of one mole of glucose by the HMP shunt. This figure would compare favorably to the yield of ATP which is derived from glycolysis and the TCA cycle (net yield of 38 moles of ATP generated per mole of glucose oxidized, Table 14.8).

It would appear, however, that most of the $NADPH_2$ produced by the phosphogluconate pathway is utilized as a reducing agent, principally for the synthesis of fatty acids and/or steroids. This conclusion is consistent with the fact that the enzymes required for these processes are present in abundance in the tissues listed above (eg, adrenal and testis), and synthetic processes of this type are negligible in striated muscle. In cells that possess the biochemical machinery for the phosphogluconate oxidative pathway, the factor that limits the rate of operation of this mechanism is the availability of NADP. This fact would indicate a coupling of glucose oxidation to the requirements of the reactions that employ $NADPH_2$ as an essential component.

Conversion of Lactate to Glucose. It has been known for many years that an isolated skeletal muscle contracting under anaerobic conditions produces lactic acid from glycogen. Furthermore, after oxygen is readmitted to such an in vitro system, the lactic acid disappears, approximately 20 percent of this compound being oxidized to carbon dioxide, whereas the remainder is reconverted back into glycogen. These facts are summarized in Figure 14.26.

A similar situation would obtain in liver that receives blood lactic acid which has been produced in, and released by, skeletal muscle provided that an adequate oxygen supply is present. ATP would be synthesized within the mitochondria by the oxidative phosphorylation of pyruvic acid to phosphoenolpyruvic acid (⑫, Fig. 14.25), and also by the oxidation of lactic acid to pyruvic acid (⑭, Fig. 14.25). If the pyruvic acid kinase reaction (⑫) were easily reversible, then the phosphoenolpyruvic acid formed could be converted readily back into glucose by the reverse of the reactions shown in Figure 14.25. The pyruvate kinase reaction, however, is reversible only at a very slow rate. Consequently the conversion of pyruvic acid back to phosphoenolpyruvic acid must take place through an alternate pathway in order that lactic or pyruvic acids (regardless of their metabolic origin) are to be converted into glucose.

This alternate pathway operates through the enzymes that underlie the process of anaplerosis. Acetyl CoA and ATP, which are generated by the intramitochondrial oxidation of some of the pyruvic acid lend direction to the process, since acetyl CoA is required for the operation of pyruvate carboxylase. The oxaloacetic acid produced by this reaction in turn is reduced by malic dehydrogenase into malate, and this latter compound then diffuses out of the mitochondria into the cytoplasm where it is reoxidized into oxaloacetic acid by cytoplasmic malic dehydrogenase. The phosphoenolpyruvate carboxykinase of the cytoplasm when activated by acetyl CoA now employs the energy of ITP or GTP in order to catalyze the transformation of oxaloacetic acid into phosphoenolpyruvic acid. The reactions and enzymes that participate in this process are summarized in Figure 14.31.

The phosphoenolpyruvic acid synthesized in this manner now is converted into fructose-1.6-diphosphate by reversal of reactions ⑪, ⑩, ⑨, ⑧, ⑦, and ⑥ (Fig. 14.25). The NADH necessary for the reduction of 1,3-diphosphoglyceric acid to glyceraldehyde-3-phosphate (⑧) is produced by the concomitant oxidation of lactic acid to pyruvic acid.

It is readily apparent from the facts presented above that glucose formation from lactate is not entirely a simple reversal of the glycolytic reactions.

From the standpoint of bioenergetics, the overall reaction for glycolysis is:

$$\text{Glucose} + 2\ \text{ADP} + 2\ P_i \rightarrow$$
$$2\ \text{Lactic acid} + 2\ \text{ATP} \qquad (1)$$

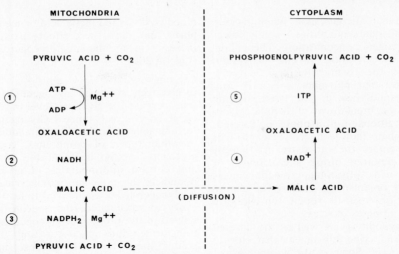

FIG. 14.31. Pathway for the metabolic formation of phosphoenolpyruvic acid. The specific enzymes that catalyze these reactions are: ①, pyruvic carboxylase; ②, malic acid dehydrogenase; ③, malic enzyme; ④, malic acid dehydrogenase; ⑤, phosphoenolpyruvate carboxykinase. (After White, Handler, and Smith. *Principles of Biochemistry*, 4th ed, 1968, The Blakiston Division, McGraw-Hill.)

The reverse of this process, ie, the formation of hexose from lactic acid is:

$$2 \text{ Lactic acid} + 6 \text{ ATP} \rightarrow$$
$$6 \text{ Glucose} + 6 \text{ ADP} + 2 \text{ P}_i \qquad (2)$$

Thus for each molecule of lactic acid that is resynthesized into hexose as shown in equation (2), a molecule of ATP is necessary in the reactions catalyzed by pyruvate carboxylase, phosphoenolpyruvate carboxykinase (in fact, ITP or GTP is utilized here), the two malic acid dehydrogenases, and by phosphoglyceric acid kinase. Since there is a net difference between the glycolytic and synthetic processes that is equivalent to four high-energy ATP bonds, then it is this difference that drives the formation of glucose from lactate. Actually the ΔF for reaction (1) is around -30 Cal/mole, whereas that for reaction (2) is about -48 Cal/mole. Consequently the energy which is derived from the hydrolysis of 6 ATP molecules is used to reverse the process.

The Physiologic Role of Glycogen; Regulation of Liver and Muscle Glycogen Stores. Normally, the tissues of the body contain large stores of chemical potential energy, principally in the form of lipids and the polysaccharide glycogen, which in turn is a high molecular weight polymer of glucose. Since lipids are insoluble in water, and glycogen is only slightly soluble, these components exert very little effect upon the overall osmotic pressure of the cells of the body. In mammals, including human beings, glycogen occurs in relatively small quantities; therefore lipids are the principal storage form for chemical energy. Glycogen is found in most tissues, including adipose tissue; however, the glycogen of liver and skeletal muscle are the stores of this material which are of major physiologic importance.

Regardless of the tissue in question, glycogen is dispersed within the cells and a considerable spectrum of molecular weights exists for this carbohydrate, the range being from around 2×10^5 to 10^7 daltons (Fig. 13.21, p. 859).

Hepatic Glycogen. In a fed mammal, the liver normally contains between two and eight percent glycogen on a wet weight basis, but the glycogen present is constantly undergoing degradation and resynthesis. That is, the hepatic glycogen undergoes rapid and continual metabolic turnover, and it is not stored as an inert substance. In contrast to a fed animal, however, fasting for less than a day will deplete the hepatic glycogen content to practically nothing, whereas feeding such an animal with glycogenic substances (for example glucose) results in a rapid increase in the liver glycogen content.

The quantity of hepatic glycogen present in a given subject obviously depends upon both the quantity as well as the type of food consumed. For example, a carbohydrate-rich diet elevates the concentration of liver glycogen, whereas a carbohydrate-poor diet generally depresses it. Strenuous exercise markedly lowers the concentration of glycogen in the liver, as this substance is degraded to glucose which then passes in the bloodstream for utilization by the contracting muscles.

The principal physiologic regulatory factors for the hepatic glycogen concentration are various hormones, and these endocrine effects on glycogen content of the liver will be considered in greater detail in Chapter 15. Certain of these factors, however, may be mentioned briefly at this point. Excess insulin tends to increase the glycogen concentration within the skeletal muscles rather than that of the liver. Insulin deficiency and excessive thyroid hormone both reduce the liver glycogen concentration. The adrenal medullary hormone epinephrine or pancreatic glucagon, when injected or secreted in adequate quantities, rapidly lower the hepatic glycogen concentration. These hormones operate by activating the enzyme system responsible for the formation of 3'5'-cyclic adenylic acid, so that phosphorylase b is converted into phosphorylase a. The latter enzyme then catalyzes the breakdown of hepatic glycogen; therefore glucose-6-phosphate is formed. Glucose-6-phosphate then is hydrolyzed into free glucose which enters the bloodstream, thereby producing an elevated blood sugar level, or hyperglycemia. This physiologic mechanism is protective to the organism as it allows an abundant supply of glucose to reach the skeletal muscles quickly in times of stress.

The adenohypophysis as well as the adrenal cortex contain endocrine substances that elevate liver glycogen levels. Some of the adrenocortical hormones act, for example, by increasing the supply of glucose that is derived from noncarbohydrate precursors by gluconeogenesis, and this de novo synthesis favors an increase in liver glycogen concentration.

It may be concluded readily from the foregoing remarks that the hepatic glycogen level normally is quite labile, hence is subject to modification by a variety of external and internal factors.

Muscle Glycogen. The usual concentration of glycogen found in mammalian skeletal muscle is around 0.5 to 1 percent. Despite the fact that this level is considerably below that found in liver, most of the total quantity of glycogen in the body is present in skeletal muscle, because of the relatively large total mass of this tissue in the body.

In marked contrast to liver glycogen, the muscle glycogen level is relatively stable, and is not depleted even by extended periods of fasting. However, convulsions, regardless of their etiology, produce a marked reduction in muscle glycogen. As noted above, the administration of insulin generally increases muscle glycogen concentration, although if hypoglycemic convulsions ensue as the consequence of too much insulin being given then the muscle glycogen content may be seriously depleted.

When epinephrine is given to a fasted animal, the formation of adenylic acid is enhanced in muscle similarly to the effect of this compound in liver. The glucose-6-phosphate thus formed, however, cannot be hydrolyzed to glucose in skeletal muscle as the phosphatase enzyme that is necessary for this process is absent from muscular tissue. Furthermore metabolism of the glucose-6-phosphate takes place via the glycolytic pathway, hence the first metabolic products that appear in the bloodstream are the highly diffusible pyruvic and lactic acids. Subsequently, rapid glycogenesis from these compounds takes place within the liver. Therefore, following the administration of epinephrine to a fasting animal, the sequence of metabolic events that takes place is: first, a decline in muscle glycogen in seen; second, an increase in blood pyruvic and lactic acid concentrations takes place; and third, these effects are followed by an increase in the hepatic glycogen concentration (Fig. 14.26).

Glycogenesis and Glycogenolysis. The synthesis of glycogen is termed glycogenesis, and nutrients that increase the glycogen level in tissues are called glycogenic substances. Glycogenic substances include several hexoses, and in addition a number of other types of compound, such as certain glycogenic amino acids (Table 14.11), glycerol, and certain intermediates of glycolysis. Compounds of this type cause an increase in the hepatic glycogen level when given to fasting animals. In every instance cited, the specific metabolite involved in the process of glycogensis first must be transformed into glucose-6- and glucose-1-phosphates, thence into UDP-glucose (uridine diphosphate-glucose) as outlined in Figure 14.32.

The breakdown of glycogen, or *glycogenolysis,* is also depicted in Figure 14.32. Within the gastrointestinal tract this process is a hydrolysis, but within living cells, only the process of glycogen phosphorolysis is physiologically significant. The metabolic product of this reaction, glucose-1-phosphate, then is converted into glucose-6-phosphate by phosphoglucomutase, and the latter hexose phosphate now may enter the major pathways for carbohydrate metabolism, ie, the glycolytic or phosphogluconate sequences.

The formation of free glucose from glucose-6-phosphate and its release into the circulating blood is possible only in liver, intestine, and kidney, because only these tissues have a glucose-6-phosphatase. Thus the physiologically important reaction, glucose-6-phosphate → glucose + P_i, can occur only within liver, intestine and kidney, but not in skeletal muscle, since the latter tissue lacks the requisite enzyme to catalyze this process.

Blood Glucose and the General Factors that Regulate Glucose Metabolism. The homeostatic regulation of the blood glucose level within relatively narrow limits is of paramount importance to the survival of the individual, as this hexose is utilized continuously throughout life by all of the tissues of the body. Normally, the concentration of glucose in human blood ranges between 70 and 90 mg per 100 ml approximately 10 hours following the last meal. Slightly higher blood sugar levels may be observed shortly after a meal has been eaten, whereas fasting for an extended period causes only a negligible decline in blood glucose concentration for several days, assuming that any decrease takes place at all.

Various tissues of the body exhibit considerable differences with regard to their dependence upon circulating glucose levels in the blood. As noted earlier, the functions of the central nervous system are quite literally at the mercy of the blood glucose level, since this hexose is the major energy source that crosses the blood–brain barrier at a rate that is adequate to maintain normal integrity

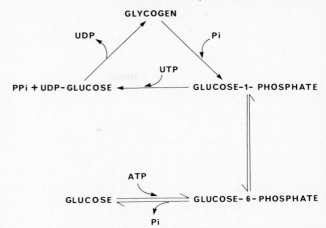

FIG. 14.32. Outline of the pathways for glycogenesis and glycogenolysis. PP$_i$ = inorganic pyrophosphate; UTP = uridine triphosphate; UDP = uridine diphosphate; P$_i$ = inorganic phosphate. (After White, Handler, and Smith. *Principles of Biochemistry,* 4th ed, 1968, The Blakiston Division, McGraw-Hill.)

and performance of the tissues involved. In fact, if the blood glucose level drops suddenly, the earliest symptoms observed are due to failure or derangement of central nervous functions.

Tissues other than those of the central nervous system (eg, skeletal muscle) obtain a large proportion of their total chemical energy requirements from substances other than glucose; consequently they are not so exacting in their demands insofar as a constant blood sugar level is concerned. Thus, the myocardium is quite efficient in extracting lactic acid and fatty acids from the coronary blood, and in utilizing these compounds as sources of chemical energy (p. 143). In fact this ability of heart muscle to derive useful energy from substances other than glucose endows the organ with a certain degree of metabolic adaptability, and thus renders it less sensitive to fluctuations in the blood glucose concentration.

Physiology of Blood Glucose. As might be expected, the regulation of blood glucose level within rather narrow limits is subject to a number of extremely delicate homeostatic controls. The various major factors contributing to this process may be considered individually, and the interplay among these factors is summarized in Figure 14.33.

Following ingestion of a high carbohydrate meal, the sugars derived from the hydrolysis of di- and polysaccharides within the gut contribute a large quantity of glucose to the bloodstream. On a daily basis, however, this process is intermittent since the dietary carbohydrate is assimilated within a few hours after a meal. However the liver, kidney, and intestine continuously supply glucose to the blood between meals via the hydrolysis of glucose-6-phosphate (Fig. 14.32), which glucose in turn is supplied either by glycogenolysis or by de novo synthesis from other metabolic precursors. Any compound capable of being transformed into an intermediate of the glycolytic pathway is a potential glucogenic compound, and as mentioned above, this includes a number of amino acids as well as glycerol in addition to a number of other substances of less quantitative significance.

The many potential metabolic pathways that glucose may follow within the body are summarized in Figures 14.33 and 14.34; however, the urine of a normal individual contains at most only a trace of glucose, regardless of diet.

As will be apparent from inspection of Figure 14.33, the blood glucose level, or concentration, at any particular time is the resultant of first, the rate of glucose formation from glycogen, amino acids, and other glucogenic compounds; second, the rate of glucose absorption from the gut; third, the rate at which glucose is utilized by all of the tissues of the body; and fourth, the rate at which glucose is lost in the urine.

The rate at which glucose is removed from the blood by the tissues is regulated partially by the glucose concentration in the extracellular fluid compartment per se. The higher the extracellular fluid glucose concentration is, the more rapidly this sugar is assimilated (or absorbed) by such tissues as liver and muscle. The contraction of skeletal muscle, in fact, is accompanied by an augmented rate of glucose entry into the muscle cells, even when the glucose level in the extracellular fluid remains constant. Probably this effect is caused by an enhanced rate of activity of the normal mechanism that is responsible for facilitated glucose transport into the cells.

If the blood glucose is maintained at elevated levels for long periods in normal individuals, this hyperglycemia favors the production of substances derived from glucose. For example, adipose tissue produces fatty acids at a greater rate, whereas glycogen as well as fatty acid synthesis are enhanced in hepatic tissue. The concept of glucose tolerance is discussed on page 1084.

LIPID METABOLISM. In the normal adult, lipid (fat) contributes a minimum of around 10 percent to the total body weight. Although lipids are found in varying quantities in all tissues and organs, the bulk of this material is concentrated in certain regions called fat depots that are composed chiefly of a specialized type of connective tissue known as adipose tissue. A large portion of the cytoplasm within the cells of adipose tissue is filled with lipid

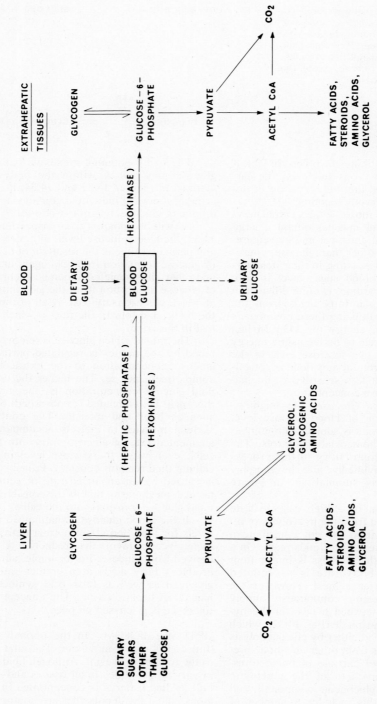

FIG. 14.33. Factors that contribute to the maintenance of the blood glucose level. (After White, Handler, and Smith. *Principles of Biochemistry*, 4th ed, 1968, The Blakiston Division, McGraw-Hill.)

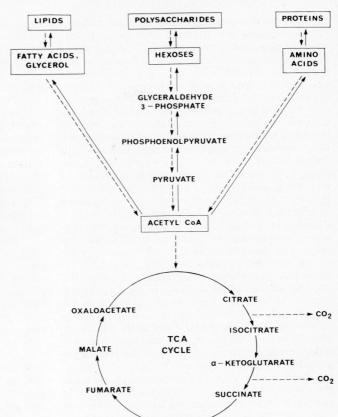

FIG. 14.34. Overall summary of catabolic reactions (dashed lines) and anabolic reactions (solid lines). The central metabolic pathways which have dual functions (ie, which have both anabolic and catabolic functions) are called amphibolic pathways. (Data from Lehninger. *Biochemistry*, 1971, Worth Publishers.)

droplets of various sizes, the amount depending upon the nutritional state of the individual.

It is important to realize that the lipid stored within adipose tissue is not inert chemically. Rather, this lipid is highly labile metabolically, even though the individual is in complete nutritional and caloric balance. This important concept will be discussed further below.

The body lipid stores are of considerable physiologic importance for several reasons: (1) the large subcutaneous fat depots prevent excessive heat loss from the body, as will be discussed later in this chapter; (2) the fat depots around certain organs (eg, the kidney) serve to cushion them against mechanical trauma; (3) most relevant to the present discussion, the lipid stores of the body provide a large reservoir of chemical potential energy that may be mobilized in times of restricted caloric intake. Lipid yields more than twice as many calories per gram upon complete oxidation (9.1 Cal) than does carbohydrate (4.1 Cal/gm) because of the presence of a greater number of oxidizable carbon bonds per gram of lipid.* Consequently the fat depots provide an extremely important source of stored energy for endergonic

Alternatively, the carbon atoms in lipids are in a more highly reduced state chemically than are the carbon atoms of glucose.

bodily processes. This fact also is especially true as the lipids that are stored within the depots throughout the body are quantitatively much larger than are the total available carbohydrate reserves. Furthermore, since lipid is stored in an almost water-free state, in contrast to carbohydrate which is hydrated to some extent, the lipids of adipose tissue represent a highly concentrated form of stored chemical energy.

General Metabolic Features of the Body Lipids. The digestion, assimilation, and transport of exogenous (or dietary) lipids were considered in Chapter 13. In this section, only the general aspects of depot lipid metabolism that are of physiologic importance will be considered.

When an individual ingests an excess of calories above his daily bodily requirements over a prolonged interval, then the quantity of total body lipid also increases, and gross obesity may ensue (p. 1215). Conversely, during extended periods of fasting (or starvation) the total quantity of lipid present within the body diminishes considerably (p. 1211).

It has been demonstrated unequivocally by the use of isotopic tracer elements that there is a continual turnover (including the synthesis, breakdown and restoration) of most bodily constituents with time, even though a constant overall body composition is maintained. The depot lipids are no exception to this general rule, as they are undergoing continuous mobilization and degradation, whereas new lipid is being synthesized and stored at the same time. The relative constancy of the lipid stores in the body of a person in caloric equilibrium is due to a rather precise balance between the rates of these two opposed processes.*

Adipose tissue exhibits two general metabolic features of physiologic importance. First, this tissue takes up carbohydrate and lipid as well as certain intermediate compounds derived from the metabolism of these nutrients for fat synthesis and storage. Second, the depot lipid is mobilized as free fatty acids (FFA), and to a lesser extent glycerol. Each of these processes is regulated largely by a number of hormones, as discussed in Chapter 15. In particular, however, insulin, thyroxine, and epinephrine, profoundly affect the synthesis and degradation of lipid by adipose tissue.

In the human body, over 99 percent of the depot lipid is present as triglycerides. Chemically, the composition of these lipids is similar, regardless of the anatomic location of the depot in the human body, but the degree of saturation or unsaturation of the particular fatty acids which are present in the triglycerides of adipose tissue may reflect dietary extremes, insofar as the type of ingested fatty acids is concerned. Generally speaking, liver lipid is richer in unsaturated fatty acids than is depot lipid. Consequently liver lipid is in the liquid state. Subcutaneous adipose tissue contains lipid that is liquid, and yet is as highly saturated as is compatible with this physical state at body temperature, 37 C. The energy yield is greater from the oxidation of saturated rather than unsaturated fatty acids, hence mammals in general, and humans in particular, deposit subcutaneously a type of lipid that has the greatest quantity of potential chemical energy; ie, the subcutaneous lipids contain a higher proportion of saturated than unsaturated fatty acids (Table 3.7, p. 68).

Since triglycerides of the fat depots provide the largest and most important single source of stored chemical potential energy in the human body, the metabolism of these compounds will be stressed in the presentations to follow.

The phrase "caloric equilibrium" or "balance" indicates that the daily average total caloric intake equals the daily average total caloric expenditure.

General Physiology and Metabolic Properties of Adipose Tissue. Adipose tissue is capable of both assimilating and storing triglycerides, and glucose administered orally has been demonstrated to stimulate the uptake of injected chylomicrons by adipose tissue.

Triglycerides may be absorbed directly from the bloodstream into fat cells; however, it is believed that maximum uptake of such lipid occurs principally through hydrolysis and reesterfication. Carbohydrate metabolism promotes such esterification, and the cells of adipose tissue contain both the glycolytic and phosphogluconate metabolic pathways (Fig. 14.25 and Table 14.9, respectively). The activity of the phosphogluconate oxidative pathway is somewhat greater than is that of the glycolytic mechanism in adipose tissue.

The glycerol moiety of stored depot triglyceride is derived chiefly from glucose at the triose phosphate step of anaerobic glycolysis (⑥, Fig. 14.25). The free alcohol arises via α-glycerophosphate, through the reversible steps shown in the same figure. Presumably, dihydroxyacetone phosphate and α-glycerophosphate are intermediates in this process. The α-glycerophosphoric acid is obtained by the reduction of dihydroxyacetone phosphate which in turn is formed during glycolysis. This process is of major importance in adipose tissue, and it provides the requisite glycerol for triglyceride synthesis in the fat depots. Conversely, free glycerol which is derived from the hydrolysis of triglyceride may be oxidized via the glycolytic pathway as shown in Figure 14.25, after phosphorylation to α-glycerophosphoric acid by glycerol kinase.

In liver, kidney, intestine, and lactating mammary gland (but not in adipose tissue!) an enzyme known as glycerokinase (or glycerol kinase) catalyzes the phosphorylation of free glycerol in the presence of ATP to α-glycerophosphoric acid. These tissues can utilize free extracellular glycerol directly, either as a substrate for oxidation, or as an intermediate for triglyceride synthesis. Lacking glycerokinase, adipose tissue is unable to utilize glycerol directly in contrast to liver, kidney, intestine, and lactating mammary gland.

Oxidation of Triglycerides. Triglycerides cannot be oxidized until they have undergone hydrolysis into their constituent fatty acids and glycerol, a reaction catalyzed by intracellular lipases. The general reaction for this process is:

$$
\begin{array}{l}
\text{R—C(=O)—O—C—H}_2 \\[4pt]
\text{R—C(=O)—O—C—H} \quad + 3H_2O \rightleftharpoons 3R\text{—COOH} + \begin{array}{l} \text{HO—C—H}_2 \\ \text{HO—C—H} \\ \text{HO—C—H}_2 \end{array} \\[4pt]
\text{R—C(=O)—O—C—H}_2
\end{array}
$$

Triglyceride   Fatty acids   Glycerol

The glycerol derived by this process from triglyceride (neutral fat) or phospholipid now enters the glycolytic pathway after it is converted into α-glycerophosphoric acid by glycerokinase and ATP as noted above (Fig. 14.25).

Hydrolysis of triglyceride occurs to a great extent in adipose tissue, and the free fatty acids (FFA) produced thereby are released into the plasma for transport to other tissues. Many tissues are able to assimilate these circulating long chain FFA and to oxidize them with the production of energy as described below. Examples of such tissues are heart, liver, kidney, skeletal muscle, lung, testis, brain, and adipose tissue itself. The ability of any given tissue to utilize glycerol depends, of course, upon whether or not the enzyme glycerokinase is present in that tissue.

Oxidation of Fatty Acids. It was proposed many years ago (Knoop, 1905) that fatty acids were oxidized under physiologic conditions by a process known as β-oxidation, and it is known today that this is the principal physiologic mechanism whereby these compounds are oxidized in the body. Since the fatty acid components of the triglyceride molecule contain the major portion of the energy which is stored in neutral fat, then energy derived from these compounds by the process of β-oxidation provides the greatest share of the total energy which is obtained from lipid metabolism.

Located on the inner membrane of the mitochondria, adjacent to the enzymes of the terminal respiratory chain, are several enzymes collectively known as the fatty acid oxidase complex. These enzymes catalyze the oxidation of long-chain fatty acids to acetyl CoA with the coupled synthesis of ATP from ADP, as shown in Figure 14.35. Inspection of this figure will reveal that the oxidation of long-chain fatty acids containing an even number of carbon atoms is accomplished by a series of reactions in which the fatty acyl chain is shortened in a stepwise fashion by the removal of two carbon atoms each time the process takes place. Note that fatty acyl CoA derivatives rather than free fatty acids are involved in these reactions. The process is termed β-oxidation simply because the second, or β-, carbon from the carboxyl group on the fatty acid is oxidized each time the process is repeated.

Like the metabolism of glucose, an initial reaction of free fatty acids with ATP is necessary in order to form an active intermediate. This intermediate now can react with the enzymes which carry out the remainder of the metabolic process. This is the only step in the β-oxidation of fatty acids that requires energy derived from ATP (Fig. 14.35, ①).

The enzyme thiokinase (or acyl-CoA synthetase) catalyzes the phosphorylation of the FFA into an active fatty acid or acyl CoA. A number of thiokinases are known, each of which is specific for a free fatty acid of specific chain length. The high-energy intermediates which are involved in oxidative phosphorylation also may serve as donors of free energy for the activation of fatty acids.

The acyl CoA formed in the first reaction of the β-oxidation sequence then loses two hydrogen atoms from the α and β carbon atoms. This oxidative process is catalyzed by acyl-CoA dehydrogenase (②), resulting in the formation of an α, β-unsaturated acyl-CoA. The coenzyme required for this process is a flavoprotein. Reoxidation of the flavoprotein by the terminal respiratory chain also requires the presence of another cofactor known as the electron-transferring flavoprotein.

During reaction ③ (Fig. 14.35) water is added and the double bond is saturated, forming β-hydroxy-acyl-CoA. This process is catalyzed by the enzyme enoyl hydrase (or crotonase).

In the next step (④), the β-carbon of the β-hydroxy-acyl-CoA is dehydrogenated (ie, oxidized) still further. This reaction takes place under the catalytic influence of β-hydroxyl-acyl-CoA dehydrogenase, and a β-keto-acyl-CoA compound results. In reaction ④ it will be noted that NAD is the coenzyme involved in the oxidation.

Lastly, in reaction ⑤, β-keto-acyl-CoA is split at the β-carbon by thiolase (or β-ketothiolase). Note that another molecule of CoA is involved in this process, so that the products of this reaction consist of, first, a fatty acid (acyl-CoA) derivative that contains two less carbon atoms than did the original molecule prior to oxidation; and second, a molecule of acetyl CoA. The acyl-CoA derivative thus formed now enters the oxidative sequence at ② as indicated by the arrow in Figure 14.35, while the acetyl CoA can be oxidized to carbon dioxide and water via the tricarboxylic acid cycle, whose enzymes also are located within the mitochondria (Fig. 14.29).

The net result of the β-oxidative process is that a long chain fatty acid ultimately is degraded completely into acetyl CoA (2-carbon units).

Bioenergetics of Fatty Acid Oxidation. The transport of electrons in the terminal respiratory chain from reduced flavoprotein and NAD will produce a minimum of five high-energy phosphate bonds of ATP, as indicated in Figure 14.35, for each of the first seven acetyl-CoA molecules generated by the β-oxidation of palmitic acid (which compound has a total of 16 carbon atoms in the chain). Thus, β-oxidation of these 14 carbon atoms into a total of 7 acetyl-CoA molecules will yield a total of 35 high-energy bonds ($5 \times 7 = 35$). Since a total of 8 mols of acetyl CoA is formed ultimately from the β-oxidation of one mole of palmitate, then the complete oxidation of this acetyl-CoA in the TCA cycle will generate 8 times 12, or 96, high-energy phosphate bonds. However, 2 molecules of ATP were required for the initial activation of the palmitate (①, Fig. 14.35); therefore the net yield of high-energy bonds for the complete oxidation of 1 mole of palmitic acid is 94. Since each of these bonds represents approximately 8 Cal/mole, a total of 752 Cal of free energy has been trapped. As the total free energy liberated by the oxidation of palmitic acid is 2,340 Cal/mole, the

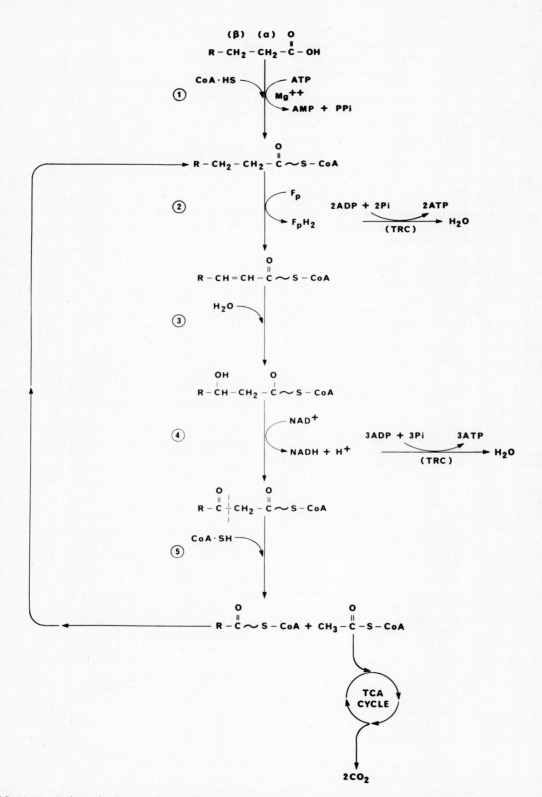

FIG. 14.35. Pathway for the β-oxidation of fatty acids. The specific enzymes that catalyze the individual reactions are: ①, thiokinase; ②, acyl-CoA dehydrogenase; ③, enoyl hydrase; ④, β-hydroxyacyl-CoA dehydrogenase; ⑤, thiolase. (After Harper. *Review of Physiological Chemistry,* 13th ed, 1971, Lange Medical Publications.)

efficiency of this process in trapping free energy as phosphate bonds in ATP is approximately 32 percent ($752/2,340 \times 100$).

The complete β-oxidation of fatty acids having a longer or shorter chain length will, of course, alter the total energy yield in proportion to the number of two-carbon units that are available in any particular fatty acid. Similarly, unsaturated fatty acids of a given chain length will yield correspondingly less free energy upon their complete oxidation, depending upon the number of double bonds present in the particular molecule, because of the fact that there are less hydrogen atoms available to enter the terminal respiratory chain.

Interrelationships Between the Metabolism of Lipids and Related Substances. Although it is beyond the scope of this text to consider in detail the biosynthetic and other metabolic pathways involving the lipids and certain related compounds, a number of these general interrelationships are summarized in Figure 14.36 because of their importance to the physiology of the endocrines (Chap. 15).

Formation of Ketone Bodies and Ketosis. Under normal circumstances, the catabolism and synthesis of fatty acids take place within the body without the accumulation of intermediates to any significant extent. During some abnormal functional states, however, three compounds derived from fatty acids, as well as certain amino acids (Table 14.11) accumulate in the blood, and these substances are commonly called ketone bodies. The compounds are acetoacetic acid, β-hydroxybutyric acid, and acetone.* All of these products arise from the metabolism of acetoacetyl CoA, a normal intermediate compound formed during the oxidation of fatty acids (Figs. 14.35 and 14.36). Furthermore, acetoacetyl CoA is generated easily by reversal of the thiolase reaction, ie, 2 Acetyl CoA $\rightleftharpoons$ Acetoacetyl CoA + CoA.

Formation of Acetoacetic Acid. In hepatic tissue, a principal metabolic fate of acetoacetyl CoA is its conversion into β-hydroxy-β-methylglutaryl CoA:

Actually only acetone is a ketone.

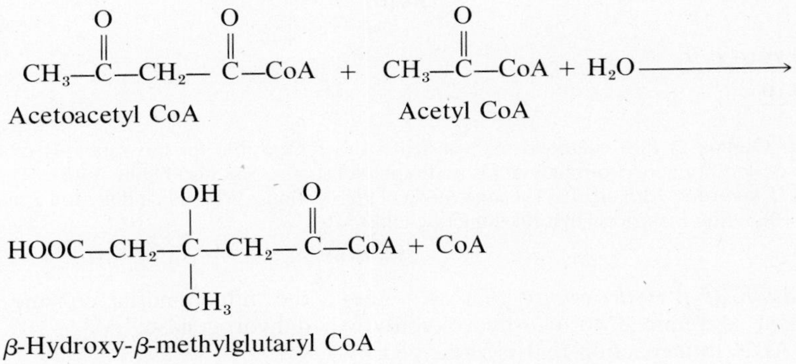

$$\underset{\text{Acetoacetyl CoA}}{CH_3-\overset{O}{\overset{\|}{C}}-CH_2-\overset{O}{\overset{\|}{C}}-CoA} + \underset{\text{Acetyl CoA}}{CH_3-\overset{O}{\overset{\|}{C}}-CoA} + H_2O \longrightarrow$$

$$\underset{\beta\text{-Hydroxy-}\beta\text{-methylglutaryl CoA}}{HOOC-CH_2-\underset{\underset{CH_3}{|}}{\overset{\overset{OH}{|}}{C}}-CH_2-\overset{O}{\overset{\|}{C}}-CoA} + CoA$$

The β-hydroxy compound is a major intermediate in the biosynthesis of cholesterol and of steroids (Fig. 14.36), as well as in the catabolism of the amino acid leucine. The cleavage of β-hydroxy-β-methylglutaryl CoA by an enzyme found in mitochondria that is distinct from the one responsible for catalyzing the above reaction is the major pathway whereby acetoacetic acid is formed in liver:

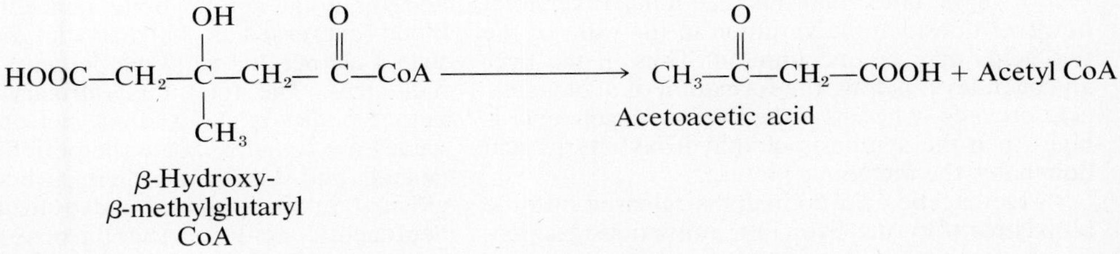

$$\underset{\substack{\beta\text{-Hydroxy-}\\\beta\text{-methylglutaryl}\\\text{CoA}}}{HOOC-CH_2-\underset{\underset{CH_3}{|}}{\overset{\overset{OH}{|}}{C}}-CH_2-\overset{O}{\overset{\|}{C}}-CoA} \longrightarrow \underset{\text{Acetoacetic acid}}{CH_3-\overset{O}{\overset{\|}{C}}-CH_2-COOH} + \text{Acetyl CoA}$$

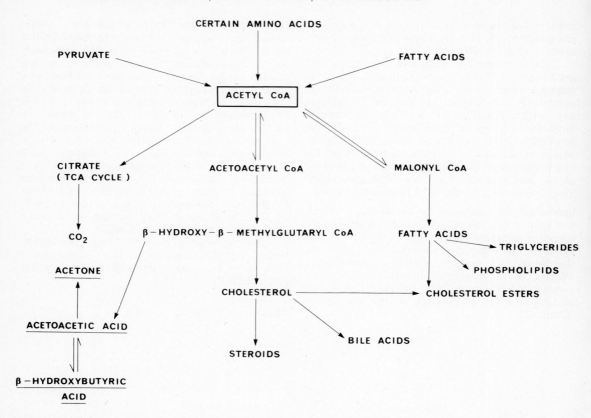

FIG. 14.36. Outline of the metabolic pathways that are responsible for the synthesis of ketone bodies (underlined compounds on left of figure) from fatty acids and other substrates. See also Figure 14.52. (Data from Harper. *Review of Physiological Chemistry*, 13th ed, 1971, Lange Medical Publications; White, Handler, and Smith. *Principles of Biochemistry*, 4th ed, 1968, The Blakiston Division, McGraw-Hill.)

Formation of β-Hydroxybutyric Acid. Acetoacetic acid is reduced to D-β-hydroxybutyric acid by NADH in a reaction that is catalyzed by the mitochondrial enzyme β-hydroxybutyric acid dehydrogenase:

$$CH_3-\overset{\overset{\textstyle O}{\|}}{C}-CH_2-COOH + NADH + H^+ \rightleftharpoons$$

Acetoacetic acid

$$CH_3-\overset{\overset{\textstyle OH}{|}}{CH}-CH_2-COOH + NAD^+$$

D-β-Hydroxybutyric acid

It will be noted that this reaction is reversible; however considerable variation in the ratio of the two acids may be encountered. Thus, if the liver glycogen level is low, the formation of acetoacetic acid prevails, whereas if the liver glycogen level is high, then the synthesis of β-hydroxybutyric acid dominates the metabolic picture.

Acetoacetic acid normally is released into the bloodstream by the liver in a continuous fashion, and the total ketone body concentration in the blood (expressed as β-hydroxybutyric acid) is less than 3 mg per 100 ml (3 mg percent) in the normal individual. The total daily urinary excretion of ketone bodies is about 20 mg per day. These low values are found because the peripheral tissues in general, and skeletal muscle in particular, are quite efficient in their uptake and utilization of blood acetoacetic acid for the purpose of obtaining energy. The metabolism of acetoacetic acid requires that it first be reconverted into its CoA derivative. This is accomplished by the transfer of a CoA moiety from succinyl CoA, a reaction that is catalyzed by a specific thiophorase, an enzyme which apparently is lacking in hepatic tissue.

The acetoacetyl CoA formed as the result of this reaction then undergoes scission into two molecules of acetyl CoA by the action of thiolase.

The acetyl CoA thus produced now may be oxidized in the TCA cycle.

Formation of Acetone. Acetone is produced from acetoacetic acid by a decarboxylation that is catalyzed by the enzyme acetoacetate decarboxylase. This reaction, which passes through several intermediate steps involving the substrate and enzyme, may be summarized as follows:

$$CH_3-\overset{\overset{\displaystyle O}{\|}}{C}-CH_2-COOH \xrightarrow[\quad CO_2 \quad \quad H_2O \quad]{} CH_3-\overset{\overset{\displaystyle O}{\|}}{C}-CH_3$$

Acetoacetic acid   Acetone

The acetone formed by this process may be converted metabolically into propanediol. The latter compound is transformed into pyruvic acid, and thus may enter into all of the metabolic pathways of pyruvate metabolism (Fig. 14.27).

Ketonemia, Ketonuria, and Ketosis. Reference to Figure 14.36 will reveal that both acetyl CoA and β-hydroxy-β-methylglutaryl CoA occupy key roles in the metabolism of lipid. When the utilization of carbohydrate decreases and/or there is an increased mobilization of fatty acids from the depots to the liver, two of the three pathways for the metabolism of acetyl CoA are inhibited. Specifically, these pathways are the TCA cycle and the synthesis of fatty acids. Under such circumstances, the fatty acid intermediates tend to accumulate at the level of acetyl CoA carboxylase, so that acetyl CoA now is funneled into its third principal metabolic pathway, viz., the formation of β-hydroxy-β-methylglutaryl CoA and the products derived therefrom, cholesterol and acetoacetic acid. The increased rate of formation of acetoacetic acid causes an elevation in the blood level of this compound together with a concomitant rise in the levels of hydroxybutyric acid and of acetone. This elevated level of ketone bodies is termed a *ketonemia*. When the blood level of ketone bodies surpasses the renal threshold for these substances they appear in the urine, resulting in the condition known as *ketonuria*. If ketonemia and ketonuria are present to any significant degree, then the expired breath of such an individual usually has the characteristic odor of acetone. The condition in which these three symptoms are present, viz., ketonemia, ketonuria, and an odor of acetone in the exhaled air is known as *ketosis*.

Insofar as the etiology of ketosis is concerned, it is important to remember that any condition that is associated with a decrease in the availability and/or utilization of the normal primary energy source, carbohydrate, will lead to an increase in the utilization of fatty acids as the principal energy source. An excellent example of this situation is to be found in starvation, wherein the body glycogen stores are depleted to such an extent that survival of the individual depends upon the mobilization and utilization of fatty acids that are obtained from the depot lipid stores as the prime energy source (p. 1212). The mobilization of depot lipids in such a circumstance is manifest biochemically by a hyperlipemia; ie, the blood lipid level is elevated above normal. The catabolism of fatty acids by the liver takes place at an increased rate; therefore, production of acetoacetyl CoA as well as its products also occurs more rapidly.

Ketosis most often is seen clinically when gastrointestinal disturbances are present, during pregnancy, in babies, and in uncontrolled diabetes mellitus. In otherwise normal individuals, ketosis also may be encountered when an average diet is replaced by one low in carbohydrate and high in fat content (p. 1220). During renal glucosuria, ketosis sometimes occurs; however, the most important clinical reason for ketosis is diabetes mellitus, ie, a lack of the hormone insulin, that prevents the utilization of glucose at an appropriate rate.

PROTEIN AND AMINO ACID METABOLISM

General Remarks. Like other bodily constituents, the protein and amino acid metabolism of the human body are in a dynamic steady state throughout the lifetime of the individual. Thus proteins and amino acids are being synthesized and degraded continuously. In turn, the absolute as well as the relative rates of both of these intracellular processes are controlled by inherent biochemical mechanisms in addition to several hormones (Chap. 15).

It is important to realize that the major processes involved in protein nitrogen metabolism within the body itself are directly concerned with amino acids or their metabolic products. The nutritive aspects of dietary proteins, and the hydrolysis

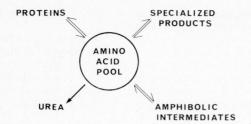

FIG. 14.37. General outline of amino acid metabolism. Urea formation is the only irreversible step in this process. The pathways, enzymes, and intermediates that participate in synthetic and degradative processes are not necessarily the same. (After Harper. *Review of Physiological Chemistry,* 13th ed, 1971, Lange Medical Publications.)

of these substances within the gastrointestinal tract were considered in Chapter 13. The principal metabolic role of exogenous and endogenous amino acids is to provide a continuing supply of precursors for the synthesis of all bodily proteins and peptides, as well as other more specialized nitrogenous constituents of physiologic impor-

tance, such as histamine, serotonin, thyroxin, epinephrine, purines, pyrimidines, and so forth.

The structural formulas of the common amino acids, classed as aromatic, acidic, basic, and neutral are depicted in Figures 14.38 through 14.43 inclusive. The amino acid residues found in proteins, together with the commonly used abbreviations for these compounds are listed alphabetically in Table 14.10.

Nitrogen Balance. Regardless of the age of an individual, the body undergoes a continual daily intake and loss of nitrogenous compounds. Proteins are quantitatively the most significant of these, although other nitrogenous substances also are assimilated, and the metabolic products are excreted in urine and feces.

When the total daily uptake of nitrogen by the body exceeds the loss, as during infancy or childhood, then a state of positive nitrogen balance is said to exist. Conversely, when the daily loss of nitrogen exceeds the intake, as during starvation or certain types of debilitating disease, then a state of negative nitrogen balance is present. Under normal circumstances, however, the daily nitrogen

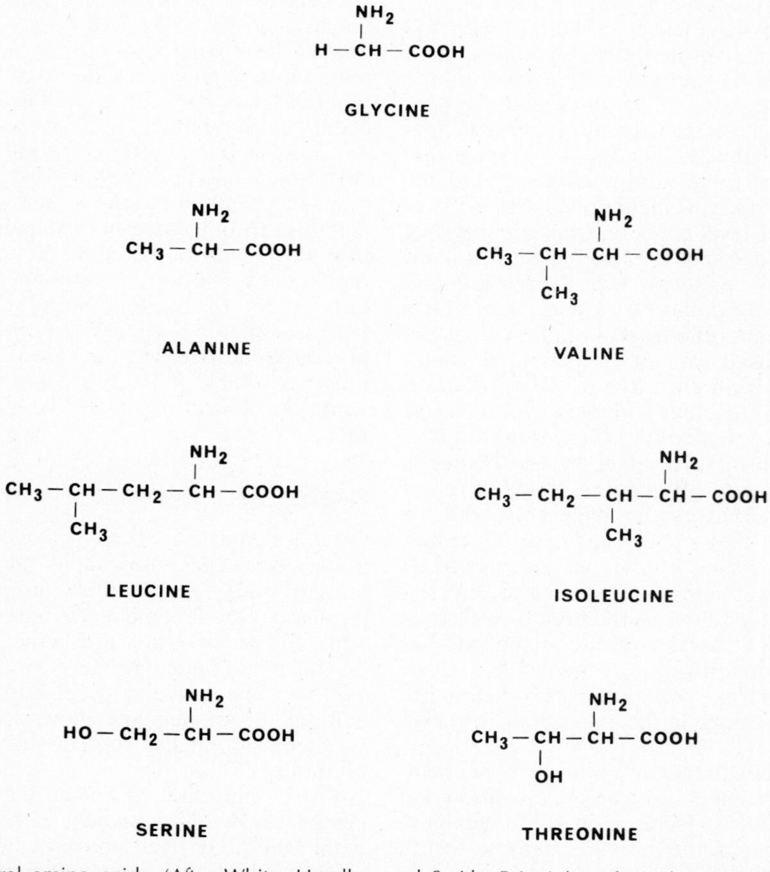

$$\underset{\text{GLYCINE}}{H-\overset{\overset{\displaystyle NH_2}{\displaystyle |}}{C}H-COOH}$$

$$\underset{\text{ALANINE}}{CH_3-\overset{\overset{\displaystyle NH_2}{\displaystyle |}}{C}H-COOH} \qquad \underset{\text{VALINE}}{CH_3-\overset{\overset{\displaystyle NH_2}{\displaystyle |}}{\underset{\underset{\displaystyle CH_3}{\displaystyle |}}{C}}H-\overset{\overset{\displaystyle NH_2}{\displaystyle |}}{C}H-COOH}$$

$$\underset{\text{LEUCINE}}{CH_3-\overset{}{\underset{\underset{\displaystyle CH_3}{\displaystyle |}}{C}}H-CH_2-\overset{\overset{\displaystyle NH_2}{\displaystyle |}}{C}H-COOH} \qquad \underset{\text{ISOLEUCINE}}{CH_3-CH_2-\overset{}{\underset{\underset{\displaystyle CH_3}{\displaystyle |}}{C}}H-\overset{\overset{\displaystyle NH_2}{\displaystyle |}}{C}H-COOH}$$

$$\underset{\text{SERINE}}{HO-CH_2-\overset{\overset{\displaystyle NH_2}{\displaystyle |}}{C}H-COOH} \qquad \underset{\text{THREONINE}}{CH_3-\overset{}{\underset{\underset{\displaystyle OH}{\displaystyle |}}{C}}H-\overset{\overset{\displaystyle NH_2}{\displaystyle |}}{C}H-COOH}$$

FIG. 14.38. Neutral amino acids. (After White, Handler, and Smith. *Principles of Biochemistry,* 4th ed, 1968, The Blakiston Division, McGraw-Hill.)

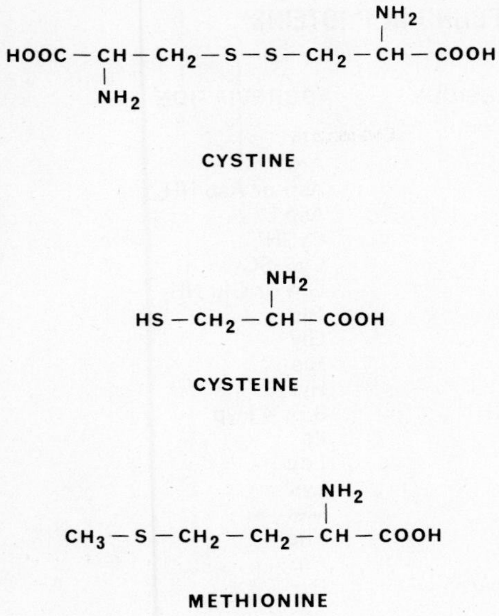

$$HOOC - CH - CH_2 - S - S - CH_2 - \overset{\overset{\displaystyle NH_2}{|}}{CH} - COOH$$
$$\underset{\displaystyle NH_2}{|}$$

CYSTINE

$$HS - CH_2 - \overset{\overset{\displaystyle NH_2}{|}}{CH} - COOH$$

CYSTEINE

$$CH_3 - S - CH_2 - CH_2 - \overset{\overset{\displaystyle NH_2}{|}}{CH} - COOH$$

METHIONINE

FIG. 14.39. Sulfur-containing amino acids. (After White, Handler, and Smith. *Principles of Biochemistry*, 4th ed, 1968, The Blakiston Division, McGraw-Hill.)

intake is approximately equal to the daily nitrogen loss, and the body is in nitrogen balance.

A General Concept Pertaining to Amino Acid Metabolism. In marked contrast to carbohydrate and lipids, proteins and amino acids are not "stored" in the body in the conventional sense of the word. Undoubtedly, however, the level of muscle, plasma, liver, kidney, and other body proteins fluctuates to some extent during any particular day in relation to the intake of protein in meals. Any surplus dietary amino acids either must be excreted, a wasteful process, or alternatively the nitrogen must be removed and eliminated, and the resulting α-keto acids then may be oxidized directly to provide energy, or else converted into glycogen or fatty acids which can be stored. The latter topics will be emphasized in the discussions to follow.

Tissue Extraction of Circulating Amino Acids. The total plasma concentration of the various α-amino acids ranges between 35 and 65 mg per 100 ml as shown in Table 14.11. Amino acids that enter the circulation via absorption from the intestinal tract or by intravenous injection are removed from the bloodstream by the tissues and organs within a matter of a few minutes.

Quantitatively the liver has the greatest capacity to extract amino acids from the circulation, but the kidneys also exhibit this property to a significant extent; the other tissues assimilate smaller quantities of amino acids. In fact, certain tissues, in particular brain, have a selective ability to extract some amino acids from the bloodstream quite readily whereas others are taken up at much slower rates.

As noted earlier (p. 864), amino acids enter cells by an active transport mechanism that requires energy, as the transport process takes place against a concentration gradient. The uptake of amino acids occurs with a concomitant increase in the water content of the cell as well as a slight elevation in the quantity of intracellular sodium ion. Potassium ion may leave the cell concurrently, especially when either arginine or lysine is absorbed, and this slight net movement of potassium has the effect of maintaining the electric neutrality of the intracellular contents. Presumably the metabolic fate of an amino acid once it enters the cellular pool is the same, regardless of its source (exogenous or endogenous).

Three general factors affect the manner in which amino acids may be apportioned among the many possible metabolic reactions that are open to them within the cell (Fig. 14.37). First, the requirement for new tissue proteins may place a great demand upon the available supply of amino acids. This effect is especially evident, for example, in infants and growing children, or during recovery from a period of starvation. The quantity of carbohydrate and lipid available in the diet markedly influence the fraction of the total caloric requirement that must be provided by the oxidation of amino acids. Second, the relative quantities as well as kinds of individual amino acids that are provided to the tissues will affect markedly their suitability as constituents of specific cellular proteins. Third, the regulatory effects of a number of hormones can alter markedly the rate and direction of specific metabolic reactions that involve protein and amino acid metabolism.

Amino Acids and Protein Synthesis. It is quite evident from the large increase in total body protein that takes place during the growth of an individual that net protein synthesis has occurred, and that dietary (ie, exogenous) amino acids must be used for this process. The adult in nitrogen balance must also synthesize protein continually to replace digestive enzymes, to elaborate protein hormones that must be replaced as they are degraded, to produce plasma proteins, and to restore the entire spectrum of proteins that are constituents of cells and which have a short life-span. A few examples of the latter are leukocytes, erythrocytes, and epithelial cells of the gastrointestinal mucosa and skin.

Experimental studies that employed isotopic tracers have shown conclusively that protein metabolism is a dynamic process, and that proteins, in common with other bodily constituents, are undergoing continual turnover (synthesis and degradation) throughout the life of the individual; and as noted earlier the synthesis and breakdown must take place at equal rates, when the individual is in nitrogen balance.

The de novo production of cells, such as those

Table 14.10 AMINO ACID RESIDUES FOUND IN PROTEINS[a]

PARENT COMPOUND; NAME OF AMINO ACID	NAME OF RESIDUE	ABBREVIATION
Alanine	Alanyl	Ala
Arginine	Arginyl	Arg
Asparagine	Asparaginyl	Asn or Asp NH$_2$
Aspartic acid	Aspartyl	Asp
Cysteine	Cysteinyl	CySH
Cystine	Cystyl	CyS-SCy
Glutamine	Glutaminyl	Gln or Glu NH$_2$
Glutamic acid	Glutamyl	Glu
Glycine	Glycyl	Gly
Histidine	Histidyl	His
Hydroxylysine	Hydroxylysyl	Hyl
Hydroxyproline	Hydroxyprolyl	3 or 4 Hyp
Isoleucine	Isoleucyl	Ile
Leucine	Leucyl	Leu
Lysine	Lysyl	Lys
Methionine	Methionyl	Met
Phenylalanine	Phenylalanyl	Phe
Proline	Prolyl	Pro
Pyrrolidone	Pyrrolidone carboxyl	Pyr
Serine	Seryl	Ser
Threonine	Threonyl	Thr
Tryptophane	Tryptophanyl	Trp
Tyrosine	Tyrosyl	Tyr
Valine	Valyl	Val

[a]*Data from White, Handler, Smith.* Principles of Biochemistry, *4th ed, 1968, The Blakiston Division, McGraw-Hill.*

types which were mentioned above, invariably is linked to net protein synthesis, and practically all of the proteins found in any given cell type must be elaborated in situ, rather than transported to the locus of cell formation. The turnover rates for different individual cellular proteins can vary markedly (on the order of from hours to months) and these rates depend only to a limited extent upon the rate at which new cells are being formed. Thus intracellular protein synthesis and degradation are

Table 14.11[a] GLYCOGENIC AND KETOGENIC ACID AMINO ACIDS[b]

METABOLIC FATE(S)		
GLYCOGEN = GLYCOGENIC AMINO ACIDS	LIPID = KETOGENIC AMINO ACIDS	GLYCOGEN AND LIPID = GLYCOGENIC PLUS KETOGENIC
Alanine (4.0)	Leucine (2.5)	Isoleucine (2.0)
Arginine (2.3)		Lysine (3.7)
Aspartic acid (0.3)		Phenylalanine (2.0)
Cystine (+ cysteine) (3.0)		Tyrosine (1.3)
Glutamic acid (0.9)[c]		Tryptophan (1.7)
Glycine (2.9)		
Histidine (2.1)		
Hydroxyproline (—)		
Methionine (0.6)		
Proline (2.6)		
Serine (1.4)		
Threonine (2.1)		
Valine (3.2)		

[a]*Data from Harper.* Review of Physiological Chemistry, *13th ed, 1971, Lange Medical Publications.*

[b]*Following removal of the amino group, the carbon skeleton of the individual amino acids forms an amphibolic intermediate that can enter one of two principal metabolic pathways. The α-keto acid either may be oxidized directly thereby providing useful energy, or it may be synthesized into glycogen or lipid thereby storing the energy for future use. Number in parentheses following the name of each compound is the plasma concentration of the amino acid following a 12 hour fast, in mg%.*

[c]*The plasma concentration of glutamine = 7.5 mg%.*

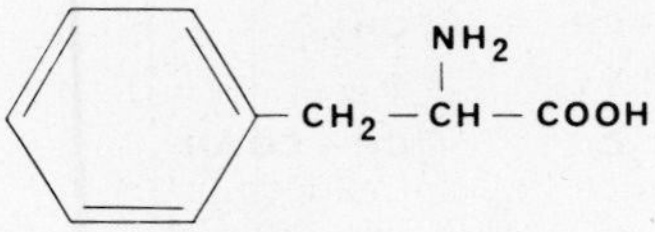

PHENYLALANINE

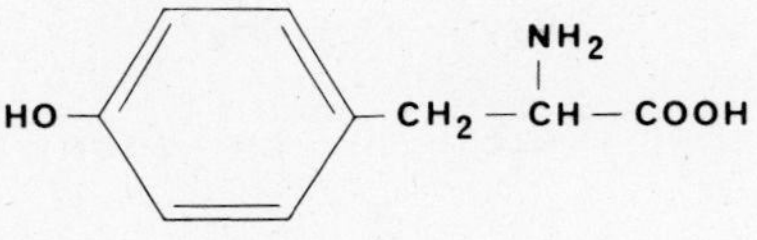

TYROSINE

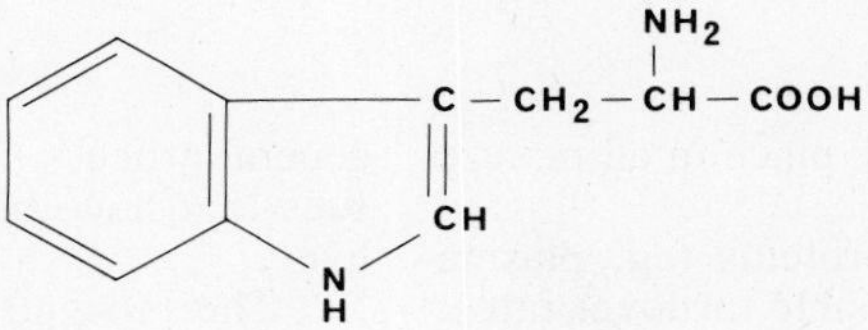

TRYPTOPHAN

FIG. 14.40. Aromatic amino acids. (After White, Handler, and Smith. *Principles of Biochemistry,* 4th ed, 1968, The Blakiston Division, McGraw-Hill.)

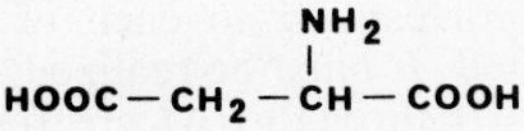

ASPARTIC ACID

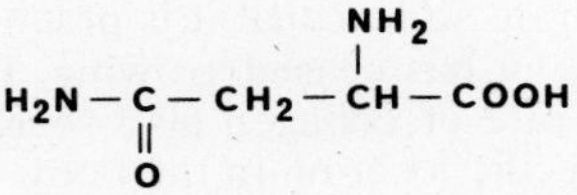

ASPARAGINE

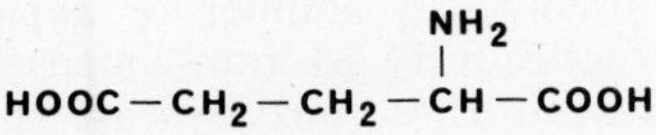

GLUTAMIC ACID

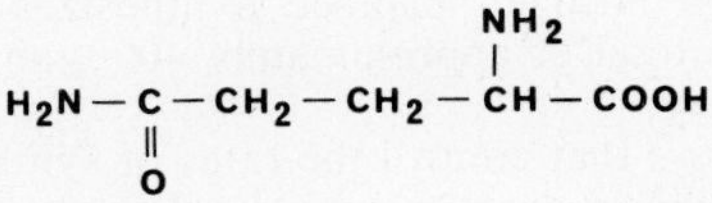

GLUTAMINE

FIG. 14.41. Acidic (monoamino dicarboxylic) amino acids. (After White, Handler, and Smith. *Principles of Biochemistry,* 4th ed, 1968, The Blakiston Division, McGraw-Hill.)

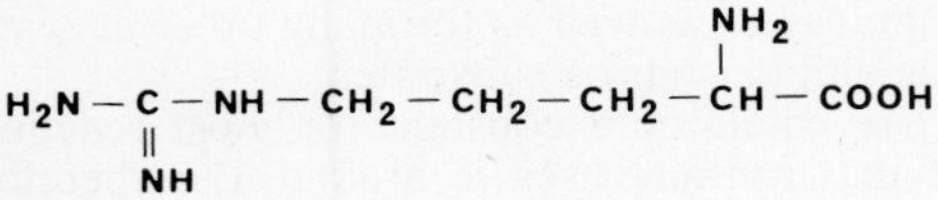

ARGININE

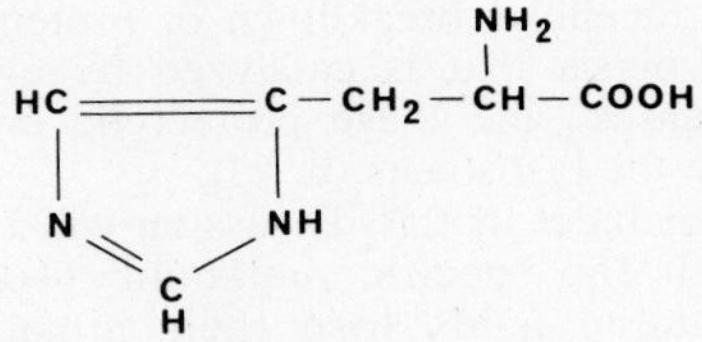

HISTIDINE

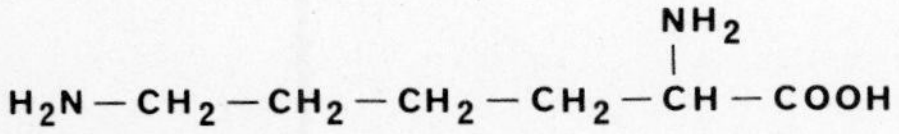

LYSINE

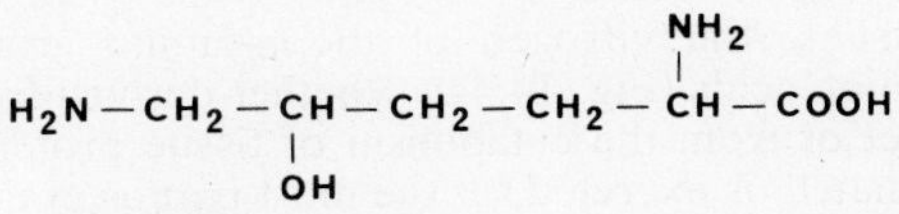

HYDROXYLYSINE

FIG. 14.42. Basic (diamino monocarboxylic) amino acids. (After White, Handler, and Smith. *Principles of Biochemistry,* 4th ed, 1968, The Blakiston Division, McGraw-Hill.)

PROLINE

HYDROXYPROLINE

FIG. 14.43. Imino acids. No amino group is present. (After White, Handler, and Smith. *Principles of Biochemistry,* 4th ed, 1968, The Blakiston Division, McGraw-Hill.)

inseparable processes that take place in all mature intermitotic cells.

Individual extracellular proteins (eg, plasma albumin) also exhibit quite variable turnover rates; however such proteins generally have a much shorter half-life than do intracellular proteins, and this is usually on the order of a few days. An exception to this statement regarding the turnover rate of extracellular protein is collagen. This important extracellular component of connective tissue has a turnover rate so low that it is practically nil once the individual has ceased growing. However, the turnover rate of collagen also seems to depend upon its specific location in the body.

The total rate of protein synthesis in man has been estimated using isotopically labeled amino acids. Using this technique, around 0.6 gm of amino nitrogen is incorporated into protein per kg of body weight per day. Consequently, a 70 kg adult male in nitrogen balance synthesizes and catabolizes a total of approximately 400 grams of protein per day.

The factors that control the rates of synthesis and degradation of protein in vivo are unclear, although the availability of free amino acids apparently is not the major rate-limiting factor. Thus, if there is an adequate supply of amino acids, increasing the quantity of these compounds does not augment the rate of net protein synthesis.

The intracellular breakdown of proteins is a simple hydrolysis that is catalyzed by enzymes called proteases, and these proteolytic enzymes are found in the lysosomes (p. 81).

The remainder of this discussion will be concerned with the specific metabolic processes whereby amino acids lose their nitrogenous groups, and with the possible metabolic fates of the remaining carbon skeletons with emphasis upon these compounds as an alternative source of energy to carbohydrate and lipid within the body.

Nitrogen Metabolism of Amino Acids. In human tissues, the nitrogen of the α-amino groups of amino acids (Fig. 14.44), whether derived from the diet or from the catabolism of tissue proteins, ultimately is excreted for the most part as urea in the urine. The removal of amino groups from amino acids takes place chiefly in the liver. This is due to the size of this organ, as well as its strategic location in the portal circulation. However, the catabolism of amino groups of amino acids is a general process. Therefore, it takes place in all tissues that have been studied and not solely in the liver.

The biosynthesis of urea involves four interrelated biochemical processes: (①) Transamination; (②) oxidative deamination; (③) ammonia transport; and (④), the chemical reactions of the urea cycle. These processes are shown in Figure 14.45 as an overall flow of nitrogen from amino acid nitrogen to urea, and the numbers in this figure correspond to each of the four processes just listed. It must be realized, however, that reactions (①) through (③) are freely reversible, and that each of the processes, which is depicted in Figure 14.45 also is capable of playing a role in the biosynthesis of amino acids.* That is, the intermediates are *amphibolic,* and thus may enter anabolic as well as catabolic pathways.

The process of transamination (①), Fig. 14.45) involves a number of hepatic enzymes known collectively as transaminases (or aminotransferases). These enzymes are specific for the reaction that takes place between α-ketoglutaric acid and the majority of the other amino acids. Transamination reactions result in the interconversion of a pair of amino acids and a pair of keto acids, usually α-amino acids and α-keto acids. Pyridoxal phosphate, derived from vitamin B_6 (p. 880), is an essential coenzyme for the action of transaminases, as well as for many other enzymes having amino acids as substrates.

The equilibrium constant for most reactions involving transaminases is around 1.0; therefore

The reactions of the urea cycle per se are irreversible (Fig. 14.50).

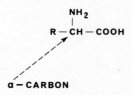

FIG. 14.44. The α-carbon of an α-amino acid is that carbon which is adjacent to the carboxyl group. α-Amino acids are depicted in Figures 14.38 through 14.42. (After Harper. *Review of Physiological Chemistry,* 13th ed, 1971, Lange Medical Publications.)

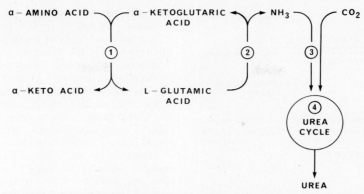

FIG. 14.45. Summary of the principal catabolic reactions that are involved in nitrogen metabolism. ① = transamination; ② = deamination of glutamic acid with the formation of ammonia; ③ = irreversible reactions of ammonia with carbon dioxide in the urea cycle, ④, results in urea formation, as depicted in detail in Figures 14.50 and 14.51. (After Harper. *Review of Physiological Chemistry,* 13th ed, 1971, Lange Medical Publications.)

the process of transamination is freely reversible. As an important biologic consequence of this fact, transaminases can function both in the synthesis as well as in the degradation of amino acids, ie, the intermediates are amphibolic. The general reaction of transamination is shown in Figure 14.46.

Two specific transaminases present in practically all mammalian cells catalyze the transfer of amino groups from most amino acids to produce alanine (from pyruvic acid) or glutamic acid (from α-ketoglutaric acid). The enzymes responsible for these transfers are, respectively, alanine transaminase (or alanine-pyruvate transaminase) and glutamic acid transaminase (also called glutamic acid-α-ketoglutaric acid transaminase). Although each of these transaminase enzymes is specific for the particular pair of amino and keto acids designated by its name, these enzymes are nonspecific insofar as the other pair of amino acids is concerned. Consequently, a broad spectrum of amino and keto acids may undergo transamination through interaction with these two specific enzymes, in accordance with the following reactions: wide variety of amino acids can be converted into glutamic acid amino nitrogen. This fact is ex-

(1) α-amino acid + pyruvic acid

$$\xrightarrow[\text{transaminase)}]{\text{(alanine}}$$

α-keto acid + alanine

(2) α-amino acid + α-ketoglutaric acid

$$\xrightarrow[\text{transaminase)}]{\text{(glutamic acid}}$$

α-keto acid + glutamic acid

However alanine also is a substrate for the glutamic acid transaminase reaction (②, Fig. 14.45). Therefore, all of the amino nitrogen from a

tremely important biologically, because L-glutamic acid is the only amino acid present in mammalian cells that can undergo oxidative deamination at a significant rate. Furthermore, the production of ammonia from α-amino groups can only take place after this nitrogen has first been converted into α-amino nitrogen groups of L-glutamic acid.

The majority of the amino acids which are listed in Table 14.11 provide substrates for transamination with the exception of lysine, threonine, proline, and hydroxyproline.

The serum levels of transaminase may rise sharply in a number of disease states (eg, following coronary thrombosis), and thus are of considerable diagnostic importance in clinical medicine.

Oxidative deamination is the next step in the metabolism of the α-amino nitrogen groups which are derived from various amino acids by transamination (②, Fig. 14.45). The two processes are closely coupled in mammalian tissues.

The enzyme L-glutamic acid dehydrogenase (or L-glutamate dehydrogenase) catalyzes the release of ammonia from L-glutamic acid which in turn is formed by transamination as described above, and in accordance with the general reaction shown in Figure 14.47.

The activity of L-glutamic acid dehydrogenase

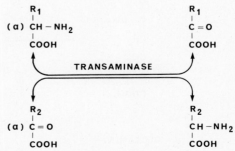

FIG. 14.46. Transamination. R_1, R_2 = different chemical groups. (After Harper. *Review of Physiological Chemistry,* 13th ed, 1971, Lange Medical Publications.)

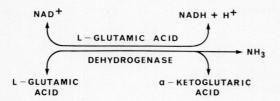

FIG. 14.47. The reversible L-glutamate dehydrogenase reaction liberates free ammonia from glutamic acid. (After Harper. *Review of Physiological Chemistry,* 13th ed, 1971, Lange Medical Publications.)

derived from hepatic tissue is altered by allosteric modifiers. Thus, ADP enhances the activity of this dehydrogenase whereas ATP, GTP, and NADP inhibit the activity of the enzyme.

Since the L-glutamate dehydrogenase reaction is reversible, it not only converts nitrogen from L-glutamic acid to urea (catabolism), but also catalyzes the formation of L-glutamic acid from α-keto glutaric acid and free ammonia, a biosynthetic process.

The L-glutamic acid dehydrogenase reaction is the major physiologic route whereby ammonia is produced in the body from the α-amino nitrogen groups of amino acids. There are, however, a number of other enzymes in mammalian liver and kidney that catalyze the oxidative deamination of amino acids, although the physiologic role of these L- and D-amino acid oxidases is unclear at present.

In addition to the ammonia produced by the dehydrogenation of glutamic acid in the tissues, another major source of ammonia is to be found within the intestines. The bacterial flora of the gut produces quantities of ammonia both from dietary protein as well as from the urea which is secreted in the fluids which enter the digestive tract. The ammonia from these sources is absorbed into the blood of the portal vein, whence it passes directly to the liver. Thus portal venous blood contains significant quantities of ammonia, whereas blood in the systemic vessels contains almost none.

Normally the liver extracts practically all of the ammonia entering it in the portal blood. Since even minute amounts of ammonia are highly toxic when present in the systemic circulation, in cases of hepatic malfunction the symptoms of ammonia intoxication may arise.

The quantity of ammonia leaving the kidneys in the renal venous blood always is greater than that found in the renal arteries. Consequently the kidneys produce ammonia, and some of this compound passes into the bloodstream. The excretion of ammonia by the kidneys into the urine, however, constitutes a far more important factor in the renal metabolism of ammonia than does the release of this compound into the bloodstream. The production of ammonia by these organs is an essential feature of the mechanisms within the renal tubules that are involved in the regulation of acid–base balance within the body as well as for the conservation of cations (Chap. 12, p. 807). As noted in this earlier discussion, the production of ammonia by the kidneys is increased sharply in metabolic acidosis and decreased markedly in metabolic alkalosis. This ammonia is not generated by the breakdown of urea; rather, it is derived from intracellular amino acids, and glutamine in particular (Fig. 14.48). This reaction is catalyzed by the enzyme renal glutaminase, which releases ammonia by breakdown of the amide group of glutamine.

Ammonia may be eliminated from the body as ammonium salts, especially when a condition of metabolic acidosis exists. However, an overwhelming proportion of the total ammonia which is produced in the body is excreted as urea, the major nitrogenous component of urine.

Ammonia is produced continuously in the tissues by the reactions described above, yet only traces of this compound usually are present in the blood (10 to 20 μg per 100 ml). Normally the liver rapidly extracts the ammonia from the blood, and then converts it into glutamic acid, glutamine, or urea. Note the considerable differences in the blood levels of free amino acids, especially

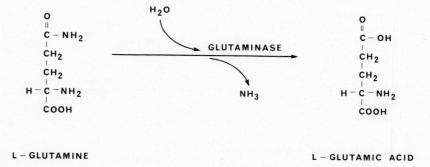

L – GLUTAMINE   L – GLUTAMIC ACID

FIG. 14.48. The glutaminase reaction is irreversible for all practical purposes in the direction shown. (After Harper. *Review of Physiological Chemistry,* 13th ed, 1971, Lange Medical Publications.)

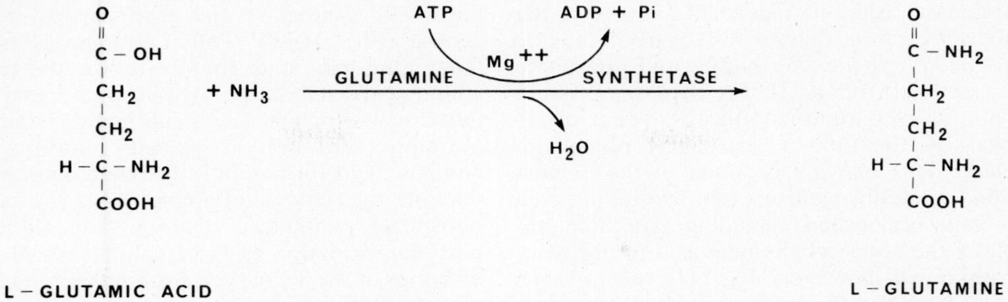

$$\text{L} - \text{GLUTAMIC ACID} \quad + NH_3 \quad \xrightarrow[\text{GLUTAMINE}]{\text{ATP} \quad \quad \text{ADP} + \text{Pi} \atop \text{Mg}^{++} \quad \quad \text{SYNTHETASE}} \quad \text{L} - \text{GLUTAMINE}$$

FIG. 14.49. The glutamine synthetase reaction proceeds strongly to the right, thus favoring the synthesis of glutamine. (After Harper. *Review of Physiological Chemistry,* 13th ed, 1971, Lange Medical Publications.)

glutamine, when compared to the trace levels of circulating ammonia (Table 14.11). These facts constitute the general basis of ammonia transport, as depicted in Figure 14.45, (③), and discussed further below.

The synthesis of glutamine takes place under the catalytic influence of glutamine synthetase (Fig. 14.49), an intramitochondrial enzyme present in highest concentration in renal tissue. Note that this reaction requires the energy that is derived from the hydrolysis of 1 molecule of ATP to ADP. This reaction proceeds far to the right; therefore glutamine formation is favored.

The amide nitrogen of glutamine is not liberated by reversal of the glutamine synthetase reaction. Rather, free ammonia is formed by the hydrolysis of glutamine under the catalytic influence of another enzyme, glutaminase, as discussed earlier (Fig. 14.48).

Thus, the two enzymes, glutamine synthetase and glutaminase, underlie the catalytic interconversion of free ammonia (or ammonium ion, NH_4^+) and glutamine in a cyclic fashion that may be outlined schematically:

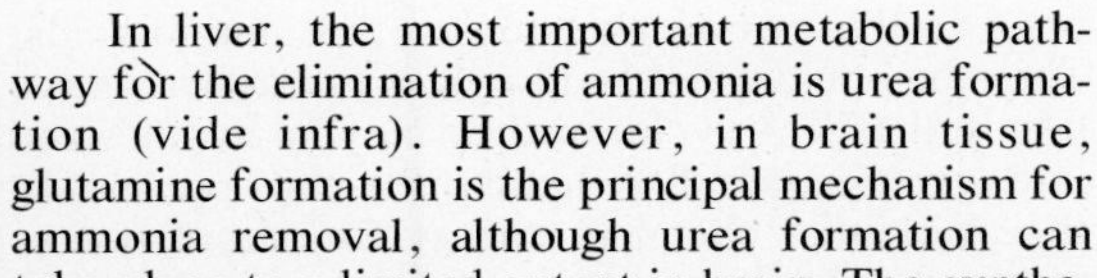

Glutamic acid + ATP

(Glutamine synthetase)

Ammonia   Glutamine

(Glutaminase)

Glutamic acid

In liver, the most important metabolic pathway for the elimination of ammonia is urea formation (vide infra). However, in brain tissue, glutamine formation is the principal mechanism for ammonia removal, although urea formation can take place to a limited extent in brain. The synthesis of glutamine in brain tissue has to be preceded by the formation of glutamic acid within this tissue, because the quantity of blood glutamic acid that is available is insufficient to permit synthesis of the increased amounts of glutamine that are formed in the brain when the blood ammonia concentration is elevated. The TCA cycle inter-

mediate, α-ketoglutaric acid, provides the direct source of glutamic acid for glutamine synthesis in brain. However, the intermediates of this oxidative cycle would be exhausted rapidly, were it not for their replacement by carbon dioxide fixation with conversion of pyruvic acid into oxaloacetic acid. In actual fact, brain tissue does fix a significant quantity of carbon dioxide into amino acids, and this process doubtless involves anaplerosis and the TCA cycle.

The synthesis of urea from ammonia (④, Fig. 14.45) involves a cyclic biochemical mechanism known as the urea cycle (Fig. 14.50). This process takes place in five individual steps as shown in Figure 14.50.

The first reaction in urea formation (①, Fig. 14.51) involves a condensation between 1 mole each of ammonia derived from the glutamic acid

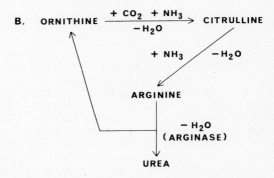

FIG. 14.50. **A.** Net reaction involved in the biosynthesis of urea. **B.** Outline of the urea cycle. (After Harper. *Review of Physiological Chemistry,* 13th ed, 1971, Lange Medical Publications.)

dehydrogenase reaction, Figure 14.47, or else the glutaminase reaction, Figure 14.48, carbon dioxide that has been activated by biotin, and phosphate which is derived from ATP. The resulting product is carbamyl phosphate, and the reaction is catalyzed by the enzyme carbamyl phosphate synthetase. This enzyme is found in the hepatic mitochondria of all organisms that synthesize urea (ie, ureotelic organisms), including man. The precise role of the N-acetyl glutamic acid in this reaction is unknown; however, it is believed to serve as an activator for the carbamyl phosphate synthetase in some manner, a role also played by the biotin (Fig. 13.31, p. 881). The high-energy terminal phosphate bonds of 2 ATP molecules supply the energy for this reaction.

Reaction ② involves the synthesis of citrulline, a process catalyzed by L-ornithine transcarbamylase, an enzyme present in the mitochondria of liver. This enzyme is highly specific for ornithine and the equilibrium is strongly toward the synthesis of citrulline.

Reaction ③ involves the synthesis of arginosuccinic acid, and this process is catalyzed by arginosuccinic acid synthetase. Note in Figure 14.51 that a molecule of aspartic acid is joined to a citrulline molecule via the amino group of the former compound. The necessary energy for this reaction involves the breakdown of 1 mole of ATP to AMP (adenosine monophosphate) plus inorganic pyrophosphate (PP_i). The equilibrium is strongly toward the formation of arginosuccinic acid.

The next reaction in urea formation, ④, involves the cleavage of arginosuccinic acid into arginine plus fumaric acid under the catalytic influence of arginosuccinase.

The final step in the urea cycle, reaction ⑤, is catalyzed by arginase, and it involves the hydrolysis of arginine into one mole each of urea and ornithine, so that the cycle can repeat indefinitely.

Arginine is present in all animal cells, as it is an essential amino acid for protein synthesis. Consequently urea can be made by any cell that has this amino acid provided that arginase also is present. In humans and other mammals the liver is the principal site of urea formation in the body as it contains not only arginase, but all of the other requisite enzymes for this process. Kidney and brain can also synthesize urea, but quantitatively the contribution of these tissues to the overall urea level in the body appears to be negligible.

It is worth noting here that the reaction sequence ornithine → citrulline → arginine is the normal biosynthetic pathway for arginine. The presence of a high arginase activity in liver, however, transforms this process from a one-way biosynthetic pathway into a cyclic process whereby the carbon skeleton of ornithine is used repeatedly with 1 molecule of urea being formed for each revolution of the cycle.

The synthesis of urea is believed to be regulated by a linked interaction between carbamyl phosphate synthetase and glutamic acid dehydrogenase (Fig. 14.47). Thus, nitrogen is funneled from glutamic acid (hence indirectly from all amino acids) into carbamyl phosphate, and thence into urea. Although the equilibrium constant of the glutamic acid dehydrogenase reaction favors glutamic acid formation rather than ammonia production, the removal of ammonia via the carbamyl phosphate synthetase reaction (Fig. 14.51, ①), plus the oxidation of α-ketoglutaric acid by the enzymes of the TCA cycle within the mitochondria (Fig. 14.29), tend to enhance glutamic acid metabolism. This effect is also stimulated by ATP, which serves not only as a substrate for the synthesis of carbamyl phosphate, but also stimulates the activity of glutamic acid dehydrogenase toward the formation of ammonia.

Fate of the Intermediates Produced Following Deamination of the Amino Acids. The present discussion is concerned with the metabolic fates of the carbon skeletons of the amino acids which result from the deamination of these compounds. Since transamination and oxidative deamination of the individual amino acids leads to the formation of the corresponding α-keto acids, then the material to be outlined here is concerned primarily with the fate of these α-keto acids.

It was noted earlier that excess dietary amino acids cannot be stored by the body in significant quantities either as such or in the form of protein. Two major pathways are open for the utilization of the α-keto acids that are derived from amino acids. First, the carbon skeletons may be converted into pyruvate, acetyl CoA, or TCA cycle intermediates, and oxidized directly to provide useful energy. Second, the carbon skeletons may be converted into these or other intermediates, and much of the chemical potential energy then may be stored by the ultimate conversion of these compounds into glucose and glycogen, or lipid. The relative extent to which either of these processes takes place, of course, depends upon the nutritional state of the body as a whole.

The potential interconversions of the individual amino acids are summarized in Figure 14.52, and the possible fates of these compounds insofar as their biochemical transformation into glycogen and/or lipid is concerned are summarized in Table 14.11. Two metabolic pathways thus are open to the carbon skeletons of the amino acids. They may be catabolized to yield energy via oxidation in the TCA cycle, or they may be synthesized into glycogen and/or lipids, thereby preserving their potential chemical energy for future use as required. These intermediates are termed amphibolic because their metabolic fate may take either of two general courses, anabolic or catabolic.

By way of summarizing the foregoing statements, it has been firmly established experimentally that carbohydrate, lipid, and protein are interconvertible. Furthermore, this work has demonstrated clearly that the carbon skeletons of 13

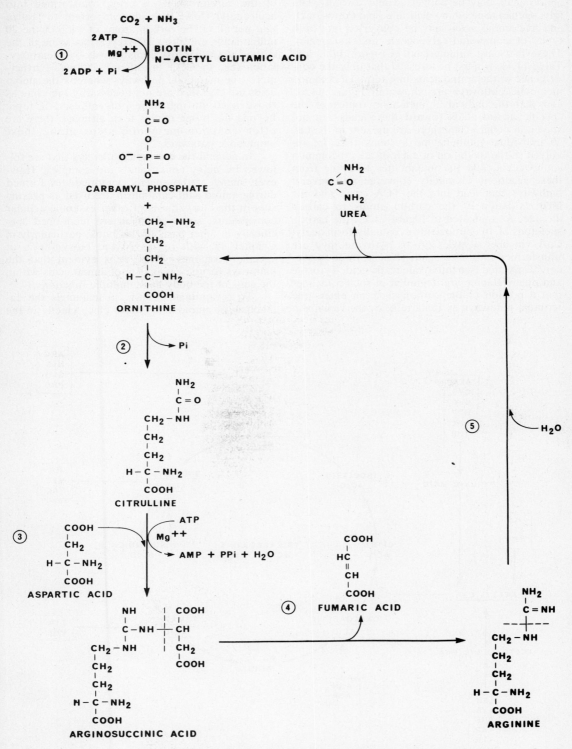

FIG. 14.51. The urea cycle. The specific enzymes that catalyze the individual reactions that are involved in urea biosynthesis are: ①, carbamyl phosphate synthetase; ②, ornithine transcarbamylase; ③, argininosuccinate synthetase; ④, argininosuccinase; ⑤, arginase. (After Harper. *Review of Physiological Chemistry*, 13th ed, 1971, Lange Medical Publications.)

amino acids may be converted into carbohydrate (glycogenic), one amino acid into lipid (ketogenic), and five amino acids may be converted into both types of compound (glycogenic and ketogenic). These facts are summarized in Figure 14.52 and Table 14.11. It must be stressed that only the convergence of the amino acids into terminal common metabolic pathways are shown in Figure 14.52, and that the individual metabolic routes of the specific amino acids toward these ends are not shown in detail. Thus there are present in the cell 20 individual multienzyme systems that are involved in the oxidation of each of the 20 common amino acids, and the metabolites resulting from these reactions ultimately converge into several pathways that lead into the TCA cycle and the final oxidation of the carbon atoms to carbon dioxide. As indicated in Figure 14.52, the carbon skeletons of 10 amino acids eventually form acetyl CoA via acetoacetyl CoA or pyruvate, five are transformed into α-ketoglutaric acid, three to succinyl CoA, and two into oxaloacetic acid. Tyrosine and phenylalanine are degraded in such a manner that a portion of the carbon skeleton enters the terminal pathways as fumaric acid, the remainder

of the carbon skeleton being transformed into acetoacetyl CoA. It is also important to realize that not all of the carbon atoms of each of the 20 amino acids enter the TCA cycle, as some of the carbon atoms are lost along the way by decarboxylation (eg, to CO_2) or other reactions. Furthermore, the metabolic pathways whereby the amino acids are degraded are not necessarily the same as those used during their biosynthesis, a topic beyond the scope of this text, although there are often common metabolic steps along these amphibolic pathways.

In general the catabolic pathways that are followed by many amino acids are complex. However, many of the specific intermediates formed during amino acid catabolism are used as precursors in the biosynthesis of other essential cellular constituents, some of which are mentioned specifically in other parts of this book, particularly in Chapter 15 with regard to the biosynthesis of specific hormones. Thus, it is evident that the pathways for the catabolism of amino acids within the cells of the body have multiple functions.

To recapitulate briefly, in mammals the catabolism of amino acids takes place chiefly in the

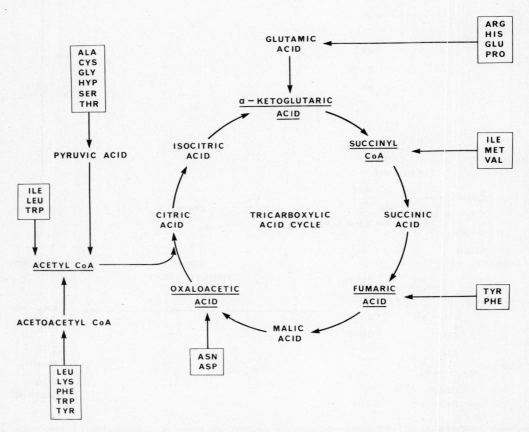

FIG. 14.52. Outline of the amphibolic intermediates which are formed from the carbon skeletons of particular amino acids. The names of the amino acids which are abbreviated in this figure are given in Table 14.10. See also Figure 14.27. (After Harper. *Review of Physiological Chemistry*, 13th ed, 1971, Lange Medical Publications.)

liver, although the kidneys participate in this process to some extent. Skeletal muscle plays a relatively minor role in amino acid catabolism; however, once the carbon skeleton of any amino acid has been incorporated into, say, a glucose or a fatty acid molecule, it then enters the common metabolic pool of the body for that particular compound and then it can be drawn upon for purposes of supplying energy for any cellular process as required. In other words, once such a metabolic interconversion has been achieved, the carbon skeleton of any particular amino acid loses its metabolic identity.

METABOLISM OF ETHYL ALCOHOL. Although ethanol hardly can be considered an essential element of human nutrition, the widespread consumption of this compound in various beverages renders the metabolism of alcohol a topic of importance in physiology.

Alcoholic beverages may be ingested for reasons that range from its use as a pleasant adjunct to everyday living all the way to the frankly pathologic ingestion of spirits by the confirmed alcoholic. This discussion is included, however, primarily because the energy content of pure ethanol is considerable. In a bomb calorimeter, the combustion of 1 gram of ethyl alcohol yields 7.13 Cal of free energy, whereas in the body the physiologic value of 7.0 Cal/gm is obtained when this compound is oxidized.

Alcohol is absorbed rapidly and completely from the gastrointestinal tract into the bloodstream, whence it is distributed throughout the body. Since alcohol is volatile, a small fraction of the compound is lost from the body in the expired air via the respiratory tract. Similarly, a small quantity is excreted directly in the urine. However, quantitatively these two fractions represent only a negligible proportion of any given quantity of ethyl alcohol that has been ingested, and the major fate of this compound in the body is its eventual oxidation, principally in the liver.

The pathways for the metabolism of ethyl alcohol at present are established only in broad outline, and may be summarized as follows:

$$CH_3\text{—}CH_2OH + NAD^+ \xrightarrow{\;\;①\;\;} CH_3\text{—}CHO + NAD^+ + H^+$$

     Ethyl alcohol   Acetaldehyde

$$CH_3\text{—}CHO + NAD^+ + CoA\text{—}SH \xrightarrow{\;\;②\;\;} CH_3\overset{\displaystyle O}{\overset{\displaystyle \|}{\text{—}C}}\text{—}S\text{—}CoA + NADH^+$$

    Acetaldehyde   Acetyl CoA

As shown in Figure 14.53, acetaldehyde can undergo spontaneous keto–enol tautomerism. Reaction ① is catalyzed by the specific enzyme alcohol dehydrogenase that is found almost exclusively in the liver. Reaction ② is carried out under the catalytic influence of an aldehyde dehydrogenase, and the resulting acetyl CoA that is formed then may be oxidized directly in the TCA cycle to provide energy (Fig. 14.29), converted into fatty acids or cholesterol (Fig. 14.27), or utilized for any other reactions that require this high-energy intermediate compound. In other words, the body does not discriminate between the acetyl CoA that has been derived from alcohol and that produced from any other nutritional source. This fact is of considerable importance clinically, as the heavy drinker may derive a considerable proportion of his total energy requirements from alcoholic beverages to the exclusion of other nutrients having a more balanced food value insofar as essential amino acids, vitamins, and other compounds are concerned. Ultimately this situation is quite detrimental to the overall nutritional state of the individual; so much so, in fact, that many investigators believe that the long-term degenerative effects of excess alcohol consumption result from a nutritional deficiency as much as from the toxic effects of the alcohol itself.

It is of interest to note that the rate of alcohol metabolism has been found experimentally to parallel the rate at which glucose, hence pyruvic acid, is oxidized. Thus increased alcohol metabolism is accompanied by an increased carbohydrate metabolism, and conversely an increased carbohydrate metabolism that results in increased pyruvic acid formation also augments the rate of alcohol oxidation.

From a practical standpoint, however, the total rate of oxidation of pure alcohol by an average adult human amounts to about 28 ml per hour.

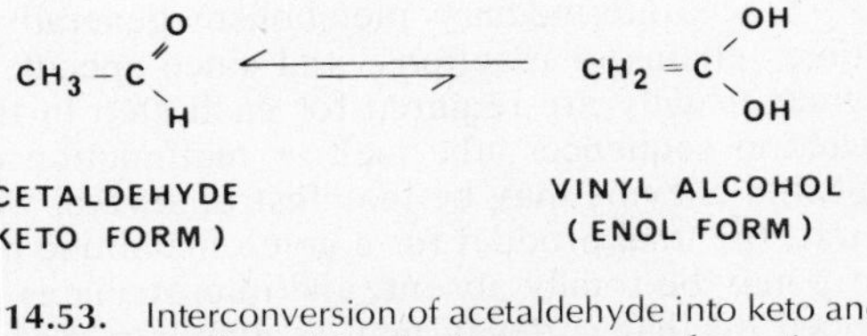

CH_3 – C (=O)(H) ⇌ CH_2 = C (OH)(OH)

ACETALDEHYDE   **VINYL ALCOHOL**
(KETO FORM)   **(ENOL FORM)**

FIG. 14.53. Interconversion of acetaldehyde into keto and enol forms. As indicated by the longer arrow, formation of the keto form is favored. (After White, Handler, and Smith. *Principles of Biochemistry*, 4th ed, 1968, The Blakiston Division, McGraw-Hill.)

The consumption of alcohol much in excess of this maximal rate leads to varying degrees of intoxication with its attendant consequences. In this regard, the blood and respiratory gas levels of alcohol, as determined chemically in a given individual, can assume considerable medicolegal significance in determining the degree of intoxication of a person under specific circumstances.

CLINICAL ASPECTS OF CERTAIN METABOLIC DEFECTS. In general, abnormalities of metabolism reflect changes in the specific rates of one or more bodily processes that may be induced by abnormal nutrition, drugs, poisons, disease, an imbalance in the normal physiologic regulatory mechanisms, or by inherited metabolic defects. These rate changes may lead in turn to qualitative or quantitative alterations in specific bodily constituents, but a clear understanding of these alterations in body composition requires a knowledge of the underlying rates for the process (or processes) which are involved. For example, it was mentioned earlier that the various endocrine glands normally exert a profound regulatory influence upon all aspects of metabolism in the human body. Abnormal functioning of these organs can lead to a broad spectrum of metabolic disorders that have a profound clinical significance. The various disease states that can arise through hypoactivity or hyperactivity of the endocrine glands are discussed in context in Chapter 15 and the concomitant metabolic derangements that appear are also stressed.

In the present section, emphasis will be placed upon a few of the many so-called "inborn errors of metabolism" that occur in humans through the inheritance of defective genes. Since the genes ultimately are responsible for controlling protein (hence enzyme) synthesis, then such genetic defects are reflected by an alteration in DNA that leads either to a complete failure to synthesize a protein or to the synthesis of a modified protein whose function is abnormal or completely absent. If the mutant gene that is concerned in an inborn error of metabolism is one that determines the biosynthesis of a specific enzyme (ie, protein) that is essential to some phase of intermediary metabolism, then the catalytic activity of that enzyme may be faulty or even totally absent. Such effects can occur, for example, if the sequence of amino acids in the enzyme, as dictated by the mutant gene (abnormal DNA), is incorrect for that particular protein moiety.

Since intermediary metabolism generally involves chains of reactions, and since specific enzymes usually are required for each step in these reaction sequences, the lack or malfunction of a specific enzyme may be manifest in several ways. First, the final product for a given metabolic pathway may be totally absent, and intermediates produced by that pathway may accumulate. Second, an intermediate metabolite may accumulate in significant quantities when no alternate pathway is available. Third, the metabolic products of a normally alternate, and minor, metabolic pathway may accumulate in large quantities because of a block in the primary pathway. Fourth, under certain conditions the presence of a particular metabolite in excess does not reflect an accumulation of that intermediate proximal to an enzymatic block, but rather it reflects an overproduction of the compound(s) which is due to the release from a normal regulatory mechanism; eg the metabolite accumulation is due to a defect in feedback inhibition. Thus, if the metabolic product of a given pathway controls the rate at which it is synthesized via that pathway by a mechanism that operates through allosteric inhibition of the first enzymatic step, then malfunction of this regulatory mechanism can lead to overproduction and accumulation of the metabolite.

In human beings, many of the metabolic defects that have been observed can be traced to the absence of activity of a single enzyme, and as one gene controls the biosynthesis of one enzyme, then the fundamental defect in such disorders is to be found in one faulty gene. A few of these metabolic disorders, together with the specific enzymes that are deficient, are listed in Table 14.12. The conditions which are listed in this table have been selected particularly to illustrate the clinical states that can arise from deficient activity of enzymes that have been discussed earlier in this text.

Certain general features of these metabolic disorders may be summarized. Each defect is hereditary, and thus is present throughout the lifetime of the individual. Furthermore, since these enzymatic defects are hereditary, the errors of metabolism observed in each case reflect the fact that the person having them is a mutant, and the mutation can be detected as the result of that person being unable to perform a specific biochemical reaction or series of reactions. In this regard, human mutants are similar to mutant microorganisms, and in both instances, the study of such subjects has revealed many previously obscure biochemical pathways.

Practically all inborn errors of metabolism are recessive characteristics, and are either autosomal or X-linked. The expression of the gene in the phenotype takes place only in the hemizygote, or in X-linked recessives, the hemizygote. In heterozygote individuals the quantity of enzyme activity normally is adequate. However, by tests in which the particular enzymatic step is subjected to stress, eg, by loading the individual with the compound or metabolite involved, the metabolic defect often is demonstrable in the heterozygote. In addition, direct assay of enzyme activity in suitable tissues often can reveal the heterozygote individual because the enzyme activity is present at a level below that for the homozygous normal person; however, in such heterozygous persons the enzyme activity is still higher than that found in the homozygous individual.

Table 14.12[a]   SOME EXAMPLES OF RECOGNIZED HEREDITARY METABOLIC DISORDERS IN MAN[b]

CONDITION	ENZYME OR OTHER COMPONENT HAVING AN ABNORMAL (DEFICIENT) ACTIVITY
Acatalasia	Catalase
Afibrinogenemia	Fibrinogen
Agammaglobulinemia	γ-Globulin
Albinism	Tyrosinase
Alkaptonuria	Homogentisic acid oxidase
Analbuminemia	Serum albumin synthesis
Fructose intolerance	Fructose-1-phosphate aldolase
Hemolytic anemia	Pyruvate kinase[c]
Maple syrup urine disease	Amino acid decarboxylase
Phenylketonuria (PKU)	Phenylalanine hydroxylase
Glycogen storage disease I	Glucose-6-phosphatase
Glycogen storage disease V	Muscle phosphorylase
Glycogen storage disease VII	Muscle phosphofructokinase
Glycogen storage disease VIII	Liver phosphorylase kinase
Gout, primary (one type)	Hypoxanthine guanine phosphoribosyl transferase (HGPRT)
Hemolytic anemia	Diphosphoglyceric acid mutase[c]
Hemolytic anemia	Hexokinase[c]
Hemolytic anemia	Hexosephosphate isomerase[c]
Hemolytic anemia	Phosphoglyceric acid kinase[c]
Hemolytic anemia	Triosephosphate isomerase[c]
Lactase deficiency, adult	Intestinal lactase
Congenital lactose intolerance	Intestinal lactase
Congenital lipase deficiency	Pancreatic lipase
Leigh's necrotizing encephalomyelopathy	Pyruvate carboxylase
Methemoglobinemia	NADH-methemoglobin reductase
Trypsinogen deficiency disease	Trypsinogen

[a] *Data from Harvey, Johns, Owens, and Ross.* The Principles and Practice of Medicine, *18th ed, 1972, Appleton-Century-Crofts.*

[b] *In each condition listed in this table, the specific protein or enzyme that is lacking or defective in activity has been identified.*

[c] *Note that a defect in any of several glycolytic enzymes (Table 14.6) can lead to hemolytic anemia.*

OVERALL ENERGY METABOLISM IN THE HUMAN BODY

In the first section of this chapter, the biochemical mechanisms whereby the body obtains useful energy from the stepwise oxidation of the three basic types of nutrients (foods) were the principal concern. In these discussions, it was explained how a fraction of the total chemical potential energy of foods is harnessed (or trapped) in specific configurations of certain high-energy compounds, notably ATP, and only in these forms can various types of work be performed by the free energy thus released. The remainder of the free energy obtained from the exergonic oxidative processes is heat energy, and as such cannot perform useful work within the body. Essentially, therefore, the heat that is produced incidentally to the transformation of free chemical energy into compounds that are able to perform useful work largely is a waste product, and thus must be dissipated from the body.

Normally, the rate at which biologic oxidations take place in vivo is governed by the rate at which free energy (ΔF) is required for the performance of useful work, and the heat energy that is lost irretrievably as a concomitant event to these processes is dissipated to the environment. Because of this continual energy loss (as heat) from the body, it is vital that new sources of energy for oxidative process be provided daily in the form of carbohydrates, lipids, and proteins, in order that a biochemical and morphologic steady state within the cells and tissues be maintained throughout the lifetime of the individual.

However, the overall process of metabolism whereby much of the free energy derived from the oxidative degradation of nutrients is lost as heat energy is not as wasteful biologically as might be imagined from the foregoing statements. This is because of the fact that man, a mammal, is a homeothermic organism; and as such he must maintain a relatively constant body temperature despite environmental temperatures that generally are below 37 C, his normal body temperature.

The study of overall energy metabolism includes a variety of subjects including the caloric value of foods, the respiratory quotient (RQ), direct and indirect calorimetry, the caloric requirements of the body, and the specific dynamic action (or calorigenic action) of foods. These topics are the concern of the present section.

The Metabolic Rate and Its Measurement

It is an important and fundamental concept of physiology that the total quantity of energy that is yielded by the catabolism of food within the body is identical with that quantity of energy liberated when the same food is burned in vitro.* In the latter instance, all of the free energy that is liberated by combustion appears as heat. Within the body, however, the oxidation of nutrients by catabolic processes yields energy that appears not only as heat, but also as external work (eg, muscular contractions, osmotic work) and energy storage. This statement may be summarized in the simple relationship: Total energy output = External work + Energy storage + Heat energy.

In a skeletal muscle performing isometric contractions, practically all of the total energy output appears as heat because very little, if any, external work is being done. This situation obtains since work is the product of a force moving a mass through a given distance. On the other hand, the same muscle contracting under isotonic conditions performs work at an efficiency of about 50 percent, and this value can be derived from the expression:

$$\text{Efficiency} = \frac{\text{Total work performed}}{\text{Total energy expended}} \times 100$$

In this situation, the quantity of energy that is used for the performance of the useful work is a large fraction of the total energy that has been expended, and the quantity of energy that appears as heat is proportionally lower.

Free energy is trapped in the tissues of the body by the formation of specific bonds in high-energy compounds.† The overall extent to which this process takes place varies considerably, depending upon the nutritional state of the individual as well as the amount of external work being performed. For example, in a fasting subject, energy storage is zero or negative, and if the subject is resting very quietly in a fully relaxed state, then practically all of the energy being used appears as heat. Under such circumstances, only a small fraction of the total energy expended is used for work, for example, the maintenance of such vital processes as the heartbeat and the bioelectric potentials in the central nervous system.

The total quantity of heat that is released per unit time by the body is defined as the metabolic rate, and this value is expressed in terms of heat energy, ie, in calories.

It will be recalled that the calorie (small c) is defined as the quantity of heat which is required to raise the temperature of 1 gm of water 1 c in the range between 15 c to 16 c. This fundamental unit also is known as the gram calorie, standard calorie, or small calorie. However, in any discussion of metabolism, the kilocalorie (kcal) or Calorie (capital C) is the unit employed. One kcal equals 1,000/cal, ie, it is the quantity of heat required to raise the temperature of 1,000 gm of water from 15 C to 16 C. The kilocalorie will be used exclusively in the discussions to follow, and the abbreviation Cal will be used to denote this quantity.

Determination of the Caloric Value of Foods and Other Substances

The caloric value (or heat of combustion) of any oxidizable substance, including purified carbohydrates, lipids, or proteins, can be determined in an apparatus such as a bomb calorimeter (Fig. 14.54). In practice, a known mass of test material is sealed within a heavy metal cylinder with a fine wire loop of an inert substance (such as platinum) inserted into the material. The closed cylinder is subjected to a moderately high pressure of pure oxygen (around 20 atmospheres), and then placed in a bath containing a known mass of water at a certain temperature, usually 15 C. When electric current is passed through the wire, combustion is started and the test material is oxidized completely, and all of the heat generated thereby is transferred to the calorimeter and water provided that suitable precautions are taken to prevent any heat transfer to any other part of the environment. Knowing the mass of the water and the temperature change, the heat produced may be calculated easily from the relationship $Cal = m(T_2 - T_1)$, in which m equals the mass of the water in gm. The caloric value per gm of test material thus may be

---

*Nitrogenous compounds do not yield an identical quantity of energy when oxidized in vivo and in vitro for reasons which will be discussed subsequently.

†Free energy is not stored to any appreciable extent in ATP and other high energy compounds because the in vivo concentration of these substances is so low.

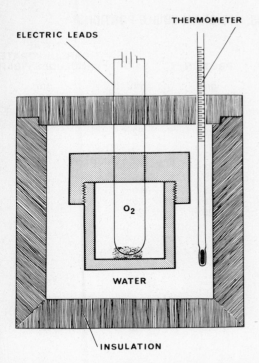

ELECTRIC LEADS

THERMOMETER

O₂

WATER

INSULATION

FIG. 14.54.   The principal features of a bomb calorimeter.

calculated from the total heat produced and the mass of the sample.

The total heat energy released by the complete oxidation of any substance is the same irrespective of the mechanism of the reaction involved, provided that the end products of the reaction are identical. Consequently, the oxidation of pure glucose, according to the equation

$$C_6H_{12}O_6 + 6O_2 \rightarrow 6\ CO_2 + 6\ H_2O + Heat$$

yields an identical quantity of heat energy whether it occurs in a bomb calorimeter or in the human body. This value is 3.75 Cal per gm (Table 14.13). Absorbed fats also yield identical caloric values when oxidized completely in vitro or in vivo; the

Table 14.13   SUMMARY OF IMPORTANT METABOLIC VALUES FOR THE BASIC NUTRIENTS[a]

NUTRIENT	RQ	CAL/G	CAL/LITER OXYGEN UTILIZED
Carbohydrate	1.00	3.75	5.01
Fat	0.70	9.4	4.75
Protein[b]	0.801	4.3	5.92

[a]*Data from Harper.* Review of Physiological Chemistry, *13th ed, 1971, Lange Medical Publications; White, Handler, and Smith:* Principles of Biochemistry, *4th ed, 1968, The Blakiston Division, McGraw-Hill.*

[b]*1 gm urinary N_2 = 5.92 liters of O_2 used, 4.75 liters of CO_2 produced, 6.25 g of protein metabolized, and a production of 26.51 Cal.*

average caloric value of these nutrients is 9.4 Cal per gm. The heats of combustion for various proteins as determined in a bomb calorimeter are somewhat higher than those determined by measurements made in vivo because the oxidation of these compounds is incomplete in the body, whereas the oxidation process is complete in a bomb calorimeter. This situation obtains since the end products of protein metabolism that are excreted in the urine are such compounds as urea, creatinine, uric acid, and so forth, and these substances still contain a certain amount of free energy in the form of oxidizable (ie, reduced) carbon and hydrogen. Hence approximately 1.3 Cal per gm of protein is lost from the body. An average value of 5.6 Cal per gm is found for mixed proteins in a bomb calorimeter. Therefore around 4.3 Cal per gm is the heat which is produced by the oxidation of absorbed protein in the body; ie, this is the physiologic value for protein oxidation. Stated another way, the nitrogenous endproducts of protein metabolism still contain a significant quantity of free energy that is unavailable to the body.

It must be realized that ingested carbohydrates, fats, or proteins are not completely absorbed from the intestinal tract. Therefore the physiologic caloric values are somewhat lower than when these are determined by bomb calorimetry. For example, cellulose is oxidized completely in a bomb calorimeter, but is neither digested in, nor absorbed from, the human gastrointestinal tract. Consequently, the caloric value of a fruit or vegetable that contains much cellulose will yield a significantly higher caloric value when determined in vitro compared to that value obtained in vivo. Furthermore the carbohydrates, lipids, and proteins which are found in the usual food materials eaten by man contain mono-, di-, and polysaccharides, as well as short- and long-chain fatty acids that may be saturated or unsaturated. In addition, a variety of amino acids of differing carbon chain lengths and nitrogen contents are obtained from protein digestion. The breakdown of complex organic compounds during the process of digestion itself reduces their caloric value somewhat. These factors, together with small differences in the average composition of natural foodstuffs, give rise to slight variations among the caloric values of various foods obtained by bomb calorimetry and those determined in vivo. Average caloric values for a few foodstuffs are given in Table 14.14. In practice, little error is introduced by rounding off the caloric values of the common substances that are essentials of human nutrition. The following values are employed commonly in nutrition and dietetics: carbohydrate, 4 Cal/gm; fat 9 Cal/gm; protein, 4 Cal/gm; ethyl alcohol, 7 Cal/gm.

Direct Calorimetry

The total heat production in the body may be measured by a technique known as direct calorimetry

Table 14.14[a]   COMPOSITION OF FOODS, 100 GRAM EDIBLE PORTIONS[b]

FOOD	WATER	ENERGY	PROTEIN	FAT	TOTAL CARBOHYDRATE (INCLUDES FIBER)
	%	Cal	gm	gm	gm
Abalone, raw	76	98	19	0.5	3
Apple, raw, not pared	84	58	0.2	0.6	15
Apricots, raw	85	51	1	0.2	13
Bacon, cured, fried, drained	8	611	30	52	3
Bananas, raw	76	85	1	0.2	22
Bass, striped, cooked	61	196	22	9	7
Beans, white, cooked	69	118	8	0.6	21
Beans, red, cooked	69	118	8	0.5	21
Beans, lima, cooked	71	111	8	0.5	20
Beef, cooked (81% lean, 19% fat)	49	327	26	24	0
Beverages					
Beer (4.5% alcohol by volume = 3.6% by weight)	92	42	0.3	0	38
Spirits (gin, whiskey, rum, vodka), 94 proof = 40% alcohol by weight	60	275	——	——	Trace
Wines, table (12% alcohol by volume, 10% by weight)	86	85	0.1	0	4.2
Breads					
Rye	36	243	9	1	52
White	36	270	9	3	51
Whole wheat	36	243	11	30	48
Cabbage, cooked	94	20	1	0.2	4.3
Carrots, cooked	91	31	1	0.2	7
Cauliflower, cooked	93	22	2.3	0.2	4
Celery, raw	94	17	1	4	1
Cheese, Cheddar	37	398	25	32	2
Chicken, cooked	64	166	32	3	0
Clams, canned	86	52	8	1	3
Cod, cooked	64	170	29	5	0
Corn, cooked	77	83	3	1	19
Crab, cooked	79	93	17	2	0.5
Cream, heavy	57	352	2	38	3
Eggs, chicken, cooked, boiled	74	163	13	12	1
Halibut, cooked	67	171	25	7	0
Heart, beef, cooked	78	108	17	4	0
Herring, canned	63	208	20	14	0
Kidneys, beef, cooked	53	252	33	12	1
Lamb, cooked (79% lean, 21% fat)	50	319	24	24	0
Lettuce, raw	96	13	1	0.1	2.9
Liver, beef, cooked	56	229	26	11	5
Macaroni, cooked	64	148	5	0.5	30
Mackerel, canned	66	180	21	10	0
Milk, cow, whole	87	65	4	4	5
Milk, goat	88	67	3	4	5
Milk, human	85	77	1	4	10
Mushrooms, canned	93	17	2	0.1	2
Muskrat, cooked	67	153	27	4	0
Onions, cooked	92	29	1	0.1	7
Opossum, cooked	57	221	30	10	0
Oranges, peeled	86	49	1	0.2	12
Oysters, raw	87	66	8	2	3
Peanuts, roasted	2	582	26	49	21
Peas, green, cooked	82	71	5	0.4	12
Pork, fresh, cooked (72% lean, 28% fat)	42	410	21	36	0

Table 14.14 (Cont'd)

Potatoes, cooked	80	76	2	0.1	17
Rabbit, cooked	60	216	29	10	0
Raccoon, cooked	55	255	29	15	0
Reindeer, cooked (91% lean, 9% fat)	67	178	22	9	0
Rice, brown, cooked	70	119	3	0.6	26
Rice, white, cooked	73	109	2	0.1	24
Salmon, canned	64	203	22	12	0
Soybeans, cooked	74	118	10	5	10
Spinach, cooked	92	23	3	0.3	4
Sugar (sucrose) granulated	0.5	385	0	0	99+
Sweet potatoes, cooked	71	114	2	0.4	26
Tongue, beef, cooked	61	224	22	17	0.4
Tuna, canned in oil	53	288	24	21	0
Turkey, cooked	55	263	27	16	0
Venison, lean, raw	74	126	21	4	0
Whale meat, raw	71	156	21	8	0

[a]*Data from Watt and Merrill.* Composition of Foods, Agriculture Handbook No. 8. *Washington, DC, United States Department of Agriculture, 1963.*

[b]*The foodstuffs listed in this table are representative only, and most of the values have been rounded off to the nearest whole digit. Slight differences will be found among the values given for similar foods which have been processed differently (e.g., if oil or sugar has been added during packing or canning), or if meat or meat byproducts contain more (or less) fat. Likewise meats, poultry, and seafoods will differ slightly in composition depending upon their particular source.*

(Fig. 14.55). In practice, the chambers used for the direct determination of total heat production by an animal or human being are quite elaborate, as no heat loss to the external environment may take place. In principle, the total heat produced by the subject may be measured by thermocouples suitably arranged to detect the heat exchange from the subject to an accurately metered quantity of water flowing through pipes surrounding a heavily insulated chamber of suitable size, while the respiratory exchange of oxygen and carbon dioxide by the subject are monitored accurately and continuously.

Direct measurements performed using large chambers of this type have given values of about 1,500 to 1,800 Cal per day for the total heat production of a resting adult human being in the fasting (postabsorptive) state. Such a direct technique also permits the effects of exercise, food consumption, environmental temperature, and other activities or factors on heat production to be studied under closely controlled experimental conditions.

Early experiments on the total heat production of human and animal subjects performed simultaneously with studies on respiratory gas ex-

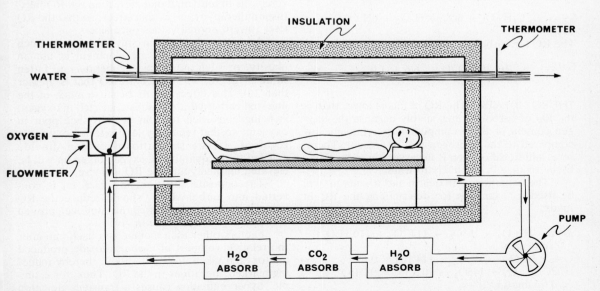

FIG. 14.55. A whole-body calorimeter. (After Ruch and Patton [eds]. *Physiology and Biophysics,* 19th ed, 1966, Saunders.)

change (ie, oxygen consumption and carbon dioxide production) soon revealed the important fact that heat production could be estimated quite accurately by measuring respiratory gas exchange alone. The application of this far simpler technique obviates the necessity for the complex apparatus required for direct calorimetry, and it yields values that are of proven validity. Therefore, indirect calorimetry is the method in widest use today for the determination of total heat production.

In order to understand how one may calculate the total heat production of the body solely from a knowledge of oxygen consumption and carbon dioxide production, it is essential to understand the concept of the respiratory quotient and its relationship to urinary nitrogen excretion.

Respiratory Quotient

The respiratory quotient (RQ) is defined as the volume of carbon dioxide produced divided by the volume of oxygen which is consumed over a definite period of time; ie, RQ is the ratio: volume of CO_2/volume of O_2. This value can be obtained for chemical reactions in vitro, and for particular organs and tissues, as well as for the entire body. In turn, the volumes of oxygen and carbon dioxide can be measured with instruments like the spirometer (Fig. 14.56).

THE RQ OF CARBOHYDRATE. The RQ for the oxidation of carbohydrate is 1.0 and this value is the same whether the combustion of this type of compound takes place in vivo or in vitro. This statement is true because the ratio of oxygen and hydrogen atoms in carbohydrates is the same as that in water. Using glucose as an example, the oxidation of this compound takes place according to the reaction:

$$C_6H_{12}O_6 + 6O_2 \rightarrow 6CO_2 + 6H_2O$$

The RQ for carbohydrate, therefore, is:

$$RQ = CO_2/O_2 = 6/6 = 1.0.$$

THE RQ OF FATS. The RQ of fats is lower than is the RQ of carbohydrates simply because the oxygen content of these compounds is relatively low compared to the hydrogen that is present. Therefore, additional oxygen is necessary in order that a fat be oxidized completely to water.

The oxidation of tristearin and tripalmitin will be used as examples for determining the RQ of lipids.

A. $2C_{57}H_{110}O_6 + 163O_2 \rightarrow 114CO_2 + 110\ H_2O$
   Tristearin
   $$RQ = CO_2/O_2 = 114/163 = 0.699$$
B. $2C_{51}H_{98}O_6 + 145O_2 \rightarrow 102CO_2 + 98H_2O$
   Tripalmitin
   $$RQ = CO_2/O_2 = 102/145 = 0.703$$

Two specific examples of lipids were given as they exemplify the very slight differences in RQ that are found among various fats because of differences in their chemical composition. For all practical purposes, however, the RQ for the oxidation of fatty materials is taken as 0.70, and any error that is introduced by the use of this average figure is negligible.

THE RQ OF PROTEINS. The RQ of proteins cannot be determined by a simple stoichiometric relationship as it can for carbohydrates and lipids as the composition and proportions of the different proteins found in the diet are in general unknown. Therefore, calculation of the RQ of mixed proteins is an indirect and complex process, but an average RQ value of 0.801 has been determined empirically for dietary proteins.

THE RQ OF OTHER COMPOUNDS. The RQ values for a few other compounds of biologic importance are:

Acetoacetic acid	1.0
β-Hydroxybutyric acid	0.89
Ethyl alcohol	0.67
Glycerol	0.86
Pyruvic acid	1.20

PHYSIOLOGIC AND OTHER FACTORS THAT CAUSE VARIATIONS IN THE RQ. The RQ of an average human subject consuming a mixed diet that contains varying proportions of carbohydrate, fat, and protein is about 0.85. If the proportion of carbohydrate that is metabolized increases, then the RQ approaches 1.0, whereas the RQ falls as the relative quantity of fat that is metabolized increases.

If carbohydrate metabolism is deranged severely, as in untreated diabetes, then the RQ falls. Insulin therapy reverses this effect, so that the RQ rises toward normal values once again.

An extremely high carbohydrate intake, such as may occur during the development of human obesity, or such as is used deliberately to fatten cattle or geese, will induce an RQ that is greater than 1.0. This effect occurs because much of the ingested carbohydrate (a substance rich in oxygen) is being converted into fat (a substance poor in oxygen), so that relatively less external oxygen is required. Hence the ratio of carbon dioxide exhaled to the quantity of oxygen taken up will be elevated markedly and the RQ will rise.

The opposite situation, in which fat is converted into carbohydrate, should reduce the RQ below 0.70, but such an effect has not been proven conclusively to take place in vivo.

Factors other than metabolism itself can alter the relative volumes of carbon dioxide produced and oxygen taken up by the body, thereby inducing marked alterations in the RQ. Thus, for example, hyperventilation causes a transient elevation in RQ merely because more carbon dioxide is exhaled. During physical exercise, the RQ may rise as high as 2.0 because of the combined effects of hyperventilation and development of an oxygen debt in the body. Following a bout of exercise the RQ may drop to 0.5 or lower because of the reversal of these processes.

In states of metabolic acidosis, the RQ is elevated because respiratory compensation for the acidosis produces an increase in the net quantity of expired carbon dioxide, and in severe metabolic acidosis the RQ may even exceed 1.0. On the other hand, during metabolic alkalosis the RQ decreases.

Since the RQ of the intact body is affected by many respiratory factors, as well as by the overall metabolism per se, some authors prefer to call the CO_2/O_2 ratio the respiratory exchange ratio (R) in preference to using the term RQ to denote this value.

THE RQ OF ORGANS. The overall oxygen consumption and carbon dioxide production of an individual organ can be obtained either in vivo or in vitro simply by multiplying the blood flow per unit time by the arteriovenous oxygen and carbon dioxide differences across the organ. The RQ then can be calculated.

Data on the RQ of single organs are of interest because one may deduce certain features pertaining to the general metabolic processes taking place. Hence, the RQ of intact brain tissue normally is found to lie between 0.97 and 0.99. This fact indicates that carbohydrate is the major substrate that is oxidized in such tissue.

Another example of the application of this technique is to the stomach during the secretion of gastric juice. While active gastric secretion is taking place the stomach actually has a negative RQ because it is taking more carbon dioxide from the arterial blood than it is eliminating in the venous blood.

Indirect Calorimetry

In order to calculate the energy production of the body from the RQ, one must know the quantity of oxygen which is needed to oxidize completely 1 mole of each of the three major nutrients, as well as the number of Cal that are yielded during this process. In other words, the caloric equivalents of each major nutrient must be known.

CARBOHYDRATE. The RQ for carbohydrate is 1.0. Again using glucose as an example, the complete oxidation of this compound yields 6 moles of CO_2 and requires 6 moles of O_2. Multiplying these figures by 22.4 (one mole of any gas occupies a volume of 22.4 liters under STP), we find that 134.4 liters of carbon dioxide are produced and 134.4 liters of oxygen are consumed during the oxidation of 1 mole of glucose ($6 \times 22.4 = 134.4$). Since the oxidation of 1 mole of glucose yields 673 Cal (p. 920), then 673 Cal/134.4 liters = 5.01 Cal per liter O_2.

In other words, the oxidation of glucose in the body yields 5.01 Cal per liter of O_2 consumed.

FAT. Similarly, the complete oxidation of 1 mole of fat yields approximately 16,300 Cal and requires 3,500 liters of oxygen. Therefore, each liter of oxygen used in the body solely for the purpose of oxidizing fat accounts for 4.65 Cal (16,300/3,500).

As noted earlier, slightly different values are found for the RQ when different fats are oxidized, hence small variations in this figure may be noted among different texts.

PROTEIN. As noted earlier the RQ for protein oxidation in the body has been calculated from empirical data to average about 0.801. Furthermore, since native proteins (ie, as ingested in foods) contain an average of 16 percent nitrogen, then it is apparent that each gram of urinary nitrogen must represent 6.25 gm of protein that has been metabolized in the body.

It has been shown experimentally that the metabolism of a quantity of protein yielding 1 gm of urinary nitrogen produces 26.51 Cal and 4.75 liters of carbon dioxide while 5.92 liters of oxygen is utilized simultaneously.

Table 14.13 summarizes the significant metabolic values that have been presented thus far for carbohydrate, fat, and protein metabolism.

CALCULATION OF TOTAL HEAT PRODUCTION IN THE BODY (INDIRECT CALORIMETRY). In order to calculate the heat production of the body per day (indirect calorimetry), one must know the total volume of oxygen utilized, the total volume of carbon dioxide expired, and the total urinary nitrogen output over the same interval. Oxygen consumption and carbon dioxide excretion can be determined with a spirometer, as illustrated in Figure 14.56.

Using a hypothetical example to illustrate the calculations involved in indirect calorimetry, the subject utilized oxygen at a rate of 410 liters per day, and excreted 350 liters of carbon dioxide during the same interval. The total urinary nitrogen excretion determined for the same period was 13.0 gm.

Since protein is metabolized incompletely, the measured gas exchange must be corrected for the quantity of protein that the subject has metabolized, so that a nonprotein RQ is obtained. The nonprotein RQ in turn represents the portion of the total RQ that is due solely to the oxidation of carbohydrate and lipid.

The nonprotein RQ is calculated first, and this value is readily obtained from the following data. One gm of urinary nitrogen represents the oxidation of a quantity of protein that requires 5.92 liters of oxygen and would result in the elimination of 4.75 liters of carbon dioxide (Table 14.13). Hence, when the quantity of urinary nitrogen (13.0 gm) is multiplied by the number of liters of oxygen that are necessary to oxidize that quantity of protein, the fraction of the total oxygen intake that

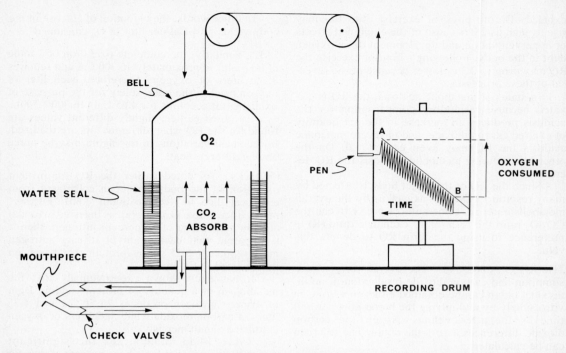

FIG. 14.56. The principal features of a spirometer. The rate of oxygen consumption per unit time is given by the slope of the line A–B. (Data from Ganong. *Review of Medical Physiology,* 6th ed, 1973, Lange Medical Publications; Ruch and Patton [eds]. *Physiology and Biophysics,* 19th ed, 1966, Saunders.)

was required for the oxidation of protein alone is obtained. Thus, $13.0 \times 5.92 = 76.9$ liters of oxygen were used by the subject to oxidize protein.

The quantity of carbon dioxide produced by the oxidation of the same quantity of protein may be calculated in a similar fashion. Thus $13.0 \times 4.75 = 61.7$ liters of carbon dioxide were produced from protein oxidation alone during the test interval.

The quantities of oxygen utilized and carbon dioxide produced by protein oxidation now are subtracted from the total values for the day of these gases. The remainders give the quantities of oxygen and carbon dioxide that were involved in carbohydrate and fat oxidation only.
Oxygen: $410 - 76.9 = 333.1$ liters
Carbon dioxide: $350 - 61.7 = 288.3$ liters

Therefore, the nonprotein RQ is:

$$\frac{\text{Volume of nonprotein } CO_2}{\text{Volume of nonprotein } O_2} = \frac{288.3}{333.1} = 0.86$$

The nonprotein RQ now must be converted into grams of carbohydrate and fat metabolized. This task has been simplified considerably, as accurate tables and graphs (Fig. 14.57) have been prepared by a number of investigators that give the relative proportions of fat and carbohydrate which are oxidized per liter of oxygen consumed. Table 14.15 is an example of this type of data, and it will be noted that when the nonprotein RQ is 0.86 [Ta-

ble 14.15(c)] then 0.622 gm of carbohydrate and 0.249 gm of fat are metabolized per liter of oxygen used by the subject. Consequently in order to determine the total quantities of carbohydrate and fat used, multiply these figures by the total liters of oxygen (obtained from the nonprotein RQ) that were utilized by the subject during the oxidation of carbohydrate and fat.

Total carbohydrate: $333.1 \times 0.622 = 207.2$ gm
Total fat: $333.1 \times 0.249 = 82.9$ gm

The total amount of protein metabolized now may be determined easily, since each gram of urinary nitrogen indicates that 6.25 gm of protein Table 14.13) has been oxidized. Thus, $13.0 \times 6.25 = 81.2$ gm protein has been oxidized by the subject during the test period.

The total heat production now may be obtained by multiplying the caloric value of each nutrient (Table 14.13) by the total quantity of that nutrient that was oxidized. This will give the heat produced by the oxidation of each foodstuff, and the sum of these individual values is the total heat production of the subject during the 24 hour test period.

Carbohydrate: 207.2×4 Cal/gm $= 828.8$
Fat: 82.9×9 Cal/gm $= 746.1$
Protein: 81.2×4 Cal/gm $= 324.8$
  Total heat production $= 1899.7$ Cal/day

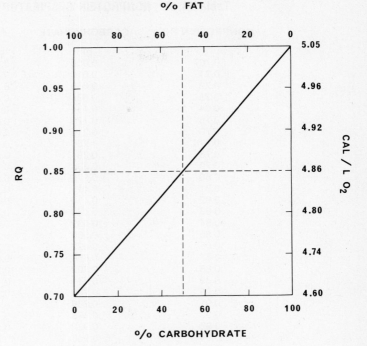

FIG. 14.57. Graph for determining the nonprotein respiratory quotient. See also Table 14.15.

Basal Metabolic Rate

The total energy expenditure of the body at any time, as indicated by the total heat production, is the sum of the energy which is required to maintain life (the basal metabolism) plus the energy required to perform any additional activities. When the basal metabolism is measured under resting conditions and in the postabsorptive state, the total heat produced by the body is the only major source of energy loss, and this heat loss in turn must reflect the utilization of stored energy as no energy is being used to do work insofar as the external environment is concerned. Rather, the basal metabolism under these conditions reflects the energy required for the maintenance of various fundamental cellular activities essential to the maintenance of life, particularly the maintenance of active transport systems in the cells. Also included in this category are the activity of brain, muscle, hepatic, renal, and other tissues, in addition to the mechanical work performed by the contraction of the respiratory, cardiac, and gastrointestinal musculature as well as the heat produced by metabolic turnover processes. The basal metabolism, therefore, is the total energy requirement for all of the individual cells, tissues, and organs of the body, and it amounts to roughly 50 percent of the total energy that is expended during the many activities carried out by an individual during the course of a normal 24-hour day.

BASAL METABOLIC RATE. The basal metabolism is affected markedly by a number of factors; consequently in order to obtain valid measurements of energy production that are comparable among individuals, certain specific and arbitrary conditions must be satisfied. When these conditions are met, the basal metabolic rate (or *BMR*) may be obtained by indirect calorimetry as described below. It should be mentioned, however, that the *BMR* does not reflect the minimal metabolism that is consistent with life, as the metabolic rate of an individual is somewhat lower during sleep that it is when determined under the conditions specified below, and may be even less, for example after a prolonged fast.

CONDITIONS ESSENTIAL TO MEASUREMENT OF THE BASAL METABOLIC RATE IN HUMANS. First, the subject must be in a postabsorptive state and have had nothing per os for 12 hours prior to determination of the *BMR*. Second, the subject is awake. Third, complete mental and physical relaxation is essential. Under ideal conditions, the test should be performed after awakening from a night's sleep but before arising. Stress, eg, anxiety over the test itself, should be avoided. Fourth, the subject must be supine. Fifth, the ambient temperature should be in the range between 20 C and 25 C.

IMPORTANT FACTORS THAT AFFECT THE BASAL METABOLIC RATE. Twelve important factors affect the basal metabolic rate directly. The first four of these factors are eliminated by determining the metabolic rate while the subject is in the basal state or can be accounted for in the calculations,

Table 14.15[a] NONPROTEIN RESPIRATORY QUOTIENTS[b]

NONPROTEIN RQ	CARBOHYDRATE	FAT	CALORIES
	gm	gm	
0.707	0.000	0.502	4.686
0.71	0.016	0.497	4.690
0.72	0.055	0.482	4.702
0.73	0.094	0.465	4.714
0.74	0.134	0.450	4.727
0.75	0.173	0.433	4.739
0.76	0.213	0.417	4.751
0.77	0.254	0.400	4.764
0.78	0.294	0.384	4.776
0.79	0.334	0.368	4.788
0.80	0.375	0.350	4.801
0.81	0.415	0.334	4.813
0.82	0.456	0.317	4.825
0.83	0.498	0.301	4.838
0.84	0.539	0.284	4.850
0.85	0.580	0.267	4.862
0.86[c]	0.622	0.249	4.875
0.87	0.666	0.232	4.887
0.88	0.708	0.215	4.899
0.89	0.741	0.197	4.911
0.90	0.793	0.180	4.924
0.91	0.836	0.162	4.936
0.92	0.878	0.145	4.948
0.93	0.922	0.127	4.961
0.94	0.966	0.109	4.973
0.95	1.010	0.091	4.985
0.96	1.053	0.073	4.998
0.97	1.098	0.055	5.010
0.98	1.142	0.036	5.022
0.99	1.185	0.018	5.035
1.000	1.232	0.000	5.047

[a]*Data from McClendon and Medes.* Physical Chemistry in Biology and Medicine, 1925, Saunders.

[b]*Each value listed in this table is related to the Cal per liter of oxygen utilized as well as to the proportions of carbohydrate and fat that are metabolized. All values are equivalent to one liter of oxygen consumed.*

[c]*Figures for a nonprotein RQ of 0.86 are used in the example given in the text for calculation of total heat production by indirect calorimetry.*

See also Figure 14.56.

whereas the others must be taken into account when interpreting the results of the test.

1. Muscular exertion performed immediately prior to or during the test will elevate the BMR significantly.
2. The ingestion of food several hours before the BMR is measured will give falsely elevated values. (See specific dynamic action of foodstuffs.)
3. The environmental temperature will affect the BMR directly. Thus an individual tested in an extremely hot room will have a lower BMR than would be found if the ambient temperature was in the range 20 C to 25 C. (See also effects of climate below.)
4. The surface area of the body is an extremely important factor, as it has been found that the basal metabolic rates of different individuals are quite comparable in general when they are expressed in terms of the total body surface area (square meters), as discussed below. An important generality in this regard, however, is that smaller individuals usually have a higher basal metabolic rate per unit surface area than do larger persons. Consequently, the height and weight of the subject must be taken into account when interpreting the results.
5. The sex of the subject is important because females normally exhibit a lower BMR than males, regardless of their age (Table 13.10, p. 874).
6. Age also is an important factor in interpreting the results of BMR measurements (Table 13.10). In newborn infants, the rate is low whereas at around the age of five years a maximum BMR is reached. Thereafter, the rate gradually and continually declines throughout the lifetime of the individual. In females, the BMR decreases more rapidly between the ages of five to 17 years than it does in males.
7. The BMR is lower in individuals living in warm climates than in persons living in temperate zones. In fact, the BMR of persons in the tropics

is related more closely to total body weight than to surface area. Conversely, Eskimos exhibit higher normal *BMR* values than are found in other groups. Whether or not this fact reflects an ethnic and/or an environmental (diet, cold?) effect is a moot question.

8. Certain variations in *BMR* have been reported among various racial groups in addition to the Eskimos. It has been reported that Chinese and Indians normally have *BMR*s that are somewhat lower than those of Caucasians.
9. Pregnancy and menstruation elevate the *BMR* somewhat.
10. The nutritional state of the subject can affect the *BMR* significantly. Thus in starvation and acute malnutrition the *BMR* is lowered.
11. Diseases of an infectious or febrile type elevate the *BMR*, and the increase generally is directly proportional to the rise in body temperature since the overall metabolic activity of the cells increases with a concomitant increase in their heat production.
12. Certain hormones have a profound effect upon the *BMR*, in particular the thyroid hormone. Thus thyroxine exerts a marked stimulatory effect upon the metabolic rate of all the cells in the body; consequently, the *BMR* is elevated in hyperthyroid states whereas it is depressed in hypothyroid conditions. In clinical practice, the principal use of calorimetry is in the diagnosis of thyroid disease. However, because of the advent of routine chemical techniques for the determination of circulating thyroid hormone levels, it should be pointed out that calorimetry has largely been discarded in favor of these more specific and accurate diagnostic tools.

Epinephrine is the only other hormone that exerts a specific effect upon heat production in the body, although in normal individuals the effects of this hormone are both rapid in onset and evanescent in time. Tumors of the adrenal medulla known as pheochromocytomas, however, induce a chronic elevation in the *BMR*.

Dysfunctions of certain other endocrine glands, eg, the pituitary, may cause alterations in the *BMR* that are secondary to derangement of the normal endocrine balance in the body.

MEASUREMENT OF THE BASAL METABOLIC RATE IN HUMANS. In actual fact, the *BMR* as determined clinically is not truly "basal," and this term merely denotes an accepted set of arbitrarily standard conditions. For practical purposes, the *BMR* can be estimated with satisfactory accuracy solely by measuring the oxygen consumption of a patient under basal conditions for two 6-minute intervals using a spirometer, such as is illustrated in Figure 14.56. The measured oxygen consumption is corrected to standard conditions of temperature and pressure, and the average volume of oxygen consumption which was recorded in liters during the two intervals is multiplied by 5 in order to

convert it to the total oxygen consumption per hour by the subject. The heat production in Cal per hour now is obtained simply by multiplying the oxygen consumption per hour by 4.825 Cal (Table 14.15). This value is the caloric equivalent of 1 liter of oxygen for an RQ of 0.82, a value that has been found empirically to lie close to the average RQ for most subjects in an approximately normal nutritional state.

This empirically determined value for heat production in Cal/hour now is divided by the surface area of the patient in square meters to give the *BMR* in Cal per square meter body surface per hour. The surface area of the body can be calculated from a number of formulas; however, in practice it is most convenient to use a nomogram that relates body weight and height to the surface area (Fig. 14.58). The accuracy of this procedure is sufficient for most purposes, as the nomogram is obtained from the classic du Bois formula for determining the surface area of the body, ie,

$$A = H^{0.724} \times W^{0.425} \times 71.84$$

In this expression, which was derived empirically from measurements on a large number of individuals, A = the surface area in cm², H = height in cm, and W = the weight in kg. The surface area in square centimeters which is obtained by use of this formula is divided by 10,000 in order to give the surface area in square meters.

Standard tables are available from which the normal *BMR* of the patient can be obtained in terms of the age and sex of an individual. Table 13.10 (p. 874) is an abbreviated presentation of such normal *BMR* values, and this table is useful as it also illustrates the effects of age and sex upon the normal basal metabolic rates.

In practice, the *BMR* as determined on a patient is expressed as a plus or minus percentage of the normal *BMR*. A *BMR* that ranges between −15 and +20 percent of the values listed in Table 13.10, is considered to lie within the normal range for that particular individual. In hyperthyroid states, however, the *BMR* may exceed +75 percent whereas in hypothyroid states, the *BMR* may fall below −60 percent.

The Specific Dynamic Action of Foods

A reduction in the total caloric intake of an individual may result in a significant fall in the rate of metabolism. On the other hand, following ingestion of foods, an increased total body heat production results, provided that all other conditions remain exactly as defined for the basal state. This increase in heat production following a meal is known as the specific dynamic action *(SDA)* or calorigenic action of foods. The biochemical and/or physiologic mechanisms which are responsible for this effect are unclear in detail, although some investigators believe that the *SDA* is caused by the extra energy which is released from

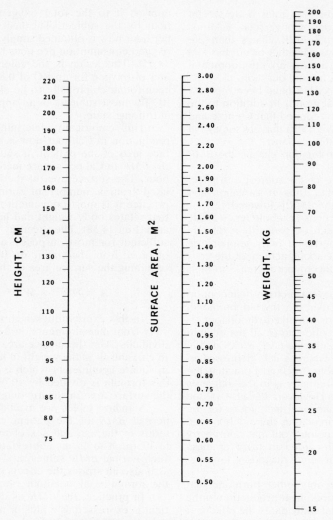

FIG. 14.58. Nomogram for determining the surfaces area of the human body from the height and weight. (Data from West, Todd, Mason, and Van Bruggen. *Textbook of Biochemistry*, 4th ed, 1966, Macmillan.)

foodstuffs as the consequence of their metabolism. Apparently the hepatic metabolism of foods is responsible to a large extent for the *SDA*, as discussed below.

The quantity of additional heat that is produced above the basal level following the ingestion of food, as well as the time interval of elevated heat production, depend upon the type and amount of food that is eaten. Protein foods cause the largest increment in heat production. Thus the *SDA* for protein alone is about 30 percent of the total caloric value of the ingested protein. Carbohydrate has an *SDA* of approximately 5 percent, and the SDA of lipid amounts to roughly 8 percent. Consequently it is essential when calculating the caloric value of a normal diet to take this factor into account in order to provide sufficient calories to compensate for those lost from the body as heat because of the specific dynamic action.

For example, the ingestion of 25 gm of protein is equivalent to 100 Cal (Table 14.13). However, this quantity of protein leads to the generation of 30 Cal of extra heat above the *BMR*, and this heat is lost from the body so that only 70 Cal of potentially utilizable chemical energy can be obtained from 25 gm of protein.

The approximate *SDA* values for each type of nutrient given above is found only when each foodstuff is eaten separately. However, when these foods are consumed in a mixed diet, the *SDA* of the diet cannot be predicted on the basis of the additive effect of the *SDA* values of the individual foods. It has been demonstrated experimentally that each foodstuff exerts a depressant effect upon the *SDA* of the others when food mixtures are ingested. Therefore, the high *SDA* of protein alone can be reduced markedly, depending upon the relative quantities of carbohydrate and fat in the diet. In fact, fat depresses the *SDA* in a mixed diet far more than either of the other two nutrients in combination.

An approximate rule-of-thumb used in calculating the total daily energy requirement for an individual is to add 10 percent to the total calories required in order to provide enough energy to compensate for the *SDA*. However, if the fat content of the diet is relatively high, then this figure may be too high. Studies performed upon human subjects have revealed that the net *SDA* following ingestion of a high-protein (37 percent) meal containing a total of approximately 1,000 Cal is around 17 percent. Similar studies using an isocaloric low-protein (7 percent), but high carbohydrate, meal gave a value around 10 percent for the *SDA*.

Consequently, there is much disagreement among various investigators as to what value should be used in estimating the caloric requirements necessary to compensate for the energy lost by the SDA. However, values between 6 and 10 percent may be deemed adequate to cover most individual cases.

Interestingly, a number of individual amino acids produce an SDA effect that is similar quantitatively, regardless of whether these compounds are ingested orally or injected intravenously. Hence it is obvious that the mechanical and chemical work that is involved in the digestion and absorption of these compounds does not contribute to the *SDA*. Furthermore, experiments performed in dogs have demonstrated that the *SDA* of amino acids does not depend upon their oxidation. In general, however, the SDA of individual amino acids appears to be related to the type of deamination they undergo. Thus oxidative deamination with transformation of the amino nitrogen into urea takes place concomitantly with the development of a large *SDA*, whereas transamination is related to production of a small *SDA*.

Glucose is converted readily into fat within the body, and thiamine is an essential cofactor in this metabolic process. Experiments performed on rats have shown that when glucose is given simultaneously with thiamine a far greater *SDA* develops (8 percent) than when glucose is administered alone (4 percent). The extra heat production in this instance may reflect the energy that is necessary to "prepare" glucose for deposition as fat, although this statement is conjectural.

Caloric Requirements of the Human Body

The total basal heat production, as given by the *BMR*, is a relatively constant value in any given individual. However, any muscular work performed results in an added caloric requirement above the basal level, and this factor is subject to enormous variation among different individuals as well as in the same person at different times. The sum of these two factors determines the total number of calories that are necessary to maintain caloric (= energy) balance (or body weight) in an adult (cf p. 873). During growth, pregnancy, or convalescence, additional calories also are necessary in order to provide the extra energy that is necessary for the biosynthetic reactions involved in the increase in total tissue mass within the body.

By way of summary, therefore, the total daily caloric requirement must be adequate to cover the needs of the individual with respect to four factors: (1) Basal metabolism, including the maintenance of body temperature; (2) The specific dynamic action of foodstuffs themselves; (3) Muscular activity; (4) Growth, reproduction, and lactation.

An intake of between 1,300 and 2,000 Cal per 24 hours has been found to be sufficient to meet the basal caloric requirements of most individuals. To this basal requirement one must add an adequate number of calories to compensate for all of the muscular activity performed during an average day; and this factor, as noted above, is subject to considerable variation. Thus it is impossible to give an adequate or accurate generalization concerning the total caloric requirements that will cover all situations in different individuals. However, a number of specific illustrative examples may be cited that will give the reader an idea of the approximate caloric requirements of individuals engaged in widely differing occupations (Table 14.16). Obviously only estimates of caloric requirements can be made unless detailed experimental studies are carried out, but the values that are cited in the discussion to follow provide an indication of the importance of muscular exercise insofar as influencing the total daily caloric requirements of the individual are concerned.

A total caloric intake ranging between 2,000 and 3,000 Cal per day is more than adequate for most individuals who are engaged in sedentary to moderately vigorous activities. If the individual is engaged in severe physical exertion, the caloric requirement becomes much higher depending upon the time spent upon, as well as the nature of, the activity. As the consequence of such vigorous activity, the total caloric requirement may reach, or even greatly exceed, 4,000 Cal per day. Several examples of this type of activity are cited in Table 14.16.

Human beings, in common with all other living organisms, obey the first law of thermodynamics insofar as caloric intake and energy expenditure are concerned. Therefore to prevent loss of body tissue mass, the total caloric intake of food must equal the total heat production of the body over a period of time, and an individual in such a state is said to be in caloric (or energy) balance. The consumption of excess calories over those which are required to maintain the body in caloric balance, if continued for a prolonged interval of time, results in obesity. This statement holds true regardless of the source of the excess calories, as carbohydrate and protein, as well as fat, are converted readily into, and stored as,

Table 14.16[a] ENERGY EXPENDITURE FOR INDIVIDUALS ENGAGED IN VARIOUS ACTIVITIES[b]

	Cal/m²/hr
REST	
Sleeping	35–65
Awake, lying still	40–80
Awake, sitting up (eg, reading)	50–100
LIGHT ACTIVITY	
Dishwashing	60
Writing, clinical work, seamstress	60–140
Typing	15–40
Standing, relaxed	85–105
Washing, dressing	100
MODERATE ACTIVITY	
Washing, dressing	100
Standing	115–120
Walking	140–260
Housework	140–150
Sweeping	110
HEAVY ACTIVITY	
Bicycling	100–300
Swimming	350
Lumbering	350–900
Skiing	500
Running	500–600
Wrestling	980
Climbing	200–960
Shivering	300–800

[a] Data from Ruch, Patton [eds]. Physiology and Biophysics, 19th ed, 1966, Saunders.

[b] The values cited in this table are to be considered as approximations only and will vary considerably depending upon the relative humidity, ambient temperature, wind velocity (Table 10.20), clothing worn, and so forth.

depot lipid. In fact, clinical obesity practically always results from excess caloric intake and excess caloric intake alone! Conversely, weight loss may be achieved merely by curtailing the total caloric intake per day over a prolonged interval. Such a decrease in total body weight may be calculated readily since depot lipid contains 9 Cal per gm. Therefore a person who consumes 1,000 Cal per day less than his total energy expenditure for the same period will lose 1 kg of fat in 9 days, as this quantity of stored lipid contains 9,000 Cal. Achieving this goal, however, is not as simple a procedure as might be suggested by this elementary fact of bioenergetics, judging by the prevalence of obesity as a clinical problem in the United States.

CALCULATION OF THE TOTAL ENERGY EXPENDITURE. In order to arrive at the approximate total daily caloric expenditure of an individual for clinical purposes, it is helpful to understand the factors and calculations involved. A hypothetical example of such a calculation is summarized in Table 14.17.

DIETARY SOURCES OF ENERGY. The energy required for all physiologic processes is derived from the oxidation of dietary foodstuffs that contain carbohydrate, fat, and protein. As noted earlier, the total daily caloric requirement is the sum of the basal energy needs of the body plus the energy necessary for performance of muscular work. From a physiologic standpoint, carbohydrates and lipids are the most economical sources of energy in the body. The major role of protein is to serve as material for tissue growth and repair, although this basic nutrient also is oxidized to provide energy, particularly when the caloric intake from carbohydrate and/or lipid is insufficient to meet the overall energy requirements. Thus, it is a fundamental nutritional truism that the presence of adequate quantities of carbohydrate and fat in the diet tend to spare protein from being oxidized, thereby allowing it to participate in a multitude of anabolic processes.

Consequently, in a well-balanced diet, 55 to 70 percent of the total calories are derived from carbohydrate sources; 20 to 30 percent are obtained from fat; and 10 to 15 percent from protein (cf. p. 872).

THERMOREGULATION

Introduction

The homeostatic regulation of normal body temperature within narrow limits (or thermoregulation) is a complex process that involves a continual interplay among a number of physiologic and environmental factors. The basic principle of thermoregulation, however, is quite simple. In order to maintain a constant body temperature, the total heat production within the body must equal the total heat loss from the body to the environment. Thus, the balance between heat production and heat loss are the major determinants of body temperature in man and other warm-blooded (or homeothermic) organisms (mammals and birds). All cold-blooded (or poikilothermic) animals, on the other hand, are unable physiologically to alter and to maintain their body temperature independently of the environment, except by certain behavioral patterns. For example, some reptiles (eg, lizards and snakes) will move deliberately into sun or shade in order to achieve some degree of optimal compensation for environmental temperature fluctuations. However, in general vertebrates such as fishes, amphibians, and reptiles, as well as all invertebrate forms of life, are unable to maintain a constant body temperature, hence these organisms and their activities are strictly at the mercy of their environment. Incidentally, the term poikilothermic or "cold-blooded" is a misnomer, because the body temperature of such animals usually is subject to considerable daily variation in accordance with the environmental temperature. Furthermore, the internal (or core) body temperature of poikilotherms is slightly above that of the ambient temperature; otherwise the heat that is generated during the course of their normal metabolic processes would not be lost, but would accumulate with lethal effects upon the organism.

Table 14.17 CALCULATION OF THE AVERAGE DAILY ENERGY REQUIREMENTS OF A HYPOTHETICAL PERSON[a]

DAILY ENERGY EXPENDITURE	TOTAL Cal/m²/hr	TOTAL Cal/2.1 m²/hr
Sleeping, 8 hr (35 Cal/m²/hr)[b]	280	588
Walking, 2 hr (150 Cal/m²/hr)	300	630
Reading, writing, sitting in class, 11 hr (110 Cal/m²/hr)	1,210	2,541
Bicycling, 1 hr (200 Cal/m²/hr)	200	420
Meals, washing, dressing, 2 hr (100 Cal/m²/hr)	200	420

Total energy expenditure/24 hrs = 4,599 Cal/day

[a]*For example, a student having a height = 180 cm; weight = 91 kg; surface area = 2.19 m² (Fig. 14.57).*

[b]*Values taken from Table 14.16.*

On the other hand, mammals as well as birds have evolved a number of reflex responses that are integrated within the hypothalamus, and these reflexes function to maintain the body temperature within a narrow range despite wide oscillations in the environmental temperature.* In a broad sense, then, the evolution of thermoregulatory mechanisms has liberated homeothermic organisms from a strict dependence upon their environment, insofar as extremes of temperature are concerned. Of more relevance, why is a constant body temperature, that usually is above that of the environment, physiologically essential to survival and proper functioning of the human body? The answer to this important question is based upon two interrelated facts. First, the rate of all of the biochemical reactions depends upon the temperature; and second, the enzyme systems that catalyze these reactions function optimally only within a very restricted temperature range. Since all of the normal functions of mammalian cells, tissues, and organs depend ultimately upon these enzymatic processes, normal function of the body as a whole is dependent upon the maintenance of a relatively constant temperature of the internal environment.

The several factors that determine heat production within, and heat loss from, the body now will be considered individually together with their interrelationships following a brief discussion of the major anatomic structures that are concerned directly with thermoregulation in the human body.

Mammals that hibernate (eg, certain ground squirrels, marmots, and hedgehogs) are an exception to this general statement when in hibernation. While awake, such animals are strictly homeothermic, but during periods of hibernation, their body temperature falls markedly in accordance with the environmental temperature. However, under such circumstances, their core body temperature still remains slightly above that of their surroundings. Bears, incidentally, are not true hibernators; they just sleep deeply for extended periods.

Anatomic Structures Concerned with Thermoregulation

In general terms, two major structures are concerned directly with the regulation of body temperature in man. These structures are the skin and the hypothalamus, and these two organs, together with appropriate sensory receptors as well as afferent and efferent nerves, constitute a remarkably efficient and integrated system for controlling body temperature.

FUNCTIONAL ANATOMY OF THE SKIN

General Properties. In Chapter 2 (p. 45) it was emphasized that the skin, far from being an inert "wrapping material" for the body, is an important and physiologically active organ. In fact, the skin is one of the largest organs in the body, and as such it comprises about 15 percent of the total body weight.

Skin has a number of important functions. First, it protects the individual from injury and desiccation as well as serving as a mechanical barrier against invasion by microorganisms and other foreign bodies. The pigment melanin, which is found in the outer layer of the skin, protects the body from ultraviolet radiation. Second, the skin serves as an important sense organ in that it receives various types of sensory inputs from the immediate environment. Third, a number of substances are excreted by the skin. Fourth, the subcutaneous fat depots have an important part in fat metabolism. Fifth, in homeothermic organisms the skin plays a highly significant role in thermoregulation; and intimately related to this function is the participation of this organ in the maintenance of the overall water balance within the body.

Anatomically the skin is composed of two principal layers, the surface epithelium, or epidermis, and the underlying connective tissue layer, the dermis, or corium (Figs. 14.59 and 14.60). Underneath there is a looser connective tissue layer, the hypodermis. In many regions the hypodermis is composed principally of subcutaneous adipose tissue. In turn the hypodermis is connected loosely to periosteum, aponeuroses, or deep fascia. Mucous membranes are continuous with the skin at several places in the body, and these regions are known as mucocutaneous junctions. Such junctions are located at the eyelids, nares, lips, prepuce, vulva, and anus.

Many functions of the skin depend primarily upon the epidermis, whose epithelium forms a continuous layer that covers the entire outer surface of the body. Local specializations of the epidermis result in the formation of such structures as nails, hair, and glands. The epidermal cells synthesize a fibrous protein, keratin, that is of major importance to the protective functions of the skin. In addition, these cells also synthesize melanin the pigment which is responsible for protection of the individual from the ultraviolet radiation of the sun.

The free surface of the skin is far from smooth; rather, it is characterized by grooves and lines along points of flexure. At the junction between the epidermis and dermis, a typical pattern of interlocking ridges and grooves is seen in vertical sections of skin. The interface between dermis and hypodermis, however, is not sharp; hence the elements of both of these regions merge and no clearly delineated boundary is present such as is found between epidermis and dermis.

Epidermis. The outermost layer of the skin is composed of stratified squamous epithelial cells that belong to two distinct functional populations. One type of epidermal cell, which is present in the greatest abundance, undergoes keratinization and thus forms the dead superficial layers of the epidermis. These cells, which form the keratinizing system of the skin, are desquamated continually from the surface and are replaced by cells that are generated by mitotic divisions in the basal layer of the epidermis. The cells migrate progressively upward as they are supplanted by new cells underneath. Keratin is synthesized as the cells progress towards the free surface of the skin, so that ultimately all of the living cytoplasm is replaced by this protein and the cell dies.

The second type of epithelial cell, the melanocyte, is concerned with the production of melanin, as discussed on page 1042. The so-called pigmentary system of the skin is comprised of melanocytes.

In certain regions of the body, notably the palms of the hands and soles of the feet, the epidermis is composed of four distinct layers (Fig. 14.59). In vertical sections, these are the stratum germanitivum (or germinative layer), the stratum granulosum (granular layer), the stratum lucidum (clear layer), and the stratum corneum (cornified or horny layer). The latter two layers form the external cornified portion of the skin. Elsewhere on the body surface, the epidermis is both thinner

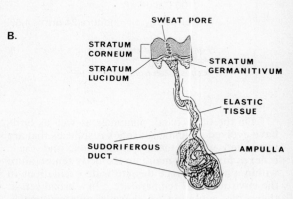

FIG. 14.59. **A.** Cross-section of human skin at low magnification illustrating the principal structures and vascular supply. **B.** Diagram of a single sweat gland. (After Bloom and Fawcett. *A Textbook of Histology,* 9th ed, 1968, Saunders.)

and simpler in structure, while the stratum lucidum generally is absent. Thus the epidermis ranges between 0.07 and 0.12 mm in thickness over most of the body surface, whereas on the palms of the hands it may be 0.8 mm thick and as much as 1.5 mm on the soles of the feet. Prolonged mechanical friction upon any body surface in postnatal life can induce a significant thickening (callus formation) of the epidermis.

Blood vessels are completely absent from the epidermis, and nutrients presumably diffuse into this region of the skin via tissue fluid which is derived from the capillaries found in the underlying connective tissue layer. Furthermore it is important to realize that the epidermis is not impervious to the passage of fluid and other substances, so that water is lost continuously from the body via this route.

Mucocutaneous junctions are transition zones between skin and mucous membranes; hence the cornified outer layer is thin or even absent. Usually these regions do not contain hairs, hair follicles, sweat glands, or sebaceous glands. Consequently, the mucocutaneous areas are lubricated by secretions of the mucous glands that are located within the body orifices. The thinness or absence of the cornified layer in the mucocutaneous regions also gives rise to a red or pinkish coloration that is due to the blood in the underlying capil-

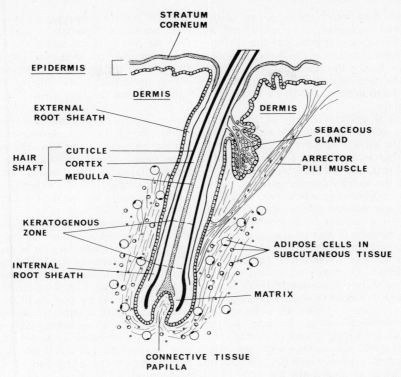

STRATUM
CORNEUM

EPIDERMIS

DERMIS

EXTERNAL
ROOT SHEATH

DERMIS

SEBACEOUS
GLAND

HAIR
SHAFT

CUTICLE
CORTEX
MEDULLA

ARRECTOR
PILI MUSCLE

KERATOGENOUS
ZONE

ADIPOSE CELLS IN
SUBCUTANEOUS TISSUE

INTERNAL
ROOT SHEATH

MATRIX

CONNECTIVE TISSUE
PAPILLA

FIG. 14.60. Cross-section of human skin showing the morphology of a single hair follicle. (After Bloom and Fawcett. *A Textbook of Histology,* 9th ed, 1968, Saunders.)

laries. Cyanosis can often be detected clinically by observing the coloration of a patient's lips.

Dermis. The dermis has a thickness that ranges between approximately 0.6 mm (on the eyelids and prepuce) to over 3 mm (on the palms of the hands and soles of the feet), although the average thickness of the dermis is between one and two mm. This layer is not clearly delineated, however, as it merges imperceptibly with the subcutaneous connective tissue (hypodermis). The outer portion of the dermis forms papillae that interdigitate with the depressions between the ridges on the innermost surface of the epidermis (Fig. 14.59), and this region is known as the papillary layer of the dermis. The reticular layer of the dermis lies beneath the papillary layer, although these two regions cannot always be distinguished clearly.

The reticular layer of the dermis is composed principally of relatively dense connective tissue whose collagenous fibers extend in all directions, although most of these fibers are oriented approximately parallel to the external surface of the skin. The collagenous connective tissue of the papillary layer is less dense, and the bundles of collagenous fibers are somewhat thinner than those which are found in the reticular layer.

Elastic connective tissue is abundant in the dermis, and bundles of this material are concentrated around the hair follicles as well as the sweat and sebaceous glands.

Situated deep within the reticular layer of the skin on the penis, scrotum, areolae, and perineum are numerous smooth muscle cells that form a loose plexus in these regions. Contraction of these muscular elements causes the skin to assume a wrinkled appearance.

Several other structures are associated with the dermis. Arrector pili muscles are smooth muscle fibers that are connected with the hairs whose follicles are found within the dermis (Fig. 14.60). Sebaceous glands also are located within the dermis, as are an abundance of nerve endings, nerves, and blood vessels.

Hypodermis. The hypodermis (or subcutaneous layer of the skin) is composed of loose collagenous and elastic connective tissue which is continuous with that of the dermis. Fat cells in varying numbers may develop in the hypodermis in most regions of the body (Fig. 14.60). For example, the abdominal fat depot that forms in this region of the skin may greatly exceed a thickness of 3 cm; however the subcutaneous layer of tissue on the eyelids and penis never develops fat cells.

The hypodermis also contains large blood vessels, nerve endings, and an abundance of nerve trunks, and these structures extend into the overlying dermis.

Hairs, Hair Follicles, and Related Dermal Structures. Hairs are thin keratinous filaments that range in length from a few mm to over a meter.

The hair follicles from which the hairs extend are tubular epidermal invaginations as illustrated in Figure 14.60. These skin appendages are distributed over the entire surface of the body, save for the palms of the hands, the soles and sides of the feet, the lips, the prepuce and glans penis, as well as the clitoris, labia minora, and internal surfaces of the labia majora.

Unlike typical epidermal cells of the keratinizing system of the skin that deposit keratin continuously, the growth of hairs is a discontinuous process. That is, phases of growth alternate with periods of rest, and the particular dermal follicular cells in the dermal papillae whence hairs originate show morphologic alterations that correspond to the intermittent phases of growth and rest. The effect of certain hormones upon the growth and distribution of hair on the body is discussed in Chapter 15 (p. 1139).

Sebaceous Glands. Each hair follicle has one or more sebaceous glands associated with it (Fig. 14.60). The secretory portion of these structures has a rounded, saclike (alveolar) appearance, that frequently appears similar to a bunch of grapes when several alveoli are present. Hence sebaceous glands are simple branched glands.

The alveoli of the sebaceous glands are composed of a basement lamina which lies adjacent to an outermost layer of thin epithelial cells. The cells within the alveoli become progressively more rounded as they fill with secretion and move toward the luminal surface of the alveolus. As these cells become engorged with secretion, they appear to be distended with fatty droplets; the nucleus then shrinks and disappears. Subsequently, the entire cell breaks down into an oily debris that is liberated as sebum into the short duct of the gland. This duct empties into the upper region of the follicular canal, and thus the oily secretion reaches the body surface. In consequence of this mode of secretion, sebaceous glands are termed holocrine, because each glandular cell degenerates and thus forms a part of the secretory product per se.

In a few regions of the body sebaceous glands are found independently of hairs. These regions are the lips and corners of the mouth, the glans penis and internal fold of the prepuce, the mammary papilla, and the labia minora. All of these areas have sebaceous glands whose ducts open directly upon the body surface rather than on a hair follicle as shown in Figure 14.60. The meibomian glands of the eyelids also belong to the class of sebaceous glands whose ducts open directly upon the external surface of the body.

The secretion of the sebaceous glands is an oily product that serves to lubricate the skin and to prevent it from excessive desiccation.

Arrector Pili. The bands of smooth muscle cells that comprise the arrector pili muscles are connected to the connective tissue sheath of the hair follicle at one end and to the papillary layer of the dermis at the other (Fig. 14.60). Normally the hair follicle lies at an acute angle to the skin surface, but when the arrector pili muscles contract in response to such stimuli as cold, fear, or anger then the hair is raised into a more erect position. Simultaneously the skin overlying the point of attachment of the muscle is depressed, whereas the region immediately surrounding the hair is elevated, causing the "gooseflesh" seen in humans.

Sweat Glands. The common eccrine sweat glands are located on the entire surface of the skin except for the margins of the lips, the nailbed and the glans penis. These structures are simple, coiled tubular glands; ie, the terminal part of the gland is merely a simple tube that is highly convoluted into a ball-like mass, and the duct is a thin, unbranched tube (Fig. 14.59B).

The principal secretory portion of the common sweat gland is situated in the dermis, and is approximately 0.3 to 0.4 mm in diameter. Certain of these structures in the axillae and perianal region are of the apocrine type, and range between 3 to 5 mm in diameter. A thick basement membrane invests the secretory portion of the sweat glands, and located just within this membrane are long spindle-shaped myoepithelial cells which resemble (but are not identical with) smooth muscle cells. The myoepithelial elements are believed to participate in the extrusion of sweat by their contraction.

Sweat is secreted by truncated pyramidal cells that are present in a single layer upon the myoepithelial cells. A large spherical nucleus is present near the base of the secretory cells, and mitochondria as well as secretory vacuoles may be seen, their abundance being in proportion to the functional state of the gland. Considerable variation is evident in the diameter and shape of the lumen of the secretory portion of the sweat gland, and once again this depends upon the functional state of the gland.

In certain regions of the skin the sweat glands exhibit peculiar morphologic and functional characteristics. For example, the cerumenous glands which are located within the skin of the external auditory meatus produce a waxy secretion known as cerumen. The secretory portions of these glands are branched and their ducts may empty, together with those of adjacent sebaceous glands, into the follicles of hairs. This is in marked contrast to the common sweat glands whose ducts exit directly upon the surface of the skin through fine pores that are independent of hair follicles (Fig. 14.59B). Furthermore, cerumenous glands in their terminal regions have a highly developed investment of ordinary smooth muscle cells.

Moll's glands are a second type of specialized sweat gland that are found in the margin of the eyelid. The terminal parts of these glands are irregularly contorted but do not form a ball, and their lumen is wide. The ducts of Moll's glands exit either upon the surface of the skin directly or into the hair follicles of the eyelashes.

Sweat is a transparent watery liquid that is

excreted principally by the small common sweat glands (Fig. 14.59B). The sweat glands of the armpits and anal region, on the other hand, produce a thicker secretion having a more complex chemical composition, and in sexually mature females, the axillary sweat glands exhibit cyclic changes that correspond with the phase of the menstrual cycle. Thus in the premenstrual interval, the epithelial cells as well as the lumens of the axillary glands enlarge, whereas these morphologic alterations are reversed during the menstrual period itself.

The differences between the eccrine and apocrine sweat glands are of physiologic importance and may be summarized briefly. Eccrine sweat glands are not attached to hair follicles; they produce a watery secretion that is of considerable importance in thermoregulation; and they are innervated by cholinergic nerves. In contrast to the eccrine sweat glands, the ducts of apocrine sweat glands are attached to hair follicles and they commence to function at puberty. The apocrine secretion is viscous, and these glands are innervated by adrenergic nerves.

All of the eccrine sweat glands are not necessarily functional simultaneously on all parts of the body surface, nor are they all functional under the same conditions. For example, anxiety causes sweating to start first on the palms of the hands and the soles of the feet. The altered skin resistance that is attendant upon this process provides one physiologic parameter which is measured in polygraph ("lie detector") tests. In contrast to this type of stimulus, exposure of the human body to markedly elevated temperatures causes sweating to begin on the forehead. This sweating spreads to the face and then to the remainder of the body.

It has been demonstrated experimentally that the concentration of electrolytes in the ducts of the eccrine sweat glands is greater than the concentration of these substances in sweat collected at the surface of the skin. Therefore a reabsorption of electrolytes must take place within the ducts of the glands themselves.

Blood and Lymphatic Vessels of the Skin. The arteries that provide blood to the skin are found in the hypodermis (or subcutaneous layer). Branches of these vessels course outward to form a network known as the rete cutaneum that is parallel to the external skin surface and which is located at the junction between the hypodermis and dermis (Fig. 14.59A). From the lower side of this vascular network, branches pass into the hypodermis, and these vessels supply arterial blood to the sweat glands, fat cells, and deeper parts of the hair follicles (Fig. 14.60). Vessels that arise from the superficial part of the rete cutaneum course into the dermis, and at the boundary between the dermis and papillary layer of the epidermis these vessels enter the dermis to form a second plexus known as the rete subpapillare (Fig. 14.59A). Thin branches from the rete subpapillare now pass to the individual dermal papillae, so that each of these structures receives an ascending artery and a descending vein, with a capillary network interposed between the two larger vessels. The veins that collect blood from the capillaries within the dermal papillae then anastomose to form a network that is located just below the papillae. At the junction between the papillary and reticular layers are located three additional, and progressively larger, networks of veins. In the center of the dermis, as well as at the junction between the dermis and the hypodermis, the venous plexus is located at the same level as the rete cutaneum. The veins which drain the sebaceous and sweat glands enter this venous network. Large subcutaneous veins pass from the deep venous network and these vessels course together with the larger arteries.

The skin also has many direct connections between arteries and veins that lack intervening capillary networks, and these direct arteriovenous shunts have a critical role in thermoregulation.

Each hair follicle has an independent blood supply. A small artery provides blood to the papilla, the rete subpapillare supplies blood to the sides of the follicle, and a number of other arterioles divide into a dense capillary network within the connective tissue that surrounds the entire follicle. Similarly, a dense capillary network surrounds the basement lamina of the sebaceous and sweat glands.

Lymphatic vessels are particularly abundant within the skin. Lymphatic capillaries originate within the papillae as networks and/or blind-ending vessels, so that within the papillary layer a dense network of these structures is present, and the lymphatics always originate somewhat deeper than do the blood vessels. Branches from this superficial lymphatic plexus run to a deeper lymphatic network that is located at the junction between the dermis and hypodermis. Beneath the rete cutaneum, the lymphatics become much larger and they are equipped with valves. The large subcutaneous lymphatic vessels take their origin from this deeper network, and these larger lymphatics now pass centrally together with the blood vessels. However, no lymphatic vessels are associated directly with the hair follicles or glands of the skin.

Nerves of the Skin; Peripheral Receptors that Participate in Thermoregulation: The entire surface of the skin is an extremely important exteroceptive sense organ as discussed in Chapter 6 (p. 179); consequently, it is hardly surprising that its structure is richly supplied with sensory nerves. Furthermore, the skin contains an abundance of motor nerves that supply the blood vessels, sweat glands, and arrector pili muscles.

Relatively thick nerve bundles are found in the subcutaneous layer of the skin, and these nerves form loose networks which are composed principally of myelineated fibers, although some nonmyelinated fibers also are present. Within the dermis, branches are given off from this network to form several additional thin plexuses.

Physiologic Correlates. In all layers of the epidermis, dermis, and hypodermis are found many different types of free and encapsulated (specialized) nerve endings. These structures were discussed in detail under cutaneous (ie, somatic) receptors in Chapter 6 (p. 179). However, it is worthwhile to reiterate here that the various sensory modalities are not associated with any particular type of specialized (ie, encapsulated) nerve endorgan, a concept that has been held by physiologists for many years. Rather, the cutaneous sensations are received by appropriate stimuli applied to physiologically specific, though morphologically indistinguishable nerve endings that terminate in the skin and there form loose plexiform networks. The nerve endings within the skin that subserve the sensations of heat and cold undoubtedly are connected to the myelinated craniospinal fibers of the dorsal spinothalamic tract (Table 7.5. p. 259).

It also must be borne in mind that "heat" and "cold" are relative terms, and that the adequate stimulus for any thermal receptor is a change in temperature. Consequently the threshold stimulus for a cold receptor is an adequate rate-of-fall in temperature whereas warmth receptors respond only to a sufficient rate-of-rise in temperature, and both types of thermal receptor exhibit rapid adaptation. On the other hand, when excessive quantities of heat are applied to the skin a sensation of pain is elicited, a sensory modality that also is mediated by bare nerve endings.

It is important to realize that the thermal sense organs in the skin do not respond to "temperature" itself. Rather these thermoreceptors record the skin temperature at the point where the nerve endings are located anatomically. Furthermore, and as a direct consequence of this fact, thermal receptors are stimulated by internal heat as well as by the heat of the external environment. On the other hand, certain of the nonmyelinated fibers that are present in the skin are the axons of motor nerves. These fibers innervate the blood vessels, smooth muscles, and glands of the skin, hence they can assume an important role in thermoregulation.

CENTRAL NEURAL STRUCTURES THAT PARTICIPATE IN THERMOREGULATION; ROLE OF THE HYPOTHALAMUS. Experimental animals whose brain stem has been transected inferior to the hypothalamus are unable to regulate their temperature, hence they become poikilothermic. More sophisticated experiments in which discrete lesions are induced in specific hypothalamic regions have revealed that the derangement in thermoregulation that follows placing such a lesion depends specifically upon the area where the damage is located. Therefore, following the placement of a lesion in the rostral (anterior) hypothalamus at the level of the optic chiasm and anterior commissure (especially in its lateral extensions), the animal be-

comes unable to regulate its body temperature in a warm environment (Figs. 7.46 through 7.49 inclusive). Similarly, human patients are unable to thermoregulate in warm environments as the result of infarcts or tumors in the anterior hypothalamus. Thermoregulation in a cold environment, on the other hand, may be relatively undisturbed following anterior hypothalamic lesions.

The defects in thermoregulation that are observed in patients or experimental animals as the result of anterior hypothalamic lesions are caused by derangement of the physiologic mechanisms that normally are activated by a warm environment, and that produce an increased heat loss from the body as a normal homeostatic adaptation to such a stimulus. In humans, these physiologic responses to a warm environment are cutaneous vasodilatation and sweating.

If a lesion is induced in the caudal hypothalamus but anterior to the mammillary bodies then the capacity of the animal to maintain a normal body temperature in either a warm or a cold environment is seriously impaired, and following such a procedure the animal is poikilothermic. The mechanisms that control heat loss from the body are defective, probably because the lesion interrupts the neural pathways that lead from the more rostrally situated heat loss centers to the periphery. The failure to maintain a normal body temperature in a cold environment following a posterior hypothalamic lesion is caused by destruction of a caudally situated hypothalamic center that controls heat production and heat conservation.

Further experiments have demonstrated conclusively that the temperature of the blood that perfuses the cranial region in general (or the temperature of the hypothalamic tissues themselves) is a major factor in determining the thermoregulatory response evoked. Thus warming the carotid blood before it enters the skull, or localized heating of the hypothalamic region by use of suitable experimental techniques (eg, diathermy, thermode probes) will elicit appropriate thermoregulatory responses. For example, heating the hypothalamus activates those mechanisms responsible for increased heat loss from the body, ie, cutaneous vasodilatation and sweating, despite the fact that no alteration in skin temperature has taken place. Conversely, cooling the hypothalamus results in vasoconstriction, piloerection, and shivering, and these effects tend to conserve body heat as well as to increase heat production. Studies such as these indicate that the hypothalamic thermoregulatory center functions somewhat like a thermostat in that the cells within this region of the brain are sensitive to alterations in the temperature in their immediate environment, so that they can elicit and integrate appropriate physiologic responses to localized heating or cooling. The thermostat analogy is not, however, entirely accurate and will be discussed further below.

Experiments such as those just cited leave no

room for doubt that the mammalian hypothalamus contains two thermoregulatory centers that function in opposite directions. Furthermore, the integrated and coordinated discharge of impulses through descending nerve pathways from these hypothalamic centers to the peripheral structures that are responsible for thermoregulation (eg, the cutaneous vasculature) plays a major role in the maintenance of a normal body temperature. It also is clear that the two hypothalamic thermoregulatory centers function in a reciprocal fashion in vivo, as they are functionally as well as anatomically interconnected by means of bilateral tracts that traverse the lateral hypothalamus. Despite these interconnections, however, the regulatory functions of these two regions are independent to a certain extent.

Normal Body Temperatures

The homeostatic mechanisms concerned with the regulation of body temperature in homeotherms tend to stabilize the temperature within narrow limits. However, these mechanisms cannot prevent small fluctuations in temperature due to external and internal thermal variations. Therefore, it is implicit in any concept of a "normal" body temperature that the specific conditions under which the temperature was measured must be stated accurately. Furthermore, the actual locus on or within the body where the temperature is measured is an important factor, as normally some variation, albeit slight, is found in the temperatures that are recorded from different regions.

ORAL TEMPERATURE. Under carefully-controlled conditions, studies which were performed on a large series of healthy young adults indicate that the mean oral temperature in the morning is 36.7C. Since the standard deviation in this group of subjects was 0.2, then 95 percent of normal young adults have an oral temperature in the morning between 36.3 and 37.1 C.

Further studies have shown that under less well-controlled conditions (ie, insofar as exercise, eating, drinking, smoking, and other factors that contribute to thermal load were concerned), the mean oral temperature of humans was 37.0± 0.5C. Thus, for all practical purposes 37C is usually considered to be the normal oral temperature of humans, although individual variations above and below this average value are encountered frequently, depending upon such miscellaneous factors as the time of day, exercise, eating, drinking warm or cold liquids, and smoking prior to taking the measurement. Usually the oral temperature is around 0.5C below rectal temperature.

RECTAL TEMPERATURE. The rectal temperature is indicative of the core body temperature, and controlled studies in a large group of normal subjects not under thermal stress indicated that this value is 37.0C ± 0.5C. Since the rectal temperature is indicative of the temperature within the

core of the body, it is least subject to variations in environmental temperature.

In normal individuals, the rectal temperature undergoes a regular diurnal variation of about 0.6C. Persons who sleep at night and are awake during the day exhibit the lowest temperature at around 6 AM, whereas the peak temperature is reached during the evening. A similar diurnal pattern is observed in patients who are hospitalized for bed rest.

Women exhibit an additional cyclic temperature fluctuation that is characterized by a slight rise in core temperature at the time of ovulation during the sexual cycle (Fig. 15.9, p. 104). This monthly pattern is, of course, superimposed upon the diurnal variations described above.

Normal ranges for oral and rectal temperatures under various conditions are summarized in Figure 14.61.

SKIN TEMPERATURES. The skin on various parts of the body surface exhibits wider temperature fluctuations than either the oral or rectal temperatures, merely because this organ is exposed to widely varying environmental temperatures. In general the extremities, especially the hands and feet, are subject to the greatest temperature differences.

The several factors that contribute to the skin temperature are discussed on page 1004.

Additional specific factors that alter body temperature will be discussed in context in the material to follow; however, in conclusion to this section it should be emphasized that an accurate clinical evaluation of an elevated temperature in a patient requires a clear understanding of the three basic factors that determine the body temperature: first, the production as well as distribution of heat within the body; second, the mechanisms whereby heat is lost from, or gained by, the body; and third, the neural mechanisms that underlie control of body temperature.

Heat Production and Distribution Within the Body

HEAT PRODUCTION. Four major factors contribute to the generation of heat within the body.

Exergonic Reactions. The exergonic chemical reactions that contribute to heat production by the body as a whole were discussed in the first major division of this chapter, and their relationship to the basal metabolic rate was considered in the second major division in conjunction with overall energy balance. These oxidative reactions liberate heat energy as a byproduct of the various specific metabolic processes, and that heat is contributed to the overall thermal pool of the body throughout the lifetime of the individual.

Ingestion of Foodstuffs. The ingestion of foods leads to an increase in basal heat production that varies in accordance with the type and quantity of

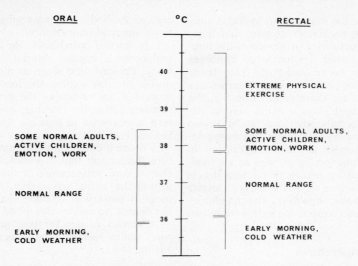

FIG. 14.61. Normal oral and rectal temperatures for a number of different physiologic states. Cf. Figure 14.65. (After Ruch and Patton [eds]. *Physiology and Biophysics,* 19th ed, 1966, Saunders.)

the foodstuffs that are consumed. This effect, called the specific dynamic action of foods, was discussed earlier.

Contraction of the Skeletal Muscles. The principal source of heat in the human body is derived from muscular work, ie, by the contraction of the skeletal muscles that accompanies various types of bodily activity. For example, during vigorous exercise the heat generated by the performance of muscular work can accumulate within the body to such an extent that the rectal temperature may rise above 40C (Fig. 14.61). Furthermore, the slight elevation in body temperature that is seen during various emotional states presumably is due, at least in part, to the contraction of skeletal muscles that may take place unconsciously as a concomitant of the emotional state itself.

Shivering is an involuntary rhythmic contraction of the skeletal muscles that serves as a physiologic mechanism to increase heat production in very cold environments.

Endocrine Factors. Totally apart from food consumption and muscular exercise, heat production within the body can be modified profoundly by certain hormones. Thus, epinephrine and norepinephrine produce an evanescent rise in heat production, whereas thyroxine causes an increase in body heat that develops slowly over a matter of days or weeks, but is of prolonged duration. Thus, in hyperthyroidism the body temperature is elevated chronically up to 0.5C above the normal range.

In addition to these four major factors, the body thermal load also may be increased somewhat by the ingestion of large quantities of very warm foods and/or beverages, although in general this factor is of minor importance physiologically.

Infants, but not adults, have a large deposit of a peculiar type of fat that is located in the scapular

regions of the body. The extremely high metabolic rate of this brown fat, as it is called, contributes significantly to the total heat production in infants, but not in adults, which have only ordinary white fat.

Thus, the total heat produced by the body at any particular time is the sum of the individual thermal contributions that are made by the several individual factors presented above.

HEAT DISTRIBUTION AND TRANSFER. Two major physical parameters underlie the distribution and transfer of heat within the body, and these factors are conduction and convection.

Conduction takes place through any material by the direct transfer of thermal energy down a temperature gradient that exists between the adjacent atoms of the material; ie, in the process of conduction, heat energy flows down a thermal gradient. Conduction of heat also takes place between two different bodies that are at different temperatures, and which are in direct contact with one another (Fig. 14.10A). The quantity of heat that is exchanged by conduction between two objects that are in direct contact and at different temperatures is directly proportional to the temperature difference between them.

Convection, on the other hand, involves the transfer of thermal energy through bulk movement of the material itself. That is, the molecules of a fluid (a gas or a liquid) at one temperature move by the process of convection to another region. A familiar example of convection is to be found in a pan of cold water that is heated slowly on a stove. As the water is heated, the warmer water rises away from the bottom of the pan, because of its increased thermal energy and reduced density. As the warmer water rises, cooler water sinks down to replace the warmer water that has risen toward the surface, and convection currents are set up within the mass of water. Ultimately, the convection currents will cause the temperature throughout the mass of water within the pan to become uniform as the water boils. Similar convection currents are set up between air masses at different temperatures.

Within the body the processes of conduction and convection act together in order to assure adequate heat transfer.

The body tissues are not especially good conductors of heat. Consequently, if heat exchange took place solely by conduction, the thermal gradients which are necessary to redistribute the heat generated by metabolism would have to be quite large. Furthermore, a rapid adaptation to internal and external thermal stresses would be impossible, as the thermal conductance of each tissue is constant.

Within the body, convective heat transport takes place largely by the bulk transport (or movement) of the body fluids, hence the cardiovascular system is of primary importance in underlying the forced convection of heat energy throughout the body. The circulatory system directly influences the distribution of heat within the body in three ways.

First, it reduces temperature differences throughout the body to a minimum. Since the distance heat must be transported solely by conduction from the cells to the capillaries is extremely minute, the temperature differences (or gradients) that are necessary to effect this transfer rapidly are on the order of hundredths of a degree Celsius. Thus in well-perfused, metabolically active tissues, the abundant capillary networks that are present provide an extremely efficient means of removing heat from the cells by forced convection with only a small increase in temperature of the blood. Conversely, in tissues that are cooler than the temperature of the blood that is coming to them, heat is transferred to the tissues.

In physical terms, this effect may be described as follows. The heat that is removed from, or added to, a body by a fluid which passes through the body is directly proportional to the difference in the inflow and outflow temperatures as well as the flow rate. Consequently, the circulatory system warms some tissues and cools others, so that wide temperature extremes in various regions of the body are minimized, and the body as a whole is maintained at a practically uniform temperature.

Second, the relative ease with which heat is transferred from the body to the environment through the skin is affected profoundly by vasomotor changes in this organ. It was noted earlier that the dermal capillaries and other vessels supplying the skin do not perfuse the epidermis; therefore, heat is transferred into this superficial layer only by conduction. If the dermal capillaries are fully dilated, heat is transferred at a sufficient rate by forced convection when a temperature gradient of only a fraction of a degree exists between the surface of the skin and the blood. Conversely, if

the cutaneous vessels constrict strongly, then heat must be transported to the surface of the skin by conduction alone, and this process requires a proportionally greater temperature difference. Vasomotor states that lie between these two extremes result in heat transfer by convection and conduction in varying proportions. The state of vasomotor tone in the arteriovenous anastomoses within the skin is also of major importance in regulating heat exchange between the core tissues and the environment as described above. Therefore, the relative facility with which heat is transferred to the surface of the body is altered significantly by vasomotor changes within the skin, and these vasomotor changes in turn markedly affect the thickness of the insulating layer that effectively separates the environment from the underlying tissues within the body.

Third, countercurrent heat exchange between the major blood vessels lying close together within the body assumes great physiologic importance under certain circumstances in regulating heat distribution and conservation within the body. Blood flowing through the extremities passes from the heart through the arterial tree, loses some heat in the capillaries of the skin, and then this cooler blood returns to mix with the warmer blood that comes from the internal organs. The net effect of this process would be to reduce the total body heat content somewhat were it not for the fact that countercurrent heat exchange takes place by conduction through the arterial and venous walls, as depicted in Figure 14.62 (see also Fig. 12.8, p. 780). The following consequences result from this heat flux from the warmer arterial blood to the cooler venous blood in certain regions of the body. The blood is somewhat precooled before it reaches the superficial areas; therefore, the capillary-to-skin thermal gradient is reduced, so that heat loss to the environment also is reduced. Lastly, the venous blood returning to the heart is somewhat prewarmed; hence the overall effect of countercurrent heat exchange is to conserve body heat.

The relative physiologic importance of countercurrent exchange of heat within the body as a mechanism for heat conservation is governed by one major factor. This factor is the relative distribution of the venous blood that is returning to the heart by way of the deep and the superficial vessels (Fig. 14.62). Thus, if blood is returned primarily via the deeper lying veins, then heat transfer by conduction to the arteries (hence conservation of heat), is increased, whereas if the blood is returned from the periphery largely in the superficial veins, then heat transfer to the arteries is minimized and body heat loss is increased somewhat. Under different physiologic conditions, venous blood can be shunted in varying degrees from superficial to deep vessels, and thereby heat loss can be enhanced or diminished.

INTERNAL BODY TEMPERATURES. The patterns of temperature distribution within the different regions of the body are quite complex in detail, and depend upon a number of highly variable factors such as local temperatures within the tissues (hence local metabolic rates), local vascular tone, and local thermal gradients. In outline however, the general features of temperature distribution within the body under different environmental extremes are quite straightforward.

Metabolic heat in the body is derived principally from exergonic chemical processes in the deep organs (eg, the brain, heart, viscera), except during muscular exercise. Therefore in order for this heat to be lost, the deep organs must be somewhat warmer than the blood that perfuses them, as well as the surrounding tissue. In this regard it is useful to conceptualize the entire body as a warmer core that is surrounded by a thin layer of insulation, as depicted in Figure 14.63. Tissues having a temperature within a fraction of a degree of the rectal temperature are considered to be at core body temperature. In this rather simplistic view, the core essentially consists of the entire body mass when cutaneous vasodilatation is at a maximum, and under these circumstances the insulating layer is reduced to the thickness of the epidermis (Fig. 14.63A).

In extremely cold environments, however, heat must be conserved by the body, and vasoconstriction sharply reduces the circulation to the peripheral regions so that the central core shrinks, thereby lowering heat transfer to the body surface (Fig. 14.63B). Indirectly, this mechanism effectively decreases heat loss from the body to the environment by reducing the surface temperature of the body. As the consequence of this effect, the temperature of the deep-lying vital organs is maintained in an extremely cold environment, although this internal protection takes place by sacrificing heat transport to the extremities to such an extent that severe thermal damage to these regions can occur.

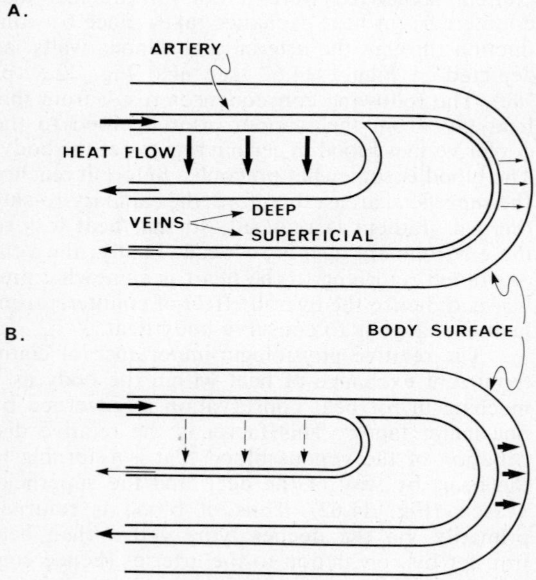

An important concept relating to the effects discussed above is the surface–volume ratio. In a cylindrical object, the diameter is inversely proportional to the surface–volume ratio. Consequently, the fingers have a greater surface area in relation to their volume than do the arms, and the arms have a greater surface–volume ratio than does the trunk. In very cold environments, excessive heat loss to the environment as well as excessive cooling of the blood takes place in the extremities, despite complete vasoconstriction of the cutaneous vessels. In this situation, heat exchange by the countercurrent mechanism allows large thermal gradients to be established parallel with the long axes of the major vessels so that the entire distal region of the extremity becomes part of the insulating layer as the body core shrinks.

In a warm environment, on the other hand, the large surface–volume ratio of the extremities is beneficial to the individual, as it permits heat loss to the environment. In fact, amputees with both legs removed can suffer acute discomfort in a warm environment, as the total surface area of the body is markedly reduced, and thus heat loss from the lower extremities no longer takes place.

FIG. 14.62. Countercurrent heat exchange in the extremities. **A.** When the arteriovenous anastomoses or other vessels in the skin are open, blood, hence heat, is diverted from the superficial vessels of the skin (large arrows) and thereby body heat is conserved. **B.** When the arteriovenous shunts or other cutaneous vessels are closed by vasoconstriction then more blood now is diverted toward the surface of the skin; hence more heat is lost from the body (large arrows). See also Figure 12.8. (Data from Ruch and Patton [eds]. *Physiology and Biophysics,* 19th ed, 1966, Saunders.)

FACTORS THAT ALTER SURFACE BODY TEMPERATURES. Under normal circumstances, the skin temperature measured upon the surface of the body lies between the core body temperature and that of the environment. Following cutaneous vasodilatation, when heat transport from the body core to the surface increases significantly, the surface temperature is elevated and thus approaches the core temperature. Conversely, vasoconstriction reduces heat transport through the skin, so that the surface temperature falls toward the environmental temperature. Two examples will serve to illustrate these points. First, alcohol not only induces a feeling of warmth due to its depressant effects upon the higher brain centers, but also because it induces a loss of vasomotor tone; hence it induces cutaneous vasodilatation. The latter effect can result in a considerable heat loss from the body, and this fact contraindicates the use of alcoholic beverages in quantity as a source of energy in extremely cold climates. Second, patients suffering from Raynaud's disease undergo intermittent and severe vasoconstriction, especially of the vessels of the extremities. An individual with Raynaud's disease may complain of being cold, although the ambient temperature is quite comfortable to a normal person. This complaint is quite real, however, as the vasospasm has caused the temperature of the thermal receptors within the skin to fall toward the environmental temperature, hence to discharge impulses that are interpreted as cold.

When the evaporation of water from the surface of the body removes large quantities of heat, then the skin temperature may become lower than either the core or environmental temperatures. Therefore, a temperature gradient is set up between the body core and the body surface down which heat can move, and this gradient can exist even when the ambient temperature is greater than 32 C.

By way of summary, therefore, the skin temperature at any point upon the body surface at any particular time is the resultant of those factors that underlie heat transport from the body core to the surface as well as the relative ease whereby heat is transferred to the immediate environment. Since these factors are capable of great variation over the body surface, the skin temperature measured upon different regions of the body surface also can vary markedly.

TISSUE INSULATION AND CLOTHING. Tissue insulation may be defined in general terms under steady state conditions as the temperature difference which is present between the body core and body surface (ie, the thermal gradient) which is necessary to produce a given heat flow per unit time per area of skin. Thus tissue insulation may be expressed according to the dimensions ° per Cal per hour per m². Thus the insulating capacity of tissues, skin, or other material, is directly proportional to the magnitude of this value.

It has been calculated that the normal value for insulation of the nude human body in "comfortable" surroundings is approximately 0.08 C per Cal per hour per m². Vasodilatation produced by a warm environment can lower this value to around 0.02 C per Cal per hour per m², whereas vasoconstriction due to exposure to cold can increase tissue insulation to 0.2 C per Cal per hour per m². On the other hand, obese persons, who have a large quantity of fat in their subcutaneous tissues, exhibit a much higher tissue insulation in a standard environment than do individuals of normal weight.

Clothing, of course, merely adds additional insulation to that which is already provided by the body tissues. The actual insulation that is provided to the body by various styles, materials used, and even the colors of clothing, are of almost infinite variety; consequently no physiologically valid generalizations can be made with regard to the insulation provided by this factor.*

Heat is conducted from the skin to the air layer that is trapped at the body surface by clothing and then some of this heat is conducted across the clothing to the surrounding air. The quantity of heat that is transported through clothing to the surroundings depends upon the texture and thickness, as well as the magnitude of the layer of air that is trapped at the body surface.

HEAT CONTENT OF THE BODY. The specific heat of the body tissues in general is around 0.83 Cal per ° per kg, and this relatively high value can be attributed to the high water content of the body. It was stressed in Chapter 1 (p. 16) that the high specific heat of water was a physical property of inestimable value to the survival of living organisms, because it permitted large changes in heat content of the body to take place without concomitant wide fluctuations in body temperature. Since the specific heat of human tissues also is quite high, thermal stresses that are imposed upon the body by wide variations in environmental temperature are buffered.

If the rate of heat loss to the environment is not balanced precisely by the rate of heat production within the body, then the difference is added to, or subtracted from, the overall body heat content. And if the thermal stress that is imposed upon the body changes rapidly, then heat production and loss are imbalanced temporarily. But because of the high heat capacity of the adult human body, the heat production and heat loss may be imbalanced for as long as several hours without deleterious effects because the body temperature changes only slightly because of the high specific heat of the tissues. The advantages of this to the individual are twofold. First, the motor responses to thermal stress do not have to take place rapidly in proportion to the degree of thermal stress that is imposed suddenly upon the body. Thus there is a time delay between the onset of the thermal stress and the development of a marked alteration in body temperature. Second, the damping effect occasioned by the time delay permits the individual to adapt gradually to a rapidly-imposed thermal stress, or if such adaptation is physiologically impossible, then adequate time may be available for escape from that environment.

Heat Exchange Between the Body and the Environment

There are four basic processes whereby heat is exchanged between the body and the environment. These processes are conduction, convection, radiation, and evaporation. The net heat loss at any

---

Since the hair covering the surface of the human body is quite scanty, erection of these appendages by contraction of the arrector pili muscles with the concomitant development of "goose flesh" (or horripilation) serves little purpose insofar as increasing the insulation of the body is concerned.

particular time is the sum of the individual quantities of heat lost through each of these four processes. The proportions contributed to the total heat loss by each of these factors can vary considerably, depending upon the body temperature, environmental temperature, degree of muscular activity, and so forth. The total surface area of the skin is a factor of major importance in heat exchange between the body and the environment, and in this regard it is well to remember that as the volume of the body (ie, its total mass) increases, there is a disproportionate increase in the total surface area that is available for heat exchange. Thus in two individuals of the same height but of quite different weights, the heavier person will have a lower surface area in relation to his body weight than the leaner person.

There are three physiologic pathways other than the skin whereby some heat is lost from the human body. These routes are through the respiratory system, and through urination, as well as defecation. Usually, however, the total amount of heat that is lost from the body through these channels is negligible insofar as the total heat exchange of the body with the environment is concerned. Some heat also may be gained or lost from the body by the ingestion of large quantities of hot or cold foods and/or beverages. Normally, the total heat gain or loss via this route also is negligible.

The relative quantities of heat which are lost from the body at 21 C by each of the processes described above are summarized in Table 14.18.

CONDUCTION. The process of heat conduction down a thermal gradient, ie, from the skin to a cooler substance or object in direct contact with the skin, is negligible under ordinary circumstances. For example, standing on a cold shower floor with bare feet causes only a slight heat loss through the soles. On the other hand, bathing or diving in an ice cold mountain lake can cause a major heat loss from the body, not only by conduction, but also by convection currents that

Table 14.18 APPROXIMATE RELATIVE QUANTITIES OF HEAT LOST FROM THE BODY AT 21 C

MECHANISM OF HEAT LOSS	HEAT LOSS, % OF TOTAL[a]
Convection	35
Radiation	34
Vaporization of insensible perspiration	27
Respiration	2
Conduction	1
Urination and defecation	1

[a]These fractions can vary considerably depending upon the specific circumstances under which they were determined, hence the values listed in this table are only approximate. Heat production in the body at any particular time is determined by the additive effects of metabolic processes, the specific dynamic action of ingested foodstuffs, and muscular activity, both voluntary and involuntary, as discussed in the text.

are set up in the water that is in contact with the body.

CONVECTION. Convection is a mode of heat exchange which involves the bulk transfer of thermal energy between fluid and solid objects or within fluids themselves. Insofar as heat loss from the body by convection is concerned, if the skin temperature is warmer than that of the environment, then heat is transferred from the body to the surrounding air. The heated air next to the body rises, and it is replaced by more dense cooler air from below. Thus the net effect of this process is a continuous flow of cool air past the body where the air is warmed, and thereby a net heat loss from the body surface takes place continuously.

In this regard it is important to mention that most ordinary clothing prevents the development of convection currents along the body surface, so that a layer of air warmed by the body heat is trapped, and thus serves to prevent excessive heat loss via the process of convection. The movement of air that is produced solely by body heat is termed natural convection. If external factors are involved in causing the air movement, eg, winds or fans, then the process is called forced convection.

As noted above, heat loss from the body by natural convection depends upon the existence of a thermal gradient between the skin and the surrounding air. Therefore, if the body surface and the ambient air are both at the same temperature no convective heat transfer can take place. On the other hand, if the thermal gradient is reversed so that the air temperature is above that of the body surface, then the body gains heat by convection.

The rate of natural convective heat transfer not only depends upon the thermal gradient described above, but also upon the total exposed surface area of the body. Clothing reduces the total surface area that is available for convective heat transfer, as does body posture. Curling up (as on entering a cold bed) reduces the surface area which is available for convective heat transfer. Normally the axillae and groin are not exposed to the surrounding air, so that the total body surface area available for heat exchange by natural convection is reduced to various degrees, depending upon the position of the arms and legs.

RADIATION. In the process of heat transfer by radiation, thermal energy is exchanged between two objects not in contact with each other but which are at different temperatures. Radiant energy is transmitted through space as discrete bits of electromagnetic energy known as photons. Upon contact with another solid or a liquid, the electromagnetic (radiant) energy of the photons is absorbed and converted into heat energy. The average energy of the photons as well as their rate of emission are directly proportional to the temperature of the surface whence they are emitted.

Since the surface of the human body is warm, it too emits radiant energy as photons, and thus some heat is lost from the body by radiation to cooler objects in the environment. Conversely, energy radiated from the sun or objects heated by the sun (eg, rocks or sand in a desert), can transfer radiant energy to the body and appreciable warming can result. Thus on a sunny day, one can be quite comfortable in light clothing even though the ambient temperature is below freezing, provided that enough sunlight is directed onto the body.

The absorption and emission of photons by a given object are but two aspects of the same physical property, so that an efficient emitter is also an efficient absorber. Thus an object that is a "perfect" emitter also is a "perfect" absorber, and this is the theoretical "black body" of the physicist. Conversely, an object that has zero emissivity also has zero absorbance and is the theoretically "perfect" reflector. In actuality, however, all surfaces that are found in the real world lie somewhere between these two extremes. In general terms more radiant energy is absorbed as well as emitted (dependent upon the temperature, of course) by dark-colored clothing, whereas less radiant energy is absorbed and emitted by light-colored garments. In this regard, it might appear from the variations in pigmentation found in human skin that emissivity of photons, hence exchange of radiant energy, would vary considerably among different ethnic groups, but this is not so in actuality. The reason for this similarity depends upon the fact that the emissivity of the skin varies depending upon the energy of the photons that impinge upon it. Photons whose energies lie within the range of the visible spectrum are indeed reflected in accordance with the cutaneous pigmentation of the individual. However, those photons primarily responsible for thermal energy exchange are found in the far infrared portion of the spectrum. Since the energies are much lower in this spectral region, most surfaces, including skins having widely varying amounts of pigmentation, are almost perfect absorbers. Very dark skin absorbs around 80 percent of incident solar radiation, while very light fair skin absorbs about 65 percent.

Under normal circumstances, radiation accounts for a large and important fraction of the thermal energy exchange between the body and the environment, and as indicated in Table 14.18, this amounts to almost half of the total quantity of heat that is lost from the body of a person in thermal equilibrium at 21 C.

VAPORIZATION (EVAPORATION) OF WATER. The evaporation of water is an extremely important mechanism insofar as heat loss from the body (hence thermoregulation) in humans is concerned. In passing from the liquid to the gaseous (or vapor) state, water requires the addition of thermal energy, and the quantity of energy that is necessary to convert liquid water into vapor at constant temperature is called the latent heat of vaporization. Within the range of temperatures that are encountered under physiologic conditions, the latent heat of vaporization of water is approx-

imately 580 Cal per liter or 0.58 Cal per gram. Consequently, the large quantity of thermal energy that is removed from the body during the evaporation of water from the skin renders this process extremely important in human thermoregulation.

Heat loss from the body through evaporation of water takes place via several routes that differ considerably in their overall physiologic significance. Thus such heat loss occurs through the mucous membranes of the oral, nasal, and respiratory passages; by passive diffusion through the skin (insensible perspiration); and by sweating.

Heat Loss from the Oral, Nasal, and Respiratory Passages. In humans, thermal exchange through the mouth and various portions of the respiratory tract is merely an incidental feature of pulmonary gas exchange. Since expired air is nearly saturated with water vapor, regardless of the rate of ventilation, some heat is being lost constantly from the respiratory passageways. Under normal resting conditions, this heat loss is approximately 9 Cal per hour, and this value increases in proportion to the ventilation rate. However, the heat required for water loss by this route is provided by the blood, and since the perfusion rate of the pulmonary tissues is extremely high, then very little change in body temperature is related to heat loss by the respiratory and related passages in human beings.

In furred mammals, on the other hand, heat loss to the environment is enhanced considerably by panting, and this mechanism is of considerable importance to thermoregulation in such species. Thermal exchange through the skin is limited to a considerable extent by the insulating layer of fur; consequently the rapid, shallow breathing at a low tidal volume that attends panting augments the total pulmonary ventilation, without, however, increasing the alveolar ventilation appreciably. Since most of the extra air transported during panting comes from the anatomic dead spaces, then heat loss from the respiratory tract is increased appreciably, but without the development of a concomitant imbalance in the blood gas concentrations such as would occur if the animal were merely hyperventilating.

Insensible Perspiration. The normal skin of human beings is relatively impermeable to the passage of water from the superficial tissues to the environment, although it is not completely impermeable. Thus a small quantity of water constantly is moving by passive diffusion through the epidermis to the body surface, whence it is evaporated, thereby causing a slight but continuous heat loss from the body. Water loss by passive diffusion through the skin is imperceptible to the individual; hence this process is often called insensible perspiration or transcutaneous diffusion.

Depending upon the ambient temperature and the relative humidity, the total water loss from the entire body surface by insensible perspiration may range between 10 and 50 ml per hour, with an attendant heat loss of between 6 and 30 Cal per hour.

The combined heat loss from the body via the respiratory passages and insensible perspiration is of rather minor importance insofar as overall thermoregulation in human beings is concerned; this total amounts to approximately 15 percent of the entire heat loss.

Sweating. Unlike insensible perspiration, sweating is an active process whereby water is secreted upon the body surface by the specialized eccrine sweat glands. In a warm and dry environment, vaporization of water from the skin becomes the major factor in thermal exchange because an active secretion of sweat markedly increases the quantity of water that is available for evaporation from the body surface. Thus the sweat glands of the skin are a physiologically important effector mechanism insofar as thermoregulation is concerned.

Sweat glands are innervated solely by the sympathetic nervous system; however, they differ markedly from other sympathetic endorgans because they are cholinergic rather than adrenergic (Table 7.6, p. 276). Furthermore, these glands actively secrete sweat only when they are stimulated via nerve impulses, and the volume of sweat that is secreted is proportional to the frequency of the efferent nerve impulses. This overall effect of the nervous system upon sweat secretion is often termed sudomotor activity.

The quantity of heat that is lost from the body surface by the vaporization of water (ie, the evaporation of sweat) depends upon two factors: first, the rate at which water is secreted by the sweat glands; and second, the ability of the ambient environment to remove water vapor.

If the air is dry and moving quickly, then heat loss from the body by evaporation is limited only by the rate at which sweat can be secreted. Conversely, if the air is humid (ie, moist) and still, then heat loss from the body is restricted by the capacity of the surrounding air to remove water from the skin surface. Thus the relative humidity of the air is a major factor in determing evaporative heat loss from the body under certain conditions.* It is common experience that one feels warmer on a hot humid day then on a hot, but dry, day.

Under most circumstances heat loss by the vaporization of sweat varies between the two extreme situations described above, so that the total heat loss from the body via the sweating mechanism can vary from 30 to over 900 Cal per hour.

The rate of sweat secretion in any individual varies markedly, of course, depending upon the

environmental temperature and degree of muscular work being performed. However, in a hot climate during a period of heavy physical work, a person can secrete over 1.5 liters of sweat per hour; and in a dry atmosphere, practically all of this sweat can be evaporated, save for that which runs off the body and is lost insofar as evaporative heat loss is concerned. In this regard, a significant water (as well as salt) depletion from the body stores can take place via the sweat glands, a fact that must be considered in relation to water (and salt) balance in individuals who are performing heavy manual labor in extremely hot environments over extended periods of time.

Physiology of the Thermoregulatory System

The various anatomic and physiologic parameters pertaining to thermoregulation that were discussed individually in the preceding sections may now be correlated in order to give a composite picture of thermoregulation in the human body. The adaptive mechanisms concerned with the control of heat production, heat conservation, and heat loss are summarized in Table 14.19.

THERMORECEPTORS. These sensory elements of the human thermoregulatory system are of two types, peripheral and central.

Peripheral Thermoreceptors. The cutaneous nerve endings contain peripheral thermoreceptor properties, and these functionally specific heat and cold receptors in turn respond to adequate thermal stimuli over an appropriate temperature range and send afferent impulses of graded frequency centrally. These afferent impulses ascend the spinal cord in the dorsal spinothalamic tracts (Table 7.5, p. 259), and ultimately these signals converge upon the hypothalamic thermoregulatory center where the impulses are integrated and appropriate effector responses are initiated.

The impulse frequency that is generated by appropriate thermal stimuli applied to the cutaneous thermoreceptors is proportional to the intensity of the stimulus, and these receptors are particularly sensitive to rapid alterations of temperature. That is, an increase in temperature within their physiologic range is the stimulus for the heat receptors, and the frequency with which these receptors fire (hence the frequency of afferent neural discharge of impulses) is in proportion to the magnitude as well as the rapidity with which the temperature changes. In a similar fashion, a decrease in temperature activates the cold receptors.

Both heat and cold receptors exhibit rapid adaptation so that after an initial fast burst of impulses that follows the application of an appropriate stimulus, the firing rate of both types of receptor decreases markedly.

Central Thermoreceptors. The central thermoreceptors are located within the hypothalamus. Thus an elevated brain temperature causes an increased heat loss due to evaporation of the sweat

---

*The relative humidity is defined as the quantity of water vapor which is contained per unit volume of air at a given temperature divided by the quantity of water vapor that the same volume of air would hold if it were saturated with water vapor at the same temperature. The relative humidity is expressed as a percent.

Table 14.19 SUMMARY OF HUMAN THERMOREGULATORY MECHANISMS

MECHANISMS ACTIVATED BY HEAT. Reflexes controlled in anterior hypothalamus; lesions in this part of brain induce hyperthermia.

    Mechanisms that increase heat loss:

        Cutaneous vasodilatation; major thermoregulatory mechanism

        Sweating; major thermoregulatory mechanism

        Increased pulmonary ventilation; minor thermoregulatory mechanism

    Mechanisms that decrease heat production:

        Anorexia (loss of appetite)

        Physical inertia, apathy

        Reduced thyrotropic hormone secretion by anterior pituitary; leads to reduced thyroxin output by thyroid gland (long term effect)

MECHANISMS ACTIVATED BY COLD. Reflexes controlled in posterior hypothalamus; lesions in this part of brain induce hypothermia.

    Mechanisms that increase heat production:

        Shivering

        Elevated level of voluntary muscular activity

        Elevated thyrotropic hormone secretion by anterior pituitary; leads to increased thyroxin output by thyroid gland (long term effect)

        Secretion of epinephrine and norepinephrine by adrenal medulla (short-term effect)

    Mechanisms that decrease heat loss:

        Cutaneous vasoconstriction; major thermoregulatory mechanism

        Behavioral adaptations; eg, depart from a cold to a warmer environment if possible, add more protective clothing, curl up to reduce surface area exposed to heat loss, use blankets, build fire, and so forth. Behavioral adaptations constitute a major thermoregulatory mechanism for humans.

        Horripilation (erection of hairs). This reaction to cold constitutes a mechanism of negligible importance to thermoregulation insofar as humans are concerned, but it is a mechanism of considerable importance to certain lower mammals.

that is secreted in response to such central stimulation, as well as peripheral vasodilatation that increases heat loss due to convection and radiation. A reduced brain temperature, on the other hand, results in the conservation of body heat due to peripheral vasoconstriction. In addition, the involuntary muscular contractions which are involved in shivering increases body heat production markedly.

CENTRAL INTEGRATION OF AFFERENT THERMOREGULATORY IMPULSES: ROLE OF THE HYPOTHALAMUS. The simple fact that body temperature is regulated at all necessitates the presence of one or more central neural mechanisms that are able to process the inputs from peripheral and central thermoreceptors. Such central integrative mechanisms also must be able to initiate suitably graded motor impulses that in turn pass to the appropriate effector systems in order to control heat loss, heat production, and heat conservation.

There are many levels of integration within the central nervous system, both conscious and unconscious, that converge ultimately upon the vasomotor and somatic motor effectors responsible for thermoregulation. For example, the functions of the human thermoregulatory system are linked inseparably to the functions of the cardiovascular and respiratory systems. Furthermore, voluntary behavior affects the process of thermoregulation, as exemplified by the act of putting on more clothes in cold weather or by seeking the shade on an extremely warm day.

The principal and critical level of integration within the central nervous system that is reponsible for the homeostatic regulation of body temperature in mammals is the hypothalamus. The two neural structures within the hypothalamus that play an integrative role in thermoregulation are connected by bilateral tracts, and these pass through the lateral hypothalamus. As noted earlier, the anterior hypothalamic region contains a heat loss center that functions reciprocally, and in conjunction with, a posteriorly situated heat production and conservation center. Afferent impulses that arise in heat thermoreceptors (whether peripherally or centrally located) stimulate the rostrally located heat loss center resulting in vasodilatation and sweating. As the anterior heat loss center is stimulated, the posterior heat production and conservation center is inhibited. Conversely, impulses that are generated in cold thermoreceptors stimulate the posteriorly located heat production and conservation center, resulting in shivering and vasoconstriction. In this instance the anteriorly situated heat loss center is inhibited.

This presentation of the reciprocal functional activities of the two hypothalamic thermoregulatory centers may be somewhat simplistic in the light of recent experimental evidence that indicates certain regions of the brain adjacent to, but not within the hypothalamus itself, also appear to have a role in temperature control of the body as a whole. Nevertheless, until the functional relationships between these structures and the hypothalamus proper are clearly elucidated, the principal integrative structures for temperature regulation in mammals must remain as described above.

PHYSIOLOGIC REGULATION OF BODY TEMPERATURE UNDER VARIOUS CIRCUMSTANCES

Adaptation to Light Exercise and Slight Thermal Stress. Alterations in vasomotor tone with an attendant change in tissue insulation are adequate to maintain the balance between heat production and heat loss when light exercise is performed or when the body is subjected to slight thermal stresses. Under such circumstances, the body core temperature remains unchanged, and the heat loss from evaporation not only is small, but is derived almost entirely from insensible perspiration. The metabolic rate is not elevated significantly, except when voluntary muscular movements are performed. The subjective feeling is one of thermal neutrality, and the skin temperatures range between 31 C and 34 C.

The vasomotor state of the individual is the primary thermoregulatory factor involved in the control of body temperature, and man generally selects clothing that is suitable to the ambient temperature and wind velocity, among other factors, in order to remain comfortable under such physiologically neutral conditions. For a subject clothed normally, an ambient temperature of around 20 C coincides with the maximum ability of the body to thermoregulate solely by alterations in vasomotor tone.

Adaptation to Vigorous Exercise and Severe Thermal Stress. A change in tissue insulation due to peripheral vasodilatation alone is inadequate to permit adequate heat loss from the body under conditions of severe muscular exertion or when the environmental temperature increases significantly. Furthermore, the metabolic rate cannot decline below that which is necessary to maintain the minimal level of bodily activity; therefore, in addition to maximal vasodilatation, additional heat must be lost from the body by the evaporation of water. Under such conditions, the rate of sweat secretion increases as the temperature rises, and concomitantly the progressive cutaneous vasodilatation readily permits transfer of heat to the surface of the body whence it can be removed by evaporation of the sweat.

There are certain physiologic drawbacks involved in the increased sudomotor activity that are coupled with maximal peripheral vasodilatation as a thermoregulatory system. First, a significant fraction of the cardiac output must be diverted from other regions of the body (eg, the skeletal muscles) solely to perfuse the dilated cutaneous vasculature. Second, sweat production and increased blood flow both require an increased metabolic energy production; hence these factors tend to elevate the overall body heat production slightly. Third, the increased temperature itself raises the tissue metabolic rate somewhat, so that this extra thermal energy is added to the heat load that must be dissipated from the body.

Adaptation to Severe Cold. If the ambient temperature falls considerably below the "neutral" range, the increased tissue insulation that results from peripheral vasoconstriction is insufficient to prevent excessive heat loss from the body. The body temperature can be maintained only by increasing metabolic heat production. This is accomplished either voluntarily by increasing muscular activity, or involuntarily by shivering. In an extremely cold environment, for example, the metabolic rate does not commence to rise until after shivering begins.

As in the case with heat loss under extremely hot environmental conditions, there are also certain physiologic disadvantages which are inherent within the adaptive mechanisms to an extremely cold environment. First, the rhythmic muscular contractions that are attendant upon vigorous shivering are quite tiring to the individual. Second, the tissue insulation is decreased significantly due to the increased blood flow to the skeletal musculature. Third, heat loss from the body by convection of surface air is enhanced by movements of the body. Consequently, heat loss to the environment is actually increased by shivering, so that as much as 30 percent of the heat which is produced by this activity is wasted insofar as increasing the overall body heat content is concerned.

With regard to heat loss from the body in an extremely cold environment, it must be borne in mind that the extremities are quite vulnerable to freezing, especially the ears, nose, fingers, and toes. As depicted in Figure 14.63, the reader will note that as the ambient temperature falls, the peripheral regions become progressively cooler, so that the deep-lying tissues at the body core remain at all times around 37 C. However, for practical purposes, the actual temperature as given by reading a thermometer does not necessarily give an accurate, or even approximate, idea of the freezing point of human tissue under different environmental conditions. For this reason, a table showing the chill factor has been included in this text, as it gives the temperature in relation to wind velocity insofar as actual cooling of the exposed skin is concerned (Table 14.20).

Convective heat loss is increased significantly by movement of the air surrounding the body. A person riding in an unenclosed vehicle can suffer severe frostbite on unprotected skin surfaces, even though no wind is blowing and the ambient temperature is actually above freezing. Naturally, if both the wind velocity is high and the movement of the individual with respect to the wind also is rapid, then these factors are additive, and accordingly the temperature is lower, as shown in Table 14.20. Consequently persons living in colder climates are vulnerable to superficial and/or deep freezing of their body tissues on exposed or underprotected extremities.

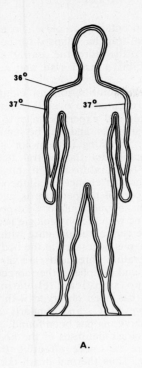

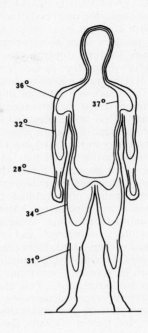

A.

B.

FIG. 14.63. Isotherms in the body. **A.** In a warm environment. **B.** In a cold environment. Note that the core body temperature is maintained at 37 C in a cold environment at the expense of the temperature of the extremities. That is, as the ambient temperature falls, the core "contracts" so that the temperature of the brain and other vital internal organs is maintained at 37 C by blood which is diverted from more superficial regions as shown in Figure 14.62. (After Ruch and Patton [eds]. *Physiology and Biophysics,* 19th ed, 1966, Saunders.)

Table 14.20[a]   **WIND CHILL INDEX**[b]

ACTUAL TEMPERATURE, °C[c]

WIND VELOCITY, km/hr	+2	−1	−4	−7	−9	−12	−15	−18	−21	−23	−26	−29	−32	−34
						APPARENT TEMPERATURE, °C[c]								
Calm	+2	−1	−4	−7	−9	−12	−15	−18	−21	−23	−26	−29	−32	−34
3	+1	−3	−6	−9	−11	−14	−17	−21	−24	−26	−29	−32	−35	−37
6	−6	−9	−13	−17	−19	−23	−26	−32	−33	−35	−38	−43	−47	−50
9	−9	−12	−17	−21	−24	−28	−32	−36	−40	−43	−46	−51	−54	−57
12	−11	−16	−20	−23	−27	−31	−36	−40	−43	−47	−51	−56	−60	−63
15	−14	−18	−22	−26	−31	−34	−38	−43	−47	−50	−55	−59	−64	−67
19	−15	−19	−24	−28	−31	−36	−41	−45	−49	−53	−57	−61	−66	−70
22	−16	−20	−25	−29	−33	−37	−42	−47	−51	−55	−58	−64	−68	−72
25[d]	−17	−20	−26	−30	−34	−38	−47	−48	−52	−56	−60	−66	−70	−74

[a] Data for this table were kindly provided by Mr. Joseph W. Barry, U.S. Climatological Service, National Oceanic and Atmospheric Administration, Denver, Colorado.

[b] The wind chill index is defined as the equivalent cooling effect on exposed skin when the bare skin is subjected to various wind velocities.

[c] The actual temperature is that measured by a thermometer. The apparent temperature is dependent upon the wind velocity, actual temperature, and relative humidity among other factors, and this is the actual temperature to which the body is exposed when out of doors. Note particularly the sharp drop in apparent temperature as the wind velocity increases due to the operation of the chill factor. For comparative purposes, the temperature of solid carbon dioxide ("dry ice") is −78.5°C.

[d] When the wind velocity exceeds 25 km/hr, little additional chilling effect takes place.

CONTROL OF THE EFFECTOR MECHANISMS IN-
VOLVED IN THERMOREGULATION. In
homeothermic animals, the deep body temperature
is maintained at a constant level by deviations in
the central temperature from some normal level;
these deviations in turn control the thermoregulatory effectors themselves in such a way as to
cause the central temperature itself to return to the
normal level. The "normal" value of this central
temperature often is called the reference temperature or set point. Thus thermoregulation in the
human body is accomplished by a negative feedback control system as shown in Figure 14.64. The
three components that are essential to such a
feedback system are present. First, there are
hypothalamic and cutaneous thermoreceptors that
sense the existing central, or controlled, temperature. Second, there are effector mechanisms that
can alter the central temperature, these effectors
being the vasomotor, sudomotor, and metabolic
systems discussed earlier. Third, there are integrative structures capable of making a comparison between the actual, or sensed, temperature and the
normal or reference temperature. This element of
the thermoregulatory system determines whether
the central temperature at any given time is too
high or too low, and then it activates the suitable
effector mechanisms in order to bring the central
temperature back to the normal level.

The thermoregulatory feedback system is
negative because a deviation of central temperature in one direction activates mechanisms that
cause the temperature to change in the opposite
(or negative) direction. Thus an increase in the
central temperature activates the physiologic
mechanisms that are responsible for heat loss
(vasodilatation, sweating), whereas a decrease in the
central temperature activates those mechanisms
which are responsible for heat conservation (vasoconstriction) and heat production (shivering).

The reference temperature (set point) should
be considered as a transition point for activation of
the mechanisms that increase or decrease body
temperature, rather than as a specific temperature
in the hypothalamus or elsewhere in the body.

The thermoregulatory system of the body is
often likened to the thermostatic control of a furnace within a building, although this analogy is
incorrect in three respects. First, the body has a
number of effector mechanisms, rather than one,
ie, turning on the furnace. Second, the body can
integrate the input signals depending upon the
degree of environmental stress, and thus grade the
effector response. A thermostat usually is an
on–off mechanism and accordingly, the response
of the furnace is all-or-nothing. Third, the thermoregulatory effectors are affected by cutaneous
temperature as well as by hypothalamic temperature, in contrast to the thermostat which is set to
respond to one temperature only.

Cutaneous thermoreception also plays a role
in the regulation of the effectors that control body
temperature. In humans, shivering can occur immediately upon entering a cold environment, and
before sufficient time has elapsed for central cooling to occur. Furthermore, the central temperature
even can exhibit a brief rise under such circumstances.

The relative contributions of cutaneous and
hypothalamic receptors to overall effector control
have been evaluated experimentally to some extent in human subjects, and may be summarized in
general terms. Sweating (sudomotor activity) is
regulated primarily by the central temperature,
and when this temperature is high shivering is inhibited. A far greater change in cutaneous temperature than in central temperature is necessary to
evoke a given thermoregulatory response. This
fact is in keeping with the physiologic necessity for
maintaining the central temperature within far narrower limits than is necessary at the body surface.
Distinct temperature thresholds are also present
for the central and peripheral thermoreceptor systems. Thus no significant elevation in the
metabolic rate has been observed with skin temperatures above 31 C, and similarly a distinct
threshold for shivering has been found.

The physiologic defenses of the body against
overheating are far more effective than those
against cooling. Thus thermal death occurs when
the body temperature exceeds 45 C, but a
hypothermia where the central temperature falls
below 24 C, is not necessarily fatal, provided that
suitable therapeutic measures are taken. For
example, the normal thermoregulatory system is
completely inactive at extremely low temperatures; hence the body will not rewarm spontaneously unless external heat is provided under such
conditions.

ABNORMALITIES OF THERMOREGULATION

Survival Limits. The core body temperature can
fluctuate around 2 C above the normal level of 37.5
C without serious derangement of body functions
(Fig. 14.65). However, as the temperature rises
above 39.5 C, a condition known as hyperthermia
exists, and the function of the central nervous system becomes progressively impaired; therefore
around 41 C to 42 C serious convulsions occur,
and if the rectal temperature remains above 41 C
for extended intervals some permanent brain damage takes place. At around 43 C, heat stroke follows, and frequently death occurs. If the temperature increases to between 44 C and 45 C, a rapid,
generalized, and irreversible denaturation of proteins (including enzymes) takes place throughout
the body, and death intervenes quickly and inevitably. Since the normal body temperature lies only
a few degrees below the upper limit for survival,
extreme hyperthermia presents a critical medical
emergency that requires immediate and drastic
treatment which is aimed toward the reduction of
body temperature.

If, on the other hand, the core body temperature falls significantly below 37.5 C, then a state of

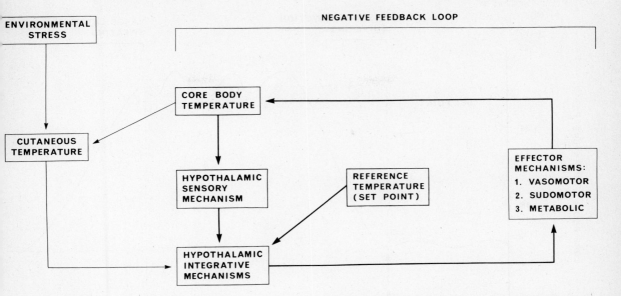

FIG. 14.64. The physiologic mechanisms which underlie thermoregulation. The temperature which is perceived by the hypothalamus as well as the cutaneous temperature both are compared to the reference temperature (or set point), and then effector mechanisms are stimulated or inhibited depending upon the extent and direction to which the ambient temperature is altered. Note also that the cutaneous temperature does not form a part of the negative feedback loop per se. (After Ruch and Patton [eds]. *Physiology and Biophysics,* 19th ed, 1966, Saunders.)

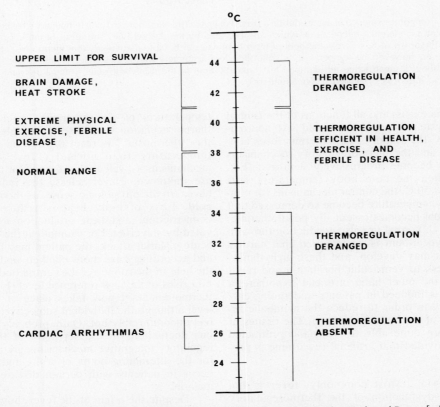

FIG. 14.65. Effects of marked alterations in body temperature. Cf. Figure 14.61. (After Ruch and Patton [eds]. *Physiology and Biophysics,* 19th ed, 1966, Saunders.)

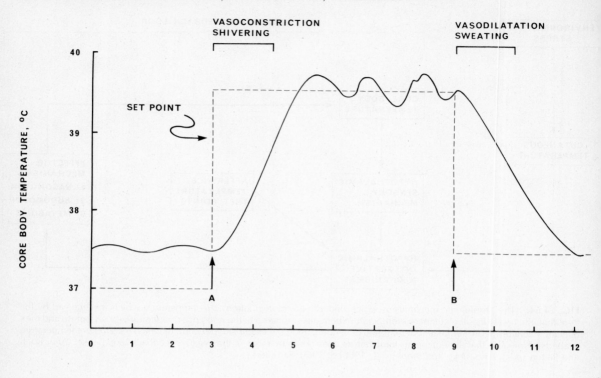

FIG. 14.66. Temporal sequence of events during a period of fever. The alterations in body temperature lag behind changes in the set point, or reference temperature, which are indicated by the dashed line. The onset of the episode of fever is indicated by the arrow at **A**, whereas the bout of fever terminates at **B**. Note particularly that thermoregulation is present during the fever, however the temperature is controlled less accurately during this state. Therefore the oscillations of temperature about a mean value are damped less critically than in the nonfebrile state. (After Ruch and Patton [eds]. *Physiology and Biophysics,* 19th ed, 1966, Saunders.)

hypothermia exists, and all functions of the central nervous system become depressed. At approximately 33 C the victim of hypothermia loses consciousness and the thermoregulatory mechanisms themselves become progressively deranged (Fig. 14.65). When the core body temperature falls much below 30 C, the central mechanisms for control of body temperature become so depressed that the individual becomes essentially poikilothermic. Around 28 C, normal impulse conduction throughout the myocardium is so impaired that cardiac arrhythmias may develop, and these arrhythmias may progress to ventricular fibrillation and rapid death. On the other hand, artificial hypothermia sometimes is induced in patients undergoing cardiac surgery in order to reduce the metabolic requirements of the body for oxygen. The tissues of the body are quite able to withstand protracted cooling, provided that artificial rewarming is provided.

Fever (Pyrexia). Most commonly, fever is not caused by dysfunction of the thermoregulatory system. In fact, this system operates normally, but it behaves as though the set point (or reference temperature) of the integrative mechanisms for thermoregulation were reset at a higher level; ie, the "thermostat" is reset to respond to a higher temperature than normal (Fig. 14.66). The mechanisms as well as the reasons for this process are unknown. Nevertheless, the individual behaves, in the physiologic sense, as though he were cold. Loss of body heat is reduced by vasoconstriction, and heat production may be elevated by shivering. For example at the onset of an acute malarial attack, the patient may feel chilled, and accordingly use more blankets and/or turn up the heat in the room. As the central body temperature rises to the new reference level (Fig. 14.66A), thermoregulation now takes place at this higher level although the individual subjectively now may feel uncomfortably hot. Quite probably, spontaneous fluctuations in the set point of the thermoregulatory integrative mechanisms are responsible for the alternating bouts of chills and fever that occur in patients with certain diseases, such as malaria.

Despite the origin of the fever, however, once the factor that caused this physiologic adaptation is removed, then the set point returns to the normal level once again (Fig. 14.65B) and physiologically the individual responds as though he were too warm. Thus sweating and peripheral vasodilatation increase heat loss, and blankets may be discarded. The body temperature returns to normal, thereby terminating the febrile episode.

Etiology of Fever. A primary neurologic disorder may be responsible for producing fever in a patient, although by far the most common cause is the presence in the bloodstream of compounds or toxins known by the nonspecific and generic name of pyrogens. At times bacterial toxins may liberate endogenous pyrogen through their action on the granulocytes of the blood, and this protein in turn acts directly upon the hypothalamic thermoregulatory centers to increase the set point as discussed above. The mechanism underlying this effect is completely unknown.

Although artificial hyperthermia has been employed clinically in the treatment of certain bacterial infections before the advent of antibiotics (eg, in the treatment of syphilis), no conclusive benefits have been demonstrated for increasing body temperature artificially. In fact, fever is a distinct handicap to recovery of the patient, so that efforts are generally directed toward reducing the severity of an existing pyrexia. In extreme cases of hyperthermia, the entire patient must be cooled physically, such as by immersion in a cold bath or by the use of ice packs. On the other hand, if a more moderate fever is present the use of antipyretic drugs is the method of choice, aspirin (acetylsalicylic acid) being used most frequently. Interestingly, the magnitude of the antipyretic effect which is produced by aspirin is proportional to the magnitude of the fever present. The higher the body temperature, the greater the fall in temperature following administration of a given quantity of aspirin. Aspirin thus has no effect upon the body temperature of a normal person, and this compound appears to block the effect of pyrogens upon the hypothalamic thermoregulatory centers, although the specific mechanism of action of aspirin is unknown.

CHAPTER 15

Endocrine and Reproductive Physiology

The two major systems of the human body that ultimately coordinate, regulate, and integrate all of its innumerable functions are the nervous system and the endocrine system. The information transmitted by the nervous system in the form of electric impulses is conducted rapidly in neurons, whereas transmission of information by the endocrine system does not involve any such specific conduits. Rather, the signals that originate within the various ductless glands comprising the endocrine system are communicated at a much slower rate by means of a number of chemical substances, or hormones, which are secreted into, and carried by, the bloodstream from their point of origin to other regions, the so-called target organs or tissues, where their messages are interpreted and acted upon. Briefly, then, coordination of the many functional activities of the body may be considered to reside in electric (neural) and chemical (endocrine) regulatory mechanisms. In many instances, there is a complex and interdependent relationship between these two systems within the body, not only from a developmental, but also from a functional standpoint. Indeed, this interrelationship is so close in some instances as to suggest that the endocrine system acts as a functional extension of the nervous system (eg, sympathetic neural reactions), although in other situations the endocrine system appears to function quite independently of any neural influences.

Since individual hormones very rarely act independently of one another, all of the endocrine glands of the body are considered to comprise an elaborate system that regulates the rates of growth, development, and function of certain bodily tissues. Underlying these processes, the rates of many specific metabolic reactions within the body in turn are regulated by various hormones. An endocrine substance, or hormone, can be defined as a specific organic chemical produced within, and secreted by, certain cells directly into the extracellular body fluids for transmission to a target organ and/or tissues, where it exerts one or more specific physiologic effects.

For convenience, endocrine substances may be divided into two broad groups, local and general hormones. Acetylcholine, liberated at synaptic junctions as well as at parasympathetic and somatic nerve endings in skeletal muscles, is an excellent example of a local hormone. This compound is synthesized, utilized, and inactivated in a circumscribed region. The important known local hormones are summarized, together with their general physiologic properties, in Table 15.1. Certain of these substances will be discussed later in this chapter whereas others have been dealt with in context as appropriate. The hormones produced by the gastrointestinal tract have been included arbitrarily in the category of local hormones.

The bulk of the material to be discussed in this chapter will deal with general hormones. These substances, in contrast with local hormones, are synthesized and secreted directly into the blood that perfuses specific ductless endocrine glands. Thus, the general hormones are transported by the cardiovascular system to other regions of the body where they produce their characteristic effects.

The endocrine glands, together with the hormones that they elaborate, their target organs, and general functions are summarized in Table 15.2. Collectively, these glands comprise the major portion of the endocrine system of the human body.

Although hormones perform a wide variety of functions, four general statements may be presented here concerning their mechanism of action.

First, hormones always act by exerting a regulatory effect, and no hormone appears to initiate metabolic reactions. In other words, the biochemical machinery (ie, the enzymes) for carrying out specific chemical reactions becomes an inherent property of the cell when it differentiates during fetal life. The need for close regulation of these intracellular reactions via turning enzymes "on"

Table 15.1 LOCAL HORMONES[a]

Acetylcholine[b]
Cholecystokinin
Erythropoietin
Epinephrine[b]
Norepinephrine[b]
Gastrin
Secretin
Renin
Histamine
Serotonin
Hypothalamic-releasing factors:
  Thyrotropin-releasing factor (TRF)
  Luteotropic hormone-releasing factor (LHRF)
  Follicle-stimulating hormone-releasing factor
    (FSHRF)
  Adrenocorticotropin-releasing factor (ARF)
  Somatotropin-releasing factor (SRF)
Kinins
Gamma-aminobutyric acid (GABA)[b]
Prostaglandins

[a]*Local hormones can be defined as chemical regulatory substances which are not produced by a specific glandular organ, but which may exert their action either locally or at a distance from the cell or tissue in which it is produced.*

[b]*These compounds also are referred to as neurohormones or neurohumoral agents.*

and "off" is critical to life, as the simultaneous and uncontrolled expression of the full enzymatic activities of all of the 3,000 different enzymes that are estimated to be present in every cell would lead to utter biologic chaos. Hormones may modify the rates of certain, presumably critical, intracellular biochemical reactions, but they do not initiate these reactions de novo, as cellular metabolic reactions can proceed autonomously in the absence of hormonal influences. Such specific rate-limiting reactions have not been identified in most instances as yet; however, it is clear that hormones do regulate the rates of a number of both general as well as specific metabolic reactions in various cells of the body. Specific illustrative examples to support this statement will be given in context.

Second, no hormone is secreted at an exactly uniform rate. For example, certain hormones are secreted according to a diurnal rhythm, eg, the adrenal hormones. Other endocrine substances, such as the ovarian hormones, are secreted in longer but precisely timed, individual cycles that cover a time span of several weeks. The secretion rates of still other hormones may depend upon internal or external environmental factors. Insulin secretion by the pancreas is dependent upon dietary carbohydrate intake, whereas thyroid hormone release is intimately associated with the environmental temperature.

Third, all hormones exert their influences at extremely low (trace or biocatalytic) concentrations; they add neither appreciable mass nor energy to the body.

Fourth, hormones are biologic catalysts, and as such these substances remain quantitatively as well as qualitatively unaltered by the reaction(s) in which they participate in carrying out their specific function (or functions). Nevertheless, hormones are being lost from the body continuously through excretion or metabolic inactivation. For example, the effect of insulin upon the regulation of blood glucose level (ie, its endocrine function) has nothing to do with the eventual degradation and inactivation of this hormone by enzymes such as proteases or disulfide reductases. Compensation for this hormonal loss must also be continuous, so that a certain minimal (or basal) quantity of hormone must be produced at all times in order that a specific deficiency of the hormone does not occur. Conversely, the clinical effects of an excess of certain hormones may reflect a defective rate of metabolic inactivation and/or excretion of the hormone itself rather than excess production by the gland or tissue involved in its synthesis.

The rate of synthesis of various hormones appears to be regulated largely by specific negative feedback mechanisms. Each endocrine gland has a tendency to secrete excessive quantities of its own specific endocrine substance(s); however, once the normal physiologic response to the secretion has been obtained, this information is fed back to the gland, and this decreases or inhibits further secretion. Conversely, if secretion is subnormal, the physiologic effects of the hormone are reduced; therefore, any inhibition of the gland that was produced by negative feedback also is diminished, and the gland now secretes the hormone at an increased rate, thereby restoring homeostatic equilibrium once again. The rate of secretion of each hormone is exquisitely balanced against the demand for that hormone in accordance with specific tissue requirements.

The general hormones belong to several fundamental classes of chemical compounds; therefore, proteins, peptides, amino acid derivatives, and steroids are all represented by specific endocrine substances. The chemical characteristics of specific hormones as well as the individual negative feedback mechanisms that regulate their secretion will be presented in context.

How do hormones exert their myriad effects at the cellular level? An important general mechanism has been described whereby certain endocrine substances mediate a number of specific cellular responses. These responses include alterations in cell permeability, as well as stimulation of the rate of degradation and synthesis of specific intracellular chemical compounds, including enzymes and other proteins (Table 15.3). The specific effects produced by a given hormone, depend, of course, upon the target cell type involved as well as on the inherent biochemical characteristics of the target cells themselves.

The effects of certain hormones appear to be mediated at the cellular level via the intracellular production of a practically ubiquitous nucleotide known as cyclic 3', 5'-adenosine monophosphate

Table 15.2 GENERAL ENDOCRINE GLANDS AND THEIR HORMONES[a]

GLAND[b]	HORMONE (SYNONYMS)	TARGET ORGAN OR TISSUE	PRINCIPAL FUNCTIONS	PRESUMED AMP-MEDIATED RESPONSE
Pituitary Gland (Hypophysis Cerebri) Adenohypophysis	Somatotropin (growth hormone, somatotropic hormone)	General	Accelerates rate of body growth, particularly growth of bone and muscle. Exerts anabolic effect upon calcium, phosphorus, and nitrogen metabolism. Affects metabolism of carbohydrate and lipid. Elevates skeletal and cardiac muscle glycogen content.	?
	Thyrotropin (thyrotropic hormone, thyroid stimulating hormone)	Thyroid gland	Synthesis and secretion of thyroid hormones	Yes
	Corticotropin (adrenocorticotropin, adrenocorticotropic hormone)	Adrenal cortex	Synthesis and secretion of adrenal cortical steroids	Yes
	Follicle-stimulating hormone	Ovary, testis	Stimulates growth of ovarian follicle in female, spermatogenesis in male.	?
	Luteinizing hormone (interstitial cell stimulating hormone)	Ovary, testis	Stimulates development of corpus luteum following ovulation and progesterone synthesis therein. In male, stimulates development of interstitial tissue of testis and secretion of androgen.	Yes
	Prolactin (luteotropic hormone, luteotropin lactogenic hormone, mammotropin)	Mammary gland	Proliferation of tissue, initiation of milk secretion.	?
Pars Intermedia	α-Melanocyte-stimulating hormone and β-melanocyte-stimulating hormone (intermedin)	Expand melanophores in lower vertebrates	Insignificant in humans.	Yes
Neurohypophysis	Antidiuretic hormone (vasopressin)	Renal tubules	Facilitates water absorption.	Yes
		Arterioles	Produces vasoconstriction, hence exerts a pressor effect.	Yes
	Oxytocin	Smooth muscle, especially of uterus	Contraction, parturition.	?

Source	Hormone	Target tissue	Physiologic action	
Thyroid Gland	Thyroxine (T₄) Triiodothyronine (T₃)	General	T_3 and T_4 accelerate the metabolic rate and oxygen consumption of all bodily tissues.	Yes
	Thyrocalcitonin (same as parathyroid calcitonin)	Skeleton	Metabolism of calcium and phosphorus.	?
Parathyroid Gland	Parathormone	Skeleton, kidney, gastro-intestinal tract	Metabolism of calcium and phosphorus	Yes
Endocrine Pancreas	Calcitonin	Skeleton	Metabolism of calcium and phosphorus	?
	Insulin	General	Regulates carbohydrate metabolism, stimulates protein synthesis.	Yes
	Glucagon, pancreatic	Liver	Stimulates hepatic gluconeogenesis, glucogenolysis	Yes
Adrenal Cortex	Adrenal cortical steroids (e.g., cortisol)	Adipose tissue	Stimulates lipogenesis	Yes
		General	Metabolism of carbohydrate	Yes
	Aldosterone	Renal tubule	Metabolism of electrolytes and water	?
Adrenal Medulla	Epinephrine	Heart muscle, smooth muscle, arterioles	Accelerates heart rate; causes arteriolar vasoconstriction, hence pressor response; stimulates contraction of most smooth muscle.	Yes
		Liver, skeletal muscle	Stimulates glycogenolysis	
		Adipose tissue	Stimulates lipolysis	
	Norepinephrine	Arterioles	Causes arteriolar vasoconstriction, hence pressor response.	Yes
Testis	Testosterone	Accessory sex organs	Stimulates normal growth, development, and functions.	?
		General	Stimulates maturation of secondary sex characteristics.	?
Ovary[c]	Estrone, estradiol	Accessory sex organs	Stimulate normal growth, development, and cyclic functions.	Yes
		Mammary glands	Development of system of ducts	
		General	Stimulate maturation of secondary sex characteristics.	
	Progesterone (from corpus luteum)	Uterus	Prepares endometrium for implantation of fertilized ovum.	?
		Mammary glands	Development of alveolar system	

[a] Data from White, Handler, and Smith. Principles of Biochemistry, 4th ed, 1968, The Blakiston Division, McGraw-Hill; Sutherland. Science 177:401, 1972; Pastan. Sci Am 227:97, 1972.

[b] The topics listed in this column follow the same sequence as the major divisions of Chapter 15.

[c] The estrogenic and other hormones that are elaborated by the placenta are discussed in the text.

Table 15.3 SOME EFFECTS OF CYCLIC AMP[a]

PROCESS OR ENZYME	TISSUE(S)	CHANGE IN ACTIVITY OR RATE OF PROCESS
Protein kinase[b]	Several	Increased
Phosphorylase	Several	Increased
Glycogen synthetase	Several	Decreased
Lipolysis	Adipose tissue	Increased
Amino acid uptake	Adipose tissue	Decreased
Amino acid uptake	Liver, uterus	Increased
Enzyme synthesis	Liver	Increased
Gluconeogenesis	Liver	Increased
Ketogenesis	Liver	Increased
Steroidogenesis	Several	Increased
Permeability to water	Epithelial cells	Increased
Permeability to ions	Epithelial cells	Increased
Calcium resorption	Bone	Increased
Renin production	Kidney	Increased
Discharge frequency	Purkinje cells of cerebellum	Decreased
Membrane potential	Smooth muscle	Increased
Tension	Smooth muscle	Decreased
Contractility	Cardiac muscle	Increased
HCl secretion	Gastric mucosa	Increased
Amylase release	Parotid gland	Increased
Insulin release	Endocrine pancreas	Increased
Thyroid hormone release	Thyroid gland	Increased
Calcitonin release	Thyroid gland	Increased
Hormone release	Adenohypophysis	Increased
Aggregation	Platelets	Decreased
Proliferation	Thymocytes	Increased

[a]Data from Sutherland. Science 177:401, 1972.

[b]An increase in the activity of protein kinase is the known mechanism whereby the effects of cyclic AMP are mediated in several systems (eg, the phosphorylase and glycogen synthetase systems). Quite possibly an enhanced protein kinase activation also is involved in most of the other effects of cyclic AMP.

(abbreviated cyclic AMP, Fig. 15.1) as shown in Figure 15.2. According to this concept, a hormone may act upon specific receptor sites for that hormone which are located within the membrane of the target cell, and the receptor specificity governs which particular hormones act upon a given cell. Following the union of intramembranous receptor and hormone, the enzyme adenyl cyclase is activated within the membrane, and the part of this enzyme that is oriented toward the cytoplasm catalyses the conversion of cytoplasmic adenosine triphosphate (ATP) into cyclic AMP (Fig. 15.1). The cyclic AMP in turn produces the distinctive cellular responses to the hormone, such as were outlined above, before it in turn becomes inactivated. The specific cellular events that are stimulated depend upon the inherent nature of the cell itself, as outlined in the first general concept relating to hormone action presented above. Thus, cyclic AMP causes adrenal cortical cells to elaborate adrenocortical hormones, renal tubular cells to alter their permeability to water in response to antidiuretic hormone, or thyroid cells to produce more thyroxine. The general hormones are summarized in Table 15.2 together with the information as to whether or not the cyclic AMP

mechanism described above and illustrated in Figure 15.2 has been implicated in their particular functions at the cellular level.

The cyclic AMP mechanism appears to be responsible for the regulation of various cellular functions when, and only when, the response is of an on–off or activation–inactivation nature. Furthermore, there is some experimental evidence that indicates that a second nucleotide known as guanosine 3', 5'-monophosphate (cyclic GMP) has an antagonistic (or reciprocal) role to that of cyclic AMP in the regulation of certain cellular functions including the responses to endocrine substances such as insulin, serotinin, oxytocin, and histamine.

Specific examples of the regulation of cellular functions by hormones that mediate changes in the intracellular concentrations of cyclic AMP and cyclic GMP will be presented in context; however, it is important to mention one more aspect of the effects produced by these compounds at the molecular level. Considerable experimental evidence is available which indicates that many native proteins, including enzymes, can undergo changes in their conformation (or shape) in response to external influences, including the presence of hormones that may alter the intracellular

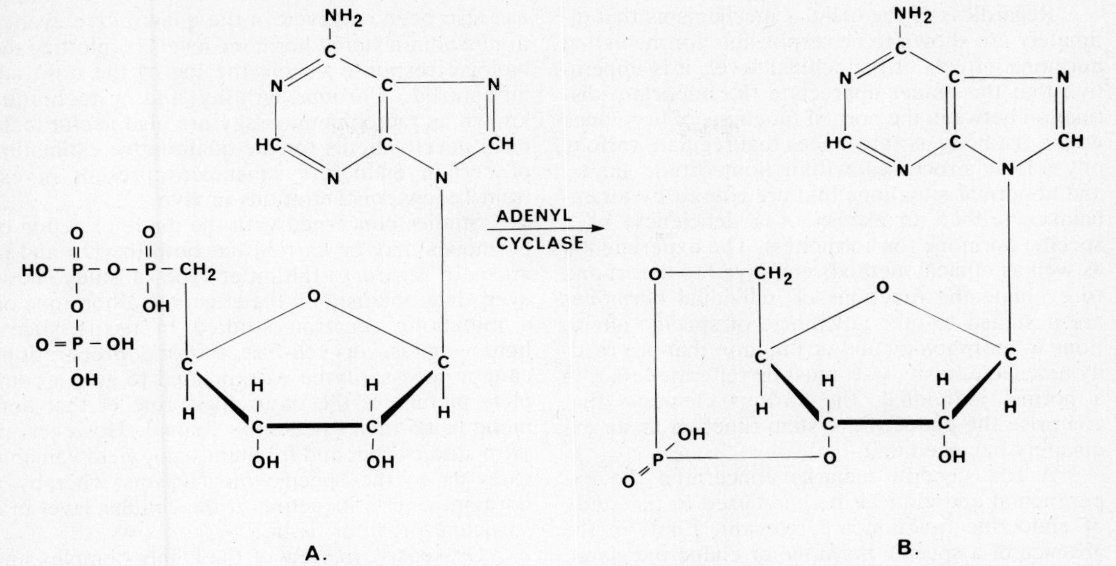

A.

B.

FIG. 15.1 Conversion of adenosine triphosphate **(A)** into cyclic AMP **(B)** takes place in the cell membrane under the influence of adenylate cyclase. (After Pastan. *Sci Am* 227:97; 1972.)

concentrations of cyclic AMP and GMP. In turn, small conformational changes in proteins are fundamental to the on–off regulation, hence control, of a host of physiologic processes at the cellular level.

Other mediators of the intracellular effects of hormones have been postulated. Included in this category is a chemically related group of miscellaneous compounds known as prostaglandins (p. 1173). In contrast to the cellular activation produced by cyclic AMP, prostaglandins often cause inhibition. However, there is also evidence that certain prostaglandins may exert their effects by altering the intracellular cyclic AMP level itself.

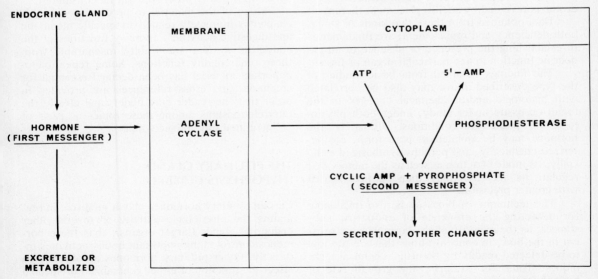

FIG. 15.2. The second messenger concept of hormone action. The hormone (first messenger) reacts with specific receptors in the cell membrane, thereby enhancing the activity of adenyl cyclase. This enzyme in turn enhances the conversion of ATP into cyclic AMP (Fig. 15.1). The latter compound in turn forms the second messenger which now accelerates specific cellular functions, eg, secretion, alterations in permeability, enzyme synthesis and other functions such as are summarized in Table 15.3. (After Sutherland. *Science* 177:401, 1972.)

Regardless of the cellular mechanisms that ultimately are shown to be responsible for mediating hormonal effects at the cellular level, it is imperative that the reader appreciate the important distinction between the normal functions of hormones within the body as substances that regulate various physiologic processes within homeostatic limits, and abnormal situations that are caused by an imbalance (either an excess or a deficiency) of a specific hormone (or hormones). The experimental as well as clinical methods employed to detect and to evaluate the functions of individual hormones are designed to take advantage of specific alterations in morphology and/or function that are readily amenable to study. It must be reiterated that, in a normal individual, the various elements that comprise the endocrine system function as an exquisitely balanced unit.

A few general remarks concerning the experimental and clinical methods used in the study of endocrine function are relevant. First, in the absence of a specific hormone or endocrine gland, a syndrome is produced that is specific for the loss of that substance or tissue. Such a condition may be observed clinically or else induced experimentally by surgical excision of individual endocrine organs. In this situation a deficiency of the hormone (or hormones) is present. Second, specific physiologic responses are observed upon the exogenous administration of the hormone, whether by injection of extracts prepared from the gland, by implantation of the glandular tissue itself, or by injection of the purified endocrine substance. In this situation an excess of the substance may produce an exaggeration of the characteristic effects of the hormone in subjects having a normal complement of the substance, or else a deficiency syndrome is reversed by such replacement therapy.

Data obtained from these two kinds of study, both deficiency and replacement, are fundamental to establishing the presence or absence of an endocrine function in any particular organ or tissue.

The findings obtained from basic studies of the type described above may also be correlated with histologic and/or chemical changes in the organ or tissue under study, under both physiologic and pathologic conditions. Ultimately, the hormone may be isolated in pure form, characterized chemically, and perhaps synthesized artificially. Its mode of action as well as the factors that regulate its secretion may then be investigated with greater precision.

The technique of bioassay is also invaluable for detecting the presence of endocrine substances, as the majority of the hormones are present in the body in concentrations that are too low to be detected readily by the usual chemical techniques. Thus, for example, the growth rate of hypophysectomized rats provides a useful index for assaying the potency of growth hormone preparations. Similarly, insulin preparations can be assayed by determining the fall in blood sugar level in depancreatized dogs. Some degree of success has also been achieved in the quantitative evaluation of administered hormone levels by plotting the biologic response versus the log of the dose administered. Chromatography and a technique known as radioimmunoassay are also useful technical developments for the quantitative estimation of certain endocrine substances present in extremely low concentrations in vivo.

Studies concerned with the mode of action of hormones may be carried out both in vivo and in vitro. In regard to the latter type of study, however, data obtained on the effects of a hormone on a metabolic reaction studied in tissue slices, homogenates, or cell-free enzyme preparations cannot necessarily be extrapolated to give a complete picture of the physiologic role of that hormone in an intact person or animal. However, in vitro studies can, and frequently do, yield valuable clues as to the specific mechanisms whereby a hormone exerts its action at the cellular level in a particular organ or tissue.

Obviously, in view of the highly complex and interrelated activities of many hormones in the body, whether these activities be synergistic or antagonistic, it is often quite difficult to determine the physiologic effects of a single hormone acting alone.

The remainder of this chapter is divided into nine major sections. Eight of these sections will present an overall survey of the endocrinology of those ductless glands that secrete the general hormones of the body as listed in Table 15.2. In the final section, a group of miscellaneous substances that are produced by various organs and tissues and which have endocrine functions of varying degrees of importance in the regulation of certain specialized bodily processes will be considered.

Much of the material to be discussed in this chapter is intimately related to specific metabolic and digestive activities, as many functions of the endocrine system are wholly inseparable from these other bodily functions. Some repetition of important material has been deemed essential for emphasis, and cross references are provided in order that the reader may understand clearly the interrelationships among these topics as parts of an integrated physiologic whole.

THE PITUITARY GLAND (HYPOPHYSIS CEREBRI)

Certain pituitary hormones play a vital role in regulating the physiologic activity of several other endocrine glands (target organs); thus these hormones control many important bodily activities indirectly. Other pituitary hormones act directly to affect the function of either generalized or specific tissues in the body. Therefore, an overall survey of the pituitary gland, its secretions, and their regulation will be presented at the outset in order that the reader may obtain a perspective on the functions of this all-important organ before a detailed consideration of the other endocrine glands and their hormones is undertaken.

Functional Morphology of the Pituitary Gland

GROSS ANATOMY. The hypophysis, located at the base of the brain, is approximately 1 cm long, 1 to 1.5 cm wide, and around 0.5 cm thick. In adult males, this structure weighs about 500 mg, and is slightly larger in adult females. The hypophysis rests in a deep depression in the sphenoid bone, the sella turcica, and is covered by a tough membrane, called the diaphragma sellae.

In actuality, the pituitary gland is composed of three distinct endocrine organs as shown in Figure 15.3 and Table 15.4. The two principal regions of the pituitary gland are the adenohypophysis (or anterior pituitary), and the neurohypophysis (or posterior pituitary). Located between these two regions is a narrow, almost avascular, zone known as the pars intermedia. The further subdivisions of these regions are indicated in the figure and the table.

The adult pituitary develops from two embryologic sources. The adenohypophysis is derived from Rathke's pouch, a dorsal outpocketing of the roof of the mouth; hence it develops from the pharyngeal epithelium. The neurohypophysis, on the other hand, originates from a downward-growing process that develops on the floor of the diencephalon; important neural connections with

Table 15.4 PRINCIPAL ANATOMIC DIVISIONS OF THE PITUITARY GLAND[a]

ADENOHYPOPHYSIS (GLANDULAR LOBE)
Pars intermedia
Pars distalis (anterior lobe)
Pars infundibularis (pars tuberalis)
NEUROHYPOPHYSIS
Infundibulum (neural stalk): median eminence + infundibular stalk
Lobus nervosus (neural lobe): infundibular stalk

[a]*Data from Bloom and Fawcett. A Textbook of Histology, 9th ed, 1968, Saunders.*
The hypophyseal stalk is composed of the median eminence, infundibular stalk, and pars infundibularis. The posterior lobe is composed of the infundibular process and the pars intermedia. See also Figure 15.3.

the hypothalamus are retained by the neurohypophysis in adults.

The pituitary is connected to the base of the brain by the slender infundibular stalk (Fig. 15.3), as well as by an elaborate vascular supply that is described below. These neural and vascular connections with the brain thus ensure the key role of the hypophysis in the integration of the nervous system with the endocrine system.

NEURAL CONNECTIONS BETWEEN PITUITARY AND HYPOTHALAMUS. Essentially all hormonal secretion by the pituitary gland is regulated by signals that are transmitted from the hypothalamus in one of two ways. First, secretion from the pos-

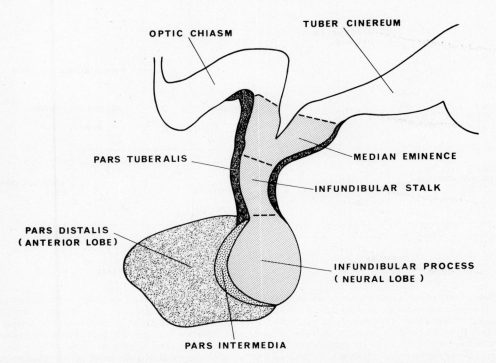

FIG. 15.3. Gross anatomy of the pituitary gland. See also Table 15.4. (After Bloom and Fawcett. *A Textbook of Histology,* 9th ed, 1968, Saunders.)

terior pituitary, or neurohypophysis, is regulated by nerve impulses originating in receptors that are located within the hypothalamus and which are transmitted along efferent effector neurons to the posterior pituitary to excite its secretion. Second, secretion by the adenohypophysis, is regulated by neurosecretory substances which are elaborated by the hypothalamus per se, and thence transmitted to the adenohypophysis via the hypothalamic–hypophyseal portal blood vessels as discussed below. The hypothalamus in turn receives information in the form of nerve impulses from practically all parts of the nervous system. For example, olfactory, psychic, pain, and osmotic stimuli all may exert excitatory or inhibitory effects upon the hypothalamus, and this collected sensory information then is integrated by the hypothalamus and relayed to the pituitary gland in order to control its secretion by one of the two mechanisms summarzied above, which are termed

collectively the neurohypophyseal relay systems (Fig. 15.4).

VASCULAR SUPPLY OF THE PITUITARY AND HYPOTHALAMUS. The blood supply to the anterior and posterior regions of the pituitary is unusual and of extreme physiologic importance in regulating the secretion of the gland (Fig. 15.5). From the internal carotid arteries, two inferior hypophyseal arteries originate. These vessels ramify extensively within the capsule of the gland and send branches to the neurohypophysis as well as to the sinusoids of the adenohypophysis. A number of superior hypophyseal arteries also arise from the internal carotids as well as from the posterior communicating artery of the circle of Willis. These vessels anastomose freely in the region of the median eminence of the hypothalamus, as well as in the base of the pituitary stalk. In turn, capillaries that comprise a network of vessels termed

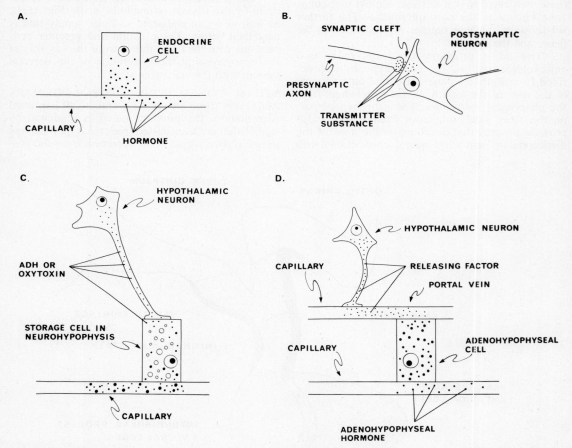

FIG. 15.4. Comparison of four endocrine secretory mechanisms. **A.** A typical endocrine cell (eg, of the adenohypophysis or adrenal cortex) secretes its product directly into the bloodstream. **B.** At a typical synapse, the presynaptic terminal releases a neurotransmitter substance (eg, acetylcholine or norepinephrine) that excites or inhibits the postsynaptic neuron or the cell. **C.** In the posterior pituitary, secretion of antidiuretic hormone (vasopressin) and oxytocin the hormones are elaborated by neurons, and then are transported through the axons of the same neurons to storage cells in the posterior pituitary. Thence the hormones ultimately are released into the bloodstream. **D.** Hypothalamic releasing factors pass from the neurons that secrete then into capillaries which transport these hormones through portal veins (Fig. 15.5) to cells of the adenohypophysis and thus stimulate secretion by these cells. (After Guillemin and Burgus. *Sci Am* 227:24, 1972.)

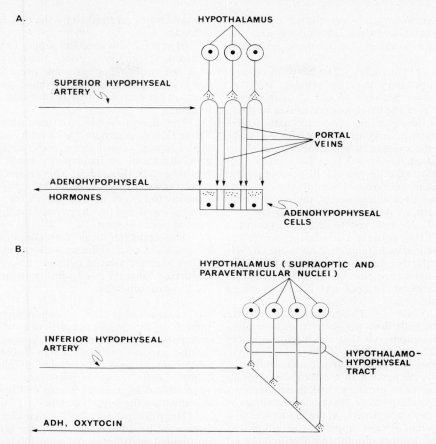

FIG. 15.5. **A.** Vascular connections between the hypothalamus and adenohypophysis. **B.** Neural connections between the hypothalamus and the neurohypophysis. (After Guillemin and Burgus. *Sci Am* 227:24, 1972.)

the primary plexus arise from the superior hypophyseal and communicating arteries. These capillaries of the primary plexus extend into the median eminence, then return to the surface where they become veins which pass downward around the hypophyseal stalk and supply the sinusoids of the adenohypophysis. Those venules that connect the capillaries in the median eminence of the hypothalamus with the sinusoidal capillaries of the anterior lobe form the hypothalamic–hypophyseal portal system.

Venous drainage of the pituitary gland takes place principally through vessels that run in the vascular layer of the capsule to the diaphragm of the sella turcica, and from there into the neighboring dural sinuses.

Thus, the hypophyseal blood supply is both abundant as well as admirably disposed for the transmission of neurosecretory materials that act as signals to regulate secretion of the adenohypophysis.

HISTOLOGY

Adenohypophysis. The pars distalis (or anterior lobe) is the largest division of the pituitary gland, as shown in Figure 15.3. This structure is comprised of irregular cords and clumps of cells that range between 20 and 180 μ in diameter, and that are found in close association with the thin-walled sinusoids of the vascular system. The pars distalis is enclosed in a dense collagenous connective tissue capsule; however, the stroma within the glandular tissue itself is not abundant.

The glandular cells themselves are classified on the basis of their affinity, or lack of affinity, for certain dyes. The two general classes of cells in the anterior pituitary are called chromophil and chromophobe. The chromophilic cells in turn are subdivided into acidophilic (or alpha) cells, whose specific granules stain strongly with acidic dyes, and into basophilic (or beta) cells, whose specific granules stain heavily with basic dyes.

Acidophils (Alpha Cells). The acidophilic cells of the anterior pituitary are rounded or ovoid, and in humans they range between 15 and 19 μ in diameter. The acidophils contain numerous refractile granules that can be visualized readily by light microscopy. Acidophils comprise approximately 40 percent of the total cell population of the anterior pituitary, and these elements are believed to secrete growth hormone (GH) and prolactin (or luteotropic hormone, LTH). There is good evidence, at least in certain species other than man, that two types of acidophils are present, and that each cell type produces one of these hormones specifically.

Basophils (Beta Cells). The basophil cells of the anterior pituitary may be angular, oval, or round, and range between 15 and 25 μ in diameter. The nuclei of the basophils are similar to those of acidophils; however, the chromatin may exhibit a coarser pattern, and in general the refractile granules in the cytoplasm are smaller than those found in acidophils.

Basophils make up around 10 percent of the total glandular cell population of the anterior pituitary, and two subclasses of these cells may also be present, as is the case with acidophils.

Basophils are believed to secrete thyrotropic hormone (TH), follicle-stimulating hormone (FSH), interstitial cell-stimulating hormone (ICSH, or synonymously, luteinizing hormone, LH). Perhaps basophils are also involved in the secretion of adrenocorticotropin (ACTH, also called corticotropin).

Chromophobes. The cytoplasm of the chromophobe cells does not stain in ordinary histologic preparations, so that this region appears clear when examined by light microscopy. Chromophobes are present at a concentration of approximately 50 percent of the total cell population in the anterior pituitary, and these elements have been postulated to be functionally neutral (or undifferentiated) precursors of the acidophils and basophils. Such a view appears too simplistic, however, as tumors composed exclusively of chromophobe cells in humans are frequently associated with excess secretion of one or more adenohypophyseal hormones.

Pars Intermedia.   In human embryos there is a distinct cleft that separates the adenohypophysis from the neurohypophysis. The neural (posterior) side of this cleft is lined by basophilic cells arranged in multiple layers, and these cells form the pars intermedia. The cleft still may be present in young children, but it is rarely seen in adult pituitary glands. If present, the basophilic cells lining the cleft make up a discrete pars intermedia; however, in the majority of adults, the pars intermedia is often represented by a region of cysts (Rathke's cysts), which generally are lined by ciliated epithelium.

If the cleft is absent, the basophilic epithelial cells of the pars intermedia blend with neurohypophyseal elements, hence this region of the pituitary loses its discrete laminar character. Within the pars intermedia the cells are polygonal, and, as noted above, are basophilic in their staining characteristics.

The only hormone known to be secreted by the pars intermedia is melanocyte-stimulating hormone (MSH). This substance actually is composed of two distinct polypeptides called α-MSH and β-MSH. Collectively these two polypeptides are sometimes referred to by the general term, intermedin.

In general, the vascular supply to the pars intermedia is relatively poor. Neural connections with the neurohypophysis are present, although these connections appear to have an inhibitory function on secretion that is of minor importance in man.

The pars tuberalis (or pars infundibularis, Table 15.4) is continuous with both the pars intermedia as well as with the adenohypophysis, and it generally forms an incomplete sheath around the pituitary stalk (Fig. 15.3). This region is the most highly vascular portion of the hypophysis, since it contains the major arterial supply for the anterior lobe as well as the hypothalamic–hypophyseal venous portal system.

The epithelial cells that comprise the pars tuberalis are acidophilic as well as basophilic; undifferentiated epithelial cells are also present. No endocrine function has been demonstrated as yet for the pars tuberalis.

Neurohypophysis.   The neurohypophysis is composed of the median eminence of the tuber cinereum, the infundibular stalk, and the infundibular process, as outlined in Table 15.4. The neurohypophysis is composed of an inherent population of cells known as pituicytes, as well as the terminations of neurons whose cell bodies are located within the hypothalamus.

The principal mass of the neurohypophysis is comprised predominantly of a large bundle of unmyelinated nerve fibers (roughly 100,000), which is known as the hypothalamo–hypophyseal tract (see Figs. 7.45 through 7.48). This tract has its origin principally in the supraoptic nucleus of the hypothalamus near the optic chiasma, as well as in the paraventricular nucleus in the wall of the third ventricle. Other regions of the hypothalamus also contribute small numbers of fibers to the hypothalamo–hypophyseal tract. All of the fibers from these regions converge upon the median eminence, and traverse the infundibular stalk into the infundibular process. In this region, the neurons terminate blindly and in very close association with the vessels comprising the abundant capillary plexus of the neural lobe. Hence, these neurons do not end directly upon either neural or other effector cells (Fig. 15.5D).

Scattered throughout the neurohypophysis are variably sized masses of deeply-staining material known as Herring bodies. These Herring bodies are localized accumulations of neurosecretory material within the axoplasm of nerve fibers of the hypothalamo–hypophyseal tract. The terminal ramifications of the hypothalamic axons within the neurohypophysis adjoin the basement layer that lines the perivascular spaces.

The endothelial cells that form the capillary walls within the neurohypophysis, similarly to those in the adenohypophysis (and other typical endocrine glands), are quite attenuated and provided with circular fenestrations covered by extremely thin diaphragms.

In addition to the dense neurosecretory granules that are present within the nerve terminals of the neurohypophysis, small (about 400 Å) agranular vescicles of unknown function are also found. These granules resemble those present within synaptic terminals in other regions of the central nervous sytem.

The pituicytes are of irregular size and shape, and are found among the nerve terminals of the neurohypophysis. Currently, the pituicytes are believed to function similarly to glial cells that act as supporting elements elsewhere in the nervous system (p. 246). No endocrine function is known for the pituicytes and it is not clear whether they function only as supporting elements, or whether they also have a role in the metabolism of the secretory elements themselves.

The neurohypophysis synthesizes, stores, and releases two important and closely related hormones, oxytocin and vasopressin (or antidiuretic hormone, ADH).

General Functions of the Pituitary Gland

Before undertaking a detailed consideration of the individual hormones that are elaborated by the pituitary gland, a general summary of the effects of total experimental extirpation of the hypophysis (hypophysectomy) will be presented and compared with the symptoms that develop clinically in total functional inactivation of this organ which occurs through disease processes (Simmonds' disease). In this way, a general perspective will be obtained on the manifold bodily processes which are affected by the pituitary gland. In both of these examples, the net effects upon the body are caused by the complete absence or a relative deficiency of the several hormones elaborated by the pituitary gland. The reader should also be aware that the age and physiologic state of the subject is of considerable importance in the development of certain of the symptoms of pituitary deficiency.

Additional manifestations of an excess and/or a deficiency of the pituitary hormones will be presented in context as each of these substances is discussed.

EFFECTS OF EXPERIMENTAL HYPOPHYSECTOMY. The several effects of total ablation of the pituitary are as follows:

First, cessation of growth occurs if the operation is performed upon a young animal. Carried out in an adult, hypophysectomy causes a loss of body tissue and a reversion of the hair to a finer, more juvenile type.

Second, the thyroid gland atrophies, and the attendant symptoms of hypothyroidism develop, including a reduced basal metabolic rate.

Third, the adrenal cortex atrophies and metabolic derangements that are attendant upon a diminished (but not abolished!) adrenal cortical function occur.

Fourth, hypophysectomy performed in juvenile subjects of either sex results in failure of the gonads to mature. In adults, such a procedure causes atrophy of the testes or ovaries, as well as of the accessory reproductive organs. In either situation, sterility ensues.

Fifth, if hypophysectomy is performed late in pregnancy (or postpartum), abortion takes place and/or lactation fails to occur.

Sixth, profound alterations in carbohydrate, fat, and protein metabolism are observed following hypophysectomy. These include: a hypersensitivity to insulin; a tendency toward hypoglycemia; a rapid loss of body glycogen stores during fasting; a decreased fat catabolism; an amelioration of diabetes mellitus, such as that induced by surgical pancreatectomy or by the injection of phlorhizin; and loss of nitrogen from the body, ie, a negative nitrogen balance develops.

Seventh, blanching of the pigment cells (chromatophores or melanophores) in the skin occurs when lower vertebrates (eg, amphibia or fish) undergo hypophysectomy. The light-adaptive changes in these cells also fail. That is, the dispersal mechanism for melanin within the chromatophores becomes inoperative so that normal skin darkening or bleaching in response to alterations in light intensity no longer occurs.

The reader should note that all of the changes induced by hypophysectomy reflect a deficiency in specific adenohypophyseal hormones. Damage to (or ablation of) the neurohypophysis, on the other hand, may result in a temporary or permanent polyuria or diabetes insipidus.

SIMMONDS' DISEASE. Pituitary insufficiency in human beings generally results from either a tumor or infarction of the blood vessels supplying this organ. The number and severity of the symptoms resulting from such pathologic situations once again reflect the relative lack of specific pituitary hormones, as well as the physiologic state of the individual at the time of onset of the disease. The possible symptoms (listed below) of clinical pituitary insufficiency are quite similar to those found after experimental hypophysectomy.

1. Dwarfism occurs if pituitary insufficiency occurs in juvenile individuals. Premature senility may be observed in adults.
2. The basal metabolic rate is low.
3. Failure of gonadal maturation occurs in young individuals. Testicular or ovarian atrophy occurs in sexually mature individuals and sterility ensues.
4. Stress, such as may be induced by surgical procedures, can induce shock quite readily.
5. There is a hypersensitivity to insulin and a tendency toward hypoglycemia.
6. The skin becomes pale and waxy in appearance.
7. An overt adrenal insufficiency may be present occasionally.

8. Cachexia (ie, a general state of malnutrition, emaciation, and debility) may develop, although taken alone cachexia is not necessarily symptomatic of pituitary insufficiency.

Hormones of the Pituitary Gland

It is evident from the material summarized in the previous section that the secretions of the pituitary gland regulate an extremely broad spectrum of complex physiologic processes. The pituitary gland synthesizes and secretes 10 chemically, and, in general, physiologically, distinct hormones; these are summarized in Table 15.2. The overall functions of the individual pituitary secretions are illustrated in Figure 15.6.

In the present section, the six adenohypophyseal hormones will be discussed first, followed by a consideration of the two melanocyte-stimulating hormones, and then by the two endocrine secretions of the neurohypophysis.

The reader should note that the sequence chosen for discussion of the general hormones in this section as well as the remainder of this chapter follows the outline presented in Table 15.2.

SOMATOTROPIN (GROWTH HORMONE [GH])
SOMATOTROPIC HORMONE [STH]

Chemistry. All of the hormones that have been isolated from the adenohypophysis thus far have proven to be small proteins or polypeptides. Human growth hormone contains about 200 amino acid residues, and it has a molecular weight of 21,500 daltons; two disulfide bridges are present in the molecule. A 50 percent reduction of the molecular weight (by hydrolysis) causes no appreciable loss in the physiologic activity of growth hormone, and recently this hormone has been synthesized artificially.

There are considerable chemical differences among the growth hormones obtained from the pituitary glands of various species of animals. For this reason, therapeutic use of growth hormone derived from bovine or other subprimate sources is useless in the treatment of growth hormone deficiencies in humans. Presumably, these foreign hormones act as antigens, as they stimulate the rapid formation of antigrowth hormone antibodies.

The basal growth hormone level in adult humans is below 3 ng per ml, and it is metabolized quite rapidly, presumably to a large extent by the liver. Since the half-life of circulating growth hormone is between 20 and 30 minutes, the total growth hormone output can be calculated to be around 4 mg per day in adults.

The circulating level of growth hormone is increased by a number of stimuli, as discussed below.

General Functions of Somatotropin. The phenomenon of growth is complex, and it is affected not only by growth hormone per se, but also by thyroxin, adrenal cortical hormones, androgens, insulin, and extrinsic influences (eg, the diet, not only with respect to nutritional quality, but also quantity), as well as inherent genetic factors. In view of the many interrelated factors that result in orderly growth and development of the individual, it is rather inaccurate to designate any single substance as "growth hormone." Hence the name "somatotropin" may be more appropriate than "growth hormone" although both of these words have been used interchangeably in the following discussion.

The long-term injection of somatotropin into young normal or hypophysectomized animals results in an acceleration of overall growth. This process includes the accumulation of both hard and soft tissues, and it has been shown that growth hormone affects all tissues of the body that are capable of exhibiting growth. Thus, the influence of somatotropin is not exerted upon any particular target organ; rather, this hormone exerts an effect upon practically all cells at one time or another throughout the life span of the individual. This effect is particularly evident, however, upon the rate of skeletal growth in young subjects.

It should be emphasized at this juncture that overall growth results from an orderly sequence of events, and not merely from an increase in the total body mass (or weight) alone. A net gain in the total quantity of protein, increases in the size of the entire body, and specific skeletal elements are additional criteria of growth.* In particular, growth hormone increases the cell size, ie, it induces hypertrophy of the cells, and it also stimulates mitosis, a process that results in an increased total number of cells, ie, hyperplasia.

General Metabolic Effects of Somatotropin. The administration of growth hormone has been shown to affect a widely diverse spectrum of integrated metabolic reactions, most of which participate in the overall process of growth.

Anabolic Effect on Protein Metabolism. Somatotropin stimulates the formation of ribonucleic acid so that protein synthesis is also stimulated. This effect takes place both in liver and peripheral tissues, and the net anabolic effect is reflected by an increased nitrogen retention (ie, a positive nitrogen balance). It has been postulated that growth hormone acts upon protein synthesis at the ribosomal level (p. 62). Somatotropin also increases the transport of both neutral as well as basic amino acids into the cells by a mechanism that probably is independent of the protein synthetic mechanism. This effect of somatotropin is, however, clearly independent of the mechanism whereby insulin facilitates the transport of amino acids into cells.

Carbohydrate Metabolism. Growth hormone exerts three major effects upon carbohydrate metabolism.

*Increased salt and water retention, as well as fat deposition, will induce weight gain without a concomitant overall growth of the individual.

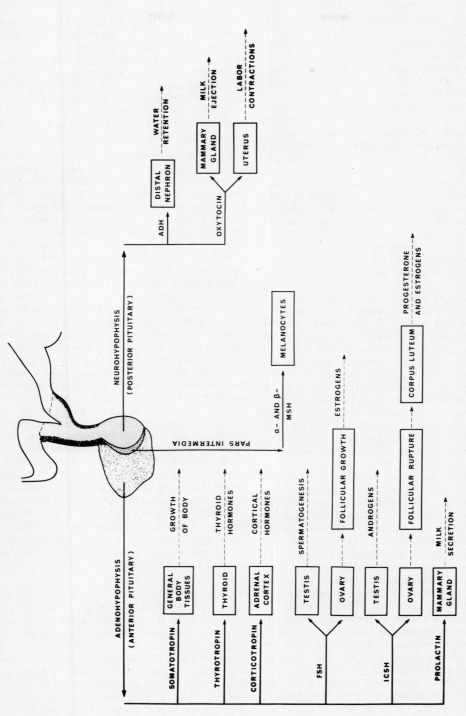

FIG. 15.6. Summary of the pituitary hormones. FSH = follicle stimulating hormone; ICSH = interstitial cell stimulating hormone; MSH = melanocyte stimulating hormone; ADH = antidiuretic hormone. See also Table 15.2.

First, the injection of somatotropin produces an elevated blood glucose level (hyperglycemia), which effect is preceded by an early hypoglycemia; this dual response is known as the pancreatotropic effect of growth hormone, since the early hypoglycemia is caused by the indirect release of excess insulin from the endocrine pancreas. Another pancreatotropic effect of somatotropin is to stimulate pancreatic release of glucagon, and this effect underlies the hyperglycemic action of growth hormone.

Second, prolonged administration of somatotropin results in a glucosuria, and the hormone may also intensify the severity of clinical diabetes in the human subject. This glucosuric effect of growth hormone is known as the diabetogenic effect of growth hormone.*

Thus, growth hormone decreases (or antagonizes) the sensitivity of the individual to insulin. Hypophysectomy ameliorates an established diabetes (Houssay preparation), and increases the sensitivity to insulin.

Briefly, growth hormone increases hepatic glucose output, resulting in a hyperglycemia, while at the same time it exerts an anti-insulin effect in skeletal muscle.

Third, and related to the effects discussed above, somatotropin also exerts a glycostatic effect in cardiac and skeletal muscle; that is, growth hormone increases the formation of glycogen in these tissues. Therefore, somatotropin tends to build up the carbohydrate reserves in certain tissues, and thus "spare" glucose, as it decreases the rate of utilization of this substrate by the body at the expense of the fat reserves as described below.

Both insulin and somatotropin, as well as exogenous carbohydrate, are all essential for normal growth. Quantitatively, however, the effect of growth hormone on protein synthesis (anabolism) is far more important in the growth and maintenance processes than is its overall role in carbohydrate metabolism.

Lipid Metabolism. Continued injection of somatotropin causes the accelerated mobilization of free fatty acids into the plasma (lipid-mobilization effect) from the fats that are stored in the body depots (eg, in the subcutaneous regions), and in turn this response may result in an increased concentration of liver lipids, as well as increased blood levels of ketone bodies (a ketonemia). A ketonuria ultimately develops. This response to growth hormone is known as the ketogenic effect. The release of nonesterified (or free) fatty acids that are stored in adipose tissue results from the direct action of somatotropin on the hydrolysis of fats within the adipose cells of the fat depots.

Since growth hormone stimulates the mobilization as well as the utilization of body fats, this hormone enables the body to use fat as an energy source preferentially over carbohydrate.

Approximately 25 percent of all patients having growth hormone secreting tumors of the adenohypophysis exhibit diabetes.

However, only when abnormally high levels of somatotropin are present are the fatty acids utilized to such an extent that an outright ketosis results.

Synthesis of Hard Tissues. Growth hormone stimulates both the formation of cartilage (chondrogenesis) as well as bone (osteogenesis). Synthesis of the bone matrix, including production of chondroitin sulfate and increased calcification, are also affected by growth hormone. Somatotropin apparently affects the formation of skeletal tissues indirectly by maintaining a serum component known as sulfation factor, which factor in turn activates chondroitin sulfate and collagen synthesis.

Miscellaneous Effects of Growth Hormone. In addition to the major influences of somatotropin upon protein, carbohydrate, and fat metabolism that were described above, this substance also exerts an influence upon several other bodily processes. First, somatotropin can be shown to stimulate the formation of reticulocytes, thus exerting an erythropoietic effect. Second, growth hormone produces an increase in both the size and the function of the kidneys. This renotropic effect is manifest by an elevated renal clearance and tubular excretion of certain substances.

In conclusion, somatotropin exerts an effect upon a wide variety of metabolic processes, but it is not clear at present just how this single hormone exerts a regulatory influence upon so many diverse tissues and types of bodily process. Three possibilities may be advanced to explain the mechanism of action of growth hormone. First, somatotropin may regulate, by a specific action, a single rate-limiting key metabolic process that is wholly unknown, but that is shared in common by practically all cells of the body at one time or another during their lifespan (cyclic AMP and/or cyclic GMP levels?). Second, somatotropin may influence cell permeability directly, thereby directly increasing the intracellular availability of various materials and substrates. Third, somatotropin may act upon a vast array of specific enzyme systems. Although it is intriguing to consider these ideas, speculation as to which one (or combination) is most nearly correct is futile in the absence of more factual evidence.

Hypothalamic Control of Anterior Pituitary Secretion. The hypothalamus plays an important regulatory role in the secretion of all of the anterior pituitary hormones, as indicated earlier in this chapter. Thus, the secretion of not only somatotropin, but also thyrotropin, corticotropin, follicle-stimulating hormone, and luteinizing hormone are all governed to a considerable extent by five chemically distinct polypeptide releasing factors produced in the hypothalamus and thence transmitted via the hypophyseal portal vessels to the adenohypophysis (Fig. 15.4D). In all of these instances, the specific hypothalamic releasing factors stimulate secretion of these five pituitary hormones and, since they share a common mechanism, may all be considered together at this point.

All of the releasing factors are small polypeptides.* Thus, any nerve fibers that pass directly from the brain to the anterior pituitary gland do not control the function of this organ. Rather, the releasing factors produced in the nerve endings within the median eminence of the hypothalamus (Fig. 15.5D), and which are secreted into the numerous closely adjacent capillary loops from which the portal vessels originate, are responsible for regulating the individual pituitary secretions.

The five stimulatory hormones elaborated by the hypothalamus are as follows: somatotropin- (or growth hormone-) releasing factor (SRF, GHRF); thyrotropin-releasing factor (TRF); corticotropin-releasing factor (CRF); follicle-stimulating hormone-releasing factor (FSHRF); and luteinizing hormone-releasing factor (LHRF).

These factors provide examples of a type of nervous control known as neurosecretion, an effect that may be defined as the secretion of a chemical substance from nerve endings directly into the bloodstream (or body fluids) so that the substance is transmitted to effector cells located at a distance from the site of origin. Neurosecretion thus differs somewhat from neurotransmission at synaptic clefts (Fig. 15.4B) as well as in situations where nerve endings directly stimulate glandular cells (such as in the adrenal medulla or posterior pituitary), resulting in the release of a stored secretion (Fig. 15.4C).

Apparently, the several releasing factors are produced in various localized regions within the median eminence of the hypothalamus. For example, TRF is secreted in the anterior portion, CRF in the middle region, and the gonadotropin-releasing factors are elaborated in the posterior region of the median eminence. The cell bodies of the neurons that produce the releasing factors have not as yet been localized, although the arcuate nucleus appears to be involved in the secretion of these substances. Doubtless, the cell bodies of the neurons responsible for secreting the various releasing factors are located in a number of regions, and are also able to respond to various stimuli.

Experimental evidence has shown that control of secretion of the releasing factors themselves is regulated by negative feedback mechanisms. In the case of the anterior pituitary hormones, when an elevated circulating level of any of these endocrine substances is present, the hypothalamus is inhibited so that the secretion of the releasing factor controlling the pituitary secretion rate of that particular hormone is also decreased. On the other hand, the pituitary gonadotropins (ie, ICSH and LH) elevate the blood level of gonadal steroids, and the concentration of these gonadal compounds in the circulation secondarily inhibits the secretion

of the specific releasing factor responsible for stimulating the secretion of that particular gonadotropin.

Specific Factors That Affect Somatotropin Release by the Anterior Pituitary. As noted in the preceeding discussion, the secretion of somatotropin by the anterior pituitary is controlled partly by the hypothalamus. In addition to the extraction of a somatotropin-releasing factor (SRF) from hypothalamic tissue, lesions of the hypothalamus or injury to the pituitary stalk both inhibit the secretion of growth hormone. Not only is growth per se inhibited under such conditions, but also the increased blood levels of growth hormone that are induced by the attendant hypoglycemia are blocked.

In adults as well as children, the rate at which somatotropin is secreted undergoes both rapid and profound fluctuations in response to a number of stimuli. The plasma concentration of growth hormone in neonatal infants is somewhat greater than in older persons; however, between older children and adults there are no demonstrably significant differences in the circulating level of this hormone. As noted earlier, the somatotropin concentration in the plasma ranges between 0 and 3 ng/ml in normal individuals, and adult pituitaries contain high concentrations of growth hormone.

The various stimuli that have been shown to increase growth hormone secretion lie in three broad groups. First, situations in which there is a real or possible deficiency in substrates required for the production of energy, such as fasting, exercise, or hypoglycemia, all stimulate somatotropin release. Second, conditions that elevate the plasma levels of certain amino acids also increase growth hormone secretion. For example, the ingestion of a high-protein meal or the intravenous infusion of certain amino acids, such as arginine, both produce such a response. Third, situations in which physiologic stress is present, or there are pyrogenic substances in the body, also augment somatotropin output by the adenohypophysis.

Furthermore, somatotropin levels may fluctuate spontaneously and markedly throughout the day without any apparent relationship to any definite stimulus that is known to affect the secretion of this hormone. Interestingly, the plasma somatotropin level rises sharply to a peak shortly after one goes to sleep; however, the physiologic significance of this fact is a mystery. The secretion of growth hormone is inhibited during the normal rapid eye movement (REM) phase of sleep in humans, but it is elevated when the subject is deprived of REM sleep.†

During fasting as well as in situations that produce physiologic stress, the elevation of free fatty acids in the plasma that attends an increased secretion of somatotropin takes several hours to occur; however, this hyperlipemic response ulti-

---

*The secretion of prolactin, on the other hand, is inhibited, rather than stimulated, by a hypothalamic chemical factor, as discussed on page 1039.

†There is some experimental evidence that supports the contention that REM sleep per se may be triggered by a deaminated metabolic product of serotonin.

mately provides an important energy source. This statement also holds true during a state of hypoglycemia. Injection of glucose solutions, on the other hand, tends to minimize the effects of exercise, and the plasma concentration of somatotropins is lowered by such treatment.

The normal basal concentration of somatotropin in the plasma is similar in males and females, but the increased growth hormone secretion observed during arginine infusion and prolonged exercise are usually greater in women than they are in men. However, after prior estrogen treatment, men exhibit a response that is similar to that normally found in women. Cortisol administration, however, inhibits the elevated secretory rate of growth hormone in response to hypoglycemia in both sexes.

The alterations in growth hormone level produced by exercise, fasting, and hypoglycemia are all reduced in obese persons, although the significance of this observation is unclear (cf. p. 1215).

Clinical Effects of Somatotropin Deficiency (or Hyposecretion): Dwarfism.   A premature arrest of skeletal development, or *dwarfism*, may result from a specific lack or absence of hypophyseal somatotropin. Unlike cretins, hypophyseal dwarfs are neither deformed, nor show mental deficiency, although they often may exhibit sexual immaturity. The relative proportions of the different portions of the skeleton are not appreciably different from normal, although the head generally is disproportionately large with respect to the rest of the body. Adult dwarfs may achieve a height of only 0.9 to 1.2 meters.

As emphasized earlier, growth is a complex event that requires the coordinated activity of several hormones as well as an adequate diet; it is also influenced by genetic factors. Since linear growth ceases normally when the epiphyses fuse to the long bones under the influence of androgens (ie, epiphyseal closure takes place), then a pituitary dwarf who is deficient in somatotropin will respond to androgen therapy by an initial stimulation of growth because of the anabolic effect of the sex hormone. However, subsequent to epiphyseal closure, growth ceases once again.

Although dwarfism may result from any of several inherent as well as environmental etiologic factors, in a syndrome known as congenital isolated growth hormone deficiency there is a specific lack of pituitary somatotropin. This defect is inherited, probably as an autosomal recessive characteristic. Such individuals are known as sexual ateliotic dwarfs, and although remaining short in stature, they do mature sexually. Women having this condition can become pregnant, deliver, and lactate in a normal fashion.

Interestingly, African pygmies have been demonstrated to possess normal plasma somatotropin levels. However, there appears to be a defect in the use of this hormone by the peripheral

tissues of their bodies such that they remain short in stature.

Clinical Effects of Somatotropin Excess (or Hypersecretion): Gigantism and Acromegaly.   Excessive secretion of somatotropin in man, before the onset of puberty (ie, prior to complete ossification of the long bones) results in a condition known as *gigantism*. This endocrine dysfunction usually is the result of an adenomatous tumor of the adenohypophysis and results chiefly in an overgrowth of the skeleton, so that individuals over 2.5 meters in height may result. In persons suffering this malady, the limbs grow to disproportionate lengths with respect to the rest of the body.

Acromegaly in man results from a cause similar to that of gigantism. In contrast to the latter condition, the onset of the tumor and the secretion of excess somatotropin take place after full skeletal growth has occurred. The four principal characteristics of acromegaly are:

1. A disproportionate overgrowth of the bones in the face, hands, and feet takes place. The jaw becomes massive, and may protrude markedly *(prognathism),* while both hands and feet are enlarged disproportionately. The hands in particular become broadened or spadelike, and the fingers are thickened. Soft tissues of the scalp, forehead, nose, and lips are thickened, giving an overall heavy or coarse appearance to the features. The body hair also overgrows.
2. The spine bows, resulting in kyphosis. If this effect is sufficiently pronounced, mechanical respiratory difficulties may ensue with an attendant dyspnea.
3. *Splanchnomegaly,* or an enlargement of the viscera also accompanies the syndrome of acromegaly. The tongue, heart, lungs, liver, thymus, and spleen are all enlarged greatly. The parathyroids, thyroid, and adrenal glands may hypertrophy or exhibit adenomatous growth. Hyperthyroidism may be present in addition to the states of hyperglycemia and glucosuria. The latter two symptoms suggest a diabetes having its origin the pancreas.
4. Early in the course of the disease, increased sexual activity may be present in acromegalics. Ultimately, however, the gonads atrophy so that sexual functions are abolished in both sexes. Impotence ensues in men and amenorrhea develops in women.

THYROTROPIN (THYROTROPIC HORMONE, THYROID STIMULATING HORMONE [TSH])

Chemistry.   Highly purified thyrotropin has been isolated from human pituitaries as well as from the glands of several animal species (eg, sheep and beef). The biologic activity of thyrotropin is related to a glycoprotein having a molecular weight of approximately 25,000 daltons.

Thyrotropin has eight or nine cystine residues per mole, and disulfide groups are present as linkages within the chain. Several sugars are present within the molecule, including galactosamine, glucosamine, fructose, mannose, and galactose. The sugars probably are combined within the thyrotropin molecule as a single oligosaccharide unit.

General Functions of Thyrotropin. As noted earlier, hypophysectomy results in a diminution of thyroid function, whereas exogenous administration of thyrotropin causes stimulation of thyroid activity. In the former situation, the thyroid involutes and atrophies, resulting in a lowered basal metabolic rate, a decreased concentration of serum protein-bound iodine, lowered iodine uptake, and a diminished overall cellular metabolic activity. Thyrotropin administration to a subject deficient in this hormone reverses all of these effects. Thus, the specific target organ for thyrotropin is the thyroid gland.

Thyrotropin affects the rate of the following reactions: First, it elevates the rate at which iodide is removed from the blood by the thyroid. Second, it accelerates the rate at which monoiodotyrosine, diiodotyrosine, and thyroxine are synthesized by the thyroid gland. Third, pinocytosis of colloid by the follicles of the thyroid gland is increased by thyrotropin. Fourth, thyrotropin increases the release of organic iodide as hormones from the thyroid gland. Thus, more thyroxine and triiodothyronine are secreted under the influence of thyrotropin. Fifth, chronic administration of excess thyrotropic hormone over prolonged intervals results in an increase in blood flow within the thyroid. Ultimately the cells hypertrophy with a concomitant increase in the total mass of the gland, a condition known as a goiter.

A hypophysectomized animal suffers a reduced rate of uptake of injected iodide from the plasma by the thyroid gland. Once the iodide enters the thyroid in such a hypophysectomized animal, it is still rapidly converted into diiodotyrosine, although the rate of conversion of the formed diiodotyrosine to thyroxine is sharply depressed in the absence of the pituitary, and, as noted above, the thyroid gland in hypophysectomized animals also undergoes atrophy.

Thyrotropin may also act directly upon the orbital tissue of the eye in producing the protruding eyeballs (or exopthalmos) that are seen clinically in cases of diffuse toxic goiter. However, preparations of thyrotropin that have been highly purified show a decrease in exopthalmos-stimulating activity, so that perhaps another distinct hypophyseal factor is responsible for this effect.

Mechanism of Thyrotropin Action. The principal mode of action of thyrotropin upon the thyroid gland appears to be mediated through an increase in the rate of synthesis of cyclic AMP. Subsequent to the increase in AMP concentration, glucose oxidation (and oxygen consumption) also increase, presumably via the hexose–monophosphate shunt (Table 14.9, p. 955). A rapid increase in the pinocytosis of colloid by the thyroid cells ensues, as well as the other observed effects of thyrotropin that were listed in the previous section.

Thyrotropin also stimulates several other metabolic reactions in the cells of the thyroid. The synthesis of NAD and NADP as well as other coenzymes necessary for the reactions involved in hormone synthesis and secretion is enhanced by the elevated glucose oxidation induced by cyclic AMP. Phospholipid and protein synthesis are also increased in the thyroid by thyrotropin, although there is no good evidence that the other effects of this hormone on the thyroid are due to an enhanced protein synthesis per se. Briefly, therefore, thyrotropin acts upon the thyroid gland by stimulating a variety of particular metabolic processes.

Based upon morphologic studies, it has also been postulated that thyrotropin affects the permeability of the membranes of the thyroid epithelial cells in addition to producing the effects noted above.

Regulation of Thyrotropin Secretion. There are two mechanisms present in the body that regulate the secretion of thyrotropin—neural and neuroendocrine.

Neural Mechanisms. It has long been recognized clinically that psychogenic (ie, neurogenic) factors can affect thyroid function and thus contribute to a condition known as thyrotoxicosis. In addition, experimental lesions that are appropriately placed in the hypothalamus inhibit the anticipated rise in thyrotropin secretion in response to reduced blood levels of circulating thyroxine. Hence it is known that a neural mechanism can participate in thyrotropin secretion, although the extent to which neural factors participate in the physiologic regulation of thyrotropin secretion is unknown.

Neuroendocrine Mechanism. The hypothalamic thyrotropic hormone-releasing factor (TRF), and its major role in controlling the secretion of this hormone by the pituitary was discussed on page 1030. The chemical feedback mechanism for the overall control of thyrotropin secretion also was presented earlier.

Clinical dysfunctions of thyrotropin secretion are reflected indirectly through impaired thyroid functions, and will be discussed in relation to the thyroid gland itself.

CORTICOTROPIN (ADRENOCORTICOTROPIN, ADRENOCORTICOTROPIC HORMONE, ACTH)

Chemistry and Metabolism. Corticotropin has been isolated, and its structure has been determined. As shown in Figure 15.7, this substance is an oligopolypeptide (a straight-chain polypeptide) having 39 amino acid residues. The first 23 amino acids, as well as their sequence, is identical in cor-

Ser	Tyr	Ser	Met	Glu	His	Phe	Arg	Trp	Gly	Lys	Pro	Val	Gly	Lys	Lys	Arg	Arg	Pro	Val
1	2	3	4	5	6	7	8	9	10	11	12	13	14	15	16	17	18	19	20

FIG. 15.7 Amino acid sequence of human corticotropin. See Table 14.10 (p. 970) for a listing of the amino acid residues found in proteins. (After White, Handler, and Smith. *Principles of Biochemistry*, 4th ed, 1968, The Blakiston Division, McGraw-Hill.)

ticotropins that have been isolated from the hypophyses of several species, including humans, pigs, cattle, and sheep.

A synthetic polypeptide (or a pepsin or acid-hydrolyzed product of natural corticotropin) that contains the first 23 amino acid residues (β-corticotropin) has an in vivo biologic activity that is similar to the same hormone (α-corticotropin) when this is isolated from natural sources. Therefore, amino acid residues 24 through 39 are not critical to the biologic action of this hormone, but apparently these residues do confer a species-specificity or individuality to the molecule. Hence corticotropin may be antigenic in species other than the one from which it was derived.

Although the chain of amino acids from 1 through 23 apparently constitutes the most active portion of the corticotropin molecule, synthetic polypeptides that have the first 16-amino-terminal (serine) residues exhibit some biologic activity. However, a synthetic polypeptide containing the first 19 residues exhibits a marked increase in activity above a peptide composed of only the first 16 residues. This fact suggests that the sequence of basic amino acids, Lys. Lys. Arg. Arg (positions 15.16.17.18), may have an important role in determining the biologic potency of the entire corticotropin molecule.

In vivo the inactivation of corticotropin is quite rapid; therefore its half-life in the human circulation is about 10 minutes, although corticotropin does not appear to be altered by the adrenals themselves when it exerts its catalytic effects within these glands.

The corticotropin level in the plasma can be determined by means of immunoassay or bioassay in test animals.

General Functions of Corticotropin. Corticotropin stimulates both the growth as well as the secretion of adrenal steroids by adrenal cortical tissue, which tissue is the principal target organ for this hormone. Thus the principal actions of corticotropin are mediated indirectly through the adrenal glands as the target organs, and the effects of corticotropin in general reflect the activity of the adrenal secretions.

In hypophysectomized humans or animals, the adrenal cortices atrophy; normal size and function are restored by the exogenous administration of corticotropin, and atrophy is prevented by postoperative injection of corticotropin preparations.

A short time following hypophysectomy, glucocorticoid hormone synthesis by the adrenal cortex falls quite rapidly (within an hour) to extremely low levels, and this effect is corrected

rapidly (within a few minutes) by the injection of corticotropin. However, in patients with chronic hypofunction of the pituitary, or some time after hypophysectomy,* a single injection of corticotropin no longer is effective in stimulating glucocorticoid secretion; therefore either repeated injections or long-term infusion of corticotropin is necessary to stimulate the adrenals. Following atrophy of the adrenals the response to corticotropin is much delayed, however secretion of adrenal hormones commences once again before any demonstrable changes in either the size or morphology of these glands takes place following exogenous corticotropin administration.

Since the injection of corticotropin stimulates the adrenal cortex directly, then the administration of this compound mimics all of the responses produced by the adrenal hormones. On the other hand, some direct responses to corticotropin that are not mediated through the adrenals are as follows. First, an augmented gluconeogenesis and a depression of protein synthesis in all tissues (except the liver) is induced by corticotropin. Second, lipid mobilization from the fat depots to the liver takes place (lypolytic or adipokinetic effect) with development of a concomitant ketonemia as well as a hypercholesterolemia. This effect can be demonstrated in vitro as well as in vivo. Third, some augmentation of salt and water retention takes place following administration of corticotropin; however this effect is much less quantitatively than that which occurs when aldosterone is given directly. Since the secretion of aldosterone depends only partly upon the hypophysis, then this reduced direct effect of corticotropin administration is not surprising. Fourth, the administration of corticotropin also results in a stimulation of erythropoiesis as well as a depression of lymphocyte and eosinophil production.

Consequently, corticotropin is of considerable value therapeutically in the management of certain clinical problems which also respond directly to adrenal cortical steroids. Addison's disease, however, provides an exception to this general statement, as in this condition there is an inadequate mass of normal adrenal tissue present to respond to exogenous corticotropin.

It must be borne in mind that the administration of corticotropin to a patient may exert different effects than those obtained by the injection of specific adrenal cortical steroids, each of which has a characteristic function (or functions), because the pituitary hormone stimulates the release

---

This effect commences around 24 hours following hypophysectomy and increases progressively.

Lys	Val	Tyr	Pro	Asp	Ala	Gly	Glu	Asp	Gln	Ser	Ala	Glu	Ala	Phe	Pro	Leu	Glu	Phe
21	22	23	24	25	26	27	28	29	30	31	32	33	34	35	36	37	38	39

FIG. 15.7. (Cont'd)

of a complex nonspecific mixture of hormones from the adrenals. Furthermore, the effects of exogenous corticotropin administration are complicated still further by inhibition of the normal endogenous pituitary secretion by the subject's own hypophysis through interference with the normal feedback regulatory mechanism discussed below. As a consequence, prolonged injection of corticotropin can induce such undesirable side effects in the patient as, for example, masculinization (in women). This response reflects hypersecretion of male sex hormones (androgens) by the adrenal cortical tissue of the patient.

In addition to the indirect, adrenal-mediated actions of corticotropin, this hormone also exerts several direct effects upon a number of target tissues as noted above. Thus, in addition to the direct effects already discussed, corticotropin stimulates the rate of pigment (melanin) formation in certain cells of the skin (melanocytes), a property that is shared in common with the two melanocyte-stimulating hormones. Since all three of these hormones also share certain similarities in chemical structure this fact may explain, at least partially, the melanin-stimulating effect of corticotropin under certain conditions. Thus, in Addison's disease the observed darkening of the skin may be due to the abnormally high blood level of corticotropin present in this condition. Corticotropin also exerts a direct action upon the pancreas, whereby it stimulates the release of insulin.

Mechanisms and Factors Involved in the Regulation of Corticotropin Secretion. The rate of corticotropin secretion by the anterior pituitary is regulated principally by the general neural and feedback mechanisms that were described earlier and are reviewed below. First, the neuroendocrine (or neurohumoral) mechanism responds to a wide variety of dissimilar stimuli, and thereby increases corticotropin release (Fig. 15.8). For example, emotional or other types of stress, such as trauma, chemical substances or bacterial toxins, drugs, or materials that are normally present in the body such as thyroxine, insulin, epinephrine, or antidiuretic hormone all may influence corticotropin secretion. The effects of stress will be considered more extensively in the discussion below. All of these stimuli share one common factor, in that they stimulate liberation of corticotropin-releasing factors (CRF), by the median eminence of the posterior hypothalamus. These humoral agents are in turn transmitted via the hypophyseal portal system to the adenohypophysis where they augment corticotropin secretion. One CRF (α_2-CRF) is structurally as well as biologically similar to

α-melanocyte-stimulating hormone. β-CRF, on the other hand, has considerably more biologic activity than α-CRF, and it also exerts a marked ADH-like action. Vasopressin (ADH) per se also exhibits a corticotropin-releasing activity.

Second, there is an inverse relationship between the plasma level of circulating adrenal cortical steroids (glucocorticoids) and corticotropin release. Therefore, an increased utilization, hence removal, of these steroids from the blood by the tissues themselves results in a lowered concentration of blood steroids, and this in turn stimulates the secretion of corticotropin directly. Conversely, the administration of exogenous adrenal cortical steroids results in an inhibition of corticotropin secretion by the anterior pituitary, so that prolonged treatment with such compounds results in adrenal cortical hypofunction and eventual atrophy in patients as well as in animals following adrenal cortical hormone injections over extended intervals of time. The blood level of steroids thus acts as the negative feedback stimulus to the hypothalamus, which in turn controls the release of CRF.

Actually, both of these mechanisms share in common the hypophyseal CRF mechanism; however, the underlying stimuli that elicit an increased or decreased corticotropin release are quite different in each case.

Since elevated circulating levels of glucocorticoids depress corticotropin secretion, this reponse provides a major physiologic mechanism for controlling the secretion of the hormone by the adenohypophysis. An inverse and linear relationship can be demonstrated between the level of circulating glucocorticoids and corticotropin secretion. There is also a parallel between the potency of the individual glucocorticoid adrenal hormones and the degree of inhibition of corticotropin secretion.

It must be reiterated here that there is an inherent danger following the cessation of therapeutic glucocorticoid administration after this therapy has been given for extended intervals, since the adrenals may have become atrophic and thus unresponsive to corticotropin treatment. In addition to this fact, the pituitary itself may not secrete normal quantities of corticotropin for up to a month, despite stimulation of the adrenals by exogenous administration of this hormone.

The level of free (or unbound) glucocorticoids in the plasma is normally quite low. Therefore, in the absence of stress the inhibition of corticotropin secretion is also normally very low. An acute drop in blood glucocorticoid levels is not a rapid and powerful stimulus to corticotropin secretion, the response only being observed after several hours. Chronic adrenal insufficiency, however, does produce a marked stimulation of corticotropin production and secretion by the adenohypophysis.

Stress and Corticotropin Secretion. The concentration of corticotropin in the body fluids can be assayed by procedures involving bioassay or immunoassay, and thus a quantitative estimate of this hormone under various physiologic states may be obtained.

It is interesting to note that under conditions of severe stress, the quantity of corticotropin that is secreted by the adenohypophysis far exceeds that quantity necessary to evoke maximal glucocorticoid secretion by the adrenal cortices. The elevated secretion of corticotropin in situations involving a physiologic crisis are mediated primarily through the hypothalamus. The median eminence and portal vessel system form the final common pathway whereby the adenohypophysis is activated by CRF, as discussed earlier. Destruction of the median eminence results in a blockage of the elevated corticotropin secretion that takes place during stress, although a minimal basal secretion of glucocorticoids by the adrenals continues following such injury, and these glands to not undergo atrophy.

A large number of afferent neural pathways from many parts of the brain converge upon the median eminence. Fibers from the amygdaloid nuclei transmit responses to emotional stresses, hence anxiety, apprehension, and fear directly stimulate corticotropin secretion markedly (Fig. 15.8).

There is a profound effect of the time of day (diurnal or circadian rhythm) upon the circulating glucocorticoid level, and this effect in turn is controlled by corticotropin secretion that is regulated by afferent impulses that arise in the limbic system and thence converge upon the median eminence.

Another source of afferent nerve impulses affecting hypothalamic CRF secretion is found in the reticular formation. As shown in Figure 15.8, these impulses stimulate corticotropin secretion via the CRF mechanism in response to injury (or trauma).

In addition to the neural factors, outlined above, which stimulate corticotropin secretion during times of stress, blood-borne humoral factors may also play a role in stimulating the hypothalamo–hypophyseal mechanism. Epinephrine and norepinephrine do not increase corticotropin secretion in man; therefore adrenocortical and adrenal medullary secretions are regulated by independent hypothalamic mechanisms.

It was believed at one time that stress resulted in increased corticotropin secretion by producing a reduction in the blood glucocorticoid level; this concept has been proven to be incorrect. It appears that the net rate of corticotropin secretion by the pituitary during periods of stress is determined by two antagonistic mechanisms. First, the summated neural (and other) stimuli that converge upon the adenohypophysis via the median eminence of the hypothalamus tend to increase the

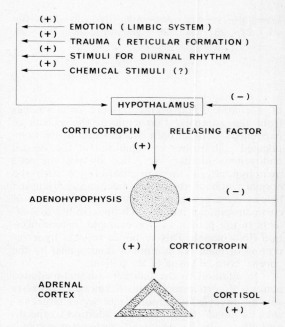

FIG. 15.8. Summary of the factors believed to regulate corticotropin secretion. (+) = stimulation; (−) = inhibition. Note that an increased circulating level of adrenal corticosteroids directly exerts a negative feedback effect upon the hypothalamus and adenohypophysis. CRF = corticotropin releasing factor. (After Ganong. *Review of Medical Physiology,* 6th ed, 1973, Lange Medical Publications.)

secretion of corticotropin. Second, and in opposition to this mechanism, the circulating blood glucocorticoids themselves tend to inhibit the secretion of corticotropin, and this curbing effect is in direct proportion to the concentration of these hormones in the plasma.

Mode of Action of Corticotropin. As might be expected, corticotropin has been shown to stimulate the synthesis of adrenal steroids in experiments conducted in vivo as well as in vitro. Furthermore, the concentration of cyclic 3′,5′-AMP in adrenal tissue is increased by corticotropin. In turn, the addition of cyclic AMP directly to adrenal tissue in vitro stimulates the biochemical reactions within the mitochondria that are involved in the hydroxylation of steroids.

A corticotropin-induced elevation in cyclic AMP concentration is observed in adrenal tissue prior to the increase in steroid synthesis. Therefore, a primary action of this pituitary hormone upon the cells of the adrenal cortex is to stimulate the synthesis of cyclic AMP via an increase in the activity of adenyl cyclase, as shown in Figure 15.2.

The increased synthesis of adrenal steroids that is produced under the trophic influence of corticotropin is not found if protein synthesis is inhibited in the adrenal tissue. This fact suggests that synthesis of one or more proteins is involved in the stimulation of steroid synthesis by corticotropin, although inhibition of protein synthesis does not affect the increase in cyclic AMP produced by corticotropin. Consequently, the protein formed de novo which is believed to be involved in steroid synthesis may act at a biochemical locus beyond the step in which adenyl cyclase is activated by corticotropin.

Clinical Effects of Excess (Hypersecretion) of Corticotropin: Cushing's Disease.   The clinical situation found in man which results from an excess secretion of adrenal glucocorticoids is known as Cushing's disease (or syndrome). This condition may result from one of several etiologic factors: the therapeutic (ie, exogenous) administration of large doses of adrenal steroids; endogenous glucocorticoid-producing tumors of the adrenal cortex; or hypersecretion of corticotropin by the adenohypophysis. In the latter instance, a defect in the regulation of corticotropin secretion itself may be present, although the mechanism which is involved in this process is not well understood. In addition to the three causative factors listed above, Cushing's disease also may be produced by corticotropin-secreting tumors of the basophil cells of the adenohypophysis; in this situation, the disease also is referred to as pituitary basophilism. Tumors arising in nonendocrine tissues also may secrete substances that are similar to, if not identical with, hypophyseal adrenocorticotropic hormone, and such substances also can produce Cushing's disease.*

The defects that may be found in a patient having Cushing's syndrome may be summarized:

1. The body fat is redistributed in a characteristic fashion. Obesity of the trunk, especially the abdominal wall, is observed. Fat also collects in the face, giving it a moon-shaped appearance, whereas on the upper back, the fat accumulation produces a typical hump. The buttocks also are enlarged by fat deposition.

   Because of the rapid distension of the skin by the process of fat redeposition, purple-red striae are observed, especially on the abdominal wall. (These scars, due to rupture of the subdermal tissues, are found normally when the skin is stretched rapidly as around the female breasts during puberty or in the skin of the abdomen during pregnancy. However, in normal persons the striae generally are faint, and the strong purple color that is present in Cushing's disease is lacking.)
2. The patient is protein-depleted in Cushing's disease owing to an increased rate of protein breakdown (catabolism). In contrast to the body proper, the muscles are poorly developed, so that the limbs are disproportionately thin compared to the size of the rest of the body. The skin

and subcutaneous tissues are also thin; wound healing is poor, and bruises and ecchymoses result easily from minor trauma.
3. Cyanosis in the feet, hands, and face may be observed, as well as pigmentation of the skin. Hair growth is abundant, and virilization of females with the attendant growth of a mustache or beard may be observed.
4. Demineralization of the bones occurs in Cushing's syndrome with the ultimate development of osteoporosis. The softening and demineralization of bone attending this process eventually leads to skeletal deformities as well as to collapse of the vertebral column.
5. Hypertension is often observed; around 85 percent of all patients having Cushing's disease exhibit this response.
6. A loss of sexual functions occurs in Cushing's disease.
7. A hyperglycemia and glucosuria are present, and frequently an insulin-resistant diabetes is precipitated, especially in persons who are predisposed to this condition. Acidosis is not severe generally although hyperlipemia and ketosis are present.
8. Severe acne may also be found in Cushing's disease.

The specific metabolic defects that are associated with Cushing's syndrome will be considered in greater detail as appropriate in later sections.

At this point, it will be helpful if the reader compares the symptoms of Cushing's syndrome (which is caused by a specific pituitary hyperfunction) with the symptoms that are induced by hypophysectomy (p. 1027) and Simmonds' disease (p. 1027). Both of the latter conditions, of course, reflect pituitary hypofunction. Furthermore, the clinical picture in Cushing's syndrome will serve as a general introduction to the functions of the glucocorticoids that are secreted by the adrenal cortex.

FOLLICLE-STIMULATING HORMONE (FSH).   The adenophypophysis secretes three so-called gonadotropic hormones that directly stimulate gonadal function in both males and females. These hormones are follicle-stimulating hormone (or FSH), which will be discussed in the present section; luteinizing or interstitial cell-stimulating hormone (LH or ICSH); and prolactin. The latter two hormones will be considered in the next two sections. In addition to the three adenohypophyseal gonadotropins, the placenta also produces a hormone known as human chorionic gonadotropin (abbreviated CG or HCG), which is discussed on page 1163, in order to distinguish it from a related gonadotropic substance that is found in the serum and urine of pregnant mares and which sometimes is designated by the acronym PMS.

The present discussion of the three pituitary gonadotropic hormones is designed merely to in-

It should be stressed that nonendocrine tumors have been found which secrete substances resembling in their actions nearly all of the known hormones.

troduce the reader to these substances. The detailed functional roles of the gonadotropins in male and female reproductive physiology will be presented in context with a discussion of the target organs for these substances, the testes and ovaries, later in this chapter.

Chemistry. Human FSH is a small protein, having a molecular weight of around 17,000 daltons, and the hormone has been prepared in a highly purified form. As the FSH molecule contains between 8 and 9 percent carbohydrate it is a glycoprotein. Hexoses, fucose, hexosamine, and a sialic acid residue constitute the carbohydrate moieties. If the sialic acid residue is removed specifically from FSH (by incubation with an appropriate enzyme), then all biologic activity of the hormone is lost.

Although little is known about the synthesis and metabolism of any of the gonadotropic hormones, FSH appears to have a half-life in the human circulation of approximately three hours.

General Functions of Follicle-Stimulating Hormone. The target organs for FSH in the male are the testes. Within the testes, FSH directly stimulates the germinal epithelium of the seminiferous tubules so that the rate of spermatogenesis is augmented. Exogenous administration of FSH induces the appearance of large numbers of spermatocytes in several stages of development, as well as mature spermatozoa.

The target organs for FSH in the female are the ovaries. In these organs, FSH causes the appearance and growth of a large number of graafian follicles concomitant with an increased total ovarian weight (or mass).

In castrates of both sexes, and following the menopause in women, the urinary excretion of FSH is increased significantly. A similar elevation of urinary FSH occurs when malignant tumors of the primary reproductive organs is present.

Mechanisms and Factors Involved in the Regulation of FSH Secretion. The two principal mechanisms concerned with regulating the secretion of FSH as well as the other pituitary gonadotropins via the hypothalamo–hypophyseal releasing factor system and feedback regulation via the products of the target organs were discussed earlier (p. 1030). It is obvious that a broad spectrum of neural stimuli can affect the physiology of sexual activity, and many of these regulatory and modifying factors have been described in both clinical as well as in experimental situations.

The specific hypothalamic releasing factor for FSH has been partially purified, and it appears to have a low molecular weight (less than 1,000).

The circulating levels of male sex hormones (androgens) or female sex hormones (estrogens) themselves constitute another regulatory mechanism for the inhibition of FSH secretion by the adenohypophysis.* The sex hormones act directly upon the hypothalamus to inhibit the secretion of specific releasing factors, and thus decrease gondotropin secretion by the anterior pituitary. In particular, the hypothalamic receptor area for androgens lies in the posterior median eminence, whereas estrogens exert their effects in the arcuate nucleus of the hypothalamus.

The pineal body has also been implicated as having a role in the regulation of gonadotropin secretion. Thus 5-hydroxytryptophane and melatonin, secreted by this structure inhibit the estrus cycle in certain species of mammal. The physiologic role of this possible third mechanism for regulating pituitary gonadotropin secretion in humans is unclear.

General Mode of Action of the Gonadotropins. The pituitary gonadotropins stimulate the synthesis of the androgen, testosterone, as well as two estrogens, estradiol and progesterone. These effects are observed both in vivo as well as in vitro in the appropriate gonadal tissues.

Cyclic AMP synthesis is stimulated by the gonadotropins in general, and this effect in turn is reflected by an increased formation of the gonadal steroid hormones. There is experimental evidence suggesting that these effects in turn are mediated secondarily following a primary regulatory effect (translational control) of the gonadotropic hormones upon protein synthesis.

INTERSTITIAL CELL-STIMULATING OR LUTEINIZING HORMONE (ICSH or LH)

Chemistry. Human ICSH, similarly to FSH, is a glycoprotein having a molecular weight of around 26,000 daltons. The carbohydrate content of this substance is around 3.5 percent.

General Functions of Interstitial Cell Stimulating Hormone. The target organs for ICSH in the male are the testes, specifically the interstitial or Leydig cells of these organs, where this hormone stimulates the synthesis and secretion of the male hormone (or androgen) testosterone.

In the female, the target organs for ISCH are the ovaries; this hormone induces three major responses. First, ICSH stimulates ripening of the ovarian follicles. Second, ICSH causes the behavioral manifestations of estrus (or heat). Third, ICSH stimulates the metamorphosis of the ruptured ovarian follicles into corpora lutea (hence the synonymous name for ICSH of luteinizing hormone).

Since ICSH acts upon the interstitial cells of testes as well as the thecal cells of the ovaries, the functions of this hormone in both sexes may be considered analogous.

Mechanisms Involved in the Regulation of ICSH Secretion. The mechanisms that control ICSH secretion are similar to those already discussed for the other adenohypophyseal hormones. However, the specific luteinizing hormone-releasing factor (LRF) has not been isolated in pure form as yet from hypothalamic tissue. Biologically active prep-

---

The same mechanism also is important in regulating the secretion of luteinizing hormone (LH or ICSH) as well as prolactin from the adenohypophysis.

arations of this releasing substance do not give a ninhydrin reaction, suggesting that this neuroendocrine substance is not a protein or peptide which contains free amino (—NH$_2$) groups.

PROLACTIN (LUTEOTROPIC HORMONE, LTH, LUTEOTROPIN, LACTOGENIC HORMONE, MAMMOTROPIN). The various names applied to prolactin are indicative not only of its functions, but also reflect the fact that it was once thought that this hormone was two discrete substances.

Chemistry. Human prolactin, which has a molecular weight of roughly 23,000 daltons, is a simple protein; thus carbohydrate is lacking in the molecule. Structurally as well as functionally, prolactin is similar to somatotropin, as discussed below.

General Functions of Prolactin. Originally prolactin was considered to be a necessary factor for the initiation of lactation following parturition in mammals. However, it was later demonstrated that this hormone also stimulates the functional activity of the corpora lutea, and consequently the secretion of progesterone. These actions of prolactin are shown diagrammatically in Figure 15.6.

Prolactin normally acts in a synergistic fashion together with estrogen in stimulating the growth (hypertrophy) of the mammary glands. This function is in addition to the ability of prolactin to initiate milk secretion in the previously hypertrophied mammary tissue. Thus prolactin has a dual role in the human, as it acts with estrogen to ready the mammary glands for milk secretion, as well as by initiating the secretion of this substance.

In young pigeons, prolactin induces a marked hypertrophy and hyperplasia of the normally thin epithelial layer of the crop sac, and this effect provides the basis of an important bioassay for prolactin.

In primates, including man, prolactin not only shares a similar general structure to somatotropin, but also has certain functional properties which are similar to those of growth hormone. In fact, at one time it was considered that prolactin and somatotropin were identical substances; however, more recent experimental work has demonstrated that there are two distinct hormones rather than one. For example, purified human somatotropin stimulates the secretion of milk (lactogenic effect) in humans as well as in dairy animals (cows). However, patients who suffer a congenital deficiency in the production of growth hormone are able to lactate normally, indicating that a lactogenic principle (ie, prolactin) is present. The plasma of such patients when tested during lactation stimulates the proliferation of crop sac epithelium in the highly sensitive pigeon bioassay for prolactin that was mentioned above. Prolactin and somatotropin undoubtedly are separate biochemical entities despite an overlap in certain of their functional properties.

Mechanisms Involved in the Regulation of Prolactin Secretion. In common with the other pituitary gonadotropins, the blood level of prolactin per se regulates the secretion of this hormone via the negative feedback control mechanism discussed on page 1031. However, in sharp contrast to FSH and ICSH, the hypothalamic control of prolactin secretion from the adenohypophysis is inhibitory rather than stimulatory. Thus the hypothalamic inhibitory factor for prolactin (PIF) exerts a suppressive rather than a stimulatory effect upon the adenohypophyseal α-cells which secrete prolactin.

It also appears that there is a reciprocal relationship between secretion of the other two pituitary gonadotropins and prolactin. When the adenohypophyseal secretion of FSH and ICSH are stimulated by their respective hypothalamic releasing factors, the concomitant secretion of prolactin is reduced through the hypothalamic release of PIF. Conversely the secretion of prolactin by the anterior pituitary, with an attendant lactogenic response, apparently results from a suppressed release of the other gonadotropins.

Mode of Action of Prolactin. In addition to the general biochemical actions common to all of the gonadotropins, prolactin also has been demonstrated to stimulate glucose uptake and lipogenesis in vitro when added to adipose tissue. In addition, prolactin when injected in vivo elicits biochemical effects that are quite similar to those produced by somatotropin (p. 1028).

Non-Pituitary Gonadotropin. During normal pregnancy, the human placenta exerts an endocrine function in that it secretes chorionic gonadotropin (HCG), a substance whose biologic effects resemble those of the pituitary hormones described above. HCG normally supplements the hypophysis in maintaining the corpus luteum during pregnancy. This hormone alone will not prevent the ovarian atrophy which follows hypophysectomy in pregnant animals, indicating that differences exist between the hypophyseal and chorionic hormones.

In addition to its role in the female, HCG when administered to males stimulates the Leydig tissue of the testes, and thus indirectly the male accessory reproductive organs are stimulated by the enhanced testosterone levels. HCG has been of some use clinically in the treatment of cryptorchidism, a condition in which the testes remain in their prepubertal location within the abdominal cavity rather than descending into the adult position in the scrotum.

Since chorionic gonadotropin appears in the urine shortly after the onset of pregnancy, ie, at around the first week subsequent to the first missed menstrual period, this substance forms the basis for two common biologic tests for detecting pregnancy. First, the Friedman test employs virgin female rabbits. Urine from the woman is injected intravenously into the animals for several successive days, and the criterion of pregnancy is the ability of the urine to produce ovulation. This effect in turn is judged by the presence of ruptured ovarian follicles following sacrifice of the rabbits. Second, in the Ascheim–Zondek test, immature female mice or rats are used as subjects. Urine, or else an alcoholic precipitate thereof, is injected into the animals for several days, and the criteria diagnostic of pregnancy are an increased ovarian weight, ripening of some ovarian follicles, and hemorrhages into some of the unruptured follicles.

In either of the tests described above, chorionic gonadotropin induces the physiologic responses noted in the animals. A more economical and rapid test for pregnancy than those described above utilizes male frogs and toads of certain species. One injection of pregnant urine into the dorsal lymph sac (ie, subcutaneously on the dorsum) of these amphibia is sufficient to cause the animal to excrete large quantities of living spermatozoa after a few hours. The presence of these sperm is determined readily by microscopic examination, and once again the response is elicited by the elevated titer of HCG in the urine.

In addition to the HCG found normally in the urine of pregnant women, certain pathologic conditions also can give rise to increased urinary quantities of gonadotropin which resembles hypophyseal gonadotropins. The presence of such a substance in the urine can not only give rise to a false positive pregnancy test, but also may be of considerable significance diagnostically. For example, such a non-pituitary gonadotropin is found in the urine of female patients suffering from a chorionepithelioma, a malignant placental tumor, or else in hyatidiform mole, a cystic disease of the chorionic tissue proper. Male patients suffering from malignant testicular tumors of the embryonic epithelial tissue also exhibit elevated urinary gonadotropin levels. Examples of such diseases are epitheliomas and teratomas. Under such conditions, the assay of urinary gonadotropin is a helpful diagnostic aid.

Gonadotropins and the Human Sexual (Menstrual) Cycle. The remarkable and close interrelationships among several endocrine substances in regulating a body function are well illustrated by their roles in the rhythmic sexual (or menstrual) cycle of the human female following the onset of puberty. A discussion of this process also will serve to illustrate and to summarize the integrated physiologic roles of the several pituitary gonadotropic substances discussed above and their influence upon the target organs (ovaries).

In prepubertal males as well as females, there is no detectable secretion of the pituitary gonadotropins; therefore no discernible morphologic changes take place in the immature testes or ovaries until the time of puberty. The physiologic mechanisms responsible for the onset of puberty are unknown. However, the rhythmic menstrual cycles initiated in human females (and other primates, eg, baboons and chimpanzees) at puberty are due to the adenohypophyseal secretion of gonadotropic hormones. The cyclic changes in the

gonadotropins as well as the ovarian estrogens during the human sexual cycle are shown in Figure 15.9, and compared with the concomitant morphologic alterations in the ovarian follicle and uterine endometrium. These alterations, hormonal as well as structural, may be considered sequentially starting with the first day of the cycle following the end of the previous menstrual period (MP in the figure). The 28-day cycle illustrated in Figure 15.9 is about average. Actually, menstrual cycles as short as 20 days or as long as 40 days have been recorded in normal females.

Follicle-stimulating hormone (FSH) is secreted by the adenohypophysis, and this hormone acts in a synergistic fashion with a small quantity of luteinizing hormone (LH) from the same source to stimulate follicular development in the ovary. Usually only one or two follicles, hence one or two ova, develop in the ovaries per sexual cycle. However, the development of multiple follicles and ova are not uncommon, especially since the recent advent of the so-called "fertility drugs."

As the follicle ripens, the cells of the theca interna also develop, and commence to secrete estrogens. The estrogens in turn induce the secretion of follicular fluid by the cells of the granulosa. These estrogens also sensitize the follicle to FSH. The stimulated ovarian follicle continues its development, whereas other follicles which were in the process of ripening undergo complete or partial atresia (or involution).

The estrogens produced by the ovarian follicle enter the blood, and now exert several effects upon the adenohypophysis. FSH secretion is inhibited, whereas the secretion of LH and also, possibly, prolactin, is stimulated. The adenohypophyseal LH causes rupture of the mature (or ripe) and sensitized follicle so that release of the mature ovum or ovulation, takes place. LH also stimulates the development of the corpus luteum (or "yellow body") from the cells lining the ruptured follicle; then prolactin initiates secretion of progesterone by the developed corpus luteum. The progesterone from the corpus luteum in turn inhibits the secretion of LH and prolactin by the hypophysis. If fertilization of the ovum and pregnancy do not ensue (p. 1190), the fall in the blood levels of the LH and prolactin now induce the onset of menstruation as the corpus luteum degenerates. When the secretion of LH and prolactin ceases, the hypophysis returns to its previous state and thereby a new cycle commences.

If fertilization and pregnancy occur, the endometrium of the uterus is already prepared for implantation of the ovum by the action of progesterone from the corpus luteum. In addition, progesterone maintains the uterine wall in a condition for the maintenance of the embryo and fetus, as the corpus luteum does not degenerate following implantation of the fertilized ovum. Progesterone from the corpus luteum also reduces the tone of the uterine smooth muscle, and it stimulates development of the mammary tissue (acting in a

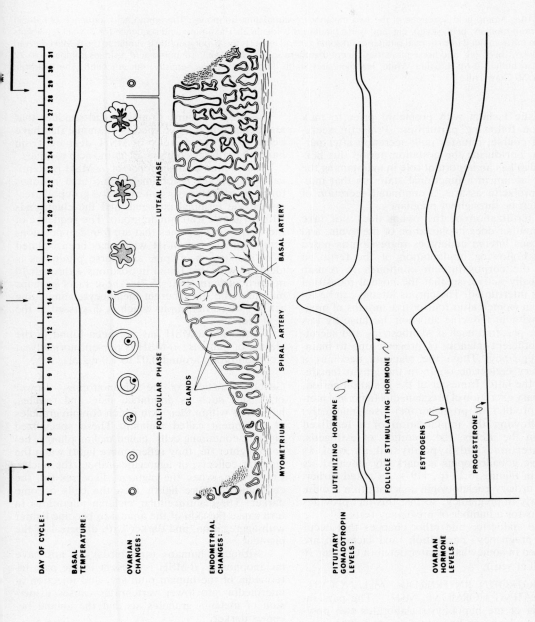

FIG. 15.9. Summary of endocrine changes and other events in the human sexual (menstrual) cycle. MP = menstrual period. (Data from Ganong. *Review of Medical Physiology,* 6th ed, 1973, Lange Medical Publications; Guyton. *Textbook of Medical Physiology,* 4th ed, 1971, Saunders; West, Todd, Mason, and Van Bruggen. *Textbook of Biochemistry,* 4th ed, 1966, Macmillan; White, Handler, and Smith. *Principles of Biochemistry,* 4th ed, 1968, The Blakiston Division, McGraw-Hill.)

DAY OF CYCLE:

1 2 3 4 5 6 7 8 9 10 11 12 13 14 15 16 17 18 19 20 21 22 23 24 25 26 27 28 29 30 31

BASAL TEMPERATURE:

OVARIAN CHANGES:

ENDOMETRIAL CHANGES:

FOLLICULAR PHASE

LUTEAL PHASE

GLANDS

MYOMETRIUM

SPIRAL ARTERY

BASAL ARTERY

PITUITARY GONADOTROPIN LEVELS:

LUTEINIZING HORMONE

FOLLICLE STIMULATING HORMONE

OVARIAN HORMONE LEVELS:

ESTROGENS

PROGESTERONE

α-MSH: Acetyl—Ser—Tyr—Ser—Met—Glu—His—Phe—Arg—Trp—Gly—Lys—Pro—Val—NH₂
             1   2   3   4   5   6   7   8   9   10   11   12   13   14

β-MSH: Ala—Glu—Lys—Lys—Asp—Glu—Gly—Pro—Tyr—Arg—Met—Glu—His—Phe—Arg—Trp—Gly—
          1   2   3   4   5   6   7   8   9   10   11   12   13   14   15   16   17

Ser—Pro—Pro—Lys—Asp
18   19   20   21   22

FIG. 15.10. Amino acid sequence of the two melanocyte-stimulating hormones. The amino acid sequence of α-MSH derived from monkey, beef, sheep, pig, and horse pituitaries is identical. The amino acid sequence for β-MSH represents this hormone as derived from human pituitaries. In some general respects this β-polypeptide is similar to the β-MSH derived from monkey, beef, pig, and horse pituitaries except that these species have only 18 amino acid residues compared to 22 residues in human β-MSH. (After White, Handler, and Smith. *Principles of Biochemistry,* 4th ed, 1968, The Blakiston Division, McGraw-Hill.)

synergistic fashion with prolactin) prior to milk secretion following parturition. Prolactin secretion, of course, initiates milk secretion after parturition, but during the gestation period this hormone also has an important role in maintaining the corpus luteum in a functional state, so that indirectly prolactin assures a continued secretion of progesterone throughout pregnancy.

If fertilization of the ovum does not take place, neither does implantation of the ovum, and the corpus luteum undergoes regression as noted above. Following implantation of the fertilized ovum, the corpus luteum continues to remain functionally active, so that the normal menstrual cycle is interrupted. The corpus luteum can be removed in women after four or five months of pregnancy without inducing abortion, because by this time the placenta itself is synthesizing and secreting a sufficient quantity of progesterone to maintain pregnancy. Thus, the placenta becomes a temporary endocrine organ in the gravid female. During the latter trimester of the gestation period, the urinary excretion of pregnanediol (a normal metabolic product of progesterone) rises markedly.

Following the implantation of a fertilized ovum in the uterus, the secretion of estrogens, progesterone, and hypophyseal as well as chorionic gonadotropins is markedly elevated, as shown in Figure 16.13 (p. 1197). As noted earlier, the net urinary gonadotropin concentration is quite high early in pregnancy, so that this fact provides the basis for a number of pregnancy tests.

The endocrine and other changes that occur during pregnancy, parturition, and lactation are discussed in somewhat greater detail in Chapter 16 (p. 1190 et seq.).

MELANOTROPIN (INTERMEDIN, MELANOCYTE-STIMULATING HORMONE, MSH). The pars intermedia of the hypophysis elaborates two polypeptides which are known as α- and β-melanotropins or α- and β-melanocyte-stimulating hormones (MSH). Collectively, these two substances are referred to as melanotropin or intermedin.

Chemistry. The amino acid sequences found in α- and β-melanotropins from several species are shown in Figure 15.10.

α-MSH contains 13 amino acid residues that are identical in the five species of animal that have been studied, whereas β-MSH derived from human pituitaries contains 22 amino acid residues.

The amino acid sequence of α-MSH is identical with the 13 N-terminal amino acid residues found in ACTH, save that an acetyl group is present on the terminal nitrogen, and the chain ends with a valine carboxamide group. The sequence of the amino acid residues that are found in positions 7 through 13 of β-MSH which has been isolated from monkey, beef, pig, and horse pituitaries is identical with that found in positions 4 through 10 in corticotropin (Fig. 15.7). These facts perhaps may underlie the inherent melanocyte-stimulating activity of corticotropin which is discussed in the next section.

Although α-MSH has not been found in the plasma of humans, β-MSH is present normally at a concentration around 0.02 to 1.0 ng/ml.

General Functions of the Melanotropins. Lower organisms such as amphibia, fish, and reptiles, have cells within their skin which contain granules of a pigment called melanin. These specialized pigment-containing cells, called malanophores, become lighter (ie, they reflect more light) when the pigment collects, or aggregates, about the nuclei. Conversely, when the pigment disperses (ie, the cells absorb more light), then the cells become darker. These intracellular melanin migrations in turn cause the skin of the animal to become lighter with aggregation, and darker with dispersal of the pigment.

Although humans (and birds) do not have melanophores, β-MSH is present in the pars intermedia of the human pituitary, and injection of intermedin into lower vertebrates causes dispersion of melanin granules so that the animal becomes darker.

Neither α- nor β-MSH exert any of the effects of adrenocorticotropic hormone, despite the common structural patterns shared by these substances which were noted earlier. Pure corticotropin, on the other hand, does exert a slight MSH activity. The melanocyte-stimulating potency that is exerted by corticotropin is around 0.5 percent of that of α-MSH and 1 percent that of β-MSH on a basis of equivalent doses of both substances.

In humans as well as in other mammals, the cells which contain melanin pigment are called melanocytes. These cells differ from melanophores in that the pigment does not migrate toward and away from the nucleus, but the granules remain stationary in the cytoplasm; hence the melanotropins exert no effect upon pigment dispersal in mammals. In fact, although a number of alleged functions of the melanotropins have been described, the normal physiologic role of these substances in the human is unknown. However, the mere presence of a substance in the body is no particular guarantee that the substance necessarily has any physiologic role at all. Oxytocin and prolactin have no conceivable functions in the human male, although both hormones are present!

Melatonin, a substance found in the pineal body of mammals and other vertebrates, reverses the skin darkening produced by melanotropin. Melatonin antagonizes the dispersal of melanin induced by melanotropin, and thereby produces pallor of the skin in amphibia and other lower vertebrates.

Possible Clinical Implications of Melanotropin Dysfunction. Although melanotropin does not appear to have any particular functional activity in normal humans, there is some evidence that his hormone may participate in certain disease states. The prolonged exogenous administration of melanotropin, whether of natural or synthetic origin, can accelerate melanin synthesis and result in a darkening of the skin of Blacks. That is, the high concentration of melanin which normally is present in these individuals is elevated still further by excess melanotropin. Unpigmented skin is unaffected by excess melanotropin.

The changes of skin pigmentation observed in a number of endocrine disorders have been ascribed to alterations in the blood concentration of corticotropin, since this compound exerts a certain MSH-like activity. Thus extreme pallor is a symptom of pituitary hypofunction, whereas excess pigmentation is found in patients suffering from tumors which secrete excess corticotropin. Patients having a primary adrenal insufficiency also exhibit hyperpigmentation, indicating that the pituitary is intact as this organ must be functional (in the absence of adrenal activity) in order that excess pigmentation be found at all. But a very high concentration of corticotropin must be present to induce such pigmentation, as the MSH-like

potency of corticotropin is quite low compared to that of the melanotropins, as noted above. It is inferred that the pigmentation which develops in certain patients cannot be caused by elevated corticotropin levels alone, but also may reflect an increased melanotropin secretion. This statement is also borne out by the observation that melanotropin secretion apparently parallels that of corticotropin. It has been found that tumors of nonendocrine origin which secrete corticotropin also secrete a substance that is quite similar chemically to melanotropin. Thus, it would appear that the changes in skin pigmentation observed clinically in some patients reflects an increased secretion of melanotropin rather than of corticotropin.

ANTIDIURETIC HORMONE (ADH, VASOPRESSIN). The posterior pituitary gland, or neurohypophysis, of mammals, including man, secretes two hormones, antidiuretic hormone (or vasopressin) and oxytocin. ADH will be considered first, and this discussion will be followed by a consideration of the properties of oxytocin.

Chemistry. As shown in Figure 15.11, ADH is a peptide comprised of eight amino acid residues, if one considers cystine as a single amino acid. Thus it is an octapeptide. The hormone has been synthesized, and artificial peptides which are even more active than the natural product have been prepared by changing particular amino acids. It also has been determined that both the chain length as well as the cyclic structure of ADH are essential for biologic activity.

In vivo ADH is inactivated rapidly, principally by the kidneys and liver, so that its biologic half-life is approximately 15 minutes. The effects of this hormone upon the kidney occur rapidly but are of a transient nature. In humans, as little as 0.1 μg of ADH will produce a maximum antidiuretic effect. Anesthetics and acetylcholine both induce an antidiuretic response by stimulation of ADH secretion as discussed further below.

General Functions of Antidiuretic Hormone. Two principal physiologic roles are ascribed to ADH. First, and of major importance, this compound increases the permeability of the distal nephron including the distal tubules and collecting ducts of the kidney to water (p. 796). Therefore, the kidneys are the principal target organs for this hormone, so that water readily enters the hypertonic interstitial fluid of the renal pyramids under the influence of ADH. Glomerular filtration apparently is unaffected by ADH. The net result is a retention

A: Cys—Tyr—Phe—Gln—Asn—Cys—Pro—Arg—Gly—NH$_2$
    1   2   3   4   5   6   7   8   9

B: Cys—Tyr—Ile—Gln—Asn—Cys—Pro—Leu—Gly—NH$_2$
    1   2   3   4   5   6   7   8   9

FIG. 15.11. Amino acid sequences in antidiuretic hormone **(A)** and oxytocin **(B)** derived from the neurohypophysis. (After Ganong. *Review of Medical Physiology,* 6th ed, 1973, Lange Medical Publications.)

of water by the body, so that the overall osmotic pressure in the extracellular fluids is decreased. Consequently, ADH plays a major role in the overall regulation of body fluid balance. If a deficiency of ADH is present, the urine volume is increased and the urine becomes hypotonic to plasma, hence there is a net loss of water from the body.

Additional functions of ADH in the regulation of renal function include a decrease of renal medullary blood flow, an increased permeability of the collecting ducts to urea, and possibly an effect upon the rate at which sodium is transported actively out of the ascending limbs of Henle's loops (see also Chapter 12, p. 788 et seq.).

Second, ADH when administered in large doses exerts a hypertensive or pressor effect by inducing a generalized peripheral vasoconstriction of the arterioles and capillaries. Although this ADH-induced vasoconstriction is seen, for example, in coronary and pulmonary vessels, the cerebral and renal vessels dilate secondarily to the rise in systemic blood pressure induced by ADH. The elevation of systemic arterial blood pressure is caused by the direct action of ADH upon the smooth muscle of the vascular walls; however, it is unlikely that this hormone exerts any effect upon the normal regulation of blood pressure in vivo.

Although hemorrhage is a powerful stimulus to ADH secretion, the magnitude or time course of the hypotension that develops following severe blood loss is not particularly affected by the absence of the neurohypophysis. At physiologic secretion rates by the neurohypophysis, ADH probably exerts a negligible pressor function. The laboratory demonstration of a pressor response to ADH has given rise to the synonymous name for this compound of vasopressin.

Mechanism and Factors which Regulate the Secretion of Antidiuretic Hormone. Unlike the adenohypophyseal hormones, the two internal secretions of the neurohypophysis are synthesized within the hypothalamus, in the supraoptic and paraventricular nuclei. The former nucleus synthesizes principally ADH, whereas oxytocin is synthesized in the latter; some overlap exists, however. These hormones, in close association with a protein having a molecular weight of about 30,000, actually migrate down the nerve fibers from the hypothalamus into the nurohypophysis where they are stored as accumulations of microscopically visible granules in the nerve endings within the neurohypophysis (Fig. 15.4C).

A number of stimuli regulate ADH secretion. Included among the factors that elevate ADH release are an increased plasma osmotic pressure, a decreased extracellular fluid volume (as in hemorrhage), severe exercise, neurogenic states such as pain and emotional states, or bodily stress in general. Certain drugs, such as nicotine, or anesthetics such as barbiturates and morphine, also increase ADH secretion. Since ADH secretion is a cholinergic response, acetylcholine and the anesthetics exert their effects through stimulating this mechanism. On the other hand, factors that diminish the secretion of ADH include a decreased plasma osmotic pressure, an increased extracellular fluid volume, sympathetic effects, and alcohol.

ADH is released into the bloodstream from the neurohypophysis by nerve impulses transmitted along the appropriate fibers from the hypothalamus. If the plasma osmotic pressure rises, the discharge rate of the supraoptic neurons increases, and conversely if the osmotic pressure falls, then the discharge rate also decreases. The osmoreceptor cells which are responsible for sensing the alterations in osmotic pressure of the extracellular fluid lie in the anterior hypothalamus, and possibly these elements are distinct from the neurons of the supraoptic nucleus per se. It also is unclear whether or not these osmoreceptor cells are the same as those responsible for regulating water intake via the thirst mechanism (p. 300).

An extremely delicate homeostatic feedback mechanism is present in the body to protect the plasma osmolality, and this mechanism operates to control ADH secretion. When the osmolality of the plasma is normal, there is a continuous and uniform secretion of ADH so that the plasma level of this hormone remains constant. If the plasma osmotic pressure increases, there is a concomitant increase in ADH secretion, and conversely, when the osmotic pressure falls, so does ADH secretion. In fact, changes in plasma osmotic pressure as low as 2 percent in either direction are sufficient to produce marked alterations in ADH secretion.

The total extracellular fluid volume itself also has an influence upon the secretion of ADH. ADH secretion is increased when this volume is low, and ADH secretion decreases when the extracellular fluid volume expands.

Hemorrhage produces the secretion of far larger quantities of ADH than does hyperosmolality of the plasma. Similarly, a decreased plasma volume (or hypovolemia) will cause ADH secretion, despite the fact that the plasma is hypotonic.

As noted earlier, alcohol depresses ADH secretion directly. This factor, plus the volume effect of the total liquid present in most alcoholic beverages, is responsible for the commonly observed diuretic effects of alcoholic drinks.

By way of summary, dehydration or increased salt intake (leading to a hypertonic extracellular fluid) elevate ADH secretion, whereas the release of this hormone is inhibited when the extracellular fluid becomes hypotonic.

Mechanism of Action of Antidiuretic Hormone. The action of ADH, in common with that of a number of other hormones, appears to be mediated through increasing the synthesis of cyclic AMP in the cells of the renal tubules. The relationship between the increased rate of cyclic AMP synthesis under the stimulus of ADH, and the increased absorption of water in the distal nephron is unknown, although presumably the permeability of these elements to water is enhanced in some way by the increased cyclic AMP level.

Clinical Effects of Excess Antidiuretic Hormone (Hypersecretion). In a number of clinical situations, the osmotic regulation of ADH secretion may be overwhelmed by a volume stimulus or by other nonosmotic stimuli, the net result being a decrease in the osmolality of the plasma combined with a retention of water. However, a relative hypersecretion of ADH is present, and water intoxication can be precipitated in such patients if the water intake is sufficiently high. For example, following surgery, there may be a decreased blood sodium level (hyponatremia) combined with water retention. This effect is explained, at least in part, by the mechanism described above.

In patients suffering from edema due to nephrosis, hepatic cirrhosis, or congestive heart failure, both water and salt retention occur. In such conditions it appears that there is a defect in the regulation of neurohypophyseal secretion of ADH combined with a reduced ability of both kidney and liver to remove ADH from the bloodstream. Thus the half-life of the hormone in the circulation is longer. The pathologic mechanism whereby hypersecretion of ADH takes place is unknown.

Clinical Effects of Antidiuretic Hormone Deficiency (Hyposecretion). A deficiency of ADH secretion may be produced by diseases which affect the supraoptic nuclei, the hypothalamo–hypophyseal tract, or the neurohypophysis itself. Surgical removal of the posterior pituitary gland may also induce hyposecretion of ADH; however, some function may be regained after the supraoptic fibers recover from the trauma and commence to secrete ADH once again.

All of these abnormal processes can induce the syndrome of diabetes insipidus. In this condition the symptoms are excretion of large volumes of a dilute urine (or polyuria), a condition which generally is combined with an intense thirst (or polydipsia) which in turn leads to the consumption of large quantities of liquid. The urine voided by patients suffering from diabetes insipidus is characterized by having a very low specific gravity (from 1.002 to 1.006), and the volume of urine eliminated each day may be far in excess of 5 liters. Without the polydipsia which stimulates fluid intake by the patient, a potentially fatal dehydration rapidly develops. Diabetes insipidus can be controlled therapeutically by the parenteral administration of purified ADH.

OXYTOCIN

Chemistry. Similarly to ADH, oxytocin is an octapeptide; the sequence of amino acid residues in this hormone are shown in Figure 15.11 in comparison with ADH. Chemically, both of the neurohypophyseal hormones are quite similar in general structure, and the disulfide linkage found in ADH also is present in oxytocin so that a ring forms part of the structure of the two neurohypophyseal hormones.

General Functions of Oxytocin. The term "oxytocin" is from the Greek and it signifies "rapid birth." This name is indicative of one function of oxytocin, as this hormone directly stimulates the smooth muscle of the uterus. The sensitivity of the uterine musculature to oxytocin is quite variable and depends upon the physiologic state. Estrogen augments the sensitivity of the uterus to oxytocin, whereas progesterone inhibits the action of oxytocin upon the uterus. During late pregnancy, the gravid uterus becomes quite sensitive to oxytocin, and this hormone is secreted at increased rates during labor. Following cervical dilatation, passage of the fetus down the birth canal undoubtedly initiates afferent nerve impulses which ultimately reach the paraventircular nuclei of the hypothalamus, and these impulses induce augmented secretion of oxytocin to stimulate labor contractions (cf. p. 1201).

Oxytocin may also have some role in precipitating labor; however, the mechanisms involved in this process are exceedingly complex, so that neuroendocrine factors alone hardly may be considered to constitute the sole mechanisms responsible for delivery.

A major physiologic role of oxytocin is in the secretion of milk from the lactating breast. Surrounding the alveoli and ducts of the mammary gland are myoepithelial cells that are stimulated directly by oxytocin. Concentration of the myoepithelial elements in turn expels milk from the alveoli of the breast into the larger ducts or sinuses; ie, oxytocin "milks the milk," so to speak, from the alveoli. From the sinuses, the milk passes from the nipple, a process known as milk ejection.

A number of hormones acting in a coordinated fashion underlie growth of the breast as well as milk production and secretion; however oxytocin is essential for milk ejection in the majority of mammals, as discussed further in Chapter 16. Briefly, the neuroendocrine reflex responsible for milk ejection acts as follows. Touch receptors are found abundantly in the breast, but especially surrounding the nipple. Afferent impulses that arise in these receptors are transmitted via the somatic touch pathways to the bundle of Schütz and mammillary peduncle. Thence the impulses pass to the paraventricular nuclei of the hypothalamus, and discharge of the paraventricular neurons in turn causes the secretion of oxytocin from the neurohypophysis. Oxytocin now is transmitted in the blood to the myoepithelial cells of the breast, causing milk ejection.

The usual stimulus to the ejection reflex is the infant suckling at the breast. The time course of this reflex is such that within one minute after the onset of suckling, milk begins to flow. In lactating women, stimulation of the external genitalia as well as emotional stimuli (eg, hearing the child cry) can also evoke an enhanced oxytocin secretion mediated by this reflex; thus occasionally the milk is ejected in spurts.

In addition to the functions of oxytocin described above, this hormone has also been shown to stimulate the smooth musculature of the gallbladder, intestine, ureters, and urinary bladder. However, the excitatory effect of oxytocin is far weaker in these organs than it is in the uterus. ADH also can stimulate contractions in the pregnant (or gravid) uterus; however, ADH is far less potent than oxytocin in this respect.

Clinical use is made of oxytocin preparations during, as well as after, parturition in order to elicit smooth muscle contractions as indicated above. On the other hand, since alcohol inhibits the secretion of oxytocin, this compound sometimes is employed clinically to decrease oxytocin secretion, hence the uterine contractions that could lead to spontaneous abortion.

Regulation of Oxytocin Secretion. The principal factors affecting oxytocin secretion were discussed in the previous section, and nerve impulses transmitted to the hypothalamus provide the major stimuli for evoking an elevated release of this hormone.

The mechanism underlying oxytocin secretion is identical to that for ADH, which was described earlier. It is not clear whether or not oxytocin mediates its effects through an influence upon the rate of cyclic AMP synthesis.

Clinical Defects in Oxytocin Secretion. No gross functional errors have been ascribed to either hyposecretion or hypersecretion of oxytocin, although it is interesting to speculate upon the possibility that such factors may be involved in difficult or prolonged labor, inadequate lactation, or perhaps even fertility problems in some patients.

MISCELLANEOUS HYPOPHYSEAL AND HYPO-PHYSEAL-LIKE COMPOUNDS. In addition to the ten specific peptide hormones that have been isolated from various regions of the pituitary gland as well as the releasing factors elaborated by the hypothalamus, it should be mentioned that a number of additional polypeptides have been isolated from these tissues, although no particular function can be ascribed as yet to these substances.

For example, three peptides have been isolated from the pituitary that stimulate fatty acid release from adipose tissue, hence these compounds are termed lipotropic or adipokinetic hormones. These substances resemble α- and β-MSH structurally, and thus also resemble corticotropin structurally as well as biologically. However, none of the lipotropic hormones is iden-

tical chemically with any of the known pituitary hormones.

In addition to the hypophyseal peptides mentioned above, the placenta produces at least four hypophyseal-like substances: first, chorionic gonadotropin; second, a corticotropinlike substance; third, a prolactinlike substance; and fourth, a somatotropinlike substance. All of these placental substances exert biologic effects like the corresponding pituitary hormones, although their normal physiologic roles, if any, are unknown at present.

CONCLUSION. It is readily evident from the foregoing material that the pituitary gland, through its individual secretions, exerts a profound regulatory influence directly upon the functions of a multitude of bodily processes. Indirectly the pituitary gland also affects the regulation of an even greater number of specific bodily functions through the effects of its hormones upon a number of other endocrine glands. These integrative and regulatory functions of the hypophysis in turn provide an additional mechanism to the nervous system (as well as an extension thereof) for supplying the individual with control systems for achieving a quite sophisticated homeostatic balance within the internal environment. The various roles of the pituitary also provide mechanisms for achieving adaptation of the individual to various external influences.

The pituitary gland is often spoken of as the "master gland" of the body. However, in view of the many complex functions performed by this organ and its various morphologic components, it would be more accurate to consider the pituitary as an ensemble of several discrete neural and glandular entities which normally function in a harmonious fashion.

The remainder of this chapter will be concerned with specific endocrine glands, their functions, and the regulatory mechanisms which control their secretion, hypophyseal and otherwise.

THE THYROID GLAND

The thyroid gland secretes two major iodine-containing hormones, thyroxine (or tetraiodothyronine, T_4) and triiodothyronine (T_3), both of which control and regulate the overall metabolic rate of the body cells and tissues (Fig. 15.12). The principal function of the thyroid gland is to sustain an optimum metabolic level that is essential to the normal function of all the tissues of the body. The general functions of the thyroid hormones are to stimulate oxygen consumption by the cells, to participate in the regulation of carbohydrate and fat metabolism, and to exert important effects in the growth and maturation of the individual.

Thyroid secretion is regulated by thyrotropic (or thyroid stimulating) hormone of the adenohypophysis, as discussed earlier. In turn, the secretion of thyrotropic hormone is regulated partly

A.

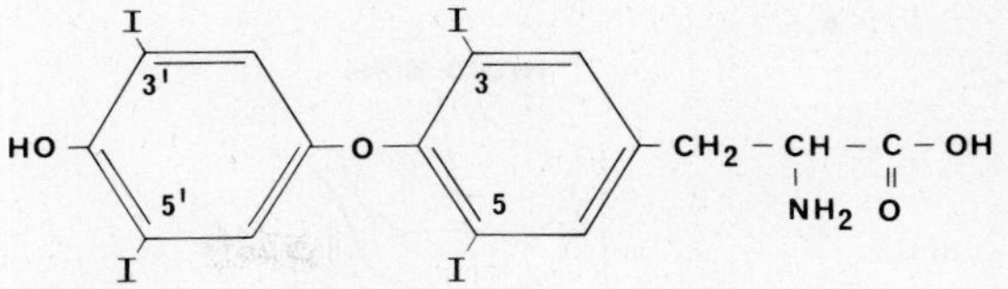

B.

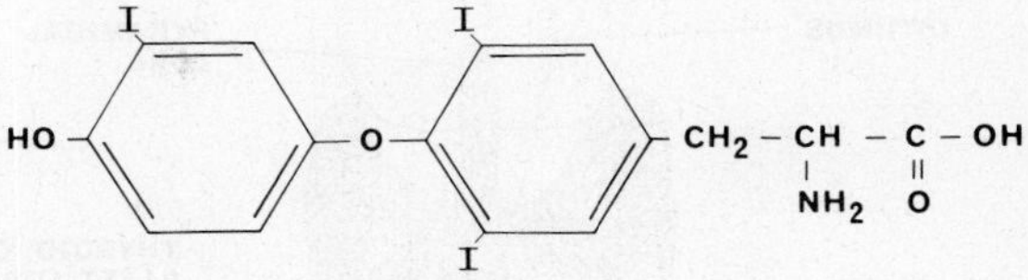

FIG. 15.12. Structural formulas of the thyroid hormones. **A.** Thyroxine (T_4) or 3, 5, 3′, 5′-tetraiodothyronine; **B.** 3, 5, 3′-triiodothyronine (T_3). (After Ganong. *Review of Medical Physiology*, 6th ed, 1973, Lange Medical Publications.)

by a direct negative feedback mechanism produced by elevated circulating thyroid hormone levels which act partly upon the anterior pituitary directly, and partly by a neural mechanism which operates through the hypothalamo–hypophyseal complex and whose effects are mediated via the thyrotropic hormone releasing factor.

Thus, homeostatic regulation of thyroid secretion is achieved in such a way that the metabolic requirements of the individual are adapted to alterations in the internal as well as external environments. In addition to thyroxine and triiodothyronine, the thyroid gland also secretes a third hormone known as calcitonin (or thyrocalcitonin), which reduces the plasma level of calcium and phosphate upon injection into experimental animals. Thus, calcitonin counteracts the hypercalcemia produced by excess parathyroid hormone secretion.

Functional Morphology of the Thyroid Gland

GROSS ANATOMY. The thyroid gland is located in the neck, anterior to the trachea (Fig. 15.13A). In normal adult humans this organ weighs roughly 32 gm (between 25 and 40 gm), and it consists of two lateral lobes which are bridged by a narrow isthmus which structure extends across the trachea immediately below the cricoid cartilages. Each lateral lobe is approximately 5 cm long, 3 cm wide and 2 cm thick. In approximately 30 percent of the population, there is also a pyramidal lobe which extends upward from the isthmus near the left lobe as shown in Figure 15.13A.

Both the size and the appearance of the thyroid gland can vary markedly depending upon a number of factors including age, the reproductive cycle in the female, diet, and the geographic area where the person lives.

Embryologically, the thyroid develops from an evagination of the pharyngeal floor. Occasionally remnants of this evagination are found in adults as the thyroglossal duct. The thyroglossal duct enables one to visualize the evagination from pharynx to neck which gave rise to the adult thyroid. In fact, the pyramidal lobe represents a remnant of this duct.

The thyroid in common with other endocrine organs is richly vascular, and it is supplied with branches of the superior and inferior thyroid arteries, which ramify ultimately into an extensive capillary network within the gland. These capillaries in turn anastomose, forming the venous plexuses from which the superior and middle thyroid veins as well as the inferior thyroid veins arise. Both superior and middle thyroid veins drain into the internal jugular veins, whereas the inferior thyroid veins empty into the innominate (or brachiocephalic) vein.

HISTOLOGY. The thyroid gland is composed of innumerable spherical follicles (or acini) which resemble cysts. The follicles range from 0.02 to 0.9 mm in diameter, and are lined with a single layer of simple epithelium (Fig. 15.13B). Each follicle is filled with a gelatinous and optically homogenous colloid that stains intensely with certain reagents; this proteinaceous material consists of thyroglobulin as well as iodinated amino acids, hence it represents the stored secretory product of the epithelial cells which line the follicle. In this regard, it should be mentioned that the thyroid gland is unique among the endocrines, since it not only synthesizes and secretes hormones, but is also provided with a well-developed mechanism for the extracellular storage of its products within the follicles.

The follicular epithelial cells show great variations in their height, but generally exhibit a squamous to low cuboidal appearance. When the gland is hypoactive, the epithelium is flattened and tends to be squamous, and when in this functional state the follicles are large and distended with abundant colloid. On the other hand, when the gland is hyperactive, the appearance of the cells ranges from a cuboidal toward a columnar appearance, and the follicles are much smaller with correspondingly less colloid. There are many exceptions to these general statements, however, so that an accurate evaluation of the functional state of the thyroid gland solely from microscopic study is difficult.

The nuclei of the glandular (or principal) cells of the follicles are centrally located, spherical, and

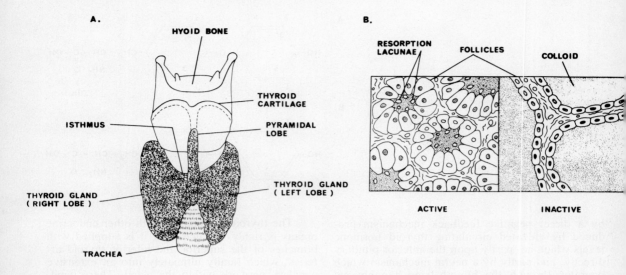

FIG. 15.13. **A.** Gross anatomy of the thyroid gland. **B.** Histology of the thyroid gland. **C.** Ultrastructure of a thyroid cell under normal conditions (left side) and following prolonged administration of thyroid stimulating hormones. (Data from Bloom and Fawcett. *A Textbook of Histology,* 9th ed, 1968, Saunders; Dawson. *Basic Human Anatomy,* 1966, Appleton-Century-Crofts; Ganong. *Review of Medical Physiology,* 6th ed, 1973, Lange Medical Publications; Goss [ed]. *Gray's Anatomy of the Human Body,* 28th ed, 1970, Lea & Febiger.)

contain little chromatin. One or several nucleoli may be present. The basophilic cytoplasm contains numerous thin rodlike mitochondria as well as an abundance of fat droplets and a highly developed endoplasmic reticulum, a feature common to most glandular cells.

Microvilli having canaliculi internally are present upon the luminal surfaces of the epithelial cells. These projections appear to have a significant role in the transport of the thyroid hormones from the follicle to the capillary blood. Lysosomes having a diameter of about 0.5 μ are abundant in the cytoplasm. These structures are shown in Figure 15.13C.

In addition to the principal cells of the follicles, there is present in the thyroid gland a much smaller population of larger cells that commonly are known as parafollicular cells, clear (C) cells, or "mitochondria-rich" cells. These structures stain less deeply in routine histologic preparations than do the principal cells, hence the name clear (or C) cells is applied to those elements. The parafollicular cells are found interspersed among the principal cells of the follicles as well as in an interfollicular location. It has been demonstrated that the parafollicular cells have a quite different biochemical structure than the principal cells. For example, the enzyme patterns are different in both cell types. The parafollicular cells have been shown to be associated with synthesis of the hormone thyrocalcitonin whereas the principal cells elaborate the other thyroid hormones, viz., thyroxine and triiodothyronine. The parafollicular cells of the thyroid gland also are similar to the calcitonin-secreting cells of the parathyroid glands.

Surrounding each follicle is an exceedingly thin basement membrane or lamina which is about 500 Å thick, as well as a delicate reticular fiber network. Each follicle is also invested with a plexus of fenestrated capillaries, and between the capillary networks of adjacent follicles are found the blind-ending terminations of lymphatic vessels. Nerve fibers are abundant along the blood vessels. These nerves represent both postganglionic sympathetic fibers and occasional preganglionic parasympathetic fibers also are found. The nerves that innervate the thyroid apparently serve a vasomotor function, as extrinsic denervation of the thyroid by transplantation to an ectopic locus in the body does not affect the secretion of hormones by this organ.

Biosynthesis and Secretion of the Thyroid Hormones Containing Iodine

IODINE METABOLISM AND IODIDE TRAPPING. An overall survey of iodine metabolism and its relationship to the biosynthesis of the thyroid hormones is presented in Figure 15.14.

An exogenous (dietary) supply of iodine (I_2) or iodide (I^-) is critical for normal synthesis of the thyroid hormones. Iodine, ingested in an elemental form, is reduced chemically to iodide within the intestinal tract and absorbed in this form. Although minute traces of iodine appear in the feces and bile, approximately 98 percent of the iodide absorbed from the gut either is taken up by the thyroid or excreted in the urine. The circulating plasma level of iodide normally is around 0.3 μg/100 ml. About 70 percent of the iodide that is filtered by the kidneys (ie, the iodine that is liberated by thyroxine degradation in the tissues) is reabsorbed. There is only a slight loss of iodide from the body via this excretory route, hence recycling of this trace element plays an important role in conserving body iodine stores. An adult requires an average of about 125 μg of iodine per day in the diet in order to replace this minimal loss.

The thyroid gland actively concentrates plasma iodide by transporting it from the circulation to the colloid within the follicles against a concentration gradient. Underlying the specific avidity of the thyroid for iodide is an active transport mechanism sometimes called the iodide pump or iodide trapping mechanism. The concentration of iodide within a normal thyroid follicle is between 25 and 50 times greater than that in the plasma. During periods of maximal activity, the thyroid can achieve a concentration of iodide that is over three hundred times greater than that in the plasma.

The thyroid cells have a resting transmembrane potential of around -50 mv with respect to the interstitial fluid; therefore iodide is pumped actively into the cell against this electric gradient. Thence it diffuses into the colloid down the electric gradient where it is oxidized into elemental iodine ($2I^- \rightarrow I_2 + 2e^-$). The operation of this mechanism has been studied extensively using tracer doses of radioactive iodine (^{131}I), and the quantities of iodine employed for such experiments are so minute that they do not increase the total body pool of iodine to any significant extent.

The oxidation of iodide to iodine within the thyroid is quite rapid. Following oxidation, the iodine then becomes bound to tyrosine, while that amino acid remains attached to the protein thyroglobulin, as discussed below (Fig. 15.14).

The active transport system for iodide within the follicular epithelium is stimulated by thyrotropin from the adenohypophysis so that the rate of clearance of this ion from the plasma is influenced directly by this pituitary hormone. On the other hand such agents as thiouracil block the binding of iodine to tyrosine, and the iodide trapping mechanism also can be inhibited by anions such as perchlorate, or drugs such as ouabain.

In addition to the thyroid gland, several other tissues are able to concentrate iodide from the plasma against a concentration gradient. Included in this category are the placenta, the ciliary body of the eye, the gastric mucosa, the choroid plexus,

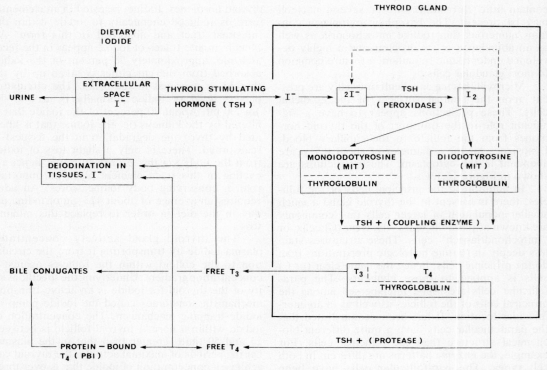

FIG. 15.14. Outline of iodine and thyroid hormone metabolism. T_4 = thyroxine; T_3 = triiodothyronine; MIT = monoiodotyrosine; DIT = diiodotyrosine. (After Harper. *Review of Physiological Chemistry,* 13th ed, 1971, Lange Medical Publications.)

and the mammary glands. However, the rate of iodide uptake by these tissues is not affected by thyrotropic hormone, and the physiologic role of this iodine concentrating process in these tissues is unknown.

THYROGLOBULIN AND THYROID HORMONE BIOSYNTHESIS. There is a continuous dynamic turnover of total iodine in the body, but most especially in the thyroid gland per se. Within this gland, the turnover of iodine has three interrelated aspects. First, iodine uptake by the gland, as discussed above; second, the incorporation of iodide into the iodine-containing thyroid proteins; and third, hydrolysis (proteolysis), and secretion of the thyroid hormones from the colloid into the circulation. The present section will deal with the second of these three aspects of thyroid function; the secretion of the thyroid hormones will be presented in the next section.

There is strong experimental evidence that the biochemical reactions involved in the synthesis of the thyroid hormones take place within the colloid of the thyroid follicles, as shown in Figure 15.14 and outlined in Table 15.5. At least three steps are involved in the sequential reactions which result in the synthesis of thyroxine (T_4), this compound being the principal hormone of the thyroid gland. The reactions that result in thyroid hormone

synthesis occur in the colloid by the iodination and condensation of tyrosine molecules bound in peptide linkage with the protein thyroglobulin. Thyroglobulin in turn is a glycoprotein consisting of four subunits (peptide chains), and has a molecular weight of about 650,000 daltons. There is approximately 10 percent carbohydrate in thyroglobulin consisting of glucosamine, fucose, sialic acid, galactose, and mannose. Hydrolysis of thyroglobulin reveals the presence of typical amino acids; however, in addition to these compounds, iodinated derivatives of histidine and tyrosine as well as thyronine are present in the protein molecule. The reactions involved in the synthesis of the thyroid hormones take place in a definite sequential fashion. The iodinated histidine accounts for only 3 percent of the total thyroid iodine, and this substance is merely a byproduct of thyroid hormone synthesis as it has no known thyroid activity. The mono- and the diiodothyronines exert the characteristic effects of thyroid hormone. The most important thyronines from a physiologic standpoint are thyroxine and 3,5,3'-triiodothyronine (Table 15.6).

The first step in the synthesis of thyroid hormone is the oxidation of iodide (I^-) to iodine (I). The iodine then is "activated" in the presence of hydrogen peroxide (H_2O_2) by iodide peroxidase, with the possible formation of iodinium ion (I^+) or

Table 15.5 OVERALL SEQUENCE OF EVENTS IN THYROID HORMONE METABOLISM[a]

1. Thyroid actively takes up blood iodide.
2. Iodide is stored as inorganic iodide in thyroid gland.
3. Iodide is oxidized to iodine, an enzyme-catalyzed process.
4. Protein-bound tyrosine is iodinated thereby forming mono- and diiodotyrosine.
5. Diiodotyrosine units coupled to form thyroxine (T_4). Mono- and diiodotyrosine units coupled to form triiodothyronine (T_3). Alternatively T_3 could be formed by removal of one atom of iodine from a T_4 molecule.
6. A thyroid protease catalyses release of T_3 and T_4 from thyroglobulin.
7. T_3 and T_4 are secreted from thyroid follicles into the blood.
8. T_3 and T_4 bind to plasma proteins.
9. The protein bound thyroid hormones are distributed to the tissues throughout the body.
10. T_3 and T_4 are metabolized, ie, deiodinated.

[a]*Data from West, Todd, Mason, and Van Bruggen.* Textbook of Biochemistry, *4th ed, 1966, Macmillan.*

See also Figures 15.14 and 15.15.

else hypoiodide, according to the following reaction:

$$I_2 + H_2O \rightleftharpoons IO^- + I^- + 2H^+$$

Within seconds the activated iodine is bound chemically to the 3-position of tyrosine residues, a reaction catalyzed by the enzyme iodinase. The iodinated tyrosine molecules remain attached via their peptide linkage to thyroglobulin (Fig. 15.15). The monoiodotyrosine (MIT) thus formed now is iodinated in the 5 position, forming diiodotyrosine (DIT). Two DIT molecules then react by oxidative condensation to form thyroxine (T_4), liberating alanine in the process. All of these steps take place while the iodinated reaction products remain attached to thyroglobulin. The formation of triiodo-thyronine (T_3) presumably occurs through the condensation of MIT and DIT, whereas reverse T_3 is synthesized by coupling of DIT with MIT (Fig. 15.15).

Although the specific mechanisms whereby the condensation reactions leading to T_4, T_3, and reverse T_3 take place are unknown in detail, it is clear that these reactions require energy and all are aerobic processes.

Changes in the stereochemical configuration of the thyroglobulin molecule may be induced experimentally by alterations in pH or salt concentration. In turn, such physical changes in the protein molecule have been shown to affect profoundly the sequential conversion of tyrosine to thyroxine residues by iodination. This fact is of potential significance in the regulation of thyroid hormone synthesis in vivo. In this regard, it is important to remember that thyroglobulin synthesis is a distinct process that is wholly independent of sequential iodination and hormone synthesis in the thyroid gland.

The functions of thyrotropic hormone from the pituitary gland in regulating various aspects of thyroid gland function, such as overall growth and development of this organ have been discussed earlier; however, it may be reiterated here that injection of thryotropin is followed within a few hours by an increased rate of iodide uptake by the thyroid. This effect is also accompanied by increased rates of synthesis of MIT, DIT, and thyroxine. These effects in turn are followed by increased pinocytosis of colloid as well as by an increased secretion of thyroid hormones as discussed below (Fig. 15.13C).

Thyrotropin mediates its effects on the cells of the thyroid chiefly by increasing the concentration of cyclic AMP in the principal cells of the thyroid follicles. Consequent to this increase in cyclic AMP concentration, the rate of pinocytosis increases rapidly, concomitant with a similarly rapid rate increase in glucose oxidation. The latter effect occurs chiefly through the hexosemonophosphate

Table 15.6 PRINCIPAL IODINATED COMPOUNDS OF THE HUMAN THYROID GLAND[a]

COMPOUND	PERCENT OF TOTAL THYROID IODINE	POTENCY COMPARED TO THYROXINE	AVERAGE DISTRIBUTION IN NORMAL THYROID
		Percent	Percent of Total
3, 5, 3′, 5′-Tetraiodothyronine (Thyroxine, T_4)[b]	35–40	100	35
3, 5, 3′-Triiodothyronine (T_3)[b]	5–8	500–1000	7
3, 3′-Diiodothyronine	1	15–75	Trace
3, 3′, 5′-Triiodothyronine (Reverse T_3)	1	5	Trace
3-Monoiodotyrosine (MIT)	17–28	0	23
3, 5-Diiodotyrosine (DIT)	Trace	0	33

[a]*Data from White, Handler, and Smith.* Principles of Biochemistry, *4th ed, 1968, The Blakiston Division, McGraw-Hill.*

[b]*The physiologic roles of T_4 and T_3 are identical qualitatively, although their potency and time course of action differ somewhat. However because of its relatively greater abundance, thyroxine, or T_4, is the principal thyroid hormone. T_3 is secreted in traces and despite its much greater biologic activity than T_4, this hormone has a relatively minor physiologic role in overall regulation of metabolic activity of the body.*

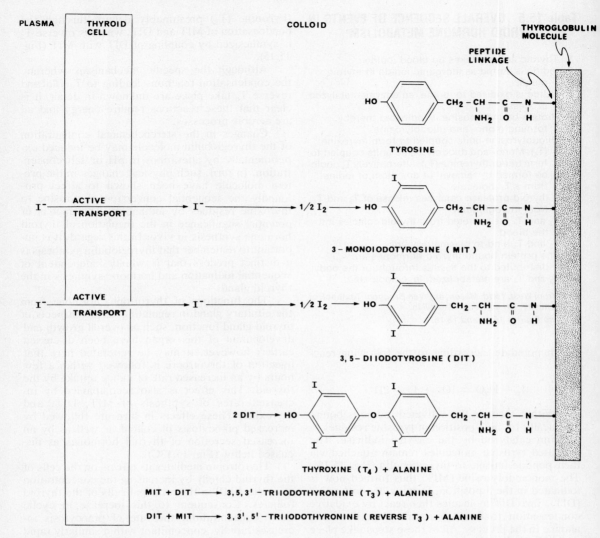

PLASMA   THYROID CELL   COLLOID   THYROGLOBULIN MOLECULE

PEPTIDE LINKAGE

TYROSINE

ACTIVE TRANSPORT

3 - MONOIODOTYROSINE (MIT)

ACTIVE TRANSPORT

3, 5 - DIIODOTYROSINE (DIT)

THYROXINE (T_4) + ALANINE

MIT + DIT ⟶ 3, 5, 3' - TRIIODOTHYRONINE (T_3) + ALANINE

DIT + MIT ⟶ 3, 3', 5' - TRIIODOTHYRONINE (REVERSE T_3) + ALANINE

FIG. 15.15. Biosynthesis of the thyroid hormones. Note that the synthesis of these compounds takes place within the colloid of the thyroid follicle while the intermediates remain attached to a thyroglobulin molecule. Only when the synthesis is complete are T_4 and T_3 released by hydrolysis of the peptide linkage which attaches them to the thyroglobulin. (After Ganong. *Review of Medical Physiology,* 6th ed, 1973, Lange Medical Publications.)

shunt. The additional functional alterations induced by thyrotropin follow, including an increased rate of synthesis of thyroglobulin.

If thyrotropin stimulation continues for some time, the thyroid gland hypertrophies perceptibly, resulting in the development of a goiter. On the other hand, hypophysectomy or pituitary hypofunction, regardless of cause, result in thyroid atrophy and a reversal of all of the effects of thyrotropin stimulation.

MECHANISM OF THYROID SECRETION. The thyroid hormones synthesized within the colloid of the follicles are stored in the bound form (ie, as part of the thyroglobulin molecule) until they are secreted. The principal cells of the thyroid gland

have two important functions. First, they concentrate and transport iodide and also synthesize and secrete thyroglobulin into the follicle. Second, the same cells free the thyroid hormones from the protein moiety by hydrolysis, and then secrete the free hormones into the circulation. As noted earlier this dual role of the thyroid gland, both as a synthetic factory for hormones as well as an extracellular storage organ for those hormones, is unique among the endocrine glands.

The thyroid cells secrete free thyroid hormone into the circulation via the mechanism which is illustrated in Figure 15.13C. The thyroid cells ingest the colloid from the follicular lumen by pinocytosis (p. 94). Once within the thyroid cell, colloid-containing pinocytotic vacuoles (or vesicles) coalesce with lysosomes thereby exposing the thyroglobulin to action by the proteolytic enzymes contained within these cytoplasmic structures. The formation of pinocytotic vescicles within the colloid adjacent to the apical border of the principal cells of the thyroid follicles thus results in vacuole formation within the colloid, and the appearance of reabsorption lacunae as shown in Figure 15.15B.

Following the union between lysosome and pinocytotic vesicle, the lysosomal proteolytic enzymes free thyroxine and other iodinated compounds from the thyroglobulin by breaking the peptide bonds. The free hormones and other iodinated compounds are liberated into the cytoplasm of the cell. MIT and DIT are deiodinated by the enzyme iodotyrosine dehalogenase found in the cytoplasmic microsomes. This enzyme does not react with thyroxine or the thyronine compounds so that these substances pass out of the cell to the interstitial fluid, and thence into the circulation. The iodide liberated from MIT and DIT by the dehalogenase is recycled and used over again.

In the normal adult, approximately 80 μg per day of thyroxine and up to 50 μg per day of triiodothyronine are secreted by the thyroid gland into the circulatory system.

FACTORS WHICH REGULATE THE SECRETION OF THYROID HORMONES. As noted in the previous section, the thyroid hormones are liberated from thyroglobulin by hydrolysis prior to their secretion into the circulation. This step is essential, as thyroglobulin itself has no thyroid hormone activity.

A number of stimuli can affect secretion of the thyroid gland. Vasoactive hormones such as antidiuretic hormone (ADH or vasopressin) and epinephrine exert a direct action upon the thyroid gland to increase hormone secretion. Exposure to a cold environment increases thyroid hormone secretion, whereas a hot environment decreases

secretion. Psychic stimuli also may exert an effect upon thyroid hormone secretion. All of these factors are of relatively minor importance, as the principal mechanism whereby thyroid hormone secretion is regulated under physiologic conditions is by the variations in the blood level of pituitary thyrotropin. Increased circulating levels of both thyroxine and triiodothyronine inhibit thyrotropin secretion through the negative feedback mechanism mediated by the hypothalamo–hypophyseal complex. Lesions produced experimentally in the hypothalamus eliminate the increased secretion of thyrotropin that occurs following the administration of such stimuli as cold or propylthiouracil.

The releasing factor for thyrotropin is a tripeptide that is secreted by the hypothalamus, and this is carried by the hypophyseal portal circulation directly to the anterior pituitary to stimulate release of thyrotropin. There also may be some direct inhibitory effect of circulating thyroid hormone levels upon the release of thyrotropin by the adenohypophysis; however, the principal control appears to be mediated via hypothalamic neural activity. This mechanism also appears to be responsible for adaptation of thyrotropin output to various environmental situations, eg, chronic exposure to a cold or warm ambient temperature.

The normal homeostatic regulation of thyroid hormone output appears to reside in the interrelationships between the circulating hormones and the hypothalamo–hypophyseal complex, as illustrated in Figure 15.16.

Transport and Metabolic Fate of the Thyroid Hormones

TRANSPORT. The free thyroxine secreted into the blood from the thyroid gland is found in the circulation principally in combination with certain plasma proteins. Plasma contains only 2 to 4 $\times$

FIG. 15.16. Feedback mechanism whereby physiologic alterations in the environmental temperature alter thyroid secretion. (+) = stimulation or enhanced secretion; (−) = inhibition or reduced secretion. Note that in similarity to the regulation of corticotropin secretion (Fig. 15.8) enhanced blood levels of T$_3$ and T$_4$ inhibit both hypothalamus and adenohypophysis. (After Ganong. *Review of Medical Phsyiology,* 6th ed, 1973, Lange Medical Publications.)

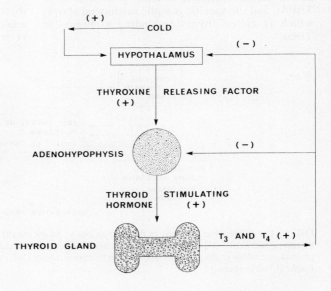

10^{-3} μg per 100 ml of free thyroxine (or about 6×10^{-11} moles per liter). This free thyroxine is in equilibrium with about 9 mg per 100 ml of thyroxine that is bound to the plasma proteins, as well as with the thyroxine which is bound to tissue proteins. These interrelationships are shown in Figure 15.17.

Free thyroxine is being secreted continuously into the circulation by the thyroid and it is this fraction which exerts the physiologic effects of the hormone to be discussed in the next section. In addition, the free thyroxine fraction also probably exerts the regulatory effect upon thyrotropin secretion by the adenohypophysis.

The addition of free thyroxine to the plasma shifts the equilibria illustrated in Figure 15.17 so that more thyroxine enters the cells. On the other hand, an increase of the thyroxine binding globulin (TBG, see below) of plasma causes the tissue uptake of thyroxine to decrease, thereby lowering the rate of thyroxine metabolism. Red blood cells also bind thyroxine; however, they do not use or metabolize the hormone.

Triiodothyronine is found in plasma at a total concentration of approximately 0.3 μg/100 ml, and around 1.5×10^{-4} μg/100 ml is free (0.5 percent) whereas 75 percent of the remainder is bound to TBG, and the rest to plasma albumin. The binding of triiodothyronine to these two proteins is considerably weaker than is that of thyroxine, so that T_3 has a much shorter half-life than T_4 and its actions are far more rapid within the tissues.

The total quantity of protein bound iodine that is found in the plasma ranges between 3.5 and 9.0 μg/100 ml (average 6.0 μg/100 ml) in normal individuals. Of this total amount, 95 percent is thyroxine; consequently the average concentration of protein-bound thyroxine is approximately 9.0 μg/100 ml of plasma.

Within the plasma, the proteins that bind the thyroid hormones are albumin, a precursor of albumin called thyroxine-binding prealbumin (or TBPA), and the specific globulin mentioned above, which is called thyroxine-binding globulin (or TBG).

The total quantities of thyroxine which these proteins can bind when saturated with hormone (ie, their binding capacity), vary widely. Albumin has the greatest capacity, TBPA is intermediate, and TBG has the least capacity for binding thyroxine. In contrast to their total capacities, however, the affinities of these three proteins for thyroxine are quite different. The specific avidity of TBG for thyroxine is such that most of this hormone in the plasma is bound to TBG, small quantities are bound to TBPA, and none, for all practical purposes, is bound to albumin.

Briefly, then, the capacity of these plasma proteins for binding thyroxine reflects their potential ability to combine with hormone, whereas their affinity denotes the actual binding of thyroxine under physiologic conditions. Even though TBG is present in the plasma at a concentration of only 0.2 μg/100 ml, this protein carries about 90 percent of the total protein-bound thyroxine in normal plasma; therefore, roughly over 30 percent of the available binding sites on this protein moiety are occupied under physiologic conditions.

METABOLISM OF THE THYROID HORMONES. Thyroxine and triiodothyronine undergo deiodination and deamination in a wide variety of tissues; consequently, only minute traces of these hormones, if any, appear in the urine. The deiodination of thyroxine to triiodothyronine has been shown to occur in man. Since T_3 is more potent biologically and exerts its effects more rapidly than does T_4, it has been postulated that thyroxine may be inactive until it has been converted to T_3. This concept is substantiated by the fact that the rate of deiodination of thyroxine in skeletal muscles is proportional to the metabolic rate. However T_3 binds more weakly to proteins than does T_4, so that this effect could be explained equally well by the greater availability of free T_3 in vivo than of T_4.

Within the tissues, the deamination of the thyroid hormones T_4 and T_3 produces pyruvic acid analogues of these compounds. Thus T_4 is converted to 3, 5, 3′, 5′-tetraiodothyropropionic acid

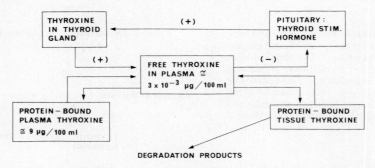

FIG. 15.17. Thyroxine fractions in the human body. An increased TSH secretion from the pituitary gland (+) stimulates thyroxine secretion, and the elevated plasma free thyroxine level in turn inhibits TSH stimulation (−) as well as increases the plasma and tissue protein bound levels of the hormone. (After Ganong. *Review of Medical Physiology*, 6th ed, 1973, Lange Medical Publications.)

(T_4 PROP), whereas T_3 yields 3, 5, 3'-triiodothyropropionic acid. Subsequent decarboxylation of these two pyruvate analogues yields 3, 5, 3', 5'-tetraiodothyroacetic acid (TETRAC), and 3, 5, 3'-triiodothyroacetic acid (TRIAC) respectively.

Both thyroxine and triiodothyronine are conjugated in the liver forming glucuronides as well as sulfates. These conjugates in turn are excreted in the bile and thus enter the intestine where they undergo hydrolysis and are reabsorbed via the enterohepatic circulation. The remainder of the conjugated T_4 and T_3 compounds are eliminated in the stools, so that roughly 15 percent of the total daily iodide loss from the body occurs through the intestinal route.

Mechanisms of Action and Physiologic Effects of the Thyroid Hormones

MECHANISMS OF ACTION. It has been demonstrated experimentally that thyroxine, as well as its chemical analogues, decrease the efficacy with which energy is converted in the cell. This is accomplished by an uncoupling of phosphorylation from oxidation within the mitochondria as described in detail on page 931, and illustrated in Figure 14.20. As the direct consequence of this effect, a lower proportion of the energy which is released by oxidation is stored in high energy phosphate bonds so that a greater proportion of the total energy that is released by oxidation appears as heat. Consequently, the thyroid hormones increase the oxygen consumption of practically all body cells and tissues (the exceptions to this general rule are noted below). This effect upon oxygen consumption is known as the calorigenic action of thyroxine. Thyroxine also stimulates mitochondrial respiration and phosphorylation, as well as the action of respiratory and other enzymes.

Since the calorigenic effect of thyroxine can be blocked by compounds that inhibit protein synthesis (eg, actinomysin D, puromycin) the possibility exists that the stimulation of metabolic rate produced by this hormone is secondary to a direct effect upon ribonucleic acid synthesis; hence the mechanism underlying the calorigenic effect would be more complex than that outlined above.

PHYSIOLOGIC EFFECTS OF THYROXINE. In general the broad spectrum of effects that are produced in the body by thyroxine are secondary to the calorigenic action of this hormone. Thyroxine also affects the processes of growth and maturation, exerts a regulatory effect upon lipid metabolism, and stimulates the absorption of carbohydrates from the gut. Thyroxine also elevates oxygen dissociation from hemoglobin by increasing the concentration of 2,3-DPG in red blood cells (p. 747).

As noted earlier, triiodothyronine is more potent than thyroxine in stimulating the metabolic processes summarized above. Reverse triiodothyronine, on the other hand, is metabolically inert for all practical purposes.

Calorigenic Effect of Thyroxine. As mentioned above, thyroxine directly exerts a calorigenic effect upon most cells of the body. Notable exceptions to this general rule are the adult brain, anterior pituitary, lymph nodes, testes, uterus, and spleen. Thyroxine, in fact, directly reduces oxygen consumption by the anterior pituitary, probably because of the inhibitory effect of the hormone upon thyrotropin secretion.

Following oral or parenteral administration of one specific dose of thyroxine there is a lag period of several hours before the metabolic rate commences to increase; thereafter, the effect of the hormone is prolonged, and it may last for over six days. The overall magnitude of the calorigenic effect produced by thyroxine depends upon two major factors. First, the metabolic rate prior to administration of the hormone. If the metabolic rate is low initially, then the percentage increase produced in the metabolic rate by a given dose of thyroxine is high. On the other hand, if the initial metabolic rate is high, then the increment of change in metabolic rate produced by a given quantity of the hormone is small. These relationships hold true in both normal (or euthyroid) individuals, as well as in hypothyroid (or athyreotic) patients. Second, the net rate of catecholamine secretion also affects the response to thyroxine administration. In general the effect of thyroxine combined with epinephrine results in a mutual potentiation of the effects of each hormone alone.

The stimulation of metabolic rate that occurs as a primary effect of thyroxine administration is accompanied by a number of secondary effects upon various metabolic processes. For example, in adults the increased metabolic rate which is produced by thyroxine is attended by an increased degradation (or catabolism) of the body proteins, urinary nitrogen excretion is enhanced, and if the dietary intake of food is not increased accordingly, the endogenous protein stores are catabolized together with fat stores and weight is lost.* In children, thyroxine is a factor that is essential for normal growth and maturation of the skeleton. It should be reiterated also that thyroxine acts synergistically together with somatotropin. In hypothyroid children, bone growth is much slower and epiphyseal closure is delayed. In such hypothyroid individuals, thyroxine administered in small doses will cause a retention of nitrogen (ie, it induces a positive nitrogen balance), since this hormone also stimulates growth. However large doses of thyroxine administered to either euthyroid or hypothyroid children produces protein catabolism as seen in adults.

The protein catabolism that follows excess thyroid hormone secretion (or administration) in either adults or children may be so pronounced that weakness of the skeletal muscles (thyrotoxic myopathy) develops. This myopathy is accom-

---

*Thyroxine administration is not a desirable way in which to control obesity (cf. p. 1215).

panied by a marked elevation in urinary creatine excretion. Furthermore, an excess of potassium is liberated during the protein catabolism, and this ion is excreted in the urine together with elevated levels of uric acid and hexosamine. Also, bone protein is degraded in abnormally great quantities in conditions where excess thyroid hormone is present, so that hypercalcemia and hypercalcuria accompanied by osteoporosis are found.

A cold environment stimulates thyroxine secretion, and the increased heat production which results from this response to cold (thyroxine thermogenesis) is important in the maintenance of body temperature. In general, large doses of thyroxine stimulate sufficient excess heat production to induce a slight rise in body temperature. This effect in turn stimulates the bodily mechanisms involved with heat loss, including cutaneous vasodilatation, and this effect results in a decreased peripheral resistance. Under such circumstances the cardiac output actually increases because of the combined actions of thyroid hormone and catecholamines (epinephrine and norepinephrine) directly upon the heart. The net effect, then, is an elevated pulse pressure combined with an increased heart rate so that the circulation time is decreased.

Relationships Between the Actions of Thyroxine and the Catecholamines. The intimate relationships among thyroxine, epinephrine and norepinephrine are quite important, both from a physiologic as well as a clinical standpoint.

Epinephrine exerts certain actions similar to those produced by thyroxine, eg, it increases the metabolic rate, increases the irritability of the nervous system, and induces effects upon the cardiovascular system which are similar to those of thyroid hormone. In general norepinephrine exerts similar effects to those of epinephrine; however, unlike thyroxine, the actions of both catecholamines are evanescent (ie, of quite short duration). In thyroidectomized subjects, the catecholamines exert no calorigenic effect, whereas in the presence of excess thyroxine the toxicity of the catecholamines is increased considerably.

In hyperthyroidism, the secretion of catecholamines is generally normal. The tremors, sweating, and cardiovascular effects produced by thyroxine are blocked by sympathectomy, drugs (eg, guanethedine, reserpine) which reduce tissue catecholamine stores, or that block β-adrenergic receptors (eg, propranolol). Sympathetic blockade may also reduce the calorigenic effect of thyroxine to some extent, although the plasma proteins which are bound to iodine remain unaltered.

As mentioned above, oxygen consumption of the brain remains unchanged in both hyperthyroid and hypothyroid states; that is, thyroxine does not influence brain metabolism in adults. Thyroxine also has no direct effect upon the cerebral blood flow or glucose utilization by the brain and may exert some effects upon the brain indirectly by increasing sensitivity to circulating catecholamines, even though only minute traces of thyroxine per se are able to penetrate the blood–brain barrier. Thus, the increased stimulation of the reticular activating system seen in hyperthyroid individuals may result from an increased catecholamine sensitivity that is caused by the presence of excess thyroxine.

In infants, on the other hand, thyroxine may stimulate oxygen consumption by the brain as well as exert other actions upon the nervous system. Presumably, these effects are due to incomplete development of the blood–brain barrier in such individuals (cf. p. 256).

The peripheral nervous system is also affected markedly by thyroxine. For example, the reaction time of phasic reflexes such as the ankle jerk (Achilles tendon reflex) is shortened in hyperthyrodism because of the increased irritability and conduction velocity of the nerves and synaptic pathways.

In summary, many of the effects of thyroxine, especially on the nervous and cardiovascular systems, are due in large measure to adrenergic effects. A clear distinction between these two endocrine factors cannot always be made. Furthermore, it is important to remember that thyroxine and the catecholamines exert mutually synergistic effects, although the mechanisms whereby these interactions take place are unknown.

Thyroid Hormones and Carbohydrate Metabolism. The increased rate of carbohydrate absorption from the digestive tract stimulated by thyroxine appears to be independent of its calorigenic activity. In hyperthyroid states the blood glucose rises quickly after a meal rich in carbohydrates. This effect may be so pronounced that the renal threshold is exceeded and glucosuria ensues. The liver glycogen concentration remains low, however, because of the elevated metabolic rate combined with the effects of circulating catecholamines upon hepatic glycogen stores. These factors cause the blood glucose level in hyperthyroid individuals to fall rapidly once again.

Thyroid Hormones and Cholesterol Metabolism. Cholesterol synthesis, as well as the hepatic mechanisms for removing this steroid from the plasma, both are stimulated by thyroxine. It has been observed that the plasma cholesterol level declines following the administration of thyroxine but before the calorigenic effect is seen. This shows that the cholesterol-lowering effect of this hormone is exerted independently of the stimulation of oxygen consumption.

Thyroid Hormones and Growth. In addition to the growth-promoting actions of thyroid hormone that are exerted together with somatotropin (eg, by stimulation of protein synthesis or anabolism, hence nitrogen retention, in young individuals), thyroxine also has been shown to activate certain inherited genetic mechanisms in the cells. Evidence for this statement is derived principally from experiments conducted on larval amphibia. However, studies of this type indicate the far-reaching effects of the thyroid gland on many cells and their mechanisms insofar as normal growth, development, and maturation are concerned.

Effects of Thyroid Dysfunction

In humans, the symptoms and consequences of hypothyroidism and hyperthyroidism result from the physiologic effects of either a deficiency or an excess of the thyroid hormones. The several clinical conditions resulting from thyroid dysfunction will be considered first. This discussion will be followed by a general presentation of certain laboratory measurements useful in the determination of thyroid function in patients. The final portion of this section will deal with several chemical agents that inhibit thyroid function, and thus find application in the therapeutic management of certain thyroid disorders.

HYPOTHYROIDISM: IODINE DEFICIENCY, CRETINISM, AND MYXEDEMA

Iodine Deficiency and Simple Goiter. If the dietary intake of iodine is grossly inadequate, eg, below 10 μg per day, the synthesis of thyroxine is impaired, and the secretion of this hormone is insufficient to meet the normal requirements of the body. As the circulating blood concentration of thyroxine falls, the secretion of thyrotropin by the pituitary increases, and the thyroid gland hypertrophies as a consequence; the result is an iodine-deficiency goiter. In this condition, the thyroid may enlarge to enormous proportions. Despite the thyroid hypertrophy, the patient is nonetheless hypothyroid because of a lack of hormone synthesis.

Iodine-deficiency goiter has been known since ancient times, and was treated prior to the routine addition of iodide to table salt by giving the patient the ash from burned seaweed, a substance rich in iodine. This condition is also called endemic goiter, because of the lack of iodine in the soil in a number of inland regions of the world, for example, in Central Europe and around the Great Lakes of North America. Leaching of the iodine from the soil in such areas results in the growth of food plants that are deficient in this element; hence the diet must be supplemented with iodine from another source.

Cretinism. Cretinism is found in children who are hypothyroid from the time of birth. Such individuals are mentally retarded to an extreme degree, are dwarfed in stature, and have enlarged, protruding tongues and abdomens. Hypothyroid infants exhibit defective myelination of the nerves, and the basal metabolic rate is low. Permanent mental retardation ensues unless the condition is recognized and therapy is instituted shortly after birth. Once the clinical syndrome of cretinism has developed treatment will not reverse the mental retardation.

Prior to the extensive use of iodized salt in the diet, maternal iodine deficiency was the most usual cause of cretinism. Other, less common, etiologic factors now are recognized, including congenital abnormalities of thyroid function that lead to hyposecretion of thyroid hormones.

Myxedema. In adults the syndrome that results from hypothyroidism is known as myxedema. It must be realized, however, that this term also denotes specifically the skin changes which are found in this syndrome (see below).

Hypothyroidism may result from a number of disease processes that directly affect thyroid gland function; therefore, thyrotropin administration has no beneficial effect upon thyroid hormone secretion. On the other hand, myxedema can result indirectly from failure of the adenohypophysis, and in this instance the thyroid gland is responsive to exogenous thyrotropin administration.

The Syndrome of Myxedema. In an adult whose thyroid gland is completely nonfunctional (athyreotic), the BMR falls to as low as -40, so that a cold environment is poorly tolerated. The hair becomes coarse and sparse; the skin is puffy, dry, and yellowish, (ie, carotenemia is present). The voice becomes hoarse and speech is slow; memory is poor, thought processes are slow, and some patients may develop a pronounced psychic derangement (or "myxedema madness"). Various degress of hypothyroidism can modify the severity of these symptoms in the syndrome of myxedema.

The Skin Condition of Myxedema. In this situation, the term myxedema denotes a derangement of metabolism in the skin, characterized by an abnormal subcutaneous accumulation of various proteins that are conjugated with polysaccharides, chondroitin sulfuric acid, and hyaluronic acid. These complex substances are normally present in the skin. In hypothyroidism they are found in excess, and since these compounds promote water retention, the skin takes on a characteristic bloated appearance which also is known as myxedema.* Exogenous thyroxine administration stimulates the metabolism of the excess proteinaceous complexes, and a diuresis ensues which eliminates the excess water. Ultimately myxedema reverts to normal following thyroxine administration.

There are several additional functional derangements that may be observed in hypothyroid patients in general. First, there is a decreased intestinal absorption of cyanocobalamin (vitamin B_{12}) combined with a reduced metabolism of the bone marrow, thus an anemia develops. This condition may be reversed by the administration of

---

In cardiac failure a "pitting edema" develops, particularly in the lower extremities. That is, when a finger is pressed firmly into the edematous region, a depression or pit remains when the finger is withdrawn. In myxedema, however, such pitting does not occur.

thyroxine so that exogenous cyanocobalamin is absorbed normally once again.

Second, the secretion of milk is decreased in hypothyroidism, and increased by exogenous thyroxine administration.

Third, although thyroid hormones do not stimulate uterine metabolism directly, thyroxine is essential for normal sexual (menstrual) cycles to take place, as well as for fertility.

Fourth, in the syndrome of myxedema, it was mentioned that the skin was yellowish. Thyroxine is essential for the conversion of carotene to vitamin A within the liver (p. 886). Consequently in hypothyroid patients there is an accumulation of carotene in the blood (carotenemia) which produces the yellowish coloration of the skin. Clinically, carotenemia can be distinguished readily from jaundice. Carotenemia does not cause the sclera of the eyes to become yellow, whereas jaundice produces a yellow tint in this structure.

HYPERTHYROIDISM OR THYROTOXICOSIS: GRAVES' DISEASE (EXOPHTHALMIC GOITER). Clinically the syndrome encountered in hyperthyroidism is characterized by an enormous increase in appetite (or hyperphagia); nervousness and tremor of the fingers; loss of weight; a warm, soft, moist skin; and a BMR of from +10 up to +100. The heart rate is elevated at rest and other cardiovascular symptoms are prominent.

A number of thyroid disorders (eg, benign or malignant tumors) can produce this clinical picture, however Graves' disease (or exophthalmic goiter) is by far the most common form of thyrotoxicosis encountered clinically. In this condition, the thyroid exhibits a diffuse enlargement and hyperplasia and the eyeballs protrude or bulge from their sockets (exophthalmos), and the patient has a "wild" or "staring" appearance.

In Graves' disease the thyroid hyperfunction is caused by a gamma globulin in the plasma unique in that it is capable of stimulating the thyroid gland directly. This protein mimics the action of thyrotropin, but the action of the globulin is more prolonged than that of thyrotropin. The protein which is responsible for producing Graves' disease has been termed long-acting thyroid stimulation (or LATS). This substance is known to be an antibody formed in response to an antigenic component of thyroid tissue, although the etiologic factors that are involved in the production of the antigen are unknown.

Neither LATS nor thyroxine are responsible directly for producing the exophthalmos of Graves' disease. In fact, thyroidectomy may exacerabate this ocular condition. Patients suffering from Graves' disease have a factor in their plasma which can induce exophthalmos in animals. This substance is not thyrotropin, but it may be a derivative of this hormone; quite possibly LATS also is of hypophyseal origin.

Thyrotoxicosis places the cardiovascular system under considerable stress, so that many pa-tients with hyperthyroidism exhibit symptoms that are predominantly, or even entirely, of this type. The heart disease produced by hyperthyroidism generally can be cured; however, early and accurate diagnosis is essential to achieve this goal.

It is axiomatic in clinical medicine that heart failure develops (and is present) whenever the cardiac output is insufficient to maintain a normal tissue perfusion. This statement is true even though the cardiac output may be elevated well above the normal range. In thyrotoxicosis the fall in peripheral resistance caused by cutaneous vasodilatation causes a marked increase in the pulse pressure (ie, the systolic pressure minus the diastolic pressure). In fact, there is a great similarity between the vasodilatation found in thyrotoxicosis, and the presence of a large arteriovenous fistula. In either situation if the compensatory mechanisms which increase the cardiac output are insufficient, then heart failure ensues despite the high cardiac output.

Prolonged and severe hyperthyroidism is also complicated by hepatic failure. The liver of the thyrotoxic patient is chronically depleted of glycogen so that it becomes highly sensitive to various types of injury. Furthermore, if the increased metabolic requirements for vitamins and calories induced by hyperthyroidism are not supplied, then the resulting malnutrition adds to the impairment of hepatic function. Lastly, hepatic dysfunction adds to the complications of long-term thyrotoxicosis because the excess circulating thyroid hormones are removed from the circulation and inactivated less rapidly.

MEASUREMENTS USED IN THE CLINICAL EVALUATION OF THYROID FUNCTION

Plasma Iodine Levels. As noted earlier, the protein-bound iodine (PBI) level in the plasma averages 9 μg per 100 ml in normal persons. In some laboratories the plasma is extracted with butanol prior to the chemical determination of the iodine concentration. This procedure effectively extracts iodine from monoiodotyrosine, diiodotyrosine, and thyroglobulin; therefore thyroxine iodine alone is measured. The plasma concentration of butanol extractable iodine (BEI) is somewhat lower than that of PBI. The BEI ranges between 2.2 and 5.6 μg per 100 ml (mean 4.0 μg per 100 ml) in normal individuals.

Both PBI and BEI levels are significantly lower than the normal mean values in plasma taken from hypothyroid patients, whereas they are higher than normal in plasma from hyperthyroid individuals. In this regard, it is important to remember that large quantities of exogenous iodine can elevate both PBI and BEI levels for months after the administration of the iodine. Falsely high values of plasma iodine are encountered in patients following the injection (or ingestion) of certain chemicals which contain iodine. For example, Lugol's solution (concentrated potassium iodide), iodopyracet (Diodrast), and other contrast media all elevate the plasma iodine level for long periods following their administration because these radiopaque iodine-containing substances bind to the plasma proteins, so that they are retained in the plasma and are eliminated very slowly. On the other hand, the plasma iodine levels may be inaccurately low in patients being treated with mercurial diuretics.

Radioactive Iodine Uptake as a Measurement of Thyroid Function. Tracer doses of ^{131}I can be administered orally to patients in order to assess thyroid function. Such doses of isotopic iodine are so small that they exert no effect upon overall thyroid function; however by placing a gamma ray counter over the neck, the rate of uptake of this element by the thyroid gland may be measured and plotted against time following administration of the isotope (Fig. 15.18A). Another counter placed

elsewhere on the body can be used to assess the total average circulating ^{131}I at the same time.

In hyperthyroidism, the quantity of radioactivity in the thyroid increases sharply, then plateaus, and may commence to fall within 24 hours (Fig. 15.18B). This is due to the abnormally rapid incorporation of iodine into, and release of, the thyroid hormones. In a normal individual, on the other hand, the uptake of ^{131}I by the thyroid gland generally is still increasing at the end of 24 hours.

In marked contrast to the picture in hyperthyroidism, a hypothyroid patient exhibits a much lower rate of iodine uptake than normal (Fig. 15.18C), and the peak levels of iodine in the thyroid always remain far below normal.

Certain aspects of iodine metabolism must be borne in mind when interpreting iodine uptake

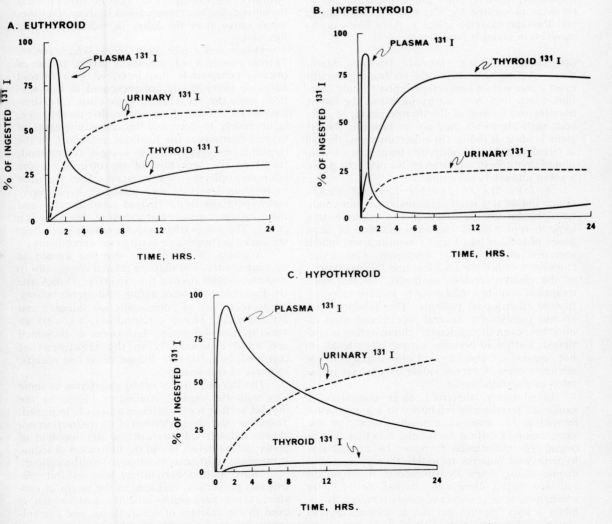

FIG. 15.18 **A.** ^{131}I uptake in normal (euthyroid), hyperthyroid **(B)**, and **(C)**, hypothyroid individuals. (After Ganong, *Review of Medical Physiology*, 6th ed, 1973, Lange Medical Publications.)

curves. A patient having a functionally normal thyroid gland, but who has a daily iodine intake that is barely adequate to prevent iodine deficiency goiter, will exhibit an ^{131}I uptake similar to that seen in a hyperthyroid individual. On the other hand, a patient whose body iodine pool is large because of an elevated dietary iodide intake will show a reduced ^{131}I uptake by the thyroid, even though his thyroid is functionally normal. This effect is caused by excessive dilution of the ^{131}I by mixing with the total body iodine; therefore, proportionally fewer radioactive atoms are taken up per unit time by the gland.

Radioactive iodine in high concentrations (nontracer doses) selectively kills thyroid tissue because the specific avidity of this gland for the element enables the radiation to be concentrated in the thyroid itself. ^{131}I has been used successfully in the treatment of thyroid carcinoma as well as in certain benign thyroid diseases. It is essential to remember, however, that gamma radiation per se (such as is emitted by ^{131}I) may be carcinogenic and this carcinogenic effect is more likely to be manifest in younger patients.

Compounds Which Inhibit Thyroid Functions. Antithyroid chemicals or drugs generally exert either one of two actions upon thyroid function. First, they may act by inhibiting the iodine trapping mechanism of the thyroid gland; or, second, such chemicals may act by blocking the organic binding of iodine. In either situation, the fall in circulating thyroid hormones leads to a stimulation of thyrotropin secretion by the pituitary gland; a goiter ultimately develops.

Anions Which Inhibit Thyroid Function. Iodide (I$^-$) itself may, under certain conditions, inhibit thyroid function. For example, in a hyperthyroid patient, the administration of large doses of iodides (eg, Lugol's solution) will inhibit secretion of the thyroid hormones. Colloid accumulates within the follicles, and the vascularity of the gland decreases markedly, so that such treatment is quite useful prior to surgery in hyperthyroid (thyrotoxic) patients. This inhibitory response persists for several weeks and then diminishes, even though iodide adminstration is continued. Euthyroid persons, on the other hand, do not respond (or else respond only slightly) to the administration of excess iodide by changes in the mass of glandular tissue.

In summary, therefore, large quantities of iodide are temporarily inhibitory to a hyperplastic thyroid gland, whereas a small quantity of the same anion is critical to normal function of this organ. The differential response by normal and hyperthyroid patients to iodide therapy is useful diagnostically when thyrotoxicosis is suspected but not yet a clear-cut clinical entity. The mechanism whereby excess iodide temporarily inhibits a hyperplastic, but not a normal, thyroid gland is unknown.

Several other monovalent anions act upon the normal thyroid gland to inhibit the active transport of iodide. These substances act by direct competition with iodide for loci on the iodine trapping (or transport) mechanism, hence the net effect is a chemically induced iodide deficiency within the thyroid tissue that is caused by competitive inhibition with iodine uptake. Examples of anions that affect iodide uptake in this manner are chlorate, perchlorate, nitrate, periodate, and biiodate. These ions are concentrated within the gland in a manner similar to iodide; however, since they cannot substitute for (or replace) iodine in the synthesis of thyroid hormone molecules, then a deficiency of thyroxine develops.

Thiocyanate ions also inhibit iodide uptake by the thyroid gland through the same mechanism of competitive inhibition. In contrast to the other anions, this substance is not concentrated within the colloid.

Perchlorate and thiocyanate are used occasionally as therapeutic agents for clinical thyrotoxicosis; the former anion is about ten times more active than the latter in reducing thyroid function.

Drugs and Other Substances Which Inhibit Thyroid Function. Examples of several classes of organic compounds that interfere with thyroid hormone biosynthesis are presented in this section. Since the net effect of these chemicals, similarly to the effects of the anions discussed above, is ultimately to decrease the circulating level of thyroid hormones, the feedback inhibition to the hypothalamo–hypophyseal system is reduced; therefore the secretion of thyrotropin by the adenohypophysis is increased. The elevated thyrotropin levels in turn stimulate hypertrophy and hyperplasia in the thyroid gland directly, and the subsequent gross enlargement of this organ is a goiter. The compounds which produce this effect are called goitrogens or goitrogenic compounds.

A group of compounds that are known as thiocarbamides, and that are related chemically to thiourea, inhibit thyroid function (Fig. 15.19A and B). Examples of potent antithyroid agents belonging to this class of compounds are thiourea and 2-thiouracil. These compounds, as well as 4-methyl and 4-n-propyl derivatives of thiouracil are used extensively in the treatment of thyrotoxicosis, Graves' disease being one specific example of this condition.

The thiocarbamides act by interfering in some way with the organic binding of iodine by the thyroid so that hormone biosynthesis is impaired. Therefore, the administration of thyroxine (but not excess iodide!) will prevent the development of goiter, as the defect lies in the utilization of iodine subsequent to its accumulation within the colloid.

Many aniline derivatives also exhibit goitrogenic properties. Included in the group of substances are para-amino-salicylic acid, which is used in the therapy of tuberculosis, and the sulfonamides used as antibacterial agents.

Sulfonylurea derivatives such as tolbutamide

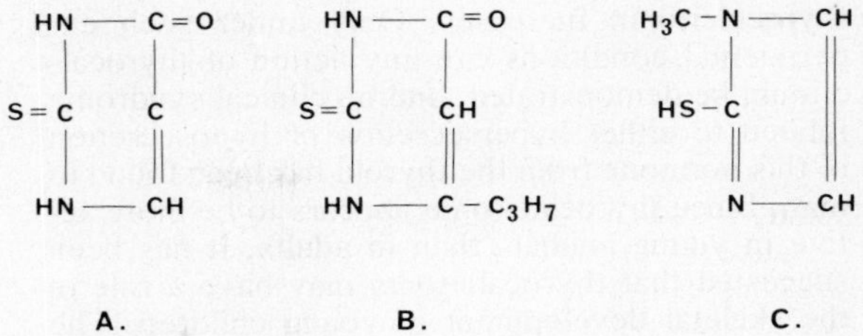

FIG. 15.19. Antithyroid compounds. **A.** Thiouracil. **B.** Propylthiouracil. **C.** Tapazole (1-methyl-2-mercapto-imidazole). (After Ganong. *Review of Medical Physiology*, 6th ed, 1973, Lange Medical Publications.)

(which is used clinically in the treatment of diabetes mellitus) also has goitrogenic properties. Likewise, mercaptoimidazoles, for example methimazole (Tapazole), are goitrogenic (Fig. 15.19C).

A group of naturally-occurring (rather than synthetic) chemical goitrogens are found in the seeds and edible portions of many plants of the cabbage family (Brassicaceae). Goitrogenic compounds also are found in various types of rutabaga and turnip.

The best known of the natural goitrogenic compounds is goitrin (Fig. 15.20). This substance is present as an inactive glycoside precursor, progoitrin, which is converted to goitrin by an enzyme (myrosinase) that is found in intestinal bacteria. This enzyme catalyses the hydrolytic reaction necessary to activate progoitrin. Cooking destroys the progoitrin activator present in the vegetable itself as the enzyme is heat labile. However, the bacteria produce the active goitrin in the intestinal tract when it is absorbed.

Normally the various compounds discussed above are not goitrogenic in usual therapeutic doses or when vegetables containing progoitrin are consumed in reasonable quantities in a well-balanced diet. A marginally adequate dietary iodide intake coupled with the presence of goitrogenic compound may be sufficient to induce a clinical thyroid deficiency and goiter, although either condition alone would be an insufficient stimulus for a goiter to develop.

The mode of action of these diverse types of goitrogenic compound at the molecular level is not clear, although several mechanisms have been postulated. First, the goitrogenic chemical may act in direct competition to tyrosine as a substrate for active iodine, thus iodination of this amino acid is diminished or cannot take place. Second, the goitrogenic compound may inhibit the action of the peroxidase enzymes. Third, the goitrogen may reduce chemically to iodide a large part of the active iodine in the colloid so that the iodination of thyrosine is inhibited (Fig. 15.14).

It appears that the reactions involved in thyroxine synthesis exhibit a graded sensitivity to goitrogenic agents. The coupling reaction is the most sensitive and it is inhibited first; diiodotyrosine formation is the next most sensitive; and monoiodotyrosine synthesis is the least sensitive as well as the most nonspecific of the reactions involved in thyroxine synthesis.

Regardless of the detailed mechanism or mechanisms that are involved in the action of goitrogenic compounds, it is clear that such substances exert their action in general by inhibiting the biosynthesis of thyroid hormones after iodide has been concentrated within the colloid of the thyroid gland.

Thyrocalcitonin (Calcitonin)

As noted earlier, the thyroid gland produces another hormonelike substance that is completely distinct from the iodine-containing hormones discussed in previous sections. This endocrine substance, known as thyrocalcitonin, causes a decrease in the circulating plasma calcium and phosphate levels following its injection. These actions are diametrically opposite to those of parathyroid hormone, which will be discussed in the next major section of this chapter.

Thyrocalcitonin is produced in the mammalian thyroid gland by the parafollicular, or clear cells.

Human thyrocalcitonin has been isolated from carcinomas of the thyroid that have an abundant population of parafollicular cells. This substance is a polypeptide having a molecular weight of about 8,700 daltons and is composed of 32 amino acid residues. Only minute traces of thyrocalcitonin are found in the normal adult thyroid.

Thyrocalcitonin exerts its effects directly upon bone to inhibit the solution of calcium as well as phosphate, and thereby the circulating level of these two ions is diminished. There is some evidence that thyrocalcitonin mediates its effects through stimulation of cyclic AMP formation, although this fact has not been established with complete certainty.

If the thyroid gland of an experimental animal is perfused with solutions containing a high concentration of calcium, then the secretion of

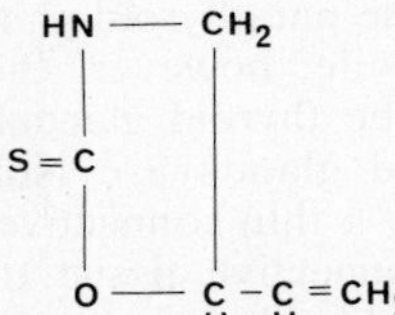

FIG. 15.20. Structural formula of goitrin (L-5-vinyl-2-thio-azolidine). (After Ganong. *Review of Medical Physiology*, 6th ed, 1973, Lange Medical Publications.)

thyrocalcitonin increases. Only under such experimental conditions can any action of thyrocalcitonin be demonstrated, and no clinical syndrome related to either hypersecretion or hyposecretion of this hormone from the thyroid has been found in man. Since thyrocalcitonin appears to be more active in young animals than in adults, it has been suggested that thyrocalcitonin may have a role in the skeletal development of young children. The role of thyrocalcitonin in relation to calcium metabolism will be discussed further in the next section.

THE PARATHYROID GLAND: OSTEOGENESIS AND THE PHYSIOLOGY OF BONE: CALCIUM AND PHOSPHATE METABOLISM

The principal function of parathyroid hormone (or parathormone) secreted by the parathyroid glands is in the homeostatic regulation of ionized calcium in the extracellular fluids, a function critical to the survival of the individual. In carrying out this major physiologic role, parathyroid hormone mobilizes calcium from bone, elevates the plasma calcium level, and increases urinary phosphate excretion. Furthermore, because of its intimate relationship to calcium and phosphate metabolism, parathyroid hormone is important in the formation of bone and teeth. Vitamin D also has an indispensible role in these processes. The present section, therefore, will present a discussion of these several interrelated physiologic processes with emphasis upon the key role of parathyroid hormone in their regulation.

Functional Morphology of the Parathyroid Glands

GROSS ANATOMY. In man, the parathyroid glands are small, oval, yellow-brown bodies that generally are found attached to the posterior surface of the thyroid gland; accessory parathyroid glands, however, are common. Occasionally the accessory parathyroid glands are found in association with the thymus, or located deep within the mediastinum, as both of these structures share a similar embryologic origin.

The total mass of parathyroid tissue ranges between 0.05 and 0.3 gm. The size of these structures varies between 3 and 8 mm in length, 2 and 5 mm in width, and 0.5 to 2 mm in thickness.

Generally the parathyroid glands are found in the thyroid capsule; however, they may be embedded within the thyroid glandular tissue itself. The parathyroid glands are isolated from the thyroid tissue by a thin connective tissue capsule. Within these connective tissue trabeculae larger ramifications of blood vessels, nerves, and lymphatics enter the parathyroids.

HISTOLOGY. Microscopic examination reveals that the parathyroids are composed of thickly packed cells that are arranged in anastomosing cords, dense masses, or occasionally as follicles containing colloid within the lumen. A loose network of reticular connective tissue fibers ramifies between the individual cells, and provides support for the abundant capillaries as well as nerve fibers. Numerous fat cells may be present within the connective tissue stroma of the parathyroids.

Two major types of epithelial cells are found in human parathyroid glands, and these are chief (or principal) cells, and oxyphilic cells.

Chief (Principal) Cells. The chief cells are polygonal, 7 to 10 μ in diameter, and have a centrally located vesicular nucleus. The cytoplasm is pale and stains very faintly with acidic dyes. The mitochondria are elongated, and a Golgi apparatus is located adjacent to the nucleus. Abundant glycogen granules as well as heavy granular deposits of lipofuscin pigment are evident with appropriate staining techniques.

Besides the intracellular constituents mentioned above, the chief cells also contain an abundance of fine granules which can be visualized with silver stains (eg, of Bodian), chrome alum hematoxylin, and iron hematoxylin. It is believed by some investigators that these structures represent secretory granules, and that the chief cells are responsible for the secretion of parathyroid hormone.

Oxyphilic Cells. The oxyphilic cells can be distinguished from the chief cells in several ways. First, the oxyphilic cells are less abundant than chief cells, and they occur singly or in small groups; second, they are much larger than chief cells; and third, they have a small nucleus which stains deeply in addition to a markedly acidophilic cytoplasm. The Golgi apparatus is smaller in the oxyphilic cells than in the chief cells, and the elongated mitochondria are relatively much more abundant in the oxyphilic cells.

The parathyroid glands undergo certain morphologic changes as the individual ages. The relative amount of connective tissue as well as fat cells increases with age. In the dense masses of glandular epithelial cells, cords and follicles appear in babies at about one year of age, and then increase subsequently. The oxyphilic cells apparently develop between the ages of four to seven years, and increase in relative abundance after puberty.

Chemistry of Parathyroid Hormone

Purified parathormone is a simple polypeptide chain comprised of 83 amino acids. It has a molecular weight of around 9,000, and cysteine is absent from the molecule. Fractional acid hydrolysis of parathyroid hormone has revealed that the biologic activity of the molecule resides in the sequence of approximately 35 amino acids which includes the terminal amino group. Further degradation of parathyroid hormone has revealed that the immunologic activity, in contrast to the biologic activity, resides in a much smaller portion of the molecule.

Metabolism of Parathyroid Hormone

In man, the half-life of parathyroid hormone in the circulation has been estimated at about 20 to 30 minutes. Following parathyroidectomy in calcium-depleted rats, the plasma calcium level commences to fall within an hour and this steady decline continues until death ensues usually within 12 hours. Data such as these are indicative of a continuous and closely-regulated secretion of parathyroid hormone in the normal individual.

Regulation of Parathyroid Hormone Secretion

The mechanism whereby parathyroid hormone secretion is controlled resides in a simple negative feedback loop. The concentration of ionized serum calcium directly affects the rate of parathyroid hormone secretion, so that an inverse relationship exists between these two factors. When a hypercalcemia is present, regardless of its cause, parathyroid secretion is inhibited, so that the excess calcium is deposited in the bones. Conversely, a hypocalcemia stimulates parathyroid hormone secretion; thereby calcium is mobilized from the bones. These responses to serum ionized calcium are relatively rapid in onset, and the net result is an exquisitely delicate homeostatic mechanism for the regulation of the circulating calcium ion level.

An increase in the plasma phosphate level will also increase the rate of parathormone secretion. This effect is indirect, since an elevated phosphate concentration reduces the plasma calcium level, and this lowered calcium level in turn acts as the direct stimulus on the parathyroid glands to enhance their secretion.

Physiologic Effects of Parathyroid Hormone

The normal regulatory effects of parathyroid hormone are manifest at three different loci in the body: the skeleton, the kidneys, and the gastrointestinal tract. It is important to note that this hormone can affect directly and independently any of these three systems.

SKELETON. Parathyroid hormone acts directly upon bone tissue to stimulate the activity of the osteoclasts; in this manner an increased quantity of ionized calcium is released directly from the bone matrix into the circulation. This calcium mobilization is accompanied by a loss of mucopolysaccharide from the bone matrix.

The calcium-mobilizing effect of exogenous parathormone may be blocked by simultaneously injecting a substance, such as puromycin, which inhibits protein synthesis. This observation suggests an interrelationship between the two physiologic processes; however, the detailed mechanism whereby parathyroid hormone exerts its effects remains obscure.

KIDNEY. An excess of parathyroid hormone, whether caused by hypersecretion of the glands or produced by exogenous administration, induces a marked increase in phosphate excretion (phosphaturia) by the kidneys, and this effect is accompanied by an increased plasma calcium level. The chief source of the calcium is, of course, the bones, which provide the principal body reservoir of this cation. On the other hand, hyposecretion of the parathyroid glands, or their surgical removal by parathyroidectomy produces a reduced urinary phosphate excretion that is coupled to a reduced plasma calcium level.

Experiments on animals (dogs) have shown that parathormone exerts a direct effect upon the renal tubule. The evidence favors the view that parathyroid hormone directly stimulates the secretion of phosphate by the distal tubule, thereby leading to the observed phosphaturia. Alternatively it has been proposed that parathormone may decrease the reabsorption of phosphate from the proximal convoluted tubule. Neither glomerular filtration rate nor renal blood flow is affected by parathyroid hormone. The net effect of the phosphaturia, however, regardless of the mechanism involved, is a decline in the plasma phosphate level.

GASTROINTESTINAL TRACT. Parathyroid hormone stimulates the absorption of calcium and phosphate from the gut. The uptake of these ions by the cells of the intestinal villi is thereby enhanced, so that more calcium and phosphate are transported into the bloodstream. This effect of parathyroid hormone depends upon the presence of an adequate dietary supply of vitamin D, as this substance is also critical for an effective active transport mechanism for calcium to operate in the intestinal (ileal) mucosa (p. 887).

Mechanism of Action of Parathyroid Hormone

The detailed biochemical mechanisms that underlie the more obvious physiologic actions of parathyroid hormone are unknown. As noted earlier, the parathyroid hormone not only stimulates the activity of the osteoclasts in bone, but also induces the conversion of osteoblasts into osteoclasts. However, these responses are far too slow to account for the rapid increase in ionized plasma calcium which follows parathormone administration. Apparently the mobilization of calcium from bone, as well as the phosphaturia, are mediated through the synthesis of cyclic AMP. DNA-dependent synthesis of RNA also may be involved in this process.

Parathyroid hormone also causes a rapid increase in blood citrate levels, and the source of this metabolite has been shown to lie in bone. It has been conjectured that the release of citrate is responsible for the calcium-mobilizing effect of parathormone. According to this concept, an elevated citrate concentration in the extracellular fluid immediately surrounding the bone binds calcium. This effect lowers the effective concentration of calcium locally so that more calcium goes into solution, in accordance with the law of mass action. Citric acid release also tends to lower the pH, again locally, thereby providing a medium in which calcium can dissolve more readily.

In vitro studies have yielded few clues concerning the physiologic mechanisms that underlie parathyroid hormone action. For example, it has been demonstrated that this hormone markedly decreases the rates of respiration and phosphorylation in isolated renal mitochondria preparations. On the other hand, renal tissue from parathyroid hormone-treated animals shows an increased rate of glucose oxidation as well as an elevated accumulation of calcium and phosphorus. Furthermore, cyclic AMP synthesis and excretion by the kidney are both enhanced by parathormone.

Studies of this type although interesting in themselves, do not explain satisfactorily the physiologic actions of parathormone at the biochemical level.

Functional Morphology and the Development of Bone

In similarity to other types of connective tissue, bone consists of cells, fibers, and an extracellular matrix (or ground substance). However, in contrast to the other connective tissues in the body, the extracellular materials are calcified in bone, so that the skeleton functions both as an essential supporting structure, as well as a protective element, for various organs and tissues of the body. These purely mechanical functions are due to the hardness of the extracellular components of the matrix. In passing, it is worth mentioning that bone has a tensile strength that is similar to cast iron, although its density is about one-third that of the latter.

Bone is hardly an inert construction material. It is also a dynamic living system which continuously is being renewed and rebuilt throughout the lifetime of the individual. Because of this active and continuous metabolism of bone, as well as its responsiveness to external mechanical, nutritional, and hormonal stimuli, bone plays a key role in the metabolism and distribution of certain important substances in the body. Notably, bone provides an important reservoir for calcium and phosphate.

It is essential at this point to consider the morphology and physiology of bone prior to a discussion of calcium metabolism, in order that these topics may be integrated with the overall regulation of calcium levels by parathyroid hormone and thyrocalcitonin.

GROSS STRUCTURE OF BONE. The principal gross anatomic features of bone to be discussed in this section are illustrated in Figure 15.21.

Macroscopic inspection reveals that two forms of bone may be distinguished. Compact bone appears as a dense continuous mass, in which spaces are visible only by use of a microscope. Spongy bone, on the other hand, consists of a delicate, three-dimensional network of branching spicules called trabeculae. The interconnecting spaces within the trabecular framework contain bone marrow. Compact and spongy bone can grade from one type to the other without any distinct boundary.

In a long bone, such as the femur or radius, the shaft or diaphysis forms a hollow cylinder having thick walls. The extensive central marrow (or medullary) cavity contains bone marrow. At each end of the shaft, the epiphyses are composed primarily of spongy bone that is surrounded by a thin layer of compact bone.

The interconnected spaces among the trabeculae of epiphyseal spongy bone of adults are continuous with the medullary cavity of the diaphysis. In juvenile individuals, the epiphysis and diaphysis are separated by a band of cartilage, the epiphyseal plate. This structure is united to the spongy bone of the diaphysis by a transitional zone known as the metaphysis. The epiphyseal plate and spongy bone adjacent to the transitional region provide a growth zone, in which elongation of the bone occurs.

A layer of hyaline cartilage invests the articular surface (or joint) of the epiphysis of long bones. This articular cartilage covers the thin layer of compact bone.

In general, bones are enclosed in a layer of specialized connective tissue called the periostium. This membranous structure is important because it has osteogenic potency, ie, it can form new bone under appropriate circumstances, such as in the healing of fractures.

The diaphyseal medullary cavity is also lined by a membrane, the endosteum, which is more delicate than the periosteum, but which exhibits erythropoeitic in addition to osteogenic properties.

Areas of bones that lack a functional periosteum (eg, loci of insertion of tendons and ligaments) heal poorly and slowly when injured, as the connective tissue in contact with these surfaces of the bone does not have osteogenic properties.

HISTOLOGY OF BONE. Compact bone is composed principally of the calcified interstitial substance, the bone matrix. This matrix is deposited in lamallae, or layers, between 3 and 7 μ in thickness, as shown in Figure 15.22.

Microscopic examination of a thin ground section of bone also reveals numerous cavities in the matrix called *lacunae,* each of which is entirely filled by a bone cell (or osteocyte). From the lacunae, exceedingly fine, branching tubular passageways, termed canaliculi, radiate in all directions. The canaliculi from a given lacuna anastomose freely with those originating from adjacent lacunae, so that the entire matrix of compact bone is traversed and interconnected by these minute tunnels. Presumably this dense network of canali-

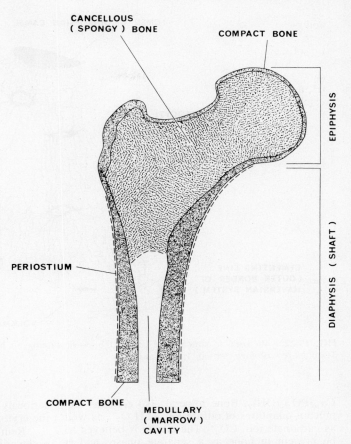

CANCELLOUS (SPONGY) BONE

COMPACT BONE

EPIPHYSIS

DIAPHYSIS (SHAFT)

PERIOSTIUM

COMPACT BONE

MEDULLARY (MARROW) CAVITY

FIG. 15.21. Gross anatomy of bone. (After Bloom and Fawcett. *A Textbook of Histology,* 9th ed, 1968, Saunders.)

culi that connects the individual lacunae, hence the osteocytes, is essential for the nutrition of the cells, unlike the situation which exists in cartilage. (The process of diffusion through the aqueous phase of the gel matrix is adequate to supply the nutritional requirements of the cells in hyaline cartilage.) The deposition of calcium salts in bone reduces permeability considerably; therefore, an elaborate network of canaliculi is necessary to connect the osteocytes lying within the lacunae with the nearest perivascular space so that metabolic exchanges can occur at reasonable rates.

The lamellae of compact bone are arranged in patterns, the commonest being the haversian system (or osteon), as shown in Figure 15.22. In cross section, these cylindrical units are composed of between 5 and 20 lamellae that are arranged concentrically around a central opening (the haversian canal) which contains one or two blood vessels. The haversian canals are between 20 and 110 μ in diameter. Generally the vessels they contain are capillaries or venules; however, an arteriole may be present occasionally. The haversian canals communicate with one another, as well as with the free surface and the medullary cavity, by a transversely oriented system of Volkmann's canals. Blood vessels from the periosteum and endosteum communicate with the vessels located within the haversian systems via the Volkmann's canals.

Spongy bone is also arranged in lamellae having lacunae within the matrix. The trabeculae of spongy bone are thin, so that generally no complete haversian systems are present together with their blood vessels. Consequently, the exchange of metabolites occurs with the endosteal surface via the system of canaliculi.

STRUCTURE AND COMPOSITION OF BONE MATRIX. The interstitial substance of bone is comprised of two principal components, an organic matrix and inorganic salts. The organic matrix of bone is composed of collagenous fibers embedded in an amorphous ground substance containing glycoproteins. During the processes of growth and development, the quantity of organic material per unit volume of bone stays relatively constant. The quantity of water increases, and the proportion of inorganic material (bone mineral) increases. In the adult up to 65 percent of the total mass of dry, fat-free bone consists of inorganic salts. When bone is poorly calcified, as in rickets or osteomalacia the total bone mineral may be as low as 35 percent.

The inorganic matrix of bone consists of submicroscopic crystals of a complex type of mineral called apatite, which in turn is composed of calcium and phosphate. In bone this mineral is similar to a form of hydroxyapatite, which has the formula

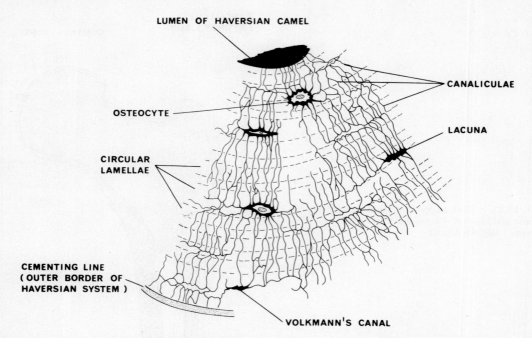

FIG. 15.22. Microscopic structure of compact bone illustrating an haversian system. (After Bloom and Fawcett. *A Textbook of Histology,* 9th ed, 1968, Saunders.)

$Ca_{10}(PO_4)_6(OH)_2$. Bone mineral also contains significant quantities of citrate ion, $C_6H_5O_7^{\equiv}$, as well as carbonate ion, $CO_3^=$. The citrate is believed to be present in bone as a separate phase, and is presumed to be concentrated upon the surfaces of the apatite crystals. The site occupied by carbonate in the ultrastructure of bone is unclear.

The apatite crystals in bone are formed as minute plates or rods that range between 8 and 15 Å in thickness, 15 to 18 Å wide, and 200 to 400 Å in length. These crystals are found embedded within the collagen fibers of the organic matrix, so that the combination of the two substances results in a very strong yet lightweight structure capable of resisting severe mechanical stress. The conditions responsible for bone mineralization will be discussed later in this section.

Sodium and magnesium are also present within the bone mineral, so that bone also serves as a storage reservoir for these cations.

Radioisotopes of calcium (^{45}Ca) and phosphorus (^{32}P) can substitute readily for the stable forms of these elements (ie, ^{40}Ca and ^{31}P), and thus are quite useful in the study of mineral metabolism and turnover in bone. Certain foreign cations also have a propensity for substituting for calcium ion in bone if they are ingested. Included in this category are lead (Pb^{++}), strontium (Sr^{++}) and radium (Ra^{++} or ^{226}Ra). Certain products of nuclear fission also are so-called bone seeking isotopes. The most dangerous of these is ^{90}Sr, because it accumulates preferentially in bone and the radiation it emits can

seriously damage the functions of bone as well as the erythropoietic tissue of the marrow.

Removal of the inorganic material of bone by a chelating agent such as ethylene diamine tetra-acetate (EDTA) results in a decalcified structure which retains its form but is no longer hard, and yet is tough and flexible. Incineration of a bone, on the other hand, removes the organic component, and although the bone retains its gross structure and hardness, it becomes extremely brittle, and has lost its tensile strength. It is clear from these facts that the proper combination and proportions of organic and inorganic materials in the bone matrix are essential to the hardness as well as the toughness and resiliency of bone. These properties in turn render bone an admirably suitable tissue for its mechanical as well as metabolic functions in the body.

BONE CELLS. Basically, actively growing bones have three types of morphologically distinct cells. These cells are osteoblasts, osteocytes, and osteoclasts. Since each of these cells can change readily and reversibly from one type to the other depending upon particular circumstances (a process that is called modulation, not differentiation), probably the bone cells represent different functional aspects of single cell type.

In addition, other body cells have latent mesenchymal properties, and can be transformed under appropriate conditions into different types of bone cells. For example, the latent osteogenic potency of the primitive spindle-shaped cells of the periosteum and endosteum may be stimulated during normal growth of bone or during healing of an injury or fracture. Subsequently, these cells may undergo mitosis and ultimately assume the morphologic and functional characteristics of osteoblasts, osteocytes, and osteoclasts.

Osteoblasts. The osteoblasts are related intimately to the formation of bone, and are always found upon the surfaces of growing or developing bones. When new bone matrix is being deposited actively, osteoblasts form a layer of cuboidal or low columnar cells which are interconnected by short, narrow processes. The spheroid nucleus usually is located at the pole of the cell farthest away from the bone surface. The cytoplasm is strongly basophilic because of the high ribonucleic acid (RNA) content. During the active formation of bone, osteoblasts give a strong reaction for alkaline phosphatase. Furthermore, the cytoplasm contains numerous small granules that are demonstrable by staining with the periodic acid–Schiff reaction, and which presumably are the precursors of bone matrix. When bone formation ceases, the alkaline phosphatase reaction declines and the cytoplasmic granules of the osteoblasts disappear as these elements revert to a spindle-shaped form.

Osteocytes. The osteocytes are the principal cellular component of mature bone. These cells, as noted above, are found within the lacunae present in the calcified interstitial substance, and each osteocyte has a flattened cell body which conforms to the shape of the cavity which it occupies. Slender processes extend from the cell body and enter the canaliculi; however, it is unknown just how far the processes extend into these tubules. The characteristics of the nuclei and cytoplasm of the osteocytes are similar in general to those of the osteoblasts, save that the cytoplasm has only a slight affinity for basic dyes. Fundamentally an osteocyte is merely an osteoblast which has become entrapped within the bone matrix, so that it no longer is actively secreting, but rather is involved in the maintenance of the interstitial substance and the canaliculi of the bone matrix.

Osteocytes may also play an important role in the transport of materials from blood to bone and vice versa. As mentioned earlier, parathyroid hormone directly stimulates the osteocytes in some manner so that the bone matrix is modified with an attendant solubilization of bone minerals. Consequently, osteocytes appear to have an important function in the homeostatic regulation of calcium in the body fluids, a function regulated by parathormone.

Similarly to osteoblasts, osteocytes apparently can undergo "metamorphosis" into the other types of bone cells. During bone resorption, many osteocytes are released from the matrix, and these now may revert to osteoblasts, or else become incorporated into multinucleate osteoclasts.

Osteoclasts. The osteoclasts are giant cells that are associated intimately with regions of bone that are undergoing active resorption. The number of nuclei within these cells is quite variable, but frequently as many as 20 nuclei are found within one osteoclast. The individual nuclei within an osteoclast are similar to these structures as found in osteoblasts and osteocytes. Hence osteoclasts are considered to be derived from the latter cell types by fusion of individual cells.

The cytoplasm of the osteoclast is faintly basophilic and highly variable in appearance so that vacuolations often are observed. The vacuolar material stains positively for acid phosphatase, in contrast to the reaction for alkaline phosphatase given by the cytoplasm of the osteoblasts and osteocytes. The significance of this observation is unknown.

There is no convincing evidence that osteoclasts are phagocytic insofar as their role in active bone resorption is concerned, although this process may play a minor role. Possibly these cells secrete proteolytic enzymes or an acid which dissolves the inorganic matrix of bone in their immediate vicinity, or else a chelating agent is released from the osteoclasts that solubilizes the bone minerals.

DEVELOPMENT OF BONE (OSSIFICATION). The development of a connective tissue "pattern" or "model" always precedes the formation of bone. The transformation of this connective tissue model into bone follows one of two different pathways of osteogenesis. If bone formation occurs directly in primitive connective tissue, it is called intramembranous ossification. On the other hand, when osteogenesis occurs within a cartilagenous model, it is termed endochondral ossification. The deposition of the bony matrix is basically similar in both types of ossification, however.

Intramembranous Ossification. During intramembranous ossification, the mesenchyme forms a highly vascular layer of connective tissue. The cells contact one another through long, slender processes, and fine bundles of collagen fibers are present in the intercellular spaces together with a matrix of a gel-like ground substance. Bone formation first becomes evident when thin bands of a highly eosinophilic material develop. These bands appear between adjacent blood vessels, and since the blood vessels form a network, then the first trabeculae of the bony matrix also exhibit a pattern of branching and anastomosis similar to that of the vessels. Concomitantly the connective tissue cells enlarge and accumulate upon the trabecular surfaces, while their shape becomes cuboidal to columnar; simultaneously they become strongly basophilic, at which point they are deemed to be osteoblasts. The cells remain connected by slender processes during these developmental changes, and as they secrete more bone matrix, the trabeculae enlarge.

Macromolecules of tropocollagen also are secreted together with the protein–polysaccharide component of the matrix, and these fibrillar structures now condense in an irregular pattern, called woven bone, to form the fibrous element of the matrix. During later remodeling of this woven bone, the layers of collagen are secreted into the lamellae in an orderly, parallel arrangement.

The ability of the matrix to become calcified is probably governed by the osteoblasts, so that when the primitive interstitial matrix develops to a certain stage, it becomes susceptible to calcification and calcium phosphate now is deposited. Thenceforth calcification occurs in all of the matrix which is formed.

As layers of bone are added progressively to the trabeculae, osteoblasts become entrapped in spaces (lacunae) within the developing calcified matrix, and these cells become the osteocytes of the mature bone. The trapped osteocytes remain connected with surface osteoblasts by means of the slender cellular processes and as calcified matrix develops around these processes, the canaliculi are formed.

In regions that are destined to remain spongy bone, the growth of the trabeculae by surface accretion of matrix ceases, and the vascular connective tissue in the interstices gradually differentiates into hemopoietic tissue. The connective tissue on the outer surface of the bone eventually condenses and ultimately becomes the periostium, hence surface osteoblasts finally assume an appearance resembling fibroblasts and come to lie in the deepest layer of the periostium as growth ceases. In response to an appropriate stimulus (eg, injury) these elements once again undergo a morphologic metamorphosis into osteoblasts.

In areas that ultimately will become compact bone, the trabeculae thicken progressively, so that the interstitial spaces surrounding the blood vessels are eliminated, and during this process, collagen fibrils are laid down upon the successive layers of bone in an orderly lamellar fashion.

Membranous bones, formed by the process of intramembranous ossification as described above, are found in the skull. These skeletal elements include the frontal, parietal, temporal, and occipital bones as well as portions of the mandible.

Endochondral Ossification. During endochondral ossification, a hyaline cartilage model of the bone develops first. In a long bone, eg, the femur, a primary center of ossification develops in the middle of the shaft, or diaphysis. This process is accompanied by profound morphologic alterations in the cartilage cells, including marked hypertrophy, cytoplasmic vacuolation, and glycogen accumulation. The lacunae within the cartilagenous matrix enlarge, so that the matrix adjacent to the center of ossification becomes reduced into thin partitions and spicules.

In the vicinity of the hypertrophied cartilage cells, the hyaline matrix becomes susceptible to calcification, so that if a properly high concentration of calcium and phosphate are present in the extracellular fluid, then calcium phosphate crystals are deposited (ie, they precipitate) in small aggregations.* The hypertrophied cartilage cells of the model then regress, die, and undergo degeneration.

While the changes in the chondrocytes that were summarized above are taking place inside of the cartilage model, concomitant alterations are occurring within those cells of the perichondrium which have osteogenic tendencies. Thus a thin layer of bone is deposited surrounding the middle region of the diaphysis; this is the periosteal collar (or band). Simultaneously blood vessels from the surrounding connective tissue, now termed the periostium, grow into the shaft. These vessels enter the cavities within the cartilagenous matrix that were produced by hypertrophy of the chondrocytes and the merging of their lacunae. The vessels now branch, and then grow toward each end of the model. Capillary loops then develop which extend into the cavities, and cells having osteogenic potency thus are transported into the cartilage along with the vessels in the connective tissue which surrounds and accompanies the blood vessels. Some of the cells then develop into hemopoietic tissue, whereas others differentiate into osteoblasts. The latter cells form an epithelioid layer upon the remaining spicules of calcified cartilage, and then secrete bone matrix upon these elements. In centers of endochondral ossification the trabeculae first have a core of calcified cartilage that is surrounded by a layer of bone.

In long bones, growth in length continues in the cartilage model after the appearance of the primary center of ossification in the diaphysis. The chondrocytes in the epiphyses become arranged in more orderly longitudinal rows, and as ossification proceeds from the primary center in the shaft, the chondrocytes undergo the same sequential changes as in the primary center of ossification, as described above.

Several growth zones are discernible in the epiphyseal cell columns. First, a zone of proliferation is present, wherein mitotic division of the cells produces a continuous elongation of the cell columns. Second, a zone of maturation develops, wherein the cells enlarge. Third, a zone of hypertrophy is seen in which the cells become vacuolated. This latter region also is called the zone of provisional calcification, because here the matrix becomes the locus for calcium deposition. At the diaphyseal end of the cell columns lies a region in which the chondrocytes are undergoing regression, and the open lacunae are invaded by capillary loops and connective tissue from the medullary space within the diaphysis. As these tissues encroach upon the ends of the columns, osteoblasts

*Stated another way, precipitation of bone matrix in the form of calcium phosphate crystals takes place if the interstitial fluid is supersaturated with calcium and phosphate ions.

accumulate upon the longitudinal bars of calcified cartilage, and bone matrix is deposited upon them.

The transitional region in which cartilage is being replaced by bone is known as the metaphysis. The spongy bone that has formed earlier in this zone undergoes reorganization as growth proceeds. However, as the entire bone elongates, spongy bone is deposited at the epiphyseal ends at the same rate at which the diaphyseal ends of the trabeculae are being resorbed. The spongy bone of the metaphysis remains approximately constant in length as growth and development of bone proceeds by endochondral ossification.

During the third month of fetal life the primary centers of ossification have appeared in each of the long bones of the skeleton. Generally, hypertrophy of the chondrocytes within the epiphysis, which indicates that ossification has commenced, is found after birth. Subsequently, the epiphyseal regions are infiltrated by blood vessels as well as osteogenic tissue, so that secondary centers of ossification develop at each end of the long bones.

Ultimately, the secondary centers expand to such an extent that all of the epiphyseal cartilage is replaced, save for the articular cartilage, and the thin epiphyseal plate which is a thin transverse disc between epiphysis and diaphysis. Growth at the epiphyseal plate now is responsible for all further elongation of the long bones of the skeleton. The cartilage cells in the proliferative zone in this region multiply at a rate equivalent to their degeneration, so that the width of the plate remains approximately constant in width as growth proceeds. Cartilage cells are continually growing away from the diaphysis, while the dead cells are being removed simultaneously at the diaphyseal end of the epiphyseal plate. The consequence of this process is a net increase in length of the diaphysis.

Closure of the epiphysis takes place at the end of the growth period when the chondrocytes decrease their rate of multiplication, and ultimately are replaced by spongy bone and marrow. After this fusion of epiphysis and diaphysis has taken place, no further elongation of the bone is possible. The important effects of pituitary somatotropin upon skeletal growth and epiphyseal closure were discussed on page 1030.

Endochondral ossification in a cartilage model is responsible for the increase in length of the long bones. Growth in diameter, however, results from the deposition of concentric lamellae of new membranous bone under the periostium; therefore, the compact bone of the diaphysis is formed almost entirely by the process of subperiosteal intramembranous ossification, although this process continues in bones which first were laid down by the endochondral ossification of a cartilage model.

Bones of the skeleton that are first modeled in hyaline cartilage prior to endochondral ossification include the bones comprising the base of the skull, the pelvis, the bones of the extremities, and the ribs.

In conclusion, it should be mentioned that throughout the development of the entire skeleton the surfaces of individual bones are constantly being remodeled, so that the external shape of the individual bones is approximately the same from early fetal stages of development through adult life. The process of remodeling involves subperiosteal deposition of bone in some regions, whereas resorption occurs in other areas.

It should be reemphasized here that bone, and the minerals which comprise its matrix, are in a constant state of dynamic flux within the body. A continuous exchange (or turnover) of minerals is taking place between the bone matrix, interstitial fluid, and plasma. Certain aspects of calcium and phosphate metabolism may now be considered in relation to these processes.

General Aspects of Calcium and Phosphate Metabolism

There are roughly 1,100 gm of calcium in the adult human body, and of this total more than 99 percent is present in the skeleton. The remaining fraction is important in a number of physiologic processes that are not related to the structure of bone. For example, calcium is an essential factor in enzyme systems, such as those underlying the mechanical contraction of skeletal and cardiac muscle. Calcium also may have a key role in the generation of the rhythmic pacemaker potentials in the sinoauricular node of the heart which initiate cardiac activity. Furthermore, calcium has a key role in the generation and transmission of nerve impulses, and in the response of muscle to such impulses (ie, neuromuscular transmission). The blood clotting mechanism also requires calcium in order to function properly.

Calcium is present normally in the intracellular fluid at a concentration of approximately 20 mg per 100 gm of tissue, and it is also is an important constituent of the intercellular cement.

INTESTINAL ABSORPTION OF CALCIUM AND PHOSPHATE. Although these topics were considered in Chapter 13, certain aspects of calcium and phosphate absorption are relevant to the present discussion, hence will be reviewed briefly.

From a nutritional point of view, the absorption of calcium is a major problem, since this cation is present in the diet mainly in the form of insoluble salts. Calcium is ingested principally as calcium phosphate. The three forms of calcium phosphate in a decreasing sequence of their solubilities are $Ca(H_2PO_4)_2$, $CaHPO_4$, and $Ca_3(PO_4)_2$. Within the stomach the pH is such that all of the calcium phosphates are readily soluble; at duodenal pH, however, the major calcium salts are $CaHPO_4$ and $Ca(H_2PO_4)_2$.

Active transport of calcium ion occurs in the ileum, and this mechanism can operate against a fivefold concentration gradient. Active calcium absorption is enhanced in the presence of vitamin D and diminished in the absence of this substance. Consequently the presence of vitamin D in the diet is essential to normal absorption of calcium (cf. p. 887).

Somatotropin elevates the urinary excretion of calcium; however, it also stimulates intestinal absorption of this cation. The enhanced absorption may be greater than the excretory effect, so that the net result is a positive calcium balance.

The intestinal absorption of phosphate is passive, in contrast to that of calcium. This anion is transported across the intestinal mucosa secondarily to the active transport of calcium.

ROUTES OF CALCIUM EXCRETION. The major normal pathway for calcium excretion is the intestinal tract so that subjects on a calcium-free diet still excrete calcium in the feces. Calcium is a normal constituent of digestive secretions, especially the bile, and the quantity eliminated is dependent upon the plasma calcium concentration.

Under normal circumstances, the kidney excretes little calcium. However, in chronic hypercalcemia, sufficient calcium may be excreted to produce renal calculi. Normally, around 99 percent of the calcium present in the glomerular filtrate is reabsorbed, even in the presence of an artificially induced hypercalcemia. This fraction decreases in a number of pathologic conditions in which mineral resorption from bone is abnormally active.

PHOSPHATE METABOLISM. Phosphate is so abundant and widespread in various biologic materials used as food that a nutritional deficiency in phosphate cannot occur when the diet is adequate in calories and protein. Phosphate is ingested generally as the orthophosphate or as organic phosphate, the latter compound being converted to orthophosphate in the gastrointestinal tract. Phosphate is not absorbed in appreciable quantities from the stomach, hence the small intestine is the principle site of absorption of this ion. As noted above, phosphate absorption is passive and secondary to active calcium ion absorption.

In plasma, the inorganic phosphate is found chiefly as orthophosphate, so that $HPO_4^=$ and $H_2PO_4^-$ are present in a ratio of about 4:1. All of the plasma phosphate is diffusible (ionized), and therefore filtrable through the glomerulus of the kidney, which is the principal excretory route for this substance.

The total orthophosphate concentration in plasma is normally 4 to 5 mg per 100 ml in children compared to 3.5 to 4 mg percent in adults.

The plasma level of phosphate is regulated by a homeostatic feedback mechanism. Similarly to calcium, the skeleton stores phosphate, and this can be mobilized when the plasma phosphate level falls, or else phosphate is deposited when the plasma concentration increases.

PLASMA CALCIUM FRACTIONS AND CONCENTRATIONS. In normal human plasma, the total calcium concentration ranges between 9 and 11 mg per 100 ml. This is equivalent to 4.5 to 6.5 mEq per liter (average 5.5 mEq/liter) or 2.5 mM/liter (Fig. 15.23).

The total plasma calcium is found in two major forms: diffusible calcium or ionized calcium (Ca^{++}), and nondiffusible (or protein bound) calcium.

Diffusible Calcium. The free or diffusible calcium fraction of the total plasma calcium amounts to approximately 1.3 mM/liter. A small quantity of the diffusible calcium also is found in chemical association (as a complex) with such anions as phosphate, bicarbonate, and citrate. Parathyroid hormone and thyrocalcitonin exert diametrically opposite effects upon the free calcium level of the extracellular fluid compartment.

Normally, a reciprocal relationship exists between serum calcium and phosphate levels. When the level of one of these components is diminished the other is elevated concomitantly. In hyperparathyroidism, however, the concentration of both of these ions is elevated, whereas in juvenile rickets, the concentrations of both calcium and phosphate may be decreased.

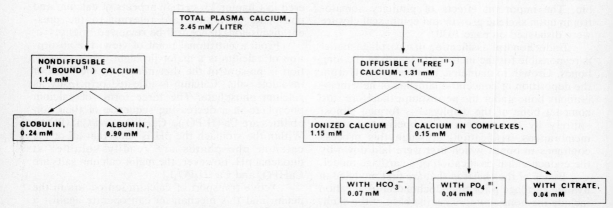

FIG. 15.23. Distribution of calcium in plasma. (Data from Ganong. *Review of Medical Physiology*, 6th ed, 1973, Lange Medical Publications.)

Nondiffusible (or Bound) Calcium. The non-diffusible calcium is chemically combined with certain plasma proteins, so that it is unable to diffuse through semipermeable membranes in contrast to ionized calcium (Fig. 15.23). The quantity of bound calcium is related directly to the total concentration of plasma proteins. If the plasma is low in protein then the total calcium level is correspondingly low. As noted above there is a direct relationship between the pH and the quantity of calcium that is bound, because of the fact that the plasma proteins are more negatively charged (anionic form) at lower H^+ concentrations.

The approximate total quantity of nondiffusible calcium in normal plasma is around 1.14 mM per liter. The greater proportion of this calcium is bound to albumin (0.90 mM per liter) while the remainder (0.24 mM per liter) is combined with the globulin fraction of the plasma proteins, as indicated in Figure 15.23.

From a practical standpoint it is difficult to obtain accurate routine clinical measurements of ionized calcium in the plasma directly. Because of the relationship between plasma protein concentration and calcium binding, it is important to determine the total protein in the sample whenever the total plasma calcium is measured. For clinical purposes, the ionized calcium can be estimated from a nomogram when the total plasma protein and calcium concentrations are known.

CALCIUM METABOLISM AND OSTEO-GENESIS. In mature bone, calcium is found in two biologically active, but not chemically different forms. First, a highly labile skeletal reservoir of calcium is present which is capable of rapid mobilization and exchange (turnover) with the interstitial fluid and plasma calcium with which it is in a dynamic equilibrium (exchangeable calcium). Second, there is a larger total pool of relatively stable calcium which also can exhibit turnover with the calcium of the extracellular fluids, but at a much slower rate. Actually these two "pools" of calcium in bone differ only in the rapidity with which they can become metabolically active.

During osteogenesis, cells of mesenchymal origin (viz, osteoblasts) secrete collagen and ground substance into the medium in their immediate vicinity. The collagen then polymerizes into fibrils. This medium also is rich in mucoproteins and mucopolysaccharides. Even though such a medium is of common occurrence in the body, normally mineralization takes place only in the presence of osteoblasts, which elements in turn are found only in regions that ultimately will become bone.

Although the details of bone mineralization are unclear, an accurate outline of this process may be presented. Calcium and phosphate ions are present in concentrations in the bone-forming medium that are sufficient to saturate the extracellular fluid locally. When the solubility product

for these ions (ie, $[Ca^{++}] \times [PO_4^=]$) exceeds a certain value, the medium now becomes supersaturated, and calcium phosphate crystals precipitate out of solution. Apparently the collagen fibrils in the matrix act as "nuclei" for this process. That is, the fibrils provide a unique surface, because of their particular stereochemical configuration, which initiates crystal formation ("seeding"). In fact, bone crystal formation has been shown to occur first within these fibrils. As bone develops, the apatite crystals grow until they entirely surround and thus entrap the collagen fibrils. Then, in turn, the crystals themselves become nucleating agents for hydroxyapatite to form in the interstices among the collagen fibrils.

Osteoblasts are quite rich in alkaline phosphatase, and this enzyme is believed by some workers to increase locally the concentration of phosphate by ester hydrolysis, so that the solubility product is exceeded and calcium phosphate precipitates more readily. It is of interest in this regard that collagen from connective tissue other than that found in the bone matrix does not induce calcification in an identical supersaturated medium under normal conditions. However, ectopic calcification is of considerable importance clinically, and several examples of this pathologic process will be presented later.

Mature fully mineralized bone is essentially water-free. Collagen is present to the extent of 20 percent by weight and it accounts for about 40 percent of the volume of bone. The mineral content of bone accounts for the remainder.

Possible Interrelationships Between Parathyroid Hormone and Thyrocalcitonin in the Regulation of Normal Calcium and Phosphate Metabolism

The hormone calcitonin was first extracted and isolated from parathyroid tissue; subsequently thyroid tissue also was shown to have an even higher total quantity of this substance, so that the calcitonin from this source was renamed thyrocalcitonin.* Thyrocalcitonin which has been obtained in a purified state from hog thyroid has been shown to be a polypeptide having a molecular weight of 4,500. No iodine is present in thyrocalcitonin.

Injection of thyrocalcitonin results in an acute hypocalcemia and hypophosphatemia. The effect of this hormone thus is directly opposite to that of parathyroid hormone, suggesting that a function of thyrocalcitonin may be to counteract the effects of hypercalcemia. Thus thyrocalcitonin increases the

The total quantity of thyrocalcitonin that is present in the thyroid gland is greater than in the parathyroids merely because of the greater mass of thyroid than parathyroid tissue. However, since the parafollicular cells of the thyroid are relatively scanty in occurrence, then the concentration of thyrocalcitonin per unit mass of thyroid tissue is somewhat lower than that found in parathyroid tissue.

transfer rate of calcium from blood to bone; ie, it acts directly upon the latter tissue. Thyrocalcitonin does not increase loss of calcium from the body, nor is its action mediated through the hypophysis, gastrointestinal tract, kidneys, or parathyroid glands. Consequently, thyrocalcitonin acts upon bone to accelerate the rate of calcium deposition, as well as to inhibit resorption of this ion from bone into the extracellular fluids. The net effect is to augment calcium retention, especially by the skeleton.

The circulating plasma level of free (ionized) calcium regulates the secretion rate of thyrocalcitonin, both from thyroid and parathyroid tissue. An elevated calcium level (hypercalcemia) acts directly upon these glands to increase the rate of thyrocalcitonin secretion, while concomitantly lowering the plasma calcium levels to normal values once again.

Clinical Correlates

The several abnormal processes that will be discussed in this section have been selected to illustrate the complex interrelationships among parathyroid gland activity, bone physiology, calcium metabolism, and certain important ancillary factors which are related to these processes.

HYPOPARATHYROIDISM. The basic function of parathyroid hormone is to maintain the plasma level of ionized calcium, a function essential to continued survival of the individual. In carrying out this function, parathyroid hormone secretion elevates (and maintains at a constant level) the plasma (hence extracellular fluid) calcium level by mobilizing this cation directly from the exchangeable pool in bone. This effect in turn results from the direct stimulation and maintenance of osteoclast activity in bone mineral resorption. Parathyroid hormone also promotes urinary phosphate excretion.

As noted earlier, the ionized (diffusible) calcium in the plasma is critical for the maintenance of normal neuromuscular activity and blood clotting, as well as cardiac and skeletal muscular contractility.

Hypoparathyroidism is of rare occurrence in man. Generally it is due to surgical procedures that involve removal of the thyroid gland, and in which the parathyroids are inadvertently removed together with the thyroid tissue. Even more rarely disease processes may decrease parathyroid function directly.

Following parathyroidectomy there is a steady decline in the plasma calcium level. Symptoms of neuromuscular hyperexcitability appear when the serum calcium level falls to 7 mg percent or lower, and these symptoms are followed by development of the syndrome of hypocalcemic tetany. It is important at this juncture to emphasize the difference between tetanus (which is either a sustained muscular contraction, or a disease caused by *Clostridium tetani*) and tetany. The syndrome of tetany, which accompanies hypoparathyroidism and a concomitant fall in plasma ionized calcium, includes the following signs. First, a sharp flexion of the wrist and ankle joints (carpopedal spasm) occurs. When the muscles of the upper extremity exhibit a spasm causing the fingers to extend, whereas the wrist and thumb flex, this response is called Trousseau's sign. In mild tetany, Trousseau's sign sometimes may be precipitated by occlusion of the circulation for a few minutes. Chvostek's sign also may be elicited in tetany, and this consists of a sudden contraction of the facial muscle following a mechanical stimulus, ie, tapping the facial nerve over the angle of the jaw on the same side of the face on which the contraction develops.

Tetany may also include twitching of the skeletal muscles (fibrillatory contractions), generalized body convulsions, cramps, and stridor (a shrill, harsh sound during respiration) which is caused by a marked contraction of the laryngeal muscles (laryngospasm). The laryngospasm may be so severe that the airway becomes obstructed and asphyxia ensues.

The tetany elicited by hypoparathyroidism in man develops from several days to several weeks following surgery as the plasma ionized calcium level gradually falls, and unless treated promptly, death inevitably results.

Tetany, such as described above, also is found in severe vitamin D deficiency, alkalosis, and following ingestion of alkaline salts. The common etiologic denominator in all of these situations is an inadequate extracellular fluid ionized calcium level.

Overall neuromuscular irritability may be expressed conveniently by summarizing the ionic variables in the plasma which contribute to this effect in the following relationship:

$$\text{Net Irritability} \cong \frac{[K^+]\,[Na^+]\,[HPO_4^-]\,[HCO_3^-]}{[Ca^{++}]\,[Mg^{++}]\,[H^+]}$$

This statement is not a precise mathematical formulation, but rather it is presented to summarize the interdependent electrolytes which collectively exert an influence upon the net cellular irritability. Thus a significant decrease in calcium ion concentration, such as is encountered in hypoparathyroidism, results in an elevated neuromuscular irritability which can lead to tetany. Conversely, a significant increase in the free calcium ion concentration may lead to functional impairment (reduced irritability) that can terminate in cardiac or respiratory failure.

Hypocalcemic tetany thus results from the effect of a decreased ionized calcium concentration acting upon the central nervous system and peripheral nerves, as well as upon the muscle fibers directly. There is evidence, however, that neuromuscular transmission per se is depressed somewhat in hypocalcemia. However, this effect is negated and overwhelmed by the hyperexcitability of the neuromuscular tissues themselves.

It also may be appreciated readily from the general expression presented above that a decreased plasma hydrogen ion concentration (eg, that produced by a respiratory alkalosis due to hyperventilation) can evoke symptoms of tetany at higher calcium ion levels. In this situation, the plasma proteins are ionized to a greater extent at an elevated pH; therefore, more protein anion is present to bind with calcium, and this binding produces a net reduction in ionized plasma calcium. From a clinical standpoint these facts are of the utmost importance in the treatment of hypocalcemic patients.

Chronic hypoparathyroidism may also be accompanied by the development of abnormally thick and heavy bones as well as cerebral calcification. These symptoms generally accompany chronic tetany.

Long-term therapy with parathyroid hormone has proven unsatisfactory because the substance is both scarce and expensive. The effect of this hormone only lasts for approximately 24 hours, and as the hormone from animal sources is antigenic, then the activity progressively decreases. Elevation of the ionized plasma calcium level must be achieved by administering this element orally in combination with vitamin D. The latter compound, in addition to promoting intestinal absorption of calcium, also has a slight action in promoting calcium and phosphate mobilization from bone (by an unknown mechanism). The vitamin D compound dihydrotachysterol has the most pronounced ability to produce this effect, so that large doses of this vitamin in combination with calcium salts are quite effective in controlling extracellular fluid calcium levels.

HYPERPARATHYROIDISM. A hyperparathyroid condition is caused by excessive secretion of parathyroid hormone and usually results from a tumor (primary hyperparathyroidism), or hyperplasia (secondary hyperparathyroidism) of the glandular tissue. Since pregnancy and lactation (as well as other conditions) may decrease calcium levels, and since parathyroid tumors are more common in women than in men, then it is possible that prolonged low calcium levels that stimulate hypersecretion of parathyroid hormone also may have a causative role in the tumorigenesis itself.

In primary or secondary hyperparathyroidism, osteoclastic activity within the bones is stimulated markedly. This in turn results in a marked elevation of the serum calcium level, while generally the phosphate level is depressed somewhat. Parathyroid hormone also elevates the rate of lactic acid production by the bone cells through the glycolytic mechanism, and the locally reduced pH produced by the lactic acid may have a role in the solubilization of bone mineral in hyperparathyroidism.

In severe and chronic primary hyperparathyroidism, osteoclastic activity far exceeds the capacity of the osteoblasts to form new bone, so that the elevation in serum calcium is accompanied by extensive decalcification and the weakened bones fracture readily owing to the net loss of apatite. X-rays reveal this decalcification together with occasional large punched-out cystic regions that are filled with osteoclasts. This cystic bone disease which accompanies chronic hyperparathyroidism is known as osteitis fibrosa cystica (or von Recklinghausen's disease). In this condition, serum phosphate is decreased, whereas the renal excretion of calcium increases markedly. The increased osteoclastic activity results in turn in a sharp elevation of plasma alkaline phosphatase.

In hyperparathyroidism, the serum total calcium level can rise to over 15 mg percent, so that depression of central and peripheral neural activity ensues. This effect is accompanied by a number of symptoms, including muscular weakness and loss of appetite, together with a prolonged diastolic relaxation phase of the cardiac cycle.

When hyperparathyroidism is severe, both calcium and phosphate levels in the serum become markedly elevated. The concentration of the latter ion rises, presumably because renal excretion of this ion is unable to keep pace with its mobilization. The body fluids become supersaturated with both of these ions, and calcium phosphate ($CaHPO_4$) crystals are precipitated within the renal tubules, alveoli of the lungs, gastric mucosa, and arterial walls. This ectopic or metastatic calcification can proceed so rapidly that death may ensue in only a few days.

Patients suffering from mild hyperparathyroidism may exhibit no gross bone lesions or defects such as were described above; however, they have a tendency to develop kidney stones (or renal calculi). This calculus formation occurs because the abnormal quantities of calcium and phosphate present in the serum tend to be excreted by the kidneys, and thus precipitate out thereby forming calcium phosphate calculi. Calcium oxalate stones may also form because of the elevated calcium ion concentration in the presence of normal oxalate levels. The urinary pH of any particular patient is, of course, an important factor in determining calculus formation.

Renal calculi are far more apt to develop in alkaline than in acid urine, because of the relatively lower solubility of calcium phosphate (or oxalate) at a higher pH. An acidotic diet or drugs that produce acidosis often are used in the therapy for renal calculi.*

Second hyperparathyroidism results from a low serum calcium ion level, regardless of its cause, because the parathyroid glands are stimu-

Oranges and other citrus fruits thus should be eliminated from such diets, because they produce an alkaline urine owing to the cations they contain, the citric acid responsible for the low pH of such fruits being oxidized in the TCA cycle to carbon dioxide and water.

lated directly to increase parathyroid hormone secretion by this relative lack of calcium, and thus to compensate for the deficiency. Such secondary hyperparathyroidism may result from a low calcium diet, impaired calcium absorption (viz, in rickets), vitamin deficiency, pregnancy, lactation, or osteomalacia.

RICKETS. Normal skeletal growth and development requires a diet that is not only adequate in calories, calcium, and phosphorous, but also in vitamins A, D, and C.

The condition of rickets (sometimes called "juvenile rickets") is found in children principally as the result of a deficiency of calcium and phosphate in the extracellular fluids. In general, the serum calcium in rickets may be only slightly lower than normal. Commonly, however, rickets is due to a lack of vitamin D, rather than to a calcium- or phosphate-deficient diet. Sunlight catalyzes the conversion of 7-dehydrocholesterol in the skin to an active form (vitamin D_3) and this substance in turn promotes the intestinal absorption of calcium, hence phosphorus.

In a patient who is suffering from rickets, the serum calcium level may be depressed only slightly; however, the phosphate concentration is reduced considerably in this disease. This situation occurs because the reduced blood calcium stimulates a secondary hyperparathyroidism in order that the serum calcium level be maintained by bone resorption. There is no effective regulatory mechanism for controlling the fall in phosphate level. Since phosphate ion is mobilized together with calcium ion during bone resorption, and since an elevated parathyroid hormone level actually increases the renal excretion of phosphate, the net effect is a fall in the serum phosphate level.

The increased parathyroid hormone secretion found in rickets tends to protect the individual against hypocalcemia, and this hypersecretion is accompanied by a marked hyperplasia of the glandular elements. The blood alkaline phosphatase level also rises considerably because of the increased osteoclastic activity. However, the chronic demands upon the bone mineral reservoir eventually weaken the bone. Although osteoblastic activity is intensified, and organic bone matrix is secreted and deposited in large amounts during rickets, this material does not become calcified. The concentrations of calcium and phosphate in the interstitial fluid remain too low to achieve an adequate concentration for hydroxyapatite crystal formation to occur. Eventually, uncalcified bone matrix (ground substance) replaces the resorbed bone.

Tetany is rare early in rickets, because the mobilization of calcium is sufficient to maintain a reasonably normal ionized calcium level in the serum. Eventually, as the disease progresses, the bone reservoir of calcium is depleted so extensively that the plasma calcium level falls rapidly and death occurs from laryngospasm and asphyx-

ia. Calcium ion, when administered intravenously, immediately reverses the tetany. The condition of rickets is prevented, or ameliorated once it has developed, simply by a diet that is adequate in calcium, phosphate, and vitamin D. The requirement for the latter nutritional factor is essential, as otherwise the ingested calcium and phosphate are not absorbed but excreted in the feces.

OSTEOMALACIA. Osteomalacia occurs in adults; this condition is essentially the same as rickets in children. Thus osteomalacia is often called "adult rickets."

Only rarely do normal adults suffer from a lack of calcium, phosphate, and vitamin D, because large quantities of these substances are not needed by the body since rapid bone growth has ceased. During chronic steatorrhea, however, which is characterized by a failure to absorb fat, bone demineralization may occur. In steatorrhea a net calcium deficiency develops because this cation forms insoluble soaps with fats which then are excreted in the stools. Furthermore, as vitamin D is a fat soluble factor and is absorbed from the gastrointestinal tract together with fats, in steatorrhea a deficiency of this vitamin acts in combination with the excess calcium excretion so that a deficiency of calcium and phosphate in the body fluids occurs. If this situation is sufficiently prolonged or severe then bone demineralization with the attendant derangements of juvenile rickets ensues.

Osteomalacia also may develop in pregnant or lactating women whose dietary intake of calcium and phosphate as well as vitamin D is marginal or actually inadequate.

RICKETS, OSTEOMALACIA, AND RENAL DISEASE. A condition sometimes called "renal rickets" can result from long-term kidney disease. The renal dysfunction leads to acidosis as there is a failure to excrete acids normally. The acidosis in turn stimulates bone resorption, and thus is one causative factor leading to the rickets and osteomalacia that ultimately develop.

A rare chronic familial disease of the kidneys which has an indefinite etiology also causes rickets and osteomalacia, but by a different mechanism than that described above. In this condition, known as congenital hypophosphatemia (or hypophosphatasia), there is an inherent failure of the renal tubules to reabsorb phosphate, hence a net phosphate deficiency ultimately develops. This condition affects primarily children. The defective bone formation that is exhibited by such individuals resembles that found in rickets.

The type of rickets encountered in congenital hypophosphatemia responds to treatment with phosphate compounds rather than to calcium and vitamin D administration; hence it is also called vitamin D-resistant rickets.

OSTEOPOROSIS. Osteoporosis differs from rickets and osteomalacia in that the defect lies in an abnormal formation of organic bone matrix (ground substance), rather than in an abnormal calcification of that matrix. In osteoporosis the osteoblastic activity is generally subnormal so that the rate of bone deposition also is subnormal, and as a consequence the skeletal structures become more fragile than normal.

Osteoporosis may be caused by a number of factors: (1) a mechanical effect because of the lack of use, such as may be caused by extended bed rest; (2) chronic malnutrition which is so extensive that the proteinaceous components of the bone matrix cannot by synthesized; (3) vitamin C deficiency, as this nutritional substance is essential to the normal secretion of intercellular materials by all cells, including osteoblasts; (4) diseases of protein metabolism in general can result in osteoporosis because of faulty matrix synthesis; (5) Cushing's disease, in which the secretion of excess glucocorticoids causes a reduced protein deposition in general, while concomitantly these hormones also stimulate increased protein breakdown (catabolism); (6) acromegaly can result in osteoporosis because of a deficiency in sex hormones, an excess of adrenocortical hormones, and sometimes a deficiency of insulin because of the diabetogenic nature of adenohypophyseal hormones, all of which defects may impair normal bone matrix formation; (7) postmenopausal decrease in estrogen secretion may result in osteoporosis, because estrogens exert an osteoblast-stimulating effect which is absent in women following the menopause; (8) in elderly (senescent) persons, protein anabolism usually is defective so that inadequate bone matrix is synthesized. The bones become brittle because of a relative lack of organic matrix, and thus fracture readily.

CALCIFICATION IN SOFT TISSUES. As noted earlier, aberrant (or ectopic) calcification and thus ossification can occur in many soft tissues of the body; renal calculi and their formation were discussed earlier as an example of this process. Myositis ossificans is another example of ectopic or metastatic calcification. In the skin, tendons, ligaments, and cardiac valves the process of calcification also is associated with collagen deposition similar to that which takes place during normal bone formation. In large blood vessels such as the aorta, however, ossification is associated with the protein elastin. Elastin is found in the intima of the vessels, and calcification and arteriosclerosis (or "hardening of the arteries") take place in an atheromatous lesion (or plaque) that previously was deposited in the subendothelial layer. The atheroma per se is an abnormal deposit of lipid material that is rich in cholesterol.

Pancreatic calcification possibly may be related to a local elevation of pH in this organ that is caused by an increased secretory activity on the part of the acinar tissue, hence the solubility of calcium and phosphate ions may be reduced locally to a point at which calcium phosphate crystals precipitate.

THE PANCREAS: ENDOCRINE FUNCTIONS

The pancreas and the testes share the distinction of being the only endocrine glands in the body whose hormones are secreted into the bloodstream in the usual fashion, but which also have ducts whereby other products elaborated by different cells that are located within the same organ are transported from their site of manufacture to their locus of action.

The endocrine properties of the pancreas reside in specialized groups of cells known as the islets of Langerhans. These cells synthesize, store, and secrete the two hormones of the pancreas, viz., insulin (antidiabetic hormone) and glucagon.

Functional Morphology of the Pancreas

GROSS ANATOMY. The gross anatomy of the pancreas was discussed on page 834 in relation to the exocrine secretion of digestive fluids by this organ into the intestinal tract.

HISTOLOGY OF THE ISLETS OF LANGERHANS. Embedded in random fashion throughout the exocrine tissue of the pancreas are small (about 75 by 175 μ), ovoid, highly vascular masses (or nests) of endocrine cells which compose the islets of Langerhans (Fig. 15.24). The islets are somewhat more abundant in the tail of the pancreas than in the body or head of the organ. The total number of islets present in the entire human pancreas has been estimated to range between 200,000 and 2,000,000; however, the total mass of the islet tissue represents only one or two percent of the entire mass of the pancreas.

The abundant blood flow from the islets of Langerhans drains directly into the hepatic portal vein.

The islets of Langerhans are delineated from the surrounding acinar tissue by a delicate layer of reticular connective tissue, which does not enter the islets to any great extent, except insofar as a reticulum is associated with the capillaries. These terminal blood vessels ramify extensively after branching off from a pancreatic arteriole, then reanastomose ultimately to form the pancreatic veins, which have a course that is parallel with the arteries. The pancreatic veins unite to join the hepatic portal vein. The capillaries found in the islets of Langerhans have the type of endothelium that is alternately thick and then punctuated by attenuated regions which have open fenestrations (or pores).

By using appropriate staining techniques, several morphologically distinct cell types can be recognized within the islet tissue of the pancreas. These cell types are distinguished by their granulation.

The islet cells generally are arranged in cords, and stain more faintly than do the surrounding

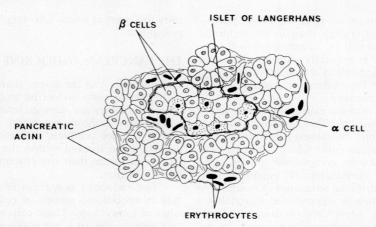

FIG. 15.24.   Islets of Langerhans, the endocrine tissue of the pancreas. (After Bloom and Fawcett. *A Textbook of Histology,* 9th ed, 1968, Saunders; Ganong. *Review of Medical Physiology,* 6th ed, 1973, Lange Medical Publications.)

pancreatic acinar cells. No granulations are visible within the islet cells in routine hematoxylin and eosin preparations; however, in tissues that have been stained by the Masson technique or with Mallory-azan, three kinds of granular cells can be distinguished within the islets of the human pancreas. These are alpha cells, beta cells, and delta cells. These cell types have been confirmed by electron microscopy.

Alpha Cells.   With the special stains mentioned above, alpha cells are seen to contain relatively large, bright red granules under the light microscope. By means of electron microscopy, the alpha cells are seen to contain large quantities of dense spherical granules that are enclosed in a membrane. A thin, clear region that separates the membrane from the granule may or may not be visualized, depending upon the type of fixation that has been employed for preparation of the tissue.

The mitochondria appear as slender rods when suitably oriented in the section. The granular endoplasmic reticulum and juxtanuclear Golgi complex of alpha cells are typical of secretory cells. Furthermore, the concentration of glucagon can be correlated with the relative abundance of alpha cells in the entire islet cell population in experimental studies that have been carried out on lower species. The alpha cells have been implicated in the secretion of glucagon, as discussed on page 1091.

Beta Cells.   The beta cells are somewhat smaller than the alpha cells, and in specially-stained sections the granules appear brown-orange. Ultramicroscopy reveals that pronounced interspecific variations exist in the granules of the beta cells. In the human as well as the canine pancreas, however, the membrane-limited beta granules contain one or more minute crystals that exhibit a polygonal or rectangular shape.

The Golgi complex is more highly developed in the beta than in the alpha cells, and the mitochondria are also larger. The endoplasmic reticulum on the other hand is less pronounced in the beta than in the alpha cells.

The beta cells produce the antidiabetic hormone insulin. This important fact can be demonstrated in several ways. First, the compound alloxan, when given in appropriate doses, selectively destroys the beta cells of the islets and diabetes is produced thereby (Fig. 15.25). Second, in clinical diabetes, the beta cells are abnormal or absent. Third, patients having beta cell pancreatic tumors exhibit a chronic hypoglycemia that results from insulin hypersecretion. Surgical extirpation of the pancreas relieves the hypoglycemia in such patients.

Delta Cells.   In the human pancreas, delta cells may contain granules that are intermediate in size between those which are present in alpha and beta cells, and some investigators consider the delta cells to be transitional forms of the other two cell types. However, this interpretation is debatable. The delta cells could also represent alpha or beta cells that are undergoing regression, or else be completely different cell types having a functional role that is undefined at present.

In general, the three types of islet cell discussed above tend to exhibit polarization with respect to the capillaries of the islets; thus, the granules are closely aggregated toward the vascular pole of the cells.

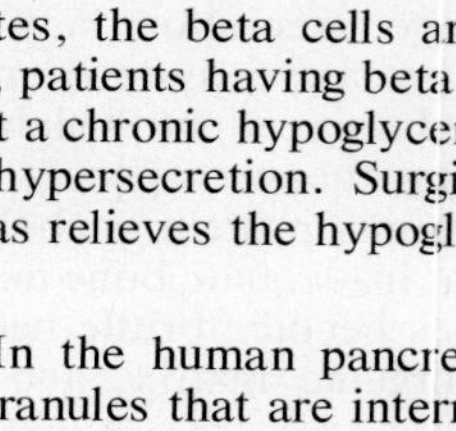

FIG. 15.25.   Structural formula of the diabetogenic compound alloxan. (After Ganong. *Review of Medical Physiology,* 6th ed, 1973, Lange Medical Publications.)

Diabetes Mellitus

In ancient Greece and Rome, the term *diabetes mellitus* was used in reference to the excretion of a large volume of urine having a sweetish taste, whereas the term *diabetes insipidus* denoted a large volume of tasteless (ie, unsweetened) urine. Today the term *diabetes insipidus* is applied solely to the disease caused by lesions in the supraoptic nuclei of the hypothalamo–hypophyseal system; hence the word "diabetes" now is used synonymously with diabetes mellitus, and thus will be used henceforth to denote the condition in which a large volume of urine containing glucose is excreted each day.

The complex nature of the diabetic syndrome is best introduced by summarizing the defects encountered in the absence of insulin, whether in an experimental animal or a human patient. Hence diabetes is a complex metabolic disorder, and all of the symptoms that are encountered in this condition may be attributed directly either to a relative or to a complete lack of insulin. Furthermore, all of these symptoms may be reversed completely by the administration of appropriate doses of exogenous insulin.

EXPERIMENTAL DIABETES MELLITUS. Following surgical removal of the entire pancreas (ie, total pancreatectomy) in an animal such as the dog, the characteristic syndrome of diabetes mellitus develops rapidly. Immediately following pancreatectomy, the animal recovers quickly and appears to be normal, but an adequate diet alone will not prevent death, which occurs at variable intervals (but usually within several weeks) following surgery. On the other hand, if the animal is provided with a diet that is adequate in the essential nutrients and supplied with insulin, it will live indefinitely. Withdrawal of insulin causes the diabetic pattern to reappear, and death will ensue unless insulin is provided once again.

The signs and symptoms encountered in experimental diabetes mellitus are as follows:

Hyperglycemia. A greatly increased blood sugar level results from an elevated hepatic glycogen breakdown so that glucose accumulates in the bloodstream due to failure of the peripheral tissues to take up and utilize this substrate properly (Table 10.7).

Glucosuria. Urinary glucose excretion occurs when the blood glucose concentration rises above the renal transport maximum for this compound. This effect usually develops when the blood glucose level rises to about 160 mg percent.

Polyuria. A marked increase in the volume of urine excreted ensues as a consequence of the hyperglycemia. The dehydrating osmotic diuresis that results in turn leads to a marked increase in thirst, so that a greatly increased ingestion of water (or polydipsia) is observed.

Nitrogen excretion. The output of urinary nitrogen is increased in diabetes because of a markedly increased catabolism of bodily protein combined with failure of protein synthesis.

Loss in body weight. Weight loss follows the defects in protein metabolism.

Metabolite levels in the blood. The concentrations of free fatty acids, neutral fats, cholesterol esters, and phospholipids in the blood all show a marked increase due to an elevated mobilization and breakdown of fats in the absence of insulin.

Ketone bodies. As a consequence of an accelerated fat catabolism, ketone bodies accumulate in the blood (ketonemia) and excessive quantities of these compounds are also excreted in the urine (ketonuria).

Metabolic acidosis. The ketonemia produces a metabolic acidosis, and this in turn causes respiratory stimulation because of the reduced blood pH. An "air hunger" develops, and the respiratory pattern that accompanies the air hunger is called Kussmaul breathing in humans.

Respiratory quotient. A reduced RQ in uncontrolled diabetes is a direct indication of the increased fat catabolism mentioned above.

Glycogen content in muscles. The glycogen level in skeletal muscles may fall or remain normal. The glycogen content of cardiac muscle actually increases in diabetes, whereas liver glycogen content falls to extremely low levels.

Phosphate excretion. Urinary phosphate excretion is increased markedly in diabetes.

In addition to the foregoing effects of insulin deficiency, diabetic animals also become highly susceptible to infection, and pathologic conditions may also develop in the eyes of pancreatectomized animals some time after the operation was performed.

If insulin is withheld, the subject eventually passes into a state of coma induced by the ketosis, and death is the inevitable consequence. However, as discussed earlier, when experimental pancreatectomy is accompanied by hypophysectomy the diabetes that develops is considerably less severe because of the absence of the anterior pituitary (Houssay animal).

HUMAN DIABETES MELLITUS. Diabetes mellitus is a pathologic condition in man encountered frequently in clinical practice. Certain etiologic factors known to be associated with this disease in patients will be discussed later. Acute insulin deficiency in a human patient is accompanied by all of the signs and symptoms that were enumerated above for animals following pancreatectomy.

In man, the air hunger produced by stimulation of the respiratory center by the acidosis is known as Kussmaul breathing, ie, paroxysms of severe dyspnea occur intermittently.

The classic triad of human diabetic symptoms is polydipsia, polyuria, and polyphagia; this triad sometimes is known as the three "P's."

It should be stressed that in chronic human diabetes, as well as in animals having experimental diabetes, there is a great excess of extracellular glucose combined with an intracellular deficiency of this substrate; hence the subject actually suffers from "starvation in the midst of plenty." In turn this intracellular inanition leads to the marked weight loss, and the patient suffers from malnutrition despite a seemingly adequate diet.

It is evident from an examination of the list of defects encountered in diabetes mellitus that insulin lack results in profound metabolic derangements in a multitude of biochemical processes in the body. Although the effects of this disease are often ascribed principally to defects in carbohydrate metabolism, such a simplistic view is quite inaccurate because insulin deficiency alters the normal metabolic patterns in protein and fat metabolism to an equally profound degree, as it does carbohydrate metabolism. The major biochemical lesions encountered in diabetes mellitus will be discussed in the sections to follow. In turn, these discussions will serve to illustrate and to emphasize the multiple roles of insulin in the normal biochemical regulation of various metabolic processes throughout the body.

Insulin

CHEMISTRY. The chemical structure of insulin from various species has been elucidated, and this compound has been synthesized in the laboratory. The structure of human insulin is shown in Figure 15.26. The hormone is a large polypeptide (or small protein), and it consists of two chains, A and B, which are linked together by disulfide bridges between two cystine residues. Human insulin has a molecular weight of 5,734 daltons, as calculated from the structural formula of the compound. In solution, the molecular weight of insulin is affected by its concentration, the pH, and the presence of other ions and proteins; therefore considerable variation in molecular weights can be obtained under different experimental conditions and when different techniques for assessing this value are employed. In general the molecular weights for insulin are found in multiples of 6,000; viz., 6,000; 12,000; 24,000; 36,000; and 48,000. This fact suggests that single insulin molecules polymerize under different conditions.

The 51 amino acid residues that comprise the two chains of insulin differ slightly among various species in the A8, A9, A10, and B30 positions. For example, insulin extracted from beef pancreas, which commonly is used in the treatment of diabetic patients, has the sequence Ala · Ser · Val in positions A8, A9, and A10, and Ala in position B30. In human insulin, the sequence in the A chain is Thr·Ser·Ile, whereas Thr is found in position B30. Usually these interspecific differences are insufficient to affect the biologic (ie, hypoglycemic) activity of the insulin when it is administered to different species; however, such differences are sufficient to make the insulin exhibit antigenic properties. A diabetic patient receiving beef insulin for extended periods of time develops antibodies against the exogenous beef insulin. Generally the titer of these antibodies is sufficiently low so that no clinical problem develops. If a patient does develop a high titer of antibodies against beef insulin, then insulin from another species (eg, pig) must be employed to control the diabetic condition.

An important chemical property of the insulin molecule resides in the presence of a number of acidic groups in its structure. This chemical property thus facilitates the combination of insulin with bases. Clinically this fact is of considerable significance, as the combination of insulin with protamine or globin results in a complex that is relatively much less soluble at tissue pH, the insulin is absorbed from the site of injection much more slowly, and thereby its action is prolonged considerably. Less frequent injections of insulin are necessary in patients for effective control of their diabetes. A number of such salts of crystalline insulin, some containing zinc, are available commercially.

A specific example of the prolongation of insulin action that takes advantage of the chemical nature of this hormone may be cited. Regular amorphous insulin starts to exert its hypoglycemic action in approximately 1 hour following injection, and the maximum hypoglycemic action of this hormone is manifest in about 3 hours, whereas the

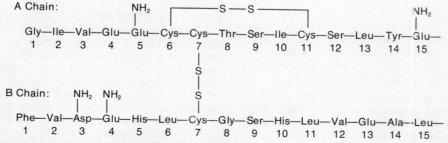

FIG. 15.26 Amino acid sequence in human insulin. Note that this hormone in its active form consists of two polypeptide chains (A and B) that are linked together by disulfide linkages. (After Ganong, *Review of Medical Physiology*, 6th ed, 1973, Lange Medical Publications.)

effect of amorphous insulin persists for about 6 hours. Protamine zinc insulin, on the other hand, exhibits a longer lag period following its injection (between 6 and 8 hours), and the maximum hypoglycemic activity is manifest between 12 and 24 hours. The duration of action of protamine zinc insulin is for 48 to 72 hours.

It should be emphasized that insulin cannot be given orally, as it is rapidly digested and thus inactivated in the gastrointestinal tract because of its peptide structure. Parenteral administration is necessary. Furthermore, it must be borne in mind that under physiologic conditions the pancreas secretes insulin into the portal vein so that it passes directly to the liver. Insulin that has been injected subcutaneously, however, must first be absorbed locally from the site of injection and then must pass through the circulatory system with an attendant dilution and partial inactivation before it reaches the portal circulation.

BIOSYNTHESIS AND SECRETION OF INSULIN. The synthesis of insulin takes place within the endoplasmic reticulum of the beta cells of the islets of Langerhans. As shown in Figure 15.27, the insulin is next transported to the Golgi complex where it forms minute spherical droplets (or granules) that are invested in a membrane, and that are visible with light microscopy. The droplets then move to the cell wall where their membranes fuse with the membrane of the cell, and the insulin is secreted into the interstitial fluid by the process of emeiocytosis (or "reverse pinocytosis"). The insulin thus released now traverses the basement membrane and fenestrated endothelial wall of an adjacent capillary in order to enter the bloodstream.

The insulin molecule is synthesized as a large, folded single chain termed proinsulin. This polypeptide consists of approximately 73 amino acid residues, and contains the two disulfide linkages between the A and B chains, as well as a peptide moiety that connects the two chains as shown in Figure 15.28. The connecting peptide (or c-peptide) linking the two chains normally is removed by proteolytic cleavage before the insulin is secreted in its biologically active form; however after prolonged beta-cell stimulation, or in tumors of the beta cells, proinsulin itself may be secreted.

Proinsulin has negligible hormonal activity compared to activated insulin (Fig. 15.26).

As shown in Figure 15.28, the amino portion of proinsulin ultimately forms the B chain, and its carboxyl portion forms the A chain of insulin. The biologically inactive c-peptide, consisting of some 22 amino acid residues, is secreted along with insulin. This initial linkage is essential to the biosynthesis of insulin, as it permits folding and disulfide bond formation in the proinsulin molecule.

The normal processes involved in the synthesis and secretion of insulin require active glucose metabolism, probably for the formation of adenosine triphosphate. Cyclic AMP also plays an essential role in the biosynthetic processes whereby insulin is formed. Calcium and potassium ions also are required for insulin secretion.

TRANSPORT AND METABOLISM OF INSULIN. Considerable research has indicated that insulin is transported in the circulation as the free molecular form, and that it is not bound to the plasma proteins as is, for example, thyroxine. On the other hand, tissues bind insulin readily. Thus, large quantities of insulin are bound in the liver and kidneys. Most other tissues can also bind insulin to varying degrees; however, the hormone is not bound either by the brain or by the erythrocytes.

Insulin is inactivated in the body by two enzyme systems which collectively are termed insulinases. One of these enzyme systems induces cleavage of the disulfide linkages which hold the A and B chains together, whereas the other enzyme system hydrolyzes the peptide chains themselves. The enzyme responsible for splitting the disulfide linkage of insulin is called hepatic glutathione insulin transhydrogenase. This enzyme splits the insulin molecule into its component A and B chains, and in this system, the tripeptide glutathione acts as a coenzyme for the transhydrogenase.

The half-life of insulin in the circulation of man is reported to be about 40 minutes. This fact is indicative of an extremely rapid physiologic turnover rate for this hormone, and it also indicates that a continuous supply of endogenous insulin is of critical importance to the individual.

It should be mentioned that determination of the insulin concentration in blood by a chemical

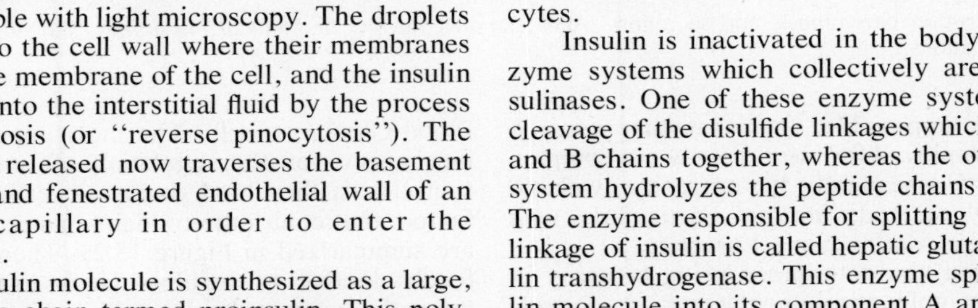

NH$_2$ NH$_2$

Leu—Glu—Asp—Tyr—Cys—Asp
16   17   18   19   20   21

             S

           S

Tyr—Leu—Val—Cys—Gly—Glu—Arg—Gly—Phe—Phe—Tyr—Thr—Pro—Lys—Thr
16   17   18   19   20   21   22   23   24   25   26   27   28   29   30

FIG. 15.26 (cont'd)

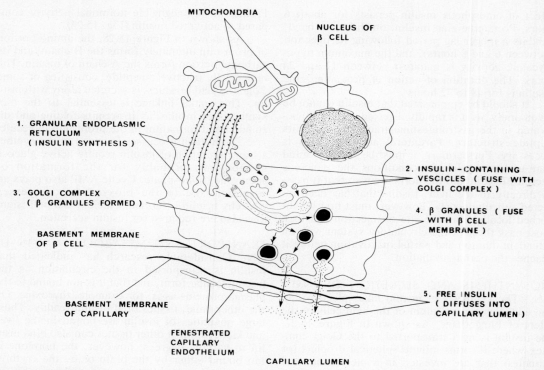

MITOCHONDRIA

NUCLEUS OF
β CELL

1. GRANULAR ENDOPLASMIC
   RETICULUM
   (INSULIN SYNTHESIS)

2. INSULIN – CONTAINING
   VESICLES (FUSE WITH
   GOLGI COMPLEX)

3. GOLGI COMPLEX
   (β GRANULES FORMED)

4. β GRANULES (FUSE
   WITH β CELL
   MEMBRANE)

BASEMENT MEMBRANE
OF β CELL

BASEMENT MEMBRANE
OF CAPILLARY

5. FREE INSULIN
   (DIFFUSES INTO
   CAPILLARY LUMEN)

FENESTRATED
CAPILLARY
ENDOTHELIUM

CAPILLARY LUMEN

FIG. 15.27. Insulin biosynthesis and secretion. (After Ganong. *Review of Medical Physiology*, 6th ed, 1973, Lange Medical Publications.)

procedure is impossible, and that the research techniques employed for insulin assay are complex and tedious. Consequently the currently available insulin assay methods are not particularly well suited to use in routine clinical practice.

Insulin may be determined using bioassay in vivo. For example, the hypoglycemia or convulsions induced in sensitized (eg, hypophysectomized) rats or mice are compared against a control (or standard) preparation of insulin in a similar group of animals. The reduction in blood sugar in each animal in each group must be determined individually and then evaluated statistically.

In vitro techniques for insulin bioassay include the quantitative effects of unknown insulin samples compared with standard insulin preparations on the rate of glucose uptake in such isolated tissues as the epididymal fat pad or mammary gland of the rat. Immunochemical assays may also be employed for insulin determination in vitro, and such procedures have largely replaced the other techniques. None of these techniques is especially precise, and all are extremely time-consuming.

FACTORS WHICH AFFECT THE PHYSIOLOGIC REGULATION OF INSULIN SECRETION. The secretion of insulin by the islets of Langerhans in a normal individual is known to be influenced by a wide variety of factors having various degrees of relative importance in the overall regulation of insulin secretion by the pancreas in vivo.

Blood Glucose Level. The most important single mechanism for the regulation of normal insulin secretion is the blood glucose level itself and the factors that contribute to maintenance of this level are summarized in Figure 15.29. There is a direct feedback effect of the blood glucose concentration upon the pancreatic islet cells, which in turn are freely permeable to this substrate regardless of the presence of insulin. In other words, the rate at which glucose enters the islet cells of the pancreas is independent of the insulin content of these tissues. An elevated glucose level in the blood perfusing the pancreas stimulates the secretion of insulin by the beta cells, so that the venous blood leaving this organ contains a higher concentration of the hormone, and this effect in turn reduces the systemic blood glucose level. Conversely, a normal or depressed blood sugar level inhibits the rate of insulin secretion so that the blood sugar level rises.

This feedback control initiated by the blood glucose concentration provides an exquisitely sensitive homeostatic mechanism whereby the secretion of insulin is regulated between narrow tolerances in normal individuals. Thus, both blood insulin as well as glucose levels are integrated closely in a dynamic fashion under physiologic conditions. The concentration of insulin in the bloodstream at any time thus depends in turn upon a balance between the rate of secretion of insulin into the blood by the beta cells of the pancreatic

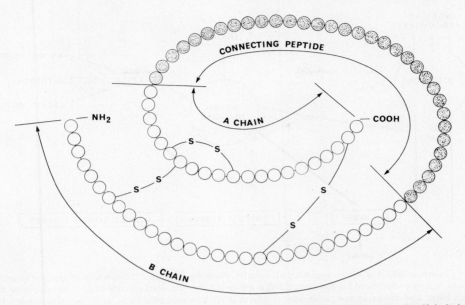

FIG. 15.28. Proinsulin. The connecting peptide (C peptide) is essential to the formation of the disulfide linkages. The C peptide is removed from proinsulin and then secreted together with active insulin in the β-granules (Fig. 15.27). (After Ganong. *Review of Medical Physiology,* 6th ed, 1973, Lange Medical Publications.)

islets and its rate of removal from the blood, and its inactivation by the various tissues of the body. In diabetes, this regulatory mechanism is absent so that homeostatic regulation of the blood sugar level is impaired, as illustrated in Figure 15.30.

Glucose normally exerts a biphasic effect upon insulin secretion in humans as well as in animals. At first the rate of secretion of the hormone increases rapidly following the administration of glucose, and then the secretory rate of insulin falls. This effect now is followed by a prolonged insulin secretion which develops more slowly and is maintained for as long as an elevated

blood sugar level also is sustained by the intravenous infusion of glucose.

The initial rapid insulin secretion that occurs in response to glucose infusion reaches its peak in a few minutes. This effect can be inhibited by drugs that block an increase in cyclic AMP formation, which response in turn is mediated via the β receptors. Propranolol, for example, exerts this effect. The second, prolonged response to glucose infusion is inhibited by compounds that inhibit protein synthesis.

Carbohydrate substrates other than glucose also exert some effect upon the pancreatic secre-

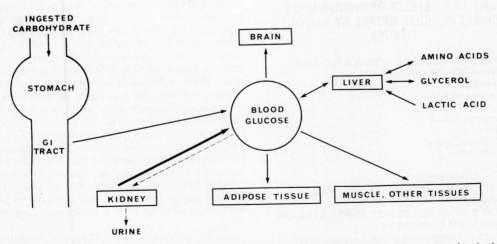

FIG. 15.29. Factors that regulate the blood glucose level in the normal individual. Note that glucose is reabsorbed strongly from the glomerular filtrate (heavy arrow). Insulin does not affect the intestinal or renal absorption of glucose nor does it affect brain uptake of this compound. See also Figures 15.30 and 15.31.

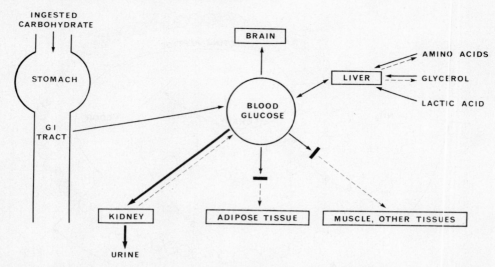

FIG. 15.30. Defects in blood glucose regulation in diabetes. Note the large quantitative loss of glucose in urine (heavy arrows) because the renal threshold for this compound is exceeded, the impaired glucose utilization in adipose tissue and muscle (black bars), and the excessive catabolism of amino acids, glycerol, and lactic acid. See also Figures 15.29 and 15.31. (After Ganong. *Review of Medical Physiology,* 6th ed, 1973, Lange Medical Publications.)

tion of insulin. Mannose and fructose also stimulate insulin secretion; however the latter compound is converted rapidly into glucose within the cells. Insulin also facilitates the entry of D-xylose, L-arabinose, and galactose into cells in general; however none of these sugars directly stimulates the secretion of insulin (Table 15.7).

Effect of Electrolytes Upon Insulin Secretion. Adequate concentrations of potassium and calcium are essential to the normal secretion of insulin by the pancreatic islets.

Effect of Amino Acids and Keto Acids Upon Insulin Secretion. Certain amino acids (eg, arginine and leucine among others), as well as β-ketoacids de-

Table 15.7 EFFECT OF INSULIN UPON NORMAL GLUCOSE UPTAKE BY VARIOUS TISSUES

UPTAKE OF GLUCOSE STIMULATED AND REGULATED BY INSULIN
Skeletal muscle
Cardiac muscle
Smooth muscle
Adipose tissue
Liver
Mammary gland
Leukocytes
Aorta
Eye, crystalline lens
Fibroblasts
Pituitary

UPTAKE OF GLUCOSE NOT STIMULATED OR REGULATED BY INSULIN
Islets of Langerhans
Brain (except a portion of the hypothalamus)
Renal tubules
Erythrocytes

rived from fat metabolism (eg, acetoacetic acid) can stimulate insulin secretion by unknown mechanisms. It is interesting to note in this connection that insulin itself stimulates the incorporation of amino acids into proteins, and in addition this hormone also inhibits the breakdown of fats which leads to the production of keto acids in the first place.

Neural Influence Upon the Secretion of Insulin. Terminal ramifications of the right vagus innervate the islets of Langerhans, and stimulation of this nerve elicits an increased insulin secretion. The physiologic role of this effect is unknown.

Intestinal Hormones and Insulin Secretion. Under physiologic conditions, ingested glucose stimulates a much greater secretion of insulin than does an equivalent dose of the same carbohydrate given intravenously. Research into the cause of this phenomenon has revealed that a number of substances that are released by the gastrointestinal mucosa can exert this effect upon insulin secretion. These substances include gastrin, secretin, cholecystokinin, and glucagon, as well as a substance resembling glucagon which is released from the intestinal mucosa. Thus, it is clear that normal insulin secretion can be enhanced by one or more endocrine factors that originate in the mucosa of the gastrointestinal tract and that are normally liberated by food entering the stomach and small intestine. It is unclear, however, just which factor (or factors) is involved under physiologic conditions in producing such a pancreatic response in vivo.

The several factors that are summarized above all have been demonstrated experimentally to be able to stimulate pancreatic insulin secretion of a greater or lesser extent. However, it should be reiterated here that the principal physiologic stimulus to insulin secretion is the blood sugar level itself; consequently, the role played by all of the other factors is relatively minor compared to this.

FACTORS THAT INHIBIT INSULIN SECRETION

Exogenous Insulin. If insulin is injected into normal animals, the content of insulin within the pancreas falls. When the exogenous insulin administration is stopped, however, the pancreatic islet cells become hyperactive and respond readily to the various normal stimuli which enhance insulin secretion that were enumerated above. A juvenile diabetic patient may exhibit a temporarily diminished insulin requirement following initial treatment with the hormone, although eventually the diabetes worsens and thus becomes permanently established as the patient grows older.

Epinephrine, Norepinephrine, and Insulin Secretion. The catecholamines act directly upon the pancreatic islets to inhibit insulin secretion. This inhibitory effect is not blocked by drugs which inhibit the β-adrenergic receptors; however, it is blocked by inhibition of the α-adrenergic receptors. Experimental evidence suggests that substances which stimulate the α-adrenergic receptors (eg, epinephrine and norepinephrine) actually act by inhibiting cyclic AMP. Conversely drugs which stimulate β-adrenergic receptors enhance insulin secretion.

Pancreatic Reserve. In common with other endocrine cells, the beta cells of the islets hypertrophy in response to appropriate stimuli, so that a prolonged and moderate hyperglycemia increases both the synthesis and secretion of insulin. If, however, the hyperglycemia is prolonged and pronounced, the cells ultimately become hyalinized and vacuolated. Subsequently, hyaline and hydropic degeneration of the islet cells ensues. If the hyperglycemia is alleviated shortly after the cells stop secreting insulin, recovery may occur spontaneously, but if the stimulus persists, then the cells atrophy completely. The pancreas appears to be the only endocrine gland which exhibits this atrophy through exhaustion. Normally the pancreas has such a large reserve of functional islet tissue that such atrophy is quite difficult to evoke in normal humans or animals. However, if this reserve is decreased through use of drugs (eg, injection of small doses of alloxan) or by partial pancreatectomy, then hyperglycemia can induce exhaustion in the remaining beta cells. Only rarely do human subjects who ingest a high-carbohydrate diet for long periods of time develop a hyperglycemia, which is associated with a glycosuria, dehydration, and coma.

Starvation and the injection of large quantities of insulin also depress insulin secretion by the islets.

Hypoglycemic and Hyperglycemic Drugs. Certain synthetic compounds reduce the blood sugar level when administered orally; however, these substances induce hypoglycemia only when functional pancreatic tissue is present. Such hypoglycemic agents include tolbutamide, chlorpropamide, and other sulfonylurea derivatives. These substances act by stimulating the release of endogenous insulin from the pancreas, and apparently this effect is exerted by mediating an increase in the quantity of cyclic AMP in the beta cells.

In contrast to the hypoglycemic drugs mentioned above, diazoxide is a drug which is diabetogenic like alloxan. This compound has been demonstrated to inhibit pancreatic insulin secretion, thereby inducing diabetes. Other drugs, such as thiazide diuretics, may also exert diabetogenic effects in susceptible individuals, once again by inhibiting normal insulin secretion.

Metabolic Effects of Insulin

The following discussion will present the overall effects of insulin upon a number of individual physiologic and metabolic processes. Emphasis will be placed upon the normal roles of this hormone, and these will be compared with the metabolic defects which arise in diabetes. In turn, the etiology of the diabetic symptoms summarized earlier will be related to these underlying defects where possible.

Despite the extensive regulatory influences that are manifest by insulin in every area of metabolism, however, the actual biochemical mechanisms whereby insulin produces these diverse regulatory effects on metabolism are largely unknown. There is little point, therefore, in attempting to present a "unified concept of insulin action." Rather, such facts as are available concerning the general effects of insulin on carbohydrate, fat, and protein metabolism will be stressed.

EFFECTS OF INSULIN ON CARBOHYDRATE METABOLISM. The factors involved in the physiologic homeostasis of blood glucose in the presence of an adequate supply of insulin are summarized in Figures 14.33 (p. 960), 15.29 and 15.31A, and the defects in this process during insulin deficiency are shown in Figure 15.30 and 15.31B.

Basically, there are two processes involved in carbohydrate metabolism that require an adequate supply of insulin in order for it to be carried out at a normal rate. These are first, the entry of glucose into the cells of a number of peripheral tissues, and second, the normal storage of blood glucose as glycogen by the liver. In diabetes, many of the biochemical abnormalities encountered, especially those of carbohydrate metabolism, can be attributed to one of these two basic factors, as discussed in the legend to Figure 15.31.

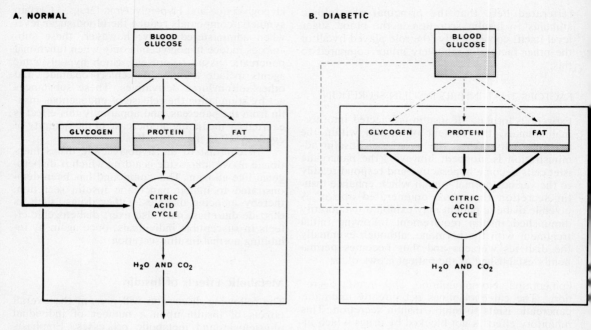

FIG. 15.31. Comparison of the metabolism in a normal and a diabetic individual. **A.** In the presence of adequate insulin, the blood glucose level is maintained at normal levels and aerobic glucose oxidation is the primary source of energy (heavy arrow at left). **B.** In insulin deficiency, the blood glucose level rises because of impaired utilization of glucose as well as defective conversion of this compound into glycogen, protein, and fat (light dashed arrows). Consequently, an excessive catabolism of the intracellular stores of these compounds takes place in order to meet the energy requirements of the body (heavy arrows). See also Figures 15.29 and 15.30. (After Ganong. *Review of Medical Physiology,* 6th ed, 1973, Lange Medical Publications.)

Insulin does not affect the intestinal absorption of ingested carbohydrate, nor does it modify the reabsorption of glucose by the nephron. Furthermore, glucose uptake by the brain in general is unaffected by insulin (Fig. 15.29). However, the defective glucose uptake into various other peripheral tissues of the body when the quantity of insulin is deficient results in a net accumulation of glucose in the bloodstream (Fig. 15.30). This impaired glucose uptake (or decreased peripheral utilization), combined with the deranged glucostatic function of the liver gives rise to the hyperglycemia observed in diabetes. Normally the hepatic uptake of glucose and its polymerization into glycogen within the liver cells is stimulated by insulin, and since this organ contains glucose-6-phosphatase, the liver also secretes glucose into the bloodstream under appropriate conditions. Insulin, however, inhibits the secretion of glucose resulting in the hepatic glucostatic effect of this hormone. When insulin is lacking there is an uncontrolled hepatic output of glucose into the bloodstream which results in development of a hyperglycemia. Under physiologic conditions, therefore, when the blood glucose level becomes greatly elevated as after a carbohydrate-rich meal, insulin secretion increases so that the net effect is a decrease in hepatic glucogenesis, combined with an augmented peripheral utilization of glucose.

In summary, then, the hyperglycemia of diabetes results from a decreased peripheral utilization of glucose, combined with an elevated rate of glucose output by the hepatic cells.

The oral glucose tolerance curve is useful clinically in the diagnosis of diabetes, since the configuration and time-course of this relationship gives a direct indication of the functional response of the pancreas to a specific glucose load. As shown in Figure 15.32 (curve 1), the ingestion of a standard dose of glucose by a fasting individual with normal pancreatic function results in a rise in blood sugar to a peak within an hour. Subsequently, the blood glucose concentration returns to the initial value, after approximately three hours. In a diabetic individual, the fasting blood glucose level may (or may not) be higher before the glucose load is administered, and thereafter the peak rise is slower, the effect is more prolonged, and the blood sugar concentration usually lies well above the normal range at the end of three hours (Fig. 15.32, curves 2, 3, and 4).

Effects of Hyperglycemia in Diabetes. By itself, a moderate hyperglycemia is harmless unless the blood sugar level rises to a value that leads to hyperosmolality of the blood. In this situation, a hyperosmolal coma may ensure because of dehydration of the central nervous system. (Note: This

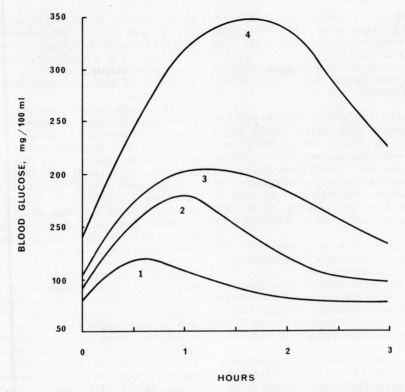

FIG. 15.32. Glucose tolerance curves obtained following ingestion of about 2 gm glucose/kg body weight at zero time. 1. Normal curve. 2. Probable or latent diabetes. 3. and 4. Overt diabetes. Note that in a normal individual (curve 1) the blood sugar level peaks in about an hour following oral administration of a glucose load, then returns to normal values in about 3 hours. (After Ganong. *Review of Medical Physiology,* 6th ed, 1973, Lange Medical Publications.)

type of coma must not be confused with the coma resulting from diabetic acidosis, as described on page 1087).

Hyperglycemia per se does not alter the reabsorption of glucose by the nephron; however, when the tubular maximum for reabsorption of this substrate is exceeded, glycosuria occurs. The excretion of large quantities of osmotically active glucose molecules in the urine leads to the loss of large volumes of water and a polyuria results because of this osmotic diuresis. The polyuria in turn causes dehydration of the body tissues, so that the water intake mechanisms are stimulated, and the intense thirst that develops leads to polydipsia.

Furthermore, about 4 kcal of energy are lost from the body for each gram of glucose excreted in the urine, therefore increasing the food intake in order to compensate for this loss of energy raises the blood glucose level still surther. The augmented food intake now increases the glycosuria still further; therefore, endogenous protein and fat stores must be mobilized in order to meet bodily energy requirements so that weight loss ensues.

Effects of Exercise. An important normal function of insulin is in the regulation of glucose uptake by skeletal muscle. However, in diabetics, during periods of exercise, glucose uptake by the muscles is enhanced even though insulin is lacking in such individuals. The cause of this phenomenon is unknown, but it may be related to a relative anoxia of the muscle cells which alone increases their permeability to glucose. Regardless of cause, in diabetics taking insulin, exercise per se can precipitate a serious hypoglycemia so that the usual insulin dosage should be decreased (or the caloric intake increased somewhat) prior to bouts of exercise in order to compensate for this effect.

Effects of Intracellular Glucose Deficiency. The extracellular fluid of a diabetic patient is overwhelmed with an abundance of glucose. However, the impaired entry of this carbohydrate into the cells of the peripheral tissues, a process which normally is regulated by insulin, results in an intracellular deficit of glucose. In effect, the metabolic machinery within the cells is starving from a lack of an important energy source. Consequently, the energy requirement in diabetes can be supplied only by drawing upon protein and fat reserves, as illustrated in Figure 15.31B. Thus, an inadequate glucose supply and utilization within the cells of the ventromedial nuclei of the hypothalamus is a possible cause of the hyperphagia that develops in diabetic patients.

A decreased glucose uptake by the peripheral

tissues in diabetes results ultimately in a depletion of their normal glycogen stores. Therefore, glycogen deficiency commonly is observed in the liver and muscles of diabetics.

All of the defects in carbohydrate metabolism encountered in diatetes that are discussed above are reversed by insulin.

EFFECTS OF INSULIN ON PROTEIN METABOLISM. In normal individuals, insulin regulates the rate at which amino acids are catabolized to water and carbon dioxide. Insulin also moderates the rate at which these compounds are converted into glucose (ie, gluconeogenesis) by the liver. In addition, insulin elevates the incorporation of amino acids into proteins, so that the synthesis of these compounds is accelerated by this hormone.

The incorporation of amino acids into proteins under the influence of insulin is independent of its effects upon glucose metabolism, and appears to be mediated in two ways. First, there is a stimulation of the ribosomes by insulin so that protein synthesis is enhanced. Second, amino acid transport into the cells is accelerated by insulin. These two effects of insulin probably are independent of one another.

Thus protein breakdown (or catabolism) is inhibited by insulin and protein synthesis is accelerated simultaneously. The anabolic effects of insulin are explained in part through the direct effects of insulin on protein synthesis via the ribosomes, but also are partly explicable on the basis of an increased availability of glucose to provide a primary energy source within the cells. This is the so-called protein sparing effect of glucose seen when this carbohydrate is present intracellularly in adequate quantities.

In diabetics, on the other hand, gluconeogenesis from amino acids is elevated sharply (Fig. 15.31) for the following reasons. First, the concentration of free amino acids in the blood and skeletal muscle is elevated markedly because in the absence of insulin, protein catabolism is increased whereas the rate of protein synthesis declines concomitantly, and the elevated amino acid level provides a ready pool of substrates for conversion into glucose. The amino acid alanine in particular is converted readily into glucose. Second, the activity of the enzymes responsible for the conversion of pyruvic acid (and other two-carbon fragments) into glucose is increased. These enzymes include phosphoenolpyruvate carboxykinase (which is responsible for the conversion of oxaloacetic acid into phosphoenolpyruvic acid); fructose-1,6-diphosphatase (which catalyses the conversion of fructose diphosphate into fructose-6-phosphate); and glucose-6-phosphatase (which controls the rate of secretion of glucose into the circulation by the liver). There is also an increase in acetyl CoA, which increase in turn augments the activity of pyruvate carboxypeptidase. Insulin deficiency increases the concentration of acetyl CoA because

of the decrease in the rate of lipogenesis as discussed below. The specific metabolic loci that are affected by insulin deficiency are shown in Figure 14.25, page 938. Certain other hormones which may contribute indirectly to the increased gluconeogenesis of diabetes include adrenal corticoids and glucagon.

In diabetic patients, the net effect of accelerated gluconeogenesis combined with a decreased protein synthesis as well as an increased protein catabolism is a profound negative nitrogen balance, protein depletion, and thus wasting of the body (Figs. 15.30 and 15.31B). In juvenile diabetics, normal growth cannot take place. Resistance to infections is considerably reduced in diabetics, possibly because of inadequate antibody formation combined with the sugar-rich intracellular body fluids which provide excellent culture media for invading microorganisms.

EFFECTS OF INSULIN ON FAT METABOLISM. Insulin decreases the catabolism of lipids and inhibits the formation of ketone bodies. This hormone also increases the synthesis of fatty acids as well as triglycerides. Insulin thus exerts a profound effect upon depot lipid metabolism, principally by facilitation of peripheral glucose uptake. The catabolism of this substrate is stimulated equally in both the hexosemonophosphate shunt (Table 14.9, p. 955), and the Embden–Myerhof glycolytic pathway (Fig. 14.25, p. 938). Particularly in adipose tissue fatty acids are synthesized from acetyl CoA under the influence of insulin, and since a high concentration of glycerol phosphate is formed by glycolysis, the fatty acids are esterified with glycerol rather than released from the cells. Therefore insulin, because of its inhibitory effect upon the hormone-sensitive lipase of adipose tissue, also inhibits the breakdown and release of free fatty acids. In turn, this effect of insulin prevents these compounds from accumulating in the bloodstream and thereby "flooding" the liver. In this manner, hepatic ketone production, as well as acetyl CoA formation, are maintained within physiologic limits.

Lipoprotein lipase is activated by insulin, and this activation assists in the maintenance of normal triglyceride levels in the circulation by removal of these fatty substances from the blood.

Insulin has three major physiologic roles in the promotion of fat storage. First, it stimulates the production of glycerol phosphate, thereby stimulating triglyceride synthesis. Second, insulin inhibits the action of the hormone-sensitive lipase of adipose tissue. Third, insulin activates lipoprotein lipase.

In a normal individual, approximately 50 percent of an ingested glucose load is metabolized to carbon dioxide and water, about 5 percent is converted to glycogen, and the remaining 30 to 40 percent is converted to fat (lipogenesis) in the adipose tissue depots. In a diabetic individual on the other hand, the derangements of lipid metabolism are so widespread and profound that diabetes could well be considered a disease of lipid, rather than carbohydrate, metabolism (Figs. 15.30 and 15.31B).

The major abnormalities of lipid metabolism found in diabetes are an uncontrolled and accelerated rate of lipid catabolism as well as a concomitant increase in the rate of ketone body formation. These effects are coupled to a decreased rate of synthesis of fatty acids and triglycerides. Therefore, in contrast with a normal individual, the fate of an ingested glucose load in a diabetic patient is roughly as follows. Less than 5 percent of the glucose is converted into fat, although the quantity metabolized to carbon dioxide and water also is decreased. The amount of glucose converted to glycogen decreases, of course, in insulin deficiency. As the consequence of these two effects, exogenous glucose accumulates in the blood and passes into the urine.

The increased fat breakdown (catabolism) in a diabetic is evidenced by an elevated circulating level of triglycerides, and the abundance of chylomicrons in the blood can give rise to a hyperlipemia which is visible to the unaided eye (ie, the plasma appears whitish, creamy, and opaque).

Insulin deficiency causes the free fatty acid level in the plasma to rise appreciably, hence these constituents increase in parallel fashion with the blood glucose level in diabetes. Normally fatty acids are metabolized to acetyl CoA in the hepatic and other tissues. Some of the acetyl CoA thereby produced then is oxidized in the citric acid cycle to yield carbon dioxide and water (Fig. 14.29, p. 948). Insulin deficiency, however, causes such an increase in acetyl CoA due to fat catabolism (Fig. 14.35, p. 964) that the capacity of the tissues to oxidize this substrate is exceeded, and an increased tissue concentration of acetyl CoA develops.

The elevated levels of free fatty acids in diabetes directly inhibit the tissue oxidation of glucose and also decrease tissue sensitivity to insulin. Under physiologic conditions, insulin decreases the release of free fatty acids and exerts a hypoglycemic effect. In diabetes, on the other hand, both of these effects are antagonized by the elevated free fatty acid levels. In view of these facts it has been postulated that a derangement of fat metabolism which elevates the free fatty acids could be the primary biochemical event in the etiology of clinical diabetes.

Under normal circumstances, a certain proportion of the acetyl CoA that is formed in the body is converted into acetoacetyl CoA, and in the liver acetoacetyl CoA is converted in turn into acetoacetic acid as well as it derivatives, β-hydroxybutyric acid and acetone (Fig. 14.36, p. 966). These ketone bodies normally form an important energy source during fasting, as discussed in Chapter 16 (p. 1213). In diabetics, the rate of ketone body formation by this mechanism is enhanced considerably because of the excess production of acetyl CoA which results from insulin deficiency. Although a diabetic without insulin can utilize a considerable proportion of the ketone bodies directly as an energy source, the rate of their production is so great than ketone bodies accumulate in the bloodstream. In fact, patients having severe diabetes actually may exhibit a decrease in their rate of ketone body utilization, a fact that worsens the ketosis. A severe metabolic acidosis (p. 808) develops in uncontrolled diabetic patients even though a considerable proportion of the total hydrogen ion liberated from acetoacetic and β-hydroxybutyric acids is buffered in the bloodstream. In turn, the low blood pH results in stimulation of the respiratory center resulting in "air hunger" or Kussmaul breathing, ie, intermittent periods of rapid, deep respiratory movements which are coupled to a severe dyspnea.

The urine becomes quite acid in metabolic acidosis and electrolyte imbalances occur when the renal capacity for exchanging plasma cations with hydrogen and ammonium ion is exceeded, and sodium and potassium are lost in the urine. The great loss of water and electrolytes during the metabolic acidosis of diabetes in turn lead to dehydration, hypovolemia, and hypotension. Ultimately these derangements reach the point where neural function is seriously imparied, consciousness is lost, and diabetic coma develops. Acidosis is the most frequent cause of death in clinical diabetes.

Diabetic coma may be due to hyperosmolality of the blood that is caused primarily by hyperglycemia (hyperosmolal coma) as well as metabolic acidosis. Furthermore, an increase in the blood lactic acid (lactic acidosis) may also complicate the ketoacidosis in diabetics, or else by itself the lactic acid accumulation may lead to coma.

Parenteral insulin administration reverses all of the defects in lipid metabolism which were discussed above.

EFFECTS OF INSULIN UPON ELECTROLYTE BALANCE

Potassium.   Under physiologic conditions, insulin causes potassium to enter body cells. The potassium concentration of the extracellular fluid is reduced significantly in normal individuals upon injection of insulin and glucose; however, in patients suffering from renal failure the hyperkalemia is relieved only briefly by such therapy. Diabetic patients in metabolic acidosis may develop a hypokalemia following insulin injection.

Insulin has been shown to increase the resting transmembrane potential of skeletal muscle and adipose cells under experimental conditions. Perhaps this increased electric gradient is responsible for the intracellular accumulation of excess potassium under the influence of administered insulin. In any event, an adequate concentration of potassium ions is essential to the normal secretion of insulin by the pancreatic islets, so that secretion of the hormone is diminished significantly in hypokalemic patients, for example, in primary hypoaldosteronism. In such a clinical situation, diabetic glucose tolerance curves develop (Fig. 15.32); however, restoration of an adequate blood potassium level causes the glucose tolerance curve to revert to a normal pattern once again.

Insulin and Overall Body Electrolyte Balance.   The osmotic diuresis is always present in diabetic patients causes significant quantities of sodium and potassium to be lost from the body secondarily to the polyuria. During metabolic acidosis when the renal capacity for exchange of these plasma cations for hydrogen and ammonium ions is exceeded, then still greater quantities of sodium and potassium are lost in the urine. Therefore, when severe metabolic acidosis accompanies diabetes, the total body sodium pool is reduced considerably. The plasma sodium concentration, on the other hand, may or may not be reduced. The total body potassium pool also is reduced under such conditions. Usually, however, the plasma potassium concentration remains at normal levels, an effect that is due partly to the fact that the acidosis reduces potassium excretion, and partly because insulin deficiency reduces potassium entry into the cells.

These facts are of considerable importance in the treatment of acidotic diabetic patients. For example, exogenous insulin causes potassium movement into the cells from the extracellular fluid in significant quantities as described above. The total body pool of potassium is depleted in such acidotic patients. A severe or fatal hypokalemia may be induced by the administration of insulin alone under such conditions.

RECAPITULATION OF THE MAJOR ABNORMALITIES PRESENT IN DIABETES.   A brief summary of the physiologic and biochemical abnormalities encountered in the diabetic patient is worthwhile at this juncture.

Insulin deficiency produces a decreased entry of glucose into many cells of the body, and this effect is accompanied by an increased hepatic release of glucose into the bloodstream. These two factors cause a hyperglycemia which in turn results in a dehydration of the body through an osmotic diuresis. Polyuria and glycosuria also result from the hyperglycemia, and these two effects in turn cause polydipsia through activation of the thirst mechanism.

The intracellular glucose deficiency that results directly from the decreased peripheral utilization of this substrate causes a hyperphagia. Gluconeogenesis from the excessive catabolism of body protein also occurs. Energy requirements of the body thus are met at the expense of body protein and fat stores both of which are consumed at abnormally high rates. The losses of body proteins and fats result in a profound loss of body weight.

The increased fat catabolism that accompanies diabetes results in an accumulation of excess triglycerides and free fatty acids in the bloodstream, whereas fat synthesis is inhibited concomitantly. The usual metabolic pathways cannot process all of the excess acetyl CoA formed by fat catabolism; therefore, in the liver this compound is converted into excessively large quantities of ketone bodies. Since the rate of production of ketone bodies exceeds their rate of utilization by the body these compounds now accumulate in the blood and a metabolic acidosis results from the ketonemia.

The body stores of sodium and potassium decrease because of the polyuria, and this loss of ions now exacerbates the dehydration produced by the osmotic effects of the hyperglycemia.

Ultimately the dehydrated, hypovolemic, hypotensive, acidotic patient enters a comatose state that rapidly terminates in death unless therapy is promptly initiated.

Insulin deficiency is responsible for every one of these defects. In emergencies, treatment of the acidosis with alkali, as well as the restoration of body stores of water, sodium, and potassium are essential adjuncts. However, only insulin can revert the fundamental metabolic abnormalities underlying the diabetic state to a normal condition once again.

COMMENT ON THE MECHANISM OF ACTION OF INSULIN.   A few remarks on the "mechanism of action" of insulin are now in order by way of conclusion to the general discussion of the metabolic effects of this hormone. Although some investigators believe that all of the myriad effects of this hormone are manifest through one fundamental physiologic mechanism that is common to all cells and tissues, there is a minimum of five aspects of insulin action that appear to be distinct and unrelated at the present time. First, insulin stimulates the rate of glucose entry into a number of cells and tissues. Second, insulin stimulates the rate of amino acid transport into many cells. Third, insulin stimulates the rate of protein synthesis by the cells. Fourth, insulin causes an increase in the resting transmembrane potential of adipose tissue and skeletal muscle fibers. Fifth, insulin inhibits the hormone-sensitive lipase in adipose tissue.

These facts argue strongly against a single common physiologic denominator that underlies all aspects of insulin action. It is clear, however, that many metabolic processes affected by insulin result from an increased glucose entry into various cells and tissues under the stimulus of this hormone, and also that many of the defects observed in diabetes may be attributed directly to an intracellular deficiency of glucose in this disease.

MISCELLANEOUS ASPECTS OF CLINICAL DIABETES AND INSULIN ACTION.   In this section, several factors pertaining to the etiology of human diabetes mellitus, certain long-term pathologic consequences of this condition, and the effects of hyperinsulinism and hypoglycemia will be discussed.

Factors Related to the Etiology of Human Diabetes. Although the injection of certain hormones can produce experimental diabetes in animals (eg, growth hormone, corticotropin), there is little evidence that diabetogenesis in man results from such a cause.

Human diabetes is divided into two broad categories, depending upon the age of the individual when the disease appears. These categories are juvenile and maturity-onset diabetes.

Juvenile Diabetes. As the name implies, juvenile diabetes appears in young persons (children and adolescents), and is characterized by a reduction in the total number of beta cells in the islets combined with a low total insulin content of the pancreas; the disease therefore is characterized by a general deficiency of insulin in the body.

Juvenile onset diabetes usually is severe, and frequently ketoacidosis accompanies and complicates the clinical picture. Overt pathology of the beta cells is often seen, although juvenile diabetics who die shortly after the disease appears may also show no (or very little) evidence of beta cell abnormality, whereas the total insulin content of the pancreas from such patients is normal.

Maturity-Onset Diabetes. Again as indicated by the name, maturity-onset diabetes develops in adults. A considerable amount of morphologically normal islet tissue may be present, and the insulin content of the pancreas may (or may not) be well within the normal range. Since the individual exhibits diabetic symptoms, however, a relative deficiency of functioning insulin must be presumed, and such a condition could result from an abnormally rapid destruction (ie, utilization) of insulin. The available clinical evidence favors the interpretation that in maturity-onset diabetics there is an abnormally low rate of release of insulin from the pancreas under the stimulus of normal blood sugar concentrations.* Consequently, maturity-onset diabetes may be considered to result from a failure of the pancreas to secrete insulin normally in response to the stimulus of an elevated blood glucose level.

Maturity-onset diabetes, in marked contrast to juvenile diabetes, generally is mild and ketoacidosis occurs but rarely. Such patients sometimes can be treated successfully merely by an appropriate diet. Weight reduction by obese patients may also ameliorate the severity of maturity-onset diabetes.

A major factor in the etiology of human diabetes that develops spontaneously is hereditary. The tendency to develop clinical diabetes probably is inherited as an autosomal recessive characteristic. This abnormal genetic property is present in roughly 20 percent of the population, and it is manifest by the development of abnormal glucose tolerance curves in such individuals (Fig. 15.32).

Approximately four percent of the entire population may exhibit some degree of diabetes mellitus during their lifetime.

Maturity-onset diabetes is related closely to obesity (p. 1215), as well as heredity; therefore, weight reduction improves glucose tolerance in such individuals. Interestingly, obese nondiabetic persons may have elevated blood insulin levels, as well as a tendency toward a somewhat abnormal glucose tolerance like that shown in Figure 15.32 (curve 2). Such subjects exhibit a resistance to the effects of insulin, and the high plasma amino acid concentrations that are also present possibly may contribute to the hyperinsulinemia.

Since clinical diabetes mellitus is caused by a relative deficiency to an absolute lack of insulin, then many gradations in the severity of the disease obviously can be manifest among different individual patients. Other possible causes of diabetes include the synthesis and secretion of a chemically abnormal inulin; a deranged storage and/or secretory mechanism for insulin; and the presence of antibodies against insulin in the circulation.

Clinical diabetes may be viewed primarily as a congenital disease that may be precipitated by obesity and other factors, including an abnormally high carbohydrate diet. Some authors consider that diabetes develops in four steps, and a gradual transition from one stage to the next occurs progressively as follows. First, prediabetes is present, and this stage represents an inherited susceptibility to diabetes present from the time of conception, although no overt pathologic or chemical changes are present. Second, chemical diabetes develops. In this stage the glucose tolerance test is abnormal, although other symptoms are lacking. Third, overt diabetes appears, in which the classic syndrome of full-blown diabetes mellitus is present. Fourth, chronic diabetes ensues, in which stage vascular lesions become evident.

Vascular Lesions Found in Diabetes. The elevated plasma lipid level present in chronic diabetics generally includes an abnormally elevated plasma cholesterol concentration and this factor may play a significant role in the genesis of the arteriosclerosis seen in diabetic patients. The rise in plasma cholesterol appears as a secondary change that is due, in part at least, to the primary increase in circulating triglycerides that is found in diabetics. The arteriosclerosis is, of course, a major complication of chronic diabetes regardless of its etiology, and this arteriosclerosis appears to develop independently of the degree of therapeutic control of the disease by insulin and diet.

In contrast to the long-term arteriosclerotic vascular changes, many diabetics exhibit vascular lesions in the retinal vessels of the eyes (retinopathies), as well as vascular lesions in the kidneys. These pathologic changes sometimes appear quite early during the clinical course of diabetes, even before other symptoms have developed. Usually, however, there is a definite lag period following

---

Obviously, if there were a failure in the insulin synthetic mechanism, the pancreatic insulin content would be lower than normal.

the onset of clinical diabetes and the development of retinopathies. Nonetheless an opthalmoscopic examination should form part of every routine physical examination.

Hyperinsulinism and Hypoglycemia. As mentioned earlier, the exogenous insulin administered parenterally to diabetic patients enters the peripheral rather than the portal circulation. It is unclear whether or not this fact is significant in the clinical management of diabetes or in the complications that arise in this disease, although it is essential that the liver receive an adequate supply of the hormone in order to carry out its normal metabolic functions properly, and this statement is true whether or not the insulin is transported to the liver by the physiologic route, ie, via the hepatic portal circulation.

Injection of insulin ultimately causes a fall in the blood sugar level in normal as well as in diabetic subjects. Commercial preparations of insulin, however, may be contaminated with the hormone glucagon so that an initial rise in blood sugar level occurs before the blood glucose commences to decline. This biphasic effect is absent in glucagon-free insulin preparations, so that the blood glucose level commences to fall shortly following injection of the hormone.

It must be emphasized at this juncture that the requirement for insulin that is necessary to control diabetes in different patients varies considerably. Furthermore, the dose of insulin that is needed by an individual patient also may vary considerably at different times, depending upon such factors as physical activity, weight gain, during periods when endogenous secretion or therapeutic administration of diabetogenic hormones increases (eg, glucocorticoids), as well as during pregnancy, infections, and febrile states. All of these situations require additional quantities of insulin.

Hypoglycemic episodes (or insulin reactions) occur frequently in juvenile diabetics and occasionally in adult diabetics. The increased physiologic uptake of glucose by skeletal muscle during bouts of exercise, an effect that is independent of insulin, is especially important in this regard. The patient on an insulin regimen must learn to compensate for this exercise effect, and to adjust the insulin dose and/or diet accordingly.

The administration of excess insulin, hyperinsulinism due to islet cell tumors (insulinomas), or hyperinsulinism caused by beta cell hyperplasia all cause a number of symptoms that may be attributed either directly or indirectly to the effects of hypoglycemia upon the nervous system. Since glucose is the only substrate used by the brain in significant quantities, a continuous supply of this substance is critical to normal function of the nervous system, as the normal reserve supply of this carbohydrate in the neural tissues themselves is extremely low.

Therefore, when the blood glucose level falls due to hyperinsulinism, regardless of its cause, the

regions of the brain having the highest metabolic rates suffer first, and thus also show evidence of signs of relative glucose deficiency first. The cerebral cortex is the first area to exhibit signs of glucose lack, and this is followed by symptoms that indicate malfunction of the centers in the diencephalon and medulla that normally respire at a somewhat lower rate than the cortex. The initial cortical symptoms of dizziness, mental confusion, weakness, and hunger are succeeded by convulsions and a comatose state ultimately develops.

Prolonged hypoglycemia results in irreversible functional damage to the brain. The sequence in which such damage occurs again reflects the intrinsic metabolic rate of the various regions of the central nervous system. Hence the cortex, diencephalon, and medulla are effected by glucose deficiency in a sequence of increasing resistance to a lack of this substrate. Death ultimately ensues from failure of the respiratory center in the medulla unless proper therapy is instituted rapidly. Glucose administration causes an immediate and dramatic remission of all of the hypoglycemic symptoms. However, if the blood sugar has been depressed for too long (around 5 minutes), then irreversible central nervous damage ranging from a slight deterioration of the intellectual faculties to a permanent comatose state results.

Hypoglycemia is a powerful stimulant to sympathetic neural discharge, and the elevated secretion of catecholamines that results from such stimulation is a probable cause of the tremors, nervousness, and palpitations that are observed in hypoglycemic patients.

The actual blood glucose level at which hypoglycemic symptoms occur is quite variable. When the concentration of blood glucose falls gradually (or remains at) about 50 mg percent, such symptoms become manifest in humans. On the other hand, a rapid fall in blood sugar to values as low as 20 mg percent elicit no symptoms until after the cortical carbohydrate stores have been utilized. Nondiabetic individuals also may develop hypoglycemia under certain conditions, and when this condition is chronic, the hypoglycemic symptoms may range in severity from motor incoordination and slurred speech to psychic aberrations and convulsions; however no coma ensues as is the situation with diabetics. The speech and coordination defects that are seen in mild symptomatic hypoglycemia resemble those seen in states of alcoholic intoxication.

If endogenous insulin secretion is chronically elevated (as, for example, in a patient having an islet cell tumor) the hypoglycemic symptoms usually are most prominent in the morning because the hepatic glycogen reservoir is depleted after fasting all night. Faulty diagnosis of this hypoglycemic condition is common, as the symptoms often resemble those of a frank psychosis or epilepsy.

Malignant tumors in some patients secrete a substance that resembles insulin in its action, so that hypoglycemia can occur in the absence of any apparent pancreatic dysfunction.

Certain forms of primary liver disease produce a diabetic-type glucose tolerange curve (Fig. 15.32); however, in contrast to diabetes, the fasting blood sugar level is low in such patients rather than elevated.

A normal rise in the glucose tolerance curve is obtained in functional hypoglycemia subsequent to administering a test dose of glucose. However, the ensuing fall in blood sugar concentration does not stabilize at normal values. Rather it drops to much lower levels, so that hypoglycemic symptoms appear within three to four hours following a meal. Occasionally this response pattern is observed in subjects who ultimately develop diabetes.

In some patients who have undergone gastrectomy or other surgical procedures on the gut, or who suffer from thyrotoxicosis, the intestinal absorption of glucose is abnormally fast. The blood glucose concentration rises to a peak very quickly following a meal. Subsequently the blood sugar concentration falls to subnormal levels, again quite rapidly, and hypoglycemic symptoms become evident within two hours after the meal.

Certain oral hypoglycemic agents, such as tolbutamide, chlorpropamide, and other sulfonylurea derivatives that are used clinically for the control of diabetes (Fig. 15.33), apparently reduce the blood glucose level by stimulating the secretion of endogenous insulin from such functional islet tissue as is present in the pancreas. This effect is probably caused by the drugs stimulating an increase in the quantity of cyclic AMP in the beta cells. Therefore, such drugs should be clinically effective hypoglycemic agents only in the early stages of juvenile diabetes, as well as in the insulin-resistant forms of maturity onset adult diabetes, in which the secretion of endogenous insulin is defective, but adequate functional islet tissue is present. Clinical studies have shown that this statement is true.

In contrast to the sulfonylurea derivatives, phenformin (DBI) and other biguanide oral hypoglycemic compounds (Fig. 15.33C) do not stimulate insulin secretion; rather these drugs accelerate glucose utilization in the body tissues by a mechanism which has not yet been satisfactorily clarified.

A word of caution must be added with regard to all of the oral hypoglycemic agents mentioned above. Recent clinical investigations have demonstrated that the mortality rate from cardiovascular disease in patients treated with oral antidiabetic agents is considerably higher than in patients whose therapy consisted of diet and/or insulin alone. Therefore, use of these compounds in the practical control of clinical diabetes entails a much greater risk to the patient than does other therapeutic measures. In fact, one clinical study indicated that oral antidiabetic agents ranged from ineffective in controlling diabetes to dangerous to the life of the patient!

Thiazide Diuretics and Diabetes. Diabetic patients who are taking thiazide diuretics in order to increase their urinary sodium and potassium excretion can suffer a decreased glucose tolerance so that the diabetic condition becomes worse. Some thiazide compounds exert this action indirectly, presumably by means of the therapeutically induced potassium depletion and its effects upon insulin secretion; however, certain other thiazide derivatives act directly upon the islet cells to induce functional and morphologic damage.

Glucagon

The second hormone that is elaborated by the islets of Langerhans within the pancreas is known as glucagon. This substance exerts a diametrically opposite effect to that of insulin upon the blood glucose level; consequently the pancreatic secretion of glucagon results in development of a

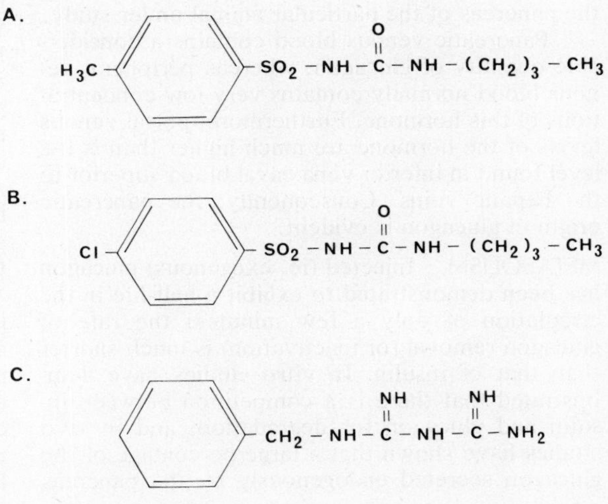

FIG. 15.33. Oral antidiabetic agents. **A.** Tolbutamide. **B.** Chlorpromide. **C.** Phenformin (phenethylbiguanide, DBI). (After Ganong. *Review of Medical Physiology,* 6th ed, 1973, Lange Medical Publications.)

His—Ser—Gln—Gly—Thr—Phe—Thr—Ser—Asp—Tyr—Ser—Lys—Tyr—Leu—Asp—Ser—Arg—Arg—Ala—Gln—
1   2   3   4   5   6   7   8   9   10   11   12   13   14   15   16   17   18   19   20

FIG. 15.34. Amino acid sequence in human glucagon. (After Ganong. *Review of Medical Physiology,* 6th ed, 1973, Lange Medical Publications.)

hyperglycemia because of increased hepatic glycogenolysis that is induced by the hormone.

CHEMISTRY. Bovine (and porcine) glucagon is a linear polypeptide containing 29 amino acid residues as shown in Figure 15.34, and it has a molecular weight of 3,485 daltons. The structural similarity of glucagon to secretin suggests a common genetic origin for both hormones.

Few data on the chemical structure of glucagons from different species of animals are available; however bovine and porcine glucagon stimulate the formation of antiglucagon antibodies in rabbits, and the rabbit antibodies have proven useful for the measurement of glucagon levels in humans as well as in experimental animals.

The walls of the stomach and duodenum secrete a material (*not* secretin) which reacts with the antiglucagon antibodies of rabbits, hence it has been called the glucagonlike immunoreactive factor (GLIF). This intestinal substance apparently is not chemically identical with pancreatic glucagon; however it is released into the gastrointestinal tract when glucose is ingested. Both pancreatic glucagon and GLIF can stimulate insulin secretion from the pancreas.

BIOSYNTHESIS AND SECRETION. Glucagon is believed to be synthesized, stored, and secreted by the alpha cells of the islets of Langerhans. Evidence for this statement is found in experiments in which the administration of cobalt chloride caused a specific destruction of the alpha cells to varying degrees, and the quantity of extractable glucagon in the pancreas was decreased accordingly. Furthermore, the quantity of glucagon that can be extracted from various animal species is related directly to the relative abundance of alpha cells in the pancreas of the particular animal under study.

Pancreatic venous blood contains a considerable quantity of glucagon, whereas peripheral venous blood normally contains very low concentrations of this hormone. Furthermore, portal venous levels of the hormone are much higher than is the level found in inferior vena caval blood superior to the hepatic veins. Consequently, the pancreatic origin of glucagon is evident.

METABOLISM. Injected (ie, exogenous) glucagon has been demonstrated to exhibit a half-life in the circulation of only a few minutes; the rate of glucagon removal (or inactivation) is much shorter than that of insulin. In vitro studies have demonstrated that there is a competition between insulin and glucagon for degradation, and in vivo studies have shown that a large percentage of the glucagon secreted endogenously by the pancreas normally is bound by the liver. Hepatic tissue also contains an enzyme that inactivates glucagon by splitting off histidyl · serine (His · Ser) from the amino terminal end of the glucagon molecules (Fig. 15.34).

REGULATION OF GLUCAGON SECRETION. Hyperglycemia suppresses the rate of secretion of glucagon, whereas hypoglycemia or fasting for several days elevates it. Under physiologic circumstances the secretory rate of this hormone is regulated by a simple feedback mechanism that operates directly through the stimulus of pancreatic blood glucose concentration upon the alpha cells of the pancreatic islets, but in a direction that is opposite to that for insulin; ie, there is an inverse relationship between the rate of glucagon secretion and the blood glucose level.

A deficiency of pancreatic glucagon may lead to clinical hypoglycemia during periods of fasting. During actual starvation in the presence of normally functional alpha cells, however, the blood glucose may fall to a certain extent, but an overt hypoglycemia is prevented by increased gluconeogenesis that results from increased glucagon secretion.

Ingestion of high-protein meals (or amino acids) also serves as a stimulus to an elevated glucagon secretion. Since amino acids can also stimulate insulin secretion, it has been postulated that the parallel rise in glucagon secretion that occurs normally following a protein meal serves to prevent the hypoglycemia that would develop if insulin alone were present at an elevated concentration.

PHYSIOLOGIC ACTIONS OF GLUCAGON. The major function of glucagon in the body is to elevate the blood glucose level; this hormone is a hyperglycemic factor. The hyperglycemia in turn is produced directly by an acceleration of glycogenolysis under the influence of glucagon, so that the liver (but not the skeletal muscles) secretes more of this substrate into the bloodstream. A second important function of glucagon in the body is to stimulate hepatic gluconeogenesis using such amino acids as are available as substrates for this process.

Glucagon also exerts a calorigenic action that is not caused directly by the hyperglycemia evoked by the hormone. Rather, the hyperglycemia appears to result from the hepatic deamination of various amino acids during the process of gluconeogenesis in the liver. Since thyroxine is necessary for the elevation in metabolic rate stimulated by glucagon, it appears that the calorigenic

Asp—Phe—Val—Gln—Trp—Leu—Met—Asn—Thr
21   22   23   24   25   26   27   28   29

FIG. 15.34. (Cont'd)

action of glucagon actually is caused by thyroxine, which hormone directly stimulates the oxidation of the deaminated amino acid residues.

The physiologic calorigenic effect exerted indirectly by glucagon also requires the presence of an adequate supply of adrenocortical hormones (in addition to thyroxine) in order to become manifest. Adrenalectomized animals die in hypoglycemia when they are fasted, since adrenal glucocorticoids may also stimulate gluconeogenesis under certain conditions. However, the role of the adrenocortical hormones in gluconeogenesis per se appears to be facultative (optional) rather than obligate (essential) under physiologic conditions, so that adrenal cortical hormone secretion is not increased during periods of actual starvation in intact persons or animals. Under physiologic conditions, glucagon may be considered as a hormone of fasting, since it stimulates the rate of hepatic gluconeogenesis, and thereby prevents development of hypoglycemia in this state by elevating the blood glucose concentration.

Glucagon also induces several other physiologic and metabolic effects in addition to those described above either indirectly or directly, and these effects can be demonstrated under appropriate experimental conditions.

First, the hepatic gluconeogenesis produced by glucagon is accompanied by an increased urea formation from the amino residues liberated during the process of deamination of amino acids. Furthermore, glucagon directly inhibits amino acid incorporation into hepatic proteins.

Second, glucagon inhibits the rate of hepatic synthesis of fatty acids and cholesterol from acetate, and this hormone also stimulates the rate of ketone body formation concomitantly. Hepatic lipase is activated during this process, presumably through the adenyl cyclase system. The net effect is an increase in the concentration of free fatty acids (derived from hepatic triglycerides) which then become available for oxidation within the liver. This effect is also accompanied by an increase in acetyl CoA, fatty acyl CoA and, of course, ketogenesis.

Third, glucagon acts directly upon adipose tissue, in which it stimulates the release of free fatty acids and glycerol. This effect results secondarily to an increase in the cyclic 3′, 5′-AMP level which occurs directly under stimulus of the hormone.

Fourth, glucagon exerts a direct effect upon the kidneys, so that renal plasma flow and the glomerular filtration rate are accelerated. These renal effects are exerted by the hormone independently of the hyperglycemia induced by the hormone, and the net result is an elevated excretion of sodium, potassium, chloride, phosphates, and total urinary nitrogen (eg, that derived from urea, uric acid, and creatinine).

Fifth, glucagon directly stimulates cardiac contractility, so that a positive inotropic effect is produced. However, glucagon does not alter myocardial irritability, so this hormone eventually may find clinical application in cardiology, as well as in the treatment of acute or chronic hypoglycemia in which area it is employed currently.

MECHANISM OF THE HYPERGLYCEMIC ACTION OF GLUCAGON. The acceleration of hepatic glycogenolysis under the influence of glucagon that results in a hyperglycemia is as follows. Glucagon increases the activity of the adenyl cyclase system which in turn increases the rate of formation of 3′, 5′-AMP. The cyclic 3′, 5′-AMP then activates phosphorylase kinase, and this enzyme now accelerates the breakdown of endogenous hepatic glycogen stores into glucose, and this carbohydrate is secreted into the portal system. This mechanism of action of glucagon is identical with that of epinephrine except that glucagon does not produce the hypertensive effect characteristic of epinephrine (see Fig. 15.48). Glucagon acts solely upon hepatic glycogen stores, and does not affect the phosphorylase system in skeletal muscle.

RELATIONSHIP AMONG GLUCAGON, INSULIN AND DIABETES. The role, if any, that glucagon plays in the pathogenesis of human diabetes mellitus is unclear at present, although certain clinical and experimental data are pertinent to this question. In diabetic patients, a glucose load does not inhibit glucagon secretion as it does in normal individuals, although protein exerts its usual stimulatory effect on the secretion of the hormone. It appears that a relative excess of glucagon is present in the blood of diabetics, and this excess contributes to the hyperglycemia. Furthermore, beta cell degeneration in the islets of Langerhans can be induced by prolonged treatment of experimental animals with glucagon, provided that the pancreatic reserve of islet tissue is first reduced (eg, by subtotal pancreatectomy). In addition, in animals that are rendered diabetic by the injection of alloxan the quantity of insulin required to control the diabetes is reduced significantly following total pancreatectomy. (Alloxan selectively destroys the beta cells of the islets, leaving the alpha cells intact). Lastly, in animals that have undergone subtotal pancreatectomy the ensuing hyperglycemia may cause functional depletion of the insulin that is present within the remaining beta cells. Under such circumstances the animal actually requires more insulin to control the hypoglycemia than does an animal which has undergone total pancreatectomy, a fact that suggests, but of course does not prove, the operation of a hyperglycemic factor.

THE ADRENAL CORTEX

The suprarenal or adrenal glands in man are paired organs that are embedded in the retroperitoneal adipose tissue at the superior pole of each kidney. As depicted in Figure 15.35, these glands are roughly triangular and flattened in appearance; the approximate dimensions are 5 cm by 3 cm by 1 cm in thickness, and the total mass of both adrenals is around 15 gm in an adult. However, the size of the glands as well as their total weight may exhibit wide variations, depending upon the age and physiologic state of the individual.

The cut surface of a transected adrenal gland reveals two discrete regions within the thick, collagenous connective tissue capsule that surrounds the entire gland. First, an outermost yellow cortex that becomes reddish-brown in its innermost layer adjacent to the thin second inner layer, the greyish medulla. Consequently, the adrenal gland consists of two distinct endocrine organs that are located within a common structure. The cortex and medulla differ markedly in their embryologic origins, morphology, hormonal secretions, and functions.

The adrenal medulla synthesizes and secretes the two catecholamine hormones, epinephrine and norepinephrine. Although these hormones are not essential to survival (as are certain of the cortical hormones), they do enable the individual to adapt physiologically to stress or emergency situations. The adrenal medulla will be discussed on page 1118 et seq.

The tissue of the adrenal cortex synthesizes and secretes all three of the general types of steroid hormones found in the body. First, glucocorticoids, which are essential to the normal metabolism of carbohydrate, protein, and fat by the body. Second, a mineralocorticoid which is critical in the regulation of normal electrolyte balance and maintenance of the proper volume of extracellular fluids in the body. Third, the adrenal cortical tissue also secretes small quantities of sex hormones, which exert minimal effects upon reproductive functions under physiologic conditions.

Functional Morphology of the Adrenal Cortex

DEVELOPMENTAL CONSIDERATIONS. The adrenal cortex develops from the coelomic mesoderm that lies medially to the urogenital ridge. In human fetuses about 10 mm in length, mesothelial cells near the cranial end of each mesonephros multiply extensively and invade the adjacent highly vascular mesenchyme. Eventually these cells form the fetal cortex, which ultimately comprises roughly 80 percent of the total mass of cortical tissue in the embryo. In the 14 mm embryo, further proliferation of the mesothelial cells forms the permanent cortex.

Following parturition, the fetal cortex degenerates rapidly, whereas the permanent cortex enlarges; the net effect of these processes is that the adrenals lose about 50 percent of their total mass during the first weeks of postnatal life.

The embryonic adrenal cortex is physiologically active, and already it is under the regulatory influence of corticotropin from the adenohypophysis.

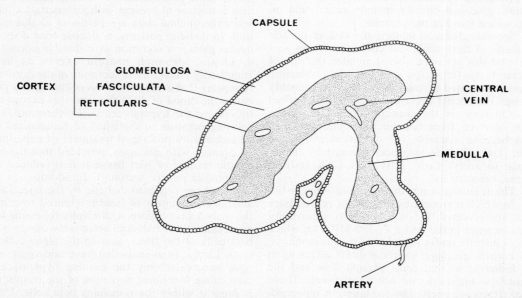

FIG. 15.35.   Gross anatomy of the adrenal gland. (After Goss [ed]. *Gray's Anatomy of the Human Body*, 28th ed, 1970, Lea & Febiger.)

HISTOLOGY. The cortex forms the major portion of the total mass of tissue found in the adrenal gland, and in adults three concentric zones are discernible within this organ, as shown in Figure 15.36: (1) a thin outermost layer, the zona glomerulosa lies just beneath the capsule; (2) a thick middle region, the zona fasciculata which lies beneath the zona glomerulosa; (3) a relatively thick inner layer, the zona reticularis which lies adjacent to the medulla. The transition between one cortical zone and the next is gradual rather than sharply delineated.

Zona Glomerulosa. The zona glomerulosa comprises about 15 percent of the total cortical mass. The cells of the zona glomerulosa are arranged in tight groups and pillars consisting of large, columnar epithelial cells that are continuous with the cells of the zona fasiculata. The individual pillars of glomerulosa cells are separated by vascular sinuses.

The spherical nuclei of the glomerulosa cells

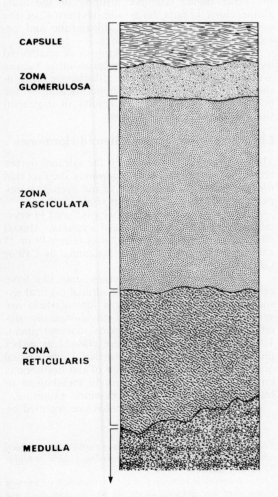

CAPSULE

ZONA
GLOMERULOSA

ZONA
FASCICULATA

ZONA
RETICULARIS

MEDULLA

FIG. 15.36. Microscopic anatomy of the adrenal gland. (After Bloom and Fawcett. *A Textbook of Histology,* 9th ed, 1968, Saunders.)

generally contain one or more nucleoli, and stain deeply; the Golgi apparatus is juxtanuclear. The cytoplasm of glomerulosa cells is more sparse than in the cells of the other two cortical zones, and is usually acidophilic, although clumps of basophilic material are often present. Filamentous mitochondria are abundant.

The cytoplasm also contains a well-developed smooth endoplasmic reticulum that is present as an anastomosing network of tubules, as well as abundant free ribosomes in the matrix. Some rough endoplasmic reticulum is also present in these cells.

The glucocorticoid hormone corticosterone is secreted by all three layers of the adrenal cortex. However, the enzymatic machinery necessary for the synthesis of the mineralocorticoid hormone aldosterone is restricted to cells of the zona glomerulosa. Furthermore, the cells of the glomerulosa that are located adjacent to, and just beneath, the capsule are able to provide cells for regeneration of all three cortical zones if necessary.

Zona Fasciculata. The zona fasciculata normally comprises the bulk of the cortical tissue, thus comprising roughly 75 percent of the total mass of the adrenal cortex. The cells in this zone are polyhedral and much larger than those present in the glomerulosa. They are aligned in long cords, and the cords are arranged radially with respect to the medulla. Usually the cords of cells that are present within the zona fasciculata are only one cell layer thick, and they are separated by sinusoidal blood vessels.

The cells of the fasciculata have one or two centrally located, spherical nuclei, the cytoplasm is weakly basophilic, and clumps of basophilic material are seen occasionally. The mitochondria are relatively more scarce and more variable in appearance than in the cells of the glomerulosa.

The cells of the fasciculata are filled with lipid droplets normally, but during routine histologic preparations, the lipid is extracted, so that the cells take on a vacuolated appearance. A narrow transitional region between the zona glomerulosa and zona fasciculata that lacks the abundant lipid droplets may also be present.

The smooth endoplasmic reticulum is even more highly developed than that in the zona glomerulosa. Cisternae of granular endoplasmic reticulum are also seen with the electron microscope, and these cisternae represent the clumps of basophilic material that can be observed with light microscopy.

As noted earlier, corticosterone is secreted by the zona fasciculata and the cells of this intermediate cortical region also secrete cortisol and small quantities of sex hormones, a functional property that is shared in common with the zona reticularis.

Zona Reticularis. The zona reticularis is the innermost region of the adrenal cortex and comprises roughly 10 percent of the total cortical tissue mass. The cells of the zona reticularis are arranged in an anastomosing network, rather than dispersed in an ordered parallel array of cords such as are found in the zona fasciculata. The transition between these two inner zones is gradual, and the cells in both regions are quite similar morphologically, save for the fact that lipid droplets are more scanty in the cytoplasm of the reticularis cells. The nongranular endoplasmic reticulum is abundant in reticularis cells like in cells of the fasciculata.

"Light" and "dark" staining nuclei are common in cells of the zona reticularis, and this effect possibly may represent degenerative changes that have taken place in some of the cells, rather than an artifact that was induced during preparation of the tissue for histologic examination.

The cells of the zona reticularis normally secrete corticosterone and cortisols as well as minute quantities of sex hormones (androgens and estrogens).

ARTERIAL BLOOD SUPPLY, CONNECTIVE TISSUE, AND LYMPHATICS OF THE ADRENAL CORTEX. The adrenal gland is supplied abundantly with blood from several arteries that enter the capsule around its periphery. The major part of the blood supply is received via the superior suprarenal arteries which originate from the inferior phrenic artery. In addition, the middle suprarenal arteries (which arise from the aorta) and the inferior suprarenal arteries (which arise from the renal artery) also provide some blood to the gland. These arteries form a plexus of vessels within the capsule, and the cortical arteries originate from this plexus, thence blood is distributed to an anastomosing network of sinusoids which surround the masses and cords of cells within the three cortical regions. The sinusoids from local areas of the cortex then converge upon a collecting vein within the zona reticularis at the junction of cortex and medulla; consequently, the cortex has no venous system.

In certain regions of the adrenal gland, the suprarenal capsule extends into the substance of the cortex as thick pillars of connective tissue or trabeculae. In addition delicate reticular connective tissue fibers form a complex framework that supports the individual cells of the cortex as well as the medullary tissue. Lymphatic vessels are limited to the capsule and cortical trabeculae, as well as to the connective tissue surrounding the large veins.

A few random nerve fibers are present in the cortex. However, morphologically as well as functionally, the innervation of the adrenals is a topic that properly belongs to a discussion of the adrenal medulla, hence will be considered later.

MORPHOLOGIC CHANGES IN THE ADRENAL CORTEX INDUCED BY ALTERATIONS IN PITUITARY FUNCTION. As mentioned earlier, normal adrenal cortical functions and morphology are dependent upon an adequate supply of the hormone corticotropin from the adenohypophysis (Fig. 15.6) Hypophysectomy or pituitary hypofunction causes marked histologic and functional alterations in the adrenal cortex. A deficiency of corticotropin, regardless of its cause, produces an almost complete atrophy of the zona fasciculata and zona reticularis, but little change is seen in the zona glomerulosa. This effect is prevented or reversed by injections of corticotropin.

Conversely, the administration of large quantities of the glucocorticoid hormone cortisol into intact animals or human patients can suppress the normal secretion of corticotropin by the pituitary so that the inner zones of the cortex atrophy.

Selective hypertrophy of the zona glomerulosa can result from inordinate demands placed upon the homeostatic mechanisms for regulating electrolyte balance. For example, this can be accomplished experimentally by placing an animal on a low sodium–high potassium diet. On the other hand, selective atrophy of the zona glomerulosa results from administration of the mineralocorticoid hormones aldosterone and deoxycorticosterone in large doses.

Possibly the zona glomerulosa is unaffected by corticotropin deficiency because other factors are involved in stimulating aldosterone biosynthesis, although persistent hypofunction of the adenohypophysis ultimately results in degeneration of this cortical region also.

Chemistry of the Adrenal Steroid Hormones

The hormones synthesized by the adrenal cortex are all steroids, and share in common the fact that they are chemical derivatives of the cyclopentanoperhydrophenanthrene nucleus (Fig. 15.37A).

About 50 steroids have been isolated in crystalline form from adrenal gland extracts. Almost all of these steroids have a total of either 19 or 21 carbon atoms,* and thus are designated as C19 or C21 steroids.†

Of the numerous steroid compounds that have been isolated chemically from adrenal cortical tissue in crystalline form, only a few of these are synthesized and secreted in physiologically significant quantities by the normal adrenal gland. These important hormones are classed as either glucocorticoid or mineralocorticoid, depending upon whether the major action of the hormone is on carbohydrate, fat, and protein metabolism or on electrolyte (sodium and potassium) excretion.

The principal glucocorticoids are secreted by

---

*The exceptions to this statement are cholesterol, which contains 27 carbon atoms, and estrone, which contains 18 carbon atoms.
†This denotation indicates the total number of carbon atoms that are present in the molecule, and does not indicate the position of specific radicals — these are lo ure 15.37A and C, and discussed further in the pages following.

A.

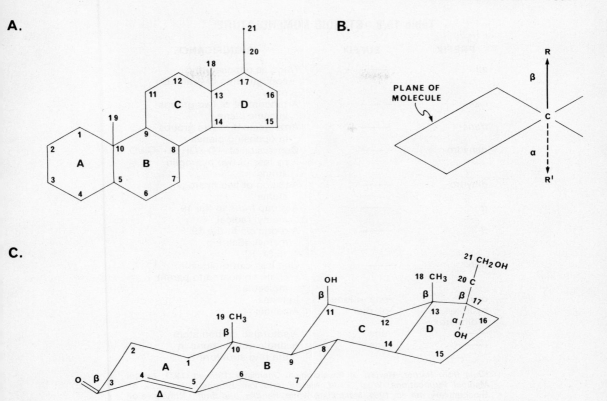

B.

C.

FIG. 15.37. **A.** Numbering and lettering of the carbon atoms and rings in the steroid nucleus (or perhydrocyclopentano-phenanthrene ring). **B.** Conventions employed in designating the β and α positions of radicals (R_1, R_2) with respect to the plane of the steroid ring in steroid compounds. **C.** Conformational structure of the steroid cholestanol illustrating the conventions shown in parts A and B of this figure. (After White, Handler, and Smith. *Principles of Biochemistry,* 4th ed, 1968, The Blakiston Division, McGraw-Hill.)

the adrenals are cortisol (synonyms: 17 α-hydroxycorticosterone, hydrocortisone, compound F) and corticosterone (synonym: compound B). The structural formulae of these hormones are shown in Figure 15.38. The principal mineralocorticoid secreted by the adrenals is aldosterone.

These three adrenocortical hormones are C21 steroids, and belong to one structural type, chemically speaking. That is, each has a two-carbon side chain attached to the 17 position on the D ring (Fig. 15.37A), and thus each has a total of 21 carbon atoms in the molecule. The C21 steroids that have a hydroxyl (—OH) group at the 17 position in the molecule are often referred to as 17-hydroxy-corticoids or 17-hydroxycorticosteroids.

A second chemical type of adrenocortical steroid hormone contains a keto (C=O) or hydroxyl (C—OH) group attached directly to the 17 position of the D ring; however only 19 carbon atoms are present in the entire molecule. The physiologically significant C19 steroids generally have a keto group at the 17 position, hence are called 17-ketosteroids. The C19 steroids exhibit androgenic activity, which effect is minimal in physiologically normal women. In adult men, however, this adrenal source of androgen appar-

ently complements the normal testicular hormone output. The principal androgenic compound secreted by the adrenal cortex normally is dehydro-epiandrosterone (Fig. 15.39).

Before proceeding into a discussion of the biosynthesis of the adrenocortical hormones, a few additional remarks on the conventions of steroid nomenclature are necessary (see also Table 15.8).

First, the individual rings, and the positions on the rings that may be occupied by various chemical groups, are indicated by letters and numbers respectively, as shown in Figure 15.37A.

Second, the chemical groups that lie above the stereochemical plane formed by the steroid ring are designated by the Greek letter beta (β) and a solid line (—R). On the other hand, the chemical groups that lie below this stereochemical plane are designated by the Greek letter alpha (α), and indicated by a dashed line (---R) in the structural formula (see also definitions of *cis-* and *trans-*, Table 15.8). These conventions are illustrated in Figure 15.37 B and C.

Third the Greek letter delta (Δ), denotes a double bond, as shown between positions 4 and 5 in Figure 15.37C.

Fourth, the methyl group (—CH$_3$) in positions

Table 15.8 STEROID NOMENCLATURE[a]

PREFIX	SUFFIX	SIGNIFICANCE
allo-	————	*Trans* as opposed to *cis:* configuration of A and B rings
cis-	————	Arrangement of two groups in same plane
trans-	————	Arrangement of two groups in opposing planes
dehydro-	————	Conversion of —C—OH to —C=O by loss of two hydrogen atoms
dihydro-	————	Addition of two hydrogen atoms
α-	————	A group *trans* to the 19 methyl radical
β-	————	A group *cis* to the 19 methyl. (See Fig. 15.37, B)
nor-	————	One less carbon in side chain compared to parent molecule
oxy-	-one, -dione	Ketones
hydroxy-, dihydroxy-	-ol, diol	Alcohols
————	-ane	A saturated carbon atom
————	-ene	A single double bond in the ring structure

[a]*Data from Harper.* Review of Physiological Chemistry, *13th ed, 1971, Lange Medical Publications; West, Todd, Mason, and Van Bruggen.* Textbook of Biochemistry, *4th ed, 1966, Macmillan; White, Handler, and Smith.* Principles of Biochemistry, *4th ed, 1968, The Blakiston Division, McGraw-Hill.*

18 and 19 of the steroid molecule usually is indicated only by short vertical lines in the structural formulae.

In the majority of the naturally occurring adrenal steroids, the 17-hydroxy groups occupy the α configuration, whereas the 3-, 11-, and 21-hydroxy groups are present in the β configuration, as shown in Figure 15.37C.

The topic of stereoisomerism was discussed briefly in Chapter 1 (p. 28), as it is of considerable importance to the biochemical activity of various metabolites in the body. In this regard, naturally occurring aldosterone is in the D-isomeric form owing to the 18-aldehyde configuration and L-aldosterone is inactive physiologically.

In the presentation to follow, the names in most common usage of the various steroids will be employed, and the synonyms for these compounds will be included as appropriate.

Biosynthesis of the Adrenal Steroid Hormones

The major synthetic pathways in the body for the adrenal cortical hormones are shown in Figure 15.39. In this outline, the most important secretory products of the gland are underlined and the structural formulae of these compounds also are given.

In vivo, the adrenal steroids are synthesized either from acetate or the cholesterol which arises from this substrate. With the exception of nervous tissue, the adrenal concentration of cholesterol is the highest found in any tissue in the body. Ad-

renal cholesterol is found almost entirely in the esterified form, and a high percentage of unsaturated, long-chain fatty acids are present in these esters.

Adrenal cortical stimulation results in a significant decrease in the concentration of cholesterol in the gland, and this effect coincides in time with the maximal release of adrenocortical steroids. This observation indicates the key role of cholesterol in the biosynthesis of the adrenal hormones. All of the biosynthetic reactions outlined in Figure 15.39 are aerobic, requiring an adequate oxygen supply in order to take place.

The adrenal glands, incidentally, contain the highest concentration of ascorbic acid of any tissue found in the body. 400 to 500 mg are present per 100 gm of fresh tissue.

Cleavage of the side chain on 20 α, 22-dihydroxycholesterol in order to form pregnenolone requires NADH, and this reaction is catalyzed by a desmolase, and is stimulated by cyclic AMP.

The 3-hydroxydehydrogenase required for the conversion of pregnenolone into progesterone as well as 17-α-hydroxypregnenolone into 17-α-hydroxyprogesterone requires NAD as cofactor. This enzyme system is found in the microsome fraction of adrenal cortical cells.

The individual enzyme systems for the hydroxylation of adrenal steroids are each specific for an individual ring position. These enzymes require for their action a supply of reduced nicotinamide adenine dinucleotide (NADH), in addition to molecular oxygen. Cyclic AMP stimulates hydroxylation because of its role in providing NADH via the phosphogluconate oxidative pathway (Table 14.9) through activation of adrenal phosphorylase.

The 11-α- and 18-hydroxylating enzyme complexes are both found in the mitochondria. On the other hand, the enzyme that catalyzes 21-hydroxylation is associated with the microsomal fraction of adrenal cortical cell preparations.

The adrenal glands are the principal organs for 11-hydroxylation among all of the steroid producing endocrine glands.

The major precursor of corticosterone and aldosterone is pregnenolone, and that of cortisol is 17-α-hydroxypregnenolone. Briefly, the sequence whereby the C21 steroid hormones are synthesized involves formation of the Δ4, 3 keto configuration in the A ring, followed by hydroxylation in the 21 position; lastly, hydroxylation in the 11 position takes place.

Aldosterone is synthesized by substitution of the methyl group in position 18 of corticosterone by an aldehyde radical. Most of the aldosterone in the body fluids appears to be in the hemiacetal configuration, as shown in Figure 15.38B.

Androgens (C19 steroids) are produced by scission of the side chain to form dehydroepiandrosterone; this compound forms testosterone and other androgens, and in turn these male hormones give rise to the estrogens, as outlined in Figure 15.39.

It is interesting to note that progesterone, the hormone of the corpus luteum, is also synthesized normally by the adrenal gland as a metabolite formed as an intermediate in adrenocortical hormone synthesis. Furthermore, the placenta also normally exhibits adrenal cortical activity that quantitatively is equal to around two percent of the total endocrine activity of the adrenal glands. Consequently, the placenta temporarily can supplement the normal supply of adrenocortical hormones, a fact that may be of significance in the amelioration of the symptoms encountered in hypoadrenal functional states in women during pregnancy. The examples just cited also serve to emphasize the fact that similar hormones can be produced by entirely different endocrine organs.

In conclusion, it should be emphasized that a defect in enzyme activity involved in any of the synthetic steps indicated in Figure 15.39 leads to an excess of the steroids in the pathway above (or before) the block, with a concomitant deficiency in the intermediates and hormones below (or beyond) the defective enzymatic step. The effects produced by certain enzyme deficiencies will be discussed subsequently.

Regulation of Adrenocortical Hormone Secretion

In normal adult humans, the adrenal glands secrete approximately 10 to 30 mg of cortisol, 2 to 4 mg of corticosterone and 0.3 to 0.4 mg of aldosterone per day. The major physiologic stimulus to synthesis and secretion of the adrenocortical hormones, particularly the glucocorticoids, is pituitary corticotropin. As shown in Figure 15.39, corticotropin catalyses the conversion of cholesterol to pregnenolone, thereby stimulating the biosynthesis of the adrenocorticoid hormones. Corticotropin increases the rate of formation of cyclic 3′,5′-AMP, and the rate of pregnenolone synthesis is probably related to the elevated cyclic AMP concentration, since the increase in steroid secretion that takes place under the influence of corticotropin follows in time the increase in cyclic AMP concentration. The augmented levels of cyclic AMP also produce increased quantities of NADH, a substance that is essential for scission of the side chain of cholesterol, as well as for the subsequent steroid hydroxylation reactions to occur. The hydroxylation reactions take place through NADH activation of adrenal glycogen phosphorylase; however, these syntheses are secondary to the principal action of corticotropin, which is upon pregnenolone synthesis from cholesterol.

Corticotropin does not stimulate adrenal steroid formation if protein synthesis is inhibited. Therefore, it appears that protein formation is necessary for the steroidogenic effect of corticotropin to be manifest. On the other hand, inhibition of protein synthesis does not hinder the increased rate of cyclic AMP formation which is induced by corticotropin, hence the protein believed to be involved in steroid synthesis must act at a locus beyond the formation of adenyl cyclase. Briefly, then, a primary action of corticotropin on the adrenal cortex lies in its ability to stimulate synthesis of cyclic AMP by increasing the activity of adenyl cyclase.

As discussed earlier in this chapter, corticotropin from the adenohypophysis regulates both the basal physiologic secretion of adrenal glucocorticoids, cortisol, and corticosterone, as well as the increased secretion of these hormones that is produced by stress. In turn, negative feedback control of corticotropin release is achieved by means of circulating blood level of glucocorticoids acting upon the hypothalamo–hypophyseal system. In man, cortisol is the most important normal regulator of corticotropin release.*

In contrast to the glucocorticoids, adrenal secretion of the mineralocorticoid hormone, aldosterone, is influenced only to a very limited extent by corticotropin. The principal factor that regulates aldosterone secretion is the adrenal blood electrolyte concentration. In particular, a reduced sodium ion concentration is the most significant single factor that stimulates aldosterone release from the adrenal cortex, although the blood level of potassium ion also has a role in this process, as elevated blood potassium causes a parallel in-

*Presumably this effect is related to the fact that cortisol also is the major free circulating adrenocortical hormone, being found at a plasma level of about 12 μg/100 ml.

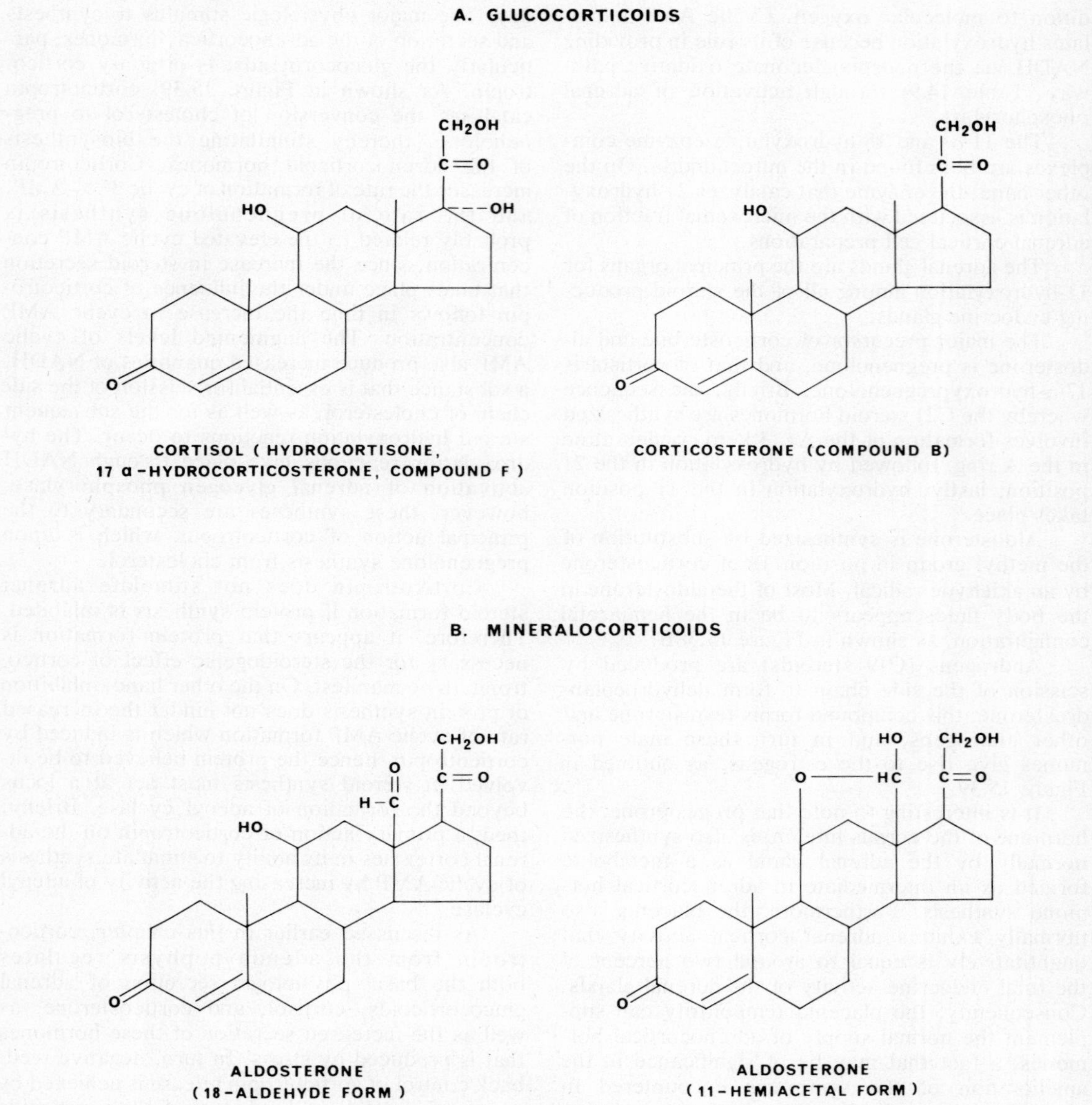

A. GLUCOCORTICOIDS

CORTISOL (HYDROCORTISONE,
17 α –HYDROCORTICOSTERONE, COMPOUND F)

CORTICOSTERONE (COMPOUND B)

B. MINERALOCORTICOIDS

ALDOSTERONE
(18–ALDEHYDE FORM)

ALDOSTERONE
(11–HEMIACETAL FORM)

FIG. 15.38. A. Structural formulas of the adrenal glucocorticoids cortisol and corticosterone. **B.** Structural formulas illustrating the aldehyde and hemiacetal forms of the mineralocorticoid aldosterone. (After White, Handler, and Smith. *Principles of Biochemistry,* 4th ed, 1968, The Blakiston Division, McGraw-Hill.)

crease in aldosterone secretion. Consequently, there is a direct and localized effect of these two ions upon the adrenal cortex; and the output of cortisol and corticosterone are regulated independently of aldosterone secretion.

Experimental evidence also has demonstrated that a specific aldosterone stimulating hormone (ASH) is present in the body, and this substance is another factor that acts directly upon the adrenal cortex to increase secretion of aldosterone.

Angiotensin II production is elevated during constriction of the thoracic vena cava or the hepatic portal vein, and such elevated angiotensin levels can also increase aldosterone secretion.

This effect of angiotensin II on aldosterone secretion is wholly independent of the pressor action of this hormone as discussed in the section "Other Organs and Substances Having Actual or Presumed Endocrine Functions" on page 1169.

Conditions that result in sodium deficiency cause the adrenal cortex to become more responsive (ie, more sensitive) to the effects of corticotropin. Hypophysectomy has no effect on the adrenal response to sodium deficiency, although aldosterone secretion is reduced in hypophysectomized subjects when they are subjected to stress stimuli.

Many other naturally occurring substances

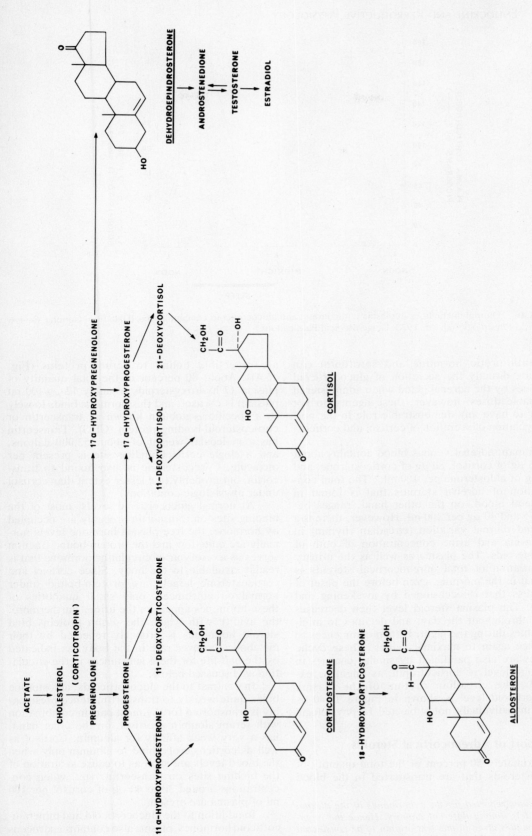

FIG. 15.39. Outline of the biosynthetic pathway for the major adrenal cortical steroids (underlined). (After White, Handler, and Smith. *Principles of Biochemistry*, 4th ed, 1968, The Blakiston Division, McGraw-Hill.)

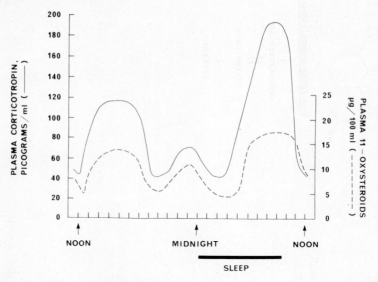

FIG. 15.40.   Diurnal fluctuations in plasma corticotropin and glucocorticoid concentrations. (Data from Ganong. *Review of Medical Physiology,* 6th ed, 1973, Lange Medical Publications.)

(eg, antidiuretic hormone and serotonin) can stimulate directly the secretion of glucocorticoid hormones by the adrenal gland when administered in suitable doses; however, these agents do not appear to have any demonstrable role in the normal regulation of secretion of cortisol and corticosterone.

In man, adrenal venous blood contains about 5 to 15 μg of cortisol, 80 μg of corticosterone, and 0.01 μg of aldosterone per 100 ml.* The total concentration of adrenal steroids that is found in peripheral blood, on the other hand, ranges between 5 and 40 μg per 100 ml. However, there is a profound diurnal variation (circadian rhythm) in the plasma and urine concentration of both of these steroids. The plasma as well as the urinary concentration of total adrenocortical steroids is maximal in the morning, even before the onset of the stress that is occasioned by awakening and arising. The plasma steroid level then decreases slowly throughout the day, and declines to minimal values during the night. It then commences to rise once again to maximal values. These cyclic changes are also paralleled by similar changes in the blood level of corticotropin, as might be expected. The circadian rhythm of the plasma glucocorticoid level is shown in Figure 15.40 as found in individuals not subjected to physiologic stress.

Transport of Adrenocortical Steroids

Approximately 50 percent of the total quantity of corticosteroids that are transported in the blood

---

The values reported for these hormones in the plasma vary widely among different authors. Hence the quantitative values given in this text are not to be considered absolute; rather they are indicative only.

are reversibly bound to serum proteins (Fig. 15.41). About 80 percent of the total quantity of plasma 17 hydroxysteroids (roughly 12 μg/100 ml plasma) is cortisol, and this hormone binds loosely to a specific α-globulin known as transcortin or corticosteroid-binding protein (CBG). Transcortin has a molecular weight of about 52,000 daltons, and a single cortisol-binding site is present per molecule. Corticosterone is also bound to transcortin, but probably to a lesser extent than cortisol under physiologic conditions.

At normal glucocorticoid levels, most of the binding sites on transcortin probably are occupied by hormone; the free plasma hormone levels normally are quite low and the protein-bound fraction serves as a reservoir of circulating hormone that is readily available to the body. Since cortisol and corticosterone largely are protein-bound under normal circumstances, only small quantities of these hormones appear in the urine. Furthermore, the avidity with which the serum proteins bind steroid hormones is probably reflected by their metabolic turnover rate in the body, as indicated by the half-life for these substances in the circulation as discussed below.

In contrast to the glucocorticoids, aldosterone binds quite weakly to transcortin, hence presumably is transported for the most part in combination with serum albumin. Cortisol, on the other hand, has a very weak affinity for albumin. Cortisol as well as corticosterone bind to albumin only when the blood levels are such as to cause saturation of the binding sites on transcortin, viz., when concentrations around 30 to 40 μg of cortisol per 100 ml of plasma are present.

In addition to the glucocorticoid and mineralocorticoid hormones, plasma also contains androgens that are elaborated by the adrenals as well as the male gonads. The plasma androgens also are present in a free state as well as in a conjugated form, the latter being formed by association of the androgens with the plasma proteins. The principal adrenal androgen, dehydroepiandrosterone (DHEA), is found in plasma at a concentration of approximately 50 $\mu g/100$ ml, whereas androsterone, the male hormone which is found in both males and females, is present at a level of about 25 $\mu g/100$ ml. Similar general relationships between free and protein-bound androgens exist as for the other adrenocortical hormones.

All of the adrenal corticoids are deemed to be physiologically inactive when they are bound to serum proteins. In this regard, it is important to emphasize here that it is the free steroid hormones in the plasma which exert a negative feedback effect upon the hypothalamo–hypophyseal axis, and thus regulate corticotropin secretion. In turn the several integrated factors that affect the free glucocorticoid hormone level in the circulation are shown in Figure 15.41 and further discussed below.

Transcortin is synthesized in the liver; therefore, its plasma concentration is reduced in hepatic disease (eg, cirrhosis), multiple myeloma, and nephrosis. The transcortin level is elevated, by thyroid hormone and estrogen, and such an effect is seen during late pregnancy. This stimulatory effect of estrogen on transcortin synthesis is mediated indirectly by hypophyseal thyrotropin. Since estrogen increases the circulating transcortin level, and since the bound steroid hormones are essentially inactive physiologically, then increased quantities of estrogen can decrease the efficacy of the glucocorticoids by reducing the concentration of these adrenal hormones that are available to the tissues in a free state.

On the other hand, progesterone has a high affinity for the single steroid binding site on transcortin so that this hormone can displace cortisol into the free, active steroid fraction by exerting a competitive effect for the transcortin-binding site.

The various equilibria that can be altered as the result of physiologic or pathologic changes in the various factors discussed above can be visualized by reference to Figure 15.41. Thus, for example, if the level of transcortin rises, more cortisol is bound, and as a consequence the free cortisol level falls, thereby stimulating the release of more corticotropin from the adenohypophysis. Eventually a new dynamic equilibrial level is achieved, at which point the free cortisol level is normal. The bound cortisol as well as the rate of adrenal cortical steroid secretion both are increased. Conversely, when the transcortin level decreases, alterations in the opposite direction take place. As a consequence of these adaptive changes, pregnant women exhibit high rates of glucocorticoid hormone secretion together with elevated total plasma 17-hydroxycorticoid levels, but they do not develop symptoms of glucocorticoid hypersecretion.

The opposite situation also holds for some patients having nephrosis. These individuals may not exhibit symptoms of glucocorticoid deficiency in spite of reduced rates of glucocorticoid secretion combined with low total plasma 17-hydroxysteroid concentrations.

Metabolism and Excretion of the Adrenocortical Steroids

The principal metabolic routes and metabolites of the major adrenocortical steroid hormones are summarized in Figure 15.42. Certain general features common to all of these hormones will be presented first, and then certain individual differences in the metabolic treatment of these hormones will be considered specifically, together with the individual metabolites that are then excreted as end-products in the urine, bile, and feces.

In contrast to most other hormones (eg, insulin) there is no significant catabolism of adrenal steroids to carbon dioxide and water. Rather, the steroid nucleus is eliminated intact after certain side groups on the molecule are chemically reduced and/or conjugated with other substances, so that the excretory products are physiologically in-

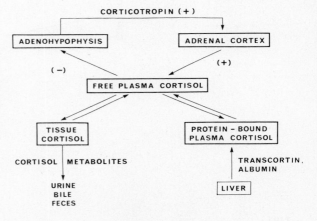

FIG. 15.41 Interrelationships among the adenohypophyseal hormone, adrenal cortex, and plasma cortisol levels. Note that a negative feedback mechanism controls the free plasma cortisol level, a mechanism similar to that involved, for example, in the regulation of thyroid hormone secretion (Fig. 15.17). (After Ganong. *Review of Medical Physiology*, 6th ed, 1973, Lange Medical Publications.)

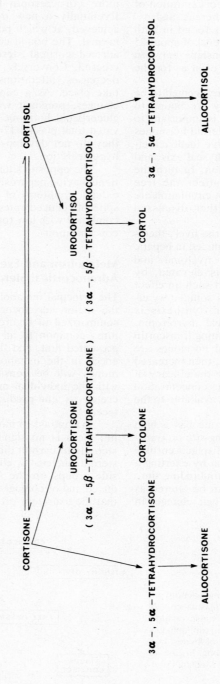

FIG. 15.42. Metabolism of cortisol. (After White, Handler, and Smith. *Principles of Biochemistry*, 4th ed, 1968, The Blakiston Division, McGraw-Hill.)

active. Furthermore, the metabolic turnover rates of steroid hormones are quite high, as discussed for individual hormones below.

The liver appears to be the principal, if not the sole, organ responsible for the conversion of active adrenal cortical hormones into inactive metabolites for elimination, because this organ contains the requisite enzyme systems and cofactors responsible for the necessary chemical alterations to the parent compounds.

CORTISOL METABOLISM. The half-life of cortisol in the circulation is approximately 60 to 90 minutes, and this hormone is metabolized chiefly by the liver.

As shown in Figure 15.42, cortisol is reduced to dihydrocortisol, next to tetrahydrocortisol, and then conjugated with glucuronic acid for excretion.

The glucuronyl transferase enzyme system that carries out the conjugation reaction of tetrahydrocortisol with glucuronic acid is also responsible for the production of glucuronides of bilirubin, and a number of other hormones and drugs. Competitive inhibition between the various substrates for this enzyme may occur.

Some cortisol also is converted into cortisone in the liver (Fig. 15.42), and although this compound is a physiologically active glucocorticoid (hence is used widely in medical practice), it is not secreted by the adrenal gland but is produced in the liver.* Under physiologic circumstances, cortisone does not enter the systemic circulation to any significant extent, since this compound is reduced and conjugated rapidly in the liver. The resulting physiologically inactive tetrahydrocortisol glucuronide derivative of cortisol is quite soluble, and it enters the circulation to be excreted rapidly in the urine, partially by tubular excretion, without binding to protein molecules.

Experiments performed in human subjects have revealed that over 90 percent of a dose of exogenous ^{14}C-labeled cortisol is eliminated within 48 hours. Approximately 70 percent of the total quantity of administered cortisol was found as reduced compounds in the urine, and the remaining 20 percent was eliminated in the feces. The fraction of free and conjugated adrenal steroid metabolites excreted in the bile is reabsorbed largely from the intestine via the enterohepatic circulation.

Roughly 10 percent of the total quantity of cortisol secreted by the adrenal cortex is converted within the liver into 17-ketosteroid derivatives of cortisol and cortisone. These derivatives then are conjugated with sulfate to form esters for excretion in the urine. Other metabolites are formed in minor quantities and this fraction includes some 20-hydroxy derivatives. However, by far the largest proportion of cortisol is eliminated in the form of urinary 17-hydroxycorticosteroids.

It is important to remember that cortisone, as well as all other steroids with an 11-keto group are metabolites of the glucocorticoid hormones that are secreted by the adrenal glands.

The normal 24-hour excretion of 17-hydroxycorticosteroids derived from cortisol and cortisone amounts to approximately 10 mg in adults males and 7 mg in females. Since most of the 17-hydroxycorticoids are excreted as the conjugated form (with glucuronic acid), a preliminary hydrolysis with glucuronidase is necessary before a quantitative determination of the steroids can be performed in the laboratory.

The measurement of urinary 17-hydroxycorticosteroids is useful as a test for adrenal function under various circumstances.

CORTICOSTERONE METABOLISM. The half-life of corticosterone in the circulation is about 50 minutes, and this shorter half-life than that of cortisol probably reflects a lesser degree of binding to the serum proteins than takes place with cortisol.

In general, the metabolism and excretion of corticosterone and its metabolites are quite similar to those for cortisol (Fig. 15.43), except that corticosterone does not form 17-ketosteroid derivatives (see next section).

ALDOSTERONE METABOLISM. The half-life of aldosterone in the circulation is about 20 minutes, as it is bound to protein only to a very limited extent. The plasma concentration of aldosterone in the circulation is quite small (about 0.006 μg/100 ml).

Within the liver, a large part of the plasma aldosterone is converted into the tetrahydroglucuronide derivative (Fig. 15.44). However some of the hormone also is converted in both liver and kidney into an 18-glucuronide, and this compound, unlike other steroid metabolites, can by hydrolyzed into free aldosterone once again by treatment with acid at pH 1.0. Consequently, the 18-glucuronide is frequently referred to as the acid labile conjugate.

Of the total aldosterone secreted by the adrenal cortex, only one percent appears in the urine in the free form; five percent appears as the acid-labile conjugate, and around 40 percent is conjugated to glucuronic acid.

URINARY 17-KETOSTEROIDS AND ADRENAL ANDROGENS. The 17-ketosteroids excreted in the urine belong to three general classes: (1) acidic compounds derived from the bile acids; (2) phenolic compounds derived from the estrogens; and (3) neutral compounds derived from the endocrine secretions of the adrenal cortex as well as the gonads. Only the neutral 17-ketosteroids of adrenal cortical origin will be considered in this section.

In a 24-hour urine sample from an adult male, a total of between 10 and 20 mg of neutral 17-ketosteroids is found, whereas an adult female excretes between 5 and 15 mg of these compounds. The urinary 17-ketosteroids reflect the overall androgenic function of the individual. In the female, these compounds are produced exclusively by the adrenal cortex, whereas in the male, both the ad-

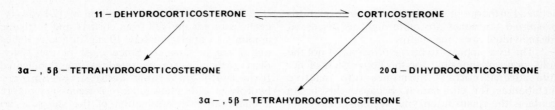

FIG. 15.43. Metabolism of corticosterone. (After White, Handler, and Smith. *Principles of Biochemistry,* 4th ed, 1968, The Blakiston Division, McGraw-Hill.)

renal cortex and the testes contribute to the total output of 17-ketosteroids. The testes are responsible for about 33 percent of the total neutral urinary 17-ketosteroids, the remaining two-thirds of the ketosteroids secreted by males are produced by the adrenals or derived from cortisol in the liver.

The 17-ketosteroid dehydroepiandrosterone originates chiefly from the adrenal cortex, and this compound is found in the urine of normal men and women; together with its derivatives, testosterone, androstenedione, and 11-β-hydroxyandrost-4-ene-3, 17-dione Fig. 15.39). The latter compound, as well as the 17-ketosteroids formed from cortisol and cortisone in the liver, are the only 17-ketosteroids having an —OH or a =O group in the 11 position of the C ring. Thus they are denoted as 11-oxy-17-ketosteroids.

Regulation of Adrenocortical Secretion

The physiologic action of pituitary corticotropin upon the adrenal cortex and certain of its secretions has been discussed earlier. In particular, the mechanisms that regulate glucocorticoid secretion were presented. The material to be discussed in this section will summarize the factors involved in the regulation of aldosterone secretion by the adrenal cortex.

A number of stimuli can increase the secretion of aldosterone by the adrenal cortex, either selectively or nonselectively, as summarized in Table 15.9. It is important to distinguish between those factors involved in the physiolgic regulation of aldosterone secretion and those stimuli which, under certain conditions (viz, stress), are able to provoke an increased secretion of this hormone. The principal physiologic regulatory factors for aldosterone secretion are adrenal blood sodium and potassium levels, corticotropin secretion, and renin from the kidney.

BLOOD ELECTROLYTE LEVELS AND ALDOSTERONE SECRETION. Regardless of the cause, a decreased blood level of sodium or an increased blood potassium level act directly upon the cells of the adrenal cortex to elevate the secretion of aldosterone. The sodium level is far more important than the potassium concentration insofar as the physiologic regulation of aldosterone output is concerned. However, there is not a corresponding and obligatory increase in glucocorticoid (cortisol and corticosterone) output by the adrenals when the aldosterone output is altered by a shift in electrolyte balance.

The adrenal glands become more sensitive to the effects of corticotropin when a relative sodium deficiency is present; however hypophysectomy does not influence the increase in aldosterone secretion that is produced by a low blood sodium concentration. This operation does reduce the augmented secretion of aldosterone that is evoked normally by noxious stimuli.

EFFECTS OF CORTICOTROPIN AND THE HYPOTHALAMUS UPON ALDOSTERONE SECRETION. Pituitary corticotropin appears to exert a nonselective, physiologic stimulatory effect upon the secretion of aldosterone in addition to its action in augmenting the output of glucocorticoids by the adrenal cortex. However, the quantity of corticotropin necessary to elicit a maximum aldosterone secretion is greater than that required to stimulate maximal secretion of the glucocorticoid hormones. Nonetheless, the level of corticotropin which is required to produce this effect on aldosterone output lies within the physiologic range of endogenous corticotropin secretion by the pituitary gland, but if the rate of corticotropin secretion remains elevated, then aldosterone secretion starts to decrease after several days; ie, the effect is transitory.

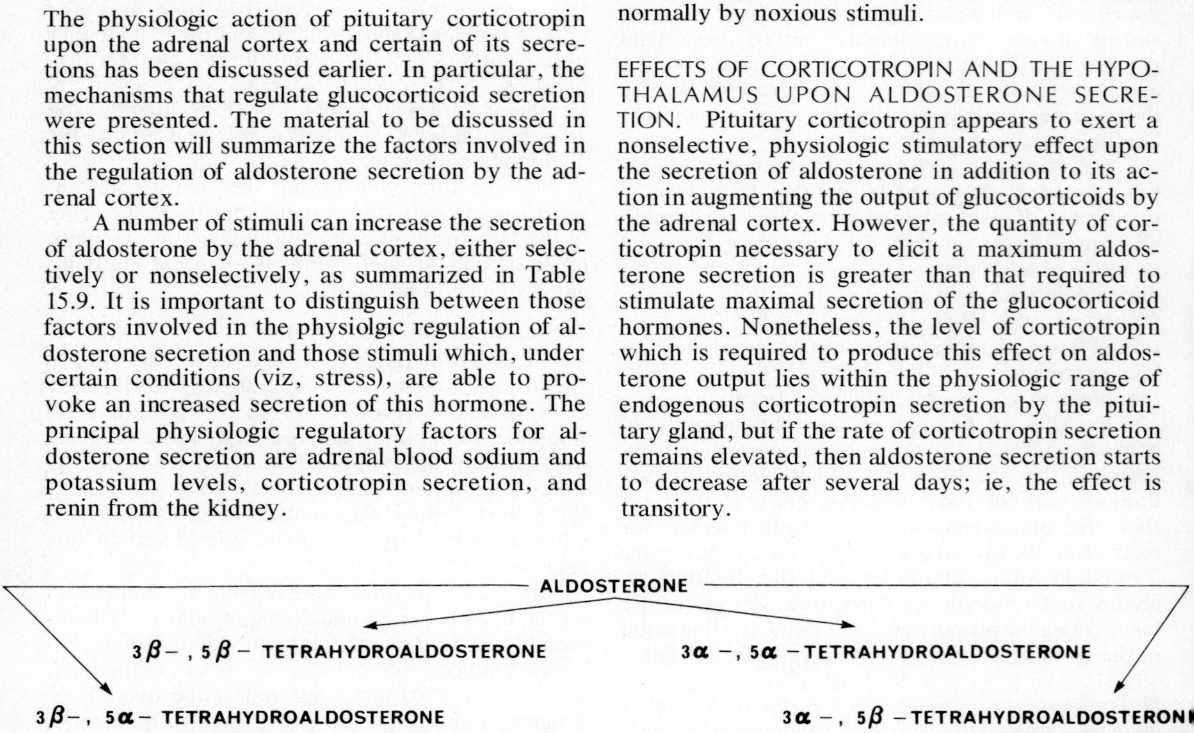

ALDOSTERONE

3β— , 5β— TETRAHYDROALDOSTERONE   3α — , 5α —TETRAHYDROALDOSTERONE

3β—, 5α— TETRAHYDROALDOSTERONE   3α —, 5β —TETRAHYDROALDOSTERON|

FIG. 15.44. Metabolism of aldosterone. (After White, Handler, and Smith. *Principles of Biochemistry,* 4th ed, 1968, The Blakiston Division, McGraw-Hill.)

Following hypophysectomy, the normal (ie, physiologic) rate of aldosterone secretion is unchanged, and the increase in secretion provoked by salt restriction is unaffected. The increased aldosterone secretion in response to noxious stimuli, eg, trauma, no longer is present however.

The hypothalamus indirectly affects the rate of aldosterone secretion through the influence of this portion of the diencephalon upon corticotropin secretion by the adenohypophysis. Therefore, any damage or injury to this region of the brain that diminishes corticotropin secretion also inhibits the increase in aldosterone secretion that is induced by stress.

EFFECTS OF RENIN AND ANGIOTENSIN II UPON ALDOSTERONE SECRETION. As discussed on page 1169, and illustrated in Figure 15.72 the octapeptide angiotensin II is produced from angiotensin I by the action of renin on an α_2 globulin present in the blood. Both renin and angiotensin II in small doses selectively can stimulate the secretion of aldosterone by their direct action upon the adrenal cortex.

Renin is secreted by the juxtaglomerular cells of the afferent renal arterioles (p. 770), and although the mechanism whereby secretion of this proteolytic enzyme is not completely resolved, it is apparent that a decrease in the mean renal arterial blood pressure increases renin secretion. It appears that a physiologic mechanism for the regulation of aldosterone secretion is operative via a negative feedback mechanism. Thus, a decline in the extracellular fluid volume (or the arterial blood volume) produces a fall in renal arterial pressure, elevating renin secretion. In turn, the increased quantity of angiotensin II produced by this enzyme acts directly upon the adrenal cortex to increase the rate of aldosterone secretion. The aldosterone expands the extracellular fluid volume by causing increased salt and water retention. The greater extracellular fluid volume in turn raises the renal arterial pressure, so that the stimulus to elevated renin secretion is removed.

Surgery, hemorrhage, and other traumatic stimuli increase corticotropin secretion, but aldosterone secretion per se also is increased by such stimuli in hypophysectomized subjects. Similarly, standing upright, and constriction of certain blood vessels likewise decrease the intraarterial vascular volume, causing a fall in the mean renal arterial blood pressure (Table 15.9). All of these factors can produce an elevated rate of aldosterone secretion via the renin–angiotensin mechanism described above.

If the dietary intake of sodium is severely restricted, the extracellular fluid volume contracts, so that once again the renin–angiotensin system is activated, and aldosterone secretion is stimulated. In this instance renin and aldosterone levels rise before any demonstrable fall in blood pressure occurs. This fact suggests that a limited sodium intake affects renin secretion by another, as yet unknown, mechanism.

Table 15.9   STIMULI TO INCREASED ALDOSTERONE SECRETION[a]

1. Decreased sodium concentration in adrenal arterial blood, regardless of cause
2. Elevated potassium concentration in adrenal arterial blood, regardless of cause
3. Standing (in) upright (position)
4. Primary hyperaldosteronism
5. Secondary hyperaldosteronism (eg, such as may be encountered in hepatic cirrhosis, congestive heart failure, and nephrosis)
6. Hemorrhage[b]
7. Physical trauma[b]
8. Anxiety[b]
9. Surgery[b]
10. Corticotropin secretion by adenohypophysis[b]
11. Renin from kidney
12. Constriction of thoracic vena cava or hepatic portal vein

[a]*Data from Harper.* Review of Physiological Chemistry, *13th ed, 1971, Lange Medical Publications; West, Todd, Mason, and Van Bruggen.* Textbook of Biochemistry, *4th ed, 1966, Macmillan; White, Handler, and Smith.* Principles of Biochemistry, *4th ed, 1968, The Blakiston Division, McGraw-Hill.*

[b]*These stimuli exert a nonselective effect upon aldosterone secretion, ie, glucocorticoid secretion also is increased concomitantly.*

It is difficult at this time to assess with accuracy the relative importance of electrolyte levels, pituitary corticotropin secretion, and the renin–angiotensin mechanism insofar as the physiologic regulation of aldosterone secretion is concerned under various conditions. Quite probably, one or more of these regulatory mechanisms may exert a dominant effect over the others under specific circumstances (ie, different types of stress), and the net basal secretion of this hormone reflects the minimal activity of all three of these systems in the normal, nonstressed individual.

Physiologic Actions of Adrenocortical Secretions

The adrenal steroids exert a profound influence upon a considerable variety of physiologic and biochemical processes. Some of these effects are undoubtedly secondary responses to a primary role (or primary action) of a given steroid hormone, whereas other responses may reflect even more indirect effects of the principal action.

For convenience, the various functions of the adrenal corticosteroid secretions may be grouped into six categories: (1) metabolic effects on carbohydrate, protein, and fat metabolism; (2) effects on electrolyte and water metabolism. The influence of the steroid hormones upon circulatory homeostasis as well as neuromuscular irritability may be included in this category; (3) hematologic effects; (4) secretory actions; (5) effects upon allergic and inflammatory reactions in the body; (6) the effects exerted by the adrenocortical hormones on bodily resistance to stress stimulation, eg, cold, or trauma.

Exogenous administration of the three principal hormones of the adrenal cortex (cortisol, corticosterone, and aldosterone) produces somewhat comparable effects that overlap to some extent (Table 15.10). Thus, cortisol exerts an effect upon all six of the categories of response summarized above, although the predominant effects of this hormone are manifest in its actions upon various metabolic activities rather than upon electrolyte and water metabolism. Aldosterone, on the other hand, exerts its major action upon electrolyte and water metabolism, but has negligible activity when compared to cortisol in other physiologic processes. Corticosterone has an effect upon all of the six actions listed; however, this hormone is significantly less potent than either cortisol or aldosterone in their respective areas of maximal efficacy.

ADRENAL INSUFFICIENCY. Prior to a discussion of the various individual functions and actions of the adrenocortical hormones, a brief survey of the signs and symptoms of acute adrenal insufficiency, together with some etiologic factors, will be presented in order to emphasize the critical importance of these hormones, and of the adrenal cortex in general, to the survival of the individual.

Humans or experimental animals suffering from a total absence of the adrenocortical hormones exhibit the following symptoms:

1. A decrease in plasma sodium and chloride is present together with a concomitant increase in serum potassium, these defects being caused by an impaired renal tubular reabsorption of sodium and chloride coupled to a defective potassium excretion.
2. Hemoconcentration occurs because of the excessive water loss through the kidney that accompanies the loss of electrolytes. The decreased extracellular fluid volume leads in turn to a diminished cardiac output. Because of failure of the sodium:potassium:hydrogen exchange mechanism in the kidney, a metabolic acidosis may develop. The renal failure and circulatory defects lead to uremia (increase in blood urea). Chronic adrenal cortical hormone deficiency leads to lymphocytosis and anemia.
3. Protein anabolism is faulty; therefore, during fasting urinary nitrogen excretion is diminished.
4. Hypoglycemia develops upon fasting owing in part to an increased carbohydrate utilization and in part to a concomitant decreased hepatic glycogen level during fasting caused by a reduced glucogenesis and gluconeogenesis. There is also a greatly increased sensitivity to insulin, because of the decreased hepatic glucogenesis and gluconeogenesis.

   Carbohydrate absorption from the gut is also impaired, possibly due to the defects in potassium metabolism, since carbohydrate and potassium move together normally in an intracellular direction, whereas in adrenal cortical insufficiency, potassium tends to leave the cells thereby contributing to the net hyperkalemia.
5. Depot lipid metabolism is faulty, for reasons unknown.
6. Anorexia, weight loss, and cessation of growth occur.
7. Muscular weakness and sensitivity to stress develop, which are due in part to potassium loss from the cells, and in part to the defective glycogenesis.
8. Individuals who exhibit chronic hypofunction of the adrenals have several additional signs, including a decreased metabolic rate, an exacerbation of the inflammatory response and hypersensitivity reactions, and excessive pigmentation. The latter effect possibly may be due to the inherent melanocyte-stimulating action of pituitary corticotropin.

Table 15.10[a] RELATIVE BIOLOGIC ACTIVITY OF NATURAL ADRENOCORTICAL STEROIDS COMPARED TO CORTISONE[b]

STEROID	GLYCOGEN DEPOSITION	SODIUM RETENTION	SURVIVAL	MUSCLE WORK	GROWTH	COLD STRESS	ANTIIN-FLAMMATORY ACTIVITY
Cortisone[c]	100	100	100	100	100	100	100
Cortisol	155	150	100	160	219	—	1,255
Corticosterone	54	255	75	46	108	9	3
Aldosterone	30	60,000	—	—	—	—	0
11-Dehydrocorticosterone	48	—	58	32	—	33	0
Deoxycorticosterone	0	3,000	400	5	—	8	0

[a]*Data from White, Handler, and Smith.* Principles of Biochemistry, *4th ed, 1968, The Blakiston Division, McGraw-Hill.*

[b]*These studies were carried out on adrenalectomized rats, except for the antiinflammatory activity studies which were conducted with adrenalectomized mice, and the muscle work performance experiments which were performed on adrenalectomized-nephrectomized rats.*

[c]*Under physiologic conditions, cortisone is not secreted by the adrenal gland, but is a metabolite of cortisol that is produced in the liver as shown in Figure 15.42.*

Unless the patient or animal suffering from adrenocortical insufficiency is subjected to appropriate therapy, the hypotension leads to cardiovascular insufficiency and fatal shock ensues because of the lack of aldosterone. The metabolic defects listed above are corrected primarily by glucocorticoids, and if these are withheld, stress stimuli can induce a rapid collapse and death.

It must be emphasized that at physiologic concentrations, the role of the adrenocorticoids is to prevent the defects outlined above. Furthermore, the responses evoked by large quantities of these hormones, regardless of their source, differ quite markedly from those which are involved in the homeostatic regulation of the various bodily processes.

The remainder of this section will deal principally with the physiologic roles of the adrenal steroids.

CARBOHYDRATE METABOLISM. The glucocorticoids cortisol and corticosterone promote glucose release from the liver as well as glycogenesis and gluconeogenesis (from amino acids), and the latter two processes result in hepatic glycogen deposition. These hormones also inhibit peripheral glucose utilization.

All of these effects can be demonstrated several hours following the injection of a dose of the hormones, especially cortisol. The elevated release of glucose following adrenal steroid administration probably reflects an increased hepatic glucose-6-phosphatase activity. In addition, the activity of pyruvate carboxylase and certain specific hepatic transaminase enzymes is also elevated, together with an increased glycogen synthesis. The net result of these several individual effects of the glucocorticoid hormones is an increase in the hepatic glycogen store.

Pyruvate carboxylase activity increases quite early following cortisol administration. This fact implies that the enzymatic conversion of pyruvic acid to oxaloacetic acid under the influence of this enzyme is an important step in the stimulation of glycogenesis by cortisol. Steroid hormones also inhibit amino acid incorporation into the extrahepatic tissues; therefore, quantities of these substrates become available as additional precursors for hepatic gluconeogenesis.

In vivo as well as in vitro, the utilization of glucose by peripheral tissues (eg, muscle, adipose, and lymphoid) is decreased by cortisol. This hormone also reduces mucopolysaccharide synthesis in connective tissue. It appears that both a transport mechanism, together with an unresolved intracellular mechanism, are involved in producing these effects.

In the fasted or fed normal individual who is given exogenous cortisol, an elevation of both blood glucose and hepatic glycogen concentrations are observed. These effects are due to the above-described effects of the hormone on carbohydrate metabolism. If, however, the administration of 17-hydroxycorticoids is prolonged, a diabetic type of glucose tolerance curve develops ultimately and glucosuria ensues. In sensitive human subjects, this response may be linked to a latent genetic tendency toward diabetes, and eventually permanent diabetes mellitus may develop after long-term steroid administration because of degeneration of the beta cells of the islets of Langerhans. Thus, under physiologic conditions both cortisol and corticosterone appear to exert a secondary effect upon the homeostatic regulation of blood glucose level during fasting, an action that is shared with glucagon.

On the other hand, during hypofunction of the adrenal cortex or following adrenalectomy, an increased sensitivity to insulin develops, an effect that is perhaps related to the reduction of glyconeogenesis within, and glucose release by, the liver. The normal insulin requirement of the individual is diminished together with the severity of any preexisting diabetic state.

The enhanced insulin sensitivity also means that the requirement of the body for this hormone is decreased significantly, so that an individual suffering from hypofunction of the adrenal cortex can rapidly develop a fatal hypoglycemia and coma during fasting, because of the exaggerated effect of the normal quantity of insulin which is being secreted by the pancreas. Therefore, in such hypoadrenal and diabetic individuals, the blood sugar is maintained at physiologic levels only as long as food intake is adequate.

Muscle glycogen levels also decrease markedly when the adrenal cortical steroids are in deficient supply. As the direct consequence of this defect the work performance of skeletal muscle is impaired severely (Table 15.10). In fact, the reduced work capacity of adrenalectomized rats may be used as a bioassay technique for adrenocortical hormones.

In summary of the effects of cortisol and corticosterone upon carbohydrate metabolism, the principal physiologic actions of these glucocorticoids are to stimulate hepatic glucose secretion, as well as glycogenesis and gluconeogenesis by the liver. In addition, these adrenal steroids also inhibit peripheral glucose utilization, an action that is antagonistic to that of insulin. Therefore, under physiologic conditions in the normal individual, there exists an exquisitely delicate and dynamic balance as well as interplay among the secretions of the adrenal cortex, the pancreas (both insulin and glucagon), and the thyroid gland insofar as the net homeostatic regulation of carbohydrate metabolism is concerned. Furthermore, the feedback mechanisms that involve the adenohypophyseal hormones are vital factors that indirectly control the secretion of adrenal cortical as well as the thyroid hormones, and this factor ultimately exerts profound effects upon carbohydrate metabolism. Under normal circumstances, the dynamic interplay among these complex factors enables the level of carbohydrate utilization and distribution throughout the body to be adapted continuously and precisely to various physiologic states, such as fasting, exercise, and so forth. But when a defect in the secretion of one or more of these hormones results in an imbalance among them, the clinical picture which emerges can range from mild symptoms to a catastrophic situation, insofar as the life of the patient is concerned.

General remarks similar to those outlined above also are pertinent in relation to the effect of various endocrine substances upon protein, fat, and electrolyte metabolism.

PROTEIN METABOLISM.   Another physiologic action of cortisol is to stimulate hepatic protein synthesis, but at the same time to inhibit this process in skeletal muscle and other peripheral tissues. Reduced amino acid transport into the cells is also found in those tissues in which protein synthesis is decreased by an excess of adrenocortical steroids. Since normal protein catabolism continues in the peripheral tissues when glucocorticoid levels are increased, the net effect is a loss of amino acids from the tissues, hence a wasting of the soft tissues occurs. Osteoporosis is also found under such circumstances.

When the blood level of cortisol is abnormally elevated, the loss of nitrogen from the cells also reflects an action of the hormone upon the free amino acid pools within the cells, and even though hepatic protein synthesis occurs at a much faster rate (ie, is stimulated) under such conditions, this process is inadequate to balance the loss of amino acids from the tissues. An elevation of the plasma total free amino acid concentration develops.

Thus one of the principal diabetogenic effects of elevated glucocorticoid levels results from an overall increase in the rate of bodily protein catabolism over the rate of protein anabolism with a concomitant augmented hepatic gluconeogenesis from the amino acid residues. This effect is, of course, combined with the increased hepatic glycogenesis as well as ketogenesis, which also develop under such conditions.

The effect of excess glucocorticoid hormones upon amino acid oxidation in turn is reflected by an elevated rate of urea synthesis and excretion when these substances are present in excess, so that the net result of excess glucocorticoids is a negative nitrogen balance.

Cortisol acts directly upon the hepatic cells to stimulate protein synthesis, and this function of the hormone has been demonstrated experimentally to affect amino acid incorporation into the protein of the ribosomes. Similar experiments performed with a peripheral tissue (eg, skeletal muscle and lymphoid cells), also have revealed a direct action of cortisol upon protein metabolism. However, in this instance there is an inhibition of amino acid transport into the cells and subsequent incorporation of the amino acids into proteins. Therefore, synthesis of RNA by liver is stimulated by cortisol, whereas in peripheral tissues, especially skeletal muscle and lymphoid cells, the hormone produces a decreased rate of RNA synthesis.

LIPID METABOLISM.   In normal subjects, the injection of excess adrenocortical steroids may increase lipogenesis in the peripheral tissues. This effect is possibly due to a related increase in insulin secretion, as this pancreatic hormone stimulates lipogenesis as discussed earlier. In vitro experiments have shown, however, that cortisol will stimulate directly the release of free fatty acids from depot adipose tissues, and also that cortisol will enhance this same activity of epinephrine. The lipid-mobilizing effect of the glucocorticoid hormones possibly contributes to the profound ketogenic effect of cortisol in human patients with Addison's disease that is combined with diabetes. A similar effect is seen in experimental animals following both adrenalectomy and pancreatectomy.

Quite possibly the lipid-mobilizing action of the adrenal cortical steroids is secondary to an inhibition of peripheral glucose utilization. Similarly to the effects of these hormones on protein synthesis that were discussed above, the effects of the glucocorticoids on liver lipid metabolism are diametrically opposite to those observed in the peripheral tissues. In hepatic tissue, glucocorticoids strongly accelerate the rate of triglyceride (triacylglyceride) synthesis, and this effect occurs before the blood fatty acid levels rise as the consequence of the lypolytic effect of the glucocorticoids on the depot lipids.

WATER AND ELECTROLYTE METABOLISM.   The adrenal steroids have a particularly important role in the homeostatic regulation of electrolyte and water balance in the body, especially in regard to the normal maintenance of sodium and potassium concentrations in the extracellular fluids.

The chief mineralocorticoid secreted by the adrenal cortex is aldosterone, although corticosterone normally is secreted in sufficient quantities to exert a slight effect upon electrolyte balance, as indicated in Table 15.10. Deoxycorticosterone is not secreted in physiologically significant amounts except under abnormal circumstances. Although this compound exerts only three percent of the activity of aldosterone, synthetic deoxycorticosterone acetate (DOCA) is used clinically as a mineralocorticoid because of its availability and low price compared to aldosterone.

The normal functions of the adrenal mineralocorticoids may best be discussed in terms of an excess and deficiency of these hormones in the body. Aldosterone and deoxycorticosterone both stimulate an increased reabsorption of sodium (hence chloride) and bicarbonate from the distal (and to some extent from the proximal) renal tubules. A similar effect of these hormones upon these electrolytes is seen in the sweat glands, salivary glands, and mucosa of the gastrointestinal tract. An elevated level of either aldosterone or deoxycorticosterone produces an increase in the extracellular sodium concentration, with an attendant expansion of the extracellular fluid volume, as well as the bicarbonate ion concentration. At the same time, the serum potassium and chloride concentrations decline as shown in Table 15.11. The sodium retention causes a net exchange of intracellular potassium for extracellular sodium; the excess potassium then is excreted in the urine. Sodium ion is also mobilized from connective tissue into extracellular fluid compartments when excess adrenocortical steroids are present.

The increased extracellular fluid volume leads secondarily to hypertension. If severe and prolonged, this elevated fluid volume ultimately can produce frank edema and congestive heart failure, although sodium excretion usually increases despite the presence of excess mineralocorticoids after a certain stage in the development of an expanded extracellular fluid volume has been reached. The elevated loss of sodium, despite the presence of continued high levels of adrenocorticoids, is due principally to a decreased renal sodium reabsorption, although the mechanism involved in this process is unknown in detail.

An elevated adrenal mineralocorticoid level may augment urinary loss of other electrolytes, particularly calcium ion. In this instance the effect is due to the retarded protein synthesis that was described earlier. Hence, osteoporosis follows the decreased osteogenesis because the calcium that would be deposited in the bone is now excreted.

A deficiency of adrenal mineralocorticoids (ie, hypoadrenal corticalism), on the other hand, results in defective renal tubular reabsorption of sodium. The concomitant elevated urinary excretion of sodium, chloride, and water in turn leads to a decrease in the total extracellular fluid (thus plasma) volume. There is also a simultaneous increase in potassium retention, hence the plasma level of this ion rises (Table 15.11). The tissue potassium concentration also increases.

Since the fall in plasma (ie, extracellular fluid) sodium is greater than the quantity which is excreted in the urine during rapidly-developing ad-

renal insufficiency, then sodium now is entering the cells of the body.

Since the plasma volume is reduced in adrenal cortical insufficiency, hemoconcentration and increased blood viscosity result. These effects in turn lead progressively to cardiovascular insufficiency, including a decreased cardiac output and hypotension. Fatal shock ultimately ensues in untreated subjects (cf. p. 1237). Augmented dietary sodium chloride intake alone may ameliorate these effects to some extent, but human patients require such large quantities of this mineral that death ensues unless mineralocorticoid therapy also is provided.

Untreated adrenal cortical insufficiency also leads to metabolic acidosis which has renal as well as extrarenal causes. The renal effects of adrenal cortical hormone deficiency result from an abnormal excretion of hydrogen and ammonium ions, so that during periods of acid loading, the physiologic acidification of the urine is below normal. The extrarenal cause of metabolic acidosis is produced by the elevation of serum potassium. This ionic imbalance causes a shift of bicarbonate into the cells resulting in hydrogen ions moving out into the extracellular fluid, and acidosis develops.

The hyperkalemia that accompanies adrenal cortical insufficiency can also result in electrocardiographic abnormalities and even cardiac arrest in diastole can occur if the hyperkalemia is sufficiently pronounced. Furthermore, the loss of extracellular sodium that occurs during adrenal cortical insufficiency may cause hyperexcitability of brain tissue, as revealed by alterations in the electroencephalogram.

As an attendant to adrenal cortical insufficiency there is a net loss of body fluid by the kidney and a concomitant decrease of total plasma volume. The attendant hypotension that develops produces a lowered kidney perfusion rate so that renal failure can develop. Following renal failure, the blood urea, calcium, potassium, and inorganic phosphate ion concentrations all become markedly elevated. However, the renal excretion of ingested water is now inhibited so that a hypotonic increase (or expansion) in the extracellular fluid volume occurs following the drinking of copious volumes of liquid. This excess water ultimately becomes osmotically distributed among all of the body fluid compartments, and since the cells of the central nervous system are involved in this effect, convulsions due to the water intoxication, and even death, may result. Clinically, these effects are reversed partially in hypoadrenal states by the administration of excess sodium chloride coupled to a low potassium intake. Such a therapeutic approach tends to correct the mineral imbalance; however, the administration of exogenous adrenal mineralocorticoids is essential in order to achieve a true physiologic regulation of the defects in salt and water balance that are present in adrenocortical insufficiency.

It is evident from the foregoing discussion that

Table 15.11　PLASMA ELECTROLYTE LEVELS AND ADRENAL CORTICAL FUNCTION[a]

	ELECTROLYTE LEVELS, mEq/liter			
	Na$^+$	K$^+$	Cl$^-$	HCO$_3^-$
Normal	142	4.5	105	25
Adrenal Insufficiency	120	6.7	85	15
Primary Hyper-aldosteronism	148	2.4	96	41

[a]*Data from Ganong.* Review of Medical Physiology, *6th ed, 1973, Lange Medical Publications.*

the homeostatic regulation of salt and water balance in the normal individual is critically dependent upon the continual secretion of aldosterone, and to a lesser extent corticosterone, by the adrenal cortex.

EFFECTS OF GLUCOCORTICOIDS UPON THE BLOOD. Cortisol and corticosterone (but not aldosterone) exert certain actions upon the quantity of formed elements in the circulation, and these effects are best demonstrated by the exogenous administration of an excess of these hormones.

Glucocorticoid administration (eg, of cortisol) stimulates erythropoiesis, and also increases the circulating level of neutrophils and platelets. Thus, moderate anemia is a feature of adrenal cortical insufficiency, although in hypoadrenal states the eyrthropoietic response to hypoxia is unaffected. Bone marrow is stimulated directly by the glucocorticoid hormones.

On the other hand, exogenous glucocorticoid administration markedly decreases the blood levels of eosinophils, lymphocytes, and basophils. Glucocorticoids decrease the eosinophil level in the circulation by increasing the number of these cells which are stored in lungs and spleen. Cortisol depresses the activity of lymphoid and thymic tissue directly by inhibiting mitosis, and ultimately may cause the involution of such tissues. These hormones also increase the rate of destruction of lymphocytes. Conversely, adrenal cortical hypofunction results in hypertrophy of the lymphoid tissue so that a lymphocytosis develops.

SECRETORY ACTION OF THE GLUCOCORTICOIDS. Cortisol and corticosterone directly stimulate the secretion of hydrochloric acid and pepsin by the stomach. Pituitary corticotropin also exerts a similar effect; however, the gastric response to this hormone is mediated indirectly via the adrenal cortex, hence increased glucocorticoid hormone release.

The glucocorticoids also reduce the quantity of gastric mucus that is secreted. Since the mucus is important in providing a physiologic barrier against the deleterious effects of acid and pepsin on the stomach itself, the presence of excess glucocorticoids can dispose a patient toward the development of gastric ulcers, and such ulcerative lesions may be encountered clinically following prolonged glucocorticoid therapy.

The adrenal glucocorticoids also stimulate trypsinogen secretion by the pancreas.

In patients suffering from Addison's disease, the pernicious anemia that is present may be caused by gastric hyposecretion with a concomitant insufficiency of secretion of intrinsic factor by the mucosa of the stomach. Thus, cyanocobalamin (vitamin B_{12}) absorption would be impaired, and as a consequence, erythropoiesis would become defective.

GLUCOCORTICOID EFFECTS ON INFLAMMATORY AND ALLERGIC REACTIONS. Cortisol and a number of synthetic steroids are effective in preventing (or inhibiting) inflammatory phenomena regardless of whether these reactions are caused by a physical, chemical, or bacterial stimulus (Table 15.12). Such an effect can be elicited by either local or systemic administration of the steroid. These antiflammatory compounds prevent the local accumulation of polymorphonuclear leucocytes as well as the local activity and destruction of fibroblasts that occur during any inflammatory process (or reaction). In addition, glucocorticoids reduce local swelling (or edema) and also inhibit the systemic effects of bacterial toxins.

The decreased local inflammatory response that is caused by steroids is due perhaps to an inhibition of kinin release within the affected tissue. Also, there is experimental evidence indicating that the glucocorticoids stabilize the membranes of the lysosomes so that the breakdown of these structures during the inflammatory tissue reaction is inhibited.

Large quantities of exogenous glucocorticoids administered to humans elevate the titer of circulating antibodies at first, and then subsequently depress this titer. The inhibition of fibroblastic activity by glucocorticoids that was noted above prevents the localization of bacterial infection as well as the formation of adhesions after surgery.

Insofar as allergic diseases are concerned, glucocorticoids depress the symptoms of those conditions caused by histamine release from the tissues. Thus, when some antibodies react with their antigens, histamine is released, and this compound in turn produces a number of allergic symptoms. The glucocorticoids neither prevent the antigen–antibody reaction nor do they interfere with the action of this substance once it is released; however, they do prevent the initial tissue release of histamine.

It should be stressed that the pharmacologic reactions of the glucocorticoids discussed above are not mediated by the adrenocortical hormones when they are secreted in physiologic quantities. Rather these actions, together with other symptoms of glucocorticoid excess, are noted following administration of large doses of natural (or synthetic) adrenal steroid compounds. Thus there are two major dangers inherent in the long-term therapeutic administration of glucocorticoids. First, these adrenal hormones inhibit the normal secretion of corticotropin by the pituitary gland to such an extent that acute adrenal cortical insufficiency can develop when therapy is discontinued. Second, in patients suffering from active bacterial infections (eg, tuberculosis or pneumonia), glucocorticoid therapy may alleviate markedly the febrile, toxic, and lung symptoms of the disease, while concomitantly permitting the bacteria to spread throughout the body. The infection spreads and can kill the patient unless appropriate antibiotic or other therapy is administered simultane-

Table 15.12   COMPARISON OF THE EFFECTS OF CORTISONE WITH CERTAIN OTHER STEROIDS[a]

NAME OF SUBSTANCE	ANTIINFLAMMATORY POTENCY	SODIUM RETENTION	GLYCOGENIC POTENCY
Naturally Occurring Compounds			
Cortisone	1	1	1
Cortisol	1.25	1.5	1.5
Deoxycorticosterone	0	30–50	0
Aldosterone	0	300–600	0
Synthetic Compounds[b]			
Δ′-Cortisol	4	slight effect	3–5
9-α-Fluorocortisol	10–15	200–400	10–15
2-Methyl-9-α-Fluorocortisol	10	1,000–2,000	10
16-α-Methyl-9-α-fluoro-Δ-cortisol	25–35	slight effect	25–30

[a]*Data from White, Handler, and Smith.* Principles of Biochemistry, *4th ed, 1968, The Blakiston Division, McGraw-Hill.*

[b]*Many synthetic steroids have been prepared in order to produce compounds having a greater potency and more selective physiologic effects than the natural adrenocortical hormones. Some of these compounds exhibit a far greater activity than cortisol, as shown above, and a degree of selectivity has been achieved. Only four examples of the many synthetic steroids have been cited in this table.*

ously with the glucocorticoids, as the latter mask the symptoms caused by the underlying disease process.

Nonetheless, glucocorticoids do have an important therapeutic role in the treatment of allergic conditions of the skin and eye, hypersensitivity reactions, certain mesenchymal tissue diseases including rheumatoid arthritis, anaphylatic shock, and other allergic and autoimmune disease states. However, the possibility of serious adverse reactions to exogenous steroid administration must be recognized clearly.

STRESS AND ADRENAL GLUCOCORTICOIDS. Normal individuals are able to adapt successfully to an extremely wide variety of noxious stimuli by an increased corticotropin secretion, which in turn stimulates an elevated adrenal glucocorticoid secretion. This homeostatic rise in circulating glucocorticoid levels is essential to survival. For example, an adrenalectomized or hypophysectomized animal or person is unable to tolerate such stimuli as physical trauma, cold, hemorrhage, antigenic substances, infection, or chemical agents at levels that would not seriously impair function of a normal subject. Such hormone-deficient individuals also exhibit a markedly abnormal sensitivity to certain hormones, such as insulin or thyroxine.

It is evident that noxious stimuli sharply increase the bodily requirements for adrenal cortical hormones, although the mechanisms whereby the effects of these hormones are mediated are unclear for the most part. An increased pituitary corticotropin secretion that follows stressful stimuli generally is accompanied by an augmentation of the activity of the sympathoadrenal medullary system and certain of the effects of adrenal glucocorticoids in stress situations perhaps may reside in the maintenance of vascular reactivity to the

catecholamines. Furthermore, the catecholamines are unable to exert their full capacity to mobilize free fatty acids as an emergency energy source in the absence of adrenal glucocorticoids, and part of the mechanism whereby the adrenal steroid hormones exert their actions may reside in their action with the catecholamines. In this regard, it is of interest that sympathectomy alone does not alter the tolerance of an animal to a number of stress conditions to as great an extent as adrenalectomy does.

Mechanism of Action of the Adrenal Hormones

The broad and complex spectrum of physiologic and biochemical responses produced by the adrenal steriods requires a clarification of the primary and secondary roles of these hormones insofar as their mechanisms of action are concerned. At present no single primary mechanism of action that underlies all of these actions can be stated with certainty. However, it is quite possible that the earliest common biochemical denominator underlying the manifold steroid effects in vivo resides in the action of the steroids in stimulating the synthesis of a specific protein within the target cell. It has been shown experimentally that aldosterone combines with the nuclei of renal cells, and there induces the synthesis of an enzyme (as yet unidentified) that facilitates sodium ion transport.

Cortisol and other glucocorticoids appear to act upon a DNA-dependent RNA synthetic mechanism, with the result that the synthetic rates of certain specific cellular proteins are modified profoundly. The earliest demonstrable biochemical change in lymphoid tissue following exposure of such tissue to a steroid is a decrease in RNA polymerase activity as well as glucose utilization by the cells. Conversely, a sharp increase in RNA polymerase activity is found in liver and other target cells under similar experimental conditions. The hepatic synthesis of other specific enzymes (eg, pyruvate carboxylase, tyrosine–glutamate transaminase) are also stimulated by cortisol. Current experimental data show that the action of adrenal steroids in the liver may be at (or following) an aminoacyl-tRNA translation step or else at the level of DNA transcription that is governed by an RNA polymerase system; an elevated rate of cyclic AMP synthesis also appears to be involved. Both of these possibilities, however, require an effect of steroids upon the rates of synthesis of specific proteins. However, as noted earlier, cortisol inhibits protein synthesis in lymphoid tissue, and enhances this process in hepatic and other tissues. At present there is no satisfactory or consistent explanation forthcoming as to how the same hormone is able to affect similar processes in two tissues in diametrically opposite directions at the same time!

Clinical Correlates

In this section several conditions that involve adrenocortical malfunction in humans will be discussed. It must be stressed that the clinical picture in such dysfunctions usually reflects the physiologic and biochemical changes that result from the principal effects of one predominant adrenocortical hormone. Thus abnormally elevated or depressed secretion of each one of the adrenal steroids results in a typical syndrome. On the other hand, if the primary etiologic factor responsible for producing adrenal hypersecretion or hyposecretion lies outside of the adrenal glands per se, ie, in the adenohypophyseal secretion of corticotropin, then the effects upon cortical secretion will be less specific insofar as the nature of the clinical picture is concerned.

This survey of certain effects of adrenal hyperfunction and hypofunction will also serve as a convenient way to summarize and review the complex actions of the hormones elaborated by the adrenal cortex.

GLUCOCORTICOID HYPERSECRETION (CUSHING'S SYNDROME). The numerous physiologic and metabolic abnormalities that result from excess cortisol production are known as Cushing's syndrome.

Etiology. The underlying defect responsible for excess cortisol secretion by the adrenal gland may reside in any of several tissues.

First, about 70 percent of the patients having Cushing's syndrome exhibit bilateral adrenal hyperplasia; in particular, the zona fasciculata of the cortex is widened. The cause of the adrenal hyperplasia generally is considered to reside in the adenohypophysis, with excessive secretion of corticotropin being the principal cause of the hyperplasia. In about 33 percent of such patients, a small to large pituitary adenoma is present (Cushing's disease*); most of the remainder of such individuals exhibit an increase in the population of basophil cells that has an obscure etiology.

Second, approximately 25 percent of the patients with Cushing's syndrome have glucocorticoid-secreting primary adrenal tumors. These may be adenomas and encapsulated, or carcinomas, and metastatic or invasive.

Third, the remaining 5 percent of patients having spontaneous Cushing's syndrome are associated with carcinomas of the lung, testis, ovary, thymus, exocrine and endocrine pancreas, as well as the chromaffin and argentaffin systems. Material exhibiting a corticotropinlike biologic activity may be extracted from some of these tumors, and it is believed that this substance in turn stimulates excess glucocorticoid production by the adrenals.

It also is important to mention that long-term therapeutic administration of cortisol can also result in the development of Cushing's syndrome.

Clinical Aspects. A patient with Cushing's syndrome presents a caricature of the physiologic actions of cortisol, and the clinical manifestations of this condition may range from quite subtle to most obvious as discussed earlier. The principal features of this condition may be reviewed briefly as follows. There is a central redistribution of fat which results in a characteristic change in the overall appearance. The extremities are thin; however, fat accumulates in the abdominal wall, face ("moon face"), and upper back; the latter gives rise to the so-called "buffalo hump" appearance. The abdominal skin is stretched by the increased mass of adipose tissue deposited in this region, and the subdermal tissues rupture, resulting in large purplish-red striae.† Hirsutism is a common finding in patients with Cushing's syndrome.

The patient suffers from increased protein catabolism; therefore, the skin and subcutaneous tissues are thin, and the skeletal muscles are poorly developed and wasted. Wound healing is poor, bruises and ecchymoses (discolorations due to extravasation of blood) are readily induced, and muscular weakness is common following the muscular wasting observed most commonly in the proximal muscle groups, especially in the gluteal region.

Osteoporosis develops as a consequence of decreased protein matrix formation combined with increased matrix catabolism in bone, so that the vertebrae collapse and the neck of the femur frac-

---

Cushing's disease *is caused by a primary adenoma of the pituitary, whereas Cushing's* syndrome *may or may not be caused by such an adenoma.*
†*Such scars occur normally when rapid stretching of the skin occurs as during pregnancy or around the breasts of girls at puberty; however, such physiologic striae are not pronounced and lack the intense coloration seen in Cushing's syndrome.*

tures readily. Skeletal deformities result. In addition, high levels of glucocorticoids also have an inhibitory effect upon the normal action of vitamin D, and elevate the glomerular filtration rate; therefore calcium excretion is increased.

An abnormal glucose tolerance test is seen in about 88 percent of patients with Cushing's syndrome, and overt, insulin-resistant diabetes is present in approximately 20 percent of such individuals. Ketosis and diabetic coma are rare, although the hyperglycemia may be refractory to the usual therapeutic measures. Thus, a frank diabetes mellitus may develop, especially in individuals who are genetically predisposed to this condition. Quantities of the amino acids released by the elevated protein catabolism are converted into glucose within the liver, and this factor combined with the depressed peripheral utilization of glucose, is sufficient to precipitate the diabetes.

In growing children, the peripheral catabolic actions of cortisol result in a failure of growth.

Malfunction of the reproductive glands can range from slight (in adrenal adenoma) to prominent (in adrenal carcinoma), although marked variations from these states are seen. Sterility may be found in adult females together with an outright amenorrhea, or else a disturbed menstrual rhythm. Occasionally masculinization (or virilization) of the female patient is sufficient to confuse the diagnosis of Cushing's syndrome with that of the adrenogenital syndrome.

Changes in electrolyte metabolism are seen, as the quantities of glucocorticoids secreted in Cushing's syndrome are sufficient to exert pronounced mineralocorticoid effects. Furthermore, if Cushing's syndrome is caused by excess corticotropin secretion because of failure of the normal feedback regulation of this pituitary hormone, then corticosterone is also secreted in abnormal quantities together with cortisol. The resulting salt (sodium) and attendant water retention in turn lead to edema, and the concomitant potassium depletion contributes to the muscular weakness. The ensuing hypokalemic alkalosis may also result in tetany. Approximately 85 percent of patients with Cushing's syndrome exhibit hypertension because of the net increase in plasma volume that ultimately can lead to congestive heart failure or cerebrovascular accident.

Accelerated electroencephalographic rhythms are also seen in Cushing's syndrome, and mental abberrations are common. These symptoms may include insomnia, increased appetite, euphoria combined with rapid changes (or swings) in mood to depression, mania, and frank psychoses with suicidal tendencies. The mental alterations found in glucocorticoid excess are more severe than those seen when a deficiency of these hormones is present.

Miscellaneous symptoms of Cushing's syndrome include peptic ulcers, skin pigmentation with a pattern resembling that found in Addison's disease, renal calculus formation, and exophthalmos.

It is sometimes difficult to distinguish between a mild state of Cushing's syndrome and simple obesity. However, in obesity, the extremities are obese in contrast with the emaciated limbs seen in Cushing's syndrome. Muscular weakness, ecchymoses, osteoporosis as indicated by radiologic evidence, and hypokalemic alkalosis are the most significant general diagnostic clues, although these findings must be confirmed by laboratory tests that show some degree of elevated autonomous cortisol secretion.

MINERALOCORTICOID HYPERSECRETION OR PRIMARY ALDOSTERONISM (CONN'S SYNDROME). Hypersecretion of the mineralocorticoid aldosterone by the adrenal cortex leads to the condition of primary aldosteronism or Conn's syndrome.

Etiology. The principal etiologic factor in primary aldosteronism is an adrenal adenoma (in about 90 percent of patients), so that the secretion of the hormone no longer is subject to the normal homeostatic regulatory mechanisms discussed earlier. Conn's syndrome is found predominantly in female patients, at a ratio of around 3:1 over males.

Clinical Aspects. Hypertension, generally but not always benign, accompanies primary aldosteronism, and this is coupled with an increase in the extracellular fluid volume. Hypernatremia is slight because of the so-called escape phenomenon described on page 801, which lowers the blood sodium content despite the elevated aldosterone secretion. For this reason also, no dependent edema usually is present in primary aldosteronism, despite the retention of excess water.

There is a severe potassium loss in primary aldosteronism, and ultimately this potassium depletion may so damage the kidney that there is a reduction of its ability to concentrate electrolytes and other substances, and this renal defect may be combined with a polyuria. Hypokalemic alkalosis with tetany, fatigue, and skeletal muscle weakness are also seen, and paralysis may eventually develop in some patients having Conn's syndrome.* The potassium deficiency also results in a slight, but demonstrable, decrease in glucose tolerance, a defect that is rectified by potassium therapy.

Electrocardiographic signs of primary hyperaldosteronism may include a normal or slight prolongation of the PR interval, a depressed ST segment, inverted T waves, and prominent U waves. All of these electrocardiographic signs revert to normal when potassium repletion is provided therapeutically.

Miscellaneous symptoms of Conn's syndrome include visual disturbances, headache, vomiting, and cerebrovascular accident. Polydipsia develops in consequence of the polyuria. Enlargement of the

---

*The alkalosis secondarily reduces the plasma ionized calcium level to such an extent that a latent or overt tetany develops.

heart *(cardiomegaly)* is found in less than 50 percent of the cases, but this condition may develop secondarily to a prolonged hypertensive condition.

ADRENOGENITAL SYNDROMES. Under physiologic circumstances, the androgenic and estrogenic substances normally secreted by the adrenal cortex (Fig. 15.39) play no significant role in the development and maintenance of typical male and female sexual characteristics. If, however, the secretion of these substances reaches abnormal levels, typical masculinizing or feminizing effects result, and these effects depend upon both sex and age of the patient, as well as on the type of hormone elaborated. It is important to stress that the adrenal secretion of these sex steroids is not regulated by gonadotropins, but rather by corticotropin. Therefore, alterations in pituitary secretion of corticotropin can affect profoundly the elaboration of sex hormones by the adrenal cortex (Table 15.13).

Estrogens. It appears that estradiol is synthesized and secreted by the adrenals. In ovarectomized females, corticotropin stimulates a rise in urinary estrogen levels, and adrenalectomy causes a decrease in the secretion of these substances. Under normal circumstances, however, the quantity of estrogens elaborated by the adrenal glands is far too small to exert any appreciable physiologic effects in individuals of either sex. Tumors that secrete feminizing quantities of estrogenic substances in males have been reported, although such a pathologic condition is of rare occurrence. Furthermore, female patients having estrogen-dependent cancer of the breast show improvement following either adrenalectomy or glucocorticoid therapy that is sufficient to inhibit corticotropin secretion. Precocious puberty caused by the secretion of abnormal quantities of estrogenic substances from the adrenals may also be seen, but this phenomenon is quite rare.

Androgens. By far the most common clinical abnormalities of adrenal sex steroidogenesis are those that produce masculinization (or virilism) in the female.

Routes of Androgen Synthesis. The pathways of steroid synthesis in the adrenal cortex are so arranged that androgen secretion is enhanced when steroid synthesis is abnormal, as depicted in Figure 15.39. As shown in this illustration, the major aspects of this metabolic scheme are such that corticotropin exerts its principal action at the initial steps that are involved in steroidogenesis. Thus, corticotropin stimulates the 20, 22-dihydroxylation of cholesterol, accelerating the conversion of this compound into pregnenolone (Δ5). Beyond this compound, several major branches of steroid synthesis can take place. The first of these branches leads to androgen synthesis. Pregnenolone and progesterone provide substrates for the 17-hydroxylase enzyme, and in turn the derivatives of these compounds are good substrates for the desmolase enzymes that split the carbon 20, 21 side chain and produce dehydroepiandrosterone (DHEA) and androstenedione. Both of these steroids in turn are weakly androgenic; however, in liver and other tissues they are readily converted into testosterone, dihydrotestosterone, and the biologically active androstanediols.

Parallel to this double pathway leading to androgen synthesis lies the sulfate pathway of the adrenal gland. The steroids conjugated with sulfur that are secreted by the adrenal are biologically inactive; however, these compounds are hydrolyzed readily in peripheral tissues to become active. Thus both pregnenolone and 17-hydroxypregnenolone are sulfated easily on the 3-hydroxyl position, and then are protected from conversion into progesterone and 17-hydroxyprogesterone. However, reactions of the carbon 20, 21 are not affected by this conjugation, so that dehydro-

Table 15.13 SOME DEFECTIVE ENZYME PATTERNS FOUND IN THE ADRENOGENITAL SYNDROME[a]

DEFECT	PRECURSOR WHICH ACCUMULATES	CLINICAL ASPECTS
Cholesterol desmolase (?)	Cholesterol	Incompatible with life
3-β-ol-dehydrogenase	Δ-5-Pregnenolone-DHEA	High mortality rate; virilism in utero
21-Hydroxylase	Progesterone 17-Hydroxyprogesterone	Mild virilism Severe virilism; salt loss
11-Hydroxylase	Desoxycorticosterone	Virilism; hypertension
18-Oxidase	Corticosterone 18-Hydroxycorticosterone	Salt loss
17-Hydroxylase	Corticosterone Desoxycorticosterone	Hypertension Defective virilization Deficiency in estrogens

[a]Data from Harvey, Johns, Owens, and Ross (eds). The Principles and Practice of Medicine, *18th ed,* 1972, Appleton-Century-Crofts.

epiandrosterone forms as an end product of the sulfate pathways, and this steroid becomes an important component of the urinary 17-ketosteroids. In pregnant women, DHEA is converted to estradiol by the placenta; however, in nonpregnant individuals this steroid may be converted into active androgens by hydrolysis in the peripheral tissues. The sulfokinase enzymes responsible for this process seem to become active only at the onset of normal puberty, which explains why certain forms of the adrenogenital syndrome appear during, or shortly following, puberty.

Since the androgenic pathways lie first upon the sequence of metabolic reactions that lead to steroid formation, any defects lying beyond these initial steps, which defects induce a rise in corticotropin levels (via the normal feedback mechanism), are potentially capable of stimulating androgen synthesis and secretion by the adrenal cortex. In actual fact, a wide variety of specific and inherited defects in steroid synthesis are recognized, and each of these results in specific symptoms that depend upon the particular biochemical locus involved in the defect. Certain of these biochemical lesions and the resulting clinical defects are summarized in Table 15.13.

Clinical Aspects of the Adrenogenital Syndromes. Only rarely will an adrenal tumor induce feminization in the male, with a consequent overdevelopment of the mammary glands (or gynecomastia) and impairment of the libido. In the sexually immature female, such an estrogen-secreting tumor may induce precocious puberty, with breast enlargement, premature menstruation, and growth of pubic hair.

By far the commonest clinical manifestation of the adrenogenital syndrome is that in which virilism of the female is seen. This masculinization is found in prepubertal as well as in adult females, and it results in the development of a number of symptoms and characteristics typical of the normal male, including a tendency to baldness and a receding hairline; growth of a beard; small breasts; heavy, muscular arms and legs; an enlarged clitoris; amenorrhea; and in general, the pattern of hair distribution on the body and extremities resembles that in the normal male.

In the genetically female fetus prior to the 12th week of gestation, various degrees of female pseudohermaphroditism may occur due to excess adrenal androgen secretion, with an attendant development of genitalia having male characteristics.

On the other hand, adult males who suffer the effects of excess adrenal androgens merely undergo an overdevelopment of the normal preexisting male characteristics. In boys, prior to the onset of puberty, however, the adrenogenital syndrome is manifest by precocious development of the secondary sex characteristics, including hypertrophy of the penis, growth of hair on the face, chest, and pubic regions, but without an attendant normal testicular growth. In addition, individuals who exhibit such precocious pseudopuberty also have

adult sexual impulses, and these manifestations of the adrenogenital syndrome may occur in male children as young as three years of age.

ADRENOCORTICAL INSUFFICIENCY (ADDISON'S DISEASE). Hypofunction of the adrenal cortices resulting in Addison's disease may be caused by idiopathic atrophy (ie, without known cause), or by specific diseases of these organs, including tuberculosis, meningococcal infections, staphylococcal septicemia, and secondary carcinomas that replace the normal cortical tissues. Furthermore, hemorrhage and necrosis are often seen in the adrenals following acute infections.

Adrenocortical failure may be chronic and develop slowly so that there is a relative insufficiency of the cortical hormones, or else the failure may be acute, in which case a critical deficiency of cortical hormones results in early death unless adequate therapy is instituted rapidly following accurate diagnosis.

It should be stressed that the symptoms of Addison's disease result from a nonspecific deficiency of all of the adrenocortical hormones, and this deficiency may range in severity from a slight decrease to a complete absence of these steroids. Consequently, the symptoms and complications themselves are of graded severity.

The clinical features of Addison's disease include the following signs and symptoms. A gradual decline in general health occurs and this is accompanied by languor, weakness, and an indisposition to perform any physical or mental exertion. Strength and work capacity decrease progressively throughout the day, so that the weakness becomes exaggerated by evening. The patient also may exhibit signs of irritability, restlessness, and drowsiness. Characteristically, weight loss occurs, together with loss of appetite in Addison's disease; nausea, vomiting, abdominal pain, and diarrhea are also of common occurrence. Prostration may follow a minor infection. All of the aforementioned signs and symptoms develop as consequences of cortisol deficiency, so that this hormone is often called the "hormone of well-being."

In Addison's disease there are deficiencies in mineralocorticoid as well as glucocorticoid secretion by the adrenal glands. A deficit of aldosterone in this condition results in a loss of salt and water from the body, and this factor, coupled with the hypoglycemia that results from cortisol deficiency thus can produce prominent symptoms. The salt deficit (or "salt wasting") may result in muscle cramps, and a salt craving often occurs, especially in children. The hypoglycemia and hyponatremia result in dizziness and syncope, and the hypoglycemia seen in Addison's disease generally, but not invariably, occurs after a period of fasting.

A characteristic hyperpigmentation of the skin is seen in chronic Addison's disease. This pigmentation probably results from the combined effects of an excess secretion of melanocyte-stimulating hormone (MSH), and a decrease in the cortisol level. A diffuse tan may be present over any region of the body, giving the skin a dirty appearance that is especially prominent at pressure points, ie, the knees and elbows. Recent scars, the areolae, the anogenital region, and skin creases all are common sites of excessive pigmentation. The mucus membranes also develop a bluish black pigmentation, eg, on the lips, gums, buccal, rectal, and vaginal surfaces. In patients having autoimmune adrenal insufficiency there also may be skin areas that are totally lacking any pigment (vitiligo).

The blood pressure is low in patients with Addison's disease (unless preexisting hypertension is present) and the heart is small. Postural hypotension is readily demonstrable in such individuals.

The wasting of muscles is often severe and weakness of these structures is a general finding. Accompanying the elevated blood potassium level an outright paralysis may occur, together with loss of tendon reflexes.

The axillary and pubic hair become sparse in females, but these secondary sex characteristics in males are not particularly affected. The testes and prostate also usually remain unaffected.

Rarely calcification of the pinnae occurs, and this finding may be related to the occasional hypercalcemia found in Addison's disease.

Laboratory findings in adrenocortical insufficiency include the electrolyte imbalances noted above, hemoconcentration, and hypoglycemia. Since the glomerular filtration rate is reduced, the patient is unable to excrete (or even to tolerate!) a water load. Thus water intoxication may be induced by using such a procedure during diagnostic evaluation of a patient. Finally, a low 24-hour excretory rate of adrenal steroids (ie, 17-ketosteroids and 17-hydroxycorticosteroids) combined with a lack of any response to corticotropin administration is useful diagnostically.

The symptoms of acute adrenocortical insufficiency may result from a lack of, or inadequate treatment of, Addison's disease. Hence a so-called Addisonian crisis may follow minor infections or surgery and this situation must be considered a clinical emergency.

An Addisonian crisis is characterized by prostration, fever, nausea, vomiting, cardiovascular collapse and shock and dehydration, cyanosis, and pulmonary edema. Death ensues rapidly unless prompt therapeutic measures are instituted. The underlying factors responsible for the precipitation of an Addisonian crisis include a marked sodium and water loss from the body, resulting in depletion of these substances. These factors are coupled with a potassium intoxication that results from hyperkalemia. In addition, acidosis and hypoglycemia also accompany the crisis. This condition is reversed rapidly by the immediate intravenous administration of glucocorticoids, glucose, and isotonic saline. Plasma may also be used to counteract the hypotension and shock state that develops as a concomitant of the Addisonian crisis.

THE ADRENAL MEDULLA

The medulla of the adrenal gland is an entirely different endocrine organ from the adrenal cortex both from the standpoint of morphology as well as physiology. Furthermore, both of these structures are derived independently and from totally different tissues in the embryo.

For these reasons, the morphologic, biochemical, and physiologic properties of the adrenal medulla are discussed separately from those of the adrenal cortex, despite the fact that both of these organs are intimately related anatomically and also share a common vascular supply.

Unlike the adrenal cortex, the mammalian adrenal medulla is not essential to life, and thus may be extirpated without producing deleterious effects, insofar as survival is concerned.

Functional Morphology of the Adrenal Medulla

GROSS ANATOMY.   The adrenal medulla in adult humans and other mammals is invested closely by the cortical tissue (interrenal tissue). However, in other vertebrates the two structures either may be entirely separate or else interrelated in a number of ways. In a transected fresh adrenal gland, the medullary tissue appears grayish in sharp contrast to the yellow cortex (Fig. 15.35).

HISTOLOGY.   Under the light microscope, the boundary between the zona reticularis and the medulla generally is irregular, and columns of cortical cells may extend for some distance into the medullary tissue, as shown in Figure 15.36.

The medulla is comprised of irregularly shaped epithelioid cells that are arranged in cords or clusters, and that are surrounded by capillaries and venules.

If adrenal medullary tissue is fixed in solutions containing bichromate ion, the medullary cells are seen to be filled with fine brown granules. This is the chromaffin reaction, and presumably this reaction is caused by the action of chromic ion on the catecholamines in the granules of the medullary cells so that oxidation and polymerization of the hormones epinephrine and norepinephrine within the cytoplasmic granules takes place. Similarly, the medulla is stained green following treatment with ferric chloride.

The chromaffin reaction is seen in other tissues of the body, principally in the argentaffin cells of the gastrointestinal tract and the mast cells. These cells give this reaction because of the 5-hydroxytryptamine and dopamine that they contain. The adrenal medullary chromaffin cells, however, originate from neural ectoderm in the embryo, secrete catecholamines, and are innervated by sympathetic preganglionic fibers.

Special histochemical techniques reveal that

two different cell types are present in the adrenal medulla, one of which contains epinephrine, the other norepinephrine. Furthermore, electron microscopy of the adrenal medulla reveals that the cells contain abundant membrane-limited dense granules, having diameters ranging between 50 and 350 nm. The granules of certain cells exhibit a rather low electron density and are considered to contain epinephrine, whereas another cell population contains very high density granules that are believed to consist of norepinephrine.

The cisternae of the granular endoplasmic reticulum are arranged in parallel systems. The Golgi apparatus is located in a juxtanuclear position, and its cisternae contain electron dense material that is believed to represent a precursor of the hormone granules.

The secretory mechanism whereby the adrenal medullary hormones are released into the bloodstream is not understood fully. Some investigators feel that this process is similar to the secretion of exocrine products, whereas other workers believe that the chromaffin granules are secreted intact, and the catecholamine then diffuses out of these "packets" of hormone through the membrane. Alternatively, an active transport process may liberate the hormone from the granules into the cytoplasm and thence the hormone passes out of the cell.

MEDULLARY BLOOD SUPPLY AND INNERVATION. Certain features of the adrenal cortical blood supply may be reviewed here briefly together with a consideration of the medullary blood supply. As discussed previously, the rich blood supply to the adrenal gland is provided by a number of arteries that penetrate the capsule of the organ, and give rise to a capsular plexus. This plexus in turn gives rise to a system of cortical arteries, and these vessels are distributed to the anastomosing network of sinusoids that invest the cords of cortical parenchymal cells. The sinusoids within a specific region of the cortex then converge upon a collecting vein at the corticomedullary junction, and as noted earlier there is no venous drainage from the adrenal cortex.

Some branches of the adrenal arteries, in contrast to the cortical vessels, pass through the trabeculae and then traverse the cortical tissue with little or no branching until they reach the medulla. Within the medulla, these vessels now branch repeatedly and thus form the abundant capillary network that surrounds the cords and clumps of medullary chromaffin cells. Consequently the adrenal medulla has a double (or dual) blood supply. First, blood comes from the sinusoids of the cortex that anastomose with the medullary capillary bed at the corticomedullary junction, and second, blood also comes from the medullary arteries that pass directly from capsule to medulla.

The medullary capillaries drain into the same venous system as do those of the cortex. Ultimately the numerous adrenal venules form large central veins within the medulla which anastomose and emerge from the capsule as the suprarenal vein.

The cells that line the medullary capillaries are composed of typical endothelial cells.

The abundant capsular nerve plexus contains some sympathetic ganglion cells; branches of these nerves pass through the cortex in the trabeculae and thence run directly, and almost exclusively, to the medulla. These preganglionic fibers terminate in multiple endings that surround individual medullary cells. In actual fact, therefore, the adrenal medulla may be considered from a functional standpoint to be a physiologic extension of the sympathetic nervous system (see also below).

EMBRYOLOGIC CONSIDERATIONS. As noted earlier, the adrenal cortex develops from coelomic mesoderm located on the medial side of the urogenital ridge. The adrenal medulla, on the other hand, arises from ectodermal tissue of the neural crest (Figs. 7.2, 7.4, and 7.5). This tissue also gives rise to the cells of the sympathetic ganglia. The sympathochromaffin cells of the neural crest that are destined to form the adrenal medulla migrate ventrally and penetrate the analgen of the adrenal cortex medially, so that ultimately they assume a central position within the primitive adrenal gland.

CHROMAFFIN SYSTEM (OR PARAGANGLIA). Within the body are a number of broadly scattered accumulations of cells that appear to share many common features with the adrenal medullary cells. These groups of cells collectively are termed the paraganglia or chromaffin system of the body. Included in this category are the organs of Zuckerkandl (which is located in the retroperitoneum) as well as similar cell groups found in the heart, liver, testis, and ovary. These paraganglia all contain cells that exhibit a typical chromaffin reaction described above, and these cells also stain with ferric ion. Furthermore, the paraganglia contain cells that are usually arranged in cords, and also possess a rich vascular supply. However, a definite endocrine role has not been demonstrated as yet for this group of tissues, although they are known to synthesize catecholamines.

Hormones of the Adrenal Medulla

CHEMISTRY. Two compounds having endocrine activity have been isolated from extracts of adrenal medullary tissue. These two hormones are the catecholamines epinephrine (adrenaline) and norepinephrine (or arterenol), and their structural formulas are given in Figure 15.45. Both of these compounds are synthesized, stored, and secreted by the adrenal medullary tissue under physiologic conditions.

There is a considerable difference among various species of animal insofar as the relative pro-

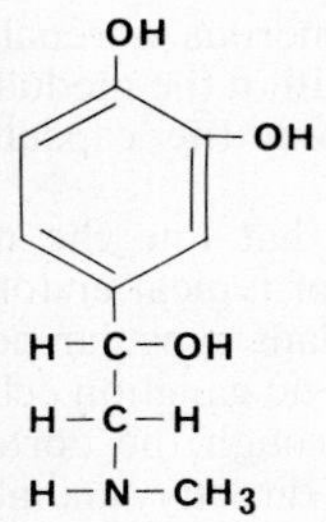

OH

OH

H—C—OH

H—C—H

H—N—CH₃

A. NOREPINEPHRINE

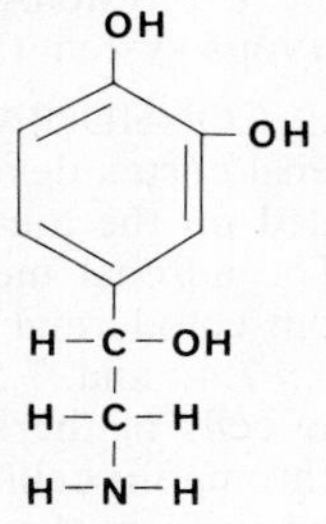

OH

OH

H—C—OH

H—C—H

H—N—H

B. EPINEPHRINE

FIG. 15.45. Structural formulas of the catecholamines. (After White, Handler, and Smith. *Principles of Biochemistry*, 4th ed, 1968, The Blakiston Division, McGraw-Hill.)

portions of these catecholamines in the adrenal medulla is concerned. For example, the adrenal gland of man and dog contain and secrete approximately 80 percent epinephrine and 20 percent norepinephrine, whereas cat adrenals secrete around 90 percent norepinephrine. On the other hand, the whale adrenal secretes norepinephrine exclusively. Norepinephrine may also enter the circulation from adrenergic nerve endings, as discussed in Chapter 6.

BIOSYNTHESIS OF THE ADRENAL MEDULLARY HORMONES. The biosynthetic pathway for the catecholamines by the adrenal medulla is summarized in Figure 15.46. The precursors of epinephrine and norepinephrine are the amino acids phenylalanine and tyrosine. The enzyme phenylalanine hydroxylase (Fig. 15.46, ①, which catalyzes the conversion of phenylalanine to tyrosine, is found in the liver but not in the adrenal gland.

The oxidation of tyrosine to 3,4-dihydroxyphenylalanine (or DOPA) in the adrenal glands is catalyzed by tyrosine hydroxylase (②) and enzyme similar to, and which uses the same cofactors as, phenylalanine hydroxylase.

The aromatic L-amino acid decarboxylase that catalyzes the decarboxylation of 3,4-dihydroxyphenylalanine to 3,4-dihydroxyphenylethylamine (hydroxytyramine or dopamine, ③) requires pyridoxal phosphate, and is found not only in adrenal medullary tissue, but also in kidney and liver.

Dopamine is converted into norepinephrine by 3, 4-dihydroxyphenylethylamine-β-hydroxylase (④). This nonspecific oxidase enzyme requires copper and catalyzes the direct addition of oxygen to the β-carbon atom. The methylation of norepinephrine to produce epinephrine is catalyzed by phenylethanolamine-N-methyl transferase (⑤). S-adenosylmethionine may (or may not) be the methylating agent in vivo, and the activity of this transferase is regulated indirectly by the pituitary–adrenocortical mechanism in that the excess adrenocortical steroid production stimulated by corticotropin depresses the activity of this enzyme. Consequently, the adrenal cortical steroids may exert some of their physiologic effects via their action upon the adrenal medulla by altering the enzymatic methylation of norepinephrine to epinephrine (Fig. 14.46, ⑤).

The hydroxylation of tyrosine to 3,4-dihydroxyphenylalanine (DOPA) is the rate-limiting step in the biosynthesis of the adrenal medullary hormones.

This pathway for the production of norepinephrine in the adrenal medulla is quite similar to the pathway for synthesis of this compound that is found in adrenergic nerve endings. Tyrosine is transported into, hence is concentrated within, the nerve endings by an active mechanism. At this site tyrosine is converted into DOPA by tyrosine decarboxylase, and then into dopamine by an aromatic L-amino acid decarboxylase. These enzymes are located within the neuronal cytoplasm. The dopamine now enters the granular vesicles and there is converted into norepinephrine. However, in distinct contrast to the adrenal medulla, the adrenergic nerve endings do not contain appreciable quantities of the enzyme phenylethanolamine-N-methyl transferase, so that only in the adrenal medulla are significant quantities of epinephrine synthesized from norepinephrine as indicated in step ⑤, Figure 15.46.

In a manner similar to the adrenals, the rate-limiting step in the adrenergic nerve endings for the synthesis of norepinephrine lies in the conversion of tyrosine into dihydroxyphenylalanine (DOPA).

Adrenergic nerve endings also take up norepinephrine as well as small quantities of epinephrine from the blood, and these catecholamines, whether from endogenous synthesis or an exogenous source such as the adrenal medulla then are concentrated in the granulated vesicles of the neurons by an active transport mechanism and stored there until released by nerve impulses.

Secretion and Transport of the Adrenal Catecholamines

Several possible mechanisms whereby the adrenal catecholamines may be secreted by the medullary cells were mentioned above. The free hormones, epinephrine and norepinephrine, are found in the

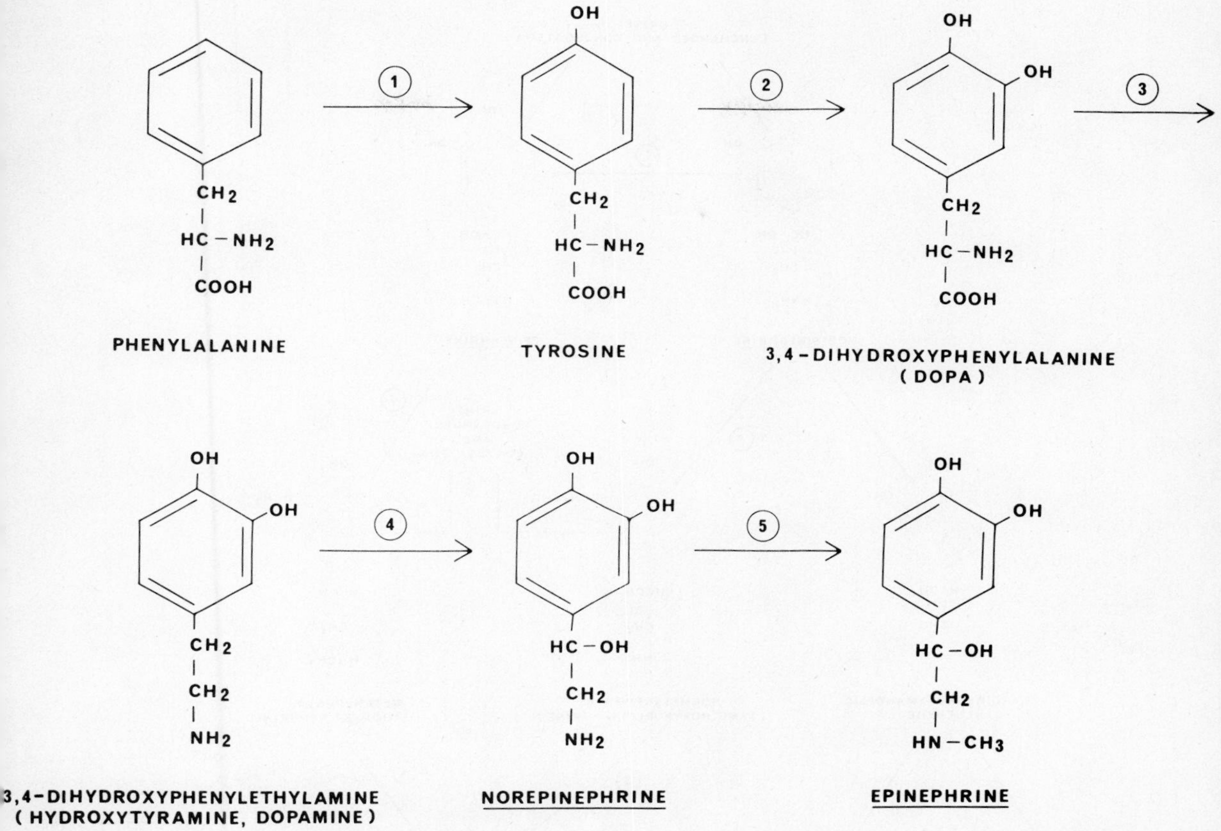

FIG. 15.46. Biosynthesis of the catecholamines. The enzymes that catalyze the specific reactions are: ①, phenylalanine hydroxylase; ②, tyrosine hydroxylase; ③, aromatic L-amino acid decarboxylase; ④, dopamine β-hydroxylase; ⑤, phenylethanolamine-N-methyltransferase plus S-adenosylmethionine. (After Ganong. *Review of Medical Physiology*, 6th ed, 1973, Lange Medical Publications.)

plasma in extremely low concentration (ie, at micrograms per liter levels), and no special transport systems are present to convey these compounds in the bloodstream. Despite the fact that human medullary tissue contains between three and ten times more epinephrine than norepinephrine, the mean plasma concentration of epinephrine is only 0.06 μg/liter compared to around 0.3 μg/liter for norepinephrine.* Adrenergic nerve endings also contribute some norepinephrine to the circulating catecholamine level, albeit the total quantity of norepinephrine that is released into the circulation from this source is negligible.

The epinephrine found in body tissues other than the adrenal medulla, and brain in general, is absorbed from the blood rather than synthesized by the tissues.

Metabolism and Excretion of the Catecholamines

The catecholamines have very short half-lives in the circulation. Hence these compounds are rapidly metabolized, principally in the liver, prior

Approximate values only.

to excretion. The pathways for the catabolism of norepinephrine and epinephrine are outlined in Figure 15.47. Each of these hormones is catabolized by three mechanisms: (1) o-methylation, a reaction catalyzed by catechol o-methyl transferase (Fig. 15.47, ②). S-adenosylmethionine is used in this reaction; (2) oxidative deamination, a reaction catalyzed by monoamine oxidase (MAO, ③); (3) conjugation reactions.

O-methylation is the chief metabolic pathway for epinephrine, and the principal metabolites of this hormone which are found in the urine are 3-methoxy-4-hydroxymandelic acid (vanilmandelic acid [VMA]), 3-methoxyepinephrine (metanephrine), and 3-methoxy-4-hydroxyphenylglycol (Fig. 15.47). The catechols 3,4-dihydroxymandelic acid, norepinephrine, and epinephrine are excreted in the urine only in minor quantities.

Roughly 50 percent of the total catecholamines secreted by the adrenal medulla are excreted in the urine as free or conjugated normetanephrine and metanephrine, and only trace quantities of free norepinephrine and epinephrine are excreted. The normal daily (24-hour) urinary output of catecholamines amounts to roughly

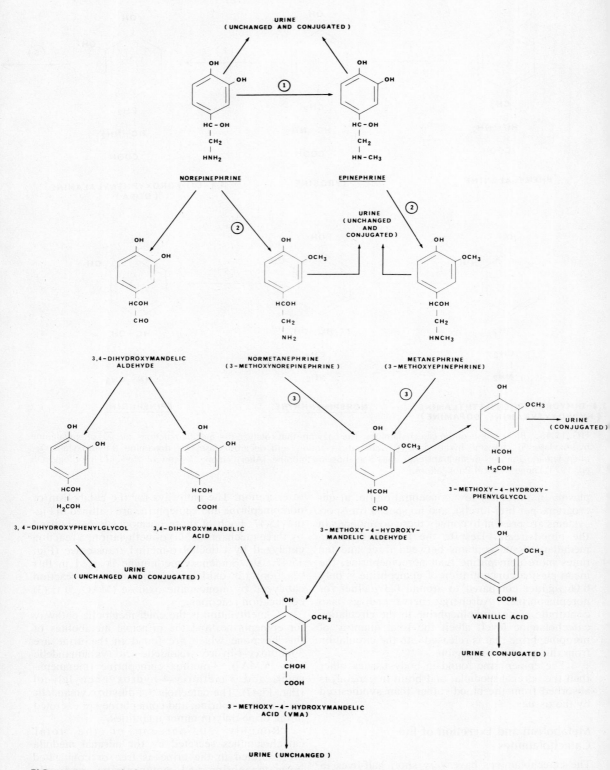

FIG. 15.47. Catabolism of the catecholamines. The enzymes which catalyze the specific reactions are: ①, phenylethanolamine-N-methyltransferase; ②, catechol-o-methyltransferase (COMT); ③, monoamine oxidase (MAO). (Data from Ganong. *Review of Medical Physiology*, 6th ed, 1973, Lange Medical Publications; White, Handler, and Smith. *Principles of Biochemistry*, 4th ed, 1968, The Blakiston Division, McGraw-Hill.)

30 μg of norepinephrine, 6 μg of epinephrine, and 700 μg of VMA.

Neural Regulation of Adrenal Medullary Secretion

Essentially, the adrenal medulla is a functional extension of the sympathetic nervous system whose postganglionic neurons have no axons, and which have evolved into specialized secretory cells. The physiologic stimuli that induce secretion by this organ are transmitted as impulses that reach the medulla via the splanchnic nerves to enhance the output of catecholamines. Although the adrenal medullary hormones are not essential to life, and no clinical condition involving hyposecretion of these organs has been recorded, the secretion of epinephrine and norepinephrine is of importance in assisting the individual to adapt rapidly to emergency situations.

Basal catecholamine secretion is low in resting states, especially during sleep. However, an increased medullary secretion of epinephrine and norepinephrine is part of the overall regulatory mechanism of the individual for meeting emergency situations. The other component of this system is, of course, the sympathetic nervous system. Furthermore, there is experimental evidence indicating that the secretion of the adrenal medullary hormones is enhanced when the cholinergic sympathetic vasodilatory system discharges at the onset of a bout of exercise (cf, p. 1177). Under these circumstances, increased adrenal medullary epinephrine secretion augments the vasodilatation produced by discharge of the sympathetic vasodilatory fibers to skeletal muscle. Thus, the principal physiologic regulatory mechanism for increasing adrenal medullary secretion operates via the sympathetic nervous system. In addition, a number of drugs act directly upon the adrenal medulla to increase secretion of the catecholamine hormones by this organ.

It is important to realize that, in general, stimulation of adrenal medullary secretion does not result in an alteration of the proportions of epinephrine and norepinephrine that are elaborated by these glands. However, certain particular circumstances can alter the ratio of epinephrine to norepinephrine secreted. Hypoxia, asphyxia and emotional stress all increase the quantity of norepinephrine secreted relative to epinephrine, whereas hemorrhage selectively increases the quantity of epinephrine secreted relative to norepinephrine. One must not conclude from these facts that the adrenal glands arbitrarily secrete the "best" hormone for a given situation, but rather that there is a "selective" release of epinephrine or norepinephrine that depends in turn upon the specific type of stimulus involved.

Physiologic and Biochemical Actions of Norepinephrine and Epinephrine

The hormones secreted by the adrenal medulla exert a broad spectrum of physiologic and biochemical effects upon a wide variety of bodily processes, as summarized in Table 15.14. The sev-

Table 15.14[a]   PHYSIOLOGIC AND BIOCHEMICAL EFFECTS OF NOREPINEPHRINE AND EPINEPHRINE[b]

PHYSIOLOGIC EFFECTS	NOREPINEPHRINE	EPINEPHRINE
Heart rate	+; −[c]	+
Cardiac output	0; −[c]	+ + + +
Systolic blood pressure	+ + + +	+ +
Diastolic blood pressure	+ +	+; 0; −
Total peripheral resistance (TPR)	+ +	− −
Action on central nervous system	0;+	+ +
Eosinopenic response	0	+

BIOCHEMICAL EFFECTS		
Release of free fatty acids	+ + + +	+ + +
Blood glucose	0; +	+ + +
Blood lactic acid	0; +	+ + +
Oxygen consumption; body heat production (ie, metabolic rate)	0; + +	+ +

[a]*Data from Ganong.* Review of Medical Physiology, *6th ed, 1973, Lange Medical Publications; Guyton.* Textbook of Medical Physiology, *4th ed, 1971, Saunders.*

[b]*A plus sign (+) denotes an increase, whereas a minus sign (−) denotes a decrease, and a zero (0) denotes no change in the function. The degree of response is indicated in approximate fashion. Thus the scale + to + + + indicates a slight to pronounced increase in the effect or process.*

[c]*Norepinephrine reduces heart rate by stimulating a reflex bradycardia, thus the cardiac output is lowered.*

eral possible effects of a single catecholamine may result from the mode of administration of the hormone (ie, endogenous or exogenous), as well as the dosage of the hormone to which the subject is exposed. These factors are important in determining the degree as well as the type of response elicited by the adrenal medullary hormones.

PHYSIOLOGIC EFFECTS OF THE CATECHOLAMINE HORMONES. Norepinephrine and epinephrine not only mimic the effects of sympathetic nervous discharge or stimulation as summarized in Table 7.6 (p. 276), but also these hormones themselves stimulate the nervous system.

The catecholamines directly augment both the rate and contractile force of the isolated heart as studied in vitro. The heart in situ, however, may respond quite differently, because the various reflexes present in the body may cause secondary alterations in heart rate because of shifts in blood pressure induced by these hormones. The catecholamines also directly increase myocardial irritability, so that extrasystoles or even fibrillation may develop under certain conditions subsequent to the administration of these compounds.

Both norepinephrine and epinephrine dilate the coronary vessels. However, norepinephrine causes vasoconstriction in other organs, whereas epinephrine dilates the vascular bed in skeletal muscles and viscera when administered in moderate doses. Thus, epinephrine when injected intravenously produces a marked rise in heart rate and cardiac output concomitantly with a sharp rise in blood pressure that is due to a generalized vasoconstriction.

As noted above, norepinephrine produces vasoconstriction in most organs other than the heart; hence it too produces a marked pressor response. However, since epinephrine causes vasodilatation of the vascular bed in skeletal muscles, this effect overcompensates for the vasoconstriction produced elsewhere; the net effect of an increased adrenal medullary secretion in vivo is a fall in the total peripheral resistance. The systolic blood pressure rises because of the direct effect of epinephrine in stimulating the heart rate increases the cardiac output sufficiently to overcome the concomitant vasodilatory effect of epinephrine in the skeletal muscles.

Infusion of norepinephrine into normal humans (or animals) results in an increase in both the systolic and diastolic blood pressures. The ensuing hypertension then stimulates the aortic and carotid pressoreceptors, with the result that reflex bradycardia occurs, and this reflex-induced bradycardia dominates the direct stimulatory effect of norepinephrine upon heart rate; hence this rate shows a net decrease and the cardiac output falls.

Epinephrine, on the other hand, causes an increased pulse pressure; however, the pressoreceptor stimulation evoked by this hormone is inadequate to obscure the direct effects of epinephrine upon the heart rate and cardiac output. The heart rate and cardiac output increase.

In addition to the effects exerted upon the nervous and cardiovascular systems by the adrenal medullary hormones, these compounds also directly affect the physiologic activity of smooth muscle. In this regard the effect of norepinephrine and epinephrine are quite variable, and their effects upon this kind of tissue depend upon the site in the body where the muscle is located.

Norepinephrine exerts a weakly inhibitory action upon the contactile activity of smooth muscle in the gastrointestinal tract, but does not relax the smooth musculature of the pulmonary bronchioles. As noted above, this hormone also elicits a marked pressor response because of its direct effect upon vascular smooth muscle, with an attendant rise in total peripheral resistance.

Epinephrine, on the other hand, markedly inhibits the contractile activity of the smooth musculature of the gastrointestinal tract, producing relaxation, while simultaneously stimulating contraction of the pyloric and ileocecal sphincters. Epinephrine also produces a strong dilatation of the bronchial musculature.

Note that these effects of the adrenal catecholamines strongly resemble the effects of adrenergic nerve stimulation to the same organs (Table 7.6).

Epinephrine decreases the rate of onset of fatigue in skeletal muscle and also increases the rate and depth of respiration. Presumably these effects are due to the enhanced glycogenolysis and lypolysis produced by this hormone as well as the calorigenic action of these amines.

In adrenalectomized animals suffering from an overall adrenal insufficiency, the vascular smooth muscle becomes unresponsive to the effects of norepinephrine and epinephrine. Capillary dilatation ensues and permeability of these vessels (eg, to dyes) increases markedly in the terminal stages. Lack of response to adrenergic stimulation (hence liberated norepinephrine) possibly hinders the compensation of the vascular bed to the reduced circulating blood volume (or hypovolemia) that attends adrenal insufficiency, and this in turn favors cardiovascular collapse. Vascular reactivity in such animals is restored by glucocorticoid administration.

METABOLIC EFFECTS OF THE CATECHOLAMINE HORMONES. Both epinephrine and norepinephrine exert certain effects upon the metabolism of carbohydrate and fats, and these actions of the catecholamines may be far more significant physiologically than their roles as endocrine extensions of the sympathetic nervous system.

Carbohydrate Metabolism. Epinephrine stimulates glycogenolysis in both liver and muscle. This results in an elevation of the blood glucose level as well as an increased production of lactic acid in muscle. Epinephrine activates phosphorylase in both of these tissues and the plasma potassium level rises coincidentally with the increased glycogenolysis.

The biochemical mechanisms whereby epinephrine mediates glycogenolysis in liver and muscle are illustrated in detail in Figures 15.48 and 15.49. Aside from their inherent physiologic significance, the mechanisms of these reactions provide examples of the first elucidation of the action of a hormone on a fundamental cellular process at the molecular level.

An increased oxygen consumption accompanies these glycogenolytic effects; this amounts to about 20 to 40 percent above the resting level in humans. The carbon dioxide production concomitantly rises to an even greater extent than does the increase in oxygen consumption, so that the respiratory quotient also increases.

Norepinephrine, on the other hand, exerts very little effect upon carbohydrate metabolism and oxygen consumption, a fact that provides the major exception to the generalization that the metabolic actions of the two adrenal catecholamines are similar.

Lipid Metabolism. Epinephrine as well as norepinephrine are almost equally potent in their effects upon lipid metabolism. Thus, both of these catecholamines exert a profound lipid-mobilizing action, and they increase the blood level of nones-terified fatty acids by stimulating the release of free fatty acids and glycerol from adipose tissue. This effect results directly from an elevated rate of lipolysis that is produced in adipose cells by these hormones, and an increased oxygen consumption (or calorigenic action) also attends these processes.

In normal subjects, the administration of epinephrine also elevates serum cholesterol and phospholipid levels. The turnover rate of the cholesterol and phospholipids, also is elevated in cardiac muscle by exogenous epinephrine.

Epinephrine directly inhibits insulin release from the pancreas, and thus serves as an emergency hormone by exerting the following metabolic actions. First, epinephrine rapidly provides a supply of fatty acids for use as fuel by skeletal muscle. Second, epinephrine mobilizes glucose by stimulating hepatic glycogenolysis as well as gluconeogenesis directly. Third, by reducing the endogenous secretion of insulin, epinephrine decreases glucose uptake by the peripheral tissues, thereby increasing the supply of this substrate that is available to the central nervous system.

MECHANISM OF ACTION OF THE ADRENAL MEDULLARY HORMONES. The general effects of norepinephrine and epinephrine have been di-

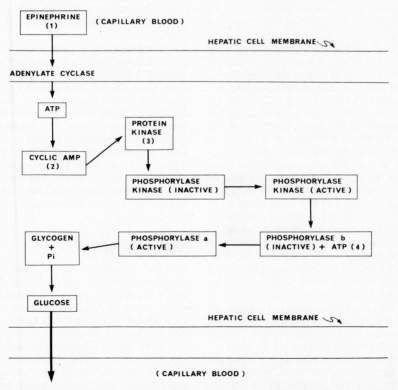

FIG. 15.48. Mechanism underlying the release of blood glucose from a hepatic cell by glycogenolysis. (1) Epinephrine, the first messenger, activates adenylate cyclase within the cell membrane so that some cytoplasmic ATP is converted into cyclic AMP (2) which is the second messenger. (3) A protein kinase then is activated by the cyclic AMP and this in turn activates a second cytoplasmic kinase. The active phosphorylase kinase now activates phosphorylase b into phosphorylase a, and the latter catalyzes the degradation of glycogen into glucose, and the glucose diffuses from the hepatic cell into the bloodstream. (After Pastan. *Sci Am* 227:97, 1972.)

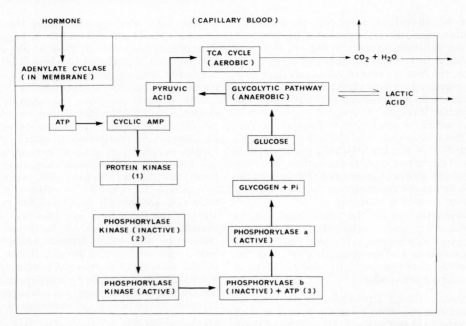

FIG. 15.49. Mechanism underlying glycogenolysis within a muscle cell. The cyclic AMP which is formed under the influence of activated adenylate cyclase, as described in the legend to Figure 15.48, activates a protein kinase (1). This kinase now activates a second kinase (2) which in turn converts inactive phosphorylase b into active phosphorylase a (3). Glycogen now is degraded into glucose and the latter compound can be metabolized either aerobically or anaerobically as shown. Note, however, that the glucose released by this process is utilized within the muscle cell, and that it does not diffuse into the bloodstream as is the situation in the liver cell (Fig. 15.48). (Data from Pastan. *Sci Am* 227:97, 1972.)

vided into two categories depending upon their sensitivity to certain drugs. These effects depend in turn upon the actions of these catecholamines upon postulated α and β receptors located within the effector organs (Table 7.6, p. 276). Drugs which inhibit α receptors (such as phentolamine) prevent such actions as the pressor effects of the catecholamines, whereas inhibition of β receptors by other compounds (eg, isoproterenol or propranolol) prevents the inotropic and chronotropic responses of the heart to the catecholamines as well as the ability of these hormones to stimulate glycogenolysis in muscle and liver and lipolysis in adipose tissue. Experimental evidence has shown that most of the β responses are due to stimulation of adenyl cyclase activity, so that an increase in the synthesis of cyclic 3′, 5′-AMP results. Epinephrine, for example, produces this effect in a number of tissues, including liver, skeletal muscle, adipose tissue, and cardiac tissue. The adenyl cyclase system in turn has been shown to be localized within the cell membranes of hepatic and other tissues as discussed in the introduction to this chapter and illustrated in Figure 15.2. Furthermore, the effects of norepinephrine and epinephrine upon glycogen metabolism have been shown to be mediated by a marked increase in the activity of the adenyl cyclase system for epinephrine as shown in Figures 15.48 and 15.49.

On the other hand, certain α responses have been demonstrated to reside in an inhibition of cyclic AMP synthesis.

It is important to realize that in certain tissues where epinephrine can stimulate both α and β types of response, the net effect of the amine depends upon the relative quantity and/or sensitivity of each type of receptor that is present in the tissue. For example, in the pancreas the α-adrenergic response to epinephrine is predominant; therefore, cyclic AMP concentration decreases, inhibiting insulin secretion. If, however, an α adrenergic blocking agent such as phentolamine (Regitine) is present, the β effect becomes predominant, and epinephrine now stimulates cyclic AMP synthesis, and insulin secretion is stimulated rather than inhibited.

Clinical Correlates

As noted earlier, no clinical condition that can be attributed directly to adrenal medullary hyposecretion is known. In fact, the adrenal medulla only rarely becomes involved in disease processes.

Certain chromaffin cell tumors of the adrenal medulla, known as pheochromocytomas, produce a clinical state resembling hypersecretion of the adrenal medulla. The symptoms resulting from such tumors include bouts of intermittent (or paroxysmal) hypertension which may develop into permanent hypertension. Death ultimately results from complications of this pathologic situation, including such long-term effects as pulmonary edema, coronary insufficiency, and ventricular fibrillation.

In general, the norepinephrine concentration in pheochromocytomas is much greater than that of epinephrine, and the hypertensive episodes seen in such patients are believed to result principally from the secretion of norepinephrine.

Laboratory tests for suspected cases of pheochromocytoma include blood and urinary catecholamine analyses, especially for VMA (vanilmandelic acid, or 4-hydroxy-3-methoxymandelic acid), since this compound is the principal urinary metabolite of the adrenal medullary catecholamines.

An interesting diagnostic test for pheochromocytoma is based upon the use of the α-adrenergic blocking agent phentolamine (Regitine). Since this compound is a specific antagonist to the effects of norepinephrine, injection of this drug into a patient with a pronounced hypertension (pressures in excess of 170/110 mm Hg) that is caused by a pheochromocytoma results in a sustained fall in blood pressure in from two to five minutes. The drop in blood pressure should be at least 35/25 mm Hg in order to implicate the adrenal medulla as the possible source of excess norepinephrine.

THE PHYSIOLOGY AND ENDOCRINOLOGY OF REPRODUCTION IN THE MALE

Ultimately, the determination of sex in higher animals, including humans, is determined at the moment of conception, and it depends upon a single chromosome called the Y chromosome. The many physiologic differences between males and females result from this genetically determined difference between the two sexes, as well as from the subsequent development of a single pair of endocrine organs, the testes in the male, and the ovaries in the female. The differentiation of primordial (or indifferent) genital tissue into testes or ovaries is determined genetically in utero. However, the development of male genitalia requires the presence of a functional testis. It also appears that the male patterns of gonadotropin secretion and male sexual behavior may result from the action of male hormones (or androgens) on the brain during early development of the individual. This fact has been demonstrated in experiments conducted on animals, although such early effects of androgens in humans have not yet been demonstrated.

In humans there are 46 chromosomes. In males, there are 22 pairs of somatic chromosomes (or autosomes) and two sex chromosomes, a large X chromosome plus a small Y chromosome; the latter determines whether or not an individual is a male. Females, on the other hand, have 22 pairs of autosomal chromosomes plus two X chromosomes.

The arbitrary arrangement of all of the chromosomes into a pattern for study is called the karyotype of the individual, and this chromosomal array is based upon the morphology of these structures. Human cells grown in vitro provide the material for such studies. After treatment of the cells with colchicine (which arrests mitosis in metaphase), the cells are exposed to a hypotonic solution that causes the chromosomes to swell, and slides can then be prepared from the treated cells which clearly illustrate the chromosomal pattern of the individual.

Since the human Y chromosome is smaller (thus lighter) than the X chromosome, it has been postulated that spermatozoa having this chromosome can reach the ovum in the female genital tract more rapidly, and this difference supposedly accounts for the fact that statistically male births are slightly preponderant over female. Basically, however, it is the male who determines the genetic sex of the offspring.

Postpartum, the gonads are relatively inactive until the onset of puberty (or adolescence), at which time they are activated by adenohypophyseal gonadotropic hormones, and the characteristic features of adult males and females appear together with the onset of sexual cycles (or menstrual cycles) in females.

The primary sex organs (or gonads) in both sexes have dual roles. First, these structures secrete steroid sex hormones. In the male these steroid hormones, known by the generic name of androgens, are masculinizing or virilizing in their effects; in the female on the other hand, the sex steroids are called estrogens, and these compounds produce feminizing effects. Second, the gonads of both sexes produce germ cells, a process known as gametogenesis. The testes produce spermatozoa, whereas the ovaries produce ova.

In the present section, three general aspects of male reproductive physiology and endocrinology will be presented: (1) spermatogenesis; (2) the male sexual act; and (3) the regulation of these functions by the androgenic hormones produced in the testes. These topics will be discussed in conjunction with, and following a review of, the pertinent anatomic considerations.

Functional Morphology of the Male Sex Organs

The male reproductive system is comprised of the testes, together with a number of accessory ducts and structures, including the epididymis, vas deferens, ejaculatory duct (or vas efferens), the distal portion of the urethra, the penis, and several glands. In the latter category are included the seminal vesicles, the glands of Littre and the bulbourethral, or Cowper's glands.

THE TESTES AND SPERMATOGENESIS

Gross Anatomy. As noted earlier, the testes, in common with the pancreas, share the distinction of being the only endocrine organs in the body which also possess a system of excretory ducts that carry the excretory products of the gland.

Each of the paired testes is oval, approximately 4 to 5 cm in length, 2.5 cm wide, and 3 cm in anteroposterior diameter. Each of these organs weighs roughly 100 gm. These compound tubular glands are enclosed in a thick fibrous capsule, the tunica albuginea; posteriorly this capsule extends into the gland proper as the rather thick mediastinum testis, and thin fibrous partitions, the septula testis, extend in a radial fashion from the mediastinum to the tunica albuginea. Thus, the organ is divided into roughly 250 compartments or lobules, the lobuli testis. Each testicular lobule in turn is composed of between one and four highly convoluted seminiferous tubules, which are about 150 to 250 μ in diameter and 30 to 70 cm in length. (The total length of the seminiferous tubules in the human testis has been estimated to be around 250 meters). The anatomic relationships just described are illustrated in Figure 15.50.

As the seminiferous tubules approach the mediastinum, they unite with adjacent tubules to form between 20 and 30 straight tubules (tubuli recti) which then pursue a direct course into the mediastinal area. In this region the straight tubules form a network of thin-walled anastomosing spaces known as the rete testis (Fig. 15.50). At the upper border of the mediastinum, the rete testis forms 12 to 15 efferent ducts which pass through the tunica albuginea. These structures become enlarged and highly convoluted as they leave the mediastinum and enter the duct of the epididymis. The seminiferous tubules comprise the exocrine portion of the testis, and the excretory product of this organ is whole cells, and is transported in this system of ducts (Fig. 15.54).

Each testis is attached at its flat posterior surface to the end of a spermatic cord. The spermatic cords are composed of the vas deferens (or ductus deferens), together with the blood vessels, and nerves on that side. The spermatic cord in turn is enclosed by a layer of longitudinal striated muscle fibers, the cremaster muscle. This muscle also invests the testes and elevates them toward the abdomen in response to fear, cold, and other stimuli, a phenomenon that is known as the cremasteric reflex.

The testes are suspended in a sac of skin, the scrotum, which has important thermoregulatory properties insofar as spermatogenesis is concerned.

The epididymis is an elongated organ that is closely attached to the posterior surface of each testis, and is composed of the convoluted proximal portion of the excretory duct system (Fig. 15.50).

On the inner surface of the tunica albuginea, the dense connective tissue blends into another, less dense layer that is called the tunica vasculosa testis. Loose connective tissue extends inward from this layer to fill the spaces between the seminiferous tubules. In addition to fibroblasts, macrophages, and other cell types, this connective tissue contains groups (clusters) of epithelioid cells called the interstitial or Leydig cells. These elements comprise the endocrine tissue of the testis.

The gonads of both males and females develop from the urogenital ridge which consists of

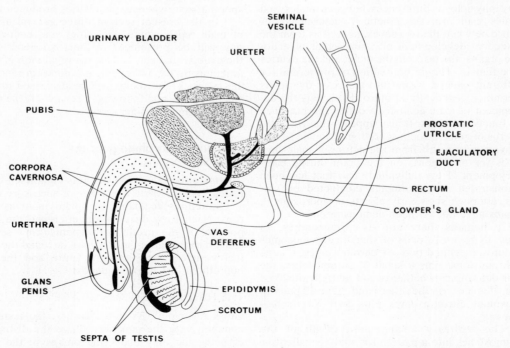

FIG. 15.50. Gross anatomy of the testis and related structures. (After Ganong. *Review of Medical Physiology*, 6th ed, 1973, Lange Medical Publications; Goss [ed]. *Gray's Anatomy of the Human Body*, 28th ed, 1970, Lea & Febiger.)

bilateral accumulations of mesodermal tissue that is situated in the dorsal wall of the peritoneal cavity near the adrenal glands. This tissue ultimately gives rise to both the kidneys and the sex glands. Until the sixth week of gestation the genital ridge is indistinguishable morphologically in either sex, and thus is called the indifferent stage of morphogenesis, even though the true sex of the individual was determined genetically at the time of conception.

The primordial gonad develops a cortex and medulla. During the seventh and eighth weeks in genetic males, the medulla develops into a testis, and the cortex disappears. Leydig cells form and androgenic hormones are secreted which stimulate development and growth of the secondary sex characteristics. In genetic females, on the other hand, the cortex develops into the ovary, whereas the medulla regresses. Again in contrast to the male, the embryonic ovary does not secrete hormones.

Histology

Seminiferous Epithelium. In adults, the seminiferous tubules are lined with a complex stratified epithelial tissue that contains two major cell types: first, supporting (sustentacular or Sertoli) cells, and second, spermatogenic cells (Fig. 15.51). Both of these cell types represent successive stages in the development of male germ cells, and they are not distinct cell populations.

The Sertoli cells are attached to the basement membranes of the seminiferous epithelium, and generally have an ovoid nucleus and a large, characteristically three-lobed, nucleolus. These cells also have an elaborate system of thin processes that extend toward the luminal surface of the tubule, and that surround the spermatogenic cells. The earliest stage in the development of the germ cells, the spermatogonia, also lie adjacent to the basal layer of the seminiferous tubules, whereas later stages in the formation of the germ cells are found progressively further away from the basal lamina, ie, toward the lumen of the tubule.

The Sertoli cells form a matrix in which are imbedded the various types of spermatogenic cells. Their form is highly irregular, and they conform closely to the irregular and ever-changing outlines of the germ cells during gametogenesis. Presumably the sustentacular cells, in addition to providing physical support for the developing spermatozoa, also have a role in the nutrition of the germinal elements.

Spermatogenesis. The complex and sequential series of events that result in the production of motile spermatozoa may be summarized briefly, and the general appearance of the cells at each stage of development are shown in Figure 15.51. The progressive series of changes that result in the formation of mature spermatozoa from spermatogonia (the process of spermatogenesis) may be divided into three stages.

First, *spermatocytogenesis*. In this phase, the spermatogonia multiply by mitotic division (Fig.

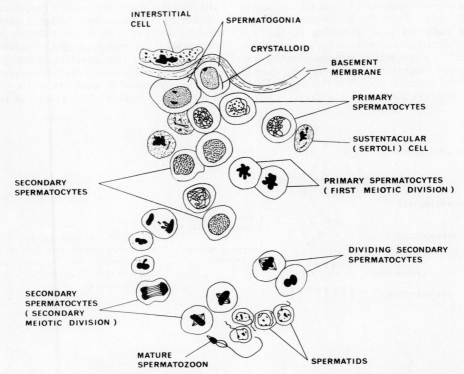

FIG. 15.51.   Seminiferous and related epithelium of the testis. (After Bloom and Fawcett. *A Textbook of Histology,* 9th ed, 1968, Saunders; Ganong. *Review of Medical Physiology,* 6th ed, 1973, Lange Medical Publications.)

3.41, p. 110), and thereby reproduce themselves and also give rise to many generations of spermatocytes. Spermatogonia have spherical nuclei, chromatin granules of either fine or irregular size, and one or more nucleoli. The pale-staining cytoplasm is homogenous. Electron microscopy has revealed the presence of two kinds of spermatogonia based upon certain minor morphologic differences, and these cells are referred to as type A and type B.

Second, *meiosis* (Fig. 15.52). Since the somatic cells contain a double, or diploid, number of chromosomes, then if sperm and ovum each had diploid chromosomes, their fusion would result in a doubling of the normal chromosomal complement. This does not happen, however, because of the process of meiosis (or reduction division) during this stage of gametogenesis in both sperm and ova. The spermatocytes undergo two maturation divisions during this process, and each spermatocyte produces a number of spermatids, each of which contains half the number of chromosomes that are found in the somatic cells. The nuclei of both mature sperm and mature ova each has a haploid set of chromosomes, since the nuclei divide twice during meiosis whereas the chromosomes divide only once, as shown in Figure 15.52.

The so-called primary spermatocytes resulting from mitosis of the spermatogonia become larger because of accumulation of cytoplasm as they move away from the basal lamina. Shortly after they are formed, these cells enter prophase of the first maturation division, and their chromatin forms the diploid number of thin chromosomes (leptotene stage). Then the homologous chromosomes of each set fuse, resulting in half the number of chromosomal strands (zygotene stage). Subsequently, these paired strands contract and appear coarse (pachytene stage); crossing over of chromosomal elements occurs at this point. These stages of prophase last for several days.

Following prophase, the nuclear membrane disappears, and the paired chromosomes become oriented on the equatorial plate of the cell (metaphase). During anaphase of the first meiotic division, entire chromosomes, instead of half chromosomes as in mitosis, move to the poles. These chromosomes may be different however, because of the exchanges of nuclear material that may have transpired during the crossing over process.

Anaphase and telophase are rapid, and the secondary spermatocytes formed during this process contain only half of the chromosomes that were present originally in the parent cells.

Since the secondary spermatocytes remain in interphase for only a very short interval they are rarely seen in histologic preparations, as they pass rapidly into the second maturation division, a process that is essentially similar to mitosis, save that a haploid number of chromosomes are involved.

The two male secondary spermatocytes that result from the first maturation division thus will contain 22 autosomal chromosomes plus a single X or a Y chromosome. Thus, union of a sperm containing the X chromosome with an ovum results in a zygote containing a complement of 44 + XX chromosomes. This union results in a genetically female offspring. On the other hand, fusion of a 22 + Y sperm with the 22 + X ovum results in a 44 + XY zygote and the offspring is a genetic male.

It is important to mention that cell division (or cytokinesis) of the spermatocytes and type B spermatogonia is incomplete, and the daughter cells resulting from this division remain connected together by a protoplasmic bridge. Consequently, the final division of each spermatocyte into eight spermatids is incomplete, a fact revealed by electron microscopy. Thus cytokinesis during the final division of the spermatogonia is incomplete; therefore pairs of joined primary spermatocytes result. In turn, these cells produce groups of four inter-

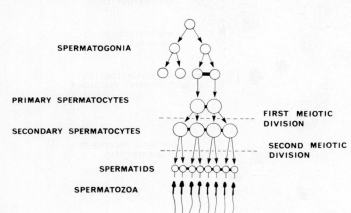

SPERMATOGONIA

PRIMARY SPERMATOCYTES

SECONDARY SPERMATOCYTES

FIRST MEIOTIC DIVISION

SECOND MEIOTIC DIVISION

SPERMATIDS

SPERMATOZOA

INTERCELLULAR BRIDGE

FIG. 15.52. Meiosis (cf. Fig. 3.41). (Data from Bloom and Fawcett. *A Textbook of Histology,* 9th ed, 1968, Saunders; DeRobertis, Nowinski, and Saez. *Cell Biology,* 5th ed, 1970, Saunders.)

connected secondary spermatocytes which then give rise to groups of eight spermatids, which also are joined together. The mechanism whereby the clusters of spermatids separate into individual spermatozoa is unknown; however, if this process were defective it could give rise to the abnormal double spermatozoa found in the ejaculate.

Third, *spermiogenesis*. During this process, the spermatids undergo a complex series of changes which result in the development of mature spermatozoa. The small spherical spermatids which result from division of the secondary spermatocytes contain a spherical nucleus about 5 μ in diameter, having pale-staining granular chromatin. The Golgi apparatus becomes granular; these proacrosomal granules then coalesce into a large globule, the acrosomal vesicle, and this structure in turn becomes attached to the outer portion of the nuclear membrane, marking the anterior tip of the future sperm nucleus. The limiting membrane of the acrosomal vesicle then increases in size as well as its contact area with the nuclear membrane, thereby becoming the head cap of the mature sperm. The acrosomal material then becomes reorganized together with the nuclear material, and the spermatid elongates. The major portion of the acrosome remains at the anterior pole of the nucleus, and forms a thin layer which fuses with the head cap, thereby forming the acrosomal cap (or acrosome). The acrosome is present in the sperm of all mammals, although considerable morphologic variations are seen among the sperm of different species.

During the early stages of acrosome formation, the centrioles migrate to the opposite end of the spermatid and the distal centriole gives rise to the slender flagellum (or tail) which grows out of the cell into the extracellular cleft located between the spermatid and the Sertoli cell which surrounds it. Ultimately, the complex structure of the flagellum comes to resemble that of a cilium (Fig. 4.39, p. 153), in that a series of longitudinal fibrillar elements are arranged about the axis of the flagellum. These are the contractile elements that lend motility to the mature sperm.

When the tail of the sperm has differentiated completely, the spermatid clusters cast off the residual cytoplasm that invests them, the sperm heads are released from the Sertoli cells, and then liberated into the lumen of the seminiferous tubule.

A mature spermatozoon (Fig. 15.53) consists of a head and tail; the latter is divided into a neck middle piece, principal piece, and end piece, depending upon differences in thickness and fine structure along its length. The head contains the chromosomal material, whereas the tail contains the contractile mechanism whereby the sperm becomes a self-propelled entity.

In conclusion to this discussion, it is important to stress that normal spermatogenesis and the production of viable sperm cannot take place at core body temperature (ie, 37.5 C). The slightly lower testicular temperature that is present when the gonads occupy an extraabdominal position in the scrotum is essential to male fertility. Cryptorchid males are sterile when both testes remain either in the abdominal cavity or within the inguinal canal.

Testicular Interstitial Tissue. The interstitial cells of Leydig, which comprise the endocrine tissue of the testis, are found in isolated clusters or rows along the smaller blood vessels. These cells are polyhedral and range between about 15 and 20 μ in diameter. The nuclei of the Leydig cells are large, peripherally located, contain a small amount of chromatin, and one or two nucleoli. The cytoplasm exhibits a strong acidophilia, and small numbers of mitochondria are present.

Crystalline structures of obscure function are characteristically seen in the cytoplasm of the Leydig cells following special staining of the tissue, eg, by azocarmine; these are called the crystals of Reinke.

Testicular Blood Vessels, Nerves, and Lymphatics. The internal spermatic artery supplies the major portion of the blood supply to the testis (Fig. 15.50), and branches of this vessel penetrate the mediastinum as well as the tunica albuginea, thence to course in the tunica vasculosa. The vessels then penetrate the individual testicular lobules and then branch into loose capillary networks which surround the seminiferous tubules. The veins exit from the testis in a pattern similar to, and parallel with, that of the arteries.

The nerve supply is derived from the spermatic plexus, and delicate fiber networks surround the blood vessels.

Networks of lymphatic capillaries are abundant in the interstitial tissue between the seminiferous tubules.

EXCRETORY DUCTS. The tubular structures that transport the sperm from the testis to the exterior of the body are, in sequence from the testis outward, the vas efferens, the epididymis, the vas (ductus) deferens, the ejaculatory duct, and the urethra. These structures are shown sequentially in Figure 15.54.

Vas Efferens (Canaliculi or Ductuli Efferentes). As noted earlier, on the upper and posterior region of each testis, approximately 12 to 15 efferent ducts arise from the rete (Fig. 15.50), perforate the tunica albuginea, and then emerge upon the testicular surface. These highly convoluted efferent ductules, or vasa efferentia, are approximately 0.6 mm in diameter, 4 to 6 mm in length, and in turn form the ductules form about 5 to 13 conical bodies called the vascular cones, which are around one cm in length. The vascular cones have their bases directed toward the upper end (or head) of the epididymis and their apices directed toward the mediastinum testis. Attached to each other by connective tissue, the vascular cones collectively comprise a portion of the head of the epididymis.

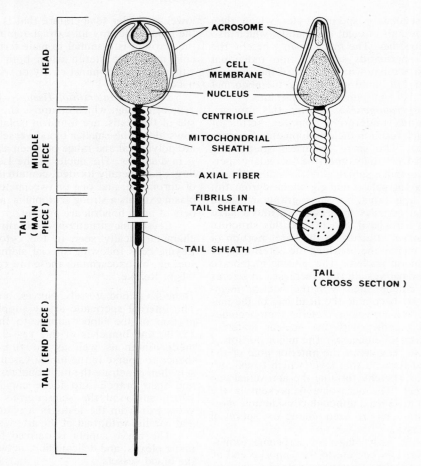

FIG. 15.53. Morphology of a human spermatozoon. (After Bloom and Fawcett. *A Textbook of Histology,* 9th ed, 1968, Saunders; Ganong. *Review of Medical Physiology,* 6th ed, 1973, Lange Medical Publications.)

The efferent ducts are lined with an epithelium that has alternate groups of low and tall cells. The low, clear, secretory cells form glandular structures, and have a brush border with a central flagellum on their free surface. The tall cells, on the other hand, are generally conical-shaped with the base directed toward the lumen of the tubule. The free surface of the tall cells of the vasa efferentia is covered with cilia that beat toward the epididymis, and thus transport the sperm that are produced in the seminiferous tubules toward that structure.

The thin basement membrane that is located around the periphery of the efferent ductules is surrounded by circularly disposed smooth muscle cells; this muscular layer thickens in the vascular cones.

Epididymis (Ductus Epididymis). The single ductus epididymis of each testis is formed by the fusion of the convoluted tubules of the vascular cones. The epididymis is a tubular structure, is quite tortuous, and when fully extended measures around 6 meters in length. However, the convolutions are such

that the epididymis is compressed (like a spring) into a structure that is approximately four cm long. This long canal, together with the surrounding connective tissues, forms the body and tail of the epididymis.

The lumen of the epididymis in its proximal region is lined by a pseudostratified columnar epithelium that is covered on its free surface by nonmotile stereocilia about 30 μ in length. A highly developed capillary network surrounds the basement membrane together with a layer of circularly arranged smooth muscle fibers that doubtless propel the sperm along the ducts by peristaltic contractions. A plexus of delicate nerve fibers as well as sympathetic ganglia are intimately associated with the vascular walls and the smooth muscle fibers.

The tail of the epididymis extends nearly to the inferior pole of the testis. In this region, the epididymal duct becomes less convoluted, its walls thicken, and its diameter enlarges. The duct leaves the tail of the epididymis and becomes the ductus deferens.

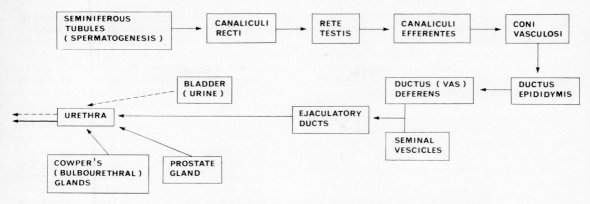

FIG. 15.54. The excretory ducts and their relationships in the male urogenital system.

Vas Deferens (Ductus Deferens) and the Ejaculatory Ducts. The ductus deferens arises from the tail of the ductus epididymis and extends to the ejaculatory duct. The vas deferens is located in the scrotum at its point of origin at the tail of the epididymis, then traverses the spermatic cord as a component of this structure to enter the abdominal cavity via the inguinal canal. In the abdominal cavity, the vas deferens leaves the other structures of the spermatic cord, traverses the pelvic cavity immediately superior to the bladder, crosses the ureter, and then terminates at the base of the prostate, where it joins the duct of the seminal vesicle to form the short (19 mm), straight, ejaculatory duct which is 0.3 mm in diameter (Fig. 15.54).

The vas deferens has an irregular outline in cross section, and the pseudostratified columnar epithelium that is found in this portion of the urogenital system is lower than that found in the epididymis. Stereocilia are present upon the luminal border of the epithelial cells.

The smooth muscle coat of the ductus deferens is highly developed (about 1 mm thick), and has clearly delineated inner and outer longitudinal layers that are separated by an intermediate circular layer. Thus duct is quite firm, and thus readily palpable through the skin of the scrotum.

After crossing the ureter in the abdominal cavity, the ductus deferens dilates into a larger ampulla. At the distal end of this ampullary dilatation there is situated a large glandular evagination of the duct, the seminal vesicle. Beyond the ampulla, the ductus deferens, now called the ejaculatory duct, passes through the body of the prostate gland which is situated at the base of the bladder. The two ejaculatory ducts now enter the lumen of the prostatic urethra by small slits that are located on the posterior wall of this structure.

The epithelial lining of the ejaculatory ducts is of the simple or pseudostratified columnar type, and these cells possibly have a secretory function. The ejaculatory ducts are covered only by a connective tissue layer, and lack the muscular layer that is found in the other tubular portions of the male genital tract.

It should be mentioned that the operation for male sterilization, known as bilateral vasectomy, involves the ligation and section of each vas deferens within the scrotum. This procedure effectively prevents the passage of sperm from the testes to the exterior, and atrophy of the seminiferous epithelium results. However, as the vascular and neural supply to the testes, which also lie within the spermatic cord, are wholly unaffected, bilateral vasectomy produces no effect upon the endocrine function of the testicular tissue. Therefore, the effects of vasectomy are quite different from those of castration, which involves removal of the testes in their entirety.

ACCESSORY GLANDS OF THE MALE REPRODUCTIVE TRACT. The principal glands associated with male reproduction are the seminal vesicles (paired), the prostate gland, and the bulbourethral or Cowper's glands (paired).

Seminal Vesicles. These glandular structures are elongated, (about 7 cm) club-shaped, and lobulated. The seminal vesicles are situated just above the base of the prostate gland between the posterior surface of the urinary bladder and the anterior surface of the rectum. Each vesicle is a tubular structure about 10 cm in total length but which is coiled upon itself, and which has numerous blind-ending lateral outpocketings (or diverticula) that arise from an irregular branching lumen. The seminal vesicles actually are evaginations of the vas deferens, and thus are quite similar to it in their structure.

The morphology of the epithelium is quite variable, depending upon the age and physiologic state of the individual; however, usually it is pseudostratified — there is a layer of round basal cells, as well as a layer of somewhat larger cuboidal or columnar cells on the luminal surface. A central flagellum is present on the cells covering the free surface of the gland, yellow pigment granules are abundant in the cells of the seminal vesicles, and these cytoplasmic inclusions appear at the onset of puberty. The seminal vesicles are surrounded by a muscular coat that is innervated by a nerve plexus as well as small sympathetic ganglia. The hypogastric plexus innervates these structures of the seminal vesicles.

The secretion that is elaborated by the seminal vesicles is clear, alkaline, yellowish, viscous, and contains globulin. It serves as a transport as well as nutrient medium for the sperm (see semen, p. 1137).

Androgenic hormones are essential to the normal growth, development, and maintenance of the seminal vesicles. Hence the glandular epithelium atrophies after castration, but is restored following exogenous testosterone administration.

Prostate Gland.   The prostate gland is the largest auxiliary gland of the male reproductive system, and in shape and size it is similar to that of a horse chestnut. This gland weighs roughly 25 gms. The prostate is located just below the urinary bladder where it completely surrounds the first segment of the urethra as its point of origin from the bladder. Hypertrophy of this gland results in urinary retention or difficulty in passing urine normally.

The prostate gland is invested in a thin and dense capsule of fibrous connective tissue and smooth muscle fibers; hence it is firm to palpation. Collagenous and elastic connective tissue as well as smooth muscle fibers form partitions (or septa) that extend into the gland from the capsule, and these septa form an abundant stroma that supports the glandular elements of the organ. Embedded within the stroma are between 15 to 30 branching tubuloalveolar (or saccular) glandular elements and their ducts. The latter empty independently into the urethra on the left and right sides near the prostatic utricle (colliculus seminalis). Surrounding the urethra, the smooth muscle fibers form a heavy ring, the internal sphincter of the bladder.

The glandular epithelium of the prostate ranges from squamous to low cuboidal, depending upon the size of the cavity. Ordinarily squamous, pseudostratified, or even low columnar epithelium is encountered. Secretory granules are numerous.

The prostate is innervated by abundant plexuses of generally nonmyelinated nerve fibers and small sympathetic ganglia. The interstitial connective tissue is richly provided with various specialized as well as unspecialized sensory nerve endings, including genital corpuscles, end bulbs, and free nerve terminations.

The epithelial elements of the prostate atrophy after castration, and, similarly to the seminal vesicles, androgens are essential to the normal growth, development, and maintenance of the prostate gland.

The prostatic secretion is a thin, slightly acid (pH 6.5) liquid, which is quite opalescent and has a characteristic seminal odor. This fluid is the principal source of the citric acid and acid phosphatase in the semen. The blood concentration of the acid phosphatase rises because of cellular necrosis when prostatic malignancy is present. Hence the

blood level of this enzyme is of considerable diagnostic importance when prostatic malignancy is suspected. The protein content of prostatic fluid is low; however, several proteolytic enzymes as well as diastase, a fibrinolysin, and a beta glucuronidase are present. The prostatic secretion is added to the seminal fluid during ejaculation and may serve to increase the motility of the spermatozoa.

Bulbourethral Glands (Cowper's Glands).   The two compound tubuloacinar bulbourethral glands are about the size and shape of a pea and are located in the urogenital diaphragm which forms the floor of the pelvis. The bulbourethral glands are found on each side of the urethra, and their irregularly shaped ducts are about 2.5 cm long, and enter the posterior portion of the cavernous urethra. The connective tissue septa that are located between the glandular lobules range from one to three mm in diameter, and have elastic tissue networks as well as thick bundles of smooth and striated muscles.

The secretory epithelium of the bulbourethral glands exhibits wide functional variations. Thus, in enlarged alveoli, the cells are flattened, whereas in other glandular regions they are cuboidal or columnar; the nuclei are located basally in the epithelial cells. Mucoid droplets are present in the cytoplasm.

The secretion of the bulbourethral glands is a clear, viscous, mucoid lubricant which has an alkaline reaction. Presumably the latter assists in the neutralization of the acid female vaginal secretions thereby enhancing the survival potentiality of sperm in the female genital tract.

PENIS AND MALE SEXUAL PHYSIOLOGY

General Morphology.   The penis, or male copulatory organ, is elongated, pendant, and is attached to the front and sides of the pubic arch anterior to the scrotum. The penis is composed of three long cylindrical masses of cavernous or erectile tissue. These are the two corpora cavernosa penis and the single corpus cavernosum urethrae (Fig. 15.55A). The corpora cavernosa penis originate from the ascending pubic rami on the left and right sides and converge at the pubic angle. Thence they pass distally side by side to their tapered ends, and form the dorsal two thirds of the shaft of the penis. A shallow longitudinal groove at the upper junction of the corpora cavernosa contains the dorsal artery and vein. On the lower surface of the corpora cavernosa lies a deep groove which contains the unpaired corpus cavernosum urethrae (or corpus spongiosum). The latter is traversed throughout its entire length by the urethra (Fig. 15.55B). The corpus spongiosum ends with a mushroom-shaped enlargement, the glans penis. The glans in turn has two convexities posteriorly which cap the conical tips of the two corpora cavernosa penis.

Each of the two cavernous bodies is invested in a thick (about 1 mm) membrane that is composed of an outer longitudinal layer of collagenous fibers and an inner circular layer of the same tissue. This structure is called the tunica albuginea

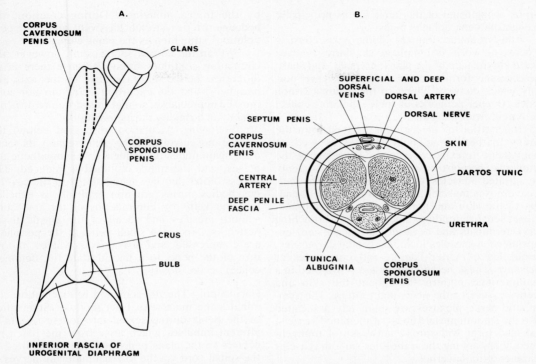

FIG. 15.55 **A.** Gross anatomy of the human penis. **B.** Anatomy of the penis in cross section. (After Goss [ed]. *Gray's Anatomy of the Human Body,* 28th ed, 1970, Lea & Febiger.)

(Fig. 15.55B). Although some elastic connective tissue networks are present, this membrane is resistant to distension beyond a certain point. In the midline between the two corpora cavernosa, the tunica albuginea forms a septum having abundant perforations that allow free communication between the cavernous spaces on each side. Efferent veins in the dense inner layer of the tunica albuginea provide numerous channels for drainage of blood from the cavernous spaces.

The erectile tissue of the corpora cavernosa consists of a spongy system of irregular vascular spaces (venous sinusoids) which are supplied with blood by afferent arteries and drained by efferent veins. These vessels are collapsed when the penis is flaccid, but during erection they form large cavities which are engorged with blood which is under very high pressure. The surface of the vascular spaces within the corpora cavernosa are lined with endothelium which is continuous with that of the arteries and veins. The partitions (or trabeculae) between these spaces are composed of dense connective tissue having collagenous and elastic fibers as well as strands of smooth muscle.

The skin that covers the penis is thin and numerous smooth muscle fibers are present in the well-developed subcutaneous layer, although adipose tissue is absent. The glans is covered by a circumferential layer of skin, the prepuce, which has the character of a mucous membrane on its inner surface adjacent to the glans.

The vascular supply to the penis is from the arteria penis. This vessel branches repeatedly into a number of large vessels, such as the arteria dorsalis penis and arteria profunda penis, among others, which supply different regions of the organ. However, all of these vessels anastomose eventually. In all of the arterial branches, the intima is thrown into thickenings which extend into the lumen before they enter the erectile tissue.

As the arterial branches penetrate the tunica albuginea to enter the corpora cavernosa, they tend to run in a longitudinal, forward direction that is highly convoluted when the penis is flaccid. These helicene arteries have a thick media, and when they are roughly 70 μ in diameter they course in the longitudinal trabeculae of the cavernous bodies and open directly into the cavernous spaces. The blood supply to the corpus cavernosum urethra is similar to that of the corpora cavernosa.

Most of the excurrent blood from the corpora cavernosa is drained via the vena profunda penis. The branches of this vessel have a thick muscular coat, and they originate beneath the tunica albuginea, particularly in the posterior regions of the three masses of erectile tissue. In turn, the branches of the vena profunda penis originate through the anastomosis of a large number of venules that are situated beneath the tunica albuginea. These postcavernous venules drain the largest spaces within the erectile tissues. Blood from the corpus spongiosum of the penis drains principally through the vena dorsalis penis.

The lymphatic drainage of the penis consists of dense networks of lymphatic capillaries that are located in the skin of the glans, prepuce, and shaft; these vessels form a superficial lymph vessel dorsally which conveys lymph to the inguinal lymph nodes. Deeper networks of such capillaries collect the lymph from the glans.

Innervation of the penis is derived from the parasympathetic (pudendal nerves) as well as sympathetic (plexus cavernosus) branches of the autonomic nervous system. These nerves supply the striated muscles of the penis (eg, the bulbocavernosus muscle), as well as the sensory endings to the skin and urethral mucous membrane. These sensory endings include free branching nerve terminals and a variety of specialized, encapsulated corpuscles including genital corpuscles, corpuscles of Vater–Pacini, and corpuscles of Meissner. These nerve endings are distributed in a characteristic pattern throughout the skin and deeper tissues of the penis and urethra. The sympathetic nerve plexuses are intimately associated with the smooth muscle fibers of the blood vessels, and also form widespread networks of unmyelinated fibers among the trabecular smooth muscles of the corpora cavernosa.

Mechanism of Erection. Various psychic as well as direct physical stimuli to the glans penis or other regions of the body can initiate the phenomenon of erection in the male.

The anatomic arrangement of the incurrent and excurrent blood vessels to the penis is responsible for the mechanism of erection. The arteries play the active role in this process, whereas the veins are passive. The degree of erection is proportional to the degree of sexual stimulation whether this is psychic, physical, or a combination of the two kinds of stimuli.

Erection is caused by afferent parasympathetic impulses which pass from the sacral region of the spinal cord via the nervi erigentes to the arteries of the penis. The smooth muscle fibers of the arteries and the principal erectile bodies (corpora cavernosa penis) relaxes; hence these structures dilate. The ensuing rise in blood pressure overcomes the elastic resistance of the tissue and stretches the media of the arteries. The longitudinal ridges in the intima that were mentioned above are believed to facilitate the rapid dilatation and filling of the vessels with blood. The lacunae of the cavernous bodies become distended with arterial blood, and as the helicene arteries expand, the peripheral spaces and thin walled veins beneath the tunica albuginea are compressed against this inelastic structure, thereby decreasing the outflow of blood from the organ. Blood thus accumulates in the corpora cavernosa under increasing pressure, and the tissue becomes rigid and elongates because of this massive vascular engorgement. Over-distension of the erectile tissues is prevented

by the tunica albuginea. During erection, the helicene arteries stretch passively and their convolutions straighten as the penis elongates.

Within the corpus spongiosum, however, the circulation continues freely, because there is no difference between the lacunae in the axis and periphery, and the excurrent veins are not compressed appreciably, so that this structure does not attain much rigidity during erection.

Following ejaculation (discussed below) the arterial musculature of the penis regains its tonus once again under the influence of sympathetic impulses, and the influx of blood is reduced. The excess blood that has accumulated in the cavernous bodies during erection gradually is expelled into the veins by contractions of the trabecular smooth muscles and recoil of the elastic tissue networks. Since the small veins in the periphery are compressed, and also possess valves, the return of the penis to a flaccid state (ie, detumescence) occurs slowly.

Ejaculation. The phenomenon of ejaculation culminates the male sexual act (coitus). Stimulation of the penis during sexual intercourse results in afferent impulses which arise from the touch receptors in the glans penis, and are transmitted to the spinal cord via the internal pudendal nerves.

Ejaculation per se is a two-stage reflex, whose centers lie in the upper lumbar centers of the spinal cord (L_1 and L_2), as well as in the sacral region (S_1 and S_2). When sexual stimulation becomes sufficiently intense, the lumbar centers begin to discharge rhythmic sympathetic impulses which leave the cord at L_1 and L_2 and are transmitted to the genitals via the hypogastric plexus. These impulses initiate emission which is the first stage in ejaculation. During this process, the smooth musculature of the epididymis, vasa deferentia, seminal vesicles, and prostatic capsule contracts so that sperm and glandular secretions are expelled into the internal urethra. During ejaculation proper, the semen is expelled from the urethra by contraction of the bulbocavernosis muscle, which in turn compresses the bulbus urethrae, under the influence of rhythmic impulses which now arise in the sacral region (S_1 and S_2) of the spinal cord. This stage of ejaculation occurs at the time of orgasm.

It should be mentioned that the cerebrum, although of considerable normal importance to the male sexual act, does not appear of critical importance to its accomplishment. Thus, decerebrate animals and humans whose cords have been transected above the lumbar region can exhibit erection and ejaculation following appropriate direct genital stimulation. Consequently, the male sexual act would appear to result from inherent reflex mechanisms which are integrated in the lumbar and sacral segments of the spinal cord, and these mechanisms are evoked by direct sexual stimulation of the genitals and/or psychic influences which arise in the cerebral cortex.

Semen. Sperm, together with the secretions of the ducts and accessory glands of the reproductive tract (ie, the seminal vesicles, prostate, and bulbourethral glands) comprise the semen which is ejaculated during the male sexual act. The approximate pH of this complex liquid is 7.5, since the alkaline prostatic and bulbourethral fluids neutralize the slightly acid secretions of the other components of the semen.

Sperm appear to be nonmotile in the seminiferous tubules, whence they are propelled by the probable action of two factors. Passive fluid pressure of the tubular secretions as well as by slow, spontaneous peristaltic contractions of the seminiferous tubules. Within the ductuli efferentes, the beat of the eipithelial cilia transports the sperm toward the epididymis, and the secretions of the nonciliated cells doubtless add fluid to the sperm.

The epididymis is traversed quite slowly; hence sperm may remain in this structure for months, especially in the tail of the organ. Rhythmic contractions of the smooth muscle surrounding the epididymis may be responsible for the slow movement of the sperm through this duct. During sexual activity, however, this circular smooth muscle exhibits vigorous rhythmic contractions, and this contractile activity is of major importance to the discharge of stored sperm during ejaculation.

As the epididymis is the major locus for sperm storage, germ cells do not accumulate in the vas deferens. The vas deferens is, in fact, merely a conduit; hence it is adapted only to the rapid propulsion of sperm to the exterior of the body during sexual activity.

The glandular seminal vesicles contribute greatly to the volume of the ejaculate with a secretion that is rich in fructose, the major carbohydrate substrate that is utilized by the sperm. This compound is used as an energy source to promote motility of the sperm after their ejaculation. Seminal fluid also contains a yellowish pigment (flavins) which gives semen a strong fluorescence in ultraviolet light, a fact which is of medicolegal significance in the detection of semen.

During ejaculation, the muscular cells of the prostate contract and the voluminous fluid that is ejected from the gland tends to dilute the thicker portions of the semen as well as to stimulate sperm motility itself.

When the semen enters the urethra, it mixes with the secretions of the bulbourethral glands and glands of Littre. Thence it is discharged from the penis by rhythmic contractions of the bulbocavernous muscle.

Mere distension of the various regions of the genital tract by fluid secretions, especially in adolescent boys, is sufficient to cause expulsion of semen and ejaculation during sleep, the so-called nocturnal emission. This phenomenon can occur without any direct genital stimulation, and as it may occur during dream states, it is indicative of a strong psychic involvement.

In man, the usual volume of ejaculate is around 3.5 ml. The sperm contribute less than 10 percent to this total volume, the remainder being the seminal plasma whose components were described above. The sperm vary in concentration in semen from around 50 to 150 million per ml; each ejaculate may contain roughly 200 to around 500 million sperm. The flagellum or tail of the sperm cells undulates rapidly and thus the sperm moves with the head forward at a rate of about 18 μ per second, while simultaneously rotating on its long axis.

Within the male, sperm can live for months and survive for some time after death of the individual. Outside of the body, these cells can survive for several days under appropriate conditions, and even for years, when frozen into a state of "suspended animation" in a suitable medium, and their capacity to fertilize ova after thawing may be retained after such treatment. Viable sperm can also be found for several days after coitus in the female genital tract, ie, within the uterus and fallopian tubes.

Some investigators are of the opinion that the various components of semen that were discussed above are discharged from the urethra in a definite sequence; viz., during erection the mucoid secretions of Cowper's and Littre's glands lubricate the urethra; at the onset of ejaculation, the prostatic secretion is discharged first, and this prostatic discharge is following in turn by expulsion of the sperm masses that are contained in the epididymis through the vas deferens; finally the ejaculate receives the secretion of the seminal vesicles.

Chemistry and Biosynthesis of the Male Sex Hormones

The male sex hormones, androgens, are synthesized by the interstitial tissue (Leydig cells) of the testes, and these substances are C19 steroids that have an —OH group at the 17 position. Testosterone is the principal testicular hormone, although other androgenic substances have been isolated from testicular tissue, as depicted in Figure 15.56. Estrogenic substances have also been isolated from the testes in small quantities, even as androgenic compounds have been found in ovarian tissue.

The biosynthetic pathways for testosterone from acetate and cholesterol in the Leydig cells as shown in Figure 15.56 should be compared with the biosynthetic pathways for androgens and for the biosynthesis of adrenocortical steroids (Fig. 15.39), as well as those involved in estrogen synthesis by the ovary (Fig. 15.68). According to current belief, the biosynthetic pathways in all endocrine tissues which synthesize steroid hormones are similar; however, these organs may differ in the specific enzyme systems present in their cells. In the Leydig cells of the testes, the 11- and

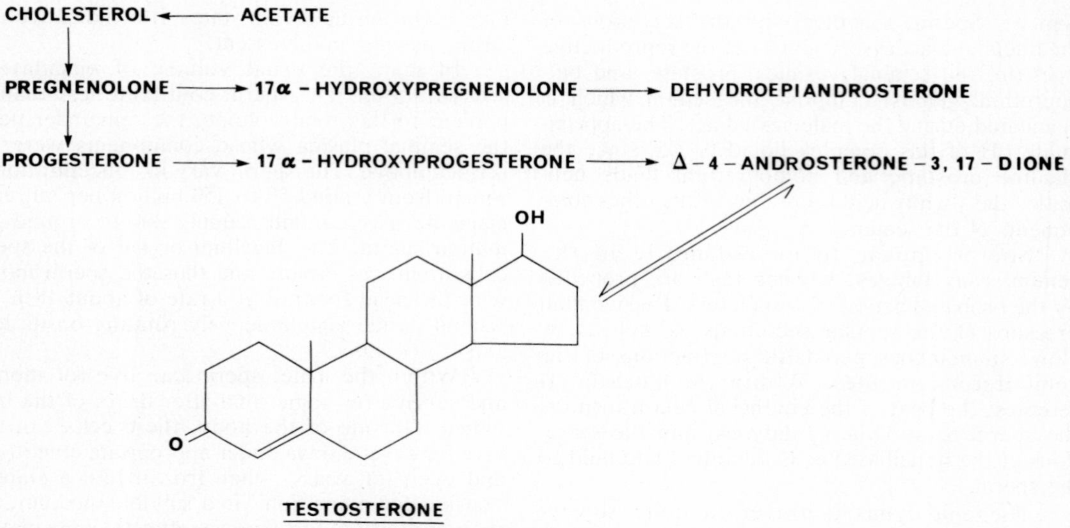

FIG. 15.56. Biosynthetic pathway for testosterone. (Data from Ganong. *Review of Medical Physiology,* 6th ed, 1973, Lange Medical Publications; White, Handler, and Smith. *Principles of Biochemistry,* 4th ed, 1968, The Blakiston Division, McGraw-Hill.)

21-hydroxylases found in the adrenal cortex are absent; therefore pregnenolone is hydroxylated in the 17 position. Scission of the side chain thus produces 17-ketosteroids which in turn are converted into testosterone.

The rate-limiting step in the synthesis of testosterone is in the 20 α-hydroxylation of cholesterol, and this reaction is regulated by the interstitial cell stimulating hormone (ICHS), which is secreted by the adenohypophysis. As discussed earlier, testosterone is also synthesized in the adrenocortical tissue of both men and women, a process that is not regulated by ICSH.

Testosterone Secretion

In normal postpubertal males, the secretion rate of testosterone ranges between four and nine mg per day. Minute quantities of this androgenic hormone are also secreted by the ovaries as well as by the adrenal cortex of normal females.

Transport, Metabolism, and Excretion of Testosterone

Roughly 66 percent of the testosterone secreted by the testes is transported in the plasma bound to proteins. Albumin and the β globulin which also binds estradiol are involved in this process. This β globulin is called gonadal steroid binding globulin (GBG).

The total plasma testosterone, both free and bound, is about 0.6 μg/100 ml in normal males and around 0.1 μg/100 ml in women. In men this value decreases with age.

The metabolism of testosterone is shown in

Figure 15.57. A minute quantity of the total secreted testosterone is converted into estrogen at an unknown site in the body. However, the major portion of this hormone is converted into 17-ketosteroids (ie, they have a ketone group at C-17) in the liver. The four principal urinary metabolites of testosterone are dehydroepiandrosterone, epiandrosterone, androsterone, and etiocholanone (etiocholane 3-α-ol-17-one) (Fig. 15.58). The latter two compounds are the most important urinary androgen metabolites from a quantitative standpoint.

Approximately one-third of the total urinary 17-ketosteroids of males are of testicular origin, whereas the remaining two-thirds are derived from the adrenal corticoids. The majority of the 17-ketosteroids are weakly androgenic, viz., they have less than 20 percent of the androgenic potency of testosterone. It should be stressed that neither are all androgens 17-ketosteroids, nor are all 17-ketosteroids androgens. Thus testosterone is not a 17-ketosteroid, and the 17-ketosteroid urinary metabolite of this hormone is not androgenic. The urinary concentration of 17-ketosteroids provides a useful clinical indication of endogenous synthesis and secretion of androgenic hormones.

In the normal adult female the daily total 17-ketosteroid excretion is between 4 to 17 mg, whereas in males the excretion of these compounds ranges between 6 and 28 mg. The excretion of 17-ketosteroids before puberty is at approximately one-third that of the adult level. These differences are caused by the precursors of 17-ketosteroids secreted by the gonads before and after puberty, as well as the contribution to the urinary 17-ketosteroids by the C19 and C21 steroids of the adrenal cortex.

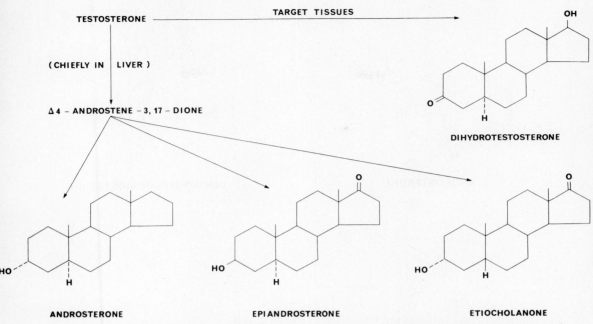

FIG. 15.57. Metabolic products of testosterone. Dihydrotestosterone, a metabolic product of testosterone, is the physiologically active male hormone in some target tissues. (Data from Ganong. *Review of Medical Physiology,* 6th ed, 1973, Lange Medical Publications; White, Handler, and Smith. *Principles of Biochemistry,* 4th ed, 1968, The Blakiston Division, McGraw-Hill.)

Regulation of Testosterone Secretion

Interstitial cell stimulating hormone (ICSH) that is secreted by the adenohypophysis under the influence of the hypothalamus acts directly upon the Leydig cells to regulate testosterone secretion. By a negative feedback mechanism the circulating testosterone level inhibits the secretion of ICSH releasing factor by the hypothalamus and thereby decreases testosterone secretion by the testes. Follicle stimulating hormone (FSH), which is also secreted by the adenohypophysis, maintains gametogenesis in the testes, a process to which androgens also contribute. A postulated scheme for the regulation of testicular function is shown in Figure 15.59. There is experimental and clinical evidence that the negative feedback effect of testosterone secretion upon hypophyseal ICSH release is exerted directly at the hypothalamic level. FSH secretion, on the other hand, appears to be regulated independently of ICSH secretion. A hypothetical inhibitory substance that is indicated by a question mark in the figure, may be secreted by the seminiferous tubules. It acts at either the hypothalamic or the pituitary level to decrease the secretion of FSH. Alternatively, the seminiferous tubules may remove circulating FSH actively from the blood and thus reduce the blood level of this hormone. This effect in turn would stimulate the further secretion of FSH. Estrogenic hormones reduce plasma testosterone levels, probably because they lower ICSH secretion; testicular estrogens also might have a role in the regulation of

FSH secretion. These concepts are largely speculative at present, however.

Physiologic Actions of Testosterone

Apart from the known action of testosterone and other androgenic compounds in the regulation of testosterone and other androgenic compounds in the regulation of ICSH secretion, such hormones also participate in the development and maintenance of the male secondary sex characteristics. In addition, androgens have an important anabolic effect in protein metabolism; consequently they exert a growth-promoting effect.

MALE SECONDARY SEX CHARACTERISTICS. In males at puberty there occur profound changes in body configuration, hair distribution, and genital size (Fig. 15.60). These and other male secondary sex characteristics will be considered individually, but all result directly from the effects of increased testicular androgen secretion (androgenic activity) of the male hormones.

Insofar as body configuration is concerned, at puberty the shoulders broaden and the skeletal muscles enlarge. A beard appears and the scalp hair line recedes somewhat. Pubic hair grows in a typical male pattern, and hair also appears in the axillae, on the chest, and around the anus together with an increase in body hair in general.

The external and internal genitalia show profound morphologic changes. The penis elongates and becomes wider. The scrotum develops pig-

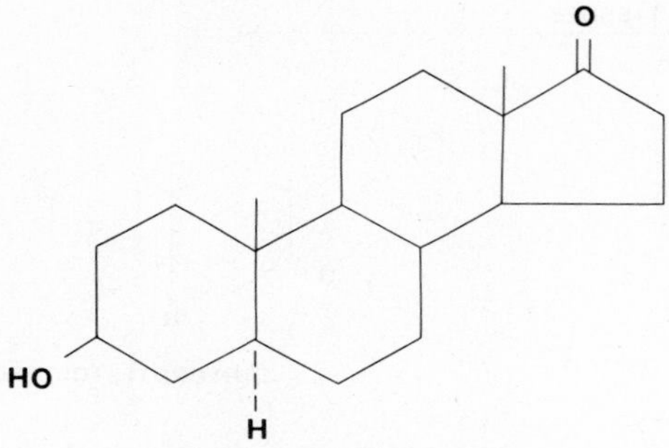

ANDROSTERONE

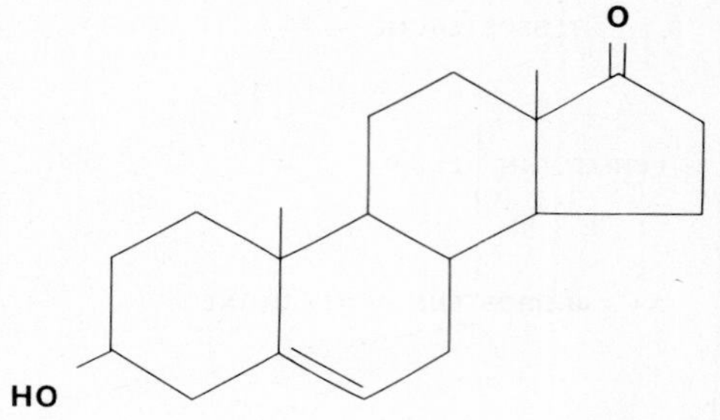

DEHYDROEPIANDROSTERONE

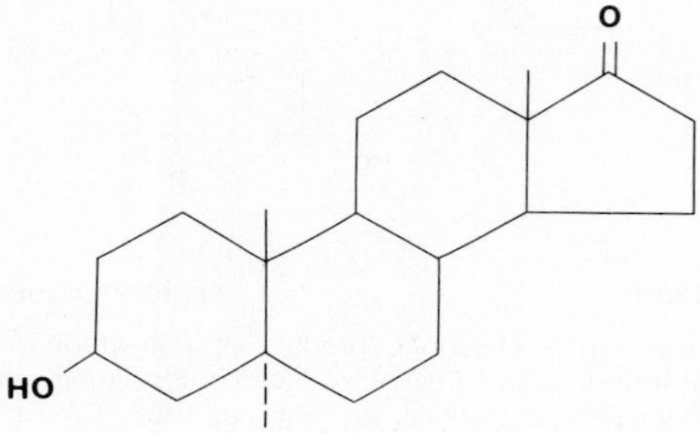

EPIANDROSTERONE

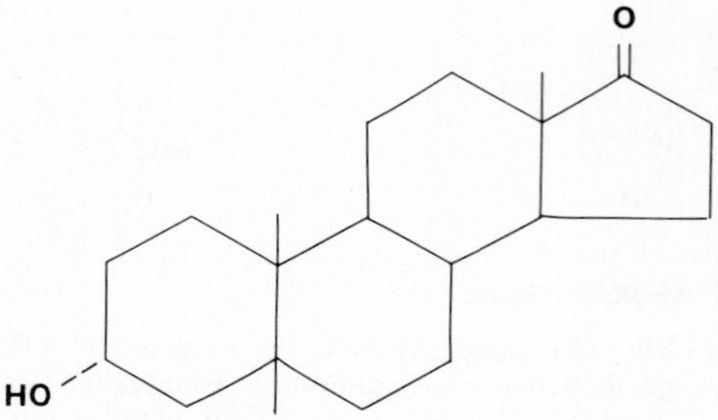

ETIOCHOLANE – 3a – OL – 17 – ONE

FIG. 15.58. Principal urinary metabolites of testosterone. From a quantitative standpoint, etiocholanone and androsterone are the most important urinary androgen metabolites. (After White, Handler, and Smith. *Principles of Biochemistry*, 4th ed, 1968, The Blakiston Division, McGraw-Hill.)

mentation and becomes rugose. The seminal vesicles enlarge, commence to secrete, and produce fructose, which provides an important substrate for the metabolic processes of the spermatozoa. The prostate and bulbourethral glands also enlarge and commence to secrete under the influence of testosterone.

The larynx and vocal cords enlarge, so that the voice becomes deeper.

The sebaceous glands of the skin secrete a more plentiful and viscous product, thereby predisposing the individual to acne.

Psychologically, the individual becomes more aggressive and active, and an interest in females appears (increased libido or sex drive).

All of these androgenic effects form the pattern of virilization or masculinization that is produced by the male hormones.

It is important to mention that androgens and estrogens exert a mutually antagonistic biologic action upon each other. Thus, the biologic actions of the male hormones are inhibited by the administration of female hormones and vice versa. Testosterone administration (or ovarectomy) sometimes are employed therapeutically for carcinoma of the breast. Similarly estrogens are used in the treatment of prostatic carcinoma, often in conjunction with castration (orchidectomy).

EFFECTS OF ANDROGENS UPON PROTEIN METABOLISM. Since androgenic compounds, particularly testosterone, stimulate the synthesis of protein and decrease the breakdown of protein in general, these hormones exert a marked anabolic effect on protein metabolism. Hence, they promote growth and result in a number of the alterations in the various secondary sex characteristics that were described above.

Androgens also stimulate epiphyseal closure in the long bones, thus causing cessation of growth.

The male hormones also produce a number of secondary effects that result from their primary anabolic action. Androgens produce a moderate retention of water, sodium, potassium, calcium, sulfate, and phosphate. Furthermore, androgens cause an increase in the size of the kidneys. However, exogenous androgens, when administered to patients with wasting diseases in doses sufficient to evoke the anabolic effect, also produce masculinization as well as an increased libido; consequently, their clinical usefulness in such conditions is limited. Synthetic androgen analogues in which the androgenic and anabolic effects of the male hormones have been separated have not proven to be especially useful. Certain synthetic androgens are discussed below.

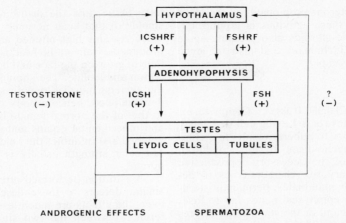

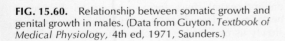

FIG. 15.59. Mechanisms underlying the regulation of testosterone synthesis and sperm production by the testis. It is unclear whether increased blood FSH levels directly inhibit the release of follicle-stimulating hormone-releasing factor (FSHRF) from the hypothalamus as indicated by the question mark or whether another humoral factor is involved in this process. (Data from Ganong. *Review of Medical Physiology,* 6th ed, 1973, Lange Medical Publications.)

Mechanism of Action of the Androgenic Hormones

Within the prostate, and presumably other organs, testosterone is converted into dihydrotestosterone (DHT, Fig. 15.57), and this metabolic product rather than testosterone per se is the physiologically active compound in such tissues.

It has been demonstrated experimentally that in prostatic tissue androgens accelerate protein synthesis. In turn, this effect appears to be related to an increased number of ribosomes which are rich in RNA. The protein-synthesizing capacity of the tissue is elevated, and this mechanism may also be responsible for the actions of testosterone in other organs.

Synthetic Androgens

Testosterone that is administered orally exhibits an androgenic potency which is only 15 percent that of injected testosterone. This is due to partial hepatic inactivation of the hormone as it is absorbed from the gut. A synthetic testosterone analogue, methyltestosterone (Fig. 15.61), has found extensive clinical use because this compound has the highest known androgenic potency when given orally.

Extensive investigations have been carried out in an attempt to dissociate the masculinizing effects of androgens from the desirable anabolic effects of these compounds, especially since testosterone is useful therapeutically in carcinoma of the breast. Steroids that lack an angular methyl group in position 10 of the steroid ring, and thus called norsteroids, have proven of interest. Examples of such compounds are 19-nortestosterone and 17-α-ethyl-19-nortestosterone, whose structural formulas are given in Figure 15.61. These substances have exhibited a relative anabolic/androgenic potency of about 20:1 in animal bioassay

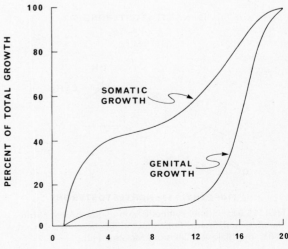

FIG. 15.60. Relationship between somatic growth and genital growth in males. (Data from Guyton. *Textbook of Medical Physiology,* 4th ed, 1971, Saunders.)

systems. For testosterone propionate, on the other hand, the ratio of these activities is around 1:1. These two synthetic steroids also promote some nitrogen retention in humans and also influence calcium balance favorably.

Clinical Correlates

CASTRATION (ORCHIDECTOMY).   Complete extirpation of the testes (or a Leydig cell deficiency which develops) prior to puberty results in a clinical picture of eunuchoidism. In this condition, there is a failure of the accessory reproductive glands and secondary sex characteristics to develop properly. In human males, permanent sterility ensues, and the epiphyseal plates fail to fuse, resulting in enlargement of the stature and disproportionately long legs. The individual is not as tall as in pituitary gigantism, however. In eunuchoidism there also is an increased amount of

fatty tissue, and the distribution of the lipids is similar to that seen normally in the female. The voice remains high-pitched and the larynx is not prominent. Hair on the head is plentiful although it fails to grow on the face and body, eg, in the pubic region and axillae. The shoulders are narrow, and the skeletal muscles remain small as well as underdeveloped, hence a body conformation similar to that of the normal female is present. The penis and other genital organs and glands remain undeveloped and infantile; the libido does not develop. Muscular strength usually is considerably below normal.

Castration performed after puberty results in similar defects to those described above. However, the effects are much less severe, ie, the difference between the two situations is in degree rather than in kind, except that sterility is permanent and the voice remains deep, as laryngeal growth at puberty is permanent.

Exogenous administration of androgenic compounds will prevent the changes that occur following castration, and will reverse to some extent the alterations that were produced by the androgen lack.

CRYPTORCHIDISM (OR CRYPTORCHISM).   In cryptorchidism, the testes develop normally within the abdominal cavity but do not migrate into the scrotum. Thus one or both testes may be retained in the abdominal cavity or inguinal canal in around 10 percent of all newborn males. Of these, spontaneous descent of the testes generally occurs, so that cryptorchidism is rare by the time that adolescence is reached. Gonadotropin therapy may accelerate testicular descent, otherwise surgical correction is necessary. Preferentially, the operation should be accomplished before puberty; otherwise the higher abdominal temperature will cause permanent and irreversible damage to the spermatogenic epithelial tissue so that the cryptorchid individual is sterile, as noted earlier.

TESTICULAR HYPOFUNCTION.   Clinically the effects of male hypogonadism depend upon whether the deficiency occurs before or after puberty, and upon whether spermatogenesis and/or endocrine function is involved. Testicular deficiency may be caused by hypothalamic and hypophyseal disease or else by a number of primarily testicular and chromosomal defects. If the gametogenic function of the testes fails, sterility ensues; whereas if the endocrine function of the testes fails, then the secondary sexual characteristics revert to the prepubertal type. However, this reversion occurs quite slowly as once these structures and functions have developed, their maintenance requires very little androgen. The voice of adult males with hypogonadism remains deep, as the laryngeal changes that accompany puberty are irreversible.

Pubic and axillary hair do appear in males with gonadal hypofunction owing to adrenocortical androgen secretion; however, the bodily hair is generally sparse and the pubic hair assumes the

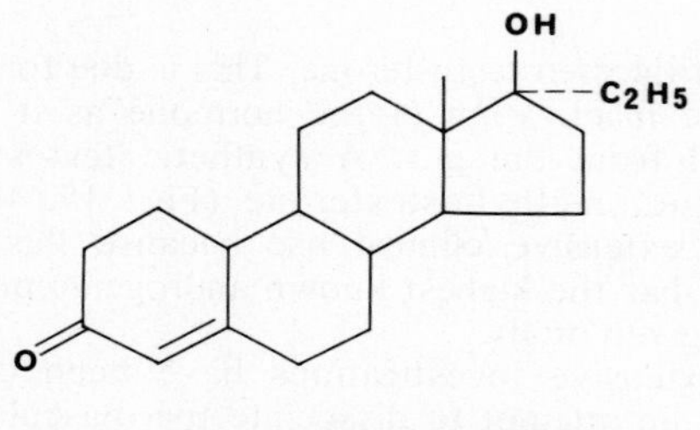

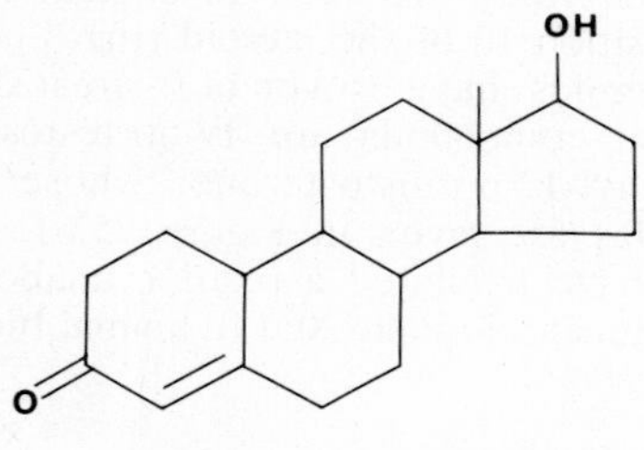

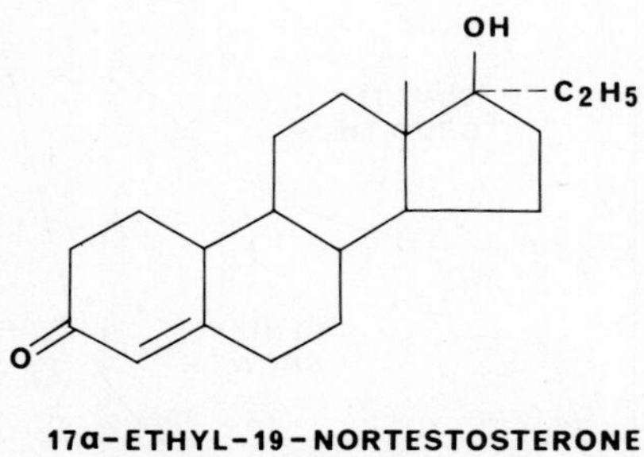

FIG. 15.61. Some synthetic androgens. (After White, Handler, and Smith. *Principles of Biochemistry,* 4th ed, 1968, The Blakiston Division, McGraw-Hill.)

inverted triangle (ie, base up) pattern of the female rather than that of the normal male.

ANDROGEN-SECRETING TUMORS. Clinically, testicular hyperfunction without the presence of a tumor is not a recognized pathologic entity. Testicular interstitial cell tumors that secrete androgens are quite rare, and endocrine symptoms of such tumors are seen only in prepubertal males. A precocious pseudopuberty develops in such patients; however, no spermatogenesis is present, although there is a prematurely early onset of most of the masculinizing characteristics that normally occur at puberty and which were enumerated earlier.

GYNECOMASTIA. Development of the breast in the male is referred to as gynecomastia. This condition, which may be unilateral, or more frequently bilateral, occurs briefly in about 70 percent of the male population at the time of puberty. Gynecomastia may also be caused directly by estrogen therapy in men, as well as by estrogen-secreting tumors of the adrenocortical tissue. It also is seen in patients with eunuchoidism that is caused by testicular disease, hyperthyroidism, hypothyroidism, cirrhosis of the liver, and during early digitalization for congestive heart failure. Because of the wide diversity in these conditions,

there are doubtless a number of primary etiologic factors involved in the development of gynecomastia, although the basic cause of gynecomastia remains obscure.

PHYSIOLOGY AND ENDOCRINOLOGY OF REPRODUCTION IN THE FEMALE

Reproductive functions in the female may be divided into two major categories: first, the physiologic changes in the body that prepare the reproductive organs for conception and gestation; and second, the period of gestation per se. Anatomically, the female reproductive system consists of the ovaries, oviducts (or fallopian tubes), uterus, vagina, and external genitalia (Fig. 15.62). In addition to these primary organs of the female genital system, the mammary glands of the skin are accessory structures which provide for nourishment of the newborn. In the sexually mature female, the ovaries, fallopian tubes, and uterus exhibit profound rhythmic morphologic and functional alterations in relation to the sexual (or menstrual) cycle and to pregnancy. Such changes are regulated in turn by complex neural and endocrine mechanisms. These and related topics will form the basis of the present discussion.

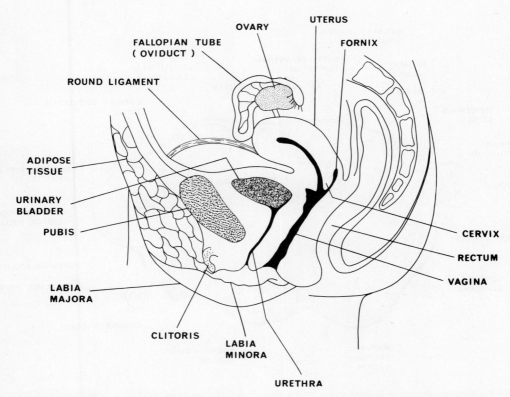

FIG. 15.62.   Gross anatomy of the female reproductive system. (After Goss [ed]. *Gray's Anatomy of the Human Body,* 28th ed, 1970, Lea & Febiger.)

Functional Morphology and Physiology of the Female Sex Organs and Mammary Glands

OVARY

Gross Anatomy.   In adult humans, the ovaries are paired, slightly flattened, roughly almond-shaped organs weighing approximately four to eight grams each. The ovary measures between 2.5 to 5 cm in length, 1.5 to 3 cm in width, and 0.5 to 1.5 cm in thickness. The hilus of the ovary is attached to the broad ligament by the mesovarum. The broad ligament in turn extends from the wall of the pelvic cavity laterally to the uterus.

Histology and Histophysiology.   In cross section, the ovary is composed of a thick cortex that surrounds the medulla or (zona vasculosa). The following layers and structures may be discerned from without inward (Fig. 15.63). On the surface, the ovary is invested in a continuous layer of squamous to cuboidal "germinal" epithelium. Beneath this layer is a dense connective tissue layer, the tunica albuginea. Embedded in the connective tissue stroma of the cortex are follicles that contain the female sex cells, the ova. The follicles in turn exhibit a broad range of sizes, a fact that reflects the various developmental stages of these structures. At maturity, the follicle ruptures at the ovarian surface, thereby releasing the ovum which

now passes into the open end of the adjacent oviduct.

The boundary between cortex and medulla is not clearly delineated in the ovary, and the medullary tissue is composed primarily of loose connective tissue as well as relatively large twisted blood vessels.

Ovarian Follicles.   The ovarian follicles are embedded within the cortical stroma beneath the tunica albuginea, and are more abundant in younger than in older individuals. Over 400,000 of these structures have been seen in sections of ovaries from young females, but the total number of follicles decreases progressively throughout life so that at, and beyond, menopause they are rare or may be absent.

The important aspects of follicular growth, development, and metamorphosis into the corpus luteum now may be discussed sequentially (see Figs. 15.9 and 15.63).

First, *primordial follicles* form the overwhelming majority of the ovarian follicles to be seen in a young ovary, and they are situated in a thick layer just beneath the tunica albuginea. Each primordial follicle is composed of a large, spherical oocyte invested in a layer of follicular cells.

The oocyte has a large, vesicular nucleus having finely granular, pale chromatin, and a prominent nucleolus. A well developed Golgi apparatus

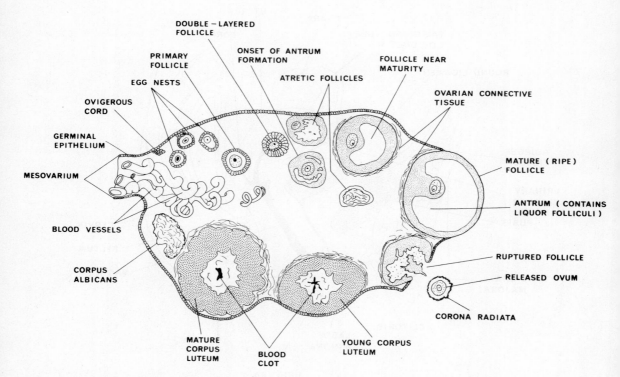

FIG. 15.63. Morphology of the human ovary. (Data from Arey. *Developmental Anatomy,* 7th ed, 1965, Saunders; Ganong. *Review of Medical Physiology,* 6th ed, 1973, Lange Medical Publications; Goss [ed]. *Gray's Anatomy of the Human Body,* 28th ed, 1970, Lea & Febiger.)

and abundant small mitochondria are also present in the oocytes. The primordial oocyte is surrounded by a single layer of flattened granulosa or follicular cells, and the surfaces of oocyte and follicular cells are closely apposed at this stage of development.

Second, *primary follicles* develop from primordial follicles. Thus, each month several primordial follicles commence to develop from previously inactive primordial follicles into mature ova. Then usually only one of these follicles commences to grow rapidly on about the sixth day of the sexual cycle, whereas the others regress for unknown reasons.

As oocyte enlarges, the single layer of follicular cells surrounding this cell becomes cuboidal or low columnar; mitotic division now converts this tissue into a stratified epithelium, and the oocyte now becomes surrounded by a clear, refractile layer called the zona pellucida which lies between the oocyte and the granulosa cells.

Concomitant with the increase in follicular size, all of these follicular structures move deeper into the cortex, and a sheath of stromal cells develops around the entire follicle (theca folliculi). Ultimately the theca folliculi differentiates into a layer of secretory cells, the theca interna, which in turn is surrounded by a connective tissue layer called the theca externa. The theca interna now develops a rich capillary plexus but the granulosa layer remains avascular throughout the growth and development of the follicle.

During growth of the oocyte, multiple and dispersed Golgi complexes develop, and these complexes eventually become located beneath the plasma membrane (or oolemma) of the ovum. The endoplasmic reticulum becomes more extensive, and the ribosomes more numerous; dense lipid granules are also present in the developing oocyte.

Third, *secondary follicles* are produced as the granulosa cells proliferate, the follicle becomes oval, and the oocyte takes on an eccentric position within the follicle. Irregular spaces that are filled with a clear liquor folliculi make their appearance within the follicle when the diameter of this structure reaches about 0.2 mm. The follicular size increases subsequently as the quantity of liquor folliculi increases. The spaces among the granulosa cells coalesce, so that a single cavity, the antrum, forms, and thereafter the follicle is called a secondary or vesicular follicle. At this stage, the ovum has reached its full size; however, the follicle continues to grow until it achieves a diameter of 10 mm or greater.

The vesicular (or secondary) follicle is lined by granulosa cells that form a stratified epithelium, and there is a thickening of these cells at one point within the ovum, the cumulus oophorus, and this region contains the ovum.

Fourth, the *mature follicle* takes between 10 to 14 days to develop from the onset of the sexual cycle. When mature, the follicle occupies the entire thickness of the ovarian cortex, and it forms a bulge like a blister upon the surface of the ovary; consequently the liquor folliculi appears to be under considerable pressure.

The membrana granulosa, which is the membrane lining the follicular cavity, now is surrounded by a basal layer which separates the granulosa from the theca interna. Mitotic division of the granulosa cells decreases and intercellular spaces among these cells become evident. The ovum gradually detaches from the granulosa by the formation of liquid-filled intercellular spaces within the cumulus oophorus. The layers of columnar granular cells which remain attached to the ovum as it becomes detached from the follicular wall are termed the corona radiata, and these cells are still present around the ovum after ovulation.

The theca folliculi is most fully developed in the mature follicule, and the internal layer (or theca interna) of these cells is located within a reticular network that is continuous with the remainder of the ovarian stroma. Despite the fact that the interna cells are derived from connective tissue cells, they have cytologic characteristics that are similar to other steroid secreting endocrine glands such as adrenocortical cells. Thus the cells of the theca interna are considered to synthesize and secrete the female sex hormones, or estrogens. The abundant capillary plexus that is present in this region substantiates this view. As noted earlier, generally only one of the many ovarian follicles that start to develop at the start of each sexual cycle reaches maturity, whereas the remainder degenerate and involute at various stages in their development.

Ovulation.   The process of ovulation liberates the ovum so that it can become fertilized after contact with a spermatozoon. Generally, only one ovum is released each month, but occasionally two, or even more, ova are liberated. Under normal circumstances, ovulation takes place on the fourteenth day of a theoretically ideal 28-day menstrual cycle (Fig. 15.9), although typical cycles sometimes may occur without ovulation (anovulatory cycles).

During ovulation, the ovum and the granulosa cells that comprise the corona radiata become detached from the remainder of the cumulus oophorus, and thenceforward float freely within the liquor folliculi. The superficial portion of the follicle on the ovarian surface now becomes progressively thinner, and fluid appears to be secreted more rapidly. The increased intrafollicular pressure causes the follicle to bulge outward in a prominent clear oval area called the macula pellucida. This process occurs rapidly (in about 5 minutes in the rat ovary), and then the macula ruptures (or bursts), thereby ejecting the ovum and the follicular fluid.

The fimbria of the oviducts lie quite close to the ovary during ovulation, and the currents developed by the cilia on their epithelial surfaces propel the ovum into the ostium of the oviduct.

Maturation of the Female Germinal Cells. The primitive germ cells of the early embryo are found in the endoderm of the yolk sac. Subsequently these cells migrate to the genital ridges and thence into the primordial ovary.* Following birth, these cells remain dormant within the primordial follicles for many years (ie, until puberty), and during this interval they are called primary oocytes. The primary oocytes thus are homologous to the primary spermatocytes of the male. In the germinal cells of both sexes, the prophase of meiosis takes considerable time and in the oocyte it takes place later in the follicular growth phase (Fig. 15.52). The first maturation or meiotic division takes place shortly prior to ovulation, and the chromatin is divided equally between the daughter cells. However, one of these cells, known as the secondary oocyte, receives nearly all of the cytoplasm, while the other daughter cell ultimately becomes the first polar body. This polar body in turn is a tiny cell with a nucleus, but which has almost no cytoplasm. The first meiotic division is complete just a few hours before ovulation, and it occurs while the ovum is still within the follicle.

After the first polar body has been extruded, the nucleus of the secondary oocyte now commences the second maturation division. This event, however, progresses only to metaphase unless fertilization intervenes, at which point the chromatin again divides equally. Once again, however, the cytoplasm is retained principally by one daughter cell, and half of the chromatin now is extruded as the second polar body.

As in spermatogenesis, the end result of the two meiotic divisions is the production of a mature ovum having a haploid chromosomal number.

Fertilization. The corona radiata that surrounds the ripe ovum undergoes a progressive dispersion following the arrival of a mass of spermatozoa, presumably due to depolymerization of the protein–polysaccharide intercellular cement, as well as to increased surface activity of the cells per se. The head of one spermatozoon penetrates the zona pellucida, a process which is unclear insofar as its detailed mechanism is concerned. Once within this clear region, the sperm stops moving and is drawn into the ooplasm, a process known as fertilization proper. At this juncture, the ovum is stimulated to complete the second meiotic division (which was arrested in metaphase), and the second polar body is expelled. Fusion of the nuclei of the egg and sperm then follows, and by this fusion a restoration of the diploid chromosome number now is achieved. Cleavage of the ovum now proceeds apace. On the other hand, an unfertilized ovum undergoes fragmentation and absorption or phagocytosis of the debris then follows. An ovum

probably remains fertile for roughly 24 hours after ovulation.

Development of the Corpus Luteum. The wall of the follicle collapses after expulsion of the ovum and liquor folliculi during ovulation, so that the granulosa becomes wrinkled (Fig. 15.9). A central blood clot may also form in this cavity because of hemorrhage from capillaries of the theca interna. The cells of the granulosa and theca interna now undergo extensive cytologic changes. They enlarge, accumulate lipid, and then metamorphose into pale-staining, polygonal lutein cells, and the follicle now is called the corpus luteum. The cells that originate from the granulosa comprise the principal mass of the lutein tissue, and are called the granulosa lutein cells. The peripheral cells derived from the theca interna are known as theca lutein cells. The mature corpus luteum now secretes the hormone progesterone.

During the process of luteinization within the follicular cells, as the capillaries of the theca interna grow and form a network about the lutein cells, the central blood clot is resorbed gradually and then is replaced by fibrous tissue.

If fertilization does not take place, a so-called corpus luteum of menstruation develops, and this structure survives for approximately 14 days. The secretion rate of progesterone falls gradually and the lutein cells ultimately degenerate, to be replaced by scar tissue (the corpus albicans) which gradually moves centrally into the ovary, a process that takes months or even years.

If fertilization takes place, however, the corpus luteum enlarges to become the corpus luteum of pregnancy. Following parturition, the degeneration and involution of this temporary endocrine organ proceeds as described above.

Follicular Atresia. In the human female, the period of reproductive capability lasts for about 30 years, which means that a total of roughly 400 ova reach maturity and are discharged during this entire interval, since usually only one ovum is released each month. Since there are approximately a half-million oocytes in the ovaries at birth over 99 percent of these elements ultimately degenerate, a process known as follicular atresia. Follicular atresia commences in intrauterine life, is pronounced at birth as well as before puberty, and then it continues at a lower rate throughout the reproductive life-span of the individual. The physiologic mechanisms underlying this process, as well as the biologic reasons for this lavish waste of germinal tissue are unknown.

Atresia can take place at any stage of follicular development, and even mature follicles can undergo this process. The oocyte shrinks and degenerates and the granulosa cells involute. Ultimately the atretic follicle resembles a degenerate corpus luteum, except for the scar tissue that surrounded the zona pellucida of the ovum, as well as the remains of other typical cellular components of the

---

The "germinal epithelium" that invests the ovary was so named because at one time it was believed that the oocytes developed from these cells, a concept that is no longer tenable. This misnomer persists, however.

follicle. Hence these structures are known as corpora lutea atretica.

Ovarian Interstitial Tissue. The interstitial tissue of the ovarian cortex is composed of spindle-shaped cells that resemble smooth muscle fibers, save that they lack myofibrillae. In addition, networks of reticular connective tissue fibers are also present, whereas elastic tissue is associated only with the ovarian blood vessels. This ovarian stroma has no endocrine functions, and ovarian follicles are scattered throughout the cortical stroma.

In contrast to the cortical stroma, the medulla of the ovary is composed of loose connective tissue, numerous elastic fibers, and smooth muscle bands that are associated with the blood vessels.

Vascular Supply and Innervation of the Ovary. The ovarian artery, which provides the principal blood supply to the ovary, arises directly from the aorta caudally to the renal vessels. The ovarian artery then anastomoses with the uterine artery, which in turn runs upward from the cervix and passes along the lateral aspect of the uterus. Vessels from the anastomosis that is produced by fusion of the ovarian and uterine arteries now enter the ovarian hilus, and these vessels branch extensively as they pass through the medulla of the ovary. These contorted vessels are known as helicine arteries, and similarly to the vessels in the corpora cavernosa penis, they have longitudinal ridges on their internal surfaces. Within the medulla, the helicine arteries form a plexus, and radial branches of the plexus then pass among the follicles into the cortex where capillary networks are formed, which networks are continuous with thecal networks of the larger follicles. The veins follow a course similar to that of the arteries, and in the medulla the veins anastomose to form a plexus in the hilus.

Lymphatic capillary networks originate in the ovarian cortex, but lymph vessels equipped with valves occur only outside of the hilus.

Nerve fibers to the ovary originate from the ovarian plexus, and the ovarian nerves enter the hilus along with the blood vessels. These nerves are principally unmyelinated, although some myelinated fibers also are present. The presence of sympathetic nerve cells in the ovary has not been demonstrated unequivocally, although sensory fibers that terminate in Pacinian corpuscles have been found in the stroma.

OVIDUCT (FALLOPIAN TUBE)

Gross Anatomy. The oviducts are paired, muscular tubes about 12 cm in length that are lined with a mucous membrane and serve to transport the ovum from ovary to uterus. The oviducts are located within the free anterior border of the broad ligament, one on each side of the uterus (Fig. 15.62).

The lumen of the oviducts is open to the abdominal cavity at its upper end, and is continuous with the uterine cavity at the lower end. There are four anatomic subdivisions of the oviduct: (1) the portion that traverses the uterine wall, called the pars interstitialis; (2) a narrow region that is located near the uterine wall, the isthmus; (3) an expanded segment, the ampulla, lying above the isthmus; and fourth, the funnel-shaped opening of the oviduct into the abdominal cavity, called the infundibulum. The latter segment has numerous long fingerlike processes, the fimbriae, which partially envelop the ovary.

Histology. Three layers are distinguishable in the wall of the various regions of the oviduct: an inner mucous membrane, a muscular layer, and an outer serous coat. The mucous membrane of the oviduct is thick and folded, so that in cross section the lumen appears as a system of branching labyrinths surrounded by masses of epithelial tissue. This folding is most highly developed in the ampulla. The oviduct is lined by simple columnar epithelium and two cell types are evident. Ciliated epithelial cells are most abundant on the fimbriae and in the ampulla, and the pattern of ciliary beat is toward the uterus. The other type of cell is nonciliated and may be secretory. Possibly each of these cell types represents different functional states of a single type of cell. Although true glands are absent in the oviduct, this structure nevertheless undergoes cyclic morphologic changes like those observed in the uterine epithelium.

The lamina propria of the mucous membrane is a meshwork of reticular connective tissue fibers and abundant fusiform cells. The muscular layer surrounding the mucosa is comprised of an inner circular to spiral layer and an outer longitudinal layer. There is no clear demarcation between these two layers, however. The bundles of smooth muscle fibers in turn are enmeshed in a framework of loose connective tissue, and these muscle bundles even extend into the broad liagment. The outermost coating of the oviduct consists of typical peritoneal serosa.

During ovulation, the oviduct undergoes active peristaltic movements and around the infundibulum, especially in the fimbriae, the large mucosal vessels and muscle fibers form a network that is essentially a kind of erectile tissue. When the vessels become engorged during ovulation, the turgor and enlargement that develop in the fimbriae, combined with contraction of their intrinsic musculature, brings the opening of the fallopian tube into contact with the ovarian surface, thereby facilitating entrance of the ovum.

The peristaltic contraction waves that pass from infundibulum to uterus at the time of ovulation doubtless facilitate transport of the ovum to the uterus; this process is facilitated by the rhythmic waves of ciliary contractions that pass down the oviduct (Fig. 4.40, p. 154). Consequently the oviduct is a highly active structure, physiologically speaking, during the ovulatory phase of the sexual cycle, rather than merely serving as a passive conduit for ovum and sperm.

Blood Vessels, Lymphatics, and Nerves of the Oviduct. A rich vascular and lymphatic supply is present in all regions of the oviduct. During intervals of vascular engorgement, the lymphatic channels are also distended with lymph, and this effect undoubtedly adds to the greater turgor of the fimbriated end of the oviduct which occurs at this time.

An abundant plexus of fine nerve bundles is found within the circular muscle layer of the oviduct that innervates the myofibrils, and branches of this plexus also pass to the mucous membrane. Even larger bundles of nerves are located in association with the blood vessels of the serous layer, as well as in the longitudinal muscle.

UTERUS. The oviduct delivers the fertilized ovum to the uterus. This organ of the female reproductive tract in turn provides the normal site for implantation of the ovum, and also develops the vascular relationships critical to the growth and maturation of the embryo (later the fetus) throughout subsequent prenatal life.

The uterus undergoes profound morphologic and physiologic alterations in preparation for implantation of the fertilized ovum, and these changes occur during a sexual (or menstrual) cycle that has a duration of approximately 28 days (Fig. 15.9). The rhythmic cycles have their onset at puberty with the first menstrual period (or menarche), and then they occur regularly through-

out the sexual life of the female, about 30 years; they cease only with the onset of the menopause.

In the event that pregnancy intervenes, the sexual cycles are interrupted temporarily until after the child has been delivered, and then they commence once more. During early pregnancy, a specialized organ develops within the uterus called the placenta, which provides an essential, though temporary, circulatory link between the maternal and the developing embryonic vascular systems. In addition, this structure also has an important, albeit temporary, endocrine function during pregnancy. The uterine changes described above, during the rhythmic sexual cycles as well as during pregnancy, are controlled by a complex balance and interplay among many neural and endocrine factors.

The aforementioned topics will form the basis of the present section; however, the processes of fertilization, implantation, pregnancy, parturition, lactation, and certain aspects of neonatal physiology will be considered in more detail and as a continuum in Chapter 16 (Figs. 16.8 through 16.20).

Gross Anatomy of the Uterus. The human uterus (or womb) is a pear-shaped hollow organ with a thick muscular wall, slightly flattened anteroposteriorly, and with its broad end, or fundus, directed upward (Fig. 15.64). In the nonpregnant

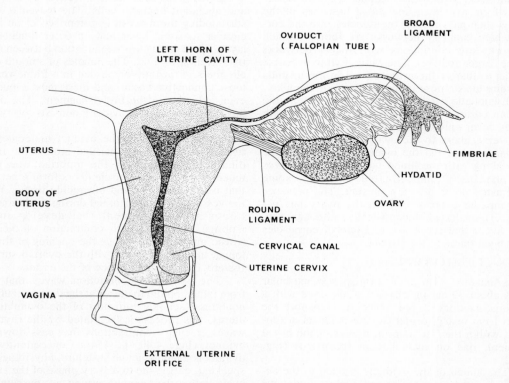

FIG. 15.64. Anatomic relationships between the human ovary, uterus, and vagina. (After Goss [ed]. *Gray's Anatomy of the Human Body,* 28th ed, 1970, Lea & Febiger.)

state, the uterus is about 7.0 cm long, 4 cm wide and 2.5 cm thick from front to back.

This organ does not occupy a vertical position within the pelvic cavity; rather, the fundic end is directed forward, upward, and slightly to the right of the midline (Fig. 15.62). The body of the uterus lies almost horizontally, and it rests upon the superior surface of the urinary bladder. The cylindrical, constricted cervix (or neck) is about 2.5 cm long and is directed downwards, and slightly posterior. The terminal, rounded portion of the cervix that protrudes into the vagina is called the portio vaginalis.

The uterine cavity is relatively small when compared to the thickness of the uterine calls. The oviducts are attached to the lateral walls of the fundus, and their lumens are confluent with the uterine cavity itself. The latter is much flattened in the body of the uterus because of apposition of the anterior and posterior uterine walls in the non-pregnant state. The uterine cavity within the fundus is continuous with the narrow cervical canal, and the constriction at the junction between the uterine body and cervix is known as the internal cervical os. The cervical canal opens into the vagina below on the portio vaginalis via a small opening, the external cervical os.

The uterus is supported within the pelvic cavity by a number of uterine ligaments. These consist of the paired broad, round, uterosacral, and cardinal ligaments, as well as the unpaired anterior and posterior ligaments.

The uterine wall is composed of three layers: (1) An outermost serous membrane, the peritoneum, covers the uterus. This membranous structure covers the fundus and most of the posterior side of the uterus, but is absent where it is reflected over the urinary bladder anteriorly and the rectum posteriorly. (2) The myometrium; the thickness of the uterine wall is due principally to this mass of smooth muscle. (3) The endometrium is a glandular mucous membrane that lines the inner wall of the uterus.

Histology of the Uterine Layers.
Myometrium. The myometrium is a thick uterine layer that consists of smooth muscle fibers arranged in bundles; the bundles in turn are separated by thin septa which also contain smooth muscle fibers. Several smooth muscle layers may be discerned within the myometrium, but these layers are not sharply delineated, since fiber bundles often pass between the adjacent layers. From the endometrium outward, these layers are: (1) the stratum submucosum, a thin layer which consists principally of longitudinal fibers, but which forms rings about the portions of the oviducts that are located within the uterus; (2) the stratum vasculare is the thickest uterine layer and it consists primarily of circular and oblique muscle fiber bundles. This layer receives its name because of the large blood vessels which are present, hence it appears spongy upon gross inspection; (3) The stratum

supravasculare consists chiefly of circular and longitudinal muscle fibers; and (4) the stratum subserosum, a thin outer layer which consists of longitudinal fiber bundles. Both the stratum supravasculare and stratum subserosum send bundles of muscle fibers into the broad and round ligaments as well as to the oviduct.

The myometrial smooth muscle fibers usually are around 50 μ in length. However, during pregnancy these cells undergo an increase in their size through hypertrophy to about 500 μ, and concomitantly the total uterine mass increases roughly 20-fold. Hyperplasia also may occur during pregnancy, so that the total number of muscle cells increases through mitosis. Following parturition, the muscle fibers diminish in size rapidly as the uterus returns to its normal size, and perhaps some of the fibers undergo complete dissolution.

Among the smooth muscle bundles of the uterus is found an abundant connective tissue network that is composed of collagenous and elastic fibers. Fibroblasts, macrophages, mast cells and other cellular and fibrillar elements also are present in this stroma.

Dense collagenous and elastic connective tissue fibers form the principal structural components of the uterine cervix. This structure has a firmer consistency than other portions of the uterus.

Endometrium. The uterine mucous membrane exhibits cyclic changes throughout the sexual life of the female, commencing with the menarche at puberty (12 to 15 years of age), and continuing until menopause (45 to 50 years of age). The changes in structure seen in the endometrium are regulated in turn by rhythmic variations in the secretion of the ovarian hormones, as summarized in Figure 15.9. At the end of each sexual cycle, the endometrium undergoes partial dissolution with an attendant blood loss. The products resulting from this process give rise to a bloody vaginal discharge, called the menstrual flow, which lasts for three to six days, and is the most obvious sign of the cyclic variations in the functional state of the uterus.

The general structure of the normal endometrium at its point of maximal development in the midpoint of the menstrual cycle will be discussed below, prior to a consideration of the morphologic variations in this tissue which accompany the menstrual cycle.

When the endometrium reaches the peak of its development during a menstrual cycle, it is about 5 mm in thickness, and is composed of a surface epithelial tissue which invaginates to form myriad tubular uterine glands (Fig. 15.9). These glands penetrate the endometrium to the level of the thick lamina propria, or endometrial stroma, as this connective tissue layer is often called.

A single columnar epithelium covers the surface of the uterine lumen, and both ciliated and secretory cells are present. The cell types within the uterine glands are similar to those of the surface epithelium; however, ciliated epithelial elements are more scanty. The ciliary beat apparently is oriented toward the endometrial surface within the glands, and toward the vagina on the uterine surface. The uterine glands generally are simple tubules, which sometimes bifurcate near the myometrium, and may enter this muscular layer for short distances.

In elderly females, the endometrium becomes attenuated and atrophies. The glandular orifices disappear and cysts may form within the glands themselves.

The stroma of the endometrium consists of irregular stellate cells with large ovoid nuclei. The processes of these cells apparently contact each other, and are closely associated with a fine meshwork of reticular connective tissue fibers. Only the uterine arterioles possess elastic fibers. Macrophages are seen, but interestingly, these cells do not phagocytize the blood which is extravasated during menstruation. The reason for the absence of mobilization of these cells during this process is not known.

The vascular supply to the endometrium is of considerable importance to the phenomenon of menstruation and to placenta formation. The uterine arteries traverse the broad ligaments along the lateral sides of the uterus, and branches of these vessels enter the stratum vasculare of the myometrium. Within the stratum vasculare, the arcuate arteries that are arranged about the circumference of the uterus pass toward the midline, where they undergo anastomosis with similar vessels from the opposite side. The arcuate arteries then send branches through the outer layers of the myometrium to the endometrium. As the junction between these two layers is traversed, small basal arteries branch off from the arcuate arteries (Fig. 15.65A) and these vessels in turn supply the deepest-lying portion of the endometrium (stratum basale) which is not cast off during menstruation. As the vessels course toward the inner uterine surface through the thicker functional layer of the endometrium, the arcuate arteries do not branch, but are markedly twisted, and these coiled arteries then branch into arterioles which form a dense capillary network near the endometrial surface (Fig. 15.65B). The uterine veins are thin-walled, and form an anastomotic meshwork which exhibits sinusoidal dilatations within all regions of the endometrium.

Throughout the major part of the menstrual cycle, the coiled arteries exhibit alternate constriction and dilatation, so that the superficial endometrial layer is, in turn, ischemic and engorged with blood. These successive fluctuations in blood supply are of importance in the process of menstruation, and will be discussed further below.

The endometrial mucous membrane of the uterine body (or corpus uteri) undergoes a gradual transition into the mucosa of the isthmus. In this region of the uterus the endometrium is thin, coiled arteries are absent, and generally no hemorrhage occurs during menstruation.

The cervical mucosa, on the other hand, is about 3 mm thick, and has folds upon its surface that exhibit branchings, called the plicae palmatae. Tall columnar epithelial cells forming the lining of the cervical canal, and the oval nuclei of these elements are located within the basal portion of the cells; the cytoplasm is generously filled with mucus. Numerous large glands are present in the cervical mucosa; however, these glands are different from those found in the uterine body and isthmus in that they exhibit extensive branching and their lining consists of elongated, mucus-secreting cells like those upon the epithelial surface. Occasional ciliated cells are observed in the cervix, and generally the cervical canal is filled with mucus. The cervical mucosa does not participate in the morphologic alterations that attend menstruation, which, as noted above, are confined largely to the corpus uteri proper.

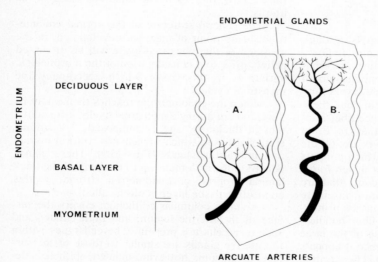

ENDOMETRIAL GLANDS

ENDOMETRIUM

DECIDUOUS LAYER

A.

B.

BASAL LAYER

MYOMETRIUM

ARCUATE ARTERIES

FIG. 15.65. Vascular supply to the endometrium of the uterus. A. Basal artery. B. Coiled or helicine artery. (After Ganong. *Review of Medical Physiology*, 6th ed, 1973, Lange Medical Publications.)

Endometrial Changes During the Menstrual Cycle. Throughout the female sexual cycle, the endometrium of the uterus exhibits a definite, regular sequence of morphologic and physiologic changes. These changes may be divided arbitrarily into three phases directly related to the functional state of the ovary, as shown in Figure 15.9.

Follicular (or Proliferative) Phase. The proliferative phase takes place concomitantly with growth of the ovarian follicles and their secretion of estrogenic hormones. Hence the follicular stage commences at the end of menstrual flow. The endometrium increases in thickness about threefold, as the consequence of frequent mitoses within the epithelial and stromal elements. The tubular glands increase both in length and in number. The coiled uterine arteries elongate somewhat, but they neither reach the uppermost third of the epithelium nor do they become appreciably contorted during this phase.

The growth of the endometrium by proliferation may persist for about one day following ovulation, which takes place on day 14 of a 28-day menstrual cycle.

Luteal (or Secretory) Phase. The luteal phase of the human sexual cycle takes place following ovulation, when the corpus luteum is functionally active and is secreting progesterone.

The endometrium may show some additional thickening during the luteal phase, principally due to the secretion within the uterine glands as well as edema in the stroma.

Typically, the nuclei of the epithelial cells exhibit a transient migration toward the free surface, an effect that is due to glycogen accumulation in their basal cytoplasm. As the glands grow they become markedly contorted. The coiled arteries lengthen and become convoluted during the luteal phase and ultimately these grow into the superficial region of the endometrium, as illustrated in Figure 15.65B.

Menstrual Phase. If fertilization fails to intervene, then approximately two weeks following ovulation endometrial stimulation by ovarian hormones decreases and profound alterations in the vascular supply to the endometrium occur. Prolonged vasoconstriction of the coiled arteries results in a sharply decreased blood flow to the superficial part of the endometrium which may last for up to several hours. Secretion by the glands of the uterine mucosa ceases, and this factor combined with interstitial fluid loss results in some endometrical shrinkage.

Following approximately two days of alternating vasoconstriction and vasodilatation, the coiled arteries shut down completely, whereas blood flow still is maintained in the basal vessels. Thus the superficial region of the endometrium becomes highly ischemic. However, after several hours the constricted vessels open once again for a short interval, and the anoxic vessels near the surface rupture. Blood now flows into the stroma and thence into the lumen of the uterus. Ultimately fragments of endometrial tissue become detached from the surface, which leaves the ruptured ends of the arteries, veins, and glands open. The uterine lumen presents the appearance of a raw wound by about the third day of menstrual flow.

The total quantity of blood that is lost during an average menstrual period amounts to less than 50 ml.

The deeper (basal) endometrial layer remains intact during the process of menstruation, and even before the vaginal discharge is complete, epithelial cells from the ends of the glands begin to move out and these rapidly generate a new surface epithelium. The circulation is restored, and the follicular phase of the next cycle commences.

During a sexual cycle in which ovulation does not occur (anovulatory cycle) no corpus luteum develops, and the proliferative phase lasts until menstruation intervenes once again.

Summary of Events During the Female Sexual Cycle. The fundamental biologic reason underlying the complex morphologic and physiologic changes of the menstrual cycle is that the primate uterine lining becomes prepared for implantation of the fertilized ovum (p. 1190). If fertilization does not occur, then the lining is removed and replaced with fresh tissue.

Clinically, detection of the various phases of the menstrual cycle from endometrial biopsy samples is of considerable importance in patients with ovarian or menstrual abnormalities. Similarly, anovulatory cycles may also be detected by this technique, a fact that is of importance in cases of infertile marriages.

The principal phases of the endometrical cycle now may be summarized briefly. First, during the *proliferative phase*, the endometrium is between 1 and 5 mm thick; the glands are generally straight, narrow, and the epithelium is tall; numerous mitotic figures are seen; coiled arteries are absent in the superficial region. Second, in the *secretory phase*, the endometrium is between 3 and 6 mm thick; the glands are waxy; the cells are tall; mitotic figures are seen only in the coiled arteries near the endometrial surface. Third, a *premenstrual phase* may be discerned during which the endometrium is 3 to 4 mm thick; both the glands and arteries are markedly twisted (or contorted); and the stroma is dense and abundant leucocytes are present. Fourth, during the *menstrual phase*, the endometrium is ½ to 3 mm thick; extravasated blood is evident; the stroma is dense, and the glands and arteries are flattened; and the epithelial surface presents a denuded as well as raw wound appearance.

Endocrine Regulation of the Female Sexual Cycle. All of the complex morphologic and physiologic changes in the uterus that take place thoughout the female sexual cycle are dependent upon the secretion of ovarian hormones. These substances, known collectively as estrogens, consist principally of the steroid hormones estradiol and estrone. In turn, ovarian activity is controlled directly, and in a cyclic fashion, by two pituitary gonadotropic hormones, viz., follicle-stimulating hormone (FSH), which stimulates follicular growth up to the point of ovulation, and luteinizing hormone (LH).* LH, acting together with FSH, is responsible for ovulation per se, as well as for the early development of the corpus luteum.

As noted earlier in this chapter, both FSH and LH secretion are regulated directly by the hypothalamus. The negative feedback mechanisms whereby the appropriate hypothalamic releasing factors are secreted to control the secretion of FSH and LH were discussed earlier (p. 1031).

The important cyclic fluctuations in the gonadotropic hormones, and concomitantly, ovarian estrogen secretion, are depicted in Figure 15.9, together with the concomitant morphologic changes that take place in the ovary and uterus.

Following ovulation, the corpus luteum develops from the thecal cells of the collapsed ovarian follicle. This temporary endocrine organ now secretes the hormone progesterone, which is responsible for the secretory and other changes in the endometrium preparatory to the arrival of a fertilized ovum. If implantation of a fertilized ovum takes place, then secretion of gonadotropins by the trophoblast (see below) helps to maintain the corpus luteum in a functional state. Otherwise, the corpus luteum degenerates and menstruation ensues. Following implantation, the corpus luteum enlarges to become the corpus luteum of pregnancy, and the secretory endometrium now persists and undergoes hyperplasia, because of the action of progesterone in preventing endometrial ischemia and subsequent menstruation. In pregnancy the secretory endometrium is maintained by the corpus luteum, and menstruation thus is inhibited until after birth of the child, by the continued active secretion of high progesterone levels by the corpus luteum (p. 1146).

As discussed earlier, the secretion of excess ovarian hormones acts directly upon the hypothalamus to inhibit gonadotropin secretion, and this simple negative feedback mechanism thus is responsible for regulation of the normal human sexual cycle. Certain drugs used for the artificial control of conception take advantage of this mechanism, since certain orally administered progresterone analogues suppress ovulation via their action upon the hypothalamic–hypophyseal system.

The effects of female sex hormones on the secondary sex characteristics, and certain other physiologic systems will be presented later.

Estrus Cycle. In mammals other than primates, menstruation does not occur, and the sexual cycle in such species is called the estrus cycle; however the uterine changes are in general similar

to those that take place in primates, even though the endometrium is not shed if fertilization and implantation do not occur. In such species, a period of heat or estrus takes place at the time of ovulation, and generally the female only accepts the male for coitus during estrus.

The endocrine changes that underlie the estrus cycle also are similar to those of the menstrual cycle; however, the days are numbered from the first day of estrus rather than the first day of menstruation, which is the convention employed for determining the onset of the sexual cycle in humans.

Usually male mammals do not exhibit cyclic reproductive activity that is similar to the estrus cycle, exceptions to this statement being the rhinoceros and elephant, which come into "rut" ("heat") or in the latter species "musth" (or "must") at periodic intervals. Certain male ground squirrels also exhibit a cyclic reproductive behavior which coincides with an actual descent of the testes into the scrotum from the abdominal cavity during their active reproductive period, ie, in the spring. The lower testicular temperature thus stimulates both sexual activity and spermatogenesis in these rodents, the animals being infertile for the remainder of the year.

Implantation. Fertilization of the ovum by a sperm (or conception) generally takes place within the upper region of the oviduct. The ovum commences to divide at once so that by the time it arrives within the uterus at around the fourth day, it is a hollow sphere of cells termed the blastocyst. Twenty-four to 48 hours after entering the uterus, the blastocyst becomes attached to the secretory endometrial surface. At this time, the cells have differentiated into an inner cell mass that lies at one pole, and a layer of primitive trophoblast cells that comprise the remainder of the blastocyst wall. The inner cell mass ultimately forms the rest of the embryo. The trophoblast cells are responsible for the attachment and implantation of the ovum, as well as for the later formation of the placenta.

Following contact with the endometrial surface, the trophoblast cells multiply rapidly, so that a multinucleate protoplasmic mass results, in which no cell boundaries are evident. This is called the syncytial trophoblast. In turn, the syncytial trophoblast produces an active dissolution (disintegration) of the epithelial surface, so that by around day 11, the blastocyst has penetrated completely into the endometrium, and the trophoblast forms a layer that surrounds the inner cell mass. On the surface, the uterine epithelium has spread over the opening caused by the invasion of the blastocyst, a process termed interstitial implantation.

Between days 9 and 11, the growing trophoblast is penetrated by a system of connected lacunae which are filled with blood that has been liberated by the erosion of the endometrial (maternal) vessels, and this blood apparently nourishes

---

The reader should note that luteinizing hormone is identical chemically with interstitial cell stimulating hormone (or ICSH) of the male.

the embryo (at least partially) prior to placental development.

At this stage, the trophoblast exhibits an inner layer of single cells, the cytotrophoblast, and an outer syncytiotrophoblast layer. Cells of the former layer divide actively by mitosis, and contribute to the latter by fusion with the syncytiotrophoblast.

By day 11, the embryo per se is a disc of epithelial cells that have been derived from the inner cell mass; hence a layer of ectoderm and a layer of endoderm are present. The ectodermal plate continues at its periphery into a layer of flattened (squamous) cells which surround a tiny amniotic cavity. In like fashion, the endoderm continues into the cells which form the yolk sac. The chorion is composed of the trophoblast which surrounds these structures.

Generally, implantation takes place in the dorsal uterine wall; however, the blastocyst can implant anywhere in this organ or even in the oviduct or abdominal cavity, giving rise to ectopic loci of implantation. The process of implantation of the fertilized ovum and its subsequent development are considered further on page 1190 et seq.

Development of the Placenta. The placenta is the organ in which the physiologic transfer of nutrients and waste products between the maternal and embryonic circulations occur. This structure is also an important, albeit temporary, source of hormones.

Between day 11 and day 16 following conception, the trophoblast enlarges rapidly and the implantation cavity enlarges continually. The maternal vascular system is progressively invaded by syncytiotrophoblast tissue, and the lacunae communicate in many areas with the endometrial venous sinuses. Beyond day 15, strands of trophoblastic cells grow outward from the chorionic surface and form the primary chorionic villi (Figs. 16.9 through 16.11.) Invasion of these structures by mesenchyme results in the formation of secondary chorionic villi; thus, a layer of syncytiotrophoblast tissue now covers a layer of cytotrophoblast tissue, at the center of which lies a column of mesenchymal cells. These structures are immersed in maternal blood which percolates slowly through a system of connected channels, the intervillous space.

Subsequently, cytotrophoblast cell columns develop across the intervillous space, and spread out upon the opposite wall to produce a sheet of tissue, the trophoblastic shell, which is perforated at intervals by maternal blood vessels. The villi absorb nutrients from the maternal blood within the intervillous space, and excrete waste products into this labyrinthine system of channels.

The vascular system of the embryo differentiates within the central mesenchymal columns of the secondary villi; later these separate spaces, which are lined with endothelium, coalesce and produce continuous vascular channels. Concomitantly, other vessels are developing in the mesenchyme of the chorion and stalk of the umbilical cord which then connects the embryonic heart with the peripheral circulation. Between day 21 and 23, fetal blood commences to circulate through the capillaries of the villi, which are now called definitive (or tertiary) chorionic villi. These structures are found radially about the periphery of the entire chorion, and the stem villi which extend to the trophoblastic shell grow branches whose terminal ends float freely in the intervillous blood.

The source of human chorionic gonadotropic (HCG) hormone has been localized. The syncytiotrophoblast cells appear to be responsible for the synthesis and secretion of this hormone. The human placenta also elaborates estrogens and progesterone in sufficient quantities beyond the third month of gestation to prevent abortion if ovarectomy is performed after, but not before, that time.

The ultimate vascular relationships that develop between the maternal and fetal blood supplies are depicted in Figure 16.11. Deoxygenated blood from the fetus (as the embryo is termed beyond the end of the second month of gestation) is transported to the placenta via the umbilical arteries (Fig. 16.19). Placental arteries radiate from the umbilical vessels at the junction of the umbilical cord with the placenta, and these run to, and branch within, the chorionic plate. Subsequently the abundant branches of the placental arteries pass into the stem villi, and then ramify still further in the branches of these villi (aborescent villi) so that ultimately capillary networks are formed in the terminal villi. In this region, oxygen-rich venous blood is collected into veins with extremely thin walls which join successively larger vessels, and which run in a course parallel with the arteries to the chorionic plate, to join still larger veins which converge upon the single umbilical vein. The latter vessel carries blood via the umbilical cord to the ductus venosus. From this structure, blood enters the inferior vena cava near its junction with the right atrium.

Maternal blood, on the other hand, comes from the arcuate branches of the uterine arteries and is transported by the coiled arteries into the intervillous space via openings in the basal plate of the placenta. The maternal arteriolar flow is pulsatile, so that the blood is ejected in intermittent spurts into the much larger intervillous space. Therefore, and in accordance with Bernoulli's principle, although the blood pressure within this space is higher, yet the blood percolates slowly among the placental villi, and metabolites are exchanged with the fetal circulation.

The diffusion barrier in the mature human placenta thus is comprised of a thin layer of syncytiotrophoblast and its basal lamina, as well as the endothelial wall of the fetal capillaries.

The incurrent pulsatile blood stream thus forces the blood back toward the basal plate, whence it is drained by an abundance of communicating channels between the intervillous space, and the decidua basalis of the maternal portion of the placenta. The decidua basalis is that portion of the placenta that develops from the endometrium that lies immediately beneath the site of implantation.

The development and several functions of the placenta are considered in somewhat greater detail in Chapter 16 (p. 1193 et seq.) in relation to the entire topic of gestation.

VAGINA. This short (about 2 cm wide by 7 cm long at rest), thin-walled, distensible, muscular tube extends from the lower margin of the uterus to the exterior of the body (vestibule of the external genitalia). The vagina is situated in the lower part of the pelvic cavity behind the bladder and in front of the rectum (Fig. 15.62). Functionally, the vagina receives the penis during coitus, and it transports unfertilized ova and menstrual products to the exterior. This organ also functions as a membranous portion of the birth canal during parturition.

The distal end of the vagina is blocked partially by an incomplete, transverse, membranous partition, the hymen, which generally is ruptured during the first sexual intercourse.

From without inward, the vagina is composed of three layers: the adventitial connective tissue, the muscular coat, and the mucous membrane. The outermost adventitial layer, composed of dense connective tissue, serves to join the vagina to surrounding organs and structures, and contains nerve fiber bundles, scattered nerve cells, and an abundant venous plexus. The bundles of smooth muscle fibers are disposed circularly as well as longitudinally, and the latter are especially prominent in the outer portion of this layer. The bulbocavernosus muscle fibers form a sphincter of striated muscle that surrounds the opening of vagina into the vestibule (ostium). The mucous membrane has a surface epithelium that consists of stratified squamous cells, and a lamina propria composed of dense elastic and other connective tissue fibers. The venous plexus is located within the deeper portion of the lamina propria.

Specific glands are not present in the vagina, and this structure is lubricated by secretions of the cervical glands. Mucus is secreted by the vaginal epithelium during sexual excitation. The actual source of this secretion is unknown, but this mucus apparently does not come from the cervix or the Bartholin's glands.

Estrogens exert a direct effect upon the vaginal epithelium during the sexual cycle, and these hormones induce a thickening and cornification of the stratified squamous epithelial lining of this organ. Progesterone, on the other hand, stimulates the secretion of a thick mucus later in the sexual cycle, and this hormone also induces a marked thickening of the vaginal epithelium which then becomes infiltrated with leucocytes. These cyclic changes are quite prominent during various phases of the estrous cycle in rats. In humans, however, such morphologic changes are not so clear-cut; therefore, accurate diagnosis of the phase of the ovarian cycle from vaginal smears alone is not possible. A determination of the particular phase of the menstrual cycle in humans must be carried out by examination of uterine biopsy samples.

The upper region of the vagina receives its blood supply from the vaginal branch of the uterine artery; the vaginal branch of the inferior vesical artery supplies the middle portion; and the middle hemorrhoidal and internal pudendal arteries distribute blood to the lower part of the vagina. Lymphatic vessels from the vagina drain into the hypogastric (internal iliac), lateral sacral, and inguinal lymph nodes.

The vagina is innervated by fibers from the hypogastric plexus, the fourth sacral, and the pudendal nerves.

During sexual excitation the vagina dilates to around 6 cm in width and lengths to about 10 cm in length; however, this structure is capable of considerably greater distension during coitus.

EXTERNAL GENITALIA. The external reproductive organs of the female include the mons pubis, the paired labia majora and labia minora, the clitoris, the bulb, and the glands of Bartholin. Collectively, these structures are known as the pudendum or vulva.

Mons Pubis. This skin-covered fatty cushion lies anterior to the pubic symphysis, and after puberty it becomes covered with coarse hair.

Labia Majora. The labia majora are thick, rounded folds of skin anatomically homologous to the male scrotum. Anteriorly they meet, forming the anterior commissure at which point they are confluent with the mons pubis. Posteriorly, the walls of the labia majora become thinner and terminate about 2.5 cm anterior to the anus in the posterior commissure. The labia majora contain large quantities of adipose tissue as well as a thin layer of smooth muscle. The latter is homologous to the tunica dartos of the scrotum.

The external surfaces of the labia majora are covered with hair, whereas the inner aspect is smooth. Sweat and sebaceous glands are abundant on both inner and outer surfaces.

Labia Minora. These small, paired, thin folds of skin enclose an area, the vestibule, where the urethra, vagina, and ducts of the greater vestibular glands (Bartholin's glands) open upon the body surface. Anteriorly the labia minora enclose the clitoris, whereas posteriorly these structures decrease in size and are connected by a transverse fold known as the frenulum pudendi.

The labia minora are invested in stratified squamous epithelium having a central connective tissue layer with delicate elastic networks. Large sebaceous glands are abundant upon both surfaces. Hairs are absent on the labia minora.

Clitoris. The clitoris is homologous to the dorsal portion of the penis; however, it does not contain the female urethra. It is comprised of two small fused, erectile, cavernous masses of tissue (corpora cavernosa of the clitoris) which terminate distally in a rounded tubercle, the glans clitoridis. The glans clitoridis is richly supplied with sensory nerve endings. The entire clitoris is covered by the vestibular mucous membrane. The entire length of this structure is about 2.5 cm, and the clitoris is supported by a suspensory ligament which extends to it from the pubic symphysis.

Bulb. This structure is comprised of two relatively large masses of erectile tissue located on each side of the vaginal orifice. Anteriorly these tissues fuse into a narrow band of tissue, the pars intermedia, and posteriorly this tissue expands and contracts the greater vestibular glands.

Greater Vestibular Glands (Bartholin's Glands). The glands of Bartholin are tubuloalveolar structures that are quite similar to the bulbourethral glands (Cowper's glands) of the male, with which they are homologous. Each vestibular gland is about one cm in diameter, and it opens upon the inner surface of the labia minora through a duct which is about two cm long. The secretion of the Bartholin's glands is a mucus which has a lubricating function during sexual activity.

PHYSIOLOGY OF FEMALE SEXUAL ACTIVITY. The female sexual act, in a manner similar to that in the male, is governed by local as well as by psychic stimulation. Local sexual stimulation of the female, as well as the male, depends upon massage or irritation of the external genital organs, perineal region, and other areas of the body (erogenous zones). The clitoris is an especially sensitive region for eliciting sexual sensations in the female. Once again in similarity to the male, the sexual sensations are transmitted via the pudendal nerve and sacral plexus to the spinal cord, and therefrom to the cerebrum.

The first sign of effective sexual stimulation in the female is that the lumen of the vagina becomes lubricated by copious volumes of mucus secreted by the vaginal epithelium within half a minute after the onset of such stimulation. Parasympathetic efferent impulses also result in secretion by the Bartholin's glands in the vicinity of the vaginal orifice (or introitus). These combined secretions normally provide effective lubrication during coitus, and result in the rhythmic massaging stimulation that is of primary significance in eliciting the appropriate spinal reflexes of both the female and male orgasm.

The erectile tissue of the bulb that surrounds the introitus, as well as the clitoris, becomes engorged with blood during sexual stimulation of the female in a manner that is quite similar to that of the penis. Efferent impulses are transmitted via parasympathetic nerves, and these impulses pass through the nervi erigentes that arise from the sacral plexus and run to the external genitalia, thereby inducing local arterial dilatation and subsequent engorgement of the tissue with blood and a concomitant erection. As the bulb enlarges, this tightens the introitus, thus facilitating stimulation of the male to the point where ejaculation occurs. During sexual stimulation, the highly distensible vagina also dilates markedly.

When local sexual stimulation reaches sufficient intensity, the lumbar and sacral spinal reflexes that underlie the female orgasm (or climax) discharge, and this process is considerably facilitated by conditioned reflexes of cerebral origin. The phenomenon of female orgasm is analogous to ejaculation in the male. There is clinical evidence that fertilization is more apt to follow normal coitus and female orgasm than is likely after artificial insemination. Consequently, it appears that fertility is enhanced to some extent by female orgasm, although the physiologic mechanisms underlying this effect in humans is not clearly defined at present.

MAMMARY GLAND. The paired mammary glands are accessory organs of the female reproductive system which are closely related to the sweat glands of the skin, both developmentally and structurally. Functionally, however, the mammary glands are regulated by the endocrine organs of the female reproductive system.

The secretion of the mammary glands (milk) provides the immature offspring with a source of nourishment.

Development of the Mammary Gland. In the embryo of about six weeks of age, the paired ectodermal mammary ridges (or milk lines) extend from the axillae to the groin on each side of the midline of the ventral surface of the body. In humans, only two mammary glands usually develop, although supernumerary nipples or masses of glandular tissue may develop anywhere along the mammary lines.

During prenatal life, the differentiation and development of the mammary glands follows a similar pattern in both males and females. In the normal male, however, there is little additional development of these structures following birth. In the female, on the other hand, the mammary glands show considerable morphologic and physiologic variation depending upon the age and functional state of the glands (ie, during pregnancy, lactation, or time of the menstrual cycle) as well as the race and nutritional state of the individual.

The human mammary gland originates in the embryo as several local thickenings within the mammary ridge in the pectoral region of the embryo at the site of the future breast. This lobular primordium becomes successively globular, bulbous, and then lobulated. During the fifth month of gestation, between 15 and 20 cords bud into the connective tissue, and form the primary milk ducts. The branching terminations of these structures then dilate into secretory acini. Lumina appear later within the milk ducts and acini.

While the above changes are taking place within the glandular tissue per se, the free surface of the primordium flattens and deepens into a pit upon which the primary ducts open. At about the time of parturition, the free surface of the mammary primordium elevates into the nipple. The areola that surrounds the nipple is discernible as a circular area free of hair primordia, but containing branched areolar glands (of Montgomery) at about five months postpartum.

Morphology of the Resting Mammary Gland. The adult mammary gland is a compound tubuloalveolar structure comprised of the 15 to 20 radiating lobes that originate from the nipple or mammillary papilla as described above (Fig. 15.66). The individual lobes in turn are separated by partitions of connective tissue and surrounded by masses of adipose tissue that are deposited at the time of puberty. The deposition of adipose tissue produces a marked gross hypertrophy of the gland at this stage of the life cycle. A lactiferous duct supplies each lobe and these ducts open upon the surface of the nipple. Beneath the pigmented areola surrounding the nipple, the lactiferous ducts exhibit a dilatation, the sinus lactiferus. The individual lobes are subdivided into lobules, and the smallest of these lobules are elongated alveolar ducts that are lined with evaginations, the alveoli. Loose connec-

tive tissue is present within the individual lobules, and this tissue permits distension of the organ during pregnancy and lactation, as well as engorgement of the breast with blood (hyperemia) and edema fluid, an event that occurs during certain phases of the menstrual cycle.

A low cuboidal epithelium is present in the secretory portions of the resting mammary gland, the alveoli, and the alveolar ducts. These cells rest upon a basement lamina as well as an incomplete layer of myoepithelial cells. These branched myoepithelial cells enclose the glandular alveoli in a loose network, and their presence gives rise to the interpretation that the mammary and sweat glands are related morphologically.

Each mammary gland has a number of ducts, and each lobe of the mammae is a separate compound tubular gland. The primary duct of each lobe unites with progressively larger ducts which ultimately drain into a lactiferous duct and these structures in turn open independently upon the surface at the tip of the nipple. The nipple and areola are covered with epidermis that has elongated dermal papillae. The capillary bed within these papillae is brought close to the surface; the skin overlying this region is pink in color in immature as well as in blonde persons. Pigmentation of the nipples and areolae develops in most persons at puberty, and this pigmentation increases during pregnancy.

Bundles of smooth muscle fibers are arranged

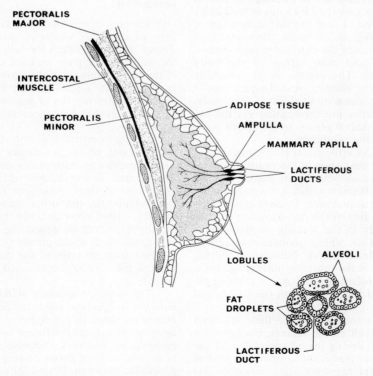

FIG. 15.66. Gross anatomy of the human mammary gland. (After Goss [ed]. *Gray's Anatomy of the Human Body,* 28th ed, 1970, Lea & Febiger.)

longitudinally along the lactiferous ducts and circularly within the nipple and its base. The nipple can become erect because of the contraction of these elements as the consequence of certain types of stimuli, eg, cold, erogenous stimuli, or suckling by a baby during nursing.

An abundant innervation is provided in the skin of both nipple and areola. Free nerve endings, Meissner's corpuscles, Merkel's discs, Krause's end bulbs, and Pacinian corpuscles are present in the dermis (p. 179 et seq.). Pacinian corpuscles are also found in the glandular tissue. This overall innervation of the mammary gland is of considerable physiologic significance, as the neural stimulus evoked by suckling is essential to the maintenance of normal lactation, as described on page 1203 and illustrated in Figure 16.17.

Morphology of the Active Mammary Gland. Marked alterations in the mammary glands occur as the direct consequence of an increase in the circulating levels of certain hormones during pregnancy as discussed in Chapter 16 (p. 1198). Briefly, however, during the first half of pregnancy, the terminal portion of the duct system of the gland exhibits rapid growth and branching, both of which processes take place concomitantly with some regression of the mammary adipose tissue. During this growth period, the interstitial tissue becomes infiltrated with plasma cells, eosinophils and lymphocytes.

Subsequently, as gestation advances, the rate of glandular hyperplasia decreases, and the parenchymal cells and alveoli become distended with a clear secretion which is rich in lactoproteins. This material is called colostrum, and it is secreted from the glands immediately following parturition. Colostrum has a laxative effect in the newborn, and it contains certain antibodies as well, which substances in turn confer a degree of passive immunity on the newborn. Shortly after childbirth, the lymphoid elements within the glandular stroma decrease, and the secretion becomes more abundant and rich in lipids, resulting in a flow of milk.

Different regions within the actively secreting mammary gland do not appear to be in the same functional state simultaneously, judging from the histologic appearance of the tissue during lactation. Some areas exhibit the presence of an abundant milk supply while the ducts are distended with secretion, whereas other areas in the secreting gland are relatively devoid of milk and the lumens of the ducts in such inactive regions are much thinner. The glandular epithelial cells also vary from flat to columnar, depending upon the secretory state of the tissue. The taller, more active cells often show a round bulging of their free ends into the lumen of the duct.

The epithelial cells of the mammary gland appear to produce two distinct secretory products which are elaborated and secreted by two mechanisms. First, the proteinaceous components of milk are synthesized upon the ribosome-covered membranes of the endoplasmic reticulum, a feature that is shared in common with other protein-secreting cells. These constituents form dense spherical vesicles or granules having a diameter of about 400 nm and which are related intimately to the Golgi complex. These vesicles ultimately fuse with the plasmalemma at the cell surface, and then discharge their granules into the acinar lumen. Second, the lipid constituents of milk are produced within the cytoplasmic matrix, and thus are not associated with the Golgi apparatus. The abundant fatty droplets pass toward the apices of the epithelial cells and enlarge simultaneously, thence they are liberated from the cell as minute droplets which are invested within a portion of the plasmalemma as well as coated by a small quantity of cytoplasm.

The neurohormonal reflex mechanism involved in continued milk production is shown in Figure 16.17 (p. 1204). Suckling, however, provides the important neural stimulus to continued lactation, so that if nursing is continued, then milk secretion continues for many months, and even may persist for several years. Cessation of nursing results in a rapid distension of the mammary glands with milk and then milk production stops rapidly. Presumably this inhibitory effect on milk production is due to an absence of the neural stimulus of suckling combined with engorgement of the breasts which decrease the availability of oxytocin to the myoepithelial cells. Shortly thereafter, the milk that remains in the alveoli and ducts is absorbed, and the gland returns gradually to the resting state, which condition persists until pregnancy stimulates mammary growth once again. Some of the alveoli which have developed during the first pregnancy do not disappear entirely, however.

The vascular supply of a functionally active mammary gland is far greater than that found in the resting state. The arterial blood supply to this gland is derived from the internal mammary artery, the intercostal arteries, and the thoracic branches of the axillary artery. These vessels course parallel with the larger ducts and then divide to form rich capillary networks upon the basement layer and secretory regions. Blood is drained via the internal mammary and axillary veins.

Capillary networks of lymphatic vessels have their origin in the connective tissue which invests the alveoli. Thence the lymphatic capillaries form a subpapillary lymphatic network whose course parallels that of the ducts of the mammary gland. Subsequently larger vessels transport lymph principally to the axillary nodes, although some lymphatic drainage to the sternum, and even the contralateral breast may occur.

During old age, the secretory epithelium and portions of the secretory ducts undergo atrophy and the mammary gland involutes, so that in general it returns to a prepubertal condition. Occasionally the epithelium is the site of pathologic changes, and in cases of carcinoma of the breast,

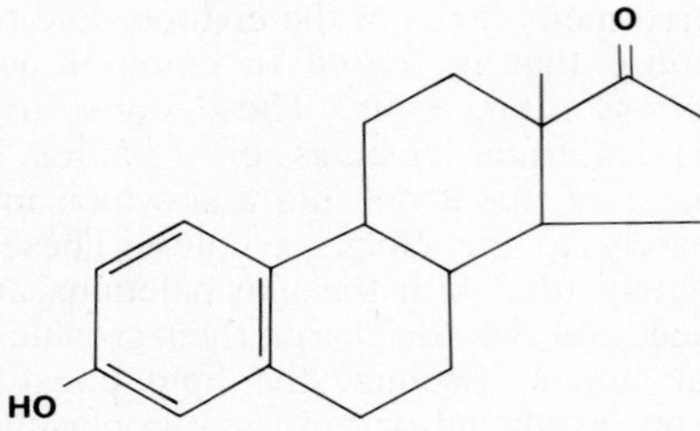

ESTRONE

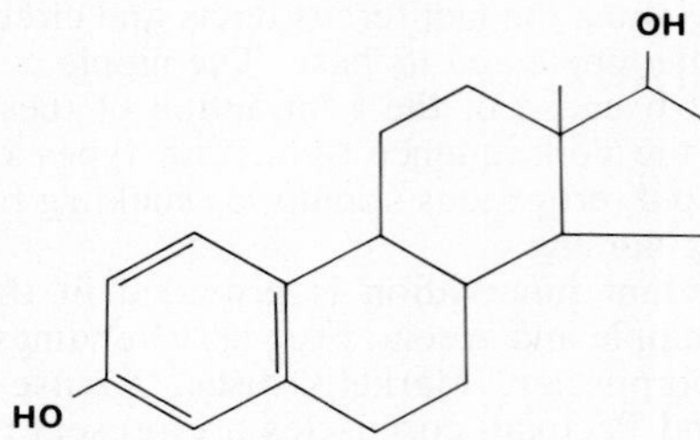

β – ESTRADIOL

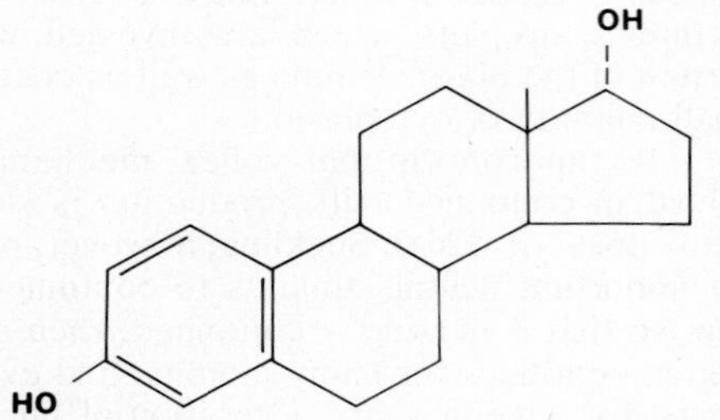

α – ESTRADIOL

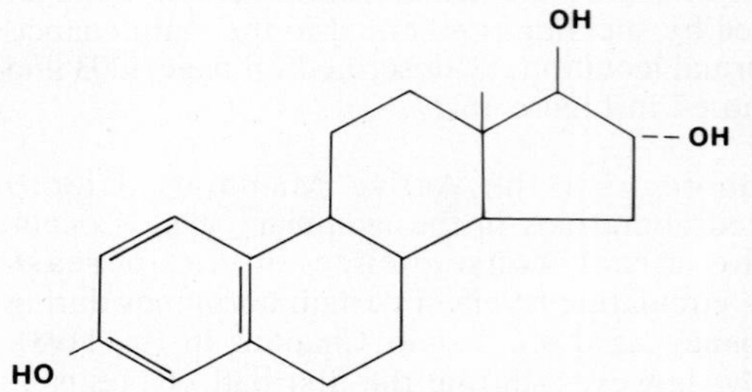

ESTRIOL

FIG. 15.67. Estrogenic hormones. β-estradiol normally is secreted by the ovarian follicle. α-estradiol as well as estrone are for the most part metabolic products of β-estradiol as shown in Figure 15.68. (After White, Handler, and Smith. *Principles of Biochemistry,* 4th ed, 1968, The Blakiston Division, McGraw-Hill.)

metastasis from the locus of the tumor can occur readily via the lymphatic channels outlined briefly above.

Estrogenic Hormones

CHEMISTRY. Four C18 steroid compounds that have an estrogenic action have been isolated from ovarian tissue and human urine (Fig. 15.67). These naturally-occurring substances are 17-β-estradiol, estrone, 17-α-estradiol and estriol. All of these compounds lack an angular methyl group on the 10 position of the steroid ring, and also they do not have a Δ4, 3-keto configuration in the A ring, which is aromatic (see Figure 15.37 for numbering of ring positions).

The relative estrogenic potency of each of these compounds depends upon the mode of administration (viz., oral or parenteral) as well as upon the type of bioassay system that is used, as shown in Table 15.15.

BIOSYNTHESIS OF ESTROGENS. The synthetic pathways of the estrogenic hormones are shown in Figure 15.68. (This illustration should be compared with Figures 15.39 and 15.56 in which the biosynthesis of adrenocortical and androgenic hormones are shown.)

Note that the pathway for estrogen synthesis involves first the formation of androgens, and that testosterone is the actual biochemical precursor of the female sex hormones. The theca interna cells of the ovarian follicles are the principal site of synthesis and secretion of estrogens in the non-pregnant female. In addition, estrogens are secreted in significant quantities by the placenta and

Table 15.15 COMPARISON OF THE ACTIVITY OF THE NATURAL ESTROGENS[a]

| | RELATIVE POTENCY | |
COMPOUND[b]	SPRAYED RAT[c]	IMMATURE MOUSE[d]
β-Estradiol	1,000	1,000
Estrone	100	100
Estriol	20	40
α-Estradiol	10	7.5

[a]*Data from White, Handler, and Smith.* Principles of Biochemistry, *4th ed, 1968, The Blakiston Division, McGraw-Hill.*

[b]*The compounds were administered subcutaneously in equivalent doses in each group of animals.*

[c]*Uterine weight gain in spayed rats was the criterion of potency.*

[d]*Uterine weight gain in intact, but immature, female mice was the criterion of potency. In both rats and mice, the mean weight gains in the uteri stimulated by the hormones were compared against the uterine weights of a similar group of control (ie, untreated) animals.*

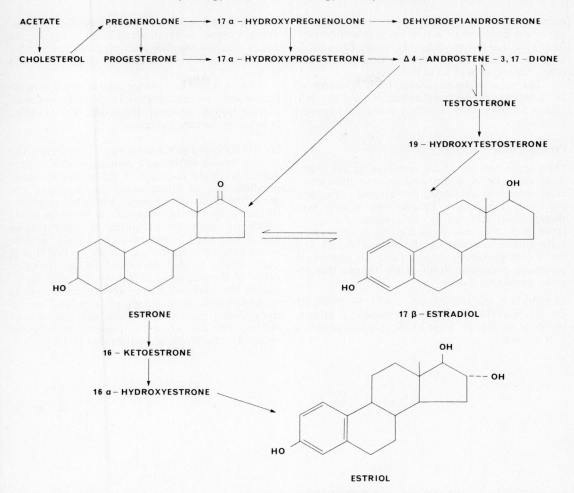

FIG. 15.68. Biosynthetic pathway for estrogens in the human ovary. Note that the androgen testosterone is the immediate precursor of 17-β-estradiol, the principal estrogenic hormone. (After White, Handler, and Smith. *Principles of Biochemistry,* 4th ed, 1968, The Blakiston Division, McGraw-Hill.)

in minute quantities by the adrenal cortex and testes.

The principal estrogenic hormone that is secreted by the ovary under normal circumstances is 17-β-estradiol. The α form of this compound, in addition to estriol and estrone, are in general products of 17-β-estradiol metabolism.

Incidentally, the richest known biologic source of estrogenic hormones is the stallion testis. Consequently, the urine of these animals contains the highest concentration of 17-β-estradiol that is found in any naturally occurring endocrine tissue.

SECRETION AND TRANSPORT OF THE ESTROGENS. The total secretion rate of estrogens has been estimated to range between 200 and 500 μg per day in nonpregnant adult females at the time of ovulation. Another peak of estrogen secretion is seen during the midluteal phase of the sexual cycle (Fig. 15.9). On the other hand, normal adult males

secrete about 50 μg of total estrogens per day. The actual values for estrogen secretion in both males and females can vary considerably according to the age of the subject as well as with the phase of the menstrual cycle in women.

Roughly two-thirds of the circulating estrogens are bound to plasma proteins; and 17-β-estradiol, the principal estogenic hormone that is secreted by the ovary, is in equilibrium in the circulation with estrone. In normal adult females, β-estradiol and estrone both are present in the plasma at a concentration of about 0.03 μg/100 ml. In adult males, on the other hand, the circulating estradiol level is approximately 0.003 μg/100 ml and that of estrone 0.02 μg/100 ml.

METABOLISM AND EXCRETION OF THE ESTROGENIC HORMONES. The liver appears to be a major organ for the metabolic interconversion of 17-β-estradiol, estradiol, and estrone, the latter two compounds being major metabolic products of estradiol. Within hepatic tissue, the estrogens either are oxidized or converted into sulfate (sulfuric acid) and glucuronide conjugates. Estradiol is also converted into estrone in the placenta and other tissues, as well as in the liver.

The urine of humans contains a minimum of 10 metabolic products of estradiol, and the bile also is an important excretory route for estrogen metabolites.

Progesterone

The corpus luteum which develops in the ovary from the cells of a ruptured ovarian follicle secretes progesterone during the last half of the sexual cycle (Fig. 15.69). Progesterone also is synthesized and secreted by the placenta during the latter part of gestation, as well as in physiologically minute amounts by the adrenal cortex and testis. Progesterone is, in fact, an important metabolic precursor in all of the tissues that secrete steroid hormones.

CHEMISTRY, BIOSYNTHESIS AND TRANSPORT OF PROGESTERONE. The structure of progesterone is shown in Figure 15.69; this hormone is a C21 steroid.

As shown in Figure 15.68, some 17-α-hydroxyprogesterone is synthesized from progesterone and is released together with ovarian estrogens by the follicle prior to luteinization (Figs. 15.39 and 15.56).

There is some evidence that secreted progesterone may be transported in combination with plasma protein; however, the details of the binding process are largely unknown.

SECRETION OF PROGESTERONE. Apparently the corpus luteum is the sole ovarian source of progesterone, as the secretion of this hormone closely parallels that of luteal development during the menstrual cycle. In fact during the follicular phase, the secretion rate of progesterone drops to levels which are similar to the normal secretion rate of men (Fig. 15.9). Thus males normally secrete about 5 mg of progesterone per day, a rate that is quite similar to that of women in the follicular phase of the menstrual cycle. On the other hand, during the midluteal phase, the secretion rate in women rises to about 32 mg per day, and toward the end of pregnancy this value can rise to over 300 mg per day when the placenta is fully developed and actively secreting (Fig. 16.13).

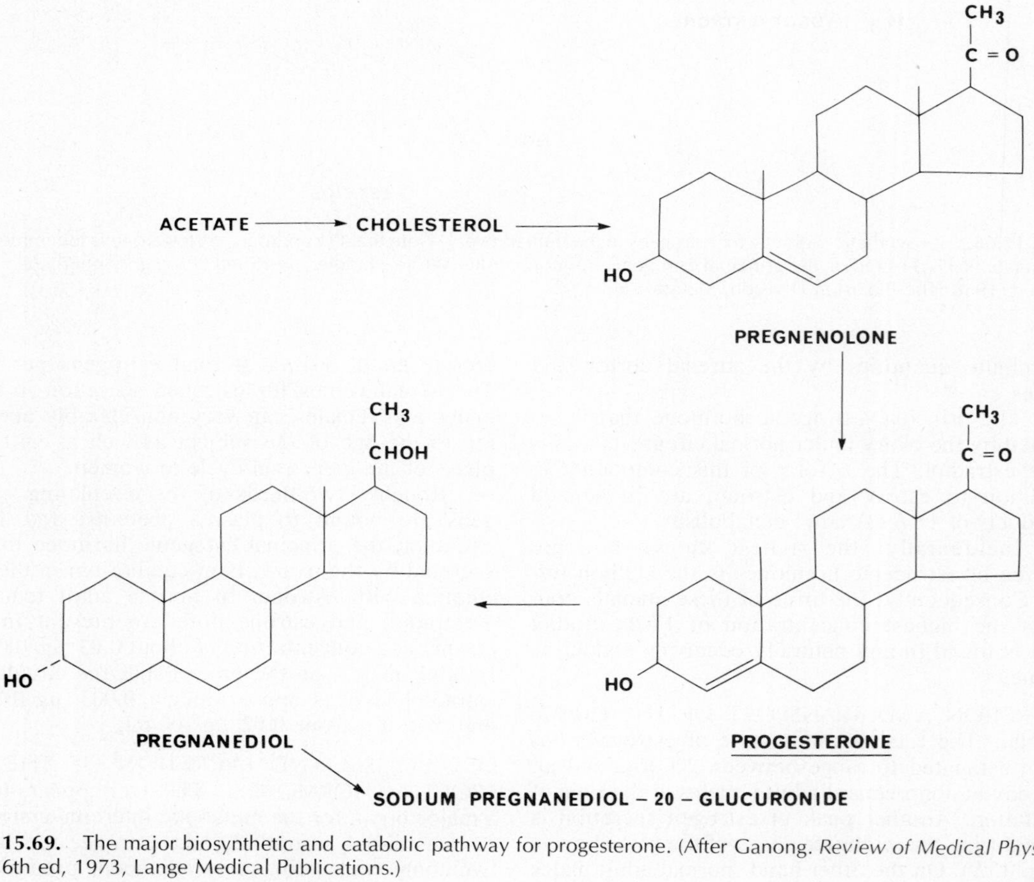

FIG. 15.69. The major biosynthetic and catabolic pathway for progesterone. (After Ganong. *Review of Medical Physiology,* 6th ed, 1973, Lange Medical Publications.)

METABOLISM AND EXCRETION OF PROGES-TERONE. As shown in Figures 15.39 and 15.56, progesterone acts as a precursor of 17-α-hydroxyprogesterone, androst-4-ene-3,17-dione, and testosterone in ovarian, testicular, placental, and adrenal tissues. These compounds, which are derived from progesterone per se, also may be considered to be metabolites of this hormone. Furthermore, estradiol and estrone are synthesized from progesterone via testosterone by ovarian tissue; thus in a sense these estrogens may be considered as progesterone metabolites.

Quantitatively, however, the reduction product of progesterone known as pregnanediol is by far the most important progesterone metabolite (Fig. 15.69). This compound is formed principally in the liver, and is neither synthesized exclusively from progesterone nor is it the only specific metabolite of progesterone. However, the urinary excretion of pregnanediol provides a clinically useful indication of progesterone secretion and metabolism. Furthermore, pregnanediol determinations are of greater significance when they are correlated with the physiologic state of the patient. For example, during the normal menstrual cycle, the chief source of progesterone is the corpus luteum, whereas during pregnancy progesterone also is synthesized by the placenta, and this latter source is of great importance later in gestation.

Relaxin

There is some experimental evidence that the corpus luteum, in addition to synthesizing and secreting the steroid progesterone, also elaborates at least three chemically-related basic polypeptides whose molecular weight is about 9,000 daltons and which cause relaxation of the symphisis pubis and uterus. The tissue extracts that contain these principles (that collectively are termed relaxin) also cause dilatation of the uterine cervix. All three of these physiologic effects can be elicited only in animals that are in normal or artificial estrus, the latter having been induced by exogenous estrogen administration. In contrast to other steroid hormones including progesterone and the estrogens, which cause relaxation of the pubic symphisis only after a long period of treatment, relaxin preparations cause such an effect within a few hours. The physiologic role of relaxin in normal parturition in human females is unclear, although such peptides have been shown to be present in the blood of pregnant women.

Regulation of Ovarian Functions

The role of the hypothalamo–hypophyseal axis is of paramount importance in the regulation of estrogen and progesterone secretion by the ovary. The factors involved in the regulation of ovarian function have been presented in context in earlier sections of this chapter and will be referred to specifically in the following discussion.

CONTROL OF ESTROGEN SECRETION AND OVULATION. Following the onset of puberty and the menarche in females, follicle-stimulating hormone (FSH), which is secreted by the adenohypophysis, is responsible for the early maturation of the ovarian follicles (Fig. 15.9). Somewhat later in the menstrual cycle, a combination of FSH and luteinizing hormone (LH or interstitial cell-stimulating hormone, ICSH) is responsible for the final maturation of the ovum. Ovulation itself takes place, in addition to the onset of development of the corpus luteum from the follicular cells following ovulation, under the impetus of a sudden increase in the pituitary secretion of LH. Subsequent maintenance of the corpus luteum in humans and secretion of progesterone by this organ may possibly be regulated by LH and prolactin (luteotropic hormone, LTH) acting together. Unlike the situation which obtains in rodents (eg, rats and mice), prolactin alone does not have a luteotropic effect in certain animal species, notably humans and other primates.

The secretion of FSH and LH by the anterior pituitary are regulated in turn by FSH-releasing factor (FSHRF) and LH-releasing factor (LHRF). These releasing factors are of hypothalamic origin, and are transported directly to the pituitary via the hypophyseal portal vascular system (Figs. 15.4 and 15.5). Consequently neural stimuli are of primary, albeit indirect, importance to the physiologic regulation of follicular development, ovarian secretion, ovulation, and perhaps luteinization. In certain species of female animals, notably the domestic cat, rabbit, and ferret, reflex ovulation is triggered directly by neural stimulation while the female is in estrus. Tactile stimulation of the external genitalia and other body regions during copulation are sufficient to stimulate ovulation via the afferent impulses which converge on the hypothalamus thereby indirectly and suddenly liberating a large concentration of LH from the pituitary. In humans, however, the neural components which influence ovulation are more subtle; and although the menstrual cycle can be influenced markedly by psychic and emotional stimuli, the physiologic role of such neural factors upon normal ovulation in women is not entirely clear.

Circulating estrogens appear to act directly upon the hypothalamus to inhibit the secretion of FSHRF by the pituitary and thus to decrease FSH secretion, a negative feedback mechanism that was considered in more detail on page 1031. In addition, it is believed by some investigators that the increase in the level of circulating estrogens which is observed immediately prior to ovulation is responsible for producing the sudden rise in LH secretion which stimulates ovulation (Fig. 15.9). And quite possibly the secretion of FSH and LH by the adenohypophysis is inhibited by the elevated levels of estrogen and progesterone found in the circulation during the luteal phase of the menstrual cycle.

Prolactin secretion, in contrast to that of FSH

and LH, is inhibited rather than stimulated by the hypothalamic releasing factor known as prolactin-inhibiting factor (PIF). There is also an approximately reciprocal relationship between the secretion of prolactin and that of LH.

It has been demonstrated experimentally that an increase in cyclic AMP accompanies the stimulating effect of LH upon progesterone secretion by the corpus luteum. Further studies have also indicated that this stimulating effect of LH (or an elevation in cyclic AMP) in turn is dependent upon the de novo synthesis of protein. Thus it appears that LH activates adenyl cyclase in the corpus luteum, so that a sequence of reactions is initiated like those which are stimulated by corticotropin in the adrenal gland (Table 15.3, Fig. 15.2). An increase in cyclic AMP stimulates biochemical reactions involving protein synthesis, and these reactions in turn result in an augmented steroid hormone synthesis.

It is not known at present what causes the spontaneous regression of the corpus luteum (or luteolysis) during the latter part of the menstrual cycle. Once luteolysis commences, however, the circulating levels of estrogen and progesterone fall, and these effects lead to an increase in the adenohypophyseal secretion of FSH and LH. New follicles are stimulated to develop, and estrogen secretion from this source shows a rapid increase. This elevation in estrogen concentration now stimulates the release of a quantity of LH, an effect mediated by a positive feedback effect upon the nervous system. Ovulation ensues, and this is followed by the development of a corpus luteum, following which event the estrogen and progesterone levels increase once again (Fig. 15.9), and these hormones inhibit FSH and LH secretion temporarily. However, luteolysis intervenes once again, and a new cycle commences assuming that pregnancy does not intervene.

General Functions and Actions of the Female Sex Hormones

A number of functions of the estrogens and progesterone have been discussed in context earlier in this chapter, especially with reference to the rhythmic morphologic and physiologic variations which occur during the menstrual cycle and which are stimulated by these endocrine substances. In the present section, certain additional physiologic roles and actions of these hormones will be discussed. In particular, the following material will deal with puberty, pregnancy, parturition, lactation, and menopause (see also Chapter 16, p. 1190 et seq.). In addition, certain other more general metabolic and physiologic actions of these hormones will be interjected as deemed appropriate in these discussions.

PUBERTY, ADOLESCENCE, AND THE MENOPAUSE. Following birth, the ovaries remain inactive until stimulated by pituitary gonadotropins at the onset of puberty, at which time both cyclic endocrine and gametogenic functions develop.* In human females, puberty with menarche are reached at approximately 12 years of age. Children between the ages of seven and 10 years exhibit a slow rise in estrogen (and androgen) secretion that heralds the sudden increase in the rate of secretion of these hormones during the next few years.

A neural mechanism apparently underlies the onset of puberty, but the actual physiologic mechanism underlying the process is unclear. Although the gonads of immature subjects are responsive to stimulation by pituitary gonadotropins, and such hormones are present in the pituitary prior to puberty, as well as appropriate releasing factors for FSH and LH in the hypothalamus, adenohypophyseal gonadotropins are not released prior to this physiologic landmark. In immature humans with lesions in the hypothalamus near the infundibulum, as well as in experimental animals with similar lesions, precocious puberty develops. These facts imply that a hypothalamic mechanism is present before puberty which actively inhibits FSH and LH release by the pituitary of both females and males. Yet such an inhibitory substance, or other type of mechanism, has yet to be demonstrated experimentally. In other words, the mechanism that underlies the actual onset of puberty is unknown, although several mechanisms, including the one described above have been postulated.

At puberty, however, the menarche indicates the onset of the rhythmic female reproductive cycles that persist until menopause intervenes. In turn this manifestation of sexual maturity is underlain by the cyclic variations in estrogen and progesterone secretion discussed previously.

The female secondary sexual characteristics which mature during adolescence under the influence of ovarian estrogens include the following general changes in the body. First, the breasts enlarge as the ducts grow and adipose tissue is deposited in these glands under the stimulus of estrogens. Adipose tissue also is deposited in the buttocks.† Pigmentation of the nipples and areolae is increased somewhat at puberty. The female larynx retains its prepubertal proportions, so that the voice remains higher-pitched in the female than in the male.

Females develop less body hair in general than do males, but more scalp hair is present. The pubic hair grows in a characteristic flat-topped pattern; however, both pubic and axillary hair growth in females result principally from the effects of adrenal androgens. The body configuration develops in such a manner that women have narrow shoul-

---

*The term adolescence refers to the period of growth and maturation (development) of the reproductive system which culminates in puberty at which point full reproductive capacity is present.
†Such a feminine pattern of fat distribution in breasts and buttocks also is seen in castrate males.

ders, broad hips, convergent thighs, and divergent arms (or a wide carrying angle). Estrogens also stimulate an increased sexual drive (or libido) in females.

Note that the secondary sex characteristics in women develop in part from the direct feminizing effects of estrogens (eg, breast development), and in part merely from the absence of a sufficient quantity of androgens (eg, absence of a beard and laryngeal growth).

In addition to these long-term effects of estrogens upon the female secondary sex characteristics, the ovarian estrogens also directly stimulate overall growth and function of the sexual organs at puberty, including the external genitalia, clitoris, vagina, uterus, fallopian tubes (oviducts), and ovarian follicles. In addition, estrogens augment uterine blood flow, and stimulate hypertrophy and hyperplasia within the myometrium. Lastly, the smooth muscle within the uterus and oviducts becomes more irritable (or excitable) and spontaneously contractile under the influence of ovarian estrogens.

In addition to the aforementioned effects of estrogens, these hormones are also responsible for stimulating epiphyseal closure in the long bones which results in cessation of overall growth of the skeleton during adolescence.

Between the ages of 45 and 55, the ovaries gradually become refractory to gonadotropic stimulation, the secretion of estrogens declines, and menstrual cycles are no longer observed. This phenomenon is known as the menopause; it may be accompanied by sensations of warmth ("hot flashes") as well as psychic symptoms of various kinds.* Estrogen therapy alleviates the hot flashes suffered by postmenopausal females and also may ameliorate to some extent the pyschic distress which may develop. Androgens sometimes are included in estrogenic preparations used for replacement therapy because of their anabolic property, insofar as protein synthesis is concerned.

The hot flashes are not confined to females, but also may occur in males who have been castrated in adulthood. The testes undergo some functional decline with advancing years; however, no distinct physiologic "male menopause" occurs.

ENDOCRINE CONTROL OF PREGNANCY AND PARTURITION. The endocrine and other aspects of pregnancy, parturition, and lactation are presented in detail in Chapter 16, but the following survey of the salient physiologic events during these processes is designed to serve as an introduction to the subsequent more extensive presentation.

Subsequent to fertilization of the ovum and implantation of the blastocyst in the endometrial wall of the uterus, the placenta commences to de-

velop. In contrast to the situation seen in a nonfertile menstrual cycle, the corpus luteum does not regress, but enlarges and develops into the corpus luteum of pregnancy in response to stimulation by a gonadotropic hormone which now is secreted by the placenta. In humans, this gonadotropic hormone is called human chorionic gonadotropin (HCG), and under the stumulus of this substance, the corpus luteum of pregnancy now secretes estrogens as well as progesterone. Following the third month (or first trimester) of pregnancy, the human placenta secretes a sufficient concentration of these hormones to maintain pregnancy, even if ovarectomy is performed. Prior to this time, ovarectomy in the human results in abortion.†

The human placenta, in addtion to secreting estrogens, progesterone, and HCG, also secretes renin as well as a hormone known as chorionic growth hormone-prolactin (CGHP) or human placental lactogen (HPL). All of the hormones that are elaborated by the placenta apparently come from the syncytiotrophoblast.

Secretion of growth hormone from the pituitary during pregnancy is not increased; however, CGHP, which is a peptide with a molecular weight of around 9,000 daltons and a structure that is similar in some respects to that of growth hormone, seems to function like growth hormone during pregnancy. Thus CGHP stimulates nitrogen potassium and calcium retention. There is a direct relationship between the placental mass and the quantity of CGHP secreted; therefore, a low plasma level of this peptide is a clinical indication of placental inadequacy.

There is little exchange of hormones between the mother and baby and vice versa. The transfer of proteinaceous endocrine substances does not take place to any great extent whereas the movement of nonprotein hormones across the placental barrier may take place to a very slight extent. There is evidence, however, that some of of sex hormones (androgens as well as estrogens) secreted by the placenta are synthesized from sulfate-confugated androgens which have been secreted by the fetal adrenal cortex.

In humans, the duration of pregnancy is approximately 270 days from the time of fertilization (or 284 days from the first day of the menstrual period which precedes conception). Despite considerable research, the physiologic mechanisms that initiate parturition are unknown, although many hypotheses have been advanced to explain this phenomenon. It is clear, however, that uterine contractions increase in frequency during the last month of gestation, although these contractions occur at irregular intervals. The myometrium becomes progessively more sensitive to the effects of oxytocin from the posterior pituitary late in pregnancy (ie, during the ninth month), and once par-

---

*The term climacteric refers to the combined phenomena that accompany the cessation of reproductive functions in the female as well as to the reduced testicular activity in the aging male.

†This fact is in direct contrast to other mammalian species, which abort following ovarectomy, regardless of the stage of pregnancy.

turition has commenced, reflex secretion of oxytocin is mediated via stimuli which arise from the genital tract itself. As the secretion of oxytocin increases, so do the uterine contractions. In effect the process of childbirth may be considered as a sort of positive feedback mechanism.

Relaxin also may participate in the process of parturition in that it may facilitate delivery by causing passive dilatation of the symphysis pubis during childbirth, although this statement is conjectural insofar as humans are concerned.

Hypothalamic lesions do not necessarily interfere with normal childbirth, a fact that suggests either that oxytocin is not entirely essential to this process, or else a sufficient quantity of this hormone is being secreted by nerve endings within the pituitary stalk per se to stimulate effective labor contractions. On the other hand, some hypothalamic lesions do cause a prolonged labor.

In addition to strong, rhythmic uterine contractions, spinal reflexes and voluntary contractions (ie, the Valsalva effect) also contribute to expulsion of the fetus, placenta, and membranes.

Following parturition, there is a general drop in the titers of circulating gonadotropic and other hormones, and the normal ovarian cycle resumes until pregnancy intervenes once again.

In certain species of mammals (eg, rats and mice) a condition known as pseudopregnancy (with an attendant luteal development, prolactin secretion, and endometrial changes that are typical of pregnancy in these species), can result from mechanical stimulation of the vagina (eg, by sterile copulation) or other means (eg, insertion of a thread or other foreign body in the uterus during estrus). Pseudopregnancy of this type is not seen in humans. However, pregnancy can be imagined so strongly that there may be cessation of the ovarian (menstrual) cycles (amenorrhea) which is combined with breast changes, abdominal distension, and morning sickness. These changes are symptoms of false pregnancy (pseudocyesis), and they stress the important influence of the emotional condition of the individual upon the secretion of hormones by the endocrine glands.

LACTATION AND MILK EJECTION.   In humans, the ducts of the mammary glands develop primarily under the stimulus of estrogens, whereas growth and development of the lobules are stimulated principally by progesterone. In some species (eg, rats and mice) the injection of prolactin causes milk production (lactation) in mammary glands that have been developed by preliminary treatment with estrogens and progesterone. A similar situation may also obtain in humans so that during pregnancy the combined actions of estrogens and progesterone prepare the breasts for milk formation, this secretory process in turn being stimulated by prolactin acting upon the fully developed mammary gland.

There is also direct evidence, at least in some species, that lactation also requires secretion of adrenocorticosteroids. It is now felt that when the levels of estrogens and progesterone in the circulation fall at the end of pregnancy, the augmented output of hypophyseal prolactin acting together with adrenocortical steroids causes milk secretion by the fully developed mammary gland.

Oxytocin from the posterior pituitary stimulates contraction of the myoepithelial cells that line the ducts of the mammary gland, and thus milk is expressed from the alveoli and ducts into the sinuses, and thence out of the nipple. This effect is known as milk ejection, a neuroendocrine process which requires the presence of oxytocin. Under normal circumstances, the milk ejection reflex is responsible for the actual initiation of milk flow. The touch receptors involved are located in the breast, especially within the nipple and areola. The afferent impulses that arise in the somatic touch receptors are transmitted via the bundle of Schutze and mammillary peduncle to the supraoptic and paraventricular nuclei, and discharge of impulses by the latter structure stimulates the secretion of oxytocin from the posterior pituitary gland.

A sustained milk flow (ie, lactation) is dependent upon a prolonged secretion of prolactin, and ejection of this secretion in turn depends upon oxytocin. Therefore, the tactile stimulus provided by active suckling by the infant is all-important to the maintenance of these interrelated processes. Emotional stimuli (eg, hearing a baby cry) or direct genital stimulation can also produce oxytocin secretion by this neuroendocrine mechanism, therefore a lactating woman occasionally will eject milk in spurts even though nursing (or suckling) is not taking place.

In humans, one to several days elapse following childbirth before milk secretion occurs in any quantity, and suckling not only stimulates and prolongs milk flow, but also increases the volume of milk which is secreted. Interestingly, a small quantity of milk normally is secreted into the ducts of the mammary glands during the fifth month of pregnancy, whereas abortion after the fourth month of gestation causes a marked increase in the volume of milk which is secreted. It would appear that ejection of the uterine contents in some way acts as a stimulus to lactation.

As noted earlier, lactation can be sustained for months, or even years, by regular suckling which both removes the milk and which also provides an important stimulus to the neuroendocrine reflex described above. Elimination of this tactile stimulus by termination of nursing causes a distension of the mammary gland with secreted milk, an effect which is followed by a rapid regression of mammary function. Oxytocin secretion is diminished and the milk present within the alveoli and ducts is resorbed shortly thereafter.

Generally women who do not nurse their babies experience a menstrual period around 6 weeks postpartum. Suckling, on the other hand, stimulates prolactin secretion, and this process also appears to inhibit FSH and LH secretion.

Ovulation is postponed, and estrogen and progesterone secretion drop markedly. The mechanism that causes the inhibition of FSH and LH secretion in a nursing female is unknown, although roughly 50 percent of such individuals do not ovulate until after the child is weaned.

METABOLIC AND OTHER EFFECTS OF ESTROGENS AND PROGESTERONE: POSSIBLE MECHANISMS OF ACTION. In view of the profound physiologic effects of estrogens, it is not surprising that these steroid hormones have been shown to influence a wide variety of biochemical processes in a number of tissues. It is quite difficult to distinguish between primary estrogenic actions and the secondary effects of these hormones. Thus estrogens markedly affect protein, nucleic acid, lipid, and inorganic ion metabolism, particularly the metabolism of calcium and phosphorus. In addition, estrogenic hormones exert an action upon the skin and related structures, in addition to exerting an antagonistic effect upon androgenic activity.

The general anabolic effects of estrogenic hormones are attested to by the fact that these compounds exert broad effects upon cellular proliferation. These effects are especially pronounced in the tissues of the uterus and mammary glands, and in turn these effects are manifest at the biochemical level by a rapid stimulation of RNA as well as protein synthesis.

Protein and Nucleic Acid Metabolism. Estradiol, when administered experimentally, is concentrated rapidly in a specific lipoprotein in normal uterine tissue as well as in mammary tumors of both humans and animals. In experimental animals (rats), less than an hour after the administration of estrogen to uterine tissue, there is a marked increase in a rapidly synthesized RNA fraction together with an increase in the activity of RNA polymerase. Both of these effects can be blocked by inhibitors of protein synthesis, a fact that suggests that a possible mechanism of estrogen action in the uterus involves an enhanced protein synthesis, concomitantly with an increased rate of synthesis of messenger RNA.

Lipid Metabolism. The effects of estrogens upon lipid metabolism are just as marked as those upon protein and nucleic acid metabolism, although their onset is somewhat later. Thus estrogens produce a marked acceleration in the turnover rate of uterine phospholipids, and when such hormones are administered to experimental animals when the diet is deficient in methyl groups, they prevent lipid accumulation in the liver. Consequently estrogens exert a lipotropic effect in hepatic tissue.

Estrogen administration in humans reduces the blood concentration of lipids, especially in subjects with hyperlipemia. This fact is of interest because of the much greater incidence of coronary arterial disease and thrombosis in men than in women.

Calcium and Phosphorus Metabolism. Experimental estrogen administration for long intervals can result in an increase in the serum calcium and phosphate levels. The attendant calcification leads to extensive ossification of the long bones, and this process may proceed to such an extent that the marrow cavity is obliterated and anemia develops. However, there is a concomitant loss of calcium from the pelvic bones.

A physiologic role for estrogens in the regulation of bone metabolism is implied by the appearance of bone decalcification and the osteoporosis that develops in women following menopause.

Estrogens also induce salt and water retention. Females gain weight just prior to menstruation, but oddly enough, this effect does not take place at the time of ovulation when the estrogen levels reach their peak (Fig. 15.9). Accompanying the salt and water retention, tenseness, irritability, and other symptoms develop and lead to the discomfort known as premenstrual tension. Antidiuretic hormone secretion may also be elevated at this time of the sexual cycle and this factor would contribute to the premenstrual fluid retention and perhaps to the other symptoms.

Miscellaneous Effects of Estrogens and Progesterone. The antagonism between the effects of estrogens and androgens was mentioned earlier. In addition, estrogen injection decreases the elevated secretion by the sebaceous glands of the skin which is stimulated by testosterone.

One further example of the antagonism between estrogens and androgens is worth mentioning. Young cockerels may be chemically castrated by the use of either natural or synthetic estrogenic compounds. For example, diethylstilbestrol has been used commercially by the poultry industry for the purpose of caponizing the fowl (Fig. 15.70A). The process involves insertion of a pellet of estrogen beneath the skin of the neck, so that a prolonged absorption of the hormone takes place and high levels of the hormone thus are maintained. If the chicken heads, containing the remains of the hormone, subsequently are fed to mink, then the males of the latter species become sterile, an effect that has caused some perturbation in the commercial mink breeding industry.

Topical application of estrogens in creams to induce breast hypertrophy sometimes is employed as these hormones stimulate duct and alveolar development in the mammary glands. Any effects of these estrogenic substances upon breast growth are due largely to the systemic absorption of the estrogen; however, only a very minor local growth stimulatory effect may also be obtained by use of such preparations. In view of the known role of estrogens in carcinogenesis, the indiscriminate usage of estrogens in dosages that are sufficient to affect breast growth for cosmetic reasons appears completely unwarranted.

Progesterone is known to stimulate respiration. During the luteal phase of the menstrual cy-

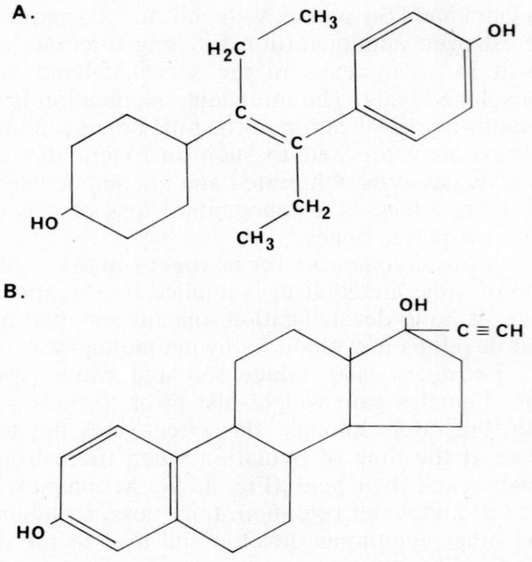

A.

B.

FIG. 15.70. Synthetic estrogens. **A.** Diethylstilbestrol. **B.** Ethinylestradiol. (After Ganong. *Review of Medical Physiology,* 6th ed, 1973, Lange Medical Publications.)

cle, the alveolar P_{CO_2} of females is lower than that of males, and this fact has been attributed directly to the effect of progesterone upon respiration.

Progesterone normally exerts its principal action upon the uterine endometrium which previously has been developed by estrogens, and it also stimulates mucus secretion during the latter half of the sexual cycle, an effect that is critical to implantation of the ovum. If pregnancy occurs, then sustained progesterone secretion is essential to normal completion of the pregnancy. In this regard, progesterone appears to inhibit uterine motility during pregnancy. Progesterone also contributes to breast development during pregnancy and this hormone also exerts an antiovulatory effect when administered during days 5 through 25 of the menstrual cycle. This important effect underlies the usage of synthetic progestins as oral contraceptive agents, as discussed below.

Synthetic Estrogens

The naturally occurring estrogens are far less potent when administered orally than when given parenterally. Presumably this decrease in efficacy is due to the instability of these hormones in the gastrointestinal tract as well as the liver, where they are largely metabolized in the hepatic parenchyma after absorption from the gut.

The synthetic compound ethinyl estradiol (Fig. 15.70B) is the most potent orally active estrogen that is known, and this steroid has about 10 times the potency of an equivalent dose of estrone when administered orally. Furthermore, ethinyl estradiol apparently is stable both in the gastrointestinal tract as well as in the liver.

Another potent synthetic estrogenic compound is diethylstilbestrol (Fig. 15.70A) which has a potency that is three to five times that of estrone.

Because of its activity when given orally, this compound also is useful therapeutically in patients suffering from estrogen deficiency.

Synthetic Progestins

Progesterone is biologically effective only when administered parenterally. Compounds that exert a progesteronelike activity are known as progestational agents (or progestins), and such substances when administered orally in appropriate doses inhibit the elevated secretion of luteinizing hormone (LH) which occurs at the midpoint of the ovarian cycle and which stimulates ovulation.

Such compounds inhibit ovulation apparently because LH release is blocked indirectly by such agents. Hence the progestins inhibit the neural factors which in turn precipitate luteinizing hormone-releasing factor from the hypothalamus.

The structural formulae of three synthetic steroid progestins are given in Figure 15.71. The potency of these compounds varies. Norethindrone given orally equals the potency of parenterally administered progesterone, wherease 19-nortestosterone has a tenfold greater activity than that of medroxyprogesterone.

The physiologic action of the synthetic progestins is enhanced when they are combined with small quantities of estrogens, and the latter compounds alone are contraceptive when administered regularly. One oral contraceptive (Enovid) is a mixture of 98.5 percent Norlutin combined with 1.5 percent of a synthetic estrogen (mestranol), and the combination of the two drugs is taken in accordance with a regular schedule to prevent ovulation, hence conception.

It should be mentioned that oral contraceptives are not completely effective, and that untoward side effects have been observed in some patients taking these drugs, including the development of massive intravascular thromboses.

Clinical Correlates

Abnormalities of the morphology and physiology of the female as well as male reproductive systems and accessory glands may run the gamut from inherited defects (ie, genetic abnormalities) to gross malfunction of the endocrine glands, which in turn produce various morphologic and physiologic effects. In the present section several of these abnormalities will be discussed; these examples were selected to illustrate typical dysfunctions of the kinds mentioned above. Endocrine disorders involving the reproductive system are not necessarily confined to that system, but may also involve other organs and functions that are quite remote from the usual target organs.

GENETIC ABNORMALITIES. Several defects involving the sex chromosomes which may occur during gametogenesis are caused by nondisjunction. In this process, a pair of chromosomes do not separate normally during meiosis; therefore both of the involved chromosomes subsequently go to

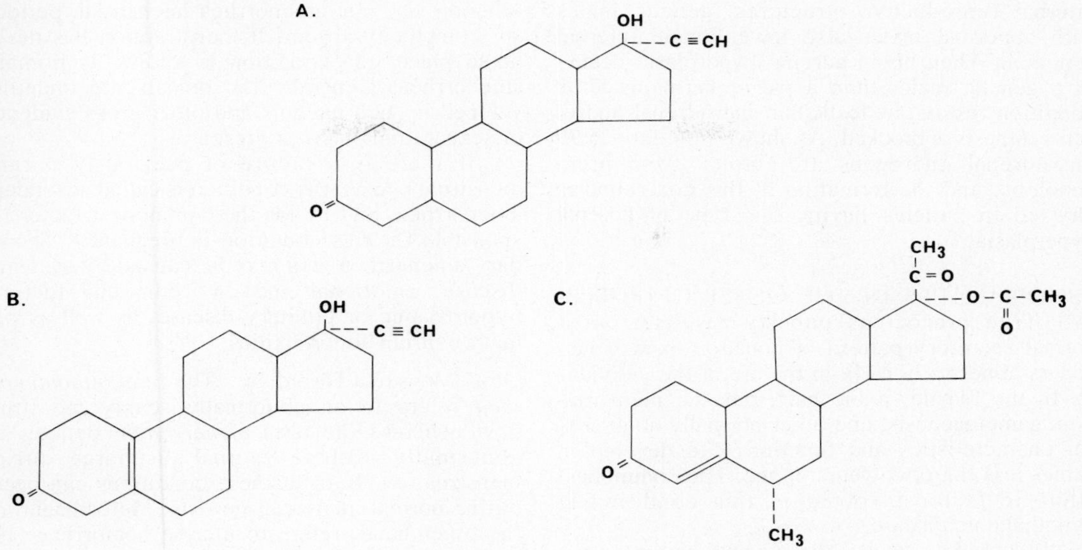

FIG. 15.71. Examples of synthetic progestational agents that are active when administered orally. **A.** Δ4(5)-17-α-Ethynyl-17-hydroxyestren-3-one (norethindrone, Norlutin). **B.** Δ5(10)-17-α-Ethynyl-17-hydroxyestren-3-one. **C.** 6-α-Methyl-17-α-acetoxyprogesterone (medroxyprogesterone, Provera). (After Ganong. *Review of Medical Physiology,* 6th ed, 1973, Lange Medical Publications.)

one daughter cell. The four possible abnormal zygotes that can result from nondisjunction during oogenesis are: XXY, leading to seminiferous tubule dysgenesis; XXX, leading to a "superfemale"; XO, in which gonadal dysgenesis develops; and YO, which is lethal.

The commonest disorder of the sex chromosomes is the XXY pattern, and individuals having this defect exhibit normal male genitalia and testosterone secretion may be sufficient at puberty for the male secondary sex characteristics to develop. The seminiferous tubules are abnormal, however, and the patient is sterile. A high incidence of mental retardation is also present in such cases. This syndrome is called seminiferous tubule dysgenesis (or Klinefelter's syndrome).

The XXX (superfemale) pattern follows the XXY distribution in frequency in the population, but apparently it does not result in any obvious abnormalities.

Individuals who inherit the XO pattern either have rudimentary gonads or else these organs are entirely absent. Female internal and external genitalia develop, the individual is short in stature, and no sexual maturation occurs at the normal time of puberty. This condition is known variously as gonadal dysgenesis, Turner's syndrome, or ovarian agenesis.

Another type of abnormality involving the sex chromosomes is that in which the individual inherits a mosaic pattern, possibly XX/XY. This pattern in turn leads to the condition of true hermaphroditism in which both ovaries and testes are present in the same individual.

The phenomenon of nondisjunction can also lead to other disease states which result in abberations in the autosomal, rather than the sex chromosomes. Down's syndrome (or mongolism), for example, is the consequence of nondisjunction of chromosome 21s during meiosis, so that three of these chromosomes (trisomy 21) are present in the offspring.

The conditions cited above represent only a few of the many known chromosomal abnormalities which result in specific disease states involving the reproductive system.

ENDOCRINE ABNORMALITIES: FEMALE AND MALE PSEUDOHERMAPHRODITISM. A pseudohermaphrodite is a person having the genetic pattern and gonads of one sex, but the genitalia of the opposite sex.

In genetic males, the external genitalia normally develop under the stimulus of androgens which are secreted by the embryonic testis. However, male genitals also may develop in genetic females who are exposed to androgens for some reason during the eighth to thirteenth week of fetal life. Such androgens may be formed endogenously (by androgen-secreting hyperplastic adrenal tissue) or due to androgens administered therapeutically to the mother. Beyond week 13, however, the female genitalia are completely developed, so that androgens do not stimulate growth (ie, hypertrophy) of the clitoris.

On the other hand, male pseudohermaphroditism can result from abnormalities in the embryonic testes, so that female external genitalia develop. Since the testes also secrete the testosterone necessary for the normal development of the male internal reproductive structures, genetic males with abnormal testes also have female internal genitalia. When lipoid adrenal hyperplasia occurs in a genetic male, than a pseudohermaphroditic condition results, as testicular and adrenal androgen synthesis is blocked. As shown in Figure 15.39 the adrenal androgens are formed from pregnenolone, and the formation of this compound is blocked in patients having this type of adrenal hyperplasia.

ABNORMALITIES IN THE ONSET OF PUBERTY.   True precocious puberty results from a normal secretory pattern of gonadotropins which occurs abnormally early in the life of the individual. In the female, pubic hair, breasts, menstruation, gametogenesis, and other normally adult sexual characteristics and functions can develop in babies less than two years of age. The commonest etiologic factor involved in this condition is hypothalamic disease.

In contrast to true precocious puberty, patients having the syndrome of precocious pseudopuberty exhibit an early development of the secondary sex characteristics, but without the concomitant onset of gametogenesis. This condition develops in young males who are exposed to abnormal quantities of androgens or in females who similarly are exposed to estrogens; neither spermatogenesis nor ovarian development are seen in either males or females who develop precocious pseudopuberty, however.

Etiologic factors related to precocious pseudopuberty include abnormalities of the adrenal gland, including congenital virilizing adrenal hyperplasia or androgen-secreting tumors in males and estrogen-secreting tumors in females. In addition, interstitial cell tumors of the testis or granulosa cell tumors of the ovary also can produce precocious pseudopuberty.

Failure of testicular development by 20 years of age or the appearance of the menarche until an age of 17 or older are considered to be abnormal. Such delayed or absent puberty may be caused by hypofunction of the pituitary or thyroid gland. Dwarfing and other gross endocrine abnormalities usually are associated with the abnormal pubertal development. Dwarfing also occurs in patients with gonadal dysgenesis that results from the XO chromosome pattern.

Some persons may exhibit delayed puberty despite the presence of apparently normal gonads and other endocrine functions. In the male this condition is termed eunuchoidism, and in females it is referred to as primary amenorrhea (see below).

OVARIAN MALFUNCTIONS

Anovulatory Cycles.   During the first year or so following menarche, and prior to the menopause, anovulatory sexual cycles are not considered abnormal.

Amenorrhea.   In amenorrhea menstrual periods are completely absent. If menstruation has never taken place, the condition is known as primary amenorrhea. Generally the breasts are underdeveloped in such patients and other gross evidence of sexual immaturity is present.

If there is a failure of previously normal menstrual cycles, the condition is called secondary amenorrhea, and by far the commonest factor responsible for this condition is pregnancy. Secondary amenorrhea also may be caused by systemic disease, emotional and environmental factors, hypothalamic or pituitary disease, as well as primary ovarian abnormalities.

Other Menstrual Disorders.   The term *oligomenorrhea* refers to an abnormally scanty menstrual flow, whereas the term *menorrhagia* denotes an abnormally profuse vaginal discharge during menstruation. Both of these conditions can occur during normal menstrual periods. Metorrhagia, on the other hand, refers to uterine hemorrhage between menstrual periods. Highly painful menstruation is called dysmenorrhea, and the cramps which are common during menstruation in younger females, often do not occur following the first pregnancy.

Ovarectomy.   In adult females, the surgical removal of the ovaries, or their destruction by disease processes, obviously results in sterility and amenorrhea. There is little alteration in libido or body conformation, although hot flashes such as are experienced following the menopause also may be present following removal of the ovaries.

Ovarian Tumors.   If an androgen-secreting tumor is present in the ovary, masculinization can result, whereas an estrogen-secreting tumor that develops in childhood can induce precocious pseudopuberty. These endocrine malfunctions are rare, however, and there is no other clinical condition that is related directly to ovarian hypersecretion.

Estrogens and Carcinoma of the Female Breast.   Roughly one-third of the breast cancers in young women are estrogen-dependent; that is, the continued growth of such tumors requires the presence of circulating estrogens. They are alleviated (but not destroyed) by ovarectomy and exacerbated by pregnancy because the circulating estrogen level increases markedly during gestation.

In addition to ovarectomy, a reduced estrogen level which causes temporary remission of the symptoms, as well as regression of the tumor per se, may also be achieved by hypophysectomy, and/or by bilateral adrenalectomy. In patients having intact adrenals, glucocorticoid administration inhibits adrenal, but not ovarian, estrogen secretion via the hypothalamo-hypophyseal axis. Furthermore, pituitary corticotropin secretion stimulates adrenal estrogen secretion but not estrogen secretion from the ovaries. Therefore hypophysectomy, by removal of the source of both corticotropin and gonadotropins, inhibits adrenal as well as ovarian estrogen secretion, and thus may inhibit the growth of the carcinoma as well as its attendant symptoms, at least temporarily.

OTHER ORGANS AND SUBSTANCES HAVING ACTUAL OR PRESUMED ENDOCRINE FUNCTIONS

In addition to the hormones that are secreted by the major endocrine glands of the body, the gastrointestinal tract, and the neuroendocrine substances elaborated by the nervous system, an endocrine function either has been demonstrated or postulated for several other organs and tissues. These organs include the kidney, spleen, thymus, and pineal body (or pineal "gland"). Furthermore, a wide variety of tissues has been shown to elaborate a number of compounds that collectively are known as prostaglandins, and which appear to exert a regulatory influence upon a number of important bodily processes. Hence prostaglandins will be considered among the endocrine substances to be discussed in this section.

The Kidney: Renin and Erythropoietin

In addition to its excretory functions (Chap. 12) the kidney also has been shown to produce two substances with hormonal actions. Renin, which is important in the regulation of aldosterone secretion and which also exerts indirectly a generalized vasopressor action; and erythropoietin, which is of importance in the homeostatic regulation of erythropoiesis under certain conditions.

RENIN. The proteolytic enzyme renin has a molecular weight of around 50,000 daltons; however little is known of the structure of this protein. Once secreted by the kidney, however, renin catalyses the hydrolysis of an α_2 globulin in the plasma which releases a decapeptide known as angiotensin I (or hypertensin I), as shown in Figure 15.72. The α_2 globulin which is acted upon by renin is synthesized within the liver, and is known variously as angiotensinogen, hypertensinogen, or renin substrate. Principally within the lungs, converting enzyme then removes histidyl leucine from the inactive decapeptide angiotensin I, thereby producing the physiologically active octapeptide angiotensin II (or hypertensin II) as shown in Figure 15.73.

The endocrine action of renin is mediated indirectly via two hydrolytic reactions.

Angiotensin II (or hypertensin II), is inactivated quite rapidly, so that the half-life of this compound in the human circulation is between one and two minutes. The group of enzymes which are involved in the inactivation of angiotensin II are known collectively as angiotensinase, and these enzymes are present in a number of tissues and include certain aminopeptidases.

Biosynthetic Origin of Renin.   Experimental evidence indicates that renin is synthesized and secreted by the juxtaglomerular cells of the kidney which were described on page 770. Briefly, the granular epithelioid juxtaglomerular cells are located within the medial layer of the afferent arteriole as it enters the glomerulus, and together with the modified tubular cells of the macula densa, they constitute the so-called juxtaglomerular apparatus.

Some Factors Involved in the Regulation of Renin Output by the Kidney.   Generally the plasma level of renin is stated in terms of nanograms of angiotensin II produced by incubation of 100 ml of plasma for three hours. In humans, normal levels of renin lie between 150 and 250 nanograms (ng), although higher values have been reported.

A number of stimuli have been shown to elicit increased renin secretion. Sympathetic stimulation via the renal nerves causes increased renin secretion as does hypoglycemia which is of sufficient magnitude to evoke generalized sympathetic discharge. Apparently the catecholamines exert a direct action upon adenyl cyclase within the membranes of the juxtaglomerular cells, although denervation of one kidney does not prevent the increased renin secretion which may be stimulated by constriction of the ipsilateral renal artery. Transplanted (hence denervated) kidneys also respond to various stimuli by increased renin secretion; therefore extrinsic innervation is not essential to an enhanced renin secretion.

Renin secretion is also increased when a fall in the mean arterial blood pressure develops (such as is induced by a massive hemorrhage), or when a person rises from a supine to the vertical position. Observations such as these have led to the hypothesis that an intrarenal pressoreceptor mechanism is present, perhaps in the juxtaglomerular apparatus, which is stimulated to increase renin secretion by a fall in mean blood pressure within the afferent arterioles.

On the other hand, renin secretion also is enhanced by a low sodium diet, hyponatremia, and certain diuretics; therefore, possibly the cells of the macula densa are the receptors and decreased sodium transport in these elements is responsible for stimulating renin secretion.

At present, the physiologic mechanisms which underlie the control of renin secretion under various conditions are unknown despite extensive research.

Physiologic Actions of Renin.
Pressor Effect.   Renin is secreted in elevated quantities in animals which have undergone certain procedures which induce chronic renal ischemia. For example, constriction of one renal artery in the rat by partial ligation of this vessel induces a sustained renal hypertension in this species. A similar effect can be induced by subtotal ligation of both renal arteries in the dog. Furthermore, the intravenous infusion of renin or

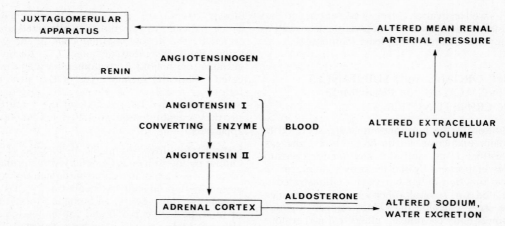

FIG. 15.72. The mechanism of action of renin in enhancing aldosterone secretion. Note that a negative feedback mechanism is involved in the regulation of this system, as an increased extracellular fluid volume increases the mean renal arterial blood pressure, and this in turn will decrease renin secretion from the juxtaglomerular apparatus. This system also appears to have a significant role in the thirst mechanism. See also Figure 15.73. (Data from Ganong. *Review of Medical Physiology,* 6th ed, 1973, Lange Medical Publications.)

angiotensin II into intact animals causes a marked, but rapidly transient, pressor response. Both systolic and diastolic blood pressures are elevated sharply in such experimentally produced renal hypertension.

Angiotensin II has been shown to be the most powerful known vasoconstrictor of the systemic arterioles, and in equivalent amounts, this octapeptide exerts a pressor effect that is between five and eight times greater than that of an equivalent dose of norepinephrine. A considerable amount of research has been conducted in this area in order to determine whether or not renin has an etiologic role in human essential hypertension. In fact some hypertensive humans (as well as animals) do have elevated circulating angiotensin levels which are coupled with high aldosterone secretion rates and hypokalemia. However, in the majority of cases, such hypertensive subjects do not exhibit any such changes in their circulating angiotensin levels. In patients with essential hypertension, unilateral renal arterial stenosis sometimes is seen, but more often such a condition is not present, so that the mechanism that is involved in the maintenance of the chronic hypertension is unclear. Nonetheless the kidney may be involved in some, as yet obscure, fashion.

In contrast to these findings, patients with idiopathic hypertrophy and hyperplasia of the juxtaglomerular apparatus exhibit elevated aldosterone secretion rates and a hypokalemia that is coupled with a concomitantly elevated circulating angiotensin II level. In such patients the blood pressure is normal.

Regulation of Aldosterone Secretion. By far the most important physiologic role of angiotensin II in the body appears to be concerned with the direct action of this hormone on the rate of secretion of aldosterone by the adrenal cortex (Fig. 15.72). Hence renin and angiotensin II play an important, though indirect, part in the regulation of extracellular fluid volume. In small quantities, this effect of angiotensin far outweighs the pressor effects.

Asp—Arg—Val—Tyr—Ile—His—Pro—Phe—His—Leu—Leu—Val—Try—Ser—PROTEIN
 1   2   3   4   5   6   7   8   9   10 11 12 13 14

Angiotensinogen

Asp—Arg—Val—Tyr—Ile—His—Pro—Phe
 1   2   3   4   5   6   7   8

Angiotensin II

FIG. 15.73. Amino acid sequence in human angiotensinogen and angiotensin II. Angiotensinogen is an α_2-globulin having the terminal amino acid sequence shown above. PROTEIN = remainder of α_2-globulin molecule. ① = peptide linkage split by converting enzyme, ie, Phe-His linkage (8–9); ② = peptide linkage split by renin, ie, Leu-Leu linkage (10–11). (Data from Ganong. *Review of Medical Physiology,* 6th ed, 1973, Lange Medical Publications.)

The negative feedback mechanism whereby the renin–angiotensin system may regulate aldosterone secretion is shown in Figure 15.72. It is well established that a decline in the renal mean arterial blood pressure stimulates aldosterone secretion. Thus a fall in the extracellular fluid volume (eg, as by a hemorrhage) causes such a fall in the renal blood pressure and increases renin secretion. As a consequence of the increased renin secretion the circulating level of angiotensin II now is enhanced. This hormone in turn acts upon the adrenal cortex directly to stimulate aldosterone secretion; therefore water and salt retention are enhanced. As the extracellular fluid volume increases, the mean renal arterial blood pressure rises once more, and the stimulus to the increased renin secretion is removed and the negative feedback cycle is completed. Conversely, an increase in the mean blood pressure within the kidney produces a decrease in renin secretion through the same mechanism, and thereby aldosterone secretion also is inhibited, and more salt and water are lost so that the blood volume is reduced, and the blood pressure falls.

ERYTHROPOIETIN.   Erythropoietin is a glycoprotein (mucoprotein) which is produced by the action of a substance that is elaborated by the kidney (renal erythropoietic factor, REF). In turn REF acts upon a plasma globulin to produce erythropoietin.

Circulating erythropoietin stimulates red blood cell production (or erythropoiesis) by stimulating an increased rate of hemoglobin synthesis, as well as enhancing the release of erythrocytes from the bone marrow. The physiologic stimuli to increased erythropoietin production are hypoxia or hemorrhage, both of which reduce the effective total circulating red cell mass. On the other hand, when the total red cell mass is increased, erythropoietin production is inhibited, and the rate of erythropoiesis declines.

Within the bone marrow, erythropoietin stimulates certain stem cells, which are called erythropoietin-sensitive stem cells, to develop into proerythroblasts, a process which is accompanied by the synthesis of mRNA.

Since antibodies against erythropoietin cause anemia, it appears that this hormone is necessary for normal erythropoiesis to be sustained.

REF in humans is synthesized principally in the kidneys, and apparently to a slight extent in an unknown extrarenal locus, and hypoxia, androgens, and cobalt salts enhance this synthetic process. The circulating globulin on which REF acts is secreted by the liver. Hypoxia apparently stimulates the synthesis of both of these substances; however, the receptors that are sensitive to a relative oxygen deficiency are unknown. In similarity to the action of renin or angiotensinogen (Fig. 15.72), the effect of REF on the plasma globulin also appears to be enzymatic, but these two endocrine substances are known to be entirely distinct from one another.

Erythropoietin is inactivated principally in the liver, and the half-life of this hormone in the blood is around five hours. In contrast, the stimulation of erythrocyte production by erythropoietin takes two or three days to become manifest since maturation of these formed elements of the blood takes place rather slowly.

Factors that increase testosterone secretion (eg, increased pituitary ICSH secretion) also stimulate erythropoietin production, whereas estrogens inhibit this process, giving still another example of the general antagonism between male and female sex hormones.

Spleen

The spleen is not essential to life, although it is a part of the reticuloendothelial system. It also contains lymphoid tissue that produces lymphoid cells and monocytes, as well as plasma cells for the manufacture of antibodies (p. 547). In man, unlike other mammals, the spleen does not function as a significant blood reservoir, although in normal infants extramedullary erythropoiesis takes place in this organ as well as in the liver.

Important effects of the spleen appear to be exerted upon the life-span of the erythrocytes, leucocytes, and other formed elements of the blood, although how these actions are exerted are not at all clear. In pathologic situations, the spleen may destroy red and white cells at an abnormal rate, whereas in other conditions the platelets may be extracted selectively from the circulation in abnormal quantities, giving rise to a thrombocytopenia that is accompanied by hemorrhages or purpura.

The effects of the spleen upon the blood elements, as well as its effects upon the toxicity of some drugs, have given rise to the postulate that the spleen exerts these actions through an endocrine mechanism. However, no convincing evidence has been presented to indicate that such a mechanism is present in splenic tissue, and any endocrine function by the spleen has yet to be proven.

Thymus

This bilobed lymphoid organ is found within the anterior mediastinum. It is large during infancy, but exhibits progressive atrophy following puberty, and this atrophy follows the general pattern which is exhibited by other lymphoid tissues and the atrophy which is produced by adrenal glucocorticoids with advancing age.

Morphologically, the thymus consists of an outer cortex of lymphoid tissue which surrounds a medulla that is composed of lymphocytes and a coarse stroma containing typical clusters of cells known as Hassell's corpuscles. The latter structures have a core of granular cells which are invested in typical epithelioid cells. Hassell's corpuscles represent the remaining vestiges of the tubules from the third branchial pouches of the embryo whence the thymus was derived.

During infancy, the thymus produces lymphoid cells which ultimately develop antibodies causing delayed hypersensitivity reactions as well as homologous tissue graft rejection. Thus humans lacking a thymus due to congenital aplasia (DiGeorge's syndrome) have a normal humoral immunity but a defective cellular immunity (p. 548). In such individuals, delayed hypersensitivity reactions are absent and tissue that is transplanted from other persons is not rejected. If thymectomy is performed on older persons, graft rejection will take place.

Experiments performed on newborn mice indicate that a chemical agent from the thymus circulates to the stem cells in the lymphoid organs throughout the body, and this agent in turn stimulates the stem cells to develop into lymphocytes with immunologic competence (Fig. 9.13, p. 549). That is, the altered cells now are capable of forming antibodies against foreign protein. A polypeptide hormonelike substance has been extracted from thymic tissue (thymosin), and this substance has been shown to increase the level of circulating lymphocytes as well as to accelerate skin graft rejection.

Enlargement (hyperplasia) of the thymus with the development of germinal centers similar to those found normally in lymph nodes has been found in a number of autoimmune diseases. In such conditions, the body seems to develop antibodies against (and thus to reject) certain of its own tissues. Included in this category of pathologic states are rheumatoid arthritis, thyroiditis, hemolytic anemia, and myasthenia gravis.

Pineal Body

GROSS ANATOMY. The pineal body or gland (epiphysis cerebri) is located above the roof of the diencephalon at the extreme posterior end of the third ventricle (Fig. 7.11, p. 247). The epiphysis lies beneath the posterior end of the corpus callosum, and is attached by a stalk to the habenular and posterior commissures. Grossly the pineal body is a flattened, conical structure, about 7 mm in length, 4 mm in width, and is covered by pia mater. Numerous connective tissue septa having an abundant vascular supply, and which originate from the pia mater, enter the tissue of the pineal body and these septa surround the epithelioid cells of the cords and follicles.

HISTOLOGY. The principal mass of the pineal body is composed of pale-staining cords of epithelioid cells having large, irregularly-shaped nuclei. These chief cells having large, irregularly-shaped nuclei. These chief cells (or pinealocytes) also exhibit long processes that are visible only after using special staining techniques, and which end in swellings about the vascular connective tissue. The cytoplasm of the pinealocyte contains vesicular cytoplasmic bodies which suggest a secretory function for this organ.

Interstitial cells also are found between the cords and groups of pinealocytes. These elements have an abundance of filaments within the cytoplasm, and are believed by some investigators to be glial cells.

Nerve fibers are found within the pineal stalk; however, such neural structures are scarce within the pineal body itself.

MORPHOLOGIC CHANGES WITH AGING. In infants and young animals the pineal body is large; however, before the onset of puberty involution commences. In man, concretions which are composed of calcium and magnesium phosphate, as well as carbonate (the so-called "pineal sand"), are found in the tissue. These concretions render the pineal body radiopaque, therefore this organ can be visualized normally in x-ray films of the adult skull.

POSSIBLE FUNCTIONS OF THE PINEAL BODY. Similarly to the posterior pituitary and area postrema, the pineal body lies outside of the blood-brain barrier; thus it accumulates dyes and other materials more rapidly than do other regions of the brain. Similarly to the area postrema, the serotonin and norepinephrine content of the pineal body are quite high, and the normal rate of phosphorus turnover is also quite high in the pineal body.

Many functions, some fanciful and some realistic, have been ascribed to the pineal body. At present it appears that the pineal body may have a neuroendocrine function, and by the synthesis of a specific methoxyindole known as melatonin from serotonin (5-hydroxytryptamine), this structure may exert a "pseudohormonal" regulatory effect upon certain endocrine glands. Specifically, melatonin has been implicated in the regulation of the reproductive (ie, gonadal) cycles of rodents and birds, an effect that in turn depends upon the diurnal light cycle. Photic stimuli are presumed to be transmitted via optic pathways and relayed by sympathetic nerves to the pineal body itself where such stimuli are converted into endocrine stimuli. The concentration of the specific enzyme required for melatonin synthesis (hydroxyindole-O-methyl transferase, HIOMT) is decreased when an animal is maintained in light of constant intensity. Phosphorus uptake and serotonin content of the pineal body also have been correlated with the diurnal rhythm of light and darkness.

In humans, clinical observation has revealed that precocious puberty may occur in males whose pineal body has been destroyed by tumors. Such facts have led to the concept that the pineal body exerts an antigonadotropic effect.

At present, however, the actual physiologic role (if any) of the pineal body in man or other mammals for that matter, remains enigmatic as well as conjectural.

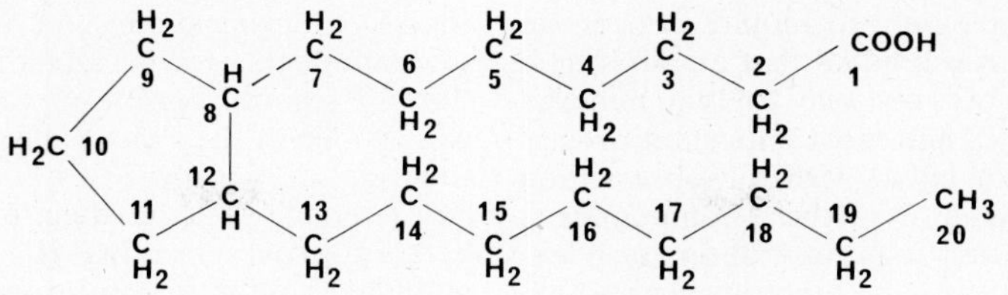

FIG. 15.74. Structural formula of prostanoic acid. (After Pike. *Sci Am* 225:84, 1971.)

Prostaglandins

Collectively the prostaglandins are a group of fatty acid compounds that exert profound and diverse physiologic effects upon a wide variety of tissues and processes in the body. Discussion of these substances has been included with the general hormones, although this is not to imply that all of the prostaglandins exert an endocrinelike regulatory influence under all circumstances.

Historically the presence of different prostaglandins in various biologic materials has been known or suspected for about 40 years, but only during the past decade or so have these compounds been subjected to extensive study for several important technical reasons, including the fact that they are present in most tissues in extremely minute quantities.

Because of the great potential importance of the prostaglandins in clinical medicine, research in this field is progressing rapidly, and the following discussion is designed to provide only a brief introduction to this field.

CHEMISTRY. The prostaglandins are derivatives of certain unsaturated fatty acids (eg, arachidonic acid, p. 69), hence they are carboxylic acids having 20 carbon atoms. All of these compounds have been shown to have the same basic chemical structure, viz., that of prostanoic acid. This compound contains two fatty acid chains attached to a 5-carbon cyclopentane ring (Fig. 15.74).

The different individual prostaglandins are formed principally by the addition of various groups or radicals to specific carbon atoms at various loci in the molecule. The presence of one, two, or three double bonds at different loci in this structure also give rise to specific differences among these compounds. The latter property gives rise to the three series of prostaglandins, ie, prostaglandins series 1, series 2, or series 3, depending upon whether one, two, or three double bonds are present in the molecule.

The primary prostaglandins are designated as PGE or PGF. PGE compounds have an oxygen atom attached to carbon 9 of the cyclopentane ring, whereas PGF prostaglandins have a hydroxyl radical on the same carbon atom (Fig. 15.74). PGA or PGB prostaglandins are formed by dehydration of a PGE molecule.

Approximately over a dozen different prosta-

glandins have been isolated and characterized, and all of these compounds exhibit biologic activity.

BIOSYNTHESIS, METABOLISM, AND EXCRETION. The prostaglandins are synthesized in the body from several polyunsaturated fatty acids (eg, arachidonic acid) by the formation of a five-carbon (cyclopentane) ring and the incorporation of three molecular oxygen atoms into a hypothetical intermediate molecule. The intermediate than is transformed into a PGE or PGF structure, and the primary prostaglandins then are converted readily into other specific, but chemically related, compounds, one example of which is PGE_1 (Fig. 15.15).

The enzyme systems that are involved in prostaglandin synthesis appear to be associated principally with the cell membranes. This fact is of considerable importance, as phospholipids are essential constituents of the plasmalemma, and thus provide a concentrated source of unsaturated fatty acids for prostaglandin biosynthesis.

The enzymes that underlie prostaglandin synthesis have been found in a great number of tissues which have been taken from various animals including man. These tissues include the lung, kidney, thymus, uterus, brain, and seminal vesicles.*

As noted earlier, prostaglandins are present in

Incidentally, semen was the first-studied source of a prostaglandin, and the secretion of this substance was originally, but erroneously, ascribed to the prostate, hence the name "prostaglandin." Subsequently, however, the seminal vesicles were shown to contribute a prostaglandin to the seminal fluid, but this misleading term has persisted.

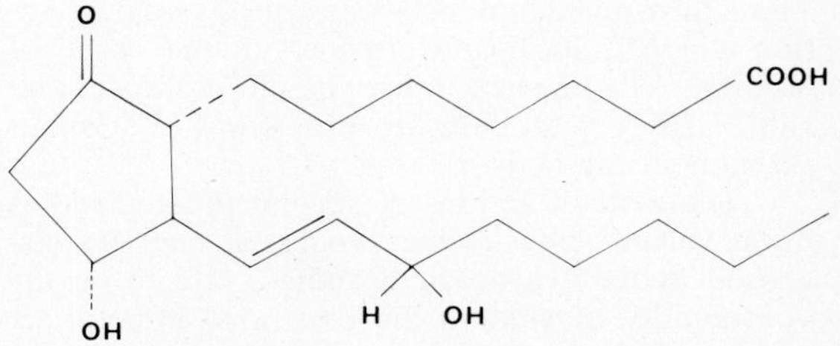

FIG. 15.75. Structural formula of the prostaglandin PGE_1. (After Pike. *Sci Am* 225:84, 1971.)

the tissue of the body only in minute concentrations, and it has been estimated that an adult man synthesizes a total of only one to two hundred micrograms per day of the most important prostaglandins. This fact, coupled with the observation that enzymatic degradation of these compounds is exceedingly rapid after their formation, also explains why their presence and functions proved exceedingly difficult to ascertain for many years. There is no evidence of prostaglandin storage, hence the tissues apparently manufacture these compounds as needed, and their survival in the body is quite evanescent. Stated another way, the turnover rate of endogenous prostaglandins in the body is exceedingly high, thus their half-life is quite short, and their concentration in any particular tissue at any moment is quite low.

Radioactive prostaglandin (PGE_1) when injected intravenously into experimental animals (mice) of both sexes is taken up rapidly by the liver, kidney, and subcutaneous tissue. Female mice also concentrated this prostaglandin in their uterus, whereas males also concentrated this material in the thoracic duct.

In man, injected tracer doses of PGE_1 were rapidly excreted by the kidney (50 percent in the urine) and liver (10 percent into the bile), so that in 20 hours about 60 percent of the exogenous compound was eliminated from the body. In similarity to the mouse experiments described above, radioautographs of both liver and kidney revealed high isotopic concentrations in these tissues.

Because of the delicate balance between the rapid synthesis and equally rapid degradation of prostaglandins in most tissues, the concentration of these compounds usually averages only one microgram per gram of wet tissue. Seminal fluid represents an exception to this general rule, so that prostaglandins are present in this source at a concentration that is about 100-fold greater than that which is present in other tissues.

PHYSIOLOGIC ACTIONS OF THE PROSTA-GLANDINS.

Several general facts concerning the action of prostaglandins are noteworthy. First, these compounds exert profound physiologic effects in extremely small doses. In fact, the prostaglandins are among the most potent known substances of biologic origin. Second, the individual prostaglandins are highly specific in their actions. Third, prostaglandins may exert a physiologic action which is local and confined to the tissue of origin, or else the action may be more diffuse, and thus affect widespread, but similar, tissues throughout the body.

The general actions of several prostaglandins upon various physiologic processes may be discussed under five broad headings: effects on the contractility of smooth muscle; renal effects; secretion of exocrine and endocrine glands; transmission by neural tissue; and miscellaneous effects.

Effects of Prostaglandins on Smooth Muscle. The several effects of prostaglandins upon the contractility of smooth muscle in various tissues and organs is one of their most important known areas of function.

Uterus. Among their most striking actions, certain prostaglandins (PGE_2 or PGF_2-alpha) stimulate rhythmic and strong contractions of the uterus of pregnant women in exceedingly small doses. Since prostaglandins are found as apparently normal constituents in uterine venous blood as well as in the amniotic fluid of women during labor, the physiologic role of these substances during parturition may be of considerable significance.

In humans, the infusion of 0.05 micrograms of PGE_2 per minute per kilogram body weight will induce delivery within several hours. Thereby this compound facilitates parturition.

The potential clinical usefulness of prostaglandins in treatment of secondary amenorrhea and for inducing abortion is the subject of extensive current investigations.

Blood Vessels. Two prostaglandins exert diametrically opposite effects upon the systemic blood pressure by their direct action on the smooth muscle that controls vascular caliber. PGE_2 infusion produces hypotension, whereas the closely-related PGF_2-alpha induces hypertension. It is interesting to speculate upon a possible role for the latter compound in the genesis of essential hypertension.

It has been demonstrated that still another prostaglandin, PGA_1, can reduce the systemic blood pressure in patients with essential hypertension. These facts imply an important physiologic role for the prostaglandins in the regulation of blood pressure.

Local application of PGE_1 to the nasal passages causes vasoconstriction of the mucosal blood vessels with concomitant enlargement of the airway. The same prostaglandin, inhaled as an aerosol, causes dilatation of the bronchial musculature, thereby facilitating respiration in patients suffering from asthma.

Renal Effects. The infusion of either of two prostaglandins (PGE_1 or PGA_1) into the renal artery of experimental animals (dogs) in minute doses produces a significant increase in the volume of urine that is excreted by the kidney. The enhanced urinary excretion is also coupled with an increase in the excretion of sodium ion. These facts, together with the effects of certain prostaglandins upon vascular tone that were noted above, substantiates the conclusion that these compounds have an important physiologic role in the homeostatic regulation of blood pressure.

Prostaglandins and Secretion. There is evidence that prostaglandins may have an important role in the regulation of both exocrine as well as certain endocrine secretions. For example, experimental work in dogs has clearly demonstrated that a sustained infusion of either PGE_1 or PGE_2 dramatically inhibits the secretion of pepsin as well as hydrochloric acid by the gastric mucosa, presumably by a direct action upon the parietal cells of the gastric glands. Furthermore, in rats, prostaglandins can prevent the development of gastric and duodenal ulcers. Since ulceration is believed to result from erosion of the mucosa by excessive quantities of gastric juice, the physiologic synthesis of prostaglandins by the stomach may protect the mucosa against ulceration by regulating its secretion. In humans who lack the normal endogenous complement of gastric prostaglandins, exogenous prostaglandins eventually may prove to be of therapeutic value in protecting the stomach lining against the corrosive action of its own secretions.

A particularly interesting example of prostaglandin action in an endocrine system has been demonstrated in monkeys. Infusion of a prostaglandin (PGF_2-alpha) into mated female monkeys causes a sharp reduction of progesterone secretion by the corpus luteum, possibly attended by regression of this structure. Normally, of course, such an event does not take place after fertilization. It is interesting to speculate upon the possibility that a prostaglandin is responsible for initiating normal luteal regression during the female sexual cycle, and thus such a compound may provide a key factor controlling the normal menstrual rhythm in women. Of more pragmatic significance, however, is the possibility that natural or artificial prostaglandins eventually may prove useful as chemical abortifacients.

Neural Effects of Prostaglandins. It has been shown experimentally (in heart and spleen) that PGE_2 infusion inhibits secretion of the physiologic neurohumoral transmitter norepinephrine when the appropriate sympathetic nerves were stimulated. Conversely, inhibition of prostaglandin synthesis within the nerve terminals by appropriate chemical agents leads to an excessively great release of norepinephrine upon stimulation of the nerves.

Results such as these clearly indicate that prostaglandins may play an important role in the negative feedback control of impulse transmission in the sympathetic nervous system.

Miscellaneous Effects of Prostaglandins. A number of actions of prostaglandins have been observed in addition to those summarized above. These include an accelerated rate of synthesis of prostaglandins by lung tissue during anaphylaxis as well as by the brain surface following stimulation of peripheral nerves. Similarly prostaglandin synthesis in skin also is increased by experimentally induced inflammatory reactions as well as in allergic eczema. Prostaglandin synthesis within the kidney also is increased during experimental renal is-

chemia which was induced by arterial constriction. In the latter instance, a possible interrelationship among prostaglandin synthesis, renin secretion, and the ischemia is intriguing, but has not yet been demonstrated.

Prostaglandins also have been implicated in the antiinflammatory action of aspirin and other drugs. Thus such compounds possibly may exert their effects in vivo by blocking the biosynthetic mechanisms for specific prostaglandins.

Evidence such as that presented above clearly suggests a probable relationship between prostaglandins and the regulation of a wide diversity of normal physiologic processes within the body. Furthermore, certain of these compounds doubtless play an important role in the etiology of certain pathologic conditions.

POSSIBLE MODE OF ACTION OF PROSTAGLANDINS: GENERAL CONCLUSIONS. It was stated in the introduction to this chapter that the two well-established major regulatory and communications mechanisms within the human body are the nervous system and the endocrine system, with an occasional overlapping of the functions which are performed by these two systems. The cell membrane with its many dynamic functions appears to be the one point at which the activity of these two systems converge, as this structure is critical to the performance of an extraordinarily large number of highly specific and complex processes. Since the cell membrane is comprised chiefly of proteins and phospholipids, and since the latter compounds provide the precursor fatty acids for prostaglandin synthesis, it is logical to assume that the prostaglandins have a major physiologic role in the regulation of membrane function per se, especially since these substances appear to be formed by the membrane.

It also is fascinating to speculate upon a possible interrelationship among prostaglandins, cyclic AMP synthesis, and various specific cellular functions, as cyclic AMP is known to have an important role in the cellular interpretation of messages which are transmitted by specific hormones as well as in other cellular processes which have been cited in context throughout this book. For example, do specific prostaglandins exert an action in certain cells insofar as their individual permeability characteristics to various substances are concerned? Do prostaglandins function to regulate intramembranous active transport systems? The answer to questions such as these must await experimental analysis; however, it is clear even at this relatively early phase of research into the field of prostaglandins that these compounds eventually will prove to have many fundamental and vital roles in normal as well as pathologic physiology, and doubtless some of these compounds eventually will assume considerable importance as therapeutic agents in clinical medicine.

CHAPTER 16

Physiologic Responses and Adaptations of the Body under Particular Conditions

"Nature gave me my model, life and thought: the nostrils breathe, the heart beats, the lungs inhale, the being thinks and feels, has pains and joys, ambitions, passions and emotions." Thus did the great sculptor Auguste Rodin express the thoughts he entertained while preparing the mold of "The Thinker," the statue that would be his life's masterwork. These words also provide a particularly apt introduction to the final chapter of this textbook, as they emphasize that the many individual functional activities of the human body are merely parts of an integrated whole, and that whole is the individual human being.

Up to this juncture the many physiologic properties of the human body have been described individually for the most part, despite the fact that in many instances these functions are inseparably interrelated, a fact of major importance to the practicing physician. In the discussions that follow, an attempt has been made to provide the reader with a composite general discussion and summary of the events taking place during certain fundamental physiologic responses and adaptations of the human body at different stages in the life cycle as well as when it is subjected to various environmental stimuli and other circumstances. Most of the examples selected for inclusion in this chapter illustrate normal processes which are (or may be) called into play during the lifetime of an individual. By way of contrast, however, several abnormal states have been presented in which the body is required to adapt to stresses forced upon it by various disease processes and other situations by using such homeostatic mechanisms as are available in order that the individual survive at all. All of these discussions will serve to emphasize that the normal as well as the diseased human body has a remarkable capacity to respond and adapt successfully to a broad range of environmental and other stresses.

Throughout this text, numerous examples of commonly encountered physiologic phenomena have been cited that have proven to be problems of considerable profundity, insofar as clarification of their underlying biologic mechanisms is concerned. Many of these seemingly obvious physiologic problems also have proven to be exceedingly refractory to solution, even by the application of the most sophisticated physical and chemical techniques that are available to the research worker today. Similarly, many of the topics to be discussed in this chapter, although understood in an overall fashion, are not yet explicable in detail. Therefore such current gaps in our knowledge will be mentioned as appropriate in the material to follow, not necessarily with the aim of overemphasizing their significance, but rather with the goal of illustrating a few of the many unresolved areas of physiology that yet may prove to be fruitfully amenable to future investigation.

The eight general topics that will be presented in this chapter are:
1. The Physiology of Exercise
2. Pregnancy, Parturition, and Lactation
3. Some Aspects of Fetal and Neonatal Physiology
4. Human Malnutrition: Starvation and Obesity
5. Survival at Decreased Ambient Pressures: High Altitude, Aviation, and Space Physiology
6. Survival at Increased Ambient Pressures: Diving Physiology
7. The Pathophysiology of Shock
8. The Pathophysiology of Chronic Cardiac Failure.

The reader will note that the cardiovascular and respiratory systems are major participants in all of the processes listed above, and thus form common biologic denominators for a comparison of the individual topics. Each of these subjects has been presented so that the discussion essentially is complete in itself. Pertinent cross-references to material discussed earlier also have been included in order to assist the reader in locating more detailed information on specific topics as may be desired for review purposes.

THE PHYSIOLOGY OF EXERCISE

Definition and General Considerations

In the broadest physiologic sense, the word *exercise* may be defined as an increased level of physical exertion (or work output) above the resting state. Yet this definition falls short of being either adequate or complete because the physical exertion involved in a given activity may be carried out at levels ranging from "very mild" to "strenuous." Thus the *type,* the *rate,* and the *duration* of the physical work involved in the performance of muscular exercise are among the many factors that enter into any definition or discussion of this subject. With regard to the duration of the exercise, it is also essential to consider whether or not the physical effort is of short duration (ie, lasting for minutes or hours) or whether it is chronic (of long duration, ie, extended over a period of days, weeks, or even months). Furthermore, in both of these instances the intermittence of the performance throughout the periods of enhanced activity plays a major role in the ultimate physiologic responses.

Other important parameters that are involved in any general consideration of exercise as a physiologic process are the *age, sex,* and *general state of health* of the individual who is performing the physical work, as well as the basic reason for undertaking the exercise.

Some general examples may be used to illustrate the foregoing concepts. Common repetitive types of exercise such as walking, jogging, running, cycling, swimming, tennis, and so forth may be performed merely for the purpose of gaining and/or maintaining a certain level of physical fitness or conditioning, whereas a far greater work output over a long period of time is demanded of an athlete in preparation for an upcoming competitive event if he wishes to achieve peak physical performance during that event. Similarly, persons engaging in different occupations are subject to enormous variations in the amount of exercise, or more accurately, physical work, that they are called upon to perform daily. Furthermore, the voluntary and involuntary muscular contractions that are associated with and accompany the various levels of sexual excitation and coitus induce physiologic responses that at times may equal the intensity of those responses that are found during periods of strenuous physical work. For this reason, a few examples of some of the bodily responses that occur during sexual activity are included in the following discussion.

In contrast to the examples cited above, the word "exercise" denotes activity at a far different level of exertion for a patient who is recuperating from the effects of a severe coronary thrombosis; thus a slow, short perambulation around the hospital ward represents, to him, the maximal work output that is feasible and appropriate to his circumstances. In fact, during certain abnormal pulmonary states, the energy expended solely for the act of breathing itself is sufficient to consume an inordinately large proportion of the total energy output of the body over the basal level (p. 722).

Consequently, the general considerations listed above must be taken into account in any discussion or evaluation of exercise or physical work.

In the strict physical sense, exercise may be considered to be the performance of work, and as such it is amenable to quantitative measurement (in the laboratory) and to expression in physical terms. As defined in Chapter 1 (p. 3), work (W) is defined as the force required for the movement of a given mass (M) through a specific distance (D); thus, $W = M \times D$. Power, on the other hand, is defined as the rate at which the work is performed, ie, $P = W/t$. The units employed for the expression of work thus may be kg-m. Power, on the other hand, is expressed in watts if the work, W, is defined in joules (1 joule $= 10^7$ ergs), and the unit for time is in seconds. More commonly, work performance by humans is expressed in kg-m/min. Actually, therefore, the work performance is defined in terms of power rather than as work per se.

Usually, however, the physical quantitation of work performance is impractical for various technical reasons. Therefore one must resort to use of such necessarily vague and subjective terms in defining the level of exercise as "light," "moderate," and "severe" when evaluating the work performance of various individuals under different circumstances. The effects of prior conditioning, if any, also must be considered in such an evaluation.

It should be emphasized that exercise (or physical work) can represent one of the greatest physiologic stresses to which the human body may be subjected during the course of a normal lifetime. The adaptive responses carried out by the body to this type of physiologic activity, far from being detrimental in the normal person, have been demonstrated amply and repeatedly to be of positive benefit to the mental as well as physical health and well-being of the individual. Furthermore, in many clinical states specific types of exercise performed at a frequency and level that are suitable for the individual patient have proven to have a marked shortening effect upon the overall convalescent period. This fact is in diametric opposition to the earlier held view that extended bed rest is the key to a successful recovery.

It must be appreciated that the entire spectrum of respiratory, cardiovascular, and other physiologic adaptations to exercise has but one basic purpose, and that is to deliver an adequate supply of oxygen to the active skeletal muscles in order that the contractile machinery can continue to function for sustained periods of time. All other responses to enhanced work output are secondary to this critical adaptive effect.

In conclusion, it should be obvious to the reader that the phenomenon known as exercise is extremely complex in its ramifications. The gamut of physiologic responses to enhanced physical activity is equally complex, and a brief presentation and discussion of these individual responses will form the remainder of this section.

Physiologic Responses and Adaptations to Enhanced Physical Activity

THE PRIMARY STIMULI. Since exercise may be defined for practical purposes as an increased level of work output above the resting state of the body due to voluntary muscular contractions, it is pertinent to inquire at the outset: What are the principal stimuli that induce the major physiologic changes that are observed in the exercise state?

The most obvious changes that occur during increased muscular exertion are those concerned with the musculo-skeletal, cardiovascular, and pulmonary systems. Consequently, a priori, it would appear that the primary stimuli for the physiologic adaptations that occur during exercise might be found in, or related to, one or more of these major systems.

In a resting individual, the respiratory drive is conditioned by the interplay of three principal factors. These factors are the arterial Po_2, the arterial Pco_2, and the concentration of hydrogen ion in the arterial blood (cH^+, p. 735). During periods of enhanced muscular activity, pulmonary ventilation and oxygen consumption increase markedly, and this increase is directly proportional to the quantitative increase in the muscular exertion performed (Fig. 16.1F). During exercise, the production of carbon dioxide and acid metabolites (eg, lactic acid) also increases concomitantly with an enhanced tissue utilization of oxygen (p. 137; p. 447). Therefore one might assume with some logic that the prime cause of the respiratory stimulation that takes place at the onset of exercise is evoked by a shift in the three factors that are concerned with the maintenance of pulmonary ventilation during rest. Experimental studies have demonstrated repeatedly that increased muscular activity is *not* necessarily or consistently associated with an immediate decrease in the arterial Po_2 or an elevation in the arterial Pco_2 or cH^+. The unequivocal demonstration of one or more such changes would, of course, be essential to provide evidence that such effects were primarily responsible for eliciting the respiratory changes occurring immediately upon the commencement of a period of exercise.

The Pco_2 of mixed venous blood is increased during strenuous exercise, and the Po_2 is decreased concomitantly. There is no conclusive evidence that any receptors that affect respiration are present anywhere in the venous portion of the systemic circulation that are comparable to the chemoreceptors found in the arterial tree (p. 674).

Some evidence exists that respiration can be stimulated reflexly via sensory impulses that arise in contracting skeletal muscles and/or within the joints that are acted upon by these muscles. The physiologic role of such reflexes in the overall pattern of adaptation to exercise is obscure at present.

It has been postulated that during voluntary muscular activity motor impulses from higher brain centers descend via certain unspecified nerve tracts to stimulate the medullary respiratory center directly. Certainly it is clear that respiratory and cardiovascular activity may increase markedly in an athlete in anticipation of an immediately forthcoming competitive event and prior to the actual start of the exercise. It is therefore obvious that cortical factors must play a role in such adaptations as take place prior to the onset of the physical activity. Such cortical stimulation is not always demonstrably present before exercise commences so that the role of such neural influences as primary stimuli in all types of exercise is hardly clear at present (see further discussion of this factor below).

It also is of interest that blood flow through the skeletal muscles can increase significantly before, as well as immediately following, the onset of a period of exercise. This fact also implies neural activity, and presumably in this case the sympathetic vasodilator system is involved (p. 671).

Both the heart rate and cardiac output increase significantly following the onset of, and during, a period of exercise (Fig. 16.1A and D). These effects result from a markedly increased venous return due to the pumping action of the skeletal muscles (p. 652). In addition, the increased movements of the thorax attendant upon an increased ventilatory response also augment venous return to the heart; hence this factor tends to elevate cardiac output. But it is impossible at present to ascribe to these physiologic changes any particular role as *primary* stimuli in the adaptations to exercise, as such alterations in the cardiovascular state of the individual usually occur shortly *after* the onset of activity. That is, the changes appear to be *secondary* rather than primary responses to exercise (see also p. 657 and Fig. 16.6).

It is evident from the brief foregoing discussion that the primary stimuli that are responsible for the respiratory and cardiovascular adaptations immediately prior to, or following, the start of a period of muscular exertion remain unclarified, although certain chemical and neural factors appear to be implicated in this process. Some investigators believe that accessory respiratory drives, as yet unidentified specifically, are present in the body. These factors presumably exert their effects upon the respiratory drive immediately, and even before, significant and readily detectable alterations in arterial Po_2, Pco_2, and arterial cH^+ take place. Such accessory respiratory drives have

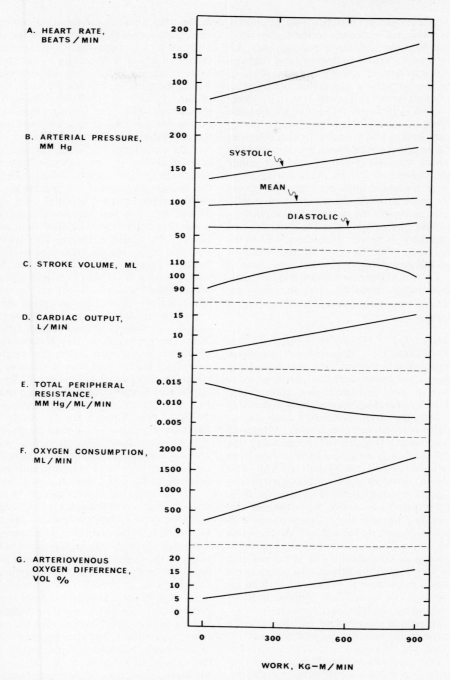

FIG. 16.1. Effects of various levels of isotonic exercise upon certain cardiovascular and respiratory functions. (Data from Ganong. *Review of Medical Physiology,* 6th ed, 1973, Lange Medical Publications.)

been referred to as the "exercise stimulus," and they operate at the onset of exercise before the three known chemical factors that regulate resting pulmonary ventilation can come into play. Probably afferent nerve impulses from a number of regions of the body also participate in this general process.

Finally, the performance of physical work is a voluntary muscular process that requires conscious initiation. Therefore it would be surprising indeed if cortical impulses did *not* play a major role in the initial as well as in the later stages of exercise, despite the fact that these impulses or their effects might not necessarily be detected readily. Once the exercise is under way, of course, the adaptive visceral responses to the higher level of muscular activity are largely automatic. Although the specific neural pathways (and perhaps certain ancillary chemical mechanisms) involved in stimulating the immediate physiologic changes at the start of exercise are unknown, it is quite likely that these stimuli ultimately will prove to originate in descending tracts that lead from the cortex to lower brain centers and thereby to the effectors which are concerned directly with regulation of the more obvious cardiorespiratory and the other more subtle changes that take place during exercise. Thus such cortical impulses and their effects quite easily could prove to be the primary stimuli necessary to activate the accessory respiratory drive and other physiologic adaptive mechanisms. It is very unlikely, however, that a single exercise stimulus is responsible for evoking the manifold graded and integrated responses that take place at the onset of physical exertion. There is a far greater likelihood that a number of exercise stimuli are responsible for the activation and coordination of these processes.

BLOOD AND LYMPH FLOW IN SKELETAL MUS-CLE DURING EXERCISE. Regardless of the origin of the primary stimuli that underlie the physiologic changes that occur at the start of a period of exercise, it is of interest that blood flow through the skeletal muscles increases immediately following, or even before, the physical exertion commences, and that there is a concomitant decline in the total peripheral resistance (Fig. 16.1E). Presumably this effect on blood flow is mediated by impulses from the sympathetic vasodilator system (p. 694), and perhaps also to some extent by a concomitant decrease in tonic vasoconstrictor discharge (p. 667). After the exercise has started, however, an elevated blood flow through the muscles themselves is maintained by local vasodilator mechanisms within the active tissue. Included among these local mechanisms responsible for maintaining a high blood flow in rhythmically contracting muscle are a decrease in tissue P_{O_2}, combined with an increase in tissue P_{CO_2} (p. 665). In addition to these factors, certain metabolites that have active vasodilatory properties accumulate in rhythmically contracting muscle, and in this category are included potassium ions and lactic acid, as well as adenosine and adenosine nucleotides.

Another factor that contributes to the increased blood flow in active muscle is the rise in temperature that accompanies the accelerated metabolic rate which occurs in the muscles as a consequence of exercise (p. 1001; p. 1002). The elevated temperature actively dilates the arterioles, metarterioles, and precapillary sphincters so that the number of patent capillaries within an exercising muscle increases between 10-fold and 100-fold, depending upon the severity of the physical effort. One important physiologic consequence of this temperature-induced vasodilatation is a decrease

in the total distance for the diffusion of oxygen to (and metabolites from) the active muscle cells (p. 643). A second important consequence of the increase in tissue temperature is an increase in the total cross-sectional area of the vascular bed within the muscle. This in turn leads to a decrease in the velocity of blood flow through the active muscle (p. 636), and thereby more time is permitted for gas and metabolite exchange between the capillaries and the muscle fibers.

As a consequence of the vascular changes described above, the net intracapillary blood pressure rises (p. 558) so that it becomes greater than the oncotic pressure of the plasma throughout the entire length of the capillaries (Fig. 18, p. 560). Therefore a marked increase in the quantity of water that enters the interstitial fluid compartment takes place during exercise. This transudation or filtration effect also is enhanced by the increased osmotic pressure of the interstitial fluid per se which results from an increased accumulation of metabolites that diffuse out of the muscle cells. The greater volume of interstitial fluid that accumulates in actively contracting muscle in turn causes a greatly increased rate of lymph flow (p. 561). The net result of this process is prevention of an excessive increase in the total volume of interstitial fluid, as well as to increase the rate at which the interstitial fluid is turned over. Thus an augmented lymph flow in exercising muscle offsets to a large extent the decreased *rate* (*not* volume!) of blood flow through the tissue, in addition to facilitating removal of some of the metabolic products released from the contracting cells. The increased formation of interstitial, hence lymphatic, fluid during vigorous exercise results in a transient decrease in the circulating plasma volume (p. 558).

It is important to remember that the blood flow through a rhythmically exercising muscle is *pulsatile* rather than continuous, as depicted in Figure 16.2. Thus when a muscle contracts so that it develops approximately 10 percent of its maximal tension, the blood flow is decreased significantly, owing to compression of the intramuscular blood vessels (p. 652). Similarly, if the muscle develops more than 70 percent of its maximal tension during contraction, then the blood flow ceases completely. Consequently, the alternate contraction–relaxation pattern that accompanies various types of rhythmic, isotonic exercise produces marked fluctuations in blood flow. On the other hand, a sustained, powerful, isometric contraction can reduce blood flow to zero, so that the onset of anaerobiosis and fatigue is quite rapid.

OXYGEN TRANSPORT AND UTILIZATION. The elevated temperature and the lowered pH that develop in active muscle cause the oxygen dissociation curve for hemoglobin to shift to the right (Fig. 11.30B, C, pp. 748, 749); therefore quantitatively more oxygen is released from the red blood cells to the muscles. It also has been shown that the quantity of 2,3-DPG (2,3-diphosphoglycerate, p.

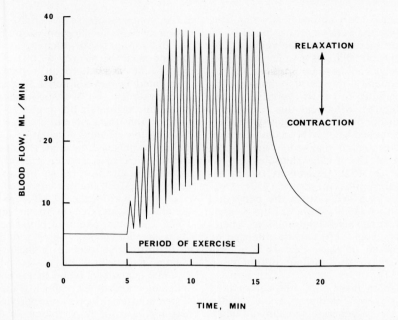

FIG. 16.2. Relationship between blood flow and rhythmic isotonic contractions in a skeletal muscle. Note that as exercise proceeds, vasodilatation increases the total blood flow through the muscle even during the contractions, as indicated by the upward trend on the bottom of the flow curve. (Data from Ganong. *Review of Medical Physiology,* 6th ed, 1973, Lange Medical Publications.)

747) increases in the red cells during exercise. Therefore an elevated level of this compound also contributes to the increased release of oxygen from hemoglobin during exercise. The net effect of these two mechanisms is to generate an arteriovenous oxygen difference that is up to threefold greater than that found during rest (Fig. 16.1G), while at the same time the transport of carbon dioxide out of the active tissue is accelerated. The physiologic mechanisms described above, together with the vascular mechanisms discussed in the previous section, exert a combined effect such that the oxygen consumption of exercising skeletal muscle can increase up to 100-fold over that observed during periods of rest. The energy output of the entire body also can increase even beyond this level during short periods of maximal physical effort. However, under such circumstances glucose is metabolized under anaerobic rather than aerobic conditions, so that an oxygen debt is incurred (p. 1184).

CARDIOVASCULAR AND RENAL ADAPTATIONS. The responses of the cardiovascular system to exercise depend upon whether the muscular contractions are principally isometric or principally isotonic (p. 122). Only in the latter situation, of course, is external work performed.

At the onset of an isometric contraction the heart rate increases, and this increase takes place even when one thinks of performing such a muscular contraction. Therefore cortical impulses doubtless exert a role in this cardioacceleratory process.

Presumably the tachycardia is mediated via the cardioregulatory center in the medulla (p. 670). The cardioacceleration observed during isometric exercise is due mainly to a decrease in vagal tone, although an increased sympathetic tone of the cardiac nerves also has a role in producing this effect. A few seconds after an isometric contraction develops, both systolic and diastolic blood pressures rise precipitously. The stroke volume of the heart is altered only slightly; however, blood flow to the muscles that are contracting isometrically is decreased sharply or completely, because of mechanical compression of the blood vessels.

Isotonic exercise also produces a marked and rapid increase in heart rate (Fig. 16.1A); however, in contrast to the situation that exists during isometric exercise, the stroke volume increases significantly (Fig. 16.1C). Owing to vasodilatation in the exercising muscles, the systolic arterial pressure may increase only to a relatively slight extent, whereas the diastolic pressure may remain at the pre-exercise level, or even show a decline (Fig. 16.1B). The mechanisms underlying these effects were discussed in detail earlier (Fig. 10.67, p. 669), but will be reviewed briefly. The increased sympathetic discharge that accompanies exercise not only stimulates heart rate, but also augments the force of the myocardial contractions, thereby increasing the efficiency of the heart as a pump. The observed rate increase is especially pronounced in normal subjects who have not undergone extensive physical conditioning (cf p. 1188).

It is a fundamental maxim of cardiac physiology that the normal heart automatically regulates its output to equal the venous return (p. 617). As the venous return increases then the venous pressure rises, the diastolic filling is greater, and therefore the ventricular end-diastolic pressure also increases. Consequently the cardiac musculature contracts with greater force, in accordance with the Starling effect (p. 616), and this response is enhanced still further by the adrenergic sympathetic discharge mentioned above. Muscular exercise increases venous return markedly due to the pumping action of the skeletal muscles on the peripheral veins (p. 652), as well as by the increased respiratory movements (p. 652). Concomitantly, there is a decrease in the peripheral resistance due to vasodilatation of the vessels within the contracting skeletal musculature. Therefore the net physiologic effect that is induced by the operation of all of these factors is a pronounced increase in the cardiac output (Fig. 16.1D; Table 16.1).

The increase in heart rate and contractility observed during exercise is due principally to the effects of increased adrenergic sympathetic activity. In addition, the cardiac rate regulatory mechanism that is an inherent property of the pacemaker tissue itself also may contribute to the net increase in frequency that occurs upon increasing the venous return during exercise, although the contribution of this mechanism to the overall tachycardia probably is slight (p. 598).

The large increase in venous return that is essential to the increased cardiac output during exercise does not induce the observed rise in cardiac output by itself. Thus the adrenergic sympathetic discharge plays a major role in increasing myocardial contractility so that the end-systolic volume is relatively lower, and consequently the stroke volume increases to a relatively higher level at a given venous filling pressure (Fig. 16.1C).

In addition to the pumping action of the contracting muscles, as well as to the increased respiratory movements, venous return to the heart is augmented by several other factors that operate during exercise. First, vasoconstriction within the splanchnic region shunts a relatively large volume of blood into the systemic circulation, thereby increasing the effective circulating blood volume in the active muscles. Second, the dilated peripheral arterioles cause a greater pressure to be transmitted through the capillary bed to the peripheral veins, an effect that is enhanced by the much larger total area of the dilated capillary bed in the skeletal muscles that is found during exercise. Third, the total volume of blood contained within the systemic veins is decreased because of the vasoconstriction induced in these vessels by adrenergic neural discharge.

The quantity of blood that is shunted from the splanchnic and other regions of the body into the systemic arterial vessels during vigorous exercise may increase the total volume in this part of the cardiovascular system by as much as 30 percent (Table 16.1).

The maximum heart rate that is attained during exercise declines with age. In children it is not abnormal to observe a frequency of over 200 contractions per minute, whereas in normal adults a heart rate of over 190 beats per minute during strenuous exercise is rare. Elderly persons exhibit an even smaller increase in heart rate during exercise over the resting state.

In this regard, it is of interest to note that during extreme sexual stimulation, both male and female human subjects achieve a tachycardia which is similar to that observed during the per-

Table 16.1[a]   CARDIAC OUTPUT AND DISTRIBUTION OF BLOOD FLOW DURING REST AND DURING EXERCISE[b]

	STANDING AT REST (liters/minute)	MAXIMAL EXERCISE (liters/minute)
Total cardiac output	5.90	24.00
Flow to[c]		
Heart	0.250	1.00
Brain	0.750	0.750
Skeletal muscle	0.650	0.300
Skeletal muscle, active	0.650	20.850
Skin	0.50	0.50
Viscera (eg, gastrointestinal tract, liver, kidneys)	3.10	0.6

[a]*Data from Mitchell and Blomquist. New Engl J Med 284: 1018, 1971.*
[b]*Values obtained on sedentary adult male at rest and during exercise at maximal oxygen uptake.*
[c]*Note particularly the enormous increase in blood flow to the exercising skeletal muscles, the marked decrease in blood flow to the viscera, and the constancy of blood flow to the brain.*

formance of vigorous physical work. During copulation, cardiac rates up to 175 beats per minute have been recorded, and during orgasm frequencies over 180 beats per minute are not uncommon. Similarly, a large elevation in blood pressure above resting values has been observed in male and female subjects during coitus, although the incremental changes in systolic and diastolic blood pressures are somewhat greater in males than in females. The elevations in systolic pressure range up to 100 mm Hg for males, and 50 mm Hg for females; the corresponding diastolic pressure increases are up to 80 mm Hg for males and to 40 mm Hg for females.

Although the figures cited above give the upper levels for the tachycardia and incremental changes in blood pressure recorded in human subjects, they are indicative of the physical stress that occurs during sexual activity, a fact that is of particular significance to the physician responsible for the management of certain cardiac dysfunctions.

Following a period of exercise, the heart rate declines rather slowly to the resting level as vagal tone increases and the sympathetic effects diminish. The blood pressure also falls, and even may reach levels slightly below normal for a short time. Probably the latter effect is due to the sustained action of vasodilator metabolites that have accumulated during the exercise, in addition to the time required for a resting vasomotor tone to become reestablished.

During strenuous exercise, the renal blood flow may be decreased by as much as 80 percent below the resting value. Although the orthostatic effect on blood flow through the kidneys may be considerable (p. 651), this factor is hardly sufficient to explain such a drastic reduction in flow during physical exertion, so that additional, and presently unknown, mechanisms doubtless are involved in this phenomenon.

Older persons undergo a considerable reduction in their capacity to perform physical work because they are unable to increase their heart rate, hence cardiac output, sufficiently to provide enough oxygen to meet the increased requirements of the exercising tissues. Thus mild exercise in such individuals results in far greater incremental changes in heart rate and blood pressure than it does in young subjects. The rate of recovery following exercise also is far slower in elderly persons than in young individuals. Thus the rates at which the heart rate, blood pressure, oxygen uptake, and carbon dioxide excretion return to resting levels are retarded in the elderly. These effects are consequences of the aging process itself. Therefore aging results in a gradual decline in the physiologic reserve of the cardiovascular and respiratory systems, as well as a gradual impairment of the efficacy of those homeostatic mechanisms that are concerned directly with cardiovascular and respiratory adaptations. In turn these changes lead to a reduced work capacity with advancing years.

RESPIRATORY ALTERATIONS. The increase in pulmonary ventilation that takes place during exercise is responsible for three important physiologic alterations. First, it provides the additional oxygen above the resting level which is essential to meet the increased metabolic demands of the active tissues. Second, the increased ventilation eliminates the extra carbon dioxide that is produced during exercise. Third, some of the metabolic heat increment generated during physical exertion is dissipated from the body by the increased ventilation (p. 1007).

The physiologic changes that accompany increased pulmonary function must, of course, act in a smoothly integrated fashion with the attendant cardiovascular responses, the local alterations in tissue blood flow, and the increased tissue oxygen extraction that were discussed in the previous sections and summarized in Figure 16.1.

It was noted earlier that pulmonary ventilation as well as oxygen consumption increase in direct proportion to the quantity of work performed by the exercising muscles; ie, the ventilation rate is proportional to the incremental change in the metabolic rate. This fact is depicted in Figure 16.3, in which the respiratory volume (or total pulmonary ventilation) in liters per minute is plotted against the metabolic oxygen consumption of the body in relation to the intensity of the work output.

During a period of exercise, the quantity of oxygen entering the blood within the pulmonary capillaries is increased because the total volume of blood flowing through the lungs per minute increases markedly and also because the volume of oxygen per unit volume of blood leaving the pulmonary capillaries increases significantly. First, the blood flow through the pulmonary circuit increases from about 5.5 liters per minute at rest to values as high as 40 liters per minute during strenuous exercise. Consequently the total volume of oxygen that enters the blood increases from around 250 ml per minute at rest to as much as 4,000 ml per minute. Second, and concomitantly with the increased rate of blood flow, the P_{O_2} of the pulmonary arterial blood that enters the lung capillaries has decreased from a resting value of 40 mm Hg to around 25 mm Hg, or even lower, an effect that is due to the enhanced tissue extraction of oxygen during exercise. The net effect due to the operation of these factors is the generation of a steeper alveolar–capillary diffusion gradient for oxygen, so that the diffusion rate for this gas is increased approximately threefold during exercise. Hence a greater volume of oxygen enters the pulmonary blood per unit time, and thus the blood in the venules from the lungs is almost completely saturated with oxygen despite the relatively enormous blood flow rates encountered in the pulmonary capillaries during exercise.

The quantity of carbon dioxide eliminated per unit volume of blood also is increased markedly during exercise. Therefore excretion of this waste product increases from around 200 ml per minute

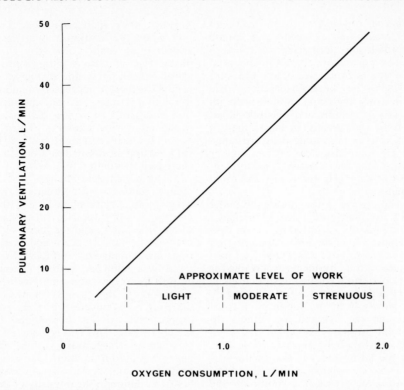

FIG. 16.3. Relationship between the level of work performance or exercise and the oxygen consumption in human subjects.

at rest to a level as high as 8,000 ml per minute during strenuous exercise.

The oxygen uptake increases to a maximal rate in exercise, and this rate is in direct proportion to the increase in work load up to a definite maximum level; beyond this point, the oxygen uptake is maximum as the physiologic mechanisms for oxygen transport are operating at their full capacity (Fig. 16.4A). During a submaximal physical effort, the metabolic production of lactic acid approximately parallels the oxygen consumption (Fig. 16.4B), and no net accumulation of lactic acid occurs; ie, the body is said to be in a *steady state*. During such a steady-state condition, the oxygen consumption (and utilization) exactly balance the aerobic metabolic requirements of the active muscles. However, if the level of exercise now is increased beyond the maximal physiologic capacity of the cardiovascular and respiratory systems to supply an adequate minute volume of oxygen to the muscles, then the aerobic metabolic processes no longer are able to keep pace with the demand for oxygen. The blood lactic acid level continues to rise as shown in Figure 16.4B. The excess lactic acid produced continues to diffuse into the blood from the muscles in which the aerobic metabolic processes are inadequate to provide resynthesis of high energy stores (eg, ATP) at a rate sufficient to equal their rate of utilization. An *oxygen debt* has been incurred, and the

energy required for muscular contractions at this new high level of physical activity now comes from anaerobic metabolism. At the termination of the strenuous exercise, an excess volume of oxygen now must be consumed that is precisely equal to the volume of the oxygen debt that was incurred during the period of anaerobiosis, in order that the excess lactic acid produced during the anaerobic work period be reoxidized aerobically to pyruvic acid once again (Fig. 16.5; pp. 936, 945). Hence, the oxygen debt must be "repaid" upon cessation of the anaerobic work.

It is apparent from the foregoing discussion that aerobic (or steady-state) exercise can be continued for long intervals. In contrast, maximal work performance of the muscles under anaerobic conditions can be sustained for only short periods, as the volume of the oxygen debt that can be incurred under such conditions is strictly limited.

The time course of the ventilatory responses to exercise is presented conceptually in Figure 16.6. The sudden increase in pulmonary ventilation that occurs immediately prior to, or at, the onset of increased activity (phase 1) presumably is related to cortical stimuli, while proprioceptive afferent impulses from the joints, tendons, and the muscles themselves probably also contribute to this effect. The gradual second rise in the ventilation curve (phase 2) presumably is due to the operation of humoral factors, although the primary

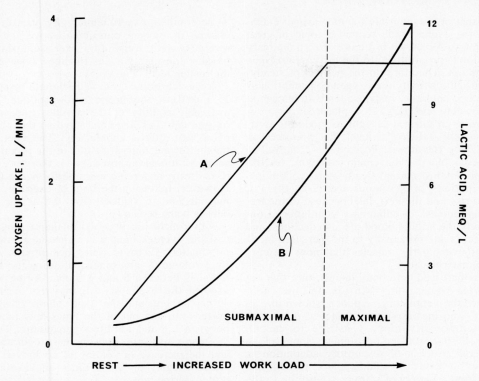

FIG. 16.4. Oxygen uptake (A) and lactic acid production (B) as related to progressively increasing work loads in human subjects. (Data from Ganong. *Review of Medical Physiology,* 6th ed, 1973, Lange Medical Publications.)

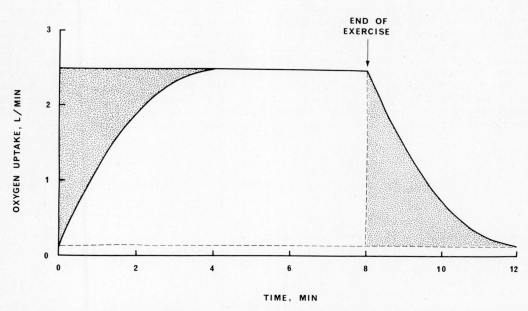

FIG. 16.5. The concept of oxygen debt. The volume of the oxygen debt incurred during a period of maximal work performance (shaded area at left) is equal to the volume of oxygen above the resting oxygen uptake (horizontal dashed line) that must be consumed at the end of the work period and before the oxygen uptake returns to the resting level (shaded area at right). In this example, the duration of work period was 8 minutes.

stimuli that are responsible for increasing ventilation during exercise still remain a topic of great controversy. As shown in Figure 16.3, pulmonary ventilation is proportional to the increased oxygen consumption. Therefore during periods of steady-state muscular activity the oxygen consumption as well as the ventilatory response tends to plateau at some elevated level (Fig. 16.6, phase 3).

During an interval of light to moderate exercise that still is below a level that produces an oxygen debt, the arterial P_{O_2} and P_{CO_2}, as well as the arterial blood pH, remain constant; ie, this situation obtains during steady-state conditions. During periods of strenuous exercise, the P_{O_2} tends to rise and the P_{CO_2} falls because of the tremendous increase in pulmonary ventilation. Concomitantly the arterial blood pH also declines, and this effect is due principally to the increased lactic acid output from the muscles that now are contracting anaerobically.

The increased core body temperature that occurs during exercise (Fig. 14.61; p. 1002) may contribute to the ventilatory response, shown in Figure 16.6, and the respiratory center also may become temporarily more sensitive to carbon dioxide. Perhaps oxygen per se plays some role in the overall process of respiratory adaptation to exercise; however, it would appear at present that stimuli from a variety of sources contribute to the increased pulmonary ventilation which occurs during periods of enhanced physical activity.

Immediately upon the cessation of exercise, the respiratory rate drops precipitously (Fig. 16.6, phase 4); however, the rate does not reach a resting level until any oxygen debt that has been incurred during the exercise has been repaid fully (Fig. 16.6, phase 5; Fig. 16.5). If a large oxygen debt is present in the body at the termination of exercise, it may take well over an hour before the basal ventilatory level is reached once again. In this situation, the stimulus to sustained ventilation at an elevated level beyond the end of the exercise period is the elevated hydrogen ion concentration in the blood that results from the lactic acidemia. Quantitatively, the magnitude of the oxygen debt

is determined experimentally by measuring the volume of oxygen consumed above the resting level between the end of the exercise, and until the oxygen consumption has declined once more to the resting level (Fig. 16.5).

The carbonic acid–bicarbonate system (p. 757) buffers the lactic acid produced during anaerobic activity of the muscles and while an oxygen debt is being contracted so that the blood pH is not altered appreciably. The operation of this buffer mechanism results in the production of some additional carbon dioxide; therefore the RQ rises during anaerobic work performance (p. 986). However, following the bout of exercise, the RQ may decline to values below 0.5 as the oxygen debt is being repaid (p. 1184).

In conclusion it is worth mentioning that at rest the average human breathes at a rate of approximately 12 to 15 times per minute, and the total volume of air exchanged during the same interval is between 6 and 8 liters. During vigorous exercise the respiratory rate may rise to as high as 40 per minute, and the total volume of air transported into and out of the lungs may increase to between 125 and 170 liters per minute. Therefore the respiratory rate may increase over threefold, and the increase in volume of air moved can increase roughly up to 30 times above that transported during rest.

As noted in the previous section, the cardiovascular responses to sexual excitation can vary from mild to quite pronounced, depending upon the degree of stimulation. Similarly, humans of both sexes exhibit a marked hyperventilatory response during coitus. Thus the maximum ventilatory rate observed during the performance of strenuous work also has been recorded in male and female subjects during periods of extreme sexual activity; that is, respiratory rates of up to 40 per minute can be found rather frequently toward the end of the preorgasmic phase of intercourse. In fact, the duration as well as the intensity of the hyperventilatory response are a clear indication of the degree of sexual tension that is developed both before and during coitus.

FIG. 16.6. The alterations in pulmonary ventilation that occur during and after a bout of physical exercise (vertical arrows). The five ventilatory phases, indicated by numbers, are discussed in the text. (After Ganong. *Review of Medical Physiology*, 6th ed, 1973, Lange Medical Publications.)

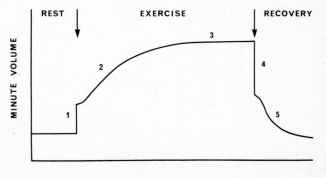

METABOLIC ADAPTATIONS TO EXERCISE. The cardiovascular and respiratory adaptations to mild as well as moderate exercise are sufficient to deliver enough oxygen to the active muscles. Therefore the aerobic metabolism of glucose via the glycolytic pathway coupled to the tricarboxylic acid cycle are able to cope with the energy requirements of the contracting muscles, as the overall rate of ATP synthesis is adequate to supply enough energy for the contractile process without an oxygen debt being incurred. If, however, the work performed is so vigorous that an oxygen debt is generated, the anaerobic glycolysis of glucose to lactic acid becomes the sole biochemical mechanism whereby ATP can be supplied for the contractile machinery of the muscles (p. 936, Fig. 14.25). As noted in Chapter 14 (p. 936), the glycolytic process is relatively inefficient from the standpoint of bioenergetics, and it also results in the production of significant quantities of lactic acid. The latter compound accumulates in the blood until the exercise ceases, at which time the increment of oxygen consumed above the resting level during repayment of the oxygen debt is utilized to reconvert the lactic acid back into pyruvic acid on a mole-for-mole basis (Fig. 16.5). The pyruvate formed from lactic acid then may be resynthesized into glucose and/or glycogen. The principal biochemical pathway whereby the lactic acid produced during anaerobic muscular work is metabolized during recovery from the oxygen debt is the *lactic acid* (or *Cori*) *cycle* (Fig. 14.26, p. 944).

In normal individuals, carbohydrate and fat are the principal sources of energy for the performance of physical work. Protein is not an important energy source.

Even when carbohydrate stores are plentiful, some fat is utilized by the muscles during exercise so that the RQ may decline progressively as exercise proceeds. As the glycogen stores are depleted during prolonged bouts of heavy physical activity, the quantity of fat oxidized increases from about 8 percent of the total energy output to around 77 percent; the quantity of carbohydrate being oxidized concomitantly shows a proportional decline.

Lipids are about 10 percent less economical as energy sources than are carbohydrates, because fatty compounds require more oxygen for their oxidation than does an equimolar quantity of carbohydrate; ie, the oxygen/carbon ratio in lipids is about 10 percent lower than that in carbohydrate. For the performance of vigorous muscular work, the bodily carbohydrate reserve is therefore an important factor; during vigorous, but short, periods of exercise carbohydrate provides the major source of energy.

During prolonged bouts of heavy exercise or physical work, especially if these are accompanied by fasting and a cold environment, the glycogen stores of the entire body can be depleted to such an extent that depot lipids then are mobilized as the primary source of energy (Fig. 14.34, p. 961). Incomplete combustion of the glycerol and fatty acids formed from the depot triglycerides during such exercise may yield an excess of ketone bodies, and a mild ketosis may ensue until the carbohydrate stores are replenished once again.

Since the metabolic rate of the body increases in direct proportion to the degree of muscular exertion, it is hardly surprising that the overall metabolic energy expenditure of the body varies considerably, depending upon the type of exercise or work performed. Some data relating total energy output of the body to the type of physical activity involved are summarized in Table 14.16, p. 994.

At the onset of physical exertion, the resting functional activities of the muscular, cardiovascular, respiratory, and biochemical mechanisms all must be "turned up" in order to function properly at higher levels. Athletes accomplish this literal, as well as figurative, warming-up process by performing certain exercises prior to active participation in an event or contest. However, there is little sound experimental evidence to indicate that ultimate physiologic performance is related to such physical warming-up activities.

THERMOREGULATION. The increased metabolic rate that occurs at all levels of muscular activity ranging from mild to strenuous is attended by a considerable increase in the production of heat energy by the body. This fact is readily attested to by the significant rise in core body temperature that accompanies physical exertion (Fig. 14.61, p. 1002). The necessity for dissipating this excess heat calls into play the several physical and physiologic mechanisms that are involved in the overall heat loss mechanism. The extent to which each of these mechanisms participates in the overall heat loss depends, of course, upon the ambient temperature, humidity, clothing, severity of the muscular effort, and so forth. Specifically, the peripheral vasodilation in the active muscles that occurs during exercise enhances bulk heat transfer to the body surface (convection, Fig. 14.62, p. 1004); sweating (p. 1008) increases heat loss by evaporation (p. 1007); and radiation (p. 1006) of body heat to the surroundings may increase somewhat owing to the elevated body temperature. Lastly, if the exercise is sufficiently prolonged and vigorous, then the hypothalamic thermoregulatory mechanism may be "reset" at a higher level so that a greater rate of heat loss is achieved (p. 1009).

Briefly, therefore, exercise results in an elevated heat production by the body which in turn stimulates a greater heat loss by activating the normal thermoregulatory mechanisms, and although the core body temperature becomes somewhat higher temporarily, nevertheless the increased heat loss is sufficient to prevent an in-

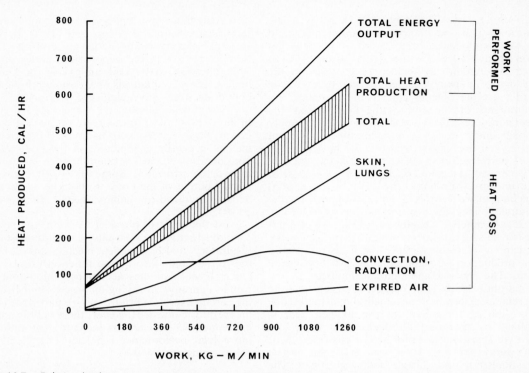

FIG. 16.7.   Relationship between work performance, energy output, and heat loss from the body during the performance of work at different levels of intensity. The shaded area indicates the excess of total heat production above the total heat loss, whereas the total energy output is equal to the total heat produced plus the total mechanical work performed. (Adapted from Ganong. *Review of Medical Physiology,* 6th ed, 1973, Lange Medical Publications.)

crease in core body temperature to such an extent that survival of the individual would be endangered.

Some Effects of Chronic Exercise

The effects of a given level of exercise upon an average subject are considerably different from the responses of an individual who has undergone a sustained period of physical training or conditioning. Specifically, there are several major physiologic differences between a sedentary individual and a trained athlete. First, the athlete has a lower resting heart rate and the maximal percent increase in heart rate during periods of exercise is considerably less than that of the untrained individual. Second, the heart of the athlete has a larger end-systolic ventricular volume. Third, the stroke volume of the heart of the athlete is much larger at rest than is that of the untrained person. Fourth, the total mass of cardiac tissue in the trained athlete is considerably greater than that of an average person of comparable height, weight, and generally similar bodily configuration (see also p. 578). Some of these differences are summarized in Table 16.2, and collectively they enable the athlete to attain a significantly greater cardiac output during exercise with a much lower incremental

change in heart rate than is possible in an untrained subject.

In addition to the four differences cited above that have been observed in comparative studies of untrained and trained individuals, it also has been found that athletes apparently have a greater arteriovenous oxygen difference during exercise than is exhibited by average subjects. The advantage of this mechanism in achieving delivery of a larger volume of oxygen to the muscles during vigorous activity is evident.

Another aspect of long-term physical conditioning is the effect of such work performance upon the structure of the muscles themselves. It is obvious that chonic exercise increases the size and strength of skeletal muscle, whereas the opposite is true of the muscles in sedentary individuals. Interestingly, however, specific training regimes for athletes can be quite selective insofar as the net effect upon the muscles themselves is concerned. Thus strength training (as for weight-lifting contests) can increase the total muscle mass as well as the number of myofibrils; speed training (as for boxing or running short races) induces a quantitative increase in the glycolytic enzymes; and endurance training (as for marathon races, or long-distance swimming) enhances the oxidative capacity of muscle, as well as the total number of mitochondria, the quantity of glycogen present,

Table 16.2 CARDIAC FUNCTION IN RELATION TO PREVIOUS TRAINING

SUBJECT	HEART VOLUME	HEART WEIGHT[a]	LEFT VENTRICULAR END–SYSTOLIC VOLUME
	ml	gm	ml
Untrained	785	300	51
Trained	1,015	350	101
Professional Athlete	1,440	500	177

[a]*Estimated.*

and the efficacy of the cardiovascular system. Following all types of training there is an increase in neuromuscular coordination which tends to minimize inefficient work output.

The greater size of a muscle that develops following prolonged exercise is due to a hypertrophy (ie, an enlargement of the individual fibers), and not a hyperplasia (ie, an increase in the total number of fibers). There is no evidence that embryonic striated muscle fibers undergo mitosis (cell division) beyond approximately four to five months of intrauterine life; hence adult muscle fibers enlarge by hypertrophy alone.

Any observed increase in the cross-sectional diameter of a muscle fiber following chronic exercise can be attributed to an increase in the volume of the sarcoplasm, indicating a greater metabolic reserve, and/or to an increase in the mass of the myofibrils themselves.

Experimental studies have indicated that prolonged isotonic exercise induces a muscular hypertrophy due to increased sarcoplasmic mass, so that the range of fiber sizes in the muscles tends to become less. That is, the smaller fibers undergo hypertrophy, and all of the fibers enlarge to some extent, although none becomes larger than the maximum size of those present in normal (ie, unexercised) muscle. The number of myofibrils does not increase in muscles that undergo hypertrophy due to isotonic exercise.

On the other hand, prolonged isometric exercise (ie, over an interval of several months) induces a quantitative increase in the total number of myofibrils present in the muscle, together with an increase in the mass of sarcoplasm within the initially smaller muscle fibers.

The practical consequences of these facts in athletic training are obvious. Since isometric exercise produces an increase in the number of myofibrils, which are, after all, the contractile machinery of the muscle, then a greater force can be generated following hypertrophy due to isometric rather than isotonic exercise. Interestingly, an isometric exercise repeated once daily for only a few (3 to 5) seconds can result in a muscular hypertrophy equal to that found after repeating an isotonic exercise several times a day for extended periods. A mere increase in muscle size is hardly the only goal in physical training, however!

It seems probable that there is a size limit that individual muscle fibers can attain through hypertrophy without developing a susceptibility to tetanic spasms or "cramps"; however, during extended periods of heavy exercise the major factor that limits performance is the ability of the cardiovascular and respiratory systems to deliver oxygen at an adequate rate to supply the active tissues.

Cardiac muscle also undergoes hypertrophy in a manner similar to that found in skeletal muscle following prolonged physical training of the individual.

Insofar as the etiology of the word "hypertrophy" in skeletal muscle is concerned, it is significant that the motor nerves to these structures exert a definite trophic function with regard to maintenance of the muscle in a good functional state (p. 129). Thus it is interesting to speculate upon the possibility that the increased frequency of nerve impulses to striated muscles that accompanies exercise conceivably might play a fundamental biochemical role in the development of hypertrophy as discussed above. On the other hand, complete motor denervation of skeletal muscle from whatever cause results in total atrophy of the contractile tissue (eg, as in poliomyelitis, p. 450); relative physical inactivity with its attendant decrease in motor nerve discharge results in atony and functional weakness of skeletal muscle, such as is obvious in persons who lead sedentary lives; and increased physical activity, or strenuous athletic training, perforce augments neural discharge to striated muscles so that good tonus and hypertrophy result as consequences of the enhanced activity. The possible physicochemical mechanisms underlying these processes are unknown at present.

Another fundamental, though transient, physiologic alteration that occurs in man following a period of vigorous exercise is a slight decrease in the total circulating plasma volume. Thus the hematocrit and the total plasma protein concentration increase to some extent; these facts are consistent with the increased net flow of water to the interstitial spaces of the active muscles (p. 558). Prolonged physical training, on the other hand, appears to cause a permanent increase in the total blood volume. However, in men who are accustomed to performing strenuous manual labor this difference, although real, is quite small when it is compared to similar values obtained from sedentary individuals of comparable height, weight, and general body configuration.

The vital capacity of the lungs exhibits much variation among comparable individuals. This measurement depends in turn upon a number of factors including age, sex, body weight, body position, pulmonary compliance, and strength of the respiratory muscles (p. 725). However, it is an established fact that a well-trained athlete normally exhibits a vital capacity that is around 30 to 40 percent greater than that of an untrained person of comparable age, sex, and body weight. This adaptation to chronic exercise provides such a trained individual with a considerable physiologic advantage, insofar as his respiratory adaptation to vigorous physical work is concerned.

Finally, it is an undisputed fact that one of the specific benefits to be derived from a program of regular daily exercise of some sort is psychologic. This statement holds true whether or not the person involved is an average individual or an athlete. Some form of physical exertion carried out on a daily basis causes the person to "feel better" than if he were completely sedentary, and mental, as well as physical, performance tend to improve.

Furthermore, a definitely long-term benefit of increased physical activity on a regular basis enhances the probability that the person will remain in better health as well as remain more active as he ages. Last, there is good epidemiologic evidence that regular exercise tends to decrease the likelihood of developing myocardial infarction, as well as to reduce the severity of such an attack should it occur.

Fatigue

The subjective phenomenon of "fatigue" is not clearly understood, insofar as its underlying physiologic mechanisms are concerned. Undoubtedly a great many factors contribute to this generalized sensation, among which may be included centripetal impulses that arise in muscle proprioceptors and the biochemical effects of prolonged acidosis on the cortex as well as on other higher centers. It is noteworthy that the subjective difficulty of performing muscular exercise is correlated with the rate at which oxygen is consumed, and not with the work performance as measured in kilogram-meters per minute.

Isometric muscular contractions that are sustained for any length of time become painful — first, because of the relative ischemia that develops in a muscle that results from the contractile tension, and second, because of the accumulation of "P substance." The latter material is unidentified as yet, although certain kinins (p. 648) and/or potassium ion may be involved in eliciting the discomfort of prolonged exercise through their chemical action on peripheral nerve endings. Intermittent bouts of rhythmic aerobic muscular con-

tractions, on the other hand, are not painful as the blood flow is not decreased for any length of time, and this rapid blood flow presumably flushes any P substance out of the muscles and thereby prevents its accumulation.

The stiff, sore muscles that develop following an unaccustomed bout of physical exertion may be caused by an accumulation of interstitial fluid in the tissues (local edema) as well as by an increase in metabolic products therein (eg, hydrogen ion, lactate, denatured protein) in addition to damage to the contractile elements themselves.

Subjective fatigue can result from boredom, ie, a deficiency of external and/or volitional cortical stimuli. Other unknown factors doubtless are involved in the etiology of muscular fatigue, and quite possibly central neurotransmitter depletion may play a role in this process, although this remark is conjectural.

PREGNANCY, PARTURITION, AND LACTATION

The endocrine and other mechanisms that underlie the sexual (reproductive) physiology of the male and female were discussed in Chapter 15 (pp. 1127, 1143). The chromosomal aspects of sex determination were presented on page 1130, and the cytologic changes that occur at the time of fertilization of the ovum by a spermatozoon were considered on page 1146.

In the present discussion, certain general physiologic aspects of conception, transport, and implantation of the fertilized ovum are presented. Subsequently some overall developmental aspects of early fetal life will be summarized in relation to the pertinent structural as well as functional characteristics of the placenta. These topics will be followed by a consideration of the major endocrine and other physiologic responses that take place in the female body during pregnancy. Last, the mechanisms involved in parturition will be discussed.

The process of lactation was dealt with in relation to endocrine physiology (p. 1157); however, certain physiologic aspects of this problem are relevant to the present discussion, and will be included in this section.

Fertilization, Transport, and Implantation of the Ovum

Following coitus, spermatozoa make the transit from the upper vagina through the uterus and fallopian tubes to the vicinity of the ovary within a few minutes to several hours. The motility of the sperm alone is inadequate to account for the rapidity with which this transport is accomplished (p. 152), but there is no evidence that rhythmic contractile movements of the uterus and tubes induced by copulatory stimuli facilitate propulsion of the semen toward the ovary. The prostaglandin found in seminal fluid conceivably might participate in stimulating sperm motility to some extent in vivo. However, there is definite experimental proof that the uterus and fallopian tubes do not undergo a "reverse peristalsis" during coitus and/or orgasm; therefore such a mechanism for accelerating sperm transport has been ruled out. In vitro, viable spermatozoa can swim at a rate of about 2.7 mm per minute.

Sperm remain viable in the normal female genital tract for approximately 24 to 72 hours, although their ability to fertilize an ovum probably is limited to about 24 hours following ejaculation. Following ovulation, on the other hand, the ovum remains viable for roughly 12 to 24 hours. Therefore fertilization obviously is possible only when coitus occurs during the interval when viable male and female gametes are present. The approximate time of fertility for women during the sexual cycle is indicated in Figure 15.9, p. 1041.

Following ovulation, the ovum enters the peritoneal cavity directly; then it must pass into the opening of the upper (superior) end of the fallopian tube. It will be recalled that this opening is surrounded by fimbriae which envelop the ovaries, and the inner surface of these fringes is lined with a ciliated epithelium (p. 1147). The pattern of ciliary beat in the fimbriae, as well as in the rest of the oviduct, is directed toward the abdominal ostium of the fallopian tube. Thus a slow liquid current flow is developed that sweeps the ovum into the fallopian tube. The ciliary beat mechanism for transport of the ovum does not explain one interesting and clinically established fact: that is, how a woman who has had one ovary together with the contralateral fallopian tube removed surgically at the same time can become pregnant! Regardless of the mechanism involved, practically all of the ova that are liberated do enter the fallopian tubes, so that abdominal (ie, ectopic) pregnancies are relatively rare events.

Only one spermatozoon is necessary for fertilization of the ovum. The entry of more than one sperm (despite the enormous numbers of these cells present in a normal ejaculate, p. 1137), is believed to be prevented by the diffusion of some chemical substance from the ovum into the zona pellucida, and this substance effectively inhibits the ingress of more than one of the male gametes. The early cytologic changes that accompany the act of fertilization are depicted in Figure 16.8. It is believed that the single sperm that fertilizes the ovum penetrates the cell wall of the egg by means of lysosomal enzymes contained in the acrosome; once this feat is accomplished, the chemical events that ensue prevent entrance of other viable sperm cells.

Generally, the ovum is fertilized in the uppermost region of the fallopian tube. Following this event, it takes 72 hours or longer for the fertilized ovum to traverse the tube to the uterus. This transport is accomplished principally by the extremely slow movement of the fluid within the fallopian tube, the current being set up by the pat-

tern of ciliary beat noted above. Possibly weak peristaltic movements of the tube passing from the fimbriated end of this structure toward the uterus facilitate the transport process.

The usual three-day lag between fertilization of the ovum and its entrance into the uterus allows time for several cell divisions to take place, so that by the time the developing embryo reaches the uterus it is known as a *blastocyst*. During its transport down the fallopian tube, the fertilized ovum is believed to be nourished by the copious secretions that are liberated from the epithelial cells lining this structure. Evidence for this statement comes from the demonstration that a fertilized ovum will continue to divide in vitro, provided that the culture medium consists solely of homogenized mucosa from the fallopian tube. Synthetic media, on the other hand, generally are not suitable for this process.

Once the blastocyst has entered the uterus, it remains free within the lumen of this organ between four to five days prior to implantation. Therefore implantation generally takes place on the seventh to eighth day following conception. During this interval, the blastocyst is nourished by the liberal secretions that are elaborated by the endometrial cells of the uterus.

Following contact with the endometrium (p. 1149), the blastocyst becomes rapidly surrounded by two layers of tissue. The outer layer, known as the *syncytiotrophoblast,* is a large multinucleate protoplasmic mass that has no clearly discernible cell walls. The inner layer, called the *cytotrophoblast,* is composed of individual cells. Implantation of the blastocyst results from the activity of the two kinds of trophoblast cells. Proteolytic enzymes are elaborated by these elements which actively digest and liquefy the endometrial tissue in the immediate vicinity of the encapsulated blastocyst so that a small crater is eroded chemically in the endometrial surface, and the blastocyst and its satellite tissue now enter this depression. Concomitant with this process of implantation, the trophoblast develops cords of cells that invade and attach to the edges of the digested endometrium. In summary, therefore, implantation is accomplished by two simultaneous processes. The blastocyst, encased within the two trophoblastic layers, chemically excavates a small depression in the endometrial surface, and then becomes attached within this depression.

Following implantation, the trophoblastic and subjacent endometrial cells multiply rapidly so that ultimately the placenta as well as the several membranes associated with the embryo are derived from these cells.

The enzymes secreted from the trophoblast cells that play such as important role in implantation also release considerable quantities of fluid and nutrients from the endometrial tissue during proteolysis. A large proportion of this fluid together with much of the soluble material contained therein is absorbed actively into the developing

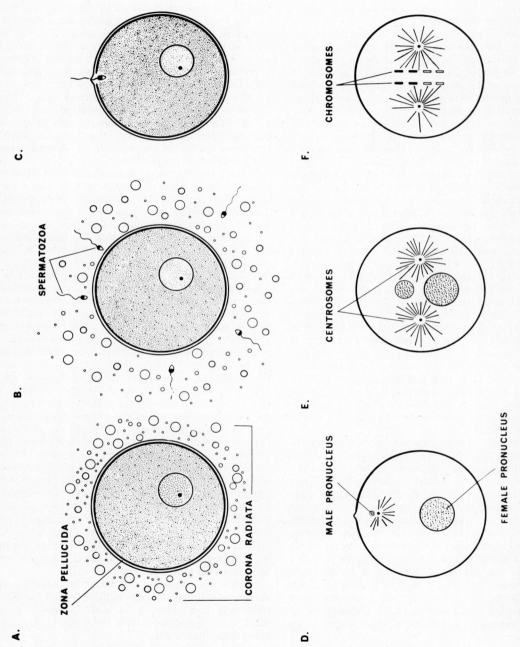

FIG. 16.8. Fertilization of the ovum. **A.** Mature ovum surrounded by the corona radiata. **B.** The corona radiata disperses. **C.** A single spermatozoon enters the ovum. **D.** and **E.** The male and female pronuclei form, then fuse to produce a diploid number of chromosomes. **F.** Onset of the first mitotic division of the fertilized ovum.

blastocyst following pinocytosis and phagocytosis (p. 94) by the trophoblast cells, and thus it serves as a "rich nutrient broth" for the developing embryo as discussed further in the next section. The dorsal wall of the uterus is the usual site of implantation for the blastocyst.

Endometrial Nutrition of the Embryo and the Subsequent Development and Functions of the Placenta

As depicted in Figure 15.9, page 1041, the progesterone that is secreted by the corpus luteum during the last half of each menstrual cycle transforms the stromal cells of the endometrium into relatively larger cells that are rich in certain nutrients (eg, glycogen, lipids, protein, certain minerals) that are essential for development of the embryo in the event that pregnancy occurs. Assuming that fertilization does take place, an orderly series of endocrine changes rapidly ensues so that even more progesterone is secreted (see also p. 1197), Fig. 16.13). This incremental secretion of progesterone during the earliest stage of pregnancy stimulates in turn an even greater hypertrophy of the endometrial cells due to the accumulation of greater quantities of nutrients in their cytoplasm, and the greatly enlarged endometrial cells now are called *decidual cells,* their mass in toto being known as the *decidua.* Since this hypertrophy occurs prior to implantation, the blastocyst actually implants into the decidual tissue that has developed while the ovum is still a free-living organism within the uterine lumen.

As mentioned earlier, while implantation is in progress, the concentrated nutrients that have been stored within the decidual cells are released locally by the proteolytic activity of the trophoblast cells with subsequent phagocytosis and pinocytosis of the released materials and fluid by the blastocyst. Thus an adequate supply of nutrients is available for the embryonic developmen-

tal processes; and as the cords of trophoblastic cells continue to grow within the decidual tissue, still more nutrients become available as more decidual cells undergo digestion. In fact, this mechanism provides the only source of nutrients for embryonic development during the first seven days following implantation. Furthermore the decidual tissue continues to supply a significant fraction of the total nutritional requirements of the embryo for the interval between 8 and 12 weeks; ie, until the placenta gradually usurps the trophoblastic nutritional responsibility. These facts are shown graphically in Figure 16.9.

The placenta may be defined as a temporary organ that develops in response to the stimulus of pregnancy, and whose principal functions are the nutrition of, and removal of waste products from, the embryo (later known as the fetus).* The endocrine and erythropoietic functions of the placenta will be discussed subsequently.

The development of the placenta takes the following general course. First, as the cords of trophoblast cells surrounding the blastocyst invade and attach to the decidua, capillaries grow into these cords from the primitive vascular system of the embryo so that blood commences to flow from the embryo through these vessels at around the sixteenth day after fertilization. Accompanying this process, maternal blood sinuses develop between the surface of the endometrium and the cords of trophoblast cells. At this point, fingerlike projections develop as the trophoblast cells continue to grow; these projections gradually become elongated so that they lie free within the maternal blood sinuses that have developed in the endometrium. Ultimately the trophoblastic projections are known as placental villi, and they contain

In the human species, the developing child arbitrarily is called an embryo *up to the end of the second month of gestation; beyond the end of the second month to parturition, the child is termed a* fetus.

FIG. 16.9. The relative contributions of trophoblastic and placental nutrition throughout a normal pregnancy. F = time of fertilization. Note that the embryo first is nourished exclusively by the trophoblast, whereas later nutrition of the fetus is carried out exclusively by the diffusion of substances through the placental membrane. (Adapted from Guyton. *Textbook of Medical Physiology,* 4th ed, 1971, Saunders.)

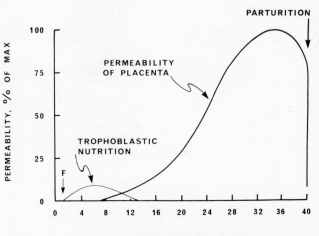

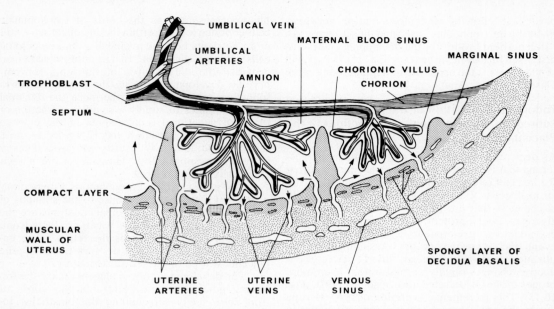

FIG. 16.10. Functional anatomy of the mature placenta in cross-section. Two placental villi are shown. The direction of blood flow in the maternal blood sinuses is indicated by arrows. The fetal arteries are unshaded, whereas the fetal veins are indicated in solid black. See also Figure 16.11.

the capillaries that grow into them from the developing embryo.

The principal anatomic features of the fully developed placenta are illustrated in Figures 16.10 and 16.11. The vascular relationships between the fetal and maternal circulations when completely established are as follows. Blood from the fetus courses through the umbilical arteries to capillary networks located within the placental villi, and then back into the fetus via the umbilical vein. The maternal blood, on the other hand, flows from the uterine arteries into the blood sinuses that surround the villi, thence back into the uterine veins (Fig. 16.11).

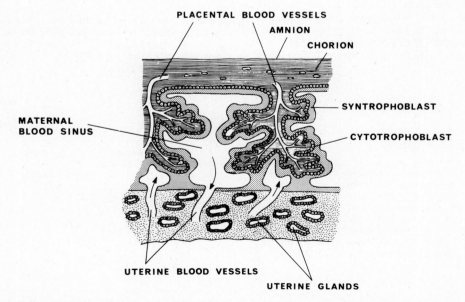

FIG. 16.11. Detailed functional anatomy of the placenta in cross-section, showing the relationship between the fetal and uterine blood vessels as well as the cell layers in the chorionic villi. The direction of maternal blood flow in the sinuses is indicated by arrows. (Cf. Fig. 16.10)

The capillaries within each villus are composed of a very thin endothelium that is surrounded by a layer of mesenchyme. This thin mesenchymal layer in turn is invested with a layer of syntrophoblastic cells. A fourth layer of cells is present surrounding the syntrophoblast cells during the first four months of gestation. The cuboidal cells that make up this layer are known as *cytotrophoblast* or *Langerhan's cells* (Fig. 16.11).*

When completely developed, the human placenta has a total surface area for diffusion that is roughly 16 square meters; this area is approximately one-quarter of the entire alveolar surface that is available for gas exchange in the adult lungs. It is important to remember, however, that even when the placenta is fully mature the barrier between the fetal blood and the materal blood is three cell layers thick. Furthermore, and in marked contrast to the alveoli, the distance between the fetal and maternal bloodstreams is considerably greater. These facts are of the utmost physiologic importance for the transfer of various substances from mother to child and vice versa.

As remarked earlier, the major functions of the placenta are to provide a temporary organ that permits diffusion of nutrients from the maternal into the fetal circulation as well as to permit fetal waste products to be eliminated in the opposite direction.

Since many substances can traverse the placental barrier simply by the process of diffusion, it is important to define the permeability of the placental membrane. This property may be expressed as the total quantity of any specific substance that passes across the placental barrier for a specific concentration difference of that substance across the membrane. As shown in Figure 16.9, the permeability of the placental membrane is low during the early stages of pregnancy because the membrane is relatively thick and the total surface area that is available for diffusion is relatively small. Subsequently, the permeability gradually increases to a maximum and then declines as the fetus reaches full term. The increase in permeability results from a progressive diminution in the thickness of the membrane (ie, from four to three layers of cells) combined with a large increase in the total surface area that is available for diffusion. The sharp decline in placental permeability that becomes evident a few weeks before parturition takes place is the result of death and degeneration of portions of this organ as it ages.

A fact that may assume considerable clinical importance (eg, in erythroblastosis fetalis, p. 573) is the rupture of areas of the placental membrane that permit fetal erythrocytes to enter the maternal circulation. (Only occasionally do maternal cells pass into the fetus.)

The placental diffusion of oxygen and carbon dioxide now may be considered. The diffusion of

These placental cells should not be confused with the islets of Langerhans of the pancreas.

oxygen across the placental barrier is quite similar to that encountered in the alveoli of the lungs (p. 740). Thus a pressure gradient for oxygen is present between the blood in the maternal sinuses and that of the fetal blood in the capillaries of the villi. Since the average Po_2 in the placental sinuses is around 50 mm Hg when the fetus is near term, and the mean Po_2 in the umbilical vessels that transport blood away from the placenta is about 30 mm Hg, then the pressure gradient for diffusion of oxygen is approximately 20 mm Hg between the maternal and fetal bloodstreams.

There are three physiologic mechanisms that enable the tissues of the developing fetus to obtain sufficient oxygen despite the fact that the Po_2 of the blood is only 30 mm Hg in the umbilical veins. First, a special type of hemoglobin called fetal hemoglobin is synthesized in utero by the fetus. As shown in Figure 16.12, the oxygen dissociation curve for fetal hemoglobin is shifted somewhat to the left of that for maternal hemoglobin. The practical consequence of this mechanism is that the fetal hemoglobin is able to transport between 20 and 30 percent more oxygen per gram than maternal hemoglobin. Second, the hemoglobin concentration (gm Hb per 100 ml) in fetal blood is approximately 50 percent greater than that of maternal blood so that volume for volume fetal blood can transport considerably more oxygen than maternal blood. This factor is of considerable importance in augmenting oxygen transport in the fetus. Third, the Bohr effect (p. 747; Fig. 11.30) also operates to increase oxygen transport by the fetal blood. Thus a given quantity of hemoglobin can transport more oxygen at a relatively higher pH than at a lower pH. Since the fetal blood that enters the placenta carries a considerable burden of carbon dioxide, loss of this carbon dioxide in the placenta causes the fetal blood to become more alkaline. Accompanying this process, the maternal blood in the uterine sinuses becomes more acidic due to the gain of the fetal carbon dioxide. Consequently these changes produce an increased oxygen combining capacity in the fetal hemoglobin, whereas that of the maternal blood is decreased. Thus the Bohr effect acts in diametrically opposite ways in fetal and maternal bloods, so this mechanism is twice as effective in fetal oxygen exchange as compared to pulmonary oxygen exchange.

The total diffusing capacity of the placenta when the fetus is at full term is approximately 1.2 ml of oxygen per minute per millimeter Hg gradient of Po_2, and this rate is quite similar to that found in the newborn child.

Carbon dioxide is produced continuously within fetal tissues by the same metabolic processes that are operative in the tissues of the mother, and this carbon dioxide can be eliminated only via the placenta. In fetal blood, the Pco_2 rises to around 41 to 46 mm Hg, in contrast to a maternal blood Pco_2 of about 40 to 45 mm Hg. Therefore a slight but quite ample diffusion gradient is present

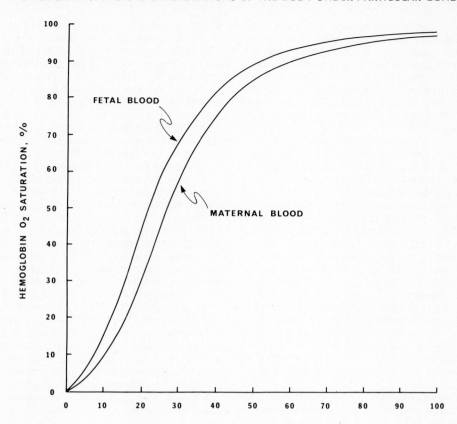

FIG. 16.12.   Dissociation curves for human hemoglobin F (fetal) and hemoglobin A (adult). This figure illustrates that at any particular P_{O_2} fetal blood can transport a significantly greater quantity of oxygen than maternal blood.

between the fetal circulation and that of the mother, and despite this low gradient, carbon dioxide can diffuse across the placental barrier into the maternal blood quite rapidly. This is true for two reasons: first, because of the extremely high solubility of this gas in the liquid phase of the placental membrane, and second, because carbon dioxide can diffuse at a rate some 20 times more rapidly than oxygen.

The many nutrients required by the fetus during its growth and development in utero diffuse into the fetal blood across the placenta by a mechanism that is identical with that for oxygen. Thus glucose is present in fetal blood at a concentration that is some 20 to 30 percent below that of maternal blood because this carbohydrate is utilized quite rapidly by the fetus. Consequently a steep diffusion gradient is present across the placental barrier for this substance; therefore diffusion of glucose into the fetal bloodstream is quite rapid.

Amino acids are relatively small molecules, and fatty acids are highly soluble in the placental membrane. Therefore both of these types of nutrient diffuse quite rapidly across the placental bar-

rier and in adequate quantities to satisfy the requirements of the fetus.

Essential minerals, including sodium, chloride, potassium, calcium, and phosphorus, also diffuse quite readily across the placental membrane into the fetal blood. Since these minerals are used rapidly by the fetus, their blood concentrations fall, and this metabolic removal tends to enhance the concentration gradients across the placental barrier thereby permitting further diffusion of these nutrients in accordance with fetal requirements.

Diffusion is, of course, a passive physical process. There is, however, some evidence that active transport of certain nutrients may take place, presumably by the action of the trophoblastic cells, at least throughout the first half of the gestation period. Thus, concentration gradients for amino acids, calcium, phosphate, and ascorbic acid have been found to be greater in fetal than in maternal blood, and this fact suggests that an active transport mechanism is operative in addition to diffusion for the transfer of these materials.

Excretory products produced by metabolism of the fetus are eliminated by diffusion down their individual concentration gradients in a manner similar to that for carbon dioxide. In particular, nonprotein nitrogenous compounds are involved. Therefore urea diffuses readily across the placental barrier, despite a low fetal blood:maternal blood concentration difference. Creatinine does not diffuse as easily as urea, making the concentration gradient for this compound much higher between the blood of the fetus and that of the mother.

In conclusion, the excretion of metabolic wastes from the fetus appears to take place solely by diffusion across the placental barrier, and once these compounds have entered the maternal circulation, they are in turn excreted from the body of the mother together with her own excretory products.

Endocrine Responses during Pregnancy

The placenta has another vital physiologic role in pregnancy in addition to the functions described in the previous section. The placenta functions as an endocrine organ which secretes a number of hormones that are essential to continuation of the pregnancy to full term. The working definition of the placenta that was presented on page 1191 must now be amplified to include its endocrine functions.

The principal hormonal substances elaborated by the placenta may be listed first, and then discussed in context with the other endocrine changes that occur in the pregnant woman. These hormones are: *human chorionic gonadotropin (HCG), estrogens, progesterone,* and *human placental lactogen* (HPL). Additional hormones that apparently are secreted in small quantities by the placenta will be mentioned in context. All of the placental hormones apparently are synthesized within, and secreted by, the syncytial trophoblast (p. 1193).

The menstrual phase of the female sexual cycle occurs at about 14 days following ovulation (p. 1151, Fig. 15.9). At this point, the secretory endometrium of the uterus sloughs off and is lost. Should menstruation take place following implantation of the fertilized ovum, then spontaneous abortion would result and the pregnancy would terminate. Menstruation is prevented during pregnancy, however, by secretion of human chorionic gonadotropin by the trophoblastic tissue (p. 1163). As shown in Figure 16.13, the secretion of HCG increases rapidly during pregnancy, so that it reaches a peak at about 7 weeks following conception; thereafter its concentration declines to a comparatively low value at about 16 weeks following conception.

HCG is a glycoprotein that contains galactose and galactosamine, and it exerts a function that is quite similar to that of the luteinizing hormone secreted by the anterior pituitary (p. 1038). The most important physiologic role of HCG, therefore, is that it prevents the regression of the cor-

pus luteum at the end of the sexual cycle. In fact, placental HCG probably stimulates the corpus luteum to secrete even greater quantities of its usual hormones, ie, estrogens and progesterone (p. 1163). The net result of these secretory processes is that the uterine mucosa continues to proliferate and quantities of nutrients thus become concentrated in the endometrium. Consequently, true decidual cells are present when the blastocyst undergoes implantation.

After a month or more has passed subsequent to fertilization of the ovum, the corpus luteum has enlarged to approximately twice its original size and the continued secretion of progesterone and estrogens from this structure ensures maintenance of the decidual cells of the endometrium. This adaptation is essential to development of the placenta and other tissues related to the fetus.

Ovarectomy performed on a pregnant female prior to the end of the third month of gestation usually induces spontaneous abortion due to loss of the luteal tissue and its hormones. If, however, ovarectomy is performed after the end of the third month, the placenta generally secretes a supply of progesterone and estrogens that is adequate to maintain the pregnancy until full term is reached.

HCG has another physiologic effect that is manifest in the genetically determined male fetus. Chorionic gonadotropin exerts an interstitial cell-stimulating effect (p. 1038), which results in the production of testosterone by the developing testes, and the small quantity of testosterone thus elaborated is of importance in stimulating development of the male sex organs (p. 1139). Furthermore, as the male fetus reaches term, the endogenous secretion of testosterone by the male fetus stimulates testicular descent into the scrotum. Chorionic gonadotropin obtained commercially from pregnant mares is often used clini-

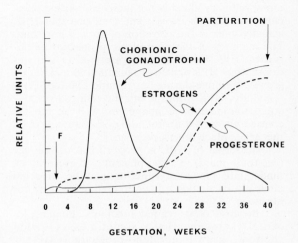

FIG. 16.13. The relationship among chorionic gonadotropin, estrogens, and progesterone during pregnancy. F = fertilization. Cf. Figure 16.16. (After Guyton. *Textbook of Medical Physiology,* 4th ed, 1971, Saunders.)

cally in patients with cryptorchidism to stimulate secretion of testosterone; the latter hormone in turn causes testicular descent.

The sharp rise in HCG that accompanies early pregnancy results in an increased urinary excretion of this hormone, a fact that forms the basis of several pregnancy tests. For example, HCG stimulates ovulation in immature female mice (Ascheim–Zondek test), and it also promotes the release of spermatozoa (or ova) from frogs and toads (Hogben and similar tests).

Secretion of human placental lactogen (or HPL) commences at approximately the fifth week of pregnancy, and the rate at which this hormone is elaborated by the placenta continues to increase gradually as gestation proceeds. It appears, by inference from animal experiments, that HPL in humans is luteotropic. Hence this hormone stimulates the secretion of large amounts of progesterone and estrogens from the corpus luteum. Maternal blood contains a high concentration of HPL; however, very little is found in the fetal circulation as the molecule is able to cross the placental barrier to only a limited extent.

Human placental lactogen also greatly stimulates growth of the mammary glands, a function that is similar to that of prolactin (p. 1039). HPL therefore appears to have a major role in breast development in the gravid female. In addition, HPL exhibits some properties that are similar to growth hormone (or somatotropin, p. 1028). In particular, HPL stimulates cell growth, and inhibits the insulin-mediated active transport of glucose into cells, the utilization of this nutrient being depressed by elevated HPL levels.

Like pituitary somatotropin (p. 1028), HPL is composed of 190 amino acid residues; other structural similarities are present in both protein molecules.

The secretion of somatotropin from the maternal pituitary gland is not increased during pregnancy; in fact, there is some evidence that HPL actually inhibits the secretion of growth hormone. The majority of the physiologic properties exhibited by HPL resemble those ascribed to somatotropin; thus HPL appears to function as a maternal growth hormone during the gestation period. Consequently, HPL brings about the reduced glucose utilization mentioned above, as well as the increased nitrogen, calcium, and potassium retention that takes place during pregnancy.

Secretion of HPL is proportional to the size (ie, total mass) of the placenta; therefore, an inadequate placental function can be detected clinically by finding a low circulating level of this hormone.

Because of its varied functions, human placental lactogen (HPL) sometimes is called chorionic growth hormone–prolactin (CGP) or human chorionic somatomammotropin (HCS).

As mentioned earlier, the placenta secretes progressively larger quantities of estrogens and progesterone throughout the course of gestation, as depicted in Figure 16.13. In this respect, func-

tionally the placenta resembles the corpus luteum.

Placental estrogen output toward the end of pregnancy may even reach levels up to 300 times those observed at the middle of a normal sexual cycle (cf Fig. 16.13 with Fig. 15.9, p. 1041). Estrogenic functions in the nongravid female were discussed in Chapter 15 (p. 1151), but the general functions of these hormones during pregnancy may be summarized as follows. First, estrogens stimulate a marked growth of the uterus in keeping with the increase in mass of the fetus. Second, estrogens stimulate proliferation of the glandular tissue of the breasts, producing an overall increase in their size. This role of the estrogens undoubtedly is shared by HPL, as discussed above. Third, the external genitalia enlarge under the influence of an elevated level of circulating estrogens (see also p. 1162). Fourth, the high circulating levels of estrogens that occur near the end of pregnancy relax the several pelvic ligaments. The sacroiliac joints tend to become more supple, and the symphysis pubis becomes more elastic. These mechanical alterations tend to render passage of the fetus through the vagina during parturition a more facile process.

At the onset of pregnancy, progesterone is secreted in moderate amounts by the corpus luteum. However, as gestation proceeds, this hormone is secreted in ever-increasing quantities by the placenta so that by the end of pregnancy progesterone may be secreted in quantities up to one gram per day, and this represents as much as a 10-fold increase in production of the hormone over the nongravid state (Fig. 16.13).

The particular and essential actions of progesterone during a normal pregnancy are as follows. First, the early nutrition of the embryo is dependent upon the development of decidual cells in the endometrium, a process that is stimulated by the direct action of progesterone upon the uterus. Second, progesterone inhibits uterine contractility, thereby preventing spontaneous abortion during pregnancy. Third, as progesterone specifically increases the secretions of the fallopian tubes and uterus, it contributes directly to the early nutrition of the morula (the earliest cleavage stages of the fertilized ovum) and of the blastocyst prior to implantation of the latter. Progesterone also appears to exert some kind of a regulatory effect upon cell cleavage in the early embryo. Fourth, progesterone has a role in the development of the mammary glands during pregnancy that serves to ready these structures for lactation (p. 1157).

By way of summary, the placenta has a critical endocrine function, since the chorionic gonadotropin, estrogens, and progesterone that are secreted by this organ are essential to completion of a normal pregnancy. Human placental lactogen also may prove to have a similarly indispensable role in this complex physiologic event.

Some of the responses seen in the maternal endocrine glands that are not directly related to reproductive functions during pregnancy may be outlined briefly. These responses are due principally to the increased metabolic stress placed upon the mother during pregnancy, but these alterations also may represent the feedback effects of certain placental hormones upon the pituitary as well as upon other endocrine glands.

During pregnancy the adenohypophysis becomes larger by approximately 50 percent. Thyrotropin and corticotropin secretions increase, although somatotropin levels remain constant or may fall somewhat due to the action of HPL. Follicle-stimulating hormone and luteinizing hormone secretions decrease sharply due to the inhibitory negative feedback effects of estrogens and progesterone that are released from the placenta.

The thyroid gland generally enlarges approximately 50 percent in pregnancy; concomitantly, thyroxin secretion is increased to a similar extent. The elevated thyroxin secretion is mediated by the increase in adenohypophyseal thyrotropin secretion noted above.

Secretion of glucocorticoids and aldosterone by the adrenal cortex is increased to some extent in the gravid female. The latter hormone increases the renal tubular reabsorption of sodium; so even during a normal pregnancy there is some fluid retention and a tendency to develop edema.

The parathyroid glands frequently enlarge during pregnancy. In particular, this effect is seen if the woman is on a diet that is calcium-deficient. The excess parathormone that is released following such parathyroid enlargement stimulates the release of calcium from the maternal skeleton. Thus a normal calcium ion concentration is maintained in the body fluids of the mother while the fetus extracts calcium for ossification of its own bones; both of these processes occur at the expense of the maternal calcium pool.

A hormone known as relaxin is found in the corpus luteum of the ovaries. The physiologic role of this substance in inducing relaxation of the ligaments of the pubic symphysis in humans is obscure, although relaxin causes this effect in rats and guinea pigs. In humans this softening effect on the appropriate ligaments is more probably due to the combined (synergistic?) actions of luteal hormones as well as placental estrogens and progesterone, as discussed on page 1198.

Maternal and fetal hormones do not exchange in significant quantities across the placental barrier. There is no transport of protein hormones, and nonprotein hormones cross the placenta to an extremely limited extent. An exception to the latter statement is dehydroepiandrosterone sulfate, which is converted by the placenta into estrogens and androgens, and these substances enter the circulation of the mother from the fetal circulation.

Maternal Physiologic Responses During Pregnancy

The entire process of gestation imposes a considerable physiologic burden upon the female body, and many of the functional adaptations that take place during this period result from this increased load. Furthermore, the hormones secreted by either the placenta or the nonreproductive endocrine glands can stimulate many responses in the mother, including an increased size of the breasts and the external genitalia, as well as an enlargement of the vagina and of the introitus. In particular, the sex hormones induce a large increase in the total mass of the uterus, from about 30 gm in the nongravid state to roughly 700 gm at term, or around a 25-fold increase in the mass of this organ.

Several important maternal cardiovascular alterations are found in pregnancy. First, approximately 750 ml of maternal blood flows through the placenta per minute in the later stages of pregnancy. Second, and as a direct consequence of this large placental blood flow, the total peripheral resistance of the maternal circulatory system is increased. Therefore, an increased venous return to the maternal heart results, and this factor operates to increase cardiac output to some extent, an effect that is enhanced by the elevated circulating blood volume noted below.

However, the overall metabolic rate also increases during pregnancy, and these two mechanisms acting in concert result in a net increase in the cardiac output that is between 30 and 40 percent above normal by around the twenty-seventh week of pregnancy. Subsequently the cardiac output declines gradually (for unknown reasons) as gestation continues; a near-normal value is reached during the final two months of gestation. Third, the total circulating blood volume of the mother increases about 30 percent above normal during pregnancy, an effect that presumably is caused principally by the elevated secretion of certain hormones, eg, estrogens, progesterone, and adrenocortical steroids. All of these hormones increase fluid retention by the kidneys, and they are secreted in relatively high quantities during the latter half of pregnancy. The approximate time course of the elevated blood volume in pregnancy is shown in Figure 16.14.

Due to the elevated fluid retention during the early stages of pregnancy, the hematocrit tends to decrease somewhat. Subsequently, however, the maternal bone marrow is stimulated to produce more erythrocytes so that by the time of parturition the hematocrit is essentially normal once again, and at that point the maternal circulation contains roughly between one and two liters of extra blood. Approximately 25 percent of this extra blood is lost during a normal parturition.

The maternal basal metabolic rate increases approximately 15 percent during the last half of the gestation period, and this effect is caused principally by the increased production of a broad spectrum of various hormones, in particular thyroxine. This increase in the maternal metabolic rate in turn produces a requirement for an elevated caloric intake during pregnancy. Furthermore, the increased weight being carried by the mother also necessitates the expenditure of larger quantities of energy than are utilized normally for the performance of muscular work.

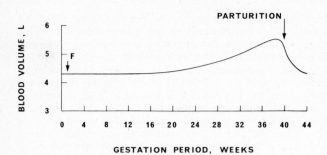

FIG. 16.14. The progressive alterations in maternal blood volume during pregnancy. F = fertilization. (After Guyton. *Textbook of Medical Physiology,* 4th ed, 1971, Saunders.)

The increase in basal metabolic rate, coupled with the progressively increasing fetal oxygen requirements, results in an increase in the total maternal oxygen utilization during pregnancy so that at term the oxygen utilized by the mother is about 20 percent above normal. Thus the respiratory rate of a pregnant woman may increase up to 50 percent above that found in the nongravid state. This effect upon the rate of ventilation is enhanced by the increased uterine size, as the pressure exerted upon the inferior surface of the diaphragm by the abdominal contents tends to limit severely the normal movements of this structure (p. 716). Thus the increased respiratory frequency compensates for the more limited tidal volume and vital capacity (p. 724) that result from fetal and uterine growth during the latter stages of pregnancy. The elevated maternal venous Pco_2 also contributes to stimulating the respiratory rate.

Throughout the entire course of an average normal pregnancy, the mother gains a total of about 11 kilograms in body weight, and most of this weight increment takes place during the last six months of gestation. Owing to the significantly increased maternal appetite during pregnancy, some women may gain considerably more weight than this approximate figure would indicate (ie, up to 34 kg). Effective prenatal counseling is essential to prevent such an undesirable weight increase.

Pregnancy imposes a severe drain upon the maternal resources insofar as specific nutritional requirements are concerned. This fact is especially true during the last third (trimester) of pregnancy, during which period the body weight of the fetus doubles. In particular, the maternal requirements for supplemental quantities of "high-quality" proteins (ie, those containing all or most of the essential amino acids, p. 874), certain vitamins, and minerals increase significantly. Thus calcium, phosphorus, iron, and vitamin D are required in excess during pregnancy.

Although the concentrations of many nutrients increase gradually within the placenta as gestation progresses, the ability of this organ to provide a completely adequate supply of nutrients to the fetus is nonetheless limited. Furthermore, it should be stressed that the fetus will continue to grow normally or almost normally even though the mother is living on a nutritionally marginal, or even frankly inadequate, diet. The fetus takes precedence for such nutrients as it needs, and will draw upon the maternal stores of these nutrients regardless of the mother's nutritional state. These enhanced requirements for specific nutrients are, of course, in addition to the increased caloric requirement discussed earlier.

The pregnant woman generally forms urine at a somewhat higher rate than when in the nongravid state. This is so because of the excretory products of the fetus that are added to the maternal circulation (p. 1193) and must be eliminated by the maternal kidneys in addition to the mother's own excretory products. Renal absorption of electrolytes (particularly sodium and chloride), as well as water, increases due to the elevated levels of corticosteroids that occur in pregnancy. However, this salt and water retention tends to be offset physiologically by the increased glomerular filtration rate that develops in the pregnant female. In fact, the GFR may increase by as much as 50 percent during late pregnancy and this mechanism increases both salt and water loss considerably.

The ureters dilate in pregnancy, compensating to some extent for the mechanical pressure exerted by uterine expansion on these structures as they pass over the rim of the pelvis. However, ureteral dilatation and the accompanying expansion of the renal pelvis can predispose the pregnant woman to urinary tract infections.

The total volume of amniotic fluid produced during pregnancy ranges from a few milliliters to several liters; normally the volume of this fluid is between 0.5 and 1.0 liter. Interestingly, the source(s) and reabsorptive loci of the amniotic fluid are unknown for the most part. However, small volumes of amniotic fluid are known to come from the fetal kidneys, which become functional and excrete in utero once they have developed sufficiently; this occurs after about 10 weeks of gestation.

Parturition

The process whereby the fetus is expelled from the uterus is known as parturition. Normally, the gestation period in humans averages 270 days (= 38.5 weeks) from the day of conception, or about 284 days from the first day of the last menstrual period preceding fertilization.

The actual physiologic mechanisms that initiate parturition are unknown in spite of considerable research done on this problem. Consequently, the underlying changes involved at the onset of childbirth remain largely conjectural. It is known, however, that during the final trimester of gestation the irritability of the uterus increases, and that weak, irregular contractions of this organ take place (Braxton–Hicks contractions). These contractions progressively increase in frequency during the final month of pregnancy. The relatively weak Braxton–Hicks contractions rapidly change within a few hours to extremely strong contractions that dilate the cervix of the uterus and subsequently expel the fetus through the widely distended vagina (or birth canal, as this structure is termed during parturition).

What specific factors have been implicated in the relatively sudden transition of the weak uterine contractions into the powerful contractions (or labor contractions) that accompany parturition, or labor, as the process commonly is termed? Two major factors are worth discussing here. First, there is a gradual alteration in hormonal balance that accompanies the increase of irritability of the uterine musculature. Second, there is a gradual increase in the response of the smooth muscle to mechanical stimulation.

Progesterone is known to reduce the irritability and contractility of the uterine musculature during pregnancy as discussed earlier (p. 1198). In marked contrast to this effect of progesterone, estrogens definitely tend to increase uterine irritability and contractility. Even though both of these hormones are secreted in gradually increasing amounts during gestation, from about the twenty-eighth week onward, the increase in estrogen secretion exceeds the relative increase in the secretion of progesterone; the estrogen/progesterone ratio rises as full-term growth of the fetus is approached. Therefore it has been postulated that an increase in the magnitude of the estrogen/progesterone ratio is one factor that participates in the increased uterine irritability and contractility that are found toward the end of the gestation period. This idea receives support from the observation that an increased level of circulating unconjugated estrogens is found in the maternal blood just before the fetus reaches term.

Several lines of experimental evidence indicate that oxytocin (p. 1045), liberated from the neurohypophysis, also participates in causing the increased uterine contractility that is seen at the end of pregnancy. It is clear, however, that dilatation or irritation of the cervix stimulates a neurogenic reflex that enhances oxytocin secretion by the neurohypophysis.

Insofar as mechanical factors are concerned in the initiation of uterine contractions, it is a well-known fact that increased physical distension of smooth muscle accelerates the inherent pacemaker frequency of this tissue, and concomitantly increases the rate of contractions thereof. Interestingly, this rate effect upon the spontaneous activity of smooth muscle is quite similar to that observed after applying mechanical stretch to cardiac pacemaker tissue (p. 659). Thus mechanical stretch or irritation applied to the uterine smooth muscle near term can increase its contractile activity considerably. Such mechanical distension or irritation of the uterine cervix also is particularly effective in stimulating contractions of the body of the uterus itself. For example, labor can be induced readily by physically dilating the cervix, and/or by rupturing the amniotic sac (or "bag of waters"). Following such procedures, the head of the fetus now is able to move still lower in the abdominal cavity and thus to dilate the cervix still more strongly. It is unclear at present how cervical stimulation induces contraction of the uterine body. Both myogenic as well as neurogenic mechanisms for impulse transmission have been postulated to explain this effect.

Undoubtedly, a number of sequential and integrated physicochemical mechanisms are operative throughout the entire process of childbirth. As depicted in Figure 16.15A and B, a positive feedback mechanism may be implicated in this process, and although this scheme is largely hypothetical at present, it does explain the sequential events taking place throughout parturition. This mechanism is particularly interesting for another reason. It provides one of the few examples in which a positive rather than a negative feedback system has a normal physiologic role in the body.

The elements of this theoretical positive feedback mechanism may be summarized briefly. First, mechanical, neural, and chemical factors increase uterine contractility toward the end of gestation (Fig. 16.15A.1), including perhaps an elevated estrogen/progesterone ratio. Second, the weak Braxton–Hicks contractions generally become stronger. Finally, a contraction of sufficient force is evoked so that uterine irritability increases to a still greater extent. Oxytocin secretion enhances uterine contractility still further (Fig. 16.15A.2). Third, positive feedback now increases uterine irritability progressively, resulting in successively more forceful contractions; ie, cyclic stimuli are induced so that as the degree of feedback increases so does the contractile response of the uterine musculature. That is, the oscillations within the feedback system become progressively greater as they are not damped, as in a negative feedback system (p. 49; Fig. 2.2).

False labor is a situation in which vigorous labor contractions gradually cease. This phenomenon may be explained on the basis of this theory merely by the fact that in order for the positive feedback mechanism to continue to operate, each contraction must be more forceful than the preceding one. Consequently, if the contractile cycles do not stimulate the uterus adequately once labor has started, then the whole pattern undergoes gradual regression and the contractile cycles disappear.

Once labor contractions have become firmly established, reflexes are initiated between the birth

A. ONSET OF PARTURITION

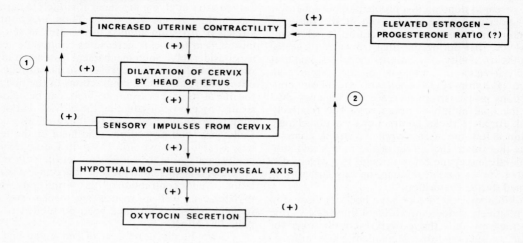

B. PARTURITION

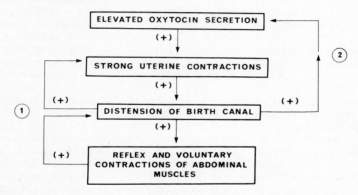

FIG. 16.15. The positive feedback mechanisms believed responsible for parturition. (+) = enhanced activity or excitation. **A.** An elevated estrogen/progesterone ratio near term may contribute to an increased uterine contractility. Impulses from the cervix, 1, as well as increased oxytocin secretion, 2, also increase uterine contractility so that positive feedback cycles which reinforce one another are induced (arrows). **B.** As parturition continues, contraction of the abdominal muscles as well as further distension of the birth canal, 1, lead to ever stronger uterine contractions. The enhanced physical distension of the birth canal also leads to enhanced oxytocin secretion, 2, and these positive feedback mechanisms tend to reinforce one another progressively until the fetus is expelled.

canal and the spinal cord; from the latter, efferent pathways conduct impulses directly to the skeletal muscles of the abdominal wall, causing powerful contractions of these structures. These abdominal contractions also may be reinforced by cortical (ie, voluntary) activity which also increases the contraction of the abdominal muscles (Fig. 16.15B), as well as coordinates these contractions with the intermittent and involuntary uterine contractions. The effects of the uterine contractions combined with those of the abdominal musculature can exert an expulsive force on the fetus that is equivalent to about 12 kg.

The uterine contractions during parturition start at the fundus, then sweep downward over the body of the organ. The force developed by each contraction becomes progressively weaker as it nears the cervix; hence the fetus is propelled downward and out through the birth canal.

In the early stages of labor, the contractions are of brief duration, and may occur only once every 30 minutes. As labor continues, however, the contractions become more frequent; ie, they occur at 1- to 3-minute intervals, and the contractions also become progressively stronger as parturition advances.

In about 95 percent of normal births, the head of the baby is expelled first, and thereby the diameter of the birth canal is forced open as the child descends. In the remaining 5 percent of births, the buttocks appear first; this position at delivery is the so-called breech presentation.

Parturition is divided arbitrarily into two stages. During the *first stage of labor,* the cervix of the uterus undergoes a gradual and progressive dilatation. This stage persists until the birth canal is of sufficient diameter to accommodate the head of the fetus. The first stage lasts roughly between 8 and 24 hours in previously nonparous women; however, the multiparous female may complete this stage in only a few minutes.

The *second stage of labor* is entered once cervical dilatation is complete, and the fetal head can enter the birth canal readily. Successive waves of uterine and abdominal contractions now propel the fetus gradually farther down the birth canal until delivery is accomplished. The second stage of labor may be extremely brief in multiparous women, or persist longer than a half hour in the previously nonparous female.

The alterations in the pelvic structures prior to childbirth were discussed in relation to the endocrine secretions involved in pregnancy (p. 1199).

The uterus contracts powerfully within a few minutes after childbirth, and this contraction effectively severs the connection between the uterus and the placenta at the original site of implantation. Hemorrhage from ruptured sinuses is limited during this process because the strong uterine contraction that shears off the placenta causes the vascular musculature of this organ to constrict. Furthermore, each uterine vessel is surrounded by smooth muscle fibers of the uterus per se, and the contraction of thses fibers in turn compresses the blood vessels much like the action of a tourniquet.

Following childbirth, the uterus involutes during the successive four to five weeks, so that by the end of this time the endometrial surface has healed and epithelial cells typical of the nongravid state have covered the luminal surface once again. Concomitantly, the size of this organ has returned to normal, and by the end of about six weeks normal sexual cycles have become reestablished once again, provided that the woman does not nurse her infant.

The endocrine changes that take place in the mother following parturition are illustrated diagrammatically in Figure 16.16.

Lactation

The growth and development of the female mammary glands at puberty were discussed in Chapter 15 (p. 1163). Briefly, these processes are linked intimately to the onset of menstrual cycles and the enhanced estrogen secretion by the ovaries that accompanies the appearance of regular sexual cycles. Estrogens stimulate growth of the duct system of the breasts as well as the supporting connective tissue, and in addition stimulate lipid deposition which lends mass to the breasts at the onset of puberty.

During pregnancy, additional growth and ramification of the secretory ducts is stimulated considerably by the high concentrations of estrogens

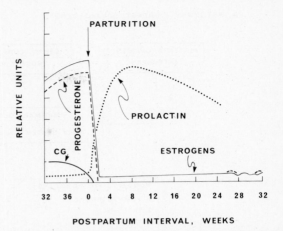

FIG. 16.16. Summary of the major endocrine changes following parturition. CG = chorionic gonadotropin. The small curved dashed lines on the estrogen curve at the lower right of the figure indicate the onset of regular sexual cycles and progesterone secretion by the ovaries following parturition. Cf. Figure 16.13. (After Guyton. *Textbook of Medical Physiology,* 4th ed, 1971, Saunders.)

that are secreted by the placenta; further increases in the mass of the stroma and of the fat deposits also take place concomitantly. In addition to these gross changes, a marked increase in protein deposition takes place, owing to the synergistic action of pituitary somatotropin and perhaps human placental lactogen (Fig. 16.17A).

However, the effects of the several hormones noted above result in only a partial development of the mammary tissue. Thus the concomitant action of progesterone is essential to produce full proliferative growth of the lobules and alveoli, as well as to cause the alveolar cells themselves to assume secretory properties. These changes induced by progesterone in the mammary glands are quite similar to those observed in the uterine endometrium during pregnancy (p. 1198).

Prolactin appears to be the hormone most intimately concerned with actual milk secretion following parturition (p. 1039). Experimental work on lower mammals has revealed that prolactin acts synergistically with estrogens and progesterone in stimulating proliferation of the alveolar system. Presumably, therefore, prolactin (and also HPL?) has a similar function in humans insofar as the final development of the mammae prior to the onset of lactation.

Thyroxine, adrenal corticoids, and insulin also are essential to lactation, as they tend to condition the overall metabolism of the mother in order that lactation can take place (Fig. 16.17A and B).

Immediately following parturition, the breasts are in a state of physiologic preparation for nursing, although only a few milliliters of a clear, almost fat-free liquid, known as *colostrum,* are secreted at this time.

Presumably during gestation the high concentrations of estrogens and progesterone exert an in-

A. PREGNANCY; BREAST DEVELOPMENT:

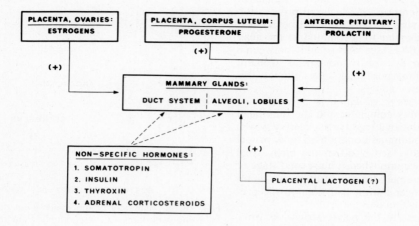

B. POSTPARTUM; MILK FORMATION:

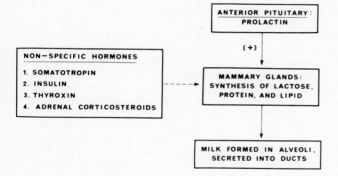

C. POSTPARTUM; MILK EJECTION:

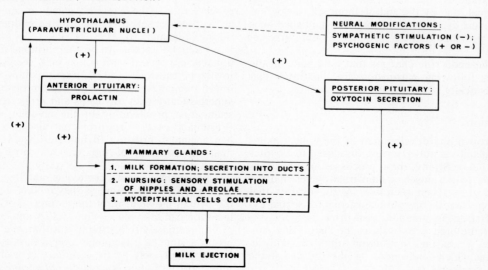

FIG. 16.17. The factors involved in the breast development which takes place during pregnancy **(A)**, as well as milk formation postpartum **(B)**, and the milk ejection reflex **(C)**. (+) = enhanced activity or stimulation; (−) = decreased activity or inhibition.

hibitory effect on prolactin secretion, hence on lactation itself. Immediately following childbirth, however, there is a sudden fall in the blood levels of estrogen and progesterone following expulsion of the placenta, and this removes any suppressive effects of these hormones on prolactin secretion. These endocrine changes are illustrated in Figure 16.16 (see also Fig. 16.13). The sudden upsurge in prolactin secretion causes the glandular tissue of the mammae to synthesize large quantities of lactose, fat, and protein (casein), as well as to secrete sizable volumes of true milk instead of colostrum (Fig. 16.17B). Normal milk secretion also requires the indirect action of somatotropin as well as adrenal cortical hormones; however, prolactin is the only hormone that acts specifically on the mammary glands to cause milk secretion (Fig. 16.17B).

Although the mammary glands produce milk continuously, the milk does not flow readily from the alveoli where it is formed (p. 1157) into the ducts (p. 1164). The process of milk ejection (or "let-down") requires the action of a reflex having both neurogenic and endocrine components (Fig. 16.17C). The act of nursing itself produces sensory stimuli that arise in the nipple and areola. Impulses from these regions of the breast are transmitted via somatic sensory nerves to the spinal cord, thence to the hypothalamus where oxytocin (p. 1045) and some vasopressin (p. 1043) are secreted. These hormones enter the circulatory system and thus are carried to the mammary tissue where they directly stimulate contraction of the myoepithelial cells that surround the alveoli. Milk is forced into the excurrent duct system of both breasts simultaneously, regardless of which one is being suckled.

An interesting example of a conditioned reflex (p. 440) can be elicited in some nursing mothers. After the baby has nursed for a period of time, say a week or so, then the sound of the infant crying and/or the sight of the child being brought into the room with the mother provide sufficient stimuli for the forcible ejection of a stream of milk, although physical contact between mother and child has not yet occurred. It is obvious that psychogenic factors, hence cortical influences, can play a major role in the success or failure of nursing. A generalized discharge of the sympathetic nervous system regardless of cause can inhibit secretion of oxytocin, and thereby curtail milk ejection, as indicated in Figure 16.17C.

Prolonged lactation while nursing an infant depends upon two principal factors. First, the breasts must be emptied of milk at regular intervals; and second, the regular suckling provides the sensory stimulus that is essential to the continued secretion of oxytocin, hence lactation. Usually the rate of milk secretion declines sharply and spontaneously about nine months following childbirth, although prolonged nursing can induce milk production for several years postpartum.

In summary, lactation is initiated by the secretion of prolactin following parturition. However suckling by the infant stimulates a neuroendocrine reflex that results in oxytocin secretion as well as milk ejection, and this sensory stimulus is necessary in order that lactation continue for any period of time. Prolactin itself does not stimulate oxytocin secretion, as was once believed.

Following parturition, the uterus involutes far more rapidly in lactating than in nonlactating women, and this effect apparently is due to prolactin secretion which inhibits the secretion of follicle-stimulating hormone (FSH, p. 1037) as well as luteinizing hormone (LH, p. 1038); thus a return to full ovarian function is delayed by lactation, so that regular sexual cycles may not reappear for several months following childbirth. Women who do not suckle their infants return to a normal pattern of sexual cycles far more rapidly, so that the first menstrual period generally occurs at about six weeks following parturition.

In some respects lactation imposes a far more serious drain upon the metabolic resources of the mother than does pregnancy. For example, up to 1.5 liters of milk are secreted each day at the peak of nursing activity, and this volume requires the formation (and loss) of about 100 gm of lactose, 50 gm of fat, and 2 to 3 gm of calcium phosphate. Thus an adequate supply of calcium, phosphorus, and vitamin D is essential in the diet of the nursing female. Otherwise the parathyroid glands become more active, and decalcification of the maternal bones as well as other metabolic disturbances can take place.

SOME ASPECTS OF FETAL AND NEONATAL PHYSIOLOGY

A few salient features of normal fetal growth and development may be summarized at the outset, as these aspects of prenatal life have important functional consequences.

The pattern of fetal growth, insofar as overall length and body weight are concerned, is depicted in Figure 16.18. It will be noted from this diagram that no appreciable gain in length of the embryo takes place during the first two or three weeks immediately following conception and implantation. In fact, the embryo remains at an almost microscopic size during this time. During the same interval, however, development of the placenta as well as the fetal membranes takes place to an appreciable extent. Following this initial lag period, the mass as well as the length of the fetus increase in almost direct proportion to the time that has elapsed since fertilization took place, so that the fetus normally weighs slightly over 3 kg at parturition.

A month after fertilization the primordia of all of the major organs have appeared, and during the next 8 to 12 weeks the detailed anatomic structures of these organs become evident. Consequently by the time 16 weeks have passed the organs of the fetus are grossly similar to those found in the newborn child, although the full his-

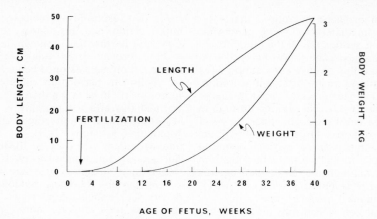

FIG. 16.18. Summary of fetal growth in utero. (Data from Guyton. *Textbook of Medical Physiology,* 4th ed, 1971, Saunders.)

tologic and cytologic development of the tissues and cells requires the additional 5 months of gestation to be completed. In fact, at the time of parturition, the nervous system, liver, and kidneys are not yet fully developed.

The general developmental sequence of certain organs and organ systems may be outlined briefly. The human heart commences beating at around the fourth week of gestation at a rate of approximately 70 contractions per minute, and the rate increases progressively with time so that the heart rate at full term is about 140 beats per minute. Details of the circulatory system before and immediately following birth will be discussed on page 1207.

Erythrocytes are formed in a number of tissues and organs that assume this function successively as development proceeds. First, nucleated red blood cells are formed in the yolk sac and mesothelial cells of the placenta (about the third week following conception). Second, nonnucleated erythrocytes are produced in the mesenchyme and vascular endothelium of the fetus (about the fourth week). Third, the liver commences to produce red blood cells (about the sixth week). Fourth, the spleen and other lymphoid tissues produce erythrocytes (around the twelfth week). Fifth, the bone marrow gradually assumes the erythropoietic and leucopoietic functions (beyond the twelfth week).

Thus during the second trimester of pregnancy the liver, spleen, and lymphoid tissues collectively are the principal sources of blood cells in the fetus. During the third trimester, the bone marrow progressively usurps the hematopoietic functions of these other tissues. The special hemoglobin produced during gestation is known as hemoglobin F (p. 749), and this respiratory pigment appears at about the time the marrow assumes a hematopoietic function (week 20). Hemoglobin F has a significantly greater oxygen affinity at a low Po_2 than does adult hemoglobin (hemoglobin A). This fact is of the utmost importance to fetal respi-

ration, as noted earlier (p. 1195; Fig. 16.12). Following parturition, 80 percent of the hemoglobin is of the F type. However, no more of this type of hemoglobin is formed in the neonate; when the infant is around four months old 90 percent of the hemoglobin found in the circulation is type A.

In utero, fetal respiratory movements are inhibited, although such movements are known to occur. It is obvious, however, that pulmonary gas exchange cannot take place in the lungs until after parturition, as the lungs are filled with alveolar fluid, and the alveoli do not expand until parturition is effected and the fluid is eliminated (p. 722).

Many of the peripheral reflexes are well established in the nervous system by the ninth to twelfth week of gestation, although many important higher functions of the central nervous system remain completely undeveloped, even when the fetus is at term. In fact complete myelinization of a number of major tracts in the CNS does not take place until a year postpartum (p. 158).

Swallowing can occur in utero; therefore the gastrointestinal tract of the fetus is exposed to large volumes of amniotic fluid, and during the third trimester of pregnancy the functions of this system are similar to that of the neonate (or newborn child). Small amounts of meconium, or fetal feces, are produced continuously in the gastrointestinal tract and eliminated via the large intestine directly into the amniotic fluid.

The kidneys of the fetus are known to excrete urine during the last half of gestation, and this urine is voided directly into the amniotic fluid. The homeostatic regulatory systems for controlling electrolyte and acid–base balance of the extracellular fluids neither become operational until about the middle of the gestation period, nor do they function to their fullest capacity until approximately one month following parturition.

Insofar as overall fetal metabolism is concerned, glucose is the primary source of energy in utero. Proteins and lipids are formed and stored rapidly in the fetus. In fact, lipid materials are derived mainly via synthesis from glucose, rather than absorbed as such from the maternal circulation.

The formation of bone was discussed in Chapter 15 (p. 1064 et seq).

The arrangement of the fetal circulatory system is of particular physiologic significance because of the profound alterations in the pattern of blood flow that take place immediately following parturition. As shown in Figure 16.19, the circulatory systems of the fetus and neonate exhibit certain major anatomic and physiologic differences when compared to the adult pattern.

In the fetus, a vessel known as the *ductus arteriosus* shunts practically all of the blood from the pulmonary artery directly to the aorta as shown in Figures 16.19 and 16.20. Thus, most of the blood (roughly 80 percent) bypasses the lungs, which, of course, are nonfunctional in utero. Since the resistance within the vascular bed of the fetal lungs is high, and since the pulmonary arterial pressure is a few mm Hg higher than that in the aorta, then blood flow directly into the aorta through the ductus arteriosus is favored.

A major difference between fetal and adult hearts is the presence of an opening, the *foramen ovale,* between the right and left atria. Most of the

blood entering the right atrium via the inferior vena cava passes directly through the patent foramen into the left atrium. Blood entering the right atrium from the superior vena cava, on the other hand, tends to enter the right atrium and then it passes into the pulmonary artery.

In the human fetus about 55 percent of the entire cardiac output flows through the placenta, and the blood leaving this organ in the umbilical veins is thought to be approximately 80 percent saturated with oxygen, in comparison with the 98 percent saturation found in the adult pulmonary veins and systemic arteries (p. 656).

As illustrated in Figure 16.20, the *ductus venosus* channels some of the oxygenated blood carried by the umbilical veins from the placenta directly into the inferior vena cava whereas the rest of this blood mingles with the portal venous blood of the fetus. Subsequent to this mixing, the portal (as well as the systemic) venous blood is approximately 25 percent saturated, while the mixed blood in the vena cava is roughly 65 percent saturated.

The overall physiologic consequences of the three fetal blood shunts discussed above are as follows. First, the head region of the fetus receives the relatively more highly oxygenated blood from

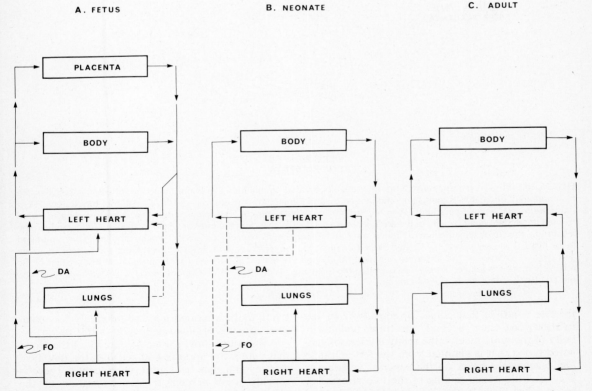

A. FETUS   **B. NEONATE**   **C. ADULT**

FIG. 16.19. The patterns of blood flow in the fetal **(A),** neonatal **(B),** and adult **(C)** circulatory systems. DA = ductus arteriosus; FO = foramen ovale. The dashed lines in **A** indicate minor pathways for blood flow in the fetus. The dashed lines in **B** indicate the alterations in blood flow that occur in the neonate immediately postpartum as the ductus arteriosus and foramen ovale close, thereby shunting blood to the adult circulatory pattern diagrammed in **C.**

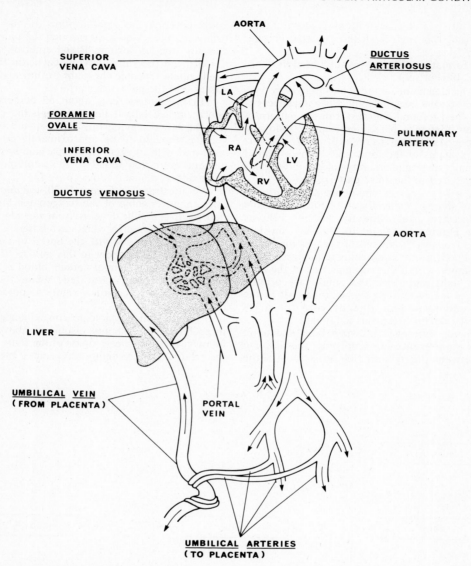

FIG. 16.20. Physiologic anatomy of the fully developed fetal circulation. The major anatomic differences between the fetal and adult circulatory systems are emphasized by underlining. The arrows indicate the direction of blood flow. RA = right atrium; RV = right ventricle; LA = left atrium; LV = left ventricle. The major proportion of the oxygenated blood passing to the right heart by way of the umbilical vein (as well as that from the inferior vena cava) is shunted directly to the left atrium, thus bypassing the lungs. Thence, most of the more highly oxygenated blood from the umbilical vein is pumped to the head, whereas the less oxygenated blood from the superior vena cava is diverted mostly through the pulmonary artery and ductus arteriosus to the lower regions of the body and the umbilical arteries. That is, a functional separation of the relatively more oxygenated and less oxygenated bloodstreams is achieved. (After Ganong. *Review of Medical Physiology,* 6th ed, 1973, Lange Medical Publications.)

the left ventricle that comes principally from the inferior vena cava. Second, the trunk and lower body of the fetus receive the relatively less oxygenated blood that is shunted from the left ventricle and reaches the heart via the superior vena cava. That is, little actual mixing of the two streams of blood takes place in the left ventricle of the fetal heart owing to the property of streamline flow exhibited by the blood in the fetal heart (see also p. 626). Thus a functional separation of the rela-

tively more saturated and the relatively less saturated streams of blood is present in this organ.

As noted above, some 55 percent of the blood that enters the aorta is pumped into the umbilical arteries and thus back into the placenta. There is roughly a 60 percent oxygen saturation of the blood within these arteries, as well as in the inferior portion of the aorta of the fetus (Fig. 16.20).

The surprising resistance of fetal and neonatal tissues to the hypoxia that normally is present in utero is not well understood. Nevertheless, and despite this resistance, the fetus might suffer irreparable hypoxic damage were it not for the greater oxygen affinity of fetal erythrocytes, compared to those of the adult, a phenomenon that has been discussed previously (p. 1195).

The cardiovascular and respiratory systems of the fetus undergo a number of profound alterations immediately following birth. As shown in Figures 16.19 and 16.20, the right and left sides of the fetal heart pump blood in parallel, rather than in series as is the situation in the adult heart. At the moment of birth, the placental circulation is arrested suddenly. Thus the total peripheral resistance in the newborn child rises sharply. The aortic pressure also rises abruptly until this pressure is above that in the pulmonary artery. At the same time the neonate becomes progressively asphyxic because of the sudden interruption of the placental blood flow, and this asphyxia leads to gasping and an attendant expansion of the lungs, an effect that is facilitated by the strongly negative intrapleural pressure that develops during the gasps (-30 to -50 mm Hg). Another factor that assists in the process of lung expansion is a substance known as surfactant. This material markedly reduces the surface tension of the alveolar fluid, particularly during expiration, so that the alveoli can inflate more readily (see also p. 722).

The first breath taken by the infant, together with the constriction of the umbilical veins immediately following delivery, forces about 100 ml of blood from the placenta into the circulatory system of the neonate, an effect that has been termed "placental transfusion."

Following expansion of the lungs, the resistance in the pulmonary blood vessels falls to below 20 percent of the level present in utero. Consequently blood flow through the lungs increases about fivefold above that present in utero. The blood that returns from the lungs to the heart now increases the left intraatrial pressure so that the valve located on the foreamen ovale is closed mechanically by pressure which forces this structure against the interatrial septum (Figs. 16.19, 16.20). The ductus arteriosus constricts strongly within a few minutes after delivery. Since the umbilical vein is occluded completely within the cord, usually by tying shortly after parturition, flow through the ductus venosus ceases completely (Fig. 16.20).

The mechanisms that underlie the functional closure of the ductus arteriosus as well as those involved in the expansion of the lungs are far from being completely understood. However, there is some evidence that an increase in arterial Po_2 and/or asphyxia both act as stimuli for constriction of the ductus arteriosus. Kinins also appear to be involved in this process (p. 648).

Ultimately, both the foramen ovale as well as the ductus arteriosus are closed by growth of fibrous connective tissue (anatomic closure) that obliterates the openings completely by the time the

infant is around one month old. An adult pattern of flow within the cardiovascular system becomes established functionally within a few days after parturition, whereas that for the respiratory system becomes established immediately after birth.

Following parturition and the attendant cessation of blood flow to and from the placenta, the ductus venosus constricts gradually. Thereby a much greater volume of blood is shunted directly through the portal vein to the liver from the gastrointestinal tract.

The drastic changes that take place in the human cardiovascular and respiratory systems at the time of birth are dramatic, but a few other important physiologic differences between the neonate and the adult should be mentioned. Basically, many of these differences result from the inherent lack of stability in the nervous and hormonal regulatory systems that underlie homeostasis in the adult body. In turn this instability results from the incomplete development of many organs at the time of birth, as well as from the time required postpartum for the control mechanisms to become fully operative.

A few of the special physiologic problems with which the neonate must cope may be summarized. The respiratory rate of the newborn is approximately 40 per minute, the tidal volume being around 15 ml; the total respiratory volume per minute is around 600 ml. This volume is about twice as great as that of an adult on a unit body weight basis. The functional residual capacity of the neonate is one-half that found in the adult when they are compared on the same body weight basis. Therefore when the normal respiratory rate of the neonate is altered significantly, the blood gas concentrations can change markedly and in accordance with whether the rate increases or decreases.

Inherent deficiencies or defects in surfactant, among other factors, can lead to a respiratory distress syndrome in the premature neonate. The respiration in such an infant may be normal at first; however, between a few minutes after birth to several days later the lungs fill with a proteinaceous fluid resembling plasma and the victim dies. At autopsy, it is found that the alveoli are filled with a material that resembles a pinkish hyaline membrane when seen in sections of the lung examined under the microscope. Hence, the alternative name for this condition is *hyaline membrane disease*. Recent investigations have revealed both the origin and the role of surfactant in preventing hyaline membrane disease in the newborn. While the embryonic lung develops as an outpocketing of the primitive gut, it undergoes a definite sequence of morphologic and physiologic changes from the trachea outward toward the most remote alveoli. Most relevant to the present discussion, however, from 16 weeks of gestation onward certain epithelial cells within the developing lung differentiate into the so-called type II cells (also called dust cells or granular pneumocytes) of the alveoli. The type II cells in turn elaborate surfactant, which is composed principally of the lipids lecithin and sphingomyelin as well as some proteins. The lecithin/sphingomyelin ratio in the surfactant increases progressively as the type II cells of the lungs develop, so that when the fetus reaches term this ratio is about 15:1.

The alveolar fluid that is present within the lungs before birth, and which was secreted by the fetal lungs, is removed by three mechanisms. First, about 30 percent of the total volume is lost passively from the upper airways as the neonate emerges from the birth canal. Second, about 50 percent of the alveolar fluid is absorbed by the lymphatic system. Third, the remainder presumably enters the infant's bloodstream owing to a favorable osmotic pressure difference as air enters the lungs and the blood flow through these organs increases immediately after parturition. At this point the surfactant normally assumes a critical role, as it tends to prevent collapse of the alveoli during expiration through its effect in lowering surface tension within the alveoli (p. 722). In premature infants, however, the surfactant is "chemically immature" and present in inadequate quantities; therefore the alveoli tend to collapse during expiration, giving rise to hyaline membrane disease of the newborn. The physically unstable alveoli of the surfactant-deficient premature infant are unable to remain open, and physical exhaustion combined with an inadequate respiratory gas exchange rapidly become incompatible with survival.

Interestingly, the injection of adrenal cortical hormones into either pregnant experimental animals or pregnant women has been found to accelerate the development of functional type II cells within the alveoli to such an extent that prematurely born offspring have a far greater chance of survival during the critical first days of postnatal life.

The neonatal blood volume is about 300 ml, and the "placental transfusion" described earlier can increase this volume to a total of around 400 ml. During the first few hours postpartum, the hematocrit rises due to the hemoconcentration occasioned by fluid loss to the interstitial spaces. Thus the circulating blood volume declines toward a normal level of around 300 ml, although the hematocrit remains elevated still longer.

The cardiac output of the newborn child is about 550 ml per minute. Similarly to the respiratory minute volume, the cardiac output is approximately twice that of an adult when expressed on a unit body weight basis.

The blood of the neonate contains about 4×10^6 erythrocytes per mm³, but this value may be elevated to around 4.75×10^6 red blood cells per mm³ by forcing blood manually from the umbilical cord at the time of parturition. However, erythropoiesis is sluggish during the first weeks of extrauterine life, and therefore, by the end of 2 to 2.5

months of age the erythrocyte count declines to approximately 3.2×10^6 red cells per mm³. Subsequently the erythrocyte concentration rises to normal values once again within another month or so as the rate of erythropoiesis accelerates. The leucocyte count of the neonatal child is roughly 4.5×10^4 cells per mm³; this level is approximately five times greater than is found in the normal adult.

The turnover rate of body water (ie, the rates of intake and excretion) in the newborn infant is about seven times more rapid per unit body weight than in an adult. Consequently, even minor shifts in fluid balance can produce serious abnormalities.

Since the metabolic rate of the neonate is double that of the adult on a per-weight basis, metabolic acids are produced far more rapidly than in an adult. Consequently, there is an inherent predilection toward the development of metabolic acidosis in the newborn infant. Furthermore, development of the kidneys is functionally inadequate until around the end of the first month of postnatal life. For example, urine is concentrated by the neonatal kidneys to an osmolality that is only 1.5 times that of the plasma, compared to the threefold to fourfold concentration which is accomplished by the adult kidneys.

Because of the inadequacy of the physiologic mechanisms described above, the two most serious problems faced by the newborn infant are dehydration (eg, as from severe diarrhea) and acidosis. Only rarely does accumulation of excess body water take place to such an extent that it presents a clinical problem.

As noted earlier, the normal metabolic rate of the neonate on a unit body weight basis is approximately twice that of the adult (p. 990), principally because of the much lower surface/volume ratio (p. 991) in the infant. This high metabolic rate underlies the doubled minute respiratory volume as well as the doubled cardiac output found in newborn infants. However, the actual heat loss per unit of surface area in the infant is somewhat less than that of the adult. Consequently, this factor is of considerable importance in temperature regulation, especially as the neonatal thermoregulatory mechanisms are poorly developed and remain so for about two weeks after childbirth. Thus wide fluctuations in body temperature commonly are seen in newborn infants. In fact, a brief postpartum temperature drop of one or two degrees is of normal occurrence.

As noted earlier, the ductus venosus closes gradually following parturition, so that hepatic blood flow increases markedly during the first few days of postnatal life, as all of the portal blood now flows through the liver. During fetal life, of course, hepatic blood flow is quite low because the patent ductus venosus shunts most of the oxygenated blood from the placenta directly to the inferior vena cava of the fetus (Fig. 16.20).

In general, the hepatic functions of the neonate are marginal or mildly deficient for a few days after birth; eg, the rate of gluconeogenesis is low, resulting in a low blood sugar; plasma protein synthesis is inadequate; and bilirubin secretion is depressed until conjugation of this compound with glucuronic acid takes place.

The newborn child has an overall capacity to digest, assimilate, and metabolize foods in a manner quite similar to that exhibited by older children except for the following specific differences. The rate of fat absorption is somewhat lower; pancreatic amylase secretion is low so that starches are not digested and adsorbed readily; and, as noted above, the blood glucose concentration is low and fluctuates markedly, since the homeostatic mechanisms for regulating the level of this compound are developed incompletely.

On the other hand, the neonatal child is quite able to synthesize proteins readily, and hence it is able to store a considerable amount of nitrogen. Thus on a complete diet that contains adequate amounts of essential amino acids, the infant can utilize about 90 percent of the ingested amino acids, a far greater proportion than is used by the adult body.

In normal infants, the endocrine glands are highly developed anatomically as well as physiologically at birth; abnormalities of the endocrines are uncommon in newborn infants. There are a number of instances, however, in which endocrine disturbances may be present in the neonate. For example, if the mother is diabetic then the infantile pancreas will exhibit considerable hyperfunction and the postpartum blood glucose level may fall to 20 mg percent or even lower. Surprisingly, this low blood glucose does not result in coma or death as it would in an adult.

Pregnancy itself may be complicated by maternal diabetes, and fetal growth often is deranged both prepartum as well as postpartum due to the metabolic abnormalities of the mother. Fetal death in utero is not uncommon in diabetic pregnant women, and there is a high degree of infant mortality (over 66 percent) in those fetuses that reach full term. Of those infants of diabetic mothers who do survive parturition, over 65 percent subsequently die of the respiratory distress syndrome discussed on page 722.

The neonatal child has a high degree of inherited immunity that is derived from the mother (p. 547). Thus the infant is protected immunologically against such common childhood diseases as measles, smallpox, diphtheria, and polio for about six months at which time the child is capable of synthesizing specific gamma globulins, hence antibodies of its own. The inherited immunity of the child at birth is due to the diffusion of gamma globulins containing antibodies from the maternal circulation into the fetal circulation across the placental barrier during prenatal life, an exception to the general rule that proteins are unable to cross the placental barrier.

HUMAN MALNUTRITION: STARVATION AND OBESITY

For the purposes of the following discussion the term *malnutrition* may be defined as a state of abnormal nutrition that is caused by an insufficient or unbalanced intake of the basic nutrients, or their impaired assimilation and/or utilization by the body. As pointed out in Chapter 13 (p. 856), there are only six basic classes of nutrients required for complete human nutrition. Fats, proteins, and carbohydrates are required in the diet for tissue synthesis and repair and to provide energy (p. 873); a number of vitamins (p. 877) and certain minerals (p. 890) perform many specific as well as nonspecific roles in cellular and extracellular structures, and in the numerous pathways that are taken by various intermediate compounds involved in the normal metabolic functions of the body; an adequate supply of water is critical to survival as this compound provides the ultimate milieu in which all of the biochemical reactions that underlie life take place. Furthermore water provides a major structural component for most of the cells, tissues, and organs of the body.

Certain defects in the metabolism and utilization of specific nutrients were mentioned in earlier chapters (ie, 13, 14, and 15) so that the present section will deal primarily with two general types of malnutrition, namely, starvation and obesity. At first glance, the reader may find it surprising that the latter condition is a type of malnutrition, because this term usually connotes a deficiency rather than an excess of certain nutrients. Yet obesity definitely is, and will be treated as, a form of malnutrition.

Starvation and Semistarvation

The human body possesses an enormous degree of resiliency as well as adaptability to a wide variety of stresses of both external and internal origin. Therefore a prolonged reduction or even a total absence of food intake is merely one type of stress that may be imposed upon an individual. War, famine, and an assortment of natural disasters (eg, floods, earthquakes, or droughts) may force starvation upon individuals as well as upon entire populations for varying periods of time.

The principal alterations that take place in the body during periods of fasting or outright starvation are of a metabolic nature. The general metabolic alterations that accompany starvation will be summarized first; subsequently the specific changes that are encountered in particular organs will be presented.

Following the onset of a period of fasting, the body first utilizes its reserves of carbohydrate (ie, glycogen), as this nutrient is metabolized most readily and rapidly for the purpose of providing energy. The carbohydrate stores of the body are quite limited; within 24 to 48 hours the metabolically available glycogen supply within the liver and muscles is completely used up, as shown in Figure 16.21. In fact, if carbohydrate were the sole energy source, and fat or protein were not catabolized simultaneously, the total energy that is available in the glycogen stores of the body would be adequate to sustain life for no longer than approximately 12 hours.

In the early period of fasting a transient hypoglycemia may occur, but the blood sugar is little affected even in the later stages of starvation, and some glycogen is always present in the liver.

After depletion of its glycogen stores, the body next shifts its metabolic emphasis to fat catabolism as the prime source of energy. Assuming that around 15 percent of the total mass of the body is fat at the onset of starvation, then by the time five to six weeks have elapsed the lipid depots have been emptied, insofar as mobilizable fat is concerned (Fig. 16.21). At this juncture the protein stores of the body, including the structural proteins of the liver, skeletal muscle, heart, and other organs (but excepting the nervous system), are catabolized rapidly to provide energy. During the earlier stages of inanition, the metabolism of fat as an energy source prevents excessive protein breakdown beyond that which is catabolized each day as the consequence of normal metabolic processes.

This general sequence of changes during starvation is accompanied by several other metabolic effects. A progressive decline in metabolic rate occurs as a concomitant to the inanition and therefore the body temperature shows a corresponding decrease. Simultaneously there is a fall in the systemic arterial blood pressure and heart rate. A ketosis develops as the RQ decreases with elevated fat utilization, and there is some retention of electrolytes and water in certain stages of inanition.

During starvation, the loss of body weight is distributed unevenly among the various tissues. Thus the subcutaneous and other body fat depots lose most of their available lipid first, and large volumes of water also are lost during the early stages of fasting as the adipose tissue is degraded.

When actual metabolic breakdown of muscular tissue occurs later in starvation, there is a corresponding increase in nitrogen and sulfur excretion in the urine, the carbon skeletons of the amino acids being oxidized to provide energy. The muscles can lose up to 35 percent of their total mass during prolonged starvation, and the heart may lose almost as much, whereas the central nervous system appears to lose only 5 percent of its total weight even during a prolonged fast. The kidneys may lose up to 20 percent of their total mass during a similar period of time.

Urinary excretion of sodium, potassium, magnesium, and chloride may decline from the onset of starvation as a compensatory reaction to the reduced mineral intake, and the blood concentration of these electrolytes is changed but little. On the other hand, the sodium bicarbonate concentration of the blood decreases progressively as a metabolic acidosis develops consequent to the ketosis that accompanies an increased fat catabolism.

An increased ketone body excretion accompanies starvation, and this results from synthesis of excessive quantities of acid metabolites, in particular acetoacetic and β-hydroxybutyric acids. These acidic compounds are produced as the consequence of the elevated fatty acid breakdown found during fasting when there is no carbohydrate available to undergo preferential degradation by the body as an energy source. During fasting, the ketosis is more pronounced in women than in men, but this metabolic difference between the sexes is unrelated to the generally larger deposits of adipose tissue found in women. Thus a lean female still excretes significantly greater quantities of ketone bodies in the urine during starvation than does a similarly fasting but obese male. Furthermore, the ketosis observed during fasting is not particularly different in an obese individual than in a person of normal weight.

The adaptive responses of several individual organs and tissues during periods of total food deprivation now may be considered. During starvation the metabolic requirements of the brain are of paramount importance; hence this organ receives top priority for available nutrients, in particular glucose. However, during a prolonged fast there is a shift in the pattern of substrate utilization. Normally the major energy source of the body tissues in general, and neural tissue in particular, is glucose. The brain has a stringent requirement for a continual supply of oxygen as well as for glucose, so that under resting conditions this organ consumes approximately 45 percent of the total oxygen supply and about 65 percent of the total circulating glucose. Consequently the human brain requires roughly 100 to 145 gm of glucose per day,

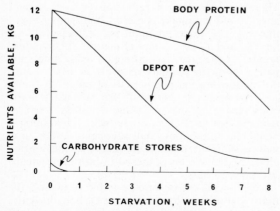

FIG. 16.21. Relative loss of metabolically available body nutrients during prolonged starvation. Note the particularly rapid utilization of carbohydrate, and also the accelerated utilization of body protein as an energy source when the depot lipids become exhausted.

an equivalent of 400 to 600 Cal/day, to meet its normal metabolic requirements. However, the total metabolically available hepatic glycogen store amounts to less than 100 gm in a fed subject. Therefore, the glycogen store of the liver in such an individual is sufficient to sustain the metabolic requirements of the brain for only a few hours. In fact, even after a short fast (ie, such as that occurring overnight between dinner and breakfast), the liver commences to utilize other substrates for gluconeogenesis as the available glycogen store becomes depleted. Thus free amino acids derived from protein breakdown as well as free fatty acids released from the fat depots begin to accumulate in the blood roughly 12 hours following the last meal.

As the fast continues and the period of early starvation ensues, the entire body can synthesize about 160 gm of glucose per day. At this stage, most of this gluconeogenesis from amino acids takes place in the liver; however, the renal cortices also contribute a significant quantity of newly synthesized glucose to the bloodstream. Somewhat later, renal cortical glucose synthesis increases so that it now exceeds the total hepatic output of this compound. As starvation continues, progressively more fatty acids are oxidized, and as these compounds yield 9 Cal/gm of energy in vivo, compared to about 4 Cal/gm for either protein or glucose, then the enhanced metabolism of fat tends to conserve the body protein stores. Therefore the rate of weight loss decreases somewhat as the length of the fast increases. When the available depot lipids are consumed, of course, the rate of protein degradation accelerates sharply.

The metabolic requirements of the brain remain unchanged, however, but the liver and kidneys together can synthesize de novo a total of only 24 gm of glucose per day after several weeks of fasting. This quantity of glucose obviously is inadequate to supply the minimal metabolic requirements of the brain, let alone the other bodily tissues. Hence ketone bodies produced by the catabolism of free fatty acids in the liver provide an alternate source of energy during prolonged starvation (p. 965). Thus the brain during starvation shifts its metabolic pattern from the practically exclusive use of glucose to one in which about 30 percent of the total metabolic energy expenditure is derived from oxidation of this carbohydrate. The remainder of the energy requirement of the brain is met after some weeks of starvation by the oxidation of amino acids (15 percent), and by the oxidation of the ketone bodies' β-hydroxybutyric acid (50 percent) and acetoacetic acid (5 percent).

Ketone bodies normally are produced in very limited quantities; however, during periods of starvation the fatty acids released from the fat depots are transported in the blood to the liver where they are converted into acetoacetic acid, acetone, and β-hydroxybutyric acid. These compounds in turn are released into the blood from the hepatic cells for transport to the various tissues where they are oxidized directly to provide energy. As noted earlier, the brain will preferentially utilize β-hydroxybutyric acid during starvation, and the magnitude of this utilization is a direct reflection of the circulating level of ketone bodies, which level in turn is dependent upon the extent to which the available carbohydrate reserve has been depleted.

The adaptive mechanism discussed above tends to conserve or spare protein stores to some extent; however, some protein still is degraded continually and the nitrogen derived therefrom is excreted as ammonia, urea, and creatinine. The nitrogen excretion that occurs throughout intervals of prolonged starvation merely reflects the fundamental and continuing turnover of all bodily proteins; however, as no exogenous amino acids are available, then extreme emaciation develops ultimately as starvation continues.

In periods of starvation some glucose is essential in order that the tricarboxylic acid cycle continue to function in the synthesis of ATP. It has been proposed on the basis of strong experimental evidence that during starvation the amino acid alanine plays a key role in the de novo synthesis of glucose in a cyclic fashion (Fig. 16.22). According to this concept, amino acids liberated by protein degradation in, for example, the skeletal muscles, transfer their amino groups to pyruvic acid thereby forming alanine. The alanine is transported in the blood to the liver. Following deamination in this organ, the pyruvic acid thus formed is reduced to glucose which then is transported to the peripheral tissues once again where it is oxidized to pyruvate and the cycle continues (cf Cori cycle, Fig. 14.26, p. 944). Thus a fixed and limited supply of glucose is recycled.

As protein is conserved through the preferential utilization of depot lipids during starvation, the total urinary output declines markedly. Consequently the requirements for water decrease to such an extent that water balance can be maintained on an intake of only a few hundred milliliters per day.

Although there has been much discussion on this point, there is some clinical as well as experimental evidence to support the contention that a reduction in the total mass of individual cells takes place during starvation rather than a decrease in the total number of cells in the body. As might be expected, this statement is best supported by histologic studies performed on biopsy samples of adipose tissue taken at intervals during long periods of enforced starvation.

The total length of time an average human being can survive total starvation, provided that water is available, is no longer than nine to ten weeks.* Obviously, the survival period would de-

The classic example in support of this general statement is that of Terence MacSwiney who died of starvation following a hunger strike of 74 days. Grossly obese patients treated in hospital by a regimen of total starvation have survived for periods of up to eight months with no apparent ill effects.

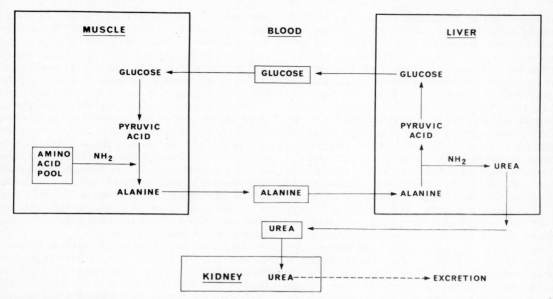

FIG. 16.22. The alanine, or Cahill, cycle. This pathway affords a mechanism whereby a limited supply of glucose can be recycled, and it also provides a way for the efficient transport of protein nitrogen to the liver following the degradation of body proteins to amino acids. (Data from Young and Scrimshaw. *Sci Am* 225:14, 1971.)

pend upon the physical state of the individual at the start of the fast; ie, upon the size of the fat and other stores of body nutrients. In this regard, it should be mentioned that the bodily supplies of certain vitamins (in particular the water-soluble B group as well as vitamin C) are limited; these factors are utilized rapidly during starvation. Following a week or more of total starvation specific vitamin deficiencies begin to appear, and these worsen progressively as starvation continues so that the general debility of prolonged starvation is complicated still further by a lack of particular vitamins, and the onset of death can be accelerated by such deficiencies. The body weight during starvation may have declined by over 50 percent by the time death intervenes.

In contrast with total starvation, undernutrition or semistarvation may permit survival for longer periods of time; however, many specific nutritional deficiencies in addition to weight loss will develop and complicate the clinical picture. Thus a prolonged nutritional regimen that provides a subminimal number of calories results in deficiencies that can involve proteins, vitamins, and essential minerals. The occurrence of particular nutritional disorders such as rickets (p. 1074), scurvy (p. 885), xerophthalmia (p. 886), and beriberi (p. 878) increases considerably as the caloric intake becomes subnormal.

The overall effects of semistarvation obviously depend upon the net caloric deficit plus the relative deficiencies in specific nutrients, eg, proteins, particular vitamins, and minerals that are either lacking or present in the diet in physiologically minimal or subminimal quantities. (See also kwashiorkor and marasmus, Chapter 13, p. 875.)

In general, six major physiologic effects result from a severe lack of adequate nutrition. These effects also are exacerbated in total starvation.

First, emaciation, or a reduction in body weight, is of universal occurrence in partial starvation. The body compensates for the inadequate caloric intake by consuming its own tissues, and the loss of weight is due chiefly to the catabolism of fat stored in the adipose tissue, although considerable amounts of body protein may be metabolized in extreme cases. In this way, the energy requirements are met. The nitrogen balance becomes negative, and overall body growth is impeded or stops completely in children.

Second, the loss of fat in the depots that normally support and cushion the visceral organs causes them to become displaced from their normal positions to some extent, and a bloated appearance may result due to some water retention as the depot fat is catabolized. A ketosis also becomes evident due to the catabolism of fat to the exclusion of carbohydrate during partial starvation.

Third, edema may develop. In certain instances this effect is due to a decreased circulating level of plasma proteins; consequently, there is a net transudation of fluid across the capillary walls into the interstitial spaces. Deranged nutrition of the capillary walls themselves also can result in edema, as the permeability of these vessels is impaired seriously during periods of inadequate nutrition.

Fourth, the basal metabolic rate declines in partial as well as in total starvation. Not only is the metabolic rate per unit surface area decreased (by up to 40 percent), but also the metabolic rate per unit mass of tissue also declines (around 20 percent). The latter effect can be explained partially on the basis of a markedly decreased cardiac work, a reduced skeletal muscle tone, and a subnormal body temperature. The subnormal temperature would, of course, result in a concomitant decline in the rates of all of the biochemical reactions in the body tissues. As a consequence of the reduction in metabolic rate, the semistarved individual exhibits an extreme sensitivity to cold because the vascular bed in the skin undergoes vasoconstriction in order to diminish heat loss via radiation as well as convection. Thus the skin temperature per se is reduced.

Fifth, fatigue occurs readily upon the performance of physical work. The accomplishment of a given amount of muscular work still requires the same energy expenditure in partial starvation as is necessary during the performance of the same activity by a well-fed subject, and this situation obtains regardless of the fact that the basal metabolic rate is reduced significantly in the malnourished individual.

Sixth, semistarvation, if sufficiently severe and prolonged, can result in an overall lack of resistance to infections in some persons. This effect is due, at least partially, to a decline in antibody production owing to a lack of dietary protein as the source of amino acids for antibody synthesis. Furthermore, as protein synthesis requires a large energy expenditure, antibody production also can be decreased by a marked caloric restriction. Obviously the severe general debility that accompanies semistarvation would render the individual more likely to succumb to an infectious disease once it has become established, in contrast to a normal person whose overall resistance to infection is much greater. In addition, the levels of certain key enzymes can decline markedly, for the reasons given above.

A number of psychologic alterations develop in partial as well as in total starvation. The severity of these symptoms is dependent upon a number of factors, including the severity of the nutritional deficit, and the circumstances surrounding the food deprivation. Frequent effects of semistarvation are: a reduced intellectual capacity; apathy; marked depression; personality changes, including introversion; and a loss of moral scruples. These effects generally are reversible upon increasing the daily caloric intake to an adequate level.

Viewed from the standpoint of general biology, the metabolic adaptations that are evoked by relatively short periods of fasting tend to minimize sharp fluctuations in the supply of essential nutrients to various tissues, in particular the brain. Man is, after all, an intermittent feeder and thus the body normally undergoes alternating periods of "feast" and "famine" in the physiologic sense. Thus the biologic mechanisms that have evolved to provide a regulated and consistent flow of nutrients to the tissues during periods of fasting are homeostatic controls in the truest sense of the term. What can be said of the mechanisms themselves that adapt the body to a discontinuous supply of exogenous nutrients? Two important endocrine factors appear to be involved, although in general the adaptive mechanisms to fasting are poorly understood. Following a meal, insulin secretion is stimulated by the absorption of glucose and amino acids from the gut into the bloodstream. The elevated blood insulin level then promotes de novo lipogenesis from glucose, particularly in the adipose tissues, and this insulin also accelerates the rate of glucose uptake as well as that of amino acids by the cells in general, but especially the muscle cells. Concomitantly the elevated insulin level inhibits hepatic gluconeogenesis. Following absorption of the meal, the circulating insulin level gradually falls to fasting values. During prolonged starvation, on the other hand, the insulin concentration in the blood may become subnormal, whereas the secretion of glucagon from the pancreas is enhanced greatly. Glucagon stimulates hepatic glyconeogenesis, promotes lipolysis in adipose tissue, and otherwise acts in a manner diametrically opposite to the actions of insulin. It would appear that an altered circulating glucagon/insulin ratio is a major factor in producing some of the primary metabolic adaptations to starvation. For example, the accelerated hepatic gluconeogenesis as well as the elevated lipid metabolism in the liver can be accounted for by the increased circulating glucagon level that occurs in starvation. Quite possibly other hormones have secondary roles in the metabolic adaptations found in starvation. In particular, the glucocorticoids of the adrenal glands as well as somatotropic hormone of the anterior pituitary may have important functions in the biochemical adaptations to starvation. The effects of somatotropin in this regard are shown in Figure 16.23 and discussed in the next section.

Obesity

Obesity is the opposite side of the biologic coin to starvation or partial starvation, and this type of malnutrition may be defined as overweight that results from an accumulation of excess body fat. Obesity also might be defined as overnutrition in contrast to starvation or semistarvation. Overweight also may occur due to an excessive accumulation of water, extreme muscular development, or in very rare instances, the growth of large tumors.

The "normal" body weights for men and women of different ages that are indicated in Table 16.3 are based upon longevity as the principal criterion for the "ideal weight" of an individual, and are grouped according to the general physical structure and sex of the subject.

There is much undisputed clinical evidence that obesity in humans accompanies and contributes significantly to the development of a number of important degenerative diseases. In particular, obesity imposes a severe stress upon the cardio-

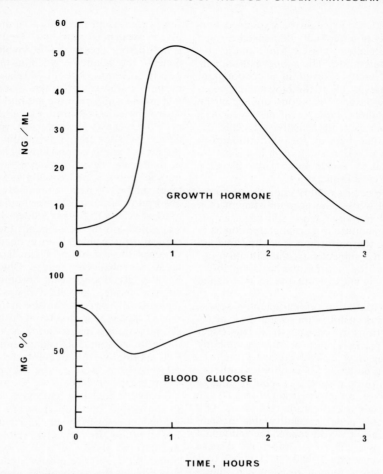

FIG. 16.23. The reciprocal effects of a single injection of insulin (at 0 time) on the plasma level of growth hormone and blood glucose in humans. (Data from Ganong. *Review of Medical Physiology,* 6th ed, 1973, Lange Medical Publications.)

vascular, respiratory, and other bodily systems. For example, obesity predisposes the individual to a lack of resistance to pulmonary infections, hypertension, congestive heart failure, coronary artery disease (including coronary thrombosis), arteriosclerosis, cerebrovascular accidents (or "strokes"), diabetes, amenorrhea, toxemia of pregnancy, and diseases of the digestive system, especially those conditions that involve the liver and gallbladder. In the United States obesity is the most commonly encountered chronic medical problem and its prevalence is increasing together with a markedly increased mortality rate from the diseases listed above.

In the clinical sense, obesity is considered to be present when the patient exceeds his ideal body weight for his height and age by 9.0 kg or more, and pathologic complications will develop eventually when the person exceeds this ideal weight by 11 kg or more. This definition of obesity assumes that any of the other possible factors mentioned earlier that contribute to overweight have been ruled out, and that the individual has only an abnormally large accumulation of body fat.

In determining the etiology of obesity, a number of physiologic as well as psychologic factors appear to be involved (Table 16.4); however, it will be noted in this table that all of these etiologic factors ultimately share one common physiologic denominator. Obesity is the result of an intake of calories that exceeds the expenditure of energy over a period of time; ie, a positive caloric balance is present. This doctrine, of course, is based upon the basic laws of thermodynamics (p. 920). Thus for each 9.3 Cal in excess of energy requirements that enter the body, 1.0 gm of fat is stored. Therefore for each 9,300 Cal ingested above the immediate energy demands of the body 1.0 kg of fat is deposited. Note that the type of nutrient is irrelevant, insofar as the exogenous source of the extra calories is concerned, and this remark may be considered a corollary to the caloric balance doctrine. These statements have for years enjoyed the status of physiologic dogma, and as such have provided the fundamental basis upon which all weight loss and weight control programs are based. Thus a sustained decrease in the total caloric intake coupled with an increased

Table 16.3[a] APPROXIMATE DESIRABLE BODY WEIGHTS[b]

HEIGHT (m)[c]	WEIGHT ACCORDING TO FRAME (kg)[d]		
	Small	Medium	Large
Men			
1.57	50.8–54.5	53.6–58.6	57.2–64.0
1.60	52.2–55.8	55.0–60.4	58.6–65.4
1.63	53.6–57.2	56.3–61.7	59.9–67.2
1.65	54.9–58.6	57.7–63.1	61.3–69.0
1.68	56.3–60.4	59.0–69.4	62.7–70.8
1.70	58.1–62.2	60.8–66.7	64.5–73.1
1.73	59.9–64.0	62.7–69.0	66.7–75.4
1.75	61.7–65.8	64.5–70.8	68.6–77.2
1.78	63.6–68.1	66.3–72.6	70.4–79.0
1.80	65.4–69.9	68.1–74.9	72.2–81.3
1.83	67.2–71.7	69.9–77.2	74.5–83.5
1.85	69.0–73.5	71.7–79.5	76.3–85.8
1.88	70.8–75.8	73.5–81.7	78.5–88.0
1.91	72.6–77.6	75.8–84.0	80.8–90.3
1.93	74.5–80.0	78.1–86.3	82.6–92.6
Women			
1.47	41.8–44.5	43.6–48.6	47.2–54.0
1.50	42.7–45.9	44.5–49.9	48.1–55.4
1.52	43.6–47.2	45.9–51.3	49.5–56.8
1.55	44.9–48.6	47.2–52.7	50.8–58.1
1.58	46.3–49.9	48.6–54.0	52.2–59.5
1.60	47.7–51.3	49.9–55.4	53.6–60.8
1.63	49.0–52.7	51.3–57.2	54.9–62.7
1.65	50.4–54.0	52.7–59.0	56.8–64.5
1.68	51.8–55.8	54.5–61.3	58.6–66.3
1.70	53.6–57.7	56.3–63.1	60.4–68.1
1.73	55.4–59.5	58.1–64.9	62.2–69.9
1.75	57.2–61.3	59.9–66.7	64.0–68.1
1.78	59.0–63.6	61.7–68.6	65.8–74.0
1.80	60.8–65.4	63.6–70.4	67.6–76.3
1.83	62.7–67.2	65.4–72.2	69.5–78.5

[a]*Data from Metropolitan Life Insurance Company.*

[b]*The criterion of longevity has been used for determining the desirable body weights. Weights are for persons 25 years of age and older.*

[c]*Heights are measured with shoes on; 2.54 cm heels for men, 5.0 cm heels for women.*

[d]*Weights are measured with person dressed in indoor clothing.*

energy output (ie, regular exercise) are considered basic features of any rational program designed to decrease, and subsequently to maintain, body weight at the ideal level for any particular individual. This statement assumes that a restriction of the total caloric intake is the primary goal, and that the diet employed for this purpose is adequate in all other respects, otherwise severe nutritional deficiencies obviously would develop. However, in the light of certain experimental evidence derived from a variety of sources, it would appear that the time has come to revise certain aspects of the traditional views underlying the management of obesity, and to discuss the pertinent endocrine and biochemical mechanisms that appear to be involved in this process. First, certain physiologic and biochemical alterations that are found in the body as accompaniments of, and perhaps contributory to, obesity should be reviewed briefly.

Since obesity basically is an increase in the total quantity of body fat, it is pertinent to examine any alterations that may be present in the adipose cells of the fat depots. Elevated fat stores could result from the development of an increased total number of adipose cells (ie, hyperplasia), an increase in the size of the individual adipose cells due to fat accumulation (ie, hypertrophy), or a combination of both of these factors. It has been found in human subjects that the total number of fat cells that are present during childhood and adolescence can increase by hyperplasia, but in adult life such hyperplasia apparently does not take place. Thus if obesity has its onset in childhood, the individual cells hypertrophy by only about 50 percent; however, the adipose tissue of obese children has been shown to contain about three times more cells than does comparable adipose tissue taken from an adult whose obesity developed subsequent to adolescence. If an adult is overfed experimentally, the total number of cells does not increase, but the hypertrophy that develops in the individual adipose cells is significantly greater than that found in the obese child. The period of life during which obesity develops underlies the relative contributory effects of cell size and cell number to adult obesity.

Experimental work has been performed on newborn rats that were overfed or underfed until they were weaned; subsequently both groups of animals were placed on an identical diet. The previously overfed animals were relatively heavier and had more adipose cells throughout their lifetimes than did the underfed group, a finding similar to that seen in human subjects.

The implications of the studies cited above are obvious. The caloric intake during childhood and adolescence well may determine to a large extent the total number of adipose cells present in the adult, especially as hyperplasia of these cells does not occur beyond adolescence. If children are overfed by zealous parents, they may be conditioned for adult obesity. This statement is substantiated by the clinically poor prognosis for sus-

Table 16.4 FACTORS RELATED TO THE ETIOLOGY OF OBESITY[a]

Increased Caloric Intake above Energy Expenditure
Cultural, social and psychologic factors
Hypothalamic damage
Depleted lipid in adipose cells (?)
Iatrogenic

Decreased Energy Expenditure below Caloric Intake
Reduced level of physical activity
Increased "thermodynamic efficiency" (?)

Gross Endocrine Involvement (Rare)
Hypothyroidism
Tumors of pancreatic islet tissue
Cushing's syndrome
Hypogonadism

[a]*Data from Harvey, Johns, Owens, and Ross (eds). The Principles and Practice of Medicine, 18th ed, 1972, Appleton-Century-Crofts.*

tained weight reduction in patients who were obese as children.

The physiologic mechanisms that underlie hunger and satiety have been presented earlier (p. 409). Briefly, it has been shown by neurophysiologic studies that there is a center in the ventrolateral hypothalamus that controls feeding, whereas a satiety center is located in the ventromedial hypothalamus. The cerebrum receives a constant stream of impulses from the feeding center that are interpreted subjectively by the individual as hunger; the intensity of this sensation can be inhibited by impulses arising in the satiety center and that are transmitted to the feeding center. Experimentally, anorexia (loss of appetite) results from damage to the ventrolateral hypothalamus while a suitably placed lesion in the ventromedial hypothalamus results in uncontrolled eating and obesity (the syndrome of hypothalamic obesity).

The satiety center may possibly be activated by any of three postulated stimuli. First, gastric distension such as follows a meal may send afferent impulses to activate the satiety center. Second, the center may be activated by increased peripheral glucose utilization, ie, by an increased arteriovenous glucose difference. Alternatively it has been postulated that the inherent activity of the hypothalamic satiety center may be regulated directly by the rate of glucose oxidation within the cells of this center themselves. These cells, called *glucostats,* are less active when the rate of glucose utilization is low (ie, when the arteriovenous blood glucose difference is slight); the activity of the feeding center is unchecked and the individual experiences a sensation of hunger. Conversely, when the rate of oxidation of glucose by the glucostats is elevated, the activity of the feeding center is inhibited, and a sensation of satiety develops. This concept, known as the *glucostatic hypothesis,* is well supported by much experimental data, including the fact that in the ventral hypothalamus the rate of glucose oxidation is proportional to the circulating insulin level, in contrast to the situation in other types of neural tissue.

Third, distension of adipose tissue cells by accumulated fat may be another major factor that controls appetite. When such cells become distended, the satiety center is activated by afferent impulses that arise in the adipose tissue itself; but when these cells are empty, this is not the case and no inhibitory impulses reach the hunger center from the periphery. Thus chronic hunger and consequent overeating result from cortical stimulation. Such an effect is observed frequently in the obese patient following a period of successful weight loss.

Metabolic factors frequently have been considered to play a role in the etiology of obesity. If the conversion of exogenous calories in excess of those required into adipose tissue were more efficient in certain individuals than in others, then the former would develop obesity more readily.

Several lines of circumstantial evidence have been offered to support this contention, namely that the thermodynamic efficiency of nutrient conversion into fat is greater in certain individuals than in others. For example, a strain of rats has been developed which has a genetic predisposition to increased rates of lipogenetic enzyme activity within their adipose tissue. Therefore obesity can be produced readily in these animals even though their total caloric intake is restricted and thus similar to that of control animals fed an identical diet, both qualitatively as well as quantitatively. Similar alterations in the enzyme activity of some humans also may possibly contribute to the development of obesity.

Direct attempts to determine an increased thermodynamic efficiency in cases of human obesity have not been successful as yet, probably because the techniques employed for determination of this effect are not sensitive enough to detect the relatively small metabolic differences that are involved. There is definite evidence, however, which indicates that there are considerable differences among different individuals insofar as their gross energy metabolism is concerned (p. 989). Furthermore, when groups of human subjects were overfed experimentally under controlled conditions, it took a considerably greater number of excess calories to produce obesity in some individuals than in others. In those subjects who proved refractory to the initial development of obesity, it took a much greater daily caloric intake (expressed on the basis of calories per square meter of body surface) than it did to maintain the excess body weight in "spontaneously" obese subjects.

Psychologic factors undoubtedly play some role in the overall genesis of obesity. For example, feeding behavior was studied in humans relative to their internal physiologic state. The eating behavior of obese and normal persons was compared in situations of satiety, food deprivation, and fear. In marked contrast to normal subjects, eating was not inhibited by fear in obese subjects; rather, food intake was increased. Thus it was concluded that the obese person was insensitive to the normal internal stimuli that regulate feeding behavior. Furthermore, obese subjects have been found to react to a much greater extent than did normal subjects when subjected to various environmental stimuli related to eating, eg, the smell, sight, and taste of food, the time of day, and the sight of other people eating. In this regard, early or late conditioning can be an important etiologic factor in the development of obesity in some persons.

A definite genetic predisposition for the development of obesity has been known for many years to be present in certain strains of animals. However, no such relationship has been proven conclusively in humans, although some studies are suggestive in this regard. For example, lean parents tend to have lean children, whereas children born to obese parents also tend to become, and to remain, obese. Studies performed upon identical twins that were separated in early childhood are inconclusive insofar as the later development or nondevelopment of obesity is concerned. It is obvious that the eating habits that are instilled in the child early in life are quite significant in this respect; ie, in humans early conditioning with regard to patterns and types of food intake in some instances may be the primary factor concerned with development of obesity.

Endocrine disturbances rarely cause obesity, although commonly such factors often are blamed by the patient for his condition. Obesity may be associated with hypothyroidism, but very few hypothyroid patients are actually obese. The obesity that occurs frequently in Cushing's syndrome (p. 1037) exhibits a characteristically different pattern of fat distribution than that found in dietary (or exogenous) obesity. Obesity rarely develops to any significant extent in patients with tumors of the islets of Langerhans, in whom the level of circulating insulin is quite high. Gonadal deficiency is a most infrequent cause of obesity, although male eunuchs and women with the Stein–Leventhal syndrome are obese. (The latter condition is associated not only with obesity, but also with virilism, hirsutism, amenorrhea, infertility, enlarged polycystic ovaries, elevated levels of luteinizing hormone, and depressed follicle-stimulating hormone levels.)

Although endocrine disturbances are an unlikely primary cause of obesity, obesity per se can nevertheless induce a number of secondary metabolic and hormonal abnormalities. Generally these effects reverse spontaneously upon a return to normal weight. Included in the abnormalities that are caused by obesity are increased glucocorticoid production by the adrenal cortex, hyperinsulinism, a diminished release of somatotropin (growth hormone) in response to appropriate stimuli, and a decreased secretion of glucagon in either the fed or fasting state following intravenous alanine administration.

A strongly increased insulin response to glucose loading invariably is observed in obesity, and since the obese subject has a decreased sensitivity to the normal peripheral effects of insulin, then a normal glucose tolerance curve is produced only by an abnormally great elevation in the circulating insulin level. In persons who have a preexisting genetic tendency to develop maturity-onset diabetes mellitus (p. 1089), the stress imposed upon the body by obesity per se probably unmasks the previously latent diabetic tendency. This response doubtless is responsible for the high correlation between maturity-onset diabetes and obesity, as well as the improved carbohydrate tolerance that is observed in obese diabetic patients following weight loss.

Obese subjects generally exhibit an abnormal carbohydrate utilization due to their excessively high insulin levels. The conversion of exogenous carbohydrate to lipid is accelerated and hypo-glycemia ensues rapidly following a meal. The attendant sensation of hunger that develops results in further eating and the cycle repeats, thereby exacerbating the original problem.

It must be borne in mind that the endocrine and metabolic malfunctions summarized above have been demonstrated clearly to be effects of obesity, rather than factors that cause this condition. These endocrine imbalances will be discussed subsequently in relation to the management of obesity.

The control of obesity is as complex a problem as the factors that may contribute to the etiology of this condition in the first place. Any rational program designed for the control of obesity has two major goals: first, reduction of body weight to the "ideal" level for that particular person, and second, to maintain that ideal weight once it is achieved for the remainder of the life of that individual.

The following discussion will emphasize the physiologic and biochemical aspects of weight loss and subsequent weight control, but with full realization that psychologic and psychiatric factors are of the utmost importance to the patient. Thus effective counseling on the part of the physician can be of major importance to the successful outcome of any regimen designed to promote weight loss.

Certain drugs are employed clinically to promote weight loss. For example, amphetamines are prescribed as appetite suppressants; however, the unwanted and even dangerous side effects of these compounds (eg, hypertension and addiction) as well as their short-term period of usefulness render them unsuitable in most instances. Thyroid hormone also is used to enhance weight loss as this compound augments energy production by the body; however, the long-term use of thyroxine and synthetic compounds having related physiologic effects also is precluded on the grounds of many untoward side effects in the patient. Diuretics, on the other hand, promote loss of water, not fat.

Mild to moderate exercise is a useful and recommended adjunct to successful weight loss and control. However, a kilogram of body fat contains about 9,300 Cal. Therefore, one would have to walk at a moderate pace, ie, consume about 240 Cal per hour, for roughly 38 hours in order to lose only 1 kilogram of fat! Exercise alone is hardly the answer to successful weight control, particularly as exercise depletes the body carbohydrate stores and stimulates hunger.

Popular "fad diets" range from the worthless to the nutritionally harmful; such diets are unworthy of serious consideration as a means to permanent weight loss.

Are there any physiologic alternatives to ameliorating the problem of obesity in a practical way?

The general consensus of medical opinion is that moderate to severe caloric restriction on an otherwise nutritionally adequate diet, combined with an increase in physical work (hence caloric output) through the performance of some form of regular exercise will induce and maintain a progressive weight loss. Parenthetically, it should be mentioned here that obesity in general is correlated with a marked decrease in overall physical activity. At the same time, the patient should be educated in order to repattern his eating habits in such a way that weight loss will be sustained. In extreme cases, a program of total starvation combined with an adequate fluid, vitamin, and mineral intake has been applied to hospitalized patients under close medical supervision. However, starvation not only causes a loss of fat, but also loss of an inordinately large quantity of body protein. Unfortunately, the prognosis for both short-term and long-term control of obesity by such regimens is extremely poor, and rebound to the predietary weight level, or even above this level, is the rule rather than the exception.

It was noted earlier that the type of nutrient, ie, carbohydrate, fat, or protein, which supplies the excess calories above those required for daily energy balance is irrelevant, insofar as the development of obesity is concerned. There is specific and convincing experimental evidence available, however, that refutes this dogma and demonstrates unequivocally that the source of the calories can be of prime importance in the problem of obesity. After all, the body is not a bomb calorimeter wherein the combustion of foods is a rapid one-step process; rather, the metabolic pathways are intricate, stepwise, and interrelated at many points.

Specifically, it was demonstrated over a century ago (1864) that a pronounced weight loss could, and did, occur on a high fat diet, a finding that was confirmed by more precise clinical studies performed approximately 25 years ago. More recently, a meticulous study was performed in which the rate of weight loss by obese subjects was shown to be in accordance with the caloric deficit when the proportions of carbohydrate, fat and protein in the diet remained constant. This finding, of course, is in strict agreement with the concept of energy balance. Further studies demonstrated that on a constant intake of energy (eg, 1,000 Cal per day), the rate at which the same subjects lost weight now depended upon the composition of the diet, and this rate of weight loss was greatest in patients maintained on high-fat diets. In fact, when subjects whose weight remained constant, or even increased, on a 2,000-Cal daily intake of a mixed diet were given a 2,600-Cal diet that contained fat and protein as the predominant energy sources, the patients actually lost weight. Control studies in these patients indicated that the observed weight losses were not caused by water loss, and that an actual loss of total body fat had occurred.

Three important general conclusions can be drawn from the studies cited above: (1) The exogenous source of the calories does make a difference insofar as gain or loss of body weight is concerned. (2) Obese patients apparently have a

defect in their carbohydrate metabolism that leads to a greater and/or more rapid conversion of this nutrient to fat, and to storage of this lipid in the fat depots of the body. (3) If the biologically available carbohydrate, ie, that which can be assimilated from the gastrointestinal tract, is eliminated completely from the diet (or even reduced significantly), then any tendency to oversynthesis of lipid from carbohydrate will be minimized.

Obviously the laws of thermodynamics are immutable so that any explanation of weight loss while consuming calories in excess of energy requirements principally in the form of fat and/or protein must be compatible with these laws. Furthermore, what explanation can be given to account for any weight loss due to lipid mobilization and utilization that occurs during the course of a high fat–low carbohydrate diet that is in accordance with known facts related to physiology, intermediary metabolism, and endocrinology? Lastly, aside from any theoretical explanation of the process, does such a diet offer hope toward a practical, long-term solution of obesity for the otherwise normal individual?

Some tentative answers to these points may be advanced at this time that are based upon firmly established experimental data. The following discussion assumes that, aside from obesity, the subject is normal and that no overt pathologic signs (eg, diabetes or cardiovascular disease) are evident.

First of all, in a normal individual who is ingesting an average mixed diet, approximately 50 to 60 percent of a given glucose load is oxidized to carbon dioxide and water, around 5 percent is transformed into glycogen, and between 30 and 40 percent is converted into lipid in the fat depots. It is clear that lipid synthesis in adipose tissue is a major pathway for carbohydrate metabolism; even a small increase in this synthetic response could result over a period of time in a considerable increase in the magnitude of the depot fat stores. In a normal subject, ketone body production is such that the blood level of these compounds is about 1 mg per 100 ml of blood, and less than 1 mg of these compounds is excreted per day in the urine. That is, the principal substrate being used as an energy source is glucose, and the rate of ketone body formation is about equal to the rate of utilization in the normal subject. Now if the exogenous carbohydrate intake suddenly is reduced to an extremely low level (biologic zero, for all practical purposes), while a sufficient quantity of calories is ingested in the form of protein and fat to ensure an adequate supply of energy, what are the biochemical consequences? Essentially they are the same as those that take place during the early stages of fasting (p. 1211). That is, the available carbohydrate stores of the body are utilized first so that by the end of 24 to 48 hours all available glycogen is depleted, and the body now must derive energy from exogenous lipid and protein, as well as by the mobilization of endogenous fat from the depot lipids. Although there is no direct pathway for the conversion of lipid into carbohydrate (Fig. 14.34, p. 961), the carbon skeletons of the exogenous amino acids supplied by the diet are converted readily into glucose (Fig. 16.22) so that the blood sugar is maintained at a physiologically adequate level.

On an average diet, the rate of metabolism of ketones is equal to their rate of formation; therefore no ketosis develops when adequate carbohydrate is present. But when the acetyl CoA formed by the β-oxidation of FFA's cannot enter the TCA cycle at the rate at which it is synthesized, then the net quantity of acetyl CoA increases. Consequently, the rate of condensation of acetyl CoA is accelerated, and larger quantities of acetoacetic acid are synthesized in the liver (Fig. 14.36, p. 966). Hence the levels of β-hydroxybutyric acid and acetone also rise. When the capacity of the tissues to oxidize the ketone bodies is exceeded a ketosis develops. The urinary output of these compounds also increases.

The principal reason that the entrance of acetyl CoA into the TCA cycle is inhibited in either a relative or absolute sense is that there is an intracellular carbohydrate deficiency in the peripheral tissues. This deficiency is caused by three factors: First, a decreased metabolism of glucose to pyruvic acid results in the production of more acetyl CoA (from FFA's) than can be condensed with the available oxaloacetic acid so that oxidation of all of the acetyl CoA cannot take place. Second, if glucose metabolism is decreased, less lipid is being synthesized de novo from acetyl CoA. This effectively adds to the net accumulation of acetyl CoA. Third, the rate of lipid oxidation is enhanced. Therefore more ketone bodies are produced as the result of the increased formation of acetyl CoA.

Starvation, a low carbohydrate–high fat diet, and diabetes mellitus all lead to a relative intracellular carbohydrate deficiency. In the first two instances, this carbohydrate deficiency occurs because no major pathway for converting fat into carbohydrate is present in the body. The magnitude of the ketosis that develops in response to fasting or a high fat and high protein diet is related to three major factors: (1) the net increase in the level of the circulating free fatty acids in the plasma; (2) the quantity of fat that is available for mobilization (as FAA's) from the depot lipids, as well as that derived from exogenous sources. These two factors are of far greater importance as sources of ketone bodies than is the amount of endogenous fatty acids that is present in the liver itself, since the liver normally extracts roughly 30 percent of the FFA's that are brought to it by the circulation. (3) As the plasma FFA level rises, proportionally greater quantities of FFA's are converted by the liver into ketone bodies, and proportionally less are oxidized via the TCA cycle. The ketosis that develops as the consequence of a relative deficiency of available carbohydrate has two principal physiologic effects. The tissues of the body, including the brain, now derive a major proportion of their energy requirement from the oxidation of ketone bodies that diffuse from the hepatic cells and are carried by the circulation to the extrahepatic tissues.* In addition, the relationship between the hepatic formation and the extrahepatic oxidation of ketone bodies is regulated in such a way that the overall energy output remains constant, although the excretory rate of ketone bodies rises as the overall production of these compounds increases and in turn their blood level rises.

Briefly, therefore, the metabolic machinery of the body has shifted its emphasis from carbohydrate metabolism to fat metabolism when exogenous calories are supplied in the form of lipid and protein. More importantly, the rates of mobilization and utilization of depot lipids are enhanced significantly as consequences of this alteration in the metabolic patterns that accompany such a dietary regimen.

It is now pertinent to examine the endocrine and other mechanisms that are involved in the mobilization and utilization of depot lipids, because, after all, these are problems that are basic to the control of obesity.

Although it has been claimed that the human pituitary gland produces a distinct fat-mobilizing hormone, at present the evidence clearly indicates that under physiologic conditions somatotropin (or growth hormone, p. 1028) is the only adenohypophyseal hormone that is directly concerned with the mobilization of depot lipids. Furthermore, it is equally clear that in normal children, as well as in adults, the level of somatotropin in the circulation can change both rapidly and significantly in response to various stimuli (p. 1031). For example, when there is a relative deficiency of energy-producing substrate such as occurs in hypoglycemia, or such as is found during strenuous exercise and while fasting, the level of somatotropin increases. Similarly, the ingestion of a high protein meal (or certain amino acids) also stimulates growth hormone secretion (p. 1031). Stress per se whether of psychologic and/or physiologic origin also will increase secretion of somatotropin.

It is particularly interesting and significant that elevated blood levels of glucose decrease the secretion of growth hormone, whereas an increased circulating level of insulin has the opposite effect (Fig. 16.23). An elevated growth hormone level per se, as well as an increased circulating level of FFA's, feed back to the pituitary directly and thus can regulate the secretory rate of this hormone. In obese subjects, the changes in the rate of secretion of somatotropin in response to the stimuli of hypoglycemia, fasting, and exercise are depressed.

Assuming that somatotropin itself is the important pituitary factor responsible for the

---

The liver per se has a negligible capacity for oxidizing ketone bodies.

physiologic mobilization of depot lipids, then the facts cited above lend credence to the view that depressed growth hormone secretion is related to the fact that obese subjects are relatively refractory to the development of a ketosis.

As noted earlier, the depot lipids are in a constant state of rapid flux, and thus normally exhibit a relatively high turnover rate (p. 961). Growth hormone, as well as the catecholamines epinephrine and norepinephrine, stimulate the activity of the hormone-sensitive lipase in adipose tissue, thereby enhancing the liberation of FFA's (and glycerol) from the depot lipids (Fig. 16.24), an effect that leads to a ketosis if sufficiently pronounced. Thus not only the anterior pituitary but also the autonomic nervous system (via the catecholamines) is involved in depot lipid mobilization, as norepinephrine released at postganglionic sympathetic nerve endings within adipose tissue can accelerate the hydrolysis of triglycerides.

It is particularly significant that increased insulin as well as carbohydrate levels can decrease the activity of the hormone-sensitive lipase of adipose tissue, thus predisposing to a deposition, rather than to a mobilization, of lipids.

Based upon the foregoing information how can one account, at least partially, for the significant weight loss that occurs on a high fat and high protein but low-carbohydrate diet? First of all, it is obvious that a significant loss of potential free energy (ie, calories) from the body could take place via urinary (and fecal?) excretion of ketone bodies, provided that the ketosis is sufficiently pronounced. Furthermore, since acetone is volatile, some added loss of calories will take place via the respiratory tract when a sufficient level of ketosis is present. However, the total quantity of energy lost by such excretory mechanisms would depend strictly upon the degree of ketosis. There is no evidence, incidentally, that a mild to moderate chronic ketosis is harmful in normal or overweight individuals, and the blood buffers significantly reduce the fall in pH which potentially might occur due to the elevated acetoacetic acid and β-hydroxybutyric acid levels in the blood. It must be emphasized that this effect is in distinct and marked contrast to the often fatal metabolic acidosis that develops in patients suffering from uncontrolled diabetes mellitus. Ketosis is a normal concomitant of prolonged, strenuous exercise, and this condition develops as the available carbohydrate reserves are depleted. In fact, however, after several weeks on a high fat–low carbohydrate regimen, the ketosis in a normal subject diminishes or disappears completely, presumably because of an increased activity of the enzyme systems that utilize fat. In this regard, Eskimos normally live on a high fat–low carbohydrate diet without obvious detriment, and individuals of this race do not exhibit a ketosis. Secondly, the increased excretion of nitrogenous carbon compounds (eg, urea) occasioned by a high protein intake also would contribute to the overall caloric loss from the body via the urine.

Third, an elevated dietary fat level could lead to a mild steatorrhea. This effect obviously would lead to a considerable loss of exogenous calories, depending upon the quantity of fat ingested.

Fourth, the endocrine changes that would occur over a period of time on a high fat and high protein but low carbohydrate diet would tend to favor hydrolysis and utilization of depot lipids, rather than their de novo synthesis from exogenous carbohydrate. Thus, insulin secretion would be suppressed, whereas that of growth hormone and glucagon would tend to be increased, and the operation of these factors would accelerate the rate of endogenous fat utilization, hence wieght loss. It is interesting to note that the hyperglycemic–hypoglycemic swings that are observed in obese subjects due to undamped insulin secretion tend to disappear on a low carbohydrate diet so that the blood sugar level exhibits a relatively more stable pattern.

Fifth, the ketosis could of itself suppress ap-

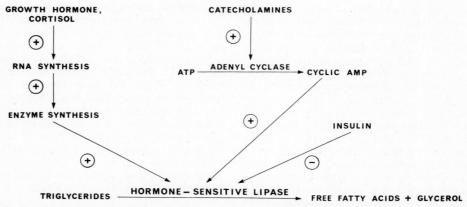

FIG. 16.24.   Effects of a number of hormones upon the lipase of adipose tissue. (+) = stimulation; (−) = inhibition. (Data from Ganong. *Review of Medical Physiology,* 6th ed, 1973, Lange Medical Publications.)

petite; thus the overall caloric intake would be reduced significantly. In this regard it is highly significant that by their very nature, fats and proteins have a very high satiety value compared to that of carbohydrates, and this factor alone would tend to limit the voluntary intake of these foods and thereby to restrict the gross caloric intake. The eventual monotony of a diet rich in fats and proteins, but of limited carbohydrate composition, also would tend to limit voluntary food intake to some extent.

Sixth, the specific dynamic action of proteins is quite high (p. 991), so that the total percentage of calories "wasted" by this effect might contribute significantly to an increased energy loss from the body. Of course the lipid component of a fat–protein diet would tend to reduce the net value of the specific dynamic action somewhat.

Seventh, it is quite possible that the metabolism of fat itself, as well as gluconeogenesis from amino acid carbon skeletons, might require a significant energy expenditure above that necessary when an unlimited carbohydrate supply was readily available.

It should be mentioned that the calorigenic effects of all nutrients are suppressed in obese subjects. Any increased energy expenditure due to the effects of the specific dynamic action while eating a high fat and protein but low carbohydrate diet would have to be verified experimentally in obese subjects. Regardless of the underlying mechanisms that are involved, however, this type of diet does afford several distinct improvements over mere caloric restriction as a means to effective weight loss and control. First, the inclusion of much fat and protein renders such a diet far more palatable and satisfying insofar as the patient is concerned. Second, the ketosis that develops early in such a regimen has distinctly beneficial psychologic as well as physiologic effects. That is, the subject "feels better" and "has more energy" in the subjective sense, while at the same time he is excreting excess calories derived from fat breakdown. Third, there appears to be a reduced tendency for the development of hypoglycemia on a ketogenic diet with attendant shifts in mood and effects on hunger pattern that lead to gross overeating. Fourth, since a high fat and protein but low carbohydrate dietary regimen is more appealing than is mere caloric restriction, the obese patient is less apt to relapse into earlier eating habits that contributed to his obesity in the first place.

It should be mentioned that persons engaged in heavy manual labor or athletic activities definitely require a certain daily ration of carbohydrate in order to perform satisfactorily, as the rate at which fat oxidation can provide sufficient energy is limited, even under optimal circumstances. Persons of this sort usually are not excessively obese in the first place so that no major dietary restrictions are necessary for their health or well-being.

In conclusion, it is evident that obesity is a clinical state of sufficient magnitude that concerted efforts should be made to study in far greater detail the effects of specific nutrients, hormones, and other factors on the etiology and amelioration of this problem. It also is apparent that the time has come to assume an iconoclastic viewpoint, insofar as certain earlier thinking and teaching in this area of physiology and medicine are concerned.

SURVIVAL AT DECREASED AMBIENT PRESSURES: HIGH ALTITUDE, AVIATION, AND SPACE PHYSIOLOGY

The human species normally occupies a relatively narrow vertical belt upon the surface of the earth that ranges from sea level up to an altitude of several thousand meters. In fact, since most of the entire world population lives below 2,000 meters, it is pertinent to examine the physiologic mechanisms that enable one to adapt to higher altitudes and to outline the limits of these adaptive mechanisms.

The principal effects of high altitude upon the human body are those related to a decreased barometric pressure (Fig. 16.25) and hence caused by a reduced partial pressure of oxygen in the inspired air (Table 16.5). The cardiovascular and respiratory adaptations to various altitudes are roughly proportional to the relative hypoxia up to a certain limit (about 6,000 meters), beyond which point the person becomes incapacitated as the extreme limit of his physiologic adaptive mechanisms has been reached. At still higher altitudes, such as are reached in aircraft or space vehicles, maintenance of a completely artificial environment is critical to survival. Such an environment must include provisions for maintaining an atmosphere that is compatible with life and efficient performance, as well as for the maintenance of suitable temperature control and protection from ultraviolet radiation. In addition, the potentially damaging effects of sudden acceleratory forces upon the body must be minimized.

As aircraft and space vehicles have become increasingly sophisticated during recent years, such problems as the effects of hypoxia, rapid acceleration, temperature, and weightlessness have assumed paramount importance insofar as adapting the human body to such conditions is concerned.

The present discussion will be concerned first with the short-term as well as chronic effects of reduced oxygen partial pressures upon the body together with a discussion of the physiologic adaptive mechanisms to such conditions; second, the effects of slow and rapid decompression will be presented; third, a number of specific problems related to aviation and space physiology will be summarized.

The general features of various types of hypoxia were discussed in Chapter 11 (p. 764). As one ascends to progressively greater elevations

Table 16.5 RELATIONSHIP BETWEEN ALTITUDE AND ALVEOLAR GAS CONCENTRATIONS[a]

ALTITUDE (m)	BAROMETRIC PRESSURE (mm Hg)	Po_2 OF AIR (mm Hg)	BREATHING AIR			BREATHING PURE OXYGEN		
			Alveolar Po_2 (mm Hg)	Alveolar Pco_2 (mm Hg)	Arterial Blood O_2 Saturation (Percent)	Alveolar Po_2 (mm Hg)	Alveolar Pco_2 (mm Hg)	Arterial Blood O_2 Saturation (Percent)
0	760	159	104	40	97	673	40	100
3,048	523	110	67	36	90	436	40	100
6,096	349	73	40	24	70	161	40	100
9,144	226	47	21	24	20	139	40	99
12,192	141	29	8	24	5	58	36	87
15,240	87	18	1	24	1	16	24	15

[a]Data from Guyton. Textbook of Medical Physiology, 4th ed, 1971, Saunders.

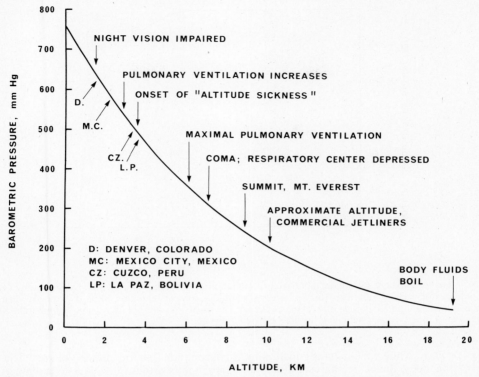

FIG. 16.25. The relationship between barometric pressure and altitude together with the altitudes of several cities and other pertinent data.

above sea level, there is a decline in the ambient barometric pressure (Fig. 16.25) which is accompanied by a proportional decrease in the atmospheric Po_2 (Table 16.5). This basic fact is the underlying cause of all of the physiologic problems related to hypoxia that are encountered at high altitudes. In this regard, it is important to remember that the Po_2 of the inspired air always is approximately 20 percent of the total barometric pressure at any altitude (see also p. 741), and that the Po_2 is altered significantly by the presence of water vapor in the air (p. 728). Although the alveolar Po_2 decreases with increased altitude, the Pco_2 does not change appreciably from 40 mm Hg until rather extreme altitudes are reached (Table 16.5). Similarly the PH_2O does not change appreciably at higher altitudes as this factor is dependent upon body temperature (Table 11.6, p. 743); the PH_2O essentially is constant, ie, about 47 mm Hg. Therefore, as one ascends to greater altitudes, the sum of the Pco_2 and PH_2O remains approximately the same as at sea level, but the barometric pressure, hence the Po_2, falls appreciably. The net effect of the combined partial pressures of carbon dioxide and water vapor at any altitude is to reduce effectively the space which is available within the lungs for such oxygen as is present so that when one breathes pure air the oxygen saturation of the blood decreases far more rapidly upon ascent than it does when breathing pure oxygen at the same ambient

pressures (Table 16.5). The consequences of these important facts are shown in Figure 16.26, in which the saturation of hemoglobin in arterial blood is plotted against the altitude.

Several physiologic mechanisms automatically become operative when a person ascends from sea level to a higher altitude, and these mechanisms tend to compensate for the hypoxia that is encountered due to such a change in environment. The types and degree to which these adaptive responses are called into play depend upon the altitude to which the person ascends, hence the relative hypoxia with which the body is challenged. Furthermore, the total length of the sojourn at a higher altitude plays an important role in the overall acclimation of the body to the hypoxia.

Immediately following exposure to a low Po_2, the chemoreceptors are stimulated by the hypoxia, so that pulmonary ventilation is increased markedly. The maximum increase in alveolar ventilation that can be produced by this mechanism is approximately 65 percent above the normal level. This effect is an immediate adaptive response to high altitude. However, the respiratory rate subsequently falls somewhat and then rises once again if the person remains at altitude for about a week. The reasons for these respiratory changes are as follows. The hyperventilation that develops immediately upon ascent to high altitude causes large quantities of carbon dioxide to be excreted thereby

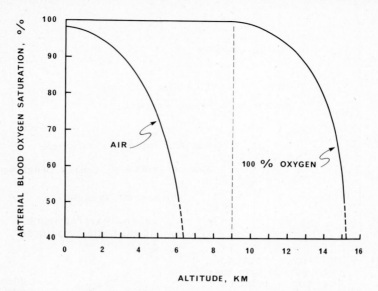

FIG. 16.26. Relationship between arterial blood oxygen saturation and altitude while breathing air and pure oxygen at the ambient pressure for each altitude (Fig. 16.25). Note that the blood oxygen saturation while breathing oxygen does not commence to decline until an altitude of 9.5 km is reached. The dashed portions of each curve represent the altitudes at which the arterial blood oxygen saturation declines below 50 percent.

lowering the Pco_2, while concomitantly increasing the pH of the body fluids; ie, a respiratory alkalosis develops (p. 759). There is a transient increase in urinary pH as a concomitant of the respiratory alkalosis because of an increase in urinary bicarbonate ion excretion which operates to compensate for the reduced blood carbonic acid level (p. 805).

The net effect of the respiratory alkalosis is a temporary inhibition of the respiratory center that tends to oppose the hypoxic stimulation. This inhibition diminishes gradually during the several days immediately following exposure to altitude, so that the respiratory center now can respond fully to the hypoxia, and ventilation increases once again about five-fold to seven-fold above the normal level. The cause of the temporarily decreased sensitivity of the respiratory center is unknown, although the lowered concentration of bicarbonate ion in the body fluids, in particular the cerebrospinal fluid, has been implicated in this process.

Like the respiratory adaptations that occur upon exposure to high altitude, certain cardiovascular alterations take place immediately upon exposure to hypoxia. Thus cardiac output can increase by 20 to 30 percent following an ascent to high altitude, and the heart rate generally increases markedly. Experimental studies performed on subjects exposed to a simulated altitude of around 3,000 m in a pressure chamber have indicated that these alterations in cardiac output result from the hypocapnia induced by the immediate hyperventilatory response to hypoxia. Subsequently, this hypocapnia causes a reduction in venous compliance (venoconstriction) so that there is a decreased venous return to the right heart with a concomitant fall in stroke volume. The cardiac output falls within a few days to essentially normal levels when the subject remains at altitude. If, however, the initial hyperventilatory response to the hypoxia is prevented — by the addition of a sufficient quantity of carbon dioxide to the air inspired at the reduced barometric pressure — then the increase in cardiac output noted above is not seen. The coronary blood flow remains unchanged at high altitudes.

On the other hand, blood flow through such organs as the skeletal muscles, heart, and brain remains elevated throughout the stay at high altitude, whereas the blood flow to organs that do not have such a large oxygen requirement, including the skin and kidneys, decreases.

As pointed out in Chapter 9 (p. 531), hypoxia is the major stimulus for erythropoiesis. Thus after complete acclimation to a low Po_2 level, the hematocrit rises from a normal value of about 45 percent to around 60 to 65 percent. Concurrently, the hemoglobin concentration of the blood increases from a normal concentration of around 15 gm percent up to around 22 gm percent. Similarly the total blood volume increases by approximately 20 to 30 percent; therefore ultimately the total increase in the circulating hemoglobin can be as great as 50 to 90 percent.

In marked contrast to the immediate respiratory and cardiovascular changes discussed above, the increase in hemoglobin and blood volume that occur upon exposure to high altitude require months to develop to a physiologically effective extent, and it may be several years before these compensatory effects are developed fully.

Another long-term adaptive response to high

altitude has been demonstrated, in animals at least. This is a marked increase in both the number and size of the capillaries in the vascular bed as a consequence of hypoxia. Such an adaptation not only would tend to compensate for the increased blood volume noted above, but it would also facilitate gas transport to the cells.

The normal capacity for diffusion of oxygen across the alveolar membrane is about 20 ml per mm Hg pressure difference per minute per cm², and this rate can become up to three times greater during exercise. Similarly, upon exposure to high altitude the rate of oxygen diffusion across the alveolar membrane shows a definite increase. This effect presumably is due to the marked increase in pulmonary blood volume that accompanies hypoxia, and this physical alteration would tend to dilate the capillaries, thereby increasing the total surface area that is available for the diffusion of oxygen. In addition, the hyperventilation that accompanies altitude would tend to increase the surface area of the alveoli themselves through mechanical distension of these structures during the forced inspirations.

In addition to the rapid and chronic physiologic adaptations that take place in the body of an unadapted individual upon ascent to high altitude, several chronic changes are found in persons and animals indigenous to altitudes ranging between 4,000 and 5,000 m.

Certain natives whose forebears have long lived at extreme altitudes have developed a "barrel chest," so that their vital capacity is significantly greater than that of comparable individuals who live at lower elevations. Furthermore, such inhabitants of high altitudes normally exhibit a greater activity of some oxidative enzyme systems than is found in their counterparts living at sea level. Perhaps this enzymatic mechanism permits a more efficient oxygen utilization, although this conclusion is largely conjectural.

Interestingly, persons moving from sea level to high altitudes for extended periods of time do not exhibit any significant alteration in their patterns of enzyme activity so that long-term genetic or other selective factors may be responsible for this effect. In this regard, it is especially pertinent that natives who are naturally acclimated to high altitudes can survive an artificially induced hypoxia equivalent to about 9,100 m for several hours without receiving any supplemental oxygen; such an environment would prove rapidly fatal to an inhabitant of lower elevations.

By way of summary, the principal physiologic adaptations to high altitudes are to be found in the cardiovascular and respiratory systems; and basically these adaptations underlie increased oxygen transport to the tissues. In certain special instances, increased metabolic utilization of the oxygen at the cellular level may provide an additional adaptive mechanism to a hypoxic environment.

The symptoms of hypoxia vary considerably in onset and severity depending upon the altitude above sea level. Thus impaired night vision apparently is the earliest physiologic symptom of hypoxia; this symptom first becomes evident at around 1,500 m and becomes progressively more acute as the altitude increases so that around 5,000 m the quantity of light needed for normal scotopic vision is approximately 150 percent above that required at sea level (Fig. 16.25).

Hyperventilation induced by high altitude generally makes its appearance at about 2,400 m (Fig. 16.25), at which point the arterial blood saturation of oxygen has declined to about 93 percent (Fig. 16.26), and this decrease is sufficient to excite the chemoreceptor mechanism. At elevations above 2,400 m, there is a progressive increase in pulmonary ventilation, and this reaches a maximum of about 65 percent above normal at elevations between 4,800 and 6,100 m. Beyond this point, the chemoreceptors are not stimulated further by additional increments in altitude; rather they undergo hypoxic depression.

At elevations around 3,600 m certain other general effects of hypoxia become manifest, and collectively these symptoms often are referred to as altitude or mountain sickness. Included in this category are lassitude, irritability, decreased mental capacity, headache, nausea, vomiting, and sleeplessness. Occasionally euphoria as well as Cheyne-Stokes respiration may be evident. The severity of these symptoms, particularly headache, generally increases in proportion to the altitude. It should be pointed out that there is considerable variation among different individuals insofar as their development of any or all of these symptoms is concerned, so that it is unlikely that any particular person will develop all of the symptoms listed above.

It is important to emphasize that memory, judgment, and discrete motor skills generally remain completely normal up to about 2,700 m; beyond this elevation, and up to around 4,600 m, such capacities may remain normal for only short periods of time; at higher altitudes, prolonged depression of these cerebral and motor functions takes place even more rapidly, a fact that is of particular importance to pilots of open aircraft or mountain climbers.

At elevations around 7,000 m, the cerebral symptoms may graduate to muscular twitches, convulsions, outright coma, and death. The latter is caused by the effects of profound hypoxia upon the respiratory center.

Figure 16.26 illustrates another important point regarding high altitude physiology. Namely, if one breathes pure oxygen the arterial blood oxygen saturation up to approximately 15,000 m is the same (50 percent) as for breathing air up to about 6,100 m. That is, in both instances the arterial blood contains the same amount of oxygen but this degree of saturation is reached at considerably different altitudes. Normally, however, a person can remain conscious even when the oxygen saturation falls to between 40 and 50 percent (dashed lines on graphs), so that the maximum ceiling an aviator can reach when breathing pure oxygen at the ambient barometric pressure is approximately 15,000 m, compared to only 6,000 m when breathing air at the same pressure. Obviously, a mask that delivers air or oxygen to the pilot under pressure, or for that matter a pressurized cabin in the aircraft, will increase the altitude that can be reached regardless of the barometric pressure. For example, commercial jet aircraft fly at an altitude of approximately 10,000 m, and the barometric pressure is about 200 mm Hg at this elevation (Fig. 16.25). The ambient Po_2 at this altitude is, accordingly, only 43 mm Hg (Table 16.5), a level that is quite incompatible with human survival. However, the cabins of such aircraft are pressurized to maintain an artificial altitude of approximately 1,000 m (barometric pressure = 675 mm Hg; Po_2 = 141 mm Hg; arterial blood oxygen saturation = 97 percent). Complete safety and comfort are ensured, insofar as the respiratory requirements of the crew and passengers are concerned.

If the cabin pressure of an aircraft flying at a high altitude falls suddenly due to a leak or other mechanical failure, there is a lag period depending upon the altitude and rate of loss of pressure from the aircraft between the fall in alveolar Po_2 and the onset of actual coma. Physiologically, this lag is caused by such oxygen as is present in the blood together with that in the tissues and other body fluids, although this oxygen store is strictly limited. From a practical standpoint this brief interval affords the crew and passengers an opportunity to switch to an alternate oxygen supply, and for the plane to dive to a lower altitude before coma intervenes. The rate of loss of pressurization

would, of course, determine the net survival time before coma or death occurred at a given altitude. For example, a reversible loss of consciousness would ensue between five to six minutes after the onset of loss of cabin pressure through a small hole (10 to 20 cm) at an altitude of about 10,000 m, assuming that no remedial steps were taken. At around 12,000 m permanent brain damage would result three to four minutes following similar damage to the aircraft. At around 15,000 m, on the other hand, death would intervene within one to two minutes following structural damage to the pressurization system. In each of these situations, however, there is sufficient time for corrective measures to be taken. But if the structural damage were so massive that the occupants were suddenly exposed to the ambient pressure surrounding the aircraft, then coma and death would be practically instantaneous, depending upon the altitude, an effect known as explosive decompression.

The two curves given in Figure 16.27 indicate the approximate times for the onset of coma after exposure to different altitudes, as well as the intervals for which a person remains conscious and able to perform effectively before losing consciousness.

In addition to the effects of hypoxia, too rapid an ascent in unpressurized aircraft or a sudden loss of cabin pressure while a plane is cruising at a moderate altitude (eg, 6,000 m) can produce a condition known as dysbarism, commonly known as the bends or caisson disease (p. 1233). This situation results from too rapid decompression so that when the ambient pressure suddenly falls below the pressure of the gases that are in solution within the body fluids, bubbles of the gases are formed in the tissues as well as in the body fluids them-

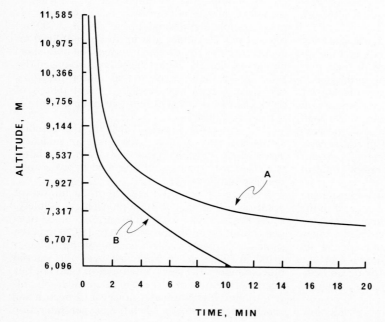

FIG. 16.27. Relationship between the time of exposure to diminished atmospheric pressures at various altitudes and the onset of coma (A) as well as the length of time consciousness is retained (B).

selves, an effect that is quite similar to that observed upon uncapping a bottle of carbonated beverage.

The effects of dysbarism can range from mild tingling sensations, to extreme pain, to death. As this condition is encountered far more frequently among deep-sea divers and caisson workers rather than among aviators, it will be discussed in greater detail in the next section of this chapter (p. 1231).

Rapid ascent to high altitudes also produces expansion of gas within the body cavities, eg, within the middle ear, and the gastrointestinal tract. Unless inflammation or other blockage of the eustachian tubes is present, the act of swallowing readily equalizes any pressure differences that may exist between the middle and external ears so that no discomfort or damage to the tympanic membranes results. Belching (eructation) and gradual expulsion of flatus through the anus generally serve to relieve any discomfort due to gas expansion within the gut caused by the decreased barometric pressure at high altitudes.

At altitudes at and above 16,000 m, a pressure suit or a pressurized cabin is mandatory for pilots or space travelers, because at a barometric pressure of 47 mm, the body fluids boil. This physical effect occurs simply because the vapor pressure of the water in the body fluids, but particularly that within the alveoli, is greater than ambient pressure. Such boiling would take place immediately in the partial vacuum that is present at such altitudes. This fact is of theoretical interest alone, because the unprotected victim exposed to such altitudes would already be dead from hypoxia.

The rapid changes in direction and velocity of aircraft and space vehicles that are encountered during their flight lead to the generation of marked acceleratory forces that can have profound effects upon the human body. In particular, linear acceleration and deceleration respectively take place at the beginning and end of each flight, whereas angular acceleration takes place during all turns made while in flight. In aircraft, angular acceleration is the force that most often causes physiologic difficulties, whereas linear acceleration and deceleration are the major force problems faced by the occupants of space vehicles.

The force (*F*) developed by an angular acceleration is directly proportional to the mass (*M*) of the object (or person) and the square of the velocity (*V*²), but inversely related to the radius of the turn (*r*). This relationship is expressed by the equation:

$$F = \frac{MV^2}{r}$$

Thus a high velocity coupled to a short turning radius results in the immediate development of extremely high forces on the body. Such acceleratory forces are measured in terms of "G"s (one G is the acceleration produced by the force of gravity

or 980 centimeters per second per second; p. 3). Thus one G is equivalent to the normal force exerted by gravitational pull on the body at sea level. A positive G force is experienced whenever a pilot comes out of a dive, ie, the mass of the body increases momentarily. On the other hand, during an outside loop or similar maneuver, a negative G force is experienced, ie, the mass of the body decreases momentarily.

The principal physiologic effects of a positive G force are upon the cardiovascular system because blood is fluid and the pattern of flow can be shifted readily during exposure to angular acceleratory forces. Thus if a person is standing, a force of 6 G will increase the hydrostatic pressure in the veins of the feet to six times that normally present, or around 540 mm Hg. When in the sitting position, a similar situation obtains so that the pressure developed is well over 300 mm Hg. Essentially, the blood is "centrifuged" to the lower regions of the body under the force of positive G, so that venous return to the heart and blood flow to brain are decreased sharply as the blood becomes pooled in the lower extremities. The decrease in venous return decreases cardiac output so that the systemic arterial pressure falls. Blackout of vision occurs as cerebral blood flow is reduced, and this is followed by unconsciousness when the angular acceleration is in the range between 4 and 6 G.

If, on the other hand, the acceleratory force is applied transversely across the anterior–posterior axis, then the subject can tolerate extremely high G forces, as they are distributed over large areas of the body. Forces of 15 to 25 G applied in this direction cause no serious physiologic dysfunction, even when maintained for many seconds.

Negative G forces of about −4 to −5 G, on the other hand, produce transient hyperemia of the head region, and as the eyes are unprotected, a so-called temporary blindness known as "redout" occurs. Little damage to the cerebral vessels occurs during exposure to negative G forces because the cerebrospinal fluid is centrifuged toward the head together with the blood and thus exerts a protective cushioning effect, so actual rupture of cerebral vessels is minimized.

Interestingly, during application of a negative G force to the body, the systemic arterial pressure at the heart level and above increases markedly, thereby causing strong stimulation of the baroreceptor reflex (p. 672), and marked vagal slowing of the heart occurs temporarily. In fact this effect may be so pronounced that total cardiac inhibition can occur for a few seconds and consciousness may be lost due to inadequate cerebral blood flow.

Various automatic mechanical devices that prevent pooling of the blood have been developed to protect the individual from exposure to the foregoing effects of excessive G forces. For example, the so-called "G-suits" are merely devices that automatically apply positive pressure to the lower extremities and abdomen in order to prevent excessive quantities of blood from being forced to the lower regions of the body. Hence the tendency toward blackout is reduced during the application of positive G forces.

In contrast to the problems encountered with angular acceleration in aviation physiology, space vehicles normally are unable to make abrupt turns. Therefore the G forces applied to the occupants of such vehicles are linear, and these occur primarily upon liftoff and during reentry.

The first-stage rocket engine supplies a thrust that develops an acceleratory force of approximately 10 G; the second-stage booster develops around 8 G; and the third-stage booster develops about 3 G. These maximal forces are sustained for only a few seconds, and since the astronauts are positioned within the vehicle in a semireclining position which is oriented transversely to the main axis of the acceleration, then these forces are well tolerated.

During reentry, on the other hand, the necessary deceleration of a space vehicle must be accomplished over a vastly greater distance than the acceleration on liftoff in order that the enormous G forces developed as the craft slows be reduced to tolerable levels. This deceleratory effect results from the fact that the spacecraft is traveling at an enormous velocity as it approaches the earth, and in accordance with the force equation presented above, the force developed is proportional to the square of the velocity. Consequently the distance required for deceleration from the velocities of space to those required for safe reentry is many thousand-fold greater than that required for liftoff.

Blind flying represents a particular situation in which instruments are essential to the pilot because the otoliths within the labyrinths of the inner ear (p. 382) only function properly when the head is inclined with respect to the vertical axis of the body. Without instruments or visual cues (ie, the horizon), however, the pilot is unable to judge the rate of ascent or descent of the aircraft, as the otoliths also undergo rapid adaptation to a constant stimulus. Therefore no perception of ascent or descent of the aircraft relative to the ground is evident at a constant velocity. Similarly, the semicircular canals perceive acceleration and deceleration for only a few seconds following their stimulation (p. 388). Upon entering a very slow turn while flying blind the pilot will be unaware that he is continuing to turn. A rate-of-turn indicator and an artificial horizon are mandatory for blind flying. The weightlessness encountered by astronauts in outer space also contributes to the necessity for adequate instruments to compensate for the absence of the usual proprioceptive cues.

The low temperatures encountered at high altitudes (eg, −55 C at about 12,000 m) require suitable clothing or heated cabins in the aircraft or space vehicle in order that the physiologic requirements of the body be met. In outer space (ie, hundreds of kilometers above the surface of the earth), the temperature rises to very high levels (ie, around 3,000 C) so that the absorption of radiant energy from the sun as well as the radiation of this absorbed energy away from the vehicle at rates that are compatible with the maintenance of physiologically acceptable temperatures within the spacecraft are engineering problems of major importance.

Although radiation poses no particular problem in conventional aviation, the advent of space missions has made this aspect of travel through outer space a significant physiologic problem. As shown schematically in Figure 16.28, cosmic rays from the sun as well as from outer space are trapped, condensed, and shaped by the earth's magnetic field into two broad bands that encircle the earth, called the Van Allen radiation belts. The innermost belt commences at an altitude of approximately 480 km, and extends outward to about 4,900 km, whereas the outermost belt commences at an altitude of approximately 9,700 km and extends to about 32,000 km. The inner belt ranges some 30° north and south of the equatorial plane, while the outer belt similarly extends through an arc of about 75°.

The radiant energy found primarily in the outer Van Allen radiation belt consists of high-energy electrons; the inner belt is a mixture of high-energy electrons and protons (Figure 1.3, p. 5). The radiant energy level of the inner belt is extremely high so that from a practical standpoint (ie, the weight involved) it is impossible to shield a space vehicle effectively from this radiation. The radiation that the human body would receive in a few hours from orbiting through either of these belts would be lethal within a short time; it is essential that space vehicles orbit below the innermost radiation belt, ie, at an altitude of 300 to 400 km. At these levels, the radiation hazard is slight.

During solar flares, the intensity of the radiation in the outer Van Allen belt rises enormously so that traveling through the belts at such a time would be extremely hazardous to the occupants of a space vehicle. However, this danger can be reduced to some extent by departing from, and returning to, the earth near one of the polar regions, at which points the radiation belts become much attenuated as shown in the figure.

Weightlessness poses an additional problem in space travel. This physical effect occurs simply because the acceleratory forces of gravity exerted upon both the space vehicle and the human body itself by any nearby planet (or other celestial body) are identical: there is no resistance to motion in space (ie, it is a vacuum for practical purposes), and both the vehicle and the body are pulled in the same direction by identical acceleratory forces. The problem of weightlessness is principally one of engineering rather than physiology, because special attention must be given to facilities for eating, drinking, waste disposal, hand holds, and so forth as the lack of gravity in the spacecraft permits any unattached person, substance, or object to float freely within the vehicle.

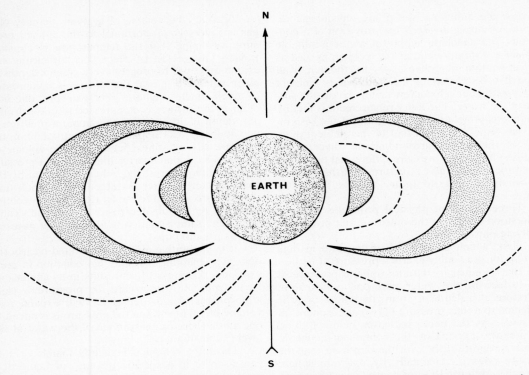

FIG. 16.28.   The Van Allen radiation belts (shaded areas to left and right of the earth's equatorial plane). N = north; S = south. The dashed curving lines represent the magnetic field of earth which shapes the Van Allen belts.

Recently, following a space mission that lasted for approximately two months, it was reported that the erythrocyte count of the astronauts was reduced slightly (about 10 percent), and that some hypotonia of the skeletal muscles, as well as a transient disequilibrium, were observed upon return to the earth's gravitational field. The reduced muscle tone probably was due to adaptive effects of the proprioceptor system in response to a lack of gravitational force acting upon the appropriate sensory receptors. Weightlessness per se, however, has not proven as yet to cause any temporary or permanent effects upon the human body.

Lastly it should be mentioned that survival of human beings in the physiologically hostile environments encountered at high altitudes and in outer space requires the maintenance of a completely artificial environment, insofar as the temperature, oxygen, carbon dioxide, and water vapor levels are concerned. The modern spacecraft usually has a total atmospheric pressure that ranges between 200 and 400 mm Hg, and sufficient nitrogen is introduced into the oxygen to minimize the great potential danger from fire that exists when pure oxygen alone is used. This low ambient pressure within the vehicle is quite adequate to provide adequate oxygen, yet at the same time it minimizes the possibility of explosion within the vehicle itself. The carbon dioxide and water vapor are kept at optimal levels by recycling the gas mixture through suitable absorbants or "scrubbers."

Temperature control within the spacecraft also must be such as to prevent any wide fluctuations that could range from incapacitating to fatal.

Incidentally, the fundamental requirements for an artificial environment in a space capsule that were enumerated above are practically identical with those necessary in a submarine that is to remain submerged for extended periods of time (ie, days, weeks, or even months). In addition, however, some concern has been expressed over, and studies have been under way to evaluate, the effects of elevated carbon monoxide levels found in nuclear submarines that are occasioned by smoking tobacco by a proportion of the crew. Thus it is essential to determine not only the tolerable levels of this gas in such a closed environment, but also any long-term effects this could have on the physical and mental performance of the crew, smokers and nonsmokers alike.

SURVIVAL AT ELEVATED AMBIENT PRESSURES: THE PHYSIOLOGY OF DIVING

Several profound physiologic alterations take place when the human body is subjected to very high alveolar gas pressures, such as those that are encountered when diving or while working in caissons that must be pressurized in order to prevent the walls of these chambers from collapsing due to the external water pressure and to prevent the ingress of water itself.

For the individual the prime physiologic requirement during diving is the provision of a continuous and adequate supply of air or a suitable artificial gas mixture containing oxygen, as well as the prevention of collapse of the thoracic wall when it is subjected to the tremendous ambient pressures that develop a few meters below the surface of the water. The latter requirement is met by continuously metering the pressure of the respiratory gas supply to equal that of the ambient pressure of the water surrounding the diver at all depths. Thereby no pressure differential exists between the intrathoracic structures and the external environment, and no injury results from this cause.

Three important physical principles must be understood clearly at the outset of any discussion of diving physiology:

First, a column of fresh water 10.0 m high (9.76 m for sea water) exerts a pressure at its bottom that is equal to the atmospheric pressure at the surface of the water plus one atmosphere. Therefore at a depth of 10 m, the entire body is subjected to a total pressure of two atmospheres at sea level. As the depth increases below 10.0 m, the pressure exerted by the column of liquid increases proportionally with the depth so that the pressure rises incrementally by one atmosphere for each additional 10 m of depth. These facts are summarized in Table 16.6. Incidentally, this effect of hydrostatic pressure is identical with that discussed earlier in relation to the cardiovascular system (p. 630; see also Pascal's third law, p. 624).

Second, the volume of a given quantity of a gas is inversely proportional to the applied pressure, assuming that the temperature is constant Boyle's law, p. 11). In effect, this law means that as one descends to progressively greater depths, a given volume of gas at the surface is compressed to much smaller volumes. For example, one liter of air at the surface is compressed to one-half liter at a depth of 10 m where the ambient pressure is two atmospheres. Similarly at a depth of 30 m, the pressure is four atmospheres, and the volume of the air now is one-quarter liter, and at a pressure of eight atmospheres, the volume occupied by the gas is one-eighth liter. In other words, each time the pressure applied to the air doubles, the volume occupied by the air decreases by one-half (Table 16.6). The same situation holds for other gases as well as gas mixtures.

It is imperative also to understand clearly the opposite situation, in which the reduction of pressure on a compressed gas results in its expansion. At a depth of 70 m where the pressure is eight atmospheres, 0.125 liter of air will expand to a volume of 1.0 liter when the pressure is reduced to one atmosphere on the surface of the water at sea level (Table 16.6).

Third, the volume of a slightly soluble gas that can dissolve in a specific quantity of liquid at a particular temperature is directly proportional to the partial pressure of that gas (Henry's law, p. 11). This physical principle is of particular significance to diving physiology in relation to the development of dysbarism.

Table 16.6 RELATIONSHIP AMONG DEPTH, PRESSURE, AND THE RELATIVE VOLUME OF A GAS[a]

DEPTH (m)	PRESSURE (atmospheres)	PRESSURE (mm Hg)	VOLUME OF GAS (liters)
0[b]	1	760	1
10	2	1,520	0.5
20	3	2,280	—
30	4	3,040	0.25
40	5	3,800	—
50	6	4,560	—
60	7	5,320	—
70	8	6,080	0.125
80	9	6,840	—
90	10	7,600	—

[a]This table illustrates the effects of depth and pressure in fresh water. Owing to its greater density, the pressure developed by an equivalent column of sea water is slightly greater; ie, a pressure of 2 atmospheres is encountered at a depth of 9.76 m compared to 10.0 m in fresh water. For practical purposes, this slight difference is negligible.

[b]Note that the atmospheric pressure to which the body is subjected at sea level is one atmosphere, and that at a depth of 10 m, this atmospheric pressure is added to that of the column of water so that the net pressure exerted now becomes 2 atmospheres. At high altitudes, the barometric pressure is lower, depending upon the particular altitude, however each 10 m column of water still exerts the same additive effect seen at sea level; ie, the increment of pressure increase is one atmosphere for each 10 m increase in depth, and this is added to the barometric pressure to give the total ambient pressure at any specific depth.

Virtually all of the major physiologic problems encountered in ordinary diving and other operations that involve breathing gases under high pressures are consequences of the three physical laws discussed above. Obviously, hypoxia is not a problem at the high alveolar oxygen pressures encountered in diving as it is during ascent to high altitudes.

One of the major problems faced by a person after exposure to elevated air pressures for any length of time is the necessity for a gradual decompression to remove the excess nitrogen that dissolves in the body fluids under high pressure. When air is breathed under pressure for an extended period of time, the total quantity of nitrogen that dissolves in the body fluids and tissues increases markedly until the body ultimately becomes saturated with this gas after several hours. This fact is in accordance with Henry's law. Oxygen is metabolized, and carbon dioxide diffuses rapidly; these gases do not pose such a problem as does nitrogen.

Thus the total volume of nitrogen that goes into solution in the body during a given dive depends upon two factors: first, the *pressure* under which the gas is breathed, ie, the depth of the dive, and second, the *time* the diver remains at that depth. Since nitrogen is not metabolized, the excess nitrogen that becomes dissolved in the body fluids and tissues during a dive must be eliminated gradually from the lungs as decompression takes place; otherwise serious consequences will occur as discussed below.

The human body at sea level contains almost precisely one liter of dissolved nitrogen. The body fluids contain slightly less than one-half of this volume, while the remainder is found predominantly in the fat depots because nitrogen is about five times more soluble in lipid than it is in water. Once the body of a diver has become completely saturated with nitrogen, the total volume of this gas that is present in his body, calculated to sea-level pressure (ie, one atmosphere), depends upon the depth at which saturation took place as shown in Table 16.7. It is important to realize, however, that a considerable span of time, usually on the order of several hours, is required for nitrogen to reach equilibrium between the alveoli and the body fluids and tissues at any given pressure, merely because nitrogen does not diffuse rapidly among the body compartments. Thus the nitrogen content of the body fluids in general reaches equilibrium in about an hour, although the depot and other lipids tend to reach the saturation point much later owing to their relatively poor blood supply. However, the greater overall capacity of the body fats for dissolving the gas results ultimately in a greater total nitrogen concentration. From a practical standpoint, if a person remains at great depths for only a few minutes, then correspondingly less nitrogen becomes dissolved in the body fluids and tissues, so that his total decompression time is shortened considerably.

Table 16.7   RELATIONSHIP AMONG DEPTH, PRESSURE, AND NITROGEN CONTENT OF THE BODY[a]

DEPTH (m)	PRESSURE (atmospheres)	TOTAL VOLUME OF DISSOLVED NITROGEN[b] (liters)
0	1	1
10	2	2
30	4	4
60	7	7

[a]*Assuming complete saturation of the body in accordance with Henry's law.*
[b]*These volumes are calculated to sea level pressure, ie, one atmosphere, or 760 mm Hg.*

If the person ascends rapidly to the surface following a relatively deep and prolonged dive the excess gases, in particular nitrogen, that have dissolved in the body fluids and tissues under pressure will come out of solution, and thus form bubbles of free gas within various regions of the body. This effect causes an abnormal condition known variously as *dysbarism, decompression sickness, the bends,** or *caisson disease.* Dysbarism thus is caused by a large gradient between the ambient pressure on the external surface of the body compared to the combined pressures of all of the dissolved gases within the body that develop upon rapid ascent from a long dive. Exercise tends to accelerate bubble formation in the body, in a manner similar to that encountered upon shaking a bottle of a carbonated beverage prior to opening it.

The pathologic consequences of dysbarism range from a very mild tingling in the arms and/or legs to coma and death, depending upon the total number, volume, and location of the bubbles formed in the body. Common symptoms of dysbarism include mild to extreme pain in the shoulders, arms, and/or legs, dizziness, paralysis, dyspnea, fatigue, collapse, and coma. The major neurologic symptoms of dysbarism are believed to be referable to bubble formation in various particular regions of the nervous system, both peripheral and central. Infrequently, bubble formation in the central nervous system will lead to permanent paralysis or psychic dysfunctions. Dysbarism generally is accompanied by severe pain, especially in the joints of the long bones, ie, the arms and legs, although this symptom may be diffuse rather than localized.

In addition to the neurogenic symptoms of dysbarism, gas bubbles also may lodge in the pul-

Some years ago men worked for many hours each day in caissons under extremely high barometric pressures, but without adequate periods of decompression. The crippling disease that resulted caused the victims to arch their backs in an attempt to alleviate the pain, hence the origin of the term "the bends."

monary capillaries and cause a severe form of dyspnea known as the "chokes." Pulmonary edema and death often follow this manifestation of dysbarism.

The normal rate of nitrogen excretion from the body via the lungs is depicted graphically in Figure 16.29. The body of the subject was saturated with nitrogen at a depth of 10 m (two atmospheres pressure), and then he was decompressed suddenly to one atmosphere pressure. Although some 60 percent of the excess nitrogen was eliminated in about two hours, note that some excess nitrogen still is present many hours later.

The rate at which a diver can ascend safely to the surface depends upon two principal factors: first, the depth to which the descent was made; and second, the total length of time that was spent at depth. Decompression tables have been prepared (eg, by the US Navy) that give the total decompression times at various depths during the ascent that are necessary to avoid dysbarism. These tables also include a generous margin of safety. Similar decompression tables also have been prepared for divers working at high altitudes in lakes and reservoirs. It is important to remember that if one carries out several deep dives during a single 24-hour period that the effects of these individual dives are cumulative so that the total decompression time must be adjusted accordingly.

Rather than undergo a slow, stepwise ascent from depth, the diver may ascend rapidly and then be placed immediately in a decompression chamber. Essentially this device is a large hermetically sealed tank within which the ambient pressure can be regulated not only over a broad range, but with considerable precision. Once in the chamber, the diver is recompressed to a pressure that is equivalent to the maximum depth of his dive, and then decompressed slowly in accordance with the standard tables. Similarly, a victim of the bends can be recompressed until he is at a pressure equal to, or even greater than, that at which his symptoms disappear, and then he is decompressed gradually as described above. In both of these instances, of course, the effect of recompression is merely to ensure that no bubbles of free gas are present in the body. The subsequent gradual decompression period is designed to eliminate the excess nitrogen at a physiologically safe rate.

One advantage to having divers live in an underwater habitat (eg, the Sealab experiments) for periods of days, or even weeks, is the fact that only one decompression is necessary at the end of their sojourn at depth. The divers remain continuously below the surface, and are able to work daily under such conditions for relatively long periods of time. As their bodies become totally saturated with nitrogen under such conditions, an extended period of decompression in a chamber is critical upon their return to the surface. This decompression may require a week, or even longer, to be accomplished successfully. It has been proven by experiments of this type that the human body can tolerate such an environment and function well without deleterious aftereffects.

The effects of working under artificially high gas pressures in a caisson, and the decompression requirements that are necessary to avoid dysbarism, are identical with those for diving.

A diver breathing compressed air normally is exposed to three gases at elevated pressures, namely oxygen, nitrogen, and carbon dioxide. Each of these gases has certain peculiar effects when respired at extremely high pressures.

When the pressure of oxygen in the respired air is extremely high, for example three atmospheres (2,280 mm Hg), certain definitely toxic effects of this gas rapidly become evident. The observed symptoms of acute oxygen toxicity (or oxygen poisoning) are dizziness, irritability, nausea, muscular twitches, visual disturbances, numbness, epileptiform convulsions, and coma. The convulsions often take place without warning and with rapidly fatal consequences. Exercise greatly enhances the tendency of a diver to develop oxygen toxicity, and there is great variability in the tolerance to an elevated P_{O_2} level in the same person from day to day, as well as among different individuals. For the reasons cited above, it is extremely dangerous to utilize pure oxygen alone while diving at depths below 6.7 m for even short periods of time. As shown in Figure 16.30, there is an inverse relationship between the time a person develops the symptoms of oxygen toxicity and the depth while breathing pure oxygen.

The underlying biologic mechanism (or mechanisms) responsible for acute oxygen toxicity is unknown, although several possible effects have been suggested. For example, severe oxygen poisoning is known to inactivate certain oxidative enzymes; it has been postulated that this inactivation results in a decreased synthesis of adequate

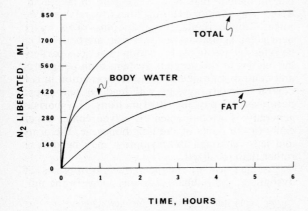

FIG. 16.29. Rates of nitrogen liberation from the human body following saturation of the body fluids and tissues by breathing air at a depth of 10 meters. (Data from Guyton. *Textbook of Medical Physiology*, 4th ed, 1971, Saunders.)

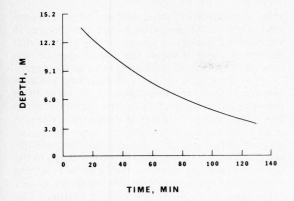

FIG. 16.30. Oxygen tolerance curve illustrating the relationship between various depths and the length of time that a person can remain safely while breathing pure oxygen. (Data from Guyton. *Textbook of Medical Physiology,* 4th ed, 1971, Saunders.)

high energy phosphate bonds. Alternatively, a decreased cerebral blood flow due to a high arterial P_{O_2} and the resulting vasoconstriction might deplete the supply of essential nutrients and/or increase the level of metabolic waste products in the brain, thereby inducing convulsions. Lastly, high tissue oxygen levels are known to produce high concentrations of free radicals that possibly could oxidize many intracellular enzymes and other components, thereby leading to a massive derangement of normal metabolic functions within neurons and other cells.

When pure oxygen is respired at one atmosphere pressure, no symptoms of acute toxicity such as those described above are seen. However, certain chronic effects develop, including pulmonary edema. This condition results from a disruption of the alveolar and bronchial epithelium, presumably by oxidation of essential cellular components due to their intimate exposure to the high alveolar P_{O_2}.

Nitrogen exerts no demonstrable physiologic effects upon the body at the normal partial pressure of this gas that is encountered at sea level (608 mm Hg). When a diver breathes nitrogen under high pressures (ie, in compressed air) for periods of an hour or more, a condition known as *nitrogen narcosis* may develop as the quantity of nitrogen dissolved in the body fluids increases progressively with time and depth. The early symptoms of nitrogen narcosis resemble those of mild alcoholic intoxication; consequently this condition has been dubbed the "rapture of the deep" or the "uglies" by experienced divers.

Mild symptoms of nitrogen narcosis often appear at depths ranging between 40 and 45 meters. At this stage, the diver becomes euphoric and tends to lose any sense of personal responsibility and judgment insofar as his immediate environment is concerned. The diver who is using a self-contained underwater breathing apparatus (SCUBA gear) may offer his mouthpiece to a passing fish, attempt to sing, or to perform other irrational acts. At depths between 45 and 60 meters, the diver becomes relaxed and drowsy; consequently, he may lose his mouthpiece and drown. Between depths of 60 and 75 meters, muscular incoordination develops and physical strength declines; therefore the diver is unable to perform any useful work, eg, swim toward the surface. At depths below 90 meters the diver becomes physically incapacitated, and coma generally ensues at depths between 105 and 120 meters. The inherent dangers of nitrogen narcosis are evident from the foregoing statements.

Presumably, nitrogen exerts a progressive narcotic effect by the same mechanism that makes gas anesthetics effective; that is, since nitrogen is freely soluble in the body fluids, it can enter the lipids of the membranes and other structural elements of the neurons in a sufficient quantity under elevated pressures to reduce their normal irritability to such an extent that neural functions are severely deranged.

The high pressures encountered in deep diving do not, by themselves, alter the carbon dioxide partial pressure within the alveoli, and provided that the diver respires normally, the alveolar P_{CO_2} remains within normal limits as the rates of production and excretion of this gas are equal when the body is submerged. These statements assume that the equipment being used for diving is well designed and is functioning properly. However, certain types of diving apparatus, notably helmets or rebreathing gear (in which pure oxygen is used and recycled after the carbon dioxide is removed by an absorbent), often permits the carbon dioxide concentration to rise to extremely high levels. When the carbon dioxide level increases up to around a 10 percent concentration in the inspired air, the diver compensates by a 6-fold to 10-fold increase in his minute respiratory volume (p. 737). But at carbon dioxide concentrations above 10 percent, an actual depression of the respiratory center occurs; therefore the hyperventilatory respiratory compensation fails. Various degrees of lethargy and narcosis develop; ultimately, anesthesia sets in as the result of acute carbon dioxide toxicity.

During prolonged dives to extreme depths, helium often is used to replace nitrogen in artificial gas mixtures, because this inert gas has four properties that are superior to those of nitrogen when respired at high pressures. First, helium exhibits practically no narcotic effect to depths around 150 meters, and even then it is only about 20 percent as toxic as nitrogen. Second, as helium has an extremely low density owing to its low atomic weight (ie, 2 compared to 14 for N_2), the airway resistance of the diver is reduced considerably (see also discussion below). Third, helium exhibits a 60 percent lower solubility in the body fluids than does nitrogen under comparable conditions of pressure and time of exposure so that the potential number of bubbles that can form during decompression is reduced markedly. Fourth, helium diffuses far more rapidly through the body tissues than does nitrogen, a fact that also is due to its low atomic weight. Thus helium is excreted far more rapidly from the body than is nitrogen. The rapid diffusion of helium combined with the low solubility of this gas in the body fluids markedly reduce the total decompression time necessary following a particular dive.

The low narcotic effect, reduced airway resistance, and reduced decompression time all are distinct advantages to using helium, but this gas exhibits two additional properties that render its usefulness limited to certain types of diving. Helium starts to form bubbles within the body fluids when the ratio of the pressures between the inside and the outside of the body is only 1.7, whereas this ratio for nitrogen is 3.0. A diver must be decompressed in many short stages, rather than with a few longer stages such as are necessary when using nitrogen. Since helium diffuses far more rapidly than nitrogen, relatively more helium than nitrogen becomes dissolved in the body fluids in a much shorter period of time. This makes nitrogen the inert gas of choice for short dives to moderate depths (eg, to 60 m for 10 min), and helium the preferred gas for prolonged dives to more extreme depths (eg, to 75 m for 2 hours). In this regard, it should be mentioned that helium accelerates the rate of onset of convulsions that are produced by acute oxygen toxicity. However, the partial pressure of oxygen in the respired gas mixture must exceed a certain threshold value for convulsions to appear at all, and helium merely potentiates this effect of oxygen. The inclusion of a small quantity of nitrous oxide in the gas mixture, on the other hand, tends to prevent the onset of convulsions due to oxygen toxicity so that practical gas mixtures for working at considerable depths can be designed to take advantage of the facts presented above.

It is essential that any compressed gas used to sustain life while diving be of the utmost purity. Even minute traces of carbon monoxide in the gas mixture could rapidly prove incapacitating or even fatal, because of the extremely high affinity of hemoglobin for this compound (p. 749). This fact is especially pertinent at the elevated gas pressures encountered while diving. Since gasoline-driven compressors commonly are used to supply air to helmet divers as well as to charge SCUBA tanks, then it is imperative that the air intake on the compressor be situated at a safe distance and upwind from the engine exhaust in order that no contamination of the compressed air by carbon monoxide occur. Furthermore, the compressor itself must operate in a water seal rather than in oil, as even small traces of oil in the compressed air supply could seriously derange the delicate alveolar membranes themselves and thereby interfere with pulmonary gas exchange.

There are a number of additional physiologic problems that may be encountered in diving other than the effects of elevated gas pressures discussed above. These problems are the changes in the physical density of the air or gas mixture that are encountered at depth; the effects of rapid descent and ascent; and the temperature of the water.

The density or viscosity of a gas or a gas mixture increases in direct proportion to the ambient pressure. At a depth of 30 m, the density of air is four times greater than it is at the surface, whereas at 60 m it is seven times greater. Since the resistance to movement of a gas mixture through the respiratory tract is also directly proportional to the density of that gas mixture, the work of breathing is enhanced considerably by breathing air or any other gas mixture at increasing depths. The maximum breathing capacity (ie, the total volume of gas moved per minute) is reduced as the density of the gas mixture increases. For example, at a depth of 30 m, the maximum breathing capacity is reduced to 50 percent of normal when breathing compressed air, whereas if an oxygen–helium mixture is used, this value is reduced to only 85 percent.

If a diver descends too rapidly, the volumes of all of the free gases within the body are reduced markedly and suddenly because of the increased ambient pressure that is applied to the entire body surface. Normally, as for example during SCUBA diving, the air that is supplied to the lungs from the tank via a demand regulator is at the same pressure as that of the surrounding water for any given depth; no pressure differential exists and no damage occurs. If, on the other hand, a person descends rapidly but without inhaling additional air, at a depth around 30 m the chest wall gradually becomes crushed owing to the increased ambient pressure; ie, 4 atmospheres at 30 m. Since the smallest lung volume that can be achieved without suffering major structural damage is about 1.5 liters, the effect of suddenly increased external pressure is known as "the squeeze." Similarly, air that is entrapped within the middle ear or sinuses is also compressed during a rapid descent, even to shallow depths. Not only does excruciating pain result, but the eardrums can rupture easily if swallowing is not employed to equalize the pressures. If a SCUBA diver does not exhale consciously into his mask as he descends, the pressure differential that develops between the air trapped within the mask and the surrounding water can result in a localized squeeze on the facial as well as ocular regions, and severe localized hemorrhages can result. On the other hand, if a "hard hat" diver suddenly loses control of the pressure within his helmet, as sometimes happens, his entire body can be forced up and into his helmet by the ambient water pressure.

During a rapid ascent from depth the precisely opposite pulmonary effects from rapid descent occur if the person does not breathe normally or exhale continuously as he rises toward the surface. Thus a spasm of the glottis, such as occurs during panic, will entrap gas within the lungs, and this gas expands during the ascent as the ambient pressure decreases (see Table 16.6). After the lungs have expanded maximally, however, the intrathoracic pressure continues to rise so that when the intra-alveolar pressure reaches between 80 and 100 mm Hg, air is forced into the pulmonary capillaries from the alveoli. This situation, known as aero-embolism,* is extremely dangerous and frequently it is lethal. In addition, the greatly elevated intra-alveolar pressure can rupture the surface of the lungs producing pneumothorax (p. 717). Rapid expansion of any gas that may have accumulated within the gastrointestinal tract during a long dive can produce serious damage to this system during a rapid ascent, although such an effect is rare.

Rapid ascent from considerable depths is a serious problem for persons attempting to escape from a disabled submarine without any diving apparatus. Such a situation also obtains when a diver is forced to rise to the surface rapidly when his suit "balloons" due to malfunction or other cause, or when a SCUBA diver exhausts his air supply at depth. In these instances aeroembolism is the most serious immediate problem. As the person rises, he must exhale continuously until the surface is reached. This is possible, even without breathing, as the physical expansion of the gas that entered the lungs during the last inhalation is sufficient to enable the person to surface. As the carbon dioxide that is produced during such an emergency ascent is eliminated by the continuous exhalation, the respiratory drive to inhale is reduced somewhat and the individual can hold his breath for a considerably longer time than usual.

In this regard, experienced skin divers (ie, persons diving without an air supply) often hyperventilate voluntarily for a minute or so prior to a dive. Such hyperventilation does not increase the total body oxygen content appreciably, but this maneuver does lower the blood P_{CO_2} somewhat so that the respiratory drive is inhibited temporarily. A "free dive" of considerably longer duration can be made than would be possible otherwise.

Diving in cold water carries with it a major stress, insofar as heat loss from the body is concerned. Since the body is immersed in a conducting medium, the major route of heat loss is by conduction (p. 1006); convection currents (p. 1006) that are caused by movements of the water close to the body surface also play a minor role in this heat loss. Usually a diver will wear some form of protective suit that cuts down this heat loss due to its insulating properties. The suit may be wet or dry, artificially heated or not, depending upon the thermal stress to which the body will be subjected under the particular conditions of the dive.

Usually the so-called "wet suit," made of sponge rubber, is used by SCUBA divers over a broad range of water temperatures throughout the world. This type of suit depends primarily upon the tiny bubbles of air that are entrapped within the rubber itself for its insulating properties. Secondarily, a thin layer of water between the skin and inner surface of the suit that is warmed by the body heat itself affords some additional insulating properties. But it must be remembered that the air bubbles contained within the material of the suit itself that provide the insulation are subject to the same gas laws as those described earlier. As one descends to greater depths, these air bubbles tend to collapse owing to the greater ambient pressures, and the effective insulating properties are decreased most at those depths where maximal insulation would be needed. For this reason, the possibility of a rapid and incapacitating heat loss from the body during a deep dive must not be ignored.

In conclusion to this discussion, it is appropriate to mention that every activity carried out by man in or around an aquatic environment carries with it the risk of drowning. In many instances of drowning (roughly 30 percent), the person does not inhale water, so that the lungs remain "dry," and the person dies of asphyxiation (p. 764). This is the case because water entering the trachea elicits a powerful reflex laryngospasm that apposes the vocal cords, thereby sealing off the lungs from the water. If, however, the drowning person inhales sea water, the osmotic effect of the salt is such that water is withdrawn from the pulmonary capillaries into the lungs, hemoconcentration ensues, and asphyxiation takes place in about eight minutes. In marked contrast to sea water, the inhalation of fresh water produces quite different physiologic effects. In this instance, the osmotic pressure of the water is far greater than that of the blood; consequently large volumes of water rapidly enter the blood of the pulmonary capillaries across the alveolar membrane. The ensuing massive hemodilution in turn causes hemolysis of large numbers of erythrocytes, so that the potassium content of these cells is added to the already-diluted plasma electrolytes. These factors, combined with the anoxia that quickly develops, result in cardiac fibrillation within about two minutes following the inhalation of fresh water.

THE PATHOPHYSIOLOGY OF SHOCK

A study of the shock syndrome provides an excellent review of the complex and interlocking mechanisms that underlie the normal homeostatic regulation of cardiovascular functions. Clinically, the term "shock" may be defined as an emergency state that is characterized by insufficient tissue perfusion due to a physiologic insult, and which results in a partial or complete depression of the normal homeostatic mechanisms that govern cardiac output. Generally the shock syndrome is defined in terms of the clinical symptoms that are produced following the physiologic insult, which include a weak or absent brachial arterial blood pressure, a rapid but weak pulse, oliguria, hyperpnea and tachypnea, pallor, cyanosis, restlessness,

*Aeroembolism must not be confused with the bends or dysbarism.

anxiety or mental dullness, nausea and/or vomiting, and a low body temperature as indicated by cool or clammy extremities. All of these symptoms can be signs of shock.

The several factors that can precipitate a shock state are summarized in Table 16.8, but despite certain obvious differences in the different shock "entities," the one common physiologic denominator that apparently is present in all cases is inadequate tissue perfusion; ie, a relative tissue ischemia that results from a decreased cardiac output (Fig. 16.31). Certain features of each type of shock now may be discussed individually.

One major cause of shock is a reduced circulating blood volume that can be produced by any of several types of injury (Table 16.8.1). This type of shock often is referred to as *hypovolemic shock* because of the pronounced reduction in the circulating blood volume. Hypovolemic shock frequently is divided into a number of subclasses depending upon the etiology of the syndrome. The designations hemorrhagic shock, wound shock, surgical shock, traumatic shock, and burn shock are useful because there are important differences in these clinical states, despite certain fundamental similarities among them. Hemorrhagic shock results directly from the loss of large volumes of whole blood, and will be considered in detail below as an illustration of the general physiologic alterations that occur in most shock states. Wound and surgical shock actually are caused by a combination of external blood loss, internal hemorrhage, and dehydration.

In traumatic shock, there is extensive damage to muscle and bone tissue so that massive hemorrhage into the injured regions can precipitate the shock state. This type of shock is often seen in victims of automobile accidents as well as in battle casualties. Myoglobin is released into the circulatory system when there is extensive muscle damage, and the precipitation of this pigment in the renal tubules can induce various degrees of kidney damage. The combination of traumatic shock and myoglobinuria sometimes is referred to as the crush syndrome, although myoglobin is only one of the factors present in shock that can produce renal insufficiency. In all forms of shock, renal failure is a possible complication.

Enormous volumes of plasma can be lost through the injured regions of the skin in burn

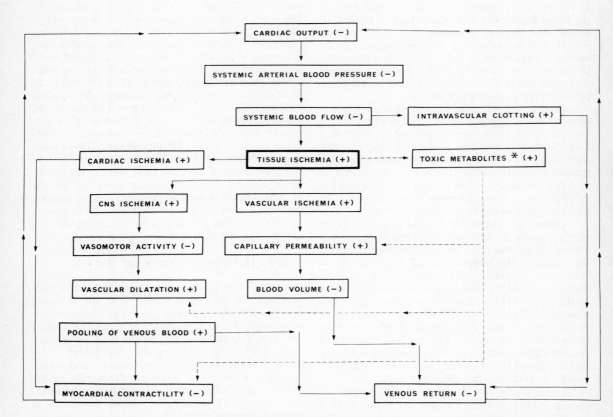

FIG. 16.31. The interlocking positive feedback mechanisms that are operative in the progressive development of shock states. Tissue ischemia (heavy box) appears to be a common factor in the development of all types of shock. (+) = increase or acceleration; (−) = decrease or inhibition. The accumulation of toxic metabolites produced by the ischemia (*) can include lactic acid and histamine as well as bacterial toxins. The effects of such agents on the genesis of shock are emphasized by dashed lines.

shock. In this situation, plasma rather than whole blood is lost in the exudate so that hemoconcentration may be pronounced. This effect is indicated by an increased hematocrit. Complex metabolic changes also develop in burn shock. Possibly these changes are related, at least in part, to the markedly reduced hepatic blood flow that is seen almost immediately following a severe burn. Thus a 50 percent rise in the basal metabolic rate follows the burn, and in some cases hemolytic anemia develops.

When myocardial contractility per se is impaired for some reason, there is a reduced cardiac output even though the circulating blood volume may be normal *(normovolemic shock)*. Hence the clinical state that develops as the result of severe cardiac malfunction is called cardiogenic or congestive shock (Table 16.8.2). The symptoms of normovolemic shock are similar to those for hypovolemic shock, but in addition a stasis of blood develops, particularly in the pulmonary and visceral regions, owing to the inability of the heart to pump out all of the blood that is returned to it. Hence a markedly increased venous pressure is a prominent feature of cardiogenic shock.

When there is extensive damage to the myocardium of the left ventricle, as through a massive thrombosis, cardiogenic shock is prominent. This condition occurs in approximately 10 percent of patients following a coronary thrombosis with severe myocardial infarction. Possibly chemical substances are liberated from the infarcted region; such substances could trigger ventricular receptors that elicit inhibition of the vasomotor center Bezold–Jarisch reflex, p. 677). The compensatory vasoconstrictor reaction to shock would be diminished and the shock state would become worsened.

Neurogenic shock, also called neurogenic hypotension, or low resistance shock, includes several conditions in which the blood volume is normal (ie, the patient is normovolemic); however the overall capacity of the vascular bed is increased greatly by a marked vasodilatation (Table 16.8.3). Fainting as a response to strong emotion, extreme fear, or grief can lead to neurogenic shock through a vasovagal reflex. The stunned reaction in severely injured persons that occurs immediately following the injury is a poorly understood shock reaction that is included in this general category of neurogenic shock.

Anaphylactic shock develops rapidly and is the consequence of an antigen–antibody reaction in a sensitized individual that leads to the release of large amounts of histamine into the bloodstream. This compound in turn causes a significant increase in the capillary permeability as well as a diffuse arteriolar and capillary dilatation throughout the body.

Shock can become a serious complicating factor during the course of various metabolic and infectious diseases, for example in diabetes mellitus (Table 16.8.2E). A decreased peripheral vascular resistance plus some hypovolemia both contribute to the etiology of shock under such conditions. Thus, adrenal insufficiency (p. 1108), diabetic ketoacidosis (p. 1077), and severe diarrhea (p. 905) all result in a serious decrease in the blood sodium ion concentration, despite the fact that the underlying mechanisms involved are completely different in each of these situations. There is a marked decrease in the circulating plasma volume, and this may be of sufficient magnitude to induce cardiovascular collapse. In certain infectious diseases, a large volume of plasma may be lost to the interstitial spaces throughout the body due to a diffuse increase in the vascular permeability. Lastly, certain microorganisms (eg, gram-negative bacteria) release chemical substances known collectively as endotoxins. The endotoxins in turn can produce dehydration or reduce the tone of vascular smooth muscle by paralysis of these contractile elements. The net result of all of the aforementioned processes, combined with the fact that cutaneous vasodilatation frequently is present in the febrile individual, tend to augment the imbalance between the volume of circulating blood and the net volume capacity of the cardiovascular system itself.

It can be seen from the brief résumé of facts cited above and summarized in Table 16.8 that the etiology of the shock state is extremely complex.

The interrelated physiologic mechanisms that are concerned with the progressive development of the shock state are summarized in Figure 16.31. Since tissue ischemia is the underlying feature that apparently is shared in common by all types of shock, this factor has been emphasized in the diagram. However, it is imperative to understand that the stimulus or injury that initiates the interlocking, progressive, and cyclic changes that take place in shock states can affect practically any one, or more than one, factor involved in this process, and tissue ischemia usually is the result, rather than the cause, of shock. A clear grasp of this concept will enable the reader to visualize from this diagram how the several etiologic factors that can induce shock will lead to a series of self-perpetuating positive feedback cycles.

If the compensatory mechanisms that are set into play following the initial physiologic insult are adequate, then the shock state is reversible, and the patient survives. If, however, these compensatory mechanisms, combined with any therapeutic measures that are applied, are inadequate then the shock state passes into an irreversible stage from which recovery is impossible.

The progressive cardiovascular and related alterations that take place in hemorrhagic shock now may be considered in detail, and will serve as a model for the events that occur during shock states in general. The immediate physiologic responses and adaptations to blood loss will be considered first (Table 16.9).

The reduction in circulating blood volume that accompanies a severe hemorrhage causes a decline

Table 16.8 FACTORS IN THE ETIOLOGY OF SHOCK[a]

TYPE OF SHOCK	MAJOR PHYSIOLOGIC AND CLINICAL ALTERATIONS
1. Reduced Circulating Blood Volume (Hypovolemic shock)	
A. Blood loss	Decreased circulating blood volume leads to a reduced venous return which leads secondarily to a reduced cardiac output. Caused by external or internal hemorrhage. Results in systemic hypotension, rapid but weak pulse, intense thirst, clammy skin, rapid shallow respiration, alterations in mental state (eg, restlessness, apathy, coma).
B. Plasma loss	As above. Caused by massive third degree burns, extensive dermatitis, intraabdominal fluid accumulation.
C. Protein-free fluid and electrolyte loss	As above. Caused by copious vomiting, diarrhea, polyuria.
2. Reduced Cardiac Output (Cardiogenic Shock)	
A. Pericardial tamponade	Rigid, fluid-filled pericardium that impairs venous return to right and left hearts. An elevated venous pressure results.
B. Myocardial disease	Myocardium damaged (eg, by thrombosis) so that the heart is unable to maintain an adequate output. Symptoms may include a gallop rhythm, abnormal heart sounds, EKG changes, bilateral ventricular failure. Chest pain is a frequent accompaniment of this condition.
C. Valvular disease; eg, thrombus in valve, rupture of aortic cusp	Obstruction to forward blood flow may be present or regurgitation of blood in a retrograde direction takes place. Cardiac murmurs are present; the patient may have a previous history of endocarditis.
D. Pulmonary embolism	Obstruction in pulmonary artery impairs venous return to left atrium and left ventricular output falls; an imbalance between pulmonary ventilation and perfusion develops. Symptoms include tachypnea, cyanosis, venous distension.
E. Metabolic derangements, eg, diabetic acidosis	Decreased myocardial contractility develops secondary to acidosis or hypoxia. Symptoms include Kussmaul respiration, cyanosis. Signs of chronic lung disease may be present.
3. Neurogenic Hypotension (Low-Resistance Shock; Normovolemic Shock)	
A. Anaphylaxis	Decreased systemic arteriolar resistance develops secondary to an antigen–antibody reaction. Rapid onset, no signs of fluid loss or heart disease, history of antigen administration, wheezing frequently is observed.
B. Vagal reflex	Bradycardia and decreased peripheral arteriolar resistance mediated by vagus nerve in response to surgical or other instrumental manipulation, visceral pain, or strong emotional states. Results in bradycardia with nausea, sweating, and occasionally syncope.
C. Pharmacologic (iatrogenic)	Reduced peripheral arteriolar resistance caused by blockade of the autonomic nervous system. Hypotension develops that frequently is not accompanied by characteristic peripheral signs of shock due to drug inhibition of sympathetic nervous system.
D. Overdose of sedative	Reduced venous return. Coma or stupor are present.

Table 16.8 (Con't.)   FACTORS IN THE ETIOLOGY OF SHOCK[a]

TYPE OF SHOCK	MAJOR PHYSIOLOGIC AND CLINICAL ALTERATIONS
4. Combinations of Above	
A. Perforated viscus	Decreased circulating blood volume combined with a reduced neurogenic vasomotor tone. Symptoms include a rigid abdomen with rebound tenderness.
B. Septicemia	Pulmonary venoarterial admixture, reduced peripheral oxygen extraction, cardiac output and peripheral vascular resistance variable. Clinical picture of bacterial blood poisoning. Results in an altered mental state. Frequently shaking chills and fever are present.

[a]*Data from Harvey, Johns, Owens, and Ross (eds). The Principles and Practice of Medicine, 18th ed, 1972, Appleton-Century-Crofts.*

in venous return; therefore the cardiac output falls rapidly. This reduced cardiac output leads in turn to a decline in the systemic arterial blood pressure as well as the systemic blood flow, so that a generalized tissue ischemia develops. This ischemia now leads to other progressive cardiovascular alterations, as depicted in Figure 16.31. The magnitude of these alterations is dependent upon the volume of the blood loss.

One particularly significant effect of the generalized tissue ischemia is that the reduced tissue perfusion leads to an elevated rate of anaerobic glycolysis; hence lactic acid production throughout the body increases markedly. The resulting lactic acidemia directly exerts a depressant effect upon myocardial contractility, reduces the vascular response to catecholamines, and may even produce coma. These general effects are emphasized by dashed lines in the figure.

From the standpoint of cardiovascular homeo-

Table 16.9   SUMMARY OF THE COMPENSATORY REACTIONS TO HEMORRHAGIC SHOCK

IMMEDIATE (SHORT-TERM) COMPENSATION	PRINCIPAL PHYSIOLOGIC EFFECT
Tachycardia	Increase cardiac output
Vasoconstriction	Increase systemic arterial blood pressure
Venoconstriction	Increase venous return to right atrium
Increased ventilation	Increase venous return to right atrium
Increased muscular contractions (Only seen in some patients)	Increase venous return to right atrium

INTERMEDIATE AND LONG-TERM COMPENSATION	
Transudation of interstitial fluid into capillaries	Restores circulating blood volume
Extravascular albumin mobilized	Helps to restore circulating blood volume
Endocrine factors. Increased secretion of:	
Catecholamines	Augments peripheral vasoconstriction and tachycardia
Vasopressin	Negligible pressor effect, if any
Glucocorticoids	Promotes retention of Na^+ thereby contributes to restoration of circulating blood volume
Aldosterone	Same as above
Renin	Pressor effect negligible if any.
Increased plasma protein synthesis	Restores plasma osmotic pressure to normal
Increased formation of erythropoietin	Stimulates erythrocyte production so that normal red blood cell mass is restored eventually in response to tissue hypoxia

stasis, any sudden reduction in the circulating blood volume represents a physiologic emergency; hence several compensatory reactions are elicited immediately (Table 16.9). The fall in systemic arterial blood pressure that develops following a massive hemorrhage decreases the tension of the arterial baroreceptors so that their tonic inhibition of the medullary vasomotor and cardioregulatory centers decreases (p. 668). This results in a reflex tachycardia as well as a generalized vasoconstriction; both of these responses tend to restore the cardiac output and blood pressure toward normal values once again.

With slight to moderate hemorrhage, the pulse pressure (p. 622) is reduced, although the mean arterial blood pressure may remain at normal levels. Nevertheless, the decline in pulse pressure alone that attends a moderate hemorrhage is sufficient to stimulate the reflex tachycardia and vasoconstriction discussed above.

The systemic arteriolar vasoconstriction induced by hemorrhage affects primarily the kidneys, viscera, and skin. Cyanosis may become evident in the latter organ due to the more sluggish peripheral blood flow. The cutaneous vasoconstriction also accounts for the pallor and coolness of the extremities observed in this type of shock. In contrast to these vasoconstrictor responses, the coronary vessels dilate, and cerebral blood pressure and flow also remain practically unchanged in the shock state.

Following hemorrhage both the afferent as well as the efferent renal arterioles constrict, although the latter vessels contract more strongly. The renal plasma flow (p. 773) is decreased more than the glomerular filtration rate (p. 783), and therefore the filtration fraction increases. That is, the GFR/RPF ratio becomes elevated (p. 786). Some blood may be shunted through the medullary regions of the kidneys; a larger proportion of the blood flow to these organs detours the cortical glomeruli (Fig. 12.5, p. 774).

In shock, the rate of urine formation falls to extremely low levels. Sodium retention becomes pronounced and nitrogenous metabolic products accumulate in the blood giving rise to uremia. If the hemorrhagic hypotension persists for any length of time, profound and irreversible renal tubular damage may develop.

A marked and widespread venoconstriction also accompanies hemorrhagic shock, and this homeostatic response tends to augment venous return to the heart.

The vasoconstriction that takes place in the splanchnic regions following hemorrhage shunts a large volume of blood from the viscera into the systemic circulation, and thereby the circulating blood volume is elevated to some extent. This effect is enhanced by a similar shunting of blood from the pulmonary and cutaneous regions of the circulation. The spleen has a negligible role as a blood reservoir in man as the volume of blood that is provided by contraction of this organ is quite small in humans (p. 692).

Hemorrhage also serves as a powerful stimulus to secretion of the adrenal medullary hormones (p. 1119). The augmented discharge of adrenergic sympathetic neurons also elevates the circulating catecholamine level to some extent (p. 1123), but this response presumably contributes to only a limited extent to the generalized peripheral vasoconstriction noted above. The elevated level of circulating catecholamines may stimulate the reticular formation of the brain, and this stimulation in turn may lead to the psychic states observed in hemorrhagic shock, ie, apathy, apprehension, restlessness, or coma.

The decrease in the total circulating mass of erythrocytes that attends hemorrhage lowers the oxygen-carrying capacity of the blood. Concomitantly blood flow in the aortic and carotid bodies declines. The relative anemia (p. 566), the stagnant hypoxia (p. 764), and the acidosis that are present in shock combine to stimulate the aortic and carotid chemoreceptors (p. 674). Presumably this stimulatory effect on the chemoreceptors is the principal mechanism whereby tachypnea occurs in shock (p. 720), and this response causes an increased pumping of venous blood to the heart (p. 652). If the patient is active or restless, rather than apathetic or comatose, the increased muscular activity can enhance this pumping effect (p. 652).

Increased chemoreceptor discharge also enhances the activity of the vasomotor center, so that the peripheral vasoconstriction noted above is reinforced to some extent.

In addition to the immediate or short-term compensatory reactions to blood loss described above, a number of long-term compensatory reactions are activated by severe hemorrhagic hypotension (Table 16.9). These effects are manifest in a matter of hours, days, or even weeks following the hemorrhage.

The loss in circulating blood volume produces arteriolar vasoconstriction and a fall in systemic arterial blood pressure. These responses, combined with a reduced venous return to the heart, produce a fall in capillary hydrostatic pressure. Interstitial fluid passes into the capillaries, and this effect restores the circulating blood volume to some extent (Fig. 9.18, p. 560). As the interstitial fluid is lost to the capillaries, the relative dehydration of the interstitial compartment causes a net movement of water out of the cells by osmosis. It has been demonstrated experimentally that the reduced extracellular fluid volume causes intense thirst, a common symptom in shock patients (Table 16.8). Following blood loss, the thirst can develop without any change in the osmolality of the plasma; therefore it is presumed that the interstitial and/or intracellular dehydration produces this sensation (see also p. 408).

The process of fluid transudation ultimately results in a restoration of the circulating plasma volume in about 10 to 72 hours following a moderate hemorrhage. Most of the tissue fluids that enter the capillaries during this interval are protein-free; however, some albumin also enters the circulation from extravascular sources, eg, the hepatic cells. The interstitial fluids tend to dilute the plasma proteins and erythrocytes so that the hematocrit gradually falls, usually in a matter of hours, following the loss of whole blood.

Once the available supply of extravascular albumin has been depleted, several days are required for the remainder of the plasma protein that was lost to be replaced by de novo synthesis; the principal source of this plasma protein is the liver.

The circulating erythropoietin level rises in response to the relative anemia (p. 1171) so that the reticulocyte count progressively increases following hemorrhage. This value rises to a maximum in around 10 days, and the total circulating red cell mass reaches a normal level in four to eight weeks following the blood loss.

The low hematocrit that develops shortly after hemorrhage is tolerated quite well because the concentration of 2,3-diphosphoglyceric acid (2,3-DPG) within the erythrocytes increases rapidly in response to the relative hypoxia. This compound enables the hemoglobin to yield more oxygen to the tissues (p. 747).

Aldosterone secretion is stimulated by hemorrhage. This effect is mediated by two factors, viz, the increased secretion of corticotropin (p. 1036) as well as the elevated secretion of renin (p. 1169). Vasopressin and angiotensin II secretion also are increased by hemorrhage, but it is doubtful that these responses play major roles in the reestablishment of a normal blood pressure following the loss of blood. The elevated circulating levels of aldosterone and vasopressin do result in a retention of water as well as sodium ion so that the net result is some increase in the total circulating blood volume. The initial decreases in the rate of urine formation as well as sodium excretion are caused mainly by the hemodynamic changes in the kidneys that occur shortly following blood loss as discussed earlier.

The severity of the shock state depends upon the total quantity of blood lost by the patient. Some patients die shortly after a massive hemorrhage, whereas others undergo a gradual recovery, provided that the short-term and long-term compensatory mechanisms are able to restore the cardiovascular functions to normal levels. Appropriate therapeutic measures, of course, assist the homeostatic mechanisms of the body to reestablish normal cardiovascular functions. Certain patients exhibit a shock state that persists for hours, and despite restoration of a normal blood volume, the patient gradually becomes progressively more refractory to the action of vasopressor and other drugs. Myocardial contractility also does not respond to appropriate therapeutic measures, so that the cardiac output continues to be depressed far below normal levels. This situation is known as *irreversible* or *terminal shock*. In such cases, the heart exhibits a progressive bradycardia, the peripheral resistance declines steadily, and ultimately the patient dies.

Although the actual mechanism(s) responsible for initiating the irreversible phase of shock are unknown in detail, it is clearly evident that a number of interlocking and highly detrimental positive feedback mechanisms are operative in the patient. For example, as shown in Figure 16.31, cardiac ischemia leads to a decline in myocardial contractility; hence cardiac output falls; this in turn adds to the systemic hypotension, the general tissue ischemia worsens, the levels of toxic metabolites increase, and so a vicious, self-perpetuating feedback cycle becomes operative. Additional positive feedback cycles are depicted in the figure.

Powerful tonic contractions of the precapillary sphincters (p. 591) and venules (p. 592) are important features of reversible shock, and these effects are of particular significance in the splanchnic region. The abnormally low capillary perfusion that results from spasm of the precapillary sphincters leads to severe hypoxic tissue damage. Thus, at the onset of irreversible shock some three to five hours after hemorrhage, the precapillary sphincters dilate, whereas the venules remain constricted. Blood now can enter the capillaries, but the flow is stagnant, consequently the tissue hypoxia remains unchanged. However, as the hydrostatic pressure within the capillaries rises following dilatation of the precapillary sphincters, plasma escapes readily from the capillaries owing to their markedly increased permeability in the hypoxic state. Ultimately these vessels undergo necrosis, and as their walls disintegrate, whole blood now pours into the tissues. When this stage of shock is reached, the administration of whole blood or other fluids to restore the circulating blood volume is worthless, as the loss of such fluids from the circulatory system to the extravascular spaces takes place almost as rapidly as the patient is transfused.

It has been claimed that bacterial toxins within the circulatory system may contribute to the failure of the precapillary sphincters to maintain a normal tonus in shock states. In this regard, it has been demonstrated that stagnant hypoxia of the gastrointestinal tract, such as occurs in advanced shock, permits bacteria from the liver and gut to invade the circulation. This effect presumably becomes manifest as the hypoxia gradually eliminates the normal barriers to such invasion.

Any therapeutic measures that are directed against shock have two principal goals: first, to eliminate the cause of the shock; and second, to assist the homeostatic mechanisms of the body to achieve a physiologic level of tissue perfusion once again. Regardless of the etiology of the shock state, it is imperative to restore an adequate systemic arterial blood pressure so that coronary blood flow will not be impaired.

In hemorrhagic, wound, traumatic, and surgical shock, since blood loss is the major etiologic factor, treatment is based upon the prompt and rapid transfusion of suitable volumes of compatible whole blood. This procedure is essential in order that an adequate circulating blood volume be established without delay, while retaining a normal hematocrit. Transfusion of saline solutions is of limited value for this purpose, as a large proportion of the water and salts that are administered rapidly enters the extracellular fluid space, as only some 25 percent of the total volume injected is retained within the circulatory system.

In situations where the primary defect is a loss of plasma (combined with hemoconcentration, eg, as in burn shock), then treatment with plasma alone is the method of choice to restore the circulating blood volume to an adequate level as well as to reduce the hematocrit.

A number of so-called plasma expanders also are useful to some extent in the treatment of shock states that are induced principally by loss of plasma from the body. These substances are non-toxic high-molecular-weight sugars (eg, dextrans) as well as related compounds. Since the molecules of plasma expanders do not cross the capillary epithelium they remain within the circulatory system, and thus increase the overall circulating blood volume in accordance with the volume of solution administered. The advantage to the use of such materials resides in their stability over long intervals during storage, hence their immediate availability under adverse conditions.

Hypertonic solutions of various substances, including human serum albumin, are of limited usefulness as emergency adjuncts in shock therapy. Although such solutions will expand the circulating blood volume due to osmotic withdrawal of fluid from the interstitial fluid compartment, they are disadvantageous because they produce additional dehydration in a patient already suffering from this abnormality.

Epinephrine is quite useful in the treatment of anaphylactic shock. In fact, it appears that this compound exerts some unknown physiologic action in addition to its known cardiotonic as well as vasoconstrictor effects upon the dilated vessels that are present in this condition. Similarly norepinephrine also is used for its vasopressor effects in shock.

Sedatives are of limited value in the treatment of shock as they may depress the physiologic compensatory mechanisms, eg, discharge from the vasomotor center. Such agents if used at all should be given in extremely small doses. In this regard, alcohol is especially dangerous in shock therapy as this compound is a central nervous system depressant (p. 943). It dilates the cutaneous vascular bed via inhibition of the vasomotor center and thereby potentiates the effects of the shock state itself.

Overheating the patient also promotes cutaneous vasodilatation so that caution must be observed to prevent such an effect. The cardiovascular effects of sitting or standing (p. 630) also will decrease the efficacy of the normal compensatory mechanisms to shock and should be avoided. The patient in shock should be supine, and by raising the level of the feet a few inches above the horizontal position not only is venous return from the lower portions of the body facilitated, but also the cerebral circulation is increased. This head-down position is helpful for only short time intervals as the abdominal viscera press upon the diaphragm, thereby increasing the difficulty encountered in breathing. This position also increases the likelihood that pulmonary complications will develop ultimately.

THE PATHOPHYSIOLOGY OF CONGESTIVE HEART FAILURE

The general syndrome of heart failure can be manifest by the onset of either acute or chronic symptoms. Acute heart failure (such as occurs in cardiogenic shock [Table 16.8], ventricular fibrillation, or aeroembolism) has a rapid onset and results quickly in the death of the patient. Chronic heart failure, on the other hand, has a gradual onset, and may result from the operation of many disease processes. Some of these processes can be defined clearly (Table 16.10), whereas others are more obscure.

The syndrome of chronic congestive heart failure was selected for discussion as this condition serves to illustrate how a number of abnormal processes can contribute to the development of a major clinical problem through derangement of certain basic physiologic mechanisms. A discussion of chronic heart failure not only will provide the reader with a useful review of certain aspects of normal cardiac physiology, but also will serve to indicate the compensatory mechanisms that are called into play and which enable the patient to survive for extended periods of time despite the stress imposed upon the body by major impairment of normal cardiac function.

The principal signs and symptoms of congestive heart failure include an inadequate cardiac output, cardiac hypertrophy, edema, an abnormally long circulation time, enlargement of the liver (hepatomegaly), dyspnea, venous engorgement that is particularly evident in the veins of the neck, and a generalized weakness. The severity of these symptoms is related to the degree of cardiac

Table 16.10   ETIOLOGIC FACTORS IN CONGESTIVE HEART FAILURE[a]

1. Systemic hypertension
2. Valvular heart disease
3. Pulmonary hypertension
4. Pericardial disease
5. Intrinsic myocardial disease states
6. Various rare conditions which include high-output states (eg, anemia, thyrotoxicosis, arteriovenous fistulas)

[a]*Data from Harvey, Johns, Owens, and Ross (eds).* The Principles and Practice of Medicine, *18th ed, 1972, Appleton-Century-Crofts.*

insufficiency as well as to the level of activity undertaken by the particular patient. The more important signs and symptoms will be considered further in context.

The term "congestive heart failure" does not specify a particular disease entity; rather, this term denotes the functional condition of a patient suffering from heart disease irrespective of the cause. The syndrome of congestive heart failure can result from a number of abnormal processes that produce derangement of cardiac function such that the heart is unable to maintain the circulation of the blood at a level that is compatible with the physical activity of the patient under different conditions. Stated another way, the normal heart pumps out all of the blood returned to it when the individual undertakes different levels of activity; ie, the heart is able to compensate for various states of physical exertion by increasing its output of blood in proportion to the requirements of the body. The failing heart, on the other hand, is unable to maintain an adequate output in accordance with the increased metabolic demands of the body caused by an increased activity level, an effect that sometimes is referred to as cardiac decompensation. In fact, patients suffering from severe congestive heart failure may be unable to maintain an adequate cardiac output even during complete rest. In failure, the heart no longer is able to adapt physiologically to the metabolic demands of the body. It may be added that many of the symptoms associated with congestive heart failure noted above are caused by the retention of fluid.

Congestive heart failure is due primarily to an abnormal myocardial contractility, and not to a deranged excitation of, or conduction of impulses through, the heart. As the contractile machinery that pumps the blood is itself defective, a review of the pertinent mechanisms that underlie normal cardiac contractility will provide a background against which the activity of the heart in chronic failure can be presented.

The cardiac output normally varies over an extremely wide range. Increased physical activity or exercise (p. 1177) is a major physiologic stimulant to increased cardiac output, and in a normal person this effect is closely related to the elevated metabolic rate, as indicated by the oxygen consumption (Fig. 16.32).

Another important cardiovascular regulatory mechanism during periods of increased physical activity is the redistribution of blood flow; eg, during periods of exercise, blood is shunted away from the gastrointestinal tract and kidneys so that the proportion of the total cardiac output that is directed to the skeletal muscles is increased markedly. In turn, this shift in the pattern of blood flow from the resting state to that which occurs during exercise is accomplished by alterations in the vasomotor control mechanisms that regulate peripheral resistance within the several vascular beds. The various regions of the systemic circulation are organized in a parallel fashion (Fig. 10.15, p. 588), and the blood flow through each of these regions is inversely proportional to the resistance of the vessels in any particular circuit (p. 624).

Cardiac output per se is regulated by a number of physiologic mechanisms. Among these, venous return to the right heart is an extremely important factor in determining cardiac output, as the normal (ie, nonfailing) heart pumps out all of the blood that is returned to it during a given interval.

The most obvious physiologic mechanism for increasing cardiac output is an increase in heart rate. If the stroke volume remains unaltered (p. 621), then a 10 percent increase in heart rate results in a 10 percent increase in the cardiac output, but such an increase in cardiac output can take place only if there is an adequate venous return.

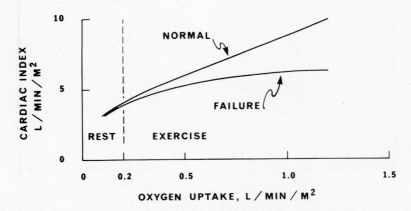

FIG. 16.32. Relationship between the cardiac index (cardiac output per unit body surface area) at rest and during exercise in a normal and a failing heart. Note that at rest there is little difference between the normal subject and the patient in failure, whereas during exercise the failing heart is unable to increase its output adequately. Note also that as the intensity of the exercise increases progressively, the divergance between the output of the normal and failing hearts increases disproportionately. (Adapted from Harvey, Johns, Owens, and Ross [eds]. *The Principles and Practice of Medicine,* 18th ed, 1972, Appleton-Century-Crofts.)

The properties of the myocardium itself also have a profound influence upon the normal cardiac output. The Frank-Starling mechanism, which specifies that the force of contraction that is produced by a muscle fiber is related directly to the length of the fiber at the beginning of the contraction, is of particular significance (Fig. 16.33; p. 616). The initial fiber length is a function of the end-diastolic volume of the ventricle in vivo, and the relationship between the fiber length and the end-diastolic volume is shown schematically in Figure 16.33. Since the end-diastolic volume is a function of venous return, then an increased venous return results directly in greater end-diastolic volume, an increased length of the individual myocardial fibers due to stretch, and a greater force of contraction. Hence, the stroke output is elevated in the normal heart by the Frank-Starling mechanism.

Roughly 50 percent of the total volume of blood that is present in the ventricles at the end of diastole is ejected during each systole, and this ejection fraction remains relatively constant despite marked variations in venous filling in the normal heart. In chronic congestive failure, however, the ejection fraction decreases markedly, and the affected ventricle dilates progressively as discussed below.

From an anatomic standpoint the functional unit of cardiac and skeletal muscle is the sarcomere in which filaments of actin and myosin (p. 116) are interdigitated (Fig. 16.34; see also p. 138). As these two sets of filaments slide past each other during activity, an overall shortening or contraction of the sarcomere results. The binding forces that develop between the actin and myosin filaments are activated by release of calcium ions from the sarcoplasmic reticulum during conduction of an impulse down the fiber, and when the calcium is sequestered once again the binding forces are inactivated and relaxation ensues, at least in skeletal muscles.

The extent to which the myosin and actin filaments overlap determines the actual number of binding sites that are in a functional state, and thus controls the force of contraction that is generated by the interdigitation of these fibrils. The sarcomere length is therefore related to the tension that is generated by the contracting muscle, as shown

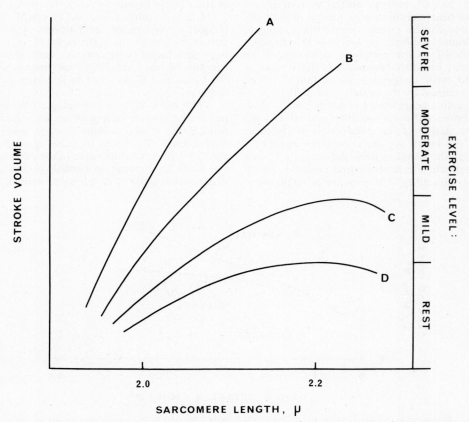

FIG. 16.33. Ventricular function (or Starling) curves illustrating the relationship between stroke volume and sarcomere length under different activity states. A. Normal heart during sympathetic stimulation. B. Normal heart, unstimulated. C. Mild heart failure. D. Severe heart failure. Note in C and D that the stroke volume plateaus or even starts to decline at sarcomere lengths at which the normal heart still is increasing its output. (After Harvey, Johns, Owens, and Ross [eds]. *The Principles and Practice of Medicine,* 18th ed, 1972, Appleton-Century-Crofts.)

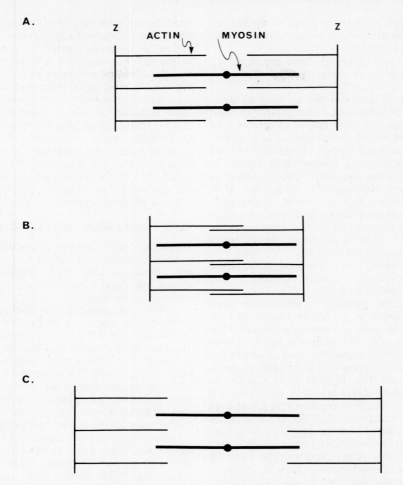

FIG. 16.34. **A.** Normal sarcomere in resting state. **B.** Normal sarcomere in contracted state. **C.** Sarcomere in overstretched state such as is encountered during congestive heart failure. The myocardial dilatation encountered in such a condition results in a fewer number of binding sites between the actin and myosin filaments, and results in the decreased stroke volume seen in curves C and D, Figure 16.33.

in Figure 10.37 (p. 616). At the bottom of this figure, the resting tension curve obtained by passive stretch of the inactive muscle is presented; and it is this resting tension curve that determines ventricular filling during the diastolic phase of the cardiac cycle.

A normal heart operates at a sarcomere length that is far below the point at which the tension developed during contraction is at a maximum (Fig. 16.33B). Enhanced contractile tension (or increased cardiac work) can be elicited by increasing the resting length of the contractile unit (the sarcomere length) as shown in Figure 16.33B. The ascending limb of the Starling curve shown in this figure is produced by such an effect. There is a definite point beyond which any further increase in sarcomere length will not result in an increased development of tension (Fig. 16.33), and at this point the sarcomere is overstretched. Since the length of the contractile unit has passed its optimal

value, the total number of binding sites that are in a functionally useful position is reduced significantly. In the intact heart this effect is indicated by the descending limb of the Starling curve (Figs. 16.33, 16.34C). Beyond this point any further increase in the end-diastolic volume does not produce an increase in the contractile tension developed during systole so that a smaller fraction of the total volume of blood that is present within the ventricle at the end of diastole will be expelled during the next systole; ie, the ejection fraction decreases. The net result of this process is additional ventricular dilatation.

Since the tension developed within the ventricular wall during systole is a function of both the radius of the chamber and the pressure within the lumen, then as the volume of the heart increases, a greater tension must be produced by the myocardium in order to generate the same pressure (Laplace's law, p. 627). The efficiency of the heart decreases when it is dilated to greater end-diastolic volumes such as are present during chronic failure (Figs. 16.32 and 16.33.)

The basic facts reviewed above are of the utmost importance to an understanding of chronic cardiac failure.

A number of morphologic as well as functional alterations have been demonstrated in heart failure, although the physiologic significance of these changes is not altogether clear. For example, a heart that has been dilated chronically beyond normal limits exhibits a phenomenon known as slippage; that is, the sarcomeres are no longer in normal alignment with respect to each other. Morphologic alterations also are found in the plasmalemma of individual myocardial cells, as well as in the T system which plays such an important role in the contractile process insofar as the release of calcium ion is concerned (p. 140).

Thus far, exhaustive biochemical studies have failed to disclose any clues related to the mechanism of cardiac failure.

The effect of sympathetic stimulation upon ventricular function is shown in Figure 16.33A. Note that during such autonomic stimulation the normal heart is able to produce a much greater stroke volume at any particular level of sarcomere length (or diastolic volume) than is the unstimulated normal heart (Fig. 16.33B). Note also that in failure the stroke volume which is delivered at a given sarcomere length is considerably lower than that in the normal heart (Fig. 16.33C and D). A failing heart is dilated, and unlike the normal heart is unable to keep its output regulated in accordance with venous return because the force of contraction developed during systole is much lower at any particular fiber length. The failing, dilated heart is working at a mechanical disadvantage, insofar as Laplace's law is concerned.

Congestive heart failure has been defined as "the inability of the cardiac output to increase in proportion to the metabolic demands placed upon the circulation." The bottom curve in Figure 16.32 illustrates this effect in a failing heart compared to the response of a normal heart under similar conditions. Note that the cardiac output is essentially normal within the lower range of oxygen consumption, ie, when the person is at rest, but as the oxygen requirements of the body increase, the failing heart does not increase its output accordingly. This lack of adaptability on the part of the failing myocardium frequently results from an inadequate contractile response of the muscle fibrils to an increased length that is caused by an elevated venous return. The reduced capacity of the failing heart to increase its output can result from myocardial disease per se, or from mechanical abnormalities such as valvular disease or systemic hypertension (Tables 16.10 and 16.11). Thus the failing myocardium no longer is able to compensate adequately to an increased metabolic requirement for blood flow throughout the body.

The principal clinical symptoms of congestive heart failure generally are divided arbitrarily into those due primarily to fluid retention (right heart failure) or to those caused by congestion of the pulmonary vascular bed (left heart failure), and this classification will be used in the following discussion. In most patients both of these conditions are present to some extent; however, one or the other situation usually predominates so that this general classification affords a functionally as well as clinically useful delineation of the two major types of failure. Nevertheless, it is imperative to remember that the functions of the entire heart, hence the circulation as a whole, are abnormal as the result of failure, regardless of which side of the organ is affected primarily.

Right Heart Failure and Fluid Retention

When the cardiac output is inadequate compared to the metabolic needs of the body, fluid retention and edema develop. The mechanisms whereby edema formation occurs in right heart failure are summarized in Figure 16.35 (cf. p. 575).

When an inadequate cardiac output develops owing to the operation of factors such as those summarized in Table 16.11, certain homeostatic mechanisms are called into play that tend to protect such vital regions of the body as the brain and heart tissue itself. Thus cerebral and coronary blood flows are maintained at normal levels, but these flows are maintained at the expense of a decreased blood flow to the skin and skeletal muscles due to vasoconstriction in these peripheral regions.

Since the right atrial pressure is elevated, the right atrial filling pressure also increases, and an elevated systemic venous pressure develops throughout the body. The liver undergoes hepatomegaly owing to the increased total volume of blood contained within the systemic venous circuit as well as the elevated systemic venous pressure.

Table 16.11   CONGESTIVE HEART FAILURE INDUCED BY VALVULAR HEART DISEASE[a]

PRINCIPALLY INVOLVING RIGHT HEART

Pulmonary hypertension that develops
  secondarily to disease of the left heart
Pulmonary stenosis
Tricuspid stenosis
Tricuspid regurgitation

PRINCIPALLY INVOLVING LEFT HEART

Aortic stenosis
Aortic regurgitation
Mitral stenosis
Mitral regurgitation

[a]*Data from Harvey, Johns, Owens, and Ross (eds). The Principles and Practice of Medicine, 18th ed, 1972, Appleton-Century-Crofts.*

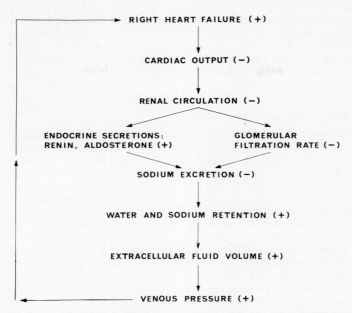

FIG. 16.35. The mechanism of edema formation in congestive heart failure. Note that positive feedback due to the increased venous pressure contributes to edema formation as this pressure causes a progressive dilatation of the right heart. (+) = increased or enhanced; (−) = decreased or reduced. It should be noted that an elevated circulating aldosterone level is not a consistent finding; rather, it is found only in some patients with congestive failure.

The difference between the oxygen content of arterial and venous blood increases subsequent to the reduced cardiac output of failure, and this effect leads to a lowered oxygen content of the body tissues in general, ie, a generalized hypoxia develops.

The circulation of blood through the kidneys is impaired seriously by the shunting of blood from the viscera that takes place when the cardiac output is reduced in heart failure. The decreased renal blood flow in turn leads to a reduced glomerular filtration rate (p. 783). The proportion of sodium that is reabsorbed from the glomerular filtrate also is increased concomitantly with the decreased glomerular filtration rate (p. 788). Not only is less sodium being filtered by the kidneys in congestive heart failure, but also more of the sodium ion that is filtered is reabsorbed. Progressively less sodium ion is excreted whereas more is retained in the body, and these processes lead to abnormal fluid retention as well. The net effect is an increased extracellular fluid volume and the development of edema.

Congestive heart failure in some patients is accompanied by an increased production of aldosterone (p. 1106) which facilitates salt (hence water) retention by the body. The circulating renin levels also are elevated in congestive failure, although other as yet unidentified endocrine factors may be involved. The increased aldosterone levels combined with the hemodynamic disturbances (principally the reduced glomerular filtration rate) are sufficient to explain most of the major distur-

bances in sodium metabolism that are found in chronic congestive heart failure.

In this regard, alterations in the tubular reabsorption of sodium that occur independently of changes in the glomerular filtration rate have been demonstrated in normal subjects. Thus mechanisms that are responsive to alterations in either the total extracellular fluid volume or the total blood volume have been postulated. Although unproven, such osmoreceptors also could play a role in the abnormal sodium retention that occurs in congestive heart failure.

Four major factors are concerned with the pattern of edema that is seen in patients with congestive heart failure. First, venous pressure is the principal factor that determines the distribution of the edema fluid in congestive heart failure, and this fact is demonstrated clearly by the effects of posture (ie, hydrostatic pressure). An ambulatory patient accumulates fluid primarily in the lower extremities because the effect of gravity produces the highest venous capillary pressures in these regions (p. 651). If the patient is supine, the edema develops principally in the sacral region. Second, the colloid osmotic pressure of the plasma proteins tends to offset to some extent the hydrostatic effect noted above; therefore edema formation takes place most prominently in patients who have a decreased concentration of circulating plasma proteins as a concomitant to heart failure. Third, tissue pressure also contributes to the distribution of edema fluid. Thus regions that contain loose connective tissue (eg, the ankle) are more prone to develop edema. Fourth, the function of the lymphatic system also determines to some extent the distribution of edema fluid (p. 561). This is so as a portion of the protein and fluid that enter the interstitial spaces from the capillaries is returned to the vascular bed via the lymphatic system.

Left Heart Failure and Pulmonary Congestion

Congestion of the pulmonary vascular bed can be due to a mechanical obstruction that prevents normal blood flow into the left heart from the lungs, to a malfunction of the left ventricle itself, or to a valvular defect (Tables 16.10 and 16.11). Thus, a mechanical obstruction to blood flow is encountered in mitral stenosis, whereas ventricular dysfunction can develop secondarily to the presence of aortic valve disease, systemic hypertension, or inherent myocardial disease. In all of these conditions, the pressure within the left atrium is elevated, and when a diseased state is present in the left ventricular myocardium then the end-diastolic pressure also is elevated in this chamber. A high pulmonary arterial pressure is necessary in order to fill the left atrium. This functional damming of the blood entering the left heart results in the development of pulmonary hypertension, and ultimately may lead to failure of the right heart in some patients. This retrograde sequence of events is, however, a rather simplistic view. It must be reiterated that the function of both ventricles is affected when any portion of the myocardium suffers any functional impairment, and the terms "right heart" and "left heart" failure are relative rather than absolute.

Since the principal defect found in left heart failure is pulmonary vascular congestion, the patient at first may exhibit dyspnea upon exertion, and this progresses to dyspnea at rest as the failure progresses. Characteristically, the patient exhibits a nocturnal paroxysmal dyspnea, in which extreme shortness of breath is manifest upon awakening from a deep sleep. Orthopnea is a related condition that necessitates the patient sleeping in a semireclining position so that the head and trunk are elevated, a posture which alleviates the dyspnea present in the supine position. In some individuals, chronic left heart failure is revealed only by cough and/or spitting of blood (hemoptysis).

Probably the same physiologic mechanism underlies both nocturnal paroxysmal dyspnea and orthopnea. As the body assumes the supine position from an erect posture, the hydrostatic pressure within the veins of the lower extremities falls (p. 651) so that the fraction of the total blood volume that had been pooled in the legs now returns to the right side of the heart to be pumped to the lungs. Since the function of the left heart is impaired, the flow of blood out of the pulmonary circuit is inadequate, and the lungs develop congestion due to the higher pulmonary intravascular pressure that results.

There is an additional mechanism that also may contribute to the genesis of nocturnal paroxysmal dyspnea, but in a more gradual fashion, ie, over a period of hours. Upon lying down, the reduction of venous hydrostatic pressure causes a shift in the fluid balance across the capillary membranes of the body in general, and fluid now enters the systemic vascular bed from the interstitial fluid compartment. The net result of this process is an increase in the total circulating blood volume, and this acts in concert with the redistribution of blood flow noted above to induce congestion in the pulmonary vessels.

Apparently the vascular congestion in the pulmonary circuit initiates reflexes that result in constriction of the bronchi and bronchioles in patients with congestive heart failure. Wheezing usually is evident, together with moist rales (see below). In some patients, particularly those having a clinical history of bronchial asthma (p. 723), the wheezing will be the principal early physical sign of heart failure. When the lung develops marked bronchial constriction as a response to pulmonary vascular congestion, the condition is known as cardiac asthma.

During the early course of left heart failure, the total volume of blood that is present within the pulmonary circulation is increased so that both arterial and venous pressures are elevated in the lungs. The congested, tumescent lung loses its normal elasticity, and pulmonary ventilation is decreased. As the disease progresses, certain other pulmonary alterations take place whose importance may overshadow the excessive volume of blood in the lungs. Thus the compliance of the lung tissue itself decreases markedly (p. 719); hence the work of respiration increases, and this effect may be partly responsible for the dyspnea observed in patients with left heart failure. Presumably reflexes that originate within the engorged vessels and/or the alveoli themselves contribute to the tachypnea as well as the "air hunger" that are present in this condition.

When the pulmonary venous pressure increases sufficiently, the normal equilibrium between the colloid osmotic pressure of the plasma and the hydrostatic pressure within the pulmonary capillaries becomes imbalanced and transudation of fluid into the alveolar lumens takes place. If this effect is pronounced and develops suddenly, then frank pulmonary edema results. On the other hand, if the transudation of fluid takes place to only a slight extent, then moist rales develop.

Etiologic Factors Related to Congestive Heart Failure

Several of the factors that can produce congestive heart failure were mentioned earlier, and certain of these now may be considered in somewhat greater detail (Table 16.10).

Various forms of hypertensive cardiovascular disease affect an estimated 20 to 25 million Americans, and contribute to one-half of the entire yearly mortality rate in the United States alone. Thus hypertension is a public health problem of major significance. Systemic hypertension generally is diagnosed when the systemic arterial blood pressure is above 140 to 150 mm Hg and the diastolic pressure exceeds 90 mm Hg. Many conditions underlie the etiology of hypertension. The major importance of obesity in this regard was stressed earlier in this chapter. Regardless of the causative factor(s), chronic hypertension lowers the life expectancy, and may lead to early death or incapacitation from congestive heart failure, a stroke (or cerebral vascular accident), coronary infarction, coronary insufficiency, or uremia.

The systemic blood pressure is maintained by the combination of cardiac output and peripheral resistance; hence any factor that enhances either or both of these parameters can induce hypertension (ie, an increased pulse rate, stroke volume, total circulating blood volume, blood viscosity, or a reduced elasticity of the vessels and vasomotor tone). In most instances, however, hypertension is attributable to an increased total peripheral resistance, and there is a tendency for the blood pressure to increase spontaneously with age, females showing a greater increase in both systolic and diastolic pressures than do males of comparable age groups (Fig. 16.36).

So-called essential hypertension is the most common form of hypertensive cardiovascular disease; the origin of this condition is unknown. All other forms of hypertension are secondary to some form of outright pathologic disease process, eg, coarctation of the aorta, primary hyperaldosteronism (p. 1115), Cushing's syndrome (p. 1114), pheochromocytoma (p. 1126), or unilateral renal disease (p. 813).

From the standpoint of the pathophysiology of chronic congestive heart failure, a prolonged and significant elevation of the systemic arterial resistance, regardless of cause, forces the heart to perform considerably more work than is normal in order to propel the blood through the systemic circulation. In particular, an elevated diastolic pressure is more damaging to the heart than is a comparable elevated systolic pressure, because at the onset of systole the already stretched myocardial fibers must immediately overcome the high arterial resistance that is present while the fibers are contracting at a mechanical disadvantage. Left ventricular hypertrophy and ultimately congestive left heart failure develop progressively as the myocardial fibers must contract against pressures that are beyond their normal morphologic and functional limits for prolonged intervals.

The hypertensive patient may exhibit shortness of breath, nocturnal paroxysmal dyspnea, edema, elevated venous pressure, and perhaps hepatomegaly.

When a significant heart murmur is evident, valvular heart disease is considered to be present. In turn, if this condition is of sufficient magnitude, it can lead to chronic left or right heart failure. As shown in Table 16.11, the predominant type of

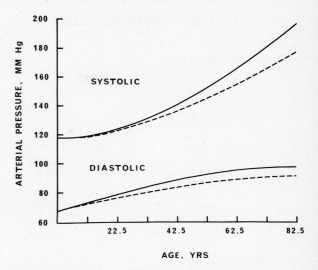

FIG. 16.36. Relationship between systolic and diastolic blood pressures in male (dashed curves) and female (solid curves) subjects at different ages. (Data from Harvey, Johns, Owens, and Ross [eds]. *The Principles and Practice of Medicine,* 18th ed, 1972, Appleton-Century-Crofts.)

failure that develops depends upon the particular valve involved.

Stenosis (or narrowing of a valvular orifice) can impose a greatly increased resistance to blood flow from that particular chamber, whereas regurgitation of blood through a leaky valve likewise increases the work that the myocardium must perform. Both of these pathophysiologic conditions if sufficiently prolonged and marked cause dilatation of the myocardium that leads eventually to chronic failure.

Most commonly pulmonary hypertension is the cause of right ventricular failure, as this condition significantly increases the work of the right ventricle in pumping blood through the vascular bed of the lungs. The mechanism underlying this type of failure is similar to that described above for systemic hypertension. Shortness of breath and fatigue are the outstanding symptoms in pulmonary hypertension, particularly if this condition is secondary to left heart failure. The common signs of right heart failure also are present, ie, an elevated venous pressure, ascites, edema, and hepatomegaly.

Chronic right heart failure also can be produced by a number of pericardial diseases, some of which are listed in Table 16.12. The pericardium becomes constrictive as the result of a disease process so that scar tissue limits the inflow of blood into the right heart. The cardiac output becomes limited as the heart cannot dilate and fill properly owing to the mechanical restriction imposed upon it. Thus the basic problem in chronic restrictive pericarditis is that the filling of the right heart is impaired. This effect is in marked contrast to myocardial failure per se in which the basic

Table 16.12   PERICARDIAL DISEASES[a]

ACUTE AND SUBACUTE

Infections
  Bacterial; eg, tuberculosis, pneumococcal,
    staphylococcal
  Protozoan; eg, amebic
  Mycotic; eg, coccidiomycosis
  Viral; eg, coxsackie
Connective tissue and allergic disease; eg,
  rheumatic fever, systemic lupus erythematosis
Chemical or metabolic; eg, uremia, myxedema
Neoplastic diseases
Pericardial disease secondary to disease of
  heart or great vessels. Related to myocardial
  infarction, pulmonary embolism
Physical injury; eg, trauma, radiation

CHRONIC DISORDERS

Chronic constrictive pericarditis
Chronic pericardial effusion; eg, myxedema,
  cholesterol pericarditis

[a]Data from Harvey, Johns, Owens, and Ross (eds). The Principles and Practice of Medicine, 18th ed, 1972, Appleton-Century-Crofts.

problem is the emptying of the heart. The clinical symptoms of restrictive failure usually are those of right heart failure, although if the constriction involves the left heart, in particular the left atrium, then congestion of the pulmonary vascular bed may develop secondarily to this mechanical limitation on the normal filling of this chamber.

Primary myocardial disease or cardiomyopathy* sometimes can be classified according to the causative factor, as shown in Table 16.13. When only the myocardium is involved, the condition is referred to as a primary cardiomyopathy, whereas if the heart is involved as a part of an overall systemic disease process, the condition is referred to as a secondary cardiomyopathy. From a clinical standpoint, the cardiomyopathies also can be classified as hypodynamic (hypotrophic) or hypertrophic.

A hypodynamic cardiomyopathy is characterized by a pronounced ventricular dilatation so that the myocardial wall is thin and the contractions are weak, as the heart is working at a distinct mechanical disadvantage. Emptying is seriously impaired during systole, ie, the ejection fraction is low. Generally the clinical picture is one of congestive left heart failure, with the patient exhibiting dyspnea and orthopnea. Right heart failure may develop subsequently to that present on the left side, and this retrograde effect is attended by an elevated venous pressure, edema, and hepatomegaly. Mitral and tricuspid murmurs are

common in hypodynamic myopathies, especially during systole, and these murmurs may lead the examining physician to conclude mistakenly that a primary valvular disease is present. The murmurs are caused by deformation of the ventricles owing to alterations in their size and shape, and not primarily to valvular disease. Thus the hypodynamic heart exhibits dilatation but hypertrophy is absent so that the ventricular wall is quite thin relative to the size of the lumen. This condition sometimes is referred to as hypotrophic cardiomyopathy. The specific etiology of this pathophysiologic situation may be unknown, but any of several of the factors listed in Table 16.13 can produce the myocardial damage leading to the failure.

In patients who recover from hypotrophic cardiomyopathy, no suitable cardiac hypertrophy takes place and a gradual dilatation of the left ventricle develops with time. The prognosis is poor for most patients who exhibit this type of condition and no specific therapy is available to ameliorate it.

Hypertrophic cardiomyopathy on the other hand differs in several important respects from the hypotrophic form; hence the ventricular wall is thick and the lumen small. The principal physiologic defect in hypertrophic cardiomyopathy is a decreased compliance (p. 629) of the hypertrophied ventricular wall which seriously impairs diastolic filling of the chamber.

The hypertrophied left ventricle usually is the predominant sign that is noted upon early clinical examination, and this major change may be accompanied by generally minor symptoms such as dyspnea on exertion or chest pain. The EKG as well as the physical examination and chest x-ray all provide evidence of the presence of left ventricular hypertrophy.

Again in contrast to the situation in hypodynamic cardiomyopathy, the hypertrophic heart contracts strongly; therefore ejection of blood is complete during each systole despite the

Table 16.13   SOME ETIOLOGIC FACTORS IN CARDIOMYOPATHY[a]

1. Ischemia: coronary atherosclerosis, syphilis
2. Infections: bacterial, protozoal, viral, rickettsial
3. Toxic agents: alcohol, sulfonamides, bacterial toxins
4. Metabolic: hyperthyroidism, hypothyroidism, beriberi, anemia
5. Collagen vascular diseases: systemic lupus erythematosis, polyarteritis
6. Neuromuscular disease: muscular dystrophy
7. Trauma: mechanical, electric, radiation
8. Postpartum
9. Cause unknown; ie, idiopathic

*The word cardiomyopathy is a general term that indicates primary myocardial disease which may have an obscure or unknown etiology.

[a]Data from Harvey, Johns, Owens, and Ross (eds). The Principles and Practice of Medicine, 18th ed, 1972, Appleton-Century-Crofts.

fact that the thickened ventricle cannot fill normally. The latter sign apparently is the major physiologic problem in hypertrophic cardiomyopathy.

The prognosis in patients having a hypertrophic cardiomyopathy is much better than that for the hypotrophic condition; in fact, roughly 50 percent of most groups of patients have no progression of symptoms when observed for long periods of time. As the disease continues, however, atrial fibrillation develops together with systemic embolism, and death may result ultimately from congestive heart failure.

Congestive heart failure occurs in patients with so-called high output states (ie, situations in which the cardiac output is increased abnormally); however, this condition is rare unless organic heart disease also is present.

There is a pronounced resting increase in cardiac output in thyrotoxicosis (p. 1058) that occurs secondarily to a fall in peripheral arterial resistance combined with the direct stimulatory effect of thyroxine on the metabolism of the heart that causes a profound tachycardia. The pulse pressure is elevated, and the heart is hyperexcitable so that atrial fibrillation is common in thyrotoxicosis. Chronic congestive failure rarely is observed, however, in young thyrotoxic patients.

Severe anemia (p. 566) produces effects similar to thyrotoxicosis in that the cardiac output must increase considerably in order to provide sufficient oxygen for the tissues. Concomitantly the heart becomes hyperreactive, a tachycardia develops, and peripheral vasodilatation is observed. Congestive failure is of rare occurrence in young persons with normal hearts, but anemia provides a stress that frequently causes failure in the hearts of older individuals who suffer from a preexisting heart disease.

High output congestive heart failure also may occur in persons with large arteriovenous fistulas such as may result from stab or gunshot wounds.

Some Aspects of the Management of Chronic Heart Failure

The foregoing brief review of the pathophysiologic and etiologic aspects of congestive heart failure should indicate to the reader the complexity of the problems that are encountered in this clinical state. It must be remembered that in congestive heart failure the normal compensatory homeostatic mechanisms underlying cardiac output usually are defective; consequently the myocardium is at all times working at a disadvantage and near the upper limit of its functional capacity compared to the normal heart. This fact, shown graphically in Figures 16.32 and 16.33, is so basic to an understanding of heart failure that it is reiterated here for emphasis.

Obviously, when the heart is no longer able to compensate to meet the minimal metabolic demands of the body, either with or without suitable treatment, then death intervenes.

As stated earlier in this text (p. 50), the physician often must play a vital role in assisting and restoring the homeostatic mechanisms of the body to such an extent that the patient can survive for a much longer time than would otherwise be possible. The management of chronic heart failure provides an excellent example of the beneficial effects of such therapeutic intervention at the physiologic level. Consequently, this discussion of chronic failure will be concluded with a brief survey of the general principles that form the basis for management of patients who suffer from chronic congestive heart failure.

In common with other disease states, an accurate diagnosis is the first step in the therapy of congestive heart failure. The etiology and the anatomy of the lesion or abnormality and its severity, together with the physiologic status of the patient, are all prime considerations in establishing the diagnosis.

It is equally important to assess the prognosis for the patient insofar as his current status is concerned with respect to the course and rate of development of the failure. Age and estimated future potential life-span are important considerations, and any therapy that is instituted should be appropriate for the patient at his particular age and stage of the disease.

The factor (or factors) that underlie the current episode of congestive failure must be recognized and removed (or ameliorated) if therapy is to prove effective. The major factors involved in the precipitation of an episode of failure are summarized in Table 16.14. From a pathophysiologic standpoint, the factors that are summarized in the first part of this table act to reduce myocardial function, ie, cardiac work, whereas those in the second group act by increasing the work load that is imposed upon the heart.

A few particularly significant factors listed in Table 16.14 may be discussed individually.

Discontinuation of prescribed digitalis medication by the patient for various reasons is a highly important causative factor in producing episodes of congestive failure.

Another factor that will produce this effect is the rapid development of an arrhythmia, particularly atrial fibrillation, in the patient with mitral stenosis.

Alcohol exerts a severe depressant effect upon the myocardium and may play an important role in cardiac decompensation in some patients, whereas if this stress is removed adequate physiologic compensation may take place spontaneously.

If the work load of the heart is increased suddenly, as by rupture of one of the chordae tendineae, acute failure can be precipitated.

More commonly, however, the ingestion of a large amount of salt or overeating in general (as during the holiday seasons) is responsible for placing an additional stress upon the heart.

Good clinical management can result in the

Table 16.14 FACTORS THAT CAN PRECIPITATE CONGESTIVE HEART FAILURE[a]

FACTORS THAT DECREASE THE CAPACITY OF THE HEART TO MEET THE METABOLIC REQUIREMENTS OF THE BODY:

Discontinuation of digitalis therapy
Arrhythmias, eg, fibrillation, tachycardia,
   bradycardia (plus) heart block
Impairment of myocardial function per se
   Infarction
   Rheumatic myocarditis
   Coronary or bacterial embolism
   Toxic substances; eg, alcohol
   Thiamine deficiency

FACTORS THAT INCREASE THE LOAD ON THE HEART:
Dietary lapse; eg, excessive salt intake,
   overeating in general
Increased physical activity
Infection
Surgical operation: transfusion, fluid therapy
Pulmonary embolism
Perforated septum or valve; rupture of chorda
   tendina
Anemia
Thyrotoxicosis
Medication that enhances salt and water
   retention; ie, iatrogenic causes

[a]*Data from Harvey, Johns, Owens, and Ross (eds).* The Principles and Practice of Medicine, *18th ed, 1972, Appleton-Century-Crofts.*

patient feeling so well that he resumes a normal pattern of activity, thereby precipitating an acute episode of failure in response to excessive physical exertion.

Pregnancy increases the work load on the heart, and frequently this state precipitates the first episode of failure in women having a disease of the mitral valve.

The effects of anemia and thyrotoxicosis in the genesis of chronic heart failure were discussed earlier and are significant because both conditions can be treated with good chances for long-term success.

In conclusion to the foregoing remarks, the precipitating factors that underlie congestive heart failure must be eliminated as part of the particular therapy for this condition; hence their proper identification assumes the utmost significance in order to preclude future episodes of failure.

Specific therapy in congestive heart failure includes rest, diet, the administration of cardiac glycosides, diuretics, mechanical removal of excess fluid from the body, and surgery (eg, to correct arteriovenous fistulas). The therapy must be tailored to the individual patient at the current stage of his disease. Overtreatment can have effects that are as deleterious as those of undertreatment.

Rest is the principal physical therapy for congestive heart failure, regardless of the factor (or factors) that has upset the equilibrium between cardiac performance and the metabolic needs of the body. The extent to which the activity of a particular patient is reduced depends entirely upon the severity of his condition. The therapeutic regimen may range from complete bed rest to a limitation of the amount of physical work that the patient may perform.

A decrease in physical activity can be achieved readily, whereas an increase in work output by the heart may not always be possible. In failure caused by myocarditis, rest has an additional effect that is beneficial; namely, it facilitates healing of the myocardium. However, venous thrombosis as well as embolism both are potential complications of congestive failure that may develop during prolonged bed rest.

Diet is another important adjunct to the overall management of patients suffering from congestive heart failure, and basically restriction of sodium intake is the goal. The severity of the failure in turn governs the extent to which salt is restricted from the diet.

In general, fluid intake need not be restricted on a low sodium intake, although concomitant diuretic medication (see below) may render it advisable to restrict the overall fluid intake in some patients, particularly if renal involvement is severe.

Cardiac glycosides are the major drugs of choice for the therapy of congestive heart failure. Digitalis and certain related compounds exert a direct action upon the failing myocardium in that they enhance the force of contraction of the heart muscle. That is, digitalis glycosides produce a positive inotropic response in the failing heart. The action of digitalis glycosides is in marked contrast to the other types of therapy that are directed toward reducing the work load of the failing heart.

The mechanism of action of the cardiac glycosides upon the failing myocardium is of particular interest. The increased contractility produced by these compounds apparently is effected by an inhibition of the sodium pump mechanism that normally maintains the intracellular sodium ion concentration well below that of the extracellular fluid (Fig. 3.31, p. 93). Inhibition of the sodium pump permits more sodium ion to diffuse into the myocardial cells, and as the intracellular concentration of this ion rises, the transport of calcium ion to the myofilaments is facilitated (p. 140). Since the force developed by the muscle during the active state is in direct proportion to the quantity of calcium that reaches the myofilaments (p. 143), the digitalis glycosides apparently act indirectly by enabling the intracellular sodium ion level to increase and thus in turn to release more calcium which then enhances the contractile process at the molecular level.

The preponderance of experimental evidence favors the mechanism of action stated above, al-

Table 16.15   SOME DIURETIC AGENTS[a]

CLASS OF COMPOUND	PREPARATION	MECHANISM OR LOCUS OF ACTION
Thiazides	Chlorothiazide	Inhibit Na^+ reabsorption in the distal tubules
Mercurials	Mercaptomerin	Inhibit Na^+ reabsorption in Henle's loop and distal tubules
Anthranilic acids	Furosemide	Inhibit Na^+ absorption in Henle's loop
Aldosterone antagonists	Spironolactone	Competitive inhibition of effects of aldosterone on renal tubules
Carbonic anhydrase inhibitors	Acetazolanide	Inhibit sodium and bicarbonate absorption in proximal and distal tubules

[a]*Data from Harvey, Johns, Owens, and Ross (eds). The Principles and Practice of Medicine, 18th ed, 1972, Appleton-Century-Crofts.*

though some investigators have advanced the view that cardiac glycosides may act by a direct effect upon the contractile proteins themselves.

The force of contraction, or the maximum rate at which pressure develops in the intact heart, can be measured experimentally; and in the left ventricle, the pressure developed during systole is a function of the active state of the myocardium (p. 660). Clinically, however, the increase in contractility of the myocardium in a digitalized patient being treated for chronic congestive failure is indicated by an increase in the cardiac output as well as by a decrease in the end-diastolic volume and end-diastolic pressure within the heart. Thus the cardiac glycosides cause the ventricular function curve of the failing heart (Figs. 16.33C and D) to shift upward toward normal (Fig. 16.33B); thus a particular end-diastolic volume (or end-diastolic fiber length) is related to a greater stroke output in the digitalized patient.

In addition to promoting enhanced contractility of the failing myocardium, cardiac glycosides exert an effect upon the conducting tissue of the failing heart (p. 584). Such drugs extend the refractory period of the atrioventricular node (p. 586) so that the rate of impulse transmission from the atria to the ventricles is reduced. This effect is of particular importance when atrial fibrillation is present, as the refractory period of the AV node is of major importance in regulating the ventricular rate (p. 701) under such conditions. The glycosides act primarily on the AV node through the vagus nerve, although a slight direct action on the conducting tissue per se is present.

In a patient having a rapid ventricular rate that is caused by atrial fibrillation, the short diastolic interval does not allow the ventricle to fill adequately. The cardiac output falls, and the onset of congestive heart failure often is accelerated. Consequently, the cardiac glycosides are particu-

larly effective in the therapy of heart failure in which a marked ventricular tachycardia is present secondarily to an uncontrolled atrial fibrillation. This is true because the effect of the bradycardia induced by the drug is added to the positive inotropic effect that is exerted directly upon the weakened myocardium.

One additional facet of digitalis therapy is worth mentioning. The magnitude of the oxygen debt that is incurred during exercise can be used as an index of the sufficiency of the cardiac output to supply the overall metabolic requirements of the body. It has been found that digitalization of the patient having valvular heart disease, but who has not yet developed congestive failure, produces a striking reduction in the oxygen debt that develops following the performance of a standardized exercise. Therefore cardiac glycosides exert a beneficial prophylactic effect in this type of patient.

As all patients who have congestive heart failure retain sodium and water, a major aim in treatment of this condition is to reduce the sodium excess in the body by increasing the urinary sodium excretion. A number of potent diuretics are available to accomplish this goal, but selection of the appropriate compound is of paramount importance to achieve sodium loss with a minimum of untoward side effects in particular cases.

Rapid diuresis rarely is desirable and is far more likely to cause major complications than is a more gradual loss of sodium which allows the body to adapt slowly to the hemodynamic alterations that are induced by the reduction in the total volume of extracellular fluid. Several diuretics are listed in Table 16.15 together with a brief summary of their principal mechanisms of action.

At times it is essential to remove edema fluid that has accumulated within the body during cardiac failure by the use of mechanical techniques (ie, a needle or drainage tube), because large amounts of such fluid, if present in the serous cavities, can seriously hinder recovery from congestive heart failure. For example, major pleural effusions hamper pulmonary ventilation as well as venous return to the right heart and should be removed. Ascites (an accumulation of fluid within the abdominal cavity), on the other hand, interferes with normal movements of the diaphragm thereby hampering normal pulmonary ventilation. The increased pressure within the abdominal cavity caused by large volumes of fluid can decrease renal venous flow markedly, inhibiting diuresis.

It will be appreciated readily from the foregoing brief summary of the factors involved that the proper evaluation and treatment of chronic congestive heart failure can be as complex as is the disease process itself. The prognosis ultimately depends not only upon the skill of the physician, but also upon whether or not a degree of physiologic compensation can be achieved that is compatible with survival of the patient.

References

GENERAL REFERENCES

Allen JM (ed): Molecular Organization and Biological Function. New York, Harper & Row, 1967

Best, CH, Taylor NB (eds): The Physiological Basis of Medical Practice, 8th ed. Baltimore, Williams & Wilkins, 1966

Bloom W, Fawcett DW: A Textbook of Histology, 9th ed. Philadelphia, Saunders, 1968

Burton BT: The Heinz Handbook of Nutrition, 2nd ed. New York, The Blakiston Division, McGraw-Hill, 1965

Chusid JG: Correlative Neuroanatomy and Functional Neurology, 14th ed. Los Altos, Lange Medical Publications, 1970

Dawson HL: Basic Human Anatomy, 2nd ed. New York, Appleton-Century-Crofts, 1974.

De Robertis EDP, Nowinski WW, Saez FA: Cell Biology. Philadelphia, Saunders, 1970

Everett NB: Functional Neuroanatomy, 6th ed. Philadelphia, Lea & Febiger, 1971

Frisell WR: Acid-Base Chemistry in Medicine. New York, Macmillan, 1968

Ganong WF: Review of Medical Physiology, 6th ed. Los Altos, Lange Medical Publications, 1973

Gatz AJ: Manter's Essentials of Clinical Neuroanatomy and Neurophysiology, 4th ed. Philadelphia, FA Davis, 1970

Goodman LS, Gilman A: The Pharmacological Basis of Therapeutics, 4th ed. New York, Macmillan, 1970

Goss CM (ed): Gray's Anatomy of the Human Body, 28th ed. Philadelphia, Lea & Febiger, 1970

Guyton AC: Textbook of Medical Physiology, 4th ed. Philadelphia, Saunders, 1971

Harper HA: Review of Physiological Chemistry, 13th ed. Los Altos, Lange Medical Publications, 1971

Harvey AMcG, Johns RJ, Owens AH Jr, Ross RS (eds): The Principles and Practice of Medicine, 18th ed. New York, Appleton-Century-Crofts, 1972

Huff BB (ed): Physicians Desk Reference, 27th ed. Oradell, NJ, Medical Economics Co. 1973

Katz B: Nerve, Muscle, and Synapse. New York, McGraw-Hill, 1966

Keele CA, Neil E: Samson Wright's Applied Physiology, 11th ed. London, Oxford University Press, 1965

Langley LL: Outline of Physiology, 2nd ed. New York, The Blakiston Division, McGraw-Hill, 1965

Lehninger AL: Bioenergetics: The Molecular Basis of Biological Energy Transformations. New York, WA Benjamin, 1965

Lehninger AL: Biochemistry. New York, Worth, 1970

Mahler HR, Cordes EH: Basic Biological Chemistry. New York, Harper & Row, 1968

Masoro EJ, Siegel PD: Acid-Base Regulation: Its Physiology and Pathophysiology. Philadelphia, Saunders, 1971

Mountcastle VB (ed): Medical Physiology, 12th ed, Vol I. St. Louis, Mosby, 1968

Mountcastle VB (ed): Medical Physiology, 12th ed, Vol. II. St. Louis, Mosby, 1968

Ruch TC, Patton HD (eds): Physiology and Biophysics, 19th ed. Philadelphia, Saunders, 1966

Selkurt E (ed): Physiology, 2nd ed. Boston, Little, Brown, 1966

Sodeman WA, Sodeman WA Jr (eds): Pathologic Physiology, 4th ed. Philadelphia, Saunders, 1967

Spector WS (ed): Handbook of Biological Data. Philadelphia, Saunders, 1956

Stecher PG, Windholz M, Leahy DS, Bolton DM, Eaton LG (eds): The Merck Index, 8th ed. Rahway, NJ, 1968

Toporek M: Basic Chemistry of Life. New York, Appleton-Century-Crofts, 1968

Toporek M: Essentials of Biochemistry. New York, Appleton-Century-Crofts, 1971

West ES, Todd WR, Mason HS, Van Bruggen JT: Textbook of Biochemistry, 4th ed. New York, Macmillan, 1966

White A, Handler P, Smith EL: Principles of Biochemistry, 4th ed. New York, The Blakiston Division, McGraw-Hill, 1968

SPECIFIC REFERENCES

Chapter 1

Breslow R: The nature of aromatic molecules. Sci Am 227:32, 1972

Gucker FT, Meldrum WB: Physical Chemistry. New York, American Book Co. 1944

Lange NA (ed): Handbook of Chemistry, 8th ed. Sandusky, Ohio, Handbook Publishers, 1952

Secrist JH, Powers WH: General Chemistry. New York, Van Nostrand, 1966

Weast RC (ed): Handbook of Chemistry and Physics, 53rd ed. Cleveland, The Chemical Rubber Co. 1972-1973

White A: Classical and Modern Physics. New York, Van Nostrand, 1940

White HE: Modern College Physics, 3rd ed. New York, Van Nostrand, 1956

Chapter 2

Cannon WB: The Wisdom of the Body, 2nd ed. New York, Norton & Co, 1939

Davson H, Eggleton MG (eds): Starling's Principles of Human Physiology, 14th ed. Philadelphia, Lea & Febiger, 1968

Koshland DE Jr: Protein shape and biological control. Sci Am 229:52, 1973

Olmsted JMD, Olmsted EH: Claude Bernard and the Experimental Method in Medicine, New York, Henry Schuman, 1952

Schoenheimer R: The Dynamic State of Body Constituents. Cambridge, Mass, Harvard University Press, 1942

Chapter 3

Anfinsen CB: Principles that govern the folding of protein chains. Science 180:223, 1973

Borek E: The Code of Life. New York, Columbia University Press, 1965

Bretscher MS: Membrane structure: some general principles. Science 181:622, 1973

Capaldi RA: A dynamic model of cell membranes. Sci Am 230:26, 1974

Coxon RV, Kay RH: A Primer of General Physiology. New York, Appleton-Century-Crofts, 1967

Crick FHC: The genetic code. Sci Am 207:66, 1962

Culliton BJ: Cell membranes: a new look at how they work. Science 175:1348, 1972

Davis RH: Metabolite distribution in cells. Science 178:835, 1972

Erikson RO: Tubular packing of spheres in biological fine structure. Science 181:705, 1973

Fox CF: The structure of cell membranes. Sci Am 226:31, 1972

Marx JL: Microtubules: versatile organelles. Science, 181:1236, 1973

Neutra M, Leblond CP: The Golgi apparatus. Sci Am 220:100, 1969

Nomura M: Ribosomes. Sci Am 221:28, 1969

Satir P: How cilia move. Sci Am 231:44, 1974

Singer SJ, Nicolson GL: The fluid mosaic model of the structure of membranes. Science 175:720, 1972

Soloman AK: Pumps in the living cell. Sci Am 207:100, 1962

Stent GS: Cellular communication. Sci Am 227:42, 1972

Tanzer ML: Cross-linking of collagen. Science 180:561, 1973

Watson JD, Crick FHC: Molecular structure of nucleic acids. Nature 171:737, 1953

Chapter 4

Brokaw CJ: Flagellar movement: a sliding filament model. Science 178:455, 1972

Eckert R: Bioelectric control of ciliary activity. Science 176:473, 1972

Florey E: General and Comparative Animal Physiology. Philadelphia, Saunders, 1966

Hurwitz L, Fitzpatrick DF, Debbas G, Landon EJ: Localization of calcium pump activity in smooth muscle. Science 179:384, 1973

Huxley HE: The contraction of muscle. Sci Am 199:2, 1958

Huxley HE: The mechanism of muscular contraction. Sci Am 213:2, 1965

Margaria R: The sources of muscular energy. Sci Am 226:84, 1972

Murakami A, Eckert R: Cilia: Activation coupled to mechanical stimulation by calcium influx. Science 175:1375, 1972

Murray JM, Weber A: The cooperative action of muscle proteins. Sci Am 230:58, 1974

Naitoh Y, Kaneko H: Reactivated triton-extracted models of paramecium. Science 176:523, 1972

Szent-Györgyi A: Chemistry of Muscular Contraction, 2nd ed. New York, Academic Press, 1951

Wessels NK: How living cells change shape. Sci Am 225:76, 1971

Worley LG, Fischbein E, Shapiro JE: The structure of ciliated epithelial cells as revealed by the electron microscope and in phase-contrast. J Morph 92:545, 1953

Chapter 5

Brimijoin S, Capek P, Dyck PJ: Axonal transport of dopamine-β-hydroxylase by human sural nerves in vitro. Science 180:1295, 1973

Eyzaguirre C: Physiology of the Nervous System. Chicago, Year Book Medical Publishers, 1969

Fischer HA, Schmatolla E: Axonal transport: simple diffusion? Science 182:180, 1973

Magid A: Axonal transport: simple diffusion? Science 782:180, 1973

Ochs S: Fast transport of materials in mammalian nerve fibers. Science 176:252, 1972

Roisen FJ, Murphy RA, Pichickers ME, Braden WG: Cyclic adenosine monophosphate stimulation of axonal elongation. Science 175:73, 1972

Chapter 6

Axelrod J: Neurotransmitters. Sci Am 230:59, 1974

Bennett JP, Logan WJ, Snyder SH: Amino acid neurotransmitter candidates: sodium-dependent high affinity uptake by unique synaptosomal fractions. Science 178:997, 1972

Berl S, Puszkin S, Niklas WJ: Actomyosin-like protein in brain. Science 179:441, 1973

Davidoff RA: Gamma-aminobutyric acid antagonism and presynaptic inhibition in the frog spinal cord. Science 175:331, 1972

Drachman DB, Witzke F: Trophic regulation of acetylcholine sensitivity of muscle: effect of electrical stimulation. Science 176:514, 1972

Eyzaguirre C: Physiology of the Nervous System. Chicago, Year Book Medical Publishers, 1969

Fambrough DM, Drachman DB, Satyamurti S: Neuromuscular junction in myasthenia gravis: decreased acetylcholine receptors. Science 182:293, 1973

Jedrzejczyk J, Wieckowski J, Rymaszewska T, Barnard EA: Dystrophic chicken muscle: altered synaptic acetylcholinesterase. Science 180:406, 1973

Katz B: Quantal mechanism of neural transmitter release. Science 173:123, 1971

Levy WB, Redburn DA, Cotman CW: Stimulus-coupled secretion of γ-aminobutyric acid from rat brain synaptosomes. Science 181:676, 1973

Marx JL: Cyclic AMP in brain: role in synaptic transmission. Science 178:1188, 1972

McAfee DA, Greengard P: Adenosine 3', 5'-monophosphate: electrophysiological evidence for a role in synaptic transmission. Science 178:310, 1972

Perisic M, Cuenod M: Synaptic transmission blocked by colchicine blockade of axoplasmic flow. Science 175:1140, 1972

von Euler US: Adrenergic neurotransmitter functions. Science 173:202, 1971

Chapter 7

Arey LB: Developmental Anatomy, 7th ed. Philadelphia, Saunders, 1965

Barcroft J: The Respiratory Function of the Blood. New York, Macmillan, 1914

Chu N, Bloom FE: Norepinephrine-containing neurons: changes in spontaneous discharge patterns during sleeping and waking. Science 179:908, 1973

DeLong MR: Putamen: Activity of single units during slow and rapid arm movements. Science 179:1240, 1973

Evarts EV: Brain mechanisms in movement. Sci Am 229:96, 1973

Freeman FR: Sleep Research: A Critical Review. Springfield, Charles C. Thomas, 1972

Gatz AJ: Manter's Essentials of Clinical Neuroanatomy and Neurophysiology, 4th ed. Philadelphia, Davis, 1970

Gordon B: The superior colliculus of the brain. Sci Am 227:72, 1972

Gould SJ: Darwin's delay. Natural History 83:68, 1974

Horn G, Rose SPR, Bateson PPG: Experience and plasticity in the central nervous system. Science 181:506, 1973

Jovanovic UJ: Normal Sleep in Man. Stuttgart, Hippokrates Verlag, 1971

Kales A (ed): Sleep: Physiology and Pathology. Philadelphia, Lippincott, 1969

Libassi PT: Where the past is present: how does memory reside in the brain? The Sciences 14:17, 1974

Lipton S: Percutaneous cervical cordotomy. Proc Roy Soc Med 66:607, 1973

Llinas RR: The cortex of the cerebellum. Sci Am 232:56, 1975

Lugaresi E: Some aspects of sleep in man. Proc Roy Soc Med, 65:11, 1972

Marx JL: Nerve growth factor: regulatory role examined. Science 185:930, 1974

Mawdsley C: Neurological complications of haemodialysis. Proc Roy Soc Med 65:871, 1972

Melzak R, Wall PD: Pain mechanisms: a new theory. Science 150:971, 1965

Merton PA: How we control the contraction of our muscles. Sci Am 226:30, 1972

Mullan, S, Harper PU, Hekmatpanak J, Torres H, Dobbin G: Percutaneous interruption of spinal-pain tracts by means of a strontium needle. J. Neurosurg 20:931, 1963

Penfield W: Memory mechanisms. Arch Neurol Psychiatry 67:178, 1952

Perlow BW: Acupuncture: its theory and use in general practice. Proc Roy Soc Med 66:426, 1973

Pessah MA, Roffwarg HP: Spontaneous middle ear muscle activity in man: a rapid eye movement sleep phenomenon. Science 178:773, 1972

Rosenzweig MR, Bennett EL, Diamond MC: Brain changes in response to experience. Sci Am 226:22, 1972

Rosomoff HL, Carroll F, Brown J, Sheptak P: Percutaneous radiofrequency cervical cordotomy: technique. J Neurosurg 23:639, 1965

Rosomoff HL: Neurosurgical control of pain. Ann Rev Med 20:189, 1969

Rosomoff HL, Sheptak P, Carroll F: Modern pain relief: percutaneous chordotomy. JAMA 196:482, 1969

Rosomoff HL: Bilateral percutaneous cervical radiofrequency cordotomy. J Neurosurg 31:41, 1969

Siffre M: Six months alone in a cave. National Geographic, 147:426, 1975

Soloman P (ed): Sensory Deprivation. Cambridge, Mass, Harvard University Press, 1961

Teyler TJ et al: Altered States of Awareness. San Francisco, Freeman, 1972

Truex RC, Carpenter MB: Human Neuroanatomy, 6th ed, Baltimore, Williams & Wilkins, 1970

Wallace RK: Physiological effects of transcendental meditation. Science 167:1751, 1970

Wallace RK, Benson H: The physiology of meditation. Sci Am 226:84, 1972

Chapter 8

Cagan RH: Chemostimulatory protein: a new type of taste stimulus. Science 181:32, 1973

Gordon B: The superior colliculus of the brain. Sci Am 227:72, 1972

Pettigrew JD: The neurophysiology of binocular vision. Sci Am 227:84, 1972

Rushton WAH: Visual pigments and color blindness. Sci Am 232:64, 1975

Chapter 9

Bishop C (ed): Overview of Blood, 1974. Buffalo, NY, Blood Information Service, 1974

Cooper MD, Lawton AR III: The development of the immune system. Sci Am 231:58, 1974

de Gruchy GC: Clinical Hematology in Medical Practice, 3rd ed. Oxford, England, Blackwell Scientific Publications, 1970

Edelman GM: Antibody structure and molecular immunology. Science 180:830, 1973

Fearnley GR: The fibrinolytic system. Proc Roy Soc Med 64:923, 1971

Frankland AW: Allergy: immunity gone wrong. Proc Roy Soc Med 66:365, 1973

Goldberg A, Brain MC (eds): Recent Advances in Haematology. London, England, Churchill Livingstone, 1971

Guyton AC, Granger HJ, Taylor AE: Interstitial fluid pressure. Physiol Rev 51:527, 1971

Marks PA, Rifkind RA: Protein synthesis: its control in erythropoiesis. Science 175:955, 1972

Notkins AL, Koprowski H: How the immune response to a virus can cause disease. Sci Am 228:22, 1973

Porter RR: Structural studies of immunoglobulins. Science 180:713, 1973

Ratnoff DD, Bennett B: The genetics of hereditary disorders of blood coagulation. Science 179:1291, 1973

Chapter 10

Ayres SM, Gregory JJ (eds): Cardiology: A Clinicophysiologic Approach. New York, Appleton-Century-Crofts, 1971

Brooker G: Oscillation of cyclic adenosine monophosphate concentration during the myocardial contraction cycle. Science 182:933, 1973

Brooks CMcC, Hoffman BF, Suckling EE, Orias O: Excitability of the Heart. New York, Grune & Stratton, 1955

Burch GE, Winsor T: A Primer of Electrocardiography, 5th ed. Philadelphia, Lea & Febiger, 1966

Burton AC: Physiology and Biophysics of the Circulation. Chicago, Year Book Medical Publishers, 1965

Collier HOJ: Kinins. Sci Am 207:111, 1962

Hoffman BF, Cranefield PF: Electrophysiology of the Heart. New York, The Blakiston Division, McGraw-Hill, 1960

Jensen D: Intrinsic Cardiac Rate Regulation. New York, Appleton-Century-Crofts, 1971

Johnson PC (ed): Autoregulation of Blood Flow. New York, The American Heart Association, Monograph 8, 1964

Mitchell JH, Wildenthal K: Analysis of Left Ventricular Function. Proc Roy Soc Med 65:542, 1972

Ross R, Glomset JA: Atherosclerosis and the arterial smooth muscle cell. Science 180:1332, 1973

Sonnenblick EH, Skelton CL: Oxygen consumption of the heart: physiological principles and clinical implications. New York, Modern Concepts of Cardiovascular Disease. American Heart Association, 1971

Warren JV: The physiology of the giraffe. Sci Am 231:96, 1974

Willius FA, Keys TE (eds): Classics of Cardiology, Vol I. New York, Dover, 1961

Willius FA, Keys TE (eds): Classics of Cardiology, Vol II. New York, Dover, 1961

Chapter 11

Comroe JH: Physiology of Respiration. Chicago, Year Book Medical Publishers, 1965

Filley GF: Pulmonary Insufficiency and Respiratory Failure, Philadelphia, Lea & Febiger, 1967

Filley GF: Acid-Base and Blood Gas Regulation. Philadelphia, Lea & Febiger, 1971

McDermott M: Interrelation between surface tension and mechanical properties of the lung: a historical review. Proc Roy Soc Med 66:381, 1973

Chapter 12

Goetz KL, Bond GC, Bloxham DD: Atrial receptors and renal function. Physiol Rev 55:157, 1975

Pitts RF: Physiology of the Kidney and Body Fluids, 2nd ed. Chicago, Year Book Medical Publishers, 1968

Smith HW: Lectures on the Kidney. Lawrence, Kansas, University Extension Division, University of Kansas, 1943

Chapter 13

Baron JH: Physiological control of gastric acid secretion. Proc Roy Soc Med 64:739, 1971

Bayliss WM, Starling EH: The mechanism of pancreatic secretion. J Physiol 28:325, 1902

Bogert LJ, Briggs GM, Calloway DH: Nutrition and Physical Fitness, 8th ed. Philadelphia, Saunders, 1966

Brady RO: Hereditary fat-metabolism diseases. Scientific American 229:88, 1973

Davenport HW: Physiology of the Digestive Tract, 2nd ed. Chicago, Year Book Medical Publishers, 1966

Davenport HW: Why the stomach does not digest itself. Sci Am 226:86, 1972

Dudrick SJ, Rhoads JE: Total intravenous feeding. Sci Am 226:73, 1972

Frieden E: The biochemistry of copper. Sci Am 218:103, 1968

Frieden E: The chemical elements of life. Sci Am 227:52, 1972

Hammond AL: Aspirin: new perspective on everyman's medicine. Science 114:48, 1971

Kassell B, Kay J: Zymogens of proteolytic enzymes. Science 180:1022, 1973

Kretchmer N: Lactose and lactase. Sci Am 227:70, 1972

Maugh TM II: Trace elements: a growing appreciation of their effects on man. Science 181:253, 1973

Myant NB: The normal physiology of lipid transport. Proc Roy Soc Med 64:893, 1971

Roy AD: True heartburn. Brit Med J 2:279, 1967

Wagner AF, Folkers K: Vitamins and Coenzymes. New York, Interscience Publishers, Division of John Wiley and Sons, 1964

Wilson ED, Fisher KH, Fuqua ME: Principles of Nutrition, 2nd ed. New York, John Wiley and Sons, 1965

Chapter 14

Baldwin E: Dynamic Aspects of Biochemistry. Cambridge, England, University of Cambridge Press, 1949

Dickerson RE: The structure and history of an ancient protein. Sci Am 226:58, 1972

Gates DM: The flow of energy in the biosphere. Sci Am 224:88, 1971

Chapter 15

Allen JE, Rasmussen H: Human red blood cells: prostaglandin E_2, epinephrine and isoproterenol alter deformability. Science 174:512, 1971

Axelrod J: Noradrenaline: fate and control of its biosynthesis. Science 173:598, 1971

Brunner HR, Kirshman JD, Sealey JE, Laragh JH: Hypertension of renal origin: evidence for two different mechanisms. Science 174:1344, 1971

Buchman MT, Peake GT. Osmolar control of prolactin secretion in man. Science 181:755, 1973

Charles MA, et al: Adenosine 3',5'-monophosphate in pancreatic islets: glucose-induced insulin release. Science 179:569, 1973

Cowie AT: Physiological actions of prolactin. Proc Roy Soc Med 66:861, 1973

Evans JI, Maclean AM, Ismail AAA, Love D: Circulating levels of plasma testosterone during sleep. Proc Roy Soc Med 64:841, 1971

Fernstrom JD, Wartman RJ: Brain serotonin content: increase following ingestion of carbohydrate diet. Science 174:1023, 1971

Guillemin R, Burgus R: The hormones of the hypothalamus. Sci Am 227:24, 1972

Harmo PG, Ojeda SR, McCann SM: Prostaglandin involvement in hypothalamic control of gonadotropin and prolactin release. Science 181:760, 1973

Itskovitz HD, Odya C: Intrarenal formation of angiotensin I. Science 174:58, 1971

Kolata GB: Cyclic GMP: cellular regulatory-agent? Science 182:149, 1973

Koshland DE Jr: Protein shape and biological control. Sci Am 229:52, 1973

Oster G: Conception and contraception. Natural History 7:46, 1972

Pastan I: Cyclic AMP. Sci Am 227:97, 1972

Pike JE: Prostaglandins. Sci Am 225:84, 1971

Schally AV, Arimura A, Kastin AJ: Hypothalamic regulatory hormones. Science 179:341, 1973

Sutherland EW: Studies on the mechanism of hormone action. Science 177:401, 1972

Tepperman J: Metabolic and Endocrine Physiology, 2nd ed. Chicago, Year Book Medical Publishers, 1968

Whitelocke RAF, Eakins KE, Bennett A: Acute anterior uveitis and prostaglandins. Proc Roy Soc Med 66:429, 1973

Chapter 16

Astrand P-O, Rodahl K: Textbook of Work Physiology. New York, McGraw-Hill, 1970

Avery ME, Wang N-S, Taeusch H Jr: The lung of the newborn infant. Sci Am 228:74, 1973

Ball MF, Canary JJ, Kyle LH: Tissue changes during intermittent starvation and caloric restriction as treatment for severe obesity. Arch Internal Med 125:62, 1970

Benoit FL, Martin RL, Watten RH: Changes in body composition during weight reduction in obesity: balance studies comparing effects of fasting and a ketogenic diet. Ann Internal Med 63-2:64, 1965

Benson AJ: Neurological aspects of disorientation in aircrew. Proc Roy Soc Med 66:519, 1973

Bradley RM, Mistretta CM: Swallowing in fetal sheep. Science 179:1016, 1972

Brierly JB: Neurological sequelae of decompression in supersonic transport aircraft. Proc Roy Soc Med 66:527, 1973

Chapman CB: Dietary factors in treating heart disease. J Am Dietetic Assoc 29:346, 1953

Cowie AT: Induction and suppression of lactation in animals. Proc Roy Soc Med 65:24, 1972.

Davies DM: The effects of extended hypercapnia. Proc Roy Soc Med 65:496, 1972

Ernsting J: Hyperbaric oxygen therapy: physiological considerations. Proc Roy Soc Med 64:873, 1971

Ernsting J: Hypoxia in the aviation environment. Proc Roy Soc Med 66:523, 1973

Jokl E: Hypoxia at the Mexico City olympic games. Bull International Soc Cardiol III:4, 1971

Karpovich PV, Senning WE: Physiology of Muscular Activity, 7th ed. Philadelphia, Saunders, 1971

Kekwick A, Pawan GLS: Caloric intake in relation to body weight changes in the obese. Lancet 2:155, 1956

Kekwick A, Pawan GLS: Metabolic study in human obesity with isocaloric diets high in fat, protein or carbohydrate. Metabolism 6:447, 1957

Kekwick A, Pawan GLS: An experimental approach to the mechanisms of weight loss. II: A comparison of the effects of Thyroxine, Fat-Mobilizing substance (FMS) and food deprivation in achieving weight loss in mice. Metabolism 12:222, 1963

Kekwick A, Pawan GLS: The effect of high fat and high carbohydrate diets on rates of weight loss in mice. Metabolism 13:87, 1964

Keul J, Doll E, Keppler D (Translated from the German by Skinner JS): Energy Metabolism of Human Muscle. Baltimore, University Park Press, 1973

Kiesow LA, Bless JW, Shelton JB: Oxygen dissociation in human erythrocytes. Its response to hyperbaric environments. Science 179:1236, 1973

Krehl WA, Lopez SA, Good EI, Hodges RE: Some metabolic changes induced by low carbohydrate diets. Am J Clin Nutrition 20:139, 1967

Lambert RJW: The nuclear submarine environment. Proc Roy Soc Med 65:795, 1972

Lightfoot NF: Chronic carbon monoxide exposure. Proc Roy Soc Med 65:798, 1972

Masters WH, Johnson VE: Human Sexual Response. Boston, Little, Brown, 1966

Newburgh LH, Johnson MS: The nature of obesity. J Clin Invest 8:197, 1930

Nicholson AN: Sleep patterns in the aerospace environment. Proc Roy Soc Med 65:192, 1972

Olesen ES, Quaade F: Fatty foods and obesity. Lancet 1:1048, 1960

Preston FS: Medical aspects of aerospace travel. Proc Roy Soc Med 65:187, 1972

Pennington AW: Pathophysiology of obesity. Am J Digest Dis 21:69, 1954

Pennington AW: Treatment of obesity: developments of the past 150 years. Am J Digest Dis 21:65, 1954

Pilkington TRE, Gainsborough H, Rosenoer VM, Carey M: Diet and weight-reduction in the obese. Lancet 1:856, 1960

Schauf GE: Diet and management of obesity. Am J Clin Nutr 24:287, 1971

Segal SJ: The physiology of human reproduction. Sci Am 231:52, 1974

Turnbull AC: Myometrial contractility in pregnancy and its regulation. Proc Roy Soc Med 64:1015, 1971

Young CM: Weight reduction using a moderate-fat diet. I. Clinical responses and energy metabolism. J Am Dietetic Assoc 28:410, 1952

Young CM: Weight reduction using a moderate-fat diet. II. Biochemical responses. J Am Dietetic Assoc 28:529, 1952

Young CM, Empey EL, Serraon VU, Pierce ZH: Weight reduction in obese young men. J Nutrition 61:437, 1957

Young CM, Scanlan SS, Im HS, Lutwak L: Effect on body composition and other parameters in obese young men of carbohydrate level of reduction diet. Am J Clin Nutrition 24:290, 1971

Young VR, Scrimshaw NS: The physiology of starvation. Sci Am 225:14, 1971

Appendix

Table 1   THE GREEK ALPHABET

SYMBOL		NAME		SYMBOL		NAME
A	α	alpha		N	ν	nu
B	β	beta		Ξ	ξ	xi
Γ	γ	gamma		O	o	omicron
Δ	δ	delta		Π	π	pi
E	ϵ	epsilon		P	ρ	rho
Z	ζ	zeta		Σ	σ, ς	sigma
H	η	eta		T	τ	tau
Θ	θ	theta		Υ	υ	upsilon
I	ι	iota		Φ	ϕ	phi
K	κ	kappa		X	χ	chi
Λ	λ	lambda		Ψ	ψ	psi
M	μ	mu		Ω	ω	omega

Table 2 ATOMIC WEIGHTS OF THE ELEMENTS BASED ON THE RELATIVE ATOMIC MASS OF ^{12}C = 12.000

ELEMENT	SYMBOL	ATOMIC NO.	ATOMIC WEIGHT
Actinium	Ac	89	(227)
Aluminum	Al	13	26.98
Americium	Am	95	(243)
Antimony	Sb	51	121.75
Argon	Ar	18	39.94
Arsenic	As	33	74.92
Astatine	At	85	~210
Barium	Ba	56	137.34
Berkelium	Bk	97	(247)
Beryllium	Be	4	9.01
Bismuth	Bi	83	208.98
Boron	B	5	10.81
Bromine	Br	35	79.90
Cadmium	Cd	48	112.40
Calcium	Ca	20	40.08
Californium	Cf	98	(251)
Carbon	C	6	12.01
Cerium	Ce	58	140.12
Cesium	Cs	55	132.90
Chlorine	Cl	17	35.45
Chromium	Cr	24	51.99
Cobalt	Co	27	58.93
Copper	Cu	29	63.54
Curium	Cm	96	(247)
Dysprosium	Dy	66	162.50
Einsteinium	Es	99	(254)
Erbium	Er	68	167.26
Europium	Eu	63	151.96
Fermium	Fm	100	(257)
Fluorine	F	9	18.99
Francium	Fr	87	(223)
Gadolinium	Gd	64	157.25
Gallium	Ga	31	69.72
Germanium	Ge	32	72.59
Gold	Au	79	196.96
Hafnium	Hf	72	178.49
Helium	He	2	4.00
Holmium	Ho	67	164.93
Hydrogen	H	1	1.00
Indium	In	49	114.82
Iodine	I	53	126.90
Iridium	Ir	77	192.22
Iron	Fe	26	55.84
Krypton	Kr	36	83.80
Lanthanum	La	57	138.90
Lawrencium	Lw	103	(257)
Lead	Pb	82	207.19
Lithium	Li	3	6.94
Lutetium	Lu	71	174.97
Magnesium	Mg	12	24.30
Manganese	Mn	25	54.93
Mendelevium	Md	101	(256)
Mercury	Hg	80	200.59
Molybdenum	Mo	42	95.94
Neodymium	Nd	60	144.24
Neon	Ne	10	20.17
Neptunium	Np	93	237.04
Nickel	Ni	28	58.71
Niobium	Nb	41	92.90
Nitrogen	N	7	14.00
Nobelium	No	102	(254)
Osmium	Os	76	190.2
Oxygen	O	8	15.99
Palladium	Pd	46	106.40

Table 2 (Cont.)   ATOMIC WEIGHTS OF THE ELEMENTS BASED ON THE RELATIVE ATOMIC MASS OF $^{12}C = 12.000$

ELEMENT	SYMBOL	ATOMIC NO.	ATOMIC WEIGHT
Phosphorus	P	15	30.97
Platinum	Pt	78	195.09
Plutonium	Pu	94	(244)
Polonium	Po	84	(~210)
Potassium	K	19	39.10
Praseodymium	Pr	59	140.90
Promethium	Pm	61	(145)
Protoactinium	Pa	91	231.03
Radium	Ra	88	226.02
Radon	Rn	86	(~222)
Rhenium	Re	75	186.2
Rhodium	Rh	45	102.90
Rubidium	Rb	37	85.46
Ruthenium	Ru	44	101.07
Samarium	Sm	62	150.35
Scandium	Sc	21	44.95
Selenium	Se	34	78.96
Silicon	Si	14	28.08
Silver	Ag	47	107.86
Sodium	Na	11	22.98
Strontium	Sr	38	87.62
Sulfur	S	16	32.06
Tantalum	Ta	73	180.94
Technetium	Tc	43	98.90
Tellurium	Te	52	127.60
Terbium	Tb	65	158.92
Thallium	Tl	81	204.37
Thorium	Th	90	232.03
Thulium	Tm	69	168.93
Tin	Sn	50	118.69
Titanium	Ti	22	47.90
Tungsten	W	74	183.85
Uranium	U	92	238.02
Vanadium	V	23	50.94
Xenon	Xe	54	131.30
Ytterbium	Yb	70	173.04
Yttrium	Y	39	88.90
Zinc	Zn	30	65.38
Zirconium	Zr	40	91.22

Index

Abadie's sign, 450
A-band, 116, 118
Abducens nerve, 245, 283–84, 286, 468
Abortifacients, chemical, 1175
Abortion, spontaneous, 1197
Absorption, 254, 562, 815, 859, 862, 865, 867, 870, 871
Absorptive cells, 825, 826
Acceleration, 3, 1223, 1229
Accommodation, 164, 472
   convergence and, 475, 476
   lens and, 472
   mechanism of, 472
   reflex, 227
Acetaldehyde, 41, 944, 979
Acetazolamide, 872
Acetic acid, 945
Acetoacetic acid, 965, 1087, 1212, 1213, 1221
Acetoacetyl CoA, 1087
Acetone, 965, 1087, 1213, 1221
Acetylcholine, 147, 195, 209, 448, 458, 484, 662, 667, 668, 672, 676, 680, 688, 693, 694, 1016
   biosynthesis of, 196
   release, mechanism of, 211
   release of during rest, 210
Acetylcholinesterase, 209, 210
Acetyl coenzyme A (acetyl CoA), 882, 945, 947, 1086–87, 1221
Acetylglucosamine, 66–67
Acetylsalicylic acid, 572
Achalasia, 837, 898, 899
Achilles tendon reflex, 1056
Achlorhydria, 861
Achromatic series, 484
Acid, 13, 18
Acid anhydride, formation of, 42
Acid-base balance, 54, 528
   erythrocyte, size in, 528
Acid labile conjugate, 1105
Acidophilic (alpha) cells, 1025
Acidosis, 674, 1074, 1190
   diabetic, 763
Acidosis, metabolic, 738, 763–64, 808, 1077, 1087, 1088, 1108, 1111, 1210, 1212, 1222
   etiologic factors and, 763
   renal compensation of, 807
   RQ and, 987
Acidosis, respiratory, etiologic factors and, 759
Acid, weak, ionization constant for, 20
Acne, 1037, 1140
Aconitase, 949–50
cis—Aconitic acid, 949
Acoustic tract, 506

Acromegaly, 1032, 1075
Acrosomal vesicle, 1131
Acrosome, 1131, 1191
Actin, 116, 117, 124, 139, 146, 616, 660
Actinomysin D, 1055
Action potential, 103–4, 120, 130, 132, 133, 162, 165, 171, 182, 482
   biphasic, 171
   compound, 173
   conduction of, 168, 184
   duration of, 617
   duration and length of contraction, relationship between, 616
   ionic fluxes during, 135
   monophasic, 171
   recording of, 163
   self-amplifying event, 135
Activation, heat of, 123
Activators, 917
Active complex, 141
Active secretory mechanisms, 794
Active site, 913
Active state, 123
Active transport, 85, 92, 97, 98, 104, 448, 779, 787, 790, 856, 860, 911
Activity, 20, 24
   coefficient, 20, 22
   carrier systems, in, 94
   mechanism of, 94, 102, 562, 864, 870, 1196
   substances reabsorbed by, 788, 791
   zones of, 831
Actomyosin, 583
Actual volume, 787
Acupuncture, 360, 464
Acyclic compounds, 33
Acyl-CoA dehydrogenase, 963
Acylglycerol, 68
Adams-Stokes syndrome, 696
Adaptation, 186, 353, 523, 673, 1000
   dark, 476, 487–88, 886
   light, 487
Addisonian crisis, 1118
Addison's disease, 1034, 1035, 1110, 1112, 1115, 1117
   blood picture in, 883
Adenine, 59
Adenohypophysis, 958, 1023, 1024, 1025
Adenoids, 821
Adenosine, 1180
Adenosine diphosphate, 78, 104, 137, 934
Adenosine triphosphatase, 94, 101, 118, 911
   activity of, 119, 139, 141, 151
   in cilia, 155